AF736094

A CALENDAR OF THE CORRESPONDENCE OF SIR JOHN HERSCHEL

This volume for the first time provides summaries, descriptions, and documentation for 14,815 letters written by or sent to Sir John Herschel (1792–1871). Herschel's numerous contributions to astronomy, as well as to mathematics, physics, chemistry (especially photochemistry and photography), meteorology, philosophy of science, and scientific organization, led his British contemporaries to regard him as the most prominent scientist of his era. Because Herschel corresponded on a remarkable array of topics and with leading figures both in Britain and beyond, this volume gives scholars access to a wealth of revealing new information. The many uses of the volume are enhanced by its Biographical Register, which identifies about 1500 of Herschel's correspondents, and by its Index, which supplies thirty thousand references. This volume is far and away the most extensive source of information on John Herschel ever published. Also included are bibliographies of Herschel's publications and of publications on him.

Michael J. Crowe is Professor in the Program of Liberal Studies and Graduate Program in History and Philosophy of Science at the University of Notre Dame.

David R. Dyck is Professor of History (Retired) at the University of Winnipeg.

James J. Kevin is Adjunct Instructor in the Curriculum and Instruction Department at East Tennessee State University.

Editor

MICHAEL J. CROWE

Associate Editors

DAVID R. DYCK   JAMES J. KEVIN

Research Assistants

KEITH A. ABNEY
GINA M. BACIGALUPI
DIANA H. BARNES
MARVIN P. BOLT
ANNE-MARIE CLAVELLI
MATTHEW F. DOWD
RYAN D. DYE
DARIN HAYTON
KATHRYN E. HIBEY
REBECCA L. LUBAS
SUSAN E. PETTI
BARBARA M. TURPIN
JAMESON M. WETMORE

Computer Consultant

A. EDWARD MANIER, JR.

A CALENDAR OF THE CORRESPONDENCE OF SIR JOHN HERSCHEL

CAMBRIDGE
UNIVERSITY PRESS

PUBLISHED BY THE PRESS SYNDICATE OF THE UNIVERSITY OF CAMBRIDGE
The Pitt Building, Trumpington Street, Cambridge CB2 1RP, United Kingdom

CAMBRIDGE UNIVERSITY PRESS
The Edinburgh Building, Cambridge CB2 2RU, UK http://www.cup.cam.ac.uk
40 West 20th Street, New York, NY 10011–4211, USA http://www.cup.org
10 Stamford Road, Oakleigh, Melbourne 3166, Australia

©Michael J. Crowe 1998

This book is in copyright. Subject to statutory exception
and to the provisions of the relevant collective licensing agreements,
no reproduction of any part may take place
without the written permission of Cambridge University Press.

First Published 1998

Printed in the United States of America

Typeset in Times 9/10 in Microsoft Word for Macintosh

A catalog record for this book is available from the British Library

Library of Congress Cataloging in Publication data

A calendar of the correspondence of Sir John Herschel /
[Michael J. Crowe, David R. Dyck, James R. Kevin, editors
p. cm.
Includes bibliographical references and index.
ISBN 0–521–63149–1
1. Herschel, John F. W. (John Frederick William), Sir, 1792–1871–
–Correspondence—Catalogs. 2. Astronomers—Great Britain–
–Correspondence. I. Crowe, Michael J. II. Dyck, David R., 1933–
. III. Kevin, James J., 1938–
QB36.H59C35 1998 98–22928
520'.92—dc21 CIP

ISBN 0 521 631491

To

Marian E. Crowe

The research for this Calendar was funded in part by the National Science Foundation under Grant No. DIR 8921705. The University of Notre Dame, including its Institute for Scholarship in the Liberal Arts, also provided extensive funding.

Advisory Board for the John Herschel Correspondence Project
PROFESSOR DAVID S. EVANS, UNIVERSITY OF TEXAS
DR. SIMON SCHAFFER, UNIVERSITY OF CAMBRIDGE
PROFESSOR REESE V. JENKINS, RUTGERS, THE STATE UNIVERSITY
PROFESSOR GREGORY A. GOOD, WEST VIRGINIA UNIVERSITY
PROFESSOR BRIAN WARNER, UNIVERSITY OF CAPE TOWN
MR. JOHN HERSCHEL-SHORLAND, DISS, NORFOLK, ENGLAND
PROFESSOR PHILLIP R. SLOAN, UNIVERSITY OF NOTRE DAME

CONTENTS

INTRODUCTION

THE IMPORTANCE OF JOHN HERSCHEL AND OF HIS CORRESPONDENCE

Contemporaries of Sir John F. W. Herschel (1792–1871) recognized that he was among the most important figures in nineteenth-century science. Writing during the late 1830s, John Ruskin remarked that through his acquaintance with William Buckland, "I met the leading scientific men of the day, from [John] Herschel downward...."[1] In an obituary notice on Herschel, T. Romney Robinson lamented: "... British science has sustained a loss greater than any which it has suffered since the death of Newton...."[2] That many of Robinson's contemporaries shared his sentiments is shown by the fact that Herschel was interred next to Newton in Westminster Abbey. The significance of Herschel for nineteenth-century British science is also suggested by the conclusion reached by W. F. Cannon in a study of Herschel's place in nineteenth-century science; after asking what it meant to be scientific in the early Victorian period, Cannon marshaled abundant evidence to support the conclusion that it meant: "Be as much like Herschel as possible."[3]

Herschel's reputation among his contemporaries derived not only from the importance of his contributions to science, but also from the breadth of his learning. His most significant achievements were in astronomy, to which area his father, Sir William Herschel, had made very notable contributions. Among the elder Herschel's achievements were the discovery of Uranus, pioneering studies in stellar astronomy, and catalogs of the nebulae and double stars of the northern heavens. John Herschel not only extended his father's studies of the northern heavens, but also published the first detailed study of the stellar and nebular objects of the southern heavens. This was based on four years of intense observation carried out by the younger Herschel at the Cape of Good Hope. John Herschel's *Outlines of Astronomy* (1849) served for decades as the definitive presentation of that science. In addition, John Herschel made important contributions to mathematics, physics, chemistry, photography, the philosophy of science, and scientific organization. The rebirth of mathematics in Britain is frequently dated from the second decade of the nineteenth century when John Herschel served as founding president of the Analytical Society, which played a major role in introducing continental methods of analysis into Britain. In physics, Herschel contributed not only to electricity but also to optics; his book-length article "Light," published in 1830 in the *Encyclopaedia Metropolitana*, was recognized as a definitive mathematical presentation of the wave theory of light not only in England, but also in France and Germany, where translations were immediately published. His discovery in 1819 of a method of fixing images formed when silver salts are exposed to light was the first of a number of important contributions he made to photochemistry. Herschel's *Preliminary Discourse on the Study of Natural Philosophy* (1830) continues to be appreciated as a classic in philosophy of science. The range and significance of Herschel's contributions were such that he can with justice be described as Britain's first modern physical scientist.[4]

Because John Herschel began his career carrying the name that his father had made famous, because the son, unlike his autodidact father, was trained at the center of British science, Cambridge University, and because the younger Herschel worked in nearly every area of science, he had almost instant access to the international leadership of science. This opportunity was not lost on Herschel, whose devotion to and diligence in correspondence were remarkable. These and other features of Herschel's life account for the fact that the correspondence of John Herschel constitutes one of the most valuable archives for nineteenth-century science. The single most remarkable aspect of his correspondence is its magnitude; this volume provides summaries of the 14,815 letters to and from Herschel that are currently known. An even more significant feature of the John Herschel correspondence is the importance of his correspondents.

Among Herschel's main correspondents (along with the approximate numbers of surviving letters exchanged with them) were George Airy (1059), Charles Babbage (422), Francis Baily (276), Francis Beaufort (172), David Brewster (61), W. R. Dawes (90), Augustus De Morgan (403), Michael Faraday (63), James David Forbes (89), William Rowan Hamilton (57), John Hind (93), William Lassell (70), Humphrey Lloyd (62), John W. Lubbock (92), Charles Lyell (57), Thomas Maclear (398), Roderick Murchison (113), George Peacock (96), Edward Sabine (528), Adam Sedgwick (45), Richard Sheepshanks (166), William Henry Smyth (182), Mary Somerville (57), James South (90), George Stokes (132), Wilhelm Struve (57), Henry Fox Talbot (130), John Tyndall (60), William Whewell (212), and Thomas Young (43). Some of the smaller collections are also of great interest, e.g., the letters Herschel exchanged with Charles Darwin (9), Maria Edgeworth (15), Carl Friedrich Gauss (23), John Stuart Mill (19), Herbert Spencer (15), and William Thomson (6). These tallies provide evidence that not only was John Herschel involved in most areas of science, but also that he was central to that crucially important group that historians of science have come to

[1] John Ruskin, *Praeterita and Dilecta* (London: George Allen, 1908), 205.

[2] T. R. R. [T. Romney Robinson], "John Frederick William Herschel," *Proceedings of the Royal Society of London, 20* (1872), xvii.

[3] Walter F. Cannon, "John Herschel and the Idea of Science," *Journal of the History of Ideas, 22* (1961), 219.

[4] M. J. Crowe, "John Herschel: Britain's First Modern Physical Scientist" in Brian Warner (ed.), *John Herschel 1792–1992: Bicentennial Symposium* (Royal Society of South Africa, 1994), pp. 57–69.

call "the Cambridge network."[1] This group has been identified as the source of much of the best of nineteenth-century British science.

The importance of Herschel's correspondence derives not only from its magnitude and range, but also from the fact that some of the individual letters were highly influential. Perhaps the most striking example of this consists of two letters Herschel wrote to Charles Lyell on the question of the origin of species, which letters many Darwin scholars describe as having played a crucial role in Darwin's investigation of that topic.[2]

Another reason why it is so important to provide increased access to Herschel's correspondence is that whereas "life and letters" volumes have been published on nearly all the leading British scientists of the nineteenth century, no such volume has ever treated Herschel. Nor has a comprehensive selected edition of his letters appeared, although Herschel scholarship has been enriched by such volumes as Brian and Nancy Warner's *Maclear and Herschel: Letters and Diaries at the Cape of Good Hope 1834–1838*.[3] Many other important documents from Herschel's years at the Cape were made available in the most significant of David Evans's many contributions to Herschel scholarship.[4] Herschel scholarship has also benefitted from a short but substantial biography published by Günther Buttmann: *In the Shadow of the Telescope: A Biography of John Herschel*.[5] It is hoped that the present Calendar will serve as a guide to the Herschel's extensive correspondence, a large portion of which is now available on microfilm.[6] The Calendar should also serve as a rich source for biographical information on Herschel by allowing scholars to follow for the first time the main lines of Herschel's activities and ideas on an almost day-to-day basis.

REPOSITORIES HOLDING CORRESPONDENCE OF JOHN HERSCHEL

Many persons have contributed to the preservation of this remarkable archive. No one, however, did more in this regard than John Herschel himself, who carefully preserved and arranged nearly all the correspondence he received and made abstracts, drafts, pressed copies, etc., for many of the letters he sent. Moreover, in the period immediately after Herschel's death in 1871, one of Herschel's children, Col. John Herschel, led a major effort to rescue from the ravages of history as many as possible of the letters his father had written and that were consequently in attics or archives of his correspondents. A document written by Col. John Herschel and preserved at the Royal Society describes in some detail the efforts that were then made.

> The question, as to a biography of Sir J. Herschel being undertaken, was raised in 1873—soon after my return from India (on furlough) in the summer of that year.
>
> It was decided to restrict operations, at that period, to what was most urgent, viz, the collection of autograph letters in the hands of correspondents and their representatives, and of copies or drafts where such could be found.
>
> This was carried out, with considerable energy and persistence, and at no small cost of time, trouble, and cash, throughout the year 1874; in November of w^h year I had to leave England again.
>
> With but few exceptions, letters thus collected were returned, after having been copied,—the copies being invariably compared, word for word, with the originals, & initialled as such. This—the finding, getting, taking verified transcripts of, & finally returning several hundred letters,—represents the bulk of the work effected up to Nov. '74.[7]

Additional documents[8] supply further details concerning these efforts as well as information as to which correspondents were contacted. Very careful comparison of these transcriptions, which number 1648, with their originals (when such still exist) shows that the transcriptions were indeed done with care and accuracy. Moreover, the most momentary glance at these transcriptions reveals that their level of legibility far exceeds that found in the letters written in Herschel's notoriously difficult hand.

Two important if frequently little appreciated points emerge from the above information. First, although Herschel preserved nearly all the letters he received, his correspondents were far less diligent. The result of this is that 9,463 of the letters treated in this volume are letters *to* John Herschel, whereas only 5,352 are letters *from* him. From the plausible assumption that Herschel wrote approximately as many letters as he received, one can infer that four thousand or more letters by Herschel were not located in the international search for Herschel letters carried on in conjunction with the preparation of this volume. No doubt many of these letters have been destroyed. The second point relates to the fact that 14,815 separate letters are treated

---

[1] Susan Faye Cannon, *Science in Culture: The Early Victorian Period* (New York: Science History Publications, 1978) and Robert Smith, "The Cambridge Network in Action: The Discovery of Neptune," *Isis, 80* (1989), 395–422.

[2] Walter F. Cannon, "The Impact of Cultural Uniformitarianism. Two Letters from John Herschel to Charles Lyell, 1836–1837," *American Philosophical Society Proceedings, 105* (June 1961), 301–14.

[3] (Rotterdam: A. A. Balkema, 1984).

[4] David S. Evans, Terence J. Deeming, Betty Hall Evans, and Stephen Goldfarb (eds), *Herschel at the Cape: Diaries and Correspondence of Sir John Herschel, 1834–1838* (Austin, Texas: Univ. of Texas Press, 1969).

[5] trans. by Bernard Pagel, edited and with an introduction by David S. Evans (New York: Charles Scribner's Sons, 1970).

[6] In 1990, University Publications of America brought out a microfilm of the entire Herschel collection at the Royal Society. This has made Herschel's correspondence far more accessible to scholars.

[7] This document appears as Frame 69 of Reel 27 of the University Publications of America Herschel microfilm.

[8] Frames 111–20 of Reel 27 of the University Publications of America Herschel microfilm.

in this volume. This fact might easily lead one to assume that this is the number of items of Herschel correspondence that have been preserved and that have been processed in the research for this volume. In fact, the situation is far otherwise. The reason is that repositories holding Herschel correspondence usually possess not only originals, but also such items as drafts, abstracts, extracts, contemporary transcriptions, and related correspondence, e.g., letters to Herschel's wife or letters sent to a correspondent who then sent the letter to Herschel. Consequently, in preparing this Calendar, it has been necessary to process not 14,815 items of correspondence but probably closer to 19,000 documents. Some of these, e.g., letters sent to correspondents and then forwarded to Herschel, have usually been excluded from consideration. In other cases, e.g., when we have located the original letter, plus an abstract, plus a draft, plus a contemporary transcription, we have listed after the summary all these items along with their locations. In most cases, however, only one or two forms of the letter have been preserved.

Col. John Herschel's efforts, so important to the preservation of his father's correspondence, failed in regard to the finding of a biographer. Seven decades later, in 1944, John Herschel's grandson, the Reverend Sir John Herschel, Bt., donated a very large portion of Herschel's correspondence as well as transcriptions of Herschel's diaries for the years 1834 to 1871 and miscellaneous other materials to the Royal Society. One document published by the Royal Society indicates that this collection contains 10,058 items,[1] but this figure is easily misunderstood if taken to be an indication of the number of original letters held by the Royal Society. On the one hand, this figure neglects the fact that various other archives in the Royal Society contain Herschel letters. On the other hand, this figure seems to be simply a sum of the number of originals plus drafts plus abstracts plus transcriptions that were part of this donation. In any case, by far the largest collection of Herschel letters is preserved at the Royal Society in London.

Large as this donation was, it by no means exhausted the number of John Herschel letters held by the Herschel family. On 3 and 4 March 1958, a sale was held at Sotheby's in London of numerous letters, manuscripts, etc. that had been owned by William, Caroline, or John Herschel, which items were at that time owned by Mrs. E. D. Shorland, niece of Rev. Sir John Herschel. The University of Texas acquired a large portion of the John Herschel items offered in this sale, including 2,555 items of correspondence and the original diaries for 1820 to 1871. Through this action, the University of Texas acquired the second largest collection of Herschel letters.

The third largest collection, preserved in the Royal Greenwich Observatory Archive, which is maintained at Cambridge University, consists of part of John Herschel's correspondence with that most prolific of correspondents George Airy. We have located 607 separate items of correspondence in this archive.

The fourth largest archive is at the Royal Astronomical Society in London. This collection, which includes letters of William and Caroline Herschel, seems to have been donated by the Herschel family. Our research in this archive resulted in the location of 440 separate items of correspondence.[2]

The Herschel family still has in its possession a very significant number of Herschel letters, most being of a personal or family nature. These have not been accessible to the public. Fortunately, we have been allowed to examine and to summarize 383 of these letters.

Despite the large size of these five collections, numerous Herschel letters are preserved in other collections throughout the world. In fact, we have located letters in a total of 73 different repositories, the names of these appearing in this volume's "Provenance List."

The number and importance of the Herschel letters held by the Royal Society as well as some special features of that collection make it useful to provide additional information on that collection, which is also now available on microfilm. The Royal Society, upon receiving the Herschel materials, carefully bound the letters in 28 volumes. Moreover, in the 1950s, a Royal Society librarian R. K. Bluhm prepared a detailed catalog of these letters, which was available at the Royal Society Library but was not published. This catalog has now been edited for publication and is available from University Publications of America.[3] The Royal Society Herschel collection can for convenience be described as consisting of five parts:

- Vols. 1–18: Original letters to or from Herschel arranged alphabetically by correspondent.
- Vol. 19: Miscellaneous items, including letters that happened not be to included in the earlier volumes.
- Vols. 20–4: Transcriptions of letters prepared under the direction of Col. John Herschel.
- Vol. 25: Primarily pressed copies of letters by John Herschel.
- Vols. 26–8: Miscellaneous items, including additional letters and the transcriptions of the diaries for 1834–71.

THE ROYAL SOCIETY SUMMARIES

Of the 14,815 summaries in this volume, 4,971 were prepared (during the 1950s, it seems) by librarians at the Royal Society in London. It appears that these librarians initially set out to summarize all Herschel let-

[1] R. K. Bluhm, "A Guide to the Archives of the Royal Society and to Other Manuscripts in Its Possession," *Notes and Records of the Royal Society of London, 12* (1956), 39.

[2] For information on materials at the Royal Astronomical Society, see J. A. Bennett, "The Manuscript Archives of the Royal Astronomical Society," *Quarterly Journal of the Royal Astronomical Society, 18* (1977), 459–63.

[3] M. J. Crowe (ed.), *The Letters and Papers of Sir John Herschel: A Guide to the Manuscripts and Microfilm* (in the series: *Collections from the Royal Society*), (Bethesda, Maryland: University Publications of America, 1991).

ters at the Royal Society, preserving these summaries on reference cards carrying the identification mark of the letter. Probably because of the immense time and effort this would have required, they terminated this ambitious and valuable project after having summarized most of the letters bound in volumes 1 to 13 of the 28 volumes of Herschel materials in the Royal Society John Herschel collection. Thus the identification marks for summaries prepared at the Royal Society always begin with the letters RS:HS followed by a number from 1 to 13 inclusive. It should not, however, be concluded that all RS:HS items followed by a number between 1 and 13 inclusive were prepared at the Royal Society. In a substantial number of cases, especially those in which the letter was in a foreign language, no summary was prepared. In such cases, the team that prepared this volume has remedied the omission. Moreover, the project team has added the physical description of all letters from volumes 1 through 13, including the designation of the number of pages for each letter. In addition, the project team prepared all summaries of letters not in volumes 1–13 of the Royal Society collection.

Information on how these Royal Society summaries became available provides background for understanding decisions made concerning the form in which the entries were included in this volume. At the time when funding for this project was being negotiated with the National Science Foundation, a representative of the Royal Society Library stated that summaries of all letters in volumes 1–13 in their Herschel collection had already been prepared. This welcome news provided a justification for a reduction in the amount of time and of funding that would be needed for the project. As our project progressed, staff members at the Royal Society Library photocopied for us the cards containing these summaries. It is important to note that the Royal Society retains copyright to these entries.[1]

In preparing these Royal Society summaries for inclusion in this volume, we had to face the issue of how much editing they should receive. Various considerations, the most important being time and funding constraints, pointed toward keeping to a minimum the level of editing applied to them. Consequently, our practice has been to retain essentially their full content, but with some seemingly necessary exceptions. For example, we have corrected all errors evident to us. Moreover, such considerations as consistency in style and economy of space have necessitated that we bring their format closer to that adopted for all other summaries in this volume. For example, we have (silently) made such alterations as substituting "JH" for "John Herschel" or "Herschel" and replaced "Royal Astronomical Society" by "R.A.S." Consistency of style led us to replace British spellings by their American equivalents and to substitute numerous semicolons and periods for the dashes used very liberally by the Royal Society librarians. Another major task resulted from the fact that the Royal Society librarians typically provided only the last names of persons mentioned in summaries. Because of the need to index all summaries for this volume, we have consequently added first names (whenever needed and possible) to the Royal Society summaries. To have applied a heavier level of editing might well have violated the integrity of the summaries prepared at the Royal Society and, in any case, would have involved the expenditure of quantities of time far beyond what had received funding.

THE CONTENTS OF THE CALENDAR

All items of correspondence treated in this Calendar fall into seven categories.

1. *Original letters.* The great majority of letters summarized in this Calendar are original letters, usually written in the hand of and signed by the sender.
2. *Copies of letters.* Nearly all these are letters from John Herschel, who sometimes wrote out a copy of an important letter or made a pressed copy of it. Also, as noted above, in the early 1870s, John Herschel's family arranged for the transcription of a large number of letters that he had written. In the vast majority of cases, these copies date from the nineteenth century. Twentieth-century photocopies are not included in this category. Wherever possible, the summaries given in this Calendar are based on examination of the original letter (or a photoduplicate of it) rather than a copy or abstract of it.
3. *Published letters.* Such letters can take either of two forms. First, these may be letters Herschel sent to journals or newspapers for publication. Usually the original is not preserved. The second type are letters that have been published, e.g., in a "life and letters" volume of one of Herschel's contemporaries, but where we have been unable to determine the location of the original on which the published form was based.
4. *Drafts.* Drafts treated in this Calendar probably all came directly or indirectly from the pen of John Herschel, who was careful to preserve drafts of his more important letters. These are precious documents not only in those cases where the final letter has been lost, but also in instances where the scholar can compare the draft with the final letter sent so as to see shifts in Herschel's thought.
5. *Third-party letters.* Repositories holding Herschel correspondence frequently contain letters not written by or sent to Herschel, but that are somehow relevant to him. We have included such items when they shed light on Herschel. One type of letter with this characteristic is a letter directed to Herschel's correspondent but containing information to be shared with Herschel. Another type of letter is written by someone close to Herschel, for example, his wife, and contains important information on Herschel, but the letter itself is directed to another party. Individual judgments (perhaps not always definable or defensible)

[1] The Royal Society has asked that the following statement applying to these summaries be included in this volume: "The Royal Society of London for Improving Natural Knowledge holds rights to portions of this work. No unauthorized reproduction of these materials is permitted. All rights reserved."

have had to be made in each such case. In the majority of cases, we have decided against inclusion of the third party letter. Nonetheless, ca. 163 such letters have been included.

6. *Memoranda.* No hard and fast line can be drawn that separates letters from memoranda, but typically the latter items are shorter than standard letters and are bereft of such formalities as salutations.

7. *Letters listed in sales catalogs.* When we have seen or been alerted to a Herschel letter appearing in a sale catalog, we have contacted the firm offering the letter, asking to receive a photocopy or a transcription of the letter. In most cases, the vendors have cooperated fully and effectively, thereby allowing us to provide information on letters that might otherwise have disappeared from the world of scholarship.

EDITORIAL METHOD

Editorial practices for this Calendar have been based in large part on those employed by the editors of the widely respected *Calendar of the Correspondence of Charles Darwin,* 1821–1882.[1] This has had a number of advantages, extending both to the present editors and to users of this Calendar. The former, in selecting format and methods, have been able to draw productively on the experience of the Darwin editors. Readers, at least those who have worked with the Darwin Calendar, will be aided by this decision because the high level of consistency in the formatting of the two volumes maximizes the ease with which users will work with our format. Differences in the collections of letters of the two scientists and other factors have suggested some departures from the Darwin Calendar format, but such instances have been rare.

Each entry in this Calendar contains the following items: a number assigned by the editors of this volume; the date of the letter; the name of the correspondent, if known; the address of the correspondent if appearing on the letter or envelope or in cases where it could be reasonably conjectured, supplied within square brackets. The summary of the contents of the letter comes next, followed by the physical description and location line. This line begins with a description of the letter as well as information on its number of pages and whether it was written in a foreign language. The location of the letter is then supplied. If the letter has been published, a reference to this publication concludes the entry.

Further information on the editorial practices followed for the parts of the entries is given below.

1. *Number.* In general, the entries have been numbered from earliest letter to the latest. Thus the earliest known letter has been assigned the number 1. Frequently, more than one letter corresponds to a particular day. In those cases, the letters for that day have been arranged in the alphabetical order of the names of the correspondent. In some instances, even these principles of ordering have proven inadequate; for example, instances have arisen where Herschel and a correspondent exchanged as many as five letters all on the same day. In such cases, efforts have been made to place the letters in the order in which they were written, responses appearing after rather than before the original letter, with the number assigned accordingly.

2. *The Date.* The date is expressed in the format year, month, day, e.g., 1863–3–27. In cases where one or more elements in the date do not appear in the letter or could not be determined with certainty, the element has been placed in square brackets to indicate its status. If no date could be assigned, this is indicated by the abbreviation "n.d." If the element was not only inferred but also significantly uncertain, it is followed by a question mark. Thus a letter bearing only the date July 15 could be assigned either of the following: [1843]–7–15 or [1843?]–7–15, the latter form indicating that the year is more uncertain than that assigned in the former case. All dates have been given in terms of the Gregorian rather than the Julian calendar. Thus a letter from Russia, where the Gregorian calendar was not instituted until 1917, may bear the date June 2/13. This would be transcribed as June 13. If a letter contains no date but was written in response to an earlier letter dated, say, 1845–9–14, then the undated later letter would be assigned the date [1845–9–14 or later]. Where it is known or conjectured that a letter was written in a particular year or month, but nothing allows the assignment of a day to the letter, the letter is placed with the letters written on the first day of that year or month. Thus a letter known to have been written sometime in 1843 has been treated in terms of placement as if it had been written on 1843–1–1.

3. *The name of the correspondent.* The name of the correspondent is preceded by a "To" or a "From", the former indicating that the letter was to that correspondent from Herschel, the latter ("From") meaning that the letter was sent from that correspondent to Herschel. The name of the correspondent is given in fairly full form, e.g., George Gabriel Stokes has been used rather than G. G. Stokes. Titles in most cases have not been included, unless meaning required such. Thus such titles as Sir, Rev., Lord, Col., Mr., Mrs., etc., have been omitted. We have departed from this rule in cases where the first name of the correspondent is not known. This has been indicated by inserting a Mr. (or Miss) where the first name would typically be placed. Where one or more elements of the name have been conjectured, this has been indicated by placing that element in square brackets. Where the evidence is significantly less than satisfactory for the conjectured addition, this is expressed by a question mark following the name in square brackets. When we have been essentially sure of the name, even though it may not appear anywhere in the letter, square brackets have not been employed.

4. *The address.* London street names stand alone (i.e., without the designation London), whereas other street names are followed by the city. Also Collingwood (the name of Herschel's home in Hawkhurst from 1840 on) usually stands alone. In naming cities outside the English-speaking world, the common English language name has been used, e.g.,

[1] Edited by Frederick Burkhardt and Sydney Smith (New York: Garland, 1985); 2nd ed. (Cambridge Univ. Press, 1994).

Vienna rather than Wien, Florence rather than Firenze. Where conjectural information has been inserted, square brackets have been employed. If more conjectural or inferential, we have inserted square brackets plus a question mark. If no location can be assigned, the abbreviation n.p. (no place) has been inserted.

5. *The summaries.* These have been designed to summarize the most significant information in the letter. Space considerations have necessitated the use of various abbreviations, e.g., "JH" for John Herschel and the first letter of the first and of the last name of the correspondent, e.g., in a letter from Charles Babbage, CB would refer the reader to the first line for the full name. Other abbreviations are listed on the page immediately before the first Calendar entries. Where a letter prominently mentions a publication, efforts have been made to provide a reference, sometimes rather cryptic, to that publication. In choosing between a readable or a space-saving telegraphic style, the tendency has been to favor the former. Most of the entries are meant to stand alone, but many will be enriched and clarified by reading adjacent summaries. Cross references (e.g., [see JH's 1842–9–16]) have frequently been provided to assist this process. Much research has been necessitated by the fact that most letters in mentioning a person give only the person's last name. Where that name has been carried over into the summary, efforts have been made to identify the person by assigning a first name. Where these efforts have not succeeded, this is indicated by prefacing the last name by an abbreviations such as Mr., Mrs., Captain, etc. Where the letter exists as, say, an original, a draft, a contemporary transcription, and an abstract, the summary has been based on the original. Quotations are indicated by the use of single quotation marks.

6. *Physical description.* Provided below (and also on the abbreviations page) are abbreviations, terms, and symbols used on the physical description line.

L	letter
D	document (other than letter)
draft	draft of letter
mem	memorandum
C	copy of letter
CC	contemporary copy of the letter (e.g., copy kept by sender)
trans	translation of letter (into or from English)
Abs	Abstract (of a letter)
A	autograph (half or more of text in sender's hand)
(A)	part autograph (any portion less than half in sender's hand)
S	signed or initialed by sender
(S)	signed with sender's name by an amanuensis
ALS	autograph letter signed by sender
LS	letter in hand of amanuensis, signed by sender
ACCS	contemporary copy in sender's hand, signed by sender
(French)	language of described item
inc	incomplete (part(s) of the original document missing)
damaged	part(s) of original document destroyed
†	annotations by recipient
†(Jones)	annotations by specified person
††	lengthy, numerous, or particularly important annotations by recipient
encl	enclosure
diag	diagram or illustration

An unusually long physical description of a letter might take the form: ALS 4pp (French) inc & damaged diag. This would mean that the item is a letter in the hand of the sender (autograph) and signed by the sender, that the letter is 4 pages long in text length, that it is written in French, is incomplete, damaged, and includes one or more diagrams. In those rare cases where it has been impossible to see the item identified or a photoduplicate of it, the fact that the length of the letter is unknown has been expressed by writing Xpp for the number of pages.

7. *Provenance.* All materials held by the Royal Society are identified by a mark beginning with the letters RS. The great majority of these begin with RS:HS, meaning that they are part of the Royal Society's Herschel collection, rather than other collections. The provenance identifications for items held by the Harry Ransom Humanities Research Center at the University of Texas begin with the letter Tx; Royal Astronomical Society materials with RAS; Royal Greenwich Observatory items with RGO. Letters held by John Herschel–Shorland, great great grandson of John Herschel, are identified by beginning with JHS. To find the full name for any provenance identification, see the "Provenance List," which follows this introduction. Persons wishing to consult individual letters will no doubt help the librarians holding the letters by supplying as much information as possible about the letter and especially its identification mark.

The atypically long provenance identification for Herschel's letter to Gauss of 1821–2–11 takes the form NSUB Cod.Ms.Gauss 99 (Draft: RS:HS 8.72 & C: 20.111). Using the Provenance List, one can determine that the original for this letter is held at the Niedersächsische Staats- und Universitäts-Bibliothek, Göttingen (identification mark being provided), that a draft is preserved at item 72 in volume 8 of the Royal Society Herschel Collection, and that a contemporary transcript occurs in volume 20, item 111 in the Royal Society Herschel Collection.

8. *Reference to published version of letters.* Although it has not been the central goal of the research directed toward the preparation of this Calendar to find all published versions of correspondence to and from John Herschel, this research has located 665 letters that appear, at least in part, in published volumes. These published letters take two forms. A minority consist of letters that Herschel sent to newspapers or journals in the successful hope that the letter be published. The majority of published letters appear either in such volumes as Brian and Nancy Warner's *Maclear and Herschel: Letters and Diaries at the Cape of Good Hope 1834–1838* or in biographies or collections of letters of Herschel's contemporaries, e.g., Charles Darwin. In every case where published letters of any of these forms have come to our attention, we have referenced the publication, in most cases using an ab-

breviated reference. These abbreviations can easily be deciphered using the "Bibliography of Works Containing Printed Herschel Correspondence," which is included in this volume.

9. *Future plans.* The research materials presented in this volume are also preserved in a database program for Macintosh computers named *4th Dimension.* This has greatly facilitated our work, for example, in determining the chronological order of the letters or in ascertaining the number that have been published. The editors of this volume have been exploring possibilities of making this database available to scholars. Its uses are numerous. With this program and data, it is easy to determine which letters were exchanged with, say, George Airy or were written in the months around the discovery of the planet Neptune or which summaries contain the word "Neptune."

The research produced in this Calendar depended very heavily on the cooperation of librarians and other persons with Herschel correspondence in their care. In making available to us photoduplicates of these letters, these persons have frequently and fully justifiably stipulated that these copies not become the basis of a competing archive of Herschel materials. Moreover, it is not the intention of the editors of this Calendar to open a service that will supply scholars with photoduplicates of letters held by repositories with professional staff responsible for these materials. Thus although the editors welcome information on the Calendar itself and will appreciate receiving corrections as well as information about newly found letters, we shall not in general be responsible for responding to requests for photoduplicates of any letters described in this Calendar. Correspondence concerning the calendar should be addressed to Professor Michael J. Crowe, Program of Liberal Studies, University of Notre Dame, Notre Dame, IN 46556.

ACKNOWLEDGMENTS

The main funding for this Calendar came from the National Science Foundation as Grant DIR 8921705, awarded in 1990 during the tenure of Dr. Ronald Overmann as Program Director for History and Philosophy of Science. An earlier National Science Foundation Grant (SES 85–21314), awarded in 1986, funded the research out of which this project arose.

The University of Notre Dame also provided extensive support for this project. For example, it funded two half-semester leaves of absence for the Editor, provided research space and computers to the staff, and in numerous other ways contributed to this project. At three crucial points in the process, Notre Dame's Institute for Scholarship in the Liberal Arts provided grants that were vital to this project. Most recently, it provided a subvention to help with the publication costs for this large volume. Special thanks are thus due to Robert P. Burke and Associate Dean Jennifer L. Warlick of the Institute for Scholarship in the Liberal Arts. In addition, Notre Dame's Graduate Program in History and Philosophy of Science provided funding for two graduate students to assist during the final stages of the project. Help also came in various ways from the Editor's department, the Program of Liberal Studies, including its chairpersons, Professors Phillip R. Sloan, Stephen M. Fallon, and F. Clark Power. Among the other Notre Dame persons who contributed significantly to this project, special mention should be made of Roger B. Skurski, Associate Dean of the College of Arts and Letters.

The very large contribution to this project made by David R. Dyck, Professor of History at the University of Winnipeg, now retired, was due partly to his great generosity and also to the very supportive leave-of-absence policy of his university. Without his involvement in the project, it is doubtful that it could have been brought to completion.

Librarians at institutions too numerous to mention were vital in securing the materials on which this Calendar is based. Special mention should be made of the librarians of the University of Notre Dame, of the Royal Society, of the Harry Ransom Humanities Research Center at the University of Texas, of the Royal Astronomical Society, of the University Library at Cambridge, and of the Royal Greenwich Observatory Archive, which is now located at Cambridge.

We express warm thanks to Margaret M. Jasiewicz for having typed the Royal Society summaries used in this volume and contributed in other important ways.

Special thanks are due to John Herschel–Shorland, great great grandson of John Herschel, who generously made available to us various Herschel family letters that are not in the public domain.

The project has benefitted very significantly from the members of the project Advisory Board. The persons who have served in the role are: Professor David S. Evans, Department of Astronomy, University of Texas; Professor Gregory A. Good, Program in History of Science and Technology, West Virginia University; Mr. John Herschel-Shorland, Crossingford Lodge, Diss, Norfolk, England; Professor Reese V. Jenkins, Department of History, Rutgers, The State University; Dr. Simon Schaffer, Department of History and Philosophy of Science, University of Cambridge; Professor Phillip R. Sloan, Program of Liberal Studies, University of Notre Dame; and Professor Brian Warner, Department of Astronomy, University of Cape Town.

In concluding these acknowledgments, the editors wish to express their very great appreciation to all the members of the team who helped prepare the materials included in this volume. The names of these persons appear on the first page of this volume as recognition of the invaluable contribution each made to it.

PROVENANCE LIST

Académie belgique: Académie Royale des Sciences, des Lettres et des Beaux-Arts de Belgique, Brussels, Belgium.
Académie des sciences (Paris): Académie des sciences, Institut de France, Paris, France.
Adler: Adler Planetarium, Chicago, Illinois.
AdWDDR Bessel: Zentrale Archiv der Akademie der Wissenschaften der DDR—Bessel Nachlass, Berlin, German Democratic Republic.
AdWDDR Encke: Zentrale Archiv der Akademie der Wissenschaften der DDR—Encke Nachlass, Berlin, German Democratic Republic.
APS: Library of the American Philosophical Society, Philadelphia, Pennsylvania.
Auckland Public Library: Auckland Public Library, Auckland, New Zealand.
Bayerische Akademie der Wissenschaften: Bayerische Akademie der Wissenschaften, Munich, Germany.
BL: British Library, London, England.
Bedgebury: Bedgebury School, Collingwood House, Hawkhurst, Kent, England.
BostonPubLib: Boston Public Library, Boston, Massachusetts.
British Astronomical Association: British Astronomical Association, London, England.
Buckinghamshire County Council: Lee Papers, Buckinghamshire County Council, Aylesbury, Buckinghamshire, England.
Cambridge University Library: Cambridge University Library, Cambridge, England.
Cape Archives: South African Archives, Cape Town, South Africa.
Columbia Univ. Library: Columbia University Library, New York, New York.
CUOL: Cambridge University Observatory Library, Cambridge, England.
DMC: Dibner Manuscript Collection of the Smithsonian Institution, Washington, D.C.
Dunedin Public Lib.: Dunedin Public Library, Dunedin, New Zealand.
Edinburgh Royal Society: Edinburgh Royal Society, Edinburgh, Scotland.
EUL: Edinburgh University Library, Edinburgh, Scotland.
Friedman, Gerald M.: Private collection.
Geological Society: Geological Society, London, England.
George Eastman House: Richard and Monay Menschel Library, International Museum of Photography at George Eastman House, Rochester, New York.
Getty Center: Getty Center for the History of Art and the Humanities, Santa Monica, California.
Gingerich, Owen: Private collection.
Harvard: Houghton: Houghton Library of Harvard University, Cambridge, Massachusetts.
Huntington Library: Huntington Library, San Marino, California.
Hydrographic Office: Hydrographic Office, Taunton, Somerset, England.
JHS: Private collection of John Herschel-Shorland.
Kasslers: Private collection of Jamie and Michael Kassler.
King's College: King's College Library, University of London, London, England.
Linnaean Society: Linnaean Society, London, England.
Lion Heart Autographs: Lion Heart Autographs (Dealer), New York, New York.
London Library, Univ. of: Senate House Library of the University of London, London, England.
Luther, Paul: Paul W. Luther Astronomy Books (Dealer), Bernardston, Maryland.
Manchester Central Library: Manchester Central Library, Manchester, England.
Massachusetts Historical Society: Massachusetts Historical Society, Boston, Massachusetts.
National Archives of Canada: National Archives of Canada, Ottawa, Canada.
National Library of Australia: National Library of Australia, Canberra, Australia.
National Maritime Museum: National Maritime Museum, Greenwich, England.
NatLibScot: National Library of Scotland, Edinburgh, Scotland.
New South Wales Library: State Library of New South Wales, Sydney, Australia.
NMPFT: National Museum of Photography Film and Television, Bradford, England
NSUB: Niedersächsische Staats- und Universitäts-Bibliothek, Göttingen, Germany.
Oxford, Bodleian: Bodleian Library, Oxford, England.
Oxford, History of Science Museum: History of Science Museum, Oxford, England.
RAS:JH Archive: Royal Astronomical Society Herschel Archive, London, England.
RGO: Archives of the Royal Greenwich Observatory, Cambridge University Library, Cambridge, England.
Rhodes Univ. Library: Cory Library of Historical Research, Rhodes University, Grahamstown, South Africa.
RI: Royal Institution, London, England.
Ross Collection, Sydney: Private collection of Sydney Ross.
Rosse Papers: Rosse Papers, Birr Scientific Heritage Foundation, Birr, Co. Offaly, Ireland.
Royal Astronomical Society of Canada: Royal Astronomical Society of Canada, Toronto, Canada.
Royal Botanic Gardens: Royal Botanic Gardens, Kew, Richmond, Surrey, England.
Royal Geographical Society: Royal Geographical Society, London, England.
Royal Mint: Royal Mint, Llantrisant, nr. Pontyclun, Mid-Glamorgan, England.
RS: Royal Society, London, England.
RS:HS: Herschel Papers, Royal Society, London, England.
Rylands, John: John Rylands University Library of Manchester University, Manchester, England.
St. Andrews: Library of Saint Andrews University, St Andrews, Fife, Scotland.
Science Museum Archives: Archives of the Science Museum, London.
South African Astronomical Observatory: South African Astronomical Observatory, Cape Town, South Africa.
St. John's College, Cambridge: Saint John's College of Cambridge University, Cambridge, England.
TC, Camb.: Trinity College of Cambridge University, Cambridge, England.
TCD: Library, Trinity College, Dublin, Ireland.
TxU: Texas University, Harry Ransom Humanities Center, Austin, Texas.
UCT: Cape Town University Library, Cape Town, South Africa.
Univ. of Witwatersrand: Library of the University of Witwatersrand, Johannesburg, South Africa.
Univ.Coll.London: Library of University College, London, England.
WT: Library of Wellcome Trust, London, England.
Yale Univ.: Yale University Library, New Haven, Connecticut.

A CALENDAR
OF THE
CORRESPONDENCE
OF
SIR JOHN HERSCHEL

ABBREVIATIONS AND SYMBOLS USED

For date and location

n.d.	no date designated
n.p.	no place designated

For physical description

L	letter
D	document (other than letter)
draft	draft of letter
mem	memorandum
C	copy of letter
CC	contemporary copy of the letter (e.g., copy kept by sender)
trans	translation of letter (into or from English)
Abs	Abstract (of a letter)
A	autograph (half or more of text in sender's hand)
(A)	part autograph (any portion less than half in sender's hand)
S	signed or initialed by sender
(S)	signed with sender's name by an amanuensis
ALS	autograph letter signed by sender
LS	letter in hand of amanuensis, signed by sender
ACCS	contemporary copy in sender's hand, signed by sender
(French)	language of described item
inc	incomplete (part(s) of the original document missing)
damaged	part(s) of original document destroyed
†	annotations by recipient
†(Babbage)	annotations by specified person
††	lengthy, numerous, or particularly important annotations by recipient
encl	enclosure
diag	diagram or illustration

For Societies and Institutions:

B.A.A.S.	British Association for the Advancement of Science
H.E.I.C.	Haileybury East India College
R.A.S.	Royal Astronomical Society
R.S.L.	Royal Society of London

For Journals:

B.A.A.S.Rep.	*British Association for the Advancement of Science Reports*
P.M.	*Philosophical Magazine*
R.A.S.Mem.	*Royal Astronomical Society Memoirs*
R.A.S.M.N.	*Royal Astronomical Society Monthly Notices*
R.S.P.T.	*Royal Society of London Philosophical Transactions*

For Publications by John Herschel:

Admiralty Manual	Herschel, John (ed.), *Admiralty Manual of Scientific Enquiry; Prepared for the Use of Her Majesty's Navy: And Adapted for Travellers in General* (1849)
Cape Results	*Results of Astronomical Observations Made during the Years 1834, 5, 6, 7, 8 at the Cape of Good Hope; Being the Completion of a Telescopic Survey of the Whole Surface of the Visible Heavens, Commenced in 1825 by Sir J. F. W. Herschel* (1847)
Essays Q.E.R.	*Essays from the Quarterly and Edinburgh Reviews, with Addresses and Other Pieces* (1857)
Familiar Lectures	*Familiar Lectures on Scientific Subjects* (1866)
Iliad	*The Iliad of Homer, Translated into English Accentuated Hexameters* (1866)
Meteorology	*Meteorology. From the Encyclopaedia Britannica* (1861)
Outlines Astr.	*Outlines of Astronomy* (1849)
Physical Geography	*Physical Geography: From the Encyclopaedia Britannica* (1861)
Prelim. Discourse	*A Preliminary Discourse on the Study of Natural Philosophy* (1830)
Telescope	*The Telescope* (1861)
Treatise Astr.	*Treatise on Astronomy* (1833)

CALENDAR OF THE JOHN HERSCHEL CORRESPONDENCE 1799–1820

1. 1799–8–9 To Mary Pitt Herschel Newbury
A thank you note for clothes, written at school [Mr. Bull's], and requesting that JH's cyphering and music books be sent.
ALS 1p JHS 2.5

2. 1799–8–9 or earlier To Mary Pitt Herschel [Newbury]
Hopes that JH's mother has had a pleasurable time in London; JH's class went to see Mrs. Mason's (?) baby.
ALS 1p JHS 2.2

3. 1799–11–28 To Mary Pitt Herschel Newbury
Writes to parents to say that JH is well, his head has healed, and he is looking forward to his holidays.
ALS 2pp JHS 2.1

4. 1807–11–11 From Alexander Rogers New Quay, Leith
Has arrived in Scotland. Recently witnessed a large comet. Thanks JH and family for hospitality at Slough.
ALS 3pp RS:HS 14.386

5. 1808–1–1 From Alexander Rogers New Quay, Leith
Thanks JH for information concerning the comet. Planned to build telescope with object glasses separated from each other. Optician said this would not correct spherical and chromatic aberrations. Wants advice.
ALS 3pp RS:HS 14.387

6. 1808–2–22 From Alexander Rogers New Quay, Leith
Wishes to teach privately. Asks William Herschel to write a recommendation certificate. Weather has been problematic. Sends greeting to the elder Herschels.
ALS 3pp RS:HS 14.388

7. 1808–3 From Alexander Rogers New Quay, Leith
Concerned over illness of William Herschel. Discusses 'Evolution of Curve Lines.' Describes [Christiaan] Huygens's terms for evolution of curves. Offers advice to avoid problems with them.
ALS 3pp RS:HS 14.389

8. 1808–5–18 To [Thomas (Sr.)] Baldwin [Slough]
Recalls JH's visit to Hammersmith. Asks TB to permit JH's cousin Mary Baldwin to visit Slough at time of Ascot races. Received letter from [TB's daughter] Sophia.
C 1p TxU:H/L-0079; Reel 1054 (C: TxU:H/M–0616.2; Reel 1086)

9. 1808–11–5 From Alexander Rogers New Quay, Leith
Grateful for letter from Brighton. Agrees with JH concerning comet. Is suspicious of those who claim it is the same comet that was seen one year previously.
L 1p inc RS:HS 14.390

10. 1809–1–2 From Alexander Rogers Bridge St., Leith
Describes the difficulty of [P. Laplace's] *Mécanique céleste*. Has a new pupil. Discusses proper boundary of a front view mirror.
ALS 3pp RS:HS 14.391

11. 1809–3–25 From Alexander Rogers Bridge St., Leith
Discusses problem of the ellipse and offers a method of solving it. States the proper boundary of a front mirror is too difficult to ascertain. Still having problems with teaching.
ALS 3pp RS:HS 14.392

12. 1809–9–20 From Alexander Rogers Bridge St., Leith
Describes lunar rainbow seen on 22 Aug. Congratulates JH on his mathematical attainments. Describes his enjoyment of mathematics. Hopes JH will continue to correspond.
ALS 3pp RS:HS 14.393

13. 1810–1–5 From [Johann] D[ietrich] Herschel Hanover
Identifies a number of the various orders of ape and comments on them.
ALS 3pp (German) TxU:H/M–0884; Reel 1085

14. 1810–3–5 From Alexander Rogers Bridge St., Leith
Found JH's resolution of a biquadratic ingenious. AR used René Descartes's method of resolution, but found it less elegant. Discusses resolution of higher order quadratics.
ALS 3pp RS:HS 14.394

15. 1810–6–19 From James A. Gordon St. John's College
Informing him of his position in the examinations, and offering congratulations.
ALS 1p RS:HS 8.158

16. 1810–10–1 From James Grahame Glasgow
Recalls time spent together; asks JH for information about Cambridge colleges, especially St. John's.
ALS 3pp RS:HS 8.233

17. 1810–12–2 From James Grahame Edinburgh
Is attending John Playfair's lectures in mathematics; talks of plans to come to Cambridge.
ALS 4pp RS:HS 8.234

18. [1811?–?]–23 To William Herschel [Cambridge]
Copy of JH's midsummer examination questions—17 on Greek drama, 22 on Mechanics—at St. John's College. Currently reading P. S. Laplace's *Système du monde*,which confirms WH's theory of galaxy formation. T. W. Hornbuckle reports that money has been received from Bruce & Co.
ALS 6pp TxU:H/L-0546.2; Reel 1053

19. 1811 2 28 From James Grahame Edinburgh
News of several lady friends and JG's progress in law studies.
ALS 4pp RS:HS 8.235

20. 1811–3–13 From James Grahame n.p.
Asks about nature of Classics examination JG must take to be considered for admission to Cambridge.
ALS 2pp RS:HS 8.236

21. 1811–6–6 From Pat Wilson 29 Kensington Square
Speculates that JH may one day rival Isaac Newton. Awaits JH's visit next Tuesday. JH's parents ordered chairs to be ready [Wednesday], when PW's sister will accompany JH to Slough.
ALS 3pp RS:HS 18.422

22. 1811–9–28 From James Grahame Birmingham
Apologizes for long delay in writing; is coming to Cambridge in October.
ALS 2pp RS:HS 8.237

23. 1811–10–4 To William Herschel Cambridge
Duties at St. John's College. James Grahame's father is arriving from Glasgow. [James] Wood is working to get Grahame into college. Ask cousin [Mary Baldwin] for address for Mr. Rogers. Recalls JH's summer vacation.
ALS 3pp TxU:H/L-0547; Reel 1053

24. 1811–10–7 From James Grahame Birmingham
Is coming to Cambridge on 10 Oct.
ALS 1p RS:HS 8.238

25. 1811–11–1 From Alexander Rogers Charlotte St., Leith
Discusses the comet that 'attracts the attention of all.' Speculates that the train of comets arises from the impulse of solar rays. Discusses shape and composition of comets.
ALS 3pp RS:HS 14.395

26. 1811–11–16 To William Herschel Cambridge
JH's hypothesis on nebular vortex and formation of solar system. Has other ideas about molecular forces and comet tails, but will wait to learn whether WH wants to hear them. JH burned over 100 pages of notes on these speculations. Will return to Slough in mid-December.
ALS 4pp TxU:H/L-0548; Reel 1053

27. [1812?] To Charles Babbage n.p.
Finding it difficult to finish his memoir. Gives some integral equations for his comments. Hopes to see him at Slough soon.
ALS 4pp RS:HS 2.361

28. [1812?] To Charles Babbage n.p.
Regarding equational problems.
draft 4pp RS:HS 2.360

29. [1812] To Charles Babbage Norfolk St.
Sends on part of letter for CB to complete and forward.
C 1p RS:HS 20.3

30. 1812 To William Nicholson n.p.
[Addressed to WN as 'Editor of the *Journal of Natural Philosophy, Chemistry, and the Arts* from 'A Lover of Modern Analysis' [JH], this letter] derives by a new method various analytical formulae for the tangent and cotangent.
L *Journal of Natural Philosophy, Chemistry, and the Arts, 31* (1812), 133–6

31. 1812–1–6 From Alexander Rogers Charlotte St., Leith
Is now a clerk for [London and Edinburgh] shipping company. Does not know if he should mention his employment to Mr. Watts.
ALS 3pp RS:HS 14.396

32. 1812–1–16 From James Grahame Whitehill, Glasgow
Has been quite ill; talks about one journey.
ALS 3pp RS:HS 8.239

33. 1812–1–19 or 26 From James Grahame Whitehill, Glasgow
Talks about JH becoming a churchman.
ALS 2pp RS:HS 8.240

34. 1812–2–5 To William Herschel Cambridge
Describes JH's public Latin examination in algebra yesterday. James Grahame and sister were dangerously ill, but Grahame returned to Cambridge. [Addendum:] Newspaper clipping naming students (including JH) who received B.A. degrees at Cambridge on 16 Jan. 1813.
ALS 3pp † TxU:H/L-0549; Reel 1053

35. 1812–3–23 To William Nicholson, n.p.
[Addressed to WN as 'Editor of the *Journal of Natural Philosophy, Chemistry, and the Arts* from 'A Lover of Modern Analysis' [JH], this letter] derives by a new method various analytical formulae for the sine and cosine.
L *Journal of Natural Philosophy, Chemistry, and the Arts, 32* (1812), 13–16

36. 1812–3–29 From James Grahame St. John's
Is coming to visit JH at Slough.
ALS 1p RS:HS 8.241

37. 1812–5–29 From James Grahame Stamford
About JG's travels to Edinburgh.
ALS 2pp RS:HS 8.242

38. 1812–6 From James Grahame Edinburgh
Note to accompany several copies of JG's thesis.
ALS 1p RS:HS 8.244

39. 1812–6–13 From James Grahame Edinburgh
Congratulates JH on having completed examinations; JG will be admitted to the bar soon.
ALS 3pp RS:HS 8.243

40. 1812–6–16 From John Playfair to William Herschel Edinburgh
Sends William and Mary Somerville to visit Herschel family. Describes and praises the couple. Hopes observatory at Edinburgh will be completed soon.
ALS 3pp RS:HS 14.169

41. 1812–6–20 From Charles Babbage Teignmouth
Sending an equation in calculus for him to give opinion.
ALS 3pp RS:HS 2.1

42. [1812–7–1] To Charles Babbage n.p.
Answer to CB's equation. Bring something to discuss at the forthcoming meeting of the Analytical Society.
ALS 5pp RS:HS 2.2 (C: 20.1)

43. 1812–7–10 From Charles Babbage [Teignmouth?]
Further comments on his original query [see CB's 1812–6–20] regarding the equation in calculus.
ALS 8pp RS:HS 2.3

44. 1812–8 or later To Charles Babbage n.p.
Giving equations.
ALS 1p RS:HS 2.358

45. [1812–8] To Charles Babbage n.p.
About the mathematical formulation for the summation of any series.
C 1p RS:HS 20.4

46. 1812–8 To John William Whittaker n.p.
Angry at self and whole world except JW. Has been drunk and travelling. Met brilliant but 'not beautiful' woman. Wonders why he left Cambridge; requests news and offers advice.
ALS 3pp St. John's College, Cambridge

47. 1812–8–6 From James Grahame Whitehill, Glasgow
Describes his experiences in court; talks about their mutual friend J. W. Whittaker [letter completed 1812–8–11].
AL 4pp inc RS:HS 8.245

48. 1812–8–14 To Charles Babbage n.p.
Further comment on the calculus equation sent by CB.
ALS 2pp RS:HS 2.4 (C: RS:HS 20.2)

49. [1812]–8–21 From Sophia Baldwin Slough
Asks JH to come to Slough to visit Alexander Herschel before he leaves.
ALS 4pp Harvard: Houghton AAH 67m–67 (42)

50. 1812–8–23 From James Grahame Whitehill, Glasgow
Philosophizes about life [letter completed 1812–9–11].
ALS 4pp RS:HS 8.246

51. 1812–10–21 From Sophia Baldwin London
Requests JH obtain a paper by Joseph Banks for William Herschel.
ALS 1p Harvard: Houghton AAH 67m–67 (42)

52. [1812]–11–2 From J[ohn] Brass Trinity College
Although nominated for the Analytical Society by Edward Bromhead, JB declines membership. Praises the society as 'highly calculated to the promotion of mathematical literature.'
ALS 2pp BM Add 37182 f5

53. 1812–12–1 To William Herschel Cambridge
Two weeks of examinations begin tomorrow. Experiments writing with glass pen. Describes glassmaker's methods. Suggests application to micrometers.
ALS 4pp TxU:H/L-0550; Reel 1053

54. 1812–12–22 From Charles Babbage Bath
Sending details of some new theorems he has come across.
ALS 4pp RS:HS 2.5

55. 1812–12–22 From James Grahame Whitehill, Glasgow
About the lady JG intends to marry; also JG's law practice.
ALS 4pp RS:HS 8.247

56. 1813–1–1 From Alexander Rogers Charlotte St., Leith
Dr. [Charles] Burney offered him a position at his academy. He will, however, remain at the shipping company.
ALS 2pp RS:HS 14.397

57. 1813–1–28 From James Grahame Edinburgh
Congratulates JH on becoming Senior Wrangler.
ALS 2pp RS:HS 8.248

58. 1813–2–8 To Charles Babbage n.p.
Regarding the equations in his memoir on the series of tangents. Queries about the Analytical Society.
ALS 4pp RS:HS 2.6 (C: RS:HS 20.5)

59. 1813–2–19 From Charles Babbage [Cambridge]
Answers to JH's questions on equations. Regarding the printing of his memoirs. Recent meeting of the Analytical Society.
ALS 4pp RS:HS 2.7

60. 1813–2–20 To John William Whittaker n.p.
Philosophizes about society, college, time passing, and the 'fog' of present life. Wants to know all that is going on at Cambridge. Sends fond greetings to friends. [Letter completed 1813–2–22].
ALS 4pp St. John's College, Cambridge

61. 1813–2–25 To Charles Babbage n.p.
Further equations discussed. Has purchased more books. Regarding the printing of the memoir. [Letter continued on 1813–3–2.]
ALS 4pp RS:HS 2.8 (C: RS:HS 20.6)

62. 1813–2–26 From John William Whittaker St. John's College, Cambridge
Hopes JH settles in Cambridge. News of their fellows. Charles Babbage's lecture popular. Includes parody on [Sir Walter] Scott.
ALS 3pp RS:HS 18.238

63. 1813–3–9 From James Grahame Edinburgh
Plans for coming to London.
ALS 2pp RS:HS 8.249

64. 1813–3–28 From James Grahame Stoke Newington
Writes to JH to encourage him to come to London.
ALS 2pp RS:HS 8.250

65. 1813–4 From Charles Babbage [Cambridge]
Comments regarding the preface to their memoir. Further comments on JH's equations.
ALS 4pp RS:HS 2.9

66. 1813–4–14 or earlier From James Grahame Edinburgh
Writes of his journey back to Edinburgh, and of his intended wife.
ALS 3pp RS:HS 8.251

67. 1813–5–1 From Charles Babbage [Cambridge]
Memoir is being printed. Gives some more strange theorems for JH's comments.
ALS 4pp RS:HS 2.10

68. 1813–5–4 To Charles Babbage n.p.
Regarding his memoir: would like to see part of CB's notes.
ALS 3pp RS:HS 2.11 (C: RS:HS 20.8)

69. 1813–5–4 From John William Whittaker St. John's College, Cambridge
Discovered JH was elected fellow. Reports vandalism at Trinity. Asks about [James] Grahame. Reports on Mr. Jephson's treatment of Charles Babbage.
ALS 3pp RS:HS 18.239

70. 1813–5–22 To John William Whittaker Slough
Warmed by drink and society of 'charming' girls, JH is thinking of Whittaker, and advises him to study less. JH is enjoying reading more than did in school. Sends other news.
ALS 3pp St. John's College, Cambridge

71. 1813–5–25 From Charles Babbage [Cambridge]
Is sending copy of part of the memoir requested by JH. Gives a further theorem for JH's consideration.
ALS 3pp RS:HS 2.12

72. 1813–5–25 From James Grahame Edinburgh
Law business is good; JG's intended desires to meet JH.
ALS 3pp RS:HS 8.252

73. 1813–6–1 or later From Charles Babbage [Cambridge]
Is sending the first 12 pages of JH's memoir. Is glad to hear that he has material for a further memoir. Gives equations for JH's comments.
ALS 3pp RS:HS 2.13

74. 1813–6–14 From Alexander Rogers Charlotte St., Leith
Congratulates JH on his progress in mathematics. Discusses forces of attraction and his employment at the shipping company.
ALS 3pp RS:HS 14.398

75. 1813–6–21 From James Grahame Edinburgh
Asks JH if he would read and criticize a manuscript of JG's.
ALS 1p RS:HS 8.253

76. 1813–6–[26?] From [John William Whittaker] St. John's College, Cambridge
Comments on JH's romance. Hints at JW's own. Notes college fervor for religious proselytizing in foreign lands. Asks about 'Catholic question.' Quotes censored epigram written on royal tombs. Charles Babbage very ill.
AL 4pp RS:HS 18.240

77. 1813–6–27 To Charles Babbage Slough
Further regarding theorems recently discussed. Regarding the preface for JH's and CB's *Memoirs of the Analytical Society.*
ALS 4pp RS:HS 2.14 (C: RS:HS 20.9)

78. 1813–6–30 From Charles Babbage [Cambridge]
Where to find further information on functional equations. Gives further equations for JH's comments. Regarding the proposed abridgement of J. B. Delambre's work on astronomy.
ALS 2pp RS:HS 2.15

79. 1813–7–1 From James Grahame Edinburgh
Note to accompany JG's manuscript [see JG's 1813–6–21].
ALS 1p RS:HS 8.254

80. 1813–7–2 To John William Whittaker Slough
Comments on decaying British domestic affairs. Deplores religion 'as established by law' in contrast with religion 'established by nature.' Has felt sombre lately. Comments on music as relief from mathematics.
ALS 4pp damaged St. John's College, Cambridge

81. [1813?]–7–7 To John William Whittaker [Slough]
Comments on unfortunately aristocratic world, JW's description of Hamlet, current investigations and perhaps too deep involvement in scientific pursuits. Is correcting friend's novel (*The Victim of Villainy* or *Travelling Tetrahedrons*). Asks JW to find mineralogical specimen.
ALS 4pp St. John's College, Cambridge

82. 1813–7–14 From James Grahame Whitehill, Glasgow
Thanks for the criticism [see JG's 1813–6–21]; is considering publication.
ALS 2pp RS:HS 8.255

83. 1813–7–25 To Charles Babbage Slough
Regarding his new researches into the theory of determining functions from given conditions. The printing of his memoir is proceeding slowly.
ALS 3pp RS:HS 2.16 (C: RS:HS 20.10)

84. 1813–7–28 From James Grahame Edinburgh
Has had manuscript [see JG's 1813–7–14] privately printed with the title *Crispin*, only six copies and not for sale.
ALS 2pp RS:HS 8.256

85. 1813–8–2 To John William Whittaker n.p.
Urges to send copies of [S. F.] Lacroix's work and École polytechnique journal. James Grahame is successfully studying law. Grahame's father is staying with JH and his presence makes JH 'better and wiser.' Has been thinking much. Hopes JW will check on printing of *Memoirs of the Analytical Society*. [Letter completed 1813–8–3.]
ALS 3pp St. John's College, Cambridge

86. 1813–8–6 To John William Whittaker n.p.
Gives further instructions on sending books [see JH's 1813–8–2]. Comments on 'refractory' nature of religious thought. Is not in love, but teases JW about frequent visits to Thackerays. Recounts current heavy involvement in music, asking [Thomas?] Greenwood for help with harmonies.
ALS 3pp St. John's College, Cambridge

87. 1813–8–16 To Charles Babbage n.p.
Has been to London and procured the books mentioned by CB. Mentions differential equations of P. S. Laplace. Regarding some of CB's theorems. Problem of law of chance. Printing of his memoir.
AL 4pp inc RS:HS 2.17 (C: RS:HS 20.11)

88. 1813–8–16 To John William Whittaker n.p.
Asks JW to check on books apparently sent [see JH's 1813–8–2 & 1813–8–6], which have not yet arrived.
ALS 1p St. John's College, Cambridge

89. [1813–9] To Charles Babbage Slough
Sends a number of theorems and integrals.
C 4pp inc RS:HS 20.12

90. 1813–9–6 From James Grahame Edinburgh
Encourages JH to get rid of his hatred of infants.
ALS 3pp RS:HS 8.257

91. 1813–9–24 or later To Charles Babbage n.p.
Orders equipment from London through CB.
C 2pp RS:HS 20.14

92. 1813–9–24 From Charles Babbage Teignmouth
Will be visiting London shortly and hopes to purchase some experimental equipment to carry out experiments with heat.
ALS 2pp RS:HS 2.18

93. 1813–10–[13] To Charles Babbage n.p.
Would like to see him as soon as possible so they can prepare the preface of their memoir for the printer.
ALS 3pp RS:HS 2.19 (C: RS:HS 20.13)

94. 1813–11–5 From James Grahame Edinburgh
Encourages JH's decision to become a barrister; comments on the joys of marriage.
ALS 3pp RS:HS 8.258

95. 1813–11–10 From Mary Pitt Herschel Slough
Please consider William Herschel's opinions on JH's career decision. MPH favors church, but will accept whatever makes JH happy. Will send money soon for [JH's tutor] T. W. Hornbuckle.
ALS 1p & C 2pp TxU:H/M–0619; Reel 1086

96. 1813–11–10 From William Herschel Slough
Responds to JH's criticism of churchmen. Criticizes law and defends church as careers. Explains why WH never discussed religious opinions with JH.
C 4pp RAS:WH Archive 1/11.1; Reel 12 Lubbock, *Herschel Chronicle*, 348–50

97. 1813–11–14 From Mary Pitt Herschel Slough
Assures JH that exchange of letters with William Herschel did not produce alienation. Encourages both to meet for discussion. JH's cousin [Mary Baldwin] anxiously awaits reply from JH.
ALS 1p (on William Herschel's 1813–11–14) RAS:WH Archive 1/11.2 verso; Reel 12

98. 1813–11–14 From William Herschel Slough
Apologies for any bitterness caused by differences of opinions. Willing to listen to JH's reason for preferring law over church as career, but must postpone arguments until later. [Typewritten copy includes annotation by unknown hand that JH's letters were not preserved in this 'only serious conflict of wills between the father & son.']
ALS 1p RAS:WH Archive 1/11.2; Reel 12 (C: 1/11.3) Lubbock, *Herschel Chronicle*, 351

99. 1813–12–4 From James Grahame 4 Shandwick Place, Edinburgh
About the opportunities in law.
ALS 4pp RS:HS 8.259

100. 1814–1–10 From James Grahame Edinburgh
Encourages JH to come to visit; more about the study of law.
ALS 3pp RS:HS 8.260

101. 1814–1–10 To John William Whittaker n.p.
Obliquely, wishes luck and offers advice on exams. Asks for news of Cambridge scandals. Is entering law school. Encloses mathematical formula for which they had earlier searched.
ALS 3pp St. John's College, Cambridge

102. 1814–1–12 To Charles Babbage n.p.
Has had the bill for the printing of their memoir. Ways of raising money to pay for the printing. Comments on one of CB's equations. Will be entering Lincoln's Inn in a week's time.
AL 4pp inc RS:HS 2.20 (C: RS:HS 20.15)

103. 1814–1–21 From Charles Babbage Peterhouse College
Regarding the proposed method of raising funds to pay for printing of the memoir. Justifies his equation queried by JH. Would he purchase certain chemicals for him.
ALS 4pp RS:HS 2.21

104. 1814–1–26 To Charles Babbage 18 Henrietta St., Covent Garden
Chemical news: decomposition of 'sulphuret of carbon' by J. J. Berzelius and Alexander Marcet. Analysis of 'azotane' and discovery of 'iode' by Humphry Davy. Congratulations on results of CB's examination.
C 3pp † TxU:H/L-0047; Reel 1054

105. 1814–1–26 To Charles Babbage n.p.
Is sending chemicals. Regarding Humphry Davy's experiments with 'azotane'. Davy's paper on iodine.
ALS 4pp RS:HS 2.22 (C: RS:HS 20.16)

106. 1814–1–28 From John William Whittaker St. John's College, Cambridge
Results of JW's exams. JH's advice in science was too modern to help. Reports local scandals among examiners. Not speaking to James Grahame. Going to 'Babylon' on Tuesday.
ALS 4pp RS:HS 18.241

107. [1814–2?] To William Herschel London
Working in law office of [F. W.?] Sanders. Attended R.S.L. last night. T. W. Hornbuckle offered JH position as subtutor at St. John's College, but JH declined. Describes paper on minerals that JH submitted through E. D. Clarke to Geological Society. Charles Babbage just made creative mathematical discovery.
ALS 3pp TxU:H/L-0546.1; Reel 1053

108. 1814–2–20 From Alexander Rogers Charlotte St., Leith
Congratulates JH on the formation of the [Analytical] Society, which will let Britain 'take the lead' in mathematics. Discusses analysis of a radical.
ALS 3pp RS:HS 14.399

109. 1814–2–26 From James Grahame Edinburgh
Thanks for essays JH sent; John Playfair was impressed by them. JG is writing material in favor of a Glasgow/Edinburgh canal.
ALS 2pp RS:HS 8.261

110. 1814–4 From Charles Babbage Rowden
Will be in town and requests addresses of shops where certain experimental apparatus can be purchased. Send [H. E.?] Roscoe's direction.
ALS 2pp RS:HS 2.23

111. 1814–4–18 From John William Whittaker Nentbury Hall, Alston, Cumberland
At home of student in highland moors along Nent River. Describes host family and barren surroundings. Greets all at St. John's. Asks about JH's chemistry lectures.
ALS 4pp RS:HS 18.242

112. 1814–6–6 From James Grahame Edinburgh
JG to be a father, has bought a new house, and urges JH to marry.
ALS 4pp RS:HS 8.262

113. 1814–7–4 From Charles Babbage Totnes
Has been working on analysis. Gives equations he has solved. Has met a mineralogist.
ALS 3pp RS:HS 2.24

114. 1814–7–19 From James Grahame Whitehill, Glasgow
Family news; encourages JH to read Jane Austen's novels.
ALS 3pp RS:HS 8.263

115. 1814–8 From Charles Babbage [Chudleigh]
Sends equations in operational calculus.
ALS 4pp RS:HS 2.27

116. 1814–8–1 From Charles Babbage Chudleigh
Their 'Analytical' has not been reviewed in the *Edinburgh Review*. Has married. Looking out for a situation. Gives an equation dealing with relative functions.
ALS 3pp RS:HS 2.25 Hyman, *Charles Babbage,* 31 inc

117. 1814–8–4 To Charles Babbage Slough
Feigned reproach for CB's return to 'the Analytics.' Inquires about CB's 'Devonshire Ghost at Chudleigh.' Results of JH's chemistry experiments. Derides scientists who promote theory of 'akasch' as the one and only form of matter. Met Alexander Marcet in Greenwich.
C 3pp TxU:H/L-0048; Reel 1054

118. 1814–8–4 To Charles Babbage n.p.
Comments on the equation sent by CB. Has been experimenting with potassium.
ALS 4pp RS:HS 2.26 (C: RS:HS 20.17)

119. 1814–8–7 To Charles Babbage Shaw
Regarding CB's marriage. Possibility of a suitable situation for him.
ALS 3pp RS:HS 2.28 (C: RS:HS 20.18)

120. 1814–8–10 From Charles Babbage Chudleigh
Temper and stinginess of CB's father. Seeking job in mining through Mr. Champernowne, local mineralogist. Describes CB's elopement, marriage by former tutor, and new residence. Considered career in Church when CB had prospect of inheriting fortune from CB's father, but that is no longer possible. Gives an equation for discussion.
ALS 8pp RS:HS 2.29 Hyman, *Charles Babbage,* 31–3 inc

121. 1814–9–4 From James Grahame Whitehill, Glasgow
Family and law business news.
ALS 3pp RS:HS 8.264

122. 1814–9–18 From James Grahame Whitehill, Glasgow
Continues to encourage JH toward law and talks about some of JG's cases.
ALS 3pp RS:HS 8.265

123. 1814–9–20 To Charles Babbage St. John's College, Cambridge
Gratitude for CB's friendship. Chemical supplies and instruments. Describes synthesis of 'Sulphurane' [S_2Cl_2] and other experiments. Theories of combustion and acids. [Letter continued on 1813–9–21:] JH avidly studying law, against father's wishes.
AL 4pp inc RS:HS 2.30 (C: RS:HS 20.19 & TxU:H/L-0049.1; Reel 1054)

124. 1814–9–25 From Susan White Shaw
Invites JH, her godson, to sit for new likeness by same artist who drew him before. Cannot accept JH's mother's invitation to Slough.
ALS 3pp † RS:HS 18.226 (Abs: TxU:H/M–0580.1; Reel 1093)

125. 1814–10–5 From Alexander Rogers Charlotte St., Leith
Describes an oil lamp that makes efficient use of all the oil placed in it.
ALS 3pp RS:HS 14.400

126. 1814–10–25 To Charles Babbage St. John's College, Cambridge
CB has founded 'a calculus totally new and immensely powerful.'
C 2pp TxU:H/L-0049.2; Reel 1054

127. 1814–10–25 To Charles Babbage St. John's College, Cambridge
Praising CB's functional equations. How to become an F.R.S. Conditions at Slough. Hears that he is coming to live in London.
ALS 4pp RS:HS 2.31 (C: RS:HS 20.20)

128. 1814–11 To Charles Babbage Cambridge
Has executed all of CB's commissions. Hopes to meet Charles Shadwell when in town. Is in the throes of exams. Thinking of taking up mathematics. Relates further chemical experiments.
ALS 3pp RS:HS 2.33 (C: RS:HS 20.21)

129. 1814–11 To [?] n.p.
Finds no experimental evidence connecting index of capillary action to index of refraction.
TC 1p inc TxU:H/L-0640; Reel 1089

130. 1814–11–14 From Charles Babbage Arundel St.
Thanking him for the paper on analysis. Has been to see Sir Joseph Banks. Hopes to see JH soon.
ALS 3pp RS:HS 2.32

131. 1814–12–12 From Charles Babbage Arundel St.
Why hasn't he written? Further regarding Charles Shadwell's advice. Is preparing his paper on functions so that JH or William Herschel can communicate it to the R.S.L.
ALS 3pp RS:HS 2.34

132. [1815 to 1829] To Miss C. [Slough]
Expresses gratitude for watching over JH's interests in his absence.
ALS 2pp BostonPubLib Ms.Eng.489(2)

133. 1815 From James Grahame n.p.
Sends best wishes on JH returning to St. John's to teach; about books JG has read and some of his cases.
ALS 3pp RS:HS 8.270

134. 1815–1–3 From James Grahame 29 Northumberland St., Edinburgh
Upbraids JH for not writing, and announces the birth of his daughter, Anne.
ALS 2pp RS:HS 8.266

135. 1815–2–11 From Charles Babbage n.p.
Explaining his new theory of the atmospheres of the sun, planets, and stars.
ALS 3pp RS:HS 2.35

136. 1815–2–16 To Charles Babbage n.p.
Regarding CB's theory of the atmosphere of the sun, planets, and stars.
ALS 4pp RS:HS 2.36 (C: RS:HS 20.22)

137. 1815–3–8 From James Grahame Edinburgh
Talks about the law, especially now that JH has begun his study of it.
ALS 3pp RS:HS 8.267

138. 1815–3–15 From Charles Babbage Lucas St.
Regarding the vacant post of Professor of Chemistry at Cambridge. Has heard that the works of Leonhard Euler are being reprinted. Would he obtain the return of one of CB's books from Mr. Higman.
ALS 3pp RS:HS 2.37

139. 1815–3–23 To Charles Babbage Cambridge
Attending E. D. Clarke's mineralogy lectures. JH's analysis of dioptase.
C 1p TxU:H/L-0050; Reel 1054

140. 1815–3–23 To Charles Babbage n.p.
Regarding the vacant professorship of chemistry at Cambridge. Recent chemical experiments. Has obtained the book from Mr. Higman. Mr. Pulling was here recently.
ALS 3pp RS:HS 2.38 (C: RS:HS 20.23)

141. 1815–5–2 From Thomas W. Hornbuckle St. John's College, Cambridge
Offering JH the position of assistant tutor at Michaelmas. [Draft from JH declining.]
ALS 2pp RS:HS 9.482

142. 1815–7–31 From William Wallace Royal Military College, Bagshot
Sends paper on fluxions written for *Edinburgh Encyclopaedia*. Uses new formulas, dispensing with binomial theorem. Brings logarithmic and exponential formulae within power of common algebraic processes.
ALS 2pp RS:HS 18.21

143. 1815–9–10 To Charles Babbage n.p.
Hopes to call on him Monday. Is just off to Brighton for medical reasons. Will return to Cambridge later.
ALS 2pp RS:HS 2.39 (C: RS:HS 20.24)

144. 1815–9–23 To William Wallace Brighton
WW's elegant presentation of doctrine of limits is best basis for elementary treatise but not for extensive work, because it involves imaginary functions. Charles Babbage is making progress in theory of functional equations.
C 2pp RS:HS 20:25

145. 1815–9–25 From Charles Babbage n.p.
Is sending a problem in analysis for his comments. Gives solution to functional equations of the second degree. Hopes to see him when he passes on his way to Cambridge.
ALS 3pp RS:HS 2.41

146. 1815–9–25 To Charles Babbage 7 Charles St., Brighton

Idling time at Brighton. Gives equations for CB to study. Regarding the equipment he needs for experiments. Will not be able to return to professional activities for some time.
ALS 4pp RS:HS 2.40 (C: RS:HS 20.26)

147. 1815–10–7 To Charles Babbage [Brighton]
Answer regarding CB's problem in analysis. Will visit him soon. Getting apparatus ready to be moved to Cambridge.
AL 4pp RS:HS 2.42 (C: RS:HS 20.27)

148. 1815–10–11 From James Grahame Edinburgh
Complains that JG had to discover from strangers that JH was ill.
ALS 1p RS:HS 8.268

149. 1815–10–28 From Charles Babbage Devonshire St.
Wished JH had visited him when in London. Gives some functional equations for his comments. Wants a form in order that he (CB) may be elected to the R.S.L.
ALS 3pp RS:HS 2.43

150. 1815–11–3 From James Grahame Edinburgh
Distressed to hear JH is giving up law; family news.
ALS 4pp RS:HS 8.269

151. 1815–11–6 To Charles Babbage n.p.
Reasons why he could not visit him when in London. Gives views on a paper of CB's. Fears he must give up the profession.
AL 4pp inc RS:HS 2.44 (C: RS:HS 20.28)

152. 1815–11–9 From Charles Babbage n.p.
Regarding his own functional theories. Let him know when next in town.
ALS 3pp RS:HS 2.45

153. 1815–11–13 From Charles Babbage Devonshire St.
Newman has sent the apparatus. Problems in calculus.
ALS 4pp RS:HS 2.46

154. 1815–11–15 From Charles Babbage Devonshire St.
Giving form for recommendation to be elected to the fellowship of the R.S.L. Has been offered space in a magazine for reviewing mathematical subjects. Suggesting anonymous work on analytics by him or JH.
ALS 4pp RS:HS 2.47

155. 1815–11–20 To Charles Babbage n.p.
Asking him to get certificate signed for his election to the R.S.L. Has had no time for mathematics recently. Will be prepared to write anonymous article against analysts.
ALS 1p RS:HS 2.48

156. 1815–11–22 To Charles Babbage [Cambridge]
Supports CB for membership in R.S.L. Little time recently for mathematics, but is quite upset with analysts.
C 1p RS:HS 20.29 (C: RS:HS 2.48)

157. 1815–11–23 From Joseph Banks Soho Square
Has received his paper, which will be laid before the R.S.L. as soon as he has an opportunity. Is obliged to him for compressing the matter to so small a compass.
ALS 1p RS:HS 5.15

158. 1815–12–8 From Charles Babbage Devonshire St.
Soon leaving for a short visit to Devon. Hopes he will call when passing through London. New means of preventing explosions in mines. Gives an equation for JH's comments.
ALS 4pp RS:HS 2.50

159. 1815–12–18 From Charles Babbage n.p.
Will see him on Thursday. Has paper ready for the R.S.L.
ALS 1p RS:HS 2.52

160. 1815–12–18 To Charles Babbage n.p.
Is in the midst of examining. Will call when in town this week. Has done nothing in analysis lately.
ALS 3pp RS:HS 2.51 (C: RS:HS 20.30)

161. 1815–12–19 From Ann Wilson 29 Kensington Square
Expects JH for dinner Friday. JH's old friend Ianetta learned that she is descended from Scotland's Queen Annabella.
ALS 2pp RS:HS 18.415

162. 1815–12–28 From Charles Babbage Totnes
Sending problems in functional analysis. Please send his manuscript on functions to Devonshire Place.
ALS 3pp RS:HS 2.53

163. [1816] From Edward Daniel Clarke n.p.
Reports account given at [Sebremnitz] of ore that JH analyzed.
C 1p TxU:H/M–0141; Reel 1087

164. [1816] To Edward Daniel Clarke n.p.
Reports on some chemical results obtained by JH.
ALS 1p Sydney Ross Collection

165. [1816] To Edward Daniel Clarke n.p.
Praises EC for 'amazing experiments' published in [William T.] Brande's journal [*Quarterly Journal of Science and Art*]. Wishes EC's publication had appeared in *R.S.P.T.* Has an experiment to show EC.
ALS 2pp Sydney Ross Collection S. Ross, 'Unpublished Letters ... to Clarke,' *Bulletin for the History of Chemistry, 11* (1991), 85

166. [1816] To Edward Daniel Clarke n.p.
Describes chemical experiments carried out by JH on corundum. Attempts to interpret these experiments.
ALS 1p Sydney Ross Collection

167. [1816] To Edward Daniel Clarke n.p.
Mentions some chemical experiments on which JH has worked; hopes EC will exhibit his experiments for R.S.L.
ALS 1p Sydney Ross Collection

168. 1816–1–6 To Charles Babbage Slough
Does not have CB's manuscript on functions in his possession. Regarding the translations of S. F. Lacroix. Has done no work while at home.
ALS 4pp RS:HS 2.54 (C: RS:HS 20.31)

169. 1816–2–7 To Charles Babbage n.p.
Is sorry that CB's paper could not be printed in the recent issue of the *R.S.P.T.* Discusses problem in analysis. [Note dated [1816–]2–7]: has been waiting for Mr. Deighton's estimate of printing.
ALS 4pp RS:HS 2.55 (C: RS:HS 20.32)

170. 1816–2–8 From James Grahame Edinburgh
JG's wife recovering from serious illness; JG writing on population.
ALS 4pp RS:HS 8.271

171. 1816–2–21 From Charles Babbage n.p.
What is he to do with his work on S. F. Lacroix. Regarding the R.S.L. and the printing of CB's paper on functions. Is sending an abridged version of this paper to JH.
ALS 3pp RS:HS 2.56

172. 1816–2–26 To Charles Babbage n.p.
Has received parcel and letter. Has not had much time to give attention to the manuscript yet. On the R.S.L. and the printing of long papers. Hopes to have CB's translation of S. F. Lacroix soon. Regarding one of his equations.
ALS 2pp RS:HS 2.57 (C: RS:HS 20.33)

173. 1816–3–9 From Charles Babbage n.p.
Is thinking of applying for the position of Professor of Mathematics at East India College. Has written to William Herschel and others for references. George Peacock is also a candidate. Regarding functional equations.
ALS 3pp RS:HS 2.58

174. 1816–3–13 To Charles Babbage n.p.
Regrets he cannot assist CB over the position at East India College. William Herschel has less influence over Sir Joseph Banks than CB thinks. Regarding his paradoxes in functional equations.
ALS 2pp RS:HS 2.59 (C: RS:HS 20.34)

175. 1816–4–1 From James Grahame Whitehill, Glasgow
Family news; agrees to forward any manuscript JH sends to John Playfair.
ALS 1p RS:HS 8.272

176. 1816–4–20 To Edward Daniel Clarke Slough
Thanks EC for his interest in a paper JH submitted to the Geological Society. JH has come to question some of his observations and to recognize that J. J. Berzelius anticipated JH's recognition of the acid nature of silica.
ALS 3pp Sydney Ross Collection

177. 1816–4–24 From Henry Warburton 45 Cadogan Place
Suggests changes for JH's paper on Swedish felspar submitted to Geological Society. Confusion between this and other silicates. Compares JH's analysis to that of Wilhelm Hisinger and others.
ALS 4pp RS:HS 18.36

178. 1816–4–27 From Charles Babbage Totnes
Regarding his application for the position of professor at East India College. Gives functional equations for JH's comments.
ALS 3pp RS:HS 2.60

179. 1816–6 From Charles Babbage Torquay
Sorry he has had all the trouble with CB's translation. Father is ill. Will move shortly to Torquay. Regarding the printing of mathematical notations.
ALS 3pp RS:HS 2.61

180. 1816–6–3 From Henry Warburton 45 Cadogan Place
Discussed JH's proposed felspar paper [see HW's 1816–4–24] with W. H. Wollaston. Suggests using Wollaston's and [recently deceased Smithson] Tennant's unpublished method to test for alkalis and give Tennant credit for it. James Smithson first coined term 'silicate'.
ALS 3pp RS:HS 18.37

181. 1816–6–9 From Charles Babbage Torquay
Regarding his translation of S. F. Lacroix. Regarding the notation of functions.
ALS 3pp RS:HS 2.62

182. 1816–7–10 From Charles Babbage Torquay
Is sending his paper on functions. Gives some functional equations and their solutions. Will be staying at Torquay until September.
ALS 3pp RS:HS 2.63

183. 1816–7–14 To Charles Babbage Slough
Has had little leisure to write. Been working on S. F. Lacroix. Regarding the appendix to Lacroix. Is going to the Isle of Wight. Hopes to write a work on analytics. Has heard that CB is writing articles for the journal of the Royal Institution. Would like to write articles for an encyclopaedia.
ALS 4pp RS:HS 2.64 (C: RS:HS 20.35)

184. 1816–7–20 From Charles Babbage Torquay
Regarding JH's plans for the third part of S. F. Lacroix. Royal Institution and CB's articles for their journal. Hopes that JH will not write his book on Newton yet. Is hoping to publish a work on the Calculus of Functions. Hopes to meet him when in Devon.
ALS 3pp RS:HS 2.65

185. [1816]–7–20 From John William Whittaker Alconbury Hill
Returning home; father dying. News of Cambridge friends. Unable to send information JH requested about university at Ryde. William Whewell upset, because Whewell's father died and was buried before Whewell was notified.
ALS 3pp RS:HS 18.243

186. 1816–[8] From Charles Babbage Torquay
Is glad to hear that he will be coming to Torquay. Regarding his new method of solving functions.
ALS 2pp RS:HS 2.66

187. 1816–8–6 From George Peacock Ferrara
News of his travels in Italy. Has lost their trunks.
ALS 4pp RS:HS 13.245

188. 1816–8–14 To [Henry] Wilkinson Bath
Gives necessary information for printing end of JH's and Peacock's translation of [S. F.] Lacroix's text on the calculus. Requests news about Cambridge. Tells of JH's travels through England, and Charles Babbage's interests.
ALS 3pp St. John's College, Cambridge

189. 1816–8–15 From James Grahame Edinburgh
Family news; is preparing his writings on population [see JG's 1816–2–8] for publication.
ALS 4pp RS:HS 8.273

190. 1816–8–22 From Thomas Smart Hughes Cambridge
Regrets to hear JH's father is not well and hopes the sea air will be of benefit. Would like to accompany him on the Continent whenever it suits him. Gives all the Cambridge news.
ALS 4pp RS:HS 10.215

191. 1816–8–29 From John William Whittaker St. John's College, Cambridge
Manuscripts [1816 translation of S. F. Lacroix's *Traité du calcul différentiel et du calcul intégral*] are ready. Send notes and missing appendix quickly, or press will stop.
ALS 1p RS:HS 18.244

192. 1816–9–2 To John William Whittaker n.p.
Describes declining health but calm mind of JH's father. Visit with Charles Babbage cheered JH's depressed state. Wants to return to joy of speculation, but lacks energy. Writes and requests news about mutual friends.
ALS 3pp St. John's College, Cambridge Hyman, *Charles Babbage,* 40 inc

193. 1816–9–11 From James Grahame Edinburgh
Family news and the law.
ALS 3pp RS:HS 8.274

194. 1816–9–17 From Richard Jones 29 Ship St., Brighton
Regrets delay in writing as he was uncertain where JH would be. Wishes JH would come and stay with him to relieve his melancholy position. He has accepted a curacy at Ferring.
ALS 4pp RS:HS 10.345

195. 1816–9–26 From Sir William Watson Dawlish
Enjoyed visit of William Herschel and JH. Sorry about accident with coach. Discusses Horne Tooke's search for metaphysics through etymology. Extensive analysis of nature of verbs. Notes [daughter] Joy's response to JH's Masonic sign.
ALS 3pp RS:HS 18.109

196. 1816–9–27 From Charles Babbage Dudmaston Hall, Bridgenorth
Temporarily at Bridgenorth as his wife's father has died. Must delay his visit to London. Has been interesting his brother-in-law in gas lights.
ALS 2pp RS:HS 2.67

197. 1816–10–10 To Charles Babbage Slough
Leaving Cambridge shortly, probably for good. Will be under his father's directions observing the stars. Has sent the appendix to S. F. Lacroix. Regarding JH's error in the printing of an equation. Gives an equation for CB to comment on.
ALS 4pp RS:HS 2.68 & 20.36

198. 1816–11 To Edward Daniel Clarke London
Sending EC a specimen that W. H. Wollaston describes as '*the most intractable substance by fire*' he has ever seen. Discusses its properties and possible chemical composition.
ALS 2pp Sydney Ross Collection

199. 1816–11–14 From George Peacock n.p.
Regrets not having notified JH of the arrival of the manuscripts. Comments on them and the other papers. An algebra is much wanted. Has been much occupied with college and public business.
ALS 3pp RS:HS 13.246

200. 1816–11–16 From Sir William Watson Bath
Thanks JH for William Herschel's gift. Wishes Horne Tooke had lived to publish third volume. E. D. Clarke's experiment with oxygen-hydrogen fuel mix to blowpipe was anticipated by Americans. Cites 1803 journal from New York.
ALS 2pp RS:HS 18.110

201. 1816–11–25 From James Grahame Edinburgh
JG announces the death of his daughter Anne, and the publication of his book on population.
ALS 1p RS:HS 8.275

202. 1816–12–1 From Edward Daniel Clarke Cambridge
Chemical experiments with EC's blowpipe. Discovered cause of blowpipe explosions. Thank Mrs. Lowry for carbon substance sent to EC. Sent metal of Baryta to W. H. Wollaston. Suspects metallic nature of carbon and diamond.
ALS 4pp RS:HS 5.256 (C: TxU:H/M–0142; Reel 1087)

203. 1816–12–3 From George Peacock Trinity College, Cambridge
Book will be published on 13 Dec. Also printing JH's notes. Will extend to 710 pages. Comments on the errors and gives ideas for future publications. Reform is afoot in the University.
ALS 4pp RS:HS 13.247

204. 1816–12–6 From John Galt Lindsey Place, Chelsea
Has dispatched the manuscripts of William Spence by coach, and encloses the key. Hopes JH will call on him when in town.
ALS 1p RS:HS 8.34

205. 1816–12–8 To Fearon Fallows Slough
Lauds FF's being chosen as mathematics examiner; hopes will be able to observe defects in Cambridge's current program and asks FF to press program of Analytical Society.
ALS? 3pp inc St. John's College, Cambridge

206. 1816–12–23 From Sir William Watson Bath
Thanks JH for cheeses and etymology letter. 'Substantive quality' of names. Agrees with [Thomas] Hobbes, questions sensible perceptions. Invites WH for visit as respite from bad health.
ALS 3pp RS:HS 18.111

207. 1816–12–24 To Charles Babbage n.p.
Has received a copy of their book. Sending a copy of his own paper printed in the *R.S.P.T.* Has prepared the way for CB's book on examples. Has been examining [John?] Spence's manuscript.
ALS 4pp RS:HS 2.69 (C: RS:HS 20.37)

208. 1816–12[?]–24 To Edward Daniel Clarke Slough
Returns with thanks EC's specimen of Island Crystal. Also discusses specimen of titanium lent by EC.
ALS 2pp Sydney Ross Collection

209. [1817 to 1822] From David Wilkie to Miss Wilson [24 Lower Phillimore Place, Kensington]
Please arrange for JH to join DW, Dr. [Thomas?] Chalmers, and possibly Captain Carnigh for dinner. Chalmers wanted to visit William Herschel in Windsor but cannot, and would at least like to meet JH before going to meet [William?] Wilberforce.
ALS 2pp RS:HS 18.250

210. 1817–1–2 From Charles Babbage Devonshire St.
Regarding a possible error by JH in one of his equations [see CB's 1816–10–10]. Proposes a plan for a new work on analysis.
ALS 3pp RS:HS 2.70

211. [1817–1–20] From Edward Daniel Clarke Cambridge
Regrets his blunder. Send him Wilson Lowry's address and he will write. His own recent experiments with platinum.
ALS 2pp RS:HS 5.257 (C: TxU:H/M–0143; Reel 1087)

212. 1817–1–21 From James Grahame 29 Northumberland St., Edinburgh
Much philosophizing about life, work, and happiness.
ALS 3pp RS:HS 8.276

213. 1817–1–25 To John William Whittaker Slough
Queries and news about friends at Cambridge. How is analytical movement proceeding and translation of [S. F.] Lacroix's treatise being received? Is writing a new algebra.
ALS 3pp St. John's College, Cambridge

214. 1817–1–30 From Charles Babbage Devonshire St.
Is glad JH is coming to town. Another son born. They must revive the Analytical. Explains a problem in functional analysis.
ALS 3pp RS:HS 2.72

215. 1817–1–30 To Charles Babbage n.p.
Will be coming to town shortly and hopes to stay with CB if

possible. They must get another volume of the transactions of the Analytical Society out.
ALS 4pp RS:HS 2.71 (C: RS:HS 20.38)

216. 1817–2–4 To William Whewell Slough
Is actively pursuing mathematical investigations. Plans to come to Cambridge soon for a short visit. Will follow this first letter by many more. Excavations by the Geological Society going on in JH's area.
ALS 4pp TC, Camb. Add. Ms.a.207[1] (draft: RS:HS 18.157 & C: RS:HS 20.39 & TC, Camb. 0.15.46[17] inc)

217. 1817–2–6 From James Grahame Edinburgh
The state of the country and friends, and please push JG's book.
ALS 3pp RS:HS 8.277

218. 1817–2–17 From Charles Babbage Devonshire St.
Has sent letters in a parcel. Regarding functional equations of the first order.
ALS 3pp RS:HS 2.73

219. 1817–2–17 From Charles Babbage Devonshire St.
Note to the effect that Dr.William Clarke of Trinity is a candidate for the Professorship of Anatomy. Hopes JH will support the Dr. [This letter is an appendix to CB's 1817–2–17]
ALS 1p RS:HS 2.74

220. 1817–2–24 From James Grahame Edinburgh
About JG's book on population and a pamphlet on the usury laws [letter completed 1817–2–26].
ALS 3pp RS:HS 8.278

221. [1817–2–25 or later] To James Grahame [Slough]
Discusses at some length the strengths and weaknesses of Lucan's *Pharsalia*. Mentions JG's pamphlet on usury laws, suggesting that the argument from analogy is especially precarious in political economy.
C 3pp RS:HS 20.52

222. 1817–2–27 From Charles Babbage Devonshire St.
Will send part of JH's manuscript and Hirsh's book. Has introduced E. F. Bromhead to the R.S.L.'s club and Sir Joseph Banks. Is about to draw up a sketch of the history of functions.
ALS 3pp RS:HS 2.75

223. 1817–3–4 From George Peacock Trinity College, Cambridge
Outlines the events at Cambridge and the defeat of the reform GP hoped to introduce in the mathematical examinations.
ALS 4pp RS:HS 13.248

224. 1817–3–6 From William Whewell Trinity College
Discusses university politics. Misses JH. Impressed by [Edward] Bromhead's mathematical abilities. Is considering translating a book by S. F. Lacroix. George Peacock's Tripos examination questions have caused controversy. Believes Peacock should have stressed applied over pure mathematics.
ALS 4pp RS:HS 18.158 Todhunter, *Whewell, 2,* 15–17

225. 1817–3–12 From Edward Daniel Clarke Cambridge
When his lectures have finished he will attend to JH's request regarding the inscription of Dr. G. H. Noehden. Hopes he will attend one of Sir Joseph Banks' Sunday evening gatherings and show the specimens he has given JH; also show them to W. H. Wollaston.
ALS 3pp RS:HS 5.258 (C: TxU:H/M–0144; Reel 1087)

226. 1817–3–17 From George Peacock Trinity College, Cambridge
Thanks for the two letters and the interesting matter they contain. Hopes JH will come when [William?] Clark[e]'s election takes place. Has put JH's name on the Syndics. Will not cease to exert himself in the cause of [mathematical] reform. Vice-Chancellor is against change.
ALS 4pp RS:HS 13.249

227. 1817–3–19 From G[eorge] C. Renouard Sidney Sussex College, Cambridge
Discusses inscriptions presented to Cambridge University by [Claudius] Bucharman. They are only facsimiles of those given to the Jews at Cochin, written in Hebrew, Arabic and Malabar.
ALS 2pp RS:HS 14.326

228. 1817–3–29 From Charles Babbage Devonshire St.
Encloses one of E. F. Bromhead's papers. Has collected all his letters relating to functions. Has he a copy of his paper 'Solutions...'? Regarding the properties of curves. His scheme for symmetrizing all functions.
ALS 3pp RS:HS 2.76

229. 1817–3–31 To Charles Babbage n.p.
Will go down to Cambridge to vote for William Clarke. Equation problem. Enquiry regarding paper on factorials by [Peter] Nic[h]olson. Regarding errors made by Clarke in his papers.
ALS 4pp RS:HS 2.77 (C: RS:HS 20.40)

230. 1817–4–3 To Charles Babbage Slough
Is sending CB all his letters dealing with functional equations. Comments on E. F. Bromhead's paper. Taylor's theorem. Is getting on well with his essay on exponential functions.
ALS 4pp RS:HS 2.78 (C: RS:HS 20.41)

231. 1817–4–10 From James Grahame Edinburgh
On JG's ignorance of mathematics.
ALS 3pp RS:HS 8.279

232. 1817–4–16 To Charles Babbage n.p.
Is returning E. F. Bromhead's paper. Matters are arising from this paper. Encloses one of his own dealing with the notation of functions. When should he set off to Cambridge to vote for William Clarke?
AL 4pp RS:HS 2.79 (C: RS:HS 20.42)

233. 1817–4–20 From John Clarke-Whitfeld Cambridge
Reports problem publishing JW's music. Relates adventures of churchmen and students at Cambridge.
ALS 3pp RS:HS 18.234

234. 1817–5–5 From James Grahame Edinburgh
Of poetry, usury laws, and the assizes [letter completed 1817–5–24].
AL 3pp RS:HS 8.280

235. 1817–5–12 From Charles Babbage Devonshire St.
Has been to Cambridge. Regarding JH's paper dealing with notation. Has been working on his history of functions.
ALS 3pp RS:HS 2.80

236. 1817–5–15 To Charles Babbage Slough
Will see him in London. Is getting on with his supplement. Has seen CB's latest paper in the *R.S.P.T.* Regarding Mr. Hirsch's theories on equations.
ALS 3pp RS:HS 2.81 (C: RS:HS 20.43)

237. 1817–5–26 From William Herschel Slough
Schedules time in London next week to cast two mirrors for 20-foot telescope. Details of mixture to be used for mirrors. Questions about casting process. Tells JH to consult 'Messr. Devey' for answers. Plans to visit Greenwich soon. Hand tremor makes writing difficult. Received letter from [Ragreneau?] inviting Herschel family to breakfast tomorrow.
ALS 3pp RAS:WH Archive 1/11.4; Reel 12

238. 1817–5–28 From Charles Babbage Devonshire St.
Come on Friday to meet E. F. Bromhead and discuss notation.
ALS 1p RS:HS 2.82

239. 1817–5–30 From George Peacock Trinity College, Cambridge
Rejoices at the prospect of seeing JH on 10 June, the day of the election. Many supporters of William Clark may be absent. Has been unwell himself lately. Hears nothing about JH's Algebra. [Richard?] Harraden is difficult to move. Tell Charles Babbage an anonymous donor has given £20,000 to Peterhouse.
ALS 3pp RS:HS 13.250

240. [1817–6] From George Peacock Trinity College, Cambridge
Has only just returned to the University. Not known at what time the election will take place. Has made inquiries about the bricks; they were given to the College by Sir J. Malcolm. Will get them copied. Just about to write to Charles Babbage who is coming down to take his degree.
ALS 2pp RS:HS 13.251

241. 1817–6–2 From George Peacock Trinity College, Cambridge
Encloses certificate that Charles Babbage sent him, with the amended signatures. Sorry it has given JH so much trouble. All rejoice to see JH on 10 June.
ALS 1p RS:HS 13.252

242. 1817–6–11 From David Brewster to William Herschel Venlaw by Peebles
Would like WH's son, JH, to write some articles for his *Edinburgh Encyclopaedia.* Would he help him on an article entitled 'Stars'?
ALS 2pp RS:HS 4.241

243. 1817–6–13 To John William Whittaker Slough
Is overwhelmed with work in algebra, studying S. F. Lacroix's work, grinding and polishing mirrors for JH's father. Sends thanks to Whitfield family for songs. Comments on controversial Cambridge news. Sends news of friends.
ALS 3pp St. John's College, Cambridge

244. 1817–6–15 From George Henry Noehden The Albany, London
Sending a query from Sir Gore Ouseley regarding some Persepolitan writing.
ALS 3pp RS:HS 13.141

245. 1817–6–18 To William Whewell Slough
Asks about WW's mathematical investigations. JH's book on algebra is progressing. Describes mathematical investigations by [Edward] Bromhead in calculus of factors. Misses Cambridge.
ALS 3pp TC, Camb. Add. Ms.a.207[2] (C: RS:HS 20.44)

246. 1817–6–19 To David Brewster Slough
Agrees to write several articles on mathematics, and one on astronomy based on William Herschel's work.
C 2pp RS:HS 20.45

247. 1817–7–7 From David Brewster Venlaw by Peebles
Is pleased that he will write articles for his *Edinburgh Encyclopaedia.* Would like him to write one on 'Stars.'
ALS 2pp RS:HS 4.242

248. 1817–7–13 From James Grahame Edinburgh
Note to announce the second edition of JG's pamphlet on the usury laws.
ALS 1p RS:HS 8.281

249. 1817–7–26 To William Whewell Slough
Is pursuing mathematical investigations with Charles Babbage. Asks WW to check whether George Peacock is progressing in printing the Supplement to Lacroix's volume.
ALS 3pp TC, Camb. Add. Ms.a.207[3] (C: RS:HS 20.46)

250. 1817–7–29 From G[eorge] C. Renouard Sidney Sussex College
Characters of Cochin inscription are common Arabic. GR, however, could not decipher them. Discusses various characters and includes a sample of some of them.
ALS 3pp RS:HS 14.327

251. 1817–8–1 To Charles Babbage n.p.
Has been trying his hand at functions; quotes some. Hopes to visit him at Torquay. Push on with the supplement.
ALS 4pp RS:HS 2.83 (C: RS:HS 20.47)

252. 1817–8–3 From George Henry Noehden The Albany, London
Is grateful for JH's assistance regarding the Persepolitan writing. G. C. Renouard's communication appears to be conclusive. Hopes he can convey the information to the Royal Society of Göttingen. Would like a copy of the whole inscription if this is not imposing too much.
ALS 3pp RS:HS 13.142

253. 1817–8–6 From Charles Babbage Devonshire St.
Is glad to hear he will be visiting Torquay. Regarding mathematical functions.
ALS 3pp RS:HS 2.84

254. 1817–8–7 From George Peacock Trinity College, Cambridge
Regrets delay in answering letter, but is setting to work and hopes to send part of his manuscript on curves in a few days. Has spoken to Smith (the printer). [Richard?] Harraden is immovable.
ALS 3pp RS:HS 13.253

255. 1817–8–8 To David Brewster Slough
Sends one mathematical article on isoperimetrical problems and asks to write the article on variations as well.
C 1p RS:HS 20.48

256. 1817–8–8 From James Grahame Edinburgh
Family news, and the sad state of current poetry.
ALS 3pp RS:HS 8.282

257. 1817–8–18 or later From Charles Babbage n.p.
Regarding the calculus of derivations.
ALS 1p RS:HS 2.85

258. 1817–8–28 To John Galt Dawlish near Exeter
Discusses plans for an edition of the mathematical writings of the late William Spence, to be edited by JH.
ALS 2pp National Archives of Canada

259. 1817–9–20 From John Galt Lindsey Place, Chelsea
Has passed JH's communication to Mrs. [William] Spence, who desires that publication may proceed as quickly as possible. Would he decide on a publisher. Will write biographical details himself. Would JH give him details of title page and contents.
ALS 2pp RS:HS 8.35

260. 1817–9–26 From John Galt Lindsey Place, Chelsea
Has received JH's letter to which he will reply fully later. Sends some more manuscripts he has received from Mrs. [William] Spence. Also encloses a copy of the unfinished essay of which only 80 copies have been printed.
ALS 1p RS:HS 8.36

261. 1817–10–3 To Charles Babbage [Slough]
Regarding his projected supplement. Calculus of derivations. Naming of periodical functions. [John?] Spence's manuscript.
ALS 4pp RS:HS 2.86 (C: RS:HS 20.49)

262. 1817–10–7 To [?] Slough
Poses mathematical problems for 'polar equations of conic sections.'
ALS 2pp TxU:H/L-0638; Reel 1089

263. 1817–10–8 From James Grahame Edinburgh
On horse back riding, the law, and the birth of a daughter.
ALS 3pp RS:HS 8.283

264. 1817–10–15 To John Galt Slough
Discusses editorial actions taken by JH in editing mathematical manuscripts of William Spence, including materials recently sent to JH by JG.
ALS 2pp National Archives of Canada

265. 1817–10–22 From Charles Babbage Torquay
Hopes to visit him in a few days. Regarding the calculus of derivations. Periodic functions. Is taking up mineralogy.
ALS 3pp RS:HS 2.87

266. [1817–10–22] To William Wallace [Slough]
Invites WW and [Thomas] Leybourn to join JH for dinner while Charles Babbage is visiting JH at Slough.
C 1p RS:HS 20.50

267. 1817–10–23 From John Galt Lindsey Place, Chelsea
Has sent JH's continuation of William Spence's 'Outlines' to press without any alteration. Gives the scheme adopted for the makeup of the volume. Hopes JH agrees; if not write at once as he is leaving London for some time.
ALS 2pp RS:HS 8.37

268. 1817–10–28 From William Wallace Royal Military College, Bagshot
Accepts invitation to visit Slough and meet Charles Babbage. [Thomas] Leybourn may accompany WW. Seeking area of conic sections.
ALS 3pp RS:HS 18.22

269. 1817–10–29 From John Galt Lindsey Place, Chelsea
Book will be arranged according to JH's wishes. Underwoods are to be the publishers. Is pleased JH has found signs of greatness in William Spence's works. Hopes JH will show this in his biographical introduction.
ALS 3pp RS:HS 8.38

270. 1817–11 From Susan White [Shaw, Berkshire]
Enjoyed 'good *dancing* account' of JH from Sophy [Sophia Baldwin?].
Abs 1p TxU:H/M–0580.2; Reel 1093

271. 1817–11–6 From Susan White Shaw
Asks JH to be trustee of SW's and her husband's will.
ALS 3pp RS:HS 18.227

272. 1817–11–8 To John William Whittaker Slough
Asks about Cambridge friends, and that JW check on various items JH left there. Sends news, especially on 'mineralogizing' with [I. H.?] Jephson and Charles Babbage. Comments on recent deaths of J. A. Deluc, and of Princess Charlotte.
ALS 3pp St. John's College, Cambridge

273. 1817–11–11 From Charles Babbage Devonshire St.
Regarding mathematical examples for their forthcoming book. Is taking up chemistry again. Comments on a paper dealing with functional equations by W. G. Horner.
ALS 3pp RS:HS 2.88

274. 1817–11–15 From George Peacock Trinity College, Cambridge
Received the parcel containing GP's manuscript just as he was intending to go into the country. Delayed his departure to prepare it for the printer, but the latter is waiting for types. Comments on JH's suggestions. Observatory was discussed at the meeting of Caput but was put off till 3 Dec. further discussion. James Wood is exerting his influence against it.
ALS 4pp RS:HS 13.254

275. 1817–11–24 To Charles Babbage Slough
Regarding W. G. Horner's paper. Printing of their book. Is preparing a paper to submit to the R.S.L. on a mathematical subject. Will be in town shortly and will visit his laboratory. Has received more sheets of [John?] Spence's essays.
ALS 4pp RS:HS 2.89 (C: RS:HS 20.51)

276. 1817–12–3 From George Peacock Trinity College, Cambridge
Syndicate appointed to consider the erection of an observatory at Cambridge. Gives details of passage through the Senate. Was obliged for J. B. Biot's letter of introduction; found his talents and candor delightful. Had estimate from Edward Troughton for instruments for the Observatory. Printing of the examples began yesterday.
ALS 3pp RS:HS 13.255

277. 1817–12–10 From George Peacock Trinity College, Cambridge
Will be a meeting of the Observatory Syndicate on Tuesday 16 Dec.; hopes that JH can come as his name has great weight.
ALS 1p RS:HS 13.256

278. 1817–12–15 From James Grahame Edinburgh
About friendship.
ALS 4pp RS:HS 8.284

279. 1818–2–18 From George Peacock Trinity College, Cambridge
Went to London, and then returned for meeting of the Syndicate. Gives details of resolutions passed and also the one to which GP objects, *viz* body responsible for appointing observer. There is to be another meeting on Friday.
ALS 4pp RS:HS 13.257

280. 1818–3–7 From George Peacock Trinity College, Cambridge
Meeting was not called by the Vice-Chancellor, so powers of the Syndicate are now at an end. Gives news of events at the University. Has had letter from J. B. Biot, who is to give a series of lectures on Newton's *Principia*.
ALS 4pp RS:HS 13.258

281. [1818]–3–9 From Thomas Catton St. John's College, Cambridge
John Hailstone is about to be married and thus the Woodwardian professorship will become vacant. Would JH be interested? If so, would he let his friends know and then they can rally support.
ALS 2pp RS:HS 5.221

282. 1818–3–10 To Charles Babbage n.p.
Would like comments on the accompanying paper on functional equations. Stephen Lee [of the R.S.L.] has revived the reading of Isaac Newton in Paris.
ALS 4pp RS:HS 2.90 (C: RS:HS 20.53)

283. 1818–3–10 From Miles Bland Cambridge
There is shortly to be a vacancy in the Woodwardian professorship at Cambridge. Would JH like to stand? If so, let him know then he can rally support.
ALS 2pp RS:HS 4.153

284. 1818–3–16 From Charles Babbage Devonshire St.
Comments on his paper sent previously [see JH's 1818–3–10]. Time has been taken up with mineralogy. Is sending some samples.
ALS 4pp RS:HS 2.91

285. 1818–4 From Charles Babbage Devonshire St.
Cannot help him with the *Commercium epistolicum* [by Pierre Fermat]. Leaving town for a few days. Regarding the kaleidoscope.
ALS 1p RS:HS 2.93

286. 1818–4–1 From George Peacock Trinity College, Cambridge
Observer for the forthcoming observatory is to be the Plumian Professor. Gives news of other professorships and gives opinion of JH's chances should he apply for any.
ALS 4pp RS:HS 13.259

287. 1818–4–10 From George Nayler to William Herschel Heralds' College
Encloses two sketches of proposed armorial ensigns, and would be pleased to receive them back with Sir William's preferences.
ALS 1p RS:HS 13.146

288. 1818–4–26 To Charles Babbage Slough
Please send the third volume of J. E. Montucla's *Histoire de mathématiques*. Is running into difficulties with his history of mathematics owing to lack of translations. Regarding kaleidoscopes. Has been reading David Brewster's experiments on Mother of Pearl.
ALS 3pp RS:HS 2.92 (C: RS:HS 20.54)

289. [1818–5–4] From George Peacock Trinity College, Cambridge
Requesting support for Adam Sedgwick in the forthcoming election for the Woodwardian Lectureship. Has been unwell himself.
ALS 3pp RS:HS 13.260

290. 1818–5–10 From David Brewster Edinburgh
There has been a delay in printing so will not require his articles until later. Will be in London in the near future and would like to see him.
ALS 2pp RS:HS 4.243

291. 1818–5–13 From Benjamin Gompertz Kennington Terrace, Vauxhall
Would like to become an F.R.S. Charles Babbage has procured a form and BG would be pleased if JH would subscribe his name. Hopes he is well.
ALS 1p RS:HS 8.136

292. 1818–5–14 From George Peacock Trinity College, Cambridge
Comments on the candidates and the process for filling the Woodwardian Lectureship at Cambridge; also some other Cambridge University matters.
ALS 3pp RS:HS 13.261

293. 1818–5–16 To Charles Babbage n.p.
Thanking him for the third volume of J. E. Montucla's *Histoire de mathématiques.* Gives functional equations for his comments. Is afraid that he will not be able to visit Cornwall this summer.
ALS 4pp RS:HS 2.94 (C: RS:HS 20.55)

294. 1818–5–26 From James A. Gordon Rome
Has been visiting Gibraltar, Malta, and the Vatican City. Comments on the sights.
ALS 6pp RS:HS 8.159

295. 1818–6–1 From James Grahame Edinburgh
JG's wife is seriously ill.
ALS 1p RS:HS 8.286

296. 1818–6–3 From George Peacock Imperial Hotel, Covent Garden
Making arrangements for a visit to JH at Slough.
ALS 2pp RS:HS 13.262

297. 1818–6–5 From James Grahame Edinburgh
Depressed over wife's serious illness.
ALS 2pp RS:HS 8.287

298. 1818–6–19 From William Whewell Trinity College
Informs JH of the place of Simon Stevin in the history of the idea of a parallelogram of forces. Has heard that JH is investigating polarized rays.
ALS 3pp RS:HS 18.159 Todhunter, *Whewell, 2,* 24 inc

299. 1818–7–24 From Charles Babbage Devonshire St.
Wishes to leave town and spend a few weeks in Torquay. Would JH join him? Has new theorem on functions.
ALS 2pp RS:HS 2.95 Hyman, *Charles Babbage*, 40 inc

300. 1818–8–9 From Johann Dietrich Herschel Hanover
Comments on a number of family matters and discusses the activities of some mutual acquaintances.
ALS 3pp (German) TxU:H/M–0885; Reel 1085

301. 1818–8–12 To William Whewell Brighton
Grateful for WW's comments on investigations of [Simon] Stevin. JH's work on polarization of light. Discovery of multiple axes of polarization was preempted by David Brewster. Diagrams polarization of niter crystals and Iceland spar.
C 4pp TxU:H/L-0496; Reel 1055

302. 1818–8–18 From James Grahame The Grove near Hamilton
JG's wife died, and he has been sick in body and spirit.
ALS 1p RS:HS 8.289

303. 1818–8–19 To William Whewell Crooked Crabtree, St. Leonard's Forest
Thanks WW for and expresses agreement with WW's analysis of Simon Stevin. Describes experiments JH is conducting on the polarization of light.
ALS 4pp TC, Camb. Add. Ms.a.207[4] (C: RS:HS 20.56)

304. 1818–10–13 From Charles Babbage Devonshire St.
Gives some analytical equations for his comments.
ALS 2pp RS:HS 2.96

305. 1818–11–1 From William Whewell Trinity College
Expresses thanks for and comments on JH's account of JH's optical experiments. Discusses efforts to improve mathematical education at Cambridge, including WW's book on mechanics.
ALS 4pp RS:HS 18.160 Todhunter, *Whewell, 2,* 28–31

306. 1818–11–5 From James Grahame Edinburgh
About letter writing, and the joy JG receives from his children.
ALS 3pp RS:HS 8.290

307. 1818–11–9 To Charles Babbage [Slough]
Welcome back to London. Explain CB's demonstration of 'theorem in transcendental arithmetic.' JH's observations of 'new acid' [hyposulfurous] and optical phenomena in 'chrystals of nitre.'
AL 4pp inc RS:HS 2.97 (C: RS:HS 20.57 & TxU:H/L-0051.1; Reel 1054 inc)

308. 1818–11–10 From Charles Babbage Devonshire St.
Wants his support for his own application for a post on the Board of Longitude.
ALS 1p RS:HS 2.98

309. 1818–11–19 From Bertie Greatheed 17 Lower Brook St.
Has seen William Artaud, who will be pleased to paint William Herschel's picture. Will undertake it in the new year as he is rather pressed for time at the moment. Hears that J. C. Ross and W. E. Parry have returned from the North.
ALS 2pp RS:HS 9.123

310. 1818–11–20 To David Brewster Slough
Further about additional mathematical articles for the *Edinburgh Encyclopaedia*; comments on DB's intention to start a new 'superior' science journal.
C 2pp RS:HS 20.58

311. 1818–11–20 From James Grahame Edinburgh
Coming to terms with the death of his wife and with his new life [letter completed 1818–12–8].
ALS 3pp RS:HS 8.291

312. 1818–11–26 From Charles Babbage Devonshire St.
When he comes up to town would he please bring Biot's work. Wants to show JH a paper on series. Has just had a visit from Edward Sabine, newly returned from Greenland.
ALS 1p RS:HS 2.99

313. 1818–12–1 From Charles Babbage Devonshire St.
Astronomer Royal's post at Greenwich vacant; hopes JH will try for it. Has not been successful in getting on the Board of Longitude. Hopes to spend a fortnight in Paris after Christmas.
ALS 3pp RS:HS 2.100

314. 1818–12–2 To Charles Babbage [Slough]
Cannot meet CB in London this week. Has verified composition of [hyposulfurous] acid. Encloses crystal of 'sulpho=sulphite of lime.'
C 1p inc TxU:H/L-0051.2; Reel 1054

315. 1818–12–2 To Charles Babbage n.p.
Thanks CB for his advice about applying for the post of Astronomer Royal when it becomes vacant, but does not think himself qualified. Will not be able to come to London this weekend. Regarding his new acid.
ALS 2pp RS:HS 2.101 (C: RS:HS 20.59)

316. 1818–12–4 From Charles Babbage n.p.
Urges JH to apply for the Greenwich post. Quotes extract from Thomas Thomson on chemistry. Gives equations for his comments. Has found a new method of producing household gas.
ALS 4pp RS:HS 2.102

317. 1818–12–6 To Charles Babbage n.p.
Fifth [1817] edition of Thomas Thomson's *System of Chemistry* shows previous discovery of JH's 'new acid,' but its reported composition is wrong.
C 2pp inc TxU:H/L-0051.3; Reel 1054

318. 1818–12–6 To Charles Babbage n.p.
Thanking CB for CB's further remarks on the Greenwich post. Further regarding his new acid.
ALS 3pp RS:HS 2.103 (C: RS:HS 20.60)

319. 1818–12–10 To [James] Moyes Slough
Sends JM material for inclusion in *Mathematical Essays by the Late William Spence, Esq.*, ed. by JH. [Includes manuscript of final paragraph of JH's Preface to this book.]
ALS 2pp inc Adler Ms. 13

320. 1818–12–12 To Charles Babbage Kensington
Remarks on CB's article on analysis and the examples used. Hopes to see him at Christmas; then they can talk about the projected visit to Paris.
ALS 4pp RS:HS 2.104 (C: RS:HS 20.61)

321. 1818–12–16 From Charles Babbage n.p.
Regarding one of JH's papers in the *Analytical*. P.S. dated 19 Dec.: Is glad to hear he would like to go to Paris.
ALS 4pp RS:HS 2.105

322. 1818–12–20 To Charles Babbage n.p.
Hopes CB will visit Slough soon to discuss the proposed trip to Paris. Gives a sketch of his investigations on theorems connected with series.
ALS 4pp RS:HS 2.106

323. 1818–12–22 From Charles Babbage Devonshire St.
Will visit him at Christmas. Regarding the theorem quoted by JH [see JH's 1818–12–20].
ALS 2pp RS:HS 2.107

324. [1819?] To Charles Babbage n.p.
Cannot accompany CB to Deptford because JH busy with James South. Mentions the 'Pearsonian dinner,' 'plagi[h]edral' crystals, and a letter by Wilhelm Olbers.
ALS 3pp RS:HS 2.362 (C: RS:HS 20.67)

325. [1819?] From Charles Babbage n.p.
Would he like to come for a trip down the river to see W. E. Parry's Expedition ships? Query regarding quartz crystals. Date for Pearsonian dinner.
ALS 2pp RS:HS 2.340

326. 1819 To Francis Baily [Slough]
Acknowledges in JH's father's name FB's notice of the upcoming annular eclipse of 1820. Stresses the importance of observing it.
ALS 2pp RS:HS 3.33 (C: RS:HS 20.62)

327. [1819 to 1826] From Samuel Young n.p.
Will show JH largest specimen of apophyllite [Argentine spar] in kingdom.
ALS 1p RS:HS 18.319

328. 1819–1–3 From James Grahame Edinburgh
Family news and other matters [letter completed 1819–2–23].
ALS 4pp RS:HS 8.292

329. 1819–1–3 From George Peacock Trinity College, Cambridge
Asks JH to write a letter of recommendation for a mutual friend, and adds some Cambridge news.
ALS 3pp RS:HS 13.263

330. 1819–1–7 From Charles Babbage n.p.
Gives equation. Is arranging the visit to Paris.
ALS 2pp RS:HS 2.108

331. 1819–1–11 To David Brewster London
Mr. Murray will bring to DB two papers by JH, one on hyposulfurous acid and one on mathematical analysis, plus requested abstract of JH's and William Herschel's 'last papers' in *R.S.P.T.*
C 1p TxU:H/L-0110; Reel 1054

332. 1819–1–11 To David Brewster London
Note to accompany the sending of some papers; JH is off to Paris.
C 1p RS:HS 20.63

333. 1819–1–11 To Mary Pitt Herschel Portland Coffee House [London]
JH arrived safely yesterday and attended event at home of Sir Joseph Banks. [Matthew] Baillie asked about Lady Mary Herschel. JH and Charles Babbage, preparing to tour France, leave tomorrow for Dover. Wishes better health for MH and William Herschel.
ALS 2pp & C 2pp TxU:H/M–0626.1; Reel 1086 (ALS? & C: TxU:H/L-0514.1; Reel 1053)

334. 1819–1–12 To Mary Pitt Herschel Dover
Will sail tomorrow for Calais. Could not find Mr. Armstrong in Dover.
ALS 1p & C TxU:H/L-0514.2; Reel 1053

335. 1819–1–17 To Mary Pitt Herschel Hotel de Bruxelles, 35 Rue du Mail
Arrival in France. Travel companions. JH and Charles Babbage will go to Institute tomorrow to visit J. B. Biot, S. D. Poisson, and other mathematicians.
ALS 4pp & C TxU:H/L-0514.3; Reel 1053

336. 1819–1–26 From William Artaud Fitzroy Square
Regarding his father's portrait. Would recommend waiting until the days become clearer and longer.
C 1p RS:HS 1.413

337. 1819–2–3 From Sir Charles Blagden n.p.
Called to wish him a good journey and to offer his compliments to Sir William and Lady Herschel.
ALS 1p RS:HS 5.75

338. 1819–2–15 To Sir William Watson Slough
JH's father and mother will call on WW during visit to Bath, scheduled after wedding of JH's cousin. Spent three weeks meeting science luminaries in Paris. Notes continuity of scientific endeavors there over three generations.
ACCS 3pp RS:HS 18.112 (C: RS:HS 20.64)

339. 1819–2–17 To Charles Babbage Slough
Would he please forward the letters and papers which he left behind.
ALS 1p RS:HS 2.109 (C: RS:HS 20.65)

340. 1819–2–19 From Charles Babbage n.p.
Relates the details of his accident. Gives equations and theorems.
ALS 4pp RS:HS 2.111

341. 1819–2–19 To Charles Babbage n.p.
Is anxious to hear how his wounds are progressing. Will not be able to come to town yet owing to various circumstances. New chemical experiments.
ALS 2pp RS:HS 2.110 (C: RS:HS 20.66)

342. 1819–3–6 From Charles Babbage n.p.
Has dispatched the tourmaline. His chemical experiments. Circulating functions solving chance problems. [Letter postmarked 1819–3–29.]
ALS 3pp RS:HS 2.114

343. 1819–3–21 From Charles Babbage Devonshire St.
Has had some more tourmaline sent. Gives some more equations. Chemical experiments.
ALS 3pp RS:HS 2.112

344. 1819–3–25 To Charles Babbage n.p.
Made apparatus for showing rings between 'two Tourmalines.' JH's process for making hyposulfurous acid. Praises W. H. Wollaston's analysis of CB's 'Tonquinate.'
C 2pp inc TxU:H/L-0051.4; Reel 1054

345. 1819–3–25 To Charles Babbage Slough
Regarding the disposal of the tourmaline. Address of his cousin. Drawing and description of the apparatus for experiments on tourmaline. Chemical experiments.
ALS 4pp RS:HS 2.113 (C: RS:HS 20.68)

346. 1819–3–25 From Robert Grahame, Sr. Walthamstow
Unable to come to dinner on Saturday but hopes to see him in the evening.
ALS 1p RS:HS 9.1

347. 1819–4–14 From Charles Babbage n.p.
Has received papers from J. B. Biot. Has been to the Customs and Excise. Functional equations.
ALS 4pp RS:HS 2.115

348. 1819–5–2 From Jean Baptiste Biot Paris
Has been very busy; otherwise he would have answered his friendly letter earlier. Observations on tourmaline. Double refraction.
ALS 3pp (French) RS:HS 4.84

349. 1819–5–7 From James Grahame Edinburgh
Family news and ramblings [letter completed 1819–5–23].
ALS 3pp RS:HS 8.293

350. 1819–5–8 To Charles Babbage n.p.
Answers CB's question on the equation. Hopes to see him in London shortly. Chemical experiments.
ALS 3pp RS:HS 2.116 (C: RS:HS 20.69 & C inc: TxU:H/L-0051.5; Reel 1054)

351. 1819–5–12 From Sir William Watson Bath
Delighted that JH proposes to use William Herschel's 20-foot telescope.
ALS 2pp RS:HS 18.113

352. 1819–5–15 To David Brewster London
Some additions to an article submitted about hyposulfurous acid; comments on work by J. B. Biot on polarization and double refraction of crystals.
C 1p RS:HS 20.70

353. 1819–6–13 To John William Whittaker Slough
Requests arrangements for JH's visit before and for Cambridge commencement. Asks whether JW would like to renew correspondence with James Grahame, and congratulates on JW's theological book.
ALS 2pp St. John's College, Cambridge

354. 1819–6–15 From James A. Gordon Brighton
Has not heard from him lately, though he wrote last year from Rome and gave him a resume of his travels. William Crackanthorpe is a good companion. The pavilion is being rapidly built. Would like JH to accept hospitality at his house. Gives news of friends.
ALS 3pp RS:HS 8.160

355. 1819–6–17 From Samuel Young Formosa Cottage
Mineral apophyllite [Argentine spar]. Offers gems and minerals from India for JH's experiments. David Brewster's papers raise questions about polarized light. Describes microscope SY found.
ALS 3pp RS:HS 18.314

356. 1819–7 To Charles Babbage n.p.
Sent JH's paper on mother-of-pearl and paper on hyposulfurous acid to David Brewster. Tell Brewster of JH's discovery of optical structure in crystals of bicarbonate of potash. Would Brewster accept paper from JH enlarging the definition of crystals?
C 2pp inc TxU:H/L-0051.6; Reel 1054

357. 1819–7 To Charles Babbage n.p.
Wishing CB success in his efforts to obtain the position of professor at Edinburgh. Relates details of his recent experiments with bicarbonate of potash.
ALS 4pp RS:HS 2.119 (C: RS:HS 20.72)

358. [1819]–7–1 From Ann Wilson 29 Kensington Square
Learned that JH has been ill since returning from Slough. Please visit soon.
ALS 1p RS:HS 18.416

359. 1819–7–14 To David Brewster Slough
Sends a paper on mother of pearl for publication in [*Edinburgh Philosophical*] *Journal.* Apologizes for confused state of paper on hyposulfurous acid.
C 2pp RS:HS 20.71 (C: TxU:H/L-0111; Reel 1054)

360. 1819–7–14 From David Brewster Edinburgh
Is anxious to receive his article on hyposulfurous acid as the volume is almost ready for the printer.
ALS 1p RS:HS 4.244

361. 1819–7–24 From Charles Babbage Devonshire St.
Has just returned from Lincolnshire. Will shortly be visiting Torquay. Would like to call on him if convenient. Has heard of a vacant professorship at Edinburgh University and may try for it. Gives an equation of W. H. Wollaston's.
ALS 3pp RS:HS 2.117

362. 1819–7–24 From Sir William Watson Dawlish
Received JH's profile. Glad William Herschel's portrait is taken. WW has one from forty years ago. Lady Watson not well. Requests information about 'the comet.' Happy for account of Beckwiths.
ALS 2pp RS:HS 18.114

363. 1819–7–25 From Charles Babbage Devonshire St.
Is going to Edinburgh to apply for the post left vacant by John Playfair's death. Would JH and his father give him certificates and references? [Note by Mrs. CB to the effect that her husband has heard that James Ivory will not be contesting the professorship at Edinburgh].
ALS 3pp RS:HS 2.118

364. 1819–7–27 To Charles Babbage Slough
JH understands CB to be a candidate for a professorial chair, and JH writes a warm letter of support.
ALS 2pp BL Aaa 32292.123

365. 1819–8–8 From Charles Babbage Edinburgh
Apologizes for the long delay in answering his letter. Relates experiences and hopes of success in his effort to obtain the position of professor at the University.
ALS 2pp RS:HS 2.120

366. 1819–8–8 From George Peacock Trinity College, Cambridge
About Charles Babbage's hopes for a position, and joint publication ventures.
ALS 4pp RS:HS 13.264

367. 1819–8–22 From David Brewster Esk Hill by Roslin
Has heard from Charles Babbage that JH has sent him the article on hyposulfurous acid, but he has not in fact received it. Has seen Babbage a lot recently. Candidates for the Mathematical chair. Regarding colored rings in Carbonate of Potash. Has reserved his paper on Series for number 3.
ALS 2pp RS:HS 4.245

368. 1819–8–29 From James Grahame Edinburgh
Of Charles Babbage's visit, family news, and lawsuits.
ALS 3pp RS:HS 8.294

369. 1819–[9] To Charles Babbage n.p.
Sorry to hear of his failure at Edinburgh. Inviting him to Slough. Enquiring about the people CB met at Edinburgh.
ALS 3pp RS:HS 2.121 (C: RS:HS 20.73)

370. 1819–9–14 From James Grahame Edinburgh
Chastises JH for misbehavior along with Charles Babbage, discusses the parlous state of the country's affairs, expresses interest in the history of astronomy, and sends family news [letter completed 1819–10–8].
ALS 4pp RS:HS 8.296

371. 1819–9–18 From David Brewster Esk Hill by Roslin
Regarding refraction in various crystals and comparison with Newton's tables. Suggests a joint paper on the subject. His parcel has never arrived. Is working on a paper on rotating phenomena.
ALS 4pp RS:HS 4.246

372. 1819–9–29 From Andrew Wilson [College St.], Glasgow
Mr. Townley cannot get B.A. degree without attending one year at University and taking usual examinations. London papers greatly magnified disturbances. Things are peaceful now, but vigorous measures are needed to suppress 'dangerous Meetings.' Expects to enjoy Beckwiths' visit.
ALS 2pp RS:HS 18.414

373. 1819–9–30 To David Brewster Slough
Thanks for, and comments on, DB's 1819–9–18.
C 5pp RS:HS 20.74

374. 1819–10–2 To James Grahame n.p.
Of Charles Babbage's trip to Scotland [letter continued 1819–10–8].
C 2pp inc RS:HS 8.295

375. 1819–10–4 From David Brewster Esk Hill by Roslin
Thanks for letter. Would he send his paper to [William?]

Blackwood. Thinks it only fair that JH's first paper on light should be written on his own, providing he acknowledges his discoveries. Comments on recent experiments. Who has written the article on polarization in the recent *Edinburgh Review*?
ALS 4pp RS:HS 4.247

376. 1819–10–13 To Charles Babbage Slough
Will not be in town until Tuesday afternoon. Has been speculating on a means of telegraphic communication between himself and CB.
ALS 2pp RS:HS 2.123 (C: RS:HS 20.75)

377. 1819–10–13 From Charles Babbage [Devonshire St.?]
Has just returned from Edinburgh and found JH's letter awaiting him. Relates various experiences in Edinburgh. Recent discoveries in his chemical experiments.
ALS 3pp RS:HS 2.122

378. 1819–10–21 From Jean Baptiste Biot Paris
Queries regarding the refraction of light in crystals.
ALS 4pp (French) RS:HS 4.85

379. 1819–10–25 From James Grahame Edinburgh
Of Charles Babbage, family, public affairs, and law suits [letter completed 1819–11–15].
ALS 4pp RS:HS 8.297

380. [1819–11] To David Brewster n.p.
Regarding terminology for polarization and refraction of light. DB's statement concerning the experiments of E. L. Malus and light through crystals.
draft 4pp RS:HS 4.273

381. 1819–11–3 From James A. Gordon 29 Hill St.
Was pleased to receive JH's farewell letter. Would be glad when JH was in town if he would call on Gordon's father and mother. Has left copies of the translation of the Memoir of Mr. De La Harpe. Gives an extract of a letter from Naples.
ALS 3pp RS:HS 8.161

382. 1819–11–5 From David Brewster Esk Hill by Roslin
Has received his papers on hyposulfurous acid and Mother of Pearl. Comments regarding latter. Agreement necessary on nomenclature for light. His own and J. B. Biot's experiments on light. Would JH read his paper of 1818 and comment on it.
ALS 3pp RS:HS 4.248

383. [1819–11–24] To David Brewster n.p.
Sends paper on deviation of tints. Comments on findings. Encloses crystals of carbonate of lime, which exhibits the system of rings. Regarding his own paper on Mother of Pearl and his indebtedness to DB. Comments on nomenclature for light.
draft 4pp RS:HS 4.249 (C: RS:HS 20.76)

384. 1819–12–1 To William Whewell 49 Charlotte St., Portland Place
Praises WW's new book [*Elementary Treatise on Mechanics*]; is working to get WW elected to the R.S.L. Criticizes a pamphlet by George Peacock. Comments on rules proposed for the new Cambridge Philosophical Society.
ALS 4pp TC, Camb. Add. Ms.a.207[5] (C: RS:HS 20.77)

385. 1819–12–5 From George Peacock Trinity College, Cambridge
About the joint publication of JH, GP, and Charles Babbage, and Babbage not doing his part.
ALS 4pp RS:HS 13.265

386. 1819–12–9 From David Brewster Edinburgh
Thanks for letters. Regarding printing of his papers. Would like to propose him Fellow of the Royal Society of Edinburgh. Biot's law. Much obliged for his opinion respecting E. L. Malus. Thanks for crystal of carbonate of lime. Comments on this.
ALS 4pp RS:HS 4.250

387. 1819–12–10 To Jean Baptiste Biot n.p.
Answers to JB's queries on the refraction of light through crystals, and Newton's rings.
draft 8pp RS:HS 4.86 (C: 20.78)

388. 1819–12–15 To David Brewster n.p.
Accepts, with thanks, DB's offer to propose JH for membership in the Royal Society of Edinburgh; some comments on DB's 1819–12–9.
C 2pp RS:HS 20.79

389. 1819–12–17 To David Brewster London
Sends signed certificate for admission to Royal Society of Edinburgh; the Cambridge Philosophical Society has just been formed.
C 1p RS:HS 20.80

390. 1819–12–29 From George Peacock London
Arrangements about a visit to Slough.
ALS 1p RS:HS 13.266

391. [1820–1] To Heinrich Wilhelm Matthias Olbers [London]
Announces new Astronomical Society. Invites WO to become associate.
AAbs 1p RS:HS 19.9

392. [1820 to 1824] From Johann Wilhelm Pastorff n.p.
Sends paper on photospheres to be read to Astronomical Society.
Abs 1p RAS:JH Archive 6/2.10 verso; Reel 8

393. [1820 to 1824] To Johann Wilhelm Pastorff Astronomical Society
JH presented JP's paper on photospheres to Society.
AAbs 1p RAS:JH Archive 6/2.10 verso; Reel 8

394. [1820]. To James Wood n.p.
Congratulates on having obtained position at St. John's. Asks whether Cambridge University Press would publish new edition of Giuseppe Piazzi's star catalog.
ALS 2pp St. John's College, Cambridge

395. 1820–1–5 From Charles Babbage Devonshire St.
Please send instructions as to how he would like the rock crystals cut so that he can pass the instructions on to Wilson Lowry. Is sending two drawings of JH. Has heard from David Brewster that Edinburgh has established a prize like that of Benjamin Thompson (Count Rumford).
ALS 2pp RS:HS 2.124

396. 1820–1–17 To Charles Babbage n.p.
Would he please send back some of his papers on astronomy as he wants them in order to draw up a prospectus.
ALS 1p RS:HS 2.125 (C: RS:HS 20.81)

397. 1820–1–22 From Francis Baily Gray's Inn
Regarding the printing and distribution of JH's *Address*.
ALS 3pp RS:HS 3.34

398. 1820–1–24 From Francis Baily Gray's Inn
Is sending copies of the *Address* and circular, and two copies

of his own pamphlets.
ALS 1p RS:HS 3.35

399. 1820–1–30 To David Brewster n.p.
Arrangements for paying dues to the Royal Society of Edinburgh; some comments about light passing through crystals.
C 2pp RS:HS 20.82

400. 1820–1–31 From Charles Babbage Devonshire St.
Recent happenings and committee decisions in the Astronomical Society.
ALS 3pp RS:HS 2.126

401. 1820–2–2 To Charles Babbage n.p.
Regarding the forthcoming meeting of the Astronomical Society. The printing of his own address. Is preparing a paper on functional equations. Is prepared to accept the office of Foreign Secretary of the Astronomical Society.
ALS 3pp RS:HS 2.127 (C: RS:HS 20.83)

402. 1820–2–3 To David Brewster Slough
Clarifies areas of disagreement with DB in JH's explanation of the results of colored tints of light passing through polarizing crystals.
C 5pp RS:HS 20.84

403. 1820–2–6 From Charles Babbage Devonshire St.
Has had a letter from Davies Gilbert regarding the presidency of the Astronomical Society. Possible alternatives. Gives a problem of the law of chance.
ALS 3pp RS:HS 2.128

404. 1820–2–11 From David Brewster Edinburgh
Thanks for the extract of his paper. Comments on this and his own discoveries on the polarization of light. His fees for the articles will pay his subscription to the Royal Society of Edinburgh.
ALS 4pp RS:HS 4.251

405. 1820–2–12 To Wilson Lowry Slough
Asks where WL procured a certain 'large crystal of Fish-eye-Stone.'
ALS 1p St. John's College, Cambridge

406. 1820–2–15 From Sir William Watson Bath
Relieved that news of death of King [George III] did not weaken William Herschel. Will see JH and Lady Mary Pitt Herschel in Bath in April. Lady Watson and Miss Joy are not well.
ALS 2pp RS:HS 18.115

407. [1820–2–16] From Wilson Lowry n.p.
Finds that the crystal of 'Fish-eye-stone' came from Sweden and was attached to a mass of oxydulous iron. Hopes that JH receives this information in time for his paper.
ALS 1p RS:HS 11.491

408. 1820–2–17 From George Peacock Trinity College, Cambridge
Business matters relating to the R.S.L. and the Astronomical Society.
ALS 2pp RS:HS 13.267

409. [1820 or 1825]–2–18 From Thomas Young 48 Welbeck St.
Gives diagram and explanation of TY's reduced parallax. JH's formula is elaborate but complicated. TY's formula is less exact but easier for public to use. Welcomes criticism.
ALS 2pp RS:HS 18.351

410. [1820]–2–19 From Fearon Fallows St. John's College, Cambridge
Pleased to hear of his election to the Astronomical Society. No further news of the projected observatory at Cambridge. Has JH any news of the Observatory to be erected at the Cape? Would JH add his name to his certificate for the R.S.L. Is he interested in the Plumian Professorship?
ALS 3pp RS:HS 7.156

411. 1820–2–22 From Francis Baily Gray's Inn
Has sent the *Address* to the printers for the corrections to be made. Committee meeting on Tuesday next. Hopes JH can come.
ALS 3pp RS:HS 3.36

412. 1820–2–26 From Adam Sedgwick Trinity College, Cambridge
Thanks JH for communication to the [Cambridge Philosophical] Society. States the reactions of Cambridge faculty to the society. Gives dates for next meetings.
ALS 2pp RS:HS 15.418

413. 1820–3–11 From Francis Baily Gray's Inn
Is sending the books. News of the Committee meeting. Relations of the Astronomical Society with Sir Joseph Banks and the R.S.L.
ALS 2pp RS:HS 3.37

414. 1820–3–18 From Charles Babbage Devonshire St.
Was unable to attend council meeting of the Astronomical Society. Difficulties over the choosing of a president. Regarding the affairs of Sir Joseph Banks.
ALS 4pp RS:HS 2.129

415. 1820–3–19 To Charles Babbage n.p.
Was unable to attend the meeting of the Astronomical Society Council. Has heard the news regarding the president. Gives a problem of chance. Regarding the printing of the supplement.
ALS 4pp RS:HS 2.130 (C: RS:HS 20.85)

416. [1820]–3–20 From Adam Sedgwick Trinity College, Cambridge
States that the [Cambridge Philosophical Society] meeting went well. Charges against council were proven unfounded with the help of a letter from [Webster?]. James Cumming has been experimenting with magneticogalvanic phenomena.
ALS 2pp RS:HS 15.419

417. 1820–3–20 From James Wood St. John's College
Please write [H. J. Temple, Viscount] Palmerston on behalf of Fearon Fallows, candidate for director of future observatory at Cape of Good Hope.
ALS 2pp RS:HS 18.291

418. 1820–3–26 From Agnes Baillie Hampstead
Invitation to dine with them; Sir Walter Scott is expected as well.
ALS 2pp RS:HS 3.7

419. 1820–3–30 From Charles Babbage n.p.
Regarding his stove for his experiments. Problem with crystals. Printing of his book in Paris.
ALS 4pp RS:HS 2.131

420. 1820–3–31 To Henry John Temple Slough
Recommends Fearon Fallows for position as astronomer at Cape of Good Hope. Praises FF for his mathematical ability.
draft 3pp RS:HS 17.343 (C: RS:HS 20.86)

421. [1820–4?] From Charles Babbage n.p.
Has been experimenting with magnetism. Two persons wish to become Foreign Members of the [Astronomical Society?]. Regarding the Lucasian at Cambridge. Would like JH to procure the medal of [Matthew Boulton?] for him.
ALS 2pp RS:HS 2.341

422. 1820–4–3 From Charles Babbage n.p.
Informing him of the death of Isaac Milner. He should try for the Lucasian Professorship at Cambridge.
ALS 4pp RS:HS 2.132

423. [1820–4–3] From Thomas Catton St. John's College, Cambridge
The Lucasian professorship is vacant. Thinks JH is worthy of this position. Would like to be proposed a member of the Astronomical Society.
ALS 3pp RS:HS 5.222

424. 1820–4–3 or 4 From George Peacock Trinity College, Cambridge
Urges JH to become an active candidate for vacant Lucasian professorship.
ALS 1p RS:HS 13.268

425. 1820–4–4 To Charles Babbage n.p.
Will be going to Cambridge to try for the vacant professorship. Remarks on his reasons for applying for the post. His recent experiments with quartz.
ALS 4pp RS:HS 2.133 (C: RS:HS 20.87)

426. 1820–4–4 From Agnes Baillie Hampstead
Invitation to spend the morning with them on the day they are expecting Sir Walter Scott.
ALS 2pp RS:HS 3.8

427. 1820–4–6 From James Grahame Edinburgh
Of family, Walter Scott's novels, and affairs in Scotland and Spain.
ALS 4pp RS:HS 8.298

428. 1820–4–9 From Joanna Baillie Hampstead
Invitation to dine with them and Sir Walter Scott.
ALS 2pp RS:HS 3.17

429. 1820–4–12 To Charles Babbage Slough
About plans for journey to visit CB in Cornwall, and to go beyond.
C 4pp RS:HS 20.88

430. 1820–4–18 To George Gretton St. John's College, [Cambridge]
Enquiry regarding the possibility of a vacancy in the position of organist at the cathedral. If there is, he knows a suitable candidate.
draft 3pp RS:HS 9.43 & 20.89

431. 1820–4–19 To David Brewster n.p.
Sending all his papers on the derivations of tints, suitably amended to give DB due acknowledgement. Queries regarding some of DB's statements. Supports his theories in a new paper he has written for the Cambridge Philosophical Society.
draft 10pp RS:HS 4.252–4 (C: RS:HS 20.90)

432. 1820–4–20 To Charles Babbage n.p.
Reasons why his father does not wish to become the president of the Astronomical Society. Has attended a meeting of the Philosophical Society. Please send three or four copies of the address of the Astronomical Society.
ALS 3pp RS:HS 2.134

433. 1820–4–20 From George Gretton Deanery, Hereford
Further news regarding the position of organist at Hereford. Hopes JH's parents are well.
ALS 3pp RS:HS 9.44

434. 1820–4–21 To Charles Babbage n.p.
Declining invitation for William Herschel to be President of the Astronomical Society.
C 2pp RS:HS 20.91 (C: RS:HS 2.134)

435. 1820–4–25 To Jean Baptiste Biot Cambridge
Fears that a letter written to him in November may not have reached him. Will send him a copy of his paper of colors. A Philosophical Society has been formed at Cambridge.
draft 2pp RS:HS 4.87 (C: RS:HS 20.93)

436. 1820–4–26 From Agnes Baillie Hampstead
Further invitation to breakfast with them and Sir Walter Scott.
ALS 2pp RS:HS 3.9

437. 1820–5–1 From Charles Babbage n.p.
No news of the Board of Longitude. Has informed Thomas Beckwith that JH will be with him on Wednesday. Gives an example in functions.
AL 3pp inc RS:HS 2.135

438. 1820–5–6 From Jean Baptiste Biot Paris
Has received his letter written in November a little late. Observations on refraction and allied subjects.
ALS 3pp (French) RS:HS 4.88

439. 1820–5–6 From George Peacock Trinity College, Cambridge
Cambridge University has agreed to build a new observatory [Madingley Road] and now the matter of collecting subscriptions for it must begin.
ALS 3pp RS:HS 13.269

440. 1820–5–6 To Henrik Johan Walbeck n.p.
Announces new Astronomical Society. Invites HW to become associate. Mentions planned observatory at Cape of Good Hope.
AAbs 1p RS:HS 19.13

441. 1820–5–8 From George Gretton Deanery, Hereford
The organist is to be judged by the Chapter. Will bear JH's suggestion in mind. Wishes JH could be considered for the Lucasian Professorship. Would like to meet him if possible. Unable to understand one of his papers.
ALS 3pp RS:HS 9.45

442. 1820–5–9 To Friedrich Wilhelm Bessel Slough
Asks if FB is willing to be proposed as an Associate member of the Astronomical Society.
ALS 2pp AdWDDR Bessel 225 (C: RS:HS 20.94)

443. 1820–5–9 To Johann Elert Bode n.p.
Rules of new Astronomical Society. Offers to propose JB as associate.
AAbs 1p RS:HS 19.1

444. 1820–5–9 To Alexis Bouvard n.p.
Announces newly formed Astronomical Society. Invites AB to join.
AAbs 1p RS:HS 19.3

445. 1820–5–9 To Carl Friedrich Gauss Slough
Announces formation of the Astronomical Society. Will be pleased to propose [CG] as an associate.
ALS 2pp NSUB Cod.Ms.Gauss 99 (C: RS:HS 20.95)

446. 1820–5–10 To Charles Babbage [Slough]
CB is 'incurable castle-builder.' Denies that JH falls easily into love. Eradicating 'one hastily conceived & silly passion' has deadened JH's heart. Any future love must follow a long acquaintance and friendship.
ALS 3pp TxU:H/L-0052; Reel 1054 (C: RS:HS 20.96)

447. 1820–5–10 From George Gretton Deanery, Hereford
There has been another fire at Hereford. [W. G.?] Hayter is to be judged next Tuesday. Hopes all are well. Own family has suffered from an attack of influenza.
ALS 3pp RS:HS 9.46

448. 1820–5–16 From George Gretton Deanery, Hereford
Has just returned from the Chapter, which has been considering the case of [W. G.?] Hayter. Will now decide on Thursday as new fires have taken place.
ALS 1p RS:HS 9.47

449. 1820–5–17 From George Peacock Observatory Committee Room
Subscriptions to the fund to build the observatory [see GP's 1820–5–6] are coming in only slowly.
ALS 2pp RS:HS 13.270

450. 1820–5–19 From John Clarke-Whitfeld Cambridge
JW to be proposed as cathedral organist by Dean of Hereford. Current organist under suspicion for starting fires at college. Thanks for support of JH's friends.
ALS 2pp RS:HS 18.235

451. 1820–5–19 From David Wilkie 24 Lower Phillimore Place, Kensington
Coming to Slough at Lady Herschel's invitation.
ALS 1p RS:HS 18.248

452. 1820–5–20 From George Gretton Deanery, Hereford
Believes that JH's candidate will get full support when the Chapter decides on the position of organist. Hopes all are well. Sends this to Slough.
ALS 2pp RS:HS 9.48

453. 1820–5–22 From Charles Babbage Devonshire St.
Regarding Prince Leopold. JH must come to town to mix with men of science. There are two vacancies on the Board of Longitude. Regarding a certain star.
ALS 3pp RS:HS 2.136

454. 1820–5–23 To Charles Babbage n.p.
Hopes to take up abode in his former lodgings. To what star does CB refer to in his last letter [see CB's 1820–5–22]?
ALS 3pp RS:HS 2.137 (C: RS:HS 20.97)

455. [1820 or 1826]–5–25 From William Hyde Wollaston Buckingham St.
John Pond's last communication to Committee is in press. Save its second printing.
ALS 1p RS:HS 18.280

456. [1820–6] To Francis Baily n.p.
Requests FB's paper on the eclipse and the reports on observations of the eclipse that JH has received. Discusses criteria to be used in deciding what to print in the Astronomical Society's journal.
ALS 3pp Bodleian Library, Oxford

457. 1820–6–1 From Alexis Bouvard Paris
Is grateful for his letter and also for sending the regulations of the Astronomical Society on which he sees the names of many distinguished men. Accepts with gratitude the offer of his book.
ALS 2pp RS:HS 5.88

458. 1820–6–1 From Jean-Baptiste Delambre Paris
Has received the rules of the Astronomical Society. Gives full details about himself.
ALS 2pp (French) RS:HS 6.482

459. 1820–6–5 From Charles L. Harding Göttingen
Is grateful for being recommended as a member of the Astronomical Society. Thinks the Society will confer great benefits in the field of astronomy.
ALS 2pp RS:HS 9.233

460. 1820–6–19 From Francis Baily Gray's Inn
Is making arrangements for the Committee meeting. Will bring some copies of the *Eclipse* for JH.
ALS 2pp RS:HS 3.38

461. 1820–7 From Charles Babbage n.p.
Humphry Davy has been canvassing. Has several Cornish introductions. Would like to visit South Wales.
ALS 2pp RS:HS 2.139

462. 1820–7–11 From George Gretton Deanery, Hereford
JH's friend Dr. J. C. Whitfield was elected organist this morning. He himself is suffering with the jaundice. Anxious to hear of Sir William's health. Does not know anyone near Lichfield.
ALS 1p RS:HS 9.49

463. 1820–7–12 To Charles Babbage n.p.
Has received the visitation letter. Work is progressing on the twenty-foot telescope. Hopes to leave for Devon soon. Places he would like to visit. Hopes CB is well versed in geology.
ALS 4pp RS:HS 2.138

464. 1820–7–12 To Carl Friedrich Gauss Slough
Has received letter of 3 June 1820 reporting on the new circle at Göttingen Observatory. Will read it to the Astronomical Society and propose CG as associate. Sends some JH publications.
ALS 2pp NSUB Cod.Ms.Gauss 99 (C: RS:HS 20.98)

465. 1820–7–15 From John Clarke-Whitfeld Cambridge
JW was elected Hereford cathedral organist. Thanks JH for letter of support and gift of poetry. Bishop promised to arrange JW's quick release from Cambridge duties. Vice-Chancellor ruled in JW's favor in case with Mrs. Deacon.
ALS 2pp RS:HS 18.236

466. 1820–7–18 From Joseph Goodall Lodge, Eton College
Would like him to come to dinner next Saturday.
ALS 1p RS:HS 8.140

467. 1820–7–22 From James South to William Herschel Blackman St., Southwark
Asks WH to contribute to a fund to place a bust of Edward Troughton in Greenwich Observatory. Introduces himself to WH.
ALS 2pp RS:HS 16.412

468. 1820–7–29 From Friedrich Wilhelm Bessel [Hamburg]
FB's observations. Happy to become associate of new Astronomical Society. Discusses observations of Halley's Comet by James Bradley and Nathaniel Bliss.
Abs(by JH) 1p RS:HS 19.1

469. 1820–7–30 From Henrik Johan Walbeck Hamburg
Reports recent eclipse observations around the world.
ALS 2pp (Latin) RS:HS 18.8

470. 1820–8–1 From John Clarke-Whitfeld Hereford
Arrived in Hereford two days before Dean's death. Lost valuable confidant. Sprained wrist and fingers before leaving Cambridge, but must play for Dean's service.
ALS 1p RS:HS 18.237

471. 1820–8–6 From Sir William Watson Dawlish
Recovering from shoulder injury. After leaving Slough, called on Joseph Banks, who died recently.
ALS 1p RS:HS 18.116

472. 1820–8–7 From Charles Babbage Dudmatson
Has been doing lots of chemical experiments. Gives one of his experiments with prisms. Regarding star observations.
ALS 4pp RS:HS 2.140

473. 1820–8–9 From Charles Babbage Dudmatson
Would JH send him the chemical formula for converting rags to sugar.
ALS 1p RS:HS 2.141

474. 1820–8–10 From Marquis Pucci London
Unable to accept JH's invitation. Will call after returning from Scotland.
L 1p RAS:JH Archive 10/4.21 verso; Reel 9

475. 1820–8–12 To Charles Babbage Slough
It looks as though their projected Cornish excursion will have to be postponed for a time. Gives the formula for the conversion of rags into sugar. A Finnish astronomer wishes to be elected a member of the Astronomical Society. Observations on the theory of parallax.
ALS 4pp RS:HS 2.142 (C: RS:HS 20.99)

476. 1820–8–[12?] To James A. Gordon Slough
Intended to write earlier but has been delayed by various causes. Their old friend the Dean of Hereford has just died. Comments on political situation in Sicily. New philosophical society started in Cambridge. Plans also for an observatory there.
draft 4pp RS:HS 8.162 & 20.100

477. [1820–8–12] From Barnaba Oriani Milan
Thanks for *Regulations* of the Astronomical Society. Is honored on being elected a member. Gives details of *Theory of the Moon* by G. A. Plana; also sends one of his own works, *Elementi di trigonometria sferoidica.*
ALS 2pp (French) RS:HS 13.200

478. 1820–8–16 From James A. Gordon London
Has read of Mr. Annesley's death. Will be staying at Stocks until November and would welcome a visit from JH. His parents are quite well and hopes JH's are as well. Is grateful for his comments on Sicily.
ALS 2pp RS:HS 8.163

479. [1820?]–9 From William Frédéric Edwards Paris
Is grateful for his paper, which he has communicated to A. J. Fresnel. Health has been bad so has confined his researches to physiology. Has won the prize of the Academy for his paper on vertebrates. Will send JH a copy. Not much happening as everyone is in the country.
ALS 2pp RS:HS 7.121

480. 1820–9–26 From Jean Baptiste Biot Marquieres pres Beauvais
Further information on his experiments with refraction in various crystals, especially topaz.
ALS 4pp (French) RS:HS 4.89

481. 1820–10–5 From James Grahame 29 Northumberland St., Edinburgh
Recollections of this house JG has just sold.
ALS 2pp RS:HS 8.299

482. 1820–10–9 To Adam Sedgwick Slough
Asks AS to send him plagiedral quartz for experimentation. Unable to commence Cornish expeditions with Charles Babbage. Requests AS discuss publication of their [JH's, George Peacock's, and Babbage's] book with Peacock.
ALS 2pp damaged—torn with piece missing RS:HS 15.420 (C: RS:HS 20.101)

483. 1820–10–17 From Johann Elert Bode Berlin
Thanks JH for news of newly formed Astronomical Society; will be pleased to be a foreign correspondent for it.
ALS 1p (German) RS:HS 4.163

484. 1820–10–20 From James Grahame Edinburgh
Of religious matters, accusations against the Queen, and of Charles Babbage [letter completed 1820–11–4].
ALS 4pp RS:HS 8.300

485. 1820–10–21 From Charles Babbage Devonshire St.
Has returned from his holiday. Thanks for the chemical periodicals. Has evolved some new schemes.
ALS 2pp RS:HS 2.143

486. 1820–10–24 To Charles Babbage Slough
Thanks for all the packages. Would like to send copies of the supplement to John Brinkley and Fearon Fallows. Regarding D. F. J. Arago's statement. Method of making a barometer clock. Wishes the president of the R.S.L. was changed every four or five years. Message for W. H. Wollaston from J. B. Biot.
ALS 3pp RS:HS 2.144 (C: RS:HS 20.102)

487. 1820–11–2 To Charles Babbage n.p.
Informing CB of the death of Mrs. Beckwith [wife of Thomas?]. W. E. Parry has returned successfully.
ALS 2pp RS:HS 2.145 (C: RS:HS 20.103)

488. 1820–11–2 To Charles Babbage Bedford Place
Gives Latin motto. Is sending his chemical book.
ALS 3pp RS:HS 2.146 (C: RS:HS 20.104)

489. 1820–11–2 From Sir William Watson Bath
William Herschel's health. Lady Watson and Miss Joy are well. Sad that JH will not visit Bath.
ALS 1p RS:HS 18.117

490. 1820–11–3 From Charles Babbage n.p.
Offering any assistance during the funeral period of Mrs. Beckwith [wife of Thomas?].
ALS 1p RS:HS 2.147

491. 1820–11–6 From James Grahame Edinburgh
Offers JH the consolations of religion to deal with his unhappiness.
ALS 3pp RS:HS 8.301

492. 1820–11–7 To Charles Babbage n.p.
Thanking him for offer of assistance. Will call on him and give him papers for the Astronomical Society.
ALS 2pp RS:HS 2.148 (C: RS:HS 20.105)

493. 1820–11–9 From Helen Wilson Glasgow
HW expected news of Mrs. Beckwith's death, following receipt of letter from Mary Baldwin describing poor health of

her sister.
ALS 2pp RS:HS 18.419

494. 1820–11–13 From Adam Sedgwick Trinity College, Cambridge
Discusses study of Oolite beds. Mother recently died. Will bring paper to publisher. Discusses last Cambridge Philosophical Society meeting, [James] Wood, and [E. D.] Clarke. Cannot locate the crystal JH requested.
ALS 3pp RS:HS 15.421

495. 1820–11–16 From George Peacock Trinity College, Cambridge
About the expenses and arrangements regarding the publication of JH's *A Collection of Examples of the Application of the Calculus of Finite Differences.*
ALS 4pp RS:HS 13.271

496. 1820–11–21 From George Dollond London
Returns the polarizing apparatus by the Windsor coach, and is grateful for the loan of it. Hopes to supply Dr. Jackson with a suitable instrument. Sending a paper on parallel plates.
ALS 2pp RS:HS 6.497

497. 1820–11–22 From James South Blackman St., Southwark
Will see JH on Friday; hoping for clear weather.
ALS 1p RS:HS 16.413

498. 1820–12–2 to 18 From George Peacock Cambridge
Urging JH to come to Cambridge to examine the plans for the new observatory.
ALS 2pp RS:HS 13.273

499. 1820–12–2 From George Peacock Trinity College, Cambridge
About the expenses of their joint publication [see GP's 1820–11–16], rumors of a vacancy in the Lucasian professorship, and about the new observatory at Cambridge.
ALS 4pp RS:HS 13.272

500. 1820–12–18 From Charles Babbage Devonshire St.
Giving him instructions arising from the recent meeting of the Astronomical Society. Gives details of JH's books he has in his possession. Has been experimenting with the game of noughts and crosses. Gives a problem in analysis. Regarding a paper by JH on numbers.
ALS 3pp RS:HS 2.149

501. 1820–12–19 To Charles Babbage Slough
Remarks on the R.S.L. Has written to J. B. Biot. Regarding the annual report. Needs a new council for the Astronomical Society. Please send him some copies of the rules. Meeting of Society of Arts. His recent chemical experiments. James South was with him last week and carried out observations.
ALS 4pp RS:HS 2.150 (C: RS:HS 20.106)

502. 1820–12–20 From George Peacock Heckfield near Reading
About observatory plans and possible vacancies at Cambridge.
ALS 4pp RS:HS 13.274

503. 1820–12–22 From E. Willes n.p.
Father died. Consoles JH on loss of Mrs. Beckwith. Will always remember kindness shown by Herschel family when EW was attending Eton.
ALS 3pp RS:HS 18.253

504. 1820–12–25 To Charles Babbage n.p.
Suggestions for the wording of the annual report of the Astronomical Society. Wishes to propose a medal to be awarded for the theory of Saturn's satellites. Regarding the Lucasian professorship at Cambridge.
ALS 4pp RS:HS 2.151 (C: RS:HS 20.107)

CALENDAR OF THE JOHN HERSCHEL CORRESPONDENCE 1821–1825

505. [1821?] From Charles Babbage n.p.
Can JH deputize for him at the Astronomical Society meeting?
ALS 2pp RS:HS 2.345

506. [1821 or later] To Michael Faraday n.p.
Discusses origin and reactions of enclosed borate of lead. Postscripts on parallel strata in glass.
ALS 3pp diag RI 499–500 F3B

507. [1821] From Mary Pitt Herschel Slough
William Herschel is glad that JH declined offer for WH to become commissioner of longitude. Hopes JH enjoys visit with James South and family.
ALS 2pp damaged & C 2pp inc TxU:H/M–0620.4; Reel 1086

508. [1821] From [J.?] Henry Royal Observatory
Compares observed times of rocket bursts at observatory and at Blackman St.
ALS 1p TxU:H/M–1002; Reel 1083

509. [1821–1822] To Henry Kater [Slough]
Sends set of William Herschel's duplicated papers. Curious results 'examining the rings,' but bad weather has hampered experiments. Correction of spherical aberration. JH is using J. d'Alembert's *Opuscules*. Image of Alpha Lyrae.
ALS 3pp RS:HS 19.17 (C: RS:HS 25.13.21)

510. [1821 to 1828] To Mrs. Mary Frances Kater Kensington
Asks MK's permission to allow a close friend to make use of the words once sent to her by JH as suitable for the music of *Cherry Ripe.*
C 1p RS:HS 25.13.23

511. [1821 to 1828] To Mrs. Mary Francis Kater [Slough]
Accidentally found some verses on a scrap of paper, and wondered if it would fit the music of *Cherry Ripe*; sends verses to MK.
C 1p RS:HS 25.13.22

512. [1821] From James South Blackman St.
Sends three chronometers set to Blackman St. time. Confirm them by a transit of Arcturus tomorrow, then return them to JS at Blackman St. Gives times for next rocket firings.
ALS 1p inc TxU:H/M–0946.1; Reel 1083

513. 1821–1–1 From Henrik Johan Walbeck Königsberg
Further eclipse observations from around the world [see HW's 1820–7–30].
ALS 3pp (Latin) RS:HS 18.9

514. [1821?] To William Whewell n.p.
Provides detailed travel suggestions for WW's planned trip to the continent. Asks WW to deliver letters to various European scientists.
ALS 6pp TC, Camb. Add. Ms.a.207[18] (C: RS:HS 21.36)

515. 1821–1–12 To Jean Baptiste Biot Cambridge
Regarding refraction in crystals of apophyllite and quartz.
draft 3pp RS:HS 4.90 & 20.108

516. 1821–1–16 To Henry Kater Slough
Collecting English observations of eclipse of 7 Sept. 1820 for H. J. Walbeck. Construction of object glasses.
ALS 3pp damaged RS:HS 19.14 (C: 20.109)

517. 1821–1–16 From G[erard] Moll Utrecht
Thanks for the kind reception afforded him at Slough. Sends a copy of [J. H.?] Schröder's paper on the discoveries of William Herschel. Also sends some observations of the late solar eclipse made in Holland.
ALS 2pp RS:HS 13.47

518. [1821 or 1827]–1–17 From William Hyde Wollaston 1 Dorset St.
Sir Edward Codrington will take charge of chronometer for Captain Copeland if delivered to Codrington at Portsmouth.
ALS 1p RS:HS 18.281

519. 1821–1–22 To Charles Babbage Slough
Hopes to call on Thursday to discuss queries. Regarding recent astronomical experiments. Alterations to his twenty-foot telescope.
ALS 2pp RS:HS 2.153 (C: 20.110)

520. 1821–1–29 From John Brinkley Observatory, Dublin
It was too cloudy for any worthwhile observations to be made during the recent eclipse. Looking forward to the publication of JH's work. Outlines his own recent and projected work. Thomas Brisbane observed the eclipse at Cork.
ALS 3pp RS:HS 4.274

521. 1821–1–30 From William Hasledine Pepys n.p.
Invites JH to the London Institution to meet Humphry Davy.
AL 1p RS:HS 13.337

522. 1821–1–31 From Charles Babbage Devonshire St.
Meeting at Dr. George Pearson's on Friday next.
ALS 1p RS:HS 2.154

523. 1821–2–5 From Charles Babbage Devonshire St.
Hopes JH will meet CB and Francis Baily in the near future to discuss arrangements for the Astronomical Society.
ALS 1p RS:HS 2.152

524. 1821–2–9 To Heinrich Wilhelm Matthias Olbers [London]
HO was elected associate of Astronomical Society. Prize for 1821 relates to Saturn's satellites.
AAbs 1p RS:HS 19.9

525. 1821–2–10 To Alexis Bouvard n.p.
AB was elected to new Astronomical Society. Extracts from 11 Feb. 1821 letter to C. F. Gauss.
AAbs 1p RS:HS 19.3

526. 1821–2–11 To Jean-Baptiste Delambre Slough
Announces election of JD as associate member of the Astronomical Society. [In a postscript,] JH states that the Society's prize for next year will be on the theory of the motions of Saturn's satellites.
ALS 3pp Académie des sciences (Paris)

527. 1821–2–11 To Carl Friedrich Gauss Slough
CG has been elected associate of the Astronomical Society. Hopes the benefits will be mutual. Details of the prize subject for the present year. Has read CG's communications on the Reichenbach circle to the Society, where it evoked great interest. Comments on this. CG's certificate has been signed by Fearon Fallows, newly appointed Cape astronomer.
ALS 4pp NSUB Cod.Ms.Gauss 99 (draft: RS:HS 8.72 & C: 20.111)

528. 1821–2–14 From William Herschel Slough
Gratitude for election to Astronomical Society. Notes on double star Zeta Orionis. Regards to James South.
LS 1p †(by ?) RAS:WH Archive 1/11.5; Reel 12

529. 1821–2–15 or earlier From James Grahame n.p.
Family news.
ALS 3pp RS:HS 8.302

530. 1821–2–17 From Francis Baily Gray's Inn
Regarding the printing of the report and sanction of council. Details several points on which he requires guidance. New observatories in India. Can JH dine with him on Thursday?
ALS 3pp RS:HS 3.39

531. 1821–2–18 To [Postmaster?] Slough
Complains of delay in receiving mail, which was addressed to Colnbrook but misdirected to Abingdon and to Marlow.
ALS 1p, encl 1p TxU:H/L-0747; Reel 1089

532. 1821–2–19 From Charles Babbage Totnes
Regarding W. T. Brande's attack on the Astronomical Society. Joseph Banks and his opposition to CB. Recent happenings in the Astronomical Society. Thanks for the volumes.
ALS 4pp RS:HS 2.155

533. 1821–2–26 To Charles Babbage Slough
Regarding W. T. Brande's recent attack on the Astronomical Society. The Board of Longitude and its dealings with CB. Has met Henry Warburton recently.
ALS 3pp RS:HS 2.156 & 20.112

534. 1821–2–26 To Henrik Johan Walbeck Slough
Reports to HW, director of the observatory at Åbo, on observations, made by a variety of British astronomers, of the solar eclipse of 7 September 1820.
ACCS 3pp (Latin) †(by JH) RS:HS 18.10a (C: RS:HS 20.113)

535. 1821–3–8 From Miles Bland Cambridge
There is an opening for a Lay fellowship. Is JH interested? News of J. W. Whittaker's appointment.
ALS 1p RS:HS 4.154

536. 1821–3–8 To James Grahame Slough
Responds to JG's consolations [see JG's 1820–11–6] in JH's loss [Miss Gwatkin?]; comments on his commitment to astronomy, together with family news [letter completed 1821–3–12].
draft 4pp RS:HS 8.303 (C: RS HS 20.114)

537. 1821–3–9 From Sir William Watson Bath
Sorry to hear JH's father and mother are ill and cannot visit Bath. Received account of new Astronomical Society of London. Honored to be listed as member with old friend William Herschel.
ALS 2pp RS:HS 18.118

538. 1821–3–10 To Miles Bland n.p.
Though he could qualify, he would rather not try for the fellowship as he does not require it so much as some other persons may.
draft 2pp RS:HS 4.154a

539. 1821–3–11 From Heinrich Wilhelm Matthias Olbers Bremen
Honored on being elected Foreign Member of the Astronomical Society of London. Finds that the comet he observed in January had already been seen by J. N. Nicollet and J. L. Pons. Saw the remarkable appearance on the dark part of the moon noticed by Henry Kater. Thinks Kater's theory is wrong. Will soon be bringing out a paper on the moon's volcanoes.
ALS 5pp (German with English trans.) RS:HS 13.174

540. 1821–3–16 From James Grahame Edinburgh
Family news, JG's reduced circumstances, and public affairs.
ALS 3pp RS:HS 8.304

541. [1821]–3–17 From Mary Pitt Herschel Slough
Sends John Flamsteed's Atlas and Catalogue. Discusses JH's intention to review all of William Herschel's double stars. Wants details of JH's appointment to committee.
ALS 2pp & C 2pp TxU:H/M–0620.5; Reel 1086

542. 1821–3–17 From Joseph Nicolas Nicollet Paris
Has received the regulations of the Astronomical Society of London—a great encouragement to astronomers. Has made observations on the orbit of the comet. Has sent two memoirs to T. M. Brisbane.
ALS 3pp (French) RS:HS 13.134

543. 1821–3–19 From Francis Baily Gray's Inn
Can he give him the address of Capt. W. E. Parry? Should he send a report to Humphry Davy?
ALS 1p RS:HS 3.40

544. 1821–3–23 To Johann Elert Bode [Astronomical Society]
JB was elected associate of Astronomical Society. Asks that Society's annual prize be announced in *Jahrbuch*. Please insert 'Reclamation' of Francis Baily against Jabbo Oltmanns, relative to the eclipse of [Thales?], into next *Ephemeris*. Send copies of JB's *Ephemeris* for T. F. Colby and for Society.
AAbs 1p RS:HS 18.10c & 19.2

545. 1821–3–23 From Mary Pitt Herschel Slough
When will JH arrive?
ALS 1p & C 1p TxU:H/M–0620.6; Reel 1086

546. 1821–3–23 To Giuseppe Piazzi [Astronomical Society]
Elected to membership in Astronomical Society. Subjects of annual prize. What is title of [C.] Brioschi's work?
AAbs 1p RS:HS 18.10d

547. [1821]–3–27 From Samuel Young Great Russell St.
Residing outside London. Will return for balloting of [William?] Bridgman at Thursday's R.S.L. meeting.
ALS 1p RS:HS 18.318

548. 1821–4 To Charles Babbage n.p.
Would CB breakfast with JH in order to meet members of the Correction Committee.
ALS 1p RS:HS 2.157 & 20.115

549. 1821–4 To Gerard Moll [Astronomical Society]
Elected to membership in Astronomical Society. Thanks for GM's paper on eclipse.
AAbs 1p RS:HS 18.10e

550. 1821–4 To Heinrich Christian Schumacher [Astronomical Society]
HS was elected associate of Astronomical Society. This year's prize relates to theory of Saturn's satellites. Does HS plan to continue his table of apparent places of fixed stars?
AAbs 1p RS:HS 18.10f & 19.11

551. 1821–4–5 From Charles Hatchett Chelsea
Invitation to dinner.
ALS 1p RS:HS 10.169

552. 1821–4–9 From William Edward Parry 3 Downing St.
Can he meet JH at the Astronomical [Society] on the 13th?
ALS 1p RS:HS 13.217

553. 1821–4–16 From Baron George Best 19 Sloane St.
Sends latest news of J. E. Bode.
ALS 2pp RS:HS 5.57

554. 1821–4–17 From Heinrich Christian Schumacher Copenhagen
Thanks for honor conferred by Astronomical Society. Mentions [F. W.] Bessel's paper on the 4th satellite of Saturn. Calculated tables of apparent places of 45 stars, which HS plans to publish annually.
ALS 3pp RS:HS 15.367

555. 1821–4–19 From Francis Baily Gray's Inn
Is sorry JH could not manage the Council meeting last Monday. Suggests a date for the next meeting.
ALS 2pp RS:HS 3.41

556. 1821–4–[19] To Joseph Nicolas Nicollet [Astronomical Society]
Thanks for JN's 'Comets Elem[ts]' and book on libration of moon. Please send JN's and J. L. Pons's first observations of comet. Invites JN to become associate of Astronomical Society.
AAbs 1p RS:HS 18.10g

557. 1821–4–20 From Alexis Bouvard Paris
Thanks for the letter informing him that he was elected corresponding member of the Astronomical Society. Comments on the prize questions. Gives details of his astronomical publications.
ALS 3pp RS:HS 5.89

558. 1821–4–29 From Mary Pitt Herschel Slough
Death of Mrs. Beckwith's baby. Warns JH against one-legged swindler 'Mr. Davis,' who may ask JH for money. Still in search of a cook.
ALS 2pp & C 2pp TxU:H/M–0620.1; Reel 1086

559. 1821–5–4 To Charles Babbage n.p.
Please forward the address of H. Meikle, the computer.
ALS 1p RS:HS 2.158 & 20.116

560. 1821–5–6 From Mary Pitt Herschel Slough
William Herschel met with Mr. Gwatkin to discuss JH's proposal of marriage to Miss Gwatkin. MH sees no impediment. WH wants JH to return soon.
ALS 1p & C 1p TxU:H/M–0620.2; Reel 1086

561. 1821–5–8 From Charles Babbage n.p.
Will call on JH tomorrow. Has had a letter from Benjamin Gompertz.
ALS 1p RS:HS 2.159

562. 1821–5–8 To Edward Pigott Slough
Sends EP a copy of the first annual report of [Astronomical Society?] and invites EP to join the 'circle.'
ALS 2pp DMC.695A.885.26

563. 1821–5–9 To Carl Friedrich Gauss London
Sends first annual report of the Astronomical Society and will be sending one of JH's papers ['On the Aberrations of Compound Lenses and Object-glasses,' *RSPT* (1821), 222–67], which JH summarizes.
ALS 3pp diag NSUB Cod.Ms.Gauss 99 (C: RS:HS 20.117)

564. 1821–5–10 Edward Pigott Bath
Declines JH's 'polite and kind offer' due to declining health.
ALS 1p RS:HS 14.162

565. 1821–5–12 From Johann Elert Bode Berlin
Thanks for letter. Will do all in his power to further the cause of the Astronomical Society. Missing parts of *Nautical Almanac* and *R.S.P.T.* Would JH look into matter. Unable to send complete set of his Astronomical Yearbook, but can supply some for a suitable sum.
ALS 3pp (German) RS:HS 4.164

566. 1821–5–14 From George Peacock Trinity College, Cambridge
Cambridge news, GP's own activities, and comments on the progress of the observatory plans.
ALS 4pp RS:HS 13.275

567. 1821–5–16 From Joseph Johann Littrow Vienna
Welcomes astronomical correspondence. Is grateful for having his name proposed as a member of the Astronomical Society. Comments on the work of the Observatory. Would like the memoir of Edward Troughton.
ALS 3pp (French) RS:HS 11.232

568. 1821–5–16 To John Rennie Blackman St.
Wishes JR to secure for the syndics of Cambridge University Observatory a copy of JR's plan for a Cape Observatory.
ALS 2pp NatLibScot MS.19830, f.210

569. 1821–5–17 To Joseph Nicolas Nicollet [Astronomical Society]
Sends packet by Mr. Sellique containing report of Astronomical Society council and James South's corrections for June to Dec. 1821.
AAbs 1p RS:HS 18.10h

570. 1821–5–19 From Basil Hall HMS Conway, Valparaiso
Sending some observations on a comet which appeared last month. Has had the pendulum up for some weeks but has been unable to carry out many observations owing to official duties. Has been working a good deal on occultations. Wishes some had been published as they are of great practical use.
ALS 4pp RS:HS 9.162

571. 1821–5–20 From Charles Babbage n.p.
Regarding JH's misunderstanding with George Dollond in the latter's quarrel with James South.
ALS 4pp RS:HS 2.160

572. 1821–5–21 From Charles Babbage n.p.
Unable to leave his wife. Has read about JH's purchases at Christie's.
ALS 1p RS:HS 2.161

573. 1821–5–29 To Francis Baily n.p.
If FB will send him his paper on the eclipse, he will write his report. Regarding the publication of information received from various members. Duties of the Foreign Secretary.
AL 3pp RS:HS 3.42 & 20.118

574. 1821–5–31 To Carl Friedrich Gauss [Astronomical Society]
Sends first report of Astronomical Society council and James South's corrections for June to Dec. 1821.
AAbs 1p RS:HS 18.10i & 19.6

575. 1821–6–4 From Francis Baily Gray's Inn
Is forwarding his paper on the eclipse. Publication of information received from members. Regarding the refereeing of Gerard Moll's paper on the eclipse.
ALS 2pp RS:HS 3.43

576. 1821–6–7 To Alexis Bouvard London
Has received AB's 1821–4–20. Thanks for report on the solar eclipse. Reports British observations of the eclipse.
ALS 2pp BostonPubLib Mss.Acc.66 (AAbs: RS:HS 19.3)

577. 1821–6–7 From George Peacock Trinity College, Cambridge
About plans for the Cambridge observatory.
ALS 4pp RS:HS 13.276

578. 1821–6–12 From Mr. Sellique Paris
Comments on J. B. Biot's reactions to JH's observations on polarization of light in apophyllite.
ALS 1p (French) RS:HS 17.187

579. 1821–6–19 From James South Blackman St.
Coordinates plans for firing rockets at Blackman St. and Slough. Problems with chronometer errors. [JH annotation: Longitudes of Slough, Greenwich, and Blackman St. Will fire rockets next on 10 July.]
ALS 3pp † TxU:H/M–0947; Reel 1083

580. 1821–6–20 To Charles Babbage Slough
No need for meeting of Correction Committee. Will be in town on Friday and hopes to meet him.
ALS 2pp RS:HS 2.162 & 20.119

581. 1821–6–22 From Sir William Watson Dawlish
Has not seen Mr. Penn's work JH mentioned. WW was chosen member of Astronomical Society. Asks JH to pay dues; WW will reimburse him.
ALS 2pp RS:HS 18.119

582. 1821–6–25 From Heinrich Christian Schumacher Altona
Sends invitation to participate in an astronomical journal. Asks JH to forward letter to [F.] Fallows at the Cape of Good Hope. Mentions [Thomas] Young's plans to travel. Requests that JH forward letters to [W. H.] Wollaston, [John] Brinkley, [David] Brewster, and [Henry] Kater. Encloses announcement of founding of *Astronomische Abhandlungen* and of *Astronomische Nachrichten*.
ALS 3pp encl RS:HS 15.368

583. 1821–6–26 From Joseph Johann Littrow Vienna
Sending his book on *Theoretical and Practical Astronomy*. Also sends books for Thomas Catton. Comments on some of the varying observations made at observatories.
ALS 3pp (German) RS:HS 11.233

584. 1821–7–2 To Charles Babbage Slough
[Theophila] Gwatkin will accept Georgiana Babbage's invitation. Will set date after Gwatkin returns from Baylin. Asks CB to be trustee in JH's marriage settlement.
ALS 2pp TxU:H/L-0053; Reel 1054 (C: RS:HS 20.120)

585. [1821–7–2 or later] To Georgiana Babbage [London]
Arranges time for GB to meet [Theophila] Gwatkin. JH leaves London tomorrow for Slough; returns Saturday for 'visitation.'
ALS 2pp TxU:H/L-0055; Reel 1054

586. 1821–7–5 From Charles Babbage Devonshire St.
Will be having a meeting with the Gwatkins. Regarding becoming a trustee of JH's and matters appertaining to it.
ALS 4pp RS:HS 2.163

587. 1821–7–9 From Mary Pitt Herschel to Charles Babbage Slough
Asks CB to to look very carefully into financial affairs relating to JH's impending marriage [to Miss Gwatkin]. Begs CB not to tell JH she has written.
ALS 1p TxU:H/M–0613; Reel 1086 (C: RS:HS 20.121)

588. 1821–7–10 From James Grahame Edinburgh
On the loss of a friend's wife and child, the death of Napoleon, and on the JH/JG correspondence.
ALS 3pp RS:HS 8.305

589. 1821–7–10 From Mary Pitt Herschel Slough
Dr. Pope examined William Herschel and reported that WH's only complaint was old age. Finances relating to JH's marriage settlement. Must decide whether to sell JH's horse. Compliments to Miss Gwatkin.
ALS 2pp & C 2pp TxU:H/M–0620.3; Reel 1086

590. 1821–7–12 To Giuseppe Piazzi n.p.
Is making arrangements with GP to republish GP's star catalogue in Cambridge.
draft 3pp RS:HS 13.394 & 20.122

591. 1821–7–23 To Mary Pitt Herschel Dieppe
JH, Charles Babbage, Richard Jones, and George Peacock arrived in France after difficult voyage.
ALS 2pp & C TxU:H/L-0515.1; Reel 1053

592. 1821–7–23 To William Herschel Dieppe
Terrible voyage across Channel. George Peacock and Richard Jones accompanied JH and Charles Babbage for part of journey.
ALS 3pp damaged & C TxU:H/L-0551; Reel 1053

593. 1821–7–31 To Mary Pitt Herschel Paris
Leaving Paris tomorrow for Geneva and Milan. Describes tour through France. Names scientific men whom JH, Charles Babbage, Richard Jones, and George Peacock met. Visited French Board of Longitude. Alexander von Humboldt was of great service.
ALS 3pp & C TxU:H/L-0515.2; Reel 1053

594. [1821–8] From [Charles Babbage] to Mary Pitt Herschel n.p.
Explains how CB responded to JH's request that CB be a trustee for marriage settlement, prior to JH's marriage to Miss Gwatkin. Will meet the Gwatkins at Friday's party. Hopes to find time to speak to JH about this in private. Declines to pursue the course requested by William and

Mary Pitt Herschel until JH confides more in CB.
C 3pp TxU:H/M–0968; Reel 1083

595. 1821–[8] To Heinrich Christian Schumacher n.p.
JH willing to communicate with HS on worthwhile matters. Subscribes Astronomical Society to HS's *Astronomische Nachrichten*. Explains 'Prize Question.' Requests copy of F. W. Bessel's essay on fourth satellite of Saturn. Society resolved to print 'daily tables of 46 stars.'
AAbs 1p RS:HS 19.11

596. 1821–8–5 To Henry Kater Paris
Delivered R.S.L. resolution on 'Junction of French & English triangles' to François Arago. Arago alone will meet HK and T. F. Colby to finish triangles this summer. J. B. Biot will collaborate next year with HK on pendulum measurements.
ALS 3pp RS:HS 19.15 (C: 20.123)

597. 1821–8–5 From Samuel Young Formosa Cottage
Laid up with gout. Thanks JH for proposing SY's eldest son [George] as member of taxonomical society. SY trying to gain captain's commission for son under Lord Melville [R. S. Dundas]. Describes crystal that SY cannot identify.
ALS 1p † RS:HS 18.315

598. 1821–8–7 From Mary Pitt Herschel Slough
Concerned about JH in his travels on Continent. Relieved to receive news from JH. William Herschel is in good health. Thomas Baldwin continues as before.
ALS 2pp † & C 3pp TxU:H/M–0620.7; Reel 1086

599. 1821–8–10 From Picot Trembley Frontenex, Geneva
Extremely formal letter of greeting to JH while traveling in Switzerland. Ends with a long postscript describing a major Swiss monument in Lucerne.
ALS 3pp (French) RS:HS 17.443

600. 1821–8–12 To William Herschel Geneva
No letters from Slough. Terrible weather. Unable to explore Alps. Purchased carriage for remainder of tour. Describes journey from Paris through Jura Mountains to Geneva. Met J. A. Deleu family, M. A. Pictet, Alexandre Marcet, [Gilbert Elliot] Lord Minto, and Italian expatriates from Neapolitan revolution. Proposes itinerary through Alps, weather permitting.
ALS 4pp & C TxU:H/L-0552; Reel 1053

601. 1821–8–15 To James A. Gordon Chamouni
Wet weather has given him time for letter writing. Spent a week in Paris. Passed through Dijon, and has spent some time at Geneva, where there is a flourishing Philosophical Society. Comments on the meetings. Weather has been bad. Gives a Neapolitan story.
draft 4pp RS:HS 8.164 & 20.124

602. 1821–8–15 To Mary Pitt Herschel Chamouni [Chamonix]
Describes scenery at Mont Blanc. Trip by mules into mountains. Barometer measurements by Charles Babbage. [Letter continued 16 Aug.:] Plans today include visit to Mar de Glace glacier. [Letter continued 20 Aug.:] Found excellent guide, Coutet, with knowledge of geology. Describes expedition to Mar de Glace glacier and JH's sketches. More barometer observations. [Letter continued 22 Aug. in Geneva:] No mail arrived from Slough. Leave tomorrow for Chambéry and Turin. Planning to leave Italy by 2 Sept.
ALS 4pp & C TxU:H/L-0515.3; Reel 1053

603. 1821–8–25 From Mary Pitt Herschel Slough
Comments on JH's European adventures. Death of MH's brother Thomas Baldwin. Mr. Grover of Eaton expects death of Mr. Vince soon. Mr. Langton is stabling JH's horse.
ALS 3pp & C 4pp TxU:H/M–0620.8; Reel 1086

604. 1821–8–28 From Johann Elert Bode Berlin
Should have received the Astronomical Yearbook. No news of its receipt or answers to letters of 12 May and 19 June. Observations which JH sent will appear in the 1824 Yearbook. Has not received the *Nautical Almanac* for 1821/2.
ALS 1p (German) RS:HS 4.165

605. 1821–8–30 To William Herschel Como
Journey of JH and Charles Babbage from Geneva via Mont Cenis to Turin, where effects of revolution are still evident. Describes Milan. Oppressive heat and torrential rain. Plan to return via Monte Rosa or 'Mont' St. Gotthard. Language problems.
ALS 4pp & C TxU:H/L-0553; Reel 1053

606. 1821–9–15 From Mary Pitt Herschel Slough
Looks forward to JH's return. Visited [E. P. T.] Babbage and Mrs. Beckwith in London.
ALS 2pp † & C 2pp TxU:H/M–0620.9; Reel 1086

607. 1821–9–15 To Mary Pitt Herschel Moudon, Vaud, Switzerland
Describes Como [Lombardy]. Visit with Pietro Configliachi [professor of physics at University of Pavia]. Crossing of alpine passes and Lake Maggiore. Ascent of Monte Rosa. JH was first Englishman to reach summit. [Letter continued 17 Sept. in Berne:] News from Slough of death of JH's uncle [Alexander Herschel]. Expects to return home in three weeks.
ALS 4pp & C TxU:H/L-0515.4

608. 1821–9–21 To Signor Configliachi Interlaken
Asks PC for barometer readings for specific days in September, as JH was at the summit of Mount Rosa with a barometer. JH would also appreciate any other barometer readings around the foot of the mountain for those days. JH is on his way back to England.
ALS 2pp WT 65659

609. 1821–10–1 From Alekis F. Orloff to William Herschel Brighton
Would like to know when he can call on the Herschels.
ALS 1p RS:HS 13.207

610. 1821–10–2 From [D. M.] Vallot Dijon
Missed JH (traveling with Charles Babbage) when they passed through Dijon. Provides some barometric observations along with a comparison of measuring instruments. Appears to be commenting on some of Charles Babbage's interests in statistics, and believes that G. W. Leibniz did some work in this also, but died before publishing it. [Parts of this letter illegible.]
ALS 3pp (French) RS:HS 17.458

611. 1821–10–3 To William Herschel Paris
Expects to be home in one week. May be delayed at Calais by astronomers' trigonometrical operations. JH's and Charles Babbage's excursion from Berne to foot of Alps. Visited [P. E.] Fellenberg at Hofwyl.
ALS 3pp & C TxU:H/L-0554; Reel 1053

612. 1821–10–5 From Heinrich Christian Schumacher Altona
Thanks for forwarding notice on journal to [F.] Fallows. Hopes to use the longitudinal differences of English observations to improve observations.
ALS 2pp RS:HS 15.369

613. 1821–10–8 To Mary Pitt Herschel Dover
Windy passage across Channel. Will arrive home in two days.
ALS 1p & C TxU:H/L-0515.5; Reel 1053

614. 1821–10–10 From James A. Gordon Stocks, Hemel Hempstead
Was pleased to receive his letter. He too has been travelling abroad. Outlines his travels. Expects JH has a good collection of sketches. Saw Louis XVIII at Paris.
ALS 3pp RS:HS 8.165

615. [1821]–10–16 From Alexis F. Orloff Brighton
Regrets to hear of the poor health of Sir William Herschel. Would like a few lines of Sir William's handwriting for his autograph collection.
ALS 2pp RS:HS 13.208

616. 1821–10–27 From Charles Babbage Devonshire St.
Has had a visit from Mr. Cox. Regarding payment of carriage for parcel from Vienna. Will be visiting him on Tuesday.
ALS 2pp RS:HS 2.164

617. 1821–10–29 To Charles Babbage Slough
Is glad he is coming on Tuesday. Regarding the parcel of books from Vienna. Recent experiments with mercury.
ALS 3pp RS:HS 2.165 & 20.125

618. 1821–10–29 To John Pond Slough
Provides a detailed critique of a mathematical paper by [J. M. H.] Wronski, which JP presented on Wronski's behalf to the R.S.L. Suggests that JP urge Wronski to withdraw the paper.
ALS 4pp RS:HS 14.30 (C: RS:HS 20.126)

619. 1821–10–31 From David William Smith Alnwick
Letter of thanks for greetings from JH.
ALS 2pp RS:HS 16.181

620. 1821–11–1 To John Pond Slough
Discussion and criticism of paper by a [J. M. H.] Wronski on methods in theory of equilibrium of fluids.
ALS 4pp RS MM.6.74

621. [1821–11] To [?] 9 Downing St.
A covering note written for an enclosed letter (not available), all of it seeming to relate to filling a Cambridge position, for which J. S. Henslow seems eminently qualified.
ALS 2pp APS B.H435p.1

622. 1821–11–5 To Charles Babbage n.p.
Would like to visit this week. Has had an attack of rheumatic fever. Gives tables of orbit of a comet.
ALS 3pp RS:HS 2.166 & 20.127

623. 1821–11–5 From Mary Pitt Herschel Slough
Inquires after JH's health. When will JH visit Slough? Efforts to find new footman to replace Mr. Brown, who leaves Thursday.
ALS 2pp & C 2pp TxU:H/M–0620.10; Reel 1086

624. 1821–11–7 From Mary Pitt Herschel Slough
Asks JH to vouch for Richard Copeland, who expects navy promotion. Directed two footmen to call on JH for interviews.
ALS 2pp & C 1p TxU:H/M–0620.11; Reel 1086

625. 1821–11–10 From Francis Baily Gray's Inn
Would like to borrow one of H. C. Schumacher's books from JH. Should they print Giuseppe Piazzi's communication on the eclipse?
ALS 2pp RS:HS 3.44

626. 1821–11–10 From Johann Elert Bode Berlin
Inquires after a case of Year-books previously dispatched. Cost £14–4–0. No answers to his letters of May, June or August.
ALS 1p (German) RS:HS 4.166

627. 1821–11–10 From Carl Friedrich Gauss Göttingen
Has been absent the whole summer and fears that his letter acknowledging his election to the Astronomical Society may have gone astray. Is grateful for the honor. Sends a paper of his own on a new method of making observations with the heliotrope. Comments on the usefulness of this instrument.
ALS 2pp RS:HS 8.73

628. 1821–11–10 From Mary Pitt Herschel n.p.
Forwards ham from Susan White and letter from Richard Copeland. Tenant farmer Mr. Davenport refuses to pay rent to William Herschel. Ask Mr. Blagrave for legal advice. Hired new footman. JH is invited to Miss De Luc's party on Monday.
ALS 2pp & C 2pp TxU:H/M–0620.12; Reel 1086

629. 1821–11–12 From George Peacock Trinity College, Cambridge
News from Cambridge, including information about the new observatory.
ALS 4pp RS:HS 13.277

630. 1821–11–14 From Sir William Watson Bath
Good health in both families. JH's letter about Alps was interesting. Will toast William Herschel's birthday tomorrow. What does WW owe JH for Astronomical Society dues?
ALS 2pp RS:HS 18.120

631. 1821–11–17 To Friedrich Wilhelm Bessel Slough
Sixteen-month delay in receipt of FB's observations. Willing to publish James Bradley's observations of Halley's Comet but questions accuracy of Nathaniel Bliss's. Sends John Brinkley's analysis [of April 1821 comet observed by Basil Hall in southern hemisphere]. Please describe Georg Reichenbach's new [transit] circle at Königsburg.
C 1p RS:HS 20.128

632. 1821–11–17 From Thomas Glanville Taylor Royal Observatory, Greenwich
Sends corrections to Polaris observations.
ALS 1p RS:HS 17.333

633. 1821–11–20 To Friedrich Wilhelm Bessel n.p.
Responds to FB's much delayed answer to JH's 1820–5–9; comments on some observational matters.
ALS 3pp AdWDDR Bessel 225 (C: RS:HS 20.128)

634. 1821–11–20 From Samuel Young Exmouth
Sends form signed by SY's eldest son [George] to JH, who will stand proxy for SY's son, still at sea. SY eager to exchange minerals with JH. Invites JH to Formosa Cottage after Christmas.
ALS 1p † RS:HS 18.316

635. 1821–11–25 To Charles Babbage Downing St., Westminster
Regarding barometric observations.
ALS 3pp RS:HS 2.167 & 20.129

636. 1821–11–29 From George Peacock Trinity College, Cambridge
News of Cambridge University.
ALS 4pp RS:HS 13.278

637. 1821–12–2 To Charles Babbage Slough
Regarding their chances of obtaining a professorship at Cambridge.
ALS 3pp RS:HS 2.168 & 20.130

638. 1821–12–2 From Thomas Catton St. John's College, Cambridge
The Plumian professorship is vacant. News of who is likely to be appointed. Congratulations on the award of the Copley medal.
ALS 2pp RS:HS 5.223

639. 1821–12–2 From George Peacock Trinity College, Cambridge.
About the university politics of filling vacant posts.
ALS 2pp RS:HS 13.280

640. 1821–12–11 From Johann Elert Bode Berlin
Acknowledges JH's receipt of Astronomical Yearbook, also payment for same. Thanks for information of Southern comets.
ALS 2pp (German) RS:HS 4.167

641. 1821–12–11 To Joseph Johann Littrow n.p.
JL was elected associate of Astronomical Society. Received JL's books and papers. Will send Society's *Transactions*. John Pond gave permission to test Robert Molyneux's clock at Royal Observatory. Questions F. G. W. Struve's transit determinations of double stars. Pond discovered errors in Greenwich transit instrument and places little dependence on its observations since late 1819. Sends John Brinkley's analysis [of April 1821 comet observed by Basil Hall in southern hemisphere]. Asks about Halley's Comet and parallax. Wants information on object glasses of 6-inch diameter or greater. Requests copy of JL's annual published observations.
AAbs 2pp RS:HS 19.8

642. 1821–12–19 From Mary Pitt Herschel Slough
Tell Mr. Blagrave to decline Mr. Ramsbottom's offer to purchase Crown Inn at Slough from William Herschel and to proceed with repairs on house and offices. Problems finding another footman.
ALS 3pp & C 3pp TxU:H/M–0620.13; Reel 1086

643. 1821–12–20 From Charles Babbage n.p.
Would like to see him and discuss plans regarding his calculating machine.
ALS 1p RS:HS 2.169

644. 1821–12–31 From James South Blackman St., Southwark
Discusses a paper JS is preparing for the Astronomical Society. Mentions his recent observations. Suspects the report of a comet is a hoax.
ALS 3pp RS:HS 16.414

645. 1821–12–31 From Thomas Young 48 Welbeck St.
Board of Longitude will meet on 3 Jan. to examine instruments and proposals, and to consider Fearon Fallows's report from Cape of Good Hope.
ALS 1p RS:HS 18.320

646. [1822?] From Charles Babbage n.p.
Has had a letter from George Peacock. Regarding the Lucasian Professorship at Cambridge.
ALS 1p RS:HS 2.346

647. [1822 or earlier] From Edward Daniel Clarke n.p.
Thanks for his kind note. Mr. Harrison wishes him to prepare a paper for the R.S.L. Will find his experiments continued in the next number of Thomson's *Annals*. As soon as his blow pipe is repaired he will let him know.
ALS 3pp RS:HS 5.260

648. [1822 or earlier] From Edward Daniel Clarke n.p.
Forgot to mention it last time they were together that another mammoth has been found at Trumpington. This agrees with G. C. L. D. Cuvier's theories. Regarding the ore JH has just analyzed.
ALS 3pp RS:HS 5.259

649. 1822 To *Edinburgh Philosophical Journal* n.p.
Lays out various practical rules for the determination of the radii of a double acromatic object glass.
L *Edinburgh Philosophical Journal, 6* (1822), 361–70

650. 1822 From Joseph Goodall n.p.
Hopes he will be more pleased with the version of the epitaph he is enclosing. Would like to discuss it with him if he has the time to spare.
ALS 4pp RS:HS 8.143

651. 1822 From Joseph Goodall n.p.
Hopes his second paper will not falsify the opinion. Would like his objections or comments.
ALS 2pp RS:HS 8.142

652. [1822 to 1829] To Henry Kater [London]
Unable to be at Katers' home for dinner, but will arrive later. JH is pleased to find HK's and JH's measurements of Mars agree with those of JH's father, William, and compares these with some other measures.
C 1p RS:HS 25.13.20

653. [1822] To Samuel Lathan Mitchill London
Sends first volume of the 'Transactions' of the Astronomical Society to the Philosophical Society of New York.
ALS 2pp Columbia Univ. Library

654. [1822] From Claude S. M. Pouillet Paris
Tells JH to meet a Monsieur Debure [?] while in Brussels; arranges a time.
ALS 1p (French) RS:HS 14.183

655. 1822 From [J. S. C.] Schweigger Halle
Sends pages from his publication. Wishes to know if JH has received previous two packages accompanied by detailed letters through Heinnemann [?] in London.
ALS 1p (German) RS:HS 15.413

656. [1822?] From James South n.p.
Regrets having missed JH. Has observed the seven satellites of Saturn and the fifth star of the Trapezium. Wishes to observe some of the objects observed by Wilhelm Struve.
ALS 4pp RS:HS 16.476

657. 1822–1–2 From Joseph Johann Littrow Vienna
Sends books and observations for various observatories. Comments on recent astronomical work. Points out some discrepancies in John Pond's works. Praises the telescopes of Josef Fraunhofer.
ALS 15pp (German with English trans.) RS:HS 11.235–6

658. 1822–1–4 From Francis Baily Gray's Inn
Has only had 100 copies of his own tables printed. Difficulties as to whom he should send them. Regarding the observations of the old astronomers. Any news regarding the printing of Giuseppe Piazzi's paper?
ALS 3pp RS:HS 3.45

659. 1822–1–10 From Franz Xaver von Zach Genoa
Thanks JH for the letter informing FZ of his nomination to the Astronomical Society. Describes how JH can send packages to Geneva via the British Consul.
ALS 1p (French) RS:HS 18.360

660. 1822–1–14 From William Pearson Kilworth near Wilford
Thanks JH for the 'kind notices respecting the tabular numbers' in a paper of JH's. Comments on the demands made of William Herschel despite age and health.
ALS 3pp RS:HS 14.150

661. 1822–1–15 From James South Blackman St., Southwark
Gives directions for some time measurements involving rockets JS was planning in conjunction with JH. Mentions various double stars.
ALS 2pp RS:HS 16.415

662. [1822]–1–20 From Mary Pitt Herschel Slough
John and William Langton will handle legal matter of Mr. Davenport's overdue rent. Sends JH's stockings. Mrs. Goodall says wonderful things about JH. Mary Baldwin leaves Slough on Wednesday. MH cannot stay alone; will seek another companion.
ALS 3pp & C 2pp TxU:H/M–0621.1; Reel 1086

663. 1822–1–20 From William Whewell Brighton
Urges JH to come to Brighton and then to return with him to Cambridge; stresses that both he and Richard Jones are anxious to talk with JH.
ALS 1p RS:HS 18.161

664. 1822–1–21 From Richard Jones Brighton
Begs JH to leave work and relax with RJ and William Whewell at RJ's house.
ALS 1p RS:HS 19.280

665. 1822–1–21 To Pierre Simon Laplace London
Advises PL that he has been elected an associate of the Astronomical Society.
ACCS 1p RS:HS 19.7 (C: 20.131)

666. 1822–1–22 To Friedrich Wilhelm Bessel n.p.
FB was elected associate of Astronomical Society.
AAbs 1p RS:HS 19.1

667. 1822–1–28 From David Brewster Edinburgh
It has been a long time since he last heard from him. Very busy with professional pursuits. Congratulations on receiving the Copley medal. Would like an abstract of his article on aberrations in chromatic lenses. Comments on this.
ALS 4pp RS:HS 4.255

668. 1822–1–28 From Mary Pitt Herschel Slough
Settlement made on Mr. Davenport's back rent. Wishes JH had been there to bargain for William Herschel. Planning trip to London. JH invited to Miss De Luc's party. William Herschel is weak, unable to walk.
ALS 3pp & C 3pp TxU:H/M–0621.2; Reel 1086

669. 1822–1–29 To D. François J. Arago London
Informs FA of his election as an associate of the Astronomical Society. Includes message for Alexis Bouvard.
ALS 2pp Columbia Univ. Library

670. 1822–1–30 To D. François J. Arago n.p.
Informing him that he (FA) has been elected an Associate of the Astronomical Society of London. Postscript : Henry Kater hopes to hear from FA soon. Thanks him for tables of Jupiter, Saturn and Uranus.
draft 3pp RS:HS 1.351 & 20.132

671. 1822–1–31 From George Peacock Trinity College, Cambridge
Of Cambridge University news, and observatory plans.
ALS 3pp RS:HS 13.281

672. 1822–2–2 From Joanna Baillie Hampstead
Would like a few of his poems so that she can publish them in an anthology.
ALS 2pp RS:HS 3.18

673. 1822–2–4 From Francis Baily Gray's Inn
Regarding a suitable date for the next meeting of Council. Construction of tables. Regulations regarding auditors.
ALS 3pp RS:HS 3.46

674. 1822–2–6 From James Grahame Edinburgh
About their relationship.
ALS 3pp RS:HS 8.306

675. 1822–2–6 To Alfred Gautier London
Announcing his election to the Astronomical Society. Would welcome any communications. Sends reports to M. A. Pictet and thanks for his communications on barometric height. Would be very pleased to be of assistance at any time.
draft 2pp RS:HS 8.79 & 20.133

676. 1822–2–7 From Thomas Young 48 Welbeck St.
Board of Longitude reluctant to pay for new building at James South's unless South and JH agree to place their observations at Board's disposal.
ALS 1p RS:HS 18.321

677. 1822–2–8 From Sir Everard Home Sackville St.
Has got [Karl/Peter?] Bauer's measurements, so if he calls Monday or Tuesday he can show them to him.
ALS 1p RS:HS 10.203

678. 1822–2–9 From Francis Baily Gray's Inn
Reasons why JH was unable to see the Atlas when he called.
ALS 2pp RS:HS 3.47

679. 1822–2–12 From Francis Baily Gray's Inn
Regarding H. C. Schumacher's *Nachrichten*.

Astronomical observations. Regarding foreign correspondents for the Astronomical Society.
ALS 4pp RS:HS 3.48

680. 1822–2–15 From Sir Everard Home Sackville St.
At the President's suggestion he is confining his Croonian lecture to the anatomical description of the eye; therefore has not sent the experiments.
ALS 1p RS:HS 10.204

681. 1822–2–16 From Mary Pitt Herschel [Slough]
Visit from Sir David Smith and wife. Family health. Mr. Davenport is ignoring all letters. Extend invitation to Slough to Mrs. Wrangham and Mrs. Beckwith.
ALS 2pp † & C 2pp TxU:H/M–0621.3; Reel 1086

682. 1822–2–16 To Charles Babbage n.p.
Regarding W. T. Brande's attack on Sir Joseph Banks, and its relation to the Council of the Astronomical Society.
ALS 3pp RS:HS 2.170 & 20.134

683. 1822–2–23 To John William Whittaker Slough
Asks for money pledged for publication of *Memoirs of the Analytical Society*, which has made no profit. Will be attending testimonial at Cambridge for JW [who has resigned position there].
ALS 2pp St. John's College, Cambridge

684. 1822–2–25 From Alfred Gautier Geneva
Thanks for his election to the Astronomical Society. Gives details of his recent activities and those of M. A. Pictet.
ALS 4pp (French) RS:HS 8.80

685. 1822–2–25 From Pierre Simon Laplace Paris
Is grateful for the news of his election to the Astronomical Society.
ALS 1p (French) RS:HS 11.99

686. 1822–3–6 From Francis Baily Gray's Inn
Regarding the plan of the tables for the places of the principal stars. Will be unable to attend the meeting of council on Friday. Will send memorandum for discussion.
ALS 2pp RS:HS 3.49

687. 1822–3–8 From Francis Baily Gray's Inn
Wishes to resign from the secretaryship of the Astronomical Society.
ALS 2pp RS:HS 3.50

688. 1822–3–11 To John William Whittaker n.p.
Shocked by death of [Professor E. D. Clarke], JH supports [J. S.] Henslow as candidate to succeed Clarke. Apologizes for incorrect charging of JW for Analytical Society *Memoirs* [see JH's 1822–2–23]. Asks JW to find out whether William Whewell will write article on physical astronomy.
ALS 2pp St. John's College, Cambridge

689. 1822–3–13 From Joseph Johann Littrow Vienna
Has he received the copies of the Vienna Observations? Encloses a small pamphlet of interest to mariners. What does he think of Josef Fraunhofer's micrometer for observing double stars? Is writing this letter on his birthday.
ALS 8pp (German with English trans.) RS:HS 11.234

690. 1822–3–14 From Schenck & Co. Berne
Mr. Schenck will consider favorably proposition made when JH passed through Berne, provided it is compatible with S's position and suitable for a married man.
C 1p inc (French) RS:HS 17.173

691. 1822–3–15 To Henry Kater n.p.
Sends the formula for finding the maximum diameter of the planets.
C 2pp RS:HS 20.135

692. 1822–3–17 From George Peacock Trinity College, Cambridge
Observatory plans are progressing well; GP also sends some Cambridge University news.
ALS 4pp RS:HS 13.282

693. 1822–3–18 To D. François J. Arago n.p.
Would like his help on observations of double stars. Comparison of data. Gives details of stars and distances of certain planets from the earth.
draft 3pp RS:HS 1.352 & 20.137

694. 1822–3–25 From Thomas Young 48 Welbeck St.
Next meeting of Board of Longitude.
ALS 2pp † RS:HS 18.322

695. 1822–3–26 From Francis Baily Gray's Inn
Regarding the method of dispatch of papers for foreign members.
ALS 2pp RS:HS 3.51

696. [1822–3–28 or later] To Henry Kater n.p.
Encloses JH's translation of J. J. Littrow's method of determining latitude by pole star. It already appeared in F. X. Zach's *Correspondance astronomique....* Send Francis Beaufort's survey of Karamania coast.
ALS 2pp RS:HS 19.16 (C: RS:HS 25.13.19)

697. 1822–3–29 From Gilbert Elliot, Second Earl of Minto Naples
Has had a letter from [J. C.?] Schenck of which he encloses a copy as it was possibly intended for JH. Will be pleased to hear about his expedition to Monta Rosa. Has found new method of estimating the mean temperature of the air. Was too late to see Vesuvius in eruption. Let him know if he wishes to say anything to Schenck.
ALS 3pp RS:HS 7.32

698. [1822–4] To Charles Babbage n.p.
Tells CB that JH has seen a calculating machine made by some 'rascally frenchman.'
C 1p RS:HS 20.135

699. [1822–4] From Thomas Young 48 Welbeck St.
Asks JH to check accuracy of *Nautical Almanac* entry for Jupiter's satellites III and IV on 20 Aug. 1819. Thanks for JH's note, but tell James South 'he is bound to give his *reasons*....'
ALS 2pp RS:HS 18.324

700. 1822–4–4 From Thomas Young 48 Welbeck St.
Fears JH misunderstood TY's remark about the injustice of James South's accusation regarding *Nautical Almanac*. TY was not trying to provoke South's hostility. Sends 'Schumacher's No. 10' for JH's perusal.
ALS 1p RS:HS 18.323

701. 1822–4–6 From James South Blackman St., Southwark
Discusses JS's efforts to reform the *Nautical Almanac*, including JS's *Practical Observations on the Nautical Almanac*. States that in this *'awful crisis'* JS has no

hostile feelings toward its superintendent, Thomas Young.
ALS 2pp RS:HS 16.416

702. 1822–4–9 From Charles Babbage n.p.
Matters relating to his calculating machine.
ALS 4pp RS:HS 2.171

703. 1822–4–10 From Pierre Simon Laplace n.p.
Please forward enclosed papers.
Abs(by JH) 1p RS:HS 19.7

704. 1822–4–19 From James Grahame Edinburgh
Comments on the need to travel, and on the boredom of the meetings of the Royal Society of Edinburgh; family news.
ALS 4pp RS:HS 8.307

705. 1822–4–20 To Joseph Nicolas Nicollet 9 Downing St.
P. L. Guinand has informed the Astronomical Society that he can make flint glass free from the usual imperfections. Would like any information from JN. One of his papers has been named as likely to receive a prize. Hears that an arithmetical machine has recently been demonstrated before the Institute. Have any details been made public?
C 2pp RS:HS 13.134bis & RS:HS 20.138

706. 1822–4–22 To Henrik Johan Walbeck [Astronomical Society]
Sends first volume of Astronomical Society's *Transactions* and JH's paper on aberrations ['On the Aberrations of Compound Lenses.' 1821]. Asks for geodesical results at Åbo.
AAbs 1p RS:HS 18.10b & 19.13

707. 1822–4–26 To Charles Babbage n.p.
Please forward enclosed note to Mrs. Roget [wife of P. M. Roget?]. Various items of interest.
ALS 1p RS:HS 2.172 & 20.139

708. 1822–4–27 From Joanna Baillie Hampstead
Thanking him for poetry. Would like some more, written on a different subject and under better conditions.
ALS 3pp RS:HS 3.19

709. 1822–5–2 From Friedrich Wilhelm Bessel Königsberg
Thanks JH for FB becoming Associate member of Astronomical Society; FB is planning to prepare a new major star catalogue.
ALS 3pp (German) RS:HS 4.66

710. 1822–5–5 From Wilhelm Struve Dorpat
WS asks that his name be given to the Astronomical Society and attaches a copy of observation of double stars made at Dorpat. WS also inquires about copies he had previously sent to William Herschel and to the R.S.L., offering to replace them if they have been lost.
ALS 2pp (Latin) RS:HS 17.103

711. 1822–5–6 From Jean-Baptiste Delambre Paris
Has received the first book of the Memoirs of the Astronomical Society but had to pay duty on the packet.
ALS 1p (French) RS:HS 6.483

712. 1822–5–17 From John Knowles Navy Office
Sends a copy of a work he has written on *Shipbuilding*. Also sends the letter of [Robert] Seppings on circular stress.
ALS 1p RS:HS 11.90

713. 1822–5–18 From Joanna Baillie Hampstead
Thanking him for his poems. Will print some; is returning the remainder.
ALS 2pp RS:HS 3.20

714. 1822–5–18 To Thomas Catton [London]
Received letter and packet from J. J. von Littrow, with request that TC order clock from Robert Molyneux for Littrow. Encloses Littrow's Vienna observations.
AAbs 1p RS:HS 19.8

715. 1822–5–18 To Henry Kater 9 Downing St.
Leaving town for a week. Will read [Francis Beaufort's] book after return. Delighted by 'two instances of good fortune' in one week at Union Place [HK's home].
ALS 2pp RS:HS 19.18

716. 1822–5–18 From Sir Robert Seppings Navy Office
Informs JH of arrangement for carriage conveyance to Deptford Yard.
L 1p RS:HS 17.190

717. 1822–5–20 From Francis Baily Gray's Inn
Regarding the tables. Would like to see J. J. Littrow's observations. Regarding the printing of various papers in the memoirs.
ALS 2pp RS:HS 3.52

718. 1822–5–20 From James Grahame Edinburgh
Comments on library facilities available to JG, followed by family news.
ALS 2pp RS:HS 8.308

719. 1822–5–20 To C[harles] Stokes 9 Downing St.
Thanks CS of Gray's Inn for being willing to introduce JH to Mr. Clift.
ALS 1p Adler Planetarium

720. 1822–5–22 To Jean-Baptise Delambre 9 Downing St.
On behalf of the Astronomical Society, JH thanks JD for a copy of JD's *Histoire de l'astronomie moderne*. JH tells JD about an interesting Arabic manuscript relating to the history of astronomy at Cambridge University, which JH will have copied for JD. In a postscript dated 15 June, JH apologizes for problems encountered in shipping printed material to JD.
ALS 3pp WT 67390

721. 1822–5–22 To Lutzardt de Wabern Slough
Has received his communication via Gilbert Elliot (2nd Earl of Minto). Comments on the chances of [J. C.?] Schenck succeeding in business in England.
C 4pp RS:HS 11.495 & 20.141

722. 1822–5–28 From Johann Elert Bode Berlin
Thanks for the *Nautical Almanac*. Still requires *R.S.P.T.* for 1817. Through Francis Baily he will receive copies of his Yearbook, one for his father and one for the R.S.L.
ALS 2pp (German) RS:HS 4.168

723. 1822–5–28 To Jean-Baptiste Delambre 9 Downing St.
Thanks him on behalf of the Astronomical Society for his *Histoire de l'astronomie moderne*. Gives details of a paper on Arabic manuscripts in the *Cambridge Philosophical Transactions*. Have D. F. J. Arago and J. N. Nicollet received his letter?
C 4pp RS:HS 6.484 & 20.142

724. 1822–5–30 From Fearon Fallows Cape Town
His health has been poor and is still not well. Outlines his surveys for the site of the Observatory, difficulties of storing instruments, lack of finance. Gives details of the observations carried out so far. First assistant is leaving and has appointed a R.C. priest. Remember him to his friends.
ALS 16pp RS:HS 7.157

725. [1822–6] From Gilbert Elliot, Second Earl of Minto n.p.
Cannot leave the instrument with Edward Troughton later than Saturday. Perhaps Troughton had better leave the instrument in the box until GE calls.
ALS 2pp RS:HS 7.34

726. 1822–6–1 From Henry Kater Union Place, Regents Park
The Miss Edgeworths will prolong their visit until Friday in order to visit Slough. George Dollond has sent the radii of the crown lens. Will JH breakfast with the Katers on Friday?
ALS 2pp RS:HS 11.7

727. 1822–[6] To the Secretary, Philosophical Society of Philadelphia London
Sends a copy of Volume 1 of the *Transactions* of the Astronomical Society, and makes arrangements for future volumes.
L(S) 2pp APS (no i.d.)

728. 1822–6–3 From Henry Kater Union Place, Regents Park
The Miss Edgeworths have changed their plans again in order to visit Portsmouth, but will arrive at Slough on Sunday morning. Hopes the Herschels can still breakfast with the Katers on Friday.
ALS 2pp RS:HS 11.8

729. 1822–6–6 From Fearon Fallows n.p.
The comet is not visible. All is ready for its observation. How is the Cambridge observatory progressing? Comet still not visible. Wishes he had a good telescope.
ALS 3pp RS:HS 7.158

730. 1822–6–10 From Charles Babbage Devonshire St.
Will not be able to pay a visit as children have whooping cough. Has been demonstrating his calculating machine. Hopes to see him at the Greenwich visitation.
ALS 3pp RS:HS 2.173

731. 1822–6–11 From Miles Bland Cambridge
JH has been appointed a Lay-fellow. It entails no duties. The present body regards it as an honor to retain JH's name.
ALS 2pp RS:HS 4.155

732. 1822–6–12 To Miles Bland Slough
Received MB's notice that JH was appointed fellow of Saint John's College. Will visit Cambridge if William Herschel's health permits.
AdraftS 2pp RS:HS 19.19

733. 1822–6–13 From Olinthus G. Gregory Royal Military Academy, Woolwich
Outlining the circumstances in which J. N. Vallot was elected an Associate of the Astronomical Society. Has only received one letter from Vallot.
ALS 3pp RS:HS 9.35

734. 1822–6–15 From [J. S. C.] Schweigger Halle
Mentions meeting [W.] Herschel in 1816. Recalls receiving catalog from Caroline Herschel, which he then gave to the Munich observatory. Looks forward to meeting JH and 'sharing love of truth.' Sends chemistry and physics papers. Discusses code for electromagnetic signals and its origin.
ALS 4pp (German) RS:HS 15.412

735. 1822–6–16 To Gilbert Elliot, Second Earl of Minto 9 Downing St.
Discusses the pros and cons of GE's proposal that [J. C.?] Schenck, a Swiss maker of optical instruments, set up a business in London. JH is not sure that JH ascended Monte Rosa, but is certain of JH's barometric reading. Interested in GE's method of estimating air temperature
ALS 4pp NatLibScot MS.12226, f.21 (draft: RS:HS 7.32b & C: 20.143)

736. 1822–6–17 To James South n.p.
Requests corrections on a paper JS was publishing with JH. Urges JS to use less excessive language.
Adraft 2pp RS:HS 16.417 (C: 20.144)

737. 1822–6–20 From John Pond Royal Observatory
Thanks JH for extract from a letter from [J. J.] Littrow. Discusses in detail positional determinations of various stars he has made, attempting to determine whether errors of either observations or computation have entered into his results. Reports that in general there is substantial agreement between determinations made by JP and those made by John Brinkley at Ireland's Royal Observatory. Mentions solar observations at Greenwich.
ALS 7pp RS:HS 14.32

738. [1822–6–20 or earlier] From John Pond n.p.
Thanks for sending your catalogues. Requests duplicates. Has made annotations indicating the 'quantity which Mr. [F. W.] Bessel's Instrument places the stars to the south of my determinations.' Cannot yet explain discrepancy. Unable to determine a parallactic shift for Alpha Lyrae, Alpha Aquila, or Gamma Draconis.
ALS 3pp RS:HS 14.31

739. [1822–6–25] From Gilbert Elliot, Second Earl of Minto Berkeley Square
Would like very much to see JH as soon as possible. Has an instrument of [J. C.?] Schenck to show him. There is one on the way to him now from Paris.
ALS 2pp RS:HS 7.33

740. 1822–6–25 To Joseph Johann Littrow Slough
Thanks for the publications sent; lists materials JH is sending JL. Asks about quality of Josef Fraunhofer's large telescopes. JH read part of JL's letter of 2 Jan. 1822 at Astronomical Society meeting. Discusses proper motion of sun, JL's work on latitude of the pole star, observatory clocks, micrometers, and the Cambridge Observatory.
ALS 4pp Columbia Univ. Library (ACCS: RS:HS 19.8 & C: 20.145)

741. 1822–6–26 From Fearon Fallows Cape Town
Has received letters from the Admiralty, which have allayed all his previous distress. Is to receive two telescopes. Weather has been unsuitable for observations.
ALS 2pp RS:HS 7.159

742. 1822–6–28 To Henry Kater [London]
Examine reflecting theodolite by [?] Schenck, now at Edward Troughton's. Arrangements for Madeira expedition. James South is willing to make observations

there. 'Bowdich (the African)' is now at Teneriffe or Madeira and needs employment. William Herschel's health.
ALS 3pp RS:HS 19.20 (C: RS:HS 20.146)

743. 1822–6–29 From Francis Baily Gray's Inn
Further regarding the transmission of papers abroad. Encloses a copy of his own paper on the formula of J. J. Littrow. Astronomical matters.
ALS 2pp RS:HS 3.53

744. [1822–6–30] From Gilbert Elliot, Second Earl of Minto Berkeley Square
[J. C.?] Schenck has reason to be grateful to JH for all his assistance. Schenck's friends can now judge whether it will be expedient to bring Schenck to London to set up in business. Awaits anxiously the verdict of Edward Troughton.
ALS 3pp RS:HS 7.35

745. 1822–7–5 To [J. S. C.] Schweigger Slough
[Responding to JS's letter dated 1822, i.e., #655,] acknowledges receipt of mail. Regrets delay but wished to examine the differences in electrical results between [J. W. A.] Pfaff and JH. Remains certain about results and request that this letter remain 'private.'
ACCS 2pp RS:HS 15:413a

746. 1822–7–6 From Friedrich Wilhelm Bessel Königsberg
Thanks JH for some papers, sends some of his own; comments about errors in Greenwich star catalogue.
ALS 4pp (German) RS:HS 4.67

747. 1822–7–13 To Charles Babbage n.p.
Has had a letter from the Secretary of the Dijon Academy offering him and CB the title of Non-Resident Associate of the Academy. Is there any chance of seeing him in the near future?
ALS 2pp RS:HS 2.174 & 20.147

748. 1822–7–15 From James Grahame Edinburgh
Invites JH to go to the Netherlands with JG.
ALS 3pp RS:HS 8.309

749. 1822–7–16 From Pierre Simon Laplace Paris
Sends one of his own memoirs, which he would like JH to compare with other theories on the same subject.
ALS 1p (French) RS:HS 11.100

750. 1822–7–23 To William Herschel Amsterdam
JH and James Grahame visited [Jan van den] Bosch's agricultural colony in Frederick's-Oort for 2,400 urban itinerants. Leave tonight for Haarlem, Hague, and Rotterdam.
ALS 3pp & C TxU:H/L-0556; Reel 1053 (C: RS:HS B28.6.c)

751. 1822–7–24 To David Brewster Slough
Accompanies a paper JH was submitting to the Royal Society of Edinburgh. Discusses motion of light rays in Apophyllite and various optical writings of DB.
C 2pp RS:HS 20.148

752. 1822–7–24 From Nicholas Fuss St. Petersburg
Gratitude from Imperial Academy of Sciences for copy of first volume of Astronomical Society's 'Transactions.' Please send future volumes to Mr. Dubatchevski, Prussian consul-general in London. Sent packet to H. J. Walbeck at Åbo.
ALS 2pp RAS:WH Archive 1/13.F.12; Reel 13

753. 1822–7–25 To David Brewster n.p.
Sends JH's 1823 paper 'Absorption of Light by Coloured Media' to Royal Society of Edinburgh. Approves DB's illustration of polarized light in mineral apophyllite. Corrects DB's quotation from JH's letter regarding crystals observed on Mont Blanc.
AAbs 1p RS:HS 19.4

754. 1822–7–25 From James Grahame Edinburgh
Further on JG's travel plans [see JG's 1822–7–15].
ALS 3pp RS:HS 8.310

755. 1822–7–30 From Francis Baily Gray's Inn
Has been absent and is sending letter he found on return, and a copy of his own remarks on J. J. Littrow's formula. Regarding the best method of procuring the Memoirs. Can lend him a copy of W. T. Brande's Journal if he would like it.
ALS 3pp RS:HS 3.54

756. 1822–7–30 To J. Henderson [9 Downing St.]
Sends payment for packet of books.
C 1p RS:HS 19.22 (C: RS:HS B28.2)

757. 1822–7–30 To Richard Jones 9 Downing St.
Invites RJ to meet James Grahame, who is joining JH for trip to Netherlands. Return *Éphémèrides du citoyen* directly to Grahame. Describes return to Cambridge by JH, George Peacock, and William Whewell after their visit with RJ at Brasted. JH's sketch of Brasted Church. Alludes to [death of Grahame's first wife and oldest daughter].
C 2pp RS:HS 19.21 (C: RS:HS B28.1)

758. 1822–7–31 To Charles Babbage n.p.
Query regarding the construction of a model of a machine for CB. CB's letter to Humphry Davy about his engines. He will be visiting the Netherlands shortly. P. S. Laplace has sent him details of his recent researches into gases. Hopes the family are well after their recent visit to Devon.
ALS 3pp RS:HS 2.175 & 20.149

759. 1822–7–31 From Richard Jones Brasted
Will come to town to meet [James] Graham[e?]. His books arrived safely for which he was very grateful. Does he know of any Dutch collections on the East Indies?
ALS 3pp RS:HS 10.346

760. [1822–8] From Richard Jones n.p.
Needs the Ephemerides for 1769. Can he ask [James?] Graham[e] for any old pamphlets. Is expecting William Whewell and George Peacock and would be glad if JH can come as well.
AL 3pp RS:HS 10.390

761. [1822–8 or earlier] From E. Wrangham n.p.
EW and daughter look forward to visiting Slough on Monday with Mr. Beckwith.
ALS 1p RS:HS 18.434

762. 1822–8–2 To Mr. Byam, Lithographic Press 9 Downing St.
Orders lithograph of enclosed letter and tables for Astronomical Society.
C 1p RS:HS 19.23a (C: RS:HS B28.3)

763. 1822–8–2 From Maria Edgeworth Edgeworthstown
Encloses list of books on Ireland as requested. Her brother is in London. Hopes that JH will visit them should he come to Ireland.
AL 4pp inc RS:HS 7.20

764. 1822–8–2 To Henry Kater 9 Downing St.
Encloses table of star declinations for 1815 and 1820, received from F. W. Bessel, who used different meridian circles for each year.
C 2pp RS:HS 19.25

765. 1822–8–2 From Henrik Johan Walbeck Åbo [Finland]
Received JH's letter [29 Apr. 1821] with 'Memoirs of Astronomical Society' and observations of recent solar eclipse. Has not published eclipse calculations. Convinced that moon's diameter is accurately determined. Latitudes, longitudes for sites in Finland. 1820–22 observations from Åbo. Design of new Åbo observatory. New instruments from Munich.
ALS 4pp RS:HS 18.11

766. 1822–8–2 To Thomas Young 9 Downing St.
Leaving for two weeks in Netherlands.
C 1p RS:HS 19.24 (C: RS:HS B28.4)

767. 1822–8–3 From Charles Babbage Teignmouth
Regarding the construction of the model machine. Things appertaining to the Netherlands. Regarding his new chemical theory.
ALS 3pp RS:HS 2.176

768. 1822–8–4 From Thomas Young 48 Welbeck St.
Has no errands for JH in Holland or Flanders. Asks for details on JH's and Charles Babbage's method of determining heights by barometrical observations. Hopes to use this to confirm TY's method by 'my formula deduced from refraction.'
ALS 1p RS:HS 18.325

769. 1822–8–6 To Mr. Byam, Lithographic Press 9 Downing St.
Orders 30 copies of lithograph.
C 1p RS:HS 19.23b

770. 1822–8–7 To Friedrich Wilhelm Bessel Astronomical Society
Thanks for the publications by FB that FB sent. JH has sent the copies to the persons FB designated. Discusses the 'enigmatical difference in the places of the stars as observed here & on the Continent.' Cannot explain this.
ALS 3pp AdWDDR Bessel 225 (C: RS:HS 19.26, 20.151, & B28.5)

771. 1822–8–7 To David Brewster [9 Downing St.]
Sends circular of F. W. Bessel's 'Declinations.' Sent paper on 25 July for Royal Society of Edinburgh, similar to DB's paper outlined in previous issue [of *Edinburgh Journal of Science*]. Argues for continuity of color spectrum versus 'per saltum' transitions.
C 1p RS:HS 19.27 (C: RS:HS 20.150 & B28.6a)

772. 1822–8–7 To William Herschel London
Leaving for Calais tomorrow. Unable to see Mr. Beckwith. Hopes [Susan] White arrived safely in Slough. Encloses Astronomical Society's notice of F. W. Bessel's determination of declinations of fixed stars.
ALS 1p, encl 2pp TxU:H/L-0555; Reel 1053

773. 1822–8–13 From Mary Pitt Herschel Slough
Received JH's letter from Calais. Visitors to Slough. Marquess of Londonderry committed suicide.
ALS 2pp † & C TxU:H/M–0621.4; Reel 1086

774. 1822–8–14 To Mary Pitt Herschel Brussels
Travels with James Grahame from Calais to Brussels. Visit with General Pierre Cambronne and tour of Waterloo. Leaving tomorrow for Antwerp. Mr. Gibson is useless as traveling companion.
ALS 3pp TxU:H/L-0516; Reel 1053 (C: RS:HS B28.6.b)

775. 1822–8–20 From Mary Pitt Herschel Slough
William Herschel is dangerously ill. Hurry home.
ALS 1p & C 1p TxU:H/M–0621.5; Reel 1086

776. 1822–8–22 From Joseph Johann Littrow Vienna
Books per N. M. Rothschild have not yet arrived. Has arranged a standing order for the Vienna Observations. The Robert Molyneux clock can be dispatched at once. Agrees with JH that double stars are the test of a good telescope.
ALS 5pp RS:HS 11.237

777. 1822–8–23 From Thomas Beckwith Slough
William Herschel is getting weaker, but is not in pain. Lady Mary Herschel is anxious for JH's arrival.
ALS 1p TxU:H/M–0079; Reel 1087

778. 1822–8–25 or earlier. From Antoinette de Luc Windsor
Invitation to JH to meet some friends of AL's.
ALS 1p RS:HS 11.493

779. 1822–8–30 From Robert Grahame, Sr. to Miss Baldwin Glasgow
Is sorry to hear the news of Sir William Herschel's death. Has no idea where JH is. His son's last letter was from Lille.
ALS 1p RS:HS 9.2

780. 1822–[9?] To Mrs. Mary Francis Kater [Slough]
Lady Mary Herschel feels great sorrow, but her health is fair. Caroline Herschel is distracted by her return to Hanover. Hopes present extraordinary season will favor Henry Kater's observations. News from [Maria] Edgeworth.
C 1p RS:HS 19.43

781. [1822–9] From Mr. March [Modmodagroan?]
JH omitted Mrs. Beckwith and [Charles] Babbage in the list given to M [undertaker] this morning. Please send their addresses.
ALS 1p TxU:H/M–0941; Reel 1083

782. 1822–9–3 From Charles Hutton 34 Bedford Row
Condolences on death of William Herschel.
ALS 1p TxU:H/M–0286; Reel 1087

783. 1822–9–3 From Richard Jones Brasted
Condolences on death of William Herschel. JH should not feel guilty for being absent.
ALS 2pp TxU:H/M–0305; Reel 1087 Lubbock, *Herschel Chronicle*, 361–2

784. 1822–9–3 From George Peacock Trinity College, Cambridge
Condolences on death of William Herschel.
ALS 3pp TxU:H/M–0414; Reel 1087

785. 1822–9–4 To James Grahame Slough
Expresses remorse for JH's inattention to William Herschel while still alive. Lady Mary Herschel's grief. Invites JG to [*missing*]. Hopes to visit before JG leaves England.
C 2pp inc RS:HS 19.29 (C: RS:HS B28.8)

786. 1822–9–4 To James South Slough
Appreciates JS's support of JH's mother at death of William Herschel. Invites JS to attend WH's remains to grave.
C 1p RS:HS 19.28 (C: RS:HS B28.7)

787. [1822–9–4?] To Commander of Steam-Packet *Talbot* Slough
JH gave passport to commander or steward with instructions to use it to retrieve mail forwarded to JH in Calais. Lack of response concerns JH. Please procure mail and send it with JH's passport to 9 Downing St. Reward.
C 1p RS:HS 19.34–35

788. 1822–9–6 From James Grahame London
JG is called home urgently. Does not want to go, but feels painful duty to do so. Cancels visit to Slough. Asks JH to retrieve prints that JG left at Downing St.
ALS 2pp TxU:H/M–0211; Reel 1087

789. 1822–9–6 To Johann Dietrich Herschel Slough
Caroline Herschel is determined to leave England before JH can arrange William Herschel's affairs and accompany her to Hanover. Please notify JH when she is safely settled. JH is too grieved now to write more.
C 1p RS:HS 19.30

790. 1822–9–6 From Joseph Planta British Museum
Extols friendship with William Herschel. Regrets that JP cannot attend funeral.
ALS 3pp TxU:H/M–0408; Reel 1087 Lubbock, *Herschel Chronicle*, 362–3

791. [1822–9–7] From James Grahame London
Hopes JH's presence is solace to JH's mother. Wants to visit Slough, but must return to Scotland before 10 Sept. Death of William Herschel and need to surrender to God's will. Prays for JH and JH's mother.
ALS 3pp TxU:H/M–0212; Reel 1087

792. 1822–9–7 To Johann Dietrich Herschel n.p.
Legacy of stock for JDH from William Herschel's will. How shall JH transfer it to Hanover?
ALS 2pp TxU:H/L-0528; Reel 1053

793. 1822–9–9 From Maria Edgeworth to Margaret Brodie Herschel Edgeworthstown
Condolences on the death of Sir William Herschel.
ALS 3pp RS:HS 7.21

794. 1822–9–10 From Charles Babbage n.p.
The account will be inserted in the *Morning Chronicle* tomorrow.
ALS 1p RS:HS 2.177

795. 1822–9–11 From Samuel Goodenough Rose Castle
Is sorry to read in the papers the news of the death of William Herschel. Had he been near he would have liked to have attended the funeral.
ALS 1p RS:HS 9.103

796. 1822–9–11 To Richard Jones 9 Downing St.
Laments loss of William Herschel's affection; grateful for that of friends. Encloses rare book on '*arcana* of Dutch commerce' that JH bought in Amsterdam. Will send list of works on same topic that JH found in library at Hague.
C 2pp RS:HS 19.31 (C: RS:HS B28.11)

797. 1822–9–11 To George Peacock 9 Downing St.
Grateful for GP's sympathy. JH's mother 'dreadfully distressed.'
C 1p RS:HS 19.32 (C: RS:HS B28.12)

798. 1822–9–12 To Postmaster at Calais [Slough]
Requests that JH's letters be forwarded to Slough. JH gave passport to steamship captain, who failed to retrieve letters and lost JH's passport.
C 1p RS:HS 19.33 (C: RS:HS B28.13)

799. 1822–9–16 To Maria Edgeworth Slough
Thanks ME for sympathies [on death of William Herschel]. Received ME's list of books on Irish history. Met [John] Brinkley and wife and ME's brother during their visit to England.
C 3pp inc RS:HS 19.36

800. 1822–9–25 From Charles Babbage Kempsey, near Worcester
Will be moving to Worchester shortly. Theory on stained glass. Regarding his recent pamphlet.
ALS 3pp RS:HS 2.178

801. 1822–9–25 To David William Smith Slough
Thanks for sympathy. Lady Herschel deeply afflicted. Caroline Herschel showing 'uncommon fortitude,' planning to return to Hanover. Heard that [*missing*] is beyond recovery; hopes she will be spared pain.
C 2pp inc RS:HS 19.37 (C: RS:HS B28.16)

802. 1822–9–26 To Mr. Davenport Slough
As executor of William Herschel's will, JH demands payment of half-year's rent on farm, due last March.
C 1p RS:HS 19.39a

803. 1822–9–26 To Sir William Watson Slough
Reluctant to intrude on WW's own sorrow, but grateful for WW's forty-year friendship with William Herschel. Caroline Herschel departing for Hanover.
C 2pp inc RS:HS 19.38 (C: RS:HS B28.17)

804. 1822–9–27 To [?] Slough
Refer to 10 Sept. *Morning Chronicle* and 7 Sept. *Literary Chronicle* for biographies of William Herschel. Submit manuscript for JH's perusal before publication.
C 1p RS:HS 19.40

805. 1822–9–27 To [?] Slough
Could not locate personal reference named Alexander Mackenzie in Cheltenham. Therefore, JH declines to employ 'F. E.'
C 1p RS:HS 19.39b

806. 1822–9–28 From James Grahame Edinburgh
Family news on JG's return from the Continent.
ALS 3pp RS:HS 8.311

807. 1822–9–28 From F. Trechsel Berne
Informs JH of delay in getting JH his instrument. Sends condolences concerning death of [William] Herschel. Informs JH of 'reunion of distinguished savants' at Gambery in August.
ALS 3pp (French) RS:HS 17.374

808. 1822–10–1 From Alfred Gautier Geneva
Sending 3 copies of a paper for his personal use and the Astronomical Society. Is grateful for the interest JH takes in his own work. Outlines recent astronomical activities of himself and M. A. Pictet. Intends visiting England.
ALS 4pp (French) RS:HS 8.81

809. 1822–10–4 To Mr. Davenport Slough
Mary Herschel will not reduce rent. Pay up or leave. Referee shall decide whether D deserves compensation.
C 1p RS:HS 19.42

810. 1822–10–4 To James Grahame Slough
Despatched JG's 'goods' this week. JH vexed by tenant [Mr. Davenport], who will neither pay rent nor leave farm. Congratulates JG's sister on change in family. Sends £10 for expenses during journey [in Netherlands]. Did Richard Jones return *Éphémèrides du citoyen*? News of JH's traveling companion Mr. Gibson.
C 2pp RS:HS 19.41 (C: RS:HS B28.17)

811. 1822–10–11 To Editor of *Morning Post* 9 Downing St.
Upset that editor printed William Herschel's will. Falsehood of accompanying editorial. Cancel JH's subscription.
C 1p RS:HS 19.44 (C: RS:HS B28.24)

812. 1822–10–11 To [stockbroker] [9 Downing St.]
Instructions for sale and transfer of stocks according to William Herschel's will, and for payment of legacy to Johann Dietrich Herschel. Meet in broker's office at noon tomorrow.
C 2pp RS:HS 19.45

813. 1822–10–12 From David Brewster 10 Coates Crescent, Edinburgh
Has been unable to answer his letters due to the King's visit. Refrained from writing on hearing of the death of W. Herschel. His paper on the absorption of light will appear in the next number of the *Transactions* of the Royal Society of Edinburgh. Comments on this. Grateful for any astronomical news. Advice welcomed on memoir of W. Herschel.
ALS 3pp RS:HS 4.256

814. [1822]–10–15 From Anne Elise Knipping Bremen
Writes to JH (AK's uncle) for the first time, asking for news. Recalls JH's visit to Hanover. Comments on the respect in the world for JH. Appears to be intending to accompany her husband to England for a business trip.
ALS 2pp (French) TxU:H/M–0323; Reel 1087

815. 1822–10–16 To Charles Babbage [9 Downing St.]
Caroline and Dietrich Herschel leave for Rotterdam in two days. CB's 'offer of the heretifical Robes' and 'sacrilegious stain.' Thank [Woolrych Whitmore] for agricultural pamphlet sent to JH and [Richard] Jones. Sent CB's letter to Humphrey Davy and others.
ALS 3pp RS:HS 2.179 (C: RS:HS 19.46 & 20.152)

816. 1822–10–17 To John Herapath London
Heard of Herapath's new powerful geometry methods. No new views of differential calculus on Continent. Refers to those of William Spence annexed to JH's *Mathematical Essays*...(1819). Explains object of combinatory analysis, popular among Germans. Has not seen M. C. T. de Damoiseau's paper on lunar theory.
C 3pp inc RS:HS 19.47

817. 1822–10–17 From William Whewell Trinity College [Cambridge]
Discusses regions of the Alps and sends sketch of Monte Rosa. Informs JH of a brilliant student, G. B. Airy, who is constructing a calculating machine. Condolences on the death of William Herschel.
ALS 3pp RS:HS 18.162 Todhunter, *Whewell, 2,* 50–51

818. 1822–10–18 From James Grahame Edinburgh
Some news from Amsterdam, and some family news.
ALS 3pp RS:HS 8.312

819. 1822–10–19 To Charles Babbage Downing St., Westminster
Destroy or return the article given him on Thursday. Has found something better.
ALS 2pp RS:HS 2.180 & 20.153

820. 1822–10–20 To Mr. Davenport Slough
Will deduct money owed for meat from D's rent, up to William Herschel's death. Rent after 24 Aug. 1822 must be paid in cash to Lady Mary Herschel.
C 1p RS:HS 19.48

821. 1822–10–21 To Mr. Styles Slough
Lady Mary Herschel asks that S referee dispute with tenant farmer Mr. Davenport, if latter should leave before lease expires.
C 1p RS:HS 19.49

822. 1822–10–28 From Charles Babbage Devonshire St.
Regarding the copying of the painting by William Artaud. Send it by coach if possible. Meeting of the Astronomical Society Council and matters arising from it.
ALS 2pp RS:HS 2.181

823. 1822–11–10 To Fearon Fallows Downing St.
Has delayed answering his letters owing to the health of his father. Grieves to hear of the health of FF. Board of Longitude are fully behind FF. Does not agree with some of FF's proposals.
draft 8pp RS:HS 7.160 & 20.154

824. 1822–11–11 To William Whewell Downing St.
Thanks WW for drawing of Monte Rosa. Comments on a recent political election.
ALS 4pp TC, Camb. Add. Ms.a.207[7] (C: RS:HS 20.155)

825. 1822–11–12 From Mary Pitt Herschel Slough
Visited Langton family. Visitors at Slough. Elizabeth Langton is staying with MH. Hopes to make peace between Elizabeth and her cousins. JH's rooms 'over the way' are almost finished. Still grieving for William Herschel. Concerned at not hearing from JH's aunt.
ALS 3pp & C 3pp TxU:H/M–0621.6; Reel 1086

826. 1822–11–13 From John Brinkley Observatory, Dublin
Is extremely sorry to hear the news of the death of William Herschel. Regarding F. W. Bessel's polar distances in relation to his own, and possible reasons for their difference. Has been working on the Greenwich observations.
ALS 3pp RS:HS 4.275

827. 1822–11–16 From George Peacock Trinity College, Cambridge
About the election to fill vacancies at Cambridge University.
ALS 4pp RS:HS 13.283

828. 1822–11–19 From Marc August Pictet Geneva
Thanks JH for the books sent, and describes the work at Geneva in making geodesic measurements.
ALS 3pp RS:HS 13.398

829. 1822–11–21 From Humphry Davy n.p.
Returning sincere thanks for the memorial inscribed with the name of JH's illustrious father. It rests with JH and his companions how posterity will judge the present age.
ALS 2pp RS:HS 6.39

830. 1822–11–26 From Matilda Parry Sion Place
Thanks Herschel family for aid given to care for Hannah Scull, a disabled orphan in the care of the Parrys. Details process of getting legal permission to find a permanent home for the child.
ALS 4pp RS:HS 14.142

831. 1822–11–27 From [Johann] D[ietrich] Herschel Hanover
Following the death of JH's father, William, JDH comments on Caroline Herschel's arrival in Hanover, and expresses his concern about the situation of JH's mother.
ALS 3pp (German) TxU:H/M–0886; Reel 1085

832. 1822–11–27 From Molyneux & Cope London
Sending their account for J. J. Littrow's astronomical clock, now packed and shipped.
ALS 1p RS:HS 13.48

833. [1822–12] To William Whewell n.p.
Gives advice on WW's planned trip to Europe. Asks WW to contact various scientists there on JH's behalf.
ALS 3pp TC, Camb. Add. Ms.a.207[6] (C: RS:HS 20.157)

834. 1822–12–5 To Charles Babbage Slough
Regarding the Cambridge affair. Has had a letter from the Edinburgh Society of Arts.
ALS 2pp RS:HS 2.182 & 20.158

835. 1822–12–5 From Joseph Goodall Lodge, Eton College
Regarding the inscription, and why he has recast it instead of altering it.
ALS 1p RS:HS 8.141

836. 1822–12–20 To Joseph Johann Littrow London
Has just heard that the ship carrying JL's clock arrived at Hamburg on the 5th. Gives the name of the Bank to which payment is to be made. Has used his *Theory of Comets* in a paper.
C 3pp RS:HS 11.239 & 20.159

837. 1822–12–20 To Heinrich Christian Schumacher 9 Downing St.
Requests that HS look for a shipment of a clock and send it on to J. J. Littrow, and inform the maker in London, Robert Molyneux, that HS has done this. JH sends along John Pond's catalogue of star position differences.
ALS 3pp† DMC.695A.885.1117

838. 1822–12–21 To Pierre Simon Laplace London
Introducing William Whewell, who will be visiting Paris.
C 3pp (French) RS:HS 11.101 & 20.159

839. 1822–12–26 From Caroline Lucretia Herschel Hanover
Sending three old letters from William Herschel. Still unpacking and organizing her books and papers.
L *Memoir ... of Caroline Herschel*, 161–2

840. 1822–12–31 From George Peacock Heckfield Place near Hertford Bridge
About intrigue and unethical dealings surrounding the election to fill vacancies at Cambridge University.
ALS 4pp RS:HS 13.284

841. 1823 To *Astronomische Nachrichten* n.p.
Gives a formula related to a table published in the *Nautical Almanac,* which table listed various stellar observations made by John Pond.
L *Astronomische Nachrichten, 1* (1823), 30–31

842. [1823] To David Brewster n.p.
Sends an abstract of his paper on observations relating to the practical construction of telescopes [JH's 'Practical Rules,' 1822]. Would like 25 copies for distribution. His recent experiments with polarized light.
draft 2pp RS:HS 4.272 (C: RS:HS 25.13.17)

843. [1823 early] From Pierre Simon Laplace n.p.
Sends paper.
Abs(by JH) 1p RS:HS 19.7

844. [1823?] From James South n.p.
Has been ill. Should we publish our observations of 340 double stars or extend the number to 400?
ALS 3pp RS:HS 16.477

845. [1823 to 1828] From William Hyde Wollaston Dorset St.
Most correct mode of proceeding will be to *suppress* incorrect corrections and not to give credit to [T. I. M.?] Forster for inferences of [Peter?] Barlow or calculations of WW.
ALS 1p RS:HS 18.290

846. [1823 to 1828] From William Hyde Wollaston n.p.
Unsure whether WW will be in town to dine with JH Monday.
ALS 1p RS:HS 18.287

847. [1823 to 1828] From William Hyde Wollaston n.p.
Previous engagement prevents WW 'joining the early party.'
ALS 1p RS:HS 18.285

848. 1823–1–3 From Joseph Johann Littrow Vienna
Has requested the government to forward payment for the clock. Regarding variations in the observations of F. W. Bessel, John Pond, etc. Has high opinion of Bessel's work. Has been sending books for Thomas Catton, but he never replies. How is Greenwich Observatory heated? Some observatories have troubles with dampness. Recommends book by K. W. Feuerbach.
ALS 13pp (German with English trans.) RS:HS 11.240

849. [1823]–1–3 From Thomas Young 48 Welbeck St.
Sends papers for R.A.S. and A. J. Fresnel's papers. Modifies analogy of infinite pipes to clarify TY's hypothesis of three principal colors. Charles Babbage sent survey of Mont Cenis. Can JH find its barometrical observations?
ALS 2pp RS:HS 18.352

850. 1823–1–6 From Joseph Goodall Lodge, Eton College
Has been very busy and has only now had a chance to work on the inscription. Suggests various forms. Would like his comments on these.
ALS 3pp RS:HS 8.144

851. 1823–1–7 From James Grahame Edinburgh
About meeting Walter Scott, lawsuits and family news.
ALS 4pp RS:HS 8.313

852. [1823–1–13] To Molyneux & Cope 13 Charlotte St., Portland Place
J. J. von Littrow delighted with clock sent by M & C and commissions JH to order another for L. A. Fallon, director of Great Austrian Trigonometrical Survey. Describes modifications. Contact Littrow for further details.
ACCS 1p RS:HS 19.8

853. 1823–1–24 From Caroline Lucretia Herschel [Hanover]
Asks JH to pay CH's debt to his mother and deduct that amount from next annuity payment to CH.
ALS 1p inc TxU:H/M–0936.4; Reel 1083

854. 1823–2–1 To George Dollond 9 Downing St.
Sends some calculations relating to the construction of lenses.
ALS 2pp RS:HS 20.160

855. 1823–2–4 From Mary Pitt Herschel Slough
Hopes that letter from Mr. Langton does not mean more trouble. Flood has made 'Complete Prisoners' of residents at Slough. Glad that JH bought good horse. Advice on JH's laundry. Caroline Herschel's finances. Concern for Georg Griesbach's health. Regards from Susan White.
ALS 3pp & C 3pp TxU:H/M–0622.1; Reel 1086

856. 1823–2–12 From Joseph Johann Littrow Vienna
Has received parcel from N. M. Rothschild. Hopes money for the Robert Molyneux clock will soon arrive. Encloses some items for him in Thomas Catton's parcel. Please accept *Analytical Geometry*. Hopes JL's two volumes on astronomy will be translated into English, and encloses copy of additions.
ALS 4pp RS:HS 11.241

857. [1823]–2–14 From Francis Legatt Chantrey Belgrave Place
If either of the designs please him, he will be pleased to advise; otherwise explain in detail what he requires.
ALS 1p RS:HS 5.243

858. 1823–2–17 To David Brewster Slough
Some comments on sun spots based on William Herschel's ideas; considering a collection of William Herschel's writings but publishers not interested.
C 2pp RS:HS 20.161

859. 1823–2–18 From Francis Baily Gray's Inn
Regarding errors and discrepancies in the tables of F. W. Bessel. Difficulty of dispatching books to foreign members.
ALS 3pp RS:HS 3.55

860. 1823–2–27 From Caroline Lucretia Herschel Hanover
Gratified that JH appreciates the astronomical legacy that she left at Slough. Explains that she returned to Hanover because she felt that 'it would be in vain to struggle any longer against age and infirmity.'
L *Memoir ... of Caroline Herschel*, 162–3

861. 1823–3–4 To Charles Babbage St. John's, Cambridge
Events at Cambridge. Has a paper by A. L. Cauchy for him. Has met Humphry Davy and discussed CB's machine. CB is celebrated in verse.
ALS 3pp RS:HS 2.183 Hyman, *Charles Babbage*, 27 inc

862. 1823–3–5 From Mary Pitt Herschel [Slough]
Birthday greetings to JH. Severe wind damage at Slough, but telescopes are safe. News of acquaintances. Condition of JH's horse.
ALS 3pp & C 3pp TxU:H/M–0622.2; Reel 1086

863. [1823]–3–10 From Francis Legatt Chantrey Belgrave Place
Bearer must show the inscription which must be corrected by JH. Regarding a price for the work.
ALS 1p RS:HS 5.244

864. 1823–3–11 From William Hasledine Pepys n.p.
Invites JH to London Institution to observe an experiment.
ALS 1p RS:HS 13.338

865. 1823–3–16 From Fearon Fallows Cape Town
Was pleased to receive JH's letter. Regrets to hear of the death of Sir William Herschel. Grateful to the Board of Longitude for their assistance. Regarding the deficiencies of one of the clocks. Work carried out. Site has been approved for the new observatory.
ALS 5pp RS:HS 7.161

866. 1823–3–19 From George Peacock Trinity College, Cambridge
About matters relating to the building of the observatory.
ALS 3pp RS:HS 13.285

867. 1823–3–22 From David Wilkie 24 Lower Phillimore Place, Kensington
Per JH's request, DW wrote to Alfred Chalon inquiring about miniature portraits. Quotes prices. Lists names of satisfied patrons.
ALS 2pp RS:HS 18.249

868. 1823–3–28 To Robert M. Patterson London
Writes to acknowledge the receipt of a volume of the *Transactions* of the Philosophical Society of Philadelphia.
L(S) 1p APS (no i.d.)

869. 1823–3–29 To Wilhelm Struve n.p.
WS was elected associate of Astronomical Society. Distributed books as WS directed. Will send JH's and James South's observations of double stars when printed. Neither Thomas Young nor William Herschel received WS's Dorpat observations.
C 3pp RS:HS 20.162

870. 1823–4–8 From Schenck & Co. Berne
Apologizes for late reply. Includes news. Hopes to see JH in Berne in summer. Has forgotten price agreed on for object.
ALS 2pp (French) JH note RS:HS 17.174

871. 1823–4–14 From Caroline Lucretia Herschel Hanover
The Germans are printing many of William Herschel's papers; comments that 'there does not pass a month but something appears in print.'
L *Memoir ... of Caroline Herschel*, 164 inc (?)

872. 1823–4–18 From Pierre Simon Laplace Paris
Sending Part XI of his own *Mécanique céleste*. Comments on this work.
ALS 2pp (French) RS:HS 11.102

873. 1823–4–25 From Wilson Lowry n.p.
JH's geological specimen is a piece of granular quartz, found in great quantities in Berks, Wilts, and Herts. JH promised to show him some of the garnets found by

[J. S.?] Henslow in Anglesea.
ALS 1p RS:HS 11.492

874. 1823–5–8 From James Grahame Edinburgh
About books JG has read including some poems by JH.
ALS 4pp RS:HS 8.314

875. 1823–5–8 From W. Herschel Griesbach Vegesack, Bremen
Hoping that JH's goodness will help WG out of difficulty.
ALS 1p (German) RS:HS 9.139

876. [1823 to 1829]–5–22 From [?] n.p.
Left 'proof plate' of William Herschel at JH's residence last week. Kept it in cottage for long time. Hopes it will find 'a better place' with JH.
AL 1p inc TxU:H/M–0912; Reel 1083

877. 1823–5–31 From Benjamin Gompertz n.p.
Is applying for the position of actuary at the Law Life Assurance Society and would be glad of a letter of recommendation from JH; please direct a letter to Messrs. Peace as soon as possible.
ALS 1p RS:HS 8.137

878. 1823–6–1 To Benjamin Gompertz Slough
Received the letter. Hopes he will be successful in his application for the position of Actuary to the Law Life Society. Has a high opinion of his mathematical papers and has used his system in his own works.
draft 2pp RS:HS 8.138 & 20.163

879. 1823–6–2 From William Hyde Wollaston n.p.
Invites JH to meet H. C. Oersted, Henry Warburton, and Henry Kater at dinner next Wednesday.
ALS 1p RS:HS 18.277

880. [1823]–6–3 To Lord Mayor of London 56 Devonshire St.
Will attend meeting of Lord Mayor's Committee to rebuild London Bridge.
draft 5pp † Harvard: Houghton AAH 67m–67

881. 1823–6–16 From Peter Barlow Royal Military Academy
Is sending him the second edition of his book on magnetic attractions. Comments on the views put forth in this work.
ALS 4pp RS:HS 3.286

882. 1823–6–22 To Marc Seguin London
Will send MS's communication about the theory of caloric to some scientific journals, rather than the Astronomical Society, where it would be inappropriate. Comments on caloric theory.
ALS 2pp RS:HS B26.61

883. 1823–6–23 To David Brewster London
Forwarding to DB, through Hans Christian Oersted, a paper on caloric by Marc Seguin.
C 1p RS:HS 20.164

884. 1823–6–24 From Caroline Lucretia Herschel Hanover
Worries that Johann Pfaff is unqualified to translate William Herschel's papers into German; hopes that JH will be the primary authority for commenting on WH's work.
L *Memoir ... of Caroline Herschel*, 165–7 inc (?)

885. 1823–6–27 From Charles Babbage Devonshire St.
Has had a meeting with the Chancellor of the Exchequer concerning his calculating machine. Is to get financial assistance from the government. Hopes to see JH on Monday night.
ALS 2pp RS:HS 2.184

886. 1823–6–30 From James South n.p.
Discusses various apparent errors JS has been detecting in observing double stars with two different telescopes. Assures JH that no serious problems exist.
ALS 5pp RS:HS 16.418

887. 1823–7–2 From Charles Babbage Devonshire St.
Is sending the agreement between the government and CB regarding payment for work on the calculating machine.
ALS 3pp RS:HS 2.185

888. 1823–7–5 To Charles Babbage n.p.
Has signed the paper regarding the machine and is returning it. Regarding its publicity with Members of Parliament.
ALS 2pp RS:HS 2.186

889. 1823–7–9 From Charles Babbage Devonshire St.
Regarding the paper on the machine for the Treasury. Will postpone his visit to Slough.
ALS 1p RS:HS 2.187

890. [1823 to 1825]–7–9 From Francis Lunn n.p.
Will follow JH's instructions. Expects to find inst[rument?] at home of Mr. Mowman or at Leicester Square.
ALS 1p TxU:H/M–0341.1; Reel 1087

891. 1823–7–14 From Caroline Lucretia Herschel Hanover
Hopes to receive more correspondence from JH. Often wishes that she could be with JH in order to ensure that he will not overwork himself like William Herschel did.
L *Memoir ... of Caroline Herschel*, 168 inc (?)

892. 1823–7–15 From Wilhelm Struve Dorpat
With the departure of [Otto von] Kotzebue, WS takes the opportunity to send JH volumes of Dorpat observations. Requests notification if volumes do not arrive.
ALS 2pp (German) RS:HS 17.104

893. 1823–7–18 From Charles Babbage Devonshire St.
There is to be a sale of pictures, including one of JH; would he like it? Has not heard anything from the Treasury yet regarding payment of his money. Plans for projected holiday.
ALS 2pp RS:HS 2.188

894. 1823–7–26 From James Grahame Bristol
Is coming to London and wants to see JH.
ALS 1p RS:HS 8.315

895. 1823–7–29 To Charles Babbage Slough
Regarding the Chancellor of the Exchequer and the calculating machine. Recent astronomical observations.
ALS 3pp RS:HS 2.189

896. 1823–7–31 To Henry Kater n.p.
'Our micrometer screws' are not in error. F. G. W. Struve's star measurements using Josef Fraunhofer's repeating micrometer agree well with JH's. So error must be in HK's observations.
ALS 3pp RS:HS 19.51 (C: RS:HS 20.165)

897. 1823–8–1 To Caroline Lucretia Herschel Downing St.
Researching double stars. Comparing his finding to William Herschel's, finds that most are compatible. The 40-ft. reflector is unusable, but hopes to try the 20-ft. Heard that WH's papers will be translated into German; wishes Johann Pfaff to undertake it. Asks CH to send him a copy of the finished translation.
ALS 4pp TxU:H/L–0568; Reel 1058 *Memoir ... of Caroline Herschel,* 168–70 inc

898. 1823–8–8 From Heinrich Christian Schumacher Altona
[J. W. A.] Pfaff plans to publish a translation and edition of all William Herschel's papers. Requests a catalogue of those papers and permission to publish a biography of his later years.
ALS 1p RS:HS 15.370

899. 1823–8–10 From Charles Babbage Devonshire St.
Has purchased the picture for £8. Will be at Birmingham tomorrow.
ALS 1p RS:HS 2.190

900. 1823–8–10 From James South n.p.
Reports on various observations of double stars that JS has made. Asks JH to check some of JS's results.
ALS 8pp RS:HS 16.419

901. 1823–8–11 From Caroline Lucretia Herschel Hanover
Assures JH that she will send Johann Pfaff's German translation of William Herschel's papers as soon as they are published.
L *Memoir ... of Caroline Herschel,* 170–1 inc (?)

902. 1823–8–11 From Wilhelm Struve Dorpat
Thanks JH for having been nominated to England's Astronomical Society. Notes that his results on double stars agree with those of JH and James South. Eagerly awaits new 14-foot achromatic telescope.
ALS 4pp (German) RS:HS 17.105

903. 1823–8–13 From C. Goring 56 Commercial Rd.
Queries regarding Sir William Herschel's telescopes, and requesting JH's advice on certain astronomical matters.
ALS 4pp RS:HS 8.224

904. 1823–8–15 To Pierre Simon Laplace n.p.
Letter of introduction to Mr. Robison, son of Edinburgh professor. Thanks PL for sending JH books 11 and 12 of LP's *Mécanique céleste.* Hopes to see PL's 'perturbations of Ceres and [J. F.] Encke's Comet.' JH's and James South's work on double stars verified views PL once held. Hopes François Arago will publish observations of double stars.
ALS 4pp National Maritime Museum (AAbs: RS:HS 19.7)

905. 1823–8–27 From H. G. Goltermann Duke St. St. James's
Has no record of payee on draft from Caroline Lucretia Herschel's account sent to HG by JH.
ALS 1p TxU:H/L-0347.1; Reel 1054

906. 1823–9–13 From James South n.p.
Reports on efforts to make measurements regarding 39 Boötes. Accepts JH's invitation for JS and wife to visit JH at Slough.
ALS 3pp damaged RS:HS 16.420

907. 1823–9–20 From Mr. Reugger Bern
Explains that because of illness, Mr. Schenck, a skilled instrument maker, will not be moving to London. [Much of letter is illegible, but it contains a draft reply from JH.]
ALS 2pp†† (French) RS:HS 14.478

908. [1823?]–9–23 From Francis Lunn Hawarden Castle
Had difficulty with actinometer observations on Etna, but obtained a few. Has just obtained a living in Somerset. Is working for the *Encyclopaedia Metropolitana* at the moment but hopes to make a set of actinometer observations soon.
ALS 4pp RS:HS 11.409

909. [1823–10 or earlier] From Heinrich Christian Schumacher n.p.
Please forward biography of WH and a list of WH's papers to J. W. A. Pfaff, who plans to translate William Herschel's works into German.
Abs(by JH) 1p RS:HS 19.10

910. [1823]–10–6 From Mary Pitt Herschel London
Wants news of visitors to Slough. Encloses papers from Thomas Young. Will send Young's plates later with Miss Delevaux.
ALS 2pp inc & C 3pp inc TxU:H/M–0622.3; Reel 1086

911. 1823–10–11 From James South Blackman St., Southwark
Discusses observations he had made of various stars, especially a star in Capricorn that William Herschel believed to be a gravitational double. Has observed 61 Cygni.
ALS 3pp RS:HS 16.422

912. 1823–10–14 From Thomas W. Hornbuckle St. John's College, Cambridge
Answers to his queries regarding fees at Cambridge and amount of living allowance necessary.
ALS 2pp RS:HS 9.483

913. [1823–10–15 or later]. To William Whewell n.p.
Praises and comments on some proposals made by WW for designating the faces of crystals. Urges him to present his method to the R.S.L.
ALS 3pp TC, Camb. Add. Ms.a.207[8] (C: RS:HS 20:168)

914. [1823–10–15 or later] To William Whewell n.p.
Comments favorably on a method developed by WW for designating the faces of crystals. Presents a method JH had worked out previously, but writes on the letter: '*Not sent*—On rereading his letter I find it is his idea—strange!'
Adraft 4pp RS:HS 18.164 (C: RS:HS 20.167 & TC, Camb. 0.15.46[18] inc)

915. 1823–10–15 From William Whewell Trinity College
Presents a system WW has devised for designating the faces of crystals. Asks JH whether it would be suitable for presentation to the R.S.L.
ALS 4pp RS:HS 18.163

916. [1823–10–20] To Johann Wilhelm A. Pfaff [London]
Heard from H. C. Schumacher that JP plans to translate William Herschel's works into German. Sends five bound volumes, three papers, and list of papers printed in *R.S.P.T.* Promises to send biography of William Herschel when it is finished.
AAbs 1p RS:HS 19.10

917. 1823–10–[20] To Heinrich Christian Schumacher n.p.
Sends packet of books and papers for delivery to J. W. A. Pfaff, copy of JH's 'Essay on Physical Astronomy,' and JH's and James South's work on double stars. Promises to send biographical notice of William Herschel. Astronomical Society received HS's 'Schreiben en Dr. Olbers.'
AAbs 2pp RS:HS 19.11

918. 1823–11–13 From James South Blackman St., Southwark
Discusses a paper by [Edward] Troughton and work that JH and JS were preparing. Remarks on observations of 61 Cygni and of Sigma Cassiopeia.
ALS 3pp RS:HS 16.423

919. [1823–11–14 or later] From James South n.p.
Will bring a transit instrument to Slough so that JH and JS can determine its longitude. Proposes joint observations of the satellites of Saturn. Has been observing 61 Cygni.
ALS 4pp RS:HS 16.475

920. 1823–11–17 From Joseph Johann Littrow Vienna
Thanks for books. Presumes Robert Molyneux has received payment for clock; this is a very good instrument, much better than [Thomas?] Young's. Encloses a work of his own, which he would like to see in the *Transactions* of the R.A.S. Sends propositions by a young Bavarian named K. W. Feuerbach.
ALS 6pp (German with English trans.) RS:HS 11.242

921. 1823–11–22 From Heinrich Christian Schumacher Altona
Thanks for forwarded essay on physical astronomy. Remarks about method for precisely determining the power of a telescope. Has [C. L.] Harding's celestial atlas. Wanted to dedicate book to Astronomical Society, but asks privately whether permission is necessary.
ALS 2pp RS:HS 15.371

922. 1823–11–23 From Charles Babbage n.p.
Enclosing an invitation from Humphry Davy to dine with him. Invitation for JH to dine with CB also.
ALS 1p RS:HS 2.191

923. 1823–11–23 From Susan White [Shaw, Berkshire]
SW's nephew Henry White is coming to Cambridge with message from [servant] Sally, who misses JH.
Abs 1p TxU:H/M–0580.3; Reel 1093

924. 1823–11–23 From Susan White Shaw
Asks JH to befriend nephew Henry White, scholar and missionary at Cambridge. Asks about Mary Baldwin. When will JH visit again? [JH annotation: Asked William Whewell to call on Henry White.]
ALS 2pp † RS:HS 18.228

925. [1823–11–24] To William Whewell [Slough]
Advises WW on how to submit WW's paper on crystallography to the R.S.L. Asks him to visit Henry White.
ALS 3pp TC, Camb. Add. Ms.a.207[9] (C: RS:HS 20.166)

926. 1823–11–25 From Joseph Johann Littrow Vienna
Has doubtless received his *Treatise on the Corrections of the Equatorials*. Col. L. A. Fallon, head of the Austrian Survey, would like a Robert Molyneux clock. Can JH arrange to order one and send it via H. C. Schumacher. Encloses some notes on practical astronomy, which may be useful.
ALS 5pp (German with English trans.) RS:HS 11.243

927. 1823–11–26 From Thomas Young 48 Welbeck St.
Board of Longitude will meet at TY's home on Saturday.
ALS 2pp RS:HS 18.326

928. 1823–11–29 To David Brewster 53 Great Portland St.
Some comments on observations by Continental observers; matters related to sending books and papers.
C 2pp RS:HS 20.169

929. 1823–11–29 From Marc August Pictet Geneva
Letter of introduction for O. F. Mossotti.
ALS 1p RS:HS 13.399

930. 1823–11–29 From F. Trechsel Berne
Sends letter by Mr. Massotti of the Milan observatory, who is leaving to find employment in England. Massotti wishes to sell his dividing machine in England. FT hopes to see JH in England at the reunion of the Societé Helvetique.
ALS 1p (French) RS:HS 17.375

931. 1823–12–2 To David Brewster n.p.
Vol. 6 and 7 of F. W. Bessel's observations. Did DB receive JH's paper?
AAbs 1p RS:HS 19.4

932. 1823–12–10 To [Jean Louis] Pons [Astronomical Society]
Society awarded JP silver medal. JH willing to accept on JP's behalf and forward it.
AAbst 1p RAS:JH Archive 6/2.12 verso; Reel 8

933. 1823–12–13 From Thomas Catton 10 Craven St.
Returning the five volumes of the Berlin yearbook with thanks. Regarding the observations of P. F. A. Méchain at Slough in 1787.
ALS 2pp RS:HS 5.224

934. 1823–12–13 From Johann Wilhelm A. Pfaff Erlangen
Is collecting and translating William Herschel's writings, projected for 4 volumes, the first one already being in press. Requests JH to provide additional biographical information about his father for one of the later volumes.
ALS 3pp RS:HS 13.378

935. 1823–12–19 From James South Blackman St., Southwark
Writing in confidence, informs JH of a conversation with Humphry Davy concerning the appointment of assistants for John Pond, Astronomer Royal. Situation complicated by Pond's precarious health and the widespread belief that JH would be the appropriate successor to Pond.
ALS 3pp RS:HS 16.424

936. 1823–12–19 To William Whewell n.p.
Strongly praises WW's completed paper on crystallography. Questions one point in it.
ALS 3pp TC, Camb. Add. Ms.a.207[10] (C: RS:HS 20.170)

937. 1823–12–21 To James South n.p.
Responding to JS's letter of 1823–12–19, advises caution in countervening John Pond's preferences in the appointment of Royal Observatory assistants. Were the

post of Astronomer Royal to become vacant, JH would be '*anything but disinclined* to offer myself for it,' but will not make a 'premature declaration' of his willingness to accept it.
Adraft 8pp RS:HS 16.425 (C: 20.171)

938. 1823–12–23 From Joseph Johann Littrow Vienna
New Observatory is being planned on the lines of the new one being built at Obo in Finland. Would like an open letter from JH on the requirements for an observatory. Is there any news of the clock for [L. A.?] Fallon?
ALS 3pp (German with English trans.) RS:HS 11.244

939. [1824 to 1848] To 'dear Sir' n.p.
Received [?]'s inquiry of 1 Feb. Lists all biographical notices pertaining to William Herschel. Comments in detail on biography and philosophy of WH in David Brewster's *Edinburgh Review*.
Adraft 4pp TxU:H/L-0743; Reel 1089

940. 1824 To David Brewster n.p.
Letter to [*Edinburgh Philosophical*] *Journal* calling attention to 1809 article by [Paul] Erman of Berlin, which some say anticipated JH's 1824 Bakerian lecture on 'Motions produced in fluid conductors when transmitting the Electric current.' JH compares both papers, admits Erman's priority on some theoretical points but claims originality for JH's experiments.
Adraft 4pp TxU:H/L-0112; Reel 1054

941. 1824–1 From Jean Baptiste Joseph Fourier Paris
Sending copies of his Éloge on William Herschel, and requesting some information regarding his publications.
ALS 2pp (French) RS:HS 7.342

942. [1824] From Mrs. Mary Frances Kater n.p.
Many thanks for his beautiful ballad.
ALS 2pp RS:HS 11.19

943. 1824 To Giuseppe Piazzi n.p.
Defends William Herschel's views against charges of hostility to religion. Describes his father as 'a sincere believer in, and worshipper of, a benevolent, intelligent, and superintending Deity....'
L Lubbock, *Herschel Chronicle*, 197

944. [1824 to 1827?] From Joseph Planta Hampstead
Regrets missing JH's visit. Refers JH to [Charles] König for admission to Department of Natural History at the British Museum.
ALS 1p RS:HS 14.166

945. [1824 or later] From William Smith Williams [65 Cornhill]
Packets will be sent Saturday by Smith, Elder & Co., so that the one for [Caroline] Herschel will not be late for Mr. Köhler.
ALS 1p inc RAS:JH Archive 6/2.1 verso; Reel 8

946. [1824 to 1829] From Thomas Young n.p.
JH is correct, and TY's 'little man is wrong.' Moon's apparent angular motion increases as spectator moves closer.
ALS 1p RS:HS 18.355

947. 1824–1–4 From William Edward Parry Admiralty
Regrets delay in answering JH's communication, but was away from Bath when the letter arrived. Interested in the extraordinary local attraction observed by Mr. Duncan in 1791. Gives locations of three such situations in his recent voyage. Comments on the use of Peter Barlow's plate. Will write to Barlow before his next voyage.
ALS 4pp RS:HS 13.218

948. 1824–1–8 To James South Charlotte St., Portland Place
Wishes JS immediately and confidentially to pass on to Humphry Davy JH's letter of 1823–12–21, making clear that JH does not favor appointment of Stephen Lee as head assistant at the Royal Observatory. Has heard that JS has felt that JH slighted JS in JH's letter. Denies this.
Adraft 5pp RS:HS 16.432 (C: 20.172)

949. 1824–1–8 To James South London
If JS has not already passed on to Humphry Davy JH's letter of 1823–12–21, JS should do so immediately. If post of Astronomer Royal should become vacant, JH would be interested in it, but would not wish to work with Stephen Lee, then under consideration for appointment as head assistant at Greenwich.
Adraft 2pp RS:HS 16.426a (C: 20.173)

950. 1824–1–15 To Alexis Bouvard London
Acknowledges letters from AB and French Board of Longitude received by Astronomical Society. Thanks for care given to JH's reflecting theodolite by Mr. Schenck. Promises copy of JH's and James South's [1824] article on double stars. Hopes to hear from J. B. Fourier regarding biography of William Herschel.
AAbsS 2pp RS:HS 19.3

951. 1824–1–15 To Heinrich Christian Schumacher London
Orders missing issues of *Astronomische Nachrichten*. Answers HS's question about JH's essay on Physical Astronomy. James South's testimony 'vindicated' Josef Fraunhofer's repeating micrometer to Edward Troughton. C. F. Gauss's mode of determining powers of telescopes. Nothing heard of Captain [Maughan?]. Directions for sending [K. L.] 'Harding's atlas' to JH. Gives J. F. Daniell's address. Astronomical Society is gratified that HS dedicated book to it. Shall direct Robert Molyneux to send L. A. Fallon's clock to HS.
AAbsS 1p RS:HS 19.11

952. 1824–1–18 To Charles Babbage Slough
Regarding the First Assistant at Greenwich Observatory, and the possibility of JH ever becoming Astronomer Royal. Hopes to see him on Wednesday.
ALS 1p RS:HS 2.192

953. 1824–1–20 From Heinrich Christian Schumacher Altona
Will forward *Astronomische Nachrichten* to JH and Astronomical Society. Comments on quality of [Joseph] Fraunhofer's micrometer, which has only the wires illuminated. [C. F.] Gauss uses a theodolite to determine the angle of the telescope.
ALS 3pp RS:HS 15.372

954. 1824–1–21 To Joseph Johann Littrow London
Receipt and shipment of various papers and letters. Robert Molyneux has not received payment for clock. JH ordered another for L. A. Fallon; it is ready to ship. Clarifies Edward Sabine's remark about pendulums. Hopes to translate JL's *Analytical Geometry*. JH translated JL's 'empirical formula of refraction.' Believes JL's method of determining latitude without knowing the time is not new. Notes J. W. A. Pfaff's translation of William Herschel's

works.
AAbs 1p RS:HS 19.8

955. 1824–1–22 From George Biddell Airy Bury St. Edmunds
Requesting information of the dispersion of rays in Crown and Flint glass.
ALS 3pp RS:HS 1.19

956. 1824–1–23 From Basil Hall 128 George St., Edinburgh
Had no report he could offer the Astronomical Society. Outlines the situation in Chile. Hopes to arrange the material on his recent voyages in the form of a book.
ALS 6pp RS:HS 9.163

957. 1824–1–25 To Joseph Johann Littrow London
Approves of the plan for the observatory. Comments on some of the features. Good equatorial is a necessity. Will try to obtain a copy of the plan of the Cambridge Observatory.
draft of Abs 2pp RS:HS 11.245

958. 1824–1–31 To George Biddell Airy Slough
Sends GA information about indices of refraction in different glasses and in different parts of the spectrum; also recommends the writings of Josef Fraunhofer.
ALS 4pp RGO 6.803.4

959. 1824–1–31 To David Brewster Slough
JH's theory of colors. Edward Troughton has DB's instrument one-third completed. When will DB want JH's encyclopedia article on 'Variation'?
C 3pp RS:HS 20.175 (C draft: 20.174)

960. 1824–2 From Jean Baptiste Joseph Fourier n.p.
Sends notes on a paper which are to be submitted to James South. Also included [?] are copies of the paper for JH. Attached is a note by JH about the papers of JH's father, William.
LS 2pp (French) † (JH note) RS:HS 19.218

961. 1824–2–4 To Mary Pitt Herschel n.p.
Mrs. Babbage, still recuperating, wishes to visit Slough. Recent publication of Charles Babbage's book should excite insurance directors. Paper by Mary Somerville was read at R.S.L. last week. Invitation for MH to visit Somervilles at Chelsea. JH admitted to Geological Society yesterday, presided over by William Buckland. Encloses JH's diploma from Royal Academy of Copenhagen. News of MH's London acquaintances.
ALS 4pp & C TxU:H/L-0517.1; Reel 1053

962. 1824–2–9 From James Grahame Edinburgh
On JG's concern for JH's well–being.
ALS 3pp RS:HS 8.316

963. 1824–2–9 From Johann Wilhelm Pastorff Berlin
Requests answer to JP's earlier letter. Observations of photospheres around planets and satellites.
Abs 1p RAS:JH Archive 6/2.10 verso; Reel 8

964. 1824–2–10 From John Brinkley Observatory, Dublin
Has just received his letter and Struve's observations. Comments on these and those of F. W. Bessel.
ALS 3pp RS:HS 4.276

965. 1824–2–10 From Joseph Johann Littrow Vienna
Has sent money for Robert Molyneux clock care of Friess, the bankers. Is grateful for JH's trouble over the clock. Do not print certain observations he sent as they form the subject of an argument with F. X. von Zach.
ALS 2pp (German with English trans.) RS:HS 11.246

966. 1824–2–13 From George Biddell Airy Trinity College, Cambridge
Description of theory of a telescope, though not practical yet.
ALS 3pp RS:HS 1.20

967. 1824–2–13 From George Biddell Airy to James South Cambridge
Concerning calculations for the parallax of Mars.
ALS 3pp RS:HS 1.21

968. 1824–2–15 To James South Charlotte St.
Urges that JS not get so emotionally involved in issues concerning astronomers. 'For my own part I mean to make Science my amusement, not my business....' Comments on JS's relations with John Pond.
ACCS 2pp RS:HS 16.426b (C: RS:HS 20.177)

969. 1824–2–20 From Francis Baily Gray's Inn
Is sending a letter from Dr. John Brinkley, and the corrected elements of the comet.
ALS 2pp RS:HS 3.56

970. 1824–2–23 To [Joseph Goodall] Slough
Comments on some ideas of J. B. Lamarck, and about large fossil remains apparently found.
ALS 2pp APS B.H435p.27

971. 1824–2–26 From Stephen Peter Rigaud Oxford
Cannot find observation of comet of 1759 in [James] Bradley's papers. Discusses eighteenth-century astronomical observations. Can send JH particulars if he wants them.
ALS 2pp RS:HS 14.336

972. [1824–3 or earlier] From Dr. [Thomas?] Hodgkins n.p.
Tour notes composed to assist JH's travels in Sicily.
AL 3pp TxU:H/M–1195; Reel 1060

973. 1824–3–3 From G. W. [Ott?] London
Acknowledges receipt by Jameson & Aders of Astronomical Society's 1824 gold medal, to be delivered to J. F. Encke in Gotha. [JH annotation: Encke confirmed receipt of medal.]
ALS 1p † RS:HS 19.214

974. 1824–3–8 From Charles Babbage [Cambridge]
Is leaving from Cambridge. Hopes that he will come. Has some new things on functional analysis to discuss.
ALS 1p RS:HS 2.193

975. 1824–3–9 From Thomas W. Hornbuckle St. John's College, Cambridge
Enclosing a draft for his dividend, which he hears will be the last. Hopes he changes his fellowship for a better one, but would be pleased if he would delay his decision for a while.
ALS 1p RS:HS 9.484

976. 1824–3–13 From David Brewster n.p.
Regarding the change of ownership and policy of the *Edinburgh Philosophical Journal*. Would like a paper from him for the first number. Marked 'Private.'
ALS 4pp RS:HS 4.257

977. 1824–3–15 To Johann Wilhelm Pastorff London
Observations of Venus by William Herschel and JH. Only J. H. Schröter claims to have seen 'deep marginal indentations' that JP claims.
AAbst 1p RAS:JH Archive 6/2.10 verso; Reel 8

978. 1824–3–16 From William Wallace 14 Argyle Square, Edinburgh
Please recommend London instrument makers to supply new Edinburgh observatory. Josef Fraunhofer was asked to make only lens for transit instrument, but wants to make entire instrument. [JH annotation: Recommended Fraunhofer for entire transit instrument and Robert Molyneux for clock.]
ALS 3pp † RS:HS 18.23

979. 1824–3–18 To Pierre Simon Laplace London
Henry Kater's triangulation survey of north Scotland. Edward Sabine's study of pendulum lengths at various latitudes. François Arago did not receive letters from JH and Francis Baily. Comments on rumor of [Isaac] Newton's madness. JH plans to reduce transit observations of fixed stars at Greenwich since [James] Bradley's time. Are French and other astronomers in agreement on common system of reductions? Will visit Paris next month. Observed curious phenomena in voltaic electricity.
ACCS 3pp RS:HS 19.7 (C: 20.178)

980. 1824–3–23 From George Peacock Trinity College, Cambridge
On the matter of helping a friend who is being maligned [Mrs. Bowdich], and about JH's projected Continental tour.
ALS 4pp RS:HS 13.286

981. 1824–3–31 From George Dollond London
Would be pleased if JH would obtain some flint-glass for him when he goes to Munich.
ALS 2pp RS:HS 6.498

982. 1824–4–1 From James Grahame n.p.
Sends blessings on JH.
ALS 1p RS:HS 8.317

983. 1824–4–2 To Mary Pitt Herschel Dover
Foul weather in Dover delays departure of JH and servant James Child for Calais. Geological formations along road to Dover. Instructions on numbering and preserving personal letters.
ALS 3pp & C TxU:H/L-0517.2 & -0517.3; Reel 1053

984. 1824–4–3 To Mary Pitt Herschel Dover
Weather is good for Channel crossing. Returns 'book of rates.' It is too cumbersome.
ALS 1p & C TxU:H/L-0517.4 & -0517.3; Reel 1053

985. 1824–4–6 From Jean Baptiste Joseph Fourier n.p.
Has received JH's note, and JF will visit JH next Wednesday, accompanied by D. F. J. Arago and [C. L.] Mathieu.
ALS 1p (French) RAS:WH Archive 1/13.F.7; Reel 13

986. 1824–4–7 To Mary Pitt Herschel Hotel de Tours, Paris
Arrival in Paris. Names chemists, mathematicians, and others whom JH met at Institute. Breakfast with D. F. J. Arago at observatory. Problems with passport. Purchase of carriage for trip to Italy. [Letter continued same day from Melun:] En route to Lyons.
ALS 4pp & C TxU:H/L-0517.5; Reel 1053

987. 1824–4–9 From Mary Pitt Herschel Brompton [Middlesex]
Correspondence with Babbages. Caroline Herschel asks JH to meet Mr. Kestner, private secretary of Hanoverian ambassador to Rome. Regards from Miss Delevaux.
ALS 2pp inc & C 3pp inc TxU:H/M–0623.1; Reel 1086

988. 1824–4–9 From Joseph Johann Littrow Vienna
[Repetition of information in JL's 1823–11–17 and JL's 1824–2–10.]
ALS 2pp (German with English trans.) RS:HS 11.247

989. 1824–4–11 To Charles Babbage Aiguebelle
Relates experiences and persons met recently in Paris. Account of Alpine scenery and travel experiences. [Continued 1824–4–20 Turin:] Events while travelling to Turin. [Continued 1824–4–22 Alessandria:] Has met Plana at Turin Observatory. Five days of discussions with him. Please order two specified books and send to Plana. View of the Alps.
ALS 4pp RS:HS 2.194

990. 1824–4–12 From [Luigi] Poletti Modena
Presents JH with his solution to a mathematical equation in finite differences; asks for JH's solution.
ALS 6pp † (Italian) RS:HS 14.173

991. 1824–4–15 From Mary Pitt Herschel Brompton [Middlesex]
Pleased to hear about JH's travels in Europe. Visit to London with Mary Baldwin. News of acquaintances. Describes a dinner with the Copelands. Death of [Susan] White's husband. Sent money and Richard Copeland's drawings to Caroline Herschel via Mr. Golterman. Misses Slough. Mr. Robinson made offer to Mrs. Morice for manuscripts.
ALS 4pp damaged & C 5pp TxU:H/M–0623.2; Reel 1086

992. 1824–4–19 To Mary Pitt Herschel Turin
Describes travels in Savoy and Sardinia. Snowstorm on Mont Cenis. Arrived in Turin on Good Friday. Meeting with G. B. Plana at Turin observatory. Leaving for Rome next week.
ALS 4pp & C TxU:H/L-0517.6; Reel 1053

993. 1824–4–26 To Mary Pitt Herschel Genoa
Describes Turin and church politics. Ill health of the Pope {Leo XII]. [Letter continued 27 April between Genoa and Modena:] Met F. X. Zach in Genoa. Incident of anti-Jewish bigotry in Genoa. [Letter continued 2 May in Florence:] Met G. B. Amici in Modena. Clarity of Amici's telescope. Describes Bologna.
ALS 4pp & C TxU:H/L-0517.7; Reel 1053

994. 1824–4–30 From Marc August Pictet Geneva
Sends some observations of Mars, and describes the functioning of some new equipment.
ALS 3pp RS:HS 13.400

995. 1824–5–2 To James Grahame Florence
Of JH's continental travels and their dangers, and of JG's possible marriage plans.
draft 4pp RS:HS 8.318

996. 1824–5–5 To Charles Babbage Florence
Recent experiences in crossing the Alps. Overturning of the coach and his narrow escape.
ALS 4pp RS:HS 2.194a

997. 1824–5–6 From Mary Pitt Herschel Slough
Received package with gifts from Caroline Herschel. [Letter continued 7 Apr.:] Forwarding mail for JH to Charles Babbage. Reminds JH to write from Rome. [Letter continued 8 Apr.:] Death of Mr. White. Georg Griesbach underwent operation. [Letter continued 9 Apr.:] Reading [August von] Kotzebue's *Travels through Italy*. Worried about dangers in JH's journey there.
ALS 3pp † damaged & C 6 pp TxU:H/M–0623.3; Reel 1086

998. 1824–5–9 To Charles Babbage Florence
Recent meeting and astronomical observations with G. B. Amici. [Repeat of his narrow escape in the Alps, related in JH's 1824–5–5]
ALS 4pp RS:HS 2.195 & 20.180

999. 1824–5–11 To Mary Pitt Herschel Monastery of Vallombrosa
Describes monastery. [Letter continued 12 May in Arezzo, Tuscany:] Journey from Bologna. Describes Florence and Florentine art. [Letter continued 16 May at Città di Londra Hotel in Rome] Received mail from home. Death of [JH's godmother] Susan White. Describes Rome. Will search for Mr. Kestner.
ALS 4pp & C TxU:H/L-0517.8; Reel 1053

1000. 1824–5–12 From Mary Pitt Herschel Slough
Anticipates news from JH in Naples. Mary Baldwin staying with MH. Glad that servant James is good companion. [Letter continued 13 May:] Advice on sending mail from Italy. Carriage repaired and repainted. [Letter continued 14 May:] Terrible weather. Miss Deluc is not expected to survive. [Letter continued 17 May:] Visitors to Slough.
ALS 3pp & C 5pp TxU:H/M–0623.4; Reel 1086

1001. 1824–5–20 To Mary Pitt Herschel Rome
Disappointed with Roman ruins and art. [Letter continued 24 May:] Hopes to receive mail from Slough in Naples. [Letter continued 27 May:] Viewed pope's benediction in St. Peter's Square. Roman weather. JH's itinerary for June and July.
ALS 4pp & C TxU:H/L-0517.9; Reel 1053

1002. 1824–5–21 From Mary Pitt Herschel n.p.
Delighted with JH's news from Italy. MH urges him to visit Englishmen on Continent, especially Lord Somerset's family. Death of Lord Byron. [Former servant] Carter was banished to Botany Bay for life. [Letter continued 22 May:] News of acquaintances. Reminder to pay wages to servant James.
ALS 2pp & C 4pp TxU:H/M–0623.5.1; Reel 1086

1003. 1824–5–24 From Mary Baldwin Oxford St.
JH's mother is well. She and MB will visit exhibition in London tomorrow. Health of acquaintances.
ALS 2pp & C 2pp TxU:H/M–0623.5.2; Reel 1086

1004. 1824–5–28 From Charles Babbage Devonshire St.
Has been appointed director of a company. Offers shares to JH. Has sent books to G. A. A. Plana. Wife and children in Devon.
ALS 4pp RS:HS 2.196

1005. 1824–6–1 To Mary Pitt Herschel Naples
Asks for news of MH's health. Lists all letters JH has sent. Problems with European post. Trip from Rome to Naples. Neopolitan countryside. [Letter continued 2 June:] At Hotel de la Città di Napoli on Bay of Naples. Describes Mount Vesuvius. Visit with Teodoro Monticelli and British minister W. R. Hamilton. [Letter continued 3 June:] Attended opera at Europe's largest theater. Prefers Naples to Rome; will stay at least one month.
ALS 4pp & C TxU:H/L-0517.10; Reel 1053

1006. [1824 to 1831]–6–3 To Committee for Rebuilding London Bridge 56 Devonshire St, Portland Place
Accepts honor of joining committee at ceremony to lay first stone of new bridge. Proposes including platina copies of standard British measures of length and weight in foundation as part of '*precious records*' preserved for posterity.
ACC 4pp TxU:H/L-0228; Reel 1054

1007. 1824–6–6 From Mary Pitt Herschel Slough
Received JH's letter from Rome. Bought [David] Cox's *Pictures of Italy* to see places JH has visited. Visits London doctor with Mary Baldwin. News of acquaintances. MH authorized seizure of Mr. Davenport's corn, because he refuses to pay rent.
ALS 3pp † & C 4pp TxU:H/M–0623.7; Reel 1086

1008. 1824–6–8 To Mary Pitt Herschel Crater of Vesuvius
Describes last letter to MH. [Letter continued 9 June in Naples:] Will write to J. H. Nelson. Problems with postal service in Naples. Met European nobility at dinner of British minister W. R. Hamilton. Ascent of Mount Vesuvius with Nicola Covelli. Barometer observations, chemical experiments, and camera lucida drawings. High society in Naples. Describes 'ridiculous' religious practices at church of Madonna dell'Arco and barbarian lower classes of Naples.
ALS 4pp & C TxU:H/L-0517.11; Reel 1053

1009. 1824–6–9 From William Richard Hamilton n.p.
The mistake was mutual; hopes to see him Friday for dinner.
ALS 2pp RS:HS 9.216

1010. 1824–6–11 From James South [London]
Writing to JH [then travelling in Europe], JS informs him that JS is moving to Paris. Comments on his plans for his observatory in Paris and on recent events in England regarding astronomers and instruments.
ALS 8pp † RS:HS 16.427

1011. 1824–6–12 From Mary Pitt Herschel Slough
Concerned about JH in Rome. Expecting visit from Mrs. [James] South and Charles Babbage, who has rented house in Passy, France. Many visitors to Slough for Ascot races next week. J. H. Nelson reminds JH to use special shipping agent in Florence to send items to England. [Letter continued 13 June:] Celebrating MH's 74th birthday. [Letter continued 14 June:] Asks for JH's itinerary.
ALS 4pp † damaged & C 6pp TxU:H/M–0623.8; Reel 1086

1012. [1824]–6–14 From Francis Lunn Chesterford, Essex
Met Mr. Smedley a few days ago and the latter has the impression that JH is to write the article on Heat for the *Encyclopaedia Metropolitana*. Comments on this and wonders if he can request the articles from JH. Sends letter via Charles Babbage.
ALS 2pp RS:HS 11.410

1013. 1824–6–15 From [William Richard] Hamilton n.p.
Invites JH to dinner.
AL 1p Harvard: Houghton AAH 67m–67 (51)

1014. 1824–6–17 To Mary Pitt Herschel Naples
Received mail from MH, Mary Baldwin, and J. H. Nelson. Death of George Gordon (Lord Byron). Dined with British minister W. R. Hamilton. News of English residents in Rome. Recent archeological discoveries in Naples. Describes Pompeii, Solfaterra, and Puzzuoli. Religious procession of king through Naples. [Letter continued 18 June:] Met British admiral H. B. Neale at Hamilton's. Problems with Dey of Algiers.
ALS 4pp & C TxU:H/L-0517.12; Reel 1053

1015. 1824–6–20 To Mary Pitt Herschel Palermo
Voyage to Sicily. Scenes of Palermo. Visit with Giuseppe Piazzi and Niccolo Cacciatore at Palermo observatory. Names scientists JH plans to visit on Sicily.
ALS 4pp & C TxU:H/L-0517.13; Reel 1053

1016. 1824–6–20 From Giovanni Antonio Plana Turin
Is working on printing GP's theory of the moon. Has received packages containing the works of a number of English scientists, and wants to know cost so as to be able to reimburse JH. Wants to be remembered to Charles Babbage.
ALS 2pp (French) RS:HS 13.411

1017. 1824–6–27 To Mary Pitt Herschel Girgente [Agrigento], Sicily
Six-day mule trip to see ruins of Girgente. Servant James Child is holding up well.
ALS 3pp & C TxU:H/L-0517.14; Reel 1053

1018. 1824–6–28 From Mary Pitt Herschel Slough
MH suffering from gout. Heavy rains are ruining crops. Miss Deluc died yesterday. Postponing trip to seaside to await visit by Parrys. Difficulty in collecting rent from Mr. Davenport. [Letter continued 27 June:] Received today JH's letter from Naples. Pleased to hear so many in Europe honor William Herschel. [Letter continued 28 June:] Reminder to use Mr. Nelson's shipping agent for packages to England.
ALS 4pp † damaged & C 7pp TxU:H/M–0623.9; Reel 1086

1019. 1824–6–30 From Mary Pitt Herschel Slough
News of family friends. Awaits JH's return. Mary Baldwin goes to London.
ALS 2pp † & C 3pp TxU:H/M–0623.10; Reel 1086

1020. 1824–6–30 To Johann Wilhelm A. Pfaff Caltagirone, Sicily
Directs JP to additional sources of biographical information about William Herschel [see JP's 1823–12–13], and comments on some other details of JP's letter.
C 1p RS:HS 13.379 & 20.182

1021. 1824–7–1 To Mary Pitt Herschel Catania, Sicily
Mount Etna marks furthest point of JH's journey. Travel is difficult. Both JH and servant James Child are in good health. Describes ruins of Girgente. Send future letters to Geneva or Paris. [Letter continued 3–4 July at Mount Etna:] Few lines from Casa Inglese, highest house in Europe, built by Carlo Gemmellaro for travelers. [Letter continued 5 July in Catania:] Ascent of Mount Etna.
ALS 4pp & C TxU:H/L-0517.15; Reel 1053

1022. 1824–7–2 To Caroline Lucretia Herschel Catania, Sicily
Describes his laborious journey to the summit of Mt. Etna; from his barometric readings, concludes that its altitude is 10,000 or 11,000 ft. Asks about the progress of Johann Pfaff's translation of William Herschel's papers; JH mentions that he wrote to Pfaff from Cattagione, Sicily. [Letter continued 20 July from Naples and 16 Aug. from Florence.]
ALS 4pp TxU:H/L–0569.1; Reel 1058 *Memoir ... of Caroline Herschel,* 172–4

1023. 1824–7–4 To Charles Babbage Etna
Is observing the heavens from Etna. [Letter continued 11 July Palermo:] Recent experiences touring Sicily.
ALS 4pp RS:HS 2.197 & 20.181

1024. 1824–7–6 From Thomas Hodgkin Rome
Morichini will be gratified to see JH when he returns through Rome. The Abate Feliciano Scarpellini is also desirous of meeting JH; he is one of the foremost makers of instruments in Italy. Weather is just becoming hot, so hopes he gets through his Sicilian tour before the heat increases.
ALS 2pp RS:HS 9.401

1025. 1824–7–8 From Carlo Gemmellaro Catania [Sicily]
Encloses letter from CG's brother regarding barometer scales. Hopes JH had pleasant trip from Catania to Palermo in company of Count [Bessa?]. [Niccolò] Cacciatore will give JH six-year collection of DG's meteorological observations, soon to be published.
ALS 1p TxU:H/M–0952; Reel 1083

1026. 1824–7–12 From Mary Pitt Herschel n.p.
Worried about safety of JH's voyage to Sicily. Mr. Tasker forced Mr. Davenport finally to pay back rent. Should telescope be painted to protect it from rain? Planning trip to Margate. Asks JH to buy watch for MH in Geneva. Declined Mr. Beckwith's request for MH to be godmother to new daughter. [Letter continued 28 June:] James South and wife will visit MH tomorrow.
ALS 4pp & C 6pp TxU:H/M–0623.11; Reel 1086

1027. 1824–7–12 To Mary Pitt Herschel Palermo
Departure delayed by week-long feast of St. Rosalie, patron of Sicily. No vessels moving. Reviews Sicilian scenery, weather, and culture. Compares Mount Etna and Vesuvius. Describes travel across center of island with companions Count Bessa Negrini of Mantua and C. G. B. Daubeny of Oxford. [Letter continued 14 July:] Oppressive heat in Palermo. [Letter continued 18 July:] Arrival in Naples.
ALS 4pp & C TxU:H/L-0517.16; Reel 1053

1028. 1824–7–16 To Mario Gemmellaro Palermo
Is grateful for the assistance received while in Catania. Comments on the divisions of the barometer. His own barometer was undamaged after travelling 500 miles over impassible roads. Would be pleased to receive communications.
draft 2pp RS:HS 8.91 & 20.184

1029. 1824–7–16 To Sir William Watson Palermo
Written from Palermo, this chatty letter describes JH's time in Italy and Sicily. JH mentions having ascended Mt. Etna and having spent the night on its summit. A second part of this letter is dated July 19 and its location is given

as Naples.
ACC 2pp TxU:H/L-0494; Reel 1055 (C: RS:HS 20.183)

1030. 1824–7–17 From Jean Louis Pons to Francis Baily Marlia
Thanks FB for informing him of his election to the Astronomical Society.
ALS 1p (French) RS:HS 14.177

1031. 1824–7–22 To Mary Pitt Herschel Naples
Lists all letters received from Slough and those sent by JH to recipients in England. News of acquaintances at home. C. G. B. Daubeny taken ill. Itinerary for return trip. Low opinion of southern Italians. English society in Naples. Describes cask of wine JH shipped to England. [P.S. 24 July:] Daubeny's illness is malaria. JH will stay with him. Grahames are not here.
ALS 4pp & C TxU:H/L-0517.17; Reel 1053

1032. 1824–7–30 To Mary Pitt Herschel Naples
Describes last leg of Sicilian tour and festival of St. Rosalie, patron of Sicily, in Palermo. [Letter continued 31 July:] C. G. B. Daubeny is recovering but not strong. JH will travel on alone. Lists letters received from England. Dinner with Lord Belvedere. Shipped case of shells and minerals to Mr. Beckwith.
ALS 4pp & C TxU:H/L-0517.18; Reel 1053

1033. 1824–8–6 To Mary Baldwin Rome
Received MB's letters of 19 and 28 May [1824]. Responds to news from Slough. Expedition to Sicily was successful, but C. G. B. Daubeny contracted malaria. Summer climate in Rome and Pontine Marshes. Danger of malaria forced JH and servant James [Child] to travel separately. [Letter continued 7 Sept.:] James arrived from Naples. JH's trip to Tivoli. Itinerary home through Germany. Send letters to Munich.
ALS 4pp damaged TxU:H/L-0061; Reel 1054

1034. 1824–8–10 From Feliciano Scarpellini Rome
Praises JH, JH's father, and their telescopes highly. Asks JH not to forget commission to FS.
ALS 1p (Italian) RS:HS 17.169

1035. 1824–8–14 To Mary Pitt Herschel Florence
Oppressive heat. Toured Rome and Florence in early morning hours only. Low regard for Romans and southerners. Glad to be in civilized Tuscany. Describes Florence. Sorry to hear of Renuel's death. Began letter in Catania to Caroline Herschel.
ALS 4pp & C TxU:H/L-0517.19; Reel 1053

1036. 1824–8–15 From Giovanni Inghirami Museo di Fisica
Invites JH to the Observatory of the Museo di Fisica.
AL 1p (Italian) RS:HS 10.265

1037. 1824–8–22 To Mary Pitt Herschel Venice
Mail and news of acquintances at home. Met Jack Fuller, fellow of St. John's College. Visited astronomer Giovanni Inghirami. Describes Bologna. [Letter continued 23 Aug.:] Deciding route for trip home. Describes servant James Child. Journey from Bologna to Venice. Dismal condition of University of Padua. Describes Venice. [P.S. 24 Aug.:] Shipment to Mr. Beckwith from Rome includes engravings for J. H. Nelson.
ALS 4pp & C TxU:H/L-0517.20; Reel 1053

1038. 1824–8–26 From Mary Baldwin n.p.
Visit to Margate, including concert and dinners.
ALS 2pp inc TxU:H/M–0623.12.1; Reel 1086

1039. 1824–8–26 From Mary Pitt Herschel n.p.
Overjoyed upon receiving two letters from JH. Try to visit Caroline Herschel despite time added to journey. Asks for 'repeating watch' from Geneva.
ALS 1p & C 1p TxU:H/M–0623.12.2; Reel 1086

1040. 1824–8–30 To Mary Pitt Herschel Neumarkt, Tyrol
Leaving tomorrow on expedition into mountains to search for minerals. Journey from Venice to Verona. Visited Giovanni Santini in Padua.
ALS 3pp & C TxU:H/L-0517.21; Reel 1053

1041. [1824–9?] From Frederick Soret n .p.
Hopes to see JH on JH's return visit to Geneva. Encloses quartz specimen, amethyst crystal. Sends regards to Charles Babbage.
ALS 2pp (French) RS:HS 17.221

1042. 1824–9–9 From Mary Pitt Herschel Slough
Returned today from Margate. Received JH's letter from Venice. Will send Mr. Ramsbottom's money to London with Mrs. Singer. New construction at Slough.
ALS 2pp & C 2pp TxU:H/M–0623.13; Reel 1086

1043. 1824–9–9 To Mary Pitt Herschel Seiss Valley near Brixen [Bressanone], Tyrol
Eight-day mineral hunting expedition in mountains, with base at Campedello. High praise for Tyrolese. [Letter continued 10 Sept. in Botzen:] Arrived safely and found servant James Child rested and waiting.
ALS 4pp & C TxU:H/L-0517.22; Reel 1053

1044. 1824–9–11 From Johann Franz Encke Seeburg
Thanks for his election to the Astronomical Society and for the award of its gold medal. Sends a proof of a new chart of the moon by W. G. Lohrmann. Sends two papers of his own. Hopes to send notice of John Pond's comet orbit in 1825.
ALS 4pp (German) RS:HS 7.82

1045. 1824–9–14 From Charles Babbage n.p.
The Insurance firm for which he acted as director has been wound up. Has taken a holiday in the country.
ALS 3pp RS:HS 2.198

1046. 1824–9–15 To Mary Pitt Herschel Innsbruck, Tyrol
Passage through Alps. Servant James Child's adventure being lost in mountains. Religious character of Tyrolese. Describes Innsbruck.
ALS 4pp & C TxU:H/L-0517.23; Reel 1053

1047. 1824–9–16 From James South n.p.
Feels he has pledged to the public to continue double star observation. Encourages JH to persevere as well. Offers JH use of his 5-foot refracting telescope. Asks about Fraunhofer telescopes.
ALS 8pp RS:HS 16.428

1048. 1824–9–17 To Caroline Lucretia Herschel Munich
Decides not to go to Switzerland because he wants to visit Hanover. On the way, hopes to see Johann Pfaff at Erlangen, Johann Encke at Seeburg, Baron Bernhard von Lindenau at Gotha, and Carl Gauss and Karl Harding at Göttingen among others.

ALS 3pp TxU:H/L–0569.2; Reel 1058 *Memoir ... of Caroline Herschel,* 175–6

1049. 1824–9–18 To James Grahame Munich
About the termination of JG's marriage plans, and the joys of travel.
draft 5pp RS:HS 8.319 (C: RS:HS 20.179)

1050. 1824–9–20 To Mary Pitt Herschel Munich
Glad that MH is vacationing with Mrs. Delevaux. Sorry about Charles Babbage's loss. Journey from Innsbruck. Meeting with Josef Fraunhofer. Georg von Reichenbach is dangerously ill. Met Captain Feilding and family. Itinerary for return journey via Hanover. Will visit caves of Bayreuth and J. W. A. Pfaff in Erlangen.
ALS 4pp & C TxU:H/L-0517.24; Reel 1053

1051. 1824–9–21 To James South Munich
Does not share JS's feeling of having pledged to continue double star observation. Reports on the high quality of the telescopes manufactured by Josef Fraunhofer.
Adraft 3pp RS:HS16.429 (C: 20.186)

1052. 1824–9–22 From Mary Pitt Herschel Slough
Caroline Herschel is anticipating JH's visit to Hanover. Expecting visit from Elizabeth Baily. New son born to Charles Babbage. Hopes that death of King of France will not impede JH's journey to Paris. Recent deaths at home.
ALS 3pp & C 4pp TxU:H/M–0623.14; Reel 1086

1053. 1824–9–25 From Caroline Lucretia Herschel Hanover
Hopes that JH will help her to finish a catalogue of 2500 nebulae that she and William Herschel began but never completed.
L *Memoir ... of Caroline Herschel*, 176–7

1054. [1824]–9–27 From Caroline Lucretia Herschel Hanover
Forwards recently received item to JH in Göttingen. [Mary] Baldwin in Slough reports good health of JH's mother. CLH will see JH soon.
ALS 1p inc TxU:H/M–0936.3; Reel 1083

1055. 1824–9–28 To Mary Pitt Herschel On road from Gotha to Göttingen
Terrible German roads. Missed J. W. A. Pfaff in Erlangen, where university is on vacation. Describes caves of Muggendorff near Bayreuth. Visited Mr. Lindeman and J. F. Encke in Gotha. [Letter continued 30 Sept. in Göttingen:] Found letter from Caroline Herschel with enclosure from Charles Babbage. Hopes to visit K. L. Harding and J. F. Blumenbach in Göttingen.
ALS 4pp & C TxU:H/L-0517.25; Reel 1053

1056. 1824–10–2 From Mary Pitt Herschel Slough
Received JH's letter from Munich. Arrangements for getting JH's belongings through Customs. Will meet JH in London. Mr. Pitt is unable to paint telescope tube due to foul weather. Scandal of Mr. White's will. Recent deaths.
ALS 3pp † & C 4pp TxU:H/M–0623.15; Reel 1086

1057. 1824–10–3 To Charles Babbage Hanover
His recent tours in Italy and Germany, and the people he met.
ALS 4pp RS:HS 2.199

1058. 1824–10–4 To Mary Pitt Herschel Hanover
News of relatives and friends in Hanover. Caroline Herschel is working on catalog of nebulae. [Letter continued 7 Oct. in Minden:] News of German scientists. Suspects that JH and H. W. M. Olbers are distantly related.
ALS 4pp & C TxU:H/L-0517.26; Reel 1053

1059. 1824–10–8 From Mary Pitt Herschel Slough
Painting of telescope begun. Thomas Baldwin Jr.'s wife soon to give birth. Charles Babbage reported being robbed.
ALS 3pp & C 3pp TxU:H/M–0623.16; Reel 1086

1060. 1824–10–10 From Charles Babbage Devonshire St.
Is glad to hear of his speedy return. Regarding JH's instruments and their passage through the customs. Is writing a book on Assurance Companies.
ALS 2pp RS:HS 2.200

1061. 1824–10–11 To Mary Pitt Herschel Steam boat between Nimeguen [Nijmegen] and Rotterdam on the Rhine
Will return earlier than planned. Comments of Mrs. Moorsom's misfortune and new children born to Baldwins and Babbages. [Letter continued in Rotterdam:] Journey from Cleves [Kleve] to Rotterdam. Report of new propulsion device for sailing vessels.
ALS 4pp & C TxU:H/L-0517.27; Reel 1053

1062. 1824–10–22 From Johann Franz Encke Seeburg
Finds no errors in JH's letter on the parallax of Mars. Sends his own calculations. Has not found the barometrical readings JH requested.
ALS 2pp (German) RS:HS 7.83

1063. 1824–10–22 To Caroline Lucretia Herschel Slough
Happy to be reunited with mother. JH thanks CH for his enjoyable stay in Hanover. Wishes CH well on her [catalogue of] nebulae.
ALS 2pp TxU:H/L–0569.3.a; Reel 1058

1064. 1824–10–25 To James Grahame Slough
About the last part of JH's continental travels, including his visit to Caroline Herschel.
draft 2pp inc RS:HS 8.320 (C: RS:HS 20.187)

1065. 1824–11–1 From Caroline Lucretia Herschel Hanover
Relieved that JH has returned safely from his European tour.
L *Memoir ... of Caroline Herschel*, 179–80 inc (?)

1066. 1824–11–4 From William Frédéric Edwards Paris
Has heard that JH has arrived back in England. Sends one of his own papers for his library. Regrets that JH did not call at Paris on his travels. Sends a parcel from P. S. Laplace. Remember him to Charles Babbage.
ALS 2pp RS:HS 7.122

1067. 1824–11–4 From Pierre Simon Laplace Paris
Sending the fourth volume of his own *Mécanique céleste*; comments on various parts. Work is being done on the figure of the earth.
ALS 4pp (French) RS:HS 11.103

1068. 1824–11–4 To Edward Sabine 2 Orchard St.
Encloses differences determined through indications of portable barometer, standard, thermometer, and various other instruments.
ALS 1p encl RS Sa.643

1069. 1824–11–4 To Thomas Young 2 Orchard St.
JH carried portable barometer by Edward Troughton and thermometer by Mr. Newman to Europe and compared these to instruments used by individual scientists and observatories. Reports results to Board of Longitude.
ALS 3pp TxU:H/L-0513; Reel 1055

1070. 1824–11–9 To Joseph Johann Littrow London
Has just returned from abroad and found his letter of April awaiting him. Hoped to visit Vienna when abroad. Has received money for the clock. Hopes to send his own paper on double stars later. Encloses some papers for distribution. Measured height of Etna when in Italy. Met Josef Fraunhofer and had an instructive discussion with him.
draft 2pp RS:HS 11.248 (C: RS:HS 20.188)

1071. 1824–11–11 To Francis Lunn 2 Orchard St.
Charles Babbage communicated [Edward] Smedley's note requesting JH to write an article on 'Heat" and an *Introduction to Physical Science*. Appears to be some misunderstanding as JH does not wish to write on Heat, no interest in the subject, but is prepared to write on *Physical* Optics. When will the latter article be required?
draft 3pp RS:HS 11.411 & 20.189

1072. 1824–11–18 From Thomas Young n.p.
Repeat computation for Jupiter's satellite IV [see TY's 1824–4]. Explains W. H. Wollaston's '*blue bow*.' See figure 422 of TY's [*Lectures on Natural Philosophy*].
ALS 1p RS:HS 18.327

1073. [1824–11–21 to 1825] From Thomas Young n.p.
Math is flimsy in John Dalton's theory of vapor dispersion, but physical grounds are firm. Sends copy of TY's volume on sound. Modules of elasticity and tension are analogies to explain TY's doctrine. Presently studying modules of elasticity that do not apply to chimney pipes investigated by Daniel Bernoulli and J. H. Lambert.
ALS 3pp RS:HS 18.331

1074. [1824–11–21 to 1825] To Thomas Young n.p.
Thanks TY for books and for taking charge of [G. A. A.] Plana's plates. JH mentioned Eilhard Mitscherlich only when no other proposals came forth. Matter was left hanging. Assures TY that JH was not offended [see TY's 1824–11–21, i.e., #1075]. Regrets 'dilapidated state of scientific matters' in R.S.L.
ACCS 2pp RS:HS 18.330 (C: 20.176)

1075. [1824–11–21 to 1825] From Thomas Young n.p.
TY would rather have JH's labor made useful another year. W. H. Wollaston does not want [Royal] medal, but medal wants Wollaston. One council member felt that TY's remarks offended JH. Assures JH it was not intended.
ALS 3pp RS:HS 18.329

1076. [1824–11–21 to 1825] From Thomas Young n.p.
As R.S.L. vice president, JH should reconsider his report favoring Eilhard Mitscherlich over W. H. Wollaston. This may be last opportunity to pay Wollaston a '*just* compliment.' Agrees with JH that much in R.S.L. '*nicknamed* science [is] sleight of hand,' and that 'great injustice is done to men of science.' But TY expects a short life and tries not to be concerned about approval of others.
ALS 3pp RS:HS 18.328

1077. 1824–11–24 From Thomas W. Hornbuckle St. John's College, Cambridge
School at Sedbergh is vacant; please let him know if he wishes to become a candidate. [Miles?] Bland is candidate for a valuable living in Cheshire.
ALS 1p RS:HS 9.485

1078. 1824–11–27 To David Brewster London
Praises old *Edinburgh Philosophical Journal.* Promises to send DB some materials for publication.
C 2pp RS:HS 20.190

1079. 1824–11–28 To William Whewell 2 Orchard St.
Reports on the reception of WW's crystallography paper at the R.S.L. Asks WW to add a comment on the paper's relation to a paper in David Brewster's journal. Asks WW to contact George Peacock.
ALS 3pp TC, Camb. Add. Ms.a.207[11] (C: RS:HS 20.191)

1080. 1824–11–29 From Mario Gemmellaro Nicolosi sull'Etna
Sending papers published by their Academy. Will be pleased to help in any way.
ALS 1p (Italian) RS:HS 8.92

1081. 1824–12–3 From John Lee No. 5. College, Doctors Commons
Invites JH to join JL in viewing William Tassie's nearly complete wax model of bust of William Herschel at Leicester Square.
ALS 1p TxU:H/MW-0246.1; Reel 15

1082. 1824–12–3 To James South n.p.
Will be pleased to correct the proof copy of JS's paper on double stars. Mentions Charles Babbage and Edward Troughton.
ACC 2pp RS:HS16.430 (C: 20.192)

1083. 1824–12–4 To Caroline Lucretia Herschel London
JH will have little time for astronomy due to election as Secretary to the R.S.L.; the appointment also has forced JH to move from Slough to London.
ALS 3pp TxU:H/L–0569.4; Reel 1058 *Memoir ... of Caroline Herschel,* 180–1 inc

1084. [1824]–12–8 From Francis Lunn Hawarden Castle
Hears a letter from JH has miscarried. Anxious that JH should write an article on Heat as FL is too busy.
ALS 1p RS:HS 11.412

1085. 1824–12–11 To John Brinkley 56 Devonshire St.
Mislaid the copy of one of JB's mss. and would be grateful if he knows of its whereabouts. In the course of his travels in Europe he has seen some astronomical instruments, which will in time surpass any British made ones and especially those of G. B. Amici and Josef Fraunhofer.
ALS 3pp RS:HS 4.277 & 20.193

1086. 1824–12–13 From James South Passy
Reports on his life in France and on details in the paper JS and JH were publishing on double stars.
ALS 5pp RS:HS 16.431

1087. 1824–12–14 From Niccolo Cacciatore Palermo
Sends pamphlet on observations they made together on Mount Cuccio. Regarding the coefficient of expansion of the atmosphere. Will repeat his observations in the coming winter. Present the other copy to the

Astronomical Society.
ALS 1p (Italian) RS:HS 5.133

1088. 1824–12–15 From John Brinkley Observatory, Dublin
Has only fragments of the paper for which JH inquires; original was left for W. H. Wollaston to amend. Thanks for congratulations on the award of the Copley Medal. Was interested to hear about the new German astronomical instruments.
ALS 3pp RS:HS 4.278

1089. 1824–12–20 From Peter Barlow Royal Military Academy
Regarding the magnetic polarity of the earth.
ALS 3pp RS:HS 3.287

1090. 1824–12–23 From William Richard Hamilton Naples
Hopes to see him next spring, in England. Teodoro Monticelli often inquires after him. The steam boat goes on with great success. Small earthquake at Messina. Has toured the crater of Vesuvius. No news of Turks or Christians.
ALS 4pp RS:HS 9.217

1091. 1824–12–31 From James Grahame Whitehill, Glasgow
Asks JH about lodgings in London for JG and his children. JG has begun his history of North America.
ALS 2pp RS:HS 8.321

1092. [1825?] To Charles Babbage n.p.
Is sorry he will be unable to see him. Is sending copies of their paper for distribution.
ALS 2pp RS:HS 2.366

1093. 1825–1 From C. Goring 56 Commercial Rd.
Query regarding the diagrams in one of JH's articles in *R.S.P.T.* Has purchased one of the compound lenses on JH's recommendation, and relates some experiments with it. Would be pleased for JH to call at his house.
ALS 4pp RS:HS 8.225

1094. [1825?] To James South n.p.
Defends Edward Sabine's transit observations, noting that JH supports them on practical rather than theoretical grounds.
ACCS 4pp RS:HS 16.482 (C: RS:HS 25.13.28)

1095. [1825] To James South n.p.
Has represented JS's interests as best JH can regarding JS's proposal. Describes himself as unskillful in such representations.
ACC 4pp RS:HS 16.481 (C: RS:HS 25.13.26)

1096. [1825?] To James South n.p.
Has examined Edward Sabine's transit measurements; finds them unproblematic.
ACC 2pp RS:HS 16.479 (C: RS:HS 25.13.27)

1097. [1825 to 1827]–[?]–22 From William Hyde Wollaston Buckingham St.
Encloses paper sent by Mr. Perkins for R.S.L, and impression just received from printer.
ALS 1p RS:HS 18.288

1098. [1825]–1–5 To Francis Lunn n.p.
Concludes original letter lost. Impossible to do any further work for the *Encyclopaedia Metropolitana*. Hopes FL will write the article.
draft 1p RS:HS 11.412verso

1099. 1825–1–14 To Giovanni B. Amici 56 Devonshire St.
Thanking him for his friendly reception when in Italy. Is sending GA journals and pamphlets likely to be of interest to him. Regarding various papers on the electrization of mercury. Please send details of Giuseppe Bianchi's barometer.
draft 4pp RS:HS 1.334 & 20.194

1100. 1825–1–14 From Caroline Lucretia Herschel Hanover
Completing the catalogue of 2500 nebulae. Impressed with a French biography of William Herschel, although CH notes a few occasions where 'too great a stress is laid on the assistance of others.'
L *Memoir ... of Caroline Herschel*, 181–2

1101. [1825]–1–14 From Francis Lunn Hawarden Castle
Their letters have missed more than once, and JH's last letter is now at the Return Office. Would like to see this letter to learn JH's views on the article on Heat. Gives own ideas how article should be written, too busy himself, so hopes JH can write it, or does he know any other person?
ALS 4pp RS:HS 11.413

1102. 1825–1–14 To Giovanni Antonio Plana n.p.
Apologizes for delay in writing. Describes JH's travels across Europe back to England. Is sending GP copies of JH's papers in physics, mathematics, and astronomy, especially double stars. Describes a new instrument, a floating collimator, described by Henry Kater at the R.S.L. [Rough notes for a later letter on the back of part of this one.]
draft 4pp (French) RS:HS 13.413 (C (inc): RS:HS 20.195)

1103. 1825–1–16 From James Hudson 30 Upper Stamford St.
Sends JH a sketch of an experiment already made.
ALS 1p RS:HS B26.26

1104. 1825–1–18 From C. Goring 56 Commercial Rd.
Is grateful for his letter and also for the brochure. Sees the folly of his own impatience. Regarding object glasses of telescopes.
ALS 4pp RS:HS 8.226

1105. 1825–1–24 To Wilhelm Struve n.p.
Asks WS for account of the performance of his new [Josef] Fraunhofer telescope. Corroborates WS's double star motion data, except for 70 Ophiuchi, where JH and [James] South find no motion. Asks for observational targets and urges care in WS's observational technique. Mentions WS's intention to undertake celestial zone with [F. W.] Bessel. Describes [Henry] Kater's 'floating collimator.'
draft 5pp RS:HS 17.106a (C: RS:HS 20.196)

1106. 1825–1–27 To Josef von Fraunhofer n.p.
JF was elected associate of Astronomical Society. Will send Society's *Transactions* and JH's 'Absorption of Light.' W. H. F. Talbot delivered JF's 'Crown glass Prism.' JH presented J. G. Soldner's observations to Society and proposed Soldner as associate.
AAbs 1p RS:HS 19.5

1107. 1825–1–27 To William Ritchie London
Presented WR's paper on a new photometer to R.S.L. Discusses how objections may be avoided. Believes [John] Leslie's photometer should be thrown aside.
draft 5pp RS:HS 14.347 (C: RS:HS 20.197)

1108. 1825–1–28 From Johann Elert Bode Berlin
Will receive 1827 Astronomical Yearbook through Mr. Hunnemann. Would be glad if he could hurry up the Parramatta Observations from K. L. C. Rümker. Various queries regarding publications.
ALS 1p (German) RS:HS 4.169

1109. 1825–1–29 From John Brinkley Observatory, Dublin
Notes and comparisons for their recent observations of stars. Was interested in the account of Henry Kater's collimator and hopes it will be adapted for Greenwich.
ALS 3pp RS:HS 4.279

1110. 1825–1–30 From William Ritchie Tain
Still anxious to become a fellow of the R.S.L. Thanks JH for offer. Would like to be elected before end of present session.
ALS 1p RS:HS 14.348

1111. 1825–1–31 To Niccolo Cacciatore n.p.
News since he left Palermo. Comments on the pamphlet NC recently sent him. Has to inform him that he has been elected an Associate of the Astronomical Society.
draft 4pp (French) RS:HS 5.134

1112. 1825–2–3 From William Ritchie Tain, [Ross-shire]
Thanks JH for introducing WR's paper on photometer to R.S.L. council. Sends drawing of instrument and alterations to WR's paper that resulted from improvements in instrument.
ALS 2pp damaged TxU:H/M–0449; Reel 1093

1113. 1825–2–5 From William Ritchie Tain
Suggested improvements to John Leslie's photometer.
AL 2pp TxU:H/M–1212; Reel 1060

1114. 1825–2–6 To Teodoro Monticelli n.p.
Outlining his Italian travels after leaving Naples. Had to return home as business pressed.
C 2pp (French) RS:HS 12.362 (C: 20.199)

1115. 1825–2–7 From William Hyde Wollaston Buckingham St.
Sends drawings associated with WW's paper and passed by [R.S.L.] Council for printing, and memoir from M. A. Pictet.
ALS 2pp RS:HS 18.278

1116. 1825–2–7 From Franz Xaver von Zach Genoa
Comments on a variety of topics about various scientific persons on the Continent, on JH's travels in Europe, on ways to send packets between England and Europe, and on FZ's copy of the *Nautical Almanac*, which FZ has not yet received. Makes comparisons between his and JH's barometer, offers some comet observations, and talks about the appropriate date for Easter. Awaits JH's double star observations.
ALS 4pp (French) RS:HS 18.361

1117. 1825–2–11 From Giovanni B. Amici Modena
Giving details of Giuseppe Bianchi's barometer. Regarding the experiments of Leopoldo Nobili on the electrization of mercury. Gives tables of star observations.
ALS 4pp (Italian) RS:HS 1.335

1118. 1825–2–14 From James Grahame Edinburgh
About lodgings in London [see JG's 1824–12–31], JG's writing, and his 'love affairs.'
ALS 3pp RS:HS 8.322

1119. 1825–2–16 From Peter Barlow Royal Military Academy
Regarding his own experiments with revolving balls.
ALS 4pp RS:HS 3.288

1120. 1825–2–19 To William Ritchie n.p.
Will present alterations of WR's paper to R.S.L. [see JH's 1825–1–27]. Asks if he may describe the similarities between WR's photometer and an instrument JH used in Italy and Sicily.
C 4pp RS:HS 20.200

1121. 1825–2–22 From James Grahame Edinburgh
More about lodgings in London [see JG's 1825–2–14], and about books JG and JH have read [letter completed 1825–3–9].
ALS 4pp RS:HS 8.323

1122. 1825–2–26 From Joshua King Queens' College, Cambridge
Has recently discovered in Cumberland a rich vein of cobalt, which is mixed with iron. Would like his opinion whether these ores can be easily separated. Can send him a specimen if required.
ALS 3pp RS:HS 11.83

1123. 1825–2–[26?] From James South n.p.
Reports on some observations from JS's observatory near Paris and on his contacts with the Paris Observatory. Relays to JH a request from Alexander von Humboldt for barometric observations. Invites JH to visit.
ALS 8pp RS:HS 16.433

1124. 1825–2–28 From Francis Baily Gray's Inn
Would like to borrow the drawing of Josef Fraunhofer's telescope. Has received an account of this telescope from Wilhelm Struve, which he is now translating. Regarding the miscarriage of the Memoirs for Struve.
ALS 2pp RS:HS 3.57

1125. 1825–3–2 From Joshua King Queens' College, Cambridge
Forwards a specimen of his cobalt. Gives price and the various methods of use and refining. Thinks the analysis did not show its true qualities. Would like to hear the results of JH's experiments.
ALS 3pp RS:HS 11.84

1126. 1825–3–3 From William Ritchie Tain
Pleased JH liked his paper on differential photometer. Discusses apparatus for experiments on light. Describes some papers he has written and some he is writing.
ALS 3pp RS:HS 14.349

1127. 1825–3–6 From [Charles] R[igobert Marie] Bonne Paris
Edward Sabine's and JH's memoir on the differences of the meridian between Greenwich and Paris. Hopes to print a notice of the same operation with similar results. Operations last year between Paris and Brest were not a success so hopes to repeat them this year.
ALS 1p (French) RS:HS 4.183

1128. 1825–3–7 From Caroline Lucretia Herschel Hanover
Wishes JH a happy birthday. Finished the catalogue of 2500 nebulae.
L *Memoir ... of Caroline Herschel*, 183–4 inc (?)

1129. 1825–3–10 From Niccolo Cacciatore Palermo
Thanks for information on the R.S.L. barometers. Would he purchase various instruments for the new observatory. Notes on star readings.
ALS 3pp (Italian) RS:HS 5.136

1130. 1825–3–10 To James South n.p.
Making, reducing, and publishing double star work is very difficult and, because of Wilhelm Struve's recent endeavors, less pressing than it had been. Mentions various doubles. Expresses interest in JS's plan to purchase a very large refractor.
CC 3pp RS:HS 16.434 (C: 20.201)

1131. 1825–3–20 To Edward Sabine n.p.
Comments on ES's measurement [in his *Account of Experiments to Determine the Figure of the Earth*] indicating 'the increase of the fraction expressing the earth's compression.' Predicts that ES's result will cause a 'sensation,' but does not commit himself to it.
C 4pp RS:HS 15.1 & 20.202

1132. [1825–3–23] To William Wallace 56 Devonshire St., Portland Place
Let Josef Fraunhofer make WW's whole transit instrument, and mural circle as well. Attests to Fraunhofer's artistry. Germans will soon leave no stars to discover. Wishes someone would import G. F. Reichenbach's meridian circle and use it on F. W. Bessel's plan. Suggests WW order clocks from [Robert] Molyneux or [William] Hardy.
C 3pp RS:HS 20.203

1133. 1825–3–24 From Giovanni Antonio Plana Turin
Is helping another scientist bring to publication a work on triangulation in the Alps. Comments on JH's idea of suspending an extremely long plumb bob at a particular precipice in the Alps [Foucault pendulum?]. GP does not believe the results would warrant the expense. Comments on the floating collimator [see JH's 1825–1–14], noting that he does not find it very helpful.
ALS 4pp (French) RS:HS 13.414

1134. 1825–3–27 From Caroline Lucretia Herschel Hanover
Explains the layout of her recently completed catalogue of 2500 nebulae.
L *Memoir ... of Caroline Herschel*, 186–7 inc (?)

1135. 1825–3–28 From Francis Baily Gray's Inn
Is sending a copy of Wilhelm Struve's paper on Josef Fraunhofer's telescope. Regarding F. W. Bessel's formula on longitude. Has copies of J. J. Littrow's paper for him.
ALS 3pp RS:HS 3.58

1136. 1825–3–29 From Niccolo Cacciatore Palermo
Is grateful for being elected an Associate of the Astronomical Society. Compares barometers. Grateful for comments on his pamphlet. News of expedition up Mont Cuccio. Local disputes regarding the results.
ALS 4pp (Italian) RS:HS 5.135

1137. 1825–3–31 From William Hyde Wollaston n.p.
Invites JH to dinner Saturday to meet J. G. S. Van Breda.
ALS 1p RS:HS 18.279

1138. [1825–4] From Edward Sabine Princes St.
Shares JH's doubts whether the Board of Longitude will fund the Teneriffe expedition, the importance of which he again stresses. Cannot personally cover the costs. Thanks JH for his support and praise.
ALS 3pp RS:HS 15.5

1139. 1825–4–3 To Edward Sabine n.p.
Responds to ES's proposal to conduct a series of meteorological observations from the peak of Teneriffe. Promises to bring it before the Board of Longitude for consideration for support.
ALS 2pp RS:HS 15.2 (C: RS:HS 20.204)

1140. 1825–4–4 From Edward Sabine Princes St.
Clarifies issues discussed in an earlier letter. Discusses whether the project of making meteorological observations will be feasible if the Board of Longitude does not offer financial support. Attempts to arrange a meeting with JH.
ALS 3 pp RS:HS 15.3

1141. 1825–4–5 To Edward Sabine n.p.
Responds to ES's letter of 1825–4–4, stating that JH now understands that ES will proceed with the Teneriffe observations only if funds are provided. Praises Sabine's previous work.
C 2pp RS:HS 15.4 & 20.205

1142. 1825–4–8 To Edward Sabine n.p.
On 1825–4–7 presented ES's proposal for experiments at Teneriffe to the Board of Longitude, which judged the experiments not sufficiently relevant to the Board's concerns. Offers to seek endorsement from the R.S.L.
ACCS 2pp RS:HS 15.6 (C: RS:HS 20.206)

1143. 1825–4–9 From Wilhelm Struve Dorpat
Illegible.
ALS 3pp (German) RS:HS 17.107

1144. 1825–4–10 From James South Passy
On the eloges of William Herschel being prepared by François Arago and J. B. Fourier. JH's and JS's work on double stars much appreciated among French astronomers. On Alexis Bouvard, Arago, and other French astronomers. On Wilhelm Struve's observations of double stars.
ALS 7pp RS:HS 16.436a

1145. 1825–4–11 From Edward Sabine Princes St.
Encourages JH to bring ES's proposal for Teneriffe experiments before the R.S.L. Council.
ALS 1p RS:HS 15.7

1146. 1825–4–16 To James South London
Agrees to purchase JS's 7-foot refractor if no other reasonable offer made, but doubts he would use it extensively. Plans to concentrate on observing nebulae with JH's reflector at Slough. Relays regards and question for P. S. Laplace.
Abs 1p RS:HS 16.435 (C: 20.207 inc)

1147. 1825–4–17 From Joseph Johann Littrow Vienna
Anticipates appearance of JH's work on double stars; asks for suggestions for employment for a young man from Germany, and offers some suggestions about refining the study of refraction.
ALS 4pp (German) RS:HS 11.249

1148. 1825–4–18 To Caroline Lucretia Herschel [Slough]
Studied 659 double stars; JH is confident that his work proves his father's conclusions. Now plans to study nebulae.
ALS 3pp TxU:H/L–0570.1; Reel 1058 *Memoir ... of Caroline Herschel,* 188–9 inc

1149. 1825–4–19 To William H. Keating London
Writes to thank the Philosophical Society of Philadelphia for the gift of the catalogue of its library.
ALS 1p APS (no i.d.)

1150. 1825–4–19 From Teodoro Monticelli Naples
Something is going badly. Comments on volcanic action observed, and on having met some English acquaintances of JH. [Many parts of this letter illegible.]
ALS 3pp (French) RS:HS 12.363

1151. 1825–4–21 To Edward Sabine n.p.
Invites ES to join him and other scientists for dinner. Requests that ES prepare a detailed description and justification of his proposed Teneriffe experiments for consideration by the R.S.L. Council.
ACCS 3pp RS:HS 15.8 (C: RS:HS 20.208)

1152. 1825–4–22 From James South Passy
On the transmission of materials between JH and JS. On their joint observations.
ALS 4pp RS:HS 16.436b

1153. 1825–4–25 From Josef von Fraunhofer Munich
Election to the Astronomical Society.
ALS 3pp (German) RS:HS 7.379

1154. 1825–5–3 From Caroline Lucretia Herschel Hanover
The Duke of Cambridge inquired about JH's research interests.
L *Memoir ... of Caroline Herschel*, 189–90

1155. 1825–5–3 To Pierre Simon Laplace n.p.
Has too long delayed thanking him for his works. Sends some memoirs of Edward Sabine, etc. Comments on work that needs doing.
draft 4pp (French) RS:HS 11.104 & 20.209

1156. 1825–5–5 From James South Passy
Has been making observations as JH requested of the star 70 Ophiuchi. Argues for the accuracy of observations of double stars that JH and JS have made.
ALS 4pp RS:HS 16.437

1157. 1825–5–6 To Edward Sabine Devonshire St.
Read your proposal on experiments at Teneriffe to the R.S.L. Council, which declined to make 'an especial recommendation' in support of them to the government.
C 1p RS:HS 20.210

1158. 1825–5–9 To Jean Baptiste Joseph Fourier 56 Devonshire St.
Copies of the Éloge, which JF sent, are still at the Customs house so he is unable to comment on it. A list of his father's works is to be found in the Index to the *R.S.P.T.* Gives a list of all the discoveries and inventions of his father.
draft 4pp RS:HS 7.343 & 20.211

1159. 1825–5–9 From Sir Charles Lyell Bedford St.
Sending this note to inform JH that he began his own serpentine paper on Friday and next Friday intends to read JH's paper on Predazzo. Would like to go over the paper with JH beforehand. Intends to publish his own paper in Brewster's Journal. Can JH meet CL here Wednesday morning?
ALS 2pp RS:HS 11.415

1160. 1825–5–9 From James South Passy
Is sending materials to JH. Reports on continued efforts to observe the star 70 Ophiuchi. Has sighted three nebulae that may be unknown to JH.
ALS 2pp RS:HS 16.438

1161. 1825–5–12 From Giovanni Inghirami Florence
Desires to present methods of occultation observation to the Astronomical Society.
ALS 1p (Italian) RS:HS 10.266

1162. 1825–5–13 From James South Passy
If you can send a copy of our paper on double stars immediately, it seems certain that the Academy will award a medal for it. Informs JH of most expeditious method of sending same.
ALS 2pp RS:HS 16.439

1163. 1825–5–14 From Sir Charles Lyell 29 Norfolk St.
Believes that JH is acquainted with R. I. Murchison, who is to accompany CL to see the works of M. I./I. K. Brunel's Thames tunnel, after they have gone through JH's paper. Would JH like to accompany them?
ALS 2pp RS:HS 11.416

1164. 1825–5–16 From James Grahame Edinburgh
More about JG's writings [see JG's 1825–5–16].
ALS 3pp RS:HS 8.325

1165. 1825–5–16 From James Grahame Edinburgh
About JG's history of the United States, the first part of which is now finished, and forwarded with this letter.
ALS 4pp RS:HS 8.324

1166. 1825–5–16 To James South n.p.
Has received materials sent by JS. Printing of paper by JS and JH on double stars is nearly completed. Urges that the issue is not whether the observations of 70 Ophiuchi made by Wilhelm Struve are better than those made by JH and JS; the issue is truth. Caroline Herschel has sent to JH a catalogue of his father's observation. Busy observing nebulae.
ACC 2pp RS:HS 16.440 (C: RS:HS 20.212 inc)

1167. 1825–5–17 From Francis Baily Gray's Inn
Enclosing vouchers relating to the stock, etc. Regarding the interpretation of a letter from F. W. Bessel. The inquiries regarding the Astronomer Royal.
ALS 3pp RS:HS 3.59

1168. 1825–5–17 From John Kidd Oxford
Much obliged for his letter. Will be glad if he would correct the press of his paper. Encloses a printed sheet, which may amuse him.
ALS 2pp RS:HS 11.82

1169. 1825–5–19 From Thomas Young Admiralty Office
Announces 2 June meeting of 'Commissioners for more effectually discovering the Longitude at Sea.' Inquired about 'excise,' but sees no reason to wait for Navy Board to act.
(A)LS 1p TxU:H/M–0612; Reel 1086

1170. 1825–5–21 To Michael Faraday Devonshire St.
Rearranges a meeting of [the R. S. L.'s Glass Committee] to take place at the Pellat and Green's Glass House in Holland St.
ALS 1p WT A-300

1171. 1825–5–24 From James South Passy
The Academy has decided to award JH and JS its astronomy prize for their paper on double stars.
ALS 4pp † RS:HS 16.442

1172. 1825–5–25 From William Ritchie Tain, [Ross-shire]
Not upset that JH withdrew WR's paper on photometer from R.S.L. Sent similar paper to David Brewster's [*Edinburgh Journal of Science*] with note on JH's method for finding relative heating power of solar rays. Wants to dedicate WR's book on light and heat to JH.
ALS 2pp TxU:H/M–0450; Reel 1093

1173. 1825–5–25 From James South Passy
Has returned from London. A preprint of JS's and JH's paper on double stars has been delivered to François Arago. Is finding many errata in their paper. Has received a letter from Josef Fraunhofer and the paper [on double stars] of Wilhelm Struve.
ALS 2pp RS:HS 16.441

1174. 1825–5–27 From Giovanni Antonio Plana Turin
Is preparing a paper on planetary perturbations, which GP would like to send to the Astronomical Society for publication. Comments on JH's ideas on the subject of perturbations. Would the paper being in French be a problem? Comments on Henry Kater's method for finding the zero point in his meridian circle.
ALS 4pp† (French) RS:HS 13.415

1175. 1825–5–28 From Giovanni Santini Padua
Compliments JH on work as Secretary of the Astronomical Society and for keeping the scientists of Europe informed of discoveries made in England. Finds that GS needs permission from Italian government to be a member of the R.S.L.; expects this will not be difficult.
ALS 2pp (Italian) RS:HS 15.342

1176. 1825–5–29 From James South Passy
Contacts with Alexis Bouvard and P. S. Laplace. Criticizes an astronomical paper by [Robert] Woodhouse.
ALS 4pp RS:HS 16.443

1177. 1825–5–30 From Mrs. Forbes Kensington
Invites JH to dinner. Expects Mr. and Mrs. [Andrew?] Spottiswoode and Mr. and Mrs. Beckwith.
ALS 1p † RS:HS 18.335a

1178. 1825–5–30 From Jean Baptiste Joseph Fourier Paris
Expressing the thanks of the Académie for JH's gift of *Observations...on 380 Double and Triple Stars.*
ALS 1p (French) RS:HS 7.344

1179. 1825–5–31 From Francis Baily Gray's Inn
Regarding the travels of a packet of Wilhelm Struve's observations. Please bring F. W. Bessel's letter with him when he comes to the Committee meeting.
ALS 2pp RS:HS 3.60

1180. 1825–5–31 To Wilhelm Struve 56 Devonshire St., Portland Place
Acknowledges receipt of WS's second volume of observations and the new telescope's results. Mentions Caroline Herschel's catalogue of nebulae.
C 2pp RS:HS 20.213

1181. 1825–6–4 From James Grahame Edinburgh
Of dreams, poetry, and JH's winning the Lalande Prize [letter completed 1825–6–6].
ALS 4pp RS:HS 8.326

1182. 1825–6–4 From James South Passy
Comments negatively on an astronomical paper by Robert Woodhouse. [J. F. A.] Gambert in Marseille has reportedly discovered a comet. Details on the medal to be awarded by the Academy to JH and JS. If, as reported, Wilhelm Struve has come to England, invite him to visit JS in Paris.
ALS 4pp RS:HS 16.444

1183. 1825–6–5 From Samuel Young 18 Gloucester Place, New Road
Returning to Cookham. Thanks for minerals. Describes shipment of new minerals from Baffin Bay. Expecting another from Rio de la Plata, courtesy of Woodbine Parish [consul-general to Buenos Aires].
ALS 2pp RS:HS 18.317

1184. 1825–6–11 From [Charles] R[igobert Marie] Bonne Paris
Concerning recent works and discussions connected with longitude.
ALS 4pp (French) RS:HS 4.184

1185. 1825–6–19 To Edward Sabine 56 Devonshire St.
Encloses R. Bonne's letter and gives further instructions regarding rockets for experiment.
ALS 3pp JH note RS Sa.644

1186. 1825–6–19 To Thomas Young n.p.
Edward Sabine declares that R.S.L. is hostile to ES's interests. If ES makes future claim for compensation, JH wants TY's opinion in writing. Need to order four night glasses for Fairlight Station. Encloses letter from [Richard] Copeland.
ACC 2pp RS:HS 18.332 (C: 20.214)

1187. 1825–6–20 To Edward Sabine n.p.
Six artillery men have been placed under ES's command. Encloses letters from Charles Bonne. Asks ES to oversee the arrival of 180 rockets to be used at Wrotham and Fairlight. Proposes that ES test whether a light at Wrotham can be seen at Fairlight in which case gunpowder will be used instead of rockets. Sets date for ES to meet Bonne in Hornoy. Urges ES to record his expenses for the Board of Longitude.
ALS 4pp RS:HS 15.9 (C: RS 20.215)DHB

1188. 1825–6–20 From Edward Sabine Princes St.
Clarifies that the six artillery men are at ES's disposal but under JH's command. Agrees to oversee arrival of the rockets. If expenses are to be paid by the Department of Ordinance, a request must be made by the Secretary of the Board of Longitude. Agrees to keep a record of expenses, not including his time spent, which he considers 'public service.'
ALS 3pp RS:HS 15.10

1189. 1825–6–21 From James South Passy
JS has received the medals awarded JS and JH by the French Academy for their paper on double stars. Reports on his new work on double stars and his observations of

[J. F. A.] Gambert's comet.
ALS 4pp RS:HS 16.445

1190. 1825–6–23 From James Grahame Edinburgh
Anxiety about the response to JG's manuscript by the publisher [see JG's 1825–5–16].
AL 3pp RS:HS 8.327

1191. 1825–6–24 From James South Passy
Laments the difficulty of transmitting publications between Paris and London. On arrangements to get JH's Lalande Prize medal to JH. JS's recent observations and contacts with Parisian scientists.
ALS 4pp RS:HS 16.446

1192. 1825–6–28 From [Charles] R[igobert Marie] Bonne Paris
Has returned to Brest, and found his letter. Is honored to correspond with JH. Further regarding pendulum observations.
ALS 3pp (French) RS:HS 4.185

1193. 1825–6–28 From Joseph Johann Littrow Vienna
About books and papers, being sent both ways.
ALS 1p (German) RS:HS 11.250

1194. 1825–7–1 From [Charles] R[igobert Marie] Bonne Paris
Has just received his letter of 27 June. Further work of Lieut. C. L. Largeteau. Longitude readings at various points.
ALS 3pp (French) RS:HS 4.187

1195. 1825–7–1 From [Charles] R[igobert Marie] Bonne Paris
Is sending him a paper dealing with the measurement of the arc of the meridian between Greenwich and Paris. Further observations along the coast by Lieut. C. L. Largeteau.
ALS 1p (French) RS:HS 4.186

1196. 1825–7–1 To James Grahame 56 Devonshire St., Portland Place
Reports to JG on the publisher's response. JH offers JG financial assistance, telling JG 'I am rich,' JH's income being significantly greater than his needs.
C 4pp RS:HS 8.329 (C: RS:HS 20.216)

1197. 1825–7–1 From James Grahame Edinburgh
Still has not heard from publisher [see JG's 1825–6–23].
ALS 1p RS:HS 8.328

1198. 1825–7–1 To Giovanni Antonio Plana n.p.
Mostly involves arrangements about exchanges of papers and printing some of GP's in the *Astronomical Society Memoirs*.
draft 1p RS:HS 13.413

1199. [1825–7] To Edward Sabine n.p.
Encloses note about transportation of rockets, telescope, and chronometer, with suggested schedule for experiments.
ALS 1p RS Sa.680

1200. 1825–7–4 From Jean Baptiste Joseph Fourier Paris
Is grateful for the information on JH's illustrious father. Was ill when the letter arrived so hopes he will be excused for his belated reply.
ALS 1p (French) RS:HS 7.346

1201. 1825–7–8 From James Grahame Edinburgh
About the publisher's offer, and JG's intent to go to Göttingen for research purposes.
ALS 2pp RS:HS 8.330

1202. 1825–7–8 From Bernard A. von Lindenau Gotha
News that he has been elected an Associate of the Astronomical Society gives him pleasure. Does not have much time at the present to devote to astronomy. Hopes to visit England to obtain materials for a life of Edmund Halley.
ALS 2pp (French) RS:HS 11.217

1203. 1825–7–12 From Edward Sabine Hornoy, France
Reports on his meeting with Charles Bonne and their subsequent failure to see any signals the first night from either Mont Javoul or La Canche. Worries that the distances are too great for measurements so close to the sea. Speaks of similar measurements between Paris and Strasburg.
ALS 2pp RS:HS 15.11

1204. 1825–7–16 From Edward Sabine Lignières
Reports his 'entire failure' with Charles Bonne to see signals from both La Canche and Mont Javoul on the same night. Tells of moving the telescopes to Tuniers on 1825–7–15. Speaks of similar experiments elsewhere, which also failed. Concludes that distance is not the only factor in distinguishing signals. Plans to visit JH on 1825–7–25.
ALS 4pp RS:HS 15.12

1205. 1825–7–19 From James South Passy
On his contacts with French scientists and on their praise for the JH's and JS's catalogue of double stars. Suggests that JH, beset by excessive commitments, cease to devote time to seeing JS's publications through the press. Comments on the personal cost to JS of his devotion to double stars. Remarks on JH's efforts to measure the difference in longitude between the Greenwich and Paris observatories. Data on double stars.
ALS 4pp RS:HS 16.447

1206. 1825–7–20 From Edward Sabine Lignières
Reports Charles Bonne's voyage to Mont Javoul and discovery that the rockets there had been sent off too early, accounting for the previous failures. Reports subsequent successes. Conveys his plans to arrive at Teneriffe.
ALS 2pp RS:HS 15.13

1207. 1825–7–23 To John Pond Hastings
Introduces [C. T.] Largeteau from France, who wishes to receive from JP the observations made at Greenwich as part of the determination of the difference in longitude between the Greenwich and Paris observatories. Recounts some difficulties encountered in making the measurements.
ACCS 4pp RS:HS 14.34 (C: RS:HS 20.217)

1208. 1825–7–24 From Hans Christian Oersted Copenhagen
Apologizes for the long delay in writing of JH's election into the Danish Royal Academy of Sciences. HO is carrying on experiments with chlorine gas and metals.
ALS 3pp (German) RS:HS 13.162 Harding, *Correspondance de ... Örsted, 2,* 386–9

1209. 1825–7–25 To Davies Gilbert Hastings
Unable to call on him as his mother wishes to return home

directly. Edward Sabine has not returned yet. Outlines some of the difficulties Sabine is experiencing. Premature to judge results at the moment.
C 4pp RS:HS 8.108 & 20.218

1210. 1825–7–28 From Giovanni Antonio Plana Turin
Pleased that Astronomical Society will publish GP's memoir (in French) [see GP's 1825–5–27]. Adds some comments for clarification on points in paper and makes arrangements for copies of the paper.
ALS 3pp† (French) RS:HS 13.416

1211. 1825–8–1 To Davies Gilbert London
Comments on JH's mother's health, some communication from Greenwich, and the weather.
C 1p RS:HS 20.219

1212. 1825–8–1 From Robert Grahame, Sr. Glenfair
Could he let him have the address of his son James as his mother died this morning.
ALS 1p RS:HS 9.3

1213. 1825–8–1 To John Herapath London
Knows nothing of the enmity of Humphry Davy towards [John] Herapath. The purpose of JH's last letter was to point out that Herapath was in error in his solution of certain equations. Must consider correspondence closed.
C 3pp RS:HS 10.181 & 20.220

1214. 1825–[8 or later] To Caroline Lucretia Herschel Hastings
Informs CH of his gold medal from the French Academy of Sciences for JH's and James South's paper on double stars. Finds CH's catalogue of nebulae in zones very useful. Heard about Johann Pfaff's death. James Grahame may be moving to Göttingen.
ALS [3?]pp TxU:H/L–0570.2.a; Reel 1058 *Memoir ... of Caroline Herschel,* 192–3 inc

1215. 1825–8–2 To James South n.p.
JH's attempts with Edward Sabine to measure the difference in longitude between Greenwich and the Paris Observatory. Supplies, for Alexander von Humboldt, JH's measurement of Mount Etna's height. Gives message for François Arago concerning paper on magnetism published by JH and Charles Babbage.
Adraft 1p RS:HS 16.448

1216. 1825–8–3 To Carl Friedrich Gauss 56 Devonshire St.
Note to accompany the sending of a number of papers; comments on the orbits of double stars.
ALS 3pp diag NSUB Cod.Ms.Gauss 99 (C: RS:HS 20.221)

1217. 1825–8–6 From [Charles] R[igobert Marie] Bonne Paris
Concerning Lieut. C. L. Largeteau's recent observations.
ALS 1p (French) RS:HS 4.189

1218. 1825–8–7 From Charles L. Harding Göttingen
Thanks for the double star catalogue of JH and James South; comments on observing Encke's and Biela's comets.
ALS 2pp (German) RS:HS 9.234

1219. 1825–8–10 To Mary Baldwin Slough
JH's mother will pay Mr. Beckwith to cover debt of MB's brother Thomas, who now owes Lady Herschel £1035. Comments on Thomas Baldwin's proposal for business partnership with JH and request for loan from JH.
ACCS 3pp TxU:H/L-0062; Reel 1054

1220. 1825–8–12 From James South Passy
Comments on various errors detected in the catalogue of double stars published by JH and JS. Scientific affairs in Paris in relation to Alexander von Humboldt, François Arago, P. S. Laplace, and others. Various cometary and stellar observations of JS.
ALS 4pp RS:HS 16.449

1221. 1825–8–15 To *Astronomische Nachrichten* Slough
Writing in response to claims made by Josef Fraunhofer, discusses the relative merits of reflecting versus refracting telescopes. Adds comments on double star observations made by JH and James South.
L *Astronomische Nachrichten, 4* (1826), 231–6

1222. 1825–8–16 To James South n.p.
Asks JS to check over and forward to *Astronomische Nachrichten* JH's paper ['Letter on Fraunhofer's Claims for the Inferiority of Reflecting as Compared to Refracting Telescopes'], which paper calls attentions to errors in the catalogue of double stars published by JS and JH.
Adraft 1p RS:HS 16.450

1223. 1825–8–22 From James South Paris
Reports observing with J. N. Nicollet two comets. Observations of Encke's comet. Gives position of nebula to be observed by JH.
ALS 3pp RS:HS 16.451

1224. 1825–8–23 From Charles Wheatstone Swan Inn, Windsor
Appreciates JH's interest in and encouragement of CW's experiments. Describes new experiments to elucidate theory of undulatory motion of sound and light. Invented 'kaleidophone' to display paths of vibrating rods. Hopes JH can confirm these results.
ALS 3pp † RS:HS 18.144

1225. 1825–8–27 From Giovanni B. Amici Modena
Is sending journals for JH and others listed. Observations on telescopes and star readings.
ALS 4pp (Italian) RS:HS 1.336

1226. 1825–8–30 From Pierre Simon Laplace Paris
Sends the last volume of his *Mécanique céleste* for the R.S.L. Comments on his various activities. Is grateful for the memoirs JH sent.
ALS 2pp (French) RS:HS 11.105

1227. 1825–9–8 From Michael Faraday 56 Devonshire St.
Has found some errors in the printing of his paper. Sorry to hear that JH has been indisposed, but believes he is progressing.
ALS 1p RS:HS 7.174 James, *Faraday Correspondence, 1,* 383

1228. 1825–9–8 From James Grahame London
Has stopped in London and learned of the death of JG's mother.
ALS 1p RS:HS 8.331

1229. 1825–9–10 From James South Passy
Reports his observations of a comet JS detected and also of Encke's comet. Pleased JH is recovering from illness.
ALS 3pp RS:HS 16.452

1230. 1825–9–13 To Michael Faraday Slough
Thanks for the copy of his paper and errata. Received a packet of specimens of glass, which he presumes came from MF. Regarding the specific gravity of some of the specimens. Comments on further experiments to be made.
draft 2pp RS:HS 7.166 & 20.222 James, *Faraday Correspondence, 1*, 383–5 & Williams, *Selected Correspondence, 1*, 149–50

1231. 1825–9–18 From James Grahame Glenfair
Mostly related to the death of JG's mother.
ALS 2pp RS:HS 8.332

1232. 1825–9–20 From Caroline Lucretia Herschel Hanover
Her social life is rather boring at the moment because CH is busy caring for her sick brother Dietrich.
L *Memoir ... of Caroline Herschel*, 191–2 inc (?)

1233. [1825]–9–30 From Mary Pitt Herschel London
Discusses JH's rent at Devonshire Place.
ALS 2pp & C 2pp TxU:H/M–0617; Reel 1086

1234. 1825–9–30 From James South Passy
Sorry JH is still sick. Will send JS's new double stars observations to JH for his comments and for publication.
ALS 1p RS:HS 16.453

1235. 1825–10–4 From Francis Baily Gray's Inn
Will be pleased to accept his invitation to spend some days at Slough.
ALS 1p RS:HS 3.61

1236. 1825–10–4 From Michael Faraday Royal Institution
Regrets that Richard Taylor has printed his own paper in the *P.M.* before it appeared in the *R.S.P.T.* Explains the reason. Is examining various pieces of glass and will forward the results.
ALS 2pp RS:HS 7.167 James, *Faraday Correspondence, 1*, 385–6 & Williams, *Selected Correspondence, 1*, 151

1237. 1825–10–5 From William Wallace 14 Argyle Square [Edinburgh]
Introduces Dionysius Lardner, author of system of algebraic geometry, visiting England.
ALS 1p RS:HS 18.24

1238. 1825–10–6 To Gregorio Barnaba Lavia Slough
Gratitude to Gioenian Academy of Natural Sciences of Catania [Sicily] for electing JH a corresponding member. [P.S. 1 Oct. 1825:] Reports elevations on Mount Etna measured by JH with mountain barometer in July 1824, with confirmation by Niccolo Cacciatore. [JH's annotations explain sources of errors in elevations.]
C 4pp †† TxU:H/L-0222; Reel 1054

1239. 1825–10–11 From James South Passy
Is sending his latest double stars observations to JH. Weather is making observation difficult. Will be in London in November; hopes to present paper at R.S.L.
ALS 2pp RS:HS 16.454

1240. 1825–10–24 From Giovanni Antonio Plana Turin
Inquires about the arrival of the memoir sent by GP to JH [see GP's 1825–7–28].
ALS 1p† (French) RS:HS 13.417

1241. 1825–10–25 From André M. Ampère Paris
Thanking him for sending two volumes. Regarding recent experiments with electricity.
ALS 4pp (French) RS:HS 1.394

1242. 1825–10–25 From James Grahame Edinburgh
Asks JH to obtain entry for JG to the British Museum library.
ALS 1p RS:HS 8.333

1243. 1825–10–31 From Francis Baily Gray's Inn
Send copies of the errata to Giuseppe Piazzi's catalogue. Cloud ruined his observation of Saturn.
ALS 1p RS:HS 3.62

1244. [1825–11] To [?] n.p.
'Messrs. [Apsley] Pellatt and Green' will fill the pots on Monday. JH will attend. Asks correspondent to attend also. Has sent for Michael Faraday's approval a formula for making [the glass].
ALS 2pp Science Museum Archives MS 1006

1245. 1825–11–2 From James South Calais
Has been delayed in France, but is sending large collection of double star observations for JH's comments and analysis. Has dismantled JS's observatory in Passy as part of return of JS to England. Comments on P. S. Laplace.
ALS 4pp RS:HS 16.455

1246. 1825–11–5 From Thomas Young 48 Welbeck St.
Ask Charles Babbage to send tables of logarithms to TY, who will compare them to [Brook] Taylor's. Bets that five errors will be found. Did JH order mountain barometer from [J. F.] Newman? Will inquire about [W. E.] Parry's instruments. Admiralty, not us, should order sextants.
ALS 3pp RS:HS 18.333

1247. 1825–11–7 From Sir John Barrow Admiralty
Richard Copeland has received such instruments as the Board thought he should have. A mountain barometer seems extravagant for his duties. If JH thinks a camera lucida is necessary, one can be supplied.
ALS 3pp RS:HS 5.21

1248. 1825–11–7 From Michael Faraday Royal Institution
Sends JH the results of the analyses of several specimens of experimental glass.
L Xpp RS:DM 3.17 James, *Faraday Correspondence, 1*, 390–1 & Williams, *Selected Correspondence, 1*, 151–2

1249. 1825–11–7 From Thomas Young 48 Welbeck St.
Questions JH's conclusion that Board of Longitude has power to interfere in Admiralty operations. Disagrees that *Nautical Almanac* errors are serious. Compares them to tables of Charles Hutton, F. X. von Zach, and H. C. Schumacher. When will TY receive Charles Babbage's table?
ALS 3pp RS:HS 18.334

1250. 1825–11–8 To Thomas Young 56 Devonshire St.
Encloses note from Sir John Barrow refusing scientific instruments and objecting to using barometers. Supplying cameras and sextants to Admiralty's marine survey is *not* interference by Board of Longitude. Sent JH's expenses to Edward Sabine; has no further word on subject. Encloses Mr. Stewart's note about damages at Wrotham.
Adraft 2pp RS:HS 18.335b (C, encl 1p: 20.223)

1251. 1825–11–11 To James South n.p.
R.S.L. Council will meet 17 Nov. to discuss awarding its Copley medal; hence JS's paper, to be presented to R.S.L. later that day, cannot be considered for the award. Other matters have prevented JH from preparing annotations of JS's latest paper on double stars.
ACCS 3pp RS:HS16.456 (C: 20.224)

1252. 1825–11–14 From Michael Faraday Royal Institution
Agrees with JH regarding the glass. Suggests that a meeting would facilitate results. Suggests that the glass maker should make several pots for the experiments.
ALS 3pp RS:HS 7.168 James, *Faraday Correspondence, 1*, 391 & Williams, *Selected Correspondence, 1*, 152–3

1253. 1825–11–[15] From Thomas Young n.p.
[Board of Longitude] must answer to Parliament, not to public opinion. Do not contest the system. Order pocket sextants; they and barometer will be sent to Captain [Richard] Copeland. Knows how JH feels about expenses sent to Edward Sabine, but please write receipt and send it to TY within week. Will write directly to Charles Babbage for tables.
ALS 2pp RS:HS 18.336

1254. [1825–11–16?] To Henry Kater [56 Devonshire St.]
Thomas Young approved supplying [Richard] Copeland with two sextants and one mountain barometer.
ALS 1p RS:HS 19.50

1255. 1825–11–16 From Edward Sabine Tunbridge Wells
An account of the rocket experiments and P. S. Laplace's letter will be left at the R.S.L. Finds in Laplace's letter encouragement for the corroboration regarding the measurement of latitudinal degrees. Dismisses Laplace's opinions that the climate of Spitzbergen is too unfavorable for experimentation and expounds on the many reasons to favor this town.
C 4pp RS:HS 15.14

1256. 1825–11–17 From Stephen Lee Royal Society
A notice of meeting of the R.S.L. committee to improve glass.
ALS 1p RS:HS B26.25

1257. 1825–11–18 From Baden Powell Clapton
Missed JH at the R.S.L. meeting. Has some experiments on which he seeks JH's opinion regarding presentation at the Society. Enclosing report.
ALS 1p RS:HS 14.47

1258. 1825–11–19 From Giovanni Antonio Plana Turin
Is very concerned because JH's recent letter did not comment on the arrival of GP's memoir [see GP's 1825–10–24]. Comments on printing and arrangements for copies.
ALS 3pp (French) RS:HS 13.418

1259. 1825–11–23 From James Grahame 1 Berkeley Square
About travel plans once he recovers his health.
ALS 1p RS:HS 8.335

1260. 1825–11–24 From Giovanni Santini Padua
Thanks profusely for two volumes of Astronomical Society's memoirs. Sends brochure containing two memoirs: one on eclipse of 1820, the other on two comets in 1823–24. Gives news of various acquaintances and friends.
ALS 3pp (French) RS:HS 15.343

1261. 1825–11–25 From James Grahame 1 Berkeley Square
Is waiting to recover his health before he leaves for Scotland.
ALS 1p RS:HS 8.334

1262. 1825–11–25 To Susan White Devonshire St.
Assures SW that slanderous reports regarding SW's nephew James White have not reached JH or Lady Herschel.
ACCS 3pp TxU:H/L-0497; Reel 1055

1263. 1825–11–27 From Gideon Mantell Castle Place Lewes
Acknowledging the letter of 25 Nov. informing GM that he has been elected to the R.S.L. Will attend at the earliest opportunity to be admitted.
ALS 1p RS:HS 13.18

1264. 1825–11–28 To David Brewster n.p.
Has received copies of reprints from Lyell. Points out errors in a printed paper of his. Has been unwell and very busy of late. Would like to add DB's name to the Optical Glass Committee.
draft 4pp RS:HS 4.258 & 20.225

1265. 1825–11–30 From Charles Babbage Worcester
Regarding [William?] French and G. B. Airy. Will breakfast with JH on Sunday. Going to Cambridge on Monday and would like his company.
ALS 1p RS:HS 2.201

1266. 1825–12–1 From William Wallace 14 Argyle Square, Edinburgh
Received volume on double stars by JH and James South. Followed JH's advice, ordered instruments from Germany. Asked Josef Fraunhofer to design transit instrument to fit pillars designed by Edward Troughton. Quotes Fraunhofer's response [in French] that pillars should be moved to accommodate his instrument. David Brewster anxious for JH to answer last letter.
ALS 3pp RS:HS 18.25

1267. 1825–12–2 From David Brewster Edinburgh
Regarding corrections requested by JH in a printed article. Ill health and overwork applies to him also. Pleased to hear of the work of the Optical Glass Committee. Comments on this.
ALS 4pp RS:HS 4.259

1268. 1825–12–3 From Giovanni Antonio Plana Turin
Is pleased that GP's memoir has finally arrived in London [see GP's 1825–11–19]. Indicates arrangements for paying for copies. Comments on GP's astronomical observations of the last several years, and on his method of making reductions.
ALS 4pp† (French) RS:HS 13.419

1269. 1825–12–11 From Thomas Young n.p.
Suggests change in JH's table of refractions. Possible error in John Pond's reductions [illustration].
ALS 1p RS:HS 18.337

1270. 1825–12–12 To William Wallace London
Edinburgh Institution's decision to order instruments from Josef Fraunhofer should send message to 'dilatory and ... abominably expensive' English artists. Axis of

Fraunhofer's 8–foot transit instrument is too long. Compares those of Cambridge, Greenwich, Paris, and James South. [Robert] Woodhouse's paper in 1825 *R.S.P.T.* eliminated JH's hope that Cambridge would devote time to general catalogue.
C 2pp RS:HS 20.226

1271. [1825–12–15 or later] To ? n.p.
Referring to the R.S.L. Royal Medals commissioned on 15 Dec. 1825 by King George IV, JH confirms that Sir Thomas Lawrence has agreed to design the medals.
ALS 4pp Adler Planetarium

1272. 1825–12–19 From Frederick Griesbach Inland Office, G.P.O.
Is grateful for the present. Did not request the money as a gift.
ALS 2pp RS:HS 9.137

1273. 1825–12–30 To Caroline Lucretia Herschel 56 Devonshire St., Portland Place
Recovering from a prolonged illness. Discusses parallax. JH reports that he now is very skilled at sweeping with the 20-ft reflecting telescope.
ALS 3pp TxU:H/L–0570.3; Reel 1058 *Memoir ... of Caroline Herschel,* 193–4 inc

1274. 1825–12–31 To Thomas John Hussey n.p.
Sending by the bearer mirror No. 3 of which he spoke. Will exchange it for Troughton's if TH can make use of JH's. Do not attempt to clean Edward Troughton's mirror with acid. Hopes he does not feel obliged to carry out any particular series of observations.
draft 4pp RS:HS 10.137 & 20.227

1275. 1826 To Hans Christian Oersted n.p.
Thanks HO for JH's election to the Royal Academy of Copenhagen. Comments on some chemical and magnetical experiments.
C 2pp RS:HS 20.228 Harding, *Correspondance de ... Örsted, 2,* 389–90

1276. [1826] To [Sir] R[ichard] Phillips n.p.
Encloses an outline of JH's paper on iron [? 'On the Separation of Iron from Other Metals,' *RSPT* (1821), 293–9].
ALS 1p Huntington Library

1277. [1826] From Edward Sabine 2 Portland Place
Thanks JH and Charles Babbage for their paper on the magnetic experiments. Asks to have 20 extra copies of the paper on the rocket experiments made when it is printed. Speaks of travel expenses. Mentions [T. F.] Colby's opinions regarding the Spitzbergen measurements.
ALS 3pp RS:HS 15.17

1278. [1826–1] From William Henry Fox Talbot 31 Sackville St.
Comments on several experiments with spectra passing through glass films.
ALS 3pp RS:HS 17.324

1279. [1826 to 1829] From Thomas Young Park Square
Resolved puzzle about sidereal year, which needs better definition. Error between true and nominal positions prevents TY adopting 'equinoctial time' for chronology investigations.
ALS 2pp RS:HS 18.356

1280. [1826 to 1829] From Thomas Young Park Square
Errors in equinoctial time calculated by John Pond. Suggests limiting decimal places in value of equinoctial time to reflect this uncertainty. Making error uniform will make it awkward to alter future almanacs.
ALS 2pp † RS:HS 18.354

1281. 1826–1–4 From William Henry Fox Talbot 31 Sackville St.
Asks JH about Dr. Balfour's account in *Asiatic Researches* about lunar effects on humans in India.
ALS 1p inc RS:HS 17.257

1282. 1826–1–12 From William Henry Fox Talbot 31 Sackville St.
Comments on use of muriate of lime on plants. Also on existence of unexplained bands on film and on presence of red light beyond normal spectrum with light originating from certain sources.
ALS 3pp RS:HS 17.258

1283. 1826–1–18 From Giovanni Antonio Plana Turin
Responds to receipt of sample impression of GP's memoir [see GP's 1825–12–3]. Comments on printing and paying arrangements, and to whom copies are to be sent. Comments on value of using an equatorial telescope as opposed to a meridian circle. [Some parts of the letter are illegible.]
ALS 4pp† (French) RS:HS 13.420

1284. 1826–1–24 To Thomas Hodgkin Slough
Is making arrangements to meet with TH in London.
ALS 2pp WT.Ho A.1

1285. 1826–1–28 From Giovanni Antonio Plana Turin
Writes to correct error JH identified in GP's memoir [see GP's 1826–1–18]. Also asks to have two additional passages added to improve memoir. Intends to come to England next winter. Is including a list of corrections for the memoir [list is missing].
ALS 3pp (French) RS:HS 13.421

1286. [1826?]–1–28 From William Henry Fox Talbot Worthing
Questions and comments regarding improvements in microscopy.
ALS 2pp RS:HS 17.323

1287. 1826–2–1 From Caroline Lucretia Herschel Hanover
Discusses nebulae. Begs JH not to overwork himself. Hopes to live long enough to see JH marry; advises JH that if he can marry a 'good-natured, handsome, and sensible young lady, pray think of it, and do not wait till you are old and cross.'
L *Memoir ... of Caroline Herschel,* 196–9 inc

1288. [1826–2] From Edward Sabine 2 Portland Place
Encloses a copy of a letter from the astronomer [George] Fisher, who strongly supports the practicability of work in Spitzbergen. Says the letter will be presented to the Council for its consideration in hopes that a special committee will be appointed. Urges JH to show the members the copy.
ALS 2pp RS:HS 15.15

1289. [1826–2–8 or later] To William Henry Fox Talbot Devonshire St.
[Responding to WT's 1826–2–7], JH thanks WT for materials sent, makes suggestions regarding WT's upcoming trip to Geneva, and asks WT to deliver a book to [Alfred] Gautier.
ALS 3pp Science Museum Talbt 1/1

1290. 1826–2–15 From James Grahame Stevenage
Of JG's journey north and the woman who nursed him back to health [letter completed 1826–2–16 at Wandsworth].
ALS 2pp RS:HS 8.336

1291. 1826–2–23 To Edward Sabine n.p.
Denies any knowledge of a R.S.L. Council meeting. Trusts the council will make its decisions based on public duty and science, not on the fact that the decision must be defended in later discussions. Has passed [George] Fisher's letter on to members as ES requested.
ACCS 1p RS:HS 15.15b

1292. [1826–2–23 or later] From Edward Sabine Margaret St.
Claims 'it never entered [his] head' that an appointed committee would be influenced solely by the need to defend its decisions. Restates his opinion that the subject should be properly considered.
ALS 3pp RS:HS 15.16

1293. 1826–2–24 From James Grahame Edinburgh
About JG's fatigue on his return home.
ALS 2pp RS:HS 8.337

1294. 1826–2–27 From William Henry Fox Talbot Worthing
Comments on Josef Fraunhofer's work on double stars, on quality and color of light from different sources, refraction of different kinds of light, and solution of a problem in conic sections. Is going to Geneva and Modena soon.
ALS 3pp RH:HS 17.259

1295. 1826–2–27 From Henry John Temple Stanhope St.
Hopes that JH will find a matter [the professorship at Cambridge] not 'incompatible with your [JH's] other avocations.'
ALS 2pp RS:HS 17.344

1296. 1826–2–28 To Wilhelm Struve London
Informs WS that the Astronomical Society has voted WS a gold medal for WS's observations of double stars. Says he is now observing nebulae. Confirms receipt of WS's fourth volume of observations sent by [E. C. F.?] Knorre. Asks WS to verify [Henry] Kater's observations on black zones in Saturn's outer ring.
C 4pp RS:HS 20.229 (Abstract: RS:HS 17.106b)

1297. 1826–3–1 From Giuseppe Piazzi Naples
Asks for JH's assistance in the acquiring of some astronomical equipment.
ALS 1p†† RS:HS 13.396

1298. 1826–3–2 From Francis Baily 37 Tavistock Place
Thanks for the loan of his paper on the parallax. Regarding Wilhelm Struve's paper.
ALS 1p RS:HS 3.63

1299. 1826–3–2 From Johann Franz Encke Berlin
Berlin Academy plans a new Astronomical Chart. Thanks for JH's work on the double stars. Will see he has changed his address.
ALS 2pp (French) RS:HS 7.84

1300. 1826–3–6 From Mario Gemmellaro n.p.
Has heard nothing from JH since he left the island. Would like to hear regarding the height of Etna and the latest comet.
ALS 1p (Italian) RS:HS 8.93

1301. 1826–3–8 From Maria Edgeworth Edgeworthstown
Hopes that JH will visit Ireland and if he does that he will visit them. David Brewster also intends visiting Ireland.
ALS 3pp RS:HS 7.22

1302. 1826–3–13 From William Samuel Stratford 3 Verulam Buildings
JH will be presented with the Gold Medal of the Astronomical Society of London.
ALS 1p RS:HS 17.80

1303. 1826–3–18 From Giovanni B. Amici Modena
Further regarding parcels of books sent to G. S. G. Santini of Padova. Details and comments on telescopes used for observations.
ALS 4pp (Italian) RS:HS 1.337

1304. 1826–3–18 From John Brinkley Observatory, Dublin
Thanks for his valuable communications. His theory regarding the method of determining the difference of parallax is ingenious. Regarding the difference in observations of certain stars between Dublin and Greenwich. Returns G. A. A. Plana's letter and comments on this. Has not yet printed his paper on fixed stars.
ALS 3pp RS:HS 4.280

1305. 1826–3–18 To Maria Edgeworth 56 Devonshire St., Portland Place
Thanks for ME's letter; JH hopes to see her in Ireland in summer.
C 2pp RS:HS 20.230

1306. 1826–3–21 To Henry Kater St. John's College, Cambridge
Asks for the use of an invariable pendulum and a clock from the Board of Longitude so that G. B. Airy and William Whewell can carry out experiments to measure the change in gravity and the mean density of the earth, in a Yorkshire coal mine.
C 3pp RS:HS 20.231

1307. 1826–3–24 From William Henry Fox Talbot Ancona
Describes observations made as he travelled and people visited on journey from London to Ancona.
ALS 3pp RS:HS 17.260

1308. 1826–3–25 From Henry Kater York Gate
Has been waiting for a letter from Basil Hall regarding the availability of his pendulums, but has heard nothing. Surprised at the error in William Lambton's calculations. What does he think of the method of finding the figure of the earth by measuring the parallax of the moon? Is writing in bed. Has been polishing a mirror for T. J. Hussey.
ALS 4pp RS:HS 11.9

1309. 1826–3–30 From Heinrich Christian Schumacher Altona
Announces a new step to all astronomers toward understanding the solar system due to work of [Thomas] Clausen of the Altona observatory. Clausen recognized that the comet discovered by Biela was the same as that of 1805 and 1772. HS gives details of circumstances.
C 2pp (French) RS:HS 15.373

1310. 1826–3–31 From Davies Gilbert Eastbourne
Has been transcribing his paper on suspensions. Comments on this and hopes it will be received by the council. Would like to know if his paper has been received.
ALS 2pp RS:HS 8.109

1311. 1826–4–1 From Francis Baily 37 Tavistock Place
Is sending his paper on the aberration so that it can be laid before the Board of Longitude. Also sending new table for the measurement of heights by means of the barometer. Regarding the distribution of the *Memoirs* to Associates.
ALS 2pp RS:HS 3.64

1312. 1826–4–2 To Francis Baily Slough
Thanks for valuable communication, which will be laid before the Board of Longitude. Has completed own paper for the Astronomical Society. J. F. Encke has sent more copies of the proof-map. Ernesto Capocci's observations of the comet similar to his own. Has been unable to use Charles Tully's or John Ramage's telescopes yet owing to

clouds.
ALS 4pp RS:HS 3.65 & 20.232

1313. 1826–4–4 From Davies Gilbert Eastbourne
Was favored with his letter. Comments on his paper and revised it. Hopes to be in town on Thursday. Believes [Stephen] Lee has the last number of *Annales de chimie*.
ALS 1p RS:HS 8.110

1314. 1826–4–6 From William Edward Parry Admiralty
Wishes to print the observations WP presented to the R.S.L. in his forthcoming Journal. Can JH assist him; has obtained permission from Humphry Davy.
ALS 1p RS:HS 13.219

1315. 1826–4–8 From Charles Louis Largeteau Paris
Expressing his gratitude for the kindness received from JH while staying in England. Thanks for the Memoir he has received.
ALS 2pp (French) RS:HS 11.463

1316. 1826–4–11 From William Whewell Trinity College
Describes plan to carry out pendulum experiments with G. B. Airy. Asks JH for help in borrowing instruments from the R.S.L.
ALS 4pp RS:HS 18.165 Todhunter, *Whewell, 2,* 64–5

1317. 1826–4–18 From George Biddell Airy Trinity College, Cambridge
Asking for introductions to certain men of science in Paris. Revision of his views on his figure of the earth, and casting doubt on Edward Sabine's pendulum observations.
ALS 2pp RS:HS 1.22

1318. 1826–4–22 To Francis Baily Devonshire St.
Remarks on FB's address on astronomy [fragment only].
AL 2pp inc RS:HS 3.66 & 20.233

1319. 1826–4–22 From Siméon Denis Poisson Paris
Preparing to give a lecture at the R.S.L., and writes seeking some information. Comments on some work of G. A. Plana and P. S. Laplace. [Much of this letter is illegible.]
ALS 3pp (French) RS:HS 14.13

1320. 1826–4–24 To George Biddell Airy n.p.
Writes to urge GA to avail himself of the opportunity of working with Basil Hall in swinging the pendulum at Greenwich; concludes with a note for William Whewell about JH's view of the 'extremely irregular' figure of the earth.
ALS 4pp RGO 6.803.7

1321. 1826–4–24 From James Grahame Whitehill, Glasgow
About JG's health, the need for JH to marry, and plans for the summer.
ALS 4pp RS:HS 8.338

1322. 1826–4–25 From George Biddell Airy Trinity College, Cambridge
Sending his paper on the 'Figure of the Earth.'
Description of work at Greenwich Observatory. Remarks on projected expedition to Falmouth and Ashbourne.
ALS 3pp RS:HS 1.23

1323. 1826–4–25 From Michael Faraday Royal Institution
As MF will be unable to attend the next meeting of the Glass Committee, he sends JH the analysis of the glass produced in one experiment, emphasizing its consistency and high quality.
L Xpp RS:DM 3.26 James, *Faraday Correspondence, 1,* 402 & Williams, *Selected Correspondence, 1,* 155–6

1324. 1826–4–27 To Caroline Lucretia Herschel 56 Devonshire St., Portland Place
Sent to CH a copy of James South's second paper on double stars and a synoptic catalog of them. Calls CH one of the first discoverers of the 'comet of 1795.' Sent to Astronomical Society a report on 300 new double stars.
ALS 4pp TxU:H/L–0571.1; Reel 1058

1325. 1826–4–29 To Johann Franz Encke 56 Devonshire St.
Asks questions about the prospectus [see JE's 1826–3–2] prepared by the Berlin Royal Academy for a map of the heavens. Asks JE to send JH a copy of [K. L.] Harding's star atlas.
ALS 4pp AdWDDR II–VII, 5

1326. 1826–4–29 To Wilhelm Struve London
Sent medal to WS. Tells WS it was awarded to WS and [James] South. Speaks of [Francis] Baily's medal presentation speech. Lists books sent to WS.
C 2pp RS:HS 20.234 (Abstract: RS: RS:HS 17.106c)

1327. [1826–5] From William Sotheby n.p.
Declines an invitation of JH's, being 'engaged sixteenfold' that day. Invites JH to a party another day.
ALS 1p RS:HS 17.222a

1328. 1826–5–3 From Wilhelm Struve Dorpat
Offers to participate in determining stellar parallaxes by finding the angle position of double stars. Apprises Henry Kater that he has never seen Saturn's rings separated. Includes micrometer measurements of Saturn and Jupiter.
ALS 3pp (German) RS:HS 17.108

1329. 1826–5–6 To Henry Kater Slough
Invites HK to dinner Saturday to meet James South and Alexis Bouvard.
ALS 3pp RS:HS 19.52

1330. 1826–5–11 From Giovanni B. Amici Modena
Has a micrometer under construction and gives details. Pietro Prandi has published a second memoir on mercury covered with sulfuric acid.
ALS 4pp (Italian) RS:HS 1.338

1331. 1826–5–22 To William Henry Fox Talbot London
[Responding to WT's 1826–3–24], JH concurs with P. S. Laplace's view about the advantage of astronomical observation from high elevations, e.g., from mountains. Comments on telescope prices, a paper by WT, Florence, and JH's plan to visit the Continent.
ALS 4pp Science Museum Talbt 1/2

1332. 1826–5[?]–24 To Charles Babbage n.p.
Encloses a letter from Dr. John Brinkley. Please send on his manuscript catalogue of nebulae. Regarding the reading and printing of CB's paper at the Analytical. Site for his pendulum experiments.
ALS 3pp RS:HS 2.202

1333. 1826–5–26 From Michael Faraday Royal Institution
Sending a paper for the R.S.L. Humphry Davy does not approve entirely of his arguments. Would be pleased if JH

would glance over it and then inform Davy that he has it.
ALS 1p RS:HS 7.169 James, *Faraday Correspondence, 1*, 409 & Williams, *Selected Correspondence, 1*, 159–60

1334. 1826–5–26 To Michael Faraday Devonshire St.
Has read his paper on vaporization with much interest and can see no objections to its arguments. Comments on some of the details.
draft 4pp RS:HS 7.170 & 20.235 & Williams, *Selected Correspondence, 1*, 159

1335. 1826–5–27 To Michael Faraday Devonshire St.
[More detailed version of JH's 1826–5–26]. Has mentioned it to Humphry Davy.
CC 4pp RS:HS 7.171 James, *Faraday Correspondence, 1*, 409–11 & Williams, *Selected Correspondence, 1*, 161–2

1336. 1826–5–27 To [Michael Faraday] Devonshire St.
Comments extensively on paper on limit of vaporization and its significance in helping to settle ideas on constitution of the atmosphere.
ALS 4pp RI 359–62 F3B

1337. 1826–5–29 From John Ramage Aberdeen
Discusses casting and construction of telescope speculae.
ALS 2pp RS:HS 14.298

1338. 1826–5–30 From Johann Franz Encke Berlin
Will send JH some barometric observations; describes the barometers with which the observations were made.
ALS 2pp (French) TxU:H/M–0180; Reel 1087

1339. 1826–5–30 From Michael Faraday Royal Institution
Was grateful for JH's letter and detailed notes. Hopes he will attach a copy to the paper when he communicates it to the R.S.L.
ALS 3pp RS:HS 7.172 James, *Faraday Correspondence, 1*, 411 & Williams, *Selected Correspondence, 1*, 162–3

1340. 1826–5–31 To [Michael Faraday] n.p.
Wrote letter to clarify ideas, without intentions of publication. Discusses light and spectrum of Thomas Drummond's lamp.
ALS 3pp diag RI 363–5 F3B

1341. [1826–6 or earlier] From Thomas Young 48 Welbeck St.
J. J. Littrow's paper in F. X. von Zach's April [*Correspondence astronomique, géologique, hydrographique et statistique*] is only 'a new hash' of [F. W.] Bessel and no improvement.
ALS 1p RS:HS 18.358

1342. [1826–6 to 1829–5] From Thomas Young Park Square
Send plan of furnace to TY or Michael Faraday before tomorrow's committee meeting.
ALS 1p RS:HS 18.359

1343. 1826–6–8 To Michael Faraday n.p.
Comments on the intent of remarks appended to a paper of MF's.
L Xpp RI:F3.B67 James, *Faraday Correspondence, 1*, 414 & Williams, *Selected Correspondence, 1*, 164

1344. 1826–6–13 From William Whewell Dolcoath Mine, Cornwall
Describes the pendulum experiments he and G. B. Airy are carrying out in Dolcoath Mine.
ALS 4pp RS:HS 18.166 Todhunter, *Whewell, 2*, 65–7

1345. 1826–6–14 From Giuseppe Piazzi Naples
Kept busy by writing. Works with Teodoro Monticelli. Expects Captain [Richard] Copeland to visit.
ALS 1p (Italian) RS:HS 13.397

1346. 1826–6–23 From Franklin Bache Philadelphia
Thanks JH for material from AS of London. Mentions 'new comet.' Informs JH of being nominated to the list of foreign members in the American Philosophical Society.
draft 1p RS:HS 17.111b

1347. 1826–6–27 From James Grahame Tidenham near Chepstow
Asks for information, on behalf of JG's father, about a Cambridge man.
ALS 1p RS:HS 8.339

1348. 1826–7 From William Henry Fox Talbot Richmond Hill
Talks about Josef Fraunhofer's failing health and his making of flint glass.
ALS 2pp RS:HS 17.261

1349. 1826–7–1 From William Whewell Fetter Lane
Reports that an accident destroyed one of the pendulums used by G. B. Airy and WW in their Dolcoath experiments. Obtained some results and believes in the general soundness of the method employed.
ALS 3pp RS:HS 18.167 Todhunter, *Whewell, 2*, 68–9

1350. 1826–7–3 From George Biddell Airy Trinity College, Cambridge
Instructions regarding the printing of his paper the 'Figure of the Earth' in the transactions. Details of the disappointing expedition to Falmouth.
ALS 3pp RS:HS 1.24

1351. 1826–7–4 To Michael Faraday Slough
Chemical constitution of meteorites. JH suspects lead-iron alloy. Believes this has not been proposed before. Send specimen for JH's analysis.
C 1p TxU:H/L-0148; Reel 1054

1352. 1826–7–6 To George Biddell Airy Slough
Comments on GA's observing in Cornwall, and on the geodetic calculations made by GA in a paper read by JH.
ALS 2pp RGO 6.803.9

1353. 1826–7–7 From George Biddell Airy Vere St.
Thanking him for his letters of introduction to Paris. Remarks on one of GA's papers; one of his calculations incorrect.
ALS 2pp RS:HS 1.25

1354. 1826–7–7 or earlier From Charles L. Harding Göttingen
A note to accompany one of CH's writings, together with some comments on comet sightings.
ALS 2pp (German) RS:HS 9.235

1355. 1826–7–9 From Giovanni Antonio Plana Turin
Has received a package of books from JH, some of which will be sent on as instructed. Is interested in the work of JH and James South on double stars. Also comments on parallax measure of the sun and on transits of Venus. Wants to respond to P. S. Laplace's theory of Jupiter and Saturn.
ALS 4pp† (French) RS:HS 13.422

1356. 1826–7–11 From James Grahame London
About JG's children and his plans to move to Hastings.
ALS 2pp RS:HS 8.340

1357. 1826–7–17 From William Somerville 12 Hanover Square
Accepts dinner invitation. Hopes no emergencies arise that would prevent their visit.
ALS 2pp RS:HS 16.383

1358. 1826–7–22 From Jean F. A. Gambart Marseilles
Has received JH's letter informing him that he has been made an associate member of the Astronomical Society. Is very grateful for this honor.
ALS 2pp (French) RS:HS 8.39

1359. 1826–7–23 To Johann Franz Encke Slough
Reports on interest at the Astronomical Society, especially of [Thomas J.] Hussey, in the plan of the Berlin Academy for a new star chart. JH cannot participate due to JH's commitment to re-examine his father's nebulae, which requires that JH reside far from London.
ALS 4pp AdWDDR II–VII, 5

1360. 1826–7–24 From Teodoro Monticelli Naples
Reports the death of Giuseppe Piazzi.
ALS 1p (Italian) RS:HS 12.364

1361. 1826–7–28 From James Grahame 54 Piccadilly St.
About plans to spend the winter at Hastings.
ALS 2pp RS:HS 8.341

1362. 1826–8–2 From William Somerville 12 Hanover Square
If possible, send a magnetic needle and the prescription for mixing chrome green paint for Mrs. Somerville. Enjoyed visiting JH's mother.
ALS 3pp RS:HS 16.384

1363. 1826–8–4 To William Somerville Slough
Sends directions for chrome solution and the magnetic apparatus and instructions for its use, as requested [in WS's 1826–8–2]. Discusses experiments with magnets.
ALS 2pp RS:HS 16.385 (C: RS:HS 20.236)

1364. 1826–8–8 From Caroline Lucretia Herschel Hanover
Discusses her will.
L *Memoir ... of Caroline Herschel*, 200–1 inc (?)

1365. [1826–8–12] From Richard Jones Brasted
William Whewell has promised to spend six weeks with him and he hopes that JH will also be able to come for a time.
ALS 2pp RS:HS 10.347

1366. 1826–8–15 From Jean F. A. Gambart Marseilles
Observed this morning, for the first time, a comet. Gives readings.
ALS 2pp (French) RS:HS 8.40

1367. 1826–8–16 From Michael Faraday Niton, Isle of Wight
Has just received his letter. Has had no opportunity of examining the iron. Several years ago he made some experiments on the alloys of lead and cast iron, but does not remember the results. Is surprised that no new metal has been found in meteorites.
ALS 2pp RS:HS 7.173 James, *Faraday Correspondence, 1*, 420–1 & Williams, *Selected Correspondence, 1*, 166–7

1368. 1826–8–17 From Charles Babbage Paris
Has been to the observatory. New comet discovered. Has met P. S. Laplace. Does he know of any tables of logarithms to 10,000?
ALS 3pp RS:HS 2.203

1369. 1826–8–24 From William Henry Fox Talbot 31 Sackville St.
Is sending on two items directed to JH through WT.
ALS 1p RS:HS 17.262

1370. 1826–8–28 To Mary Pitt Herschel Calais
Returns house keys. Smooth passage across Channel. Met James Grahame's brother-in-law, Mr. Robison, who is going to Paris to be married. James South arranged carriage for JH's travels in France. Servant James Child is well and observed four criminals executed in London before departing. News of Beckwiths.
ALS 3pp TxU:H/L-0518.1; Reel 1053

1371. 1826–8–28 To William Whewell Calais
Asks WW to watch over the printing of JH's 'Light' for the *Encyclopaedia Metropolitana.* Describes that work as not especially original and as free of historical information. Discusses JH's trip to Calais.
ALS 3pp TC, Camb. Add. Ms.a.207[13] (C: RS:HS 20.237)

1372. [1826–9] To Richard Jones n.p.
JH plans to explore volcanic districts of Auvergne and Vivarais, then along Rhine to Bonne and Andermatte. Has cancelled plan to cross to Staffa [Scotland] and explore there.
ALS 1p inc TxU:H/L-0205; Reel 1054

1373. 1826–9–3 To Mary Pitt Herschel Montargis
Journey from London and French countryside. Towns and Paris are vastly improved since 1801 visit. Charles Babbage and wife are in Paris. Received letters of introduction from Georges Cuvier, J. N. Nicollet, and J. P. Delessert for JH's expedition to Auvergne. Alexis Bouvard is recovering from fall. Georgianna Babbage has recovered and promised to write to MH. [Letter continued 4 Sept. in Moulins:] Traveled more than 100 miles today. [Letter continued 5 Sept. in Clermont-Ferrand:] Massive flooding in Riom. Impressive view of volcanoes. [Letter continued 6 Sept.:] Will ascend Puy de Dôme today with two companions.
ALS 4pp TxU:H/L-0518.2; Reel 1053

1374. 1826–9–6 From G. W. Müller London
May choose one of each of the new specimens of Boracite crystals; the remainder send to C. D. E. König at the British Museum, together with the enclosed letter. Is now recovering from his sickness.
ALS 2pp RS:HS 13.56

1375. 1826–9–7 From Giovanni B. Amici Modena
Comments on articles recently published on microscopy.
ALS 6pp (Italian) RS:HS 1.339

1376. 1826–9–11 From G. C. L. D. Cuvier Paris
Has received the gift of books for which he is grateful. They have been placed in the library.
ALS 1p (French) RS:HS 5.384

1377. 1826–9–11 To Mary Pitt Herschel Issoire
Two unwanted companions caused JH to waste two days in Clermont-Ferrand. Describes volcanoes and high winds on

peaks. Actinometer experiments on Puy de Dôme. Hunting fossil bones of extinct animals.
ALS 4pp TxU:H/L-0518.3; Reel 1053

1378. 1826–9–12 From James Grahame 3 Cavendish Place, Hastings
About the weather and people in Hastings.
ALS 3pp RS:HS 8.342

1379. 1826–9–17 To Caroline Lucretia Herschel Montpellier
Studying the volcanoes of Auvergne; hopes to use his newly-created actinometer at the Puy de Dôme.
ALS 4pp TxU:H/L–0571.2; Reel 1058 *Memoir ... of Caroline Herschel,* 201–3 inc

1380. 1826–9–17 To Mary Pitt Herschel Montpellier
Describes region around St. Flour, where servant James Child was seriously ill for two days. Low opinion of French physicians. Journey to Montpellier.
ALS 4pp TxU:H/L-0518.4; Reel 1053

1381. 1826–9–17 To William Henry Fox Talbot Montpellier
Thanks WT for bringing to JH Josef Fraunhofer's publications and Franz Gruithuisen's plates. Recounts JH's geologizing in France and JH's efforts to measure solar radiation.
ALS 3pp Science Museum Talbt 1/3

1382. 1826–9–22 To Mary Pitt Herschel Montélimart on the Rhône
Tour delayed at least one week. Clouds have restricted JH's actinometer experiments on mountains. Describes active volcano and vineyards near Agde. Ancient ruins in Nismes [Nîmes].
ALS 4pp TxU:H/L-0518.5; Reel 1053

1383. 1826–9–24 From Robert Jameson Edinburgh
Acknowledging the letter informing him of his election to the R.S.L.; unable to attend on 9 Nov. but will attend at the first opportunity after that date.
ALS 1p RS:HS 10.433

1384. 1826–9–25 To Charles Babbage Entraygues
Geological remarks on the surrounding area and excursions made.
ALS 4pp RS:HS 2.204

1385. 1826–9–29 From Caroline Lucretia Herschel Hanover
Pleased that JH supports CH's plans for her will. Worries about JH's safety at Puy de Dôme.
L *Memoir ... of Caroline Herschel,* 203–4 inc (?)

1386. 1826–9–30 From Giovanni Antonio Plana Turin
Is sending JH some of GP's work, some of it to be forwarded. Refers to a note by GP in some of F. X. von Zach's writings, which GP would like to have added to his memoir being published by the Astronomical Society. GP comments on the status of some of his own astronomical projects.
ALS 3pp (French) RS:HS 13.423

1387. 1826–9–30 From John Ramage Aberdeen
Discusses construction of his 20-foot reflecting telescope, which he modeled after the 40-foot reflector at Slough. Discusses differences of gallery and asks for advice.
ALS 3pp RS:HS 14.299

1388. [1826–10] To George Biddell Airy n.p.
A letter of introduction for a Mr. Thomson, who intends to go to South America and establish schools, and wishes to talk to GA about university education.
ALS 2pp RGO 6.807.22

1389. 1826–10–1 To Mary Pitt Herschel Lyon
Death of James Grahame's relative. Explored volcanoes in Vivarais Mountains. Describes Mont Blanc at sunset. Regrets bypassing St. Bernard. Expects to reach London by 13 Oct.
ALS 4pp TxU:H/L-0518.6; Reel 1053

1390. 1826–10–5 To Mary Pitt Herschel Hotel de Castiglioni, Paris
Plans leisure sightseeing in Paris. Visited Louvre. [P.S. 7 Oct.:] Called on Alexis Bouvard yesterday. Met W. H. Wollaston, returning from Tyrol. Wrote to Caroline Herschel from Montpellier. Will view A. M. Ampère's electrical experiments today.
ALS 4pp TxU:H/L-0518.7; Reel 1053

1391. 1826–10–8 From James Grahame 3 Cavendish Place, Hastings
JG urges JH to marry. JG has begun to write history again.
ALS 3pp RS:HS 8.343

1392. 1826–10–10 From Thomas W. Hornbuckle St. John's College, Cambridge
The Lucasian Professorship will shortly become vacant. Hopes JH will become a candidate for this chair, once occupied by Newton.
ALS 1p RS:HS 9.486

1393. [1826–10]–11 From William Hyde Wollaston Hotel de Rivoli, No.2
Previous engagement prevents WW witnessing experiment of A. M. Ampère tomorrow. Waits for JH's report. Hopes to see JH at [Georges] Cuvier's tomorrow.
ALS 2pp RS:HS 18.286

1394. [1826–10–13] To Thomas W. Hornbuckle [56 Devonshire St.]
Declines being considered for the Lucasian professorship at Cambridge, stating his convictions 'that I shall do more for science as an independent amateur than as a Professor of any particular branch or department of it.' Proposes Charles Babbage as candidate. [Also includes draft of letter to George Peacock requesting his support for Babbage.]
draft 4pp RS:HS 9.487 (C: RS:HS 20.238)

1395. 1826–10–13 From William Whewell Trinity College
Tells JH that the Lucasian Professorship has become available. Believes most Cambridge people favor JH. Urges JH to write the electors, i.e., the heads of the Cambridge colleges.
ALS 3pp RS:HS 18.168 Todhunter, *Whewell, 2,* 73–4

1396. 1826–10–[13 or later] To James Wood n.p.
Learned of vacancy in Lucasian Professorship at Cambridge. Wrote to [T. W.] Hornbuckle explaining why JH declines. Asks JW's support for Charles Babbage.
C 2pp RS:HS 20.239

1397. 1826–10–16 From James Wood St. John's College
Wanted [JH] to become Lucasian Professor of Mathematics, but electors favor one of Heads. This will also operate against Charles Babbage.
ALS 2pp RS:HS 18.292

1398. 1826–10–17 To William Whewell n.p.
Declines to become a candidate for the Lucasian professorship. Explains his reasons. Has written in support of Charles Babbage. Thanks WW for seeing JH's 'Light' through the press. Describes JH's travels on the continent, including his 'geologising.'
ALS 4pp TC, Camb. Add. Ms.a.207[12] (C: RS:HS 20.240)

1399. 1826–10–20 From Giovanni Antonio Plana Turin
Still concerned about possible changes to GP's memoir [see GP's 1826–1–28] due to the work of P. S. Laplace. Comments on illness [?] of F. X. von Zach.
ALS 2pp (French) RS:HS 13.424

1400. 1826–10–31 From James Grahame 3 Cavendish Place, Hastings
Shares some gossip, and asks advice about a dedication for JG's book.
ALS 2pp RS:HS 8.344

1401. 1826–11–1 From Caroline Lucretia Herschel Hanover
The first volume of Johann Pfaff's German translation of William Herschel's papers has been published; asks JH to inform her whether he wants a copy. Commenting on her boredom, CH laments that her 'sole employment consists in keeping myself in good humor and not be disagreeable to others.'
L *Memoir ... of Caroline Herschel*, 204–5 inc

1402. 1826–[11] From C. I. Parnell [London]
Thanks JH for advice on a legal matter.
ALS 1p RS:HS 14.139

1403. 1826–11–3 From James Grahame Hastings
Worried about JH's love life.
ALS 3pp RS:HS 8.345

1404. 1826–11–4 From Wilhelm Struve Dorpat
Mentions a four month absence from Dorpat. Commends [James] South on his work on double stars yet states that South's distances for nearby stars are too large. Considers his measurements, previous to his [Josef] Fraunhofer telescope, significant, even if South refuses to accept them. Thanks JH for the difference in the longitudinal measurements between Paris and Greenwich.
ALS 4pp (German) RS:HS 17.109

1405. 1826–11–12 To John Ramage London
Discusses the construction of JR's telescope. Mentions a number of material changes that must be made if it is to be used for sweeping [see JR's 1826–9–30].
draft? 4pp RS:HS 14.300 (C: RS:HS 20.241)

1406. 1826–11–14 From James Grahame Hastings
About plans for the Christmas season, the weather, history writing, and book publishing.
ALS 4pp RS:HS 8.346

1407. 1826–11–14 To James South [London]
Looked for JS to talk with him about the secretaryship of the R.S.L., specifically discussing Charles Babbage's position. JH would prefer to have JS nominate Babbage.
ACCS 1p RS:HS B27.1

1408. 1826–11–17 To Charles Babbage London
Tells CB that everyone is anxious to have CB allow his name to stand for the Secretaryship of the R.S.L. JH lists many names of people who have all said they would vote for CB.
ALS 3pp RS:HS B27.2

1409. 1826–11–17 From Georgiana Babbage Thernsy[?] Lodge
Writing on behalf of her husband, Charles, who is away, GB expresses her thankfulness for JH's kindness, and believes Charles will agree to be nominated for secretary of the R.S.L.
ALS 2pp RS:HS B27.3

1410. 1826–11–18 To Caroline Lucretia Herschel Slough
[Addition to Mary Baldwin's letter.] Health of JH's uncle. Travels through volcanic country of Vivarais. Asks CH to purchase for JH 10 copies of the first volume of William Herschel's works, recently translated into German. Sent to CH six copies of JH's 'Account of a Series of Observations Made with a 20-feet Reflecting Telescope' (1826) via Mr. Golterman, and six copies of JH's 'On the Parallax of the Fixed Stars' (1826) via Capt. Müller. James South received Copley Medal this year. Asks about comet near sun on 18 Nov.
ALS 3pp TxU:H/M–0691.2; Reel 1084

1411. 1826–11–20 To George Biddell Airy n.p.
Comments on a paper on achromatic eyeglasses submitted by GA to the R.S.L.; also on a paper of William Whewell's on the operation of a pendulum.
ALS 4pp RGO 6.803.11

1412. 1826–11–22 From George Biddell Airy Trinity College, Cambridge
Information on a Mr. Atkinson. Remarks on dispersion of rays in glass.
ALS 4pp RS:HS 1.26

1413. 1826–11–22 From Jean F. A. Gambart Marseilles
Giving further details and readings of the comet he has been observing.
C 3pp (French) RS:HS 8.41

1414. [1826]–11–23 From Francis Baily Somerset House
FB has seen P. M. Roget and Stephen Groombridge, who both give their support to Charles Babbage's election. FB is very concerned about the lack of an R.S.L. Council meeting when most council members would attend.
ALS 2pp RS:HS B27.7

1415. 1826–11–23 To Humphry Davy Devonshire St.
Is not able to accede to HD's suggestion that JH interview J. G. Children. JH now feels that given HD's views, JH must withdraw Charles Babbage's name from nomination for secretary of the R.S.L.
ACCS 2pp RS:HS B27.5

1416. [1826–11–23 or earlier] From William Henry Fitton n.p.
Has heard a report that [Humphry] D[avy] is not prepared to nominate Charles Babbage and that D[avy] would be glad if J. G. Children would become Secretary of the R.S.L. Nonetheless, WF hopes that Davy will change his mind.
ALS 3pp RS:HS B27.13

1417. 1826–11–23 From Jean F. A. Gambart Marseilles
Sending a copy of his letter he has sent to J. B. J. Fourier [see JG's 1826–11–22]. Comments on readings and

observations of this comet.
ALS 2pp RS:HS 8.42

1418. 1826–[11–23] From Edward Ryan 30 Upper [Green?] St.
Is anxious to talk to JH about the matter of Charles Babbage's election. Believes Babbage being out of town is best, thereby leaving matters in the hands of his friends.
ALS 2pp RS:HS B27.6

1419. 1826–11–24 From Charles Babbage Worcester
Has just returned from collecting information on machines. Regarding the Lucasian professorship at Cambridge. Has met a clockmaker who is able to work on his machine.
ALS 3pp RS:HS 2.205

1420. 1826–11–[24 to 29] To James South Slough
Urges JS not to insist on a minor point of dispute about the title of a prospective member. JH feels it is inappropriate and will simply reflect against Charles Babbage, who proposed this person for membership.
ACC 2pp RS:HS B27.15

1421. 1826–11–25 To Charles Babbage [London]
In describing the way in which J. G. Children was elected to the secretaryship, JH says 'the matter was irregularly brought by the President before a meeting of eight or nine persons whom he chose to call a council' to meet for a few minutes 'to transmit some business connected with the Treasurer's accounts.' JH feels Humphry Davy's actions will not be excused even by Davy's best friends.
draft 4pp RS:HS B27.9

1422. 1826–11–25 To Francis Baily [London]
On 23 Nov. in the evening, at what JH calls an irregular meeting of the R.S.L., J. G. Children was elected Secretary, to serve with JH, who had been secretary for some years. JH urges that the R.S.L. and its members must now rise beyond this shabbiness and move ahead.
ACCS 4pp damaged RS:HS B27.14 (C: RS:HS B27.8 inc)

1423. 1826–11–26 From James Hudson Royal Society
Informs JH of the need for a short R.S.L. Council meeting.
ALS 1p RS:HS B27.4

1424. 1826–11–30 To James Wood London
Dr. [William] French withdrew candidacy for Lucasian Professorship. JH again attests to Charles Babbage's qualifications for that position, which includes seat on Board of Longitude.
ACCS 3pp RS:HS 18.293 (C: 20.242)

1425. [1826–12] To Charles Babbage n.p.
Regarding certificates at the R.S.L. Relates recent occurrences over form of certificates. Would like to see him at Christmas.
AL 4pp inc RS:HS 2.397

1426. 1826–12–2 From James Wood St. John's College
Restriction to reside constantly [at Cambridge] may deter Charles Babbage from position of Lucasian Professor.
ALS 1p RS:HS 18.294

1427. 1826–12–5 From Caroline Lucretia Herschel Hanover
Describes her role in William Herschel's work as 'a mere tool which he had the trouble of sharpening and to adapt for the purpose he wanted it.' Sending to JH the first volume of Johann Pfaff's translation of William Herschel's papers in three weeks; believes that JH will be unhappy with it.
L *Memoir ... of Caroline Herschel*, 205–7 inc

1428. 1826–12–7 From James Grahame 3 Cavendish Place, Hastings
Of plans to go to London.
ALS 1p RS:HS 8.347

1429. 1826–12–8 From Adolphe Quetelet Brussels
Although not known to JH, is writing under the auspices of Charles Babbage. Sends various works. Has honor of having proposed JH as correspondent to Royal Academy of Science of Brussels, unanimously supported. Would JH please seriously consider?
ALS 3pp (French) RS:HS 14.194

1430. 1826–12–9 From Wilhelm Struve Dorpat
Interested in JH's catalogue of dimmer double stars. Worries about determining parallax of very close stars. Points out mistake in JH's formula for determining 'the maxima and minima of the position angles.' Informs JH of German translation of William Herschel's writings, *Wilhelm Herschels sämmtliche Schriften.*
ALS 4pp (German) RS:HS 17.110

1431. 1826–12–12 From Niccolo Cacciatore Palermo
Thanks for instruments and books. Death of Prof. Giuseppe Piazzi. Returns instruments for repair. Books have not reached him. Sends first volume of one of his own works.
ALS 2pp (Italian) RS:HS 5.137

1432. 1826–12–14 From Charles Babbage Dudmaston
Would like to spend Christmas with him. Does he want a Newfoundland puppy?
ALS 1p RS:HS 2.206

1433. 1826–12–14 To Charles Babbage [London]
Tells CB of the mistake on the certificate for admission [to the R.S.L.?] for Mr. Moreau, which meant it had to be withdrawn. Doubts that reapplication would be successful. Other news from London.
ALS 4pp BL Aaa 32293.384

1434. 1826–12–14 From William Ritchie Tain
Heard election for position in London will not take place for several weeks. If JH can secure situation in Virginia for WR, WR will accept it.
ALS 1p inc RS:HS 14.350

1435. 1826–12–15 From James Grahame 3 Cavendish Place, Hastings
About JH's mother's illness, and an invitation to JH to visit Hastings.
ALS 2pp RS:HS 8.348

1436. 1826–12–16 From Charles Babbage Dudmaston
Regarding Mr. Moreau and his certificate. Recent happenings at the R.S.L. Has been pheasant shooting.
ALS 3pp RS:HS 2.207

1437. 1826–12–16 From Johann Franz Encke Berlin
Regrets he has taken so long to answer his letter. Gives reasons why the Commission does not allocate the same zone to two astronomers. Readings of J. Grooby and T. J. Hussey. Thanks for JH's work. J. E. Bode's work will be

continued.
ALS 3pp (French) RS:HS 7.85

1438. 1826–12–17 From Joseph Johann Littrow Vienna
Bad weather has prevented him from observing the transit of the comet. Has purchased a refractor similar to the one at Dorpat.
ALS 4pp (German) RS:HS 11.251

1439. 1826–12–24 From Caroline Lucretia Herschel Hanover
Sending a copy of Johann Pfaff's German translation of William Herschel's papers.
L *Memoir ... of Caroline Herschel*, 207–8

1440. [1827 to 1829] To Charles Babbage n.p.
Regarding various publications for distribution.
ALS 3pp RS:HS 2.369

1441. [1827?] From Charles Babbage n.p.
Understands his intellectual pursuits. Gives address of his son [Adelaide].
ALS 3pp RS:HS 2.347

1442. [1827] To Georgiana Babbage n.p.
Announcement of Baden Powell's election to the Savillian chair of Geometry.
ALS 1p RS:HS 2.367

1443. [1827] From William Henry Fitton n.p.
Regarding the affairs of the R.S.L. and JH's intentions.
ALS 3pp RS:HS 7.241

1444. [1827 to 1829?] To Davies Gilbert Slough
Will not attend council meeting. JH comments regarding the nomination for foreign membership [of R.S.L.?] of Teodoro Monticello.
C 2pp RS:HS 25.14.12

1445. [1827] From C. I. Parnell [London]
Details forgery case on which JH gave legal advice.
ALS 3pp RS:HS 14.140

1446. [1827] From [Sir] R[ichard] Phillips Birmingham
Receives copy of JH's *Light*. Invites him to Birmingham.
ALS 1p RS:HS 14.159

1447. [1827?] From John Pond n.p.
Asks that list of errata be published before a committee meets to investigate 'the subject of Mr. Lee's [?] animadversions.'
ALS 1p RS:HS 14.33

1448. [1827?] To [a member of the Glass Committee of the R.S.L.?] n.p.
It appears that J. T. Cooper would be willing to work in the performing of glass experiments. If this seems acceptable, then a meeting of the Glass Committee should be called.
draft 2pp RS:HS B26.46

1449. [1827?] From Charles Rümker n.p.
Sends [F. G. W.] Struve's observations on William Herschel's double stars because of JH's expressed interest in the work.
ALS 2pp RS:HS 14.508

1450. 1827–1–3 To Wilhelm Struve Slough
May not be able to separate double stars with the twenty foot reflector. Is reviewing nebulae. Regrets [Josef] Fraunhofer did not live to see WS's work. Speaks of standard catalogue of 2881 stars as joint labor of all observatories. Interested in bright stars with minute companions.
ALS 5pp RS:HS 17.111a (C RS:HS 20.243)

1451. 1827–1–4 From Niccolo Cacciatore Naples
The state of the observatory after the death of Giuseppe Piazzi. Has been put on a permanent basis now. Instruments and books have been received. Send some mainsprings for the chronometers.
ALS 1p (Italian) RS:HS 5.138

1452. 1827–1–9 From William Ritchie Paris
Thanks JH for proposing WR for R.S.L. Discusses construction of air pump. Trying to simplify construction of vacuum. Will be in London in March.
ALS 1p RS:HS 14.351

1453. 1827–1–11 To Mario Gemmellaro London
Sends his reasons for not writing. Includes results of his readings of Etna and observations of the comet.
draft 2pp (Italian) RS:HS 8.94 & 20.244

1454. 1827–1–16 To Davies Gilbert n.p.
Some R.S.L. business matters.
C 2pp RS:HS 20.245

1455. 1827–1–17 From Davies Gilbert Eastbourne
Talks about the glass experiments of John Dollond and Mr. Stanwood[?].
ALS 2pp RS:HS B26.27

1456. [1827–1–20] To William Whewell n.p.
Asks WW to sign a certificate on behalf of the election of [William] Ritchie to the R.S.L. and to ask Adam Sedgwick to sign also.
ALS 1p TC, Camb. Add. Ms.a.207[14] (C: RS:HS 20.246)

1457. 1827–1–24 From James Grahame Hastings
About JG's book publisher. Requests support in the application of Granville Sharp Pattison for a professorship at London University [letter completed 1827–1–25].
ALS 3pp RS:HS 8.349

1458. 1827–1–25 To David Brewster London
Asks DB to sign a certificate for the candidature of William Ritchie for admission to R.S.L.
C 1p RS:HS 20.247

1459. 1827–1–28 From Francis Baily n.p.
Further regarding his theory of measuring heights by means of the barometer.
ALS 3pp RS:HS 3.67

1460. 1827–1–29 From Georgiana Babbage n.p.
Urges JH to miss the next R.S.L. Council meeting, as many friends have been to see Charles Babbage, and are proposing to make the meeting difficult. GB suggests that JH remain at Slough because of his mother's illness.
ALS 2pp RS:HS B27.10

1461. 1827–1–30 To Charles Babbage Slough
Argues that neither CB nor his friends should in any way act to create difficulty at future R.S.L. meetings, as that will allow those attacked to resign with dignity. Rather elect someone else at the next anniversary meeting.
ALS 4pp RS:HS 2.208 (ACCS: RS:HS B27.11)

1462. 1827–[2] From Georgiana Babbage 1 Dorset St., Manchester Square
Thanking him for his letter and giving him news concerning the recent death of her father-in-law.
ALS 3pp RS:HS 2.353 Hyman, *Charles Babbage,* 62–63 inc

1463. 1827–2–1 From James Grahame Hastings
About JH's mother, and the behavior of Charles Babbage.
ALS 2pp RS:HS 8.350

1464. [1827–2–2] To Johann Franz Encke n.p.
Pleased that JE will continue the Berlin Ephemeris. Makes various recommendations regarding it. Hopes that someone will reduce Thomas Brisbane's observations of southern heavens.
ALS 3pp AdWDDR II–VII, 5

1465. 1827–2–3 To William Whewell n.p.
Has burned the certificate signed by WW and JH on behalf of election of [William] Ritchie to the R.S.L., because JH, as a R.S.L. secretary, should not endorse candidates. Discusses recent Geological Society meeting.
ALS 3pp TC, Camb. Add. Ms.a.207[15] (C: RS:HS 20.248)

1466. 1827–2–9 From George Biddell Airy Trinity College, Cambridge
Asking for the procedure for bringing business before the Board of Longitude. Has found solar tables incorrect.
ALS 2pp RS:HS 1.27

1467. 1827–2–11 To James Grahame n.p.
Comments on behavior of Charles Babbage [see JG's 1827–2–1].
draft 3pp RS:HS 8.351 (C: RS:HS 20.249)

1468. 1827–2–12 To George Biddell Airy n.p.
Talks at length about the operation of the Board of Longitude and of procedures GA should know if he wishes to present proposals to the Board; JH seems to be trying to calm down GA.
ALS 4pp RGO 6.803.13

1469. 1827–2–12 From Georgiana Babbage 1 Dorset St., Manchester Square
Views as to whether her husband would be interested in applying for the professorship at Oxford.
ALS 2pp RS:HS 2.210 Hyman, *Charles Babbage*, 63 inc

1470. 1827–2–12 To Georgiana Babbage n.p.
Regarding support for Charles Babbage at Oxford.
ALS 1p RS:HS 2.209

1471. [1827–2–13 or later] From Georgiana Babbage 1 Dorset St., Manchester Square
Gives JH news of Charles Babbage's plan to apply for the Savilian chair of mathematics at Oxford.
ALS 1p RS:HS 2.357 Hyman, *Charles Babbage,* 63 inc

1472. 1827–2–13 From James Grahame Hastings
More about Charles Babbage [see JH's 1827–2–11]; requests JH's assistance to gain entry to the British Museum library.
AL 3pp RS:HS 8.352

1473. [1827–2]–13 From William Hyde Wollaston n.p.
[Charles Babbage] should understand that no effort will be made for 'alien' unless that candidate offers himself. Since William Buckland suggested Babbage for the position, let Buckland take the initiative.
ALS 2pp RS:HS 18.283

1474. 1827–2–13 To William Hyde Wollaston n.p.
Notified Georgiana Babbage that two or three electors for professorship at Oxford favor Charles Babbage. JH not free to name them. Death of father may alter Babbage's desire for position. Meanwhile Mrs. Babbage asks that we keep 'any one favourable to C. on neutral ground.'
ACCS 3pp RS:HS 18.282

1475. 1827–2–19 From George Biddell Airy Trinity College, Cambridge
Requesting permission to borrow Christiaan Huygens's telescope. Postscript of further remarks on the solar tables.
ALS 3pp RS:HS 1.28

1476. [1827 to 1829]–2–21 From Mary Pitt Herschel Kensington
Regrets what MH told JH yesterday about Slough estate. If JH receives good offer, he may do what he judges best.
ALS 1p & C 1p TxU:H/M–0618; Reel 1086

1477. 1827–2–22 From James Grahame Hastings
About the state of JG's health, and his feelings about slavery.
ALS 2pp RS:HS 8.353 (C: RS:HS 20.250)

1478. 1827–2–27 From William Ritchie Paris
Requests JH make any necessary corrections in WR's paper on 'Diff[erential] Thermometer.' Thanks JH for his efforts on behalf of WR.
ALS 1p RS:HS 14.352

1479. 1827–3–4 To George Biddell Airy n.p.
Asks GA about the purpose of the request for the use of Christiaan Huygens's telescope in the possession of the R.S.L.; JH comments on the accuracy of James South's astronomical observations.
ALS 3pp RGO 6.803.15

1480. 1827–3–5 To John Thomas Cooper 56 Devonshire St.
Outlines experiments for JC to carry out on 'agglutination of Earths for living Crucibles.'
ALS 3pp diag inc RS DM.3.36

1481. [1827–3–5] From Richard Jones [Brasted]
The valley has been swept with an inflammatory rheumatism. Is much pleased with [James] Graham[e?]'s history. Comments on his views. Who is to be the new president of JH's society? Hopes JH will visit him in the Spring.
ALS 3pp RS:HS 10.348

1482. 1827–3–7 From George Biddell Airy Trinity College, Cambridge
Giving his reasons for wanting to borrow Christiaan Huygens's telescope. Postscript on recent observations with a prism.
ALS 4pp RS:HS 1.29

1483. 1827–3–15 To Thomas Makdougall Brisbane Devonshire St.
Has had a collection of Astronomical Observations from K. L. C. Rümker. Should these be printed and has Rümker the sanction of TB for this kind of observation from the Observatory?
draft 3pp RS:HS 5.101 & 20.251

1484. 1827–3–16 From Thomas John Hussey and [Ira?] Dawson Chislehurst
Reports on various strange changes in the position and color of Jupiter repeatedly observed by local persons known to be trustworthy. Can provide no explanation.
ALS 3pp Univ.Coll.London MS Graves 23(1)

1485. 1827–3–16 To Michael Faraday n.p.
Proposes an adjustment in the proportions of the ingredients in the next glass-making experiment.
L Xpp RI:F3.B395 James, *Faraday Correspondence, 1*, 430

1486. 1827–3–16 From Thomas Thomson Glasgow
Requests JH to read his paper on the properties of chromium at R.S.L. meeting. Discusses the experiments described in the paper.
ALS 2pp RS:HS 17.435a

1487. 1827–3–17 From Thomas Henderson to Basil Hall Edinburgh
Has looked into JH's paper and found the reason for the discrepancy in the observations. Comments on this. Thinks the enclosed paper should be sent to JH. Would like to obtain cheap copies of the Greenwich observations if possible.
ALS 2pp RS:HS 9.282

1488. 1827–3–20 From Thomas W. Hornbuckle St. John's College, Cambridge
Encloses draft. On the point of leaving College for Staplehurst, where he will be pleased to see JH whenever he is in the neighborhood.
ALS 1p RS:HS 9.488

1489. 1827–4 To George Biddell Airy n.p.
A letter accompanying several other letters that might be of interest to GA; encourages GA to act quickly if he means to try for 'the appointment.'
ALS 1p RGO 6.807.30

1490. 1827–4 From Caroline Lucretia Herschel Hanover
Discussing William Herschel's financial difficulties, CH confides that 'she never felt satisfied with the support your father received toward his undertakings, and far less with the ungracious manner in which it was granted.' Regrets WH was not able to do more work with the 40-ft. reflecting telescope.
L *Memoir ... of Caroline Herschel*, 208–11 inc (?)

1491. [1827–4] To ? n.p.
George Airy has written JH to say that he does not at this time wish to go to Dublin to offer himself as a candidate for the Andrews Professorship at Trinity College, Dublin. Asks advice on aspects of Airy's candidacy.
ALS 4pp BostonPubLib Ms.Eng.488(1)

1492. 1827–4–2 From Maria Edgeworth Edgeworthstown
Regrets that JH did not visit Ireland last year. Hopes he will come this year. Has he heard of W. R. Hamilton? David Brewster intends visiting Ireland, though he is busy with his Encyclopaedia.
ALS 4pp RS:HS 7.23

1493. 1827–4–3 From John Thomas Cooper 9 Paradise St., Lambeth
Describes glass experiments he is performing.
ALS 1p RS:HS B26.28

1494. 1827–4–5 From George Biddell Airy Trinity College, Cambridge
Questions concerning a vacant professorship at Dublin Observatory.
ALS 2pp RS:HS 1.30

1495. 1827–4–7 To George Biddell Airy n.p.
Offers information, and strategy, which GA may find useful if he is seriously considering a position at Dublin Observatory; much of the information JH provides comes from Francis Beaufort.
ALS 3pp RGO 6.803.19

1496. 1827–4–8 From George Biddell Airy Trinity College, Cambridge
Proposed visit to Dublin and letters of introduction from JH.
ALS 2pp RS:HS 1.31

1497. 1827–4–9 From James Grahame 20 Strand, Dawlish
About reviews of JG's book, and of a painting of Galileo that JG saw.
ALS 1p RS:HS 8.354

1498. 1827–4–10 From Leopold Nobili Reggio
Sends copies of a paper to JH and D. F. J. Arago on a chemical experiment being conducted at the Institute of France. Asks JH to present it to the R.S.L.
ALS 1p (Italian) RS:HS 13.155

1499. 1827–4–17 To Davies Gilbert 56 Devonshire St.
On the poor functioning of the R.S.L. Glass Committee.
C 2pp RS:HS 20.252 (ACCS: RS:HS B26.30)

1500. 1827–4–17 From James Grahame 20 Strand, Dawlish
Of JH's health, of JG's daughter Matilda, and the people he has met in Dawlish.
ALS 4pp RS:HS 8.355

1501. 1827–4–19 From Davies Gilbert Eastbourne
Thanks for his letter. Can JH or Charles Babbage suggest someone to experiment with glass? Michael Faraday? Regarding the £100 voted for John Newman or his assistant. Has to attend a local meeting on Wednesday.
ALS 2pp RS:HS 8.111

1502. 1827–4–20 From George Biddell Airy Dublin
Observations on the position of professorship at Dublin, conditions of service, income, etc.
ALS 3pp RS:HS 1.32

1503. 1827–4–20 To Dionysius Lardner Slough
As G. B. Airy has chosen not to stand as a candidate for the astronomical position in Dublin [Astronomer Royal of Ireland], JH is strongly supporting DL for that position.
ALS 3pp APS B.H435p.2

1504. 1827–5–1 From [George Biddell Airy] Trinity College, Cambridge
Dionysius Lardner is candidate for Dublin Observatory. Please send Edward Sabine's paper on longitude.
AL 1p inc TxU:H/L-0397; Reel 1055

1505. [1827–5] To William Henry Smyth 56 Devonshire St.
On the questions of the usefulness of occultations and the use of mean time in astronomy, JH would much value WS's

opinions.
ALS 3pp APS B.Sm98.1

1506. 1827–5–3 To George Biddell Airy n.p.
Comments on GA's response to an offer from Dublin [see JH's 1827–4–7]; also about the discussion in the Council of the R.S.L. of a report of experiments from William Whewell and GA; JH makes some disparaging remarks about the work of John Pond.
AL 4pp inc RGO 6.803.21

1507. 1827–5–4 From George Biddell Airy Trinity College, Cambridge
No news of Dublin. Remarks on a series of experiments. Observations on JH's paper on the *Nautical Almanac*.
ALS 3pp RS:HS 1.33

1508. [1827–5–4?] To Caroline Lucretia Herschel n.p.
Received the first volume of Johann Pfaff's German translation of William Herschel's papers. Just completed a second catalog of double stars; review of nebulae going slowly. JH comments on his precise sweeps.
ALS 4pp TxU:H/L–0572; Reel 1058 *Memoir ... of Caroline Herschel*, 212–13

1509. 1827–5–7 From David Brewster Edinburgh
Would have answered his last letter sooner but has been confined to his house. Encloses specimen of calcareous spar. Has in the past carried out many experiments on this substance. Is he satisfied with a statement in the Optical Glass report? Missing Transactions for the library.
ALS 4pp RS:HS 4.260

1510. 1827–5–7 To Thomas Thomson Slough
Informs TT that the R.S.L. will print his paper on chromium. Discusses results obtained in TT's experiments.
ACCS 4pp RS:HS 17.435b (C: RS:HS 20.253)

1511. 1827–5–8 From Giovanni B. Amici Lyons
Regarding the telescope of the Rev. [T. J.?] Hussey of Chislehurst.
ALS 2pp (Italian) RS:HS 1.340

1512. 1827–5–8 From Caroline Lucretia Herschel Hanover
She 'can only think of what is past, and is for ever forgetting the present.'
L *Memoir ... of Caroline Herschel*, 211–12 inc

1513. 1827–5–12 From William Hyde Wollaston Dorset St.
Corrects yesterday's mistake about 'Russian Platina.' It is chemically pure, but mechanically the worst.
ALS 1p RS:HS 18.284

1514. 1827–5[?]–16 To Michael Faraday n.p.
Suggests change in composition of a pot to be made by [Apsley] Pellatt for an experiment.
ALS 2pp RI 395–6 F3B

1515. 1827–5–17 From George Biddell Airy Trinity College, Cambridge
Further observations on JH's paper on the Nautical Almanac.
ALS 3pp RS:HS 1.34

1516. 1827–5–17 To Joseph Johann Littrow London
Discusses JH's efforts to send various publications to JL and to receive publications from JL. Hopes Franz von Gruithuisen, whose 'strange' lunar observations are causing controversy, will come to England with his telescope. Discusses JH's progress in preparing a catalog of nebulae.
ALS 4pp NatLibScot MS.582, no.667

1517. 1827–5–18 To George Biddell Airy n.p.
Explains to GA the disposition, by the Committee on Papers of the R.S.L., of GA's experimental results [see JH's 1827–5–3].
ALS 2pp RGO 6.803.22

1518. 1827–5–18 From James Grahame Dawlish
A silly imaginary dialogue among JH's servants, and about JG's travels in Devonshire.
ALS 3pp RS:HS 8.356

1519. 1827–5–22 From William Henry Fox Talbot A 3, Albany
Describes prismatic spectrum of flame of cyanogen and offers to repeat experiment with Michael Faraday for JH's benefit.
ALS 2pp RS:HS 17.263

1520. 1827–5–24 From William Prout 40 Sackville St.
Unable to attend [R.S.L.] council meeting.
ALS 3pp RS:HS 14.191

1521. 1827–5–25 To [Edmund R.] Daniell 56 Devonshire St.
Thanks managers of Royal Institution for generosity in offering facilities [to Optical Glass Committee] for erection of a furnace.
ACCS 2pp draft RS DM.3.42

1522. [1827–5–27?] From William Hyde Wollaston 1 Dorset St.
Will be out of town for next [R.S.L.?] council meeting. Comments on glass subcommittee report.
ALS 1p RS:HS B26.36

1523. 1827–5–29 From Olinthus G. Gregory Royal Military Academy, Woolwich
Regrets he will be unable to dine with him as he will be engaged in Huntingdonshire. Outlines the regulations for foreigners to visit the Dockyard.
ALS 2pp RS:HS 9.36

1524. 1827–5–29 From Joseph Nicolas Nicollet Paris
Introducing his friend Mr. De Lavigne, who is visiting England. Mentioned him in his memoir on the *Measure of the Arc....* Gives news of his own astronomical work. Edward Sabine arrives at the end of the month. Has obtained remarkable results with his two pendulums.
ALS 2pp (French) RS:HS 13.135

1525. 1827–5–29 To Adolphe Quetelet London
Received letters of 8 Dec. through [Charles] Babbage, including interesting work on statistical returns for Netherlands. Rejoices at prospect of new observatory in Brussels. Promises to do his utmost to help. Honored to be correspondent of Academy of Brussels. Requests AQ address further letters to JH as 'President of Astronomical Society.'
ALS 3pp Académie belgique: #1

1526. 1827–5–30 From Adolphe Quetelet Brussels
Asks where AQ can procure best instruments in England. Desires to view the interiors of English observatories. Wishes to have [Alexis] Bouvard, who is in need of a

change of scene after [P. S.] Laplace's death, accompany him. Sees translations of English scientific works, especially JH's, as a desirable project.
ALS 3pp (French) RS:HS 14.195

1527. 1827–6–2 To Leonard Horner n.p.
Outlining his scheme for election of candidates to a post [appears to be an astronomer for an observatory].
draft 4pp RS:HS 9.492 & 20.254

1528. 1827–6–3 To Thomas Henderson 56 Devonshire St.
Has received TH's alterations for TH's publication. Astronomical Society's council is offering TH some volumes of Greenwich observations.
ALS 2pp NatLibScot MS.3419, f.86

1529. 1827–6–4 From Giovanni B. Amici Paris
Relates the events of his journey to London.
ALS 2pp (Italian) RS:HS 1.341

1530. 1827–6–5 From Jean Baptiste Joseph Fourier Paris
Sending to the R.S.L. *Memoires* of the Académie for 1823.
ALS 1p (French) RS:HS 7.347

1531. 1827–6–11 From W. Herschel Griesbach Vegesack, Bremen
Is in financial difficulties and would appreciate £80.
ALS 3pp (German) RS:HS 9.140

1532. 1827–6–11 From Teodoro Monticelli Reale Accademia di Scienze
Invites JH to Royal Academy of Science's museum.
(A)LS 1p (Italian) TxU:H/M–1028; Reel 1083

1533. 1827–6–12 To John Goldingham n.p.
Comments on the observations in two of JG's papers.
draft 3pp RS:HS 9.100 & 20.255

1534. 1827–6–18 From Jean Baptiste Joseph Fourier Paris
Expressing the thanks of the Académie for the Discourse pronounced on the award of the medals at the Astronomical Society Apr. 1827.
ALS 1p (French) RS:HS 7.348

1535. 1827–6–23 From Giovanni B. Amici London
Has sent Mr. Talbot's Microscopy. Has received a letter from Leopoldo Nobili. Observations on Nobili's experiments on the electrification of mercury.
ALS 2pp (Italian) RS:HS 1.342

1536. 1827–6–23 To Mary Pitt Herschel [London]
Invitation from Duchess of Clarence [Adelaide] calls JH to London. Appointments with [Davies] Gilbert and R. H. Inglis on Monday. Will reschedule Board of Longitude meeting. Sent containers for Mr. Atherton. Paid MH's bills. [Agnes] Kater is recovering. News of Beckwiths and James Grahame.
ALS 2pp TxU:H/L-0519; Reel 1053

1537. 1827–6–25 From John Goldingham Knighton, Radnor
Is grateful for his letter and also the address to the Astronomical Society. Comments on his own paper on satellites.
ALS 4pp RS:HS 9.101

1538. 1827–6–27 To Charles Babbage Devonshire St.
Announcing his decision of resigning from the Secretaryship of the R.S.L.
ALS 2pp RS:HS 2.211

1539. 1827–6–27 To Francis Beaufort Devonshire St.
Informs FB that it is JH's intention to resign as secretary of the R.S.L. at the next anniversary meeting.
ALS 1p WT 68115 (C: RS:HS 20.256)

1540. 1827–6–27 To Davies Gilbert [56 Devonshire St.]
Informs DG that JH is resigning as secretary of R.S.L. [at the anniversary date?] and expresses great regret that some regard the position of secretary not as an elected position, but in the patronage of the President.
ACCS 3pp RS:HS B27.12 (C: RS:HS 20.257)
Philosophical Magazine, 4th ser., 29 (1865), 246–7

1541. 1827–6–27 To Henry Kater Devonshire St.
Informs HK that JH intends to resign as secretary of the R.S.L.
ALS 1p RS:HS B26.62

1542. 1827–6–28 From Francis Beaufort Manchester St.
FB pleased by JH's note yesterday announcing JH's resignation [as secretary of] R.S.L. Hopes JH will accept 'another office' in R.S.L. at future date.
ALS 2pp TxU:H/M–0060; Reel 1087

1543. 1827–6–28 From John George Children British Museum
Regrets JH's resignation of R.S.L. secretaryship.
ALS 3pp TxU:H/M–0923; Reel 1083

1544. 1827–6–28 From Davies Gilbert London
Regrets JH's retirement as secretary to R.S.L. Names possible successors.
ALS 3pp TxU:H/M–0930; Reel 1083

1545. 1827–6–28 From James Hudson Royal Society Apartments
JH's resignation as R.S.L. secretary. Reports finding papers. Will confirm that Greenwich observations reached Thomas Henderson. Thermometers and [John] Dalton's *Elements* arrived safely. Thanks for invitation to visit Slough.
ALS 3pp TxU:H/M–0282; Reel 1087

1546. 1827–6–28 From Henry Kater York Gate
Regrets JH's resignation as secretary of R.S.L.
ALS 2pp TxU:H/M–0314; Reel 1087

1547. 1827–6–28 To William Henry Smyth London
Astronomical Society will lend instruments, including a Beaufoy Clock and a transit circle, to WS for his observations.
C 2pp RS:HS 20.258

1548. 1827–6–30 From Henry Peter Brougham Hill St.
Offering him the position of Professor of Higher Mathematics at London University.
ALS 4pp RS:HS 4.303

1549. [1827–7] From Charles Babbage n.p.
News of the illness of his son.
ALS 1p RS:HS 2.342

1550. 1827–7–1 To Henry Peter Brougham Slough
Declines the offer of the Professorship of Higher

Mathematics at the University of London. Wishes to have time to devote to research.
ALS 4pp Univ.Coll.London College Correspondence (draft: RS:HS 4.304 & C: 20.259)

1551. [1827–7–1 or later] To Henry Peter Brougham n.p.
Further reasons why he cannot accept the position of Professor at London University.
draft 2pp RS:HS 4.306 (C: 20.260)

1552. [1827–7] From Henry Peter Brougham Hill St.
Has laid his letter before the Council of London University and hopes he will reconsider his refusal [of professorship].
ALS 4pp RS:HS 4.305

1553. 1827–7–3 From William Henry Fox Talbot 31 Sackville St.
Asks JH to clarify what goods are his from a joint order with WT from a German optical firm.
ALS 1p RS:HS 17.264

1554. 1827–7–10 From Caroline Lucretia Herschel Hanover
Feels distant from her family in England. Notes that the French occupation has changed Hanover from what it was when CH left in 1772.
L *Memoir ... of Caroline Herschel*, 214–15 inc

1555. 1827–7–14 From James Grahame Woodford
Family news, and marriage prospects for JG and JH [letter completed 1827–8–15 at Dawlish].
ALS 3pp RS:HS 8.357

1556. 1827–7–18 From William Henry Fox Talbot 31 Sackville St.
Queries and comments about telescopes and microscopes; some thoughts about a diffraction grating.
ALS 2pp RS:HS 17.265

1557. 1827–7–19 From Wilhelm Struve Dorpat
Announces eight copies of his *Catalogus novus stellarum duplicium et multiplicium* are being sent. Has delegated some work on double stars to his aid [E. W.] Preuß.
ALS 5pp (German) RS:HS 17.112

1558. [1827]–7–22 From George Pearson George St.
Expresses regret over JH's resignation of R.S.L. post [as Secretary].
ALS 1p RS:HS 14.149

1559. 1827–7–25 From Joseph Johann Littrow Vienna
About observing comets, problems with book sellers and shipping books. JL has written a popular astronomy in which he deals with perturbations, precession, and other difficult topics. Would JH like a copy?
ALS 4pp (German) RS:HS 11.252

1560. 1827–7–28 From Peter Barlow Woolwich
Has completed one of his telescopes. Sends a paper which he hopes JH will accept.
ALS 2pp RS:HS 3.289

1561. 1827–7–28 From William Ritchie Edinburgh
Heard JH declined mathematics professorship at London University. WR is a candidate for the position. Asks for JH's support.
ALS 1p RS:HS 14.353

1562. 1827–7–29 To Charles Babbage Slough
Letter of condolence on the death of CB's son Charles.
ALS 4pp RS:HS 2.212

1563. 1827–8–8 From Ottaviano F. Mossotti London
Is sailing for Buenos Ayres in a few days' time. Would like to thank JH for his kindness; would also like to see him, if convenient, before he departs.
ALS 1p RS:HS 13.54

1564. 1827–8–8 From William Henry Fox Talbot Lacock Abbey
Renders an accounting of expenses for optical goods received from Germany.
ALS 3pp RS:HS 17.266

1565. 1827–8–9 From Joseph Nicolas Nicollet Paris
Has received the actinometer safely by the hand of Mr. De Lavigne, also the instructions. Hopes to use it in the Alps, and will send some observations. De Lavigne was enchanted with England.
ALS 3pp (French) RS:HS 13.136

1566. [1827–8–9 or later] To William Henry Fox Talbot n.p.
Thanks WT for optical equipment sent. Notes death of Augustin Fresnel. Has purchased a telescope from James South.
ALS 3pp Science Museum Talbt 1/4

1567. 1827–8–16 From Caroline Lucretia Herschel Hanover
Thanks JH for sending his second catalog of double stars; remarks that 'by the manner in which you gentlemen now attack the starry heavens, it seems that there will soon remain nothing to be discovered.'
L *Memoir ... of Caroline Herschel*, 215–16

1568. 1827–8–21 From Michael Faraday Royal Institution
Has been trying to see him, unsuccessfully, about the affairs of the Optical Glass Committee. Hopes that JH will not resign from the Committee now that he is no longer Secretary at the R.S.L.
ALS 2pp RS:HS 7.175 James, *Faraday Correspondence, 1*, 437–8 & Williams, *Selected Correspondence, 1*, 167–8

1569. 1827–8–26 From John Pond Royal Observatory
Lacks information on Augustin Fresnel. Praises JH's drawing as rendering unnecessary annexing one to his paper. Has not yet unpacked the new telescope sent by John Ramage.
ALS 1p RS:HS 14.35

1570. 1827–8–27 To Michael Faraday Slough
Hopes to meet JH in London, and to finish building for R.S.L. soon. Discusses various prisms for experiments. Mentions 'neat' process for drawing uranium from iron.
ALS 3pp RI 397–9 F3B

1571. 1827–8–28 From Francis Baily London
Will be traveling to Hampshire. Agrees with Richard Sheepshanks over the papers on the Comet. Hopes JH will be able to visit him in Hampshire. Remarks on J. B. Delambre's *History of Astronomy*.
ALS 3pp RS:HS 3.68

1572. 1827–8–28 From James Grahame 20 Strand, Dawlish

Trivia about life in Dawlish [letter completed 1827–8–31].
ALS 4pp RS:HS 8.358

1573. 1827–8–30 To Elizabeth [Betty] Plumleigh Babbage n.p.
Is sorry to hear the distressing news. Offers suggestions to relieve Charles Babbage of his sorrows.
ALS 4pp RS:HS 2.213

1574. 1827–9–3 From Elizabeth [Betty] Plumleigh Babbage 5 Devonshire St., Portland Place
Regarding the funeral of Charles Babbage's son.
ALS 1p RS:HS 2.214

1575. 1827–9–3 From James Holman Sierra Leone
Has arrived safely after a delightful voyage from England. Regrets JH was not with them. Will sail on to Fernando Po after a week with the transport. Will be writing soon to the Provost of Eton.
ALS 3pp RS:HS 10.201

1576. 1827–9–6 From André M. Ampère Paris
Has received the new observations. Expresses his gratitude on being nominated as a Foreign Member of the R.S.L. Further regarding his experiments with electricity and magnetism.
ALS 4pp (French) RS:HS 1.395

1577. 1827–9–6 To Davies Gilbert Royal Society
Business matters for the R.S.L.
C 1p RS:HS 20.261

1578. 1827–9–7 From William Henry Fitton Tanhurst near Dorking
Has written to Charles Babbage to express sympathy on the death of his wife, but fears there is not much can be done for him. Has given up the Dorset operations. Comments on JH's theory of sulphurate of iron. May be one of his neighbors soon as he has been offered a residence at Beaconsfield.
ALS 3pp RS:HS 7.225

1579. [1827]–9–7 To William Henry Smyth Boughton, Worcester
Discusses WS's comparison of results with transit circle and Astronomical Society catalogue.
ALS 3pp APS B.Sm98.2 (C: RS:HS 20.262)

1580. 1827–9–7 From Sarah Wrangham n.p.
Regrets not thanking JH for his kindness during SW's visit to Slough.
ALS 1p RS:HS 18.435

1581. 1827–9–8 From Elizabeth [Betty] Plumleigh Babbage 5 Devonshire St., Portland Place
Thanks for JH's sympathy and help during Charles Babbage's bereavement following the death of his wife.
ALS 2pp RS:HS 2.215 Hyman, *Charles Babbage,* 65 inc

1582. 1827–9–11 From Adolphe Quetelet London
Has missed seeing JH, who must be in the countryside with [Charles] Babbage. Hopes to see JH and ask advice for observatory at Brussels.
ALS 1p (French) RS:HS 14.196

1583. 1827–9–15 From Elizabeth [Betty] Plumleigh Babbage 5 Devonshire St., Portland Place
Thanks, and hopes for the success of his help and consolation to Charles Babbage.
ALS 3pp RS:HS 2.216

1584. 1827–9–15 From Joseph Johann Littrow Vienna
Sends some additions to an earlier paper on object glasses of telescopes; also some observations of the August comet, made by other astronomers.
ALS 2pp (German) RS:HS 11.253

1585. 1827–9–25 From Caroline Lucretia Herschel Hanover
Writing her memoirs, CH sends for JH's perusal an account of her youth.
L *Memoir ... of Caroline Herschel*, 217–19

1586. 1827–10–1 From James Grahame London Coffee House, Ludgate Hill
Looking for winter lodgings.
ALS 2pp RS:HS 8.359

1587. 1827–10–3 From Mary Ann Gilpin Somers Town
Her mother has died and she would like him to use his influence at the R.S.L. to raise a subscription for the benefit of her and her brothers and sisters.
ALS 2pp RS:HS 9.97

1588. 1827–10–4 From Joseph Nicolas Nicollet Paris
Has just returned from the Alps, where he worked on the glacier at Chamouny and in the St. Bernard. Comments on his work and on the work of the actinometer. Regrets to hear of the death of Mrs. Babbage.
ALS 3pp RS:HS 13.137

1589. 1827–10–5 To Davies Gilbert Slough
Comments on geodesical measurements being made in Ireland.
C 2pp RS:HS 20.263

1590. 1827–10–6 From James Grahame London
Has taken lodgings at 5 Kensington Square, and now needs introductions to friends of the Herschels in the area [letter completed 1827–10–8].
ALS 2pp RS:HS 8.360

1591. 1827–10–9 From [Aimé] Guinand 17 Quai Voltaire, Paris
Would like to establish himself in England for the purpose of making high quality glass using a process developed by his father [P. L. Guinand].
ALS 2pp (French) RS:HS B26.29a

1592. 1827–10–11 To Thomas John Hussey London
Has been able to destroy a large portion of the color of brown sugar by the use of chlorine. If these hints are of use to TH's friend, TH may convey them.
ALS 3pp Univ.Coll.London MS Graves 23(1) (draft: RS:HS 10.138 & C: 20.264)

1593. 1827–10–12 From William Rowan Hamilton Dublin
Sorry that he could not meet JH when JH was in Dublin; hopes to be introduced to JH soon. Answers JH's queries about WH's work; explains that he is studying 'the general properties of systems of rays, and of the surfaces with which they are connected.'
L Graves, *Hamilton, 1,* 279

1594. 1827–10–18 To George Biddell Airy Slough
Comments on GA's work on the solar tables, GA's pending paper on eyepieces, and on the quality of observations made at Greenwich and Paris; GA is intending to repeat the experiment of swinging a pendulum in a

mine, and JH believes the Board of Longitude can provide the equipment.
ALS 4pp RGO 6.803.29

1595. 1827–10–19 From Vincenzo Amici Modena
His father is ill. Wishes to thank everyone for their kindness to his father on his recent visit to London.
ALS 2pp (Italian) RS:HS 1.393

1596. 1827–10–20 From Thomas Colby n.p.
Is still looking for suitable people to work in the glass experiments.
ALS 3pp RS:HS B26.31

1597. 1827–10–20 To [Aimé] Guinand Slough
Tells AG that JH is a member of a committee looking for ways of making better glass, so AG should not reveal any secrets to JH. JH provides the names of the leading telescope-makers in London, whom AG may wish to contact.
ACCS 2pp RS:HS B26.29b

1598. 1827–10–22 From Jean Baptiste Joseph Fourier Paris
Expressing the thanks of the Académie for *Observations* made with a 20-foot telescope.
ALS 1p (French) RS:HS 7.349

1599. 1827–10–24 From George Biddell Airy Trinity College, Cambridge
Requesting a copy of JH's paper on 'Object planes.' Encloses with this letter his own paper on 'Eyepieces.' Requesting that Greenwich Observations be sent to the University Library, Cambridge.
ALS 2pp RS:HS 1.35

1600. 1827–10–24 From William Henry Fitton 53 Upper Harley St.
Is having some of JH's friends to dinner and would be pleased if he would join them. Affairs of the Geological Society. Has notice from Capt. P. P. King with specimens from the Straits of Magellan.
ALS 3pp RS:HS 7.226

1601. 1827–10–24 From Adolphe Quetelet Kensington Observatory
Thanks JH for letter. Regrets having missed possibility of seeing JH while in England. Asks JH to convince [Charles] Babbage to pass through Brussels instead of Rotterdam.
ALS 2pp (French) RS:HS 14.197

1602. 1827–10–25 From Charles Tulley Islington
Comments on [Pierre] Guinand's glass-making process. Believes the glass made Guinand's way will be too expensive [see JH's 1827–10–20].
ALS 2pp RS:HS B26.32

1603. 1827–10–27 To William Rowan Hamilton London
Thanks for, praises, and comments on WH's 'Systems of Rays' paper. Says WH will have a distinguished career. Mentions other works and hopes to forward them with others he receives to WH.
ALS 4pp TCD 1493:15 Graves, *Hamilton, 1,* 279–81

1604. 1827–10–27 From Claude Louis Mathieu Paris
Thanks for the works. How can he send some for the R.S.L. and Astronomical Society?
ALS 1p (French) RS:HS 13.26

1605. 1827–10–28 From William Whewell Trinity College
Hopes to draw up 'a comparative view of the chemical and mineralogical classification of fossils....' Asks JH's view of J. J. Berzelius's system of chemical notation.
ALS 2pp RS:HS 18.169

1606. 1827–10–29 From William Henry Fox Talbot Lacock Abbey.
Writes in favour of a London University professorship for Thomas Key. Refers to some optical experiments of Michael Faraday, and extends solution of conic section problem first raised in WT's letter of 1826–2–27.
ALS 3pp RS:HS 17.267

1607. 1827–10–29 From Sarah Wrangham n.p.
Encloses parcel for JH's mother. Mr. Beckwith suffered paralytic attack. SW did not report this to Lady Mary Herschel, but JH may if he deems it wise.
ALS 2pp RS:HS 18.436

1608. [1827–11 or later] To William Henry Fox Talbot n.p.
Responds to WT's 1827–10–29. Sending JH's *Light,* which uses some of WT's publications.
ALS 3pp Science Museum Talbt 1/5

1609. [1827–11] To [?] 9 Downing St.
Offers the correspondent's friend the information that JH has received a letter indicating that John Henslow is a formidable candidate for the professorship of botany.
ALS 2pp BostonPubLib Ms.Eng. 476

1610. 1827–11–3 From Michael Faraday Royal Institution
A furnace for glass making is under construction at the Royal Institution.
ALS 1p RS:HS B26.33 James, *Faraday Correspondence, 1*, 443

1611. 1827–11–3 From [Mario Gemmellaro] Nicolosi, [Sicily]
Awaits JH's instructions on duplicating barometer observations taken at Mt. Etna. Francesco Lunn visits, will verify observations made by MG. MG congratulates JH on the success of his efforts to measure the brightness of light with precision.
ALS 3pp (Italian) TxU:H/M–0399; Reel 1087

1612. 1827–11–5 or earlier From James Grahame 5 Kensington Square
Arrangements about meeting JH.
ALS 1p RS:HS 8.361

1613. 1827–11–6 From Francis Baily 37 Tavistock Place
Is sending him a proofsheet of the first of his charts of the zodiac. Does he think the work worthwhile, and is it suitable for printing in the *Memoirs*?
ALS 3pp RS:HS 3.69

1614. 1827–11–6 To Michael Faraday Devonshire Place
Indicates the best way for the glass subcommittee [of the R.S.L.] to deal with the results of glass experiments at the Royal Institution [see MF's 1827–11–3].
ACCS 2pp RS:HS B26.34

1615. 1827–11–9 To Michael Faraday [Devonshire Place]
Informs MF that JH has been called to serve on a grand jury

and so may have to miss some glass subcommittee meetings. [This note enclosed the letter written by JH on 1827–11–6.]
ACCS 1p RS:HS B26.35

1616. 1827–11–9 From Richard Gwatkin St. John's College, Cambridge
The health of Robert Woodhouse is very precarious and wonders if JH would be interested in the Plumian Professorship. Is sure of support.
ALS 1p RS:HS 9.78

1617. 1827–11–9 or later To Richard Gwatkin n.p.
JH says he has too much work to do to consider the Plumian Professorship [see RG's 1827–11–9].
draft 1p RS:HS 9.78 verso

1618. 1827–11–15 To Wilhelm Struve n.p.
Communicates to WS the award of the Royal Gold Medal for his 'magnificent work on the New Catalogue of 3112 Double Stars.' Informs WS that award is for research completed within 5 years of presentation.
C 2pp RS:HS 20.265

1619. 1827–11–17 From Charles Babbage Aachen
Has visited Louvain and also a steam engine factory near Liége. Account of various excursions.
ALS 3pp RS:HS 2.217

1620. 1827–11–21 To Claude Louis Mathieu Slough
Congratulates CM for completing Jean-Baptiste Delambre's *Histoire de l'astronomie du XVIIIe siècle* [1827]. Explains how to send copies for distribution to England.
ALS 3pp (French) Académie des sciences (Paris)

1621. 1827–11–23 From William Whewell Trinity College
Pleased to learn that they share the same view of the reform of chemical nomenclature. Discusses various professorships at Cambridge and the question of who should head the R.S.L. Praises Richard Jones's ideas in political economy.
ALS 4pp RS:HS 18.170 Todhunter, *Whewell, 2,* 85–7

1622. 1827–11–29 From James South Kensington
Attended inquest regarding the 'calamitous' fate of [James] Grahame. Is sending JH inquest report, which contains important information.
ALS 1p RS:HS 16.457

1623. [1827]–11–30 From Richard Gwatkin St. John's College, Cambridge
Further regarding the Lucasian Professorship. Only two candidates left, Charles Babbage and G. B. Airy, and the former is not in a good position. Outlines the duties of the position.
ALS 2pp RS:HS 9.79

1624. 1827–12–5 From Joseph Clement [London]
Needs money to employ men working on Charles Babbage's calculating machine.
ALS 1p RS:HS B27.43

1625. 1827–12–5 From William Rowan Hamilton Observatory (?)
Thrilled by JH's encouragement concerning WH's 'Theory of Systems of Rays.' Enthusiastic about his career prospects at the University of Dublin.
L Graves, *Hamilton, 1,* 281–2

1626. 1827–12–6 From William Ritchie Tain
Cannot at present accept the situation [at University of Virginia] discussed in JH's letter. Is still a candidate for position at London University [see WR's 1827–7–28]. Thanks JH for honor bestowed.
ALS 2pp RS:HS 14.354

1627. [1827]–12–7 From Richard Gwatkin St. John's College, Cambridge
Robert Woodhouse is in immediate danger. Urges him to think again before he finally turns down the offer. Would like to visit him and discuss the matter. Can JH come up during the Christmas holiday?
ALS 2pp RS:HS 9.81

1628. 1827–12–7 From Alexander Rogers Bridge St., Leith
Left his job because of precarious health. Is retired. Discusses achromatic telescopes and his ideas for correcting spherical and chromatic aberrations.
ALS 3pp RS:HS 14.401

1629. [1827]–12–10 From Richard Gwatkin St. John's College, Cambridge
Sending a clause from Dr. Thomas Plume's will which he hopes will remove JH's objections to the professorship at Cambridge. There should be no difficulty in traveling between Slough and Cambridge with steam carriages.
ALS 2pp RS:HS 9.82

1630. 1827–12–12 To William Rowan Hamilton n.p.
Expresses regrets that George Airy's memoir on spherical aberration in telescope eyepieces and WH's 'Theory of Systems of Rays' reached JH too late for adequate treatment in JH's 'Light.'
ALS 1p TCD 1493:17 John Herschel, 'Light,' *Encyclopaedia Metropolitana,* 2nd Div., vol. 2 (1830), 582

1631. 1827–12–17 From Joseph Nicolas Nicollet Paris
Outlines the difficulties he has had to return JH's actinometer. Gives details of his observations in the Alps. Comments on the chronometer used.
ALS 2pp (French) RS:HS 13.138

1632. 1827–12–17 From William Henry Smyth Crescent, Bedford
Apologizes for controversy at meeting with James South over the transit circle.
ALS 2pp RS:HS 16.236

1633. 1827–12–19 To William Henry Smyth Slough
Apologizes for incident at meeting; feels partially at fault for encouraging WS to speak on the subject. Advises WS on his microscope order. Stays at Slough because of good observing conditions.
ALS 4pp APS B.Sm98.3 (C: RS:HS 16.237 & 20.266)

1634. 1827–12–19 To Wilhelm Struve Slough
Has more leisure after resigning as Secretary to the Royal Society. Is sending his and [James] South's magnitude scales and positions of observation for WS to compare. Lists common double stars and offers means of comparison. Remarks on 'new star,' the fifth star in the trapezoid of Orion.
ALS 5pp RS:HS 17.113 (C: RS:HS 20.267)

1635. 1827–12–20 To Francis Baily Slough
Has just had a letter from W. S. Stratford announcing his decision to resign from the secretaryship of the

Astronomical Society. Comments on this. He himself is also thinking of resigning the Presidency at the Anniversary meeting. Regarding the microscopes.
ALS 4pp RS:HS 3.70 & 20.268

1636. 1827–12–22 To Charles Babbage n.p.
News of his family. Joseph Clement's work on the machine and alterations proposed. Clement wants payment of £150. Recent happenings at the R.S.L.
ALS 4pp RS:HS 2.218

1637. 1827–12–22 From Caroline Lucretia Herschel Hanover
Discusses the poor health of Heinrich Olbers, and Johann Encke's inability to acquire the necessary instruments. Wants JH to send to CH a few of her 'Indexes.'
L *Memoir ... of Caroline Herschel*, 219–20

1638. 1827–12–22 From Joseph Nicolas Nicollet Paris
Sending observations of his barometrical observations made during the summer at Chamouny. Comments on them and the work of [Alfred?] Gautier.
ALS 2pp (French) RS:HS 13.139

1639. 1827–12–22 From Wilhelm Struve Dorpat
Surprised and grateful for JH's encouragement concerning his *Stellarum duplicium*. Informs JH that the observatory survived the fire at Abo. The university is moving to Helsingfors, Finland. [F. W. A.] Argelander will remain and is working on determining delineations of the foundation stars.
ALS 3pp (German) RS:HS 17.114

1640. 1827–12–24 From Thomas Catton St. John's College, Cambridge
Plumian professorship is vacant owing to the death of Robert Woodhouse. How does JH feel about applying for it? The University would welcome him.
ALS 2pp RS:HS 5.225

1641. 1827–12–24 From William Henry Fitton 53 Upper Harley St.
Regarding the suitable persons for the Observatories of Greenwich and Cambridge, is JH interested? Has been discussing R.S.L. Council matters with W. H. Wollaston.
ALS 4pp RS:HS 7.227

1642. [1827]–12–24 From Richard Gwatkin St. John's College, Cambridge
Robert Woodhouse has died and if JH intends applying for the position at Cambridge, he should lose no time in making his application.
ALS 1p RS:HS 9.83

1643. 1827–12–24 From Johann Wilhelm A. Pfaff Erlangen
Thanks JH for his assistance with edition of William Herschel's writings [see JH's 1824–6–30]. JP indicates volume two is ready for submission to printer.
ALS 1p RS:HS 13.380

1644. 1827–12–26 From Arthur Judd Carrighan St. John's College, Cambridge
Hopes that JH will apply for the vacant position at the University.
ALS 3pp RS:HS 5.326

1645. 1827–12–26 From Davies Gilbert Eastbourne
Sending an appointment [probably V-P of R.S.L.], which will please every member of the R.S.L. Unable to be in London on the 10th, when Henry Kater will be at the breakfast at the Club. Hopes to attend on the 17th.
ALS 1p RS:HS 8.112

1646. 1827–12–26 To Richard Gwatkin n.p.
Was glad to see the clause in Dr. Thomas Plume's will. Sees many advantages in the professorship at Cambridge, but still adheres to his former decision not to apply for the position.
draft 4pp RS:HS 9.80 & 20.269

1647. [1827]–12–27 From Thomas Young Park Square
'Our volume is nearly ready.' Problem dating vernal equinoxes before Christ's birth.
ALS 1p RS:HS 18.338

1648. 1827–12–28 To Arthur Judd Carrighan Slough
Declines being considered for Cambridge's Plumian Professorship. Comments on his plans for the future, including JH's work on JH's father's nebulae.
ALS 3pp St. John's College, Cambridge *Journal for the History of Astronomy, 25* (1994), 142–3

1649. 1827–12–28 To Davies Gilbert Slough
Responds to having been chosen to be the Vice–President of the R.S.L.
C 1p RS:HS 20.270

1650. 1827–12–28 From Heinrich Wilhelm Matthias Olbers [Bremen]
Comments on the observations JH has made with his 20-foot telescope, especially nebulae. HO provides some of his own observational data of Comet Encke.
ALS 3pp (German) RS:HS 13.176

1651. 1827–12–28 From Heinrich Wilhelm Matthias Olbers Bremen
Discusses comet observations, together with variations from different observers. HO offers some historical fragments about the planet Venus, and about the Lilienthal telescope.
ALS 3pp (German) RS:HS 13.175

1652. 1827–12–29 From Francis Baily London
Regarding the possibility of losing the services of W. S. Stratford, and also the president (JH). Lord High Admiral will be balloted for at the next meeting.
ALS 3pp RS:HS 3.71

1653. 1827–12–30 From George Peacock Trinity College, Cambridge
Is afraid JH will decline the invitation to be a candidate for the Lucasian professorship, so it is assumed Charles Babbage will be elected.
ALS 3pp RS:HS 13.287

1654. 1827–12–31 From William Whewell Trinity College
Asks JH for efforts on behalf of [William] Ritchie. Urges JH to apply for vacant Lucasian professorship. This failing, hopes Charles Babbage will be a candidate.
ALS 2pp RS:HS 18.171

1655. [1828–1] From William Buckland Christ Church, Oxford
The Murchisons and Somervilles will be visiting him in February and he would be glad if JH would come and meet them then.
ALS 1p RS:HS 4.322

1656. 1828 To Niccolò Cacciatore n.p.
Letter of introduction for Charles Babbage who wishes to visit NC's observatory in Palermo.
ALS 2pp? (French) C. A. Chant, 'Two Letters of a Famous Astronomer,' *Royal Astronomical Society of Canada Journal, 38* (1944), 225–7

1657. [1828] To Davies Gilbert n.p.
Captain [Henry] Foster states that the Admiralty must very shortly receive directions from the R.S.L. concerning the scientific objectives for Foster's voyage.
C 2pp RS:HS 21.1

1658. 1828 To James Grahame n.p.
Writes to complain that JH has not seen or heard from JG for some time.
C 5pp RS:HS 21.3

1659. [1828] To John Thomas Graves n.p.
Has had some difficulty accepting JG's arguments about logarithms in a paper JG sent to JH.
ALS 2pp TCD 10047/30 (C: RS:HS B26.63 & TCD 4015:113)

1660. [1828?] From Charlotte Murchison 3 Bryanston Place
Sorry they will not have a sight of the Herschels before they depart. Has arranged for the camera to be delivered to Kensington. The Murchisons are off to Dover tomorrow.
ALS 2pp RS:HS 13.59

1661. [1828] From Thomas Romney Robinson Armagh
Introduces [Dionysius] Lardner, who is a mathematician. Asks how [John] Brinkley's collimators succeeded.
ALS 3pp RS:HS 14.382

1662. [1828] To James South n.p.
Thanks JS for materials sent. Laments the decline of science in England. Believes decline has deep roots.
ACC 2pp RS:HS 16.480 (C: RS:HS 25.13.1)

1663. 1828–1–7 From Francis Baily 37 Tavistock Place
Regrets he was unable to dine with him on Monday. Regarding the secretary of the Astronomical Society. Will meet him at the R.S.L. on Tuesday.
ALS 2pp RS:HS 3.72

1664. 1828–1–7 From Joseph Bianchi Modena
Wishes to obtain a compensating pendulum clock for the new observatory of Archduke Maximilian. Can JH advise on one and the price? Has started several series of observations.
ALS 3pp (French) RS:HS 5.59

1665. 1828–1–11 From William Ritchie Edinburgh
WR's testimonials must be sent to America before further action can be taken. Asks JH for recommendation.
ALS 1p RS:HS 14.355

1666. [1828–1–11 or later] To ? Slough
Requests particulars concerning the electors for position at University of Virginia for which William Ritchie has applied. Encloses Ritchie's letter of 1827–12–6.
ALS 1p RS:HS 14.355b

1667. 1828–1–15 To Caroline Lucretia Herschel Slough
Hopes to send to CH a catalog of new double stars soon. JH is very disappointed with Johann Pfaff's German translation of William Herschel's papers.
ALS 3pp TxU:H/L–0573.1; Reel 1058

1668. 1828–1–22 From William Buckland Oxford
Went to see the bones of the mastodon last Friday. R. I. Murchison and W. H. Wollaston will be coming to Oxford on the 18 Feb.; can JH arrange to meet them there?
ALS 1p RS:HS 4.321

1669. 1828–1–24 From Adolphe Quetelet Brussels
Translation of JH's article *Light* begun. Publisher is in Paris. Gives name of collaborator as [P. F.] Verhulst. Has heard of others translating JH's work. As director of Brussels observatory, AQ needs best works of astronomy. Would JH help?
ALS 3pp (French) RS:HS 14.198

1670. 1828–1–26 From Francis Baily 37 Tavistock Place
Regarding the possibility of the Astronomical Society providing one or two pendulums. Remarks on pendulums. F. W. Bessel has been experimenting with pendulums.
ALS 3pp RS:HS 3.73

1671. 1828–1–29 From Peter Barlow n.p.
Regarding his own forthcoming experiments with his telescopes.
ALS 3pp RS:HS 3.290

1672. 1828–2–1 From Joseph Clement [London]
Needs more money to continue with Charles Babbage's calculating machine.
ALS 1p RS:HS B27.44

1673. 1828–2–4 From Peter Barlow Woolwich
Regarding his own experiments with light.
ALS 3pp RS:HS 3.291

1674. 1828–2–6 To Davies Gilbert n.p.
Accepts an invitation to dine with DG; invites DG to the anniversary meeting of the Astronomical Society; comments on scientific voyage being undertaken on behalf of R.S.L., and paid for by the government.
C 2pp RS:HS 21.2

1675. 1828–2–8 From James Grahame 5 Kensington Square
Hoping to meet JH again.
ALS 1p RS:HS 8.362

1676. 1828–2–12 To Charles Babbage n.p.
Family is well. News of Joseph Clement's work on the machine. Details of the new drawings made. Recent events at the R.S.L. Is giving up lodgings in town and returning to Slough. Sorry to hear of G. B. Amici's health. Regarding Amici's gifts to the R.S.L.
ALS 4pp RS:HS 2.219

1677. 1828–2–13 From Martin Davy Caius Lodge
The Lucasian Professorship is vacant and he would be pleased if JH would consider standing for the position.
ALS 2pp RS:HS 6.476

1678. 1828–2–15 To Michael Faraday n.p.
Sends a note with a letter from W. H. Wollaston to JH, dealing with glass-making experiments.
L Xpp RI:F3.B401 James, *Faraday Correspondence, 1*, 449

1679. 1828–2–17 From James Grahame Kensington
News of mutual friends.
ALS 1p RS:HS 8.363

1680. 1828–2–18 From George Griesbach Villa Place, Walworth
Is in difficult circumstances and would welcome any assistance from JH in obtaining employment.
ALS 2pp RS:HS 9.50

1681. 1828–2–22 From James Grahame 5 Kensington Square
News of friends and of travel.
ALS 2pp RS:HS 8.364

1682. 1828–2–24 From George Biddell Airy Trinity College, Cambridge
Asking JH to express his willingness to becoming Lucasian Professor at Cambridge University.
ALS 2pp RS:HS 1.36

1683. 1828–2–24 To Martin Davy Slough
Declining to become a candidate for the Lucasian Professorship. Has written to Charles Babbage and informed him of the vacancy.
draft 3pp RS:HS 6.477

1684. 1828–2–25 From Elizabeth [Betty] Plumleigh Babbage 5 Devonshire St., Portland Place
Books are safely lodged in Charles Babbage's library. Recent travels and persons met by her son.
ALS 3pp RS:HS 2.220

1685. 1828–2–26 To George Biddell Airy Slough
Informs GA that JH turned down the offer of the Lucasian Professorship at Cambridge, and suggests that Charles Babbage be offered the job.
ALS 4pp RGO 6.803.41

1686. 1828–2–26 From Alexander Rogers Bridge St., Leith
Sends copy of his account of the achromatic telescope because JH did not receive the original. Constructing an achromatic telescope of 5–foot focal length.
ALS 3pp RS:HS 14.402

1687. 1828–3–1 From Sir Edward Ryan Calcutta
Introduces botanist [Nathaniel] Wallich in letter. Any kindness shown him will be appreciated. Discusses some matters of geology.
ALS 3pp RS:HS 14.449

1688. 1828–3 To William Henry Smyth 56 Devonshire St.
Will visit WS after attending two R.S.L. committee meetings. Asks WS's opinion of usefulness of occultation information in the *Nautical Almanac.*
C 1p RS:HS 21.5

1689. 1828–3–6 From Martin Davy Caius Lodge
Babbage has been elected Lucasian Professor by a large majority. Is pleased by the unselfish service of JH to scientific research.
ALS 2pp RS:HS 6.478

1690. 1828–3–6 From George Peacock Trinity College, Cambridge
About the election of Charles Babbage to the Lucasian professorship.
ALS 4pp RS:HS 13.288

1691. 1828–3–6 From William Whewell Trinity College
Charles Babbage has been elected to Cambridge's Lucasian professorship. Thanks JH for his efforts. Hopes Babbage will fill it effectively.
ALS 3pp RS:HS 18.172 Todhunter, *Whewell, 2,* 90–1

1692. 1828–3–7 From George Griesbach Villa Place, Walworth
Was away from home when JH's letter came, which accounts for his belated reply. Would welcome a situation as clerk at the Bank of England should JH have any influence there.
ALS 2pp RS:HS 9.51

1693. 1828–3–8 From Elizabeth [Betty] Plumleigh Babbage 5 Devonshire St., Portland Place
Thanks for JH's efforts on behalf of her son (Charles Babbage) on his election to the Lucasian professorship at Cambridge.
ALS 2pp RS:HS 2.221

1694. 1828–3–9 To Charles Babbage Slough
Writes CB at Naples to inform CB that CB has been elected Lucasian Professor at Cambridge. Urges CB to write to thank the electors.
ALS 2pp Royal Astronomical Society of Canada C. A. Chant, 'Two Letters of a Famous Astronomer,' *Royal Astronomical Society of Canada Journal, 38* (1944), 225–7

1695. 1828–3–9 To Charles Babbage Slough
Informing him that he has been elected to the Lucasian professorship at Cambridge. Offers congratulations and gives views of other friends of his success. He should return to take up his appointment.
ALS 4pp RS:HS 2.222

1696. 1828–3–12 To Francis Baily Slough
Would FB take the chair at the next meeting as he is busy nebulae-hunting. Regarding a printer for the papers of the Astronomical Society. Would like his own paper printed before June. Discusses J. J. Littrow's optical paper. Sends article on Light, which he has written for the *Encyclopaedia Metropolitana.*
ALS 3pp RS:HS 3.74 (C: RS:HS 21.4)

1697. [1828]–3–18 From Mrs. Mary Frances Kater York Gate
Has just had a visit from James Graham[e], who came to apologize for a misunderstanding. Sends a Hindoo myth, which she is sure he can utilize for a poem. Her husband was pleased with JH's paper on light. Has himself written a paper on the collimator.
ALS 4pp RS:HS 11.20

1698. 1828–3–20 From Francis Baily 37 Tavistock Place
Thanks for his paper on light. Regarding the meeting on Friday. Pendulums. Invitation to dine with him. Geological Society has been offered room; would like him to think about the possibility of one for the Astronomical Society.
AL 4pp RS:HS 3.75

1699. 1828–3–20 From James Grahame Kensington
Of people JG has met.
ALS 3pp RS:HS 8.366

1700. 1828–3–20 From James Grahame London Coffee House, Ludgate Hill

Invites himself and his brother to JH's for tea.
ALS 1p RS:HS 8.365

1701. [1828–3–20?] To Mrs. Mary Francis Kater [London?]
Confusion around [James] Grahame's possible use of the verses that JH had offered to MK as suitable for the music *Cherry Ripe*.
C 2pp RS:HS 25.13.24

1702. [1828–3–22] To Mrs. Mary Frances Kater [Slough]
Many thanks for the Hindoo myth. May attempt to put it in verse when he can spare the time from star gazing and computing. The South American Indians imagine their ancestors are hunting among the constellations. Is glad James Grahame called. Sorry to hear a poor account of Captain Kater's health.
ALS 3pp RS:HS 11.20bis (C: RS:HS 21.6)

1703. 1828–3–24 From Sarah Bowdich 27 Burton St.
There may be a delay in supplying the copy of her book on fishes as she has to wait for a sufficient number of subscribers.
ALS 3pp RS:HS 4.204

1704. 1828–3–26 From Thomas Makdougall Brisbane Makerstoun
Has not heard from Thomas Young about the reduction of the Parramatta Observations. James Dunlop will not carry out this work; best to leave it to the one named. Will be coming to London soon as the New South Wales Bill comes up, so will call on JH.
ALS 3pp RS:HS 5.102

1705. 1828–3–27 From Joseph Clement 21 Prospect Place, Southwark
Needs more money to proceed with Charles Babbage's calculating machine.
ALS 1p RS:HS B27.45

1706. 1828–3–28 From Thomas Young 9 Park Square
Board of Longitude will accomplish nothing by meeting before T. M. Brisbane comes to town, but TY suggests informal discussion, either at TY's or Henry Kater's.
ALS 1p RS:HS 18.340

1707. 1828–3–29 From Jean Baptiste Joseph Fourier Paris
Sending the *Memoires* of the Académie.
ALS 1p (French) RS:HS 7.350

1708. 1828–3–30 From Thomas Young Park Square
Henry Kater not well [enough to travel] and requests committee meet at Kater's house. TY urges JH to verify W. H. Wollaston's measurement of refraction before republishing JH's essay. Diagrams Christiaan Huygens's principle of refraction. TY is making forms of aplanatic lenses.
ALS 2pp RS:HS 18.341

1709. 1828–4 From Thomas Young Park Square
Hopes JH will stay for Sunday dinner at TY's before JH visits Dr. Litten. If so, please invite T. M. Brisbane and Francis Baily to join them. Writes formula TY discovered for discontinuous spectra. TY can now communicate with H. W. M. Olbers and T. J. Seebeck via H. C. Schumacher.
ALS 1p RS:HS 18.342

1710. 1828–4–2 From Charles Babbage Rome
His feelings on hearing of his being elected to the Lucasian professorship at Cambridge. Regarding Joseph Clement and the work on his machine.
ALS 4pp RS:HS 2.224

1711. 1828–4–3 From William Whewell Trinity College
Encourages JH's efforts in writing a treatise on sound. Asks JH's advice on a system of mineralogical classification that WW, as Cambridge's new professor of mineralogy, had drawn up.
ALS 3pp RS:HS 18.173

1712. 1828–4–5 From Sarah Bowdich 27 Burton St.
Thanks for his generous letter and enclosures.
ALS 1p RS:HS 4.205

1713. 1828–4–8 From R. Baldwin Paternoster Row
Conveys the best thanks from the directors of the *Encyclopaedia Metropolitana* for the liberal settlement over his article 'Light.' He will receive a fine paper copy of the encyclopaedia in due course.
ALS 1p RS:HS 5.10

1714. 1828–4–9 To Prussian Academy of Sciences Slough
Thanks for JH having been elected a Corresponding Associate of the Prussian Royal Academy of Sciences.
ALS 1p AdWDDR II–III, 115

1715. 1828–4–12 From Alexander Rogers Bridge St., Leith
Recently received telescope from the optician. Some adjustments are necessary. [Robert] Blair has a similar idea regarding telescopes. Asks JH to mention his idea to the Astronomical Society.
ALS 3pp RS:HS 14.403

1716. 1828–4–13 To [Mary Baldwin] Slough
Health of JH's mother. Recommends *History of George Godfrey*. Asks if *Waterloo*, recently lost, was ship that 'C. White sailed in.' Gift for Wesleys. Letter from JH's aunt. Declining many invitations in order to continue star observations. Remembrances to the Grahames.
ALS 4pp TxU:H/L-0063; Reel 1054

1717. 1828–4–14 From James Grahame Kensington
Invites JH to London to meet a friend of JG.
ALS 1p RS:HS 8.367

1718. 1828–4–15 From Elizabeth [Betty] Plumleigh Babbage 5 Devonshire St., Portland Place
Gives views of her son (Charles Babbage) on hearing of his election to the Lucasian professorship at Cambridge.
ALS 2pp RS:HS 2.223

1719. 1828–4–17 To Elizabeth [Betty] Plumleigh Babbage n.p.
Recent work and money expended on CB's machine. News of town. James South intends to sell his instruments.
ALS 4pp RS:HS 2.225

1720. 1828–4–17 From Francis Baily 37 Tavistock Place
Regarding letters from J. L. Tiarks concerning Saturn, and Dr. K. L. C. Rümker concerning J. C. Borda's pendulum at Parramatta.
ALS 3pp RS:HS 3.76

1721. 1828–4–18 From James Grahame Kensington
JH's mother has been ill. Is JH coming into town?
ALS 2pp RS:HS 8.368

1722. 1828–4–19 From James South n.p.
Attended meeting of the Astronomical Society and made observations of Saturn. Invites JH to dinner party. Activities at the Board of Longitude. Contacts with French scientists.
ALS 4pp RS:HS 16.458

1723. 1828–4–19 To William Henry Smyth [London]
Observed two known satellites of Uranus, confirming William Herschel's observations. States, 'I am sure there are more than two.'
C 1p RS:HS 21.7

1724. 1828–4–21 To Francis Baily n.p.
Remarks on Saturn and measurable discs of stars.
ALS 4pp RS:HS 3.77 & 21.8

1725. 1828–4–25 From Alexander Rogers Bridge St., Leith
Gives permission to release his telescope proposal to others. Discusses the progress of the telescope. Considers modifying the arrangement of lenses.
ALS 3pp RS:HS 14.404

1726. 1828–4–27 From James Dunlop Makerstown, Kelso
Regrets the error over the double stars. Is grateful to the Astronomical Society for the honor conferred on him. Regarding the observations of stars that he would like to deposit in a library.
ALS 2pp RS:HS 6.515

1727. 1828–4–29 From Maria Edgeworth Edgeworthstown
Is grateful for the gift of his essay on Light, though she cannot comprehend all of it. Regarding her brother Francis and his work at Cambridge. Would be pleased to see JH at Edgeworthstown.
ALS 6pp RS:HS 7.24

1728. 1828–5–1 From Alfred Gautier Geneva
Is about to leave for England with Alphonse de Candolle. Gives news of his latest astronomical observations. Lists people he has met or hopes to meet.
ALS 3pp (French) RS:HS 8.82

1729. [1828–5 or earlier] From Thomas Young n.p.
Board of Longitude would approve providing Lieutenant Forbes with apparatus JH suggested. Inappropriate for any member to address Admiralty separately. Payment of expenses to committee members.
ALS 1p RS:HS 18.357

1730. 1828–5–2 From Giovanni Antonio Plana Turin
Is sending copies of a memoir on the triangulation of Savoy, prepared by GP and [Mr. ?] Carlini. Is including copies for other scientists and organizations.
ALS 2pp† (French) RS:HS 13.425

1731. 1828–5–3 From William Ritchie Tain
Has not heard from [University of Virginia]. Position of lecturer in mathematics and natural philosophy has opened at Dollar Institution in Edinburgh. WR asks for recommendation for the position from JH.
ALS 1p † RS:HS 14.356

1732. 1828–5–4 From Thomas Young Park Square
Abandons TY's argument on Mr. Roberts's object glass. Asks JH's help in understanding A. J. Fresnel's calculations of diffraction and Josef Fraunhofer's discontinuous functions.
ALS 1p † RS:HS 18.343

1733. 1828–5–5 To Caroline Lucretia Herschel [Slough]
Sending the Gold Medal awarded to CH by the Astronomical Society and also a pair of bracelets from his mother; will send the indexes later. Just erected James South's large equatorial.
ALS 4pp TxU:H/L–0573.2; Reel 1058 *Memoir ... of Caroline Herschel,* 221–2

1734. 1828–5–6 From Wilhelm Struve Dorpat
Accepts JH's extension of Class I double stars. Comments on comparison of these and suggests comparison of respective scales might be mutually useful. Includes continuation of work on micrometer measurements of Saturn. Notices strange eccentricity of Saturn's rings. Is considering comparing his degree-measurements with Finland and Lapland.
ALS 3pp (German) RS:HS 17.115

1735. 1828–5–9 From Charles Babbage Santa Lucia, Naples
Regarding Joseph Clement and the work on his machine. Account of his recent excursions.
ALS 4pp RS:HS 2.226

1736. 1828–5–9 From William Edward Parry Admiralty
Believes it was not *our* Captain Foster to whom the advertisement alluded. Singular way in which Mr. Barrowcliff discharges his trust.
ALS 2pp RS:HS 13.220

1737. 1828–5–10 From Francis Baily 37 Tavistock Place
Regarding the meeting, and the papers read last Friday. Invitation to dine with him and Sir Thomas Brisbane. Regarding Brisbane's observations. Various Astronomical Society affairs.
ALS 4pp RS:HS 3.78

1738. 1828–5–10 From William Ritchie Tain
Has not heard from America. Discusses paper on electric conduction. Wishes to move further south. Thanks JH for securing WR's election to R.S.L.
ALS 2pp RS:HS 14.357

1739. 1828–5–12 To Francis Baily Slough
Is glad Thomas Brisbane was in the chair at the meeting. Will be pleased to join the party on Friday. The Parramatta observations. Someone to supervise their publication. His own recent observations.
ALS 3pp RS:HS 3.79 & 21.9

1740. 1828–5–13 From Thomas Young Park Square
T. M. Brisbane is in town. Can JH meet him? JH's computations of definite integrals are simpler, more direct than P. S. Laplace's and more conclusive than Leonhard Euler's. Comments on W. H. F. Talbot's letters demonstrating Josef Fraunhofer's theorem.
ALS 2pp RS:HS 18.339

1741. [1828]–5–14 To William Henry Fitton Slough
Accepts invitation to WF's party on 19th, but will miss reading of Adam Sedgwick's and R. I. Murchison's paper.

Chemical composition of 'Fish limestone.'
ALS 2pp RS:HS 19.53 (C: RS:HS 21.10)

1742. 1828–5–15 From James Grahame Tours
Of JG's travels and the people he meets [letter completed 1828–5–17 in Nantes].
ALS 3pp RS:HS 8.369

1743. 1828–5–17 To Maria Edgeworth Kensington
Response to a positive letter from ME, which was sent in response to JH's sending her portions of his essay on light.
ALS 4pp Cambridge University Library 7656/TR59

1744. 1828–5–20 From Maria Edgeworth Edgeworthstown
Is grateful for JH's invitation to her brother to visit him. Would like advice on a suitable mathematics tutor for her brother.
ALS 4pp RS:HS 7.25

1745. 1828–5–21 To David Brewster Kensington
Sending DB a draft of a paper; comments on quality of a chromatic lens (that of [Charles] Tulley's telescope using Pierre Guinand's glass).
C 2pp RS:HS 21.11

1746. 1828–5–25 To William Rowan Hamilton Kensington
Thanks for remainder of WH's essay on 'Systems of Rays.' Recalls error-ridden copy of JH's *Light* given to WH; sends correction. Wishes to propose WH for membership in Astronomical Society.
ALS 3pp TCD 1493:20

1747. 1828–5–28 To Caroline Lucretia Herschel Kensington
JH played no role in the awarding of an Astronomical Society medal to CH. Sent to CH a few copies of JH's *Light.* His catalogs of double stars and nebulae are progressing well.
ALS 4pp damaged TxU:H/L–0573.3; Reel 1058 *Memoir ... of Caroline Herschel,* 227 inc

1748. 1828–6–1 From Charles L. Harding Göttingen
On the apparent unevenness in the space between Saturn and its rings.
ALS 2pp (German) RS:HS 9.236

1749. 1828–6–3 From Caroline Lucretia Herschel Hanover
Thanks JH for Gold Medal from the Astronomical Society. Asks if JH knows the Imperial Astronomer [J. J. von] Littrow.
L *Memoir ... of Caroline Herschel*, 228–9

1750. 1828–6–5 From Maria Edgeworth Edgeworthstown
Is grateful for his kindness. Has addressed a letter to her brother c/o JH's house but destroy it if he has returned to Ireland.
ALS 1p RS:HS 7.26

1751. 1828–6–5 From James Grahame Kensington
About JG's just completed travels to Europe.
ALS 3pp RS:HS 8.370

1752. 1828–6–6 To Michael Faraday Kensington
Describes the tests JH made on a small glass prism of not very high quality, which JH has left at the Royal Institution.
ACCS 2pp RS:HS B26.39 James, *Faraday Correspondence, 1*, 455–6

1753. 1828–6–8 From Johann Franz Encke Berlin
Sending the first volume of his Ephemerides for the Astronomical Society, also for Francis Baily and W. H. Smyth. Comments on various aspects of the Ephemerides. Thanks for JH's excellent work on *Light.*
ALS 2pp (French) RS:HS 7.86

1754. 1828–6–10 From Joseph Clement 21 Prospect Place, Southwark
Needs more money to continue work on Charles Babbage's calculating machine.
ALS 1p RS:HS B27.46

1755. 1828–6–13 From [J. Richards] Chapel St.
Spoke yesterday to His Royal Highness [Duke of Sussex], who would be flattered to be elected. Friday will be last chance until November.
ALS 1p RAS:JH Archive 12/1.2.5 verso; Reel 10

1756. 1828–6–14 From William Ritchie Tain
[Edward] Sabine's letter confirmed WR's membership in R.S.L. Feel he was ill-used by Dollar Institution. Apologizes for trouble he caused JH.
ALS 1p RS:HS 14.358

1757. 1828–6–17 From Francis Baily n.p.
Invitation to dine with him. Regarding Richard Taylor the printer.
ALS 2pp RS:HS 3.80

1758. 1828–6–21 To 'Sir' Kensington
Declines offer by Archbishop of Canterbury to join provisional committee to govern newly proposed King's College.
ALS 1p TxU:H/L-0749.1; Reel 1089

1759. 1828–6–23 From Caroline Lucretia Herschel Hanover
Describes a terrible hail-storm that broke all the windows in her house. Will send her Astronomical Society Medal for JH to store alongside his and William Herschel's medals.
L *Memoir ... of Caroline Herschel*, 229–31

1760. 1828–6–26 From Friedrich Wilhelm Bessel Königsberg
Letter of thanks for a copy of JH's thoughts on light.
ALS 1p (German) RS:HS 4.68

1761. 1828–6–29 To Francis Baily Kensington
Charles Babbage has every reason to be grateful for FB's letter to the *Times*. Richard Sheepshanks and the observations. The ring of Saturn. Has been observing lots of Wilhelm Struve's double stars.
ALS 2pp RS:HS 3.81 & 21.12

1762. 1828–6–30 From Jean Baptiste Joseph Fourier Paris
Expressing the thanks of the Académie for JH's article on *Light.*
ALS 1p (French) RS:HS 7.351

1763. 1828–6–30 From E. Willes Goodrest Lodge, near Reading
Invites JH, his mother, and cousin to visit EW anytime.
ALS 3pp † RS:HS 18.254

1764. 1828–7–1 From Elizabeth [Betty] Plumleigh Babbage 5 Devonshire St., Portland Place
Giving news of the intentions of her son (Charles Babbage) and his experiences on his travels.
ALS 3pp RS:HS 2.227

1765. 1828–7–1 To Henry Kater Kensington
Has found some errors in the sheets of computations HK sent. Lists these. Comments on astronomical work of [William] Richardson and on the reading of 'The repeal of the Longitude Acts Bill.'
ALS 3pp New South Wales State Library

1766. [1828–7] To Stephen Lee n.p.
There is no truth in SL's statement to James Ivory regarding work on the figure of the earth belonging exclusively to JH. Has written to Ivory stating the facts of the misunderstanding.
draft 1p RS:HS 11.187 & 21.18

1767. 1828–7–1 From William Henry Smyth Crescent, Bedford
Microscopes installed successfully. Describes observation of second satellite of Jupiter transiting the disk of the planet and how it seemed to linger at the planet's edge.
ALS 2pp RS:HS 16.238

1768. 1828–7–6 To Wilhelm Struve n.p.
Lists contents of a letter sent on 1828–7–6. JH's *Light* and third catalogue of double stars sent. Details of nebulae included.
draft 1p RS:HS 17.115b

1769. 1828–7–6 To Wilhelm Struve n.p.
Sends WS JH's 'Light' published in the *Encyclopaedia Metropolitana.* Mentions recalibration of his magnitude measurements of small stars near other small ones. Comments on results in WS's paper concerning Saturn's ring. Confirms sighting of 'well defined, equable, planetary disk.'
C 4pp RS:HS 21.13

1770. 1828–7–8 To William Henry Fitton n.p.
Discusses atmospheric refraction, Francis Bacon's distinction between argument and testimony, and the manner in which the Board of Longitude was dissolved.
ALS 4pp Gerald M. Friedman

1771. 1828–7–10 From James Ivory 20 Southampton St., Pentonville
Informs JH that JI had been warned off, by Stephen Lee, from writing about the figure of the earth because that was JH's territory.
ALS 3pp RS:HS 10.254a

1772. 1828–7–11 To Francis Baily Slough
Further regarding Richard Taylor the printer. Reports on observations of Mars by William Pearson. JH's new micrometer is aiding his observations of double stars.
ALS 4pp RS:HS 3.82 & 21.14

1773. 1828–7–11 To Francis Beaufort Slough
[Richard] Taylor is very anxious that the Astronomical Society reconsider its decision to discontinue his services as printer of the Society's publications. Suggests presentation of a gift to William Stratford for his services as secretary of the Astronomical Society.
ALS 4pp Huntington Library (C: RS:HS 21.15)

1774. 1828–7–11 To William Henry Smyth Slough
Suggests that phenomenon in observing Jupiter satellite was an optical illusion. Asks WS to observe spots on Mars to confirm observations of William Pearson. Asks WS to contribute to W. S. Stratford memorial.
ALS 2pp APS B.Sm98.4 (C: RS:HS 21.16)

1775. 1828–7–12 To James Ivory n.p.
Assures JI that JH could not, and would not, utter such threats, and has sent Stephen Lee a copy of this letter [see JI's 1828–7–10].
Cdraft 1p RS:HS 21.17

1776. 1828–7–14 From Francis Baily 37 Tavistock Place
Regarding the printing for the Astronomical Society. Errors in FB's astronomical tables.
ALS 4pp RS:HS 3.83

1777. 1828–7–14 From Thomas John Hussey Leckhampstead
Did not pass near JH's place when he came home; otherwise he would have returned JH's mirror. Returns next Monday and will be pleased to call if JH can offer him a bed. The storm did no great damage here, but at Chislehurst it turned a house inside out.
ALS 1p RS:HS 10.139

1778. 1828–7–14 From Stephen Lee British Museum
When he last saw James Ivory at the British Museum, no mention was made of JH's name. Thinks the remarks of Ivory are caused by mental disease.
ALS 3pp RS:HS 11.188

1779. 1828–7–14 From William Henry Smyth Crescent, Bedford
Agrees to contribute to W. S. Stratford memorial presentation. Agrees with JH that Jupiter satellite observation was an optical illusion; sends sketches of more Jupiter observations. Sends data obtained with new instruments.
ALS 2pp RS:HS 16.239

1780. 1828–7–15 From Henry Coddington n.p.
Mr. Knowles has left college and taken the plates of quartz that illustrated the experiment so well. Sends further explanations of the experiment he previously mentioned to JH.
ALS 2pp RS:HS 5.354

1781. 1828–7–16 From J. K. Eduard Schmidt Göttingen
Requests permission to translate JH's work on light into German.
ALS 1p RS:HS 17.176

1782. 1828–7–17 To Henry Kater Slough
JH very upset at the government abolishing the Board of Longitude; has many questions as to what will happen now.
C 3pp RS:HS 21.19

1783. 1828–7–18 From Henry Kater York Gate
Sends Knowles's paper for his comments. The Board of Longitude has now been officially dissolved. Hopes the R.S.L. will not take on its work unless well paid by the government. Answers to JH's queries regarding the Nautical Almanac and other works of the Board of Longitude. Present time not good for state recognition of science.
ALS 4pp RS:HS 11.10

1784. [1828]–7–18 From Dionysius Lardner 33 Percy St., Bedford Square
Is engaged in preparing a new popular cyclopaedia, and would be pleased if JH would contribute an article on Light.
ALS 2pp RS:HS 11.106

1785. 1828–7–18 From Alexander Rogers Bridge St., Leith
Asks JH to report how James South's telescope performs. There was a defect in the settings of the telescope. Hopes JH's method of eliminating chromatic aberration improves the microscope.
ALS 3pp RS:HS 14.405

1786. 1828–7–20 From Francis Beaufort Brighton
Regarding Richard Taylor the printer and the printing of the papers of the R.A.S.
ALS 4pp RS:HS 3.312

1787. 1828–7–21 From Joseph Clement 21 Prospect Place, Southwark
Wants to meet with JH, and needs more money to continue work on Charles Babbage's calculating machine.
ALS 1p RS:HS B27.47

1788. 1828–7–21 To Michael Faraday Slough
Hears that the Board of Longitude is to be dissolved; it will therefore be necessary to seek a new source of income for the continuance of the Optical Glass experiments. Will be guided by MF's wish whether to continue the experiments. Charles Anderson has showed him curious experiments with the fusion of Fluor with clay.
draft 2pp RS:HS 7.176 & 21.20 James, *Faraday Correspondence, 1*, 461

1789. 1828–7–21 To Davies Gilbert Slough
Understands the Board of Longitude has been dissolved. Wonders what the position now is regarding the Optical Glass Committee. Outlines his own views and would welcome DG's instructions.
C 4pp RS:HS 8.113 (C: RS:HS 21.21)

1790. 1828–7–21 From Joseph Goodall Lodge, Eton College
Would like him to come to dinner next Saturday.
ALS 1p RS:HS 8.145

1791. 1828–7–23 From Robert Everest Devereux Court Temple
Has given away all his drawings; all he has left is a small oil-color of Christiana. Comments on the scenery of Norway. Hopes the paintings will stimulate interest in a country of which he is very fond.
ALS 3pp RS:HS 7.136

1792. 1828–7–23 From Giovanni Santini Padua
Hopes to meet with Charles Babbage in Vienna. Inquires if JH received a set of books from University of Padua for the Astronomical Society. Praises JH's double star work; uses it for own observations.
ALS 1p (Italian) RS:HS 15.344

1793. 1828–7–24 To Joseph Clement London
Is concerned about the rapid rate of expenditure of the money on Charles Babbage's calculating machine. Recommends a reasonable rate of expenditure until Babbage returns home, which should be in about four months.
ACCS 2pp RS:HS B27.48

1794. 1828–7–24 From Michael Faraday Royal Institution
The furnace is breaking away at the top, and he encloses Joshua Ramsay's estimate for its repair. Thinks the present a suitable time for the repairs. Would JH speak to the Council?
ALS 2pp RS:HS 7.177 James, *Faraday Correspondence, 1*, 458 & Williams, *Selected Correspondence, 1*, 171

1795. 1828–7–24 To Joseph Johann Littrow Slough
Has been overwhelmed with work, which accounts for the delay in writing. Thanks for communications. Both are now printed. Comments on some of the points raised.
C 3pp RS:HS 11.254 & 21.22

1796. 1828–7–25 From Davies Gilbert Eastbourne
Has just returned from an exhausting meeting. Has no doubt funds will still be available for the Optical Glass experiments. Has written to Michael Faraday.
ALS 2pp RS:HS 8.114

1797. 1828–7–25 To Dionysius Lardner Isle of Wight
Has been contemplating a larger work on Light, but thinks he may be able to write a shorter, more popular work for the *Cabinet Cyclopaedia.* Wonders who the other contributors will be.
ALS 3pp APS B.H435p.3 (C: RS:HS 11.107 & 21.23)

1798. 1828–7–26 From Francis Baily n.p.
Thanks for the two communications. J. F. Encke's work on the Ephemeris. Francis Beaufort's remarks concerning Richard Taylor the printer. Hoping for fine weather for observations.
ALS 1p RS:HS 3.84

1799. [1828]–7–28 From Dionysius Lardner 33 Percy St., Bedford Square
Sees no objection to JH publishing a larger work on Light later if desired. Gives the names of some of the contributors to the *Cyclopaedia.*
ALS 4pp RS:HS 11.108

1800. 1828–7–28 From Giovanni Antonio Plana Valley of Lanzo
Is sending the letter with the British ambassador [Augustus Foster] to the Court of Turin; Foster is much interested in science and in meeting JH. GP will be sending additional work that GP has done, and indicates further plans.
ALS 3pp† (French) RS:HS 13.426

1801. 1828–7–29 From Davies Gilbert Eastbourne
Has had a note from J. W. Croker stating his willingness to help. Would like a sketch and details from Michael Faraday of the experiments now going on.
ALS 2pp RS:HS 8.115

1802. 1828–7–29 From Nathaniel Wallich 6 Princes St., Hanover Square
Encloses letter from Edward Ryan. NW departed Bengal on 10 Mar. and brought home [8,000+ plant specimens] for H.E.I.C.
AL 2pp RS:HS 18.28

1803. 1828–7–31 From James Grahame Kensington
JG is worried about the health of his daughter.
ALS 2pp RS:HS 8.371

1804. 1828–8–1 To Henry Kater Southampton
Comments on a paper about differential calculus, prepared by Mr. Knight, which JH confesses he does not

understand. Adds some astronomical observations JH made at Slough before leaving there, and noting the appearance of J. F. Encke's *Berlin Ephemeris* concludes with 'England is fated to lag in the rear of all her neighbours—at least in science.'
ALS 4pp DMC.695A.885.6 (C: RS:HS 21.25)

1805. 1828–8–1 To Dionysius Lardner Slough
Is quite prepared to contribute an article on Light by 1830 for DL's *Cyclopaedia.* Would like William Whewell as the contributor on Sound. Suggests other names. Would not like to use colored plates as they are so poor. Gives him an idea of the style of engraving.
C 2pp RS:HS 11.109 & 21.24

1806. 1828–8–2 To Thomas John Hussey Portsmouth
Gives references to good sources on astronomical projections. Discusses this subject and also logarithmic tables.
ALS 3pp Univ.Coll.London MS Graves 23(1)

1807. 1828–8–4 From Jean Baptiste Joseph Fourier Paris
Expressing thanks for his third paper on Observations made with a 20-foot telescope.
ALS 1p (French) RS:HS 7.352

1808. 1828–8–5 From William Henry Fitton 53 Upper Harley St.
Regarding a suitable book on the Isle of Wight. Regarding Dionysius Lardner and his projected *Cabinet Cyclopaedia.* Urges JH not to reject idea until he has made further inquiries. Affairs of the Board of Longitude.
ALS 4pp RS:HS 7.228

1809. 1828–8–5 From William Henry Smyth Crescent, Bedford
Mr. Stocken[?] has discovered a comet; have been unable to observe it due to poor weather.
ALS 2pp RS:HS 16.240

1810. 1828–8–12 From Elizabeth [Betty] Plumleigh Babbage 5 Devonshire St., Portland Place
Regarding an anonymous letter concerning her son (Charles Babbage), which has appeared in *The Record.*
ALS 2pp RS:HS 2.228

1811. 1828–8–14 To Elizabeth [Betty] Plumleigh Babbage Slough
Thanking her for the information on the anonymous letter. Would like a printed copy. Will write an answering letter to the *Times.*
ALS 3pp RS:HS 2.229

1812. 1828–8–14 From William Henry Fitton 53 Upper Harley St.
Regarding the projected *Cabinet Cyclopaedia* and its authors. State of Ireland in 1828. Feels hard against the Saxons.
ALS 4pp RS:HS 7.229

1813. 1828–8–15 From Elizabeth [Betty] Plumleigh Babbage 5 Devonshire St., Portland Place
Unable to send the cutting from *The Record* yet. Will try to obtain it from her daughter. Views of friends on the attack.
ALS 2pp RS:HS 2.230

1814. 1828–8–15 To the London *Times* Slough
In response to an article in *The Record,* JH writes to *The Times* to explain what has happened to date regarding the expenditure of public funds for the calculating devices of Charles Babbage, who was then out of England.
ACCS 3pp TxU:H/L-0229; Reel 1054 (L: RS:HS B27.54b & Cdraft: RS:HS 25.15.7) London *Times* (19 August 1828), p. 3, col. 2

1815. 1828–8–18 From Elizabeth [Betty] Plumleigh Babbage 5 Devonshire St., Portland Place
Is sending the cutting.
ALS 1p RS:HS 2.231

1816. 1828–8–19 From Elizabeth [Betty] Plumleigh Babbage 5 Devonshire St., Portland Place
Thanking him for refuting the attack on Charles Babbage.
ALS 1p RS:HS 2.232

1817. 1828–8–21 From Caroline Lucretia Herschel Hanover
Commenting on the praise that she received upon the awarding of her Astronomical Society Medal, she notes that 'whoever says too much of me says too little of your father!' Soon will send to JH 'every scrap of paper' that has assisted CH while writing her memoirs.
L *Memoir... of Caroline Herschel,* 231–3

1818. 1828–8–22 From Francis Baily London
Further invitation to visit him at Tytherley. Will be using a new telescope. William Wallace's paper for the *Monthly Notices.*
ALS 2pp RS:HS 3.85

1819. 1828–8–22 From James Grahame Kensington
Going to live at Nantes for awhile; asks JH's help with Customs.
ALS 2pp RS:HS 8.372

1820. 1828–8–24 From Thomas Young Park Square
Sudden demolition of Board of Longitude. TY is left as superintendent [of *Nautical Almanac*]. Notes quality of J. F. Encke's almanac for 1830. Should English compete with German astronomers, or discontinue almanac supplements? TY's paper on 'Astronomical Chronology' in W. T. Brande's [*Quarterly Journal of Science and Art*] illustrates JH's equatorial time.
ALS 2pp † RS:HS 18.344

1821. 1828–8–28 To Dionysius Lardner Slough
Is now prepared to write a signed article on light requested by DL [see JH's 1828–7–25]; JH encourages DL to ask William Whewell, W. H. Fitton, and George Peacock to write as well; recommends against use of color printing of pictures as the reproduction is too poor.
ALS 3pp APS B.H435p.4

1822. 1828–8–30 To Thomas Young Slough
Encourages TY to publish 1829 supplement to *Nautical Almanac*. JH originally proposed supplement to introduce improvements that would bring *N. A.* to be like *Encke's Ephemeris.* That is all ended now. English are dropping behind French and Germans in science. JH refuses to accept money from government 'which treats its agents so cavalierly.' Reduction of Parramatta [Australia] observations. Did not see TY's paper in W. T. Brande's [*Quarterly Journal of Science and Art*], but objects to substitution of true for mean equinox.
AdraftS 4pp RS:HS 18.345 (C: 21.26)

1823. [1828–8–31] From Thomas Young Park Square
Will not hold JH responsible for accuracy of William Richardson's computations. No instructions yet from Admiralty about *Nautical Almanac*, which goes to press

soon. TY will insert JH's equinoctial time. Notes two works by A. F. C. Kollman on harmony and [Beck] on 'thorough-bass.'
ALS 1p † RS:HS 18.346

1824. 1828–9–1 To Davies Gilbert Slough
Comments on two business matters in the R.S.L., namely, the work of the Glass Committee and the question of calculation of observations made at Parramatta, Australia.
C 5pp RS:HS 21.27

1825. 1828–9–1 From Thomas Young Park Square
Feels responsible for JH refusing to accept Navy's £25 reimbursement. Used JH's [equinoctial] time [in *Nautical Almanac* supplement], but with different notation. Disagrees that England's scientific reputation depends upon its government. William Richardson is justified in continuing work of late Board of Longitude. Reports meeting in Paris with Jacques Babinet and François Arago.
ALS 2pp RS:HS 18.347

1826. 1828–9–2 To Davies Gilbert Slough
Giving his own and Michael Faraday's views concerning the continuation of the Glass experiments. Regarding the reduction of the Parramatta Observations. Does not wish to undertake it himself.
draft 4pp RS:HS 8.116

1827. 1828–9–2 From Joseph Johann Littrow Vienna
Thanks for comments on JL's papers [see JH's 1828–7–24]; Charles Babbage had visited JL.
ALS 2pp (German) RS:HS 11.255

1828. 1828–9–2 From William Henry Smyth Crescent, Bedford
Sends information on occultation of Alpha Tauri. Describes meteors sighted during the night of 12 Aug.
ALS 2pp RS:HS 16.241

1829. 1828–9–9 From Sir Charles Lyell Milan
Arrived here yesterday with the Murchisons and starts tomorrow for the Vicentini. Would welcome JH's advice on what to see and read when in Sicily.
ALS 4pp RS:HS 11.417 Katherine M. Lyell, *Lyell, 1,* 200–2

1830. 1828–9–9 From Nathaniel Wallich Tavistock Hotel, Covent Garden
Found house in Turnham Green Terrace. Specimen collection still in baggage warehouse.
AL 2pp RS:HS 18.29

1831. 1828–9–10 From Elizabeth [Betty] Plumleigh Babbage 5 Devonshire St., Portland Place
Is taking her grandchild to Southend. Charles Babbage is at Berlin.
ALS 1p RS:HS 2.233

1832. 1828–9–17 To Davies Gilbert Slough
Last night he discovered Encke's Comet. Gives readings.
C 1p RS:HS 8.117 & 21.28

1833. 1828–9–18 From Francis Baily East Titherley House
Instructions on how to reach Tytherley. Has not seen Encke's Comet yet.
ALS 3pp RS:HS 3.86

1834. 1828–9–18 From James South Observatory, Kensington
Thanks JH for information on JH's detection of Encke's Comet. JS's contacts with various astronomers, including Peter Barlow, John Pond, and François Arago.
ALS 4pp RS:HS 16.459

1835. 1828–9–19 To Davies Gilbert Slough
Hastens to correct his own error; the comet he observed is no comet, but one of his father's nebulae. Regrets the errors.
C 1p RS:HS 8.118 & 21.29

1836. 1828–9–19 To William Henry Smyth Slough
Informs WS that JH located Encke's Comet the previous night, and gives the location.
ALS 1p WT B.1

1837. 1828–9–20 From William Whewell Trinity College
Will sponsor JH for the Cambridge Philosophical Society. Discusses WW's published report of WW's Dolcoath mine experiments. Comments on mineralogical classification. Promises to send books to JH on sound and music.
ALS 5pp RS:HS 18.174

1838. 1828–9–23 From [George Augustus Friederick] Fitzclarence 3 Cloisters
GF and family plan to visit JH at Slough on 29 Sept. for 'observation.'
ALS 2pp RAS:WH Archive 1/13.F.6; Reel 13

1839. 1828–9–[26 or earlier] From Charles Wesley No. 2 Paul St., Kingsdown, Bristol
Describes sister's ill health. Mentions meeting mutual friends. Inquires about welfare of JH's acquaintances.
ALS 1p † RS:HS 18.142a

1840. 1828–9–26 To Charles Wesley Slough
Heard from [James] Grahame of CW's sister's death. Invites CW to seclusion at Slough and to occupy himself arranging William Herschel's musical compositions. Tells of JH's narrow escape today in accident aboard Southampton Coach.
ACCS 2pp RS:HS 18.142b

1841. 1828–9–29 From Caroline Lucretia Herschel Hanover
Sends packet of CH's personal papers to JH. Explains why CH destroyed her diary, which reminded her of painful last years in England before William Herschel's death.
ACC 1p TxU:H/M–0678 (p. 11); Reel 1084

1842. 1828–9–29 From Charles Wesley Bristol
Thanks for invitation to Slough, but CW will remain home with relatives and friends.
ALS 1p RS:HS 18.143

1843. 1828–9–30 To Charles Babbage Slough
Not much news. Regarding Joseph Clement and his work on the machine. William Whewell and G. B. Airy's pendulum experiments. Has had a letter from J. R. Ryan.
ALS 4pp RS:HS 2.234

1844. 1828–10–3 From James Grahame Nantes
About JG's journey to Nantes.
ALS 4pp RS:HS 8.373

1845. 1828–10–4 From William Reid Chatham
Sends letter from [George] Gipps. WR wants to assists Gipps in astronomical pursuits. Asks JH for assistance.
ALS 2pp † RS:HS 14.321

1846. 1828–10–6 From Elizabeth [Betty] Plumleigh Babbage 5 Devonshire St., Portland Place
Has returned from Southend. Has letter from Charles Babbage in which he relates events and people met at the Berlin Academy.
ALS 3pp RS:HS 2.235

1847. 1828–10–8 From Alexander von Humboldt to JH [?] Berlin
Requesting the opinion of JH's late father on the nebulae in Orion.
ALS 1p (French) RS:HS 10.63

1848. 1828–10–14 From William Whewell Trinity College
Asks about JH and WW collaborating on a volume for Dionysius Lardner's *Cabinet Cyclopaedia.* Discusses aspects of mineralogical classification and crystal structure.
ALS 4pp RS:HS 18.175 Todhunter, *Whewell, 2,* 95 inc

1849. 1828–10–15 From James Grahame Nantes
Asks JH to communicate to some friends for him.
ALS 1p RS:HS 8.374

1850. 1828–10–21 From Francis Baily London
His own paper on moon culminating stars for 1829 is now printed. Time Encke's Comet was visible. Foreign encouragement for science. Remarks on pendulum experiments.
ALS 3pp RS:HS 3.87

1851. 1828–10–23 To Adolphe Quetelet Kensington
A letter of introduction to AQ on behalf of John Franklin [famous explorer of the Northwest Passage].
ALS 2pp DMC.695A.885.10

1852. 1828–10–26 To Michael Faraday n.p.
Sends MF some older samples of imperfect glass. Talks about the way in which different kinds of glass should be layered in lenses.
ACCS 2pp diag RS:HS B26.41

1853. 1828–10–27 To James Grahame Slough
Informs JG about mutual friends and JG's daughter.
ALS 4pp RS:HS 8.375

1854. 1828–10–29 From George Biddell Airy Observatory, Cambridge
Asking JH's opinion on how he should print his catalogue of observations of transits. Requests information on any forthcoming comets.
ALS 3pp RS:HS 1.37

1855. 1828–10–31 From Carl Friedrich Gauss Göttingen
Many thanks for his communications, especially for the one on light. This may be translated into German by J. E. E. Schmidt, and he would be grateful if JH would authorize this to be done.
ALS 2pp RS:HS 8.74 (C: RS:HS 21.30)

1856. 1828–11 From James Ivory 20 Southampton St., Pentonville
Would like to know the position regarding the Royal medal, which he understands he has been awarded.
ALS 2pp RS:HS 10.255a

1857. 1828–11 or later To James Ivory n.p.
In reply to JI's 1828–11 states that the delay over sending the medal has arisen because there is no die in existence so that no medal has been struck.
Cdraft 1p RS:HS 10.255b

1858. 1828–11–6 From James Grahame Nantes
Of life in Nantes and mutual friends in England [letter includes a sketch of General Cambourne by JG's son Robert].
ALS 4pp RS:HS 8.376

1859. 1828–11–7 To Davies Gilbert 26 Lower Phillimore Place
Should JH prepare a report for the R.S.L. Council on the work of the Glass Committee, including Michael Faraday's important results? W. H. Wollaston's health is failing. On recent sunspot and comet observations.
C 2pp RS:HS 21.31

1860. 1828–11–11 From James Grahame Nantes
Advice to JH, who is considering marriage.
ALS 2pp RS:HS 8.377

1861. 1828–11–15 To George Biddell Airy 26 Lower Phillimore Place
States JH's position on the question of publishing all the observations of an observatory, or working out results and publishing only those; notes clearness of Encke's Comet; comments on the return of Charles Babbage from abroad.
ALS 4pp RGO 6.803.58

1862. 1828–11–17 From Thomas Leybourn Royal Military College, Bagshot
Is just about to publish the 20th number of the *Mathematical Repository* in which he would like to include a short biography of William Herschel. Can JH supply him with an account of his father, and also inform him if a portrait of Sir William has been published?
ALS 1p RS:HS 11.481

1863. 1828–11–19 From Elizabeth [Betty] Plumleigh Babbage 5 Devonshire St., Portland Place
Is expecting her son Charles Babbage home and would be pleased if JH would be there to meet him.
ALS 3pp RS:HS 2.236

1864. 1828–11–19 From Francis Beaufort Manchester St.
Regarding the climate in Northern Australia.
ALS 2pp RS:HS 3.313

1865. 1828–11–19 To James Grahame 12 Charlotte St., Portland Place
JH describes his meeting with Margaret Stewart's mother to acquaint her with JH's intentions towards her daughter.
ALS 4pp RS:HS 8.378

1866. 1828–11–19 To John Thomas Graves 12 Charlotte St., Portland Place
Has found JG's paper [on logarithms] and will give it to the R.S.L. President and Council.
ALS 1p TCD 10047/30

1867. 1828–11–20 To Emilia Stewart [London]
Is writing to ES, JH's future mother-in-law, to request an opportunity for a private conversation with ES, on a matter 'that concerns me deeply.'
ALS 1p JHS 2.15

1868. [1828]–11–20 From Nathaniel Wallich 8 Turnham Green Terrace
Invites JH to join NW and Robert Brown for midday meal.
ALS 3pp RS:HS 18.33

1869. 1828–11–21 To Johann Franz Encke London
Informs JE that the R.S.L. is awarding JE a Royal Medal for JE's work on Encke's Comet.
ALS 3pp AdWDDR II–VII, 5

1870. 1828–11–21 From Emilia Stewart [London]
Has kept her promise to JH not to speak to the subject of their conversation [see JH's 1828–11–20]; she is pleased with JH's honorable behavior.
ALS 3pp JHS 2.13

1871. 1828–11–21 To Emilia Stewart 12 Charlotte St.
Accepts invitation for tea at the Stewart's house; has been trying to resolve the problem [which appears to have been a social *faux pas* committed by someone against a lady].
ALS 3pp JHS 2.16

1872. [1828]–11–22 From Nathaniel Wallich Turnham Green [Terrace]
Regretted JH's absence. Hopes Mary Herschel recovers. Robert Brown attended. NW's wife also ill.
ALS 4pp RS:HS 18.34

1873. 1828–11–24 From Emilia Stewart 17 Nottingham Place
A note accompanying a volume [unidentified], which ES feels JH must read.
ALS 1p JHS 2.14

1874. 1828–11–24 To Emilia Stewart 12 Charlotte St.
In view of the Stewart family concerns about the illness of Isabella Stewart, JH will reduce the frequency of his visits, although he would like to come oftener. Thanks ES for the book she sent [see ES's 1828–11–24].
ALS 3pp JHS 2.19

1875. 1828–11–25 From Michael Faraday Royal Institution
Has made a piece of glass 5" square, which George Dollond reports is good. Proposes making a furnace at the Royal Institution. Does JH have any objections?
ALS 2pp RS:HS 7.178 James, *Faraday Correspondence, 1*, 470 & Williams, *Selected Correspondence, 1*, 173

1876. 1828–11–25 From James Grahame Nantes
Sends happiest wishes to JH.
ALS 2pp RS:HS 8.379

1877. 1828–11–25 To Emilia Stewart 12 Charlotte St.
A note accompanying some panoramic views of the Alpine scenery in Piedmont, sent to JH by G. A. Plana. JH sends them along in the hope that they may provide some amusement for the invalid Isabella Stewart.
ALS 2pp JHS 2.20

1878. 1828–11–26 From William Henry Smyth Crescent, Bedford
John Lee has obtained transit circle for Astronomical Society. WS suggests someone other than Thomas Maclear operate it so Maclear has time for occultation project. Details observations of aurora.
ALS 3pp RS:HS 16.242

1879. 1828–11–27 From George Biddell Airy Observatory, Cambridge
Asking JH to get in touch with Charles Babbage. Thanking him for his observations on his catalogue of stars and transits.
ALS 3pp RS:HS 1.38

1880. 1828–11–27 To Emilia Stewart [12 Charlotte St.]
A note to accompany some drawings and prints for the amusement of the invalid, Isabella Stewart; was happy about reports of Isabella's improvement.
ALS 3pp JHS 2.21

1881. 1828–11–27 From Emilia Stewart 17 Nottingham Place
Writes to report continuing improvement in Isabella's health, and to thank JH for the prints and drawings [see JH's 1828–11–27].
ALS 4pp JHS 2.22

1882. 1828–11–28 To William Henry Smyth [London]
John Lee donation pleases JH; concerned over who to appoint to take charge of the instrument. Asks WS to write account of aurora sighting for a scientific journal.
C 1p RS:HS 21.32

1883. 1828–11–28 To Emilia Stewart [12 Charlotte St.]
Writes to offer to convey a friend of the Stewart family to their house, offers to join with Margaret Stewart to make music for Isabella, and sends some more prints and an illustrated Milton for Isabella's amusement.
ALS 4pp JHS 2.23

1884. 1828–11–29 From James Grahame to Peter Stewart Nantes
Commends JH to PS as an excellent man and ideal husband for PS's sister Margaret.
ALS 3pp JHS 2.26

1885. 1828–11–29 From Alexander Rogers Bridge St., Leith
Has received a letter from the Astronomical Society secretary. Worries AR may have offended JH and requests guidance. Describes performance of his telescope.
ALS 3pp RS:HS 14.406

1886. 1828–12–2 From Stephen Groombridge Blackheath
Regarding the testimonial for W. S. Stratford, would like to be included among the subscribers. Hopes Stratford has found beneficial employment.
ALS 1p RS:HS 9.144

1887. [1828]–12–3 From Thomas Young Park Square
Problems calculating sun's mean longitude due to movement of perigee. Questions S. D. Poisson's value for sidereal year.
ALS 2pp RS:HS 18.350

1888. 1828–12–4 To Emilia Stewart [12 Charlotte St.]
Writes to initiate JH's suit for the hand of Margaret Stewart, and place his cause before ES. Enclosed is a letter to Margaret [see JH's 1828–12–4 to Margaret Stewart] and also a letter from James Grahame to ES.
ALS 4pp encl JHS 2.24

1889. 1828–12–4 To Margaret Brodie Stewart Slough
JH declares his undying love for MS. [Letter very correctly sent through MS's mother; see JH's 1828–12–4 to Emilia Stewart].
ALS 3pp JHS 2.25

1890. 1828–12–5 To William Rowan Hamilton London
Acknowledges receipt of papers sent by WH for Astronomical Society and R.S.L. and will deliver them, but currently has not time to say more. Looks forward to enrollment of WH's name among members at next Astronomical Society meeting.
ALS 2pp TCD 1493:30

1891. 1828–12–5 To Emilia Stewart [12 Charlotte St.]
A note of thanks for the ambassadorial services performed by ES. Enclosed is a letter to Margaret Stewart [see JH's 1828–12–5].
ALS 1p JHS 2.28

1892. 1828–12–5 From Emilia Stewart [17 Nottingham Place]
Responds to JH's letters of 1828–12–4, and invites JH to visit them on 6 Dec.
ALS 3pp JHS 2.27

1893. 1828–12–5 From Margaret Brodie Stewart [17 Nottingham Place]
A reply to JH's declaration of love [see JH's 1828–12–4 to MS; seems intended to encourage JH].
C 2pp JHS 2.27a

1894. 1828–12–5 To Margaret Brodie Stewart [12 Charlotte St.]
A letter of thanks for MS's hopeful letter [see MS's 1828–12–5] in response to JH's declaration of love.
ALS 3pp JHS 2.29

1895. 1828–12–5 or later To Margaret Brodie Stewart London
Informs MS that JH will probably be arriving late as the sudden death of a friend's sister may delay him; includes many expressions of JH's affection for MS.
ALS 1p JHS 2.17

1896. 1828–12–6 From David Brewster Allerly by Melrose
Thanks for paper on light, which he has only just read as it was mislaid. Grateful for the acknowledgement of his scientific labors. Comments and queries regarding the paper in the light of his own experiments. Wishes to reprint parts of it so would be grateful for several plates.
ALS 7pp RS:HS 4.261

1897. 1828–12–7 To George Biddell Airy 26 Lower Phillimore Place
Has prepared a paper on the doctrine of sound; expects to be up to spend several days with GA at the Cambridge Observatory.
ALS 2pp RGO 6.803.73

1898. 1828–12–7 From Mary Pitt Herschel Slough
Wishes JH well on his visit to Stewart family and to Cambridge. Personal finances. Arrangements for JH's new servant.
ALS 2pp & C 2pp TxU:H/M–0624; Reel 1086

1899. 1828–12–7 To Mary Pitt Herschel n.p.
JH's love for Margaret Stewart. Financial plans for future. Isabella Stewart is still ill. Burn this letter. P.S.: Do not burn it. JH may want to read it in later years. Mrs. Stewart gave JH her husband's journal recording Margaret's sixth year. Regards from James Grahame to MH.
ALS 2pp & C TxU:H/L–0520.1; Reel 1053

1900. 1828–12–8 From Nathaniel Wallich 8 Turnham Green Terrace
Encloses letter from Edward Ryan to Charles Babbage that JH agreed to forward.
ALS 3pp RS:HS 18.30

1901. 1828–12–8 From George Hyde Wollaston Dorset St.
Seeking JH's address. Encloses dispatch. W. H. Wollaston getting weaker.
AL 1p RS:HS 18.273

1902. 1828–12–9 To Caroline Lucretia Herschel London
Acknowledges receipt of CH's diary. JH apologizes for mishandling one aspect of the awarding of the Astronomical Society's Gold Medal to CH.
ALS 3pp TxU:H/L–0573.4; Reel 1058 *Memoir ... of Caroline Herschel,* 233 inc

1903. 1828–12–10 From William Samuel Stratford 3 Verulam Buildings
Thanks JH for award and dinner invitation given to him by Astronomical Society.
ALS 1p RS:HS 17.82

1904. 1828–12–12 From George Hyde Wollaston Dorset St.
Asks JH to insert full name of Peter Dollond into P & J Dollond firm, who constructed W. H. Wollaston's telescope in 1771. W. Wollaston is feeling better.
AL 1p RS:HS 18.274

1905. 1828–12–13 From James Grahame Nantes
Of mutual friends in England and life in Nantes.
ALS 4pp RS:HS 8.380

1906. 1828–12–15 To Mary Pitt Herschel n.p.
Enamored of Margaret Stewart. Orders MH to secrecy about JH's engagement. Praises fiancée's musical talent.
ALS 2pp & C TxU:H/L–0520.2; Reel 1053

1907. 1828–12–17 From Stephen Groombridge Blackheath
Regrets his health prevents him from attending the dinner in honor of W. S. Stratford. Will discharge his debt when called upon.
ALS 1p RS:HS 9.145

1908. 1828–12–19 From Peter Mark Roget Bernard St.
Discovered a mistake in Mr. Perkins's paper published in the *Annals of Philosophy*. Perkins has not yet corrected the error, although he was requested to do so.
ALS 2pp RS:HS 14.500

1909. 1828–12–19 From George Hyde Wollaston Dorset St.
Will convey JH's note to W. H. Wollaston, who is sinking gradually.
ALS 2pp RS:HS 18.275

1910. 1828–12–24 To Margaret Brodie Stewart Slough
A love letter on Christmas Eve from JH, at home with his mother, to his 'attached,' MS; comments about their mutual friend, James Grahame.
ALS 4pp JHS 2.18

1911. 1828–12–29 To Eilhard Mitscherlich n.p.
EM has just missed being awarded a Royal Medal; it was

given to W. H. Wollaston. Thinks it advisable for EM to draw up a brief resume of his own scientific discoveries.
C 1p inc RS:HS 13.43 & 21.33

1912. [1828]–12–30 From Dionysius Lardner 33 Percy St., Bedford Square
Encloses a new prospectus of the *Cabinet Cyclopaedia.* Intends to write physical astronomy himself. William Whewell cannot write Sound. Has a stock in hand for three years.
ALS 2pp RS:HS 11.110

1913. 1828–12–31 From James Grahame Nantes
Seasons greetings and thoughts on JH's upcoming marriage [letter completed 1829–1–4].
ALS 4pp RS:HS 8.382

1914. [1829 to 1832] To [Mary Baldwin] n.p.
JH's health good. Expects fatiguing day tomorrow. Will meet MB for dinner. Give JH's love to Margaret. William brought good news of JH's mother.
AL 2pp TxU:H/L-0059; Reel 1054

1915. [1829 or 1830] From Francis Legatt Chantrey n.p.
Suggesting suitable painters of miniatures.
ALS 2pp RS:HS 5.247

1916. [1829] From [Davies Gilbert ?] 45 Bridge St., Westminster
Seems to be about an arrangement to travel together.
ALS 1p (almost illegible) DMC.695A.885.33

1917. [1829 to 1848] From Caroline Lucretia Herschel [Hanover]
Encloses notes from years of breakfast conversations with CH's brother William Herschel. Asks JH 'to take the burden of the [Zemata?] M.S. Secretaryship of [*sic*] my shoulders.' Recalls that JH did the same for 'Mr. Pickering some years ago.' Wishes JH and Margaret a happy new year. CH's notebook [79pp in English] including: necrology of astronomers and mathematicians; notes on plane and spherical geometry; star observations; various tables regarding measurement conversion, planetary systems, geography, and algebra.
ALS 2pp inc, encl Amem 79pp TxU:H/M–0672; Reel 1084

1918. [1829?] From Thomas John Hussey Chislehurst
Wants to know when he can visit JH.
ALS 1p Harvard: Houghton AAH 67m–67 (107)

1919. [1829?] To Francis Lunn n.p.
Hears that the editors of *Encyclopaedia Metropolitana* have taken exception to JH's plan to write a work on Light for the *Cabinet Cyclopaedia.* Gives his views why the works should not conflict.
draft 3pp RS:HS 11.414 & 21.37

1920. [1829?] From William Prout 40 Sackville St.
Sends abstract of a paper; wants JH's opinion.
ALS 1p RS:HS 14.192

1921. 1829 From Adolphe Quetelet Brussels
Has just received invitation. Regrets delaying JH's dinner. AQ is in the countryside. Asks to be allowed to visit toward the end of the dinner.
ALS 1p (French) RS:HS 14.201

1922. 1829–1–1 From Alexander Rogers Bridge St., Leith
Proposes to send telescope to JH to have a new lens made. Desires JH's opinion of the telescope's utility. Is hesitant to join Astronomical Society.
ALS 3pp RS:HS 14.407

1923. 1829–1–10 From James Grahame Nantes
Response to JH's wedding being set for March, and some news from Nantes.
ALS 3pp RS:HS 8.381

1924. 1829–1–13 To John Thomas Graves Slough
JG's paper was read to the R.S.L. on 13 Dec. [1828]. An assessment of it is now being made concerning its publishability, the report to be given to the R.S.L. Council.
ALS 1p TCD 10047/30

1925. 1829–1–13 To William Rowan Hamilton Slough
Was pleased to recently admit Capt. [Francis] Beaufort as WH's proxy in Astronomical Society.
ALS 1p TCD 1493:34

1926. 1829–1–14 To Caroline Lucretia Herschel 26 Lower Phillimore Place, Kensington
JH asks for a large oil portrait of CH, the size of his father's.
ALS 3pp TxU:H/L–0574.1; Reel 1058 *Memoir ... of Caroline Herschel,* 234

1927. 1829–1–17 To Thomas Henderson London
Rejoices to hear TH is a candidate for the Professorship of Astronomy at Edinburgh University. His astronomical work has been of great assistance to JH, especially his detection of the reasons for the error in the Greenwich Observations.
C 2pp RS:HS 9.283 & 21.34

1928. 1829–1–21 To William Samuel Stratford n.p.
Wishing to maintain good opinion of public, JH cautiously recommends that a certain passage be omitted from a monthly notice in the *Nautical Almanac.*
ACCS 2pp RS:HS 17.83 (C: RS:HS 21.35)

1929. 1829–1–23 From Charles Babbage Trinity Lodge
Will endeavor to obtain information about the fellowships at St. John's.
ALS 1p RS:HS 2.237

1930. [1829–1–13 or earlier] To [John Thomas Graves] n.p.
Remains unconvinced by JG's explanatory note on imaginary logarithms, but will pass paper on to 'more capable hands' in the R.S.L.
ALS 2pp TCD 4015:114 (C: RS:HS B26.64)

1931. 1829–1–24 To John Thomas Graves 12 Charlotte St., Portland Place
The Council of the R.S.L. has asked JH to be the referee of JG's paper on logarithms, and JH disagrees with JG in a fundamental way. JH therefore asks JG to withdraw the paper and publish it elsewhere.
ALS 4pp RS:HS B26.65 & 66 (C: TCD 4015:112))

1932. 1829–1–25 To William Buckland London
Regrets will be unable to join WB's party at Oxford.
ALS 1p EUL:LaII643/25

1933. 1829–1–26 To Charles Babbage n.p.
Can CB come down on Sunday to see about the machine?
ALS 1p RS:HS 2.238

1934. 1829–1–26 From John Thomas Graves Keppel St.
Is grateful for his comments. Will withdraw his paper as requested; at the same time he points out various aspects of how he arrived at the results.
ALS 3pp RS:HS 9.119

1935. 1829–1–31 To William Henry Fox Talbot 12 Charlotte St., Portland Place
Asks about a servant JH is considering hiring to accompany JH on JH's trip to the Continent.
ALS 3pp Science Museum Talbt 1/6

1936. [1829–2] To William Whewell n.p.
Invites WW to JH's wedding. Asks him to come early to meet the bride's family.
ALS 2pp TC, Camb. Add. Ms.a.207[16] (draft: RS:HS 18.176 & C: 21.37)

1937. [1829?–2?]. To [Thomas] Y[oung] n.p.
Appears to have refused an offer of a payment or position.
ALS 2pp APS B.H435p.29

1938. 1829–2–2 From James Grahame Nantes
Inconsequential chatter [letter completed 1829–2–7].
ALS 3pp RS:HS 8.383

1939. 1829–2–3 From William Wallace Edinburgh
Thomas Henderson will visit London and inform JH of events at Edinburgh Observatory. Will WW's paper on instrument for multiplication interfere with Charles Babbage's proposed machine?
ALS 3pp RS:HS 18.26

1940. 1829–2–4 From Giovanni Antonio Plana Turin
Is concerned to discover if the various packages of memoirs that GP sent have been received by JH. Talks about coming to London to present several papers.
ALS 2pp† (French) RS:HS 13.427

1941. 1829–2–5 To Caroline Lucretia Herschel [London]
A note of praise for Margaret Stewart [JH's fiancée] as a covering note for an enclosed 6pp letter from JH's cousin, Mary Baldwin, to CH, describing Margaret and her family.
ALS 1p encl JHS.HCEJ.502

1942. [1829]–2–6 To Isaac Lyon Goldsmid 12 Charlotte St., Portland Place
Pays bill for a palladium chain. May wish to purchase another object [another palladium chain?].
ALS 1p Univ.Coll.London MS MOCATTA

1943. 1829–2–6 From Henry Warburton [45 Cadogan Place]
Apologizes for delay in writing out what JH requested. He will have it tomorrow.
ALS 1p RS:HS 18.38

1944. 1829–2–7 From Joseph Bianchi Modena
Will be receiving a clock for Robert Molyneux. Comments on this and gives details of his latest observations.
ALS 3pp (French) RS:HS 5.60

1945. 1829–2–7 From Basil Hall Edinburgh
Has requested his friend Thomas Henderson of Edinburgh University to call on JH. Would be pleased if JH can help him in any way.
ALS 1p RS:HS 9.164

1946. 1829–2–11 From James Wood St. John's College
JW has been at St. John's for fifty years. Wishes JH well in preparing for marriage. JH is entitled to quarters dividend from [feast of] St. Thomas 'to Lady day next.' Notify JW when wedding occurs. JH's fellowship will then be vacant. Please bring bride to visit Cambridge.
ALS 2pp RS:HS 18.295

1947. 1829–2–12 From E. Willes Goodrest Lodge, near Reading
Compares hotels in Leamington. Recommends Regent for JH's friends. EW will soon visit JH and invalid Lady Mary Herschel.
ALS 3pp RS:HS 18.255

1948. 1829–2–13 From Alexander Rogers Bridge St., Leith
Thinks his telescope was not well executed. Would be pleased if James South tried another. Advises how to improve performance of telescope.
ALS 3pp RS:HS 14.408

1949. 1829–2–15 From Francis Baily 37 Tavistock Place
Regarding F. W. Bessel's instrument for his zone observations. Who was the first to publish lunar distances from the planets?
ALS 2pp RS:HS 3.88

1950. 1829–2–15 From William Whewell Trinity College
Will attend JH's marriage, but cannot be present for some preliminaries. Discusses relations between chemistry and mineralogy.
ALS 3pp RS:HS 18.177 Todhunter, *Whewell, 2,* 97–8 inc

1951. 1829–2–16 To Dionysius Lardner Slough
Has some difficulty in producing his article on Light for the *Cabinet Cyclopaedia* as the publishers of the *Encyclopaedia Metropolitana* consider it would be harmful to their own publication.
draft 2pp RS:HS 11.111 & 21.39

1952. [1829]–2–16 From Thomas Young Park Square
JH's determination of length of equinoctial year will remain for long time. Mrs. Young is delighted to hear of JH's new discovery.
ALS 2pp RS:HS 18.348

1953. 1829–2–18 From Dionysius Lardner 33 Percy St., Bedford Square
Regrets to hear of JH's difficulties. Will consult the proprietors and let him know the answer later.
ALS 1p RS:HS 11.112

1954. 1829–2–18 From Thomas Young Park Square
Invites JH to join TY to meet Thomas Henderson at [Davies] Gilbert's party Saturday. T. F. Colby accepted payment from Board of Longitude up to Christmas; Henry Kater would not accept payment beyond dissolution of Board. TY believes JH should accept payment at least from March to June.
ALS 1p RS:HS 18.349

1955. 1829–2–19 From Few & Co. Covent Garden
Sending the marriage settlement, which should be deposited with a banker.
ALS 1p RS:HS 7.199

1956. [1829]–2–19 From Dionysius Lardner 33 Percy St., Bedford Square
Has met the proprietors of the *Cabinet Cyclopaedia*. Would JH be agreeable to write an article on Astronomy or Optics, instead of Light? Will approach David Brewster regarding Light.
ALS 2pp RS:HS 11.113

1957. 1829–2–25 From William Rowan Hamilton Observatory
Expresses views on nature of exponential functions and defines terms that may have caused confusion in [J. T.] Graves's paper, which JH found unsatisfactory. Requests JH's reply.
ALS 2pp TCD 1493:36a

1958. 1829–2–25 From John Lee Doctor's Commons
A note accompanying a list of mathematical and astronomical works, to which JL asks JH to add names of others to make the collection more complete.
ALS 1p† BL Aaa 37184.226

1959. 1829–2–27 From William Whewell Trinity College
Where should WW meet JH on Tuesday before JH's wedding?
ALS 1p RS:HS 18.178

1960. 1829–2–28 To Dionysius Lardner 12 Charlotte St.
Accepts arrangement proposed by DL. JH will write a volume on astronomy [*Treatise Astr.*], but will need time.
ALS 2pp BostonPubLib Ms.Eng.530 (C:RS:HS 11.114 & 21.40)

1961. 1829–3 To James South n.p.
Explains that he does not wish to be a member of the new Board of Longitude; wishes instead to devote his energies to his research. Is leaving to travel for some months [with his bride]; cannot attend committee meetings during this period.
Adraft 3pp RS:HS 16.461 (C: 21.41)

1962. 1829–3–2 To Davies Gilbert 12 Charlotte St., Portland Place
Is leaving London for many months and would therefore like his name removed from the Council of the Royal Society, and also his Vice-presidency to be terminated. Would like his name removed from the glass sub-committee.
draft 1p RS:HS 8.119 & 21.42

1963. 1829–3–3 From Davies Gilbert 45 Bridge St.
Congratulates him on the cause of his temporary absence from London. Hopes he will reconsider his views on resignation from the offices of the R.S.L. at the end of the year. His resignation from the Glass sub-committee is a compliment to Michael Faraday.
ALS 1p RS:HS 8.120

1964. 1829–3–3 From Caroline Lucretia Herschel Hanover
Congratulates JH on his marriage to Margaret Brodie Stewart.
L *Memoir ... of Caroline Herschel*, 236–7

1965. 1829–3–3 From Mary Baldwin to Caroline Lucretia Herschel 26 Lower Phillimore Place
Describing the wedding of Margaret Brodie Stewart and JH on this day.
ALS 4pp JHS.HCEJ.503

1966. 1829–3–5 From James A. Gordon 29 Hill St.
Sends thanks for the wedding cake, and congratulations on JH's marriage.
ALS 1p RS:HS 8.166

1967. 1829–3–5 To Mary Pitt Herschel Regent Hotel, Leamington
JH's and Margaret's honeymoon journey through Buckinghamshire. Concern for MH's confinement.
ALS 4pp & C TxU:H/L-0521.1; Reel 1053

1968. 1829–3–6 From Charles Babbage Dorset St.
Difficulties over his machine. Joseph Clement threatens strike action. Would like to become acquainted with JH's friend. Advice on marriage given.
ALS 4pp RS:HS 2.239

1969. 1829–3–8 From William Whewell Trinity College
Asks JH to return a musical instrument he had borrowed. Mentions dining with Mrs. Alexander Stewart [JH's new mother-in-law]. Asks JH to look over a sheet of chemical elements to be used as a basis for a mineralogical classification. Invites JH and his bride to visit WW in Cambridge.
ALS 3pp RS:HS 18.179

1970. 1829–3–9 From S. Jey to Margaret Brodie Herschel Bath
Lady [Watson] would like JH to have a silver cake basket. Can she send it to Lady Herschel for JH? Lady Herschel should try the Bath waters.
ALS 2pp RS:HS 10.444

1971. 1829–3–11 From John Penn Spring Gardens
Congratulates JH on 'alteration of state' [marriage to Margaret Brodie Stewart].
ALS 2pp RS:HS 14.152

1972. 1829–3–12 From James Grahame Nantes
Best wishes to JH on his marriage.
ALS 3p RS:HS 8.384

1973. 1829–3–13 To John Thomas Graves Leamington
Is pleased that the R.S.L. has decided to publish JG's paper [on logarithms, which JH earlier recommended JG withdraw from consideration; see JH's 1829–1–24]. Please convey JH's thanks to W. R. Hamilton for his letter supporting JG's paper.
ALS 3pp TCD 10047/30

1974. 1829–3–18 From Mary Pitt Herschel Kensington
Poor weather. Visits from Stewart family. James South will write to JH regarding telescope.
ALS 2pp & C 2pp TxU:H/M–0625.1; Reel 1086

1975. 1829–3–18 To Alexander von Humboldt Leamington
In response to AH's 1828–10–8, JH writes to tell him of William Herschel's ideas on nebulae and the Magellanic Clouds.
C 2pp RS:HS 21.43

1976. 1829–3–18 From Stephen Peter Rigaud Oxford
Has found information regarding [James] Bradley's observation of 1759 comet [see SR's 1824–2–26]. Will send copies if JH wants them. Discusses [James] South's concern for Bradley's observations.
ALS 2pp RS:HS 14.337

1977. 1829–3–19 To Caroline Lucretia Herschel Leamington Priors, Warwickshire
Honeymooning, JH admits that he is happier than he has ever been. Describes Leamington.
ALS 4pp TxU:H/L–0574.2; Reel 1058

1978. 1829–3–19 From James South Observatory, Kensington
Reports on meeting of Francis Baily, Francis Beaufort, and JS with Lord Melville on the subject of a new Board of Longitude. Discusses proposed voyage of [James] Ross. Activities at the Astronomical Society. Congratulations to JH on his marriage.
ALS 4pp RS:HS 16.460

1979. 1829–3–24 From Mary Pitt Herschel Kensington
Feeling poorly. Finances for employees at Crown Inn. Encloses letter from [James] Grahame.
ALS 2pp & C 2pp TxU:H/M–0625.2; Reel 1086

1980. 1829–3–25 To Stephen Peter Rigaud Leamington, Warwick
Thanks SR for letter announcing SR's discovery of observation of comet of 1759 [see SR's 1829–3–18]. Asks SR to communicate information directly to [F. W.] Bessel. Will keep SR's letter confidential.
ALS 2pp Bodleian Library, Oxford (C: RS:HS 21.44)

1981. 1829–3–25 From William Whewell Trinity College
Asks JH for advice on a table designed to show the relation between the chemical elements and the known minerals. Asks JH about various problems relating to determining chemical formulas.
ALS 4pp RS:HS 18.180

1982. 1829–3–27 To James Grahame Leamington, Warwick
Of JH's happiness on his honeymoon.
ALS 4pp RS:HS 8.385

1983. 1829–3–27 From H. H. Hughes St. John's
Has directed £167–13–6 to be paid to JH's account at Drummond's. Gives list of donors to the fund for new buildings [at Cambridge].
ALS 1p RS:HS 10.214

1984. 1829–3–30 From Caroline Lucretia Herschel Hanover
Sending a portrait of herself; CH comments that receiving word of JH's marriage made her look 'a dozen years younger all at once.'
L *Memoir ... of Caroline Herschel*, 237 inc

1985. 1829–3–31 From Mary Somerville Royal Hospital, Chelsea
Thanks JH for his suggestions for her paper. Will accept any corrections he makes. Sends warm regards to JH and his new wife.
ALS 2pp RS:HS 16.327

1986. 1829–4–1 To Matthew Robinson Boulton Royal Hotel, Birmingham
JH and wife, passing through Birmingham, want to visit manufactory at Soho. Recalls friendship of MB's father and William Herschel and friendship of JH and late James Watt.
C 1p RAS:WH Archive 1/6.10; Reel 12

1987. 1829–4 To Davies Gilbert n.p.
Asks DG to respond to a note from George Rümker.
C 1p RS:HS 21.45

1988. 1829–4–1 From Mary Pitt Herschel Kensington
Plans to see JH at Slough. Visit by Mrs. Stewart and daughter. Declines money offered by JH.
ALS 1p & C 1p TxU:H/M–0625.3; Reel 1086

1989. 1829–4–2 From Matthew Robinson Boulton Soho
Regrets he cannot carry out JH's request. Hopes to see him and Mrs. Herschel at dinner.
ALS 4pp RS:HS 5.86

1990. 1829–4–2 To Matthew Robinson Boulton Royal Hotel, Birmingham
Apologizes for request to visit MB's manufactory. Declines dinner invitation. Plans to visit other local manufactories.
C 1p RAS:WH Archive 1/6.11; Reel 12

1991. 1829–4–3 From Alexander Rogers Bridge St., Leith
Anxious to hear the opinion of his telescope. Notices best resolution with a 3.5-inch aperture. Also, distinctness is best with correcting lenses in certain diametric position.
ALS 2pp RS:HS 14.409

1992. 1829–4–5 From William Henry Fitton 53 Upper Harley St.
Hopes the Herschels are well. Is worried about the state of Charles Babbage's engine. Thinks an approach should be made to Arthur Wellesley (Duke of Wellington).
ALS 7pp RS:HS 7.230

1993. 1829–4–6 From Mary Pitt Herschel Kensington
Visit by Patrick Stewart. Painting of 40-foot telescope completed. William Langton and two daughters are staying with MH. Somervilles report that Charles Babbage bought R. I. Murchison's house and plans to remarry.
ALS 2pp & C 2pp TxU:H/M–0625.4; Reel 1086

1994. 1829–4–6 From William Ritchie Tain
Will not be going to America. Will advertise for boarders to assist them in the study of mathematics and physical science. Sent paper to R.S.L.
ALS 3pp † RS:HS 14.359

1995. 1829–4–14 From Mary Pitt Herschel Slough
Returns to Slough. Visits by Stewart family and Somervilles. Charles Babbage visited and denied rumor of his remarriage.
ALS 2pp & C 2pp TxU:H/M–0625.5; Reel 1086

1996. 1829–4–17 To Mary Pitt Herschel Chepstow
JH and Margaret will see MH on Monday. Rode along Wye River and visited ruins of Tintern Abbey by full moonlight last night.
ALS 2pp & C TxU:H/L-0521.2; Reel 1053

1997. 1829–4–27 To Carl Friedrich Gauss Slough
Note regarding the translation into German of JH's *Light*, which JH notes has already appeared in French.
ALS 3pp NSUB Cod.Ms.Gauss 99 (C: RS:HS 21.46)

1998. 1829–4–29 From George Griesbach St. George's St., Camberwell
Is grateful for JH's assistance with his request. Would he lend him £5 until his first payment falls due.
ALS 1p RS:HS 9.52

1999. [1829–5 or earlier] From Thomas Young n.p.
Measurement of moon's diameter and distance. Expects JH will soon have '*official* authority over me' and [*Nautical Almanac*]. Astronomers probably regulate sidereal clock by variable equinox because they cannot find a fixed one. Errors in JH's equinoctial time.
ALS 2pp RS:HS 18.353

2000. 1829–5–14 To Mary Pitt Herschel [London]
Margaret's cold developed into serious fever, similar to those of Isabella and Mrs. Stewart. Financial transactions. Visited Duke of Wellington yesterday with Wolryche Whitmore on behalf of Charles Babbage. Fears Babbage's case will not go well. James Grahame is in London. Mrs. Stewart asks if Isabella and her maid Catherine may stay at Slough. Received visits from Somervilles, Katers, and Dunlops. Death of Thomas Young.
ALS 4pp & C TxU:H/L-0521.3; Reel 1053

2001. 1829–5–15 From Margaret Brodie Herschel to Mary Pitt Herschel London
Isabella is eager to visit Slough. Called today on Charles Babbage's mother, who is anxious about her son. JH believes Babbage is reconciled to loss of government support. Lists social visitors. James Grahame plans to take daughter Matilda back to Nantes. Sir Edward Parry leaves for Australia for five years. [JH note:] Do not sell [horse] Zephyr until JH returns.
ALS 4pp TxU:H/L-0521.4; Reel 1053

2002. 1829–5–15 From Mary Pitt Herschel Slough
Concern for Margaret Brodie Herschel's illness. Income from family holdings. Regards to James Grahame and his father.
ALS 2pp & C 2pp TxU:H/M–0625.6; Reel 1086

2003. 1829–5–17 To William Rowan Hamilton London
Thanks WH for letter on [J. T.] Graves's paper [see WH's 1829–2–25]. Admits JH could be in error concerning Graves's doctrines, but will let mathematical world form its own opinion.
ALS 2pp TCD 1493:51

2004. 1829–5–17 To William Whewell 17 Nottingham Mew
Recommends G. B. Airy for superintendence of the *Nautical Almanac*. Expresses hope that Richard Jones's book will be published. Laments lack of recognition given to British scientists.
ALS 4pp TC, Camb. Add. Ms.a.207[17] (C: RS:HS 21.47)

2005. 1829–5–18 From Mary Pitt Herschel [Slough]
Bring Isabella Stewart to Slough on Wedesday.
ALS 1p & C 1p TxU:H/M–0625.7; Reel 1086

2006. 1829–5–20 From Adolphe Quetelet Brussels
Regrets tremendously not being home when JH visited. Wishes to present himself at JH's hotel this evening. Would like to spend tomorrow together. Will discuss translation of *Light*.
ALS (French) 2pp RS:HS 14.199

2007. 1829–5–24 To Mary Pitt Herschel Hotel de Quillae, Calais
JH's and Margaret's voyage from London with John Stewart and Mrs. Tucker. Met R. H. Inglis on ship. Plan to go to Cassell or St. Omer tomorrow. News of Stewart family.
ALS 4pp & C TxU:H/L-0521.5; Reel 1053

2008. [1829]–5–25 From James South Observatory, Kensington
Reports unfavorably on the telescope of A. Rogers. Information regarding Charles Babbage and his engines. Controversy concerning the *Nautical Almanac*.
ALS 4pp RS:HS 16.478

2009. 1829–5–26 From Stephen Peter Rigaud Oxford
Releases JH from silence. Oxford University will publish portions of [James Bradley's] observations of comet of 1759 [see SR's 1829–3–18, 1829–3–25]. Never expected SR would trace and recover them.
ALS 1p RS:HS 14.338

2010. 1829–5–30 To Mary Pitt Herschel Hotel de Bellevue, Brussels
Scenes in France and Flanders. Hotel is adjacent to king's palace. Made drawings of cathedral yesterday. Visited Waterloo today.
ALS 4pp & C TxU:H/L-0521.6; Reel 1053

2011. 1829–6–2 From Baron Charles L. G. J. Keverbey de Kessel Brussells
Greetings to Mr. and Mrs. Herschel.
ALS 1p (French) RS:HS 11.79

2012. 1829–6–2 From Adolphe Quetelet Brussels
Is pleased to write letter to [G. P.] Dandelin for JH. JH should visit Mr. Van Rees at Liege, for which AQ will also write a letter of introduction. Nevertheless AQ believes JH's name suffices as introduction. If JH visits Namur, he should see the governor of the province, Mr. D'Omolius[?].
ALS 3pp (French) RS:HS 14.200

2013. 1829–6–7 To Mary Pitt Herschel Aix-la-Chapelle [Aachen]
Tour of Brussels. Music festival in progress. Met Belgian artists F. J. Navez and J. D. Odevaere. Dined with L. A. J. Quetelet. Met C. J. Keverberg, cofounder [with Johannes van den Bosch] of colonies at Frederiksoord and Hoogstraeten. Day in forest of Soignies. Journey from Brussels through Liege. Industrial activity at Seraing. [Letter continued 8 July:] Describes city of Aix-la-Chapelle.
ALS 4pp & C TxU:H/L-0521.7; Reel 1053

2014. 1829–6–13 From Mary Pitt Herschel Slough
Visit to Slough by Mrs. and Miss Stewart. Mr. Secker obtained part payment of Mr. Davenport's overdue rent.
ALS 1p & C 2pp TxU:H/M–0625.8; Reel 1086

2015. 1829–6–15 To Mary Pitt Herschel [Remagen]
Scenery along Rhine river. Visit to Bonn, where party was received by A. W. Schlegel, translator of Shakespeare into German, and geologist J. J. Nöggerath. [Letter continued in Andernach:] John Stewart is becoming good artist. [Letter continued 17 June in Koblenz:] Received mail from Slough. Itinerary for Germany and Switzerland.
ALS 4pp & C TxU:H/L-0521.8; Reel 1053

2016. 1829–6–24 From Mary Pitt Herschel Slough
Comforted by JH's and Margaret's good health. Plans to visit Margate. [James] Grahame will visit Slough soon. Mrs. and Miss Stewart are enjoying Slough.
ALS 2pp & C 1p TxU:H/M–0625.9; Reel 1086

2017. 1829–6–24 To Mary Pitt Herschel Carlsruhe
Journey from Koblenz to Bingen and Mannheim. Diagram of Grand Ducal Gardens at Carlsruhe. [Note by Margaret Herschel:] Enjoys travel but looks forward to seeing family again.
ALS 4pp & C TxU:H/L-0521.9; Reel 1053

2018. 1829–6–24 From Giovanni Antonio Plana Turin
Pleased to hear that JH received GP's memoirs. Comments on the nature of GP's response to an error on one of GP's memoirs on planetary perturbations. Expresses sorrow at the deaths of W. H. Wollaston, Humphry Davy, and Thomas Young.
ALS 3pp† (French) RS:HS 13.429

2019. 1829–6–29 From James Grahame Slough
Informs JH that his mother has been ill but is improving [letter completed 1829–6–30].
ALS 2pp RS:HS 8.386

2020. 1829–7–2 To Mary Pitt Herschel Baden Baden
Received mail from Slough. Margaret is superb traveler. Itinerary for Switzerland. Beauty of Baden Baden.
ALS 4pp & C TxU:H/L-0521.10; Reel 1053

2021. 1829–7–6 To Mary Pitt Herschel Strasbourg
Received Mary Baldwin's report of MH's illness. JH and Margaret are cutting short their tour and returning home in haste.
ALS 3pp & C TxU:H/L-0521.11; Reel 1053

2022. 1829–7–7 From Heinrich Wilhelm Matthias Olbers Bremen
About JH's work in reducing the observations of many nebulae. HO also speaks about JH's aunt Caroline.
ALS 1p (German) RS:HS 13.177

2023. 1829–7–8 To [Mary Baldwin] Metz [France]
JH and Margaret will arrive at Verdun tonight. Concerned for health of JH's mother and competence of doctors. Gratitude for care given by MB and Mrs. Stewart.
ALS 3pp TxU:H/L-0064; Reel 1054

2024. 1829–7–16 To Mary Pitt Herschel [London]
JH and Margaret arrived safely today. Will reach Slough tomorrow.
ALS 2pp & C TxU:H/L-0521.12; Reel 1053

2025. 1829–7–20 From James Grahame Wanstead
Of travel and holiday plans.
ALS 2pp RS:HS 8.387

2026. 1829–7–25 From James Grahame Wanstead
Of changes in JG's plans.
ALS 1p RS:HS 8.388

2027. 1829–7–29 From James Grahame Wanstead
Has no letters about General Cambourne, which JH apparently had requested.
ALS 1p RS:HS 8.389

2028. 1829–8–1 From James Grahame Wanstead
Asks JH to be trustee for JG in a legal matter.
ALS 2pp RS:HS 8.390

2029. 1829–8–5 From James Grahame Wanstead
Of plans to come to Slough.
ALS 1p RS:HS 8.391

2030. 1829–8–8 From Alexander Rogers Bridge St., Leith
Discusses further adjustments for the telescope. If JH returns to England soon, AR will come to Slough with the telescope.
ALS 3pp RS:HS 14.410

2031. 1829–8–19 From John Henry Griesbach London
Has only just had the opportunity of studying JH's paper. Comments on JH's theories regarding the musical scales and includes some of his own.
ALS 2pp RS:HS 9.53

2032. 1829–8–31 To James Hudson Slough
Asks that JH allow Jacques Babinet to see the experimental equipment given to the R.S.L. by Josef Fraunhofer.
ALS 1p WT B.2

2033. 1829–9 To Mary Pitt Herschel Slough
Lists MH's property and annual rental fees that JH arranged with each tenant. Projects earnings. Contract with Mr. Nash was witnessed by Mr. Trumper. Expects trouble getting Mr. Davenport to pay back rent that was awarded to MH.
ALS 4pp & C TxU:H/L-0521.13; Reel 1053

2034. 1829–9–3 To Roderick Impey Murchison Slough
Urges RM to read *The Times* for the previous day.
ALS 1p WT B.3

2035. 1829–9–17 From Thomas John Hussey Chislehurst
Felicitations on JH's marriage.
ALS 1p RS:HS 10.140

2036. 1829–10–7 To John Henry Griesbach Slough
Is grateful for his comments on his own theories regarding the flat 7th as a harmonic. Enlarges on this theme.
draft 4pp RS:HS 9.54 & 21.48

2037. 1829–10–7 To William Somerville Slough
Invites the Somervilles to dinner the following week.
ALS 2pp RS:HS 16.386

2038. 1829–10–11 From Giovanni Antonio Plana Turin
Is preparing to observe the occultation of Aldebaran. Seeks JH's comments on GP's position on the perturbations of Jupiter and Saturn compared to P. S. Laplace's work on the same subject. Comments further on the sending and receiving of packages of memoirs.
ALS 3pp† (French) RS:HS 13.428

2039. 1829–10–18 From Joseph Johann Littrow Vienna
Is sending papers for the *Astronomical Society Memoirs*, on telescope lenses and the construction of telescopes, planetary observations, and the use of equatorial telescopes. Is anxious to see JH's writings on light and his nebulae catalogue.
ALS 3pp (German) RS:HS 11.256

2040. 1829–10–20 From Mary Somerville Royal Hospital, Chelsea
Wants JH to return the last 8-10 pages of her manuscript.

She wishes to correct and improve it.
ALS 1p RS:HS 16.328

2041. 1829–10–22 From Mary Pitt Herschel Kensington
If Margaret Brodie Herschel is too ill, do not travel to London for MPH's sake. Will send Mrs. Moorsom to Slough.
ALS 1p & C 1p TxU:H/M–0625.10; Reel 1086

2042. 1829–11–2 From George Biddell Airy Observatory, Cambridge
Remarks about his recent tour of Italy including his meeting with G. A. A. Plana and other Italian astronomers How to observe the Moon's mass.
ALS 2pp RS:HS 1.39

2043. 1829–11–4 To Charles Babbage Slough
Hears he is at Dudmaston. Has a brother-in-law who will shortly be travelling to India and would like suggestions of books and instruments for him to take. Also letter of introduction to Edward Ryan.
ALS 4pp RS:HS 2.240

2044. 1829–11–7 To Caroline Lucretia Herschel Slough
Informs CH of his wife's pregnancy; will name the child Caroline if a girl. Reports of repairs on Slough.
ALS 4pp TxU:H/L–0574.3; Reel 1058

2045. 1829–11–8 From James Grahame Kensington
Of the health of Peter Stewart, JH's brother–in–law.
ALS 2pp RS:HS 8.392

2046. 1829–11–10 From Francis Baily 37 Tavistock Place
Regarding apartments for the Astronomical Society at Somerset House. G. B. Airy called and left a parcel for JH.
ALS 3pp RS:HS 3.89

2047. 1829–11–10 From Mary Pitt Herschel Kensington
Concerned that JH is having trouble dealing with Mr. Davenport. [Stewart family's servant] Mrs. Tucker visited MH overnight. Mrs. Moorsom will come to Slough. Visit by [James] Grahame last evening.
ALS 2pp & C 1p TxU:H/M–0625.11; Reel 1086

2048. 1829–11–11 To Francis Baily Slough
Will be unable to attend the Council Meeting of the Astronomical Society. Gives his views on accommodation for the Astronomical Society at Somerset House. J. Soldner's tables. Will send for G. B. Airy's parcel.
ALS 4pp RS:HS 3.90 (C: RS:HS 3.91 & RS:HS 21.49)

2049. 1829–11–12 From Sir Robert Harry Inglis Manchester Bldgs.
Would be pleased to know when the Herschels are coming to London so that Lady Inglis may call on Mrs. Herschel.
ALS 1p RS:HS 10.224

2050. 1829–11–13 From Nathaniel Bowditch Boston (Mass.)
Sending the first volume of his translation of P. S. Laplace's *Mécanique céleste*.
ALS 1p RS:HS 4.199

2051. 1829–11–17 From Davies Gilbert 45 Bridge St.
Time is approaching when the new members of council have to be considered. Has JH changed his mind about resigning. Hopes he will continue.
ALS 2pp RS:HS 8.120bis

2052. 1829–11–19 To Davies Gilbert n.p.
Received his letter of the 17th today. Does not think he can change his mind about resigning from the council.
draft 1p RS:HS 8.121 & 21.51

2053. 1829–11–19 To Henry Kater Slough
About the use of the collimator, together with some family news.
C 3pp RS:HS 21.50

2054. 1829–11–20 From Davies Gilbert 45 Bridge St.
Regrets the R.S.L. is to lose JH's services. Eilhard Mitscherlich has been nominated for a medal on JH's recommendation; can JH give DG some details of Mitscherlich's discoveries?
ALS 1p RS:HS 8.122

2055. 1829–11–21 From Henry Kater York Gate
Has been searching for the regulations regarding the Military College. William Fitton and HK have proposed Eilhard Mitscherlich for the Royal Medal and JH is to be asked to prepare a statement. Is pleased J. J. Littrow has adopted the collimator. Has heard of their visit to Gadesburgh. Still at the R.S.L. though he really wished to resign. Michael Faraday was voted the Copley Medal for his glass experiments.
ALS 3pp RS:HS 11.11

2056. 1829–11–24 To Davies Gilbert Slough
His library is in chaos at the moment due to house decoration, so he is unable to lay his hands on the note about Eilhard Mitscherlich. Gives a long survey of the work of Mitscherlich and states why in his opinion he should receive a medal.
C 8pp RS:HS 8.123 & 21.52

2057. 1829–11–27 From Sir Charles Lyell 9 Crown Office Row
Carlo Gemmellaro measured elevation of Benedictine monastery on Mount Etna at 294 canne [6,688 feet].
ALS 1p † TxU:H/M–0343; Reel 1087

2058. 1829–12–6 To George Biddell Airy Slough
Comments on several astronomical topics, viz., the publication of unreduced observations, the strange behavior of some of GA's clocks, the problem of no significant stars in some areas when sweeping the sky, and the effects of Venus and the Moon on each other's behavior; asks GA for confirmation of some star positions.
ALS 4pp RGO 6.803.118

2059. 1829–12–7 To Thomas Romney Robinson Slough
Congratulates TR on his volume of astronomical observations. Skeptical about alleged changes in Orion nebula. Feels better telescopes and drawings would be needed to confirm this. Discusses own double star observations.
AL 3pp Cambridge University Library 7656/TR60 (C: RS:HS 21.53)

2060. 1829–12–[10 or earlier] To Margaret Brodie Herschel Slough
Is reporting on the state of renovations at Slough.
ALS 2pp JHS.HCEJ.504a

2061. 1829–12–11 From Charles Babbage Dorset St.
Regarding the monetary affairs of his sister-in-law. Has had discussions regarding his machine. Has just commenced writing a pamphlet on the state of science in England. Everyone has been asking for JH.
ALS 3pp RS:HS 2.241

2062. 1829–12–11 To Margaret Brodie Herschel Slough
About the renovations at Slough [see JH's 1829–12–10 or earlier], and about a poem by Ovid JH had been reading.
ALS 4pp JHS.HCEJ.504b

2063. 1829–12–15 To Charles Babbage Slough
Will be in town on Thursday to execute the deed of attorney. Is glad to hear that Arthur Wellesley (1st Duke of Wellington) has taken an interest in the machine. Remarks on his pamphlet *State of Science in England.* Alterations at Slough.
ALS 4pp RS:HS 2.242

2064. 1829–12–15 To Margaret Brodie Herschel Slough
More about the renovations at Slough [see JH's 1829–12–11], and about JH's travels from London.
ALS 3pp JHS.HCEJ.504c

2065. 1829–12–[16] To Margaret Brodie Herschel Slough
Has been suffering with a cold; sends a few instructions for MH to carry out; hopes the renovations will be done by 6 Jan. [see JH's 1829–12–15].
ALS 3pp JHS.HCEJ,504d

2066. 1829–12–16 From Thomas Smart Hughes Cambridge
When he met him in Paris some years ago JH was on his way to Etna. Could he direct him to a published account of his observations?
ALS 1p RS:HS 10.216

2067. 1829–12–18 To Roderick Impey Murchison Slough
Thanks RM for information RM sent on fossils. JH's brother-in-law, due to an accident, has, however, ended geological investigations.
ALS 2pp Geological Society

2068. 1829–12–20 From James Grahame Kensington
Needs a copy of Euclid from JH; quotes some Latin poetry of Nicholas Copernicus.
ALS 2pp RS:HS 8.393

2069. 1829–12–[21] To Margaret Brodie Herschel Slough
About the sermon in church on the day before, and some difficulties encountered in the renovations [see JH's 1829–12–[16]].
ALS 4pp JHS.HCEJ.504e

2070. 1829–12–22 From Joseph von Utzschneider Munich
Has seen JH's article on light in an encyclopedia, and writes to ask if JH would send JU a copy of it.
ALS 1p (German) RS:HS 17.457

2071. [1830 or later] To Charles Babbage n.p.
Please send enclosed to Mr. Tukinor [?]. Gives brother-in-law's statement concerning Mrs. Urquhart and Mr. Deguise.
ALS 2pp RS:HS 2.399

2072. [1830] To Mary Baldwin n.p.
JH shocked by MB's note. Will tell Mrs. Moorsom tomorrow.
AL 1p TxU:H/L-0058; Reel 1054

2073. 1830 To James Grahame n.p.
Silly story asking 'what has become of JG?'
AL 4pp RS:HS 8.401

2074. [1830 or later] To Mr. Roberts n.p.
Wanted [Michael] Underwood's book on the diseases of children [*Treatise on the Diseases of Children*], not [Astley P.] Cooper's book on surgery [*Lectures on the Principles and Practices of Surgery*]. Is an exchange possible?
ALS 3pp Yale Univ.

2075. 1830–1–1 From Alexander Rogers Bridge St., Leith
Sent some Edinburgh ale to the Herschels. Will not patent telescope. Discusses Giovanni Santini's article on telescopes in the *Quarterly Journal.*
ALS 2pp RS:HS 14.411

2076. [1830] From Richard Sheepshanks 30 Woburn Place
Responds to JH's request for RS's opinion on [H. C.] Schumacher's contributions to astronomy. Praises Schumacher extensively.
ALS 4pp RS:HS 16.1

2077. [1830] From Mary Somerville n.p.
Used a counter spell to keep the moon at a 'due distance.' Her previous computations were careless. Asks JH to return part of her manuscript [*Mechanism of the Heavens*] for corrections.
ALS 2pp RS:HS 16.340

2078. [1830] From Mary Somerville n.p.
Asks JH to send the astronomical part of her manuscript [*Mechanism of the Heavens*] by Mr. Richards on Thursday.
ALS 1p RS:HS 16.339

2079. [1830–1?] To Mary Somerville n.p.
Admires her manuscript [*Mechanism of the Heavens*]. Wishes [P. S.] Laplace would have lived to see it. Notes a problem with the principle of virtual velocity.
ALS 3pp RS:HS 16.382 (C: RS:HS 25.13.29b) *Personal Recollections ... of Mary Somerville* (London, 1873), p. 167 inc

2080. 1830–1–3 To Margaret Brodie Herschel Slough
About the state of the renovations [see JH's 1829–12–[21]], and JH's plans for coming to London, where he needs to do some business both for himself and his mother.
ALS 4pp JHS.HCEJ.505a

2081. 1830–1–[5] To Margaret Brodie Herschel Slough
There are some problems with some of the renovations [see JH's 1830–1–3].
ALS 4pp JHS.HCEJ.505c

2082. 1830–1–6 To Margaret Brodie Herschel Slough
The renovations are complete [see JH's 1830–1–3], and JH sends details of his coming to London.
ALS 2pp JHS.HCEJ.505b

2083. 1830–1–6 From J. W. Morton Navy Office, Admiralty
JH's Board of Longitude salary [£75] from June through

Dec. 1828 has not been picked up. If JH does not claim it soon, it will be cancelled.
ALS 1p RS:HS 19.54a

2084. 1830–1–8 From Mrs. Mary Frances Kater York Gate
Capt. Kater is too unwell to write, but he would like to know whether JH is likely to claim the Navy Bills for his term as Commissioner of Longitude, as Kater will do whatever JH does in this matter.
ALS 3pp RS:HS 11.21

2085. 1830–1–11 From Caroline Lucretia Herschel Hanover
Thrilled to be named godmother of JH's first child.
L *Memoir ... of Caroline Herschel*, 239–40 inc

2086. [1830–1–11] From Joseph Langton n.p.
Would like to rent the piece of land at Upton, lately occupied by Mr. Davenport.
ALS 1p RS:HS 11.458

2087. 1830–1–12 To J. W. Morton, Navy Office Slough
Rejects offer of payment of £75 for serving as a Commissioner of Longitude, explaining that JH has 'ceased to consider myself a public functionary from the moment I became acquainted with the intentions of Government respecting the dissolution of the Board of Longitude.'
ACCS 1p RS:HS 19.54b (C: RS:HS 21.54)

2088. 1830–1–15 To Thomas Smart Hughes Slough
Replies to TH's 1829–2–16, giving details of observations. Have not been printed yet. Household in a state as they have just moved.
draft 2pp RS:HS 10.216 verso (C: RS:HS 21.55)

2089. 1830–1–17 To Robert Molyneux Slough
Is ordering a pocket chronometer from Molyneux and Co. on behalf of Joseph Johann Littrow of Austria.
ALS 1p RS:HS 25.1.18

2090. 1830–1–19 From Dionysius Lardner Royal Chambers, St. James
Intends to preface each great division of his Cyclopaedia with a preliminary discourse by an author of repute. Would JH contribute one on Natural Philosophy? Can offer £250. Would also like an article on Sir William Herschel later on.
ALS 3pp RS:HS 11.115

2091. 1830–1–24 From George Biddell Airy Observatory, Cambridge
Giving observations on stars. Discussion of clock rates and sky sweeps.
ALS 3pp RS:HS 1.40

2092. 1830–2 From James Grahame Kensington
Of a romance that JG has written.
ALS 4pp RS:HS 8.394

2093. 1830–2–1 To Dionysius Lardner Slough
Has considered DL's request to the best of his ability and thinks that he will be able to write the preliminary essay in about six months' time. Regarding his friend Thomas Greenwood of the Inner Temple, will be able to put DL in touch with him.
draft 2pp RS:HS 11.116 & 21.56

2094. 1830–2–3 From Francis Beaufort Admiralty
Commander Richard Copeland wishes to be remembered to him. Sends his own congratulations to Mrs. Herschel (Margaret).
ALS 2pp RS:HS 3.314

2095. 1830–2–3 From James Grahame Kensington
Asks JH to translate a letter of JG's into French.
ALS 2pp RS:HS 8.395

2096. 1830–2–7 To Margaret Brodie Herschel Kensington
Sends news of JH's mother's recovery of health, and of his intentions of returning to Slough with his mother.
ALS 4pp JHS.HCEJ.505d

2097. 1830–2–12 From Basil Hall Paris
Would like his opinion on the merits of the camera lucida of G. B. Amici and W. H. Wollaston. Half dead with idleness here at Paris, but hopes to return to London soon.
ALS 2pp RS:HS 9.165

2098. 1830–2–14 To D. François J. Arago n.p.
Thanking him for his letter and announcement that he has been made a foreign correspondent of the Royal Academy of Sciences of France. [Also contains part of a letter to J. B. J. Fourier which is copied out fully in RS:HS.21.58.]
draft 4pp RS:HS 1.353 & 21.57

2099. 1830–2–15 To Jean Baptiste Joseph Fourier Slough
Expresses elation and gratitude at being elected corresponding member of the astronomy section of the Académie des sciences. Believes that JH's name has led to JH's researches gaining more praise than their merits justify.
ALS 3pp Académie des sciences (Paris) (draft: RS:HS 7.353 & C: 21.58)

2100. 1830–2–15 From Henry Kater York Gate
Has been confined to bed with pleurisy. Davies Gilbert called yesterday and gave startling news of the wonders of James South's new telescope. Can JH inform him of its true capabilities? Gilbert talks of having an object glass of 2" diameter made of Michael Faraday's glass.
ALS 3pp RS:HS 11.12

2101. 1830–2–18 To Henry Kater Slough
Laments indiscretion of R.S.L. president [Davies Gilbert], attributing to JH statements JH never made about astronomical observations and James South's telescope. Clarifies what JH actually saw.
ALS 4pp RS:HS 19.55

2102. 1830–2–20 From Mrs. Mary Frances Kater York Gate
She understands Davies Gilbert's report was from James South. JH's description of the telescope meets Kater's ideas. Kater has been much struck by William Parsons' (3rd Earl of Rosse) experiments on the formation of large mirrors. Wishes JH had called him when he was in town.
ALS 3pp RS:HS 11.22

2103. 1830–2–21 From James South Observatory, Kensington
Has confirmed the existence of a sixth star in the Nebular Trapezium [of Orion]. Discusses interest among leaders of the Astronomical Society in securing a royal charter. Asks for JH's views.
ALS 4pp RS:HS 16.462

2104. 1830–2–23 To Mary Somerville Slough
Sends 40 pages of her manuscript [*Mechanism of the Heavens*] along with suggestions for improvement. Suggests clarifying the principles of the first chapter and a fuller explanation of various other principles.
ALS 4pp RS:HS 16.329 (C: RS:HS 21.60) *Personal Recollections ... of Mary Somerville* (London, 1873), pp. 168–70 inc

2105. 1830–2–23 From William Somerville Chelsea College
Thanks JH for attention to Mrs. Somerville's work [*Mechanism of the Heavens*]. Please send any part of the work that he has finished.
ALS 3pp RS:HS 16.387

2106. 1830–2–24 From Johann Franz Encke Berlin
Is carrying out a study of double stars, comments on some readings, and would welcome any of JH's observations. Is planning to build up the library commenced by J. E. Bode at the Observatory. The chart of T. J. Hussey is being engraved.
ALS 3pp (French) RS:HS 7.87

2107. 1830–2–26 To Emilia Stewart Slough
A note about ES being with JH and Margaret on the anniversary of their wedding. Enclosed is a 2pp letter from Margaret to her mother.
ALS 2pp encl JHS.HCEJ.506

2108. 1830–2–28 To Charles Babbage Slough
Has heard the news about Government support for the machine; is this correct?
ALS 2pp RS:HS 2.243

2109. 1830–3–1 From George Biddell Airy Observatory, Cambridge
Informing JH that he and GA have just been appointed to a committee to consider the extension of the Astronomical Society's catalogue. Observations on this catalogue.
ALS 3pp RS:HS 1.41

2110. 1830–3–1 From Mary Somerville Royal Hospital, Chelsea
Thanks JH for criticisms of her manuscript. Will outline principles more carefully. Hopes Lady Herschel is 'gaining strength.'
ALS 2pp RS:HS 16.330

2111. 1830–3–6 From Mary Somerville Royal Hospital, Chelsea
Sends first revised pages [of MS's *Mechanism of the Heavens*] back for further opinion. Highly respects JH's comments and believes he will be a truthful critic as a friend.
ALS 3pp RS:HS 16.331

2112. 1830–3–8 From Charles Babbage Dorset St.
Latest news of his machine. His book is going to press in 10 days. Would like proofsheet of last sheet of JH's 'Sound.'
ALS 2pp RS:HS 2.244

2113. 1830–3–[8 or later] To Charles Babbage n.p.
Regarding CB's attitude to the machine. Is sending rough proof of the page on 'Sound.'
ALS 3pp RS:HS 2.245

2114. 1830–3–8 To Nathaniel Bowditch Slough
Thanks NB for sending volume I of NB's translation into English of P. S. Laplace's *Mécanique céleste.* Praises the translation and notes that it shows the high level of science in the U.S. Will send NB some of JH's publications.
ALS 2pp BostonPubLib E2210.19 v.2 (39)

2115. 1830–3–8 From Thomas Drummond Ordnance Map Office
Would like JH's drawing of the lake [?] measuring operation.
ALS 2pp RS:HS 6.453

2116. 1830–3–9 To George Biddell Airy Slough
Sets out JH's preferences with regard to developing a system of reductions of astronomical observations that could then be easily applied.
ALS 3pp RGO 6.803.130

2117. 1830–3–9 To [George Biddell Airy?] Slough
Has received and comments on materials sent, including Nathaniel Bowditch's translation of P. S. Laplace's *Mécanique céleste* and a paper by J. W. Lubbock.
AL 2pp inc RS LUB.H.346

2118. [1830–3–9 or later] From Mary Somerville n.p.
Thanks JH for his communications. Anxious to get into the proof JH criticized. Sends [Joseph] LaGrange's variations of constant quantities.
ALS 2pp RS:HS 16.333

2119. 1830–3–9 To Mary Somerville Slough
Sends new translation of [P. S. Laplace's] *Mécanique céleste.* Says her revision [of MS's *Mechanism of the Heavens*] has been effective. Still doubts the derivation of the fundamental equation. Discusses force, resistance, and reaction with regard to this equation.
ALS 4pp RS:HS 16.332 (C: RS:HS 21.61)

2120. 1830–3–10 From Wilhelm Struve Dorpat
Hopes to meet JH and [James] South in summer. Announces emperor's decision to support measurement of the 3.5° latitude meridian throughout the entire Russian empire. WS is responsible for directing work in the north. Emphasizes the importance of maintaining correct linear standards uniform with both France and England. Requests a copy of the English standard so as to compare it with the Russian measurements of degrees.
ALS 3pp (German) RS:HS 17.116

2121. 1830–3–10 From Eliza Young Park Square
Asks help in summarizing scientific labors of her late husband, Thomas Young, for memoir by Hudson Gurney. [Addressed to 'William Herschale, Esqr.']
ALS 3pp † RS:HS 18.309

2122. 1830–3–12 To Eliza Young Slough
JH agrees to assist Hudson Gurney in writing biography of Thomas Young. Please send catalogue of Young's works. Ask Hudson Gurney to write JH.
C 1p RS:HS 18.310

2123. 1830–3–14 From Peter Barlow Woolwich
Plan for the comparison of the strengths of variously sized telescopes.
C 1p RS:HS 3.292

2124. 1830–3–16 From Robert Spottiswoode London
Reports circumstances surrounding death of a Mr. Beckwith, for whom JH serves as an executor.
ALS 2pp RS:HS 17.227

2125. 1830–3–18 From Davies Gilbert 45 Bridge St.
Sending a letter he has received, and would like JH's opinion on it [probably about his telescope; see P. Barlow's 1830–3–14].
ALS 1p RS:HS 8.125

2126. 1830–3–19 From Charles Babbage Dorset St.
Regarding his forthcoming pamphlet and his attitudes to scientific societies. Sent him the wrong page of 'Sound.' Has just returned from Leicestershire. David Brewster sent a problem.
ALS 4pp RS:HS 2.246

2127. 1830–3–19 To Johann Franz Encke Slough
Thanks for his letter, which had been forwarded to his old address. Is pleased he intends working on double stars, though 70 Ophiuchi may prove difficult. Will send some of his own readings for double stars. Like him, he is astonished at the acrimony of the attack on Thomas Young and the *Nautical Almanac*.
draft 4pp RS:HS 7.88 & 21.63

2128. 1830–3–19 To Davies Gilbert Slough
Suggests Peter Barlow should test his telescope by carrying out a series of observations.
C 3pp RS:HS 8.126 & 21.62

2129. 1830–3–21 To Davies Gilbert Slough
On behalf of [J. F.] Encke, asks that Greenwich observations be given to library of Berlin Observatory.
ALS 3pp RS MC.1.211

2130. 1830–3–22 From Peter Barlow Woolwich
Outlining his scheme for the comparison of the strengths of various telescopes, which JH seems to have misunderstood.
ALS 4pp RS:HS 3.293

2131. 1830–3–23 From Charles Babbage Dorset St.
Regarding his projected biography of Dr. Thomas Young. His own book on the decline of science is in the press. Regarding the life and manners of Dr. Young.
ALS 3pp RS:HS 2.247 (C: RS:HS 2.248)

2132. 1830–3–23 From William Somerville Chelsea College
Mrs. Somerville cannot decide what to call her book [*Mechanism of the Heavens*]. Please advise so publishing can commence.
ALS 3pp RS:HS 16.388

2133. 1830–3–24 To Charles Babbage Slough
Is sorry to note the tone of CB's last letter concerning his projected biography of Dr. Thomas Young, but he means to carry out his project. Edward Sabine received £1000 from the Board of Longitude.
ALS 3pp RS:HS 2.249 (C: RS:HS 2.250)

2134. 1830–3–24 To Hudson Gurney Slough
Does not wish to see HG's biography of Thomas Young until its publication. Anything JH submits must remain unaltered, with JH's name attached. Will not give reasons. Requests copy of Young's article 'Tides' in *Encyclopaedia Britannica*.
C 1p RS:HS 19.56

2135. 1830–3–27 From Charles Babbage Dorset St.
Further remarks on Dr. Thomas Young and JH's projected biography. Has had a copy of a letter from J. F. Encke about an Astronomical Ephemeris for the Society for Promoting Useful Knowledge.
ALS 2pp RS:HS 2.251

2136. [1830 to 1839]–3–27 From [Isaac] Gosset n.p.
Accepts JH's invitation to visit next Thursday. Expecting IG's brother today.
ALS 1p RAS:JH Archive 4/2.15 verso; Reel 7

2137. 1830–3–30 To Henry Kater n.p.
Has had a letter from Wilhelm Struve regarding the Standard. Could HK bring it to the notice of the R.S.L., and advise JH?
AL 1p inc RS:HS 11.13 & 21.64

2138. 1830–3–31 To Peter Barlow Slough
Not interested in any contest of telescopes; will view objects desired by PB as they fit with JH's observational plans.
C 2pp RS:HS 21.65

2139. 1830–3–31 To Hudson Gurney Slough
Send *Encyclopaedia Britannica* volumes with Thomas Young's article on tides and T. R. Robinson's article on sound. When will HG's [biography of Young] be published?
C 1p RS:HS 19.57

2140. 1830–3–31 From Henry Kater York Gate
Has never seen the fair copy of William Herschel's polishing book. Will be pleased to assist Wilhelm Struve regarding the Standards. Will ask George Everest to call on him and discuss the Indian operations. How did Herschel perfect the adjustment of his object glass? Has been experimenting with the Josef Fraunhofer spots.
AL 4pp inc RS:HS 11.14

2141. 1830–3–31 To Mary Somerville Slough
Says manuscript [*Mechanism of the Heavens*] cannot be improved materially except perhaps on one or two small points. Will look at it again when it is together in a whole. Announces birth of JH's daughter [Caroline] that morning.
ALS 2pp RS:HS 16.334 (C: RS:HS 21.66)

2142. 1830–3–31 To Emilia Stewart Slough
Writes to announce the birth of JH's and Margaret's first daughter, Caroline Emilia Mary.
ALS 1p JHS.HCEJ.507

2143. 1830–4–3 To Henry Kater [Slough]
Misplaced papers are now found. HK must be 'first authority' on scale constructed for Wilhelm Struve. JH improved telescope by separating eyepiece lenses. JH's new baby girl.
AL 2pp damaged RS:HS 19.58 (C: RS:HS 21.67)

2144. 1830–4–5 From Peter Barlow Woolwich
Further regarding his projected comparison of telescopes. Did not realize that JH was still working on double stars.
ALS 3pp RS:HS 3.294

2145. 1830–4–5 From Mary Somerville Royal Hospital, Chelsea
Offers congratulations on arrival of JH's first child. Thanks him for communicating his thoughts on her manuscript [*Mechanism of the Heavens*] at such a time.
ALS 2pp RS:HS 16.335

2146. 1830–4–5 To Emilia Stewart Slough
Relieves ES's concern about the health of Margaret Brodie Herschel—all is well, and all around them are very helpful;

wishes recovery for Margaret Herschel's sister Isabella.
ALS 4pp JHS.HCEJ.508

2147. 1830–4–9 From Francis Beaufort n.p.
Appointment to an unnamed post. Is to meet John Lee to discuss the instruments; would JH like to join them?
ALS 4pp RS:HS 3.315

2148. 1830–4–10 To Wilhelm Struve Slough
Glad WS is coming to England but cautions that London empties in June. Congratulates WS on measurement of a 'great Northern arc.' Complains about the division of the Royal Society into 'parties'. Adds [James] South's results showing a sixth star, nearer than WS's fifth, in trapezium of Orion.
C 3pp RS:HS 21.68

2149. 1830–4–14 From E[neas] MacKintosh London
Will be pleased to act as sponsor to JH's child. Could not stay as an inmate but will visit them from Windsor from Friday till Tuesday.
ALS 2pp RS:HS 12.20

2150. 1830–4–14 To Peter Stewart Slough
About the upcoming christening, the health of Margaret, and JH's concern about the tone of Charles Babbage's upcoming book.
ALS 3pp JHS.HCEJ.510

2151. 1830–4–15 To Emilia Stewart Slough
About arrangements for the christening of daughter Caroline, and about the health of the family.
ALS 3pp JHS.HCEJ.509

2152. 1830–4–21 From Robert Grahame, Sr. London
Has been laid up with a cold and unable to answer his letter. Congratulations on the new baby. Hopes mother and child are doing well. Will wait for confirmation of his son's marriage before sending the news to the papers.
ALS 1p RS:HS 9.4

2153. 1830–4–30 To Margaret Brodie Herschel London
Still making arrangements for the christening [see JH's 1830–4–15]; concerned that it should be done soon.
ALS 2pp JHS.HCEJ.511

2154. 1830–5–1 From Francis Beaufort Admiralty
Has had a letter from Henry Foster regarding his pendulum experiments at Ascension.
ALS 3pp RS:HS 3.316

2155. 1830–5–1 From Wilhelm Struve Dorpat
Requests advice in determining optimal time for visit to England. Thanks for invitation to Slough. Mentions completion of his sixth volume of observations, the introduction of which could prove interesting to JH. Fascinated by JH's observation of a sixth star in the trapezoid of Orion.
ALS 3pp (German) RS:HS 17.117

2156. 1830–5–2 From Sarah Bowdich 27 Burton St.
Sending two volumes of her *Fishes* and explaining the reason for the delay.
ALS 3pp RS:HS 4.206

2157. 1830–5–3 From Adolphe Quetelet Brussels
Takes a moment to write. Sends various brochures and observations on magnetic intensity in Germany. Asks JH to give a copy of it to [Charles] Babbage and [Henry] Kater. Congratulates JH on being named correspondent to the Institute of France. Lost his father-in-law. Has asked [James] South to press construction of AQ's instruments, but no luck. Requests JH visit Edward Troughton and the instruments.
ALS 3pp (French) RS:HS 14.202

2158. 1830–5–8 To Peter Barlow Slough
Sending a list of the stars to be observed to test the strength of his telescope.
ALS 4pp RS:HS 3.295 & 21.69

2159. 1830–5–9 From Eliza Young Park Square
Will send engraving of late husband, Thomas Young, to JH. Hudson Gurney nearly finished with memoir. When will JH be done? Wishes to announce biography of Young.
ALS 2pp † RS:HS 18.311

2160. 1830–5–10 To Adolphe Quetelet Slough
Tells AQ of 'negligent inattention' regarding Belgium's instruments. Has extracted promise from [William?] Simms that the equatorial should be finished by Christmas and mural circle within the following year. Mentions seeing Colonel Blake and [G. P.] Dandelin. Regrets [P. F.] Verhulst's illness. Discusses lichen.
ALS 4pp Académie belgique: #2 (fragment (French) Académie belgique: #3)

2161. 1830–5–15 From Mary Somerville Royal Hospital, Chelsea
Sending back part of manuscript [*Mechanism of the Heavens*] for criticism. Hopes Mrs. Herschel and baby are well.
ALS 3pp RS:HS 16.336

2162. 1830–5–16 From Eilhard Mitscherlich n.p.
Has received the *R.S.P.T.* Also comments on several chemical matters, including isomorphism and crystallization.
ALS 4pp (German almost illegible) RS:HS 13.44

2163. 1830–5–18 From Adolphe Quetelet Brussels
Announces second volume of translation of JH's *Light* is out in Paris. Is sending a copy. Work progresses slowly. Problems because of revolution. AQ's collaborator [P. F.] Verhulst is very ill. Will leave for Italy soon. Discusses his previous trip to Germany. Saw [William] Whewell. Has received Astronomical Society's memoirs.
ALS 3pp (French) RS:HS 14.203

2164. 1830–5–19 To Caroline Lucretia Herschel n.p.
Reports on good health and behavior of JH's new daughter, Caroline Emilia Herschel. Comments on Greek music and on recent results of the R.S.L. Glass Committee.
AL 2pp inc TxU:H/L–0575.1; Reel 1058

2165. 1830–5–22 To Charles Babbage Slough
Has read his book on the decline of science and thinks that the book may do some good. Comments on various parts, and outlines things that could happen as the result. Has replied to Encke's letter about Ephemeris.
ALS 2pp RS:HS 2.252 (ACCS: RS:HS 25.1.1)

2166. 1830–5–22 To Dionysius Lardner Slough
Is reluctant to sign any certificates for memberships in the R.S.L. just now [does not explain reason].
ACCS 1p RS:HS 25.1.2

2167. 1830–5–23 From George Biddell Airy Observatory, Cambridge
Wants JH's views on a 20-foot reflector. Inviting him to

Cambridge. With this letter he encloses Volume 2 of the Cambridge Observatory's Observations and Lectures in Optics.
ALS 1p RS:HS 1.42

2168. 1830–5–24 From Thomas Hodgkin New Broad St.
Sending a paper by J. J. Lister on the microscope and some crystals of Borate of Soda. Carlo Gemmellaro has been examining temperature of air in the subterranean passage near Nicolosi. Hopes JH will visit Tottenham; would give him great pleasure to see him.
ALS 3pp RS:HS 9.402

2169. [1830–5–24 or later] To Thomas Hodgkin [Slough]
Thanks TH for some crystals, including borax, as well as a paper by J. J. Lister. JH has been cutting the borax crystals preparatory to grinding and polishing them; JH has used these crystals to observe a phenomenon previously seen with a mass of niter.
ALS 2pp WT.Ho A.3

2170. 1830–5–25 To George Biddell Airy Slough
About the prices of telescope mirrors, and specifically the one of T. J. Hussey; recommends use of certain stars to test the optical quality of the mirror.
ALS 2pp RGO 6.803.137

2171. 1830–5–27 To William Henry Smyth Slough
Thanks WS for excerpt from Niccolo Cacciatore's letter; comments on the awarding of medals for merit, and points out some necessary conditions which must be considered if the observations of the satellites of Jupiter are to be used in determining the constant of aberration.
ALS 4pp APS B.Sm98.5 (C: RS:HS 21.70)

2172. [1830 to 1833]–5–31 To Henry Kater n.p.
Arrangements for R.S.L. annual banquet. Lord High Admiral may attend.
ALS 2pp RS:HS 19.76

2173. 1830–6–3 From James South Observatory, Kensington
Reports on discussion in Astronomical Society council whether to favor JH's recommendation of a new nomenclature regarding the angular positions of double stars, or to support JS's preference for William Herschel's method. Other activities concerning the Astronomical Society.
ALS 4pp RS:HS 16.463

2174. 1830–6–5 From Charles Henry Knowles Old Windsor
Returns the *Edinburgh Review*. Has written to the *Morning Chronicle* regarding G. B. Rodney's (1st Baron Rodney) naval maneuver. Sends three pamphlets.
AL 1p RS:HS 11.89

2175. 1830–6–5 To Eliza Young Slough
Received Thomas Young's portrait. Will send summary to Hudson Gurney soon. Suggests correction in proposed title.
C 1p RS:HS 18.312

2176. 1830–6–6 To Henry Kater [Slough]
F. G. W. Struve will arrive in July. Before HK and wife leave for Continent next week, please leave directions how Struve may obtain Struve's scale.
ALS 2pp RS:HS 19.59

2177. 1830–6–7 To James South Slough
Is defending a change in notation for the angles of position system that JH introduced in a paper. [Postscript dated 1830–7–7 marked 'Private' and deals with business relating to the *Athenaeum* and a comment on Charles Babbage's book.]
ACCS 3pp RS:HS 25.1.3

2178. 1830–6–8 From Henry Kater York Gate
Does he want a scale ordered form George Dollond for Wilhelm Struve? Will not be in London when Struve arrives. Has written a paper on the Standards for the R.S.L. and will send JH a copy when printed. Comments on the comparison of Standards.
ALS 3pp RS:HS 11.15

2179. 1830–6–15 From Benjamin Gompertz Vauxhall
His brother-in-law, Moses Montefiore, desires to become a member of the Athenaeum and he would be grateful for JH's assistance in this matter.
ALS 1p RS:HS 8.139

2180. 1830–6–15 To Mary Somerville Slough
Will bring her manuscript [*Mechanism of the Heavens*]. Apologizes for holding it so long. Critiques various passages concerning permanent rotation and equations of stability. Sends condolences on Dr. Somerville's loss.
ALS 4pp RS:HS 16.337 (C: RS:HS 21.71)

2181. 1830–6–18 From Caroline Lucretia Herschel Hanover
Sending a box to JH, containing chiefly books.
L *Memoir ... of Caroline Herschel*, 240–1 inc

2182. 1830–6–22 From Wilhelm Struve Altona
Will arrive in England in mid-August. Mentions convention of German scientists on 18 September and invites JH.
ALS 2pp (German) RS:HS 17.118

2183. 1830–6–25 To Hudson Gurney Slough
Provides HG [who was writing a biography of Thomas Young] with a detailed evaluation of Young's contributions to physical science, e.g., Young's development of the principle of interference in optics.
L *Letter from J. F. W. Herschel [Containing Remarks on Dr. Young's Mathematical and Physical Investigations]* (London, 1830), 15 pp.

2184. 1830–6–25 To Hudson Gurney Slough
Submits JH's manuscript on Thomas Young's mathematical works for HG's biography of Young. Notes their excessive obscurity. Reasons for insisting that JH's name be added no longer exist. Asks HG's votes for two candidates for Athenaeum.
C 2pp RS:HS 19.60

2185. 1830–6–25 From William Henry Smyth Crescent, Bedford
Has difficulty observing sunspots. Thomas Maclear sends WS information for next year's occultation of Alpha Tauri. Completes rotating roof for personal observatory.
ALS 2pp RS:HS 16.243

2186. 1830–6–27 To Dionysius Lardner Slough
Some comments about the [Vignette] together with its return.
ACCS 1p RS:HS 25.1.4

2187. 1830–6–29 From Davies Gilbert 45 Bridge St.
Regarding authors for the Bridgewater Treatises. Would he

write on astronomy? Would like his opinion on a paper by John Herapath on optics.
ALS 3pp RS:HS 8.127

2188. 1830–7–1 To Davies Gilbert Slough
Declines to write the Bridgewater Treatise on astronomy; also declines to comment on a paper due to lack of time and difficulty with the author.
ACCS 3pp RS:HS 25.1.5 (C: RS:HS 21.72) W. H. Brock, 'The Selection of the Authors of the Bridgewater Treatises,' *Notes and Records of the Royal Society, 21* (1966), 167 inc

2189. 1830–7–10 To Caroline Lucretia Herschel Slough
Been working on a new mirror for the 20-ft. telescope; very pleased with its performance. Asks CH to distribute his fourth catalog of double stars among the Prussian and German astronomers.
ALS 3pp TxU:H/L–0575.2.a; Reel 1058

2190. 1830–7–21 From George Biddell Airy Observatory, Cambridge
Sends volume of observations. Remarks on A. J. Fresnel's experiments. Lists errors in JH's treatise on light. Requests information on circular double refraction.
ALS 2pp RS:HS 1.43

2191. 1830–7–22 or earlier To Margaret Brodie Herschel Slough
Is hesitating about a decision to come in to London, because his mother seems very ill.
ALS 3pp JHS.HCEJ.512

2192. 1830–7–23 To Margaret Brodie Herschel Slough
About JH's mother's health, his coming to London, and daughter Caroline's vaccination. Is thinking about writing a 'discourse on science,' and complains about sitting up several nights, waiting for two stars to appear and now means 'to break the telescopes & melt the mirrors.' Found passage in Cicero to serve as motto for JH's *Prelim. Discourse.*
ALS 4pp JHS.HCEJ.513 (C: TxU:H/L-0537.1; Reel 1053)

2193. 1830–7–24 To Hudson Gurney 22 Beaumont St.
Inadvertently opened draft of Thomas Young's biography sent by HG, but sealed it immediately. Adamant that JH not see this before it is published. Will not change one word of JH's critique of Young's mathematics. If criticism offends Mrs. Young, then burn JH's manuscript.
C 2pp RS:HS 19.61

2194. 1830–7–27 To George Biddell Airy Slough
Thanks GA for the second volume of his observations; wishes the publications of the Astronomer Royal were as good; adds some comments on polarization of light by liquids, and then talks about observing Georgium Sidum [Uranus], and deals with the orbits of satellites of planets generally.
ALS 3pp RGO 6.803.147

2195. 1830–7–28 To Margaret Brodie Herschel Slough
Expressing his grief over Isabella Stewart's approaching death. JH's mother is bearing the heat quite well. JH is busy observing again and his new mirror is excellent. Declines to sign [F. W.] Hope's certificate for R.S.L. because JH objects to council's current policies. Tell Peter Stewart that JH insists on paying for chinaware.
ALS 4pp JHS.HCEJ.514 (C: TxU:H/L-0537.2; Reel 1053)

2196. [1830–8] To Margaret Brodie Herschel Slough
Isabella Stewart's death seems imminent. Should JH bring baby Caroline to London for funeral? Mrs. Moorsom reports that Lady Mary Herschel needed help of physician Mr. Fowler. Baby is fine.
C 1p TxU:H/L-0537.4; Reel 1053

2197. [1830–8] To Margaret Brodie Herschel n.p.
Wrote and asked F. L. Chantrey to write to MH. Baby Caroline is fine. Condolences as MH attends to sister Isabella.
C 1p TxU:H/L-0537.3; Reel 1053

2198. 1830–8 To Margaret Brodie Herschel Slough
Is trying to help arrange for a cast to be made of Isabella Stewart's face on her death; MH's daughter Caroline is doing very well.
AL 2pp JHS.HCEJ.516

2199. 1830–8 To Margaret Brodie Herschel Slough
About Isabella Stewart's health, MH's toothache, JH's mother's health, and MH's daughter Caroline's good health.
ALS 3pp JHS.HCEJ.515

2200. 1830–8–6 From Charles Babbage Dudmaston
Will he be at home later this week as he would like to call on him. Wishes to discuss the *Nautical Almanac.* Gives political news.
ALS 1p RS:HS 2.253

2201. 1830–8–7 From Wilhelm Struve Paris
Briefs JH on his productive Paris visit and offers details concerning his upcoming England visit.
ALS 2pp (German) RS:HS 17.119

2202. 1830–8–8 To William Henry Smyth Slough
Advises WS on best ways to make a 'darkening glass' for solar observations. Also suggests method for delineating sunspots. Hopes WS will attend *Nautical Almanac* meeting and support his points. Completes two telescope mirrors and observes Uranian satellites. Asks WS to look for a double star between $Beta^1$ and $Beta^2$ Capricorni.
ALS 4pp APS B.Sm98.6 (C: RS:HS 21.73)

2203. 1830–8–10 To Charles Babbage n.p.
Will be very pleased to see him and hopes he will spend longer with him. Has just finished a new 20-foot mirror for his telescope. Regarding alterations in the *Nautical Almanac.*
ALS 3pp RS:HS 2.254

2204. 1830–8–18 From James Grahame L'Eperonniere
About the French Revolution and an anticipated civil war [letter completed 1830–8–25].
ALS 4pp RS:HS 8.397

2205. 1830–8–24 From Peter Barlow n.p.
Has examined the stars listed by JH with his telescope and gives the results of his observations.
ALS 3pp RS:HS 3.296

2206. 1830–8–27 From Francis Legatt Chantrey to Margaret Brodie Herschel Belgrave Place
Has received JH's letter requesting a bust, but is unable to assist him at present. Suggests [Samuel?] Joseph may be

able to help him.
ALS 1p RS:HS 5.245

2207. 1830–9 To Margaret Brodie Herschel Slough
John [Stewart] arrived with news that JH's basket never reached MH. JH will come to London tomorrow. Baby Caroline is at Windsor with grandmother Mary Herschel.
AL 2pp inc JHS.HCEJ.517a (C: TxU:H/L-0537.5; Reel 1053)

2208. [1830–9] To Dionysius Lardner n.p.
Gives reason why he wishes to omit the F.R.S. from the title page. Gives layout of entry for the title page. Prints the slips into a book so would like them printed on one side only of the page.
draft 4pp RS:HS 11.118 & 21.74

2209. 1830–9–4 From Wilhelm Struve Kensington
Deeply regrets having missed JH before departure. Thanks for priceless gift of William Herschel's works and for the glimpses of nebulae. Requests his catalogue on noteworthy nebulae when completed.
ALS 1p (German) RS:HS 17.120

2210. 1830–9–6 From Stephen Peter Rigaud Oxford
Discusses [James] Bradley's observations of double stars. Asks JH for assistance.
ALS 2pp RS:HS 14.339

2211. 1830–9–6 To William Henry Smyth Slough
Discusses WS's observations of double stars and method of averaging positional determinations. Believes he has verified the two Uranian satellites observed by William Herschel.
AL 4pp APS B.Sm98.7 (C: RS:HS 21.75)

2212. 1830–9–7 To Stephen Peter Rigaud Slough
[Responding to SR's 1830–9–6], provides information on James Bradley's observation; also notes a 'striking' anomaly in observation. Mentions study JH is doing on the Castor star system and JH's observations of Uranus's two moons.
ALS 4pp Bodleian Library, Oxford

2213. 1830–9–8 From Stephen Peter Rigaud Oxford
Thanks JH for letter [see SR's 1830–9–6]. Discusses publication of [James] Bradley's observations.
Congratulates JH on confirming satellites of Uranus seen by William Herschel.
ALS 3pp RS:HS 14.340

2214. 1830–9–10 From Thomas John Hussey Chislehurst
Dawson and TH think it would be a good idea to republish William Herschel's papers together with a biographical memoir. Has JH thought of doing this; if not would he allow Dawson to do it? Has not heard if the foreigners engaged on the Berlin maps have relinquished their portions. The Munich instrument is a most vicious instrument.
ALS 3pp RS:HS 10.141

2215. 1830–9–11 From Basil Hall Bulstrode Park
Giving information on his voyage, which he may use as he desires. In the course of 89 days he passed through the Pacific, rounded Cape Horn, and crossed the South Atlantic without seeing any land. Would be pleased to provide any further details.
ALS 4pp RS:HS 9.166

2216. 1830–9–20 From John Leslie Coates House
Accept his best thanks for his article on *Sound.* Would be grateful for his discourse on *Light.* Will soon send his own dissertation on the *Progress of Science.*
ALS 3pp RS:HS 11.473

2217. 1830–9–21 From F. W. A. Argelander Åbo
South's double stars. Regarding Francis Baily's observations and catalogue of stars.
ALS 2pp (German) RS:HS 1.368

2218. 1830–9–21 From Dionysius Lardner 122 Regent St.
Encloses slips of the mss. as far as he has sent it and three extra copies. Will JH decide how to arrange it in chapters. Gives suggestions for the title page. Does he wish the letters F.R.S. to be added?
ALS 2pp RS:HS 11.117

2219. 1830–9–21 From Stephen Peter Rigaud Oxford
Discusses [James] Bradley's observations and possibility of double stars [see SR's 1830–9–6]. Seems no double stars are readily apparent. Discusses another experiment.
ALS 3pp RS:HS 14.341

2220. 1830–9–22 From Daniel Birkett Trinity Square
Sending a book of Henry Moseley for JH's comments, and would be pleased if he would supply a testimonial for Moseley's use.
ALS 2pp RS:HS 4.97

2221. 1830–9–26 or later To Charles Babbage n.p.
Sending books for Wilhelm Struve. Would like to see the latter before he returns to Germany.
ALS 3pp RS:HS 2.371

2222. 1830–9–30 From Daniel Birkett Trinity Square
Thanks for his reply, which he will pass on to Dr. Henry Moseley. Mr. Atkinson is in town and it may be advisable to bring Thomas Beckwith's affairs to a conclusion.
ALS 1p RS:HS 4.98

2223. 1830–10 To Margaret Brodie Herschel Slough
About the editing of JH's writings by W. H. Fitton, and about some errands to be done for JH.
ALS 3pp JHS.HCEJ.517b

2224. 1830–10–2 From Dionysius Lardner 122 Regent St.
Encloses proofs of the copy of JH's *Prelim. Discourse.* JH's essay is giving him great pleasure. Comments on the view of understanding physics without the use of mathematics.
ALS 3pp RS:HS 11.119

2225. 1830–10–5 From Daniel Birkett London
Atkinson is in London and would JH meet them on Friday to effect the necessary business.
ALS 1p RS:HS 4.99

2226. 1830–10–6 To Thomas John Hussey Slough
[Responding to TH's 1830–9–10], JH notes that John Murray declined to publish an edition of William Herschel's papers. Believes if republished his father's writings should be corrected and annotated, which JH could not at present undertake.
ALS 2pp Univ.Coll.London MS Graves 23(1) (ACCS: RS:HS 25.1.7)

2227. 1830–10–15 To Charles Babbage Slough
Has not seen the letter in the *Herald* about the possibility

of himself standing for the presidency of the R.S.L. Raises six queries regarding manufactures and other scientific facts.
ALS 2pp RS:HS 2.255 (ACCS: RS:HS 25.1.8)

2228. 1830–10–16 From Joseph Nicolas Nicollet Paris
Introducing a Mr. Lerebours, son of the optician at the Bureau des Longitudes, who is on a visit to England. Hopes JH has received the copy of JN's work on geometry, written in a new way.
ALS 2pp (French) RS:HS 13.140

2229. 1830–10–16 From Wilhelm Struve Dorpat
Reminds WS to select certain nebulae for 'communal observation.' Received good news from Finland. Gave [James] South the list of 36 double stars to be observed. Stars compare except result in smaller distances than JH's and South's observations.
ALS 2pp (German) RS:HS 17.121

2230. 1830–10–17 From Charles Babbage Dorset St.
Agrees with his opinions held about the R.S.L. Is going out of town immediately; so unable to answer the queries completely. Regarding the magnetic muzzle used by needle makers.
ALS 2pp RS:HS 2.256

2231. 1830–10–18 To William Henry Fitton Slough
About the sad state of the R.S.L. under the presidency of Davies Gilbert, and what is to be done to improve it.
ACCS 6pp RS:HS 25.1.9

2232. 1830–10–18 To John Herapath Slough
Declines an opportunity to consult [with JH?].
ACCS 1p RS:HS 25.1.10

2233. 1830–10–23 From Mary Somerville Chelsea
Sends a method of estimating the variations in the elements of orbits during the perturbations in longitude and distance. Used it in manuscript when examining Saturn and Jupiter.
ALS 2pp RS:HS 16.338

2234. 1830–10–24 To Stephen Peter Rigaud Slough
Apologizes that JH has been to busy to study James Bradley's observations of double stars sent by SR. Has been busy writing on another subject.
ALS 2pp Bodleian Library, Oxford

2235. 1830–10–27 From Caroline Lucretia Herschel Hanover
Wishes to learn if Isabella Stewart's health has declined.
L *Memoir ... of Caroline Herschel*, 242 inc

2236. 1830–10–27 To Margaret Brodie Herschel Slough
About the pressure on JH to complete his book, *Prelim. Discourse*, and JH's plans about coming to London.
ALS 4pp JHS.HCEJ.517c

2237. 1830–10–27 To Richard Sheepshanks Slough
Cannot attend meeting [concerning the *Nautical Almanac*]. Makes numerous specific recommendations concerning its format and contents. Raises the theoretical question 'Does sidereal time flow uniformly?'
ACS 4pp RS:HS 25.1.11

2238. 1830–10–28 or later To Margaret Brodie Herschel Slough
A note sent with a parcel to be delivered; JH's book is progressing well.
ALS 1p JHS.HCEJ.517d

2239. 1830–10–30 From Stephen Peter Rigaud Oxford
Discusses eighteenth-century observations from Greenwich. Describes curious observations of stars in Gamma Virginis and Polaris.
ALS 1p RS:HS 14.342

2240. 1830–11–1 From James Grahame L'Eperonniere
Family and travel news.
ALS 3pp RS:HS 8.398

2241. 1830–11–1 To Margaret Brodie Herschel Slough
Arrangements about JH coming up to London, and the health of his mother.
ALS 3pp JHS.HCEJ.517e

2242. 1830–11–20 From James Grahame L'Eperonniere
Of lost letters and the death of Isabella Stewart.
ALS 2pp RS:HS 8.399

2243. 1830–11–22 From Robert Saunders Dundas [Lord Melville], Henry Hotham, and John Barrow Admiralty
By order of lords of Admiralty, notice is sent of royal appointment of JH and others to Board of Visitors to Greenwich Observatory. [Letter forwarded 29 Nov. from R.S.L. to JH by Davies Gilbert with compliments.]
LS 2pp †(by Davies Gilbert) TxU:H/M–0942; Reel 1083

2244. 1830–11–26 To Charles Babbage Slough
His views on hearing that he has been nominated for the position of President of the R.S.L. Does not wish to become the president, but will do so if elected.
ALS 2pp RS:HS 2.257 (ACCS: RS:HS 25.1.12)

2245. 1830–11–26 To Francis Beaufort Slough
With great reluctance due to his plans for the future, JH agrees to let his name stand in nomination for the Presidency of the R.S.L.
ALS 3pp Huntington Library (ACCS: RS:HS 25.1.13)

2246. 1830–11–27 From Robert Peel Court of St. James
Royal warrant appointing Board of Visitors to Royal Observatory, Greenwich.
D 4pp † TxU:H/M–1225; Reel 1060

2247. 1830–11–29 From Joseph Johann Littrow Vienna
Asks JH to advise about where in London to buy a good chronometer for a friend.
ALS 2pp (German) RS:HS 11.257

2248. 1830–11–30 From Joseph Johann Littrow Vienna
Details about the chronometer order [see JL's 1830–11–29].
ALS 1p (German) RS:HS 11.258

2249. 1830–12–3 From William Rowan Hamilton Observatory
Sorry to hear that JH was not elected President of the R.S.L. Sending a supplement to WH's 'Theory of Systems of Rays' essay. Recommends [Edwin Richard Windham Wyndham-Quin,] Lord Adare for consideration as a member

of the R.A.S.
L Graves, *Hamilton, I,* 408–10

2250. 1830–12–5 From James Grahame L'Eperonniere
Comments on JH's lost election for President of the R.S.L.
ALS 3pp RS:HS 8.400

2251. 1830–12–6 From Dionysius Lardner 122 Regent St.
Has received the parcels of corrected slips and directed the printer to continue composing from the corrected slips. Has been requested by the editor of the *Edinburgh Review* to prepare an article on Humphry Davy. Can JH supply such an article as DL has no time himself?
ALS 2pp RS:HS 11.120

2252. 1830–12–6 From Alexander Rogers Bridge St., Leith
Pleased that JH enjoyed the ale. Asks whether comets could be portions of the luminous atmosphere surrounding the sun.
ALS 2pp RS:HS 14.412

2253. 1830–12–7 From George Biddell Airy Observatory, Cambridge
Encloses paper 'Figure of the Earth.' Wants information on rays in quartz. Has tried A. J. Fresnel's experiments.
ALS 3pp RS:HS 1.44

2254. 1830–12–9 To Charles Babbage n.p.
Would he please send back the proofs he sent the other day as the printers have requested their return. Regarding affairs at the R.S.L. Has he heard from the Treasury?
ALS 2pp RS:HS 2.258

2255. [1830–12–16] From Richard Jones University Club
Wants to print some of JH's observations in his own appendix. Congratulates him on the progress and issue of the R.S.L. business. One of JH's papers was read at the Geological Society last evening and created a good impression, especially from Adam Sedgwick.
ALS 3pp RS:HS 10.349

2256. 1830–12–18 From William Henry Fitton 53 Upper Harley St.
JH's paper was read at the Geological Society and received great approbation; it will be printed in the *Proceedings* or *Transactions*. Was unable to be present as Mrs. Fitton gave birth to a son. Sorry state of the R.S.L. affairs. Intends to have nothing to do with the R.S.L. at present.
ALS 3pp RS:HS 7.231

2257. 1830–12–18 From James Hudson Royal Society
Informs JH that he has been put onto an R.S.L. committee to consider the continuation of observations in the Southern Hemisphere, and to announce a meeting of that committee and of the Glass Committee.
ALS 1p RS:HS B26.42

2258. 1830–12–19 To Francis Baily Slough
Concerning the necessity for two observatories in the Southern Hemisphere.
ALS 3pp RS:HS 3.92 & 21.76

2259. 1830–12–21 From Roderick Impey Murchison 3 Bryanston Place
Regarding the printing of JH's memoir, which William Fitton has now taken to read: suggests it be printed in abstract first of all.
ALS 4pp RS:HS 12.383

2260. 1830–12–22 To Peter Mark Roget Slough
Will not be coming to either meeting [see James Hudson's 1830–12–18], because JH has fully expressed himself previously on the matter of Southern Hemisphere observations, and had already asked to be excused from further association with Glass Committee and any other connection with 'that subject.'
ALS 3pp RS DM.4.106 (ACCS: RS:HS B26.43)

2261. 1830–12–27 To William Henry Fitton Slough
Sending two copies of JH's *Prelim. Discourse*; keep one and pass one to Charles Babbage. JH tried to cancel the motto for the book but it was too late.
ALS 1p Gerald M. Friedman

2262. 1830–12–28 To Hudson Gurney Slough
Regrets that JH's statements about Thomas Young have offended Young's wife. Appreciates HG's dilemma. Hopes Mrs. Young will find competent person to write Young's biography. Please return JH's manuscript and destroy copies.
ACCS 2pp RS:HS 25.1.14

2263. 1830–12–28 To William Henry Fox Talbot Slough
Signing certificates for election to fellowship in the R.S.L. at this time [JH having been nominated for President] would be very awkward for JH, so he will not sign for WT.
ACCS 1p RS:HS 25.1.15

2264. 1830–12–29 To George Biddell Airy Slough
Comments on GA's and Edward Sabine's work on the figure of the earth; describes some results JH obtained from using quartz for polarizing light, and is unhappy with the proposed recipient of the R.A.S. medal, M. C. T. Damoiseau, whose lunar theory JH cannot understand.
ALS 3pp RGO 6.807.142 (draft: RS:HS 1.45 & 21.77)

2265. 1830–12–29 To William Rowan Hamilton Slough
Regrets lack of time for WH's paper but has been occupied with nebulae and double stars. Finds it difficult to concentrate on one subject. Has proposed Edward Quin for Astronomical Society membership. Mentions [Francis] Beaufort's penchant for astronomy and the recent R.S.L. elections, as well as JH's wish to meet WH personally.
ALS 3pp TCD 1493:68

CALENDAR OF THE JOHN HERSCHEL CORRESPONDENCE 1831–1835

2266. [1831–1] To Miles Bland n.p.
Takes a lively interest in his application for a professorship at Durham. Remembers him gratefully as his coach for mathematics while at Cambridge.
draft 1p RS:HS 4.157 & 21.76

2267. [1831 or 1832] From Sir Gilbert Blane n.p.
Has had great satisfaction in reading JH's *Prelim. Discourse*, but would like to correct his statement regarding the date of the introduction of lemon juice as a treatment for scurvy.
ALS 4pp RS:HS 4.162

2268. [1831] From William Vernon Harcourt Wheldrake near York
Would like JH's views on the proposed meeting at York of the Friends of Science (later the B.A.A.S.).
ALS 7pp RS:HS 9.222

2269. [1831] To Mary Pitt Herschel [Slough]
Plans to arrive tomorrow for visit.
ALS 1p TxU:H/M–0784.2; Reel 1084

2270. [1831 or later] From Elise Rufenacht n.p.
Thanks Herschel family for their 'kind interest' in her behalf.
ALS 1p Harvard: Houghton AAH 67m–67 (107)

2271. 1831 To William Henry Smyth n.p.
Wants WS to make a choice between the several skeleton forms JH has devised for recording astronomical observations. Also comments on orbit of Gamma Virginis and 30 Scorpii, as well as Encke's Comet.
ALS 3pp WT 91800.1

2272. [1831] To William Henry Smyth [Slough]
Recommends WS pay attention to planetary nebulae, solid clusters, and double stars in his observations. JH lists some of each.
ALS 3pp APS B.Sm98.10

2273. [1831] To ? n.p.
Expresses thanks for extract on cholera. Sends a copy of a paper on double stars. Hopes soon to be able to send supplement on double star orbits.
ALS 1p TC, Camb. Add. Ms.a.207[21]

2274. 1831–1–2 To Hudson Gurney Slough
Surprised that HG misinterpreted JH's comments to suggest disrespect for Thomas Young. Will not defend them. If HG does not want to include these in Young's biography, then return manuscript to JH and destroy all copies of it.
ACCS 1p RS:HS 25.1.16

2275. 1831–1–3 From Miles Bland Ramsgate
Has had many melancholy afflictions since they last met. Would like JH's support for his intended application for Professorship of Mathematics at the forthcoming University of Durham.
ALS 2pp RS:HS 4.156

2276. 1831–1–5 From George Biddell Airy Woburn Place
Opinion on M. C. T. Damoiseau's theory and tables; worth a medal.
ALS 3pp RS:HS 1.46

2277. 1831–1–9 To Fearon Fallows Slough
Brings FF up-to-date on astronomical news in England, and suggests that JH may come to the Cape.
ACCS 4pp RS:HS 25.1.17

2278. [1831–1–10] From Richard Jones Brasted
Wanted to read JH's book before writing. Thinks it a useful production and should prove popular. William Fitton has been preparing a statement on the late struggle for the chair, but William Whewell has advised him to withdraw it. Hopes he will receive RJ's book in a fortnight. Comments on this and his recollections.
ALS 4pp RS:HS 10.350

2279. 1831–1–11 To James Calder Stewart Slough
Sorrow of Mrs. Stewart [at death of Isabella]. JH has not seen much of Captain [John] Franklin, who arrived ill. JH's contest with Duke of Sussex for presidency of R.S.L. JS's invitation to visit India has rekindled JH's desire to study southern skies. Wanted once to move to Paramatta. Is considering Cape of Good Hope, and possibly visiting India to view Himalaya mountains. Will send copies of JH's *Prelim. Discourse*. Answers JS's questions about Babbage's Engine.
ALS 4pp TxU:H/L-0412; Reel 1055 (ACCS: TxU:H/L-0522.2; Reel 1053)

2280. [1831–1–12] From Richard Jones n.p.
Is writing to request JH not to let William Fitton know that RJ informed JH about Fitton's statement.
ALS 2pp RS:HS 10.351

2281. 1831–1–15 To Caroline Lucretia Herschel London
Sends to CH his new book, *Prelim. Discourse*. Writes that he 'took little part in the proceedings about the Royal Society. Under any circumstances I would not have held the office of President more than a year and should have felt it a grievous evil to have held it at all.'
ALS 4pp TxU:H/L–0576.1; Reel 1058

2282. 1831–1–17 From John Robison Edinburgh
Describes experiment of [Walter] Trevelyan with heated metals and lead, which produced a variety of sounds. New experiment will be described in [David] Brewster's journal.
ALS 2pp RS:HS 14.383

2283. 1831–1–20 From Roderick Impey Murchison 3 Bryanston Place
JH's request for one hundred copies of the abstract was discussed at the Geological Society yesterday and it was decided to print the whole paper in the *Transactions*.
ALS 3pp RS:HS 12.384

2284. 1831–1–23 To Roderick Impey Murchison Slough
Gives permission for publication of JH's paper ['On the Astronomical Causes Which May Influence Geological Phenomena'] in Geological Society's *Transactions*. Mentions report of a giant fossil found in United States.
ALS 2pp Geological Society

2285. 1831–1–27 From George Biddell Airy Observatory, Cambridge
Remarks on Greenwich observations.
ALS 2pp RS:HS 1.47

2286. 1831–1–30 To George Biddell Airy n.p.
An extensive discussion of the best way of presenting the results produced from the observations made at the Royal Observatory at Greenwich, together with comments about the duties and responsibilities of the Astronomer Royal.
ALS 4pp RGO 6.804.20

2287. 1831–2–2 From James South Observatory, Kensington
Comments on efforts toward getting a royal charter for the Astronomical Society, a paper by Peter Barlow, and activities of the R.S.L. Correspondence with Wilhelm Struve. Difficulties in getting his new telescope constructed.
ALS 8pp RS:HS 16.464

2288. 1831–2–7 From James Grahame L'Eperonniere
Of the health of JH's brother–in–law, John Stewart, and the publication of JH's *Prelim. Discourse* [letter completed 1831–2–8].
ALS 3pp RS:HS 8.402

2289. 1831–2–10 To Francis Baily Slough
Provides FB, Chairman of the Greenwich Visitation Committee, with detailed recommendations for the preparation of catalogues coming out of the Greenwich observations.
ACCS 6pp RS:HS 25.1.19

2290. 1831–2–10 From William Whewell Trinity College
Sends JH a copy of Richard Jones's new book [*Essay on the Distribution of Wealth*]. Congratulates JH on the publication of his new book [*Prelim. Discourse*].
ALS 2pp RS:HS 18.181 Todhunter, *Whewell, 2,* 115

2291. 1831–2–11 From Mary Somerville Chelsea
Wants to use method of indeterminate coefficients to integrate the differential equations of the planets' perturbations. [Originally written to W. Fairfax.]
(S) 1p RS:HS 16.341

2292. 1831–2–15 From D. François J. Arago Paris
Is sending Volume 10 of the Memoirs of the Royal Academy of Sciences.
ALS 1p (French) RS:HS 1.354

2293. 1831–2–15 From Sir Gilbert Blane Sackville St.
Has recently sent him a book and letter and wonders if he has received it. Congratulations on his recent honors. If he has the book perhaps he would look at certain passages.
ALS 2pp RS:HS 4.160

2294. [1831–2–15] From Richard Jones n.p.
Regarding the affair of Charles Babbage and William Fitton's pamphlet. Would like JH to dine with him and William Whewell to discuss various matters. JH's book is popular at Cambridge.
ALS 4pp RS:HS 10.352

2295. 1831–2–15 To William Whewell [Slough]
Thanks WW for sending Richard Jones's new book and comments on it. Comments on WW's 'Science of the English Universities' and WW's review of Charles Lyell's *Principles of Geology.* Has heard that WW is reviewing JH's *Prelim. Discourse.*
ALS 4pp TC, Camb. Add. Ms.a.207[19] (C: RS:HS 21.79)

2296. [1831–2–17] To Richard Jones Slough
Praises and comments on RJ's book on economics [*Essay on the Distribution of Wealth*]. Laments overly strong language used by William Fitton and Charles Babbage in criticizing R.S.L. Comments on JH having lost the election for R.S.L. presidency.
C 5pp RS:HS 21.80

2297. 1831–2–19 From Francis Baily 37 Tavistock Place
Has been obtaining information for his report on the observatories. Is JH's paper ready to be printed? Will be able to economize on the printing of the Greenwich Observations.
ALS 3pp RS:HS 3.93

2298. 1831–2–19 To Hudson Gurney Slough
Received HG's memoir of Thomas Young. JH's manuscript must be destroyed and all printer type taken down. Any deeper critique of Young's writings would be improper and would only become stronger, further hurting Mrs. Young's feelings.
ACCS 3pp RS:HS 25.1.20 (C: 19.62)

2299. 1831–2–19 From Alexander Rogers Bridge St., Leith
Sending more ale. Abandoning AR's theory of comets coming from solar volcanoes, although he thinks it is somewhat consistent with observed phenomena of solar spots.
ALS 3pp RS:HS 14.413

2300. 1831–2–21 From Charles Babbage Dorset St.
Has just found out that the government has approached the R.S.L. as to the usefulness and chances of success of his machine. Hopes that if JH is appointed he will not become involved in the committees of the R.S.L. Has been deciphering a cipher.
ALS 3pp RS:HS 2.259

2301. 1831–2–21 To Francis Baily Slough
More on the observations and publication of results of the Greenwich Observatory [see JH's 1831–2–10].
ACCS 4pp RS:HS 25.1.22

2302. 1831–2–21 To Dionysius Lardner Slough
Would prefer not to write a popular account of William Herschel and his work which DL has requested; JH has always intended to produce a major volume on the life and work of his father.
ACCS 3pp RS:HS 25.1.21

2303. 1831–2–22 To E[dward] Smedley Slough
Is writing to inquire about the copyright status of some earlier writings for the *Encyclopaedia Metropolitana*; JH is considering collecting some of these early writings, together with others, for a separate publication.
ACCS 2pp RS:HS 25.1.23

2304. 1831–2–23 To Peter Mark Roget Slough
JH writes to excuse himself from a meeting the following day at which Charles Babbage's calculating engine will be discussed.
ACCS 1p RS:HS 25.1.24

2305. 1831–2–24 From Dionysius Lardner 122 Regent St.
Would not write anything to make JH change his mind over the proposed biography of Sir William Herschel. Thinks there is room for a popular and more serious study of Sir William. Has received good reports of JH's *Discourse*. Has not had a chance of seeing William Whewell yet.
ALS 3pp RS:HS 11.121

2306. 1831–2–24 From Joseph Jackson Lister n.p.
Describes JL's self education in grinding compound lenses and devising trials for testing these for spectacles.
L *DNB 11*: 1228

2307. 1831–2–25 To Charles Babbage Slough
Has noted his remarks about attending the committees of the R.S.L. Received one to discuss the engine, but was unable to attend as he had other arrangements. Views on present attitude of the R.S.L. Council to the machine. Intends to

revise and enlarge his books *Light* and *Sound.*
ALS 3pp RS:HS 2.260

2308. 1831–2–25 To Augustus De Morgan Slough
Requests the return of his manuscript paper on double stars so that he may prepare his presentation to the next meeting of the Astronomical Society.
ACCS 2pp RS:HS 25.1.25

2309. 1831–2–25 To James Hope Slough
Comments on proposed heat experiments of [James] Hall.
ACCS 3pp RS:HS 25.1.26

2310. [1831]–2–25 From J[ames] MacKintosh Great Cumberland St.
Expressing appreciation of JH's book on philosophy.
ALS 1p RS:HS 12.26

2311. 1831–2–25 From James Wood St. John's College
Received JH's 'Fourth Series of Observations...' [1831] on double stars. Donated it to College library.
ALS 1p RS:HS 18.296

2312. 1831–2–28 To Augustus De Morgan London
Has received a copy of JH's paper on double stars, but is confined to bed by a severe attack of rheumatism, so is not sure when he can finish the corrections.
ALS 1p WT 67597.1

2313. 1831–3–1 From Francis Baily n.p.
Notifying him of a forthcoming meeting of the Council of the R.A.S. Report is ready.
ALS 2pp RS:HS 3.94

2314. 1831–3 From Caroline Lucretia Herschel Hanover
Thanks JH for sending a copy of his *Prelim. Discourse;* notes that it has been translated into German from a French translation. Laments that she is 'decaying.'
L *Memoir ... of Caroline Herschel*, 243–4

2315. 1831–3–2 To Dionysius Lardner Slough
Is upset by some statement attributed to JH in an advertisement for the encyclopedia [*Cabinet Cyclopaedia*], which DL edits.
ACCS 2pp RS:HS 25.1.27

2316. 1831–3–2 From William Samuel Stratford n.p.
Asks JH to join Sir James South in recommending that WS succeed John Pond as superintendent of the *Nautical Almanac.* [Marked 'Confidential'.]
ALS 3pp RS:HS 17.84

2317. 1831–3–3 From Mary Somerville Royal Hospital, Chelsea
Needs advice on integrating equations. Praises JH's book [*Prelim. Discourse*]. Says James Mackintosh and [Henry] Hallam also praise it. Note from William Somerville, which needs a reply, is enclosed.
ALS 2pp RS:HS 16.342

2318. 1831–3–3 From William Samuel Stratford n.p.
Discusses confusion with regard to possible separation of *Nautical Almanac* from Astronomer Royal and ensuing confusion regarding who will succeed John Pond as superintendent of *Nautical Almanac.* Asks JH to send letter recommending WS to the Duke.
ALS 3pp RS:HS 17.85

2319. 1831–3–3 To William Samuel Stratford Slough
Refuses to write letter recommending WS as superintendent of *Nautical Almanac*. Lists reasons, among them that he will try to attend the meeting of [Board of Visitors] personally.
ACCS 2pp RS:HS 17.86

2320. 1831–3–3 To William Samuel Stratford Slough
Writes a second letter to clarify the first [see JH's 1831–3–3] about a matter involving the Greenwich Observatory and the Astronomical Society.
ACCS 2pp RS:HS 25.1.29

2321. 1831–3–3 To William Samuel Stratford Slough
Is disturbed by a suggestion of how JH should behave with respect to the Duke of Sussex.
ACCS 1p RS:HS 25.1.28

2322. 1831–3–5 To Francis Beaufort Slough
Expresses his views to FB about the meeting of the Admiralty, Greenwich Observatory, [and members of the Astronomical Society?] held the previous day.
ACCS 3pp RS:HS 25.1.31

2323. 1831–3–5 To Peter Mark Roget Slough
Declines to serve on a committee to assist the Council of the R.S.L. to revise its charter and statutes.
ACCS 1p RS:HS 25.1.30

2324. 1831–3–5 To William Samuel Stratford n.p.
Believes that if position of Astronomer Royal should become vacant, G. B. Airy would be interested; JH would gladly support such a candidacy.
ACCS 2pp RS:HS 25.1.32

2325. 1831–3–7 From Francis Beaufort Admiralty
Thanks for his comments on the *Nautical Almanac* in relation to the report of the Board of Visitors. W. S. Stratford and his appointment to the *Nautical Almanac*.
ALS 4pp RS:HS 3.317 & ACC: Hydrographic Office Letter Book

2326. 1831–3–7 From William Samuel Stratford 3 Verulam Buildings
Appreciates JH's candid reply. Explains his reasons for requesting a recommendation, but understands that JH might not give it.
ALS 3pp RS:HS 17.87

2327. 1831–3–8 To Hudson Gurney Slough
Agrees to let HG keep one copy of JH's manuscript, but it must not circulate. Justifies strong terms in JH's description of Thomas Young.
C 1p RS:HS 19.63

2328. 1831–3–8 To Mary Somerville Slough
Explains to MS P. S. Laplace's method of indeterminate coefficients in mathematics. Thanks MS for her praise of his book [*Prelim. Discourse*], which JH values more highly than newspaper reviews.
ALS 4pp RS:HS 16.343 (ACCS: RS:HS 25.1.33)

2329. 1831–3–18 From R.S.L. Council or James Hudson Royal Society
Regrets that JH will be unable to serve on charter revision committee. Asks for any suggestions.
draft 1p encl RS DM.1.26

2330. 1831–3–18 From Joseph Johann Littrow Vienna
Is grateful for news. Please order a chronometer from Robert Molyneux. Is pleased to hear JH has resumed work on the nebulae again. Would like agreement on stating latitude and time. Would like to send some more memoirs for the R.A.S. No news from Hamburg.
ALS 7pp (German with English trans.) RS:HS 11.259

2331. 1831–3–26 From Daniel Birkett Trinity Square
The next time he is in town would he call at the Imperial Fire Office to sign share transfer.
ALS 1p RS:HS 4.100

2332. 1831–4–1 To William S. Stratford Slough
Offers to smooth the way for WS in communicating with the various astronomers on the Continent whom JH knows.
ACCS 1p RS:HS 25.2.1

2333. 1831–4–6 From Francis Baily London
Amended report of Committee [Board of Visitors?] was 'bitter pill' for John Pond. [P.S.:] Meeting went quietly. 'Sir James' [South?] was not there.
ALS 2pp TxU:H/M–0038; Reel 1062

2334. 1831–4–8 To Francis Baily Slough
Recommends some changes in committee report being prepared that deals with the use to be made of the Royal Observatory's observational data.
ACCS 1p RS:HS 25.2.3

2335. 1831–4–8 To Stephen Peter Rigaud Slough
Has now studied James Bradley's observations of the Castor double star system and found them very valuable. Please send Bradley's observations regarding Gamma Virginis. What is best method of returning Bradley's observations to SR?
ALS 1p Bodleian Library, Oxford (ACCS: RS:HS 25.2.2)

2336. 1831–4–13 From Maria Edgeworth to Margaret Brodie Herschel n.p.
Relating her experiences and places visited while on a recent visit to England.
ALS 4pp RS:HS 7.27

2337. 1831–4–13 From James Grahame L'Eperonniere
Family news and news of JG's plans for travel.
ALS 2pp RS:HS 8.404

2338. 1831–4–13 From Stephen Peter Rigaud Oxford
Received packet from JH. Will attend meeting at Greenwich in June. Describes [James] Bradley's observations of stars in Gamma Virginis.
ALS 3pp RS:HS 14.343

2339. 1831–4–15 From John Robison Edinburgh
Mentions earlier letter [RS:HS 14.383]. Discusses proposed meeting of the Cultivators of Science, which would occur that summer. Asks if JH would be interested in attending.
ALS 2pp RS:HS 14.384

2340. 1831–4–18 From D. François J. Arago Paris
Thanking him for his publication on the sun, and stating that it will be placed in the library of the Institute.
ALS 2pp (French) RS:HS 1.355

2341. 1831–4–24 From Charles Babbage Committee Room
Encloses a letter sent to his mother in Devonshire St.. Regarding JH's votes. His plan for removal prospers.
ALS 1p RS:HS 2.261

2342. 1831–4–24 To Mr. Richardson Slough
Is advising Mr. R about a number of considerations to be taken into account while considering an invitation to apply for the position of Professor of Mathematics at the College of William and Mary in the United States.
ACCS 3pp RS:HS 25.2.4

2343. 1831–5–1 To William Rutter Dawes Slough
Is very pleased with the results of WD's double star observations of Zeta Cancri, which William Stratford forwarded to JH.
ACCS 3pp RS:HS 25.2.5

2344. 1831–5–12 From Henry Kater York Gate
Sends details of the measurement of Wilhelm Struve's standard. Is pleased that JH has given an invitation to [E. de?] Beaumont, who is anxious to become an astronomer and is about to spend £1,200 on instruments. Wishes that JH appeared in town more often.
ALS 3pp RS:HS 11.16

2345. [1831–5]–12 From Mary Somerville to Margaret Brodie Herschel Royal Hospital, Chelsea
Accepts her invitation.
ALS 1p RS:HS 16.346

2346. 1831–5–13 To Francis Baily Slough
Suggests an improvement in measuring techniques for astronomical observations, by making the angular measurement by direct and then reflected vision. JH wonders whether this idea might be appended to a paper currently under consideration by the R.A.S.
ACCS 3pp RS:HS 25.2.6

2347. 1831–5–16 To William Whewell [Slough]
Is sending WW a large collection of writings on optics and light.
ALS 2pp TC, Camb. Add. Ms.a.207[20] (C: RS:HS 21.81)

2348. 1831–5–17 To Wilhelm Struve Slough
Discusses a variety of observations related to double stars, and asks FS to make some specific observations to compare with JH's work. Reveals JH's dream of going to the Southern Hemisphere for a number of years, but asks FS to keep it quiet, as 'many ties' stand in the way.
ACCS 4pp RS:HS 25.2.7

2349. 1831–5–19 From Francis Baily n.p.
Vacant rooms at Somerset House and possibility of the R.A.S. being allocated some. Suggests a meeting with Frederick Augustus (Duke of Sussex) to press their cause.
ALS 2pp RS:HS 3.95

2350. 1831–5–21 From Francis Baily London
Further remarks on the possibility of rooms for R.A.S. at Somerset House. Will be pleased to see him at the visitation. Is working on a paper on the proper motion of the stars.
ALS 2pp RS:HS 3.96

2351. 1831–5–22 To Caroline Lucretia Herschel Slough
Has determined the periods of four double stars. JH considers the determination of the elliptical orbits of double stars as 'one of the greatest [discoveries] ever made.'
ALS 4pp TxU:H/L–0576.3; Reel 1058

2352. 1831–5–23 From James David Forbes 23 Suffolk St., Pall Mall
When they last met, JH had promised to show him some of the wonders of his 20-ft. reflector. Is in town for a few days and would welcome the chance of visiting him.
ALS 2pp RS:HS 7.275

2353. 1831–5–26 From James Grahame L'Eperonniere
Of JG's travels in the south of France, fighting in France, and family news.
ALS 4pp RS:HS 8.405

2354. 1831–5–28 From Mary Somerville Royal Hospital, Chelsea
Sends preliminary observations that will be prefixed to her book [*Mechanism of the Heavens*]. Wants JH to make any criticisms of it before it is published.
ALS 2pp RS:HS 16.344

2355. 1831–5–30 To William Rutter Dawes Slough
Comments with great satisfaction on WD's observations, and their close agreement with JH's for many of the same double stars.
ACCS 4pp RS:HS 25.2.8

2356. 1831–6 From Thomas Hodgkin New Broad St.
Would like to see him when in town and introduce him to J. J. Lister and his uncle Luke Howard. Thanks for JH's *Prelim. Discourse*. Baron Etten is dead.
ALS 2pp RS:HS 9.403

2357. 1831–6–3 From Basil Hall Putney Heath
Borrowed JH's book (*Prelim. Discourse*) but is so valuable that he intends to purchase a copy. Comments on various passages.
AL 4pp inc RS:HS 9.167

2358. 1831–6–3 From William Whewell Trinity College
Returns the collection of optical writings that JH had sent. Comments on the controversy over the nature of light. Criticizes paper by Richard Potter attacking JH and other wave theorists of light.
ALS 2pp RS:HS 18.182

2359. 1831–6–4 From Caroline Lucretia Herschel Hanover
Regrets that William Herschel could not see JH finish WH's catalogue of double stars; thrilled that JH has revived WH's name.
L *Memoir ... of Caroline Herschel*, 436–7 inc

2360. 1831–[6–5 or later] From Margaret Brodie Herschel to Mary Pitt Herschel n.p.
Must postpone visit with MPH until tomorrow. Need to say farewell to Guthries, who leave Balstrode soon. Both babies are fine. [JH note:] Regrets postponement.
ALS 4pp TxU:H/L-0522.1; Reel 1053

2361. 1831–6–6 To Joseph Langton [Slough]
Sorry to hear of JL's financial misfortune. Must delay responding to JL's request until JH confers with JH's wife.
ACCS 2pp TxU:H/L-0219; Reel 1054

2362. [1831]–6–6 From Joseph Langton Cippenham Court
Is in financial difficulties, about to quit the farm, and would be grateful for a loan of £300.
ALS 3pp RS:HS 11.459

2363. 1831–6–6 To Emilia Stewart Slough
Announces the birth of JH's second daughter, Isabella, of whom JH draws a picture; all are well.
ALS 3pp diag JHS.HCEJ.518a

2364. 1831–6–7 From William Henry Smyth Crescent, Bedford
Prof. Barlow's paper on 'fluid refractor' leaves WS doubting JH's report of Beta Capricorni.
ALS 1p TxU:H/M–0490.1; Reel 1093

2365. 1831–6–8 To Francis Baily n.p.
His views on seeing Frederick Augustus (Duke of Sussex) about rooms for the R.A.S., and the actions of the Duke.
ALS 3pp RS:HS 3.98 (draft: RS:HS 3.97 & C: 21.83)

2366. 1831–6–8 To Joseph Langton Slough
Offers to loan JL £300 without interest or collateral to meet 'unjustificable severity' of JL's landlord. Tell JL's son to call on JH to discuss loan and unsatisfactory condition of the Merke [Herschel family's farm property].
ACCS 3pp TxU:H/L-0220; Reel 1054

2367. [1831–6–8] From Joseph Langton Cippenham Court
Is grateful for his help. Encloses note.
ALS 2pp RS:HS 11.460

2368. 1831–6–8 To Emilia Stewart Slough
Margaret Herschel's health is good, and baby Isabella is blooming; JH draws another sketch of Isabella.
ALS 3pp diag JHS.HCEJ.518b

2369. 1831–6–9 To Basil Hall Slough
Was pleased to receive his letter and comments on his book (*Prelim. Discourse*). Much of the book was written in great haste, but will bear his comments in mind should a new edition be required.
C 3pp RS:HS 9.168 & 21.82

2370. 1831–6–10 From Francis Baily 37 Tavistock Place
Has met Frederick Augustus (Duke of Sussex), who would be pleased to receive a deputation. The Council of the R.A.S. have appointed a deputation, including JH. F. W. Bessel's innovation with observations.
ALS 2pp RS:HS 3.99

2371. 1831–6–11 To Thomas Hodgkin Slough
Is arranging to spend an evening with TH and [J. J.] Lister.
ALS 3pp WT.Ho A.2

2372. 1831–6–11 To William Henry Smyth Slough
Sends double star observations and periods. Asks WS if he has met W. R. Dawes. JH believes JH's eyes to be lacking in light sensitivity.
ALS 4pp APS B.Sm98.8 (C: RS:HS 21.85)

2373. 1831–6–11 To Mary Somerville Slough
Returns proofs and sends a few remarks concerning them. Discusses difficulty of measuring Eta Coronae. Anxious to see her [*Mechanism of the Heavens*]. Has another daughter [Isabella].
ALS 3pp RS:HS 16.345 (C: RS:HS 21.84)

2374. 1831–6–15 From Francis Legatt Chantrey Belgrave Place
Introducing a friend, Lieut. Lister Shaw, who has traversed the course of the River Amazon.
ALS 2pp RS:HS 5.246

2375. 1831–6–15 From Joseph Johann Littrow Vienna
Sends this letter by an Armenian minister from Vienna, who would like to become acquainted with prominent men and institutions of England. He will also carry back the gold chronometer for His Excellency. Will send payment for the chronometer as soon as he knows the amount due. JH's work on Light has made a great impression. Will any of JH's countrymen be attending the Vienna meeting in September?
ALS 4pp (German with English trans.) RS:HS 11.260

2376. 1831–6–15 From Francis Maria Wilson North Audley St.
Pleased by account of JH's wife and new baby [Isabella]. Copies passage and diagram from letter of Welsh clergyman

trying to construct sound amplifier for pulpit. Asks JH to offer solutions. Husband L. P. Wilson sends regards. Forwarded JH's letters to FW's sister [Maria Edgeworth] in Ireland.
ALS 3pp † RS:HS 18.417

2377. 1831–6–16 To Walter Forman Slough
A stiff and formal letter telling WF very clearly that JH fully agrees with the Newtonian system [as WF clearly does not], and that JH would not consider trying to apply pressure to the editors of philosophical magazines to publish WF's writings, in spite of threats from WF.
ACC 2pp RS:HS 25.2.9

2378. 1831–6–20 From Charles Babbage Dorset St.
David Brewster is in town. Has arranged a series of meetings and excursions for him. Hopes JH will come to town. His mother and daughter are at Brighton.
ALS 2pp RS:HS 2.262

2379. 1831–6–20 From Caroline Lucretia Herschel Hanover
Observations on [double] stars. Congratulations [on birth of Isabella Herschel].
ALS 2pp RAS:JH Archive 1/10.2; Reel 4

2380. 1831–6–23 To Charles Babbage [Slough]
Will come up to town and meet David Brewster. Effect of violet light on chemical compound. Gives some of his recent astronomical observations.
ALS 3pp RS:HS 2.263

2381. 1831–6–24 To William Rowan Hamilton Slough
Thanks for sending extension of JH's theorem and expresses its significance. Will send copy of paper on differential and integral calculus. Includes results of calculations of orbits of double stars and mentions orbits of satellites.
ALS 3pp TCD 1493:75

2382. 1831–6–26 From Adolphe Quetelet Brussels
Sends works for JH, the Astronomical Society, and several members. Worries about [P. F.] Verhulst. Asks for JH's recent observations. Has spent much time lately on the human laws. Thanks JH for seeing Edward Troughton and instruments. Has just heard that his friend Alexis Bouvard, who was like a father, is seriously ill.
ALS 3pp (French) RS:HS 14.204

2383. 1831–6–[27/28] To James Hudson Slough
Writes to tell James Hudson where to find tables that will allow comparison between Italian and English measurements.
ALS 3pp APS B.H435p.5

2384. 1831–7–2 From James Grahame L'Eperonniere
Family news.
ALS 3pp RS:HS 8.406

2385. 1831–7–7 From William Somerville Chelsea College
Sir James Mackintosh wishes to meet JH. Invites the Herschels to a dinner party for this purpose.
ALS 1p RS:HS 16.389

2386. 1831–7–13 From John Wallis Albany Road, Camberwell
JW's observations of Jupiter on 15 Apr. 1826, with all satellites hidden. This confirms similar observation by [Samuel] Molyneux quoted in JH's [*Treatise Astr.*]. Has JH's letter of 29 Aug. 1819 relating to spots on Jupiter's disc. [JH annotation: 'To be added to my *Astronomy*.']
ALS 2pp † RS:HS 18.371

2387. 1831–7–16 From George Biddell Airy Observatory, Cambridge
Informs JH that he has published a new edition of his tracts (to turn Cambridge mathematics to physical applications). Wants his permission to use his theory on biaxial crystals. Remarks on David Brewster's theories and experiments.
ALS 4pp RS:HS 1.49

2388. 1831–7–16 From William Somerville Chelsea College
Fears JH did not receive dinner invitation because he has not replied. Hopes he will be able to come.
ALS 2pp RS:HS 16.390

2389. 1831–7–27 From Francis Baily London
John Brinkley (Bishop of Cloyne) is in town and is to dine with him next week. Hopes JH can join him.
ALS 1p RS:HS 3.100

2390. 1831–7–27 From V. Joseph London
Would like to visit JH at Slough and bring him some communications from J. J. Littrow at Vienna Observatory, where he has been recently visiting.
ALS 1p (Latin) RS:HS 10.450

2391. 1831–7–28 To Walter Forman Slough
WF's manuscript, which JH returned with his letter of 1831–6–16, has not been received by WF. JH suggests some possibilities as to what might have happened, and deplores the tone of WF's letters, indicating that this tone causes JH to decline further correspondence on the subject.
ACCS 2pp RS:HS 25.2.10

2392. 1831–7–29 To William Somerville Slough
JH was in Isle of Wight and did not receive dinner invitation. Apologizes for not coming. Asks when Mrs. Somerville's book [*Mechanism of the Heavens*] will be published.
ALS 1p RS:HS 16.391

2393. 1831–7–30 To John Phillips Slough
Other engagements preclude JH from accepting JP's invitation to present a paper at the 'general Scientific meeting' [inaugural meeting of the B.A.A.S.] to be held at York.
C 1 RS:HS 21.86

2394. 1831–8–2 To Thomas John Hussey Slough
Responds to TH's complaint that the government and the R.A.S. are not liberal enough with grants for science. JH rejects both complaints and explains why.
ACCS 2pp RS:HS 25.2.11

2395. 1831–8–5 From Francis Baily 37 Tavistock Place
Regrets that JH could not dine with him and John Brinkley (Bishop of Cloyne). Has had a letter from the Admiralty about a grant to [T. J.?] Hussey.
ALS 4pp RS:HS 3.101

2396. 1831–8–7 To Thomas John Hussey Slough
TH has been hurt by the tone of JH's letter [see JH's 1831–8–2] and felt he was writing to JH in confidence. JH assures TH that he has spoken to no one about any private details in TH's letter.
ACCS 2pp RS:HS 25.2.13

2397. 1831–8–7 To William Meadows Slough
Explains to WM that in the present chaotic state of scientific responsibilities [the Board of Longitude has been abolished and the Royal Observatory at Greenwich's sphere of activity has been severely restricted], the Cape observations have not been published, but JH hopes that will soon occur. JH then

asks WM to write when he arrives at the Cape, telling JH about the possibilities for an 'Amateur Observer' to live and observe at the Cape.
ACCS 4pp RS:HS 25.2.12

2398. 1831–8–9 To William Henry Smyth Slough
Notes that 36 Ophiuchi and 30 Scorpii have parallel proper motion; believes that James South overestimates the importance of this fact. Suggests double stars for WHS to observe.
ALS 4pp APS B.Sm98.9 (C: RS:HS 21.87)

2399. 1831–8–14 From James Grahame L'Eperonniere
About JG's travel plans, and about JH's *Prelim. Discourse.*
ALS 3pp RS:HS 8.407

2400. 1831–8–15 To Stephen Peter Rigaud Slough
Has correlated James Bradley's observations of the Castor and the Gamma Virginis double stars systems, which observations SR sent to JH, with later observations. Agreement is excellent. JH lays out elements of orbits for each system.
ALS 3pp Bodleian Library, Oxford (ACCS: RS:HS 25.2.14)

2401. 1831–8–18 To Mary Pitt Herschel Marine Hotel, Worthing
JH and Margaret arrived safely. Changes in town since 1826 visit. [Letter continues 19 Aug.:] Unable to find lodgings. Both babies [Caroline and Isabella] enjoy sea.
ALS 3pp & C TxU:H/L-0522.3; Reel 1053

2402. 1831–8–23 From James McCullagh Trinity College Dublin
Sending some papers that will appear in the Irish Transactions, on the intensity of light. If JH agrees would he send it on to one of the scientific journals.
ALS 2pp RS:HS 12.9

2403. 1831–8–25 From Francis Beaufort Admiralty
Would like his opinion regarding a matter concerning [Stephen] Groombridge's catalogue. Would he care to write the preface?
ALS 2pp RS:HS 3.318

2404. 1831–8–29 To Francis Beaufort Little Hampton
In reply to FB [see FB's 1831–8–25], JH believes that [Henry] Taylor, the person who had done all the reductions of the observations in Groombridge's catalogue, should be asked to write the preface, both by reason of his familiarity with the material, as well as having earned the right to do so.
ACCS 4pp RS:HS 25.2.15

2405. 1831–9–2 From Francis Beaufort Admiralty
Is much obliged for his advice regarding [Stephen] Groombridge's catalogue. Will send him the proof sheets as available. Has he seen Gerard Moll's *Decline of Science in England*?
ALS 3pp RS:HS 3.319 & ACC: Hydrographic Office Letter Book

2406. 1831–9–2 To Eilhard Mitscherlich Slough
Of isomorphic crystals and their varying behavior in response to heat, and of double stars.
ALS 4pp RS:HS 25.2.16

2407. 1831–9–2 To William Samuel Stratford n.p.
Comments on a number of astronomical matters including double stars, awarding of medals, and on the elements of [Halley's] comet.
ACCS 4pp RS:HS 25.2.17

2408. 1831–9–5 To Francis Beaufort Slough
Remarks by JH on proof sheet of Henry Taylor's catalog of circumpolar stars observed by Stephen Groombridge. Why is Gerard Moll, a foreigner, engaged in purely English controversy? JH wants no part of Charles Babbage and James South in this matter. Regrets JH's earlier statements. Sends extract from [Richard] Copeland's letter.
ACCS 2pp RS:HS 19.174 (C: RS:HS 21.88)

2409. 1831–9–5 To William Vernon Harcourt Slough
Favors concept of organization [B.A.A.S.] to guide scientific research and disseminate discoveries, but doubts that one is possible. Particular societies promoting special subjects are more likely to succeed. Anyone devoted to promoting such societies will have little time for research. Prefers to follow JH's own pursuits. Does not know James South's opinion.
ACCS 3pp RS:HS 25.2.18 Morrell and Thackray, *Early Correspondence*, 55–7

2410. 1831–9–9 From Francis Beaufort Admiralty
Is grateful for his observations on the method of printing [Stephen] Groombridge's catalogue. Fearon Fallows is seriously ill; does JH know of a suitable successor?
ALS 4pp RS:HS 3.320 & ACC: Hydrographic Office Letter Book

2411. 1831–9–13 To Charles Babbage Slough
Is sending mirror in perfect state for Vincenzio Antinori at Florence, together with his letter. Regarding Gerard Moll's book. Has no time to spare to attend the [*B.A.A.S.*] meeting at York.
ALS 3pp RS:HS 2.265

2412. 1831–9–13 From Charles Babbage n.p.
Is sending a query from Florence, also a letter from Vincenzio Antinori, which please return. Additional facts supporting his views expressed in the *Decline of Science*. Frederick Augustus (Duke of Sussex) has been challenged over government of the R.S.L. Engine progresses well. [Also contains draft of reply which is the same as JH's 1831–9–13.]
ALS 4pp RS:HS 2.264

2413. 1831–9–13 To Francis Beaufort Slough
Grieved to hear of Fearon Fallows broken state of health; in response to FB's request, JH reviews a number of possible candidates to succeed Fallows.
ACCS 3pp RS:HS 25.2.19

2414. 1831–9–13 From William Ritchie Royal Institution
Introduces Miss Hope, who has been studying mathematics with WR. WR has made some curious discoveries in Voltaic electricity.
ALS 1p RS:HS 14.360

2415. 1831–9–14 From William Somerville Chelsea College
Brought a copy of Mrs. Somerville's book [*Mechanism of the Heavens*] to Chancellor [Henry Brougham]. Wants JH to review it for the *Quarterly Review*. The Chancellor wishes to know if JH would like to be knighted.
ALS 3pp RS:HS 16.392

2416. 1831–9–16 To William Somerville Slough
Discusses feelings about offer of knighthood. Is ambivalent. Praises the Chancellor's [Henry Brougham's] interest in science.
ALS 3pp RS:HS 16.393 (ACCS: RS:HS 25.2.20)

2417. 1831–9–18 From William Whewell Trinity College
Comments negatively on whether *photonomy* is a proper

name for the study of light. Has suggested that the B.A.A.S. invite eminent scientists to survey their specialties. Discusses whether B.A.A.S. should meet at Cambridge. Has published his review of JH's *Prelim. Discourse.*
ALS 4pp RS:HS 18.183 Morrell and Thackray, *Early Correspondence,* 63–4 inc

2418. [1831–9–20] To Walter Forman [Slough]
Still relating to the manuscript of WF's not yet returned to him [see JH's 1831–7–28], JH explains very carefully how it was sent back and where WF should look for it.
[Correspondence books contain a shorter version, partly torn out.]
ACCS 1p RS:HS 25.2.22 (damaged portion of letter at RS:HS 25.2.20b)

2419. 1831–9–20 To William Whewell Slough
Thanks WW for his favorable review of JH's *Prelim. Discourse.* Vernon Harcourt has asked JH for his views on the establishment of a B.A.A.S. Expresses some support but numerous reservations. Responds favorably to WW's suggestion that prominent scientists prepare surveys of their specialties for the meeting.
ALS 5pp TC, Camb. Add. Ms.a.207[22] (C: RS:HS 25.2.21) Morrell and Thackray, *Early Correspondence,* 66–8 inc

2420. 1831–9–21 To William Henry Fitton Slough
Is sending sketches, etc., from his field book made while visiting the Isle of Wight. Disagrees with some of WF's representations.
ALS 4pp Gerald M. Friedman

2421. 1831–9–24 From Francis Beaufort Admiralty
His views on a successor to the Cape Astronomer.
ALS 5pp RS:HS 3.321 & ACC: Hydrographic Office Letter Book

2422. [1831]–9–25 From William Somerville Saplain [?] Lodge
Invites Herschels to dinner at Mrs. Turno's. Wishes to take a look at sky when JH gets home.
ALS 4pp RS:HS 16.411

2423. 1831–9–27 From William Somerville Bognor
Received word from the Chancellor [Henry Brougham] that he mentioned knighting JH to King [William IV]. If JH does not want this honor, he should write the Chancellor immediately.
ALS 3pp RS:HS 16.394

2424. 1831–9–29 From William Somerville Bognor
Says JH virtually has no choice concerning accepting knighthood, because it is the will of the Chancellor [Henry Brougham] and King [William IV]. Urges JH to acquiesce for the benefit of science.
ALS 3pp RS:HS 16.395

2425. [1831–9–30] To Francis Beaufort [Slough]
Expresses JH's preferences with respect to the appointment of the Cape Observatory astronomer in terms of the choices available [Follow-up letter to JH's 1831–9–13].
ACCS 3pp RS:HS 25.2.23

2426. 1831–9–30 To William Somerville Slough
Will not oppose the the wishes of the Chancellor [Henry Brougham] concerning JH being knighted, especially if this might benefit science. Will visit Chancellor to thank him.
ALS 2pp RS:HS 16.396 (draft: RS:HS 16.397; C: RS:HS 21.89)

2427. 1831–10 To Francis Beaufort n.p.
Is returning the proof of [Stephen] Groombridge's catalogue. Comments on the expense of printing it. Is pleased that Thomas Henderson has been appointed Cape Astronomer.
draft 2pp RS:HS 3.322bis

2428. 1831–10–3/4 To Gilbert Elliot, Second Earl of Minto Slough
Sends JH's observations and measurements made during JH's Sept. 1821 ascent of a peak near Mt. Rosa. Also sends JH's observations of Mt. Etna made in June 1824.
ALS 4pp NatLibScot MS.12226, f.114 (Adraft; TxU:H/L-0637; Reel 1089)

2429. 1831–10–4 To Gilbert Ellio, Second Earl of Minto Slough
Provides detailed observations made by JH on his ascent up Monte Rosa in the Alps.
ACCS 4pp RS:HS 25.2.24

2430. 1831–10–4 To William Somerville Slough
Has written the Chancellor [Henry Brougham] to accept the knighthood offered him. Sent observations from Monte Rosa to Lord Minto. Thanks WS for his help with the Chancellor.
ALS 2pp RS:HS 16.398 (C: RS:HS 21.90)

2431. 1831–10–6 From Thomas Hodgkin New Broad St.
Sending some crystals on behalf of J. Howard. Comments on them.
ALS 1p RS:HS 9.404

2432. 1831–10–8 From William Henry Fitton Highwoodhill, Hendon
Comments on some of JH's geological papers. Has received a paper on Yorkshire geology from R. I. Murchison.
ALS 4pp diag RS:HS 7.233

2433. 1831–10–10 From Augustus De Morgan 90 Guilford St.
His paper on double stars is required for press so would be grateful if it could be sent along. The society is short of papers for reading so any JH can produce would be very welcome.
ALS 1p RS.HS 6.174

2434. 1831–10–10 From Gilbert Elliot, Second Earl of Minto Admiralty
Is grateful for the copies of his notes on his observations. Has compared them with his own and found them remarkably similar. JH should see Mr. Hildyard of Eton if he wishes to compare notes on their expeditions. Regarding JH's measurement of Etna.
ALS 2pp RS:HS 7.36

2435. 1831–10–13 From William Ritchie Royal Institution
Is a candidate for Rectorship of Ayr Academy. Asks for testimonial from JH. Having financial difficulties and must leave London.
ALS 1p RS:HS 14.361

2436. 1831–10–14 From James Ivory 20 Southampton St., Pentonville
Packet sent to Scotland has now been delivered. Would like JH's advice on how to express thanks in the proper quarter. Thinks science would benefit by the presence of G. B. Airy or W. R. Hamilton at Greenwich.
ALS 3pp RS:HS 10.256

2437. 1831–10–15 From Francis Beaufort Admiralty
Is sending the first sheet of [Stephen] Groombridge's

catalogue. Regarding the costs of printing this catalogue. Fearon Fallows has died and Thomas Henderson has been appointed in his place. Congratulations on JH's knighthood.
ALS 4pp RS:HS 3.322 & ACC: Hydrographic Office Letter Book

2438. 1831–10–15 To Caroline Lucretia Herschel London
JH awarded the Guelphic Order of Knighthood by King William IV, an honor earlier bestowed on William Herschel.
ALS 3pp TxU:H/L–0576.6; Reel 1058

2439. 1831–10–16 To William Henry Smyth Slough
Sends data concerning double stars and new form for making observations of them.
C 3pp RS:HS 21.91

2440. 1831–10–17 From Charles Babbage Dorset St.
Please could he obtain through Miss Caroline Herschel several copies of the *Hannover Gazette* which includes his own name when he was awarded the Guelphic Order. There is talk that Frederick Augustus (Duke of Sussex) is going to Ireland.
ALS 1p RS:HS 2.266

2441. 1831–10–18 To Francis Beaufort Slough
Comments on the work of T. G. Taylor in reducing the observations for Stephen Groombridge's catalogue, and on an appropriate remuneration for the work. Enclosed are comments for the benefit of Taylor.
ALS 3pp enc Hydrographic Office Misc F32F3.10(enc: F32F1)

2442. 1831–10–19 From William Rowan Hamilton Observatory
Encouraged that a calculus can be constructed out of his exponential series findings.
L Graves, *Hamilton, 1,* 481–3

2443. 1831–10–19 From James Grahame L'Eperonniere
About JG's father-in-law's precarious position in the Protestant church of France.
ALS 3pp RS:HS 8.408

2444. 1831–10–21 To John George Children Slough
Declines election to R.S.L. Council.
ALS 2pp TxU:H/L–0125; Reel 1054 (ACCS: RS:HS 25.2.25)

2445. 1831–10–22 To Charles Babbage Slough
Will contact his aunt for the *Hannover Gazette.* Wishes to see Joseph Clement about moving the telescope.
ALS 1p RS:HS 2.267

2446. 1831–10–23 From William Henry Smyth Crescent, Bedford
Pleased that JH has received knighthood. Observes Alpha^1 Capricorni in attempts to see a double star. Thanks JH for double star observation outlines; asks for a 'large supply' of them.
ALS 3pp RS:HS 16.244

2447. 1831–10–25 From Caroline Lucretia Herschel Hanover
Received news from the Duke of Cambridge of JH's appointment to the Royal Hanoverian Guelphic Order.
L *Memoir ... of Caroline Herschel,* 249–50 inc

2448. 1831–10–27 To Georges C. L. D. Cuvier Slough
Writes asking that James Grahame's father-in-law, Rev. Wilson, be retained as pastor of the Protestant church at Nantes.
ACCS 3pp RS:HS 25.2.26

2449. [1831–10–29] From Charles Babbage Dorset St.
One of Joseph Clement's workmen will be coming down to work on JH's telescope. Gives rates of pay, etc. Has just got the Treasury to give their order for the completion of the buildings.
ALS 2pp RS:HS 2.268

2450. 1831–10–30 To Gilbert Blane Slough
Thanks GB for the gift of a work on medicine, and then JH proceeds to comment at some length on various aspects contained in it.
ACCS 4pp RS:HS 25.2.27

2451. [1831–11] From Sir Gilbert Blane Sackville St.
Is grateful for JH's approbation. Includes another of his works for his perusal.
ALS 2pp RS:HS 4.161

2452. 1831–11–1 From G. C. L. D. Cuvier Paris
Regarding the position of Mr. Wilson at Nantes [see JH's 1831–10–27].
ALS 1p (French) RS:HS 5.385

2453. 1831–11–2 From Augustus De Morgan 90 Guilford St.
Encloses the proof of the first part of his paper, with queries. Regarding the seal used. Volume of the memoirs have been published. There are important matters for the next Astronomical Council.
ALS 3pp RS:HS 6.175

2454. 1831–11–3 From James Grahame L'Eperonniere
Family news, and congratulations to JH on his knighthood.
ALS 3pp RS:HS 8.409

2455. 1831–11–14 From David William Smith Alnwick
Congratulates JH on his knighthood.
ALS 4pp RS:HS 16.182

2456. 1831–11–24 To John F. Daniell Slough
Asks that in JD's lectures at King's College, London, he remove an unfair and incorrect reference to JH.
ACCS 2pp RS:HS 25.2.29

2457. [1831–11–25] To [David Brewster] Slough
In an article commenting on a pamphlet about the alleged decline of science in England, JH's name has frequently been taken in vain. JH writes to protest.
ACCS 3pp RS:HS 25.2.28

2458. 1831–11–25 To [David Brewster] Slough
Wants to ensure that a letter of JH's on the decline of science controversy is to be published in Brewster's journal.
ACCS 1p RS:HS 25.2.30

2459. 1831–11–26 From John F. Daniell Kings College London
Regrets exceedingly that JH should think he was referring to him in his lecture. Regrets his work on *Sound* has been dragged into the controversy.
ALS 3pp RS:HS 6.470

2460. 1831–11–26 From William Somerville Chelsea College
Sir James Macken wishes the Herschels to come to dinner. Hopes they will attend.
ALS 3pp RS:HS 16.399

2461. 1831–11–28 From Sir Robert Harry Inglis Manchester Bldgs.
Congratulations on his knighthood. Regrets they were unable to call on the Herschels. Would JH like to become a candidate for the Literary Society?
ALS 3pp RS:HS 10.225

2462. 1831–11–28 From James South Observatory, Kensington
Has received JH's communication. Asks that JH not write JS unless he can write in a 'different manner.'
ALS 1p RS:HS 16.465

2463. 1831–11–29 To John F. Daniell Slough
Thanks JD for his letter [see JD's 1831–11–26].
ACCS 1p RS:HS 25.2.31

2464. 1831–11–29 To James South Slough
Surprised at JS's response [see 1831–11–28] to his offer. Because the offer was well intentioned, repeats it and hopes that their mutual interests will lead to continued correspondence.
ALS 2pp RS:HS 16.466 (C: RS:HS 21:93)

2465. 1831–11–30 From George Biddell Airy Observatory, Cambridge
Wants 100 of JH's star forms. Wants any information on a collection of observations of Venus.
ALS 1p RS:HS 1.48

2466. 1831–11–30 From James South Observatory, Kensington
Regrets the tensions that arose between JH and JS. Congratulates JH on his knighthood. Laments the costs and difficulties involved in the construction of JS's new observatory.
ALS 4pp RS:HS 16.467

2467. [1831–12] From Charles Babbage n.p.
Regarding J. F. Daniel[l]'s attack on JH and CB.
ALS 1p RS:HS 2.339

2468. 1831–12 To Francis Beaufort n.p.
Comments on the instructions for the Cape Astronomer and adds some of his own.
draft 4pp RS:HS 3.323b

2469. 1831–12–1 From David Brewster Allerly by Melrose
Regarding JH's letter concerning an anonymous article, which appeared in the *Edinburgh Review* and was actually by DB himself.
ALS 5pp RS:HS 4.262

2470. 1831–12–3 From William Rutter Dawes Ormskirk
Has been from home. Hopes to restore his instruments when he moves to a new house. Would welcome an exchange of information between the members of the R.A.S. Is much obliged for his hints and comments on these.
ALS 3pp RS:HS 6.49

2471. 1831–12–8 To David Brewster Slough
Having now learned that the offensive article in the *Edinburgh Review* was by DB [see DB's 1831–12–1], JH writes in surprise that DB could have written what he did, and requests that JH's letter [see JH's 1831–11–25] be returned unpublished. JH will decide how to proceed.
ACCS 4pp RS:HS 25.3.1

2472. 1831–12–8 From Mary Ann Fallows Portsmouth
Just before the death of her husband he received a letter announcing JH's intention of visiting the Cape. Would be pleased to provide him with any information if he desires it.
ALS 2pp RS:HS 7.162

2473. 1831–12–9 To Charles Babbage Slough
Had to return without seeing him owing to his mother's health. Regarding the present controversies. Is annoyed by the conduct of David Brewster.
ALS 2pp RS:HS 2.269 & 21.95

2474. [1831–12–10] From Charles Babbage Dorset St.
Regarding J. F. Daniel[l]'s lecture. His own views on the validity of the attacks. Regarding JH's opinion on David Brewster.
ALS 4pp RS:HS 2.270

2475. 1831–12–11 To Mary Ann Fallows Slough
Sends condolences on the death of MF's husband [Fearon Fallows]. JH thanks MF for her kind offer of information about the Cape; JH is so unsure of ever being able to go there that he hopes that MF will not mention this to anyone.
ALS 2pp WT A-298

2476. 1831–12–11 To William Vernon Harcourt Slough
Explains at some length the behavior of colored light, tying this to several articles that JH has written on the subject.
ACCS 4pp RS:HS 25.3.2

2477. 1831–12–12 From Mary Ann Fallows London
Grateful for JH's kind letter, but will be unable to visit him as her time is taken up visiting relations. Will send some Cape bulbs if Lady Herschel would like them.
ALS 2pp RS:HS 7.163

2478. 1831–12–15 From Basil Hall 4 St. James's Place
Has just contacted Henderson who will be very pleased to call on JH any time that suits him. Awaits his reply.
ALS 2pp RS:HS 9.169

2479. 1831–12–17 From Francis Beaufort Admiralty
Would like his comments on the instructions for the new Cape Astronomer.
ALS 3pp RS:HS 3.323a

2480. 1831–12–17 From David Brewster Allerly by Melrose
Further regarding his recent statements. Concerning JH's non-use of F.R.S. in his work.
ALS 6pp RS:HS 4.263

2481. [1831–12–18 or later] To Francis Beaufort [Slough]
Responds to FB's request for advice regarding instructions for Thomas Henderson upon taking the position at the Cape Observatory. Suggests Henderson supervise tidal observations, but not get involved in measurement of arc of meridian. Hopes to see Henderson before his departure, but JH's mother now gravely ill.
C 4pp RS:HS 21.94

2482. 1831–12–19 To Francis Baily n.p.
Sends the orbit of 70 Ophiuchi. Is progressing with the reductions of his sweeps. Justifies the method of printing adopted for [Stephen] Groombridge's catalogue.
ALS 4pp RS:HS 3.102 & 21.96

2483. 1831–12–19 To Adolphe Quetelet Slough
Apologizes for being poor correspondent. Interested in AQ's idea of taking magnets along on trips. JH used to bring his barometer. Is busy reducing sweeps. Has little time for chemistry and optics. Discusses Borax. Has worked with orbits of double stars. Mentions Mrs. Somerville's *Celestial*

Mechanics.
ALS 4pp Académie belgique: #4 *Correspondance mathématique et physique, 7* (1832), 77–9 & A. Quetelet, 'Notice sur Sir John Fred. Will. Herschel,' *Académie royale des science, des lettres et des beaux-arts annuaire* (1872), 165–6

2484. 1831–12–20 To Francis Beaufort Windsor
Offers opinions about the instructions to be furnished to the Cape astronomer, especially about tide and meteorological observations, and about surveying an arc. JH hopes to see Thomas Henderson before Henderson leaves for the Cape to assume the post of astronomer there.
ALS 4pp Hydrographic Office Misc F33F1.10a

2485. 1831–12–21 From Francis Beaufort Admiralty
Thanks for his comments on the Instructions for the newly appointed Cape Astronomer. Is sorry to hear of the failing health of his mother.
ALS 3pp RS:HS 3.324

2486. 1831–12–24 From Francis Baily 37 Tavistock Place
Thanks for the paper on Ophiuchi. On the value of JH's works. Printing of [Stephen] Groombridge's catalogue. Candidates for medals awarded by the R.A.S.
ALS 4pp RS:HS 3.103

2487. 1831–12–25 From Caroline Lucretia Herschel Hanover
Losing her strength and eyesight. Describes a conversation with the Duke and Duchess of Cambridge.
L *Memoir ... of Caroline Herschel*, 250–1 inc (?)

2488. 1831–12–30 To James South Windsor
Commiserates with JS's problems in the construction of JS's new observatory. Seeks to smooth over a misunderstanding between them. Failing health of JH's mother.
ACCS 2pp RS:HS 16.468 (C: 21.97)

2489. 1831–12–31 From James Grahame L'Eperonniere
Responds to the news that JH's mother is seriously ill.
ALS 1p RS:HS 8.410

2490. [1832 or earlier] To Charles Babbage London
Is arranging a meeting with CB.
AL 1p inc BL Aaa 37184.161

2491. 1832–1–1 To Francis Baily Windsor
Thanks for his remarks on his own work on double stars. Regarding award of the medals of the R.A.S. Method of printing [Stephen] Groombridge's catalogue.
ALS 4pp RS:HS 3.104 (draft: RS:HS 3.105; C: RS:HS 21.127)

2492. 1832–1–1 To David Brewster Windsor
Is pleased with Mr. Johnston's reply. Regarding a figure used by JH in his paper and its true significance. Is not certain yet whether to join the B.A.A.S.
draft 8pp RS:HS 4.264–5 (C: RS:HS 21.98)

2493. [1832?] From John Dalton n.p.
Answers JH's queries on visual perception. Describes trials conducted by JD. Concludes that many people see as JD sees. Sends colored silks submitted by [Richard] Potter to test color vision.
ALS 5pp TxU:H/M–1187; Reel 1060

2494. [1832?] To Augustus De Morgan n.p.
Returns manuscript and corrected proofs of one of JH's papers on double stars, with the rest of the paper going to Francis Baily.
C 1p RS:HS 25.14.6

2495. [1832?] To Mary Ann Fallows n.p.
Having now decided on going to the Cape would like a few queries answered regarding travel, housing, servants, dress, and medical facilities.
draft 4pp RS:HS 7.164 & 21.126

2496. [1832 or earlier] From James A. Gordon n.p.
Is grateful for JH's congratulations on his marriage. Revolution is in the air. Gives account of his marriage at St. Albans Abbey.
AL 4pp inc RS:HS 8.205

2497. [1832 or 1833] From James A. Gordon The Archimedes, Bristol
Pleased to find JH and family are well. Will shortly be visiting the Wye valley. Wishes JH could be induced to come and live in the Bristol area as it is so beautiful and yet convenient to London.
ALS 3pp RS:HS 8.204

2498. [1832] From James A. Gordon n.p.
Was grateful for JH's hospitality. Mrs. Gordon is on the way to Stocks. Would like to pair off with someone in order to save the journey to Cambridge for J. W. Lubbock's election. Encloses a letter he has had printed.
ALS 3pp RS:HS 8.203

2499. [1832] From Thomas John Hussey n.p.
Came with the intention of paying JH money, but finds he is out; what shall he do? Can he also advise him about the map for the R.A.S.?
ALS 1p RS:HS 10.144

2500. [1832] To Thomas John Hussey [London]
Because TH's letter has not reached JH, JH cannot reply. Is taking up residence in Slough.
ALS 3pp Univ.Coll.London MS Graves 23(1)

2501. [1832] To [John William Lubbock?] n.p.
Will leave Slough, if necessary, to vote in Parliamentary election for JL as representative for Cambridge.
ALS 2pp RS LUB.H.349

2502. [1832] To Roderick Impey Murchison [Slough]
Writes to inform RM of the arrival of Mr. Simonoff, director of the astronomical and magnetic observatory of Kasan, and indicates Mr. Simonoff's likely interests.
ALS 3pp APS 509.L56.14

2503. [1832] To ? n.p.
Gives permission to do anything with JH's letter; asks to be excused to J. W. Lubbock for not having written back.
ALS 1p RS LUB.H.350

2504. 1832–1–2 From Francis Beaufort Admiralty
Has asked Thomas Henderson to inform JH of his movements so that they can meet one another.
ALS 3pp RS:HS 3.325

2505. 1832–1–6 To Susan White Windsor
Writes to inform SW of the death of JH's mother.
ALS 2pp BL Aaa 37951.31

2506. [1832]–1–7 From Mary Harcourt St. Leonards [near Windsor]
Considers his letter a particular favor [possibly giving the news of the death of JH's mother]. May be able to meet Lady

Herschel in the Spring.
ALS 3pp RS:HS 10.155

2507. 1832–1–9 To Charles Babbage Windsor
Recent loss [death of his mother] has put the controversies temporarily aside. Gives further news of various matters which affect him. Feels angry at David Brewster's manner.
ALS 4pp RS:HS 2.271 & 21.99

2508. 1832–1–9 To William Henry Slough
Comments on a letter from WH in which he discussed experiments WH had made on the infection of cowpox, typhoid, and scarlatina.
ACCS 3pp RS:HS 25.3.3

2509. 1832–1–10 From Charles Babbage Dorset St.
Has just heard of the death of Lady Herschel. Will be at Slough for the funeral on Friday.
ALS 1p RS:HS 2.272

2510. 1832–1–10 From Francis Beaufort Admiralty
Would like his opinion on a political matter *viz* disenfranchising of boroughs.
ALS 3pp RS:HS 3.326

2511. 1832–1–10 From David William Smith Alnwick
Letter of condolence on death of JH's mother.
ALS 2pp RS:HS 16.183

2512. 1832–1–11 From James Wood St. John's College
Delighted by JH's discovery of orbit of Gamma Virginis. JH has conclusively established 'the indefinite extent and uniformity of the principle of gravitation.' Invites JH and wife to visit Cambridge.
ALS 2pp RS:HS 18.297

2513. 1832–1–13 To Francis Beaufort n.p.
Responds to FB's request for advice on a political document prepared by Thomas Drummond analyzing parliamentary representation and taxation in mathematical terms.
draft 6pp RS:HS 3.326bis (C: RS:HS 21.100)

2514. 1832–1–13 From Thomas Henderson Old Palace Yard, Westminster
Spent a pleasant time with G. B. Airy at Cambridge. The 'Melville' ship sails from Portsmouth on Thursday and he has been promised a passage to the Cape in it. Proposes to call on him tomorrow to discuss certain things.
ALS 2pp RS:HS 9.284

2515. 1832–1–14 From Augustus De Morgan 90 Guilford St.
Returns his paper and would be glad to receive an abstract. JH has been removed from the Council of the R.A.S. in deference to his wishes. King of Denmark has offered a gold medal for the discovery of a comet.
ALS 3pp RS:HS 6.176

2516. 1832–1–17 To John Gore Slough
Thanks MF for his kindness and attention to JH's mother in her final illness, and asks JG for his bill.
ALS 1p WT A-299

2517. 1832–1–20 From Caroline Lucretia Herschel Hanover
Saddened at the death of JH's mother; knows that 'it can't be long before I shall follow the dear departed.'
L *Memoir ... of Caroline Herschel*, 252 inc

2518. 1832–1–25 To Roderick Impey Murchison n.p.
Attempts explanation of curious subterranean acoustic phenomenon at Nakoos. Explains other natural acoustic phenomena and eruptions of geysers in Iceland. P.S.: Do not nominate JH to council of Geological Society.
(A)L 6pp TxU:H/L-0267; Reel 1054

2519. 1832–1–30 From Thomas Hodgkin New Broad St.
Was pleased to receive the elements of Gamma Virginis. Congratulations on his knighthood. J. J. Lister called on him recently and would like to see JH.
ALS 2pp RS:HS 9.405

2520. 1832–2–9 From George Biddell Airy Observatory, Cambridge
Has been making optical experiments. Hopes to see JH in the Spring. Hopes JH will be able to observe Encke's comet this year.
ALS 2pp RS:HS 1.50

2521. [1832–2–9 or earlier] To Thomas Hodgkin [Slough]
Makes arrangements for a visit to TH to see some of J. J. Lister's experiments. JH is sending TH some gelatinous matter that JH found in JH's telescope.
ALS 3pp WT.Ho A.4

2522. 1832–2–9 From Thomas Hodgkin New Broad St.
The time he proposes to visit them is very suitable and J. J. Lister will be there. The substance found in JH's telescope is very interesting. Hopes his mother is still well.
ALS 3pp RS:HS 9.406

2523. 1832–2–14 From Margaret Brodie Herschel to Caroline Lucretia Herschel Slough
Lovingly describes various events regarding JH's children and relatives.
ALS 6pp TxU:H/L-0577.1; Reel 1058

2524. 1832–2–16 To Margaret Brodie Herschel Atheneum
Brief note about JH's travel to Slough and the need to come back to London soon.
ALS 1p JHS.HCEJ.521

2525. 1832–2–22 From Susan White [Shaw, Berkshire]
[SW's servant] Sally celebrates JH's birthday every year and longs to see JH again.
Abs 1p TxU:H/M–0580.4; Reel 1093

2526. 1832–2–22 From Susan White Speenhamland [near Newbury, Berkshire]
Thanks JH for ring of deceased Mary Herschel. Would like to see JH's children, but age confines SW.
ALS 2pp RS:HS 18.229

2527. 1832–2–23 To Margaret Brodie Herschel Slough
JH and the baby are getting along well; JH talks about cholera; is 'almost certain I had the comet in the [JH's telescope's] field.'
ALS 3pp JHS.HCEJ.519

2528. 1832–2–25 From Miles Bland Ramsgate
The postal packet that he forwarded to Durham has been lost, so would be grateful if JH would send him another recommendation. Must not speak too humbly of himself. Finds Cambridge much changed.
ALS 2pp RS:HS 4.158

2529. 1832–2–25 To Charles G. B. Daubeny Slough
Sending a paper for CD and one for S. P. Rigaud and William Buckland. Would like Buckland's support for W. H. Mill, a candidate for the Boden professorship at Oxford. Has been

speculating on the effect of snow on the heights of mountains. Thinks David Brewster has carried his joke about the decline of chemistry too far.
draft 3pp RS:HS 6.18 & 21.102

2530. 1832–2–25 To Stephen Peter Rigaud Slough
Asks SR to vote for [William Hodge] Mill for the new Boden Professorship of Oriental Literature at Oxford. Thanks SR for gift of SR's book on James Bradley.
ALS 2pp Bodleian Library, Oxford (C: RS:HS 21.101)

2531. 1832–2–26 To Margaret Brodie Herschel Slough
All is well in the household; JH is to chair the cholera committee for the parish, which includes inoculation for all committee members and an oath to treat the sick if nurses cannot be found; JH is receiving political pamphlets [which he ridicules], after Lord John Russell cited JH as an authority in parliament.
ALS 4pp JHS.HCEJ.520

2532. 1832–2–29 To Mr. Briarly Slough
Has received a pamphlet from B about the apportionment of boroughs in a manner to eliminate the 'rotten' boroughs; JH critiques the pamphlet, especially its attempt at mathematical logic.
ACCS 4pp RS:HS 25.3.5

2533. 1832–2–29 To Thomas Drummond Slough
Comments on comparison of boroughs based on assessed value and on taxation.
C 5pp RS:HS 21.103 McLennan, *Thomas Drummond*, 154

2534. 1832–2–29 To Mr. McIntyre Slough
Arguing a statistical principle related to assessed values of several different boroughs.
ALS 4pp RS:HS 25.3.4

2535. 1832–2–29 To an Unknown Pamphleteer Slough
Responds to an unidentified mathematician who had written a critique of ideas put forward by Thomas Drummond, regarding whether to estimate the importance of various boroughs primarily in terms of population or in terms of the wealth of the population.
L John F. McLennan, *Memoir of Thomas Drummond* (Edinburgh, 1867), 157–9 inc

2536. 1832–2–29 To [?] Slough
Compares 'Products' principle with 'Sums' principle for numerical analysis of taxation in several boroughs. JH's opinion favoring Sums was quoted 'in the Assembly.'
C 4pp RS:HS 19.64

2537. 1832–3–1 From Stephen Peter Rigaud Oxford
Thanks JH for paper on micrometrical measurement of double stars. Has already promised to vote for someone else for Boden Professorship [see JH's 1832–2–25].
ALS 1p RS:HS 14.344

2538. 1832–3–2 To Joseph Trumper Slough
Thanks JT for loan of [John] Bullar's lectures. Stresses the value of good literature, especially by John Milton and Miguel de Cervantes.
C 2pp RS:HS 21.104

2539. 1832–3–6 From Thomas Romney Robinson Armagh
Introduces Edward Cooper, who has second largest telescope in the world [after JH]. Asks JH to assist Cooper. Discusses various astronomical issues and instruments.
ALS 3pp RS:HS 14.374

2540. 1832–3–12 From Charles Babbage Dorset St.
Regarding the situation on the Cape of Good Hope. Is sending a paper of Dr. [William?] Henry. David Brewster suffered last Thursday.
ALS 1p RS:HS 2.273

2541. 1832–3–12 From William Henry Fitton Geological Society
Has not received an invitation from Frederick Augustus (Duke of Sussex) for his soiree nor been introduced to him. Charles Babbage has likewise had no invitation. JH's case is different.
ALS 4pp RS:HS 7.232

2542. 1832–3–13 From Francis Beaufort Admiralty
A note about procedures for forwarding JH's papers to the Admiralty.
ACC 1p Hydrographic Office Letter Book

2543. 1832–3–15 From Francis Beaufort Admiralty
A note accompanying a letter about meteorological and atmospheric investigations.
ACC 1p Hydrographic Office Letter Book

2544. 1832–3–16 From Robert Grahame, Sr. Whitehall
Regrets to hear of the death of JH's mother. Has lost many friends and relatives during the last eight years. Staying with his two sons and has retired so would welcome a visit from the Herschels.
ALS 2pp RS:HS 9.5

2545. 1832–3–16 From Richard Jones Brasted
His wife is ill and is now at Brighton. Has just received JH's parcel. Is pleased JH likes William Whewell's article. Adam Sedgwick has refused a living.
ALS 3pp RS:HS 10.353

2546. 1832–3–17 From Wilhelm Struve Dorpat
Asks JH to thank Henry Kater for his comparison of WS's scale to that of the standard-yard. Mentions breakout of cholera as reason for delay in responding. Includes many measurements of year's work.
ALS 8pp (German) RS:HS 17.122

2547. 1832–3–18 To Francis Baily Slough
Is to dine with Frederick Augustus (Duke of Sussex). Regarding pendulums and French theorists. Will be pleased to see his paper on proper motion when ready. Regarding new amateur astronomers.
ALS 3pp RS:HS 3.107–8 (C: RS:HS 21.105)

2548. 1832–3–18 From Francis Baily 37 Tavistock Place
Is sending a copy of N. L. Lacaille's catalogue of principal stars. New observer in Yorkshire, a Mr. Beaumont. Pendulum theories known in France, and printed, but not apparently generally known. Is working on new paper for the R.S.L. on various kinds of pendulums.
ALS 4pp RS:HS 3.106

2549. 1832–3–27 To William Somerville Slough
Asks WS to forward enclosed note to Professor Schlegel. Disappointed that his celestial mechanics lecture [review] did not appear in last issue of the *Quarterly Review*.
ALS 1p RS:HS 16.400

2550. 1832–4–3 From Rummohun [?] Roy 48 Bedford Square
Must decline invitation to dinner due to poor health.
ACCS 2pp RS:HS 14.504

2551. 1832–4–5 From Dionysius Lardner 122 Regent St.
Wonders how JH's projected volume on Astronomy is progressing. Would also like a short biography of Francis Bacon.
ALS 3pp RS:HS 11.122

2552. 1832–4–10 From Richard Potter Smedley Hall
Discusses study of light, JH's theories on the subject, and his own photometer observations. Asks JH for further explanations of his theory.
ALS 3pp RS:HS 14.181

2553. 1832–4–12 From James Grahame L'Eperonniere
JG is revising his diary.
ALS 3pp RS:HS 8.411

2554. 1832–4–13 From Thomas Hodgkin New Broad St.
Has received the chronometer and will return it to Lister, together with his remarks on apertures. Charles May has found another individual suffering from color blindness. A friend has successfully made a reflecting microscope.
ALS 2pp RS:HS 9.407

2555. 1832–4–17 To Caroline Lucretia Herschel Slough
Wilhelm Struve's observations support JH's findings concerning the rapid revolution of Eta Coronae. In acknowledgement of the discovery that double stars are a 'revolving binary system,' JH changed the inscription on William Herschel's monument.
ALS 3pp TxU:H/L–0577.2.a; Reel 1058 *Memoir ... of Caroline Herschel,* 281 inc

2556. 1832–4–20 From Caroline Lucretia Herschel Hanover
Reminds JH to send his catalog of double stars.
L *Memoir ... of Caroline Herschel,* 254 inc

2557. 1832–4–22 From Francis Baily 37 Tavistock Place
Passing on the offer of assistance by Frederick Augustus (Duke of Sussex) for JH's projected visit to the Cape of Good Hope.
ALS 3pp RS:HS 3.109

2558. 1832–4–22 To Richard Potter Slough
Responds to RP's concerns about the nature of light and the interpretation of some interference experiments. JH believes that the undulatory theory of light is the best supported by the experiments at this time.
ACCS 3pp RS:HS 25.3.6

2559. 1832–4–24 To Editors of the *Annales de chimie et de physique* Slough
Discusses a method of separating iron oxide from the oxides of other metals and a process for the purification of uranium oxide.
L *Annales de chimie, 49* (1832), 306–11

2560. 1832–4–24 To Francis Baily Slough
Is grateful for the offer of assistance from Frederick Augustus (Duke of Sussex), but would not care to avail himself of public funds as his intended visit is for his own private reasons.
ALS 3pp RS:HS 3.110 & 21.106

2561. 1832–4–24 From Mary Ann Fallows to Margaret Brodie Herschel Bridekirk
Has received Sir John's letter. Gives details of the necessities for a voyage to the Cape. Conditions at the Cape, housing, food, servants, etc.
ALS 3pp RS:HS 7.165

2562. 1832–4–[26] To Editors of the *Annales de chimie et de physique* Slough
JH's notes on separation of iron oxide and a new procedure for complete purification of uranium.
C 4pp † (French) TxU:H/L-0036; Reel 1054

2563. 1832–4–26 To [François J. D. Arago] Slough
Sends letter, inspired by J. von Liebig's paper on separation of metallic oxides, to *Annales de Chimie et de Physique*. Fears JH's scientist friends may fall victim to the 'terrible malady...now raging' [in Paris].
ACCS 1p TxU:H/L-0035; Reel 1054

2564. 1832–4–27 From George Biddell Airy Observatory, Cambridge
Will he be visiting Cambridge this Spring?
ALS 1p RS:HS 1.51

2565. 1832–4–28 From James Grahame L'Eperonniere
Encourages JH to undertake a proposed trip to the Southern Hemisphere [letter completed 1832–5–1].
ALS 2pp RS:HS 8.412

2566. 1832–5–4 To George Biddell Airy Slough
Is trying to clean up JH's nebulae observations preparatory to leaving for the Cape; comments on communications from F. G. W. Struve about double stars.
ALS 4pp RGO 6.804.93

2567. 1832–5–6 From George Biddell Airy Observatory, Cambridge
Observations on the various ways of measuring angular positions (of double stars). Opinion on JH publishing his catalogue of nebulae. Remarks on the low standard of astronomical observations in England compared with the Continent.
ALS 4pp RS:HS 1.52

2568. 1832–5–6 From William Henry Smyth Crescent, Bedford
Will observe transit [of Mercury] hoping it will provide good meridian data. Encloses sketch of telescope in Bedford.
ALS 2pp encl RS:HS 16.245

2569. 1832–5–10 or 1832–5–16 From Francis Beaufort Admiralty
Regarding instructions for James Dunlop at Parramatta Observatory. Richard Copeland has been reappointed to the Meteor, now renamed Beacon.
ALS 2pp RS:HS 3.327 & ACC: Hydrographic Office Letter Book

2570. 1832–5–10 From James David Forbes Greenhill, Edinburgh
Is proposing to visit the continent after the Oxford meeting to make observations and intends taking some meteorological instruments. Would like JH's advice on some points regarding the large thermometer filled with a blue liquid.
ALS 3pp RS:HS 7.276

2571. 1832–5–12 To Wilhelm Struve Slough
Is now committed to the Cape journey, and will be occupied most of the time till then with the reduction of JH's nebulae observations. Comments on several other astronomical matters. [Letter is finished on board a steam packet on the way to Hamburg to visit JH's aunt, Caroline, dated 1832–6–9.]
ACCS 4pp RS:HS 25.3.7

2572. 1832–5–15 From James South Observatory, Kensington

Has heard that JH is publishing an attack on JS's double star observations made in France. Is this correct?
ALS 2pp RS:HS 16.469

2573. 1832–5–16 To James South Slough
Responds to JS's fear [see 1832–5–15] that JH has been preparing an attack on JS's double star observations by explaining in detail and providing extensive evidence that JH has no such intention.
ACCS 4pp RS:HS16.470 (C: 21.107)

2574. 1832–5–17 To Francis Baily Slough
Would he inform him of the date of the next visitation meeting. Will bring his own catalogue of double stars for the next R.A.S. meeting. Thanks for his paper on proper motion.
ALS 3pp RS:HS 3.111 & 21.108

2575. 1832–5–18 From James David Forbes Greenhill, Edinburgh
In preparation for foreign travel, asks JH about instruments and method for measuring radiation.
ALS 3pp St. Andrews LB I, 678–81

2576. 1832–5–21 From Francis Baily 37 Tavistock Place
Giving the dates of the visitation day. Also the date of the next meeting of the R.A.S. Has been experimenting with pendulums.
ALS 2pp RS:HS 3.112

2577. 1832–5–21 From Francis Beaufort Admiralty
Is sending letters from Thomas Henderson, the Cape Astronomer, also observations from St. Helena.
ALS 2pp RS:HS 3.328

2578. 1832–5–22 From William Henry Smyth Crescent, Bedford
Sends Gamma Virginis observations. Devotes time to nautical astronomy; experiments with formulae for calculating occultations. Takes interest in eclipse of Jovian satellites.
ALS 4pp RS:HS 16.246

2579. 1832–5–23 To James David Forbes Slough
Sends thanks and congratulations for paper verifying electric origins of all magnetism. Will give JF an actinometer with complete instructions. Includes some instructions for obtaining and using it. Sends regards to Mario Gemmellaro, should JF go to Catania.
ALS 4pp St. Andrews 73 (C: RS:HS 21.110)

2580. 1832–5–23 To William Henry Smyth Slough
Sends observations of Gamma Virginis, which differ from those of WS and James South. Observed transit of Mercury. Has abandoned plans of going to Cape of Good Hope this year. Plans to visit Caroline Herschel in Hanover.
C 2pp RS:HS 21.109

2581. 1832–5–25 To George Biddell Airy Slough
Is grateful for GA's encouragement to prepare JH's nebulae observations for publication before leaving for the Cape; it now seems clear to JH that he cannot leave before 1833; JH has had further communications from F. G. W. Struve about double stars [see JH's 1832–5–4].
ALS 4pp RGO 6.804.105

2582. 1832–5–26 From James David Forbes Greenhill, Edinburgh
Is grateful for JH's letter and offer of assistance. Will come to London earlier than expected so that he can call on him and discuss his problems before foreign travel. Would be glad of an introduction to Mario Gemmellaro and other scientists JH thinks would help him in his studies.
ALS 4pp St. Andrews LB I, 684–7 (C: RS:HS 7.277)

2583. 1832–6–1 From Rummohun [?] Roy 48 Bedford Square
Aware that letter of 3 April did not reach the Herschels' home until after the engagement. Wishes to explain the apparent neglect.
ALS 2pp RS:HS 14.505

2584. 1832–6–4 From William Rutter Dawes Ormskirk
Regrets belated reply to JH's pamphlet on double stars. Comments on the readings and compares them with some of his own.
ALS 3pp RS:HS 6.50

2585. 1832–6–4 From Alexander Rogers Baltic St., Leith
Discusses the ale he sent. Hopes his hypothesis of the origin of the solar system did not offend JH.
ALS 2pp RS:HS 14.414

2586. 1832–6–6 From Nathaniel Bowditch Boston (Mass.)
Has received copies of his papers that he communicated to the R.S.L. for which he thanks him. JH's study on Natural Philosophy has been reprinted in the States. Has forwarded the second volume of *Mécanique céleste,* one for him and one for the R.A.S.
ALS 1p RS:HS 4.200

2587. 1832–6–7 From David Brewster Adelphi
His son is shortly traveling to Calcutta and will be able to take any letter or parcel for Mr. Calder.
ALS 2pp RS:HS 4.266

2588. 1832–6–8 To David Brewster n.p.
Excusing JH for returning to Slough suddenly before leaving for Hanover.
C 1p RS:HS 21.111

2589. 1832–6–8 From James Wood St. John's College
Commissioned to ask JH to select painter and to sit for half-length portrait to be placed in Old House at St. John's. Received JH's paper on the measures of double stars.
ALS 2pp † RS:HS 18.298a

2590. 1832–6–11 To Margaret Brodie Herschel Belvedere Hotel, Hamburg
Recounts JH's travels to date, together with stories about some of the people he has met; also comments on cholera in various cities [letter completed 1832–6–13].
ALS 4pp JHS.HCEJ.522a

2591. 1832–6–12 To Caroline Lucretia Herschel Hamburg
On his German travels, JH left Margaret Herschel and the children at Slough because he feared their exposure to the cholera epidemic.
ALS 4pp TxU:H/L–0577.3; Reel 1058

2592. 1832–6–14 From Thomas Drummond Whitehall
Capt. Beaufort has just put two of JH's letters into his hands. Is conscious of the kindly way in which JH entered into the problem. Would like the chance of a conversation with him on the subject.
ALS 4pp RS:HS 6.454

2593. 1832–6–18 From James Grahame L'Eperonniere
Wants to hear more specifics of JH's travel plans.
ALS 2pp RS:HS 8.413

2594. 1832–6–18 From Heinrich Christian Schumacher Altona
Sends news of a cholera epidemic to help JH make his summer plans. Mentions a number of new cases from Altona, Hamburg, and Berlin. Asks JH not to spread information.
ALS 2pp RS:HS 15.374

2595. 1832–6–19 To Carl Friedrich Gauss Hanover
Unable to visit, but has sent two papers on the orbits of double stars. Discusses double stars, noting that four double stars are now known to orbit with periods less than a century long. Wilhelm Olbers is seriously ill.
ALS 3pp NSUB Cod.Ms.Gauss 99 (C: RS:HS 21.112)

2596. 1832–6–19 To Margaret Brodie Herschel Hanover
Reports of further travels, and JH's visit to his aunt Caroline Herschel; interesting stories about his stay in Hanover, including a report on JH's attendance at the Waterloo banquet.
ALS 4pp JHS.HCEJ.522b *Memoir ... of Caroline Herschel*, 254–6

2597. 1832–6–22 From Heinrich Christian Schumacher Altona
Sends two bottles of Claret and a box to be forwarded for [Francis] Baily.
ALS 2pp RS:HS 15.375

2598. [1832 to 1841]–6–23 From John Murray, [Sr.?] 50 Albemarle St.
Compliments to 'Sir John Hershall.' Sends four copies of next issue of *Quarterly Review*. Will send twelve copies of JH's article tomorrow.
L 1p RAS:JH Archive 3/1.5.7; Reel 7

2599. 1832–6–26 To Margaret Brodie Herschel Cuxhaven
Warns MH about delays in crossing the North Sea, due to storms; will search out the earliest alternative that will carry JH with some comfort.
ALS 3pp JHS.HCEJ.522c

2600. 1832–6–27 To Adolphe Quetelet Cuxhaven
Requests that AQ forward letter to London explaining JH's delay in returning home. Promises to send latest paper on orbits of double stars. Mentions [Wilhelm] Struve's confirmation of data. States that [Wilhelm] Olbers is dying. [H. C.] Schumacher is recovering.
ALS 3pp Académie belgique: #5

2601. [1832–6–28?] To Mario Gemmellaro n.p.
Recalls with delight experiments in solar radiation at MG home and puts MG in communication with James Forbes of Edinburgh.
ALS 2pp St. Andrews 113

2602. 1832–6–29 From Charles L. Harding Göttingen
Thanks JH for his recent writings on double stars, and comments on a number of observational issues related to the same topic.
ALS 2pp (German) RS:HS 9.237

2603. 1832–6–30 To Margaret Brodie Herschel [London]
More about JH's difficulties in reaching England [see JH's 1832–6–26] because of storms at sea; will now try to clear Customs House and come to MH this day.
ALS 4pp JHS.HCEJ.522d

2604. [1832–7 or later] To [?] Slough
Please send vol. 2 of Nathaniel Bowditch's translation of P. S. Laplace's *Mécanique céleste.*
ALS 1p BostonPubLib Ms.196(1)

2605. 1832–7–8 From D. François J. Arago Paris
Wants information for the journal. Apologizes for the delay in writing the letter.
ALS 1p (French) RS:HS 1.356

2606. 1832–7–12 From Joseph Johann Littrow Vienna
Sends two articles for the R.A.S. Gained much pleasure from reading JH's *Prelim. Discourse.* If JH has written any other books for Lardner's *Cabinet Cyclopaedia,* he would be glad to receive copies. Work on Light has made great impression. Has Charles Babbage finished his *Logs.* yet? Cholera prevalent in Vienna.
ALS 8pp (German with English trans.) RS:HS 11.261

2607. 1832–7–14 To Caroline Lucretia Herschel Slough
JH and Margaret Herschel visited the Isle of Wight.
ALS 4pp TxU:H/L–0577.4; Reel 1058

2608. 1832–7–16 To Henry Kater Slough
About the need to have the standard pound carefully measured so it can be compared with other standards of weight.
C 2pp RS:HS 21.113

2609. 1832–7–17 From James David Forbes Spa
Regarding JH's actinometers, intends to compare their results when in Switzerland and would be pleased if JH has any determinations on them. Outlines the plan of his experiments. Thinks JH should publish an account of them before JF publishes the results of his experiments. Met L. A. J. Quetelet in Brussels. Intends visiting the Pyrenees.
ALS 3pp RS:HS 7.278

2610. 1832–7–17 To William Henry Smyth Slough
Returned from Hanover visit. Observed the star WS had discovered between 36 Ophiuchi and 30 Scorpii. Requests WS to observe a nebula in which JH has spotted a triple star. H. C. Schumacher is recovering from illness, while H. W. M. Olbers is very ill and not expected to recover.
ALS 4pp APS B.Sm98.11 (C: RS:HS 21.114)

2611. 1832–7–18 From Augustus De Morgan 90 Guilford St.
Has sent off his observations by the coach. Account of the Oxford meeting is in Tilloch's Magazine, now amalgamated with David Brewster's Edinburgh Journal. Will be able to include the new catalogue of stars in the forthcoming volume if JH desires it.
ALS 2pp RS:HS 6.177

2612. 1832–7–19 From Walter Forman Pilton
Thanks for his angry effusion. His non-reply to WF's paper proves that his theories must be correct.
ALS 3pp RS:HS 7.376

2613. 1832–7–22 From Basil Hall Portsmouth
Regrets a hasty opinion should have evoked so long a letter. Regarding the status of cholera. Would be grateful for a specimen of W. Herschel's handwriting. Sun has been showing symptoms of decay and several have been trying to observe it with limited success.
ALS 3pp RS:HS 9.170

2614. 1832–7–26 To James Wood Slough
Flattered by JW's request [see JH's 1832–6–8] for portrait. Business matters delay JH selecting artist.
ACCS 1p RS:HS 18.298b

2615. 1832–7–31 From James Grahame L'Eperonniere
Of travels and family health.
ALS 2pp RS:HS 8.414

2616. 1832–8–4 From Henry Kater Tunbridge Wells
Has been prevented from answering his letter earlier by increased illness. If he returns to town he will have pleasure in meeting H. C. Schumacher. Believes the pound must be made of bell metal. Thought JH was on his way to the Cape, but sees his letter was written from Slough.
ALS 2pp RS:HS 11.17

2617. 1832–8–5 To James David Forbes Slough
Suggests that JF use his instruments to compare the force of solar radiation at high and low elevations. Delineates possible methods for this experimentation, and lists results of similar experiments. Includes further information on the actinometer.
ALS 3pp St. Andrews 123 (C: RS:HS 25.3.8)

2618. 1832–8–14 From Francis Beaufort Admiralty
Sending letter for him which was delivered to the Admiralty. Has had a letter from James South who is pleased with his reception in Copenhagen.
ALS 2pp RS:HS 3.329

2619. 1832–8–15 To [?] Slough
As offered in [?]'s letter of 9 Aug. 1832, JH accepts position as member of B.A.A.S. Council.
ALS 2pp TxU:H/L-0746; Reel 1089

2620. 1832–8–16 To William Buckland Slough
Accepts position on Council of B.A.A.S. Approves of next meeting's being held at Cambridge.
ALS 2pp St. John's College, Cambridge

2621. 1832–8–20 From Mario Gemmellaro Nicolosi sull'Etna
Sending three memoirs by his brother, two on Pantellaria and one on the forms of mountains. Would like a copy of JH's observations made in Sicily.
ALS 1p (Italian) RS:HS 8.95

2622. 1832–8–25 From Jacques Babinet Paris
Introduces Mr. A. Perrey. Remarks on tourmaline. Astronomical matters.
ALS 3pp (French) RS:HS 3.1

2623. 1832–8–25 From Alexander Rogers Baltic St., Leith
Has no intentions of making solar volcano theory public. Offers one more possibility for its validity. Pleased that Caroline Herschel is well and remembers him.
ALS 3pp RS:HS 14.415

2624. 1832–8–27 From Thomas John Hussey Hayes
Queries regarding a corroded clock face, and description of harvest bugs. Has an intermittent river in their district. Has now set up the Munich telescope and would be pleased to show it to JH. Does he want Amici's mirrors?
ALS 4pp RS:HS 10.142

2625. 1832–9–3 From Basil Hall Portsmouth
Describes a new form of sketch book, which may interest him. Encloses a paper from a friend, which can be answered by means of algebra, of which he himself is ignorant. JH's affairs were mentioned by Sir James Graham, who suggested that the Admiralty may be able to assist JH with a passage to the Cape.
ALS 3pp RS:HS 9.171

2626. 1832–9–3 From William Henry Smyth Crescent, Bedford
Sends micrometer observations of stars suggested by JH.
ALS 3pp RS:HS 16.247

2627. 1832–9–10 From Thomas John Hussey Hayes
The black broth that the Spartans liked must have been mushroom ketchup. Suggests way of obtaining a portable dome. Sends drawing and description of the harvest bug. Is pleased to hear of his projected Cape expedition. Is carrying out observations of the moon.
ALS 4pp RS:HS 10.143

2628. 1832–9–12 To Jacques Babinet Slough
Responds to JB's 1832–8–25 with comments about the optical properties of tourmaline. JH also talks about his astronomical work, including the collimator on his telescope and his double star work.
ACCS 4pp (French) RS:HS 25.3.9

2629. 1832–9–14 From William Whewell Trinity College
Sends JH a report by Mrs. Thomas Malthus on experiments relating to vision. Mentions T. Malthus's views of books on political economy by Thomas Chalmers and Richard Jones. Praises T. Malthus.
ALS 4pp encl TxU:H/M-0564

2630. 1832–9–16 To Basil Hall Slough
Gives a reply to the query on life assurance, but is not clear what the question is really about. Is flattered by the suggestion for a passage to the Cape in a Naval vessel, but has made up his mind to travel privately.
C 4pp RS:HS 9.172 & 21.115

2631. 1832–9–23 To Wilhelm Struve Slough
Compares in detail observations of several double stars made by several astronomers, and comments on differences; JH also describes a sighting of Biela's comet.
ACCS 4pp RS:HS 25.3.10

2632. 1832–9–23 To the London *Times* Slough
Gives data necessary for astronomers to observe Biela's Comet.
L London *Times* (Sept. 26, 1832), p. 2, col. 6

2633. 1832–9–24 From Pierre L. Dulong Paris
Acknowledging receipt of JH's book on double stars, which has been placed in the library.
ALS 1p (French) RS:HS 6.510

2634. 1832–9–24 To London *Times* Slough
Discusses the history, nature, and appearance of Biela's Comet.
L London *Times* (Sept. 27, 1832), p. 2, col. 6

2635. 1832–9–25 To William Henry Smyth Slough
Sends positions of Biela's Comet. Has much success with repolished telescope mirrors and collimator. Designs equatorial revolving roof for Cape observations; sends sketches. Compares WS's, W. R. Dawes's, and JH's observations of Gamma Virginis.
ALS 4pp APS B.Sm98.12 (C: RS:HS 21.116)

2636. 1832–9–28 From William Henry Smyth Crescent, Bedford
Unable to spot Biela's Comet. Outstanding observing conditions allow WS to sight normally difficult objects. Makes suggestions concerning JH's rotating roof plans.
ALS 3pp RS:HS 16.248

2637. 1832–10–2 From Augustus De Morgan 5 Upper Gower St.
Hopes to have JH's catalogue for the first meeting of the R.A.S. Anything about the comet would be welcome. Regarding one of Franciscus Vièta's mathematical theories.
ALS 3pp RS:HS 6.178

2638. 1832–10–2 From James David Forbes Locarno
Has postponed answering his letter until he has some worthwhile results to transmit. Local savants are unwilling to take much trouble but has met a German meteorologist, L. F. Kämtz, and together they have carried out observations on some of the mountains. Encloses some of the actinometer readings. Has 80 series of magnetic observations.
ALS 4pp RS:HS 7.279

2639. 1832–10–2 From William Jones St. John's College
Is gratified that JH has consented to sit for his portrait. Let him know the artist he has selected and they can arrange sittings.
ALS 2pp RS:HS 10.411

2640. 1832–10–6 From Basil Hall Portsmouth
Sent JH's last letter to Sir James Graham to show him JH's feelings in the matter of transport to the Cape. Sir James replied to the effect that he would be very pleased to arrange a passage for JH. Hall feels in an awkward situation and thinks that JH should visit the Admiralty and have conversation with Sir James.
ALS 3pp RS:HS 9.173

2641. 1832–10–10 To Basil Hall Slough
Is grateful for BH's interest over the matter of travel to the Cape. Gives his own views on the intentions of Sir James Graham. Intends to see Graham personally and thank him for his interest.
draft 1p RS:HS 9.174 & 21.117

2642. 1832–10–15 From D. François J. Arago Paris
Thanking him for sending the Memoirs.
ALS 1p (French) RS:HS 1.357

2643. 1832–10–16 From Augustus De Morgan 5 Upper Gower St.
Has received the catalogue and it will be sent to the press. Is grateful for JH's offers. Does he know anything of Bullialdi's (Ismael Boulliau) *Astronomica philolaica*? Refers to Franciscus Vièta's *Harmonicum coeleste*. 25 copies of JH's latest paper has been forwarded.
ALS 2pp RS:HS 6.179 S. De Morgan, *Memoir of Augustus De Morgan*, 75–6

2644. 1832–10–20 To John George Children Slough
Declines invitation to serve on next R.S.L. Council, being busy with preparations for Cape trip.
ALS 2pp RS MC.2.51

2645. 1832–10–21 To Francis Baily Slough
Is sending, for communication to the R.A.S., observations on Biela's comet and a speculation on comets.
C 1p RS:HS 21.118

2646. 1832–10–24 From Giovanni Antonio Plana Turin
Still worrying about planetary perturbations of Jupiter and Saturn [see GP's 1829–10–11]. Seeks advice on the matter from JH, and asks JH to make any necessary corrections on papers GP has sent.
ALS 4pp (French) RS:HS 13.430

2647. 1832–10–25 From James South Hanover
Has visited Caroline Herschel, who is well. Discusses whether JH will make magnetic observations at the Cape of Good Hope.
ALS 3pp RS:HS 16.471

2648. 1832–10–26 From John Gibson Lockhart Sussex Place, Regents Park
Sending check in payment of JH's article on Celestial Mechanism in the *Quarterly Review*. Would be pleased if JH would now write an article on comets.
ALS 2pp RS:HS 11.301

2649. 1832–10–26 From William Whewell Trinity College
Is sending JH some writings by Mr. (Mrs.?) Thomas Malthus on idiotic people and a book by WW [*Astronomy and General Physics*]. Mentions Richard Jones. Asks whether WW has seen 'the comet.'
ALS 2pp TxU:H/M-0565

2650. [1832–11] To David Brewster n.p.
Regarding DB's views on support for JH in the presidency of the R.S.L.
draft 1p RS:HS 4.271 (C: RS:HS 25.13.16)

2651. [1832–11] From John Elliot Drinkwater-Bethune British Coffee House
Sends a printed circular setting forth the claims of J. W. Lubbock as a candidate for the forthcoming election.
ALS 1p RS:HS 6.445

2652. 1832–11–1 To John Gibson Lockhart Slough
Did not expect payment for JH's [review of Mary Somerville's *Mechanism of the Heavens*] in *Quarterly Review*. Returns John Murray's remittance. François Arago's 'comet paper' and ensuing panic. Condolences on [death of JL's son].
ALS 3pp RS:HS 19.65 (C: RS:HS 21.119)

2653. 1832–11–4 To William Henry Smyth Slough
As no one seemed to be finding Biela's Comet, JH sat up several nights looking for it and now has located it.
ALS 2pp WT 91800.2

2654. 1832 11 5 From James Pillans Edinburgh
Requests JH consider taking Chair of Natural Philosophy at University of Edinburgh due to John Leslie's illness.
ALS 7pp RS:HS 14.163

2655. 1832–11–6 From John Learmonth Edinburgh
Inviting JH to become a candidate for the Professorship of Natural Philosophy at Edinburgh University.
ALS 3pp RS:HS 11.466

2656. 1832–11–7 From John S. Forbes Edinburgh
His brother intends to apply for the professorship at Edinburgh University and would be glad of a testimonial from JH.
ALS 2pp RS:HS 7.319

2657. 1832–11–8 From William Ritchie University of London
Will offer himself as candidate for Professor of Natural Philosophy at University at Edinburgh. Asks JH for recommendation.
ALS 1p RS:HS 14.362

2658. 1832–11–[9 or earlier] To Michael Faraday n.p.
Writes a letter of introduction for a friend [Dr. Robinson] of a friend to MF. Comments favorably on MF's electromagnetic research.
C 1p RS:HS 25.14.9

2659. 1832–11–9 To James Pillans Slough
Expresses shock at death of John Leslie. Declines with thanks offer to be considered for professorship of natural philosophy at University of Edinburgh.
C 3pp RS:HS 14.164

2660. 1832–11–9 To William Ritchie Slough
Supports WR's pursuit of Edinburgh professorship. Will not, however, write to the Lord Provost for WR. Feels public steps should be taken to fill position [see WR's 1832–11–8].
C 3pp RS:HS 14.363 (C: 21.120)

2661. 1832–11–10 From Michael Faraday Royal Institution
Has received his letter introducing Dr. Robinson, but thinks that the latter is not a worthwhile person to know. Is pleased that JH likes his recent experiments. Owes a debt to JH's *Prelim. Discourse.*
ALS 2pp RS:HS 7.179 & Williams, *Selected Correspondence, 1*, 235

2662. 1832–11–11 To Michael Faraday Slough
Apologizes for sending a letter of introduction for someone recommended to JH, without first getting to know the third party.
C 2pp RS:HS 21.121

2663. 1832–11–12 To John S. Forbes Slough
Will not publicly recommend JSF's brother, James Forbes, for Chair of Natural Philosophy (at Edinburgh), but praises him, his enthusiasm, and his brother's determination that he obtain the position.
ALS 4pp St. Andrews 149

2664. 1832–11–13 From Robert Jameson Edinburgh
Has probably already received a letter from the Lord Provost inviting JH to become a candidate for the chair of Natural Philosophy at Edinburgh. Outlines the duties. Hopes he will also accept the presidency of the Royal Society of Edinburgh.
ALS 3pp RS:HS 10.434

2665. 1832–11–16 From Thomas Galloway Edinburgh
Is a candidate for the Chair of Natural Philosophy at Edinburgh and would be glad if JH would supply him with his views on any of TG's papers.
ALS 3pp RS:HS 8.4

2666. 1832–11–19 From David Brewster Allerly by Melrose
Regarding his request for JH's support for his candidature for the chair of Natural Philosophy at Edinburgh University. Reasons why he thinks JH should give him a testimonial.
ALS 3pp RS:HS 4.267

2667. 1832–11–20 to 1832–12–4 To *Astronomische Nachrichten* Slough
Reports observations of Delta Cygni, 4 Aquarii, Alpha^2 Capricorni, and Biela's Comet.
L *Astronomische Nachrichten, 10* (1833), 317–20

2668. 1832–11–20 From William Jones St. John's College
Entirely approves of his choice of Thomas Phillips for his portrait. Has already painted Hugh Percy (3rd Duke of Northumberland).
ALS 1p RS:HS 10.412

2669. 1832–11–20 From Giovanni Santini Padua
Praises JH's work on double star orbits. Translated *'Fisica Celeste'* [JH's *Physical Astronomy?*] for use of University of Padua. Receives fourth volume of *R.S.P.T.* of R.S.L. R.A.S. awards GS for astronomical work. Wants to purchase a Barlow telescope for Padua Observatory. Observes Biela's Comet at Rome Observatory. Makes study of perturbations in the orbit of Jupiter.
ALS 2pp (Italian) RS.HS 15.345

2670. 1832–11–22 From James A. Gordon Clifton
Sorry to hear that JH is contemplating a long voyage and stay from England. Hoped JH would have succeeded Professor John Leslie. Would like to know his views on reform. Has been making heavy going with JH's book on Light. Hopes to bring Mrs. Gordon to see Lady Herschel soon.
ALS 2pp RS:HS 8.167

2671. 1832–11–22 From Henry Kater York Gate
Obliged for the present of his papers. Will keep them until the catalogue arrives. The improvements in his 20' telescope are interesting; comments on this. Will obtain the troy pound as soon as Parliament meets.
ALS 3pp RS:HS 11.18

2672. 1832–11–23 From Charles May Ampthill
J. J. Lister also made set of experimental glasses similar to JH's [see TxU:H/L–0255]. Lister's experiments on color-blind persons. Will forward Lister's glasses to [John] Dalton. CM plans similar experiments with JH's glasses; will forward results to JH at Cape of Good Hope. Wishes to see JH's grinding and polishing machinery.
ALS 2pp TxU:H/M–0351; Reel 1087

2673. 1832–11–23 From Siméon Denis Poisson Paris
Seems to be writing about the awarding of the R.S.L.'s Copley Medal. [Letter almost totally illegible.]
ALS 3pp (French) RS:HS 14.14

2674. [1832–11–23 or earlier] To Charles May n.p.
Sends set of experimental spectacles for CM to use in testing persons for color-blindness, or 'Dichromic Vision.' Proposes several tests to be conducted with these.
AdraftS 4pp & AdraftS 3pp TxU:H/L-0254 & -0255; Reel 1054

2675. [1832]–11–29 From Sir Charles Bell 30 Brook St.
Hopes that JH will reconsider his decision regarding the professorship at Edinburgh University.
ALS 3pp RS:HS 5.40

2676. 1832–11–29 From John William Lubbock St. James's Place
Has been induced to become a candidate to represent the University of Cambridge in Parliament and encloses his address. Would like JH's support.
ALS 2pp RS:HS 11.358

2677. 1832–11–29 From Heinrich Wilhelm Matthias Olbers Bremen
Reports the elements of a comet's orbit [Biela's ?] based on the observations of F. B. G. Nicolai of Mannheim. Based on these observations together with some made by F. W. Bessel, WO gives the locations for the comet during the coming month. Also comments on astronomical observations made elsewhere.
ALS 3pp (German) RS:HS 13.178

2678. 1832–11–30 From William Rutter Dawes Ormskirk
His astronomical work has been delayed by ill health and he is behind with Wilhelm Struve's list of stars. Gives some errors he has noticed. A friend of his, Dr. Thomas Raffles, would like autographs of any famous men; can JH oblige?
ALS 4pp RS:HS 6.51

2679. 1832–11–30 To John Elliot Drinkwater-Bethune Slough
Responds favorably to JD's request that JH vote for J. W. Lubbock in a forthcoming election [to represent the University of Cambridge in Parliament].
ACCS 1p RS:HS 19.66 (C: RS LUB.H.347 & RS:HS 21.123)

2680. 1832–12–1 To Sir Charles Bell n.p.
Indebted to CB and friends for their encouragement, but JH declines Provost's invitation to be candidate for professorship at Edinburgh University.
AdraftS 2pp RS:HS 19.67

2681. [1832–12] From John Elliot Drinkwater-Bethune n.p.
Hopes that JH will not be absent from the Poll as great harm will be done if he is. Has circulated his letter.
ALS 3pp RS:HS 6.446

2682. 1832–12–1 From John Elliot Drinkwater-Bethune British Coffee House
Is gratified by the way in which JH has spoken of J. W. Lubbock's suitability as a Parliamentary candidate. Sends details of the resolutions passed at Cambridge.
ALS 3pp RS:HS 6.444

2683. [1832–12] From John Elliot Drinkwater-Bethune n.p.
The altered state of events makes his last letter redundant. J. W. Lubbock has retired from the contest.
ALS 1p RS:HS 6.447

2684. 1832–12–1 From James David Forbes Edinburgh
Returned suddenly from the continent and found JH's letter to his brother awaiting him. Outlines his reason for applying for the vacant post at the University and would welcome a testimonial from JH. Brewster is his chief rival. Gives a few observations on his recent trip to the continent.
ALS 4pp RS:HS 7.280

2685. 1832–12–1 From James Hudson Royal Society
Sends items by F. W. Bessel and G. A. Plana. Will be dispatching boxes to various European towns next week and will be pleased to include any items from JH. His observations made recently show interesting results. Comments on these results and the influence of the bore.
ALS 3pp RS:HS 10.32

2686. 1832–12–2 From George Biddell Airy Observatory, Cambridge
Observations on the gravimeter.
ALS 3pp RS:HS 1.53

2687. 1832–12–2 From D. François J. Arago Paris
Has asked L. F. Debure to send to JH the third volume of the *Memoirs* of the Foreign Correspondents of the Academy of Sciences.
ALS 1p (French) RS:HS 1.358

2688. [1832?]–12–3 From William Vernon Harcourt Wheldrake near York
Has great hopes for the success of the [B.A.A.S.]. Would like JH's advice on the phenomenon of radiant matter to the laws of common matter, which he is preparing for a future meeting.
ALS 4pp RS:HS 9.223

2689. 1832–12–5 From John William Lubbock St. James Place
A circular letter to the electors of the University of Cambridge, asking for their support in the upcoming poll for representative to Parliament [see JH's 1832–11–30].
LCC 1p DMC.L.928A.2

2690. 1832–12–6 To William R. Dawes [Slough]
Sends some autographs requested by WD. Discusses observations of comets, especially Biela's, and of various stars, especially double stars.
ALS 2pp Columbia Univ. Library

2691. 1832–12–7 To James David Forbes Slough
Gives JF permission to use JH's letter to JF's brother in any way JF sees fit to obtain the Chair of Natural Philosophy (at Edinburgh), but still does not wish directly to promote JF's selection. Briefly discusses results of JF's experiments with the actinometer.
ALS 3pp St. Andrews 171 (C: RS:HS 21.124)

2692. 1832–12–8 From James A. Gordon 29 Hill St.
Regarding J. W. Lubbock's Committee and JH's letter. Will probably meet at the Cambridge election. Unable to vote for Lubbock.
ALS 2pp RS:HS 8.168

2693. 1832–12–9 To ? Slough
Hopes for conveyance to Cambridge in covered vehicle for meeting of J. W. Lubbock's committee.
AL 1p inc RS LUB.H.348

2694. 1832–12–10 From James Grahame London
Has just put his son, Robert, into an institution for 'persons of unsound mind.'
ALS 2pp RS:HS 8.415

2695. 1832–12–11 To John S. Forbes Slough
Feels it difficult to supply a formal testimonial but has great admiration for the work of his brother and his prospects.
draft 3pp RS:HS 7.320 (C: RS:HS 21.122)

2696. 1832–12–11 To Thomas Henderson Slough
Proposes to come to the Cape in the ensuing season and would be pleased to bring anything TH may require. Enquires about conditions at the Cape and suitable sites for JH's observations.
ALS 4pp National Maritime Museum (draft: RS:HS 9.285 & C: RS:HS 25.13.3)

2697. 1832–12–13 From James David Forbes Edinburgh
Is sorry to trouble him again but would like a letter from JH that he can print with his testimonials, as David Brewster has now printed JH's letter, also [William?] Ritchie. Awaits the results before setting out for Europe.
ALS 3pp RS:HS 7.281

2698. 1832–12–14 To Charles Babbage Slough
Regarding CB's failure in political matters. Sees cause for congratulation. Would he tell him where he can see the manufacture of springs.
ALS 4pp RS:HS 2.274

2699. 1832–12–15 From Charles Babbage n.p.
Regarding the results of the Finsbury election. Has been invited to a public dinner and would like him to come, together with some of his scientific friends. Part of his machine is ready.
ALS 2pp RS:HS 2.275

2700. 1832–12–16 To Charles Babbage Slough
Will attend CB's dinner on Wednesday. Can JH bring his brother-in-law? Will call at CB's house before the dinner.
ALS 2pp RS:HS 2.276

2701. 1832–12–17 From George Biddell Airy Observatory, Cambridge
Requests that the engravings for the catalogue of nebulae be speeded up. Further observations on the gravimeter. Observations on W. R. Hamilton's experiments on biaxial crystals using A. J. Fresnel's wave surface theory.
ALS 3pp RS:HS 1.54

2702. 1832–12–17 To James David Forbes [Slough?]
Is drawing up a new actinometer and comments on JF's ideas about it. Encloses a letter to JF's brother.
ALS 2pp St. Andrews 183 (C: RS:HS 21.125)

2703. 1832–12–17 To John S. Forbes n.p.
Further letter regarding the qualifications of James D. Forbes for the position at Edinburgh University.
draft 2pp RS:HS 7.321 (C: RS:HS 21.125)

2704. 1832–12–18 From Charles Babbage Dorset St.
Is delighted that JH is coming to dinner.
ALS 1p RS:HS 2.277

2705. 1832–12–18 From Francis Baily London
Has forwarded the parcel as requested. Is unable to send O. F. Mossotti's parcel as he does not know his address. Wilhelm Olbers' letter is being translated. Attended a committee meeting of the R.S.L. on a new telescope. Printing of the *Memoirs* has been delayed.
ALS 3pp RS:HS 3.113

2706. 1832–12–22 From Roderick Impey Murchison 3 Bryanston Place
Would like JH to have his portrait painted by H. W. Pickersgill, at no cost, before JH departs for the Cape. RM's wife is in Hampshire.
ALS 4pp RS:HS 12.386

2707. 1832–12–25 From James Grahame L'Eperonniere
On the health of the family.
ALS 3pp RS:HS 8.416

2708. 1832–12–25 From William Henry Smyth Crescent, Bedford
Sights bright patch on moon; asks JH to confirm the observation.
ALS 1p RS:HS 16.249

2709. 1832–12–28 From Sir Edward Ryan Stetterbury
Introduces JH to a Sanskrit professor. Has heard JH may travel to the Cape, but thinks family matters may deter him from going.
ALS 3pp RS:HS 14.451

2710. 1832–12–30 From George Biddell Airy Observatory, Cambridge
Requesting a copy of G. A. A. Plana and Francesco Carlini's 'Lunar Theory.' Further remarks on the practicability of the gravimeter.
ALS 3pp RS:HS 1.55

2711. 1833–1–1 From William Jones St. John's College
Has had no communication with Thomas Phillips regarding JH's portrait and they would approve of JH's choice of H. W. Pickersgill. Will call on Pickersgill when he is in town and arrange size of the portrait.
ALS 3pp RS:HS 10.413

2712. [1833] To Henry William Pickersgill n.p.
Must rearrange date of sitting for HP's preparation of a portrait of JH.
ALS 1p British Astronomical Association

2713. [1833] To William Henry Smyth n.p.
Encourages WS to spend time observing nebulae and double stars, especially the latter, 'since [James] South has given up observing, and [W. R.] Dawes is in habitual ill health.'
C 2pp RS:HS 25.14.31

2714. 1833–1–2 From E. Beckwith 18 Russell Square
In her absence JH has called on her. She would like to see him when convenient to arrange a new trustee for the marriage settlement.
ALS 3pp RS:HS 5.33

2715. 1833–1–4 From Wilhelm Struve Dorpat
Mentions [James] South's visit to Dorpat. Includes his observations of the past year. Concedes that JH saw Biela's comet first.
ALS 4pp (German) RS:HS 17.123

2716. 1833–1–7 From Richard Sheepshanks 30 Woburn Place
Asks JH to review a publication RS was preparing. Comments on James South's claim concerning supposed defects in the telescope made for him by Edward Troughton. Suggests JH intercede with the government to get it to support G. B. Airy's reduction of the Greenwich planetary observations.
ALS 4pp RS:HS 16.2

2717. [1833]–1–8 To Charles Wheatstone n.p.
Postpones visit of Lady [Margaret] Herschel to CW. Wishes to take CW's Symphonion to Slough. Has made favorable mention of CW's tuning fork experiments in paper on absorption [JH's 'On the Absorption of Light by Coloured Media...'].
ALS 3pp Kasslers

2718. 1833–1–9 From Roderick Impey Murchison 3 Bryanston Place
Further regarding the projected portrait of JH by H. W. Pickersgill.
ALS 3pp RS:HS 12.387

2719. 1833–1–12 To the *Philosophical Magazine* n.p.
Reports observing curious depositions of ice around the decaying stems of vegetables during frost. Speculates on the possible source.
L *P.M.*, n.s. *2* (1833), 110–12

2720. 1833–1–13 To Caroline Lucretia Herschel Slough
JH reports the birth of his son William James Herschel.
ALS 2pp TxU:H/L–0578.1; Reel 1058

2721. 1833–1–14 From Sarah Bowdich 27 Burton St.
Sending the Eloge on G. C. L. D. Cuvier. Plans to write a biographical memoir of Cuvier herself.
ALS 3pp RS:HS 4.207

2722. 1833–1–14 From Basil Hall Athenaeum
Congratulations on his addition to the family; hopes all are well. Would like JH to meet T. T. Grant, who has a new invention, which may prove of great utility.
ALS 3pp RS:HS 9.175

2723. 1833–1–14 From William Whewell Brasted
Comments on the optical experiments with 'Mr [Thomas] Malthus's eyes.' Reports that Richard Jones has received a professorship at King's College. Mentions WW's intention to investigate tides.
ALS 3pp RS:HS 18.184 Todhunter, *Whewell*, *2*, 152–3

2724. 1833–1–15 To William Rowan Hamilton Slough
Thanks for sending results about conical polarization and introductory lecture on astronomy communicated through Francis Beaufort. Had hoped to send WH JH's yet to be printed catalogue of double stars.
ALS 2pp TCD 1493:88 (C: TCD 7762–72:321)

2725. 1833–1–15 To William Jones Slough
Expresses some concerns about the proposed painting of his portrait [see WJ's 1833–1–1].
C 1p RS:HS 21.128

2726. 1833–1–15 To John St. John Slough
H. W. Pickersgill proposed to paint JH's portrait for free, to be given to St. John's College. Now finds JH must pay for picture. Will delay until another artist is chosen.
Adraft 1p RS:HS 19.68

2727. 1833–1–16 From Francis Beaufort Admiralty
Passing on E. J. Cooper's information that he has found a new star. Congratulations on another addition to the family.
ALS 2pp RS:HS 3.330

2728. 1833–1–16 From Alexander Rogers Baltic St., Leith
Apologizes for delaying in getting ale to the Herschels. A new cask is being sent.
ALS 2pp RS:HS 14.416

2729. 1833–1–22 From James Grahame L'Eperonniere
Of ill health and medical stories [letter completed 1833–2–2].
ALS 3pp RS:HS 8.417

2730. 1833–1–22 From Richard Jones n.p.
Will be at his service any time after the 29th. Congratulations for the boy. JH's political economy is quite good. Wrote out the verses of Homer. He can take them back with him when he comes. Is glad that [James] Grahame is satisfied.
ALS 3pp RS:HS 10.354

2731. 1833–1–23 From Giovanni Antonio Plana [Turin]
Excerpts from a letter to the R.S.L. that accompanied copies of GP's work on lunar theory.
ALS 2pp inc (French) RS:HS 13.431

2732. 1833–1–25 From James A. Gordon 29 Hill St.
Congratulations on the birth of a son. J. W. Lubbock had no chance at Cambridge. Is pleased he did not have to make the journey. Mr. Lacy, a curate at Tring, would like an introduction to JH as he has several interesting communications to make.
ALS 2pp RS:HS 8.169

2733. 1833–1–29 From Francis Baily 37 Tavistock Place
Must see him on an important matter.
ALS 1p RS:HS 3.114

2734. 1833–1–29 From William Rowan Hamilton Observatory
Rev. Humphrey Lloyd's paper on conical refraction agreed with WH's conclusions.
L Graves, *Hamilton, 1,* 632–4

2735. 1833–1–30 To Roderick Impey Murchison [Slough]
Unsatisfied with H. W. Pickersgill as the person to paint JH's portrait. JH wants to consider further what to do.
ALS 3pp BL Aaa 46126.400

2736. 1833–1–30 From William Whewell Trinity College
Sending JH various publications. Reports that Richard Jones is giving his introductory letter. Congratulates JH on birth of an 'experimental philosopher' [JH's son William James].
ALS 2pp RS:HS 18.185 Todhunter, *Whewell, 2,* 154

2737. 1833–2–1 From James David Forbes Edinburgh
Sending a copy of his testimonials and one of his reports on meteorology. Was elected to the chair of Natural Philosophy by a majority of 21 to 9. His cousin Leslie Melville hopes to be able to provide JH with letters before JH's departure for the Cape.
ALS 2pp RS:HS 7.282

2738. 1833–2–2 From Stephen Peter Rigaud Oxford
Sends paper for *Philosophical Magazine* with reference to JH's communication in the last issue. JH may annex comments, unless he will cut SR 'to pieces.'
ALS 1p RS:HS 14.345

2739. [1833]–2–2 From Eliza Young Park Square
Sends copy of [Hudson Gurney, *Memoir of Thomas Young, M.D.* (1831)]. Plans larger work. Does JH have any letters fit for publication?
ALS 4pp RS:HS 18.313

2740. 1833–2–3 From George Biddell Airy Observatory, Cambridge
Inviting JH to his house if he attends the meeting of the B.A.A.S. Proposes to visit JH in London.
ALS 1p RS:HS 1.56

2741. 1833–2–4 From Roderick Impey Murchison 3 Bryanston Place
Further comments on the painting of H. W. Pickersgill. Hopes that JH will sit for Pickersgill. Also hopes that JH will attend the Geological Society Anniversary on the fifteenth.
ALS 3pp RS:HS 12.388

2742. 1833–2–13 From James Grahame L'Eperonniere
Of health and reminiscences of the past [letter completed 1833–2–22].
ALS 4pp RS:HS 8.418

2743. 1833–2–15 To D. Robertson Slough
Has sought, and is agreeing with, advice given about JH's involvement with several trusts [?] during JH's time in South Africa.
ALS 2pp APS B.H435p.6

2744. 1833–2–17 From Siméon Denis Poisson Paris
Best wishes to JH on his voyage to the Cape, and comments on what wonders JH may observe there. [Letter largely illegible.]
ALS 4pp (French) RS:HS 14.15

2745. 1833–2–20 To James David Forbes Slough
Congratulates JF on obtaining, after some difficulty, the Chair of Natural Philosophy at Edinburgh. Accepts offer to discuss meteorology with JF's relation Leslie Melville[?], and hopes to mention his actinometer.
ALS 3pp St. Andrews 17 (C: RS:HS 21.129)

2746. 1833–2–23 From Sarah Bowdich 27 Burton St.
Is grateful for his letter and offer to show her any G. C. L. D. Cuvier letters. Was pleased to see his letter as she heard he was at the Cape in pursuit of constellations. What shall she do with the forthcoming copies of her book on fishes he ordered?
ALS 3pp RS:HS 4.208

2747. 1833–2–23 To Henry Kater Slough
Shocked to read in paper of HK's loss. Offers condolences.
ALS 2pp RS:HS 19.69 (C: RS:HS 21.130)

2748. 1833–2–23 From Thomas Maclear Biggleswade
Hears that JH is contemplating a rotatory roof for his telescope at the Cape. Sends details of a method he has devised for his own observatory. Will send a drawing of it next week. Will be pleased to show it to JH if he is that way.
ALS 4pp RS:HS 12.34

2749. 1833–2–23 To James Calder Stewart Slough
JS is most welcome to visit Slough to see new baby [William James]. [Margaret Herschel note: Invites JS to christening.]
ALS 4pp TxU:H/L-0413; Reel 1055

2750. [1833–2–25] From Richard Jones Brasted
Exhibits on Wednesday at 2 o'clock. Sends tickets. Hopes to dine at the Athenaeum.
ALS 1p RS:HS 10.355

2751. 1833–2–28 From John Dalton Manchester
Thanks for sending him his paper on sound. Comments on this and gives some of his own findings on this subject.
C 2pp RS:HS 6.466

2752. 1833–3–1 From Francis Beaufort Admiralty
Unable to help him as Richardson's Brisbane Observations have not yet come to hand. Feels the same as JH does regarding the presentation of the true colors of his parents.
ALS 3pp RS:HS 3.331

2753. 1833–3–4 From William Henry Fox Talbot 31 Sackville St.
Asks about rumor regarding JH going to Cape. Describes some electrical experiments WT saw at Royal Institution and refers to some WT carried out years earlier.
ALS 3pp RS:HS 17.268

2754. 1833–3–6 From Charles Babbage Dorset St.
Has glanced over the papers and made a few corrections. Suggests one point for alteration.
ALS 2pp RS:HS 2.278

2755. 1833–3–7 From Charles May Ampthill
Has received the glasses from Dr. John Dalton with very minute answers. Dalton gave a paper on this subject in 1794 to the Manchester Society. Will bring them to London tomorrow. Dr. Simms is in a critical condition.
ALS 2pp RS:HS 12.310

2756. 1833–3–9 From William Henry Fox Talbot 31 Sackville St.
Asks for seeds and roots of native plants of the Cape. Refers to JH's writings on beating of the heart and comments on some optical experiments of Charles Wheatstone and David Brewster.
ALS 3pp RS:HS 17.269

2757. 1833–3–11 To William Somerville Slough
Will write [J.-B.] Biot, but does not know to which enquiries Biot wishes JH to respond. Also wants to thank Biot for his kindness. S.–D. Poisson in a statement before the Académie des sciences has misdescribed JH's planned voyage.
ALS 2pp RS:HS 16.401 (C: RS:HS 21.131)

2758. 1833–3–12 To Wilhelm Struve Slough
The observational data from FS regarding Gamma Virginis are so significant that JH begs FS to continue to observe and send JH the results. Comments on a number of other astronomical observational matters.
ACCS 4pp RS:HS 25.3.11

2759. 1833–3–14 From Adolphe Quetelet Brussels
Concerned about payment for instruments. Has read in newspaper that JH is preparing to study the southern hemisphere. Asks to keep in touch. Will soon go to Paris for transit instrument and another publisher for the translation of JH's work. Problems on account of Revolution.
ALS 3pp (French) RS:HS 14.205

2760. 1833–3–[16] to [24] To Caroline Lucretia Herschel Slough
JH preparing for his journey to Cape Town. Compiling a catalog of nebulae and clusters. Sending to CH a comparison of a new set of elements of Gamma Virginis and a set of observations of it from 1720–1833; JH asks CH to send a copy to Carl Gauss.
ALS 4pp TxU:H/L–0578.3; Reel 1058

2761. 1833–3–16 To M. Perrey Slough
Gently rejects MP's offer to accompany JH to the Cape to act as JH's assistant. JH claims the voyage to be a 'pure family party of pleasure,' and if astronomy can not be well practiced at the Cape, the family may go on to India to visit the relatives of JH's wife, Margaret.
ACCS 2pp RS:HS 25.3.12

2762. 1833–3–18 From William Smith Blandford Square
Praises JH's *Prelim. Discourse* and work in general. Wishes well at the Cape.
ALS 3pp RS:HS 17.217

2763. [1833]–3–21 From Edward Joshua Cooper Markree, [Sligo, Ireland]
Thanks JH for clues to interesting star 'near Eta Coronae.' Sends sketches of April [1832?] observations of nebulae.
ALS 4pp † RAS:JH Archive 3/2.18; Reel 7

2764. 1833–3–27 From William Henry Fox Talbot London
Comments on JH's paper on potash in *Annales de chimie*. Has observed reaction products in microscope. Concerned over delay in publication by David Brewster of his experimental results in optics and light, especially as WT not convinced of legitimacy of rumored results.
ALS 4pp RS:HS 17.270

2765. 1833–3–30 From Caroline Lucretia Herschel Hanover
Feeling simply 'dismal.'
L *Memoir ... of Caroline Herschel*, 256–7 inc

2766. 1833–4–2 To William Somerville Slough
Has received payment from [John] Murray [for review of Mary Somerville's *Mechanism of the Heavens* ?]. Cannot answer Mrs. Somerville's question about poles of maximum cold.
ALS 5pp RS:HS 16.402 (C: RS:HS 19.71 & 21.132)

2767. 1833–4–3 From Caroline Herschel to Carl Friedrich Gauss Kramer Strasse, Hanover
Sends on a copy of JH's orbital elements for Gamma Virginis.
C 2pp RS:HS 21.133

2768. 1833–4–4 To Edward Joshua Cooper Slough
Thanks EC's for EC's observations made with EC's excellent telescope. Discusses Eta Coronae, Gamma Virginis, EC's drawing of M51, and whether Orion changes, stating that JH is 'no believer in *rapid* changes in Nebulae yet there *may* be

changes....'
C 4pp RS:HS 21.134

2769. 1833–4–5 From Thomas Henderson Cape of Good Hope
Has taken some time to consider the climate at the Cape in response to JH's request. Gives details of the climate, suitability of site for observatory, and general conditions at the Cape. There is a reasonable library at Cape Town.
ALS 7pp RS:HS 9.286

2770. 1833–4–6 From James David Forbes Edinburgh
Thanks for his last letter. Assures him he has no bitter feelings over the spirit of David Brewster after his disappointment of the election. May make any use he wishes of his actinometer results. Intends going over to France in a month's time, but is very busy at the present time.
ALS 2pp RS:HS 7.283

2771. 1833–4–8 From George Biddell Airy Observatory, Cambridge
Observations on the work of S. D. Poisson and P. G. le D. Pontecoulant. Is JH's 20 -ft. telescope any use for measuring angles and distances? Discussion of solar spots.
ALS 2pp RS:HS 1.57

2772. 1833–4–12 From Carl Friedrich Gauss to Caroline Herschel Göttingen
Thanks for her communication on her nephew's new calculations for Alpha Virginis. Comments on these and gives some of his own. Sends best wishes to her nephew on his projected visit {to the Cape] and hopes he will study the magnetic force of the earth while he is there.
ALS 2pp RS:HS 8.75

2773. 1833–4–16 From Charles Babbage Dorset St.
The first evening party is on Saturday, and hopes JH will be able to come.
ALS 1p RS:HS 2.279

2774. 1833–4–16 From Charles May Ampthill
Sends some specimens of aloes and calocynth with some remarks on them. Also sends some sugar made by evaporating the cane juice under diminished pressure. Inquired from J. Lister regarding the elasticity of hardened steel wire.
ALS 2pp RS:HS 12.311

2775. 1833–4–16 To Giovanni Santini Slough
Asks GS to provide hospitality to J. C. Stewart, who is traveling through Italy. Sent package of books. Hopes GS will write to JH at Cape of Good Hope.
ALS 4pp TxU:H/L-0381; Reel 1054

2776. 1833–4–17 To Friedrich Wilhelm Bessel Slough
Requests FB to send JH any observations of Gamma Virginis made by FB. Anxious for these because of differences between JH's and Wilhelm Struve's measurements of Gamma Virginis. Provides JH's new determination of the orbital elements of this object. Thanks FB for publications FB sent. JH's brother-in-law, J[ames Calder] Stewart, may visit FB.
ALS 3pp AdWDDR Bessel 225 (ACCS: RS:HS 25.3.13)

2777. [1833]–4–17 To James Calder Stewart Slough
Letters of introduction for JS and suggestions whom to visit, especially F. W. Bessel, during JS's trip to Continent and Russia.
ALS 3pp TxU:H/L-0414; Reel 1055

2778. 1833–4–18 To Charles Babbage n.p.
Is trying to prepare his book for the press, and is suffering from a severe attack of influenza.
ALS 2pp RS:HS 2.280

2779. 1833–4–24 To Charles Babbage n.p.
Will not be in town until early June owing to pressure of work. Thanks for information on Van Dieman's Land, but they have decided on the Cape.
ALS 2pp RS:HS 2.281

2780. 1833–4–24 From Basil Hall St. James's St.
Sending his sextant and instructions for its use, which he will doubtless find useful. Also a copy of James Horsburgh's *Directions*. Has not heard from Captain Hardy regarding the rope. Advises boxes covered in mackintosh. Plans to publish his book on Saturday.
ALS 4pp RS:HS 9.176

2781. 1833–4–25[?] To Caroline Lucretia Herschel [Slough]
JH explains how CH's money will be allotted to her while JH is in Cape Town. Margaret Herschel writes CH a note describing JH's new son William.
ALS 3pp TxU:H/L–0578.4.a; Reel 1058

2782. 1833–4–27 From Thomas Henderson Cape of Good Hope
Has resolved to leave the Cape and soon after JH receives this letter he hopes to be in England. Will bring the barometer and the exterior thermometer. Saw Dr. Duncan Stewart yesterday.
ALS 2pp RS:HS 9.287

2783. 1833–4–28 From James David Forbes Edinburgh
Has acquired a 7-ft. Newtonian telescope of William Herschel made in 1802. Is well preserved but the speculum appears to be tarnished. Would like to bring it to London to obtain JH's advice on its shortcomings. Is going to France to collect the actinometers and the results of L. F. Kämtz.
ALS 2pp RS:HS 7.284

2784. 1833–4–28 From Caroline Lucretia Herschel Hanover
Formal note and stamp acknowledging receipt of £50 annuity from estate of William Herschel, sent by executor JH, followed by note to JH on arrangements for sending such payments. Asks for news about Stewart family.
ALS 3pp TxU:H/M–0936.1; Reel 1083

2785. 1833–4–29 To Roderick Impey Murchison Slough
Wishes to nominate Mr. Hardin [?], who has written on Indian geology, for membership in the Geological Society.
ALS 2pp Geological Society

2786. 1833–5–1 From William Rutter Dawes Ormskirk
Is grateful for the autographs and Dr. Thomas Raffles sends him thanks. Regarding various star readings. Has a problem connected with the satellites of Saturn. Hears that JH is contemplating a visit to the Cape to make observations.
ALS 4pp RS:HS 6.52

2787. 1833–5 To Peter Stewart n.p.
Today's newspapers confirm shocking report heard last night. Imagines loss and ruin to members. Confident that exalted character of PS's uncle will result in uncle being found blameless. Certain that James [C. Stewart], as new partner, could not bear any shame for magnitude of this long-standing problem. Hopes PS's mother will bear up well. Offers to assist Mrs. Stewart financially. Grateful that James had not yet left for Munich before this happened. Margaret is in

shock.
ALS 3pp TxU:H/L-0393; Reel 1054

2788. 1833–5–3 From William Jones St. John's College
Has been laid up with influenza, which delayed his answer. Has seen H. W. Pickersgill, who will be very pleased to paint JH's portrait. Can he arrange a meeting?
ALS 4pp RS:HS 10.414

2789. 1833–5–4 To William Rutter Dawes Slough
Comments on, and compares, WD's observations of several double stars with the observations of JH and other astronomers. In a postscript JH says 'The papers have told the truth for once—we sail sometime between Sept. and Dec.'
ACCS 3pp RS:HS 25.3.14

2790. 1833–5–6 To James David Forbes Slough
Offers JF advice about the telescope mirror [?] and looks forward to seeing JF on his way to France.
ALS 1p St. Andrews 28

2791. 1833–5–13 From James David Forbes Royal Society, London
Sending the speculum to Slough in order that JH may examine it. Departs for Paris tomorrow. Hopes to see him at the Cambridge meeting.
ALS 2pp RS:HS 7.285

2792. 1833–5–15 To Thomas John Hussey Slough
Will check to see whether JH has a spare telescopic mirror JH can give TH. Comments on observing Gamma Virginis and other double stars.
ALS 4pp Univ.Coll.London MS Graves 23(1)

2793. 1833–5–15 From John William Lubbock Royal Society
Will JH be able to dine with the other visitors at Greenwich on Saturday? JH has been awarded a Royal Medal. Thinks JH is wrong in not desiring a passage to the Cape in a Man of War.
ALS 2pp RS:HS 11.359

2794. 1833–5–16 To John William Lubbock Slough
About observing binary stars. JH has decided not to accept the offer of passage on a war ship to the Cape of Good Hope, as he wishes to be beholden to no one for the results of the expedition.
ALS 4pp RS LUB.H.351 (C: RS:HS 21.136)

2795. 1833–5–20 To John Dalton Slough
Having been provided with answers by JD to some questions by JH about color-blindness, JH now comments on them and offers some theoretical considerations.
ACCS 4pp RS:HS 25.3.15

2796. 1833–5–20 To Margaret Brodie Herschel Slough
It appears that Peter Stewart, MH's brother, is fighting some kind of battle, the outcome of which will affect the whole family. JH wants to help and offers £5000 through MH to Peter. JH has just completed writing his *Treatise Astr.*
ALS 4pp JHS.HCEJ.523

2797. [1833?]–5–22 From Alfred Gautier n.p.
Recounts details of his visit to England. Gives his own recent astronomical observations. Is sending this via G. B. Airy in case JH has left.
ALS 2pp (French) inc RS:HS 8.90

2798. 1833–5–28 From William Henry Fox Talbot 29 Albemarle St.
Has performed some Newton's rings interference experiments. Asks whether such experiments have already been communicated to the R.S.L.
ALS 3pp RS:HS 17.271

2799. 1833–5–[30] To Margaret Brodie Herschel Slough
JH is 'immersed in my nebulae ... working day and night for dear life'; also found another double star. Urges MH to come to Slough with her mother and brothers.
AL 2pp inc JHS.HCEJ.524

2800. 1833–5–31 From William Henry Fox Talbot 29 Albemarle St.
Trying to understand David Brewster's results on absorption of light in gaseous state, and the dark lines in solar spectrum, the cause of which he misunderstands.
ALS 4pp RS:HS 17.272

2801. 1833–6 To Francis Beaufort n.p.
Answer to FB's queries regarding chronometers.
draft 3pp RS:HS 3.332bis (C: RS:HS 21.138)

2802. 1833–[6] To Margaret Brodie Herschel Slough
Is pressed by work, but will come to London if needed; domestic matters brought up to date.
ALS 3pp JHS.HCEJ.525

2803. 1833–6–3 To William Henry Smyth 37 Tavistock Place
Intends to visit WS after attending Cambridge B.A.A.S. meeting.
ALS 4pp APS B.Sm98.13 (C: RS:HS 21.137)

2804. 1833–6–5 From William Henry Smyth Crescent, Bedford
Will be able to receive JH during Cambridge trip.
ALS 2pp RS:HS 16.250

2805. 1833–6–8 To Henry William Pickersgill Slough
Inquires about the time when WP will paint JH's portrait and about the costs involved.
AL 2pp Yale Univ.

2806. 1833–6–10 From Charles Babbage Feldhausen, Cape of Good Hope
Outlines possible program for Wednesday.
ALS 1p RS:HS 2.282

2807. 1833–6–11 From Francis Beaufort Admiralty
Would like his opinion on the enclosed plan regarding chronometers.
ALS 4pp RS:HS 3.332

2808. 1833–6–13 From Robert Grahame, Sr. London
His son Thomas has set out for France and he himself is in London to see the passage of a bill through Parliament. Would be grateful for letters of introduction to M.P.'s with which JH is acquainted.
ALS 1p RS:HS 9.6

2809. 1833–6–15 From Charles Babbage Dorset St.
Regarding Society matters. Small parcel has arrived for JH.
ALS 2pp RS:HS 2.283

2810. 1833–6–19 From James Horsburgh East India House
Has sent to Francis Baily a copy of his *Sailing Directions* and a roll of charts. Hopes JH will point out any errors he may note.
ALS 1p RS:HS 9.498

2811. 1833–6–20 From Jacques Babinet Paris
Introducing A. T. D'Abbadie, one of his scientific friends.
ALS 1p (French) RS:HS 3.2

2812. 1833–6–22 From James Grahame L'Eperonniere
Sharing some good and some bad news.
ALS 1p RS:HS 8.419

2813. 1833–6–25 From Thomas Drummond Manchester
As he leads a roving life he did not receive JH's letter until a few days ago. Answers JH's problem concerning some method of measurement in his surveying.
ALS 12pp RS:HS 6.455

2814. 1833–6–29 From Robert Grahame, Sr. London
James's daughter has improved in health and there are hopes for her recovery. Bill passed through the House of Commons successfully. Would be glad to receive letters of introduction to any members of the House of Lords with which JH may be acquainted.
ALS 1p RS:HS 9.7

2815. 1833–6–29 From William Henry Smyth Crescent, Bedford
Meets Lord Bute at Oxford. Requests JH observe Gamma Virginis, Castor, and Polaris with WS's 10-ft. achromatic telescope.
ALS 2pp RS:HS 16.251

2816. [1833–7–3] From Richard Jones Brasted
Regrets he did not see more of him at the Cambridge meetings. [Herbert?] Mayo has promised to write a full report to JH of young [Robert?] Graham[e?].
ALS 3pp RS:HS 10.356

2817. 1833–7–4 From William Thomas Brande Royal Institution
Sending him books and expressing his admiration for JH's *Prelim. Discourse.*
ALS 1p RS:HS 5.94

2818. 1833–7–4 To James South Slough
A friend has requested JH to ask JS whether JS has a large Cauchoix objective lens that he would be willing to sell. JH has dismantled his 20-ft. telescope for voyage to Cape Town.
AdraftS 1p RS:HS16.472 (C: 21.139)

2819. 1833–7–5 From William Henry Smyth Cardiff Castle
Sends best wishes for JH's journey to Cape. Regrets missing him at Bedford during Cambridge trip.
ALS 2pp RS:HS 16.252

2820. 1833–7–6 To Charles Babbage n.p.
Let him know when the next Levee is.
ALS 2pp RS:HS 2.284

2821. 1833–7–6 From James South Observatory, Kensington
Does the person wishing to purchase JS's lens have a worthy objective in mind [see 1833–7–4]?
ALS 2pp RS:HS 16.473

2822. 1833–7–7 From Charles Babbage n.p.
Giving the date of the next Levee.
ALS 1p RS:HS 2.285

2823. 1833–7–8 To Hugh Percy (3rd Duke of Northumberland) Slough
Is unable to get a firm answer from James South as to the availability of a large telescope object glass that James South owns.
draft 2pp RS:HS 13.341 & 21.140

2824. 1833–7–10 To Caroline Lucretia Herschel Slough
Sent to CH JH's fifth catalog of double stars. JH will send to CH his new nebulae observations very soon.
ALS 2pp TxU:H/L–0578.5.b; Reel 1058

2825. 1833–7–10 From Hugh Percy (3rd Duke of Northumberland) Northumberland House
JH is to pursue the matter of acquiring an object glass further [see JH's 1833–7–8], without indicating that it may go to Cambridge Observatory.
AL 3pp RS:HS 13.342

2826. 1833–7–11 To James South n.p.
Assures JS that the person interested in purchasing JS's lens has worthy intentions for it, i.e., its use in a public institution. Regrets that JS distrusted JH's good intentions.
CC 3pp RS:HS16.474 (C: 21.141)

2827. 1833–7–17 From Thomas Maclear Biggleswade
Wife has been ill, which accounts for his belated reply. Comments on the pros and cons of taking furniture to the Cape. May leave the infant at home. Has received the plan of the observatory from Francis Beaufort.
ALS 3pp RS:HS 12.35 Warners, *Maclear & Herschel Letters,* 6–7

2828. 1833–7–18 From James Grahame Jersey
Bad news about the health of JG's daughter Matilda.
ALS 2pp RS:HS 8.420

2829. 1833–7–27 From John Ramage Aberdeen
Discusses construction of achromatic and reflecting telescopes, and stands for large telescopes. Asks JH to sign a subscription list to raise funds to construct telescope.
ALS 4pp RS:HS 14.301

2830. 1833–7–28 To William Henry Smyth Slough
Reports observations with WS's equatorial during Bedford visit. Packs instruments for voyage. Pleased that Thomas Maclear will take post of Astronomer Royal at the Cape Observatory.
ALS 4pp APS B.Sm98.14 (C: RS:HS 21.142)

2831. 1833–8 From C. De La Vane to Margaret Brodie Herschel Datchet
Is Miss Macqueen expected with him tomorrow and would she give the enclosed memo to JH [query as to whether the Duke of Gloucester was in Metz in 1776].
ALS 2pp RS:HS 6.488

2832. 1833–8–1 From Caroline Lucretia Herschel Hanover
Wishes JH well on his upcoming trip to the Cape of Good Hope. Asks JH to study the lower part of the Scorpion, for William Herschel was mystified by the 'uncommon appearance of that part of the heavens.' [Written as a postscript to a letter by Margaret Herschel.]
L *Memoir ... of Caroline Herschel*, 258

2833. 1833–8–3 To Francis Baily n.p.
Has received a note from J. C. Spencer (Lord Althorp) stating that a grant of £500 is available for them [R.A.S.?]. He has acknowledged it.
ALS 2pp RS:HS 3.115 & 21.143

2834. 1833–8–3 To Adam Sedgwick Slough
Received £500 from J. C. Spencer (Lord Althorp) for reduction under the direction of G. B. Airy of earlier

planetary observations. Althorp conferred with Lord Grey. Conjectures possible need of a larger grant.
ALS 3pp RS:HS 15.422 (C: RS:HS 21.144) Morrell and Thackray, *Early Correspondence*, 174–5

2835. 1833–8–9 To Hugh Percy (3rd Duke of Northumberland) Slough
JH has written to James South again [see JH's 1833–7–8] but has received no answer; is suggesting that it may well be that [R. A.] Cauchoix, the French optician, can supply object glasses as good as those James South has.
C 2pp RS:HS 13.343 & 21.145

2836. 1833–8–11 To D. François J. Arago n.p.
[Extract from a letter of this date to James South:] Tells FA that FA's magnetic experiments have been successfully repeated in England by Peter Barlow, Samuel Christie, Charles Babbage, and JH. Gives details. Publication in process.
CC 2pp Académie des sciences (Paris)

2837. 1833–8–11 From Thomas Maclear Biggleswade
Hears that Thomas Henderson is in town and thinks of bringing him down to Slough. Intends to take some iron bedsteads to the Cape; can JH advise him where they may be obtained?
ALS 2pp RS:HS 12.36 Warners, *Maclear & Herschel Letters*, 7–8

2838. 1833–8–12 From Thomas Henderson London
Has written two letters from the Cape. Arrived three days ago. Will shortly proceed to Edinburgh. Has left the barometer on the ship. Compared this barometer with the one at St. Helena.
ALS 2pp RS:HS 9.288

2839. 1833–8–12 From Hugh Percy (3rd Duke of Northumberland) Cowes
Agrees with JH to go direct to the French optician [see JH's 1833–8–9] for a large object glass; HP will provide a telescope with such a lens if G. B. Airy agrees such a large telescope is appropriate to Cambridge Observatory.
AL 2pp RS:HS 13.344

2840. 1833–8–13 From Wilhelm Struve Altona
Sends letter from 'chronometer expedition' in Baltic Sea under the leadership of [T. F.] Schubert and supported by the Prussian, Danish, and Swedish governments. Mentions talk written for a larger audience in French publication. Envies JH's opportunity to observe the southern hemisphere.
ALS 3pp (German) RS:HS 17.124

2841. 1833–8–14 From William Rutter Dawes Liverpool
Business matters calls him to London and would be happy to make his acquaintance. Will visit Slough if necessary.
ALS 1p RS:HS 6.53

2842. 1833–8–15 To George Biddell Airy Slough
Encourages GA to consider favorably the offer of Hugh Percy [Duke of Northumberland] of donating a telescope to Cambridge Observatory.
ALS 3pp RGO 6.157.162 (C: RS:HS 25.3.16)

2843. 1833–8–15 To Thomas Henderson Slough
Arranges to meet TH, probably at Slough, and also to get a barometer from TH.
ALS 3pp NatLibScot MS.3134, no.51

2844. 1833–8–18 From John Narrien Royal Military College
Hopes that JH will accept a copy of JN's work [on the history of astronomy]. Wishes JH all success in his forthcoming mission [to the Cape].
ALS 2pp RS:HS 13.145

2845. 1833–8–19 To D. François J. Arago Slough
Offers a paper ['Note sur la manière d'agir de l'acide nitrique sur le fer,' *Annales de chimie*, *54* (1833), 87–94] for publication in *Annales de chimie*. Informs FA how to correspond with JH while JH is at the Cape.
ALS 1p encl (4pp) Académie des sciences (Paris)

2846. 1833–8–20 From George Biddell Airy Edensor, Bakewell
Copy of letter to Hugh Percy (3rd Duke of Northumberland) stating that the Duke's offer of a refracting telescope of great power, to Cambridge Observatory would be a valuable gift. To JH: Remarks on gift of telescope, planet reduction money, and on his geological exploration of the Charnwood Forest.
ALS 3pp RS:HS 1.58

2847. 1833–8–21 To Charles Wheatstone Slough
Invites CW to Slough to demonstrate CW's 'elegant experiments on light & sound.' Will need time to get JH's electric machine in order. Asks for copy of CW's paper on Jew's harp.
C 2pp damaged RS:HS 21.146

2848. 1833–8–22 From William Rutter Dawes London
Has just arrived in London and will proceed to Slough tomorrow in the company of William Lassell.
ALS 1p RS:HS 6.54

2849. [1833?]–8–23 From William Vernon Harcourt Bishopsthorpe, York
Pleased to hear of JH's success with his application to the Treasury. Would like a statement from him on the nature of his communication to the Cambridge meeting concerning the actinometer. Hopes he is not too busy preparing for his voyage.
ALS 3pp RS:HS 9.226 Morrell and Thackray, *Early Correspondence*, 179

2850. 1833–8–23 From Charles Wheatstone 20 Conduit St, London
Accepts JH's invitation to demonstrate [on 28 Aug.] CW's portable optical and acoustical experiments and the 'whirling machine.' Asks JH to supply 'an electrical machine and a good sized jar.' Encloses copy of CW's paper on Jew's harp. Will bring another paper and new 'Symphonion' for Mary Herschel.
ALS 1p RS:HS 18.145

2851. 1833–8–24 From Thomas Maclear Biggleswade
Has been residing with [Robert?] Lee at Hartwell for a few days. Hears that JH and Benjamin D'Urban will be travelling to the Cape on the *Elphinstone*; is this true?
ALS 1p RS:HS 12.37 Warners, *Maclear & Herschel Letters*, 8

2852. 1833–8–26 From George Biddell Airy Edensor, Bakewell
Has visited Hugh Percy (3rd Duke of Northumberland) at Buxton, who suggests that JH approaches R. A. Cauchoix for the glass for the new telescope.
ALS 2pp RS:HS 1.59

2853. 1833–8–27 To George Biddell Airy Slough
A note along with letters from R. A. Cauchoix, the telescope maker, and [J. C. Spencer,] Lord Althorp, about the B.A.A.S.; JH is trying to clean up observation reports before leaving for

the Cape.
ALS 1p RGO 6.157.164

2854. 1833–8–28 From James Grahame Falmouth
Preparation to go to Madeira for health reasons.
ALS 2pp RS:HS 8.421

2855. 1833–8–29 To George Biddell Airy n.p.
Further about the telescope to be donated by Hugh Percy [see JH's 1833–8–15], along with a promise that JH will write to R. A. Cauchoix about the telescope.
ALS 3pp RGO 6.157.167

2856. 1833–9–1 To Robert A. Cauchoix Slough
Asks for reply to specific questions regarding telescope quality and price [see JH's 1833–8–29].
L(S) 4pp RGO 6.157.169 (C: RS:HS 25.3.17)

2857. 1833–9–1 From Hugh Percy (3rd Duke of Northumberland) Buxton
Wants JH to negotiate with [R. A.] Cauchoix [see HP's 1833–8–12], keeping HP's name out of the transactions.
ALS 4pp RS:HS 13.345

2858. 1833–9 From Henry White n.p.
Notice of Susan White's death.
Abs 1p TxU:H/M–0580.5; Reel 1093

2859. 1833–9–2 From James Gye 1 Charing Cross
Informs JH that JH will be solicited on behalf of a 'great national undertaking.' JG will call on JH in this regard.
ALS 1p Harvard: Houghton AAH 67m–67 (88)

2860. 1833–9–6 From Thomas Henderson London
Proceeds to Edinburgh in the course of a few days and intends to compute the star declinations. Has some suggestions to offer before JH sails for the Cape. Would be glad to hear of the comparison of the barometer and thermometer. Left a [Thomas] Jones thermometer at the Cape.
ALS 2pp RS:HS 9.289

2861. 1833–9–6 From Thomas Maclear Biggleswade
Visited the *Elphinstone* and estimated the amount of room required for his family. Offered £300 but offer was refused.
ALS 2pp RS:HS 12.38 Warners, *Maclear & Herschel Letters,* 8

2862. 1833–9–8 From Johann Franz Encke Berlin
Has been informed that JH will be going to the Cape. Comments on JH's information on double stars, and on a passage in one of his memoirs. Regarding JH's *Prelim Discourse.* Work of Fearon Fallows and Thomas Henderson at the Cape.
ALS 3pp (French) RS:HS 7.89

2863. 1833–9–9 or 1833–9–19 From Francis Beaufort Admiralty
Offering assistance for JH's forthcoming voyage to the Cape.
ALS 2pp RS:HS 3.333

2864. 1833–9–15 From William Henry Smyth Crescent, Bedford
Very impressed after meeting W. R. Dawes and discussing double stars with him. Asks JH to send Gamma Virginis observations from Cape.
ALS 3pp RS:HS 16.253

2865. 1833–9–15 From J[ames] William White Speenhamland [Berkshire]
Reports death of JW's aunt Susan on Friday [13 Sept.].
ALS 2pp RS:HS 18.399

2866. 1833–9–17 From Edward Joshua Cooper Markree
Would like to know where he can obtain JH's observations on double stars. His own recent observations of planets.
ALS 4pp RS:HS 5.278

2867. 1833–9–17 From William Smith Blandford Square
On learning that JH's trip to the Cape will be delayed somewhat, invites JH to dine with R.S.L. Club in October.
ALS 3pp RS:HS 17.218

2868. [1833–9–18 or later] To [William Smith] 20 Charlotte St.
Excuses himself from a dining engagement; JH is too busy preparing for a voyage.
ALS 2pp APS B.H435p.30

2869. 1833–9–18 To Wilhelm Struve Slough
Informs JH that WS has sent fifth catalogue of double stars by [H. C.] Schumacher. Provides African address. Departure by 20 October on 'Mt. Stewart Elphinston—a superb India ship.' Notifies of [Thomas] Maclear's succeeding [Thomas] Henderson at Cape observatory.
C 3pp RS:HS 21.147

2870. 1833–9–18 To [?] Slough
Describes sighting while at Slough a remarkably bright occurrence of the aurora borealis.
draft 2pp inc Harvard: Houghton AAH 67m–67 (103)

2871. 1833–9–19 From Francis Baily 37 Tavistock Place
Is at present in town. Sorry he was out when JH called. What shall he do with JH's barometer? Thanks for the telescopic tests.
ALS 2pp RS:HS 3.117

2872. 1833–9–21 From Francis Beaufort Admiralty
Has found a cheap passage for Thomas Maclear. Has JH anything that Maclear can do for him?
ALS 2pp RS:HS 3.334

2873. 1833–9–22 From Thomas Maclear Bedford
Has given up the attempt to travel on the same ship with JH as it proved too expensive. Will now travel on the *Andromache* next Saturday; has only a few days to obtain the necessary stores.
ALS 2pp RS:HS 12.39 Warners, *Maclear & Herschel Letters,* 9

2874. [1833–9–23] From Richard Jones Ditchling, Lewes
Hopes to visit JH ere now but has been laid up with tic-douloreux. Moves to Weymouth in November. Glad to see the decided tone JH took on causation in astronomy. Cannot quite give up the Portsmouth Expedition.
ALS 3pp RS:HS 10.357

2875. 1833–9–25 From Francis Beaufort Admiralty
Will be pleased to distribute his papers. Has he made any arrangements for forwarding of his letters? If not, can he do this for him?
ALS 4pp RS:HS 3.335

2876. 1833–9–25 To Margaret Brodie Herschel Slough
Reports on hysterical fit by one of the young servants, and comments on the appropriateness of recommending employment elsewhere for another servant (a request for a recommendation is enclosed). JH alludes to some matter troubling his relationship with MH; Thomas Maclear is about

to leave for the Cape.
ALS 4pp JHS.HCEJ.526b

2877. 1833–9–25 From James Grahame Funchal, Madeira
On the state of Matilda's health [letter completed 1833–10–2].
ALS 2pp RS:HS 8.422

2878. 1833–9–25 To William Henry Smyth Slough
Unable to send Gamma Virginis observations because papers are packed. Declines offer of WS's transit circle for use at Cape.
ALS 4pp APS B.Sm98.15 (C: RS:HS 21.148)

2879. 1833–9–26 To Margaret Brodie Herschel Slough
Painful reflections on the problem troubling the relationship between JH and MH [see JH's 1833–9–25], namely, money, and how it is spent; JH offers some advice with much love.
ALS 4pp JHS.HCEJ.527

2880. 1833–9–[27] To Margaret Brodie Herschel Slough
Mostly about what to pack where; JH concludes with a loving apology for what he said and wrote [see JH's 1833–9–26]; question about re-arranged sitting for JH with H. W. Pickersgill.
ALS 4pp JHS.HCEJ.526c

2881. 1833–9–28 or later To Margaret Brodie Herschel Slough
Packing and domestic arrangements related to Cape voyage preparations.
ALS 4pp JHS.HCEJ.526a

2882. 1833–9–28 From Hugh Percy (3rd Duke of Northumberland) Alnwick Castle
The deal has been made [see HP's 1833–9–1] and HP is most grateful for JH's work.
ALS 4pp RS:HS 13.346

2883. 1833–9–29 To George Biddell Airy Slough
More about arrangements with R. A. Cauchoix and the telescope for Cambridge Observatory; JH also comments on the state of his nebula work.
ALS 4pp RGO 6.157.171

2884. 1833–9–30 From James Hudson Royal Society
Acknowledging the receipt of two boxes of manuscripts to be held at the R.S.L. until the return of JH from the Cape.
ALS 1p RS:HS 10.33

2885. 1833–10–1 From George Biddell Airy Observatory, Cambridge
Summary of the letter of same date sent to JH at Slough (HS.1.60).
ALS 2pp RS:HS 1.61

2886. 1833–10–1 From George Biddell Airy Observatory, Cambridge
Further discussion of the telescope being donated by Hugh Percy (3rd Duke of Northumberland).
ALS 3pp RS:HS 1.60

2887. 1833–10–3 To George Biddell Airy n.p.
Hugh Percy has approved the arrangements, which JH had made with R. A. Cauchoix [see JH's 1833–9–29].
ALS 2pp RGO 6.157.173

2888. 1833–10–3 From Thomas Henderson Edinburgh
Memorandum on the site of N. L. Lacaille's Observatory.
ALS 4pp RS:HS 9.290

2889. 1833–10–4 From James David Forbes Edinburgh
Sending a proof of a paper he is now having printed. Wishes him all the best for his forthcoming expedition.
ALS 1p RS:HS 7.286

2890. 1833–10–4 From Thomas Murdoch to L. D. Webster Gordon London
Letter introducing JH.
ALS 1p RS:HS 13.60

2891. 1833–10–5 From Wilhelm Struve Dorpat
Promises to write JH more often at the Cape. Is waiting for JH's work on nebulae. Promises to send his work on 3000 double stars. Plans to begin work on parallax of fixed stars.
ALS 5pp (German) RS:HS 17.125

2892. [1833–10–6] From Richard Jones Ditchling, Lewes
Cannot move until Saturday due to medical reasons.
ALS 1p RS:HS 10.358

2893. 1833–10–7 From James Hudson Royal Society
Acknowledging receipt of tin box containing the papers of the late Sir William Herschel.
ALS 1p RS:HS 10.34

2894. 1833–10–10 To Augustus De Morgan Slough
Arrangements related to JH's going to the Cape.
C 1p RS:HS 21.149

2895. 1833–10–10 From Mario Gemmellaro Nicolosi sull'Etna
Was pleased to receive JH's letter and his observations of Etna. Would like his memoir on the Nebulae of the Southern Hemisphere. Wishes him a prosperous time at the Cape. Would like some of the copper coins from Madagascar.
ALS 3pp (Italian) RS:HS 8.96

2896. [1833]–10–10 From Martha Langton Farnham Lodge
Would be grateful for the loan of £10 to meet urgent needs.
ALS 2pp RS:HS 11.461

2897. 1833–10–11 To Miss C. Langton Slough
Preparations for leaving England make it impossible for JH to [give money] to [Joseph] Langton, but JH has paid entire bill for fencing in the Merke, part of which Langton should have paid, and now dismisses Langton's £300 debt to JH, incurred on 8 June 1831.
ALS 2pp TxU:H/L-0218; Reel 1054

2898. 1833–10–14 From Francis Beaufort Admiralty
Called on him yesterday to tell him that [Robert?] Maconochie has a copy of a work that JH wants and he would most likely let him have it.
ALS 2pp RS:HS 3.336

2899. 1833–10–14 From William Rutter Dawes Ormskirk
Hopes JH is still in England as the barometer indicates a gale. Gives some of his recent star observations. Was disappointed with the state of James South's instruments at Kensington. Met W. S. Stratford and Thomas Henderson there. William Lassell is an ingenious man.
ALS 3pp RS:HS 6.55

2900. 1833–10–15 To Francis Baily Slough
Thanks for the gift of pheasants. Would he receive JH's medal from the R.S.L. for him and keep it until he can collect it. Will call on him next week.
ALS 2pp RS:HS 3.116 (C: RS:HS 21.150)

2901. 1833–10–15 From Thomas Maclear Ship *Tam O'Shanter*
Have embarked on the *Tam O'Shanter* but had to put back because of the violent gales. All have been seasick except TM. Outlines domestic necessities for a long voyage. Will JH bring a collimator from [Thomas] Jones of Charing Cross. Hopes JH will interest D'Urban in celestial matters. Many cockroaches on board.
ALS 4pp RS:HS 12.40 Warners, *Maclear & Herschel Letters,* 13–14

2902. 1833–10–15 From Sir James McGrigor Campden Hill
Hears that JH wishes to be known to Dr. John Murray, so encloses a letter of introduction. Will find him a pleasant and amiable man.
ALS 1p RS:HS 12.16

2903. 1833–10–16 To James David Forbes Slough
Wishes JF well in [Edinburgh] and leaves addresses where JH may be reached [at the Cape].
ALS 3pp St. Andrews 48 (C: RS:HS 21.151)

2904. 1833–10–18 From Alexander Rogers Baltic St. Leith
Discusses problem with transporting the ale JH wanted. Wishes him a safe journey to the Cape of Good Hope.
ALS 2pp RS:HS 14.417

2905. [1833–10–19] From Thomas Maclear ship *Tam O'Shanter*
Have had a dreadful time since Wednesday. Gives list of instructions for a voyage; plenty of candles. Finds the poop cabins very pleasant.
ALS 3pp RS:HS 12.41 Warners, *Maclear & Herschel Letters,* 16

2906. 1833–10–20 To Thomas Henderson n.p.
Details about JH's upcoming voyage from Portsmouth [to Cape of Good Hope] and about an expedition going north from the Cape into Central Africa.
ALS 2pp EUL:LaII425/24

2907. 1833–10–21 From George Biddell Airy Observatory, Cambridge
Has not been requested to settle the affair about Henry Taylor and the Stephen Groombridge business. Would JH send him what he has ready for printing, with directions? Observations on the 4th satellite of Jupiter.
ALS 1p RS:HS 1.62

2908. 1833–10–21 To Francis Baily Slough
Thanks for the pheasants. Has a communication for the R.A.S.
ALS 1p RS:HS 3.118 & 21.152

2909. 1833–10–21 From Francis Beaufort Admiralty
Has laid JH's letter about the Cape expedition before the Board. Captain J. C. Ross has arrived safely.
ALS 2pp RS:HS 3.337

2910. 1833–10–25 To Johann Franz Encke Slough
Hopes JE will write to JH at Cape. Responds to JE's comments on JH's method of dealing with double stars. Thanks for generous comments on JH's *Prelim. Discourse.*
ALS 4pp AdWDDR Encke 49

2911. 1833–10–25 From Charles May Ampthill
Hopes this letter will reach him before his departure for the Cape. Wishes someone in the proper climate would make a study of the best kind of Aloe for medicinal purposes. Encloses letter and plates on this subject and would be pleased if JH would show them to Thomas Maclear. Has not been able to obtain any further information from dichromic persons.
ALS 2pp RS:HS 12.312

2912. 1833–10–26 To Margaret Brodie Herschel Slough
More about packing and domestic arrangements [see JH's 1833–9–28 or later]; JH asks MH to write to James South to thank him for forwarding a packet to JH.
ALS 3pp JHS.HCEJ.528

2913. 1833–10–28 To Margaret Brodie Herschel Slough
Still further to packing and domestic arrangements [see JH's 1833–10–26], which seem to be more frenetic and nearing finality.
ALS 3pp JHS.HCEJ.529

2914. 1833–10–29 To Wilhelm Struve Slough
Confirms receipt of WS's last letter from Hamburg. Advises of his departure to Cape. Requests 'measure of 3000 double stars.' Directs WS's attention to 'remarkable stars' and urges WS to procure JH's *Astronomy.* Asks WS to solicit [F. W.] Bessel's correspondence for JH. Adds [W. H.] Smyth's address.
C 3pp RS:HS 21.153

2915. [1833–10–30] From Richard Jones Ditchling, Lewes
Expected another letter from JH. Awaits this letter before leaving Portsmouth.
ALS 1p RS:HS 10.359

2916. 1833–10–31 From Francis Beaufort Admiralty
Admiralty not able to lend JH any further equipment to take on JH's expedition to the Cape.
ACC 1p Hydrographic Office Letter Book

2917. 1833–11–2 To Caroline Lucretia Herschel London
Mentions to CH that 2000 of the nebulae and all of the engravings are printed.
ALS 1p TxU:H/L–0578.7.b; Reel 1058

2918. 1833–11–3 To William Somerville Slough
Regrets he will not get to visit Chelsea before departing for the Cape of Good Hope. Sends regards to family.
ALS 2pp RS:HS 16.403

2919. [1833–11–4] From Richard Jones to Margaret Brodie Herschel Ditchling, Lewes
Mrs. Jones is laid up with a cold, but hopes to go to Portsmouth on the 7th to arrange the lodgings. Is becoming more reconciled to the Herschels' expedition.
ALS 4pp RS:HS 10.360

2920. 1833–11–6 From James Hudson Royal Society
Has now completed his examination of the mode of reading off the barometer of J. W. G. Bu[t]zengeiger, made under the direction of H. C. Schumacher. Comments on the readings. Has now sent off all the Italian papers. Hopes JH will have a prosperous voyage.
ALS 3pp RS:HS 10.35

2921. 1833–11–10 To Caroline Lucretia Herschel Portsmouth
Ready to embark to Cape Town; JH expects to arrive at the end of January. Sent out the last proof sheet of JH's nebulae catalog last night.
ALS 3pp TxU:H/L–0578.8; Reel 1058 Evans, *Herschel at the Cape,* 18–19

2922. 1833–11–12 To Francis Baily Portsmouth
Waiting for the ship to be repaired. Has sent him his paper on satellites. Would like 100 offprints. Just about to go on board ship.
ALS 2pp RS:HS 3.119 & 21.153

2923. 1833–11–12 To [Charles] May Portsmouth
Thanks CM for his communication about the aloe. JH and family are about to embark for the Cape, and JH will keep CM's interests in mind while there.
ALS 2pp WT 56482

2924. 1833–11–12 To Adolphe Quetelet Portsmouth
Addresses last words before departing to Cape. Informs AQ that results concerning absorption of light will be in next *London and Edinburgh Philosophical Journal.* One will be sent to AQ. Reminds AQ of promise to write JH at Cape.
ALS 3pp (French) Académie belgique: #6 A. Quetelet, 'Notice sur Sir John Fred. Will. Herschel,' *Académie royale des science, des lettres et des beaux-arts annuaire* (1872), 166–7

2925. 1833–11–13 To Emilia Stewart Off Needles, Isle of Wight
Last letter to come off the ship with the pilot; next communication will come after landing at the Cape; all are well and comfortable.
ALS 3pp JHS.HCEJ.530

2926. 1833–11–23 From J. Peyrot 3 Place du Panthéon, Paris
Has translated some of JH's astronomical works, and is asking JH to comment on the accuracy of JP's translation.
ALS 1p (French) RS:HS 13.376

2927. 1833–11–25 From D. François J. Arago Paris
JH has won the gold medal of the Academy of Sciences, and 635 francs, for his work on double stars. Sends his congratulations. The medal will be presented by the president.
ALS 1p (French) RS:HS 1.359

2928. 1833–12–8 To Caroline Lucretia Herschel *Mountstuart Elphinstone*
Sailing to Cape Town, JH reports sighting an English ship.
ALS 2pp TxU:H/L–0578.9.b; Reel 1058 Evans, *Herschel at the Cape*, 19–20

2929. 1833–12–27 From Augustus De Morgan 5 Upper Gower St.
Thanks for his paper on nebulae. Scarlet fever has been very prevalent this season. R.S.L. in financial straits. Paper on John Flamsteed has caused friction between James Ivory and the R.A.S. Government has promised financial aid to Edinburgh Observatory.
ALS 4pp RS:HS 6.180 S. De Morgan, *Memoir of Augustus De Morgan*, 76–7

2930. [1834 or later] From Catherine Bell n.p.
Is sorry they were unable to see the Herschels but would be happy to dine at Feldhausen.
ALS 3pp RS:HS 4.39

2931. 1834–1–1 From James Grahame Madeira
Poem in honor of JH's arrival at Cape of Good Hope.
C 1p TxU:H/M–0213; Reel 1087

2932. 1834–1–1 From James Grahame Madeira
Of family news, health, and an eclipse of the moon [letter completed 1834–1–8].
ALS 3pp RS:HS 8.423

2933. [1834] From Thomas Best Jervis Cape Town
Hopes he received his *Essay on the Primitive Standard.* Arrived at Simon's Town on Tuesday and hopes to call on JH before he departs.
ALS 2pp RS:HS 10.316

2934. [1834 to 1835] To Thomas Maclear [Feldhausen]
Comparisons of the functioning of JH's barometer and that of TM.
C 1p inc RS:HS 25.14.26

2935. [1834 to 1838] To James Prinsep [Cape of Good Hope]
Recent article in *Journal of Asiatic Society of Bengal* confirms JH's notion about transfer of atmospheric pressure between hemispheres. Meteorological journal from 1828 to 1833 of Captain Barnes, port officer of Cape Town, provided JH with formula for determining annual maxima and minima of pressure.
Adraft 1p TxU:H/L-0317; Reel 1054

2936. [1834]. From [Duncan Stewart] n.p.
Recognizing the limitations of DS's medical education and DS's location in India as well as JH's superior learning, DS requests that his brother-in-law (JH) recommend to him a 'course of study' suitable to DS's situation and needs.
AL 7pp UCT JH Letters A1.46

2937. [1834 to 1838] To [?] [Feldhausen]
Describes in detail many of the plants at the Cape, especially flowering and variously scented ones.
AL 4pp WT 65667

2938. 1834–1–16 From James Grahame Madeira
Of dreams and illness.
ALS 2pp RS:HS 8.424

2939. 1834–1–17 To George Richardson Cape Town
Includes temperature readings for waters near the Cape. Sends 'Thomson's Lunar Tables' and [Thomas] 'Lynn's Horary Tables.' Servant's health is failing. May have to send her home immediately.
ALS 4pp RS:HS 14.483 (RS:HS 21.155)

2940. 1834–1–21 From Christian C. J. Bunsen n.p.
Leaving part of his manuscript on alphabets. Will be a meeting at his house and would be pleased if JH would come and give his views.
ALS 3pp RS:HS 5.124

2941. 1834–1–21 To Caroline Lucretia Herschel Cape Town, Cape of Good Hope
Reports that they are 'safely landed and comfortably housed' in Africa. The Herschels were at sea for nine weeks and two days. Required several days to unload the instruments and luggage from the ship. Has chosen a house, about five miles from town, called 'The Grove.' Excited with the calm, clear nights, which will be excellent for observations.
ALS 4pp TxU:H/L–0579.1; Reel 1058 Evans, *Herschel at the Cape,* 39–41 & *Memoir ... of Caroline Herschel,* 262–3

2942. 1834–1–22 From Thomas Maclear Royal Observatory [Cape of Good Hope]
Feels uneasy giving response to members of expedition into interior, who requested government instruments, until TM hears from Admiral [Frederick] Warren and judges expedition's chances for success. Asks JH's opinion.
ALS 1p TxU:H/M–0940 & *Maclear*; Reel 1083

2943. 1834–1–23 To Francis Baily Cape Town
Has arrived safely and unpacked his instruments. Has seen a

suitable house for his observatory. Difficulty with the female servant brought from England. Has met Thomas Maclear. Difficulties of the latter.
ALS 4pp RS:HS 3.120 (C: RS:HS 21.154 & 25.8.1)

2944. 1834–1–23 To Thomas Maclear n.p.
Matters relating to providing observatory instruments to an expedition into Central Africa.
C 1p RS:HS 21.156 Warners, *Maclear & Herschel Letters*, 33–4

2945. 1834–1–27 From Christian C. J. Bunsen 9 Carlton Terrace
Would like him to attend a small dinner party on Monday next.
D 1p RS:HS 5.123

2946. 1834–1–28 To William Whewell Cape Town
Has arrived at Cape Town; finds it an 'Earthly Paradise.' Describes its climate and noteworthy features of the southern heavens. JH's views on geology transformed by having read volume III of Charles Lyell's *Principles of Geology.*
ALS 4pp TC, Camb. Add. Ms.a.207[23] (CC: RS:HS 25.3.18)

2947. [1834–2–4] To Thomas Maclear Feldhausen
Oppressive heat prevents JH from visiting. Sending a list of furniture from Major John Ross's house, which JH definitely will purchase.
L Warners, *Maclear & Herschel Letters*, 34

2948. 1834–2–4 From Thomas Maclear Royal Observatory, Cape of Good Hope
Heat has driven Mrs. Maclear to leave a card with Lady D'Urban. Sends the state of the barometer and thermometer for the last three days. Stars were in capital form for the telescope last night. Has retained the list of furniture of Major Ross.
ALS 2pp RS:HS 12.44 Warners, *Maclear & Herschel Letters,* 35

2949. 1834–2–9 To Charles Babbage Cape of Good Hope
Is settled near Cape Town. Experiences on the voyage. Details of his telescope arrangements, and observations carried out.
ALS 4pp RS:HS 2.286

2950. 1834–2–12 From Francis Baily 37 Tavistock Place
Hopes he is safely landed. The government have agreed to give financial assistance to set up an observatory at Edinburgh. Financial difficulties at the R.S.L. Dispute between James South and Edward Troughton. Admiralty has referred the question of the printing of [Stephen] Groombridge's catalogue to FB and G. B. Airy. Has received the Royal Medals on behalf of JH. JH has also been awarded the Lalande medal.
ALS 3pp RS:HS 3.121

2951. 1834–2–14 From Augustus De Morgan 5 Upper Gower St.
Sending a letter by Francis Baily. Anniversary of the R.A.S. takes place on Friday. Capt. [Walter?] Forman has had a rise in the world. Observations of Henry Foster fills volume VII of the memoirs.
ALS 4pp RS:HS 6.181 S. De Morgan, *Memoir of Augustus De Morgan*, 77–8

2952. 1834–2–15 From Adolphe Quetelet Brussels
JH's last letter before departure has brought AQ great pleasure. Has not received JH's memoir on absorption of light. Has mentioned JH's name in an attempt to explain the idea of Cambridge Reunion. Disappointed at not yet having received equatorial and mural circle. Asks for news of JH's observations.
ALS 3pp (French) RS:HS 14.206

2953. 1834–3–1 To John Fairbairn Cape Town
Comments on a visit to the infant school, including comments on the equality of black and white children.
C 2pp RS:HS 21.157 (draft RS:HS 7.141verso)

2954. 1834–3–1 From John Fairbairn Cape Town
Would like a copy of JH's speech he made at the Infant School Society, for his report.
draft 4pp RS:HS 7.141

2955. 1834–3–2 To Thomas Maclear n.p.
Informs TM that JH's 20-ft. telescope is fully in operation; invites TM to come and observe with it.
C 1p RS:HS 21.158 Warners, *Maclear & Herschel Letters*, 7–8

2956. 1834–3–3 From Thomas Maclear Royal Observatory, Cape of Good Hope
Points out errors in his own chronometer. Gives thermometer and barometer readings. The work of the bricklayers is very slow. Is obliged for the drawing of the nebula.
ALS 2pp RS:HS 12.45 Warners, *Maclear & Herschel Letters,* 38

2957. 1834–3–13 From Thomas Maclear Royal Observatory, Cape of Good Hope
The chronometer can remain at the Grove for the time being; it seems to preserve a more uniform rate than the transit clock.
ALS 2pp RS:HS 12.46

2958. 1834–3–22 From Thomas Maclear Royal Observatory, Cape of Good Hope
Invites Herschels and Dr. Duncan Stewart to dine with TM. Would like to compare their barometers. Would like also to show JH his own investigation on the mural circle. Has procured a chameleon.
ALS 3pp RS:HS 12.47 Warners, *Maclear & Herschel Letters,* 40–1

2959. 1834–3–28 To Caroline Lucretia Herschel Wynberg, Cape of Good Hope
Reviews his voyage to Africa in case CH did not receive JH's 1834–1–21 letter. Repairing the house that they bought outside of Cape Town, called 'Feldhausen' by the Dutch and 'The Grove' by the English; they are staying at another home close by. Happy that Table Mountain near the house shelters Feldhausen from southeast gales.
ALS 4pp TxU:H/L–0579.2; Reel 1058 Evans, *Herschel at the Cape*, 52–4

2960. 1834–4 To [Duncan] Stewart [Feldhausen]
After a very candid analysis of his own strengths and limitations in learning, JH surveys mathematics and empirical science, noting the works of highest quality and greatest accessibility. JH then provides an analysis of historical inquiry and the allied area of political philosophy, stressing the importance of statistically based studies. Concludes by a discussion of philosophy, especially logic and methodologies of inquiry.
draft 25pp UCT JH Letters A1.47 (C: Univ. of Witwatersrand) *South African Libraries, 7* (1940), 138–144; 145–154

2961. 1834–4–2 To Thomas Maclear Wynberg
Saw a 'small round perfectly defined body' last night that

looks like a planet.
L Warners, *Maclear & Herschel Letters*, 41

2962. 1834–4–4 To Thomas Maclear Wynberg
Found a 'very fine' planetary nebula. Unable to learn anything conclusive from TM's microscope readings. JH's observation tower is erected and is currently being plastered.
C 2pp RS:HS 21.159 Warners, *Maclear & Herschel Letters*, 41–2

2963. 1834–4–5 From Edward Joshua Cooper Markree
Should have written earlier to thank him for the useful catalogue of stars. Weather has been fine, but he is still waiting for his eye tube from William Simms. Remarks on Orion's nebulae.
ALS 4pp RS:HS 5.279

2964. 1834–4–5 From Thomas Maclear Royal Observatory, Cape of Good Hope
Congratulations on his success. With JH's information he soon detected the fellow in the mural circle. Gives readings. As the sky is clear he will have another night at the mural circle.
ALS 3pp RS:HS 12.48 Warners, *Maclear & Herschel Letters*, 42–3

2965. 1834–4–8 From James Hudson Royal Society
All his friends are pleased to hear of his safe arrival at the Cape. Sends a cutting from the *Athenaeum*, which announces the fact. Is anxious to know if he obtained any observations with his barometer.
ALS 4pp RS:HS 10.36

2966. 1834–4–13 From Thomas Maclear Royal Observatory, Cape of Good Hope
Giving details of his facial paralysis.
AL 5pp inc RS:HS 12.49 Warners, *Maclear & Herschel Letters*, 44–5

2967. 1834–4–13 To James Calder Stewart Wynberg
Describes Wynberg's climate, plants, scenery, and social customs. Margaret's collection of flower paintings. 20-ft. telescope became active on 26 Feb. Duncan Stewart departed today aboard *Claudine*. Thank Peter [Stewart] for papers sent to JH. French Academy awarded medal to JH.
ALS 4pp TxU:H/L-0415; Reel 1055 Evans, *Herschel at the Cape*, 55–8

2968. [1834]–4–14 From Thomas Maclear n.p.
Gives news of his seizure. Gives readings for JH's nebulae. William Meadows will point out the observations in the transit book.
ALS 3pp RS:HS 12.203 Warners, *Maclear & Herschel Letters*, 45–6

2969. 1834–4–15 To Thomas Maclear [Feldhausen]
Happy that TM is handling his facial paralysis so well; JH believes that it is due primarily to anxiety and overwork.
L Warners, *Maclear & Herschel Letters*, 46

2970. [1834–4–15] From Thomas Maclear n.p.
Has taken the liberty of sending James Fayrer with his levers for JH's advice.
ALS 1p RS:HS 12.187 Warners, *Maclear & Herschel Letters*, 46

2971. 1834–4–18 From Sir John Bell Colonial Office, Cape Town
Meteorological Observations and Journal appears to be defunct. Fears that he will be unable to recover any of the papers.
ALS 2pp RS:HS 4.49

2972. 1834–4–20 From Feliciano Scarpellini Rome
Has heard from Charles Babbage about JH, JH's father, and their telescopes. Describes telescope commission. Hopes to get to know JH better.
ALS 2pp (Italian) RS:HS 17.170

2973. 1834–4–24 From George Biddell Airy Observatory, Cambridge
Observations on the weather of the past winter. 42 Virginis. Printing of [Stephen] Groombridge's catalogue. Has written article for the *Penny Cyclopaedia*. Refracted light in prisms. Machine for polarization of the sky. Petition to Parliament to abolish religious tests for degrees at Cambridge.
ALS 4pp RS:HS 1.63

2974. 1834–5–1 From Caroline Lucretia Herschel Hanover
Thrilled that JH has arrived safely in Cape Town; notes that JH's Cape Town trip has captivated the intellectual world.
L *Memoir ... of Caroline Herschel*, 264–6 inc

2975. 1834–5–14 From Niccolo Cacciatore Palermo
Has heard of his safe arrival at the Cape. Recent personal happenings and events at Palermo Observatory. Accuracy of G. B. Airy's observations.
ALS 3pp (Italian) RS:HS 5.139

2976. 1834–5–15 To William Henry Smyth Feldhausen, Cape of Good Hope
Describes voyage to Cape as being uneventful. 20-ft. reflector operating since late February; has already examined much of the southern sky. Believes the Milky Way is not equidistant from the earth at all points; and that the earth is between Canopus and Alpha Centauri. Amazed at clusters such as Omega Centauri. Describes richness of Magellanic Clouds. Details progress in erecting instruments. Sends observations of Alpha Crucis, which he believes is a double star.
C 3pp RS:HS 21.160

2977. 1834–5–19 From C. Hope Maclean 4 St. Martin's Place
[Form letter] Announces next two meetings of Statistical Society.
LS 1p RS:HS 19.93a

2978. 1834–5–28 From Basil Hall Malta
Is glad Mrs. Hall has written to Lady Herschel. Would like a private yacht so that he could travel round the world making scientific observations. Is pleased JH had a good voyage to the Cape and hopes things are progressing.
AL 4pp inc RS:HS 9.177

2979. 1834–5–29 From Thomas Maclear Royal Observatory, Cape of Good Hope
Has decided to put the dinner party off until Monday due to the wet weather. Is working hard at the reductions. Is indebted to the Herschels for the use of their carriage.
ALS 2pp RS:HS 12.50 Warners, *Maclear & Herschel Letters*, 49

2980. [1834–6–1] To Thomas Maclear n.p.
JH will break his rule about never going out when it is possible to sweep the sky, and dine with TM; will offer some help over mural circle problems.
C 1p RS:HS 21.161 Warners, *Maclear & Herschel Letters*, 50

2981. 1834–6–5 From Charles T. Beke Finsbury Circus
Is sending a volume dealing with primeval history, which he hopes JH will accept. Is in some way related to JH.
ALS 2pp RS:HS 4.26

2982. 1834–6–6 To Caroline Lucretia Herschel Feldhausen
Sweeping progressing quickly; has used the 20-ft. reflecting telescope since February. Discovered two planetary nebulae. Studying Scorpio closely, as CH suggested; has found gorgeous globular clusters there. The equatorial was erected recently.
L Evans, *Herschel at the Cape*, 71–2 & *Memoir ... of Caroline Herschel*, 266–7 inc

2983. 1834–6–6 From Thomas Hodgkin Finsbury Circus
Rejoices to hear of his safe arrival at the Cape, and his interest in Cape educational matters. Hears he has already paid attention to the native race. Sends letter by Dr. J. D. Lang of New South Wales.
ALS 3pp RS:HS 9.408

2984. 1834–6–10 From William Henry Smyth Crescent, Bedford
Sends observations of the double stars, which JH had suggested.
ALS 2pp RS:HS 16.254

2985. 1834–6–10 To Wilhelm Struve Feldhausen
Describes enthusiastically the journey to the Cape, the beauties of the country, and the joys of astronomical observing there.
ACCS 4pp RS:HS 25.3.19

2986. 1834–6–16 From Thomas Henderson London
Is happy to hear of his safe arrival at the Cape and comfortably housed. Comments on two stars Alpha^1 and Alpha^2 Centauri.
ALS 2pp RS:HS 9.291

2987. 1834–6–20 From George Biddell Airy Observatory, Cambridge
Summary of GA's 1834–4–24, with additional information that he has written a book entitled *Gravitation*. Confusion over 42 Virginis. Affair between James South, and Edward Troughton and William Simms has been referred to arbitration. Experiments on Newton's rings by transmission. Letter to Thomas Maclear about his mural circle. Saw the eclipse of 16 July. Tarnish of JH's mirrors. R. A. Cauchoix's delay with telescope. [The final two pages of this letter are from Richarda Airy to Margaret Brodie Herschel.]
ALS 12pp RS:HS 1.64 (C: RS:HS 1.65)

2988. [1834–7–3] To Thomas Maclear [Feldhausen]
Gives location of a planetary nebula.
L Warners, *Maclear & Herschel Letters*, 50

2989. 1834–7–4 From Friedrich Wilhelm Bessel Königsberg
Wishes JH well at the Cape; describes some of his current observations.
ALS 2pp (German) RS:HS 4.69

2990. 1834–7–5 From Francis Beaufort Admiralty
Thanks for the packet of letters. Thomas Henderson has been appointed to the Edinburgh Observatory. Richardson has just completed Brisbane's Catalogue of Southern Stars. Has observed a fortnight's tides for William Whewell.
ALS 4pp RS:HS 3.338

2991. 1834–7–11 To Francis Baily Feldhausen
Describes the departure of an expedition north of the Cape, as well as the pleasures of the country and the beautiful skies for observing. The delivery of mail is, however, so erratic as to be vexatious.
ACCS 4pp RS:HS 25.3.20

2992. 1834–7–11 From Richarda Airy to Margaret Brodie Herschel Greenwich
Thanks for letter. Boy born on 12 May. Intends to go to Keswick for a few weeks. [This is written as part of the letter from G. B. Airy to JH, dated 1834–6–20.]
ALS 3pp RS:HS 1.64

2993. 1834–7–17 To James Calder Stewart Feldhausen
Settling in. Regrets departure of Duncan Stewart. Ignore James South's increasing attacks; they cannot harm JH. Hunted Cape 'wolf' last night. All observing instruments are up and active. Laments poor local workmanship. Praises John Fairbairn's editorials.
ALS 4pp TxU:H/L-0416; Reel 1055 Evans, *Herschel at the Cape*, 78–82

2994. [1834]–7–19 To Peter Stewart [Feldhausen]
Describes enclosed sketches of Cape Flats and mountains, viewed from Feldhausen. Names ships that carry JH's letters, sketches, and bird skins to friends in England.
ALS 2pp TxU:H/L-0394; Reel 1054

2995. 1834–7–22 From John Rennie London
Introduces Captain James Alexander, who has explored Africa. Will hail JH's return from the Cape with pride.
ALS 2pp RS:HS 14.476

2996. 1834–7–24 From Thomas Maclear Royal Observatory, Cape of Good Hope
Has received dispatches from Bedford at last; hopes that JH has been equally fortunate. Encloses a letter from Mrs. Smyth, which gives amusing details of the anniversary of the R.A.S. Comments at length on the character of Sir James South.
ALS 3pp RS:HS 12.51 Warners, *Maclear & Herschel Letters*, 51–2

2997. 1834–7–25 From Thomas Maclear Royal Observatory, Cape of Good Hope
Regarding some queries of N. L. Lacaille's Catalogue of Southern Stars. He was severe on Edmund Halley. Can JH inform him if the acorns sent fell off the tree or were beaten off.
ALS 4pp RS:HS 12.52 Warners, *Maclear & Herschel Letters*, 52

2998. [1834–8–12] To Thomas Maclear [Feldhausen]
Discusses JH's recent observations of Gamma Virginis.
C 1p RS:HS 21.166 Warners, *Maclear & Herschel Letters*, 53

2999. 1834–8–14 From Thomas Maclear Royal Observatory, Cape of Good Hope
Thanks for the acorns; will put them into the ground instantly. Received dispatches from the Admiralty yesterday; also a bag from Bedford. Cannot find the Greenwich Observations for 1830 in the Observatory Library.
ALS 2pp RS:HS 12.53 Warners, *Maclear & Herschel Letters*, 53

3000. 1834–8–18 From Francis Baily London
R.A.S. has been allocated rooms at Somerset House. The James South affair has been submitted to referees. Has discovered a large collection of John Flamsteed's papers at

the Greenwich Observatory.
ALS 3pp RS:HS 3.122

3001. 1834–8–23 From Adolphe Quetelet Brussels
Has JH's memoir on elliptical orbits of Saturn. Asks for JH's indulgence with lack of astronomy in annals. AQ has no big instruments. Has planted thermometers for experiments. Finally received transit instrument from [H. P.] Gambey.
ALS 3pp (French) RS:HS 14.207

3002. [1834–8–25] To Thomas Maclear [Feldhausen]
Predicting poor weather, JH advises TM not to go to the Grove tomorrow.
L Warners, *Maclear & Herschel Letters*, 53–4

3003. 1834–9–1 From Thomas Maclear Royal Observatory, Cape of Good Hope
Resumed the N. L. Lacaille operation on Saturday. Discusses the position of Lacaille's Observatory. Has obtained an old plan of the area.
ALS 4pp RS:HS 12.54 Warners, *Maclear & Herschel Letters*, 54–5

3004. 1834–9–9 From Giovanni B. Amici Florence
Concerning Capt. Basil Hall. GA is going to live at Florence as Director of the Florence Observatory.
ALS 3pp (Italian) RS:HS 1.343

3005. 1834–9–9 From Giovanni Antonio Plana Turin
In his travels, GP met James Stewart (Margaret Herschel's brother); gives GP's assessment of James Stewart. Comments on JH's intended voyage to the Cape, and wishes him well.
ALS 3pp (French) RS:HS 13.432

3006. 1834–9–11 From Caroline Lucretia Herschel Hanover
Asks about JH's children. Remarks that JH's discovery of globular clusters in the Scorpion is not what she remembered William Herschel being mystified about; remembers that WH exclaimed that there seemed to be a 'Loch im Himmel' ('hole in heaven') there.
L *Memoir ... of Caroline Herschel*, 268–9 inc

3007. 1834–9–11 From Thomas Maclear Royal Observatory, Cape of Good Hope
Approached Col. R. Thom[p]son on the subject of some rockets, and he advises an official letter to the Officer of Ordnance. Is inclined to use gunpowder instead if JH is agreeable. Hopes Lady Herschel and the new infant are progressing.
ALS 3pp RS:HS 12.55 Warners, *Maclear & Herschel Letters*, 55–6

3008. 1834–9–18 From Charles Babbage Dorset St.
Regarding his machine. Suggested modifications for it. Recent meeting of the B.A.A.S. Political events.
ALS 3pp RS:HS 2.287

3009. 1834–9–18 From John Philip Grahamstown
Is greatly exercised over the passing of the Vagrants Act, which JP believes is intended to, and will, discriminate against the blacks.
ALS 5pp RS:HS 13.381

3010. 1834–9–21 To William Whewell Feldhausen near Wynberg
Reports on the meteorological and tidal conditions at the Cape. Praises climate and observing conditions. Comments negatively on James Dunlop's catalogue of southern nebulae.
ALS 3pp TC, Camb. Add. Ms.a.207^{24} (CC: RS:HS 25.3.21)

3011. 1834–10–1 From William Henry Smyth Crescent, Bedford
Thanks JH for Gamma Virginis observations, which are finally in accord with his. Sends observation of the double stars JH suggested watching. Works with Richard Sheepshanks. Francis Baily delivers details on Edinburgh meeting of philosophers. Sends news from Niccolo Cacciatore, who has returned to the Observatory in Naples.
ALS 4pp RS:HS 16.255

3012. [1834–10–9] To [Sir] Edward Ryan [Feldhausen]
Extract from JH's journal on voyage to Cape of Good Hope, showing barometer and thermometer readings from 23 Nov. 1833 (Lat. 33°20' N) to 5 Jan. 1834 (Lat 31°30' S). Compares these to ER's thermometer and R.S.L. standard thermometer. Will accompany ER to [Cape] observatory next week. Please conduct barometer and temperature observations on ER's return trip [to India].
ALS 3pp TxU:H/L-0765.2; Reel 1089

3013. 1834–10–12 To Thomas Maclear n.p.
Comments on social matters, and on what JH has been observing.
C 2pp RS:HS 21.163 Warners, *Maclear & Herschel Letters*, 60

3014. 1834–10–13 From James Hudson [Royal Society]
Sends some packages and his *Transactions*. Sent through [Peter?] Stewart the first volume of Col. Mark Beaufoy's *Observations*. Sends copies of the *Athenaeum*, which contains details of the Edinburgh meeting. Will be placed on the list to receive free copies of the *Athenaeum*.
ALS 2pp RS:HS 10.37

3015. 1834–10–14 To Richard Jones Feldhausen
Thanks RJ for having helped J[ohn] Stewart, [JH's brother-in-law], to a job; offers to try to get some South African statistics for the Statistical Society, and along the way comments on South African political life. JH describes life at the Cape, and the beauties of the skies.
ACCS 4pp RS:HS 25.4.1

3016. 1834–10–16 From Thomas Maclear Royal Observatory, Cape of Good Hope
Thanks for the perusal of J. A. Lloyd's paper. Has written to him on the subject of the pendulum. Had only one letter from the *Elizabeth*. No satisfactory news of [Andrew?] Smith's expedition. Unable to make any circumpolar observations due to the deviation of the transit instrument.
ALS 3pp RS:HS 12.96 Warners, *Maclear & Herschel Letters*, 60–1

3017. 1834–10–17 From William Rowan Hamilton n.p.
Developed a method that greatly simplifies Joseph Lagrange's integration of the 3n differential equations of motion of the second order by reducing all calculations to the application of 'one principal function.' Applies the method to planetary motion.
L Graves, *Hamilton*, *2*, 112–7 inc

3018. 1834–10–19 From George Biddell Airy Observatory, Cambridge
Abstracts of barometrical pressure in various parts of the earth. Has used James South's equatorial with Richard Sheepshanks's clockwork attached. Constitution of Paris Observatory being changed. D. F. J. Arago's experiments on glass and metal. Humphrey Lloyd's experiments on reflected light. Edinburgh's Observatory site likely to be changed. John Pond (Astronomer Royal) has been asked to resign. Thomas Maclear's Cape circle. Reasons for the delay in

delivery of R. A. Cauchoix's telescope to Cambridge. House of Lords and Commons burnt down.
ALS 7pp RS:HS 1.66 (C: RS:HS 1.67)

3019. 1834–10–22 To Francis Baily Feldhausen
Situation at the Cape of Good Hope. JH's astronomical observations.
ALS 4pp RS:HS 3.123 (C: RS:HS 25.4.3 & 25.8.2)
R.A.S.M.N., 3 (1835), 75–7 (inc)

3020. 1834–10–22 To Peter Stewart Feldhausen
Acknowledges receipt of some parcels, and then discusses in some detail JH's financial situation, including the fact that JH's aunt, Caroline, has refused the annual portion of rent from Slough. JH expresses concern for the safety of Mr. Calder, and concludes with some family news.
ACCS 4pp RS:HS 25.4.2

3021. 1834–10–24 From Sir John Bell n.p.
Has just received the Stellenbosch diary, which he is sending on to JH. Hopes this will be of some use, but is unable to make up the deficiency.
ALS 1p RS:HS 4.51

3022. 1834–10–24 From Sir John Bell n.p.
Would like JH's comments on a plan for a water supply for Cape Town.
ALS 2pp RS:HS 4.50

3023. 1834–10–26 From Thomas Maclear Royal Observatory, Cape of Good Hope
Sending some barometric comparisons. Had a narrow escape from Table Mountain fog. Has been quail shooting. P.S. If JH would like to see some quail shooting, join him this or tomorrow evening.
ALS 2pp RS:HS 12.56–57 Warners, *Maclear & Herschel Letters,* 62

3024. 1834–10–28 To [Sir] Edward Ryan Feldhausen
Compares readings of JH's and ER's barometers to those of Cape Observatory and of R.S.L. Observations of sea surface temperatures made by JH during voyage to Cape of Good Hope. Returns manuscripts.
ALS 3pp TxU:H/L-0765.1; Reel 1089

3025. 1834–10–31 From [J.?] Watermeyer 3 Orange St. [Cape of Good Hope]
JW's 'tabular scheme,' employing Price's method, is only a first attempt and yields life probabilities too low for Cape Colony, where births exceed deaths. Will submit recent data next week.
ALS 7pp RS:HS 18.378

3026. [1834–11] To Sir John Bell [Cape Town]
Thanks for the Stellenbosch diaries, which will no doubt be useful. Comments on the prospects for water supply for Cape Town.
draft 2pp RS:HS 4.52 & 21.162

3027. [1834–11?] From [J.?] Watermeyer n.p.
Regrets missing appointment yesterday. Can JH arrive early tomorrow and meet before Council convenes? Wishes to submit more data on Colonial population. Suggests appointing subcommittee on statistical research.
ALS 3pp RS:HS 18.379

3028. 1834–11–4 From Sir Edward Ryan Calcutta
Wants advice on reading thermometers and barometers. Asks for some temperature readings JH made. Will send some Calcutta newspapers. Invites Herschel's to dinner.
ALS 2pp RS:HS 14.452

3029. 1834–11–8 From Mario Gemmellaro Nicolosi sull'Etna
Hears that JH has arrived safely at the Cape and discovered two new planets already. Will shortly receive Niccolo Cacciatore. The shape of Etna has changed. His own brothers are away at the moment.
ALS 3pp (Italian) RS:HS 8.97

3030. 1834–11–11 From [J.?] Watermeyer Cape Town
Announces 13 Nov. meeting of Committee of South African Infant Schools.
(A)LS 1p RAS:JH Archive 3/6.28; Reel 7

3031. 1834–11–15 From James Grahame L'Eperonniere
Of the return to France and the family health.
ALS 2pp RS:HS 8.425

3032. 1834–11–15 From James Horsburgh East India House
Delayed acknowledging JH's letter until the return of [James?] Hudson from Derbyshire as he knew he would be interested in the atmosphere and temperature of the ocean. Comments on this and quotes a review of Captain Foster's voyages in the *Athenaeum*.
ALS 2pp RS:HS 9.499

3033. 1834–11–17 To Thomas Maclear Feldhausen
Invites TM to a meeting of the Meteorological Committee, which James Adamson, Abraham Cloete, and James Bance hope to attend.
L Warners, *Maclear & Herschel Letters*, 64

3034. 1834–11–24 To John Philip Feldhausen
JH encourages JP's compassionate approach to the problem of racial relationships [see JP's 1834–9–18].
C 4pp RS:HS 13.382 & 21.164

3035. 1834–12–2 To James Prinsep Feldhausen
Comparing meteorological records from Bengal with those at the Cape, JH is led to some theoretical considerations about air movements, supporting his ideas with reference to other meteorological observations.
ACCS 5pp RS:HS 25.4.4

3036. 1834–12–4 From Thomas Maclear Royal Observatory, Cape of Good Hope
T. W. Bowler came and apologized after JH had gone. May get less trouble from him in the future. Sends the book for JH to look over.
ALS 2pp RS:HS 12.58 Warners, *Maclear & Herschel Letters,* 66

3037. [1834–12–5] To Thomas Maclear [Feldhausen]
Wants to learn more about JH's book-keeping system.
L Warners, *Maclear & Herschel Letters*, 66

3038. 1834–12–6 To James Calder Stewart Feldhausen
JS's visit with G. A. A. Plana. Ship *Mercury* was found safe but weather beaten. New baby [Margaret Louisa] is healthy. Stars are very clear. Commercial panic in Cape colony.
ALS 4pp TxU:H/L-0417; Reel 1055 Evans, *Herschel at the Cape,* 126–8

3039. 1834–12–13 From Thomas Hodgkin Finsbury Circus
Sends letter by Jonathan Stander, who is to work in the educational field at the Cape. Is pleased JH serves on the management committee of the Infants schools. Wishes success to the expedition to explore the interior of S. Africa. Thanks for his paper on nebulae. Enjoyed the B.A.A.S. meeting at Edinburgh. Sends one of his own pamphlets on

African colonization. Thomas Bell wishes to be remembered to him.
ALS 5pp RS:HS 9.409

3040. 1834–12–13 From James Prinsep Calcutta
Offers honors to JH from Asiatic Society in return for information and observations sent there. Oversees care of Asiatic Society telescope. Anxious to hear of JH's work at the Cape of Good Hope.
ALS 3pp RS:HS 14.66

3041. 1834–12–21 To Thomas Maclear [Feldhausen]
Thanks JH for reading Andrew Smith's letter.
L Warners, *Maclear & Herschel Letters*, 67

3042. 1834–12–22 From Teodoro Monticelli Naples
TM's nephew will deliver the letter to JH during his travels. Writes of his geological studies of volcanic rock and rock from the lower parts of the earth's crust.
ALS 2pp (Italian) RS:HS 12.365

3043. 1834–12–23 To Charles Babbage n.p.
Regarding CB's machine. Edward Ryan's visit. Events at the Cape. Failure of his astronomical observations due to cloud conditions. Has been trying to stir up the South African Philosophical Society. Recent South African expedition.
ALS 4pp RS:HS 2.288 & 25.4.5

3044. 1834–12–24 To Basil Hall Feldhausen
Provides an account of JH's observatory arrangements, some interesting observations, and the effect of the weather on observing.
ACCS 4pp RS:HS 25.4.6

3045. 1834–12–26 To William Henry Smyth Feldhausen
Discusses contribution of WS's son, C. P. Smyth, as Thomas Maclear's assistant. Discusses observations of Halley's Comet, Gamma Virginis, other celestial objects, and the use of a double image micrometer.
ALS Xpp South African Library David S. Evans, 'Sir John Herschel to Captain Smyth,' *Quarterly Bulletin of the South African Library, 25* (September, 1970), 11–14

3046. 1834–12–29 From Basil Hall Heinfeld near Graz
Forwarding two letters via Francis Beaufort. Has no political or scientific news as they are at present residing in a remote Austrian castle which they find very refreshing after the noise of Vienna. Longs to hear of JH's labors with the constellations.
ALS 4pp RS:HS 9.178

3047. [1835 to 1842] From Charles Ewan Law n.p.
As MP for Cambridge, CL promises to extend influence of ministers of Church of England and uphold privileges of University. Asks for JH's confidence.
ALS 1p inc RS:HS 18.413a

3048. [1835 or later] From Thomas Maclear n.p.
Urging JH to recommend the services of Lieut. James Bance to Gilbert Elliot (2nd Earl of Minto).
ALS 1p RS:HS 12.215

3049. 1835 From Teodoro Monticelli Naples
Writes of geological work on terrain of Sicily.
ALS 3pp (Italian) RS:HS 12.366

3050. 1835–1–5 From Francis Beaufort Admiralty
JH's last letter has given him great pleasure. Thanks for his hints about observations. The instruments at St. Helena Observatory. Payment of carriage on books sent abroad. Difficulties over the printing of the Cape observations. [François] Arago has sent him a paper on [John] Brinkley. Progress with the triangulation of the U.S.A.
ALS 8pp RS:HS 3.339

3051. [1835–1–7] To Thomas Maclear Feldhausen
Too ill to attend the South African Institution meeting; asks JH to turn in the packet containing TM's and JH's meteorological observations.
L Warners, *Maclear & Herschel Letters*, 69 & in Thomas Maclear, 'Sir John Herschel at the Cape,' *Monthly Magazine*, new ser., *3* (Sept., 1871), 128–40:136

3052. [1835]–1–10 From Thomas Maclear n.p.
Expects an occultation of Jupiter this evening. Hopes JH enjoyed the fine cool breeze of last evening. Another English vessel reported. N.S. Wales acacias are showing signs of ambition.
ALS 2pp RS:HS 12.42 Warners, *Maclear & Herschel Letters,* 70

3053. [1835]–1–11 From Thomas Maclear n.p.
Readings for the occultation of Jupiter.
ALS 2pp RS:HS 12.43 Warners, *Maclear & Herschel Letters,* 70–1

3054. 1835–1–12 From James Holman to Margaret Brodie Herschel Windsor
Sending a copy of his *Travels in the Cape*. Niece has just married a Mr. Kirbey, who is on his way now to Madras. Hopes JH is well pleased with the Cape.
ALS 1p RS:HS 10.202

3055. [1835–1–15] To Thomas Maclear n.p.
Describing JH's observations and comparing notes.
C 2pp RS:HS 21.165 Warners, *Maclear & Herschel Letters*, 71

3056. 1835–1–17 From Thomas Maclear Royal Observatory, Cape of Good Hope
Wind is likely to be higher in the evening. Would like JH's views on the effect that the motion of the center of the mural circle has on the readings of the microscopes. Wind was too high last night to observe Zeta Octantes.
ALS 2pp RS:HS 12.59 Warners, *Maclear & Herschel Letters,* 72

3057. 1835–1–20 To Charles Babbage Feldhausen
Commiserating on CB's calamity.
ALS 2pp RS:HS 2.289

3058. 1835–1–20 From James Grahame L'Eperonniere
Matilda becomes weaker and weaker [letter completed 1835–3–10].
ALS 2pp RS:HS 8.426

3059. 1835–1–21 From Hugh Percy (3rd Duke of Northumberland) Alnwick Castle
Has placed £1000 at JH's disposal for whatever use JH and Lady Herschel see fit.
ALS†† 2pp RS:HS 13.347

3060. 1835–1–22 To Thomas Maclear Feldhausen
Responds to TM's query in his 1835–1–17 about the effect that motion of the center of the mural circle would have on the microscopic readings.
C 3pp RS:HS 21.169 Warners, *Maclear & Herschel Letters*, 72–3 inc

3061. 1835–1–25 From Francis Beaufort Admiralty
Believes that the copies of the Nautical Almanac he sent him

have gone astray. Is sending him the sheets of the Parramatta catalogue.
ALS 4pp RS:HS 3.340 & ACC: Hydrographic Office Letter Book

3062. 1835–2–4 From William Henry Smyth Crescent, Bedford
Adds clock to telescope, enhancing distance measuring power. Search for Halley's Comet begins. Reports double star data; observes nebulae in JH's *Catalogue of Nebulae.* R.A.S. location moved to Somerset House. Quarrel between James South and Richard Sheepshanks being arbitrated.
ALS 4pp RS:HS 16.256

3063. 1835–2–5 From James David Forbes Edinburgh
Mentions letter from William Whewell sending JF JH's interesting but puzzling meteorological results. Will communicate them to a Dr. [John?] Davy at Malta. Has been busy preparing lectures. Made a short tour in England last summer. G. B. Airy has been awarded the Lalande Medal.
ACCS 3pp St. Andrews LB II, 215–17 (C: RS:HS 7.287) J. C. Shairp, P. G. Tait, and A. Adams-Reilly, *Life and Letters of James David Forbes, F.R.S.* (London, 1873), 116–17

3064. [1835–2–5] To Thomas Maclear n.p.
Suggests a means of resolving TM's problem with the mural circle.
draft 11pp RS:HS 12.179 (C: RS:HS 21.170) Warners, *Maclear & Herschel Letters,* 73–4

3065. 1835–2–6 From J. Peyrot 3 Place du Panthéon, Paris
Is considering issuing a translated edition of JH's astronomical work. All of France awaits the valuable results that are sure to come out of JH's journey to the Cape, and JP will be pleased to translate it.
ALS 2pp (French) RS:HS 13.377

3066. 1835–2–7 From Mario Gemmellaro Nicolosi sull'Etna
Would like to hear of any new discoveries. Is pleased he will see JH on his way home from the Cape. Nothing happening at the moment to Etna.
ALS 1p (Italian) RS:HS 8.98

3067. 1835–2–7 To Thomas Maclear Feldhausen
Additional approaches to resolving TM's mural circle problem [see JH's 1835–2–5].
C 13pp RS:HS 21.171

3068. 1835–2–7 To William Whewell Feldhausen near Wynberg
Discusses tidal observations made at the Cape and a paper by WW on tides. Praises viewing conditions at the Cape. Describes political struggles at the Cape.
ALS 4pp TC, Camb. Add. Ms.a.207[25] (CC: RS:HS 25.4.9)

3069. 1835–2–11 From John Philip Church Square
About race relations in South Africa, and anti-slavery movements in England.
ALS 4pp RS:HS 13.383

3070. 1835–2–12 From Thomas Maclear Royal Observatory, Cape of Good Hope
Has just finished the observations of Mars and the circle is free from any particular use. Further comments on his query regarding the motion of the center of the mural circle. Would like JH's assistance on the examination. Has had some brass ingots stolen from the observatory.
ALS 7pp RS:HS 12.60 Warners, *Maclear & Herschel Letters,* 74–5

3071. 1835–2–12 From Francis Beaufort to Mr. Stewart Admiralty
Intends to dispatch a parcel of books to JH and has he anything he would like to include in the parcel?
ALS 3pp RS:HS 3.341

3072. 1835–2–14 From George Biddell Airy Observatory, Cambridge
Has just been made president of the R.A.S. No news of John Pond's retirement. R. A. Cauchoix's telescope has arrived. Printing of his 1834 observations are well advanced.
ALS 2pp RS:HS 1.68

3073. 1835–2–18 From Thomas Maclear Royal Observatory, Cape of Good Hope
Sends the projection of Monday's readings and one of those from the lever of contact for comparison; comments on this.
ALS 2pp RS:HS 12.61 Warners, *Maclear & Herschel Letters,* 76

3074. 1835–2–20 To George Biddell Airy Feldhausen
Describes the tests and adjustments JH has recommended to Thomas Maclear to try to remove anomalies from the functioning of the transit circle at the Royal Observatory, Cape of Good Hope; also comments on the beauty of some southern nebulae.
ALS 4pp RGO 6.368.596 (C: RS:HS 25.4.7)

3075. [1835–2–21] From Thomas Maclear n.p.
Has replied to Richard Sheepshanks in the *Memoirs of the Astronomical Society.* Is surprised how near he got to the general figure described by the Circle Centre. Comments on calculations he intends to carry out. Cannot say decidedly that Zeta Octantes is double; it is doubtful seen through the 8-ft. transit.
ALS 3pp RS:HS 12.226 Warners, *Maclear & Herschel Letters,* 76–7

3076. 1835–2–22 To Caroline Lucretia Herschel Feldhausen
JH describes his sweeps of Scorpio. Writes that the Milky Way is 'extremely curious and interesting.' Has collected a large catalog of southern nebulae, many never before detected. JH's search for Halley's Comet has been unsuccessful thus far.
ALS 4pp TxU:H/L–0580.1.a; Reel 1058 *Memoir ... of Caroline Herschel,* 270–1

3077. 1835–2–22 From Margaret Brodie Herschel to Caroline Lucretia Herschel Feldhausen
MH discusses her determination to educate the Herschel children herself.
ALS 1p TxU:H/L–0580.1.b; Reel 1058 Evans, *Herschel at the Cape,* 143–5 & *Memoir ... of Caroline Herschel,* 270–1 inc

3078. 1835–2–23 To Thomas Maclear Feldhausen
More on the mural circle [see JH's 1835–2–7].
C 1p RS:HS 21.172

3079. 1835–2–24 To *Astronomische Nachrichten* Cape of Good Hope
Reports erecting his telescopes at the Cape and on his observations, especially of double stars, during his first year there.
L *Astronomische Nachrichten, 12* (1835), 273–6

3080. 1835–2–24 To Heinrich Christian Schumacher Feldhausen
Describes JH's location at the Cape and his observational sweeps, lists nebulae and double stars, as well as comet and eclipse observations. [Letter finished on 1835–3–5.]
ACCS 5pp RS:HS 25.4.8

3081. 1835–2–25 From John Fairbairn Cape Town
Would like a note of JH's speech to the Supporters of Infant Schools, last week. Comments on the educational facilities at the Cape.
ALS 2pp RS:HS 7.142

3082. [1835–3 or later]. From Adolphe Quetelet Brussels
Even though JH has not written, AQ has received news from JH's compatriots. Asks about JH's health. Has not really used equatorial. Continues work on terrestrial magnetism and includes observations. Asks if JH has received last part of translation of JH's work on optics.
ALS 2pp (French) RS:HS 14.277

3083. 1835–3–3 To Thomas Maclear Feldhausen
Found two planetary nebulae not listed in any catalog; hopes to show them to JH tonight.
L Warners, *Maclear & Herschel Letters*, 78

3084. 1835–3–5 From Heinrich Christian Schumacher Altona
[Otto] Rosenberger has calculated perturbations of Halley's Comet by the inner planets. [Wilhelm] Beer and [J. H.] Mädler have a map of the moon and have located two moving spots on Jupiter. F. W. Bessel's assistant has reduced James Bradley's observation to constants of aberration and nutation.
ALS 2pp RS:HS 15.376

3085. 1835–3–7 From Thomas Maclear Royal Observatory, Cape of Good Hope
Has been unable to locate the planetary nebulae; did JH give their places correctly, or is the instrument unequal to the task? Sends a comparison of John Pond's declinations and the Cape Zenith distances. Comments on the latitude of the Cape. The *Hungerford* got off yesterday.
ALS 3pp RS:HS 12.62 Warners, *Maclear & Herschel Letters*, 79

3086. [1835–3–7 or 8] To Thomas Maclear [Feldhausen]
Apologizes for his erroneous reduction. Thanks TM for his list of 'Zenith points.'
L Warners, *Maclear & Herschel Letters*, 79–80

3087. [1835–3–8 or 9] To Thomas Maclear [Feldhausen]
Reports on location of a planetary nebula. Expresses thanks for book by [Robert] Woodhouse.
C 1p RS:HS 21.168 Warners, *Maclear & Herschel Letters*, 80

3088. 1835–3–8 To Mrs. Kenneth Murchison Feldhausen
After returning to Britain, please deliver the annexed letter to her brother-in-law, Roderick I. Murchison.
ALS 1p Geological Society

3089. 1835–3–8 To Roderick Impey Murchison Feldhausen
Describes various geological aspects of the Cape area and mentions JH's astronomical work. Regrets that RM's brother is leaving the Cape.
ALS 3pp Geological Society

3090. 1835–3–10 To Francis Baily Feldhausen
This letter is being delivered by a Capt. William Edie of the 98th, recently returned from Andrew Smith's expedition up the Caledon River, in the Orange Free State.
ALS 2pp RS:HS 3.124 (C: RS:HS 25.8.3)

3091. 1835–3–10 To Francis Beaufort Feldhausen
Sent as a letter of introduction for Captain [William] Edie, who was hurt on an expedition in southeastern Africa, and who knows much about the region.
ALS 2pp Hydrographic Office Misc F23F5.1

3092. 1835–3–12 To Francis Baily Feldhausen
Sends papers of astronomical observations by a Capt. J. A. Lloyd of Mauritius. G. T. Napier's projected plan for an observatory at Canton.
ALS 2pp RS:HS 3.125 (C: RS:HS 25.8.4)

3093. 1835–3–12 To [John Augustus] Lloyd Feldhausen
Comments on importance of 'geographical observatories' like JL's as complementary to astronomical observatories, and outlines their chief objects of enquiry. Suggests that JL have observational results examined and published, and recommends taking many hourly observations.
ALS 4pp RS Sa.645

3094. 1835–3–16 To Thomas Maclear Feldhausen
Some comments about constant error in TM's observations.
C 2pp RS:HS 21.173 Warners, *Maclear & Herschel Letters*, 80–1

3095. [1835–3–17?] From Thomas Maclear n.p.
Much obliged for JH looking over the observations. Comments on these and the accuracy of the Cape circle and mural circle. Will send his observations on *Crucis*.
ALS 3pp RS:HS 12.205 Warners, *Maclear & Herschel Letters*, 81

3096. 1835–3–17 From Sir Edward Ryan Calcutta
Sends the thermometers he made at sea. Made them as accurately as he could. Introduces Lady William Bentinck.
ALS 1p RS:HS 14.453

3097. 1835–3–25 From Mario Gemmellaro Nicolosi sull'Etna
Congratulations on the birth of an infant. Niccolo Cacciatore is still working with the meridian. There has been no fall of aereolites in Sicily in spite of reports in the papers.
ALS 1p (Italian) RS:HS 8.99

3098. [1835]–3–30 From Thomas Maclear n.p.
Encloses the Brisbane observations. Appears to be a discrepancy in some of the stars.
ALS 1p RS:HS 12.207 Warners, *Maclear & Herschel Letters*, 81–2

3099. 1835–4–11 From Francis Beaufort Admiralty
Thanks him for his letter of sympathy. Gives news of financial awards made, and forthcoming to Mary Somerville, W. S. Stratford, and Michael Faraday.
ALS 4pp RS:HS 3.342

3100. [1835–4–11] To William Edie [Feldhausen]
Asks WE to take charge of a packet of JH's astronomical papers to be given to Francis Baily in England. Wishes WE restored health.
ALS 1p South African Astronomical Observatory

3101. 1835–4–13 From Thomas Maclear Royal Observatory, Cape of Good Hope
Has arranged with Lady Herschel for Sir John to spend a night at the Observatory. Come early; then he can reverse the transit instrument. Is not satisfied with the performance of the transit clock. Intends to clean it. Is ready to adopt JH's

suggestions for measuring the tides. Curious that JH and [T. G.?] Taylor should suggest similar plans for measuring the divisions of a mural circle.
AL 3pp RS:HS 12.63 Warners, *Maclear & Herschel Letters,* 84–5

3102. 1835–4–14 To Thomas Maclear Feldhausen
About the difficulty of measuring high and low tide with the instruments available to TM; offers the design of an instrument that would measure the mid–point between high and low tides.
C 6pp RS:HS 21.174 Warners, *Maclear & Herschel Letters,* 82–4

3103. 1835–4–21 From Johann Franz Encke Berlin
Sending his observations of the recent comet. JH's journey has aroused great interest in Germany. Taking over the Observatory and expecting a visit from F. W. Bessel. Wilhelm Struve has planned a new observatory for St. Petersburg. Has met astronomer Manuel Johnson from St. Helena. C. F. Gauss has produced a method of observing the magnetic needle.
ALS 3pp (German) RS:HS 7.90

3104. 1835–4–24 From Basil Hall Feistritz near Graz
Found his recent stay at the remote German castle very restful. Gives details of a large flight of crows and compares it to nebulae. Cure for sore throats. Hopes the South Easters do not disturb him. Intends spending the winter in Paris. Hopes that JH is preparing a two volume work on his scientific experiences.
ALS 4pp RS:HS 9.179

3105. 1835–4–24 From Andrew Smith Kuruman, near Lattakoo [Cape Province]
Prospects good for crossing Kalahari Desert and visiting Musalacatzie [Moselekatse], '*dreaded tyrant of the interior*.' Heard reports of extraordinary animals and great inland lake. Describes planned itinerary and John Barrow's observations.
ALS 3pp RS:HS 19.295

3106. 1835–4–25 To James Hudson n.p.
Reports on JH's meteorological observations at the Cape of Good Hope, including observations of barometric fluctuations and of the intensity of the sun's rays.
L *Athenaeum, 391* (April 25, 1835), 320 inc

3107. 1835–4–26 To Peter Stewart Feldhausen
Arranges sale of stocks and transfer of funds through Drummond's Bank for use by J. C. Stewart and John Stewart.
ACCS 2pp TxU:H/L-0395; Reel 1054

3108. 1835–5–1 From James Prinsep Calcutta
Thanks JH for meteorological observations at Cape. Wants to publish discoveries for scientists working in India. Will collect meteorological data with a special focus on atmospheric activity from all over Indian peninsula.
ALS 4pp RS:HS 14.67

3109. [1835–5–8 or 9] From Thomas Maclear Royal Observatory, Cape of Good Hope
Has received letters and books from England. Leave of absence of William Meadows and nomination of Piazzi Smyth. Is preparing a packet of observations for England; has JH anything to send?
ALS 3pp RS:HS 12.64 Warners, *Maclear & Herschel Letters,* 86

3110. 1835–5–8 From Robert Wauchope Simons Bay [near Cape Town]
Sorry to involve JH in dispute with H. H. Gird over 120 dollars for horse. Will settle out of court. Sends two mineral specimens from Ascension [Island] and barometric observations.
ALS 2pp RS:HS 18.121

3111. 1835–5–9 To William Whewell Feldhausen
Describes a work on ethics with an empirical orientation, which JH plans to write. Discusses the meteorology and geology of the Cape region.
ALS 4pp TC, Camb. Add. Ms.a.207[26]

3112. [1835–5–10] From Thomas Maclear n.p.
Acknowledges Lady Herschel's two notes and will communicate with her in the morning. Encloses the note of Mrs. Smyth [wife of W. H. Smyth?]. Hears that Dr. [John] Lee has been contesting Aylesbury, without success. Estimates for stable have been cut out.
ALS 2pp RS:HS 12.186 Warners, *Maclear & Herschel Letters,* 86–7

3113. 1835–5–10 To William Henry Smyth Feldhausen, Cape of Good Hope
C. P. Smyth appointment as assistant to Thomas Maclear pleases JH. Suggests method of determining right ascension of double stars with collimating telescope. Sends brochure proposing worldwide meteorological observations.
C 5pp RS:HS 21.175

3114. [1835]–5–12 From Robert Wauchope Simons Bay [near Cape Town]
Discusses barometer behavior during coastal gales and tornadoes. Will send JH a better specimen of oxide in granite fissure [illustration]. RW mistakenly assumed it was basalt. Regrets JH involved in RW's transaction with H. H. Gird.
ALS 3pp RS:HS 18.123

3115. 1835–5–13 From Charles Wheatstone 20 Conduit St.
Returns Mary Herschel's repaired concertina. Sends CW's recent paper on velocity of electricity [*R.S.P.T. 124* (1834)]. Electrical experiments CW is planning. Appointed professor of experimental philosophy at King's College.
ALS 2pp RS:HS 18.146

3116. 1835–5–16 To Friedrich Wilhelm Bessel Feldhausen
FB's letter of 4 July 1834 reached JH in March 1835. Discusses the observability of the sixth and seventh Saturnian satellites. Meteorological observations at sea, at Cape, and from Calcutta lead JH to suspect 'annual transfer of atmospheric pressure' between hemispheres and permanent depression at equator. Describes JH's work at the Cape. Praises Thomas Maclear.
ALS 3pp AdWDDR Bessel 225 (ACCS: RS:HS 19.70 & C: RS:HS 25.4.10)

3117. 1835–5–17 To James Calder Stewart Feldhausen
Excerpts from JH's letter [n.d.] to Peter Stewart regarding financial investments. Asks for meteorology and tide data in Canton, China.
ALS 4pp TxU:H/L-0418; Reel 1055 Evans, *Herschel at the Cape,,* 165–7

3118. 1835–5–19 To Francis Baily Feldhausen
Introducing a Mr. Fraser, who is interested in the setting up of an observatory at Inverness.
ALS 2pp RS:HS 3.126 (C: RS:HS 25.8.5)

3119. 1835–5–20 From John Philip Cape Town
Is much concerned about the conflict between the races.
ALS 4pp RS:HS 13.384

3120. 1835–5–25 To Abraham Josiah Cloete Feldhausen
Gives AC notes for Francis Beaufort and for Giovanni Plana. Praises both. Mentions various books.
ALS 3pp BostonPubLib Ms.Eng. 360

3121. 1835–5–27 From James David Forbes London
Has just come from Scotland on his way to the continent for two or three months. Pleased to hear JH's work is making good progress. Does he know that David Brewster made observations for two days of the year? Has sent him his paper on polarization. Has been occupied by the preparation of his paper on optics. Hopes to reduce his observations made on the continent in 1832. Would like T. C. Robinson, the instrument maker, to see JH's instructions.
ALS 3pp RS:HS 7.288

3122. 1835–5–29 To Thomas Maclear Feldhausen
Excited about JH's observations, and has shown them to several people.
C 2pp RS:HS 21.176 Warners, *Maclear & Herschel Letters*, 87

3123. [1835]–5–29 From Thomas Maclear Royal Observatory
Last night was a splendid night for observing. Will be pleased to place the barometer at Captain Richard Wolfe's disposal. Eclipse was interesting. Will call if he goes to Simon's Bay.
ALS 3pp RS:HS 12.206 Warners, *Maclear & Herschel Letters,* 87–8

3124. 1835–5–30 To James David Forbes Feldhausen
Discusses reduction of barometric observations. Praises the Cape's climate.
ALS 4pp St. Andrews 16 (C: RS:HS 25.4.11)

3125. [1835–6] From Thomas Maclear Royal Observatory, Cape of Good Hope
Has received some papers and newspapers. Four parcels are on their way to JH via H.M.S. Rattlesnake. Encloses a letter from William Wallace regarding the Edinburgh Observatory.
ALS 2pp RS:HS 12.66 Warners, *Maclear & Herschel Letters,* 89

3126. 1835–6–2 To John Philip Feldhausen
JH is very disturbed by the fighting between the South African military and the Kaffirs.
C 4pp RS:HS 13.385 & 21.177

3127. 1835–6–5 From Thomas Maclear Royal Observatory, Cape of Good Hope
Sends his two mountain barometers; comments on them. Method of interpolating the clock errors graphically answers well.
ALS 2pp RS:HS 12.65 Warners, *Maclear & Herschel Letters,* 88–9

3128. 1835–6–9 From John Philip Church Square
Tends to agree with most of what JH says [see JH's 1835–6–2].
ALS 3pp RS:HS 13.386

3129. 1835–6–11 To Thomas Maclear Feldhausen
William Whewell asked JH to compare their June tidal observations with those of the British Admiralty; JH will urge John Deas Thompson to be extra precise in his findings.
L Warners, *Maclear & Herschel Letters*, 89 & in Thomas Maclear, 'Sir John Herschel at the Cape,' *Monthly Magazine,* new ser., *3* (Sept., 1871), 128–140:135

3130. 1835–6–12 From Thomas Maclear Royal Observatory, Cape of Good Hope
Captain John Stevens was at the Observatory yesterday and wishes to superintend the tidal observations himself. TM intends to make the observations personally, or with the assistance of T. W. Bowler. The lunar eclipse was uninteresting. Has had copies of G. B. Airy's papers; they are excellent.
ALS 2pp RS:HS 12.67 Warners, *Maclear & Herschel Letters,* 90

3131. 1835–6–12 To Hugh Percy (3rd Duke of Northumberland) Cape of Good Hope
Responds to HP's generous offer [see HP's 1835–1–21] by saying JH does not need the money, and then suggests that HP might want to use it to support the publication of JH's results from his stay at the Cape.
C 4pp RS:HS 13.348bis & 13.348 inc & 21.178

3132. 1835–6–13 To William Rowan Hamilton Feldhausen
Gives incomplete report on observations of nebulae and double stars and theories on the Milky Way and Magellanic Clouds based thereupon. Thanks WH for his explanation of WH's new dynamical method, but JH admits that he understands only its 'general scope.'
ALS 7pp TCD 1493:109 J. Herschel, '[Extract of a Letter to William Rowan Hamilton],' *Athenaeum,* #462 (Sept. 3, 1836), 627–8, M. Hoskin, 'Astronomical Correspondence of William Rowan Hamilton,' *Journal for the History of Astronomy, 15* (1984), 69-73, and Graves, *Hamilton, 2,* 127–9 inc

3133. 1835–6–13 From Thomas Maclear Royal Observatory, Cape of Good Hope
Has received the box from H.M.S. Rattlesnake and lists the parcels for JH. T. W. Bowler commenced the tidal observations yesterday. Outlines the method used. Was beset by dogs on the way home. Expects to hear from the first ship of an Assistant.
ALS 4pp RS:HS 12.68 Warners, *Maclear & Herschel Letters,* 90–1

3134. 1835–6–13 From Jean Baptiste Joseph Fourier to JH and to James South Paris
Announcing that JH and James South have been awarded the J. J. Lalande prize for their *Double and Triple Stars.*
ALS 1p RS:HS 7.345

3135. 1835–6–19 From Mario Gemmellaro Nicolosi sull'Etna
Thanks for JH's observations and notes on nebulae. Hopes JH will come to Etna to see the changes in the crater.
ALS 1p (Italian) RS:HS 8.100

3136. 1835–6–19 From James Horsburgh East India House
Hudson has printed JH's observations on the temperature of the oceans in the *Athenaeum.* Sends a chart of the east coast of China. Edward Troughton died last Friday.
ALS 1p RS:HS 9.500

3137. [1835]–6–21 From Sir John Bell n.p.
Has been appointed to Government House. Family news.
ALS 1p RS:HS 4.58

3138. 1835–6–24 From Bernard A. von Lindenau Dresden
Sends a few lines via Capt. Basil Hall. Remembers with pleasure JH's visit to Gotha. His visit to the Cape inspires interest. C. F. Gauss is occupied with magnetic observations.

Lists some problems he would like resolved if JH has time while at the Cape. Capt. J. C. Ross has found the magnetic pole.
ALS 3pp (French) RS:HS 11.218

3139. 1835–6–24 From Thomas Maclear Royal Observatory, Cape of Good Hope
Encloses a copy of the hourly observations. Comments on the tides of Table Bay.
ALS 1p RS:HS 12.69 Warners, *Maclear & Herschel Letters,* 91–2

3140. [1835–6–25] From Thomas Maclear n.p.
Sent the barometric readings with Mr. Curnan's paper. Comments on the tidal observations. Send the tide book. William Lamb (2nd Viscount Melbourne) is reinstated.
ALS 3pp RS:HS 12.221 Warners, *Maclear & Herschel Letters,* 92–3 & Evans, 'The Astronomical Work of Sir John Herschel at the Cape,' *Quarterly Bulletin of the South African Library, 12* (1957), 51

3141. [1835–6–25] To Thomas Maclear [Feldhausen]
Made a 'mess' of the tidal observation and meteorological observations that he took earlier in the week.
L Warners, *Maclear & Herschel Letters*, 92

3142. [1835–6–25] To Thomas Maclear [Feldhausen]
Attempting to correct the faulty tidal and meteorological observations that JH made earlier in the week.
L Warners, *Maclear & Herschel Letters*, 93

3143. 1835–6–28 From Thomas Maclear Royal Observatory
Urges JH to observe Ceres. Describes TM's observations of it.
ALS 1p Sydney Ross Collection

3144. [1835–7] To Thomas Maclear [Feldhausen]
Has interpolated the hourly tidal observations for Simon's Bay.
C 1p RS:HS 21.192 Warners, *Maclear & Herschel Letters*, 94–5

3145. 1835–7–4 To William Whewell Feldhausen
Describes in detail efforts, only partially successful, made by Thomas Maclear and JH to record tidal data at Simon's Bay and Table Bay. Gives some interpretation of the data.
ALS 10pp TC, Camb. Add. Ms.a.207[27] (C: RS:HS 21.179)

3146. 1835–7–5 From Thomas Maclear Royal Observatory, Cape of Good Hope
Will be obliged for the Tide Observations book of Table Bay. Intends to enter into an agreement with the Tide Master.
ALS 1p RS:HS 12.70 Warners, *Maclear & Herschel Letters,* 93

3147. 1835–7–5 To Thomas Maclear Feldhausen
Busy interpolating the Simon's Bay tidal series; JH then intends to study Table Bay and to compare the results. Asked William Whewell to back an application to the British Admiralty for a tide clock to be erected at Simon's Bay as a permanent tide station.
L Warners, *Maclear & Herschel Letters*, 93–4

3148. 1835–7–6 From Sir Charles Lyell 16 Hart St.
Hears that JH expressed to William Whewell great pleasure in parts of CL's book. Has been altered and improved since the 1st edition. When in Copenhagen last year [H. C.] Oersted was reading JH's paper on double stars. Murray has sold 1750 copies of CL's book.
ALS 3pp RS:HS 11.418 Katherine M. Lyell, *Lyell, 1,* 449

3149. 1835–7–8 From Basil Hall Berlin
Is in Berlin together with the astronomers J. F. Encke and F. W. Bessel. Relates their researches. The new Berlin Observatory is a magnificent affair. May have to travel home sooner than expected owing to the illness of his mother-in-law.
ALS 7pp RS:HS 9.180

3150. 1835–7–8 From Thomas Maclear Royal Observatory, Cape of Good Hope
Sends the Tide observations made in Table Bay. These are now at a standstill due to the expense involved. Tomorrow he intends to engage the Tide Master. Do they wind up with a dinner?
ALS 2pp RS:HS 12.71 Warners, *Maclear & Herschel Letters,* 95

3151. 1835–7–8 To Thomas Maclear Feldhausen
Sends TM some tide observations, and some calculations done on them.
C 2pp RS:HS 21.180 Warners, *Maclear & Herschel Letters,* 94

3152. 1835–7–9 From Heinrich Christian Schumacher Berlin
Visits F. W. Bessel in Berlin to discuss pendulum experiments. [C. A.] Steinheil invented a photometer to measure relative light of different stars.
ALS 2pp RS:HS 15.377

3153. 1835–7–10 From George Biddell Airy Observatory, Cambridge
Sends his 1834 observations. Has received JH's weather observing instructions. Likely to accept the post of Astronomer Royal when John Pond resigns. Further remarks on new telescope and of E. J. Cooper's (near Sligo) - planetary disks - errors of divisions of circles. Possibility of small observatory in Upper Canada.
ALS 7pp RS:HS 1.69

3154. 1835–7–12 From William Henry Smyth Crescent, Bedford
Describes factors that led to C. P. Smyth's appointment to Cape. Uses rock crystal prism for double star observations. Continues observing JH's nebulae with equatorial clock. Approves of proposal for worldwide meteorological observations; will distribute brochures.
ALS 4pp RS:HS 16.257

3155. 1835–7–14 From Francis Baily London
Regarding meteorology. John Pond has resigned as Astronomer Royal and G. B. Airy been appointed in his place. Has finished his own account of John Flamsteed and sent it to the press. Is drawing up a report for the new standard scale.
ALS 4pp RS:HS 3.127

3156. 1835–7–17 To William Somerville [Feldhausen]
Kaffirs dispossessed by colonists. Condemns 'Colonial insolence.' Editor John Fairbairn, detested by colonists, exposed this in *South African Commercial Advertiser.*
ACC 2pp RS:HS 19.71

3157. 1835–7–17 To William Somerville Feldhausen
Glad the government has recognized Mrs. Somerville's work. Discusses the hostility toward natives at the Cape and criticizes actions of colonists. Astronomical observing conditions at Cape are very good.
ALS 4pp RS:HS 16.404 (C: RS:HS 21.181) *Personal Recollections ... of Mary Somerville* (London, 1873), pp. 216–18 inc

3158. 1835–7–21 From Francis Beaufort Admiralty
The Series of Brisbane catalogues has been interrupted. Regarding the appointment of a new Astronomer Royal. Reprint of one of JH's papers. Various scientific news.
ALS 4pp RS:HS 3.343

3159. 1835–7–21 From Sir John Bell Colonial Office Cape Town
Further queries regarding the new water supply for Cape Town.
ALS 3pp RS:HS 4.53

3160. 1835–7–22 From Richard Sheepshanks 30 Woburn Place
On JH's portrait, which RS is trying to send to JH at the Cape. Asks JH to forgive the 'liberty' he has taken with JH's 'physiognomy,' but RS wants to show off JH as an 'ornament' of Cambridge. Updates JH on Francis Baily's progress and the status of the St. Helena Observatory.
ALS 4pp RS:HS 16.3

3161. 1835–7–25 To William Whewell Feldhausen near Wynberg
Reports on efforts made in June 1835 by JH and Thomas Maclear to have tidal observations made at Simon's Bay and Table Bay. Reports briefly on JH's astronomical observations. Describes increasing tensions with the [Kaffirs] in the Cape region.
ALS 2pp TC, Camb. Add. Ms.a.207[28] (C: RS:HS 21.182 & 25.4.12)

3162. [1835–7–26] To Thomas Maclear [Feldhausen]
Describes his comparison of the tidal series at Simon's Bay and Table Bay. Sending his barometer to TM.
L Warners, *Maclear & Herschel Letters*, 95–6

3163. 1835–7–31 From Thomas Maclear Royal Observatory, Cape of Good Hope
Only made two copies of the Tide Observations, one for the Observatory and one to be forwarded to the Admiralty. Is engaged instrumentally with the transit instrument. Has engaged a new carpenter for the observatory.
ALS 4pp RS:HS 12.72 Warners, *Maclear & Herschel Letters*, 96–7

3164. 1835–8–1 From Thomas Henderson Edinburgh
Received his February letter and is pleased to hear that things are progressing. No one has seen Halley's comet yet. His work on the reduction of the Cape observations is proceeding slowly. Had no idea he gave such a poor impression of the Cape climate. David Brewster and J. D. Forbes have not visited Edinburgh recently.
ALS 3pp RS:HS 9.292

3165. 1835–8–3 From John Augustus Lloyd Mauritius
Was pleased to receive his flattering letter. Now two years since he sent copies of his observations to London and so far has had no reply. Would like to see them printed. Outlines some of the work he has been able to do with his very limited means. Hopes to be transferred soon.
ALS 10pp RS:HS 11.293

3166. 1835–8–6 From Johann Heinrich Mädler and Wilhelm Beer Berlin
The departure of two young missionaries for Africa provide the occasion for JM and WB to comment on the present status of work in preparing for printing the *Mappa Selenographica,* and they ask JH's opinion on some suggestions for printing. Have heard from F. W. Bessel of JH's sighting of the two inner moons of Saturn.
ALS 2pp (French) RS:HS 12.269

3167. [1835–8–15] To Thomas Maclear [Feldhausen]
Reports having read the record JH sent of [tidal] observations and comments on it.
C 1p inc. RS:HS 21.167 Warners, *Maclear & Herschel Letters*, 97

3168. 1835–8–15 From Thomas Maclear Royal Observatory, Cape of Good Hope
Sorry to have forgotten the curves; now encloses them. A. Walter has been here and brought the scrap he now encloses. Walter's tidal observations have been useless so far.
ALS 3pp RS:HS 12.73 Warners, *Maclear & Herschel Letters,* 97–98

3169. 1835–8–17 To Francis Baily Feldhausen
Sending some observations of double stars. Has had some of the sheets of the Brisbane Catalogue. His recent observations. Domestic affairs in the Cape of Good Hope.
ALS 4pp RS:HS 3.128 (C: RS:HS 25.4.13 & 25.8.6)

3170. 1835–8–19 From John Fairbairn n.p.
Would like a complete report of last evening's proceedings.
ALS 2pp RS:HS 7.143

3171. [1835–8–19] From Thomas Maclear n.p.
Will long remember the triumphant meeting last night. Has heard nothing from England. Hopes C. P. Smyth may have come in the *Romney,* which arrived in Simon's Bay last evening. The traversing floor for the reflector will be finished tomorrow. Has been told that JH foretells that Halley's comet will not be visible.
ALS 3pp RS:HS 12.210 Warners, *Maclear & Herschel Letters,* 98–9

3172. [1835–9–1] To Thomas Maclear n.p.
Believes that TM is the first to see the return of Halley's comet [see TM's 1835–9–1].
C 1p inc RS:HS 21.183 Warners, *Maclear & Herschel Letters*, 100

3173. 1835–9–1 From Thomas Maclear Royal Observatory, Cape of Good Hope
Has been observing since 3 a.m. and his fingers are powerless with the cold. Observed a star group, which may be Halley's comet. Comments on the working of his telescope. Has JH placed his 20-ft. telescope in position for the Blueberg Mountains?
ALS 4pp RS:HS 12.74 Warners, *Maclear & Herschel Letters,* 99–100

3174. 1835–9–3 From Hugh Percy (3rd Duke of Northumberland) Syon
Accepts JH's suggestion about using the offered money for publication costs [see JH's 1835–6–12]; is concerned about the telescope at Cambridge, as G. B. Airy has left for Greenwich; final comments on the comet's effects and the weather.
ALS 3pp RS:HS 13.349

3175. 1835–9–6 From Thomas Maclear Royal Observatory, Cape of Good Hope
Sorry to hear he had such wretched weather for their jaunt. The violent winds overturned his 14-ft. reflector and split the tube. Has obtained a capital tent for the Cape. No news from England. Commences the solar observations tomorrow, if fine.
ALS 3pp RS:HS 12.75 Warners, *Maclear & Herschel Letters,* 100–1

3176. 1835–9–8 To Charles Wheatstone Feldhausen
Comments on CW's work on light and on the velocity of electricity. JH's ruminations lead him to the idea of motion pictures.
ACCS 2pp RS:HS 25.4.14 (C: RS:HS 25.11.1)

3177. 1835–9–11 From William Rutter Dawes Ormskirk
Has been unable to do any observations owing to ill-health and lack of an observatory owing to change of residence. Sends his paper on double stars. Pleased to hear everything progresses at the Cape. Will be happy to hear from him.
ALS 4pp RS:HS 6.56

3178. [1835–9–12] From Thomas Maclear Royal Observatory
Received a newspaper from England, which gave news of a comet. A ship entered Table Bay in the night; she may be from England.
ALS 2pp RS:HS 12.218 Warners, *Maclear & Herschel Letters,* 101

3179. [1835–9–13] To Thomas Maclear n.p.
Sends information about where to look for Encke's comet.
C 2pp RS:HS 21.185 Warners, *Maclear & Herschel Letters,* 101–2

3180. [1835–9–13?] From Thomas Maclear n.p.
Sending a copy of the memorandum he made of the first mirage he has ever seen. Describes various points. Much obliged for J. F. Encke's Ephemeris; his writing reminds TM of Isaac Newton.
ALS 2pp RS:HS 12.217 Warners, *Maclear & Herschel Letters,* 102–3

3181. [1835–9–13?] From Thomas Maclear n.p.
Is grateful for J. F. Encke's letter. JH's suggestion of the 14-ft. reflector was anticipated for Halley's comet. Wishes that JH's aunt could visit the Cape.
ALS 3pp RS:HS 12.216 Warners, *Maclear & Herschel Letters,* 102

3182. 1835–9–18 From Mario Gemmellaro Nicolosi sull'Etna
Has received the printed copy of JH's observations on nebulae from Naples. Hopes JH will return to England via Sicily.
ALS 1p (Italian) RS:HS 8.101

3183. 1835–9–19 To George Biddell Airy South Africa
Congratulates GA on being honored by H. M. Government [GA was offered a knighthood, which he declined], and of being appointed Astronomer Royal, and further tells GA about some of JH's observations, and the instruments in use [letter finished 11 Oct.].
ALS 5pp RGO 6.368.598 (drafts RS:HS 1.70 & 21.187)

3184. 1835–9–20 From Thomas Maclear Royal Observatory, Cape of Good Hope
On the evening on which JH was at the Observatory, he believes he observed Encke's comet; gives details. Searched this morning for Halley's comet, without success. Rejoices to hear that G. B. Airy has been secured for the public service. No news from England.
ALS 3pp RS:HS 12.76 Warners, *Maclear & Herschel Letters,* 103–4

3185. [1835–9–20 or 21] To Thomas Maclear n.p.
Has not had any success in finding comets.
C 2pp RS:HS 21.186 Warners, *Maclear & Herschel Letters,* 104

3186. [1835–9–22] To Thomas Maclear n.p.
JH is chagrined to note that he forgot to observe the equinox, after having impressed on any and all the need to do so.
C 2pp RS:HS 21.190 Warners, *Maclear & Herschel Letters,* 104–5

3187. [1835]–9–24 From Thomas Maclear Royal Observatory, Cape of Good Hope
Wind prevents the measurement of Encke's comet from the small stars in the neighborhood. Nothing nebulous in Halley's predicted place. Admiral Patrick Campbell is expected to advance the money for the enclosure. Laborer is employed digging a trench round the Observatory.
ALS 4pp RS:HS 12.77 Warners, *Maclear & Herschel Letters,* 105–6

3188. [1835]–9–26 To Thomas Maclear Royal Observatory
Found Encke's comet the night before last. Has not had a chance to see Halley's comet. Will JH come to the Observatory? There are no fleas.
ALS 3pp RS:HS 12.212 Warners, *Maclear & Herschel Letters,* 106

3189. 1835–10–11 To Francis Beaufort Feldhausen
Has received the letters and parcels safely. Is starting reducing his sweeps. Observations in Dublin Observatory. Comments on the various functions of observatories. News of scientific events at the Cape and various expeditions in progress.
draft 8pp RS:HS 3.343bis (C: RS:HS 21.188)

3190. 1835–10–11 From James Grahame London
Of time spent with Peter Stewart, JH's brother-in-law.
ALS 1p RS:HS 8.427

3191. 1835–10–11 From Basil Hall London
Breakfasted today with G. G. Stokes, met Charles Lyell, [W. J.?] Broderip, and [Robert] Brown and had a fine time. Halley's comet is visible to the naked eye.
ALS 1p RS:HS 9.181

3192. 1835–10–12 From William Henry Smyth Crescent, Bedford
Details sighting of Halley's Comet; sends chart of dates, times, right ascension, and declination. Receives letter from Niccolo Cacciatore, which 'gives hopes of a new planet.'
ALS 4pp RS:HS 16.258

3193. [1835–10–14?] From Thomas Maclear Royal Observatory
Only brought two portraits away on Sunday. Will lend JH his own portrait. No slates are available, but a shipload is expected daily.
ALS 3pp RS:HS 12.232 Warners, *Maclear & Herschel Letters,* 107

3194. [1835]–10–18 To Thomas Maclear Royal Observatory
Is glad JH has given his opinion to Francis Beaufort on the subject of observations of the first class. Encke's comet is invisible in the 14-ft. within the range of 10 degrees. Last two nights were very favorable. Has seen Encke's comet three times, and it appears just as he saw it in England.
ALS 2pp RS:HS 12.230 Warners, *Maclear & Herschel Letters,* 107

3195. 1835–10–19 From Thomas Best Jervis Bombay
Sending a little volume he has written on Native Standards to show their usefulness. Designed to influence the Indian Commission on Standards. Hopes to see JH at the Cape next

February or March.
ALS 2pp RS:HS 10.317

3196. 1835–10–22 To Francis Baily Feldhausen
C. P. Smyth has arrived. Remarks on the Brisbane Catalogue. Thanks for his meteorological observations. His own meteorological observations. Affairs in the Cape.
ALS 4pp RS:HS 3.129 (C: RS:HS 25.4.16 & 25.8.7)

3197. 1835–10–23 To Thomas Maclear Feldhausen
Is intrigued by the apparent displacement of several planetary nebulae from the positions found in the catalogues of other observers. Does not know what to think.
ACCS 2pp RS:HS 25.4.15 (C: RS:HS 21.189) Warners, *Maclear & Herschel Letters,* 108–9

3198. 1835–10–24 To Caroline Lucretia Herschel Feldhausen
Applauds CH elected an [honorary] member of the R.A.S. Missed seeing Halley's Comet and Johann Encke's comet; Thomas Maclear, however, viewed Encke's three times. Has studied nearly the entire Southern sky.
ALS 3pp TxU:H/L–0580.3.a; Reel 1058 Evans, *Herschel at the Cape,* 191–3 & *Memoir ... of Caroline Herschel,* 280–1 inc

3199. 1835–10–[24] To Richard Sheepshanks [Feldhausen]
Provides list of desiderata for activities for observatories of the '2nd and 3rd classes.' Reports his progress from the Cape, and his inability as yet to see Halley's or Encke's comet.
ALS 2pp inc RS:HS 16.4 (C: RS:HS 25.9.1)

3200. [1835–10–25 or later] From [Charles Davidson] Bell [Cape of Good Hope]
Quotes newspaper article from 25 Oct. [1835] describing two tails of Halley's Comet.
ALS 1p RAS:JH Archive 7/3.20; Reel 8

3201. 1835–10–25 From James David Forbes Edinburgh
Unable to assist JH's friend Major [Abraham] Cloete as JF was away from home. Comments on results of JH's table and JF's own papers and scientific occupations. B.A.A.S. would welcome JH's ideas for meteorological work. S. D. Poisson has published a new work on heat, in which he attacks J. B. Fourier.
ACCS 4pp St. Andrews LB II, 233–6 (C: RS:HS 7.289)

3202. [1835–10–25] From Thomas Maclear Royal Observatory, Cape of Good Hope
JH's letter is curious regarding the planetary nebulae. Intended to ride over this morning but the heat was too great. Encloses estimate for printing the observations; comments on this. Would like to send it to Francis Beaufort on Tuesday. Returns some papers and journals. C. P. Smyth will call today.
ALS 4pp RS:HS 12.99 Warners, *Maclear & Herschel Letters,* 109–10

3203. [1835–10–25] To Thomas Maclear [Feldhausen]
Discusses printing costs. Heard that Halley's Comet was seen with the naked eye in Cape Town; JH is excited about his own prospects of seeing the comet.
L Warners, *Maclear & Herschel Letters,* 110

3204. [1835–10–26] From Thomas Maclear n.p.
The figure is 7. Will see [F. W.?] Grey again. Hears from Miss Geard that a Mr. Beck has seen the comet with a long tail. Prepared the 14-feet reflector on Saturday but the evening was unfavorable.
ALS 3pp RS:HS 12.211 Warners, *Maclear & Herschel Letters,* 110

3205. [1835–10–26] To Thomas Maclear n.p.
Still struggling with the problems inherent in using the mural circle; the comet has apparently been seen by JH's mechanic, John Stone, prior to having been sighted by TM.
C 2pp RS:HS 21.184 Warners, *Maclear & Herschel Letters,* 110–1

3206. [1835–10–26] From Thomas Maclear Royal Observatory, Cape of Good Hope
Mr. Stone has informed about the discovery of the comet. This may cause complications. Sends the ledger of Fearon Fallows, where JH will find other comparisons.
ALS 3pp RS:HS 12.79 Warners, *Maclear & Herschel Letters,* 111

3207. [1835–10–26] To Thomas Maclear [Feldhausen]
Busy taking the means of TM's reduced star observations. Angry that John Stone and James Rance saw Halley's Comet and never told JH.
L Warners, *Maclear & Herschel Letters,* 110–11

3208. 1835–10–26 To William Henry Smyth Feldhausen, Cape of Good Hope
C. P. Smyth arrived at Cape on 10 October; Thomas Maclear seems pleased with his work. Asks WS to continue observing Gamma Virginis; JH has observed it only as a single star in recent attempts. Thanks WS for distributing meteorological brochures. George Eden visits and will send meteorological observations from India.
C 3pp RS:HS 21.195

3209. [1835–10?–27] To Thomas Maclear [Feldhausen]
Warns that 'a share in a new planet is advertised and that those who will bid must be on the alert.'
L Warners, *Maclear & Herschel Letters,* 111–12

3210. [1835–10–29] From Thomas Maclear n.p.
Can JH spare a few minutes on Friday to be present when the errors of the transit instrument are being determined. Outlines the investigations necessary. The comet is a splendid fellow. Is busy measuring the diameter of the ring micrometer.
ALS 2pp RS:HS 12.220 Warners, *Maclear & Herschel Letters,* 112

3211. 1835–11–5 From James Grahame London
JG is ill in London.
ALS 1p RS:HS 8.428

3212. 1835–11–11 To Thomas Maclear [Feldhausen]
Hopes that TM can explain the errors in JH's star reductions. Saw Halley's Comet last night; comments that 'its tail was again fine and head very bright.'
L Warners, *Maclear & Herschel Letters,* 113

3213. 1835–11–13 To Thomas Maclear Feldhausen
Asks TM to measure the position of some stars that are difficult for JH to observe.
C 3pp RS:HS 21.191 Warners, *Maclear & Herschel Letters,* 113–4

3214. 1835–11–15 From James Grahame London
Has completed his revision of his historical work.
ALS 1p RS:HS 8.429

3215. 1835–11–17 From Olinthus G. Gregory Royal Military Academy, Woolwich
Introducing a friend, Lieut. R. J. Nelson of the Royal

Engineers, who is a lover of science. Trusts JH is enjoying good health.
ALS 2pp RS:HS 9.37

3216. [1835–11–19] From Thomas Maclear Royal Observatory
Intended to call on Sunday but wind and other interruptions interfered. Saw the Comet, probably for the last time on Thursday. Sends the working sheet of the R.A.'s of the Brisbane catalogue. Comments on various aspects of this catalogue. Has written to *Commercial Advertiser* requesting information from anyone who experienced the recent earthquake shock.
ALS 4pp RS:HS 12.233 Warners, *Maclear & Herschel Letters,* 114–15

3217. 1835–11–20 To Thomas Maclear [Feldhausen]
In some places, Halley's Comet will be seen in or near its perihelion during the solar eclipse.
L Warners, *Maclear & Herschel Letters*, 115

3218. 1835–11–21 To [James Constantine Adamson] Feldhausen
Sets out JH's views concerning the education that should be offered at the South African College. Stresses, among other points, that preference should be given to positive knowledge rather than to systems. Recommends which languages should be studied. Favors a broad education. Advises on many details.
draft 18pp UCT JH Letters A1.1 JH, 'Views on Scientific and General Education, Applied to the Proposed System of Instruction in the South African System of Instruction in the South African College,' *P.M., 8* (1836), 432–8 & in Ferguson and Immelman, *Herschel and Education*, 40–51

3219. [1835–11–21] To Thomas Maclear [Feldhausen]
Actinometer observations during the eclipse.
L Warners, *Maclear & Herschel Letters*, 115–16

3220. 1835–11–25 To James Calder Stewart Feldhausen
News about Italy, Haileybury College, and JH's family. Describes colds [disease] at Cape. Enjoyed journal of JS's travels in Europe. James Grahame was author of epigrams [in JH's 1834–7–17]. Family finances. Halley's comet and solar eclipse of 20 Nov. Asks again for meteorology and tide data in Canton, China.
ALS 4pp TxU:H/L-0419; Reel 1055 Evans, *Herschel at the Cape,* 198–201

3221. 1835–12 From Thomas Maclear Royal Observatory, Cape of Good Hope
JH cannot lay on too much of this kind of work, the closer in AR [Right Ascension] the better. Comments on the list of stars. The comet should be preceded by certain stars, some of which he cannot identify. If JH cannot see the comet for the trees come to the Observatory.
ALS 3pp RS:HS 12.80 Warners, *Maclear & Herschel Letters,* 112

3222. [1835–12 or earlier] To [N.] Spahr n.p.
Concern for health of Madame Gerlach.
Adraft 1p inc RAS:JH Archive 3/6.7; Reel 7

3223. 1835–12–4 To Francis Baily Feldhausen
Has reduced almost the whole catalogue of Southern stars. Comments on these. Is proceeding with his sweeps. Has seen Halley's comet. Phenomenon of shooting stars. Has had a slight earthquake. Is thinking of returning via Rio de Janeiro.
ALS 5pp RS:HS 3.130 (C: RS:HS 25.4.17 & 25.8.8)

3224. 1835–12–5 To John Augustus Lloyd Feldhausen
Comments on the possible loss of some mail going to England, the longitude of Mauritius, and the tides at the Cape. Wonders if anyone was able to observe Halley's Comet during the recent solar eclipse.
ACCS 4pp RS:HS 25.4.18

3225. [1835–12–15] From Thomas Maclear Royal Observatory
Captain [F. W.?] Grey has just been to the Observatory as he wished to ascertain the error of his chronometers. Outlines the signalling devices adopted. Fine definition last evening. Will call on Saturday if convenient.
ALS 3pp RS:HS 12.204 Warners, *Maclear & Herschel Letters,* 116–7

3226. [1835–12–15] To Thomas Maclear [Feldhausen]
Asks TM to come over and to select the trees for the Time Ball.
L Warners, *Maclear & Herschel Letters*, 116

3227. 1835–12–16 To Thomas Maclear Feldhausen
Asks TM to inquire of Captain F. W. Grey of the *Jupiter* about N. L. Lacaille's mural quadrant, which was supposedly aboard the ship two years ago.
L Warners, *Maclear & Herschel Letters*, 117

3228. 1835–12–17 From Thomas Maclear Royal Observatory, Cape of Good Hope
N. L. Lacaille's quadrant appears to have been shipped to England and landed at Portsmouth Dockyard. Will put the ball on the new flagstaff to be erected. Has spoken to George Eden (1st Earl of Auckland) about the Simon's Bay tide machine.
ALS 3pp RS:HS 12.78 Warners, *Maclear & Herschel Letters,* 118

3229. 1835–12–20 From Sir Edward Ryan Gardens
Discusses readings made by his barometer and that sent from the Observatory by Charles Babbage. Sends publications concerning a new fossil and on geology.
ALS 3pp RS:HS 14.454

3230. 1835–12–25 To [Francis Beaufort] Feldhausen
Tells FB what JH has been able to find out as to the status of N. L. Lacaille's quadrant. JH comments on his own astronomical activity, and comments on the family's delight in [Charles] Piazzi Smyth.
ALS 4pp Hydrographic Office Misc 566

3231. 1835–12–29 From Alfred Miles Hydrographic Office
Understands from Francis Baily that JH has received the first 88 pages of the Parramatta Catalogue. The remainder has been sent and now encloses the final part.
ALS 2pp RS:HS 13.39

3232. [1835]–12–31 From Thomas Maclear Royal Observatory
Will retain JH's observations for a day to make a copy. Only knew of the arrival of the Governor [?] yesterday by the sound of the artillery. Intends to visit Cape Town, leave his card, and call back on JH. Expects new outburst of party spirit.
ALS 3pp RS:HS 12.214 Warners, *Maclear & Herschel Letters,* 119

3233. [1835–12–31] To Thomas Maclear [Feldhausen]
Is TM planning to go to Paarl?
L Warners, *Maclear & Herschel Letters*, 118–19

3234. [1836] To François Arago Cape of Good Hope
Discusses the 'Moon Hoax' and JH's observations of Halley's Comet.
L *Comptes Rendus, 3* (1836), 505 (inc)

3235. 1836 To William Henry Harvey Feldhausen
Is sending WH a plant sent earlier, but this one has a bulb attached.
ALS 1p WT 69294.2

3236. [1836 ?] To William Henry Harvey Feldhausen
JH and his family went on an excursion and collected plants. JH is now sending some that survived well and asking WH to identify them.
ALS 2pp WT 69294.1

3237. [1836 early?] From Basil Hall to Margaret Brodie Herschel n.p.
P.S. Has received her letter regarding travel home via Rio and will endeavor to find out the answers and send letters of recommendation later; is now in Paris. Hopes to send a copy of a small book he has written. Hoped to include something from D. F. J. Arago but it has not arrived yet. The French are unpunctual.
ALS 3pp inc RS:HS 9.196

3238. [1836] To William Henry Fox Talbot Feldhausen
Sending WT some seeds. Had hoped to send some bulbs.
ALS 1p Science Museum Talbt 1/7

3239. 1836–1–5 To Captain C[aldwell?] Feldhausen
Thanks for alerting JH to extraordinary hoax in New York Journal of Commerce (2 Sept. 1835). Invites [Caldwell] to visit JH at Feldhausen.
AL 2pp TxU:H/L-0120; Reel 1054 (C: RS:HS 25.15.1)

3240. 1836–1–7 To John I. Taylor Feldhausen
Compares JT's observations of stars very favorably with most others available to JH; explains how JH had helped Thomas Maclear improve the mounting on a telescope at the Cape observatory, together with comments on other equipment.
ACCS 4pp RS:HS 25.4.19

3241. [1836–1–8] To Thomas Maclear Feldhausen
Uncertain when they are traveling to Paarl.
L Warners, *Maclear & Herschel Letters*, 120

3242. 1836–1–10 From James Prinsep Calcutta
Formally thanks JH on behalf of the Asiatic Society for astronomical memoirs and observations of the satellites of Uranus.
ALS 1p RS:HS 14.68

3243. 1836–1–14 From Thomas Maclear Royal Observatory, Cape of Good Hope
Many thanks for the seeds. Will record in the minutes the source of the plants. Breakfast will await JH's arrival. Has been comet hunting every morning to no purpose.
ALS 1p RS:HS 12.81 Warners, *Maclear & Herschel Letters,* 120

3244. 1836–1–15 From Roderick Impey Murchison 3 Bryanston Place
Charles Lyell was gratified to receive JH's letter, which RM has now perused. Sees that JH has some Trilobites for RM, and he would be pleased to receive them as soon as possible so that he can incorporate the new information into his forthcoming book. Affairs at R.S.L. in a sad way, though useful papers are being printed. About to visit Devon with Adam Sedgwick.
ALS 4pp RS:HS 12.389

3245. 1836–1–18 From Francis Baily London
Has sent copies of various star catalogues. Regarding the Brisbane Catalogue and its accuracy. R.A.S. has awarded JH its medal. J. J. Littrow and the appearance of the comet.
ALS 4pp RS:HS 3.131

3246. 1836–1–19 From Alfred Gautier Geneva
Alphonse de Candolle would like some information on the region where JH is now residing [at the Cape]. Has read about JH's voyage and his work on the constellations. Gives his own recent astronomical observations.
ALS 3pp (French) RS:HS 8.83

3247. 1836–1–20 From James Prinsep Calcutta
Requests meteorological observations from JH for compilation of data for 1835. Trouble getting refracting telescope for Asiatic Society.
ALS 3pp RS:HS 14.69

3248. 1836–1–22 To Thomas Maclear [Feldhausen]
Finishing his copy of the panorama taken from the summit of Paarl rock. Needs TM's star reductions as soon as possible.
L Warners, *Maclear & Herschel Letters*, 120–1

3249. 1836–1–24 From Thomas Maclear Royal Observatory
Found comet in Scorpio this morning. Gives map and description. Something wrong with reflector. 'Paarl expenses.' Describes puff adder captured by Dr. Guerd. Glad that Stone is doing well.
ALS 3pp RAS:JH Archive 7/3.6; Reel 8 Warners, *Maclear & Herschel Letters,* 121–2

3250. [1836–1–25] To Thomas Maclear n.p.
Invites TM to come see Halley's Comet in JH's large reflector. JH reports his observations of the comet.
L Warners, *Maclear & Herschel Letters*, 122

3251. 1836–1–26 From Niccolo Cacciatore Palermo
His recent work. Star readings. JH has been elected a member of the Academy of Palermo.
ALS 3pp (French) RS:HS 5.140

3252. 1836–1–26 From Alphonse Candolle Geneva
Botanical and meteorological queries. Sends meteorological observations.
ALS 4pp (French) RS:HS 5.320

3253. 1836–1–26 From Thomas Maclear Royal Observatory, Cape of Good Hope
As the error of the African chronometer is only $0^m\ 1.1^s$ fast on Cape mean time, he sends it. James Adamson is with TM or TM would avail himself of JH's invitation. The comet was superb last night. A most interesting phenomenon last night was the motion of the comet's nucleus over a small star.
ALS 3pp RS:HS 12.82 Warners, *Maclear & Herschel Letters,* 122

3254. [1836–1–27] From Thomas Maclear n.p.
Lack of rest prevented TM calling this morning, but would like to come this evening and stay the night. Comet did not appear so round last night. There was a terrible wind. Happy

with how the wall and the building protected JH's 20-ft reflecting telescope from high winds.
ALS 2pp RS:HS 12.208 Warners, *Maclear & Herschel Letters,* 122–3

3255. 1836–1–31 To Thomas Maclear n.p.
Sends a list of additional stars to be measured for JH to use as standards.
C 1p inc RS:HS 21.196 Warners, *Maclear & Herschel Letters,* 123

3256. 1836–1–31 From Thomas Maclear Royal Observatory, Cape of Good Hope
Gives measurements for the nucleus noted in his letter of 26 Jan. Regrets the delay of the last list of stars. Wind was tremendous last evening. Feared for the safety of the dome. Finds there is a quantity of sand in the box chronometer.
ALS 3pp RS:HS 12.83 Warners, *Maclear & Herschel Letters,* 123–4

3257. 1836–2–1 From William Clift Royal College of Surgeons
Regarding the collections of anatomical specimens assembled by Mr. Verreaux and the possibility of its disposal if sent to London.
ALS 2pp RS:HS 5.352

3258. 1836–2–1 From Carl F. H. von Ludwig n.p.
Dr. Andrew Smith arrived yesterday in the Bay and landed in the evening; has not seen Smith himself yet.
ALS 1p RS:HS 11.406

3259. 1836–2–3 To Francis Beaufort Feldhausen
Has been observing Halley's Comet —'altogether the most beautiful thing I ever saw in a telescope.' Comet has tripled in diameter during the last week.
ALS 1p inc Science Museum Archives MS 1130

3260. [1836–2–5] From Thomas Maclear n.p.
Congratulations on the birth of JH's son Alexander. Has a man preparing the old Ramsden Equatorial as the calculations from the repeating circle are so troublesome. Observations of the comet this morning were excellent.
ALS 3pp RS:HS 12.219 Warners, *Maclear & Herschel Letters,* 124

3261. 1836–2–6 From James Grahame L'Eperonniere
Of health, family news, and the reception of JG's latest book [letter completed 1836–2–17].
ALS 4pp RS:HS 8.430

3262. 1836–2–7 From George Biddell Airy Royal Observatory, Greenwich
Past difficulties and faults with running of Royal Observatory. Hopes Halley's Comet has been observed. Further information on the telescope presented to Cambridge Observatory by Hugh Percy (3rd Duke of Northumberland). James Challis has succeeded GA at Cambridge. Proposed new railway near Royal Observatory and its effect on observations.
ALS 7pp RS:HS 1.71 (C: RGO 6.368.601)

3263. 1836–2–7 To Thomas Maclear Feldhausen
About an anomaly in one of JH's lists of observations.
C 2pp RS:HS 21.198 Warners, *Maclear & Herschel Letters,* 124–5

3264. 1836–2–8 To Thomas Maclear Feldhausen
Calculates the location of the comet [Halley's].
L Warners, *Maclear & Herschel Letters,* 125

3265. 1836–2–9 From Adolphe Quetelet Brussels
Wonders if JH has received AQ's letters and works sent to the Cape. Sends more. Has been doing meteorological observations. Will send some if JH wishes. Mural circle and transit instrument finally in place. Problems with equatorial; therefore missed Halley's Comet.
ALS 3pp (French) RS:HS 14.208

3266. [1836]–2–10 From John Augustus Lloyd Mauritius
Has just received JH's two letters of 5 and 25 Dec. Will prepare duplicates of his letters and pendulum observations and send them to Francis Beaufort. Has still heard nothing about the Observatory.
ALS 2pp RS:HS 11.294

3267. 1836–2–10 From Wilhelm Struve Dorpat
Reports the death of family members. Reveals news of new Pulkowa observatory. Has completed his work on micrometer measurements, resulting in 10,500 observations. Includes lists of aspects of his works.
ALS 4pp (German) RS:HS 17.126

3268. 1836–2–12 From William Henry Smyth Royal Astronomical Society
Presents JH with R.A.S. medal for *Catalogue of Nebulae.*
ALS 1p RS:HS 16.259

3269. 1836–2–13 To William Rowan Hamilton Slough
Replies to WH's letter on 'Scotodynamics.' Cannot follow WH's analysis of velocities of vibrations and disturbances, but finds it symbolically beautiful and powerful. Notes that diploma [?] arrived and was sent to Hanover. Thanks WH on behalf of Caroline Herschel, who sent a letter acknowledging the honor bestowed on her by Royal Irish Academy.
ALS 4pp TCD 1493:176

3270. 1836–2–14 From John Augustus Lloyd Mauritius
Forwarding meteorological and magnetic observations for Sept. and Dec. 1835. Has had no letter from the Governor, but is given to understand he may spend £100 towards the erection of a tide gauge. Have had two violent floods.
AL 2pp inc RS:HS 11.300

3271. 1836–2–17 From Thomas Maclear Royal Observatory, Cape of Good Hope
Many thanks for the portrait; it shall soon have a resting place. Will not rest until the comet observations are en route to England. T. G. Taylor does not observe himself. There was a beautiful occultation of three stars last night.
ALS 3pp RS:HS 12.84 Warners, *Maclear & Herschel Letters,* 125

3272. 183[6]–2–17 To John Philip Feldhausen
Comments on the papers of evidence placed before the Aborigines Committee, and expresses himself vehemently about the sin of having penal colonies.
ALS 4pp Univ. of Witwatersrand (draft: RS:HS 13.387 & 21.197)

3273. 1836–2–19 From Caroline Emily Mary Herschel to Caroline Lucretia Herschel Feldhausen
Reports that JH has been looking at Halley's Comet 'every night.'
ALS 3pp TxU:H/L–0581.2.a; Reel 1058 Evans, *Herschel at the Cape,* 214

3274. 1836–2–19 To Caroline Lucretia Herschel Feldhausen
Reports the birth of his fifth child, Alexander Stewart. Has made some excellent observations of Halley's Comet. Asks CH to inform Carl Gauss that JH has not received Gauss's

magnetometer nor is JH certain of how to use it.
ALS 3pp TxU:H/L–0581.2.b; Reel 1058 Evans, *Herschel at the Cape*, 214–15

3275. 1836–2–19 From John Philip Church Square
Comments on the accusations made about himself before the Aborigines Committee.
ALS 3pp RS:HS 13.388

3276. 1836–2–20 To Charles Babbage Feldhausen
Speculates on extinction of species by natural (not miraculous) causes intermediate to Creator. JH's theory of shifting crust over Earth's surface and its subsequent transfer of pressure. Seeks geological *primum mobile* for volcanoes based on continual deposition and melting of strata. 'All records of former worlds must ultimately perish.'
L *9th Treatise*, 202–17

3277. 1836–2–20 From Francis Beaufort n.p.
Unable to send the Parramatta observations as the vessel is leaving too soon. Has just returned from a meeting to consider [William?] Ritchie's new optical glasses. George Dollond speaks well of them.
ALS 3pp RS:HS 3.344

3278. 1836–2–20 To Sir Charles Lyell Feldhausen
Thanks CL for the gift of a new edition of CL's *Principles of Geology*. JH comments on a number of issues, in a very long letter, including the replacement of extinct species by other species, CL's solution of the problem of the variation of climate, and the evolution of language. JH speculates at length on the isothermal activity under the earth's crust that may result in volcanoes or the formation of new land masses, and describes a considerable number of geological examples that JH sees as supporting CL's ideas.
ALS 12pp APS B.D25.L.1 (C: RS:HS 25.4.20) Cannon (1961), 304–11; Babbage, *Ninth Bridgewater Treatise*, 202–17 inc

3279. 1836–2–20 To Thomas Maclear [Feldhausen]
Having read John Barrow's book, JH proposes procedure for analyzing masses of data from [African] Expedition. Will volunteer to interpolate chronometer errors if TM will undertake computations.
AdraftS 4pp TxU:H/L-0241; Reel 1054 (Cdraft: RS:HS 25.15.2) Warners, *Maclear & Herschel Letters*, 125–6

3280. 1836–2–21 To Thomas Maclear Feldhausen
About the principles whereby the nomenclature of the constellations might be reformed.
C 3pp RS:HS 21.199 Warners, *Maclear & Herschel Letters*, 126–8

3281. 1836–2–23 From Thomas Maclear Royal Observatory, Cape of Good Hope
Would welcome changes in the nomenclature of the constellations. Hesitates to compute the observations of John Burrow.
AL 4pp inc RS:HS 12.85 Warners, *Maclear & Herschel Letters*, 128–9

3282. [1836–2–26] From Thomas Maclear n.p.
When in Cape Town he heard there was a parcel, probably of books, for JH, so has arranged for it to be delivered. Has had newspapers, etc., from England but not a word about the comet's northern tour. Encloses a copy of Thomas Henderson's *Cape Declinations*.
ALS 1p RS:HS 12.213 Warners, *Maclear & Herschel Letters*, 129

3283. 1836–2–27 To Francis Baily Feldhausen
Regarding the double star Gamma Virginis.
ALS 3pp RS:HS 3.132 *R.A.S.M.N., 3* (1836), 197–8 (inc)

3284. 1836–3–3 To Thomas Maclear Feldhausen
Comments concerning Francis Baily's book on John Flamsteed.
L Warners, *Maclear & Herschel Letters*, 129

3285. 1836–3–4 From Thomas Maclear Royal Observatory
Problems observing Halley's Comet. Sends first batch of TM's reduction of Thomas Brisbane's stars. Does JH approve J. C. Adamson's and J. R. Innes's offer to help calculate and tabulate Expedition data? Francis Baily's proposal to rename constellations. Renegotiating price for repairs on equatorial. Describes dream resulting from reading Baily's life of John Flamsteed.
ALS 4pp RAS:JH Archive 7/3.8; Reel 8 Warners, *Maclear & Herschel Letters*, 129–30

3286. [1836–3–4] To Thomas Maclear [Feldhausen]
Thanks TM for the right ascension reductions. Saw the comet [Halley's] last night, but it was dimmer than before.
L Warners, *Maclear & Herschel Letters*, 130–1

3287. 1836–3–5 To Thomas Maclear Feldhausen
JH's observations of the comet [Halley's], together with comments on constellation nomenclature and stellar magnitudes.
C 3pp RS:HS 21.203 Warners, *Maclear & Herschel Letters*, 132

3288. [1836–3–5 to 23] To Thomas Maclear n.p.
Has compared TM's observational measurements with those of the Brisbane catalogue, and finds TM's observations to be better; is working on a star photometer.
C 2pp inc RS:HS 21.201 Warners, *Maclear & Herschel Letters*, 134–5

3289. 1836–3–5 From Thomas Maclear Royal Observatory
Observation of Halley's Comet. Sends servant to pick up information on star 'Constants.'
ALS 3pp RAS:JH Archive 7/3.10; Reel 8 Warners, *Maclear & Herschel Letters*, 131

3290. 1836–3–5/6 To Thomas Maclear [Feldhausen]
Calculates the location of [Halley's] comet.
L Warners, *Maclear & Herschel Letters*, 132–3

3291. 1836–3–8 To Caroline Lucretia Herschel Cape of Good Hope
Is too busy to use Carl Gauss's magnetometer. Viewing Halley's Comet has interrupted JH's sweeps. Reports that Gamma Virginis is a single star in both the 20-ft. reflector and 7-ft. equatorial.
ALS 3pp TxU:H/L–0581.1.b; Reel 1058 Evans, *Herschel at the Cape*, 219–20 & *Memoir ... of Caroline Herschel*, 283–84 inc

3292. 1836–3–8 From Carl F. H. von Ludwig n.p.
Just arrived home from the meeting (which was convened to arrange landing of articles from the Expedition) and found JH's note. Is unable to return to England yet for various reasons. Sure Capt. Frederick Warren will be pleased to take any message to England for JH.
ALS 2pp RS:HS 11.407

3293. [1836–3–10/11] To Thomas Maclear [Feldhausen]
Star within 20 seconds of the nucleus of the comet [Halley's].
L Warners, *Maclear & Herschel Letters*, 133

3294. 1836–3–11 From Basil Hall Paris
Encloses a queer pamphlet, which may amuse him, but it has put D. F. J. Arago and others into a towering passion. The public tends to forget one's existence if too long away from London, so hopes that JH will bear this in mind.
ALS 3pp RS:HS 9.182

3295. [1836–3–11] From Thomas Maclear n.p.
Encloses the list in JH's own handwriting. Has placed two silver wires in the eye piece of the micrometer. Will JH secure him some cobweb for placing in the micrometer.
ALS 2pp RS:HS 12.238 Warners, *Maclear & Herschel Letters*, 133–4

3296. 1836–3–20 From Mario Gemmellaro Nicolosi sull'Etna
Pleased to hear of the good health of JH and family. Everyone hopes JH will make new discoveries in the physiological and anatomical fields as a result of his observations. Etna's crater is slowly changing. A fireball has been observed; also Halley's Comet as correctly predicted by Niccolo Cacciatore.
ALS 2pp (Italian) RS:HS 8.102

3297. [1836–3–24] From Thomas Maclear Royal Observatory
Sends the original of the barometer observations as he cannot spare the time to copy them. Had hoped to have the Brisbane stars ready on Monday. Comet was invisible last night.
ALS 2pp RS:HS 12.231 Warners, *Maclear & Herschel Letters*, 135–6

3298. [1836–3–25 or 26] To Thomas Maclear n.p.
Has installed his astrometer to help judge star magnitudes; also makes a comparison of barometric observations with those of TM.
C 1p inc RS:HS 21.200 Warners, *Maclear & Herschel Letters*, 136–7

3299. 1836–3–28 From Thomas Maclear Royal Observatory, Cape of Good Hope
Dr. Andrew Smith called yesterday. Has succeeded in establishing a postal system for the Observatory. The comet was faint last evening. Had a parcel from the Admiralty but no word about the tide gauge. Sends the Simon's Bay observations, also the P. P. Barraud chronometer, which loses time at a great rate.
ALS 4pp RS:HS 12.86 Warners, *Maclear & Herschel Letters*, 137–8

3300. 1836–3–31 To Thomas Maclear Feldhausen
Sending some equipment back and forth, and some astrometer readings.
C 1p inc RS:HS 21.204 Warners, *Maclear & Herschel Letters*, 138–9

3301. [1836–4] To Charles Grant [Feldhausen]
Following up on an earlier communication, JH expresses to CG [Lord Glenelg] his views on a variety of areas, e.g., curriculum, discipline, administration, and methods of teaching for the Government Free Schools at the Cape of Good Hope.
draft 12pp UCT JH Letters A1.2

3302. 1836–4–2 To Thomas Maclear Feldhausen
Arranges to have tea with TM tomorrow at the Royal Observatory.
L Warners, *Maclear & Herschel Letters*, 139

3303. [1836–4–5] From Thomas Maclear Royal Observatory
Returns the hourly observations of the last equinox. There will be a meeting of the Institute tomorrow. Has removed the silver lines from the mural circle and replaced them by Bermuda cobwebs. Would like to insert these in JH's micrometer. Has been busy calculating a large table of precessions in R.A. Capt. Alexander should desist from newspaper writing.
ALS 3pp RS:HS 12.185 Warners, *Maclear & Herschel Letters*, 139–40

3304. 1836–4–7 From Harrison Watson Weltevreden [Cape of Good Hope]
Purchased property near Feldhausen. Will JH sell strip of land between HW's property and main road?
ALS 1p RS:HS 18.382

3305. 1836–4–9 From William Whewell Trinity College
Comments on JH's proposals to James Adamson regarding education at the Cape. Criticizes Francis Bacon's philosophy. Plans to devote his efforts to philosophy and has begun a history of the sciences. Describes his work on the tides. Richard Jones's professorship at the East India College is in jeopardy.
ALS 6pp RS:HS 18.186 Todhunter, *Whewell, 2,* 233–6

3306. 1836–4–12 From Francis Beaufort Admiralty
Encloses Annual Report of the R.A.S. Hopes he has been amused by the French pamphlets on JH's discoveries relating to the moon. Capt. George Back will be proceeding this year in the *Terror* to investigate the discoveries of J. C. Ross and John Franklin on the N. coast of America. Read JH's last note about the comet to the Society [R.A.S.?].
ALS 4pp RS:HS 3.345

3307. 1836–4–13 From James [Struthers] Stewart Wynberg
Gives JH an estimate for the value of his property [Feldhausen] at £50 per acre.
ALS 2pp RS:HS 17.9

3308. 1836–4–15 From John Augustus Lloyd Mauritius
Sends the meteorological observations for March; unable to send any magnetic observations as he has been unwell. The Observatory escaped serious damage during the hurricane. Has been observing the disappearance of the comet. Comments on nebulae seen near its tail.
AL 4pp inc RS:HS 11.295

3309. 1836–4–15 From Thomas Maclear Royal Observatory
Letters from W. H. Smyth. Notes on Halley's Comet and possible planet sighting.
ALS 1p RAS:JH Archive 7/3.12; Reel 8 Warners, *Maclear & Herschel Letters,* 140

3310. 1836–4–18 From D. François J. Arago Paris
The Academy has received JH's 'Instructions for making ... meteorological observations' in South Africa. Thanks him for this.
ALS 1p (French) RS:HS 1.360

3311. [1836–4–21] To Thomas Maclear n.p.
Sends observations of several more standard stars, and of the comet.
C 4pp RS:HS 21.205 Warners, *Maclear & Herschel Letters*, 140–2

3312. 1836–4–23 From Thomas Maclear Royal Observatory, Cape of Good Hope
Believes his acquaintance with the comet is drawing to a close. A 30" transit was placed in the meridian last night. Returns the Pulkowa observations. Believes the bad weather may have upset Lady Herschel's journey.
ALS 3pp RS:HS 12.87 Warners, *Maclear & Herschel Letters,* 142

3313. 1836–4–[24] To James Calder Stewart [Feldhausen and] Hout Bay
Advice on how best to survive business world of Canton. [Letter continues from Hout Bay:] Family finances. [Letter continues 9 May at Feldhausen:] Asks help in developing JH's 'Phonetic Alphabet' of Chinese language. Considers writing book for Chinese. Comments on Chinese literature and culture. JH has changed shipping agents in Cape Town. Poem by JH.
ALS 10pp TxU:H/L-0420; Reel 1055 Evans, *Herschel at the Cape,* 230–5

3314. 1836–4–26 From William Hawkins Rio de Janeiro
Since his arrival he has explored the suburbs and found some pretty parts. Thinks it would appeal to JH. Houses are rather expensive but could find one to his liking. Will inquire about the landing of his instruments.
ALS 3pp RS:HS 9.256

3315. 1836–4–28 To Edward Joshua Cooper Hout's Bay
About observations of Halley's Comet. [In postscript dated 1836–5–3], describes further observations of the comet.
C 4pp RS:HS 21.202

3316. 1836–4–29 From William Henry Smyth Crescent, Bedford
Reports conflicting observations of Gamma Virginis. Uses JH's method of measuring double star positions.
ALS 4pp RS:HS 16.260

3317. 1836–5–2 From Francis Baily London
Regarding the Brisbane Catalogue. JH's discoveries relating to the moon. The James South lawsuit still hangs on. Is about to depart for Scotland. W. S. Stratford has finished his ephemeris of Halley's Comet.
ALS 4pp RS:HS 3.133

3318. [1836–5]–10 From Thomas Maclear Royal Observatory
Mural and transit are at JH's command. Will have pleasure in joining him on Saturday. Has a good supply of balls and will procure some goose shot.
ALS 2pp RS:HS 12.228 Warners, *Maclear & Herschel Letters,* 143

3319. 1836–5–10 To Thomas Maclear Feldhausen
Asks TM to accompany himself and Daniel Cloetes in hunting a leopard.
L C: RS:HS 21.206 inc Warners, *Maclear & Herschel Letters,* 143

3320. 1836–5–12 From Giovanni Antonio Plana Turin
Comments on communications with the Cape. James Stewart (Margaret Herschel's brother) is well liked by the family. GP is sending on his meteorological observations, noting that he has reduced his emphasis on astronomy. Is working on a long memoir on integral calculus. [Many parts of the letter are illegible.]
ALS 8pp (French) RS:HS 13.433

3321. 1836–5–13 From Francis Beaufort Admiralty
Received his last letter regarding Alpha Virginis in time to read it at the council. Has sent him all Stratford's Ephemerides. Regarding the sounding of the Brazil Banks. Back goes to Wager River the first week in June.
ALS 4pp RS:HS 3.346

3322. 1836–5–14 From Mario Gemmellaro Nicolosi sull'Etna
Discusses observations of Mt. Etna since its last eruption in 1832 and the work of geologist Teodoro Monticelli.
ALS 1p (Italian) RS:HS 8.103

3323. 1836–5–22 From Basil Hall Paris
Has been busy correcting proofs of a little book to be printed by Cadell in Edinburgh which has occasioned the delay. Observed the eclipse at the Observatory with D. F. J. Arago. Comments on the work of Arago. French science and literature at a low ebb.
ALS 4pp RS:HS 9.183

3324. [1836–5–24] From Thomas Maclear Royal Observatory
Has asked Dr. Andrew Smith, Capt. J. E. Alexander, and their friends for Saturday; hopes JH can join them. Has recognized fifteen of the November cometary stars; they are in Harding's map with one exception. The Ephemeris will require a small correction.
ALS 2pp RS:HS 12.184 Warners, *Maclear & Herschel Letters,* 143–4

3325. 1836–5–24 To Thomas Maclear [Feldhausen]
Does not want to compute precessions for each individual star.
L Warners, *Maclear & Herschel Letters*, 144

3326. 1836–5–24 From Robert Treat Paine Boston, Mass.
Wonders if people at the Cape observed the large fall of meteors during the night of 12 Nov. 1833; comments on this and gives theories of one professor. JH's name was used for newspaper article on discoveries on the moon; the person responsible was R. A. Locke. Encloses one of his own memoirs on the eclipse.
ALS 3pp RS:HS 13.209

3327. [1836–5–30] From Thomas Maclear n.p.
Is proceeding at a respectable speed with the Brisbane list. Has received [Andrew?] Smith's box safely. Saw Dr. Smith last night; he is more reconciled. Many thanks for the trees; the ground is suitable to receive them.
ALS 2pp RS:HS 12.234 Warners, *Maclear & Herschel Letters,* 144

3328. 1836–5–31 To William Henry Smyth Feldhausen, Cape of Good Hope
Thanks WS for having accepted his medal from R.A.S. Finds that the [Thomas] *Brisbane Catalogue* contains insufficient right ascension information, complicating reduction observations. Observed Halley's Comet.
C 3pp RS:HS 21.207 David S. Evans, 'Sir John Herschel to Captain Smyth,' *Quarterly Bulletin of the South African Library, 25* (1970), 15–16

3329. 1836–6–1 From Sir Charles Lyell 16 Hart St.
Preparing fifth edition of *Principles of Geology*, using JH's facts about boiling seeds. Mathematical error in CL's climate theory. CL's first speech as president of Geological Society. Criticizes R.S.L. JH's volcanic theory was too complex; will incorporate it into later editions. Benefits to science of Charles Babbage's social parties. Latest publications in geology. P.S. Glad to learn that JH thinks origin of new species may occur through intermediate causes; Germans criticized this. CL chooses not to engage in controversies.

Speculates on successive extinction of species. Notes on African hot springs.
ALS 4pp RS:HS 11.420 inc Katherine Lyell, *Lyell, 1,* 464–9 inc & Hyman, Charles Babbage, 130 inc

3330. 1836–6–1 From Thomas Maclear Royal Observatory, Cape of Good Hope
Believes the wires are 0.2" thicker than the old ones. Dust has settled on the parallel wires of the telescope. Advises how adjustment may be made to JH's micrometer for parallelism.
ALS 4pp RS:HS 12.88 Warners, *Maclear & Herschel Letters,* 144–5

3331. [1836–6–3] From Thomas Maclear Royal Observatory
Has returned the drawings to [Andrew?] Smith not having received *carte blanche* to look at them. Bent one of the thick wires of the micrometer when taking out the plate, so has inserted another. The micrometer will be ready tomorrow. Clean Bermuda webs and two silver wires of 6" diameter.
ALS 3pp RS:HS 12.236 Warners, *Maclear & Herschel Letters,* 145

3332. [1836–6–4] To Thomas Maclear [Feldhausen]
Thanks TM for restoring JH's micrometer.
L Warners, *Maclear & Herschel Letters*, 145–6

3333. 1836–6–7 From Sir Charles Lyell 16 Hart St.
Charles Lyell wrote letter to JH recently. Now encloses with it letters of introduction to persons in Rio and abstract of paper by Charles Babbage. [Henry] Brougham's attacks on Thomas Young.
ALS 7pp RS:HS 11.419 Katherine M. Lyell, *Lyell, 1,* 470

3334. 1836–6–10 From William Whewell University Club
WW's results regarding tidal observation and theory. Requests further data from the Cape. Has nearly completed his history of the inductive sciences.
ALS 4pp RS:HS 18.187 Todhunter, *Whewell, 2,* 242–4

3335. 1836–6–11 From Francis Beaufort Admiralty
Is enclosing some accounts of the eclipse of the sun. The St. Helena instruments have arrived. Some observations are going to be made in Upper Canada. G. B. Airy proposes some extensive magnetic observations at Greenwich. Is going to Greenwich to see and plan extension to the observatory.
ALS 4pp RS:HS 3.347

3336. 1836–6–16 To Peter Stewart Feldhausen
Comments on financial arrangements for John [Stewart] together with other business matters, and asks for an estimate for printing the results of JH's Cape observations.
ACCS 4pp RS:HS 25.5.1

3337. 1836–6–19 From James Prinsep Calcutta
Thanks JH for meteorological observations and tells how he will apply them to predictions for Calcutta. Describes need to correct barometer observations. Compiles and compares barometer reading from points of India and Central Asia. Thanks JH for double star observations, which he sends to Madras Observatory.
ALS 3pp RS:HS 14.70

3338. 1836–6–20 From Jean Baptiste Biot Paris
Discussion of his latest experiments on polarization.
ALS 3pp (French) RS:HS 4.91

3339. 1836–6–20 To Sir James McGrigor [Cape of Good Hope]
Regarding the affair of Dr. Andrew Smith, who has done good work with an expedition. Thinks it desirable that Smith should be given leave of absence to return to England and write up the expedition.
draft 4pp RS:HS 12.17 & 21.208

3340. 1836–6–21 From Andrew Smith Cape Town
Note of thanks to JH for interceding on AS's behalf.
ALS 2pp RS:HS 16.175

3341. 1836–6–29 From Caroline Lucretia Herschel Hanover
Has recently been too ill to write. Sent portraits of William Herschel to Wilhelm Struve, Heinrich Schumacher, Karl Gauss, Friedrich Bessel, and others.
L *Memoir ... of Caroline Herschel*, 284–5 inc

3342. 1836–6–29 From Harrison Watson Cape Town
Returns title deeds JH sent to HW for transfer of land [see HW's 1836–4–7]. Forwards cancelled 'Bond in favor of Neethling' for £150.
ALS 1p RS:HS 18.383

3343. 1836–6–29 From Robert FitzRoy to Margaret Brodie Herschel HMS Beagle, at sea
Making such slow progress that he fears the paper may not reach her in time, so will send a duplicate. The paper is longer than intended but it may be altered as desired. Remembers his days at the Cape with pleasure.
ALS 4pp RS:HS 7.245

3344. 1836–6–30 From Francis Beaufort Admiralty
Basil Hall has just brought in the accompanying work of J. B. Biot. Sends another batch of meteorological observations. Is pleased with the success of JH's lunar discoveries.
ALS 3pp RS:HS 3.348 & ACC: Hydrographic Office Letter Book

3345. 1836–6–30 From Hugh Percy (3rd Duke of Northumberland) Syon
Comments on observation of comets, the position of Southern Hemisphere stars, and the observation of an annular eclipse.
ALS 3pp RS:HS 13.350

3346. [1836–7] To John Fairbairn n.p.
Feels unfitted for the appointment of Commissioner offered in JF's letter of 13 July.
draft 2pp RS:HS 7.145

3347. 1836–7–1 From Robert FitzRoy HMS Beagle, at sea
Sending some curious magnetic observations made at the Cape. Regarding chronometer measurements. Changes of temperature over sea and land compared.
ALS 12pp RS:HS 7.246

3348. [1836–7?] To Thomas Maclear [Feldhausen]
Believes that JH and TM can compile a better comet catalog together than they could separately.
L Warners, *Maclear & Herschel Letters*, 147

3349. 1836–7–3 To Francis Beaufort Feldhausen
Has received Francis Baily's *Account of the Rev. John Flamsteed* [1835]. Comments that its best part is Baily's restoration of the *British Catalogue.* JH now thinks less of Flamsteed.
C(extract) 1p Univ. of London Library

3350. 1836–7–8 From Francis Baily London
Regarding his own observations of the recent solar eclipse. Is sending G. B. Airy's report on Alexander von Humboldt's letter on magnetic observations.
ALS 4pp RS:HS 3.134

3351. 1836–7–8 From Robert FitzRoy HMS Beagle, St. Helena
Sending a packet. Weather and winds have been unsuitable. Will send Thomas Maclear's packet by the next vessel.
ALS 4pp RS:HS 7.247

3352. 1836–7–10 From Augustus De Morgan 5 Upper Gower St.
Sending him the *Calculus of Functions,* which he may retain, and also some maps of the stars published by Society for the Diffusion of Useful Knowledge. The R.A.S. is preparing a picture of Francis Baily and has added JH's name to the subscription list. Supposes he has fixed a date for his return. Regarding 'Warren's jet blacking.'
ALS 4pp RS:HS 6.182 S. De Morgan, *Memoir of Augustus De Morgan*, 82–4

3353. 1836–7–10 From Thomas Glanville Taylor [Madras]
JH's reaction to *Madras Results Volume III* pleases TT. Discusses his observing methods in the volume.
ALS 4pp RS:HS 17.334

3354. 1836–7–12 From Thomas Maclear Royal Observatory, Cape of Good Hope
Was necessary to reduce the whole of the sixth months' observations. More scanty than he desired but much of his time was taken up observing the comet stars. Is now busy with the comet stars. Does not send the precession as JH has them in his list.
ALS 2pp RS:HS 12.89 Warners, *Maclear & Herschel Letters,* 146

3355. 1836–7–13 From John Fairbairn Commercial Hall
JH has been nominated a member of the newly formed commission to inquire into the suitability of Table Bay as a port.
ALS 2pp RS:HS 7.144

3356. 1836–7–19 To Thomas Maclear Feldhausen
Notes a discrepancy in TM's latest observations and those of Thomas Brisbane.
L Warners, *Maclear & Herschel Letters*, 146–7

3357. [1836–7–20] From Thomas Maclear Royal Observatory
Observations of 'Comet stars.' Sends original 'Transit book.' Take care of it.
ALS 2pp RAS:JH Archive 7/3.13; Reel 8 Warners, *Maclear & Herschel Letters,* 147

3358. 1836–7–21 From Francis Beaufort Admiralty
Is sending Captain W. F. Owen's Quarterly Meteorological statement.
ALS 2pp RS:HS 3.349

3359. 1836–7–21 From Basil Hall 71 Harley St.
Enclosing some letters of introduction for Rio. Advises him to be there in the winter. Makes a duplicate set of his observations in case the ship sinks.
AL 4pp inc RS:HS 9.184

3360. 1836–7–23 From William Henry Smyth Crescent, Bedford
JH letter on Gamma Virginis read at R.A.S. meeting. Niccolo Cacciatore appointed to new position at Palermo Observatory, where he plans to sell the *Almanack.* Many R.A.S. members experimenting with telescopic lenses.
ALS 3pp RS:HS 16.261

3361. [1836–7–25] From Thomas Maclear to Margaret Brodie Herschel Royal Observatory
Mrs. Maclear is much better and has given birth to a son, their first.
ALS 1p RS:HS 12.235 Warners, *Maclear & Herschel Letters,* 147

3362. [1836–7–25] To Thomas Maclear [Feldhausen]
Congratulates TM on the birth of his son, George.
L Warners, *Maclear & Herschel Letters*, 147–8

3363. 1836–7–28 From James Grahame L'Eperonniere
News of family and friends.
ALS 3pp RS:HS 8.431

3364. 1836–[8] To Caroline Lucretia Herschel [Cape of Good Hope]
Work nearly finished here. Has documented the sixth satellite of Saturn and has seen possibly a seventh.
ALS 2pp TxU:H/L–0581.3.b; Reel 1058 Evans, *Herschel at the Cape*, 237–8

3365. 1836–8–1 From William Henry Smyth Crescent, Bedford
Plans study of Joseph Bianchi's *New Sidereal Catalogue.* Sends Bianchi excerpt describing double star of Gamma Virginis.
ALS 2pp RS:HS 16.262

3366. 1836–8–4 From Mario Gemmellaro Nicolosi sull'Etna
Sees Niccolo Cacciatore. Mt. Etna 'continues its silence.'
ALS 1p (Italian) RS:HS 8.104

3367. 1836–8–12 From James David Forbes London
Encloses the second part of his paper on heat. Has procured some meteorological readings from Malta. Trusts everything is proceeding well at the Cape. Met G. B. Airy recently at Greenwich, where C. F. Gauss's apparatus is now installed. Mr. Gaubert is dead, also A. M. Ampère and G. C. F. M. R. De Prony.
ALS 4pp RS:HS 7.290

3368. 1836–8–16 From Francis Baily London
Further regarding the phenomenon of the shooting stars in America in November. Operation on Frederick Augustus (Duke of Sussex). Is preparing a paper on the recent solar eclipse.
ALS 3pp RS:HS 3.135

3369. [1836–8–22] To Thomas Maclear n.p.
A note along with a request from Niccolo Cacciatore for some observations that JH says he cannot make.
C 1p RS:HS 21.215 Warners, *Maclear & Herschel Letters,* 148

3370. 1836–8–24 From Thomas Maclear Royal Observatory, Cape of Good Hope
Sends the list of comet stars; the constants will be computed at the Observatory and will send a copy. Encloses letters from [W. H.] Smyth and others. Lead is being removed from the roof of the lantern.
ALS 2pp RS:HS 12.90 Warners, *Maclear & Herschel Letters,* 148

3371. [1836–8–26] From Thomas Maclear Royal Observatory
Sends annotations to JH's 'sweeping sheet.' Complains of errors in ring micrometer.
ALS 2pp RAS:JH Archive 7/3.16; Reel 8 Warners, *Maclear & Herschel Letters,* 149

3372. 1836–8–26 To Thomas Maclear Feldhausen
Inquires about a star not found in TM's list of comet stars. Believes that their comet collaboration is going well.
L Warners, *Maclear & Herschel Letters*, 149–50

3373. 1836–8–26 To Robert Treat Paine Feldhausen
JH's theory of meteors. Will convey to England RT's recent observations of eclipses. Received New York paper claiming discoveries in moon by JH. Invites American observers to join international system for simultaneous meteorological observations.
ALS 4pp † TxU:H/L-0291; Reel 1054

3374. 1836–8–26 To Robert Treat Paine Feldhausen
Mostly about periodic meteors [see RP's 1836–5–24], which JH observed in November 1835. JH offers some theoretical explanation for such a phenomenon.
C 6pp RS:HS 25.15.3

3375. 1836–9–2 From Francis Beaufort Admiralty
Is sending a paper which has barometrical information of interest. Comments on sea and atmospheric pressure. William Whewell and J. W. Lubbock are working on the tides.
ALS 3pp RS:HS 3.350

3376. [1836–9–3] To Thomas Maclear Feldhausen
Sends formulae for dealing with the apparent differences, in viewing the comet and compared star, due to the effects of parallax and refraction.
C 3pp RS:HS 21.209 Warners, *Maclear & Herschel Letters,* 150 inc

3377. [1836–9–5] From Thomas Maclear [Royal Observatory]
Sorry for surprise in church yesterday. Last minute request for JH and wife [to be godparents] was unplanned. Notes on 'Parallax & refraction.' Will soon observe all stars in [F. W.] 'Bessel's zone.'
ALS 3pp RAS:JH Archive 7/3.18; Reel 8 Warners, *Maclear & Herschel Letters,* 150

3378. [1836–9–8] To Thomas Maclear [Feldhausen]
Discusses their table of comet stars.
L Warners, *Maclear & Herschel Letters*, 151

3379. [1836–9–10] To Thomas Maclear n.p.
Much obliged for F. W. Bessel's zone. Thinks his own extra meridian observations of the comet will be respectable. Feels sure Fearon Fallows could not have approved of the supports of the dome. Has been promised the pleasure of inspecting the expenses of the Observatory. Captain J. E. Alexander and Dr. John Murray will call this morning for a parting cup of tea.
ALS 3pp RS:HS 12.237 Warners, *Maclear & Herschel Letters,* 151

3380. 1836–9–12 From Sir James McGrigor London
JH has no need to apologize for bringing before him the case of Dr. Andrew Smith. Heard the news from Dr. John Murray. Will press Smith's request in the appropriate quarter. JH should address his request to Benjamin D'Urban.
ALS 4pp RS:HS 12.18

3381. 1836–9–15 From Francis Beaufort Admiralty
Sending the meteorological reports. Hopes JH will let him know when to cease sending them. William Whewell has produced an interesting instrument. The sensation at the B.A.A.S. meeting was the manufacture of diamonds by a Frenchman.
ALS 4pp RS:HS 3.351

3382. 1836–9–19 To Thomas Maclear Feldhausen
Calms TM about not having his comet observations published yet; those published so far are not of much value.
C 2pp RS:HS 21.210 Warners, *Maclear & Herschel Letters*, 152

3383. 1836–9–20 From Thomas Maclear Royal Observatory, Cape of Good Hope
Intends to dispatch the packet to the Admiralty in a few days and await Francis Beaufort's instructions regarding the printing. TM's son would be grateful for the Herschels' gift could he express thanks himself. Would like the loan of the actinometer. Sends the dome observations.
ALS 4pp RS:HS 12.91 Warners, *Maclear & Herschel Letters,* 152

3384. [1836]–9–24 From Thomas Maclear Royal Observatory, Cape of Good Hope
Returns copies of the Comet observations and the formula for parallax and refraction. The 21st was a busy circle day. The mural circle is in better trim than ever. Sends a chronometer.
ALS 2pp RS:HS 12.92 Warners, *Maclear & Herschel Letters,* 153

3385. 1836–9–24 From John Henry Nelson n.p.
Received the print of JH with great pleasure. The picture of H. W. Pickersgill seems a truthful likeness. The bulbs have been planted and are thriving. Has just returned from the continent, where he visited Caroline Herschel, who is still well in mind, but weak in body. Church building progresses at Slough.
ALS 4pp RS:HS 13.92

3386. 1836–9–28 From Thomas Maclear Royal Observatory, Cape of Good Hope
Encloses the sweeps for June 1836, which JH lent him. Not a word about printing, or the Tide gauge in the letters he received. Will forward the letters from G. B. Airy and Francis Baily. Thomas Henderson is about to be married. Cannot procure white deal without knots. Will try an old mast.
ALS 2pp RS:HS 12.94 Warners, *Maclear & Herschel Letters,* 154–5

3387. 1836–9–28 From Thomas Maclear Royal Observatory, Cape of Good Hope
Barometer is swinging without encountering any accident. Believes there is to be a new clergyman at Rondebosch, the old Observatory housekeeper, John Fry, shortly to land from England. Hopes he is more adept at sermons than he was arranging the catalogue of instruments.
ALS 2pp RS:HS 12.93 Warners, *Maclear & Herschel Letters,* 154

3388. [1836–9–28] To Thomas Maclear [Feldhausen]
Wants to reset his barometer by TM's.
L Warners, *Maclear & Herschel Letters*, 153–4

3389. 1836–9–30 To The Albany Institute (Albany,New York) Feldhausen
Is very thankful for the meteorological observations sent to JH, and makes some suggestions for additional and continued

observations.
ALS 3pp APS B.H435p.7

3390. [1836–10] To Thomas Maclear [Feldhausen]
Asks for TM's barometer and thermometer readings for 20–28 September.
L Warners, *Maclear & Herschel Letters*, 154

3391. 1836–10–3 From John Fairbairn n.p.
Comments on the wise measures introduced for the governing of the Cape Colony. Has taken over the editorship of a periodical and intends to introduce some of the best local writers. Would welcome JH's advice.
ALS 3pp RS:HS 7.146

3392. 1836–10–3 To Robert FitzRoy Feldhausen
Sends a copy of the printed paper by RF. Regarding chronometer measurements. Has just received a communication from Alexander von Humboldt in which he seems to have anticipated both JH's and RF's ideas. Comments on a new machine for sounding.
draft 10pp RS:HS 7.248 & 21.216 & 25.5.2

3393. 1836–10–3 To Caroline Lucretia Herschel Feldhausen
Confident that 1837 will be his last year in Africa. JH's catalog of nebulae and double stars is nearly complete. Has a series of observations of Halley's Comet from 1836–1–25 to 1836–5–5. Although JH is too busy to undertake any magnetic studies, he has been providing information about Carl Gauss's magnetometer to the new observatory in Bombay. Recalls that Charles Babbage mentioned the 'principle' of Gauss's method 'at least 10 or 12 years' ago.
ALS 3pp TxU:H/L–0581.4.b; Reel 1058 Evans, *Herschel at the Cape*, 254–6

3394. [1836–10–12] To Thomas Maclear [Feldhausen]
Compares JH's and TM's barometric readings. Criticizes Pierre Morin's work.
L Warners, *Maclear & Herschel Letters*, 155

3395. 1836–10–13 To Thomas Maclear [Feldhausen]
Reports on JH's actinometer observations. Explains why he subscribed to 'Col. [Harry] Smith's piece of plate.'
C 1p RS:HS 12.95 (C: RS:HS 21.212) Warners, *Maclear & Herschel Letters*, 156

3396. 1836–10–15 From John Fairbairn n.p.
Sending a few more of the 'examinations.' Also sends an extract of a letter from Mr. Philips. The English papers have got hold of Charles Grant's (Baron Glenelg) dispatches.
ALS 1p RS:HS 7.147

3397. 1836–10–18 To the R.A.S. Feldhausen
Reports in detail JH's observations of Halley's Comet as observed from the Cape of Good Hope.
L *R.A.S.Mem., 10* (1938), 325–36

3398. 1836–11–1 From James Prinsep Calcutta
Sends *Journal of the Asiatic Society*. Calcutta Museum wants a hippopotamus skeleton; asks JH for leads.
ALS 3pp RS:HS 14.71

3399. 1836–11–1 From Nathaniel Wallich [Calcutta] Botanic Gardens
Completed tour of Upper Assam searching for [wild tea shrub]. Wife and seriously ill children leaving for England. Lord Auckland [governor-general of India] urges sending consignments of plants and seeds to JH and Baron C. F. H. Ludwig.
ALS 3pp RS:HS 18.31

3400. 1836–11–2 From John Augustus Lloyd Mauritius
Sends meteorological observations for September. Has ceased making a charge at the observatory for the checking of chronometers. Has lost income but gained work. Comments on the effects of gravity on the balance wheel of a chronometer.
AL 6pp inc RS:HS 11.296

3401. 1836–11–3 From George Biddell Airy Royal Observatory, Greenwich
Another son born. Has given lectures at Cambridge University. Remarks on routine at Royal Observatory. St. Helena Observatory closed. Printing of the Greenwich Observations. 'Northumberland' telescope not ready yet. Observations on the Greenwich zenith tube. Estimates for the Observatory. Suggests that the Observatory should be moved to a new site. Railway plan near Greenwich altered. James MacCullagh's new theory of optics.
ALS 8pp RS:HS 1.72 (C: RGO 6.368.608)

3402. 1836–11–3 From Thomas Maclear Royal Observatory, Cape of Good Hope
Cornerstone and tree answer admirably. Gives the points in the present triangulation. Lieut. John Williams of the Engineers has promised assistance. There is a French frigate in the harbor, which he intends to visit.
ALS 3pp RS:HS 12.97 Warners, *Maclear & Herschel Letters*, 156–7

3403. [1836]–11–10 From Thomas Maclear Royal Observatory
Has succeeded in fixing an 18 pounder gun firmly in the ground to mark the western termination of the base-line query on how to measure the meridian of a mountain at sea level. Has requested permission from the Governor to inspect the military orders for 1751–2.
ALS 2pp RS:HS 12.243 Warners, *Maclear & Herschel Letters*, 157

3404. 1836–11–10 From Peter Stewart 65 Cornhill
Stewart & Co., printers, and publishers Smith, Elder & Co. would be honored to print JH's proposed book. Estimates costs. Printing would require three years while JH is in Africa, but only four months after JH returns to England. Anticipates problems if [Hugh Percy] Duke of Northumberland purchases entire edition for free distribution.
ALS 8pp RAS:JH Archive 10/1.5; Reel 9

3405. 1836–11–14 From James David Forbes Edinburgh
Pleased to hear JH's work is progressing. Disappointed with Dr. [John] Davy's meteorological observations at Malta. Comments on other scientific news, including recent eclipse, [Andrew] Crosse's controversial artificial crystals at Bristol B.A.A.S. meeting, JF's 1832 observations in Switzerland, and William Whewell's and William Buckland's books.
ALS? 4pp St. Andrews LB II, 391–4 (C: RS:HS 7.291)

3406. 1836–11–15 To James David Forbes n.p.
Discusses methods of magnetic observation, JH's actinometric readings in the Cape, and the need for more physical observatories.
ALS 4pp St. Andrews 60 (C: RS:HS 21.213)

3407. 1836–11–15 To Roderick Impey Murchison Feldhausen
Some geological samples have been lost on the way to England; JH comments on his ideas of the effect of the earth's hot core on the construction of the earth's crust.
ACCS 4pp RS:HS 25.5.3

3408. 1836–11–17 From Thomas Hodgkin Finsbury Circus
Is pleased JH takes an interest in the natives. Comments on the necessity of promoting the welfare of the native races. Hopes to found an International Society for providing assistance and advice to foreigners.
ALS 4pp RS:HS 9.410

3409. 1836–11–20 From Basil Hall Edinburgh
Outlines the election of J. D. Forbes and attributes his success to JH's letter. Regarding the work of Thomas Henderson. Pity JH has to waste time doing his own reductions. Paris observations. Saw the comet in Valparaiso. Magnitude of stars compared to the moon. Hopes he liked his own recent book. Pleased the sextant is of use.
ALS 7pp RS:HS 9.185

3410. [1836–11–20] Thomas Maclear [Feldhausen]
JH is frustrated with his inaccurate barometric readings.
L Warners, *Maclear & Herschel Letters*, 157–8

3411. [1836]–11–21 or 22 From Thomas Maclear Royal Observatory
Sends C. P. Smyth with the observing lists of JH's zero stars. Comments on these lists. Would like to see the Cedar Mountains, but cannot spare the time at present. Wished to visit Riebeck's Castle to verify one of N. L. Lacaille's remarks. Did not notice any air bubbles in JH's barometer. The Observatory barometer is similar to the one of Sir John Franklin. Goes to the Block House tomorrow to commence the triangulation.
ALS 3pp RS:HS 12.227 Warners, *Maclear & Herschel Letters,* 158

3412. 1836–11–22 From John Skirrow Rondebosch
Asks to borrow chain for tomorrow's survey.
ALS 1p RAS:JH Archive 7/3.25 verso; Reel 8

3413. 1836–11–25 To Hugh Percy (3rd Duke of Northumberland) Feldhausen
Comments on inaccuracies in earlier Southern Hemisphere star catalogues, along with an extended description of comet observations.
draft 4pp RS:HS 13.351 & 21.214

3414. 1836–11–29 From Thomas Henderson Edinburgh
Pleased to hear that all is progressing well with JH. Regrets the discordance in the Parramatta Right Ascensions. Observed an occultation of Gamma Virginis in April. Has seen the method of determining the nadir point of a circle. Cape reductions suffered a setback last season as he had to lecture on mathematics.
ALS 4pp RS:HS 9.294

3415. 1836–11–30 To Augustus De Morgan Feldhausen
Comments about AD's work in mathematical functions, and then refers to JH's reductions and other astronomical matters, including the need for reform of stellar nomenclature.
ACCS 3pp RS:HS 25.5.4

3416. 1836–12–1 From Francis Baily London
Has received the Royal medal from the R.S.L. on JH's behalf. Shooting star phenomenon seen in France and Scotland. The Americans are preparing an expedition to the South Pole. Has constructed two pendulums.
ALS 3pp RS:HS 3.136

3417. 1836–12–4 From William Whewell Trinity College
Reports on his tidal studies and requests further observations from South Africa. WW's history of the inductive sciences is in press and WW has begun a philosophy of the inductive sciences. Discusses Richard Jones's career and Charles Darwin's return to England.
ALS 4pp RS:HS 18.188 Todhunter, *Whewell, 2,* 247–50

3418. 1836–12–6 To Augustus De Morgan n.p.
Please convey thanks to the Society for the Diffusion of Useful Knowledge for its star maps. Finds them preferable to [Johann] Bode's maps, 'which are full of egregious errors.'
C 1p inc Univ.Coll.London SDUK Papers

3419. 1836–12–8 From Francis Baily London
Regarding the gales which swept England at the end of November.
ALS 3pp RS:HS 3.137

3420. 1836–12–17 To Thomas Maclear [Feldhausen]
Setting a unit of measurement for solar radiation (the 'actine'); proposes to 'take for a unit of solar heat that which if <u>all</u> employed in heating a cubic inch of water exposing a horizontal surface of one square inch, to a vertical Sun during one minute would produce a dilation of one thousandth part of its volume.'
L Warners, *Maclear & Herschel Letters*, 159

3421. 1836–12–17 From Sir Edward Ryan Calcutta
Introduces [Charles?] Cameron. Received angry and excited letter from Charles Babbage. Discusses barometers. Heard Herschels may be travelling to Rio de Janeiro.
ALS 7pp RS:HS 14.450

3422. 1836–12–18 From Thomas Maclear Royal Observatory, Cape of Good Hope
Definition at the Observatory last night was abominable; therefore no comparisons can be made between the stations. Much obliged for the details of the actinometer. Will call on JH later.
ALS 1p RS:HS 12.98 Warners, *Maclear & Herschel Letters,* 159

3423. 1836–12–19 From William Henry Fitton Geological Society
Regrets the time that has elapsed since receiving JH's letter. Charles Babbage does not contemplate finishing his machine and is already planning another. Richard Jones appointed one of the Tithe Commissioners. News of the activities of Robert Brown, Francis Baily, Charles Lyell, William Buckland, J. W. Lubbock, and others. Railroads expanding in England. Many built and projected.
ALS 8pp RS:HS 7.234

3424. [1836–12–20] To Thomas Maclear [Feldhausen]
Testing whether an actinometer with a glass back is more accurate than one without.
L Warners, *Maclear & Herschel Letters*, 159–60

3425. [1836]–12–23 From Thomas Maclear Royal Observatory
From a newspaper received today sees that John Pond is dead and buried. Affair of James South and Edward Troughton and William Simms is as before. Sends the observations of the 21st; rain was not good for astronomical observations. Comments on these readings.
ALS 3pp RS:HS 12.242 Warners, *Maclear & Herschel Letters,* 160

3426. 1836–12–28 From William Rutter Dawes Ormskirk
Ill health has prevented him from working in astronomy. Talk of a Liverpool Observatory. Details of the large display of meteors seen on 12 November. Faults in his instruments. Queries regarding Saturn. Hopes to see him at the B.A.A.S.

meeting in Liverpool next year.
ALS 4pp RS:HS 6.57

3427. [1836–12–30] From Thomas Maclear Royal Observatory
Thinks Monday 2 Jan. a suitable day for taking down the Circle. Has engaged J. K. Gibbs the carpenter for the coming quarter. No official notice has been taken of his Colonial Measure of Reference.
ALS 4pp RS:HS 12.241 Warners, *Maclear & Herschel Letters,* 161

3428. [1837 late or 1838 early] To John Fairbairn [Feldhausen]
Has been impressed by the new governor and thinks the colony will benefit from his prudent judgment.
draft 3pp RS:HS 7.155 (C: RS:HS 35.13.3)

3429. [1837 or earlier] To [?] n.p.
Comments on letter received. Regrets replies about 'the Muriate of Chrome.'
Adraft 1p inc RAS:JH Archive 6/2.59 verso; Reel 8

3430. [1837–1–3] From Thomas Maclear n.p.
Proceedings with the Circle went on as well as expected; fixed the additional microscopes with plaster of paris. Sends the *P.M.* that contains F. W. Bessel's paper on the repeating circle. Opposition of Mars will require the use of the circle in the last of the present month.
ALS 3pp RS:HS 12.240 Warners, *Maclear & Herschel Letters,* 162

3431. 1837–1–10 To Caroline Lucretia Herschel Cape of Good Hope
Reports that he viewed Halley's Comet from 1836–1–25 to 1836–5–10. Now studying sunspots and solar radiation. Comments that the sun has more spots than JH previously thought. Has been busy during the December solstice determining the constant of solar radiation.
ALS 3pp TxU:H/L–0582.1.a; Reel 1058 Evans, *Herschel at the Cape,* 280–2 & *Memoir ... of Caroline Herschel,* 286–7 inc

3432. 1837–1–17 To Johann Franz Encke Feldhausen
Tried unsuccessfully to observe Encke's Comet; succeeded in observing Halley's Comet, which JH discusses. Mentions JH's graphical method of treating orbits of double stars, JH's plans to return to England, moon maps, and sunspots.
ALS 4pp AdWDDR Encke 49 (C: RS:HS 25.5.5)

3433. 1837–1–19 From James David Forbes Edinburgh
Encloses observations made by some of his pupils who have now formed a physico-mathematical society.
ALS 2pp RS:HS 7.292

3434. 1837–1–29 To William Henry Fox Talbot Feldhausen
Is sending WT some botanical specimens through Andrew Smith, whose scientific abilities JH praises. Asks WT to assist Smith in England. Encourages WT's work in mathematics and in '<u>photology</u>.' Reports on JH's astronomical observations and theorizes about sunspots.
ALS 4pp Science Museum Talbt 1/8

3435. 1837–1–31 From Sir Charles Lyell 16 Hart St.
Has been requested by Gideon Mantell to write JH a letter of introduction for a Lieut. H. I. Thomas, a friend of Davies Gilbert. Thomas wishes to be the bearer of any letter or parcel CL may have. Believes that a copy of the President's speech he sent may be lost in the post. Thinks that JH's theory of subterranean isothermal lines interesting. Dined with the Somervilles last Sunday.
ALS 4pp RS:HS 11.421

3436. 1837–2–1 To Thomas Maclear Feldhausen
Comments on disagreements in stellar observations and on sun spot activity; and requests observations of a specific star.
C 2pp inc RS:HS 21.217 Warners, *Maclear & Herschel Letters,* 162–3

3437. 1837–2–6 To Thomas Maclear Feldhausen
Wants to borrow TM's carpenter. Found a faint planetary nebula.
L Warners, *Maclear & Herschel Letters,* 163

3438. 1837–2–8 From James Prinsep Calcutta
Miss Pattle and C. R. Prinsep, Secretary of Government Proceedings, are going to Cape of Good Hope and will bring JP's *Journals of the Asiatic Society* to JH.
ALS 3pp RS:HS 14.72

3439. 1837–2–13 To Thomas Maclear n.p.
Cannot send the Brisbane list by the end of the week. Will not expect J. K. Gibbs until the cart is finished. Returns solar spots.
ALS 2pp RS:HS 12.225 Warners, *Maclear & Herschel Letters,* 163

3440. 1837–2–14 From James David Forbes Edinburgh
Sending a letter via Capt. Basil Hall. Is surprised JH did not mention the starry showers of 12 Nov. in his last letter, D. F. J. Arago is investigating them. J. B. Biot has published an extension of [Denison?] Olmsted's theory. Humphrey Lloyd has produced an interesting book on the undulatory theory. What measure does JH employ for the force of the wind? Has got William Whewell's anemometer. Influenza prevalent. William Farish has been succeeded by Robert Willis.
ALS 8pp RS:HS 7.293

3441. [1837–2–15] To Thomas Maclear Feldhausen
Notes discrepancies between TM's and Thomas Brisbane's observations.
L Warners, *Maclear & Herschel Letters,* 163–4

3442. 1837–2–16 From Basil Hall Edinburgh
Sends a small dispatch from J. D. Forbes. The barometer seems to have no connection with the weather this winter. Had a snow shower but could see no clouds in the sky. Preparing for the occultation of Mars.
ALS 4pp RS:HS 9.186

3443. 1837–2–16 From Thomas Maclear Royal Observatory, Cape of Good Hope
Regrets having to send the observations on so many sheets; those in red ink will be re-observed. John Steuart has given him a note about spots on the sun. Sends Mr. Stebbing's letter and would like JH's comments when he sees him.
ALS 3pp RS:HS 12.101 Warners, *Maclear & Herschel Letters,* 164

3444. 1837–2–16 From Matthew Henry Webster Albany [New York]
Communicated JH's letter of 30 Sept. [1836] to the Albany Institute. Hopes JH received meteorological register of observation points in U.S., which are increasing in number. Sent meteorological reports to societies and individuals JH requested. Describes barometer and temperature measurements in wells and barometer extremes in great December storm. W. C. Redfield of New York says duration of observation periods recommended by JH is too short.
ALS 3pp † RS:HS 18.128

3445. 1837–2–20 From Augustus De Morgan 5 Upper Gower St.
Will be a great reception for JH when he returns. Is working on a theory of probabilities. Francis Baily is the man for mapping the entire heavens. G. B. Airy is to test JH's topsy-turvy observations of Orion.
ALS 8pp RS:HS 6.183

3446. 1837–2–20 From John Lee London
Thanks JH for drawing of JH's house. Sends drawing of JL's Hartwell Observatory and publications from the observatory.
C 2pp Buckinghamshire County Council

3447. 1837–2–22 From William Henry Smyth Crescent, Bedford
Thanks JH for kindness to C. P. Smyth. Francis Baily re-elected president of R.A.S. W. S. Stratford prints a comet ephemeris.
ALS 3pp RS:HS 16.263

3448. 1837–2–23 From Francis Beaufort Admiralty
Sending more barometric observations. Interesting articles by S. D. Poisson and D. F. J. Arago. Peculiar crimson light seen in the sky recently. The 'Beagle' will sail in May and will call on JH.
ALS 4pp RS:HS 3.352

3449. 1837–2–28 From Roderick Impey Murchison Petersfield
Received JH's letter with pleasure; delayed replying as hoping to comment. Theories have been shown to Charles Lyell, William Whewell, etc. Gives his own comments on JH's theory. Installed Whewell as President of the Geological Society. Pleased JH is to become a subscriber to RM's book, which moves slowly along. Received a collection of fossils from Fernando Po recently.
ALS 4pp RS:HS 12.390

3450. 1837–3–1 To Francis Baily Feldhausen
Has been very busy observing so has had little time to spare for writing letters. Remarks on the recent star catalogues which he has received. Would FB bring Capt. J. A. Lloyd's case to Frederick Augustus (Duke of Sussex). Sun spots. Meteors in N. America and those at the Cape.
ALS 7pp RS:HS 3.138 (C: RS:HS 25.5.6 & 25.8.9)

3451. 1837–3–1 From Adolphe Quetelet Brussels
Still has not heard from JH. Has put much effort into horary observations of solstices and equinoxes. Sends JH articles by AQ on temperatures of earth. Will send magnetic observations.
ALS 3pp (French) RS:HS 14.209

3452. 1837–3–14 To James Duff Watt Feldhausen
Appointment of John Fry as pastor at Wynberg and Rondebosch. JW prefers external authority over selection by congregation. Prefers to retain Edward Judge as pastor. Unable to attend tomorrow's meeting.
ACCS 4pp RS:HS 18.387 (C: 21.218)

3453. 1837–3–16 From John Lee Doctors' Commons
Earlier sent JH some pencils, etc., and copies of the first two numbers of the 'Terror of the Northern Hemisphere.' Now sends third number.
C 1p Buckinghamshire County Council

3454. 1837–3–18 From Robert Treat Paine Boston, Mass.
Acknowledges receipt of JH's letter of last August. Sends a copy of the American Almanac containing a catalogue of all the solar eclipses made by himself. Had a magnificent display of red aurora on 25 Jan.
ALS 2pp RS:HS 13.210

3455. 1837–3–19 From John Fairbairn n.p.
Sending papers connected with the recent ruling in the Supreme Court, which affected JF, and may show him in an unfavorable light to JH.
ALS 3pp RS:HS 7.148

3456. 1837–3–22 To Thomas Maclear Feldhausen
Exhausted from his meteorological observations, JH will not keep his appointment for tomorrow with TM.
L Warners, *Maclear & Herschel Letters*, 165

3457. [1837–3–23] From Thomas Maclear Royal Observatory, Cape Town
The mica will do beautifully. Sorry to hear of the return of JH's complaint. Will send barometric observations tomorrow.
ALS 2pp RS:HS 12.180 Warners, *Maclear & Herschel Letters,* 165

3458. 1837–3–28 From John Fairbairn Cape Town
Returns the *Edinburgh Review* with many thanks. Intends making use of the article on Advocates. Comments on articles in the *United Services Journal.* Regarding the situation at the Cape between the natives and the settlers.
ALS 4pp RS:HS 7.149

3459. [1837–3–29] From Thomas Maclear Royal Observatory
Call on Col. G. G. Lewis; the latter expressed a desire to see the rods laid off. Can JH come and breakfast with them to see the operation? JH needs a rest from astronomy to restore his health.
ALS 3pp RS:HS 12.222 Warners, *Maclear & Herschel Letters,* 165–6

3460. 1837–3–30 From Caroline Lucretia Herschel Hanover
Experiencing 'a daily increase of pain and feebleness.'
L *Memoir ... of Caroline Herschel,* 287–8 inc

3461. 1837–4–3 To D. François J. Arago n.p.
Regarding the meteor seen in New York. Gives tables of meteors seen during a specified time. Comments on the phenomenon of meteors. Sun spots.
draft 8pp RS:HS 1.361 & 21.219

3462. 1837–4–6 To James Calder Stewart [Feldhausen]
JS's serious illness, recovery, and return to England. JH dismisses hope of devising phonetic alphabet for Chinese. Poem by Michelangelo. Notes on classic literature.
ALS 4pp TxU:H/L-0421; Reel 1055 Evans, *Herschel at the Cape,* 291–4

3463. 1837–4–7 To Francis Baily Feldhausen
Would FB convey his thanks to the R.S.L. for the award of their medal. Has observed the whole of the southern hemisphere twice. Is pleased to hear of the American support for science.
ALS 4pp RS:HS 3.139 (C: RS:HS 25.8.10)

3464. 1837–4–7 From Thomas Henderson Edinburgh
Sending a few lines via Captain Basil Hall. Has sent his observations of the refractions of stars at lower altitudes than 5°, to the R.A.S. Started work on determining the position of the stars from Baily's zodiacal catalogue. Government has authorized the printing of the observations.
ALS 3pp RS:HS 9.295

3465. [1837–4–7?] From Thomas Maclear n.p.
Sends the *Comptes rendus* in which he will find the notice TM alluded to. Beautiful definition last night. Gauze diminishes the light without further mischief.
ALS 1p RS:HS 12.192 Warners, *Maclear & Herschel Letters,* 166–7

3466. 1837–4–8 From Basil Hall Edinburgh
Encloses a dispatch from Thomas Henderson, who is very pleased with his work and a word from JH would spur him on. Is more out of touch with science at Edinburgh than JH is at the Cape. Young engineer, J. S. Russell, has idea for swift towing canal boats. Doubtless he has seen Francis Baily's supplement on John Flamsteed.
ALS 8pp RS:HS 9.187

3467. 1837–4–18 From Basil Hall Edinburgh
Further information about boats in canals. Sends proof sheets of the article in question.
ALS 2pp RS:HS 9.188

3468. 1837–4–23 From Thomas Maclear Royal Observatory, Cape of Good Hope
Intended to ride over today, but urgent mail from England claimed his attention. No news from the Admiralty. No news of the tide machine. The tripods are finished. Gathers that the Bonita cannot perform all D. F. J. Arago wished.
ALS 3pp RS:HS 12.102 Warners, *Maclear & Herschel Letters,* 166

3469. [1837–5] From Thomas Maclear n.p.
Returns JH's book of estimated magnitudes and would like it again when preparing his catalogue. Sends letters he received from Francis Beaufort and [Richard?] Jones. Is oppressed with business and an awful cough.
ALS 1p RS:HS 12.190 Warners, *Maclear & Herschel Letters,* 169

3470. [1837–5] To Thomas Maclear [Feldhausen]
Asks TM to return JH's book of star magnitudes.
L Warners, *Maclear & Herschel Letters*, 168

3471. 1837–5–5 To Thomas Maclear n.p.
About the observation of sun spots and the second satellite of Saturn.
C 2pp inc RS:HS 21.220 Warners, *Maclear & Herschel Letters*, 167

3472. 1837–5–5 From Thomas Maclear Royal Observatory, Cape of Good Hope
Received the enclosed communications by the steamer. Can JH send him a list of unobserved comet stars? Work on N. L. Lacaille's Observatory gives satisfaction. Regarding Capt. Mackonochee. Lacaille's angular measurements are not above suspicion. Observed the two rings of Saturn distinctly.
ALS 4pp RS:HS 12.103 Warners, *Maclear & Herschel Letters,* 167–8

3473. 1837–5–7 To Caroline Lucretia Herschel Feldhausen
Has decisively mapped Saturn's sixth satellite; doubts, however, that he will ever see the seventh. Asks CH to inform Friedrich Bessel of his observation.
L Evans, *Herschel at the Cape*, 300–3 & *Memoir ... of Caroline Herschel*, 288–9 inc

3474. 1837–5–15 From John Augustus Lloyd Mauritius
Is grateful for JH's kindness conveyed in his letter brought to him by Mohammed Ismail Khan. The latter has been attempting to persuade JL to go to Lucknow as an engineer, but has declined pending offer from the King of Oude. Will send his observations to Francis Beaufort after JH has left the Cape. Answers some of JH's queries.
ALS 4pp RS:HS 11.297

3475. 1837–5–19 From James Grahame L'Eperonniere
Family news update.
ALS 2pp RS:HS 8.432

3476. 1837–5–19 From Thomas Maclear Royal Observatory, Cape of Good Hope
Will call on JH tomorrow as he intends visiting Simon's Bay. Is taking a proof sheet to Cape Town. There appears to be a great discrepancy between the Captain's statement and N. L. Lacaille's station. The latter's station at Rubeck Castel remains perfect.
ALS 2pp RS:HS 12.104 Warners, *Maclear & Herschel Letters,* 169

3477. 1837–5–24 From Sir Charles Lyell 16 Hart St.
Charles Babbage will publish JH's volcanic theory in *Ninth Bridgewater Treatise*, together with JH's letter to R. I. Murchison. William Whewell read both at Geological Society. Babbage wants to compare JH's theory with passage from Bishop [Samuel] Butler. Whewell's notion of new species. Charles Darwin's new theory of coral islands is forcing CL to give up CL's volcanic crater theory. [Letter continues 28 May.] Sends copy of Babbage's *Treatise* with quotes from JH. News of Charles Wheatstone's submarine telegraph and David Brewster's proof that diamond is of vegetable origin.
ALS 4pp RS:HS 11.422 & 450 Katherine Lyell, *Lyell, 2,* 11–13

3478. 1837–5–29 From William Richard Hamilton King's Rd., Chelsea
Introducing Franklin Lushington who is going to New Holland on a surveying expedition. The ship carrying him will call at the Cape for instructions. Has written to the Governors here to induce them to adopt the Greek style, not the Gothic for the Houses of Parliament.
ALS 4pp RS:HS 9.218

3479. 1837–5–29 From Thomas Maclear Royal Observatory, Cape of Good Hope
The examination of the standard deal rod was commenced yesterday. Comments on the procedure and gives readings. The base line was levelled on Wednesday, Thursday, and Friday and he is now ready to measure it. Would like JH to be present if possible when he measures it. Had 25 visitors in one morning recently.
ALS 4pp RS:HS 12.105 Warners, *Maclear & Herschel Letters,* 169–72

3480. [1837–6 or 7] From Richard Jones n.p.
Relating the events of the last three years. Charles Babbage has published his Bridgewater treatise. Charles Lyell is just off to Norway. Mr. D'Arblay dies of a fever. Hopes JH is not working too hard. Very pleasant garden at Haileybury.
ALS 20pp RS:HS 10.410

3481. 1837–6–1 From J. H. Rutherford George St.
Encloses list of subscriptions raised to offset debt incurred by John Fairbairn [Editor of *Commercial Advertiser*] in law actions brought by [John?] Chase. Desires JH's support.
L 3pp RS:HS 14.443

3482. 1837–6–1 To J. H. Rutherford Feldhausen
Donates £25 to fund for John Fairbairn, whose editorial enthusiasm led to Fairbairn's court conviction. Advises JR to frame this support on grounds that cannot be interpreted as challenging court's decision.

Adraft 4pp TxU:H/L-0355; Reel 1054 (Cdraft: RS:HS 25.15.4)

3483. 1837–6 To Harry George Wakelyn Smith [Feldhausen]
Comments on HS's dealings with native tribal groups and returns a volume of related material.
draft 3pp RS:HS 16.199 & 21.225

3484. 1837–6–2 From J. H. Rutherford George St.
Wants JH's opinion of what to do with the subscription list. Will wait to proceed until he has an answer.
ALS 2pp RS:HS 14.445

3485. 1837–6–2 From J. H. Rutherford George St.
Has JH's letter and check. Wants to avoid improper use of JH's name. Sent list of first names and contributions to John Fairbairn. List will be returned to JR and then sent to JH.
ALS 2pp RS:HS 14.444

3486. 1837–6–3 From Roderick Impey Murchison Petersfield
JH's letter has appeared in print in Charles Babbage's Bridgewater Treatise. Comments on the books of Babbage and William Whewell. C. R. Darwin has read his paper on Coral Reefs. RM is to be General Secretary of the next B.A.A.S. meeting. [William?] Ayrton was elected F.R.S.
ALS 4pp RS:HS 12.391

3487. 1837–6–3 To Harry George Wakelyn Smith Feldhausen
Comments on the shooting death in 1835 of Hintsa, leader of the Xhosa people. JH believes that no wrongdoing occurred.
draft 2pp RS:HS 16.197 (C: RS:HS 21.221 & Rhodes Univ. Library inc)

3488. 1837–6–4 From Harry George Wakelyn Smith Charlie's Hope
Thanks for kind thoughts JH sent, after a difficult time about the Kaffir wars.
ALS 3pp RS:HS 16.198

3489. 1837–6–6 From J. H. Rutherford George St.
Discusses further corrections for the statement concerning the *Commercial Advertiser.*
ALS 1p RS:HS 14.447

3490. 1837–6–6 From J. H. Rutherford George St.
Discusses legal problems of the *Commercial Advertiser.* Has rewritten the list of subscribers. Sends documents to JH for approval. JH's response is included. JH states that slight alterations of the form are still necessary.
ALS 3pp RS:HS 14.446

3491. 1837–6–6 To J. H. Rutherford Feldhausen
Wants some rewording in the statement that is to go with the monetary support for John Fairbairn [see JR's 1837–6–6].
Cdraft 4pp RS:HS 25.15.5

3492. 1837–6–6 To J. H. Rutherford Feldhausen
Declines to get involved in political or judicial disagreements. JH donated to fund for John Fairbairn solely on grounds of benefits rendered to Cape Colony by Fairbairn's newspaper.
ACCS 3pp TxU:H/L-0356; Reel 1054

3493. 1837–6–7 From J. H. Rutherford George St.
Will finish up documents concerning the *Commercial Advertiser.* Thanks JH for his help and interest in the matter.
ALS 3pp RS:HS 14.448

3494. 1837–6–8 To Thomas Maclear n.p.
About a variety of observations that JH has made.
C 2pp inc RS:HS 21.222 Warners, *Maclear & Herschel Letters*, 172

3495. 1837–6–8 To Adolphe Quetelet Feldhausen near Wynberg
Expresses regret for not writing earlier. Thanks AQ for meteorological observations. Has complete series of observations from March 1835 to December 1836. Thanks AQ for reports on papers sent. Will depart in early 1838 and requests duplicates of all observations sent: one to England, one to the Cape. Apprises AQ of JH's work on double stars and nebulae.
ALS 4pp Académie belgique: #9 (ACCS: RS:HS 25.5.7)

3496. 1837–6–11 From Caroline Lucretia Herschel Hanover
Hopes that JH will receive Friedrich Bessel's paper called 'On the Influence of the Irregularities of the Earth on Geodetic Operations, and their Comparison with Astronomical Determinations.'
L *Memoir ... of Caroline Herschel*, 290–1 inc

3497. 1837–6–12 To Sir Charles Lyell Feldhausen
Ideas expressed by JH in his previous letter to CL (see JH's 1836–2–20) about secular variations of isothermal surfaces of the interior of the earth were similar to ideas expressed by Charles Babbage in a paper to the Geological Society. This was drawn to JH's attention by CL and by R. I. Murchison. JH argues that he did not know of Charles Babbage's paper, and goes on to point out how JH's ideas differ. Concludes with mention of having re-discovered the sixth satellite of Saturn, not seen since its discovery by JH's father, William.
ALS 5pp inc APS B.D25.L.2 (C: RS:HS 25.5.8) Cannon (1961), 311–14 inc)

3498. [1837–6–12] From Thomas Maclear n.p.
The flood has done much mischief to his baseline; there are still some holes to be filled in. Has written to Col. G. G. Lewis. Hopes to see JH at breakfast on Thursday.
ALS 2pp RS:HS 12.239 Warners, *Maclear & Herschel Letters,* 172–3

3499. 1837–6–13 From Thomas Maclear Royal Observatory, Cape of Good Hope
Weather was so unpromising that he sent off the men who were to repair the ravages to the base line levelings. Will JH therefore come on Wednesday instead of Monday provided the weather remains fine. Sorry he missed JH's request for barometric comparisons.
ALS 1p RS:HS 12.106 Warners, *Maclear & Herschel Letters,* 172

3500. [1837–6–14] From Thomas Maclear n.p.
Weather is most provoking. Thinks it best to delay the action until Friday.
ALS 2pp RS:HS 12.188 Warners, *Maclear & Herschel Letters,* 173

3501. [1837–6–14?] To Thomas Maclear Feldhausen
Asks TM whether he plans to work today.
L Warners, *Maclear & Herschel Letters*, 173

3502. [1837–6–18] To Thomas Maclear [Feldhausen]
Feeling very ill. Discusses adjustments to TM's transit instrument.
L Warners, *Maclear & Herschel Letters*, 173–4

3503. 1837–6–19 From John Lee Doctors' Commons
Is sending JH a lithographic print of Isaac Newton.
C 1p Buckinghamshire County Council

3504. 1837–6–20 From Thomas Maclear Royal Observatory, Cape of Good Hope
Had an unfortunate accident with the rods yesterday; one was smashed by the force of the wind. Intends to stake the whole line before the actual measurement takes place. Will calculate JH's Brisbane stars.
ALS 3pp RS:HS 12.107 Warners, *Maclear & Herschel Letters*, 174

3505. [1837–6–20] To Thomas Maclear [Feldhausen]
Not surprised that the measuring rods broke; knew that they were top heavy. Discusses improvements. Very busy. A horse stepped on JH's foot, but no broken bones.
L Warners, *Maclear & Herschel Letters*, 174–5

3506. 1837–6–22 From Charles G. B. Daubeny Oxford
Has received the samples of mineral waters JH sent from the Cape. His own pamphlet on this is now printed. Would be pleased if JH would obtain some bulbs of *massonia lutea* for him. Rev. Sandys in the area is an intelligent man.
ALS 4pp RS:HS 6.19

3507. 1837–6–23 From Thomas Maclear Royal Observatory, Cape of Good Hope
Sends the observations of the 21st copied by his daughter Mary. The leak in the center room has been repaired. Comments on JH's plan for measuring the base line. The Brisbane stars are nearly finished.
ALS 3pp RS:HS 12.108 Warners, *Maclear & Herschel Letters*, 176

3508. [1837–6–23 or 24] To Thomas Maclear [Feldhausen]
Discusses various matters relating to telescopic and to tidal observations. Relays request from William Whewell that TM continue tidal observations.
C 2pp RS:HS 21.193 Warners, *Maclear & Herschel Letters*, 176–7

3509. 1837–6–26 From Francis Baily London
Frederick Augustus (Duke of Sussex) has still not visited the R.S.L. Will discuss J. A. Lloyd's case with the Duke at the first opportunity. Home news. Will continue to forward pamphlets.
ALS 3pp RS:HS 3.140

3510. 1837–6–26 From William Buckland Oxford
Sending a copy of W. D. Conybeare's most recent publication; also his own Bridgwater treatise. Is grateful for his gift of luminous animals, which he has deposited in the College of Surgeons. Richard Owen has been giving the best lectures ever given. Dr. [William?] Smith is to exhibit his collection at Bulloch's museum. Is anxious to know what kind of fossil quadrupeds are to be found at the Cape.
ALS 4pp RS:HS 4.323

3511. 1837–7–4 To J. H. Rutherford Feldhausen
Continues to be concerned that the support for John Fairbairn not be seen as a political act, nor as one intended to impugn the decision of the South African Court of Justice. [Letter completed with a lengthy postscript on 1837–7–10.]
Cdraft 4pp RS:HS 25.15.6

3512. 1837–7–4 From Robert Wauchope Simons Town [near Cape Town]
J. D. Thompson plans to write JH about problem of positioning tide gage in dockyard. Admiral [Patrick Campbell] may move it to end of jetty. RW suggests omitting night observations in winter; jetty dangerous in bad weather. Will soon be ordered home. Amused by JH's report on Ascension [Island] sulfur. Sorry to hear Mary Herschel is ill. RW's son William is well again, but weak.
ALS 3pp RS:HS 18.122

3513. 1837–7–4 To J. H. Rutherford Feldhausen
Prefers that JH's name not appear on HR's list of subscribers in support of John Fairbairn. Gives reasons. Suggests appropriate wording for HR's public statement of support for Fairbairn.
ACCS 4pp TxU:H/L-0357; Reel 1054

3514. 1837–7–6 To Thomas Maclear n.p.
About some observations, the weather, and the difficulty of establishing tide observations, given the slowness of the Admiralty.
C 3pp RS:HS 21.223 Warners, *Maclear & Herschel Letters*, 178–9

3515. [1837–7–6] From Thomas Maclear Royal Observatory, Cape Town
Awful night of thunder, wind, and rain; floods around the Observatory. Gives rainfall since 22 June. Has written to [J. D.?] Thompson to observe tides at Simon's Bay. Has no one at Table Bay. Measured Alpha Centauri on the meridian. Found JH's planetary nebulae with the mural circle.
ALS 4pp RS:HS 12.199 Warners, *Maclear & Herschel Letters*, 177–8

3516. [1837–7–7?] To Thomas Maclear [Feldhausen]
Measurement of h Centauri.
L Warners, *Maclear & Herschel Letters*, 179

3517. [1837–7–15] To Thomas Maclear [Feldhausen]
Comments on stellar observations made by TM and reports JH's observations on sunspots. Notes error in [J. E.] Bode's celestial map.
C 2pp RS:HS 21.194 Warners, *Maclear & Herschel Letters*, 163–4

3518. 1837–7–17 To Stephen Peter Rigaud Cape Town
Gives altered directions for making meteorological observations to be sent to the South African Literary and Philosophical Institution.
AL 1p Bodleian Library, Oxford

3519. [1837–7–20] To Thomas Maclear n.p.
Found the 'real' h Centauri; unsure what star is the 'false' h Centauri. Saw the largest sunspot that JH has ever seen. Saddened to learn that TM's son [George] is seriously ill.
L Warners, *Maclear & Herschel Letters*, 179–80

3520. [1837–7–21] From Thomas Maclear n.p.
Thanks for JH's interest in the 'little fellow'; he has now passed all danger. Is glad JH has unravelled the mysteries of h Centauri. Wrote to Thomas Brisbane pointing out imperfections in the catalogue. Has started to publish an abstract of the Met[eorological] J[our]n[a]l in the *Government Gazette*. Will not attend the glittering assembly at Government House tonight.
ALS 3pp RS:HS 12.191 Warners, *Maclear & Herschel Letters*, 180

3521. 1837–7–22 To William Whewell Feldhausen
Reports on JH's efforts to secure tidal observations for WW, the Duke of Northumberland's offer to finance the printing of JH's Cape astronomical observations, and JH's sighting of Saturn's sixth satellite. Encourages WW's efforts to write a

philosophy of the inductive sciences.
ALS 6pp TC, Camb. Add. Ms.a.207[29] (C: RS:HS 21.224)

3522. 1837–7–25 From Johann Heinrich Mädler and Wilhelm Beer Berlin
The departure of missionaries [again; see JM's and WB's 1835–8–6] allows JM and WB to send along further material on the *Mappa Selenographica.*
ALS 1p† (French) RS:HS 12.270

3523. 1837–7–28 From William Somerville Royal Hospital, Chelsea
Introduces Frederic Smith and asks the Herschels to be kind to him when he arrives at the Cape. Is anxious to see JH again.
ALS 4pp RS:HS 16.405

3524. 1837–7–30 To Thomas Maclear Feldhausen
Sends TM his nebulae observations and drawings of what he saw.
C 2pp RS:HS 21.226 Warners, *Maclear & Herschel Letters,* 181

3525. [1837–7–31] From Thomas Maclear n.p.
Is pleased to have a look at JH's labors, but is not at home with the observations. Sorry to hear of JH's bodily sufferings. Has another sheet printed off, but no copies at the moment. Transit clock has increased its daily rate. South end of the mural circle has heaved up. Land to west of the Observatory is being discussed with view to purchase.
ALS 4pp RS:HS 12.189 Warners, *Maclear & Herschel Letters,* 181–2

3526. 1837–8–1 To Alphonse Candolle Feldhausen
A long letter on climate and vegetation and his experiences whilst at the Cape.
draft 12pp RS:HS 5.320bis (C: RS:HS 21.227)

3527. 1837–8–12 From James William Grant Westbrook
Is grateful for the loan of the book. Comments on some of [Alexander?] Stewart's works, especially his Gaelic ones. Queries regarding light in William Buckland's Bridgewater treatise; would like to borrow the volume if possible.
ALS 4pp RS:HS 9.20

3528. 1837–8–20 To William Whewell Feldhausen
Thanks for WW's *History of the Inductive Sciences* and for dedicating it to JH. Has read volume I, being especially impressed by WW's stress on hypothesizing. Recounts difficulties in getting tidal observations for WW. Is sending back a large number of JH's astronomical observations.
ALS 4pp TC, Camb. Add. Ms.a.207[30] (C: RS:HS 21.228)

3529. [1837–8–24] From Thomas Maclear Royal Observatory
Check proofs, including first announcement of JH's 'Planetary Nebulae.' Instrument inconsistencies produce immense labor for TM and C. P. Smyth. Tempted to omit one-third of readings. Praises Smyth; happy to be rid of William Meadows. Speculates on causes of TM's and JH's headaches. Will propose that Fearon Fallows's observations be published.
ALS 5pp RAS:JH Archive 6/2.34; Reel 8 Warners, *Maclear & Herschel Letters,* 182–3

3530. 1837–8–27 From Andrew Smith Egyptian Hall, Piccadilly
Tells JH about artifact exhibition AS has set up, and about preparation for publication of some astronomical observations.
ALS 4pp RS:HS 16.176

3531. 1837–8–27 From William Henry Fox Talbot Lacock Abbey
About plant seeds sent by JH from the Cape, WT having sent him some Mexican seeds in return. Halley's Comet only clearly seen on two nights.
ALS 3pp RS:HS 17.273

3532. [1837–9] To Stephen Peter Rigaud Cape Town
Thanks for meteorological observations sent by SR. Discusses other meteorological observations JH has been receiving. Will soon be returning to England.
ALS 1p Bodleian Library, Oxford

3533. 1837–9 To James Calder Stewart [Feldhausen]
Concern for JS's trip to South America. Plans for JH's return to England and visit to Hanover. Asks JS to seek new residence for JH and family. [Note from Margaret Herschel on criteria for new house.]
ALS 4pp TxU:H/L-0422; Reel 1055 Evans, *Herschel at the Cape,* 316–19

3534. 1837–9–3 To Adolphe Quetelet Wynberg
Attached circular from Meteorological Committee of South African Literary and Philosophical Institution stating a change from 36 to 24 hours of hourly observation at solstices and equinoxes. Letter to AQ thanks for great work and requests AQ continue.
ALS 4pp Académie belgique: #10

3535. 1837–9–7 To Caroline Lucretia Herschel Cape of Good Hope
Eliminated plans to visit Rio de Janeiro on the Herschels' return home, which will enable them to visit Hanover. Notes that JH's rheumatism attacks have interfered with his observations. JH considers his sweeping work finished. Has sent to the R.S.L. the first six hours of JH's 'Southern Catalogue of Nebulae and Double Stars' along with the micrometric measures of 400 southern double stars made with the equatorial.
ALS 4pp TxU:H/L–0582.2.a; Reel 1058 Evans, *Herschel at the Cape,* 313–16 & *Memoir ... of Caroline Herschel,* 291–2 inc

3536. [1837–9–12] To Thomas Maclear n.p.
Sees little benefit in trying to triangulate between stations, certainly at this time; JH's mirror polishing has been very successful.
C 3pp inc RS:HS 21.236 Warners, *Maclear & Herschel Letters,* 56–7

3537. 1837–9–13 From Harry George Wakelyn Smith Cape Castle
Note accompanying copy of dispatch relating to Kaffir wars.
ALS 2pp RS:HS 16.200

3538. 1837–9–15 From Mr. Lichtenberg London
Has communicated information from JH about JH's aunt Caroline to Mary Baldwin. Mary Baldwin has decided to go to visit Caroline Herschel.
ALS 2pp (German) TxU:H/M–0332; Reel 1087

3539. [1837?]–[9?]–18 To [James Edward Alexander] n.p.
Comments on a mining prospectus and on copper ore JH analyzed.
ALS 3pp APS B.H435p.28

3540. 1837–9–20 From George Robert Waterhouse London
Received JH's collection of African insects, etc. Describes its luminous molluscs, crustaceans, Hottentot fish.

Government granted £1500 to [William?] Smith and £1000 to C. R. Darwin to publish discoveries, and £500 to R.A.S. [Elizabeth Ann Waterhouse adds 1.5 pages to Margaret Brodie Herschel about birth of EW's second daughter, deaths in family.]
ALS 4pp RS:HS 18.94

3541. [1837]–9–22 From Thomas Maclear n.p.
Shall have the barometers and thermometers (readings?) tomorrow. Was disappointed with his observations. Admiralty has allowed the extra expense. Comments on the report on the death of 'poor young Musgrave.' Does JH know a particular problem in the damming of water?
ALS 3pp RS:HS 12.194 Warners, *Maclear & Herschel Letters,* 183

3542. 1837–9–24 To Thomas Maclear [Feldhausen]
Wants to invite members of the *Beagle* to dinner.
L Warners, *Maclear & Herschel Letters*, 184

3543. [1837–9–24] To Thomas Maclear n.p.
Has received books and papers by the *Beagle* but only one letter and nothing from Francis Beaufort. Expects a party of visitors from the *Beagle*. If JH intends to visit the Bay, TM would like to accompany him.
ALS 2pp RS:HS 12.196 Warners, *Maclear & Herschel Letters,* 184

3544. 1837–9–25 To James David Forbes Feldhausen
Thanks JF for papers and requests copies of those JH has borrowed; comments on JH's observations on solar heat, light, and lines in spectra.
ALS 4pp St. Andrews 40 (C: RS:HS 21.229)

3545. [1837–9–26] From Thomas Maclear n.p.
News of Wickham [Commander of the *Beagle*] and J. L. Stokes, who are still at Simon's Bay. Called on George Grey and Franklin Lushington yesterday. *Childe Harold* from Bombay will sail about Saturday. Did JH see Wilhelm Beer's and J. H. Mädler's paper in the *Comptes rendus* on the 6th and 7th satellites of Saturn?
ALS 4pp RS:HS 12.198 Warners, *Maclear & Herschel Letters,* 184–5

3546. 1837–9–30 From Sir John Franklin Van Diemen's Land
Sends a parcel received from England and the meteorological and tidal registers kept at Port Arthur. Has he heard of the grievous loss to Sydney by the death of Dr. Burnett? Hopes he is progressing with his labors.
ALS 7pp RS:HS 7.358

3547. 1837–10–1 From Adolphe Quetelet Brussels
Welcomes JH back to England. Proof that AQ never ceased thinking of JH is that AQ never quit his horary observations. Has sent books to R.S.L. for JH. Has restarted his notices on meteors. Is determining their longitude in relation to Greenwich. [Richard] Sheepshanks brought AQ JH's portrait and bust.
ALS 3pp (French) RS:HS 14.211

3548. 1837–10–2 To Charles Babbage Feldhausen
Comments on some parts of CB's *Ninth Bridgewater Treatise,* especially on CB's mathematical estimating of the credibility of miracles. JH shows that such a process cannot work. [Letter finished 1837–10–25.]
ALS 6pp RS:HS 2.290a (ACCS: RS:HS 25.5.9)

3549. 1837–10–3 From William Hawkins Cape Town
Has secured a passage home for JH on the *Windsor*. Thinks he will be pleased with the accommodation. Will call on him shortly.
ALS 2pp RS:HS 9.257

3550. 1837–10–3 From Dionysius Lardner Great Queen St.
Introducing his son George, who is to be stationed at the Cape for some time. Would like JH to advise him when necessary. DL has become co-editor with Bulwer Lytton of a new quarterly and would like an article from JH on his work in the Southern Hemisphere. Work on the Cyclopaedia is progressing.
ALS 7pp RS:HS 11.123

3551. 1837–10–4 To John Fairbairn Feldhausen
Has been astonished by the resolution of the meeting of the Kirk. Hopes JF will do something to calm down the parties.
draft 4pp RS:HS 7.150 & 21.230

3552. [1837–10–5] From John Fairbairn n.p.
Thanks for his note about the resolution of the Kirk. Gives his own views on the subject, which agree somewhat with those of JH.
ALS 3pp RS:HS 7.154

3553. 1837–10–7 From James Prinsep Calcutta
Calcutta Museum will trade an elephant skeleton to the Cape Museum for a hippopotamus, rhinoceros, or tiger. Work on ancient Indian inscriptions and at the mint leaves JP little time for meteorological work. Miss Pattle draws Cape scenes and sends them to JP.
ALS 3pp RS:HS 14.73

3554. [1837–10–9] To Thomas Maclear [Feldhausen]
Agrees with George Airy that Mercury should be studied more closely. Discusses TM's researches of N. L. Lacaille's Northern Extremity.
L Warners, *Maclear & Herschel Letters*, 185

3555. [1837–10–10] From Thomas Maclear n.p.
If the weather clears up would like to compare his observations of the eclipse with those of JH. Thinks of riding over on Sunday so that JH can question him on Klypfonteyn and N. L. Lacaille's station.
ALS 1p RS:HS 12.197 Warners, *Maclear & Herschel Letters,* 185

3556. 1837–10–10 From William Henry Smyth Crescent, Bedford
Sends early nineteenth-century astronomical observations. Francis Baily receives £500 from Parliament to enlarge stellar catalogue. Otto Struve's observations of Gamma Virginis agree with JH's.
ALS 4pp RS:HS 16.264

3557. [1837–10–13] From Thomas Maclear Royal Observatory
Problems determining exact times of lunar eclipses. Cites *B.A.A.S.Rep.* article by T. R. Robinson on change of color. Base line covered by water after recent storm.
ALS 3pp RAS:JH Archive 6/2.38; Reel 8 Warners, *Maclear & Herschel Letters,* 186

3558. [1837–10–14] From Thomas Maclear Royal Observatory
Compares TM's observations of 'spots' and colors [on lunar surface] to observations by others. At next eclipse, TM will use drawing of moon's face with spots labeled on it. Did JH note occultation? Chronometer is slow.
ALS 2pp RAS:JH Archive 6/2.39; Reel 8 Warners, *Maclear & Herschel Letters,* 186–7

3559. 1837–10–14 To Thomas Maclear Feldhausen
Sending the chronometer.
L Warners, *Maclear & Herschel Letters*, 186

3560. [1837–10–21 or later] To William Henry Harvey Feldhausen
A note to accompany some more plant specimens for WH to identify.
ALS 1p WT 69294.4

3561. 1837–10–21 To William Henry Harvey Feldhausen
Sends WH some plant specimens, and will send on any unique plants. JH has been asked for a description of the 'botanical climate' of the Cape, and asks WH's assistance with this.
ALS 3pp WT 69294.3

3562. 1837–10–24 From William Henry Harvey [Cape Town]
Many thanks for the beautiful specimen of Satyrium, which he has named in honor of Lady Herschel. Discusses various plants and the effect of climate. Is sorry to hear the Herschels are contemplating leaving the Cape.
ALS 8pp RS:HS 9.242

3563. 1837–10–26 From William Henry Harvey [Cape Town]
Found ten species of orchids on Table Mountain yesterday. Gives details of some and discusses their distribution.
ALS 4pp RS:HS 9.243

3564. 1837–10–26 From Robert Treat Paine Boston, Mass.
Sends hourly meteorological observations for March, June and September. Comments on these and the climate of New England. Further details of the fall of meteors. Francis Baily has sent his article on the eclipse of 1836. Delivered JH's message to Nathaniel Bowditch. Other scientific news.
ALS 6pp RS:HS 13.211

3565. 1837–10–27 From Thomas Glanville Taylor [Madras]
Observing 2000 stars in a check of Giuseppe Piazzi's catalogue.
ALS 4pp RS:HS 17.335

3566. 1837–10–28 To Mary Maclear Feldhausen
Reports the birth of John Herschel, JH's sixth child and third son.
L Warners, *Maclear & Herschel Letters*, 187

3567. 1837–11 To William Buckland Feldhausen
Thanks for paper. Discusses fossils, 'singular' land formation in Cape area. Asks WB to remember JH to 'geological friends' and tell [Charles] Daubeny of JH's specimen of Daubeny's root.
ALS 3pp diag RS Bu.83

3568. [1837–11] From Thomas Maclear n.p.
Hard at work mapping the polar stars. Rejoices at Dr. Andrew Smith's success. The results of tonight will be delivered by J. K. Gibbs in the morning.
ALS 2pp RS:HS 12.224

3569. [1837–11–1] To Thomas Maclear [Feldhausen]
Sending the spherometer. Discusses stars and nebulae observed near the South Pole.
L Warners, *Maclear & Herschel Letters*, 187–8

3570. 1837–11–2 From Thomas Maclear Royal Observatory, Cape of Good Hope
Gives a list of the close Polar stars. Started last night to observe the stars on JH's list but was put off by JH's right ascensions. Will observe each star again this night. John Fairbairn cannot divest himself of the apprehension of danger to Andries Stockenström.
ALS 4pp RS:HS 12.109 Warners, *Maclear & Herschel Letters*, 188–9

3571. [1837–11–2 or 3] To Thomas Maclear [Feldhausen]
Reports that Dr. Andrew Smith received a grant of £1500 from the British government.
L Warners, *Maclear & Herschel Letters*, 189

3572. 1837–11–2 From James Prinsep Calcutta
Sends September 1837 *Asiatic Journal.* T. E. Cantor leaves for the Cape. JP asks if the Cape Museum would want an elephant or rhinoceros skeleton.
ALS 2pp RS:HS 14.74

3573. [1837–11–3] From Thomas Maclear n.p.
Sends the results of one good night's observations entirely reduced. More transit observations are in process of being reduced.
ALS 1p RS:HS 12.181 Warners, *Maclear & Herschel Letters,* 190

3574. [1837–11–4] From Thomas Maclear n.p.
Polar region was thick this evening so that very minute stars were invisible. Set C. P. Smyth to the transit and himself at the circle to observe JH's list of stars. Received the spherometer safely. Intends to stake out the base line again on Tuesday. [Andrew?] Sm[y]th's expedition will be a serious undertaking in the Eastern direction.
ALS 3pp RS:HS 12.229 Warners, *Maclear & Herschel Letters,* 190–1

3575. 1837–11–6 From William Henry Harvey n.p.
Returns the specimens with the names. Sends a magnificent specimen of Orobanche, which may be a new species. Comments on the scent of plants. Will be glad to hear the results of his experiments on the temperature of bulbs. Comments on changes due to removal. Sleep in plants.
ALS 6pp RS:HS 9.244

3576. 1837–11–7 From John Augustus Lloyd Mauritius
Has delayed writing in order to send the correspondence about the Observatory. Has received a letter from the Colonial Office praising his work and increasing his salary, also taking over financial responsibility for the instruments. Is not satisfied with the life at Mauritius.
ALS 12pp RS:HS 11.298

3577. 1837–11–7 From John McClelland Calcutta
Sending his pamphlet on the tea plant. Comments on the geological features of the Chittagong area. Hopes that JH will have time to look at the fossils.
ALS 3pp RS:HS 12.5

3578. [1837–11–8] To Thomas Maclear n.p.
Compares his observations for some stars with the observations of TM.
C 1p RS:HS 21.237 Warners, *Maclear & Herschel Letters,* 191

3579. [1837–11–9] From Thomas Maclear Royal Observatory
Examined the base line on Tuesday; gives a resume of the operations. On Monday night observed the Polar stars; comments on the disappearance of certain stars in the Brisbane catalogue. Thinks a chain is preferable to deal rods in measuring base lines.

ALS 4pp RS:HS 12.223 Warners, *Maclear & Herschel Letters,* 191–2

3580. 1837–11–16 To Caroline Lucretia Herschel Cape of Good Hope
[Postscript to a letter by Margaret Herschel:] JH notes that he has finished the reduction of the first 9 hours in Right Ascension of his southern nebulae and double stars.
L Evans, *Herschel at the Cape*, 328

3581. [1837–11–18] To Thomas Maclear Feldhausen
Polishing telescope mirrors.
L Warners, *Maclear & Herschel Letters*, 192

3582. 1837–11–19 From Adolphe Quetelet Brussels
Thanks for two letters. Updates JH on observatories participating in horary observations. Proposes to reduce observations. Sends second part of Brussel's annals and other works. Discusses a phenomenon on 10 August. Mentions several observations of the Aurorae Borealis.
ALS 4pp (French) RS:HS 14.210

3583. 1837–11–21 Thomas Maclear Feldhausen
Asks TM for some trigonometric information regarding a particular set of points.
L Warners, *Maclear & Herschel Letters*, 193–4

3584. [1837–11–22] From Thomas Maclear n.p.
Sending the mean of all the observations on the polars. Comments on the base line, and gives readings.
ALS 10pp RS:HS 12.182 Warners, *Maclear & Herschel Letters,* 194–8

3585. [1837–11–23] From Thomas Maclear n.p.
Sends two drawings of C. P. Smyth and a portrait of Isaac Newton sent by Dr. [John?] Lee; also a plan of the triangulation.
AL 1p RS:HS 12.200 Warners, *Maclear & Herschel Letters,* 198–200

3586. 1837–11–23 From Thomas Maclear Royal Observatory, Cape of Good Hope
Encloses a plan of the triangulation and comments on his method. Thinks of making the baseline on the parade a permanent standard and has written to the Governor to obtain permission, as there is friction between the Board of Ordnance and the Cape Authorities over the ownership of the Parade. Has had no reply from the Governor as yet.
AL 4pp inc RS:HS 12.109bis Warners, *Maclear & Herschel Letters,* 198–200

3587. 1837–11–29 To Richard Jones Cape of Good Hope
Comments on William Whewell's and Charles Babbage's writings, together with family news.
C 3pp RS:HS 21.231

3588. [1837–11–30] From Thomas Maclear n.p.
Thanks for the perusal of [W. H.?] Smyth's letter; also had one, which he will send later when he has considered certain points. The zenith sector is on board the *Wellesley*, expected daily. Benjamin D'Urban has granted all TM's wishes regarding the survey. Re-surveyed the ground this morning.
ALS 2pp RS:HS 12.202 Warners, *Maclear & Herschel Letters,* 200

3589. [1837]–11–30 To Thomas Maclear [Feldhausen]
Asks TM to attend a meeting of the Cape of Good Hope Association for Exploring Central Africa.
L Warners, *Maclear & Herschel Letters*, 200

3590. 1837–11–30 To [?] Feldhausen
Is trying to expedite the shipping back to England of part of the zoological collection of Andrew Smith.
ALS 2pp APS B.H435p.8

3591. 1837–12–3 To Thomas Maclear Feldhausen
Talks about the very hot weather, and some matters of business that JH needs to work out with TM.
C 4pp inc RS:HS 21.232 Warners, *Maclear & Herschel Letters*, 201–2

3592. [1837–12–3] From Thomas Maclear Royal Observatory
Availed himself of the calm weather to finish the parade measurement. Gives an account of the work.
ALS 3pp RS:HS 12.201 Warners, *Maclear & Herschel Letters,* 200–1

3593. [1837–12–4] From Thomas Maclear Royal Observatory, Cape Town
Thanks for the perusal of Capt. J. A. Lloyd's papers. Comments on Lloyd's work at Mauritius. Does not think a mural circle necessary for the observatory at Mauritius. Regarding the Parade base line at the Cape. Not easy to put the guns into the ground.
AL 4pp inc RS:HS 12.195 Warners, *Maclear & Herschel Letters,* 202–3

3594. 1837–12–12 From Thomas Maclear Royal Observatory, Cape of Good Hope
The zenith sector was brought by the *Wellesley* and is now at the Observatory. Outlines some observations he hopes to make with it. Is about to write to R. Thom[p]son. Has some important matters to discuss with JH when he calls.
ALS 2pp RS:HS 12.110 Warners, *Maclear & Herschel Letters,* 204

3595. 1837–12–13 From Basil Hall Galashiels
Is pleased to hear the Herschels are contemplating their return but regrets it will not be via Rio. J. D. Forbes is progressing well, may be in danger of overwork. Thomas Henderson too is doing good work at the Observatory. Gives results of lunar eclipses of Oct. 1837.
ALS 8pp RS:HS 9.189

3596. [1837–12–13] To Thomas Maclear n.p.
Some minor matters of astronomical business followed by an extended list of principles on which a constellation reform might be undertaken.
C 2pp inc RS:HS 21.238 Warners, *Maclear & Herschel Letters*, 204–6

3597. 1837–12–15 To James Dunlop Feldhausen
JH's observations disagree with JD's 'Catalogues of Nebulae and Double Stars' [observed at Paramatta, Australia]. Believes errors lie in JD's work. Will send copy of JH's catalogue as soon as it is printed. [8pp of notes extracted by JH from JD's catalog; 7 pp of JH's comments, written in another hand, about these notes.]
ACCS 18pp TxU:H/L-0145; Reel 1054

3598. [1837–12–16] To *Astronomische Nachrichten* n.p.
Reports observations of the 'remarkable' increase in magnitude of the 'nebulous' star Eta Argus.
L *Astronomische Nachrichten, 15* (1838), 311–12

3599. 1837–12–16 To Francis Baily Feldhausen
Observations regarding the nebulous star 1281.
ALS 3pp RS:HS 3.141 (C: RS:HS 25.8.11)

3600. 1837–12–16 To Thomas Maclear n.p.
About the remarkably bright variable star, Eta Argus.
C 3pp RS:HS 21.233 Warners, *Maclear & Herschel Letters*, 207–8

3601. 1837–12–16/17 To Hugh Percy (3rd Duke of Northumberland) Feldhausen
Describes his observational activities, especially the discovery of the variable nature of Eta Argo, and the rediscovery of the sixth satellite of Saturn.
draft 6pp RS:HS 13.352 & 21.234

3602. 1837–12–16 to 1837–12–17 To the R.A.S. Cape of Good Hope
Reports observations of the 'remarkable' increase in magnitude of the 'nebulous' star Eta Argus.
L *R.A.S.M.N.*, *4* (1838), 121–2

3603. 1837–12–16 From Andrew Smith Fort Pitt, Chatham
More about lack of success of exhibition, and about publication of observations [see AS's 1837–8–27].
ALS 3pp RS:HS 16.177

3604. [1837–12–17?] From Thomas Maclear n.p.
Thanks for the interesting account of the star whose variability has now been established. Used the 3 1/2 foot last night and definition was good. Returns JH's letter to [James?] Dunlop and thinks there should be no offense taken. Opening for the Zenith is being enlarged. Measured JH's azimuth East and West yesterday.
ALS 4pp RS:HS 12.193 Warners, *Maclear & Herschel Letters*, 208–9

3605. 1837–12–18 From Francis Beaufort Admiralty
Is sending him a letter of Basil Hall. William Whewell is preparing another paper on the tides and is hoping for data from JH when he returns. Hears that JH will be home in June. Looks forward to seeing him again.
ALS 3pp RS:HS 3.353

3606. 1837–12–26 To George Gipps Feldhausen
Intricate enquiries at Paramatta observatory are a waste of time. Instruments are not first class. Energy would be better spent cataloguing smaller magnitude stars and measuring double stars. Urges triangulation survey, important for Asian and Pacific geography, and investigation of weights and measures, to avoid litigation as Australia develops.
AdraftS 11pp RS:HS 19.72

3607. 1837–12–27 To George Gipps Feldhausen
An extended statement of the importance of the observatory at Parramatta, Australia.
C 12pp RS:HS 21.235

3608. [1837]–12–28 From Elizabeth Gipps Cape Town
Is grateful for his gift of flower roots; she will take great care of them on her arrival at Sydney.
AL 2pp RS:HS 9.99

3609. 1837–12–28 From George Gipps Cape Town
Is grateful for the paper that JH has sent; he is sure it will promote scientific knowledge in New South Wales.
ALS 2pp RS:HS 9.98

3610. [1838] From William Henry Fitton n.p.
Regarding a successor to Frederick Augustus (Duke of Sussex) at the R.S.L. Is glad JH intends to devote his time to scientific labor. Pleased to hear his book meets with JH's approval.
ALS 4pp RS:HS 7.243

3611. [1838] From William Hawkins n.p.
Not convenient for R. J. Jones to take transfer of the property until Friday week. Sends a parcel which has just arrived.
ALS 1p RS:HS 9.260

3612. [1838] From William Hawkins n.p.
R. J. Jones will be ready to receive transfer on Friday next.
ALS 1p RS:HS 9.259

3613. [1838 or later] To [Margaret Brodie Herschel] n.p.
Remembers [wedding] anniversary.
ALS 1p TxU:H/L-0536.7; Reel 1053

3614. [1838 or later] From Thomas Maclear [Royal Observatory]
Gives latitude of 'this observatory' used in reduction of 'Comet stars.' Will accept systematic correction in all data to compensate for differences between TM's and Thomas Henderson's data. Instruments are erratic since telescope was moved. Using transit instrument only for sun, planets, and [reduction of Thomas] Brisbane's stars.
ALS 4pp RAS:JH Archive 7/3.15; Reel 8

3615. [1838] To ? n.p.
Announces F. W. Bessel's success in measuring stellar parallax.
AL 1p inc TxU:H/L-0410; Reel 1055

3616. [1838 or later] To ? [Slough?]
Appears to be commenting on JH's availability for a position.
ALS 2pp inc DMC.695A.885.1

3617. 1838–1–6 From Giovanni Antonio Plana Turin
Has made meteorological observations as desired by JH, and has sent and received various sets of such observations. Would like to come to England to visit JH, but first wants to find a husband for his only daughter. GP's assistant at the observatory died and he is looking for a replacement. [Many parts of the letter are illegible.]
ALS 3pp† (French) RS:HS 13.434

3618. 1838–1–12 From Sir John Bell n.p.
Herewith is the reply to his memorial. Do not inform people in England of the amount of rent paid for a piece of land near Cape Town.
ALS 2pp RS:HS 4.54

3619. 1838–1–13 From Thomas Maclear Royal Observatory, Cape of Good Hope
The zenith sector seems to perform very well, and he is now ready to move it to the stations. Has been troubled with cloudy weather. C. P. Smyth is working the mural circle on the stars. Thomas Jones has not prepared the diagonal scale promised by Francis Baily. Has not been well lately. No news from England except that Admiral George Elliott has permission from the Admiralty to settle the land question. Encloses a catalogue of Zenith stars.
ALS 4pp RS:HS 12.111 Warners, *Maclear & Herschel Letters*, 211–12

3620. 1838–1–13 From James Prinsep Calcutta
Sends drawings of India landscapes. Speaks of JH's return to England.
ALS 2pp RS:HS 14.75

3621. 1838–1–16 To John William Lubbock Feldhausen
Thanks for work on lunar theory. Sends news from Feldhausen, noting 'latest Astronomical novelty,' the sudden growth of star Eta Argus.
ALS 4pp RS LUB.H.352

3622. 1838–1–16 To Thomas Maclear n.p.
JH is preparing to bring his observations at the Cape to a close; has polished a mirror for TM.
C 2pp RS:HS 21.239 Warners, *Maclear & Herschel Letters*, 212–3

3623. 1838–1–16 From Robert Treat Paine Boston, Mass.
Sending hourly observations of 21 and 22 Dec. last. Comments on the weather during the winter. Performance of the shooting stars in November was poor. Hopes that JH will visit the United States on his way from Brazil to England.
ALS 1p RS:HS 13.212

3624. 1838–1–19 To *Astronomische Nachrichten* Slough
Reports on his observations of the variable stars Eta Argus, Alpha Hydrae, and Alpha Cassiopeiae, and on the comparative brightness of the stars. Mentions that ill health is forcing him to curtail astronomical observing.
L *Astronomische Nachrichten, 16* (1839), 187–90

3625. 1838–1–21 To Thomas Maclear Feldhausen
On the variability of Alpha Hydrae and some other stars.
C 2pp RS:HS 21.240 Warners, *Maclear & Herschel Letters*, 213

3626. [1838–1–24] From Thomas Maclear n.p.
Thanks for information in his letter. Agrees with JH about the variability of Hydra. Argus is returning to its original brightness. The zenith sector is down and in parts and hopes to get it re-erected on Friday. Results at the Observatory are very good.
ALS 2pp RS:HS 12.183 Warners, *Maclear & Herschel Letters,* 213

3627. 1838–1–26 From George Thomas Napier Government House, Capetown
Regrets that the Herschels were not allowed in when they called; he had given strict instructions that they should be shown in immediately. Grieves to hear of the short stay of the Herschels. Would like to talk to him privately on the State and the policy pursued.
ALS 5pp RS:HS 13.63

3628. 1838–1–27 From Sir John Bell n.p.
Is returning the packet and would JH send the envelope to Mr. Crozier. Inform Sir John Franklin that any further letters should be sent c/o of JB's private address. Includes two short verses.
ALS 3pp RS:HS 4.55

3629. 1838–1–27 From Lord John Russell Court of St. James
Royal warrant appointing Board of Visitors to Royal Observatory, Greenwich.
D 4pp TxU:H/M–1224; Reel 1060

3630. 183[8]–1–28 From Thomas Maclear Royal Observatory, Cape of Good Hope
C. P. Smyth will deliver the micrometer series JH lent him for examining the values of the transit level scale; comments on this. Had another peep at the stars JH mentioned and still finds Argus larger than Rigel. The tent fills Mrs. De Witt's yard. She is a friend and correspondent of Charles Grant (Baron Glenelg).
ALS 4pp RS:HS 12.100 Warners, *Maclear & Herschel Letters*, 213–14

3631. [1838]–2–1 From Thomas Maclear Royal Observatory, Cape of Good Hope
Will not be time to take the zenith sector to Klypfonteyn before JH's departure. Must finish work with the pendulum as it has to be returned to Francis Baily via JH. Will ride over on Sunday morning.
ALS 2pp RS:HS 12.112 Warners, *Maclear & Herschel Letters*, 215–16

3632. 1838–2–3 To Thomas Maclear n.p.
Urges against carrying up Table Mountain the zenith sector sent to TM to check N. L. Lacaille's measurement of a degree of terrestrial longitude. Comments on JH's skill with precision instruments.
C 2pp RS:HS 21.241 Warners, *Maclear & Herschel Letters*, 216

3633. [1838–2–4] From Thomas Maclear Royal Observatory, Cape of Good Hope
Intended calling on JH today but legs and feet were too swollen. Has postponed his journey to Klypfonteyn until the first week in March, to make some alterations to the instrument. Is glad JH will visit the mountains with him. Believes the *Thalia* is about to sail soon. Formalhoet is now useless at noon.
ALS 3pp RS:HS 12.115 Warners, *Maclear & Herschel Letters,* 217

3634. [1838]–2–7 From George Thomas Napier Government House, Capetown
Inviting the Herschels to dinner on Friday or Saturday.
ALS 3pp RS:HS 13.72

3635. 1838–2–8 From Adolphe Quetelet Brussels
Sends latest publications. Has received highly surprizing letter on meteors by Mr. Wood. Discusses 'famous apparitions.'
ALS 2pp (French) RS:HS 14.214

3636. 1838–2–9 To Thomas Maclear London
Received meteorite specimen. Michael Faraday will conduct analysis. Admiralty agrees to give TM an assistant. Finds a theodolite telescope for Cape Observatory.
ALS 4pp Cape Archives/Maclear Papers File 100

3637. 1838–2–11 To Wilhelm Struve Feldhausen
Pleased with the observations and astronomical literature JH received from FS; JH is packing up for the return to England; comments on causes of variations in stellar magnitude and several other astronomical topics.
ACC 6pp RS:HS 25.5.10

3638. 1838–2–17 To Sir John Bell [Feldhausen]
Presents, in response to a memorandum from JB, JH's views on such matters as the administration, staffing, and funding of the Government Free Schools at the Cape of Good Hope.
draft 23pp UCT JH Letters A1.4 (C: Rhodes Univ. Library inc) Ferguson and Immelman, *Herschel and Education*, 14–25 & C. J. F. Bunbury, *Journal of a Residence at the Cape of Good Hope* (London, 1848), 281–97

3639. 1838–2–20 From Sir John Bell n.p.
Thanks JH for his long and insightful memorandum of 1838–2–17 on the system of Government Free Schools at the Cape of Good Hope. Accepts many of JH's ideas and requests that JH, after returning to England, present JH's ideas to Lord Glenelg.
ALS 3pp UCT JH Letters A1.5

3640. 1838–2–28 To Dionysius Lardner Feldhausen
Due to his impending departure from the Cape he has had little opportunity of seeing DL's son George. Does not think he will be in a position to write a popular article on his work in the Southern Hemisphere yet awhile. Several errors in his

Treatise Astr., which he would like to correct.
draft 4pp RS:HS 11.124 (C: RS:HS 21.243)

3641. 1838–2–28 To William Samuel Stratford Feldhausen
JH, in preparing to leave Cape, thanks WS for having sent *Nautical Almanacs* to him there. Discusses motions of a comet appearing to move whimsically, the periodical star alpha Hydrae, other stars, and the re-discovery of Saturn's sixth satellite.
ALS 4pp RS:HS 17.88 (C: RS:HS 21.242)

3642. [1838–3–1] To Thomas Maclear [Feldhausen]
Asks TM to set JH's chronometer and barometer.
L Warners, *Maclear & Herschel Letters*, 217–18

3643. 1838–3–6 To Charles Grant n.p.
[Writing to Charles Grant, Baron Glenelg, British Secretary of State for the Colonies], JH advises on such aspects as the administration, staffing, and the form and content of the instruction at the newly established Government Free Schools at the Cape.
L W. T. Ferguson and R. F. M. Immelman, *Sir John Herschel and Education at the Cape 1834–1840* (Cape Town, 1961), 27–34

3644. 1838–3–6 To James Calder Stewart [Feldhausen] and aboard *Windsor Castle*
Packing to leave Cape. Desires only furnished lodgings, not new house, for month of May in London. [Letter continues 6 May 1838 on board *Windsor*:] Becalmed off France. Expect to land in Portsmouth and visit Anstey on way to London.
ALS 4pp TxU:H/L-0423; Reel 1055 Evans, *Herschel at the Cape,* 363–4

3645. 1838–3–[7] To Thomas Maclear n.p.
Final arrangements before JH and family depart for England.
C 2pp RS:HS 21.244 Warners, *Maclear & Herschel Letters*, 218

3646. 1838–3–8 From John Fairbairn Cape Town
JH is leaving the Cape at an unfortunate time as the Governor has to rely on the support of an unsympathetic party. Comments on the political situation at the Cape. The Herschels have done much good for the Cape since their residence there.
ALS 4pp RS:HS 7.151

3647. 1838–3–18 To William Henry Fox Talbot Slough
Thanks for specimens of [light] 'sensitive paper.' Praises it. JH has handed over all his specimens of photography to R.S.L. Recommends a paper by the chemist Henri Regnault.
ALS 3pp Science Museum Talbt 1/14

3648. 1838–3–27 To William Henry Fox Talbot Slough
Thanks for photographic specimens WT sent. Reports on JH's recent experiments, including some using lenses, in photography; comments on WT's experiments and on the process of patenting.
ALS 3pp Science Museum Talbt 1/15

3649. 1838–3–29 From George Peacock 9 [Artiste] Parade, Hastings
Forwards to JH letters sent by Humphrey Lloyd and G. B. Airy to William Whewell. Concurs with Lloyd's recommendations [for Canada survey] and with Airy's suggestion of photographic registers to determine simultaneity of distant observations. [JH note: Routing list for letter among committee members.]
ALS 2pp, JH note 1p TxU:H/M–0424; Reel 1087

3650. 1838–4–3 To Thomas Maclear [*Windsor Castle*]
Describes observations made while aboard the *Windsor Castle.* Mentions stopover at St. Helena.
L Warners, *Maclear & Herschel Letters*, 218–19

3651. 1838–4–9 From Joseph Bianchi Modena
Is pleased he will accept his book on the observatory. Has read with pleasure the account of JH's work at the Cape, which was sent to Wilhelm Struve.
ALS 2pp (French) RS:HS 5.61

3652. 1838–4–9 From Roderick Impey Murchison Reading
Has now received JH's letter of the 5th. John Phillips will give JH the exact terms of the grant from the B.A.A.S. Will call at Slough to see the Herschels on his way to town. Deeply regrets that JH is not interested in the position at Cambridge, but M. J. Johnson will go far.
ALS 3pp RS:HS 12.392

3653. [1838–5 or later] From George Biddell Airy Royal Observatory, Greenwich
[Form Letter] Government committee appointed to replace standards of length and measure destroyed in Parliament fire. List of questions seeking recipients' opinion on subject.
LS 1p RAS:JH Archive 5/4.7 verso; Reel 8

3654. 1838–5–7 From Mary Smith Alnwick
Waxes poetic in reply to JH's note of sympathy for MS's husband's death. Hopes for JH's safe return from the Cape and that JH will visit when in town. Updates on family news.
ALS 4pp RS:HS 17.216

3655. 1838–5–14 To James Grahame Slough
On John Stewart's marriage to JG's daughter Matilda. JH supports Stewart's suggestion that Matilda continue to live in France and that Stewart will spend time there without neglecting his London business [as partner in Murray's Printing Works]. JH offers to put £1000 into Murray's business if JG will consent to Stewart's proposal and give Matilda an annual income of £150.
C 4pp & Summary 1p TxU:H/L-0169.1 & -0169.2; Reel 1054

3656. [1838–5–15 or later] From Sarah Bowdich 27 Burton St.
An eleventh number of her series on fishes is to be published. Hopes he will subscribe to it. Congratulations on his return.
ALS 2pp RS:HS 4.209

3657. 1838–5–17 To Francis Baily n.p.
Arrangements about meeting with FB.
C 1p RS:HS 21.245

3658. 1838–5–20 To Caroline Lucretia Herschel Batt's Hotel, Dover St., Piccadilly
Herschels safely landed from Africa and in fine spirits.
ALS 3pp TxU:H/L–0583.1; Reel 1058 *Memoir ... of Caroline Herschel,* 293 inc

3659. 1838–5–22 From Rowland Hill 2 Burton Crescent
Sending a copy of his pamphlet on the Post Office. Hopes he will give his opinion to the Parliamentary Committee now sitting.
ALS 2pp RS:HS 10.190

3660. 1838–5–23 From William Henry Fox Talbot Lacock Abbey
Writing to welcome JH home from the Cape.
ALS 2pp RS:HS 17.274

3661. 1838–5–24 From Thomas Spring-Rice Downing St.
Invites JH to join committee to replace standards of weight and measure destroyed in 1834 Parliament fire. Happy that JH returned safely to England.
ALS 2pp TxU:H/M–0375; Reel 1087

3662. 1838–5–24 To George Biddell Airy Bath Hotel
Has been asked to serve on a Standards Commission, to reestablish standards destroyed by the fire in the Houses of Parliament [1834]; is willing to serve.
ALS 3pp RGO 6.340.270

3663. 1838–5–26 From William Rowan Hamilton Observatory
Congratulates JH on his successful Cape Town enterprise.
L Graves, *Hamilton, 2,* 256–7 inc

3664. [1838–5–28] To William Henry Fox Talbot n.p.
Says that for the next four to five months will be in residence at 10 Hanover Crescent, Regent's Park. Offers various botanical specimens to WT, including a *Satyrium* named after JH. Mentions Eta Argus. Believes the 'Southern Sky has been pretty well rummaged.'
ALS 3pp Science Museum Talbt 1/9

3665. 1838–5–29 From Alexander von Humboldt Berlin
Calling his attention to the works of J. C. Poggendorf and Guillaume Weber. Does he know of the ingenious work carried out at the Cape by C. F. Gauss and Weber?
ALS 1p (French) RS:HS 10.64

3666. 1838–5–31 From Johann Heinrich Mädler and Wilhelm Beer Berlin
Are looking forward to JH's trip to the Continent. Comment on recent observations of the moon's surface.
ALS 1p (French) RS:HS 12.271

3667. [1838–6 to 1840–5] To Mrs. Joseph Goodall Slough
Thanks for loan of letters of Captain [Richard] Copeland.
AL 1p Gerald M. Friedman

3668. 1838–6 To Frederick Augustus, Duke of Sussex n.p.
Thanking him for his offer of a baronetcy. Matters relating to this.
draft 3pp RS:HS 1.421 & 20.247

3669. 1838–6 To Frederick Augustus, Duke of Sussex n.p.
Giving his reasons for not wanting to become a Baronet.
draft 2pp RS:HS 1.420 & 21.246

3670. 1838–6–4 From Francis Beaufort Admiralty
Thursday will suit Charles Wood and FB is writing to G. B. Airy to meet JH in (FB's) room.
ALS 1p RS:HS 3.354

3671. 1838–6–5 From Friedrich Wilhelm Bessel Königsberg
Welcomes JH back from Cape; includes results of FB's work on geodesical measurements in East Prussia.
ALS 2pp (German) RS:HS 4.70

3672. 1838–6–6 To Henry Bunbury [London]
Informs HB that a cask of bulbs is waiting for him at the Customs House. From JH's experience of his own bulbs, urges HB to get them quickly as they will deteriorate rapidly.
ALS 1p WT B.4

3673. 1838–6–6 From James Grahame 36 Roubrard, Nantes
Welcome home to the Herschels.
ALS 1p RS:HS 8.433

3674. 1838–6–7 From George Biddell Airy Royal Observatory, Greenwich
A notice of meeting of the Standards Commission.
LS 1p TxU:H/M-0626.3; Reel 1084

3675. 1838–6–7 From John Landeman 3 Melrose St., Glasgow
Asks JH for assistance in obtaining financial aid and new instruments from English government in support of meteorological observatory in British Guiana, where JL's son is stationed. Can JH suggest employment opportunities as astronomer in U.K. for JL's son?
ALS 3pp damaged TxU:H/M–0325; Reel 1087

3676. 1838–6–8 From Roderick Impey Murchison 2 Eccleston St.
Regarding tickets for Lady Herschel's gallery. There has been some fracas with Charles Babbage over the distribution of these tickets.
ALS 4pp RS:HS 12.393

3677. 1838–6–9 From George Biddell Airy n.p.
Asks JH to compile list of questions to be sent to 'commercial bodies' regarding prospective changes in standards of weights and measures. To be presented at June 1838 meeting of Commission of Standards [see JH's 1838–5–24].
ALS 1p TxU:H/M–0626.1; Reel 1084 (ACCS: RGO 6.340.272)

3678. 1838–6–9 From John Elliot Drinkwater-Bethune [Home Office] Chambers, Downing St.
Lists 11 British cities with active Chambers of Commerce and two societies that might come 'within the Spirit of our resolution.' Will make further inquiries.
ALS 2pp TxU:H/M–0080; Reel 1087

3679. 1838–6–9 From Humphrey Lloyd Trinity College, Dublin
Would have liked to have welcomed JH on his return from the Cape, but academic duties prevented it. Royal Irish Academy progresses under W. R. Hamilton's presidency. His late father established a magnetic observatory at Dublin and HL is in charge.
ALS 4pp RS:HS 11.263

3680. 1838–6–11 To Jean Baptiste Biot Slough
Thanks JB for letter to JH at the Cape. Requests JB to provide introductions for Mr. Stanford, who is making statistical inquiries in France.
ALS 3pp Columbia Univ. Library

3681. 1838–6–11 To Humphrey Lloyd Slough
Some personal and some general astronomical news.
C 2pp RS:HS 21.248

3682. 1838–6–11 From William Henry Fox Talbot Lacock Abbey
About Cape bulbs and seeds, the new travel by railway from Slough to London, and concern about interest in animal magnetism.
ALS 4pp RS:HS 17.275

3683. 1838–6–12 From Siméon Denis Poisson Paris
Seems to be commenting on someone's wonderful work, and describing some of SP's travels. [Letter almost totally

illegible.]
ALS 3pp (French) RS:HS 14.16

3684. 1838–6–15 To William Rowan Hamilton n.p.
Comments and expands on James MacCullagh's paper on laws of reflection and polarization in crystals.
ALS 4pp TCD 4015:96 Graves, *Hamilton, 2,* 262

3685. 1838–6–18 William Rowan Hamilton Royal Irish Academy
Pleased that the Royal Irish Academy awarded its Science Medal to James MacCullagh for his paper on the 'Laws of Crystalline Reflexion and Refraction;' he deems MacCullagh's essay superior to WH's 'On Algebra as the Science of Pure Time.'
L Graves, *Hamilton, 2,* 263–4 inc

3686. 1838–6–18 From John Richard Datchet
Thanks JH for gift of roots from the Cape. Discusses Testimonial dinner for JH, at which JR's Royal Master [Augustus Frederick] spoke. Congratulates JH.
ALS 4pp RS:HS 14.330

3687. [1838–6–19] To Charles Babbage n.p.
Has had a long letter from Frederick Augustus (Duke of Sussex), who wishes him to accept a baronetcy. JH's comments on this.
ALS 3pp RS:HS 2.400 (draft marked '*Most confidential'*: RS:HS 2.290a dated 1838–6–19; C: RS:HS 21.252)

3688. 1838–6–19 to 21 From George Peacock to Margaret Brodie Herschel [Athenaeum]
A further letter saying Augustus Frederick (Duke of Sussex) can wait an extra day for a response from JH.
ALS 4pp RS:HS 13.292

3689. 1838–6–19 From George Peacock to Margaret Brodie Herschel 1 Dorset St.
A note accompanying a letter from Augustus Frederick (Duke of Sussex) asking JH to reconsider his refusal of the offer of a baronetcy [see JH's 1838–6 to AF].
ALS 2pp RS:HS 13.289

3690. 1838–6–19 From James Ivory 20 Southampton St., Pentonville
Congratulations on his return. Sends some papers on the equilibrium of fluids on which he would like JH's opinion.
ALS 2pp RS:HS 10.257

3691. 1838–6–19 To George Peacock n.p.
JH's reasons for wishing to decline the baronetcy remain, but has decided to accept it because of Augustus Frederick's (Duke of Sussex) 'really noble conduct as respects me.'
C 1p RS:HS 21.251

3692. 1838–6–19 From Hugh Percy (3rd Duke of Northumberland) Syon
A note of thanks for sending HP a full account of the welcoming dinner given for JH.
ALS 3pp RS:HS 13.353

3693. 1838–6–19 To Frederick Augustus, Duke of Sussex n.p.
Regarding a baronetcy about to be conferred on JH.
draft 3pp RS:HS 1.419 & 21.246

3694. 1838–6–19 To Frederick Augustus, Duke of Sussex Hanover Terrace
In hesitating to accept the offer of a baronetcy, JH did not wish to be perceived as arguing. Now gratefully accepts that generous offer.
C 1p RS:HS 21.250

3695. 1838–6–20 From Jons J. Berzelius Stockholm
Last year he sent the diploma of the Academy of Sciences of Stockholm of which JH had been elected a member. Introduces J. A. Wahlberg, who is coming to the Cape.
ALS 2pp (French) RS:HS 5.56

3696. 1838–6–20 To Thomas Maclear 10 Hanover Terrace
Arrived in England on 19 May after a nine week voyage. Made suggestions to Lord Minto [Gilbert Elliott] concerning Cape Observatory, including Magnetic Observatory, mural circle repair, and the addition of a theodolite telescope.
ALS 3pp Cape Archives/Maclear Papers File 100

3697. 1838–6–20 To Roderick Impey Murchison Athenaeum
Has read RM's note to the *Morning Post* correcting error in the report on the dinner [in JH's honor]. Wishes money had not been mentioned in RM's reply. Is grateful for all RM's work in arranging for the dinner.
ALS 2pp EUL:Gen523/4/26 (C: RS:HS 12.394)

3698. 1838–6–20 From George Peacock Athenaeum
GP informs JH how gratified Augustus Frederick (Duke of Sussex) was by JH's letter [see JH's 1838–6–19].
ALS 2pp RS:HS 13.291

3699. 1838–6–20 From Wilhelm Eduard Weber London
WW sends JH some of the results of his magnetic observations. WW comments on the involvement of Carl F. Gauss in the work on terrestrial magnetism and urges JH, on his upcoming visit to Hanover, to make a point of seeing Gauss at Göttingen.
ALS 3pp (German) RS:HS 18.390

3700. 1838–6–21 From Roderick Impey Murchison 2 Eccleston St.
Explaining why he wrote the letter to the *Morning Post.* Has heard that JH is about to be offered a baronetcy. Must also ask him about the future Presidency of the B.A.A.S.
ALS 7pp RS:HS 12.395

3701. 1838–6–21 To George Peacock Hanover Terrace
A note to GP accompanying a letter to Augustus Frederick (Duke of Sussex) in which JH indicates a change of heart and an acceptance of the baronetcy [see JH's 1838–6–19].
ALS 1p RS:HS 13.290 & 21.251

3702. 1838–6–22 From Francis Beaufort Admiralty
The Admiralty would like JH's opinion regarding the observatory at Mauritius and any suggestions for its improvement.
ALS 3pp RS:HS 3.355 & ACC: Hydrographic Office Letter Book

3703. 1838–6–22 From Stephen Peter Rigaud Oxford
Asks JH to come to Oxford to receive honorary degree of D.C.L. Invites Herschels to Oxford Observatory.
ALS 3pp RS:HS 14.346

3704. [1838–6–23 or later] To Francis Beaufort n.p.
Responds to request for advice on operation of the observatory established at Mauritius by John A. Lloyd. Urges it can profitably make meteorological observations and positional determinations for southern stars. Stresses the need for prompt reduction of observations.
C 5pp RS:HS 21.249

3705. [1838]–6–24 From Henry Hart Milman Westminster
Offering tickets for the nave of the abbey for the Coronation as he hears that JH has not been provided with any. Elected [Thomas?] Greenwood into the Athenaeum.
ALS 3pp RS:HS 12.350

3706. 1838–6–25 From James Ivory 20 Southampton St., Pentonville
Would be obliged for the return of the papers when he has perused them.
ALS 1p RS:HS 10.258

3707. 1838–6–26 From George Biddell Airy Royal Observatory, Greenwich
Encloses official instructions to Thomas Maclear. Prices of telescopes. Remarks on caustics.
ALS 4pp RS:HS 1.73 (C: RGO 6.368.616)

3708. 1838–6–26 From W. C. Wallace 381 Broadway, New York
Sends calf eye as sample to demonstrate anatomy of eye. Disputes theories of Charles Bell and John Dalrymple.
ALS 3pp † RS:HS 18.370

3709. 1838–6–29 To Francis Beaufort London
Fulfills request for advice on the operation of the Cape Observatory. Urges, among other points, that Thomas Maclear be given a second assistant and an improved telescope and that the observatory be assigned to undertake tidal, meteorological, and magnetic observations.
draft 12pp RS:HS 3.355bis (C: RS:HS 21.253)

3710. 1838–6–30 From Francis Beaufort Admiralty
Many thanks for his report on the Cape Observatory, which he will lay before the Lords of the Admiralty. Also thanks for the promised opinion on the Mauritius Observatory.
ALS 2pp RS:HS 3.356 & ACC: Hydrographic Office Letter Book

3711. [1838–7] From Sophie Beckedorff to Caroline Herschel Hanover
Invitation to Caroline, Sir John, and William Herschel to dine with her on Friday.
ALS 2pp RS:HS 4.8

3712. 1838–7–2 From George Biddell Airy Royal Observatory, Greenwich
Announces 7 July meeting of Standard Commission at R.A.S. apartments.
LS 1p † TxU:H/M–0626.2; Reel 1084

3713. 1838–7–2 From William Plackett Whitehall
Will send a representative to JH to discuss some matters.
L 1p RS:HS 14.165

3714. 1838–[7–4] To [?] Hanover Terrace, Regent's Park
Is leaving England very shortly, so cannot meet to discuss the correspondent's Indian Survey. Provides advice on various metrological matters.
ALS Xpp Paul Luther

3715. 1838–7–5 From Sir Robert Harry Inglis 7 Bedford Square
Sends congratulations on his safe return and also on the award of his baronetcy. Would like to submit JH's name as a candidate for the Literary Society; gives names of those already members.
ALS 3pp RS:HS 10.226

3716. 1838–7–5 From Teodoro Monticelli Naples
Congratulates JH on return to England, his success at the Cape, and his ascension to the baronetcy. Praises JH for his contributions to many areas of science: 'The name of Herschel is ... the most spoken in this century.' Describes eruption of Mt. Vesuvius in 1834 and his studies of the event. Makes general comments on the progress of science in Italy.
ALS 4pp (Italian) RS:HS 12.367

3717. 1838–7–7 From Andrew Smith Fort Pitt, Chatham
Note accompanying one of AS's publications on South African zoology.
ALS 3pp RS:HS 16.178

3718. 1838–7–8 From William Henry Fox Talbot Lacock Abbey
Thanks for more Cape bulbs.
ALS 2pp RS:HS 17.276

3719. 1838–7–[9] To Margaret Brodie Herschel Alster Hotel, Hamburg
JH and son William arrived safely. H. C. Schumacher is out of town but JH will visit Altona anyway. Son William's reactions to travel. JH tried unsuccessfully to understand Immanuel Kant and J. G. Fichte during voyage.
ALS 4pp TxU:H/L-0538.1; Reel 1053

3720. 1838–7–11 To Margaret Brodie Herschel Hanover
Fatiguing trip from Hamburg. Mr. Groskopff is 'preparing' aunt Caroline for arrival of JH and son William. Comments on German philosophers whom JH is reading on trip. Visit with Caroline Herschel. Miss Beckedorff is out of town. Architecture and language of Hanover. Groskopffs named their country home 'Slough' near Hameln.
ALS 4pp TxU:H/L-0538.2; Reel 1053

3721. 1838–7–12 To Margaret Brodie Herschel Hanover
JH and son William made surprise visit to aunt Caroline Herschel. Mrs. Beckedorff is recovering from long illness. All Detmering family, who resided at Cumberland Lodge, are dead except Mrs. Detmering. Dr. Groskopff reports that Caroline, age 89, can still 'put her foot behind her back and scratch her ear in imitation of a dog.'
ALS 3pp TxU:H/L-0538.3; Reel 1053

3722. 1838–7–16 To Margaret Brodie Herschel Einbeck
Hot temperatures killed all fish in rivers and moats of Hanover. JH's and son William's journey from Hanover through Ammensen to Göttingen. Describes outdated architecture, poor agricultural practices, and sad plight of peasant women. Attended church in Göttingen then visited C. F. Gauss at observatory. J. F. Blumenbach was too ill to receive JH and Gauss.
ALS 4pp TxU:H/L-0538.4; Reel 1053

3723. 1838–7–17 From Heinrich Christian Schumacher Altona
Received JH's note from Belvedere Hotel. Mentions various arrangements regarding JH's European journey, including JH's planned visit to Wilhelm Olbers.
ALS 2pp RS:HS 15.378

3724. 1838–7–19 From Heinrich Christian Schumacher Altona
Unless other responsibilities interfere, HS will meet JH in Bremen.
ALS 1p RS:HS 15.379

3725. 1838–7–21 To George Biddell Airy Hanover
About JH's visit to Carl Friedrich Gauss, and Gauss's well-

devised magnetic measuring instruments.
ALS 3pp RGO 6.676.46 (draft RS:HS 1.74 & 21.254)

3726. 1838–7–22 To Francis Beaufort n.p.
Admiralty wants to adapt observatory at Mauritius for rating ship chronometers in southern hemisphere. Rigorous attention to time determination would benefit meteorological, hydrographic, and astronomical observations, making Mauritius 'a National Center of Reference' for global research.
AdraftS 4pp RS:HS 19.74

3727. 1838–7–22 To Margaret Brodie Herschel Hanover
Received three letters from MH. Hopes 'awful visitation' left no permanent effect on MH. JH is not impressed by German philosophers. Journey to Hameln to meet JH's cousins Mrs. Groskopff and Mrs. Knipping. Describes country. Visit to Pyrmont and Lackern. Visited aunt Caroline Herschel and Mrs. and Miss Beckedorff in Hanover. Convinced Madame Witte to build third moon model that JH will give to F. L. Chantrey. News of son William Herschel. Composes two hymns. Departure from Hamburg delayed; may miss H. W. M. Olbers in Bremen.
ALS 4pp TxU:H/L-0538.5; Reel 1053

3728. 1838–7–23 From Fr[iedrich ?] Herschel Baden
Appears to be an attempt to establish a family connection to JH.
ALS 3pp†† (German almost illegible) TxU:H/M–0808; Reel 1085

3729. 1838–7–23 From Heinrich Wilhelm Matthias Olbers Bremen
Is making arrangements for JH's visit to WO, during a continental tour by JH.
ALS 2pp (German) RS:HS 13.180

3730. 1838–7–24 From William Benjamin Carpenter 22 Park St., Bristol
Has a book on physiology in preparation and would like to dedicate it to JH in recognition of the help he has received from his works.
ALS 3pp RS:HS 5.183

3731. 1838–7–24 From Heinrich Christian Schumacher Altona
In compliance with JH's request, HS did not go to Bremen to meet JH. Hopes JH will visit him in Altona and stay at his house.
ALS 4pp RS:HS 15.380

3732. 1838–7–25 From Mrs. Witte Hanover
Please send J. H. Mädler's sketch of fortresses that F. P. Gruithuisen claims to have discovered on moon. Formed another 'moon ball of 12 inches.' Will send piece to JH after he returns to England.
AL 2pp RS:HS 18.430

3733. 1838–7–31 To Alexander von Humboldt Altona
Regrets he was unable to meet him when in Hanover. Took no part in Guillaume Weber's operations at the Cape, but has become acquainted with them through Weber's volumes. Has pressed the Admiralty on the necessity of Magnetic Observatories at various places. Regarding AH's observations on the barometer.
C 3pp RS:HS 10.65 & 21.255

3734. 1838–8–1 From Alphonse Candolle Geneva
Thanks for his long letter on the climate and vegetation of the Cape. Congratulations on the success of his mission.
ALS 2pp (French) RS:HS 5.321 *Bibliothèque universelle des sciences, belles-lettres, et arts, 14* (1838), 118–29

3735. [1838–8 or later] From Roderick Impey Murchison n.p.
Urges JH to attend next Geological Society meeting, meet [Carlo Andrea, Conte de] Pozzo di Borgo, and see RM's new ordnance maps.
ALS 3pp RS:HS 19.292

3736. 1838–8–2 From Thomas Emerson Headlam to Margaret Brodie Herschel Newcastle/Tyne
Will reserve a room for JH at the time of the B.A.A.S. meeting.
ALS 2pp RS:HS 9.261

3737. 1838–8–3 From Heinrich Christian Schumacher Altona
Asks if he can publish the finding that the star in Argo is dimming in *Astronomische Nachrichten*. Questions whether barometric information should be in feet or yards. Asks JH about a publication that HS and [Alexander von] Humboldt are planning.
ALS 4pp RS:HS 15.381

3738. 1838–8–4 From Humphrey Lloyd Trinity College, Dublin
Rejoices that JH is in favor of a South Polar expedition. Three years since a memorandum was presented to the government and thinks the B.A.A.S. should now present a further memo. Hopes JH will give it his support.
ALS 3pp RS:HS 11.264 Morrell and Thackray, *Early Correspondence,* 270–1

3739. 1838–8–4 From Edward Sabine Dublin
Urges JH to present to the B.A.A.S. a proposal regarding research in terrestrial magnetism.
C 3pp RS:HS 15.18

3740. 1838–8–5 To Thomas Emerson Headlam 10 Hanover Terrace
Has just returned from Europe and responds to TH's invitation to stay with them while JH is at the Newcastle meeting of the B.A.A.S.
ALS 2pp WT 67597

3741. 1838–8–6 To Caroline Lucretia Herschel [London]
Reports a joyous reunion with family and friends.
ALS 4pp TxU:H/L–0583.3; Reel 1058 *Memoir ... of Caroline Herschel,* 297–8 inc

3742. 1838–8–6 To Humphrey Lloyd Hanover Terrace
JH reiterates his support of a South Polar expedition.
C 1p RS:HS 21.256

3743. 1838–8–6 To John David Roberton n.p.
Hesitant to sign application to become general advisor to Court of Directors of the East India Company. Describes concerns.
C 4pp RS:HS 14.364 (C: RS:HS 21.257)

3744. 1838–8–8 From George Jeans Tetney near Grimsby
Has been reading JH's volume on astronomy in the Cabinet Cyclopaedia and would like the answers to a few queries that have arisen. Is interested in double stars and would like a suitable telescope in which to observe them successfully. Gives description of his present telescope.
ALS 3pp RS:HS 10.279

3745. [1838]–8–8 From Roderick Impey Murchison n.p.
Charles Babbage believes that RM prevented Babbage from

becoming president of B.A.A.S. Denies this. Hopes JH can keep Babbage 'from overloading the cup of misfortune.'
ALS 4pp RS:HS 12.441 Morrell and Thackray, *Early Correspondence*, 271–2

3746. 1838–8–[10] To Margaret Brodie Herschel Newcastle
Reports on persons and meetings at B.A.A.S. Elected officers for 1839 today. Charles Babbage gave up his office of trustee in disgust. JH dislikes speaking in public.
C 4pp TxU:H/L-0538.6; Reel 1053

3747. 1838–8–15 To Charles Babbage n.p.
Regarding his party. Difficulties and accidents.
ALS 2pp RS:HS 2.291

3748. 1838–8–16 To Margaret Brodie Herschel Liverpool
Arrived at Liverpool on the way to Newcastle; some details about the journey.
ALS 1p JHS.HCEJ.531a

3749. 1838–8–18 To Margaret Brodie Herschel Newcastle
Informs MH that JH arrived safely at Newcastle for the B.A.A.S. meeting.
ALS 1p JHS.HCEJ.531b

3750. 1838–8–19 From William Reid Newcastle
Discusses WR's recent work on nature of storms. [Edward] Sabine suggested WR's paper be put into JH's 'Section [at B.A.A.S.] as Meteorology.' Sends copy of work on storms.
ALS 3pp RS:HS 14.322

3751. 1838–8–19 From Frederick Augustus, Duke of Sussex Kensington Palace
Stating his reasons for his resignation from the presidency of the R.S.L.
ALS 6pp (Lithographed circular) RS:HS 1.422

3752. 1838–8–20 From William Henry Harvey [Cape Town]
Sending a copy of a little work on Cape botany for local circulation. Intends to attempt a larger Flora later. Starts on a three months' excursion next month. Sees that JH has arrived in England.
ALS 3pp RS:HS 9.245

3753. 1838–8–20 From Abraham Follett Osler Birmingham
Sending copy of a resolution passed by the Birmingham Philosophical Society and requesting JH to let his name be included as an honorary member.
ALS 1p RS:HS 13.201

3754. 1838–8–20 or later From John Phillips Newcastle on Tyne
Informs JH that he has been appointed member of a B.A.A.S. committee to translate and publish in England foreign scientific memoirs.
C 1p RS:HS B26.1

3755. 1838–8–22 From Albert Hopkins Williamstown, Mass.
Sending an address delivered at the opening of the Observatory at Williams College. Comments on the building and gives ideas for forthcoming work. Would be happy to communicate news to the new Meteorological Association formed in London.
ALS 4pp RS:HS 10.207

3756. 1838–8–23 To Margaret Brodie Herschel Newcastle
Details about the B.A.A.S. meetings, as well as comments about JH's hosts, and comments about the evening meetings.
ALS 4pp JHS.HCEJ.531c

3757. 1838–8–27 To Spencer J. A. Compton [Newcastle]
Encloses JH's letter to William Lamb, Lord Melbourne. Asks SC, as B.A.A.S. president, to sign 1838 resolutions then forward these and JH's letter to Lord Melbourne.
ACC 1p TxU:H/L-0257.2; Reel 1054

3758. 1838–8–27 To Margaret Brodie Herschel Newcastle
A paean of praise about the B.A.A.S. meetings, especially some of the unscheduled activity, such as Adam Sedgwick giving a lecture on the seashore to 3000 or more 'colliers and rabble'; reproduces a lengthy poem given by W. R. Hamilton.
ALS 4pp JHS.HCEJ.531d

3759. 1838–8–27 To William Lamb Newcastle
Encloses 1838 B.A.A.S. resolutions. B.A.A.S. committee [for proposed Antarctic expedition] will be in London after 16 Sept. and wants to visit WL [Earl of Melbourne] to explain [purpose of expedition].
ALS 1p APS B.H435p.9 (ACCS: TxU:H/L-0257.1; Reel 1054)

3760. 1838–8–30 From Francis Baily 37 Tavistock Place
Is on the point of leaving town for a few days. Is preparing reductions of the observations of N. L. Lacaille.
ALS 2pp RS:HS 3.142

3761. [1838–9] From Karl Kreil n.p.
Enquiring about some magnetical observations. Sends a memoir of his own work on magnetism.
ALS 1p (French) RS:HS 11.69

3762. 1838–9–2 To Francis Baily 10 Hanover Terrace
Regarding the catalogue of N. L. Lacaille. Has sent the resolution regarding the Polar expedition to Hugh Percy (3rd Duke of Northumberland). Must postpone introducing the business concerning the Cape Town Observatory.
ALS 3pp RS:HS 3.143 (C: RS:HS 25.8.12)

3763. 1838–9–2 From Thomas Emerson Headlam Newcastle/Tyne
Sending results of the experiment on the lens of a large animal. Hopes he was not fatigued by the meeting.
ALS 2pp RS:HS 9.262

3764. 1838–9–2 To Caroline Lucretia Herschel 10 Hanover Terrace, Regent's Park
Copies the inscription on a vase awarded to JH to commemorate his work at Cape Town.
ALS 1p TxU:H/L–0583.6.b; Reel 1058

3765. 1838–9–3 From William Lamb Windsor Castle
Will meet with B.A.A.S. committee after WL returns to London.
ALS 1p TxU:H/M–0354; Reel 1087

3766. 1838–9–3 From Charles Babbage to Hugh Percy (3rd Duke of Northumberland) Newcastle
Proposes JH for several positions in B.A.A.S. [Annotations by JH explain reasons for declining.]
C 3pp †(by JH) TxU:H/M–0969; Reel 1083

3767. 1838–9–7 From James Pulman Herald's College
Sends designs of the coat of arms to be assigned to the

Herschel family.
ALS 1p RS:HS 14.193

3768. 183[8]–9–10[?] From William Whewell Lancaster
Can leave Lancaster immediately for interview with William Lamb, Lord Melbourne regarding the magnetic survey. Congratulates JH for escaping presidency of the B.A.A.S.
ALS 3pp RS:HS 18.189

3769. 1838–9–12 From John Fairbairn Cape Town
Outlining the political situation at the Cape.
ALS 4pp RS:HS 7.152

3770. 1838–9–13 To William Whewell Slough
William Lamb, Lord Melbourne, will receive 'our committee.' Has heard that the R.S.L. Council expects JH to become the next R.S.L. president. Requests WW to put an 'absolute negative' to any such claim.
ALS 2pp TC, Camb. Add. Ms.a.207[31] (C: RS:HS 21.258)

3771. 1838–9–[16?] To Francis Baily London
Forwards some material from William Lamb [Lord Melbourne]; JH wants to stop any rumors that he is available for the presidency of the R.S.L.
C.2pp RS:HS 25.8.13

3772. [1838–9–17] To Francis Baily n.p.
Has had a letter from William Lamb (2nd Viscount Melbourne) regarding the Polar expedition. Is not interested in becoming president of the R.S.L. in succession to Frederick Augustus (Duke of Sussex). Has he the case of H. C. Schumacher's snuff box?
ALS 3pp RS:HS 3.144

3773. 1838–9–17 To William Whewell 10 Hanover Terrace
Alerts WW about the upcoming meeting with William Lamb, Lord Melbourne. Asks WW to squelch any reports that JH will be the next R.S.L. president.
ALS 4pp TC, Camb. Add. Ms.a.207[32] (C: RS:HS 21.259)

3774. 1838–9–20 From Palm Heinrich Ludwig Boguslawski Breslau
Has been looking for Encke's Comet. Compares his own and JH's readings concerning Halley's Comet. Law of comets in relation to Biela's Comet. Will shortly be visiting Wilhelm Olbers.
C 3pp RS:HS 4.170

3775. 1838–9–21 From Thomas Maclear Royal Observatory, Cape of Good Hope
Sends report on [N. L.] Lacaille's Observatory Stations. Tells of expeditions to Klipfontein. Lack of observing due to absence at Cape. Mural circle must be corrected or replaced. Needs an equatorial telescope. Agrees theodolite telescope would be useful.
ALS 12pp RS:HS 12.113

3776. 1838–9–23 To William Radcliffe Birt Slough
Checked June observations of Mr. Robertson in *Athenaeum*. Found WB's statement correct [see WB's 1831–9–11]. Wrote to Meteorological Society for explanation. Thanks for pointing out difference in March observations. JH is honorary member of [Meteorological] Society.
ALS 3pp RS:HS 19.75

3777. 1838–9–23 To Edward Joshua Cooper Slough
Thanks for meteorological records sent by EC, and asks for some additional information.
C 2pp RS:HS 21.260

3778. 1838–9–26 To Richard Jones n.p.
Invites RJ to join Andrew Smith and [Damein?] for dinner at JH's tonight.
ALS 1p TxU:H/L-0207; Reel 1054

3779. 1838–9–27 From Nathaniel Bowditch Boston (Mass.)
Sending him an engraving, and the *Rules of the Bowditch Library*. Would like copies of JH's works for this library.
ALS 2pp RS:HS 4.201

3780. 1838–9–27 From William Henry Fitton Bloomfield Lodge
Wondered what JH intended doing about the Presidency of the R.S.L. until he heard from Francis Baily that JH would not accept the offer to become a candidate. Has to review Charles Lyell's book on geology, which has been dedicated to himself. Where does he intend to settle down? Has he seen H. D. de Blainville's paper in the *Comptes rendus*?
ALS 8pp RS:HS 7.235

3781. 1838–9–30 To Margaret Brodie Herschel Slough
On the state of decorating at the house, the flowers blooming in the garden, and domestic financial arrangements.
ALS 3pp JHS.HCEJ.532

3782. 1838–9–30 To John Pringle Nichol Slough
Thanks for sending JN's *Architecture of the Heavens* and *Phenomena and Order of the Solar System.* Critiques the illustrations of nebulae given in these books.
ALS 4pp NatLibScot MS.3650, f.78

3783. 1838–10–1 From Charles Richardson Tulse Hill
Working on a new English dictionary. Writing about sound. Asks permission to use JH's work on sound.
ALS 1p RS:HS 14.333

3784. 1838–10–3 To Michael Faraday Slough
Asks for copies of some of MF's writings to complete JH's set.
C 1p RS:HS 21.261

3785. 1838–10–3 From Matthew Henry Webster Albany [New York]
Received JH's two letters last autumn. Delayed sending observations until JH returned to England. Compared actinometers with those of Mr. Danberry of England and J. S. McCord of Montreal. Encloses letter of introduction to Harmanus Bleecker [American chargé d'affaires to the Hague] .
ALS 2pp †, encl ALS 1p RS:HS 18.129

3786. 1838–10–4 From William Hawkins Cape Town
Pleased to hear he had a good voyage home and met with a great reception on his arrival. Sends some casks of sweet wine. Has paid the installment due on the shipping shares.
ALS 2pp RS:HS 9.258

3787. 1838–10–8 From John Lindley 21 Regent St.
Is grateful for the flowers. The Satyriums are most interesting. Comments on the variations in this species.
AL 3pp RS:HS 11.220

3788. 1838–10–9 From Friedrich Wilhelm Bessel Königsberg
Announces the measurement of the annual parallax of 61 Cygni.
ALS 6pp (German) RS:HS 4.71

3789. 1838–10–9 From Michael Faraday Royal Institution

Esteems it a great honor to receive JH's request. Has searched through all his papers but has a spare copy only of the Third series. Is thinking of reprinting the whole series in one volume.
ALS 2pp RS:HS 7.180 Williams, *Selected Correspondence, 1*, 318–19

3790. 1838–10–9 To the London *Times* Athenaeum
[Responding to a suggestion for a fund campaign aimed at refurbishing William Herschel's 40-foot reflector], JH suggests that it would not be practical to reestablish this telescope.
L London *Times* (Oct. 10, 1838), p. 5, col. 4

3791. 1838–10–12 From William Henry Smyth Crescent, Bedford
Suggests Moyes Publishing Company to JH. Observes Encke's Comet. Visited by W. R. Hamilton and John Russell.
ALS 2pp RS:HS 16.265

3792. 1838–10–15 From Francis Beaufort Admiralty
Forwarding a letter from C. P. Smyth and a batch of observations from J. A. Lloyd, who intends to visit England next Spring.
ALS 1p RS:HS 3.357 & ACC: Hydrographic Office Letter Book

3793. [1838]–10–15 To George Peacock Slough
Sends copy of B.A.A.S. resolutions relative to South Polar expedition. Will wait for GP to set date for deputation [to meet government officials].
ALS 1p TxU:H/L-0297; Reel 1054

3794. 1838–10–17 From John Lindley 21 Regent St.
Exhibited JH's gift of the Satyriums at the meeting of the Horticultural Society and he has been awarded the Banksian medal. To whom shall the medal be ascribed? Any further information on the orchidaceae would be very welcome.
ALS 2pp RS:HS 11.221

3795. 1838–10–17 To William Henry Smyth Slough
Chooses a publisher and printer for *Cape Results*. Decides against getting portrait painted. Makes suggestions for observations.
ALS 4pp APS B.Sm98.16(C: RS:HS 21.262)

3796. 1838–10–18 From George Thomas Napier Government House, Capetown
Has just returned from a long sojourn on the frontier and has much information on the political situation. Comments on the case of Andries Stockenström. Has sent a mission to the Boers.
ALS 9pp RS:HS 13.64

3797. 1838–10–18 From E. Willes Stewarts Hotel, 34 Albemarle St.
Found house and 60 acres near Worcester. Needs to know soon if JH is interested. Asks about copper mines on Orange River described by J. E. Alexander [*An Expedition of Discovery into the Interior of Africa...*(1831)]. Believes missionaries may help Negroes there.
ALS 3pp † RS:HS 18.256

3798. 1838–10–20 To Gilbert Elliot, Second Earl of Minto Slough
Conveys to GE nine resolutions passed at the B.A.A.S. meeting supporting a southern hemisphere expedition of discovery, especially concerned with magnetism, and also the establishment of a series of magnetic observatories.
ALS 7pp encl NatLibScot MS.12051, f.49 (C: TxU:H/L-0263; Reel 1054)

3799. [1838]–10–21 From Edward Joshua Cooper n.p.
Will send a corrected set of Meteorological Observations later. Has not seen the comet, but has seen many shooting stars. Family is in Paris. Will leave 'planches' of his discoveries at this house and they will be delivered to Lady Herschel.
ALS 4pp RS:HS 5.280

3800. 1838–10–22 From Francis Beaufort Admiralty
Forwarded JH's letter to Lord Minto [Gilbert Elliot]. John Russell sent Antarctic resolution of B.A.A.S. to Minto at Admiralty six weeks ago. Glad that JH finds Lord Melbourne [William Lamb] favorable to this enterprise.
ALS 2pp TxU:H/M–0061; Reel 1087 (ACC: Hydrographic Office Letter Book)

3801. 1838–10–22 From Samuel B. Howlett 83 1/2 Pall Mall
Has made a barometer for his own use, which requires no correction for zero or temperature. Would like JH to see it.
ALS 2pp RS:HS 10.210

3802. 1838–10–22 To Humphrey Lloyd Slough
Is anxious to hear from HL about his ideas for a major expedition for magnetic observations being considered by the B.A.A.S.
C 2pp RS:HS 21.263

3803. 1838–10–22 To William Whewell Slough
Had an encouraging conversation with William Lamb, Lord Melbourne, about funding for the South Polar expedition. Asks WW and George Peacock to prepare a proposal regarding the expedition, JH doing the same.
ALS 4pp TC, Camb. Add. Ms.a.207[33] (RS:HS 21.264)

3804. 1838–10–23 To Francis Baily Slough
Has received a long letter from F. W. Bessel regarding the discovery of the parallax of a fixed star.
ALS 3pp RS:HS 3.145 (C: RS:HS 25.8.14)

3805. 1838–10–23 From Edward Sabine Torington, Arundel
Received JH's letter. Has been in Scotland working on magnetic observations. Eager to meet with JH at any time. Humphrey Lloyd has set up the three magnetic instruments, a 'great advance on Wilhelm Weber's investigations.'
C 4pp RS:HS 15.19

3806. 1838–10–24 From Francis Baily 37 Tavistock Place
Would he send more details of F. W. Bessel's communication so that it can be read at the next meeting of the R.A.S.
ALS 2pp RS:HS 3.146

3807. 1838–10–24 To Friedrich Wilhelm Bessel Slough
Thanks FB for letter in which FB announces the discovery of annual stellar parallax; JH questions whether effects of temperature have been adequately considered.
ALS 3pp AdWDDR Bessel 225 (C: RS:HS 21.265)

3808. 1838–10–24 From Edward Sabine Torington, Arundel
Plans a meeting in London with JH, James Ross, Humphrey Lloyd, and himself. Says an earlier cost estimate was too low and worries that the government will not fund the 'special' magnetic project.
ALS 3pp RS:HS 15.20

3809. 1838–10–24 From Gilbert Elliot, Second Earl of Minto [Admiralty]
[Marked 'Private.'] Will meet with JH next month. Will assist B.A.A.S. in research on terrestrial magnetism, but is not convinced that global expedition is most effective way; perhaps current survey teams could perform as well. Compares relative costs.
ALS 7pp TxU:H/M–0372; Reel 1087

3810. 1838–10–24 From Gilbert Elliot, Second Earl of Minto Minto
Expresses general interest in supporting research in terrestrial magnetism and possibly am expedition of discovery. Will be happy to discuss these matters further with JH.
C 3pp NatLibScot MS.12051, f.53

3811. 1838–10–25 From Humphrey Lloyd Trinity College, Dublin
Agrees on the general plan for the expedition with the views of Edward Sabine. Gives some of his own comments for magnetic observations.
AL 4pp inc RS:HS 11.265

3812. 1838–10–25 To Adolphe Quetelet Slough
Thanks for duplicates of horary observations. Requests that AQ and his associates cease them after December 1838. Questions 'whether any and what' corrections are applied to observations. Has successfully seen second satellite of Saturn.
ALS 4pp Académie belgique: #12 A. Quetelet, 'Notice sur Sir John Fred. Will. Herschel,' *Académie royale des science, des lettres et des beaux-arts annuaire* (1872), 175–6

3813. 1838–10–26 To Duncan MacFarlane Slough
Sees that JH has been named as a candidate for the Rectorship of Glasgow University. Though honored wishes it to be known that he will not be a candidate.
draft 2pp RS:HS 13.2

3814. [1838]–10–28 To William Whewell n.p.
Invites WW to a meeting with James Ross and Humphrey Lloyd on the South Polar expedition. Asks WW to inform George Peacock. Encourages WW to defend the expedition, which Lord Minto reports is in jeopardy.
ALS 3pp TC, Camb. Add. Ms.a.207[34] (C: RS:HS 21.266)

3815. 1838–10–29 From Sophie Beckedorff to Margaret Brodie Herschel Hanover
Regarding a governess for the Herschels' children. News of the family.
ALS 3pp RS:HS 4.1

3816. 1838–10–29 From Thomas Maclear Royal Observatory, Cape of Good Hope
Mr. Borcherd's son, who is contemplating work in the educational field in South Africa, would like to call on JH to obtain his advice. Question of the land about the Observatory has been settled. Has had a copy of G. B. Amici's paper on a prismatic reflecting circle; comments on this.
ALS 3pp RS:HS 12.114

3817. 1838–10–29 From Thomas Maclear Royal Observatory, Cape of Good Hope
Property of Royal Observatory will be extended.
C 1p Cape Archives/Maclear Papers File 100

3818. 1838–10–30 From Palm Heinrich Ludwig Boguslawski Hanover
Is visiting JH's aunt Caroline. Was at [Wilhelm] Olbers's 80th birthday in Bremen. Talks about some of PB's other travels.
ALS 2pp (German) RS:HS 4.171

3819. 1838–10–30 From Sir Robert Harry Inglis 7 Bedford Square
Finds Francis Beaufort has forwarded his note to Slough. Reminds him that honorary members of the Literary Society are not resident in London.
ALS 1p RS:HS 10.228

3820. 1838–10–30 From Sir Robert Harry Inglis 7 Bedford Square
As JH has now returned from abroad can he submit his name as a candidate for the Literary Society.
ALS 2pp RS:HS 10.227

3821. 1838–10–30 From Duncan MacFarlane Glasgow
The nomination of JH for the position of Rector at Glasgow University did not originate with DM or his friends. Will bring JH's views to the knowledge of his colleagues, though the students are responsible for the election of the rector.
ALS 2pp RS:HS 13.3

3822. 1838–10–30 From Edward Sabine Torington, Arundel
Writes to change the London meeting to Thursday to accommodate James Ross.
ALS 3pp RS:HS 15.21

3823. 1838–11 To Thomas Maclear [Slough]
Lord Normanby stalling education system program. Government recognizes Land Observatories. Writes letter to Admiralty requesting TM's assistant remain at Royal Observatory.
ALS 3pp Cape Archives/Maclear Papers File 100

3824. 1838–11–1 From Francis Beaufort Admiralty
Has prepared a room for him and Edward Sabine to discuss Australian affairs on Tuesday next.
ALS 2pp RS:HS 3.358 & ACC: Hydrographic Office Letter Book

3825. 1838–11–1 From Edward C. Herrick New Haven, Conn.
Is interested in meteoric showers and shooting stars. Encloses a letter [Copy preserved at Yale University] to be sent to someone at the Cape for observations from the Southern Hemisphere.
ALS 1p RS:HS 10.185

3826. 1838–[11–2] From Charles Robert Darwin 36 Gt Marlborough St.
Has promised Alexander Macleay that he would send details of the nets JH used to catch pelagic animals.
ALS 3pp RS:HS 6.10

3827. 1838–11–2 From Sir John Franklin Van Diemen's Land
Was pleased to hear of his great reception on his return to England. Endeavoring to start a small scientific society and would be glad of any communications from JH. Sends a series of tidal and meteorological observations made at Port Arthur.
ALS 4pp RS:HS 7.359

3828. 1838–11–2 To James Calder Stewart Slough
Urgent request for JS to visit and help JH and Margaret during 'these desperate nights.'
ALS 1p TxU:H/M–0777.2; Reel 1084

3829. 1838–11–2 From Margaret Brodie Herschel to James Calder Stewart Slough
Urgent request for information about good hospitals in London. Neighbor with typhus fever needs immediate care.
ALS 2pp TxU:H/M–0777.1; Reel 1084

3830. 1838–11–3 To George Biddell Airy Slough
Asks a number of questions about a projected South Polar magnetic expedition.
ALS 3pp RGO 6.675.211

3831. 1838–11–4 From Friedrich Wilhelm Bessel Königsberg
Responds to concerns about effect of temperature on measurement of stellar parallax [see JH's 1838–10–24]; comments on other observations.
ALS 2pp (German) RS:HS 4.72

3832. 1838–11–4 From John Pringle Nichol Glasgow
Has just returned from a German tour and has seen JH's letter to Dr. D. MacFarlan. Outlines the procedures that led to JH being proposed as the new Rector of the University. Comments on the illustrations of two of his own works. Resolution passed urging the free examination of foreign astronomical instruments in public institutions.
ALS 4pp RS:HS 13.131

3833. 1838–11–5 To Gilbert Elliot, Second Earl of Minto Slough
Is prepared to see GE anytime about resolutions [see JH's 1838–10–20]. Outlines the requirements of the Cape Observatory for more efficient service. Eloquently pleads the case for government support of research in the physical sciences.
ALS 6pp NatLibScot MS.12051, f.55 (draft: RS:HS 7.36b & C: 21.268)

3834. 1838–11–5 To Humphrey Lloyd Slough
Further comments in preparation for discussion of major expedition [see JH's 1838–10–22] with representatives of the government.
C 2pp RS:HS 21.267

3835. 1838–11–6 From Adolphe Quetelet Brussels
Has received JH's letter of 29 October. No need to thank for meteorological observations. Done in interest of science. Sends measurements of December 1837 and other bulletins. Asks JH to send his extra copy of AQ's *Correspondence mathématique* of July 1838 to [J. D.] Robertson. Asks more about satellites of Saturn. Thanks for information on meteors.
ALS 4pp (French) RS:HS 14.212

3836. 1838–11–8 From Gilbert Elliot, Second Earl of Minto Minto
Has not returned to town yet, but will be happy to discuss the proposals in JH's last letter. Attaches more importance to JH's words than any resolutions of the B.A.A.S.
ALS 4pp RS:HS 7.37 (C: NatLibScot MS.12051, f.59)

3837. 1838–11–8 From William Francis Cowper Downing St.
Prime minister Lord Melbourne [William Lamb] will receive JH's deputation and Chancellor of Exchequer [Thomas Spring-Rice] whenever convenient.
ALS 2pp † TxU:H/M–0153; Reel 1087

3838. 1838–11–8 From John Phillips B.A.A.S., York
B.A.A.S. requests that JH head a committee to work on the reduction of observations of N. L. Lacaille's stars. G. B. Airy and Thomas Henderson will also be assigned to the committee.
(A)LS 1p TxU:H/M–0428.1; Reel 1093

3839. 1838–11–11 To James Calder Stewart Slough
John Stewart's dilemma regarding marriage to James Grahame's daughter Matilda, and JH's letter to Grahame. Margaret Herschel is recovering from mumps.
ALS 4pp & Adraft 2pp TxU:H/L–0424 & -0404; Reel 1055

3840. 1838–11–12 From George Biddell Airy Royal Observatory, Greenwich
No one on committee has taken sufficient initiative to obtain funding for magnetic instruments. Hopes S. H. Christie and JH will act in this regard. Discusses doubts about South Polar magnetic expedition.
ALS 4pp RS MM.16.147 (ACCS: RGO 6.675.213)

3841. 1838–11–12 From Charles Richardson Tulse Hill
Grateful for permission to use JH's work on sound for the dictionary [see RS:HS 14.333]. Suggests JH write a synopsis.
ALS 2pp RS:HS 14.334

3842. 1838–11–12 From Edward Sabine London
Humphrey Lloyd will attend the meeting [announced in ES 1838–10–24].
ALS 1p RS:HS 15.22

3843. 1838–11–12 From James South Observatory, Kensington
Informs JH that in a letter to *The Times* signed by Richard Sheepshanks JH's name appears. JS wants to know if JH is aware of this fact.
ALS 1p RS:HS B27.16a

3844. 1838–11–13 From Humphrey Lloyd Morley's Hotel
Has been considering with Edward Sabine that morning the requirements for the fixed observatories. Gives lists and also estimate of finances required for a three-year period.
ALS 4pp RS:HS 11.266

3845. 1838–11–13 To James South Slough
JH has seen the letter in *The Times* [see JS's 1838–11–12]. As JS only asked if JH's name was there, JH responds to the question of fact and does not wish further communication on the matter.
ACCS 1p RS:HS B27.16b

3846. 1838–11–14 From George Biddell Airy Royal Observatory, Greenwich
The Standards Commission will need to meet soon; JH is asked to indicate most convenient, and most inconvenient, days. [See GA's 1838–6–9].
ALS 1p TxU:H/M–0626.4;.Reel 1084 (ACCS: RGO 6.340.273)

3847. 1838–11–14 From Edward Sabine Kew
Speaks of translations being made of papers from *Resultate* [*aus den Beobachtungen der Magnetische Vereins*], some of which papers indirectly pertain to ES's simultaneous observations. Lists expenses.
ALS 4pp RS:HS 15.23

3848. 1838–11–15 To George Biddell Airy Slough
Details about proposals for magnetic expeditions, as they are to be presented to British government [see GA's 1838–11–12]. Also included is a copy of the resolutions passed at a B.A.A.S. meeting on 25 Aug. 1828.
ALS 5pp RGO 6.675.217 (C: TxU:H/L–0010)

3849. 1838–11–15 To Adolphe Quetelet Slough
Thanks AQ for meteorological observations for September 1838 and other materials sent. Singles 12 and 13 November as extraordinarily clear. Observed a 'paucity of shooting stars' and an Aurora Borealis.
ALS 4pp Académie belgique: #11

3850. 1838–11–15 To James Calder Stewart [Slough]
Encloses JH's [letter to James Grahame] for JS to edit. Use discretion as to whether John [Stewart] should see this. Margaret is continually nauseous.
ALS 1p TxU:H/L-0425; Reel 1055

3851. 1838–11–16 To George Biddell Airy Slough
Comments on the work of the Standards Commission, hoping that it will move quickly to a conclusion [see GA's 1838–11–14].
ALS 3pp RGO 6.340.274

3852. 1838–11–16 To Humphrey Lloyd Slough
Would like a statement from Edward Sabine and HL outlining the probable expense and materials required for the proposed expedition so that he can hand it to the government at the interview. Has written to G. B. Airy, who is not in favor of the proposed expedition, so will write again. Thinks it necessary for men of science to agree before presenting petitions to the government.
draft 4pp RS:HS 11.267 & 21.270

3853. 1838–11–17 From George Grey Downing St.
As JH has taken so much interest in education at the Cape, he would like some converse with him before drafting the final instructions to Sir G. Napier.
ALS 3pp RS:HS 9.132

3854. 1838–11–17 From Henry Holland Brook St.
Inquires about unusual weather at Cape of Good Hope in late 1836. Was there an epidemic of influenza at Cape early in 1837?
ALS 3pp † RS:HS 9.428

3855. 1838–11–17 From John Lindley 21 Regent St.
Sends the silver Banksian medal for JH's exhibit of 16 Oct.
ALS 1p RS:HS 11.222

3856. 1838–11–17 From Humphrey Lloyd Morley's Hotel
Illness forces him to make a quick return home. Agrees with JH that scientific men should act unanimously when approaching government. Thinks the R.S.L. should go along with the B.A.A.S. as well. Hoped to consult G. B. Airy. Will see Edward Sabine to draw up the memorial.
ALS 3pp RS:HS 11.268

3857. 1838–11–18 To George Biddell Airy Slough
Replies to GA's request for meeting dates [see GA's 1838–11–14], and suggests ways to shorten the work of the commission [see JH's 1838–11–16].
ALS 3pp RGO 6.340.276

3858. 1838–11–19 From James Grahame Nantes
Of family troubles, and Matilda's impending marriage.
ALS 2pp RS:HS 8.434

3859. 1838–11–19 From William Rowan Hamilton Royal Irish Academy
Miss Caroline Herschel has just been elected an honorary member of the Academy. How shall they forward the diploma?
ALS 1p RS:HS 10.152

3860. 1838–11–19 From E. Willes 34 Albemarle St.
Sending missionary to Cape of Good Hope in ten days to investigate mines and Christianize inhabitants of Namaqualand. Does JH or Margaret Herschel have letters or messages to be taken? Lady [Byron?] wants to meet JH.
ALS 3pp † RS:HS 18.257

3861. 1838–11–20 From George Biddell Airy Royal Observatory, Greenwich
Trying to organize a meeting time for the Standards Commission that will suit everybody; GA wants to make very few changes in the standard relationships [see JH's 1838–11–18].
ALS 2pp TxU:H/M-0626;.5 Reel 1084 (ACCS: RGO 6.340.278)

3862. 1838–11–21 To George Biddell Airy Slough
Agrees to GA's suggested dates [see GA's 1838–11–20]; comments on several approaches to defining a standard of weight.
ALS 3pp RGO 6.340.279

3863. 1838–11–21 From Francis Beaufort Gloucester Place
Thanks for the congratulations. News of his daughter's marriage. Mrs. Beaufort would like to renew her acquaintance at some future date.
ALS 3pp RS:HS 3.359

3864. 1838–11–22 To George Thomas Napier n.p.
Regrets the delay in his letters but has been occupied with various matters ever since he returned from the Cape. Visited his Aunt Caroline in Germany. Visited Charles Grant (Baron Glenelg) and gave him his own views on education. Later met Sir George Grey, who outlined the new plan for education at the Cape.
draft 6pp RS:HS 13.64bis & 21.269

3865. 1838–11–23 From George Biddell Airy Royal Observatory, Greenwich
A notice of meeting of the Standards Commission.
LS 1p TxU:H/M-0626.6; Reel 1084

3866. 1838–11–23 From Alexander Bridport Becher Admiralty
Sending papers from Francis Beaufort on hurricanes.
ALS 2pp RS:HS 3.360

3867. 1838–11–23 From Edward Sabine Woolwich
Reports Francis Beaufort's response to providing needed equipment and funds [for the proposed magnetic survey]. James Ross has not yet arrived. Discusses publishing an article in the *Resultate* [*aus den Beobachtungen der Magnetische Vereins*].
ALS 3pp RS:HS 15.24

3868. 1838–11–23 From James South Observatory, Kensington
Someone has written to *The Times* as JH's 'champion' and has impugned the conduct of JS towards the R.S.L. To clear his name, JS wants to make public JH's letter to JS of 1826–11–14, which JS quotes.
ALS 3pp RS:HS B27.17a

3869. 1838–11–24 To James South Slough
As JH is not aware that JS has been attacked because of the incident referred to in JH's 1826–11–14, JH is unwilling to give permission for the publication of that letter. It seems to JH that JS is being attacked for unsubstantiated attacks he made on the President and Council of the R.S.L.
ACCS 2pp RS:HS B27.17b

3870. 1838–11–26 To Caroline Lucretia Herschel Slough
Impressed with the new President of the R.S.L., Spencer J. A. Compton. Has learned that Carl Gauss was awarded the Copley Medal for his research in terrestrial magnetism.
ALS 4pp TxU:H/L–0583.7; Reel 1058 *Memoir ... of Caroline Herschel,* 300–1 inc

3871. 1838–11–26 From James South Observatory, Kensington
As JH will not allow the publication of the letter [see JH's 1838–11–24], JS will provide a plain narrative of the facts to justify his conduct.
ALS 2pp RS:HS B27.18

3872. 1838–11–28 From Theo E. Cantor Alfred St., Bedford Square
Sending book from H. C. Oersted; also the best wishes from H. C. Schumacher.
ALS 1p RS:HS 5.322

3873. 1838–11–28 From William Henry Harvey [Cape Town]
Sending some bulbs of Daubenya in which he believes JH was interested.
ALS 2pp RS:HS 9.246

3874. 1838–12[?] To Roderick Impey Murchison n.p.
Writes in great haste, thanking RM on behalf of Lady Herschel for taking 'trouble,' especially for 'Literary Tributes' apparently sent for Lady H's album.
ALS 2pp EUL:[?]

3875. 1838–12–3 From Sir John Bell Cape
Since JH's departure the weather has been bad. A stone is to be erected to commemorate JH's work at the Cape. Political news of the Cape. Is anxious to know views of the educational scheme.
ALS 4pp RS:HS 4.56

3876. 1838–12–3 To Carl Friedrich Gauss Slough
Informs CG of the awarding to CG of the Copley medal for CG's magnetic researches; hopes that the British government will support such magnetic researches around globe.
ALS 3pp NSUB Cod.Ms.Gauss 99 (C: RS:HS 21.271)

3877. 1838–12–3 From Edward Sabine Athenaeum
Describes his meeting with Francis Beaufort and James Ross. Strongly urges that the Southern magnetic research begin in the spring. Speaks of efforts to have the expedition approved by the R.S.L., which is favored by the Board.
ALS 4pp RS:HS 15.25

3878. 1838–12–3 From William Whewell Trinity College
Entreats JH, who had recently rebuffed efforts to head the R.S.L. and the B.A.A.S., to become president of the Geological Society. Specifies responsibilities involvéd.
ALS 4pp RS:HS 18.190 Todhunter, *Whewell, 2,* 273–4

3879. 1838–12–5 To Edward Sabine Slough
Writes about arrangements for an R.S.L. council meeting.
ALS 3pp RS:HS B26.67

3880. 1838–12–5 From Wilhelm Struve Dorpat
Has spent much time in Petersburg dealing with Pulkowa affairs. Mentions his works that he has recently sent to JH. Thanks for comments on his *Mensurae micrometricae*. Looks forward to publication of JH's work at the Cape. Discusses suitability of the Russian winter for astronomical work. Requests JH's opinion of the instruments made by [J. G.?] Repsold and designated for Pulkowa. Comments on the heliometer. Invites JH to opening of Pulkowa Observatory.
ALS 6pp (German) RS:HS 17.127

3881. 1838–12–5 To William Whewell Slough
Despite respecting the distinguished company in the Geological Society, JH must refuse its presidency.
ALS 4pp Harvard: Houghton fMS Am 1301.1 (17)

3882. 1838–12–7 From C. Munby Downing St.
Writing on behalf of Lord Glenelg, invites JH to visit Lord Glenelg at the Colonial Office to present JH's views on education at the Cape of Good Hope.
ALS 2pp UCT JH Letters A1.6

3883. 1838–12–7 From George Thomas Napier Government House, Capetown
Relating recent political events at the Cape. The Apprentices have been freed. Port Natal occupied. Kaffirs at peace. Regrets that JH would not stand as president of the R.S.L. in succession to Frederick Augustus (Duke of Sussex).
ALS 12pp RS:HS 13.65

3884. 1838–12–10 To Margaret Brodie Herschel White Horse Livery, Gray's Inn Road
Has bought some horses, and needs the coachman to come to bring them home.
ALS 1p JHS.HCEJ.533

3885. 1838–12–10 From Thomas Glanville Taylor Madras Observatory
Madras Observatory established by James Caldecott. TT gives JC some ideas of JH's to aid the state of the observatory. Re-observes the *Brisbane Catalogue*.
ALS 2pp RS:HS 17.336

3886. 1838–12–17 From Caroline Lucretia Herschel Hanover
Grateful to be made an honorary member of the Royal Irish Academy. Wants 'one or two' copies of her index to John Flamsteed's catalog of omitted stars.
L *Memoir ... of Caroline Herschel*, 301 inc

3887. 1838–12–17 From Edward Sabine Torington
Writes of finding barometric data from some 1836 meteorological observations. Is sending two diagrams. Offers to help JH write a report for the R.S.L.
ALS 2pp RS:HS 15.26

3888. 1838–12–18 From Marianne Compton Castle Ashby
Her father has burnt his hand so she is writing to inform him that there will be a joint committee meeting on Saturday. When is JH coming to visit them?
ALS 2pp RS:HS 5.359

3889. 1838–12–18 From William Cook City of London School
Declines offer to teach at the Cape of Good Hope. Offers to recommend another candidate. Thanks JH for an earlier testimonial on WC's behalf.
ALS 3pp UCT JH Letters A1.7

3890. 1838–12–19 To John Timbs Slough
Apologizes for delay in responding to JT's wish to publish JH's portrait as the frontispiece of JT's *Arcana of Science*; adds that the only portrait of JH of which JH is aware is that at St. John's College, Cambridge.
ALS 2pp Yale Univ.

3891. [1838]–12–19 To William Whewell n.p.
Asks WW, and through him George Peacock, to judge

whether B.A.A.S. funds should provide some instruments for the Breslau Magnetic Observatory.
ALS 4pp TC, Camb. Add. Ms.a.207[35] (C: RS:HS 21.272)

3892. 1838–12–21 From Charles Richardson Tulse Hill
Long discussion of the sounds and etymologies of various words. [Probably in reference to the dictionary CR was preparing; see RS:HS 14.333].
ALS 3pp RS:HS 14.335

3893. 1838–12–21 From Andries Stockenström 3 Wakefield St.
Sends JH a letter from Lord Glenelg [concerning his position as Lieutenant Governor of the Eastern Province of the Cape Colony].
ALS 1p RS:HS 17.14

3894. 1838–12–21 To Charles Wheatstone [London]
Arranges a meeting with CW and Edward Sabine to prepare a report on magnetic proposals.
C 1p RS:HS 25.11.2

3895. 1838–12–22 From Francis Beaufort Admiralty
Is sending him the plans and report of Klip Fonteyn [?]. He will forward them later to the Astronomer Royal.
ALS 2pp RS:HS 3.361 & ACC: Hydrographic Office Letter Book

3896. 1838–12–23 To Spencer J. A. Compton Slough
Urges SC to arrange meeting of R.S.L. deputation with Lord Melbourne as quickly as possible to discuss funding for magnetic observatories.
ACCS 4pp TxU:H/L-0283; Reel 1054 (C: RS:HS 21.273)

3897. 1838–12–23 From Roderick Impey Murchison Stoke, near Chichester
The *Silurian System* is now at the binders, and he is now relaxing and doing Christmas visiting. Proposes to write to the editor of the *Annals of Philosophy* giving details of the vase to be presented to JH to mark his Cape sojourn. Can also send the literary effusions to Lady Herschel if she desires them.
ALS 4pp RS:HS 12.396

3898. [1838]–12–24 From Francis Beaufort Admiralty
Concerned that JH has not received 'Cape papers' sent with FB's letter. Estimates not yet called for. FB will see that Cape [necessities?] are not forgotten. When will JH present R.S.L. recommendation of 'the Expedition'? That would be good opportunity to 'fire another shot' at Lord Minto [Gilbert Elliot] or Mr. Wood.
ALS 3pp † TxU:H/M–0059; Reel 1087

3899. 1838–12–26 From Edward Sabine Torington
Encloses maps from the Magnetic Department and a paper published in 1836.
ALS 1p RS:HS 15.27

3900. 1838–12–27 To Roderick Impey Murchison Slough
Glad to see RM's geological labors on *The Silurian System* completed.
AL 1p EUL:523/4/27

3901. 1838–12–28 From Adolphe Quetelet Brussels
Discusses observation of meteors. Sends observations of last solstice and details of that night's constellations. Offers results of observations of occultation of Gamma[?] Tauri[?].
ALS 4pp (French) RS:HS 14.213

3902. 1838–12–31 From Spencer J. A. Compton Castle Ashby
Putting pressure on Lord Melbourne [William Lamb] regarding Antarctic expedition. Asked Melbourne to communicate with J. W. Lubbock, who is in London, instead of with JH.
ALS 2pp TxU:H/M–0400; Reel 1087

3903. 1838–12–31 To Caroline Lucretia Herschel Slough
Sent to CH her diploma from the Royal Irish Academy on account of her election as an honorary member.
ALS 2pp TxU:H/L–0583.9.b; Reel 1058

3904. [1839 or 1840] To Admiralty n.p.
Regarding Thomas Maclear's Cape Meridian.
draft 2pp RS:HS 1.14

3905. [1839?] To Francis Baily n.p.
Observations of [Kensa?] received. The George Everest-Thomas Jervis affair.
ALS 6pp RS:HS 3.283

3906. 1839 From Madame Biot College de France
Offering her services while JH is in Paris.
ALS 2pp (French) RS:HS 5.66

3907. [1839?] To [John William Lubbock?] n.p.
Advice on obtaining magnetical and meteorological apparatus for Van Diemen's Land observatory. Need to clear up finances.
ALS 4pp RS LUB.H.353

3908. [1839] To Thomas Maclear n.p.
Has heard from G. B. Airy that T. F. Colby has written to Ireland for the compensation bars. An additional assistant has been designated but no appointment made yet. Received a letter from their good friend but does not feel justified in bringing it to the notice of a minister. Admiralty has borrowed Fuller's theodolite (from R.A.S.) and the mural circle should soon arrive. Gives Michael Faraday's analysis of the meteorolite.
draft 2pp RS:HS 12.178

3909. 1839 To My dear Sir n.p.
Will send two photometers and three-year supply of paper to [?]. One goes to J. C. Ross and one to F. Eardley Wilmot at Cape of Good Hope. Instructions for using photometers. Hopes [?] Robinson forwarded actinometers to Woolwich as JH directed. These should be checked at St. Helena. Instruction for using actinometers. Sent report to Humphrey Lloyd at B.A.A.S. Returns [?]'s Daguerre newspaper containing 'impudent notice about M. Pambour and his great Wheels.'
Adraft 3pp TxU:H/L-0642; Reel 1089

3910. [1839?] From Harriet Ouseley Hall Barn, Beaconsfield
Is grateful for the gift to her son and for JH's time devoted to her son for his edification. Sir Gore hopes to call on the Herschels on the first fine day.
ALS 3pp RS:HS 13.206

3911. [1839] To Edward Sabine n.p.
Discusses Harvard University offer to make magnetic measurements. Requests ES to read Admiralty extracts. Asks questions regarding JH's upcoming review [in *Quarterly Review*] of several works on terrestrial magnetism.
ALS 3pp draft RS:HS 15.333

3912. [1839?]. From Edward Sabine n.p.
Cabinet found R.S.L. recommendations unsatisfactory. Merchants are urging the value of good geographical and magnetic charts of southern ocean. Geographical Society is

preparing a document. Wolverly Attwood, M.P., questioning ministers in Parliament on this matter.
ALS 3pp RS:HS 15.322

3913. [1839] To [Edward Sabine] n.p.
Submits payment to J. D. Roberton for 'Association' bills. Glad that ES approves of JH's draft article on terrestrial magnetism for *Quarterly Review*. Fears R.S.L. was premature in proposing experimental institutions; public was not prepared for such large expenditures for science. Prefers ES's suggestion of temporary magnetical and meteorological observatory. Strong disagreement from [C. J.] Riddell in America.
Adraft 3pp TxU:H/L-0632; Reel 1089

3914. [1839?] From J. Smith n.p.
Changes the date of an invitation to JH.
ALS 1p UCT JH Letters A1.8a

3915. [1839?] To Charles Wheatstone [Slough]
Too ill to attend a demonstration by CW of one of CW's instruments [?].
C 1p RS:HS 25.11.12

3916. [1839] To [?] n.p.
Fourth observatory, at Van Diemen's Land [Tasmania], will be conducted by J. H. Kay, to be landed from one of vessels [*Terror* and *Erebus*] bound for Antarctica. This vessel will also carry observers and instruments for stations at St. Helena and Cape.
Amem 1p inc RAS:JH Archive 5/4.6 verso; Reel 8

3917. 1839 To [?] n.p.
JH's experiments to find compounds suitable for photography. Accidental discovery of effect of 'nitrate of Silver.' Note of 22 Jan. from Francis Beaufort alerted JH to L. J. M. Daguerre's secret processes and W. H. F. Talbot's experiments.
Adraft 4 fragments RAS:JH Archive 4/2.3, 2.9, 2.10, and 2.13 (all verso); Reel 7

3918. 1839–1–5 From Charles Babbage Paddington Station
Will bring CB's 'Engine' to Slough tomorrow, together with Dionysius Lardner's apparatus, which CB has 'taught to write.'
ALS 2pp RS:HS 2.292 Hyman, *Charles Babbage,* 161

3919. 1839–1–7 From Caroline Lucretia Herschel Hanover
Recounts a letter from William Rowan Hamilton, President of the Royal Irish Academy, commending CH on her service to astronomy.
L *Memoir ... of Caroline Herschel*, 302 inc (?)

3920. 1839–1–12 From Francis Baily 37 Tavistock Place
At the recent council meeting of the R.A.S., JH was nominated as president. Is returning the letters of F. W. Bessel.
ALS 3pp RS:HS 3.147

3921. 1839–1–12 From Martha Langton Farnham Royal
Please accept her heartfelt thanks for his kindness.
ALS 1p RS:HS 11.462

3922. 1839–1–14 To Francis Baily Slough
Is honored by being nominated as president but would prefer someone else as he is very busy and can ill spare the time.
ALS 4pp RS:HS 3.148 (C: RS:HS 25.8.15)

3923. 1839–1–14 From James Grahame Nantes
Encourages JH to visit at Nantes.
ALS 2pp RS:HS 8.435

3924. 1839–1–16 From H. Moreton Dyer Devonshire Place House
Sent wrong paper to JH in Nov. 1838. Now sends complete meteorological journal from Mr. Lees, HD's stepson. [JH annotation: Answered 18 Jan. 1839.]
ALS 1p † TxU:H/M–0172.1; Reel 1087

3925. 1839–1–19 From Heinrich Christian Schumacher Altona
Forwards a request from C. F. Gauss for magnetic observations made in England.
ALS 2pp RS:HS 15.382

3926. 1839–1–19 To Heinrich Christian Schumacher Slough
Compares magnitudes for some specific stars and then comments more generally on that question; JH believes he will no longer be able to observe in winter in England as he is too severely afflicted by rheumatism.
ACCS 4pp RS:HS 25.5.11

3927. 1839–1–20 From George Biddell Airy Royal Observatory, Greenwich
Any news of Field Measuring bars—alteration in size of Observatory at Cape Town? Observations on use of magnetic needle. Has JH found a suitable country house yet?
ALS 2pp RS:HS 1.75 (C: RGO 6.368.618)

3928. 1839–1–20 From Francis Baily 37 Tavistock Place
Regrets that he cannot change the mind of the council of the R.A.S. over its nomination of JH as president.
ALS 1p RS:HS 3.149

3929. 1839–1–21 From E. Willes 12 A, Great Cumberland Place
EW and wife plan trip to Eton soon to see sons. Hope to visit Herschels in Slough then. Despairs of Africa venture. Received details from JH's agent at the Cape of Good Hope.
ALS 4pp RS:HS 18.259

3930. 1839–1–22 To James Calder Stewart Slough
Willing to write to [secretary of state for War and the Colonies, Charles Grant] Lord Glenelg on behalf of JS's application. JH searching for new residence.
ALS 4pp TxU:H/L-0426.1; Reel 1055

3931. 1839–1–23 From Francis Beaufort Admiralty
Has mentioned to Dr. T. R. Robinson their hopes of supplying the Cape Observatory with an equatorial telescope and theodolite. Encloses Robinson's comments. George Dollond would take two years to mount such a telescope. Curious paper in the *Comptes rendus*.
ALS 2pp RS:HS 3.362

3932. 1839–1–24 To [John] Lindley Slough
Sends JL various roots from the Cape, including *Satyriums* and *Disa.* Describes each type in detail.
ALS 3pp BostonPubLib Mss.Acc. 67

3933. 1839–1–24 From E. Willes 12 A, Great Cumberland Place
Recalls hospitality at Slough during EW's Eton days. Sets date to visit JH.
ALS 3pp RS:HS 18.258

3934. 1839–1–25 From William Henry Fox Talbot 44 Queen Anne St.

Will give a paper on fixing the image formed by the camera obscura to R.S.L. and wants to review it with JH first.
ALS 1p RS:HS 17.277 Buckland, *Talbot*, 39

3935. 1839–1–27 To James Calder Stewart n.p.
Sends letter of recommendation for JS and returns draft of JS's letter to [Charles Grant] Lord Glenelg, with recommended changes. Promise from Glenelg's father to JS's father [A. Stewart] is grounds for JS's application.
ALS 3pp TxU:H/L-0427; Reel 1055

3936. 1839–1–27 To William Henry Fox Talbot Slough
[Responding to WT's 1839–1–25], JH reports that he cannot come to London to see WT's 'curious process of fixing the image formed by a Camera obscura.' Invites WT to come to Slough.
ALS 3pp Science Museum Talbt 1/10 Buckland, *Talbot*, 39–40 inc

3937. 1839–1–28 To [Hugh Percy (3rd Duke of Northumberland)] Slough
Grieved that J. C. Stewart desires schoolmaster position at Cape of Good Hope. Attests to Stewart's many business talents, and asks HP to find 'more prominent station' for Stewart.
C 3pp TxU:H/L-0287; Reel 1054

3938. 1839–1–28 From William Henry Fox Talbot London
JH is indisposed, so WT will come to Slough to review paper [see WT's 1839–1–25]. Asks JH about appropriate curve on lenses for camera obscura for 2-foot focus.
ALS 2pp RS:HS 17.278 Buckland, *Talbot*, 40

3939. 1839–1–28 To William Whewell [Slough]
Invites WW to Slough before JH travels to Midhurst [to consider property for sale].
ALS 2pp Harvard: Houghton fMS Am 1301.1 (18)

3940. 1839–1–29 From John Lindley 21 Regent St.
Is greatly obliged for the present of flowers. Has introduced another Herschellia into the world as he has used the name for a *Disa*.
ALS 2pp RS:HS 11.223

3941. 1839–1–29 From John Pringle Nichol Glasgow
Has had orders from the Treasury for full remission of duty on their astronomical instruments. Observatory will be founded early in March. Gives list of instruments added or to be added, in case JH's opinion is asked.
ALS 4pp RS:HS 13.132

3942. 1839–1–29 From William Henry Fox Talbot 44 Queen Anne St.
Has received urgent request from *Athenaeum* to allow them to publish paper on 'Photogenic Drawing' before it is read to R.S.L. in light of announcement of the 'Parisian invention.'
ALS 2pp RS:HS 17.279a Buckland, *Talbot*, 40 inc

3943. 1839–1–30 To William Henry Fox Talbot Slough
[Responding to WT's 1839–1–29], urges WT to find some way of giving at least slightly different papers to R.S.L. and *Athenaeum*. Comments on process of fixing image, referring to a trial of his own.
ALS 3pp Science Museum Talbt 1/11 (draft 'much altered' RS:HS 17.279b & C: RS:HS 22.1) Buckland, *Talbot*, 40–1 inc

3944. [1839]–2–1 From Francis Beaufort Admiralty
Inquires about appropriate person to add to the Cape Observatory staff, if FB can persuade the Admiralty of the need. Thanks JH for his comments on chloride of silver; were much better than W. H. F. Talbot's.
ALS 3pp RS:HS 3.404 (ACC: Hydrographic Office Letter Book)

3945. 1839–2–2 From Charles Grant London
CG (Lord Glenelg) raises questions about the suitability of Mr. Train for a position teaching in the government schools at the Cape of Good Hope.
ALS 2pp UCT JH Letters A1.24

3946. 1839–2–2 From William Henry Fox Talbot 44 Queen Anne St.
More about paper being published in both the *R.S.P.T.* and *Athenaeum*.
ALS 2pp RS:HS 17.280

3947. [1839–2–2] To [?] n.p.
Mentions visit of 1 Feb. 1839 from W. H. Fox Talbot. Reports that JH has now 'accomp[lishe]d the whole problem [of photography].' Describes JH's recent results.
L R. S. Schultze, 'Re-discovery and Description of Original Material on the Photographic Researches of Sir John F. W. Herschel, 1839–1844,' *Journal of Photographic Science, 13* (1965), 62 inc

3948. 1839–2–3 From George Biddell Airy Close, Norwich
Further observations on the alterations at Cape Town Observatory. Any news of the measuring bars?
ALS 2pp RS:HS 1.76

3949. 1839–2–7 To Heinrich Christian Schumacher Slough
Sends HS's observations of magnetic declination and dip near London for forwarding to Carl F. Gauss. Encloses a letter from Edward Sabine.
ALS 2pp NSUB Cod.Ms.Gauss 110

3950. 1839–2–8 From William Rowan Hamilton Observatory
Asks JH to keep him updated on English research of the dynamics of light. Currently considering the 'propagation' of light waves, as distinguished from their mere 'preservation'; asks whether this is a new study.
L Graves, *Hamilton, 2,* 289–90 inc

3951. 1839–2–8 From William Reid 5 Maddox Street
Discusses own efforts in meteorology. Must wait to print second edition of work on storms. Appreciates nomination for R.S.L., but has little time for such pursuits.
ALS 3pp RS:HS 14.323

3952. 1839–2–8 From William Henry Fox Talbot 44 Queen Anne St.
Questions JH's intent to display 'photogenic drawings' to R.S.L.
ALS 2pp RS:HS 17.281a Buckland, *Talbot*, 45 inc

3953. 1839–2–9 To William Henry Fox Talbot Slough
Considerable comment on procedure for submitting paper to R.S.L. Urges WT to mention JH's method of washing with hyposulfite. Comments on some photographic experiments of WT.
draft 'much altered' 4pp RS:HS 17.281b (C: RS:HS 22.2)

3954. 1839–2–10 To William Henry Fox Talbot Slough
JH welcomes WT's mentioning JH's method of fixing with hyposulfite. JH has put some of JH's images on display at R.S.L. Mentions JH's recent experiments relating to photography. Comments regarding a paper submitted to

R.S.L. by WT and on some experiments of WT.
ALS 4pp Science Museum Talbt 1/12 (draft 'much altered': RS:HS 17.281b & C: RS:HS 22.2) Buckland, *Talbot,* 45–6 inc

3955. 1839–2–11 From Roderick Impey Murchison 2 Eccleston St.
Has a few pounds surplus from the subscription for JH's vase and wonders if it could be used to supply a plinth for the pedestal. Does JH agree? Perhaps Lady Herschel can supply the wording.
ALS 2pp RS:HS 12.397

3956. 1839–2–11 From William Henry Fox Talbot 44 Queen Anne St.
Wants to withhold description of 'developing' process until it is perfected to obviate others using it. Also refers to second process to make 'positive' and 'negative.'
ALS 3pp RS:HS 17.282

3957. 1839–2–12 From Heinrich Christian Schumacher Altona
Asks permission to use information from JH's letter in *Astronomische Nachrichten.* Believes [Alexander von] Humboldt can explain observational anomalies that JH notes.
ALS 2pp RS:HS 15.383

3958. 1839–2–12 To William Henry Fox Talbot Slough
JH will as WT requests stop mentioning JH's hyposulfite fixing process and await publication of WT's process. Reports in detail on some of JH's experiments relating to photography and to spectrum analysis. [In a P.S. dated 1839–2–13,] JH states: 'I have discovered your secret of fixing [or its] equivalent.'
ALS 4pp Science Museum Talbt 1/13 (ACCS: RS:HS 25.5.12 & C: TxU:H/L–0474; Reel 1055)

3959. 1839–2–13 To William Rowan Hamilton Slough
In reply to WH's 1839–2–8 inquiry, JH believes that 'Skotodynamics' (the propagation of light waves) is a new line of research. Caroline Herschel is ecstatic about her Royal Irish Academy medal.
L Graves, *Hamilton, 2,* 290–1

3960. 1839–2–14 From Michael Faraday Royal Institution
Saturday's experiments have brought a severe inflammation of the eyes so has been unable to use the aerolite. Returns it, but will try some experiments later if JH sends it back.
ALS 1p RS:HS 7.181 Williams, *Selected Correspondence, 1*, 335

3961. 1839–2–14 To William Swainson Slough
Thanks WS for his pamphlet on the 'Statistical Zoology of Britain.' Comments on the dynamics of animal and plant ecology.
ALS 3pp Linnean Society

3962. 1839–2–14 From William Swainson n.p.
Sends JH pamphlet on intimate connection between a country's zoology and its political statistics. Regrets that JH wished not to be nominated for presidency of R.S.L. on returning from Cape.
ALS 2pp RS:HS 17.250

3963. 1839–2–14 From William Henry Fox Talbot 44 Queen Anne St.
Some comments on JH's light and photography experiments, especially effects of different kinds of glass and fixing with ferrocyanate.
ALS 4pp RS:HS 17.283

3964. 1839–2–15 To Francis Baily Slough
Approves of the arrangements made by FB regarding the loan of the theodolite.
ALS 2pp RS:HS 3.151 (C: RS:HS 25.8.16)

3965. 1839–2–15 From Francis Baily n.p.
Unable to see him as he has another meeting to attend. Regarding the loan of the R.A.S.'s theodolite to Thomas Maclear. Two foreign members have been proposed for the R.A.S.
ALS 3pp RS:HS 3.150

3966. 1839–2–15 From Sir John Franklin Van Diemen's Land
Introducing John Montague, the Colonial Secretary of Van Diemen's Land.
ALS 2pp RS:HS 7.360

3967. 1839–2–16 From John Lindley 21 Regent St.
Will be pleased to examine the dried plants. Send them to the Horticultural Society's Garden, Turnham Green.
ALS 1p RS:HS 11.224

3968. 1839–2–19 From William Henry Fox Talbot 44 Queen Anne St.
Describes mediocre results obtained using ferrocyanate of potash for fixing image. Mentions some other methods, including his favorite, common salt. Asks JH to keep these secret.
ALS 2pp RS:HS 17.284 Buckland, *Talbot,* 47 inc

3969. 1839–2–20 From William Henry Fox Talbot 44 Queen Anne St.
Finally had success washing with ferrocyanate. Wants to present note of 'his' processes to R.S.L. and so to the world.
ALS 2pp RS:HS 17.285 Buckland, *Talbot,* 47–8 inc

3970. [1839]–2–22 To William Whewell n.p.
Requests WW and George Peacock to judge the appropriateness of expending B.A.A.S. funds for some instruments for the Breslau Magnetic Observatory. Mentions other matters relating to the meteorology committee.
ALS 3pp TC, Camb. Add. Ms.a.207[36] (C: RS:HS 22.3)

3971. 1839–2–23 From George Phipps (Lord Normanby) n.p.
Has been requested by Lord Glenelg [Charles Grant] to propose candidates for teaching positions in the Government Free Schools at the Cape of Good Hope. Proposes a candidate. Asks for clarification concerning the salaries that will be offered.
draft 2pp UCT JH Letters A1.8 Ferguson and Immelman, *Herschel and Education*, 59–60

3972. [1839?]–2–24 From John Pringle Nichol Westminster Hotel
Along with the Lord Advocate and other Scotch members of the House of Commons had an interview with [Thomas Spring-?] Rice concerning the enclosed memorial (Observatory of the University). Comments on their proposed plan for the work of the university.
AL 5pp inc RS:HS 13.133

3973. 1839–2–25 From George Biddell Airy Royal Observatory, Greenwich
Thomas Maclear's assistant and equipment arranged. Observations on copper horse-shoe bars applied to the end of his magnet.
ALS 2pp RS:HS 1.77 (C: RGO 6.368.619)

3974. 1839–2–25 From Francis Baily 37 Tavistock Place
Will compare the standards later. Has sent the theodolite to the Admiralty and received a letter of thanks. Thomas Maclear's paper on N. L. Lacaille's star positions will be read at the next meeting of the R.A.S.
ALS 3pp RS:HS 3.152

3975. 1839–2–25 From Andries Stockenström 3 Wakefield St.
Awaits decision on his fate as Lieutenant Governor at the Cape.
ALS 2pp RS:HS 17.15

3976. 1839–2–25 From Charles Wheatstone 20 Conduit St.
Invites JH to Kew Green to see Francis Bauer's plates etched by J. N. Niepce's photographic process. Perhaps L. J. M. Daguerre assumes more credit than is due. Heard Niepce visited London in 1826, but his paper was rejected by R.S.L. at W. H. Wollaston's urging.
ALS 2pp RS:HS 18.147

3977. 1839–2–25 From Lt. Col. [Charles] Yorke Colonial Office
Writing on behalf of George Phipps, Lord Normanby, thanks JH for his letter of 1839–2–23. Promises that Normanby will give it consideration.
ALS 2pp UCT JH Letters A1.9

3978. 1839–2–26 To Francis Baily Slough
Will come up to town so that they can work on Thomas Maclear's four foot scale standard. News of the Glasgow Observatory.
ALS 4pp RS:HS 3.153 (C: RS:HS 25.8.17)

3979. 1839–2–27 To George Biddell Airy Slough
Thanks GA for all his efforts on Thomas Maclear's behalf [see GA's 1839–2–25]; comments on some developments in photography, including the work of Nicephore Niepce done in approximately 1826.
ALS 3pp RGO 6.368.620

3980. 1839–2–27 From William Henry Fox Talbot 44 Queen Anne St.
Wishes to communicate to J. B. Biot, and so to Académie des sciences, JH's two 'beautiful' fixing methods, by hyposulfite and ferrocyanate.
ALS 1p RS:HS 17.286

3981. 1839–2–28 To William Henry Fox Talbot n.p.
Invites WT to mention JH's hyposulfite process to J. B. Biot. Recounts JH's recent experiments.
L Xpp Buckland, *Talbot,* 49 inc

3982. [1839?]–3–1 From Francis Beaufort Admiralty
Is grateful for the information on photography. Encloses president's card.
ALS 3pp RS:HS 3.405

3983. 1839–3–1 To Edward C. Herrick Slough
Has sent on one of EH's two letters to Cape Town. Reports on meteor showers, especially meteors seen on 14 Nov. 1837. P.S. adds that JH has received a number of papers by EH.
ALS 4pp Yale Univ.

3984. 1839–3–1 To Caroline Lucretia Herschel Slough
Prepared to send to CH copies of the Index to John Flamsteed.
ALS 3pp TxU:H/L–0584.1; Reel 1058

3985. 1839–3–1 From William Henry Fox Talbot 44 Queen Anne St.
Enlarges upon his method of using salt as photographic fixer; then asks JH to explain the chemistry thereof. Has tried Louis Daguerre's fixing process without success.
ALS 2pp RS:HS 17.287

3986. [1839–3] To William Whewell n.p.
Asks WW to approve cost of report forms to be sent to the Breslau Observatory. Has heard that WW's book is nearing completion and contains 'heresy.' Has just submitted a paper on 'photographical matters' to the R.S.L.
ALS 3pp TC, Camb. Add. Ms.a.207[37] (C: RS:HS 22.41)

3987. 1839–3–2 From Charles Babbage Paddington Station
Making his last experiments on the morrow. Travels to and from Slough in the course of his experiments. Hopes JH will join him.
ALS 2pp RS:HS 2.293

3988. 1839–3–2 From Francis Baily London
Regarding the four foot scale standard of Thomas Maclear. Will be pleased to see him on Thursday. R.A.S. has had a letter from the Treasury regarding a grant for Glasgow Observatory.
ALS 2pp RS:HS 3.154

3989. 1839–3–2 From Samuel B. Howlett 83 1/2 Pall Mall
Thanks for his letter. His barometer is not intended as a mountain one. Though a slight error does appear now and then it is no worse than other barometers, excepting that at the R.S.L. Would like JH to communicate the drawing and description to the R.S.L.
ALS 4pp RS:HS 10.211

3990. 1839–3–2 From Andries Stockenström 3 Wakefield St.
Outraged over a [London] *Times* report of a massacre at Fort Natal in the Cape colony, which AS believes cannot be true; backs up his belief with a letter from a soldier at Fort Natal. Will take a business trip to Holland after which he 'will be ready for anything the Government may please.'
ALS 4pp RS:HS 17.16

3991. 1839–3–3 To William Henry Fox Talbot Slough
Sent WT's specimens to J. B. Biot and F. J. D. Arago in Paris. Developed paper sensitive to heat rays of sun 'beyond the spectrum.' Diagram of three spectra: 'luminous, chemical, and thermic.'
C 2pp TxU:H/L-0475; Reel 1055

3992. [1839]–3–5 From Francis Beaufort n.p.
Is pleased to hear that JH has taken steps to avoid the carelessness at the Post Office. Sends original of R. W. Fox's paper.
ALS 1p RS:HS 3.406

3993. [1839–3–5] From Richard Jones Tithe Commission
The belladonna JH gave him is throwing up vigorous leaves. Has a scheme for making it leaf and flower in the same season. Has some ingenious machines to show JH when he calls, which may be useful to JH.
ALS 4pp RS:HS 10.361

3994. 1839–3–6 From Charles Grant London
Would like to assist James Stewart as JH asked, but CG resigned recently as Colonial Secretary and cannot help. [JH annotation: 'Called on Lord G. [Glenelg, i.e., Charles Grant]

with Js St. [James Stewart] May 1/39.']
ALS 4pp † RS:HS 19.273

3995. 1839–3–6 From John Lindley 21 Regent St.
Has opened the box of dried specimens, but they are unsuitable for his purpose as they are not flat. Suggests that King's College or University College may be a more suitable home for them.
ALS 2pp RS:HS 11.225

3996. 1839–3–7 From William Cranch Bond Boston
Read JH's letter to [Felix] Booth regarding magnetic observations in Europe. Describes WB's building and instruments for magnetic observations. Wants information about German observations.
ALS 2pp TxU:H/M–0112; Reel 1087

3997. 1839–3–9 From Johann Franz Encke Berlin
JH has been elected to the Prussian Academy of Science; comments on JE's astronomical observations.
ALS 3pp (German) RS:HS 7.91

3998. 1839–3–9 From James Hudson 7 Foxley Road, Kensington Common
Thanks JH for the interest JH has taken in him, but declines the offer to teach in the Government Free Schools at the Cape of Good Hope.
ALS 3pp UCT JH Letters A1.10

3999. 1839–3–11 From Spencer J. A. Compton Piccadilly
Lord Melbourne [William Lamb] will meet next Wednesday at 4 o'clock with deputation [from R.S.L. regarding Antarctic expedition]. Duke of Sussex is willing to meet with R.S.L. council next Saturday at SC's house.
ALS 1p † TxU:H/M–0401; Reel 1087

4000. 1839–3–12 From Spencer J. A. Compton Piccadilly
Lord Melbourne [William Lamb] changed meeting [with R.S.L. deputation] to 4:30.
ALS 1p TxU:H/M–0402; Reel 1087

4001. 1839–3–12 From Thomas Galloway Serjeants' Inn
Sending the resolution of the Council relating to the Glasgow Observatory. Would be pleased if JH would sign it and forward it to the Treasury.
ALS 1p RS:HS 8.5

4002. 1839–3–14 From Thomas Maclear to Margaret Brodie Herschel Royal Observatory, Cape of Good Hope
Will send observations to JH. Granite column will be constructed at Feldhausen as a memorial [to JH]; will forward plans.
ALS 3pp RS:HS 12.116

4003. 1839–3–14 From Carlo Matteucci Ravenna
Is working on a memoir concerning animal electricity. Would like to present it to the R.S.L. and wonders if it would qualify for a prize.
ALS 2pp (French) RS:HS 12.304

4004. 1839–3–[15 or earlier] To Humphrey Lloyd [Slough]
Passes on the suggestion, from G. B. Airy, of the use of copper rings for damping the movements of measuring instruments in measuring terrestrial magnetism.
C 2pp RS:HS 25.14.20

4005. 1839–3–15 From Edward Sabine Torington
Asks JH to read enclosed item and send it to [J. F.] Daniell, who in turn will send it to John Phillips, so that they might be ready to reply to the Council.
ALS 3pp RS:HS 15.29

4006. 1839–3–15 From Edward Sabine Torington
Strongly urges JH to support employment of artillery men in the St. Helena and Canada fixed observatories, citing the approval of the private secretary. Reports that four sets of magnetic instruments have been ordered.
ALS 4pp RS:HS 15.28

4007. 1839–3–15 From William Henry Fox Talbot 44 Queen Anne St.
Further chemicals used for fixing, including bromide potash.
ALS 1p RS:HS 17.288

4008. 1839–3–16 From Humphrey Lloyd Trinity College, Dublin
Was pleased to hear of the success of JH's exertions. Comments on the instruments required. Thanks for his communication on the copper ring; will try it with the bifilar. W. H. F. Talbot talks of applying the photographic method to meteorological and magnetical instruments. Encloses a paper.
ALS 3pp RS:HS 11.269

4009. 1839–3–18 To William Henry Fox Talbot Slough
Thanks for specimen of light sensitive paper. Provides hint on the chemical process that Louis Daguerre may have used in his photographic work. Sends specimens of JH's own work.
ALS 3pp NMPFT 1937–4840

4010. 1839–3–20 From William Buckland Oxford
Three positions are vacant at Oxford due to the death of S. P. Rigaud. Would JH be interested in applying for them as it would be a great honor to have him at Oxford.
ALS 3pp RS:HS 4.324

4011. [1839–3–20] To William Whewell [Slough]
Informs WW about various matters relating to James Ross's expedition. Is considering purchasing a house in Kent.
ALS 3pp TC, Camb. Add. Ms.a.207[38] (C: RS:HS 22.20)

4012. 1839–3–21 From Edward Sabine Torington
Humphrey Lloyd has asked for magnetometers, assuming that the 'Authority' approves of the instruments. Remarks that Lloyd's observatory will be ready for visitors in late June and that Lloyd plans to go to the continent in July. States nothing more can be done regarding the selection of employees until the Authority acts. Believes an application will have to be made to the R.S.L.
ALS 4pp RS:HS 15.31

4013. 1839–3–21 From Edward Sabine Athenaeum
Speaks of his meeting with the acting general of artillery and the master general secretary, who approved of employing men and engineers in the fixed observation work in St. Helena and Canada. Encloses the remainder of the translation of C. F. Gauss's first volume.
ALS 4pp RS:HS 15.30

4014. 1839–3–21 From William Henry Fox Talbot London
Comments on Louis Daguerre's use of muriatic ether, and encloses a photograph of hand writing.
ALS 2pp RS:HS 17.289 Buckland, *Talbot,* 50 inc

4015. 1839–3–22 From George Phipps (Lord Normanby) Downing St.
Transmits a letter from Mr. McMaster offering to teach at the

Cape of Good Hope.
L 1p † UCT JH Letters A1.11

4016. 1839–3–22 From John Wrottesley Somerset House
Applying for astronomical observer and Savilian professorship of astronomy at Oxford. Asks JH to convey this to R.A.S. Council and testify on JW's behalf.
ALS 2pp † RS:HS 18.299

4017. 1839–3–23 To James Calder Stewart n.p.
[Richard] Jones's view is correct. Will wait to hear what Mr. [Meade?] has to say, before JH writes to Mr. Hodges.
ALS 3pp TxU:H/L-0428; Reel 1055

4018. 1839–3–24 To Johann Franz Encke Slough
Asks JE to express JH's gratitude to the Royal Academy of Berlin for electing JH a foreign member. Discusses JE's observations of divisions in the ring of Saturn, JH's sightings of Saturn's satellites, the British Antarctic expedition, and comets.
ALS 4pp damaged AdWDDR Encke 49

4019. 1839–3–24 To John Wrottesley Slough
Commends JW's application for Savilian professor of astronomy at Oxford [see JW's 1839–3–22]. But it is improper for JH to propose JW unless asked to do so by R.A.S. council.
ACCS 3pp TxU:H/L-0509; Reel 1055 (C: RS:HS 25.15.8)

4020. 1839–3–25 From Manuel John Johnson 42 Chapel St.
Is applying for the position of Radcliffe Observer in the place of S. J. Rigaud and would like to use JH's opinion of MJ's star catalogue he used at the Cape.
ALS 1p RS:HS 10.332

4021. 1839–3–26 From William Edward Parry Admiralty
Enclosing letter just received from Mr. Stanley regarding [James Duff?] Watt's son. WP's wife and children are at Tunbridge Wells.
ALS 2pp RS:HS 13.221

4022. 1839–3–26 From John Phillips 23 Norfolk St.
Discusses need for barometric observations; asks if JH has plans to build a barometer.
ALS 4pp † RS:HS 14.158

4023. 1839–3–27 From Roderick Impey Murchison 2 Eccleston St.
The Bishop of London favors the idea of JH becoming Savilian Professor of Astronomy. What are JH's views on the subject? William Buckland would rejoice.
ALS 3pp RS:HS 12.398

4024. 1839–3–27 From Edward Sabine Torington
Encloses a paper. Says C. F. Gauss will make six magnetometers in two months.
ALS 1p RS:HS 15.32

4025. 1839–3–27 To William Henry Fox Talbot Slough
Describes method of varying the size of photographic images. Praises Talbot's light sensitive paper and suggests improvements regarding it. Complains about persons too ready to seek a patent.
ALS 3pp NMPFT 1937–4842

4026. 1839–3–27 From John Wrottesley Blackheath
Sorry to have made request contrary to R.A.S. policy. Asks permission to transmit JH's reply to electors for [Radcliffe] Observatory.
ALS 2pp † RS:HS 18.300

4027. 1839–3–28 From Francis Beaufort Admiralty
Has been requested in confidence to ascertain whether JH would be interested in accepting the Savilian professorship and the Radcliffe Observatory.
ALS 2pp RS:HS 3.363

4028. 1839–3–28 From George Thomas Napier Government House, Capetown
Thanks for the letter delivered by Mrs. Freeman. Has now had authority from Charles Grant (Baron Glenelg) to put JH's educational scheme into practice. Professor J. R. Innes has been put in charge of the educational system. News of the Andries Stockenström investigation will please JH. Comments on the political scene at the Cape. Hopes to assist [Alexander?] Hutchinson.
ALS 20pp RS:HS 13.66

4029. 1839–3–28 From James Clark Ross 11 St. George Terrace, Hyde Park near Cumberland Gate
Admiralty assigned *Erebus* and *Terror* to Antarctic expedition. Asks JH and Lord Northampton to request approval from Treasury for procurement of scientific instruments. Wants to consult with JH soon.
ALS 2pp † TxU:H/M–0461; Reel 1093

4030. 1839–3–28 From William Henry Fox Talbot London
Comments on a number of chemicals used in photography, as well as referring to JH's process of making 'enlargements.'
ALS 3pp RS:HS 17.290

4031. 1839–3–28 From [J. E.] Benjamin Valz Marseilles
[Letter appears to be a response to some criticisms by JH of BV's paper on Encke's Comet submitted to the R.A.S.] Deals with the density of the comet's nebulosity and with the effect of the comet's approach to perihelion, gives estimates of the comet's magnitude, and comments on the behavior of gases.
ALS 4pp (French) RS:HS 17.459

4032. 1839–3–29 From Roderick Impey Murchison 2 Eccleston St.
JH's letter settles the case of the Savilian chair. Gives news of suitable houses for the Herschels that are likely to come up for sale.
ALS 3pp RS:HS 12.399

4033. 1839–3–29 To Roderick Impey Murchison Slough
Declines RM's request that JH become a candidate for Oxford's Savilian Professorship [see RM's 1839–3–27]. JH plans to move farther from London.
ALS 3pp Geological Society

4034. 1839–3–29 To James Clark Ross Slough
On instruments for fixed stations. Fears that naval expedition will be detained.
ALS 3pp RS Sa.646

4035. 1839–3–29 From William Henry Fox Talbot London
Consists mostly of a transcription of two short notes from J. B. Biot to JH by way of WT.
ALS 2pp RS:HS 17.291

4036. 1839–3–29 From Acton Tindal Aylesbury
Informs JH that JH has been recommended to the Queen for the Commission of Peace.
ALS 1p RS:HS 17.438

4037. 1839–3–30 To James David Forbes Slough
Discusses prospects for meteorological committee of

B.A.A.S., on which they both will serve. Lacks time for it. Does not need hourly observations made by Committee of Physico-Mathematical Society of Edinburgh.
ALS 4pp St. Andrews 24 (C: RS:HS 22.5)

4038. 1839–3–30 To Hugh Percy (3rd Duke of Northumberland) Slough
About arrangements for the publication and distribution of JH's Cape observations.
draft 3pp RS:HS 13.354 & 22.4

4039. 1839–3–30 From James Clark Ross n.p.
Wants to order three complete sets of magnetic and meteorological instruments for observations at St. Helena, at Cape of Good Hope, and aboard expedition ships. Admiralty cannot furnish instruments. Asks JH to continue efforts to obtain authorization from Treasury.
ALS 4pp † TxU:H/M–0462; Reel 1093

4040. 1839–3–31 From Carl Friedrich Gauss Göttingen
Thanks for his letter informing him of his having been awarded the Copley medal. Was pleased to receive the observations of the magnetic declination and inclination of London. Projected expedition of Captain J. C. Ross pleases him. C. F. Gauss is now preparing the third volume of the *Resultate*. Would like JH's opinion on one of the papers in this volume. Gives his own views on the magnetic poles.
ALS 2pp RS:HS 8.76

4041. 1839–3–31 To William Whewell n.p.
Thanks WW for some verses. Informs him that James Ross's expedition has been approved. Discusses various aspects of the expedition, e.g., the placing of instruments on Van Diemen's Land [Tasmania].
ALS 4pp TC, Camb. Add. Ms.a.207[39] (C: RS:HS 22.6)

4042. 1839–4–1 From William Rutter Dawes Manchester
Has been unable to carry out many observations so has not written. Hears Kew Observatory is vacant with the death of S. P. Rigaud, and would like JH's support for his application for this appointment. Has been approached about taking charge of the projected Manchester Observatory. Has written to G. B. Airy for support.
ALS 3pp RS:HS 6.58

4043. [1839–4] From Richard Sheepshanks n.p.
Suggests that JH publish JH's portrait along with his book [*Cape Results*]. Wants JH to approach the Duke of Northumberland with this idea, and if he has no objections, RS will put the touched up plate at the Duke's disposal. Does not like the idea of selling prints of it.
ALS 4pp RS:HS 16.9

4044. 1839–4–2 From John Robison Edinburgh
Sends extracts of solar spots observations by [E.] Capocci. JR will be in London on his way to Paris. Asks if JH would like to accompany him to an exposition.
ALS 2pp RS:HS 14.385

4045. 1839–4–3 To Margaret Brodie Herschel The Bell, Maidstone
Has been getting information about the area with a view to buying a house there; met J. W. Lubbock and was very shocked by Lubbock's unhealthy appearance.
ALS 3pp JHS 3.16

4046 1839–4–3 From Thomas Maclear Royal Observatory, Cape of Good Hope
Presents suggestions for improving and extending Land Station observatories. Will study parallax of Alpha[1] and Alpha[2] Centauri. Government will survey the colony. Quarantine due to measles outbreak. Boers hostile.
ALS 5pp RS:HS 12.117

4047. 1839–4–4 To James Clark Ross Slough
Concerns about equipment for fixed observatories, for which aid will come from B.A.A.S., and for [naval] expedition. Expresses vital importance of meteorological observatory on Van Diemen's Land.
ALS 4pp RS Sa.647

4048. 1839–4–4 From Edward Sabine Admiralty
Asks JH if he would like the Meteorological Committee to meet before R.S.L. Council meets. Having received news that the wife of [Thomas] Spring Rice, the person in charge of the fixed observations, is 'dangerously ill,' ES proposes that responsibility of the project be moved to one of the executive departments.
ALS 3pp RS:HS 15.33

4049. 1839–4–4 From ? 18 Wood St., Cheapside
Asks JH if he has thought of compiling William Herschel's works.
AL 1p inc Harvard: Houghton

4050. 1839–4–5 To George Biddell Airy Slough
About the question of an astronomer for a vacant position at Kew Observatory.
ALS 3pp RGO 6.245.63

4051. 1839–4–6 To William Henry Harvey [Slough]
Thanks WH for the bulbs, seeds, and a book, which he sent from South Africa. JH describes how well the tubers that JH brought back from South Africa are doing. JH complains about the weather in England. Sends WH a specimen of a plant 'photograph' made with W. H. F. Talbot's newly devised system.
ALS 4pp enc WT 69294

4052. 1839–4–6 To John Phillips Slough
Sends plans to construct barometer that he never built; invites JP to use the plans.
C 3pp RS:HS 22.7

4053. 1839–4–6 From Edward Sabine 3 St. James's Place
Informs JH that a meteorological committee has been called. Discusses financial matters concerning the fixed observatories. Has written to [Thomas] Spring Rice.
ALS 3pp RS:HS 15.34

4054. 1839–4–9 To Francis Baily Slough
Finds that he is unable to attend the Council Meeting so would FB take the chair. Business to be brought before council. Ernesto Capocci has seen a spot on the sun.
ALS 4pp RS:HS 3.155 (C: RS:HS 25.8.18)

4055. 1839–4–9 From William Rutter Dawes Ormskirk
JH's letter has gone a long way to destroying his hopes for success at Kew. A great pity that the Observatory could not be put to good use. Must soon obtain employment more congenial to his tastes and this would seem an ideal position.
ALS 2pp RS:HS 6.59

4056. 1839–4–9 From Hugh Percy (3rd Duke of Northumberland) Northumberland House
Agrees to arrangements suggested by JH [see JH's 1839–3–30] regarding publication of JH's *Cape Results*.
ALS 4pp RS:HS 13.355

4057. 1839–4–10 From Edward Sabine 3 St. James's Place
Reports that the precept has been sent for putting the fixed

magnetic observatories at St. Helena, the Cape, and Canada into execution, and that the instruments for three Indian observatories have been ordered. Writes that Humphrey Lloyd has made plans for the portable buildings and fixed stations. Discusses expenses of the observatories.
ALS 3pp RS:HS 15.35

4058. 1839–4–10 From William Henry Fox Talbot Lacock Abbey
Sends a sample picture fixed by use of iodine [potash].
ALS 1p RS:HS 17.292 Buckland, *Talbot,* 52 inc

4059. 1839–4–11 From Edward Sabine Somerset House
Reports his successful meeting with Hussey Vivian, whose full support of the fixed observatories is to be communicated to the R.S.L. Council by ES. Says that Vivian desires to read the memorials of the B.A.A.S. and the R.S.L. Will look into the changes in design of the magnetometers. Suggests budgets for the observatories.
ALS 8pp RS:HS 15.36

4060. 1839–4–11 From William Henry Sykes Royal Society
Offers JH professorship of astronomy at Oxford and hopes to see him at meeting of Meteorological Committee.
ALS 3pp RS:HS 17.139

4061. 1839–4–12 To Edward Sabine Slough
Expresses great pleasure that the fixed observatories have received official approval and financial support. Approves of ES's cost estimates. Encloses copy of the B.A.A.S. memorial. Agrees to tell Humphrey Lloyd to order instruments as requested by ES.
ALS 1p RS:HS 15.37 (C: RS:HS 22.9)

4062. 1839–4–12 To William Henry Sykes [Slough?]
Thanks WS for offer of professorship of astronomy at Oxford, but declines it, partly because of other pursuits and health reasons, and considers his astronomical career terminated.
ACCS 1p RS:HS 17.140 (C: RS:HS 22.8)

4063. 1839–4–13 From Francis Baily London
Is sending a packet and several letters. Events at the last council meeting. Will read Ernesto Capocci's letter at the next meeting of the R.A.S.
ALS 2pp RS:HS 3.156

4064. 1839–4–13 From Comte Jarsana de Briguola 6 Raymond Buildings, Gray's Inn
Encloses statutes of General Shipwreck Relief Society, which JH is interested in joining.
LS 1p RAS:JH Archive 12/1.2.11 verso; Reel 10

4065. 1839–4–13 From Edward Sabine London
Encloses a copy of financial estimates concerning the fixed observatories. Expects bills will be paid by the Treasury. Has conveyed Hussey Vivian's support of the project to the Council of the R.S.L. Clarifies that no vertical magnetometers have been or will be ordered until Humphrey Lloyd is satisfied with their design. Reports that the three officers have been selected for the project. Mentions praise from C. F. Gauss for the project.
ALS 6pp RS:HS 15.38

4066. 1839–4–15 From Francis Baily London
Is sending some drawings of the comet, which had been mislaid. Thomas Maclear's paper is being referred to JH for his corrections. Will discuss which drawings are to be included.
ALS 3pp RS:HS 3.157

4067. 1839–4–15 From R. W. Howard-Vyse Stoke
Is grateful for JH's goodness regarding the pyramids. Gives measurements of the pyramids and would be grateful for any comments JH can make regarding the significance of these measurements.
ALS 2pp RS:HS 10.9

4068. 1839–4–15 From Manuel John Johnson 42 Chapel St.
Was grateful for JH's communication and is pleased his catalogue met JH's approval.
ALS 2pp RS:HS 10.333

4069. [1839?]–4–15 From Edward Sabine Somerset House
Sends documents with earliest suggestion of Antarctic magnetic variation. Also sends letter from [A. T.] Kupffer. Materials received from [Adolphe] Quetelet, Prague, and Milan. All sent to [Humphrey] Lloyd.
ALS 2pp RS:HS 15.327

4070. 1839–4–16 From George Biddell Airy Royal Observatory, Greenwich
Knows nothing about Kew Observatory, but has heard of the name of the present office holder there.
ALS 1p RS:HS 1.78 (C: RGO 6.245.65)

4071. 1839–4–16 From George Biddell Airy Royal Observatory, Greenwich
Encloses graph of C. F. Gauss's magnetic observations.
ALS 3pp diag RS MM.16.148

4072. 1839–4–17 From Thomas Grubb Dublin
Encloses description of two magnetometers approved by Humphrey Lloyd, and proposes to make these for £70.
ALS 2pp TxU:H/M–1194; Reel 1060

4073. 1839–4–17 From Humphrey Lloyd Trinity College, Dublin
Has ordered the instruments for the Van Diemen's Land observatory. Is uncertain regarding the third magnetometer; T. C. Robinson's has not yet been tested. Will order a dip circle from Paris.
ALS 3pp RS:HS 11.270

4074. 1839–4–19 To James Booth Slough
Translation of a work by C. F. Gauss and Wilhelm Weber will provide the information requested. Hopes [William Cranch] B[ond] will participate in magnetic observations. Explains 'Copper Dampers.'
ALS 3pp diag BostonPubLib Ch.I.6.34

4075. 1839–4–19 From Edward Sabine 3 St. James's Place
Suggests changes in the memorial regarding fixed observatories.
ALS 2pp RS:HS 15.40

4076. 1839–4–19 From Edward Sabine n.p.
Reports that the original cost estimate of the magnetometers was low and that other instruments have been ordered. Asks to borrow clocks from the R.S.L. and the R.A.S. Has written to Major Jarvis for observers for the East Indies observatories but expects none will be found.
ALS 6pp RS:HS 15.39

4077. 1839–4–21 To Humphrey Lloyd Slough
Responds positively to HL's 1839–4–17, and adds some additional comments.
C 2pp RS:HS 22.10

4078. 1839–4–22 From Thomas Galloway Royal Astronomical Society
Forwarding Thomas Maclear's paper on N. L. Lacaille's Stations for JH's Report. Would he also mark any unnecessary passages or drawings.
ALS 1p RS:HS 8.6

4079. 1839–4–22 From Edward Sabine 3 St. James's Place
Favors C. F. Gauss's theory of two magnetic poles, provided it satisfies the data. Office of the Artillery at Woolwich plans to 'furnish a set of instruments by private subscription.' Suggests a staff officer be appointed for the correspondence between observatories. Worries that the East Indies observatory will fail.
ALS 3pp RS:HS 15.41

4080. 1839–4–22 To William Henry Fox Talbot Slough
Comments on several approaches to fixing images on paper, especially using pure water as a fixing agent. Has developed a new sensitive paper but is keeping it secret until perfected. Is leaving for Continent on 1 May.
ALS 2pp Science Museum Talbt 1/16 (C: RS:HS 22.11)

4081. [1839]–4–26 From Thomas Best Jervis 22 Notting Hill Square
In R.S.L.'s request to H.E.I.C. court of directors for support for magnetic survey in India, Lord Northampton omitted to stress need for training TJ and other officers in use of special instruments.
ALS 2pp † TxU:H/M–0298.1; Reel 1087

4082. 1839–4–26 From Hans Christian Oersted Copenhagen
This letter will be delivered by Prof. [C. A.?] Jensen of the Royal Danish Academy of Fine Arts, a distinguished painter. Would like him to paint a portrait of JH if agreeable.
ALS 2pp RS:HS 13.163 Harding, *Correspondance de ... Örsted, 2,* 390–1

4083. 1839–4–26 From William Henry Smyth Crescent, Bedford
Wants to know if JH has a larger telescope in operation; if so, will he confirm some observations. Asks JH to recommend an epoch.
ALS 2pp RS:HS 16.266

4084. 1839–4–27 To William Henry Fox Talbot Slough
JH reports his discovery of a highly sensitive paper. Tint continues to deepen with time, as bromine does not. Not prepared to reveal process at present. Is off to Paris.
ALS 3pp Science Museum Talbt 1/17 (draft 'altered': RS:HS 17.293a & C: RS:HS 22.12)

4085. 1839–4–27 From William Henry Fox Talbot Lacock Abbey
More photography comments. Problems with use of hyposulfite fixer. Superiority of English over Daguerreotype in being able to make copies.
ALS 3pp RS:HS 17.293 Buckland, *Talbot,* 53 inc

4086. 1839–4–28 To Francis Baily Slough
Has marked Thomas Maclear's paper and returned it to Thomas Galloway.
ALS 2pp RS:HS 3.159 (C: RS:HS 25.8.19)

4087. 1839–4–28 To Francis Baily Slough
On the variability of the star Alpha Cassiopeia.
ALS 2pp RS:HS 3.158 (C: RS:HS 25.8.20) *R.A.S.M.N., 4* (1839), 195–6 (inc)

4088. 1839–4–28 To Thomas Best Jervis n.p.
Let India Board move at its own pace, and do not call on Lord Northampton. JH will write privately to 'Major J.' to suggest proper training for appointed observers.
ACC 1p inc TxU:H/M–0298.2; Reel 1087

4089. 1839–4–28 To Thomas Maclear Slough
Sends Michael Faraday's analysis of TM's meteorite specimen. Cape Observatory will receive theodolite and sidereal telescopes. Government orders Antarctic Magnetico-Explorative Expedition. JH's photography work results in unfixed images. Will report on TM's paper concerning the [N. L.] Lacaille's Stations.
ALS 4pp Cape Archives/Maclear Papers File 100

4090. 1839–4–28 To William Henry Smyth Slough
Does not have operational telescope. Suggests WS set 1840 as epoch.
ALS 3pp APS B.Sm98.17 (C: RS:HS 22.13)

4091. 1839–4–29 To Richard Sheepshanks Slough
Discusses arrangements for the printing of a portrait of JH in some copies of a forthcoming volume [JH's *Cape Results*].
ALS 3pp † RS:HS 16.5 (C: RS:HS 25.9.2)

4092. 1839–4–30 To Richard Jenkins Slough
Describes three-year magnetic survey proposed by R.S.L. Asks H.E.I.C. court of directors to provide support for three observing stations: at Madras, at Bombay, and in Himalayas. Humphrey Lloyd will supervise training of observers.
C 4pp TxU:H/L-0202; Reel 1054

4093. [1839]–4–30 To John William Lubbock Slough
Has just done photographic experiments using bromine paper, and is very pleased with them. JH notes in a postscript that he had just received a letter from W. H. Fox Talbot stating that Fox Talbot had just discovered bromine paper as well.
C 1p RS:HS 25.14.21

4094. [1839–5] From Madame Biot College de France
Will meet him at 10 a.m. to conduct him to the Botanical Gardens.
ALS 1p (French) RS:HS 5.67

4095. 1839–5–4 From Pompilio Decuppis Rome
Discusses elements of telescopic mirror making, including the curving the glass. Inquires about methods of telescope, lens, and mirror makers in England.
ALS 4pp (Italian) RS:HS 6.481

4096. 1839–5–4 From Henry Raper Admiralty
Writing at suggestion of [Francis] Beaufort. Needs information concerning [J. A.] Lloyd's observations concerning longitude of his observatory at Mauritius. Encloses prospectus of forthcoming work.
ALS 2pp RS:HS 14.302

4097. 1839–5–6 From Nathaniel Bowditch Boston (Mass.)
Letter and parcel of papers has reached them safely for which they thank him. Third volume of *Mécanique céleste* will be sent later together with the fourth volume.
ALS 3pp RS:HS 4.202

4098. 1839–5–6 From Thomas Maclear Royal Observatory, Cape of Good Hope
JH's letter of 9 Feb. has left him in high spirits. Is also pleased TM's paper was communicated to the R.S.L. Will make arrangements to collect parts of the meteorite and send the 4 lb. part to JH. G. B. Airy consents that Greenwich Mural Circle be sent to Cape Observatory. Will request more

engineering personnel for Colonial Survey.
AL 4pp inc RS:HS 12.118

4099. 1839–5–6 From Henry Perigal 5 Smith St., Chelsea
A note accompanying some machine-made mathematical curves.
ALS 1p RS:HS 13.364

4100. 1839–5–7 From Jean Baptiste Biot Paris
Concerning polarization of light rays in crystals of quartz.
ALS 3pp (French) RS:HS 4.92

4101. 1839–5–7 From Carl Friedrich Gauss Göttingen
Sends two copies of his own *General Theory of Terrestrial Magnetism*, one for JH and one for the R.S.L.
ALS 1p RS:HS 8.77

4102. 1839–5–9 To William Henry Fox Talbot Paris
Just visited L. J. M. Daguerre. Praises his images as 'miraculous' and also Daguerre's times of exposure. Daguerre puts his images on 'thin sheets of plated copper.'
ALS 3pp Science Museum Talbt 1/18 Buckland, *Talbot*, 53–4

4103. 1839–5–11 From [Michel] E[ugène] Chevreul n.p.
Concerning JH's projected visit to the Gobelin establishment.
ALS 2pp (French) RS:HS 5.338

4104. 1839–5–18 From James Elliot Brownlow St., Liverpool
Sending an extract from one of his articles on the *Theory of the Spinning Top*. Relates this to the theory of Saturn's rings.
ALS 3pp RS:HS 7.127

4105. 1839–5–18 From Richard Sheepshanks 30 Woburn Place
Suggests that JH print the portrait of JH with but not in JH's [*Cape Results*]. Has been working on Lt. Murphy's observations in Syria. Is pleased with M. J. Johnson's appointment to the Radcliffe Observatory. Asks JH for suggestions on projects for Johnson.
ALS 3pp RS:HS 16.6

4106. 1839–5–20 From [Étienne] Geoffroy Saint-Hilaire Paris
Comments on some work of JH. Disagrees with James South, and urges consideration of other ideas on a physical, [synthetic?] theory. [Much of this letter completely illegible.]
ALS 2pp† (French) RS:HS 17.167

4107. 1839–5–22 From William Reid Bermuda
Bermuda *Gazette* will publish weekly report of weather. Will send reports to JH. Discusses work of others on storms.
ALS 3pp RS:HS 14.324

4108. 1839–5–25 From Thomas Macdougal 4 Minerva Terrace, Islington
Transmits a letter from Mr. Train, who is interested in securing a teaching position in the schools at the Cape of Good Hope.
ALS 3pp † UCT JH Letters A1.12

4109. 1839–5–27 From John Cam Hobhouse India Board
Asks JH to comment on draft of letter from H.E.I.C. court of directors to Marquis of Northampton in response to R.S.L.'s request for support of magnetic survey.
ALS 2pp TxU:H/M–0277; Reel 1087

4110. 1839–5–30 To James Elliot Slough
Thanks JE for the account of an experiment performed by JE using an iron ring and a magnet, the results of which JE finds illustrative of the phenomena of Saturn's rings.
ALS 3pp DMC.695A.885.24

4111. 1839–5–30 To Hugh Percy (3rd Duke of Northumberland) Slough
[Extract] Suggests change in proposed distribution [of *Cape Results*]. Richard Sheepshanks offered to subsidize portrait of JH, to be distributed with complimentary copies [of *Cape Results*].
ACCS 2pp inc TxU:H/L-0288; Reel 1054

4112. 1839–5–31 From Lord John Russell Whitehall
Asks permission to add JH's name to the body of fellows of London University.
ALS 2pp RS:HS 14.436

4113. 1839–6 To Thomas Maclear Slough
Urges TM to observe Comet Encke. Anxious about *Cape Results* completion. Announces move to Collingwood. Requests another meteorite sample. Five schoolmasters to be sent to South Africa. Unable to send refractor.
ALS 4pp Cape Archives/Maclear Papers File 100

4114. 1839–6–4 To John Cam Hobhouse Slough
Received Hobhouse's letter of 27 May with reply from H.E.I.C. board of directors regarding establishment of magnetic observatories in India. D. F. J. Arago recommends additional station on Socotra Island [in Gulf of Aden].
C 3pp TxU:H/L-0194; Reel 1054

4115. 1839–6–4 From Hans Christian Oersted Copenhagen
Recommending Capt. Zahrtman of the Royal Danish Navy. Hopes to hear good news of JH.
ALS 1p RS:HS 13.164

4116. 1839–6–5 From Adolphe Quetelet Brussels
Sends continuation of Belgian meteorological observations. Just finished magnetic observations. Plans to continue certain ones. Regrets being alone, without aids, 'in front of the transit instrument, the mural circle and the equatorial.'
ALS 3pp (French) RS:HS 14.215

4117. 1839–6–5 To Lord John Russell Slough
Reticent to accept fellowship of London University if obligations of time and attention are too great. Requests more time to consider responsibilities of the post.
Copy 2pp RS:HS 14.437 (C: RS:HS 22.14)

4118. 1839–6–6 To James David Forbes Royal Society
Writes 'in haste' to thank JF for a pair of papers. Comments on Louis Daguerre's pictures in Paris.
ALS 2pp St. Andrews 32 (C: RS:HS 22.15)

4119. 1839–6–6 To [?] London
Departure of J. C. Ross for Antarctic expedition is approaching. Hopes that Admiralty will soon authorize R.S.L. to assist Ross in preparing instruments and coordinating global observations.
Adraft 4pp TxU:H/L-0641; Reel 1089

4120. 1839–6–9 From Nathaniel Bowditch Boston (Mass.)
Have obtained an act of incorporation for their library. Their father did not finish his translation of the fifth volume of the *Mécanique céleste*.
ALS 2pp RS:HS 4.203

4121. 1839–6–9 From Thomas Galloway Serjeants' Inn
Sending William Simm's paper, which was read at the last meeting. Would be glad of JH's views on whether it should

be printed in the *R.A.S. Memoirs.*
ALS 2pp RS:HS 8.7

4122. 1839–6–10 From Thomas Maclear Royal Observatory, Cape of Good Hope
Mr. Zahn and John Truter in search of more meteorite specimens. Truter obtains a specimen near site of impact.
ALS 1p RS:HS 12.119

4123. 1839–6–15 From Thomas Maclear Royal Observatory, Cape of Good Hope
Sends another meteorite specimen. Mr. Zahn sends group to Bokkerveld to seek more specimens. John Truter describes specimens in possession.
ALS 2pp encl † RS:HS 12.120

4124. 1839–6–15 From James Cosmo Melvill East India House
Chairman of H.E.I.C. granted permission for Major [Tenis] to delay departure to India for one month in order to become better acquainted with instruments to be used in magnetic observatories there.
ALS 1p TxU:H/M–0951; Reel 1083

4125. 1839–6–16 To Julia Margaret Cameron Slough
Thanks for gifts sent to JH's children. Gives information on various persons including John Wrottesley and Sir Edward Ryan.
ALS 4pp Getty Center

4126. 1839–6–16 To Luke Howard Slough
LH's meteorological circular register was laid before the Meteorological Committee, which considers it worthy of notice. Comments on this, and would like to discuss the matter further.
C 4pp RS:HS 10.1 & 22.16

4127. 1839–6–17 To Carl Friedrich Gauss Slough
Comments on CG's paper on the south magnetic pole, and on British efforts now underway to make magnetic observations.
ALS 3pp NSUB Cod.Ms.Gauss 99 (C: RS:HS 22.17)

4128. 1839–6–17 From William Henry Harvey Ludgate Hill
Must have received a long rambling letter written in February; hopes he will ignore it as he was suffering from a fever at the time. Returned on leave of absence for 12 months. Has introduced J. A. Wahlberg to Col. John Bell. P. H. Polemann died in April.
ALS 3pp RS:HS 9.247

4129. 1839–6–18 To Johann Franz Encke Slough
Asks if JE or the Academy of Sciences wish to participate in the British plan to make worldwide magnetic observations.
ALS 3pp AdWDDR Encke 49

4130. 1839–6–19 From George Biddell Airy Royal Observatory, Greenwich
Has sent to JH papers relating to London University. Hopes JH will accept a senatorship.
ALS 2pp RS:HS 1.79

4131. 1839–6–19 To D. François J. Arago Royal Society
Requests FA's assistance in efforts toward establishing magnetic observatories. Especially recommends an observatory at Algiers. Communication from the R.S.L. will arrive shortly. Comments on biography of William Herschel written by J. B. J. Fourier.
ALS 4pp Académie des sciences (Paris)

4132. 1839–6–19 To William Cranch Bond London
Invites WB to participate in making magnetic observations. Is sending relevant materials to WB.
ALS 2pp BostonPubLib Ms.Eng. 190 (1)

4133. 1839–6–21 To Thomas Maclear [Slough]
J. C. Ross goes to Cape; will establish the proposed Magnetic Observatory for three years. TM should seek a site for the Magnetic Observatory.
ALS 3pp Cape Archives/Maclear Papers File 100

4134. 1839–6–21 From Edward Sabine Woolwich
Dublin University is available on 2 July for instruction of the officers to be employed in the magnetic observatories. Suggests a note be sent to the R.S.L. stating that the changes in instrumentation are only slight modifications.
ALS 2pp RS:HS 15.42

4135. 1839–6–24 From C. A. Jensen 36 Great Marlborough St.
Will be leaving London on 16 July and would like to meet JH at some convenient time to carry out H. C. Oersted's commission.
AL 1p RS:HS 10.442

4136. 1839–6–24 To William Henry Fox Talbot Slough
Admires L. J. M. Daguerre's process, but hopes to see paper become dominant medium for photographs. Describes new paper JH developed. [Letter continues 6 July.] Experiment with fixed lines in chemical spectra; shows sensitivity of ultraviolet light.
ALS 3pp Science Museum Talbt 1/19 (ACCS: RS:HS 25.5.13 & C: TxU:H/L–0476; Reel 1055)

4137. 1839–6–25 To D. François J. Arago Slough
Regarding L. J. M. Daguerre's photographic experiments.
draft 2pp RS:HS 1.362 & 22.18

4138. 1839–6–26 From Friedrich Wilhelm Bessel Königsberg
Sending JH some writings on geodesic measurements; reports the second measurement of annual parallax of 61 Cygni; comments on some other observations.
ALS 2pp (German) RS:HS 4.73

4139. 1839–6–26 To Caroline Lucretia Herschel Slough
Has enclosed with the letter 'a letter sketch of the 40 feet [telescope] (as it now stands) made without hands, by Photography.'
ALS 2pp TxU:H/L–0584.4.b; Reel 1058

4140. 1839–6–28 From Andries Stockenström to Margaret Brodie Herschel 3 Wakefield St.
Thanks JH and MH for their support. Admits that before leaving the Cape he told the governor that he had no intention of returning. Ascribes difficulties of his term as Lieutenant Governor to problems inherent in having a military man in office.
ALS 10pp RS:HS 17.17

4141. 1839–6–28 From Henry Holland Brook St.
Sending HH's *Medical Notes and Reflections* (1839).
ALS 2pp † RS:HS 9.429

4142. 1839–6–28 From Edward Sabine Woolwich
Supports furnishing a wooden building at St. Helena despite the high temperatures of the tropical climate. Considers the expenses involved.
ALS 3pp RS:HS 15.43

4143. 1839–6?–29? To Adolphe Quetelet Slough
JH's brother-in-law John Stewart, who is delivering this letter, seeks admission to a printing establishment. Asks AQ to aid JS. Lists recently received letters. Glad that AQ has [C. F.] Gauss's apparatus. Discusses an experiment in photography.
ALS 4pp Académie belgique: #13

4144. 1839–6–30 To Andries Stockenström Slough
Discusses the tensions between AS and Cape Governor George Napier.
ACCS 3pp RS:HS 17.18 (C: RS:HS 22.19)

4145. 1839–7 To [Prince Albert] n.p.
[Two copies of] Printed letter from R.S.L. announcing proposed system for global magnetic observations. [One copy edited by JH for submission to 'Your Highness' with] Request for assistance in adding magnetical observatory to John Caldecott's astronomical observatory in Travancore [India].
L 2pp & (A)draftS 2pp TxU:H/L-0768.1 & -0768.2; Reel 1089

4146. 1839–7–2 From Edward Sabine Woolwich
Suggests JH include in his report a discussion of the 'dividing line of the positive and magnetic fluids' and 'separating line between the northern and southern intensities.' Presents C. F. Gauss's ideas on this topic. Suggests a trip plan for James Ross in order to obtain absolute measures of magnetic intensity on the globe.
ALS 5pp RS:HS 15.44

4147. 1839–7–2 From Andries Stockenström Devonshire St.
Discusses Lord Glenelg's role in the decision to keep him as Lieutenant Governor of the Cape and his support during the inquiry of AS's actions in a confrontation with the natives. Writes of Glenelg's retirement.
ALS 6pp RS:HS 17.19

4148. 1839–7–2 From William Henry Fox Talbot 31 Sackville St.
Sends JH a positive copy and a transparency. Remarks on JH's use of muriate of lead and asks JH to explain chemistry of it.
ALS 2pp RS:HS 17.294

4149. 1839–7–4 From Edward Sabine Woolwich
Relays the opinions of a returning officer from St. Helena that a wooden building need not be built there.
ALS 2pp RS:HS 15.45

4150. 1839–7–5 To D. François J. Arago Slough
Is sending FA information about magnetic proceedings.
ALS 1p WT 67390.1

4151. 1839–7–7 To William Henry Fox Talbot n.p.
Recounts experiments with chemicals, spectra, and photographic paper.
ALS 2pp Science Museum Talbt 1/20 (ACCS: RS:HS 25.5.14 & C: TxU:H/L–0477; Reel 1055)

4152. 1839–7–9 From James David Forbes Bourg d'Oysans
Regrets that JH had to write another letter regarding the two papers he forwarded. Missed him when in London. Saw L. J. M. Daguerre's specimen and also those of J. N. Niepce, the latter undoubtedly the originator. Has made an excursion to the volcanic region of the South of France.
ALS 2pp RS:HS 7.294

4153. 1839–7–10 To John Frederick Goddard Slough
Suggests arrangements for submission of papers, reports, models, etc., to the B.A.A.S. meeting.
ALS 4pp Manchester Central Library

4154. 1839–7–11 From Luke Howard Ackworth
Will not be able to come to town for some time so sends answers to JH's comments on his Circular Register. Hopes JH can call to see him; then he can explain more fully.
ALS 4pp RS:HS 10.2

4155. 1839–7–11 From John P. Gassiot to [John] Stewart Clapham Common
Intends erecting his large constant battery in the course of a few days and wonders if JH or Charles Babbage would be interesting in seeing it.
ALS 1p RS:HS 8.50

4156. 1839–7–12 From Edward Sabine Woolwich
Has presented the draft of the report to [T. R.] Robinson, [Charles] Wheatstone, and [J. F.] Daniell. Points out mistakes in the report. Remarks that the dipping needle has greatly improved over the years, thus increasing instrumental accuracy. Encloses Humphrey Lloyd's contribution to the report.
ALS 5pp RS:HS 15.46

4157. 1839–7–14 To Richard Sheepshanks Slough
On the difficulties of writing his book [*Cape Results*]. Feels he has been at everyone's disposal but his own and is finding notes made at the Cape difficult to decipher. Asks RS to report to him from Germany on the state of telescope manufacturing there and on the progress of F. G. W. Struve's great refractor.
ALS 3pp RS:HS 16.8 (C: RS:HS 25.9.3)

4158. 1839–7–14 From Richard Sheepshanks n.p.
Sends JH the name and address of the printer who has an engraving arranged for by RS of JH's portrait. Requests that JH give thirty copies to Lady Herschel for her own use.
ALS 1p encl RS:HS 16.7

4159. 1839–7–14 To John Stewart Slough
Is willing to be a witness to some battery experiments by J. P. Gassiot, but JH is busy, mostly with house hunting, and so may not be available.
ALS 4pp APS B.H435p.10

4160. 1839–7–16 From George Biddell Airy Royal Observatory, Greenwich
Has lost two sons with scarlet fever. Description of house at Playford, Sussex.
ALS 4pp RS:HS 1.80

4161. 1839–7–16 From Edward Sabine Woolwich
States he has received the list of instruments and their prices sent by JH.
ALS 1p RS:HS 15.47

4162. [1839]–7–17 From Francis Beaufort Admiralty
On Friday the compensating bars will be embarked for the Cape; William Mann will be in charge of them at the Tower. Dr. T. R. Robinson is anxious to have some of the instruments from the Kew Observatory; does JH know how they may be obtained? Regarding an actinometer made by Robinson.
ALS 3pp RS:HS 3.408 (ACC: Hydrographic Office Letter Book)

4163. 1839–7–17 From James Calder Stewart 11 Bloomsbury Square

Advises JH to have Foreign Office send letter to Col. Campbell, consul in Alexandria, to accompany R.S.L.'s letter to Pasha. Otherwise, R.S.L. letter may be ignored. Good news from [Duncan Stewart]. JS is not going to Scotland; may visit Slough instead.
ALS 3pp † TxU:H/M–0542; Reel 1093

4164. 1839–7–19 To James Clark Ross n.p.
Describes method of suspending furniture in a ship such that the furniture is less influenced by the ship's motion.
draft 4pp RS:HS 14.432 (C: RS:HS 25.13.5)

4165. 1839–7–19 To Edward Sabine Maidstone
Sends plans for and detailed explanation of device to suspend cot or couch in ship so as to 'destroy' ship's motion and alleviate seasickness.
ALS 4pp diag RS Sa.648

4166. 1839–7–19 From William Henry Fox Talbot Lacock Abbey
Confused by an 'unfixed' photo from JH in which the image shows different effects from different colors of the spectrum. Also other comments on photography.
ALS 2pp RS:HS 17.295

4167. 1839–7–20 From Francis Beaufort n.p.
Encloses two meteorological reports of J. A. Lloyd (who has been ill with fever), and a paper of Dr. T. R. Robinson.
ALS 1p RS:HS 3.364 & ACC: Hydrographic Office Letter Book

4168. 1839–7–20 To Margaret Brodie Herschel Ticehurst
About JH's travels and viewing of prospective homes to which to move; saw Collingwood.
ALS 4pp JHS 3.17

4169. 1839–7–21 From Joseph Goodall Lodge, Eton College
Hopes JH will honor his election by attending his dinner on Saturday.
ALS 1p RS:HS 8.146

4170. [1839]–7–22 From Francis Beaufort [Admiralty]
Is dismayed that Thomas Maclear is unable to obtain manual assistance at the Cape for the compensating bars. Bars were shipped on Friday. Next time JH passes the Admiralty he will show him the debris of the actinometer.
ALS 4pp RS:HS 3.407 (ACC: Hydrographic Office Letter Book)

4171. 1839–7–23 From Andrew Smith Fort Pitt, Chatham
Declines an invitation to visit JH; comments on political situation in South Africa.
ALS 3pp RS:HS 16.179

4172. 1839–7–26 From John Phillips B.A.A.S., Birmingham
A report by JH, George Peacock, William Whewell, and Humphrey Lloyd on magnetic observations will be read at the B.A.A.S. meeting in August.
(A)LS 1p TxU:H/M–0428.3; Reel 1093

4173. 1839–7–26 From John Phillips B.A.A.S., Birmingham
The report by G. B. Airy, Thomas Henderson, and JH on N. L. Lacaille's stars will be brought forth at the next meeting of the B.A.A.S. JP requests a copy in advance of the meeting.
(A)LS 1p TxU:H/M–0428.2; Reel 1093

4174. 1839–7–26 From John Phillips B.A.A.S., Birmingham
Reminder that JH must report on 'Reduction of Meteorological Observations' at next meeting.
(A)LS 1p RAS:JH Archive 4/9.15 verso; Reel 7

4175. 1839–7–26 From Edward Sabine Woolwich
Encloses a letter from Humphrey Lloyd. Anticipates Lloyd's arrival. Worries he will not be ready with the necessary forms before James Ross sets sail. Includes many calculations.
ALS 4pp RS:HS 15.48

4176. 1839–7–27 To James Clark Ross n.p.
Because of R.S.L. council meeting on Thursday, suggests Friday meeting with JL and several others.
ALS 3pp RS Sa.649

4177. 1839–7–29 To Hans Christian Oersted Slough
Apologizes for having been too busy to sit for a portrait as proposed, thanks HO for his poems, and describes JH's current experiments related to photography.
C 2pp RS:HS 22.21 Harding, *Correspondance de ... Örsted, 2,* 391–2

4178. 1839–7–29 From Edward Sabine Woolwich
Reports that the ships to be used for James Ross's voyage will not be available for a week and that Humphrey Lloyd is busy preparing instructions. Proposes that the ships receive their own sets of instruments.
ALS 3pp RS:HS 15.49

4179. 1839–7–30 From Francis Beaufort Admiralty
Would like his opinion on whether an officer should be sent to the Cape to be in charge of the compensation bars sent there recently by the Board of Ordnance.
ALS 3pp RS:HS 3.365 & ACC: Hydrographic Office Letter Book

4180. 1839–7–30 From Edward C. Herrick New Haven, Conn.
Thanks for JH's letter and for the references to meteors. The Austrian observations are important. Gives news of American observers and their work.
ALS 3pp RS:HS 10.186 (C: Yale University)

4181. 1839–7–30 From Edward Sabine Woolwich
Accepts JH's dinner invitation, regretting that James Ross is unable to attend. Encloses a letter from Humphrey Lloyd.
ALS 3pp RS:HS 15.50

4182. 1839–7–31 From James Clark Ross Woolwich
Preparations for departure prevented JR from dining with JH at Slough. Working on pendulum with Edward Sabine. Thanks JH for sending gimbals.
ALS 3pp TxU:H/M–0463; Reel 1093

4183. 1839–8–1 To George Biddell Airy Slough
Sympathizes with GA in the family sorrow [deaths of GA's sons Arthur and George]; JH complains that house hunting, and other matters, are keeping him from important work.
ALS 3pp RGO 6.368.622

4184. 1839–8–1 To Francis Beaufort Slough
Agrees with FB and others about the need for adequately trained men to be sent out to remeasure N. L. Lacaille's arc.
ALS 3pp APS B.H435p.11

4185. 1839–8–1 To Spencer J. A. Compton Royal Society
Problem of resolution of R.S.L. Council not transmitted to

SC.
C 1p RS:HS 22.22

4186. 1839–8–1 To Louis Daguerre Royal Society
On behalf of the Council of the R.S.L., JH writes to LD to request a camera and a hundred plates of Daguerre's about to be announced process, so that the R. S. L. may equip the James Ross expedition to the Antarctic with the means of making a photographic record of the voyage.
ALS 4pp WT 67390.2

4187. [1839–8 or later] To Mr. Neve n.p.
As agent for Sarah Collingwood's property, JH requests receipts for 'tithe and modus' of Moorhouse property, to protect from overpayment to 'Mr. Morrel,' whose past collections were 'irregular.' [JH annotation: Not sent.]
ACCS 1p RAS:JH Archive 4/11.35 verso; Reel 7

4188. [1839–8] From Edward Sabine Woolwich
Supports JH's idea of having one simultaneous observation each day. Suggests that there might be two a day. Has suggested to Humphrey Lloyd that hourly observations take place once a fortnight. Says Lloyd wishes to speak about supplies.
ALS 2pp RS:HS 15.52

4189. 1839–8–1 From Edward Sabine Woolwich
Asks JH whether he presented to the Council of the R.S.L. the resolution of the Meteorological Committee regarding references made by the 'Colonial Office, Trinity[?] Board, etc.'
ALS 1p RS:HS 15.51

4190. 1839–8 From John P. Gassiot to [John] Stewart Mark Lane
Is grateful for the trouble he has taken. Would he forward the enclosed note to JH. If JH can call on him he would make arrangements to show him the battery.
ALS 1p RS:HS 8.52

4191. 1839–8–2 From John P. Gassiot Clapham Common
Thanks for letter and also for the experiments suggested. Has been working on the tension spark. Would be pleased to show JH or Charles Babbage the battery at any time.
ALS 3pp RS:HS 8.51

4192. 1839–8–3 From Humphrey Lloyd Trinity College, Dublin
Has just had a letter from C. F. Gauss regarding the hours for observations; comments on this.
ALS 4pp RS:HS 11.271

4193. 1839–8–3 From Lionel Rodwell 10 Duke St., St. James'
Sends copy of statutes of Société general des naufragés [General Shipwreck Relief Society] in response to JH's visit yesterday. Which class of members does JH wish to join?
ALS 1p RAS:JH Archive 12/1.3.21 verso; Reel 10

4194. 1839–8–4 To John William Lubbock Slough
Asks JL to take on the preparations for the magnetic observations, including the instructing of ships' personnel.
C 3pp RS:HS 22.23

4195. 1839–8–5 From Adolphe Quetelet Brussels
Addresses a few last words before departing for Italy with his wife. Sends 3 memoirs just published. Will be in Paris to observe meteors. Proposes to redo terrestrial magnetism observations in Italy.
C 1p RS:HS 14.216

4196. 1839–8–5 From Edward Sabine Woolwich
Encloses a letter from Humphrey Lloyd. Sees no problem in adding an officer to the staff of each observatory, if necessary. Remarks on a cheap but fragile magnetometer.
ALS 3pp RS:HS 15.54

4197. 1839–8–5 From Edward Sabine Woolwich
Explains that the purpose of his trip with Humphrey Lloyd is to 'establish the cooperation of the continental magnetic observatories.' Desires that the trip be authorized by the R.S.L.
ALS 2pp RS:HS 15.53

4198. 1839–8–7 To Caroline Lucretia Herschel Slough
The 20-ft. reflecting telescope used at Cape Town was delivered safely to Slough.
ALS 3pp TxU:H/L–0584.5.a; Reel 1058

4199. 1839–8–7 To Humphrey Lloyd Slough
About times at which magnetic observations should be made at each observation station.
C 7pp RS:HS 22.25

4200. 1839–8–7 To William Whewell n.p.
WW, JH, and George Peacock must prepare a report for the B.A.A.S. on the magnetic expedition and observations; offers to draft it. Laments great loss of time in working on their committee. Is involved in buying a house in Hawkhurst.
ALS 4pp TC, Camb. Add. Ms.a.207[40] (C: RS:HS 22.24)

4201. 1839–8–9 From John William Lubbock Mitcham Grove
J. T. Boileau and L. E. O. Ludlow were appointed by H.E.I.C. to direct Indian observatories. R.S.L. is concerned that it has no choice in selecting directors. Instruments for H.E.I.C. observatories. Should JL request additional staff for all 16 observatories, or only for nine permanent ones?
ALS 3pp TxU:H/M–0340; Reel 1087

4202. 1839–8–9 From Edward Sabine Woolwich
Is grieved by the R.S.L. report, which supports a 'voyage of Discovery to the Antarctic' rather than 'research in Physical Sciences in the Southern Hemisphere.' Speaks of the discoveries made by expeditions to the Antarctic, including great changes in magnetic variations.
ALS 4pp RS:HS 15.55

4203. 1839–8–10 From James Clark Ross St. George's Terrace
Saturday the 17th will be convenient because it gives him time to finish his pendulum experiments.
ALS 1p RS:HS 14.427

4204. 1839–8–10 From John George Charles Fox Strangeways Foreign Office
Asks about letter that Lord Northampton requested Lord Palmerston to give to Mr. Stewart, who leaves for Egypt soon.
ALS 1p TxU:H/M–0510; Reel 1093

4205. 1839–8–11 From Humphrey Lloyd Trinity College, Dublin
Finds that he misunderstood Edward Sabine's statement and may have misled JH. Comments on the observations desired from the Antarctic Expedition.
ALS 4pp RS:HS 11.272

4206. 1839–8–11 From Edward Sabine Woolwich
Admiralty is dissatisfied at still being responsible for the Van Diemen's Land Observatory. Suggests responsibility be shifted to the Ordinance. Treasury will no longer finance the

building of observatories when other buildings are available. Describes new procedures whereby officers observing in the colonies must present colonial governors with proper letters upon request.
ALS 4pp RS:HS 15.56

4207. 1839–8–12 To Francis Baily Slough
Has had a letter from [John?] Phillips regarding star nomenclature revision. The display of meteors on the 10th was very fine.
ALS 3pp RS:HS 3.160 (C: RS:HS 25.8.21)

4208. 1839–8–12 From John Lindley 21 Regent St.
Is working on horticultural considerations of Cape plants and would welcome information on the temperature just below the surface of the soil in relation to the temperature of the air.
ALS 3pp RS:HS 11.226

4209. [1839]–8–12 From Thomas Maclear Royal Observatory, Cape of Good Hope
J. M. Craig requests TM to send meteorite to British Museum. TM asks JH to transfer the specimen sent earlier to the Museum.
ALS 2pp encl RS:HS 12.122

4210. 1839–8–12 From Thomas Maclear Royal Observatory, Cape of Good Hope
Encloses letters concerning [William] Reid's *Laws of Storms*. Discusses problems obtaining money for Meteorology Society from government. Reports mural circle working well. Receives another meteorite specimen.
ALS 4pp RS:HS 12.121

4211. 1839–8–14 From William Edward Parry Admiralty
Requesting letters of introduction for Lieut. F. M. Eardley Wilmot, who is to undertake magnetic observations at the Cape.
ALS 2pp RS:HS 13.222

4212. 1839–8–15 From George Biddell Airy Walton-on-the-Naze
Regarding C. F. Gauss's magnetical observations. Domestic happenings.
ALS 2pp RS:HS 1.81

4213. 1839–8–16 From Richard Bourke Downing St.
Will visit C. R. Fox at Ordnance to speed up official letter to Treasury requesting additional officer on magnetic expeditions. Will inform JH when letter reaches [S. H.] Christie [at Royal Military Academy, Woolwich].
ALS 3pp TxU:H/M–0114; Reel 1087

4214. 1839–8–17 From Francis Baily 37 Tavistock Place
Returns reports on Cape Observatory and reform of star nomenclature. FB was not included in committee to report reduction of stars in N. L. Lacaille's *Coelum australe stelliferum*, but sends FB's own report on it, which JH may send in.
ALS 2pp TxU:H/M–0039.1; Reel 1062

4215. 1839–8–17 From Henry Perigal 5 Smith St., Chelsea
More mathematical curves made by machine [see HP's 1839–5–6].
ALS 1p RS:HS 13.365

4216. 1839–8–17 To John Phillips n.p.
Report to B.A.A.S. on progress and expenses in reduction of stars in N. L. Lacaille's *Coelum australe stelliferum.*
C 1p TxU:H/M–0039.2; Reel 1062

4217. 1839–8–17 From Edward Sabine Woolwich
Anticipates Humphrey Lloyd's arrival at London this night. Announces that the pendulum experiments are finished. Interested in inspecting the magnetic instruments in preparation for the Antarctic journey. Plans JH's visit.
ALS 2pp RS:HS 15.57

4218. 1839–8–20 From Edward Sabine Athenaeum
Disappointed not to have received a response from JH to his invitation for a visit. Goes ahead with plans.
ALS 3pp RS:HS 15.58

4219. 1839–8–21 From Giovanni Antonio Plana Turin
Reports on sending and receiving barometric and magnetic observations. GP's 20 year old daughter is still not married [see GP's 1838–1–6], but GP believes her chances are improving.
ALS 3pp (French) RS:HS 13.435

4220. 1839–8–22 To Thomas Maclear London
F. E. Wilmot arrives in Cape Town for direction of the Magnetic Observatory.
ALS 3pp Cape Archives/Maclear Papers File 100

4221. [1839]–8–25. From Edward Sabine Birmingham
[A. T.] Kupffer, [Humphrey] Lloyd, and ES will meet [C. F.] Gauss at later date. Desires cooperation with Russia. Birmingham good for B.A.A.S., but the meeting lacks distinguished scientists.
ALS 2pp RS:HS 15.316

4222. 1839–8–26 To John Lindley Slough
In response to a request, JH provides such details as he can about atmospheric and ground temperatures at the Cape; JH goes on to describe how his flowers brought from the Cape are doing.
ACCS 4pp RS:HS 25.5.15

4223. 1839–8–27 From Francis Beaufort Admiralty
Is forwarding a note about an instrument of [Louis] Daguerre's.
ACC 1p Hydrographic Office Letter Book

4224. 1839–8–28 To William Whewell Slough
Reports to the B.A.A.S. [through WW as section president] on experiments JH has conducted concerning the photochemical effects of the rays at the far red end of the spectrum.
L *B.A.A.S. Report for 1839* (1840), pt. 2, 9–11 & *Athenaeum,* #621 (Sept. 21, 1839), 724–5

4225. 1839–8–28 To [William Whewell] Slough
Wishes to have mentioned to the Physical Section of the B.A.A.S. that JH has discovered that the extreme red rays [infra-red?] of the sun do not darken photographic paper, but tend to whiten it.
ACCS 4pp RS:HS 25.5.16

4226. 1839–8–30 From Francis Beaufort Admiralty
Sending a letter of Basil Hall's, which please return when he has read it.
ALS 1p RS:HS 3.366 & ACC: Hydrographic Office Letter Book

4227. [1839]–9–1 From Edward Joshua Cooper Liverpool
Would like JH to pay them a visit at Markree to see his 'circle' instrument erected. Hospitality will be rough owing to their lack of furniture. Hopes he is satisfied with his own photographic results.
ALS 4pp RS:HS 5.281

4228. 1839–9–1 From Roderick Impey Murchison Birmingham
The *Times* has published an attack on the B.A.A.S. and stated that JH has left the B.A.A.S. Would JH write to the *Times* to contradict this statement. RM is to continue as General Secretary for a further year. [Note from Herschel to see *Times* of yesterday (3–3–39).]
ALS 3pp RS:HS 12.400

4229. 1839–9–4 From Edward Sabine Woolwich
Announces Charles Riddell's departure for Canada and his own trip schedule in Germany. Has read JH's report. Informs JH that ES has been made a member of the Committee. Glad that [Vernon] Harcourt's speech covered magnetic observatories. Eager to promote the public image of the fixed observatories. The translation committee was reappointed and includes JH, but ES will spare JH any work.
ALS 6pp RS:HS 15.59

4230. 1839–9–5 To [Mr. Smith?] Slough
Is very insistent that JH receive clear title to Collingwood, with no encumbrances.
ACCS 2pp RS:HS 25.5.17

4231. 1839–9–6 From Richard Jones Tithe Commission
Whewell and RJ plan to visit South Wales and would like to stop at Slough en route.
ALS 1p RS:HS 10.362

4232. 1839–9–6 From Giovanni Santini Florence
Makes micrometer observations for a star catalogue to be published in 1840 for the University of Padua. Padua Observatory offers meridian circle to Vienna Observatory. Searches for comets.
ALS 2pp (Italian) RS:HS 15.346

4233. 1839–9–8 From William Henry Fox Talbot 31 Sackville St.
Sends some samples of photography made with a solar microscope, and comments briefly on the Daguerreotype.
ALS 2pp RS:HS 17.296 Buckland, *Talbot,* 55

4234. [1839]–9–9 From Edward Sabine Woolwich
Recommends that a fellow scientist be properly cited in the R.S.L. report for his method of measuring the depth of the ocean. Encloses observations.
ALS 3pp RS:HS 15.60

4235. 1839–9–10 From [Humphrey] Lloyd n.p.
Cannot accept JH's invitation to dinner. HL and Edward Sabine depart tomorrow for [Batind?].
ALS 1p † TxU:H/MW-0246.6; Reel 15

4236. 1839–9–10 To William Henry Fox Talbot Slough
Has succeeded in producing a 'photograph on glass.' Describes process and also JH's experiment photographing spectra.
ALS 3pp Science Museum Talbt 1/21 (ACCS: RS:HS 25.5.18 & C: TxU:H/L–0478; Reel 1055) Buckland, *Talbot,* 55

4237. 1839–9–12 From William Henry Fox Talbot Lacock Abbey
Advises JH not to publish any new processes in photography as there are patent problems. Comments on the Daguerreotype, and on the difficulty of preserving images in some circumstances.
ALS 3pp RS:HS 17.297 Buckland, *Talbot,* 55–6 inc

4238. [1839]–9–13 From Francis Beaufort n.p.
Enclosing letters from Thomas Maclear so that he may see all he says about his meteorites. Will keep the meteorite until he tells him what to do with it.
ALS 2pp RS:HS 3.409

4239. 1839–9–14 From Thomas Best Jervis Livorno [Tuscany]
Attended [French] Institute and heard D. F. J. Arago read account of L. J. M. Daguerre's process, which is very different from those of JH and W. H. F. Talbot. Has not yet visited Francesco Carlini to give him JH's letter. Introduces J. T. Boileau, who will participate in magnetic survey.
ALS 2pp TxU:H/M–0299; Reel 1087

4240. 1839–9–16 From Frederick A. G. Ouseley Hall Barn, Beaconsfield
Is delighted with the photogenic apparatus JH has sent; it will afford him great amusement. Anxiously awaits some sunshine.
ALS 2pp RS:HS 13.203

4241. 1839–9–16 From Christian F. Schönbein Stanmore
Sorry he did not see JH at B.A.A.S. meeting in Birmingham. Asks if he may call at Slough.
ALS 1p RS:HS 15.354

4242. 1839–9–17 From J. H. Tredgold 41 Wellclose Square
Discusses the dismissal of Andries Stockenström from his post of Lieutenant Governor of South Africa; expresses worry about the future of the natives of the colony.
ALS 2pp RS:HS 17.376a

4243. 1839–9–18 To J. H. Tredgold Slough
Discusses Andries Stockenström's dismissal. Suspects he will be appointed to some other post.
ACCS 3pp RS:HS 17.376b (C: 22.26)

4244. 1839–9–20 From Abraham Follett Osler Birmingham
Sends a pamphlet containing an account of AO's anemometer, also an account of the great storm of January last. Comments on his own and William Reid's theories of storms.
ALS 1p RS:HS 13.202

4245. 1839–9–24 From Francis Baily 37 Tavistock Place
Thomas Brisbane's application to the government for increase of power at the Parramatta Observatory. Can JH dine with him?
ALS 2pp RS:HS 3.161

4246. 1839–9–24 From J. H. Tredgold 41 Wellclose Square
Regrets missing JH's visit. Discusses Andries Stockenström.
ALS 2pp RS:HS 17.377

4247. 1839–9–27 To Francis Baily n.p.
Regarding the Parramatta Observatory. Has produced some colored photographic images.
ALS 3pp RS:HS 3.162 (C: RS:HS 25.8.22)

4248. 1839–9–27 From Andries Stockenström Devonshire St.
Colonial Secretary Lord Normanby dismissed AS Lieutenant Governor at the Cape, citing the feelings toward him in the colony. AS declares, 'I rejoice in having done with the Cape.' Declines an offer of a government position in the West Indies. Claims his future depends on how much he will be given as a pension.
ALS 3pp RS:HS 17.20

4249. 1839–9–30 From George Thomas Napier Government House, Capetown
Would JH communicate with C. H. Phipps (1st Marquis of Normanby) to speed up the dispatch of the four teachers selected for the Cape educational system. Frontier is quiet except for some cattle stealing. Safe arrival of GN's daughter in England.
ALS 3pp RS:HS 13.67

4250. [1839–10] To Marshall Hall n.p.
Pressure of business prevented him from replying earlier. Reserves his advice for matters that fall within his own researches. Thinks it would be advisable to obtain the opinions of several competent Fellows, *not* on the Council.
C 4pp RS:HS 9.201 & 22.28

4251. 1839–10–1 From Marshall Hall 14 Manchester Square
Regarding the attitude of the Council of the R.S.L., particularly that of P. M. Roget, to the publication of his papers, and the award of medals to other persons who have done similar work to his own.
ALS 12pp RS:HS 9.198

4252. 1839–10–3 From George Biddell Airy Royal Observatory, Greenwich
George Everest-Thomas Jervis affair.
ALS 3pp RS:HS 1.82

4253. 1839–10–5 From William Henry Fitton Bloomfield Lodge
Intends that his second son shall enter business and would like him to study in Hanover. Can JH recommend any person with which he can place him. Went to Boulogne with some members of the Geological Society to meet French geologists. Quite a successful meeting, though the French Geologists are split into two opposing factions.
ALS 4pp RS:HS 7.236

4254. 1839–10–8 From Sir Gore Ouseley Hall Barn, Beaconsfield
As JH likes music invites him to dinner on the 15th, when he should find something to enjoy. Regrets he was out when JH called.
ALS 1p RS:HS 13.204

4255. 1839–10–9 To George Biddell Airy n.p.
George Everest-Thomas Jervis affair. Observations on the production of color by chemical rays.
draft 3pp RS:HS 1.83 & 22.27

4256. 1839–10–11 From John William Lubbock London
Has he succeeded in obtaining a Daguerrotype for J. C. Ross? Thinks one or two would be desirable. Has got one himself and obtained good results. Two of his family are ill, which brings him to town. Hears that JH has got colors, which is a great step.
ALS 3pp RS:HS 11.360

4257. 1839–10–12 From John Theophilus Boileau Upper Seymour St.
Has been appointed by the East India company to supervise the instruments at the Bengal Observatory and finds there is no actinometer. Would like details of a good one and JH's views on his suggested modifications.
ALS 2pp RS:HS 5.81

4258. 1839–10–14 From Marshall Hall 14 Manchester Square
Sending a letter and some volumes for his perusal. No need to reply.
ALS 2pp RS:HS 9.199

4259. [1839–10–16] To Marshall Hall n.p.
Returns his packet of papers together with the letters of M. J. P. Flourens, [J. P.?] Müller, etc. Does not feel competent to judge papers such as MH's, which lie outside his field of knowledge. Should rid his mind of bias on the part of Council. Submit papers as from himself.
draft 2pp RS:HS 9.200 & 22.29

4260. 1839–10–18 From Charles D. E. König British Museum
Thanks for the gift of the meteorite to the British Museum; an official letter of thanks will follow later.
ALS 1p RS:HS 11.92

4261. 1839–10–20 From James Watt, Jr. Cox's Hotel, Jermyn St.
Sends English version of D. F. J. Arago's eulogy of JW's father, James Watt. Calls attention to friendship of JW's father and William Herschel. Missed seeing JH in Paris.
ALS 2pp RS:HS 18.386

4262. 1839–10–22 To [Richarda Airy] Slough
Announces the birth of JH's daughter Maria Sophia, and that all are doing well.
ALS 1p JHS 2.7

4263. 1839–10–22 To Elizabeth [Betty] Plumleigh Babbage Slough
Birth of a daughter [Maria Sophia].
ALS 2pp RS:HS 2.294

4264. 1839–10–22 From Karl Kreil Prague
Sending magnetic observations. Would like a description of A. F. Osler's Anemometer. Regarding the periodical return of the magnetic disturbances on 18 Oct.
ALS 4pp (German) RS:HS 11.70

4265. 1839–10–22 To James Calder Stewart and Mrs. Rose Stewart Slough
Birth of [Maria Sophia Herschel]. Awaiting JS's news about Bombay. John [Stewart] and Matilda [Grahame] sailed for France yesterday.
ALS 4pp (on Margaret Brodie Herschel's 1839–10–18)
TxU:H/L-0429.2; Reel 1055

4266. 1839–10–23 To Caroline Lucretia Herschel Slough
Reports the birth of JH's seventh child, Maria Sophia. Met Adolphus Frederick, the Duke of Cambridge, who asked about CH.
ALS 3pp TxU:H/L–0584.7; Reel 1058 *Memoir ... of Caroline Herschel*, 307–8 inc

4267. 1839–10–23 To Thomas Maclear Slough
Sends meteorite and analysis to British Museum. Refers to work on Alpha Centauri; predicts occultation in fifteen years. Will move to Collingwood in spring. Announces birth of JH's daughter [Sophia].
ALS 4pp Cape Archives/Maclear Papers File 100

4268. 1839–10–23 From M. Tunno Amport, Andover
Congratulates JH on the birth of a daughter [Maria Sophia].
ALS 4pp RS:HS 17.448

4269. 1839–10–28 From Edward Joshua Cooper Carlton Club
Not sure where JH is now residing. Sending the 13 sets of Meteorological Observations he has made. Will shortly be residing at Brighton for a few months and hopes to meet JH

then. The Munich circle is a beautiful instrument.
ALS 4pp RS:HS 5.282

4270. [1839]–10–28 From Edward Sabine Liège [Belgium]
Reports meetings of ES and Humphrey Lloyd with scientists in Berlin and Göttingen. Alexander von Humboldt requests information from JH about southern hemisphere. Humboldt and Gauss are pleased by British respect for their work. ES will visit JH in Slough next week. Wants to see magnetic observations made by Navy lieutenant who accompanied Sir John Franklin to Van Diemen's Land.
ALS 3pp TxU:H/M–0517.10; Reel 1093

4271. 1839–10–29 From Thomas Maclear Royal Observatory, Cape of Good Hope
William Mann arrives with compensation bars. Receives six men for transportation of equipment to Land Observatories. Military possession of Natal ends. Foundation laid for college at Government Gardens.
ALS 4pp RS:HS 12.123

4272. 1839–10–30 To D. François J. Arago Somerset House
Did Académie des Sciences agree to R.S.L. request for magnetic observatory in Algiers? Stresses importance of multinational cooperation. Details of planned global survey of earth's magnetic field. Encloses list of observations needed from Paris.
ALS 8pp DMC.695A.885.30 (Adraft: TxU:H/L-0037; Reel 1054)

4273. 1839–10–31 From George Biddell Airy Royal Observatory, Greenwich
Concerning JH's experiments with photogenic light. Encloses paper of the comparisons between JH's and the Observatory's barometers. R.A.S. and R.S.L. want a subject for medals in astronomy, R.S.L. for a Copley medal in anything.
ALS 2pp RS:HS 1.84 (C: RGO 6.368.628)

4274. 1839–11–2 To Edward Joshua Cooper Slough
Thanks for observations; comments on one of his own observations; notes imminent move to Collingwood and birth of seventh child.
C 1p RS:HS 22.30

4275. 1839–11–3 From William Richard Hamilton Chelsea
Received the enclosed in August from Teodoro Monticelli but did not know how to forward it to JH. Would he return the letter when finished with.
ALS 1p RS:HS 9.219

4276. 1839–11–4 To Robert Treat Paine Slough
Just received RP's Dec. 1837 observations, forwarded from Cape of Good Hope. JH is now president of R.S.L. joint committee of Physics and Meteorology. Meteor shower of 10 Aug. 1839. Describes global magnetic survey conducted on Göttingen Mean Time. Hopes U.S. will join survey. [Enclosure: Printed notice of global magnetic survey, listing participating governments, purposes, and instruments.]
ALS 2pp damaged, encl 2pp † TxU:H/L-0292; Reel 1054 (C: RS:HS 25.15.9)

4277. 1839–11–5 From Edward Sabine 3 St. James's Place
Speaks of arrangements for magnetic observatories in India, in the Himalayas, and at Bombay. Discusses the locations and buildings of the sites. Wishes to add Singapore as an observatory station.
ALS 3pp RS:HS 15.61

4278. [1839]–11–7 From Edward Sabine Teignmouth
Americans plan to investigate laws of North American storms. U.S. government should apply to British if cooperation is desired. Thanks JH for letter recommending ES for royal medal. Hopes someone will work on scientific data collected by different governments.
(S) 7pp RS:HS 15.318

4279. 1839–11–9 From Edward Sabine 3 St. James's Place
Encloses a letter from Alexander von Humboldt to Lord Minto endorsing the idea of an international magnetic survey. Stresses the value of the letter.
ALS 3pp RS:HS 15.62

4280. 1839–11–10 From Francis Beaufort n.p.
Thomas Maclear has sent him another Cape meteorite. Believes that JH gave part of a meteorite to the B.M. so does not see why this one should be sent there.
ALS 2pp RS:HS 3.367 & ACC: Hydrographic Office Letter Book

4281. 1839–11–11 From Francis Baily London
Has written to G. B. Airy about the Cavendish experiment and encloses his diagram and letter.
ALS 1p RS:HS 3.163

4282. 1839–11–11 From Edward Sabine Admiralty
Encloses notes. Has received JH's message about the action of the R.S.L. Council.
ALS 1p † RS:HS 15.63

4283. 1839–11–12 From Edward Sabine Admiralty
Encloses Humphrey Lloyd's plans for [magnetic] observation. Requests a reply.
ALS 1p RS:HS 15.64

4284. 1839–11–13 To [Agnes Baillie] Slough
Accepts invitation to party next Tuesday. Regrets not visiting more often at Cavendish Square or Hampstead. JH's wife and 'little one' are well.
ALS 2pp TxU:H/L-0749.2; Reel 1089

4285. 1839–11–14 From Sir Gore Ouseley Hall Barn, Beaconsfield
Hopes to hear a good account of Lady Herschel and their little boy. If JH comes to luncheon on the 22nd he will be able to see the drawings of the Caucasus.
ALS 1p RS:HS 13.205

4286. 1839–11–15 From Francis Baily London
Encloses a letter to the Admiralty for JH's signature.
ALS 1p RS:HS 3.164

4287. [1839–11–15] To Sir Gore Ouseley [Slough]
[Responding to GO's 1839–11–14], JH will come to the luncheon and meet GO's 'Illustrious Guest.' Will bring additions for Miss Ouseley's autograph collection. Margaret Herschel is recovering well [from birth of Maria Sophia].
ALS 2pp BostonPubLib Ms. Eng. 204

4288. 1839–11–15 From Edward Sabine 3 St. James's Place
Announces that copies of the observations are being sent by post and states that additions and corrections will be approved and added before the full report is submitted to the R.S.L. Council for authorization for printing. Agrees to write

up the Meteorological Committee's annual report.
ALS 3pp RS:HS 15.66

4289. 1839–11–15 From Edward Sabine 3 St. James's Place
Claims the name 'meteorological committee' is misleading for a R.S.L. committee that handles all branches of observational physics. Suggests that physics be divided into 'Experimental and Terrestrial.'
ALS 3pp RS:HS 15.65

4290. 1839–11–15 From William Henry Fox Talbot Lacock Abbey
Transmits a message from J. B. Biot, comments on the Daguerreotype, although WT has not tried it.
ALS 3pp RS:HS 17.298

4291. 1839–11–17 To Francis Baily Slough
Further regarding the Cavendish experiments.
ALS 4pp RS:HS 3.165–6 (C: RS:HS 25.8.23)

4292. 1839–11–18 From Edward Sabine 3 St. James's Place
Agrees that in the R.S.L., Geology belongs under Terrestrial Physics, and would be there had it not its own committee. Says Terrestrial Magnetism no longer belongs under the head of Atmospheric Physics. Claims he plans to discontinue his membership in the R.S.L. if his committee is no longer in charge of terrestrial magnetism. Refers to the awarding of medals.
ALS 4pp RS:HS 15.67

4293. 1839–11–19 From Edward Sabine 3 St. James's Place
Compliments Georg Erman's map of the declination lines made from observations, which confirms C. F. Gauss's map derived by theory. Asks JH whether the map should be printed and attached to the R.S.L. report. Plans to see JH at Slough. Thinks the idea of there being two auroras, one in the west and one in the east, is interesting.
ALS 4pp RS:HS 15.68

4294. 1839–11–20 From Gilbert Elliot, Second Earl of Minto Admiralty
[Marked 'Private.'] Encloses letter from Alexander von Humboldt containing suggestions to R.S.L. regarding J. C. Ross's expedition.
ALS 1p † TxU:H/M–0373.1; Reel 1087

4295. 1839–11–20 From Thomas Maclear Royal Observatory, Cape of Good Hope
Went to meteorite site in Bokkeveld, interviewed witnesses. Describes site in detail.
ALS 5pp Cape Archives/Maclear Papers File 100

4296. 1839–11–21 From Marshall Hall 14 Manchester Square
Some time ago he sent JH a parcel of books and letters. Wonders if they ever reached him. Now sends some letters that shows how his work is appreciated on the continent.
ALS 4pp RS:HS 9.202

4297. 1839–11–22 From Edward Sabine 3 St. James's Place
Talks of new plans for an observatory in Egypt. Encloses two letters. Has begun to compare data of the last three years from five or six observatories all over the world.
ALS 3pp RS:HS 15.69

4298. 1839–11–23 From Joseph Laurie Edinburgh
Recommends John McNaughton for a position as a master in the Government Free Schools at the Cape of Good Hope. Asks for information on the salary, pension, etc., for persons hired.
ALS 2pp UCT JH Letters A1.13

4299. 1839–11–24 To Gilbert Elliot, Second Earl of Minto n.p.
Suggestions by Alexander von Humboldt will be added to R.S.L. report regarding J. C. Ross's expedition. Send copy of Humboldt's letter so that it too may be included.
ACCS 1p TxU:H/M–0373.2; Reel 1087

4300. [1839]–11–25 From Francis Beaufort Admiralty
Sends hourly meteorological register for 21 Sept. in Bahama Islands. [Partial sentence about] Equipment reaching J. C. Ross at Madiera.
ALS 1p damaged RAS:JH Archive 4/2.4; Reel 7 (ACC: Hydrographic Office Letter Book)

4301. 1839–11–25 From John Phillips B.A.A.S., York
JH appointed to head a committee including George Peacock, Humphrey Lloyd, and Edward Sabine to study meteorological instruments. They will be granted £400. The grant will be presented in Glasgow next September.
(A)LS 1p TxU:H/M–0428.5; Reel 1093

4302. 1839–11–25 From John Phillips B.A.A.S., York
JH appointed to committee with G. B. Airy and Thomas Henderson to study the reductions of the calculations on N. L. Lacaille's stars. They will be granted £189.
(A)LS 1p † TxU:H/M–0428.4.1; Reel 1093

4303. 1839–11–25 From John Phillips York
Informs JH that the B.A.A.S. has again placed him on a committee to translate and publish foreign scientific memoirs.
(A)LS 1p RS:HS B26.2

4304. 1839–11–26 From George Biddell Airy Royal Observatory, Greenwich
Double image micrometer has arrived; limitations and use of it. Wants list of stars for regular observations.
ALS 1p RS:HS 1.85

4305. [1839]–11–26 From Edward Sabine n.p.
Speaks of [John] Caldecott, Astronomer to the Rajah of Travancore, who plans to build an observatory. Needs a letter from the R.S.L. for the Rajah and also for Johann Lamont, who is working in Germany.
ALS 1p RS:HS 15.70

4306. 1839–11–27 From Gilbert Elliot, Second Earl of Minto Admiralty
Has returned Alexander von Humboldt's letter to Edward Sabine for alterations before it is printed.
ALS 1p RS:HS 7.38

4307. 1839–11–27 From Marshall Hall 14 Manchester Square
Regrets his letter should have given offense but wrote it with the intention of claiming his rights at the R.S.L.
ALS 4pp RS:HS 9.203

4308. 1839–11–28 From Francis Baily Royal Society
Invitation to spend the night with him after the R.S.L. dinner on the Saturday.
ALS 1p RS:HS 3.167

4309. 1839–11–29 To George Biddell Airy Slough
About astronomical matters, such as parallax and variable stars.
ALS 4pp RGO 6.368.629 (ACCS RS:HS 25.5.19)

4310. 1839–11–29 To Francis Baily Slough
Fears that he will be unable to attend the Anniversary meeting. Has written to G. B. Airy concerning Orionis.
ALS 3pp RS:HS 3.168 (C: RS:HS 25.8.24)

4311. 1839–11–29 From William Benjamin Carpenter 22 Park St., Bristol
Sending his work for perusal. Draws his attention to the new advances in the treatment of physiology.
ALS 4pp RS:HS 5.184

4312. 1839–12–1 From Caroline Lucretia Herschel Hanover
Details her daily routine.
L *Memoir ... of Caroline Herschel*, 302–3 inc

4313. 1839–12–2 From Francis Baily London
Is sorry JH was unable to be with them at the Anniversary meeting. Has had a further letter from the Admiralty regarding a pendulum.
ALS 2pp RS:HS 3.169

4314. 1839–12–2 From Adolphe Quetelet Brussels
Is comparing 'points and measures' of new stations with [François] Arago. Also busy with daguerreotypes. Thanks for encouragements. Has necessary instruments for meteorological and magnetical observations. Mentions work done and astronomers met on Italy trip. Discusses Prince Albert.
ALS 4pp (French) RS:HS 14.217

4315. 1839–12–4 To William Henry Fox Talbot Slough
Discusses L. J. M. Daguerre's patent application, JH's experiments in making light sensitive paper, and JH's determination to give up photographic researches so as to return to preparation of JH's *Cape Results*.
ALS 4pp Science Museum Talbt 1/22

4316. 1839–12–4 From E. Willes Shanklin, near Newford, Isle of Wight
Still invalid. Offers to be godfather for JH's new child [Maria]. When does JH plan to move from Slough?
ALS 3pp RS:HS 18.260

4317. 1839–12–5 From Edward Sabine Worcester Park, Ewell
Approves of the R.S.L. letter written to the Rajah of Travancore. Notes that the Court of Directors has approved the change in observatory locations from Bombay to Singapore and from Doon[?] to Simla. The site of the Canada observatory in Toronto has been chosen. Discusses a committee meeting time and expenses. Mentions Mr. Clark's work in atmospheric electricity.
ALS 3pp RS:HS 15.71

4318. 1839–12–6 From Francis Baily London
Has had no further tidings of the pendulum so has drafted the enclosed letter.
ALS 1p RS:HS 3.170

4319. 1839–12–7 From William Henry Fox Talbot Lacock Abbey
Comments on many different processes in photography. Still has not tried the Daguerreotype, although WT has had the equipment for a considerable time.
ALS 4pp RS:HS 17.299 Buckland, *Talbot*, 589 inc

4320. 1839–12–9 From Robert Hunt Devonport
Has been experimenting with photographic papers. Outlines some of his experiments. Has prepared some papers for sale.
ALS 4pp RS:HS 10.80

4321. 1839–12–10 To Lord John Russell[?] n.p.
Discusses JH's efforts to secure applications for teaching positions at the Government Free Schools at the Cape of Good Hope. Lists a number of candidates.
draft 2pp UCT JH Letters A1.14 Ferguson and Immelman, *Herschel and Education*, 60–2

4322. 1839–12–10 From Lord John Russell Bowood
Replies to JH's letter of 1839–12–10. States that George Napier has informed him that four positions are to be filled. Asks JH which candidates he would recommend.
ALS 2pp UCT JH Letters A1.16

4323. 1839–12–11 To William Rowan Hamilton Slough
Requests another copy of WH's paper on light, recently read at Royal Irish Academy, having sent JH's own copy to Prague's [Karl] Kreil for description of [Humphrey] Lloyd's vertical magnetometer contained therein. Has WH yet found the three axes of the universe? Comments on WH's sister's 'charming' poetry.
ALS 3pp TCD 1493:182 Graves, *Hamilton, 2*, 312–13 inc

4324. 1839–12–11 From Margaret Hall to Margaret Brodie Herschel 30 Portland Place
In a postscript, Basil Hall says that he plans to bring his camera when the Hall's come to visit so that he may photograph what remains of the great 40-foot reflector.
ALS 3pp RS:HS 9.190

4325. 1839–12–11 To Karl Kreil Slough
In response to KK's 1839–10–22, JH sends information on [A. F.] Osler's anemometer, and on magnetic measurements to be carried out.
C 2pp RS:HS 22.31

4326. 1839–12–11 To George Thomas Napier Slough
Is working towards having the government hire the school masters [see GN's 1839–9–30] that GN asked JH to find. JH has also spoken to Lord John Russell about Andries Stockenström's plight. [Letter finally completed on 1840–2–11.]
Cdraft 5pp RS:HS 25.15.10

4327. 1839–12–13 From Edward Sabine London
To remedy the negligent omission of Humphrey Lloyd's name and the work of the observatories in a recent R.S.L.'s presidential speech, special communications will be sent to all foreign scientific institutions. Requests a special meeting of either the Council or the Committee be called to discuss the many foreign communications.
ALS 4pp RS:HS 15.73

4328. 1839–12–13 From Edward Sabine London
Announces that the R.S.L. Council has passed the resolution to reorganize previous committees into one entitled 'Physics and Meteorology.'
ALS 2pp RS:HS 15.72

4329. 1839–12–14 To Robert Hunt Slough
Intrigued by RH's experiments with photographic papers; will buy six sheets of RH's specimens. Admits having conducted numerous experiments himself and asks to include some of RH's results in a forthcoming paper on the subject.
ALS 3pp George Eastman House 1.17 (draft: RS:HS 10.81 & C: 22.32)

4330. 1839–12–14 From Thomas McBean 35 Bedford St., Covent Garden
Requests an appointment to see JH in regard to TM's candidacy for a position teaching at the Government Free

Schools at the Cape of Good Hope.
ALS 1p † UCT JH Letters A1.15

4331. 1839–12–15 To Edward Sabine Slough
Regrets deeply the little notice given to the observatories and to Humphrey Lloyd in the R.S.L. President's recent speech. Urges ES to comfort Lloyd. States R.S.L. 'is no longer the sole arena in which a scientific name can be acquired.'
ALS 7pp RS:HS 15.74 (C: RS:HS 22.33)

4332. 1839–12–16 to 1840–1–10 To George Biddell Airy Royal Observatory, Greenwich
Brief note supporting GA's medal recipient ideas [see GA's 1839–12–16], and adding some of JH's ideas.
ALS 3pp RGO 6.368.632

4333. 1839–12–16 From George Biddell Airy Royal Observatory, Greenwich
Sends suggestions for recipient of R.A.S. medal and comments on JH's 1839–11–29.
ACCS 1p RGO 6.368.631

4334. 1839–12–16 From Edward Sabine Worcester Park
Writes of recent data. Addition of the Prague and Milan magnetic observatories will bring the total to 18 sites. In presenting to the Council a map ready for publication, has found out that the other committees have not been informed about the need to revise their own reports. Appreciates JH's letter of regret about the lack of appreciation in the recent R.S.L.'s President's speech.
ALS 4pp RS:HS 15.76

4335. 1839–12–16 To Edward Sabine [Slough]
Declines to take steps to call a special committee or council meeting of the R.S.L. regarding foreign communications of the observatories, noting his own wish to stay out of R.S.L. affairs.
ACCS 1p inc RS:HS 15.75 (C: RS:HS 22.34)

4336. [1839 to 1854]–12–16 To [J. T.?] [Collingwood]
Write to Richard Sheepshanks for permission to copy engraving.
C 3pp (JT note) RS:HS 19.163

4337. 1839–12–18 From Robert Hunt Devonport
Is grateful for his letter. Sends some of his photographic sheets. Now uses a different chemical to prepare his papers. Will be pleased to send JH his results and to receive a copy of the memoir in due course.
ALS 3pp RS:HS 10.82

4338. 1839–12–18 From Thomas McBean 35 Bedford St., Covent Garden
Cannot afford to visit JH and has no testimonials to support his candidacy for a teaching position at the Cape. Describes his attainments and pleads for consideration.
ALS 3pp UCT JH Letters A1.17

4339. 1839–12–18 From Edward Sabine Worcester Park
Concurs with JH in his desire to stay out of R.S.L. affairs but feels it is important to promote terrestrial magnetism in the R.S.L., especially because many foreign countries with observatories like to deal officially with the R.S.L. instead of private individuals. Asks if JH wishes to continue receiving information about magnetism after JH moves to Collingwood.
ALS 5pp RS:HS 15.77

4340. 1839–12–21 From Roderick Impey Murchison Up Park, Petersfield
Requesting JH to write the next Anniversary address for the meeting of the B.A.A.S. as there is no local person capable of producing one.
ALS 4pp RS:HS 12.401

4341. 1839–12–24 From Francis Beaufort Admiralty
Enclosing a page and a drawing, which may be of interest to him. Best wishes for Christmas.
ALS 2pp RS:HS 3.368

4342. [1839]–12–24 From Edward Sabine 3 St. James's Place
[John] Boileau brought drafts of reporting forms from [Humphrey] Lloyd. Wants JH to examine these books. Will obtain supplies for observatories.
ALS 3pp inc RS:HS 15.295

4343. 1839–12–29 To Friedrich Wilhelm Bessel Slough
Thanks FB for a variety of observational material, including establishment of standards of measurement [letter completed 1840–1–3].
ALS 4pp AdWDDR Bessel 225 (C: RS:HS 22.35)

4344. 1839–12–30 From Edward Sabine n.p.
Ready to propose Hammerfest to the Council as a new observatory site. Says the Christiania station is lost due to oversized equipment.
ALS 2pp RS:HS 15.78

4345. [1840?] To [Francis Baily?] n.p.
A series of 'resolutions to be proposed as fundamental principles for a reform of the Southern constellations.'
draft 4pp RS:HS 25.14.1

4346. [1840 or later] From Joseph J. Hart n.p.
Asks for appointment with JH and directions from Etchingham station [to JH's residence].
ALS 1p inc RAS:JH Archive 13/3.94 verso; Reel 10

4347. [1840 or later] From Henry Anthony Jeffreys [Parsonage, Hawkhurst]
Thanks JH for his support.
ALS 1p Harvard: Houghton AAH 67m–67 (63)

4348. [1840 or later] From Henry Anthony Jeffreys [Parsonage, Hawkhurst]
Thanks JH and Lady Herschel for their kindness to his family [in time of his brother's illness].
ALS 1p inc Harvard: Houghton AAH 67m–67 (63)

4349. [1840 or later] From Henry Anthony Jeffreys [Parsonage, Hawkhurst]
Notifies JH that a number of visitors will be coming to the Parsonage.
ALS 1p † Harvard: Houghton AAH 67m–67 (107)

4350. [1840 or later] From Henry Anthony Jeffreys [Parsonage, Hawkhurst]
Reports on HJ's brother's health.
ALS 1p inc Harvard: Houghton AAH 67m–67 (63)

4351. [1840 or later] From Henry Anthony Jeffreys [Parsonage, Hawkhurst]
Tells JH who will be at a gathering at the parsonage.
ALS 1p inc Harvard: Houghton AAH 67m–67 (63)

4352. [1840 or later] From Henry Anthony Jeffreys [Parsonage, Hawkhurst]
Thanks JH for a favor.
ALS 1p † Harvard: Houghton AAH 67m–67 (107)

4353. [1840 or later] From Henry Anthony Jeffreys [Parsonage, Hawkhurst]

Invites JH to dinner.
ALS 1p Harvard: Houghton AAH 67m–67 (63)

4354. [1840 to 1855] To Richard Jones n.p.
Sorry that JH missed RJ at W. H. Fitton's. Offers proposal to modify property taxation procedures to yield 'taxation by payments downward' and to decrease load on poorer classes.
C 2pp TxU:H/L-0206; Reel 1054

4355. [1840 or later] From Charles Knight [13 Pall Mall?]
Compliments JH on 'greatly improved edition' of a book adding to 'every department of knowledge.'
ALS 1p inc Harvard: Houghton AAH 67m–67 (63)

4356. [1840] To John William Lubbock n.p.
On the 'misfortunes & mishaps' of the Aden magnetic observatory and on the proper care of magnetical instruments.
ALS 4pp RS MM 11.191

4357. [1840 or later] From Sir Charles Lyell n.p.
Is sorry JH did not inform him of the paper he was to read on the astronomical influence on climate. Would like an abstract when published. Would also like a copy of the verses on the great telescope. Encloses a paper of his own.
ALS 2pp RS:HS 11.448

4358. [1840 to 1846] To Roderick Impey Murchison Tavistock Place
Did not vote for RM's friend at the R.S.L. because JH missed the meeting. JH will also miss the next meeting of the Geological Society, as he is returning to Collingwood.
ALS 2pp WT 67430.1

4359. [1840?] To Hugh Percy [3rd Duke of Northumberland] n.p.
Having heard that HP had expressed pleasure at some of JH's colored photographs, JH sends some more recent ones.
Cdraft 1p RS:HS 25.13.4

4360. [1840 to 1847] To [Hugh Percy (3rd Duke of Northumberland)] n.p.
Lady [Pleasance] Smith conveyed Duke of Northumberland's interest in JH's color photographs. Encloses others, and explains use of light-sensitive juices of flowers to produce colors.
Adraft 1p RS:HS 19.164

4361. [1840 or later] From Jesse Piper Conghurst, [Kent]
Asks JH to loan map to JP.
ALS 1p TxU:H/M–0407.1; Reel 1087

4362. [1840–1] To Adolphe Quetelet n.p.
Herschelian Telescope Song in English and Latin, translated by [T. I. M.?] Forster into Latin. Requiem of 40-foot reflector sung at New Year's Eve 1839–40.
Poem 1p Académie belgique: #16

4363. [1840] To Recipients of Report of R.S.L. Joint Committee of Physics and Meteorology n.p.
Letter designed to accompany report of R.S.L. Joint Committee of Physics and Meteorology. Discusses plans for magnetic observatories and for James Ross's expedition.
draft 5pp RS MM 11.190

4364. [1840] To Edward Sabine n.p.
On what is to happen at the expiration of a three year period assigned for the undertaking of certain magnetic observations.
ALS 2pp RS MM 11.192

4365. [1840 or earlier?] From Christian F. Schönbein Blackfriar's Road
Thanks JH for invitation to Slough. Will come to visit.
ALS 1p RS:HS 15.365

4366. [1840 or earlier] To Mr. Wagner [Slough}
Invites Mr. Wagner at Eton to bring his pupils to Slough to observe an occultation [of Jupiter by the moon?].
ALS 2pp DMC.695A.885.15

4367. [1840 to 1841] To ? n.p.
Note on diurnal changes in [magnetic] variation at St. Helena and on estimated cost (£950) of observatory at St. Helena.
Amem 1p TxU:H/M–0407.2; Reel 1087

4368. [1840?] To [R.S.L. Council] n.p.
Reports on and enthusiastically recommends for publication G. B. Airy's paper on light polarity [see *RSPT, 130* (1840), 225–], which JH believes contains true explanation for phenomenon of spectra bands.
CDS 3pp RS RR.1.3

4369. [1840?] To [R.S.L. Council] n.p.
Reports on and suggests improvements in but recommends publication of a paper [see *RSPT, 130* (1840), 325–] by Robert Hunt on iodine's effect in rendering 'argentine paper' sensitive to light and thus useful for photographic purposes.
CD 3pp RS RR.1.110

4370. [1840?] To [R.S.L. Council] n.p.
Reports on William Parson's paper [see *RSPT, 130* (1840), 503–] on large mirrors for reflecting telescopes. Suggests omitting comments on William Herschel's mode of polishing, but recommends paper for publication.
CD 7pp RS RR.1.211

4371. 1840–1–3 From Francis Baily London
Invitation to spend the night with him when he comes to town for the council meetings of the Royal and Astronomical Societies.
ALS 2pp RS:HS 3.171

4372. 1840–1–3 From William Henry Fitton Bloomfield Lodge
Has just returned from a trip to Hanover, and has decided not to place his son there in accordance with advice received while he was there. Met Miss Herschel. Travelled via Belgium and met A. H. Dumont, a very original and intelligent geologist.
ALS 4pp RS:HS 7.237

4373. 1840–1–4 To Francis Baily [Slough]
Will be pleased to avail himself of the offer of a bed when the council meets. Will prepare a paper on Alpha Orionis for the Friday meeting. Can FB give him any ideas for the Anniversary address?
ALS 4pp RS:HS 3.172 (C: RS:HS 25.8.25)

4374. 1840–1–4 From Thomas Hodgkin Lower Brook St.
Affairs at the Cape; thinks Andries Stockenström is the man to be sent out to negotiate with the Boers. Would like JH's views on the subject.
ALS 4pp RS:HS 9.411

4375. 1840–1–7 From Francis Baily 37 Tavistock Place
Will be expecting him on Thursday.
ALS 1p RS:HS 3.173

4376. 1840–1–7 From Francis Beaufort Admiralty
Thanks for sending him the requiem.
ALS 2pp RS:HS 3.369

4377. 1840–1–7 From Humphrey Lloyd Trinity College, Dublin
Has not heard of any results of JH's letter to D. F. J. Arago, but Georges Aimé has been appointed to direct an observatory in Algiers. Great interest at the Institute when JH's letter was read. Glad he did not waste a second letter to the *Times*.
ALS 3pp RS:HS 11.273

4378. 1840–1–8 From Basil Hall Streatham
Francis Beaufort has shown him his copy of the 'Requiem' and would be pleased if JH would send him a copy. Mr. Harvey and himself have been observing nebulae on the clear nights. Is having a Munich object glass mounted. Will be in town towards the end of the month.
ALS 3pp RS:HS 9.191

4379. 1840–1–8 From Edward Sabine n.p.
Sends the corrections of the Meteorological Instructions to JH. Presents a new plan for recording data.
ALS 4pp RS:HS 15.79

4380. 1840–1–9 From Francis Beaufort Admiralty
Sending him the last *Comptes rendus* that can go by post. Hopes he can arrange some other method. Is always pleased to receive his publications.
ALS 3pp RS:HS 3.370 & ACC: Hydrographic Office Letter Book

4381. 1840–1–9 From James Hayler Moorhouse, Hawkhurst, Kent
Long-time servant at Collingwood, first of Jesse Gregson, then to G. N. Collingwood. Able to care for poultry, dairy, or gardens. Received 'orders to leave Elpick' in March but begs permission to remain as JH's servant.
ALS 1p RAS:JH Archive 2/1.2; Reel 5

4382. 1840–1–10 To Charles Babbage n.p.
Regarding communications from Paddington. Gives verses written in honor of the 40-foot telescope. Has some strange results in photographic work.
ALS 2pp RS:HS 2.295

4383. 1840–1–11 From Robert Vernon Smith Downing St.
Writing on behalf of Lord John Russell, asks JH to notify the persons JH has selected to teach at the Cape of Good Hope to visit RS at the Colonial Office.
ALS 2pp UCT JH Letters A1.18

4384. 1840–1–12 To William Rowan Hamilton Slough
Notice on the 'final laying up' of the Old Telescope, with a 'Requiem of the Forty-Foot Reflector,' sung by JH's family on New Year's Eve.
ALS 1p encl 1p TCD 1493:184 Graves, *Hamilton, 2,* 313–15

4385. 1840–1–13 From Philip Melvill India House
Sending a list of the principal instruments possessed by Lt. Col. George Everest. Also the last report made by Everest of his observations.
ALS 2pp RS:HS 13.33

4386. 1840–1–13 To William Whewell Slough
On 1 Jan. 1840, JH's family laid William Herschel's 40-foot reflector to rest by singing a song (enclosed). Mentions Hammerfest instruments.
ALS 1p encl TC, Camb. Add. Ms.a.207[42]

4387. 1840–1–15 From James Grahame Nantes
Comments on JH's laying to rest of William Herschel's 40-foot telescope.
ALS 2pp RS:HS 8.436

4388. 1840–1–16 From John William Lubbock n.p.
Has read over all JH's report. Makes suggestions for alterations and omissions to bring it into line with the wishes of Council.
ALS 4pp RS:HS 11.361

4389. 1840–1–17 From Andries Stockenström to Margaret Brodie Herschel Devonshire St.
Apologizes for not visiting Slough. Tells of difficulties of attempting to move his family to London.
ALS 4pp RS:HS 17.21

4390. 1840–1–17 From Edward Sabine n.p.
Discusses corrections and title page of a report. Discusses the damage to Charles Riddell's instruments.
ALS 3pp RS:HS 15.80

4391. 1840–1–17 To John Taylor, F.R.S. Slough
Requests aid in obtaining materials for experiments, including metal ores.
ALS 4pp RS:HS 17.427b (C: RS:HS 22.37)

4392. 1840–1–18 From [Edward Joshua Cooper] Leamington Priors
Recalls earlier letter in which [EC] made suggestions to JH for improving Daguerreotype.
AL 1p inc † RAS:JH Archive 5/4.8 verso; Reel 8

4393. 1840–1–18 From Michael Faraday Royal Institution
As JH has tried many varnishes he wonders if he has ever tried shellac. Has tried it on specimens of brass. Many thanks for his invitation.
ALS 2pp RS:HS 7.182 Williams, *Selected Correspondence, 1*, 358–9

4394. 1840–1–18 From William Rowan Hamilton Observatory
Confides that the past year has been 'a sombre one ... but not unhappy.'
L Graves, *Hamilton, 2,* 315

4395. 1840–1–19 From Robert Vernon Smith Downing St.
Writing on behalf of John Russell, asks that JH inquire about the suitability of T. J. Paterson for a position teaching in the Government Schools at the Cape of Good Hope. Mentions Thomas McBean.
ALS 2pp UCT JH Letters A1.37

4396. 1840–1–21 From James Grahame Nantes
On the effect of JH's visit to Nantes.
ALS 1p RS:HS 8.437

4397. 1840–1–21 From Andries Stockenström Devonshire St.
Thanks JH for offer of financial assistance which he declines. Expresses concern over the impending decision concerning his pension. Discusses arrangements for his family's voyage to London.
ALS 2pp RS:HS 17.22

4398. 1840–1–22 To Philip Melvill Slough
Returns, with many thanks, the report of George Everest. Gives adverse comment on the alterations that Everest has made to some of the instruments.
draft 4pp RS:HS 13.34 & 22.38

4399. 1840–1–24 From George Biddell Airy Royal Observatory, Greenwich
A note, sent with the manuscript papers of Stephen Groombridge, to JH as President of the R.A.S.
ACCS 1p RGO 6.192.395

4400. 1840–1–24 From Charles Few Covent Garden
Hopes to send him the lease tomorrow. If JH is coming to town will be able to see him after 10 a.m. Has written to Mr. Forbes.
ALS 2pp RS:HS 7.200

4401. 1840–1–25 From Giovanni Antonio Plana Turin
Very pleased to hear that GP received the gold medal of the R.A.S. Sends instructions for receiving. Packet from JH with a portrait of JH has not arrived; could JH send another?
ALS 4pp† (French) RS:HS 13.436

4402. 1840–1–27 From John Theophilus Boileau Upper Seymour St.
The magnetometer will not be in position tomorrow. J. C. Melvill wants to know if JH has written since his last packet. Thanks for adding name to his own certificate.
ALS 2pp RS:HS 5.82

4403. 1840–1–27 From John William Lubbock London
Sends a copy of his work on the tides. S. D. Poisson's theory is noticed near the end of the 2nd volume of *Mécanique*. Thinks some experiments are necessary to find the effect of the sun on climate. Has received a note from the Treasury.
ALS 3pp RS:HS 11.362

4404. 1840–1–27 From William Simms Fleet St.
Instruments JH referred to may not be best for JH's purposes.
ALS 1p RS:HS 17.206

4405. 1840–1–28 To Hans Christian Oersted Slough
Offers HO a small requiem, which JH's family sung at the beginning of the astronomical year at the memorial of the 40-foot reflector.
C 2pp RS:HS 22.57 Harding, *Correspondance de ... Örsted, 2*, 392–3

4406. 1840–1–28 From Henry Perigal 5 Smith St., Chelsea
Invites JH to visit HP to see the machine drawings being made [see HP's 1839–8–17].
ALS 1p RS:HS 13.366

4407. 1840–1–29 To John William Lubbock Slough
Has received JL's memoir on the tides. JH would like to amalgamate certain portions with a memoir from William Whewell. JH adds some comments on photographic experiments he has made.
C 2pp RS:HS 25.14.22

4408. 1840–1–30 To Edward Belfour Slough
Declines the invitation to the annual Hunterian Oration and dinner. Now that JH is moving even farther away from London, he feels he must reluctantly resign as a member of the Trustees of the Hunterian Collection.
ALS 3pp WT 91800.3

4409. 1840–1–30 From James Rose Innes Cape Town
Writing as the recently appointed Superintendent of Government Schools at Cape Colony, JI recounts his findings from a tour of the schools at the Cape, reporting many problems.
ALS 8pp † UCT JH Letters A1.20

4410. 1840–1–31 From Few & Co. Covent Garden
Forwarding the draft of the proposed lease. The deeds are to be sent to Dover for registration.
ALS 2pp RS:HS 7.201

4411. 1840–1–31 From John William Lubbock London
Would like JH to amalgamate William Whewell's and JL's memoirs. Thought the principle of universal gravitation was conceded by all. Treasury have sent back the bills and want the Antarctic ones separated from the Observatories ones. Should take notice of the effect of iron steams [on instruments?].
ALS 3pp RS:HS 11.363

4412. 1840–2–1 From Charles Few Covent Garden
Check for the purchase money satisfactory.
ALS 1p RS:HS 7.202

4413. 1840–2–1 To Margaret Brodie Herschel Slough
JH is concerned about MH's health, especially her sore teeth; JH sends along some letters, and reports that all are well at Slough.
ALS 3pp JHS 4.19

4414. 1840–2–1 From Thomas McBean 35 Bedford St., Covent Garden
Is sending further testimonials as a candidate for teaching at the Cape of Good Hope. Laments his financial condition. Thanks JH for his encouragement.
ALS 4pp UCT JH Letters A1.22

4415. 1840–2 To Lord John Russell Slough
Appeals to JR on behalf of Andries Stockenström, whose dismissal from Cape Colony created grief. Stockenström is ready to resign that office if he can do so in a manner that permits him to continue his career in government.
AdraftS 3pp TxU:H/L-0346; Reel 1054

4416. 1840–2–1 From Lord John Russell Downing St.
Has decided to appoint five teachers for the Government Schools at the Cape. Three have visited him. Asks JH about the remaining two.
ALS 2pp UCT JH Letters A1.21 (C: RS:HS 22.39)

4417. 1840–2–2 To Lord John Russell [Slough]
Is in communication, as requested, with Mr. Train and Thomas McBean about their candidacy for teaching at the Government Schools at the Cape of Good Hope.
draft 2pp UCT JH Letters A1.23 Ferguson and Immelman, *Herschel and Education*, 64

4418. 1840–2–3 From Francis Baily 37 Tavistock Place
Has had a letter from H. C. Schumacher announcing his discovery of a new comet in Draco. There is to be a R.S.L. Council on Thursday.
ALS 1p RS:HS 3.174

4419. 1840–2–3 From F. M. Eardley-Wilmot St. Helena
Weather has been unsuitable for using the actinometers. Sends a small parcel of geological specimens. J. H. Lefroy is in a very good situation at Longwood.
ALS 3pp RS:HS 7.1

4420. 1840–2–3 To Margaret Brodie Herschel Slough
About buying horses, and closing the deal to buy Collingwood. JH refers to an enclosure about anti-slavery.
ALS 2pp JHS 3.18

4421. 1840–2–3 From Thomas McBean 35 Bedford St., Covent Garden
Sends further testimonials on behalf of his candidacy for a

position teaching in the Government Schools at the Cape of Good Hope. Describes himself as 'an unfortunate man of genius.'
ALS 4pp UCT JH Letters A1.25

4422. 1840–2–5 To Lord John Russell Slough
Attests to the honorable personal character of Captain Andries Stockenström. Wishes the circumstances of his dismissal could be remedied.
Extract (S) 4pp RS:HS 14.438

4423. 1840–2–5 From Francis Tudhope Edinburgh
Inquires as to the salary he would receive if selected to teach in one of the new Government Schools at the Cape of Good Hope. Supplies further testimonials.
ALS 3pp UCT JH Letters A1.26

4424. 1840–2–6 From Edward Sabine London
Believes Treasury will cover all expenses. The application for a Cairo observatory looks promising and H. C. Oersted is eager to set up a station in Copenhagen. Writes of the instrument requests of other stations.
ALS 5pp RS:HS 15.81

4425. 1840–2–7 To Thomas Hodgkin n.p.
Has written a letter to Lord John Russell outlining the case of Andries Stockenström. Does not wish to become involved in Cape politics or the affairs of the Anti-Slavery Society much as he appreciates their importance.
C(draft) 2pp RS:HS 9.412 & 22.40

4426. 1840–2–7 To Thomas McBean n.p.
Tactfully explains to TM why he cannot recommend him for a teaching position in the Government Schools at the Cape of Good Hope.
draft 3pp UCT JH Letters A1.28

4427. 1840–2–7 From Mrs. A. W[auchope?] n.p.
Writes favorably on behalf of the qualifications of T. J. Paterson for a teaching position at the Government Schools at the Cape of Good Hope.
ALS 4pp UCT JH Letters A1.31

4428. 1840–2–8 From Francis Baily 37 Tavistock Place
JH did not miss anything by being absent from the council meetings of the R.S.L. and R.A.S.
ALS 1p RS:HS 3.175

4429. 1840–2–8 From William Benjamin Carpenter 22 Park St., Bristol
Sending copies of the reviews of his work, which has been attacked in some quarters. Would like JH's views on two of the chapters.
ALS 4pp RS:HS 5.185

4430. 1840–2–8 From John William Lubbock Royal Society
Sends a copy of the report, which he has looked over with P. M. Roget. S. H. Christie seems averse to altering it. Thinks it important to make observations of the temperature with barometric readings in balloon ascents. Only readings by J. L. Gay-Lussac at present.
ALS 3pp RS:HS 11.364

4431. 1840–2–9 To Francis Baily Slough
Will attend R.A.S council and stay with FB; question of expulsion of three members.
C 1p RS:HS 22.42

4432. 1840–2–11 From Philip Jennings Hampstead
Much indebted for his kind letter, but is unable to interfere in the affairs of Sir Francis Knowles as he has no contact with him. Encloses a list of instruments belonging to the late W. C. Holford and would be pleased if JH would show it to his astronomical friends.
ALS 4pp RS:HS 10.441

4433. 1840–2–11 To George Thomas Napier Slough
[Letter begins 11 Dec. 1839] Wrote to John Russell and Charles Grant in support of school system [at Cape of Good Hope]. [Letter continues 6 Jan. 1840:] JH sent to Colonial office testimonials for four applicants [for schoolmaster] approved by Russell. Notes devastating effect of dismissal on Andries Stockenström. Took liberty to inform Russell of JH's and GN's views on harshness of this dismissal. [Letter continues 12 Jan.:] Russell will appoint four schoolmasters picked by JH. [P.S. 11 Feb.:] JH is moving to Collingwood. Russell indicates sympathy for Stockenström. JH hopes this means new assignment for Stockenström.
AdraftS 6pp TxU:H/L-0281; Reel 1054

4434. 1840–2–11 From Heinrich Christian Schumacher Altona
Asks JH to petition King Christian VIII of Denmark to support the Altona observatory.
ALS 3pp RS:HS 15.384

4435. 1840–2–13 From Thomas McBean 35 Bedford St., Covent Garden
Makes another plea for consideration for a teaching position in the Government Schools at the Cape of Good Hope. Attempts to answer some of JH's objections as given in JH's letter of 1840–2–7.
ALS 3pp UCT JH Letters A1.32

4436. 1840–2–15 To the *Philosophical Magazine* Slough
Reports on a simple mode of obtaining from a common Argand oil lamp a greatly increased quantity of light.
L *P.M.*, 3rd. ser., *16* (1840), 194–5

4437. 1840–2–15 From Lord John Russell Downing St.
Acknowledges letter regarding Captain Andries Stockenström. Sorry Stockenström could not be reappointed to Cape. Offered to recommend him to the Queen for other positions. Agrees with JH's sentiments.
ALS 4pp RS:HS 14.439

4438. 1840–2–15 From Edward Sabine 12 Park Place
Plans to order equipment needed for observatories. Asks JH to examine the bills. Announces that both Bavaria and Denmark are planning observatories.
ALS 3pp RS:HS 15.82

4439. 1840–2–15 From Robert Vernon Smith Downing St.
Writing on behalf of John Russell, asks whether JH wishes to recommend Thomas McBean for a position teaching in the Government Schools at the Cape of Good Hope.
ALS 3pp UCT JH Letters A1.33

4440. 1840–2–17 From Edward Sabine 12 Park Place
Is of the opinion that plans for the Bavarian observatory station should continue. Announces that the report is finally being printed.
ALS 4pp RS:HS 15.83

4441. 1840–2–18 From Mrs. N. Collingwood Dover
Is grateful that he has given her father's name to his recently acquired property. Hopes he will enjoy many years there.
AL 2pp RS:HS 5.357

4442. [1840]–2–18 From A. W[auchope?] 6 Maria Place
Writes to support the candidacy of Francis Tudhope for a position teaching in the Government Schools at the Cape of Good Hope.
ALS 4pp UCT JH Letters A1.34

4443. 1840–2–18 From Francis Tudhope to Robert Wauchope Stockbridge
Sends further testimonials regarding his candidacy for a position teaching in the Government Schools at the Cape of Good Hope. Asks about the terms of the appointment.
ALS 2pp UCT JH Letters A1.35

4444. 1840–2–19 From Francis Baily London
They should not refuse H. C. Schumacher's request. Regarding a diploma for the King of Denmark.
ALS 1p RS:HS 3.176

4445. [1840 or later]–2–19 From Henry Anthony Jeffreys Parsonage, Hawkhurst
Asks JH to be a member of a parish committee.
ALS 1p Harvard: Houghton AAH 67m–67 (63)

4446. 1840–2–19 From Thomas Maclear Royal Observatory, Cape of Good Hope
Work will begin on obelisk at Feldhausen. Occultation of $Gamma^1$ and $Gamma^2$ Virginis observed. Triangulation survey will take TM away from Royal Observatory for extended time. Slave ship carries small pox and causes outbreak. Parallax observations show brightest stars, not nearest.
ALS 8pp RS:HS 12.124

4447. 1840–2–19 From Adolphe Quetelet Brussels
Sends publications from Royal Academy and Brussels Observatory. Asks 'particular indulgence' for one about vapors. Is now able to do observations required by R.S.L. A Gauss instrument has been mounted. Gives times of observations each day. Continues horary observations.
ALS 3pp (French) RS:HS 14.218

4448. 1840–2–20 From Robert Fox Mosebattle near Kelso
Devotes his spare time to optics and has made several telescopes. Would welcome his advice on where to obtain good quality flint glass.
RS:HS 7.378

4449. [1840]–2–20 From Edward Sabine 12 Park Place
Will send information from observatories where aurora are visible. Discusses occurrences at various observatories and a letter from [C.] Kreil to [A. T.] Kupffer regarding his observations.
ALS 4pp RS:HS 15.303

4450. 1840–2–20 From Robert Vernon Smith Downing St.
Writing on behalf of John Russell, asks that JH inform T. J. Paterson of his appointment to a position teaching in the Government Schools at the Cape of Good Hope. Mentions travel arrangements for those appointed.
ALS 2pp UCT JH Letters A1.38

4451. 1840–2–21 From John Taylor, F.R.S. Duke St., Adelphi
Promises to help JH obtain materials for experiments including mercury and silver ore.
ALS 3pp RS:HS 17.428

4452. 1840–2–22 From Luke Howard Ackworth
Has been applying his mode of calculation for the barometrical heights for the years 1815–23 and sent the results to the R.S.L. As JH may not see them he sends a few comments on them. Would like a suitable colleague to share his labors.
ALS 4pp RS:HS 10.3

4453. 1840–2–22 From Robert Vernon Smith Downing St.
Writing on behalf of John Russell, asks that JH inform Francis Tudhope of his appointment to a position teaching in the Government Schools at the Cape of Good Hope.
ALS 2pp UCT JH Letters A1.39

4454. 1840–2–22 From William Whewell Trinity College
Discusses what instruments should be given to Palon Boguslawski. Discusses tidal observations, noting the complexity of the results obtained.
ALS 6pp RS:HS 18.191–2

4455. 1840–2–24 From John Theophilus Boileau Gravesend
Loading scientific instruments aboard ship [*Anshimy*?] tomorrow prevents JB from visiting JH before departure. Missed annual R.A.S. meeting. Book of 'Observations' is nearly finished. Told [J. C.] Melvill that instruments in India were not adequate for observations of astronomical refractions.
ALS 2pp † TxU:H/M–0108; Reel 1087

4456. 1840–2–25 To [Humphrey Lloyd?] Slough
Both William Whewell and George Peacock agree to using the magnetometers, forms, books, etc, but do not agree to the provision of actinometers [for the equipping of terrestrial magnetism stations??].
ALS 1p DMC.695A.885.27

4457. [1840–2–25] To Richard Sheepshanks n.p.
Alerts RS to the fact that there is an 'underhand sale' of the prints of [H. W.] Pickersgill's portrait of JH going on, as JH's cousin bought one.
ALS 2pp RS:HS 16.10 (C: RS:HS 25.9.4)

4458. 1840–2–26 To William Henry Fox Talbot n.p.
Waited until after 14 Feb. to submit JH's paper on photography ['On the Chemical Action of the Rays of the Solar Spectrum...,' *R.S.P.T.* (1840), 1–59]. Will send WT a copy.
ALS 1p Science Museum Talbt 1/23

4459. 1840–2–27 From Richard Sheepshanks Trinity College
Assures JH that engravers typically take for themselves a few copies of any print they have made. If a large number of these appear for sale, then it is fraud. Suggests JH not act because the engraver is 'insane' and the exposure would hurt his family.
ALS 2pp RS:HS 16.11

4460. 1840–2–28 From Henry Anthony Jeffreys Hawkhurst Parsonage
Hears that JH is to be one of his parishioners. Outlines the local educational establishments. Should be grateful to know if JH would be subscribing towards their upkeep.
ALS 4pp RS:HS 10.282

4461. 1840–2–28 From Heinrich Christian Schumacher Altona
Thanks JH for suggesting that the R.A.S. write to King Christian VIII, but requests that JH include his own name and mention *Astronomische Nachrichten.*
ALS 2pp † RS:HS 15.385

4462. 1840–2–28 From William Henry Fox Talbot Lacock Abbey
Concerned about the Daguerreotype patent in England, and questions at some length a paper on the origin of freemasonry.
ALS 3pp RS:HS 17.300

4463. 1840–2–29 From Henry Perigal 5 Smith St., Chelsea
More on machine-made curves [see HP's 1840–1–28].
ALS 2pp RS:HS 13.367

4464. 1840–3–2 From Richard Sheepshanks 30 Woburn Place
Needs the name of the print seller from whom JH's cousin bought the copy of the print of JH's portrait, so RS can learn how many copies are available. RS will then know whether to 'treat with [the printer] as an honest man or a rogue.'
ALS 3pp RS:HS 16.13

4465. 1840–3–2 To Richard Sheepshanks n.p.
JH concerned not so much that a few copies of the print of JH's portrait may have been sold, but that the plate itself may have been stolen and sold. Seems relieved to know that the plate was left with the printer, not the engraver. Asks RS to make an inquiry with the printer.
ALS 2pp RS:HS 16.12 (C: RS:HS 25.9.5)

4466. 1840–3–3 To James David Forbes Slough
Outlines theory behind and method for preparing thermographic paper, which allows observers to determine 'what share each ray of the spectrum bears in producing the observed total effects.'
ALS 2pp St. Andrews 12 (C: RS:HS 25.6.1)

4467. 1840–3–3 To William Henry Fox Talbot Slough
Tells WT about JH developing the thermographic paper which is susceptible to 'calorific' and 'chemical' rays of light.
ACCS 1p RS:HS 25.6.2

4468. 1840–3–4 From Francis Baily n.p.
Is the R.A.S. to send an address to the Queen? The diploma is ready for the King of Denmark.
ALS 1p RS:HS 3.177

4469. [1840]–3–4 From Edward Sabine 12 Park Place
Sends enclosure to JH. [Humphrey] Lloyd also has a copy and Lloyd will send comments directly to JH.
ALS 1p RS:HS 15.297

4470. 1840–3–4 From Robert Vernon Smith Downing St.
Writing on behalf of John Russell, informs JH that Francis Tudhope has accepted the offer of a position teaching in the Government Schools at the Cape of Good Hope. Mentions the travel arrangements for Tudhope and for the other teachers hired.
ALS 2pp UCT JH Letters A1.42

4471. 1840–3–4 From Robert Vernon Smith Downing St.
Writing on behalf of John Russell, sends JH a letter to Francis Tudhope regarding the offer to Tudhope of a position teaching in the Government Schools at the Cape of Good Hope.
ALS 2pp UCT JH Letters A1.40

4472. [1840 or later]–3–5 From Henry Anthony Jeffreys Parsonage, Hawkhurst
Asks JH's opinion on a candidate for a committee position.
ALS 1p Harvard: Houghton AAH 67m–67 (63)

4473. 1840–3–5 From Christopher Johnson Lancaster
Has just seen JH's communication in the *P.M.* regarding an improvement to a lamp. Points out that the late Charles Seward obtained a patent for this improvement some thirty years ago.
ALS 4pp RS:HS 10.446

4474. 1840–3–6 From Francis Baily 37 Tavistock Place
Regarding an address to the Queen. Hopes to see him on Thursday.
ALS 3pp RS:HS 3.178

4475. 1840–3–9 From James David Forbes Edinburgh
Would like to know the degree of sensibility to heat of JH's paper. The paper of W. H. F. Talbot is useless for his purpose as only violet heat affects it. Did JH use a flint glass prism for his spectrum experiments? Has he ever used photography to show the impression of polarization and diffraction?
ALS 2pp St. Andrews LB III 71, 72 (C: RS:HS 7.295)

4476. 1840–3–9 To John William Lubbock Slough
Comments on several different reports; also on some experiments that lead JH to think much of the radiant energy of the Sun is absorbed in the atmosphere.
ALS 4pp RS LUB.H.354 (C: RS:HS 22.43)

4477. 1840–3–10 From Thomas Buchanan 65 Cornhill
Having accepted a position teaching in the Government Schools at the Cape of Good Hope, has found himself lacking important information and necessary funds for the impending voyage.
ALS 3pp UCT JH Letters A1.43

4478. 1840–3–10 From John William Lubbock n.p.
Has received letter and report. Will he consider by Thursday whether to refer the report back to the Committee. Does not think the balloons should be too large. Regarding the sun's effect on climate. Presumes Council cannot alter report.
ALS 3pp RS:HS 11.365

4479. 1840–3–13 To Christian VIII, King of Denmark London
Announces King's enrollment as patron of R.A.S. Recalls Danish monarches who promoted astronomy. Praises Altona observatory and work of H. C. Schumacher.
AdraftS 4pp RS:HS 19.77

4480. 1840–3–13 To James David Forbes n.p.
On the preparation of paper to exhibit a thermal spectrum.
ALS 4pp St. Andrews 13 (C: RS:HS 22.44)

4481. 1840–3–13 From Andries Stockenström London
The Treasury awards AS a pension.
ALS 2pp RS:HS 17.23

4482. 1840–3–14 From John William Lubbock 29 Eaton Place
Regrets that JH in his address to the R.A.S. should have omitted to mention the names of P. G. le D. Pontecoulant and JL in connection with the calculation of the lunar and planetary perturbations.
ALS 3pp RS:HS 11.366

4483. 1840–3–15 To [John William Lubbock] Slough
JH has apparently insulted JL by crediting JL's colleagues but not JL with improvements in lunar and planetary theory. JH explains how this happened, that JH's good intentions went awry.
ALS 3pp RS LUB.H.355 (ACCS: RS:HS 25.6.3)

4484. 1840–3–16 To Thomas Maclear Slough
Four appointed directors sent to Cape schools. Will leave Slough in two weeks for Collingwood. Laid William Herschel's telescope to rest at Slough.
ALS 4pp Cape Archives/Maclear Papers File 100

4485. [1840]–3–17 From Edward Kater Parthenon Club
Sending a paper, compiled from his late father's notes, for an escapement for an astronomical clock. Would be pleased if JH would communicate it to the R.S.L.
ALS 3pp RS:HS 11.71

4486. 1840–3–20 From Johann Franz Encke Berlin
Comments extensively on the orbits of recent comets.
ALS 3pp (German) RS:HS 7.92

4487. 1840–3–20 From John William Lubbock n.p.
Thanks for his letter. Will communicate to P. G. le D. Pontecoulant what JH has to say about him. It was certainly unwelcome that JH should omit JL's name when referring to the lunar theory.
ALS 2pp RS:HS 11.367

4488. 1840–3–21 From Francis Baily 37 Tavistock Place
Has sent the letter and diploma to the King of Denmark. Regarding the printing of Thomas Maclear's paper on N. L. Lacaille. H. C. Schumacher has seen a third comet.
ALS 2pp RS:HS 3.179

4489. 1840–3–21 From Luke Howard Tottenham
Encloses some more calculations of height by means of the barometric mean temperature. Comments on these. Staying at Ackworth until April.
ALS 8pp RS:HS 10.4

4490. 1840–3–25 From Thomas Maclear Royal Observatory, Cape of Good Hope
Acknowledges letter requesting land for the Magnetic Observatory. Has a portion of land available in the S.E. corner of the grounds, but it will be necessary to obtain the permission of Rear Admiral George Elliot.
C 2pp RS:HS 12.126

4491. [1840?]–3–26 From Edward Sabine n.p.
Reports on instruments being sent to Hamburg and to Egyptian observatories. Encloses portion of [K.] Kriel's letter to [A. T.] Kupffer. Reports on [Charles] Riddell's securing a site for a meteorological observatory.
ALS 2pp RS:HS 15.324

4492. 1840–3–27 From John Henry Lefroy Longwood, St. Helena Island
Progress in construction of observatory. Completion expected in May. Poor weather hampers actinometer observations.
ALS 3pp TxU:H/M–0330; Reel 1087

4493. 1840–3–28 From Richard Jones Tithe Commission
Sends two lithographs. Has seen the Archbishop and expects to launch his scheme with the approval of the leading politicians.
ALS 1p RS:HS 10.363

4494. [1840]–3–28 From Edward Kater Parthenon Club
Obliged for his kind note and also for forwarding the paper to the R.S.L. Would like to belong to the R.S.L. and would be pleased if JH would sign a certificate for him.
ALS 4pp RS:HS 11.72

4495. 1840–3–28 From John William Lubbock Eaton Place
Sent an extract of JH's letter to P. G. le D. Pontecoulant. Now gives an extract from Pontecoulant's reply. Is sorry they are not to see JH at S. J. A. Compton's (2nd Marquis of Northampton) tonight.
ALS 2pp RS:HS 11.368

4496. 1840–3–29 From Thomas Maclear Royal Observatory, Cape of Good Hope
Erebus and *Terror* have arrived in Simon's Bay. Discussed Magnetic Observatory with J. C. Ross, Captain Croper, and F. E. Wilmot, and wrote government requesting funds. Obelisk has not yet reached the Cape. Small pox rages in Cape Town.
ALS 4pp encl RS:HS 12.127

4497. 1840–3–31 From T. S. Redman Peckham
Describes the process of preparing photogenic paper, taking drawings, and fixing them.
ALS 3pp RS:HS 14.474

4498. [1840–4 or earlier] To 'Dear Sir' [Slough]
JH is preparing to leave residence at Slough, and all notes on William Herschel's life are packed away. Refers [?] to published biographies, but adds some corrections to those.
AL 1p TxU:H/L-0758; Reel 1089

4499. [1840–4] To Francis Baily [Collingwood]
Indicates travel plans, including a visit to FB, with a side trip for business to Slough. JH comments on some astronomical observations he has received.
C 2pp RS:HS 25.8.65

4500. [1840]–4? To John William Lubbock n.p.
Thanks for '*small*, but *intense*' volume sent.
ALS 1p RS LUB.H.356b (C: RS:HS 22.46)

4501. 1840–4 To Richard Sheepshanks Slough
Thanks RS for all the trouble he took with the problem of his portrait. Is anxious to see the specimens of the engraved nebulae of which RS spoke, but confesses little hope in photographic 'reverses' of pencil drawings. Has installed his family at Collingwood.
ALS 4pp RS:HS 16.14 (C: RS:HS 25.9.6)

4502. 1840–4–3 From Marshall Hall 14 Manchester Square
Was not certain how to reply to his last obliging letter. Has resolved to write to Professor [J. P.?] Müller. Thought it right to inform JH of his plans.
ALS 3pp RS:HS 9.204

4503. [1840–4–5] To Robert Hunt n.p.
Hopes that his paper does not anticipate any of RH's work; will share with the R.S.L. any of RH's results. Laments that he is unable to fix the spectral colors on photographic paper. Asks if RH has studied the bromines.
ALS 3pp George Eastman House, 2.17

4504. 1840–4–[7?] To Caroline Lucretia Herschel [Slough]
Packing up the 7-ft. and 10-ft. reflecting telescopes for their trip to Collingwood in Hawkhurst in Kent, JH's new home.
ALS 3pp TxU:H/L–0585.3; Reel 1058

4505. 1840–4–7 To Margaret Brodie Herschel Slough
About the choices to be made in moving furniture from Slough to Collingwood; problems about the election of a minister at Eton.
ALS 4pp JHS 3.19a

4506. 184[0]–4–8 From J[ames] R[osse] I[nnes] n.p.
Anthony Oliphant has died. The plan to establish a system of meteorological observations at the Cape Colony and in Caffreland has failed.
ALS 1p UCT JH Letters A1.44

4507. 1840–4–10 To Margaret Brodie Herschel Slough
About a tragic fire nearby, and more about moving goods to Collingwood.
ALS 2pp JHS 3.19e

4508. 1840–4–10 To Margaret Brodie Herschel Slough
Mostly about boxes packed and shipped to Collingwood, along with questions about items of equipment there.
ALS 3pp JHS 3.19b

4509. 1840–4–10 From George Thomas Napier Camp Ground, Cape of Good Hope
Apologizes for his long silence. Outlines recent events at the Cape concerning Andries Stockenström, who has been replaced by J. Hare, and Major C. C. Michell, who has resigned. J. R. Innes has been all over the Colony and will shortly be coming to England. F. M. Eardley-Wilmot has dined with the Napiers.
ALS 17pp RS:HS 13.68

4510. 1840–4–10 From Edward Sabine 12 Park Place
Discusses Charles Riddell's desire to serve at a North American observatory and the expenses involved.
ALS 4pp RS:HS 15.84

4511. 1840–4–12 To William Whewell Slough
Sends WW some verses, asks WW's and George Peacock's advice on a request by Edward Sabine for funds, and invites WW to visit Hawkhurst, where the Herschels were then in the process of moving.
ALS 3pp TC, Camb. Add. Ms.a.207[44] (C: RS:HS 22.48)

4512. 1840–4–13 To Margaret Brodie Herschel Slough
Complains about the loneliness now that the family is gone from Slough, as are most of their goods.
ALS 3pp JHS 3.19c

4513. [1840–4–14] To Francis Baily [Collingwood]
Received a 5-ft. Newtonian reflector from JH's aunt Caroline; is offering it to the R.A.S. JH is on the verge of moving to Collingwood.
ALS 2pp RS:HS 3.180 (C: RS:HS 25.8.26)

4514. [1840 or later]–4–15 From A. Beale Moor
Discusses glass manufacturing.
ALS 1p Harvard: Houghton AAH 67m–67

4515. 1840–4–15 From Robert Hunt Devonport
Would like JH's account of his own method of fixing a photograph as noted in his memoir. Singular phenomenon of the Daguerreotype. Hopes to produce a photographic paper to equal the silver plate before the month has expired. Will let him know when his process is complete.
ALS 3pp RS:HS 10.83

4516. 1840–4–15 From Henry Piddington Calcutta
Sends JH a pamphlet about meteorological observations in India.
ALS† 2pp RS:HS 13.401

4517. 1840–4–17 To Margaret Brodie Herschel Slough
Trying to establish which of the boxes of household effects that have been shipped from Slough have arrived at Collingwood; comments on the weather.
ALS 3pp JHS 3.19g

4518. 1840–4–17 From Edward Sabine n.p.
Suggests a book of expenses be kept at the R.S.L.
ALS 1p RS:HS 15.85

4519. 1840–4–18 From Friedrich Wilhelm Bessel Königsberg
Describes additional measurement of annual parallax.
ALS 3pp (German) RS:HS 4.74

4520. 1840–4–19 To Robert Hunt Slough
Exchanging further information on several aspects of the photographic process [see RH's 1840–4–15].
C 2pp RS:HS 22.49

4521. 1840–4–19 To Royal Academy of Sciences in Göttingen Collingwood
Honored to be elected as a foreign member of the Academy.
ALS 1p NSUB Cod.Ms.Hist.Lit.116, Band III, Nr. 3

4522. 1840–4–19 To William Henry Fox Talbot Slough
Sending specimens sent to JH by [Robert] Hunt. Praises Hunt. Has WT received JH's paper ['On the Chemical Action of the Rays of the Solar Spectrum...,' *R.S.P.T.* (1840), 1–59]? JH moving to Collingwood.
ALS 2pp Science Museum Talbt 1/24

4523. 1840–4–20 From Francis Baily 37 Tavistock Place
Should think that the R.A.S. would be pleased to receive the 5-ft. telescope. Regarding the printing of Thomas Maclear's paper. H. C. Schumacher has sent a paper by Wilhelm Olbers on the constellations.
ALS 2pp RS:HS 3.181

4524. 1840–4–21 From Thomas Galloway Serjeants' Inn
Has received [Temple?] Cheval[l]ier's paper and JH's letter informing him of his valuable gift. G. B. Airy is anxious to have Thomas Maclear's paper printed in the *Memoirs* as soon as possible without waiting for the government.
ALS 2pp RS:HS 8.8

4525. 1840–4–22 From Edward Sabine 12 Park Place
Discusses plans for a U.S. magnetic observatory and of [Karl] Kreil's paper about predicting earthquakes. Discusses the affairs of other observatories.
ALS 4pp RS:HS 15.86

4526. 1840–4–23 To Margaret Brodie Herschel Slough
The packing and shipping of goods to Collingwood is complete; JH will now attend to a few final details and then come to Collingwood, having spent the most horrible 'fortnight in my life.'
ALS 3pp JHS 3.19f

4527. 1840–4–23 From Robert Hunt Devonport
Is grateful for the handsome manner in which JH has acknowledged RH's work. Would be willing to prepare a memoir for the R.S.L. Does Louis Daguerre's patent affect RH's manufacture of photographic papers? Has not studied the bromnets so far, but has studied the effects of bromine on plates. Comments on his results.
ALS 5pp RS:HS 10.84

4528. 1840–4–24 To Margaret Brodie Herschel [Slough]
Mostly about damage to furniture during shipment from Slough to Collingwood.
ALS 3pp JHS 3.19d

4529. [1840]–4–24 From Edward Sabine n.p.
Has received materials from [A. D.] Bache indicating that the U.S. government has proposed to establish 5 magnetic

observatories.
ALS 1p RS:HS 15.87

4530. 1840–4–25 From Thomas Galloway Serjeants' Inn
When the case containing the stand of the telescope was opened it was found to contain a number of books. Were these intended for the R.A.S. or should they be returned to JH?
ALS 2pp RS:HS 8.9

4531. [1840]–4–25 or later. From Edward Sabine n.p.
Hopes [Humphrey] Lloyd will reconsider V.F. Magnetometer based on JH's remarks. Discusses Canadian observatories. States who will perform U. S. magnetic survey. Discusses micrometer use.
ALS 8pp inc RS:HS 15.331

4532. 1840–4–26 From Joseph Lovering Cambridge, [Mass.]
Asks JH's advice in establishing new astronomical, meteorological, and magnetic observatory at Harvard University. Describes instruments already owned and those ordered from Europe.
ALS 4pp TxU:H/M–0339; Reel 1087

4533. 1840–4–30 From Edward Sabine Royal Society
Writes of a magnetic survey of a portion of India and the inaccuracy of the equipment used in magnetic observations at sea.
ALS 4pp RS:HS 15.88

4534. 1840–4–30 From William Henry Fox Talbot Lacock Abbey
Waxes lyrical about nature, and comments on a new 'positive' paper JH has sent. Refers to JH's move to Hawkhurst.
ALS 3pp RS:HS 17.301 Buckland, *Talbot,* 15 inc

4535. 1840–5–1 From Edward Sabine 12 Park Place
R.S.L. Council approved publication of ES's paper on magnetic observation at sea. Discussions for North American magnetic observatories continue.
ALS 3pp RS:HS 15.89

4536. 1840–5–3 To William Henry Fox Talbot Collingwood
Thanks for excellent pictures WT sent. JH wishes JH had time for further experiments, especially on the 'thermic spectrum.' Praises [Robert] Hunt's results. Recommends experiment to WT.
ALS 3pp Science Museum Talbt 1/25

4537. 1840–5–6 To Francis Baily n.p.
Would FB deputize for him at the council meeting on Friday as he is unable to attend. Letter from F. W. Bessel regarding 61 Cygni. Has written to G. B. Airy regarding Thomas Maclear's paper.
ALS 2pp RS:HS 3.182 (C: RS:HS 25.8.27)

4538. 1840–5–9 From Robert Hunt Devonport
Sends his paper for communication to the R.S.L. Will send some more examples later as his camera has developed a defect.
ALS 2pp RS:HS 10.85

4539. [1840]–5–10 From Edward Sabine Woolwich
[John] Lefroy should replace [Charles] Riddell in Canada. Discusses proposed Norwegian observatory and proposals for various instruments. Navy officers may be helpful observers in colonies.
ALS 6pp RS:HS 15.308

4540. 1840–5–12 To Elizabeth Drummond [Collingwood]
Just moved [to Collingwood]. Will send information about ED's late son, Thomas Drummond. Recalls Drummond's fame and triangulation work on Irish Survey. Describes invention and first demonstration of Drummond Light.
Adraft 4pp RS:HS 19.78 R. Barry O'Brien *Thomas Drummond*, 28–9 & John F. McLennan, *Memoir of Thomas Drummond*, 437–42

4541. 1840–5–12 To Robert Hunt Collingwood
Will forward RH's paper to the R.S.L.; intrigued by RH's results using mercury vapor to produce the image on exposed paper.
ALS 3pp George Eastman House, 3.17 (C: RS:HS 22.15)

4542. [1840 or later]–5–13 To [John Ellor Taylor] Collingwood
Thanks for series of portraits of members of Ipswich Museum.
ALS 2pp TxU:H/L-0748.2; Reel 1089

4543. 1840–5–16 To George Biddell Airy [London]
Has a problem with the explanation of interference bands produced by mica being placed between the eye and prism-produced dispersion; the explanation is Baden Powell's.
ALS 4pp RGO 6.463.70

4544. [1840–5–16 or later] To George Biddell Airy 28A Devonshire St., Portland Place
Writes to supply GA with an address for a reply to JH's 1840–5–16.
ALS 1p RGO 6.463.76

4545. 1840–5–16 From Robert Hunt Devonport
Returns the photographic drawings of W. H. Fox Talbot with many thanks. Fears for their permanence. Has produced similar pictures himself acting on bromide of silver with hydrocholoric acid. Has had trouble with his periscope lens.
ALS 2pp RS:HS 10.86

4546. 1840–5–17 From George Biddell Airy Royal Observatory, Greenwich
Regarding Prof. Baden Powell's paper on dark bands. Is making catalogues of the moveables of the observatory.
ALS 7pp RS:HS 1.86 (C: RGO 6.463.72)

4547. 1840–5–18 From George Biddell Airy Royal Observatory, Greenwich
Considers paper by [Baden] Powell to be worthless. Sent it to JH today care of Mrs. Stewart.
ALS 1p TxU:H/L-0760.2; Reel 1089 (ACCS: RGO 6.340.283)

4548. 1840–5–18 From George Biddell Airy Royal Observatory, Greenwich
Responds to JH [see JH's 1840–5–16 or later] indicating to what address GA had sent his reply [see GA's 1840–5–17].
ACCS 1p RGO 6.463.77

4549. 1840–5–19 To George Biddell Airy n.p.
Thanks GA for the time and trouble GA took in writing about interference bands [see GA's 1840–5–17]; JH's wife, Margaret, is making many visits to the dentist.
ALS 3pp RGO 6.463.78

4550. 1840–5–19 From Edward Sabine n.p.
Hopes to meet with JH to discuss some data. Announces that at the Newcastle B.A.A.S. meeting, JH, Humphrey Lloyd, and ES were appointed to a subcommittee. Discusses what James Ross will do on his return from the Cape. Announces

the possibility of ten new observatories in Russia.
ALS 4pp RS:HS 15.90

4551. 1840–5–20 From George Biddell Airy Royal Observatory, Greenwich
Concerning prism refraction.
ALS 1p RS:HS 1.87 (C: RGO 6.463.80)

4552. [1840–5–20 or later] To George Biddell Airy n.p.
A note agreeing with GA about prism interference [see GA's 1840–5–20].
ALS 1p RGO 6.463.81

4553. 1840–5–21 To [R.S.L. Council] London
Formally reports on paper by Baden Powell on theory of dark bands found in solar spectrum. G. B. Airy agrees with JH's report.
AD 3pp RS RR.1.188

4554. 1840–5–22 To William Radcliffe Birt 28 A Devonshire St., Portland Place
How to reduce barometer observations and project them in curves. R.S.L. will provide table by H. C. Schumacher for this purpose. How to obtain 'law of decimal oscillations' for each station. WB's name will be affixed to results. JH needs receipt for observations, which belong to South African Literary and Philosophical Society.
AdraftS 7pp & ACCS 5pp RS:HS 19.79–80 (C: RS:HS 22.52)

4555. 1840–5–22 From Noel Paymal Lerebours Paris
Father has died and he has taken over the business. Would like references as to where he can find information on instruments.
ALS 3pp (French) RS:HS 11.470

4556. 1840–5–23 From Robert Hunt Devonport
Was gratified to receive his memoir and to find many of his doubts corrected by JH's experiments. Hoped to enclose some sulphuretted papers but the chemicals went astray in the post. Intends to follow JH's example and precipitate a glass.
ALS 2pp RS:HS 10.87

4557. 1840–5–25 From Giovanni Antonio Plana Turin
Gives new instructions about receiving medal; asks about the timing of the award.
ALS 1p (French) RS:HS 13.437

4558. 1840–5–26 From Edward Sabine 12 Park Place
Relays communications from [Charles] Riddell in Toronto, who is making progress on the observatory there and has selected an officer of the artillery for an assistant. Suggests the meeting of the Physical Committee be moved. Discusses other matters of various observatories.
ALS 6pp RS:HS 15.91

4559. 1840–5–27 From George Jeans Tetney near Grimsby
Is puzzled by JH's use of a word in his volume on astronomy; can he explain its meaning. Points out further misprints. Hopes JH will edit the works of his father [Sir William Herschel].
ALS 4pp RS:HS 10.280

4560. 1840–5–27 From Edward Sabine 12 Park Place
Encloses a letter from Boston, which may host an observatory. Has sent Boston information.
ALS 1p RS:HS 15.92

4561. 1840–5–28 From Wilhelm Struve Pulkowa
Rejoices at welcome JH received on returning to England from the Cape. Excellent resources at his disposal at Pulkowa. Mentions coin, commemorating opening of the observatory, sent to JH by means of [H. C.] Schumacher. Refers to [Otto] Struve's work on double stars. Lauds capacity of large refractor at Pulkowa Observatory.
ALS 4pp (German) RS:HS 17.128

4562. 1840–5–29 From Edward Sabine n.p.
Encloses a letter from [John] Lefroy about an 'outbreak of [solar] spots' observed at a certain time. Speaks of purchasing instruments and of writing to the secretary of the Amer[ican Philosophical?] Society.
AL 4pp inc RS:HS 15.93

4563. 1840–5–31 To John William Lubbock Collingwood
R.S.L. business about an observatory at the North Cape in Sweden.
C 3pp RS:HS 22.53

4564. 1840–5–31 To [John William Lubbock] Collingwood
Will be unable to attend Council meeting, but comments on issue to be discussed of whether R.S.L. should do anything respecting observatory at Hammerfest (North Cape).
ALS 3pp RS LUB.H.356a

4565. 1840–5–31 From George Thomas Napier Government House, Capetown
Thanks for his long letter regarding education at the Cape. The five teachers have now arrived at the Cape. Comments on education at the Cape. Further regarding the Andries Stockenström affair.
ALS 8pp RS:HS 13.69

4566. [1840–6] From Edward Sabine n.p.
Announces that the *Quarterly Review* has approved JH's article for immediate publication. Relays the message that JH should finish it as soon as possible.
ALS 3pp RS:HS 15.100

4567. [1840–6] To William Whewell n.p.
Gives JH's new address in Hawkhurst. Bring Mr. Bunt's planetarium when WW comes to Hawkhurst. Announces new photographic results.
ALS 2pp TC, Camb. Add. Ms.a.207[41] (RS:HS 22.47 & 22.36)

4568. 1840–6–1 From Edward Sabine 12 Park Place
Tells of James Ross's broken instrument at the Cape and of efforts to send new equipment to him. Speaks of [Francis] Beaufort's magnetic observations at sea. Has looked for the memorial in the *Athenaeum* but does not believe it was published there. Does not know if Humphrey Lloyd is getting married.
ALS 4pp RS:HS 15.94

4569. [1840]–6–2 From Francis Beaufort [Admiralty Office]
Encloses seven notes from J. C. Ross to Edward Sabine and FB.
ALS 1p RAS:JH Archive 4/3.5 verso; Reel 7

4570. 1840–6–3 From Adolphe Quetelet Brussels
Still waits for the instrument that Edward Sabine had built. Wishes to make actinometer observations but has no instrument. Asks JH to have one made. Sends observations from Italy trip to Sabine to present to R.S.L. Continues horary observations.
ALS 3pp (French) RS:HS 14.219

4571. 1840–6–4 From James David Forbes Athenaeum Club, London
Explains results obtained by R.S.L. subcommittee considering state of meteorological journal published in *R.S.P.T.* Wonders about feasibility of establishing 'simple but accurate meteorological register' under public authority.
ALS 4pp RS MM.16.137

4572. 1840–6–5 From Augustus De Morgan 69 Gower St.
Possible to purchase 12 postage covers for 11d in the Strand. Points out an apparent error in one of JH's books on sound. How does an open pipe vibrate? Regarding the private character of J. L. Lagrange.
ALS 2pp RS:HS 6.184

4573. 1840–6–6 From Edward Sabine n.p.
Says the needle of a magnetic instrument has been made heavier as requested by ES. Mentions other instruments and their performances. Remarks on the *Quarterly* [*Review*].
ALS 2pp RS:HS 15.95

4574. 1840–6–8 To Francis Baily Collingwood
Invitation to Hawkhurst. Has plenty of R.S.L. work to keep him from his observational work. Question of a new observatory. Has had letter from Wilhelm Struve regarding the fifteen-inch object glass.
ALS 4pp RS:HS 3.183 (C: RS:HS 25.8.28)

4575. 1840–6–8 From Edward Sabine n.p.
Desires that JH read the enclosed work by Mrs. Sabine about Antarctica.
ALS 2pp RS:HS 15.97

4576. 1840–6–8 From Edward Sabine n.p.
Gentleman from Holland, sent by his King, has arrived and is eager to learn about and set up a magnetic observatory in Holland and two or three more in the Dutch colonies. Is excited about the possibility of a South American station. Believes it possible that in a few years there will be maps of all the oceans.
ALS 3pp RS:HS 15.96

4577. 1840–6–10 From William Henry Fitton Bloomfield Lodge
Is grateful for his paper on the solar spectrum. Hopes he will continue his experiments with light so that posterity may benefit, as would have been the case if Smithson Tennant had persisted with his efforts. Regarding some of the geological theories of the times. Hopes JH will come and stay with them. R. I. Murchison is on the continent.
ALS 6pp diag RS:HS 7.238

4578. 1840–6–10 To James David Forbes n.p.
R.S.L. will deviate from its intellectual purpose whenever it meddles in practical observations. Proper role for R.S.L. is to suggest goals and methods and to reward discoveries. Recommends discontinuing meteorological observations at Somerset House; other stations now provide better data. If R.S.L. insists on establishing a meteorological register, JH suggests one like that of J. F. Daniell.
ALS 3pp St. Andrews 34 (Adraft: TxU:H/L-0158 & –0145.2; Reel 1054 & C: RS:HS 22.44)

4579. 1840–6–10 From Edward Sabine 12 Park Place
Encloses some comments of Archibald Smith about C. F. Gauss's General Theory. Believes Smith plans to go on an expedition. Mentions two new magnetometers, the building of a Cape observatory, and the recommendation to the Parliament of Holland to build an observatory.
ALS 3pp RS:HS 15.98

4580. 1840–6–11 From James David Forbes Athenaeum Club, London
Describes difficulties in attempting to produce good meteorological register and establish national meteorological observatory.
ALS 4pp RS MM.16.138 (C inc: TxU:H/M–0145.3; Reel 1087)

4581. 1840–6–12 From Noel Paymal Lerebours Paris
Has received no reply to his previous letters so probably gave an incorrect address.
ALS 2pp RS:HS 11.471

4582. 1840–6–12 From Edward Sabine 12 Park Place
Thanks JH for reading Mrs. Sabine's work. Discusses the work of C. F. Gauss and his observations regarding the southern hemisphere.
ALS 3pp RS:HS 15.99

4583. 1840–6–15 From Francis Beaufort n.p.
No objection to JH publishing soundings of J. C. Ross. FB will send JH detailed records of these. Any letters to Ross must arrive by 18 June for dispatch to Hobart Town. FB has copies of Thomas Maclear's correspondence about magnetic observations.
ALS 3pp RAS:JH Archive 4/1.9; Reel 7 & ACC: Hydrographic Office Letter Book

4584. 1840–6–15 From Thomas Galloway Serjeants' Inn
Sending officially the thanks of the Council of the R.A.S. to JH and Caroline Herschel for their gift of the telescope. Hopes JH will convey these thanks to Miss Herschel.
ALS 2pp RS:HS 8.10

4585. 1840–6–15 From William Henry Fox Talbot London
Sends a few photographs and says he has read JH's recent memoir. Will comment later.
ALS 2pp RS:HS 17.302a

4586. 1840–6–15 To [?] Collingwood
Has no objection to the recipient's republication of the 'Examples of the Differ. & Integl Calculus by Mr. Peacock,' to which JH contributed.
ALS 2pp BostonPubLib Ms.Eng. 397

4587. 1840–6–16 From John Sedgwick Hawkhurst Moor
Asks for JH's support at next week's vestry meeting. JH is entitled to six votes.
ALS 1p RAS:JH Archive 4/3.22 verso; Reel 7

4588. 1840–6–18 From John David Roberton Royal Society
If JH wishes, will continue barometric [?] observations at Royal Society Apartments, which observations JH had requested while at Cape.
ALS 2pp RS MM.16.140

4589. 1840–6–18 From Edward Sabine n.p.
Physical Committee yesterday amended report by J. D. Forbes and recommended that R.S.L. establish a magnetical and meteorological observatory in vicinity of London.
C 1p inc TxU:H/M–0145.4; Reel 1087

4590. 1840–6–18 From Edward Sabine n.p.
Discusses difficulties with establishing public magnetical and meteorological observatory. Mentions other business.
ALS 3pp RS MM.16.139

4591. [1840]–6–19 From Edward Sabine n.p.
[Charles] Riddell's observations suggest Europe and North

America experience magnetic disturbances at similar times, but in opposite directions. Suggests course of action for [Charles] Riddell and Lieutenant Younghusband in North America.
ALS 2pp RS:HS 15.326

4592. 1840–6–19 To William Henry Fox Talbot Collingwood
Praises various photographs by WT and notes changes in some. Recounts optical experiments, recent and planned, by JH on rock-salt.
ALS 4pp diag Science Museum Talbt 1/26

4593. 1840–6–20 From Edward Sabine 12 Park Place
Payment for instruments sent to Breslau observatory. Received proofs of JH's paper, which does justice to Humphrey Lloyd. Spoke to Lord Melbourne about establishing a magnetical and meteorological observatory in vicinity of London. Compares expenses for permanent and temporary observatories.
ALS 3pp TxU:H/M–0518; Reel 1093

4594. 1840–6–21 From John William Lubbock 25 Eaton Place
Has talked to William Lamb (Lord Melbourne) about permanent magnetic and meteorological observatory. Was received 'as well as we could have anticipated,' but without definite answer. Thinks head for observatory should be appointed and give lectures.
ALS 3pp RS MM.16.141

4595. 1840–6–23 From John David Roberton Royal Society
Announces 30 June meeting of Committee of Physics including Meteorology.
(A)LS 1p RAS:JH Archive 4/3.13 verso; Reel 7

4596. 1840–6–24 From John Sedgwick National School, Hawkhurst
Committee will meet next Monday to appoint 'a Master and Mistress.'
ALS 1p RAS:JH Archive 4/3.14 verso; Reel 7

4597. 1840–6–[25 or earlier] To Humphrey Lloyd n.p.
Sends to HL a sheet of curves laid down by [Charles] Riddell, comparing Brussels declination observations with those made in Toronto. JH frequently refers to HL's work on the subject of terrestrial magnetism.
C 1p RS:HS 25.14.19

4598. 1840–6–25 From John David Roberton Royal Society
Announces meeting of R.S.L. Council on Saturday 27 June. [JH annotation: Received on '27th June!']
(A)LS 1p † RAS:JH Archive 4/3.23 verso; Reel 7

4599. 1840–6–26 To Augustus De Morgan Collingwood
Thanks AD for a correction in one of JH's writings; tries to explain the partial echo that occurs when blowing across the end of an open pipe.
C 2pp RS:HS 22.55

4600. 1840–6–26 From Humphrey Lloyd Trinity College, Dublin
Has received JH's letter with the sheet of curves laid down by [Charles] Riddell. Comments on these. Is honored by the notice JH has taken of his share in the magnetic project. Encloses note on the wet bulb [thermometer].
ALS 3pp RS:HS 11.274

4601. [1840?]–6–26 To John William Lubbock Collingwood
Believes that R.S.L.'s request for £3000 grant to establish permanent magnetic and meteorological observatory is too hasty. Urges caution. Points out potential problems.
AdraftS 3pp TxU:H/L-0233; Reel 1054

4602. 1840–6–27 From John William Lubbock n.p.
Council passed 'with great unanimity' resolutions regarding permanent magnetic observatory to be established in London.
ALS 4pp RS MM.16.142

4603. 1840–6–28 To James O. Halliwell Collingwood
Feels unable to become a member of the new society proposed by Halliwell, but wishes it well.
draft 4pp RS:HS 10.147 & 22.56

4604. 1840–6–29 To Samuel Hunter Christie n.p.
Recommends printing all three papers sent to JH—G. B. Airy's on spectral bands, [Robert] Hunt's on influence of iodine on photographic papers, and that by Lord Oxmantown [William Parsons] on large reflecting telescopes.
AdraftS 8pp TxU:H/L-0126; Reel 1054

4605. 1840–6–29 From John David Roberton Royal Society
Prince Albert will meet R.S.L. Council next Thursday at Buckingham Palace.
ALS 1p RAS:JH Archive 4/3.20 verso; Reel 7

4606. 1840–6–30 To John William Lubbock Collingwood
In light of minutes of R.S.L. Council meeting, outlines vision of permanent national physical observatory and, more importantly, an experimental institute or college. Hopes Edward Sabine will be interested.
ALS 4pp RS LUB.H.357 (ACCS: RS:HS 25.6.4)

4607. 1840–6–30 From John David Roberton Royal Society
Formal statement that JH received parcel of manuscripts. Announces R.S.L. Council meeting for 9 July.
(A)LS 1p RAS:JH Archive 4/3.26 verso; Reel 7

4608. 1840–6–30 From Matthew Henry Webster Albany [New York]
Received JH's letter of 4 Nov. [1829] in February. Sends report by university regents [to New York state legislature] and MW's [meteorological] observations. Will send more U.S. observations later.
ALS 2pp † RS:HS 18.130

4609. 1840–7 To Spencer J. A. Compton Atheneum
Organizational matters relating to sending two vessels on an Antarctic expedition.
C 2pp RS:HS 22.74

4610. 1840–7–2 From John William Lubbock n.p.
Further remarks on council meeting about permanent magnetic observatory and G. B. Airy's answer [to invitation to head it?]. Awaits communication from William Lamb (Lord Melbourne).
ALS 2pp RS MM.16.143

4611. 1840–7–3 To John William Lubbock Collingwood
Is displeased by R.S.L. Council's seemingly rash resolution with regard to idea of permanent magnetic observatory.
ACCS 2pp RS MM.11.147 (ACCS: RS:HS 25.6.5)

4612. 1840–7–6 To George Biddell Airy Collingwood
GA's correspondence with [R.S.L. president] Lord Northampton [S. J. A. Compton] will be read before Physical

Committee. JH favors establishing national physical observatory and experimental institute, but wishes plan had been discussed more fully. Proposes sites, procedures, and instruments for magnetic and meteorological observations in Britain. Ideas for experimental institute.
ALS 8pp TxU:H/L-0011; Reel 1054

4613. 1840–7–6 To George Biddell Airy Hawkhurst
Comments on a proposed request for a government grant to establish a magnetic and meteorological institute.
ALS 12pp RGO 6.675.239

4614. 1840–7–6 From Caroline Lucretia Herschel Hanover
Gives reasons why CH rejoiced at news of Mrs. Papendick's death, and grieves that two daughters of Papendick are gaining influence at Court.
C 1p TxU:H/M–0936.2; Reel 1083

4615. 1840–7–6 From Caroline Lucretia Herschel Hanover
Hopes to introduce to Margaret Herschel some of her acquaintances, including the family of General Baron Hugh Halkett.
L *Memoir ... of Caroline Herschel,* 312–3 inc

4616. 1840–7–8 To Francis Baily Collingwood
Dissatisfied with R.S.L. procedures for establishing magnetic and meteorological observatory. Called meeting of Physical Committee tomorrow and invited G. B. Airy to attend. Busy reducing JH's observations from Cape of Good Hope.
C 3pp TxU:H/L-0056; Reel 1054

4617. 1840–7–8 To [Spencer J. A. Compton] Collingwood
The Physical [?] Committee is meeting the next day to deal with the question of establishing a magnetic and meteorological observatory. JH [Chair of the Physical Committee?] cannot attend the meeting and sends his supportive, but cautioning, comments.
ACCS 4pp RS:HS 25.6.6

4618. 1840–7–9 To Spencer J. A. Compton Collingwood
Has called special meeting of Physical Committee to discuss SC's response to R.S.L.'s proposal to establish permanent observatory. Expresses JH's opinion and recommends temporary observatory instead. Keep this issue separate from funding for publication of present observations.
C 4pp TxU:H/L-0284; Reel 1054 (C: RS:HS 22.58)

4619. 1840–7–10 From Francis Baily 37 Tavistock Place
Thanks for his invitation. N. L. Lacaille's observations are in process of reduction and work is continuing on the constellations. Regarding the project for a new observatory. Greenwich observation could probably cover a wider field if staff were increased. Has received a silver medal for JH from Wilhelm Struve. Government is to print Thomas Maclear's paper. T. G. Taylor has arrived from Madras bringing his fifth volume of observations.
ALS 3pp RS:HS 3.184

4620. 1840–7–12 From John Gibson Lockhart Sussex Place, Regents Park
JH must have read William Whewell's *History of the Inductive Sciences*. John Murray thinks JH's opinion of this work would form a valuable article for the *Quarterly Review*.
ALS 4pp inc RS:HS 11.302

4621. 1840–7–13 To Wilhelm Struve Collingwood
Apologizes for not having written, but has been 'harassed with desultory and importunate calls on [his] name.' Commends WS on accomplishments at Pulkowa and performance of Repsold instruments. Congratulates WS on his son Otto Struve. Received WS's figures on Halley's Comet. Comments on bright stars of variable nature. Estimates 35 or 36 magnetic observatories in the world.
C 4pp RS:HS 22.59

4622. 1840–7–14 From George Biddell Airy Royal Observatory, Greenwich
Regarding the printing of Thomas Maclear's Survey papers. Work on magnetical and meteorological observations.
ALS 3pp RS:HS 1.88

4623. 1840–7–18 To William Radcliffe Birt Collingwood
L. A. J. Quetelet's observations follow centigrade thermometer and French meter. Latitude and longitude for five sites. Three sites are unknown to JH. Hopes WB completes [reduction of barometer observations] before B.A.A.S. meeting.
ALS 3pp RS:HS 19.81

4624. 1840–7–20 From John David Roberton Royal Society
At direction of J. W. Lubbock, encloses letter addressed to William Lamb (Lord Melbourne) by R.S.L. president [Spencer Compton] [regarding magnetic observatory].
ALS 1p encl RS MM.16.144

4625. 1840–7–23 From John R. Hind 7 Spencer Place Brixton
Drawing attention to his prediction of the storms of 1838 and 1839.
ALS 2pp RS:HS 9.319

4626. 1840–7–23 From James Rose Innes 9 Craven St., Strand
Arrived in England two days ago after a tedious voyage. Would like to visit JH. Off to Scotland next week. Called at the Colonial Office and had a friendly reception. G. T. Napier's dispatch has been received.
ALS 2pp RS:HS 10.269

4627. 1840–7–24 To William Radcliffe Birt [Collingwood]
Advice on how to project barometric observations onto curves.
ALS 3pp RS:HS 19.82

4628. 1840–7–25 From George Biddell Airy Royal Observatory, Greenwich
Copies of report are available if JH's friends on Board [of Visitors] want to see it. R.S.L. selected GA's polarity paper for Bakerian Lectures. Thanks for JH's influence.
ALS 1p RAS:JH Archive 14/3.9.11 verso; Reel 11 (ACCS: RGO 6.368.638)

4629. 1840–7–25 From George Biddell Airy Royal Observatory, Greenwich
Sends JH a copy of extracts from unpublished papers on the question of standards, and asks for JH's comments on these extracts.
LS 1p TxU:H/M-0627.10; Reel 1084

4630. 1840–7–28 From John William Draper New York University
Sending a heliographic portrait taken from life. Comments on his methods.
ALS 2pp RS:HS 6.501

4631. 1840–7–29 From George Biddell Airy Royal Observatory, Greenwich

Requests the views of JH, and all other Standards Commission members, on the various subjects before the Commission; GA will compile these and circulate them to all Commission members to expedite discussions at future meetings.
LS 3pp TxU:H/M-0627.12; Reel 1084

4632. 1840–7–29 To George Biddell Airy Collingwood
Responds to GA's circulated material for the Standards Commission; suggests time to digest the information.
ALS 4pp RGO 6.340.281

4633. [1840–8?] To [William Radcliffe Birt] n.p.
Must report progress of meteorological reductions to B.A.A.S. Send specimen of 'Curves' before first week in September. Gives latitudes and longitudes for North American stations.
ALS 3pp RS:HS 19.165

4634. 1840–8–1 To Henry Bunbury Collingwood
Requests documents about the disagreement between George Napier and Andries Stockenström, which developed at the Cape. In the course of this request, JH expresses his feelings on the matter.
ALS 8pp DMC.695A.885.5 (C: RS:HS 5.121 & 22.60)

4635. 1840–8 To Caroline Lucretia Herschel Collingwood
Reports that there is no bust of William Herschel at the R.S.L., as JH once believed.
ALS 3pp TxU:H/L–0585.8.a; Reel 1058

4636. 1840–8–1 From Robert Treat Paine Boston, Mass.
Magnetic observations have been made regularly for a year and a half at the college at Cambridge by W. C. Bond. After JH's letter another set of instruments has been obtained to make another series of observations. Money from Hugh Smithson has been loaned to Arkansas. Comments on red auroral light seen. With the more frequent steamship service now hopes for quicker communications from JH.
ALS 3pp RS:HS 13.213

4637. 1840–8–5 From John Murray, [Sr.?] Albemarle St.
Sending check and a note from the editor of the *Quarterly Review.*
ALS 1p RS:HS 12.445

4638. 1840–8–6 To William Whewell Collingwood
Asks WW to write letters of introduction for [James] Innes, an educator from the Cape. Comments on WW's *Philosophy of the Inductive Sciences,* expressing reservations about its a prioristic approach.
ALS 4pp TC, Camb. Add. Ms.a.207[45] (C: RS:HS 22.61)

4639. 1840–8–7 From Charles T. Beke St Martin's Place
Has received his letter from the Cape acknowledging his book. Is about to travel to the eastern coast of Africa and would be pleased to do anything for JH if he requires it in that direction. Is leaving England by the end of the month.
ALS 2pp RS:HS 4.27

4640. 1840–8–8 From Robert Hunt Devonport
Is grateful for the suggested emendations to his paper. Comments on the action of iodine on silver and its oxide. When can he expect publication of his paper? Has noted a peculiar property in the calorific rays when copying engravings. Encloses some Daguerreotype papers.
ALS 3pp RS:HS 10.88

4641. 1840–8–10 To James David Forbes Collingwood
Explains efforts at using a makeshift apparatus for 'fixing the spectrum.' Discusses current confusion and ultimate hope surrounding JF's meteorological recommendations. In a postscript, notes observations of meteors in Persei.
ALS 3pp St. Andrews 42 (C: RS:HS 22.62)

4642. 1840–8–10 To Caroline Lucretia Herschel Collingwood
JH has given away his sweeping telescope to [Johann] Hausmann and the 5-ft. Newtonian reflector to the R.A.S. to be preserved 'long after I and all the little ones are dead and gone.' The skies have been excellent for JH to observe variable stars and to connect the northern with the southern magnitudes. Proved that Alpha Orionis is both a variable star and a periodical star.
ALS 7pp TxU:H/L–0585.7; Reel 1058 *Memoir ... of Caroline Herschel*, 314–15 inc

4643. 1840–8–13 From George Biddell Airy Royal Observatory, Greenwich
The Royal Observatory wants to make magnetic observations at Greenwich, and it appears will need to make some building alterations. Estimates are required immediately by the Admiralty/Treasury, and GA seeks JH's assistance.
ALS 3pp TxU:H/M-0627.14; Reel 1084

4644. 1840–8–13 From George Biddell Airy Royal Observatory, Greenwich
Asks for JH's help in trying to determine the likely cost of making magnetic observations at Greenwich.
ACCS 3pp RGO 6.693.94

4645. 1840–8–13 To [Sir Henry Bunbury] Collingwood
Returns papers dealing with the treatment of Andries Stockenström by Sir George Napier.
ALS 3pp WT 67624.75 (C: RS:HS 5.122 &22.63)

4646. 1840–8–14 To George Biddell Airy Collingwood
Provides information [see GA's 1840–8–13] based on equipping the Breslau Observatory; also comments on lawyers, meteor showers, and photography.
ALS 4pp RGO 6.693.96

4647. 1840–8–14 From Charles Few Covent Garden
Have deposited the title deeds for the Hawkhurst estate, and also the lease to Mr. Grover in a tin box lodged with Messrs. Drummond.
ALS 2pp RS:HS 7.203

4648. 1840–8–14 To Humphrey Lloyd Collingwood
G. B. Airy needs HL to send him the information used to set up colonial magnetic observatories.
C 2pp RS:HS 22.64

4649. 1840–[8–14 or earlier] To George Thomas Napier n.p.
Is trying to reduce the ill feeling between GN and Andries Stockenström relating to the governing of the Cape colony; JH feels hampered by limited information.
C 5pp RS:HS 22.75

4650. 1840–8–15 To George Thomas Napier Collingwood
Conduct of GN [governor at Cape of Good Hope, 1837–43] toward Andries Stockenström was unimpeachable. Stockenström's letters to JH show favorable view of GN. GN was not responsible for Stockenström's removal.
C 8pp RS:HS 19.83 (C: RS:HS 22.75)

4651. 1840–8–17 From James David Forbes Edinburgh
Thanks for JH's letters and the paper on photography with special reference to the effects of heat. Was present at the

R.S.L. during consideration of a national observatory. Would like JH's opinion. Is working on a supplementary report on meteorology.
ALS 3pp St. Andrews LB III, 120–2 (C: RS:HS 7.296)

4652. 1840–8–17 To Robert Hunt Collingwood
Thanks RH for a sample of Daguerreotype paper. Comments on some aspects of the action of 'thermic' rays and 'chemical' rays in photography.
ALS 3pp RS:HS B26.68

4653. 1840–8–17 From Andries Stockenström n.p.
Statement describing 'The conduct of Sir G. Napier in relation to Sir Andries Stockenström and his former office as Lieutenant Governor.'
C 5pp RS:HS 17.24

4654. 1840–8–17 To Andries Stockenström Collingwood
Congratulates AS on AS being made baronet; wishes him well on re-joining his family at the Cape. Hopes that the problems between AS and Governor George Napier were due only to 'incompatibility of positions.'
ALS 3pp RS:HS 17.25 (C: 22.66)

4655. 1840–8–20 From George Biddell Airy Royal Observatory, Greenwich
Regarding equipment at Royal Observatory. Comparison of planets.
ALS 3pp RS:HS 1.89

4656. 1840–8–20 From George Biddell Airy Royal Observatory, Greenwich
Thanks JH for his help [see JH's 1840–8–14], and tells JH about a visit to Greenwich by Charles Wheatstone.
ACCS 2pp RGO 6.693.98

4657. 1840–8–21 From Adolphe Quetelet Brussels
Royal Academy of Sciences and Fine Arts acknowledges receipt of JH's 'On the Chemical Action of the Rays of the Solar Spectrum.'
ALS 1p (French) RAS:JH Archive 5/2.1.24; Reel 8

4658. [1840?]–8–23 From Edward Sabine Dunraven Castle
Reports on letter from Humphrey Lloyd and latest information from [J. C.] Ross's voyage. Has received JH's meteorological plates.
ALS 4pp RS:HS 15.320

4659. 1840–8–27 To Edward C. Herrick Somerset House
Acknowledges receipt of meteorological observations from EH. P.S. [dated 1840–8–28] reports on August meteor showers observed by JH.
ALS 3pp Yale Univ.

4660. 1840–8–28 From Edward Sabine Interlaken
Will give Humphrey Lloyd's and his own latest magnetic data to JH for use in JH's report to the B.A.A.S. Observations continue in Milan. Year-long observations will soon be completed at Dublin Observatory. Proposes 400–page publication of simultaneous observations. Mentions Lloyd's insistence on processing incoming data himself.
ALS 3pp RS:HS 15.101

4661. 1840–8–30 To William Henry Fox Talbot Collingwood
Sends WT a copy of the solar spectrum in which each color is represented clearly in a 'positive' picture. JH sees this as giving hope for color photography.
ALS 2pp Science Museum Talbt 1/27 (draft 2pp RS:HS 17.302b & C: RS:HS 22.67)

4662. 1840–8–31 From Richard Jones Tithe Commission
Is not well and is planning a holiday. Will be in the vicinity of Hawkhurst next Saturday and would like to visit JH.
ALS 2pp RS:HS 10.364

4663. 1840–9–1 From Caroline Lucretia Herschel Hanover
Asserting that her memory remains sharp, CH promises that she will periodically describe 'what passed in old times.'
L *Memoir ... of Caroline Herschel*, 315 inc

4664. 1840–9–1/13 From Adolph Theodor Kupffer St. Petersburg
Sending the fourth volume of the magnetic observations; comments on these.
ALS 2pp (French) RS:HS 11.94

4665. [1840]–9 From John Murray, {Sr.?} Albemarle St.
Has forwarded JH's note regarding a review of William Whewell's book to J. G. Lockhart, who will be very pleased to receive such an article.
ALS 1p RS:HS 12.451

4666. 1840–9–1 From William Henry Fox Talbot Lacock Abbey
Concerned with developing photography for tourist use. Doubts that colors of the spectrum can be readily displayed on photographic paper. The 'chemistry' of photography is not understood.
ALS 3pp RS:HS 17.303

4667. 1840–9–5 From Johann Franz Encke London
Writing to try to set up a visit to JH.
ALS 1p (German) RS:HS 7.93

4668. 1840–9–5 From Few & Co. Covent Garden
Obliged for his remittance. Requires the key for his title deeds at Drummonds. Cancellation of his will will vest the estate in his eldest son.
ALS 1p RS:HS 7.204

4669. 1840–9–8 From Thomas Henderson Edinburgh
Activities of Mr. Jones and [J. C.] Adamson in Edinburgh. Finished reducing all stars in N. L. Lacaille's *Coelum australe stelliferum*, but catalog will not be finished before B.A.A.S. meeting. Will send it in portions to Francis Baily.
ALS 2pp TxU:H/M–0271; Reel 1087

4670. 1840–9–8 From Edward Sabine 3 St. James's Place
Encloses letters from Heinrich Boguslawski and others about magnetic observatories at various foreign locations, e.g., St. Helena, Toronto, and Hammerfest, Norway. Is going to Glasgow for the B.A.A.S. meeting. Mentions seeing Humphrey Lloyd in Berne.
ALS 2pp RS:HS 15.102

4671. 1840–9–9 To William Radcliffe Birt [Collingwood]
WB's tables [of barometer observations] are beautifully executed. Will submit them to B.A.A.S. Notes dissimilarity of British and Irish data.
ALS 3pp RS:HS 19.84

4672. 1840–9–10 From Heinrich Christian Schumacher Pulkowa
Has been busy with geodetic work. Pulkowa instruments are excellent. [F. W.] Struve has sent zenith distances of Ursa Majores. [G. von] Reichenbach's death dealt blow to precision of Munich instruments.
ALS 4pp RS:HS 15.386

4673. 1840–9–10 From Edward Sabine 3 St. James's Place
Excited about brilliant aurora in Toronto and the possibility of comparing magnetic data from observatories in North America, Europe, and St. Helena.
ALS 1p RS:HS 15.103

4674. 1840–9–13 From Johann Franz Encke London
Thanks for JH's hospitality; tells of JE's further travel plans.
ALS 2pp (German) RS:HS 7.94

4675. 1840–9–15 To B.A.A.S. n.p.
Reduction of all stars in N. L. Lacaille's *Coelum australe stelliferum* is finished. Thomas Henderson's assistant is arranging these into catalogue. This data was sent to Francis Baily.
ACCS 1p TxU:H/M–0428.4.2; Reel 1093

4676. 1840–9–16 From Edward Sabine Glasgow
Encloses letter from Sanchez Cerquero about meteorological observations. Has not yet received JH's report on the magnetic observatories.
ALS 1p RS:HS 15.104

4677. 1840–9–18 From Adolphe Quetelet Brussels
Refers to meteors of August. JH's account similar to [Edward] Herrick's. More meteors seen in America. Sends magnetic and meteorological works. Had hoped to go to Glasgow.
ALS 4pp (French) RS:HS 14.220

4678. 1840–9–24 From Edward Sabine Glasgow
Received and corrected JH's magnetism [B.A.A.S.] report. Mentions interesting account by Dr. [Johann] Lam[ont] about a vertical force magnet used in Munich. ES read earlier letter from Heinrich Boguslawski proposing a special set-up of magnets. Humphrey Lloyd acknowledges it might be an improvement.
ALS damaged 1p RS:HS 15.105

4679. 1840–9–26 From George Jeans Tetney near Grimsby
Would like permission to make a copy of one of the plates from JH's book on astronomy in order to print it in one of his own works intended for amateur astronomers.
ALS 3pp RS:HS 10.281

4680. 1840–9–28 From William C. Redfield New York
Thanks JH for his attention to WR's earlier letter. Discusses communication sent to B.A.A.S. and problems with its delivery to JH.
ALS 2pp RS:HS 14.309

4681. 1840–10–2 From C[harles Louis] Chevalier Paris
Thanks for his letter regarding his own new object glass. As he has shown interest he sends further details of his latest modifications.
ALS 2pp (French) RS:HS 5.336

4682. 1840–10–3 From James Grahame Nantes
Note accompanying a sketch of an earlier time in the lives of JG and JH (sketch by Jane Grahame).
ALS 1p RS:HS 8.438

4683. 1840–10–8 From Edward Sabine Somerset House
Actinometers for [Louis] Agassiz and John Caldecott are ready. Asks if JH wants to compare them to his standard instrument. Excessive magnetic disturbances in May were found simultaneously in Göttingen and Toronto, but the curves do not correspond.
ALS 2pp RS:HS 15.106

4684. 1840–10–10 To James Challis Collingwood
A note of thanks for the eleventh volume of the Cambridge Observatory observations; JH is sorry that he missed JC in Oxford during the summer.
AL 2pp CUOL 1840.19

4685. 1840–10–12 From Edward Sabine 3 St. James's Place
Has requested John Caldecott to do what is needful to his own and [Louis] Agassiz's actinometers. Mentions great magnetic disturbances in June are not correlated in Toronto, Greenwich, and other observatories. Mentions proposal for correlation of observations between U.S. and Britain. ES leaves soon for Guernsey.
ALS 3pp RS:HS 15.107

4686. 1840–10–13 From George Biddell Airy Royal Observatory, Greenwich
Asks JH for comments on GA's proposals for the Standards Commission [see JH's 1840–7–29]; GA comments on some time spent recently in Glasgow and surrounding regions.
ALS 1p TxU:H/L-0760.1; Reel 1089 (ACCS: RGO 6.340.283)

4687. 1840–10–17 From Richard Jones to Margaret Brodie Herschel Tithe Commission
Has been troubled with his face again. Regarding the Scotch ale, which can be obtained to order in London. Saw J. F. Encke in town and he has now sailed. Francis Jeff[re]y (Lord Jeffrey) has been staying at Haileybury and speaks much of JH.
ALS 4pp RS:HS 10.365

4688. 1840–10–17 From R. Wilcox Lucknow [India]
New magnetic observatory to be established in India. Instrument, made by [T. R.] Robinson and Mr. Newman, will be sent to R.S.L. for examination and testing before shipment. Questions best times for daily barometer observations.
ALS 2pp † RS:HS 18.407

4689. 1840–10–19 From John Phillips B.A.A.S., York
JH, William Whewell, George Peacock, Humphrey Lloyd, and Edward Sabine are appointed by the B.A.A.S. to study systems of simultaneous magnetical and meteorological observations. They will be granted £50.
(A)LS 1p TxU:H/M–0428.7; Reel 1093

4690. 1840–10–19 From John Phillips B.A.A.S., York
JH, G. B. Airy, and Thomas Henderson are appointed to oversee the publication of the reductions of the calculations of N. L. Lacaille's stars. They will have £184 at their disposal.
(A)LS 1p TxU:H/M–0428.6; Reel 1093

4691. 1840–10–19 From John Phillips York
Informs JH that the B.A.A.S. has again placed him on the committee to supervise the translation and publication of foreign scientific memoirs.
(A)LS 1p RS:HS B26.5

4692. 1840–10–19 From John Phillips York
Informs JH that the B.A.A.S. has placed him on a committee to study the possibility of using balloon ascents to study the upper atmosphere.
(A)LS 2pp † RS:HS B26.4

4693. 1840–10–19 From John Phillips York
Informs JH that the B.A.A.S. has placed him on a committee to provide two actinometers for observing high in the Alps.
(A)LS 1p RS:HS B26.3

4694. 1840–10–21 From William Wallace 6 Lauriston Lane, Edinburgh
Notes loss of many old friends. Three years seclusion with bad health produced WW's treatise on conic sections, translated to Russian. Sends book on geometrical theorems and two memoirs. Discovered JH stated in 1813 what WW thought was recent discovery about catenary.
ALS 3pp RS:HS 18.27

4695. 1840–10–27 To John David Roberton Collingwood
Meteorolite given to JH by [Thomas] Maclear has been lost. Asks JR for assistance in locating it.
C 2pp RS:HS 14.365 (C: RS:HS 22.68)

4696. 1840–10–30 To George Biddell Airy Collingwood
Is sending comments on GA's proposals [see GA's 1840–10–13]; JH has been asked about obtaining a heliometer from Munich; seeks GA's opinion on the quality of the brass workmanship in Munich.
ALS 4pp RGO 6.340.284

4697. 1840–10–30 To George Biddell Airy Collingwood
Provides detailed comments on GA's circulated material on standards [see JH's 1840–10–30]; tends toward maintaining basic units but finding decimal relationships between them.
LS 18pp RGO 6.340.286

4698. 1840–10–30 From Humphrey Lloyd 2 Ryder St., St. James's
Has just received JH's letter of August as he has been traveling on the continent. Will see G. B. Airy tomorrow. Comments on Johann von Lamont's observatory at Munich. Discussed the Report of the Magnetic arrangements with Edward Sabine. Encloses letters. L. A. J. Quetelet would like support from the R.S.L. for additional observers in his observatory.
ALS 3pp RS:HS 11.275

4699. 1840–10–30 From Hugh Percy (3rd Duke of Northumberland) Northumberland House
Elected today as Chancellor of Cambridge University. Thanks for JH's support.
ALS 1p RAS:JH Archive 5/2.1.21 verso; Reel 8

4700. 1840–10–30 From William Henry Smyth Edenhall
Disappointed with format of *Nautical Almanac*. Arguments over this have caused many hard feeling among R.S.L. members. Details the controversy.
ALS 4pp RS:HS 16.267

4701. 1840–10–31 From Johann Franz Encke Berlin
About the calculations of JE, and others, on the motion of the solar system.
ALS 3pp (German) RS:HS 7.95

4702. 1840–10–31 From Humphrey Lloyd 2 Ryder St., St. James's
Regarding the publication of the magnetical and meteorological observations. Wonders if a letter from JH to the appropriate bodies would prove useful. Will give early consideration to the plan.
ALS 3pp RS:HS 11.276

4703. 1840–10–31 From John David Roberton Royal Society
Questioned porter concerning meteorolite, but received no additional information concerning its whereabouts. Hopes JH recovers it soon.
ALS 4pp RS:HS 14.366

4704. 1840–11–1 To Humphrey Lloyd Collingwood
About the urgent need to prepare a report on the magnetic observations, and that HL is the most logical person to give direction to it.
C 4pp RS:HS 22.69

4705. [1840–11] To John William Lubbock n.p.
Thanks for letting JH know about honor accorded at 'the Anniversary' [Meeting]. Offers condolences on JL's 'late severe loss.'
ALS 3pp RS LUB.H.358

4706. 1840–11–2 From George Biddell Airy Royal Observatory, Greenwich
Further observations on David Brewster's experiments. On heliometers. Observations on suggestions for Standards.
ALS 1p RS:HS 1.90 (C: RGO 6.340.295)

4707. 1840–11–3 From Basil Hall Portsmouth
Wishes JH would lay hands on the memorandum concerning the Zenith observatory on Etna. Explains why JH's theories regarding a point in gunnery do not work in actual practice. Queries the measurement given by JH for the height of Etna.
ALS 4pp RS:HS 9.192

4708. 1840–11–5 To Caroline Lucretia Herschel Collingwood
JH and family are beginning to feel at home at Collingwood.
ALS 4pp TxU:H/L–0585.14; Reel 1058

4709. 1840–11–6 From Edward Sabine 3 St. James's Place
Consults JH before responding to Wilhelm Weber's proposal for publication of observations. ES proposes to publish German data with data from Toronto, St. Helena, the Cape, and Van Diemen's Land. The Physical Committee of the R.S.L. suggests strengthening Adolphe Quetelet's application to the Belgian government for magnetic observations. The Hammerfest [Norway] magnetic curve does not harmonize with European ones.
ALS 4pp RS:HS 15.108

4710. 1840–11–7 From Humphrey Lloyd Trinity College, Dublin
Took the chair at a meeting of the Committee for Physics at the R.S.L. Is to bring the matter before Council, who in turn will request the Master of Ordnance to publish the observations. Material from each observatory will require about half of a quarto volume.
ALS 4pp RS:HS 11.277

4711. 1840–11–8 To Francis Baily Collingwood
Encloses observations on a comet by Carl Bremiker for the next meeting of the R.A.S. Will be unable to attend many of the meetings for this session. F. W. Bessel has lost his only son. C. F. Gauss's researches into the motion of the sun.
ALS 4pp RS:HS 3.185 (C: RS:HS 25.8.29)

4712. 1840–11–9 From Thomas Galloway Serjeants' Inn
Council of the R.A.S. will have to consider claims for the medal award at its next meeting. Has JH anyone in mind so that his claims may be considered.
ALS 2pp RS:HS 8.11

4713. 1840–11–9 From Basil Hall Portsmouth
Regarding the correct figure for the height of Etna, W. H. Smyth favors the half foot measurement. Regarding falls of stones from Vesuvius. Encloses calculations made by professor at Naval College. Remembers large balls of lava at Teneriffe.
ALS 4pp RS:HS 9.193

4714. 1840–11–10 From George Biddell Airy Trinity College, Cambridge
On the possibility of obtaining a heliometer from Munich.
ALS 1p RS:HS 1.91

4715. 1840–11–10 To Robert Hunt n.p.
Does not know anywhere that such processes as RH uses are used for registry work, although the idea is frequently raised.
ALS 3pp George Eastman House, 4.17 (C: RS:HS 22.70)

4716. 1840–11–11 From Francis Baily 37 Tavistock Place
Has sent his communication on the comet to various observatories and the *Times*. Will consider candidates for medals at the next council meeting.
ALS 1p RS:HS 3.186

4717. [1840?]–11–12 To 'dear Sir' Collingwood
Thanks for [?]'s note and enclosed sonnet. Sorry that old church was replaced during JH's absence from England.
ALS 2pp TxU:H/L-0749.5; Reel 1089

4718. 1840–11–12 From Edward Sabine 3 St. James's Place
Has JH any special instructions for [Louis] Agassiz's actinometers? Has sent Wilhelm Weber recent data for Toronto and St. Helena observatories while requesting Göttingen lunar observations. Discusses publication of the German Magnetic Association. Seeks advice on the inclusion of a translation of JH's paper on photometric apparatus. Has left at the R.S.L. a copy of C. F. Gauss's and Weber's magnetic atlas.
ALS 4pp RS:HS 15.109

4719. [1840 or later]–11–12 From F. A. Young Hall House, Hawkhurst
Cannot remember all four lines of Latin maxim. Will ask Dr. Ferguson for correct version.
ALS 1p RAS:JH Archive 13/3.92 verso; Reel 10

4720. 1840–11–13 From John Gibson Lockhart Sussex Place, Regents Park
John Murray transmitted JH's note regarding his review of William Whewell's book. Wonders if this letter has been acknowledged. Would like the paper as soon as possible.
ALS 3pp RS:HS 11.303

4721. [1840 or later]–11–14 From F. A. Young Hall House, Hawkhurst
Has not heard from Dr. Ferguson. Quotes what TY can recall of Latin maxim.
ALS 1p damaged RAS:JH Archive 13/3.90–91 verso; Reel 10

4722. 1840–11–15 From John Gibson Lockhart Sussex Place, Regents Park
Wishes he had communicated direct with JH regarding his article on William Whewell's *History of the Inductive Sciences*. Could not put it in the present number, but if any part of the article has been written please send it to their printer.
ALS 4pp RS:HS 11.304

4723. 1840–11–15 To John Gibson Lockhart Collingwood
Hopes to have JH's review of William Whewell's *History of the Inductive Sciences* ready for the next issue of the *Quarterly Review*.
ALS 1p NatLibScot MS.929, no.11

4724. 1840–11–15 From Thomas Maclear Base Line, Zwartland
Arrived in Zwartland in September. Surveyed grounds; measured baselines. Will be absent from observatory until Zwartland work finished.
ALS 4pp RS:HS 12.128

4725. 1840–11–16 From Henry Piddington Calcutta
Writes about HP's observations relating to the 'Law of Storms.'
ALS 4pp RS:HS 13.402

4726. 1840–11–17 From Sir John Richardson Haslar Hospital
Thanks JH for letter introducing Mr. Burnet at the Cape. The Hudson Bay Company is recording the temperature of the subsoil at JR's request. Sends some findings, which he will publish.
ALS 4pp RS:HS 14.485

4727. 1840–11–20 From Francis Baily 37 Tavistock Place
It was a quiet meeting at the R.A.S. Medals have been appropriated by the R.S.L. Presumes he has had notice of the next visitation. Will be pleased to see him when in town. A new comet is to be seen.
ALS 2pp RS:HS 3.187

4728. 1840–11–20 From Samuel Hunter Christie Woolwich
Council of the R.S.L. has awarded JH the Royal Medal for his research on the solar spectrum. Hopes to have a communication from him soon.
ALS 3pp RS:HS 5.340

4729. 1840–11–21 From Edward Sabine n.p.
Has received papers from Colonel Sykes. Describes (and includes a portion of) Lieutenant [Henry] Yule's report on broken instruments, which have prevented operations at Aden [?]. Suggests appointing a director there. Notes safe arrival of instruments at Toronto.
ALS 3pp encl RS:HS 15.110

4730. 1840–11–23 To Francis Baily n.p.
Regarding the award of the Royal Medal of the R.S.L. Is unable to come to town. Regarding the appointment of visitors to the Royal Observatory. Is preparing a paper on celestial reform.
ALS 4pp RS:HS 3.188–9 (C: RS:HS 25.8.30)

4731. 1840–11–24 From Francis Baily 37 Tavistock Place
Will communicate JH's remarks concerning the Royal Medal to the R.S.L. council. Regarding the constitution of the Board of Visitors. He is now printing the star catalogues of Edmund Halley, Tycho Brahe, etc.
ALS 3pp RS:HS 3.190

4732. 1840–11–25 From Francis Baily 37 Tavistock Place
Correcting a statement he made in his letter of 24 Nov. [see JH's 1840–11–24], regarding the members of the R.A.S. and R.S.L. who are also members of the Board of Visitors.
ALS 1p RS:HS 3.191

4733. 1840–11–25 From Edward Sabine 3 St. James's Place
Encloses letter to JH on magnetic survey in Canada and two letters from Leipzig, as well as proposals for reforms in B.A.A.S. Hopes to meet JH at the upcoming anniversary of

the R.S.L.
ALS 1p RS:HS 15.111

4734. 1840–11–26 From Humphrey Lloyd Trinity College, Dublin
Forwards papers relating to the Aden Observatory; please return, with any comments, to Col. W. H. Sykes. Submits suggestions for monthly information and comments on various topics.
ALS 4pp RS:HS 11.278

4735. 1840–11–27 From Francis Baily 37 Tavistock Place
Regarding the award of the R.S.L.'s Royal Medal.
ALS 1p RS:HS 3.192

4736. 1840–11–28 From Henry Raper J. U. S. Club
Sending copy of first volume of HR's work on navigation [*Practice of Navigation*].
ALS 2pp RS:HS 14.303

4737. [1840–12?] To Francis Baily Collingwood
Has not had time to work on the constellations paper because of interference due to tithe surveys and land re-assessments, which JH fears may involve a lawsuit.
C 1p RS:HS 25.14.5

4738. 1840–12–1 To S. I. Osborne Collingwood
About the desirability of closing a restricted public footpath across JH's property at Slough.
ALS 4pp RS:HS 25.6.7

4739. 1840–12–4 From Edward Sabine Margate
Encloses own copy of correspondence with Humphrey Lloyd and C. R. Fox about publication of magnetic observations. Mentions the affirmation of Lieutenant Younghusband in Canada. Will send the information to Charles Wheatstone and members of the Committee of Physics. Includes comments by Hew Ross and [Hussey] Vivian. Advises about cost of publication of data and sources for funding.
ALS 4pp RS:HS 15.113

4740. 1840–12–4 From Edward Sabine Margate
B.A.A.S. defers letter from [John] Washington to Magnetic Committee. Asks for JH's approval for funding [Robert] Schomburgk's portable magnetic observatory. JH should send Washington's letter to George Peacock. William Whewell, Humphrey Lloyd, and ES have contingently approved. Offers to order the appropriate instrument from Wilhelm Weber and seek partial funding from the Geographical Society. JH appends notice of his approval.
ALS 4pp † RS:HS 15.112

4741. 1840–12–10 To Francis Baily Collingwood
Regarding the arrangement of the constellations.
AL 4pp inc RS:HS 3.193 (C: RS:HS 25.8.31)

4742. [1840]–12–11 From Edward Sabine 12 Park Place
Pleased JH will be in London soon. George Peacock and William Whewell will become members of Magnetical Committee. Received information from [J. C.] Ross at Sydney.
ALS 4pp RS:HS 15.298

4743. 1840–12–13 To Henry Raper Collingwood?
Received HR's book on navigation. Says it will be very useful. Makes several remarks and objections concerning the list of principal fixed stars.
draft 3pp RS:HS 14.304 (C: RS:HS 22.71)

4744. 1840–12–13[?] From Edward Sabine Astronomical Society
Encloses letter from Humphrey Lloyd, whose opinion of himself ES finds unpalatable. Asks for letter's return via Francis Beaufort.
ALS 2pp RS:HS 15.114

4745. 1840–12–14 To William Radcliffe Birt Collingwood
Sends barometer observations from Flushing series, Elias Loomis in U.S., and [J. S.] McCord in Montreal. Can WB prepare these before next B.A.A.S. meeting?
ALS 3pp RS:HS 19.85

4746. 1840–12–15 From Francis Baily 37 Tavistock Place
Regarding the planning of the constellations. Is having success with his own Cavendish experiments.
ALS 3pp RS:HS 3.194

4747. 1840–12–16 To William Wallace Collingwood
Gratified by WW's volume. Recalls JH's catenary theorems. Writes function to complete theory of exponential transcendents. WW's applications have greater practical bearing than JH's early efforts, which JH now finds difficult to decipher. Sorry that WW declared Mr. Gilbert's tables defective.
C 2pp RS:HS 22.72

4748. 1840–12–17 From James Constantine Adamson Burr St., Smithfield
Difficulties over affairs of South African College, Cape Town. Observations on tornados.
ALS 4pp RS:HS 1.10

4749. 1840–12–17 To John Phillips Collingwood
Requests JP send JH's projected barometer observation to W. R. Birt.
C 1p RS:HS 22.73

4750. 1840–12–17 From Edward Sabine London
Requests recommendations on the instructions proposed by ES for the magnetic expedition to Africa. Encloses meteorological observations by [Edward] Belcher, and other data related to magnetism in North America.
ALS 3pp RS:HS 15.115

4751. 1840–12–20 To Charles Wheatstone Collingwood
Discusses when and how often R.S.L. committees, particularly the physical committee, should meet.
ALS 3pp RS MM.11.152

4752. [1840]–12–21 From Francis Beaufort [[Admiralty]]
Apologizing for the delay. Sends various map locations. Also extract from Captain F. W. Beechy's letter regarding magnetic operations.
ALS 2pp RS:HS 3.403 (ACC: Hydrographic Office Letter Book)

4753. 1840–12–24 From William C. Redfield New York
Glad papers were received [see WR's 1840–9–28]. Discusses observations of storms and other meteorological phenomena.
ALS 4pp RS:HS 14.310

4754. 1840–12–28 To William Radcliffe Birt Collingwood
Returns WB's 'Curves,' which JH presented to B.A.A.S. Sends results of 1837–1838 [barometer] observations by Captain [Edward] Belcher along Pacific coast of North and

Central America.
ALS 2pp RS:HS 19.86

4755. 1840–12–29 From Giovanni Santini Venice
Sends stellar catalogue. Requests *R.S.P.T.* from 1800 to 1840 for University of Padua library.
ALS 2pp (Italian) RS:HS 15.347

4756. 1840–12–30 To Francis Baily Collingwood
Giving news of his own recent work on the grouping of the constellations.
ALS 4pp RS:HS 3.195 (C: RS:HS 25.8.32)

4757. 1840–12–30 From Edward Sabine Margate
Encloses letter from Mr. Meyerstein concerning the sending to England of a magnetometer with improvements introduced by Wilhelm Weber. Asks JH to arrange for the payment of the bill.
ALS 4pp RS:HS 15.117

4758. 1840–12–30 From Edward Sabine Margate
Relates arrangements for the magnetic observations at Lucknow in India. Proposes using Göttingen mean time for observations at Lucknow. Humphrey Lloyd relates that Captain Boileau has begun observing at Sinnlaw [Sikkim?] after preliminary moisture–related difficulties with the instruments. ES proposes solution. Charles Riddell reports safe arrival of magnets from Lloyd. Riddell and Lieutenant Younghusband approve of survey of the American possessions, but R.S.L. has not yet considered it.
ALS 4pp RS:HS 15.116

4759. 1840–12–31 From Richard Jones Tithe Commission
Has just returned from Ely, where George Peacock is very happy and prosperous. Gives news of other scientific friends. Hawkhurst tithes have been peaceably settled.
ALS 4pp RS:HS 10.366

4760. [1841?] To Charles Babbage n.p.
The Ryans are in London.
ALS 1p RS:HS 2.365

4761. [1841–1?] To [William Radcliffe Birt] n.p.
[WB's] sheets [of barometer observations] are well done. Sends European latitudes and longitudes. Will send 'Engraved Squares' soon. Notes errors in observations.
ALS 2pp RS:HS 19.169

4762. 1841 To Margaret Brodie Herschel n.p.
JH is captivated by science of photography. Has blundered onto 'Calotype *leaving out* [W. H. F.] Talbot's principal ingredient!!' [Note added by daughter Isabella Herschel:] Papa is in hurry. Greetings to family.
ALS 3pp TxU:H/L-0539; Reel 1053

4763. [1841] To Margaret Brodie Herschel [London]
Compliments MH on her poetry; JH is trying to clarify when various friends are to visit Collingwood.
ALS 4pp JHS 3.21d

4764. [1841 to 1848?] From Robert Hunt [Falmouth]
Asks to borrow a book.
ALS 1p Harvard: Houghton AAH 67m–67 (107)

4765. [1841 to 1847?] From Robert Hunt [Falmouth]
Wishes to discuss an urgent matter.
ALS 1p Harvard: Houghton AAH 67m–67 (63)

4766. [1841] From Henry Anthony Jeffreys Hawkhurst Parsonage
Sending some papers he should have returned earlier. Anxiety regarding his brother is relieved by the acquittal of Captain Levick. Will possibly return to England now.
ALS 3pp RS:HS 10.306

4767. [1841?] From Edward Sabine 12 Park Place
Sends letter from [A. T.] Kupffer regarding magnetic observations. Discusses Great Toronto Disturbance of 18 November and observation of it at Greenwich, St. Helena, and St. Petersburg.
ALS 4pp RS:HS 15.300

4768. [1841] From Richard Sheepshanks n.p.
Argues that John Hind is sick from too much observing, and should be given funds to take an extended vacation.
ALS 4pp RS:HS 16.124

4769. 1841 From F[rancis] Tudhope Cape of Good Hope
Reports that the lectures on various subjects that JH proposed are underway in Cape Town. Sends transcripts of the lectures.
ALS 4pp RS:HS 17.447

4770. 1841–1–2 From Francis Baily 37 Tavistock Place
Latest news of his Cavendish experiments. Will get the star maps printed for him. There is to be a meeting of the Council of the R.A.S. to consider the award of medals.
ALS 2pp RS:HS 3.196

4771. 1841–1–2 From Andries Stockenström Cape of Good Hope
Express gratitude for JH's role in recommending him for baronet. Reports on the political situation in the colony.
ALS 2pp RS:HS 17.26

4772. 1841–1–3 To Caroline Lucretia Herschel Collingwood
The Herschels bought a Christmas tree this year; JH reports that his children loved it and that 'they will be sure to keep up the custom which is a very merry one.'
ALS 4pp TxU:H/L–0586.1; Reel 1058

4773. 1841–1–5 From Henry Raper J. U. S. Club
Thanks JH for letter [RS:HS 14.304]. Discusses the list of principal fixed stars and its problems. Mentions several corrections that must be made.
ALS 4pp RS:HS 14.305

4774. 1841–1–5 From George Peacock Deanery, Ely
About some aspects of the magnetic survey, and who shall pay for it.
ALS 2pp RS:HS 13.293

4775. 1841–1–8 · From Wilhelm Struve Pulkowa
Regrets that his work at Pulkowa eliminates the possibility of being absent for longer periods of time and hopes to see JH in Russia. Mentions his and his son Otto Struve's most recent work. Wishes JH well in his new home at Collingwood.
ALS (German) 4pp RS:HS 17.129

4776. 1841–1–9 From Francis Baily 37 Tavistock Place
Is sorry to hear of his illness. The R.A.S. has awarded its medal to F. W. Bessel. Robert Main's paper on parallax is being printed. Is to be a meeting of the Standards Committee in February. Has arranged the printing of the star maps. John Wrottesley is to be the next president of the R.A.S.
ALS 3pp RS:HS 3.197

4777. 1841–1–9 From Edward Sabine Margate
Discusses problem involving JH, George Peacock, Humphrey Lloyd, and Wilhelm Weber regarding the price of magnetometers. Also mentions reports of magnetic observations in Russia and at the Cape of Good Hope.
ALS 4pp RS:HS 15.118

4778. 1841–1–11 From Richard Jones Haileybury
Enjoying a couple of days idleness. Regrets to hear about the proposed new road. Suggests ways of preventing it if on JH's land. Proposes a February visit.
ALS 4pp RS:HS 10.367

4779. 1841–1–11 From Edward Sabine Margate
Asks JH's advice on the publication of various observations. Is also awaiting advice from Humphrey Lloyd.
ALS 3pp RS:HS 15.119

4780. 1841–1–12 From S. B. P. Micklethwaite Iridge Place, Hurstgreen
Requests permission to submit JH's name to the Lieutenants of the Counties of Kent and Sussex as a Magistrate for the two counties.
ALS 4pp RS:HS 12.324

4781. 1841–1–13 To S. B. P. Micklethwaite Collingwood
Giving reasons why he would not like his name put forward as a magistrate.
C 3pp RS:HS 12.325 & 22.76

4782. 1841–1–14 From S. B. P. Micklethwaite Iridge Place Hurstgreen
Thanks for his kind and flattering letter. Fully understands why JH cannot spare the time to be a Magistrate, so will not

submit his name.
ALS 3pp RS:HS 12.326

4783. 1841–1–15 To Edward Sabine Collingwood
Comments on the operational and organizational details of the magnetic and meteorological observatories, and on the supervisory committee's [the Physical Committee of the R.S.L.] responsibilities.
ACCS 6pp RS:HS 25.6.9

4784. 1841–1–16 From George Biddell Airy Royal Observatory, Greenwich
Sending two copies of the 'Extracts relating to Standards.' Remarks on these extracts. Requests JH to fix a date for a meeting to discuss the report.
ALS 2pp RS:HS 1.92 (C: RGO 6.340.296)

4785. 1841–1–16 From Francis Beaufort Admiralty
Is preparing an estimate for Parliament and would like his views on the usefulness of re-measuring N. L. Lacaille's base, etc., at the Cape.
ALS 4pp RS:HS 3.371 & ACC: Hydrographic Office Letter Book

4786. 1841–1–16 From Macedonio Melloni Bde du Temple, Paris
MM has been selected by the King of Naples to be Director of the Central Board of Meteorological Observations for Naples. Would like any of the publications of the R.S.L. relative to the Fixed Observatories.
ALS 3pp (French) RS:HS 12.316

4787. 1841–1–17 From Robert Hunt Falmouth
Is grateful for referring him to Capt. Ludlow. Has now moved himself to Falmouth. What does JH think of the chances of success of a popular book on photography? He is contemplating the idea. Hears that J. B. Biot considers the production of colored drawings impossible, but is not convinced himself. Encloses description and drawing of a very clever instrument.
ALS 3pp RS:HS 10.89

4788. 1841–1–17 To Heinrich Christian Schumacher n.p.
Discusses a letter in which JH explains why JH cannot, as a British subject, accept the foreign order offered him [by the Danish king]. Asks HS to read over the letter and to decide how it should be delivered.
ALS 3pp RS:HS 15.387 (C: RS:HS 22.77)

4789. 1841–1–19 From Adolphe Quetelet Brussels
Says JH's interest in AQ's work has been source of motivation. Notes that on 21 September and 21 December, aurora borealis and magnetic perturbations have been spotted. Asks where to procure an actinometer in England.
ALS (French) 3pp RS:HS 14.221

4790. 1841–1–19 From Edward Sabine Margate
Relates Humphrey Lloyd's complaint that JH sent Mr. Meyerstein's letter by mistake. Asks JH to forward [A. T.] Kupffer's letter along with those enclosed from Charles Riddell on the magnetism survey in America. Asks for JH's advice on the publication of the observations.
ALS 2pp RS:HS 15.120

4791. 1841–1–20 To George Biddell Airy Collingwood
Agrees to proposed meeting date [see GA's 1841–1–16], and encourages GA to come with a draft final report prepared.
ALS 3pp RGO 6.340.297

4792. 1841–1–20 To Francis Beaufort Collingwood
Supports the remeasurement of N. L. Lacaille's geodetic arc in South Africa, giving a number of reasons for his support.
ACCS 3pp RS:HS 25.6.8

4793. 1841–1–26 To Francis Baily [Collingwood]
Is sending him a circumpolar projection of the Southern stars. Regarding the star maps being printed by John Arrowsmith. Will visit him the following week. Has received Robert Main's paper.
ALS 5pp RS:HS 3.198 (C: RS:HS 25.8.33)

4794. 1841–1–27 From George Biddell Airy Royal Observatory, Greenwich
A notice of meeting of the Standards Commission.
LS 1p TxU:H/M-0627.1; Reel 1084

4795. 1841–1–28 From Francis Baily 37 Tavistock Place
Will be pleased to accommodate him. Star map is being printed. Encloses a paper on Sound by Dr. R. J. Kane, for his comments.
ALS 1p RS:HS 3.199

4796. 1841–1–28 From George Thomas Napier Graham's Town
Has dispatched a military force to St. John's River. Outlines recent political events at the Cape. The educational scheme progresses well. Has had an acrimonious letter from Andries Stockenström.
ALS 4pp RS:HS 13.70

4797. 1841–1–28 To Edward Sabine [Collingwood]
Insists on inclusion of all data and graphs in the publication of the accumulated magnetic observations, made as part of 'perhaps the very greatest scientific operation ... ever ... undertaken....' Urges that funding be sought from the government.
C 3pp RS:HS 15.121 (C: RS:HS 22.78)

4798. 1841–1–29 From Edward Sabine Margate
Assures JH that he will publish magnetic observations as JH desires. [Hussey] Vivian awaits the estimate of cost of publication. Encloses circular sent to the British magnetic observatories. Informs JH that the Niger and Guiana magnetometers have arrived safely. They should provide interesting comparisons with other observatories.
ALS 4pp RS:HS 15.122

4799. 1841–1–30 From Francis Beaufort Admiralty
Is sending him Thomas Maclear's account of taking the field and of commencing his base operations.
ALS 1p RS:HS 3.372

4800. [1841–2?] To [William Radcliffe Birt] n.p.
Explains 'discordances' as contraction of damp paper. Use 'Engraved Squares' to avoid this error. Possible error in barometer observations from Montreal. Diurnal 'undulation' appearing in data. American observations arrived, may fill 'missing intervals.' Keep record of expenses incurred. Postscript on color variation in stars.
ALS 3pp RS:HS 19.166

4801. 1841–2–1 From Luke Howard Tottenham
Has been considering the reasons for the great depression of the barometer in November 1840. Thinks JH may be interested in the enclosed essay. Would be pleased if it could be read to the Royal Society.
ALS 1p RS:HS 10.5

4802. 1841–2–2 To Robert FitzRoy Collingwood
Returning some papers that should have been returned earlier. Hopes all are well.
draft 2pp RS:HS 7.249

4803. 1841–2–2 From Edward Sabine 12 Park Place
Encloses letter from G. B. Airy. Suggests new method of magnetic observation.
ALS 2pp † RS:HS 15.139

4804. 1841–2–5 To Julian Jackson Collingwood
A portable magnetometer purchased by the B.A.A.S. is now available as a loan for use in magnetic survey of British Guyana.
ALS 3pp Royal Geographical Society

4805. 1841–2–5 From Edward Sabine 12 Park Place
Presents data on various disturbances. Awaits advice from C. F. Gauss. Discusses determination of intervals between observations. Mentions proposal to continue observations for 3 years.
ALS 4pp RS:HS 15.140

4806. 1841–2–5 From Edward Sabine 12 Park Place, St James
Has learned of the availability of a house at Kew, which could be used as a magnetic and meteorological observatory. Dr. Robinson has received the instruments.
ALS 4pp RS:HS 15.123

4807. [1841?]–2–7 From Edward Sabine Admiralty
Will be happy to send particulars from [C. F.] Gauss's information. [Humphrey] Lloyd's instrument records changes in horizontal intensity best. Wants to employ magnetometers in New York.
ALS 3pp inc RS:HS 15.319

4808. 1841–2–8 From George Biddell Airy Royal Observatory, Greenwich
Sends JH a copy of draft report [see JH's 1841–1–20], and indicates procedure for coming meeting.
ACCS 1p RGO 6.340.298a

4809. 1841–2–8 From George Biddell Airy Royal Observatory, Greenwich
A note to accompany a draft of the report of the Standards Commission.
ALS 1p TxU:H/M-0627.3

4810. 1841–2–8 From George Biddell Airy Royal Observatory, Greenwich
A notice of meeting of the Standards Commission.
LS 1p TxU:H/M-0627.2

4811. 1841–2–9 From Augustus De Morgan 69 Gower St.
Regarding JH's telescope. Includes a rhyme on it.
ALS 2pp RS:HS 6.185

4812. 1841–2–9 To Edward Sabine Collingwood
Advice for standardizing observations, to be given to the Physical Committee of the R.S.L.
C 4pp RS:HS 15.141 (C: RS:HS 22.111)

4813. 1841–2–10 To Robert Hunt Collingwood
Has not lately done much work in photography, except to work with paper treated with vegetable substances.
C 2pp RS:HS 22.79

4814. 1841–2–11 To Margaret Brodie Herschel 37 Tavistock Place
It will be difficult for JH to get away to Collingwood as planned as discussions [?] are slow; some family news.
ALS 3pp JHS 3.21e

4815. 1841–2–12 From George Biddell Airy Royal Observatory, Greenwich
A notice of meeting of the Standards Commission.
LS 1p TxU:H/M-0627.4

4816. 1841–2–17 From George Biddell Airy Royal Observatory, Greenwich
Asks JH to put in writing proposals JH had made at a Standards Commission meeting, especially those relating to nomenclature [see GA's 1841–2–8].
ALS 2pp TxU:H/M-0627.5; Reel 1084 (C: RGO 6.340.298b)

4817. 1841–2–17 or later To George Biddell Airy n.p.
A note to accompany proposals for the Standards Commission to consider [see GA's 1841–2–17]; twenty pages of proposals are appended.
ALS 1p RGO 6.340.299

4818. 1841–2–19 From Elizabeth [Betty] Plumleigh Babbage 5 Devonshire St., Portland Place
Thanks and appreciation for past and present friendship.
ALS 2pp RS:HS 2.296

4819. 1841–2–20 From Francis Beaufort Admiralty
Edward Sabine has asked him to forward J. C. Ross's notes. If JH has finished with Thomas Maclear's letter would he return it.
ALS 2pp RS:HS 3.373 & ACC: Hydrographic Office Letter Book

4820. 1841–2–20 From Robert FitzRoy 25 Lowndes St.
Was pleased to receive the letter and old papers. Remembers his time at the Cape with pleasure. Has been interested, but not convinced, by J. C. Ross's soundings.
ALS 4pp RS:HS 7.250

4821. 1841–2–20 To Robert Hunt Collingwood
Unable to give any specifics concerning Louis Jacques Mandé Daguerre's new photographic process. Discovered that old paper specimens produced a much better representation of the spectrum in its natural colors than those obtained at the date of JH's paper; these results are 'light on a dark ground,' which makes JH more hopeful that colored photography will someday be perfected. Has experimented with vegetable substances.
ALS 4pp George Eastman House, 5.17

4822. 1841–2–20 From Edward Sabine Margate
Francis Beaufort will send JH sheets from James Ross covering trip from the Cape to Van Diemen's Land. Laments Charles Riddell's leave of absence but acknowledges replacement Lieutenant Younghusband's skill. States that C. F. Gauss has informed him by letter of the location of the south magnetic pole. Informs of [Edward] Belcher's letter concerning various magnetic stations in the South Pacific.
ALS 6pp RS:HS 15.124

4823. 1841–2–22 To Richarda Airy Collingwood
Writes to announce the birth of daughter Amelia, and to assure RA that both JH's wife, Margaret, and new daughter are doing well.
ALS 3pp JHS 2.6

4824. 1841–2–22 From John Taylor, F.R.S. Duke St., Adelphi
Sends JH a quantity of silver ore from Mexico.
ALS 1p RS:HS 17.429

4825. 1841–2–23 From Few & Co. Covent Garden
Has deposited the new conveyance with Messrs. Drummonds and returns the key.
ALS 1p RS:HS 7.205

4826. 1841–2–23 To Macedonio Melloni n.p.
Will do all in his power to ensure that MM receives copies of the publications relating to the Fixed Observatories. Does not think MM could have seen a copy of JH's paper on photographic action when he wrote the notice in *Annales de chimie*. Comments on this paper and more recent experiments in this field.
C 4pp RS:HS 12.317 & 22.80

4827. 1841–2–23 To Peter Mark Roget Collingwood
Notes interest expressed by Macedonio Melloni in meteorological observations. Urges that R.S.L. Council give its attention to the mass of meteorological observations that have accumulated.
ALS 3pp Lion Heart Autographs

4828. 1841–2–24 From Francis Beaufort Admiralty
Admiralty wished to present him with another part of the Cape Observations, and will hold it until someone is deputed to call at the office.
ALS 1p RS:HS 3.374

4829. 1841–2–24 From Caroline Lucretia Herschel Hanover
Angry that many call William Herschel's 40-ft. reflecting telescope 'useless.'
L *Memoir ... of Caroline Herschel*, 317–8. inc

4830. 1841–2–24 From Edward Sabine 12 Park Place
On discrepancy of data from G. B. Airy and Humphrey Lloyd. Captain Boileau and John Caldecott exchange instruments to normalize data. Will send letters from Airy, [A. T.] Kupffer, Lloyd, ES, and JH to John Lubbock and the R.S.L. Council concerning the discrepancies.
ALS 4pp RS:HS 15.142

4831. [1841–3?] To [William Radcliffe Birt] n.p.
Received WB's packet. Will send sheets tomorrow. Needs all 'Curves and data' before end of March for final report to B.A.A.S. JH's nerves 'shattered with night work.'
ALS 3pp RS:HS 19.167

4832. 1841–3–2 From Macedonio Melloni Paris
Is grateful for JH's obliging offer. Regrets his own unfortunate article regarding JH's experiments; comments on this. Is pleased JH is to continue with his on light.
ALS 4pp (French) RS:HS 12.318

4833. [1841?]–3–3 From Edward Sabine n.p,
Sends JH some actinometer observations and a letter from G. B. Airy to forward to Humphrey Lloyd. Also some observations of shooting stars made by [James] Farquharson.
ALS 2pp RS:HS 15.325

4834. 1841–3–4 From Edward Sabine 12 Park Place
Congratulates JH on his latest child. Notes that data from James Ross show great disturbances at the Kerguelen station simultaneously with Toronto and Greenwich. Has sent various data to C. F. Gauss. Wilhelm Weber reports a new magnetic station at Havana recording by the British system. Notes Ross's dip data supports Gauss's theoretical expectations.
ALS 4pp RS:HS 15.125

4835. 1841–3–5 To William Rowan Hamilton Collingwood
Requests autographed copy of WH's sister's poems, wanting to send them to an acquaintance who translated some of JH's work. Expresses a theory about cause and effect. Mentions minutes of Royal Irish Academy, a reform of the constellations, and WH's three axes of the universe.
ALS 7pp TCD 1493:192

4836. 1841–3–7 From F. M. Eardley-Wilmot Cape of Good Hope
Magnetic observatory is still not operational one year after FW's arrival. Thomas Maclear is now in 'Zwartsland' [Zwartkops?]. FW shares quarters with C. P. Smyth and William Mann. Hopes to begin observations next week at Cape Town and Simon's Bay; G. G. Lewis will make observations at Grahamstown. Problems with meteorological instruments. Proposes magnetic survey of entire colony. [JH annotation: Enquire what FW meant about measurement conversions, malfunctioning needle, and broken tube in 19 Feb. 1841 letter to Humphrey Lloyd.]
ALS 4pp †† TxU:H/M–0174; Reel 1087

4837. 1841–3–7 To Caroline Lucretia Herschel Collingwood
Reports the birth of Amelia Herschel, JH's eighth child. Forwarding to CH an article describing the telescope of William Parsons.
ALS 3pp TxU:H/L–0586.7.a; Reel 1058

4838. 1841–3–8 From Thomas Maclear Royal Observatory, Cape of Good Hope
Came back from Zwartland for a day, but returns tomorrow. Relates news of the measuring of the base line. Visited magnetic observatory; work in Zwartland not yet complete. Magnetic Observatory satisfactory despite slow construction. No news of the obelisk yet.
ALS 4pp RS:HS 12.129

4839. [1841]–3–8 From Edward Sabine 12 Park Place
Thinks enclosed letter should be appended to [James] Farquharson's report on aurora. Wants to bring magnetometer to Alford. Hopes JH has [Humphrey] Lloyd's article regarding instruments.
ALS 3pp RS:HS 15.301

4840. 1841–3–9 From Eliza Mary Hamilton Dublin
Her brother has sent JH's letter and she will be only too pleased to send her poems for whatever purpose he requires them.
ALS 1p RS:HS 10.148

4841. 1841–3–11 From Edward Sabine 12 Park Place
Discusses plans for publishing magnetic storm data. Mentions transportable magnetic instruments and their use for a travelling observatory on an expedition to the Falkland Islands.
ALS 4pp RS:HS 15.143

4842. 1841–3–13 From Robert Warington King Villa, South Lambeth
Invites JH to join new Chemical Society. Committee proposed JH as first president.
ALS 1p, encl RS:HS 18.376

4843. 1841–3–16 To William Radcliffe Birt [Collingwood]
Progress on barometer curves. Invites WB to send report, to be read at B.A.A.S. meeting in July.
ALS 3pp RS:HS 19.88

4844. 1841–3–16 To William Henry Fox Talbot Collingwood
Praises WT for invention of 'Kalotype' process; predicts it will be called 'Talbotype.' Notes excellence for photography of the weather during past year. Mentions possibility of cases

of excessively rapid photographic change.
ALS 4pp Science Museum Talbt 1/28

4845. 1841–3–17 From William Richard Hamilton 12 Bolton Row
Has just received the enclosed communication from Sir Hilgrove Turner, lately Governor of Bermuda. Believes JH is the only person who can deal with it.
ALS 2pp RS:HS 9.220

4846. 1841–3–17 From William Henry Fox Talbot Lacock Abbey
Announces his invention of the 'calotype', and his application for a patent. Questions JH on his address to R.A.S., objecting to idea that universe is infinite.
ALS 3pp RS:HS 17.304a Buckland, *Talbot,* 66 inc

4847. 1841–3–18 From S. Jey to Margaret Brodie Herschel Bath
Has been confined to the house and only goes out in a wheelchair. Would JH like the bust of Sir Joseph Banks?
ALS 3pp RS:HS 10.445

4848. 1841–3–18 From Edward Sabine 12 Park Place
Norwegian government plans to support a magnetic observatory. Discusses details concerning R.S.L. logistical support and sources for instruments and training. Encloses Wilhelm Weber's letter concerning the instruments. Notes Charles Riddell's return and Lieutenant Younghusband's continued good work.
ALS 2pp RS:HS 15.126

4849. 1841–3–18 From William Henry Fox Talbot Lacock Abbey
Preparing his calotype paper for publication. Looking for a summer house, somewhere interesting to photograph.
ALS 3pp RS:HS 17.305a

4850. 1841–3–19 From William Rowan Hamilton Observatory
Considering motion as a 'successive excitement of powers.'
L Graves, *Hamilton, 2,* 337–8 inc

4851. 1841–3–20 From George Biddell Airy Royal Observatory, Greenwich
Sending draft of report and appendix. Specimen of a gauging scale.
ALS 1p RS:HS 1.93 (C: RGO 6.340.313)

4852. 1841–3–20 or later To George Biddell Airy n.p.
Has received material sent by GA [see GA's 1841–3–20]; comments on the report and JH's unavailability for meetings of the Standards Commission in the near future.
ALS 4pp RGO 6.340.314

4853. [1841]–3–22 From Edward Sabine 12 Park Place
Discusses proposed temperature experiments involving balloons. Prince Edward Island would be good observatory but funds are lacking. [C. F.] Gauss's article translated for *Scientific Memoirs.*
ALS 6pp RS:HS 15.302

4854. 1841–3–22 To William Henry Fox Talbot Collingwood
Congratulates WT on the calotype invention and the patent. Responds at some length to WT's concerns about the infinity of the universe. In postscript, provides a 4-page dialogue on the subject.
ALS 8pp Science Museum Talbt 1/29 (C: RS:HS 17.304b & 22.81) Buckland, *Talbot,* 66 inc

4855. 1841–3–24 To William Rowan Hamilton Collingwood
On mode of conveyance of WH's sister's verses and the autographs for JH's acquaintance.
ALS 2pp TCD 1493:194

4856. 1841–3–25 From George Biddell Airy Royal Observatory, Greenwich
A note to accompany additions, comments, and emendations relating to the report of the Standards Commission.
LS 1p TxU:H/M-0627.6; Reel 1084

4857. 1841–3–25 To William Henry Fox Talbot Collingwood
A further two dialogues elaborating on JH's earlier arguments [see JH's 1841–3–18] on the infinity/finitude of the universe.
ALS 4pp Science Museum Talbt 1/30 (C: RS:HS 17.305b & RS:HS 22.82)

4858. 1841–3–26 From James David Forbes Edinburgh
Sending a notice about the polarization of the sky. Also some practical papers he set at Durham. Has the obelisk ready for shipment to the Cape. Has asked Richard Taylor to send him a copy of his report on Meteorology.
ALS 2pp RS:HS 7.297

4859. 1841–3–31 From Caroline Lucretia Herschel Hanover
Lists those who visited CH on her birthday.
L *Memoir ... of Caroline Herschel,* 318–19 inc (?)

4860. 1841–3–31 From Macedonio Melloni Paris
Is about to leave Paris and hopes to receive the meteorological information at Naples. Awaits with impatience JH's work on the rays of light announced in his last letter.
ALS 3pp (French) RS:HS 12.319

4861. 1841–4 To Caroline Lucretia Herschel Collingwood
Has not unpacked William Herschel's letters that JH secured from the R.S.L.
ALS 2pp TxU:H/L–0586.8.b; Reel 1058

4862. 1841–4–2 To Francis Baily Collingwood
Regarding the payment of John Arrowsmith's bill for the printing of the star maps, which seems excessive in relation to the low standard achieved. Would like FB's comments.
ALS 5pp RS:HS 3.201 (C: RS:HS 25.8.34)

4863. 1841–4–2 To Francis Baily n.p.
Sends letter on the subject of the star maps.
ALS 2pp RS:HS 3.200 (C: RS:HS 25.8.35)

4864. 1841–4–4 From Thomas Maclear Base Line Camp, Zwartland
Measuring work of the base line is virtually completed and levelling of the line will commence forthwith. Has devoted every day to the work except Sundays. Comments on various aspects of the work. F. E. Wilmot wishes he had not returned the rusty dip bar. Obelisk is on board the *Romeo.*
ALS 4pp RS:HS 12.130

4865. 1841–4–5 From John Gibson Lockhart London
Is delighted at the prospect of JH's article being included in the next number of the *Quarterly Review.* Is not surprised it will extend to fifty pages, but hopes he will remember the 'general reader.'
ALS 3pp RS:HS 11.305

4866. 1841–4–6 From Francis Baily 37 Tavistock Place
Has communicated JH's remarks to John Arrowsmith. Is sorry for the inconvenience caused.
ALS 2pp RS:HS 3.202

4867. 1841–4–6 To William Henry Fox Talbot Collingwood
Doubts that isomerism of all elements can be verified experimentally, but notes a number of groups with similar atomic weights that might well fit. Further on vegetable colors in photography. Refers to enclosure of a further dialogue on infinity.
ALS 4pp Science Museum Talbt 1/31 (C: RS:HS 17.306b & 22.83)

4868. 1841–4–6 From William Henry Fox Talbot Lacock Abbey
Reports receipt of a communication, which argues that all elements are merely isomers of one substance. Expresses support for such a view.
ALS 4pp RS:HS 17.306a

4869. 1841–4–[9] To Francis Baily Collingwood
Further about the map printing problem [see FB's 1841–4–6].
C 1p RS:HS 25.8.36

4870. [1841–4–10] To Francis Baily n.p.
Would like some dry prints of the star map, if no further expense is involved. Is sorry that he has caused him all the bother.
ALS 2pp RS:HS 3.203

4871. 1841–4–12 From Caroline Lucretia Herschel Hanover
List of papers JH took from CLH on his 'last visit' to [Hanover]. Miscellaneous notes about relatives and acquaintances.
Amem 2pp TxU:H/M–0684; Reel 1084

4872. 1841–4–15 From Francis Baily 37 Tavistock Place
John Arrowsmith has undertaken to make further copies of the star maps on dry paper, free of charge.
ALS 1p RS:HS 3.204

4873. 1841–4–15 From F. M. Eardley-Wilmot Cape of Good Hope
M. F. B. Dausse visited FW several times to study instruments and copy meteorological observations. Dausse urged FW to share these with D. F. J. Arago, who is concerned at delay in publishing survey results. Please notify R.S.L. that Arago's official request will arrive soon. Encloses copy of FW's 15 Apr. 1841 letter to Arago. New rain gauge. Problems with anemometer and dipping needle. Will visit Cape Agulhas with Thomas Maclear to make magnetic observations.
ALS 4pp †, encl 2pp TxU:H/M–0175; Reel 1087

4874. [1841–4–16] To William Whewell [Collingwood]
Asks WW to be godfather to JH's new daughter, Amelia. In reviewing WW's [*Philosophy of the Inductive Sciences*] has found many of JH's views changing, but cannot accept his dear friend's a prioristic approach.
ALS 4pp TC, Camb. Add. Ms.a.207[46] (draft: RS:HS 18.193 & C: RS:HS 22.84 & C: TC, Camb. 0.15.46[19] inc)

4875. 1841–4–20 To William Radcliffe Birt Collingwood
Sends missing barometer observations from Montreal, Flushing, and Albany. B.A.A.S meeting postponed.
ALS 1p RS:HS 19.89

4876. 1841–4–20 From William Whewell Trinity College
Will be pleased to be godfather of JH's new daughter. Pleased also that JH is reviewing his [*Philosophy of the Inductive Sciences*]. Presents an argument to support his a prioristic view of the nature of science.
ALS 3pp RS:HS 18.194 Todhunter, *Whewell, 2,* 297–9

4877. 1841–4–22 To William Whewell [Hawkhurst]
Is moving the date for christening JH's daughter [Amelia] and WW's goddaughter to 3 May to accommodate the godmother, Maria Edgeworth.
ALS 1p TC, Camb. Add. Ms.a.207[47]

4878. 1841–4–24 From Edward Sabine Woolwich
Forwards request of the Secretary of the Treasury for justification of public monies for expenses incurred for magnetic and meteorological observations of the Antarctic expedition. Brief note from JH should suffice for House approval. Note should mention G. B. Airy's request concerning the reduction of the Greenwich lunar and planetary observations and Francis Beaufort's concerning navigation. ES promoted to Lieutenant Colonel.
ALS 4pp RS:HS 15.127

4879. 1841–4–27 To Edward Sabine Collingwood
Provides examples to justify governmental support of various magnetic and meteorological observations. States that 'Terrestrial magnetism ... of all sciences of observation holds out the greatest present prospect of decisive reduction under ... exact theory.' Mentions work by N. L. Lacaille and Thomas Maclear at the Cape.
LS 5pp RS:HS 15.128 (C: RS:HS 22.85)

4880. 1841–4–28 From Edward Sabine Woolwich
Quotes passage from C. E. Trevelyan (Treasury Office) to [Hussey] Vivian concerning the number of copies of published data and the number of magnetic observatories. Requests JH's perusal and advice. Includes letter from [A. T.] Kupffer on latest observations.
ALS 4pp RS:HS 15.129

4881. 1841–5–1 From Robert Treat Paine Boston, Mass.
Sends pamphlet for JH from Philadelphia. Author [E. P.] Mason died at the early age of 22. Believes he used a Herschelian telescope. Has been a mild winter in Boston.
ALS 2pp RS:HS 13.214

4882. 1841–5–1 From Edward Sabine Woolwich
Relates discussion with [Hussey] Vivian on magnetic survey of Africa and Treasury support thereof. Discusses administration of various stations.
ALS 4pp RS:HS 15.130

4883. 1841–5–4 To [?] Collingwood
Suggests a resolution of a small amount of money having to do with the exchange of some chairs; also asks about a dining room table.
ALS 2pp WT B.5

4884. 1841–5–11 To [George Biddell Airy?] Collingwood
Outlines advantages of placing magnetometers for Hammerfest Observatory at Edward Sabine's disposal, so that he can instruct those going to Toronto and the Cape in their use.
ALS 4pp RS MM.11.155

4885. 1841–5–12 From George Biddell Airy Royal Observatory, Greenwich
Seeks clarification from JH about a number of matters in the draft report of the Standards Commission [see JH's 1841–5–

12 or earlier].
ALS 3pp TxU:H/M-0627.7; Reel 1084 (C: RGO 6.340.317)

4886. 1841–5–12 or earlier To George Biddell Airy n.p.
Comments on additional proposals for consideration by the Standards Commission [see JH's 1841–3–20 or later].
ALS 4pp RGO 6.340.315

4887. 1841–5–12 From [J. S. C.] Schweigger Halle
Sends book on Samothracian mysteries. Mentions article by [C. B.] Hafe. Suggests JH ask [William] Whewell to review the article in *Quarterly Review*.
ALS (German) 4pp RS:HS 15.414

4888. 1841–5–12 From Andrew Smith Chatham
Apologizes for illness preventing him from coming to Collingwood; comments on the South African political picture, and AS's intention to retire.
ALS 4pp RS:HS 16.180

4889. [1841–1858]–5–12 From Henry Holland 25 Brook St.
Invites JH to dinner if possible, or later if not at present possible.
ALS 2pp RS:HS 9.441

4890. 1841–5–13 From E. Mühry London
Is in London studying English criminal law. EM was pleased with his visit with JH, and will be telling JH's aunt Caroline about it when he returns to Hanover.
ALS 3pp (German) RS:HS 19.290

4891. 1841–5–14 To William Henry Fox Talbot Collingwood
Has been experimenting with paper prepared with light-sensitive vegetable juices. Sends two specimens of JH's results.
ALS 3pp Science Museum Talbt 1/32

4892. 1841–5–18 From George Biddell Airy Royal Observatory, Greenwich
A summons to a meeting of the Standards Commission, together with some proposals for amending the report.
ALS 1p TxU:H/M-0627.8; Reel 1084 (C: RGO 6.340.319)

4893. 1841–5–19 From William Henry Fox Talbot Sackville St.
Some details of improvements in the calotype method.
ALS 3pp RS:HS 17.307

4894. 1841–5–21 To George Biddell Airy Collingwood
Does not expect to come to the meeting [see GA's 1841–5–18], but sends comments on altered sections of the draft report.
ALS 4pp RGO 6.340.320

4895. 1841–5–21 To Robert Hunt Collingwood
Thanks RH for his 'extensive and instructive view on the present state of photography.' Found many interesting results in experimenting with vegetable substances. Surprised by RH's discovery that bi-chromate of potash is a photographic substance; JH nearly achieved this result.
ALS 4pp George Eastman House, 7.17 (C: RS:HS 22.87)

4896. 1841–5–21 To John William Lubbock Collingwood
Is unable to come to London for the Standards Commission meeting and thus will also not be able to come to the dinner party. Encloses some recent photographs.
ALS 2pp DMC.695A.885.9

4897. 1841–5–22 From Francis Baily 37 Tavistock Place
Has seen John Arrowsmith, who has sent the maps to JH. There is to be a meeting of the Standards Commission.
ALS 1p RS:HS 3.206

4898. 1841–5–24 From George Biddell Airy Royal Observatory, Greenwich
A note to accompany a number of papers on the 'toleration of inaccuracy,' a question of concern to JH in relation to the setting up of standards.
ALS 1p TxU:H/M-0627.9; Reel 1084 (C: RGO 6.340.322)

4899. 1841–5–25 From George Biddell Airy Royal Observatory, Greenwich
Informing JH of a meeting of the Commission on Standards at Somerset House on 28 May 1841.
ALS 1p RS:HS 1.94

4900. 1841–5–26 To George Biddell Airy Collingwood
Believes no error should be legally tolerated [see GA's 1841–5–24] in selling goods; thus the seller always provides a trifle over the due quantity.
AL 1p inc RGO 6.340.323

4901. 1841–5–26 To Francis Baily n.p.
Has received the maps from John Arrowsmith and sends a blank check for FB to fill in correct amount to pay the bill. Is unable to attend the Standard Commission meeting. Hopes FB can visit him in the summer.
ALS 3pp RS:HS 3.207 (C: RS:HS 25.8.37)

4902. 1841–5–27 From James Joseph Sylvester 22 Doughty St.
Sends paper on 'elimination.' Requests recommendation from JH for professorship of mathematics at University of Virginia.
ALS 4pp RS:HS 17.160

4903. 1841–5–31 To Augustus De Morgan Collingwood
Thanks AD for his theorems; comments on the value of such abstract speculations.
C 1p RS:HS 22.86

4904. 1841–6–1 From Richard Jones Tithe Commission
Sends back the slips. Does not approve of the phrase *transcendentalists*. Regrets that the subject forms a review; it is worth a treatise. Is working hard on his book. Would like to raise some rose trees for Lady Herschel.
ALS 4pp RS:HS 10.368

4905. [1841–6] To William Whewell [Hawkhurst]
Margaret Herschel and JH congratulate WW on his engagement to marry. JH sending WW an imperfect copy of JH's review of WW's *Philosophy of the Inductive Sciences*.
ALS 4pp TC, Camb. Add. Ms.a.207[48] (C: RS:HS 22.90)

4906. 1841–6–4 To William Radcliffe Birt Collingwood
Received WB's report on barometer observations. Admires perfect finish of work.
ALS 2pp RS:HS 19.90

4907. 1841–6–5 From Robert Hunt Falmouth
Returns with grateful thanks the example of photographs recently sent him. They open up a new field of curious processes. Will not have much time this summer for photographic experiments as he is carrying out a series of investigations into the air of mines. Has been experimenting with the cyanurets.
ALS 3pp RS:HS 10.90

4908. 1841–6–8 To Francis Baily n.p.
Is sending him his chart of the stars.
ALS 2pp RS:HS 3.208 (C: RS:HS 25.8.38)

4909. 1841–6–11 From Hans Christian Oersted Copenhagen
Will be receiving the official news soon that JH has been nominated Knight Commander of the Order of Dannebrogg. New magnetic observatory to replace the old one. Is grateful for JH's excellent song about the telescope.
ALS 2pp RS:HS 13.165 Harding, *Correspondance de ... Örsted, 2,* 393–4

4910. 1841–6–11 From Adolphe Quetelet Brussels
Is thrilled about 4 aides to help with meteorological and magnetic observations. Points out that number of stations participating in horary observations increased to 22. Lists them. Hopes to go to Plymouth [B.A.A.S. meeting].
ALS (French) 3pp RS:HS 14.222

4911. 1841–6–12 From Francis Baily 37 Tavistock Place
JH's paper was read last evening and was well received. Astronomical Council has decided to print it immediately. Would like his opinion on the printing of various parts. Has paid John Arrowsmith's bill.
ALS 3pp RS:HS 3.209

4912. 1841–6–12 From William Henry Fox Talbot Sackville St.
Sends JH sample of chemicals and paper to make calotype himself.
ALS 3pp RS:HS 17.308

4913. 1841–6–14 From Jean Baptiste Biot Paris
Regarding his recent experiments with light in crystals of alum etc.
ALS 2pp (French) RS:HS 4.93

4914. 1841–6–15 To Francis Baily Collingwood
Regarding the printing of the star map. Is pleased to hear that the FBs are coming to spend a few days at Hawkhurst.
ALS 4pp RS:HS 3.210 (C: RS:HS 25.8.39)

4915. 1841–6–20 To Hans Christian Oersted Collingwood
Expresses his great appreciation for informing JH of the honor being conferred by making JH a Knight Commander of the ancient and venerable order of Danneborg.
C 2pp RS:HS 22.88 Harding, *Correspondance de ... Örsted, 2,* 394–5

4916. 1841–6–22 To John William Lubbock Collingwood
Comments on a paper on the calotype by W. H. F. Talbot, adding some experiences of his own on the subject.
C 2pp RS:HS 22.89

4917. 1841–6–24 From Richard Sheepshanks n.p.
Reports on Francis Baily's accident and his progress. JH's 'stars' were brought to the R.A.S. and examined by Baily. Mrs. G. B. Airy's health improving. Notes William Whewell's planned marriage.
ALS 2pp RS:HS 16.15

4918. 1841–6–25 From John Gibson Lockhart Sussex Place, Regents Park
Sends check for £100 being payment for JH's article on William Whewell's book. Will always be pleased to receive suggestions.
ALS 1p RS:HS 11.306

4919. 1841–6–25 To Richard Sheepshanks Collingwood
Thanks RS for the news of Francis Baily; requests an update on his condition. Criticizes 'reduced map' prepared by [James] Basire. Has drawn up a catalogue to go with his paper, as RS requested, arranged in regions. Praises William Whewell's choice of a bride.
ALS 3pp RS:HS 16.16 (C: RS:HS 25.9.7)

4920. 1841–6–26 From Richard Sheepshanks 30 Woburn Place
Updates Francis Baily's medical condition, noting that it is more serious than RS had thought. Thinks [James] Basire missed JH's meaning. Suggests how to improve the engraving, perhaps using pewter. Explains that he wants the catalogue to serve the needs of expeditions.
ALS 3pp RS:HS 16.17

4921. 1841–6–26 From William Whewell Trinity College
Thanks for JH's review of WW's *Philosophy of the Inductive Sciences.* Admits that too much of it was written in a 'spirit of needless pugnacity.' Believes that JH's and WW's views are closer than may appear.
ALS 4pp RS:HS 18.195 Todhunter, *Whewell, 2,* 299–300

4922. 1841–6–28 To Julia Margaret Cameron Collingwood
[Composite letter written by JH, Margaret Herschel, and their daughter Caroline.] JH sends thanks to JC for gifts for the children and also sends JC some specimens of photography.
ALS 4pp Getty Center

4923. 1841–7–1 From John Phillips B.A.A.S., Plymouth
Asks whether JH will be prepared to present the committee report on N. L. Lacaille's stars at the B.A.A.S. meeting on 29 July.
(A)L 1p TxU:H/M–0428.8; Reel 1093 [JH annotation on verso]

4924. 1841–7–1 From John Phillips B.A.A.S., Plymouth
Asks whether JH will be prepared to give the committee report on systems of simultaneous magnetical and meteorological observations at the B.A.A.S. meeting on 29 July.
(A)L 1p TxU:H/M–0428.9; Reel 1093

4925. [1841–7] To Edward Sabine n.p.
Encloses letter on and discusses extent of government aid for proposed Hammerfest magnetic observatory and various alternatives with regard to it.
ALS 5pp RS DM.3.88

4926. 1841–7–1 From William Henry Fox Talbot Sackville St.
Has read WT's paper before the R.S.L. and now complains they will not print it in the *R.S.P.T.* as they understand it has appeared elsewhere.
ALS 3pp RS:HS 17.309a Buckland, *Talbot,* 70–1 inc

4927. 1841–7–2 From Jean Baptiste Biot Paris
Possesses JH's tract on light. Would like his comments on various phenomena.
ALS 3pp (French) RS:HS 4.94

4928. 1841–7–3 From George Biddell Airy Royal Observatory, Greenwich
Regarding JH's examination of the observations of equinoxes and solstices; GA assures him that these occurrences are observed by Cambridge and Greenwich, though JH does not mention it. Hopes he has received the report of the last 'Standards' meeting.
ALS 2pp RS:HS 1.95 (C: RGO 6.693.99)

4929. 1841–7–4 From Hans Christian Oersted Copenhagen
Explaining how JH may express his thanks at being made a Knight Commander. HO has also been made a Knight Commander. Will see that he now spells his name Ørsted.
ALS 2pp RS:HS 13.166 Harding, *Correspondance de ... Örsted, 2,* 395–6

4930. [1841]–7–5 From Henry Holland Brook St.
Thanks for mentioning HH's book [*Medical Notes and Reflections* (1839)] in JH's review of William Whewell in latest issue of *Quarterly Review*. Sends copy of its second edition.
ALS 3pp RS:HS 9.430

4931. 1841–7–5 To William Henry Fox Talbot [Collingwood]
Explains the operation of the Council of the R.S.L. and hopes that WT will treat their unintended slight as a 'gaucherie'.
ALS 4pp Science Museum Talbt 1/33 (draft: RS:HS 17.309b & C: RS:HS 22.91)

4932. 1841–7–9 From George Biddell Airy Royal Observatory, Greenwich
Returning photographs. Hopes Lady Herschel is well. Remarks on Admiralty printing Planetary reductions.
ALS 3pp RS:HS 1.96 (C: RGO 6.368.636)

4933. 1841–7–9 or earlier To George Biddell Airy n.p.
Comments on the reduction of observations, on enclosed photographs, and the health of Margaret Brodie Herschel.
ALS 3pp RGO 6.368.639

4934. 1841–7–9 From Adolphe Quetelet Brussels
Announces results of last observations, both magnetic and meteorological. Hopes to see JH.
ALS (French) 2pp RS:HS 14.223

4935. 1841–7–9 From Edward Sabine Woolwich
Discusses the reductions and publication of magnetic observations from various stations. Describes report of the Antarctic expedition and the determination of the magnetic pole. Endorses proposal for Canadian Survey. Discusses survey of South Africa and British Guiana.
ALS 6pp RS:HS 15.131

4936. [1841]–7–12 From Edward Sabine Woolwich
Wants JH to include [J. C.] Ross's instructions in his forthcoming report. Encloses tract on station errors. Discusses [Elias] Loomis's findings regarding errors in measurement.
ALS 4pp RS:HS 15.309

4937. 1841–7–16 From Alfred Gautier Geneva
Introducing Professor [L. F.] Wartmann of Lausanne University. Observations on the comet of Edmond Halley. Has had many afflictions during the winter. Has been working in the meteorological field lately. Death of [S.] L'Huil[l]ier in March. Would like Wartmann to see G. B. Airy.
ALS 4pp (French) RS:HS 8.84

4938. 1841–7–17 From Humphrey Lloyd Trinity College, Dublin
Gives brief details of the work of the magnetic observations since JH's last record. Arrangements have been completed for the publication of some of the observations.
ALS 4pp RS:HS 11.279

4939. 1841–7–19 To William Radcliffe Birt Collingwood
Fully examined WB's report [on barometer observations]. Agrees to combine British and Continental observations. Curves clearly show atmospheric processes. Likes WB's suggestion of tracing atmospheric wave beyond 24 hours; will propose this to B.A.A.S. Irregularity of Asiatic and South African data.
ALS 6pp RS:HS 19.91

4940. 1841–7–19 From Richard Jones Tithe Commission
Willy [Herschel] is doing well and is looking much healthier. Has hired a pony for him to ride. The family doctor has given him a good examination.
ALS 3pp RS:HS 10.369

4941. 1841–7–19 From John David Roberton Royal Society
Forwards from J. W. Lubbock letters from [R.S.L.] foreign office regarding establishment of magnetic observatory at Hammerfest and informs JH that Adolphe Quetelet will be in town.
ALS 2pp RS MM.16.149

4942. 1841–7–20 To John William Lubbock Collingwood
Comments on Louis Daguerre's work, and on the question of a North Cape observatory.
C 1p RS:HS 22.93

4943. 1841–7–20 To Adolphe Quetelet Collingwood
Will not attend B.A.A.S. meeting at Plymouth. Invites AQ to visit, preferably before 29 July, the date of the meeting. Sends thanks to [P. F.] Verhulst for elliptic functions.
ALS 4pp Académie belgique: #14

4944. 1841–7–21 From Thomas Henderson Edinburgh
All N. L. Lacaille's observations have been reduced and a complete catalogue sent to Francis Baily to be incorporated into the catalogue of the R.A.S. Hopes Baily is recovering his health.
ALS 2pp RS:HS 9.296

4945. [1841?]–7–22 From Edward Sabine n.p.
Corrects information from an earlier letter. [John] Caldecott stated that diurnal curve of declination changed with monsoon at Trivandum. Discusses [William] Allen and observation with transportable magnetometer.
ALS 4pp RS:HS 15.321

4946. 1841–7–24 From Francis Baily 37 Tavistock Place
Is slowly recovering from his accident. Query regarding the reduction of N. L. Lacaille's stars, and the revision of the nomenclature of the stars.
ALS 2pp RS:HS 3.211

4947. 1841–7–25 To Francis Baily Collingwood
Is very pleased to hear that FB's health is improving. Must take it very easy. Regarding the finance available for the printing of the star catalogue.
ALS 3pp RS:HS 3.212 (C: RS:HS 25.8.40)

4948. 1841–7–25 or later To William Henry Fox Talbot n.p.
Expresses reservations about Samuel Brown's work. Apologizes for delay in trying calotype; has done only some vegetable substance trials.
draft 4pp RS:HS 17.310b (C: RS:HS 22.92)

4949. 1841–7–25 From William Henry Fox Talbot 31 Sackville St.
Reports on paper by Samuel Brown to the Royal Society of Edinburgh on the conversion of carbon to silicon, and other transmutations. Reports on discovery of new element, ozone,

and complains further about R.S.L. Council.
ALS 4pp RS:HS 17.310a

4950. 1841–7–25 From William Whewell University Club
Asks JH whether he knows of any projects that the B.A.A.S. should fund. Makes further comments on JH's review of WW's *Philosophy of the Inductive Sciences.*
ALS 4pp RS:HS 18.196 Todhunter, *Whewell, 2,* 300–1

4951. 1841–7–26 From Francis Baily 37 Tavistock Place
Further regarding the printing of the star catalogues.
ALS 3pp RS:HS 3.213

4952. 1841–7–26 To William Radcliffe Birt n.p.
Cambridge barometer observations. Postponed WB's detailed report until 1842 B.A.A.S. meeting. This year JH will report generally on atmospheric curves and WB's notes. Hopes WB will attend to give explanations.
ALS 3pp RS:HS 19.92

4953. 1841–7–26 From Richard Jones Haileybury
Will bring Willy [Herschel] to town tomorrow. Comments on his health. Has been trying to buy the Shetland pony for him.
ALS 4pp RS:HS 10.370

4954. 1841–7–27 From George Biddell Airy Royal Observatory, Greenwich
Has sent JH's report on N. L. Lacaille's star computations to William Whewell.
ALS 1p RS:HS 1.97 (C: RGO 6.203.302)

4955. 1841–7–27 From George Biddell Airy Royal Observatory, Greenwich
A letter accompanying a proof copy of the Report on Standards, asking JH to read it carefully and annotate as completely as necessary.
ALS 1p TxU:H/M-0627.11; Reel 1084 (ACCS: RGO 6.340.327)

4956. 1841–7–27 or earlier To George Biddell Airy n.p.
GA and JH are on a committee (with Thomas Henderson) of the B.A.A.S. to supervise the reduction of N. L. Lacaille's star observations; JH encloses a draft report.
ALS 2pp RGO 6.203.301

4957. 1841–7–27 From Francis Baily 37 Tavistock Place
Further regarding the printing of N. L. Lacaille's catalogue.
ALS 2pp RS:HS 3.214

4958. 1841–7–28 To Francis Baily Collingwood
Regarding the printing of scientific information from government funds, with special reference to N. L. Lacaille's star catalogue.
ALS 4pp RS:HS 3.215 (C: RS:HS 25.8.41)

4959. 1841–7–28 From Adolphe Quetelet Brussels
Found letter sent at the R.S.L. Left with William Whewell for Plymouth. Results are forthcoming on meteorological observations that JH suggested. Regrets not seeing JH in England.
ALS (French) 2pp RS:HS 14.224

4960. 1841–7–30 To William Whewell n.p.
Discusses various items regarding the B.A.A.S., e.g., reduction of meteorological observations, Edward Sabine's proposal for a 'Council of 24,' and areas of science needing funding.
ALS 5pp TC, Camb. Add. Ms.a.207[49] (draft: RS:HS 18.197)

4961. 1841–7–31 From Georg Merz & Son Munich
Grateful for kind reception that JH gave to GM's son, who attended Exhibition to demonstrate GM's telescopes.
ALS 1p TxU:H/M–0362; Reel 1087

4962. 1841–7–31 To Physical Section of the B.A.A.S. n.p.
Submits for inspection fifteen colored photographic copies of engravings and mezzotintos prepared by casting luminous rays on substances prepared from vegetable sources.
L *B.A.A.S. Report for 1841* (1842), pt. 2, 40

4963. 1841–8–6 To Francis Baily Collingwood
Encloses catalogue of 1677 stars. Has sent his own report on star nomenclature to Plymouth. Will call at his house later in the week.
ALS 3pp RS:HS 3.216 (C: RS:HS 25.8.42)

4964. 1841–8–10 To George Thomas Napier Collingwood
Hears with pleasure of the peace reigning at the Cape between the Caffres. The Boers are a different proposition and JH agrees with GN's policy. J. R. Innes called in on his way to Scotland. Weather in England is gloomy, like the political events.
draft 4pp RS:HS 13.70bis & 22.95

4965. 1841–8–10 From [?] n.p.
Requests JH's opinion on five specific points in 'Report on Standards.'
AL 2pp inc TxU:H/M–0914; Reel 1083

4966. 1841–8–11 From George Biddell Airy Royal Observatory, Greenwich
Asks for JH's comments on George Peacock's assessment that the Report of the Standards Commission [see GA's 1841–7–27] is directed too much to scientific men, and others will not understand it; suggests an appendix to show the relationship between the proposed measures.
ALS 1p TxU:H/M-0627.13; Reel 1084 (C: RGO 6.340.324)

4967. 1841–8–11 From Adolphe Quetelet London
Presents best wishes to Lady Herschel. Regrets not having met on account of meeting with [G. B.] Airy at Greenwich. Is keeping four pictures of JH and has offered 2 additional ones to Prince Albert.
ALS (French) 2pp RS:HS 14.225

4968. 1841–8–13 To George Biddell Airy Collingwood
Reliability of measuring instruments used in scientific and public surveys. Approves George Peacock's suggestion to revise survey tables for use by non-scientist surveyors. Schools should teach only legalized systems of weights, measures, and money.
Adraft 4pp TxU:H/L-0012; Reel 1054

4969. 1841–8–13 To George Biddell Airy Collingwood
Comments on the report [see GA's 1841–8–11]; wants the standards legislation to be demanding of rigor in application.
ALS 4pp RGO 6.340.325

4970. 1841–8–14 From George Biddell Airy Royal Observatory, Greenwich
Has received JH's 'Report' proof regarding JH's description of specific gravity. Captive balloons. Would JH refer him to D. F. J. Arago's experiments on the shadow of a disk.
ALS 1p RS:HS 1.98 (C: RGO 6.368.636)

4971. 1841–8–14 or earlier To George Biddell Airy n.p.
Is confused by French system of recording specific gravity and seeks GA's assistance; also GA and JH are named to a

committee that JH thinks only wastes money.
ALS 3pp RGO 6.368.641

4972. 1841–8–15 To George Biddell Airy Collingwood
Complains further about French specific gravity designation [see JH's 1841–8–14 or earlier], and describes observing colored fringes on the edge of clouds.
ALS 3pp RGO 6.368.644

4973. 1841–8–15 To the *Athenaeum* Collingwood
Discusses meteor showers of 9 August 1840, emanating from a point in the constellation Perseus. Calls attention to a report on these meteors as seen from near Vesuvius in 1779 by William Hamilton.
L *Athenaeum,* #721 (Aug. 21, 1841), 643

4974. 1841–8–17 From George Biddell Airy Royal Observatory, Greenwich
Comments specifically on the accuracy of surveyors' chains [see JH's 1841–8–13], together with brief comments on a few other matters.
ACCS 2pp RGO 6.340.337

4975. 1841–8–17 or later To George Biddell Airy n.p.
Is fully occupied reducing Cape observations, and is producing a star list by magnitudes, resulting in some interesting anomalies.
ALS 4pp RGO 6.368.634

4976. 1841–8–19 From George Biddell Airy Royal Observatory, Greenwich
Outlines suggestions from [Drinkwater-]Bethune for improving 'Report of Standards.' Bethune distinguished Exchequer, Local, and Parliamentary (or National) standards, and preserved unit called 'land chain' equal to twenty yards.
LS 2pp TxU:H/M–0662; Reel 1084

4977. 1841–8–21 From Humphrey Lloyd Falmouth
Thinks that there should be more definite instructions for the Directors of the Magnetical Observatories. Has drawn up a paper, which if JH agrees with, can be signed and printed for distribution. Points out the significant parts. There should be a common nomenclature for the weather.
ALS 4pp RS:HS 11.280

4978. 1841–8–22 To 'Dear Sir' Collingwood
Replies to [?]'s questions about 'Mr. Bethune's suggestions' concerning terminology to be used for standards of weight and measure. Agrees with G. B. Airy that term 'land' should be used with 'land chain' unit.
ACCS 3pp TxU:H/L–0643; Reel 1089

4979. 1841–8–22 To George Biddell Airy Collingwood
Further comments on the Standards Commission Report [see GA's 1841–7–27].
ALS 3pp RGO 6.340.338

4980. 1841–9–2 From Edward Sabine Athenaeum
Expresses his approval of James Ross's Antarctic survey, providing some of the details of it and his determination of the south magnetic pole. Gives details of Ross's plans.
ALS 4pp RS:HS 15.132

4981. 1841–9–10 From John Phillips B.A.A.S., York
JH, William Whewell, George Peacock, Humphrey Lloyd, and Edward Sabine are appointed by the B.A.A.S. to study systems of simultaneous magnetical and meteorological observations. They will be granted £100.
(A)L 1p TxU:H/M–0428.11; Reel 1093

4982. 1841–9–10 From John Phillips B.A.A.S., York
JH, G. B. Airy, and Thomas Henderson are appointed to oversee the publication of the reductions of the calculations of N. L. Lacaille's stars. They will have £105 at their disposal.
(A)L 1p TxU:H/M–0428.10; Reel 1093

4983. 1841–9–10 From John Phillips York
Informs JH that the B.A.A.S. has placed him on a committee to conduct experiments by capture balloon on the atmosphere.
(A)L 1p RS:HS B26.7

4984. 1841–9–10 From John Phillips York
JH continues to serve on the committee to supervise and translate foreign scientific memoirs [see JP's 1840–10–19].
(A)L 1p RS:HS B26.6

4985. 1841–9–10 From Edward Sabine Dunraven
Mentions a deletion of part of James Ross's dispatch. Clerk is receiving instructions from Charles Riddell before sailing to the Cape of Good Hope, from which observatory he expects a report in eight months. Asks whether the R.S.L. should request government funding for Cape observations and experiments. Wonders if astronomical and magnetic observers there could assist each other. Requests information on the atmosphere.
ALS 9pp RS:HS 15.133

4986. 1841–9–20 From Adolphe Quetelet Brussels
Regrets not having seen JH in England. Discusses horary observations. Lists current projects. Talks about C. F. Gauss's reactions to magnetic observations by JH in two hemispheres and difference of lines for Europe and America. Gives results for August meteors.
ALS (French) 4pp RS:HS 14.226

4987. 1841–9–29 To Charles Babbage Collingwood
Regarding a missing picture of his father.
ALS 4pp RS:HS 2.297

4988. 1841–9–30 From Francis Baily 37 Tavistock Place
Is now completely recovered from his accident. Has resumed work on his Cavendish experiments, which he will soon present to the public. Hopes his new residence at Hawkhurst suits his health.
ALS 2pp RS:HS 3.217

4989. 1841–9–30 From John Hartnup Somerset House
Received JH's enquiry today. Sent 100 copies of JH's paper on the 'Nomenclature of the Stars' to Peter Stewart.
ALS 1p RS:HS 19.93b

4990. 1841–10 To Mrs. Gregory n.p.
A testimonial on behalf of Olinthus Gilbert Gregory, deceased, to assist Mrs. Gregory in obtaining a Civil List pension.
Adraft 3pp inc RS:HS 19.93c (C: RS:HS 22.99)

4991. 1841–[10]–1 To [?] Collingwood
Explains the phenomenon known as the 'harvest moon.'
ALS 2pp WT 52331

4992. 1841–10–4 From Élie François Wartmann 3 Leicester Place
Has visited scientific institutions in the British Isles, and is writing to make arrangements to see JH. Talks about Halley's comet.
ALS 2pp (French) RS:HS 18.92

4993. 1841–10–8 From Zanobi Zucchini Florence [Tuscany]
Charles Babbage will bring ZZ's Italian translation of Madame Somerville's *On the Connection of the Physical Sciences* to JH. Hopes to translate JH's *Treatise Astr.* into Italian soon. May ZZ list JH among ZZ's formal associates?
ALS 1p (Italian) RS:HS 18.364

4994. [1841]–10–9 From Edward Sabine 12 Park Place
Sends [James Clark] Ross's journal and revised B.A.A.S. magnetic report. Discusses proposals for observation in Africa. ES will send various instruments to Africa.
ALS 4pp RS:HS 15.299

4995. 1841–10–10 From Augustus De Morgan 69 Gower St.
Points out an error JH has made concerning J. E. Bode's law. What is the proper notation for the satellites?
ALS 3pp RS:HS 6.186

4996. 1841–10–10 To John William Lubbock Collingwood
As the Committee is to meet on Thursday he thinks it should reconsider its decision not to publish W. H. F. Talbot's paper on the Calotype process. Gives reasons as stated in a letter he has received from Talbot.
draft 4pp RS:HS 11.369 & 22.94 & 22.97

4997. 1841–10–10 To Hans Christian Oersted Collingwood
Asks HO if he can discover, without mentioning JH's name, what it might cost JH to be formally enrolled in Denmark for the honor he has received [see JH's 1841–6–20].
C 2pp RS:HS 22.98 Harding, *Correspondance de ... Örsted, 2*, 396–7

4998. 1841–10–10 To Wilhelm Struve Collingwood
Thanks WS for Pulkowa Medal. Problematic health and large family make visit to WS impossible. Mentions [Francis] Baily's hope of terminating 'Cavendish experiment.' Sends copy of 'Revision of Southern Constellations.'
C 4pp RS:HS 22.96

4999. 1841–10–11 To Francis Baily Collingwood
Is delighted to hear that the Cavendish experiments are proving successful. Is progressing slowly with the reductions of his own southern sweeps. Regarding a state pension for Mrs. Gregory (widow of O. G. Gregory); would FB give him some idea of Mrs. Gregory's circumstances.
ALS 4pp RS:HS 3.218 (C: RS:HS 25.8.43)

5000. 1841–[10–12] To Hugh Percy (3rd Duke of Northumberland) [Collingwood]
Recommends that William Whewell be appointed Master of Trinity College.
draft 4pp RS:HS 13.357 & 19.164 (C: RS:HS 25.13.8)

5001. 1841–10–13 From Edward Sabine 12 Park Place
Speculation about land masses and climate in Arctic regions. Extols James Ross's talents. Discusses John Dalton's pension and who should be devoting efforts for approving pensions.
ALS 5pp inc RS:HS 15.134

5002. 1841–10–14 From Francis Baily 37 Tavistock Place
Regarding the circumstances of Mrs. Gregory (widow of O. G. Gregory) and the possibility of support for a state pension for her.
ALS 3pp RS:HS 3.219

5003. 1841–10–15 From John William Lubbock n.p.
Placed JH's letter before the Committee of Papers but the Committee did not change its views regarding the publication of W. H. F. Talbot's paper. Is pleased JH is promoting his experiments so successfully.
ALS 3pp RS:HS 11.370

5004. 1841–10–15 From Richard Jones Tithe Commission
The Master of Trinity has resigned and there are intrigues on foot to bypass William Whewell. Outlines the people interested in the appointment. Not much they can do but JH might pass a word to Hugh Percy (3rd Duke of Northumberland).
ALS 4pp RS:HS 10.371

5005. 1841–10–17 To [Hugh Percy] Collingwood
Writing HP as Chancellor of the University of Cambridge, JH strongly recommends that William Whewell be named Master of Trinity College, Cambridge.
C 2pp †(JH) TC, Camb. Add. Ms.5858

5006. 1841–10–19 From Richard Jones Tithe Commission
William Whewell is Master of Trinity and he will write to Hugh Percy (3rd Duke of Northumberland).
ALS 1p RS:HS 10.372

5007. 1841–10–19 To William Whewell Collingwood
Congratulates WW on becoming Master of Trinity College, Cambridge.
ALS 2pp TC, Camb. Add. Ms.a.58[43]

5008. 1841–10–20 From Henry Raper 11 Albion Street
Asks if JH has published maps of Southern Hemisphere. Discusses [C. H.] Paravey's work on the history of constellations and ancient conceptions of the Zodiac in China, India, Egypt, and Greece.
ALS 4pp RS:HS 14.306

5009. 1841–10–22 From Heinrich Christian Schumacher Altona
Has not received parcels from Royal Society. Wants JH to send notice of proposal to *Astronomische Nachrichten*. Remarks on JH's requiem for William Herschel's large telescope. Asks JH to thank King of Denmark directly for the order conferred.
ALS 4pp RS:HS 15.388

5010. 1841–10–24 From William Whewell Ambleside
[Writing shortly after his marriage (12 Oct. 1841) and acceptance a few days later of the Mastership of Trinity College], WW thanks JH for his expressions of good will. Invites the Herschels to Trinity Lodge.
ALS 3pp RS:HS 18.198

5011. 1841–10–28 From Thomas I. M. Forster Tunbridge Wells
Wrote some three months ago, but heard nothing so presumes he was away. Did he receive the parcel of books? Has taken a house here and would be pleased to see him. Has just recovered from a serious brain fever. Was elected a member of the Association but has not paid his fees; is he still a member?
ALS 3pp RS:HS 7.329

5012. 1841–10–29 From Charles Hippolyte de Paravey London
Commenting on ancient star maps, and noting that not all ancient stars can be identified now. Further talks about ancient meanings of heavenly bodies together with much lore

including astrology, astronomy, and mythology. [Many parts of this letter illegible.]
ALS 8pp (French) RS:HS 13.215

5013. 1841–11 From Richard Jones Tithe Commission
The health of Mrs. Jones is slowly improving. Picked up the enclosed medal a little while ago. Believes the reverse depicts the scaffolding of Sir William Herschel's telescope.
ALS 2pp RS:HS 10.373

5014. [1841–11] To William Whewell [Hawkhurst]
Congratulates WW on moving into Master's Lodge at Trinity College. Asks WW to look after a new Trinity student, the son of JH's friend Mr. Hartnell. Is making progress in reducing his Cape observations and is teaching his sons Latin and Greek. Adds some comments on happiness.
ALS 7pp TC, Camb. Add. Ms.a.207[50] (C: RS:HS 22.100 inc)

5015. 1841–11–2 From Henry Raper J. U. S. Club
Encloses [C. H.] Paravey's letter [see HR's 1841–10–20].
ALS 2pp RS:HS 14.307

5016. 1841–11–7 To Caroline Lucretia Herschel Collingwood
The reductions for JH's *Cape Results* are progressing. JH's mapping work has been 'carried over the whole surface of the heavens' this year.
ALS 2pp TxU:H/L–0586.19.b; Reel 1058

5017. 1841–11–7 To Caroline Lucretia Herschel Collingwood
Summarizes James Clark Ross's expedition to reach the Southern Magnetic Pole; JH reports that Ross has discovered that the pole lies several degrees more south than Carl Gauss had calculated.
ALS 4pp TxU:H/L–0586.20; Reel 1058

5018. 1841–11–9 From Francis Baily 37 Tavistock Place
Regarding the method of printing the R.A.S.'s Star Catalogue. Which star distances are they to use, North Polar or in declination?
ALS 3pp RS:HS 3.220

5019. 1841–11–11 To Francis Baily Collingwood
Has no hesitation in adopting North Polar distances for his stars. Has two queries regarding the R.A.S.'s catalogue of stars. Weather has upset his own observations. Regarding a volume of T. G. Taylor's Madras Observations.
ALS 8pp RS:HS 3.221 (C: RS:HS 25.8.44)

5020. 1841–11–12 From Adolphe Quetelet Brussels
System of meteorological observation works well but no English station. AQ is willing to publish results in England if need be. Points out additional observers for September.
ALS (French) 3pp RS:HS 14.227

5021. 1841–11–13 From Francis Baily 37 Tavistock Place
Finds that JH has not changed his mind on star distance terminology. Progress on the preparation of the star catalogue. Regarding T. G. Taylor's Madras Observations.
ALS 4pp RS:HS 3.222

5022. 1841–11–16 To Samuel Hunter Christie Collingwood
Recommends awarding Royal Medal to [William Parsons] Lord Oxmantown for paper on large reflecting telescopes.
C 3pp TxU:H/L-0127; Reel 1054

5023. 1841–11–16 To [Spencer J. A. Compton?] Collingwood
Recommends that William Parsons be awarded Royal Medal for 'Account of Experiments on the Reflecting Telescope.'
ALS 4pp RS MM.11.158

5024. 1841–11–16 From Hans Christian Oersted Copenhagen
Glad to inform JH that there are no fees or expenses involved in being made a Knight Commander. Has had little leisure himself for scientific pursuits due to being Rector of the University.
ALS 1p RS:HS 13.167

5025. 1841–11–17 From Charles Babbage Dorset St.
Regarding the missing picture [RS:HS.2.297].
ALS 1p RS:HS 2.298

5026. 1841–11–17 From Roswell Park Philadelphia
Sends his work *Pantology, or a Systematic Survey of Human Knowledge* for JH's inspection.
ALS 2pp RS:HS 14.136

5027. 1841–11–20 To Charles Babbage Collingwood
Please send on the picture. Hopes he had a good time in Italy.
ALS 3pp RS:HS 2.299

5028. 1841–11–20 From Theo E. Cantor Calcutta
Has found a Chinese map of the heavens in Chusan, which he forwards. Comments on the flora and fauna of Chusan.
ALS 3pp RS:HS 5.323

5029. 1841–11–27 From Charles Hippolyte de Paravey Paris
Comments on the death of a mutual [?] acquaintance. Adds additional information about ancient constellations [see CP's 1841–10–29]. Complains about editors, publishers, and others who do not take CP's work seriously. [Many parts of this letter illegible.]
ALS 6pp (French) RS:HS 13.216

5030. 1841–11–29 From Adolphe Quetelet Brussels
Discusses meteorological observations. Stations increased from 26 to 28. Talks about 36 hours versus 24 hours periodical observations and asks for JH's opinion.
ALS (French) 4pp RS:HS 14.228

5031. 1841–11–30 From John David Roberton Royal Society
Forwards from Edward Sabine resolution of council regarding JH's appointment to committee considering use of building formerly occupied by Kew Observatory.
ALS 2pp RS MM.16.146

5032. 1841–12–1 To William Radcliffe Birt Collingwood
Heard of WB's bout with typhus. Interest in atmospheric waves should ensure publication of WB's article in *Athenaeum*. Will write editor on WB's behalf. Awaits completion of atmospheric curves for B.A.A.S. Congratulations on recently acquired office.
ALS 4pp RS:HS 19.94

5033. 1841–12 To *Astronomische Nachrichten* Collingwood
Discusses the question of whether the position of the moon has an influence on weather and whether either JH or his father have made this claim.
L *Astronomische Nachrichten, 19* (1842), 175–8

5034. 1841–12 To John William Lubbock n.p.
Comments on balloon ascent report, photographic paper made with vegetable colors, and the need to provide a polarizing crystal to J. B. Biot.
C 2pp RS:HS 22.106

5035. [1841?–12] From R[obert] A[nnesley] Ogilvie Lloyd Square, Islington
Due to the death of George Birkbeck, the Committee of Management of the London Mechanics Institute would like JH to become President. [Note by JH: declined.]
ALS 2pp RS:HS 13.199

5036. 1841–12–1 From Edward Sabine 12 Park Place
Discusses recent magnetic disturbances and their publication, as well as sending reports on them to G. B. Airy, Humphrey Lloyd, C. F. Gauss, and others.
ALS 3pp RS:HS 15.135

5037. 1841–12–3 To Adolphe Quetelet Collingwood
Discusses advantages and disadvantages of 24 versus 36 hour series in meteorological observations. R.S.L. officially endorses 24 hours, but AQ should continue 36 hours if best for him. Does not possess the *Comptes rendus* of French Institute and cannot therefore see No. 413. Will attempt to procure sight of it. Asks AQ to return [P. F.] Verhulst's work to him. Expresses respect for AQ's work. Asks for understanding if JH does not answer every letter promptly.
ALS 4pp Académie belgique: #15 (Adraft: RS:HS 14.229 & C: RS:HS 22.103)

5038. 1841–12–4 To Hans Christian Oersted Collingwood
Thanks HO for his effort on JH's behalf [see JH's 1841–10–10], and then describes some of his photographic experiments.
C 2pp RS:HS 22.102 Harding, *Correspondance de ... Örsted, 2,* 397–8

5039. 1841–12–6 From Francis R[awdon] Chesney Down St.
Encountered storm and lost the Tigris and 20 men. Would like a preface on this subject from JH for his own forthcoming book.
ALS 3pp RS:HS 5.335

5040. 1841–12–9 To Charles Wheatstone Collingwood
Arranges a meeting with CW and Edward Sabine to discuss Kew [magnetic] observations.
C 1p RS:HS 25.11.3

5041. 1841–12–13 From Charles Wheatstone 20 Conduit St.
Agrees to meet JH and Edward Sabine on Saturday. Kew Observatory was given to R.S.L., but R.S.L. has not yet taken possession of it. [JH annotation: election of General Morrison to R.S.L.]
ALS 2pp † TxU:H/M–0145.1; Reel 1087

5042. 1841–12–15 To [Spencer J. A. Compton?] Tonbridge
Wishes not to be re-elected as chair of R.S.L. Physical Committee, feeling that JH's distance from London and other occupations impede Committee's efficiency.
ALS 4pp RS MM.11.159

5043. 1841–12–22 From Richard Jones Tithe Commission
Sends some extracts from the Acts. Regarding the enforcement of the rates. [To Lady Herschel:] Should have let him know when they were in town. Believes Mrs. Jones is going on well at Brighton. Sets out for Cambridge to see William Whewell tomorrow. Hears rooms of Henry VIII have been opened for parties.
ALS 4pp RS:HS 10.374

5044. 1841–12–23 To J. E. Bidwell Collingwood
Regarding the advisability of purchasing a patent in the photographic process.
draft 4pp RS:HS 5.62 & 22.104

5045. 1841–12–24 From Heinrich Christian Schumacher Altona
Asks about the claim that JH professes that the moon affects the atmosphere and about possibly related research. Received paper on constellations for *Astronomische Nachrichten.*
ALS 3pp RS:HS 15.389

5046. 1841–12–25 From George Biddell Airy Royal Observatory, Greenwich
Points out an error in the instructions for magnetic observers provided for the Antarctic expedition.
ACCS 4pp RGO 6.675.268

5047. 1841–12–25 From George Biddell Airy Royal Observatory, Greenwich
A letter accompanying a 'fair copy' of the Report of the Standards Commission; GA indicates he has incorporated the views of all members of the Commission.
ALS 1p TxU:H/M-0627.15; Reel 1084 (C: RGO 6.340.340)

5048. 1841–12–25 From George Biddell Airy Royal Observatory, Greenwich
Asks the Physical Committee (Chair, JH) of the R.S.L. to note the concurrent disturbances of magnetometers and the appearance of aurora borealis, and to make a long term study of this relationship.
draft 2pp inc RGO 6.675.265 (C: RGO 6.675.267 inc)

5049. 1841–[12–25 or later] To Heinrich Christian Schumacher Altona
JH replies (to RS:HS 15.389) that neither JH nor his father held that the moon influenced the atmosphere, though JH does connect a full moon with a clear sky. [Eliza] Hamilton, sister of W. R. Hamilton, has published some poetry that is not well known, but deserving. Cannot accept foreign order.
ALS 3pp RS:HS 15.411 (C: RS:HS 22.101)

5050. 1841–12–27 To George Grey Collingwood
GG's book [*Journals of Two Expeditions of Discovery in the North–West and Western Australia,* 2 vols. (London: Boone, 1841)] is creating a sensation in London. Pleased to learn from it of the range of mountains named after JH. Hopes GG will carry out some study into the Aboriginal dialects. Standards of weight and measure should be sent to all Colonies. Important to have a magnetic survey of all colonized and colonizable parts of Australia.
ALS 8pp RS:HS 9.130 (C: RS:HS 22.105 & Auckland Public Library)

5051. 1841–12–29 From Francis Beaufort Admiralty
Sends JH a copy of Thomas Maclear's base measurements, and will transmit JH's letter to Maclear.
ACC 1p Hydrographic Office Letter Book

5052. 1841–12–29 From Thomas Makdougall Brisbane Makerstoun
Has heard that J. D. Forbes will be leaving Edinburgh to work in Glasgow, where the salary is higher. Would like JH's support for a movement to increase the salary so that

Forbes may be retained at Edinburgh.
ALS 4pp RS:HS 5.103

5053. 1841–12–30 From James David Forbes Edinburgh
About various testimonials, from JH and others, for JF's application for a government pension.
ALS 3pp St. Andrews LB III, 292– (C: RS:HS 7.298)

5054. 1841–12–31 From Adolphe Quetelet Brussels
As 'perpetual Secretary' of the Academy, AQ expresses thanks for JH's work 'On the Advantages to be Obtained by a Revision and Re-arrangement of the Constellations.'
ALS (French) 1p RS:HS 14.240

5055. [1842] To George Biddell Airy n.p.
A letter of introduction, asking GA to show the observatory to a Mr. Knipping from Hanover, a relative of JH's aunt Caroline.
ALS 3pp RGO 6.706.250

5056. [1842?] To Charles Babbage [London]
Tells CB about a visit by JH and Wolryche Whitmore to the Duke [of ?] to present the case for more government funds for Charles Babbage's calculating machine.
draft 2pp RS:HS B27.52

5057. [1842?] To [Francis Baily?] n.p.
A revised set of proposals for Southern constellation reform [see JH's 1840] to be presented to the next B.A.A.S. meeting.
draft 4pp RS:HS 25.14.2

5058. [1842?] To Alexander von Humboldt n.p.
Convey his thanks to the King for the Order of Merit. Regrets he did not see the King or AH when they visited London recently. Gives astronomical information.
draft 2pp RS:HS 10.75

5059. [1842] From Roderick Impey Murchison n.p.
Pleased that JH can come on the occasion of F. W. Bessel's visit, as W. R. Hamilton, G. B. Airy, and Francis Baily cannot come. Comments regarding Foreign Orders.
ALS 4pp RS:HS 12.440

5060. 1842 To Roderick I. Murchison n.p.
Offers an explanation of the phenomena of the freezing cave of Illetzkaya Zatchita in Russia.
L *Geological Society Proceedings, 3* (1842), 697–9

5061. [1842 or later] To Secretary of R.S.L. Physical Committee n.p.
Offers to provide committee with plans for and to supervise construction of new instrument, devised by JH, for study of solar spectrum.
Adraft 2pp TxU:H/L-0722; Reel 1089

5062. [1842]. From John Ruskin n.p.
Thanks JH for sending copy of [JH's translation of Friedrich Schiller's] 'The Walk.' Although he has no knowledge of German, he treasures this translation and appreciates JH's kindness very much.
ALS 2pp RS:HS 14.512

5063. [1842 to 1845] From Edward Sabine Woolwich
Hopes royal observatory at Cape of Good Hope will relieve Ordnance department. U.S. government agreed to pay for publication of observations there. A. D. Bache favors continuance of hourly meteorological observations in U.S. for one year. G. A. Erman will attend meeting at Cambridge. German translations by ES's wife.
ALS 2pp inc TxU:H/M–0517.3; Reel 1093

5064. [1842?] From Edward Sabine n.p.
Discusses report of R.S.L. regarding observatories. [Charles] Wheatstone agrees with report on Kew except for tax paragraph. Discusses November 18 [1841] magnetic disturbance at Toronto.
ALS 5pp RS:HS 15.323

5065. 1842 To Edward Sabine n.p.
Discusses term and extra magnetic observations with regard to [Humphrey] Lloyd's letter. Would prefer to keep them separate, but feels no harm will come from keeping them together.
ALS 2pp RS:HS 15.148 (C: RS:HS 22.144)

5066. [1842] To Edward Sabine n.p.
Encloses [G. B.] Airy's letter, which he thinks should accompany [Humphrey] Lloyd's reply regarding magnetism. Thinks all Physical Committee members must be made aware of new developments.
L 2pp † inc RS:HS 15.147

5067. 1842 or later From Richard Sheepshanks n.p.
About some unfortunate private business, and difficulty with [Francesco] De Vico's star catalogue.
ALS 3pp RS:HS 16.125

5068. 1842–1 From Charles Piazzi Smyth Royal Observatory, Cape of Good Hope
Construction of Herschel Obelisk at Feldhausen underway. Describes items to be included at the monument.
CC 3pp inc RS:HS 16.224

5069. [1842?] From Charles Piazzi Smyth Royal Observatory, Cape of Good Hope
Describes rock formations in Kamiesberg region seen while working on Arc of Meridian project. Thinks Bokkeveld would make ideal site for a meteorological observatory.
ALS 4pp inc RS:HS 16.235 B. Warner, *Charles Piazzi Smyth*, 54–5

5070. [1842] To [?] n.p.
Submit revisions to instructions for conducting and reporting magnetic and meteorological observations.
C 3pp TxU:II/L-0301; Reel 1054

5071. 1842–1–2 To Thomas Makdougall Brisbane Collingwood
A strong letter on behalf of the provision of financial support for Professor [James David] Forbes, along with JH's personal statement of belief in the need to free distinguished scientists from teaching duties.
ALS 4pp BL Aaa 32565.164 (draft: RS:HS 5.104 & C: 22.107)

5072. 1842–1–2 From William Whewell Trinity Lodge
Describes life in Trinity Lodge and invites JH to visit him there. Is writing on ethics and puzzling over the nature of causation.
ALS 4pp RS:HS 18.199 inc Todhunter, *Whewell, 2*, 303–5 inc

5073. 1842–1–3 From Marie J. P. Flourens Paris
Acknowledging a gift of one of JH's works.
L 1p RS:HS 7.371

5074. 1842–1–3 From Alfred Smee 7 Finsbury Circus
Thanks JH for a copy of a paper on prismatic study of the sun's rays, and for accompanying photographs.
AL 1p RS:HS 16.150

5075. 1842–1–4 From George Biddell Airy Royal Observatory, Greenwich
Another copy of the Report of the Standards Commission is being circulated [see GA's 1841–12–25] and needs JH's signature; J. E. Drinkwater is sending in a personal report complaining about the Commission's conservatism.
ALS 1p TxU:H/M-0628.1; Reel 1084 (C: RGO 6.340.340a)

5076. 1842–1–4 From Heinrich Christian Schumacher Altona
JH's return of order offered him was painful. Will report in *Astronomische Nachrichten* that JH denies meteorological theory attributed to him and that the 'Wiener Wirtschaftskalender für 1840' is based on JH's authority.
ALS 4pp RS:HS 15.390

5077. 1842–1–5 From George Biddell Airy Royal Observatory, Greenwich
Sending two papers, one for the Physical Committee.
ALS 1p RS:HS 1.99 (C: RGO 6.693.100)

5078. 1842–1–5 or later To George Biddell Airy n.p.
Has passed on GA's note to Edward Sabine; cannot write more because his hand shakes too much, having just made a snow-man for the children.
ALS 2pp RGO 6.693.101

5079. 1842–1–6 To William Whewell Collingwood
Comments by words and a drawing on WW's statement that WW was trying to 'puff himself out' to fill his new office and house. Also discusses moral philosophy and the theory of causation.
ALS 2pp TC, Camb. Add. Ms.a.207[51] (C: RS:HS 22.108)

5080. 1842–1–11 From Heinrich Christian Schumacher Altona
Asks JH to thank King Christian VIII of Denmark for the order, even though JH cannot accept it. Sends a mathematical puzzle.
ALS 2pp RS:HS 15.391

5081. 1842–1–12 From James David Forbes Edinburgh
Is grateful for JH's letter and commendation regarding application for a government pension; will endeavor to carry out his recommendations.
ALS 2pp St. Andrews LB III, 318–19 (C: RS:HS 7.299)

5082. 1842–1–12 To Richard Sheepshanks [Collingwood]
Is insistent that JH cannot perform the functions of Foreign Secretary [of the R.A.S.] as he has too much to do already, and has had to give up much of his personal scientific correspondence.
ACC 2pp TxU:H/L-0384; Reel 1054 (Cdraft: RS:HS 25.15.11)

5083. 1842–1–13 From Edward Sabine 12 Park Place
Will give JH's report on Kew Observatory to Charles Wheatstone. Sends JH actinometer reports. Questions methods of observation and quality of observers. Discusses government support of scientific research.
ALS 6pp RS:HS 15.136

5084. [1842]–1–14 From John Elliot Drinkwater-Bethune [Home Office] Chambers, Downing St.
Encloses JB's letter [21 Dec. 1841] to Chancellor of Exchequer [H. Goulburn] advocating adoption of metric system. Hopes [members of Standards Commission] will not consider JB obstinate. Admits that JH's opposition to system has advantage of authority and conservatism.
ALS 1p †† TxU:H/M–0081; Reel 1087

5085. 1842–1–15 To William Robert Grove Collingwood
Sorry that JH can only send 'very poor' photographic process specimens.
ALS 1p RI

5086. 1842–1–17 To George Biddell Airy n.p.
Sends a letter from Humphrey Lloyd for GA's comments.
ALS 1p RGO 6.693.102

5087. 1842–1–17 To Christian VIII, King of Denmark Collingwood
Grateful for being made Knight Commander of the Illustrious Order of Dannebrog, but British rules forbid JH accepting foreign title.
AdraftS 3pp RS:HS 19.87

5088. 1842–1–17 From Edward Sabine n.p.
Transmits message from [A. T.] Kupffer about setting up an observatory at Hammerfest and details about various apparati. Notes proposal for observatory in northern Siberia. Has ordered 250 copies of the Physical Committee Report.
ALS 4pp RS:HS 15.137

5089. 1842–1–18 From George Biddell Airy Royal Observatory, Greenwich
Provides extensive comments, about GA's magnetic instruments and observations, in reply to Humphrey Lloyd's letter [see JH's 1842–1–17].
ACCS 4pp RGO 6.693.103

5090. 1842–1–18 or later To George Biddell Airy n.p.
Thanks GA for his comments [see GA's 1842–1–18]; hopes that GA would be willing to have such statements, and other similar statements, made public.
ALS 7pp RGO 6.693.105

5091. 1842–1–19 From Francis Baily 37 Tavistock Place
Requesting any information on J. J. Littrow for his obituary notice.
ALS 2pp RS:HS 3.223

5092. 1842–1–19 From Edward Sabine 12 Park Place
Discusses at length and rejects JH's recommendation that communication in scientific projects take place solely by letter and theorization solely by individuals. Mentions subsequent variation in intervals between magnetic observations and disputes between G. B. Airy and Humphrey Lloyd, as well as other difficulties.
ALS 11pp RS:HS 15.138

5093. 1842–1–20 To Thomas Maclear Collingwood
Congratulates TM on work at Zwartland. Fears *Cape Results* will not be ready within the year.
ALS 4pp Cape Archives/Maclear Papers File 100

5094. 1842–1–21 To George Biddell Airy n.p.
Concerning Airy's papers recently submitted [see GA's 1842–1–5]. Hopes he will continue to send a report of his experiments and observations.
draft 2pp RS:HS 1.100 & 22.110

5095. 1842–1–21 To Francis Baily Collingwood
Regrets that he knows nothing about J. J. Littrow. A new astronomical instrument. Weather has been most unsuitable for observations.
ALS 4pp RS:HS 3.224 (C: RS:HS 25.8.45)

5096. 1842–1–21 From William Benjamin Carpenter 22 Park St., Bristol
Would like JH to write a review of his recent book for the

Quarterly Review.
ALS 7pp RS:HS 5.186

5097. 1842–1–22 From George Biddell Airy Royal Observatory, Greenwich
Agrees to having GA's views made public [see JH's 1842–1–18 or later] as JH sees fit; comments on aurora and magnetic storm observations.
ACCS 2pp RGO 6.693.107

5098. 1842–1–22 or later To George Biddell Airy n.p.
As many Russian observatories are now involved in magnetic observations, JH thinks the British observations should continue [see GA's 1842–1–22].
ALS 4pp RGO 6.675.271

5099. 1842–1–22 From George Biddell Airy n.p.
Comments on whether magnetic observations should be continued.
ACC 1p inc RGO 6.675.270

5100. 1842–1–23 To Edward Sabine Collingwood
JH's reasons for requesting three-year extension of magnetic and meteorological survey from British government and H.E.I.C. Preliminary results from global stations.
C 3pp TxU:H/L-0361; Reel 1054

5101. 1842–1–27 To Edward Sabine Collingwood
Clarifies role and methods of R.S.L. Council and scientific committees, suggesting that members of Physical Committee need to be better informed about its activities.
ALS 4pp RS MC.3.204

5102. 1842–1–28 To Edward Sabine Collingwood
Agrees with suggestions made by ES concerning the publication of meteorological observations.
ALS 4pp RS MM 11.166

5103. 1842–1–[29] From Richard Jones Tithe Commission
Should request Mr. Wildes to wait until JH gives his orders. Is gratified by JH's suggestions, but will not regret being passed over. Did not find Lady Herschel's letter at once for various business reasons.
ALS 3pp RS:HS 10.375

5104. 1842–1–31 From Thomas Henderson Edinburgh
Has received a series of observations from Thomas Maclear on Alpha^1 and Alpha^2 Centauri. Comments on these observations and gives results made by the old circle.
ALS 5pp RS:HS 9.297

5105. [1842–2?] From Edward Sabine 12 Park Place
[Hussey] Vivian supports E.S.'s efforts regarding a magnetic survey of North America. Has examined magnetic measurements made at Boston and Toronto.
ALS 4pp RS:HS 15.296

5106. 1842–2–2 To William Benjamin Carpenter Collingwood
Declines to review WC's recent work, and then JH justifies his position for the rest of the letter.
ALS 4pp WT 67597.2

5107. 1842–2–3 To John Lindley Collingwood
Asks about partially white leaves and about droplets of water forming at the tops of leaves. Has been studying the effects of spectral rays on vegetable colorings.
ALS 4pp diag Royal Botanic Gardens

5108. 1842–2–4 To Thomas Henderson Collingwood
Congratulates him on his discovery of the parallax of Alpha Centauri. Comments on this and makes suggestions for further examination.
C 2pp RS:HS 9.298 & 22.112

5109. 1842–2–4 From John Lindley 21 Regent St.
Not much is known about the circumstances that affect the color of plants. JH's researches will be valuable if they throw any light on the subject. Comments on the various theories affecting the color of plants.
ALS 5pp RS:HS 11.227

5110. [1842]–2–6 From Ann Smyth 3 Cheyne Walk, Chelsea
Thanks JH for reporting on C. P. Smyth's progress. Has read JH's translations of Schiller's 'The Walk.' Sends copy of C. P. Smyth's work at Baseline Camp Zwartland on the remeasurement of N. L. Lacaille's meridian of arc.
ALS 4pp RS:HS 16.217

5111. 1842–2–7 From Adolphe Quetelet Brussels
Thanks for letter of 3 December and for encouragement. Informs JH that 30 observatories participated in observation of winter solstice. Is beginning to study humans in their different relations.
ALS (French) 3pp RS:HS 14.241

5112. 1842–2–10 To Caroline Lucretia Herschel Collingwood
Is confident that by summer his sweeps will all be reduced and arranged in three catalogs for JH's *Cape Results.*
ALS 2pp TxU:H/L–0587.2.b; Reel 1058

5113. 1842–2–11 From Thomas Henderson Edinburgh
Thanks for letter and printed address and for the favorable notice of his investigations into Alpha Centauri. Comments on these and some of his observations on other stars. Encloses a note of the position of seven circumpolar stars.
ALS 6pp RS:HS 9.299

5114. 1842–2–13 To Thomas I. M. Forster Collingwood
Has been ill for two months. Will be visited by Dr. Andrew Smith shortly. Hopes TF can come to Collingwood to meet Smith.
ALS 3pp Bodleian Library, Oxford

5115. 1842–2–13 To Roderick Impey Murchison Collingwood
On RM's stepping down as president of Geological Society. Where exactly abroad will RM pursue enquiries?
ALS 4pp EUL:Gen523/4/28

5116. 1842–2–17 To the R.S.L. Physical Committee Collingwood
Discusses how exceptional observations of meteorological and magnetical phenomena should be recorded and what significance should be attached to them. Cannot attend next meeting of committee.
ALS 4pp RS MM 11.171 & 173

5117. 1842–2–27 To George Biddell Airy Collingwood
Thanks GA for exercising GA's usual discretion in the matter of the funding of Charles Babbage's calculating machine [see GA's 1842–9–26].
ALS 2pp RGO 6.427.74

5118. [1842–3?] To Francis Baily n.p.
Proposes to come to London to visit him. Has just received the proofs of Robert Maine's paper on parallax.
ALS 3pp RS:HS 3.285

5119. 1842–3–2 From William Rutter Dawes Alsop Place Regents Park
Has received from William Lassell some of the new observations he carried out with his new 9" speculum. Comments on these and various clusters of stars changing their position.
ALS 2pp RS:HS 6.60

5120. 1842–3–5 From Roderick Impey Murchison 16 Belgrave Square
Sends a memoir he has scribbled off for next Wednesday's meeting. Would like JH's advice on several points.
ALS 4pp RS:HS 12.402

5121. [1842–3–9] To Caroline Lucretia Herschel Collingwood
Remarks that he is 50 years old, and that he and CH have 'seen something of that odd and most changeable compound called Human Nature.'
ALS 2pp TxU:H/L–0587.4; Reel 1058

5122. 1842–3–9 From Thomas I. M. Forster to Margaret Herschel Tunbridge Wells
Will be pleased to come to Collingwood at any time and hopes the Herschels will also visit them. Her son was right about the word *quinque.* Comments on the various ways different people observe the colors of stars. Will be pleased to see the poetic scraps.
ALS 4pp RS:HS 7.330

5123. 1842–3–10 From George Biddell Airy Royal Observatory, Greenwich
Regarding JH's receipt of the 'Standard' reports. Remarks on the electrometer and barometer in the recent storm.
ALS 1p RS:HS 1.101 (C: RGO 6.340.341)

5124. 1842–3–10 From William Lassell Starfield, Liverpool
Sending a description of the observatory he has erected, containing an equatorially mounted Newtonian reflector. Has high opinion of its efficiency. Has detected a sixth star near the trapezium of *Orionis*.
ALS 3pp RS:HS 11.125

5125. 1842–3–10 From Roderick Impey Murchison 16 Belgrave Square
Many thanks for JH's explanations, which gave satisfaction to the geologists. Gives two queries of his own for which he would like clarification. Can they print JH's Alpine phenomena memoir?
ALS 3pp RS:HS 12.403

5126. 1842–3–11 To George Biddell Airy Collingwood
Arranging a visit for GA to Collingwood.
ALS 3pp RGO 6.368.646

5127. 1842–3–14 From Adolphe Quetelet Brussels
Sends works on meteorology and instructions for natural periodic phenomena. More proselytes in horary observations. Asks for JH's recommendations for those observing meteors in the other hemisphere.
ALS (French) 3pp RS:HS 14.242

5128. 1842–3–15 From George Biddell Airy Royal Observatory, Greenwich
Regarding his letter writing. Has a promise of a trip on the Dover Railway.
ALS 1p RS:HS 1.102 (C: RGO 6.368.648)

5129. 1842–3–15 To Robert Hunt Collingwood
Believes that the operative rays in JH's thermographic process are neither 'calorific' nor 'thermal'.
ALS 4pp George Eastman House, 12.17

5130. 1842–3–16 From 'Q in a Corner' Kensington
Encloses newspaper clipping that reports JH is member of commission of inquiry on weights and measures. Protests that the poor are defrauded more by false *scales* than by nonstandard weights. Proposes remedy.
AL 4pp TxU:H/M–0908; Reel 1083

5131. 1842–3–18 To George Biddell Airy Collingwood
Final arrangements for GA's visit to JH [see GA's 1842–3–15].
ALS 3pp RGO 6.368.649

5132. 1842–3–18 From Alfred Gautier Geneva
Regarding the publication of his own observations on Halley's comet; comments on these. [L. F.] Wartmann was very pleased with the results of his recent visit to England. Has been occupied with political problems. Looks forward to seeing the completion of JH's Cape observations.
ALS 4pp (French) RS:HS 8.85

5133. 1842–3–20 To William Lassell Collingwood
Comments on WL's finding a sixth star in the trapezium of Orion [see WL's 1842–3–10]; lists other catalogues that identify that star.
C 2pp RS:HS 22.114

5134. 1842–3–20 To Roderick Impey Murchison Collingwood
Responds to RM's request for an analysis of the cause of the difference in climate between Orenburg and Catherineburg in Russia. Agrees that Geological Society can print JH's Alpine memoir.
ALS 3pp Geological Society

5135. 1842–3–21 From George Biddell Airy Royal Observatory, Greenwich
Hopes to meet JH on 29 March.
ALS 1p RS:HS 1.103 (C: RGO 6.368.651)

5136. 1842–3–23 From Thomas I.M. Forster Tunbridge Wells
Will be pleased to visit them before they move from Tunbridge Wells. Comments on the Arabic names of stars.
ALS 8pp RS:HS 7.331

5137. 1842–3–25 From George Peacock Deanery, Ely
About arrangements to visit JH at Collingwood.
ALS 2pp RS:HS 13.294

5138. 1842–3–25 To [?] Collingwood
Appreciates receiving the curious fossil, even though JH knows little of such matters.
ALS 1p BostonPubLib Ms. Eng. 204

5139. 1842–3–26 From Richard Jones Tithe Commission
Has just confirmed the Hawkhurst apportionment; comments on this. Encloses a copy of the notice.
ALS 3pp RS:HS 10.376

5140. 1842–3–28 From Thomas Maclear Royal Observatory, Cape of Good Hope
Received JH's constellation reform paper. Reports on triangulation at Zwartland and observations of possible variable stars. Obelisk at Feldhausen erected; relates controversies over the site.
ALS 8pp RS:HS 12.131

5141. 1842–3–28 to 1842–4–9 From Minna Mädler Dorpat
Has only just received JH's parcel dispatched some six months ago, due to lack of transport during the winter months. JH's poem on the *Old Telescope* has been in great demand. Pass the enclosed to Miss Hamilton. Would like an autograph of JH for her album.
ALS 4pp RS:HS 13.15

5142. [1842–4] To Johann Heinrich Mädler n.p.
Is gratified that JM is taking an interest in double stars. Comments on some of the observations and time of their orbits.
C 2pp RS:HS 12.273 & 22.144

5143. [1842–4]. From Stephen J. Rigaud 4 St. Gile's, Oxford
Is applying for vacant professorial chair in astronomy. Wants testimonial from JH in his support. Attached note of JH is negative in this regard.
ALS 3pp † RS:HS 14.490

5144. 1842–4–3 To William Henry Smyth Collingwood
Plans to calculate Gamma Virginis orbit. Received WS paper on astronomical observations. Suggests WS look for a house in Kent after learning plans to move near London.
C 2pp RS:HS 22.116

5145. 1842–4–4 From Thomas Henderson Edinburgh
Gives the parallaxes of a few southern stars obtained from his own mural circle. Has sent his memoir on Alpha Centauri to the R.A.S.
ALS 3pp RS:HS 9.300

5146. 1842–4–6 From James David Forbes Edinburgh
Would like to know the composition of the fluid in JH's actinometers.
ALS 1p RS:HS 7.301

5147. 1842–4–8 From Johann Heinrich Mädler Dorpat
Observations on double stars.
ALS 2pp (German) RS:HS 12.272

5148. 1842–4–9[?] To James David Forbes n.p.
Instructions for preparing liquid for the actinometer. Describes 'very strange' experiments on the spectrum as well as an 'unusual' cloud formation.
ALS 3pp St. Andrews 21? (C: RS:HS 22.115)

5149. 1842–4–10 To George Biddell Airy Collingwood
Sends GA a spectrum photograph, and explains how JH obtained it.
ALS 3pp RGO 6.463.300

5150. 1842–4–16 From William Rutter Dawes Alsop Place, Regents Park
Sends all his observations of Alpha Virginis. Has not done much to this recently. Regarding JH's observations of the fifth and sixth stars in Trapizio. Has informed James Challis of William Lassell's observations.
ALS 3pp RS:HS 6.61

5151. 1842–4–18 From William Henry Fox Talbot Lacock Abbey
Sends JH samples of his calotypes. Off to Germany on a photographic expedition.
ALS 1p RS:HS 17.311

5152. 1842–4–19 To [F. M. Eardley]-Wilmot n.p.
[JH's outline of letter] Compares Wilmot's actinometer observations with JH's. Acknowledges receipt of St. Helena specimens.
AAbs 1p RS:HS 18.413b

5153. 1842–4–19 To Joseph Henry Kay Collingwood
Impressed by accuracy of actinometer measurements made by JK, Joseph Dayman, and P. A. Scott from Feb. to Sept. 1841 at JK's observatory at Rossbank [Tasmania]. Suggests methods to improve readings and simplify reductions.
(A)CCS 4pp TxU:H/L-0211; Reel 1054

5154. 1842–4–21 From George Biddell Airy Royal Observatory, Greenwich
Returns photographs of spectrum, with observations on the results.
ALS 1p RS:HS 1.104 (C: RGO 6.463.302)

5155. 1842–4–21 To William Henry Fox Talbot [Collingwood]
Thanks for calotypes involving persons engaged in actions. Gives encouragement to WT's planned photographic tour in Germany. JH has been experimenting regarding color photography.
ALS 4pp Science Museum Talbt 1/34

5156. [1842 to 1843]–4–23 From Edward Sabine Woolwich
Urged by G. B. Airy, ES withdraws earlier letter to JH and encloses letter with ES's opinions regarding continuance of [magnetic and meteorological] observations.
ALS 2pp TxU:H/M–0517.4; Reel 1093

5157. 1842–4–24 To George Biddell Airy Collingwood
Sends GA two more spectrum photographs [see JH's 1842–4–10], and believes that these may lead the way to color photography.
ALS 3pp RGO 6.463.303

5158. 1842–4–25 From Alfred Smee 7 Finsbury Circus
Sends several samples of ferrocyanide JH had requested.
ALS 3pp RS:HS 16.151

5159. 1842–4–28 From George Biddell Airy Flamsteed House, Greenwich
Returning photographs, with observations. Regarding the transport of loads in hilly country. Remarks on the transcription and pronunciation of foreign languages.
ALS 3pp RS:HS 1.105 (C: RGO 6.368.652)

5160. 1842–4–29 To Adolphe Quetelet Collingwood
Stresses importance of development and phases of periodic phenomena on earth's surface. Magnitudes of stars are the only observations JH is doing. Asks for help in planning trip on continent for three of his daughters and their governess.
ALS 4pp Académie belgique: #17

5161. [1842–4–29]. From Stephen J. Rigaud 4 St. Gile's, Oxford
Still pursuing vacant professorial chair in astronomy. Wants JH's testimonial. Sends his qualifications and description of recent activities in astronomy.
ALS 3pp RS:HS 14.491

5162. [1842–5] To James David Forbes n.p.
Proposes meeting times in town. Suggests that JF observe the upcoming total eclipse astronomically. Agrees with JF's seemingly 'cold &c' circular on scientific partnerships, while regretting that it needed to be written.
ALS 3pp St. Andrews 43?

5163. 1842–5 To Robert Hunt n.p.
Some information about Charles Piazzi Smyth. JH needs

information about some of RH's photographic paper, which JH has tried but without success.
ALS 4pp George Eastman House, 10.17 (C: RS:HS 22.117)

5164. 1842–5–2 From Thomas Henderson Edinburgh
Sending the results of his calculations of the parallaxes of the southern stars. Has recommended these stars to Thomas Maclear to observe.
ALS 3pp RS:HS 9.301

5165. 1842–5–5 From Adolphe Quetelet Brussels
Thanks JH for interest taken in periodic phenomena. Discusses train information and preferred hotel accommodations for JH's visit.
ALS (French) 3pp RS:HS 14.243

5166. 1842–5–10 From John Phillips B.A.A.S., Manchester
Asks whether JH will be prepared to present the committee report on N. L. Lacaille's stars at the B.A.A.S. meeting on 23 June.
L 1p † TxU:H/M–0428.13; Reel 1093

5167. 1842–5–10 From John Phillips B.A.A.S., Manchester
Asks whether JH will be prepared to give the committee report on systems of simultaneous magnetical and meteorological observations at the B.A.A.S. meeting on 23 June.
L 1p TxU:H/M–0428.12; Reel 1093

5168. 1842–5–10 From John Phillips B.A.A.S., Manchester
Will JH be prepared to present report on 'Nomenclature of Stars' at Manchester meeting in June?
(A)LS 1p † RAS:JH Archive 4/9.9 verso; Reel 7

5169. 1842–5–10 From Alfred Smee 7 Finsbury Circus
Has been trying to make other metallo–cyanides at JH's request, but without success.
ALS 3pp RS:HS 16.152

5170. 1842–5–12 From James David Forbes 3 Bolton St., Piccadilly
Would like to see him for a few minutes to discuss a paper he is preparing for the R.S.L. on actinometer observations.
ALS 2pp RS:HS 7.302

5171. 1842–5–13 From George Biddell Airy Royal Observatory, Greenwich
Sends JH the current position of Gamma Virginis.
ALS 1p TxU:H/M-0628.2; Reel 1084 (C: RGO 6.368.654)

5172. 1842–5–13 From C. A. von Steinheil Munich
Suggests new method of gold plating the telescopic mirrors to prevent oxidation. Mentions having tried to galvanize the mirrors with 'Galvana plastik.'
ALS (German) 3pp RS:HS 17.4

5173. 1842–5–14 From James David Forbes 3 Bolton St., Piccadilly
No meeting of the R.S.L. on the 19th but could arrange to see him before dinner. Hopes to spend the summer amongst the glaciers.
ALS 3pp RS:HS 7.303

5174. [1842–5–16?] To James David Forbes n.p.
Sets up approximate meeting time, depending on his family's timely departure to Antwerp, with JF at the R.S.L. apartments.
ALS 2pp St. Andrews 49

5175. 1842–5–18 From Bernard A. von Lindenau Dresden
It is a long time since he thanked him for the assistance in becoming a member of the R.S.L. and the R.A.S. Doubtless his work at the Cape was a great success. Hoped to visit England again but seems improbable now. Sends J. F. Encke's memoir for the R.S.L., R.A.S. and JH. Is very interested in his memoir on the variable light of Orionis. C. F. Gauss continues his magnetism.
ALS 3pp RS:HS 11.219

5176. 1842–5–19 To Adolphe Quetelet London
Thanks for assistance in preparing trip. JH's daughters will probably not stay in Brussels long. Mentions his first engraving for Cape observations of nebulae is to be engraved this very day. Three quarters of his reductions of the southern sweeps are completed.
ALS 4pp Académie belgique: #18

5177. [1842 or 1845]–5–22 From George Peacock Deanery of Ely
GP's views on continuation of magnetic observatories in colonies.
ALS 3pp TxU:H/M–1206; Reel 1060

5178. 1842–5–26 To William Henry Smyth Collingwood
Thanks WS for C. P. Smyth drawings of Cape monument and Gamma Virginis observations. Notes J. H. Mädler's work on double stars and the doubt that they obey the inverse square law. Discusses use of photography in astronomy.
C 2pp RS:HS 22.118

5179. 1842–5–28 From Robert Hunt Falmouth
Was hoping to send him some positive paper, but has been unable to prepare this due to a medical problem. Did make a sheet or two today. Returns some of the specimens JH sent him but would like to keep those made by JH himself.
ALS 3pp RS:HS 10.91

5180. 1842–5–31 From George Biddell Airy Royal Observatory, Greenwich
About the length of a Board of Visitors' meeting, and about arrangements afterward.
ACCS 1p RGO 6.8.544

5181. [1842–6] To Friedrich Wilhelm Bessel [Collingwood]
Offers to meet FB at Manchester B.A.A.S. meeting and to bring FB to Collingwood for a quiet visit.
ALS 3pp AdWDDR Bessel 225 (C: RS:HS 22.119)

5182. 1842–[6] To Margaret Brodie Herschel Manchester
Has stayed in Manchester an extra day, and has managed to avoid becoming involved in the organization of next year's B.A.A.S. meeting at Cork; talks about plans for F. W. Bessel to visit Collingwood, and about JH's plans to return home.
ALS 3pp JHS 6.3

5183. [1842–6] To William Whewell n.p.
Alerts WW that F. W. Bessel, in very poor health, will attend the Manchester B.A.A.S. meeting. JH plans to bring Bessel to Collingwood after the meeting.
ALS 3pp TC, Camb. Add. Ms.a.207[52] (C: RS:HS 22.120)

5184. 1842–6–1 To [R.S.L. Council?] Collingwood
Discusses JH's spectrometer and possibility of using surplus from R.S.L. Donation Fund for its construction and enquiry with it into nature of light and heat.
ALS 4pp RS MC.3.218

5185. 1842–6–2 To James David Forbes Collingwood
Compares actinometric measurements of radiation. Explains interest in total eclipse relative to actinometry. Discusses the possibility of making actinometric measurements by photographic means.
ALS 3pp St. Andrews 55 (C: RS:HS 22.122)

5186. 1842–6–2 To Robert Hunt n.p.
Sorry to hear RH has been ill; JH sends some photographic examples.
ALS 3pp George Eastman House, 9.17 (C: RS:HS 22.121)

5187. 1842–6–2 From Richard Sheepshanks n.p.
Gives his opinion of [George] Merz, stating that he has not done anything to 'distinguish himself from a mere practical follower of [Josef] Fraunhofer.' Requests that JH send encouragement to James Challis.
ALS 3pp RS:HS 16.18

5188. 1842–6–2 From William Henry Fox Talbot Munich
Carl Steinheil has proposed making telescope mirrors by plating silver on glass. Is writing to JH to tell him that somewhat earlier he had included this process in one of his patents.
ALS 4pp RS:HS 17.312

5189. 1842–6–3 From James Grahame 5 Place de Launcy, Nantes
Of JG's intention to write a pamphlet against slavery.
ALS 2pp RS:HS 8.439

5190. 1842–6–5 From Friedrich Eduard Beneke Berlin
Letter to accompany a complimentary copy of FB's new book on logic.
ALS 1p (German) RS:HS 4.59

5191. 1842–6–5 From Wilhelm Struve Pulkowa
Thanks JH for his letter [1841–10–10]. Discusses difficulty of measuring distance between a star of primary largeness and a small telescopic neighbor star. Rejects theory that temperature influences observation results. Appreciates JH assessment of the correspondence between WS's work and that of [F. W.] Bessel. Complains confidentially that Bessel neglects to acknowledge WS's work on parallax, begun a year before Bessel's own. Assures JH of WS's continual efforts in observing new phenomena. Comments on JH's 'On the Revision and Rearrangement of the Constellations.'
ALS (German) 8pp RS:HS 17.130

5192. 1842–6–7 To Caroline Lucretia Herschel Collingwood
Enjoyed CH's last letter.
ALS 1p TxU:H/L–0587.9; Reel 1058

5193. 1842–6–8 To [William Radcliffe Birt] Collingwood
Received WB's atmospheric curves yesterday, but too late to report them [at 1842 B.A.A.S. meeting]. Final report will be presented at next [1843] meeting.
ALS 1p RS:HS 19.95

5194. 1842–6–8 From [Peter Mark Roget] 39 Bernard St., Russell Square
With R.S.L. Council, is glad that compliance with JH's wishes regarding photographical inquiries has pleased JH.
AL 1p inc RS MM.16.151

5195. 1842–6–8 To [Sir] Edward Ryan n.p.
Welcomes ER back to England. Met ER's host, Charles Babbage, at Greenwich yesterday. Wants to visit before ER departs for Geneva, but JH's wife is ill. Based on accounts of Geneva by ER's wife, JH sent three oldest daughters and their governess to spend summer there.
ALS 7pp TxU:H/L-0765.3; Reel 1089

5196. 1842–6–9 From William Benjamin Carpenter 22 Park St., Bristol
Has been working on the structure of mother of pearl and has come to different conclusions than those of JH. Would be pleased if he could send a sample to him to investigate further.
ALS 4pp RS:HS 5.187

5197. 1842–6–9 From Sir Edward Ryan 24 Manchester St.
Hopes Lady Herschel is well soon. Meeting of JH and ER in England must be postponed. Visiting his family soon. Hopes to see Italy and Germany.
ALS 4pp RS:HS 14.455

5198. [1842–6–11] From James Grahame Tavistock Hotel, London
Note to let JH know that JG is in England.
ALS 1p RS:HS 8.440

5199. [1842 to 1843]–6–12 From Edward Sabine Woolwich
Asks JH to inspect first magnetic and meteorological observations from [Tasmania] and Toronto. Hopes North American observations will be increased.
ALS 4pp TxU:H/M–0517.5; Reel 1093

5200. 1842–6–13 To William Radcliffe Birt Collingwood
Reviews WB's assistance in JH's meteorological investigations as testimonial on behalf of WB's application for position of curator at Kew Observatory.
ALS 3pp RS:HS 19.96

5201. 1842–6–13 From Charles Wheatstone 20 Conduit St.
Encloses B.A.A.S. statement of objectives for Richmond Observatory, for which B.A.A.S. Council asks JH to organize goals and support annual subsidy.
ALS 2pp RS:HS 18.148

5202. 1842–6–15 To Alfred Smee n.p.
On the basis of some compounds referred to by AS [see 1842–5–10], JH has now developed a new form of photography using gold as a stimulant; JH calls it the 'Argyrotype'.
draft 2pp RS:HS 16.153

5203. [1842–6–16] From James Grahame n.p.
A note to arrange a meeting with JH.
ALS 2pp RS:HS 8.441

5204. 1842–6–16 From Roderick Impey Murchison 3 Bryanston Place
Would like JH to come to Manchester on 23–25 to give support for the Kew Observatory. F. W. Bessel and G. A. Erman are coming.
ALS 4pp RS:HS 12.385

5205. 1842–6–16 To C. A. von Steinheil Collingwood
Thanks for suggestion of gold plating the telescope mirrors. Mentions idea of mounting a reflector horizontally but finds its execution 'insufferable'. Thanks for work on photometric measurement of light of stars. Disagrees on some results. Encloses new photographic specimens.
ACCS 4pp RS:HS 17.5 (C: RS:HS 22.124)

5206. 1842–6–17 From Edward Sabine Woolwich
Encloses bill from W. H. Allen & Co. regarding Breslau

observatory. [JH annotation: Details of bill.]
ALS 1p † TxU:H/M–0519; Reel 1093

5207. 1842–6–17 To Charles Wheatstone Collingwood
Received prospectus of Kew observatory. Its objectives seem incompatible. It appears better suited as experimental institution. Refer Francis Beaufort to JH's 11 Oct. 1835 letter to Beaufort defining physical observatories.
ACCS 4pp RS:HS 18.149 (C: 22.125)

5208. 1842–6–21 From Alfred Smee 7 Finsbury Circus
Note accompanying a sample of iridium.
ALS 1p RS:HS 16.154

5209. [1842–6–22 or later] From James Grahame 5 Kensington Square
A note to arrange dinner with JH.
ALS 1p RS:HS 8.442

5210. 1842–6–24 From William H. Allen 7 Leadenhall St.
Acknowledges JH's payment for stationery for Breslau magnetic observatory.
ALS 1p TxU:H/M–0004; Reel 1062

5211. 1842–6–25 From Friedrich Wilhelm Bessel Manchester
Informs JH of FB's arrival in Manchester and of further travel plans.
ALS 1p (German) RS:HS 4.75

5212. [1842–7] To Friedrich Wilhelm Bessel [Collingwood]
Informs FB on how to visit Woolwich Arsenal and Dock Yard.
ALS 2pp AdWDDR Bessel 225

5213. [1842–7] From Edward Sabine Woolwich
Sorry [F. W.] Bessel's health is 'so indifferent.' Glad JH is coming to [B.A.A.S. meeting].
ALS 1p RS:HS 15.305

5214. [1842?–7?] To William Whewell [Collingwood]
Hopes to arrange a meeting with WW at the invitation of a Mr. Hope. Discusses JH's recent photographic works and sends some photographs of improved quality.
ALS 4pp inc? TC, Camb. Add. Ms.a.207[43] (C: RS:HS 22.126)

5215. 1842–7–5 From Friedrich Wilhelm Bessel Edinburgh
About further travel plans, and people FB is meeting as he travels in Britain.
ALS 2pp (German) RS:HS 4.76

5216. 1842–7–6 From [Alexander J. B. Hope?] 13 Connaught Place
A proposal to complete the calculating machine has been made by Charles Babbage, and AH wants JH's opinion of it.
AL 2pp inc RS:HS B27.53

5217. 1842–7–8 To Friedrich Wilhelm Bessel Collingwood
Detailed arrangements for FB's visit [see JH's 1842–6].
ALS 4pp AdWDDR Bessel 225 (C: RS:HS 22.127)

5218. 1842–7–10 From Francis Baily Milan
Account of the eclipse seen at Pavia.
ALS 4pp RS:HS 3.225

5219. 1842–7–10 From T. Dyson Gainsborough
Does JH plan to publish 'Treatise on Light' in separate form?
ALS 1p TxU:H/M–0173.1; Reel 1087

5220. 1842–7–10 From John Phillips B.A.A.S., York
JH, William Whewell, George Peacock, Humphrey Lloyd, and Edward Sabine are appointed by the B.A.A.S. to study systems of simultaneous magnetical and meteorological observations. They will be granted £89.
L 1p TxU:H/M–0428.14; Reel 1093

5221. 1842–7–10 From John Phillips York
Informs JH that the B.A.A.S. has placed him on a committee to supervise the translation and publication of foreign scientific memoirs.
(A)L 1p RS:HS B26.9

5222. 1842–7–10 From John Phillips York
Informs JH that the B.A.A.S. has place him on a committee to conduct experiments by capture balloon on the atmosphere.
(A)L 1p RS:HS B26.8

5223. 1842–7–11 To Caroline Lucretia Herschel Collingwood
Met Friedrich Bessel at the Manchester B.A.A.S. meeting; invited him to Collingwood, where he expects Bessel in a few days. Enclosed with the letter a specimen of a new photographic process called 'Chrysotype.' Marvels at traveling from Hawkhurst to Manchester round trip (420 miles) in under 23 hours!
ALS 4pp TxU:H/L–0587.10; Reel 1058

5224. 1842–7–12 To Edward Sabine 12 Park Place
Discusses time intervals for [magnetic] observation. Sorry a change was ever contemplated. Asks that his opinion be sent to [Humphrey] Lloyd and that ES make the final decision on how to proceed.
draft 7pp RS:HS 15.145 (C: RS:HS 22.128)

5225. 1842–7–15 From Thomas John Newbold Bombay
On his last passage down the Red Sea the Captain informed TN of the singular alteration of his chronometers. Comments on this and would like JH's views. Magnetic station at Aden has never been set up and the instruments are still at Bombay. Examined the ancient canal while at Suez. Will be pleased to assist JH in any way.
ALS 4pp RS:HS 13.121

5226. 1842–7–16 From Friedrich Wilhelm Bessel 32 Norfolk St.
Thank you note after FB visited Collingwood.
ALS 1p (German) RS:HS 4.77

5227. 1842–7–19 From Friedrich Wilhelm Bessel 32 Norfolk St.
Warm thanks for JH's kind hospitality; some comments about Prussian Order of Merit.
ALS 1p (German) RS:HS 4.78

5228. 1842–7–19 From M. Tunno Amport, Andover
Thanks JH for a letter of introduction.
ALS 3pp RS:HS 17.449

5229. 1842–7–20 To F. W. A. Argelander Collingwood
Letter of introduction for some friends of JH, to FA in Bonn.
C 2pp RS:HS 22.130

5230. 1842–7–20 To Friedrich Wilhelm Bessel Collingwood
Comments on expected honor from Prussian government and

on British rules against officially accepting it.
ALS 4pp AdWDDR Bessel 225 (C: RS:HS 22.129)

5231. 1842–7–21 To William Whewell Collingwood
[Same message as JH's letter of 1842–7–21, but sent to a second location.]
ALS 4pp TC, Camb. Add. Ms.a.207[54] (C: RS:HS 22.131)

5232. 1842–7–21 To William Whewell Collingwood
Invites WW for a short visit, JH's wife being ill. Thanks WW for paper on cause and effect. Clouds kept JH from photographing a recent solar eclipse.
ALS 4pp TC, Camb. Add. Ms.a.207[53] (C: RS:HS 22.131)

5233. 1842–7–22 To [Mr.] Collen Collingwood
Goes into considerable detail in providing information about the business organization of the Munich lens makers, and the high quality of their lenses. Urges the benefits of achromatic lenses, and sends MC some specimens of JH's photography.
ALS 4pp DMC.695A.885.31

5234. 1842–8–4 From Caroline Lucretia Herschel Hanover
Promises in her next letter to comment about JH's 'chrysotype' photographic process.
L *Memoir ... of Caroline Herschel,* 326–7 inc

5235. 1842–8–8 To J. Coates n.p.
[To JC as Secretary, Society for the Diffusion of Useful Knowledge:] Thanks for offer to send a copy of Augustus De Morgan's *Differential and Integral Calculus.*
ALS 1p Univ.Coll.London SDUK Papers

5236. 1842–8–9 To Caroline Lucretia Herschel Collingwood
Describes the total solar eclipse seen by Francis Baily at Pavia and George Airy at Turin. They were thrilled to witness three purple flames from the blocked sun emerge around the edge of the moon. Thirty more Cape Town sweeps remain to be reduced.
ALS 2pp TxU:H/L–0587.12.b; Reel 1058 *Memoir ... of Caroline Herschel*, 327–8

5237. 1842–8–10 To the *Athenaeum* Collingwood
Seeks to clear up confusions about various photographic processes, e.g., the chrysotype process, developed by JH, about which confusions arose from earlier reports on them in the *Athenaeum.*
L *Athenaeum,* #773 (Aug. 20, 1842), 748

5238. 1842–8–12 From F. M. Eardley-Wilmot Cape of Good Hope
Has been unable to make many readings with the actinometer. Is expecting some irregular readings with the magnetometers. Survey of South Africa appears very distant. Relates recent events in Natal and with the Boers.
ALS 4pp RS:HS 7.2

5239. 1842–8–15 From Benjamin Peirce Cambridge, Mass.
Sends Joseph Cranch to consult with JH over an equatorial telescope for Harvard College Observatory.
ALS 2pp RS:HS 14.151

5240. 1842–8–17 To Mr. Collen Collingwood
Would be pleased to be visited by C to discuss the application regarding glass [lenses] of various algebraic formulae devised by JH. Will also show C some of JH's recent work in photography.
ALS 3pp BostonPubLib Ch.J.6.21

5241. [1842]–8–17 From Henry Holland Brook St.
Compliments JH's translation (1842) of J. C. F. Schiller's *The Walk.* Is going to Scotland, then Switzerland. Has seen enough of America.
ALS 2pp RS:HS 9.432

5242. [1842]–8–17 From Edward Sabine Woolwich
Magnetic survey of Austrian dominions to be undertaken. Sending portable apparatus for survey in Mediterranean along with instructions from [Charles] Riddell. Completed sea observations for R.S.L.
ALS 4pp RS:HS 15.306

5243. 1842–8–19 To George Biddell Airy Collingwood
Asks GA's opinion as to whether Ordinance observatories should make their magnetic observations on a 3-minute or 5-minute system.
ALS 4pp RGO 6.693.108

5244. 1842–8–19 From Alexander von Humboldt Sans Souci
Recounts details of his recent visit to England. Recent astronomical observations. Has JH heard from C. K. J. Bunsen about his award?
ALS 3pp (French) RS:HS 10.66

5245. 1842–8–22 From Humphrey Lloyd Caernarvon
Thinks it necessary to revise the instructions for the magnetic observations. Makes some suggestions for the Report of the Committee.
ALS 4pp RS:HS 11.281

5246. 1842–8–24 From George Biddell Airy Royal Observatory, Greenwich
Returning packet of papers on magnetism. Encloses copy of a letter he has sent to Edward Sabine. Has seen the recent eclipse of the sun.
ALS 2pp RS:HS 1.106 (C: RGO 6.368.655)

5247. [1842–8–24 or later] To Alexander von Humboldt [Collingwood]
Writes to send thanks by way of AH to the King of Prussia for the award of the Order of Merit. JH also comments on some astronomical matters.
Cdraft 3pp RS:HS 25.14.14

5248. 1842–8–24 To Humphrey Lloyd Collingwood
Regarding the revision of the instructions for the magnetic observatories.
C 4pp RS:HS 11.282 (C: RS:HS 22.137)

5249. 1842–8–25 To George Biddell Airy Collingwood
Is organizing a committee to revise the instructions for observers at magnetic observatories; GA is on it.
ALS 4pp RGO 6.693.110 (C: RGO 6.675.327; draft not in JH's hand RS:HS 1.107 & 22.132)

5250. 1842–8–25 To George Biddell Airy n.p.
A note to accompany some letters being returned to GA.
ALS 1p RGO 6.675.329

5251. 1842–8–26 From Francis Baily 37 Tavistock Place
Further remarks concerning the recent eclipse and his own and G. B. Airy's papers on the subject.
ALS 2pp RS:HS 3.226

5252. 1842–8–27 To Francis Baily Collingwood
Thanks for the information on the eclipse. Comments on this. Events at the B.A.A.S. meeting. News of Wilhelm Struve's activities.
ALS 2pp RS:HS 3.227 (C: RS:HS 25.6.10 & 25.8.46)

5253. 1842–8–29 From Francis Baily 37 Tavistock Place
Has read G. B. Airy's paper on the eclipse. Would JH send on FB's remarks on the eclipse to Airy. Comments on the phenomenon of the eclipse.
ALS 4pp RS:HS 3.228

5254. 1842–8–31 From George Biddell Airy Royal Observatory, Greenwich
Encloses copy of a letter written to Edward Sabine. Wants instruments of quicker vibration.
ALS 1p RS:HS 1.108 (C: RGO 6.675.330)

5255. 1842–8–31 From Humphrey Lloyd Beaumaris
Has just received JH's circular. Was not fully aware of the labor involved in drawing up the Report of the Council. Needs a new edition of the part concerning the fixed observatories. Outlines method for new edition. Sabine should be responsible for the material on Magnetic Surveys.
ALS 5pp RS:HS 11.283

5256. 1842–8–31 From Edward Sabine Dunraven
Asks JH to consult [C. F.] Gauss concerning magnetical instructions. [Charles] Wilkes made mistake in distinguishing Balleny's[?] Land from 'American discoveries.' This has caused dispute.
ALS 4pp RS:HS 15.146

5257. 1842–8–31 From Alfred Smee 7 Finsbury Circus
Note accompanying a work detailing the study of voltaic circuits.
ALS 2pp RS:HS 16.155

5258. 1842–9–1 To George Biddell Airy Collingwood
About eclipse observations of 1842, together with JH's speculations about the existence of an atmosphere around the sun. Enclosed is an eclipse drawing by Francis Baily.
ALS 2pp RGO 6.112.75 (ACCS: RS:HS 25.6.11)

5259. 1842–9–1 To Francis Baily Collingwood
Has sent the papers on the eclipse to G. B. Airy.
ALS 1p RS:HS 3.229 (C: RS:HS 25.8.47)

5260. 1842–9–1 From J[oseph] Cranch Islington
Would like JH's views on the suitabilities of the various makes of telescopes for their observatory.
ALS 2pp RS:HS 5.370

5261. [1842–9–1 or later] To J[oseph] Cranch n.p.
Does not feel qualified to judge the merits of the various makes of telescopes. Suggests that the Astronomer Royal may be able to help or one of the other observatories.
draft 2pp RS:HS 5.371 (C: RS:HS 25.13.18)

5262. 1842–9 To Robert Hunt n.p.
Thanks for RH's paper on mineralogy. JH has explored use of mercury together with iron in photography.
ALS 2pp George Eastman House, 8.17 (C: RS:HS 22.133)

5263. 1842–9–2 To Humphrey Lloyd Collingwood
Urges some revision in both the report and instructions for the magnetic observations.
C 3pp RS:HS 22.134

5264. 1842–9–4 To Carl Friedrich Gauss Collingwood
The instructions for making magnetic and meteorological observations are to be revised. JH asks CG's advice about this.
ALS 3pp NSUB Cod.Ms.Gauss 99 (ACCS 2pp RS:HS 25.6.12)

5265. 1842–[9–4] To Caroline Lucretia Herschel Collingwood
Reports the erection of an obelisk at Feldhausen to commemorate the site of JH's 20-ft. reflector. Back at Cape Town, Thomas Maclear is measuring N. L. Lacaille's Arc of the Meridian. JH received the Prussian Order of Merit.
ALS 4pp TxU:H/L–0587.13; Reel 1058

5266. 1842–9–6 To [George Biddell Airy ?] Collingwood
Remarks on the significant increase in brightness of the star Eta Cygni, and comments on several other variable stars.
ACCS 1p RS:HS 25.6.13

5267. 1842–9–6 To Francis Baily n.p.
Calls attention to the increase in magnitude of the star Eta Cygni. Mentions some other variable stars.
L *R.A.S.M.N.*, 6 (1842), 23–4

5268. 1842–9–9 From F. W. A. Argelander Bonn
Regarding the Misses Tunros and their residence at Bonn. Stating where his memoir on the motion of the sun is printed. Observations on stars and object glasses.
ALS 4pp (German) RS:HS 1.369

5269. 1842–9–13 From Francis Baily 37 Tavistock Place
Comments on the various reports of the recent eclipse. Regarding the star constellations.
ALS 3pp RS:HS 3.230

5270. 1842–9–16 From George Biddell Airy Royal Observatory, Greenwich
Asking JH for his views on Charles Babbage's calculating machine, so that he can pass on the information to the Chancellor of the Exchequer.
ALS 2pp RS:HS 1.109 (C: RGO 6.427.68)

5271. 1842–9–[16 or later] From Elizabeth [Betty] Plumleigh Babbage 5 Devonshire St., Portland Place
Congratulations on the birth of his daughter.
ALS 1p RS:HS 2.300

5272. 1842–9–16 To Mary Baldwin n.p.
Birth of JH's ninth child [Julia].
ALS 1p TxU:H/L-0065; Reel 1054

5273. [1842–9–16 or later] To Henry Goulburn [London]
Writes to the Chancellor of the Exchequer to give JH's opinion about the probable utility of continuing to spend money on Charles Babbage's calculating machine.
draft 11pp RS:HS B27.51

5274. 1842–9–17 To Bridget Taylor Collingwood
Informs BT of the birth of JH's daughter Julia. Mother and child are both doing well.
ALS 2pp WT 64710

5275. 1842–9–18 To Caroline Lucretia Herschel Collingwood
Reports the birth of JH's ninth child, Julia.
ALS 4pp TxU:H/L–0587.14; Reel 1058

5276. 1842–9–20 To George Biddell Airy Collingwood
Needs some clarification of GA's letter [see GA's 1842–9–16] before JH is ready to reply about the calculating engine.
ALS 3pp RGO 6.427.69

5277. 1842–9–21 To George Biddell Airy n.p.
Reply to GA's letter on Charles Babbage's calculating machine.
draft 2pp RS:HS 1.110 & 22.135

5278. 1842–9–21 From Richard Jones Haileybury
Grateful for the account of Lady Herschel. Feels sorry for James Stewart; hopes to visit him later. Enjoyed himself in Holland; then went on to Germany, which he hopes to revisit when he has learned more of their language. Brother of Mrs. Jones has died.
ALS 3pp RS:HS 10.377

5279. 1842–9–22 From George Biddell Airy Royal Observatory, Greenwich
Further remarks on his letter concerning Charles Babbage's calculating machine (see GA's 1842–9–16) in order to clarify JH's mind.
ALS 2pp RS:HS 1.112 (C: RGO 6.427.71)

5280. 1842–9–22 From George Biddell Airy Royal Observatory, Greenwich
Robinsons (Devonshire St.) have offered their standard Troy pound, together with its history for £20–0–0. [JH has added notes from his reply regarding the present whereabouts of the various Troy pounds.]
ALS 3pp RS:HS 1.111 (C: RGO 6.340.342)

5281. 1842–9–22 To [William Radcliffe Birt] Collingwood
Must complete graphs of barometer curves before B.A.A.S. meeting. Did JH mistakenly include 1837 *Bogota Almanac* in barometer observations?
ALS 2pp RS:HS 19.97

5282. 1842–9–23 To George Biddell Airy Collingwood
Gives GA some accounting of standard weights in Britain, and feels that it would be worth acquiring Robinson estate weight [see GA's 1842–9–22].
ALS 4pp RGO 6.340.343

5283. 1842–9–26 From George Biddell Airy Royal Observatory, Greenwich
Encloses the Chancellor of the Exchequer's letter concerning Charles Babbage's calculating machine.
ALS 2pp RS:HS 1.113 (C: RGO 6.427.72)

5284. 1842–9–26 To [William Radcliffe Birt] Collingwood
Sends two dozen more papers on barometer graphs. Must have all papers on subject by January. Pleased with WB's lunar observations.
ALS 2pp RS:HS 19.98

5285. 1842–9–26 From John William Draper New York University
Sending a Daguerreotype impression of the spectrum. Comments on this. Is publishing a work on the chemical rays.
ALS 2pp RS:HS 6.502

5286. 1842–9–27 From Richard Sheepshanks R.A.S.
F. W. A. Argelander has requested the R.A.S. to pledge itself to publish one of his writings. RS notes that it is a rule of the R.A.S. not to do this, but rules can be broken if there is 'sufficient advantage.' Suggests Argelander's request be granted and that JH write him.
ALS 4pp RS:HS 16.19

5287. 1842–9–29 From Alfred Smee 7 Finsbury Circus
Explains how AS first learned of various compound salts of iron from medical writings.
ALS 3pp RS:HS 16.156

5288. 1842–9–29 To William Whewell Collingwood
Congratulates WW on a testimonial held in WW's honor in Lancaster. Reports the birth of JH's sixth daughter [Julia].
ALS 3pp TC, Camb. Add. Ms.a.207[55] (C: RS:HS 22.136)

5289. 1842–10–4 From Thomas Henderson Edinburgh
Has had a reply from Thomas Maclear stating that all work has ceased at the Cape Observatory until the work on calculating the Arc of the Meridian has finished. F. W. Bessel has expressed a wish that *Centaure* should be observed with a heliometer. Wishes a good heliometer could be sent to the Cape.
ALS 3pp RS:HS 9.302

5290. 1842–10–12 To Caroline Lucretia Herschel Collingwood
JH finished his catalog of stars for his *Cape Results*; hopes to be finished with his nebulae and double star catalogs soon.
ALS 3pp TxU:H/L–0587.15.a; Reel 1058

5291. 1842–10–12 From Robert Hunt Falmouth
Is grateful for the specimens sent him. Has been carrying out a series of experiments on the Calotype process with extraordinary results, especially the crimson skies. What does he think of the idea of publishing a treatise on 'Light' considered as a chemical agent?
ALS 2pp RS:HS 10.92

5292. 1842–10–13 From Richard Sheepshanks Reading
Urges JH to plead to Lord Palmerston on behalf of [H. C.] Schumacher. Suggests arguments JH might use. States that JH as 'chief of British astronomy by inheritance, by *age* & by service' is the ideal person to make this plea.
ALS 4pp RS:HS 16.20

5293. 1842–10–16 From Alfred Gautier Geneva
Shortly coming to England with daughter and Madame Gerlach. Gives family news. Mr. Muller has some observations for the Memoirs. Emile Plantamour has succeeded him at the Observatory. Met G. B. Airy and his wife on their journey to Turin. Has also conversed with J. D. Forbes.
ALS 3pp (French) RS:HS 8.86

5294. 1842–10–17 From Richard Sheepshanks Reading
Thanks JH for the letter to Lord Palmerston on [H. C.] Schumacher's behalf. Comments on British politics.
ALS 3pp RS:HS 16.21

5295. 1842–10–24 From Carl Friedrich Gauss Göttingen
Pleased to hear that the magnetic observations are to be continued for a further three years. Mode of conducting observations in some foreign observatories would seem to need improvement. Has made a series of observations of magnetic inclination, which will be published in the sixth volume of *Resultate*.
ALS 2pp RS:HS 8.78

5296. 1842–10–28 From F. W. A. Argelander Bonn
Giving his results respecting Cassiopeia. Requesting any details of observations of Orionis since April.
ALS 3pp (German) RS:HS 1.370

5297. 1842–10–28 To Mrs. Grover Collingwood
Mother and baby [Margaret and Julia] are doing well; JH asks questions relating to the Grovers giving up the lease on JH's property at Slough.
ACCS 2pp RS:HS 25.6.14

5298. 1842–10–29 From Richard Taylor Red Lion Court
Sends packet from John William Draper containing material for an issue of *P.M.* Asks JH's advice on publication of

articles for *Scientific Memoirs*.
ALS 2pp RS:HS 17.337

5299. 1842–10–31 From George Grey Govt. House, Adelaide
Has only just received his letter of Dec. 1841. Has introduced a Bill on Standards into the Legislative Council for S. Australia with the aid of JH's letter and report. Has been collecting Natural History specimens and would be pleased to send anything for JH. Making new geological discoveries daily.
ALS 5pp RS:HS 9.131

5300. 1842–11–1 To F. W. A. Argelander n.p.
Has passed on the papers sent by FA. Tables of chief stars. Observations on FA's letter of 28 October 1842 [see FA's 1842–10–28]. Sends observations of Orionis.
draft 4pp RS:HS 1.371 & 22.139

5301. 1842–11–2 To Editor of the [*R.S.P.T.?]*
Collingwood
Comments on some papers on vision and photography by Mr. Moser, and compares this with some of JH's own writings on the subject.
ALS 3pp RS:HS B26.69

5302. 1842–11–5 From George Biddell Airy Royal Observatory, Greenwich
Has arranged meeting to discuss N. L. Lacaille's observations. Sends a toy purchased in the streets of London.
ALS 2pp RS:HS 1.114 (C: RGO 6.368.656)

5303. 1842–11–7 To George Biddell Airy Collingwood
Unwilling to come to a meeting as JH anticipates the return of some of his children from the continent; seems considerable confusion as to who is to do what about star catalogues [see GA's 1841–7–27].
ALS 4pp RGO 6.203.315

5304. 1842–11–7 From Francis Baily 37 Tavistock Place
Has seen G. B. Airy. No meeting at Airy's. Further regarding reports of the recent eclipse.
ALS 2pp RS:HS 3.231

5305. 1842–11–9 From Robert Hunt Falmouth
Has been working at [L. F.] Moser's experiments and believes he is wrong in considering latent light as the agent that produces the photographic image. Read a paper last evening on the subject, which will be printed. Has been producing images by the action of heat and electricity and proposes to call this process Thermography.
ALS 3pp RS:HS 10.93

5306. 1842–11–10 To Francis Baily [Collingwood]
Has written to G. B. Airy. Regarding the various reports of the recent eclipse.
ALS 2pp RS:HS 3.232 (C: RS:HS 25.8.48)

5307. 1842–11–10 From William Rutter Dawes 9 Park Place, Regents Park
Enclosing his observations of Alpha Virginis for the past year. Has been attempting some improvements in the double refracting crystal micrometer. Comments on this and his findings. Would like JH's comments.
ALS 4pp RS:HS 6.62

5308. 1842–11–10 To Adalbert Adolf Mühry
Collingwood
Is anxious about his aunt Caroline Herschel and would like news of her health.
C 4pp RS:HS 12.371 & 22.141

5309. 1842–11–12 From Adolphe Quetelet Brussels
Continues horary observations suggested by JH. Also magnetic observations. Discusses observations of migration of birds. November has been cloudy so far and meteors cannot be seen.
ALS (French) 3pp RS:HS 14.244

5310. 1842–11–14 From Friedrich Wilhelm Bessel
Königsberg
More about Prussian Order of Merit; further annual parallax observations, together with comments on other observations.
ALS 3pp (German) RS:HS 4.79

5311. 1842–11–15 To James David Forbes Collingwood
Received JF's paper on 'thermographic process.' Believes that thermic rays were responsible for experimental results of JF, J. W. Draper, and Ludwig Moser. JH's theory of and experiments with thermic rays, which are different from 'calorific' and visible rays, in solar spectrum.
ACCS 4pp RS:HS 25.6.15 (C: TxU:H/L-0159; Reel 1054)

5312. 1842–11–16 From F. W. A. Argelander Bonn
Readings of comets.
ALS 3pp (German) RS:HS 1.372

5313. 1842–11–17 To Augustus De Morgan Collingwood
Thanks AD for the corrections to JH's writing on astronomy; congratulates AD on his book on differential calculus.
C 3pp RS:HS 22.142

5314. 1842–11–17 From Alfred Smee 7 Finsbury Circus
A note accompanying a sample of high quality ferrocyanide of [iron?].
ALS 3pp RS:HS 16.157

5315. 1842–11–21 From Robert Hunt Falmouth
Was pleased to receive JH's letter and to read that he had really proved the existence of [invisible light?]. One of his students has produced images on copper plates by sulfur fumes. Has just seen Macedonio Melloni's memoir, which is now governed by JH's views.
ALS 3pp RS:HS 10.94

5316. 1842–11–22 From Augustus De Morgan 69 Gower St.
Thanks for the gift of the book. Thomas Longman is now the editor of Lardner's *Cyclopaedia*. Latitude of the pole star incorrect in Charles Knight's publications. Regarding truth.
ALS 8pp RS:HS 6.187 S. De Morgan, *Memoir of Augustus De Morgan*, 145–7

5317. 1842–11–23 From J. William Griesbach St. Omer
Would like JH's advice how to obtain a place in Christ's Hospital for his brother-in-law. Queries regarding the firing of shot guns and on the spectrum.
ALS 3pp RS:HS 9.138

5318. 1842–11–30 From Richard Taylor Red Lion Court
Thanks JH for advice on the issue of *P.M.* that has just been published.
ALS 1p RS:HS 17.338

5319. 1842–12–2 To Robert Hunt Collingwood
Sends RH a packet of photographs with a description of each type.
ALS 3pp George Eastman House, 11.17 (C: RS:HS 22.143)

5320. 1842–12–2 From Alfred Smee 7 Finsbury Circus
About chemical solutions used in photography, and problems produced by impurities in photographic paper.
ALS 3pp RS:HS 16.158

5321. 1842–12–4 To Charles Wheatstone [Collingwood]
Sends CW a specimen of JH's 'mercurial photographic process.' JH is pleased that W. H. Fox Talbot received the Rumford medal.
C 1p RS:HS 25.11.4

5322. 1842–12–5 To John William Draper Collingwood
Comments on JD's work in photography; JH goes on to talk about JH's ideas of 'chemical' rays of light.
ACCS 4pp RS:HS 25.6.16

5323. 1842–12–8 To Caroline Lucretia Herschel n.p.
Finished the reductions of all of the nebulae and double stars recorded at Cape Town; JH soon hopes to prepare for the publication of his *Cape Results.*
L *Memoir... of Caroline Herschel*, 328

5324. 1842–12–8 From Robert Hunt Falmouth
Was very pleased with the specimen JH sent him; it is still in good condition. Has he read [John William] Draper's paper? Comments on some of the points. Outlines some of his own proposed experiments.
ALS 4pp RS:HS 10.95

5325. [1842–12–9 or later] To Robert Hunt n.p.
Is anxious for some means to be developed to get a measurable quantity [preferably by weight] from the action of a beam of light on a surface. JH has been experimenting in photography with 'flouric' compounds.
C 2pp RS:HS 25.14.15

5326. 1842–12–10 From Richard Taylor Red Lion Court
Asks for advice on a number of scientific papers.
ALS 3pp RS:HS 17.339

5327. [1842–12–12?] To Robert Hunt n.p.
Read John William Draper's papers; although he believes that Draper's instruments are inconsistent, JH feels that they are still important because they are measurable. Decries [L. F.] Moser's skepticism of photography's value, calling it a 'blindfolding to some of the most interesting physical relations that have ever been discovered.'
ALS 4pp George Eastman House, 6.17

5328. 1842–12–15 To George Dollond Collingwood
Sends description and drawings of JH's actinoscope. Requests estimate for time and expense of constructing it. Questions about clock mechanism and prisms.
AAbs 2pp TxU:H/L-0137; Reel 1054

5329. 1842–12–15 From Robert Hunt Falmouth
Much indebted for his paper on the influence of the solar spectrum on vegetable colors. Has a series of experiments ready that he hopes will resolve some of his own doubts. Does not like the term J. W. Draper uses for the new rays.
ALS 3pp RS:HS 10.96

5330. 1842–12–19 From George Dollond St. Paul's Church Yard
Papers arrived safely. Will report GD's opinion on these soon.
ALS 1p TxU:H/M–0164; Reel 1087

5331. 1842–12–22 From Alfred Smee 7 Finsbury Circus
Thanks JH for a specimen of 'mercury type,' which AS describes as he examined it under a microscope.
ALS 3pp RS:HS 16.159

5332. 1842–12–22 From Richard Taylor Red Lion Court
Discusses matter touching *P.M.* and *Scientific Memoirs.*
ALS 1p RS:HS 17.340

5333. 1842–12–23 From George Dollond St. Paul's Church Yard
Sees too many difficulties in optics of JH's proposed telescope. Can build mechanical portion, but not by March as JH requires. Cannot give estimate, because design is too novel.
ALS 2pp TxU:H/M–0165; Reel 1087

5334. 1842–12–23 To [John Frederick Goddard?] Collingwood
JH mislaid the request for specimens of photography, but now sends them; describes those sent. Explains how to get a copy of JH's paper on photography.
ALS 3pp Manchester Central Library

5335. 1842–12–24 From Robert Hunt Falmouth
Sends a few samples of plates made by the cyanotype process; they are poor because the sun has not been very strong of late. Comments on this process and wishes JH would make a few experiments with his more perfect apparatus. Inclined to agree with him about the mechanism of the eye.
ALS 3pp RS:HS 10.97

5336. 1842–12–25 To George Dollond Collingwood
Regrets that GD cannot build JH's entire actinoscope. If GD can make at least mechanical components, JH will supply optics and clockwork. Can GD complete these [by end of March]? Funds are limited; please estimate cost.
AdraftS 4pp TxU:H/L-0138; Reel 1054

5337. 1842–12–25 From John McClelland Calcutta
Sending, care of Dr. Grant, the first part of a series of experiments on the magnetic influence of light, which appeared recently in the *Calcutta Journal*. Has been unable to obtain a characteristic specimen of Kankar in Bengal. Dr. Hugh Falconer is an expert on the Kankar in Bengal and he will try to obtain information from Falconer.
ALS 3pp RS:HS 12.6

5338. 1842–12–27 From George Dollond St. Paul's Church Yard
Will begin building components that JH identified as most immediately needed. Confirm scale of JH's drawings. Will send samples of prism glass for JH to inspect. Cannot yet estimate cost.
ALS 3pp TxU:H/M–0166; Reel 1087

5339. 1842–12–27 From William Whewell Trinity Lodge
Comments on JH's translation of Friedrich Schiller's poem 'The Walk.' Invites the Herschels to Trinity Lodge, then under repair. Mentions James South's odd behavior.
ALS 4pp RS:HS 18.200

5340. 1842–12–29 To George Dollond [Collingwood]
Confirms scale of drawings [for JH's actinoscope]. Has found simple contrivance to replace clockwork. Will order lenses and prisms from Munich; all English glass has color aberrations.
ACCS 2pp TxU:H/L-0139; Reel 1054

5341. 1842–12–29 From Samuel Rogers St. James's Place
Thanks JH for the poem [JH's translation of Friedrich Schiller's 'The Walk'] he sent. Requests a song sung by 'faery elves' in William Herschel's telescope.
ALS 2pp RS:HS 14.418

5342. 1842–12–30 From Joanna Baillie Hampstead
Thanks for JH's translation of Frederick Schiller's poem.

Hopes that they had a pleasant Christmas. Family news.
ALS 4pp RS:HS 3.21

5343. 1842–12–30 From Augustus De Morgan 69 Gower St.
Thanks for his reductions of Schiller's observations. Many people believe the claims made in the 'Moon Hoax'. Is writing a book on the use of the globes.
ALS 4pp RS:HS 6.188 S. De Morgan, *Memoir of Augustus De Morgan*, 147–8 inc

5344. 1842–12–30 From William Rowan Hamilton Observatory
Thanks JH for JH's translation of Friederich Schiller's poem 'The Walk.'
L Graves, *Hamilton, 2*, 399 inc

5345. 1843–1–1 To [William Radcliffe Birt] n.p.
Reduction of barometer curves is nearly complete. Must have all papers within two weeks to prepare for B.A.A.S. meeting.
ALS 2pp RS:HS 19.99

5346. [1843 to 1850 ?] To [William Henry Fitton] [Collingwood]
Comments on the intention of a friend of WF to write a work on geography. JH's son William was ill and at home and had commented on WF's son.
ALS 2pp inc WT 62824.2

5347. 1843 To Margaret Brodie Herschel [Manchester]
Is visiting Dr. [Richard?] Hobson, where JH has met some European scientists, such as F. W. Bessel and G. A. Erman; JH anticipates they will come to visit at Collingwood.
ALS 2pp JHS 3.32

5348. [1843] To Francis Hodgson n.p.
Returns with many thanks Edward Craven Hawtrey's translations from Homer and Kallinos. Both are beautifully done. Comments on these and the meters suitable for English ears.
draft 3pp RS:HS 10.197 & 22.173

5349. 1843 From John David Roberton Royal Society
Asks JH to report on whether and why JH thinks J. D. Forbes's paper on transparency of atmosphere and solar range worthy of Royal Medal.
ALS 2pp (encl?) RS MM.16.152

5350. 1843 or later From Richard Sheepshanks Athenaeum
About organization of the executive of the R.A.S.
ALS 3pp RS:HS 16.134

5351. 1843–1–2 From Francis Baily 37 Tavistock Place
Encloses a publication. Thanks for the gift to Miss Elizabeth Baily.
ALS 1p RS:HS 3.233

5352. 1843–1–2 From G. A. Rovell Oxford
Sends excerpt of letter from Mr. Weekes [?] stating an hypothesis concerning electricity and clouds. Asks JH to test it. Sends own ideas about electricity as well.
ALS 3pp RS:HS 14.503

5353. 1843–1–3 From George Dollond St. Paul's Church Yard
Suggests modifying square base of JH's telescope to round base. Estimates cost at £80 to £100.
ALS 3pp TxU:H/M–0167; Reel 1087

5354. 1843–1–4 To George Dollond Collingwood
Approves GD's modification of driving wheel [for JH's actinoscope]. But GD's cost estimate exceeds total R.S.L. grant of £100. JH cannot approve more than £70; must have funds left for optics and clockwork mechanism.
C 2pp † TxU:H/L-0140; Reel 1054

5355. 1843–1–4 From Thomas I. M. Forster Bruges
Has taken a house in Bruges. Comments on some of his own recent work on stars. Strange reports in French newspapers of James South and his instruments. Does he know a cure for double vision? Sends drawing of a colored meteor. Has another paper by L. A. J. Quetelet on falling stars with which he disagrees.
ALS 4pp RS:HS 7.332

5356. [1843]–1–4 From Henry Holland Brook St.
Grateful for JH's translation of J. C. F. Schiller's 'The Walk.' Comments on several points in the translation.
ALS 4pp RS:HS 9.442

5357. 1843–1–6 From George Dollond St. Paul's Church Yard
Refuses to risk GD's funds by giving cost estimate for unusual telescope, especially if R.S.L. expects GD to pay for any expenses over estimate. Regretfully stops manufacture of telescope.
ALS 2pp TxU:H/M–0168; Reel 1087

5358. 1843–1–6 From Samuel Rogers St. James's Place
Thanks JH for sending the verses he requested. His sister wishes she were still a neighbor of the Herschels.
ALS 2pp RS:HS 14.419

5359. 1843–1–7 From John Richard to Margaret Brodie Herschel Datchet
Everyone loves JH's translation of [F. Schiller's] 'The Walk.' Discusses Shakespeare, literature, JH's literary endeavors, and translations.
ALS 3pp RS:HS 14.331

5360. 1843–1–9 From George Biddell Airy Royal Observatory, Greenwich
Sending copy of his 'Tides and Waves' for *Encyclopaedia Metropolitana*; would be glad of JH's comments. Also encloses GA's eclipse paper.
ALS 3pp RS:HS 1.115 (C: RGO 6.368.657 & LACC: RGO 6.503.244)

5361. 1843–1–9 To George Biddell Airy Collingwood
Needs a good artisan to construct the framework of a solar spectrum photographic apparatus.
ALS 4pp RGO 6.462.37

5362. 1843–1–9 To [?] Collingwood
Sorry to have caused trouble. Please send JH's 'description and sketches' to Peter Stewart at 65 Cornhill.
AdraftS 1p TxU:H/L-0430; Reel 1055

5363. 1843–1–10 From George Biddell Airy Royal Observatory, Greenwich
Answer to one of JH's on 9 January [apparently on equatorial telescopes].
ALS 3pp RS:HS 1.116 (C: RGO 6.462.39)

5364. 1843–1–17 To Richard Jones Collingwood
Looks forward to the possibility of a visit from RJ.
C 1p inc RS:HS 22.145

5365. 1843–1–18 From George Biddell Airy Royal Observatory, Greenwich

Regarding William Cubitt (Engineer of the Dover Railway) and the proposed blasting down of the cliffs between Dover and Folkestone.
ALS 2pp RS:HS 1.117 (C: RGO 6.368.659)

5366. 1843–1–19 To George Biddell Airy n.p.
Intends to accept offer [see GA's 1843–1–18]; JH questions the correctness of P. S. Laplace's theory of capillary action.
ALS 4pp RGO 6.368.660

5367. 1843–1–19 To [William Radcliffe Birt] Collingwood
WB's packet arrived safely today.
ACCS 2pp RS:HS 19.100

5368. 1843–1–19 To James David Forbes n.p.
Protesting about the refusal of the Royal Society of Edinburgh to publish Thomas Brisbane's magnetic observations; hopes it will be reconsidered. Thinks JF's theory on glaciers has good points, but comments on some of its shortcomings.
C 2pp inc RS:HS 7.304 & 22.146

5369. 1843–1–20 To William Radcliffe Birt Collingwood
Ask Edward Sabine for authority to insert additional notes [on barometric observations] into report already at printer's. Does not advise publishing observations. Asked Howard Elphinstone to send observations to WB.
ALS 3pp RS:HS 19.101

5370. 1843–1–20 From H. Cubitt 6 Gt. George St.
Is planning to blow down part of the cliffs near Dover for the railway and gives details of how he plans to do it.
ALS 7pp RS:HS 5.380

5371. 1843–1–23 From George Biddell Airy Royal Observatory, Greenwich
Regarding projected committee for the construction of new Standards for weights and measures. Any observations or further instructions for captain visiting the China coast?
ALS 2pp RS:HS 1.119 (C: RGO 6.340.345)

5372. 1843–1–23 From George Biddell Airy Royal Observatory, Greenwich
Further news of his proposed visit to Dover to see the blasting of the cliffs?
ALS 1p RS:HS 1.118

5373. 1843–1–23 From H. Cubitt London
Details of arrangements for reaching Dover in preparation for the blasting of the cliffs.
ALS 4pp RS:HS 5.381

5374. 1843–1–23 From James David Forbes Edinburgh
Is grateful for his letter and photographs. Will send him his paper on glaciers, which he is pleased to see interests him. Thomas Brisbane must be in error about the refusal of the Royal Society of Edinburgh to print his magnetic observations.
ALS 2pp RS:HS 7.305

5375. 1843–1–23 From William O'Shaughnessy Brooke 14 Bernan St.
Has now heard that the explosion at Dover is to take place on Thursday. Hopes to see JH there. If not, will call on him at Hawkhurst.
ALS 2pp RS:HS 13.184

5376. 1843–1–25 From Herbert Mayo 56 Wimpole St.
Would like JH to peruse HM's work and publicize its considerable use to Robert Graham.
ALS 4pp RS:HS 13.31

5377. 1843–1–25 From Edward Sabine Woolwich
Sends letter of [Johann] Lamont, whose suggestions have been adopted by the Physical Committee. Discusses size of instruments, particularly needles.
ALS 4pp RS:HS 15.149

5378. 1843–1–26 From Edward Sabine Woolwich
Discusses plans for printing of 4 maps of terrestrial isotherms prepared by [H. W.] Dove along with Dove's commentary.
ALS 4pp RS:HS 15:150

5379. 1843–1–30 From John Richard Datchet
Thanks JH for poem [F. Schiller's 'The Walk']. Sends greetings to Lady Herschel.
ALS 3pp RS:HS 14.332

5380. 1843–1–31 To the *Athenaeum* Collingwood
Reports his observations on the explosion at Dover of 19,000 pounds of gunpowder.
L *Athenaeum,* #797 (Feb. 4, 1843), 111

5381. 1843–2 To John Stewart n.p.
Agrees to provide Josiah Quincy with extracts from letters by James Grahame. Highlights of Grahame's life.
C 3pp †(by E. S. Quincy) RS:HS 19.102

5382. 1843–2–2 From Richard Taylor Red Lion Court
Has 50 copies of JH's paper ['On the Action of the Rays of the Solar Spectrum on the Daguerreotype Plate'] for *P.M.* printed exclusively for JH. Asks opinion of papers for next issue of *Scientific Memoirs*.
ALS 2pp RS:HS 17.341

5383. 1843–2–4 From Robert Hunt Falmouth
Has been reading JH's communication to the *P.M.* [on solar spectrum] and wishes to point out that RH was the first to carry out those experiments, which are recounted in the *P.M.* for 1840.
ALS 3pp RS:HS 10.98

5384. 1843–2–6 From William O'Shaughnessy Brooke 73 Newman St.
Has ventured on the risks of an election to the R.S.L. and would like JH's support and influence with those on the Council. Was prevented from visiting Dover. Sails for Canada on 4 Mar.
ALS 4pp RS:HS 13.185

5385. 1843–2–12 From Charles Babbage Dorset St.
Regarding his brother-in-law's sale of property. Sorry he was unable to meet him at Dover recently. Mother in good health.
ALS 2pp RS:HS 2.301

5386. 1843–2–14 To Charles Babbage Collingwood
Regarding the deeds for CB's brother-in-law. Is sorry that CB could not see the blasting of the cliffs at Dover.
ALS 3pp inc RS:HS 2.302

5387. 1843–2–14 From Edward Sabine Woolwich
Sent JH's letter concerning actinometers to all observatories. [Johann] Lamont must be informed of recent work. ES comparing fluctuations of direction and force at Toronto and Van Diemen's Land. Mentions [G. B.] Airy and air currents.
ALS 4pp RS:HS 15.151

5388. 1843–2–17 From Friedrich Wilhelm Bessel Königsberg
Note of thanks for paper of JH's; replies with a printing of a

talk by FB about William Herschel.
ALS 1p (German) RS:HS 4.80

5389. 1843–2–17 To Robert Hunt Collingwood
Clarification of priority to the prismatic analysis of the Daguerreotype photograph; comments on the location of a limiting diaphragm in a camera obscura.
ALS 4pp George Eastman House, 15.17 (C: RS:HS 22.147)

5390. 1843–2–17 From William Henry Smyth 3 Cheyne Walk
Extract from observations of 70 Ophiuchi by F. W. Bessel. Results agree with WS's measurements by different method and increase WS's confidence. Hopes for more information on Gamma Virginis before finalizing WS's manuscript.
ALS 2pp, encl 2pp TxU:H/M–0491; Reel 1093

5391. 1843–2–20 From G. W. Müller Hanover
Reports on GM's health and immediate travel plans.
ALS 1p (German) RS:HS 13.57

5392. 1843–2–23 To [Robert Hunt?] Collingwood
Thanks for the 'exquisite specimen of Daguerreotype.' Speculates on the possibility of making Daguerreotype portraits small enough to be set in rings or in shirt pins.
ALS 3pp Manchester Central Library

5393. 1843–2–24 To Francis Baily Collingwood
Thanks for his paper on the Earth. Wishes his own Cape work was finished and of the same standard as FB's work. Regarding support for Dr. W. B. O'Shaughnessy when his election to the R.S.L. is being considered.
ALS 4pp RS:HS 3.234 (C: RS:HS 25.8.49)

5394. 1843–2–24 From Christian Ludwig Gerling Marburg
A note accompanying some papers on the geodesic work carried out in Germany.
ALS 1p (German) RS:HS 9.96

5395. 1843–2–25 From Francis Baily 37 Tavistock Place
Regarding his own support for W. B. O'Shaughnessy. Star reform in the Southern hemisphere.
ALS 4pp RS:HS 3.236

5396. 1843–2–[25 or later] To Francis Baily n.p.
Will start work on the Southern Constellations without delay. Thanks for his remarks on W. B. O'Shaughnessy.
ALS 4pp RS:HS 3.235 (C: RS:HS 25.8.50)

5397. 1843–2–27 From Benjamin Silliman Boston, Mass.
Thanks JH for favorable comments on BS's biography of [Charles?] Mason. Has been given part of an unusual meteorite fallen in the Cape. If JH has a piece of it, hopes to obtain JH's opinion.
ALS 3pp RS:HS 17.205

5398. 1843–2–28 From Alfred Smee 7 Finsbury Circus
Describes experiments in the electrochemical decomposition of metallic salts.
ALS 2pp RS:HS 16.160

5399. [1843–3] To Francis Baily n.p.
Has sent note to the *Times* about the comet.
AL 3pp RS:HS 3.281

5400. 1843–3 To Francis Baily Collingwood
Describes his view of the Great Comet of 1843.
C 1p RS:HS 25.8.53

5401. 1843–3–1 From John Cruickshank Marischal College, Aberdeen
Announces election of Marquis of Breadalbane as lord rector of Marischal College for coming year. Thanks for JH's advice to college senate.
ALS 1p TxU:H/M–0154.1; Reel 1087

5402. 1843–3–4 From Francis Baily 37 Tavistock Place
Further regarding the method to be adopted for the printing of the star catalogue.
ALS 4pp RS:HS 3.237

5403. 1843–3–5 From R. W. Howard-Vyse Stoke
Is grateful for his note and the book; has been ill, which accounts for the delay in his acknowledgement. The verses appear to give a good idea of the original.
ALS 2pp RS:HS 10.10

5404. 1843–3–8 To Caroline Lucretia Herschel Collingwood
Wishes CH a happy 93rd birthday. Occupied with remodelling the Southern Constellations, which 'are all in confusion;' admits that not everyone will be satisfied with his constellation reforms.
ALS 4pp TxU:H/L–0588.4; Reel 1058

5405. 1843–3–9 From George Biddell Airy Royal Observatory, Greenwich
A new committee on standards is to be created to oversee the preparation and measuring of appropriate standard measures; seeks JH's opinion about asking Francis Baily and W. H. Miller to do the accurate weighing and measuring.
ALS 2pp TxU:H/M-0629.1; Reel 1084 (C: RGO 6.340.346)

5406. 1843–3–9 To George Biddell Airy Collingwood
Would be pleased if Francis Baily and W. H. Miller would agree to serve [see GA's 1843–3–9]; believes sufficient remuneration should be provided for these responsibilities, and comments on the desirability of a continuing standards committee.
ALS 4pp RGO 6.340.348

5407. 1843–3–9 From Charles Piazzi Smyth Royal Observatory, Cape of Good Hope
Sends JH observations of Comet of 1843; disappointed with Cape Observatory's records of the comet. Oversees Observatory in absence of Thomas Maclear, who is at Zwartland for meridian survey. Lists and describes instruments in use at Cape Observatory. Discusses sources of error.
ALS 6pp RS:HS 16.225

5408. 1843–3–9 To [R.S.L. Secretary] Collingwood
Requests on behalf of Richard Taylor, editor of *Philosophical Magazine,* use of plate from *R.S.P.T.*
ALS 2pp RS MC.3.270

5409. 1843–3–11 From William Radcliffe Birt Upper Seymour St.
Has forwarded a parcel to JH, but wonders if he has received it as he has had no reply.
AL 1p inc RS:HS 4.101

5410. [1843–3–13?] To Francis Baily n.p.
Invitation to come to Collingwood to discuss star arrangements.
ALS 4pp RS:HS 3.280

5411. 1843–3–13 From Alfred Smee 7 Finsbury Circus
Further on the electrochemical decomposition of metallic salts [see AS's 1843–2–28], and its relationship to photography.
ALS 3pp RS:HS 16.161

5412. 1843–3–15 To Francis Baily Collingwood
Further regarding the constellations.
ALS 4pp RS:HS 3.238 (C: RS:HS 25.8.51)

5413. 1843–3–16 From Adalbert Adolf Mühry Hanover
About the health of Caroline Herschel, and other news from Hanover.
ALS 3pp (German) RS:HS 12.372

5414. 1843–3–17 From Francis Baily 37 Tavistock Place
Further regarding the method to be adopted for the printing of the star catalogue.
ALS 4pp RS:HS 3.239

5415. 1843–3–17 To Samuel Hunter Christie Collingwood
Reports observing on 17 March a long, thin cloudy streak that moved with the stars. [P.S. of 18 March concludes the object is a comet [Great Comet of 1843].]
L *Royal Society Proceedings, 4* (1843), 450–3

5416. 1843–3–18 To George Biddell Airy n.p.
Tells GA about a new comet becoming visible.
ALS 3pp RGO 6.92.173

5417. 1843–3–19 To the London *Times* Collingwood
Reports that a very prominent comet [Great Comet of 1843] is coming into view.
L London *Times* (March 21, 1843), p. 5, col. 3

5418. 1843–3–20 From Francis Baily 37 Tavistock Place
Further regarding the constellations.
ALS 4pp RS:HS 3.240

5419. 1843–3–21 To the *Morning Herald* Collingwood
Reports where to observe the comet [Great Comet of 1843] just becoming visible.
L *Morning Herald* (March 24, 1843), p. 5, col. 3

5420. 1843–3–21 To the London *Times* Collingwood
Gives latest observations of the comet [Great Comet of 1843]. Predicts head will soon be visible.
L London *Times* (March 23, 1843), p. 6, col. 2

5421. 1843–3–22 From Francis Baily 37 Tavistock Place
Encloses a draft of a letter that it is proposed should be sent to Sir Robert Peel.
ALS 1p RS:HS 3.241

5422. 1843–3–22 From William Henry Fox Talbot Lacock Abbey
JH's account of a comet reminds WT of a strange sighting many years ago. Comments further on improvements in photography, both those made and those needed.
ALS 3pp RS:HS 17.313

5423. 1843–3–24 From George Biddell Airy Royal Observatory, Greenwich
Covering letter to one by Edward Cooper regarding appearance and observations on new comet.
ALS 1p RS:HS 1.120

5424. 1843–3–[23?] To Francis Baily Collingwood
Invites FB and his sister for a visit to Collingwood, where FB and JH can most easily complete the work of charting the southern stars.
C 2pp RS:HS 25.8.52

5425. 1843–3–23 From Few & Co. Covent Garden
Sending the deeds and the schedule. Please send notice of receipt.
ALS 2pp RS:HS 7.214

5426. 1843–3–24 From Robert Hunt Falmouth
Enlarging on the luminous phenomena he has seen in the S.W. and giving some comments from a naval officer who had observed it.
ALS 3pp RS:HS 10.100

5427. 1843–3–24 To the *Morning Herald* Collingwood
Reports where to observe the comet [Great Comet of 1843], its head having now become visible.
L *Morning Herald* (March 29, 1843), p. 6, col. 4

5428. 1843–3–24 From William Henry Fox Talbot Lacock Abbey
Describes observing of comet that evening.
ALS 2pp RS:HS 17.314

5429. 1843–3–25 From Francis Baily 37 Tavistock Place
Accepts his invitation to visit him in order to discuss various matters.
ALS 4pp RS:HS 3.242

5430. 1843–3–25 To Robert Hunt Collingwood
Some comments about photographic processes and about RH's observations of the heavens [see RH's 1843–3–24].
ALS 3pp George Eastman House, 13.17 (C: RS:HS 22.150)

5431. 1843–3–25 To William Henry Fox Talbot Collingwood
Comments on confusion between observing comet and zodiacal light. Extensive comments on various materials used in a photographic process where negatives change to positives over long time.
ALS 4pp Science Museum Talbt 1/35 (C: RS:HS 22.149)

5432. 1843–3–26 From E. Willes Lausanne [Switzerland]
Newspaper noted Caroline Herschel's 93rd birthday. Requests copy of William Herschel's portrait. Gives address of London relative, Charles Stonhouse, who can copy it. Would JH sit for portrait there also? What did JH decide about 'our comet'?
ALS 4pp RS:HS 18.261

5433. 1843–3–27 From D. François J. Arago Paris
Official acknowledgement of JH's work on photographic reproduction.
ALS 1p (French) RS:HS 1.363

5434. 1843–3–29 From William Lassell Starfield, Liverpool
Sending a letter he has received from a friend regarding a curious observation of the moon. Has seen the comet and encloses diagrams of its position.
ALS 3pp RS:HS 11.126

5435. 1843–3–29 From William Henry Fox Talbot Lacock Abbey
Suggests 'amphitype' as name for JH's process whereby same picture can go from negative to positive. Further comments on comet sighting, and on some improvements to Daguerreotype.
ALS 3pp RS:HS 17.315

5436. 1843–3–31 To William Henry Fox Talbot n.p.
Zodiacal light is a real phenomenon. Further comments on negative to positive process. Questions whether WT can define one cometary orbit with two observations.
ALS 5pp Science Museum Talbt 1/36 (C: RS:HS 22.151)

5437. 1843–3–31 To the London *Times* Collingwood
Argues against the view that what JH has taken to be the tail of a comet [Great Comet of 1843] is actually due to the

zodiacal light.
L London *Times* (April 4, 1843), p. 6, col. 1

5438. [1843-4] To Francis Baily [Collingwood]
Sends along Frederik Kaiser's observations of Halley's comet; [Robert] Everest is upset with JH because JH had signed a letter of support for a grant for T. B. Jervis.
C 2pp RS:HS 25.8.62

5439. 1843–4–1 From Francis Baily 37 Tavistock Place
Has copied out JH's map showing the constellations as named by JH. Is moving to Richmond for a few weeks. Gives a message from H. C. Schumacher.
ALS 3pp RS:HS 3.243

5440. 1843–4 From Caroline Lucretia Herschel Hanover
Describes the visit of the Crown Prince and Princess [of Hanover?] on her 93rd birthday (16 March).
L *Memoir ... of Caroline Herschel,* 330–1 inc

5441. 1843–4–2 From Joseph Hartnell Elfords
Has just received the news that Hartnell's son has passed the exams at Trinity College, Cambridge.
ALS 2pp RS:HS 10.167

5442. 1843–4–5 From Thomas Henderson Edinburgh
Sending readings from his observations of the comet.
AL 2pp RS:HS 9.303

5443. 1843–4–5 From William Henry Fox Talbot Lacock Abbey
Comments on JH's positive calotype. Suggests comet may be return of (J.-D.) Cassini's comet of 1668. Gives theorem for finding parabolic orbit of comet if two observations are made in the ecliptic.
ALS 2pp RS:HS 17.316

5444. 1843–4–7 From James Kemplay Kings Bench Walk
Sends a paper of his brother, Christopher Kemplay, on a theory respecting comets. Has already submitted it to James Challis, who objects to some points. Would be glad of JH's comments.
ALS 4pp RS:HS 11.78

5445. 1843–4–8 From Octavian Blewitt 73 Great Russell St.
Invites JH to 10 May anniversary dinner of Literary Fund Society.
ALS 1p RAS:JH Archive 3/8.13; Reel 7

5446. 1843–4–8 From William Whewell Trinity Lodge
Asks JH about a comet JH had been observing. Notes the publication of J. S. Mill's *System of Logic,* describing it as closer to JH's than to WW's position.
ALS 4pp RS:HS 18.201 Todhunter, *Whewell, 2,* 315

5447. 1843–4–12 To Caroline Lucretia Herschel [Collingwood]
Concerning the great comet of 1843, JH remarks that both Thomas Henderson and Johann Encke noted that the comet bounced off the sun's atmosphere.
ALS 1p TxU:H/L–0588.8; Reel 1058

5448. 1843–4–13 From H. Cubitt 6 Gt. George St.
There is to be another cliff-blasting. Gives details of travel arrangements.
ALS 3pp RS:HS 5.382

5449. 1843–4–15 From René J. H. Dutrochet Paris
Has received the two memoires on photographic processes and rays of the solar spectrum. Presumes JH has now received his own work on Physical researches. Comments on various electrical experiments.
ALS 3pp RS:HS 6.519

5450. 1843–4–21 From Thomas Henderson Edinburgh
Regarding the recent comet, believes it to be the one of 1668. Does not feel competent to comment on the proposal to change the names and boundaries of the constellations.
ALS 4pp RS:HS 9.304

5451. 1843–4–21 To Thomas Andrews Collingwood
Thanks TA for papers on heat of acid-base reactions.
C 1p RS:HS 22.152

5452. 1843–4–22 From George Biddell Airy Royal Observatory, Greenwich
Offers JH information about, and maps of, Switzerland, as GA understands JH is going there.
ALS 2pp RS:HS 1.121 (ACCS: RGO 6.368.662)

5453. 1843–4–23 From Robert Hunt Falmouth
Sends some specimens of photographs he has made by using bichromate of potash and an acid solution of nitrate of mercury. Gives details of a luminous phenomena he has seen in the S.W. and would like an explanation of it from JH.
ALS 3pp RS:HS 10.99

5454. 1843–4–24 From James Prescott Joule New Baily St., Salford, Manchester
Refers JH to page 18 of memoir of Manchester Literary and Philosophical Society, in which JJ, using 'the magnetic electrical machine,' succeeded in proving 'that heat may be generated by mechanical power without accompanying chemical change.'
ALS 1p RAS:JH Archive 3/7.13; Reel 7

5455. 1843–4–25 From Francis Baily 37 Tavistock Place
Gives his proposal for the amendment of nomenclature for the northern constellations. F. W. A. Argelander's new celestial atlas. Regarding JH's southern map. The letter from the B.A.A.S. has been sent to Sir Robert Peel.
ALS 4pp RS:HS 3.244

5456. 1843–4–26 From Alfred Gautier Geneva
Introducing Professor Emile Plantamour, who is visiting England. Read JH's article on the rays of the solar spectrum with pleasure. Sends some articles by Plantamour from recent journals.
ALS 2pp (French) RS:HS 8.87

5457. 1843–4–26 From Antoine C. G. Jobert 13 Portland Crescent, Leeds
Met JH when he was residing in the Auvergnes. Since then he has had to leave France and now resides in England, where he is making a living teaching French and writing books. Would be grateful if he could add JH's name as a subscriber to a forthcoming work.
ALS 7pp RS:HS 10.329

5458. 1843–4–27 To William Hopkins Collingwood
Thanks WH for his drawing of the comet, which JH will communicate to the R.A.S.
ALS 3pp WT B.6

5459. 1843–4–29 From Edward Joshua Cooper Paris
Unable to write before as he was packing for his journey home. Regrets none of his northern friends had seen the comet. Its orbit has been calculated as 35.3 years. Remarks on the tail of the comet.
ALS 8pp RS:HS 5.283

5460. 1843–5–1 From John Stuart Mill India House
Acknowledges with pleasure JH's kind note. Owes much to JH's book in the chapter on experimental methods. JH will doubtless find Auguste Comte's book worth reading. Would like a note of any errors in JM's book.
ALS 4pp RS:HS 12.328

5461. 1843–5–4 From Caroline Lucretia Herschel Hanover
Urges JH to study a temporary index made by William Herschel that contains observations about types of zodiacal light.
L Memorandum *Memoir ... of Caroline Herschel,* 331 inc

5462. 1843–5–5 From M. P. Edgeworth 1 North Audley St.
Has been able to obtain some deodar seeds from Ireland. Sends them, together with some seeds of the Himalayan cypress. Hopes they will prove a success.
ALS 2pp RS:HS 7.120

5463. 1843–5–5 From Antoine C. G. Jobert 13 Portland Crescent, Leeds
Wishes JH had added his name at the head of the list of subscribers. Thinks he is being too cautious. AJ is responsible for the whole of the work though it appears under two authors.
ALS 3pp RS:HS 10.330

5464. 1843–5–7 To the *Athenaeum* Collingwood
Describes a particularly brilliant aurora, which JH observed from Collingwood on the night of 6 May.
L *Athenaeum,* #811 (May 11, 1843), 465

5465. 1843–5–8 From Christian F. Schönbein Basle
Thanks JH for earlier letter. Discusses JH's work on chemical action of light, with which he is familiar. Sends some papers for JH's comments.
ALS 3pp RS:HS 15.355

5466. 1843–5–9 To Francis Baily Collingwood
Has received observations of the comet from C. P. Smyth, which he may like to communicate to the R.A.S. Comments on the deplorable state of the Cape Town [?] Observatory
ALS 4pp RS:HS 3.245 (C: RS:HS 25.8.54)

5467. 1843–5–9 From William Henry Smyth 3 Cheyne Walk, Chelsea
Sends comet observations from W. S. Jacob. Obtains 'another excellent epoch of Gamma Virginis.'
ALS 3pp RS:HS 16.268

5468. 1843–5–10 From George Biddell Airy Royal Observatory, Greenwich
Regarding Captain [Charles ?] Gray's letter. Robert Peel has written asking Francis Baily and W. H. Miller to do the 'Standards.'
ALS 1p RS:HS 1.122 (C: RGO 6.340.347)

5469. 1843–5–10 or earlier To George Biddell Airy n.p.
Concerned that the Colonial governors were not provided with copies of the Standards Commission Report; will not be able to get to Switzerland this year.
ALS 2pp RGO 6.340.351

5470. 1843–5–10 To Francis Baily Collingwood
Regarding FB's paper on the revision of the boundaries of the constellations.
C 10pp RS:HS 3.245bis

5471. 1843–5–13 From John Stevens Henslow Cambridge
Recommends a method for destroying wasps' nests. Refers him to the *Gardener's Chronicle* for information on flowering Cape bulbs. Has no idea where he can obtain deodar seeds. Would like to see his paper on the influence of light on vegetable colors.
ALS 5pp RS:HS 9.309

5472. 1843–5–15 From William Henry Smyth 3 Cheyne Walk, Chelsea
Francis Beaufort to publish comet information; also wants JH's opinion as to what data should be sent to Cape Observatory. James South will lecture at Royal Institution. South's requests for several historical astronomical instruments refused.
ALS 3pp RS:HS 16.269

5473. 1843–5–17 From Francis Baily 37 Tavistock Place
Regarding JH's answer to his own paper on the revision of the boundaries of the constellations.
ALS 3pp RS:HS 3.246

5474. 1843–5–18 From S. B. P. Micklethwaite 48 Albemarle St.
Hopes that JH will maintain the road from the canals to Hawkhurst as SM possesses a right of way over it.
ALS 1p RS:HS 12.327

5475. 1843–5–21 To Francis Baily Collingwood
Will be at the Visitation and intends to spend a few days in town. Has got the catalogue of objects for the star maps.
ALS 3pp RS:HS 3.247 (C: RS:HS 25.8.55)

5476. 1843–5–21 To Francis Beaufort Collingwood
His views on the imperfections and lack of suitable instruments at the Cape Observatory.
draft 4pp RS:HS 3.374bis & 22.153

5477. 1843–5–22 From Francis Baily 37 Tavistock Place
Can JH come to the congress on Friday?
ALS 1p RS:HS 3.248

5478. 1843–5–23 From Francis Beaufort Admiralty
Thanks JH for his letters about the Cape equatorial instrument.
ACC 1p Hydrographic Office Letter Book

5479. 1843–5–26 From Francis Baily 37 Tavistock Place
Has invited some astronomical friends to the meeting on Friday.
ALS 1p RS:HS 3.249

5480. 1843–5–26 To Alexander John Ellis Collingwood
Encourages AE to consider translating some mathematical works; additional comments on mathematics in general.
C 1p RS:HS 22.154

5481. 1843–5–26 From Edward Sabine Woolwich
Received a stray dispatch, dated at sea, from [James] Ross. Assumes Ross is on his way home.
ALS 2pp RS:HS 15.152

5482. 1843–5–26 From Sherwood & Co. 23 Paternoster Row
A note accompanying a new book on etymology and requesting JH comment on it.
ALS 1p RS:HS 16.144

5483. 1843–5–27 From John Lindley 21 Regent St.
Has just heard that JH would like some deodar seeds; so he

encloses some. Could also supply some young plants if he would like them.
ALS 2pp RS:HS 11.228

5484. 1843–5–28 To John Lindley Collingwood
Thanks for the deodar seeds sent; accepts JL's offer of some young deodar plants. Needs instructions on planting these. Does JL have a copy of JH's 'On the Action of the Rays of the Solar Spectrum on Vegetable Colours, and on Some New Photographic Processes'?
ALS 3pp Royal Botanic Gardens

5485. 1843–5–29 To [Robert Hunt?] Collingwood
JH sends some examples of a unique kind of latent photograph, and adds other comments about various aspects of photography. JH has seen a good aurora. He now agrees with A. C. Becquerel's theory of the spectrum.
ALS 4pp WT 68037.1

5486. [1843–5–30] From John Lindley n.p.
Twelve deodar plants are being dispatched by coach. If JH plants the seeds now he should have nice young plants by the autumn. Would be obliged for his paper on vegetable photography.
ALS 2pp RS:HS 11.229

5487. 1843–5–31 From W. H. Hunt London Custom House
Sending some pages from the *Mechanics Magazine* containing details of clock escapements, superior to those at present in use.
ALS 1p RS:HS 10.218

5488. [1843–6] From William Keleher [Cork]
Proposed travel arrangements for JH.
ALS 1p inc Harvard: Houghton AAH 67m–67 (63)

5489. 1843–6–5 To George Biddell Airy n.p.
Asks GA on behalf of F. W. Bessel, for unreduced positions of a star that FB wishes to examine for annual parallax.
ALS 2pp RGO 6.368.663

5490. 1843–6–5 To Friedrich Wilhelm Bessel London
Thanks FB for some papers; raises questions relating to remodelling the constellations of the southern sky.
ALS 4pp AdWDDR Bessel 225 (C: RS:HS 22.155)

5491. 1843–6–5 From Thomas I. M. Forster Bruges
The colored meteors were too low on the horizon for JH to see them. Comments on the number of falling stars when comets approach the earth. Sends copies of his essay on comets and a book for children. Is working on various papers including one on 'Deafness in Balloons.' Is co-operating with L. A. J. Quetelet in a *Register of Periodical Phenomena.*
ALS 4pp RS:HS 7.333

5492. 1843–6–5 From Richard Taylor Keswick
Decides not to print G. Reichenbach's papers on JH's advice. Will consider Reichenbach's papers in the future.
ALS 3pp RS:HS 17.342

5493. 1843–6–6 From George Biddell Airy Royal Observatory, Greenwich
Regarding H. C. Schumacher and the Stephen Groombridge observations.
ALS 2pp RS:HS 1.123 (C: RGO 6.368.664)

5494. 1843–6–6 From William Henry Smyth 3 Cheyne Walk, Chelsea
Requests copy of JH's observations made in 1811. Asks latest opinion of JH on Gamma Virginis before the publication of his *Cycle of Celestial Objects for the Use of Naval, Military, and Private Astronomers.*
ALS 4pp RS:HS 16.270

5495. 1843–6–7 From Robert Hunt Falmouth
Is grateful for the very interesting photograph; will keep the preparation secret if he discovers it. Has been trying many experiments with benzoates with curious results, but bad weather has stopped progress so has now turned to thermography to test the truth of Mr. Prater's conclusions. Comments on these.
ALS 3pp RS:HS 10.101

5496. 1843–6–9 From William Keleher Cork
Invitation from the Local Council to visit Cork for the B.A.A.S. meeting.
ALS 1p RS:HS 11.75

5497. 1843–6–10 To George Biddell Airy Slough
Apologizes for confusion about F. W. Bessel's request [see GA's 1843–6–6], but JH had neglected to check date on FB's letter.
ALS 3pp RGO 6.368.665

5498. 1843–6–10 From Antoine Lombardi Modena
President of the Società Italienne des Sciences has received the invitation for members to attend the B.A.A.S. meeting at Cork. Regrets the length of the journey prevents members attending, but have nominated Michael Faraday to attend.
ALS 2pp (French) RS:HS 11.488

5499. 1843–6–10 To William Henry Smyth Slough
Will review all Gamma Virginis observations before revising orbit. Catalogue of major nebulae complete; will work on minor nebulae next.
C 1p RS:HS 22.156

5500. 1843–6–12 From George Biddell Airy Royal Observatory, Greenwich
Regarding a letter from F. W. Bessel. On holiday on the Isle of Wight.
ALS 2pp RS:HS 1.124 (C: RGO 6.368.666)

5501. 1843–6–13 From F. M. Eardley-Wilmot Blackheath
Is on the point of returning to the Cape and will be pleased to deliver anything to Thomas Maclear. Has borrowed three chronometers from the Admiralty for the magnetic survey. His stay was too short for him to visit JH.
ALS 3pp RS:HS 7.3

5502. [1843–6–15] To Caroline Lucretia Herschel London
Completed 'one of the most curious products' of JH's *Cape Results*: the map and catalog of the 'Nubecula major' region.
ALS 2pp TxU:H/L–0588.10.b; Reel 1058

5503. 1843–6–16 To Sherwood & Co. Collingwood
Comments on book on etymology forwarded by Sherwood & Co. [see 1843–5–26].
draft 4pp RS:HS 16.145 (C: 22,158)

5504. 1843–6–16 To William Henry Smyth Collingwood
Sends JH's 1817 observations; cannot locate any he made in 1811. Tells WS Gamma Virginis measurements will take more time.
C 2pp RS:HS 22.157

5505. 1843–6–19 From Arthur Swain Howard Lowe Nottingham
Sending details of a singular appearance in the heavens on 16

June.
ALS 2pp RS:HS 11.490

5506. 1843–6–20 From Friedrich Wilhelm Bessel Königsberg
Answers questions raised in JH's 1843–6–5; comments on other astronomical matters.
ALS 4pp (German) RS:HS 4.81

5507. 1843–6–21 To [Spencer Compton (Marquis of Northampton)] Collingwood
Discusses deficiencies in astronomical equipment at Cape's observatory, and troubles in observing comets, double stars, and other phenomena. Hopes equipment can be updated.
ALS 3pp RS MC.3.291 ((A)CCS: TxU:H/L-0285; Reel 1054)

5508. 1843–6–22 From George Clerk Treasury Chambers
Commissioners of Treasury invite JH to join committee to oversee construction of 'New Parliamentary Standards of Weight and Measure.'
ALS 1p TxU:H/M–0924; Reel 1083

5509. 1843–6–22 From F. M. Eardley-Wilmot n.p.
Has received the two parcels of books. Regarding imperfections in the actinometer.
ALS 2pp RS:HS 7.4

5510. 1843–6–22 To William Keleher Collingwood
Regrets he will be unable to visit Cork for the B.A.A.S. meeting, but is conscious of the honor of the invitation.
C 2pp RS:HS 11.76 & 22.159

5511. 1843–6–23 From Few & Co. Covent Garden
Legal matters concerning the lease.
ALS 2pp RS:HS 7.206

5512. 1843–6–23 From Sherwood & Co. London
Thanks for JH's 1843–6–16.
ALS 1p RS:HS 16.146

5513. 1843–6–23 From William Henry Smyth 3 Cheyne Walk
JH's theory of orbits of Castor and Gamma Virginis. Observations of Gamma Virginis by W. R. Dawes and by WS.
ALS 3pp TxU:H/M–0492; Reel 1093

5514. 1843–6–24 From George Biddell Airy Royal Observatory, Greenwich
Regarding a meeting of the new 'Standard' committee.
ALS 1p RS:HS 1.125 (C: RGO 6.340.350)

5515. 1843–6–24 From Few & Co. Covent Garden
Have received his draft and done as instructed.
ALS 1p RS:HS 7.207

5516. 1843–6–24 From Alexander Henning 10 Russell Square
A friend has inquired when Pisces became the sign of the Vernal Equinox; can JH help him? Regarding the comet they observed on their passage home.
ALS 4pp RS:HS 9.305

5517. 1843–6–26 To George Biddell Airy Collingwood
Is expecting to be summoned to Slough, about repossessing JH's [and William Herschel's] old premises; this will affect JH's availability for a meeting [see GA's 1843–6–24].
ALS 1p RGO 6.340.352

5518. 1843–6–26 From Few & Co. Covent Garden
Have now completed the surrender on the lease and enclose the counterpart.
ALS 1p RS:HS 7.208

5519. 1843–6–28 From Francis Baily 37 Tavistock Place
Has obtained maps from the Diffusion Society and marked in the new stars and the constellations. Has arranged to send some to him.
ALS 3pp RS:HS 3.250

5520. 1843–6–29 From George Biddell Airy Royal Observatory, Greenwich
Agenda items for next meeting of 'Committee for Superintending the Construction of Standards.' Compares English, French, Danish, and Prussian standards.
C 3pp †(by JH) TxU:H/M–0659; Reel 1084

5521. 1843–6–29 From George Biddell Airy Royal Observatory, Greenwich
A notice of meeting of the Standards Committee.
LS 1p TxU:H/M-0629.2; Reel 1084

5522. 1843–6–29 To William Henry Smyth [Collingwood]
Reaches conclusion about Gamma Virginis. States in regard to shape of orbit, 'We are all wrong.' Believes the orbit is less than 150 years. Claims that many errors exist in British measurements made between 1829 and 1834.
C 1p RS:HS 22.160

5523. 1843–[6]–29 To Peter Stewart Collingwood
Encloses bill, which JH signed, received from J. C. Stewart. JH and family visited Dover with G. B. Airy's family, where JH witnessed noiseless explosion of 18,000 pounds of powder distributed over 18 acres.
ALS 3pp TxU:H/L-0396; Reel 1054

5524. 1843–7–2 From James Heywood Ryde, I.o.W.
Sending a copy of the *English Universities*. Has quoted JH in the third volume. Hopes that university reform will be a popular topic.
ALS 3pp RS:HS 10.187

5525. 1843–7–2 From William Henry Smyth 3 Cheyne Walk, Chelsea
Agrees with JH opinions of Gamma Virginis orbit. Sends JH a drawing by C. P. Smyth of Halley's Comet.
ALS 3pp RS:HS 16.271

5526. 1843–7–4 To Francis Baily Collingwood
Has received the maps. Gives F. W. Bessel's views on the re-arrangement of the constellations which has been proposed. Comments on FB's memorandum.
ALS 4pp RS:HS 3.251 (C: RS:HS 25.8.56)

5527. 1843–7–6 From Robert Hunt Falmouth
Enclosing the specimens of photographs by different methods; comments on the methods used. Encloses a paper by Were Fox.
ALS 3pp RS:HS 10.102

5528. 1843–7–9 From William Parsons (Lord Rosse) The Castle, Parsonstown
Should JH be attending the B.A.A.S. meeting at Cork, hopes he will visit WP and spend a few days with him. Gives news of the progress of his great telescope.
ALS 4pp RS:HS 13.227

5529. 1843–7–9 To William Henry Smyth Collingwood
Sends calculations of Gamma Virginis and determines its eccentricity. Believes the orbit is highly elongated, like that

of a comet.
C 3pp RS:HS 22.161

5530. 1843–7–11 From William Henry Smyth 3 Cheyne Walk, Chelsea
Pleased with JH's Gamma Virginis results. Mr. Visconti made head of Military Geographical Institute at Naples.
ALS 3pp RS:HS 16.272

5531. 1843–7–13 From George Biddell Airy Royal Observatory, Greenwich
Provides JH with the measurements from a series of observations of Gamma Virginis.
ALS 2pp † TxU:H/M-0629.4; Reel 1084 (C: RGO 6.368.667)

5532. 1843–7–13 From James William Grant London
Had to leave Calcutta owing to his health and was entrusted with a packet for JH from a Mr. McClelland, and has only now succeeded in tracking down JH's new address. Will remain in England for a few years. Hopes that JH's Cape labors are progressing.
ALS 4pp RS:HS 9.21

5533. 1843–7–15 From Francis Hodgson Eton College
Is much obliged for letter and pretty specimen of English hexameters. Is sad that there has been no success with the classical hexameter in our language. Comments on its use.
ALS 5pp RS:HS 10.198

5534. [1843]–7–15 From Edward Sabine Woolwich
Will send JH's report to JH in August. J. C. Ross will not be at Cork meeting, but JH may announce there that first year of Ross's observations at sea will appear in R.S.P.T.
ALS 1p TxU:H/M–0517.6; Reel 1093

5535. 1843–7–16 From James A. Gordon Naish, Bristol
If not already invited, he would be very welcome to be their guest at Bristol when the new iron screw ship is named by the Prince Consort. Outlines method of reaching Bristol.
ALS 3pp RS:HS 8.170

5536. 1843–7–17 To William Parsons (Lord Rosse) Collingwood
Declines WP's invitation to attend Cork B.A.A.S. meeting and afterwards to visit WP and to see his reflecting telescopes. Informs WP that [C. A. von] Steinheil had written JH about a process of 'Electro-gilding' mirrors for telescopes.
ALS 3pp damaged Rosse Papers K2.1

5537. 1843–7–17 From John Phillips B.A.A.S., Cork Library, South Mall, Cork
The committee report on simultaneous magnetical and meteorological observations will be read at the B.A.A.S. meeting in Cork on 17 Aug.
(A)L 1p TxU:H/M–0428.15; Reel 1093

5538. 1843–7–18 To George Biddell Airy Collingwood
Finds GA's measurements of Gamma Virginis far away from JH's own, and instructs GA in the best way to measure double star positions.
ALS 4pp RGO 6.193.203

5539. 1843–7–18 To Edward Joshua Cooper Collingwood
Comments on calculation of orbit of Gamma Virginis; requests any observations EC has of that star.
C 2pp RS:HS 22.162

5540. 1843–7–18 To William Henry Smyth Collingwood
Devises and suggests mode of micrometric measurement of angles.
C 1p RE:HS 22.163

5541. 1843–7–18 To [?] Collingwood
Refers to a variety of salts and their reaction to light in the production of photographs. JH hopes that [?] still intends to publish his work in this area.
ALS 4pp DMC.695A.885.28

5542. 1843–7–19 From James William Grant London
Was gratified to receive JH's letter and intends to avail himself of the invitation to visit Hawkhurst. Will be visiting General [Bell]. Gives observations of the comet.
ALS 4pp RS:HS 9.22

5543. 1843–7–20 From George Biddell Airy Royal Observatory, Greenwich
Regarding the measurement of angles. Observations on a problem set by JH.
ALS 2pp RS:HS 1.126 (C: RGO 6.193.205)

5544. 1843–7–20 From William Radcliffe Birt Bethnal Green
Since he forwarded the parcel he has changed his address.
ALS 1p RS:HS 4.102

5545. 1843–7–20 To [William Radcliffe Birt] Collingwood
Received copies of barometer observations. Searching for original data (lists stations) that JH sent to WB. Invites additions to JH's report. Please send account of expenses.
ALS 3pp RS:HS 19.103

5546. 1843–7–20 From James Heywood Ryde, I.o.W.
Is pleased JH has such a high opinion of the new work on the English universities; it has hardly been noticed by the public. V. A. Huber belongs to the Conservative party but tries to be impartial as a historian. Hopes that the amount of reading at universities will be reduced.
ALS 4pp RS:HS 10.188

5547. 1843–7–21 To William Radcliffe Birt Collingwood
Yesterday's letter was mistakenly sent to WB's former address. Sends copy.
ALS 3pp RS:HS 19.104

5548. 1843–7–21 From William Rutter Dawes 9 Park Place, Clarence Gate, Regent's Park
WD's 1843 measurements for Gamma Virginis. Used crystal micrometer for all observations, to compare with parallel wire micrometer used in past.
ALS 3pp TxU:H/M–0157; Reel 1087

5549. 1843–7–22 From Francis Baily 37 Tavistock Place
Is returning the report to the B.A.A.S., with one alteration. Will forward the map to Mr. Jensen with JH's remarks.
ALS 2pp RS:HS 3.252

5550. 1843–7–[23?] To Francis Baily Collingwood
Was unable to visit FB; JH is suggesting some options for FB dealing with the star map, and whether it needs to go to the B.A.A.S. for approval.
C 3pp RS:HS 25.8.57

5551. 1843–7–24 From Francis Baily 37 Tavistock Place
Will be unable to attend the meeting of the B.A.A.S. His sister is having an operation to her eyes. Is sending an account of the comet. Mr. Jensen will be coming down with

the map.
ALS 4pp RS:HS 3.253

5552. 1843–7–25 From Francis Baily 37 Tavistock Place
Mr. Jensen called to say that he will be visiting Collingwood on Monday with the map.
ALS 2pp RS:HS 3.254

5553. 1843–7–26 From Charles Few Covent Garden
Regarding a nuisance created at Slough by a steam engine.
ALS 3pp RS:HS 7.209

5554. 1843–7–28 To William Radcliffe Birt Collingwood
JH's theory of barometric waves. Comparison with ocean tides. Non-trade winds follow laws of periodicity and are predictable, accounting for previously unexplained phenomena. Attributes rotating storms to interference of two or more wave trains. Will propose WB as director of new B.A.A.S. project to discover laws of weather behavior. Sends payment for WB's expenses. Never saw anything more beautiful than WB's 'Symmetrical Barometric Curve.'
ALS 6pp †(by JH) RS:HS 19.106 (C: 19.105, 22.166, & 22.167)

5555. 1843–7–28 From Edward Sabine Woolwich
Will undertake magnetic report except parts for which research is not yet completed. Needs [Johann] Lamont's recommendation regarding experiments. Sends [James] Ross's observations for perusal.
L inc 4pp RS:HS 15.153

5556. 1843–7–29 From Charles Few Covent Garden
Does not recommend him to write to Mr. Cooper, but if any further work starts advises a visit to the spot.
ALS 1p RS:HS 7.210

5557. 1843–7–31 From William Henry Smyth 3 Cheyne Walk
Problems with computation of Gamma Virginis.
ALS 4pp TxU:H/M–0493; Reel 1093

5558. 1843–8–1 From G. Schwinck Berlin
Humbly requests that JH accept his 'celestial map' and his infinite esteem.
ALS (German) 1p RS:HS 17.181

5559. 1843–8–2 To [William Radcliffe Birt] Collingwood
Has check for WB's expenses.
ALS 3pp RS:HS 19.107

5560. 1843–8–5 From Francis Baily 37 Tavistock Place
Has received official communication from the Treasury that the government will provide financial assistance to print the catalogues of stars. Both these catalogues and the one to be printed by the B.A.A.S. await their decision on star nomenclature.
ALS 4pp RS:HS 3.255

5561. 1843–8–6 To [William Radcliffe Birt] Collingwood
Sends check. Asks for latitude and longitude of four American observation stations.
ALS 2pp RS:HS 19.108

5562. 1843–8–6 From Edward Sabine Woolwich
Discusses state of science in France. Magnetic instruments were ordered for Algiers, but ES is uncertain whether they left Paris. Will send portion of magnetic report and letter from [A. D.] Bache among other things.
ALS 4pp RS:HS 15.154

5563. 1843–8–7 From John Henry Griesbach Carlton Villas, Maida Vale
Amuses himself in his spare time with astronomy. Would like advice on the composition of the metal for the speculum. Hears that Miss Caroline Herschel is in good health.
ALS 3pp RS:HS 9.55

5564. 1843–8–13 From George Biddell Airy Gresham's Hotel, Dublin
Regarding a telescope for Thomas Maclear.
ALS 1p RS:HS 1.127

5565. 1843–8–14 To Francis Baily Collingwood
Regarding the forthcoming publication of the star catalogues and matters arising. Has written a report on meteorological observations. Sends paper by J. H. Kay on the comet.
ALS 4pp RS:HS 3.256 (C: RS:HS 25.8.58)

5566. 1843–8–16 To George Biddell Airy Collingwood
Has received a request from the Admiralty to order a new telescope for the Cape observatory; at the same time, JH has received an unsolicited offer of a lens. JH seeks advice and information from GA.
ACCS 3pp RS:HS 25.6.17

5567. 1843–8–19 To Eneas Mackintosh Collingwood
Peter Stewart informed JH of EM's offer to assist JH's son William in obtaining writership appointment in India. JH and wife are deeply grateful. JH's desires for William's future.
ACCS 3pp TxU:H/L-0238; Reel 1054

5568. 1843–8–22 From Francis Baily 37 Tavistock Place
Has received the parcel. Papers regarding the comet have been sent to the R.A.S. John Hartnup is to copy out the catalogue of stars. Will have to wait awhile for the maps of F. W. A. Argelander and Mr. Schwinck. Is preparing for the printing of the catalogues.
ALS 4pp RS:HS 3.257

5569. 1843–8–22 From E[neas] MacKintosh Richmond
Is happy to hear that JH approves of the suggestion to send his eldest son to seek his fortune in the East. Will request that the appointment be conferred on William Herschel [JH's son]. Is spending some days with the Bryants.
ALS 3pp RS:HS 12.21

5570. 1843–8–24 From Francis Baily 37 Tavistock Place
Believes 'Mr. Maclean,' who reported from Africa observing increasing brightness of Eta Argus is not [Thomas] Maclear, but [George Maclean], the 'Governor of Cape Coast Castle [Ghana] & the husband of the unfortunate [Letitia Elizabeth Landon], whose singular death caused such a sensation some time ago.'
ALS 2pp RAS:JH Archive 3/1.1.1; Reel 7

5571. 1843–8–24 To [William Howley] Collingwood
[Writing in response to Archbishop [of Canterbury, William Howley]], accepts the honor of being named a Trustee of the British Museum with much gratitude.
draft 2pp RS:HS B27.55

5572. 1843–8–26 From George Booth Tyndale Hayling Island
Congratulates JH on his election as Trustee of the British Museum.
ALS 3pp RS:HS 17.453

5573. 1843–8–28 From Jonathan Pereira Finsbury Square
Sends diagrams of woodcuts representing planes of polarization. Chooses not to publish chemistry lectures.

Asks JH's opinion of a failed polarized light experiment.
ALS 2pp diag RS:HS 14.156

5574. 1843–8–30 From George Biddell Airy Royal Observatory, Greenwich
Provides JH with information and advice about large lenses, which JH is considering for a Cape of Good Hope Observatory equatorial telescope.
ALS 2pp TxU:H/M-0629.5; Reel 1084

5575. [1843–9] To George Biddell Airy n.p.
Regarding custody of the Standards.
draft 3pp RS:HS 1.333 (C: RS:HS 25.13.12)

5576. 1843–9–2 To Georg Merz & Son Collingwood
Is inquiring, on behalf of the Admiralty, about ordering an equatorially mounted achromatic telescope for the Royal Observatory at the Cape of Good Hope.
ALS 4pp RS:HS 25.6.18

5577. 1843–9–4 From Sir Robert Harry Inglis Milton Bryan, Woburn
Will find one friend at the R.S.L. Club dinner next Thursday.
ALS 1p RS:HS 10.229

5578. 1843–9–6 From Charles Babbage n.p.
Sends deeds already signed by J. J. Guest. JH should sign and return them. [Annotation by unknown hand: Mrs. Hollier's marriage and settlement.]
ALS 1p †(by ?) TxU:H/M–0027; Reel 1062

5579. [1843]–9–7 From Edward Sabine Tulla, Co. Clare
Sends letter from [Palm Heinrich Ludwig] Boguslawski for JH's magnetic report. [John] Lefroy has found line of greatest intensity further south in Canada than ES expected.
ALS 4pp RS:HS 15.317

5580. 1843–9–8 From John David Roberton Royal Society
[J. C.] Ross has returned safely. Ross has a box for JH.
ALS 2pp RS:HS 14.367

5581. 1843–9–8 To William Henry Smyth Collingwood
Offers condolences to Ann Smyth. WS's calculation of Gamma Virginis orbit surprises JH. Concerned about accurate methods of calculation. Remarks on excavation of Chelsea mammoth.
C 2pp RS:HS 22.167

5582. 1843–9–9 From George Biddell Airy Royal Observatory, Greenwich
Regarding the storage of the Standard weights and measures.
ALS 3pp RS:HS 1.128 (C: RGO 6.340.353)

5583. 1843–9–10 To George Biddell Airy Collingwood
Makes some suggestions about possible storage places for the United States' standards [see GA's 1843–9–9].
ALS 1p RGO 6.340.354

5584. 1843–9–11 From Robert Hunt Falmouth
Enclosing specimens of his chromatype; comments on how these were produced and their characteristics. Is very busy with the Annual Exhibition of the Polytechnic Society.
ALS 3pp RS:HS 10.103

5585. 1843–9–11 From Adolphe Quetelet Brussels
Has succeeded in increasing stations around Alps for meteorological observations. [G. B.] Airy is very obliging. Hopes one day to collect all meteorological observations in *Sur le climat de la Belgique*.
ALS (French) 3pp RS:HS 14.245

5586. 1843–9–12 From George Biddell Airy Royal Observatory, Greenwich
Describes some available glass discs, which might do for making lenses for a large refracting telescope [see GA's 1843–8–30].
ALS 2pp † TxU:H/M-0629.6; Reel 1084

5587. 1843–9–12 From Francis Baily 37 Tavistock Place
Sends specimens of works which are to go to press. Has had an estimate for the printing. Nomenclature has to be settled first.
ALS 4pp RS:HS 3.258

5588. 1843–9–12 From William Radcliffe Birt Bethnal Green
Could he loan him the Russian observations. Recent readings of the barometer.
ALS 2pp RS:HS 4.103

5589. 1843–9–13 To Francis Baily Collingwood
Regarding the cost of printing the catalogues, and matters pertaining to them. Can FB give him any general table of processions. Has started work on the letter press of his own book.
ALS 3pp RS:HS 3.259 (C: RS:HS 25.8.59)

5590. 1843–9–13 To Caroline Lucretia Herschel Collingwood
Enjoyed reading CH's account of some significant events from her life. Reports that James Clark Ross has returned safely from his South Pole expedition in which Ross discovered the true position of the South Magnetic Pole.
ALS 3pp TxU:H/L–0588.13.a; Reel 1058 *Memoir ... of Caroline Herschel*, 333–4 inc

5591. 1843–9–15 To William Radcliffe Birt Collingwood
B.A.A.S. approved WB's employment by Magnetical and Meteorological Committee to explore atmospheric waves. Will send Russian observations tomorrow. Howard Elphinstone agrees to send his observations [at Ore, near Hastings]. Lists meteorology books.
ALS 3pp RS:HS 19.109

5592. 1843–9–16 From George Biddell Airy Royal Observatory, Greenwich
Requires information on an actinometer.
ALS 1p RS:HS 1.129

5593. 1843–9–17 To Robert Hunt Collingwood
Comments on RH's chromatype photographic process and other processes tried by JH.
ALS 3pp George Eastman House, 14.17 (C: RS:HS 22.168)

5594. 1843–9–17 To Sir Charles Lyell Collingwood
Inquires of CL about the suitability of George Robert Waterhouse for a natural history post in the British Museum. Congratulates CL on his successful trip to the United States.
ALS 3pp APS B.D25.L.3 (ACCS: RS:HS B27.57)

5595. 1843–9–18 From Francis Baily 37 Tavistock Place
Regarding bringing N. L. Lacaille up to a modern period. Regarding a table of precessions. FB's sister.
ALS 2pp RS:HS 3.260

5596. 1843–9–18 From John Henry Griesbach Carlton Villas, Maida Vale
Thanks for letter and advice. Has been able to get copies of William Parsons' (3rd Earl of Rosse) papers. Would like advice on the size of the grinding tool. Hopes to avail himself of JH's invitation to visit Collingwood.
ALS 3pp RS:HS 9.56

5597. 1843–9–18 From George Robert Waterhouse 57 Pall Mall
Explains how it comes about that GW is looking for a position at the British Museum.
ALS 4pp RS:HS B27.58

5598. 1843–9–20 From William Radcliffe Birt Bethnal Green
Thanks for letter and loan of the books. Regarding the movement of waves of the atmosphere.
ALS 2pp RS:HS 4.104

5599. 1843–9–20 From Georg Merz Munich
Provides prices and delivery times for a variety of different sizes of refracting telescopes, as well as some mounting advice.
ALS 2pp (German) TxU:H/M–0356; Reel 1087

5600. 1843–9–21 From Sir Charles Lyell Kimmondy, Kirriemuir
Talks about British Museum business, including the purchase of fossils and the hiring of George R. Waterhouse.
ALS 4pp RS:HS B27.59

5601. 1843–9–23 From William Henry Smyth 3 Cheyne Walk, Chelsea
W. R. Dawes observes sub-division of Saturn's rings. WS continues observing double stars recommended by JH.
ALS 3pp RS:HS 16.273

5602. 1843–9–25 To George Biddell Airy n.p.
Gives formula for defining the measure of the scale of an actinometer.
LCC 1p inc RGO 6.701.106

5603. 1843–9–27 To the *Athenaeum* Collingwood
[Replying to remarks by T. R. Robinson reported in *Athenaeum,* #830 (Sept. 23, 1843), 866–7,] JH argues that William Herschel's 40-foot reflector was not a failure, noting, for example, the discovery by it of the sixth and seventh Saturnian satellites.
L *Athenaeum,* #831 (Sept. 30, 1843), 884

5604. 1843–9–28 To [William Howley] Collingwood
[Writing to the Archbishop [of Canterbury, William Howley]], recommends George R. Waterhouse as a candidate for employment at the British Museum.
draft 2pp RS:HS B27.60

5605. 1843–9–29 From James Clark Ross St. George's Terrace
Cannot at present accept Lady Herschel's invitation to Collingwood. Is getting married.
ALS 4pp RS:HS 14.428

5606. [1843]–10 To George Biddell Airy n.p.
Recommends buying optics for 7.5-inch telescope from Metz and Mahlers in Munich and having those mounted in London.
C 2pp TxU:H/L-0013.2; Reel 1054

5607. 1843–10–3 To William Radcliffe Birt [Collingwood]
Sends Howard Elphinstone's barometer observations [at Ore, near Hastings].
ALS 1p RS:HS 19.110

5608. 1843–10–4 From George Biddell Airy Royal Observatory, Greenwich
Encloses copy of a paper for JH's comments. Regarding reviews of GA's paper on 'Tides.'
ALS 2pp RS:HS 1.130 (C: RGO 6.401.91)

5609. 1843–10–8 To Francis Baily Collingwood
Has to come to town for a Trustee meeting of the British Museum and wonders whether it would be convenient for him to visit Tavistock Place to consult him on various matters.
ALS 3pp RS:HS 3.261 (C: RS:HS 25.8.60)

5610. 1843–10–9 From Francis Baily 37 Tavistock Place
It will be quite convenient for JH to visit him on Friday. Sister's health is improved. Delay in the celestial maps ordered.
ALS 3pp RS:HS 3.262

5611. 1843–10–10 From John Phillips B.A.A.S., York
Names committee, including JH, to superintend 'translation & publication of scientific memoirs.' [Printed note regarding] grants of money.
(A)LS 1p RAS:JH Archive 4/9.10 verso; Reel 7

5612. 1843–10–11 From William Radcliffe Birt Bethnal Green
Has received Howard Elphinstone's observations. Has had an interview with Edward Sabine who has arranged for observations to be forwarded to WB. Would like JH's views on publication of some of these observations.
ALS 2pp RS:HS 4.105

5613. 1843–10–12 From George Robert Waterhouse 57 Pall Mall
Expresses GW's concern if the British Museum were to require him to pledge to do no remunerative work outside what they would pay GW to do. GW, however, is prepared to make that pledge, if necessary.
ALS 3pp RS:HS B27.61

5614. 1843–10–13 To George Biddell Airy Collingwood
Urges doubling number of lunar observations.
C 3pp TxU:H/L-0013.1; Reel 1054

5615. 1843–10–14 To Peter Mark Roget n.p.
Construction of JH's new instrument for photographing spectral lines is still incomplete, and A. E. Becquerel has already announced discovery. JH will return £100 granted to JH by R.S.L.
C 2pp TxU:H/L-0340; Reel 1054

5616. 1843–10–17 From Ann Smyth Cheyne Walk
History of husband's [W. H. Smyth] exchange of 1836 letters with T. R. Robinson regarding Gamma Virginis and JH's confirmation of W. H. Smyth's discovery. Sends copy of Augustus De Morgan's letter on renaming constellations.
ALS 3pp RS:HS 19.296

5617. 1843–10–19 To [R.S.L. Secretary?] Collingwood
Relinquishes R.S.L. Donation Fund for spectrometer [see JH's 1842–6–1], because [A. C.] Becquerel's research makes it obsolete.
ALS 4pp RS MC.3.306

5618. 1843–10–20 To N. Spahr Collingwood
Both JH and his wife, Margaret, are very concerned about the health of Madame Gerlach, the aunt of NS. Details are provided.
ACCS 3pp RS:HS 25.6.19 (draft RAS:JH 3/6.7)

5619. 1843–10–23 To George Biddell Airy Collingwood
Would support improvements, both in number and quality, of lunar observations at the Royal Observatory; current practices produce unacceptably large predictive errors.
ALS 4pp RGO 6.8.72

5620. 1843–10–24 From Henry John Rose Haughton Conquest
Edited the last volumes of the *Encyclopaedia Metropolitana*. Needs JH to summarize main points of his treatises on light and sound for the general preface.
ALS 4pp RS:HS 14.502

5621. 1843–10–26 From Josiah Quincy Harvard University
Writing memoir of James Grahame. Asks JH to send list of all Grahame's publications, copies of his work, and any available reviews.
ALS 2pp RS:HS 14.291

5622. 1843–10–29 From Edward Joshua Cooper Dublin
Thought JH had given up interest in Meteorological Observations until he saw his report in the B.A.A.S. meeting. Would like his comments on Meteorological Observations to be made at Nice to where he is now going. Is taking the big object-glass and eye pieces.
ALS 4pp RS:HS 5.284

5623. 1843–10–29 From E. Mühry Stade
Was pleased to have had the opportunity of learning more of England and its people. EM visited JH's aunt Caroline in Hanover, and reports the death of Mrs. [Sophia] Beckedorff at age 87. Sends greetings to Madame Gerlach.
ALS 3pp (German) RS:HS 19.291

5624. 1843–10–30 From Heinrich Christian Schumacher Altona
Saw article in the [London] *Times* about Lord Rosse's telescope. Has received a letter from [James] South. Asks for information from JH for private use. [Francesco] del Vico has observed triple stars, but HS thinks the telescope by [R. A.] Cauchoix is flawed.
ALS 2pp RS:HS 15.392

5625. 1843–10–31 From Alexander Wilson Great New St., Gough Square
Asks JH to prevail on Lord Ashley [A. A. Cooper] to revise money laws in House of Commons for benefit of working classes. Encloses tracts [issued by Society for the Emancipation of Industry].
ALS 3pp RS:HS 18.262

5626. 1843–11–2 To the *Athenaeum* Collingwood
Requests publication of what will hopefully be JH's last letter on the Slough telescope [see JH's 'Reply to Dr. Robinson [on the Reflecting Telescopes of the Late Sir William Herschel],' *Athenaeum*, #836 (Nov. 4, 1843), 983–4].
ALS 1p WT 69293.1

5627. 1843–11–2 To the *Athenaeum* Collingwood
Continues the debate with T. R. Robinson [see *Athenaeum*, #830 (Sept. 23, 1843), 866–7; #831 (Sept. 30, 1843), 884; and #834 (Oct. 21, 1843), 945–6] on the quality of William Herschel's 40-foot reflecting telescope.
L *Athenaeum*, #836 (Nov. 4, 1843), 983–4

5628. 1843–11–3 From George Biddell Airy Royal Observatory, Greenwich
A notice of meeting of the Standards Committee, together with an indication of business to be conducted at that meeting.
LS 2pp † TxU:H/M-0629.7; Reel 1084

5629. 1843–11–3 From George Biddell Airy Royal Observatory, Greenwich
Changes the date of a Board of Visitors' meeting and urges JH to be there.
ACCS 1p RGO 6.8.87

5630. [1843–11–7] To George Biddell Airy Collingwood
Will try to attend meeting called by GA, although JH is plagued by rheumatism; comments on the periods of Saturn's satellites.
ALS 4pp RGO 6.368.669

5631. 1843–11–8 From Richarda Airy [Greenwich]
G. B. Airy will be happy to receive JH tomorrow.
ALS 2pp RS:HS 19.312

5632. 1843–11–8 From Francis Baily 37 Tavistock Place
Has a copy of F. W. A. Argelander's *Uranometria*. Comments on this in relation to the proposed revision of the constellations.
ALS 3pp RS:HS 3.263

5633. 1843–11–12 From Mary Somerville Rome
Asks JH to tell her where to find his newly published works on the action of light on chemical substances. Rome is not intellectually stimulating. Describes summer in Venice.
ALS 4pp RS:HS 16.347

5634. 1843–11–16 To *Astronomische Nachrichten* Collingwood
Reports observations of the comet of March 1843 [Great Comet of 1843] made from India by a Mr. Clerihew, who reported seeing a bifurcation of the comet's tail.
L *Astronomische Nachrichten, 21* (1844), 199–202

5635. 1843–11–17 From Thomas I. M. Forster Bruges
Would like a copy of JH's pamphlet on *Education* written at the Cape. Did he ever receive the Perennial Calendar? Comments on recent falling stars. Does he know where Stephen Lee's *Catalogue of Changing Stars* is? Mr. Slegg's brother is in a mental home.
ALS 4pp RS:HS 7.334

5636. 1843–11–17 To Thomas Phillips Collingwood
Comments on the value of meteorological observations that TP proposes to undertake.
C 1p RS:HS 22.169

5637. 1843–11–19 From Robert Hunt Falmouth
Has received from the *P.M.* proof of an article replying to the attack of [L. F.] Moser on RH, also details of a new instrument by J. W. Draper. Would like JH's opinion on the instrument. Has had some good results with the fluorates recently.
ALS 3pp RS:HS 10.104

5638. 1843–11–20 From William Henry Smyth 3 Cheyne Walk, Chelsea
Asks JH for description of JH's 'little sweeper' telescope; plans to give a sketch of it in WS's book.
ALS 2pp † RS:HS 16.274 David S. Evans, 'Sir John Herschel to Captain Smyth,' *Quarterly Bulletin of the South African Library, 25* (1970), 16–17

5639. 1843–11–20 From George Robert Waterhouse 10 Gloucester Grove West, Old Brompton
Was appointed assistant in [mineralogical and geological branch of] British Museum. Thanks for JH's support.
ALS 2pp RS:HS 18.95

5640. 1843–11–20 To Alexander Wilson Collingwood
Disagrees with AW. Declines to intercede. [See AW's 1843–10–31].
ACCS 2pp RS:HS 18.263

5641. 1843–11–21 From Charles Babbage Dorset St.
Is sending him some new additions for the memoir by L. F. Menabrea. Mother keeps well.
ALS 3pp RS:HS 2.303

5642. 1843–11–21 To William Henry Smyth Collingwood
Sends diagram of the 'little sweeper,' which is now at Hanover with 94-year-old Caroline Herschel.
C 1p RS:HS 22.170

5643. 1843–11–22 From Francis Baily 37 Tavistock Place
Has received his letter and incorporated the amendments in the paper.
ALS 2pp RS:HS 3.264

5644. 1843–11–23 From Robert Grahame Jr. Glasgow
Does not have a complete list of his father's publications. Some he wished to suppress. Approves of JH's suggestion to send the manuscript of the history to Harvard. Is gratified by Dr. [Edmund?] Quincy's letter and also JH's impression of his father's character.
ALS 3pp RS:HS 9.10

5645. 1843–11–24 To William Morison Collingwood
WM is not the first person to write to JH on the application of photography to astronomy. Comments on the question and gives some of his own views on the subject.
draft 4pp RS:HS 13.53 & 22.171

5646. 1843–11–29 To Maria Edgeworth n.p.
Would she read the enclosed letter and show the passage marked to Mrs. F. M. Wilson. This morning he dispatched to her a book and gloves. Will send the photographs and picture by some other means later.
draft 2pp RS:HS 7.28

5647. 1843–11–29 To William Whewell Collingwood
Recounts visit to Collingwood by the novelist Maria Edgeworth. Thanks WW for information on Charles Prichard and his school.
ALS 4pp TC, Camb. Add. Ms.a.207[56] (C: RS:HS 22.172)

5648. 1843–12–1 To Charles Babbage Collingwood
Has had visitors which accounts for the delay in answering letter. Thanks for memoir additions. Is sending paper dealing with use of CB's work. Somervilles are still in Rome. Is glad to hear good news of his mother.
ALS 4pp RS:HS 2.304

5649. 1843–12 From Francis Baily 37 Tavistock Place
Has compared all the stars in N. L. Lacaille's catalogue to the new observations and plan for the constellations.
ALS 2pp RS:HS 3.267

5650. 1843–12–1 To William Rowan Hamilton Collingwood
Has been delayed in congratulating WH on pension by a delightful visit from Maria Edgeworth. Praises highly verses sent by WH's sister to Lady Herschel.
ALS 3pp TCD 1493:227 Graves, *Hamilton, 2,* 452 inc

5651. 1843–12–1 From Heinrich Christian Schumacher Altona
Sends copy of announcement from [Hervé] Faye at Paris observatory of a new comet in Orion. Requests use of JH's letter about double tail on [Halley's] comet in *Astronomische Nachrichten.* Note from Faye has a gross error in location of comet.
ALS 4pp RS:HS 15.393

5652. 1843–12–2 From Charles Babbage Dorset St.
Sending him papers on his machine. Would like to see him when in town to discuss various matters. His youngest son has arrived in Calcutta.
ALS 4pp RS:HS 2.305

5653. [1843–12?–2] To William Rowan Hamilton n.p.
Highly recommends a [William H.] Harvey (formerly Government Treasurer at the Cape) for the professorship of botany at Dublin.
ALS 3pp TCD 1493:1391

5654. 1843–12–4 From William Henry Harvey Limerick
Owing to ill health has had to quit H.M. service at the Cape and has been residing with his brother. Intends applying for the position of curator of the herbarium at Trinity College, Dublin, and would welcome a testimonial from JH. Sending this letter c/o W. J. Hooker as he does not know JH's address.
ALS 4pp RS:HS 9.248

5655. 1843–12–4 From William Henry Smyth 3 Cheyne Walk, Chelsea
Thomas Henderson discovered a comet on 2 December. Thanks JH for diagram of JH's comet-sweeping telescope.
ALS 2pp RS:HS 16.275

5656. 1843–12–6 From Sir William J. Hooker Royal Botanic Gardens, Kew
Believes the enclosed letter from W. H. Harvey refers to his candidature for a position in Dublin College. Thinks him a worthy candidate. Would like to show the Herschels round Kew anytime.
ALS 3pp RS:HS 9.448

5657. 1843–12–8 To [William Henry Harvey] Collingwood
Is glad to have been put in touch with WH again. Supports WH for the Professorship of Botany in Dublin. Reminisces about the good times at the Cape.
ALS 4pp WT 69294.5

5658. 1843–12–8 To Sir William J. Hooker Collingwood
[Responding to WH's 1843–12–6], JH has written to [William H.] Harvey and, in support of Harvey, to William R. Hamilton at Trinity College, Dublin. Thanks for invitation to visit Kew Gardens.
ALS 4pp Royal Botanic Gardens

5659. 1843–12–9 From George Biddell Airy Royal Observatory, Greenwich
Regarding glass for the telescope for the Cape Town Observatory?
ALS 1p RS:HS 1.131

5660. 1843–12–10 From Sir William J. Hooker Royal Botanic Gardens, Kew
Regarding the habits and flowering of Cape bulbs. Has just had a good collection from the Cape. The Dublin post is still vacant.
ALS 6pp RS:HS 9.449

5661. 1843–12–[11] To Margaret Brodie Herschel London
Saw Charles Pritchard and his school [Clapham Grammar School] and was favorably impressed; has enrolled son William James beginning in January.
ALS 3pp JHS LMJF.45.1a

5662. 1843–12–11 From Thomas John Newbold Kurnoal
Gives his own ideas concerning the origin of diamonds and would like JH's comments. Would also like JH's views

regarding the phosphorescent light that plays at night over the magnesite excavations.
ALS 4pp RS:HS 13.122

5663. 1843–12–12 From William Rowan Hamilton Observatory
Jokes about WH's recent 'astronomical insignificance.'
L Graves, *Hamilton, 2,* 452 inc

5664. 1843–12–12 From Robert Hunt Falmouth
Would like his views on a name for the combination of hydrogen and chlorine. Comments on the various theories. Has replied to [L. F.] Moser.
ALS 4pp RS:HS 10.105

5665. 1843–12–13 To Caroline Lucretia Herschel Collingwood
Assures CH that he has preserved everything that she ever wrote to JH or that he found in William Herschel's library.
ALS 4pp TxU:H/L–0588.17 & Extract 3pp TxU:H/L–0588.20; Reel 1058

5666. 1843–12–14 To Francis Baily Collingwood
Further regarding the revision of the constellations.
ALS 4pp RS:HS 3.265 (C: RS:HS 25.8.61)

5667. 1843–12–14 To Sir William J. Hooker Collingwood
Discusses his success in growing in England various flowers JH brought back from the Cape.
ALS 4pp Royal Botanic Gardens

5668. 1843–12–[16 or earlier] From Francis Baily n.p.
More on confused systems of star magnitudes used by Thomas Henderson.
C 1p inc TxU:H/M–0040.2; Reel 1062

5669. 1843–12–16 From Francis Baily 37 Tavistock Place
Comments regarding the proposed revision of the constellations.
ALS 4pp RS:HS 3.266 (C inc: TxU:H/M–0040.1; Reel 1062)

5670. 1843–12–17 From Thomas Maclear Royal Observatory, Cape of Good Hope
Recent travels. Dangers and difficulties of triangulation survey. Observations of variable brightness of Eta Argus. [Letter continues 29 December] Returned from 'Pequet Brig,' leaving C. P. Smyth with theodolite. Events in colony. Sending observations of Encke's Comet for R.A.S. Observations of 'great comet of March.' News of Nathaniel Wallich and family. Sends 'plan of the Herschel obelisk drawn by Smyth.' Details of obelisk. Progress on observations of 'Parallax stars.' Organized system to prevent mistakes in calculating and copying meridian work.
ALS 14 pp RAS:JH Archive 3/1.1.2; Reel 7

5671. 1843–12–20 To Charles Babbage Collingwood
Requests advice on the setting up of a sick benefit society. Remarks on some of CB's queries.
ALS 4pp RS:HS 2.306

5672. 1843–12–22 From Charles Babbage Dorset St.
Referring him to someone who can advise him on the setting up of a sick benefit society. Regarding CB's automan.
ALS 2pp RS:HS 2.307

5673. 1843–12–22 From Sir Robert Harry Inglis 7 Bedford Square
May he submit JH's name as a candidate for the Honorary membership of the Literary Society; gives list of the other candidates.
ALS 3pp RS:HS 10.230

5674. 1843–12–25 From Augustus De Morgan 69 Gower St.
Sending two formulae.
ALS 1p RS:HS 6.189

5675. 1843–12–26 From Friedrich Eduard Beneke Berlin
Thanks JH for sending FB several of his works, including the translation of Schiller.
ALS 3pp (German) RS:HS 4.60

5676. 1843–12–27 To N. Spahr Collingwood
A letter of even greater urgency about the very serious state of health of Madame Gerlach, the aunt of NS [see JH's 1843–10–20]. Madame Gerlach is living with the Herschels [as tutor? or governess?].
ACCS 4pp RS:HS 25.6.20

5677. 1843–12–31 To N. Spahr Collingwood
As JH has not heard from NS, he has arranged for Madame Gerlach to stay with Mrs. Mason, now living in part of JH's former residence at Slough. But under no circumstances can Madame Gerlach return to her duties with JH's family.
ACCS 2pp RS:HS 25.6.21

5678. 1844 From Charles Babbage n.p.
She did wrong in letting him have JH's letter.
ALS 1p RS:HS 2.350

5679. 1844 To Augustus De Morgan n.p.
Government intention to reduce interest in savings banks may well affect JH's benefit society.
C 2pp RS:HS 22.216

5680. [1844 or earlier] To Augustus De Morgan n.p.
Thanks AD for having pointed out an error in one of JH's papers.
C 1p RS:HS 25.14.3

5681. [1844–1 or later] From W. S. Mackay General Assembly's Mission, Calcutta
Extract [in JH's hand] from WM's letter describing increasing brightness of Eta Argus. [JH annotation: See '*Ast. Soc. Notices* Jan 12/44']
AAbs 1p RAS:JH Archive 3/1.1.16; Reel 7

5682. [1844] To Thomas Maclear [Collingwood]
States the case for the accuracy of the observations and reports of the discoveries of the sixth and seventh satellites of Saturn made by JH's father, William. [This letter seems to be tied to JH's 1844–7–22 in terms of date of writing.]
Cdraft 7pp RS:HS 25.15.14

5683. [1844] From Charles Pritchard Clapham Grammar School
JH to visit CP's house. CP suggests they both visit J. P. Gassiot, where they might also meet with Michael Faraday. Asks JH if there are experiments he would like to see during visit. Reports observing Fraunhofer lines through a prism.
ALS 3pp RS:HS 14.107 Ada Pritchard, *Charles Pritchard: Memoirs of His Life,* 219–20

5684. [1844 to 1845] From Edward Sabine n.p.
Funding for magnetic and meteorological instruments for Ceylon, which would be valuable addition to H.E.I.C. observatories. Compares [barometric] curves from America and Europe. Magnetometers for [Nossrey?] were paid for out

of R.S.L.'s Donation Fund.
ALS 4pp TxU:H/M-0517.1; Reel 1093

5685. [1844 to 1845] To Edward Sabine n.p.
Wishes ES had expressed opinion on requesting government money to maintain 'establishment' at R.S.L. committee meeting.
Adraft 2pp TxU:H/L-0360.1; Reel 1054

5686. [1844] To Richard Sheepshanks n.p.
Put no confidence in [Francesco] DeVico's drawing of the nebula of Orion. JH has a better one now at the engravers. Notes that DeVico has left out the 'real' 6th star in 'Trapezio.'
ALS 3pp RS:HS 16.37 (C: RS:HS 22.215 & 25.9.17)

5687. [1844] To Richard Sheepshanks n.p.
Requests RS to let him know how much JH owes him for the books by Francis Baily, which RS purchased for him. Has received RS's pamphlet [on the Liverpool Observatory].
ALS 2pp RS:HS 16.36 (C: RS:HS 25.9.16)

5688. [1844] To Richard Sheepshanks n.p.
Has mislaid RS's letter and forgotten the question RS asked concerning Baily's print. Asks him to refresh his memory. Could not get to the sale [of Francis Baily's books]. Gives the works of Baily that he wants to purchase.
ALS 3pp RS:HS 16.35 (C: RS:HS 25.9.15)

5689. [1844] To Richard Sheepshanks n.p.
Discusses controversy between [George] Bishop and [W. R.] Dawes. Recommends that the R.A.S. publish a paper by Dawes correcting errors in the published work. Wants to purchase some of Francis Baily's books. Mentions RS's work on the standard of length.
ALS 4pp RS:HS 16.34 (C: RS:HS 25.9.13)

5690. 1844 or later From Richard Sheepshanks n.p.
About the question of a bust to be made of the late Francis Baily and how the proposal might be presented.
ALS 12pp RS:HS 16.126

5691. 1844 To Richard Sheepshanks Collingwood
Very upset at an attack launched against JH in a manner that denigrates JH's father, William Herschel, in the process.
C 2pp RS:HS 25.9.12

5692. [1844] From Ann Smyth 3 Cheyne Walk, Chelsea
Warington Smyth returns from mineral gathering expedition on the Continent for Cambridge and will present JH with a specimen of Herschelite. C. P. Smyth has sent sketches he made at the Cape.
ALS 3pp RS:HS 16.218

5693. [1844 or later] To James Calder Stewart n.p.
Letter from Charles Pritchard regarding [William James Herschel's] examination next Friday. Arranges for 'Willy' to come with 'Johnnie' to India House on Tuesday to see Mr. Prinsep.
ALS 3pp TxU:H/L-0401; Reel 1055

5694. [1844 or later] From Hensleigh Wedgwood 16 Gower St.
Gives reason for writing [*Principles of Geometric Demonstration* (1844)]. Plans later edition in which HW will cite JH's statement in [*Quarterly Review*]. Role of sense experience in axioms.
ALS 12pp RS:HS 18.136

5695. [1844?] From [?] n.p.
Refers to 17 July 1827 communication from W. J. Burchell in Brazil to J. S. Duncan, brother of Phillip Duncan, noting increased brightness of Eta Navis, and that Alpha Centauri and Alpha Crucis are double stars.
Amem 1p RAS:JH Archive 3/1.1.15; Reel 7

5696. 1844–1–2 To Adalbert Adolf Mühry Collingwood
Regrets to hear of AM's sad losses. Would like him to withdraw quarterly amounts and give them to Caroline Herschel.
draft 4pp RS:HS 12.373 & 22.174

5697. 1844–1–3 From Sir Robert Harry Inglis 7 Bedford Square
Was pleased to receive his letter and the enclosure; the latter he would like to retain for a few days to show it to interested persons.
ALS 2pp RS:HS 10.231

5698. 1844–1–5 From William Henry Harvey Limerick
Delayed thanking him for the testimonials until he could give news of the outcome of his application. Position is still undecided, but he cannot be made professor as he does not possess an M.D.
ALS 3pp RS:HS 9.249

5699. 1844–1–5 From Sir Robert Harry Inglis 7 Bedford Square
Showed JH's marvelous production to Sir Robert Peel, who was much struck with it.
ALS 1p RS:HS 10–232

5700. 1844–1–6 To John Lindley Collingwood
[Responding to JL's 1844–12–30], declines JL's request that JH write a series of articles on meteorology, because of JH's need to work on the manuscript for JH's *Cape Results.* Hopes eventually to write on meteorology.
ALS 3pp Royal Botanic Gardens (draft: RS:HS 11.231 & 22.221)

5701. 1844–1–11 From William Radcliffe Birt Bethnal Green
Thanks for publishing his curves. Is proceeding with his reductions. G. B. Airy has forwarded the Greenwich observations. Would he obtain Howard Elphinstone's observations for him.
ALS 3pp RS:HS 4.106

5702. 1844–1–11 From William Rutter Dawes 9 Park Place, Regents Park
Intends moving into the country after midsummer and would like JH's advice on a suitable spot for an observatory. Regrets that the atmosphere of London is not suitable for astronomical work now.
ALS 3pp RS:HS 6.63

5703. 1844–1–11 From James A. Gordon 16 Suffolk St.
Returned home full of health after his stay with JH. Visited the Pirie's at Maidstone. Someone suggested JH was the author of *Vestiges*. Mr. Pirie agrees with their project. Experiments by electricity for inducing plant growth. Mrs. Pirie has been giving him details of mesmerism at Maidstone.
AL 4pp inc RS:HS 8.171

5704. 1844–1–12 From Edward Sabine Woolwich
Sends actinometer returns for remarks and suggestions. Mentions public dinner for scientific community.
ALS 3pp RS:HS 15.155

5705. 1844–1–13 From William Rutter Dawes 9 Park Place, Regents Park
Is extremely obliged for his prompt reply to his query

regarding a suitable site for his observatory. Outlines the kind of residence he would like.
ALS 2pp RS:HS 6.64

5706. 1844–1–16 From Francis Baily 37 Tavistock Place
Sends proof sheets of the preface to the Star catalogue of the B.A.A.S., together with a few amendments. Has not sent the catalogue to press yet.
ALS 2pp RS:HS 3.268

5707. 1844–1–19 To [Robert Hunt ?] Collingwood
Letter writing has been delayed by the crush of other business. Comments on names for various aspects of heat, light, color, and phlogiston. JH describes the actions in his 'amphitype' process. Is looking forward to RH's forthcoming work on photography.
ALS 4pp WT 68037.2

5708. 1844–1–22 From Francis Baily 37 Tavistock Place
Justification for his nomenclature in the star catalogue.
ALS 4pp RS:HS 3.269

5709. 1844–1–22 From Friedrich Wilhelm Bessel Königsberg
Comments about the uses of a meridian telescope.
ALS 3pp (German) RS:HS 4.82

5710. 1844–1–22 From Paul M. Fuss St Petersburg
Sending copies of his *Correspondence mathématique* for the R.S.L. Will send some notices of the Arctic expedition.
ALS 2pp (French) RS:HS 7.388

5711. 1844–1–25 From William Rutter Dawes 9 Park Place, Regents Park
Has had to terminate his appointment at George Bishop's Observatory and is determined to leave London and settle in the country. Has heard of a residence near Hawkhurst and would like to visit JH this coming Monday.
ALS 2pp RS:HS 6.65

5712. 1844–1–25 From Edward Sabine Woolwich
Sends actinometric returns and drawing of comet from Hobarton Magnetic Observatory. [John] Lefroy's expedition in U.S. and Canada is going well. Received the observations from Algiers.
ALS 4pp RS:HS 15.156

5713. 1844–1–26 From George Biddell Airy Royal Observatory, Greenwich
Regarding electrometers and their efficiency.
ALS 4pp RS:HS 1.132

5714. 1844–1–26 From Robert Hunt Falmouth
Thanks for his instructive letter. Comments on the terms suggested for the chemical change. Has now improved his chromatype process. Unable to continue his researches on light owing to lack of finance.
ALS 4pp RS:HS 10.106

5715. 1844–1–27 From Adalbert Adolf Mühry Hanover
About the financial arrangements for Caroline Herschel [see JH's 1844–1–2], and news of her health.
ALS 3pp (German) RS:HS 12.374

5716. 1844–1–27 From Martyn Roberts Leeds
Describes an experiment that may explain optical 'delusion' created when stars pass behind moon. Will publish it if JH finds it worthy of publication.
ALS 3pp RS:HS 14.495

5717. 1844–1–29 From Francis Baily 37 Tavistock Place
Regarding the arrangement of 'Argo.'
ALS 4pp RS:HS 3.270

5718. 1844–1–29 To Augustus De Morgan Collingwood
Wants to set up a benefit club for mutual insurance of artisans, etc., and is asking AD for actuarial information.
C 3pp RS:HS 22.175

5719. [1844–1]–29 From Peter Stewart 65 Cornhill
If publication costs exceed £1000 available, PS suggests publishing JH's [*Cape Results*] in two separate volumes, subsidizing 'drier' part for 'men of Science,' and selling 'interesting' part. Explains how PS applied [limited] government grants to works of [Andrew] Smith and Charles Darwin.
ALS 3pp RAS:JH Archive 10/1.9; Reel 9

5720. 1844–1–30 From Augustus De Morgan 69 Gower St.
Regarding the workings of Insurance Companies and Benefit Societies. Recommends an actuary. Hopes JH will come down for the Anniversary meeting.
ALS 4pp RS:HS 6.190

5721. 1844–1–31 From George Biddell Airy Royal Observatory, Greenwich
Further regarding glass for telescope. Postscript relating experiments with electricity.
ALS 3pp RS:HS 1.133

5722. 1844–2–1 From Thomas Gardner London
Is hoping to establish a practical school of chemistry in London, and would like to add his name to the list of supporters.
ALS 2pp RS:HS 9.91

5723. 1844–2–5 To Augustus De Morgan Collingwood
Questions about actuarial information in AD's 1844–1–30.
C 2pp RS:HS 22.176

5724. 1844–2–5 From Henry Anthony Jeffreys Hawkhurst Parsonage
Finds the surplus is about to be divided among the members of the Strand Friendly Institution. Comments on this and gives reasons for the surplus.
ALS 3pp RS:HS 10.283

5725. 1844–2–5 From Adolphe Quetelet Brussels
Has worked hard to augment number of stations for horary observations. Now close to 80. Asks JH's advice whether the R.S.L. or B.A.A.S. would print his observations.
ALS (French) 3pp RS:HS 14.246

5726. 1844–2–6 From Augustus De Morgan 69 Gower St.
Has just had a tract on Friendly Societies from J. W. Woolgar. Answers some queries regarding the work of such societies.
ALS 3pp RS:HS 6.191

5727. 1844–2–8 From Mrs. N. Collingwood York Place, Portman Square
Does not know the answer to his query regarding the pew, but has written to her lawyer about it.
AL 1p RS:HS 5.358

5728. [1844]–2–8 From Peter Stewart to Margaret Brodie Herschel [65 Cornhill]
Sends sample of paper to show JH. Postpones visit [to Collingwood] until PS has more information. Anxious that PS's last letter [see PS's 1844–1–29] did not clarify how

government grants were applied to publications of [Andrew] Smith and Charles Darwin. News of Stewart family.
ALS 2pp RAS:JH Archive 10/1.11; Reel 9

5729. 1844–2–8 From Henry Anthony Jeffreys Hawkhurst Parsonage
Returns the documents with many thanks. Augustus De Morgan's letter seems very satisfactory. Regarding the expenses of management. Will call on him later.
ALS 3pp RS:HS 10.284

5730. 1844–2–10 From Francis Baily 37 Tavistock Place
Would like to hear from him regarding 'Argo.'
ALS 1p RS:HS 3.271

5731. 1844–2–10 From Augustus De Morgan 69 Gower St.
All the tables in Charles Ansell's work are expressed in decimals. Gives advice on type of insurance for sickness and death. Regarding the true inventor of J. E. Bode's law. Very few at the Anniversary dinner.
ALS 4pp RS:HS 6.192

5732. 1844–2–10 From Sir Richard Owen Royal College of Surgeons
At the request of Mr. Clift, acknowledges JH's communication of the case of the Mare with the intestinal calculi. The specimens are very acceptable.
ALS 3pp RS:HS 13.187

5733. 1844–2–11 To Francis Baily Collingwood
Gives his own views regarding 'Argo' and the renaming of the constellations.
ALS 4pp RS:HS 3.272 (C: RS:HS 25.8.63)

5734. 1844–2–11 To Adolphe Quetelet Collingwood
Will attempt to arrange publication and reception of AQ's work in England. Grieves that AQ has suffered attacks because of his zeal in meteorological observations.
ALS 4pp Académie belgique: #19 (draft Académie belgique: #20) A. Quetelet, 'Notice sur Sir John Fred. Will. Herschel,' *Académie royale des science, des lettres et des beaux-arts annuaire* (1872), 180

5735. 1844–2–12 From Francis Baily 37 Tavistock Place
Further regarding the arrangement of 'Argo.'
ALS 3pp RS:HS 3.273

5736. 1844–2–13 From William Radcliffe Birt Bethnal Green
Can sympathize with JH as he has just lost a close relation. Has been unable to contact Edward Sabine. The Dublin and Munich curves have been engraved. Thanks for the *Annalen*. Howard Elphinstone has sent his [?].
ALS 2pp RS:HS 4.107

5737. 1844–2–15 From Francis Baily 37 Tavistock Place
Regarding the arrangement of 'Malus.'
ALS 2pp RS:HS 3.274

5738. 1844–2–17 From Charles D. E. König British Museum
Regarding the disposal of A. G. Bain's fossil remains from the Cape.
ALS 4pp RS:HS 11.93

5739. 1844–2–18 To Augustus De Morgan Collingwood
On poor attendance at R.A.S. dinner, and on the results of unrestricted population growth.
C 2pp RS:HS 22.177

5740. 1844–2–21 From William Radcliffe Birt Bethnal Green
Is concerned to hear that L. A. J. Quetelet's system of observations is in danger of being broken up. Requires funds from the B.A.A.S. to publish his observations. Meteorological Society not likely to support his aims. Will consult Edward Sabine.
ALS 3pp RS:HS 4.108

5741. 1844–2–21 From William Rutter Dawes 9 Park Place, Regents Park
Has taken Camden Lodge and is making preparations for moving there. Has adapted his Amician micrometer to an excellent telescope. Comments on its faults and virtues.
ALS 3pp RS:HS 6.66

5742. 1844–2–22 From Edward Sabine Woolwich
B.A.A.S. should not assume publication of meteorological observations if L. A. J. Quetelet gives these up. Johann Lamont in Munich is better qualified. Received from John Caldecott five years of meteorological observations at Trevandrum [India]. George Peacock has invited B.A.A.S. to meet at Cambridge in 1845.
ALS 4pp TxU:H/M–0520.1; Reel 1093

5743. 1844–2–24 From William James Herschel Clapham Grammar School
Describes activities at Charles Pritchard's school, including lessons in botany. Also reports that one fellow student is very ill.
ALS 2pp (German) TxU:H/M–0803

5744. 1844–2–24 From Robert Hunt Falmouth
Encloses some specimens of photographs made by the use of chloride of gold and bichromate of potash. Comments on the process.
ALS 2pp RS:HS 10.107

5745. 1844–2–27 From [I. I.] Boye Paris
Would like a suitable entry for JH to be printed in their edition of the *Archives Historiques.*
ALS 3pp RS:HS 5.93

5746. 1844–3 To Augustus De Morgan Collingwood
Details of printing the policy for the benefit society.
C 1p RS:HS 22.178

5747. 1844–3 To Augustus De Morgan Collingwood
More details of the benefit society.
C 2pp RS:HS 22.181

5748. 1844–3 To Augustus De Morgan Collingwood
More details of the benefit society.
C 3pp RS:HS 22.180

5749. 1844–3 To Augustus De Morgan Collingwood
More details of the benefit society.
C 2pp RS:HS 22.179

5750. 1844–3 From Augustus De Morgan n.p.
Advice on how to start a Friendly Society. Calculated the sun's pull on the Earth.
ALS 4pp RS:HS 6.193

5751. 1844–3 From Charles Piazzi Smyth Elands Bay, Cape of Good Hope
Received JH's packet of books and papers mailed a year earlier. Living alone for nine months, participating in Thomas Maclear's survey. Quotes J. C. F. Schiller's 'The Walk.' Ideas on natural theology. Experiments with

photography.
AL 4pp inc RS:HS 19.268

5752. 1844–3–7 To Augustus De Morgan Collingwood
Discussion of actuarial tables for the benefit society.
C 2pp RS:HS 22.182

5753. 1844–3–8 From Augustus De Morgan 69 Gower St.
Is working on tables A and B. Further regarding Charles Ansell's life tables.
ALS 4pp RS:HS 6.194

5754. 1844–3–12 From George Biddell Airy Royal Observatory, Greenwich
Has JH received Thomas Maclear's observations on Encke's comet?
ALS 2pp RS:HS 1.134 (C: RGO 6.92.120)

5755. 1844–3–13 From Thomas Galloway Serjeants' Inn
Sending a notice, which JH may be pleased to see, though he suspects the subject did not fall into able hands.
ALS 1p RS:HS 8.12

5756. 1844–3–14 From Thomas I. M. Forster Bruges
Sends a small College album of poems and snaps. Wishes JH would visit him at Bruges as they have ample room to entertain him. Hopes to publish meteorological observations kept by his grandfather, father, and himself. Comments on Saturday moons and wet and windy weather. Is he working on changes in the apparent magnitude of stars?
LS 4pp RS:HS 7.335

5757. 1844–3–14 To Margaret Brodie Herschel London
Is spending time in court, and all the rooms are filled with tobacco smoke, which bothers JH.
ALS 2pp JHS 3.23

5758. 1844–3–15 To [James Calder] Stewart Maidstone
Congratulations on JS's new position at Bank. Received JS's account from Malta of comet. JH's son William is at Clapham grammar school under Charles Pritchard. Eneas Mackintosh promised to procure job as 'writer' in India for Willy. News from India and Europe. Teaching Latin to Alexander and three oldest daughters.
ACCS 4pp TxU:H/L-0431; Reel 1055

5759. 1844–3–16 From Palm Heinrich Ludwig Boguslawski to Caroline Lucretia Herschel Breslau
A note to CH on her 95th birthday.
ALS 3pp (German) RS:HS 4.173

5760. 1844–3–17 To [M. Bosse?] Collingwood
Reacts strongly to substantial errors in unauthorized biography of JH. Suggests it be burned. [JH annotation: 'Not sent. Answered none of their letters and refused to take in 2 or 3. A regular piece of French *insolence*.']
ALS 3pp †(by JH) RS:HS 19.111

5761. 1844–3–17 To Mary Somerville Collingwood
Sending papers on actinochemistry (photography). Discusses [James] Forbes's viscosity theory, nebulae of Southern Hemisphere, great refractor at Collegio Romano, and Lord Rosse's telescopes.
ALS 5pp RS:HS 16.348 (C: RS:HS 22.148) *Personal Recollections ... of Mary Somerville* (London, 1873), pp. 265–8

5762. 1844–3–19 To George Biddell Airy Collingwood
Has not heard from Thomas Maclear [see GA's 1844–3–12]; has the Cape equatorial object glass been dealt with?
ALS 1p RGO 6.92.125 (C: RGO 6.92.124 inc)

5763. 1844–3–19 To George Biddell Airy n.p.
Comments on the large number of stars identified in one square degree of sky.
AL 2pp inc RGO 6.193.55

5764. 1844–3–20 From Thomas Lawrence General Post Office
Please reseal letter and send to TL, who will investigate JH's complaint.
ALS 1p RAS:JH Archive 3/4.7; Reel 7

5765. 1844–3–21 From George Biddell Airy Royal Observatory, Greenwich
Regarding Cape Town telescopes. Has written to Thomas Maclear for observations on Encke's comet.
ALS 1p RS:HS 1.135

5766. 1844–3–22 To J. Park Harrison Collingwood
Has not received Tables A and B of the endowment tables. Comments on the use of these tables and would welcome his views on a possible alteration in policy.
draft 1p RS:HS 10.164

5767. 1844–3–22 From Robert Hunt Falmouth
Outlines his domestic history. If JH knows of a suitable position vacant that may suit RH he would be pleased to hear of it. Cannot understand JH's lack of success with the photographic process RH uses. His own book is printed and is waiting for an engraving. Will send him a copy.
ALS 4pp RS:HS 10.108

5768. 1844–3–23 From Edward Sabine Woolwich
Encloses letter from Dr. [John] Locke. Wants to loan him a magnetometer to set up a station at Cincinnati. Received [John] Caldecott's volume of JH's observations. Thinks publishing costs will be high.
ALS 4pp RS:HS 15.157

5769. 1844–3–24 From Augustus De Morgan 69 Gower St.
Sending more [life] tables. No age 65 in Charles Ansell's tables. Gives further advice on the running and financing of Friendly Societies.
ALS 4pp RS:HS 6.195

5770. 1844–3–25 From Francis Baily 37 Tavistock Place
Returns F. W. Bessel's letter, part of which has been read to the R.A.S. Has he any communications for the R.A.S. as they are short of suitable material?
ALS 1p RS:HS 3.275

5771. 1844–3–25 From Augustus De Morgan 69 Gower St.
Has just received his letter. Further advice on Friendly Societies.
ALS 2pp RS:HS 6.198

5772. 1844–3–25 From Augustus De Morgan 69 Gower St.
Just dispatched the letter when JH's arrived. Regarding benefit societies and their workings. Will write to J. W. Woolgar.
ALS 6pp RS:HS 6.197

5773. 1844–3–25 From Augustus De Morgan 69 Gower St.
Further regarding life tables and their workings. Francis Baily has a sheet of his grand catalogue on hand.
ALS 4pp RS:HS 6.196

5774. 1844–3–26 To Augustus De Morgan Collingwood
Corrections to the actuarial tables.
C 2pp RS:HS 22.184

5775. 1844–3–26 From Augustus De Morgan 69 Gower St.
Their Tidd Pratt tables crossed. Comments on Charles Ansell's and J. T. Pratt's tables.
ALS 2pp RS:HS 6.200

5776. 1844–3–26 To Augustus De Morgan Collingwood
Under pressure of time, JH is anxious to resolve the question of appropriate actuarial tables.
C 2pp RS:HS 22.183

5777. 1844–3–26 From Augustus De Morgan 69 Gower St.
Has already sent him [William?] Tidd's tables as he requested. Regarding the printing by 5th April. Has sent tables to Farley.
ALS 2pp RS:HS 6.199

5778. 1844–3–28 From Augustus De Morgan 69 Gower St.
Is sending all the tables to the printer tomorrow. Has given tables A and B a heading. J. T. Pratt's table is similar to his own though not accurate in decimals to the pound.
ALS 4pp RS:HS 6.201

5779. 1844–3–28 From Thomas Galloway Serjeants' Inn
Would be grateful if JH would look at the translation of F. W. Bessel's letter and make any comments. Expresses thanks for his kind remarks concerning [William] Wallace.
ALS 1p RS:HS 8.13

5780. 1844–3–29 To James Calder Stewart n.p.
Newspaper accounts of JS's first success for 'the Bank' in India. Approves JS's plan to present personal report from JH on recent twin-tailed comet to Pasha of Egypt. Family news. Margaret Herschel's health. JH's son William is at Clapham grammar school under Charles Pritchard. Visit by John Stewart. Matilda Grahame's finances. Outbreak of class struggles in England.
ALS 4pp TxU:H/L-0432; Reel 1055

5781. 1844–3–30 From Augustus De Morgan 69 Gower St.
Printer unable to send proof to Richard Farley so JH will not be able to have proof by the 5th. Has requested printer to forward proof to JH.
ALS 2pp RS:HS 6.202

5782. 1844–3–30 From William Henry Harvey Dublin
Has been appointed Curator of the Herbarium at Trinity College, and his friend Dr. G. J. Allman professor. Is very pleased, and the use of library and society makes up for the loss in salary compared with the Cape. JH's testimonial had great weight with the authorities.
ALS 4pp RS:HS 9.250

5783. 1844–3–31 To Augustus De Morgan Collingwood
Further emendations regarding printing the policy of the benefit society.
C 1 p RS:HS 22.185

5784. 1844–4–1 From Thomas Galloway Serjeants' Inn
Sending back a note, which JH had included by mistake with the proof of F. W. Bessel's letter. Thanks for his corrections to the proof.
ALS 1p RS:HS 8.14

5785. [1844–4] To Sir Charles Lemon n.p.
Would like CL to use his influence to obtain a place in Christ's Hospital for one of Robert Hunt's sons.
draft 3pp RS:HS 11.191 & 22.186

5786. [1844–4] To Roderick Impey Murchison n.p.
Willing to chair B.A.A.S. meeting next year, provided no other duties are imposed on JH's time and that no one else desires that office.
ACCS 2pp RS:HS 19.116 (C: 22.214)

5787. 1844–4–3 To F. W. A. Argelander n.p.
Thanking him for his Uranography; comments on this.
draft 4pp RS:HS 1.373 & 22.187

5788. 1844–4–3 To Jean Baptiste André Dumas Collingwood
Thanks JD for sending JD's *Mémoires de chimie,* which JH praises lavishly as a contribution to organic chemistry of 'great importance and originality.'
ALS 3pp Académie des sciences (Paris)

5789. 1844–4–3 To Longman and Co. Collingwood
Is submitting a list of corrections for the astronomy section of the *Cabinet Cyclopedia.* Proposes a major new work on astronomy.
ACCS 2pp RS:HS 25.6.22

5790. 1844–4–7 To Augustus De Morgan Collingwood
Thanks for all the help with the benefit society material.
C 2pp RS:HS 22.188

5791. 1844–4–8 From Augustus De Morgan 69 Gower St.
Hopes Richard Farley's proof was in order. Gives details of steel pens he uses. Picture he sends is by W. H. Smyth. Is moving to new address.
ALS 3pp RS:HS 6.203

5792. 1844–4–8 From Thomas Galloway Serjeants' Inn
Advice on how to obtain admission to Christ's Hospital for one of JH's sons.
ALS 3pp RS:HS 8.15

5793. 1844–4–10 From Thomas Galloway Serjeants' Inn
Name of Prince Albert's private secretary is G. E. Anson. He is abroad at the present but sees he will be soon returning.
ALS 2pp RS:HS 8.16

5794. 1844–4–11 To James Challis Collingwood
A note of thanks upon receipt of the thirteenth volume of the observations of the Cambridge Observatory.
AL 1p CUOL 1844.25

5795. 1844–4–11 From James [Struthers] Stewart Cape of Good Hope
Sends JH bulbs of Cape flowers to plant. Thanks him for sending JH's translation of Friedrich Schiller's 'The Walk.' Describes the obelisk placed on JH's telescope site at Feldhausen. Reports of effort to install improved roads in the colony. Construction of St. George's Church completed. Describes planned Botanical Garden at the Cape.
ALS 11pp RS:HS 17.10

5796. 1844–4–13 From Robert Hunt Falmouth
Is most grateful for JH's offer of assistance for one of his sons. Suggests the second one would be most suitable. Is certain Sir Charles Lemon, their president, would support him.
ALS 3pp RS:HS 10.109

5797. 1844–4–13 From Roderick Impey Murchison 16 Belgrave Square
B.A.A.S. have been trying to arrange their 1846 meeting at Cambridge. William Whewell alone is opposed to this suggestion. Everyone agrees that JH should be the next president when the Dean of Ely retires. Can they have JH's views.
ALS 5pp RS:HS 12.404

5798. 1844–4–17 To Roderick Impey Murchison Collingwood
Appreciates foresight, but cannot commit to be president of B.A.A.S. meeting at Cambridge in summer of 1846. Date is too far away; personal and domestic commitments are pressing.
ALS 4pp EUL:Gen 1425/392 (C: RS:HS 12.405 & 22.189)

5799. 1844–4–18 From Alexander von Humboldt Berlin
Sending a letter via Ca'colia Gibson raising various queries regarding nebulae.
ALS 3pp (French) RS:HS 10.67

5800. 1844–4–18 From Roderick Impey Murchison 16 Belgrave Square
Regrets he made a mistake in his letter; he meant 1845, not 1846. William Whewell has been overruled and the B.A.A.S. are to meet at Cambridge in June 1845.
ALS 4pp RS:HS 12.406

5801. 1844–4–20 To Roderick Impey Murchison Collingwood
Can raise no objections to being president of the B.A.A.S. for 1845, but will be unable to be in York this year.
C 4pp RS:HS 12.407 & 22.190

5802. 1844–4–22 To R. Custance Collingwood
Provides a list of some of the better works for RC's library in the fields of astronomy and natural philosophy. In a postscript JH refers briefly to a few of JH's own writings.
ALS 4pp WT 89267

5803. 1844–4–22 To [Robert Hunt ?] [Collingwood]
Thanks RH for the papers of researches on light just received.
ALS 2pp WT 68037.3

5804. 1844–4–24 From Roderick Impey Murchison 16 Belgrave Square
Rejoiced to receive his friendly letter, even though they have no hopes of seeing him in 1846. They are still on friendly terms with William Whewell, though they disagree on the Cambridge meeting.
ALS 4pp RS:HS 12.408

5805. 1844–4–25 From Edward Joshua Cooper Nice
Is shortly leaving for the Tyrol. Has been engaged in unsuccessful comet hunting. Has found stars not in JH's catalogue and comments on these.
ALS 4pp RS:HS 5.285

5806. [1844]–4–25 From Sir Charles Lemon Charles St., Berkeley Square
Will act as JH suggests and call to see the Prince Consort's secretary. Has no power himself to secure a place at Christ's hospital.
ALS 5pp RS:HS 11.192

5807. 1844–4–26 To [William Radcliffe Birt] n.p.
Delighted by Prince Albert's benefaction to [WB]. Curious results by [Robert Hunt] may lead to 'chemical discoveries of a new order.' Gratified by attention given JH's testimonial for Hunt.
Adraft 4pp RS:HS 19.112

5808. 1844–4–26 From Edward Sabine Woolwich
Encloses bills for the Committee. [Johann] Lamont will be in England that year to correct continental surveys. Glad JH will be president [of B.A.A.S. meeting] at Cambridge.
ALS 6pp † RS:HS 15.158

5809. 1844–4–27 From Charles Richard Weld Royal Society
Committee on Physics will meet 2 May.
(A)LS 1p RAS:JH Archive 7/1.6 verso; Reel 8

5810. 1844–4–29 To Augustus De Morgan Collingwood
The benefit society is launched and all is well.
C 1p RS:HS 22.191

5811. 1844–4–29 To Mrs. Wrench Collingwood
Thanks MW for the beautiful picture created by her out of one of JH's sketches made on a journey to Europe. At MW's request, JH sends along some astronomical work in the handwriting of JH's father, William.
ALS 4pp enc WT 68134.1

5812. 1844–4–30 From Augustus De Morgan 7 Camden St.
Send money for Richard Farley to the Nautical Almanac Office. Has just moved house and is in the process of moving thousands of books.
ALS 3pp RS:HS 6.204

5813. 1844–5–1 From Robert Hunt Falmouth
Charles Lemon has communicated to him the result of an application made by JH to Prince Albert on behalf of one of RH's children. Is deeply grateful for his assistance.
ALS 4pp RS:HS 10.110

5814. 1844–5–1 To William Henry Smyth [Collingwood]
Skeptical of reports of seeing Jupiter's satellites with the naked eye.
C 1p RS:HS 22.192

5815. 1844–5–3 From Robert Hunt Falmouth
Sends a copy of G. R. Anson's letter to Charles Lemon regarding a scholarship for one of RH's children. Is deeply grateful to JH for his nomination and support.
ALS 2pp RS:HS 10.112

5816. [1844]–5–3 From Sir Charles Lemon Charles St., Berkeley Square
Encloses a copy of a letter received from Prince Albert's secretary; has sent the original to Robert Hunt. Encloses a note received from Hunt this morning.
ALS 2pp RS:HS 11.193

5817. 1844–5–5 From William Ivory Edinburgh
Is grateful for the interest JH takes in their late friend's orphan daughter. Gives summary of her financial position. A pension would be welcome. Outlines the arrangements for the sale of Thomas Henderson's books. Would JH mark the ones he thinks should be purchased for the Observatory.
ALS 3pp RS:HS 10.260

5818. 1844–5–5 From Wilhelm Struve Pulkowa
Sends this letter through his son Otto Struve. Informs JH of his upcoming visit to England and desires to renew the acquaintance at Collingwood.
ALS (German) 1p RS:HS 17.131

5819. 1844–5–7 To Augustus De Morgan Collingwood
Forwards a copy of the rules and regulations of the benefit society, with much thanks.
C 1p RS:HS 22.193

5820. 1844–5–9 To Caroline Lucretia Herschel Collingwood
Happy that his *Cape Results* are nearly finished. After reviewing his work, JH concluded that the amount of error per observation is no more than 30 or 35 seconds.
ALS 4pp TxU:H/L–0589.5; Reel 1058

5821. 1844–5–11 From Augustus De Morgan 7 Camden St.
Thanks for book. Has written three books on the foundation of algebra. Comments on these.
ALS 12pp RS:HS 6.205

5822. 1844–5–15 To Adam Sedgwick Collingwood
Discusses the meeting of the B.A.A.S. and William Whewell's view of the proceeding. Will assume the chair and expects support from AS, George Peacock, and probably Whewell.
ALS 4pp RS:HS 15.424 (C: RS:HS 22.195) J. W. Clark, *Life and Letters of the Reverend Adam Sedgwick*, vol. 2 (Cambridge, England, 1890), 73

5823. 1844–5–15 To William Whewell Collingwood
Has been converted to some of WW's philosophical views. Accepts WW's invitation to stay at Trinity Lodge during the 1845 B.A.A.S. meeting in Cambridge. W. R. Dawes has taken a house in Kent.
ALS 4pp TC, Camb. Add. Ms.a.207[57] (C: RS:HS 22.194)

5824. 1844–5–17 From Robert Hunt Falmouth
Sending some specimens of photographs by a new process that he thinks is unrivalled for simplicity and sensibility. Comments on his method. Has sent an account of another curious process to the *P.M.* His little boy will go to Christ's Hospital in July or September.
ALS 3pp RS:HS 10.113

5825. 1844–5–17 From Adam Sedgwick London
Discusses prior meeting of B.A.A.S. to determine if [Cambridge] University would receive it. Says meeting went well. Asks if JH will agree to chair the B.A.A.S. meeting.
ALS 5pp RS:HS 15.423

5826. 1844–5–21 From Benjamin Herschel Babbage Genoa
Would like JH to be a godfather. Regarding his own work in Italy. Preparing the plans for the railway from Genoa to Milan under the direction of I. K. Brunel.
ALS 2pp RS:HS 2.308

5827. 1844–5–23 From Robert Wauchope Lisieux, Calvados, France
Death of son William on [21 May 1844].
ALS 3pp RS:HS 18.124

5828. 1844–5–27 From George Biddell Airy Royal Observatory, Greenwich
Regarding JH's forthcoming visit to the Greenwich Observatory.
ALS 1p RS:HS 1.136 (C: RGO 6.8.621)

5829. 1844–5–30 To George Biddell Airy n.p.
Whether a repulsive force from the sun affects the tails of comets and thus the constancy of their orbits.
draft 1p RS:HS 22.196

5830. 1844–5–30 To George Biddell Airy n.p.
Arrangements about attending a meeting, together with some thoughts on the behavior of comets.
ALS 4pp RGO 6.368.671

5831. 1844–5–31 From Catherine Bell n.p.
Family news of the Wauchopes.
ALS 4pp RS:HS 4.36

5832. 1844–5–31 From Josiah Quincy Harvard University
Discusses possibilities for republication of James Grahame's [*History of the United States of North America*]. Memoirs of Grahame are complete.
ALS 2pp RS:HS 14.292

5833. 1844–6–4 From Caroline Lucretia Herschel Hanover
Remarks that the citizens of Hanover 'are all out of their senses' over the newly completed railway. Has been confined to the upstairs of her home since 3 February.
L *Memoir ... of Caroline Herschel*, 334–5 inc

5834. 1844–6–8 From H. T. Winther [Neuwegen?]
Has developed a new process for making photographs that are better than the Daguerreotype. HW would like to sell the process [patent?] to someone in England.
ALS 2pp (German) RS:HS 18.426

5835. 1844–6–10 or [1844–5–6?] To Alexander von Humboldt Collingwood
Extensive letter answering the many queries about nebulae raised by AH in his 1844–4–18 [letter contains notes made by AH].
C 6pp RS:HS 22.197

5836. 1844–6–11 From Edward Sabine Woolwich
Add to [ES's] enclosed letter, which ES will explain more fully at committee meeting next week.
ALS 1p TxU:H/M–0520.2; Reel 1093

5837. 1844–6–12 To Chairman of the R.S.L. Committee of Physics (including Meteorology) Collingwood
Encloses letter from Edward Sabine and supports proposal contained therein for excursion southward from Cape of Good Hope to complete magnetic data.
ALS 4pp RS MC.4.32

5838. 1844–6–12 From Edward Sabine Woolwich
Colonial governor at Ceylon will grant £150 annually for operation of magnetic observatory at Colombo, directed by Royal Artillery officers already there, if [R.S.L.] will supply instruments. Proposes seeking money for instruments from Wollaston Fund.
ALS 4pp TxU:H/M–0520.3; Reel 1093

5839. 1844–6–13 To Thomas Emerson Headlam Collingwood
When he was staying with Headlam he arranged to show him round a chemical works. Wonders if the sketch JH made at the time of a method for reducing chemical fumes is still extant. Would be glad to receive a copy of it, and the name of the works.
C 2pp RS:HS 9.263 & 22.198

5840. 1844–6–13 To Thomas Thomson Collingwood
Sends TT some specimens of the plant *tritonia uvaria.*
ALS 2pp APS B.H435p.12

5841. 1844–6–15 From Wilhelm Struve Greenwich
Notifies JH of his and [P. A.] Hansen's arrival in Greenwich

from Gotha. Plans on visiting JH the following day via Staplehurst.
ALS (German) 2pp RS:HS 17.132

5842. 1844–6–18 To George Biddell Airy Collingwood
Regarding object glasses for telescopes. Relative merits of Munich and French object glasses.
draft 2pp RS:HS 1.137 & 22.199

5843. 1844–6–18 To Robert Grahame Jr. Collingwood
If he will send him what notes he has on his father he will forward them to [Edmund?] Quincy. Can he let him have definite news as to the disposal of the *History*?
C 2pp RS:HS 9.10bis

5844. 1844–6–18 From William A. Baillie-Hamilton Surrey Place
Details regarding the Admiralty Manual of Instruction. Number of copies and compensation for JH.
ALS 5pp RS:HS 9.206

5845. 1844–6–18 To Edward Sabine Collingwood
Wollaston Fund should be reserved for furtherance of private individual research, not for vast public projects. Prefers to ask assistance from secretary for Colonies.
C 3pp TxU:H/L-0362; Reel 1054

5846. 1844–6–18 To Wilhelm Struve n.p.
Mortified at *Punch* cartoon libeling WS's sovereign. Enjoyed WS's visit and invites Otto Struve again. Request WS to give [G. B.] Airy his 'full impressions' of Munich glass as opposed to British glass. Supports British glass for Cape's colonial instruments.
ALS (German) 1p RS:HS 17.133 (C: RS:HS 22.200)

5847. 1844–6–20 From Edward Sabine Woolwich
Appropriate use of Wollaston Fund for supporting research. Possible sources of instruments for new magnetic and meteorological observatory at St. John's, Newfoundland.
ALS 2pp TxU:H/M–0520.4; Reel 1093

5848. 1844–6–21 To Spencer J. A. Compton Collingwood
Encloses letter from Edward Sabine suggesting establishment of magnetic and meteorological observatory in colony of Ceylon.
ALS 3pp RS MC.4.34

5849. 1844–6–26 From Augustus De Morgan 7 Camden St.
Hopes the Friendly Society goes on well.
ALS 1p RS:HS 6.206

5850. 1844–6–26 From Robert Grahame Jr. Glasgow
Has been occupied in London with Railway business and so has been unable to make a copy of his father's *History*. Will obtain the services of an amanuensis to speed things up.
ALS 2pp RS:HS 9.11

5851. 1844–6–28 From Wilhelm Struve Greenwich
Introduces [Wilhelm] Döllen, who wishes to meet JH. WS recommends Munich glass for the Cape telescope.
ALS (German) 2pp RS:HS 17.134

5852. 1844–6–29 From George Biddell Airy Royal Observatory, Greenwich
Further observations on object glasses.
ALS 4pp RS:HS 1.138

5853. 1844–7 From John Henry Griesbach Carlton Villas, Maida Vale
Has now succeeded in grinding his 20" speculum and it is ready for polishing. Would like to see the mounting of JH's telescope as it is not clear from the engraving.
ALS 2pp RS:HS 9.57

5854. 1844–7–5 From Thomas Emerson Headlam Newcastle/Tyne
Remembers taking him round the chemical works of Losh, Wilson and Bell, but is unable to give any news of the manuscript as Mr. Losh is away on the continent. Hopes he will visit them again in the future.
ALS 5pp RS:HS 9.264

5855. 1844–7–8 From Edward Sabine Woolwich
A. T. Kupffer proposes international meeting of authorities in terrestrial magnetism. ES suggests combining this with 1845 meeting of B.A.A.S. at Cambridge. Plans for publishing observations before Cambridge meeting.
ALS 4pp † TxU:H/M–0520.5; Reel 1093

5856. 1844–7–8 From Robert Grahame Jr. Glasgow
Hopes to have the corrected copy of his father's *History* ready in about 10 days. If JH will give him the style and address of Dr. [Edmund?] Quincy he will send it direct.
ALS 2pp RS:HS 9.12

5857. 1844–7–11 From Elizabeth Drummond to Margaret Brodie Herschel Portobello, Edinburgh
Now a long time since she has written but she has happy memories of their last meeting. Is anxious about the posthumous fame of her son. Would be pleased to hear news of the family.
ALS 4pp RS:HS 6.448

5858. 1844–7–12 From Heinrich Christian Schumacher Altona
Announces the discovery of a new comet by [Heinrich] D'Arrest, a university student. Wilhelm Struve has seen it.
ALS 1p RS:HS 15.394

5859. 1844–7–13 From Joseph Jopling 29 Wimpole St.
Hoping to publish a register of examples of curves based on his own Septenary system of generating lines. Would JH be interested?
ALS 3pp RS:HS 10.415

5860. 1844–7–15 To Mary Baldwin Collingwood
Birth of JH's daughter [Mathilda Rose]. JH's 'little travelers' will arrive soon to visit Baldwins at Anstey.
ALS 3pp TxU:H/L-0066; Reel 1054

5861. 1844–7–17 To Caroline Lucretia Herschel Collingwood
Reports the birth of JH's 9th child, Mathilda Rose Herschel.
ALS 3pp TxU:H/L–0589.10; Reel 1058

5862. 1844–7–17 To Edward Sabine Collingwood
Believes that rather than asking the government for more money to continue making magnetic observations, the six years of observations should be studied to see if further observations can be justified.
ACCS 2pp RS:HS 25.6.23

5863. 1844–7–18 From Francis Beaufort Admiralty
When his packet is ready for the Cape, forward it to him at the Admiralty and he will arrange for it to be sent. Congratulations on the new addition to the family. When next in town perhaps he would be interested in seeing the new object glass for the Cape, which William Simms is preparing.
ALS 3pp RS:HS 3.375

5864. 1844–7–19 From Edward Sabine Woolwich
JH is premature in concluding that international meeting of magnetic researchers would result in appeal to British government for funds to continue B.A.A.S. observatories beyond 1845. ES has not reviewed later observations [1842–43] and cannot judge whether observations have met original goals of R.S.L. and B.A.A.S. Agreement by Robert Peel and F. I. Brunnov on need for longer period of observations. Affirms value of JH's role in evaluating results of survey.
ALS 7pp TxU:H/M–0520.7; Reel 1093

5865. 1844–7–19 From Edward Sabine Woolwich
Ship is ordered for Lt. Clark's excursion. Feels certain that Lord Stanley has consulted colonial governor.
ALS 1p TxU:H/M–0520.6; Reel 1093

5866. 1844–7–20 From William Rutter Dawes Camden Lodge
Has traced the comet notified in J. F. Encke's circular. Gives his readings for its position. Has compared the stars near it. Hopes JH's family are all well.
ALS 4pp RS:HS 6.67

5867. 1844–7–21 To Thomas Maclear Collingwood
Tries to obtain equatorial telescope for Cape Observatory. Works only on *Cape Results* and his children's education. Mentions reforms in southern constellations. Announces birth of seventh daughter [Mathilda Rose].
ALS 4pp Cape Archives/Maclear Papers File 100

5868. 1844–7–22 To George Biddell Airy Collingwood
Will try to organize an international magnetic conference at the B.A.A.S. meeting in 1845, if GA agrees.
ALS 2pp RGO 6.675.354

5869. 1844–7–22 To Humphrey Lloyd Collingwood
Informs HL of the intention of holding a magnetic congress at the meeting of the B.A.A.S. in 1845, and JH seeks HL's opinion.
ACCS 1p RS:HS 25.6.25

5870. 1844–7–22 To Thomas Maclear n.p.
'Penny Post' is deluging JH with nuisance mail. Reviews JH's efforts to obtain object glass for equatorial telescope at Cape of Good Hope. Trying to convince G. B. Airy to purchase from Merz and Mahler in Munich, rather than from William Simms. Russia is using TM's 'Cape Arc' in survey of its entire country.
Adraft 2pp TxU:H/L-0242; Reel 1054 (Cdraft: RS:HS 25.15.13 inc)

5871. 1844–7–22 To Edward Sabine Collingwood
Having been assured by ES that the government might receive a request for more money for magnetic observations, JH retreats from his earlier position [see JH's 1844–7–17], and is willing to accept the possibility of a magnetic conference, which JH does not want to lead in any way. JH believes his scientific life is almost over and he has much to do.
ACCS 4pp RS:HS 25.6.24

5872. 1844–7–22 To William Whewell Collingwood
Seeks WW's advice on a proposal that an international magnetic congress be held in conjunction with the 1845 B.A.A.S. meeting in Cambridge.
ALS 1p TC, Camb. Add. Ms.a.207[58] (C: RS:HS 22.201)

5873. 1844–7–23 From Francis Beaufort Admiralty
Will send his letter to Thomas Maclear with one of his own. Agrees with him regarding the object glass, but feels that English artists should be encouraged.
ALS 3pp RS:HS 3.376 & ACC: Hydrographic Office Letter Book

5874. 1844–7–23 From Robert Main Royal Observatory, Greenwich
As G. B. Airy is traveling in Ireland for his health, RM acknowledges receipt of JH's 1844–7–22, with a promise to deal with it as JH wishes.
ACCS 1p RGO 6.675.356

5875. 1844–7–23 From Edward Sabine Woolwich
ES agreed to superintend publication of magnetic observations only because Humphrey Lloyd refused to do so any further. Format intended for each volume. Gratitude for JH's contributions to magnetic research. [P.S.: Provision of instruments to U.S. observers.]
ALS 5pp TxU:H/M–0520.8; Reel 1093

5876. 1844–7–24 To Robert Main Collingwood
Asks RM to forward letter [see JH's 1844–7–22] immediately to G. B. Airy.
ALS 1p RGO 6.675.357

5877. 1844–7–24 From George Peacock Deanery of Ely
Agrees to host congress at Cambridge to assess data from [magnetic and meteorological] observations. Suggests adding D. F. J. Arago to list. Will include this in GP's address to B.A.A.S at York. [JH annotation: List of five members.]
ALS 3pp † TxU:H/M–0415; Reel 1087

5878. 1844–7–25 From William Whewell 41 U. Grosvenor St.?
Approves the idea of a magnetic conference sponsored by the B.A.A.S. Assumes Edward Sabine will lead the way.
ALS 3pp TxU:H/M-0566

5879. 1844–7–28 To Friedrich Wilhelm Bessel Collingwood
Talks about stiffening a vertical circle, or tube; explains that southern constellation remodelling has been very minimal.
ALS 4pp AdWDDR Bessel 225 (C: RS:HS 22.202)

5880. 1844–7–28 From Johann Heinrich Mädler Hanover
JM is traveling in Germany for some time and intends to end up at a scientific conference in Bremen. He encourages JH to try to attend that conference so they may become acquainted.
CCS 3pp (German) RS:HS 12.274

5881. 1844–7–30 From George Biddell Airy Cruise's Hotel, Limerick
Does not believe that a magnetic congress held at this time would be productive or representative; a better job could be done by letter.
ALS 2pp † TxU:H/M-0630.1; Reel 1084 (C: RGO 6.675.360 & RS:HS 1.139)

5882. 1844–7–31 From Mary Somerville Lower Belgrave St.
Will be in London and hopes to see JH. Thanks for his letter and the promise of sending his papers on light. Also wants information on [Louis] Daguerre's work.
ALS 3pp RS:HS 16.349

5883. 1844–8 To Mary Baldwin [Collingwood]
Suggestions for travel [to Collingwood] by railroad and coach.
ALS 4pp TxU:H/L-0067.2; Reel 1054

5884. 1844–8–1 From Edward Sabine Woolwich
Informed A. T. Kupffer that JH will preside at 1845 B.A.A.S.

meeting and invited Kupffer to attend. Activities of [James] Renwick and A. D. Bache in trigonometrical survey of U.S.
ALS 3pp TxU:H/M–0520.9; Reel 1093

5885. 1844–8–2 From William Henry Fitton Wandsworth
One of the sections of the beds below the chalk, which JH sent him in 1834, has become the subject of interesting discussion *viz* the Neocomian beds and their appearance in the British Isles. Comments on the various theories regarding these beds.
ALS 8pp RS:HS 7.239

5886. 1844–8–2 From John Henry Griesbach Carlton Villas, Maida Vale
Is grateful for JH's offer of assistance and proposes to visit him next Wednesday. Congratulations on the addition to his family.
ALS 1p RS:HS 9.58

5887. 1844–8–9 From Thomas I. M. Forster Bruges
Hopes he will look out for the periodical meteors and give him his observations. Has seen some curious falling stars. Is constructing a new anemometer and used it to detect surging gales. Visited a monastery yesterday to see a new instrument for perpetual motion.
ALS 4pp RS:HS 7.336

5888. 1844–8–9 From E. C. Hawtrey Downgate
Enjoying the beautiful scenery, but the weather has not been good. Many thanks for the beautiful translation of Frederick Schiller. Has he added anything to his photographic discoveries? Saw some calographic pictures at Charles Wheatstone's, which improved on the Daguerreotype.
ALS 7pp RS:HS 10.172

5889. 1844–8–11 To [William Radcliffe Birt] n.p.
Parcel arrived safely. Will examine it with interest.
ALS 1p RS:HS 19.113

5890. 1844–8–13 From William Samuel Stratford 3 Verulam Buildings
Expresses the ill Francis Baily's appreciation for JH's letter. Writes of suffering of Baily and distress of Miss Baily [his sister].
ALS 4pp RS:HS 17.89

5891. 1844–8–17 From William Samuel Stratford 3 Verulam Buildings
Updates JH on declining health of Francis Baily, reporting Baily realizes he has few days to live but remains calm.
ALS 3pp RS:HS 17.90

5892. 1844–8–18 From Robert Grahame Jr. Glasgow
Has sent a copy of his father's *History* by the Boston steamer to Dr. Quincy. Believes the latter was hurt by the delay in his reply. Has been engrossed with railway business. Intends sending the original manuscript to Quincy.
ALS 4pp RS:HS 9.13

5893. 1844–8–20 To Mary Baldwin Collingwood
MB's intended visit to Collingwood conflicts with arrival of other guests.
AdraftS 4pp TxU:H/L-0068; Reel 1054

5894. 1844–8–22 From William Samuel Stratford 3 Verulam Buildings
Apologizes for not having informed JH of Francis Baily's further declining health. Praises Miss Baily's character and devotion to her brother.
ALS 4pp RS:HS 17.91

5895. 1844–8–23 From Giovanni Santini Padua
Sends medical writings of Philip Spongia. Sends observations of Mauvais's Comet from Berlin, Milan, and Padua. Plans to present personal observations to R.A.S. in the following year.
ALS 3pp (Italian) RS:HS 15.348

5896. 1844–8–23 From William Samuel Stratford 3 Verulam Buildings
Reports on the dying Francis Baily's increasing weakness, calm acceptance of death, gratitude for friends, and freedom from pain.
ALS 2pp RS:HS 17.92

5897. 1844–8–24 From Edward Sabine n.p.
Let George Peacock take care of inviting foreign magnetic researchers to B.A.A.S meeting at Cambridge. ES could prepare report on results of British government observatories by next February.
ALS 3pp TxU:H/M–0520.10; Reel 1093

5898. 1844–8–24 From William Samuel Stratford 3 Verulam Buildings
Attests that the dying Francis Baily remains alive and free from pain.
ALS 1p RS:HS 17.93

5899. 1844–8–25 To George Biddell Airy Collingwood
About a suggested magnetism conference, Francis Baily's illness, and some remarkable observations by F. W. Bessel.
ALS 4pp RGO 6.368.673

5900. 1844–8–26 From George Biddell Airy Royal Observatory, Greenwich
Informing JH that he certainly wrote to him from Limerick.
ALS 1p RS:HS 1.140 (C: RGO 6.675.360)

5901. 1844–8–26 From George Biddell Airy Royal Observatory, Greenwich
Cannot explain why GA's letter did not reach JH.
ACCS 1p RGO 6.675.360

5902. 1844–8–26 From William Samuel Stratford 3 Verulam Buildings
The dying Francis Baily, although unable to sit up, has calmly survived another day.
ALS 1p RS:HS 17.94

5903. 1844–8–27 From William Samuel Stratford 3 Verulam Buildings
The dying Francis Baily remains alive, 'sustained by the amazing quantity of food he takes.'
ALS 2pp RS:HS 17.95

5904. 1844–8–27 To William Whewell Collingwood
Has WW received a packet of letters for the magnetism committee, which packet was circulating.
ALS 1p TC, Camb. Add. Ms.a.207[59] (C: RS:HS 22.203)

5905. 1844–8–28 From George Biddell Airy Royal Observatory, Greenwich
Regarding F. W. Bessel's letter on the irregularity of proper motion in stars. Is it worth printing Nevil Maskelyne's observations on star places?
ALS 2pp RS:HS 1.141 (C: RGO 6.193.118)

5906. 1844–8–29 From William Samuel Stratford 3 Verulam Buildings
Although the dying Francis Baily sleeps all day and has not eaten, he remains calmly alive.
ALS 2pp RS:HS 17.96

5907. 1844–8–30 To George Biddell Airy Collingwood
Urges GA to print any of John Pond's and Nevil Maskelyne's observations that seem free from serious error [see GA's 1844–8–28].
ALS 4pp RGO 6.193.119

5908. 1844–8–30 From William Samuel Stratford 3 Verulam Buildings
To the surprise of all, the dying Francis Baily still lives.
ALS 1p RS:HS 17.97

5909. 1844–8–31 From William Samuel Stratford 37 Tavistock Place
Reports gentle death of Francis Baily.
ALS 1p RS:HS 17.98

5910. 1844–8–31 From William Samuel Stratford 3 Verulam Buildings
Requests JH's opinion on whether to attempt to obtain a grave in Westminster Abbey for the deceased president of the Astronomical Society, [Francis Baily].
ALS 1p RS:HS 17.99

5911. 1844–8–31 From William Whewell Penrith
Have received the packet of 'magnetic letters' and passed them to George Peacock. Will attend the York B.A.A.S. meeting.
ALS 2pp TxU:H/M-0567

5912. 1844–9–1 To [William Radcliffe Birt] Collingwood
WB's wave models show that barometer observations will reveal atmospheric dynamics like astronomical observations revealed planetary orbits. Encourages WB to seek simple method of temperature corrections applicable to all barometers. Offers possible mechanism. Send results to B.A.A.S quickly. Will send WB's letter to [Magnetical and Meteorological] Committee.
ALS 4pp RS:HS 19.114

5913. 1844–[9 or later] To Augustus De Morgan Collingwood
JH agrees to write a biographical sketch of Francis Baily, and hopes he can count on others for information.
C 2pp RS:HS 22.217

5914. 1844–[9 or later] To Augustus De Morgan Collingwood
Continues to work on biographical sketch of Francis Baily [see JH's 1844–[9 or later]]; is working through a lot of material to do it.
C 1p RS:HS 22.218

5915. 1844–9–2 From George Biddell Airy Royal Observatory, Greenwich
Expands on GA's proposed letter writing scheme in place of holding a magnetic congress [see GA's 1844–7–30].
ALS 3pp TxU:H/M-0630.2; Reel 1084 (ACCS: RGO 6.675.363 & C: RGO 6.675.361)

5916. [1844–9–2?] To [William Radcliffe Birt] [Collingwood]
Omitted important detail yesterday in JH's description of self-compensating barometer. Columns must be variable, not constant.
ALS 3pp RS:HS 19.168

5917. 1844–9–6 To Augustus De Morgan Collingwood
Has sent some of AD's algebra papers to Mary Somerville; comments on death of Francis Baily.
C 2pp RS:HS 22.204

5918. 1844–9–7 From Alexander Bridport Becher Admiralty
Sends JH a letter from Thomas Maclear, in line with Francis Beaufort's instructions.
ACC 1p Hydrographic Office Letter Book

5919. 1844–9–8 From Augustus De Morgan 7 Camden St.
Will send the information to Mary Somerville. Death of Francis Baily is most grievous. Hopes JH will write an obituary. Society has to elect a new president. Regarding his own algebraic papers.
ALS 4pp RS:HS 6.207

5920. 1844–9–8 From William Henry Fox Talbot Lacock Abbey
Some comments and questions on mathematics.
ALS 2pp RS:HS 17.317

5921. 1844–9–9 From William Radcliffe Birt Bethnal Green
Has just examined the area between Dublin, Paris, and Heligoland for waves. Self registering barometers would be useful.
ALS 4pp RS:HS 4.109

5922. 1844–9–9 From Edward Sabine Woolwich
Sends copy of ES's memorandum to George Peacock describing progress in publication of magnetic and meteorological observations. Encloses copies of observations from Antarctic survey and Ceylon. Reasons for H.E.I.C.'s hesitation to expand survey by C. M. Elliott. Praises C. J. B. Riddell's book, which described portable magnetic apparatus that is now requested by observers worldwide. Recent changes in other instruments. Reimbursement for instruments provided to observatory in Newfoundland. Recent results of surveys on Continent, in Canada, and by Capt. Blackwood in Torres Straits.
ALS 8pp † TxU:H/M–0520.11; Reel 1093

5923. 1844–9–12 or earlier From Richard Sheepshanks 14 London Road, Reading
Having difficulty drawing the Great Nebula in Orion using both JH's and [Francesco] De Vico's observations; prints of Francis Baily are almost ready.
ALS 4pp RS:HS 16.133

5924. [1844]–9–12 or later From Richard Sheepshanks n.p.
About trying to convert some surveys to a base using RS's standard yard.
ALS 3pp RS:HS 16.130

5925. 1844 or later–9–12 From Richard Sheepshanks Trinity College, Cambridge
Left prints of Francis Baily for JH, and will send a pamphlet about a long standing controversy in which RS was involved.
ALS 2pp RS:HS 16.128

5926. 1844–9–13 From B. Baily Threadneedle St.
Doubtless JH has heard of the death of the writer's cousin Francis Baily. There is a legacy of £1,000 due from Baily's estate.
ALS 2pp RS:HS 5.6

5927. 1844–9–14 To Charles Babbage Collingwood
Regarding the health of his aunt and other family matters.
ALS 4pp RS:HS 2.309

5928. 1844–9–14 To B. Baily Collingwood
Is deeply affected by the news of the death of Francis Baily,

and also at the news that he is to receive £1,000 under the terms of the will. Hopes to put it to a use of which Baily would have approved.
draft 3pp RS:HS 5.7 & 22.205

5929. 1844–9–14 To [William Radcliffe Birt] Collingwood
Sent WB's paper and models to [B.A.A.S. meeting in] York, to be included in report of [Magnetical and] Meteorological Committee.
ALS 4pp RS:HS 19.115

5930. 1844–9–14 From John Henry Griesbach Carlton Villas, Maida Vale
Has been unable to return the papers as he has been so busy. Proposes making working drawings as he suggests. Will take the papers to the Dover Railway to ensure their safety.
ALS 2pp RS:HS 9.59

5931. 1844–9–14 From Edward Sabine Woolwich
Forwarded to George Peacock JH's letter of 11 Sept. Karl Kreil will send self-registering instruments that ES ordered for Kew observatory. Influence of vapor pressure on barometer readings. Introduction of hygrometers.
ALS 7pp † TxU:H/M–0520.12; Reel 1093

5932. 1844–9–16 From Edward Sabine Woolwich
George Peacock returned JH's papers, with letter by Peacock. ES forwards these to JH.
ALS 3pp TxU:H/M–0520.13; Reel 1093

5933. 1844–9–16 From Mary Somerville London
Thanks JH for papers on light and photography. Enjoyed her visit to Collingwood. Feels privileged to be godmother to JH's daughter.
ALS 3pp RS:HS 16.350

5934. 1844–9–16 From William Henry Fox Talbot 31 Sackville St.
Comments and questions about a 'cone of the second order' in mathematics.
ALS 3pp RS:HS 17.318

5935. 1844–9–17 From Robert Hunt Falmouth
Has just returned from London where he has taken his son, ready to go to Christ's Hospital. Would have called on JH but believed he was not at Collingwood. Hopes to meet him at York. Comment on his recent photographic work. Would like to obtain a position in London.
ALS 3pp RS:HS 10.114

5936. 1844–9–17 From Thomas Maclear Zwart Kop near Simons Town [Cape of Good Hope]
Magnitude changes in Eta Argus. Compares methods for measuring brightness. Difficulties completing 'sector observations.' C. P. Smyth's problems with theodolite at Cape Agulhas. Mrs. [Mary] Maclear recovering after serious illness. New roads and bridges in colony.
ALS 3pp RAS:JH Archive 3/1.1.13; Reel 7

5937. 1844–9–17 From Thomas Maclear Baseline Camp Zwartland, near Simons Town
Attempts to determine light curve of Eta Argus; describes methods for determining magnitude. C. P. Smyth uses R.A.S. theodolite telescope. TM will triangulate Bushman Flat. Parallax work at Cape Observatory continues. Reports progress of new government Secretary, [John] Montague.
ALS 4pp Cape Archives/Maclear Papers File 100

5938. 1844–9–18 From Palm Heinrich Ludwig Boguslawski Breslau
His answer to the queries has not reached him; sends duplicate. Will receive *Trans. of the Silesian Soc.* Forthcoming map of stars and revision by JH. Does he require the observations of the meteorological terms be fixed?
ALS 3pp RS:HS 4.174

5939. 1844–9–19 From W. L. Newman York
Offering hospitality to JH should he attend the B.A.A.S. meeting at York. Has been studying the theory of the manufacture of achromatic telescopes. Comments on the quality of glass available. Swiss seems to be best, though expensive. Deplores the loss of Francis Baily.
ALS 3pp RS:HS 13.153

5940. 1844–9–20 From Edward Sabine Woolwich
Received letter from J. H. Lefroy in Athabasca [Canada]. Extract of C. J. B. Riddell's winter magnetic operations and summer plans along Red River.
ALS 2pp TxU:H/M–0520.14; Reel 1093

5941. 1844–9–23 From George Biddell Airy Royal Observatory, Greenwich
Hopes JH will write a memoir of Francis Baily.
ALS 1p RS:HS 1.142 (C: RGO 6.229.192)

5942. 1844–9–23 From George Biddell Airy Royal Observatory, Greenwich
A note accompanying a draft of the report on magnetic observations to date at British stations, and a circular to be sent out.
ALS 1p TxU:H/M-0630.3; Reel 1084 (C: RGO 6.675.368)

5943. 1844–9–23 From Augustus De Morgan 7 Camden St.
Has been unable to reply to JH's note as he was ill. There will be a special General meeting of the society in November to hear the memoir of Francis Baily. Will send all the information as soon as possible. Has invented a new algebraic symbol.
ALS 3pp RS:HS 6.208

5944. 1844–9–23 From Thomas I. M. Forster Bruges
Has sent a pamphlet on Education to the R.S.L. to be transmitted to JH. Explains the mechanism of his new anemometer. Is most interested in the causes of falling stars and their relation to meteorites. Has he seen [Robert?] Hunt's book on *Light*? Does he know where a description of William Parsons' (3rd Earl of Rosse) large telescope is to be found?
ALS 5pp RS:HS 7.337

5945. 1844–9–24 From W. L. Newman York
Glad to hear that there is a possibility that JH will visit York. Another expected visitor will be T. Donkin, brother of Bryan Donkin.
ALS 2pp RS:HS 13.154

5946. 1844–9–26 From Edward Sabine York
Magnetic phenomena move in opposite directions in the northern and southern hemispheres. This contradicts a report made by JH. Asks permission to omit JH's statements concerning this issue.
ALS 4pp RS:HS 15.159

5947. 1844–9–26 From Alexander Wilson Great New St. [Gough Square]
Begs JH to read enclosed [*Essay on the Standard and Measure of Value*, 2nd ed. 1832] by John Taylor and reconsider assisting the working classes.
ALS 2pp RS:HS 18.264

5948. 1844–9–27 To [B.A.A.S.] Collingwood
Encourages [B.A.A.S.] to accept offer in enclosed letter by W. S. Stratford, volunteering to assume duties of late Francis Baily in publishing star catalogs of N. L. Lacaille and J. J. L. Lalande.
C 4pp TxU:H/L-0644; Reel 1089

5949. 1844–9–28 From Josiah Quincy Harvard University
Discusses publication of James Grahame's [*History of the United States of North America*] revised. Includes circular describing Grahame, his project and the cost of the books.
ALS 2pp (plus 2pp circular) RS:HS 14.293

5950. 1844–9–30 From George Peacock Bishopthorpe, York
Advises that JH has been elected President of the B.A.A.S. for its meeting at Cambridge in 1845; also some comments on the magnetic survey.
ALS 3pp RS:HS 13.295

5951. [1844–10] From P. H. Holland Grosvenor St., Manchester
Is grateful for his letter concerning the proposed professorship. Outlines some of the alterations he proposes to submit to the Council. Hopes to raise a subscription of £5,000 to £6,000. Offer a salary of £200 per annum. Encloses the newspaper articles.
ALS 2pp RS:HS 9.446

5952. 1844–10–2 From Robert Hunt York
Returns his interesting specimens. Has read his communication in the Physical and Chemical sections. Regarding the use of certain terms. Had long discussion with W. H. Fox Talbot on matters of Light. Has been offered a position at the Birmingham Institution. Is there any chance of obtaining a position in London?
ALS 2pp RS:HS 10.115

5953. 1844–10–2 To Richard Sheepshanks Collingwood
Requests information from RS on the life of Francis Baily for a biographical sketch JH was asked to draw up for a special meeting of the R.A.S.
ALS 4pp RS:HS 16.23 (C. RS:HS 25.9.8)

5954. 1844–10–3 From Augustus De Morgan 7 Camden St.
Sends rough notes on Francis Baily's interests and works. Has not heard from W. S. Stratford lately.
ALS 1p RS:HS 6.209

5955. 1844–10–4 From John Burke 89 Sloane St.
Revising JB's *Dictionary of the Peerage and Baronetage*. Requests details of JH's family.
ALS 1p RAS:JH Archive 1/10.8; Reel 4

5956. 1844–10–5 To [Robert Hunt ?] Collingwood
Does not think that RH would be satisfied with the position of Assistant Secretary of the R.S.L., as it is largely clerical, and would not allow much time for chemical or photographical experiments. JH is not aware whether there is a position available at the Royal Institution, but would be willing to speak to Michael Faraday.
ALS 4pp WT 68037.4

5957. 1844–10–7 From Augustus De Morgan 7 Camden St.
Mr. Malby has obtained copyright of 36' globe and intends to lay down nebulae, also double stars. Is JH's work [on double stars] likely to be finished soon? Should wait for this work.
ALS 3pp RS:HS 6.210 S. De Morgan, *Memoir of Augustus De Morgan*, 148

5958. 1844–10–7 From Edward Sabine Scarborough
Received notes on omission of passage [1844–9–26] in good time. Asks JH to return proof of report. Regaining strength before going back to study Toronto volume.
ALS 4pp RS:HS 15.160

5959. 1844–10–8 To *Astronomische Nachrichten* Collingwood
Discusses critically a theory put forth by J. C. Houzeau on the 'effect of aberration arising from proper motion in changing the apparent orbits of double stars.'
L *Astronomische Nachrichten, 22* (1845), 249–54

5960. 1844–10–8 From Humphrey Lloyd Trinity College, Dublin
Approves of proposal by A. T. Kupffer and Edward Sabine for international magnetic congress under JH's leadership. [Reply to JH's 1844–7–22 to HL.]
ALS 2pp TxU:H/M–0334; Reel 1087

5961. 1844–10–8 To Roderick Impey Murchison Collingwood
Gratefully accepts role as President Elect for 1845 B.A.A.S. meeting.
ALS 1p EUL:Gen 1425/393

5962. 1844–10–8 From Adolphe Quetelet Brussels
Thanks for letter regarding equinoxes and solstices. Intends to work on physical phenomena of globe. Resumes meteorological observations. Awaits *Cape Results*.
ALS (French) 3pp RS:HS 14.247

5963. 1844–10–9 From Frederick S. Archer 18 Tavistock St., Covent Garden
Offering JH an opportunity to buy a copy of a bust of Mary Somerville.
ALS 1p RS:HS 1.404

5964. 1844–10–10 From Friedrich Wilhelm Bessel Königsberg
Responds to JH's 1844–7–28 about stiffening a vertical circle; comments on other astronomical matters.
ALS 5pp (German) RS:HS 4.83

5965. 1844–10–10 From John Phillips B.A.A.S., York
JH, William Whewell, George Peacock, Humphrey Lloyd, and Edward Sabine are appointed by the B.A.A.S. to study systems of simultaneous magnetical and meteorological observations. They will be granted £50.
(A)L 1p TxU:H/M–0428.17; Reel 1093

5966. 1844–10–10 From John Phillips B.A.A.S., York
JH appointed to a committee with G. B. Airy and William Stratford on the reduction of the calculations of N. L. Lacaille's stars.
(A)L 1p TxU:H/M–0428.16; Reel 1093

5967. [1844]–10–11 From Agnes Baillie Hampstead
Requesting autographs for her niece.
ALS 1p RS:HS 3.16

5968. 1844–10–11 From Richard Taylor Red Lion Court, Fleet St.
Francis Baily contracted with RT to print N. L. Lacaille's catalog of stars and J. J. L. Lalande's *Histoire céleste*. Account was suspended after Baily's death until B.A.A.S. meeting in York approved its continuation. Requests £300

from JH, chairman of grant committee.
ALS 1p † TxU:H/M–0548; Reel 1093

5969. 1844–10–12 To Richard Sheepshanks Collingwood
Sent his letter of 2 Oct. to Woburn Place. Has since learned that RS moved to Reading. Requests again information on the life of Francis Baily, especially on Baily's pendulum experiments.
ALS 3pp RS:HS 16.24 (C: RS:HS 25.9.9)

5970. 1844–10–12 To Alexander Wilson Collingwood
Critiques John Taylor's essay [see AW's 1844–9–26]. Views AW's position on currency measures as too vehement; urges caution. JH conceived similar plan years before. Gives reasons for disowning it. Protests JH's being painted as perpetrator of injustice simply because JH rejected AW's arguments.
C 3pp RS:HS 18.265 (C: 22.206)

5971. 1844–10–13 To George Biddell Airy Collingwood
Has been put (with GA and W. S. Stratford) on a B.A.A.S. committee to publish the J. J. L. Lalande and the N. L. Lacaille star catalogues [see GA's 1842–11–7]; sends draft request for a grant.
ALS 2pp RGO 6.203.329

5972. 1844–10–14 From George Biddell Airy Royal Observatory, Greenwich
Has sent on the draft report on magnetic observations to W. S. Stratford for completion [see JH's 1844–10–13].
ALS 1p TxU:H/M–0630.4; Reel 1084 (C: RGO 6.203.330)

5973. 1844–10–14 From George Everest Bury St., St James's
Went to the York meeting of the B.A.A.S. hoping to see JH, but was disappointed to hear he was not there. Would assure him of his friendship and esteem for his work.
ALS 4pp RS:HS 7.135

5974. 1844–10–14 From Edward Sabine Scarborough
Edinburgh [*Review*] named someone else as the person who found the true correction of the rate of pendulums on account of the atmospheric medium. ES actually discovered the correction first. Discusses [James] Ross's intensity ovals.
ALS 4pp RS:HS 15.161

5975. 1844–10–14 From Richard Sheepshanks 14 London Road, Reading
Responds in detail to JH's request [1844–10–12] for RS's assessment of Francis Baily's character and contributions.
ALS 6pp RS:HS 16.25

5976. 1844–10–15 To George Biddell Airy Collingwood
Nine conditions under which JH will accept and test instruments constructed for Cape [of Good Hope] observatory, including the assistance of Mr. Davies in evaluating these.
C 4pp TxU:H/L-0014; Reel 1054 (C: TxU:H/M-0646; Reel 1084)

5977. 1844–10–15/27 From Adolph Theodor Kupffer St. Petersburg
Sending his Magnetic and Meteorological Annual for 1842. Any publications in exchange should be addressed to the Institute of Mines, St. Petersburg.
ALS 2pp RS:HS 11.95

5978. 1844–10–16 From George Biddell Airy Royal Observatory, Greenwich
Regarding object planes for the telescope.
ALS 2pp RS:HS 1.143

5979. 1844–10–17 From Thomas Watson 16 Henrietta St., Cavendish Square
Seeks explanation for apparent oscillation of patterns on mat when viewed only by peripheral vision.
ALS 4pp RS:HS 18.384

5980. 1844–10–18 From Augustus De Morgan 7 Camden St.
Takes great interest in Mr. Malby's globes as he is writing an article on the uses of globes. Comments on the work of Malby.
ALS 4pp RS:HS 6.211 S. De Morgan, *Memoir of Augustus De Morgan,* 148–9

5981. 1844–10–19 From Agnes Baillie Hampstead
Many thanks for the autographs, which she will pass on to her niece.
ALS 1p RS:HS 3.10

5982. 1844–10–19 From P. H. Holland Grosvenor St., Manchester
Would like JH's advice on the setting up of a professorship in memory of John Dalton.
ALS 3pp RS:HS 9.444

5983. 1844–10–19 From Edward Sabine Scarbro'
J. H. Lefroy returned from McKenzie River [Canada] having established at least 350 magnetic stations between there and Red River, 'a fine pendant for [J. C.] Ross's survey' in southern hemisphere.' If expedition continues through winter, it will provide excellent survey of North American disturbances.
ALS 4pp TxU:H/M–0520.15; Reel 1093

5984. 1844–10–19 From Edward Sabine Scarborough
Sending proof copies of the report on the Toronto meteorological observations. Asks JH to make suggestions for corrections and improvements.
ALS 4pp RS:HS 15.162

5985. 1844–10–19 To Richard Sheepshanks Collingwood
Thanks RS for his letter on Francis Baily. Wants a copy of the engraving of Baily's portrait that RS is preparing, and thinks RS ought to make arrangements for its distribution among R.A.S. members.
ALS 3pp RS:HS 16.26 (C: RS:HS 25.9.10)

5986. 1844–10–21 From Alexander Wilson Letterfoundry, near Hemel Hempstead
Regrets JH was dissuaded by arguments grounded on human pettiness. Explains why AW took strong stand on currency issue. Assurance that AW respects JH and did not paint him as evil.
ALS 3pp RS:HS 18.266

5987. 1844–10–22 From Edward Sabine Scarborough
Obliged for JH's comments regarding pendulum apparatus. Discusses the article that wrongly attributes ES's discovery to [F. W.] Bessel and [Francis] Baily.
L 4pp inc RS:HS 15.163

5988. 1844–10–23 To P. H. Holland Collingwood
Comments on the pros and cons of a statue or professorship in memory of John Dalton. Would favor a professorship. Thinks the rules should be amended. Knows a good man for the position.
draft 3pp RS:HS 9.445 & 22.207

5989. 1844–10–23 From William Samuel Stratford 3 Verulam Buildings
Lists supplies sent to JH. Reports on Miss Baily's need for

quiet repose but generally good recovery after her brother's death. (She looks forward to JH's visit.) Outlines Francis Baily's biography and lineage.
ALS 4pp RS:HS 17.100

5990. 1844–10–25 From Richard Sheepshanks Reading
Supplements earlier letter [1844–10–19] giving JH his views on Francis Baily
ALS 4pp RS:HS 16.27

5991. 1844–10–26 From William Henry Smyth Edenhall
Sends copy of *Cycle of Celestial Objects*. Meets an excellent optician named [Richard?] Gwatkin, who shows him four self-made telescopes.
ALS 3pp RS:HS 16.276

5992. 1844–10–27 To William Henry Smyth Collingwood
Compliments WS on his *Cycle of Celestial Objects*. Inquires who instigated the move for reform of the *Nautical Almanac*.
C 2pp RS:HS 22.208

5993. 1844–10–28 To George Biddell Airy n.p.
Seeks GA's help to provide information to complete the obituary notice for Francis Baily [see GA's 1844–9–23].
ALS 4pp RGO 6.229.193

5994. 1844–10–29 From George Biddell Airy Royal Observatory, Greenwich
Giving abridged history of the establishment of the new Board of Visitors to the Royal Observatory. Details concerning the Nautical Almanac. Standards commission.
ALS 4pp RS:HS 1.144 (C: RGO 6.229.195)

5995. 1844–10–29 From John Henry Nelson Stoke Damerell
Gives news of the death of JN's father.
ALS 4pp RS:HS 13.93

5996. 1844–11 To 'Sir' London
[Form letter] Asks recipients to consider value of magnetical research supported by British government and submit opinions on whether B.A.A.S. should seek continuation of this support after present agreement with government expires at end of 1845.
ACCS 4pp TxU:H/L-0645; Reel 1089

5997. 1844–11–1 To George Biddell Airy Collingwood
Changing a date of meeting for the magnetic committee of the B.A.A.S., and inviting GA to be there.
ALS 2pp RGO 6.675.369

5998. 1844–11–1 To Humphrey Lloyd Collingwood
Asks HL for comments about the question of continuing the magnetic observations.
C 2pp RS:HS 22.209

5999. 1844–11–1 From George Peacock Deanery of Ely
GP prefers to meet on Saturday at Somerset House, rather than on Friday as JH proposed.
ALS 1p TxU:H/M-0416; Reel 1087

6000. 1844–11–1 From William Whewell Trinity Lodge
Sorry that WW cannot go to London to meet [with the magnetic committee]. Regrets missing JH at York B.A.A.S. meeting.
ALS 2pp TxU:H/M-0568

6001. 1844–11–2 From George Biddell Airy Royal Observatory, Greenwich
Accepts JH's invitation to attend the meeting [see JH's 1844–11–1], and offers JH accommodation for the night.
ALS 1p TxU:H/M-0630.5; Reel 1084 (C: RGO 6.675.370)

6002. 1844–11–2 or later To George Biddell Airy n.p.
Thanks GA for offer of accommodation [see GA's 1844–11–2], but JH says he must stay with Miss [Elizabeth] Baily at Tavistock Place.
ALS 1p RGO 6.675.371

6003. 1844–11–2 From Augustus De Morgan Tavistock Place
JH has forgotten to name the day of his arrival. Can he let Elizabeth Baily know? Thinks he did the right thing in staying at Tavistock Place.
ALS 2pp RS:HS 6.212

6004. 1844–11–2 From Edward Joseph Lowe Highfield House near Nottingham
Is publishing a work on rare atmospheric phenomena. Lists names of persons who have assisted with observations. Would like to inscribe the book to JH.
ALS 4pp RS:HS 11.311

6005. 1844–11–2 From Thomas Maclear Royal Observatory, Cape of Good Hope
Observed Comet Mauvais. Is grateful for JH's intervention on Observatory matters. Extra Meridian Instruments defective. Continues observations of variables. F. E. Wilmot's land survey cancelled. Magnetic survey progressing. Mary Maclear ailing.
ALS 8pp RS:HS 12.132

6006. 1844–11–6 From Humphrey Lloyd Trinity College, Dublin
Not acquainted with proposal by A. T. Kupffer. Replies to queries by B.A.A.S. committee regarding continuation of magnetic and meteorological observatories. Expects self-registering instruments to supersede human observers soon.
ALS 4pp TxU:H/M–0333; Reel 1087

6007. 1844–11–8 From F. M. Eardley-Wilmot Cape of Good Hope
Outlines the recent activities connected with the magnetic survey of the Cape and comments on the progress. Gives details of difficulties under which Thomas Maclear works at the Cape Observatory.
ALS 6pp RS:HS 7.5

6008. 1844–11–9 From Edward Sabine Woolwich
Sends actinometer observations of 1841 and 1842, which he will publish. Asks JH to look them over and make comments.
ALS 4pp † RS:HS 15.164

6009. 1844–11–11 From Edward Sabine Woolwich
Discusses ES's experiments with convertible and travelling pendulums. Asks how 'für vaterländische Cultur' is to be translated for the list of societies.
ALS 4pp RS:HS 15.165

6010. 1844–11–12 From John William Draper New York University
Sends papers and new book on 'Chemical influence of Light' by JD. Received JH's papers. Apologizes for failing to give JH discovery credit. Notes on light interference.
ALS 2pp RS:HS 19.270

6011. 1844–11–12 From Thomas Galloway Serjeants' Inn
Has received the Memoir of Francis Baily and JH's note. Will instruct the printer to follow JH's wishes. Query

regarding the printing of the catalogue of Baily's works.
ALS 3pp RS:HS 8.17

6012. 1844–11–14 From Richard Jones to Margaret Brodie Herschel Tithe Commission
Is not certain whether Mrs. Jones has written, but wishes to let her know his health is much better. Has heard good reports of JH's memoir of Francis Baily. Frightful tragedy for the Youngs. Is James well again? Hopes JH's family members are well.
ALS 4pp RS:HS 10.378

6013. 1844–11–14 From Edward Sabine Woolwich
Pleased that JH agrees on removal of tail pieces and moveable weights from pendulums. Discusses pendulum experiments in air and vacuums.
ALS 4pp RS:HS 15.166

6014. 1844–11–15 To Sir Charles Lyell Collingwood
Regarding L. F. Menabrea's papers, thinks he is a good mathematician. Is not so sure about Augusta A. King's (Lady L[ovelace]) notes.
C 2pp RS:HS 11.423 & 22.210

6015. 1844–11–15 From Richard Sheepshanks Reading
[John] Lee has a silhouette of Caroline Herschel as a young lady; would JH like it? Discusses whether JH should include Francis Baily's relationship to his family in JH's memoir on Baily. Names those to whom he plans to give copies of Baily's portrait.
ALS 4pp RS:HS 16.28

6016. 1844–11–16 From George Biddell Airy Royal Observatory, Greenwich
Regrets his delay in not returning F. W. Bessel's letter. Observations arising from this letter.
ALS 3pp RS:HS 1.145 (C: RGO 6.193.124)

6017. 1844–11–17 To Richard Sheepshanks n.p.
Does not think Francis Baily's relatives should take part in JH's presentation on Baily to the R.A.S., but if they have asked to do so, they should not be refused. Would like a copy of Caroline Herschel's silhouette. Is 'facing the engraver's hands' with his nebula, and must get the plate of JH's portrait, which RS placed at the disposal of the Duke of Northumberland.
ALS 3pp RS:HS 16.29 (C: RS:HS 25.9.11)

6018. 1844–11–18 From Edward Sabine Woolwich
Discusses the best pendulum models inside and outside a vacuum apparatus. ES will be given credit for improving the manipulation of pendulums. Makes several other comments regarding pendulums.
ALS 4pp RS:HS 15.167

6019. 1844–11–18 From Thomas Galloway Serjeants' Inn
Sending manuscript and slips received today.
ALS 1p RS:HS 8.18

6020. 1844–11–20 From Jesse Piper \, [Kent]
Apologizes for blunder in account. Encloses £1.
ALS 1p RAS:JH Archive 7/1.2 verso; Reel 8

6021. 1844–11–21 To George Biddell Airy n.p.
Inquires about the possibility of holding a Standards Commission meeting on the Royal Observatory's Visitation Day.
AL(S) 1p RGO 6.340.355

6022. 1844–11–21 To Augustus De Morgan Collingwood
Sends a list of James Dunlop's nebulae.
C 2pp RS:HS 22.211

6023. 1844–11–21 To Richard Sheepshanks Collingwood
Adds remarks to his previous comments [JH 1344–11–17] on Francis Baily's relatives participating in the memorial to Baily.
ALS 3pp RS:HS 16.30 (Adraft 3pp RS:HS 16.31; C: RS:HS 25.9.14)

6024. 1844–11–23 From Richard Sheepshanks Athenaeum
Clarifies a point raised in an earlier letter about JH being the proper person to deal with the relatives of Francis Baily. On the mishandling of arrangements for Baily's portrait by the R.A.S. On [W. S.] Stratford's poor leadership in this matter.
ALS 4pp RS:HS 16.32

6025. 1844–11–24 To George Biddell Airy n.p.
Making arrangements to stay with GA when JH comes for a meeting.
ALS 3pp RGO 6.368.675

6026. 1844–11–25 From Edward Sabine Woolwich
Will attend to JH's wishes regarding the correspondence. Discusses the buoyancy correction for particular types of pendulums. Asks if JH has received the magnetic results from Geneva.
ALS 5pp RS:HS 15.168

6027. 1844–11–27 To Edward Sabine Collingwood?
Responds to ES's request for JH's remarks on the mode of registering and printing the actinometer observations.
draft 4pp RS:HS 15.169 (C: RS:HS 22.212)

6028. 1844–11–27 From William Henry Sykes India House
Regrets that he cannot offer JH the information he needs regarding the stature of service-men, but suggests places for further searching. Shares observations about relationships between dew point and barometer.
ALS 4pp RS:HS 17.141

6029. 1844–11–28 From Thomas Galloway Serjeants' Inn
Informing him of the death of Thomas Henderson at Edinburgh. Robert Main is anxious to obtain the post. Outlines facts concerning Main and the great benefit it would be if he were appointed.
ALS 3pp RS:HS 8.19

6030. 1844–11–28 From James A. Gordon Naish, Bristol
Would like his opinion on which paper on electricity relating to the growth of vegetables, he is to give the £30 reward, as it has to be decided by 3 Dec.
ALS 4pp RS:HS 8.172

6031. 1844–12–1 From Augustus De Morgan 7 Camden St.
Is obliged for his list of James Dunlop. Has been reading JH's memoir of Francis Baily. Comments on this.
ALS 3pp RS:HS 6.213

6032. 1844–12–2 From Robert Hunt Falmouth
Received the enclosed silver plate on which is an impression made by the moon; has he any comments. Has been experimenting with solution of bichromate of potash and sulphate of copper exposed to actinic influence. Has been giving his attention to ferrotypes recently.
ALS 2pp RS:HS 10.116

6033. 1844–12–5 To George Biddell Airy n.p.
A copy of the question circular, signed by JH, sent to GA [see GA's 1844–12–17].
LS 3pp RGO 6.675.374

6034. 1844–12–5 To Alexander Dallas Bache n.p.
Requests AD's views on the benefits deriving from the worldwide system of magnetic and meteorological observatories. Various publications coming from these observations will be sent to AD.
LS 3pp Huntington Library

6035. 1844–12–5 To Carl Friedrich Gauss n.p.
[Writing as an addendum to a letter on magnetic observatories circulated by Edward Sabine], JH states that various publication concerning magnetic observations will be sent to CG.
ALS 2pp NSUB Cod.Ms.Gauss 99

6036. 1844–12–5 From Charles Moxon 4 Mt. Pleasant Lane, Upper Clapton
As secretary of Wernerian Club, CM requests information from JH on atmospheres of moon and planets.
ALS 1p TxU:H/M–0350.1; Reel 1087

6037. 1844–12–5 To Adolphe Quetelet n.p.
Requests information on behalf of the Magnetic and Meteorological Committee of the B.A.A.S. Desires AQ's assessment of past achievements by 1845–3–10. Lists works to be forwarded before end of year.
LS 3pp Académie belgique: #21

6038. 1844–12–6 From Thomas Galloway Serjeants' Inn
Querying two statements of JH in his memoir of Francis Baily. Gives details of Thomas Henderson's position at Edinburgh.
ALS 4pp RS:HS 8.20

6039. 1844–12–6 From Edward Sabine Woolwich
Sends 30 copies of JH's letters by coach. Will charge to the B.A.A.S. account.
ALS 2pp RS:HS 15.170

6040. 1844–12–7 To James David Forbes Collingwood
Requests papers relative to a preface to the catalogue of [N. L.] Lacaille's stars, so that the catalogue may be printed.
ALS 3pp St. Andrews 77

6041. 1844–12–12 From Edward Sabine Woolwich
Sent out copies of JH's remarks on Toronto actinometer. Expects that observations recommended by JH will be made next summer. Replacement for J. H. Kay's broken actinometer will arrive [in Tasmania] next June. Possible remedies to declination readings.
ALS 4pp TxU:H/M–0520.16; Reel 1093

6042. 1844–12–16 From Hugh Falconer 23 Norfolk St., Strand
Is grateful for his interest. Met the Trustees who promised to apply to the government for a grant of £1000 to assist his work.
ALS 3pp RS:HS 7.365

6043. 1844–12–17 From George Biddell Airy Royal Observatory, Greenwich
Magnetic questions have been sent to foreign observers but not the British. Why?
ACCS 2pp RGO 6.675.372

6044. 1844–12–19 To George Biddell Airy n.p.
A note of acknowledgment for receipt of books, which JH has forwarded.
ALS 1p RGO 6.368.677

6045. 1844–12–19 From Thomas Galloway Serjeants' Inn
Council of the R.A.S. have decided to print 250 extra copies of the Memoir of Francis Baily. How many would JH like and how shall he send them?
ALS 3pp RS:HS 8.21

6046. 1844–12–24 From Augustus De Morgan 7 Camden St.
Sending the note for JH's son on how to decimalize fractions of a pound. JH's memoir has gone to press. Would like details of JH's work on the Southern stars for the Annual Report.
ALS 3pp RS:HS 6.214

6047. 1844–12–25 From William Ivory Edinburgh
Has forwarded some of the late Professor Thomas Henderson's manuscripts to JH. An application may be made to the Government for a pension for the daughter. The books are to be sold and he will send a catalogue.
ALS 3pp RS:HS 10.261

6048. [1844–12–27 or earlier] To John William Lubbock [Collingwood]
Mostly news of JH's family; JH inquires as to the state of health of JL's mother.
ALS 2pp Oxford History of Science Museum 58/5

6049. 1844–12–28 To Augustus De Morgan Collingwood
Still working on his Cape observations.
C 1p RS:HS 22.213

6050. 1844–12–28 From John William Lubbock n.p.
J. F. Daniell informs JL that the table was calculated by William Galbraith. James Apjohn states everyone is in agreement that the absolute heat of steam is constant, but JL disagrees. Intends raising the matter of John Southern's experiments at the next meeting of the B.A.A.S.
ALS 3pp RS:HS 11.371

6051. [1844–12–29 or later] To John William Lubbock n.p.
JH expresses some concern about the accuracy of a wet bulb thermometer, as he is not clear about the nature of the experimental base for its use.
ALS 3pp Oxford History of Science Museum 58/7

6052. 1844–12–30 From John Lindley 5 Charles St., Covent Garden
Wonders if JH could be persuaded to write a series of articles on meteorology for the *Gardeners' Chronicle*. If unable, could he suggest a suitable substitute?
ALS 4pp RS:HS 11.230

6053. 1844–12–30 From Ann Smyth 3 Cheyne Walk, Chelsea
Requests that JH write a recommendation for C. P. Smyth's nomination to the position of Astronomer Royal for Scotland.
ALS 3pp RS:HS 16.215

6054. 1844–12–31 From Thomas Galloway Serjeants' Inn
Enclosing catalogue of Thomas Henderson's library. Printers have not yet delivered the separate copies of Francis Baily's Memoir. Points out two small errors.
ALS 3pp RS:HS 8.22

6055. 1844–12–31 From John William Lubbock n.p.
Obliged by JH's letter. Would he let him know what he thinks of raising the matter of repeating John Southern's experiments. His own eldest son has gone to Langley; he shows little aptitude for figures but a decided turn for Natural History.
ALS 3pp RS:HS 11.372

6056. [1845] To James Basire n.p.
Errata and alterations to be made in [star] magnitude symbols before publication of JH's chart.
Adraft 1p inc RAS:JH Archive 4/1.5; Reel 7

6057. [1845?–1?] To Margaret Brodie Herschel London
Asks to have carriage sent to fetch JH, as he is trying to get home in a hurry. JH has to go to a mesmeric exhibition.
ALS 2pp JHS 6.36

6058. [1845] From Margaret Brodie Herschel St. Leonards
Describes activities of MH and daughters at St. Leonards.
ALS 4pp JHS 6.26

6059. [1845] To Margaret Brodie Herschel [Collingwood]
Brings MH up to date on happenings at home; JH hopes to get away to join MH at St. Leonards.
ALS 3pp JHS 6.20

6060. [1845] To Margaret Brodie Herschel 37 Tavistock Place
Is cooling his heels in London waiting for George Peacock and others to commit themselves to meeting.
ALS 2pp JHS 6.19

6061. [1845?] To Margaret Brodie Herschel London
Met John Franklin, who is to lead a polar expedition; JH makes disparaging comments about mesmerists, such as W. E. Parry; mentions having seen some other prominent men.
ALS 4pp JHS 3.20

6062. 1845–1 To Margaret Brodie Herschel London
About the health of Elizabeth Baily, and carpet advice from Miss Sheepshanks.
ALS 4pp JHS LMJF.45.1c

6063. [1845] From Richard Jones Tithe Office
Is starting for Weisbaden to take the waters in hope of curing his rheumatism. Unable to send the rose he promised Lady Herschel. Sends some sheets of his own lectures.
ALS 4pp RS:HS 10.400

6064. [1845?] From Henry William Pickersgill 18 Soho Square
Discusses the copying of HP's portrait of JH [to be distributed with *Cape Results*].
L 2pp RS:HS 14.161

6065. [1845?] To William Henry Smyth [Collingwood]
Thanks WS on behalf of JH's brother-in-law [Peter Stewart?] for acting on [Peter Stewart's?] behalf to make him a member [?]. Comments on some comet sightings.
C 2pp RS:HS 25.13.25

6066. [1845 ?]. To Mary Somerville n.p.
Augustus DeMorgan and [Henry Fox] Talbot are sending their papers to her. Family is well. Outlines a process of photography.
ALS 3pp RS:HS 16.381 (C: RS:HS 25.14.4)

6067. [1845 to 1850?] To William Whewell [Collingwood]
Comments on WW's publication on education [*Liberal Education,* 1845 ?], especially as it applies to the teaching of mathematics. JH has been teaching JH's son William mechanics by WW's book on the subject.
C 1p RS:HS 25.14.33

6068. [1845–1–1 or later] To [?] n.p.
JH began popular series on meteorology to 'suit your readers' over one year ago, but discovered excellent treatise by L. F. Kämtz and chose instead to translate that. 'Your letter' revived the idea.
Adraft 1p (on J. Piper's 1844–11–20) RAS:JH Archive 7/1.2 verso; Reel 8

6069. 1845–1–7 From Augustus De Morgan 7 Camden St.
The memoirs are cut enough already. Hopes his health is better.
ALS 1p RS:HS 6.215

6070. 1845–1–7 From James A. Gordon Naish, Bristol
Thanks for his note. Will avail himself of the offer of hospitality and Mr. Forster and himself will call on Friday. Quotes example of electricity applied to agriculture.
ALS 3pp RS:HS 8.173

6071. 1845–1–7 From Edward Sabine Woolwich
Objection to G. B Airy's suggestion unexpected. Will see Dean of Ely [George Peacock] soon. Must review foreign letters to be sure none of consequence were omitted. Sends copy of the Toronto meteorology.
ALS 3pp RS:HS 15.171

6072. 1845–1–9 To James A. Gordon Collingwood
Will be happy to receive him but doubts the wisdom of bringing Mr. Forster as JH is not interested in joining any more societies. If Forster's experiments succeed he will be a benefactor to agriculture.
draft 4pp RS:HS 8.174 & 22.219

6073. 1845–1–11 To James David Forbes Edinburgh
Believes JH expressed support for pension for surviving child of Thomas Henderson. Comments on this, stating his own view that she is already well provided for.
ALS 2pp St. Andrews LB III, 602–3 (C: RS:HS 7.306)

6074. 1845–1–11 To Margaret Brodie Herschel Collingwood
It is pouring rain at Collingwood; JH writes a long but reasonably temperate review of the visit of two mesmerists; JH opposed to mesmerism because he sees it as fraudulent.
ALS 4pp JHS 3.24

6075. 1845–1–12 From H. Fraser Halle Lambeth
Sends a copy of the *Britannic Censor.* Outlines his aims in producing this new periodical.
ALS 5pp RS:HS 10.146

6076. 1845–1–12 or later To H. Fraser Halle n.p.
Replies to HF's 1845–1–12 that JH has no time to comment on HF's *Britannic Censor.*
draft 1p RS:HS 10.146 verso

6077. 1845–1–12 To Margaret Brodie Herschel Collingwood
Has had a sleepless night and a headache after yesterday's mesmerists [see JH's 1845–1–11]; talks about the children with JH, and tells a silly joke about mesmerists.
AL 4pp JHS 3.25

6078. [1845–1–12 or earlier] To John William Lubbock n.p.
Promises to become more knowledgeable about [heat?] theory in order to be able to discuss it with JL. Cautions that the subject is an intricate one.
ALS 2pp Oxford History of Science Museum 58/4

6079. 1845–1–13 To James A. Gordon Collingwood
Regarding Mr. Forster's experiments, there have been many on record. They are cheap to carry out and so it should be easy to arrive at a decision soon.
draft 4pp RS:HS 8.175 & 22.220

6080. 1845–1–13 From John William Lubbock n.p.
Will find the reference to S. D. Poisson in his own paper on the *Heat of Vapours*. The subject is in a state of confusion. John Southern's experiments are detailed in John Robison's *Mechanical Philosophy,* which can be sent from the R.S.L. library if JH does not possess it already.
ALS 4pp RS:HS 11.373

6081. 1845–1–14 From James A. Gordon n.p.
Thanks for his three invaluable letters. His own grandfather put up the first steam engine in London in 1770. His own knowledge is limited but he is expert on some things. Hopes he will keep his letter for a year.
AL 1p inc RS:HS 8.176

6082. 1845–1–15 To Margaret Brodie Herschel Collingwood
MH has gone for a rest to St. Leonards, and JH is expecting to go there, which will allow them both to avoid a ball invitation.
ALS 4pp JHS 3.26a

6083. 1845–1–16 From William Ivory Edinburgh
Thanks for returning the catalogue marked for the books suitable for the Observatory. J. D. Forbes has purchased some to the value of £127. Intends to apply for a pension for the orphan.
ALS 1p RS:HS 10.262

6084. 1845–1–18 To Margaret Brodie Herschel Collingwood
Details about JH's travel to St. Leonards to be with MH [see JH's 1845–1–15]; comments on the building of a 'Free Church' at Exeter.
ALS 4pp JHS 3.26b

6085. 1845–1–19 To Richard Sheepshanks Collingwood
Cannot attend the testimonial committee meeting, but recommends that for a R.A.S. testimonial, a well written parchment would be far more effective than a bound book.
C 1p RS:HS 23:5

6086. 1845–1–20 To Edward Sabine 17 Montague Square
Will write to Court of Directors to ask H.E.I.C. to publish observations from Indian observatories. Needs to know where observations are stored, whether Humphrey Lloyd has reduced them, ES's reasons for making this request, best method for preparing data, and probable expense. Must go to Slough tomorrow to do business with former tenant.
ACCS 4pp TxU:H/L-0363; Reel 1054

6087. 1845–1[?]–22 To James David Forbes Collingwood
Requests information about actinometers in JF's possession, and provides further information regarding their use.
ALS 2pp St. Andrews 7

6088. 1845–1–22 From Alfred Smee Bank of England
On the usefulness of 'Bishmuth' candles, and some necessary additions to a future edition of AS's book.
ALS 3pp RS:HS 16.162

6089. 1845–1–25 From William Radcliffe Birt Bethnal Green
Has received the communication and the proof, and returned the latter to the printer. Found a section had been omitted but cannot trace the manuscript at present.
ALS 3pp RS:HS 4.110

6090. 1845–1–29 From Peter Mark Roget 15 Upper Bedford Place
Will forward JH's paper ['Ἀμόρφωτα ...,' *R.S.P.T.* (1845), 143–53] to the R.S.L.
ALS 1p Harvard: Houghton AAH 67m–67 (107)

6091. 1845–1–30 From B. Baily Threadneedle St.
Is now in a position to pay to JH Francis Baily's legacy and would be glad to know in which way JH would like it paid.
ALS 1p RS:HS 5.8

6092. 1845–1–31 To Margaret Brodie Herschel [London]
Speaks of JH's visits on the previous day; later this day, JH is to visit Mr. Plowden about a position for son William James.
ALS 4pp JHS 4.21

6093. 1845–1–31 To Margaret Brodie Herschel 17 Montague Square
Describes how JH arranged for an opening [in the Indian Civil Service] for JH's son William James at the time when he will be the proper age [see JH's 1845–1–31].
ALS 4pp JHS 4.20

6094. 1845–1–31 From Sir Robert Harry Inglis 7 Bedford Square
Has invited Charles Grant (Baron Glenelg) to breakfast tomorrow; will JH join them?
ALS 1p RS:HS 10–233

6095. 1845–2–1 From B. Baily n.p.
Sends the printed receipt for £1,000 being the amount of Francis Baily's legacy. Please sign and send it to his own bankers.
ALS 1p RS:HS 5.9

6096. 1845–2–1 From William Rutter Dawes Camden Lodge
Introducing a Dr. Warwick, who has given lectures on the Elements of Chemical Science and who would like some authoritative advice on some astronomical matters.
ALS 2pp RS:HS 6.68

6097. 1845–2–1 From Edward Sabine Woolwich
Observations from H.E.I.C. stations were supposed to be sent to foreign secretary of R.S.L. Prospects for persuading H.E.I.C. to reduce and publish these. Best methods of publication. Toronto volume should be available next week. Report by W. R. Birt was delayed.
ALS 11pp TxU:H/M–0521.1; Reel 1093

6098. 1845–2–4 From William Radcliffe Birt Bethnal Green
Has received the missing manuscript and models from York, yesterday.
ALS 1p RS:HS 4.111

6099. 1845–2–5 To Humphrey Lloyd Collingwood
Question of procedure about considering the magnetic observations for publication.
C 1p RS:HS 22.222

6100. 1845–2–7 From John F. Daniell King's College, London
Rejoices that JH plans to persuade H.E.I.C. to publish results of magnetic and meteorological observations. No one at R.S.L. has been in charge of these observations since Humphrey Lloyd gave it up. Submits JD's list of stations and observations. Encloses specimens of JD's pure 'anchovia' and 'Sabrina.'
ALS 8pp TxU:H/M–0155; Reel 1087

6101. 1845–2–7 From Henry Anthony Jeffreys Hawkhurst Parsonage
Sends the Ticehurst Account books. Comments on the various funds. Regarding the local rates.
ALS 4pp RS:HS 10.285

6102. 1845–2–8 From Henry Anthony Jeffreys Hawkhurst Parsonage
Sends the amended books. Regarding payment by the week or day. Hopes that the dispute with Mr. Morrell can be amicably settled.
ALS 4pp RS:HS 10.286

6103. 1845–2–10 To Robert [Peel?] Collingwood
Is horrified at the wanton destruction of the Portland Vase in the [British] Museum. JH suggests flogging as appropriate for such acts of vandalism [see JH's 1845–3–18].
ALS 3pp DMC.695A.885.25

6104. 1845–2–12 From Edward Sabine Woolwich
How to obtain copies of H.E.I.C. observations. Humphrey Lloyd never intended to process data from H.E.I.C. observatories; these were added to survey after Lloyd's original (1838) offer to superintend B.A.A.S. observations. Sends copy of Toronto volume to JH to use as model for H.E.I.C. publications. Asks for JH's suggestions for Arctic observations next winter. Encloses note from [T. R.] Robinson in Parsonstown.
ALS 9pp TxU:H/M–0521.2; Reel 1093

6105. 1845–2–12 From Sir Robert Harry Inglis 7 Bedford Square
Is grateful for his letter regarding the Portland vase; has shown it to Peel, who has agreed to introduce a short bill to protect such properties. Will read JH's letter to the Trustees.
ALS 4pp RS:HS 10.234

6106. [1845]–2–12 From Humphrey Lloyd Trinity College, Dublin
Reviews establishment of H.E.I.C. observatories and comments on quality of magnetic and meteorological observations at Simla, Madras, and Singapore. Has heard nothing from A. B. Orlebar, director of Bombay observatory.
ALS 8pp † TxU:H/M–0335; Reel 1087

6107. 1845–2–12 From Humphrey Lloyd Trinity College, Dublin
Activities at H.E.I.C. magnetic and meteorological observatories in Simla, Madras, and Singapore. Results from Simla, elevation 8000 feet, will complement high altitude observations in Europe. Vertical force instrument at Singapore is better than at all other stations, but high humidity there makes actinometer unreliable.
L *B.A.A.S. Report 1845* (1846), 3–5 inc

6108. 1845–2–13 From John William Lubbock n.p.
Much obliged for JH calling but regrets he was out. The question [on steam] they have been discussing seems to be answered by a paper by a Prof. Karl Holtzmann he has just come across; comments on this. If JH is coming to S. J. A. Compton's (2nd Marquis of Northampton) soiree will he come and dine at St. James' Place?
ALS 4pp RS:HS 11.374

6109. 1845–2–15 From Henry Barrow 26 Oxendon St.
Thanks for JH's valuable information on actinometers. Encloses particulars that HB sent to [J. H.] Kay along with instruments.
ALS 1p TxU:H/M–0043; Reel 1062

6110. 1845–2–16 To John William Lubbock Collingwood
JH comments on JL's 'memoir on the heat of vapours,' indicating JH's dissatisfaction with the experimental basis (the experimental work of J. L. Gay–Lussac) for the conclusions. JH offers improvements in the construction and operation of the ice calorimeter.
ALS 4pp diag Oxford History of Science Museum 58/1

6111. 1845–2–18 From W. L. Newman Insurance Office, York
Based on JH's 1821 formula in *R.S.P.T.*, WN calculated aberrations of compound double achromatic object glasses, to be used in determining radii for grinding lenses. Explains theory. Attempting to construct 4.2-inch lens.
ALS 3pp RAS: MSS Add.29.1

6112. 1845–2–19 From John William Lubbock n.p.
Was not aware when he wrote his paper that doubt existed regarding the invariability of [the absolute heat of steam?]. Comments on the variations in the theories of S. D. Poisson and Karl Holtzmann. The experiments JH suggests would be highly interesting. Would he like to see Holtzmann's paper?
ALS 4pp RS:HS 11.375

6113. 1845–2–24 From John William Lubbock n.p.
Has no doubt that any improvements JH suggests should be published. [John?] Taylor has borrowed Holtzmann's pamphlet, which will be sent as soon as returned. Karl Holtzmann's formula for air appears quite out. Gabriel Lamé evades the question. If JH writes anything on his calorimeter hopes he will mention the various questions raised.
ALS 4pp RS:HS 11.376

6114. 1845–2–24 From Edward Sabine Woolwich
Lists 17 foreign and six English scientists to whom ES sent JH's letter and Toronto volume.
ALS 3pp TxU:H/M–0521.3; Reel 1093

6115. 1845–2–24 From Richard Sheepshanks Reading
Judges the unfinished portrait of Francis Baily a faithful rendering, although he is portrayed without the glasses, which he usually wore. Fears the intractability of [W. S.] Stratford and its implications for the workings of the R.A.S. Council. On the dispute between [W. R.] Dawes and [George] Bishop and how to resolve it.
ALS 4pp RS:HS 16.38

6116. 1845–2–25 From Adolph Theodor Kupffer St. Petersburg
Reports in considerable detail on magnetical and meteorological observation in Russia. Supports the idea of international conference of leaders in such work.
L *B.A.A.S. Report 1845* (1846), 18–19 (French)

6117. 1845–2–25 From Richard Sheepshanks London
[*Detached postscript*] Asks JH's opinion of new apparatus that [William] Simms will soon put on paper.
AL 1p RS:HS 19.294

6118. [1845–3] To John William Lubbock n.p.
Has received the books. Began with Karl Holtzmann's and finds the reasoning at fault. Comments on this. [Marked 'not

sent.']
draft 3pp RS:HS 11.405 (C: RS:HS 25.13.11)

6119. 1845–3–1 From Edward Sabine Woolwich
Forwards reply from C. R. Weld explaining why Greenwich magnetic and meteorological observations were not sent to English and foreign scientists involved in global survey.
ALS 4pp TxU:H/M–0521.4; Reel 1093

6120. [1845–3?] To William Whewell n.p.
Is sending a 'Memoir' [on Francis Baily?] to WW. JH and wife but not daughter Caroline will be at Trinity Lodge in June. Comments on a paper by William Hopkins on glaciers.
ALS 3pp TC, Camb. Add. Ms.a.207[63] (C: RS:HS 22.224)

6121. 1845–3–4 From Augustus De Morgan Tavistock Place
Elizabeth Baily would like him to write an inscription for her brother's memorial tablet. Francis Baily was buried in the land of his fathers at Thatcham.
ALS 2pp RS:HS 6.216

6122. 1845–3–6 From John William Lubbock n.p.
Further regarding the various theories on the latent heat of steam. Agrees with JH regarding the magnetic observations. Hudson's Bay Company has sent a large bill for J. H. Lefroy's expenses, which the R.S.L. declines to pay. Fears that Edward Sabine is out of humor about something.
ALS 16pp RS:HS 11.377

6123. 1845–3–7 To Richard Sheepshanks n.p.
Will help with the dispute between [W. R.] Dawes and [George] Bishop, provided certain things are understood by all the parties involved; enumerates these. Will speak as discretely as he can with [W. S.] Stratford.
ALS 4pp RS:HS 16.39 (C: RS:HS 25.9.18)

6124. 1845–3–8 From Humphrey Lloyd Trinity College, Dublin
Responds to questions on magnetic circular sent out by JH [see JH's 1844–12–5]. Comments on continuing government funding for global system of magnetic and meteorological observatories. Suggests goals for future observers.
LCC 4pp RGO 6.694.349 *B.A.A.S. Report 1845* (1846), 35–7 (inc)

6125. 1845–3–8 From Edward Sabine Woolwich
JH's influence worked; J. W. Lubbock asked ES for list of recipients for Greenwich magnetic and meteorological observations. Preparations for new Arctic expedition. Will give copies of JH's letter to F. M. E. Wilmot and J. H. Kay. New observatory at Colombo [Ceylon]. Problem defining seasons for observations at St. Helena. Reply from W. E. Weber [to JH's circular].
ALS 4pp TxU:H/M–0521.5; Reel 1093

6126. 1845–3–9 To William Parsons (Lord Rosse) Collingwood
Thanks WP for sending drawings and descriptions of various nebulae observed with WP's giant reflecting telescope. Comments on various nebulae. Expresses the idea that some planetary nebulae are 'hollow elliptical or spherical shells with or without a central Star.'
ALS 4pp Rosse Papers K2.2

6127. 1845–3–10 From Edward Sabine Woolwich
Actinometers are 'counterordered.' Reply from H. W. Dove to JH's circular. Problem defining seasons for tropical stations. Concerned that some authorities are reaching decision to extend or to terminate magnetic observations *before* receiving and reading Toronto volume. Failure of R.S.L. to send Greenwich magnetic and meteorological observations to appropriate recipients. Details of J. H. Lefroy's magnetic survey for Hudson's Bay Company, and misunderstanding over who would pay for Lefroy's transportation. Efforts by F. M. E. Wilmot to obtain ship for survey of [Cape of Good Hope]. ES concurs with Treasury's suggestion that copy of Toronto volume be sent to each colony. Observations of nebulae by T. R. Robinson.
ALS 16pp, Sabine note 1p TxU:H/M–0521.6; Reel 1093

6128. 1845–3–10 To William Whewell Collingwood
On behalf of Lady Herschel, JH thanks WW for the copy of WW's book [*Indications of the Creator*]. Chides WW for having written an elementary work.
ALS 3pp TC, Camb. Add. Ms.a.207[64] (C: RS:HS 22.225)

6129. 1845–3–11 From Georg Adolph Erman Berlin
Expresses thanks for receipt of Toronto magnetical and meteorological observations. Encourages continuation of British observations. Makes proposal for future work.
L *B.A.A.S. Report 1845* (1846), 38–42

6130. 1845–3–11 From [Georg Adolph] Erman Berlin
Response to questions on magnetic circular sent out by JH [see JH's 1844–12–5].
LCC 5pp RGO 6.694.351

6131 1845–3–11 From Palm Heinrich Ludwig Boguslawski to Caroline Lucretia Herschel Breslau
A note to CH on her 96th birthday.
ALS 2pp (German) RS:HS 4.175

6132. 1845–3–11 From John William Lubbock n.p.
Now sees JH's point regarding the theory of steam. Comments on various experiments. Glad to see JH approves of repeating John Southern's experiments. Unable to go to Cambridge; his banking business requires unceasing attention. Doing all he can about the observations. Council will take up the East India observations.
ALS 20pp RS:HS 11.378

6133. 1845–3–12 From Richard Sheepshanks Reading
More on the matter of [W. R.] Dawes and [George] Bishop. Dawes deserves credit for the observation and reduction, Bishop for being 'founder & proprietor.' On JH's ability to mediate such situations, because he is so well liked. Confident in Bishop's 'good sense,' once he knows that he is being placed in JH's hands.
ALS 4pp RS:HS 16.40

6134. 1845–3–12 From William Whewell Trinity Lodge
Explains to JH that WW's *Indications of the Creator* was written to counteract [Robert Chambers's] *Vestiges of Creation.* Comments critically on William Hopkins's glacial theory.
ALS 4pp † RS:HS 18.211 Todhunter, *Whewell, 2,* 325–6 inc

6135. 1845–3–13 To Caroline Lucretia Herschel Collingwood
Wishes CH a happy 96th birthday. Reports that when Margaret Herschel's brother John Stewart was in Egypt, he saw a comet. JH remarks that 'there seems to be no end of the comets.'
ALS 3pp TxU:H/L–0590.3; Reel 1058

6136. 1845–3–13 From Edward Sabine Woolwich
Encloses letters from H. W. Dove, A. T. Kupffer, and L. A. J. Quetelet. Dove is not familiar with methods used in Toronto.
ALS 1p TxU:H/M–0521.7; Reel 1093

6137. 1845–3–14 From Carl Friedrich Gauss Göttingen
Has not seen enough magnetical and meteorological observations to judge their effectiveness. Encourages British government to continue these, but perhaps reduce frequency.
L *B.A.A.S. Report 1845* (1846), 42–4 (German); trans. 44–5

6138. 1845–3–14 From Carl Friedrich Gauss Göttingen
Response to questions on magnetic circular sent out by JH [see JH's 1844–12–5].
LCC 3pp RGO 6.694.353

6139. 1845–3–14 To William Henry Chichele Plowden Collingwood
Encourages board of directors to pay for publication of magnetic and meteorological observations made at four H.E.I.C. observatories. Names countries that participated in global survey.
ACCS 4pp TxU:H/L-0314; Reel 1054

6140. 1845–3–15 To John William Draper Collingwood
Comments on writings of JD on the chemistry of plants [letter completed 1845–5–7].
C 4pp RS:HS 22.226

6141. 1845–3–16 To George Biddell Airy n.p.
Has been informed that William Parson's telescope resolves all the nebulae in JH's catalogue into stars.
AL 1p inc RGO 6.368.678

6142. 1845–3–17 From Augustus De Morgan 7 Camden St.
Suggesting alterations in the wording for the memorial tablet to Francis Baily.
ALS 2pp RS:HS 6.217

6143. 1845–3–17 From Sir Robert Harry Inglis 7 Bedford Square
Solicitor General is to introduce a Bill for the better protection of 'things' in Museums. Does not include the National Gallery. The culprit in the Portland vase incident was an Irish student. Is JH dining at the R.S.L. Club on the 27th?
ALS 3pp RS:HS 10.235

6144. 1845–3–18 From Augustus De Morgan 7 Camden St.
Encloses the corrections by his friend T. H. Key to the wording for the memorial to Francis Baily.
ALS 2pp RS:HS 6.218

6145. 1845–3–18 From John William Lubbock n.p.
Is JH aware that the magnetic and meteorological observations now being made at Greenwich are to cease at the end of the year? As JH was zealous for these observations he may care to write to S. J. A. Compton (2nd Marquis of Northampton). It escaped his notice that S. D. Poisson and J. B. Biot referred to saturated steam. Return John Robison and Karl Holtzmann to Mansion House when he has read them. Has consulted Thomas Graham about repeating John Southern's experiments.
ALS 4pp RS:HS 11.379

6146. 1845–3–18 To Robert [Peel ??] Collingwood
Is pleased that a bill [JH's comments seem to hint at corporal punishment for offenders] is to be brought to Parliament [see JH's 1845–2–10].
ALS 3pp DMC.695A.885.23

6147. 1845–3–18 From Edward Sabine Woolwich
Receiving replies [to JH's circular] from English and foreign correspondents. George Peacock sent them invitations to B.A.A.S. meeting at Cambridge. Suggests defining tropical 'seasons' by monsoon activity. Board of Ordnance sent copy of Toronto volume to each British colony.
ALS 4pp TxU:H/M–0521.8; Reel 1093

6148. 1845–3–19 From Robert Hunt Falmouth
Has applied to the Council for the position and Michael Faraday has promised him support. Has also heard of a forthcoming vacancy in the Museum of Economic Geology. Is undecided which to apply for and would welcome JH's views.
ALS 5pp RS:HS 10.117

6149. 1845–3–20 From Thomas Crick St. John's College, Cambridge
[Form letter] Asks friends of Mr. Griffin, candidate for Librarian, to return to Cambridge for election vote on 3 Apr.
LS 1p RS:HS 19.122a

6150. [1845–3–20 or later] To James Grahame n.p.
C. P. Smyth, chief assistant at Cape observatory for past nine years, is capable observer and qualified to succeed Thomas Henderson as Royal Astronomer at Edinburgh.
Adraft 2pp RS:HS 19.122b

6151. 1845–3–20 To Charles Wheatstone [Collingwood]
Recommends Robert Hunt to fill the professorship at King's College, left vacant by the death of John F. Daniell.
C 2pp RS:HS 25.11.5

6152. 1845–3–22 From John William Lubbock n.p.
Agrees with JH that the Government Observatories at the Cape and St. Helena require mature consideration. Greenwich observations are a different matter and were suggested by the R.S.L. Committee of Physics. Has directed the minute to be sent to JH. Has been studying the wet bulb question.
ALS 4pp RS:HS 11.380

6153. 1845–3–22 From Edward Sabine Woolwich
JH misunderstood ES's method for organizing St. Helena meteorological observations. Problem defining seasons for tropical stations. Received more replies to JH's circular. Difficulty of making hourly observations in European observatories, where civilian assistants are employed. Does not agree with Humphrey Lloyd that self-registering instruments will soon replace observers. Invited [L. F.] Kämtz to [B.A.A.S. meeting at] Cambridge. Believes that Elias Loomis will become leading meteorologist in U.S.
ALS 12pp TxU:H/M–0521.9; Reel 1093

6154. 1845–3–22 From Charles Wheatstone 20 Conduit St
Agrees that vacancy in chemistry professorship left by J. F. Daniell's death at King's College should be filled by Robert Hunt, but W. A. Miller appears likely to get job.
ALS 2pp RS:HS 18.150

6155. 1845–3–23 From Thomas I. M. Forster Bruges
How has he fared during the spell of cold weather? What does he think of A. J. Ellis's colored circles? Would like the last two volumes of the *R.S.P.T.* containing the account of the comet. Encloses some impromptus.
ALS 3pp RS:HS 7.338

6156. 1845–3–24 From Augustus De Morgan 7 Camden St.
Further regarding the inscription for the memorial to Francis Baily. Has forwarded it to the executors and translated it for Elizabeth Baily. Wire pens satisfy him.
ALS 2pp RS:HS 6.219

6157. 1845–3–25 From Edward William Brayley London Institution
Queries relating to the actinometer. Is a candidate for the position of Lecturer in Chemistry at the East India Company's College.
ALS 4pp RS:HS 4.215

6158. 1845–3–26 From John William Lubbock n.p.
The magnetic and meteorological observatories were never part of any other system. Is not certain whether the Visitors had anything to do with these observatories. Captain Herbert in an Indian journal has proposed a formula for the wet bulb thermometer. Has arranged for the notice to be sent to JH.
ALS 3pp RS:HS 11.381

6159. 1845–3–27 From Edward Sabine Woolwich
Lord Minto [Gilbert Elliot] forgot about publication arrangements. Lt. [Home?] and Lt. [?] Clark left Cape of Good Hope for Antarctica on 9 Jan. Reply from C. F. Gauss. German copies of Toronto observations were delayed in Cologne.
ALS 4pp TxU:H/M–0521.10; Reel 1093

6160. 1845–3–28 From James Robert George Grahame Whitehall
Chair of Astronomy in the University of Edinburgh is vacant. Has had the son of Capt. W. H. Sm[y]th recommended as a suitable candidate. Would like JH's views on the suitability of Sm[y]th for the position.
ALS 4pp RS:HS 9.113

6161. 1845–[3–28 or later] To [?] [Collingwood]
Testimonial to C. P. Smyth's dependability and competence as an astronomer.
Adraft 1p RS:HS 19.172d

6162. 1845–3–29 From Augustus De Morgan 7 Camden St.
The epitaph has been sent to be drawn onto the stone. Comments on some of the wording. Sends some Alderton's blotting paper and the compliments of the season.
ALS 5pp RS:HS 6.220

6163. 1845–3–30 To William Henry Smyth Collingwood
Has written recommendation supporting C. P. Smyth for position of Astronomer Royal for Scotland. Suggests that C. P. Smyth, after returning to Britain, seek a teacher to help overcome his stutter. Encloses letter from P. H. L. Boguslawski for WS to present to R.A.S.
C 1p RS:HS 22.227

6164. 1845–3–31 From William Henry Smyth [3 Cheyne Walk, Chelsea]
Thanks JH for recommendation of C. P. Smyth. James South returns from Ireland; reports great observing conditions there.
ALS 4pp RS:HS 16.277

6165. 1845–4 From Augustus De Morgan 7 Camden St.
Regarding pens and suitable ink for various types. Further comments on the inscription for Francis Baily's memorial tablet.
ALS 8pp RS:HS 6.221

6166. 1845–4–1 From Edward Sabine Woolwich
Reply from G. A. Erman to JH's circular is best one yet received. Erman recommends calculating magnetic constants of 1829. ES suggests grant from B.A.A.S. to fund this. Alexander Humboldt will reply soon. Charles Wheatstone should describe new apparatus in letter to JH that can be sent to observatories.
ALS 4pp TxU:H/M–0521.11; Reel 1093

6167. 1845–4–1 From George Biddell Airy Royal Observatory, Greenwich
Regarding magnetic establishments and observations.
ALS 4pp RS:HS 1.146 (C: RGO 6.675.387)

6168. 1845–4–2 From George Biddell Airy Royal Observatory, Greenwich
Comments on JH's observing a 'falling star.'
ACCS 1p RGO 6.699.231

6169. 1845–4–2 From Thomas Galloway Serjeants' Inn
Sending the manuscript of James Dunlop's paper.
ALS 2pp RS:HS 8.23

6170. 1845–4–2 From Alfred Smee 7 Finsbury Circus
Responds to comments about Robert Hunt [see JH's 1845–4–2]; talks about making an artificial eye.
ALS 3pp RS:HS 16.164

6171. 1845–4–2 To Alfred Smee n.p.
About a letter of recommendation JH is preparing for Robert Hunt.
draft 2pp RS:HS 16.163

6172. 1845–4–3 To George Biddell Airy Collingwood
Wants to know if GA's opinions [see GA's 1845–4–1] are available for public information, or only for JH.
ALS 3pp RGO 6.675.389

6173. 1845–4–3 From Charles Babbage Dorset St., Manchester Square
As trustee of marriage settlement of CB's sister [Mary Anne Hollier], JH is asked to sign enclosed order. Reports distribution of money to sister and to H. J. Hollier.
ALS 2pp † TxU:H/M–0028; Reel 1062

6174. 1845–4–4 From George Biddell Airy Royal Observatory, Greenwich
Wants information on dispersion/separation of light in plate or crown glass.
ALS 2pp RS:HS 1.147 (C: RGO 6.463.324)

6175. 1845–4–4 or later To George Biddell Airy n.p.
Provides the best values available for dispersion and separation, which will allow GA to calculate what he needs [see GA's 1845–4–4].
ALS 3pp RGO 6.463.325

6176. 1845–4–4 From George Biddell Airy Royal Observatory, Greenwich
Last letter to JH was private [see GA's 1845–4–1].
ACCS 1p RGO 6.675.391

6177. 1845–4–4 To Charles Babbage Collingwood
Regarding the deed for CB's brother-in-law. Is reading G. W. Leibniz's metaphysical works.
ALS 3pp RS:HS 2.310

6178. 1845–4–4 From Edward Sabine Woolwich
Received reply from Karl Kreil in letter from Count Senftenberg, who owns private magnetic and meteorological observatory east of Prague run by Kreil's former assistant. ES asks JH to write letter to Clemens von Metternich, as requested by Senftenberg. High regard for Kreil's work.
ALS 4pp TxU:H/M–0521.12; Reel 1093

6179. 1845–4–5 From John William Lubbock n.p.
The President was requested to write to the East India Company about the observatories. Continuance of the magnetic observations at Greenwich was referred to the Committee of Physics. Cannot find where Nicolas Clement-

Desormes's observations are published. Has been studying the wet bulb problem. Capt. Herbert has denounced the article on hygrometry in Brewster's *Encyclopaedia.*
ALS 4pp RS:HS 11.382

6180. 1845–4–5 From William Henry Smyth [3 Cheyne Walk, Chelsea]
C. P. Smyth receives appointment as Astronomer Royal of Scotland. Again thanks JH for recommendation.
ALS 2pp RS:HS 16.278

6181. 1845–4–5 From Peter Stewart [65 Cornhill]
Sending proofs and costs of four plates from engraver James Basire. Plate of large nebulae in progress. PS located old style type that JH prefers. Finances prohibit including all plates. JH must select which ones to include.
ALS 3pp RAS:JH Archive 10/1.12; Reel 9

6182. 1845–4–6 To [John William Lubbock] Collingwood
Discusses procedures that should be followed when the Committee on Physics is summoned to deal with a 'momentous' matter, i.e., the discontinuation of the magnetic observatories.
ALS 2pp RS MM 14.131

6183. 1845–4–6 To Charles Wheatstone Collingwood
The continuation of magnetic and meteorological observatories is under question, and CW's instrument for observing atmospheric electricity should be part of the discussion.
C 2pp RS:HS 25.11.6

6184. 1845–4–7 From George Biddell Airy Royal Observatory, Greenwich
Discourages the continuation of the magnetical and meteorological observations organized by the British.
L *B.A.A.S. Report 1845* (1846), 52–3

6185. 1845–4–7 From George Biddell Airy Royal Observatory, Greenwich
Asking him to read GA's medal-speech.
ALS 1p RS:HS 1.148 (C: RGO 6.227.367)

6186. 1845–4–7 From George Biddell Airy Royal Observatory, Greenwich
Gives detailed, official, replies to the questions in the magnetic committee circular [see JH's 1844–12–5].
ACCS 7pp RGO 6.675.393 (C: RGO 6.694.357)

6187. 1845–4–7 From Friedrich Eduard Beneke Berlin
Letter to accompany a gift of part of FB's latest work on psychology and knowledge, urging JH to encourage England to study psychology.
ALS 2pp (German) RS:HS 4.61

6188. 1845–4–7 From Edward Sabine Woolwich
Received thirteen replies to JH's circular; nine remain unanswered. Suggests printing these, before they are published by B.A.A.S, and distributing these along with second circular by JH.
ALS 4pp TxU:H/M–0521.13; Reel 1093

6189. [1845]–4–7 From Edward Sabine Woolwich
Will invite G. A. Erman to act as B.A.A.S. commissioner in reduction of 1829 Gaussian constant. Will try to remedy misconceptions of G. B. Airy. T. M. Brisbane will not get involved. Testing new magnetic instrument by Humphrey Lloyd for Arctic expedition. Invite C. F. Gauss to Cambridge.
ALS 2pp TxU:H/M–0517.2; Reel 1093

6190. 1845–4–7 From Peter Stewart [65 Cornhill]
James Basire to proceed with copper engraving of JH's 'Chart plate.'
ALS 1p RAS:JH Archive 10/1.13; Reel 9

6191. 1845–4–8 From Charles Babbage Dorset St., Manchester Square
Received JH's directive to send dividend to Mary Ann Hollier. Other money is invested in property at Cardiff, owned by Mr. Hollier. CB in bad health. Questions authorship of *Vestiges* [*of Creation*].
ALS 2pp TxU:H/M–0029; Reel 1062

6192. 1845–4–9 From George Biddell Airy Royal Observatory, Greenwich
Illegible.
ACCS 2pp RGO 6.675.397

6193. 1845–4–9 From Robert Hunt Falmouth
Has been appointed keeper of Mining records at the Museum of Economic Geology. Hopes to meet him personally and thank him for his assistance.
ALS 2pp RS:HS 10.118

6194. 1845–4–10 To Carl Friedrich Gauss Collingwood
CG having now received the Greenwich and Toronto observations, JH invites CG to write further material for the forthcoming publication.
ALS 2pp NSUB Cod.Ms.Gauss 99 (C: RS:HS 22.228)

6195. 1845–4–11 From Edward Sabine Woolwich
H. U. Addington will send JH's letter to Clemens von Metternich via Robert Gordon, English minister to Vienna. JH may include George Peacock's signature also. Printer will send proofs of replies to JH's circular to JH for corrections. JH's estimate for expenses for observatories is too high. Sent copy of JH's circular to Charles Wheatstone.
ALS 4pp TxU:H/M–0521.14; Reel 1093

6196. [1845–4–11 or later] From Adam Sedgwick n.p.
Asks various questions about P. S. Laplace's nebular hypothesis and about Auguste Comte's discussion of it. AS comments: 'Your Father and you have given ample materials for observations for a 1000 years to come.'
ALS 4pp inc RS:HS 15.460

6197. 1845–4–11 From Adam Sedgwick Cambridge
Discusses meetings of B.A.A.S. Describes [Robert Chambers's] *Vestiges of Creation* as a 'singularly shallow work.' Asks various astronomical questions of JH. Says the author misunderstands P. S. Laplace and Auguste Comte. AS is quite harsh in criticism.
ALS 8pp RS:HS 15.425 & RS:HS 15:460

6198. 1845–4–12 From Ernst F. Apelt Jena
Asks JH to accept a paper EA has written.
ALS 3pp (German) RS:HS 1.402

6199. 1845–4–12 From Alfred Smee 7 Finsbury Circus
Thanks for JH's certificate concerning AS's battery, now finding increasing support.
ALS 2pp RS:HS 16.165

6200. 1845–4–13 To Robert Hunt Collingwood
Expresses his pleasure at RH's new position at the Museum of Economic Geology.
ALS 3pp George Eastman House, 16.17 (C: RS:HS 22.229)

6201. 1845–4–14 From Thomas Graham University College, London
Sends specimen of esculine, or polychrome, extracted from

horse chestnut. It exhibits optical properties that JH recently investigated. Offers to send specimen of colophene. Asks JH's recommendations for foreign secretary of R.S.L., following death of J. F. Daniell. This officer must supervise magnetic and meteorological observatories. [JH annotation: 'By *all means* appoint Sab(ine).']
ALS 8pp † TxU:H/M–0217; Reel 1087

6202. 1845–4–14 From Edward Sabine Woolwich
Published opinions on magnetic and meteorological survey should be restricted to correspondents who were directly involved in observations, particularly foreign participants. Most others may express their views at B.A.A.S. meeting in Cambridge. Recommends omitting M. F. Maury's letter and including that of J. M. Gilliss.
ALS 4pp TxU:H/M–0521.15; Reel 1093

6203. 1845–4–15 From Henry Unwin Addington Foreign Office
Will direct JH's 14 Apr. letter to Prince Metternich.
AL 1p TxU:H/M–0003; Reel 1062

6204. 1845–4–15 To George Biddell Airy Collingwood
Seeks GA's assistance in solving a puzzle in the mathematics of rotating bodies.
ALS 2pp RGO 6.245.569

6205. 1845–4–15 To Adam Sedgwick Collingwood
[Robert Chambers's] *Vestiges of Creation* is shallow and full of errors. Tells AS not to spare it when he reviews it. Discusses various topics including the nebular hypothesis, P. S. Laplace, central heat in gravitation, and the density of planets. Also, the 'nonsensical calculations' of Auguste Comte's astronomy.
ALS 4pp RS:HS 15.427 (C: RS:HS 22.230 & draft: RS:HS 15.426) Lubbock, *Herschel Chronicle*, 231 inc

6206. [1845]–4–15 From Richard Taylor Red Lion Court, Fleet St.
Asks JH's opinion on a paper on magnetism by Karl Reichenbach, which has some 'very extraordinary' statements.
ALS 1p RS:HS B26.10

6207. 1845–4–15 To Charles Wheatstone [Collingwood]
Makes some amendments in JH's request for information [see JH's 1845–4–6].
C 2pp RS:HS 25.11.7

6208. 1845–4–16 From George Biddell Airy Royal Observatory, Greenwich
Regarding the acceleration of retention of a shrinking sphere.
ALS 3pp RS:HS 1.150 (C: RGO 6.245.570)

6209. 1845–4–16 From George Biddell Airy Royal Observatory, Greenwich
Is enclosing two transcripts of letters from William Simms relating to the object-glass. a. Hopes to give account next week. b. One glass good, other has failed.
ALS 1p RS:HS 1.149

6210. 1845–4–16 From George Biddell Airy Royal Observatory, Greenwich
Comments on the various responses to the circulated questions about magnetic observations [see JH's 1844–12–5].
ALS 1p † TxU:H/M-0631.1; Reel 1084 (C: RGO 6.675.417)

6211. 1845–4–16 To Richard Taylor n.p.
Is very critical of a series of three foreign memoirs that JH believes certainly should not be published. Concludes with a strange postscript on catalepsy.
ALS 3pp RS:HS B26.11

6212. 1845–4–17 From John William Lubbock n.p.
Has been with Philip Melvill of the East India Co. discussing the printing of the observations. Would it be a good use of JH's photographic papers to take a reverse of William Parsons' (3rd Earl of Rosse) drawings of nebulae? Cannot obtain Nicolas Clement-Desormes's or John Sharpe's papers. Doubts if the mode of measuring latent heat hitherto used can be recommended for accuracy.
ALS 4pp RS:HS 11.383

6213. 1845–4–17 From W. Robarts Burnham, Bucks
[William] Penn produced letter written on WR's behalf by JH. Thanks JH for letter, which increased Mr. Slingsby's [?] confidence.
ALS 4pp RS:HS 14.493

6214. 1845–4–18 To George Biddell Airy Collingwood
Mostly about the health of JH and his wife, Margaret.
ALS 3pp RGO 6.368.679

6215. 1845–4–18 From Edward Sabine Woolwich
ES will write to M. F. Maury. Concurrence of both R.S.L. and B.A.A.S. is needed before congress is announced. Proofs of foreign responses to JH's circular will be ready next week.
ALS 3pp TxU:H/M–0521.16; Reel 1093

6216. 1845–4–18 To William Henry Smyth Collingwood
Sends P. H. L. Boguslawski's address. Asks WS to send J. H. Mädler's chart to P[eter] Stewart.
C 1p RS:HS 22.231

6217. 1845–4–19 To the London *Times* Collingwood
Responds to recently printed letter by James South. Denies South's assertion and confirms that William Herschel *did* discover two interior satellites of Saturn with 40-feet reflector telescope in 1789.
ACCS 4pp TxU:H/L-0230; Reel 1054

6218. 1845–4–20 From Richard Sheepshanks 14 London Road, Reading
George Bishop refuses permission for W. R. Dawes to edit Dawes's own observations. Portrait of Francis Baily should be nearly finished. Auction of Baily's books. Plans to start over 'de novo' and not use Baily's method for evaluating standards of measure. Convinced that standards must be connected with Ordnance survey of U.K. and T. F. Colby's units.
ALS 8pp TxU:H/M–0530; Reel 1093

6219. 1845–4–21 From Edward Sabine Woolwich
Evaluates in a detailed manner the magnetical and the meteorological observations organized by the British at various overseas locations. Makes many specific recommendations, generally supporting their continuation.
L *B.A.A.S. Report 1845* (1846), 53–7

6220. 1845–4–21 From Edward Sabine Woolwich
Response to questions on magnetic circular sent out by JH [see JH's 1844–12–5].
LCC 5pp RGO 6.694.358

6221. [1845–4–22 or later] To Thomas Maclear n.p.
Privately acknowledges lack of clarity in William Herschel's report of discovering Saturn's inner rings, but JH was hurt by James South's public refutation of that discovery and by [T. R.] Robinson's abrupt refusal to communicate. Reviews WH's papers to demonstrate falsity of South's claim. Asks

TM's opinion.
Adraft 4pp TxU:H/L-0240; Reel 1054

6222. [1845–4–22 or later] To Richard Sheepshanks n.p.
On Francis Baily's titles and on how to get a print with Baily's picture, autograph, and name in regular capitals. Explains remarks he made in a controversy with T. R. Robinson and published in the *Athenaeum* [Autumn, 1843] about JH's father's telescopes.
ALS 3pp RS:HS 16.33

6223. 1845–4–22 From Richard Sheepshanks Athenaeum
Hopes to inscribe a copy of Francis Baily's autograph on the proofs of his portraits. On the plates his full titles should be inscribed; asks JH's help in determining these. How RS can combine this with the autograph. On JH's dispute with James South.
ALS 3pp RS:HS 16.41

6224. 1845–4–27 From George Biddell Airy Royal Observatory, Greenwich
Passes on information received from Thomas Maclear regarding stars. Has tried to use William Simm's object-glass.
ALS 1p RS:HS 1.151 (C: RGO 6.193.56)

6225. 1845–4–30 From William Henry Smyth Hartwell House
R.A.S. making exception in letting J. H. Mädler's chart be removed for JH. Continues observation of Gamma Virginis.
ALS 2pp RS:HS 16.279

6226. 1845–4–30 From William Henry Smyth [3 Cheyne Walk, Chelsea]
Sends correction to data for Gamma Virginis. Asks if JH has 'considered the puzzle of Aristarchus [lunar crater]' discussed in WS's *Cycle of Celestial Objects*.
ALS 2pp RS:HS 16.280

6227. 1845–5–1 From George Biddell Airy Royal Observatory, Greenwich
Giving the plan of the Observatory.
ALS 2pp RS:HS 1.152 (C: RGO 6.699.232)

6228. 1845–5–1 To George Biddell Airy n.p.
Seems clear that JH's 'falling star' [see JH's 1845–4–29] was the same one reported by a correspondent of GA's as having been seen in Nottingham; in a postscript JH is not so sure.
ALS 4pp RGO 6.699.234

6229. [1845]–5 To George Peacock Collingwood
Accepts GP's invitation to visit him at Ely Cathedral after the B.A.A.S. meeting at Cambridge. Asks advice on whether it would be proper for JH to read his address at the meeting.
ALS 4pp Dunedin Public Lib.

6230. [1845–5 or later] To William Whewell [Collingwood]
Alerts WW, who was coming to visit the Herschels, that scarlet fever has arrived at JH's household. Comments onWW's writings about glacial theory.
ALS 4pp TC, Camb. Add. Ms.a.207[60]

6231. [1845–5] To William Whewell [Collingwood]
Invites Whewells to visit for a luncheon. Thanks WW for and comments on WW's publications on glaciers. Comments on Richard Jones.
ALS 4pp TC, Camb. Add. Ms.a.207[64] (C: RS:HS 22.232)

6232. 1845–5–2 To Augustus De Morgan Collingwood
On the subsuming of one scientific society by another; comments favorably on the *Memoirs* of the Cambridge University Philosophical Society and asks AD to help fill in JH's missing items.
C 2pp RS:HS 22.234

6233. 1845–5–4 To Thomas Maclear Collingwood
Remarks on observation of new comet. Discord in R.A.S. follows Francis Baily's death. C. P. Smyth will leave Cape after being appointed Astronomer Royal to Scotland. Glass for Cape equatorial tested. JH has no success appealing for continuation of Colonial Survey. Southern constellation reform compromised.
AL 4pp inc Cape Archives/Maclear Papers File 100

6234. 1845–5–6 From George Biddell Airy Royal Observatory, Greenwich
Describes testing William Simms's object glass and then gives details of the equatorial mounting GA used; offers to send carpenter to JH to construct mounting so JH can test object glass [diagram].
ALS 3pp TxU:H/M-0631.2; Reel 1084

6235. 1845–5–7 To John William Draper Collingwood
Critiques JD's paper on plant chemistry, contrasting 'luminous, colorific, calorific, & chemical or photographic' rays. Physics of light. [Letter continues two months later.] JH is done with photography. Has no desire to claim priorities in that field. Entrusts future to JD and others.
ACCS 2pp RS:HS 19.117

6236. 1845–5–8 To Margaret Brodie Herschel London
Comments about engaging a gardener, concerns about Peter Stewart, and waiting for council and committee meetings [of the R.S.L.?]. The Physical Committee meeting was deferred because of a transit of Mercury.
ALS 3pp JHS 6.11

6237. 1845–5–8 From Robert H. Schomburgk 19 Golden Square
Note sent with package from [J. F.] Encke of Berlin.
AL 1p RS:HS 17.178

6238. 1845–5–8 From William Henry Smyth [3 Cheyne Walk, Chelsea]
Sends Gamma Virginis data. Will report Aurora Borealis observations to R.A.S.
ALS 2pp RS:HS 16.281

6239. 1845–5–[8] To William Henry Smyth [Collingwood]
Doubtful over WS's angle for Gamma Virginis; urges precision in measurement. Hopes to obtain moon map from R.A.S. during next visit to London. Unable to solve Aristarchus puzzle.
C 1p RS:HS 22.235

6240. 1845–5–9 From Roderick Impey Murchison 16 Belgrave Square
Hears from the Dean of Ely that JH is busy writing his B.A.A.S. oration. Sent a new volume yesterday. Wished JH had been at RM's soiree on Wednesday.
ALS 4pp RS:HS 12.410

6241. 1845–5–9 From George Peacock Deanery, Ely
Several items of R.S.L. business.
ALS 4pp RS:HS 13.296

6242. 1845–5–9 From [Paul Sinnis?] B.A.A.S., 2 Duke St., Adelphi
Announces change in meeting time of Council on 14 May.
(A)LS 1p TxU:H/M–0921.7; Reel 1083

6243. 1845–5–9 From Charles Richard Weld R.S.L. Apartments
Announces 15 May meeting of Physics Committee.
ALS 1p TxU:H/M–0561; Reel 1093

6244. 1845–5–10 From George Biddell Airy Royal Observatory, Greenwich
Encloses transcript of letter dealing with object-glasses. Has given notice that he has withdrawn from the Physical Committee of the R.S.L.
ALS 1p RS:HS 1.153 (C: RGO 6.385.748)

6245. 1845–5–10 From George Peacock Deanery, Ely
Provides detailed information about JH's responsibilities as President of the B.A.A.S.
ALS 8pp RS:HS 13.297

6246. 1845–5–11 To George Biddell Airy 37 Tavistock Place
Gives reasons why telescope for Cape Observatory should be as perfect as possible. Doubts that its 'finder' is adequate.
ACCS 2pp TxU:H/L-0015; Reel 1054

6247. 1845–5–11 To Roderick Impey Murchison 37 Tavistock Pl.
On JH's Cambridge oration for B.A.A.S. Asks RM about weaknesses in paper, and character of people to be addressed, so that JH will not offend them. JH hopes to read speech, distrusting own extemporaneous speaking abilities.
ALS 3pp EUL:1999/1/9

6248. 1845–5–13 or earlier To Margaret Brodie Herschel London
Comments about spending time with JH's son William James; then describes dinner at the Archbishop's palace.
ALS 4pp JHS 3.21b

6249. 1845–5–13 To Margaret Brodie Herschel London
News about JH's son William James, JH's doings, and JH misses MH and Collingwood very much.
ALS 4pp JHS 3.21a

6250. 1845–5–15 To Carl Friedrich Gauss London
Writing on behalf of the B.A.A.S. Magnetic and Meteorological Committee, JH invites CG to submit a reply to the materials sent to him.
C 2pp NSUB Cod.Ms.Gauss 99

6251. 1845–5–15 To Adolphe Quetelet London
Includes copies of AQ's reply to circular sent previously [1844–12–5]. Requests additional information by 1845–6–15 and, if possible, AQ's presence at the meeting on 19 June.
ALS 2pp Académie belgique: #22

6252. 1845–5–15 From Karl Emil Schaffheütl Academy of Science, Munich
Updates and requests opinion on KS's newly devised 'universal photometer.'
ALS 3pp RS:HS 17.172

6253. 1845–5–17 To [William Radcliffe Birt] Collingwood
B.A.A.S. meets 18 June at Cambridge. Send contributions to report of Magnetical and Meteorological Committee to JH by 13 June.
ALS 1p RS:HS 19.118

6254. 1845–5–18 From James William Grant London
Returned home and found JH's welcome letter awaiting him. Relates his experiences while traveling in Belgium and Holland. Gives his views on the beauties of Antwerp and Ghent. Saw a Frenchman ascend in a balloon. Starts for Scotland in two days' time for two months.
ALS 12pp RS:HS 9.23

6255. 1845–5–19 From Augustus De Morgan 7 Camden St.
W. H. Smyth, Thomas Galloway, and AD went down and visited the library of the Mathematical Society and found it in very good state.
ALS 2pp RS:HS 6.222 S. De Morgan, *Memoir of Augustus De Morgan,* 150

6256. 1845–5–20 To Count Paul Edmund de Strzelecki Collingwood
Thanks for PS's *Physical Description of New South Wales.* Praises that publication.
ALS 3pp New South Wales State Library

6257. 1845–5–20 From Charles Richard Weld Somerset House
Announces meeting of R.S.L. Committee of Mathematics on 22 May.
LS 1p RS:HS 19.122c

6258. [1845–5–21 or later] To [?] n.p.
Membership dues, voting privileges, and need for discussions after reading of papers [probably at B.A.A.S. meetings]. Suggests announcing at each adjournment what the next topic will be. Considers prohibiting any discussion following medical papers.
Adraft 1p inc RS:HS 19.122d

6259. [1845]–5–22 From Thomas Graham Athenaeum
Delayed answering his letter until he had carried out the observations upon pure Colophene. Gives the results of his experiments. H. E. Sainte-Claire Deville's original account is in *Annales de chimie, vol.* 75. Council of the R.S.L. recommended Edward Sabine to be the new Foreign Secretary.
ALS 4pp RS:HS 8.232

6260. 1845–5–23 From George Peacock Deanery, Ely
Further arrangements about the B.A.A.S. meeting [see GP's 1845–5–10].
ALS 4pp RS:HS 13.298

6261. [1845–5–26] To Spencer J. A. Compton 5 Lowndes St.
Comments on SC's suggestions regarding the proposed changes in the projected new charter for the R.S.L.
draft 2pp RS:HS 5.266bis (C: RS:HS 22.263)

6262. 1845–5–26 From Spencer J. A. Compton Lowndes St.
Suggesting points to be considered for the proposed new charter for the R.S.L. Remarks on the recent meeting of the Trustees of the British Museum.
ALS 4pp RS:HS 5.266

6263. 1845–5–26 From Sir William J. Hooker Royal Botanic Gardens, Kew
A friend of his would like JH's advice on a suitable telescope.
ALS 3pp RS:HS 9.450

6264. 1845–5–27 From George Bancroft Washington D.C.
Sends a series of magnetical and meteorological observations made at Washington.
ALS 1p RS:HS 5.14

6265. 1845–5–28 From Augustus De Morgan 7 Camden St.
Regarding the by-laws of the Society. Does not know anything about the Cambridge Transactions. Recommends various mathematical papers.
ALS 5pp RS:HS 6.223 S. De Morgan, *Memoir of Augustus De Morgan,* 150–1

6266. 1845–5–30 From George Biddell Airy Royal Observatory, Greenwich
Giving notice of a meeting of the Standards Commission.
ALS 1p RS:HS 1.155

6267. 1845–5–30 From George Biddell Airy Royal Observatory, Greenwich
Hopes JH will be staying with him. Would like statement from JH regarding the Magnetic Observatory.
ALS 1p RS:HS 1.154

6268. 1845–5–30 To Humphrey Lloyd Collingwood
Is trying to organize a meeting of the Magnetic Committee before the meeting of the B.A.A.S. in Cambridge, to deal with reporting and committee membership.
C 3pp RS:HS 22.236

6269. 1845–5–31 From William Radcliffe Birt Cambridge Road, Bethnal Green
Discusses work on barometric observations.
ALS 4pp Harvard: Houghton AAH 67m–67 (86)

6270. [1845?–6?] From Charles Spring-Rice 17 Eaton Place South
Happy JH is back safely from Cambridge. Thanks JH for letter brought by [Adam] Sedgwick. Desires an account of JH.
ALS 2pp RS:HS 14.480

6271. 1845–6–2 From James A. Gordon Naish, Bristol
The paper for the prize essay has been withdrawn. Tring Agricultural Association has published the results of Mr. Forster's experiments, which so far have been very successful. Was grateful for his poetry. Has been ill as well as JH. Hopes he is improving.
ALS 4pp RS:HS 8.178

6272. 1845–6–2 From Edward Sabine Woolwich
Suggests time and place for [committee] meeting. Good report from observatory in British Guiana. Problems in establishing observatories in colonies. Changes occurring at Toronto. Time is not right to propose turning observatories over to colonial governments.
AL 4pp inc † TxU:H/M–0521.17; Reel 1093

6273. 1845–6–2 To William Whewell Collingwood
Informs WW when JH and wife will arrive at Trinity Lodge for B.A.A.S. meeting, before which JH proposes a meeting of the magnetic committee.
ALS 4pp TC, Camb. Add. Ms.a.207[61] (C: RS:HS 22.237)

6274. 1845–6–4 From Humphrey Lloyd Trinity College, Dublin
Proposed date for B.A.A.S. meeting creates difficulty for HL. Assurance that no one other than JH could have provided such vital leadership for global magnetic survey. Doubts that [James] McCullagh could serve committee well; suggests [J. D.] Forbes instead.
ALS 3pp † TxU:H/M–0336; Reel 1087

6275. 1845–6–5 To Humphrey Lloyd Collingwood
Further to arrangements about meeting [see JH's 1845–5–30].
C 2pp RS:HS 22.238

6276. 1845–6–5 To William Whewell Collingwood
Urges WW either to be present for a meeting in preparation for the Magnetic Congress or to send written recommendations.
ALS 3pp TC, Camb. Add. Ms.a.207[60] (C: RS:HS 22.239)

6277. 1845–6–6 From Edward Joseph Lowe Highfield House near Nottingham
Received yesterday a letter from G. B. Airy informing him that JH also observed the meteor of 24 April; sends an account of it. Would also like JH's account to compare the two. Is obliged to JH for trying to get his printed paper filled.
ALS 2pp RS:HS 11.312

6278. 1845–6–8 From Alexander von Humboldt Sans Souci
Sending the first volume of *Kosmos*. Queries regarding the *Magellanic Clouds*.
ALS 1p (German) RS:HS 10.68

6279. 1845–6–9 To Michael Faraday Collingwood
Comments on MF's experimental results when an electromagnetic field affects polarized light.
C 4pp RS:HS 22.240

6280. 1845–6–9 From Sir William J. Hooker Royal Botanic Gardens, Kew
Has received the enclosed from James Graham and would be pleased if W. R. Dawes would choose an instrument.
ALS 1p RS:HS 9.451

6281. 1845–6–10 To James David Forbes Collingwood
Hopes to add JF's name to B.A.A.S. Meteorological Committee, and to see JF at Cambridge meeting of the committee.
ALS 3pp St. Andrews 73 (C: RS:HS 22.241)

6282. 1845–6–10 From James A. Gordon 16 Suffolk St.
JH does not appear to have received his last letter. Has not been well, but hopes to be better for tomorrow. Own brother has been behaving in a peculiar manner. Daniel has been dealing rude blows. Hopes to visit JG's laboratory. Has now visited the laboratory and spent two agreeable hours.
ALS 7pp RS:HS 8.179

6283. 1845–6–10 From Heinrich Christian Schumacher Altona
[J. C.] Houzeau has sent a paper on the determination of parabolic elements of a comet. Relays a copy of his method, which HS will examine more closely before publishing.
ALS 2pp RS:HS 15.395

6284. [1845]–6–10 From Ann Smyth 3 Cheyne Walk, Chelsea
Mentions how helpful JH's name would be for C. P. Smyth's recommendations to the position of Astronomer Royal for Scotland. C. P. Smyth working on zodiacal light observations.
ALS 4pp RS:HS 16.219

6285. 1845–6–10 To William Whewell Collingwood
Because of the 'severe loss' recently suffered by Mrs. Whewell, Margaret Herschel will not stay at Trinity Lodge during the B.A.A.S. meeting. JH plans to go to Ely after the meeting.
ALS 2pp TC, Camb. Add. Ms.a.207[65] (C: RS:HS 22.242)

6286. 1845–6–11 From William Rutter Dawes Camden Lodge
The comet is a splendid one. Observed it last night with his

Munich 32 inch.
ALS 2pp RS:HS 6.69

6287. 1845–6–12 To George Biddell Airy n.p.
Arranging a meeting of the Magnetic Committee to set the agenda for the magnetic conference to be held during the B.A.A.S. meetings in Cambridge.
ALS 1p RGO 6.675.509

6288. 1845–6–12 From James David Forbes Edinburgh
Accepts JH's invitation to join Magnetical and Meteorological Committee of B.A.A.S. Will see JH at Cambridge next week. T. M. Brisbane's magnetic observations will be published in *Transactions* [of Royal Society of Edinburgh]. JH was responsible for making these known.
ALS 2pp TxU:H/M–0202; Reel 1087

6289. 1845–6–12 To Sir William J. Hooker Collingwood
Cannot supply information requested about telescope until WH reminds him of the detailed information necessary for decision.
ALS 3pp Royal Botanic Gardens

6290. 1845–6–12 To Hugh Percy (3rd Duke of Northumberland) Collingwood
Sends copies of plates and details about them, and expects. printing of JH's *Cape Results* to begin early in 1846.
C 2pp RS:HS 13.356 inc & 22.243

6291. 1845–6–13 From George Peacock Deanery, Ely
Still more about B.A.A.S. meeting arrangements, especially about the magnetic survey {see GP's 1845–5–23].
ALS 6pp RS:HS 13.299

6292. 1845–6–13 From John Phillips B.A.A.S., Philosophical Society, Cambridge
Informs JH that 'It is necessary to ask you as President of the British Association to preside at the General Advisory' to be held on 16 June. JH should also inspect the Cambridge Observatory.
ALS 2pp TxU:H/M–0428.18; Reel 1093

6293. 1845–6–13 From Arthur-Auguste De La Rive Geneva
Invites B.A.A.S. members to Geneva meeting of Helvetic Society of Natural Sciences.
ALS 2pp (French) TxU:H/M–1012; Reel 1083

6294. 1845–6–14 To [William Radcliffe Birt] London
Forwarded WB's report to B.A.A.S. at Cambridge. Sends payment for expenses. Two German meteorologists will be at B.A.A.S. meeting.
ALS 2pp RS:HS 19.119

6295. 1845–6–14 From George W. Hearn Royal Military College, Sandhurst
Has submitted a paper on the projection of a star on the moon's disc to the R.A.S. Wishes to draw JH's attention to it. Comments on some of his observations.
ALS 4pp RS:HS 9.265

6296. 1845–6–15 From Christian F. Schönbein Dover
Has arrived in England. Would like to call on JH.
ALS 1p RS:HS 15.356

6297. 1845–6–17 To Margaret Brodie Herschel Cambridge
Has arrived safely in Cambridge, and explains arrangements for MH to come some days later.
ALS 3pp JHS 6.8

6298. 1845–6–18 To Margaret Brodie Herschel Cambridge
The meetings of the B.A.A.S. council have gone well so far; JH sends more details on travel arrangements for MH [see JH's 1845–6–17].
ALS 4pp JHS 6.7

6299. 1845–6–18 From George Peacock Downing College, Cambridge
Brief note alerting JH to the presence of a foreign visitor.
ALS 1p RS:HS 13.300

6300. 1845–6–18 From John Phillips Cambridge
Invites JH to come to the magnetic conference meeting being held at the B.A.A.S. meeting.
L 1p RS:HS B27.66

6301. 1845–6–19 To Margaret Brodie Herschel Cambridge
The weather is great in Cambridge for B.A.A.S. meeting; JH was at a small dinner party given by William Whewell, where the talk was mostly of architecture. JH refers to a battle in the Section A meetings between David Brewster and G. B. Airy.
ALS 4pp JHS 6.10 (C: TxU:H/L-0540.1; Reel 1053)

6302. 1845–6–21 To Margaret Brodie Herschel Cambridge
The B.A.A.S. meetings are continuing, and JH has had an opportunity to talk to Charles Pritchard, who feels that their son William is doing very well at Pritchard's school; more about MH's travel arrangements [see JH's 1845–6–18].
ALS 4pp JHS 6.6

6303. 1845–6–22 To Margaret Brodie Herschel Cambridge
At the B.A.A.S. meeting, JH spoke and was thanked by the most eloquent speaker he has ever heard, an American minister by the name of Everett. Everett is 'worthy of a much better country.' JH has also been busy in the magnetic committee.
ALS 2pp JHS 6.2 (C: TxU:H/L-0540.2; Reel 1053)

6304. 1845–6–23 To Margaret Brodie Herschel Cambridge
Writes at the end of a very busy day of meetings, which was capped by a meeting in the Senate House, where Roderick Murchison spoke very eloquently, and Samuel Wilberforce thanked him even more eloquently, supporting the findings of geology in so doing. In the last part of the letter marked 'Private', JH expresses concern about the conduct of 'W. W.' [William Whewell?].
ALS 4pp JHS 6.5

6305. 1845–6–24 From Edward Everett London
Had to leave Cambridge suddenly and was thus unable to take his leave of JH. Will always remember his visit. Would like the manuscript of JH's address.
ALS 3pp RS:HS 7.114

6306. 1845–6–25 From James A. Gordon Naish, Bristol
Returning JH's [?] to keep for him. Does not think [Forster] is wrong in his assumptions. Mesmerism is better than electricity.
AL 4pp inc RS:HS 8.177

6307. 1845–6–25 To Margaret Brodie Herschel Cambridge
There has been a great disagreement between W. W.[William Whewell?] and S.[Adam Sedgwick?]. JH, together with others, has been instrumental in bringing the two back

together again, at least for the time being.
AL 5pp JHS 6.4

6308. 1845–6–28 From H. C. Jennings Calais
Has found a method of making a gum equal to gum arabic. Would like the support of the Royal Society and JH to enable him to take out a patent.
ALS 3pp RS:HS 10.438

6309. 1845–6–30 From John Henry Griesbach Carlton Villas, Maida Vale
Encloses the tickets for the Philharmonic concerts and gives details of the program. Is going to cast another 20" speculum in a few days.
ALS 1p RS:HS 9.60

6310. 1845–7 To Chairman, H.E.I.C. Court of Directors Collingwood
B.A.A.S. resolutions include closing H.E.I.C. magnetic and meteorological observatories at Simla and Singapore but continuing those in Bombay and Madras. Propose magnetic survey of India Sea.
ACCS 1p TxU:H/L-0133; Reel 1054

6311. 1845–7 From John Phillips B.A.A.S., York
Arrangements made for copies of the B.A.A.S. *Catalogue of Stars* to be distributed to public institutions and to individuals.
L 1p TxU:H/M–0428.19; Reel 1093

6312. 1845–7 From John Phillips York
Informs JH that the committee to deal with the publication of *Fauna Antiqua Sivalensis* has been struck and is to prepare a request to the government for aid in publishing.
L 1p RS:HS B27.68

6313. 1845–7 From John Phillips York
Sends JH a copy of the resolution dealing with the publication of *Fauna Antiqua Sivalensis.*
L 1p RS:HS B27.67

6314. 1845–7 From John Phillips York
Informs JH that JH together with several others will be responsible for the distribution of the B.A.A.S. catalogue of stars.
(A)L 1p RS:HS B26.14

6315. 1845–7 From John Phillips York
Expands on the work of the coal map committee [see JP's 1845–7] by sending along the resolution that created it.
AL 2pp RS:HS B26.13

6316. 1845–7 From John Phillips York
Informs JH that he is to work with a committee of the B.A.A.S. to prepare a map of the distribution of coal around the world.
(A)L 1p RS:HS B26.12

6317. 1845–7–3 From George Bishop Regents Park
The Moon map arrived safely on Saturday. Is pleased it was of some use. Includes J. R. Hind's corrections for double star Alpha Virginis. Will forward later ones.
ALS 2pp RS:HS 4.146

6318. 1845–7–3 From Robert Hunt Museum of Pract. Geology
Sends an analysis of the salts he sent him at Cambridge.
ALS 3pp RS:HS 10.119

6319. 1845–7–3 To Robert Peel London
As president of B.A.A.S., JH forwards annual resolutions for prime minister's review. Especially requests continuing support for magnetic observations.
LCC 1p RGO 6.694.365v *B.A.A.S. Report 1845* (1846), 67

6320. [1845]–7–5 From Edward Sabine Woolwich
Encloses resolutions and report for Sir Robert Peel for JH's signature. Other information to follow.
ALS 3pp RS:HS 15.304

6321. 1845–7–7 From William Henry Sykes India House
Reduction of observations will begin immediately.
ALS 2pp TxU:H/M–0514; Reel 1093

6322. 1845–7–9 From John Stuart Mill India House
Is gratified by the mention of JM's name in JH's B.A.A.S. address, but thinks that JH has made a grave error in his castigation of the work of Auguste Comte. Comments on his theories.
ALS 8pp RS:HS 12.329

6323. 1845–7–9 To Robert Peel Collingwood
Informs of B.A.A.S. proposal to encourage by 'specific pecuniary reward,' improvement of 'self-recording magnetical and meteorological apparatus.' Outlines advantages of proposal.
ALS 3pp RS Sa.650 (C: TxU:H/L-0298; Reel 1054)

6324. 1845–7–9 From Edward Sabine Clifton
[Roderick] Murchison, [John] Phillips, and ES will take care of present business. Will not convene B.A.A.S. council until autumn.
ALS 1p RS:HS 15.172

6325. 1845–7–9 To [Franc Sadleir] Collingwood
Sends copy of B.A.A.S. resolutions to fellows of Trinity College, Dublin, containing recommendation to continue magnetic and meteorological observatory there.
ACCS 2pp TxU:H/L-0379; Reel 1054

6326. 1845–7–10 From Edward Everett London
Is grateful for JH's note and the manuscript. Will treasure it. Sails on 4 Sept. Has heard who his successor will be. Would be very pleased to assist JH in any way when he returns to America.
ALS 4pp RS:HS 7.115

6327. 1845–7–10 To John Stuart Mill Collingwood
Comments on JH's disagreement with the nebular hypothesis of Auguste Comte, and points out to JM that JH disagrees with some of JM's writings on physical science as well.
ACCS 3pp RS:HS 25.6.26

6328. 1845–7–12 From James David Forbes Edinburgh
Sends new edition of JF's *Travels [through the Alps of Savoy]*. Hopes JH will examine the theoretical part.
ALS 1p St. Andrews LB III, 700 (C: RS:HS 7.307)

6329. 1845–7–13 To [John Stuart Mill] n.p.
Is sending to Auguste Comte JH's *Athenaeum* comments on Comte. Asks JM not to mention that JH sent them to JM. Does not wish to offend Comte.
ALS 2pp Yale Univ. (RS:HS 12.330 & 22.244)

6330. 1845–7–14 From John Stuart Mill India House
Would like to send the *Athenaeum* to Comte together with a copy of JH's last note. Comments on the theory of Auguste Comte. Will be thankful to receive comments on JM's own book.
ALS 5pp RS:HS 12.331

6331. 1845–7–14 From Edward Sabine Woolwich
Selling price of B.A.A.S.'s catalogue must be decided. Sends JH's report for revision.
ALS 3pp RS:HS 15.173

6332. 1845–7–14 From Christian F. Schönbein Upper Goucester Place
Will call on JH at Hawkhurst before leaving England.
ALS 1p RS:HS 15.357

6333. 1845–7–16 To John Stuart Mill Collingwood
Proceeds to show JM in detail that Comte's nebular hypothesis is arguing in a 'vicious circle' [see JH's 1845–7–10].
ACCS 4pp RS:HS 25.6.27

6334. [1845–7–16?] To John Stuart Mill n.p.
Do as JM suggests. Further comments on JH's dispute with the theories of Auguste Comte.
draft 4pp RS:HS 12.332 & 22.245

6335. 1845–7–16 From Edward Sabine Woolwich
Not satisfied with arrangement of subjects in account of magnetic proceedings. Suggests alternative arrangement. [Francis] Beaufort appreciates suggestions regarding colonial establishments.
ALS 2pp RS:HS 15.174

6336. 1845–7–17 From Friedrich Eduard Beneke Berlin
Letter to accompany FB's complete work on psychology and knowledge.
ALS 2pp (German) RS:HS 4.62

6337. 1845–7–17 From John Stuart Mill India House
Has nothing more to do than to thank him for his explicit letter. Will forward a copy of this letter to Auguste Comte.
ALS 1p RS:HS 12.333

6338. 1845–7–18 To John Stuart Mill Collingwood
Is willing to have JM show Comte JH's letter of 1845–7–13, but not JH's 1845–7–16. The latter of these two was intended to show JM the nature of the argument from JH's perspective.
ACCS 1p RS:HS 25.6.28

6339. 1845–7–21 From Augustus De Morgan 7 Camden St.
Hopes he has rested after his Cambridge labors. Comments on Thomas Wright of Durham and his book on the Universe, 1750.
ALS 3pp RS:HS 6.224 S. De Morgan, *Memoir of Augustus De Morgan*, 152

6340. 1845–7–23 From C. A. von Steinheil London
Tells JH that his travels will take him to London. Requests time and day when they can meet personally in Kent.
ALS (German) 1p RS:HS 17.6

6341. 1845–7–25 From George Biddell Airy Royal Observatory, Greenwich
Asking him to arrange distribution of the R.A.S. catalogues as he is too busy. When is JH going to try William Simm's object planes?
ALS 2pp RS:HS 1.156 (C: RGO 6.203.339)

6342. 1845–7–28 From Mr. McLachlan Stellenbosch
Criticizing the present system of Education at the Cape.
ALS 5pp RS:HS 13.9

6343. 1845–7–29 From Sir Robert Harry Inglis 7 Bedford Square
Will find certain friends at the R.S.L. Club on Thursday, including Edward Everett, the American Minister in England.
ALS 1p RS:HS 10.236

6344. 1845–7–30 From Josiah Quincy Harvard University
Sends two copies of memoirs of James Grahame. More copies may be purchased with proceeds benefiting the Massachusetts Historical Society.
ALS 1p RS:HS 14.295

6345. 1845–7–30 From Josiah Quincy Harvard University
Asks JH to distribute engravings of James Grahame among Grahame's friends and family. Discusses memoir of Grahame and Grahame's *History of the United States.*
ALS 3pp RS:HS 14.294

6346. 1845–8 To Margaret Brodie Herschel Collingwood
Writes mostly about arrangements for meeting MH's train from London.
ALS 3pp JHS 4.23

6347. 1845–8–1 From James Nasmyth Patricroft, Manchester
Offers to make drawings of the moon's surface for illustrating JH's paper on that subject to be presented to the R.A.S.
ALS 2pp RS:HS 13.76

6348. 1845–8–2 To George Biddell Airy n.p.
No news yet about the request to Robert Peel for money for magnetic observations.
LCC 1p inc RGO 6.675.520

6349. 1845–8–4 From Thomas Ayern Newcastle
Received JH's payment for coal. Estimates cost increases for next year.
ALS 1p RAS:JH Archive 4/9.22 verso; Reel 7

6350. 1845–8–4 To Margaret Brodie Herschel Collingwood
Mostly about family activity while MH is away; comments on the review by Adam Sedgwick of [Robert Chambers's] *Vestiges of a Natural History of Creation* in the *Edinburgh Review* in which the author receives 'a reasonable currycombing.'
ALS 4pp JHS 6.1

6351. 1845–8–5 To Caroline Lucretia Herschel Collingwood
JH expresses pleasure in receiving and reading extracts from CH's biography. Expects to begin printing his *Cape Results* by Christmas. In finalizing his *Cape Results*, JH has found that several Southern double stars moved in the five-year span of his observations.
ALS 4pp TxU:H/L–0590.9; Reel 1058

6352. 1845–8–5 From Edward Sabine Woolwich
[John] Stevelly wants ES to write about him. ES sends Stevelly's letter directly to JH instead. Says he is a conscientious and painstaking preceptor.
ALS 1p RS:HS 15.175

6353. 1845–8–5 From John Taylor, [F.R.S.] Penmear near Falmouth
Agrees with JH that the price of the B.A.A.S. catalogue of stars should be kept moderate to achieve the widest useful distribution.
ALS 2pp RS:HS B26.15

6354. 1845–8–6 From Agnes Baillie Hampstead
Informing him of the death of her sister.
ALS 2pp RS:HS 3.11

6355. 1845–8–7 From George Biddell Airy Royal Observatory
Encloses a copy of JH's 'letter of conditions.' Has been extremely busy.
ALS 1p BostonPubLib Ms.Eng.232(2)

6356. 1845–8–7 From Thomas Gardner London
At a recent meeting of the Council of the College of Chemistry it was resolved that JH should be invited to become a Vice-President. Hopes he will agree to this request.
ALS 1p RS:HS 9.92

6357. 1845–8–11 To James David Forbes Collingwood
Thanks JH for second edition of work on the Alps. Praises first edition.
ALS 3pp St. Andrews 111 (C: RS:HS 22.246)

6358. 1845–8–11 From William Samuel Stratford Notting Hill Square
Has communicated with [John] Phillips, and indicates that the committee will take responsibility for its actions [in distributing star catalogues?].
ALS 3pp RS:HS B26.16

6359. 1845–8–18 From Augustus De Morgan 7 Camden St.
When JH comes to town he will send him the book. Comments on this book [probably one by Wright of Durham] and the works on 'Harmonics' by Robert Smith.
ALS 4pp RS:HS 6.225

6360. 1845–8–20 From William Whewell Hastings
Comments on glacial theory. Asks JH for his views on mathematical education at Cambridge. WW favors concentration on mathematical classics.
ALS 4pp RS:HS 18.202 Todhunter, *Whewell, 2,* 328–9

6361. 1845–8–21 From Sir William J. Hooker Royal Botanic Gardens, Kew
His letter possibly did not reach him as his friend is still inquiring about a telescope for Lima.
ALS 2pp RS:HS 9.452

6362. 1845–8–21 From Richard Jones Wiesbaden
Regarding the use of the Carlsbad and Seltzer water for rheumatism. Thinks JH should come to Wiesbaden to effect a cure. A good place to study the economics of the Duchy. Much religious agitation.
ALS 4pp RS:HS 10.379

6363. 1845–8–23 To Sir William J. Hooker Collingwood
[Responding to WH's 1845–8–21], JH states that he has not yet received the information needed from [William] Simms. Has sent a further communication to Simms.
ALS 3pp Royal Botanic Gardens

6364. 1845–8–25 From John Chapman 121 Newgate St.
Received parcel from America containing 'Portraits of [James] Grahame.'
ALS 1p RAS:JH Archive 4/9.19 verso; Reel 7

6365. 1845–8–27 From Sir William J. Hooker Royal Botanic Gardens, Kew
Many thanks for the trouble he has taken over the matter of the telescope for Lima. His own son Joseph is a candidate for the chair of botany at Edinburgh and would be grateful for a testimonial.
ALS 4pp RS:HS 9.453

6366. 1845–8–28 From John Stevelly Belfast
Thanks extensively for JH's expression of approval. Rejoices in JH's acquaintanceship.
ALS 3pp RS:HS 17.232

6367. 1845–9–1 From Henry Anthony Jeffreys 47 Marina, St. Leonards
JH will hear in due course of the resignation of Mr. Young, the Medical Officer for Hawkhurst. Thinks something must be done quickly to improve matters.
ALS 4pp RS:HS 10.287

6368. 1845–9–2 From William Henry Harvey Cork
Chair of Botany at Edinburgh is now vacant and Joseph Hooker intends applying for it. Would be grateful if JH would supply a testimonial for Hooker. There is only one other candidate for the position.
ALS 8pp RS:HS 9.251

6369. 1845–9–2 To Henry Anthony Jeffreys Collingwood
Received his note this morning, the first tidings he had heard of the matter. Gives his own opinion of the matter. Would have liked to have conferred with him. His neighbors are away so has been unable to consult them.
draft 3pp RS:HS 10.288 & 22.248

6370. 1845–9–3 From Edward Sabine Brussels
Consulting with [Adolphe] Quetelet regarding some magnetic instruments. Discusses reply of Treasury regarding various endeavors. Discusses shipping charges on Russian materials.
ALS 4pp RS:HS 15.176

6371. 1845–9–3 From Henry Piddington Calcutta
Is sending JH some powder that HP thinks might be a good compound for polishing telescope specula.
ALS 3pp RS:HS 13.403

6372. 1845–9–4 From Henry Anthony Jeffreys 47 Marina, St. Leonards
Was at Hawkhurst for a few days but was unable to call on JH. Further regarding the case of the Medical Officer for Hawkhurst, Mr. Young.
ALS 4pp RS:HS 10.289

6373. 1845–9–9 From William Henry Smyth n.p.
Thanks for inspecting WS's [manuscript] and noting errors. Computations by Mr. Hinde [J. R. Hind?] from observations of Gamma Virginis by WS.
ALS 4pp TxU:H/M–0494; Reel 1093

6374. 1845–9–11 From John R. Hind 7 Allsop's Place, Regents Park
Sends a copy of an ephemeris of Biela's comet for November and December.
ALS 3pp RS:HS 9.320

6375. 1845–9–13 From Ottaviano F. Mossotti Pisa
Sending one of his own books for the R.A.S. and one of his memoirs, also one for JH. Is lacking part of his set of the *Memoirs of the Astronomical Society.*
ALS 2pp (French) RS:HS 13.55

6376. 1845–9–14 To Richard Sheepshanks Collingwood
Thanks RS for the engraving [of JH's portrait?]. Who should receive copies? Suggests that the 'Knight of C. Hill' [Sir James South and his Campden Hill Observatory] 'is not worth your powder.'
ALS 3pp RS:HS 16.42 (C: RS:HS 25.9.19)

6377. 1845–9–19 From George Biddell Airy Royal Observatory, Greenwich
Is not clear what the statement from Robert Peel means [see JH's 1845–9–29 or earlier].
ACCS 3pp RGO 6.675.530

6378. 1845–9–19 To Sir William J. Hooker Collingwood
[Responding to WH's 1845–8–27], has been slow in writing recommendation [see JH's 1845–9–19] for WH's son, Joseph D. Hooker, because, not having met WH's son, wished to get judgment from James Ross. Enclosing Ross's very complimentary letter.
ALS 3pp Royal Botanic Gardens

6379. 1845–9–19 To Sir William J. Hooker Collingwood
At WH's request [see WH's 1845–8–27], JH writes in support of WH's son, Joseph D. Hooker, as a candidate for the chair of botany at Edinburgh University.
ALS 3pp Royal Botanic Gardens

6380. 1845–9–20 From Mary Somerville Rome
Discusses light experiments, which 'amused' her during summer. Has heard no scientific news lately. Describes summer in Rome.
ALS 4pp RS:HS 16.351

6381. [1845]–9–21 From Sir William J. Hooker Royal Botanic Gardens, Kew
Is grateful for JH's testimonial for his son Joseph. Wishes the Herschels would come and look around Kew. Regarding the Cape bulbs. Has no official person to collect seeds.
ALS 3pp RS:HS 9.454

6382. 1845–9–24 From John R. Hind 7 Allsop's Place, Regents Park
Has just completed the orbit for the binary star Mu^2 Bootis. Sends details. Could not send his notice of Biela's comet to W. R. Dawes as he did not know his address.
ALS 3pp RS:HS 9.321

6383. 1845–9–25 From William Henry Smyth [3 Cheyne Walk, Chelsea]
J. R. Hind determined orbit of Mu^2 Boötes using JH method.
ALS 2pp RS:HS 16.282

6384. 1845–9–27 From Josiah Quincy Boston, Mass.
Sends memoir of James Grahame. Asks JH to distribute copies to JH's friends. If JH thinks it appropriate, he may place copies at bookseller's as well.
ALS 3pp RS:HS 14.396

6385. 1845–9–27 From Alexander F. Ridgway 42 Leicester Square
[Printed notice] Latest developments at Royal Agricultural and Commercial Society of British Guiana, including plans to form branch in London. Observer for Astronomical and Meteorological Society recently sailed for British Guiana with full complement of instruments.
L 1p TxU:H/M–0444; Reel 1093

6386. 1845–9–29 From George Biddell Airy Royal Observatory, Greenwich
Is anxious to know whether JH has used William Simms's object glass, and seeks JH's opinion.
ALS 1p TxU:H/M-0631.3; Reel 1084

6387. 1845–9–29 or earlier To George Biddell Airy n.p.
Robert Peel has suggested that the Admiralty has money available to pay for improvements in scientific instruments of value to navigation.
ALS 3pp RGO 6.675.528

6388. 1845–9–29 From Johann Heinrich Mädler Dorpat
About a number of astronomical matters, including double stars. Provides observations and calculations for Gamma Virginis. Also comments on lunar models.
ALS 3pp (German) RS:HS 12.275

6389. 1845–9–30 To George Biddell Airy n.p.
Asks for new observational data on Gamma Virginis [see JH's 1843–7–18].
AL(S) 1p inc RGO 6.193.210

6390. 1845–9–30 To George Biddell Airy n.p.
Encourages GA to ask the Admiralty about the implications of Robert Peel's statements [see JH's 1845–9–29 or earlier].
ALS 2pp RGO 6.675.532

6391. 1845–9–30 From James Basire Red Lion Square
Sent proofs of JH's plates and drawings. Will send proofs of 'large Nebula.'
LS 2pp RAS:JH Archive 4/9.38 and verso; Reel 7

6392. [1845–10] To George Biddell Airy n.p.
Regarding measure of Virginis.
draft 8pp RS:HS 1.332

6393. 1845–10–1 From George Biddell Airy Royal Observatory, Greenwich
Provides JH with the latest Royal Observatory values for Gamma Virginis [see JH's 1845–9–30], and advises JH to change his method of measuring double stars.
ALS 2pp TxU:H/M-0631.4; Reel 1084 (C: RGO 6.193.211)

6394. 1845–10–1 From Edward Sabine Woolwich
Discusses magnetic experimentation with regard to naval officers. [A. T.] Kupffer's report is based on proceedings at Cambridge. Discusses scientific conference and foreign governments.
ALS 6pp RS:HS 15.177

6395. [1845–10?] From Peter Stewart 65 Cornhill
Refused several requests from Lieutenant Sutherland in Elgin to publish 'extraordinary astronomical work.' Judges him to be 'half crack'd.' Forwards to JH Sutherland's latest letter.
AL 2pp RAS:JH Archive 4/9.20 and verso; Reel 7

6396. 1845–10–2 To George Biddell Airy Collingwood
Believes GA's measurements [see GA's 1845–10–1] seriously in error, and strenuously defends JH's method.
ALS 5pp RGO 6.193.212 (C: RS:HS 25.13.6)

6397. 1845–10–2 From George Robert Waterhouse British Museum
Sends to JH's children first part of [GW's?] new book on quadrupeds for young readers.
ALS 3pp RS:HS 18.96

6398. 1845–10–3 From George Biddell Airy Royal Observatory, Greenwich
Thanking JH for his letter on double star orbits. Observations on angles of position of Gamma Virginis.
ALS 1p RS:HS 1.158 (C: RGO 6.193.215)

6399. 1845–10–3 From George Biddell Airy Royal Observatory, Greenwich
Requesting JH to jog Sir Robert [Peel's?] memory.
ALS 1p RS:HS 1.157 (C: RGO 6.675.521)

6400. 1845–10–5 To William Whewell Collingwood
Working on double star orbits. Responds to WW's question [see WW's 1845–8–20] about proper mathematical education at Cambridge, for the most part agreeing with

WW's preference for classical works.
ALS 4pp TC, Camb. Add. Ms.a.207[67] (C: RS:HS 22.249)

6401. 1845–10–6 To John R. Hind n.p.
Rejoices that RH is studying orbit of Mu^2 Bootis. Suggests way to simplify calculations. Compares orbits for Gamma Virginis calculated by various observers. Notice of systematic errors.
ACCS 4pp TxU:H/L-0192; Reel 1054

6402. 1845–10–7 From John R. Hind 7 Allsop's Place, Regents Park
Is grateful for details of his method of calculating the orbits of double stars. Comments on some of his observations.
ALS 3pp RS:HS 9.322

6403. 1845–10–8 To George Biddell Airy n.p.
How does one jog the memory of the First Lord of the Treasury [see GA's 1845–10–3]?
LCC 1p inc RGO 6.675.522

6404. 1845–10–8 To Spencer J. A. Compton Collingwood
Concerned about delay in government decision about magnetic observatories.
C 1p RS:HS 22.250

6405. 1845–10–9 From George Biddell Airy Royal Observatory, Greenwich
Still trying to solve the problem of jogging Robert Peel's memory [see JH's 1845–10–8].
ALS 2pp TxU:H/M-0631.5; Reel 1084 (C: RGO 6.675.523)

6406. 1845–10–9 From Spencer J. A. Compton Castle Ashby
Has received JH's letter and written to Robert Peel to urge him to come to an early decision. Was not aware an answer had not been given. Will let him know immediately when a reply has been received from Sir Robert.
ALS 1p RS:HS 5.267

6407. 1845–10–9 To William Whewell Collingwood
Letter of introduction to WW for the son of Sir Jeremiah Bryant, then entering Trinity College.
ALS 4pp TC, Camb. Add. Ms.c.89[32]

6408. 1845–10–10 To George Biddell Airy [Collingwood]
Results of testing at Collingwood by JH and W. R. Dawes of two 8-inch object glasses made by William Simms for Cape of Good Hope observatory.
AdraftS 21pp TxU:H/L-0016; Reel 1054

6409. 1845–10–10 To George Biddell Airy n.p.
S. J. A. Compton [Lord Northampton] has addressed a private letter to Robert Peel [see GA's 1845–10–9].
LCC 1p inc RGO 6.675.525

6410. 1845–10–10 From William Rutter Dawes 'Starfield', Liverpool
Has received his letter. Gives his recent observations of Alpha Virginis, and other stars.
ALS 5pp RS:HS 6.70

6411. 1845–10–10 From Edward Sabine Sandgate
Will neither affirm nor deny that he is Foreign Secretary of R.S.L. Discusses importance of Foreign Secretary.
ALS 4pp RS:HS 15.178

6412. 1845–10–14 To George Biddell Airy Collingwood
Supplement to JH's report [see JH's 1845–10–10] on object glasses for Cape Observatory. Different results at lower powers when observing planets.
ACCS 2pp TxU:H/L-0017; Reel 1054

6413. 1845–10–14 From George Biddell Airy Royal Observatory, Greenwich
Thanks JH for his report on William Simms's object glass trials [see GA's 1845–9–29], together with some other astronomical matters, including reference to a letter from F. W. Bessel.
ALS 2pp TxU:H/M-0631.6; Reel 1084

6414. 1845–10–14 From Edward Sabine Sandgate
Wants JH to write the communication to Sir Robert Peel regarding the science conference. Foreign members are particularly anxious that JH do this.
ALS 3pp RS:HS 15.179

6415. 1845–10–15 To John R. Hind Collingwood
Has been considering what principle should be followed in the computation of the orbits of double stars. Gives his views.
C 4pp RS:HS 9.323 & 22.251

6416. 1845–10–15 From Edward Sabine Sandgate
Received JH's note saying Lord Northampton already wrote to Sir Robert Peel. Wants to explain how R.S.L. was left without a foreign secretary.
ALS 2pp RS:HS 15.180

6417. 1845–10–17 From John R. Hind 7 Allsop's Place, Regents Park
Will apply JH's principle at the first opportunity. Has made some successful observations of the star Epsilon Arietes. Intends observing Castor on every fine evening.
ALS 2pp RS:HS 9.324

6418. 1845–10–17 From Edward Sabine Sandgate
Does not think JH should write Sir Robert Peel pending his reply to Lord Northampton. Preparing [John] Lefroy's Canadian Survey for publication.
ALS 4pp RS:HS 15.181

6419. 1845–10–18 From Edward Sabine Sandgate
Sends communication on meteorology in Bombay for comments. [A. T.] Kupffer's communication makes ES think large scale magnetical surveys are being planned. Discusses possibilities for such experiments in Russia.
ALS 1p RS:HS 15.182

6420. 1845–10–18 From James [Struthers] Stewart Cape of Good Hope
Sends Cape roots and bulbs. Reports that plans for the Cape Botanical Garden progress slowly. Road improvement underway. Minor outbreak of Boer activity dispersed.
ALS 5pp RS:HS 17.11

6421. 1845–10–19 William Bateman Fridge Place
Unable to accept Herschels' invitation for 23 Oct.
AL 1p RAS:JH Archive 4/9.32 verso; Reel 7

6422. 1845–10–19 From W. Robarts Burnham, Bucks
Thanks Herschels for allowing his daughter to visit. [William] Penn died suddenly. Will mention JH's sentiments to the family.
ALS 5pp RS:HS 14.494

6423. 1845–10–20 From George Biddell Airy Royal Observatory, Greenwich
Is sending letters by W. R. Dawes and F. W. Bessel, the latter occasioned by a volume of 'Planetary reductions' presented

to him by GA.
ALS 1p RS:HS 1.159

6424. 1845–10–20 or later To George Biddell Airy n.p.
Thanks GA for the information about the presentation made to F. W. Bessel [see GA's 1845–10–20], and for sharing Bessel's reply.
ALS 2pp RGO 6.530.84

6425. 1845–10–21 From Adolphe Quetelet Brussels
Deprived of news from JH. Works on physical phenomena of globe. Calls attention to [J. C. A.] Peltier's work on electricity of air of barometer. Asks JH to please write soon.
ALS (French) 3pp RS:HS 14.248

6426. 1845–10–21 From Charles Richard Weld Royal Society
Committee of Physics will meet 27 Oct. to recommend foreign members and candidates for medals.
(A)LS 1p RAS:JH Archive 4/9.3 verso; Reel 7

6427. 1845–10–22 From Spencer J. A. Compton Castle Ashby
Has received a letter from Henry Goulburn regarding his own letter to Robert Peel. Delay has been due to contracts with foreign governments. Has referred Goulburn to Edward Sabine for details.
ALS 2pp RS:HS 5.268

6428. 1845–10–22 To [Robert Hunt ?] Collingwood
Thanks for the paper on the photographic effect of light on drugs. JH also comments on the photographic experiments of Mary Somerville.
ALS 3pp WT 68037.5

6429. 1845–10–23 From William Henry Smyth [3 Cheyne Walk, Chelsea]
Discusses importance of Gamma Virginis: 'I consider it a grand key to several important physical points.' W. R. Dawes claims to have observed a third Uranian satellite.
ALS 3pp RS:HS 16.283

6430. 1845–10–24 From George Biddell Airy Royal Observatory, Greenwich
Has received a satisfactory response from the Admiralty [see JH's 1845–9–30].
ALS 2pp TxU:H/M-0631.8; Reel 1084 (C: RGO 6.675.537)

6431. 1845–10–24 From George Biddell Airy [Royal Observatory, Greenwich]
A note to accompany the return of a letter to JH.
ALS 1p TxU:H/M-0631.7; Reel 1084

6432. 1845–10–24 From Sackville Phelps Tongswood
SP and wife accept Herschels' invitation for 28 Oct.
ALS 1p RAS:JH Archive 4/9.48 verso; Reel 7

6433. 1845–10–28 From George Biddell Airy Royal Observatory, Greenwich
Asks JH to write to George Merz for an object glass for a telescope for the Cape Observatory.
ALS 2pp TxU:H/M-0631.9; Reel 1084

6434. 1845–10–28 From Edward Sabine Sandgate
Sent Lord Northampton a statement describing the manner in which foreign communications have taken place. Returning to Woolwich. [Henry] Goulburn may see him there.
ALS 1p RS:HS 15.183

6435. 1845–10–29 From George Biddell Airy Royal Observatory, Greenwich
Mrs. Airy has given birth to a son.
ALS 1p RS:HS 1.160

6436. 1845–10–29 To Josiah Quincy Collingwood
Received 50 copies of memoirs of James Grahame. Will delete one passage in the copies he will distribute. Congratulates JQ on his literary accomplishment.
C 2pp RS:HS 22.252

6437. 1845–10–29 From Josiah Quincy Boston, Mass.
Sends new edition of James Grahame's *History of the United States*. Pleased to have been in correspondence with JH. Describes events at Harvard Observatory.
ALS 3pp RS:HS 14.297

6438. 1845–10–29 From Edward Sabine Sandgate
Saw [Henry] Goulburn, who will ask Lord Aberdeen to notify foreign governments that the British government will continue the magnetic observatories. Hopes observatories will continue in those countries too.
ALS 2pp RS:HS 15.184

6439. 1845–10–31 From William Henry Smyth [3 Cheyne Walk, Chelsea]
Pleased with JH Gamma Virginis observations. 'Lunar affair' disappoints WS. *Cycle of Celestial Objects* being translated into Italian.
ALS 2pp RS:HS 16.284

6440. [1845–11] From J[oseph] Hartnell Hawkhurst Dispensary
[Form letter] Requests parishioners to attend meeting on 1 Dec. to determine future of dispensary, which affords medical relief to 89 families in parish.
LS 1p RAS:JH Archive 4/9.31 verso; Reel 7

6441. 1845–11–1 From Edward Sabine Woolwich
Communication from [Henry] Goulburn is official. Government will continue observatories and acquaint other governments with them. Admiralty will furnish meteorological instruments for coast stations.
ALS 4pp RS:HS 15.186

6442. 1845–11–1 From Edward Sabine Woolwich
Thinks Dean of Ely [George Peacock] should be appointed treasurer of R.S.L. Discusses George Rennie and Robert Brown as other possible candidates.
ALS 4pp RS:HS 15.185

6443. 1845–11–1 From Charles Richard Weld Royal Society
Committee of Physics will meet 6 Nov.
(A)LS 1p RAS:JH Archive 4/9.4 verso; Reel 7

6444. [1845–11?] To William Whewell [Collingwood]
Thanks WW for his book [*Of a Liberal Education,* 1845]. Agrees with WW on the educational value of the calculus and on the importance of Isaac Newton's *Principia.* Is teaching mechanics to his son William.
ALS 4pp TC, Camb. Add. Ms.a.207[66]

6445. 1845–11–2 From George Biddell Airy Royal Observatory, Greenwich
Is writing about further arrangements about the telescope to go to the Cape Observatory; JH is negotiating with George Merz for it [see GA's 1845–10–28].
ALS 3pp TxU:H/M-0631.10; Reel 1084

6446. 1845–11–2 To Spencer J. A. Compton Collingwood
Hopes extracts from Mary Somerville's letters on her

experiments with 'Coloration of Light,' specifically, 'action of rays on vegetable juices,' can be read to R.S.L.
ALS 3pp RS MC.4.92

6447. 1845–11–2 To Alexander von Humboldt Collingwood
Sends thanks and very high praise for AH's first volume of *Kosmos*. Also comments on a number of other astronomical matters.
C 4pp RS:HS 22.253

6448. 1845–11–2 To Mary Somerville Collingwood
Discusses MS's light experiments. JH will present the results to Royal Society. Describes apparatus that may help her. Recently attended a meeting concerning continental magnetics and meteorology.
ALS 3pp RS:HS 16.352 (C: RS:HS 22.254) *Personal Recollections ... of Mary Somerville* (London, 1873), pp. 278–80 inc

6449. 1845–11–4 From William Henry Sykes India House
Expresses satisfaction with results of continued magnetic and meteorological observations in various locations.
ALS 4pp RS:HS 17.142

6450. 1845–11–5 From Field, Wardell, etc. 22 Henrietta St., Covent Garden
Encloses invoice for stout ordered by Mr. Stewart and sent to Mrs. Waterhouse of Old Brompton.
AL 1p RAS:JH Archive 4/9.27; Reel 7

6451. 1845–11–5 From William Radcliffe Birt Bethnal Green
Regarding recent barometric readings and atmospheric waves.
ALS 4pp RS:HS 4.112

6452. 1845–11–6 From James A. Gordon Naish, Bristol
Crops from the land that has been electrically heated have not been exceptional. Scotch newspapers mention some more. Is pleased that JH devoted so much time to him last Spring.
ALS 4pp RS:HS 8.180

6453. 1845–11–7 From William Carr Beresford Bedgebury Park, [Kent]
Invites JH and Margaret Brodie Herschel to dinner on 11 Nov.
AL 1p RAS:JH Archive 4/9.53 verso; Reel 7

6454. 1845–11–7 To [William Radcliffe Birt] Collingwood
Perhaps periodic surges in atmospheric waves noticed by WB will repeat. Speaks of 'new climate' for globe. Hopes WB can attend [B.A.A.S.] meeting at Southampton.
ALS 3pp RS:HS 19.120

6455. 1845–11–7 To Georg Merz & Son Collingwood
Is now finally empowered to order the telescope he inquired about [see JH's 1843–9–2].
ALS 3pp RS:HS 25.6.29

6456. 1845–11–7 From John Phillips St. Mary's Lodge, York
Sends copy of JH's Cambridge B.A.A.S. address. Printer will send proofs for JH's inspection.
ALS 1p RAS:JH Archive 4/9.39 verso; Reel 7

6457. 1845–11–7 To William Henry Smyth Collingwood
Will give lecture on moon in which he will discuss its composition, the craters Kepler and Gassendi, plus generalities.
C 1p RS:HS 22.253

6458. 1845–11–8 From James [Struthers] Stewart Cape of Good Hope
Urges JH to seek funding for Cape Botanical Garden. Notes that Cape Colonial Government proposes that superintendent of the Garden would also serve as professor of chemistry and botany at the South African College.
ALS 4pp RS:HS 17.12

6459. 1845–11–9 To Michael Faraday Collingwood
Has just seen mention of MF's latest experiments showing the connection of light and magnetism. Comments on these. Proves some of his own theories to be correct.
draft 4pp RS:HS 7.183 Williams, *Selected Correspondence, 1*, 461–3

6460. 1845–11–9 To William Hasledine Pepys Collingwood
Asks WP to recall the particulars of an experimental session [see WP's 1823–3–11] where JH tried to demonstrate the polarization of light by an electrical current in a coil of wire.
draft 2pp RS:HS 13.339 & 22.256

6461. 1845–11–10 From William Carr Beresford Bedgebury Park, [Kent]
Invites JH and Margaret Brodie Herschel to dinner on 27 Nov.
AL 1p RAS:JH Archive 4/9.52 verso; Reel 7

6462. 1845–11–10 From John R. Hind 7 Allsop's Place, Regents Park
Communicating the elements of Castor and Corona by JH's method. Comments on these readings.
ALS 2pp RS:HS 9.325

6463. 1845–11–10 From Charles Richard Weld Royal Society
[Form letter] Reminder that JH is expected to attend election of Council and officers on 1 Dec.
LS 1p RAS:JH Archive 4/9.8 verso; Reel 7

6464. 1845–11–11 From Spencer J. A. Compton Castle Ashby
Mary Somerville's paper has been given to P. M. Roget and it will be read at an early meeting. Interesting communications from Michael Faraday. Museum has received the head of a mastodon from India. East India Co. has sent to India in conformity with the Cambridge recommendation.
ALS 2pp RS:HS 5.269

6465. 1845–11–11 From Charles Pritchard Clapham
Looking forward to visit to Collingwood.
ALS 1p RAS:JH Archive 4/9.50 verso; Reel 7

6466. 1845–11–12 From George Biddell Airy Flamsteed House, Greenwich
Introducing B. A. Gould, who wishes to study observatories and astronomy in Europe. GA doubts whether he will be able to attend the meeting of the R.A.S.
ALS 2pp RS:HS 1.161

6467. 1845–11–12 From William Hasledine Pepys Kensington
WP identifies date and people present at the time of the experiment referred to by JH in his 1845–11–9.
ALS 1p RS:HS 13.340

6468. 1845–11–13 From Thomas Ayern Newcastle
Acknowledges JH's orders for coal.
ALS 1p RAS:JH Archive 4/9.43 verso; Reel 7

6469. 1845–11–13 From Michael Faraday Brighton
His letter was a great encouragement and pleasure. Comments on his own and JH's experiments with light and magnetism. Contemplates a further series of experiments but requests JH not to mention it. Is at Brighton for a rest.
ALS 3pp RS:HS 7.184 Williams, *Selected Correspondence, 1*, 463–5

6470. 1845–11–13 From R. Young Horley Peckham
Asks for JH's autograph.
ALS 1p RAS:JH Archive 4/9.47 verso; Reel 7

6471. 1845–11–16 From William Henry Smyth [3 Cheyne Walk, Chelsea]
Asks JH to write an abstract of the discourse JH presented to R.A.S. Moon model is on display at R.A.S.
ALS 4pp RS:HS 16.285

6472. 1845–11–17 From Edward Sabine Woolwich
Encloses copy of letter to Lord Northampton that Northampton was to send on to [Henry] Goulburn. JH may send it to [Henry] Addington or to the states that have cooperated.
ALS 2pp RS:HS 15.187

6473. 1845–11–18 From Augustus De Morgan 7 Camden St.
Has compared tables A and B and can find no identity between payments. Regarding the moon.
ALS 3pp RS:HS 6.226

6474. 1845–11–18 From Georg Merz Munich
Provides more specific prices and information for the refractor that GM believes will meet the specified needs [see GM's 1843–9–20].
ALS 2pp (German) TxU:H/M–0357.1; Reel 1087

6475. 1845–11–24 To Augustus De Morgan Collingwood
On insurance tables for the benefit society; comments on AD's punning humor; JH has a chest cold.
C 1p RS:HS 22.257

6476. 1845–11–24 From Josiah Forshall British Museum
Announces 29 Nov. meeting of subcommittee to consider [Anthony] Panizzi's report on state of library.
ALS 1p RAS:JH Archive 4/9.54 verso; Reel 7

6477. 1845–11–25 To Georg Merz Collingwood
Approves GM's price for making 6.5-inch telescope. Proceed with work, which must be completed in five or six months. JH must consult others before approving mounting and clockwork [see GM's 1845–11–18].
ACCS 1p TxU:H/M–0357.2; Reel 1087

6478. 1845–11–26 To Ann Smyth Collingwood
Details attempts to purchase Mme. Witte's lunar model for the 'Museum of Oconomic Geology.'
C 2pp RS:HS 22.258

6479. 1845–11–29 From Augustus De Morgan 7 Camden St.
Has undertaken the Annual Report again. Has JH anything to report? Hopes he is recovering from his cold. Should read Richard Sheepshanks's pamphlet for a good laugh. Further regarding the insurance tables.
ALS 4pp RS:HS 6.227S. De Morgan, *Memoir of Augustus De Morgan*, 153

6480. 1845–11–29 From James William Grant s.s. Great Liverpool
Attended the General Assembly of the Church in Edinburgh in May last. Comments on its proceedings. Travelled around Scotland. Is now on his way to India. Has had good weather for the most part. Comments on Gibraltar, Malta, and North Africa.
ALS 16pp RS:HS 9.24

6481. 1845–11–29 To William Henry Smyth Collingwood
Cannot attend R.A.S. meeting to explain Mme. Witte's lunar model; will send notes of explanation. Busy with the publication of *Cape Results*. Will write paper on improved method of calculating double star orbits. Discusses relationships of periods of Saturnian satellites.
C 3pp RS:HS 22.259 R.A.S.M.N., 7 (1847), 24 inc

6482. 1845–12–1 From George Biddell Airy Royal Observatory, Greenwich
Has requested Admiralty funding for the proposed telescope [see GA's 1845–11–2]; GA does not like the mounting the telescope is to be given; has gathered, from Otto Struve, that it does not work well at Pulkowa Observatory.
ALS 2pp TxU:H/M-0631.11; Reel 1084

6483. 1845–12 To Margaret Brodie Herschel Collingwood
A little news, including some comments about JH's health.
ALS 1p JHS 4.25

6484. 1845–12 To Margaret Brodie Herschel Collingwood
Writes to ask MH's advice about ordering flowers for planting.
ALS 4pp JHS 4.24

6485. 1845–12–3 From George Biddell Airy Royal Observatory, Greenwich
Is returning JH's report on William Simms's object glass [see GA's 1845–10–14]; GA reiterates Otto Struve's statement about the mounting at Pulkowa Observatory [see GA's 1845–12–1].
ALS 1p TxU.H/M-0631.12, Reel 1084

6486. 1845–12–3 From William Radcliffe Birt Bethnal Green
The great atmospheric wave has returned; observations on this.
ALS 4pp RS:HS 4.113

6487. 1845–12–6 From William Alexander Baillie Hamilton to George Biddell Airy Admiralty
Has requested of Admiralty £485 for equatorial telescope, from Merz in Munich, for Cape of Good Hope Observatory. Ask JH to estimate cost for 'next year's Estimates.'
C 1p TxU:H/M–0996.1; Reel 1083

6488. 1845–12–8 From George Biddell Airy Royal Observatory, Greenwich
A note enclosing a copy of a letter from the Admiralty, on which JH is requested to comment.
ALS 1p RS:HS 1.162a

6489. 1845–12–8 From George Biddell Airy Royal Observatory, Greenwich
Passing on information received from a O. M. Mitchel[l] of Cincinnati regarding Antares, and the surface of the sun.
ALS 2pp RS:HS 1.162 (C: RGO 6.193.233)

6490. 1845–12–8 To [William Radcliffe Birt] Collingwood
JH suffering chest pains. Delighted that WB again observed 'great symmetrical wave.' Sun's effect on atmosphere is like the wake of 'Leviathan going round & round in a confined ... harbour.'
ALS 3pp RS:HS 19.121

6491. 1845–12–8 To Caroline Lucretia Herschel Collingwood
George Airy, George Peacock, and [Adam] Sedgwick are visiting Collingwood for Christmas. Airy writes JH that an American astronomer named [Ormsby] 'Mitchell' has seen Antares double. JH reports that William Lassell and W. R. Dawes have observed the 7th Saturnian satellite and also another of the six satellites of the 'Georgium Sidus' [Uranus]. JH is confident that his *Cape Results* will go to the presses in January.
ALS 4pp TxU:H/L–0590.13; Reel 1058

6492. 1845–12–8 From Edward Sabine Woolwich
Elected Foreign Secretary of R.S.L. Discusses a shipment of books from Russia and a shipment of releases for fellows, the public, and the R.S.L. library.
ALS 4pp RS:HS 15.188

6493. 1845–12–9 To George Biddell Airy n.p.
Is skeptical of the American observations [see GA's 1845–12–8], and reports other observations communicated to JH.
AL 3pp inc RGO 6.193.234

6494. 1845–12–10 From George Biddell Airy Royal Observatory, Greenwich
Regarding estimates and payment for a parallectical mounting.
ALS 1p RS:HS 1.163

6495. 1845–12–10 To Adam Sedgwick Collingwood
Invites AS to Collingwood for Christmas. JH has been ill.
ALS 4pp RS:HS 15.428 (C: RS:HS 22.260)

6496. 1845–12–13 From Charles Richard Weld Royal Society
Announces 18 Dec. meeting of Committee of Mathematics, Astronomy, and Physics.
(A)LS 1p RAS:JH Archive 4/9.41 verso; Reel 7

6497. 1845–12–15 From Edward Sabine Woolwich
Discusses shipping charges for books sent by [Adolphe] Quetelet from Russia. Discusses position of Admiralty regarding observatories.
ALS 5pp RS:HS 15.189

6498. 1845–12–15 From William Samuel Stratford *Nautical Almanac* Office, Verulam Buildings
Requests advance payment by B.A.A.S. to Mr. Harris, assistant secretary to *Nautical Almanac*.
ALS 1p TxU:H/M–0544.1; Reel 1093

6499. 1845–12–16 From James Ranald Martin Grosvenor St.
Regrets to learn that JH is still suffering. Advises him to come to London for examination.
ALS 3pp RS:HS 12.295

6500. 1845–12–19 From William Alexander Baillie Hamilton to George Biddell Airy Admiralty
Admiralty has approved equatorial telescope for Cape of Good Hope Observatory. Advise JH to order it from Munich.
C 1p TxU:H/M–0996.2; Reel 1083

6501. 1845–12–19 From Charles Babbage Dorset St.
Has just seen Lady Herschel and heard the news of his health. Urges him to come to town for treatment.
ALS 2pp RS:HS 2.311

6502. 1845–12–19. To George Bellas Greenough Collingwood
Sending statement [written by C. White on Turkish coal fields] relating to the Coal Committee on which JH has been a 'sleeping partner.'
ALS 1p encl(2pp) Univ.Coll.London Greenough

6503. 1845–12–19 From Mrs. C. Harris London
Acknowledges receipt of JH's order from B.A.A.S. for £30. Mr. Harris is much worse and cannot respond personally.
ALS 1p TxU:H/M–0247; Reel 1087

6504. 1845–12–19 To Margaret Brodie Herschel Collingwood
About travel arrangements for MH's return home, along with the health of the family, and JH's progress on his latest book [*Cape Results?*].
ALS 4pp JHS 4.27

6505. 1845–12–19 From John Stuart Mill India House
Hears from his publisher that a second edition of JM's *System of Logic* is being prepared for the press, so would welcome any suggestions from JH regarding alterations or amendments.
ALS 3pp RS:HS 12.334

6506. 1845–12–20 From George Biddell Airy Royal Observatory, Greenwich
The Admiralty has approved the purchase of the telescope for the Cape Observatory [see GA's 1845–12–1].
ALS 1p TxU:H/M-0631.13; Reel 1084

6507. 1845–12–20 To Charles Babbage Collingwood
Health has improved. Regarding an experiment carried out at the Royal Institution.
ALS 4pp RS:HS 2.312

6508. 1845–12–20 From William Radcliffe Birt Bethnal Green
Is sorry to hear of his indisposition. Regarding movements connected with the sun.
ALS 3pp RS:HS 4.114

6509. 1845–12–20 To Michael Faraday Collingwood
Sends details of an experiment he would like MF to carry out. Cannot do it himself as he has no suitable glass.
draft 3pp RS:HS 7.185 & 22.261 Williams, *Selected Correspondence, 1*, 473–4

6510. 1845–12–22 From Michael Faraday Royal Institution
Has received his letter and suggestions for the experiment. Has ordered the glass from Newman. Is very busy so may not be able to carry out the experiment for a week or more. JH must keep the results secret for the time being.
ALS 2pp RS:HS 7.186 & 22.262 Williams, *Selected Correspondence, 1*, 475

6511. 1845–12–22 To John Stuart Mill Collingwood
Sorry for the delay in response; JH has been ill for some time. JM has objected to P. S. Laplace's theory of probabilities; JH strongly supports Laplace.
ACCS 4pp RS:HS 25.6.30

6512. 1845–12–22 From Edward Sabine Woolwich
Wants JH to sign certificate for [Joseph] Kay. Mrs. Sabine

will resume her translation of [Alexander von Humboldt's] *Cosmos*. Mr. Gyde has not received list of presentation copies of the B.A.A.S. catalogue.
ALS 4pp RS:HS 15.190

6513. 1845–12–23 From George Bellas Greenough Regents Park
Has collected information about coal deposits, but GG is very troubled that the committee [coal map committee of the B.A.A.S] has not yet met and has established no procedures. Describes some of the information GG has at hand.
ALS 5pp RS:HS B26.17

6514. 1845–12–23 From Heinrich Christian Schumacher Altona
Announces the discovery of a new 'planet' by [K. L.] Hencke of Prussia, of the class of 'the 4 small ones' [the asteroids].
ALS 2pp RS:HS 15.396

6515. 1845–12–24 or later To George Bellas Greenough [Collingwood]
AS JH understands the matter [see GG's 1845–12–23], GG is the only knowledgeable member of the committee, and so it will be up to GG to present information and propose some means of moving ahead. JH makes some critical remarks about the nature of B.A.A.S. committees.
draft 4pp inc RS:HS B26.18

6516. 1845–12–24 From James Ranald Martin Grosvenor St.
Was gratified to hear of the improvement in JH's health. Hopes it will progress.
ALS 4pp RS:HS 12.296

6517. 1845–12–24 To Georg Merz & Son Collingwood
Some modifications to the telescope ordered [see JH's 1845–11–7].
ALS 2pp RS:HS 25.6.31

6518. 1845–12–27 From George Biddell Airy Royal Observatory, Greenwich
Regarding the condition for determining the best possible orbit for a double star.
ALS 1p RS:HS 1.164 (C: RGO 6.193.216)

6519. 1845–12–27 From George Biddell Airy Royal Observatory, Greenwich
A note of instructions about paying for the Cape telescope [see GA's 1845–12–20].
ALS 1p † TxU:H/M-0631.14; Reel 1084

6520. 1845–12–29 From John Henry Griesbach Carlton Villas, Maida Vale
Did not realize that JH would leave town so quickly. Cast a 10" speculum successfully after William Parsons' (3rd Earl of Rosse) plan. Is having difficulty with his grinding and polishing and would welcome JH's advice.
ALS 3pp RS:HS 9.61

6521. 1845–12–29 From Robert Hunt Museum of Pract. Geology
Has just been showing Michael Faraday some of his experiments on the influence of magnetism of bodies slowly precipitating; comments on this and will be pleased to carry out any experiments for JH.
ALS 4pp RS:HS 10.120

6522. 1845–12–29 From John Stuart Mill India House
Is grateful for JH's letter. Has already had criticisms of the chapter, which JH mentions. Will reconsider it.
ALS 2pp RS:HS 12.335

6523. 1845–12–29 From William Samuel Stratford Notting Hill Square
Returns to JH, and comments on, the distribution lists [for the B.A.A.S. star catalogue].
ALS 4pp enc RS:HS B26.19

6524. 1845–12–30 From Friedrich Eduard Beneke Berlin
Writes in thanks for JH's kind words on several occasions about FB's ideas on psychology.
ALS 3pp (German) RS:HS 4.63

6525. 1845–12–30 From John R. Hind 7 Allsop's Place, Regents Park
Sending him the elements of the new planet Astrea. Orbit is based on the Berlin observations. Differs from those given by J. F. Encke.
ALS 2pp RS:HS 9.326

6526. 1845–12–30 From Edward Sabine Ramsgate
Please notify ES of date for the council [of B. A. A. S?] to meet. Oxford is most eligible of Southampton invitations. Sir Robert [Inglis] is best candidate for president.
ALS 4pp RS:HS 15.191

6527. 1845–12–31 To George Biddell Airy n p
Will consider GA's objections [see GA's 1845–12–27] when JH works further on double star orbits.
AL 1p inc RGO 6.193.217

6528. 1845–12–31 To Thomas Maclear Collingwood
Karl Hencke discovers the asteroid Astrea. Admiralty orders complete equatorial for Cape Observatory; JH explains delay. Complains of poor health.
ALS 4pp Cape Archives/Maclear Papers File 100

6529. 1846 To Dr. [Hugh?] Falconer Collingwood
Needs information on [H?]F's government aid, publication plans, costs, to be presented at upcoming B.A.A.S. committee meeting for obtaining further funds.
ALS 4pp EUL:Gen523/4/30

6530. [1846 to 1850] To Margaret Brodie Herschel [Collingwood]
About the health of several family members, and a desire to see MH return home.
ALS 2pp JHS 5.31

6531. [1846] To Margaret Brodie Herschel London
Arrangements for JH's son William James and for travel to Collingwood.
ALS 3pp JHS 3.14d

6532. 1846 To Margaret Brodie Herschel [Collingwood]
JH is in poor health and asks MH to obtain more pills for him.
ALS 4pp JHS 3.14c

6533. [1846?] To Margaret Brodie Herschel St. Paul's Coffee House, St. Paul's Churchyard
Resolution to a problem of Peter Stewart seems to be occurring, so JH will not interfere; JH has seen a shower bath in which he is interested.
AL 2pp inc JHS LMJF.45.1g

6534. 1846 To Margaret Brodie Herschel Collingwood
MH and Isabella are away; JH reports on how things are at home.
AL 4pp JHS 1.58

6535. [1846 to 1847] To Roderick Impey Murchison n.p.
W. S. Stratford, G. B. Airy, and JH agree that B.A.A.S. should publish star catalogs of N. L. Lacaille and J. J. L. Lalande without using government money. Proposes distribution to observatories. [JH annotation: Committee added two more observatories to list.]
ACCS 3pp † TxU:H/L-0268; Reel 1054

6536. 1846 To Roderick Impey Murchison n.p.
Finally encloses note to [Christian?] Schönbein. Says S will come to Southampton. Wonders whether can recommend purchase of S's secret or military adoption of it without overstepping limits. Comments on [H. C.] Oersted's discovery.
ALS 4pp EUL:Gen523/4/29

6537. [1846?] From Charles Pritchard Clapham Grammar School
Congratulates JH and Margaret Herschel on birth of a daughter [Francisca?]. Asks JH's opinion of the ideas of 'lines of force' and 'lateral shakes.' Conducts a force experiment with J. P. Gassiot.
ALS 4pp RS:HS 14.108

6538. [1846] To the R.A.S. n.p.
Reports dimming of Beta Ursae Minoris since ca. 1840. States that it is a slowly variable star.
L *R.A.S.M.N., 7* (1847), 95–6 inc

6539. [1846 or later] From Edward Sabine [Collingwood]
[*Detached postscript*] Ask president of 'B. I. Academy' to inform council of Math and Physics section how to handle [Robert] Mallet's paper and Sir W. Hamilton's letter.
AL 1p inc RS:HS 19.293

6540. [1846?] From Christian F. Schönbein London
Will call at Collingwood. Wants to show JH some experiments and asks JH to have phosphorous and starch ready.
ALS 2pp RS:HS 15.363

6541. [1846 to 1855] From Richard Sheepshanks 24 Rutland Gate enroute from Reading
Reminds JH of new format in *Nautical Almanac* to represent symbols of late F. W. Bessel. Thomas Henderson was working with unknown quantity [in preparing N. L. Lacaille's star catalog].
ALS 1p TxU:H/M–0529.1; Reel 1093

6542. 1846 From Richard Sheepshanks Reading
Explains to JH the R.A.S. Council's intent in asking JH to be Foreign Secretary, and the arrangements that have been made when JH declined.
ALS 4pp RS:HS 16.131

6543. 1846 or later From Richard Sheepshanks n.p.
Asks JH to prepare an eloge of F. W. Bessel.
ALS 4pp RS:HS 16.127

6544. 1846–1 To William Henry Smyth Collingwood
Discusses the purchase price of Mme. Witte's lunar model.
C 1p RS:HS 22.266

6545. [1846 to 1855] To Henry Venn n.p.
Hopes Sir R. H. Inglis explained JH's delay in answering. Missionaries have obligation to preserve unwritten languages. Suggests 'Phonotypic' system of [Isaac] Pitman, with some alterations.
Adraft 3pp TxU:H/L-0486; Reel 1055

6546. [1846] From Wilhelm Eduard Weber Leipzig
Sends WW's *Elektrodynamische Massbestimmungen* (1846) via J. E. Wappäus.
AL 1p RS:HS 18.391

6547. [1846 to 1847] To William Whewell n.p.
Comments on proofs of WW's German translations. JH recommends that the magnetic observations already made should be reduced to show some results, rather than adding more observations.
C 2pp RS:HS 25.14.35

6548. 1846 To [?] n.p.
Box that [Thomas] Maclear indicated would contain Cape bulbs from Klapmuts, arrived yesterday containing common European roots. Perhaps wrong box was sent from London warehouse.
AL 2pp inc RS:HS 19.123

6549. 1846–1–2 From Edward Sabine Ramsgate
Has the opinion regarding Oxford of S. J. A. Compton (Lord Northampton). This will make his absence less consequential if he cannot attend the Council meeting.
ALS 1p RS:HS 15.192

6550. 1846–1–5 To Robert Hunt Collingwood
RH's experiments are extremely interesting and deserve to be pursued. Comments on experiments made by Michael Faraday and others relating to the formation of crystals in a magnetic field and the effect such crystals have on polarized

light.
ALS 4pp WT 68037.6 (C: RS:HS 10.121 & 22.264)

6551. 1846–1–6 From Barthes & Lowell 14 Great Marlborough St.
Received P. S. Laplace's [book] from Paris and sent it to Mr. Stewart as JH directed.
ALS 1p RAS:JH Archive 4/7.11; Reel 7

6552. 1846–1–7 From Edward Sabine Woolwich
Council will meet Friday 16 January. Does not think [Roderick] Murchison has returned. [Alexander von Humboldt's] *Cosmos* was reviewed well.
ALS 3pp RS:HS 15.193

6553. 1846–1–7 To Richard Sheepshanks Collingwood
Thanks RS for his brochure [on the affairs of the Liverpool Observatory?].
ALS 2pp RS:HS 16.43 (C: RS:HS 25.9.24)

6554. 1846–1–8 From Adam Sedgwick to Lady Herschel London
Thanks her for her hospitality. Discusses letter read at Geological Society the previous night. Asks that Mrs. [G. B.] Airy's picture be returned.
C 2pp RS:HS 15.429

6555. 1846–1–9 From Thomas I. M. Forster Bruges
Sends his last year's observations of falling stars and comments on them. Has he seen the new planet [Astrea]? Sends some new songs of his own composition.
ALS 4pp RS:HS 7.339

6556. 1846–1–10 From J[ohn] Gould Burwash
Sending some extra large trout caught by the miller, together with the 'local' trout so that he can compare them.
ALS 4pp RS:HS 9.110

6557. 1846–1–10 From Richard Sheepshanks 14 London Road, Reading
Has entered JH's name for Foreign Secretary of the R.A.S. and explains the circumstances. On his observations and method of observation of the recently discovered asteroid Astrea.
ALS 6pp RS:HS 16.22

6558. 1846–1–12 To Richard Sheepshanks Collingwood
Is too busy to accept the office of Foreign Secretary of the R.A.S. Says he cannot even keep up with his correspondence at home. Has been forced to give up foreign correspondence altogether.
ALS 2pp RS:HS 16.44 (C: RS:HS 25.9.19)

6559. 1846–1–13 From Georg Merz Munich
Asks for some specifics of the mounting and proceeds to explain how the mountings will work [see JH's 1845–11–25].
CC(S) 2pp (German) TxU:H/M–0358; Reel 1087

6560. 1846–1–13 From Leonard Romer 2 Bedford Place
A notice of meeting of the B.A.A.S. coal distribution map committee.
ALS 1p RS:HS B26.23

6561. 1846–1–14 To Sir William J. Hooker Collingwood
[James C.] Stewart has written to suggest the establishment of botanical gardens at the Cape. JH supports this idea. Would WH also?
ALS 3pp Royal Botanic Gardens

6562. 1846–1–15 From Michael Faraday Royal Institution
Has procured some glass rod fit for the experiment. Comments on the results. [Robert] Hunt has been working on crystallization and magnets and has told him of JH's interest. Grieves to hear of his illness. His paper in JH's hands will be left for the time being until he has more time.
ALS 3pp RS:HS 7.187 Williams, *Selected Correspondence, 1*, 483

6563. 1846–1–15 From Sir William J. Hooker Royal Botanic Gardens, Kew
Regarding a botanical garden at the Cape and suitable staff. Would like to join with JH in a memorial to the Colonial Office.
ALS 10pp RS:HS 9.455

6564. 1846–1–15 From Thomas Maclear Royal Observatory, Cape of Good Hope
Observatory work slowed since William Mann's absence due to head injury. C. P. Smyth's replacement not yet arrived. Regrets lack of assistants; wants to complete triangulation at Bushman Flat.
ALS 3pp RS:HS 12.133

6565. 1846–1–15 From A. B. Orlebar Bombay Magnetic Observatory
Circular Magnetic results. Observations of solar spots. Determining Longitude.
ALS 2pp RS:HS 13.181

6566. 1846–1–16 From William C. Redfield New York
Sends paper on storms. JH's advocacy of jointly progressive and rotary character of storms has been supported by new evidence.
ALS 2pp RS:HS 14.311

6567. 1846–1–16 From Edward Sabine Woolwich
Describes mistakes made by R.S.L. in sending letters to [Christopher] Hansteen through Stockholm.
ALS 2pp RS:HS 15.195

6568. 1846–1–17 To Margaret Brodie Herschel Montague Square
The goods from Collingwood arrived safely; JH has seen and called on various people; Emma Stewart is 'poorly.'
ALS 1p JHS 6.43

6569. 1846–1–19 To Richard Sheepshanks Collingwood
Clarifies his position with regard to becoming R.A.S. Foreign Secretary. Sets conditions under which he would allow his name to stand. Has been 'harassed' for the last two months by 'stomach derangements.' 'It is old Paracelsus's "Archaeus" in a fit of the Sulks.' On the disposal of some of Francis Baily's remaining manuscripts.
ALS 4pp RS:HS 16.45 (C: RS:HS 25.9.19)

6570. 1846–1–22 To Michael Faraday Collingwood
Has opened the sealed paper MF sent, and JH is now returning it.
C 2pp RS:HS 22.265 Williams, *Selected Correspondence, 1*, 486

6571. 1846–1–26 From John R. Hind 7 Allsop's Place, Regents Park
Regarding the strange appearance of the comet of Biela. Will observe the double star as soon as he has an opportunity.
ALS 2pp RS:HS 9.327

6572. 1846–1–27 To Margaret Brodie Herschel 17 Montague Square
Sent son William James to Clapham by himself; JH much concerned about the future of Peter Stewart as Mr. Smith of

Smith, Elder & Company is very ill. Brief report on Emilia Stewart's health.
ALS 4pp JHS LMJF.45.1d

6573. 1846–1–27 From Sir William J. Hooker Royal Botanic Gardens, Kew
Has heard nothing from him regarding the Cape garden. Wrote to the Colonial Office advising that the Cape garden should be similar to that at Ceylon. Hopes he agrees that Wynberg is the most likely spot.
ALS 2pp RS:HS 9.456

6574. 1846–2–2 From George W. Hearn Royal Military College, Sandhurst
Sent him a paper on differential equations about a month ago and wonders if it has gone astray as he has heard nothing from JH.
ALS 3pp RS:HS 9.266

6575. 1846–2–2 From James Ranald Martin Grosvenor St.
Truly grateful to hear such a good report of JH's health. Should continue the medicine. Hopes to avail himself of JH's kindness when he has leisure.
ALS 3pp RS:HS 12.297

6576. 1846–2–3 To Augustus De Morgan Collingwood
Asks AD for the return of a paper JH had sent him.
C 1p RS:HS 22.267

6577. 1846–2–5 From Augustus De Morgan 7 Camden St.
Should have returned the paper before, but has been very busy with the Annual Report and other matters. Paper seems worthy of publication.
ALS 2pp RS:HS 6.228

6578. 1846–2–8 From John R. Hind 7 Allsop's Place, Regents Park
Is obliged for his notice of the separation of Biela's comet; has sent extracts to H. C. Schumacher. Sends some observations made that evening. [Francesco] de Vico announces a new comet; so far he has been unable to trace it.
ALS 2pp RS:HS 9.328

6579. 1846–2–10 From George W. Hearn Royal Military College, Sandhurst
Regrets any inconvenience he has caused but is grateful for JH's act and offer to communicate his paper to the R.S.L.
ALS 3pp RS:HS 9.267

6580. 1846–2–10 To [Robert Hunt?] Collingwood
Cannot provide any specimens of photography or the spectrum good enough for an exhibition, as most have faded. A French chemist just announced as a discovery a fact long since known by JH.
ALS 3pp WT 68037.7

6581. 1846–2–12 From Georg Merz Munich
Reports on progress of refractor [see GM's 1846–1–13], and deals with finances.
ALS 1p (German) TxU:H/M–0359; Reel 1087

6582. 1846–2–14 From George W. Hearn Royal Military College, Sandhurst
Wishes he had seen JH's paper before submitting his own; then he could have rectified his errors. Has been carrying out some electrical experiments. Sent abstract to Michael Faraday, who does not understand them. Would be pleased to send JH further scientific communications.
ALS 4pp RS:HS 9.268

6583. 1846–2–14 From Sir William J. Hooker Royal Botanic Gardens, Kew
Thinks it a good idea to send a memo to the Colonial Office on the question of a Cape botanical garden. Has consulted W. H. Harvey on the question. What a curious plant is found on the coasts of Africa.
ALS 5pp RS:HS 9.457

6584. 1846–2–15 From A. B. Orlebar Bombay Magnetic Observatory
Circular: Magnetic and meteorological observations.
ALS 2pp RS:HS 13.182

6585. 1846–2–17 To Sir William J. Hooker Collingwood
Notes a small error in materials WH sent.
ALS 1p Royal Botanic Gardens

6586. 1846–2–19 From William Henry Smyth 3 Cheyne Walk, Chelsea
Asks JH's advice on the purchase of the lunar model. Anxious to see JH's new method of calculating double star orbits.
ALS 2pp RS:HS 16.286

6587. 1846–2–20 From John Stuart Mill India House
Has JH any comments on the first volume of JM's *Logic,* which the publisher wishes to prepare for a second edition.
ALS 2pp RS:HS 12.336

6588. 1846–2–20 To William Henry Smyth Collingwood
Suggests trustees of British Museum make an offer for Mme. Witte's lunar model. *Cape Results* at press. Observed double comet.
C 3pp RS:HS 22.268

6589. 1846–2–23 From William Henry Smyth 3 Cheyne Walk, Chelsea
Double star orbits not ready; R.A.S. waiting for JH's new calculating method. Reads accounts of Biela's Comet. Describes Edinburgh Observatory in detail.
ALS 4pp RS:HS 16.287

6590. 1846–2–23 From Mary Somerville Rome
Delighted that her paper was read at Royal Society. Discusses discovery of new planet [Neptune], and the future. Asks about his work.
ALS 3pp RS:HS 16.353

6591. 1846–2–28 From John Stuart Mill India House
Thanks for his remarks, which will be attended to. Comments on the various alterations he will make in line with JH's criticisms.
ALS 4pp RS:HS 12.337

6592. 1846–3–2 To Hugh Falconer Collingwood
The committee dealing with applying to the government for aid should meet soon, and so JH wants some questions answered from one of the authors of *Fauna Antiqua Sivalensis.*
ACCS 4pp RS:HS B27.71

6593. 1846–3–10 To Roderick Impey Murchison Collingwood
Has unsuccessfully tried to get in touch with [Hugh] Falconer and Falconer's publisher about obtaining further government aid for Falconer's research. Hopes RM, JH's fellow B.A.A.S. committee member, can help.
ALS 3pp EUL:Gen523/4/31

6594. 1846–3–12 From Roderick Impey Murchison 16 Belgrave Square

Thanks for his congratulations. Gives a resume of the events leading up to the knighthood. Saw Hugh Falconer yesterday. Sister-in-law has just died.
AL 4pp inc RS:HS 12.409

6595. 1846–3–13 From Hugh Falconer 23 Norfolk St., Strand
As HF is likely to be immediately ordered back to India, he is in no position to prepare himself to appear before the government seeking funds to publish *Fauna Antiqua Sivalensis.* Encloses a brief statement to the above effect to be read to the B.A.A.S. committee.
ALS 3pp enc RS:HS B27.72 & 73

6596. 1846–3–16 From John R. Hind 3 Allsop's Terrace, New Rd.
Has not been able to secure any angles of Gamma Virginis due to poor weather. Perplexed over the difference between J. H. Mädler and W. R. Dawes. George Bishop will be pleased to send any measures he may require.
ALS 4pp RS:HS 9.329

6597. 1846–3–17 From Michael Faraday Royal Institution
Has a sufficient quantity of heavy glass at his disposal and offers some to JH.
ALS 1p RS:HS 7.188

6598. 1846–[3–18 or later] To Caroline Lucretia Herschel Collingwood
A collection of birthday greetings from JH, his wife, Margaret, and seven of their children, six of whom wrote their greetings in German. JH refers to Biela's Comet having thrown off a portion of itself, producing what JH calls a 'Double Comet.'
ALS 3pp (part German) TxU:H/L–0591.3; Reel 1058

6599. 1846–3–20 From Hans Christian Oersted Copenhagen
Printed notice of thanks for memoir presented by JH.
ADS 1p RS:HS 13.168

6600. 1846–3–22 From Heinrich Christian Schumacher Altona
Circular announcing that [Theodor] Brorsen in Kiel has discovered a new comet. HS adds note that F. W. Bessel has died.
(A)L 1p RS:HS 15.397

6601. 1846–3–23 From William Rutter Dawes Camden Lodge
Thanks for the Cometen Circular. Observations of comets he has recently carried out. Thanks for the book on the telescope of William Parsons (3rd Earl of Rosse).
ALS 4pp RS:HS 6.71

6602. 1846–3–24 From George Bishop South Villa, Inner Circle, Regent's Park
Received JH's check for £300 for Harris family. Measurements of Gamma Virginis made by W. R. Dawes under GB's direction. Will send future observations to JH.
ALS 3pp TxU:H/M–0105; Reel 1087

6603. 1846–3–24 From Christian F. Schönbein Basle
Discovered an agent that renders his paper tough and waterproof. Sends sample for JH's inspection. Treated paper is highly explosive and can discharge a gun.
ALS 3pp RS:HS 15.358

6604. 1846–3–26 To Richard Sheepshanks Collingwood
Has received the prints of the portrait of Francis Baily. Finds them impressive.
ALS 3pp RS:HS 16.46 (C: RS:HS 25.9.22)

6605. 1846–3–27 From Michael Faraday Royal Institution
Will be showing some experiments in the laboratory next Monday to G. B. Airy, François Marcet, and a few others; JH would be welcome.
ALS 1p RS:HS 7.189

6606. 1846–3–28 From Thomas Maclear Royal Observatory, Cape of Good Hope
William Mann returns to England still in poor health. Expresses concern over running of Edinburgh Observatory by C. P. Symth. Regrets poor data from Cape Observatory on Comet of 1843. Observed Comet Gambert. Situation with Caffres hostile.
ALS 4pp RS:HS 12.134

6607. 1846–3–30 From John Stuart Mill India House
The first volume of JM's *Logic* has been printed and the printers require what is ready of the second volume. In Book Three would like comments from JH regarding the difference with P. S. Laplace's theories.
ALS 3pp RS:HS 12.338

6608. 1846–3–31 From Edward Everett Boston, Mass.
Sending a letter for a Benjamin Peirce, one of the professors at the University of Cambridge (Mass.). EE has just assumed the presidency of this university.
ALS 1p RS:HS 7.116

6609. 1846–4–1 From William Rutter Dawes Camden Lodge
Returns the account of Lord Rosse's telescope with many thanks. Weather has not been good for observations, but gives some observations on the comets.
ALS 3pp RS:HS 6.72

6610. 1846–[4] To Caroline Lucretia Herschel Collingwood
[Written as a family letter.] JH has corrected the first proof sheet of his *Cape Results.* Reports that Biela's Comet has been found to be a double comet.
ALS 3pp (part German) TxU:H/L–0591.3; Reel 1058

6611. 1846–4–2 To John Stuart Mill Collingwood
Continues JH's defense of Laplace's writings on probabilities [see JH's 1845–12–22].
ACCS 3pp RS:HS 25.6.32

6612. 1846–4–3 To John Stuart Mill Collingwood
A note to accompany a working out in convenient form an example of Laplace's probability ideas. [Enclosure not found.]
ACCS 1p RS:HS 25.6.33

6613. 1846–4–4 From William Whewell Trinity Lodge
Planning to edit a volume of hexameter verse translations. Wishes to include JH's translation of Friedrich Schiller's 'The Walk.'
ALS 3pp RS:HS 18.203 Todhunter, *Whewell, 2,* 333–4

6614. 1846–4–6 From Robert Hunt Museum of Pract. Geology
Has been pursuing his magnetic experiments with crystals and will be publishing a memoir on them. Comments on these experiment. Is there a difference in the mass of the earth at the poles?
ALS 4pp RS:HS 10.122

6615. 1846–4–8 To Caroline Lucretia Herschel Collingwood
Reports that Biela's Comet split into two comets; JH has observed it several times. 30–40 pages of JH's *Cape Results* have been printed.
ALS 4pp TxU:H/L–0591.4; Reel 1058

6616. 1846–4–8 From John Stuart Mill India House
JH's second letter has convinced him. Has now written a new conclusion to the chapter showing the principles of P. S. Laplace in a new light. Does not think Laplace's example is a fair type. Has already rewritten the chapter on the Doctrine of Chances.
ALS 4pp RS:HS 12.339

6617. 1846–4–8 From Hans Christian Oersted Copenhagen
Sending a German translation of an unfinished work of HO. Regrets that so few people can read the Danish language. Is grateful for JH's papers.
ALS 1p RS:HS 13.169

6618. 1846–4–8 From Charles Piazzi Smyth Hillside Crescent, Edinburgh
Describes conditions and instruments at Edinburgh Observatory. Observatory assistant Alexander Wallace works on reduction of Thomas Henderson's transit observations.
ALS 2pp RS:HS 16.226

6619. 1846–4–11 To Richard Sheepshanks n.p.
Does not possess enough knowledge of F. W. Bessel's writing to compose more than a brief obituary on him. Would like RS to help [A. B.] Orlebar of the Bombay Observatory by providing a corresponding observation of the occultation of Alpha Virginis on 19 January.
ALS 4pp RS:HS 16.47 (C: RS:HS 25.9.22)

6620. 1846–4–12 To Adolphe Quetelet Collingwood
Looks to reap fruits of observations. AQ's treatise on atmospheric waves has not arrived. R.S.L. has refused to 'receive and retain works communicated through them.'
ALS 4pp Académie belgique: #23 (C: Académie belgique: #24)

6621. 1846–4–16 To [?] M. Collingwood
Compliments of JH and his wife.
ALS 1p inc APS B.H435p.13

6622. 1846–4–16 From William Henry Smyth 3 Cheyne Walk, Chelsea
Receives report on poor state of Parramatta Observatory. Encloses letter from Thomas Brisbane, who believes James Dunlop should be removed from his post of authority there.
ALS 2pp encl RS:HS 16.288

6623. 1846–4–16 To William Henry Smyth Collingwood
Believes Beta Ursae Minoris to be a slow variable star. Remarks on diminution of the star's brightness between 1840 and 1841.
C 1p RS:HS 22.269

6624. 1846–4–20 To William Henry Smyth Collingwood
Suggests method to check the effectiveness of James Dunlop's work at Parramatta Observatory.
C 2pp RS:HS 22.271

6625. 1846–[4–22 or later] To Thomas Maclear [Collingwood]
Describes controversy over his father's discovery of two satellites of Saturn in detail. Hopes to publish *Cape Results* before end of year.
ALS 4pp inc Cape Archives/Maclear Papers File 100

6626. 1846–4–22 From William Henry Smyth 3 Cheyne Walk, Chelsea
Sends copy of James Graham's letter concerning transfer of Parramatta Observatory to government authority. Thanks JH for advice on the matter. Discusses methods of magnitude estimation.
ALS 2pp encl RS:HS 16.289

6627. 1846–4–22 From William Smith Williams 65 Cornhill
Smith, Elder & Co. received JH's revised proofs. Encloses page of 'Reduced Observations' for JH's approval. Details of printing process.
ALS 1p RAS:JH Archive 5/2.1.5 verso; Reel 8

6628. 1846–4–25 From Roderick Impey Murchison 16 Belgrave Square
Giving days on which the Council could be called. Who shall be president of Section 'A'. Could JH think of suitable persons for Chairmen and Secretaries? Sir J. Richardson is on the spot and will be president of the zoologists.
ALS 3pp RS:HS 12.411

6629. 1846–4–26 To Roderick Impey Murchison Collingwood
Approves date of B.A.A.S. meeting. Will consult with [Charles?] Wheatstone about date of Kew committee meeting.
ALS 3pp EUL:Gen523/4/32

6630. [1846–4–26 or earlier] To Roderick Impey Murchison n.p.
Suggests dates for upcoming B.A.A.S. meeting that will allow time to get report on Kew Observatory and committee meeting.
AL 2pp EUL:Gen523/4/44

6631. 1846–4–29 To George Biddell Airy Collingwood
A note to arrange a meeting to discuss the propriety of continuing the Kew Observatory.
ALS 1p RGO 6.401.193

6632. 1846–4–29 To Roderick Impey Murchison Collingwood
Probably cannot attend Kew meeting after 10th; has suggested 6th or 7th. Wishes to invite Michael Faraday as 'Physical Chair,' but knows of Faraday's poor health.
ALS 4pp EUL:Gen523/4/33

6633. 1846–4–30 From Charles Wheatstone 20 Conduit St.
Gives JH directions to Kew Observatory.
ALS 2pp RS:HS 18.151

6634. 1846–5–1 From John Knightsee Bank of England
Procedure to follow in case of missing bank notes.
(A)LS 1p RAS:JH Archive 5/2.1.4; Reel 7

6635. 1846–5–1 To Roderick Impey Murchison Collingwood
Accepts dinner invitation.
ALS 1p EUL:Gen523/4/34

6636. [1846–5] To William Whewell n.p.
Agrees to WW's request [that JH's translation of Friedrich Schiller's 'The Walk' be included in WW's planned volume?]. Regrets missing [C. K. J.] Bunsen's speech. Busy

editing N. L. Lacaille's catalogue.
ALS 3pp TC, Camb. Add. R.18.14[2] (C: RS:HS 25.13.29a)

6637. 1846–5–[7?] To Margaret Brodie Herschel London
Mostly about JH's return from London. [Seems to have been written after JH's 1846–5–8.]
ALS 2pp JHS 3.14b

6638. 1846–5–8 To Margaret Brodie Herschel 37 Tavistock Place
JH talks about visits he has made in London, and about plans to return to Collingwood.
ALS 4pp JHS 3.14a

6639. 1846–5–9 From John Galloway Kew Observatory
JH's letter has been forwarded via Charles Wheatstone as his town address was unknown.
ALS 2pp RS:HS 9.87

6640. 1846–5–11 From Sir Robert Harry Inglis 7 Bedford Square
Thanks for his kind letter. Congratulations on another addition to his family.
ALS 1p RS:HS 10.237

6641. 1846–5–17 To William Whewell Collingwood
Thanks for WW's *Lectures on Systematic Morality.* Announces the birth of another daughter [Francisca]. Working on *Cape Results*.
ALS 3pp TC, Camb. Add. Ms.a.207[68] (C: RS:HS 22.272)

6642. 1846–5–18 To Charles Babbage Collingwood
Urges a visit to Collingwood by CB. Refers to the development of a cypher.
ALS 3pp BL Aaa 32303.279

6643. 1846–5–20 From John R. Hind 3 Allsop's Terrace, New Rd.
Sends his observations of [Gamma] Cygni. Comments on the differences in the readings of various observers. Still unable to observe the double star JH requested, but the weather has been poor.
ALS 4pp RS:HS 9.330

6644. 1846–5–20 From Sir Robert Harry Inglis 7 Bedford Square
Hopes all his family are well. If JH proposes to attend the Trustees meeting of the British Museum on Saturday, will he breakfast with RI and Dr. T. R. Robinson of Armagh?
ALS 1p RS:HS 10.238

6645. 1846–5–24 From Thomas Maclear Royal Observatory, Cape of Good Hope
Suggests JH take trip to Continent to recover. Training a naval officer and the son of John Montague, Colonial Secretary, as observatory assistants. Describes outbreak of war with Caffres in detail.
ALS 7pp RS:HS 12.135

6646. 1846–5–25 From Francis Beaufort Admiralty
Encloses a letter of Thomas Maclear. William Mann will be a great loss to the Observatory. Hopes Lady Herschel and new infant are doing well.
ALS 2pp RS:HS 3.377

6647. 1846–5–27 From William Rowan Hamilton Admiralty
Sends copies of three letters by G. B. Airy on proposed railway through Greenwich Park. Asks JH's opinion on possible effects on observatory.
ALS 2pp TxU:H/M–0934; Reel 1083

6648. 1846–6–1 To 'Sir' Collingwood
JH's opinions about disruptive effect of tremors on observatory if proposed railway tunnel is built under Greenwich Park. Proposes method for adjusting train schedule to reduce interference with observations.
C 4pp TxU:H/L-0646; Reel 1089

6649. 1846–6–1 From Robert Hunt Museum of Pract. Geology
Is giving a series of lectures at the Collegiate Institution. Has made some interesting discoveries recently and has deposited a paper with Faraday for communication to the R.S.L. Comes to the conclusion that chemical action can be suspended by magnetic force. Comments on this.
ALS 4pp RS:HS 10.123

6650. 1846–6–3 From John R. Hind 3 Allsop's Terrace, New Rd.
Obtained first rate measures of the double star on 31 May, but since then the atmosphere has been hazy and so unable to send any further measures. Thinks it worthwhile to observe all the stars in *Cassiopeia* and *Ophiuchus*.
ALS 3pp RS:HS 9.331

6651. 1846–6–10 From Sophie Beckedorff Gloucester House
News of Caroline Herschel. Would be pleased to see any of the family if they are coming to town.
ALS 2pp RS:HS 4.2

6652. 1846–6–11 From Thomas I. M. Forster [Woodford]
Would a paper on falling stars be suitable for reading to the R.A.S.? Has much material on this subject and wonders where he can publish it.
ALS 4pp RS:HS 7.340

6653. 1846–6–15 From Henry T. De La Beche Geological Survey Office
Unable to afford the purchase of the model of the moon.
ALS 2pp RS:HS 6.128

6654. 1846–6–15 From Sophie Beckedorff Gloucester House
Thanks for the invitation to visit him. Will avail herself of it, if feasible, on her way to Dover.
ALS 3pp RS:HS 4.3

6655. [1846–6–15 or later] To Mrs. Witte n.p.
Replies that he knows no one who would be willing to purchase the model of the moon.
draft 1p RS:HS 6.128bis

6656. 1846–6–16 From Robert Hunt Museum of Pract. Geology
Has often thought of the inactive condition of iron and has made some notes before he received JH's comments. Has overcome all JH's objections to the intrinsic difference in the two poles. Comments on some of these experiments.
ALS 4pp RS:HS 10.124

6657. 1846–6–17 From Thomas Maclear Royal Observatory, Cape of Good Hope
Relating recent events in the latest Caffre war. [George] Childe has arrived and is engaged in preparing his rooms. Has had a letter from C. P. Smyth at Edinburgh. Price of commodities are rising rapidly at the Cape. Hopes that JH has taken expert medical advice.
ALS 5pp RS:HS 12.136

6658. 1846–6–20 From Sophie Beckedorff Gloucester House

Will leave by the train he suggests and hopes to see him and his family on Wednesday 24 June.
ALS 2pp RS:HS 4.4

6659. [1846]–6–20 From E. L. B[?] Farnbuck Place
Many thanks for his help. Hopes to hear good news of his invalids.
ALS 3pp RS:HS 5.1

6660. 1846–6–20 From Edward Sabine Woolwich
Proceedings of R.S.L. sent to recently regenerated Poudois[?] Society. [Joseph] Kay's observations will be sent to all observatories. Discusses election of fellows to R.S.L., the successor to S. J. A. Compton (Lord Northampton), proposal for single building to house all scientific societies.
ALS 8pp RS:HS 15.196

6661. 1846–6–24 From George Biddell Airy Royal Observatory, Greenwich
Letter of introduction for Professor [J.?] Hoffer of Vienna.
ALS 1p RS:HS 1.165

6662. 1846–6–24 From Sir John Bell Clarifield
Astronomical news.
ALS 4pp RS:HS 4.57

6663. 1846–6–26 From Antoine C. G. Jobert 7 Warwick Place, Leeds
Is applying for the French Mastership at the Belfast Academy and would be pleased to receive a testimonial from JH. Is also inviting subscribers for a new work to be published later.
ALS 3pp RS:HS 10.331

6664. 1846–6–27 To James McCullagh Collingwood
Has just received his remarkable paper on the dynamical theory of crystalline reflection and refraction. Comments on this. The subject has not lately occupied his attention. Would like clarification on some points.
draft 5pp RS:HS 12.10 & 22.274

6665. 1846–6–29 From John E. Gray British Museum
Saw a curious effect of some drawings on colored paper at Oxford. Thought JH may be interested.
ALS 2pp RS:HS 9.122

6666. 1846–6–29 From Owen Stanley to Isabella [?] Herschel 38 Lower Brook St.
Thanks and suggests date for dinner invitation. Reports on trip to Torres Straits and New Guinea. Updates on progress of OS's ship *Rattlesnake* and lack of midshipmen, and health of a Catherine (his sister?).
ALS 4pp RS:HS 17.229

6667. 1846–6–29 From Richard Jones Tithe Commission
Wants to visit JH at Collingwood this weekend.
ALS 1p TxU:H/M–0306.1; Reel 1087

6668. 1846–6–29 To William Whewell Collingwood
Presents a detailed discussion of the state of double star astronomy, including number known and number that are probably gravitationally linked. Gives an extended critique of WW's *Lectures on Systematic Morality,* arguing that WW's system is ultimately a happiness-of-mankind system and consequently rests on expediency considerations.
ALS 9pp TC, Camb. O.15.48[2] (C: RS:HS 18.204 inc; C: TC, Camb. 0.15.46[20] inc; C: RS:HS 22.273 inc)

6669. 1846–7–3 From William Whewell Lowestoft
Responds to JH's argument that WW's system of morality rests on expediency considerations. Argues that it rests more on empirical considerations than JH recognizes.
ALS 4pp RS:HS 18.205 Todhunter, *Whewell, 2,* 337–9

6670. 1846–7–4 From Roderick Impey Murchison 16 Belgrave Square
Agrees with JH that Charles Lyell is the man to give an evening at Southampton. Richard Owen is giving the other evening lecture. Suggests that Lyell lecture on the United States. Regarding leaders for the other sections.
ALS 4pp RS:HS 12.412

6671. 1846–7–4 From Owen Stanley 38 Lower Brook St.
Accepts invitation to dine at Collingwood, and will also accept a Mr. Jeffrey's invitation.
ALS 1p RS:HS 17.230

6672. 1846–7–5 From Sir Charles Lyell 16 Hart St.
When he received JH's request to give a lecture on American geology he did not treat it seriously. Now R. I. Murchison has made a similar request so thinks he could work up a popular lecture on the Mississippi Delta. Will keep off his old ground.
ALS 3pp RS:HS 11.424

6673. 1846–7–7 From Antoine Lombardi Modena
H. C. Oerstedt has been elected to the Society to replace F. W. Bessel, who has died. The Society is lacking one part of the *Memoirs* of the R.A.S.
ALS 1p (Italian) RS:HS 11.489

6674. 1846–7–8 To Edward Sabine Collingwood
[Charles] Lyell, [Roderick] Murchison, and JH will speak at B.A.A.S. [T. R.] Robinson should be notified that he will not need to speak at the meeting. Sees few advantages to combining all existing scientific societies into one.
draft 3pp RS:HS 15.194 (C: RS:HS 22.275)

6675. 1846–7–10 From Edward Sabine Woolwich
Spoke to [T. R.] Robinson, who will not speak at meeting [see JH's 1846–7–8]. Housing all scientific societies in one building would not impede their individual endeavors and would have some advantages.
ALS 3pp RS:HS 15.197

6676. 1846–7–13 From George Biddell Airy Royal Observatory, Greenwich
Regarding the printing of JH's 'countings of field-falls of stars.'
ALS 2pp RS:HS 1.166 (C: RGO 6.245.33)

6677. 1846–7–15 To George Biddell Airy Collingwood
About family health; JH's star counting has led him to believe that all stars are part of one sidereal system [the Milky Way].
ALS 4pp RGO 6.193.239

6678. 1846–7–15 From Edward William Brayley London Institution
Luke Howard wants to borrow from W. R. Birt JH's German work describing Karl Kreil's 'Barometrograph.'
L 1p † TxU:H/M–0130.1; Reel 1087

6679. 1846–7–21 From Augustus De Morgan 7 Camden St.
By which of JH's photographic processes can drawings be made permanent? Hopes printing goes on well.
ALS 2pp RS:HS 6.229

6680. 1846–7–21 From Humphrey Lloyd Trinity College, Dublin
Did not realize his question would require so much of JH's

valuable time to answer. Hopes to introduce some safe German Philosophy into the appropriate departments at Trinity College. Will look into the *Lehrbuch*.
ALS 4pp RS:HS 11.284

6681. 1846–7–23 To Augustus De Morgan Collingwood
Describes several photographic processes.
C 2pp RS:HS 22.276

6682. 1846–7–23 From John James Waterston London
Several journals rejected JW's paper on theory of gases. Hopes JH might read it at next B.A.A.S. meeting. Outlines theory of heat based on motion and conservation of *vis viva*.
ALS 3pp RS:HS 18.380

6683. 1846–7–24 From William Henry Smyth 3 Chenye Walk, Chelsea
Sends copy of J. R. Hind's observations of Gamma Virginis; believes they are 'too slow.' Also sends report of the Charter Committee of the R.S.L. Believes R.S.L. not selective enough. Has difficultly using JH's method of magnitude determination.
ALS 3pp RS:HS 16.290

6684. 1846–7–25 From George Peacock Bolton Percy, York
GP accepts a position which JH offers him.
ALS 3pp RS:HS 13.301

6685. 1846–7–29 From George Biddell Airy Royal Observatory, Greenwich
Informs JH that GA has stated that he will no longer work with James South on the Board of Visitors of the Royal Observatory; wants JH to be prepared for possible trouble.
ALS 2pp TxU:H/M-0632.1; Reel 1084

6686. 1846–7–29 From George Biddell Airy Royal Observatory, Greenwich
Expresses some concerns about GA's having indicated support for a particular candidate for the Board of Visitors. [Very faint.]
ACCS 2pp RGO 6.140.567

6687. 1846–7–31 To George Biddell Airy Collingwood
Has still not received F. W. Bessel's original letter [see GA's 1844–11–16].
ALS 3pp RGO 6.193.130 (drafts RS:HS 1.167 & 22.277)

6688. 1846–8–1 From John R. Hind 3 Allsop's Terrace, New Rd.
Elements of the new comet observed at Regent's Park.
ALS 1p RS:HS 9.332

6689. 1846–8–1 From William Henry Smyth 3 Cheyne Walk
Encloses copy of 29 July 1846 letter from G. B. Airy, who refuses to act on any further recommendations from Board of Visitors so long as James South is member of Board. Insinuations by South in R. H. Inglis's motion before House of Commons are deemed malevolent by Airy, who no longer considers South competent to guide scientific inquiry.
ALS 1p, encl 2pp TxU:H/M–0495; Reel 1093

6690. 1846–8–3 To John James Waterston Collingwood
Not surprised that scientific journals decline simple mechanical explanation of heat based on principle of conservation of *vis viva*. Admits that it has advocates, but JH has no time to devote to it. [JH's annotation: Not sent. Waterston gave no address but 'London'.]
Adraft 4pp †(by JH) RS:HS 18.381 (C: 22.278)

6691. 1846–8–4 From Asa Gray Boston, Mass.
Has sent to the R.S.L. Vol. 2 of the *Memoirs of the American Academy*.
ALS 1p RS:HS 9.121

6692. 1846–8–4 To William Henry Smyth Collingwood
Received WS's note about G. B. Airy's action regarding new warrant for Board of Visitors. Feels that Airy's cooperation is necessary in light of anomalous position of one board member. Recommends referring matter to Admiralty.
ACCS 1p TxU:H/L-0388; Reel 1054

6693. 1846–8–5 From George Biddell Airy Royal Observatory, Greenwich
Cannot find F. W. Bessel's letter [see JH's 1846–7–31]; believes GA returned it to JH a long time ago.
ACCS 2pp RGO 6.193.133 (C: RGO 6.193.134)

6694. 1846–8–7 From Christian F. Schönbein Basle
Discusses his explosive matter [see CS's 1846–3–24], which can discharge rifles and cannons. Four ounces of his gun cotton demolished an old tower in town.
ALS 3pp RS:HS 15.359

6695. 1846–8–7 From Richard Taylor and John Edward Taylor Red Lion Court, Fleet St.
Received £450 today from B.A.A.S. treasurer for printing N. L. Lacaille's catalog of stars and J. J. L. Lalande's *Histoire céleste*.
ALS 1p TxU:H/M–0549; Reel 1093

6696. [1846–8–8 or later?] From Christian F. Schönbein Golden Square
In London to demonstrate gun cotton. Wants JH to mention his name to influential members of government. Promises JH will be surprised at its 'wonderful effects.'
ALS 3pp RS:HS 15.364

6697. 1846–8–8 From William Samuel Stratford Feltham near Hounslow
Unable to publish catalogs in time for meeting.
ALS 2pp TxU:H/M–0545.1; Reel 1093

6698. 1846–8–12 From Robert Main Royal Observatory, Greenwich
Acknowledges, in G. B. Airy's absence, receipt of a letter from JH.
ACCS 1p RGO 6.368.681

6699. 1846–8–14 From William Samuel Stratford Feltham [near Hounslow]
Thanks for JH's encouragement. Sends three bills for publication expenses [regarding J. J. L. Lalande's *Catalogue of Those Stars in the Histoire céleste française*. Letter completed 1846–8–16.]
ALS 3pp † TxU:H/M–0545.2; Reel 1093

6700. 1846–8–15 To Julia Margaret Cameron Collingwood
Thanks JC for the writing material [?] JC sent. Pleased at the prospect of JH's eldest son [William] serving in the East India Company. Discusses developments in India and some chemical processes.
ALS 4pp Getty Center

6701. 1846–8–17 From E. Russell Pig's Quay, Blackfriars
Received JH's order for £21.
ALS 1p TxU:H/M–0469; Reel 1093

6702. 1846–8–20 From William Henry Smyth 3 Cheyne Walk, Chelsea
Discusses magnitude determination. Refers to incident in which one of WS's sons will be removed from his position but not given poor references. W. S. Jacob sends double star measurements.
ALS 3pp RS:HS 16.291

6703. 1846–8–23 To Roderick Impey Murchison Collingwood
Wonders how [Christian?] Schönbein, who is in London, could best exhibit explosives. Hopes RM will attend upcoming meeting; [William] Stratford, who is ill, cannot. [Francis] Baily, [Thomas] Henderson, and [R.] Harris have died working on 'fatal' catalogue Stratford is doing.
ALS 4pp EUL:Gen523/4/35

6704. 1846–8–23 From Christian F. Schönbein Golden Square
Thanks JH for kind letter. Asks JH to write letters of introduction to various people. Thinks [Roderick] Murchison will give good advice.
ALS 2pp RS:HS 15.360

6705. 1846–8–24 To Charles William Pasley Collingwood
The chemist C. F. Schönbein has invented a form of 'explosive cotton' and is looking for an opportunity to demonstrate it. JH is asking CP to help arrange matters.
ALS 3pp BL Aaa 41964.271

6706. 1846–8–24 From Christian F. Schönbein Golden Square
Thanks JH for interest in his gun cotton. Will demonstrate its use next week. Includes list of items he will need to perform experiments.
ALS 3pp RS:HS 15.361

6707. 1846–8–24 To John T. Towson Collingwood
Discusses JT's method for Great Circle sailing.
ALS 4pp RS:HS17.358 (C: RS:HS 22.279)

6708. 1846–8–25 From Roderick Impey Murchison Buxton
Received JH's letter without the enclosure from C. P. Schönbein, but has now received a letter from Schönbein. Has advised him to come to the Southampton meeting. Has written his discourse and would welcome a paragraph from JH on physical science and on some of the foreigners expected at the meeting. What can he say about Carlo Matteucci? Had letter from William Whewell, who wishes to meet H. C. Oersted.
ALS 8pp RS:HS 12.414

6709. 1846–8–26 From William Lassell Starfield, Liverpool
Has seen the seventh satellite of Saturn. Encloses diagrams and comments on them.
ALS 4pp RS:HS 11.127

6710. [1846]–8–27 From Carlo Matteucci Leicester Square
Happy to hear that JH is to preside over the Physical and Mathematical Section. Will present Miss Robina Young to Lady Herschel before leaving for Southampton.
ALS 3pp (French) RS:HS 12.306

6711. 1846–8–28 To William Lassell Collingwood
Comments on WL's observation of the seventh satellite of Saturn as noted in WL's 1846–8–26.
C 1p RS:HS 22.280

6712. 1846–8–28 From William Lassell Starfield, Liverpool
Was in error when in his last letter he stated he had seen the seventh satellite of Saturn; he has yet to see it.
ALS 2pp RS:HS 11.128

6713. 1846–8–29 From Edward Sabine Woolwich
Discusses cotton that allegedly gives equal projectile with less recoil than gunpowder. May have important implications for breech loading rifle.
ALS 8pp RS:HS 15.198

6714. 1846–8–30 From Roderick Impey Murchison Buxton
Thanks for his letter and C. F. Schönbein's. Will go on board the *Excellent* via rail to Gosport. Encloses what he has written on Michael Faraday for JH's amendments. Regarding the dinners at the meeting.
ALS 3pp RS:HS 12.415

6715. 1846–8–30 To [?] Collingwood
Thanks [?] for the honor of having her book on astronomy dedicated to JH.
ALS 1p DMC.695A.885.19

6716. 1846–8–31 From William Rutter Dawes Camden Lodge
Would like details of the approximate position of the planet. Has heard nothing about his equatorial. Would like it to pass through the Customs unopened to save possible damage.
ALS 4pp RS:HS 6.73

6717. 1846–9 To Margaret Brodie Herschel London
JH has expressed strongly his concern about P[eter Stewart]'s health; JH hopes he did not speak out too strongly.
AL 4pp JHS 3.13f

6718. 1846–9 To Margaret Brodie Herschel Southampton
Describes how time is spent at a conference, with a large section being directed to JH's daughter Isabella and written in Latin.
ALS 4pp JHS 1.2

6719. 1846–9–1 From John T. Towson Edward St., Blackfriars
Thanks JH for suggestions on his methods and on publishing his work.
ALS 4pp RS:HS 17.359

6720. 1846–9–2 From Charles T. Beke 6 St Mildred's Court
Hopes that he will accept the accompanying pamphlet. Would like to make his acquaintance at some future date.
ALS 1p RS:HS 4.28

6721. 1846–9–2 To Roderick Impey Murchison Collingwood
On fitness of Michael Faraday, as a chemist and physicist, to chair B.A.A.S. section. Remarks on [H. C.] Oersted's identifying electricity and magnetism, and connection between Oersted's and Faraday's work. Further comments on Faraday's work, and on instances of rash observation ultimately proving correct.
ALS 4pp EUL:Gen523/4/37 (C: RS:HS 12.416 & 22.281)

6722. 1846–9–2 To William Whewell Collingwood
Pleased that WW is giving a paper at Southampton B.A.A.S. meeting. JH will attend. Has completed the printing of most of his *Cape Results*. Smoke-drifts are besetting Kent.
ALS 4pp TC, Camb. Add. Ms.a.207[69] (C: RS:HS 22.282)

6723. 1846–9–3 To Roderick Impey Murchison Collingwood
To help RM, who must speak about [F. W.] Bessel, writes of Bessel's astronomical accomplishments.
ALS 2pp EUL:Gen523/4/48

6724. 1846–9–5 From Robert Mallet Dublin
Sends results of RM's experiment on wave motion and earthquakes that RM will report at B.A.A.S. meeting in Southampton. Proposes new committee to implement RM's project.
ALS 4pp TxU:H/M–0346; Reel 1087

6725. 1846–9–5 From Charles William Pasley London
Would like to see C. F. Schönbein's invention tried but has not the means to afford any assistance at the moment. Understands that Edward Sabine is to introduce Schönbein to the Master General of Ordnance, so he will be in good hands.
ALS 4pp RS:HS 13.234

6726. 1846–9–5 From Charles Piazzi Smyth Hillside Crescent, Edinburgh
Thanks JH for spotting error in a publication about the Edinburgh Observatory. Discusses 'the planet that is expected to be found beyond Uranus.' Asks JH to help him send astronomical news to Thomas Maclear.
ALS 4pp RS:HS 16.227

6727. 1846–9–7 From William Lassell Oakwood Hall near Stockport
Details of WL's observations of the seventh satellite of Saturn.
ALS 4pp RS:HS 11.129

6728. 1846–9–8 From [S?] Bold Belvedere Villa
Invites JH to join SB and Mr. Bold at 'our stable' tomorrow evening. Hopes [J. B. A.] Dumas will arrive from Paris then.
ALS 1p TxU:H/M–0109.1; Reel 1087

6729. 1846–9–8 From Robert John Kane Dublin
Is about to leave for a journey through Belgium and Germany for the purpose of studying collegiate education on behalf of the Irish government. Would welcome any letters of introduction from JH
ALS 3pp RS:HS 11.1

6730. 1846–9–[9 or 16] To Margaret Brodie Herschel Southampton
Tells MH about his journey to the B.A.A.S. meetings in Southampton, and comments on some of the people there, including a lady about whom JH finds his 'heart a little in danger.' George Peacock is very ill.
ALS 4pp JHS 3.13c

6731. 1846–9–9 From William Samuel Stratford Feltham [near Hounslow]
WS has been quite ill for nearly a month. Sends the rest of the proof sheets of the catalogue [J. J. L. Lalande's *Catalogue of Those Stars in the Histoire céleste française...* or of N. L. Lacaille's *Catalogue of 9766 Stars in the Southern Hemisphere.*]
ALS 3pp TxU:H/M–0545.3; Reel 1093

6732. 1846–9–10 From Charles T. Beke 6 St Mildred's Court
Thanks for his letter and friendly sentiments. Hopes to meet him at Southampton, though there are difficulties.
ALS 2pp RS:HS 4.29

6733. 1846–9–10 To Margaret Brodie Herschel Southampton
Is very busy at the B.A.A.S. meetings; reports on some humorous incidents.
ALS 4pp JHS 3.13d

6734. 1846–9–10 From Luke Howard Ackworth
Hopes his friends will support the publication of his *Barometrographia,* which will be exhibited at the forthcoming B.A.A.S. meeting.
ALS 3pp RS:HS 10.6

6735. 1846–9–12 To Margaret Brodie Herschel Southampton
Writing from B.A.A.S. meeting; all is going well; makes special reference to lectures of Roderick Murchison and Charles Lyell. JH honored by sitting next to Prince Albert.
ALS 4pp JHS 3.13a

6736. 1846–9–13 To Caroline Emily Mary Herschel Bishopstoke
Describes a visit to a friend of JH's youth and her husband, the Dean of Winchester.
AL 4pp JHS 1.59

6737. 1846–9–13 To William Lassell Southampton
Believes that the observations WL has made are quite sufficient to establish the existence of the seventh satellite of Saturn.
C 1p RS:HS 22.283

6738. 1846–9–14 To Luke Howard Southampton
Will distribute the prospectuses of LH's book as desired, but regrets he will be unable to subscribe for a copy himself as its price is too expensive for his means. Pleased to hear of his good health.
draft 2pp RS:HS 10.7 & 22.284

6739. 1846–9–16 To Margaret Brodie Herschel Southampton
Further about the B.A.A.S. meeting [see JH's 1846–9–12]; reports on some misunderstandings as a result of statements made by Dr. [T. R.] Robinson about JH.
ALS 4pp JHS 3.13b

6740. 1846–9–17 To Margaret Brodie Herschel 17 Montague Square
JH has 'slain' a dragon of a bore; he is concerned about Peter Stewart's being ill at ease in his mother's [Emilia Stewart's] presence, and JH declares his love for MH.
ALS 4pp JHS 3.22

6741. 1846–9–[18] To Margaret Brodie Herschel 17 Montague Square
Describes a tour of ships and dockyards at Southampton, including a view of James Nasmyth's steam hammer; got the last train for London, where JH is spending a few days with E. MacKintosh before returning home.
ALS 3pp JHS 3.13e

6742. 1846–9–19 To the London *Times* London
Protests erroneous impression in today's report of JH's conduct at closing meeting of B.A.A.S. Does not know what gave rise to T. D. Morris-Stirling's remarks regarding electric telegraph. Witnesses confirm that JH did not malign British science. [JH annotation: Printed 21 Sept. 1846.]
ACCS 1p † TxU:H/L-0231; Reel 1054 London *Times*, (Sept. 21, 1846), p. 6

6743. 1846–9–21 From Henry T. De La Beche Geological Survey Office
Sending the first volume of the *Memoirs* for Great Britain.
ALS 2pp RS:HS 6.129

6744. [1846]–9–21 From T. D. Morris-Stirling Union Club, Trafalgar Sq.
[Marked 'Private.'] Encloses copy of TM's letter to the *Times* responding to news report of TM's altercation with JH and [T. R.] Robinson at B.A.A.S. meeting.
AL 3pp, encl 5pp TxU:H/M–0381; Reel 1087

6745. [1846–9–21] From [?] n.p.
Challenges JH's letter in the *Times* today justifying JH's remarks at B.A.A.S. that Charles Wheatstone invented electric telegraph. Believes Wheatstone 'pilfered' invention from Mr. Baines.
AL 3pp inc TxU:H/M–0913; Reel 1083

6746. 1846–9–22 From Johann Franz Encke Berlin
Asks for permission to look at R.S.L. letters from G. W. Leibniz to Henry Oldenburg; comments on dioptrics, and several astronomical topics.
ALS 3pp (German) RS:HS 7.96

6747. 1846–9–[22 or later] To William Rowan Hamilton n.p.
Returns two papers, which had been missent, thus preventing JH from communicating them at a meeting.
ALS 3pp TCD 1493:1390

6748. 1846–9–22 To T. D. Morris-Stirling Collingwood
Received copy of TM's letter to the *Times* and sends copy of JH's letter to the *Times*. Recounts words that JH and TM used; wishes 'foam' had been used instead of 'scum,' to prevent misunderstanding [at B.A.A.S. meeting].
C 2pp TxU:H/L-0266; Reel 1054

6749. 1846–9–24 From John Bullar Bembridge, Ryde
Is obliged for his hints on saving heat in his ventilating apparatus. Comments on this and encloses leaflets on baths and wash-houses.
ALS 3pp RS:HS 4.325

6750. 1846–9–25 From William Rutter Dawes Camden Lodge
Thanks to JH's assistance the telescope was delivered safely and unopened. It is beautifully made.
ALS 2pp RS:HS 6.74

6751. 1846–9–26 To Richard Roberts Collingwood
Talks about considerations to be taken into account when building telescope specula, especially large ones.
ACCS 4pp RS:HS 25.6.34 & 25.7.1

6752. 1846–9–27 From John Hingston London
Regarding Benjamin Peirce's *Elementary Treatise on Sound.* Can lend him his own copy if JH has no copy.
ALS 2pp RS:HS 10.194

6753. 1846–9–30 From John R. Hind 3 Allsop's Terrace, New Rd.
In case he has not received a letter from Berlin he sends details of the planet of U. J. J. Leverrier.
ALS 1p RS:HS 9.333

6754. 1846–10–1 To the *Athenaeum* Collingwood
[The discovery of Neptune having been just announced], JH calls attention to JH's recent suggestion to the B.A.A.S. that such a discovery was imminent; states that in 1842 JH had discussed the idea of a trans-Uranian planet with F. W. Bessel, and that [J. C.] Adams of Cambridge had carried out an investigation comparable to that of U. J. J. Leverrier.
L *Athenaeum,* #988 (Oct. 3, 1846), 1091

6755. 1846–10–1 From Spencer J. A. Compton Belton
Will place J. F. Encke's request before the Council, but sees no difficulty in copying the letters at Somerset House. Is waiting for the return of George Peacock (Dean of Ely) before summoning council.
ALS 1p RS:HS 5.270

6756. 1846–10 or later To Richard Jones n.p.
A long statement against a very nasty article in the *Mechanics Magazine* on the priority controversy surrounding the discovery of Neptune.
C 5pp RS:HS 22.295

6757. 1846–10–1 To William Lassell Collingwood
Announces, to WL, the discovery of a new planet beyond Uranus, gives co–ordinates, and urges WL to look for satellites 'with all possible expedition.'
C 1p RS:HS 22.285

6758. 1846–10–2 From James McCullagh Trinity College Dublin
Has been on the Continent, which accounts for the delay replying to JH's letter of June. Regrets he cannot give an answer to JH's queries regarding the constitution of the ether. Fears his hypothesis on the density of the ether is not clear as it has misled JH. Has read Michael Faraday's papers with eagerness.
ALS 4pp RS:HS 12.11

6759. 1846–10–4 or later From Wilhelm Struve n.p.
Sends paper on naming of Neptune. Denounces elimination of [J. C.] Adams's part versus U. J. J. Leverrier's role regarding Neptune. Will forward Otto Struve's paper on the expedition between Altona and Greenwich and his own *Positiones mediae stellarum fixarum.*
ALS 4pp (German) RS:HS 17.136;

6760. 1846–10–4 From Wilhelm Struve Pulkowa
Description of Pulkowa observatory sent fourteen days before. Included a listing of the library. Otto Struve works on micrometer measurements of double stars. Complains about [J. H. von] Mädler's unprofessional attitude at Dorpat. Announces 'Neptune' as new planet's name.
ALS 4pp(German) RS:HS 17.135

6761. 1846–10–5 From Richard Roberts Manchester
Thanks JH for suggestions concerning optics. Will soon build a model telescope. Asks a series of questions concerning the construction of telescopes.
ALS 4pp RS:HS 14.496

6762. 1846–10–6 From William Rutter Dawes Camden Lodge
Thanks for sending J. R. Hind's letter about the new planet. Erected his equatorial on Saturday. Gives some of his observations on Saturn. Has sent William Lassell a prism by Georg Merz.
ALS 4pp RS:HS 6.75

6763. 1846–10–7 From John Phillips B.A.A.S., York
The B.A.A.S. grants £70 for the publication of the catalog of J. J. L. Lalande and N. L. Lacaille.
(A)L 1p TxU:H/M–0428.20; Reel 1093

6764. 1846–10–8 From John Hingston London
Has sent the book. Gives an outline of his own work and ideas he has for the improvement of the piano. Would like JH's comments.
ALS 4pp RS:HS 10.195

6765. 1846–10–8 From John Stevelly Belfast
Asks whether JH's committee will consider Belfast as a location for B.A.A.S.'s catalog of stars. Remembers clearly JH's quote about Neptune. Thinks JH should not yield to the French in calling Uranus so.
ALS 4pp RS:HS 17.233

6766. 1846–[10]–9 To [William Samuel] Stratford n.p.
Do not let catalogues make WS uneasy. Government money is available to pay printer. Will try to bring out N. L. Lacaille's work before 25 Sept. if JH can obtain [Thomas] Henderson's papers and WS's notes for the preface. Will not interfere with [T.R.] Robinson's committee.
ALS 4pp TxU:H/L-0473; Reel 1055

6767. 1846–10–12 From William Lassell Starfield, Liverpool
Details of WL's observations of a ring and satellite of U. J. J. Leverrier's planet.
ALS 4pp RS:HS 11.130

6768. 1846–10–13 From George Biddell Airy Royal Observatory, Greenwich
Is thinking of writing an account of the new planet.
ALS 4pp RS:HS 1.169

6769. 1846–10–13 From George Biddell Airy Flamsteed House, Greenwich
Just returned from nine weeks on the continent. Met Miss Caroline Herschel, who mentioned a catalogue of John Flamsteed's stars to which she had attached some notes.
ALS 4pp RS:HS 1.168

6770. 1846–10–17 From Humphrey Lloyd Trinity College, Dublin
Received his letter enclosing Edward Sabine's a few days ago on returning home. Busy at College; otherwise would have written. Could have attended a meeting of Committee of Physics had he known. Discovery of the new planet is a triumph. Has he heard of the attempt to destroy the telescope of William Parsons (3rd Earl of Rosse)?
ALS 3pp RS:HS 11.285

6771. 1846–10–18 To Richard Roberts Collingwood
Questions related to JH's 1846–9–26 having been asked by RR, JH writes to answer these questions.
ACCS 3pp RS:HS 25.7.2

6772. 1846–10–20 From James Constantine Adamson Cape Town
Requesting help in procuring person for the staff of the South African College.
ALS 4pp RS:HS 1.11

6773. 1846–10–22 To George Biddell Airy Collingwood
There is a need for a meeting of the Committee of Physics [R.S.L.] to discuss the current method of making temperature corrections for magnetic observations.
ALS 4pp RGO 6.676.451

6774. 1846–10–24 From George Biddell Airy Royal Observatory, Greenwich
Detailed response to JH's 1846–10–22 about the matter of temperature corrections for magnetic observations.
ACCS 4pp RGO 6.676.453

6775. 1846–10–26 From George Biddell Airy Royal Observatory, Greenwich
Suspects that the original complainant about temperature corrections [see JH's 1846–10–22] was in fact working with a faulty magnetometer.
ACCS 1p RGO 6.676.458

6776. 1846–10–26 From George Biddell Airy Royal Observatory, Greenwich
Regarding the slow motion screw in declination.
ALS 1p RS:HS 1.170 (C: RGO 6.159.370)

6777. [1846]–10–29 From E. L. B.[?] n.p.
Family news.
ALS 4pp RS:HS 5.2

6778. 1846–10–29 From James Prescott Joule Salford
Has been carrying out a series of experiments with spiders immersed in acids and finds their bodies are not protected in any way. Has also been experimenting with elastic tissues to see if there is a change of temperature.
ALS 4pp RS:HS 10.428

6779. 1846–10–30 From George Biddell Airy Royal Observatory, Greenwich
Regarding his paper about the new planet. Attacks by article in the *Athenaeum*.
ALS 2pp RS:HS 1.171

6780. 1846–10–30 From William Henry Smyth 3 Cheyne Walk, Chelsea
Discusses controversy over the discovery of Neptune. Astronomical committee of R.S.L. advised to award U. J. J. Leverrier the Copley Medal.
ALS 2pp RS:HS 16.292

6781. 1846–10–31 From William Rutter Dawes Camden Lodge
Thanks for the copy of the Guardian containing his own letter about the discovery of the new planet. Award of the Copley medal to U. J. J. Leverrier should allay French jealousies. Gives details of his own observations.
ALS 4pp RS:HS 6.76

6782. 1846–11–2 To George Biddell Airy Collingwood
Has obtained puzzling results when reducing JH's observations of the sixth satellite of Saturn.
ALS 3pp RGO 6.79.336

6783. 1846–11–2 To George Biddell Airy n.p.
A committee meeting is being called [see JH's 1846–10–22].
ALS 1p inc RGO 6.676.459

6784. 1846–11–2 From Spencer J. A. Compton Castle Ashby
Regrets his lateness in writing his request, but would JH write an obituary notice of F. W. Bessel? Does not expect it to be ready before the Anniversary meeting.
ALS 2pp RS:HS 5.271

6785. 1846–11–2 From Leonard Horner 2 Bedford Place
When on board the *Excellent* at Portsmouth he was deafened by the firing of a gun and has since suffered from 'singing' in his ear. Can JH recommend a relief? His daughter would like an engraving of JH.
ALS 3pp RS:HS 9.493

6786. 1846–11–2 From William Henry Smyth 3 Cheyne Walk, Chelsea
Lunar model will be placed in Somerset House.
ALS 2pp RS:HS 16.293

6787. 1846–11–3 From William Whewell Lowestoft
Informs JH about WW's forthcoming collection of hexameter verse translations. Laments that the discovery of the new

planet [Neptune] did not occur at Cambridge.
ALS 4pp RS:HS 18.206

6788. [1846–11–4?] To William Whewell [Collingwood]
Gives WW permission to print JH's translation of 'The Walk' by Friedrich Schiller. Comments further on the acceptance of pentametric and hexametric verse by the English, and includes a few verses of JH's own. JH well along in having his *Cape Results* in print.
ALS 4pp TC, Camb. Add. Ms.R.18.14[1] (C: RS:HS 25.13.7)

6789. [1846]–11–5 To Johann Franz Encke London
Has secured the permission JE requested to allow [Karl I.] Gerhardt to get transcripts of letters held by R.S.L. from G. W. Leibniz to Henry Oldenburg. [Continued 1846–11–6] Mentions controversy over discovery of Neptune. Thanks JE for publications sent.
ALS 4pp AdWDDR Encke 49

6790. 1846–11–6 To William Whewell Collingwood
Gives corrections to the proof copy of JH's translation of Friedrich Schiller's 'The Walk.' Comments on the failure in England to discover the new planet [Neptune] and on his own and James Challis's role in the controversy that erupted.
ALS 4pp TC, Camb. Add. Ms.R.18.14.4[4(1&2)] inc (C: RS:HS 22.286 & C: RS:HS 18.207 inc †)

6791. 1846–11–7 From John Webb Woollgar Lewes
Sends communication from R.A.S.
ALS 1p Harvard: Houghton AAH 67m–67 (62)

6792. 1846–11–11 To R.S.L. Chairman of Committee of Physics Collingwood
Being unable to attend committee meeting to consider recommendations for Royal Medal, elaborates reasons for thinking Edward Sabine's memoirs on terrestrial magnetism worthy of award.
ALS 4pp RS Sa.651 (C: RS Sa.652, RS Sa.653; ACCS: RS:HS 25.7.3)

6793. 1846–11–15 From Roderick Impey Murchison 16 Belgrave Square
Thanks for his letter and the enclosure of Dr. Andrew Smith. Will now write to C. R. Darwin. Can now wait on the Chancellor of the Exchequer.
ALS 1p RS:HS 12.417

6794. 1846–11–16 From William Rutter Dawes Camden Lodge
Gives comparisons between some of Wilhelm Struve's observations of double stars and his own. Would like to purchase JH's lens if he does not require it himself. Regrets the weather was so bad when JH's daughters visited him.
ALS 3pp RS:HS 6.77

6795. 1846–11–18 To George Biddell Airy Collingwood
Has made some minor adjustments in JH's satellite of Saturn results [see JH's 1846–11–2], but the major discrepancy still exists.
ALS 2pp RGO 6.79.338

6796. 1846–11–18 To George Biddell Airy Collingwood
Proposes a plan for the distribution of printed copies of the star catalogues [see GA's 1845–7–25].
ALS 2pp RGO 6.203.344

6797. 1846–11–18 From William Rutter Dawes Camden Lodge
Is grateful for the lens JH sent. Puzzled by the discrepancies in JH's results. They were pleased to receive his children.
ALS 3pp RS:HS 6.78

6798. 1846–11–19 From George Biddell Airy Royal Observatory, Greenwich
Regarding F. W. Bessel's table for the Huygenian satellite. On early chronology.
ALS 2pp RS:HS 1.172 (C: RGO 6.79.339)

6799. 1846–11–19 From George Biddell Airy Royal Observatory, Greenwich
Essentially agrees with JH's plan for distribution of the star catalogues [see JH's 1846–11–18].
ALS 2pp TxU:H/M-0632.2; Reel 1084 (C: RGO 6.203.345)

6800. 1846–11–19 From William Rutter Dawes Camden Lodge
Could he help with respect to the readings for the satellite of Saturn?
ALS 2pp RS:HS 6.79

6801. 1846–11–19 To William Rutter Dawes Collingwood
Comments on calculation of star locations, and how to make special observations.
C 2pp RS:HS 22.287

6802. [1846?–11–19] To William Rowan Hamilton Collingwood
WH's theorems on ellipsoids are new to JH, but JH is unfamiliar with the field, so they may not actually be new. Wishes to master mathematics. Still working on Cape observations. Believes Neptune was almost an English discovery, and would have been well-named Minerva.
ALS 4pp TCD 1493:374 Graves, *Hamilton, 2,* 532–3 inc

6803. 1846–11–20 To George Biddell Airy Collingwood
GA's suggestions [see GA's 1846–11–19] have helped JH solve his problem; now JH has a problem with observations of the second satellite of Saturn.
ALS 4pp RGO 6.79.340

6804. 1846–11–21 From George Biddell Airy Royal Observatory, Greenwich
Regarding Johann Lamont's numbers.
ALS 2pp RS:HS 1.173 (C: RGO 6.79.342)

6805. 1846–11–21 From Edward Sabine Woolwich
JH must inform Lt. Dayman of the actinometer observations to be done. Thanks JH for his actions regarding the Royal Medal. Wants to show Lord Angbury[?] a copy of JH's letter.
ALS 4pp RS:HS 15.199

6806. 1846–11–22 From D. François J. Arago Paris
Has forwarded JH's letter to U. J. J. Leverrier. Thinks that the time has come for an analysis of the work of William Herschel.
ALS 2pp (French) RS:HS 1.364

6807. 1846–11–22 From Urbain J. J. Leverrier Paris
François Arago has communicated JH's letter informing him that UL has been awarded the Copley Medal.
ALS 1p (French) RS:HS 11.195

6808. 1846–11–23 From William Rowan Hamilton Observatory
Elated that WH's account of the generation of an ellipsoid is an original result.
L Graves, *Hamilton, 2,* 533–5 inc

6809. 1846–11–23 From William Whewell Trinity Lodge
Suggests final revisions to JH's hexameter translation of Friedrich Schiller's 'The Walk.' Informs JH of the other authors of verses in the volume.
ALS 4pp RS:HS 18.208

6810. 1846–11–25 To Spencer J. A. Compton Collingwood
Informing SC that U. J. J. Leverrier has asked JH to accept the Copley Medal on his behalf.
C 1p RS:HS 22.289

6811. 1846–11–25 To Urbain J. J. Leverrier Collingwood
Will put the matter of UL's medal into the hands of Edward Sabine, the Foreign Secretary, who will hand it to the Ambassador for delivery to France.
draft 2pp (French) RS:HS 11.196 & 22.288

6812. 1846–11–25 From Edward Sabine Woolwich
U. J. J. Leverrier has written and asked that JH receive the [R.S.L.] medal for him.
ALS 2pp RS:HS 15:200

6813. 1846–11–26 From F. M. Eardley-Wilmot HMS Apollo
Forwarding a letter of Thomas Maclear. Has just left the Cape. Comments on the conditions there and outlines some of the difficulties.
ALS 2pp RS:HS 7.6

6814. 1846–11–28 From Urbain J. J. Leverrier Paris
Is pleased with the arrangements for the forwarding of his medal. Is sending for the R.S.L. five examples of his researches.
ALS 2pp (French) RS:HS 11.197

6815. 1846–11–29 From Augustus De Morgan 7 Camden St.
Has been planning the new duties for officers of the R.A.S. now that Francis Baily is dead. Hopes JH will consent to become president now, in order to give the prestige of the Society a boost.
ALS 4pp RS:HS 6.230 S. De Morgan, *Memoir of Augustus De Morgan*, 154–6

6816. [1846–12] To William Whewell n.p.
Returns, with comments, proofs of WW's hexameter translation. Favors encouraging theoretical more than observational work regarding terrestrial magnetism.
ALS 4pp TC, Camb. Add. Ms.R.18.14[3]

6817. 1846–12–3 To Richard Sheepshanks Collingwood
Agrees reluctantly to serve as Foreign Secretary of R.A.S., but will be unable to attend most meetings. JH's [*Cape Results*] is nearing publication; needs to know where he can get the plate [of JH's portrait]. Urges that if J. C. Adams 'is medallized, it should be *most cautiously* worded so as not to bear the least allusion to that ugly *word* "priority."'
ALS 4pp RS:HS 16.48 (C: RS:HS 25.9.25)

6818. 1846–12–4 To George Biddell Airy Collingwood
Asks a number of questions that need answers before considering the purchase of a large new telescope for the Royal Observatory.
ALS 4pp RGO 6.9.79

6819. 1846–12–4 To Augustus De Morgan n.p.
Would not like to become President and shirk the duties it involves, but if the R.A.S. cannot find someone else for the President then he will consent to let his name go forward. Has had a letter from W. H. Smyth on the same subject.
draft 3pp RS:HS 6.231 & 22.290

6820. 1846–12–8 To Caroline Lucretia Herschel Collingwood
JH is working hard on his *Cape Results*.
ALS 4pp TxU:H/L–0591.10; Reel 1058 *Memoir ... of Caroline Herschel*, 341–2 inc

6821. 1846–12–10 From Dr. Groskopf Hanover
Appears to be providing JH with the most recent news about JH's aunt Caroline.
ALS 2pp (German almost illegible) RS:HS 9.146

6822. 1846–12–10 From Johann Lamont Munich
Does not have *Astronomische Nachrichten 316* to check error. Refers [JH] to Sept. 1841 issue of *Gelehrte Anzeigen* of 'our Academy' for correct determination of 'elements of the 8th Satellite of Saturn.'
AL 3pp RAS:JH Archive 7/6.6; Reel 8

6823. 1846–12–11 From James Constantine Adamson Cape Town
Regarding appointment of staff at South African College and remarks on educational facilities. Last part of letter missing.
ALS 4pp RS:HS 1.12

6824. 1846–12–12 From Thomas Galloway 45 Torrington Square
Has written an obituary notice of F. W. Bessel for the R.S.L. Has arranged for the printers to send JH a proof and would be glad to receive any comments as it was hurriedly written.
ALS 3pp RS:HS 8.24

6825. 1846–12–14 From Thomas Galloway Serjeants' Inn
Is obliged for his encouraging note. G. B. Airy has sent him two useful pamphlets on F. W. Bessel, which JH may care to see. Duties of the President of the R.A.S. should not prove too difficult as W. H. Smyth can do most of the duties.
ALS 4pp RS:HS 8.25

6826. 1846–12–17 To Richard Sheepshanks Collingwood
Would not object if the [R.A.S.] council resolved to give J. C. Adams a medal, but the wording must be carefully considered, especially regarding the word 'discovery.' Presents a 'string of resolutions,' on how to deal with Adams and other contenders in the discovery of Neptune. Leave the question of priority for posterity. If the council sets up Adams as a competitor to U. J. J. Leverrier, JH will reserve to himself 'the privilege of protest.'
ALS 4pp RS:HS 16.50 (C: RS:HS 25.9.27)

6827. 1846–12–17 To Richard Sheepshanks Collingwood
Does not want the R.A.S. to make any assertions with respect to priority regarding the discovery of Neptune: '...though Neptune ought to have been born an Englishman and a Cambridge man every inch of him,' you will never make 'an English Discovery of it do what you will.' States that [J. C.] Adams is owed much. If the R.A.S. throws down the gauntlet, this will do much 'mischief' and demoralize Adams.
ALS 3pp RS:HS 16.49 (C: RS:HS 25.9.26)

6828. 1846–12–20 To George Biddell Airy n.p.
Seeking information to help JH prepare an obituary notice, for the R.A.S., of F. W. Bessel.
ALS 3pp RGO 6.56.171

6829. 1846–12–21 From George Biddell Airy Royal Observatory, Greenwich

Lists the contents of a parcel being sent.
ALS 1p RS:HS 1.174 (C: RGO 6.56.228)

6830. 1846–12–21 From George Biddell Airy Royal Observatory, Greenwich
Sends some books in response to JH's request for information about F. W. Bessel [see JH's 1846–12–20]; GA's daughter Hilda is very ill.
ACCS 1p RGO 6.56.173

6831. 1846–12–21 From Augustus De Morgan 7 Camden St.
Is beginning to work on the Annual Report. Has he any information about F. W. Bessel or J. J. L.Lalande's catalogue?
ALS 3pp RS:HS 6.232

6832. 1846–12–22 To Richard Sheepshanks Collingwood
Clarifies his position regarding the resolutions on the R.A.S. medals relating to the discovery of Neptune. Is unwilling to urge that they be accepted by the R.A.S, but can live with them because they contain the truth and nothing 'improper or offensive.' Suggests that if the R.A.S. does adopt these, it may 'prove a more fatal apple of discord than any that has been thrown down among us for years.'
ALS 4pp RS:HS 16.51 (C: RS:HS 25.9.28)

6833. 1846–12–22 To William Henry Smyth Collingwood
Believes council should decide the handling of the medal over the discovery of Neptune. Thinks that if J. C. Adams receives a medal, U. J. J. Leverrier should receive 'the more prominent distinction.'
C 2pp RS:HS 22.291

6834. 1846–12–24 To Augustus De Morgan n.p.
Some comments related to several astronomical publication problems; difficulties with naming the planet [Neptune].
C 2pp RS:HS 22.292

6835. 1846–12–25 To Richard Sheepshanks Collingwood
Clarifies some details in his letter to RS of 1846–12–25 on the R.A.S. deliberations regarding giving medals for the discovery of Neptune.
ALS 3pp RS:HS 16.53 (C: RS:HS 25.9.30)

6836. 1846–12–25 To Richard Sheepshanks Collingwood
Has received the list of six persons proposed by the R.A.S. for its medal, along with who nominated each. Has harsh words for the RAS's actions [regarding the discovery of Neptune] and says it has been badly mismanaged in England. Urges RS to exert his influence with G. B. Airy to get some names removed. At end writes 'Burn this.'
ALS 3pp RS:HS 16.52 (C: RS:HS 25.9.29)

6837. 1846–12–26 From George Biddell Airy Flamsteed House, Greenwich
Domestic news. Further regarding telescopes and slow motion in declination.
ALS 1p RS:HS 1.175 (C: RGO 6.159.371)

6838. 1846–12–26 From Thomas Maclear to Margaret Brodie Herschel Royal Observatory, Cape of Good Hope
Expresses concern over health of JH.
C 1p inc Cape Archives/Maclear Papers File 100

6839. 1846–12–26 To Charles Moxon Collingwood
Indicates that JH is unaware of any such constellation as that to which CM has referred.
ALS 1p APS B.C831

6840. 1846–12–26 To Richard Sheepshanks Collingwood
Did not understand the steps necessary by the R.A.S. rules and that an appeal to a general meeting was necessary [regarding awarding medals, e.g., for the discovery of Neptune]. Assures him that W. H. Smyth did not write anything of which RS could disapprove. Believes the council has the discretion to deal with anything that comes before it.
ALS 2pp RS:HS 16.54 (C: RS:HS 25.9.31)

6841. 1846–12–26 From Richard Sheepshanks n.p.
Responds in detail to JH's concerns expressed in his letters of 1846–12–25. Explains that in awarding R.A.S. medals regarding the discovery of Neptune, the Council wished to avoid slighting either U. J. J. Leverrier or J. C. Adams. Gives views of G. B. Airy, James Challis, and Augustus De Morgan.
ALS 16pp RS:HS 16.55

6842. 1846–12–27 To [François J. D. Arago] Collingwood
Is very pleased about FA's biography of JH's father, William. JH comments on what pictures of William Herschel might be available to be included in this biography.
ALS 4pp WT 67390.3

6843. 1846–12–27 To Wilhelm Struve Collingwood
Thanks WS for the description of the Pulkowa Observatory. Remarks on the discovery of the new planet [Neptune], 'a most spirit stirring event.' Questions whether a small distant satellite of Jupiter could have escaped detection.
C 3pp RS:HS 22.293

6844. 1846–12–29 From Mary Maclear to Margaret Brodie Herschel Royal Observatory, Cape of Good Hope
Nearly a year has passed since they received letters from the Herschels. Hopes JH's health is improving. Gives news of her own children, and the work of her husband at Orange River. The Equatorial has arrived safely.
ALS 8pp RS:HS 12.246

6845. 1846–12–29 From Edward Sabine Woolwich
Discusses possible meteorological and magnetic research possibilities in the colonies, particularly at Bermuda.
ALS 11pp RS:HS 15.201

6846. 1846–12–29 To William Whewell Collingwood
Responding to a published statement by WW regarding the discovery of Neptune, JH presents in detail a different interpretation, giving more credit to U. J. J. Leverrier than WW favored. Suggests that some observatory make a photographic record of sunspots.
ALS 4pp TC, Camb. Add. Ms.a.207[70] (C: RS:HS 18.209 †(JH) & RS:HS 22.294)

6847. 1846–12–29 From Francesco Zantedeschi Venice
Discusses issues raised in JH's 'On the Chemical Action of the Rays of the Solar Spectrum on Preparations of Silver....'
ALS 2pp (Italian) RS:HS 18.362

6848. [1847?] To Francis Baily n.p.
Regarding the revision of the nomenclature of the constellations. Its effect on the printing of the star catalogues.
ALS 5pp RS:HS 3.282

6849. [1847] From William Empson East India College, Hertford
Longmans reckons on 60 pages for the next number so would JH have his article ready on time.
ALS 2pp RS:HS 7.57

6850. [1847] From J[ohn] Henry Griesbach n.p.
Presents to the R.A.S. a drawing of solar spots made during 1843.
L *R.A.S.M.N., 8* (1847), 14 inc

6851. [1847?] To William Robert Grove n.p.
Encloses post office order; not knowing WG's 'Christian' name; hopes that WG can sign 'William.' Apologizes for being unable to attend 'the Dinners.'
ALS 3pp RI

6852. 1847 To Margaret Brodie Herschel London
Reports on a visit to Grandma [Emilia Stewart]; may go to Cambridge; sends MH a draft beginning of a review of Alexander von Humboldt's *Kosmos.*
ALS 3pp JHS 3.15b

6853. [1847] From John Lee n.p.
Discusses whether John Couch Adams should receive the Copley Medal for his work in the attempted discovery of Neptune.
ALS 1p inc Harvard: Houghton AAH 67m–67 (61)

6854. [1847 or later] From Charles Pastorff n.p.
Wishes to obtain copy of JH's observations of sun spots at Cape of Good Hope.
C 1p inc TxU:H/M–1031.1; Reel 1083

6855. [1847] To the R.A.S. n.p.
Accepts with thanks J. Henry Griesbach's drawings of solar spots. Encourages others to observe the sun. Suggests the use of photography in recording solar spots. Pledges to give the R.A.S. all JH's drawings of solar spots.
L *R.A.S.M.N., 8* (1847), 14–15 inc

6856. [1847] From Edward Sabine n.p.
Elizabeth Sabine's translation of [Alexander von Humboldt's] *Cosmos* is nearly complete. Hopes JH will accept invitation to review it in *Edinburgh Review.* Encloses account of great disturbance of December.
ALS 3pp RS:HS 15.328

6857. [1847] From Richard Sheepshanks n.p.
Reports organization of N. L. Lacaille's star catalog. Suggests method by which to compare this with Thomas Henderson's figures to determine Henderson's method of computation.
ALS 7pp TxU:H/M–0529.3; Reel 1093

6858. 1847 or later From Richard Sheepshanks n.p.
Is trying to straighten out the origin of the R.A.S., and giving appropriate credit.
ALS 2pp inc RS:HS 16.138

6859. 1847–1–1 From William Whewell Paris
Recommendations on how best to make tidal observations.
ALS 4pp RS:HS 18.210

6860. [1847] To William Whewell n.p.
Thanks WW for the supplemental dedication to JH in the second edition of WW's *Philosophy of the Inductive Sciences.* Is making slow progress correcting JH's *Cape Results.*
ALS 3pp TC, Camb. O.15.48[1]

6861. [1847] To William Whewell n.p.
Agrees to inclusion of his initials in WW's *Verse Translations* and to vote for Prince Albert in some Cambridge election. Comments on the 'mess' in the R.A.S. over awarding its medal for 1848.
ALS 3pp TC, Camb. Add. Ms.a.57[51]

6862. 1847–1–6 From George Biddell Airy Royal Observatory, Greenwich
Sends extracts of a letter from H. C. Schumacher. Does not agree with Schumacher regarding F. W. Bessel's 'specific attraction.'
ALS 1p RS:HS 1.176

6863. 1847–1–6 From George W. Hearn Royal Military College, Sandhurst
Has been investigating the Cavendish experiments recently repeated by Francis Baily. Thinks G. B. Airy has neglected the resistance of the air in his mathematical theory. Would like to send an abstract of his investigations so that JH can judge their merits.
ALS 4pp RS:HS 9.269

6864. 1847–1–6 From Charles Johnson General Post Office
Explains combined British and foreign postage rate applied to JH's letter, which has been forwarded to France. [JH annotation on verso about actinometer.]
ALS 1p † TxU:H/M–0301; Reel 1087

6865. 1847–1–8 To Augustus De Morgan Collingwood
Needs someone to translate German/Latin letters; about James Ross's polar expedition.
C 2pp RS:HS 22.296

6866. 1847–1–8 From Richard Sheepshanks Athenaeum
Recounts discussion in R.A.S. council regarding how the Society's medal should be awarded in response to the discovery of Neptune. U. J. J. Leverrier led J. C. Adams in the voting, but did not receive enough votes.
ALS 4pp RS:HS 16.56

6867. 1847–1–11 From Augustus De Morgan 7 Camden St.
Will do what he can for J. C. Gerhardt. Appears to be unfair regarding Isaac Newton and G. W. Leibniz. Is working on a list of writings in arithmetic.
ALS 4pp RS:HS 6.233

6868. 1847–1–11 From George W. Hearn Royal Military College, Sandhurst
Is grateful for his letter on the Cavendish experiment. Gives outline of his own theories regarding certain aspects of the Cavendish experiments, and would like his opinion on these before they are incorporated into a paper.
ALS 6pp RS:HS 9.270

6869. 1847–1–12 From Alfred Smee Bank of England
Sends a note along with a work on vision.
ALS 2pp RS:HS 16.166

6870. 1847–1–14 To Richard Sheepshanks Collingwood
Congratulations on how the deliberations concerning awarding the R.A.S. medal came off 'well & quietly.' Recommends how JH's notice on F. W. Bessel should be distributed.
ALS 3pp RS:HS 16.57 (C: RS:HS 25.9.32)

6871. 1847–1–17 From Augustus De Morgan 7 Camden St.
Has seen the papers, which consist of some 30 pages, the longest being in the *Commercium Epistolicum.* Knows someone who would copy them accurately and at a reasonable price.
ALS 2pp RS:HS 6.234

6872. 1847–1–21 From Thomas Galloway 45 Torrington Square

Sending a revised copy of his own memoir on F. W. Bessel. Council of the R.A.S. have not awarded a medal this year.
ALS 2pp RS:HS 8.26

6873. 1847–1–22 From George W. Hearn Royal Military College, Sandhurst
Sends his manuscripts. Discusses some of Michael Faraday's experiments, in particular his theory on the non-magnetism of blood. Has sent Faraday a paper on the action of voltaic currents.
ALS 4pp RS:HS 9.271

6874. 1847–1–28 To Augustus De Morgan Collingwood
Thanks for the pains about a translator [see JH's 1847–1–8]; preparing biographical information about F. W. Bessel.
C 2pp RS:HS 22.297

6875. 1847–1–29 To George Biddell Airy Royal Observatory, Greenwich
Seeks GA's opinion about including a specific passage in F. W. Bessel's obituary notice for the R.A.S. Also inquires about the printing of the star catalogues of N. L. Lacaille and J. J. L. Lalande.
ALS 3pp RGO 6.229.309

6876. 1847–2–1 From George Biddell Airy Royal Observatory, Greenwich
Regarding F. W. Bessel's doubt concerning the law of gravitation.
ALS 3pp RS:HS 1.178 (C: RGO 6.229.311)

6877. 1847–2–1 From George Biddell Airy Royal Observatory, Greenwich
Regarding W. S. Stratford and the printing of the catalogues. F. W. Bessel's star corrections.
ALS 3pp RS:HS 1.177

6878. 1847–2–1 From George Biddell Airy Royal Observatory, Greenwich
Concerned about W. S. Stratford, who has not responded to letters; GA has a problem with the German notation of star positions used by Thomas Henderson.
ACCS 3pp RGO 6.203.347

6879. 1847–2–1 From George W. Hearn Royal Military College, Sandhurst
Thanks for prompt reply. Does know something about blood and circulation, but may not have expressed himself clearly. Thinks Michael Faraday has made important discoveries in magnetism and electricity.
ALS 4pp RS:HS 9.272

6880. 1847–2–1 To William Lassell Collingwood
Ask WL's permission to make reference to WL's sighting of the seventh satellite of Saturn in JH's observations of all the satellites of Saturn.
C 1p RS:HS 22.298

6881. 1847–2–2 To George W. Hearn Collingwood
Is little conversant in physiology so is unable to discuss GH's theories. Suggests some alterations to the text of the article.
C 3pp RS:HS 9.273 & 22.299

6882. 1847–2–3 To Augustus De Morgan Collingwood
Sends what JH has written on F. W. Bessel.
C 1p RS:HS 22.300

6883. 1847–2–3 From James David Forbes Edinburgh
Comments, after delays due to busyness, on JH's queries on the calibration of actinometers and gives results of experiments.
ALS 3pp St. Andrews LB IV, 193–5 (C: RS:HS 7.308)

6884. 1847–2–4 From William Lassell Starfield, Liverpool
About observations of the seventh satellite of Saturn, and using William Parsons' (3rd Earl of Rosse) 'machine'.
ALS 3pp RS:HS 11.131

6885. 1847–2–5 From George W. Hearn Royal Military College, Sandhurst
Agrees wholeheartedly with JH's recommendation for his paper. Will rewrite it if necessary. Hopes JH will produce a second edition of his work on finite differences.
ALS 3pp RS:HS 9.274

6886. 1847–2–6 From H. S. Wigginton Albert Villa, Slough
As secretary of newly formed 'Mechanic's Institute' in Slough, HW invites JH to preside at 18 Feb. meeting.
ALS 2pp, Wigginton note RS:HS 18.406

6887. 1847–2–7 To Charles Babbage Collingwood
Is thinking of publishing a new edition of *Examples in the calculus of finite differences*. Would like his views on this.
ALS 4pp RS:HS 2.313

6888. 1847–2–7 From John R. Hind 3 Allsop's Terrace, New Rd.
Discovered a telescopic comet in Cepheus. Gives readings.
ALS 1p RS:HS 9.335

6889. 1847–2–7 To Richard Sheepshanks Collingwood
Has heard that the R.A.S. council's 'no-medal resolution' has produced much dissatisfaction, which will not disappear by a general vote supporting the acts of the council. Thinks that the course adopted is wrong. If the council will not reconsider in favor of a positive conclusion, JH suggests that the issue be raised in a general meeting of the R.A.S. and that its decision be accepted with 'alacrity.' Holds himself free, as President, from any obligation to uphold 'in speaking or writing' whatever decision results.
ALS 2pp RS:HS 16.58 (C: RS:HS 25.9.33)

6890. 1847–2–7 To William Henry Smyth [Collingwood]
Discusses Neptune discovery controversy; believes both scientists should receive medals.
ALS 2pp inc RS:HS 16.294 (C: RS:HS 22.301)

6891. 1847–2–8 To George Biddell Airy Collingwood
Comments on the unsatisfactory [to JH] resolution regarding the awarding of the annual R.A.S. medal.
ALS 4pp RGO 6.229.80

6892. 1847–2–8 To Augustus De Morgan Collingwood
About the dispute in the R.A.S. over awarding its medal(s) after discovery of Neptune.
C 2pp RS:HS 22.302

6893. 1847–2–8 From Sir Francis Ronalds Chiswick
Wishes to use remarks JH made at Kew Observatory in 1846 for a Royal Society publication. Asks JH to read transcript for accuracy.
L 2pp RS:HS 14.421

6894. 1847–2–8 From Edward Sabine Woolwich
Encloses application for meteorological and magnetic observatories from governor of New Brunswick. May establish observations in all colonies for limited time. Puzzled over [John] Lefroy's results from Athabasca.
ALS 5pp RS:HS 15.202

6895. 1847–2–8 To Richard Sheepshanks Collingwood
Has learned of more dissatisfaction with the R.A.S. council's decision regarding awarding its medal. Because the council rejected the proposal to give more than one medal, the decision to medalize no one was a logical consequence. Expects the matter will be raised in the general meeting. Wants some positive conclusion, not a negative one. Restates his original position that medals ought to go to both, with care taken in the wording so as to avoid national rivalry.
LS 4pp RS:HS 16.59 (C: RS:HS 25.9.35)

6896. 1847–2–9 From George Biddell Airy Royal Observatory, Greenwich
Is sending papers, from an American, for JH's opinion.
ALS 1p RS:HS 1.179 (C: RGO 6.56.248)

6897. 1847–2–9 From George Biddell Airy Royal Observatory, Greenwich
As disturbed about the medal situation as JH [see JH's 1847–2–8]; the solution seems to be to award two medals.
ACCS 4pp RGO 6.229.87

6898. 1847–2–9 From Ralph Tatham St. John's College
Asks JH to suggest candidates for a professorship at the South African College.
ALS 3pp RS:HS 17.427a

6899. 1847–2–11 To George Biddell Airy Collingwood
Proposes that R.A.S. award medals regarding the discovery of Neptune to both U. J. J. Leverrier and J. C. Adams, giving first honors to Leverrier, and specifying distinctly the grounds for awarding the second. Hopes his resolution will not be seen as blaming anyone and that it will have a 'healing and a friendly effect.'
ACS 4pp RS:HS 16.61 (C: RS:HS 25.9.37)

6900. 1847–2–11 To George Biddell Airy Collingwood
Continuing to try to resolve the medal problem [see GA's 1847–2–9].
ALS 4pp RGO 6.229.96

6901. 1847–2–11 To Richard Sheepshanks n.p.
Has written G. B. Airy stating his approval of having JH's statement favoring awarding two R.A.S. medals read. Wants to be fair to U. J. J. Leverrier, and does not want to 'stultify' the existing council. Is enclosing a copy of his note to Airy.
ALS 4pp † encl RS:HS 16.60 (C: RS:HS 25.9.36)

6902. 1847–2–11 From George Wagstaffe Yapp 68 Cheapside
As secretary, GY requests JH's presence at first soirée of Whittington Club [for journalists]. Will JH accept election as vice president?
ALS 1p RS:HS 18.442

6903. [1847]–2–[12 or later] To Richard Sheepshanks n.p.
Has written to G. B. Airy requesting that at the R.A.S. meeting, Airy show RS the note JH wrote to him. Announces 'LV's' [U. J. J. Leverrier's] election as an associate of the R.A.S.
ALS 2pp RS:HS 16.62 (C: RS:HS 25.9.34)

6904. 1847–2–13 From George Biddell Airy Royal Observatory, Greenwich
Reports to JH on the discussion, about the medal problem, at an R.A.S. meeting [see JH's 1847–2–11].
ACCS 3pp RGO 6.229.99

6905. 1847–2–13 From Urbain J. J. Leverrier Paris
Regarding his Copley medal, has written a letter of thanks to the Society. Does not doubt that the new planet will be discovered.
ALS 3pp (French) RS:HS 11.198

6906. 1847–2–14 From George Biddell Airy Royal Observatory, Greenwich
More information on the activity of the R.A.S. in trying to resolve the medal problem [see GA's 1847–2–13].
ACCS 4pp very faint RGO 6.229.101

6907. 1847–2–17 To Johann Franz Encke Collingwood
Informs JE what would be the cost for transcription of the G. W. Leibniz letter requested by JE [see JH's [1846]–11–5]. Mentions controversy over discovery of Neptune.
ALS 3pp AdWDDR Encke 49

6908. 1847–2–17 From Alfred Smee 7 Finsbury Circus
Has been preparing famine food, such as hay biscuits.
ALS 3pp RS:HS 16.167

6909. 1847–2–20 To William Henry Fitton Collingwood
Expresses opinions about the medal fiasco at the R.A.S., and in so doing, also defends G. B. Airy. JH is very concerned about the mean attacks being made on various people over the Neptune discovery controversy. Makes especially strong statements in a PS written on Feb. 21.
ACCS 3pp RS:HS 25.7.5

6910. 1847–2–20 To Edward Sabine Collingwood
Has discovered a flaw in the theory of actinometer observations. Realizes that JH's actinometer observations at the Cape may be useless and that observations made at Government and East India observatories should be suspended.
ALS 4pp RS:HS 15.203 (ACCS: RS:HS 25.7.4; C: RS:HS 22.303; RS MM 11.188)

6911. 1847–2–22 From William Samuel Stratford Notting Hill Square
Discusses the cost of printing and publishing J. J. L. Lalande's *Catalogue of Those Stars in the Histoire céleste française* and of N. L. Lacaille's *Catalogue of 9766 Stars in the Southern Hemisphere* [for which JH wrote the Preface]. Hopes to have both books published by the end of April.
ALS 3pp TxU:H/M–0546.2; Reel 1093

6912. 1847–2–22 From William Samuel Stratford *Nautical Almanac* Office
Apologizes to JH for worrying him about the printing of J. J. L. Lalande's *Catalogue of Those Stars in the Histoire céleste française....*
ALS 2pp TxU:H/M–0546.1; Reel 1093

6913. 1847–2–23 From Johann Franz Encke Berlin
Comments relating to G. W. Leibniz correspondence [see JE's 1846–9–22]; comments on naming the new planet Neptune.
ALS 2pp (German) RS:HS 7.97

6914. 1847–2–23 From Edward Sabine Woolwich
Copies of 1847–2–20 letter will be sent to observatories utilizing actinometers. Earlier series will be replaced with accurate figures. Suggests method to measure temperature in actinometers.
ALS 2pp RS:HS 15.204

6915. 1847–2–24 From John P. Gassiot Clapham Common
W. R. Grove and JG would like an opportunity of discussion with JH on a scientific matter. Will visit him if necessary.
ALS 3pp RS:HS 8.53

6916. 1847–2–24 From William Samuel Stratford Notting Hill Square
Nothing more has been done with J. J. L. Lalande's *Catalogue of Those Stars in the Histoire céleste française...;* WS hopes to spend more time on it henceforth. WS also has been too busy to oversee the progress on the bust of Francis Baily.
ALS 3pp TxU:H/M–0546.3; Reel 1093

6917. 1847–3–1 To James David Forbes Collingwood
Believes all actinometric measurements made are useless because they did not take into account the rate of expansion of the liquid, which is not constant.
ALS 4pp St. Andrews 10a (C: RS:HS 22.304)

6918. 1847–3–1 From [John] Lee Doctors Commons
Plans to send parcel to observatory at Cape of Good Hope; does JH have anything to include?
ALS 1p † TxU:H/M–0329; Reel 1087

6919. 1847–3–2 From John P. Gassiot Clapham Common
Regrets he was unable to see JH when in town last week. Outlines the scheme for a new club.
ALS 4pp RS:HS 8.54

6920. 1847–3–2 To William Henry Smyth Collingwood
Plans to write John Russell in an effort to secure a pension for Thomas Maclear; asks for R.A.S. and WS's support. Asks WS to obtain some facts relevant to pension request.
C 2pp RS:HS 22.305

6921. 1847–3–3 From Robert Hunt Museum of Pract. Geology
Is desirous of procuring impressions of prismatic spectra from all parts of the globe. These should show the chemical influence of the solar rays. The B.A.A.S. may be able to help or Colonel Edward Sabine. Can JH use his influence?
ALS 2pp RS:HS 10.125

6922. [1847?]–3–3. From Edward Sabine Woolwich
Glad JH is writing the meteorology. Collecting materials for charts of ocean. Forwards some of this to JH for comments.
ALS 4pp RS:HS 15.307

6923. 1847–3–4 From James David Forbes Edinburgh
Sympathizes with JH's feelings of regret over actinometer readings. Has procured a new thermometer and carried out experiments toward correction of actinometers.
ALS 3pp diag St. Andrews LB IV, 214–16 (C: RS:HS 7.309)

6924. 1847–3–4 To William Henry Smyth Collingwood
Asks WS if possible to enclose a note by Francis Beaufort to John Russell concerning request for pension for Thomas Maclear. Lays aside double star observations; continues work on orbit equation.
C 2pp RS:HS 22.306

6925. 1847–3–4 To [?] Collingwood
In response to a request, JH is not willing to part with any letters from his father, William, but will send one of William Herschel's notes to his sister, Caroline.
ALS 2pp WT 67597.3

6926. 1847–3–5 From John P. Gassiot Clapham Common
Thanks for his note. Is writing to remove any misconceptions JH may have about the proposed new club. Michael Faraday will be joining on condition that there will be no dinners.
ALS 2pp RS:HS 8.55

6927. 1847–3–6 From Charles Piazzi Smyth Hillside Crescent, Edinburgh
W. S. Jacob sends Alpha Centauri observations to R.A.S.; uses JH's *Cape Results* as a guide to double stars.
ALS 2pp RS:HS 16.230

6928. 1847–3–8 From Augustus De Morgan 7 Camden St.
Giving details of Thatcham Church and of the tablet, which is to be erected in memory of Francis Baily. Sculptor will be sending proof of the inscription.
ALS 2pp RS:HS 6.235

6929. 1847–3–9 To William Henry Smyth Collingwood
Asks WS to take the chair at the next meeting of the R.A.S [probably concerning the Adams-Leverrier dispute].
C 2pp RS:HS 307 (C: RS:HS 22.307)

6930. 1847–3–11 To William Henry Sykes Collingwood
Notes that actinometer observations prove faulty due to problem with instruments. Asks that WS notify Indian observatories to join others in terminating such observations until JH finds method of getting around problem.
C 3pp RS:HS17.143 (C: RS:HS 22.308)

6931. 1847–3–12 To James P. Muirhead Collingwood
Thanks for sending his book on the *Correspondence of James Watt.* Hopes to give it the attention it deserves, but meanwhile comments on the water controversy of Henry Cavendish, Joseph Priestley, and James Watt.
C 8pp RS:HS 12.381 & 22.309

6932. 1847–3–13 From John Forbes, D.D. Glasgow
Believes he has succeeded in integrating elliptic and hyperbolic functions in finite terms, and sends a paper in which this is discussed.
ALS 2pp RS:HS 7.372

6933. 1847–3–13 To Charles Richard Weld Collingwood
George III defrayed all costs (£4000) for William Herschel's 40-feet reflecting telescope. JH began dismantling it in Dec. 1839. Lenses and equipment are in storage. Tube's internal structure of corrugated iron and framework's diagonal bracing were originated by WH.
ACCS 2pp RS:HS 18.392 (C: 22.310)

6934. 1847–3–17 To William Henry Smyth Collingwood
Proposal of giving medals to both J. C. Adams and U. J. J. Leverrier by William Whewell turned down by R.A.S.
C 2pp RS:HS 22.311

6935. 1847–3–19 From Edward Joseph Lowe Highfield House near Nottingham
Sends a work for JH; would he also present the other copy to the R.S.L.
ALS 1p RS:HS 11.313

6936. 1847–3–20 From [Karl] Herman Knoblauch Berlin
Sending treatises on radiant heat for the R.S.L. and JH.
ALS 1p RS:HS 11.86

6937. 1847–3–23 To Augustus De Morgan Collingwood
Thanks for AD's paper on probability of arguments and on the syllogism.
C 1p RS:HS 22.312

6938. 1847–3–23 To James P. Murchison Collingwood
Thanks JM for a copy of the *Correspondence of the Late J. Watt.* In commenting on it, JH discusses at some length the doctrine of phlogiston as it relates to the discovery of the

composition of water.
ACCS 4pp RS:HS 25.7.6

6939. 1847–3–29 From Alfred Gautier Geneva
Introducing his nephew, Emile Gautier, Doctor of Mathematics. Sends a memoir of J. H. Maedler.
ALS 2pp (French) RS:HS 8.88

6940. 1847–3–29 From Manuel John Johnson Radcliffe Observatory, Oxford
Would be pleased to accommodate JH if he is attending the B.A.A.S. meeting.
ALS 2pp RS:HS 10.334

6941. 1847–3–30 From George Biddell Airy Royal Observatory, Greenwich
Regarding pamphlets concerning F. W. Bessel. Altitude instrument finished.
ALS 2pp RS:HS 1.180 (C: RGO 6.56.184)

6942. 1847–3–30 or later To George Biddell Airy n.p.
The books are on their way [see GA's 1847–3–30].
ALS 2pp RGO 6.56.185

6943. 1847–4–1 From William Rowan Hamilton Observatory
Sends some results concerning undisturbed parabolic motion. Laments the Irish famine.
L Graves, *Hamilton, 2,* 565

6944. 1847–4 From Anne Elise Knipping Hanover
Comments on the reports of the birthday celebrations for JH's aunt Caroline.
ALS 2pp (German) RS:HS 11.59

6945. 1847–4–1 From Sir Charles Shadwell Barnes
Presents 'tables for facilitating the approximate prediction of occultation and eclipses for any particular place,' so that seamen without specialized backgrounds in mathematics can observe and improve hydrography.
ALS 2pp RS:HS 17.198

6946. 1847–4–3 From William Radcliffe Birt Bethnal Green
Is very grateful for his assistance and encloses the page of the report in which he publicly acknowledges this. Further results since the return of the great atmospheric wave of 1845.
ALS 3pp RS:HS 4.115

6947. 1847–4–4 To [William Radcliffe Birt] Collingwood
Returns WB's proofs. Compliments WB's successes in meteorology. Barometric waves may be caused by something 'complementary' in opposite seasons or hemispheres.
ALS 3pp RS:HS 19.124

6948. 1847–4–4 To F. A. Herschel Griesbach n.p.
Has signed the certificate, and stated that FG is known to him.
draft 1p RS:HS 9.135

6949. 1847–4–4 From Marian Koller Kremsmunster
Sending a series of magnetical and meteorological observations made at the observatory. Comments on these.
ALS 4pp (French) RS:HS 11.91

6950. 1847–4–7 To [Robert Hunt] Collingwood
Attached letter from Edward Sabine [see Sabine's 1847–4–7] will allow RH to communicate directly with Sabine.
ALS 1p BostonPubLib Ms.Eng.406

6951. 1847–4–7 From Edward Sabine Woolwich
If given precise instructions, will attempt to obtain the prismatic spectra from Toronto and St. Helena that [Robert] Hunt requested [see Hunt's 1847–3–3].
ALS 3pp BostonPubLib Ms.Eng.406

6952. 1847–4–12 From William Rutter Dawes Camden Lodge
Is reading JH's book and marvelling at the condensed amount of information it contains. Has been observing some of Wilhelm Struve's more difficult objects. Regarding the double star *Antares*. Has been testing his telescope for errors and found it accurate.
ALS 4pp RS:HS 6.80

6953. 1847–4–17 From John P. Gassiot Clapham Common
Meeting for the formation of a new club took place last Monday. Gives list of persons who attended. To be called the Philosophical Club. Outlines its rules and regulations.
ALS 3pp RS:HS 8.56

6954. 1847–4–17 From William Henry Smyth 3 Cheyne Walk, Chelsea
Studies two possible orbits of Gamma Virginis; sends complete descriptions of both, including calculation of the apparent and actual ellipse. The first orbit combines the data of other astronomers.
ALS 7pp RS:HS 16.295

6955. 1847–4–20 To William Rowan Hamilton Collingwood
On WH's 'Hodograph' and theorems of parabolic motion and the relation between velocities, initial velocities, and time. Praises WH's son. *Cape Results* nearly finished. Revising book on astronomy. Plans to 'attack' quaternions. Mentions parabolic functions and Benjamin Peirce's claim that the discovery of Neptune was accidental.
ALS 4pp TCD 1493:387 Graves, *Hamilton, 2,* 565–6

6956. 1847–4–21 From William Rutter Dawes Camden Lodge
Will be pleased to accept his offer of the glass wedges. Bad weather has upset his observations. Gives some of his recent observations. James Challis doubts the existence of a ring round Neptune. Is not impressed by Capt. W. S. Jacob's observations.
ALS 4pp RS:HS 6.81

6957. 1847–4–[21] To William Henry Smyth [Collingwood]
Skeptical over plan for galvanic illumination of wires. Has not yet seen J. B. Biot article; states he has always held Biot in high regard. Discusses naming of new planet.
C 2pp RS:HS 22.313

6958. 1847–4–21 From William Henry Smyth 3 Cheyne Walk, Chelsea
Sends angular results from various astronomers using similar epoches.
ALS 2pp RS:HS 16.296

6959. 1847–4–22 To John P. Gassiot n.p.
Asks JG to allow JH to join the Philosophical Club [see JG's 1847–4–17] on a trial basis for a year or two.
draft 1p RS:HS 8.56 verso

6960. 1847–4–22 To William Robert Grove Collingwood
Asks to be allowed to join the new Philosophical Club on a trial basis [see letter of same date to John P. Gassiot].
ALS 2pp RI (C: RS:HS 22.314)

6961. 1847–4–22 From Georg Merz & Son Munich
Are preparing an objective for JH, which they expect to send in August.
ALS 2pp (German) RS:HS 13.36

6962. 1847–4–23 From John P. Gassiot Clapham Common
Will forward JH's note. Is pleased he will be joining the new club. Wishes him happiness in his retirement.
ALS 2pp RS:HS 8.57

6963. 1847–4–26 From William Rutter Dawes Camden Lodge
Returning the sheets of JH's *Cape Results* together with a list of errors he has noted. Regarding further reading of star *Antares*. Thanks for the glass wedges. Hopes weather will improve for JH's journey.
ALS 3pp RS:HS 6.82

6964. [1847–4–26 or later] To William Robert Grove n.p.
Asks time and place of next meeting of Philosophical Club, having received invitation lacking that information.
ALS 1p RI

6965. 1847–4–26 From William Robert Grove 4 Hare Court Temple
Announces next meeting of Philosophical Club.
ALS 1p RI

6966. 1847–4–26 From William Rowan Hamilton Observatory
Sends some new theorems concerning undisturbed parabolic motion; believes that much remains to be discovered in this field.
L Graves, *Hamilton, 2,* 566–70 inc

6967. 1847–4–26 From Edward Craven Hawtrey Eton College
Would like some suggestions for books to be awarded for the Maths prize at Eton.
ALS 4pp RS:HS 10.173

6968. 1847–4–27 From William Robert Grove 4 Hare Court
Has received JH's letter from Gassiot and submitted it to the committee dealing with members of the Philosophical Club.
ALS 1p RS:HS 9.148

6969. 1847–4–28 From William Smith Williams 65 Cornhill
Drawings of comets and sun spots are finished and ready for JH's inspection. Name and address of Mr. Ford, lithographer.
ALS 1p RAS:JH Archive 10/6.11 verso; Reel 9

6970. 1847–4–30 From James Clark Ross Aston House, Aylesbury
Wants JH to name the circle of uniform temperature of the ocean near the equator. Explains the circle of uniform temperature.
ALS 4pp RS:HS 14.429

6971. 1847–[5] To John S. Forbes n.p.
Comments on the paper he has sent him. Finds it difficult to give an opinion as so much verification is required.
C 3pp RS:HS 7.322 (C: RS:HS 22.315)

6972. [1847–5] From Richard Sheepshanks n.p.
Arranges concurrent meetings of Standards Committee and Visitation [to Greenwich observatory]. Possible explanation for notations in [Thomas Henderson's] 'tables at June 11.' Please review and edit 'Preface to Lalande.' Will write to [Elizabeth] Baily.
ALS 3pp TxU:H/M–0531; Reel 1093

6973. 1847–5–3 From William Samuel Stratford *Nautical Almanac* Office
Asks if the recent discussion of what name, 'Leverrier' or 'Neptune,' should be used for the new planet changed JH's opinion on whether the name used in the *Nautical Almanac* for JH's father's planet should be 'Georgian Planet' or 'Uranus.'
ALS 2pp TxU:H/M–0546.4; Reel 1093

6974. 1847–5–4 From Dr. Groskopf Hanover
Is searching for a Robert Georg Twaddel; asks JH whether he knows anything of him.
ALS 1p (German) RS:HS 9.147

6975. 1847–5–4 To William Samuel Stratford 37 Tavistock Place
Approves naming the new planet Neptune and continued designation of JH's father's planet as Uranus. Suggests reasons and alternatives.
ALS 2pp RS:HS 17.101 (C: RS:HS 22.316)

6976. 1847–5–4 From W. Witte Hanover
About the lunar model made by WW, and being considered for purchase [see JH's 1846–2–20].
ALS 4pp (German) RS:HS 18.431

6977. 1847–5–5 To James Clark Ross 37 Tavistock Place
Speculates on name for JR's 'hot and cold basins' and on causes of equatorial heat and polar cold.
ALS 4pp RS Sa.654

6978. 1847–5–5 From Richard Sheepshanks n.p.
Encloses letter from W. S. Stratford at R.A.S. How RS plans to distribute works on J. J. L. Lalande and N. L. Lacaille left unfinished at death of Francis Baily.
ALS 3pp, encl ALS 1p TxU:H/M–0944; Reel 1083

6979. 1847–5–5 From William Samuel Stratford *Nautical Almanac* Office
Looking forward to learning JH's opinion of Richard Sheepshanks.
ALS 2pp TxU:H/M–0546.5; Reel 1093

6980. 1847–5–6 To Richard Sheepshanks 37 Tavistock Place
On a puzzling aspect of [Thomas] Henderson's papers. One column in his tables seems to have been altered in someone else's handwriting. Wants to know the origin of these corrections, and how Henderson's tables ought to be printed. Offers several options.
ALS 4pp RS:HS 16.64 (C: RS:HS 25.9.39)

6981. [1847–5–6 or earlier] To Richard Sheepshanks [37 Tavistock Place]
Has the papers by [Thomas] Henderson but they do not contain much information.
ALS 2pp RS:HS 16.63 (C: RS:HS 25.9.38)

6982. 1847–5–7 From Heinrich Christian Schumacher Altona
[H. J.] Anderson of New York wishes to be introduced to JH. Number 600 of *Astronomische Nachrichten* will finish the 25th volume of the journal. Wilhelm Struve will be present.
ALS 3pp RS:HS 15.398

6983. [1847]–5–7 From Richard Sheepshanks Reading
Cannot explain how Thomas Henderson arrived at special

figures in N. L. Lacaille's star catalog. Will examine Lacaille's volumes tomorrow.
ALS 3pp TxU:H/M–0529.2; Reel 1093

6984. 1847–5–7 From William Samuel Stratford Notting Hill Square
Replies to some of the queries from JH's 1847–5–6 letter.
ALS 2pp TxU:H/M–0546.8; Reel 1093

6985. 1847–5–7 From William Samuel Stratford *Nautical Almanac* Office
Discusses the printing of J. J. L. Lalande's *Catalogue of Those Stars in the Histoire céleste française....*
ALS 3pp TxU:H/M–0546.7; Reel 1093

6986. 1847–5–7 From William Samuel Stratford *Nautical Almanac* Office
Arranges a meeting the next morning with JH.
ALS 1p TxU:H/M–0546.6; Reel 1093

6987. 1847–5–8 To George Biddell Airy 37 Tavistock Place
Believes he has worked out the German notation of star positions [see GA's 1847–2–1].
ALS 3pp RGO 6.203.349

6988. 1847–5–[8?] To Richard Sheepshanks 37 Tavistock Place
Has figured out why [Thomas] Henderson's tables [see JH's 1847–5–6] appear as they do.
ALS 3pp RS:HS 16.65 (C: RS:HS 25.9.40)

6989. 1847–5–8 From William Samuel Stratford *Nautical Almanac* Office
Reports that the RAS has an available balance of £77. He calculates that to complete the printing of J. J. L. Lalande's *Catalogue of Those Stars in the Histoire céleste française...* and N. L. Lacaille's *Catalogue of 9766 Stars in the Southern Hemisphere,* however, will require £83 pounds; WS hopes to cut costs so that they do not overspend their grant.
ALS 4pp TxU:H/M–0546.9; Reel 1093

6990. 1847–5–8 From Alexander Wallace Edinburgh
Can offer no explanation of how Thomas Henderson computed tables from data in N. L. Lacaille's star charts. Henderson did all his work at home. Suggests explanations for Henderson's corrections in red ink.
ALS 6pp TxU:H/M–0558; Reel 1093

6991. 1847–5–9 To Charles Babbage Collingwood
Asks CB's advice on how correctly to avoid attending a Queen's ball JH has been commanded to attend.
ALS 2pp RS:HS 2.314

6992. 1847–5–10 From George Biddell Airy Royal Observatory, Greenwich
Concerning Thomas Henderson's reductions of N. L. Lacaille's observations.
ALS 2pp RS:HS 1.181 (C: RGO 6.203.351)

6993. 1847–5–10 To George Biddell Airy n.p.
Uses GA's comments [see GA's 1847–5–10] finally to achieve clarification of Thomas Henderson's calculations.
ALS 3pp RGO 6.203.352

6994. 1847–5–11 From William Samuel Stratford Notting Hill Square
Sending to JH some information about 'L's catalogue' [either N. L. Lacaille's *Catalogue of 9766 Stars in the Southern Hemisphere* or J. J. L. Lalande's *Catalogue of Those Stars in the Histoire céleste française....*]
ALS 1p TxU:H/M–0546.10; Reel 1093

6995. 1847–5–12 From William Samuel Stratford Notting Hill Square
Although WS is suffering from gout, he promises to work on the layout of J. J. L. Lalande's *Catalogue of Those Stars in the Histoire céleste française....*
ALS 1p TxU:H/M–0546.11; Reel 1093

6996. 1847–5–13 From Richard Sheepshanks Reading
Congratulates JH on finding solution [to Thomas Henderson's notations]. Sends extract of 11 Mar. 1841 letter from Henderson, stating that Henderson used computations in Francis Baily's edition of N. L. Lacaille's catalog. RS's notes on observation and reductions of J. J. L. Lalande. Sent skeleton of RS's 'Preface' to W. S. Stratford.
ALS 5pp TxU:H/M–0532; Reel 1093

6997. 1847–5–13 From William Samuel Stratford Notting Hill Square
Discusses the layout of J. J. L. Lalande's *Catalogue of Those Stars in the Histoire céleste française....*
ALS 2pp TxU:H/M–0546.12; Reel 1093

6998. 1847–5–15 From George Biddell Airy Royal Observatory, Greenwich
Asks JH for measures of several double stars.
ACCS 1p RGO 6.193.242

6999. 1847–5–15 From George Biddell Airy Royal Observatory, Greenwich
About a time conflict involving the R.S.L. and Visitation Day at the Royal Observatory.
ACCS 2pp RGO 6.9.574

7000. 1847–5–15 From George Biddell Airy Royal Observatory, Greenwich
A note enclosed with the printed observations made with the Northumberland equatorial telescope.
ACCS 1p RGO 6.158.369

7001. 1847–5–15 From Luke Howard Tottenham
Has directed the printers, Taylors, to send JH's booksellers the parts of the *Barometrographia* as issued, free of all costs. The science of meteorology attracts little attention at the moment.
ALS 4pp RS:HS 10.8

7002. 1847–5–16 From Richard Sheepshanks n.p.
Believes that RS's suggested correction, '*when translated into its equivalent value in terms of the inclination,*' will provide sought-for values.
ALS 1p TxU:H/M–0533; Reel 1093

7003. [1847–5–17] To Algernon Percy (4th Duke of Northumberland) n.p.
JH's *Cape Results* is ready for publication. Gratitude for support of late Hugh Percy. Regrets HP did not live to see completion of *Cape Results*. Trusts that AP will continue support. Asks for interview. Encloses correspondence between HP and JH to show how arrangement originated.
AdraftS 2pp RAS:JH Archive 10/4.1; Reel 9

7004. 1847–5–18 From Algernon Percy (4th Duke of Northumberland) Northumberland House
Suggests JH visit tomorrow at noon.
ALS 1p RAS:JH Archive 10/4.2; Reel 9

7005. 1847–5–18 From William Samuel Stratford Notting Hill Square

Reports that the printing of J. J. L. Lalande's *Catalogue of Those Stars in the Histoire céleste française...* and N. L. Lacaille's *Catalogue of 9766 Stars in the Southern Hemisphere* (which includes a preface by JH) is 'progressing satisfactorily.'
ALS 2pp TxU:H/M–0546.13; Reel 1093

7006. 1847–5–19 From George Biddell Airy Royal Observatory, Greenwich
A notice of meeting of the Standards Commission.
LS 1p TxU:H/M-0633.1; Reel 1084

7007. 1847–5–19 To George Biddell Airy London
Further to resolving the time conflict [see GA's 1847–5–15] and sending out invitations for Visitation Day.
ALS 2pp RGO 6.9.578

7008. 1847–5–20 From William Samuel Stratford Notting Hill Square
Sends a proof of the preface that JH wrote for N. L. Lacaille's *Catalogue of 9766 Stars in the Southern Hemisphere*; WS also gave a proof to G. B. Airy and to Richard Sheepshanks and requested that they comment on it.
ALS 2pp TxU:H/M–0546.15; Reel 1093

7009. 1847–5–20 From William Samuel Stratford Notting Hill Square
The alterations to N. L. Lacaille's *Catalogue of 9766 Stars in the Southern Hemisphere* are nearly finished.
ALS 1p TxU:H/M–0546.14; Reel 1093

7010. 1847–5–21 From William Samuel Stratford Notting Hill Square
Is concerned about the location of packages relating to the new edition of N. L. Lacaille's *Catalogue of 9766 Stars in the Southern Hemisphere.*
ALS 2pp TxU:H/M–0546.17; Reel 1093

7011. 1847–5–21 From William Samuel Stratford Notting Hill Square
Wants Thomas Henderson's ambiguous reductions mentioned in N. L. Lacaille's *Catalogue of 9766 Stars in the Southern Hemisphere.*
ALS 2pp TxU:H/M–0546.16; Reel 1093

7012. 1847–5–21 From William Samuel Stratford Notting Hill Square
WS has given instructions for forwarding JH's letters to Laplace [?] Lodge on Sunday and to Alton afterwards.
ALS 1p TxU:H/M–0546.18; Reel 1093

7013. 1847–5–22 From Richard Sheepshanks Reading
Dispute with W. S. Stratford over proof sheets. Efforts to appease [T. R.] Robinson. RS's progress on works of N. L. Lacaille and J. J. L. Lalande.
ALS 4pp TxU:H/M–0534; Reel 1093

7014. 1847–5–22 To Richard Sheepshanks Taplow Lodge near Maidenhead
Sends RS his address so that RS or G. B. Airy can send proofs.
ALS 3pp RS:HS 16.66 (C: RS:HS 25.9.41)

7015. 1847–5–23 To George Biddell Airy n.p.
Will send measurements of the requested stars [see GA's 1847–5–15] as soon as JH can consult the appropriate papers.
LCC 1p inc RGO 6.193.243

7016. 1847–5–23 From Richard Sheepshanks n.p.
RS will disclaim editorial responsibility for 'Preface' to J. J. L. Lalande's work if W. S. Stratford modifies it substantially. G. B. Airy is upset by omissions in 'Preface.' Usefulness of J. B. J. Delambre's work in understanding observations of N. L. Lacaille.
ALS 4pp TxU:H/M–0535; Reel 1093

7017. 1847–5–23 From William Samuel Stratford Notting Hill Square
Discusses how the late Francis Baily should be credited in J. J. L. Lalande's *Catalogue of Those Stars in the Histoire céleste française....*
ALS 3pp TxU:H/M–0546.19; Reel 1093

7018. 1847–5–24 or earlier To George Biddell Airy Taplow Lodge near Maidenhead
Gives GA an address for the next while, and tells GA what JH has done about invitations for Visitation Day [see JH's 1847–5–19].
ALS 4pp RGO 6.9.594

7019. 1847–5–24 From Julia Margaret Cameron [Calcutta], India
Has been unable to write previously owing to an attack of fever. Pleasant recollections of life at the Cape. Is sorry to hear the news of JH's son William. News of Hardinge [J. C.'s child].
ALS 8pp RS:HS 5.150

7020. 1847–5–24 From William Samuel Stratford Notting Hill Square
Asks JH to retain all proofs of the preface that JH wrote for N. L. Lacaille's *Catalogue of 9766 Stars in the Southern Hemisphere.*
ALS 3pp TxU:H/M–0546.20; Reel 1093

7021. 1847–5–25 From Richard Sheepshanks n.p.
Aggravated by W. S. Stratford's refusal to send proofs [of 'Preface' to J. J. L. Lalande's work] to RS.
ALS 1p TxU:H/M–0536; Reel 1093

7022. 1847–5–26 From Richard Sheepshanks n.p.
Repeats yesterday's protests against W. S. Stratford's refusal to send revised proofs to RS. Hopes G. B. Airy will not blame JH for delay.
ALS 4pp TxU:H/M–0537; Reel 1093

7023. 1847–5–26 From William Samuel Stratford Notting Hill Square
Discusses a possible measurement error in N. L. Lacaille's *Catalogue of 9766 Stars in the Southern Hemisphere.* Richard Sheepshanks is certain that WS has the proofs of JH's preface to Lacaille's book, but an annoyed WS contends that they are with JH.
ALS 4pp TxU:H/M–0546.21; Reel 1093

7024. 1847–5–26 To Richard Sheepshanks Anstey Cottage near Alton
Answers questions raised by RS concerning the preface JH was writing for N. L. Lacaille's *Catalogue of 9766 Stars.* Objects to JH being seen as responsible for the contents of the book itself, even though he served on the first two B.A.A.S. committees dealing with the reduction of Lacaille's stars.
ALS 4pp RS:HS 16.67 (C: RS:HS 25.9.42)

7025. 1847–5–27 From William Samuel Stratford Notting Hill Square
Received G. B. Airy's proof of JH's preface of N. L. Lacaille's *Catalogue of 9766 Stars in the Southern Hemisphere* from JH, but still awaits his, JH's, and Richard Sheepshanks's copies. Completing the index and the preface to J. J. L. Lalande's *Catalogue of Those Stars in the Histoire*

céleste française....
ALS 2pp TxU:H/M–0546.22; Reel 1093

7026. 1847–5–28 From Anne Elise Knipping Hanover
Thanks JH for birthday wishes for JH's aunt Caroline Herschel. Goes on to tell JH about Caroline's health, both physical and mental, and about the birthday celebration for Caroline.
AL 4pp inc (German) RS:HS 11.60

7027. 1847–5–28 From Heinrich Christian Schumacher Altona
We will drink to your health on June 9, the day on which #600 of the *Astronomische Nachrichten* will appear.
ALS 1p RS:HS 15.399

7028. 1847–5–28 From William Samuel Stratford Notting Hill Square
Elated to learn from JH that JH has the remaining proofs of JH's preface to N. L. Lacaille's *Catalogue of 9766 Stars in the Southern Hemisphere,* as WS had believed; he also is thrilled that Richard Sheepshanks, who asserted that WS held the proofs, will now have to eat his words!
ALS 4pp TxU:H/M–0546.23; Reel 1093

7029. 1847–5–29 From William Samuel Stratford Notting Hill Square
Has supplied Richard Sheepshanks with revisions of J. J. L. Lalande's *Catalogue of Those Stars in the Histoire céleste française....* Because Sheepshanks previously accused WS of misplacing the proofs of JH's preface to N. L. Lacaille's *Catalogue of 9766 Stars in the Southern Hemisphere* (which later were found to be where WS had asserted), WS quips that he 'returned every scrap of paper' to Sheepshanks.
ALS 1p TxU:H/M–0546.24; Reel 1093

7030. 1847–5–30 From Thomas Maclear Bushman Flat Sector Sta.
Has just read W. R. Dawes's description of the observatory at Camden Lodge and the telescope by Georg Merz makes TM envious. Hopes JH will press claims for one for the Cape. Is at present engaged sinking the Sector for shelter. TM has suffered much from the effects of the heat. Dr. Gall has died, aged 98. What a drubbing G. B. Airy has given R. H. Inglis.
ALS 4pp RS:HS 12.137

7031. 1847–5–31 From William Samuel Stratford Notting Hill Square
Discusses the layout of J. J. L. Lalande's *Catalogue of Those Stars in the Histoire céleste française....*
ALS 3pp TxU:H/M–0546.26; Reel 1093

7032. 1847–5–31 From William Samuel Stratford Notting Hill Square
Busily preparing sections of J. J. L. Lalande's *Catalogue of Those Stars in the Histoire céleste française...* and N. L. Lacaille's *Catalogue of 9766 Stars in the Southern Hemisphere* for the printer.
ALS 1p TxU:H/M–0546.25; Reel 1093

7033. [1847–6 or earlier] From Benjamin Morgan Cowie Putney
[Form letter] Invitation to attend annual awards ceremony at College for Civil Engineers.
LS 1p † TxU:H/M–0152.1; Reel 1087

7034. 1847–6–1 To John William Lubbock Collingwood
Comments on JL's improvements to planetary theory.
C 2pp RS:HS 22.317

7035. 1847–6–4 To Algernon Percy (4th Duke of Northumberland) Collingwood
JH's servant died in accident yesterday. Encloses papers addressed to AP but intended for [Count Jarnac?]. Suggests how to obtain help from [H. J. Temple] Lord Palmerston without appearing to violate rules. [G. H. Gordon] Lord Aberdeen shows no interest. Purpose of JH's 'memorial' to Lord John Russell was to obtain pension for Thomas Maclear.
Adraft 2pp RAS:JH Archive 10/4.3; Reel 9

7036. 1847–6–6 From William Samuel Stratford Notting Hill Square
Asks JH to compare the stars in Thomas Henderson's reductions to those in N. L. Lacaille's *Coelum Australe Stelliferum.*
ALS 3pp TxU:H/M–0546.27; Reel 1093

7037. 1847–6–8 To Charles Babbage Collingwood
Regarding the physical features of the moon.
ALS 2pp RS:HS 2.315

7038. 1847–6–8 To Spencer J. A. Compton Collingwood
JH's opinions regarding agreement between R.S.L. and Hudson Bay Company for survey of northern Canada. Transportation expenses should be paid by Company, officer's salary by Ordnance department, but R.S.L. may choose to bear costs in order to avoid dishonoring science. Reviews magnetic data obtained by survey.
C 6pp †(by JH) TxU:H/L–0286; Reel 1054

7039. [1847–6–8?] To Algernon Percy (4th Duke of Northumberland) [Collingwood]
Encloses proof of Dedication to be inserted in presentation copies [of JH's *Cape Results*]. Book will be ready soon. Forwarding copies would result in great expense and half would be lost. Let recipients be notified and send for their copies. Sends three lists of individuals and institutions to be so favored. Expects unavoidable omissions.
ALS 2pp RAS:JH Archive 10/4.4; Reel 9

7040. 1847–6–8 From Sir Edward Ryan Garden Lodge Kensington
Sorry he did not see JH in London. Hopes JH will support John Lefevre for a position at the University.
ALS 3pp RS:HS 14.456

7041. 1847–6–9 To [Josiah Forshall?] 24 Berkeley Square
Reports on the changes to be made to the British Museum heating system to improve heating and ventilation in the manuscripts department.
C 4pp inc RS:HS B27.62 (draft: RS:HS B27.64 same topic but significantly different)

7042. 1847–6–9 From George Woolcott Kent Ophthalmic Hospital, Maidstone
Requests permission to include JH on list of '[Five?]' presidents of hospital.
ALS 1p RAS:JH Archive 10/4.14.2; Reel 9

7043. 1847–6–10 From Charles Babbage Dorset St.
How to evade attending the Queen's ball. Thanks for information on the moon.
ALS 3pp RS:HS 2.316

7044. 1847–6–10 From Algernon Percy (4th Duke of Northumberland) Whitehall Gardens
Received JH's lists of suggested recipients [of presentation copies of JH's *Cape Results*]. Wishes to discuss them with

JH.
ALS 2pp RAS:JH Archive 10/4.5; Reel 9

7045. 1847–6–11 From George Biddell Airy Royal Observatory, Greenwich
Regarding the final payments for the Liverpool Telescope.
ALS 2pp RS:HS 1.182 (C: RGO 6.155.364)

7046. 1847–6–11 From Heinrich Christian Schumacher Altona
[In explanation of HS's 1847–5–28,] reminds JH of all he wrote for *Astronomische Nachrichten* previously. Cannot attend B.A.A.S. meeting because of daughter's wedding, but would have declined anyway because of foreseen arguments over the merits of J. C. Adams and U. J. J. Leverrier. Thanks JH for standing invitation to visit England.
ALS 3 pp RS:HS 15.400

7047. 1847–6–12 From Algernon Percy (4th Duke of Northumberland) Whitehall Gardens
James Hayday is recommended as best bookbinder [for *Cape Results*]. Explanation of how to bind copies for royal patrons. Encloses list of eleven royal recipients and Hayday's address. What shall AP do with JH's correspondence with late Hugh Percy?
ALS 2pp RAS:JH Archive 10/4.6; Reel 9

7048. 1847–6–12 From W. W. Rundell Falmouth
Sends ideas about southern deviation of bodies falling from the perpendicular. Desires corrections. Is investigating methods of launching bodies so they fall without deflection.
ALS 3pp RS:HS 14.433

7049. 1847–6–13 To Algernon Percy (4th Duke of Northumberland) Collingwood
Will send 13 copies [of JH's *Cape Results*] for royal recipients to bookbinder James Hayday. Appropriate to bind all other presentation copies 'in Boards.' Please return letters from late Hugh Percy. Return incomplete copy of book to Smith, Elder & Co. JH suggests changes in epigraph and preface to express better HP's and AP's contribution to *Cape Results*.
AdraftS 2pp RAS:JH Archive 10/4.8; Reel 9

7050. 1847–6–14 To George Biddell Airy Collingwood
Provides the measurements for Beta Aquarius and Alpha 2 Capricorni [see JH's 1847–5–23].
ALS 2pp RGO 6.193.244

7051. 1847–6–14 To George Biddell Airy n.p.
Agrees with GA [see GA's 1847–6–11] about paying up accounts.
AL 1p RGO 6.155.366

7052. 1847–6–14 To Thomas Collins Collingwood
Refuses permission for JH's portrait to appear in Collins's *Portraits & Memoirs of Eminent Conservatives* (1846). JH has never identified publicly with any political party.
ACCS 1p TxU:H/L-0130; Reel 1054

7053. 1847–6–14 From Edward Sabine Woolwich
Should ES send thermometers to observatories? JH and [Francis] Beaufort to consider debt owed to Hudson Bay Company. Sends letter describing photometer of [John] Ball for JH's opinion.
ALS 4pp RS:HS 15.205

7054. 1847–6–14 From William Samuel Stratford Notting Hill Square
Promises to send to JH copies of J. J. L. Lalande's *Catalogue of Those Stars in the Histoire céleste française...*, and of N. L. Lacaille's *Catalogue of 9766 Stars in the Southern Hemisphere* by week's end. WS has tried to connect Francis Baily with JL's *Catalogue* in a 'formal' manner.
ALS 3pp TxU:H/M–0546.28; Reel 1093

7055. [1847]–6–15 From Thomas P. Beckwith Christ Church, Oxford
Offers his rooms to JH when he attends the B.A.A.S. meeting.
ALS 3pp RS:HS 5.35

7056. 1847–6–16 From Augustus De Morgan 7 Camden St.
Does he know anything about E. J. Lowe's book [*Treatise on Atmospheric Phenomena*, 1846] on atmospheric phenomena? Looks good but cannot trust his own judgement. Could not dine with him as he had another appointment.
ALS 2pp RS:HS 6.236

7057. 1847–6–16 From Sir William J. Hooker Royal Botanic Gardens, Kew
The Professor of Botany at the Royal Dublin Society has died. Would like to propose their friend W. H. Harvey for the position. Would JH write a letter of commendation to the Society?
ALS 4pp RS:HS 9.458

7058. 1847–6–17 To Sir William J. Hooker Collingwood
[Responding to WH's 1847–6–16], JH doubts that JH's endorsement of [William H.] Harvey's candidacy for professorship of botany would have much credibility. Does hold Harvey in high esteem.
ALS 3pp Royal Botanic Gardens

7059. 1847–6–17 To W. W. Rundell Collingwood
Has examined southern deviation of falling bodies. His results differ greatly from WR's. Presents formula and asks WR to reconsider. Will not present these ideas to the B.A.A.S.
(S) copy 4pp RS:HS 14.434

7060. 1847–6–17 From William Samuel Stratford Notting Hill Square
Exultant that J. J. L. Lalande's *Catalogue of Those Stars in the Histoire céleste française...* and N. L. Lacaille's *Catalogue of 9766 Stars in the Southern Hemisphere* will be published tomorrow. Asks whether it would be appropriate for the B.A.A.S. to present copies of these two catalogs to J. C. Adams and to U. J. J. Leverrier; discusses other people who deserve copies. In printing and publishing the catalogs, they have spent all of the grant money.
ALS 4pp † TxU:H/M–0546.29; Reel 1093

7061. [1847]–6–18 From Thomas P. Beckwith Eaton Place
Has no one else to offer his rooms so they are at JH's disposal. Has notified Mr. Hill that JH will be requiring them.
ALS 2pp RS:HS 5.36

7062. 1847–6–18 From Algernon Percy (4th Duke of Northumberland) Whitehall Gardens
AP's name should not appear in dedication of JH's *Cape Results*. Consulted friend; awaiting reply. But will accept whatever JH wishes.
ALS 3pp RAS:JH Archive 10/4.9; Reel 9

7063. 1847–6–18 From William Samuel Stratford Notting Hill Square
Now has in hand published copies of J. J. L. Lalande's *Catalogue of Those Stars in the Histoire céleste française...*

and N. L. Lacaille's *Catalogue of 9766 Stars in the Southern Hemisphere*.
ALS 1p TxU:H/M–0546.31; Reel 1093

7064. 1847–6–18 From William Samuel Stratford Notting Hill Square
Sends the revised financial report that records the final printing and publication costs of J. J. L. Lalande's *Catalogue of Those Stars in the Histoire céleste française...* and N. L. Lacaille's *Catalogue of 9766 Stars in the Southern Hemisphere*.
ALS 1p TxU:H/M–0546.30; Reel 1093

7065. 1847–6–19 From William Radcliffe Birt Bethnal Green
Does JH think it worthwhile to extend his atmospheric wave investigation? Will be at the B.A.A.S. meeting at Oxford.
ALS 3pp RS:HS 4.116

7066. 1847–6–19 To Algernon Percy (4th Duke of Northumberland) Collingwood
Considers it proper that AP's name be associated with [*Cape Results*]. Explains importance of public support of science by prominent persons.
Adraft 2pp RAS:JH Archive 10/4.10; Reel 9

7067. 1847–6–20 From Joanna Baillie Hampstead
Regarding his observations of the heavens and its relation to theological matters.
ALS 3pp RS:HS 3.22

7068. 1847–6–20 To [William Radcliffe Birt] Collingwood
Encourages barometer observations in Mediterranean and Atlantic. Admiralty should support B.A.A.S. request for shipboard observations. Will probably not see WB in Oxford.
ALS 4pp RS:HS 19.125

7069. 1847–6–20 From Hans Christian Oersted Copenhagen
Sends two of his papers. R. I. Murchison has now received the Order of Dannebrogg. If JH thinks he can utilize the order once conferred on him he should write to the King. P.S. King is now dead, but could pass the letter to the appropriate body.
ALS 2pp RS:HS 13.170

7070. 1847–6–21 From Algernon Percy (4th Duke of Northumberland) Whitehall Gardens
Satisfied with JH's decision [to include AP's name in *Cape Results*].
ALS 2pp RAS:JH Archive 10/4.11; Reel 9

7071. 1847–6–21 From W. W. Rundell Falmouth
Has re-examined work on southern deviation. Still cannot see his mistake. Re-presents his ideas for further comments.
ALS 3pp RS:HS 14.435

7072. 1847–6–22 To Joanna Baillie Collingwood
Surprised that JB sees theological doctrines in JH's astronomical observations.
C 2pp RS:HS 22.321

7073. 1847–6–22 To James David Forbes Collingwood
Asks for details of several actinometers.
C 1p RS:HS 22.320

7074. 1847–6–22 To W. Herschel Griesbach London
Sends £5 note. This is all he can spare. Has no recollection of ever meeting him and their relationship is very remote. Wishes a speedy improvement in his prospects.
draft 1p RS:HS 9.140a

7075. 1847–6–22 From Charles Richard Weld R.S.L., Somerset House
Sends paper for JH to read. Should it be published in *R.S.P.T.*?
ALS 1p TxU:H/M–0562.1; Reel 1093

7076. [1847]–6–25 From Joanna Baillie Hampstead
Thanking him for his letter in answer to hers on theological doctrine and its relation to JH's observations.
ALS 3pp RS:HS 3.25

7077. 1847–6–25 To John William Lubbock Collingwood
JH has arranged for U. J. J. Leverrier, J. C. Adams, and Wilhelm Struve to visit JH together at Collingwood. JH is writing to invite JL to join them.
ALS 2pp Oxford History of Science Museum 58/2

7078. 1847–6–25 To [R.S.L. Council] Collingwood
Reports on and recommends for publication experiments in paper [see *RSPT, 137* (1847), 253–] by [A. F. J.] Claudet on effect of solar radiation on silver plate's mercury deposit, but questions conclusion about triplicity of photographic action.
ALS 4pp RR.1.42

7079. 1847–6–26 From C. A. von Steinheil Munich
Tells JH of [P. L. von] Seidel's success in comparing luminosity of stars by means of a photometer. The results match those of JH. Sends JH a copy requesting that the form be ignored and it be communicated to the R.A.S.
ALS 1p (German) RS:HS 17.8

7080. 1847–6–26 From C. A. von Steinheil Ostend
Tells JH that his invitation to visit came too late. Regrets deeply not having the opportunity to meet, but expects to do so next year. Hopes to present two works to JH at that time. Mentions working with the talented P. L. Seidel, his assistant.
ALS 2pp (German) RS:HS 17.7

7081. 1847–6–28 From George Peacock Oxford
A resolution passed by members of the B.A.A.S. meeting at Oxford urges the publication of William Herschel's papers in one volume.
ALS 4pp RS:HS 13.302

7082. 1847–6–28 To Richard Sheepshanks Collingwood
Asks RS to check the background of Samuel B. Lamb, who wants to lease JH's house at Slough.
ALS 2pp RS:HS 16.68 (C: RS:HS 25.9.43)

7083. 1847–6–29 From Augustus De Morgan 7 Camden St.
Will send his man to C. R. Weld. F. W. Bessel will be in good time for council. Is glad he is on the last chapter.
ALS 3pp RS:HS 6.237

7084. 1847–6–29 From Richard Sheepshanks Reading
Responds to JH's request [1847–6–28] that RS inquire about Samuel B. Lamb. Has visited [W. R.] Dawes and his telescope; thinks he will be productive. Lefevre [George Shaw-Lefevre?] has started for the university, adding 'what chance he has of success I can't guess.' On JH's contacts with Cambridge University.
ALS 4pp RS:HS 16.69

7085. 1847–6–29 To [Richard Sheepshanks] n.p.
Is still working on the obituary of F. W. Bessel; JH will not be at the R.A.S. anniversary dinner; comments on the usage

of various fixed points for producing standard thermometers.
ALS 2pp inc RGO 6.356.158

7086. 1847–6–29 From William Smith Williams 65 Cornhill
As instructed, sent 14 copies of JH's book to James Hayday, bookbinder. Confirms style of binding, lettering, and folding of plates requested by JH. One hundred copies will be finished by end of week.
ALS 1p RAS:JH Archive 10/3.1; Reel 9

7087. 1847–6–30 From James Hayday 31 Little Queen St.
Has not yet received copies [of JH's *Cape Results*] from Smith, Elder & Co. [Algernon Percy] Duke of Northumberland called to discuss binding but made no decisions. Awaiting JH's instructions by return post.
ALS 3pp RAS:JH Archive 10/3.2; Reel 9

7088. 1847–6–30 To George Peacock Collingwood
Replying to GP's 1847–6–28, is worried about the amount of time that will be required of JH in preparing a suitable biography; does not want it to interfere with other work on hand.
C 3pp RS:HS 13.303 & 22.322

7089. 1847–7 To Margaret Brodie Herschel Cambridge
JH has organized something [?] special, and he believes it will come off.
AL 1p JHS 6.40

7090. 1847–7 To Margaret Brodie Herschel Cambridge
Met U. J. J. Leverrier, F. G. W. Struve, and George Peacock at the gate of Trinity College; has arranged for Leverrier and Struve to visit Collingwood.
AL 4pp JHS 6.41

7091. 1847–7 To Margaret Brodie Herschel London
JH is apologizing for harsh words he inadvertently used, whereby he hurt MH.
AL 3pp JHS 3.15a

7092. 1847–7–1 To Algernon Percy (4th Duke of Northumberland) Collingwood
Bookbinder James Hayday assured JH that two copies [of JH's *Cape Results*] for royal recipients would be ready Monday. Proposes presenting these to [Queen Victoria] and Prince Albert during Albert's installation at Cambridge University.
ALS 2pp RAS:JH Archive 10/4.12; Reel 9

7093. [1847–7–1?] To Richard Sheepshanks [Collingwood]
Suggests RS visit Hawkhurst ca. 12 July as U. J. J. Leverrier, F. G. W. Struve, and J. C. Adams will be visiting. Suggests RS come with [W. R.] Dawes. JH's book [*Cape Results*] is nearly ready for private distribution. Asks RS that JH be allowed to tell parties, to whom a copy of the engraving of JH's portrait will be given, that it is a gift from RS, lest JH be considered immodest for giving his own 'effigy.'
ALS 4pp RS:HS 16.72 (C: RS:HS 25.9.44)

7094. 1847–7–[1?] To Richard Sheepshanks Collingwood
Invites RS to dinner in Collingwood on 10 [July], as F. G. W. Struve will be there. U. J. J. Leverrier may also come, but G. B. Airy and John Lubbock cannot attend.
ALS 2pp RS:HS 16.71 (C: RS:HS 25.9.45)

7095. 1847–7–2 From Charles T. Beke 6 St Mildred's Court
Sending a copy of his Essay on the Nile and its tributaries. Has received evidence from Africa, which corroborates his theories.
ALS 2pp RS:HS 4.30

7096. 1847–7–2 From James Hayday 31 Little Queen St.
Before JH's letter of yesterday arrived, [Algernon Percy] Duke of Northumberland returned and selected binding pattern for thirteen copies [of JH's *Cape Results*]. Two presentation copies will be ready and be sent to Peter Stewart Monday.
ALS 1p RAS:JH Archive 10/3.3; Reel 9

7097. 1847–7–2 To Adam Sedgwick Collingwood
States he must get to Cambridge to present his book [*Cape Results*] to the Queen and Prince. Seeks advice from AS on handling the matter.
ALS 2pp RS:HS 15.430 (C: RS:HS 22.323)

7098. 1847–7–4 From Algernon Percy (4th Duke of Northumberland) Albury
Was unable to confer with Chancellor about JH's proposal [1847–7–1]. Suggests arrangements JH should make before presenting copies [of JH's *Cape Results*] to Queen Victoria and Prince Albert.
ALS 4pp RAS:JH Archive 10/4.13; Reel 9

7099. 1847–7–5 From Maria Edgeworth Edgeworthstown
Sending a volume of poetry by D. P. Starkey and would like JH's views on its merits. Is recovering slowly from an illness.
ALS 4pp RS:HS 7.29

7100. 1847–7–5 From John Hymers n.p.
Has placed Mr. Yorke's room at JH's disposal; gives times of meals.
AL 1p RS:HS 10.222

7101. 1847–7–5 From Richard Sheepshanks Athenaeum
People do not consider JH vain. If JH wants to give his [portrait of JH; see JH's 1847–7–1] to people, this would not be considered vanity on his part. Gives impressions of U. J. J. Leverrier after meeting him at the Royal Observatory.
ALS 4pp RS:HS 16.70

7102. [1847–7–11] To Caroline Lucretia Herschel n.p.
JH is sending with the letter a finished printing of his *Cape Results,* which JH describes as 'the completion of my Father's work—the survey of the Nebulous Heavens.' Also mailing a copy of the *Cape Results* to the King of Hanover. Wilhelm Struve, John Couch Adams, and U. J. J. Leverrier are visiting Collingwood.
ALS 4pp TxU:H/L–0592.7; Reel 1058 & Reel 1089
Memoir ... of Caroline Herschel, 342–3

7103. 1847–7–8 From Charles Richard Weld R.S.L., Somerset House
Please send papers relating to 'the matter in agitation' between Hudson's Bay Company and R.S.L. CW will forward these to Admiral [Francis] Beaufort.
ALS 1p TxU:H/M–0563; Reel 1093

7104. 1847–7–14 To Josiah Forshall [Collingwood]
Again writes [see JH's 1847–6–9] about the heating system at the British Museum.
draft 2pp RS:HS B27.63

7105. 1847–7–14 From Charles Piazzi Smyth Hillside Crescent, Edinburgh
Asks if JH observed Gamma Scorpii as a double star; sends observations of it from W. S. Jacob in India. W. S. Jacob observes Neptune.
ALS 2pp RS:HS 16.228

7106. 1847–7–15 or later To Charles Piazzi Smyth [Collingwood]
Responds to CS's [see CS's 1847–7–14] request for observations of Gamma Scorpii as a double star. JH comments on CS having omitted J. C. Adams from the list of Neptune discoverers.
C 3pp RS:HS 25.14.30

7107. 1847–7–17 To Algernon Percy (4th Duke of Northumberland) Collingwood
[First draft] Just returned from presenting copies [of JH's *Cape Results*] to Queen and Prince Albert at Cambridge. Sends two copies to AP and 'Duchess Dowager.' Exploring ways to have copies delivered overseas by foreign ambassadors. Encloses system of graphs arranging recipients according to priority of delivery. Best way to package and address copies.
Adraft 5pp (p. 5 on George Woolcott's 1847–6–9) RAS:JH Archive 10/4.14.1; Reel 9

7108. 1847–7–18 To Algernon Percy (4th Duke of Northumberland) Collingwood
[Second draft] Just returned from presenting copies [of JH's *Cape Results*] to Queen and Prince Albert at Cambridge. Sends two copies to AP and 'Duchess Dowager.' Exploring ways to have copies delivered overseas by foreign ambassadors.
Adraft 1p inc RAS:JH Archive 10/4.15; Reel 9

7109. 1847–7–19 To Algernon Percy (4th Duke of Northumberland) Collingwood
[Third draft] Just returned from presenting copies [of JH's *Cape Results*] to Queen and Prince Albert at Cambridge. Sends two copies to AP and 'Duchess Dowager.' Exploring ways to have copies delivered overseas by foreign ambassadors. Kings of Prussia and Denmark were omitted on JH's list. Wilhelm Struve will carry copies to Altona and St. Petersburg. May JH begin distributing copies to recipients in Great Britain?
Adraft 5pp RAS:JH Archive 10/4.16; Reel 9

7110. 1847–7–21 From Algernon Percy (4th Duke of Northumberland) Whitehall Gardens
Gratitude for receipt [of JH's *Cape Results*]. To assure delivery of copies to foreign sovereigns, JH must first write letter to minister of each legation asking for royal acceptance. Lists nine ambassadors with whom AP has already communicated.
ALS 4pp RAS:JH Archive 10/4.17; Reel 9

7111. 1847–7–22 From John Wrottesley 1 Albemarle St
Invites JH to join royal commission to investigate May [1847] collapse of railway bridge over Dee River by designing experiments to test engineering limits of cast iron. Names other members. All expenses will be paid.
ALS 8pp RS:HS 18.301

7112. 1847–7–24 From Algernon Percy (4th Duke of Northumberland) Whitehall Gardens
Encloses answers from three foreign ambassadors. Hopes JH does not have trouble.
ALS 2pp RAS:JH Archive 10/4.18; Reel 9

7113. 1847–7–27 From Richard Jones Tithe Commission
Threatened at Cambridge with [C. E.?] Law and Mr. Fielding. Wishes JH would write a letter of support for [J. G. Shaw-?] Lefevre. Is short of helpers but not overworked. William Empson is anxious for a review of *Kosmos*. Goes on Thursday to vote for [Henry?]Goul[d]burn and Lefevre.
ALS 4pp RS:HS 10.380

7114. 1847–7–28 To Margaret Brodie Herschel Tavistock Place
Mostly about the health of Elizabeth Baily; JH ends his letter with a silly riddle.
ALS 4pp JHS 6.38

7115. 1847–7–28 From Algernon Percy (4th Duke of Northumberland) Northumberland House
JH must come to London to supervise distribution [of JH's *Cape Results*] to foreign sovereigns. Received more replies from ambassadors.
ALS 2pp RAS:JH Archive 10/4.19; Reel 9

7116. 1847–7–30 To Algernon Percy (4th Duke of Northumberland) London
Sends form letter to accompany AP's presentation of copies of JH's *Cape Results* to six foreign sovereigns, with six individualized paragraphs to be inserted as appropriate for each sovereign.
AdraftS 3pp RAS:JH Archive 10/6.14; Reel 9

7117. 1847–7–31 To George Bishop London
Invites GB to accept copy of JH's *Cape Results*.
ALS 1p RAS:JH Archive 10/6.5; Reel 9

7118. 1847–8 To John Hartnup Collingwood
Sending a copy of JH's *Cape Results* to Hartnup for deposit in the Liverpool Observatory.
ALS 1p British Astronomical Association

7119. 1847–[8] To Margaret Brodie Herschel Collingwood
Expresses considerable anger at Parliament having changed property laws to allow enormous speculation to occur.
AL 4pp JHS 6.33

7120. 1847–8–1 From Wilhelm Struve Greenwich
Has arrived from Paris and will continue to Hamburg. Regrets not being able to revisit Collingwood. Requests JH's work concerning the southern skies.
ALS 1p (German) RS:HS 17.137

7121. 1847–8–2 From Thomas Galloway 45 Torrington Square
Gratitude for receipt of 'splendid volume' [JH's *Cape Results*].
ALS 1p RAS:JH Archive 10/5.17; Reel 9

7122. 1847–8–2 From Charles Piazzi Smyth Hillside Crescent, Edinburgh
Thanks JH for Gamma Scorpii observations. Plans to have Edinburgh Observatory record observations of 2–3000 stars over 30 years. Edinburgh observations for 1841 published.
ALS 3pp RS:HS 16.229

7123. 1847–8–3 From Algernon Percy (4th Duke of Northumberland) Northumberland House
Copy [of JH's *Cape Results*] sent to French king. Kings of Hanover and Holland have accepted. AP today paid £1000 to Drummonds [for JH's publishing costs].
ALS 1p RAS:JH Archive 10/4.20; Reel 9

7124. 1847–8–4 From John William Lubbock n.p.
Thanks JH for the receipt of a report of JH's work.
ALS 1p DMC.695A.885.36

7125. 1847–8–4 From Robert Main Greenwich
Received JH's *Cape Results*.
ALS 1p RAS:JH Archive 10/5.18; Reel 9

7126. 1847–8–5 From Charles Manley 25 Great George St.
[Form letter] Institution of Civil Engineers acknowledges receipt of JH's *Cape Results*.
(A)LS 1p RAS:JH Archive 10/5.20; Reel 9

7127. 1847–8–5 From Ralph Tatham St. John's College, Cambridge
Received personal copy of JH's *Cape Results*, and three for College Library, J. C. Adams, and Cambridge Philosophical Society. Please advise on propriety of sending letter of thanks to Duke of Northumberland [Algernon Percy]. [9 Aug.postscript:] Visited [T. W.] Hornbuckle last week.
ALS 3pp RAS:JH Archive 10/5.19; Reel 9

7128. 1847–8–6 From William Lassell Starfield, Liverpool
Note to accompany details of WL's observations of the satellite of Neptune. [Details are found in WL's letter to H. C. Schumacher of 1847–8–6.]
ALS 3pp RS:HS 11.132

7129. 1847–8–6 To Algernon Percy (4th Duke of Northumberland) [Collingwood]
Publishing expenses from Smith, Elder & Co. [for *Cape Results*] are nearly covered by AP's grant [£1000]. Bookbinder James Hayday will dispatch to Northumberland House copies for kings of Holland and Hanover. Details of printing and publishing costs.
Adraft 2pp (on Marquis Pucci's 1820–8–10) RAS:JH Archive 10/4.21; Reel 9

7130. 1847–8–6 From William Lassell to H. C. Schumacher Starfield, Liverpool
Details of steps in his discovery of the satellite of Neptune. Gives details of the telescope he is using.
C 7pp RS:HS 11.133

7131. 1847–8–7 From John Lee Hartwell House
Gratitude for *Cape Results*. Placed it next to account of the Northumberland Equatorial and Dome.
ALS 2pp RAS:JH Archive 10/5.21; Reel 9

7132. 1847–8–8 From William Empson Craig Crook, Edinburgh
Is grateful to JH for undertaking to review Alexander von Humboldt's *Kosmos*. Regarding copyright and printing of this book.
ALS 4pp RS:HS 7.43

7133. 1847–8–8 From Algernon Percy (4th Duke of Northumberland) Stanwick, Darlington
Sent copies [of JH's *Cape Results*] to royal sovereigns in seven nations. Encloses letter from Russian ambassador, F. I. Brunnov. Agrees with JH about 'large Commission' on account for publishers Smith, Elder & Co. AP owes letter to former U.S. minister Edward Everett in Boston; will send [*Cape Results*] with it.
ALS 3pp † RAS:JH Archive 10/4.22; Reel 9

7134. 1847–8–8 From William Henry Smyth 3 Cheyne Walk, Chelsea
Gratitude for [*Cape Results*]. Feels JH 'let off [James] Dunlop in a softer stile than he deserves.'
ALS 2pp RAS:JH Archive 10/5.22; Reel 9

7135. 1847–8–8 To Mary Somerville Collingwood
Has noticed a resemblance between [Alexander von Humboldt's] *Kosmos* and MS's chapter, 'On Man,' in her manuscript. Advises her to revise. Sends results of observations at the Cape.
ALS 4pp RS:HS 16.354 (C: RS:HS 22.324)

7136. 1847–8–9 To Francis Beaufort Collingwood
[JH's abstract] Destinations for copies of JH's *Cape Results* addressed to FB.
AAbst 1p inc RAS:JH Archive 10/6.6; Reel 9

7137. 1847–8–9 From John R. Hind 3 Allsop's Terrace, New Road
Gratitude for *Cape Results*. Special comments on JH's 'method of sequences' and cometary theory. Refers to article on Halley's Comet in *Astronomische Nachrichten*. Reports observations of Saturn's satellites and orbit of Gamma Virginis.
ALS 2pp RAS:JH Archive 10/5.23; Reel 9

7138. 1847–8–10 From Spencer Joshua Alwyne Compton Piccadilly
Thanks for [*Cape Results*].
ALS 1p RAS:JH Archive 10/5.26; Reel 9

7139. 1847–8–10 From James Hayday 31 Little Queen St.
Sent eleven copies [of JH's *Cape Results*] to Northumberland House to be packed and sent to destinations chosen by [Algernon Percy,] Duke of Northumberland. Encloses JH's account.
ALS 2pp RAS:JH Archive 10/3.4; Reel 9

7140. 1847–8–10 From Charles Piazzi Smyth 3 Cheyne Walk, Chelsea
Received JH's note of 8 [Aug 1847] and went immediately to Smith, Elder & Co., but no copies [of JH's *Cape Results*] were ready. Will gladly transmit them to recipients designated by JH.
ALS 3pp RAS:JH Archive 10/5.24; Reel 9

7141. 1847–8–10 From William Henry Sykes 47 Albion St., Hyde Park
Assures JH that [*Cape Results*] will receive warm welcome in libraries of India. Suggests other recipients there. Send copies with dispatches from India House.
ALS 3pp RAS:JH Archive 10/5.25; Reel 9

7142. 1847–8–11 From Richard Jones Tithe Commission
Asked [William] Empson to write to JH about 'republication.' Thanks for [Cape *Results*]. Invites JH to visit. Glad that JH voted for [J. G. Shaw-]Lefevre [for M.P. from Cambridge].
ALS 4pp RAS:JH Archive 10/5.27; Reel 9

7143. 1847–8–11 To Algernon Percy (4th Duke of Northumberland) Collingwood
Will send Edward Everett's copy [of JH's *Cape Results*] to Northumberland House. Everett was elected principal of Harvard College but resigned. Returns letter from Russian ambassador, F. I. Brunnov. Now that AP has sent copies to foreign sovereigns, JH will send to their ministers those copies destined for [observatories and institutions in] their nations.
Adraft 3pp RAS:JH Archive 10/4.23; Reel 9

7144. 1847–8–13 From Francis Beaufort Admiralty
Will be pleased to dispose of the copies of the *Reports* when they arrive. Thanks for his own copy. Maria Edgeworth will be pleased with hers.
ALS 2pp RS:HS 3.378

7145. 1847–8–13 or earlier To Margaret Brodie Herschel London
Elizabeth Baily is not in good health; JH had his pocket

picked that day, but nothing of consequence was taken.
ALS 4pp JHS 6.37

7146. 1847–8–13 From John R. Hind 3 Allsop's Terrace, New Rd.
Has discovered another member of the group of planets between Mars and Jupiter. Gives readings. May be the lost planet of Niccolo Cacciatore.
ALS 1p RS:HS 9.336

7147. 1847–8–14 From John Couch Adams St. John's College, Cambridge
Thanks for [*Cape Results*] and copy of JH's portrait. Attributes JA's passion for astronomy to JH's influence. Recalls happy visit to Collingwood.
ALS 3pp RAS:JH Archive 10/5.28; Reel 9

7148. 1847–8–14 To Margaret Brodie Herschel London
Has been to Cambridge to support a friend in a vote; JH comments on the dry state of the countryside, and responds to a number of family matters, including one in which JH is trying to get the message across that he is not interested in personal advancement.
ALS 4pp JHS 6.39

7149. 1847–8–14 To John R. Hind Collingwood
Obviously a ring of planets is being discovered. Congratulations on the discovery. Is the reading based on one night's observations only? Gives it a female name, possibly Iris.
draft 3pp RS:HS 9.337 & 22.325

7150. 1847–8–14 From William Lassell Starfield, Liverpool
Grateful for JH's Cape observations. Has nothing new to report on the ring of Neptune. Has been troubled with dew on his prism, but has now succeeded in overcoming this difficulty.
ALS 4pp RS:HS 11.134

7151. 1847–8–14 From William Smith Williams 65 Cornhill
Bookbinder is inserting epigraphs into selected copies [of JH's *Cape Results*] for distribution Monday to addresses in London. Took liberty to include four copies for members of Royal Society in Edinburgh. Waiting for JH's instructions for dispatch to Cape of Good Hope.
ALS 1p RAS:JH Archive 10/3.5; Reel 9

7152. [1847–8–14?] From John Wrottesley Birmingham Philosophical Institution.
Gratitude on behalf of Institution for receipt of [*Cape Results*].
ALS 1p RAS:JH Archive 10/5.30; Reel 9

7153. 1847–8–14 From John Wrottesley Wrottesley House
Received Cape *Results*. Compliments JH's industry.
ALS 2pp RAS:JH Archive 10/5.29; Reel 9

7154. 1847–8–15 From Edward Sabine Woolwich
Grateful for receipt of [*Cape Results*].
ALS 2pp RAS:JH Archive 10/5.31; Reel 9

7155. 1847–8–16 From William Empson Craig Crook, Edinburgh
Encloses Thomas Longman's letter. Is very pleased that JH is undertaking the great work.
ALS 4pp RS:HS 7.44

7156. 1847–8–16 From John R. Hind 3 Allsop's Terrace, New Rd.
Giving the details of his planetary discovery. Likes JH's name, but several more have been suggested. K. L. Hencke may have observed the new planet, but sees no reason why it should not have an English name.
ALS 4pp RS:HS 9.338

7157. 1847–8–16 To Phillipe Jarnac Collingwood
Asks instructions for shipping copies of JH's *Cape Results* for distribution to French institutions and individuals. Gratitude for delivering copy to French sovereign.
Adraft 1p (on W. S. Williams's 1847–4–28) RAS:JH Archive 10/6.11; Reel 9

7158. 1847–8–16 From Edward Sabine Woolwich
Sent new actinometers and thermometers to various observatories along with JH's instructions.
L inc 2pp RS:HS 15.206

7159. 1847–8–16 From Benjamin Lewis Vulliamy London
Sends Parliament papers relating to clock proposed for tower of Westminster Palace. Asks JH to comment on letters from G. B. Airy to [Stratford] Canning and to [E. J.] Dent.
ALS 1p RAS:JH Archive 14/3.7.3 verso; Reel 11

7160. 1847–8–17 From James Hudson Royal Agricultural Society
Was pleased to see one of JH's letters with his mail that morning. Recalls his first meeting with JH. Has arranged for JH's name to be added to those who receive free copies of the Journal.
ALS 4pp RS:HS 10.38

7161. 1847–8–17 From Adolphe Quetelet Brussels
Thanks JH for introducing him to Robert Kane. Hears that JH will publish Cape results. Impatient to see them. Mentions JH's philosophical work.
ALS 3pp (French) RS:HS 14.249

7162. 1847–8–18 From James Challis Cambridge Observatory
Observatory received *Cape Results*. Congratulates JH on completing project. Keeping busy with '*new Planets*.' Gives credit to J. R. Hind for diligent searching [for Iris].
ALS 3pp RAS:JH Archive 10/5.32; Reel 9

7163. 1847–8–19 From Charles Babbage Dorset St.
Remarks on JH's Southern astronomy volume [*Cape Results*].
ALS 6pp RS:HS 2.317

7164. 1847–8–19 From Temple Chevallier Esh Parsonage
Thanks from curators of Durham Observatory for *Cape Results*.
ALS 1p RAS:JH Archive 10/5.33; Reel 9

7165. 1847–8–19 From Phillipe Jarnac Tunbridge Wells
Will distribute copies [of JH's *Cape Results*] if JH will forward them to French embassy.
L 2pp RAS:JH Archive 10/5.84; Reel 9

7166. 1847–8–19 From James Cosmo Melvill East India House
Gratitude for copy of JH's *Cape Results* received by HEIC library. Will forward copies to HEIC observatories in Madras and Bombay and to library of Asiatic Society of Bengal.
ALS 1p RAS:JH Archive 10/5.98; Reel 9

7167. 1847–8–19 From Count [Frederik] Reventlow Danish Legation
Will forward copies of JH's *Cape Results* to recipients in Denmark.
L 2pp RAS:JH Archive 10/5.14; Reel 9

7168. 1847–8–20 From Count [Gerrit] Schimmelpenninck 5 Great Cumberland St.
Will forward copies of JH's *Cape Results* to Netherlands on Sunday, if received at embassy by tomorrow.
L 1p RAS:JH Archive 10/5.93; Reel 9

7169. 1847–8–20 From Richard Sheepshanks 14 London Road, Reading
Congratulations for *Cape Results*. Excellent memorial to William Herschel. [Engravings of JH's] portrait were distributed to famous astronomers in northern Europe. Fragile nature precludes shipping it to Italy, India, and America. Believes J. R. Hind and George Bishop have right to discovery of new planet [Iris].
ALS 3pp RAS:JH Archive 10/5.34; Reel 9

7170. [1847–8–20?] To William Smith Williams [Collingwood]
Lists ten recipients in England to whom JH sent circular announcing presentation copies [of JH's *Cape Results*] and portrait of JH.
Adraft 1p RAS:JH Archive 10/3.23; Reel 9

7171. 1847–8–21 From George Dollond St. Paul's Church Yard
Gratitude for receipt of JH's *Cape Results*.
ALS 1p RAS:JH Archive 10/5.35; Reel 9

7172. 1847–8–21 From Thomas Maclear Royal Observatory, Cape of Good Hope
Has had to leave his work with the Sector due to ill health. Found [George] Child overworked at the Observatory. Comments on some observations. Pleased to hear that JH's health has improved. Mr. Watermeyer has died and C. F. H. von Ludwig remarried. Gives latest news of Kaffir War. Comments on the quarrel between G. B. Airy and James South.
ALS 8pp RS:HS 12.138

7173. 1847–8–23 From Francis Beaufort n.p.
Two unsuccessful trips to Kensington to carry out JH's commission. Unable to meet 'Sir J.' Please write 'Private' on letters addressed to FB.
ALS 2pp RAS:JH Archive 10/6.7; Reel 9

7174. 1847–8–23 From Temple Chevallier Esh Parsonage
Apologizes for not noticing JH's offer 'written upon the second leaf' [of JH's *Cape Results*] to send engraving of JH's portrait. Assures JH that TC would treasure it.
ALS 2pp RAS:JH Archive 10/5.39; Reel 9

7175. 1847–8–23 From Michael Faraday Royal Institution
Gratitude for receipt of JH's *Cape Results*. Congratulates Margaret Brodie Herschel for her part in it.
ALS 1p RAS:JH Archive 10/5.37; Reel 9

7176. 1847–8–23 From C. Klingemann 4 Hobart Place, Eaton Square
Count Kielmansegg is absent. Sent copy of JH's *Cape Results* to king this week. Will distribute other copies to institutions [in Netherlands] if JH will forward these to Embassy.
ALS 3pp RAS:JH Archive 10/5.91; Reel 9

7177. 1847–8–23 From James Clark Ross Aston House [Aylesbury]
Congratulations on completion of 'Herculean labors' represented in *Cape Results*.
ALS 3pp RAS:JH Archive 10/5.38; Reel 9

7178. [1847]–8–23 From Edwin Richard Windham Wyndham-Quin Dover
Gratitude for receipt of JH's *Cape Results*.
ALS 1p RAS:JH Archive 10/5.36; Reel 9

7179. 1847–8–24 To William Rowan Hamilton Collingwood
Seeks London bookselling agent through whom to direct copies of JH's *Cape Results* for Dublin scientific groups. Discusses pretty names for new planets. Thanks for WH's memoir on quaternions.
ALS 2pp TCD 1493:406 Graves, *Hamilton, 2,* 590 inc

7180. 1847–8–24 From Henry Lawson 7 Lansdown Crescent, Bath
Gratitude for receipt of JH's *Cape Results*. Recalls JH's parents. Delighted to own portrait of JH.
ALS 1p RAS:JH Archive 10/5.40; Reel 9

7181. 1847–8–25 From John Hartnup Observatory, Liverpool
Thanks JH for having sent JH's *Cape Results* to Hartnup's Liverpool Observatory. Would appreciate receiving a picture of its author.
ALS 2pp British Astronomical Association

7182. 1847–8–25 From Mr. Van De Meyer 50 Portland Place
Sent *Cape Results* to Belgian king. Will forward other copies to recipients in Brussels. Names universities that would appreciate receiving copies.
L 2pp RAS:JH Archive 10/5.8; Reel 9

7183. 1847–8–26 From Smith, Elder & Co. 65 Cornhill
Received JH's instructions. Six copies [of JH's *Cape Results*] and 125 epigraphs were sent by rail to JH. If JH will inscribe and return epigraphs, binder can substitute these for those already bound in books.
ALS 1p RAS:JH Archive 10/3.6; Reel 9

7184. 1847–8–26 From John Webb Woollgar Lewes
Believes no representation of northern portion of Milky Way exists to match JH's plate XIII [in *Cape Results*] showing southern portion. Would JH construct one?
ALS 1p RAS:JH Archive 10/5.42; Reel 9

7185. 1847–8–27 From W. E. Maxwell South Kilworth [Leicestershire]
On behalf of uncle William Pearson, who is ill, WM thanks JH for *Cape Results*.
ALS 1p RAS:JH Archive 10/5.43; Reel 9

7186. 1847–8–27 From [Benjamin Lewis] Vulliamy London
Sends papers to JH as president of the RAS.
ALS 3pp RS:HS 17.465

7187. 1847–8–28 From John Hartnup Liverpool Observatory
Gratitude on behalf of Observatory committee for receipt of JH's *Cape Results*. Has ordered JH's portrait. Hopes to see 'our Equatorial mounted in a few weeks.'
ALS 1p RAS:JH Archive 10/5.41; Reel 9

7188. [1847?]–8–28 From Edward Sabine Woolwich
Forwarded letters to William Whewell, as JH requested. Will show copies of them to G. B. Airy. Completing volume of J. C. Ross's second crossing. Willing to take charge of W. R. Birt's models.
ALS 3pp TxU:H/M–0517.7; Reel 1093

7189. [1847–8–28] From William Smith Williams 65 Cornhill
J. P. Nichol of Glasgow purchased copy [of JH's *Cape Results*] before JH's presentation copy reached Nichol. WW found eleven more epigraphs.
ALS 1p damaged RAS:JH Archive 10/3.22; Reel 9

7190. 1847–8–29 To [Benjamin Lewis] Vulliamy n.p.
Acknowledges receiving papers [see MV's 1847–8–27].
ALS 1p RS:HS 17.465a

7191. 1847–8–30 From William Empson Craig Crook, Edinburgh
Encloses sheet from Mr. Moriarty. Assumes he is to make no more inquiries regarding the volumes of *Kosmos*. Translation keeps abreast of the original. Returns to Haileybury about the 15th.
ALS 4pp RS:HS 7.45

7192. 1847–8–30 From Smith, Elder & Co. 65 Cornhill
Received JH's instructions. Fifty cloth copies [of JH's *Cape Results*] with epigraphs inserted are ready.
ALS 1p RAS:JH Archive 10/3.7; Reel 9

7193. 1847–9–1 To George Bancroft Collingwood
Asks help in forwarding copies of JH's *Cape Results* to U.S. List of 19 recipients and addresses. Asks GB's help in correcting errors before JH inscribes copies.
AdraftS 2pp RAS:JH Archive 10/6.12; Reel 9

7194. 1847–9–1 From William Rowan Hamilton Observatory
Wants WH's quaternion mathematics to be challenged; also worries that quaternions will become merely 'a private and personal skill' instead of a method that can be taught.
L Graves, *Hamilton, 2,* 591–2 inc

7195. 1847–9–1 From Count [Gerrit] Schimmelpenninck 5 Great Cumberland St.
King of Netherlands received copy of JH's *Cape Results* and wishes to award gold medal to JH. Does JH accept?
ALS 1p RAS:JH Archive 10/5.94; Reel 9

7196. 1847–9–2 From John Bullar to Lady Margaret Brodie Herschel Southampton
Thanks MH for fulfilling promise. Hails JH's [*Cape Results*] as grand testimony to God's infinite power.
ALS 3pp RAS:JH Archive 10/5.44; Reel 9

7197. 1847–9–2 From William Parsons (Lord Rosse) 19 East Cliff, Dover
Will call for JH's [*Cape Results*] when in London. One of WP's assistants reported ring around Neptune last winter, but WP is skeptical. Instruments idle for nine months. WP too busy with [Irish famine] relief. Wrote to T. R. Robinson to find competent assistant.
ALS 4pp RAS:JH Archive 10/5.102; Reel 9

7198. 1847–9–2 From Count Revel 4 Grosvenor St.
In response to JH's letter to the late Count [Polton?], Revel will forward two copies of JH's *Cape Results* to Turin for distribution to academic institutions.
L 2pp RAS:JH Archive 10/5.118; Reel 9

7199. 1847–9–4 From George Peacock Brighton
Will soon marry [Frances] Selwyn, sister of Bishop of New Zealand. Favorable comments on [*Cape Results*].
ALS 4pp RAS:JH Archive 10/6.8; Reel 9

7200. [1847]–9–4 From [C. M. Wilke] Eastbourne
Thanks JH for citing CW's 'endeavours in connexion with Mr. [Sherman?].'
ALS 2pp RAS:JH Archive 10/5.45; Reel 9

7201. 1847–9–5 From Georg Merz & Son Munich
Thanks for the address of the Admiralty; will send the instrument away in three weeks. Would JH send his observations to them via Williams Norgate (bookseller).
ALS 1p RS:HS 13.37

7202. 1847–9–6 From William Smith Williams 65 Cornhill
Received 112 epigraphs inscribed by JH for insertion into presentation copies [of JH's *Cape Results*], to be distributed as directed.
ALS 1p RAS:JH Archive 10/3.8; Reel 9

7203. 1847–9–7 From James Constantine Adamson Cape Town
Relating to the appointment of Langham Dale to the staff of the South African College.
ALS 4pp RS:HS 1.13

7204. 1847–9–7 From James William Grant Calcutta
Has been appointed Apothecary General. His own wife's health has not been good. Hopes all is well at Collingwood. What is he doing about William? Comments on the field open to chemistry students.
ALS 4pp RS:HS 9.25

7205. 1847–9–7 From Manuel John Johnson Radcliffe Observatory, Oxford
Gratitude on behalf of observatory trustees for receipt of JH's *Cape Results*. Distributed copies as directed.
ALS 2pp RAS:JH Archive 10/5.46; Reel 9

7206. 1847–9–7 From Élie François Wartmann Geneva
Comments on electrical equipment, and then refers to a variety of experiments, including shocking chickens awake that have been given ether, and the study of other electrical effects on gases and various liquid substances. Wonders about the use of ether/electric shock on humans in surgery. Congratulates JH on his *Cape Results*.
ALS 4pp (French) RS:HS 18.93

7207. 1847–9–8 From Edward Joshua Cooper Mount Eagle, Cabinteeley [Ireland]
Praises JH's *Cape Results*. Sends EC's paper read at Royal Irish Academy, three engravings of EC's 'Circle,' and EC's recent work dabbling 'in other than Astronomical affairs.'
ALS 4pp RAS:JH Archive 10/5.103; Reel 9

7208. 1847–9–8 From Adolphe Quetelet Brussels
[Form letter] Royal Academy of Science and Humanities acknowledges receipt of JH's *Cape Results*.
(A)LS 1p (French) RAS:JH Archive 10/5.9; Reel 9

7209. 1847–9–9 From Sir Francis Ronalds Kew Observatory
Has had a compensating barometer constructed. Requests thoughts concerning compensating barometers. Sends some memos concerning pendulums and other topics.
ALS 3pp RS:HS 14.422

7210. 1847–9–10 From Paul Partsch Vienna
Has received package in Turkey from JH. Sends thanks to JH through Sir Robert Brown. Values highly the gift of fragments from a meteor fallen near Cold-Bokkovold on 13 October 1838.
ALS 1p RS:HS 14.143

7211. 1847–9–10 From William Yolland Ordnance Map Office, Southampton
As successor to T. F. Colby, WY invites JH to send offered copy of JH's *Cape Results* to library of Ordnance Survey.
ALS 2pp RAS:JH Archive 10/5.47; Reel 9

7212. 1847–9–11 From Thomas Bell Elcock Fletcher Birmingham
Will announce JH's offer to give [*Cape Results*] to Philosophical Institution's library.
ALS 2pp RAS:JH Archive 10/5.48; Reel 9

7213. [1847]–9–11 From William Jerdan Literary Gazette Office
Sends copy of *Literary Gazette*, containing notice of [JH's *Cape Results*], which WJ saw while visiting mutual friend J. C. Ross.
L 1p RAS:JH Archive 10/6.10; Reel 9

7214. 1847–9–11 From John Pringle Nichol Glasgow Observatory
Gratitude for receipt of JH's *Cape Results*, donated to university library. David Brewster is head of St. Andrews College, and Daniel Dewar is head of Marischal College in Aberdeen. Hopes observatory will be complete when B.A.A.S. meets again in Glasgow.
ALS 5pp RAS:JH Archive 10/5.119; Reel 9

7215. 1847–9–13 From Edward Sabine Spa
Discusses Mrs. Sabine's translation of [Alexander von] Humboldt's *Cosmos,* particularly some theological passages.
ALS 5pp RS:HS 15.207

7216. 1847–9–16 From William Radcliffe Birt Bethnal Green
Informing JH that he has sent a letter to John Lee [RS:HS 4.117] and that Lee agrees with WB and wonders if it would be possible to obtain JH's support. Encloses drawings and details of Francis Ronalds's self registering barometer.
ALS 4pp RS:HS 4.118

7217. 1847–9–16 From Edward Joshua Cooper Cabinteeley
Encloses copies of configurations of nebulae clusters and explains his readings. Has he heard of Macedoine Melloni's water eye piece? [Andrew] Graham has invented a new micrometer.
ALS 6pp RS:HS 5.286

7218. 1847–9–16 To Robert John Kane Collingwood
Asks for latest chemical information on platina groups, or 'Platinoids.' In trying to remove platina from mineral specimens JH may have found new metal. Describes its characteristics. P.S.: Keep this information private.
ACCS 2pp TxU:H/L-0209; Reel 1054

7219. 1847–9–17 From Francis Hodgson and G. R. Green Eton College
Gratitude for library's receipt of JH's *Cape Results*.
ALS 1p RAS:JH Archive 10/5.49; Reel 9

7220. 1847–9–17 From William Smith Williams 65 Cornhill
Fifteen copies [of JH's *Cape Results*] sent as JH directed. Remainder of 112 copies due today from binder. Awaiting JH's instructions for distribution.
ALS 1p RAS:JH Archive 10/3.9; Reel 9

7221. 1847–9–18 From James David Forbes Eastertyne by Dunkeld, [North Britain]
Gratitude for gift of JH's *Cape Results*. Just returned from geological tour of highlands with [Bernhard] Studer from Bern [Switzerland]. Hopes to purchase Madame Witte's 'model of the Moon...for Edinburgh,' if it has not been bought in London. Observed Neptune at Oxford in company with U. J. J. Leverrier, J. C. Adams, and Wilhelm Struve.
ALS 4pp RAS:JH Archive 10/5.120; Reel 9

7222. 1847–9–19 From Christian C. J. Bunsen Carlton Terrace
Copy of JH's *Cape Results* was sent to king of Prussia. Other copies will be forwarded to astronomers in Germany. Note on [Alexander von] Humboldt's itinerary.
ALS 2pp RAS:JH Archive 10/5.105; Reel 9

7223. 1847–9–20 To Edward Joshua Cooper Collingwood
Comments on EC's observations of some nebulae and comets.
C 2pp RS:HS 22.327

7224. 1847–9–20 From Edward Riddle Greenwich Hospital School
Gratitude for receipt of JH's *Cape Results*.
ALS 1p RAS:JH Archive 10/5.50; Reel 9

7225. 1847–9–20 From ? Prospect Row, Cork
Asks JH about calculations of arcs of meridian.
AL 1p inc Harvard: Houghton

7226. 1847–9–21 From David Brewster St. Leonard's College, St. Andrew's University
Gratitude for gift of JH's *Cape Results*, not yet received. Send copy for [St. Andrews] University through Smith, Elder & Co., booksellers to the university. Concurs with JH on names for asteroids.
ALS 2pp RAS:JH Archive 10/5.121; Reel 9

7227. 1847–9–22 To Georg Merz Collingwood
Happy to learn that Cape equatorial [telescope] is ready. Address it to Francis Beaufort at Admiralty Office. Discusses payment. Sent [*Cape Results*] as GM instructed.
Adraft 1p TxU:H/M–0109.2; Reel 1087

7228. 1847–9–23 From William Smith Williams 65 Cornhill
Copies [of JH's *Cape Results*] have been distributed as JH instructed. U.S. government does not permit envoy to carry private parcels on official travel. Ambassador George Bancroft is out of town. Suggests writing to each American recipient and ask directions for delivery.
ALS 2pp † RAS:JH Archive 10/3.10; Reel 9

7229. 1847–9–24 From Edward Charlesworth Yorkshire
Gratitude on behalf of Yorkshire Philosophical Society for JH's offer to send *Cape Results*.
ALS 3pp RAS:JH Archive 10/5.51; Reel 9

7230. 1847–9–24 From Robert John Kane 51 Stephens Green, Dublin
Only received JH's letter yesterday as he does not visit Cork very often. Is not surprised at JH's findings on platinum. Gives some of the more important memoirs on platinum. Is pleased that JH is interested in chemical research into

platinum.
ALS 4pp RS:HS 11.2

7231. 1847–9–24 From William Smith Williams 65 Cornhill
Delivery of copies [of JH's *Cape Results*] to Munich as directed by JH will result in charge. Does JH want to apply to Bavarian minister to transmit these free?
ALS 1p RAS:JH Archive 10/3.11; Reel 9

7232. 1847–9–25 From Francis Beaufort Admiralty
Completion and transfer of telescope for Cape [of Good Hope]. Grateful for receipt of [*Cape Results*].
ALS 3pp TxU:H/M–0062; Reel 1087

7233. 1847–9–25 From William Yolland Ordnance Map Office, Southampton
Acknowledges receipt of JH's *Cape Results* by Ordnance Survey library.
ALS 1p RAS:JH Archive 10/5.52; Reel 9

7234. 1847–9–26 From [John Henry Griesbach] 3 Carlton Villas, Maida Vale, Edgware Road
Gratitude for receipt of JH's *Cape Results*. Offers suggestions from painter [John] Martin on how to alter engravings on steel plates, like those with which JH was dissatisfied.
ALS 1p inc RAS:JH Archive 10/5.53; Reel 9

7235. 1847–9–27 From George Biddell Airy Royal Observatory, Greenwich
Wants a copy of his *Cape Results* for his assistant. Would he continue to observe the duration of annularity in the coming eclipse?
ALS 2pp RS:HS 1.183 (C: RGO 6.56.271)

7236. 1847–9–27 From George Fisher Greenwich Hospital School
Gratitude for receipt of JH's *Cape Results*.
L 1p RAS:JH Archive 10/5.54; Reel 9

7237. 1847–9–28 To George Biddell Airy Collingwood
Will be happy to supply a copy of the *Cape Results* [see GA's 1847–9–27]; the next asteroid should be called Flora, to provide 'young goddesses for Mars and Jupiter to admire.'
ALS 3pp RGO 6.56.272

7238. 1847–9–28 From William Smith Williams 65 Cornhill
Smith, Elder & Co. wrote to all American intended recipients [of JH's *Cape Results*; see SE's 1847–9–23]. Still trying to get cooperation from American envoy. Asks JH's permission to publish now.
ALS 1p RAS:JH Archive 10/3.12; Reel 9

7239. 1847–9–29 From George Biddell Airy Royal Observatory, Greenwich
Regarding a letter from the Newcastle Philosophical Society.
ALS 1p RS:HS 1.184 (C: RGO 6.56.274)

7240. 1847–9–29 From George Bancroft 90 Eaton Square
Legation does not ordinarily forward parcels to America, but JH's unique reputation leads GB to set a precedent. Instructions for packaging copies [of JH's *Cape Results*]. S. C. Walker is now at National Observatory in Washington, D.C.
ALS 2pp RAS:JH Archive 10/5.133.1; Reel 9

7241. 1847–9–30 From William Empson East India College, Hertford
Has been here some 10 days. Received a copy of JH's great work before leaving Scotland. Hopes of a review of it from George Peacock. Longmans is now arranging for the printing of the review.
ALS 4pp RS:HS 7.46

7242. 1847–9–30 To Baron [Karl Alexander von] Hügel Collingwood
Presents a copy of JH's *Cape Results* for transmission to the observatory of the University of Tübingen.
AL 1p Science Museum Archives 523.89 Q

7243. 1847–9–30 From William Samuel Stratford Notting Hill Square
Please send latitude and longitude of Hawkhurst. Family benefited by trip to Southampton. Gratitude for receipt of JH's *Cape Results*.
ALS 2pp RAS:JH Archive 10/5.55; Reel 9

7244. 1847–10–1 From George Biddell Airy Royal Observatory, Greenwich
Concerning the method of observation by direct and reflected vision.
ALS 3pp RS:HS 1.185 (C: RGO 6.368.684)

7245. 1847–10–1 or later To George Biddell Airy n.p.
Was too cloudy to observe the eclipse; very worried about the health of Thomas Maclear.
ALS 3pp RGO 6.368.686

7246. 1847–10–1 or earlier To George Biddell Airy n.p.
About observing stars with the transit circle, and timing the annulus [of an eclipse?].
ALS 3pp RGO 6.368.682

7247. [1847–10–1] To George Bancroft n.p.
Thanks GB for making exception. Will send copies [of JH's *Cape Results*] immediately to American legation. Offers personal copy to GB. Notes national and private efforts at astronomy in America.
AdraftS 2pp (on G. Bancroft's 1847–9–29) RAS:JH Archive 10/5.133.2; Reel 9

7248. 1847–10–1 From John Simson Consulate of Baden, London
Will forward *Cape Results* to F. B. G. Nicolai at Mannheim Observatory.
ALS 1p † RAS:JH Archive 10/5.4; Reel 9

7249. 1847–10–1 From William Smith Williams 65 Cornhill
American ambassador George Bancroft gave permission for envoy to transmit official copies [of JH's *Cape Results*; see SE's 1847–9–23 and 9–28]. Smith, Elder & Co. retrieved all letters written to American recipients before these were sent from American consulate. Gratified to receive JH's permission to publish [JH's *Cape Results*]. Already have several orders. Encloses Bancroft's letter.
ALS 1p RAS:JH Archive 10/3.13; Reel 9

7250. 1847–10–1 From Wesley Stoker Barker Woolhouse 26 Cornhill
Gratitude for receipt of JH's *Cape Results*.
ALS 1p RAS:JH Archive 10/5.56; Reel 9

7251. 1847–10–2 From William Rowan Hamilton Observatory
The equatorial on the Dublin Observatory dome needs improvement. In the last fifty-two months, WH has seen three new planets: Neptune, Iris, and Flora.
L Graves, *Hamilton, 2,* 592–3 inc

7252. 1847–10–2 From Henry Raper 5 Chapel St., Grosvenor Square
Gratitude for receipt of JH's *Cape Results*. Congratulates JH on completion of many years' labor.
ALS 3pp RAS:JH Archive 10/5.57; Reel 9

7253. 1847–10–2 From William Smith Williams 65 Cornhill
Just learned that American ambassador George Bancroft is disappointed at not receiving presentation copy [of JH's *Cape Results*].
ALS 1p RAS:JH Archive 10/3.14; Reel 9

7254. 1847–10–3 From Algernon Percy (4th Duke of Northumberland) Alnwick Castle
Sending last Royal copy [of JH's *Cape Results*] today. Edward Everett's copy goes soon. Suggests sending copy to Prof. Robinson of New York.
ALS 2pp † RAS:JH Archive 10/4.24; Reel 9

7255. 1847–10–3 From William Whewell Trinity Lodge
Received JH's *Cape Results* with joy. JH gives new interest to 'our old friend the Great Bear.' WW thinking of setting up self registering photographic magnetic observatory at Lowestoft.
ALS 3pp RAS:JH Archive 10/5.58; Reel 9

7256. [1847–10–4] From Baron de Cello Bavarian Legation [London]
Asks JH to send [*Cape Results*] to Bavarian Legation.
AL 1p inc † RAS:JH Archive 10/5.5; Reel 9

7257. 1847–10–4 From Daniel Dewar Craiglochie, Errol
Absence from Aberdeen delayed DD's thanks for *Cape Results*. Official acknowledgement from [Marischal] College senate will be sent soon.
ALS 4pp RAS:JH Archive 10/5.122; Reel 9

7258. 1847–10–4 From William Henry Fitton 53 Upper Harley St.
Gratitude for receipt of JH's *Cape Results*. Marvels at scope of JH's work.
ALS 3pp RAS:JH Archive 10/5.59; Reel 9

7259. 1847–10–4 From John Henry Royal Observatory
Gratitude for receipt of JH's *Cape Results*.
L 1p RAS:JH Archive 10/5.60; Reel 9

7260. 1847–10–4 From Baron de Linden 40 Portman Square
Agrees, in Baron K. A. Hügel's absence, to forward copy [of JH's *Cape Results*] to Tübingen. Please send it to Württemberg legation.
L 1p † RAS:JH Archive 10/5.150; Reel 9

7261. 1847–10–4 From L. Craeyvanger Utrecht
Gratitude from Council of University of Utrecht for copy of JH's *Cape Results* donated to library.
ALS 1p RAS:JH Archive 10/5.95; Reel 9

7262. 1847–10–5 From John Davies and James Prescott Joule George St., Manchester
Acknowledgement of receipt of JH's *Cape Results* by Literary and Philosophical Society of Manchester.
ALS 1p RAS:JH Archive 10/5.62; Reel 9

7263. 1847–10–5 From John Davies and James Prescott Joule George St., Manchester
[Form letter] Literary and Philosophical Society of Manchester acknowledges receipt of JH's *Cape Results*.
(A)LS 1p RAS:JH Archive 10/5.61; Reel 9

7264. 1847–10–5 From George Thomas Napier Nice
One of GN's relatives, Henry Bunbury, is writing a book on the Cape and would like information on the educational system. Has advised Bunbury to write to JH. Would like a copy of the speech JH made at Cambridge in which he praised GN's government at the Cape. Comments on the present sorry state at the Cape.
AL 4pp inc RS:HS 13.71 & 75

7265. 1847–10–6 From George Biddell Airy Royal Observatory, Greenwich
Regarding H. C. Schumacher and the intended distribution of J. J. F. Lalande's catalogue.
ALS 3pp RS:HS 1.186 (C: RGO 6.203.381)

7266. 1847–10–6 or later To George Biddell Airy n.p.
Has dealt with the problem of thanking H. C. Schumacher [see GA's 1847–10–6].
ALS 3pp RGO 6.203.383

7267. 1847–10–6 To William Rowan Hamilton Collingwood
Thanks WH for 'plan' of [asteroid] Iris. Family is happy to have WH's son (JH's son William's friend) with them for holidays. Recounts their playful and adventurous activities.
ALS 3pp TCD 1493:411

7268. 1847–10–6 To Baron de Linden Collingwood
Agrees to a request to forward some work to BL.
AL 1p WT 63313.5

7269. 1847–10–7 To William Radcliffe Birt Collingwood
JH avoids participating in meteorological enquiry. Suggests improvements on WB's plan to establish government office to collect and publish meteorological reports from worldwide network of military stations. Has read Francis Ronalds's papers.
C 4pp RS:HS 19.126 (C: RS:HS 22.328)

7270. 1847–10–7 From William Rutter Dawes Camden Lodge
Observed a bright nebulous mass this evening. Gives readings for it. The comet is proceeding North. Has notified the *Times*.
ALS 3pp RS:HS 6.83

7271. 1847–10–7 From Algernon Percy (4th Duke of Northumberland) Alnwick Castle
Delighted that JH was awarded Gold Medal by King of Holland [for *Cape Results*].
ALS 3pp RAS:JH Archive 10/4.25; Reel 9

7272. 1847–10–8 From George Bancroft 90 Eaton Square
Devotes all leisure time to reading JH's gift copy of JH's *Cape Results*. Reports distribution of twenty copies to America.
ALS 1p RAS:JH Archive 10/5.134; Reel 9

7273. 1847–10–8 From Heinrich Louis d'Arrest Berlin
Gratitude for receipt of JH's *Cape Results*.
ALS 1p RAS:JH Archive 10/5.106; Reel 9

7274. 1847–10–8 From Edward Sabine Woolwich
R.S.L. committee for distribution of star catalogs should notify H. C. Schumacher of its work. Ordered additional copies to meet requests by foreign corresponding members. Deciding whom to include on distribution list. Concern by Treasury that distribution not be done for profit. Encloses letter from F. G. W. Struve.
ALS 4pp TxU:H/M–0521.18; Reel 1093

7275. 1847–10–8 From Edward Sabine Woolwich
Further discusses the translation of [Alexander von] Humboldt's *Cosmos* [see ES's 1847–9–13].
ALS 4pp RS:HS 15.208

7276. 1847–10–9 From William Empson East India College, Hertford
Is sorry that JH has been introduced to the law on Trusts under unfavorable circumstances. Longmans wants the review [JH's review of Humboldt's *Cosmos*] printed and bound by Christmas day. Let him know as soon as his paper is ready.
ALS 4pp RS:HS 7.47

7277. 1847–10–9 To Edward Sabine Collingwood
Agrees that H. C. Schumacher and foreign institutions should be added to list of recipients for catalogs. Please submit recommended names to B.A.A.S. Needs addresses of [W. S.] Stratford and Parramatta observatory. Complications in mailing B.A.A.S. catalog to foreign recipients.
C 2pp TxU:H/L-0364.2; Reel 1054

7278. 1847–10–9 To Edward Sabine Collingwood
Discusses use of 'force' versus use of 'power' in translating [Alexander von] Humboldt's *Cosmos*. Says science has outgrown 'force.' [See ES's 1847–10–8]
ALS 1p RS:HS 15.209 (C: RS:HS 22.329)

7279. 1847–10–9 From [Heinrich Christian Schumacher] [Altona]
Thanks JH for a copy of his *Cape Results* and comments on several other astronomical topics.
ALS 2pp (German) RAS:JH Archive 10/5.107; Reel 9

7280. 1847–10–9 From Alfred Smee Bank of England
About the eye's means of adjusting in vision; asks JH's permission to dedicate AS's writings on vision to JH.
ALS 4pp RS:HS 16.168

7281. 1847–10–10 To Margaret Brodie Herschel Collingwood
Offers to assist Peter S[tewart] in his financial difficulties; also comments on E. M[ackintosh]'s financial problems. JH expects more funds soon from dividends.
ALS 4pp JHS 6.34

7282. 1847–10–10 From Adam Sedgwick Cambridge
Disappointed not to have seen Sir Harry Smith or JH's family while AS was in Norwich. Grateful for receipt of JH's *Cape Results*. Congratulates JH on this 'grand harvest,' a monument to William Herschel's honor. JH is 'now again a free man.'
ALS 4pp RAS:JH Archive 10/5.63; Reel 9

7283. 1847–10–11 From James [W.?] Bell Tuscan Consulate, London
[Cavr.?] Peruzzi departed for Florence Wednesday and took Duke of Northumberland's letters. At next opportunity, JB will forward copies of JH's *Cape Results* to Florence observatory and to G. B. Amici.
L 2pp † RAS:JH Archive 10/5.131; Reel 9

7284. 1847–10–11 To Margaret Brodie Herschel Collingwood
Is concerned about a domestic [?] who is having fits; asks MH to get some information on the rights of domestics, and then JH asks MH to get him some scientific information that would be at the R.S.L. library.
ALS 4pp JHS 6.35

7285. 1847–10–11 To Robert John Kane Collingwood
Encloses JH's letter of 11 Oct., which was not mailed to RK. Discovered traces of titanium in Astraeum. Puzzled by saline compounds that appear in JH's experiments with 'Platinoids.'
(A)CC 2pp TxU:H/L-0210.2; Reel 1054

7286. 1847–10–11 To Robert John Kane Collingwood
Thanks for information [on latest platina research]. Has encountered 'a *totally new metallic chemistry*.' Searching for papers by Mr. Gros and Mr. Beiset. Describes JH's experiments with 'Platinoids,' which JH named like asteroids: Heberum, Astraeum, and Neptunium.
C 4pp † TxU:H/L-0210.1; Reel 1054

7287. 1847–10–13 From George Biddell Airy Royal Observatory, Greenwich
Did he see the eclipse? Agrees with him over the H. C. Schumacher proposals.
ALS 2pp RS:HS 1.187 (C: RGO 6.112.273)

7288. 1847–10–13 From John Adamson and John Fenwick Newcastle upon Tyne
Acknowledge receipt of JH's *Cape Results* by Literary and Philosophical Society.
ALS 1p RAS:JH Archive 10/5.64; Reel 9

7289. 1847–10–13 To Richard Sheepshanks Collingwood
Gives information from W. R. Hamilton on the location of the asteroid Iris. Proposes R.A.S. begin collecting observations of sunspots, the goal being a complete series of 'the sun's aspect for *every day* of *every year*.' Offers the observations of H. Griesbach for 1846 as a 'nest egg' as well as some of his own. Requests address of G. Schwinck, who prepared a star atlas. Reports on F. G. W. Struve's Neptune observations.
ALS 4pp damaged RS:HS 16.73 (C: RS:HS 25.9.46)

7290. 1847–10–14 From Charles James Fox Bunbury Mildenhall
Would like information on the system of education used at the Cape, for his forthcoming book.
ALS 3pp RS:HS 5.118

7291. 1847–10–14 From William Empson East India College, Hertford
Has just received his note. Would like to print [JH's review of] *Kosmos* at Christmas, so let him know if there is a chance of it being ready. Mrs. [Richard] Jones has not been feeling well. Richard Jones is visiting Alexander Baring (1st Baron Ashburton).
ALS 4pp RS:HS 7.48

7292. 1847–10–14 To Caroline Lucretia Herschel Collingwood
Answered CLH's question in JH's letter to Sophia Beckedorff. Expects that CH has received the copy of his *Cape Results* that JH sent recently.
ALS 1p TxU:H/L–0592.9; Reel 1058 & Reel 1089

7293. 1847–10–14 From Edward Sabine Woolwich
Problems with expensive postage for [W. S.] Stratford's notices could have been avoided if these had been sent to R.S.L. for mailing. Explains R.S.L. system for foreign mailings. Believes that council's list of recipients for star catalogs should include many institutions that are not associated with observatories. Advises sending more copies to French institutions. Problems at Parramatta Observatory.
ALS 8pp TxU:H/M–0522.1; Reel 1093

7294. 1847–10–15 From Filipp Ivanovich Brunnov Ashburnham House

Has received a note from JH and will look after sending it on, but navigation to St. Petersburg will be questionable for the next while. FB enjoyed JH's work [*Cape Results*] as did the Emperor. JH is given directions how to send a copy to F. G. W. Struve.
L 2pp (French) RAS:JH Archive 10/5.113; Reel 9

7295. 1847–10–15 To Adolphe Quetelet Collingwood Hawkhurst, Kent
Thanks AQ for package No. 32. Asks AQ to acknowledge receipt of JH's *Cape Results.* Commends AQ for outstanding work on magnetic meteorological and periodic phenomena.
ALS 4pp Académie belgique: #25

7296. 1847–10–16 From Georg Merz & Son Munich
The refractor [see GM's 1846–2–12] has now been shipped, and the account is rendered. As well, GM provides two pages of instructions about setting up the telescope.
ALS 2pp (German) †, encl 2pp (German) TxU:H/M–0360; Reel 1087

7297. 1847–10–17 To [Alexander V. Bewley] Collingwood
Tells AB about the state of education in South Africa when JH was there, comments on his part in such reforms as were made, and gives all the credit to Sir George Napier.
ALS 3pp WT B.7

7298. 1847–10–17 To Charles James Fox Bunbury Collingwood
Sending him details of the system of Education used at the Cape.
C 3pp RS:HS 5.119 & 22.330

7299. 1847–10–17 From Thomas Jones Chetham's Library, Manchester
Brief history of library and its 20,000 volumes, many donated. Asks JH to donate copy of JH's *Cape Results*. Refers JH to *Gentleman's Magazine*, June 1847, for further information on library.
ALS 2pp † RAS:JH Archive 10/5.65; Reel 9

7300. 1847–10–17 To Anne Elise Knipping Collingwood
Grateful that AK withheld 'fanatical letter' addressed to Caroline Herschel. JH authorized Drummonds Bank in London to honor any check from [A. A.] Mühry up to £50 to pay for Caroline's needs. Tell Miss Beckedorff and Mühry to use this channel, rather than writing to JH directly.
Adraft 3pp TxU:H/L-0214; Reel 1054

7301. 1847–10–18 From William Rutter Dawes Camden Lodge
If not too busy JH would have doubtless seen the recent comet on several evenings. Gives latest observations of Saturn and its satellites.
ALS 4pp RS:HS 6.84

7302. 1847–10–18 From Count Dietrichstein Chandos House
Forwarded *Cape Results* to Emperor. Will forward other copies to observatories, libraries, and individuals in Austro-Hungarian empire as requested.
L 2pp RAS:JH Archive 10/5.1; Reel 9

7303. 1847–10–18 From John R. Hind 3 Allsop's Terrace, New Rd.
Has discovered another new planet [Flora]. Gives readings.
ALS 1p RS:HS 9.339

7304. 1847–10–18 From Baron Koller Chandos House
Count Dietrichstein is out of town. What is Duke of Northumberland's [Algernon Percy's] connection to *Cape Results*? Embassy wants to inform superiors in Vienna.
ALS 2pp RAS:JH Archive 10/5.2; Reel 9

7305. 1847–10–18 From Karl Vasilevich Nesselrode St. Petersburg
Thanks JH for a copy of JH's *Cape Results,* and indicates that it was also appreciated by the Emperor.
LS 1p (French) RAS:JH Archive 10/5.114; Reel 9

7306. 1847–10–19 From James [W.?] Bell Tuscan Consulate, London
Mr. Henderson left London today taking copies of JH's *Cape Results* to Florence.
L 1p RAS:JH Archive 10/5.132; Reel 9

7307. 1847–10–19 From Heinrich Christian Schumacher Altona
Wilhelm Struve stayed several days with HS enroute from England to Pulkowa, but forgot to leave copies of [JH's *Cape Results*] for HS and K. L. C. Rümker. Later, Struve forgot to give these to G. B. Airy, who was visiting Pulkowa, for delivery to HS. HS has not yet seen *Cape Results*, or J. J. Lalande's catalogue to be sent by B.A.A.S. Happy to publish R.A.S. invitation for observations of solar spots. Short poem about Rümker losing a comet.
ALS 3pp RAS:JH Archive 10/5.15; Reel 9

7308. 1847–10–19 From Henry Warburton 45 Cadogan Place
Sends copy of HW's paper on combinations of 'Plural Elements,' printed in *Transactions* of Cambridge Philosophical Society, based on the method of continued subtraction suggested by Augustus De Morgan. Shows solution for Bernoulli's fifth number.
ALS 4pp RS:HS 18.39

7309. 1847–10–20 From George Biddell Airy Royal Observatory, Greenwich
Is prepared to report on GA's trip to see Pulkowa Observatory, if JH does not need the meeting time [see JH's 1847–10–28].
ACCS 2pp RGO 6.230.189

7310. 1847–10–20 From George Bishop Regents Park
Thanks for his congratulatory letter. The name of J. R. Hind will become famous. Would JH suggest a name for the new planet. Saw another new planet last week.
ALS 3pp RS:HS 4.147

7311. 1847–10–20 From Charles James Fox Bunbury Mildenhall
Is grateful for the detailed answer to his query on the system of Education used at the Cape. Is very busy and hopes to go to Italy when his book has been seen through the press.
ALS 3pp RS:HS 5.120

7312. 1847–10–20 From Robert Phelps Sidney Lodge, Cambridge
Grateful for receipt of JH's *Cape Results.*
ALS 1p RAS:JH Archive 10/5.67; Reel 9

7313. 1847–10–20 From William Henry Fox Talbot Athenaeum Club
Grateful for receipt of JH's *Cape Results.* Sending photographs by several artists to show that 'knowledge of the art is spreading.'
ALS 2pp RAS:JH Archive 10/5.66; Reel 9

7314. 1847–10–21 From Richard Sheepshanks Reading
Hopes JH will attend the November R.A.S. meeting to stress

the importance of recording solar observations [drawings of sunspots]. On J. R. Hind's work. Says Hind is not strong enough for the work he gets through. Cannot find G. Schwinck's address; suggests contacting H. C. Schumacher.
ALS 4pp RS:HS 16.74

7315. 1847–10–22 From William Samuel Stratford Notting Hill Square
Communicates with others JH's choice of name for new planet [Iris]. Sends position of Hawkhurst Church tower obtained from map department. Describes rainy-day eclipse and reactions to it.
ALS 3pp RS:HS 17.102

7316. 1847–10–23 From William Rowan Hamilton Observatory
WH notes that quaternions derive some interesting results in problems with three moving bodies.
L Graves, *Hamilton, 2,* 594 inc

7317. 1847–10–23 From John William Lubbock High Elms, Farnborough
Thought his letter had miscarried. Thanks for the gift of his recently published work [*Cape Results*]. Will be glad to supply any information from his own work on comets, which he hopes to send in the course of a few days. Is studying U. J. J. Leverrier's method of development by elimination.
ALS 4pp RS:HS 11.384

7318. 1847–10–23 From Alfred Miles Admiralty
Francis Beaufort is on leave. Does JH know of Beaufort's intentions regarding instruments assigned to [William] Mann, who departs aboard *Arab* in two days?
ALS 3pp TxU:H/M–0366; Reel 1087

7319. 1847–10–23 From A[?] M[?] Admiralty
In Francis Beaufort's absence, AM is trying to clarify the matter of an astronomical instrument, apparently obtained from George Merz & Son.
ACC 1p Hydrographic Office Letter Book

7320. 1847–10–23 From Edward Sabine Woolwich
R.S.L. policy for paying postage on scientific communication by members to and from foreign correspondents. Addresses of proposed recipients for *Cape Results*. ES's illness and problem with eyes.
ALS 13pp TxU:H/M–0522.2; Reel 1093

7321. [1847–10–23] To William Henry Fox Talbot n.p.
Thanks for excellent Talbotypes. Notes how few photographs are 'perspective representations on a vertical plane' and suggests how to correct this.
ALS 3pp Science Museum Talbt 1/37

7322. 1847–10–23 From William Smith Williams 65 Cornhill
Unable to explain how [Robert] Phelps received copy [of JH's *Cape Results*]. His name not on any list, and printer Peter Stewart did not send him one. Stewart requests copies be sent to John Ruskin and James Thomson. Will carry out JH's instructions for copy to Wilhelm Struve.
ALS 1p RAS:JH Archive 10/3.15; Reel 9

7323. 1847–10–24 From George Bishop Regents Park
Thanks for his early reply and for his name and symbol for the new planet (Flora). J. R. Hind has informed H. C. Schumacher. Hopes that K. L. Hencke in his new charts is not merely enlarging the Berlin maps.
ALS 3pp RS:HS 4.148

7324. [1847]–10–24 From William Empson East India College, Hertford
Regarding progress with the printing and translation of *Kosmos*. Richard Jones has visited Alexander Baring (1st Baron Ashburton).
ALS 4pp RS:HS 7.49

7325. 1847–10–25 From John R. Hind 3 Allsop's Terrace, New Rd.
Thanks for the neat name and symbol he has given to the new planet; thinks H. C. Schumacher has read Hora for Flora. The next one will have to be called Thetis as so many have suggested that name.
ALS 1p RS:HS 9.340

7326. 1847–10–25 To William Henry Smyth Collingwood
Thanks WS for copy of W. S. Jacob paper on American telescope. Hopes that Jacob will attempt Talbotypes of sun spots while at Aden.
C 2pp RS:HS 22.331

7327. 1847–10–26 From Thomas Makdougall Brisbane Greenock
Gratitude for receipt of JH's *Cape Results*. Glad that Parramatta catalogue was helpful. Laments government decision to do away with Parramatta observatory. TB suffering from rheumatism.
ALS 2pp RAS:JH Archive 10/5.123; Reel 9

7328. 1847–10–26 From Richard Sheepshanks R.A.S.
Suggests alteration in list of observatories that JH sent to W. S. Stratford. Advises JH to include private astronomers on list also. News of [Elizabeth] Baily. Needs to borrow Francis Baily's apparatus for measuring length.
ALS 4pp TxU:H/M–0538; Reel 1093

7329. 1847–10–26 From William Henry Fox Talbot Athenaeum
Comments on astronomical symbols, and position of camera when photographing buildings.
ALS 2pp RS:HS 17.319

7330. 1847–10–28 To George Biddell Airy Collingwood
Clarifies the agenda for the next R.A.S. meeting, and reports that the large refracting telescope is on the way to the Cape.
ALS 4pp RGO 6.230.187

7331. 1847–10–28 From Prince di Castelcicala 15 Princes St., Cavendish Square
Will forward [copies of JH's *Cape Results*] to Naples.
L 1p RAS:JH Archive 10/5.104; Reel 9

7332. 1847–10–28 From Urbain J. J. Leverrier Paris
Thanks for his visit to Collingwood. Regarding JH's great work. Would like to exhibit them to the King of France.
ALS 4pp (French) RS:HS 11.199

7333. 1847–10–28 From Adolphe Quetelet Brussels
Thanks JH for a copy of JH's *Cape Results.* Is pleased by JH's encouragement of AQ's meteorological observations, about which AQ comments further. Speaks of a family of comets related to Mars and Jupiter and asks JH's opinion about this.
ALS 3pp (French) RAS:JH Archive 10/5.10; Reel 9

7334. 1847–10–28 From Peter Stewart London
Received JH's note to present copies [of JH's *Cape Results*] to [James] Thomson, John Ruskin, and Chetham's Library [Manchester]. Of 350 copies allotted for gratis presentation, 289 copies have been sent and 61 remain. [Alexander]

Henning paid for a copy, but PS notified him that he was on JH's list of recipients. Sends letter that PS received from James [Stewart] before James's recent arrival in York Terrace. Comments on proper way to handle failing businesses.
ALS 4pp RAS:JH Archive 10/3.16; Reel 9

7335. 1847–10–29 From George Biddell Airy Royal Observatory, Greenwich
Regarding the forthcoming meeting. Concerning the telescope used at Pulkowa.
ALS 4pp RS:HS 1.188 (C: RGO 6.192.372 inc)

7336. [1847]–10–29 From William Empson East India College, Hertford
Hopes no inconvenience will be caused by booking JH's paper for 10 Dec.; its length has been estimated as 50 pages. Will send him the text of *Kosmos* as it comes from the printers. Richard Jones has returned in triumph. Agrees with Alexander Baring (1st Baron Ashburton) on the currency.
ALS 4pp RS:HS 7.50

7337. 1847–10–30 From William Cranch Bond to George Bancroft Harvard Observatory, Cambridge, Mass.
Acknowledges receipt of JH's *Cape Results*.
ALS 1p RAS:JH Archive 10/5.137; Reel 9

7338. 1847–10–30 From Henry Ingersoll Bowditch Boston
Gratitude from proprietors of Bowditch Library for receipt of JH's *Cape Results* and for JH's memory of their late father [Nathaniel Bowditch].
ALS 2pp RAS:JH Archive 10/5.136; Reel 9

7339. 1847–10–30 From Asa Gray Boston
[Form letter] Acknowledges receipt of JH's *Cape Results* by library of American Academy of Arts and Sciences.
(A)LS 1p RAS:JH Archive 10/5.135; Reel 9

7340. 1847–10–30 From William Rowan Hamilton Dublin
Describes the funeral of James MacCullagh; mystified why he committed suicide.
L Graves, *Hamilton, 2,* 595–6 inc

7341. 1847–10–30 From John R. Hind 3 Allsop's Terrace, New Rd.
Sends his first approximation of the elements of the asteroid Flora.
ALS 1p RS:HS 9.341

7342. 1847–10–31 From Royal Academy of Science (Berlin) Berlin
A note of thanks for the gift of a copy of JH's *Cape Results*.
L(S) 1p (German) RAS:JH Archive 10/5.108; Reel 9

7343. 1847–11–1 From George Biddell Airy Royal Observatory, Greenwich
Regarding magnetic disturbances.
ALS 3pp RS:HS 1.189 (C: RGO 6.693.113)

7344. 1847–11–1 From Benjamin Peirce to George Bancroft Harvard University
Received *Cape Results*. Praises JH's drawings of nebulae. American astronomers will be honored by JH's tribute to memory of E. P. Mason.
ALS 2pp RAS:JH Archive 10/5.138; Reel 9

7345. 1847–11–1 From Spencer J. A. Compton Castle Ashby
As Hugh Percy (3rd Duke of Northumberland) has died, it would be very fitting if JH would contribute an obituary notice for the R.S.L.
ALS 1p RS:HS 5.272

7346. [1847–11] From Thomas John Hussey Hayes Rectory, Kent
Grateful for receipt of JH's *Cape Results*. Describes TH's observations of thread-like 'bridges' among solar spots and questions their cause. Describes 'velocipede' constructed by local mechanic.
ALS 4pp RAS:JH Archive 10/5.73; Reel 9

7347. 1847–11–2 To George Biddell Airy Collingwood
Suggests to GA that his observations [see GA's 1847–11–1] might rather support than destroy the theory of the thermo-electric origin of terrestrial magnetism.
ALS 4pp RGO 6.693.115

7348. 1847–11–2 From John Russell Bartlett New York University
Gratitude for receipt of JH's *Cape Results* by New York Historical Society.
ALS 1p RAS:JH Archive 10/5.139; Reel 9

7349. 1847–11–2 From William Beattie Hampstead
Is printing letters of the poet Thomas Campbell and would like JH to look at the proofs of those from his father to Campbell.
ALS 2pp RS:HS 5.29

7350. 1847–11–2 From Sophie Beckedorff Hanover
Withdrew £25 from Drummonds [Bank] for Caroline L. Herschel. Describes Caroline's deteriorating condition. JH's [*Cape Results*] arrived three months ago. Despite efforts by friends to convince her that it was hers, [Caroline] presented it to king of Hanover, who already had copy. Received manuscript from JH's daughter Louisa.
ALS 4pp TxU:H/M–0077; Reel 1087

7351. 1847–11–2 From John Cruickshank Aberdeen
Senate of Marischal College acknowledges gift of JH's *Cape Results*.
ALS 1p RAS:JH Archive 10/5.124; Reel 9

7352. 1847–11–2 From Walter R. Johnson South Broad St., Philadelphia, Penn.
[Form letter] Acknowledges receipt of JH's *Cape Results* by Academy of Natural Sciences of Philadelphia. [Personal note on verso] Reports Association of American Geologists and Naturalists has extended its scope of interest and changed name to 'American Association for the Promotion of Science,' hoping to attract foreign scientists. As secretary, WJ invites JH to attend September meeting in Philadelphia.
(A)LS 1p, WJ note RAS:JH Archive 10/5.141.1 and 141.2; Reel 9

7353. 1847–11–2 From Duncan Macfarlane Glasgow College
Gratitude for receipt of JH's *Cape Results* by University of Glasgow.
ALS 3pp RAS:JH Archive 10/5.125; Reel 9

7354. 1847–11–2 From Matthew Fontaine Maury National Observatory, Washington, D.C.
Gratitude for receipt of JH's *Cape Results*. Outlines MM's project to catalogue all ninth and tenth magnitude stars by duplicate sweeps with different instruments and observers. Can JH offer suggestions?
ALS 2pp RAS:JH Archive 10/5.140; Reel 9

7355. 1847–11–3 From Joseph Lovering Cambridge University, [Cambridge, Mass.]
Gratitude for receipt of JH's *Cape Results*. Recently resolved Orion nebula with new refractor from Munich. Working now on annular nebula in Lyra.
ALS 3pp RAS:JH Archive 10/5.142; Reel 9

7356. 1847–11–3 From William Simms London
Grateful for receipt of JH's *Cape Results*.
ALS 1p RAS:JH Archive 10/5.68; Reel 9

7357. 1847–11–3 From William Smith Williams 65 Cornhill
Regrets letting one copy [of JH's *Cape Results*] go out containing 'cancelled sheet.' Gives printer's explanation for error. Assures JH that no other copies were affected.
ALS 1p RAS:JH Archive 10/3.17; Reel 9

7358. 1847–11–4 From George Biddell Airy Royal Observatory, Greenwich
Currents in telegraph wires.
ALS 2pp RS:HS 1.190 (C: RGO 6.693.117)

7359. 1847–11–4 From Edward Joshua Cooper Cabinteeley
Delayed writing in the hope of obtaining some interesting news. Would like a name for some asteroids. Andrew Graham's micrometer works well. Further observations on various stars.
ALS 5pp RS:HS 5.287

7360. 1847–11–4 From William John Hamilton Somerset House
[Form letter] Acknowledges receipt of JH's *Cape Results* by Geological Society of London.
(A)LS 1p RAS:JH Archive 10/5.69; Reel 9

7361. 1847–11–4 From Digby P. Starkey Dublin
In flowery prose thanks JH for 'feeling with me in what I have felt deeply myself,' perhaps referring to one of JH's poems. Is delighted to have learned through W. R. Hamilton that JH is a poet.
ALS 4pp RS:HS 17.231

7362. 1847–11–6 From E[ugène Charles] Catalan Paris
Writes, on behalf of the Societé philomatique, to thank JH for the magnificent volume [JH's *Cape Results*].
ALS 1p (French) RAS:JH Archive 10/5.85; Reel 9

7363. 1847–11–7 To Edward Joshua Cooper Collingwood
Talking about the possibility of discovering an 'Irish Asteroid' or perhaps an exterior planet; suggests names for them.
C 2pp RS:HS 22.333

7364. 1847–11–8 From William Beattie Hampstead
Is greatly obliged for his prompt attention to the proof sheets. Has nearly 1,000 letters and 50–60 ms. poems to edit and wishes someone could be found to write Thomas Campbell's life, as he does not feel equal to it.
ALS 3pp RS:HS 5.30

7365. 1847–11–8 From Thomas Galloway 45 Torrington Square
Has received JH's letter and also the one of [J. H.?] Griesbach. These will be read at the next meeting of the R.A.S. Thinks enough has been written about E. J. Cooper's planets.
ALS 3pp RS:HS 8.27

7366. 1847–11–9 From Johan Gothard, Baron de Rehausen 14 Halkin St. West, Belgrave Square
Received JH's letter to JG's late predecessor at Swedish legation, Count Björnstjerna. Will forward copies of JH's *Cape Results* to Swedish institutions named by JH.
ALS 2pp † RAS:JH Archive 10/5.128; Reel 9

7367. 1847–11–10 From John Couch Adams St. John's College, Cambridge
Accepting JH's invitation to spend a few days with him. Remarks on planets in general. Information on academic success of Hugh Percy (3rd Duke of Northumberland).
ALS 4pp RS:HS 1.1

7368. 1847–11–10 From William Rutter Dawes Camden Lodge
Sending paper by Andrew Ross, who would be glad of any comments. [William] Simms informs him that he has improved his 8" object glasses. Gives recent observations for *Enceladus*.
ALS 3pp RS:HS 6.85

7369. 1847–11–11 From William Rutter Dawes Camden Lodge, Cranbrook
Measurements and diagrams of WD's 8.5-foot telescope, dome, and other structures at Camden Lodge [to be duplicated for shipment to Cape of Good Hope observatory]. Laments deaths of Julia Cornwallis and J. L. Felix Mendelssohn-Bartholdy.
ALS 4pp TxU:H/M–0158; Reel 1087

7370. 1847–11–12 From Filipp Ivanovich Brunnov n.p.
Encloses letter from K. V. Nesselrode.
L 1p (French) RAS:JH Archive 10/5.115; Reel 9

7371. 1847–11–12 From Thomas Galloway 45 Torrington Square
Outlining the circumstances leading to JH's name being proposed for the new council of the R.S.L. Hopes he will agree to his name being proposed.
ALS 4pp RS:HS 8.28

7372. 1847–11–12 From William Rutherford Royal Military Academy
Grateful for receipt of JH's *Cape Results*.
L 1p RAS:JH Archive 10/5.70; Reel 9

7373. 1847–11–13 From Edward Everett Cambridge, Mass.
Has received JH's volume and letter from Algernon Percy (4th Duke of Northumberland). Regards it as a valuable contribution to his library. Would like JH's opinion on the papers of Benjamin Peirce and his views on the planet Neptune.
ALS 4pp RS:HS 7.117

7374. 1847–11–13 From John Ruskin Dover
Grateful for receipt of JH's *Cape Results*. Was aesthetically inspired by JH's descriptions and drawings.
ALS 3pp RAS:JH Archive 10/5.71; Reel 9

7375. 1847–11–14 To George Biddell Airy Collingwood
Asks GA for details about the scientific support provided by Hugh Percy; also asks about the availability of star charts on the Berlin Academy scheme.
ALS 4pp RGO 6.368.688

7376. 1847–11–14 From Henry Warburton 45 Cadogan Place
Close to finding expeditious, symmetric method for computing Bernoulli's numbers. Resolves four formulas that

HW sent to JH earlier today.
ALS 3pp RS:HS 18.40

7377. 1847–11–15 From George Biddell Airy Royal Observatory, Greenwich
Replies to JH's 1847–11–14, with respect to Hugh Percy and star charts.
ACCS 2pp RGO 6.368.690

7378. 1847–11–16 From Francis Beaufort [Trim?]
[Received JH's 12 Nov. letter.] Send bill for Munich telescope to Admiralty. Order parts for dome construction, which [Thomas] Maclear cannot procure at Cape [of Good Hope].
ALS 3pp TxU:H/M–0063; Reel 1087

7379. 1847–11–17 To George Biddell Airy Collingwood
Is still working on the distribution list for star catalogues [see GA's 1846–11–19].
ALS 3pp RGO 6.203.389

7380. 1847–11–17 To William Henry Smyth Collingwood
Believes that if R.S.L. gives medals that year, J. R. Hind should receive one. John Lubbock is another possible candidate for his work on perturbation theory.
C 2pp RS:HS 22.334 D. Evans, 'Sir John Herschel to Captain Smyth,' *Quarterly Bulletin of the South African Library, 25* (1970), 17

7381. 1847–11–17 To William Henry Smyth Collingwood
Believes that Thomas Galloway's paper on solar motion deserves a medal.
C 1p RS:HS 22.335

7382. 1847–11–17 From William Samuel Stratford Notting Hill Square
Discusses persons who deserve copies of J. J. L. Lalande's *Catalogue of Those Stars in the Histoire céleste française...* and N. L. Lacaille's *Catalogue of 9766 Stars in the Southern Hemisphere* (?).
ALS 4pp TxU:H/M–0546.33; Reel 1093

7383. 1847–11–17 From William Samuel Stratford Notting Hill Square
Advises that the catalogs [J. J. L. Lalande's *Catalogue of Those Stars in the Histoire céleste française...* and N. L. Lacaille's *Catalogue of 9766 Stars in the Southern Hemisphere*] not be bound for another five or six months because some proof-sheets are still quite wet.
ALS 2pp TxU:H/M–0546.32; Reel 1093

7384. 1847–11–18 From George Biddell Airy Royal Observatory, Greenwich
Offers his suggestions for any trimming of the catalogue distribution list that may be necessary [see JH's 1847–11–17].
ALS 2pp TxU:H/M-0633.2; Reel 1084 (C: RGO 6.203.391)

7385. [1847]–11–18 From William Empson East India College, Hertford
Does he mean his paper will be ready for the printer by 10 Dec.? Should be at the binders by the 19th. Political signs need watching. Richard Jones is not well. George Peacock will be home next week.
ALS 3pp RS:HS 7.51

7386. 1847–11–19 From George Eden Admiralty
Outlining his plan for a Scientific Manual for Naval officers. Would be glad if he would superintend its publication.
ALS 8pp RS:HS 7.9

7387. [1847]–11–20 To George Eden Collingwood
Agrees to gather and edit contributions to *Admiralty Manual* and supervise its printing. Will correspond next month with all [scientists] chosen by GE.
AdraftS 2pp RS:HS 19.137 (RS:HS 23.51)

7388. 1847–11–20 From Richard Sheepshanks Reading
Has received a letter from [W. C.] Bond stating that Maria Mitchell observed the comet on 1 Oct., several days before it was seen by [Francesco] DeVico (3 Oct.) and by W. R. Dawes (7 Oct.). Proposes calling it 'Miss M. Mitchell's comet' and asks if JH has any objections. Knows DeVico will 'growl.'
ALS 2pp RS:HS 16.75

7389. 1847–11–22 From George Eden Admiralty
Is obliged for JH's readiness to assist in the preparation of the *Admiralty Manual*. Will put matter in hand at once.
ALS 1p RS:HS 7.10

7390. [1847]–11–22 From William Empson East India College, Hertford
Everything safe for Longmans. Let him know if he requires any separate copies. Richard Jones is better. William Whewell is mortified at being passed over by the Windsor invitation. George Peacock does not care; he is too happy.
ALS 4pp RS:HS 7.52

7391. 1847–11–22 From Johann Franz Encke Berlin
The excellence of JH's *Cape Results* has humbled JE. Much of the letter is taken up with more in the same vein, citing specific examples.
ALS 3pp (German) RAS:JH Archive 10/5.109; Reel 9

7392. [1847]–11–22 From Edward Sabine Teighmouth
Send presentation copies [of JH's *Cape Results*] to foreign recipients in name of R.S.L. Treasury charges duty only if books are presented by individuals. Will present JH's list [of recipients] at next meeting of B.A.A.S. council. Congratulates JH on winning Copley Medal. Will transmit copy [of *Cape Results*] to Naples.
ALS 4pp RAS:JH Archive 10/6.9; Reel 9

7393. 1847–11–22 From Henry Warburton 45 Cadogan Place
Thanks for comments on HW's paper and work on properties of Bernoulli's numbers. Claims HW's method of continued subtraction is convenient process for determining numerical coefficients. Found error, long perpetuated in literature, regarding Bernoulli's 13th number. Offers correct solution.
ALS 4pp RS:HS 18.41

7394. 1847–11–23 From Francis Beaufort Admiralty
[Reply to JH's 19 Nov. note.] FB may be able to get Admiralty to have roof [for telescope at Cape of Good Hope] built at Cranbrook under JH's direction. Submit plan or estimate of time and expenses. Just saw [Thomas] Maclear's letter telling of adventures in 'northern expedition.'
ALS 3pp TxU:H/M–0064; Reel 1087

7395. 1847–11–23 To Richard Sheepshanks Collingwood
The comet should bear Maria Mitchell's name. If [Francesco] DeVico 'grumbles,' RS could note that Madame Rümker also discovered the comet on [Oct] 3rd. 'What was there in it to attract the eyes of the female part of the creation so especially?'
ALS 2pp RS:HS 16.76 (C: RS:HS 22.336 & 25.9.47)

7396. 1847–11–23 From Henry Warburton 45 Cadogan Place
More calculations showing continued subtraction method in solution of Bernoulli's 13th number.
ALS 1p RS:HS 18.42

7397. 1847–11–24 From Spencer J. A. Compton Piccadilly
Is obliged for JH's letter and enclosure. Pleased that he will be at the Anniversary meeting to receive the medal [Copley Medal] recently awarded to him.
ALS 1p RS:HS 5.273

7398. 1847–11–24 From Paul Auguste Ernest Laugier n.p.
Thanks JH for a copy of JH's *Cape Results* and makes admiring comments on it.
ALS 2pp (French) RAS:JH Archive 10/5.86; Reel 9

7399. 1847–11–24 From Henry Warburton 45 Cadogan Place
More calculations using continual subtraction method to test Bernoulli's 15th number.
ALS 3pp RS:HS 18.43

7400. 1847–11–25 From George Biddell Airy Royal Observatory, Greenwich
Sending account of the magnetic disturbances of 24 September 1847.
ALS 1p RS:HS 1.191

7401. 1847–11–25 From William Rutter Dawes Camden Lodge
Will certainly help him regarding the Cape dome. Gives details of his mounting. Sends a copy of his observations.
ALS 4pp RS:HS 6.86

7402. 1847–11–25 From John William Lubbock n.p.
Regarding his own method of calculating the perturbations of the comets. Expense of computation is high for an individual and a mention by JH would bring support for his scheme. Would like to explain his method at an ordinary meeting in January. Every branch of science is facilitated by tables and those he suggests seem most desirable.
ALS 15pp RS:HS 11.385

7403. 1847–11–25 From Henry Warburton 45 Cadogan Place
More continual subtractions to resolve Bernoulli's 17th number. Error in official value given in *Encyclopaedia Metropolitana*. Also working on Bernoulli's 18th number.
ALS 3pp RS:HS 18.44

7404. 1847–11–26 From Samuel Hunter Christie Royal Society of London
[Form letter] Acknowledges receipt of JH's *Cape Results* by Royal Society of London.
(A)LS 1p RAS:JH Archive 10/5.72; Reel 9

7405. 1847–11–26 From Edward Joshua Cooper Cabinteeley
Would be pleased if JH would look out for 'shooting stars' on certain specified evenings.
ALS 2pp RS:HS 5.288

7406. 1847–11–26 To Thomas Maclear [Collingwood]
Complains of lack of time for correspondence. Equatorial for Cape Observatory completed. Advises TM to take relaxed observatory schedule to recover health. *Cape Results* published. Works on *Outlines Astr.* New governor arrives in Cape Town.
ALS 4pp Cape Archives/Maclear Papers File 100

7407. [1847]–11–26 From Richard Sheepshanks Reading
Maria Mitchell 'shall have her due honor,' despite [Francesco] DeVico. Discusses problems with the R.A.S. awarding medals and honors, the loss to the R.A.S. of Francis Baily, and the issue of proposing foreign associates. Proposes a list of such to JH. Objects to DeVico, but is the head of a publishing observatory and has found comets. Urges caution in appointing Americans. On publishing *R.A.S.M.N.*
ALS 8pp RS:HS 16.77

7408. 1847–11–27 From Theodore Dwight Woolsey Yale College, New Haven, Conn.
Gratitude for receipt of JH's *Cape Results* by Yale library. Pleased by references to recently deceased alumnus E. P. Mason.
LS 1p RAS:JH Archive 10/5.143; Reel 9

7409. [1847]–11–29 From William Empson East India College, Hertford
Do not strike out anything he has written. Has been suffering with nose-bleed.
ALS 4pp RS:HS 7.53

7410. 1847–11–29 From Félix Victor Mauvais Paris Observatory
Thanks JH for a copy of JH's *Cape Results* and compliments JH and his illustrious family.
ALS 1p (French) RAS:JH Archive 10/5.87; Reel 9

7411. 1847–11–29 To Richard Sheepshanks Collingwood
Likes RS's R.A.S. associate list. Must give the Americans a year or two more before it can be adequately ascertained who their top astronomers are. Sends a letter JH received from John Lubbock, who wants the opportunity in January to present new methods in perturbation theory. Responds to RS on the means of the R.A.S. recognizing significant contributions to astronomy.
ALS 4pp RS:HS 16.78 (C: RS:HS 25.9.48)

7412. 1847–11–29 From Henry Warburton Welwyn, Hertfordshire
Will submit paper to Cambridge [Philosophical] Society revising all [Leonhard] Euler's values for Bernoulli's numbers. Doubts accuracy of values up to B=49 quoted by George Peacock.
ALS 2pp RS:HS 18.45

7413. 1847–11–30 From George Biddell Airy Royal Observatory, Greenwich
Seeks opinions on discontinuing use of 25-foot zenith tube and altering form of printed observations for transit instrument and mural circle.
PrintedS 3pp TxU:H/L-0019; Reel 1054

7414. 1847–12–1 From Francis Beaufort Admiralty
Sends JH a copy of sailing tables, and asks JH's opinion as to the value of printing them.
ACC 2pp Hydrographic Office Letter Book

7415. 1847–12–1 From Richard Sheepshanks Reading
John Lubbock entitled to bring his proposal before R.A.S., and to the best advice they can give, but the idea of explaining new views on physical astronomy strikes RS as 'objectionable' in principle. No one will understand his presentation. How to proceed with this. RS's work on the standard yard.
ALS 8pp RS:HS 16.79

7416. 1847–12–4 From Francis Beaufort Admiralty
Thanks for JH's reply about 'Towson Tables.' Please return copy of JH's letter to FB. Encloses Georg Merz's account. [Telescope] is at Custom House awaiting shipment to Cape of Good Hope. Does JH have plan and estimate yet for dome, so FB can present it to Admiralty? Received JH's 24 Oct. note about Mr. Morris's account.
ALS 4pp TxU:H/M–0065; Reel 1087 (ACC: Hydrographic Office Letter Book)

7417. 1847–12–4 To John William Lubbock Collingwood
Has been considering a reply to JL's letter of the 25th. Outlines two courses to take and prefers the latter himself. Communicate with Richard Sheepshanks regarding the *viva voce*.
C 2pp RS:HS 11.386 & 22.336

7418. 1847–12–4 To John William Lubbock Collingwood
Advice as to best way to present JL's method of calculating the orbits of comets to R.A.S. If JL desires R.A.S. to undertake calculations, might be some difficulty.
ALS 4pp RS LUB.H.359

7419. 1847–12–4 From Richard Sheepshanks R.A.S.
John Lubbock wishes only to explain his system, so there is no reason why the R.A.S. should not listen to him. Certain conditions must be met. Troubles with the *R.A.S.M.N.*
ALS 3pp RS:HS 16.80

7420. 1847–12–5 From Henry Warburton 45 Cadogan Place
Error in Leonhard Euler's value for Bernoulli's 21st number. Promotes HW's method of continued subtraction as a reliable means of proving accuracy of values for higher Bernoulli numbers.
ALS 4pp RS:HS 18.46

7421. 1847–12–6 From James Renwick Columbia College, New York
Gratitude for receipt of JH's *Cape Results* [transmitted via George Bancroft].
ALS 1p RAS:JH Archive 10/5.144; Reel 9

7422. 1847–12–7 From [Elihu] Rich Hatton Garden
Sends *Principles of Natural Philosophy* from the Swedenborg Association. Published nearly one century earlier, its contents should interest JH. Contains a nebular theory commonly attributed to [P. S.] Laplace.
ALS 2pp RS:HS 14.482

7423. 1847–12–7 To Adam Sedgwick Collingwood
Invites AS to Collingwood for Christmas. The Somervilles and the Adamses will also be present.
ALS 3pp RS:HS 15.431 (C: RS:HS 22.339)

7424. [1847]–12–9 From William Empson East India College, Hertford
Has been reading as far as page 36 of the proofs with great pleasure. Regarding choice of words for a phrase. Richard Jones is uninfluenced as yet.
ALS 4pp RS:HS 7.54

7425. 1847–12–10 From George Richardson Cambridge Terrace
Now commanding the *Herefordshire*. All navigators must prove their competency. Asks JH for such a certificate.
ALS 2pp RS:HS 14.484

7426. [1847]–12–12 From William Empson East India College, Hertford
Congratulations on finishing his paper. Would he check over the proofs. Richard Jones is well, though not well enough to attend at the College chapel. J. E. D. Bethune and John Lefevre are staying with Jones. Bethune starts for India in February.
ALS 3pp RS:HS 7.55

7427. 1847–12–13 From Henry Warburton London
Destroy all HW's letters claiming errors in Leonhard Euler's values for Bernoulli's numbers. Mr. Hensley of Trinity College found weakness in HW's continued subtraction method. Clarifies symbols HW used to represent P. S. Laplace's expressions of Bernoulli's numbers.
ALS 11pp RS:HS 18.47

7428. 1847–12–14 From Henry Warburton 45 Cadogan Place
Continues explicating P. S. Laplace's formula for determining Bernoulli's numbers. Concerned that JH sees HW claiming too much for HW's continued subtraction method.
ALS 6pp RS:HS 18.48

7429. 1847–12–14 From Élie François Wartmann Geneva
Thanks JH for the gift of the *Cape Results*, and comments on his impending move to Lausanne.
ALS 2pp (French) RAS:JH Archive 10/5.129; Reel 9

7430. 1847–12–15 From William Rutter Dawes Camden Lodge
Has repaired the damage to his equatorial. Observed another of Wilhelm Struve's double stars last evening. Compares some of their star readings.
ALS 3pp RS:HS 6.87

7431. 1847–12–16 To John William Lubbock Collingwood
Arrangements for JL's presentation to be made to the R.A.S.
C 1p RS:HS 22.338

7432. 1847–12–16 To John William Lubbock Collingwood
Asks whether JL wishes to present method [see JH's 1847–12–4] at next R.A.S. meeting.
ALS 3pp RS LUB.H.360

7433. 1847–12–16 From Georg Merz & Son Munich
Some small disagreement about the bill for the refractor [see GM's 1847–10–16].
ALS 1p (German) TxU:H/M–0361; Reel 1087

7434. 1847–12–16 To William Henry Smyth Collingwood
Suggests John Lubbock's improvements in calculating planetary perturbations be discussed at an R.A.S. meeting. Plans to bring subject before the Council.
C 1p RS:HS 22.337b

7435. 1847–12–18 From John William Lubbock High Elms, Farnborough
Is obliged by his note and will accordingly write to Richard Sheepshanks. Spot on the sun was very visible to the naked eye recently. Looks forward to seeing JH.
ALS 3pp RS:HS 11.387

7436. 1847–12–20 From Henry Warburton 45 Cadogan Place
Verified Leonhard Euler's values for Bernoulli's numbers up to B=25 by method previously outlined and to be stated fully in this letter. Mr. Hensley [of Trinity College] proposes to revise incorrect values given in *Penny Cyclopaedia*. Computing a Bernoulli number by P. S. Laplace's formula

takes three days. HW's method of continued subtraction requires only 24 hours.
ALS 6pp RS:HS 18.49

7437. 1847–12–21 From Drummond's Bank London
Received three bills for £316 from Georg Merz and Sons, but JH's balance is only £19.
ALS 2pp TxU:H/M–0170; Reel 1087

7438. 1847–12–22 To Francis Beaufort Collingwood
Requests reimbursement for £316 JH paid to [Georg] Merz for two bills owed by Astronomer Royal.
AdraftS 2pp TxU:H/L-0087; Reel 1054

7439. 1847–12–22 From William Rutter Dawes Camden Lodge, Cranbrook
Should Mr. Morris, carpenter at Cranbrook, include canvas for covering dome, or will that be available at Cape Town? Improvements in WD's dome should be incorporated into Cape dome. Apologizes for WD's carelessness in managing telescope when JH visited.
ALS 4pp TxU:H/M–0159; Reel 1087

7440. 1847–12–22 From Samuel Heinrich Schwabe Dessau
Thanks JH for the gift of JH's *Cape Results.*
ALS 2pp (German) RAS:JH Archive 10/5.110; Reel 9

7441. 1847–12–22 From John T. Towson Devonport
Thanks JH for suggestion on his method of Great Circle sailing. Further enquires over some cartographic matters.
ALS 3pp RS:HS 17.360

7442. 1847–12–23 To George Biddell Airy Collingwood
Has been requested to prepare a manual of scientific instructions for the use of ship's officers, and is asking GA to prepare a section on astronomy.
LS 2pp RGO 6.2.135

7443. 1847–12–23 From Francis Beaufort Admiralty
Admiralty lords paid £316 to JH's account at Drummond's [Bank] for dome to house Cape of Good Hope telescope.
ALS 1p TxU:H/M–0066; Reel 1087 (ACC: Hydrographic Office Letter Book)

7444. 1847–12–23 To William Radcliffe Birt Collingwood
Lord Auckland [George Eden] asked JH to edit [*Admiralty Manual*] for officers on surveying and exploring expeditions. Charles Wheatstone will prepare instructions for meteorological observations. Asks WB to write instructions on detecting 'atmospheric waves and barometric fluctuations.'
(A)LS 3pp RS:HS 19.127

7445. 1847–12–23 From Johann Friedrich Ludwig Hausmann Göttingen
Acknowledges receipt of JH's *Cape Results* by Royal Society of Sciences of Göttingen.
ALS 1p (German) RAS:JH Archive 10/5.92; Reel 9

7446. 1847–12–23 To Sir William J. Hooker [Collingwood]
Lord Auckland [George Eden] has asked JH to edit *Admiralty Manual* and proposed that WH write section on botany. Would WH do this and when could WH complete it?
LS 2pp Royal Botanic Gardens

7447. 1847–12–23 To Adam Sedgwick Collingwood
Will edit a manual [*Admiralty Manual*] of Scientific Desiderata and Queries as requested by [George Eden,] Lord Auckland. Is AS preparing any material for this work? What is its content and when will it be ready?
LS 2pp RS:HS 15.432

7448. 1847–12–23 From Henry Warburton 45 Cadogan Place
Proved accuracy of Leonhard Euler's values for Bernoulli's 27th number. Plans to test number 29. Will investigate Thomas Clausen's theorem, the generality of which JH appears to have disproved.
ALS 4pp RS:HS 18.50

7449. 1847–12–23 To William Whewell Collingwood
Is editing a manual for surveying and exploring expeditions. Asks WW to do section on tides.
LS(A) 3pp TC, Camb. Add. Ms.a.207[71] (C: RS:HS 22.340)

7450. [1847?]–12–25 From Anne Elise Knipping Hanover
Caroline Herschel, JH's aunt, is very ill, and not expected to live much longer. AK expresses some concern about some financial matters.
ALS 4pp (German) TxU:H/M–0322; Reel 1087

7451. 1847–12–25 From Edward Sabine Woolwich
Preparing instruments and instructions for various expeditions. Discusses ES's wife's translation of [Alexander von] Humboldt's *Cosmos* and the [R.S.L.] Council.
ALS 4pp RS:HS 15.210

7452. [1847–12–25 or later]. From Adam Sedgwick n.p.
Thanks JH for letter of 7 December, inviting him to Collingwood for Christmas. Apologizes for not coming. Received bad domestic news and was robbed. Will testify at Old Bailey.
AL 4pp RS:HS 15.461 inc

7453. 1847–12–26 From George Biddell Airy Playford near Ipswich
Is willing to assist JH, but feels that the Admiralty must instruct GA to do so [see JH's 1847–12–23].
ALS 2pp RGO 6.2.136

7454. [1847?]–12–25 From Richard Sheepshanks Athenaeum
Informs JH that RS will present a case for more than one medal to be awarded.
ALS 4pp RS:HS 16.129

7455. 1847–12–26 From George Biddell Airy Playford near Ipswich
Has sent out a number of circulars to JH as a member of the Royal Observatory's Board of Visitors.
ALS 1p RGO 6.9.137

7456. 1847–12–27 From Francis Beaufort Admiralty
Will look into the affair of the overpayment for the telescope. Regarding geographical and hydrographical desiderata for the *Admiralty Manual.*
ALS 3pp RS:HS 3.379 & ACC: Hydrographic Office Letter Book

7457. 1847–12–27 From William Radcliffe Birt Bethnal Green
Is honored at George Eden's (2nd Baron Auckland) request for meteorological directions for atmospheric waves and barometric fluctuations. Will there be any engravings in the *Admiralty Manual*? The three annual barometric movements have occurred this year.
ALS 4pp RS:HS 4.119

7458. 1847–12–27 From [George] Buist St. Andrews University
Gratitude for receipt of JH's *Cape Results* by university library.
ALS 1p RAS:JH Archive 10/5.126; Reel 9

7459. 1847–12–27 From Sir William J. Hooker Royal Botanic Gardens, Kew
Has been requested by George Eden (1st Earl of Auckland) to contribute the botanical section to the manual to which JH alludes. Comments on this and the requirements for a manual of this type.
ALS 4pp RS:HS 9.459

7460. 1847–12–28 From Henry T. De La Beche Geological Survey Office
Has just seen his letter [apparently requesting an article on mineralogy for the *Admiralty Manual*] on returning to his office. Is very busy at the moment with a survey of coal for George Eden (Earl of Auckland), but could probably help, given time.
ALS 4pp RS:HS 6.130

7461. 1847–12–28 To George Eden Collingwood
All scientists named by GE, except [Adam] Sedgwick, will contribute to *Admiralty Manual*, but only one has a paper already prepared.
AdraftsS 4pp RS:HS 19.128 (C: RS:HS 22.341)

7462. 1847–12–28 From Sir Richard Owen Royal College of Surgeons
Introducing Benjamin Travers, who intends visiting the Cape of Good Hope. Would welcome advice from JH.
ALS 4pp RS:HS 13.188

7463. 1847–12–28 From Sir John Richardson Haslar Hospital
Asks: for name of instrument maker JH recommends for construction of actinometers; which hydrogrameters should be taken to Mackenzie's River; and whether JH has any experiments he would like done in America.
ALS 2pp RS:HS 14.486

7464. 1847–12–28 From Sears Cook Walker Washington, D.C.
Gratitude for receipt of JH's *Cape Results* by observatory of Central High School of Philadelphia.
ALS 1p RAS:JH Archive 10/5.146; Reel 9

7465. 1847–12–28 From Charles Wheatstone Lower Mall, Hammersmith
[George Eden] Lord Auckland seeks meteorological 'reports.' CW will contribute instructions on atmospheric electricity, photometry, and optical meteorology. W. R. Birt will report on meteorological instruments and wind observations.
ALS 3pp † RS:HS 18.152

7466. 1847–12–29 From George Eden Admiralty
Has now sent the letters and circulars to the various people in connection with the *Admiralty Manual.* Has not sent one to W. R. Birt, but will do so if JH thinks it advisable.
ALS 2pp RS:HS 7.11

7467. 1847–12–29 From Sir Robert Harry Inglis Milton Bryan, Woburn
Wishes him the compliments of the season. Can JH dine with RI and Edward Sabine next Wednesday?
ALS 2pp RS:HS 10.239

7468. 1847–12–30 To [Augustus De Morgan?] Collingwood
Is trying to establish the nature of C. I. Gerhardt's request about letters of G. W. Leibniz.
ALS 2pp APS B.H435p.14

7469. [1847]–12–30 From William Empson East India College, Hertford
Francis Jeffrey (Lord Jeffrey) is the writer of *Plaidoyer* in the last number; intends it to be a judicial deliverance. Has sent JH's review of the translation of *Kosmos* to Lord Jeffrey and he praised it highly. Gives some of his comments.
AL inc 4pp RS:HS 7.56

7470. [1847–12–30 or later ?] To [Sir John Richardson] [Collingwood]
Provides details about where to buy actinometers, and what type to buy.
ALS 2pp WT 64710.309

7471. 1847–12–30 To Richard Sheepshanks Collingwood
Adds list of names for consideration as foreign associates [of the R.A.S.], including Americans. Changes in John Lubbock's requirements for his project.
ALS 4pp RS:HS 16.81 (C: RS:HS 22:342 inc & 25.9.49)

7472. 1847–12–30 From Henry Warburton Ore near Hastings
Demonstrates property of 'Coefficients of the terms of the Numerators of the Fractions which generate the odd powers of the Numbers of the Natural Series.' Claims this method can compute Bernoulli's numbers 37, 39, and 41.
ALS 5pp RS:HS 18.51

7473. 1847–12–30 To Sir John [Richardson] Collingwood
[Responding to JR's 1847–12–28], JH gives advice where to buy actinometers and what features to look for. JH cannot advise about hygrometers as these are to be used in very cold climates.
ALS 3pp WT 67597.4

7474. 1847–12–31 From John Couch Adams St. John's College, Cambridge
Thanking him for his hospitality.
ALS 3pp RS:HS 1.2

7475. 1847–12–31 From Henry T. De La Beche Geological Survey Office
It appears that Mineralogy has been assigned to him in the new manual. Comments on this and would like to know who will be writing the chapter on Geology as it is likely to overlap.
AL 3pp inc RS:HS 6.131

7476. 1847–12–31 From Sir Richard Owen Royal College of Surgeons
Has received JH's request regarding the *Admiralty Manual.* Will be pleased to contribute a section on collecting specimens.
ALS 3pp RS:HS 13.189

7477. 1847–12–31 From Algernon Percy (4th Duke of Northumberland) Stanwick, Darlington
Congratulations on success [of JH's *Cape Results*], and happy new year. Encloses letter from Austrian ambassador with Emperor's testimonial to JH. Exchanged private notes with Count Jarnac regarding French king's intention to reward JH. Hopes that British government will permit JH to accept special honor from Louis Phillipe.
ALS 4pp RAS:JH Archive 10/4.26; Reel 9

7478. 1847–12–31 From Edward Sabine Woolwich
Received memorandum from JH concerning magnetic instruments and instructions for their use. Happy to comply with Lordship's request for general instructions for magnetic instruments. Received meteorological observations from around the world.
ALS 4pp RS:HS 15.212

7479. 1847–12–31 From Edward Sabine Woolwich
Discusses the meteorological observations made on board the *Pagoda*, particularly the effect of latitude and longitude on barometric readings.
ALS 4pp RS:HS 15.211

7480. 1848 To D. François J. Arago n.p.
Glad to learn of the Academy's interest in his Cape Observations. Regarding his father. Comments on the 'Equalizations of stars.' [Written on an experimental leaf photographic plate.]
draft 4pp RS:HS 1.366 (C: RS:HS 25.13.9)

7481. [1848?] To Charles Babbage n.p.
Thanks for his offer of help in the trusteeship. Also for his pamphlet on income tax. Is about to become involved in Chancery proceedings. Comments on tax.
ALS 4pp RS:HS 2.388

7482. 1848 or later From Julia Margaret Cameron n.p.
Health of family.
ALS 6pp inc RS:HS 5.178 & 179

7483. [1848] From William Empson Craig Crook, Edinburgh
Has he made a decision regarding writing a review for the second volume of *Kosmos*? Longmans is contemplating a second edition. George Peacock is much better.
ALS 4pp RS:HS 7.62

7484. [1848 to 1852] To Margaret Brodie Herschel n.p.
JH learned from daughters that JH and MBH *must* appear in drawing room for birthday celebration after girls return from Ball.
ALS 1p TxU:H/L-0536.3; Reel 1053

7485. 1848–1–1 To Margaret Brodie Herschel Collingwood
Is glad to see that P[eter] S[tewart]'s prospects have improved [see JH's 1847–10–10]; JH promises MH a Daguerreotype, and comments on family activities.
ALS 4pp JHS 6.31

7486. 1848 To Adam Sedgwick Collingwood
Family will arrive at Norwich on 16 May. Discusses Norfolk strata and Cray fossils.
ALS 3pp RS:HS 15.441 (C: RS:HS 22.233)

7487. [1848 to 1855] To Richard Sheepshanks n.p.
Asks RS to summarize T. G. Taylor's contributions to astronomy. Hopes to assist Taylor's widow and family.
ACCS 1p TxU:H/L-0383; Reel 1054

7488. [1848?] From Hensleigh Wedgwood 7 Lancaster Place, Strand
Sends small work on foundations of thought, with citation of JH's article in *Quarterly* [*Review*]. Note on geometry of planes.
ALS 2pp RS:HS 18.135

7489. 1848–1–2 From Adam Sedgwick Trinity College, Cambridge
Is ill and cannot write instructions. Directs JH to instructions written earlier by the Council of the Geological Society.
ALS 3pp RS:HS 15.433

7490. 1848–1–2 From Otto Wilhelm Struve Pulkowa
Sends JH a copy of his paper on the satellites of Uranus, including comments on the work of JH's father, William. OS is amazed at what JH has accomplished in the last several years.
ALS 2pp (German) JH note RS:HS 17.245

7491. 1848–1–3 From Admiralty Admiralty Office
Advising JH that the Admiralty will be sending him information which they have received from G. B. Airy, Edward Sabine, and Richard Owen.
ALS 1p RS:HS 1.15

7492. 1848–1–3 From William Rutter Dawes Camden Lodge
Mr. Morris, the Cranbrook carpenter, has the plan and specification ready. Would JH like to see it before sending it to the Admiralty, presumably through Francis Beaufort?
ALS 4pp RS:HS 6.88

7493. [1848–1–3?] To Algernon Percy [Collingwood]
Thanks for New Year's wishes. Honored by French king's award, but asks AP not to press this matter. British government will not permit JH to accept it, and JH does not want to draw attention to himself while trying to arrange memorial to Thomas Maclear.
Adraft 2pp RAS:JH Archive 10/4.28; Reel 9

7494. [1848–1–3?] To Algernon Percy (4th Duke of Northumberland) [Collingwood]
Returns letter from French ambassador, Count Jarnac. Deeply honored by French king's desire to bestow 'Cross of the Legion of Honour' on JH, but rules of British government do not permit him to accept it.
Adraft 2pp RAS:JH Archive 10/4.27; Reel 9

7495. 1848–1–4 To Adam Sedgwick Collingwood
Because AS cannot visit JH, JH has sent a circular. Assumes circular of [George Eden,] Lord Auckland will also be received. Will wait for response to query.
ALS 1p RS:HS 15.434 (C: RS:HS 23.1)

7496. 1848–1–5 To William Radcliffe Birt Collingwood
Encloses memorandum on *Admiralty Manual*. General meteorology is assigned to Charles Wheatstone, but special section [on barometer observations] will be attributed to WB.
ALS 2pp RS:HS 19.129

7497. 1848–1–5 To Sir William J. Hooker Collingwood
Likes what WH has outlined on botany for JH's *Admiralty Manual*. Suggests some additions.
ALS 2pp Royal Botanic Gardens

7498. 1848–1–5 From Edward Sabine Woolwich
R.S.L. selecting foreign members. Asks JH's opinion of a number of foreign scientists and also his opinion of what he, as foreign secretary, should do in this matter.
ALS 4pp RS:HS 15.213

7499. 1848–1–5 To William Whewell Collingwood
Suggests how WW should write his contribution giving instructions for tidal observations for JH's *Admiralty Manual*. Notes the death of JH's butler.
AL 3pp TC, Camb. Add. Ms.a.207[72] (C: RS:HS 23.2)

7500. 1848–1–6 From George Biddell Airy Playford near Ipswich
Responds to some of JH's concerns [see JH's 1847–12–30]

about changes in instrumentation at the Royal Observatory.
ALS 3pp TxU:H/M-0634.1; Reel 1084 (C: RGO 6.9.141)

7501. 1848–1–7 From Admiralty Admiralty Office
Sending copies of Admiralty Memorandum of 7 December, and requesting JH to request papers from Edward Sabine.
ALS 1p RS:HS 1.16

7502. 1848–1–7 From George Biddell Airy Playford near Ipswich
A note accompanying a printed circular about the proposed changes in instrumentation at the Royal Observatory. [Encloses GA's circular to Board of Visitors.]
ALS 1p, encl L 2pp TxU:H/M-0634.2; Reel 1084

7503. 1848–1–8 From William Radcliffe Birt Bethnal Green
Encloses the manuscript of the directions for atmospheric waves and barometric fluctuations. Suggests that it would be useful if the vessel about to travel to New Zealand would be utilized to supply barometric readings en route.
ALS 4pp RS:HS 4.120

7504. [1848]–1–8 From William Empson 29 George St., Hanover Square
Many thanks for his replies to the queries of Francis Jeffrey's (Lord Jeffrey); they will be appreciated. Use of the word 'vindicated' not happy. Mr. Moriarty called yesterday; he is not appreciated in England. Will send copy of JH's review to Alexander von Humboldt. Hopes he will review the 2nd volume. The Joneses are at Pall Mall, and William Whewell in Paris.
ALS 7pp RS:HS 7.58

7505. 1848–1–8 From William Keddie Glasgow
[Form letter] Acknowledges receipt of JH's *Cape Results* by library of Philosophical Society.
(A)LS 1p RAS:JH Archive 10/5.127; Reel 9

7506. 1848–1–8 From James P. Muirhead Aston Hall near Birmingham
Acknowledges JH's letter on the water controversy. Draws his attention to an article on the subject in the current *Edinburgh Review*. Comments on the various theories.
ALS 7pp RS:HS 12.382

7507. 1848–1–8 From Denison Olmsted Yale College
Gratitude for receipt of JH's *Cape Results*. Sends copies of notice of JH's *Cape Results* prepared by DO for *American Journal of Science*.
L 1p † RAS:JH Archive 10/5.147; Reel 9

7508. 1848–1–9 From George Biddell Airy Playford near Ipswich
A note accompanying some manuscript pages of instructions for ship's officers [see JH's 1847–12–23].
ALS 1p RGO 6.2.143

7509. 1848–1–10 To George Biddell Airy Collingwood
Thanks GA for his contributions to some material JH is compiling; asks about date of Board of Visitors meeting.
ALS 3pp RGO 6.9.151

7510. 1848–1–10 To George Biddell Airy n.p.
Has received GA's manuscript instructions for ship's officers [see GA's 1848–1–9]; there is no hurry on the remaining pages.
LCC 1p RGO 6.2.153

7511. [1848–1–10 or later] From William Somerville to Lady Herschel n.p.
Condolences regarding sad event [death of Caroline Herschel]. Somervilles enjoyed their visit to Collingwood. Has sent a box of gifts.
ALS 4pp RS:HS 16.410

7512. 1848–1–11 To Richard Sheepshanks Collingwood
Praises RS for work on *R.A.S.M.N.* Will arrive in time for the R.A.S. council meeting. Has heard nothing of John Lubbock's paper or lecture, except through J. C. Adams. Fears that before the R.A.S. anniversary, it will lose one of its female associates, his aunt Caroline.
ALS 2pp RS:HS 16.82 (C: RS:HS 25.9.50)

7513. 1848–1–11 To William Henry Smyth Collingwood
Suggests WS take the chair at the next R.S.L. meeting. Discusses a project of John Lubbock's. Fears Caroline Herschel will not survive the year.
C 1p RS:HS 23.3

7514. 1848–1–12 To Sophie Beckedorff Collingwood
Gratitude for news [of Caroline Herschel], whose death seems imminent. Notes on burial, disposition of property, and transfer of papers to JH. Sends obituary written by JH for use at appropriate time. Will send inscription for headstone later.
AdraftS 4pp TxU:H/L-0088; Reel 1054

7515. 1848–1–12 From Thomas Galloway 45 Torrington Square
Has sent the review of George Everest's *Indian Survey* to Longmans. Fears it is hard and dry after JH's review of *Kosmos*. If JH decides to publish it, would he let him have it back for revision.
AL 2pp inc RS:HS 8.29

7516. 1848–1–14 From [E. W.] L. Kirchoff Hanover
Writes about matters relating to the death of JH's aunt Caroline.
ALS 4pp (German) TxU:H/M–0317; Reel 1087

7517. 1848–1–15 From William Radcliffe Birt Bethnal Green
Omitted a detail concerning observations in his last letter. Further regarding the proposed observations in New Zealand.
ALS 5pp RS:HS 4.121

7518. 1848–1–16 From James Constantine Adamson St. Stephen's Square, Cape Town
Thanks for personal copy of JH's *Cape Results*.
ALS 2pp RAS:JH Archive 10/5.12; Reel 9

7519. [1848]–1–16 From William Empson 29 George St., Hanover Square
[George] Peacock asks for more time for his review. Enough science in vol. 2 to tempt JH. JH should be the new editor for Brown. Will send Alexander von Humboldt a copy of the review unless he thinks it should come from the author and not the editors. Returns to Haileybury on Thursday.
ALS 4pp RS:HS 7.59

7520. 1848–1–17 From William Henry Sykes India House
Requests letter ascertaining William Mann's competence, assuring his hiring as astronomer at Madras.
ALS 1p encl RS:HS 17.144a

7521. 1848[–1–17 or later] To William Henry Sykes n.p.
JH writes to the best of his knowledge about William Mann's competence in various areas.
A 1p RS:HS 17.144b

7522. 1848–1–18 From James Constantine Adamson South African College
Expresses gratitude from Council of Directors for library copy of JH's *Cape Results*.
ALS 1p RAS:JH Archive 10/5.13; Reel 9

7523. [1848?]–1–18 From Anne Elise Knipping Hanover
Writes of the death of JH's aunt Caroline. Is happy that Caroline is at rest. Describes funeral. Deals with some matters related to servants and other business matters.
ALS 4pp (German) TxU:H/M–0320; Reel 1087

7524. 1848–1–19 To William Lassell Collingwood
Urges WL to make public his observations of the seventh satellite of Saturn as Otto Struve is about to announce his observation of the same body.
C 1p RS:HS 23.4

7525. 1848–1–19 From Thomas Maclear Royal Observatory, Cape of Good Hope
Illness in the family has prevented him from acknowledging JH's splendid work, which arrived some weeks ago. Is grateful for JH's appreciation of TM's work. Wife has been very ill. C. F. H. Ludwig has died. H. G. W. Smith has said there must be a Botanic Garden at the Cape. New treaty regarding Kaffir land may increase the chance of peace.
ALS 4pp RS:HS 12.139

7526. 1848–1–19 To Richard Sheepshanks Collingwood
Recommends that the form that a testimonial should take would be the presentation of a parchment bearing the seal of the R.A.S. rather than a bound book.
C 1p RS:HS 23.5

7527. 1848–1–19 To Richard Sheepshanks Collingwood
RS will receive two communications for the R.A.S., one from S. C. Walker on Neptune's elements, the other from Otto Struve on the interior satellite of Uranus. Walker's could go into the *R.A.S.M.N.*, whereas Struve's ought be read at a regular meeting. Recommends reading of William Lassell's work on Neptune's satellite as it predates Struve's. Caroline Herschel died on the 9 [Jan.] at age 98.
ALS 2pp RS:HS 16.83 (C: RS:HS 23.6 & 25.9.51)

7528. 1848–1–20 From Francis Beaufort Admiralty
Cannot find any record of his having written. Is sending an official request from the Admiralty for JH to superintend the dome of the Cape telescope. Advice regarding his overpayment of money to the Astronomer Royal.
ALS 4pp RS:HS 3.380 & ACC: Hydrographic Office Letter Book

7529. 1848–1–20 From Palm Heinrich Ludwig Boguslawski Breslau
Letter of condolence after death of Caroline Herschel; comments on Herschel family accomplishments in astronomy.
ALS 4pp (German) RS:HS 4.176

7530. 1848–1–20 To William Howley Collingwood
Asks Archbishop of Canterbury [William Howley] to accept JH's resignation. Impossible to give adequate attention to JH's duties as trustee [of British Museum].
AdraftS 2pp RS:HS 19.130 (C: RS:HS 23.7)

7531. 1848–1–20 From Sir Richard Owen Royal College of Surgeons
The packet from the Admiralty was not the manuscript; that will follow speedily. Regrets any anxiety caused.
ALS 4pp RS:HS 13.190

7532. 1848–1–20 From Henry George Ward Admiralty
Permission is granted for JH to order construction of dome for Cape of Good Hope observatory.
ALS 1p TxU:H/M–0559; Reel 1093

7533. 1848–1–21 From George Biddell Airy Playford near Ipswich
Proposes changes in the buildings at the Royal Observatory to accommodate the shifting of one of the telescopes [see GA's 1848–1–6].
ALS 3pp TxU:H/M-0634.3; Reel 1084 (C: RGO 6.9.172)

7534. 1848–1–22 To George Biddell Airy Collingwood
Thinks the building changes proposed [see GA's 1848–1–21] are fine; explains JH's position on the whole matter.
ALS 3pp RGO 6.9.174

7535. 1848–1–22 To George Biddell Airy n.p,
Comments on all of GA's instructions for ship's officers [see JH's 1848–1–10], mostly suggesting some additions about meteor showers.
ALS 2pp inc RGO 6.2.154

7536. 1848–1–22 From Thomas Galloway 45 Torrington Square
Sending particulars concerning George Everest's Arc of Meridian. Would be glad if they could be returned when he has read them as he intends writing an article on them for Taylor's Journal.
ALS 2pp RS:HS 8.30

7537. 1848–1–22 From William Lassell Starfield, Liverpool
Has sent his observations on Saturn and Neptune to the R.A.S. and encloses a copy for JH. Weather has been unsuitable for many observations recently. Has not heard of any observations of a second satellite of Neptune.
ALS 2pp RS:HS 11.135

7538. 1848–1–22 To Adam Sedgwick from M. B. Herschel Collingwood
Looks forward to seeing AS at Norwich. Hopes he is recovering. Caroline Herschel has died.
ALS 4pp RS:HS 15.135

7539. 1848–1–23 From Augustus De Morgan 7 Camden St.
Enclosing details of the proposed testimonial for meritorious work in astronomy. Would like confirmation of the date of birth of JH's aunt.
ALS 4pp RS:HS 6.238

7540. 1848–1–23 To Adam Sedgwick Collingwood
Accepts invitation to Norwich. For Geological Instructions, JH will wait to hear from [George Eden,] Lord Auckland. If necessary, he will consult [Charles] Darwin regarding the Instructions.
ALS 2pp RS:HS 15.436

7541. 1848–1–24 From Edward Sabine Woolwich
[S. J. A. Compton,] Lord Northampton will resign as president of R.S.L. ES discusses proposed changes in office of president, and asks JH to consider taking the office under the newly proposed changes.
ALS 4pp RS:HS 15.214

7542. 1848–1–24 To Edward Stanley Collingwood
Informs ES [President of the Royal College of Surgeons] that JH is unable to attend the Hunterian Oration and the dinner.
ALS 1p BL Aaa 42240.19

7543. 1848–1–25 To Edward Sabine Collingwood
Regrets to hear of retirement of Spencer Compton (Marquis of Northampton) from R.S.L. presidency. Is grateful that ES and others wish him to hold office temporarily, but regards it as 'impossible.'
ALS 4pp RS Sa.655 (draft: RS:HS 15.215 & C: RS:HS 23.8)

7544. 1848–1–27 To Augustus De Morgan Collingwood
Some comments about dates related to JH's aunt Caroline Herschel.
C 1p RS:HS 23.9

7545. 1848–1–27 From Augustus De Morgan 7 Camden St.
The memoir of JH's aunt is forthcoming. Comments on old and new style in calendars.
ALS 4pp RS:HS 6.239

7546. 1848–1–29 From George Biddell Airy Royal Observatory, Greenwich
Further regarding the distribution of the J. J. F. Lalande catalogue.
ALS 2pp RS:HS 1.192 (C: RGO 6.203.397)

7547. 1848–1–29 From George Biddell Airy Royal Observatory, Greenwich
Responds to a number of JH's concerns about buildings, instruments and printing at the Royal Observatory [see JH's 1848–1–22 & 1847–12–30].
ALS 4pp TxU:H/M-0634.4; Reel 1084 (C: RGO 6.9.183

7548. 1848–1–29 From George Biddell Airy Royal Observatory, Greenwich
Responds to suggestions by JH [see JH's 1848–1–22]; is worried about giving too much information.
ACCS 2pp RGO 6.2.155

7549. 1848–1–31 From Augustus De Morgan 7 Camden St.
Encloses article for JH to work on. Can he inform him of Encke's Christian names?
ALS 1p RS:HS 6.240

7550. 1848–1–31 From George Eden Admiralty
Is sorry to hear that they are not to have the services of Adam Sedgwick. Would be pleased if JH would write to C. R. Darwin. Thinks that 200 pages will be quite long enough for the *Admiralty Manual*.
ALS 2pp RS:HS 7.12

7551. [1848?]–1–29 From Anne Elise Knipping Hanover
About legal matters related to the death of Aunt Caroline, as well as identifying friends and dignitaries who paid their respects. Deals with bills to be paid, and with the question of a memorial tablet in the church.
ALS 7pp (German) TxU:H/M–0321; Reel 1087

7552. 1848–1–29 To Anne Elise Knipping Collingwood
Waited for legal documents from L. Kirchoff before responding to AK's letter announcing death of JH's aunt Caroline. Named AK's son Adolph as JH's attorney. Authorizes disposition of CLH's papers and possessions. Approves tombstone inscription by Mr. Müller. Donates portrait of William Herschel to Mrs. Groskopff.
ACCS 4pp & AdraftS 4pp TxU:H/L-0216 & -0215; Reel 1054

7553. 1848–1–29 From Edward Sabine Woolwich
Sends own article on magnetic surveys at sea. Requests comments. Discusses foreign distribution of star catalogues approved by the council and election of foreign members of R.S.L.
ALS 4pp RS:HS 15.216

7554. 1848–1–[31?] From Sophia Beckedorff Hanover
Caroline Herschel's death and funeral. Awaiting JH's instructions for disposition of property and headstone inscription. Erroneous biographies appearing in French and German papers. Encloses copy of response from Mr. Hahn, bookseller, to manuscript by Margaret Louisa Herschel. [P.S., 3 Feb. 1848] Accidentally burned JH's Latin inscription for Caroline Herschel. Does JH have copy?
ALS 6pp, encl 2pp (German) TxU:H/M–0078; Reel 1087

7555. 1848–1–31 From Thomas Galloway 45 Torrington Square
The paragraph he sent JH was not intended for his address. Augustus De Morgan must be under some misapprehension. Has asked De Morgan to send JH the revised form.
ALS 3pp RS:HS 8.31

7556. 1848–1–31 From William Lassell Starfield, Liverpool
Has received a letter from W. C. Bond of the Cambridge Observatory (U.S.A.), who speaks of the satellite of Neptune. Bond also sends details of the stars around Orion. Relates his own work on the specula of telescopes. Intends to submit a paper on this to the R.A.S.
ALS 6pp RS:HS 11.136

7557. 1848–2 To George Biddell Airy n.p.
JH comments on quality of Georg Merz object glasses.
draft 1p inc RS:HS 23.12

7558. 1848–2–1 From George Biddell Airy Royal Observatory, Greenwich
Regarding spots on the sun and [T. J.?] Hussey and [A.?] Dawson's observations.
ALS 2pp RS:HS 1.193 (C: RGO 6.230.443)

7559. 1848–2–1 From Francis Beaufort Admiralty
JH must write official letter to Admiralty stating completion of [Cape of Good Hope] telescope and explaining reason for additional £170.
ALS 1p † TxU:H/M–0067; Reel 1087

7560. 1848–2–2 To George Biddell Airy Collingwood
Is aware of T. J. Hussey's drawings [see GA's 1848–2–1]; needs information about P. A. Hansen's lunar theory for an appreciation JH is to provide.
ALS 2pp RGO 6.230.444

7561. 1848–2–2 To George Biddell Airy n.p.
Agrees that some of JH's suggestions about meteor showers were probably unnecessary [see GA's 1848–1–29].
ALS 1p inc RGO 6.2.156

7562. 1848–2–2 From Eduard Heis Aix la Chapelle [Aachen]
Gratitude for receipt of JH's *Cape Results*. Owning no instruments, EH appreciated JH's chapter on 'naked eye' astronomy. Notes and drawings of variable stars.
ALS 2pp † RAS:JH Archive 10/5.111; Reel 9

7563. 1848–2–3 From George Biddell Airy Royal Observatory, Greenwich
Concerning P. A. Hansen's observations of Lunar irregularities.
ALS 4pp RS:HS 1.194 (C: RGO 6.230.221)

7564. 1848–2–3 From William Rutter Dawes Camden Lodge
Has found that *Orionis* is a double star. Compares William Lassell's and Wilhelm Struve's observations of various stars.
ALS 3pp RS:HS 6.89

7565. 1848–2–3 To Augustus De Morgan Collingwood
Some additional details about JH's aunt Caroline [see JH's 1848–1–27].
C 1p RS:HS 23.10

7566. 1848–2–3 To Richard Sheepshanks Collingwood
JH must arrange for the celebration of Bishop's Observatory. Needs information on its equipment, achievements, and staff, e.g., on W. R. Dawes's and J. R. Hind's observations. Will contact Hind.
ALS 4pp RS:HS 16.84 (C: RS:HS 25.9.52)

7567. 1848–2–4 From Charles Robert Darwin Down, Farnborough
As JH has accepted the superintendency of the *Admiralty Manual* he will try to undertake the geological part, but it will take time as he is busy on another project.
ALS 2pp RS:HS 6.11 Darwin, *Correspondence, 4,* 107

7568. 1848–2–4 From Augustus De Morgan 7 Camden St.
Sending something from Airy relating to the testimonial. Anything he does not use can be returned to Somerset House for AD.
ALS 2pp RS:HS 6.241

7569. 1848–2–4 From William Henry Fox Talbot Lacock Abbey
Has performed experiments measuring the deviation of spectra through an obliquely placed diffraction grating. Believes they strongly support the wave theory of light.
ALS 4pp RS:HS 17.320

7570. 1848–2–5 From Francis Beaufort Admiralty
Received yesterday JH's receipts for payment of Cape telescope. Forwarded these to accountant general of Navy.
ALS 1p TxU:H/M–0068; Reel 1087 (ACC: Hydrographic Office Letter Book)

7571. 1848–2–5 From Francis Beaufort n.p.
Informing him of the recent death of Mrs. Lestock Wilson.
ALS 3pp RS:HS 3.381

7572. 1848–2–7 To Augustus De Morgan Collingwood
Needs clarification about John Lubbock's planetary theory; thanks for G. B. Airy's lunar reductions.
C 1p RS:HS 23.11

7573. 1848–2–7 From M. Macann 56 Upper Seymour St.
Has been carrying out researches on the reflection of light. Would like JH to communicate his paper on this subject to the R.S.L.
ALS 4pp RS:HS 12.1

7574. 1848–2–7 From [?] Hanover
[P.S. on verso of cover. Rest of letter is missing.] Notified Mrs. Knipping of kind wishes sent by JH and wife. Will notify JH when personal effects of Caroline Lucretia Herschel have been inventoried and distributed according to JH's directions. Questions fate of [telescope] given to [J. F. L.] Hausmann by Caroline Herschel now that Hausmann is dead. Suggests donating it to observatory at Göttingen or to 'Archiv' in Hanover.
AL 1p TxU:H/M–0916; Reel 1083

7575. 1848–2–8 From Pierre Daussy Paris
Thanks JH for the gift of a copy of JH's *Cape Results,* and admires it greatly.
ALS 1p (French) RAS:JH Archive 10/5.88; Reel 9

7576. 1848–2–8 To Richard Sheepshanks Collingwood
Returns [Max] Weisse's manuscript. He must say something about it and so has swallowed 'whole hog' what RS had written about it. Will make a mess [in his R.A.S. address] of John Lubbock's work on perturbations.
ALS 3pp RS:HS 16.85 (C: RS:HS 25.9.53)

7577. 1848–2–9 From Alfred Gautier Geneva
Informs JH that JH has been made an honorary member of the Physical and Natural History Society of Geneva; the Society thanks JH for a copy of his *Cape Results.* Comments on F. V. Mauvais's discovery of a comet and on the illness and death of Caroline Herschel. Will send comments on some of JH's astronomical work later.
ALS 3pp (French) RAS:JH Archive 10/5.130; Reel 9

7578. 1848–2–9 From Alfred Miles Admiralty
Paid £5 into JH's account at Drummond's Bank today.
ALS 1p TxU:H/M–0367; Reel 1087

7579. 1848–2–10 From Algernon Percy (4th Duke of Northumberland) Whitehall Gardens
Arrangements for a meeting with JH.
ALS 2pp RS:HS 13.358

7580. 1848–2–11 From Francis Beaufort Admiralty
Financial matters are dealt with relating to the purchase of a telescope [for Royal Observatory, Cape of Good Hope].
ACC 1p Hydrographic Office Letter Book

7581. 1848–2–11 From George Eden Admiralty
Sends the pamphlet by W. Snow Harris on lightning conductors. May find some extracts useful for the *Admiralty Manual.* Also extracts from William Reid and Mr. Liddington's work on hurricanes.
ALS 2pp RS:HS 7.13

7582. 1848–2–11 From Michael Faraday Royal Institution
Leopold Gmelin's *Handbuch der Chemie* contains all JH requires. If he does not have access to it will send it to the R.S.L. for him.
ALS 1p RS:HS 7.190

7583. 1848–2–11 To Margaret Brodie Herschel 37 Tavistock Place
Lists a number of people JH called on but all were out. Received a gold medal with the head of the King of the Netherlands on it. Also received a 'very superb and truly Imperial' ring from 'the Duke.' [This, presumably, is the diamond ring he received from the Emperor of Austria.]
ALS 3pp JHS 6.32

7584. 1848–2–11 From Sir Robert Harry Inglis 7 Bedford Square
JH's letter grieves him, but he will convey his decision to the Trustees. Has just heard of the death of the Archbishop [of Canterbury, William Howley].
ALS 2pp RS:HS 10.240

7585. 1848–2–12 From Admiralty Admiralty Office
Informing JH that J. C. Pritchard [Prichard] has been asked to send him material for chapter on Ethnology for 'Scientific Manual' for naval officers, which JH is preparing.
ALS 1p RS:HS 1.17

7586. 1848–2–[12] From M. Macann 56 Upper Seymour St.
Is grateful for JH's kind offer to look over his paper. Draws JH's attention to certain points. Thinks that the reasoning will please him.
ALS 3pp RS:HS 12.2

7587. 1848–2–16 To Charles Babbage Collingwood
Regarding the originator of the R.A.S. Regarding his duties as trustee and other legal matters.
ALS 4pp RS:HS 2.318

7588. 1848–2–16 From George Eden Admiralty
Suggesting a title for the *Admiralty Manual.* Thinks it early to commence printing as papers may be delayed. Wording on the circular may do for the preface.
ALS 4pp RS:HS 7.14

7589. 1848–2–16 To William Whewell Collingwood
Gives directions for and raises questions concerning WW's contribution to JH's *Admiralty Manual.*
ALS 3pp TC, Camb. Add. Ms.a.207[73] (C: RS:HS 23.13)

7590. 1848–2–17 From George Biddell Airy Royal Observatory, Greenwich
Regarding the proposed eight-inch object glass. Supplementary estimate for the Admiralty. Extract of a letter from Otto Struve.
ALS 3pp RS:HS 1.195 (C: RGO 6.9.192 & 6.368.691)

7591. 1848–2–17 To Richard Sheepshanks Collingwood
Parramatta Observatory has been broken up and is now defunct.
ALS 3pp RS:HS 16.86 (C: RS:HS 25.9.54)

7592. 1848–2–18 From Thomas Williams Northumberland House
Requested letters from [Hugh Percy] late Duke of Northumberland to JH offering to pay for publishing [*Cape Results*]. Hopes to include that debt in late Duke's estate, rather than as obligation upon [Algernon Percy] new Duke. Encloses receipt for £1,000. [JH note (20 Feb 1848): Sent receipt to TW. Expenses came within £50 of Duke's gift.]
ALS 3pp, JH note 1p RS:HS 18.412

7593. 1848–2–20 From Thomas Maclear Royal Observatory, Cape of Good Hope
Gives recent observations with the 46" achromatic. G. B. Airy has sent the elements of the recently discovered planets. Latest news of the Governor of the Cape and his diplomatic missions. Mrs. Maclear's sister has come out to visit them and her health and Mrs. Maclear's is improving. R. J. Jones has decided to keep Feldhausen in the family and not sell it.
ALS 8pp RS:HS 12.140

7594. 1848–2–23 To [William Radcliffe Birt] Collingwood
Received drafts from WB and Charles Wheatstone for *Admiralty Manual.* JH must edit and connect them. Promises to preserve WB's intent and to give full credit.
AL 2pp RS:HS 19.131

7595. 1848–2–24 From George Biddell Airy Royal Observatory, Greenwich
Regarding modifications for the actinometer. Has written to Georg Merz regarding the object-glass. Regarding the eight-inch object-glass. Name of the maker of his actinometer. [JH has written a rough draft of his reply to this letter on the same sheet.]
ALS 2pp RS:HS 1.196 (C: RGO 6.701.112)

7596. 1848–2–24 From Smith, Elder & Co. 65 Cornhill
Sends statement of JH's account [for publishing *Cape Results*]. Sales of book are less than anticipated but continually increasing.
ALS 1p, encl 6pp RAS:JH Archive 10/3.18; Reel 9

7597. 1848–2–26 To George Biddell Airy Collingwood
Surprised by GA's unhappiness with George Merz [see GA's 1848–2–17]; suggests GA seek information from Thomas Maclear at the Cape, who is getting a new telescope from George Merz.
ALS 3pp RGO 6.9.195

7598. 1848–2–26 From William Radcliffe Birt Bethnal Green
Will be happy to abide by JH's judgement. Would like to know if the publication of the instructions in the manual would interfere with their separate publication. Will have to obtain financial assistance if his observations are to be continued.
ALS 3pp RS:HS 4.122

7599. 1848–2–26 From Charles Wheatstone Lower Mall, Hammersmith
Seriously ill. Will forward manuscripts to JH by next week.
ALS 1p RS:HS 18.153

7600. 1848–2–27 From George Biddell Airy Royal Observatory, Greenwich
Georg Merz's practice regarding the sale of object-glasses. Is preparing a model of a transit circle.
ALS 2pp RS:HS 1.197 (C: RGO 6.9.197)

7601. 1848–2–29 From Charles Babbage Dorset St.
Regarding the origins of the R.A.S. Will try and find a substitute trustee.
ALS 2pp RS:HS 2.319

7602. 1848–2–29 To William Henry Fox Talbot Collingwood
Refers WT to reference work to check WT's experiments [see WT's 1848–2–4]. JH says he will perform no more experiments in physical optics.
ALS 3pp Science Museum Talbt 1/38 (C: RS:HS 23.14)

7603. [1848–3?] To [William Radcliffe Birt] n.p.
Admiralty Manual must be succinct. WB's specialized notes [on atmospheric waves] are too extensive.
ALS 3pp RS:HS 19.170

7604. 1848–3–1 To Augustus De Morgan Collingwood
On the origin of the R.A.S.
C 1p RS:HS 23.15

7605. [1848–3] To Sir William J. Hooker [Collingwood]
Extent of WH's manuscript for JH's *Admiralty Manual* is fine. Please send it as soon as possible.
ALS 2pp Royal Botanic Gardens

7606. 1848–3–1 From Robert Treat Paine Boston, Mass.
Acknowledges receipt of JH's *Cape Results.* Offers map of Massachusetts, constructed from RP's triangulation and astronomical observation. Zeal for astronomy is increasing in America. Reports discoveries by W. C. Bond and total solar eclipse of Nov. 1834.
ALS 2pp RAS:JH Archive 10/5.148; Reel 9

7607. 1848–3–3 From Sir Robert Harry Inglis 7 Bedford Square
Has written a letter to Mrs. James Maclaren and enclosed

three orders of admission to the British Museum.
ALS 2pp RS:HS 10.241

7608. 1848–3–3 From Edward Sabine Woolwich
Describes R.S.L. Council meeting regarding succession of presidency. Most members wish to nominate JH to serve for a few years. ES urges JH to accept. [Marked 'private.']
ALS 4pp RS MM.16.134

7609. 1848–3–4 From Charles Davidson Bell Woolwich
Is off to Scotland and has not been able to collect the parcel. News of the Cape. Dr. J. R. Innes's letters. Is sorry to hear of Lady Herschel's slow recovery.
ALS 4pp RS:HS 5.41

7610. 1848–3–5 To Edward Sabine Collingwood
Returns Robert Peel's letter. JH has not changed opinion. Will not accept office [of R.S.L. president], even for one year.
ACCS 2pp TxU:H/L-0365; Reel 1054

7611. 1848–3–6 To Richard Sheepshanks Collingwood
Unable to chair the [R.A.S.] meeting on Friday. On various persons proposed as R.A.S. associates or honorary members. Has recently been busy reading newspapers: 'There seem to be 2 worlds going on at once and it is not easy to keep up with both.'
ALS 4pp RS:HS 16.87 (C: RS:HS 25.9.55)

7612. 1848–3–6 To William Somerville Collingwood
Thanks WS for W. H. Wollaston's crystal models and other gifts. Comments on the 'strange and exciting events' [Revolution of 1848] that are in progress in political affairs.
ALS 4pp RS:HS 16.406 (C: RS:HS 23.16)

7613. 1848–3–8 From de Belime. St Germain-en-Laye
Sending a Latin poem he has written on the death of Caroline Herschel. Comments on the troubled state [Revolution of 1848] of Paris at the time.
ALS 4pp RS:HS 5.39

7614. 1848–3–9 From Sir Charles Lyell Royal Society, Somerset House
Asks when JH can receive CL and Edward Sabine at Collingwood, that they may explain to JH why R.S.L. Council desires JH to serve as president.
ALS 5pp RS MM.16.135

7615. 1848–3–10 From Josiah Forshall British Museum
The [British Museum] board accepts, with regret, JH's resignation as a Trustee.
ALS 1p RS:HS B27.65

7616. 1848–3–10 To Sir Charles Lyell [Collingwood]
Begs CL and Edward Sabine to cancel their plan to interview JH for office [Presidency] in R.S.L. To continue will cause JH 'inexpressible pain.'
ACCS 1p TxU:H/L-0236; Reel 1054 (C: RS:HS 23.17)

7617. 1848–3–10 To Sir Charles Lyell Collingwood
'Confidential' Insists that CL honor JH's decision regarding R.S.L. office and drop subject.
ACCS 1p TxU:H/L-0235; Reel 1054

7618. 1848–3–10 To Edward Sabine Collingwood
Asked [Charles Lyell] to communicate JH's response to ES.
ACCS 1p TxU:H/L-0366; Reel 1054

7619. [1848]–3–10 From Edward Sabine Woolwich
Discusses ES's visit with Charles Lyell to try inducing JH to accept nomination for R.S.L. presidency. In light of JH's earlier letter, ES did not propose JH as nominee, but others regard JH's nomination as exceedingly important.
ALS 3pp RS MM.16.136

7620. 1848–3–13 From Henry T. De La Beche Geological Survey Office
May alter his own article on mineralogy as he pleases.
ALS 4pp RS:HS 6.133

7621. 1848–3–13 From Henry T. De La Beche Geological Survey Office
Sending him the material on mineralogy for the new manual. Comments on the difficulty of writing this sort of work and gives an outline of how he has written it.
ALS 4pp RS:HS 6.132

7622. 1848–3–13 From Edward Sabine Woolwich
Sorry JH will not undertake R.S.L. presidency. Discusses meteorology and magnetism. [J. D.?] Robinson's instrument works well.
ALS 3pp RS:HS 15.217

7623. 1848–3–15 From George Eden Admiralty
Is prepared to enlarge the *Admiralty Manual* provided the extra material is of good quality and not too bulky. Will contact John Murray [Jr.] about printing and publishing rights. Will give W. R. Birt every facility for a separate publication.
ALS 4pp RS:HS 7.15

7624. 1848–3–15 From James Cowles Prichard 1 Woburn Place
Submits a paper [on ethnology] for JH's *Admiralty Manual.* Welcomes information from JH on the subject.
AL 2pp RS:HS 14.61

7625. 1848–3–16 To William Radcliffe Birt Collingwood
Returns WB's draft with JH's editorial corrections. Admiralty has no objection to WB publishing it as separate article. JH will write instructions on 'general' meteorological observations to fill gaps left by WB's and Charles Wheatstone's specialized instructions. Devote time to analyzing existing observations, not to making more.
ALS 4pp RS:HS 19.132

7626. 1848–3–17 From Robert Hunt Museum of Pract. Geology
Has been working for a long time on the supposed influence of magnetism on chemical action. Outlines an experiment he has used to show whether magnetism affects chemical action; would like his comments.
ALS 4pp RS:HS 10.126

7627. 1848–3–17 From Georg Merz & Son Munich
Expresses thanks at considerable length for gift of JH's *Cape Results.*
ALS 2pp (German) RAS:JH Archive 10/5.6; Reel 9

7628. 1848–3–17 From Edward Sabine Woolwich
Sends astronomical observations of the *Rattlesnake* received from Mr. Dayman.
ALS 3pp RS:HS 15.218

7629. 1848–3–18 From Francis Beaufort Admiralty
Has had a note from the carpenter stating that the dome is ready; will notify him when the ship in which it is to sail is decided. The two papers of the *Admiralty Manual* are in progress by W. J. Hamilton and F. W. Beechey.
ALS 3pp RS:HS 3.382 & ACC: Hydrographic Office Letter Book

7630. 1848–3–18 From James Cowles Prichard 1 Woburn Place
Submits a hastily written paper for JH's *Admiralty Manual.* Hopes it will suit JH's purpose. Comments in response to JH's remarks on systems of orthography.
ALS 4pp RS:HS 14.62

7631. 1848–3–20 From James Cowles Prichard 1 Woburn Place
Explains orthographical details (nasal ng, etc.) and suggests individual letter values. Invites JH to make alterations in JP's paper for JH's *Admiralty Manual.*
ALS 4pp RS:HS 14.63

7632. 1848–3–21 From Spencer J. A. Compton Piccadilly
Hopes JH will be able to come to town to attend the next Council meeting. Hopes William Parsons (3rd Earl of Rosse) will be the next president.
ALS 2pp RS:HS 5.274

7633. [1848–3–21] From Charles Robert Darwin Down, Farnborough
Has forwarded JH's note to Robert Mallet, who no doubt will compress his part. Encloses his own part,which he fears may be too long. Would like a chance to see the proofs. When completed would like several copies.
ALS 3pp RS:HS 6.14 Darwin, *Correspondence, 4,* 123–4

7634. 1848–3–23 From Francis Beaufort Admiralty
Encloses W. J. Hamilton's paper to show its extent. Please send proofs for Hamilton to correct.
ALS 2pp RS:HS 3.383

7635. 1848–3–23 To Spencer J. A. Compton Collingwood
Regrets unable to dine with SC; explains why he declines to be a candidate for the presidency of R.S.L.
C 1p RS:HS 23.18

7636. 1848–3–23 From Edward Sabine Woolwich
Will correct his contribution to JH's scientific manual. Discusses [Humphrey] Lloyd's apparatus for measurement of total force.
ALS 4pp RS:HS 15.219

7637. 1848–3–23 From William Somerville 3 Chester Terrace
Sending a large specimen of rock salt. Wants JH to make a lens for Mrs. Somerville, if possible. Discusses the political transitions that are occurring.
ALS 3pp RS:HS 16.407

7638. 1848–3–24 From Elias Loomis New York University
Heard from George Bancroft that he was forwarding JH's book on the Southern Nebulae. Has not received it yet so presumes it must have been lost in transit. Sends some of his own astronomical papers for JH. Has made observations of Halley's comet; gives details as he believes they are from a period for which JH lacks information.
ALS 4pp RS:HS 11.307

7639. 1848–3–24 From Edward Joseph Lowe Highfield House near Nottingham
Thanks for the account of the large meteor. Has not heard of its observation from anyone else. Has seen a remarkable solar phenomenon of which he encloses sketch and details. Intends building a larger observatory and taking a partner.
ALS 4pp RS:HS 11.314

7640. 1848–3–24 From Heinrich Christian Schumacher Altona
The 'psychical epidemic' that is pervading Europe has invaded HS's country, and reason flies to the detriment of science.
ALS 1p RS:HS 15.401

7641. [1848]–3–25 From Francis Beaufort Admiralty
'Enclosed' will answer JH's question. Will deliver dome [for Cape of Good Hope telescope] to Deptford Dock Yard.
ALS 1p TxU:H/M–0058; Reel 1087 (ACC: Hydrographic Office Letter Book)

7642. 1848–3–26 To William Somerville Collingwood
Will use rock salt to study heat spots. Discusses views of [Macedonio] Melloni regarding solar heat. If England can avoid the mistakes of her neighbors, she will enjoy a great triumph.
ALS 4pp RS:HS 16.408 (C: RS:HS 23.19)

7643. 1848–3–27 From Spencer J. A. Compton Piccadilly
Is sorry to hear that he will be unable to be present next Saturday. Is desired by Council to express their regret that he will not become president of the R.S.L. Hopes William Parsons (3rd Earl of Rosse) will accept the presidency.
ALS 2pp RS:HS 5.275

7644. 1848–3–27 From John Stevens Henslow Mitcham, Hadley, Sufk
Sending a paper in which he recommends JH's *Prelim. Discourse;* it is addressed chiefly to mechanics.
ALS 2pp RS:HS 9.310

7645. [1848–3–27?] To Richard Sheepshanks [Collingwood]
Read enclosure and judge what changes it will necessitate regarding 'S.'s [H. C. Schumacher?] position'. Wrote to 'S.' to tell him that as far as JH is concerned, JH is ready to act on any suggestion 'S' may make regarding the king.
ALS 1 p RS:HS 16.88 (C: RS:HS 25.9.58)

7646. 1848–3–28 From Alexander Dallas Bache to George Bancroft Washington, D.C.
Asks GB to convey gratitude to JH for AB's receipt of JH's *Cape Results.*
ALS 1p RAS:JH Archive 10/5.149; Reel 9

7647. 1848–3–28 From William Radcliffe Birt Bethnal Green
Has received his parcel and will pay particular attention to the points raised. Regarding *using* and *corrections* for the barometer. Will send manuscript to the Admiralty.
ALS 4pp RS:HS 4.123

7648. 1848–3–28 From James Cowles Prichard 1 Woburn Place
Discusses problems with his paper on ethnology for JH's *Admiralty Manual.* Asks JH for fuller information on what is needed in JP's paper.
ALS 3pp RS:HS 14.64

7649. 1848–3–28 From Edward Sabine Woolwich
Sorry JH's family has influenza. Encloses letter from [John] Lefroy regarding actinometers.
ALS 2pp RS:HS 15.220

7650. 1848–3–29 From Anne Elise Knipping Hanover
Wants to send JH a remembrance which his Aunt Caroline wanted JH to have. The obituaries in some of the papers are inadequate. The astronomical journals need to be provided

with suitable information about Caroline Herschel.
ALS 4pp (German) RS:HS 11.61 & 61a

7651. 1848–3–30 From James Cowles Prichard 1 Woburn Place
Glad JH finds CP's paper suitable for *Admiralty Manual.* JP allows question is physiological but also ethnological; admits problems but sticks to his thesis. Would JH be willing to assume a leading role in a planned new journal somewhat similar to the *Athenaeum,* but giving more attention to science?
ALS 4pp RS:HS 14.65

7652. 1848–3–31 From Sir Richard Owen Royal College of Surgeons
Will resume work on his 'Instructions' for the *Admiralty Manual* after the conclusions of his lectures. Will be able to forward it before the end of May.
ALS 1p RS:HS 13.191

7653. 1848–4–1 From William Radcliffe Birt Bethnal Green
Has forwarded the manuscript, with notes, to the Admiralty and JH should receive it in a few days. Leaves the arrangement to him. Did he receive his letter of 15 Jan. and a copy of his Southampton report?
ALS 3pp RS:HS 4.124

7654. [1848–4?] To [William Radcliffe Birt] n.p.
Returned [WB's] manuscript with Admiralty notes last week. Concerned, because postal service lost C. R. Darwin's manuscript when JH returned it.
ALS 2pp RS:HS 19.171

7655. 1848–4–2 From D. François J. Arago Paris
Has received the Cape of Good Hope observations. Regarding work on the Board of Longitude. Has received the portrait of William Herschel.
ALS 3pp (French) RS:HS 1.365

7656. 1848–4–3 To [William Radcliffe Birt] Collingwood
Received [WB's] packet. JH has influenza, will read manuscript after recovery.
ALS 1p RS:HS 19.133

7657. 1848–4–3 From [Jacob Gijsbertus Samuel van] Breda Harlem
Acknowledges receipt of JH's *Cape Results* by Holland Society of Sciences.
ALS 2pp (French) RAS:JH Archive 10/5.96; Reel 9

7658. 1848–4–3 To William Rowan Hamilton Collingwood
Wonderingly admires WH's quaternions. Lady Herschel has not yet thanked Eliza Hamilton (WH's sister) for the poetry because of serious illness. Except for influenza, would wish WH's son to visit for Easter. Mentions 'political extravaganzas.'
ALS 4pp TCD 1493:424 Graves, *Hamilton, 2,* 599 inc

7659. 1848–4–4 From Sophie Beckedorff to Margaret Brodie Herschel Hanover
Death of her brother-in-law. Riots in Germany. Thanks for the gift.
ALS 4pp RS:HS 4.5

7660. 1848–4–4 From Adam Sedgwick Cambridge
Needs clarification of the relationship between the sun's rotation and the nebular hypothesis. Do the planets between Mars and Jupiter create problems for the hypothesis?
ALS 3pp RS:HS 15.437

7661. 1848–4–5 From George Biddell Airy Royal Observatory, Greenwich
Concerning the new meridional instrument. Is JH coming to town soon?
ALS 2pp RS:HS 1.199 (C: RGO 6.9.203)

7662. 1848–4–5 From George Biddell Airy Royal Observatory, Greenwich
Concerning object-glasses.
ALS 4pp RS:HS 1.198 (C: RGO 6.9.201)

7663. 1848–4–5 To George Biddell Airy Collingwood
Willing to let GA choose the best objective lens [see GA's 1848–4–5]; then JH offers another possibility; all are ill at Collingwood.
ALS 4pp RGO 6.9.204

7664. 1848–4–5 To Adam Sedgwick Collingwood
Family has been ill. Discusses possibilities for the formation of the sun and their effects on the law of area and the nebular hypothesis.
ALS 4pp RS:HS 15.438 (C: RS:HS 23.20)

7665. 1848–4–7 From Charles Davidson Bell Woolwich
Has found the plan of Sussex very useful on his travels. Comments on Brighton. Has seen his aunt and uncle. Has had no chance of visiting Clapham and so unable to send his sketch.
ALS 4pp RS:HS 5.42

7666. 1848–4–8 From Sir William J. Hooker Royal Botanic Gardens, Kew
Has finished his paper for JH's *Admiralty Manual* and it only requires copying. Can reduce it if necessary.
ALS 2pp RS:HS 9.460

7667. 1848–4–9 To Charles Wheatstone [Collingwood]
Asks for CW's contribution to the Admiralty's scientific manual.
C 1p RS:HS 25.11.8

7668. 1848–4–10 From William Somerville 3 Chester Terrace
Informs JH that before JH's [Cape] expedition, WS recommended it to [H. R. V. Fox,] Lord Holland, who recommended it to James Graham, then First Lord of the Admiralty. Sends JH a letter from Lord Holland to WS praising JH's efforts.
ALS 1p RS:HS 16.409

7669. 1848–4–12 From Robert Hunt Museum of Pract. Geology
Encloses a copy of his own paper from the *P.M.* Has not found time to do the experiments suggested by JH, but hopes to shortly. Is revising some papers on the physical powers of the universe for re-publication.
ALS 4pp RS:HS 10.127

7670. 1848–4–12 From Edward Sabine Woolwich
Sends meteorological register contracts from [John] Lefroy. Discusses observations of aurora in North America. Discusses recent magnetic disturbances and similar disturbances in 1841 and 1847.
ALS 8pp RS:HS 15.221

7671. 1848–4–13 From Frederik Kaiser Leiden
Gratitude for receipt of JH's *Cape Results.* Offers to send FK's complete series of micrometrical observations of [Neptune].
ALS 2pp RAS:JH Archive 10/5.97; Reel 9

7672. 1848–4–14 From Francis Beaufort Admiralty
Encloses a letter from Thomas Maclear to show how enthusiastic he is for the great telescope.
ALS 1p RS:HS 3.384

7673. 1848–4–15 From Henry Liddell Slough Mechanics' Inst.
Officers and Members of the Slough Mechanics Institute would like JH to become their president and also officiate at the laying of the stone of the new building later this month.
ALS 4pp RS:HS 11.214

7674. 1848–4–16 From Augustus De Morgan 7 Camden St.
Requesting that JH would send notices to all the new honorary members. Eleven testimonials are ready for his signature; would he sign them when next in town. Where shall he send the Cape newspapers?
ALS 4pp RS:HS 6.242

7675. 1848–4–17 From John Forbes, D.D. Glasgow
Last year JH agreed to review JF's paper [see JF's 1847–3–13] and to forward it to R.S. L. if it is accurate. JF now submits paper to JH.
ALS 3pp, encl 25pp RS:HS 7.373 & TxU:H/M-1146; Reel 1060

7676. 1848–4–18 From Frederick W. Beechey to Francis Beaufort n.p.
Has been prevented from working for the past ten days due to pains in the head, but will send his manual to JH as soon as he can get it arranged and copied.
ALS 1p RS:HS 4.9

7677. 1848–4–19 From Anne Elise Knipping n.p.
Thanks Margaret Brodie Herschel for a gift. Talks about some legal matters related to the death of JH's aunt Caroline Herschel. AK is much concerned with political turmoil around her. Apparently Dr. Mühry died soon after the death of Caroline Herschel.
AL 4pp inc (German) RS:HS 11.62

7678. 1848–4–19 From M. Macann 56 Upper Seymour St.
Has heard no more news of his paper; shall he write to the R.S.L. for news? Comments on the claims of James MacCullagh for theories regarding light. Hopes to get something published by the London publishers so has written a paper on trigonometry, which he hopes JH will present to the R.S.L. Doubtless JH is aware that there is no general proof for John Napier's rules.
ALS 5pp RS:HS 12.3

7679. 1848–4–20 From Francis Beaufort Admiralty
Encloses a note from F. W. Beechey explaining why there has been a delay with his paper for the *Admiralty Manual.*
ALS 1p RS:HS 3.385

7680. 1848–4–[20?] From Adam Sedgwick Norwich
Tells JH to arrive 16 May. Will send train schedule. Sends regards to family.
ALS 4pp RS:HS 15.439

7681. 1848–4–21 To Henry Liddell n.p.
Regrets he cannot attend the meeting to lay the stone of the new lecture room, nor can he take an active part in the lectures due to an attack of influenza. Would like his name added to the ordinary members [of the Slough Mechanics Institute].
draft 2pp RS:HS 11.215 & 23.21

7682. 1848–4–21 From M. Macann 56 Upper Seymour St.
Has read G. B. Airy, James MacCullagh, and JH on light and several other minor writers. His own paper has been read at the R.S.L. Regarding the standard of teaching at Trinity College, Dublin. Intends to induce Parliament to set up a commission of the great scientific men of England.
ALS 6pp RS:HS 12.4

7683. 1848–4–23 From James Clark Ross Athenaeum
Going to the Arctic. Wants suggestions for experiments. Will pursue barometric experiments. Royal Society has not published his Antarctic research. Will send those documents to JH first for revision.
ALS 4pp RS:HS 14.430

7684. 1848–4–24 To Augustus De Morgan Collingwood
Organizational matters related to R.A.S.; intrigued by AD's partial differential.
C 2pp RS:HS 23.22

7685. 1848–4–24 From Augustus De Morgan 7 Camden St.
Agrees with his proposals. Will meet him at Barclay the printers or if unable to manage then, will forward the testimonials to Somerset House for his signature.
ALS 2pp RS:HS 6.243

7686. 1848–4–25 To Augustus De Morgan Collingwood
About authorship of an encyclopedia article, meeting arrangements, and the mathematical cleverness of his son William.
C 1p RS:HS 23.23

7687. 1848–4–25 To Adam Sedgwick Collingwood
Says the family will arrive at Norwich on 16 or 17 May.
ALS 3pp RS:HS 15.440

7688. 1848–4–25 From Mary Somerville to Lady Herschel Chester Terrace
Sends Lady Herschel her book [*Physical Geography*] as a token of friendship. Discusses the current political turmoil. Regards to family.
ALS 3pp RS:HS 16.355

7689. 1848–4–28 From Augustus De Morgan 7 Camden St.
Thinks Herschel is getting on very well with his mathematics. Boys learn mathematics at a much earlier age now. Will have testimonials ready at Barclays if he will name the day. Will bring Cape newspapers to the Society.
ALS 3pp RS:HS 6.244

7690. 1848–4–29 From Edward Joshua Cooper Watford
Andrew Graham has found another planet [Metis]. Gives readings made with the meridian circle of latest observations. Returns to Ireland soon.
ALS 1p RS:HS 5.289

7691. 1848–5 To Augustus De Morgan n.p.
Arrangements for meeting with AD.
C 1p RS:HS 23.24

7692. 1848–5–1 To Edward Joshua Cooper Collingwood
Congratulates EC on 'Thetis' [proposed name for asteroid Metis]; J. R. Hind has discovered a new star.
C 1p RS:HS 23.25

7693. 1848–5–2 From Sarah Lee 27 Regent Square
Is applying for a Government pension and would value a

supporting testimonial from JH.
ALS 3pp RS:HS 11.189

7694. 1848–5–6 From George Biddell Airy Royal Observatory, Greenwich
Further information on the zenith tube. Description of a zenith tube.
ALS 2pp RS:HS 1.201 (C: RGO 6.160.30)

7695. 1848–5–4 From George Biddell Airy Royal Observatory, Greenwich
Regarding the reinstatement of the zenith tube.
ALS 3pp RS:HS 1.200 (C: RGO 6.160.29)

7696. 1848–5–6 To George Biddell Airy Collingwood
Does not believe that GA's proposal [see GA's 1848–5–4] will work.
ALS 4pp RGO 6.160.32

7697. [1848]–5–7 From Charles Robert Darwin Down, Farnborough
Sends his manuscript. P.O. has been unable to trace the missing parcel. Comments on J. B. Elie de Beaumont's method of measuring lava streams; emphasizes the importance of repeated observations. Would like copies of the manual when completed.
ALS 3pp RS:HS 6.15 Darwin, *Correspondence, 4,* 136–7

7698. 1848–5–8 To George Biddell Airy Collingwood
Apologizes for having questioned GA's proposal [see GA's 1848–5–6], which is very good and should work well.
ALS 1p RGO 6.160.34

7699. 1848–5–8 From Henry Liddell Slough Mechanics' Inst.
Sends copy of the rules of the Institute. At a Special General meeting JH was elected president. This is subject to JH's approval. Would like him to be present at the opening of the new Institute on the 25th of the next month.
ALS 2pp RS:HS 11.216

7700. 1848–5–9 From William A. Baillie-Hamilton Maiden Newton
The Manual appears to have lost time. Outlines the general form to be taken, but leaves the details to JH. Size might be related to Alexander von Humboldt's *Kosmos* volume. Returns to the Admiralty next week.
ALS 6pp RS:HS 9.207

7701. 1848–5–10 From Edward Joshua Cooper Dublin
'Thetis' has yet to be discovered. Discusses names for new planet.
ALS 4pp RS:HS 5.290

7702. 1848–5–10 From Augustus De Morgan 7 Camden St.
Returns the papers of John Forbes (on mathematical subject) and comments on them. Suggests what the R.S.L. should do in such cases.
ALS 4pp RS:HS 6.245

7703. 1848–5–11 From Charles Robert Darwin Down, Farnborough
As JH is interested in colors he sends him details of the color changes that the genus *Ibla* undergoes. Will send him a specimen if desired.
ALS 3pp RS:HS 6.12 Darwin, *Correspondence, 4,* 142

7704. 1848–5–11 To John Taylor, F.R.S. 37 Tavistock Place
Received JT's letters regarding Francesco De Vico's possession of R. A. Cauchoix's object-glass. Will follow JT's example in this delicate matter and withhold judgment.
ALS 3pp BostonPubLib Ms.Acc. 68 (ACCS: RS:HS 19.134 & RS:HS 23.26)

7705. 1848–5–15 From George Bishop Regents Park
Is returning the letters sent him by JH. Regarding Francesco de Vico and the instruments belonging to the Roman College and [Richard?] Taylor's letter concerning them.
ALS 4pp RS:HS 4.149

7706. [1848]–5–[16] To John Forbes, D.D. Collingwood
Extensive comments on JF's mathematical paper [see JF's 1847–3–13].
C 3pp RS:HS 22.315

7707. [1848–5–16] To Richard Sheepshanks n.p.
Is leaving for Norwich to return on the 25th. Leave the forms with [John] Williams and JH will sign them. Will leave his letters to have the R.A.S. seal affixed. [Thomas] Taylor has died; can RS tell him who is likely to 'offer for the place' [Madras Observatory] and what its 'appointments' are.
ALS 2pp RS:HS 16.90 (C: RS:HS 25.9.57)

7708. 1848–5–17 From John Couch Adams St. John's College, Cambridge
Regarding the elements of Uranus and Neptune, and Alexis Bouvard's 'Elements of Uranus.'
ALS 4pp RS:HS 1.3

7709. 1848–5–17 From Sarah Lee 27 Regent Square
Is grateful for JH's testimonial, which she has forwarded with others to Lord John Russell.
ALS 4pp RS:HS 11.190

7710. 1848–5–19 From Richard Jones Tithe Commission
Presumes they are at Adam Sedgwick's. Supposes they will come to Broxbourne by the 11 o'clock train from Norwich and will arrange for it to be met. Expects a new theory of the earth's crust from JH. Mrs. Jeff[re]y is seriously ill, which delays the Jeff[re]ys at Haileybury.
ALS 4pp RS:HS 10.381

7711. 1848–5–20 From Thomas Galloway 45 Torrington Square
Has been extracting some paragraphs from JH's *Cape Results* for Taylor's *Philosophical Magazine.* Hopes he approves. Unable to get to the last meeting of the Astronomical.
ALS 1p RS:HS 8.32

7712. 1848–5–22 From Benjamin Lewis Vulliamy Pall Mall
Objects to appointment of G. B. Airy as *sole* referee in determining contract for making Great Clock at Westminster Palace. Forwards copies of letters and papers that demonstrate Airy's prejudice and ill will toward BV and Mr. Whitehurst.
ALS 1p TxU:H/M–0557; Reel 1093

7713. 1848–5–23 From Henry Piddington Calcutta
Uses the barometer for measuring both vertical and horizontal distances, and believes the rotation of cyclones may be connected to the sun spot activity.
ALS 4pp RS:HS 13.404

7714. 1848–5–24 From Frederick W. Beechey Westbourne Crescent
Regarding his own publication on hydrography and welcoming any suggestion from JH on various points.
ALS 4pp RS:HS 4.10

7715. 1848–5–24 From William Radcliffe Birt
Cambridge House Academy, Cambridge Road, Bethnal Green
Francis Ronalds stated that observations at Kew Observatory may be discontinued by B.A.A.S. for budget reasons. WB volunteers to supervise observations at no additional cost. Hopes JH has recovered from influenza.
ALS 3pp † TxU:H/M–0101.1; Reel 1087

7716. 1848–5–24 From William A. Baillie-Hamilton
Admiralty
Acknowledges the receipt of his letter, together with the preface and list of contents for the *Admiralty Manual*. Comments on this. Will arrange for copies to be sent to those designated. Admiralty is grateful for JH's work on the manual.
ALS 4pp RS:HS 9.208

7717. 1848–5–24 From [Sir] Francis Ronalds Cambridge House [Academy, Cambridge Road, Bethnal Green]
Urges JH to persuade B.A.A.S. to hire W. R. Birt to continue observations at Kew.
ALS 1p TxU:H/M–0101.2; Reel 1087

7718. 1848–5–25 From E. Beckwith 52 Eaton Place
Has not seen him for a long time, but is sure he would like to know that her daughter is to be married.
ALS 4pp RS:HS 5.34

7719. 1848–5–25 From John Forbes, D.D. Glasgow
JH's letter of 16 May 1848 was delayed by wrong title for JF, who is now minister of Free Church. Earlier letter from JH never reached JF. Happy to learn that JH received JF's mathematics paper. JF prefers to submit it to R.S.L., but when JF received no reply from JH, JF sent paper to Philip Kelland in Edinburgh, who now expects JF's paper.
ALS 3pp TxU:H/M–0203; Reel 1087

7720. 1848–5–30 From Frederick W. Beechey Weston-super-Mare
Is pleased to hear that his paper is of some use. Various queries regarding the work in question.
ALS 4pp RS:HS 4.11

7721. 1848–5–31 From William Radcliffe Birt
Cambridge House Academy, [Cambridge Road, Bethnal Green]
Following interview with Edward Sabine, WB compiled reasons for continuing Kew Observatory. Sends copy to JH.
ALS 2pp TxU:H/M–0102; Reel 1087

7722. 1848–5–31 From Christian C. J. Bunsen 4 Carlton House Terrace
Relays gratitude of king of Prussia and royal librarians for copies of JH's *Cape Results*.
L 1p RAS:JH Archive 10/5.112; Reel 9

7723. 1848–5–31 From John Forbes, D.D. Glasgow
JH's letter of 26 Apr. went to wrong person with same name as JF. Grateful for JH's interest in JF's mathematical paper. Hopes to submit it to R.S.L. Replies to error noticed by JH. Encloses note from Philip Kelland.
ALS 3pp TxU:H/M–0204; Reel 1087

7724. 1848–6–1 To [William Radcliffe Birt] Collingwood
Agrees that Kew Observatory should remain open. Will urge B.A.A.S. to continue observations. *Admiralty Manual* going slowly. WB may use Admiralty's printer for WB's separate article. Did not tell Lord [Auckland, i.e., George Eden] of WB's offer to go to New Zealand.
ACCS 8pp RS:HS 19.135

7725. 1848–[6] To William A. Baillie-Hamilton n.p.
Thanks for the offer of 25 copies of the *Admiralty Manual*. Thinks contributors should receive 6 copies instead of the 2 copies originally intended. Some misunderstanding of his proposals regarding returns of information. Printer has made a blunder in the *Manual* and JH has given orders for new pages to be substituted.
draft 2pp RS:HS 9.209 & 23.29

7726. 1848–6–1 To Edward Sabine Collingwood
Believes B.A.A.S. should continue to maintain Kew Observatory, but worries about finances. Encloses letter from William R. Birt, whose aid will maintain it for a while.
ALS 4pp RS Sa.656

7727. 1848–6–2 From Edward Sabine Woolwich
Members of R.S.L. committee to study Kew observatory. Copies of correspondence between ES and W. R. Birt. If Birt is retained, expenses of Kew observatory will exceed money granted by B.A.A.S. Before applying for government assistance, JH and G. B. Airy must agree on plan.
ALS 4pp TxU:H/M–0523.1; Reel 1093

7728. 1848–6–2 To Edward Sabine n.p.
Agrees with W. R. Birt on advantages of continuing Kew observatory if B.A.A.S. can afford it. Encloses Birt's letter to JH. Understands difficult financial position of B.A.A.S.
AdraftS 2pp TxU:H/L–0367; Reel 1054

7729. [1848]–6–3 From Richard Napier 4 Chesham Place
RN's brother [G. T. Napier] is travelling in Italy, but RN opened JH's *Cape Results* when Charles Bunbury explained its contents. Reports recovery of 'Cecilia's eyes.'
ALS 4pp RAS:JH Archive 10/5.74; Reel 9

7730. [1848]–6–5 From Frederick W. Beechey
Westbourne Crescent
Is sending a sketch of methods to make a scale to meet increasing pressure at depths. Perhaps JH could suggest a suitable instrument.
ALS 3pp RS:HS 4.12

7731. 1848–6–5 From Edward Sabine Woolwich
JH may meet with Kew committee in London on 16 June. William Mann will probably replace late T. G. Taylor in Madras. Suggests that W. R. Birt replace Mann at Cape of Good Hope. Position at Kew would not be suitable to Birt.
ALS 4pp TxU:H/M–0523.2; Reel 1093

7732. 1848–6–6 To Frederick W. Beechey Collingwood
Sending suggestions for a sympiesometer.
draft 4pp RS:HS 4.13 & 23.27

7733. 1848–6–6 From Charles Robert Darwin Down, Farnborough
Encloses samples of the *Ibla*. Outlines the method of making it change color. Hopes he will not have to re-write his geological chapter for the *Admiralty Manual*.
ALS 3pp RS:HS 6.13 Darwin, *Correspondence, 4,* 149

7734. 1848–6–6 From William Smith Williams 65 Cornhill
Per JH's directions, sent copies [of JH's *Cape Results*] to [Francesco] De Vico, Chetham's Library [Manchester], and Wilhelm Struve.
ALS 1p RAS:JH Archive 10/3.19; Reel 9

7735. 1848–6–7 To Frederick W. Beechey Collingwood
Further suggestion for the sympiesometer.
draft 4pp RS:HS 4.14 & 23.28

7736. 1848–6–7 To Humphrey Lloyd Collingwood
Asks questions about how to proceed for the continuance of magnetic observations.
C 1p RS:HS 23.30

7737. 1848–6–7 To ? Collingwood
Writes to tell of arrangements whereby the recipient of the letter may pick up a copy of JH's *Cape Results*.
ALS 3pp DMC.695A.885.4

7738. 1848–6–9 From Francis Beaufort Admiralty
Will be sending on 10 June all the equipment for Thomas Maclear, but the tracing paper is not forthcoming. Has JH anything for the Cape?
ALS 2pp RS:HS 3.386

7739. 1848–6–9 From Frederick W. Beechey n.p.
Thanks for his suggestions and gives a possible modification.
ALS 4pp RS:HS 4.15

7740. 1848–6–10 From William Rutter Dawes Camden Lodge
Admiralty have applied to Mr. Morris for the drawing of the stone pillar for the Cape equatorial by Georg Merz. Suggests that the type in use at Pulkowa be used. Weather prevented outing to Bodiam.
ALS 2pp RS:HS 6.90

7741. 1848–6–10 From Edward Sabine Woolwich
William Mann's application to replace late T. G. Taylor [at Madras observatory] has been approved by H.E.I.C. B.A.A.S. cannot offer good position at Kew to W. R. Birt. Goals of B.A.A.S. committee to report on Kew observatory. Please return [T. R.] Robinson's account of William Parson's telescope.
ALS 4pp TxU:H/M–0523.3; Reel 1093

7742. 1848–6–12 From Frederick W. Beechey Weston-super-Mare
Is going to get one of the instruments that JH suggested manufactured. Would he therefore send the dimensions of the instrument.
ALS 3pp RS:HS 4.16

7743. 1848–6–12 To Thomas Maclear Collingwood
William Mann applies for Madras Observatory post; JH believes TM should be consulted for recommendation. [Andrew] Graham, observer for E. J. Cooper, discovered the asteroid Metis. Describes children's education.
ALS 4pp Cape Archives/Maclear Papers File 100

7744. 1848–6–13 From Francis Beaufort Admiralty
Regarding a successor to William Mann should the latter be appointed to Madras.
ALS 2pp RS:HS 3.387

7745. 1848–6–13 From F. M. Eardley-Wilmot Woolwich
T. G. Taylor of the Madras Observatory has just died. Would like JH's support for William Mann for the position.
ALS 2pp RS:HS 7.7

7746. 1848–6–13 From Edward Sabine n.p.
Encloses letter to ES from Mr. Hutton.
ALS 1p TxU:H/M–0523.4; Reel 1093

7747. 1848–6–14 To William Rowan Hamilton Collingwood
Asks WH to forward Dr. [Mortimer] O'Sullivan thanks for two sermons on preparation of the world for the reception of Christianity.
ALS 2pp TCD 1493:426

7748. 1848–6–14 From Sir William J. Hooker Royal Botanic Gardens, Kew
Sending his notes for JH's *Admiralty Manual*. Would have been earlier but he dislocated his left shoulder and had to have the services of an amanuensis.
ALS 2pp RS:HS 9.461

7749. 1848–6–14 To Sir William J. Hooker Collingwood
WH's manuscript on botany for JH's *Admiralty Manual* has arrived and deserves praise. JH awaits three final manuscripts.
ALS 2pp Royal Botanic Gardens

7750. 1848–6–17 From Frederick W. Beechey Weston-super-Mare
Further queries about the manufacture of the sympiesometer. Will be moving to Shirehampton in a few days time.
ALS 3pp RS:HS 4.17

7751. 1848–6–17 From Edward Sabine Woolwich
Discusses reduction of observations, continuation of various observatories, and the necessary financial arrangements for each. JH's actinometer instructions were misunderstood at Toronto.
ALS 8pp RS:HS 15.222

7752. 1848–6–18 To Frederick W. Beechey Collingwood
Answers to FB's queries concerning the sympiesometer.
draft 4pp RS:HS 4.18 & 23.31

7753. 1848–6–20 To Frederick W. Beechey Collingwood
Further points regarding the sympiesometer.
draft 8pp RS:HS 4.19 & 23.32

7754. 1848–6–21 From [Paul Auguste Ernest] Laugier Paris
Please convey to the R.A.S. his thanks at the honor of being elected an Associate.
ALS 1p (French) RS:HS 11.454

7755. 1848–6–23 From Edward Sabine Woolwich
Hopes JH can attend council meeting on 7 July. Thinks Kew Observatory could be made an important magnetic and meteorological station.
ALS 4pp RS:HS 15.223

7756. 1848–6–26 From William A. Baillie-Hamilton Admiralty
Comments on JH's estimate for the number of pages for the *Admiralty Manual*. Auckland does not mind the manual being enlarged should JH think it necessary.
ALS 3pp RS:HS 9.210

7757. 1848–6–28 From Frederick W. Beechey Shirehampton
Thanks for his suggestions on the instrument which E. A. L. Negretti is working on. Gives some more details of the instrument.
ALS 4pp RS:HS 4.20

7758. 1848–7–3 From Edward Sabine Woolwich
Hoping JH receives [Francis] Ronalds's and [W. R.] Birt's statements regarding Kew Observatory. Asks JH to bring William Parson's note to their [JH and ES's] upcoming meeting. Discusses finances of Kew.
ALS 4pp RS:HS 15.224

7759. 1848–7–7 From Hervé A. E. A. Faye Paris
Thanks for being elected an Associate of the R.A.S. Astronomy binds the continent of Europe together. Presents his compliments to Richard Sheepshanks, George Bishop,

and J. R. Hind.
ALS 3pp RS:HS 7.366

7760. 1848–7–7 From William A. Baillie-Hamilton Admiralty
Regarding an appendix to W. J. Hooker's article for the *Admiralty Manual* by [Augustin] Creuze if it is not already covered by Hooker.
ALS 3pp RS:HS 9.211

7761. 1848–7–7 From Félix Victor Mauvais Paris
Feels honored at receiving the diploma of the R.A.S. Will redouble his efforts in the field of astronomy.
ALS 1p (French) RS:HS 13.29

7762. 1848–7–8 From Pierre Daussy Paris
Expressing thanks for being elected an Associate of the R.A.S.
ALS 1p (French) RS:HS 6.473

7763. [1848]–7–8 From Charles Pritchard [Clapham Grammar School?]
Receives object glass from Munich for double star observations.
ALS 2pp inc RS:HS 14.109

7764. [1848]–7–11 From William Empson Craig Crook, Edinburgh
Has not heard of the results of the Rolfe's visit to Hawkhurst. The Sabines are translating *Kosmos* and all would be pleased if JH would look at the proofs. Intended stopping at York but had to hurry on to Edinburgh when Francis Jeffrey (Lord Jeffrey) was seriously ill.
ALS 4pp RS:HS 7.60

7765. 1848–7–11 From Green & Co. Cockspur St.
Going out of business. Offer to sell gold plate and jewelry to preferred customers at reduced prices.
LS 1p TxU:H/M–0219.1; Reel 1087

7766. 1848–7–11 From William A. Baillie-Hamilton Admiralty
George Eden (1st Earl of Auckland) concurs in JH's view that [Augustin] Creuze's letter should form an appendix to the article by W. J. Hooker.
ALS 2pp RS:HS 9.212

7767. 1848–7–11 From Sir Robert Harry Inglis 7 Bedford Square
Sending a letter on the Orthography of the African language from a friend of his, the Rev. Henry Venn. Would be grateful if JH could help him.
ALS 2pp RS:HS 10.242

7768. 1848–7–11 From Henry Venn Church Missionary House, Salisbury Square
Seeks interview with JH on subject of orthography of unwritten languages. Lists six African languages into which Church Missionary Society has translated Scriptures.
ALS 3pp TxU:H/M–0556; Reel 1093

7769. 1848–7–13 From Edward Joseph Lowe Highfield House near Nottingham
Sends a description of a large meteor he observed last night. Since his letter he has erected a meteorological observatory.
ALS 3pp RS:HS 11.315

7770. 1848–7–13 From Eliza Susan Quincy Boston, Mass.
Requests JH forward [James] Grahame's last letter to the Stewarts. Enjoyed JH's *Cape Results*. Discusses Harvard Observatory and a recent bequest.
ALS 4pp RS:HS 14.287

7771. 1848–7–14 From Sir Richard Owen Royal College of Surgeons
Has now completed his manuscript for the *Admiralty Manual*. Runs to 90 pages. To what Office shall he forward it?
ALS 1p RS:HS 13.192

7772. 1848–7–15 From George Biddell Airy Royal Observatory, Greenwich
Regarding Mary Somerville's proposals for extraordinary terrestrial refractions—*Admiralty Manual*.
ALS 4pp RS:HS 1.202

7773. 1848–7–15 or later To George Biddell Airy n.p.
Agrees that no requests for additional observations should be made of Thomas Maclear; *Admiralty Manual* is about to be printed [see GA's 1848–7–15].
ALS 4pp RGO 6.2.160

7774. 1848–7–15 From George Biddell Airy Royal Observatory, Greenwich
Received proof pages of *Admiralty Manual* [see JH's 1848–2–2], and wants to add a paragraph on the zodiacal light [which GA attaches].
ACCS 2pp RGO 6.2.157

7775. 1848–7–16 To [William Empson] Collingwood
Declines to review C. R. Weld's book on Royal Society [*A History of the Royal Society*, 2 vols., London, 1848]. Fears will be unable to review Alexander von Humboldt's *Kosmos* to own satisfaction.
ALS 3pp St. John's College, Cambridge

7776. 1848–7–17 From George Biddell Airy Royal Observatory, Greenwich
A note accompanying JH's copy of *Admiralty Manual* [see JH's 1848–7–15 or later].
ACCS 2pp RGO 6.2.161

7777. 1848–7–17 From Francis Beaufort Admiralty
Captain F. W. Beechey is in town and will remain until Wednesday evening. He received the enclosed this morning.
ALS 1p RS:HS 3.388

7778. 1848–7–17 From William A. Baillie-Hamilton Admiralty
Accompanying packet has just arrived from [Richard?] Owen. Appears that it will make the *Admiralty Manual* larger. Would like JH's views.
ALS 2pp RS:HS 9.213

7779. 1848–7–17 From Sir Richard Owen Royal College of Surgeons
Has been condensing his contribution to the *Admiralty Manual*, now reduced to forty pages; comments on this. The Westminster fever delayed the Admiralty manuscript.
ALS 4pp RS:HS 13.193

7780. [1848]–7–20 From William Empson Craig Crook, Edinburgh
Francis Jeffrey (Lord Jeffrey) may be able to go on holiday. Does JH know anything about a Capt. J. T. Smith who has been named as a reviewer for C. R. Weld's *History of the R.S.*? Sorry to hear he has misgivings about the second part of *Kosmos*; may change his mind later.
ALS 4pp RS:HS 7.61

7781. 1848–7–20 From Edward Sabine Woolwich
Meeting of B.A.A.S. committee with JH. JH's report was

included with extracts from Francis Ronalds's report.
ALS 4pp TxU:H/M–0523.5; Reel 1093

7782. 1848–7–21 From [Heinrich Louis] d'Arrest Leipsic
On various astronomical matters.
ALS 1p RS:HS 1.412

7783. 1848–7–21 From Richard Jones Tithe Commission
Gathering mercantile statistics spoken of by John Stewart and preparing these for House of Commons.
ALS 1p TxU:H/M–0307.1; Reel 1087

7784. 1848–7–21 From Thomas Jones Chetham's Library, Manchester
Gratitude for donation of JH's *Cape Results* [see TJ's 1847–10–17]. James Nasmyth, library patron, also 'gratified.' Sent separately a local newspaper announcement of this acquisition. Encloses prospectus of index of library's holdings.
ALS 2pp, encl 2pp RAS:JH Archive 10/5.75; Reel 9

7785. 1848–7–23 From Manuel John Johnson Radcliffe Observatory, Oxford
Encloses a letter from W. J. Burchell, the African traveller, showing the variability of Argus. Has added a 10 foot telescope to the instruments at the Observatory.
ALS 4pp RS:HS 10.335

7786. 1848–7–25 From Charles William Pasley London
Thanks for the letter concerning William Mann; has no doubt now he will be appointed to Madras. Met Lady Herschel and Miss Herschel at Guthries. Is going to Burnsall or Hurst Green to see a mine exploded by CP's son. Wonders if JH would like to come and witness the explosion.
ALS 4pp RS:HS 13.235

7787. 1848–7–25 From Adolph C. Petersen Altona
A note of thanks for being elected to membership in the R.A.S.
ALS 1p (German) RS:HS 13.372

7788. 1848–7–25 From Edward Sabine Woolwich
Deliberations on continuance of Kew Observatory, and ES's willingness to step aside for another member. Successes and failures of B.A.A.S. objectives at Kew and prospective goals for future. Increasing government involvement in science.
ALS 8pp TxU:H/M–0523.6; Reel 1093

7789. 1848–7–26 From Charles William Pasley Bonsell
Does not think JH will be interested in the exploding of the mine by CP's son. Gives details of the arrangements.
ALS 4pp RS:HS 13.236

7790. 1848–7–27 From Edward Joseph Lowe Highfield House near Nottingham
Thanks for the account of the thunderstorm. Comments on the phenomenon of the reversal of wind with a thunderstorm. Has been experimenting with wet and dry bulb thermometers. Also with the effect of electricity on the temperature of jars.
ALS 3pp RS:HS 11.316

7791. 1848–7–27 From James Nasmyth Patricroft
Grateful for receipt of JH's *Cape Results*.
ALS 1p RAS:JH Archive 10/5.76; Reel 9

7792. 1848–7–29 To Margaret Brodie Herschel Collingwood
MH is having dental trouble; JH talks about changing some of his plans saying 'The fact is I am a prisoner for life and know it.' But JH goes on to say that companionship means that he is not in solitary confinement. [It seems JH is giving up reading and research plans because of the needs of his family, but he does not regret it.]
ALS 3pp JHS 6.30

7793. 1848–7–29 From Edward Sabine Woolwich
[John] Murray is wrong; ES has no more manuscripts. Admiralty orders many ships to perform meteorological observations. Old R.S.L. meteorological forms out of date.
ALS 3pp RS:HS 15.225

7794. [1848–7–30 or earlier]. From Edward Sabine n.p.
[Francis] Ronalds's method of automatic registry is more useful than [Charles] Brooke's in the colonies. Brooke's required camphine[?], which is difficult to obtain.
ALS 4pp inc RS:HS 15.332

7795. 1848–7–30 From Edward Sabine Woolwich
Received intelligence from [W. H.] Sykes that [William] Mann will have appointment at Madras. Knows [W. R.] Birt wants to succeed Mann [at Cape Observatory].
ALS 1p RS:HS 15.226

7796. 1848–7–31 To Edward Sabine Collingwood
Believes Kew Observatory will be ineffective if not private observatory of B.A.A.S., so wishes not to attend meeting to draw up application for government support. Approves, however, of committee revising report.
ALS 2pp RS Sa.657

7797. 1848–7–31 To [Charles Wheatstone] Collingwood
Declines to attend a meeting to prepare a proposal for government support for Kew Observatory.
ALS 2pp King's College London Wheatstone 5/36

7798. 1848–7–31 To [B.A.A.S. Committee] Collingwood
Unwilling to support request for government grant to Kew observatory. Wants to see improvements in report by B.A.A.S. committee first.
ACCS 3pp TxU:H/L-0647; Reel 1089

7799. 1848–8 From Richard Clark and William Shepherde 447 West Strand
Increasing importation of [whale oil] allows Metropolitan Light Company to reduce [candle] prices.
ALS 1p RS:HS 19.136a

7800. 1848–8–2 From Francis Beaufort Admiralty
Has sent to India House to find out if William Mann has been appointed to Madras. Thomas Maclear's views on a new assistant. Unable to get a copy of Collet's Trigonometrical tables.
ALS 3pp RS:HS 3.389 & ACC: Hydrographic Office Letter Book

7801. 1848–8–2 To Henry Venn(?) Collingwood
Responds to a letter received long ago concerning sounds. Is convinced that a 'new printed and written character' is needed. Suggests consulting [Isaac] Pitman's phonotype system. Believes joint use of Roman, italic, and Greek characters would create a more complete system. Stresses that he is willing to correspond, but is unable to meet in person.
C 5pp RS:HS 17.461b (C: RS:HS 23.33)

7802. 1848–8–2 To Henry Venn Collingwood
Lists difficulties with the [Isaac] Pitman phonetic system for writing Oriental languages in the Roman alphabet. Problems include inaccurate vowel usage. Suggests a revised system.
ALS 5pp RS:HS 17.461c

7803. 1848–8–3 From George Biddell Airy Royal Observatory, Greenwich
Forwarding some misdirected proofs of ship's officers instructions to JH [see GA's 1848–7–17].
ACCS 1p RGO 6.2.166

7804. 1848–8–3 From William Radcliffe Birt Bethnal Green
Is grateful for JH's support for his application for a post in the Cape Observatory. Outlines his practical experience in astronomy. Does not mind what post is offered him.
ALS 4pp RS:HS 4.125

7805. 1848–8–4 From Francis Beaufort Admiralty
Will keep W. R. Birt in view for any vacancy that may occur. Will leave Thomas Maclear to appoint Mr. Montague. Will be obliged for the trigonometrical tables.
ALS 2pp RS:HS 3.390 & ACC: Hydrographic Office Letter Book

7806. 1848–8–4 From Alexander Bridport Becher Admiralty
A note accompanying a letter from Thomas Maclear, forwarded to JH.
ACC 2pp Hydrographic Office Letter Book

7807. 1848–8–5 From William Radcliffe Birt Bethnal Green
Does not mind being considered for the alternative position at the Cape. Has seen Edward Sabine about the prospects.
ALS 4pp RS:HS 4.126

7808. 1848–8–8 From William A. Baillie-Hamilton Admiralty
Forwards William Burnett's paper on medicine and medical statistics for the *Manual.* It exceeds the desired length but JH may prune it as he thinks fit.
ALS 2pp RS:HS 9.214

7809. 1848–8–10 From Richard Sheepshanks Reading
RS's progress in establishing standards of measurement. Lists objections to earlier methods of measuring. Compares various materials for use in making bars for measurement standards.
ALS 4pp TxU:H/M–0539; Reel 1093

7810. 1848–8–15 From George Eden Admiralty
Sending a publication similar to the one which JH is editing. Hardly seems worthy of the names of G. C. L. D. Cuvier and A. L. Jussieu.
ALS 2pp RS:HS 7.16

7811. 1848–8–18 To [Richard Sheepshanks] Collingwood
Comments on the quality of comparisons made on bars for standard lengths, and on the possible metals of which such standard bars might be made.
ALS 4pp RGO 6.356.159

7812. 1848–8–19 From George Eden Admiralty
Is obliged for his note. Does not object to the increased size of the *Admiralty Manual,* but should keep it within reasonable limits.
ALS 2pp RS:HS 7.17

7813. 1848–8–22 From Adolphe Quetelet Brussels
Hopes JH has received letter thanking JH for *Cape Results.* Sends meteorological work. Hopes to quit barometric wave observations of atmospheric waves. Has worked with aides on meteorological variations. AQ's mother died.
ALS 3pp (French) RS:HS 14.250

7814. 1848–8–26 To [Robert Hunt] Collingwood
Wants to be supplied with a copy of RH's work on the 'poetry of science' [RH's *Poetry of Science,* 1848].
ALS 2pp WT 68037.8

7815. 1848–8–28 From Robert Hunt 6 Craig's Court
Sends more copies of prospectus of RH's book (Poetry of Science) for JH to distribute.
ALS 1p TxU:H/M–0285.1; Reel 1087

7816. 1848–8–30 From Henry Venn Church Missionary House
Considers JH's suggestions for correcting the phonetic alphabet. Plans to arrange a meeting of the missionary committee on the subject.
ALS 3pp RS:HS 17.462

7817. [1848–9 or later] To Patrick Francis Robertson n.p.
Questions about PR's proposals regarding gold currency. Proposal 6, if unlimited, means that government will always buy [Consale?] at 90.5 and sell at 91.5, regardless of commercial market price.
Adraft 1p inc TxU:H/M–0285.2 & Adraft inc: TxU:H/M–0295.2; Reel 1087

7818. [1848]–9–2 From William Empson Craig Crook, Edinburgh
Francis Jeffrey (Lord Jeffrey) is slowly recovering. He has stated several times that JH is the man to review the 2nd volume of *Kosmos*. Sends prospectus of the Physical Atlas. Returns to Haileybury on the 18th. Richard Jones is in Argyle.
ALS 2pp RS:HS 7.63

7819. [1848]–9–3 From Anne Elise Knipping Hanover
Concerned about the sending and receiving of some of Caroline Herschel's bequests to JH. AK also talks about arrangements for some other people following Caroline Herschel's death.
ALS 5pp (German) RS:HS 11.63 & 11.63 bis

7820. 1848–9–4 From Thomas Maclear Royal Observatory, Cape of Good Hope
Pleased to receive JH's letter of 12 June. Has to visit Agulhas to view the erection of the lighthouse. Gives details of the history of this new lighthouse. Pity François Arago does not continue his experiments with gas. Governor is in the North of the Country visiting the emigrant Boers.
ALS 4pp RS:HS 12.141

7821. 1848–9–4 From William Whewell Trinity Lodge
Discusses WW's essay on tides for JH's *Admiralty Manual,* aspects of JH's *Outlines Astr.,* and planetary astronomy in general.
ALS 8pp RS:HS 18.212

7822. 1848–9–5 From William Rutter Dawes Camden Lodge
Returning JH's proof sheet with errors noted in pencil. Gives some of his recent observations on the satellites of Saturn.
ALS 4pp RS:HS 6.91

7823. 1848–9–6 From John R. Hind 11 Hanover Cottages
Sending notices of several variable stars he has observed recently. Observed Encke's comet twice. Newly discovered variable stars will be discussed in his forthcoming notes to the Ecliptic charts. Sends a small pamphlet on the expected comet.
ALS 4pp RS:HS 9.342

7824. 1848–9–9 From Edward Sabine Valencia Island
[Charles?] Brooke received £500 for his photographic apparatus. Thinks [Francis] Ronalds should receive similar compensation for his device. Discusses apparatus and award.
ALS 4pp RS:HS 15.227

7825. 1848–9–11 From Johann Galle Berlin
Thanks for the diploma of the R.A.S. and for JH's *Cape Results*.
ALS 1p (German) RS:HS 9.85

7826. 1848–9–11 From W. Leslie Melville Edinburgh
Delayed replying to JH's letter until he had more leisure. Will be pleased to nominate JH's son for a cadetship at Addiscombe.
ALS 2pp RS:HS 13.35

7827. 1848–9–12 From Frederick W. Beechey Portishead
Francis Beaufort thinks that FB's essay in the *Admiralty Manual* should include a few lines on the height of the waves of the ocean. FB has asked Beaufort to write these lines.
ALS 2pp RS:HS 4.21

7828. 1848–9–13 From George Biddell Airy Royal Observatory, Greenwich
Asks JH for his opinion of F. W. A. Argelander's letter concerning the publication of his work by the R.A.S.
ALS 2pp RS:HS 1.204 (C: RGO 6.230.453)

7829. 1848–9–13 From Augusta Ada King Ashley-Combe, Somerset
Lord Lovelace [William King] is collecting all works on moon's influence on weather. Asks for JH's suggestions.
ALS 6pp RS:HS 19.281

7830. 1848–9–17 To Eliza Susan Quincy Collingwood
Discusses state of affairs in France. Hopes celestial objects in Southern Hemisphere will continue to be studied. Praises astronomical work done in U.S.
C 4pp RS:HS 14.278 (C: RS:HS 23.34)

7831. 1848–9–18 From John Herapath Kensington
Sends JH a mathematical theorem.
ALS 1p RS:HS 10.182

7832. 1848–9–19 From Urbain J. J. Leverrier Paris
Can expect a visit from Jacques Babinet. Regarding the planet Neptune and its discovery.
ALS 2pp (French) RS:HS 11.200

7833. 1848–9–20 To William Henry Smyth Collingwood
Sends paper by F. W. A. Argelander, which G. B. Airy wishes to have published. Revising chapter on perturbations in his *Treatise Astr.* Will include theory of motions of the apsides and the variations of the eccentricities.
C 2pp RS:HS 23.35

7834. 1848–9–20 To Wilhelm Struve Collingwood, Hawkhurst, Kent
Uneasy at no confirmation of arrival of sent *Cape Results.* Lists observatories to have received copies. Is preparing new edition of *Treatise Astr.*, now titled *Outlines Astr.*
C 3pp RS:HS 23.36

7835. 1848–9–21 From George Biddell Airy Royal Observatory, Greenwich
Wants more information about a Venus parallax expedition.
ACCS 2pp RGO 6.245.120

7836. 1848–9–22 From Richard Jones Tithe Commission
Louisa and Johnny [Herschel] are well. Called at India House to see Hooper. Gives requirements of the various examinations and thinks Willy [Herschel] would do well to stay with [Charles] Pritchard, while Johnny can remain and study at home.
ALS 4pp RS:HS 10.382

7837. 1848–9–22 To William Lassell Collingwood
Is pleased to hear of the discovery of further satellites of Neptune.
C 1p RS:HS 23.37

7838. 1848–9–23 From Maria Edgeworth Edgeworthstown
Was pleased to receive his note and to hear that he is in good health and spirits. Urges him to come to Ireland.
ALS 4pp RS:HS 7.30

7839. 1848–9–24 To George Biddell Airy Collingwood
Personally would like to have F. W. A. Argelander's work printed [see GA's 1848–9–13], but will need to get a wider opinion; also asks about sharing money available from the Admiralty [see GA's 1845–10–24].
AL(S) 3pp RGO 6.230.454 (LCC: 6.675.578 inc)

7840. 1848–9–24 To George Biddell Airy n.p.
Reports that William Lassell has written to say he has discovered the 8th satellite of Saturn.
ALS 2pp inc RGO 6.245.153

7841. 1848–9–24 To George Biddell Airy n.p.
Forwards a letter about the Venus parallax expedition [see GA's 1848–9–21], and adds some comments.
LCC 1p RGO 6.245.121

7842. 1848–9–24 From Francis Beaufort Charmouth
Is sending on the proofs as he does not know the printer.
ALS 2pp RS:HS 3.391

7843. 1848–9–25 From George Biddell Airy Royal Observatory, Greenwich
Concerning the eighth satellite. Observations on U. J. J. Leverrier's paper.
ALS 2pp RS:HS 1.205

7844. 1848–9–25 From George Biddell Airy Royal Observatory, Greenwich
Comments about the proposed Venus parallax expedition and their method of measurement [see JH's 1848–9–24].
LACC 2pp RGO 6.245.122

7845. 1848–9–25 From George Biddell Airy Royal Observatory, Greenwich
Explains how the money from the Admiralty came to be awarded as it was [see JH's 1848–9–24].
ACCS 3pp RGO 6.675.579

7846. 1848–9–25 From George Biddell Airy Royal Observatory, Greenwich
Sends comments on the pages of JH's account of the moon's motions [see JH's 1848–9–25].
ACCS 3pp RGO 6.370.39

7847. 1848–9–25 To George Biddell Airy n.p.
JH has revised his account of the moon's motions given in JH's *Treatise Astr.* in preparation for JH's *Outlines Astr.*
ALS 4pp inc RGO 6.370.36

7848. 1848–9–25 From George Biddell Airy n.p.
Comments on observational quality obtainable with the great telescope of William Parsons [Earl of Rosse].
ACCS 2pp RGO 6.159.171

7849. 1848–9–26 From William Rutter Dawes Starfield, Liverpool
Has experienced bad weather since the discovery of Saturn's 8th satellite. William Lassell proposes to call it *Hyperion.* Wishes Starfield Observatory was situated in a clearer atmosphere.
ALS 4pp RS:HS 6.92

7850. 1848–9–28 From George Biddell Airy Royal Observatory, Greenwich
Queries regarding the perturbation sheets.
ALS 3pp RS:HS 1.206

7851. 1848–9–30 To William Whewell Collingwood
Discusses revisions in WW's and Frederick Beechey's papers in JH's *Admiralty Manual.* Also comments on drafting *Outlines Astr.,* the controversy over the discovery of Neptune, and William Lassell's discovery of a new Saturnian satellite.
ALS 7pp TC, Camb. Add. Ms.a.207[75] (C: RS:HS 23.38)

7852. 1848–10–2 From Richard Jones Tithe Commission
Col. Ouseley had a son at Addiscombe. Has ascertained the relative value of the various branches of study in the final examination. Gives table of values. Hears that military drawing puzzles some. Children are well. If Johnny [Herschel] draws as well as Louisa [Herschel] at 14 he will do well.
ALS 4pp RS:HS 10.383

7853. 1848–10–4 From George Biddell Airy Royal Observatory, Greenwich
Encloses perturbation sheets for checking.
ALS 2pp RS:HS 1.207 (C: RGO 6.370.41)

7854. 1848–10–5 To George Biddell Airy Collingwood
Remarks in response to GA's 1848–10–4.
ALS 4pp RGO 6.370.42

7855. 1848–10–5 From Augusta Ada King Ashley-Combe, Somerset
Thanks for answers about moon and proof of JH's next book. Seeks evidence to support claim that clouds disappear when moon is full. Enquires about lunar formulas.
ALS 9pp RS:HS 19.282

7856. 1848–10–6 From John Couch Adams Launceston
Discusses U. J. J. Leverrier's paper on Neptune. Difference between distance of real and theoretical planets. News that JH is going to publish *Outlines Astr.* Hopes to see him at Cambridge on the day the Public Orator is being elected.
ALS 8pp RS:HS 1.4

7857. 1848–10–8 From Augusta Ada King Ashley-Combe, Somerset
Asks JH to review Lord Lovelace's [William King's] paper on moon's influence on weather.
AL 4pp inc RS:HS 19.283

7858. 1848–10–11 To William Lassell Collingwood
Some question of Harvard University observer having observed the satellite [see JH's 1848–9–22] one day before WL.
C 2pp RS:HS 23.39

7859. 1848–10–11 To General [?] Collingwood
Comments on memo on refraction sent to Thomas Maclear.
ALS 2pp St. John's College, Cambridge

7860. [1848]–10–12 From Augusta Ada King Ashley-Combe, Somerset
Quotes passage from A. E. P. Gasparin comparing JH's actinometer to other such instruments. Seeks JH's opinion of Adolphe Quetelet's hypothesis of heat. Question on photographic chemicals and 'calorific rays.'
ALS 13pp RS:HS 19.284

7861. 1848–10–12 From William King Ashley-Combe, Somerset
Returns JH's diaries with observations of moon's effect on cloud cover. Quotes citation of these observations to appear in WK's new paper.
AL 4pp inc RS:HS 19.285

7862. 1848–10–13 From George Biddell Airy Royal Observatory, Greenwich
Has received further proof sheets of the perturbations from the printer. Regarding Isaac Newton's theory of the motion of the moon's apse.
ALS 4pp RS:HS 1.208a (C: RGO 6.370.44)

7863. 1848–10–13 From William Lassell Starfield, Liverpool
Outlining the apparent controversy between himself and W. C. Bond over the discovery of the satellites.
ALS 6pp RS:HS 11.137

7864. 1848–10–14 From William Lassell Starfield, Liverpool
Further regarding W. C. Bond's discovery of the Neptunian satellites.
ALS 3pp RS:HS 11.138

7865. 1848–10–15 To Augusta Ada King n.p.
Returned proof of Lord Lovelace's [William King's] review of [A. E. P.] Gasparin. Notes and sketch regarding actinometers and other instruments.
AAbs 1p RS:HS 19.136b

7866. 1848–10–15 To William Lassell n.p.
Criteria for deciding who can claim to be the discoverer of the satellite. [This letter marked 'not sent on second thoughts; see RS:HS 23.41 for letter sent.]
draft 4pp † RS:HS 11.139 & 23.40 & 23.41

7867. 1848–10–16 To Henry John Temple Collingwood
Asks HT to appeal for a pension for H. C. Schumacher, director of the Royal Danish Observatory.
C 3pp RS:HS 17.345

7868. 1848–10–16 From Henry Venn Church Missionary House
Lists problems with JH's proposal for reform of the phonetic alphabet found by the missionary committee.
ALS 6pp RS:HS 17.463

7869. 1848–10–17 From George R. Porter Board of Trade
Replies to JH's letter of [9]–11 stating that he will do the article on statistics for JH's *Admiralty Manual,* and outlines what he expects it to include and to exclude. Asks JH for directions.
ALS 3pp RS:HS 14.42

7870. 1848–10–17 From W. H. B. Webster Carr St., Ipswich
Requests donation of JH's *Cape Results* to library of Ipswich Philosophical Society.
ALS 1p RAS:JH Archive 10/5.77; Reel 9

7871. 1848–10–18 From Henry John Temple Carlton Gardens
Gives JH's 1848–10–16 letter to the Danish Minister as a

direct appeal for H. C. Schumacher.
ALS 2pp RS:HS 17.346

7872. 1848–10–19 From George Biddell Airy Royal Observatory, Greenwich
A note accompanying the return of a paper.
ACCS 1p RGO 6.230.460

7873. 1848–10–19 From Maximilian Weisse Cracow
Thanks for the honor of being elected a member of the R.A.S. MW is afraid his star catalogue has too many errors to suit him. Tells JH about the many disruptions, both personal and astronomical, over the last few years. MW has recently been doing some meteorological and magnetic observations.
ALS 4pp (German) RS:HS 18.137

7874. 1848–10–20 To Richard Sheepshanks Collingwood
Sends letter of Lord P [Palmerston, i.e., Henry John Temple] on 'Ch. Reventlow's [Count de Reventlow, Danish Ambassor to London?] case.' Comments on the present political situation and the causes for conflict, citing William Shakespeare.
ALS 3pp RS:HS 16.89 (C: RS:HS 25.9.56)

7875. 1848–10–21 To George Biddell Airy Collingwood
Regarding Isaac Newton's theory of the motion of the moon's apse. Concerning the reliability of J. H. Seyfforth's theories. Status of the Ipswich Philosophical Society.
draft 4pp RS:HS 1.208b & 23.43

7876. 1848–10–21 To John Couch Adams Collingwood
Giving his views on planetary and lunar perturbations.
draft 4pp RS:HS 1.5 & 23.42

7877. 1848–10–22 To George Biddell Airy Collingwood
JH begs to differ with GA and Isaac Newton on the moon's motions [see GA's 1848–10–13]; warns GA against accepting J. H. Seyffert's work as accurate.
ALS 4pp RGO 6.370.67

7878. 1848–10–23 From William Rutter Dawes Camden Lodge
Recently observed *Hyperion* very faintly. Gives observation and drawings. The recent bad weather has prevented important observations.
ALS 4pp RS:HS 6.93

7879. 1848–10–23 From Thomas Galloway 45 Torrington Square
Sending a complete copy of the excerpts from JH's book [*Cape Results*]. One or two of the pictures are not by JH. Thinks a fuller abstract would have been useful.
ALS 2pp RS:HS 8.33

7880. 1848–10–24 From John Couch Adams St. John's College, Cambridge
Answer to JH's letter (HS 1.5) on planetary and lunar perturbations.
ALS 10pp RS:HS 1.6

7881. 1848–10–24 From George Biddell Airy Royal Observatory, Greenwich
Further regarding GA's defense of Isaac Newton's theory on the moon. J. H. Seyfforth's theories seem plausible. Ipswich Philosophical Society.
ALS 4pp RS:HS 1.209 (C: RGO 6.370.69)

7882. 1848–10–24 To George Biddell Airy Collingwood
Still arguing that Isaac Newton is not correct concerning the moon's motions [see GA's 1848–10–24].
ALS 4pp RGO 6.370.71

7883. 1848–10–24 From [William Henry Smyth] 3 Cheyne Walk, Chelsea
Asking JH to sign an enclosed item.
AL 1p RS:HS 1.398

7884. 1848–10–25 From George Biddell Airy Royal Observatory, Greenwich
Regarding the annual equation of the moon's motion.
ALS 4pp RS:HS 1.211 (C: RGO 6.370.73)

7885. 1848–10–25 To George Biddell Airy Collingwood
Further regarding his dispute with Isaac Newton's theory on the moon.
draft 4pp RS:HS 1.210 (C: RS:HS 23.44 & C: RGO 6.370.75)

7886. 1848–10–26 To George Biddell Airy Collingwood
Will now work with S. J. A. Compton, the President of the R.S.L., to try to get fair treatment for another applicant for the Admiralty money [see GA's 1848–9–25].
ALS 4pp RGO 6.675.581

7887. 1848–10–26 To George Biddell Airy Collingwood
Further clarification to the moon's motions [see JH's 1848–10–25].
ALS 4pp RGO 6.370.77

7888. 1848–10–26 From Christian August Friedrich Peters Pulkowa
Thanks for the gift of a copy of JH's *Cape Results.*
ALS 1p (German) RAS:JH Archive 10/5.116; Reel 9

7889. 1848–10–27 To George Biddell Airy Collingwood
Regarding the moon's orbit.
draft 4pp RS:HS 1.212 & 23.45

7890. 1848–10–27 To Spencer J. A. Compton Collingwood
Regrets Admiralty's giving whole of grant to [Charles] Brooke, when [Francis] Ronalds's photographic processes seem at least as satisfactory. Will join SC in supporting Ronalds's claims.
ALS 3pp RS Sa.659

7891. 1848–10–27 To William Rowan Hamilton Collingwood
Is glad that WH is trying in his *Lectures on Quaternions* to clarify principles of quaternions. Suggests that WH define highly abstract terms.
ALS 2pp TCD 1493:472

7892. 1848–10–28 To George Biddell Airy Collingwood
Further clarification to JH's 1848–10–26.
ALS 2pp RGO 6.675.582

7893. 1848–10–28 To George Biddell Airy Collingwood
Thanks GA for the time and trouble he took in the drawn-out debate over the moon's motions [see 1848–9–25 to 1848–10–26].
ALS 2pp RGO 6.370.79

7894. 1848–10–30 From George Biddell Airy Royal Observatory, Greenwich
Regarding the annual equation of the moon's orbit.
ALS 3pp RS:HS 1.213 (C: RGO 6.370.80)

7895. 1848–10–30 From George Biddell Airy Royal Observatory, Greenwich
Sends GA's correspondence on the subject of the Admiralty grant, with further explanatory comments [see JH's 1848–10–

28].
ACCS 4pp RGO 6.675.583

7896. 1848–10–31 From John Herapath Kensington
On the deviation from the perpendicular of falling bodies.
ALS 3pp RS:HS 10.183

7897. 1848–11–1 To Spencer J. A. Compton Collingwood
Comments unfavorably on Admiralty's attitude to matters concerning magnetic observations.
C 1p RS:HS 23.46

7898. 1848–11–1 From John Simson 1 Riches Court, Lime St.
Encloses letter received today from Mannheim. [JH note on verso] Enclosed letter of 26 Dec. 1847 was from [J. C.] Schröder to John Simson. Gratitude for receipt of [*Cape Results*]. F. B. G. Nicolai died last year. Schröder was appointed provisional director of Mannheim Observatory.
L 1p †, JH note RAS:JH Archive 10/6.2; Reel 9

7899. 1848–11–2 From George Biddell Airy Royal Observatory, Greenwich
A note to accompany further copies of correspondence to the Admiralty [see GA's 1848–10–30].
ACCS 1p RGO 6.675.585

7900. 1848–11–7 To Augustus De Morgan Collingwood
About a number of new associate members of the R.A.S.
C 2pp RS:HS 23.47

7901. 1848–11–7 From Robert Hunt Museum of Pract. Geology
Is grateful for JH's request to add his name as a subscriber to his book. Sends a volume entitled the *Poetry of Science* for JH's acceptance.
ALS 2pp RS:HS 10.128

7902. 1848–11–7 To William Henry Smyth Collingwood
Will miss R.A.S. meeting; asks WS to take chair. William Lassell announces discovery of Hyperion; believes he may have also observed an eighth satellite of Saturn.
C 1p RS:HS 23.48 David S. Evans, 'Sir John Herschel to Captain Smyth,' *Quarterly Bulletin of the South African Library, 25* (1970), 17

7903. 1848–11–8 To [Robert Hunt] Collingwood
Thanks RH for a copy of RH's *Poetry of Science,* and comments on an experiment in growing seeds under glass and in darkness.
ALS 4pp WT 68037.9

7904. 1848–11–8 From Roderick Impey Murchison Coolhurst, Hersham
Returned from Continent in September, discovered JH's gift of JH's *Cape Results*. Grateful for this memento of friendship. Hopes JH received memoirs from [Giovanni] Plana, which RM sent from Turin. Notes on transition from Cretaceous to Miocene in Alps and Apennines.
ALS 4pp RAS:JH Archive 10/5.78; Reel 9

7905. 1848–11–9 From Robert Hunt Museum of Pract. Geology
Is pleased he approves of his style. Regarding the experiments on the germination of seeds. His own son is progressing satisfactorily.
ALS 6pp RS:HS 10.129

7906. 1848–11–9 From George R. Porter Board of Trade
States he has complied with Lord Auckland's and JH's wishes on the statistical section of the *Admiralty Manual* but fears such restrictions leave it little more than 'dry bones.'
ALS 1p RS:HS 14.43

7907. 1848–11–10 From Charles Wheatstone King's College
R.S.L. Physical Committee considering giving Copley medal to W. E. Weber for work in electromagnetism. Does JH concur? Other candidates are F. E. Neumann or A.L. Cauchy in physical optics, and Michael Faraday on condensation of gases. Rumford medal was awarded to H. V. Regnault for work on steam engine.
ALS 3pp RS:HS 18.154

7908. 1848–11–11 To the *Athenaeum* Collingwood
Reports that JH will discuss the discovery of Neptune in his forthcoming *Outlines Astr.* Mentions Benjamin Peirce's calculations regarding Neptune.
L *Athenaeum,* #1100 (Nov. 25, 1848), 1176

7909. 1848–11–11 To Augustus De Morgan Collingwood
Asks for reference to date of discovery of the moon's variation by Aboul Wefa.
C 1p RS:HS 23.50

7910. 1848–11–12 To Charles Wheatstone Collingwood
Strongly supports W. E. Weber as the best candidate for the receipt of the Rumford Medal [see CW's 1849–11–10].
C 2pp RS:HS 25.11.9

7911. 1848–11–13 To the *Athenaeum* Collingwood
Reports witnessing a lunar rainbow, including a secondary rainbow.
L *Athenaeum,* #1099 (Nov. 18, 1848), 1149

7912. 1848–11–13 From Augustus De Morgan 7 Camden St.
The tract JH means is by L. A. Sédillot on the history of astronomy of the Arabs. Comments on this. Possesses it but it is bound in with a lot of other papers so is too heavy to send. Would he give John Williams permission to open letters addressed to the President of the R.A.S.
ALS 3pp RS:HS 6.246

7913. 1848–11–13 From George R. Porter Board of Trade
Writes about 'floss[?]-growers of Kent' in response to JH's letter. Characterizes the activity as a form of lottery and profit-making in the context of the free-trade/protectionist debate.
ALS 3pp RS:HS 14.44

7914. 1848–11–15 To [Charles] Wentworth Dilke Collingwood
Regarding the letter on the planet Neptune, which JH submitted to the *Athenaeum* [see Athenaeum, #1100 (Nov. 25, 1848), 1176], JH requests suppression of the final paragraph 'referring to certain peculiar ideas of Professor [Benjamin] Peirce.'
ALS 2pp BostonPubLib Ms.KF, v.3, No. 296

7915. 1848–11–15 To Edward Sabine Collingwood
Supports proposed course of action to procure for [Francis] Ronalds half of the grant and recognition given to [Charles] Brooke alone for similar photographic invention.
ALS 4pp RS Sa.658

7916. 1848–11–16 From George Eden Admiralty
Would like to hear how the *Admiralty Manual* is progressing.
ALS 2pp RS:HS 7.18

7917. 1848–11–16 From Sir Richard Owen Royal College of Surgeons
Received the seven leaves of his manuscript from [John?] Murray yesterday. Comments on matters relating to the *Admiralty Manual* and RO's contribution.
ALS 4pp RS:HS 13.194

7918. 1848–11–18 From Francis Beaufort Admiralty
The Board of Admiralty has inquired whether a paper on earthquakes by Robert Mallet has been received for the *Admiralty Manual*. Can JH advise?
ALS 2pp RS:HS 3.392

7919. 1848–11–18 From George Eden Admiralty
Thanks for letter. Is grateful for all the trouble JH is taking over the *Admiralty Manual*. Presumes JH has dealt with rotatory storms; otherwise he would have asked Col. William Reid to write a few pages.
ALS 2pp RS:HS 7.19

7920. 1848–11–[20] To the *Athenaeum* n.p.
Provides a rule for dividing vanishing lines in perspective drawing. [Written by 'Geometer', i.e., JH.]
L *Athenaeum*, #1100 (Nov. 25, 1848), 1179

7921. 1848–11–21 From George Biddell Airy Royal Observatory, Greenwich
Wants a meeting of the Board of Visitors to determine the extent of time to be given to magnetic and meteorological observations at the Royal Observatory.
ALS 3pp TxU:H/M-0634.5; Reel 1084 (C: RGO 6.9.241)

7922. 1848–11–21 To James David Forbes Collingwood
Thanks for medal of the Royal Society of Edinburgh.
ALS 1p St. Andrews 87

7923. 1848–11–22 From Alexander Bridport Becher Admiralty
Would JH authorize him to call a meeting of the Board of Visitors in the absence of S. J. A. Compton (2nd Marquis of Northampton)?
ALS 2pp RS:HS 3.393

7924. 1848–11–23 From Marshall Hall 14 Manchester Square
Sending a second edition of his letter to the President-elect of the R.S.L. Ten years ago he wrote to JH on this subject. Thinks the R.S.L. has its duties as well as its rights.
ALS 3pp RS:HS 9.205

7925. 1848–11–23 From Edward Joseph Lowe Highfield House near Nottingham
Thanks for the account of the lunar rainbow. Encloses account of Aurora Borealis. Has been residing at Bath and working with Henry Lawson. Observed a peculiar solar appearance on the 18th.
ALS 3pp RS:HS 11.317

7926. 1848–11–23 From W. H. B. Webster Carr St., Ipswich
Requests donation of JH's *Cape Results* to library of Ipswich Philosophical Society. Explains goals of Society. Note on 'lunar rainbows.'
ALS 3pp RAS:JH Archive 10/5.79; Reel 9

7927. 1848–11–26 To William Henry Smyth Collingwood
Gives suggestions for Edinburgh Board of Visitors. Gives no specific names, but says it should include theoretical and practical astronomers, geographers, and meteorologists.
C 1p RS:HS 23.52

7928. 1848–11–27 From George Biddell Airy Royal Observatory, Greenwich
Observations on JH's chapters [of *Admiralty Manual*?] received from the printers.
ALS 3pp RS:HS 1.214 (C: RGO 6.370.82)

7929. 1848–11–27 From George Biddell Airy n.p.
Urges JH to appear at the next meeting of the Board of Visitors as JH's opinion is very important.
CC 1p RGO 6.9.254

7930. 1848–11–27 From Sir Robert Harry Inglis 7 Bedford Square
Does JH intend visiting the R.S.L. on Thursday and probably the R.S.L. Club dinner on that day? If so, will he breakfast with him?
ALS 2pp RS:HS 10.243

7931. 1848–11–28 To George Biddell Airy Collingwood
Thanks GA for his letter [see GA's 1848–11–27] and brings the whole episode to a close.
ALS 3pp RGO 6.370.84

7932. 1848–11–28 To Augustus De Morgan Collingwood
Arrangements about meetings and signing of the R.A.S. testimonials.
C 1p RS:HS 23.53

7933. [1848]–11–28 From John Murray, [Jr.] Albemarle St.
Has received the papers of Richard Owen and H. T. De La Beche for paging. Encloses notices of a portable chemical apparatus; is it entitled to a place in the Manual?
AL 1p RS:HS 12.452

7934. 1848–11–28 To William Whewell Collingwood
Comments on a philosophical paper by [Robert Leslie] Ellis, sent him by WW. Discusses his own and Ellis's views on physical causality.
ALS 4pp TC, Camb. Add. Ms.a.207[74] (C: RS:HS 23.49)

7935. 1848–11–29 From Joseph Baxendell Princess St., Manchester
Regarding his observations of certain stars.
ALS 3pp RS:HS 3.304

7936. 1848–11–29 From [J. Yinsly?] Munich
[Form letter] Royal Baverian Academy of Science acknowledges receipt of JH's *Cape Results*.
(A)LS 1p (German) RAS:JH Archive 10/5.7.1; Reel 9

7937. 1848–11–30 From Augustus De Morgan 7 Camden St.
Has been hunting JH's proposition through the books and finds that it has dropped out of use, though the older geometrical books include it.
ALS 1p RS:HS 6.247

7938. 1848–11–30 From John Stevens Henslow Mitcham, Hadley, Sufk
Received his letter just as he was about to lecture. Regarding the Institution, there are five or six telescopes in use and G. B. Airy is to deliver lectures, so JH's work may be useful.
ALS 3pp RS:HS 9.311

7939. 1848–12–2 From George Smith London
W. S. Williams visited Russian embassy and ascertained only that copies [of JH's *Cape Results*] were shipped. Suggests writing to [Russian ambassador F. I.] Brunnov for more exact information.
ALS 1p RAS:JH Archive 10/6.1; Reel 9

7940. 1848–12–3 To [?] Collingwood
Writes to a lady at Slough to thank her for her kindness to JH's son [?].
ALS 2pp WT 75871

7941. 1848–12–4 From Henry T. De La Beche Geological Survey Office
Returning the proofs of his article for the *Admiralty Manual.* 25 copies will be enough for him.
ALS 2pp RS:HS 6.134

7942. 1848–12–10 From Robert John Kane Dublin
Some of his friends are wishing to put his name forward as a candidate for the fellowship of the R.S.L. Would JH sign his certificate?
ALS 3pp RS:HS 11.3

7943. 1848–12–12 From Georg A. Erman Berlin
Sending one of his own books, which he hopes JH will study.
ALS 1p RS:HS 7.102

7944. 1848–12–12 From Thomas Graham 4 Gordon Square
Sending a certificate for R. J. Kane and would be pleased if JH would put his name at the head of the list.
ALS 2pp RS:HS 8.227

7945. 1848–12–12 From W. H. B. Webster Carr St., Ipswich
Acknowledges receipt of JH's *Cape Results* donated to library of Ipswich Philosophical Society. Explains public features of library. Invites JH to lecture there. Note on 'lunar rainbow.'
ALS 3pp RAS:JH Archive 10/5.80; Reel 9

7946. 1848–12–19 From Wilhelm Struve Pulkowa
Acknowledges receipt of JH's *Cape Results.*
ALS 3pp (German) RS:HS 17.138

7947. 1848–12–20 From H. B. Maskew Public Library, Cape Town
Grateful for *Cape Results*, received through Thomas Maclear.
ALS 2pp RAS:JH Archive 10/5.11; Reel 9

7948. 1848–12–21 From Royal Bavarian Academy of Science Munich
An official certificate and a note of appreciation for the gift of JH's *Cape Results.* Also comments on JH's *Prelim Discourse.*
AL 3pp (German) RAS:JH Archive 10/5.7.1&2; Reel 9

7949. 1848–12–23 To F. W. A. Argelander Collingwood
JH, president of R.A.S., is writing to FA to discover the quantity of results that FA wants to present for printing.
C 2pp RS:HS 23.54

7950. 1848–12–24 From Thomas Maclear Royal Observatory, Cape of Good Hope
While TM was at Agulhas the large equatorial arrived, perfectly safe. Dome and railway have not arrived yet. Had a letter from W. R. Dawes. Found lighthouse at Agulhas well made and has sent sample of the material to Francis Beaufort. Not much news of the war in the Colony. Will take care over the mounting of the equatorial.
ALS 3pp RS:HS 12.142

7951. 1849 From Jacques Babinet Paris
Presenting A. T. D'Abbadie, who has made a voyage to Brazil. Remarks on JH's calorimeter.
ALS 3pp (French) RS:HS 3.3

7952. [1849] To Francis Beaufort n.p.
JH's article on 'Meteorology' [for the *Admiralty Manual*] is finished; JH will make some comments on aneroid barometers in the preface.
ALS 2pp Hydrographic Office Misc F24F1.3

7953. 1849–1–2 From John R. Hind 11 Hanover Cottages
Sends a copy of the *Astronomical Remembrancer* to which he referred. The garnet star is BAC 7582. Encloses a list of variable stars and comments on the proximity of two of them.
ALS 2pp RS:HS 9.343

7954. 1849–1–3 From Sophie Beckedorff to Margaret Brodie Herschel Hanover
Recent political events in Germany. Family news. Hopes Lady Herschel's health has improved.
ALS 5pp RS:HS 4.6

7955. 1849–1–3 From Edward Sabine Woolwich
Discusses letter to [G. B.] Airy and JH's response to it. Discusses instruments at Toronto Observatory. Describes 18 November 1841 magnetic disturbances. Working on Atlantic declination maps.
ALS 4pp RS:HS 15.228

7956. 1849–1–4 From Richard Sheepshanks Reading
H. C. Schumacher wants an application from the R.A.S. RS complied, giving reasons. 'Miss B.' [Elixabeth Baily] reluctant to write even the shortest formal note. Suggests JH announce her gift on taking the chair. Wants to 'jump' at [F. W. A.] Argelander's offer [to complete the British Catalogue]. Progress on his 'apparatus.' Has asked J. C. Adams to give the R.A.S. notice of the controversy about Neptune, in which U. J. J. Leverrier ('a very cantankerous fellow') has engaged. If Adams refuses, would JH respond?
ALS 6pp RS:HS 16.91

7957. 1849–1–4 From William Henry Smyth 3 Cheyne Walk, Chelsea
Compiles published work, unpublished work, and lectures; asks JH to proof the compilation. Discusses meetings of R.S.L. concerning charter by-laws.
ALS 3pp RS:HS 16.297

7958. 1849–1–5 From Richard Jones Tithe Commission
Has mislaid a piece of paper on the Addiscombe examinations that he intended sending to JH. Appear to be some changes in policy. Has taken a house in Grosvenor place for one or two months. The Commission is not yet out.
ALS 4pp RS:HS 10.384

7959. 1849–1–5 From Robert John Kane 51 Stephens Green, Dublin
Would like JH's opinion on a memoir by M. McCann on the reflection of light from binaxal crystals. McCann is a candidate for a mathematical chair in Queen's College [Cork].
ALS 2pp RS:HS 11.4

7960. 1849–1–8 From Frederick W. Beechey Westbourne Crescent
Has at last received his letter and withdrawn the index.
ALS 2pp RS:HS 4.22

7961. 1849–1–8 To Robert John Kane Collingwood
M. McCann sent his paper to JH for communication to the R.S.L., which he did, but does not know the outcome of it. Was unable to report on it as his own knowledge was not modern enough.
C 4pp RS:HS 11.5

7962. 1849–1–9 From C[onstantine] E[stlin] Prichard Bristol
Informs JH of the death of CP's father, James Cowles Prichard.
ALS 4pp RS:HS 14.187

7963. 1849–1–13 From F. W. A. Argelander Bonn
Has notes on John Flamsteed. Shall he send them? Gives notices of two variable stars.
ALS 7pp (German) RS:HS 1.374

7964. 1849–1–13 From Richard Sheepshanks R.A.S.
R.A.S. voted to give its medal to William Lassell. Recounts the discussion of this among the members. G. B. Airy will be the next president. Applauds JH for the work he has done in this position. An 'olive branch' tendered to [W. S.] Stratford; he was made vice president.
ALS 4pp RS:HS 16.92

7965. [1849]–1–15. From Mary Grantham 37 Tavistock Place
Elizabeth Baily will be pleased to see him on the day mentioned. Also agrees with him to present the bust of her brother to the R.A.S. JH may have casts of the bust.
ALS 4pp RS:HS 9.118

7966. 1849–1–15 From Robert John Kane 51 Stephens Green, Dublin
Thanks for his letter regarding M. McCann. Does he know anything of the capabilities of Parkinson Wilson, Mr. Hobson, and Mr. Sandiman? Had intended to communicate his own researches on platina a long time ago, but was hoping to conclude them first. Thomas Graham will send down the form of certificate for him to sign.
ALS 3pp RS:HS 11.6

7967. 1849–1–15 From Richard Sheepshanks Royal Astronomical Society
RS, Vice President of R.A.S., writes to JH, President, about a series of Council and membership matters.
ALS 5pp RS:HS 16.93

7968. 1849–1–17 To Richard Sheepshanks Collingwood
JH replies to RS's 1849–1–15
ALS 4pp RS:HS 16.94 (C: RS:HS 25.9.59)

7969. 1849–1–19 From Francis Beaufort Admiralty
The Equatorial has arrived safely at the Cape. William Mann has declined the position at Madras and G. R. Smalley offered his services. What does JH think? Encloses letter from Thomas Maclear and one for Lady Herschel.
ALS 3pp RS:HS 3.394 & ACC: Hydrographic Office Letter Book

7970. 1849–1–19 From John R. Hind 11 Hanover Cottages
As JH is compiling a complete catalogue of variable stars, wonders if he has noticed the one listed in the Berlin maps. Has recently made a cometary summary.
ALS 2pp RS:HS 9.344

7971. 1849–1–20 To Richard Sheepshanks Collingwood
Says he will guardedly write in his *Outlines Astr.* what he has to say about Neptune and the controversy surrounding its discovery. States that he eschews formulae.
ALS 2pp RS:HS 16.95 (C: RS:HS 25.9.60)

7972. 1849–1–21 To F. W. A. Argelander Collingwood
Further to reception of results from FA [see JH's 1848–12–23].
C 1p RS:HS 23.55

7973. 1849–1–21 From William Radcliffe Birt Bethnal Green
Has just heard from Edward Sabine that there is still a vacancy at the Madras Observatory. Would like JH's support if he applies to the East India Company.
ALS 2pp RS:HS 4.127

7974. 1849–1–22 From John R. Hind 11 Hanover Cottages
May use his cometary summary by all means; believes that J. G. Galle's is the most complete but his method requires improvement. Thinks he has traced the history of Halley's comet and will be writing a paper on the subject.
ALS 2pp RS:HS 9.345

7975. 1849–1–22 From Sir Robert Harry Inglis 7 Bedford Square
Archbishop of Canterbury is to take his seat at the R.S.L. on Thursday next and is to dine with RI beforehand; will JH join them?
ALS 1p RS:HS 10.244

7976. 1849–1–23 From Charles T. Beke 6 St Mildred's Court
Sending an extract from a letter dealing with magnetic and electrical instruments recently sent to Egypt for purposes of exploration.
ALS 1p RS:HS 4.31

7977. 1849–1–27 From William Radcliffe Birt Bethnal Green
Thanks for the letter and his interest in the matter. Has written to W. H. Sykes relative to his astronomical qualifications.
ALS 2pp RS:HS 4.128

7978. 1849–1–29 From William Parsons (Lord Rosse) The Castle, Parsonstown
Has never seen the great nebula of Andromeda satisfactorily resolved; comments on this and gives some of his observations. Will be going to Brighton and then take his family to London for Easter.
ALS 12pp RS:HS 13.228

7979. 1849–1–31 From [Adolph Theodor Kupffer?] St. Petersburg
Thanks to JH for the copy of JH's *Cape Results.*
ALS 1p (French) RAS:JH Archive 10/5.117; Reel 9

7980. 1849–1–31 From Prof. Keely to Edward Sabine Waterville College
Encloses note to be directed to JH, pointing out discrepancies in JH's article on sound in first edition of *Encyclopaedia Metropolitana.*
L 2pp TxU:H/M–1008; Reel 1083

7981. 1849–2–1 To Augustus De Morgan Collingwood
Arrangements about how the R.A.S. meeting will proceed.
C 1p RS:HS 23.56

7982. 1849–2–2 From Edward H. Baily 17 Newman St.
Will be pleased to supply the casts [of the bust of Francis Baily; see M. Grantham's [1849]–1–15]; charge is five guineas each.
ALS 1p RS:HS 3.27

7983. 1849–2–2 From Augustus De Morgan 7 Camden St.
Would he trouble himself about the address only; all the rest will be arranged for him.
ALS 1p RS:HS 6.248

7984. 1849–2–4 From Thomas Maclear Royal Observatory, Cape of Good Hope
Sends details of the progress of erecting the equatorial. William Mann is engaged on the Simon's Bay tide observations. The Madras Observatory post was offered to Mann but he declined it for health reasons. Comments on the affairs of [G. R.Smalley], who is in financial difficulties. Not much news from the Cape. Comments on some of the educational news.
ALS 6pp RS:HS 12.143

7985. 1849–2–8 From Urbain J. J. Leverrier Paris
Regarding the work of Yvon Villarceau on double stars. Would like a reference from Sir William Herschel's paper.
ALS 3pp (French) RS:HS 11.201

7986. 1849–2–12 To William Whewell 3 Henrietta St., Covent Garden
Asks WW, as one of the trustees of JH's marriage settlement, to sign some legal documents.
ALS 4pp TC, Camb. Add. Ms.a.207[76] (C: RS:HS 23.57)

7987. 1849–2–13 From William Radcliffe Birt Bethnal Green
Wishes to embody in his reports on atmospheric waves details of JH's work in this field. Would like his comments on his suggested wording. Can E. W. Brayley borrow Karl Kreil's book? Nothing from Madras yet.
ALS 4pp RS:HS 4.129

7988. 1849–2–14 From John Edwin Mayall 433 West Strand
Sends the proof of the lithograph from the Daguerreotype for JH's comments and additions.
ALS 3pp RS:HS 13.30

7989. 1849–2–15 From William Parsons (Lord Rosse) The Castle, Parsonstown
Examined the nebula of Andromeda the night before last with his three-foot instrument. Comments on his observations and the sketch made by [W. C.?] Bond. Would require several good nights to make a sketch of value.
ALS 3pp RS:HS 13.229

7990. 1849–2–18 To Augustus De Morgan Collingwood
Returns H. P. Brougham's (Baron Brougham and Vaux) paper. Does not recall any similar theories, but does not think his own theories are shaken by those of Brougham. Finds that AD has already done extensive work on the calendar.
draft 2pp RS:HS 6.249 & 23.58

7991. 1849–2–21 From William Radcliffe Birt Bethnal Green
Has received his letter of the 20th and is anxious to know if E. W. Brayley can borrow Karl Kreil's book.
ALS 2pp RS:HS 4.130

7992. 1849–2–27 From Edward Joseph Lowe Bath
Encloses some sketches to show some remarkable changes in solar spots. Would he place the observations before the R.A.S. when finished with them.
ALS 2pp RS:HS 11.318

7993. 1849–3–2 From William Cranch Bond Observatory, Cambridge, Mass.
Expresses appreciation for being elected an associate of the R.A.S.
C 2pp Univ. of London Library

7994. 1849–3–3 From Augustus De Morgan 7 Camden St.
Finds that some of the pages from the paper JH recently sent him are missing. Please send if he still has them. H. P. Brougham (Baron Brougham and Vaux) does not mean to attack the undulating theory. Regarding solar spots and the surface of the moon.
ALS 4pp RS:HS 6.250

7995. 1849–3–4 From Richard Sheepshanks Reading
Shall send a catalogue of double stars observed by [W. S.] Jacob. Gives some information about several measurements to establish standards.
ALS 4pp RS:HS 16.96

7996. 1849–3–4 To [James Calder] Stewart Collingwood
Received news today of JS's arrival in Bombay. Describes conflict as 'Armageddon.' Family news. John Stewart returned from China and leaves soon to meet Matilda [Grahame] in Paris. [Richard] Jones's concern for JS is genuine. Eneas Mackintosh obtained cadetship for JH's son John at Addiscombe.
ALS 4pp TxU:H/L-0433; Reel 1055

7997. 1849–3–6 From Edward Joseph Lowe Bath
Has just received JH's letter giving details of an interesting auroral phenomenon. Has been experimenting with wire insulates. Henry Lawson has observed streams of red light resembling aurora.
ALS 2pp inc RS:HS 11.319

7998. 1849–3–7 From George Biddell Airy Royal Observatory, Greenwich
Regarding the 'chilled' casting of a telescope.
ALS 2pp RS:HS 1.215 (C: RGO 6.160.77)

7999. 1849–3–7 To Augustus De Morgan Collingwood
On return of a paper, and approval of AD's method of resolving fractions.
C 1p RS:HS 23.59

8000. 1849–3–8 From John Johnstone New Prison, Wandsworth, Surrey
Sends sample of photographic lenses ground by JJ. Asks JH for formula of curvature to admit more light.
ALS 3pp TxU:H/M–0304; Reel 1087

8001. 1849–3–8 To George Biddell Airy Collingwood
Tells GA about the internal counterpoise system in use in some German telescope mountings, and includes a diagram; JH knows nothing about chilling speculum metal [see GA's 1849–3–7].
ALS 5pp RGO 6.160.78

8002. 1849–3–9 To Richard Sheepshanks Collingwood
About an astronomer for the Madras observatory. Suggests ways to reduce glare and so improve measurements [see RS's 1849–3–4].
ALS 3pp RS:HS 16.97 (C: RS:HS 25.9.61)

8003. 1849–3–17 From Spencer J. A. Compton Piccadilly
It is now time that they speak to John Russell (1st Earl Russell) concerning [Francis?] Ronald[s]. Would next Friday suit him?
ALS 2pp RS:HS 5.276

8004. 1849–3–18 From Augustus De Morgan 7 Camden St.
Regarding the early history of the calendar. Comments on the various versions.

ALS 7pp RS:HS 6.251 S. De Morgan, *Memoir of Augustus De Morgan*, 204–6

8005. 1849–3–19 To Spencer J. A. Compton Collingwood
Arrangements about JH coming to London to go with SC to see John Russell (1st Earl Russell) [see SC's 1849–3–17].
C 1p RS:HS 23.60

8006. 1849–3–19 To Urbain J. J. Leverrier Collingwood
Has just been able to consult Sir William Herschel's papers. Regarding Yvon Villarceau's method of measuring the orbits. May be like his own, which he briefly outlines.
C draft 4pp RS:HS 11.202 & 23.61 JH, 'Extrait d'une lettre à M. Le Verrier sur sa méthode pour la détermination de l'orbite des étoiles doubles,' *Comptes Rendus, 28* (1849), 421–2 inc

8007. 1849–3–22 From Urbain J. J. Leverrier n.p.
Would like to ask a few more questions regarding JH's method outlined in his last letter. Can UL submit it to the Institute?
ALS 3pp (French) RS:HS 11.203

8008. 1849–3–25 To Augustus De Morgan Collingwood
Sends AD part of JH's paper on double stars; concerned about AD's dating it on arrival.
C 1p RS:HS 23.62

8009. 1849–3–25 From William King East Horsley Pr., Ripley, Surrey
Delay in printing WK's meteorology paper.
ALS 3pp RS:HS 19.286

8010. 1849–3–25 To Urbain J. J. Leverrier Collingwood
Found his letter on return from London. Would be pleased for UL to submit part of his letter of 19 Mar. to the Institute. Further comments on part of this letter.
C 2pp RS:HS 11.204

8011. 1849–3–26 To Augustus De Morgan Collingwood
Correction to JH's paper on double stars [see JH's 1849–3–26 & 1849–3–25].
C 1p RS:HS 23.64

8012. 1849–3–26 To Augustus De Morgan Collingwood
Another part of JH's double star paper; question of priority between JH and Yvon Villarceau.
C 1p RS:HS 23.63

8013. 1849–3–26 From Maximilian Weisse Cracow
Thanks JH for *Cape Results* sent through [F. G. W.] Struve. Further grateful for the N. L. Lacaille and Jérôme Lalande works sent to the observatory. Appreciates reduction of the moon results as well as the magnetic observation of 1845. Reveals new development in double star zones (?). Comments on making magnetic observations.
ALS 3pp (German) RS:HS 18.138

8014. 1849–3–27 To Augustus De Morgan Collingwood
Correction to paper on double stars following up JH's 1849–3–26; a terrible pun sent on by JH's wife, Margaret.
C 1p RS:HS 23.65

8015. 1849–3–27 From Augustus De Morgan 7 Camden St.
Has received three packets, dated and initialed. Further comments regarding Julius Caesar and the calendar.
ALS 2pp RS:HS 6.252

8016. 1849–3–28 From George Biddell Airy Royal Observatory, Greenwich
Asks JH's opinion about GA's plan not to print a separate volume of magnetic and meteorological observations for 1848.
ACCS 2pp RGO 6.694.248

8017. 1849–3–29 To George Biddell Airy n.p.
Asks GA if he knows of any rules or pitfalls in calculating time from earlier times [even B.C.] to the present.
ALS 2pp inc RGO 6.465.26

8018. 1849–3–29 To George Biddell Airy n.p.
Some concerns about GA's intent of attaching magnetic and meteorological observations to the R.A.S.'s *Astronomical Observations* [see GA's 1849–3–28].
ALS 2pp RGO 6.694.250

8019. 1849–3–30 From George Biddell Airy n.p.
Has little advice to offer JH about time calculation [see JH's 1849–3–29], except for fairly standard reminders.
ACCS 2pp inc RGO 6.465.27

8020. 1849–3–30 From George Biddell Airy Royal Observatory, Greenwich
Will reconsider the proposal about printing observations [see JH's 1849–3–29], and get back to JH.
LACC 1p inc RGO 6.694.251

8021. 1849–4 To Augustus De Morgan n.p.
Needs to revise *Outlines Astr.*, and will include Ernesto Capocchi's announcement of the discovery of another planet; some further corrections of JH's double star paper.
C 2pp RS:HS 23.66

8022. 1849–4–1 From [A. J. F.] Yvon Villarceau Paris
Asks for JH's comments on YV's specific double star observations. YV presents the mathematical method he uses to calculate double star orbits, which he had presented to the Paris Académie in 1847, but it was not published.
ALS 5pp (French) RS:HS 17.418

8023. 1849–4–2 From Henry Anthony Jeffreys Hawkhurst Parsonage
Describes HJ's formal petition against Stuart Watley's bill for reform of marriage laws.
ALS 4pp TxU:H/MW-0025.1; Reel 14

8024. [1849–4–3] From Henry Anthony Jeffreys Hawkhurst Parsonage
Thanks for JH's corrections to HJ's formal petition [against marriage reform bill].
ALS 1p TxU:H/MW-0025.2; Reel 14

8025. 1849–4–3 From Urbain J. J. Leverrier n.p.
Read part of JH's letter to the Institute and Yvon Villarceau also presented his own method. Will see the letter printed in the *Comptes rendus*.
ALS 3pp (French) RS:HS 11.205

8026. 1849–4–4 From George Biddell Airy Royal Observatory, Greenwich
Glad to hear that Margaret Louisa [JH's daughter] is coming to visit GA's family; could JH send along the key to the R.A.S. strong box?
ACCS 1p RGO 6.231.395

8027. 1849–4–[5] To George Biddell Airy n.p.
Responds to GA's 1849–4–4, and sends the key.
ALS 2pp RGO 6.231.396

8028. 1849–4–5 From Francis Charles Knowles Beaumont St.
Looks forward to the new edition of JH's *Treatise Astr.* Comments on the atmosphere and gases on the surface of the moon. What does JH think of these comments?
ALS 4pp RS:HS 11.64

8029. 1849–4–7 From George Biddell Airy Royal Observatory, Greenwich
Some responses to the question of copyright and of revision of articles prepared for the *Encyclopaedia Metropolitana.*
ACCS 4pp RGO 6.371.106

8030. 1849–4–9 To George Biddell Airy Collingwood
Is making a case for JH's priority claims with regard to the means of determining double star orbits, in conflict with Yvon Villarceau.
ALS 4pp RGO 6.231.176

8031. 1849–4–10 From William Radcliffe Birt Cambridge Road, 2 Sydney Place
Has received the proof sheets of his contributions to the *Admiralty Manual* and would like to make a few alterations. Has heard no news of the Madras appointment.
ALS 4pp RS:HS 4.131

8032. 1849–4–10 From Edward Joseph Lowe Highfield House near Nottingham
Has just published a small work to remove popular superstitions. Presents a copy to JH as a mark of respect.
ALS 1p RS:HS 11.320

8033. 1849–4–11 From George Biddell Airy Royal Observatory, Greenwich
A note about forwarding papers related to the double star question [see JH's 1849–4–9].
ACCS 1p RGO 6.231.178

8034. 1849–4–17 From George Biddell Airy Royal Observatory, Greenwich
Returning papers of Yvon Villarceau and an abstract of them, together with comments on them.
ALS 2pp RS:HS 1.216 (C: RGO 6.231.180

8035. 1849–4–19 To George Biddell Airy Collingwood
Extended comments about some of GA's statements in GA's abstract of Yvon Villarceau's papers on double stars [see GA's 1849–4–11].
ALS 8pp RGO 6.231.181

8036. 1849–4–19 To [A. J. F.] Yvon Villarceau Collingwood
Has passed YV's memoirs on the calculation of orbits on to G. B. Airy for publication; copies will be sent [see YV's 1849–4–1]. Notes that YV's methods are similar to those of U. J. J. Leverrier.
ACCS 2pp (French) RS:HS 17.418a

8037. 1849–4–21 From Francis Beaufort Admiralty
Identifies the positive features of the aneroid barometer as in use by the Navy.
ACC 1p Hydrographic Office Letter Book

8038. 1849–4–21 From Robert Hunt Museum of Pract. Geology
Is thinking of withdrawing his son from Christ's Hospital owing to his frequent epileptic fits. Should he inform the Prince Consort? Outlines his recent photographic work.
ALS 5pp RS:HS 10.130 & 10.136

8039. [1849]–4–24 From Robert Hunt Museum of Pract. Geology
Since writing to JH he has taken medical advice and decided to leave his son at Christ's Hospital for the time being. Gave a lecture at the Royal Institution on Friday evening.
ALS 5pp RS:HS 10.131 & 10.136

8040. 1849–5–4 From Jacob Hagen Adelaide
Regarding tides in the atmosphere. Suggests a possible way of measuring these tides. Has never seen the subject mentioned.
ALS 2pp RS:HS 10.145

8041. 1849–5–4 From Sir Francis Ronalds Kew Observatory
Grateful for JH's help in making a successful claim upon Her Majesty's Government.
ALS 1p RS:HS 14.423

8042. 1849–5–5 From Augustus De Morgan 7 Camden St.
Many thanks for his astronomy [*Outlines Astr.*]. Will read it later when he has more time.
ALS 3pp RS:HS 6.253

8043. 1849–5–7 From George Biddell Airy Royal Observatory, Greenwich
A note of thanks for a copy of JH's *Outlines Astr.*
ACCS 1p inc RGO 6.371.108

8044. 1849–5–12 From Alfred Smee Bank of England
Note to present JH with a copy of AS's latest work.
ALS 2pp RS:HS 16.169

8045. [1849?]–5–13 From Sara Prinsep 9 Chesterfield Street
Thanks JH for his support of Henry Prinsep in his attempt to be elected [to seat of directors of the East India House].
ALS 3pp RS:HS 14.190

8046. 1849–5–13 To William Henry Smyth Collingwood
Approves of James Nasmyth's application to join R.S.L. Describes good points of Nasmyth's telescope. Supposes that if he had drawing talent, he would sketch the members of the R.S.L.
C 2pp RS:HS 23.68

8047. 1849–5–14 From Sir Charles Shadwell Barnes
Describes usual method for interpolation of altitudes at sea. JH's graphic method probably would be practicable.
ALS 2pp RS:HS 17.199

8048. 1849–5–18 From Algernon Percy (4th Duke of Northumberland) Whitehall Gardens
Is looking forward to receiving a copy of JH's *Outlines Astr.*; describes an aurora AP had seen at Alnwick Castle.
ALS 4pp RS:HS 13.359

8049. 1849–5–18 From [A. J. F.] Yvon Villarceau Paris
Is sending a communication to JH for insertion in the R.A.S.M.N. Seems to be some problem with understanding YV's calculations of the orbits of double stars, so YV explains his method.
ALS 2pp (French) RS:HS 17.419

8050. 1849–5–22 From James Challis Cambridge Observatory
Is grateful for his *Outlines Astr.* Comments on its quality and usefulness. Sends the first report of the Board of Mathematical Studies.
ALS 3pp RS:HS 5.235

8051. 1849–5–22 From Augustus De Morgan 7 Camden St.
Sends extracts from book by Mr. Hopkins on the solar system. Has his own theories on the Augustan calendar proved to his own mind. No authority for giving the name Saras to the period of 223 lunations.
ALS 4pp RS:HS 6.254

8052. 1849–5–29 From James David Forbes Trinity Lodge, Cambridge
Regarding the possibility of a new edition of JH's *Treatise Astr.* originally published in Lardner's Cyclopaedia. Has found it very useful for his students.
ALS 4pp RS:HS 7.310

8053. 1849–5–29 From William Galbraith Edinburgh
Pointing out some errors in JH's *Outlines Astr.*
ALS 3pp RS:HS 8.1

8054. 1849–6–1 To James David Forbes Collingwood
Comments on JH's *Treatise Astr.* in the Cabinet Cyclopaedia, mentioning its continued publication, misstatements he wishes to correct, and his addition of A. de Gasparis's newly-discovered planet [Hygeia].
ALS 3pp St. Andrews 48

8055. 1849–6–9 From John Murray, [Jr.] Albemarle St.
Admiralty Manual is now published and an early copy has been sent by post. Remaining copies will be sent later. Regarding the distribution of other copies, can JH assist with some of the addresses?
AL 2pp RS:HS 12.446

8056. 1849–6–12 From William Galbraith Edinburgh
Thanks for his letter. Was pleased to hear about the corrections. Points out another error in JH's figures in JH's *Outlines Astr.*
ALS 3pp RS:HS 8.2

8057. 1849–6–12 From Richard Jones Tithe Commission
Before he can exchange appointment of Willy [Herschel] he must find out what his present appointment is. Still thinks of going to St. Leonards and hopes to call on the Herschels. Would like Maria Edgeworth's last letter; misses her very much. Should cultivate Angelica.
ALS 4pp RS:HS 10.385

8058. 1849–6–12 From Heinrich Christian Schumacher Altona
Alexander von Humboldt reports reading that Lord Rosse's telescope has resolved the nebula of Orion. Humboldt requests JH's view on the matter. Describes new government.
ALS 3pp RS:HS 15.402

8059. [1849]–6–13 From Charles Robert Darwin The Lodge, Malvern
Thanks for his obliging note. Has already had one from John Murray. Hopes the *Admiralty Manual* will bear fruit. Has been at Malvern for three months and his health has improved enormously.
ALS 4pp RS:HS 6.16 Darwin, *Correspondence, 4,* 238–9

8060. 1849–6–13 To Richard Jones Collingwood
Family news, questions of son Willy's position, frost in the garden, and JH met Louis Philippe.
C 3pp RS:HS 23.69

8061. 1849–6–15 From William Radcliffe Birt Cambridge Road
Has received the copies of the *Admiralty Manual.* Discussions of the electrical observations at Kew are almost finished and will be able to forward the manuscript if desired.
ALS 3pp RS:HS 4.132

8062. 1849–6–15 From Henry Raper 5 Chapel Street
Thanks JH for subscription to the Astronomical Club. Discusses JH's objection to use of symbols in works on navigation. Notes that symbols are becoming more common. Requests advice.
ALS 4pp RS:HS 14.308

8063. 1849–6–16 From William Rutter Dawes Camden Lodge
Was grateful for C. R. Darwin's note. Glad to hear of the improved health of Sir J. Dundas. Gives recent observations of Alpha Virginis.
ALS 3pp RS:HS 6.94

8064. 1849–6–22 From George Biddell Airy Royal Observatory, Greenwich
Is passing on a request for a copy of JH's *Cape Results* from C. L. Littrow of Vienna.
ACCS 2pp RGO 6.57.149

8065. 1849–6–25 To George Biddell Airy Collingwood
A number of copies of the *Cape Results* were sent out, including one to Charles Ludwig Littrow [see GA's 1849–6–22]; JH will now try to trace the books whereabouts.
ALS 4pp RGO 6.57.151

8066. 1849–6–26 From William Cranch Bond Cambridge, [Mass.]
WB's latest observations of Gamma Virginis.
ALS 1p † TxU:H/M–0113; Reel 1087

8067. 1849–6–26 From Heinrich Christian Schumacher Altona
Relays debate about the use of names of Roman gods for 'planets.' [Annibal] de Gasparis proposes to name his discovery in Italian, 'Igia,' rather than the Roman Hygia [Hygeia]. Which version will JH adopt?
ALS 3pp RS:HS 15.403

8068. 1849–6–29 To Charles Wheatstone Collingwood
Invites CW to come out to Collingwood next week when some other friends are coming, too.
C 1p RS:HS 25.11.10

8069. 1849–6–30 From Charles Wheatstone Lower Mall, Hammersmith
Will visit JH next week. Describes machine CW will bring to illustrate undulatory theory of light.
ALS 3pp RS:HS 18.155

8070. 1849–6–30 To Charles Wheatstone Collingwood
Is pleased CW is coming [see JH's 1849–6–29]; JH asks CW to bring some of his apparatus to do polarized light experiments.
C 2pp RS:HS 25.11.11

8071. [1849]–7–2 From E[neas] MacKintosh Montagu Square
Has judged wisely for Willie [JH's son] in the present uncertain state of his health. Theodore Hope has just returned from Madeira to take his place at College. Will visit the Herschels before they set out for the Pyrenees.
ALS 2pp RS:HS 12.24

8072. 1849–7–13 To Eneas Mackintosh n.p.
Efforts by Duncan Stewart to provide new career for brother Peter Stewart, whose mischief created economic ruin, left

Duncan Stewart in difficult financial situation. JH is willing to relinquish EM's appointment to Haileybury College for JH's son John and to seek position for John at [Royal Military Academy,] Woolwich, if EM would transfer recommendation to Haileybury from JH's son John to Duncan Stewart's second son.
Adraft 2pp TxU:H/L-0239; Reel 1054

8073. 1849–7–15 To Augustus De Morgan Collingwood
Thanks for present of Thomas Wright's books [*An Original Theory of the Universe,* 1750]. Finds it curious. Suggests that it is conceptually as distant from 1849 as the ancients were from it. Praises printing of its plates.
ALS 3pp Univ. of London Library

8074. 1849–7–16 From E[neas] MacKintosh Ramsgate
Was not at home when JH's letter arrived. Appointment of John [JH's son] is to all intents JH's own, and Leslie Melville will sanction any appropriation JH may desire. Thinks that JH and his wife should consider the situation before embarking further.
ALS 4pp RS:HS 12.22

8075. 1849–7–22 From Thomas Maclear Royal Observatory, Cape of Good Hope
Has had four men working on the Dome to complete it before the bad weather. Has not heard who has been appointed to Madras. The candidate from the Cape is in financial difficulties. Tide reductions are progressing. Outlines the disturbances caused by the expected arrival of the convicts.
ALS 4pp RS:HS 12.144

8076. 1849–7–26 From Charles Wheatstone Lower Mall, Hammersmith
Lists seven French and German works [1837–1847] on optical science. Comments on one. Quotes copyright law on reproducing passages from these.
ALS 3pp RS:HS 18.156

8077. 1849–7–27 From James A. Gordon South Saxon Hotel
Favorable comment from all quarters on JH's latest book of poems. Draws attention to an article on Scottish abbeys and cathedrals. Has just paid his respects to the King [of France], who remembers JH warmly.
ALS 4pp RS:HS 8.181

8078. 1849–8–1 To [William Radcliffe Birt] Collingwood
Will forward WB's report on Kew Observatory to [John] Taylor. Praises WB's thoroughly inductive discussion of atmospheric electricity.
ALS 3pp RS:HS 19.138

8079. 1849–8–1 To Edward Sabine Collingwood
Comments favorably on report by W. R. Birt on Kew meteorological observations. Sees special significance concerning atmospheric electricity.
C inc RS MM

8080. 1849–8–3 From V. Tracy Paris
Thanks JH for a copy of a scientific manual (*Admiralty Manual*?).
1p (French) RS:HS 17.442

8081. 1849–8–5 From James A. Gordon University Club
Outlining the advantage of a visit to Scotland. Hopes JH will have time to visit Knockspock as some of his friends will be there.
AL 4pp inc RS:HS 8.182

8082. 1849–8–7 From William Reid Woolwich
Appreciates JH's praise of his meteorological work. Describes curious waterspouts in the sea and their boiling appearance.
ALS 4pp RS:HS 14.325

8083. 1849–8–10 From James Hudson Royal Agricultural Society
Has arranged for the missing parts of their journal to be forwarded. His own health is good.
ALS 4pp RS:HS 10.39

8084. 1849–8–11 To [William Radcliffe Birt] Collingwood
Certain that JH did not receive WB's 'Sheet of Curves.' They may be in Edward Sabine's or WB's own possession.
ALS 4pp RS:HS 19.139

8085. 1849–8–15 From William Radcliffe Birt Cambridge Road
Has received from Edward Sabine the manuscript of his own report on the electrical observations. Is pleased that his labors have been useful.
ALS 2pp RS:HS 4.133

8086. 1849–8–16 From Francis Beaufort Admiralty
The Dome has been returned to the Cape and is now in process of erection as the enclosed will show.
ALS 1p RS:HS 3.395

8087. [1849]–8–17 From William Empson Craig Crook, Edinburgh
Expects he has received the proofs by now. Is pleased he will review L. A. J. Quetelet's book on probabilities. Returns to Haileybury next month.
ALS 4pp RS:HS 7.76

8088. 1849–8–17 From James A. Gordon Knockspock
Is sorry to hear that JH will not be visiting them this year. Hopes he will come another year. Sir George Grey should read JH's translation of Frederick Schiller. Thomas Maclear ought to be Scottish. The Queen has arrived at Balmoral.
ALS 4pp RS:HS 8.183

8089. 1849–8–19 To Francis Beaufort Collingwood
Returns a letter from Thomas Maclear to FB, and then JH expresses his concern about the state of health and work of Maclear.
ALS 4pp Hydrographic Office Misc F24F1.1

8090. 1849–8–27 From William Radcliffe Birt Cambridge Road
Has altered the words in accordance with JH's suggestions. Encloses extract dealing with clouds.
ALS 4pp RS:HS 4.134

8091. [1849]–8–28 From George Thomas Napier Brussels
Has been wandering over Europe for the last two years and has only just heard of JH's gift of his *Cape Results*. Is grateful for the gift and the memories it invokes. Comments on the present political scene at the Cape.
ALS 4pp RS:HS 13.74

8092. [1849]–8–29 From William Empson Craig Crook, Edinburgh
Thanks for his letter concerning L. A. J. Quetelet; trusts he will be able to prepare the article for the Christmas number. Cholera is at Haileybury.
ALS 4pp RS:HS 7.64

8093. 1849–8–29 From Hans Christian Oersted Copenhagen
Has not heard from him for a long time or his letter was not delivered and the unrest in Europe has not encouraged him to write letters. Sends articles, also part of a book, which he has been trying to get published by an English publisher. Would like the advice of JH on this effort.
ALS 3pp RS:HS 13.171 Harding, *Correspondance de ... Örsted, 2,* 402–4

8094. 1849–8–31 To Augustus De Morgan Collingwood
Raises an objection to a method of calculating probabilities set out by AD in his *Formal Logic* [1847]. Otherwise, praises AD's discussion.
ALS 4pp Univ. of London Library

8095. [1849]–9–1 From Eneas MacKintosh Montagu Square
Has seen Leslie Melville who can be relied on to give John [JH's son] the Addiscombe appointment.
ALS 2pp RS:HS 12.23

8096. [1849–9] To Thomas Maclear n.p.
Encloses star observations. Comments on his instrumental system. Cannot find the rule for position when two stars are of equal magnitude. *Neptune* with the convicts is anchored in Simon's Bay, awaiting the answer from H. G. Grey (3rd Earl Grey).
ALS 2pp RS:HS 12.245

8097. 1849–9–1 From Thomas Maclear Royal Observatory, Cape of Good Hope
Sends observations made with the nine-foot equatorial, with comments. Received JH's *Astronomy* and the *Logs.* of Charles Babbage; the latter aroused protests in the family. [Archibald] Robertson sold five copies of the *Astronomy* instantly and has ordered further copies.
ALS 2pp RS:HS 12.145

8098. 1849–9–2 From Augustus De Morgan 7 Camden St.
Regarding their various points of view on a certain question.
ALS 4pp RS:HS 6.255

8099. 1849–9–7 or later To Richard Sheepshanks n.p.
Appreciates additional time made available by RS to allow JH to finish paper properly.
ALS 3pp RS:HS 16.140

8100. 1849–9–7 From Richard Sheepshanks Reading
Suggests some means of improving a paper submitted by JH; more on standard measures of length [see RS's 1849–3–4], and other R.A.S. business.
ALS 8pp RS:HS 16.98

8101. 1849–9–9 From Thomas Maclear Royal Observatory, Cape of Good Hope
Has had some splendid evenings, and sends the observations made. Comments on the work of the new equatorial. W. R. Dawes would like to compare the measures of *Antares* with the enclosed. Admiral Reynolds is being succeeded by Commodore Christopher Wyvill, a queer person. Sends Cape newspapers to show events in the convict question. All public servants are to be deprived of food, etc.
ALS 3pp RS:HS 12.146

8102. 1849–9–10 To Francis Beaufort [Collingwood]
A note expressing JH's thanks for allowing JH to read a letter from Thomas Maclear to FB; JH again expresses his concern for the welfare of Maclear.
ALS 2pp Hydrographic Office Misc F24F1.2

8103. 1849–9–13 From Edward H. Baily London
Cast of the late Francis Baily is ready, also the one of JH. These will be forwarded as arranged.
ALS 2pp RS:HS 3.28

8104. 1849–9–14 From Francis Beaufort Admiralty
Encloses letters from Thomas Maclear and returns one of JH's.
ALS 2pp RS:HS 3.396

8105. 1849–9–16 To Spencer J. A. Compton Collingwood
Asking SC for a letter of introduction for a friend to see J. H. Cust (Viscount Alford).
C 1p RS:HS 23.70

8106. 1849–9–17 From Francis Beaufort Admiralty
Encloses more letters from Thomas Maclear. Admiration for Maclear's energy. Cape Town is in a state of semi-rebellion.
ALS 2pp RS:HS 3.397

8107. 1849–9–18 From Spencer J. A. Compton Birmingham
Has written letter of introduction to J. H. Cust (Viscount Alford) and sent it to [Robert?] Grahame. The Association has voted £250 for Kew Observatory. Had interesting session on nebulae from William Parsons (3rd Earl of Rosse). Meeting has been successful in all respects except the pecuniary one.
ALS 2pp RS:HS 5.277

8108. [1849]–9–21 From William Empson Craig Crook, Edinburgh
Hopes his paper on L. A. J. Quetelet will be ready for the Easter number. Richard Jones has returned from Switzerland, and dined with Quetelet in Brussels. Has been suffering himself with congestion of the brain and consequently unable to write.
ALS 3pp RS:HS 7.65

8109. 1849–9–25 From George Biddell Airy Royal Observatory, Greenwich
Informing JH that they have been appointed members of a committee for proving a three foot reflector for the Cape Town Observatory.
ALS 4pp RS:HS 1.217 (C: RGO 6.248.130)

8110. 1849–9–26 To Adolph Knipping Collingwood
Gratitude for AK's services as executor for affairs of Caroline L. Herschel. Did not receive AK's Sept. 1848 letter. Give remaining money to [Caroline's servant] Betty, whose pending marriage surprised JH. Thank AK's mother for her letter of 3 Sept. to JH. Happy that AK's brother Georg returned safely from war in Holstein with honors. Hopes peace comes to Europe soon.
ALS 4pp TxU:H/L-0217; Reel 1054

8111. 1849–10–3 From Edward H. Baily 17 Newman St.
The two casts will be forwarded tomorrow.
ALS 1p RS:HS 3.29

8112. 1849–10–5 From [Edward] Champnes Slough
Would like JH's views on the enlarging of the local church.
ALS 4pp RS:HS 5.331

8113. 1849–10–10 From Edward H. Baily 17 Newman St.
Is glad to hear that the casts arrived safely. Is flattered by Lady Herschel's remarks.
ALS 1p RS:HS 3.30

8114. 1849–10–10 From William Henry Smyth 3 Cheyne Walk, Chelsea
Discusses the site of Cold Harbor. Visits the small personal observatory of J. G. Barclay, who WS thinks will produce good work.
ALS 2pp RS:HS 16.298

8115. 1849–10–11 From William Sharp Rugby
Asks JH's opinion on a passage from [John] Williams's *Narrative of Missionary Enterprises in the South Seas* regarding unusual tides in Tahiti and Society Islands.
ALS 3pp JH note RS:HS 17.203

8116. 1849–10–12 From Hans Christian Oersted Copenhagen
Thanks for promise of kind intervention on his behalf, but he believes that JH would not advise him to publish the work in England. Thinks it would be more welcome in Germany. Would like comments and corrections.
ALS 3pp RS:HS 13.172 Harding, *Correspondance de ... Örsted, 2,* 404–5

8117. 1849–10–14 To William Henry Smyth Collingwood
Discusses Cold Harbor. Suggests the Duke of Bedford [F. C. H. Russell] could help secure a pension for Thomas Maclear.
C 1p RS:HS 23.71

8118. 1849–10–19 To William Whewell Collingwood
Thanks WW for two papers, one dealing with G. F. W. Hegel. Discusses a problem in calculating double star orbits.
ALS 3pp TC, Camb. Add. Ms.a.207[77] (C: RS:HS 23.72)

8119. 1849–10–21 From Thomas Romney Robinson Castle, Parsonstown
Discusses B.A.A.S. resolution regarding the Cape observatory. Wants accurate drawings of nebulae to attempt to ascertain change. Wants JH to petition government to get reflector at Cape.
ALS 3pp RS:HS 14.375

8120. 1849–10–22 From Edward H. Baily 17 Newman St.
Is sorry his letter [see EB's 1849–10–10] has not been received, but he replied by return of post.
ALS 1p RS:HS 3.31

8121. 1849–10–23 From George Biddell Airy Royal Observatory, Greenwich
Is William Dawes equatorial telescope available for purchase?
ACCS 1p RGO 6.142.149

8122. 1849–10–24 To George Biddell Airy Collingwood
JH's wife, Margaret, is writing to Mrs. Dawes and will ask about the telescope [see GA's 1849–10–23].
ALS 2pp RGO 6.142.149a

8123. 1849–10–25 From Edward H. Baily 17 Newman St.
Many thanks for the letter and check. Will require no further sitting yet, but would like one later to give the finishing touches.
ALS 1p RS:HS 3.32

8124. 1849–10–25 From Isaac Fletcher Tarn Bank near Cockermouth
Has JH any of his papers he wrote on the micrometrical measures of the double stars as he does not possess the volumes that contain them. Has erected an equatorial for the purpose of measuring the double stars. Gives details.
ALS 2pp RS:HS 7.271

8125. 1849–10–27 To Thomas Romney Robinson Collingwood
Will not urge government to construct a great reflector at the Cape; believes it should be done through private funding; does look forward to the time when his work on southern skies will be reexamined. [See TR's 1849–10–21.]
ALS 3pp Cambridge University Library 7656/TR61 (AC: RS:HS 14.376 & C: RS:HS 23.73)

8126. 1849–10–28 To George Biddell Airy Collingwood
Has heard nothing beyond what GA has said [see GA's 1849–9–25], but agrees with GA's assessment of the situation.
ALS 4pp RGO 6.248.132 (drafts RS:HS 1.218 & 23.74)

8127. 1849?–10–29 From William Parsons (Lord Rosse) The Castle, Parsonstown
About adjudication of R.S.L. medals.
ALS 2pp RS MM.16.153

8128. 1849–11–1 From Julia Margaret Cameron 10 Chesham Place
Regrets that she does not receive so many letters from him, even though she lives so much nearer. Comment on literature. News of Hardinge [J. C.'s child].
ALS 8pp RS:HS 5.151

8129. 1849–11–1 From Edward Joseph Lowe Beeston
Sending printed notice appealing for funds to establish a Midland Observatory. Would like a word of encouragement from JH.
ALS 1p RS:HS 11.321

8130. 1849–11–1 From Thomas Romney Robinson Edgeworthstown
Did not wish JH to join application to government [see TR's 1849–10–21 & JH's 1849–10–27]. Committee hoped JH would help carry plan out if funding were secured. Doubts private funding will be given.
ALS 4pp RS:HS 14.377

8131. 1849–11–1 To Edward Sabine Collingwood
Supports proposal made by C. M. Elliot to execute 'running survey' of India with instruments from observatories.
ALS 3pp RS MC.4.296

8132. 1849–11–5 To William Parsons (Lord Rosse) n.p.
Recommends Edward Sabine's work on terrestrial magnetism for Royal Physical Medal, and joint award of Copley Medal to WP and William Lassell for improvements in reflecting telescopes.
C 3pp TxU:H/L-0341; Reel 1054

8133. 1849–11–6 From Alfred Smee Bank of England
Encloses sequel to AS's volume on electro-biology, which AS dedicated to JH. Both are being translated into French.
ALS 1p TxU:H/M–0484; Reel 1093

8134. 1849–11–8 or later To Robert V. Dixon n.p.
Reply to RD's 1849–11–8.
draft 2pp RS:HS 6.495verso

8135. 1849–11–8 From Robert V. Dixon Trinity College Dublin
Sending a copy of the first part of his work on *Heat*.
AL 1p RS:HS 6.495

8136. 1849–11–9 To Edward Joseph Lowe Collingwood
Giving his views on the desirability of establishing a Midland Observatory.
C 3pp RS:HS 11.322 & 23.76

8137. 1849–11–10 From John Stevens Henslow Mitcham, Hadley, Sufk
Well-wishers of the Ipswich Museum hope to raise a fund for a lithograph portrait of G. Ransome; wonders if JH is interested in becoming a subscriber. Richard Owen is to deliver a lecture at the anniversary meeting next month.
ALS 3pp RS:HS 9.312

8138. 1849–11–12 From Ebenezer Henderson Newferry near Birkenhead
Is having an astronomical clock constructed and would like his advice on the calculation of some of the movements. Gives a brief description of the clock.
ALS 4pp RS:HS 9.279

8139. 1849–11–12 From Edward Joseph Lowe Beeston
Thanks for his letter regarding the proposed observatory. Points out some provisions he intends making for the observatory. Encloses extracts from letters favoring the scheme.
ALS 4pp RS:HS 11.323

8140. 1849–11–13 To George Gabriel Stokes Collingwood
Thanks GS for a number of his papers on railway bridge fractures, gravitation, and the spectrum.
C 1p RS:HS 25.10.1

8141. 1849–11–15 From Julia Margaret Cameron Tunbridge Wells
Has shown his letter to John Taylor and quotes the latter's opinion of JH. Views on the poetry of Taylor.
ALS 4pp RS:HS 5.152

8142. 1849–11–15 To Edward Joseph Lowe Collingwood
JH is now very supportive of EL's intention of establishing a Midland Observatory following EL's 1849–11–12.
C 1p RS:HS 11.324 & 23.77

8143. 1849–11–16 From [A. J. F.] Yvon Villarceau Paris
Thanks JH for his interest in YV's work on double stars. YV sends a copy of a note on double stars published in the Académie des Sciences for consideration by the R.A.S. Raises a problem of agreement between Newtonian theory and observational data. YV feels astronomical studies in France are slowing down.
ALS 2pp (French) RS:HS 17.420

8144. [1849]–11–17 From William Empson Craig Crook, Edinburgh
Is much better but has to take care. Francis Jeffrey (Lord Jeffrey) has taken over the editorial duties of the [*Edinburgh*] *Review*. Pleased to hear he is working on his paper. Will inquire about the average Englishman. W. H. Sykes and S. D. Poisson have published papers on this subject.
ALS 4pp RS:HS 7.66

8145. 1849–11–18 From Richard Jones Haileybury
Mrs. Jones has lately lost a sister-in-law, which accounts for her laxness in correspondence. Will search his index for an article on the average Englishman. Tour through France and Switzerland was a great success. Feels much better for it.
ALS 4pp RS:HS 10.386

8146. 1849–11–19 From George Biddell Airy Royal Observatory, Greenwich
Responds to a JH question about the accuracy of the calculations for the figure of the earth in GA's article in the *Encyclopaedia Metropolitana.*
ACCS 3pp RGO 6.418.78

8147. 1849–11–21 From Ebenezer Henderson Newferry near Birkenhead
Thanks for his letter but it does not allude to the calculation he sent. Has seen astronomical clocks previously.
ALS 2pp RS:HS 9.280

8148. [1849–11–21 or later] From George R. Porter n.p.
Returns the proofs of [W. J.] Hamilton's paper, and hopes to send his contribution to the *Admiralty Manual* soon.
ALS 1p RS:HS 14.46

8149. 1849–11–21 From George R. Porter Board of Trade
Must consider and enquire before replying to JH's letter of yesterday. Sends volumes of the 'tables of revenue' for 1846 and 1847 to JH, plus a supplement relating to foreign countries. If this is not enough, let GP know.
ALS 2pp RS:HS 14.45

8150. 1849–11–23 From William Rutter Dawes Furze Park Torquay
Sending a copy of the Torquay directory where Edward Vivian's comparison of thermometer readings are given. The climate of Torquay does not suit them; Mrs. Dawes is going north for a few weeks. Would like a comparison of the state of the weather for various parts of the country.
ALS 4pp RS:HS 6.95

8151. 1849–11–23 From James David Forbes Edinburgh
B.A.A.S. and Royal Society of Edinburgh have been approached to apply to the government for finance to send an observer and reflector to the Cape Observatory. Does JH think this advisable, with his knowledge of conditions at the Cape?
ALS 3pp RS:HS 7.311

8152. 1849–11–24 From William Galbraith Edinburgh
Has received his note enclosing G. B. Airy's paper. Regrets any trouble he may have caused. Sends copies of his own paper for measuring heights by the barometer. Took some observations on the tides at Brodick and comments on William Whewell's statements concerning tides.
ALS 4pp RS:HS 8.3

8153. 1849–11–[26] To James David Forbes n.p.
Agrees that it would be desirable to make further observations at the Cape, but foresees difficulties, especially liaison with the Astronomer Royal at the Cape.
draft 2pp RS:HS 7.312 & 23.75

8154. 1849–11–27 To James David Forbes Collingwood
[Responding to JF's 1849–11–23], JH cautiously discusses various considerations bearing on the idea of sending an astronomer and a large reflecting telescope to the Cape.
ALS 4pp Edinburgh Royal Society

8155. 1849–11–27 From James A. Gordon Stocks, Tring
Wrote to Francis Russell (10th Duke of Bedford) about Thomas Maclear and encloses his reply. Has just framed JH's prints. Sends a few scraps. Hopes to visit G. G. Waldegrave (2nd Baron Radstock).
ALS 2pp RS:HS 8.184

8156. 1849–12–2 From Leonard Horner Rivermead, Hampton Wick
Queries regarding the statement by William Herschel on the time taken for the light from the stars to reach the earth.

Have there been any new developments in this theory?
ALS 3pp RS:HS 9.494

8157. 1849–12–3 To Thomas Maclear Collingwood
Thanks TM for double star observations. Refers to convict incident in South Africa.
ALS 2pp Cape Archives/Maclear Papers File 100

8158. 1849–12–4 From Heinrich Christian Schumacher Altona
Sends account of rainbow-like atmospheric phenomena; asks JH for an explanation. Remarks on American empirical discoveries. Too many hypotheses being put forward, including [Daniel] Kirkwood's. Comments on [B. A.] Gould and [Stephen] Alexander. Americans think they are absolute judges in scientific matters. *Astronomische Nachrichten* of 3 November contains improved method for computing planetary orbits.
ALS 8pp RS:HS 15.404

8159. 1849–12–6 From Thomas Maclear Royal Observatory, Cape of Good Hope
Sends observations of Lambda Toucani and Zeta Eridani.
ALS 1p Cape Archives/Maclear Papers File 100

8160. 1849–12–8 From John William Lubbock n.p.
Even though he was Vice-Chancellor [of London University], he had no opportunity of becoming acquainted with the internal regulations of the colleges connected with the University. Regarding the education of young children. Sends some music for Lady Herschel. Would like them to pay him a visit.
ALS 7pp RS:HS 11.388

8161. [1849?]–12–11. From Elizabeth Sabine [to Margaret Brodie Herschel] n.p.
Edward Sabine will recover. Please tell JH how grateful ES is for JH's positive remarks regarding his report. Discusses several scientific matters. Contains message from ES for JH on instrumentation at Kew Observatory.
ALS 6pp RS:HS 15.337

8162. 1849–12–14 From John Fletcher Miller Whitehaven
Sends copies of his own papers and would be pleased to receive any of JH's in exchange. Is a candidate for the R.S.L. and would welcome JH's support in any way. Curious phenomenon seen on the surface of the moon in the Lake District.
ALS 9pp RS:HS 12.341

8163. 1849–12–17 To Charles May Collingwood
Learned of John Gorrie's mechanical ice-making method. Asks CM to send letter to JH recounting conversation of 11 Feb. 1848 in which JH told CM about JH's theory for making ice by compressing air. JH does not want to interfere with Gorrie's claim, but will submit CM's letter to *Athenaeum* and put on record, for sake of 'scientific history,' that JH was '*certain*' that such a principle would work. Condolences for death of CM's partner.
AdraftS 4pp TxU:H/L-0256; Reel 1054

8164. 1849–12–17 From Arthur Cleveland Wigan Benenden [Kent]
Thanks for JH's candor and observations of W. H. Wollaston. A. L. Wigan revised book [*Duality of the Mind* (1844)] before death, but revision is still under review. Offensive passages will be removed, and JH's suggestions incorporated.
L 3pp RS:HS 18.404

8165. 1849–12–18 From Heinrich Christian Schumacher Altona
Thanks JH for explanation of phenomena [see HS's 1849–12–4]. B. A. Gould [?] has begun to publish American version of *Astronomische Nachrichten*, but without strong financial backing.
ALS 2pp RS:HS 15.405

8166. 1849–12–20 From Charles May Ipswich
Sends note with needed facts [related to JH's early discovery of principle for ice-making machine], plus paragraph that JH may wish to insert [in notice to Athenaeum].
ALS 1p TxU:H/M–0352; Reel 1087

8167. 1849–12–20 From Hans Christian Oersted Copenhagen
Gratitude for JH's *Cape Results* sent to Royal Society of Sciences.
ALS 1p RAS:JH Archive 10/5.16; Reel 9

8168. 1849–12–21 From William Radcliffe Birt Kew Observatory
Charles Wheatstone has offered to apply for an R.S. grant for him to continue his atmospheric wave investigations. His *Hurricane Guide* will soon be issued by Murray.
ALS 3pp RS:HS 4.135

8169. 1849–12–21 From John Fletcher Miller Whitehaven
Very kind of JH to agree to sign JM's certificate. Has written to C. R. Weld to forward the certificate to JH, if this is allowable. Will be pleased to receive copies of any astronomical papers JH has for disposal.
ALS 2pp RS:HS 12.342

8170. 1849–12–26 From Antoine Thompson d'Abbadie Chateau d'Audaux near Navarre
Thanks JH for the copies of JH's *Cape Results*; some are being sent on as directed, as are the other papers forwarded to AA.
ALS 1p (French) RAS:JH Archive 10/5.89; Reel 9

8171. 1849–12–26 To Augustus De Morgan Collingwood
Seeking a tutor for his young nephews. Yvon Villarceau has sent printed copies of Villarceau's double star papers; JH has dealt with them.
C 2pp RS:HS 23.78

8172. 1849–12–27 From Sir Edward Ryan Garden Lodge, Kensington
Will gather information JH requested. It may be difficult to change the Indian appointment to Bengal of JH's son [Wiliam].
ALS 3pp RS:HS 14.457

8173. 1849–12–28 From Hans Christian Oersted Copenhagen
Thanks for the trouble that JH has taken in searching for a publisher for HO's work upon the natural philosophy of beauty. [Ludwig] Colding, Inspector of Waterworks, Copenhagen, who is coming to England, will collect the manuscript from JH. Direct them to 70 Old Bond St.
ALS 1p RS:HS 13.173

8174. 1849–12–31 From William Radcliffe Birt Kew Observatory
Thanks for his letter, and will arrange with John Williams concerning L. A. J. Quetelet's books. Regarding recent barometrical readings.
ALS 3pp RS:HS 4.136

8175. [1850] To Henry Peter Brougham n.p.
Praises in very strong terms and in detail the contributions to astronomy made by Thomas Maclear. Believes Maclear deserves a pension.
ALS 4pp Univ.Coll.London Brougham

8176. [1850] From Henry Peter Brougham Grafton St.
Would like JH's view on Thomas Maclear.
ALS 2pp RS:HS 4.308

8177. 1850–1 To Augustus De Morgan n.p.
Comments on calculation by Frédéric Petit of the hyperbolic orbit of the meteor of 19 Aug. 1847.
C 1p RS:HS 23.80

8178. [1850 to 1852] From John Graham n.p.
Suggests changes in statutes governing colleges.
L 1p inc TxU:H/M–0915; Reel 1083

8179. [1850 or later ?] To Isabella Herschel n.p.
Talks about the concepts of thinking and consciousness.
ALS 3pp WT 91327

8180. [1850–1855] To Margaret Brodie Herschel London
Part of a letter about not having received any hampers from Collingwood, and about some household staff arrangements.
AL 1p inc JHS 5.35 Note DRD

8181. [1850] To Sir William J. Hooker 37 Tavistock Place
Requests favor from WH on behalf of Richard Jones and his wife in light of her move to Richmond.
ALS 2pp Royal Botanic Gardens

8182. [1850] To William Parsons (Lord Rosse) n.p.
Discusses possible recipients of Royal Medals, continuing to favor Edward Sabine.
ALS 4pp RS Sa.681

8183. [1850] To Charles Pritchard n.p.
Inability of JH's son [William] to pass Greek. Concerned that this may delay [William's] entry into Haileybury College. Notes similar inability with languages in JH's other two sons.
ACCS 5pp TxU:H/L-0318; Reel 1054

8184. 1850–1–1 To Elizabeth Sabine Collingwood
Regrets to hear of Edward Sabine's serious illness but praises ES's accomplishment and devotion; hopes ES will rest and soon recover good health.
ALS 4pp RS Sa.660

8185. 1850–1–1 To Adam Sedgwick Collingwood
Has heard of AS's broken arm. Sends regards from the family.
ALS 3pp RS:HS 15.442 (C: RS:HS 23.79)

8186. 1850 To Richard Sheepshanks [Collingwood]
Has received interesting observations of double stars from Thomas Maclear at the Cape observatory. The new equatorial lens sent to the Cape arrived safely, in spite of sloppiness by the English custom-house.
ALS 3pp RS:HS 16.105 (C: RS:HS 25.9.69)

8187. [1850 to 1851] To John Shepherd n.p.
Petition for pension for widow of T. G. Taylor, H.E.I.C. astronomer at Madras who published error-free observations of southern stars.
AdraftS 2pp RS:HS 19.162b (C: RS:HS 25.10.10)

8188. 1850–1–3 From Elizabeth Sabine Hastings
Thanks JH for letter and concern shown during Edward Sabine's illness. He is recovering.
ALS 8pp RS:HS 15.335

8189. 1850–1–5 From John R. Hind 11 Hanover Cottages
On the evening of 4 Jan. he found a nebulosity in or near Draco that he has been unable to discover in any catalogue. Has discovered another red star in Taurus, which is not given in JH's list.
ALS 2pp RS:HS 9.346

8190. 1850–1–5 From Henry Piddington Calcutta
A note accompanying another of HP's meteorological publications.
ALS 2pp RS:HS 13.405

8191. [1850]–1–7 From Henry Holland Brook St.
Delayed answering JH's letter until HH could find details of works by 'Göttingen Professor,' but has had difficulty in locating them. Any treatise connecting epidemics with fungous origins would be of interest. Will bring distressing medical case to attention of one of HH's committees. HH's *Medical Notes and Reflections*, 3rd edition (1855).
ALS 3pp RS:HS 9.439

8192. [1850?]–1–8 From William Empson East India College
His manuscript arrived and has been dispatched to the printers with the necessary instructions. Regrets to hear that JH's labors are impeded by ill health.
ALS 4pp RS:HS 7.67

8193. 1850–1–11 To [Robert Hunt ?] Collingwood
Inquires whether Charles Wheatstone ever fulfilled his intention to carry out experiments to measure the speed of light in the atmosphere by means of 'revolving machinery.'
ALS 3pp WT 68037.10

8194. 1850–1–15 From William Radcliffe Birt Kew Observatory
His engagement at Kew will soon terminate. Regarding the grant for Kew and the position he holds there. Can JH inform him of the zodiacal light.
ALS 6pp RS:HS 4.137

8195. [1850?]–1–17 From William Empson East India College
Would be happy to hear of any improvement in JH's health. Encloses a letter from Francis Jeffrey (Lord Jeffrey). J. S. Mill edited Jeremy Bentham's works in two volumes.
ALS 4pp RS:HS 7.68

8196. 1850–1–19 From William Radcliffe Birt Kew Observatory
Regarding the luminous rings recently observed in the solar system.
ALS 4pp RS:HS 4.138

8197. 1850–1–19 From Alexander von Humboldt Berlin
Queries regarding JH's star lists. Sending magnetic information from A. T. Kupffer. Regarding the distribution of nebulae.
ALS 4pp (German) RS:HS 10.69

8198. 1850–1–21 From Augustus De Morgan 7 Camden St.
Regarding the delicate case of priority of investigation. Quotes example of Henry Warburton and John Brinkley.
ALS 4pp RS:HS 6.256

8199. 1850–1–21 From Lord Francis Jeffrey Edinburgh
Returns 'slips' with JH's corrections. Asks explanation of

probability theory.
ALS 4pp RS:HS 19.279

8200. 1850–1–21 From Sir Edward Ryan Garden Lodge, Kensington
Has no information to send yet. JH should not be too hopeful of changing the appoint to Bengal of his son [William James].
ALS 3pp RS:HS 14.458

8201. [1850?]–1–23 From William Empson East India College
Returns the slips received from Francis Jeffrey (Lord Jeffrey); would JH forward them to the printer with his instructions. His mathematical demonstration will be allowed enough room. Pleased to hear his deafness has been relieved.
ALS 2pp RS:HS 7.69

8202. 1850–1–24 To Augustus De Morgan Collingwood
About Henry Warburton's theorem, and the Gregorian calendar.
C 1p RS:HS 23.81

8203. 1850–1–24 From Henry Anthony Jeffreys Hawkhurst Parsonage
Sends drawing and statement showing what they are about at the church. Would welcome any assistance from JH.
ALS 2pp RS:HS 10.290

8204. 1850–1–24 To Henry Anthony Jeffreys Collingwood
Regrets he is unable to support his project for the church as he has many more worthy commitments.
C draft 1p RS:HS 10.291 & 23.82

8205. [1850–1–25] To the *Athenaeum* n.p.
[Writing anonymously], offers a solution to a Latin problem discussed earlier in the *Athenaeum.*
L *Athenaeum,* #1162 (Feb. 2, 1850), 132

8206. [1850–1–25] To the *Athenaeum* n.p.
[Writing anonymously], asks a question about the public understanding of the Gregorian calendar.
L *Athenaeum,* #1162 (Feb. 2, 1850), 132

8207. 1850–1–25 From Augustus De Morgan 7 Camden St.
Further regarding the Gregorian calendar and comments on some of the questions involved in its interpretation.
ALS 2pp RS:HS 6.257

8208. 1850–1–26 To Lord Francis Jeffrey Collingwood
Explaining his own idea of probability. [This appears to be the original, which was never sent as Lord Jeffrey died 1850–1–25.]
RS:HS 10.437 & 23.83

8209. 1850–1–[28] From William Empson 24 Moray Place, Edinburgh
Reports death of Francis Jeffrey. JH's letter to Jeffrey arrived too late. Mrs. Jeffrey will return to [Haileybury?] with WE.
ALS 3pp RS:HS 19.271

8210. 1850–1–28 From Henry Warburton 45 Cadogan Place
Thanks for citing (18 months ago) John Brinkley's paper on 'General Term.' Heard that Augustus De Morgan notified JH of deductions HW drew from Brinkley's theorems, extending them into permutations and combinations. Re-read JH's ['On the Development of Exponential Functions' (1816)]. Asks where to buy JH's *Examples in Finite Differences* [1820] for HW's great-nephew [Howard Elphinstone].
ALS 9pp RS:HS 18.52

8211. 1850–1–29 From E. W. Payne Plymouth
Asks questions relevant to a series of books presenting science for juveniles.
ALS 3pp RS:HS 14.147

8212. 1850–1–31 From Henry Warburton 45 Cadogan Place
Thanks for sending *Examples in Finite Differences* [1820] to WH's great-nephew Howard Elphinstone. John Brinkley's paper opened up new possibilities for permutations and combinations.
ALS 4pp RS:HS 18.53

8213. 1850–2–4 From Augustus De Morgan 7 Camden St.
Would he drop him a few lines on the orbit of the meteor for the R.A.S. Robert Potts of Trinity is to publish by subscription a translation of Robert Simson's *Porisms.*
ALS 4pp RS:HS 6.258

8214. 1850–2–4 From Sophie Beckedorff to Lady Herschel Hanover
Family and political news.
ALS 6pp RS:HS 4.7

8215. 1850–2–4 From John R. Hind 11 Hanover Cottages
In looking through Lamont 1845–6 he found two observations of Neptune not previously recognized. Thanks for his memoir on the orbits of double stars.
ALS 2pp RS:HS 9.347

8216. 1850–2–4 To Sir Henry Taylor Collingwood
Returning with gratitude HT's *Statesman,* which stimulated much political and social reflection in JH.
AL 4pp inc Bodleian Library, Oxford

8217. 1850–2–6 From Charles Piazzi Smyth Hillside Crescent, Edinburgh
A. G. Melville will start Meteorological Observatory at Queen's College, Galway. CPS discusses JH article on plans for building an ice-making machine.
ALS 4pp RS:HS 16.231

8218. 1850–2–7 From E. W. Payne Plymouth
Discusses points of lunar and solar eclipse observations.
ALS 2pp RS:HS 14.148

8219. [1850]–2–9 From Henry Holland Brook St.
Marriage of HH's eldest daughter. Thanks for JH's letter of introduction to George Bishop and J. R. Hind.
ALS 3pp RS:HS 9.438

8220. 1850–2–11 To Alexander von Humboldt Collingwood
Answers questions raised by AH in his 1850–1–19 [letter contains notes made by AH].
C 3pp RS:HS 23.84

8221. 1850–2–14 From William Radcliffe Birt Kew Observatory
Has received an interesting communication from the Rev. J. B. Reade of Stone, near Aylesbury, on the zodiacal light. Comments on this.
ALS 4pp RS:HS 4.139

8222. [1850]–2–14 From Henry Holland Brook St.
Thanks for congratulations on marriage of HH's daughter.

Notes HH's review of J. C. Prichard's *Natural History of Man* in December issue of *Quarterly Review*, written while on holiday in Armenia.
ALS 3pp RS:HS 9.437

8223. 1850–2–15 To Roderick Impey Murchison Collingwood
Proposes priorities for allocating £1000 granted to R.S.L. by government. Disagrees with RM and Edward Sabine; R.S.L. should *not* assume responsibility for observatory or any other permanent institution.
ALS 3pp EUL:Gen523/4/39 (ACCS: TxU:H/L-0269; Reel 1054)

8224. 1850–2–15 From Edward Sabine St. Leonards-on-Sea
Glad to have seen Lady Herschel and children. Is recovering slowly. Thanks JH for interest in [Francis] Ronalds's work. Discusses paper ES has submitted for *R.S.P.T.*
ALS inc 8pp RS:HS 15.336

8225. 1850–2–15 From Heinrich Christian Schumacher Altona
[A. C.] Petersen [?] claims to have found a new comet near JH's nebula #379. Look for it, but do not announce this discovery.
ALS 1p RS:HS 15.406

8226. [1850]–2–17 From William Empson East India College
Would JH agree that his review of L. A. J. Quetelet's book on probabilities be allowed to stand until the June number as there is a large paper on observations for the next number [of the *Edinburgh Review*]?
ALS 4pp RS:HS 7.70

8227. 1850–2–20 From R. Courtney New Street Square
Notes discrepancy in date for discovery of planet Hygeia. Please furnish correct date and return to Spottiswoodes & Shaw, printers.
ALS 1p TxU:H/M–0151.1; Reel 1087

8228. 1850–2–24 From Henry Warburton 45 Cadogan Place
Admires chimes analogy in JH's 'On Circulating Functions' [1818]. Resumed interest in partitions of numbers in development of periodic functions. Finds JH's method superior. Encourages him to publish it. WH and Augustus De Morgan developed original method, which WH erred in attributing to Leonhard Euler. De Morgan's paper to Cambridge [Philosophical] Society demonstrates theorem leading to proof of Euler's theorem.
ALS 8pp RS:HS 18.54

8229. 1850–2–25 From Edward Sabine St. Leonards–on–Sea
Discusses shipping some of JH's books, with a shipment of other books for the Royal Society, to the R.S.L. in order to save shipping charges.
ALS 3pp RS:HS 15.229

8230. 1850–2–25 From Henry Warburton 45 Cadogan Place
Explains HW's method for periodic functions.
ALS 3pp RS:HS 18.55

8231. 1850–2–26 From Thomas Maclear Royal Observatory, Cape of Good Hope
Pleased to receive JH's letter, which he hastens to answer to remove any misconceptions regarding the equatorial. Comments on some of his observations of stars. Received [T. R.] Robinson's proposal for a large reflector. William Mann has just finished 8 years of tide gauge measures. Grieves to hear of the illness of W. R. Dawes. [George] Smalley's financial affairs have come to a crisis. Convict question has been settled; they are to be sent to Van Diemen's Land.
ALS 7pp RS:HS 12.147

8232. 1850–2–27 From Henry Warburton n.p.
'Private & Confidential.' Urges JH to accept University of London's senate invitation to join Board, which JH previously declined. Names others invited.
ALS 4pp RS:HS 18.56

8233. 1850–[2–27 or later] From Henry Warburton n.p.
'*Private & Confidential.*' All seven invited to join [University of London's] senate have agreed. Urges JH to reconsider.
ALS 3pp RS:HS 18.58

8234. 1850–2–28 From John R. Hind 11 Hanover Cottages
Sends an interesting result of the comet computed by Mr. Barber of Etwell. W. R. Dawes's observations are about to be printed. Wilhelm Struve has some curious optical equations. Will be pleased to see Dr. [Henry?] Holland when he arrives.
ALS 6pp RS:HS 9.348

8235. 1850–3–1 From Frederick Pollock Guildford St.
Announces R.S.L. committee meeting concerning a government grant. Invites JH to dinner after the meeting.
ALS 3pp RS:HS 14.174

8236. 1850–3 To Richard Sheepshanks n.p.
Writes about possible astronomers to send to Trivandrum in India.
draft 2pp RS:HS 16.99

8237. [1850]–3–3 From William Empson East India College
When JH sends his paper to Mr. Shaw would he inform WE as he wishes to start the next number off with it. Hopes Mrs. [Richard] Jones's health has improved by her visit to the college.
ALS 4pp RS:HS 7.72

8238. 1850–3–4 To Roderick Impey Murchison Collingwood
Has been requested to serve on committee on R.S.L. government grant [see JH's 1850–2–15] and to draw up report to council for adoption on next day. JH declines request but offers opinions, suggestions, and a resolution.
ALS 4pp EUL:Gen523/4/40

8239. 1850–3–4 From Henry Warburton 45 Cadogan Place
'Private & Confidential.' Secretary of State George Grey disappointed that JH declined to join University [of London] Board. Extended offer again. HW offers conjecture on JH's non-periodic numbers.
ALS 4pp RS:HS 18.57

8240. 1850–3–5 From John Fletcher Miller Whitehaven
Has received the papers JH sent and values them highly. Can JH advise him on making regular observations of sun spots. Explains his apparatus.
ALS 4pp RS:HS 12.343

8241. 1850–3–6 From Edward Sabine St. Leonards–on–Sea

Glad JH received gelatine paper. Discusses magnetic experiments at Toronto and Hobarton. Discusses R.S.L. committees and preparation of instruments for Arctic expedition.
ALS 8pp RS:HS 15.230

8242. 1850–3–8 From Edward Sabine St. Leonards-on-Sea
Hypothesizes that annual variation of Dip and Total Force at Toronto may be caused by greater proximity of earth and sun. Discusses annual variations.
inc 4pp RS:HS 15.231

8243. 1850–3–10 To Edward Sabine Collingwood
Thanks to [Francis] Ronalds for supply of paper. Comments on ES's discovery about direction of change in earth's magnetism, relating it to sun's and other electric currents. In this light, discusses auroras. Mentions abolishing of R.S.L. scientific committees and possible successor to John Caldecott.
ALS 7pp RS Sa.661–2

8244. 1850–3–13 From Reginald Stuart Poole 25 White Rock Place, Hastings
Asks JH about RP's astronomical deductions concerning the ancient dates of five celestial phenomena, e.g., heliacal rising of Sirius, a new moon.
ALS 5pp † RS:HS 14.36

8245. 1850–3–13 To [?] Collingwood
Thanks for [?]'s kindness shown to JH's son [?] in visit to Mr. Peter's observatory. [?]'s friends 'the Wrenches' are well.
ALS 2pp TxU:H/L-0749.3 (C: –0749.4); Reel 1089

8246. 1850–3–14 To Roderick Impey Murchison Collingwood
Has received report of R.S.L. grant committee, noting that it did not adopt three of his recommendations. Asks questions. Is glad that RM and Charles Babbage are again on friendly terms.
ALS 3pp EUL:Gen523/4/41 (ACCS: TxU:H/M–0151.2; Reel 1087)

8247. 1850–3–15 To Edward Sabine 32 Harley St.
Thanks Mrs. Sabine for translation of third volume of Alexander von Humboldt's *Kosmos*. Admires its author. Questions references in notes. Discusses meeting for Wintringham Committee.
ALS 4pp RS Sa.664

8248. 1850–3–15 To Richard Sheepshanks Collingwood
About an error in the annual report of the R.A.S.; would RS give some thought to a suitable candidate for the post of astronomer at Trivandrum, made vacant by the death of John Caldecott.
ALS 3pp RS:HS 16.116 (C: RS:HS 25.9.62)

8249. 1850–3–16 To Augustus De Morgan Collingwood
About JH's poor health; some problems in perspective. Having read a book on Egyptology, JH wanders off in flights of fanciful numerology.
C 4pp RS:HS 23.85

8250. 1850–3–16 To Roderick Impey Murchison Collingwood
Thanks for clarifications on grant comments [see JH's 1850–3–14]. JH's work is plentiful and time is short, so wishes not to undertake public projects, like working on R.S.L. grant committee. Received RM's enclosure regarding Russian Academy. Has heard nothing about medal committee.
ALS 4pp EUL:Gen523/4/42 (ACCS: TxU:H/L-0270; Reel 1054)

8251. 1850–3–19 From Reginald Stuart Poole 25 White Rock Place, Hastings
Clarifies questions and includes his calculations of number of days since the five phenomena (see RP 1850–3–13). Will send some calculations to Astronomer Royal.
ALS 4pp RS:HS 14.37

8252. 1850–3–22 From William Henry Sykes India House
Regrets missing JH's visit; discusses influences of atmosphere and solar position on Indian weather and difficulty it causes in obtaining meteorological laws.
ALS 8pp RS:HS 17.145

8253. 1850–3–25 From George Biddell Airy Royal Observatory, Greenwich
Apologizes for not being at home when JH called, and thanks JH for his family's kindness to GA's children during the past winter.
ACCS 4pp RGO 6.372.219

8254. 1850–3–25 From Charles Henry Parry Belgrave Place, Brighton
Recommends F. E. Wilmot for election to R.S.L.
ALS 3pp RS:HS 14.141

8255. 1850–3–25 From Frédéric Petit Toulouse
Thanks for the gift of a book [*Cape Results*?] made to the observatory at Toulouse. Comments favorably on the book.
ALS 1p (French) RAS:JH Archive 10/5.90; Reel 9

8256. [1850?]–3–26 To [Mary Baldwin] [Haileybury?]
Thanks for birthday gift. Health of friends.
ALS 4pp TxU:H/L-0069; Reel 1054

8257. 1850–3–26 From Augustus De Morgan 41 Chalcot Villas NW
Has never heard the polar axis approximation. JH's treatise on perspective must be very complete. John Taylor is his old publisher. Has got 64 more syllogisms symbolized.
ALS 4pp RS:HS 6.259 S. De Morgan, *Memoir of Augustus De Morgan*, 209

8258. 1850–3–28 From John R. Hind 11 Hanover Cottages
Has just received a memoir from Bastiaan Bomme of Middelburg on the comet of 1264 and 1556. Sends the chief results. Comments on the findings and future appearances of the comet.
ALS 3pp RS:HS 9.349

8259. 1850–3–29 To George Biddell Airy Collingwood
Asks GA for help in identifying an astronomer for the observatory at Trivandrum, India.
ALS 4pp RGO 6.143.7

8260. 1850–3–29 From Johann Franz Encke Berlin
Sends a book on optics; if JH likes it, then tell others about it. Comments on some aspects of JH's *Outlines Astr.*
ALS 4pp (German) RS:HS 7.98

8261. 1850–3–[30 or earlier] To Richard Sheepshanks Collingwood
More on prospects for the vacant astronomer post at Trivandrum.
ALS 3pp RS:HS 16.104 (C: RS:HS 25.9.63)

8262. 1850–3–31 To Richard Sheepshanks n.p.
Other possibilities for post at Trivandrum [see JH's 1850–3–[30 or earlier]].
ALS 3pp RS:HS 16.100 (C: RS:HS 25.9.64)

8263. 1850–4–1 From George Biddell Airy Royal Observatory, Greenwich
Comments on a number of possibilities for the post at Trivandrum [see JH's 1850–3–20].
ACCS 4pp RGO 6.143.9

8264. 1850–4–1 From Thomas Maclear Royal Observatory, Cape of Good Hope
Giving his views on the proposed supply of a large reflector for the Cape Observatory.
C 6pp RS:HS 12.148

8265. 1850–4–2 To George Biddell Airy n.p.
Thanks GA for his 1850–4–1, and provides details of a likely candidate.
ALS 3pp RGO 6.143.11

8266. 1850–4–5 To L. H. Robson Collingwood
Sends some post office stamps to pay for a 'certificate'.
ALS 1p DMC.695A.885.2

8267. 1850–4–6 From Agnes Baillie Hampstead
Letter of thanks and good wishes.
ALS 2pp RS:HS 3.23

8268. 1850–4–6 From Joanna Baillie Hampstead
Letter of thanks and good wishes.
ALS 2pp RS:HS 3.23a

8269. 1850–4–6 From C. Gubbins Meruth, Bengal
Has just received the *Admiralty Manual*. Thinks it would be helpful if some tables for determining the heights of mountains by means of the barometer could be included in the next edition. Points out an error in the *Nautical Almanac*.
ALS 3pp RS:HS 9.150

8270. 1850–4–7 To Richard Sheepshanks Collingwood
Selection procedures for the vacant astronomy post at Trivandrum [see RS's 1850–3–31].
ALS 1p RS:HS 16.101 (C: RS:HS 25.9.65)

8271. 1850–4–10 From Sir Robert Harry Inglis 7 Bedford Square
Found his letter and petition waiting for him on his return from East Kent, and had already presented it to the House of Commons. Outlines the events.
ALS 4pp RS:HS 10.245

8272. 1850–4–10 To [William Parsons (Lord Rosse)?] Collingwood
Discusses changes in system of awarding R.S.L. Royal Medals.
ALS 4pp RS MC.4.340

8273. 1850–4–11 From Friedrich Eduard Beneke Berlin
Letter to accompany a gift of the first part of FB's popular psychology work.
ALS 3pp (German) RS:HS 4.64

8274. 1850–4–12 From William Radcliffe Birt Kew Observatory
Is sending for his perusal his own observations on the zodiacal light for the first quarter of the year. Comments on this.
ALS 2pp RS:HS 4.140

8275. 1850–4–13 From Samuel Rogers St. James's Place
Thanks JH for the volume of verse. Hopes the Herschels will visit if they come to London.
ALS 2pp RS:HS 14.420

8276. 1850–4–16 From Heinrich Christian Schumacher Altona
Asks JH to accept sunspot observations made over 14 years by the late [J. W.] Pastorff of Altona Observatory. Accompanying micrometrical measurements are worthless due to mounting of telescope.
ALS 4pp RS:HS 15.407

8277. [1850]–4–17 From William Empson East India College
Would be pleased to receive JH's paper. His own house is still sad and silent.
ALS 4pp RS:HS 7.71

8278. 1850–4–18 To [William Radcliffe Birt] Collingwood
Returns [WB's] paper on the zodiacal light. Notes error and offers JH's ideas about light's source.
ALS 3pp RS:HS 19.140

8279. 1850–4–18 To [?] Collingwood
Cannot accept the invitation to stay during the B.A.A.S. meeting in Edinburgh, as JH will be unable to come.
ALS 3pp APS B.H435p.15

8280. 1850–4–19 From William Henry Sykes 47 Albion St.
Clarifies results of pressure oscillations on two coasts of India as similar despite different weather conditions, and deems results thus unsatisfactory; explores reasons and suggests solutions for lack of success.
ALS 9pp RS:HS 17.146

8281. 1850–4–20 From John Henry Lefroy Toronto Observatory
Believes that Edward Sabine has brought J. H. Lefroy's paper on the Aurora Borealis to JH's notice. Now presents another report on this subject, with comments.
ALS 4pp RS:HS 11.469

8282. 1850–4–20 From Thomas Lewin n.p.
Thanks JH for some historical information on events of the first century A.D. [for TL's study on the life of St. Paul].
ALS 1p Harvard: Houghton AAH 67m–67 (91)

8283. 1850–4–22 From William Morison Savile Row
Sends a copy of the notice in the *Times* of a fall of ice in Rossshire, also an account of a similar fall in India in 1826. Comments on these occurrences. [James] Dalmahay has constructed a slide rule for computing the dew-point.
ALS 2pp RS:HS 13.51

8284. 1850–4–23 From Charles Hay Cameron Upper East Sheen Lodge
Has addressed a letter to the East India Company, which has raised a storm.
ALS 4pp RS:HS 5.145

8285. 1850–4–23 To Richard Sheepshanks Collingwood
About some difficulty in trying to determine [Andrew] Graham's interest in the Trivandrum post.
ALS 3pp RS:HS 16.102 (C: RS:HS 25.9.66)

8286. 1850–4–[24 or earlier] To Richard Sheepshanks Collingwood
[John Russell] Hind is being considered for the post of

astronomer at Trivandrum in India.
C 1p RS:HS 25.9.68

8287. 1850–4–25 From John Thomas Briggs to Francis Beaufort Somerset House
Please ask JH to send receipts to account for £316 spent on equatorial telescope from Munich.
ALS 3pp TxU:H/M–0977; Reel 1083

8288. 1850–4–25 To Richard Sheepshanks Collingwood
[Andrew] Graham is not interested in the post [see JH's 1850–4–23].
ALS 2pp RS:HS 16.103 (C: RS:HS 25.9.67)

8289. 1850–4–26 From Francis Beaufort Admiralty
Encloses note from accountant general of Navy about payment of Cape telescope. Send receipt to FB to complete transaction.
ALS 2pp TxU:H/M–0069; Reel 1087 (ACC: Hydrographic Office Letter Book)

8290. 1850–4–26 To William Rowan Hamilton Collingwood
A Mr. Patton has applied for a position as astronomer. Is he 'conversant with practical observing'?
ALS 4pp TCD 1493:478

8291. 1850–4–26 From Heinrich Christian Schumacher Altona
Sends [J. W.] Pastorff's solar observations [see HS's 1850–4–16] as JH's own property.
ALS 4pp RS:HS 15.408

8292. 1850–4–27 From Sir Robert Harry Inglis 7 Bedford Square
Hopes to present the petition on Monday. Had a letter from Mary Somerville at Turin introducing Count and Countess Grizzo.
ALS 1p RS:HS 10.246

8293. 1850–4–27 From William Henry Sykes 47 Albion St.
Discusses balloon experiments to explain fall of barometer with humidity, and explains more discrepancies in vapor-pressure observations. Mentions new 'Hygrometric condenser' as an alternative to wet bulb experiments.
ALS 8pp RS:HS 17.147

8294. 1850–4–29 From Agnes Baillie Hampstead
Packet and documents are ready. When can he call?
ALS 2pp RS:HS 3.12

8295. 1850–4–29 To Francis Beaufort Collingwood
Reimbursement from Navy for bills paid by JH to Georg Merz for 'Cape Equatorial.'
ACCS 3pp †(by JH) TxU:H/L-0759; Reel 1089

8296. 1850–4–30 From William Radcliffe Birt Kew Observatory
Outlining the influence of the recent aurora on the zodiacal light.
ALS 3pp RS:HS 4.141

8297. 1850–4–30 From Humphrey Lloyd Trinity College, Dublin
Before he sends the 'Instructions' to the press would like to know if there is any objection to the use that he has made of JH's *Admiralty Manual.*
ALS 3pp RS:HS 11.286

8298. 1850–5–1 To Charles Hay Cameron Collingwood
Thanks for sharing ideas of CC and Sir Edward Ryan regarding taking a step toward making India self-governing. Discusses these ideas.
ALS 4pp Getty Center

8299. 1850–5–2 To George Bellas Greenough Collingwood
Thanks GG for a publication on mensuration. Praises it, but having never favored the French metrical system, JH doubts the usefulness of a system that departs from it.
ALS 3pp Univ.Coll.London Greenough

8300. 1850–5–2 From William Whewell Trinity Lodge
Sending JH a paper by WW on the nature of induction. Reformulating Aristotle's view. Discusses a proposed Royal Visitation, which WW opposes.
ALS 3pp TxU:H/M-0569

8301. 1850–5–4 From George Bellas Greenough Regents Park
The metrical system is decaying. Does not see any future for the standard yard. A universal language is a necessity.
ALS 3pp RS:HS 9.125

8302. 1850–5–5 To William Whewell Collingwood
Thanks for WW's paper on Aristotle. Discusses the philosophical meaning of the term *conception.* Mentions a visitation to the universities involving the Royal Commission on the Universities of Oxford and Cambridge [on which JH eventually served].
ALS 4pp TC, Camb. Add. Ms.a.207[78] (C: RS:HS 23.86)

8303. 1850–5–9 To [Julia Margaret] Cameron Collingwood
Thanks JC for the gift of her work *The Virgin* [*Widow?*], which both JH and his wife, Margaret, are anxious to read. JH also comments that he would like to discuss the Indian situation with JC's husband.
ALS 4pp DMC.695A.885.20

8304. 1850–5–9 From Ebenezer Henderson St. Helen's
Sends his fractional approximation of a sidereal revolution. Would like his opinion on its accuracy. Clock will not be finished for several years.
ALS 4pp RS:HS 9.281

8305. 1850–5–10 From Reginald Stuart Poole 25 White Rock Place, Hastings
Astronomer Royal [G. B. Airy] has responded to RP's queries. JH was right about him being off one day in his calculation.
ALS 2pp RS:HS 14.37bis

8306. 1850–5–11 To Thomas Romney Robinson [Collingwood]
Discusses the possibility of a large reflecting telescope being constructed for observation of the southern heavens.
draft 2pp RS:HS 14.378

8307. 1850–5–13 From Annibal de Gasparis Naples
Announcing his discovery of a new planet. Gives readings.
ALS 1p RS:HS 8.43

8308. 1850–5–16 From John Hymers St. John's
Names mentioned in his note probably refers to F. D. Barker and W. H. Smith. Persons bearing JH's name have recently graduated; will be pleased to make inquiries.
ALS 2pp RS:HS 10.223

8309. 1850–5–20 From Ebenezer Henderson St. Helens
Requests JH's opinion on his calculation on 'solar and sidereal ratios.'
ALS 1p Harvard: Houghton AAH 67m–67 (51)

8310. 1850–5–22 To Edward Sabine Collingwood
Mentions cirrus cloud formations foreshadowing aurora. Is unable to attend R.S.L. committee meetings and concerned that 'outlying' members be able to relay opinions through letters.
ALS 4pp RS Sa.663

8311. 1850–5–23 From Edward Joseph Lowe Beeston
Henry Lawson has nominated EL for the R.S.L. and would be grateful if JH would add his name to the certificate. Has just received an 11' refracting telescope from Lawson. Has doubtless heard of the newly formed British Meteorological Society. S. C. Whitbread is president, and they already have one hundred members. Has had several gifts of books for the Midland Observatory.
ALS 3pp RS:HS 11.325

8312. 1850–5–23 From Edward Sabine Woolwich
Proposes date for next and perhaps final R.S.L. government grant committee meeting. Wishes there were more funding. Predicts results of meeting. Updates on ES's health.
ALS 8pp RS MM.16.131

8313. 1850–5–27 To the *Athenaeum* Collingwood
Announces the discovery by Annibal de Gasparis of the asteroid Parthenope; gives its location.
L *Athenaeum,* #1179 (June 1, 1850), 589

8314. 1850–5–27 From Joseph Hume Bryanston Square
Sending a note from M. R. Gubbins of the Bengal Civil Service, which may be of interest to JH.
AL 1p RS:HS 10.217

8315. 1850–5–27 From Thomas Romney Robinson Edgeworthstown
Members of B.A.A.S. doubt private funding for telescope will be forthcoming soon [see see TR's 1849–10–21, JH's 1849–10–27, & TR's 1849–11–1]. Will apply to government for funds. Asks JH's advice regarding telescope design.
ALS 7pp RS:HS 14.379

8316. 1850–5–28 To George Biddell Airy n.p.
Sends on announcement of 'new planet Parthenope' from Annibal de Gasparis at Naples.
ALS 2pp RGO 6.97.135

8317. 1850–5–28 From Heinrich Christian Schumacher Altona
Reports that [Annibal] de Gasparis has discovered a new asteroid, Parthenope. Gasparis credits JH with the discovery because JH had proposed the name Parthenope when AG had discovered Hygeia.
ALS 2pp Some French RS:HS 15.409

8318. 1850–5–29 From George Biddell Airy Royal Observatory, Greenwich
Acknowledges receipt of JH's 1850–5–28; GA will pass on the news to some others.
ACCS 1p RGO 6.97.137

8319. 1850–5–30 To Karl von Martius Collingwood
Thanks to the Royal Bavarian Academy of Science for electing JH a 'Foreign Member.' Trusts that the Academy received a copy of JH's *Cape Results.*
ALS 3pp Bayerische Akademie der Wissenschaften 68, S. 116

8320. 1850–5–29 From Lord John Russell 37 Chesham Place
Invites JH to join Bishop of Chester [John Graham] and George Peacock as member of royal commission of enquiry into state and revenues of Cambridge University.
ALS 4pp TxU:H/M–0470; Reel 1093

8321. 1850–5–30 To Lord John Russell Collingwood
Received JR's letter as JH was preparing to leave home. Will consult with George Peacock and reply to JR soon.
AdraftS 1p (on H. Goltermann's 1823–8–27) TxU:H/L-0347.2; Reel 1054

8322. 1850–5–30 From Alfred Smee 7 Finsbury Circus
Note to present JH with a copy of AS's work on 'Instinct and Reason.'
ALS 1p RS:HS 16.170

8323. 1850–5–31 To Thomas Romney Robinson Collingwood
Continues opposed to application to government to support the 'Great Southern Reflector' [see JH's 1849–10–27]; responds to a number of specific queries about the construction of such a telescope.
ALS 4pp Cambridge University Library 7656/TR62 (C: RS:HS 23.87)

8324. 1850–6–1 From Henry Warburton 45 Cadogan Place
R.S.L. sent JH's ['On the Algebraic Expression of the Number of Partitions...' (1850)], asking HW to judge its merits for publication in *R.S.P.T.*
ALS 3pp RS:HS 18.59

8325. [1850]–6–2 From William Empson East India College
Sending proofs. Will see that he has altered a few words here and there. Mrs. Jones's health has improved and she talks of going to Richmond.
ALS 4pp RS:HS 7.73

8326. 1850–6–2 To Margaret Brodie Herschel London
George Peacock is seriously ill; Charles Lyell urged JH to see a hippopotamus that had been brought to London; news of family and friends.
ALS 4pp JHS 3.21c

8327. 1850–6–2 From William Whewell Trinity Lodge
Comments on the Royal Commission on the Universities of Oxford and Cambridge and the possibility, favored at Cambridge, that JH would serve on that committee.
ALS 4pp TxU:H/M-0570 Todhunter, *Whewell, 2,* 359–60

8328. [1850 or later]–6–3 To Margaret Louisa Herschel Royal Mint
Gratitude that Louisa sustained her resolution and resisted temptation. Trust mother [MBH] if in doubt.
ALS 4pp TxU:H/L-0532; Reel 1053

8329. 1850–6–3 From William Parsons (Lord Rosse) 19 Great Cumberland St.
Invitation to dinner next Wednesday.
ALS 1p RS:HS 13.230

8330. 1850–6–4 To Lord John Russell 37 Tavistock Square
Agrees to serve on proposed commission of enquiry into state of Cambridge University, provided that JH may resign if enquiry lasts more than one year.
ACCS 1p TxU:H/L-0348; Reel 1054

8331. 1850–6–4 To Richard Sheepshanks 32 Harley St.
Describes JH's decimal coinage system as he would like to see it introduced.
ALS 3pp RGO 6.356.161

8332. 1850–6–5 From Henry Warburton 45 Cadogan Place
Favorable report to R.S.L. on JH's paper. Suggests minor changes. Verified only methods, not numerical values. Shows more general treatment of partitions than one that JH developed.
ALS 7pp RS:HS 18.61

8333. 1850–6–5 From Henry Warburton 45 Cadogan Place
Review of JH's paper [see HW's 1850–6–1] renewed HW's interest in 4- and 5-partitions of numbers. Will complete paper on combinations during vacation.
ALS 2pp RS:HS 18.60

8334. 1850–6–6 From Lord John Russell Pembroke Lodge
Does not want JH to enquire into the University of Oxford. JH may resign at the end of the year.
ALS 2pp RS:HS 14.440

8335. 1850–6–7 From Joanna Baillie Hampstead
Has the two copies of the book reserved for JH and Lady Herschel. Will send them later.
ALS 3pp RS:HS 3.24

8336. [1850–6–7] From William Empson East India College
Would like JH's opinion on the layout of his page.
ALS 4pp RS:HS 7.74

8337. 1850–6–9 From Algernon Percy (4th Duke of Northumberland) Syon
Invites JH to come to see the flowering plants in Syon gardens.
ALS 3pp RS:HS 13.360

8338. 1850–6–9 or later To Algernon Percy (4th Duke of Northumberland) [Collingwood]
Margaret Herschel will avail herself of the offer [see AP's 1850–6–9]; inquires what is happening to the letters from AP's father that JH had given to AP.
draft 3pp RS:HS 13.361

8339. 1850–6–11 To Margaret Brodie Herschel Collingwood
Has received an invitation from Algernon Percy and JH has accepted for MH and a few friends; JH has received a letter offering some lost spoons with JH's arms and crest. JH says this must be an attempt at a swindle as he has lost no such spoons, since they never existed. Urges some checking by MH. [Letter contains a note by MH asking someone to investigate on JH's behalf.]
ALS 4pp JHS 6.29

8340. 1850–6–14 From William Radcliffe Birt Kew Observatory, Old Deer Park
Stunned to learn that WB's letter to Edward Sabine was interpreted by B.A.A.S. committee as WB's resignation from Kew Observatory. Encloses copy of resolution terminating WB. Plans to fight it.
ALS 7pp † TxU:H/M–0103; Reel 1087

8341. 1850–6–14 To Margaret Brodie Herschel Collingwood
Great distress about a letter from Charles Pritchard to the effect that JH's son William may not be ready for the entrance examination to Haileybury [to prepare for the Indian Civil Service]. JH is considering working with William to teach him Greek.
ALS 4pp JHS 6.28

8342. 1850–6–15 From Algernon Percy (4th Duke of Northumberland) Whitehall Gardens
Note accompanying return of letters [see JH's 1850–6–9 or later].
ALS 1p RS:HS 13.362

8343. 1850–6–16 To William Radcliffe Birt Collingwood
Sorry about misunderstanding between WB and Edward Sabine at Kew Committee. Suggests WB write statement to clear things up.
ACCS 4pp TxU:H/L-0100; Reel 1054

8344. 1850–6–16 To Margaret Brodie Herschel Collingwood
Following another letter from Charles Pritchard [see JH's 1850–6–14], it now appears that son William's Latin is deficient; JH will work with William during William's holidays, but is concerned about interruptions, mentioning an invitation to a ball at Buckingham Palace and 'Mrs. [Julia] Cameron's Artist,' who is coming to paint JH.
ALS 4pp JHS 6.27

8345. 1850–6–19 From William Radcliffe Birt Kew Observatory, Old Deer Park
Failed to obtain interview with Edward Sabine. B.A.A.S. committee considers WB finished [see WB's 1850–6–14], yet claims rights to WB's report on luminous meteors, which WB submitted *after* involuntary termination. WB wants report returned to him. Plans no more work in science. Thanks JH for ten years of assistance and kindness.
ALS 4pp TxU:H/M–0104; Reel 1087

8346. 1850–6–19 From Charles Pastorff Zulichem
Thanks JH for remarks on his father's astronomical contributions. Graciously receives JH's *Cape Results.*
ALS 1p RS:HS 14.144

8347. 1850–6–20 To [William Radcliffe Birt] Collingwood
Attests to W. R. Birt's scientific capabilities and dependability. Regrets Birt's departure from previous position.
ALS 3pp RS:HS 19.141

8348. 1850–6–21 From William Parsons (Lord Rosse) 19 Great Cumberland St.
Asks JH's opinion on candidates for Copley Medal to be presented in November.
ALS 2pp RS MM.16.154

8349. 1850–6–24 To Edward Sabine Collingwood
Cannot attend meeting of grant committee tomorrow. Favors grant to print observations made at Armagh Observatory by T. R. Robinson, but with conditions. Protests any grants to public institutions. Mr. Forbes's proposition is too premature to act on it.
ACCS 4pp †(by JH) TxU:H/L-0368; Reel 1054

8350. 1850–6–26 From Edward Sabine Woolwich
Encloses letters pertaining to committee decision, proposed by G. B. Airy and regarding which JH wrote a letter, to grant aid to Armagh Observations.
ALS 3pp RS MM.16.130

8351. 1850–6–27 From William Rutter Dawes Wateringbury, Maidstone
Regarding a certain Mr. Ray who applied for the position at Ormskirk. Is thinking about the removal of his observatory. His illness has upset his plans.
ALS 4pp RS:HS 6.96

8352. 1850–7–1 From James A. Gordon Stocks, Tring
Sending details of a case of mesmerism known to him. Met one of his sons the other day. Pleased to hear JH is to become a Crown Commissioner. Wishes Mr. Naish was accessible to JH.
ALS 4pp RS:HS 8.185

8353. 1850–7–6 From Giovanni Antonio Plana Turin
Asks for some of JH's memoirs to complete GP's collection. Has observed the comet recently discovered. Comments on one of GP's recent memoirs, and asks if JH has a copy.
ALS 4pp (French) RS:HS 13.438

8354. 1850–7–6 To [R.S.L. Council] Collingwood
Reports on paper by William Thomson on mathematical theory of magnetism. Approves paper, but believes JH lacks expertise to report properly on it.
ALS 5pp RS RR.2.237

8355. 1850–7–9 From William Henry Smyth 3 Cheyne Walk, Chelsea
Sends JH copy of his address to Royal Geographical Society. Inquires about a means to make astronomical instruments steadier with poles devised by JH.
ALS 2pp RS:HS 16.299

8356. 1850–7–15 To William Henry Smyth Collingwood
Compliments WS on a written address. Describes how to set up a telescope to avoid unnecessary vibration; includes diagram.
C 2pp RS:HS 23.88

8357. 1850–7–18 From Friedrich Eduard Beneke Berlin
Letter to accompany the second part of FB's popular psychology, together with more of FB's constant urging of the need for psychology in England.
ALS 2pp (German) RS:HS 4.65

8358. 1850–7–19 From David Brewster St. Leonard's College, St. Andrews [University]
Received Dr. [T. R.] Robinson's memorial to John Russell. Believes six-foot reflector could have been obtained as easily as a three-foot reflector.
ALS 2pp TxU:H/M–0127; Reel 1087

8359. 1850–7–20 or later From Thomas Maclear Royal Observatory, Cape of Good Hope
TM's attention has been drawn to the question raised by Joseph Hume in the House of Commons regarding the use of the Cape Observatory. Will forward a copy of the weekly register and later a copy of the general report.
C 2pp RS:HS 12.149bis

8360. [1850–7–20 or later] From Thomas Maclear Royal Observatory, Cape of Good Hope
Encloses copy of the letter he has written [1850–7–20] to the Secretary of the Admiralty.
ALS 1p RS:HS 12.149

8361. 1850–7–20 From Alfred Smee 37 Old Jewry
Comments on some of AS's writings, and on the effect of strong coffee on vision.
ALS 2pp RS:HS 16.171

8362. 1850–7–26 From John Phillips 4 Castle St., Edinburgh
Asks JH to send copy of [T. R.] Robinson's memorial to John Russell on the southern telescope. Robinson's statement 'is wanted' for upcoming council meeting and has not yet arrived.
ALS 1p RS MM.16.132

8363. 1850–7–27 To William Empson Collingwood
[A. J.] Ellis was correct; JH used earlier version of proof by [C. F.] Gauss in *Theoria Motus Corporum Coelestium* [1809] and was not aware of newer proof. But Ellis's objection to JH's proof applies only to properties of space.
C 1p inc TxU:H/L-0146; Reel 1054

8364. 1850–7–27 To James David Forbes Collingwood
Calls JF's attention to an article in which an answer is attempted to JF's objection to 'the argument from probability of a physical connection between individuals of a double star.'
ALS 3pp St. Andrews 56 & C: TxU:H/L-0146; Reel 1054 inc J. C. Shairp, P. G. Tait, and A. Adams-Reilly, *Life and Letters of James David Forbes, F.R. S.* (London, 1873), 469–70

8365. 1850–7–27 From James A. Gordon Naish, Bristol
Sends a copy of the *Globe* newspaper with the critique on Karl Reichenbach. Sorry to hear JH will be travelling southwards. The West country is very attractive. Is pleased to hear JH is a Commissioner.
ALS 4pp RS:HS 8.186

8366. 1850–7–31 From Alexander von Humboldt to JH [?] Sans Souci
Astronomical queries, with particular reference to the Coal Sack.
ALS 2pp RS:HS 10.70

8367. 1850–8–1 From John George Children Halstead Place
Has sent the aneroid barometer to [Edward] Dent's, who will give it to the messenger when he calls.
ALS 2pp RS:HS 5.339

8368. 1850–8–2 From Heinrich Christian Schumacher Altona
Alexander von Humbolt requests that JH answer questions about 'coal sacks (plur.)' [Coal Sack of the Southern heavens].
ALS 2pp RS:HS 15.410 (C: TxU:H/M–1031.2; Reel 1083)

8369. 1850–8–7 From William Empson T.W. [Tunbridge Wells?]
Death of WE's child. Will return to [Haileybury]. Encloses letter from [A. J.] Ellis. Let Ellis know if JH wants to see Ellis's works.
ALS 4pp TxU:H/M–0179; Reel 1087

8370. 1850–8–7 To Alexander von Humboldt Collingwood
Offers some information on 'Coal Sacks' in Southern Cross [see AH's 1850–7–31]. Happy to hear that volume 3 of AH's *Kosmos* is near completion.
C 3pp RS:HS 23.89 (C: TxU:H/M–1031.3; Reel 1083)

8371. 1850–8–7. From Christian F. Schönbein n.p.
Forwards results of experiment on influence of solar light on the chemical action of oxygen. Asks JH to experiment further and to give CS his views on the subject.
ALS 2pp RS:HS 15.362

8372. 1850–8–9 From Giovanni Santini Padua
Gets copy of *Cape Observations* through ministry. May visit England to meet with JH and G. B. Airy. Sends observations of Biela's Comet.
ALS 3pp (Italian) RS:HS 15.349

8373. 1850–8–11 From Alexander von Humboldt Sans Souci
Is grateful for his constructive letter on the Coal Sack. Regarding the work of the Schlaginweit brothers.
ALS 1p (French) RS:HS 10.71

8374. 1850–8–12 From J. Riedl-Sevenstern Vienna
Has made a globe of J. H. Mädler's lunar observations. Would JH accept one as a gift; if so, how can he send it?
ALS 2pp RS:HS 11.474

8375. 1850–8–14 From William Lassell to G. B. Airy Starfield, Liverpool
Believes he has discovered a second satellite of Neptune; gives readings.
C 4pp RS:HS 11.140

8376. 1850–8–18 From George Biddell Airy Aviemore
Committee set up by the B.A.A.S. to consider methods of observing the forthcoming eclipse of the sun; observations on this.
ALS 3pp RS:HS 1.219 (C: RGO 6.119.188)

8377. 1850–8–18 or later To George Biddell Airy n.p.
Is just leaving for France, and has time only for a few thoughts in response to GA's 1850–8–18.
ALS 3pp RGO 6.119.190

8378. 1850–8–20 From John Murray, [Jr.] Knock, Largs
JH's obliging note concerning corrections to the maps in JM's *Handbook for France* has been sent to him at Knock. Would like to present the appropriate handbook should JH or his friends visit the Continent. *Admiralty Manual* has now been all sold and a new edition is required; has JH any alterations or suggestions?
ALS 2pp RS:HS 12.447

8379. 1850–8–21 From James Challis Cambridge Observatory
Is glad to hear that he can retain the Adams prize essay for a further period. Is pleased that William Lassell has discovered a second satellite of Neptune. Regarding shooting stars seen on 9 Aug.
ALS 3pp RS:HS 5.236

8380. 1850–8–21 To William Whewell Collingwood
Discusses JH's chemical experiences with sulphuret of lead. Leaves for the continent 'on Monday next.' Has conditionally accepted a position on the Royal Commission on the Universities of Oxford and Cambridge.
ALS 4pp TC, Camb. Add. Ms.a.207[79] (C: RS:HS 23.90)

8381. 1850–8–22 To [?] Collingwood
Declines offer of support. JH does not wish to become involved in politics, representing Cambridge.
ALS 3pp BostonPubLib Ms.196(1)

8382. 1850–8–26 From Augustus De Morgan 7 Camden St.
Guessed the name of his friend. Has not read the article attentively yet. If JH is in the south of France he may meet H. P. Brougham (Baron Brougham and Vaux) opticizing. Picked up an old book on the reformation of the calendar owned by C. Clavius 1556. Has sent to Rome for signature.
ALS 3pp RS:HS 6.260

8383. 1850–8–26 From James A. Gordon Knockspock
Congratulations on his success. Outlines persons to call on when in France. Met H. S. Fox-Strangways (3rd Earl of Ilchester) at Exeter. Has met a friend who has read Karl Reichenbach's book and believed it.
ALS 4pp RS:HS 8.187

8384. 1850–8–26 From John Henry Griesbach Carlton Villas, Maida Vale
Has been unable to carry on his astronomical pursuits for the past three years. Would like to visit JH to see the construction of his telescope.
ALS 3pp RS:HS 9.62

8385. 1850–8–27 From James A. Gordon Knockspock
Thinks JH is the man for Cambridge.
ALS 4pp RS:HS 8.188

8386. 1850–8–31 From George Grey Court of St. James
Royal warrant appointing John Graham, George Peacock, JH, John Romilly, and Adam Sedgwick to constitute Cambridge University Commission.
D 3pp TxU:H/M–1223; Reel 1060

8387. [1850–9] To William Whewell n.p.
The publisher [John] Murray wants a new edition of JH's *Admiralty Manual.* Suggests revisions for WW to make in WW's paper in that volume.
ALS 2pp TC, Camb. Add. Ms.a.207[80]

8388. 1850–9–3 From Thomas Wilson Crimble House, Leeds
As president, TW asks JH to chair next annual soirée of Leeds Mechanics' Institution and Literary Society. Names distinguished predecessors. Describes members and activities.
ALS 3pp RS:HS 18.423

8389. 1850–9–5 To James A. Gordon Paris
Has no intention of standing as MP for Cambridge. Thanks him for his generous offer. Would be pleased to pass on £200 to his cousin Thomas Baldwin.
draft 4pp RS:HS 8.189 & 23.91

8390. [1850–9–8?] From Louisa Herschel to Thomas Wilson [Collingwood]
JH is touring France and may not return before November. Presumes JH will not accept TW's invitation [see TW's 1850–9–3] to preside at annual soirée.
ACCS 2pp RS:HS 18.424

8391 1850–9–17 From Charles Eyres Caius College, Cambridge
CE is candidate for office of secretary to Commission of Inquiry into state of University of Cambridge.
ALS 3pp TxU:H/M–0186; Reel 1087

8392 1850–9–25 From John Graham Camaroon
Received royal warrant to appoint commission of inquiry into state of Cambridge University. Asks JH to attend between 16 and 22 Oct. Also inviting Adam Sedgwick and John Romilly. Wants joint meeting with Oxford inquiry commission, headed by Bishop of Norwich.
ALS 3pp TxU:H/M–0214; Reel 1087

8393. [1850]–9–26 From Joanna Baillie Hampstead
Thanks for his encouragement. Regarding the Hoggerts.
ALS 3pp RS:HS 3.26

8394. 1850–9–30 From James A. Gordon Knockspock
Was pleased with JH's reply to his offer. Has just returned

from Poole. Remembers Mary Baldwin. Long discussion on politics.
ALS 12pp RS:HS 8.190

8395. 1850–10–[5 or earlier] From Margaret Louisa Herschel to John Graham Collingwood
JH is traveling in France and cannot respond to summons from [John Graham,] Bishop of Chester to attend October meeting of Cambridge University Commissioners. John Russell was notified in advance of JH's absence.
ACCS 1p †(by JH) TxU:H/M–0801; Reel 1085

8396. 1850–10–10 To John Graham Avignon
JH is traveling in Europe and unable to attend meeting of Cambridge University Commissioners on 16 or 22 Oct.
AdraftS 2pp (†by JH) TxU:H/L-0170; Reel 1054

8397. 1850–10–15 From Augustus De Morgan [University College, London]
Raises a question regarding the size of an infinitely small quantity. Argues that 1850 is the last year of the first half of the nineteenth century.
L S. De Morgan, *Memoir of Augustus De Morgan*, 213

8398. 1850–10–25 From James A. Gordon University Club
Hopes he is on the way back from his travels. Has been corresponding about the Black Prince. Nothing to add about mesmerism, but biology is strong at Glasgow. Agricultural matters seem to be improving. Has left Scotland and is travelling around.
ALS 4pp RS:HS 8.191

8399. 1850–10–25 From Arthur Cleveland Wigan 26 Osnaburgh St., Regent's Park
Requests testimonial on behalf of AW's application for position of undersecretary at St. George's Hospital.
ALS 2pp † RS:HS 18.405

8400. 1850–10–29 From George Peacock Cambridge
First regular meeting of Cambridge [University] Commission will be tomorrow at Treasury. W. H. Bateson was elected secretary at private meeting earlier. GP caught terrible cold on unlucky trip to Germany.
ALS 3pp † TxU:H/M–0417; Reel 1087

8401. 1850–10–30 From John Graham Euston Hotel
Cambridge University Commission met formally for first time today at Treasury Chambers.
ALS 2pp TxU:H/M–0216; Reel 1087

8402. 1850–10–30 To Thomas Maclear Collingwood
Returns from vacation in France. Offers to be a trustee for the Cape Observatory. J. R. Hind has discovered asteroid Victoria.
ALS 3pp Cape Archives/Maclear Papers File 100

8403. 1850–10–31 To James David Forbes Collingwood
Describes travels through a volcanic district in France, highlighting beautiful scenery, various rock and lava formations, analyzing their causes, and comparing JH's observations with JF's.
ALS 7pp St. Andrews 93 A/a (C: RS:HS 23.93)

8404. 1850–10–31 From George Peacock Deanery of Ely
Reports results of first regular meeting of [Cambridge University Commission] yesterday. Health of Adam Sedgwick and [Richard] Jones.
ALS 2pp TxU:H/M–0418; Reel 1087

8405. 1850–11–1 From James A. Gordon St. George's Hotel
Hopes he will do what is best with the sum of money. Is most anxious he should have the letter.
ALS 3pp RS:HS 8.192

8406. 1850–11–4 From Henry Piddington Calcutta
Sends JH some publications, including one that tends to show that cyclones are electrical phenomena.
ALS 3pp RS:HS 13.406

8407. 1850–11–6 From James David Forbes Edinburgh
Was pleased by JH's favorable opinion of his paper on the Vivarais. Finds JH's arguments concerning [John] Michell's theory of probabilities unconvincing. Will soon write a paper on the problems.
ALS 2pp St. Andrews LB IV, 522–3 (C: RS:HS 7.313)

8408. 1850–11–9 From Lord John Russell Chesham Place
Annual salary of Master of Mint was reduced to £1000 upon R. L. Sheil's resignation, and office was declared non-political. Proposes to submit JH's name to Queen as next master.
ALS 4pp TxU:H/M–0471; Reel 1093

8409. 1850–11–11 From John R. Hind 22 Grove Rd., St. John's Wood
On Nov. 4 observed a red star, which does not appear in any of the recent maps or charts. Gives readings. JH may be able to throw some light on it.
ALS 2pp RS:HS 9.350

8410. 1850–11–12 From William Henry Bateson St. John's College, Cambridge
Bishop of Chester [John Graham] has called meeting of Cambridge University Commissioners for 19 Nov.
ALS 2pp & CS TxU:H/M–0055.1 & M–0054; Reel 1087

8411. 1850–11–14 From William Henry Stephenson Downing St.
Chancellor of Exchequer [Charles Wood] will see JH at noon tomorrow.
ALS 1p TxU:H/M–0499; Reel 1093

8412. [1850]–11–16 From Edward Sabine 11 Old Burlington St.
Received series of actinometer observations from [Joseph] Dayman aboard *Rattlesnake*. JH's trip to Continent. Completion of Francis Ronalds's magnetographs. Improved access to Kew Observatory by railway. Next committee meeting. Bakerian lecture by Michael Faraday.
ALS 4pp TxU:H/M–0524; Reel 1093

8413. 1850–11–17 From Thomas Maclear Royal Observatory, Cape of Good Hope
Was agreeably surprised by the communication in Lady Herschel's letter; did not expect any pension. Outlines his ideas for careers for some members of his own family. India and the Services seem the best propositions. Does not expect much from the observations of Venus in Chile unless the definition is good.
ALS 3pp RS:HS 12.150

8414. 1850–11–17 To Lord John Russell Collingwood
Accepts position as Master of Mint. Asked Chancellor of Exchequer [Charles Wood] to specify housing arrangements for JH. Will reside with friend [Eneas Mackintosh] at 17 Montague Square until JH hears from JR.
ACCS 2pp TxU:H/L-0349; Reel 1054

8415. 1850–11–18 To Margaret Brodie Herschel 17 Montague Square
Went to meet Lord John Russell. JH considers [Mastership of Mint] settled, but do not speak openly of it yet. Carried out MBH's errands in London. Encloses note for daughter Louisa from Miss A. Bryant.
ALS 4pp TxU:H/L-0541.1; Reel 1053

8416. 1850–11–19 To Margaret Brodie Herschel 17 Montague Square
Cannot determine if mastership of Royal Mint will be vacant. R. L. Sheil is in Dublin. Rumors about Sheil's new appointment. Will attend Cambridge University Commission meeting today. [G. G.] Gordon congratulated JH on Mint position. JH's host, Eneas Mackintosh, suggests that JH's son John enter Addiscombe College at age 16, but JH prefers 17. Searching for house in London. P.S.: Mr. Stroud had not heard of Mrs. Jeffrey's notice. Tell JH what to wear for 'Royal presence.'
ALS 4pp TxU:H/L-0541.2; Reel 1053

8417. 1850–11–19 From William Parsons (Lord Rosse) Castle Parsonstown
Sends sketches of nebulae. It is time to do something about Toronto magnetic observatory. Should it be continued?
ALS 2pp TxU:H/M–0465; Reel 1093

8418. [1850–11–20] From William Henry Bateson [No. 10 Downing St.]
[Cambridge University] Commission will continue its sittings for whole week.
ALS 2pp TxU:H/M–0053; Reel 1087

8419. 1850–11–20 From William Thomas Brande Royal Mint
Congratulates JH for appointment as next Master of Mint. WB at Mint for 25 years. Offers any assistance JH may need.
ALS 2pp TxU:H/M–0117; Reel 1087

8420. 1850–11–[20] To Margaret Brodie Herschel London
Learned from H. D. Harness today that entire Mint is undergoing reorganization. JH expects to devote considerable time to this transition. Admires Harness's capability. Read report yesterday by commission to reform Mint. Will see Lord John Russell tomorrow about Mastership of Mint.
AL 4pp TxU:H/L-0541.4; Reel 1053

8421. 1850–11–20 To Margaret Brodie Herschel 17 Montague Square
JH still in suspense about Mastership of Mint. Wants to find house for MBH in London. Describes work that faces University Committee.
AL 4pp TxU:H/L-0541.3; Reel 1053

8422. 1850–11–20 From Charles Wood Downing St.
Asks JH to call on CW tomorrow.
ALS 1p TxU:H/M–0589; Reel 1086

8423. 1850–11–21 To Margaret Brodie Herschel London
Chancellor of Exchequer [Charles Wood] confirmed JH's appointment as Master of Mint today. R. L. Sheil will resign on 1 Dec. JH will be installed on 3 Dec. as first non-political appointee.
ALS 3pp TxU:H/L-0541.5; Reel 1053

8424. 1850–11–21 From Lord John Russell Pembroke Lodge
Glad that JH accepted Mastership of Mint. Will provide annual housing allowance of £500. George Grey will take steps necessary to put JH into office.
ALS 2pp TxU:H/M–0472; Reel 1093

8425. 1850–11–21 To Lord John Russell 17 Montague Square
Government allowance for JH's residence [in London] is sufficient. Will make necessary arrangements for housing while awaiting official notice to assume Mastership of Mint.
ALS 2pp TxU:H/L-0350; Reel 1054

8426. 1850–11–22 To James A. Gordon London
Has accepted the offer of the Mastership of the Mint and has to find security for £20,000; can JG assist him in finding this sum as his own resources are not enough?
draft 3pp RS:HS 8.193

8427. 1850–11–23 From James A. Gordon Stocks, Tring
Is pleased he asked his help in obtaining security, but could not offer more than £3,000 as his lands are heavily mortgaged.
ALS 4pp RS:HS 8.194

8428. 1850–11–23 To Richard Lalor Sheil 17 Montague Sq.
Please convey RS's views of changes in organization of Mint.
ACCS 1p TxU:H/L-0385; Reel 1054

8429. 1850–11–24 To George Grey Collingwood
Requests permission to post £20,000 of JH's funds as security bond for JH's position as Master of Mint, rather than calling on JH's friends to act as surety. Will JH also need to pay 'Office Fee,' which Royal Commission of 1849 recommended be abolished?
AdraftS 4pp TxU:H/L-0175; Reel 1054

8430. [1850]–11–25 From James A. Gordon Stocks, Tring
Was pleased to receive JH's reply concerning the surety. His trip to London did him good. Curious JH should come under Sir George Grey. He has a mortgage on his property of £20,000. There is a society that will insure any sum of money.
ALS 4pp RS:HS 8.210

8431. 1850–11–25 To James A. Gordon Collingwood
Is grateful for JG's offer. Did not realize what he was asking. In a state of chaos at the moment due to their impending removal.
draft 4pp RS:HS 8.195

8432. 1850–11–25 From Sir Robert Harry Inglis 7 Bedford Square
Congratulations on JH's appointment to Mastership of Mint. Still wishes JH had been president of R.S.L.
ALS 1p TxU:H/M–0287; Reel 1087

8433. 1850–11–25 From Richard Lalor Sheil Templemore, Long Orchard, Ireland
Will contact JH when RS returns to London.
ALS 2pp TxU:H/M–0480; Reel 1093

8434. 1850–11–25 From William Henry Stephenson Downing St.
Encloses copies of papers from R. L. Sheil. Originals will be sent later.
ALS 2pp TxU:H/M–0500; Reel 1093

8435. 1850–11–26 From Alexander Cornewall Duff-Gordon Treasury Chambers
Directions for entering into JH's position as Master of Mint will be sent to JH from Treasury. George Maule approved

JH's method for posting security bond. Committee has recommended abolishing official fees, but this has not yet been done.
ALS 4pp TxU:H/M–0171; Reel 1087

8436. 1850–11–26 From John Stevelly Belfast
Encloses note from [W. P.] Wilson explaining need for astronomical instruments for instructing advanced students at Belfast. Asks JH's advice. Congratulations on JH's appointment to Mastership of Mint. [JH annotation: Suggests appropriate instruments.]
ALS 3pp † 1p TxU:H/M–0503; Reel 1093

8437. 1850–11–27 From William Henry Fox Talbot Lacock Abbey, Chippenham
Congratulates JH on appointment to Mastership of Mint. Hopes JH, as member of Cambridge University Commission, will introduce needed reforms.
ALS 3pp TxU:H/M–0547; Reel 1093

8438. 1850–11–27 From Nathaniel Wallich 5 Upper Gower St., Bedford Square [Bloomsbury]
Trying to produce quality flax from Bengal's china grass *Urtica*. JH's suggestions were helpful. Will see H. C. Oersted in Copenhagen. Account of Copenhagen festivities was bungled by editor.
ALS 8pp RS:HS 18.32

8439. 1850–11–28 To William Henry Stephenson 37 Tavistock Place, Tavistock Sq.
Received papers sent by WS. JH has changed addresses; no longer resides at Montague Square.
AL 1p TxU:H/L-0391; Reel 1054

8440. 1850–11–29 From Charles Edward Trevelyan Treasury
JH's security as Master of Mint may be met with deposit of £10,000 of stock plus personal bond of £10,000, or with £20,000 of stock.
LS 2pp TxU:H/M–0550.1; Reel 1093

8441. 1850–11–30 To William Thomas Brande n.p.
Thanks for congratulations, but JH is not yet appointed [Master of Mint]. Sorry to hear that JH inherits painful disputes. Plans to rely on WB's long experience [as superintendent of Die department].
Adraft 1p TxU:H/L-0102; Reel 1054

8442. 1850–11–30 To Charles Edward Trevelyan 37 Tavistock Place
Agrees to deposit £10,000 of stock for JH's security and give JH's own bond for £10,000.
ACCS 1p TxU:H/M–0550.2; Reel 1093

8443. [1850–12] To Mary Baldwin 37 Tavistock Place
JH's new duties at Royal Mint. JH and Margaret staying with [Elizabeth] Baily while searching for residence in London. JH's son William to be examined at Haileybury College in Jan. 1851. Son John is [attending Clapham grammar school] with Charles Pritchard.
ALS 3pp TxU:H/L-0070; Reel 1054

8444. 1850–12–2 From Richard Hankins Treasury Chambers
C. E. Trevelyan asks JH to describe stock that JH plans to post as security bond for office of Master of Mint. Deposit stock in names of Treasury officials.
ALS 3pp TxU:H/M–0239; Reel 1087

8445. 1850–12–2 To [Henry Drury Harness] 37 Tavistock Square
Returns 1837 report. Expects announcement of JH's appointment soon.
ACCS 1p TxU:H/L-0654; Reel 1089

8446. 1850–12–2 From Henry Pearse St. John's Rectory, Bedford
Thanks JH for his work on behalf of HP's sister, Mrs. Mary Maclear [wife of Thomas Maclear] in obtaining a government pension for her.
ALS 4pp RS:HS 13.328

8447. 1850–12–2 From Hermann Schlagintweit Leicester St.
Sends letter from [Alexander] von Humboldt and HS's new book on the geography of the Alps. Would like to visit JH at Collingwood.
ALS 3pp RS:HS 15.350

8448. 1850–12–2 To William Henry Fox Talbot London
Appreciates WT's letter of congratulation on JH's appointment [as Master of the Mint, to the Cambridge Commission?]. Has begun work with Cambridge.
ALS 2pp Science Museum Talbt 1/39

8449. 1850–12–3 From Henry Drury Harness Royal Mint
Received documents of 1837 evidence from JH. Forwarded letter to JH's former address, 17 Montague Square, before learning of JH's new address [32 Harley St.].
ALS 3pp TxU:H/M–0250.1; Reel 1087

8450. 1850–12–3 From William Henry Sykes India House
Thanks for signing certificate for [C. M.] Elliott. Sends extracts from George Buist's history of heat storms in India, with evidence of great masses of ice dropped by these storms.
ALS 2pp TxU:H/M–0515; Reel 1093

8451. 1850–12–4 From William Henry Bateson St. John's College, Cambridge
Bishop of Chester [John Graham] wishes to convene 10 Dec. meeting of Cambridge University Commissioners. Will JH attend?
ALS 2pp TxU:H/M–0057.1; Reel 1087

8452. 1850–12–4 From Richard Lalor Sheil 73 Eccleston Square
Returned from Ireland. Invites JH to call on RS.
ALS 2pp TxU:H/M–0481; Reel 1093

8453. 1850–12–5 From William Henry Bateson St. John's College, Cambridge
Queen will receive University address next Tuesday. Bishop of Chester [John Graham] wishes to convene meeting of Cambridge University Commissioners on 11 Dec.
ALS 2pp TxU:H/M–0057.2; Reel 1087

8454. 1850–12–5 From Sir John Bell Guernsey
Offers congratulations on JH's new position [Master of Mint]. Recalls 'our water-pipe operations at Cape Town fifteen years ago' [1835]. Speculates on water wave theory. Pity for 'poor Charles Bell.'
ALS 4pp TxU:H/M–0909; Reel 1083

8455. 1850–12–5 From George Peacock Deanery of Ely
Compliments JH's list of questions [for Cambridge University Commission to discuss]. Issues that GP wants to discuss.
ALS 4pp TxU:H/M–0419; Reel 1087

8456. 1850–12–6 From Henry Drury Harness Royal Mint
Regrets JH's not stopping at the Mint today. HH will not be

in tomorrow.
ALS 2pp TxU:H/M–0250.2; Reel 1087

8457. 1850–12–7 From Charles Edward Trevelyan Treasury Chambers
Asks JH for suggested improvements in R. L. Sheil's plan for reforming Mint.
LS 1p *Return*, 18

8458. 1850–12–8 From George Peacock Deanery of Ely
Changes to questions prepared by GP and JH for tutors [at Cambridge University].
ALS 3pp TxU:H/M–0420; Reel 1087

8459. 1850–12–9 From Henry Drury Harness Royal Mint
Mistakenly opened enclosed letter, addressed to JH.
ALS 1p TxU:H/M–0250.3; Reel 1087

8460. [1850]–12–10 To Charles Edward Trevelyan Royal Mint
JH's review of report to Treasury on Mint reforms.
Adraft 1p & AdraftS 1p TxU:H/L-0480 & -0480 verso; Reel 1055

8461. 1850–12–11 To Charles G. B. Daubeny n.p.
Introducing two German scientists who have written an elaborate work on the physical geography of the Alps. Thinks they may both benefit by their knowledge of each other. Has just returned from a ramble over volcanic country.
draft 3pp RS:HS 6.20

8462. 1850–12–11 From Henry Drury Harness Royal Mint
Forwarded letter as JH requested.
ALS 1p TxU:H/M–0250.4.1; Reel 1087

8463. 1850–12–12 From Alexander Wilson 3 Granville Square
Requests permission to publish JH's and AW's 1843–1844 correspondence in '*Journal of Industry*.'
ALS 2pp RS:HS 18.267

8464. 1850–12–14 From George Biddell Airy Royal Observatory, Greenwich
Congratulates JH on his important new position [Mint]; wants to start seriously to prepare for July 1851 solar eclipse [see JH's 1850–8–18 or later].
ACCS 4pp RGO 6.119.192

8465. 1850–12–15 To George Biddell Airy Collingwood
Will be very busy learning his new job for a while, so GA should start working with other committee members; JH makes some suggestions about eclipse observation [see GA's 1850–12–14].
ALS 4pp RGO 6.119.194

8466. 1850–12–15 To George Biddell Airy n.p.
A note accompanying a letter that seems to come from François Arago, and that seems to JH to show a disturbed state of mind; asks GA for his opinion.
ALS 3pp inc RGO 6.372.221

8467. 1850–12–15 From Richard Lalor Sheil 73 Eccleston Square
Needs to know date of JH's [letters] patent, to compute salary for RS's final pay period as Master of Mint.
ALS 2pp TxU:H/M–0482; Reel 1093

8468. 1850–12–16 From George Biddell Airy n.p.
Acknowledges the receipt of JH's suggestions [see JH's 1850–12–15]; urges JH to consider going to Norway or the Baltic to see the eclipse.
ACCS 1p inc RGO 6.119.196

8469. 1850–12–16 From George Biddell Airy Royal Observatory, Greenwich
Is confident the letter [see JH's 1850–12–15], which is supposedly from François Arago, is a forgery.
ACCS 2pp RGO 6.372.223

8470. 1850–12–17 From Henry Drury Harness Royal Mint
Encloses a letter needing JH's consideration. Believes that it should be in Mint records.
ALS 1p TxU:H/M–0250.5; Reel 1087

8471. 1850–12–17 From Lacy Rumsey Treasury Chambers
Letters patent for JH's appointment as Master of Mint were officially sealed yesterday. Will send statement for fees soon.
L 1p TxU:H/M–0466; Reel 1093

8472. 1850–12–18 From Henry Drury Harness Royal Mint
Minutes regarding contracts with the Moneyers and Melter show that these were temporary arrangements. R. L. Sheil's accounts should be closed at Audit office by next Monday.
ALS 2pp TxU:H/M–0250.6; Reel 1087

8473. 1850–12–18 From Henry Drury Harness Royal Mint
HH learned only this morning that JH will attend today's meeting of Mint board. Lists all formalities and business at R. L. Sheil's first board meeting. Does JH want to follow this same procedure? If so, meeting should be postponed until Friday to prepare documents.
ALS 4pp TxU:H/M–0250.7; Reel 1087

8474. 1850–12–18 From Charles Edward Trevelyan Treasury
Fees paid on appointment of private individuals were reduced but not abolished. JH's fee upon appointment as Master of Mint will be two-thirds of former fee for that office.
LS 2pp TxU:H/M–0550.3; Reel 1093

8475. 1850–12–19 From Lacy Rumsey Treasury Chambers
Please return Queen's Warrant for Coining for correction of erroneous date.
L 1p † TxU:H/M–0467; Reel 1093

8476. 1850–12–20 From Lacy Rumsey Treasury Chambers
Corrected Queen's Warrant for Coining and returns it to JH.
ALS 1p TxU:H/M–0468; Reel 1093

8477. 1850–12–20 To William Whewell 32 Harley St.
Refers to WW's work on tides and on the mathematics of political economy. Fears that JH's duties [as Master of the Mint] will keep him from science. Has taken a house on Harley St.
ALS 4pp TC, Camb. Add. Ms.a.207[82]

8478. 1850–12–21 From Robert John Kane Queen's College, Cork
Grateful for JH's approval of RK's efforts on behalf of 'enlightened education.' Congratulations on JH's appointment to Mastership of Mint.
ALS 3pp TxU:H/M–0313; Reel 1087

8479. 1850–12–21 From Adolph Schlagintweit 15 Leicester St., Leicester Square
Grateful for JH's letters of introduction. Met Prince Albert. Edward Sabine arranged for meteorological observations in Alps by AS and brother Hermann to be published by Longmans. Plans more observations of geology and atmospheric chemistry. Alexander Humboldt advised AS to ask JH where to apply for financial support.
ALS 4pp † TxU:H/M–0477; Reel 1093

8480. 1850–12–21 From J. Riedl-Sevenstern 436 Landstrasse, Vienna
Has not heard from JH since sending JR's moon globe to JH by way of Paul Mendelsohn. Asks for JH's opinion of globe and whether it would be proper for JR to send others to royalty in England.
ALS 1p TxU:H/M–0445; Reel 1093

8481. 1850–12–23 From Lord John Russell Woburn Abbey
Directs JH to investigate international supply and demand for gold and silver and to make projection for next twenty years. Refer to [J. R.] Macculloch and [James?] Wilson for information.
ALS 3pp TxU:H/M–0473; Reel 1093

8482. 1850–12–24 From William Wyon Royal Mint
Encloses two letters from WW, one to R. L. Sheil in Oct. 1850, and one to JH, as Sheil's successor, dated 17 Dec. 1850. Both letters state WW's response to Sheil's announcement of proposed changes in organization of Mint. Appeals to JH to reconsider plan to deprive WW of use of Mint machinery for WW's private engraving contracts. Protests abolition of office of probationer engraver, for which WW's son has been training.
ALS 1p, encl 6pp, encl 4pp TxU:H/M–0609 & TxU:H/M–0607 & TxU:H/M–0608; Reel 1086

8483. 1850–12–25 To William Wyon Collingwood
Received copies of WW's letters of 17 Dec. and 17 Oct. to R. L. Sheil. Will study them.
ACCS 1p TxU:H/L-0512; Reel 1055

8484. 1850–12–26 From Sir Robert Harry Inglis 7 Bedford Square
Invites JH to dinner next Monday to meet African missionary J. L. Krapf.
ALS 2pp TxU:H/M–0288; Reel 1087

8485. 1850–12–27 To Margaret Brodie Herschel n.p.
Richard Jones is wrong in attacking Mr. Tucker. JH must inform W. H. Sykes that [JH's son] William is going to Madras and not to Bombay. Believes that [W. H. C. Plowden] will change William's assignment. Letter from Lord John Russell has increased JH's work load. John Barlow suggests that MBH, not JH, should become member of Royal Institution. Encloses draft of letter to E. Mackintosh.
ALS 3pp TxU:H/L-0541.6; Reel 1053

8486. 1850–12–27 To Lord John Russell 32 Harley St.
Will report on recent changes in supply and demand for precious metals after JH contacts persons named in JR's letter.
ACCS 2pp TxU:H/L-0351; Reel 1054

8487. 1850–12–29 To William Parsons (Lord Rosse) 32 Harley St.
Believes Toronto magnetic observatory should somehow be continued, at least temporarily.
ALS 4pp RS MC.4.373

8488. 1850–12–30 To Margaret Brodie Herschel 32 Harley St.
Son William is adapting to life and study in London. Dined with R. H. Inglis today and met African explorer J. H. Krapf, who discovered source of Nile. MBH will soon be member of Royal Institution. Please send copy of letter from F. M. Eardley-Wilmot to JH regarding admission to R.S.L. JH blundered in signing it. JH's note to W. H. C. Plowden was not worded correctly. Richard Jones is arranging lodging for JH's son William at Haileybury College.
ALS 4pp TxU:H/L-0541.7; Reel 1053

8489. 1850–12–31 From Alexander von Humboldt Berlin
News of the astronomical activities of his colleagues. Astronomical queries.
ALS 4pp (German) RS:HS 10.72

8490. 1850–12–31 From Adolph C. Petersen Altona
A note to inform JH of the death of H. C. Schumacher.
ALS 1p RS:HS 13.373

8491. [1851 to 1853?] To Charles Babbage n.p.
Will he and David Brewster dine with him on Tuesday.
ALS 1p RS:HS 2.396

8492. 1851–1 From Antoine C. Becquerel Jardin des Plants
Comments on JH's work on the spectrum and photography.
ALS 4pp (French) RS:HS 5.37

8493. [1851 or 1852?] From William Rutter Dawes n.p.
Concerning his recent visit to Starfield and the observations made by William Lassell with his 20 foot. Regarding the rings of Saturn. Has not heard of the Craig telescope.
ALS 4pp RS:HS 6.127

8494. [1851] From Edward Enfield [Royal Mint]
Returns letter from JH to C. E. Trevelyan. It correctly reflects position of EE and Robert Rintoul, as stated in letter from Jasper Atkinson to JH.
ALS 2pp TxU:H/M–0181; Reel 1087

8495. [1851 to 1853] From Caroline Emily Mary Herschel [London]
Recovering from cold. Wants to visit suburbs while Lou [Margaret Louisa Herschel?] is there.
ALS 3pp TxU:H/M–0792.2; Reel 1084

8496. [1851 to 1854] From James A. Gordon n.p.
First time he has used a pen since his illness. Approves of leaving the Education question in abeyance. Thanks for the Mint return.
AL 2pp RS:HS 8.208

8497. [1851 to 1855] To George Hamilton-Gordon n.p.
Sends copy of letter from JH to Chancellor of Exchequer on state of Royal Mint. Fears JH's position there may be misunderstood.
AdraftS 2pp TxU:H/L-0001; Reel 1054

8498. 1851–1 From Henry Drury Harness Royal Mint
Asks JH's approval for conditions that HH will set for [William] Wyon to use Mint equipment to conduct private business as medallist.
ALS 4pp TxU:H/M–0251.8; Reel 1087

8499. 1851–1–1 From Henry Drury Harness Royal Mint
Sends check for salary due to JH for previous quarter. Itemizes taxes and fees deducted. As first non-political appointee to Mastership, JH is eligible for superannuation fund, but HH needs instructions from Treasury about this. Will arrange to receive dies for Waterloo Medal from Benedetto Pistrucci.
ALS 3pp TxU:H/M–0251.1; Reel 1087

8500. [1851 to 1853–7] To Henry Drury Harness n.p.
Notes on weighing and coining silver. Payment of wages in Coining Room.
AdraftS 3pp TxU:H/L-0183.1; Reel 1054

8501. [1851 to 1852] To Margaret Brodie Herschel [32 Harley St.]
Tell Miss [Rufeneals?] that JH will take care of money owned by Miss Brandon. JH and son William will come to Collingwood for weekend. Will write to Colonial Office on behalf of [William?] Mann. Thomas Maclear will rejoice to have daughter return to Cape.
ALS 4pp TxU:H/L-0536.14; Reel 1053

8502. [1851 to 1855] To Margaret Brodie Herschel [Royal Mint]
Describes 'First Dinner' at Billingsgate. Adjourned to [Richard] Jones's rooms on Suffolk Street for tea and herrings. Charles Trevelyan is out of town.
ALS 3pp TxU:H/L-0536.11; Reel 1053

8503. [1851 to 1855] To Margaret Brodie Herschel Royal Mint
Home remedies for JH's insomnia. Mr. Gibsone returned from Berlin and visited JH last night, causing JH to miss seeing comet again. It is not J. R. Hind's missing comet.
ALS 4pp TxU:H/L-0536.9; Reel 1053

8504. [1851 to 1852] To Margaret Brodie Herschel n.p.
JH and daughters attended German minister C. C. J. Bunsen's dinner, met distinguished guests. Describes delicate Etruscan vases at Bunsen's. Has MH received money from [Thomas] Ayern for timber? James [C. Stewart] was in London yesterday. Mr. M [Murchison?] says that 'G. is in *very* bad hands,' but JH's daughters report that she was well yesterday. Encloses note from G. B. Airy, who was not impressed by [?]'s test results. Bills for [son William's] tuition at Haileybury.
ALS 4pp TxU:H/L-0536.8; Reel 1053

8505. [1851 to 1855] To Margaret Brodie Herschel Royal Mint
Distressing train trip back to London. JH protested at Tunbridge Wells and had third class passengers removed from second class carriage. MH's medicine works well for JH. Will go to Charley Street this evening.
ALS 4pp TxU:H/L-0536.6; Reel 1053

8506. [1851 to 1852] To Margaret Brodie Herschel n.p.
No other 'letter of agents' here. MBH is right about son William's camera. JH longs for Collingwood.
ALS 3pp TxU:H/L-0536.5; Reel 1053

8507. [1851] To Margaret Brodie Herschel London
Will be home a few days later than expected as JH is involved in a court case to be heard immediately after the one he is sitting through, and which he describes.
ALS 3pp JHS 6.16

8508. [1851] To Margaret Brodie Herschel London
As the Mint is closing for a week, JH is trying to arrange a long weekend at Collingwood.
ALS 1p JHS 6.42

8509. [1851] To Margaret Brodie Herschel London
Mostly about JH's health and instructions about the madeira and claret JH is sending home.
ALS 4pp inc JHS 6.60

8510. [1851–1855] To Margaret Brodie Herschel [London]
About their son Alexander, and the striking of 60,000 medals.
ALS 4pp JHS 1.15

8511. [1851 or 1852–1–1] To Margaret Brodie Herschel 32 Harley St.
Mostly family matters, especially the education of several of their sons.
ALS 4pp JHS 1.1

8512. [1851 or later] To Thomas Hodgkin 32 Harley St.
Apologizes for delay in answering TH's letter. JH makes arrangements to visit TH at his home.
ALS 2pp WT.Ho A.5

8513. [1851?] To [Robert Hunt ?] [Collingwood]
Comments on different washes to be used in photography, and clarifies the matter of a supposed Daguerreotype of the nebula in Orion.
ALS 3pp WT 68037.12

8514. [1851]–1–1 To Richard Jones 32 Harley St.
RJ's letter to William Whewell helped JH understand why public panic creates pressure for more money. Asks for RJ's opinion about concurrent circulation of gold and silver. Gives example of this 'compound tender' or 'binary standard' that haunts JH.
ALS 2pp TxU:H/L-0208; Reel 1054

8515. [1851] To Charles Shaw-Lefevre n.d.
Recommending Robert Waterhouse to succeed C. D. E. König as Keeper of the Mineralogical Collection at the British Museum.
C 2pp RS:HS 11.468 & 23.100

8516. 1851–1 To Roderick Impey Murchison n.p.
Requests copy of RM's lecture on distribution of gold, and further information on value of it and silver. Asks specific questions about Russian gold.
AL 3pp inc? EUL:Gen 523/6

8517. [1851 to 1852] To William Parsons (Lord Rosse) [Collingwood]
Discusses W. H. Fox Talbot's patent application, commenting that JH wishes all patents were abolished. Regrets that duties at the Mint and for the Cambridge University Commission have hindered JH's ability to carry out JH's responsibilities to WP.
ALS 4pp Rosse Papers K2.12

8518. [1851 to 1852] To Théophile Jules Pelouze [Royal Mint]
Asks advice on how to purify gold of residual silver.
ACCS 2pp TxU:H/L-0726; Reel 1089

8519. [1851 to 1855] To Benedetto Pistrucci [Royal Mint]
Received BP's letter of 'the 9th.' Concerned that BP's words were colored by strong emotion.
draft 4pp TxU:H/L-0306; Reel 1093

8520. [1851] To Privy Council [Royal Mint]
Advises substantial changes in organization and operation of Royal Mint.
Adraft 1p inc TxU:H/L-0484.15 verso; Reel 1055

8521. [1851] From Henry Rich Treasury
Chancellor of Exchequer [Charles Wood] approved annual £1200 salary for G. F. G. Mathison.
ALS 2pp TxU:H/M–0443.14; Reel 1093

8522. [1851] From Henry Rich 19 Charles St., Berkeley Square
Committee duty at House of Commons prevented HR visiting JH today. Plans to visit JH at Harley Street tomorrow to discuss 'compensation question.'
ALS 3pp TxU:H/M–0443.3; Reel 1093

8523. [1851] From Henry Rich House of Commons
Arranges for meeting with JH to decide on adjustments in Mint offices. Wants to discuss Henry Bingley's salary request.
ALS 4pp TxU:H/M–0443.2; Reel 1093

8524. [1851] From Henry Rich Treasury
Proposes compromise that might settle dispute resulting from abolition of company of Moneyers. Make revisions if JH chooses before sending document to Chancellor of Exchequer [Charles Wood].
ALS 6pp TxU:H/M–0443.1; Reel 1093

8525. [1851 to 1855] To Henry Rich n.p.
Discusses sketch of new coin to be struck. C. E. Trevelyan will attempt to get Chancellor of Exchequer [Charles Wood] to grant money to Mint. Discusses calculations.
Adraft 3pp TxU:H/L-0321; Reel 1054

8526. [1851?] To James Calder Stewart n.p.
JS's 'great book' on Augustine arrived safe and is testimonial to 'Sir Joshua' and JS. Offers informal critique. Will send magnetic circulars to Pasha of Egypt as JS asked. Tell Peter [Stewart] that telescope arrived safely.
ALS 4pp damaged TxU:H/L-0434; Reel 1055

8527. [1851 to 1854] To Charles Edward Trevelyan [Royal Mint]
Encloses letter from moneyer contractors claiming compensation.
AdraftS 2pp TxU:H/L-0479.5; Reel 1055

8528. [1851 to 1854] To Charles Edward Trevelyan Harley St.
[Marked 'Confidential.'] JH will confer again with CT on 'this paper.'
ALS 1p TxU:H/L-0479.4; Reel 1055

8529. [1851 to 1854] To Charles [Edward Trevelyan] Harley St.
Comments on proposed gold coinage.
AdraftS 6pp TxU:H/L-0479.7; Reel 1055

8530. [1851 to 1854] To Charles [Edward Trevelyan] n.p.
Chemical training needed for prospective assayers at Royal Mint. Workload of John Field, resident assayer, does not allow time to train apprentices.
Adraft 2pp TxU:H/L-0479.1; Reel 1055

8531. [1851] To [Charles Edward Trevelyan] [Royal Mint]
Regrets trouble created by Benedetto Pistrucci's obnoxious refusal to accept new title as Mint's chief medallist. Asks [CT] to seek assistance from Chancellor of Exchequer [Charles Wood].
Adraft 5pp TxU:H/L-0635; Reel 1089

8532. [1851 to 1855] To [Charles Edward Trevelyan] [Royal Mint]
Asks Treasury commissioners for permission to increase pay of workmen in melting rooms. Describes living and working expenses incurred by regular and by temporary employees.
Adraft 2pp TxU:H/L-0628; Reel 1089

8533. [1851 to 1855] To [Charles Edward Trevelyan] [Royal Mint]
Two severe accidents recently at Mint. Charitable Fund covers only permanent employees. Asks permission to establish account with nearby hospital to cover temporary workers also.
Adraft 2pp TxU:H/L-0626; Reel 1089

8534. [1851 to 1855] To James Wilson n.p.
Asks for information about foreign gold.
Adraft 1p inc TxU:H/M–0169.2; Reel 1087

8535. [1851–1?] To [?] n.p.
Can [?] write memorial to Treasury that embodies main objects of proposed national institution? Indicate number and kind of instrumental trials needed to prepare for future expeditions. JH will be in London next Thursday; call committee meeting then. Send reply to JH care of E. Mackintosh at 17 Montague Square by Wednesday.
ACCS 2pp inc TxU:H/M–0219.2; Reel 1087

8536. [1851 to 1854] To [?] [Royal Mint]
Testimonial to professional service rendered to Mint by G. F. G. Mathison, contractor for melting and refining gold and silver bullion.
Adraft 1p TxU:H/L-0750.3; Reel 1089

8537. [1851 to 1855] To [?] [Royal Mint]
[Last page only] Estimates time to create dies for [?] medal. Reports progress in production of medal.
Adraft 2pp TxU:H/L-0616; Reel 1089

8538. [1851 to 1855] To [?] [Royal Mint]
[Last page only; continues article VIII and begin article IX.] Mint rules for accounting for precious metals. W. T. Brande is responsible for enforcing these in Coining department.
ACCS 1p inc TxU:H/L-0612; Reel 1089

8539. [1851 to 1855] To [?] [Royal Mint]
[Unidentified Mint employee] was not properly informed that his duties included instructing persons not employed at Mint.
ACCS 1p inc TxU:H/L-0610; Reel 1089

8540. [1851 to 1855] To [?] [Royal Mint]
Will forward [?]'s petition to W. H. Barton when Barton returns next week.
ACCS 1p TxU:H/L-0599; Reel 1089

8541. [1851 to 1855] To [?] n.p.
Encloses copy of note from [W. D.] Haggard regarding production of quicksilver in California. Use of gold bullion coins will be influenced by issue of equilibrium of gold and silver as 'concurrent measures of value.' [JH note on verso on copper alloy ratios.]
Adraft 1p, JH note 1p TxU:H/L-0508.2; Reel 1055

8542. [1851 to 1854] To [?] [Royal Mint]
Received letter of 10th. Understands that there is no need to forward any more copper.
Adraft 1p inc TxU:H/L-0490.2; Reel 1055

8543. [1851 to 1855] To [Bank of England] n.p.
Government wants to know if public has accepted half crown and florin, introduced as part of transition to decimal coinage.
Adraft 2pp inc TxU:H/L-0608; Reel 1089

8544. 1851–1–2 From Benedetto Pistrucci Flora Villa, Bishopsgate, Chertsey
Will go to the Royal Mint to discuss the Medal of Waterloo with H. D. Harness. Apologizes for not writing in English.
ALS 2pp (Italian) TxU:H/M–0429.1; Reel 1093

8545. 1851–1–2 To Charles [Edward Trevelyan] 32 Harley St.
Needs more time to consider R. L. Sheil's proposal for Mint reforms.
ACCS 1p TxU:H/L-0481.1; Reel 1055

8546. [1851–1–3] To Margaret Brodie Herschel [32 Harley St.]
Received formal notice that MH will be voted to membership in Royal Institution on 3 Mar. John Barlow will reserve seats for daughters Caroline and Isabella. JH has tickets to Institution lectures. Son William's exams with W. T. Hooper [at Haileybury College] yesterday. Family finances.
ALS 4pp TxU:H/L-0536.2; Reel 1053

8547. 1851–1–3 From Hermann Schlagintweit Leiceister St.
Thanks JH for letter. Would like to call on JH Sunday morning.
ALS 2pp RS:HS 15.351

8548. [1851]–1–4 To George Biddell Airy 32 Harley St.
Makes proposals about decimal coinage, based on one thousand units in the sovereign.
ALS 3pp RGO 6.349.5

8549. 1851–1–5 To Margaret Brodie Herschel Collingwood
Describes son William's entrance exam to Haileybury College yesterday and procedure of announcing results. Assured by [H.E.I.C. director] W. H. C. Plowden that William's assignment can be changed from Madras to Bombay. W. H. Fitton delivered 'Umbrella and [Panel?]' last night. *Daily News* criticized JH for inaction. Promoting decimal coinage. Mint is in crisis.
ALS 4pp TxU:H/L-0542.1; Reel 1053

8550. 1851–1–5 From Benedetto Pistrucci Flora Villa, Bishopsgate, Chertsey
Will go to the Mint the next Friday to confer with H. D. Harness. States what a great honor it is to work with the Mint.
ALS 1p (Italian) TxU:H/M–0429.2; Reel 1093

8551. 1851–1–6 To [William A. Baillie-]Hamilton Royal Mint
Asks about policy at Royal Dockyards for paying workmen on major holidays. [JH annotation: Summary of WH's reply, with procedures for specific holidays or leaves.]
AdraftS 2pp †(by JH) TxU:H/L-0178; Reel 1054

8552. 1851–1–6 From Henry Drury Harness Royal Mint
Received JH's draft letter concerning Engraving department. Noticed newspaper article that could come only from source inside Mint. Advises JH to warn employees against discussing Mint operations with public press.
ALS 2pp TxU:H/M–0251.2; Reel 1087

8553. 1851–1–6 From Edward Hawkins British Museum
Wishes to confer with JH on Mint business.
ALS 1p TxU:H/M–0249; Reel 1087

8554. 1851–1–7 From George Biddell Airy Playford near Ipswich
Likes JH's ideas about decimal coinage [see JH's [1851]–1–4]; will send suggestions for eclipse observations.
ACC 2pp inc RGO 6.349.7

8555. 1851–1–7 From George Biddell Airy n.p.
About the possible use of decimal subdivisions of the pound weight on the docks and at the Customs House.
ACCS 1p inc RGO 6.341.4

8556. 1851–1–7 From George Biddell Airy Playford near Ipswich
Wants to determine JH's whereabouts for mailing eclipse

observation suggestions [see GA's 1850–12–16].
ACCS 1p RGO 6.119.197

8557. 1851–1–7 To Augustus De Morgan 32 Harley St.
About JH's poor health; mentions discovery of two asteroids and new rings of Saturn.
C 1p RS:HS 23.94

8558. 1851–1–8 From Henry Drury Harness Royal Mint
Suggests alterations to JH's proposal to the Treasury concerning production of medals.
ALS 4pp TxU:H/M–0251.4; Reel 1087

8559. 1851–1–8 From Henry Drury Harness Royal Mint
Found copy of JH's report that can be sent to Treasury. Complaint about copper coins by local shopkeeper, Mr. [Armstrong?], appears to be attempt for publicity, not authentic problem. Hopes to avoid enquiry by Treasury. Encloses list of Mint officers. Returns letter from G. B. Airy. Benedetto Pistrucci will be at Mint on Friday.
ALS 4pp TxU:H/M–0251.3; Reel 1087

8560. 1851–1–8 To Charles Edward Trevelyan Royal Mint
Approves reforms suggested by R. L. Sheil. Suggests using Assay department to assay foreign coins during quiet periods. Recommends introducing new coin weighing machine by Mr. Cotton to inspect every coin, not just random samples. Expects difficulty in eliminating private artistic work of William Wyon, chief engraver.
LS 2pp *Return*, 18–19

8561. 1851–1–10 From E. R. Dovrell Duke St., Stramford St., [Blackfriars Road]
Should slips of *Admiralty Manual* now be sent from William Clowes & Sons, printers, to Collingwood?
ALS 1p TxU:H/M–0169.1; Reel 1087

8562. [1851–1–10 or earlier] From William Henry Smyth n.p.
Excellent weather for observations of variable colors [in stars].
ALS 1p inc TxU:H/M–0496.1; Reel 1093

8563. 1851–1–10 To ? n.p.
Record of JH and H. D. Harness witnessing turning of matrix of Waterloo Medal by Benedetto Pistrucci. Harness affirmed R. L. Sheil's official release of Pistrucci from responsibility if matrix breaks.
Amem 1p TxU:H/M–0496.2; Reel 1093

8564. 1851–1–11 From George Biddell Airy Royal Observatory, Greenwich
Further comments on JH's coinage ideas [see JH's [1851]–1–4].
ACCS 2pp RGO 6.349.8

8565. 1851–1–11 To George Biddell Airy 32 Harley St.
Wants GA's opinion about applying for a pension for the widow of the late astronomer at Madras, T. G. Taylor.
ALS 3pp RGO 6.143.254

8566. 1851–1–11 From Richard Sheepshanks R.A.S.
Advises JH to limit appeals on behalf of worthy causes, such as [T. G.] Taylor's widow, or public may abuse JH's kindness.
ALS 4pp TxU:H/M–0540; Reel 1093

8567. 1851–1–13 From Henry Drury Harness Royal Mint
Sends coinage statistics from Royal Mint for 1848–50. Similar statistics from foreign mints may be obtained through Foreign Office. Records of silver prices are kept only by Lloyds.
ALS 3pp TxU:H/M–0251.5; Reel 1087

8568. 1851–1–14 From George Biddell Airy Playford, near Ipswich
Has made some efforts at suggesting places of employment to T. G. Taylor's son [see JH's 1851–1–11].
ACCS 2pp RGO 6.143.257

8569. 1851–1–14 From Henry Drury Harness Royal Mint
Board meeting tomorrow does not require JH's attendance.
ALS 1p TxU:H/M–0251.6; Reel 1087

8570. 1851–1–15 From George Biddell Airy Royal Observatory, Greenwich
Believes the quality of the astronomical work of T. G. Taylor was certainly high [see JH's 1851–1–11].
ALS 2pp RGO 6.143.256

8571. 1851–1–15 From Julia Margaret Cameron Sheen Lodge
Wrote to JC's brother-in-law, H. T. Prinsep, who promised that JH's 'nephew' will receive direct appointment when nephew comes of age. Gratitude for JH's friendship of 14 years. JC's husband is in Ceylon. Invites JH and wife to visit JC. Names other recent visitors.
ALS 4pp TxU:H/M–0133; Reel 1087

8572. 1851–1–15 From William Rutter Dawes Wateringbury, Maidstone
Observed increased breadth of Saturn's ring on 25 and 29 Nov. 1850, confirmed by William Lassell on 3 Dec. Independent discovery of inner ring at Harvard University on 15 Nov. was announced in the *Times* on 3 Dec. Sent WD's notes to R.A.S. to establish independent discovery, but *R.A.S.M.N.* reported this erroneously.
ALS 4pp TxU:H/M–0160; Reel 1087

8573. 1851–1–15 From Henry Drury Harness Royal Mint
Obtained silver prices for 1838–39 and 1848, but prices for 1849–50 are not yet available.
ALS 2pp TxU:H/M–0251.7; Reel 1087

8574. 1851–1–17 From Adolph C. Petersen Altona
Is very appreciative of JH's letter of condolence at the death of H. C. Schumacher; also clarifies how Schumacher's responsibilities are being carried out.
ALS 2pp (German) TxU:H/M–0406; Reel 1087

8575. 1851–1–17 From Henry James Prescott Bank of England
Gold from America is not identified by its local source, so quantity of California gold can only be estimated. Contact W. D. Haggard or Matthew Marshall for information.
ALS 2pp TxU:H/M–0409; Reel 1087

8576. 1851–1–17 To Henry James Prescott 32 Harley St.
Needs to know how much gold from California has entered England.
ACCS 2pp TxU:H/L-0316; Reel 1054

8577. 1851–1–17 From James Wilson [Fontainville?], Westbury, Wiltshire
Happy to assist JH in researching fluctuations in global supplies of gold and silver. Lists sources of information.
ALS 4pp TxU:H/M–0583; Reel 1093

8578. [1851]–1–18 To Julia Margaret Cameron Harley St.
Asks that JC hold JH excused from appearing tomorrow,

because JH has a severe cold.
ALS 2pp Getty Center

8579. 1851–1–18 From William Rutter Dawes Wateringbury, Maidstone
Wishes WD had heeded JH's advice [in JH's 1851–1–18] regarding report of WD's discovery to R.A.S. Description of Saturn's inner ring.
ALS 4pp TxU:H/M–0161; Reel 1087

8580. 1851–1–19 To Robert Hunt 32 Harley St.
Does not believe JH can help RH in some delicate matter. JH talks about photographic chemistry, including passing on information from A. C. Becquerel.
ALS 4pp WT 68037.11

8581. 1851–1–20 From Benedetto Pistrucci Flora Villa, Bishopsgate, Chertsey
Will continue to write in Italian so JH can censor the translation of the letters when giving information to other officials at the Mint.
ALS 1p (Italian) TxU:H/M–0429.4; Reel 1093

8582. 1851–1–20 From Benedetto Pistrucci Flora Villa, Bishopsgate, Chertsey
Receives workload from H. D. Harness. Discusses the striking of the commemorative Waterloo Medal.
ALS 2pp (Italian) TxU:H/M–0429.3; Reel 1093

8583. 1851–1–21 From William Debonaire Haggard Bank of England
Sends information on imported California gold as JH requested.
L 1p TxU:H/M–0223; Reel 1087

8584. 1851–1–22 From William Debonaire Haggard Bank of England
Sent papers to Mint in response to JH's questions. Will soon send information on circulation of Bank of France.
ALS 1p TxU:H/M–0224; Reel 1087

8585. 1851–1–22 To Henry Drury Harness [Royal Mint]
Letter from Benedetto Pistrucci incorrectly quotes JH as stating that no Mint officer except Pistrucci is capable of handling dies. W. T. Brande may interpret this as lack of JH's confidence in Brande's abilities.
AdraftS 2pp TxU:H/M–0490.2; Reel 1093

8586. 1851–1–22 From Henry Drury Harness Royal Mint
Gave permission for F. M. Russell [wife of prime minister] to tour Mint on 30 Jan. Do not be concerned over Benedetto Pistrucci's remarks; JH's note accompanying Pistrucci's letter will be sufficient for Treasury to understand the situation. P.S.: Suggests letting Lady Russell see studio of [William] Wyon.
ALS 4pp TxU:H/M–0251.9; Reel 1087

8587. 1851–1–24 From Henry Drury Harness Royal Mint
Sends 1849–50 silver prices.
ALS 1p TxU:H/M–0251.10; Reel 1087

8588. 1851–1–24 To Benedetto Pistrucci 32 Harley St.
Wants BP's opinion on the number of casts that should be made for medal.
ACCS 2pp TxU:H/L-0307; Reel 1093

8589. 1851–1–24 From Charles Wood Downing St.
Asks JH and H. D. Harness to meet CW on Monday.
ALS 1p TxU:H/M–0591; Reel 1086

8590. 1851–1–25 From Henry Drury Harness Royal Mint
Two books arrived for the Mint library. When are JH and HH scheduled to visit Chancellor of Exchequer [Charles Wood]?
ALS 2pp TxU:H/M–0251.11; Reel 1087

8591. 1851–1–27 From Frances A. M. Russell Chesham Place
Obliged to JH for his note. Will be at the Mint next Thursday [2 Feb].
ALS 1p RS:HS 14.513

8592. 1851–1–28 From William Debonaire Haggard Bank of England
WH's report on California gold refers only to gold imported to London by Bank, not to gold arriving at other ports or gold imported by individuals.
ALS 2pp TxU:H/M–0225.1; Reel 1087

8593. 1851–1–28 From Henry Drury Harness Royal Mint
Benedetto Pistrucci insists on purchasing special press for medals. Does JH approve 'indulging the artist' and spending £5 for this equipment?
ALS 3pp TxU:H/M–0251.12; Reel 1087

8594. 1851–1–28 From Benedetto Pistrucci Flora Villa, Bishopsgate, Chertsey
Thankful for help in organizing his office. Receives instructions from H. D. Harness.
ALS 1p (Italian) TxU:H/M–0429.5; Reel 1093

8595. 1851–1–28 From James Wilson [Fontainville?], Westbury, Wiltshire
Encloses latest report from Bank of France.
ALS 2pp TxU:H/M–0584; Reel 1093

8596. 1851–1–29 From Thomas Maclear Royal Observatory, Cape of Good Hope
Congratulations on JH's appointment to mastership of Mint. New Kaffir war. Sent report of parallax of Alpha Centauri to R.A.S.
ALS 2pp TxU:H/M–0345; Reel 1087 (ALS?: Cape Archives/Maclear Papers File 100)

8597. 1851–1–31 From Henry Drury Harness Royal Mint
Family matters take HH away. Postpones discussion of Henry Bingley's letter.
ALS 1p TxU:H/M–0251.13; Reel 1087

8598. 1851–1–31 From Henry Drury Harness Royal Mint
Forwards letter from Henry Bingley. Will supply JH with records relative to Bingley before JH responds to it.
ALS 1p TxU:H/M–0251.14; Reel 1087

8599. 1851–2–1 From William Henry Bateson St. John's College, Cambridge
Bishop of Chester [John Graham] requests that JH attend 5 Feb. meeting of Cambridge University Commissioners.
ALS 2pp TxU:H/M–0057.3; Reel 1087

8600. [1851–2]. To William Whewell n.p.
Seeks to coordinate WW's and Frederick Beechey's contributions to JH's *Admiralty Manual.*
ALS 3pp TC, Camb. Add. Ms.a.207[81] (C: RS:HS 23.97)

8601. 1851–2–2 To William Whewell 32 Harley St.
Suggests how to reconcile WW's directions for tidal observation with those provided by Frederick Beechey for JH's *Admiralty Manual.* Thanks WW for and comments on his publication on political economy. Asks WW for views on

decimal coinage.
ALS 4pp TC, Camb. Add. Ms.a.207[83] (C: RS:HS 23.95)

8602. 1851–2–4 To Henry Bingley 32 Harley St.
Assures HB that transitional state of Assay department will soon be replaced by more permanent conditions.
Adraft 1p TxU:H/L-0092; Reel 1054

8603. 1851–2–4 To [John Tyndall?] 32 Harley St.
Asks that Louisa Herschel be admitted to attend James Forbes's lectures.
ALS 1p encl RI 164

8604. 1851–2–4 To William Whewell 32 Harley St.
Suggests revisions WW should make in his paper in the 2nd ed. of JH's *Admiralty Manual* so as to make it more consistent with Frederick Beechey's.
ALS 4pp TC, Camb. Add. Ms.a.207[84] (C: RS:HS 23.96)

8605. 1851–2–5 From James A. Gordon Naish House, Wrangle, near Bristol
JG's health. Rejoices that JH and family are residing at Harley Street. Doubts that Alfred Smee's 'reasoning machine' is any more attractive than was Smee's electro-biology. Results of recent tests for magnetic power in human body. Insufficient data to answer question of gold standard. JG's groom was injured in accident.
ALS 4pp TxU:H/M–0931; Reel 1083

8606. 1851–2–5 From Henry Drury Harness Royal Mint
As directed by Charles Wood, HH gave formal notice [of contract termination] to Moneyers and sent Wood's letter to G. F. G. Mathison. Forwarded JH's note to Henry Bingley.
ALS 1p TxU:H/M–0252.1; Reel 1087

8607. 1851–2–5 To Margaret Brodie Herschel 32 Harley St.
About the family, and the need for wine from the cellar at Collingwood.
ALS 4pp JHS 1.61

8608. 1851–2–5 From Charles Edward Trevelyan Treasury Chambers
Treasury commissioners have approved reform plan for Mint First step will be to terminate contract with company of Moneyers. Other reforms can be delayed. Make necessary preparations for Mint staff to take over those duties.
LS 1p *Return*, 19

8609. 1851–2–5 From Charles Wood Downing St.
Calls meeting for Friday of JH, H. D. Harness, the Moneyers, and [G. F. G. Mathison].
ALS 2pp TxU:H/M–0590; Reel 1086

8610. 1851–2–6 From Henry Drury Harness Royal Mint
Encloses copy of 5 Feb. 1851 letter from C. E. Trevelyan containing Treasury approval for 'part of' Mint reforms recommended by R. L. Sheil and JH. Treasury solicitors will serve notice of termination to Moneyers. HH sees difficulties if proposed changes are not all adopted simultaneously.
ALS 1p, encl C 1p TxU:H/M–0252.2; Reel 1087

8611. 1851–2–6 From Sir Robert Harry Inglis 7 Bedford Square
JH was elected member of Literary Society last night. Identifies other members. Invites JH and wife to dinner next Saturday.
ALS 2pp TxU:H/M–0289; Reel 1087

8612. 1851–2–7 To James A. Gordon 32 Harley St.
Expresses deep appreciation for JG's loan of £500 to JH in difficult period. Sorry to hear of JG's suffering.
ACCS 4pp TxU:H/L-0168; Reel 1054

8613. 1851–2–8 From Michael Faraday Royal Institution
Finds he made an error last night due to his bad memory. The Duke was asking for JH; hopes he met him after Richard Owen's discourse.
ALS 1p RS:HS 7.191

8614. 1851–2–8 From Francis Napier Millwall
Questions a statement in JH's *Outlines Astr.* regarding the light of Sirius. Also repeated in Alexander von Humboldt's *Cosmos*.
ALS 1p RS:HS 13.143

8615. 1851–2–9 From James A. Gordon Naish, Bristol
Obedience is better than sacrifice. Is progressing slowly. Hopes to visit JH at Harley St. soon.
ALS 2pp RS:HS 8.196

8616. 1851–2–10 From Henry Drury Harness Royal Mint
Legality of reforms at Mint may be challenged if Mint indenture is treated as an obsolete document. Details of appointments of Mint officers. Recommends establishing schedule of salaries for all officers and employees.
ALS 4pp † TxU:H/M–0252.4; Reel 1087

8617. 1851–2–10 From Henry Drury Harness Royal Mint
Forwards notices from George Maule [terminating contract with Moneyers and G. F. G. Mathison] for JH's signature. Will call meeting with Moneyers for Thursday.
ALS 1p TxU:H/M–0252.3; Reel 1087

8618. 1851–2–10 To Company of Moneyers Royal Mint
Official notice that Mint contract with Moneyers, begun on 8 Nov. 1820, will terminate three months from today.
LS 1p Return, 20

8619. 1851–2–11 To Samuel Hunter Christie 32 Harley St.
Believes Edward Sabine's paper [*R.S.P.T., 141* (1851), 123–] on periodicity of irregular magnetic disturbances important and certainly fit for publication.
ALS 2pp RS RR 2.213

8620. 1851–2–11 From Henry Drury Harness Royal Mint
Has read JH's letter to Charles Wood. Will notify Moneyers and George Maule.
ALS 2pp TxU:H/M–0252.5; Reel 1087

8621. 1851–2–14 From Charles Babbage Dorset St., Manchester Square
Sends deciphered code. Explains how CB solved it. [CB note dated 24 Mar. 1846: Technique for deciphering code.]
ALS 2pp, Babbage note 1p TxU:H/M–0030; Reel 1062

8622. 1851–2–14 From William Henry Bateson St. John's College, Cambridge
Bishop of Chester [John Graham] calls meeting of Cambridge University Commissioners for 18 Feb.
ALS 1p TxU:H/M–0057.4; Reel 1087

8623. 1851–2–14 From Charles Edward Trevelyan Treasury
Bring H. D. Harness and [J. W.] Morrison to meet CT next week to discuss matter of [Moneyers] Edward Enfield and Robert Rintoul and to fix salaries for Mint officers. Must refer JH's legal question to George Maule, Treasury solicitor.
LS 3pp TxU:H/M–0551.1; Reel 1093

8624. 1851–2–14 To Charles [Edward Trevelyan] Royal Mint
Salaries for Chief and Assistant Coiners, if these appointments are accepted by Edward Enfield and Robert Rintoul, respectively.
Adraft 3pp TxU:H/L-0481.2; Reel 1055

8625. 1851–2–15 To Charles Babbage 32 Harley St.
CB has 'decyphered' JH's puzzle. JH will not try to puzzle CB again.
ALS 2pp BL Aaa 37194.478

8626. 1851–2–16 To Charles Wood Royal Mint
Potential problems with establishing refinery at Mint. Remodeling coining department. Approval by Treasury is required before proposed changes can be made. Asks for formal schedule of salaries for Mint employees.
ACCS 2pp TxU:H/L-0481.3; Reel 1055

8627. 1851–2–19 From William Richard Hamilton 12 Bolton Row
Left at JH's door recent specimen of fossil breccia from [Aingo?] Island, where WH was detained 50 years earlier. Grateful for manner in which JH handled affair with Benedetto Pistrucci. Attests to Pistrucci's honesty and genius.
ALS 4pp TxU:H/M–0228; Reel 1087

8628. 1851–2–19 From Henry Pearse St. John's Rectory, Bedford
More about the arrangements for the pension for Mary Maclear [see HP's 1850–12–2].
ALS 4pp RS:HS 13.329

8629. 1851–2–20 From Henry Drury Harness Royal Mint
Informed George Maule of legal complications of old Mint indenture as applied to JH. Maule's solution was similar to what JH proposed to Charles Wood. Advises against telling Edward Enfield and Robert Rintoul about difficulties. Has not yet received statement from G. F. G. Mathison.
ALS 4pp TxU:H/M–0252.6; Reel 1087

8630. 1851–2–21 From Henry Drury Harness Royal Mint
G. F. G. Mathison wishes to consult colleagues before making a decision.
ALS 2pp TxU:H/M–0252.7; Reel 1087

8631. 1851–2–22 From Henry Drury Harness Royal Mint
Sends proposals for Mint reform.
ALS 1p TxU:H/M–0252.8; Reel 1087

8632. 1851–2–22 To Charles Edward Trevelyan Royal Mint
Legal notice of contract termination was served to company of Moneyers by Treasury solicitor on 13 Feb. Requests that reform of Mint Board, Mint office, Assay office, and Melting department be implemented *before* Mint staff assume duties of Moneyers in May. Encloses JH's plan for reorganizing departments. Offered positions in new structure to four Moneyers.
LS 1p, encl 1p *Return*, 20–1

8633. 1851–2–24 From Henry Drury Harness Royal Mint
Royal patent and warrant for Master to coin were discovered in R. L. Sheil's cellar. These are now safely at Mint.
ALS 1p TxU:H/M–0252.9; Reel 1087

8634. 1851–2–25 From Charles Edward Trevelyan Treasury
Chancellor of Exchequer [Charles Wood] wants to see JH and H. D. Harness tomorrow regarding arrangement with G. F. G. Mathison.
LS 2pp TxU:H/M–0551.2; Reel 1093

8635. 1851–2–26 From Henry Drury Harness Royal Mint
Forwards correspondence. Will meet JH at Chancellor of the Exchequer today.
ALS 1p TxU:H/M–0252.10; Reel 1087

8636. 1851–2–27 From Gilbert Farquhar Graeme Mathison Royal Mint
Offers arguments and evidence to support GM's protest to Charles Wood, Chancellor of Exchequer, on 7 Feb. 1851, against plan to abolish Mint refinery.
LS 2pp *Return*, 24–31

8637. 1851–2–27 From Charles Edward Trevelyan Treasury Chambers
Treasury commissioners will not authorize changes to Mint Board, mint office, or assay office until solicitors can determine legal ramifications. Changes in Melting department must await counter-proposal from G. W. G. Mathison. JH's plan to maintain other departments on present footing is approved.
LS 1p *Return*, 22

8638. 1851–2–27 To William Whewell 32 Harley St.
Asks approval for proposed changes in WW's revised contribution to JH's *Admiralty Manual.*
ALS 4pp TC, Camb. Add. Ms.a.207[85] (C: RS:HS 23.98)

8639. 1851–2–28 From Frederick W. Beechey Westbourne Crescent
Considers JH's form preferable to his own so will omit his. As JH is in London perhaps he would call and see his tidal stream theory, as he is free most days after 3 p.m.
ALS 4pp RS:HS 4.23

8640. 1851–2–28 From Henry Drury Harness Royal Mint
Encloses Treasury's remarks on statement by Charles Wood. British Museum requests coins from pyx chest for their collections. [JH annotation: Replied to Edward Hawkins on 28 Feb.]
ALS 3pp † TxU:H/M–0252.11; Reel 1087

8641. 1851–[3] From Jasper Atkinson 24 Portman Square
Will visit JH at Mint today. [JH annotation: JA wanted to know if government persists in plan to terminate Moneyers' contract.]
ALS 1p † TxU:H/M–0014; Reel 1062

8642. [1851]–[3] From Henry Rich Treasury
Spoke to Chancellor of the Exchequer [Charles Wood] about JH's memorandum.
ALS 2pp TxU:H/M–0443.4; Reel 1093

8643. 1851–3–2 To Mary Somerville Harley St.
JH's new position [Master of Mint] took him by surprise. People seem 'wild' over renewed agitation of Papal aggression issue. Describes the discovery of Saturn's new ring.
ALS 4pp RS:HS 16.356 (C: RS:HS 23.99)

8644. 1851–3–4 From Jasper Atkinson 24 Portman Square
Asks for appointment with JH tomorrow.
ALS 1p TxU:H/M–0015; Reel 1062

8645. 1851–3–4 From Hermann Schlagintweit Französische Strasse 7, Berlin
Sends regards from scientific men in Berlin and Munich, where HS and brother Adolph lectured after returning from

England. Alexander Humboldt will write to Duke of Northumberland, as JH suggested, to request financial aid for HS's research. May Humboldt use JH's name in this appeal?
ALS 3pp † TxU:H/M–0478.1; Reel 1093

8646. 1851–3–4 From Charles Edward Trevelyan Treasury
George Maule will help JH prepare draft of new organization for Mint to present to Privy Council. Inform Mint officers of their need to post security for their positions.
LS 2pp TxU:H/M–0551.3; Reel 1093

8647. 1851–3–5 From Henry Drury Harness Royal Mint
W. T. Brande has only one recommendation for vacant position in Die department. If JH approves this person, send note to HH.
ALS 2pp † TxU:H/M–0253.1; Reel 1087

8648. 1851–3–6 From Frederick W. Beechey Westbourne Crescent
Is enclosing the revised sheets to save time. Suggestions regarding these. Is sending his tidal paper to Airy for his perusal.
ALS 3pp RS:HS 4.24

8649. 1851–3–6 From Henry Drury Harness Royal Mint
Edward Enfield postponed decision until Moneyers' contract is settled. HH will forward G. F. G. Mathison's letter to Treasury and arrange for leasing refinery.
ALS 3pp TxU:H/M–0253.3; Reel 1087

8650. 1851–3–6 From Henry Drury Harness Royal Mint
Hopes to prepare necessary papers before Privy Council meets tomorrow.
ALS 1p TxU:H/M–0253.2; Reel 1087

8651. 1851–3–7 From George Biddell Airy Royal Observatory, Greenwich
The eclipse observation suggestions are now ready for printing [see GA's 1851–1–7]; would JH proof read them?
ACCS 2pp RGO 6.119.198

8652. 1851–3–7 From Charles C. F. Greville Buckingham Palace
[Formal proclamation] JH's request was read to Queen Victoria. Former warrant to JH, issued 17 Dec. 1850, is suspended. Effective this day, Treasury Commissioners and JH may take steps necessary to reform Mint.
LS 2pp *Return*, 22–3

8653. 1851–3–7 From Henry Drury Harness Royal Mint
Needs JH's signature for Mint letters. Needs copy of a statement by G. F. G. Mathison.
ALS 2pp TxU:H/M–0253.4; Reel 1087

8654. 1851–3–7 To Charles Edward Trevelyan Royal Mint
Forwards proposal from G. F. G. Mathison to retain refinery as part of Mint. JH still concurs with R. L. Sheil that refinery should be sold or leased. Mr. Haggard's report in 1848 shows that purified gold may be purchased more cheaply than it can be processed in Mint. Again requests permission to proceed with reforms to Mint Board and three offices.
LS 2pp *Return*, 23–4

8655. 1851–3–8 To George Biddell Airy 32 Harley St.
Will be happy to proofread the material [see GA's 1851–3–7].
ALS 3pp RGO 6.119.199

8656. [1851]–3–8 From Edward Enfield Royal Mint
Asks for written confirmation from JH that Treasury intends to abolish company of Moneyers.
ALS 4pp † TxU:H/M–0182; Reel 1087

8657. 1851–3–8 From Henry Drury Harness Royal Mint
Order [for Mint reform] has arrived from [Privy] Council. Royal warrant will follow soon.
ALS 1p TxU:H/M–0253.5; Reel 1087

8658. 1851–3–8 To Hermann Schlagintweit 32 Harley St.
As stated before, JH does *not* want JH's name used in Alexander Humboldt's appeal to Duke of Northumberland on behalf of HS and brother Adolph. Feels hurt by HS's refusal to accept this.
ACCS 1p TxU:H/M–0478.2; Reel 1093

8659. 1851–3–9 From George Biddell Airy Royal Observatory, Greenwich
Will send the eclipse observation suggestions in print, not manuscript [see JH's 1851–3–8]; the path of the eclipse has been misrepresented by J. F. Encke's carelessly identified latitudes.
ACCS 1p RGO 6.119.201

8660. 1851–3–10 From Adolphe Quetelet Brussels
Comments on exchange of papers with JH over the years, especially on electricity. Thanks JH for his assistance in AQ's research. Notes the loss of H. C. Schumacher. Hopes to come to England for the Great Exposition.
ALS 4pp (French) TxU:H/M-0910.2; Reel 1083

8661. 1851–3–10 To John Bird Sumner, Archbishop of Canterbury 32 Harley St.
The death of [Charles] König having created a vacancy at the British Museum, JH recommends that not only Richard Owen but also George Waterhouse be given very serious consideration for the position.
C 2pp RS:HS 23.101

8662. 1851–3–11 From [Benedetto Pistrucci] Royal Mint
Notes on the production of the Waterloo Medal, including the quality of medal, the color, and the design. Suggests a forger in Birmingham, John Robinson, be brought in to work on the production.
Benedetto Pistrucci note 3pp (Italian) TxU:H/M–0429.6; Reel 1093

8663. 1851–3–11 From Charles Edward Trevelyan Treasury
George Maule will prepare draft of Privy Council's warrant to reorganize Mint. Sends draft of Treasury minutes enabling JH to make all arrangements. Will return G. F. G. Mathison's statement when minutes have been approved. [JH annotation: Returned minutes with note that it and Council warrant invest all powers that JH needs.]
(A)LS 3pp † TxU:H/M–0551.4; Reel 1093

8664. 1851–3–12 From Alexander John Ellis 7 Apsley Place, Redland, Bristol
Points out error in JH's scale of solar system in *Outlines Astr.* and in JH's article on solar system in *Penny Cyclopaedia*. Returns papers borrowed from JH two years ago. [JH annotation: No error; AE misinterpreted it.]
ALS 4pp † TxU:H/M–0176; Reel 1087

8665. 1851–3–12 To Michael Faraday n.p.
Thanks MF for sending a 'healing liquid.'
C 1p RS:HS 23.102

8666. 1851–3–12 From Michael Faraday Royal Institution
Enclosing the tickets. Hopes the whiskey will be good.
ALS 1p RS:HS 7.192

8667. 1851–3–13 From Jasper Atkinson Moneyers' Office
Junior Moneyers Edward Enfield and Robert Rintoul decline offer from Chancellor of Exchequer [Charles Wood] for new positions in reorganized Mint. It would put them into inferior status.
ALS 2pp TxU:H/M–0016; Reel 1062

8668. 1851–3–13 From Henry Drury Harness Royal Mint
Jasper Atkinson refuses to take charge of the Coining Room. HH believes that Moneyers intend to subvert Mint reforms. Hopes to recruit Captain [Turner] for this Coining Room.
ALS 3pp TxU:H/M–0253.6; Reel 1087

8669. 1851–3–14 From Charles Edward Trevelyan Treasury Chambers
Treasury commissioners have read G. F. G. Mathison's letter. JH is now authorized to proceed with reforms to Mint Board, Mint office, Assay office, Melting department, and Refinery. Offer Refinery to Mathison for private refining business.
LS 1p *Return*, 32

8670. [1851]–3–14 To William Whewell 32 Harley St.
Suggests a slight change in WW's paper in JH's *Admiralty Manual* (2nd ed.).
ALS 2pp TC, Camb. Add. Ms.a.207[86]

8671. 1851–3–15 To Charles Edward Trevelyan Royal Mint
JH offered position of chief coiner to Moneyer Edward Enfield, and position of assistant coiner to Moneyer Robert Rintoul. Both declined these offers.
ACCS 2pp TxU:H/L-0481.4; Reel 1055 *Return*, 32

8672. 1851–3–15 To Edward Enfield Royal Mint
Encloses copy of JH's letter to C. E. Trevelyan announcing decisions by EE and Robert Rintoul to decline offices of Chief Coiner and Assistant Coiner, respectively.
AdraftS 2pp TxU:H/L-0147; Reel 1054

8673. 1851–3–17 From Richard Franklyn Royal Mint
Asks JH to 'modify the office offered to' RF's nephew, who will otherwise lose privileges and position under new organization of Mint.
ALS 5pp TxU:H/M–0925; Reel 1083

8674. 1851–3–18 From Charles Edward Trevelyan Treasury
Agrees to JH's division of Mint clerks into two classes according to salaries, with additional pay for those selected to be Melter and Assistant Coiner.
LS 2pp TxU:H/M–0551.5; Reel 1093

8675. 1851–3–19 From Henry Bingley Royal Mint
Withdraws HB's letter of 30 Sept. [1850] regarding Assay department, as JH requested.
ALS 1p TxU:H/M–0089; Reel 1087

8676. 1851–3–19 From Henry Bingley Royal Mint
Requests additional time to study JH's offer to HB of new position in reorganized Mint.
ALS 2pp TxU:H/M–0088; Reel 1087

8677. 1851–3–19 From Henry Bingley Royal Mint
Acknowledges JH's plan to change office of resident assayer.
ALS 1p TxU:H/M–0087; Reel 1087

8678. 1851–3–19 From Charles Edward Trevelyan Treasury
Chancellor of Exchequer [Charles Wood] wishes to see Moneyers Edward Enfield and Robert Rintoul in person on Saturday. Asks JH to be present also. P.S.: Adopted JH's suggestion to send Moneyers' memorial back to JH.
ALS 2pp TxU:H/M–0551.6; Reel 1093

8679. 1851–3–19 From Charles Edward Trevelyan Treasury Chambers
CE returns joint petition submitted by Moneyers. Treasury commissioners do not accept Moneyers' claim to be a corporation. Each Moneyer must submit an individual petition for compensation.
LS 1p *Return, 76* (1852), 44

8680. 1851–3–20 From W. Gibson Craig and R. M. Bellew Osborne House
[Formal proclamation] Queen Victoria revokes warrant of 17 Dec. 1850 to Master of Mint and issues new warrant directing JH to follow Queen's Order of 7 Mar. 1851 pertaining to reform of Mint.
LS 1p *Return*, 23

8681. [1851]–3–20 From Edward Enfield [Royal Mint]
EE and Robert Rintoul will meet JH at Downing St. next Saturday. Was this appointment with Chancellor of Exchequer [Charles Wood] arranged *before* EE requested one, or as a result of EE's request?
ALS 3pp TxU:H/M–0183; Reel 1087

8682. 1851–3–20 To Richard Franklyn 32 Harley St.
Difficulties arising from reform at Mint. Hopes new equality of rank for junior clerks will satisfy RF and [F. R.] Brande.
ALS 1p TxU:H/L-0756; Reel 1089

8683. 1851–3–20 From Henry Drury Harness Royal Mint
Letters from Treasury and from JH were sent to five Moneyers. Two notes were also sent to Edward Enfield and Robert Rintoul.
ALS 1p TxU:H/M–0253.8; Reel 1087

8684. 1851–3–20 To Company of Moneyers Royal Mint
Forwards letter from C. E. Trevelyan, requiring company of Moneyers to submit detailed statement to Treasury commissioners regarding Moneyers' claim to compensation.
LS 1p *Return*, 45

8685. 1851–3–20 From William Henry Stephenson Downing St.
Chancellor of Exchequer [Charles Wood] changed time of meeting on Saturday with JH, Robert Rintoul, and Edward Enfield.
ALS 2pp TxU:H/M–0501; Reel 1093

8686. 1851–3–20 From Charles Edward Trevelyan Treasury
[Privy Council's] warrant to reorganize Mint and letter regarding compensation for Moneyers were sent to JH yesterday. Copy of Henry Bingley's letter of 30 Sept. 1850 came to CT with note that Bingley had withdrawn it. [JH annotation: Returned Bingley's letter to Bingley.]
ALS 1p † TxU:H/M–0551.7; Reel 1093

8687. 1851–3–21 From Henry Bingley [Royal Mint]
Needs explanation of plan [to abolish Assay Office]. Define terms on which JH offers HB position as non-resident assayer.
ALS 3pp TxU:H/M–0090; Reel 1087

8688. 1851–3–21 From Alexander John Ellis 7 Apsley Place, Redland, Bristol
Apologizes for citing nonexistent error in JH's *Outlines Astr.* [see AE's 1851–3–12].
ALS 2pp TxU:H/M–0177; Reel 1087

8689. 1851–3–22 From Henry Drury Harness Royal Mint
Forwards papers from the Mint Board meeting.
ALS 1p TxU:H/M–0253.10; Reel 1087

8690. 1851–3–22 From Henry Drury Harness Royal Mint
Urges JH to advertise in London papers for candidates for Chief Coiner. This would show Mint officers that 'there is no difficulty in filling vacant places quickly with efficient people.' Writes advertisement describing position and benefits.
ALS 3pp TxU:H/M–0253.9; Reel 1087

8691. 1851–3–24 From Jasper Atkinson 24 Portman Square
Forwards letter signed by all Moneyers, with assurance that they will cooperate with Treasury plan.
ALS 1p TxU:H/M–0017; Reel 1062

8692. 1851–3–24 To Edward Everett 32 Harley St.
Thanks for EE's 'Orations and Speeches.' Recalls first meeting EE at Cambridge. Praises Harvard Observatory and W. C. Bond's discovery of an interior Saturnian ring. Notes that William R. Dawes had independently discovered this ring. Encourages EE to attend the Great Exhibition.
ALS 4pp Massachusetts Historical Society

8693. 1851–3–24 From Company of Moneyers Royal Mint
Disagree with Treasury commissioners. Offer several examples to support Moneyers' claim to corporate status. List each member's age, time of entering Mint, and money received from coining activities.
LS 1p *Return*, 45

8694. 1851–3–25 To Jasper Atkinson 32 Harley St.
Suggests change in procedure for 'applicants' to Royal Mint.
AdraftS 3pp TxU:H/L-0039; Reel 1054

8695. 1851–3–25 From Jasper Atkinson 24 Portman Square
Will take JH's note to Mint tomorrow.
ALS 1p TxU:H/M–0018; Reel 1062

8696. 1851–3–25 From Richard Franklyn Royal Mint
Corrects error in yesterday's note to JH reporting annual share of profits earned by each [Moneyer].
ALS 3pp TxU:H/M–0927; Reel 1083

8697. 1851–3–25 From Richard Franklyn Royal Mint
Grateful for JH's offer to place all junior clerks in Mint on equal salary, but this will still result in salary loss for RF's nephew.
ALS 4pp TxU:H/M–0926; Reel 1083

8698. 1851–3–25 From Henry Drury Harness Royal Mint
JH's presence is not necessary today. Encloses letter from G. F. G. Mathison.
ALS 1p TxU:H/M–0253.11; Reel 1087

8699. 1851–3–25 From Robert Rintoul Royal Mint
Encloses RR's individual claim for compensation.
ALS 2pp TxU:H/M–0446; Reel 1093

8700. 1851–3–26 To Gilbert Farquhar Graeme Mathison Royal Mint
Received GM's letter of 20 Mar., stating GM's position for official record. Regrets that GM interpreted conversations with JH and H. D. Harness in manner that led to so many misunderstandings. Asks for early reply from GM regarding office of Melter.
ACCS 3pp TxU:H/L-0249; Reel 1054

8701. 1851–3–26 From William Parsons (Lord Rosse) Castle Parsonstown
Happy to assist by writing to John Russell on behalf of [pension for J. R.] Hind. Results of WP's recent observations. Recommends Thomas Grubb as candidate for Chief Coiner at Mint.
ALS 7pp TxU:H/M–0464; Reel 1093

8702. 1851–3–26 To Robert Rintoul Royal Mint
Treasury commissioners insist on dealing with Moneyers as individuals, not as corporation. Returns RR's memorial and suggests that RR delete reference to Company of Moneyers.
AdraftS 2pp TxU:H/L-0338; Reel 1054

8703. 1851–3–27 From William Brockedon Devonshire St.
Thanks for JH's suggestions, which will be incorporated. Comments on various patents.
ALS 4pp RS:HS 5.105

8704. 1851–3–27 From Gilbert Farquhar Graeme Mathison Royal Mint
Concerned that GM misinterpreted meaning of JH's 22 Mar. letter. Apologizes to JH and H. D. Harness. Troubled by fundamental changes in Mint. Withdraws GM's earlier letter.
ALS 3pp TxU:H/M–0389; Reel 1087

8705. 1851–3–27 To Robert Rintoul 32 Harley St.
Advises RR to keep individual claim for compensation ready, but wait to see if Treasury accepts joint claim by Company of Moneyers.
ACCS 1p TxU:H/M–0447.2; Reel 1093

8706. 1851–3–27 From Robert Rintoul Royal Mint
Made changes, as JH suggested, to RR's individual claim for compensation, but other Moneyers advised RR to participate in their joint memorial to Treasury. Asks JH's opinion of best course.
ALS 3pp TxU:H/M–0447.1; Reel 1093

8707. 1851–3–28 From Henry Bingley Royal Mint
Has considered JH's proposal [of new position in reorganized Mint] and encloses HB's official answer.
ALS 2pp TxU:H/M–0091; Reel 1087

8708. 1851–3–28 From Carlo Matteucci Villa di Corliano, Pisa
Is sending JH a book recounting some of CM's researches. Comments on various factors influencing conductance in carrying out experiments in electricity and telegraphy. [Letter illegible in parts.]
ALS 8pp (French) RS:HS 12.305

8709. 1851–3–29 From George Biddell Airy Royal Observatory, Greenwich
About the location of Julius Caesar's landing in Britain and the location of subsequent battles.
ACCS 4pp RGO 6.466.197

8710. 1851–3–29 To Henry Bingley [Royal Mint]
Changes offer that JH made to HB in note to H. D. Harness.
AmemS 1p TxU:H/L-0093; Reel 1054

8711. 1851–3–29 From Henry Drury Harness Royal Mint
Sends proposed letter expressing regret that Henry Bingley refuses to accept the office of resident Assayer. Will arrange meeting with Mr. Blake. Received two testimonials for William Buckle. Professor Johnson is 'one of the assayers.' Another £100,000 will be sent to Cape of Good Hope. [JH annotation: Adapted HH's letter and sent it to Bingley on 29 Mar.]
ALS 4pp † TxU:H/M–0253.12; Reel 1087

8712. 1851–3–29 To Richard Sheepshanks 32 Harley St.
Writing in support of an application for a pension for John Hind.
ALS 3pp RS:HS 16.106 (C: RS:HS 25.9.70)

8713. 1851–3–29 From Charles Edward Trevelyan Treasury Chambers
Treasury commissioners appoint JH and Henry Rich to investigate Moneyers' claim to compensation for loss of earnings due to reform at Mint.
LS 2pp *Return*, 45–6

8714. 1851–3–31 From George Biddell Airy Royal Observatory, Greenwich
Letter of clarification about the enclosed proof of the eclipse suggestions [see GA's 1851–3–9].
ACCS 3pp RGO 6.119.202

8715. 1851–4–1 From John Couch Adams St. John's College, Cambridge
Stating he has approached certain Government officials in order to obtain a pension for John Hind.
ALS 2pp RS:HS 1.7

8716. 1851–4 To Henry Bingley Royal Mint
JH cannot give definite date for HB's termination in Assay Office. This will be delayed for as long as gold is being imported for coinage.
ACCS 3pp TxU:H/L-0094; Reel 1054

8717. 1851–4 From William Brockedon Devonshire St.
Will have some power punches made. Gives Mr. Hodges's notes on JH's points. Very busy at the moment with the exhibition.
ALS 4pp RS:HS 5.106

8718. 1851–4–2 To George Biddell Airy n.p.
Note to accompany return of printed text [see GA's 1851–3–31], with one annotation enclosed.
LACC 1p RGO 6.119.204

8719. 1851–4–2 To George Biddell Airy n.p.
Encourages GA to write a letter in support of time off for John Hind.
ALS 3pp RGO 6.248.6

8720. 1851–4–3 From George Biddell Airy Royal Observatory, Greenwich
Needs clarification of JH's annotation [see JH's 1851–4–2].
ACCS 2pp RGO 6.119.206

8721. 1851–4–3 From Henry Drury Harness Royal Mint
Encloses draft of proposed letter to Treasury.
ALS 1p, JH note 1p TxU:H/M–0254.1.1; Reel 1087

8722. 1851–4–4 To George Biddell Airy n.p.
Responds, clearing up the misunderstood instruction [see GA's 1851–4–3].
ALS 2pp RGO 6.119.207

8723. 1851–4–4 From Edward Enfield Sheffield House, Kensington
Encloses official response by EE to offer from Chancellor of Exchequer [Charles Wood]. Asks JH to see that memorial from EE [and Robert Rintoul] is read by [Henry Rich].
ALS 3pp TxU:H/M–0184.1; Reel 1087

8724. 1851–4–5 To Edward Enfield 32 Harley St.
Received EE's letters [see EE's 1851–4–4] and will forward these to Henry Rich.
ACCS 1p (on E. Enfield's 1851–4–4) TxU:H/M–0184.2; Reel 1087

8725. 1851–4–5 To William Parsons (Lord Rosse) n.p.
[Extract] Appeals [to R.S.L.] for financial support to J. R. Hind, whose health is failing. Encloses letter and sketch from Hind, certificate for Dr. Hall, and letter from Richard Sheepshanks.
ACC 1p inc TxU:H/L-0342; Reel 1054 (C: RS:HS 25.15.15)

8726. 1851–4–5 To Henry Rich Royal Mint
Sends proposals for HR's approval. Encloses excerpt from Edward Enfield's letter concerning salary raise.
AdraftS 2pp TxU:H/L-0322 & 0323; Reel 1054

8727. 1851–4–5 To Robert Rintoul 32 Harley St.
If RR still wishes, RR may now submit individual claim for compensation to JH and Henry Rich for consideration.
ACCS 1p TxU:H/M–0447.3; Reel 1093

8728. 1851–4–5 To Charles Edward Trevelyan Royal Mint
Forwards letter from Edward Enfield, who feels that Enfield's compensation as terminated Moneyer should not be diminished for declining offer of position in new Mint establishment.
ACCS 1p TxU:H/L-0481.5; Reel 1055 *Return*, 32

8729. 1851–4–6 To Margaret Brodie Herschel London
Unhappy with matters at the Mint; JH talks about the activities of the children, and about his involvement with the Great Exhibition. [Most of the letter written on 1851–4–7.]
ALS 4pp JHS 6.15

8730. 1851–4–7 To [Jasper Atkinson] Royal Mint
Landed estates held in common by Company of Moneyers may affect their claims to compensation.
AdraftS 1p TxU:H/L-0040; Reel 1054

8731. [1851]–4–7 From Henry Rich Treasury
Will take JH's memorandum to Chancellor of Exchequer [Charles Wood] for consideration.
ALS 2pp TxU:H/M–0443.5; Reel 1093

8732. 1851–4–8 From Michael Faraday Royal Institution
No doubt he is aware of F. A. Nobert's work on glass. Has received from him a paper, which he intends communicating to the R.S.L. Thinks JH may like to see it.
ALS 1p RS:HS 7.193 Williams, *Selected Correspondence*, 2, 620

8733. 1851–4–8 From [William] Law Downing St.
Would JH forward the enclosed letter to [J. R.?] Hind as he cannot find Hind's address.
AL 3pp RS:HS 11.465

8734. 1851–4–8 From Robert Rintoul Royal Mint
Encloses RR's individual claim for compensation, to be forwarded to Treasury.

ALS 2pp, encl 9pp TxU:H/M–0448; Reel 1093 & TxU:H/M–1211; Reel 1060

8735. 1851–4–8 From Charles Edward Trevelyan Treasury Chambers
Treasury commissioners will not make exception in case of Edward Enfield, who must accept position in reformed Mint or lose compensation for termination of Moneyers' contract.
LS 1p *Return*, 32

8736. 1851–4–9 From George Biddell Airy Royal Observatory, Greenwich
Has sent testimonial regarding John Hind to William Parsons (3rd Earl of Rosse).
ALS 3pp RS:HS 1.220 (C: RGO 6.248.11)

8737. 1851–4–9 To George Biddell Airy 32 Harley St.
John Hind has been granted time off with pay to recover his health [see JH's 1851–4–2].
ALS 1p RGO 6.248.12

8738. [1851]–4–9 From Henry Rich Treasury
Returns document to JH. Has not yet discussed it with Chancellor of Exchequer [Charles Wood].
ALS 1p TxU:H/M–0443.6; Reel 1093

8739. 1851–4–10 To William Buckle Royal Mint
WB is accepted as new assistant Coiner of Mint. Official appointment will follow.
ALS 3pp TxU:H/L-0114; Reel 1054

8740. 1851–4–10 From Henry Drury Harness Royal Mint
Gave enclosed letter to William Buckle. Schedule, as altered by Mr. Kerr, will return from printer today.
ALS 1p TxU:H/M–0254.2; Reel 1087

8741. 1851–4–10 From John R. Hind 22 Grove Rd., St. John's Wood
Is grateful for JH's efforts in bringing his case before the government. Hopes to prove useful to the cause of science if his health permits.
ALS 2pp RS:HS 9.351

8742. 1851–4–10 To Mary Somerville n.p.
Poem, entitled 'Intelligence,' celebrating creation of mankind.
C 1p RS:HS 19.183

8743. 1851–4–11 From C. P. Barrett Eton
Wishes to improve Slough and encloses a map of proposed alterations for JH's comments. Would like to purchase a portion of JH's garden for a new road.
ALS 3pp RS:HS 5.20

8744. 1851–4–12 From Henry Drury Harness Royal Mint
Charles Sterry accepted office [as assistant to assay master].
ALS 1p TxU:H/M–0254.4; Reel 1087

8745. 1851–4–12 From Henry Drury Harness Royal Mint
H. W. Field expects to receive Charles Sterry's answer today. Mr. Johnson's reply has not arrived. Percival Johnson stopped to inquire about position as assayer. Forwards letter sent from Mr. Stirling to W. H. Barton.
ALS 2pp TxU:H/M–0254.3; Reel 1087

8746. 1851–4–14 To Charles Edward Trevelyan Royal Mint
Reports reforms made so far in Coining, Melting, and Assaying departments. Hired William Buckle, engineer from Birmingham, as assistant coiner. Robert Mushet will accept position of senior clerk and melter after G. F. G. Mathison's contract is terminated. Office of resident assayer, declined by Henry Bingley, was accepted by H. W. Field, who will assume duties after J. M. Beckwith, Master's assayer, retires. Position of junior clerk and assistant assayer, offered to Henry Bingley and declined, was accepted by Charles Sterry. Duties of nonresident assayers were accepted by W. A. Miller, Thomas Graham, and T. Johnson.
LS 1p *Return*, 33

8747. 1851–4–15 To C. P. Barrett 32 Harley St.
Will not sell any portion of Slough. Disapproves CB's plan [see Barrett's 1851–4–11] to put road through its garden.
AdraftS 2pp † RS:HS 19.142

8748. 1851–4–15 From Henry Bingley Royal Mint
Accepts JH's decision to abolish Assay Office. Will not compress HB's response to this 'distressing and anxious subject.'
ALS 3pp TxU:H/M–0092; Reel 1087

8749. 1851–4–15 From Michael Faraday Royal Institution
The accompanying book was addressed to MF but inscribed to JH, so he considers it belongs to JH.
ALS 1p RS:HS 7.194

8750. 1851–4–17 From Henry Bingley Brighton
Will rewrite HB's claim to Treasury as JH recommended.
ALS 1p TxU:H/M–0093; Reel 1087

8751. 1851–4–17 To Henry Bingley Royal Mint
Regrets abolition of Assay office. Will transmit to Treasury HB's letter claiming compensation, but JH feels HB should provide more precise information. HB misunderstood purpose of JH's letter.
AdraftS 4pp TxU:H/L-0095; Reel 1054

8752. 1851–4–18 From James A. Gordon 8 Park St.
Regrets he was unable to dine at Harley St. Regarding the stars and the earth. Is grateful for the poetry JH sent. Hopes he will visit them this Spring.
ALS 4pp RS:HS 8.202

8753. 1851–4–22 From Richard Franklyn Moneyers' Office
Objects to presence of visitors in Moneyers' department at Mint. Proposes one day each week when coinage could be suspended and visitors permitted.
ALS 4pp TxU:H/M–0928; Reel 1083

8754. 1851–4–22 To George Maule Royal Mint
Prepare legal notice for termination of G. F. G. Mathison's contract as Melter.
ACCS 1p TxU:H/L-0253; Reel 1054

8755. [1851 to 1855]–4–22 From George Peacock Deanery, Ely
Forwards letter from William Whewell. Referees seem to agree with Humphrey Lloyd's proposals. Busy with [Cambridge University] Commission business. Colleges are resisting proposed changes.
ALS 3pp TxU:H/M–0425; Reel 1093

8756. 1851–4–22 From Charles Edward Trevelyan Treasury
Consider CT's official letter of 16 Apr. suspended. Fee will probably be abolished and H. W. Field's salary increased to £500. Received today's letter from JH with enclosure by H. D. Harness regarding conduct of Moneyers. Will forward these to Chancellor of Exchequer [Charles Wood].
ALS 2pp TxU:H/M–0551.8; Reel 1093

8757. 1851–4–22 To Charles [Edward Trevelyan] Royal Mint
JH needs someone in higher authority to admonish moneyers and to confirm Mint master's and deputy's right to enter any part of Mint.
ACCS 3pp TxU:H/L-0481.6; Reel 1055

8758. 1851–4–23 To Samuel Hunter Christie n.p.
Reports and offers suggestions on a paper [R.S.L. *Proceedings, 6*, 96–] by W. H. Sykes on depressions of wet bulb thermometer at Ahmednaggur.
ALS 4pp RS RR.2.235

8759. 1851–4–24 To Gilbert Farquhar Graeme Mathison Royal Mint
Announces that Robert Mushet accepted office of Melter, which GM declined. When does GM want to terminate contract as Melter and Refiner? Draw up GM's claims for compensation and JH will forward these to Treasury.
AdraftS 1p TxU:H/L-0250; Reel 1054

8760. 1851–4–25 From Gilbert Farquhar Graeme Mathison Royal Mint
Just received JH's official letter. Will meet with H. D. Harness next week. Prefers to have GM's contract legally end on 30 Apr.
ALS 2pp TxU:H/M–0390; Reel 1087

8761. 1851–4–26 From William Henry Bateson St. John's College, Cambridge
Cambridge University Commissioners will meet on 13 and 14 May.
ALS 2pp TxU:H/M–0057.5; Reel 1087

8762. 1851–4–26 From Henry Bingley n.p.
HB's memorial, detailing effect on HB and son of JH's plan to abolish Assay Office, should be sent to Treasury.
Abs (by JH) 2pp TxU:H/M–0094; Reel 1087

8763. 1851–4–26 From Henry Bingley Royal Mint
Claims greater compensation from Treasury for loss of earnings, residence, and future employment of HB's son, due to abolition of office of Queen's assay master, held by HB's family for more than 100 years. Considers earlier offers of compensation unacceptable.
LS 2pp *Return*, 48–9

8764. 1851–4–26 To Henry Bingley Royal Mint
JH forwarded HB's claim for compensation to Treasury.
AdraftS 2pp TxU:H/L-0096; Reel 1054

8765. 1851–4–26 From Charles Edward Trevelyan Treasury
Decision by Chancellor of Exchequer [Charles Wood] regarding admissions to Coining rooms. Fee for assayer H. W. Field will now come from Mint accounts, not from Field personally.
LS 3pp TxU:H/M–0551.9; Reel 1093

8766. 1851–4–26 From Charles Edward Trevelyan Treasury Chambers
Treasury commissioners approved JH's appointments and retirement of J. M. Beckwith [see JH's 1851–4–14]. Give legal notice of contract termination immediately to G. F. G. Mathison. Henry Bingley's claim for compensation will not be considered until JH's evaluation of it is completed.
LS 1p *Return*, 34

8767. 1851–4–28 From Henry Drury Harness Royal Mint
Sent W[illiam] Lalor home, because JH did not return. Encloses proposed letter to Richard Franklyn.
ALS 1p TxU:H/M–0254.5; Reel 1087

8768. 1851–4–28 To Charles Edward Trevelyan [Royal Mint]
Encloses two letters. JH reached agreement with Richard Franklyn and Edward Enfield regarding Master's right of access to moneyers' machinery in future.
AdraftS 2pp TxU:H/L-0481.7; Reel 1055

8769. 1851–4–29 From George Biddell Airy Royal Observatory, Greenwich
Comments on a contract to be used by the Mint, for its coin manufacturers.
ACCS 4pp RGO 6.3.171

8770. [1851–4]–29 To Archdeacon Julius Charles Hare Mint
Enclosed materials will admit JCH and his friends to the Mint.
ALS 1p Bodleian Library, Oxford

8771. 1851–4–30 From George Biddell Airy Royal Observatory, Greenwich
Reports on observations made of one of the Mint's potential contractors [see GA's 1851–4–29].
ACCS 2pp RGO 6.3.174

8772. 1851–4–30 To Gilbert Farquhar Graeme Mathison Royal Mint
Formal notice of termination of GM's contracts as first clerk, melter, and refiner at the Mint.
LS 1p *Return*, 34

8773. [1851–5 to 1851–10] From William Thomas Brande Royal Mint
Grateful for JH's reply. Will wait for [Henry] Rich's answer to JH's references.
ALS 1p TxU:H/M–0116.3; Reel 1087

8774. 1851–5–1 From Sir Robert Harry Inglis 7 Bedford Square
Asks JH to pay brief call at RI's home on way to London.
ALS 1p † TxU:H/M–0290; Reel 1087

8775. [1851–5 or later] To [?] [Royal Mint]
P.S.: Henry Bingley asks Treasury for further clarification on compensation for Bingley's son.
Adraft 1p inc TxU:H/L-0609; Reel 1089

8776. 1851–5–2 To William Buckle Royal Mint
After their contract expires on 13 May, Moneyers will continue their operations at Mint until importation of bullion ceases, at which time WB's duties as assistant Coiner will commence. Living accommodations are not yet available at Mint.
ALS 4pp TxU:H/L-0115; Reel 1054

8777. 1851–5–5 To Jasper Atkinson Royal Mint
JH received request for employment from workman employed by JA. Forwards this to JA.
AdraftS 1p TxU:H/L-0041; Reel 1054

8778. 1851–5–5 To Charles Edward Trevelyan Royal Mint
Forwards Henry Bingley's appeal to Treasury for compensation for loss of Bingley's office as Queen's assay master. Notes previous offers that Bingley declined, including position as resident assayer in reorganized Mint. Points out that Bingley's son worked in Mint for past two years without pay in preparation for succeeding his father.

Compliments Bingley's past work in raising Mint standards for purity of gold.
LS 2pp *Return*, 49–50

8779. 1851–5–6 To Charles Babbage n.p.
Would like a reference from him regarding a candidate for a position at the mint. Would like him and David Brewster to dine with him.
ALS 3pp RS:HS 2.320

8780. 1851–5–6 To Margaret Brodie Herschel n.p.
Describes JH's day at Mint and at home. Reading H. V. Regnault's 'Chemistry.' Met with Richard Grenville about 'miserably knotty affair' with Benedetto Pistrucci. Inspected [W. A.] Miller's improved assay process at R.S.L. Arranged with Sir [Charles] Fremantle to employ twenty Custom House clerks at Mint temporarily. Arranged lease of refinery to Rothschild. Too busy to join daughters and Miss Stewart for dinner and opera. Dined at Duke of Somerset's with Charles Babbage, [Holland?], and David Brewster.
ALS 4pp TxU:H/L-0542.2; Reel 1053

8781. 1851–5–6 From George Maule Treasury
Returns copy of JH's letter to Queen, with suggested alterations. Sign and return this, so GM can officially present matter to Privy Council tomorrow.
ALS 2pp TxU:H/M–0349; Reel 1087

8782. 1851–5–6 or earlier From William Henry Fox Talbot [Lacock Abbey]
Refers to process which produces pictures that are negative or positive depending upon the light. Hopes to use it to photograph lunar landscape. Wants to call it 'amphitype' if JH agrees.
ALS 3pp RS:HS 17.321

8783. 1851–5–6 To William Henry Fox Talbot 32 Harley St.
Agrees to WT's use of name 'amphitype' [see WT's 1851–5–6 or earlier].
ALS 3pp Science Museum Talbt 1/40 (C: RS:HS 23.103)

8784. 1851–5–9 From Georg A. Erman Berlin
Sends a paper. Comments on temperature at low altitudes and listing papers known on this subject.
ALS 3pp (German) RS:HS 7.103

8785. 1851–5–10 From George Biddell Airy Royal Observatory, Greenwich
A note to accompany the forwarding of a packet of 'Eclipse Observation Suggestions' [see JH's 1851–4–4].
ACCS 1p RGO 6.119.208

8786. 1851–5–12 To Richard Franklyn [Royal Mint]
Assurance that compensation for Moneyers, whose contract terminates tomorrow, will not be affected by Moneyers' continuing to coin on temporary basis.
ACCS 2pp (†by JH) TxU:H/L-0160; Reel 1054

8787. 1851–5–13 From William Henry Bateson [No. 10] Downing St.
Cambridge University Commissioners meet tomorrow.
AL 1p TxU:H/M–0056; Reel 1087

8788. 1851–5–13 From George Cornewall Lewis Treasury Chambers
Treasury commissioners appoint Henry Rich to assist JH in investigating claim for compensation by Henry Bingley, due to abolition of office of Queen's assay master.
LS 1p *Return*, 50

8789. 1851–5–16 From George Biddell Airy Royal Observatory, Greenwich
Describes GA's journey to Gothenburg (Sweden), the purpose of which was to observe a solar eclipse.
ACCS 3pp RGO 6.373.167

8790. 1851–5–16 To George Biddell Airy London
Is trying to arrange to have GA join JH and some friends for dinner.
ALS 4pp RGO 6.373.165

8791. 1851–5–16 From William Benjamin Carpenter 6 Regents Park Terrace
Would like JH's opinion on an article in the *R.S.P.T.* Has had good reports of his own physiological works. New editions are forthcoming and he will send them to JH.
ALS 4pp RS:HS 5.188

8792. 1851–5–16 From Henry Drury Harness Royal Mint
Encloses letter from collector of customs at Cork and HH's proposed letter to Treasury regarding Cork memorial. Charles Elouis could not find information on 'token coinage' for Bank [of England] in the early part of the century.
ALS 2pp TxU:H/M–0255.1; Reel 1087

8793. 1851–5–16 To Henry Rich 32 Harley St.
Calculates coinage and gold use by Mint between 1801 and 1850. Asks HR to check HR's calculations.
ACCS 3pp TxU:H/L-0324; Reel 1054

8794. 1851–5–17 From Henry Drury Harness Royal Mint
Encloses letter regarding claim by Henry Bingley. Charles Elouis found new information about 'token coinage' for Bank of England.
ALS 1p TxU:H/M–0255.2; Reel 1087

8795. 1851–5–19 From John R. Hind 22 Grove Rd., St. John's Wood
Has discovered another planet [Irene]. Gives readings. Suggests source of name.
ALS 3pp RS:HS 9.352

8796. 1851–5–19 To Henry Rich 32 Harley St.
Discusses calculations of Mint gold usage from 1801 to 1850. Also plans to calculate amount of silver used from Bank of England.
ACCS 4pp TxU:H/L-0325; Reel 1054

8797. 1851–5–20 To George Bishop 32 Harley St.
Suggesting the name 'Irene' for the new planet.
draft 4pp RS:HS 4.150 & 23.104

8798. 1851–5–20 To Henry Rich [32 Harley St.]
Proposals for salaries of Edward Enfield and Robert Rintoul.
Adraft 4pp TxU:H/L-0326; Reel 1054

8799. 1851–5–21 From Henry Drury Harness Royal Mint
Discussion of contract with [Joshua] Field.
ALS 1p TxU:H/M–0255.3; Reel 1087

8800. 1851–5–21 From Charles Edward Trevelyan Treasury
Tenders [for bids] should be let soon. Chancellor of Exchequer [Charles Wood] will see JH and H. D. Harness this afternoon.
ALS 2pp TxU:H/M–0551.10; Reel 1093

8801. [1851–5–21 or later] To [Charles Edward Trevelyan] [Royal Mint]
Under old system of political appointments, master of Mint was not eligible for superannuation fund. Asks Treasury to

modify Treasury minutes of 20 May 1851 to permit JH to receive this benefit. Compares superannuation rates before and after 5 Aug. 1829.
Adraft 2pp TxU:H/M–0353.2; Reel 1087

8802. 1851–5–22 From John R. Hind 22 Grove Rd., St. John's Wood
Thanks for the name [Irene]; will lose no time in making it known. Observed a faint nebulous light around the planet.
ALS 3pp RS:HS 9.353

8803. [1851–5–22] To William Whewell 32 Harley St.
Invites WW to meet with the two representatives of French science in the Jury of Philosophical Instruments [for the Great Exhibition] and to dine with Lord Rosse [William Parsons].
ALS 2pp TC, Camb. Add. Ms.a.207[88]

8804. 1851–5–23 From Jasper Atkinson Royal Mint
Moneyers are extremely anxious. Chancellor of Exchequer [Charles Wood] announced in Apr. 1850 that government would abolish Company of Moneyers, whose contract terminated on 13 May. But still there is no arrangement for compensation.
ALS 3pp TxU:H/M–0019; Reel 1062

8805. 1851–5–23 From George Bishop Regents Park
Thanks for the name for the new planet [Irene]. J. R. Hind has written to the *Times* concerning it.
ALS 2pp RS:HS 4.151

8806. 1851–5–23 To William Whewell n.p.
'[J. R.] Hind and [George] Bishop are delighted with your name Eirene (vulgo Irene) for the new planet[oid]....'
ALS 1p TC, Camb. Add. Ms.a.207[87] (C: RS:HS 23.201)

8807. 1851–5–26 From Annibal de Gasparis Naples
Announcing his discovery of another new planet. Gives readings.
ALS 1p (French) RS:HS 8.44

8808. 1851–5–26 From Henry Drury Harness Royal Mint
Silver was purchased for Mint on basis of [P. N.] Johnson's assay. Such purchases usually require assay by Henry Bingley. Records show that Bingley entered Mint in 1820.
ALS 2pp TxU:H/M–0255.4; Reel 1087

8809. 1851–5–26 From Henry Anthony Jeffreys Hawkhurst Parsonage
Has purchased a pendulum from London and would like JH's permission to suspend it in his barn. Was not entirely satisfied with the experiment at the Polytechnic.
ALS 3pp RS:HS 10.292

8810. 1851–5–27 To [Henry Bingley] 32 Harley St.
Please clarify [HB]'s claim for compensation by distinguishing income derived from bullion assays for coinage and that for general market. By whose authority were constructed those stables and coach house for which [HB] claims compensation?
AdraftS 3pp TxU:H/L-0615 & L-0648; Reel 1089

8811. 1851–5–27 From William Benjamin Carpenter 6 Regents Park Terrace
Sending a copy of the 3rd edition of his treatise [*Principles of General and Comparative Physiology*]. Points out differences in the editions and draws attention to some novelties.
ALS 3pp RS:HS 5.189

8812. 1851–5–27 From Henry Drury Harness Royal Mint
[Joshua] Field appears satisfied with Mint contract.
ALS 1p TxU:H/M–0255.5; Reel 1087

8813. 1851–5–27 To R. L. \ Harley St.
Regrets not having had time to examine George Cayley's views on figure of the Earth.
ALS 2pp St. John's College, Cambridge

8814. 1851–5–28 To Jasper Atkinson 32 Harley St.
Does not yet have word from government regarding claims for compensation made by Company of Moneyers.
AdraftS 1p TxU:H/L-0042; Reel 1054

8815. 1851–5–29 From Jasper Atkinson Royal Mint
Thanks for JH's assurance that arrangements for Moneyers' compensation will soon be made. Hopes that government will be liberal.
ALS 2pp TxU:H/M–0020; Reel 1062

8816. 1851–5–29 From Henry Bingley Royal Mint
There are no records comparing private and official assays performed at Mint. History of coach house and stable now in HB's possession.
ALS 3pp TxU:H/M–0095; Reel 1087

8817. 1851–5–29 From Henry Drury Harness Royal Mint
Sends document from Maudslay and Field for JH to examine and forward to the Chancellor of the Exchequer.
ALS 1p TxU:H/M–0255.6; Reel 1087

8818. 1851–5–30 To William Benjamin Carpenter 32 Harley St.
Thanks WC for the work he has completed on JH's behalf.
ALS 2pp WT 67597.5

8819. 1851–5–31 From Henry Drury Harness Royal Mint
Mint holiday for Queen's birthday. Will arrange with Mr. Peligot for inspection.
ALS 1p TxU:H/M–0255.7; Reel 1087

8820. 1851–6–1 From William Richard Hamilton 12 Bolton Row
Do not make copies of C. K. J. Bunsen's letter to WH regarding Waterloo Medal, except for JH's use or at request of cabinet ministers. If Benedetto Pistrucci had been treated properly by JH's predecessors at Mint, Waterloo Medal could have been finished 15 years ago.
ALS 3pp TxU:H/M–0229; Reel 1087

8821. 1851–6–2 From George Biddell Airy Royal Observatory, Greenwich
Asks for JH's assistance in inviting foreign astronomers to Visitation Day at the Royal Observatory.
ACCS 2pp RGO 6.10.339

8822. 1851–6–2 From Henry Drury Harness Royal Mint
Benedetto Pistrucci should not exhibit [Waterloo] medal on his own, but 'by permission of the Master of the Mint.' Mr. Peligot did not show up at Mint on Saturday. Forwards two letters, one from [J. D.] Forbes.
ALS 3pp TxU:H/M–0256.1; Reel 1087

8823. 1851–6–2 To Henry Rich 32 Harley St.
Henry Bingley wants coach house and stables owned by Bingley's father, who occupied position of King's Assayer before Bingley. Bingley's private assaying practice for Bank of England. Distribution of gold bullion by Bank of England.
ACCS 4pp TxU:H/L-0327; Reel 1054

8824. 1851–6–3 From Henry Drury Harness Royal Mint
Maudslay, Sons, and Field will not lower their bid offer [for contract coinage work].
ALS 1p TxU:H/M–0256.2; Reel 1087

8825. 1851–6–3 To Henry Rich 32 Harley St.
Calculates salary budget for Jasper Atkinson, Edward Enfield, and Robert Rintoul.
AdraftS 3pp TxU:H/L-0328; Reel 1054

8826. 1851–6–4 From Henry Drury Harness Royal Mint
Informed C. E. Trevelyan that Maudslay, Sons, and Field will not lower their bid [for contract coinage work]. HH believes that Mint should no longer seek outside contractors. No need to meet at Treasury today.
ALS 2pp TxU:H/M–0256.3; Reel 1087

8827. 1851–6–5 To Jasper Atkinson [Royal Mint]
JH is not authorized to communicate officially with Moneyers. Will do so as soon as authority is granted.
ACCS 1p TxU:H/M–0021.2; Reel 1062

8828. 1851–6–5 From Jasper Atkinson Moneyers' Office
Moneyers still have not received confirmation about compensation for loss of their centuries-old contract. Urges JH to arrange speedy settlement.
ALS 2pp TxU:H/M–0021.1; Reel 1062

8829. 1851–6–6 From Jasper Atkinson Moneyers' Office
Did JH ever report Moneyers' situation to Treasury? Or are JH and [Henry] Rich still giving it 'consideration'?
ALS 2pp TxU:H/M–0022; Reel 1062

8830. 1851–6–6 From Henry Drury Harness Royal Mint
Encloses HH's semi-official letter regarding staff for Coining department. JH may forward it to Chancellor of Exchequer [Charles Wood] with JH's own letter. Encloses monthly accounts for JH's signature.
ALS 2pp TxU:H/M–0256.4; Reel 1087

8831. 1851–6–6 From John R. Hind 22 Grove Rd., St. John's Wood
Has received the elements of Irene from J. F. Encke, which places the planet near Egeria. Eduard Heis has asked for JH's address as he has prepared a paper on variable stars.
ALS 3pp RS:HS 9.354

8832. 1851–6–7 To James Ramsey Macculloch 32 Harley St.
Gratefully returns [Michel] Chevalier's work, which contained numerical data that JH sought.
ALS 1p TxU:H/L-0237; Reel 1054

8833. 1851–6–9 To Jasper Atkinson n.p.
Assurance that JH's lack of response is due not to discourtesy but to lack of information.
AdraftS 1p TxU:H/L-0043; Reel 1054

8834. 1851–6–9 To Henry Rich [32 Harley St.]
Delays decision over assigning Henry Bingley as Chief Coiner.
ACC 2pp TxU:H/L-0329; Reel 1054

8835. 1851–6–10 From William Henry Bateson St. John's College, Cambridge
Cambridge University Commissioners will meet on 17 June.
ALS 1p TxU:H/M–0057.6; Reel 1087

8836. 1851–6–12 From Henry Drury Harness Royal Mint
Nothing to report from Treasury. Hopes JH is recovering from cold.
ALS 1p TxU:H/M–0256.5; Reel 1087

8837. 1851–6–12 From Henry Pearse St. John's Rectory, Bedford
Still more on the pension for Mary Maclear [see HP's 1851–2–19].
ALS 3pp RS:HS 13.330

8838. 1851–6–13 From Henry Drury Harness Royal Mint
Messrs. Rennie submitted estimate [for contract coinage work] yesterday that HH considers preposterous.
ALS 1p TxU:H/M–0256.6; Reel 1087

8839. 1851–6–14 From George Biddell Airy Royal Observatory, Greenwich
About a strange mixture of coinage in Prussia.
ACCS 3pp RGO 6.349.9

8840. 1851–6–16 From Charles Edward Trevelyan Treasury Chambers
Treasury commissioners rejected bids from Mr. Nutting, George Rennie, and firm of Maudslay, Sons, & Field for contracts to coin gold, silver, and copper coins for Mint. JH should develop a plan for conducting coinage in some government agency.
LS 1p *Return*, 38–9

8841. 1851–6–17 From Henry Drury Harness Royal Mint
Treasury rejected all bids [for contract coinage work] and called for establishing coining facilities within Mint. Sends copy of such plan that HH already prepared for JH's signature.
ALS 1p TxU:H/M–0256.7; Reel 1087

8842. 1851–6–18 From William Henry Bateson [No. 10] Downing St.
Cambridge University Commissioners will meet on 11 July.
ALS 1p TxU:H/M–0057.7; Reel 1087

8843. 1851–6–20 From William Wyon Royal Mint
Thanks for timely warning of date when dies for new florin will be needed. Confirms inscription that JH wants on florin.
ALS 2pp TxU:H/M–0610; Reel 1086

8844. 1851–6–21 To Jasper Atkinson n.p.
Mint must resume gold coinage and needs to take possession of Moneyers' equipment quickly. This will relieve Moneyers of additional responsibility.
AdraftS 2pp TxU:H/L-0044; Reel 1054

8845. 1851–6–23 To Henry Rich 32 Harley St.
Calculates budget for Mint salaries.
ACCS 4pp TxU:H/L-0331; Reel 1054

8846. 1851–6–23 To Henry Rich 32 Harley St.
Compensation for Henry Bingley [former Sovereign's Assay Master whose office was recently terminated]. Bingley refused offers of £500 and £750.
ACCS 4pp TxU:H/L-0330; Reel 1054

8847. 1851–6–24 From George Phillips Bond Paris
Unable to avail himself of JH's invitation. Thanks for his interest in American astronomy. Met U. J. J. Leverrier, J. B. Biot, and others since he has been in Paris.
ALS 4pp RS:HS 4.177

8848. 1851–6–25 From Jasper Atkinson Moneyers' Office
Thanks JH for offer to relieve Moneyers of responsibility for coinage, but Moneyers would rather have some information

about what Treasury is planning to *do* with them.
ALS 2pp TxU:H/M–0023; Reel 1062

8849. 1851–6–26 To Henry Rich [32 Harley St.]
Discusses the operation and equipment of Coining Room.
Adraft 4pp TxU:H/L-0332; Reel 1054

8850. 1851–6–27 To Lord John Russell Royal Mint
Official report on international supply and demand for gold and silver, with projection of supply for next twenty years.
Adraft 16pp & AdraftS 28pp & Adraft 7pp TxU:H/L-0352 & -0353 & -0354; Reel 1054

8851. 1851–6–28 From Gail Borden, Jr. 34 Bury St., St. James's
At request of Ashbel Smith of Texas, GB, inventor of the 'Meat Biscuit,' sends sample to JH with instructions for cooking.
ALS 1p TxU:H/M–0919.2; Reel 1083

8852. 1851–6–28 From Henry Drury Harness Royal Mint
Sends note from J. T. Pratt arranging meeting. [JH annotation: 'National Debt Office, Old Jewry, Wednesday.']
ALS 1p † TxU:H/M–0256.8; Reel 1087

8853. 1851–6–30 From Henry Drury Harness Royal Mint
Will search for papers and be at Treasury at appointed time. Found Oxford more beautiful than expected.
ALS 1p TxU:H/M–0256.9; Reel 1087

8854. 1851–7–1 From Henry Drury Harness to Joshua Field [Royal Mint]
JH asks that JF recommend someone to value articles at Mint claimed by Moneyers as their property.
ALS 1p TxU:H/M–0998; Reel 1083

8855. 1851–7–2 To Henry Rich Royal Mint
Wants to release records of Mint salaries as public documents. Discusses pensions.
AdraftS 4pp TxU:H/L-0333; Reel 1054

8856. [1851–7–3] To Jasper Atkinson [Royal Mint]
Final adjustments to Moneyers' claims for compensation is in progress. Expects these to be ready when Moneyers' work terminates next week.
ACCS 2pp TxU:H/M–0024.2; Reel 1062

8857. 1851–7–3 From Jasper Atkinson Moneyers' Office
Moneyers will stop Mint operations next Thursday [10 July]. But do not expect Moneyers to leave if compensation for their machinery, tools, and lost income has not been settled by then.
ALS 3pp TxU:H/M–0024.1; Reel 1062

8858. 1851–7–3 From Joshua Field to Henry Drury Harness Lambeth
Recommends T. Cooke as someone who 'could value some Chemicals' at Mint. [JH annotation: seeking someone as 'Valuer of the Moneyers Rollers and tools.']
ALS 1p †(by JH) TxU:H/M–0987.1; Reel 1083

8859. 1851–7–4 From Henry Drury Harness to Joshua Field [Royal Mint]
JF misunderstood HH's note. Mint needs someone to value tools, not chemicals. Thought Field's firm [Maudslay, Sons, & Field] could recommend one of its employees.
ALS 1p TxU:H/M–0987.2; Reel 1083

8860. 1851–7–6 To Margaret Brodie Herschel London
Of JH's activities with some of the children; about an invitation to the Mansion House ball, and of the health and activities of several friends.
ALS 4pp JHS 6.14

8861. 1851–7–6 From William Henry Sykes India House
Glad that Dr. [Duncan] Stewart received promise from Mr. Prinsep of cadetship [in H.E.I.C.] for Stewart's son. WS cannot promise same for young [George] Maclear but will add Maclear to list of applicants for WS's patronage.
ALS 4pp TxU:H/M–0516; Reel 1093

8862. 1851–7–8 From Charles Edward Trevelyan Treasury Chambers
Encloses minutes of Treasury commissioners, approving compensation by Paymaster-general to company of Moneyers for loss of earnings due to reform at Mint.
LS 1p, encl 2pp *Return*, 46

8863. 1851–7–8 To [?] Royal Mint
Anxiously awaiting reply to JH's letter of 4 July requesting [?]'s recommendation of a London firm that can quickly evaluate Moneyers' equipment.
ACC 1p inc TxU:H/L-0750.2; Reel 1089

8864. 1851–7–9 From Jasper Atkinson Moneyers' Office
Requests copy of Treasury's minutes relating to Moneyers' claims for compensation.
ALS 2pp TxU:H/M–0025; Reel 1062

8865. [1851]–7–9 From Henry Rich Treasury
Settled matter of Moneyers' compensation with Chancellor of Exchequer [Charles Wood]. Wants to discuss Henry Bingley's position at Mint. Will be in House of Commons that week.
ALS 3pp TxU:H/M–0443.7; Reel 1093

8866. 1851–7–9 From Lord John Russell Downing St.
Sends Mr. Benme's [?] letter and prospectus. Wants JH's opinion on whether the Queen should dedicate it.
ALS 3pp RS:HS 14.441

8867. 1851–7–10 To Charles Wood Royal Mint
Henry Finch and F. R. Brande declined offices of junior clerks at Mint. JH needs to fill these positions quickly with qualified young men.
AdraftS 4pp TxU:H/L-0507; Reel 1055

8868. 1851–7–11 From Jasper Atkinson Moneyers' Office
Evaluation of Moneyers' tools and machinery will be completed by Monday. Asks for written assurance that Treasury will pay Moneyers what Mr. Richards, the evaluator, determines their equipment to be worth.
ALS 3pp TxU:H/M–0026; Reel 1062

8869. 1851–7–11 From Charles Edward Trevelyan Treasury
Please return Treasury minutes regarding compensation for Moneyers so CT may alter it. JH may give copy of it to Moneyers. Returns Jasper Atkinson's application.
LS 2pp TxU:H/M–0551.11; Reel 1093

8870. 1851–7–12 To Jasper Atkinson Royal Mint
Accuses JA and Moneyers of ignoring agreement by refusing to vacate rooms and relinquish tools at Mint, and of purposely slowing down Mint officers' preparation to resume coining money. [Memorandum on verso: Edward Enfield and H. D. Harness were present at time of agreement.]
ACCS 4pp TxU:H/L-0045; Reel 1054

8871. 1851–7–12 From William Henry Bateson [No. 10] Downing St.

John Romilly, invited by John Russell, will attend 14 July meeting of Cambridge University Commissioners to discuss Modern History professorship.
ALS 3pp TxU:H/M–0057.8; Reel 1087

8872. 1851–7–12 From Augustin Reslhuber Kremsmünster Observatory
Observatory recently received *Cape Results* from Marian Koller, former director until 1847 and now minister of public instruction in Vienna.
ALS 2pp RAS:JH Archive 10/5.3; Reel 9

8873. 1851–7–15 From Henry Bingley [Royal Mint]
Urges JH to obtain HB's retirement allowance and compensation quickly.
ALS 2pp † TxU:H/M–0096; Reel 1087

8874. 1851–7–17 From [Sir] Edward Ryan Garden Lodge
[Richard Jones?] agreed to prepare letter to Lord John [Russell?] but not to send it. Now learns that meeting is scheduled for tomorrow. Agrees with JH and [William] Empson that [Thomas] Maclear's work should be rewarded, but wants to move more deliberately.
ALS 3pp TxU:H/M–0475; Reel 1093

8875. 1851–7–17 From A[lexander] Spiers Paris
Has used 'elegant' extracts of JH in book [*Prelim. Discourse*] to familiarize students 'on the continent' with Francis Bacon's work. Is sending copy of book to JH.
ALS 2pp RS:HS 17.226

8876. 1851–7–18 From Sir Edward Ryan Board of Trade
Saw [Richard] Jones and stopped letter he was sending to Lord John [Russell]. Has consulted [John] Lefevre. Needs to tell JH what passed between [Sir William] Hodges and Lord John.
ALS 3pp RS:HS 14.459

8877. 1851–7–18 From Charles Edward Trevelyan Treasury
Returns JH's original schedule for Mint reform, with amendments.
LS 1p TxU:H/M–0551.12; Reel 1093

8878. 1851–7–19 From Charles Edward Trevelyan Treasury
Please send copy of R. L. Sheil's letter of 10 May 1849 regarding retirement of John Beckwith.
ALS 1p TxU:H/M–0551.13; Reel 1093

8879. 1851–7–20 From Henry Anthony Jeffreys Hawkhurst Parsonage
Regarding the Tithe question. Thinks there are some in the parish who would sign a document for JH. Would he let him know the form to adopt.
ALS 3pp RS:HS 10.293

8880. 1851–7–20 From [A. J. F.] Yvon Villarceau Paris
Appears to be presenting YV's calculations for the orbits of double stars. [Much of this letter is illegible.]
ALS 4pp (French) RS:HS 17.421

8881. 1851–7–21 From Mr. Medwell Maryland Point, Stratford, Essex
Requests permission for M's parents to tour Royal Mint.
ALS 1p TxU:H/M–0353.1; Reel 1087

8882. 1851–7–23 From Henry Drury Harness Royal Mint
Will meet JH tomorrow. Mint staff is taking over rooms from Moneyers. Transition is going smoothly. Joseph Nicoll is dying and will probably not draw much compensation.
ALS 2pp TxU:H/M–0257.1; Reel 1087

8883. 1851–7–24 From Sir Edward Ryan Board of Trade
Sends proposed letter to Lord John [Russell]. Wants it signed by George Peacock, William Whewell, [John] Lefevre, JH, and the archbishop. [William] Hodges will deliver it.
ALS 4pp RS:HS 14.460

8884. 1851–7–25 To Charles Edward Trevelyan Royal Mint
Submits JH's reformed organizational plan for Mint and new personnel assignments [see JH's 1851–7–26] for approval by Treasury commissioners.
LS 1p *Return*, 39

8885. 1851–7–25 To William Whewell 32 Harley St.
Asks WW to sign a letter to Lord John Russell being circulated on behalf of Richard Jones.
ALS 4pp TC, Camb. Add. Ms.a.207[89] (C: RS:HS 23.105)

8886. 1851–7–26 From Henry Drury Harness Royal Mint
Sends W[illiam] Lalor with letters for JH. Will go to Treasury at four o'clock.
ALS 1p TxU:H/M–0257.2; Reel 1087

8887. 1851–7–26 To Mint Staff Royal Mint
[Formal notice] Promulgation of positions and duties within reorganized Mint establishment.
D 2pp *Return*, 39–40

8888. [1851]–7–26 From Henry Rich 19 Charles St. [Berkeley Square]
Sends report on Henry Bingley for JH's approval.
ALS 1p TxU:H/M–0443.8; Reel 1093

8889. 1851–7–26 From Charles Edward Trevelyan Treasury Chambers
Treasury commissioners have approved JH's new organizational plan and assignments for Mint.
LS 1p *Return*, 40

8890. 1851–7–27 From Henry Drury Harness Royal Mint
Recovering from illness.
ALS 1p TxU:H/M–0257.3; Reel 1087

8891. 1851–7–27 To William Henry Fox Talbot 32 Harley St.
Wants recommendation from WT for teacher of the 'art of Talbotyping.'
ALS 3pp Science Museum Talbt 1/41 (C: RS:HS 23.106)

8892. 1851–7–29 From Sir Edward Ryan Board of Trade
Hopes archbishop will sign the letter. [John] Lefevre will have a good copy made for presentation to Lord John [Russell]. Will arrange delivery with [William] Hodges when all signatures are received.
ALS 4pp RS:HS 14.461

8893. 1851–7–29 From Charles Edward Trevelyan Treasury
Treasury commissioners admit merits of claims by Mint officers, whom JH supports, but does not agree to pay raises. Calculating salaries for more years than officers have actually served would set bad precedent for other public offices.
LS 3pp TxU:H/M–0551.14; Reel 1093

8894. [1851–8] To Margaret Brodie Herschel London
Is awaiting notice that he needs to go to see [Richard] Jones, who is very ill with a carbuncle on his neck. JH also comments on arrangements to be made related to Jones's

illness.
ALS 4pp JHS 6.17

8895. [1851–8] To Margaret Brodie Herschel [London]
About the expected death of [Richard] Jones; JH is about to leave to visit him and Mrs. Jones.
ALS 3pp JHS 6.25

8896. 1851–8–1 From Richard Jones Tithe Commission
JH's exertions and friendship are a treasure. Will visit them on Monday next.
ALS 3pp RS:HS 10.387

8897. 1851–8–2 From Francis Beaufort Admiralty
Anxious to have JH meet the new assistant, Pierce Newton, to be sent to the Cape Observatory.
ACC 1p Hydrographic Office Letter Book

8898. 1851–8–2 From Sir Robert Harry Inglis 7 Bedford Square
Has forwarded his letter regarding Dr. [O. W.?] Fiedler to Sir Henry Ellis. C. D. E. König is on vacation at the moment so is unable to estimate how much the British Museum will offer for the fulgurite. Perhaps JH would call and inspect the specimens.
ALS 3pp RS:HS 10.247

8899. [1851]–8–4 From Henry Rich Treasury
Sends final version of report on Henry Bingley's position at Mint. Awaits decision from Chancellor of Exchequer [Charles Wood]. Would like to have Bingley's approval of the report, but believes that Chancellor should see it first.
ALS 3pp TxU:H/M–0443.9; Reel 1093

8900. 1851–8–5 From Henry Drury Harness Royal Mint
Awaits assay report from W. A. Miller. [F. R.] Brande is absent again when most needed [during transition to new Mint system]. George Arbuthnot sent list of coins needed for colonies. HH's daughter returns German book lent by JH.
ALS 3pp TxU:H/M–0258.1; Reel 1087

8901. 1851–8–7 To Charles Edward Trevelyan Royal Mint
Forwards report by Mint Board evaluating claims from company of Moneyers to compensation for equipment taken over by Mint during reform. Mr. Richards has evaluated Moneyers' equipment, and JH submits recommended compensation for various articles.
LS 2pp *Return*, 53–4

8902. 1851–8–8 From S. M. Seake Treasury Chambers
Henry Rich asks JH to sign and return enclosed report.
ALS 1p TxU:H/L-0723; Reel 1089

8903. 1851–8–9 From Henry Drury Harness Royal Mint
Regrets JH's poor health. Forwards message from Henry Rich at Treasury. Official appointment of [C. B. Piers] arrived. Letter from [W. G.] Hayter named Mr. Aberdein as new clerk in Mint office.
ALS 1p TxU:H/M–0258.2; Reel 1087

8904. 1851–8–11 From George Phillips Bond Pulkowa
Description of the recent total eclipse of the sun as seen at Pulkowa Observatory.
ALS 3pp RS:HS 4.178

8905. 1851–8–12 To Margaret Brodie Herschel London
It appears JH is just back from Paris, and so brings MH up to date on the health of various individuals and on JH's activities. Describes the procedure for making fused muriate of lime, and the costs of insurance and income taxes.
ALS 4pp JHS 6.13

8906. [1851]–8–12 From Henry Rich 19 Charles St. [Berkeley Square]
Chancellor of Exchequer [Charles Wood] rejects proposal to allow Henry Bingley an annual salary of £750. Bingley must accept salary offered by Treasury.
ALS 4pp TxU:H/M–0443.10; Reel 1093

8907. 1851–8–12 From James Wilson India Board
Large amounts of gold are arriving from U.S., but Bank of England has little. Is gold accumulating at Royal Mint, or is it passing on to France?
ALS 3pp TxU:H/M–0585; Reel 1093

8908. 1851–8–12 To James Wilson 32 Harley St.
Suspension of operations at Mint is only temporary, due to introduction of new systems for assaying and coining. Mint is storing bullion for resumption of gold coinage in a few days. JH's opinion on continuing drain of bullion from Bank [of England].
ACCS 4pp TxU:H/L-0501; Reel 1055

8909. 1851–8–13 From Henry Bingley [Royal Mint]
Unable to vacate Assay Office and HB's residence by dates that JH set. Requires extension. Still has not seen terms, in writing, of HB's retirement from Mint.
ALS 3pp TxU:H/M–0097; Reel 1087

8910. 1851–[8?]–13 To William Rowan Hamilton 32 Harley St.
Expresses deep sorrow upon hearing of death of WH's sister [Eliza Hamilton]. Is anxious to see WH's finished work on quaternions. Hopes WH has made it clear and simple.
ALS 4pp TCD 1493:489

8911. 1851–8–13 From Henry Drury Harness Royal Mint
Sent JH's note to G. F. G. Mathison today. Leaving for Woods and Forests office to expedite this matter. Describes where HH put keys to coin cases for JH's sons, who will probably arrive while HH is away.
ALS 2pp TxU:H/M–0258.3; Reel 1087

8912. 1851–8–13 From Alexander von Humboldt Sans Souci
Introducing Adolphe Barth. Astronomical news and queries.
ALS 2pp (French) RS:HS 10.73

8913. 1851–8–13 From Gilbert Farquhar Graeme Mathison Richmond
Grieved by delay in leasing of refinery. GM completed inventory of refinery seven weeks ago and needs to leave for Germany soon. Urges JH to finalize lease.
ALS 10pp TxU:H/M–0391; Reel 1087

8914. 1851–8–13 From Henry Drury Harness to Gilbert Mathison Royal Mint
Delay in converting offer made to GM into legal form is not JH's fault, but Treasury's. Rent cannot be stated until Mr. [Pusnithorne?] reports his valuing of refinery equipment.
ALS 3pp TxU:H/M–1000; Reel 1083

8915. 1851–8–13 From Benedetto Pistrucci Flora Villa, Bishopsgate, Chertsey
Pleased that JH returns to health.
ALS 1p (Italian) TxU:H/M–0429.7; Reel 1093

8916. 1851–8–13 From Adolphe Quetelet Brussels
Thanks JH for his hospitality when AQ visited England. Sending some copies of AQ's writings to be passed on to W. R. Birt.
ALS 3pp (French) TxU:H/M-0910.1; Reel 1083

8917. 1851–8–13 To Henry Rich 32 Harley St.
Discrepancies in Mint documents. Error might have occurred in copying.
AdraftS 4pp TxU:H/L-0334; Reel 1054

8918. 1851–8–14 To Henry Bingley Royal Mint
Office of Queen's Assayer will terminate 18 Aug. HB must vacate his house by 13 Oct. JH cannot extend HB's occupancy to Christmas.
ACCS 3pp TxU:H/L-0097; Reel 1054

8919. 1851–8–14 From James Wilson India Board
Thanks for explaining disposition of gold belonging to Bank of England but stored at Mint. Suspected that importers of U.S. gold were not selling it to Bank but were privately coining it themselves.
ALS 4pp TxU:H/M–0586; Reel 1093

8920. 1851–8–15 To James Wilson [Royal Mint]
Profits that Bank [of England] makes from assay operations at Royal Mint.
Adraft 2pp TxU:H/L-0506; Reel 1055

8921. 1851–8–19 To Charles Babbage n.p.
Please let him have David Brewster's address.
ALS 1p RS:HS 2.321

8922. 1851–8–19 From Henry Drury Harness Royal Mint
Waiting for official assay report is creating delay that gives bad appearance at Mint. HH suspects that someone is creating distrust of Mint's assaying arrangements. Proposes to begin work without reports.
ALS 2pp TxU:H/M–0258.4; Reel 1087

8923. 1851–8–19 From George Cornewall Lewis Treasury Chambers
Treasury commissioners received report on JH's and Henry Rich's investigation of claim by Henry Bingley for compensation due to abolition of office of Queen's assay master. Treasury grants Bingley annual pension of £750.
LS 1p *Return*, 50

8924. 1851–8–20 To Henry Rich 32 Harley St.
Awaits decision from Treasury on salary for Henry Bingley.
ALS 1p TxU:H/L-0335; Reel 1054

8925. 1851–8–20 From Charles Edward Trevelyan Enmore [Bridgewater]
Asked Chancellor of Exchequer [Charles Wood] to answer JH's letter regarding two apprentices. Thanks for gift of third coin struck under new Mint system. Congratulates JH and H. D. Harness for successful reform.
ALS 2pp TxU:H/M–0551.15; Reel 1093

8926. 1851–8–20 To Charles [Edward Trevelyan] Royal Mint
Encloses JH's letter to commissioners of Treasury requesting official permission to meet Bank [of England's] demand for silver.
AdraftS 2pp TxU:H/L-0481.8; Reel 1055

8927. 1851–8–21 To Charles Babbage n.p.
Will show his friends round the Mint.
ALS 3pp RS:HS 2.322

8928. 1851–8–21 From William Thomas Brande Royal Mint
Moneyers failed to fulfill conditions of apprenticeship for WB's son [F. R. Brande], yet they refuse to refund £1,000 apprenticeship fee that WB paid them five years ago. Asks advice from JH, who is member of committee that awarded compensation to WB's son.
ALS 3pp TxU:H/M–0126; Reel 1087

8929. 1851–8–21 To Henry Rich [Royal Mint]
Discusses the matter of pensions for Mint workers.
Adraft 4pp TxU:H/L-0336; Reel 1054

8930. 1851–8–22 From Edward Sabine Woolwich
Wants advice regarding recalculation of magnetic lines using [C. F.] Gauss's method and new numerical data. This will be a provisional calculation.
ALS 3pp † RS:HS 15.232

8931. 1851–8–22 From S. M. Seake Treasury
Responds to JH's letter of 21 Aug. to Henry Rich, who is out of town. Finds nothing that would affect claim made by Mr. Bingley's son. Advises JH to proceed as JH thinks proper.
ALS 4pp TxU:H/M–0943; Reel 1083

8932. 1851–8–23 To Margaret Brodie Herschel London
Is enjoying son William's stay with JH; comments on the poetry of MH's letters as opposed to the lack of poetry in London. JH and son William are to go to dinner, but JH complains that he would rather have his Sundays in London free of engagements.
ALS 4pp JHS 6.12

8933. 1851–8–23 To Edward Sabine Harley St.
Cannot offer advice regarding geo-magnetism. Suggests ES ask [C. F.] Gauss for assistance.
(S) 1p RS:HS 23.107

8934. 1851–8–25 From Benedetto Pistrucci Flora Villa, Bishopsgate, Chertsey
BP states problems concerning his office, salary, residence, and pension.
trans (by JH) 3pp TxU:H/M–0431; Reel 1093

8935. 1851–8–25 From Benedetto Pistrucci Flora Villa, Bishopsgate, Chertsey
Believes it might be possible to remain at Bishopsgate and commute to London instead of moving to the Mint. Discusses a medal commemorating the coronation of Queen Victoria.
ALS 5pp † (Italian) TxU:H/M–0429.8; Reel 1093

8936. 1851–8–26 From Henry Bingley Royal Mint
What steps has JH taken regarding [HB's claim for compensation] for HB's son?
ALS 1p TxU:H/M–0098; Reel 1087

8937. 1851–8–26 To [Henry Bingley] Royal Mint
Henry Rich has left England, so JH presented HB's question to C. E. Trevelyan. Believes that [Treasury's] earlier response will be considered official. Doubts that HB's son will receive compensation for loss of future profession.
ALS 3pp TxU:H/L-0649; Reel 1089

8938. 1851–8–26 From William Henry Fox Talbot Lacock Abbey
Describes observing solar eclipse at Marienburg in eastern Prussia. Gives advice about whom to see for photography training [see 1851–7–27].
ALS 3pp RS:HS 17.322

8939. 1851–8–27 To Benedetto Pistrucci Royal Mint
Discusses changes in the duties of the office of Chief Medalist. Claims his duty as Master of the Mint is to put 'high talent' where it can be useful.
Adraft 1p TxU:H/L-0308; Reel 1093

8940. 1851–8–28 From Henry Drury Harness Royal Mint
Mint is 'striking shillings fast.'
ALS 1p TxU:H/M–0258.5; Reel 1087

8941. 1851–8–28 To [Henry Drury Harness] 32 Harley St.
Suspects that JH sent unsigned check to HH yesterday. Please send messenger William Lalor to tell if JH is needed at Mint today.
ACC 2pp TxU:H/L-0650; Reel 1089

8942. 1851–8–28 From Charles Wood Hickelton
Received specimen coin from JH. It is as good as anything produced by Moneyers.
ALS 2pp TxU:H/M–0595; Reel 1086

8943. 1851–8–29 From George Biddell Airy Royal Observatory, Greenwich
Regarding the observations on the recent eclipse of the sun.
ALS 3pp RS:HS 1.221

8944. 1851–8–29 To George Biddell Airy London
Bringing GA up to date about JH's family news and commenting on news of GA's family.
ALS 3pp RGO 6.373.169

8945. 1851–8–29 To Margaret Brodie Herschel London
Has been to see [Richard] Jones, who has undergone some 'abominable' treatment; Mrs. Jones believes Jones is recovering; JH talks about a plague of boils in England. JH comments on the activities of a little kitten, on Annibale de Gasparis having found another planet [Eunomia?], and on a strange poem being written about the Crystal Palace by a man named [?] Warren.
ALS 8pp JHS 6.21

8946. 1851–8–31 To Margaret Brodie Herschel Harley St.
Most of the letter is taken up in replying to MH's letters; JH talks about Charles Wheatstone's idea of 'binocular Daguerreotypes,' and about a chart of all balloon ascents, prepared in Paris, and brought to JH by a friend coming from there.
ALS 4pp JHS 6.22

8947. [1851]–9–1 From Elizabeth Ann Waterhouse to Lady Margaret Herschel Kensington
Make JH aware that G. R. Waterhouse desires position at British Museum vacated by C. D. E. König's death. GW is trying to return quickly from Paris, but delay might suggest that GW is not interested.
ALS 4pp RS:HS 18.377

8948. 1851–9–3 From B. Hebeler Royal Prussian General Consulate, 106 Fenchurch St.
Requests permission for Mr. Knoblauch and five friends to inspect Mint.
ALS 1p TxU:H/M–0270.1; Reel 1087

8949. 1851–9–3 From Charles Edward Trevelyan Enmore, Bridgewater
Forwards decisive response by Chancellor of Exchequer [Charles Wood] to letter from [W. T.] Brande. CT will not get involved.
ALS 1p TxU:H/M–0551.16; Reel 1093

8950. 1851–9–3 To Charles Edward Trevelyan Royal Mint
Forwards evaluation of melting and refining equipment taken over from G. F. G. Mathison. Portions of refinery were evaluated by Mr. Abud, named by Woods and Forests commissioners. Submits recommended compensation for various articles.
LS 1p *Return*, 54

8951. 1851–9–6 To William Thomas Brande Royal Mint
Awards made by master [of Mint] to Moneyers should not interfere with Moneyers settling accounts among themselves.
ACCS 1p TxU:H/L-0103; Reel 1054

8952. 1851–9–8 To My dear Sir Royal Mint
Approves of [?]'s memorial to Treasury, but returns it to [?] for minor corrections before forwarding it.
ACCS 2pp TxU:H/L-0651; Reel 1089

8953. 1851–9–9 From Henry Drury Harness Royal Mint
Complains that H. W. Field seems to have forgotten that, as officer in new Mint, Field should be available whenever required.
ALS 2pp TxU:H/M–0259.1; Reel 1087

8954. 1851–9–10 From Henry Bingley Royal Mint
Has JH heard from C. E. Trevelyan regarding claim by HB's son for compensation? If not, HB asks JH to compensate HB's son for one and a half years of unpaid services at Mint.
ALS 3pp TxU:H/M–0099; Reel 1087

8955. 1851–9–10 From Charles Edward Trevelyan Treasury
Read enclosed memoranda, and inform CT whether JH was aware of these before JH made recommendations regarding compensation for Moneyers' tools.
ALS 1p TxU:H/M–0551.17; Reel 1093

8956. 1851–9–11 To David Brewster 32 Harley St.
Asks DB to write a paragraph explaining the novelty of the photographic process of '[Andrew?] Ross and [Thomas?] Thomson' as distinct from [Nicéphore] Niépce's albuminous process.
ALS 4pp Science Museum Archives MS 1174

8957. 1851–9–11 To Charles Edward Trevelyan Royal Mint
Proposes budget for repair of moneyers' machinery.
Adraft 1p TxU:H/L-0481.9; Reel 1055

8958. 1851–9–12 To Henry Bingley Royal Mint
Suggests how HB should request government compensation for HB's son for former services as probationer assayer and for loss of future employment.
AdraftS 3pp TxU:H/L-0098; Reel 1054

8959. 1851–9–12 From Charles Edward Trevelyan Treasury
JH's explanation of discrepancy in allowance for repairs of [Moneyers'] machinery is satisfactory. Please return memoranda.
ALS 1p TxU:H/M–0551.18; Reel 1093

8960. 1851–9–15 From Annibal de Gasparis Naples
Gives details of the new planet discovered 29 July. Proposes to call it *Eunomia*.
ALS 2pp (French) RS:HS 8.45

8961. 1851–9–16 From Henry Bingley Royal Mint
Hopes accompanying official letter is adequate.
ALS 1p TxU:H/M–0100; Reel 1087

8962. 1851–9–17 To Henry Bingley Royal Mint
Will forward HB's letter to Treasury but with note that JH did not regard HB's son as actual probationer assayer. Denies HB's claim to deserve new accommodations after HB vacates quarters at Mint.
ACCS 1p TxU:H/L-0099; Reel 1054

8963. 1851–9–17 To Charles Edward Trevelyan Royal Mint
Forwards new claim by Henry Bingley for compensation for work performed by Bingley's son as probationary assayer. JH suggests £300. Believes that Bingley's further claim for reimbursement for coachhouse and stables, built at Mint residence by Bingley, was covered by Treasury's earlier compensation to Bingley.
LS 1p *Return*, 50

8964. 1851–9–18 From George Robert Waterhouse British Museum
Embroiled in Museum politics with Richard Owen over position and salaries. Sends copies of testimonials GW collected to take to Archbishop [of Canterbury, J. B. Sumner].
ALS 4pp RS:HS 18.97

8965. 1851–9–20 From Charles Edward Trevelyan Treasury Chambers
Treasury commissioners approve compensation claims for equipment belonging to company of Moneyers and to G. F. G. Mathison.
LS 1p *Return*, 55

8966. 1851–9–24 From Henry Drury Harness Royal Mint
Runs daily Mint business so JH can turn attention to the Exhibition papers. Will propose to William Buckle that books in Coining department be incorporated into general records of Mint.
ALS 2pp TxU:H/M–0259.2; Reel 1087

8967. 1851–9–25 From Charles Edward Trevelyan Treasury Chambers
Treasury commissioners approve payment of £300 to son of Henry Bingley for work performed as probationary assayer.
LS 1p *Return*, 51

8968. 1851–9–27 To George Biddell Airy Royal Mint
Some high quality sets of weights are becoming available. Should they be acquired for use with the standards work?
ALS 4pp RGO 6.341.5

8969. 1851–9–27 From Gilbert Farquhar Graeme Mathison Richmond
Asks JH to endorse GM's enclosed memorial to Treasury. Did not include financial accounts, which are still incomplete. Encloses letters to JH and H. D. Harness.
ALS 2pp TxU:H/M–0392; Reel 1087

8970. 1851–9–28 To Gilbert Farquhar Graeme Mathison [Royal Mint]
[P.S. to letter of 28 Sept. 1851:] JH wrote to J. W. Morrison about GM's claim to melting pots in refinery. Asks GM to clarify Mint records regarding pots. [JH note: Excerpts from Mint records, 1820–1844, outlining history of stopped pots in Mint refinery.]
ACC 2pp †, JH note 2pp TxU:H/L-0251 & -0252; Reel 1054

8971. 1851–9–29 From George Biddell Airy Royal Observatory, Greenwich
Is inquiring further into the usefulness of the weights [see JH's 1851–9–27].
ACCS 2pp RGO 6.341.7

8972. 1851–9–29 From Thomas Maclear Royal Observatory, Cape of Good Hope
Hears that JH is using his influence to obtain a transit circle for the Cape. Hopes the application will be successful. Comments on the siting of the instruments. Cape politics are deplorable. Kaffir war still continues and appears likely to be expensive.
ALS 3pp RS:HS 12.151

8973. 1851–10–1 From George Biddell Airy Royal Observatory, Greenwich
Now believes that at least one specific weight should be acquired [see GA's 1851–9–29].
ACCS 2pp RGO 6.341.8

8974. 1851–10 To James David Forbes 32 Harley St.
Asks JF to deliver papers about professorships and a council position at a new university at Sydney to interested, qualified parties.
ALS 4pp St. Andrews 87 (C: RS:HS 23.108)

8975. 1851–10 To Mint Staff Royal Mint
[Formal notice] Promulgation of regulations for coinage of gold bullion at Mint.
L 1p *Return*, 55

8976. 1851–10–2 From John Hearn Northshields
Fears Hearn's first letter (13 Sept. 1851) to JH was lost. Anxiously asks again for testimonial from JH to governor at Sandhurst on behalf of Hearn's son, who 'gets daily worse.'
ALS 1p † TxU:H/M–0269; Reel 1087

8977. 1851–10–2 From Gilbert Farquhar Graeme Mathison Royal Mint
Encloses GM's memorial to Treasury, with financial accounts included, for JH's approval.
ALS 1p TxU:H/M–0393.1; Reel 1087

8978. 1851–10–4 From [Jean]-D[aniel] Colladon London
Leaves tomorrow for Geneva. Expresses thanks for all JH's help and presents his compliments to Lady Herschel.
ALS 1p (French) RS:HS 5.356

8979. 1851–10–4 From Edward Enfield Sheffield House, Kensington
All Moneyers have signed receipt for £3095. Where should EE pick up check?
ALS 2pp TxU:H/M–0185; Reel 1087

8980. 1851–10–4 To Charles Edward Trevelyan Royal Mint
Forwards, with JH's approval, G. F. G. Mathison's claim for compensation due to termination of office of Melter and Refiner at Mint.
ACCS 1p TxU:H/L-0481.10; Reel 1055

8981. 1851–10–4 To ? [Royal Mint]
Note summarizing G. F. G. Mathison's memorial of 2 Oct. 1851, outlining GM's career in public service, earnings while at Mint, and grievances against reforms at Mint. Forwarded GM's memorial to C. E. Trevelyan on 4 Oct.
Amem 2pp TxU:H/M–0393.2; Reel 1087

8982. 1851–10–7 From George Biddell Airy Royal Observatory, Greenwich
Informs JH about the receipt of important papers from the Provost of the new college [Sydney University] at Sydney,

Australia.
ACCS 1p RGO 6.337.63

8983. 1851–10–7 From Edward Sabine Woolwich
Saw [J. B.] Listing, who may see JH soon. Sends volume of Cape magnetic observations. Thermometer at Kew is a great success. New thermometers compare favorably with [Henri] Regnault's.
ALS 4pp RS:HS 15.233

8984. 1851–10–8 From Henry Drury Harness Royal Mint
Forwards letters from Robert Mushet and Mr. Hill. Suspended Mr. Fowler, lamplighter, for drunkenness; another gas man who neglected his duty was referred to Mr. [Louis?]. Gave JH's papers and gas company's report to W. T. Brande, who will report on future gas supplies for Mint. Reports accuracy of 27 melting pots. [JH annotation: Recommends getting rid of gas production facility. Response to report on melting pots.]
ALS 4pp † TxU:H/M–0260.1; Reel 1087

8985. [1851]–10–8 From Henry Rich Treasury
Decisions made concerning compensation for Moneyers. HR objects to G. F. G. Mathison's request [for higher salary].
ALS 3pp TxU:H/M–0443.11; Reel 1093

8986. 1851–10–9 To [Sir] Edward Ryan Collingwood
Mr. Hodges will arrive at Benenden tonight. JH is visiting Collingwood for weekend to conduct parish business. Explains efforts to get assignment of JH's son William changed to Bengal. Asks ER to help.
ALS 4pp TxU:H/L–0765.4; Reel 1089

8987. 1851–10–10 To Henry Rich Collingwood
Plans to obtain pension for G. F. G. Mathison. Mathison feels moral duty to [Richard] Lynn, Head Melter to Mathison, who has no legal claim to compensation.
C 4pp TxU:H/L–0337

8988. 1851–10–12 From James A. Gordon Naish, Bristol
His own health does not improve very rapidly. Has he heard how Owen Jones is progressing? Has heard good reports of Dr. J. A. Symonds of Bristol.
ALS 4pp RS:HS 8.197

8989. 1851–10–13 To Charles Edward Trevelyan Royal Mint
Lists wages for assistant workers in Mint.
ACC 2pp TxU:H/L–0481.11; Reel 1055

8990. 1851–10–14 To George Biddell Airy 32 Harley St.
Has received the papers [see GA's 1851–10–7]; Sydney University seeks assistance in finding professors. What should the response be?
ALS 2pp RGO 6.337.64

8991. 1851–10–16 To George Biddell Airy 32 Harley St.
An invitation sent out by JH to invite all search committee members [Sydney Trust] to JH's house [see GA's 1851–10–16].
ALS 3pp RGO 6.337.67

8992. 1851–10–16 From George Biddell Airy Royal Observatory, Greenwich
Is trying to organize a meeting of all involved in the new Sydney University project [see JH's 1851–10–14].
ACCS 3pp RGO 6.337.65

8993. 1851–10–16 To Margaret Brodie Herschel London
Son William's report from the school at Haileybury is very good; JH attended the closing of the Crystal Palace Exhibition, which 'was very stupid.' A basket of grapes from Cornwall has been sent for MH, and JH is eating them.
ALS 4pp JHS 6.18

8994. 1851–10–16 To [Charles Edward Trevelyan] Royal Mint
Encloses form completed by G. F. G. Mathison, together with second submission of Mathison's memorial to Treasury and letter regarding Mathison's compensation.
ACCS 2pp TxU:H/L–0652; Reel 1089

8995. 1851–10–18 From A. Fyers Mauritius
Has established a Meteorological Society at Port Louis. What type of barometer would JH advise them to adopt? Hopes to extend their activities to magnetic observations later.
ALS 4pp RS:HS 7.389

8996. 1851–10–20 From Henry Drury Harness Royal Mint
Bank of England wants more gold coined. If JH can obtain discretionary form requested in JH's 13 Oct. letter to Treasury, then coining can begin next week.
ALS 1p TxU:H/M–0260.2; Reel 1087

8997. [1851 to 1855]–10–20 To My dear Sir Royal Mint
Gave [?]'s instructions to policeman at [Mint] gate. No silver bullion is available on market. Coinage may have to cease. Gives quantities of gold and silver delivered last week.
ACC 3pp TxU:H/L–0621; Reel 1089

8998. 1851–10–21 From Charles Edward Trevelyan [Treasury]
Will respond tomorrow to JH's important note.
ALS 1p TxU:H/M–0551.19; Reel 1093

8999. 1851–10–22 To Margaret Brodie Herschel London
About JH's health and news of friends and relatives.
ALS 4pp JHS 6.23

9000. 1851–10–22 From Charles Edward Trevelyan Treasury
Before Treasury commissioners approve pay scale for Mint workmen proposed by JH, they want to know pay scale previously used by Moneyers. If JH or H. D. Harness can call on Chancellor of Exchequer [Charles Wood] today, matter can be settled immediately.
ALS 2pp TxU:H/M–0551.20; Reel 1093

9001. [1851–10–23 or later] To Charles [Edward Trevelyan] n.p.
Difficulties in pay system for moneyers.
ACC 1p inc TxU:H/L–0479.3; Reel 1055

9002. 1851–10–24 To John Lubbock 32 Harley St.
Cannot accept JL's kind invitation. Predicts a storm regarding the jury awards for the Great Exhibition.
ALS 4pp Sydney Ross Collection

9003. 1851–10–[24] From Charles Wood Downing St.
Expresses pleasure at results of CW's visit to Mint today. Particularly impressed by capabilities of JH and H. D. Harness.
C 3pp & C 1p TxU:H/M–0592 & TxU:H/M–0593.1; Reel 1086

9004. 1851–10–25 From Charles Edward Trevelyan Treasury
Proposal to transfer books and papers of Mint Commission from Treasury Chambers to Mint vaults.
LS 1p TxU:H/M–0551.21; Reel 1093

9005. 1851–10–25 To Charles [Edward Trevelyan] Harley St.
JH and H. D. Harness were deeply gratified by CT's note of [24 Oct.].
AdraftS 2pp TxU:H/L-0481.12; Reel 1055

9006. 1851–10–26 From William Henry Bateson St. John's College, Cambridge
George Peacock was unable to attend meeting of Cambridge University Commissioners. Quorum needed to conclude 'Inquiry' before end of next session of Parliament.
ALS 4pp TxU:H/M–0057.9; Reel 1087

9007. 1851–10–26 To Charles Wood 32 Harley St.
Grateful for CW's compliments. Attributes most of credit for new system at Mint to H. D. Harness.
C 1p TxU:H/M–0593.2; Reel 1086

9008. 1851–10–28 From Charles Davidson Bell Cape Town
Provides JH a sketch of the geological study and mapping being carried on in South Africa by Andrew Bain. Asks JH to bring a geological map and collection made by Bain to the attention of geologists.
C 4pp Univ. of Witwatersrand

9009. 1851–10–29 To George Biddell Airy n.p.
Is having some difficulty in acquiring the single weight [see GA's 1851–10–1].
AL(S) 4pp RGO 6.341.9

9010. 1851–10–29 To George Biddell Airy n.p.
Note with copies of an advertising circular about the Sydney University professorships [see JH's 1851–10–16].
LCC 1p inc RGO 6.337.69

9011. 1851–10–29 From Charles Babbage Dorset St.
Regarding his plans for lighthouses. Had a visit from L. F. Menabrea in his absence. Who manufactures the specified telescopes?
ALS 3pp RS:HS 2.323

9012. 1851–10–29 From Henry Drury Harness Royal Mint
JH is needed for tomorrow's meeting with Mr. [Hooter] regarding wrought iron melting pots.
ALS 1p TxU:H/M–0260.3; Reel 1087

9013. 1851–10–29 From Leonard Charles Wyon 12 Bedford Place, Brighton
LW's father, William Wyon, died today.
ALS 1p TxU:H/M–0601; Reel 1086

9014. 1851–10–30 To Charles Babbage Harley St.
Further regarding lighthouses.
ALS 3pp RS:HS 2.324

9015. 1851–10–30 To Mary Baldwin 32 Harley St.
Curiosity about quill-like 'pen' sent to JH's son John by MB. Glad that MB moved to Brighton for medical care.
ALS 3pp TxU:H/L-0071; Reel 1054

9016. 1851–10–30 From Sir William Snow Harris Plymouth
Sends a new edition of his own work on electricity. Has been studying the manuscripts of Henry Cavendish and thinks they should be published. JH in one of his articles attributes an invention to C. F. Gauss, whereas WH suggested the idea seven years before Gauss.
ALS 4pp RS:HS 10.163

9017. 1851–10–31 From Charles Edward Trevelyan Treasury
Prepare general report of Mint reforms for submission to Parliament.
(A)LS 2pp TxU:H/M–0551.23; Reel 1093

9018. 1851–10–31 From Charles Edward Trevelyan Treasury
Concurs with JH's proposal for reorganizing Engraving department. Will inform Chancellor of Exchequer [Charles Wood].
LS 2pp TxU:H/M–0551.22; Reel 1093

9019. 1851–10–31 To Charles Edward Trevelyan Royal Mint
Discontinuance of Gas department of Mint. Superintendent, Mr. [Lorde?], may claim compensation. Historic item discovered during sale of Henry Bingley's tools and machinery.
Adraft 4pp TxU:H/L-0481.14; Reel 1055

9020. 1851–10–31 To Charles Edward Trevelyan [Royal Mint]
Death of William Wyon offers opportunity to reorganize engraving department.
ALS 3pp † (by 'M') †(by Charles Wood) TxU:H/L-0481.13; Reel 1055

9021. 1851–11–1 From William Richard Hamilton 12 Bolton Row
Death of William Wyon. Suggests returning Benedetto Pistrucci to office of Chief Engraver, to which WH recommended Pistrucci in 1816 but from which Pistrucci was removed because he was Catholic.
ALS 3pp TxU:H/M–0230; Reel 1087

9022. 1851–11–1 To [Emilia Stewart] [32 Harley St.]
JH's wife sent two pheasants [from Collingwood]. JH forwards one to 'Grandm'a.'
ALS 1p TxU:H/L-0727; Reel 1089

9023. 1851–11–1 From Charles Edward Trevelyan Treasury
JH and H. D. Harness may call on CT at Treasury whenever convenient.
LS 1p TxU:H/M–0551.24; Reel 1093

9024. 1851–11–3 From George Biddell Airy Royal Observatory, Greenwich
Responds with a suggestion of a reasonable price for all the weights [see JH's 1851–10–29].
ACCS 2pp RGO 6.341.11

9025. 1851–11–3 From George Biddell Airy n.p.
A note to clarify receipt of a letter transmitted by Margaret Brodie Herschel.
ACCS 1p inc RGO 6.373.171

9026. 1851–11–3 From William Lassell Starfield, Liverpool
Has discovered two new satellites of Uranus; gives observations. Would JH suggest names for them.
ALS 3pp RS:HS 11.141

9027. 1851–11–4 From George Biddell Airy Royal Observatory, Greenwich
Still pursuing Julius Caesar's path in Britain [see GA's 1851–3–29]; wants local information from JH.
ACCS 4pp RGO 6.466.291v

9028. 1851–11–4 From George Biddell Airy Royal Observatory, Greenwich
A note accompanying testimonials for a candidate for a Sydney University professorship [see JH's 1851–10–29].
ACCS 1p RGO 6.337.70

9029. 1851–11–4 From James H. Cole Downing St.
Chancellor of Exchequer [Charles Wood] wants to know value of alloy presently used in sovereign sterling.
ALS 2pp TxU:H/M–0146; Reel 1087

9030. 1851–11–4 To William James Herschel n.p.
Encourages his son Willie to show that he deserves the favors WH has received from W. H. Sykes.
C 2pp RS:HS 23.111

9031. 1851–11–4 To William Lassell 32 Harley St.
Congratulations to WL on his discovery [see WL's 1851–11–3]; naming should occur once the various satellites have been described and verified.
C 1p RS:HS 23.109

9032. 1851–11–[4 or later] To William Lassell [Collingwood]
Suggests names for the satellites of Uranus, in response to a request from WL [see WL's 1851–11–3].
C 1p RS:HS 25.14.18

9033. 1851–11–4 To Charles Edward Trevelyan Royal Mint
With death of William Wyon, chief engraver, JH wants to leave position of medalist vacant and eliminate Engraving department, turning to outside artists when needed. Proposes to deprive Benedetto Pistrucci, chief medalist, and L. C. Wyon, probationer engraver, of their residences at Mint and appoint them outside 'modellers and engravers' to Mint, retaining James Wyon, presently assistant to chief engraver, as probationary resident engraver under direction of Die department.
LS 1p *Return*, 40–2

9034. 1851–11–4 To Charles [Edward Trevelyan] Harley St.
Analysis of copper alloy in sovereign coin.
ALS 3pp TxU:H/L-0481.15; Reel 1055

9035. 1851–11–5 To George Biddell Airy [Collingwood]
Invites GA to stay at Collingwood and explore the area for evidence of Roman occupation [see GA's 1851–11–4], and talks about a friend of JH's family. Indicates a price offered for weights, but has had no response [see GA's 1851–11–3].
ALS 5pp RGO 6.466.295 & 6.341.12

9036. 1851–11–5 From George Biddell Airy Royal Observatory, Greenwich
Are there any earthworks in the wood near Hawkhurst [see GA's 1851–11–4]?
ACCS 1p RGO 6.466.294

9037. 1851–11–5 To Richard Sheepshanks 32 Harley St.
As the Kew Observatory needs a standard yard, JH is proposing that they be allowed to use one of RS's as a provisional standard until Parliament acts.
ALS 4pp RS:HS 16.107 (C: RS:HS 25.9.71)

9038. 1851–11–6 From Richard Sheepshanks Reading
About the establishment of a standard yard, and how the Kew Committee should proceed [see JH's 1851–11–5]; RS also plans to prepare standard thermometers. Received notice of William Lassell's discovery of more satellites of Uranus.
ALS 4pp RS:HS 16.108

9039. 1851–11–7 To Margaret Brodie Herschel Harley St.
Is extremely depressed at living in London, a 'howling wilderness of brick and mortar;' JH compares at some length the personality of MH and Richarda Airy; brings MH up to date on some family and acquaintance happenings, and mentions in passing that William Lassell has discovered two interior satellites of Uranus [Ariel and Umbriel].
AL 8pp inc JHS 6.24

9040. 1851–11–8 To George Biddell Airy Royal Mint
Is not very knowledgeable about local terrain that might help GA; adds comments about weight buying, and offers a solution [see GA's 1851–11–5].
ALS 4pp RGO 6.466.297 & 6.341.13

9041. 1851–11–8 To Alexander von Humboldt London
JH has been delayed in replying to AH's 1851–8–13 because of official duties at the Great Exhibition. Comments further on several astronomical matters.
C 3pp RS:HS 23.110

9042. 1851–11–8 From John Tyndall Queenwood College
Letter in application for Professorship of Experimental Philosophy and Chemistry at University College, Sydney.
ALS 2pp RS:HS 17.381 (C: RI 493)

9043. 1851–11–8 From John Tyndall Queenwood College
Sends memoirs on magnetism of crystalline bodies. Notes that remarks on atomic mechanism of Iceland spar contrast with JH's views expressed in JH's *Light*.
ALS 2pp RS:HS 17.382 (C: RI 494)

9044. 1851–11–9 From George Biddell Airy Royal Observatory, Greenwich
Will send an assistant to bid for the weights at the auction [see JH's 1851–11–8].
ACCS 1p RGO 6.341.14 (C: RGO 6.341.15)

9045. 1851–11–10 From Christian C. J. Bunsen 9 Carlton Terrace
JH's letter has already gone to Alexander von Humboldt, who has been waiting to hear from him. The fourth volume of *Cosmos*.
ALS 1p RS:HS 5.125

9046. 1851–11–[10] To Robert Mushet Royal Mint
Send names of persons who will provide bonds for RM's new office. JH will forward these to Treasury. Explains purpose of bond.
AdraftS 2pp TxU:H/L-0273; Reel 1054

9047. 1851–11–10 From Edward Sabine Woolwich
Read JH's *Admiralty Manual* meteorology article. Disagrees with statement regarding barometer and pressure of dry atmosphere. Discusses atmosphere and barometer at some length.
ALS 4pp RS:HS 15.234

9048. 1851–11–10 To John Bird Sumner, Archbishop of Canterbury 32 Harley St.
Strong testimonial to Archbishop of Canterbury on behalf of George Waterhouse, over Richard Owen, for late C. D. E. König's position at British Museum.
ACCS 2pp RS:HS 19.143

9049. 1851–11–11 From Charles Edward Trevelyan Treasury
Will seek approval from Chancellor of Exchequer [Charles

Wood] for JH's new arrangement to pay Mint police.
LS 2pp TxU:H/M–0551.25; Reel 1093

9050. 1851–11–12 From Henry Drury Harness Royal Mint
Forwards books for JH.
ALS 1p TxU:H/M–0261.2; Reel 1087

9051. 1851–11–12 From Henry Drury Harness Royal Mint
Chancellor of Exchequer [Charles Wood] agreed to police arrangements at Mint. Asked Sir R. [Manque?] to state when he 'will receive the gates.' Treasury did not approve Bank of England's request. Reviews procedure by which Commissariat department obtains money from Mint through Bank of England. [JH annotation: There was *no time* to count money before it was sent to Commissariat.]
ALS 3pp † TxU:H/M–0261.1; Reel 1087

9052. 1851–11–12 From Gilbert Farquhar Graeme Mathison Richmond
Doctor urges GM to move to warmer climate. Please finalize terms for GM's retirement so GM may leave London.
ALS 3pp TxU:H/M–0394; Reel 1087

9053. 1851–11–13 From George Biddell Airy Royal Observatory, Greenwich
GA's assistant bought the weights at the auction for less than JH had offered [see GA's 1851–11–9].
ACCS 1p RGO 6.341.16

9054. 1851–11–13 From Henry Drury Harness Royal Mint
Received letter from Robert Mushet regarding Mushet's securities. Advises JH to consult George Maule.
ALS 1p TxU:H/M–0261.4; Reel 1087

9055. 1851–11–13 From Henry Drury Harness Royal Mint
Forwards two letters, one from Thomson Hankey.
ALS 1p TxU:H/M–0261.3; Reel 1087

9056. 1851?–11–13 From William Henry Sykes India House
Reports on and discusses applications of hourly wet bulb observations in ocean.
ALS 4pp RS MM.16.164

9057. 1851–11–13 To Charles Wood [Royal Mint]
Reports decision by governors of Bank [of England] concerning which coins are best for exportation. JH suggests alloy of copper and gold for Canadian circulation.
Adraft 2pp TxU:H/L-0508.1; Reel 1055

9058. 1851–11–14 From Michael Faraday Royal Institution
Did as JH wished with the papers. Is busy with magnetism and lines of force. Sends a couple of printings.
ALS 1p RS:HS 7.195

9059. 1851–11–14 From Charles Edward Trevelyan Treasury
Annual pension for G. F. G. Mathison was fixed by Chancellor of Exchequer [Charles Wood] at 'Fourteen Hundred Pounds (1200£).' JH and Henry Rich should submit formal report on this matter.
ALS 2pp TxU:H/M–0551.26; Reel 1093

9060. 1851–11–14 To [Charles Edward Trevelyan] [Royal Mint]
Asks for reconsideration of £2000 security required of melter Robert Mushet, whose wages are 'very moderate.' Now that all workmen are under jurisdiction of Mint Police, JH believes that dishonesty is less likely and that securities should be reduced.
ACCS 3pp TxU:H/L-0653; Reel 1089

9061. 1851–11–15 From Gilbert Farquhar Graeme Mathison Richmond
Thanks JH for arranging terms of retirement for GM, who is now free to leave England. Presumes that GM's pension began as of last July.
ALS 3pp TxU:H/M–0395; Reel 1087

9062. 1851–11–15 From Charles Wood Downing St.
Asks JH to join CW at meeting with governor of Bank of England [Thomson Hankey] on Monday to discuss coining matters. Thanks for JH's letters on quicksilver and on coinage.
ALS 3pp TxU:H/M–0596; Reel 1086

9063. 1851–11–16 To George Biddell Airy 32 Harley St.
Information about some candidates for Sydney University professorships [see JH's 1851–10–29].
ALS 3pp RGO 6.337.71

9064. 1851–11–16 To George Biddell Airy 32 Harley St.
Partly identical to JH's 1851–11–16, but with different information about some candidates.
ALS 3pp RGO 6.337.73

9065. 1851–11–17 From William Keith Murray Stonehaven
Is thinking of erecting an observatory and would welcome JH's advice on telescopes and object-glasses, particularly those of William Simms.
ALS 4pp RS:HS 13.61

9066. 1851–11–17 From H. R. Reynolds Treasury
HR assumed duties of George Maule, who died Friday. Please send papers relating to Robert Mushet, so HR can process Mushet's security bond for office of Melter.
ALS 4pp TxU:H/M–0433; Reel 1093

9067. 1851–11–18 From George Biddell Airy Royal Observatory, Greenwich
Comments about some candidates for Sydney University professorships [see JH's 1851–10–29].
ACCS 3pp RGO 6.337.75

9068. 1851–11–18 From James H. Cole Downing St.
Chancellor of Exchequer [Charles Wood] postponed tomorrow's meeting with governor and deputy governor of Bank of England.
ALS 2pp TxU:H/M–0147; Reel 1087

9069. 1851–11–18 From Charles Edward Trevelyan Treasury
CT delays answering JH's letter of 3 Nov. 1851 until JH submits report on financial impact of changes in Engraving department after death of William Wyon.
LS 2pp TxU:H/M–0551.27; Reel 1093

9070. 1851–11–19 From Henry Drury Harness Royal Mint
Passage in Report of Mint Commission referred to by Robert Mushet, regarding securities for Mint officials, does not pertain to Mushet. Details of Mushet's appointment as senior clerk and melter. Thanks for extract from [Michel] Chevalier's book. Cost of [mercury] constitutes 18 percent of expense for reducing ore.
ALS 3pp TxU:H/M–0261.5; Reel 1087

9071. 1851–11–19 From H. R. Reynolds Treasury
Send copies of security bonds for former Melters, so HR may compare these to provisions in Robert Mushet's bond.
ALS 3pp TxU:H/M–0434; Reel 1093

9072. 1851–11–20 From Charles Edward Trevelyan Treasury
What public service will L. C. Wyon render that justifies Wyon's annual salary of £400? Compares this with duties and salary of Benedetto Pistrucci. Suggests retaining Wyon at present salary of £200 and paying separately for piecework.
LS 4pp TxU:H/M–0551.28; Reel 1093

9073. 1851–11–21 To Charles [Edward Trevelyan] Royal Mint
Case of L. C. Wyon, promoted to chief engraver, is different from those of Henry Bingley, whose office [Queen's assay master] was abolished, and of Mr. Field, who succeeded J. M. Beckwith [as resident assayer].
ACCS 3pp TxU:H/L-0481.16; Reel 1055

9074. 1851–11–22 From William Henry Sykes India House
Noting discrepancy between dew points obtained by wet bulb and direct methods, WS cannot say which is nearer the truth; suspects that both have inaccuracies.
ALS 4pp RS:HS 17.148

9075. 1851–11–23 To Edward Sabine 32 Harley St.
Remarks on a correction in manual regarding some fluctuation of heating and cooling.
ALS 3pp RS Sa.665

9076. 1851–11–24 From Charles Wood Downing St.
Accepts JH's argument against coining gold dollar, but feels that two-dollar gold piece is needed. Reports plan to build mint in New South Wales. U.S. will establish mint in California.
ALS 3pp TxU:H/M–0597; Reel 1086

9077. 1851–11–25 From George Biddell Airy Royal Observatory, Greenwich
Is sending some papers on standards prepared by Richard Sheepshanks.
ACCS 2pp RGO 6.341.17

9078. 1851–11–25 From James H. Cole Downing St.
Chancellor of Exchequer [Charles Wood] invites JH to tomorrow's meeting with governor and deputy governor of Bank of England.
ALS 2pp TxU:H/M–0148; Reel 1087

9079. 1851–11–25 From H. R. Reynolds Treasury
Sends draft of security bond for Robert Mushet that names master of Mint as sole judge of what constitutes reasonable waste in melting process. Insert names of persons, including banker, who will act as surety for Mushet.
ALS 3pp TxU:H/M–0435; Reel 1093

9080. 1851–11–26 To George Biddell Airy n.p.
A note about the receipt of papers from Richard Sheepshanks.
ALS 2pp RGO 6.337.79

9081. 1851–11–26 From George Biddell Airy Royal Observatory, Greenwich
Does not believe the Australian gold rush should affect the selection process for Sydney University professors [see GA's 1851–11–18].
ACCS 3pp RGO 6.337.77

9082. 1851–11–26 From Sir William Snow Harris Plymouth
Encloses and comments on abstract of important manuscripts by the late English scientist [Henry] Cavendish.
ALS 7pp encl RS MM.16.124

9083. [1851]–11–26 From Henry Rich Treasury
Chancellor of Exchequer [Charles Wood] agrees to give G. F. G. Mathison annual allowance of £1200. Suspects mistake in report on Mathison.
ALS 4pp TxU:H/M–0443.12; Reel 1093

9084. 1851–11–27 From James David Forbes Edinburgh
Regrets he has not been able to find anyone suitable for the chair of chemistry and natural philosophy at Sydney University.
ALS 2pp St. Andrews LB V, 103–4

9085. 1851–11–27 From William Lassell Starfield, Liverpool
Giving his recent observations of the satellites of Uranus. Has not detected more than four satellites.
ALS 4pp RS:HS 11.142

9086. 1851–11–27 From Charles Edward Trevelyan Treasury Chambers
Treasury commissioners approved JH's plan to eliminate Engraving department [see JH's 1851–11–4]. Benedetto Pistrucci and L. C. Wyon will receive fixed salaries and housing allowances as nonresident modelers and engravers to Mint.
LS 1p *Return*, 42

9087. [1851]–11–27 To [Charles Wood] [Royal Mint]
Accepts responsibility for error in amount awarded to [G. F. G. Mathison]'s pension by JH and Henry Rich. Sum of £1200 should have read £1400.
AdraftS 4pp TxU:H/L-0675; Reel 1089

9088. 1851–11–28 From Thomas Maclear Royal Observatory, Cape of Good Hope
Has recommended [Andrew] Bain use JH's patronage to present the geological map of South Africa to the Geological Society. Comments on Bain's visit. Affairs have not improved in Kaffir Land.
ALS 2pp RS:HS 12.152 (C: Univ. of Witwatersrand)

9089. 1851–11–28 To Robert Mushet Royal Mint
Needs names of banks that can verify credit of persons who will provide RM's bond. Clause that names master of Mint as sole judge is normal practice.
ALS 2pp TxU:H/L-0274; Reel 1054

9090. 1851–11–29 From Charles Edward Trevelyan Treasury
Grateful for JH's correction of CT's error in pension for G. F. G. Mathison [see CT's 1851–11–14]. This was CT's first such error in 25 years of public service.
ALS 2pp TxU:H/M–0551.29; Reel 1093

9091. 1851–12–1 From George Biddell Airy Royal Observatory, Greenwich
May not be able to attend next Sydney Trust meeting [see GA's 1851–12–1], as GA has family commitments.
ACCS 1p RGO 6.337.81

9092. 1851–12–1 From George Biddell Airy Royal Observatory, Greenwich
A note accompanying testimonials for candidates for Sydney University professorships [see GA's 1851–11–18].
ACCS 1p RGO 6.337.80

9093. 1851–12–1 From William Henry Bateson St. John's College, Cambridge
Meeting of Cambridge University Commissioners postponed to 17 Dec.
ALS 2pp TxU:H/M–0057.10; Reel 1087

9094. 1851–12–1 To [William Snow Harris] 32 Harley St.
Thanks for WH's book on electricity, including experiments by Henry Cavendish. Wishes that these had been published by R.S.L. Asks for clarification on experiments by C. A. Coulomb. [JH annotation: 'Not Sent. Met Sir WH at R.S. and transmitted verbally.']
ACC 3pp †(by JH) TxU:H/L–0655; Reel 1089

9095. [1851–12–1 or later] To [R.S.L. Council] n.p.
Reports on a paper by Baden Powell read at the R.S. L. and relating to the wave theory of light and to a paper by Henry Brougham.
ADS 3pp RS RR.2.92

9096. 1851–12–4 From George Biddell Airy Royal Observatory, Greenwich
Indicates a preferred date for a meeting of the Sydney Trust [see GA's 1851–12–1].
ACCS 1p RGO 6.337.82

9097. 1851–12–4 To George Biddell Airy 32 Harley St.
Still arranging the next meeting of the Sydney Trust [see GA's 1851–12–4].
ALS 3pp RGO 6.337.83

9098. [1851]–12–4 From Henry Rich Treasury
Believes information is missing from report on G. F. G. Mathison.
ALS 7pp TxU:H/M–0443.13; Reel 1093

9099. 1851–12–5 From George Biddell Airy Royal Observatory, Greenwich
Agrees to JH's meeting date [see JH's 1851–12–4].
ACCS 2pp RGO 6.337.85

9100. 1851–12–5 From Richard Sheepshanks Athenaeum
Asks JH to sign certificate for membership in R.S.L. for William Simms.
ALS 4pp RS:HS 16.111

9101. 1851–12–5 From Richard Sheepshanks Athenaeum
Tells JH of RS's dealings with the Kew Observatory Committee; comments further on standard thermometers.
ALS 4pp RS:HS 16.110

9102. 1851–12–5 From Charles Edward Trevelyan Treasury
G. F. G. Mathison called on CT yesterday to discuss error in Mathison's pension. Will finalize this after CT receives report from JH and Henry Rich.
ALS 2pp TxU:H/M–0551.30; Reel 1093

9103. 1851–12–6 To Arthur Hugh Clough 32 Harley St.
Acknowledges the safe arrival of AC's testimonials.
ALS 1p Bodleian Library, Oxford

9104. 1851–12–6 From William Richard Hamilton 12 Bolton Row
Conferred with Benedetto Pistrucci before sending WH's letter to JH on Pistrucci's behalf. Asks JH to review JH's decision regarding Pistrucci.
ALS 2pp TxU:H/M–0231; Reel 1087

9105. 1851–12–7 From Benedetto Pistrucci Royal Mint
The Royal Commission discusses BP's position at the Mint. BP will meet with H. D.Harness to confer over BP's position.
ALS 4pp (Italian) TxU:H/M–0429.9; Reel 1093

9106. 1851–12–7 To Richard Sheepshanks [32 Harley St.]
Agrees to sign certificate for William Simms, but as JH is at present on the Council of the R.S.L., his signature may not be allowed.
ALS 2pp RS:HS 16.112 (C: RS:HS 25.9.72)

9107. 1851–12–8 From William Richard Hamilton 12 Bolton Row
Grieved to hear that JH's recommendation to Treasury does not do justice to Benedetto Pistrucci's public service. Reviews Pistrucci's appointment by Lord Maryborough [William Wellesley-Pole] in 1815. Lists Pistrucci's objections to present appointment at Mint.
ALS 4pp TxU:H/M–0232; Reel 1087

9108. 1851–12–8 From Gilbert Farquhar Graeme Mathison 10 Dover St.
Delighted to learn by JH's letter of 6 Dec. that JH plans to provide for men at Mint who worked under GM. Will prepare statements of their services and pay scales.
ALS 1p TxU:H/M–0396; Reel 1087

9109. 1851–12–8 To My dear Sir 32 Harley St.
Just received [?]'s letter, but believes it was written before JH visited [?] today. Assumes that all points in letter were answered at meeting.
ACCS 3pp TxU:H/L–0656; Reel 1089

9110. 1851–12–9 From Benedetto Pistrucci Royal Mint
Speculates that there might be an increase in duties for BP, now that W. C. Wyon is dead. Considers living at the Mint.
ALS 3pp (Italian) TxU:H/M–0429.10; Reel 1093

9111. 1851–12–9 To [Benedetto Pistrucci] Royal Mint
Death of William Wyon necessitated reorganization of Engraving department, which Treasury commissioners have already sanctioned. Sorry that [BP] finds new title objectionable. Hopes that [BP] will not focus on 'painful recollections over which the grave has closed.'
AdraftS 5pp TxU:H/L–0657; Reel 1089

9112. 1851–12–9 From Richard Sheepshanks Royal Astronomical Society
Thanks JH for his attention to his request. Has inquired about the etiquette of members of the Council signing certificates: it is done, though not regularly.
ALS 4pp RS:HS 16.113

9113. 1851–12–10 From Sir Charles Lyell 11 Harley St.
JH should speak privately to Archbishop [J. B. Sumner] on behalf of G. R. Waterhouse. Richard Owen's supporters are strong, but as governor of British Museum, [Sumner] is obliged to speak favorably for staff members who expect promotion. Will tell [Sumner] that Waterhouse is JH's relative.
ALS 3pp TxU:H/M–0344; Reel 1087

9114. 1851–12–11 To George Biddell Airy n.p.
Still trying to arrange a meeting of the Sydney Trust [see JH's 1851–12–4].
ALS 3pp RGO 6.337.86

9115. 1851–12–11 From George Biddell Airy Royal Observatory, Greenwich
Agrees to meeting date [see JH's 1851–12–11] and asks JH to

make some additional arrangements.
ACCS 3pp RGO 6.337.88

9116. 1851–12–11 From Benedetto Pistrucci Royal Mint
Requests JH to pursue the matter of BP's retirement pension. Compares BP's salary with those of other Mint officials, including the Modeller and the Engraver. Discusses the medal designs that the Royal Commission charged BP to create over the years.
ALS 7pp (Italian) TxU:H/M–0429.11; Reel 1093

9117. [1851]–12–11 From Henry Rich Treasury
More considerations concerning G. F. G. Mathison. HR believes mistakes in report were due to error made by C. E. Trevelyan.
ALS 3pp TxU:H/M–0443.15; Reel 1093

9118. 1851–12–12 From William Henry Bateson St. John's College, Cambridge
Sorry for mistake that led JH to show up [on 10 Dec.] for postponed meeting. University senate authorized vice chancellor to address queries raised by Cambridge University Commissioners regarding college revenues.
ALS 4pp TxU:H/M–0057.11; Reel 1087

9119. 1851–12–12 From William Rutter Dawes Wateringbury, Maidstone
Suggesting someone suitable as a Greek scholar. News of his relations. Is pleased she likes his description of the solar eclipse. Regarding the priority of the discovery of the faint rings of Saturn.
ALS 11pp RS:HS 6.97

9120. 1851–12–12 From William Richard Hamilton 12 Bolton Row
Clarifies attitude of Benedetto Pistrucci toward probationary medallist [L. C.] Wyon. Pistrucci objects to JH giving same title to both artists. If JH desires, WH will accompany JH to see John Russell about this matter.
ALS 4pp TxU:H/M–0233; Reel 1087

9121. 1851–12–12 From Henry Drury Harness Royal Mint
Thanks JH for book. Asks approval for HH's letter about medals drafted to Leonard Wyon.
ALS 1p TxU:H/M–0262.1; Reel 1087

9122. 1851–12–12 To Sir Robert Harry Inglis n.p.
Bringing the position of G. R. Waterhouse to his notice and giving list of persons who have testified in his favor.
draft 3pp RS:HS 10.248

9123. 1851–12–13 From George Biddell Airy Royal Observatory, Greenwich
Is working on a short list of candidates for the Sydney University professorships [see GA's 1851–12–11].
ACCS 1p RGO 6.337.91

9124. 1851–12–13 From Sir Robert Harry Inglis 7 Bedford Square
Thanks for his communication enclosing the letter concerning G. R. Waterhouse, which has been forwarded to the Archbishop of Canterbury [J. B. Sumner] with a covering note.
ALS 2pp RS:HS 10.249

9125. 1851–12–16 From Charles Edward Trevelyan Treasury
Bring report by R. L. Sheil when JH and H. D. Harness visit CT to discuss amalgamation of Coining and Machinery departments. Congratulations on 'remarkable distinction' obtained by JH's son [William] at Haileybury College, CT's alma mater.
LS 3pp TxU:H/M–0551.31; Reel 1093

9126. 1851–12–17 From Henry Drury Harness Royal Mint
Will go to the Treasury as requested by JH. Sees nothing in letters from Benedetto Pistrucci or W. R. Hamilton that justifies JH's reconsidering this case. Leonard Wyon requested copy of Treasury letter that changed arrangements for Engraving department.
ALS 3pp, JH note 2pp TxU:H/M–0262.2; Reel 1087

9127. [1851–12–17] To [Benedetto Pistrucci] [Royal Mint]
Sorry that [BP] refuses to accept new position in reorganized Engraving department. Admires [BP]'s artistic qualities, but will not reinstate distinction between medallist and engraver.
Adraft 4pp TxU:H/L–0634; Reel 1089

9128. 1851–12–17 From Leonard Charles Wyon Royal Mint
Seeking appointment as engraver to Goldsmith's Company. Needs to learn status of LW's position as engraver to Mint.
ALS 2pp TxU:H/M–0602; Reel 1086

9129. 1851–12–17 To Leonard Charles Wyon Downing St.
Assures LW that title of 'Modeller and Engraver at the Royal Mint,' conferred on LW, is '*no "inferior appointment."*'
ACCS 3pp TxU:H/L–0511; Reel 1055

9130. 1851–12–18 From George Biddell Airy Flamsteed House, Greenwich
Regarding his recent tour around the Kent and Sussex borders.
ALS 4pp RS:HS 1.222

9131. 1851–12–19 From John Couch Adams St. John's College, Cambridge
Concerning the recent eclipse of the sun, seen in Norway. Observations on candidates for Mathematical Professorship at Sydney.
ALS 8pp RS:HS 1.8

9132. 1851–12–19 From Leonard Charles Wyon Royal Mint
Thanks for JH's information about status of LW's position as engraver to Mint. LW was today appointed engraver to Goldsmith's Company.
ALS 1p TxU:H/M–0603; Reel 1086

9133. 1851–12–21 From George Biddell Airy Royal Observatory, Greenwich
Sends GA's choices for Sydney University professorships to JH [see GA's 1851–12–13].
ACCS 3pp RGO 6.337.92

9134. 1851–12–21 To George Biddell Airy 32 Harley St.
Comments and further material on some candidates for the Sydney University professorships [see GA's 1851–12–21].
ALS 3pp RGO 6.337.94

9135. 1851–12–21 From Benedetto Pistrucci Royal Mint
Repeats his desire for a pension. Criticizes the efficiency of some of the Mint officials. Complains many others have been granted residences and pensions. Needs a competent worker to harden the dies for the Waterloo Medal.
ALS 7pp (Italian) TxU:H/M–0429.12; Reel 1093

9136. 1851–12–22 From Francis Beaufort Admiralty
Thomas Maclear has asked for a transit circle and therefore the Admiralty would like JH's comments on this.
ALS 3pp RS:HS 3.398 & ACC: Hydrographic Office Letter Book

9137. 1851–12–22 From Charles Edward Trevelyan Treasury
Has returned to Treasury. JH and H. D. Harness may call whenever convenient.
LS 1p TxU:H/M–0551.32; Reel 1093

9138. 1851–12–22 To Charles [Edward Trevelyan] Royal Mint
JH and H. D. Harness will meet CT tomorrow as planned.
ACCS 1p TxU:H/L-0481.17; Reel 1055

9139. 1851–12–23 From George Biddell Airy Royal Observatory, Greenwich
A note about papers sent and received [see GA's 1851–12–21].
ACCS 1p RGO 6.337.96

9140. 1851–12–23 From Charles Edward Trevelyan Treasury Chambers
Treasury commissioners approve compensation by Paymaster-general to G. F. G. Mathison for loss of earnings due to abolition of refinery at Mint.
LS 1p *Return*, 53

9141. 1851–12–23 From George Robert Waterhouse British Museum
Received appointment [as keeper of mineralogical and geological branch of British Museum]. Grateful for JH's support. Celebrating birthday of GW's wife.
ALS 2pp RS:HS 18.98

9142. 1851–12–24 From Henry Drury Harness Royal Mint
Asks JH to sign enclosed letter concerning estimate called for by Treasury. Needs decision from Benedetto Pistrucci in order to pay Pistrucci this month. Will send copy of Treasury letter to G. F. G. Mathison.
ALS 2pp TxU:H/M–0262.3; Reel 1087

9143. 1851–12–24 From Thomas Maclear Royal Observatory, Cape of Good Hope
Now forwards [Andrew] Bain's Geological map, which had been delayed by the Agent. Has heard nothing definite regarding the transit circle but would appreciate JH's recommendation, together with those of Francis Beaufort and G. B. Airy. Gives details of the work at the Observatory. Estimates the cost of mounting the transit circle.
ALS 4pp RS:HS 12.154

9144. 1851–12–24 To Charles Edward Trevelyan Royal Mint
Encloses draft of letter, resulting from yesterday's meeting, for submission to Chancellor of Exchequer [Charles Wood].
AdraftS 3pp TxU:H/L-0481.18; Reel 1055

9145. 1851–12–25 From William Henry Bateson St. John's College, Cambridge
Meeting of Cambridge University Commissioners for 13 Jan. is changed to 20 Jan.
ALS 2pp TxU:H/M–0057.12; Reel 1087

9146. 1851–12–25 To [James Calder] Stewart Hawkhurst
Lists letters and papers to be sent to JH from Collingwood.
AL 3pp TxU:H/L-0435; Reel 1055

9147. 1851–12–26 From George Biddell Airy Playford near Ipswich
Some business relating to the Sydney Trust [see GA's 1851–12–23].
ACCS 1p RGO 6.337.97

9148. 1851–12–26 From Charles Edward Trevelyan Treasury
Chancellor of Exchequer [Charles Wood] will not need to discuss JH's draft of 'Estimate' until after New Year's Day.
LS 2pp TxU:H/M–0551.33; Reel 1093

9149. 1851–12–27 To George Biddell Airy Collingwood
Differences of opinion about the best candidates for the Sydney University professorships [see GA's 1851–12–26].
ALS 4pp RGO 6.337.98

9150. 1851–12–27 To Francis Beaufort Collingwood
Report for the Lords of the Admiralty on the desirability of supplying Thomas Maclear with a transit circle.
ALS 4pp RS:HS 3.398a (C: RS:HS 23.112)

9151. 1851–12–28 From George Robert Waterhouse British Museum
Richard Owen retiring, recommending GW for post of C. D. E. König [died 6 Sept. 1851].
ALS 2pp RS:HS 18.99

9152. 1851–12–30 To George Biddell Airy Collingwood
About a procedure to make a final decision about two of the Sydney University professorships without having another meeting [see JH's 1851–12–27].
ALS 4pp RGO 6.337.100

9153. 1851–12–31 To George Biddell Airy Collingwood
Still not finished with the process [see JH's 1851–12–30].
ALS 3pp RGO 6.337.104

9154. 1851–12–31 From George Biddell Airy Playford near Ipswich
Has some difficulty with JH's proposal [see JH's 1851–12–30], but accepts it and makes his choices for the Sydney University professorships.
ACCS 3pp RGO 6.337.102

9155. 1852–1–1 To George Biddell Airy Collingwood
Is circulating JH's votes for the remaining Sydney University professorships [see JH's 1851–12–31].
ALS 2pp RGO 6.337.106

9156. [1852?] To Charles Babbage n.p.
Can he lend him the Cambridge Calendar or tell him the examiners in the mathematical tripos.
ALS 2pp RS:HS 2.379

9157. [1852–1857] From Ernesto Capocchi n.p.
Presents JH with a work [*Lezioni di astronomia* by François Arago and translated by EC] on the foundations of astronomy. Discusses the beauty of the science.
ALS 3pp (Italian) RS:HS 5.325

9158. [1852 or later] To Robert FitzRoy Collingwood
Is trying to help RF understand the difference between the dynamics of solid bodies moving over each other and the dynamics of moving fluids, both water and air.
C 5pp RS:HS 25.14.11

9159. [1852] To George Hamilton-Gordon n.p.
Attributes reorganization of Mint to H. D. Harness. Recommends Harness for its mastership.
Adraft 2pp TxU:H/L-0002; Reel 1054

9160. 1852 To Margaret Brodie Herschel 32 Harley St.
Clarifies family finances, including the sale of stock to pay living expenses.
ALS 4pp JHS 1.85

9161. 1852 To Margaret Brodie Herschel [London]
Is explaining why JH will not be able to meet MH's train.
ALS 2pp JHS 1.84

9162. 1852–1–1 or later To Margaret Brodie Herschel [London]
Complains of overwork and of other people trying to involve JH in their quarrels; is glad MH has company.
ALS 4pp JHS 1.8

9163. [1852?]–[?]–20 To William James Herschel [London]
Is giving WH advice about buying books to take to India. JH stresses the need for economy and discernment; urges WH to retain some interest in science, although WH will not be 'a scientific man.'
ALS 4pp JHS 5.34

9164. [1852] From John Stewart Pau, Pyrenees
Explains in detail his photographic method for producing landscapes on paper.
L *Athenaeum,* #1311 (Dec. 11, 1852), 1363–4

9165. [1852?] From William Henry Sykes India House
Suggests possible times to meet with JH.
ALS 2pp RS MM.16.165

9166. [1852 to 1854] To [Charles Edward Trevelyan] [Royal Mint]
Requests Treasury commissioners' approval for increasing salary of C. W. Goodwin, fireman in Die department. Lists new orders for dies that have increased Goodwin's workload.
Adraft 2pp TxU:H/L-0607; Reel 1089

9167. 1852–1–2 To George Biddell Airy 32 Harley St.
The process of selection of the two remaining Sydney University professors is complete [see JH's 1851–12–31].
ALS 3pp RGO 6.337.108

9168. 1852–1–2 From George Biddell Airy Playford near Ipswich
A note accompanying some letters related to the Sydney Trust [see GA's 1851–12–31].
ACCS 1p RGO 6.337.107

9169. 1852–1–2 To My dear Sir [Royal Mint]
JH must know if [?] accepts or declines new arrangements for [?]'s office, already approved by Treasury commissioners.
AdraftS 2pp TxU:H/L-0659; Reel 1089

9170. 1852–1–2 From William Digby Sadleir Trinity College, Dublin
Apologizes for unwarranted familiarity used in previous letter requesting JH's evaluation of the scientific work of [Humphrey] Lloyd.
ALS 3pp RS:HS 17.166

9171. 1852–1–3 To George Biddell Airy 32 Harley St.
Many of the testimonials sent to the Sydney Trust are missing [see JH's 1852–1–2].
ALS 3pp RGO 6.337.110

9172. 1852–1–3 From Benedetto Pistrucci Royal Mint
Expects a salary increase now that BP has the title of Chief Medalist.
ALS 1p (Italian) TxU:H/M–0430.1; Reel 1093

9173. 1852–1–4 From George Biddell Airy Playford near Ipswich
Thankful the process is over and pleased with the results; has no testimonials [see JH's 1852–1–3].
ACCS 1p RGO 6.337.112

9174. 1852–1–5 From Benedetto Pistrucci Royal Mint
Finds it scandalous that BP has been promoted without an increase in salary: '... in this Establishment, I have been degraded from what was my due as an Artist.' Wishes to take up the matter with the Treasury.
ALS 5pp TxU:H/M–0430.2; Reel 1093

9175. 1852–1–6 From Sir Robert Harry Inglis Milton Bryan, Woburn
Will the Herschels come to dinner on Thursday the 13th with one of their daughters?
ALS 1p RS:HS 10.250

9176. 1852–1–7 From Henry Drury Harness Royal Mint
Information about Sydney branch mint. Agrees with JH on price of assays for Bank of England. Salary dispute in Mint.
ALS 3pp TxU:H/M–0263.1; Reel 1087

9177. 1852–1–7 From Benedetto Pistrucci Royal Mint
Asks JH to give him a few days before making a reply to the proposed pension offer.
ALS 1p TxU:H/M–0430.3; Reel 1093

9178. 1852–1–8 From Benedetto Pistrucci Royal Mint
Thanks JH for proposing a memorial and retirement allowance for him to the Treasury, but insists on a £500 allowance (his old salary) and a residence at the Mint.
ALS 4pp TxU:H/M–0430.4; Reel 1093

9179. 1852–1–9 To George Biddell Airy 32 Harley St.
One of the candidates chosen for a Sydney University professorship may not be free to go on time; still missing testimonial letters [see JH's 1852–1–3].
ALS 4pp RGO 6.337.113

9180. 1852–1–9 To Margaret Brodie Herschel [London]
Worried about war and a French invasion; JH has been treating his cold with laudanum and mustard blisters.
ALS 4pp JHS 1.75

9181. 1852–1–9 To Benedetto Pistrucci Royal Mint
Denies BP a request concerning his acceptance of the office of Chief Medalist. Discusses BP's reluctance to accept the office without being awarded a place of residence by the Treasury.
ACC 4pp TxU:H/L-0309; Reel 1093

9182. 1852–1–9 To Richard Sheepshanks 32 Harley St.
Is in urgent need of William Lassell's address in Malta.
ALS 2pp RS:HS 16.121

9183. 1852–1–10 From John Tyndall Queenwood College
Asks JH to keep memoirs JT sent with regard to JT's application for position at University of Sydney.
ALS 1p RS:HS 17.383a (C: RI 497)

9184. 1852–1–12 To Margaret Brodie Herschel 32 Harley St.
Consoles MH on 'horrid affair.' Thankful that consequences were not worse. Arranging transportation for son William to Paris. Dinner with [Margaret Seymour,] Duchess of Somerset. Encloses letter from Richard Sheepshanks. Honored by Society of Arts request to include [JH's]

photography in display to be shown to Prince Albert.
ALS 4pp TxU:H/L-0543.1; Reel 1053

9185. 1852–1–12 From Charles Edward Trevelyan Treasury
Sends altered letter for JH's records. Received JH's altered letter for Treasury records.
LS 1p TxU:H/M–0552.2; Reel 1093

9186. 1852–1–12 From Charles Edward Trevelyan Treasury
Memorandum from Treasury's Financial Room showing quarterly installments paid to Master of Mint. Please explain unusually large expenditures. Previous Mint administrators seldom expended entire grant from Parliament as JH has done. This will be last installment until Parliament approves grant for 1852–53. CT wants JH's assurance that this installment will last until end of fiscal year [31 Mar.].
LS 2pp, CT note 1p TxU:H/M–0552.1; Reel 1093

9187. 1852–1–12 To Charles [Edward Trevelyan] Royal Mint
W. T. Brande is willing to give up professorship [of chemistry] at Royal Institution to accept combined duties of superintendent of coining and die departments.
AdraftS 8pp TxU:H/L-0482.1; Reel 1055

9188. 1852–1–12 To Charles [Edward Trevelyan] Royal Mint
Explains large disbursement of silver to Moneyers and to Melter.
ACCS 3pp TxU:H/L-0482.2; Reel 1055

9189. 1852–1–13 From Henry Drury Harness Royal Mint
Asks JH to approve two letters regarding arrangements with W. T. Brande, whose salary is divided between Mint and paymaster general.
ALS 1p TxU:H/M–0263.2; Reel 1087

9190. 1852–1–13 From Charles Edward Trevelyan Treasury
Explain £47,694 excess of Mint expenses in fiscal year 1851, relative to 1850, shown in enclosed statement.
LS 2pp, CT note 1p TxU:H/M–0552.4; Reel 1093

9191. 1852–1–13 From Charles Edward Trevelyan Treasury
Replace old title of 'Melter' for Robert Mushet's office with new title used in reorganized Mint. Why does Mushet call his annual pay of £100 'inadequate'? Send JH's 'draft letter & Estimate' so CT can explain to Chancellor of Exchequer [Charles Wood] difference between JH's original and present proposals regarding [W. T.] Brande.
LS 3pp TxU:H/M–0552.3; Reel 1093

9192. 1852–1–13 To Charles Edward Trevelyan Royal Mint
Proposes to amalgamate Coining department and Machinery and Die department, with W. T. Brande as superintendent of new Coining and Die department.
LS 1p *Return*, 42

9193. 1852–1–14 From George Biddell Airy Royal Observatory, Greenwich
Has found the missing testimonials; if the chosen candidate cannot go, send the next one in line [see JH's 1852–1–9].
ACCS 3pp RGO 6.337.115

9194. 1852–1–14 To Charles Edward Trevelyan Treasury
Requests raise in salary for W. H. Barton, whose prior service was not properly credited during reorganization of Mint.
ACCS 1p TxU:H/M–0552.6; Reel 1093

9195. 1852–1–14 From Charles Edward Trevelyan Treasury
Chancellor of Exchequer [Charles Wood] approved new position of W. T. Brande and compensation for Brande's loss of income at Royal Institution, but declined to raise salary immediately for W. H. Barton, new deputy master. Please send to CT details of Brande's former salary. Wood will submit to John Russell [Prime Minister] JH's recommendation to appoint Brande's son [F. R. Brande] to position of supernumerary clerk.
LS 3pp † TxU:H/M–0552.5; Reel 1093

9196. 1852–1–14 From Charles Edward Trevelyan Treasury Chambers
Treasury commissioners approved JH's plan to amalgamate Coining department and Machinery and Die department. W. T. Brande is approved as superintendent of new Coining and Die department.
LS 1p *Return*, 42

9197. 1852–1–14 To Charles Edward Trevelyan [Royal Mint]
Returns Robert Mushet's note. Changed title of Mr. Reynolds's office from melter to "Senior Clerk and Melter.' Adjustments to salary of W. T. Brande.
ACC 2pp TxU:H/L-0482.3; Reel 1055

9198. 1852–1–15 From George Biddell Airy Royal Observatory, Greenwich
Writes to inquire as to the whereabouts of the Richard Sheepshanks's papers [see GA's 1851–11–25].
ALS 1p TxU:H/M-0635.1; Reel 1084 (C: RGO 6.341.18)

9199. 1852–1–15 From George Biddell Airy Royal Observatory, Greenwich
Apologizes for the business of the missing testimonials [see GA's 1852–1–14].
LACC 2pp inc RGO 6.337.119

9200. 1852–1–15 From Charles Pritchard to Margaret and John Herschel Ramsgate
Reports on progress of the three Herschel sons at Clapham. Asks JH's advice on his plan to construct a telescope. Looks forward to visit from JH.
ALS 8pp RS:HS 14.76

9201. 1852–1–15 From William Henry Sykes India House
Agrees with papers demonstrating falsity of present methods of correcting barometer for moisture and 'dogma of one gas being a vacuum to another.'
ALS 2pp RS:HS 17.149

9202. 1852–1–15 From Charles Edward Trevelyan Treasury
Returns drafts and estimate. Expects decision soon on claims by Robert Mushet and Mr. Hill.
LS 1p TxU:H/M–0552.7; Reel 1093

9203. 1852–1–16 To George Biddell Airy n.p.
Faster steam ships available may solve the time problem for the Sydney University professor [see GA's 1852–1–14].
ALS 4pp RGO 6.337.117

9204. 1852–1–16 To Margaret Brodie Herschel 32 Harley St.
Believes that war is imminent [see JH's 1852–1–9], that the south of England may well be attacked, and that MH and the

girls should be prepared to go north.
AL 4pp JHS 1.72

9205. 1852–1–16 From H. R. Reynolds Treasury
Security bond has been approved for Robert Mushet. Will send bond to JH after Mushet signs it.
ALS 3pp TxU:H/M–0436; Reel 1093

9206. 1852–1–16 To Richard Sheepshanks 32 Harley St.
Because of strong feeling about the procedure in the R.S.L. Council, JH has decided he will not sign the certificate for William Simms, although JH will support his candidature.
ALS 4pp RS:HS 16.114 (C: RS:HS 25.9.73)

9207. 1852–1–17 From Reginald Stuart Poole 1 West Terrace, Worthing
JH having liked RP's book on Egyptian chronology, asks JH for a testimonial for position in Department of Antiquities at British Museum.
ALS 2pp RS:HS 14.38

9208. 1852–1–17 From Sir Edward Ryan Garden Lodge Kensington
Has sent three letters of introduction for JH's son [William J. Herschel?]. Says he will be accepted in the best society because of his accomplishments and name.
ALS 3pp RS:HS 14.462

9209. 1852–1–18 From Richard Sheepshanks Reading
Argues cogently against the policy of the Council [see JH's 1852–1–16]; is continuing his work on the standard yard.
ALS 4pp RS:HS 16.115

9210. 1852–1–20 or later To Margaret Brodie Herschel 32 Harley St.
Son Willy is getting better but the doctor is coming to see him anyway; JH sympathizes with MH and daughter Caroline.
ALS 3pp JHS 1.76

9211. 1852–1–20 To Margaret Brodie Herschel [London]
Son Willy is ill and JH is treating him.
ALS 3pp JHS 1.74

9212. 1852–1–20 From Sir Robert Harry Inglis 7 Bedford Square
Will he join them at dinner tomorrow; regrets he was prevented from being with them on the 13th due to a cold.
ALS 1p RS:HS 10.251

9213. 1852–1–20 From Benedetto Pistrucci Royal Mint
Desires more time before making an official response to the Treasury's retirement offer.
ALS 1p (Italian) TxU:H/M–0430.5; Reel 1093

9214. 1852–1–20 To Edward Sabine 32 Harley St.
Disagreement with W. H. Sykes over whether to publish observations by General [Cullen?] that Sykes claims have never been reviewed or read [to R.S.L.]. Encloses copy of JH's reply to Sykes.
AdraftS 1p, encl ACC 2pp TxU:H/L-0369; Reel 1054

9215. 1852–1–20 From Charles Edward Trevelyan Treasury
Sends memorial by G. F. G. Mathison, 1850 report by R. L. Sheil, and Privy Council warrant of 7 Mar. 1851 superseding former Mint indenture. Please return these.
LS 2pp TxU:H/M–0552.8; Reel 1093

9216. 1852–1–21 From Charles Edward Trevelyan Treasury
Chancellor of Exchequer [Charles Wood] wants to avoid coining new silver money for Commissariat and for Colonies if these can be obtained through Bank [of England]. JH will receive such request today from Western Australia.
LS 2pp TxU:H/M–0552.9; Reel 1093

9217. 1852–1–22 To Edward Joshua Cooper 32 Harley St.
Congratulates EC on the first volume of *Stars near the Ecliptic*: JH expects never to observe again.
C 1p RS:HS 23.113

9218. 1852–1–22 From Charles Edward Trevelyan Treasury
Chancellor of Exchequer [Charles Wood] authorizes increasing salary of W. H. Barton to £700. Write official letter to this effect with full details, as if no previous correspondence had occurred. [Prime Minister] John Russell approved appointment of W. T. Brande's son as supernumerary clerk in Mint. Treasury is now ready to receive JH's budget for 1852–53.
LS 3pp TxU:H/M–0552.10; Reel 1093

9219. 1852–1–23 To Charles [Edward Trevelyan] Royal Mint
Changes in engraving department are delayed by Benedetto Pistrucci's 'extreme reluctance' to alter title of Chief Medallist. Suggests offering Pistrucci official assurance of continued employment.
AdraftS 2pp TxU:H/L-0482.4; Reel 1055

9220. 1852–1–23 From Charles Edward Trevelyan Treasury
Does not understand difficulty with Benedetto Pistrucci. Expects BP to be available for employment whenever needed. Whether BP should execute a particular work is JH's decision, not BP's.
LS 2pp TxU:H/M–0552.11; Reel 1093

9221. 1852–1–24 From Benedetto Pistrucci to the Treasury Royal Mint
Outlines his history with the Mint, including his promotion to Chief Engraver in 1817. Claims he was promised a pension at this time. Reports that the Treasury declared there were 'insurmountable difficulties as [BP] was a foreigner and a Catholic ...' in obtaining the original pension granted. The pension was then reduced from £500 to £350. Complains that his son was not granted the apprenticeship promised him. Outlines all pension plans proposed by the Treasury and why they are inadequate.
ACCS 6pp TxU:H/M–0430.6; Reel 1093

9222. 1852–1–26 To Margaret Brodie Herschel [London]
His concern [see JH's 1852–1–16] was not to suggest less bravery on the part of women and girls, but because Collingwood might well be in the middle of any fray. Agrees with MH's precautions about revaccination.
ALS 4pp JHS 1.73

9223. 1852–1–29 To George Biddell Airy 32 Harley St.
Proposal for dividing travel money for the Sydney University professors. Does GA agree [see JH's 1852–1–9]?
ALS 3pp RGO 6.337.120

9224. 1852–1–30 To [Robert Hunt ?] London
Has no photographs with him and when next JH goes to Collingwood, he will look to see if there are any left.
ALS 2pp WT 68037.13

9225. 1852–1–30 From Benedetto Pistrucci Royal Mint
Sends JH a copy of a request to the Lords of the Treasury for

a pension settlement [see BP's 1852–1–24].
ALS 1p (Italian) TxU:H/M–0430.7; Reel 1093

9226. 1852–1–30 From H. R. Reynolds Treasury
Encloses completed security bond for Robert Mushet, with additional title of Senior Clerk added, as suggested by JH. [JH annotation: Procedure for processing and storing Mushet's bond.]
ALS 1p † TxU:H/M–0437; Reel 1093

9227. 1852–1–31 From George Biddell Airy Royal Observatory, Greenwich
Agrees with JH's proposal regarding travel money for Sydney University professors [see JH's 1852–1–29].
ACCS 2pp RGO 6.337.122

9228. 1852–1–31 From George Biddell Airy n.p.
A note to add to a previous letter.
ACCS 1p RGO 6.374.125

9229. [1852 or earlier]–1–31 From John Barlow Royal Institution
Encloses item that is not to be distributed.
ALS 1p RS:HS 19.162a

9230. 1852–1–31 To Margaret Brodie Herschel [London]
MH is suffering [teeth?] and is urged to come to London; a few items of news and household accounting.
ALS 4pp JHS 1.78

9231. 1852–2–2 From Thomas Maclear Royal Observatory, Cape of Good Hope
Has forwarded four boxes of geological specimens by the *Bosphorus*; these are the illustrations for [Andrew] Bain's map. Peace is about to be declared in the Kaffir War. There have been several bush fires and one almost damaged Feldhausen.
ALS 4pp RS:HS 12.155

9232. 1852–2–2 To William Henry Sykes n.p.
Suggests course of action regarding 'Gen'l. Cullen's [?]' hydrometric observations submitted to R.S.L.
draft 1p inc RS MM 16.171

9233. 1852–2–3 From William Rutter Dawes Wateringbury, Maidstone
Thanks for his help regarding his nephew. Hopes mild weather has not upset JH. Doubtless he has seen notes of the various observations of Saturn in *R.A.S.M.N.*
ALS 3pp RS:HS 6.98

9234. 1852–2–4 To William Hallowes Miller Royal Mint
The balances in the Mint are not adequate for the work WM wants to do.
C 1p RS:HS 23.115

9235. 1852–2–4 From Reginald Stuart Poole 1 West Terrace, Worthing
Thanks JH for recommendation for position in Department of Antiquities at British Museum.
ALS 3pp RS:HS 14.39

9236. 1852–2–4 To Charles Edward Trevelyan Royal Mint
How should JH deal with Benedetto Pistrucci's sudden and temperamental resignation [as chief medallist]?
Adraft 2pp TxU:H/L-0482.5; Reel 1055

9237. 1852–2–5 To Henry Drury Harness 32 Harley St.
C. E. Trevelyan's corrections to report. Arrangement for disposing of refinery is officially ratified and may be mentioned. Include payments for Moneyers' equipment, but omit reference to standards. Trevelyan knows nothing of Benedetto Pistrucci's memorial.
ACCS 4pp TxU:H/L-0184.1; Reel 1054

9238. 1852–2–5 From Lord John Russell Downing St.
Cannot nominate [Richard] Jones for position at Fenly. Has already given another name to the Queen.
ALS 2pp RS:HS 14.442

9239. 1852–2–6 From William Henry Bateson St. John's College, Cambridge
Forwards copy of memorial sent to all Cambridge University Commissioners by town clerk of Cambridge.
ALS 1p TxU:H/M–0057.13; Reel 1087

9240. 1852–2–6 From Francis Beaufort Admiralty
Encloses recently received letters and a geological map. Shall he send it to Harley St.? There is a strong possibility that the Cape will get a transit circle. William Simms has a good object glass; does JH know anything about it?
ALS 2pp RS:HS 3.399 & ACC: Hydrographic Office Letter Book

9241. 1852–2–7 From George Biddell Airy Royal Observatory, Greenwich
Would like the brief version of Richard Sheepshanks's paper to send to William Parsons.
ALS 1p † TxU:H/M-0635.2; Reel 1084

9242. 1852–2–7 From George Biddell Airy n.p.
A note about how to send mail to Richarda Airy in Madeira.
ACCS 1p RGO 6.374.126

9243. 1852–2–10 To Edward Joshua Cooper Harley St.
Returns EC's manuscript with some comments.
C 1p RS:HS 23.116

9244. 1852–2–10 To Charles Edward Trevelyan Royal Mint
Assignment of residences to Mint officials.
AdraftS 2pp TxU:H/L-0482.6; Reel 1055

9245. 1852–2–11 To George Biddell Airy n.p.
Returns Richard Sheepshanks's papers with some comments [see GA's 1852–1–15].
ACCS 2pp RGO 6.341.20

9246. 1852–2–11 To [Richard Madox] Bromley Royal Mint
RB's certificates were issued in 1825 by now-abolished Mint Board. Explains reorganized administration of Mint.
AdraftS 4pp TxU:H/L-0113; Reel 1054

9247. 1852–2–11 To Margaret Brodie Herschel [London]
About JH's health and MH's problems with her teeth, followed by a series of brief news items.
ALS 4pp JHS 1.77

9248. 1852–2–11 To William Henry Smyth 32 Harley St.
Discusses the naming of a Plateau of Brussels. Discusses names to go on list; gives JH's and Michael Faraday's recommendations.
C 1p RS:HS 23.117

9249. 1852–2–12 From Hugh Lee Pattinson Newcastle-upon-Tyne
Thanks JH for recommendation to R.S.L.; asks if JH can write a note so HP can continue for candidacy.
ALS 2pp RS:HS 14.145

9250. 1852–2–12 To Charles Edward Trevelyan Royal Mint
Does not know how to reply to Robert Mushet's claim for compensation. Can this case be reconsidered?
AdraftS 3pp TxU:H/L-0482.7; Reel 1055

9251. [1852]–2–12 To Charles Edward Trevelyan 32 Harley St.
Sends duplicate of report to Parliament that JH submitted to Chancellor of Exchequer. W. T. Brande's plan to resign professorship.
ACCS 4pp TxU:H/L-0479.6; Reel 1055

9252. 1852–2–13 From Charles Edward Trevelyan Treasury
Does not know how to reply to Robert Mushet's claim for compensation for loss of succession to G. F. G. Mathison's office. No such succession was ever granted before. Returns Mushet's letter.
(A)LS 3pp TxU:H/M–0552.14; Reel 1093

9253. 1852–2–13 From Charles Edward Trevelyan Treasury
Will present JH's report on Mint immediately. [Parliamentary printer] Hansard will be informed that JH is responsible for all corrections. Advises including two notices in report: that Mint refinery was leased to Anthony de Rothschild, and that Mastership is no longer a Parliamentary office.
LS 4pp TxU:H/M–0552.13; Reel 1093

9254. 1852–2–13 From Charles Edward Trevelyan Treasury
Suggests putting notice of new weighing machine into JH's general report on Mint.
LS 1p TxU:H/M–0552.12; Reel 1093

9255. 1852–2–13 From Henry Drury Harness Royal Mint
Instruct Mr. [Housard?] not to 'twist off and distribute' until we have more proof. Anthony de Rothschild visited Mint today to learn when he can take possession [of refinery]. Treasury solicitors should arrange this lease soon.
ALS 2pp TxU:H/M–0263.3; Reel 1087

9256. 1852–2–14 From Edward Joshua Cooper 10 Chesham St.
Contrasts EC's catalogue of comets with that of J. F. Encke.
ALS 4pp RS MM.16.159

9257. 1852–2–16 From Charles Edward Trevelyan [Treasury]
Asks for JH's opinion on answering Benedetto Pistrucci's letter.
ALS 1p TxU:H/M–0552.17; Reel 1093

9258. 1852–2–16 From Charles Edward Trevelyan Treasury
Chancellor of Exchequer [Charles Wood] approved JH's report on Mint reform. CT noted some corrections. JH may make other corrections up to time report is read to Parliament.
LS 3pp TxU:H/M–0552.16; Reel 1093

9259. 1852–2–16 From Charles Edward Trevelyan Treasury
Suggests form for announcing new character of Mastership of Mint in JH's report to Parliament. Let Chancellor of Exchequer [Charles Wood] approve it before JH inserts it into report.
LS 2pp TxU:H/M–0552.15; Reel 1093

9260. 1852–2–16 To Charles Edward Trevelyan Royal Mint
Added CT's suggested paragraphs to report. There is no documentary evidence to support [non-political] basis of Mastership of Mint. Suggests issuing a warrant to that effect.
ALS 3pp TxU:H/L-0482.9; Reel 1055

9261. 1852–2–17 From William Henry Bateson St. John's College, Cambridge
Meeting of Cambridge University Commissioners for 25 Feb. is postponed to 26 Feb.
ALS 2pp TxU:H/M–0057.14; Reel 1087

9262. 1852–2–17 From Francis Beaufort Admiralty
Is pleased to inform him that the Admiralty has ordered a transit circle for the Cape.
ALS 1p RS:HS 3.400

9263. 1852–2–17 From Charles Edward Trevelyan Treasury
Chancellor of Exchequer [Charles Wood] approved JH's change to Mint report regarding office of Master. P.S.: Who authorized annual housing allowance of £300 for Master's residence?
(A)LS 1p TxU:H/M–0552.18; Reel 1093

9264. 1852–2–18 To Thomas Maclear 32 Harley St.
Congratulates TM on engineering the construction of transit circle. Sent letter to Francis Beaufort concerning transit circle.
ALS 3pp encl Cape Archives/Maclear Papers File 100

9265. 1852–2–18 To Charles Edward Trevelyan [Royal Mint]
[Marked 'Private.' Non-political basis of Mastership of Mint] rests solely on Charles Wood's verbal communication. Regrets Benedetto Pistrucci's effort to resign [as chief medallist].
Adraft 4pp TxU:H/L-0482.10; Reel 1055

9266. 1852–2–19 To Charles Babbage 32, Harley St.
Does not wish to become a Vice-Patron of the College of Preceptors.
ALS 3pp RS:HS 2.325

9267. 1852–2–19 From Henry Drury Harness Royal Mint
Asks JH to review a document. HH will look for errors in other copy of it.
ALS 1p TxU:H/M–0263.4; Reel 1087

9268. 1852–2–[19] To My dear Sir Royal Mint
Has reconsidered and dismissed JH's previous plan. Will explain reason to [?] in person later today. Please change wording in document regarding Master's salary.
ACC 1p TxU:H/L-0660; Reel 1089

9269. 1852–2–19 From Charles Edward Trevelyan Treasury
Does Mint provide official housing for Master? If not, JH's annual housing allowance of £300 should be extra, and not deducted from JH's salary of £1500. Thanks for JH's hints on responding to Benedetto Pistrucci.
LS 3pp TxU:H/M–0552.19; Reel 1093

9270. 1852–2–21 From Hugh Lee Pattinson Newcastle-upon-Tyne
Thanks JH for note of recommendation for his continued candidacy to the R.S.L.
L 1p inc RS:HS 14.146

9271. [1852–2–21 or earlier] To Charles Edward Trevelyan n.p.
In Mr. Rich's absence, JH does not know how to process claims by apprentice moneyers F. R. Brande and Henry Finch. Ask for advice from Chancellor of Exchequer [Charles Wood]. Henry Bingley's concern for apprenticeship of Bingley's son.
AdraftS 2pp TxU:H/L-0479.2; Reel 1055

9272. 1852–2–22 From William Henry Bateson St. John's College, Cambridge
Meeting of Cambridge University Commissioners is again postponed. Dean of Ely [George Peacock] unable to attend.
ALS 2pp TxU:H/M–0057.15; Reel 1087

9273. 1852–2–23 To Charles Edward Trevelyan Royal Mint
Returns letters of Jasper Atkinson. Issue of moneyers appears to have been closed by decision of lords of Treasury on 8 July 1851. Sees no reason to reconsider it. [JH Note on verso: Abstract of two letters from Jasper Atkinson to Sir John Russell, prime minister, requesting compensation to moneyers for lost income at Mint.]
ACCS 1p [JH note] TxU:H/L-0482.11; Reel 1055

9274. 1852–2–26 To Margaret Brodie Herschel [London]
Encloses son Willy's school report; adds some news items of family and friends.
ALS 4pp JHS 1.81

9275. 1852–2–27 From Sir Robert Harry Inglis 7 Bedford Square
Sending the first fruits of the enterprise of Dr. J. L. Krapf.
ALS 1p RS:HS 10.252

9276. 1852–2–28 From William Henry Bateson St. John's College, Cambridge
Cambridge University Commissioners will meet on 16 Mar.
ALS 1p TxU:H/M–0057.16; Reel 1087

9277. 1852–2–28 From Charles Richard Weld R.S.L., Somerset Place
Announces meeting of R.S.L. council on 11 March.
(A)LS 1p TxU:H/MW-0246.9; Reel 15

9278. 1852–3–1 From George Biddell Airy Royal Observatory, Greenwich
A notice of meeting of the Standards Committee.
LS 1p TxU:H/M-0635.3; Reel 1084

9279. 1852–3–1 From James A. Gordon St. George's Hotel
Many thanks for the correction to his book. Pleased to find that the Russians had not escaped him. Suggests subjects for drawings for the Herschels.
ALS 3pp RS:HS 8.198

9280. 1852–3–2 To Margaret Brodie Herschel [London]
Sends home son Alexander's school report, and includes some news items of relatives and friends.
ALS 4pp JHS 1.83

9281. 1852–3–[2?] To Margaret Louisa Herschel 32 Harley St.
Louisa's studies and her tutors. Wedding in which Louisa will take part. Activities and studies of daughters Caroline and Isabella in London. Lists book and papers to be sent to JH and MBH in London.
ALS 4pp TxU:H/L-0533; Reel 1053

9282. 1852–3–2 To Benedetto Pistrucci Royal Mint
On the importance of record keeping in the Mint to attest to the labor and talent of those working there.
draft 2pp TxU:H/L-0310; Reel 1093

9283. 1852–3–2 From Reginald Stuart Poole 1 West Terrace, Worthing
Being able to conclude only that J. B. Biot was wrong in his conclusion, RP cannot make computation (per JH's request) about heliacal rising himself. Has applied to G. B. Airy to do it. Expresses difficulty in attaining accuracy about any ancient author's recording of a rising.
ALS 6pp RS:HS 14.40

9284. 1852–3–4 From Benedetto Pistrucci Royal Mint
Awaits a response to his pension proposal.
ALS 1p (Italian) TxU:H/M–0430.8; Reel 1093

9285. 1852–3–5 To George Biddell Airy 32 Harley St.
Wants to begin using decimal fractional weights of pounds and ounces at the Royal Mint, rather than pennyweights and grains. Would like to extend this method beyond the Mint. Asks for date of heliacal rising of Sirius at Athens in the time of Euripides.
ALS 4pp RGO 6.341.21

9286. 1852–3–6 From Edward Sabine 11 Old Burlington St.
Suggests postponing meeting of Wintringham Bequest Committee. Will bring up [Edward] Cooper's letter about comet memoir at next council meeting. Has worked out cost reductions for printing of *R.S.P.T.*
ALS 4pp RS MM.16.160

9287. 1852–3–8 From George Biddell Airy Royal Observatory, Greenwich
Is asking JH to bring the long version of Richard Sheepshanks's paper to the next meeting of the Standards Committee [see JH's 1852–2–11].
ALS 1p TxU:H/M-0635.4; Reel 1084 (C: RGO 6.341.23)

9288. 1852–3–8 From Benedetto Pistrucci Royal Mint
Rejects the Treasury's offer as 'dishonorable and degrading to my character as an artist.' Disagrees with combining the offices of Clerk of Irons, Superintendent of Machinery, and Chief Coiner. Lists eleven names of Mint officers, comparing years of service in relation to their retirement settlements. Proposes to accept £400 annual salary with the title of Her Majesty's Chief Medalist.
ALS 7pp TxU:H/M–0430.9; Reel 1093

9289. [1852?]–3–9 To 'My Lord' Royal Mint
Submits to Treasury commissioners a design by L. C. Wyon for Kaffir medal.
Adraft 2pp (on Jones's and Smith's 1853–12–5) TxU:H/L-0719.2; Reel 1089

9290. 1852–3–11 To [Robert Hunt ?] [London]
Says JH will not join the society to which he has been invited, because he has done nothing on the subject for a long time and JH's attendance would not be productive. Thanks for some lectures and papers sent to him.
ALS 3pp WT 68037.14

9291. 1852–3–13 To Margaret Brodie Herschel London
Describes music making evenings with daughters Caroline and Isabella; has made arrangements for the sale of a carriage.
ALS 4pp JHS 1.82

9292. 1852–3–15 From George Biddell Airy Royal Observatory, Greenwich
Agrees with JH's idea of decimalizing the pound weight, but the stone should not then be used [see JH's 1852–3–5].
ALS 2pp TxU:H/M-0635.5; Reel 1084 (C: RGO 6.341.24)

9293. 1852–3–15 To William Hallowes Miller 32 Harley St.
Identifies WM as a committee member for the R.S.L.; raises the question of standard weights in relation to weighing gold bullion.
C 1p RS:HS 23.118

9294. 1852–3–16 To Edward Joshua Cooper Harley St.
Offers a variety of solutions to stopping 'offensive light' during solar observations.
C 2pp RS:HS 23.119

9295. 1852–3–16 From Edward Sabine 11 Old Burlington St.
Discusses regulations of bequest, which will benefit chemistry and physics. Discusses relationship between solar spot period and magnetic disturbances. Wants JH's comments on paper regarding this topic.
ALS 4pp RS:HS 15.235

9296. 1852–3–17 From William Thomas Brande Royal Mint
Asks JH for introduction to C. E. Trevelyan for WB and Mr. Jenkins, representative for Mrs. Finch, to discuss their claims against Company of Moneyers regarding apprenticeship fee.
ALS 2pp TxU:H/M–0118; Reel 1087

9297. 1852–3–18 To Charles Edward Trevelyan Royal Mint
Notifies commissioners of Treasury that Mastership of Mint is not included among official government offices listed in Superannuative Act.
ACCS 1p TxU:H/L-0482.14; Reel 1055

9298. 1852–3–20 From Charles Edward Trevelyan Treasury
Asks for JH's opinion on enclosed note from accountant W. G. Anderson.
LS 2pp TxU:H/M–0552.20; Reel 1093

9299. 1852–3–21 From Annibal de Gasparis Naples
Has discovered another new planet [Psyche]. Gives readings.
ALS 1p (French) RS:HS 8.46

9300. [1852–3–21 or later] To Annibal de Gasparis n.p.
[JH note:] Annibal de Gasparis acknowledged receipt of JH's *Cape Results* [sent in 1847] 'by letter of March 21/52!!'
Amem 1p RAS:JH Archive 10/6.3; Reel 9

9301. 1852–3–21 To Roderick Impey Murchison 32 Harley St.
Gives RM a letter from Charles Bell [see Bell's 1851–10–28] on a geological map and collection prepared by Andrew Bain. Suggests the British Museum might wish to purchase Bain's collection of fossils.
C 1p Univ. of Witwatersrand

9302. 1852–3–23 To Francis Beaufort 32 Harley St.
The theoretical considerations that JH would take into account if he were going to study the subject of 'increasing and conveying to great distances and in given directions the reports of guns or other explosive instruments, as signals.'
ALS 11pp Hydrographic Office Misc F26F1.1

9303. [1852]–3–23 To Francis Beaufort 32 Harley St.
Suggestions and diagrams for projecting sound great distances by means of small arms and parabolic reflectors.
CS 8pp & Adraft 4pp TxU:H/L-0086 & L-0085; Reel 1054

9304. 1852–3–24 To Francis Beaufort [32 Harley St.]
A note probably accompanying JH's 1852–3–23. Comments concerning someone's collection.
ALS 3pp Hydrographic Office Misc F26F1.2a

9305. 1852–3–24 From Francis Beaufort Admiralty
JH's paper will 'infuse new life into' sound experiments now being conducted at Woolwich [Academy].
ALS 2pp TxU:H/M–0070; Reel 1087 (ACC: Hydrographic Office Letter Book)

9306. 1852–3–24 From William Lassell Starfield, Liverpool
Is well satisfied with the names JH has suggested for the satellites, but wonders whether the continental astronomers will accept them. Is puzzled by the orbits of two satellites.
ALS 4pp RS:HS 11.143

9307. 1852–3–24 From H. R. Reynolds Treasury
Altered draft security bond to make Mr. Shaw obliged to Queen instead of to master of Mint. Left penalty amount blank for JH to fill in. [JH annotation: Completed draft of bond and returned it.]
ALS 4pp † TxU:H/M–0438; Reel 1093

9308. 1852–3–25 From William Lassell Starfield, Liverpool
Found he had made an error in his letter of yesterday. Encloses an amended table.
ALS 3pp RS:HS 11.144

9309. 1852–3–26 From Benedetto Pistrucci Royal Mint
Refuses JH's proposal that BP accept the title of Modeller and Engraver with a £400 salary. Also requests that the monthly coal allowance for BP's workshop be restored.
ALS 2pp (Italian) TxU:H/M–0430.10; Reel 1093

9310. 1852–3–27 From H. R. Reynolds Treasury
Returns completed draft of security bond for Mr. Shaw.
ALS 1p TxU:H/M–0439; Reel 1093

9311. 1852–3–28 To [Charles Edward Trevelyan?] Royal Mint
After Treasury commissioners returned [Benedetto Pistrucci's]'s first memorial, to be transmitted through proper channels, JH hoped that [Pistrucci] would alter its content. But because [Pistrucci] insisted, JH forwarded it to Treasury again.
ALS 2pp TxU:H/L-0661; Reel 1089

9312. 1852–3–29 From Thomas Maclear Royal Observatory, Cape of Good Hope
Has received two notes, one from G. B. Airy and one from Francis Beaufort informing him that the Admiralty has promised to supply a transit circle; the estimates have now been passed. The Magnetic Observatory building at the Cape has been destroyed by fire. Andries Stockenström is taking an active part in politics. H. G. W. Smith has been recalled.
ALS 3pp RS:HS 12.156

9313. 1852–3–29 To W. C. Peters 32 Harley St.
Explains algebraic equation.
ALS 3pp TxU:H/L-0300; Reel 1054

9314. 1852–3–29 To [?] 32 Harley St.
As part of a discussion about decimal coinage, JH states that

an early issue of florins is contemplated.
ALS 3pp WT 76066

9315. 1852–3–31 From Henry Walter Haselbury Bryan, Blandford [Dorset]
Was JH's examiner forty years ago. Please explain why northern winters, at perigee, are not colder than southern winters, at apogee. Gratified that JH is in office held by Isaac Newton.
ALS 7pp † RS:HS 18.373

9316. 1852–4–1 From Henry Drury Harness Royal Mint
JH's illness. Accounts are not ready for JH's signature.
ALS 1p TxU:H/M–0263.5; Reel 1087

9317. 1852–4–2 From William Henry Bateson St. John's College, Cambridge
Meeting of Cambridge University Commissioners for 6 Apr. is postponed to 20 Apr.
ALS 2pp TxU:H/M–0057.17; Reel 1087

9318. 1852–4–3 From Charles Edward Trevelyan Treasury
Wants to discuss three topics when JH has time to call on CT. [JH annotation: 'Pistrucci; Superannuation; Deficits in the Exchequer credit accounts.']
LS 1p † TxU:H/M–0552.21; Reel 1093

9319. 1852–4–4 From William Richard Hamilton 12 Bolton Row
Asks JH to grant permission for WH's son and friends to tour Mint. Hopes business with Benedetto Pistrucci will be settled soon. [JH annotation: JH also wants Pistrucci business settled, but latest developments lead JH to expect anything *but* a satisfactory conclusion.]
ALS 3pp † TxU:H/M–0234; Reel 1087

9320. 1852–4–6 To Benedetto Pistrucci Royal Mint
Informs BP that the Treasury has approved his pension request. JH declines to mention BP's refusal to accept a salary from the Treasury, a statement made in haste.
ACCS 1p TxU:H/L-0311; Reel 1093

9321. 1852–4–6 From Edward A. Seymour Park Lane
Thanks for work recently sent and generally for astronomical discoveries. Sends two 'little works' for JH's inspection.
ALS 3pp RS:HS 17.193

9322. 1852–4–6 From George Gabriel Stokes Pembroke College Cambridge
Asks JH questions about a paper on the dispersion of light from beyond the visible violet end of spectrum.
ALS 2pp RS:HS 17.27

9323. 1852–4–7 From Charles Edward Trevelyan Treasury
Call on CT today.
ALS 1p TxU:H/M–0552.22; Reel 1093

9324. 1852–4–8 From William Lassell Starfield, Liverpool
Intends to use JH's names for the satellites of Uranus in a communication unless JH has any objections.
ALS 2pp RS:HS 11.145

9325. 1852–4–12 From Edward Sabine Woolwich
Believes similarity of period of solar spots and that of terrestrial magnetic variation is more than coincidental.
ALS 2pp RS:HS 15.236

9326. 1852–4–15 From John P. Gassiot Mark Lane
Is grateful for his prompt reply to his request for friend to visit the Mint.
ALS 1p RS:HS 8.58

9327. 1852–4–16 To Margaret Brodie Herschel [London]
Discusses the political atmosphere in London and some family matters.
ALS 4pp JHS 1.86

9328. 1852–4–16 From Benedetto Pistrucci Royal Mint
Ceases receiving coal supply to his office, an allowance made to BP for 35 years. Mint owes him four months' salary. Makes a reference to plight of G. B. Airy, who might be turned out of his residence at the Royal Observatory.
ALS 7pp TxU:H/M–0430.11; Reel 1093

9329. 1852–4–16 From Edward Sabine Woolwich
Discusses magnetic observations from Toronto and Hobarton from the 1840s. Attempting to determine when range of variation was greatest.
ALS 2pp RS:HS 15.237

9330. 1852–4–17 From Alexander von Humboldt Potsdam
Thanks JH for JH's review of AH's *Cosmos*. Recommends E. H. DuBois-Reymond to JH's attention.
C 1p (German) Sydney Ross Collection

9331. 1852–4–17 To George Gabriel Stokes 32 Harley St.
Replies to several queries about papers of JH. JH describes the accidental production of some interesting copper salt crystals.
C 2pp RS:HS 25.10.2

9332. [1852]–4–19 From Benedetto Pistrucci [Royal Mint]
Clarifies BP's 24 Jan. request for pension. Prefers to remain active with current salary and title. If required to leave Mint, BP will continue to execute dies for free, if compensated for loss of BP's home and studio. History of BP's contract with Mint.
C (by JH) 3pp TxU:H/L-0766.1; Reel 1089

9333. 1852–4–20 From Francis Beaufort Admiralty
Sends a note to JH that William Lassell [may use as a letter of introduction to the Governor of Malta]. FB hopes that Lassell will determine the longitude of Malta.
ACC 1p Hydrographic Office Letter Book

9334. 1852–4–20 From Robert Grant Solely Terrace, Lloyd Square
Sends a copy of RG's *History of Physical Astronomy* as a token of his admiration for JH's work.
ALS 1p RS:HS 9.15

9335. 1852–4–20 From William Richard Hamilton 12 Bolton Row
Suggests that JH authorize special payment from Audit office to Benedetto Pistrucci. If JH will not come to Pistrucci's relief, then WH will assist Pistrucci. Is JH trying to reduce Pistrucci into submission?
ALS 2pp TxU:H/M–0235.1; Reel 1087

9336. [1852–4–20] To Charles Edward Trevelyan Royal Mint
Sent Treasury's retirement form to Benedetto Pistrucci, who returned it blank. Encloses copy of Pistrucci's reply.
ACCS 1p TxU:H/L-0766.2; Reel 1089

9337. 1852–4–22 From Francis Beaufort Admiralty
Copy of Mr. Stewart's note to JH was sent to Baldwin Walker, surveyor of Navy, to evaluate new 'stopper' it recommended. Returns original to JH.
ALS 1p TxU:H/M–0071; Reel 1087 (ACC: Hydrographic Office Letter Book)

9338. 1852–4–22 From Thomas Cole 42 Manchester St., Argyle Square, New Road
Requests permission to visit Mint as personal favor to TC, whose father knew JH's father personally.
ALS 2pp TxU:H/M–0149; Reel 1087

9339. 1852–4–22 To William Richard Hamilton Royal Mint
Private. Benedetto Pistrucci appealed to Lord Derby [E. G. G. S. Stanley]. If consulted, JH will recommend that former distinctions within Engraving department be abolished. If Pistrucci insists on using his former title, he may no longer do so as an officer of the Mint. Will approve pension if Pistrucci chooses to retire.
AdraftS 4pp TxU:H/L–0181; Reel 1054

9340. 1852–4–22 From Johann Lamont Munich
Bookseller at Stuttgart has requested JL to make a translation of JH's *Outlines Astr.* Would like JH's views on the matter.
ALS 3pp RS:HS 11.452

9341. 1852–4–22 From Benedetto Pistrucci Royal Mint
Thanks JH for having the kindness to warn BP of construction on the wall adjacent to BP's studio. Asks for BP's salary again.
ALS 4pp TxU:H/M–0430.12; Reel 1093

9342. 1852–4–22 To [James Calder] Stewart Harley St.
Enclosed items just arrived from Francis Beaufort. [Richard?] Jones is in good health.
ALS 1p TxU:H/L–0436; Reel 1055

9343. 1852–4–23 To Robert Grant 32 Harley St.
Thanks RG for a copy of the *History of Physical Astronomy*; comments on the behavior of cometary tails.
draft 3pp RS:HS 9.15 verso (C: RS:HS 23.120)

9344. 1852–4–23 From William Richard Hamilton 12 Bolton Row
Benedetto Pistrucci reports that JH refuses to pay Pistrucci. Admonishes JH to treat Pistrucci kindly and prevent Pistrucci from having to beg for a living. Pistrucci also worries that new construction in Weighing Room at Mint will ruin Pistrucci's furnaces and prevent completion of Waterloo Medal. WH criticizes latest changes in Mint policies as injurious to great artists like Pistrucci for the sake of 'free Trade.'
ALS 6pp TxU:H/M–0236; Reel 1087

9345. [1852]–4–24 To William Richard Hamilton Royal Mint
Private. If JH *did* try to pay Benedetto Pistrucci as 'Chief Medallist,' Audit office would refuse to approve it, because that position no longer exists at Mint. No one opened walls of Pistrucci's studio for repairs; workmen simply requested permission to examine them to see if repairs were needed. Pistrucci was never denied access to dies for Waterloo Medal; he asked once for them when keys were not available and never came back.
ACCS 4pp TxU:H/L–0180; Reel 1054

9346. 1852–4–24 From John R. Hind 22 Grove Rd., St. John's Wood
Sending a table of variable stars. Appears to have seen the new planet [Psyche] on 29 Jan.
ALS 3pp RS:HS 9.355

9347. 1852–4–25 To Thomas Maclear [32 Harley St.]
Andrew Bain's geological work received well. JH has little time for personal correspondence.
ALS 4pp Cape Archive/Maclear Papers File 100

9348. 1852–4–27 From William Henry Bateson St. John's College, Cambridge
Cambridge University Commissioners will meet on 11 May.
ALS 1p TxU:H/M–0057.18; Reel 1087

9349. 1852–4–27 To William Robert Grove n.p.
Offers assistance as needed.
ALS 1p RI

9350. 1852–4–27 From William Lassell Starfield, Liverpool
Thinks the climate of this country is unsuitable for good observations and is thinking of setting up his telescope in Malta. What does JH think of this idea? Can JH provide an introduction to the Governor of Malta.
ALS 7pp RS:HS 11.146

9351. [1852]–4–28 To Francis Beaufort 32 Harley St.
William Lassell wishes to go to Malta to observe the heavens, and would like a letter of introduction to the governor there. Can FB assist with this?
ALS 3pp Hydrographic Office Misc F26F1.3

9352. 1852–4–28 From William Richard Hamilton 12 Bolton Row
Lists injustices suffered by Benedetto Pistrucci under JH's administration. Chancellor of Exchequer [Charles Wood] stated that Pistrucci's salary and duties would not change. Urges JH to let Pistrucci keep former title of Chief Medallist as well; otherwise JH will sign Pistrucci's 'Death-Warrant.'
ALS 4pp TxU:H/M–0237; Reel 1087

9353. 1852–4–28 To William Lassell n.p.
Is willing to assist in obtaining letters of introduction for WL's projected trip to Malta.
C 1p RS:HS 23.121

9354. 1852–4–29 From Thomas Cole 42 Manchester St., Argyle Square, New Road
Gratitude for JH's permission for TC to visit Mint.
ALS 1p TxU:H/M–0150.1; Reel 1087

9355. 1852–4–29 To William Richard Hamilton 32 Harley St.
Delay in resolving matter with Benedetto Pistrucci is not JH's fault. JH will not dignify WH's last question with a response [see WH's 1852–4–20].
ACCS 1p TxU:H/M–0235.2; Reel 1087

9356. 1852–4–29 From John Drew Salmon Wenham Lake Ice Company, 164 Strand
Please pay overdue bill for ice delivered in Feb. 1851.
LS 1p TxU:H/M–0476.1; Reel 1093

9357. 1852–4–30 From Thomas Maclear Royal Observatory, Cape of Good Hope
Is grateful for JH's letter to the Admiralty recommending a transit circle for the Cape. Comments on a suitable room for it in the Observatory. Has just had a peep at Encke's comet. Still trouble with the Kaffirs. Will be pleased to see Willy [Herschel] when he passes that way.
ALS 3pp RS:HS 12.157

9358. 1852–4–30 To the Treasury [Royal Mint]
There being greater need of gold coinage than the Mint can presently meet, JH asks for additional staff and calls attention to a possible reason for the shortage.
L *Report from the Royal Commission on International Coinage; Together with the Minutes of Evidence and Appendix* (London, 1868), Appendix: pp. 324–5

9359. 1852–5–1 To Charles Edward Trevelyan Royal Mint
Information for Lord Chandos [Richard Grenville, junior lord of Treasury]. History of Benedetto Pistrucci's employment at Mint.
AdraftS 4pp & Adraft 2pp TxU:H/L-0482.12 & -0482.13; Reel 1055

9360. 1852–5–2 To Thomas Bell Harley St.
Agrees to serve on and hopes will be able to attend meeting of Wintringham Bequest Committee. Comments on interpretation of will. P.S. stresses keeping to the time period specified in the will.
ALS 4pp RS DM.6.140 & P.S.: RS MM 16.161

9361. 1852–5–2 From Margaret Seymour Park Lane
Invites JH to dinner.
ALS 1p RS:HS 17.194

9362. 1852–5–4 From Richard Grenville Treasury
Call on RG Thursday to discuss Benedetto Pistrucci.
ALS 2pp TxU:H/M–0135.1; Reel 1087

9363. 1852–5–4 To Margaret Brodie Herschel n.p.
Hopes daughter Maria is better. Arrangements for Ball. Family finances: lists income and expenses at Collingwood and London residence. Describes carriage JH wants to buy. Ugly accident at [wedding?]; warns MH to avoid borrowing carriages. Anticipating six months of intense work following departure of H. D. Harness, after which JH intends to resign from Mint. Describes life as '*unendurable*.' Misses children.
ALS 4pp TxU:H/L-0543.2; Reel 1053

9364. 1852–5–4 From Joanna B. Horner 4 the Grove Highgate
Has had a letter from Miss Mathilde Oersted, who was gratified by JH's speech regarding her father. Her father's sudden death has been a great blow to her. Does not know what will happen to H. C. Oersted's books.
ALS 5pp RS:HS 9.489

9365. 1852–5–4 From William Henry Sykes India House
Asks JH to look at a letter by John Forbes Watson, whose paper the R.S.L. recently refused to read.
ALS 3pp Harvard: Houghton AAH 67m–67 (99)

9366. 1852–5–5 From Charles Edward Trevelyan Treasury
Forwards confidential memorandum by George Arbuthnot concerning JH's letter. Will discuss this with JH tomorrow.
LS 2pp TxU:H/M–0552.23; Reel 1093

9367. 1852–5–6 To My dear Sir 32 Harley St.
Must meet Lord Charles [?] this morning. Cannot join [?] and C. E. Trevelyan at Downing Street. Will try to see Charles Fremantle today.
ACCS 1p TxU:H/L-0663; Reel 1089

9368. 1852–5–8 From Francis Beaufort Admiralty
Forwards a letter to JH from Thomas Maclear, and FB comments on several accidents that have occurred at the Cape.
ACC 1p Hydrographic Office Letter Book

9369. 1852–5–8 From Robert Hunt Museum of Pract. Geology
Regarding the absence of his own name from the candidates list for the R.S.L.; thinks there is a clique working against him.
ALS 4pp RS:HS 10.132

9370. 1852–5–8 From William Lassell Starfield, Liverpool
Illness has prevented an earlier reply thanking him for his introduction to the Governor of Malta. Expects good results in Malta.
ALS 3pp RS:HS 11.147

9371. 1852–5–8 From Charles Edward Trevelyan Treasury
Prefers to send entire £50,000 to Cape in gold coins. Will authorize JH next week to obtain another £20,000 in gold coin from Bank [of England] for other shipments.
LS 2pp TxU:H/M–0552.24; Reel 1093

9372. 1852–5–9 To Margaret Brodie Herschel 32 Harley St.
Daughters Caroline and Isabella left London yesterday. J. M. Cameron will pick up Julia today. JH stays late at Mint each night, leaving 'Miss R' [Rennie?] at home alone. Mr. Bulstrode agreed to let JH rent residence from month to month.
AL 4pp TxU:H/L-0543.3; Reel 1053

9373. 1852–5–10 From Thomas Bell n.p.
Hopes JH can attend meeting of committee on Wintringham Bequest.
ALS 2pp RS MM.16.162

9374. 1852–5–11 From William Henry Bateson [No. 10] Downing St.
Cambridge University Commissioners met today, made progress on issue of Fellowships, and saw copy of [Bessed?] Commission report. Hopes JH can attend tomorrow.
ALS 3pp TxU:H/M–0057.19; Reel 1087

9375. 1852–5–12 From Henry Drury Harness Royal Mint
No assays today. Notified Mint staff that tomorrow is holiday for Queen's birthday.
ALS 1p TxU:H/M–0263.6; Reel 1087

9376. 1852–5–12 From John R. Hind 22 Grove Rd, St. John's Wood
Annibal de Gasparis's new planet Psyche. Alexander von Humboldt selected its name. Other new planet, found by [C. Robert] Luther, was named Thetis by F. W. A. Argelander. Detected new nebula on 26 Apr. Nervous trouble interrupting Hind's search for planets.
ALS 4pp RAS:JH Archive 12/1.6.2; Reel 10

9377. 1852–5–13 From Charles Edward Trevelyan Treasury
Inform CT of present pay scale for workmen in Coining and Melting departments. JH was correct in giving assistance and information to two French visitors. Wants to discuss F. R. Brande and Henry Finch.
LS 2pp TxU:H/M–0552.25; Reel 1093

9378. 1852–5–14 From Richard Sheepshanks Athenaeum
Suggests that JH assist William Lassell's travels to Malta by asking for a letter of introduction from the Duke of Northumberland; is concerned about the precarious position of the observatory at Altona, and of the *Astronomische Nachrichten* published there.
ALS 4pp RS:HS 16.117

9379. 1852–5–14 To James Calder Stewart Harley St.
Sends draft for insurance papers. Description of Cornhill shops.
ALS 3pp TxU:H/L-0437; Reel 1055

9380. 1852–5–15 To Joanna B. Horner n.p.
Regrets a week has elapsed before replying to her letter, but is pleased that Miss Mathilde Oersted had so high a regard for his speech concerning her illustrious father.
draft 3pp RS:HS 9.490 & 23.122

9381. 1852–5–15 To Sir John Potter London
Cannot attend the opening of the Manchester Free Library. Praises the institution.
ALS 2pp Manchester Central Library

9382. [1852–5–17] From Richard Grenville 12 Grosvenor Square
[Marked 'Private.'] Thanks for calling RG's attention to error in Treasury minutes. Hopes to correct it before these reach Parliament.
ALS 3pp TxU:H/M–0134; Reel 1087

9383. 1852–5–17 To Richard Grenville Royal Mint
[Marked 'Private.'] Found written evidence of the denial in 1825 by Lord Maryborough [William Wellesley-Pole] of lifetime tenure at Mint for Benedetto Pistrucci. Suggests contacting [W. R.] Hamilton for clarification.
AdraftS 4pp TxU:H/L-0122; Reel 1054

9384. 1852–5–17 To Benedetto Pistrucci [Royal Mint]
Urges BP to accept the Treasury's offer.
ACCS 1p TxU:H/L-0312; Reel 1093

9385. 1852–5–18 From Benedetto Pistrucci Royal Mint
Settlement reached concerning BP's salary.
ALS 1p (Italian) TxU:H/M–0430.13; Reel 1093

9386. 1852–5–19 From Richard Grenville Treasury
Learned that Treasury minutes have not yet been read to Parliament. Will incorporate JH's amendments into these.
ALS 3pp TxU:H/M–0136; Reel 1087

9387. 1852–5–19 To Adolph C. Petersen London
Is concerned about the possibility that the observatory at Altona will be discontinued or moved.
draft 4pp RS:HS 13.374 & 23.123

9388. 1852–5–19 To Richard Sheepshanks 32 Harley St.
Will speak to the Duke [of Northumberland, Algernon Percy] at the earliest opportunity and encloses a draft note to Adolph Petersen at Altona [see RS's 1852–5–14].
ALS 2pp RS:HS 16.118 (C: RS:HS 25.9.74)

9389. 1852–5–19 To Charles Edward Trevelyan Royal Mint
[Matthew] Marshall from Bank [of England] indicates that great demand for gold coins will continue. Mint will be expected to increase deliveries.
AdraftS 3pp TxU:H/L-0482.15; Reel 1055

9390. 1852–5–20 From George Arbuthnot Treasury
Encloses GA's memorandum on exporting large quantities of coins, especially to Australia. Will see JH and C. E. Trevelyan tomorrow.
ALS 4pp TxU:H/M–0007; Reel 1062

9391. [1852–5–20] To Richard Grenville Royal Mint
[Marked 'Private.'] JH found passage in Treasury's letter to Benedetto Pistrucci that could contradict the denial by Lord Maryborough [William Wellesley-Pole] of Pistrucci 's claim to lifetime job and salary at Mint.
ACCS 4pp TxU:H/L-0121; Reel 1054

9392. 1852–5–20 From Charles Edward Trevelyan Treasury
JH and H. D. Harness may call on CT tomorrow.
LS 1p TxU:H/M–0552.26; Reel 1093

9393. 1852–5–25 To Richard Grenville Royal Mint
Returns letter from C. E. Trevelyan. Awaits final conclusion in difficult case of Benedetto Pistrucci.
ACCS 1p TxU:H/L-0123; Reel 1054

9394. 1852–5–25 From Henry Piddington Calcutta
A note accompanying another publication on turbulent winds.
ALS 2pp RS:HS 13.407

9395. 1852–5–25 From Richard Sheepshanks Reading
Make a few suggestions about JH's letter to Adolph Petersen [see JH's 1852–5–19].
ALS 4pp RS:HS 16.119

9396. 1852–5–26 From Henry Drury Harness Royal Mint
Treasury solicitors responded to JH's request for date on which to lease Mint refinery to Anthony de Rothschild. Found estimates for 1850 that JH wanted.
ALS 3pp TxU:H/M–0263.7; Reel 1087

9397. 1852–5–27 To Charles [Edward Trevelyan] Royal Mint
Explains difference in 'Estimates' for years 1850 and 1852.
Adraft 1p TxU:H/L-0482.16; Reel 1055

9398. 1852–5–28 To Thomas Bell Harley St.
Expresses opinion regarding Wintringham Bequest.
ALS 4pp RS DM.6.141

9399. 1852–5–28 From Alvan Clark Boston (Mass.)
Has a copy of JH's book on astronomy, which AC values. Has been carrying out experiments with the grinding of object glasses. Remarks on various points relating to telescopes. Intends to search for double stars.
ALS 4pp RS:HS 5.341

9400. 1852–5–28 From Henry Drury Harness Royal Mint
Forwards papers for JH to study.
ALS 1p TxU:H/M–0263.8; Reel 1087

9401. 1852–5–29 To Margaret Brodie Herschel 32 Harley St.
H. D. Harness leaves Mint today, but may return later if assignment to colonies is delayed. JH cannot go to park with daughters today. Completed report on Irish coinage. Duties regarding Cambridge University Committee.
ALS 4pp TxU:H/L-0543.4; Reel 1053

9402. 1852–5–29 From Charles Edward Trevelyan Treasury
Does monetary part of enclosed document require more attention than political part? [JH annotation: 'monstrous proposal' concerning counterfeit copper.]
LS 2pp † TxU:H/M–0552.27; Reel 1093

9403. 1852–6–1 From Charles Edward Trevelyan Treasury
Asks clarification of JH's plans for countersigning drafts from Mint.
LS 2pp TxU:H/M–0552.28; Reel 1093

9404. 1852–6–1 To Charles Edward Trevelyan Harley St.
H. D. Harness went on leave with only unofficial authorization from CT. Requests that official confirmation

be sent to JH.
Adraft 1p inc TxU:H/L-0482.17; Reel 1055

9405. 1852–6–2 To 'Sir' [Royal Mint]
If enclosed report meets [?]'s approval, please show it to Chancellor of Exchequer. Parts of it should not be seen by JH's subordinates.
AdraftS 1p inc TxU:H/L-0622; Reel 1089

9406. [1852–6–2] To Charles Edward Trevelyan [Royal Mint]
Comptroller W. H. Barton is senior officer at Mint and is signatory to Mint drafts. JH's earlier letter to CT included officers empowered to countersign drafts when senior officers are absent. Plans to have Charles Elouis initial all drafts.
ACCS 2pp TxU:H/M–0552.29; Reel 1093

9407. 1852–6–3 From Francis Beaufort Admiralty
Indicates that trials have been made with a 'trumpeted cannon,' and JH is welcome to request further trials to be made at Woolwich Arsenal.
ACC 1p Hydrographic Office Letter Book

9408. 1852–6–4 To George Biddell Airy n.p.
Thanks GA for his description of Madeira; JH will try to get to Greenwich next day.
ALS 3pp RGO 6.10.23

9409. 1852–6–4 To Francis Beaufort [32 Harley St.]
Reports quite unsatisfactory results from gun firing sound experiments [see JH's 1852–3–23].
ALS 2pp Hydrographic Office Misc F26F1.4

9410. 1852–6–4 From Henry Drury Harness Sherborne, Dorsetshire
Encloses copy of HH's answer to George Arbuthnot, who agrees with HH but raised interesting questions. Describes Dorsetshire.
ALS 2pp TxU:H/M–0263.9; Reel 1087

9411. 1852–6–4 From Rowland Hill General Post Office
Letter will receive immediate attention.
ALS 1p RS:HS 10.191

9412. 1852–6–5 From George Arbuthnot Treasury
Understands that JH has seen GA's letter to H. D. Harness on 'Seigneurage' question. Encloses GA's response to Harness's most recent letter.
ALS 1p TxU:H/M–0008; Reel 1062

9413. 1852–6–5 From Harry George Wakelyn Smith 1 Albemarle St.
Note of thanks to JH for his kind wishes.
ALS 2pp RS:HS 16.201

9414. 1852–[6?]–6 To Benedetto Pistrucci [Royal Mint]
JH ready to support BP if he decides to accept office at the Mint. Assures BP that a pension can be negotiated.
ACCS 3pp TxU:H/L-0313; Reel 1093

9415. 1852–6–7 From Francis Beaufort Admiralty
Encloses a letter from Thomas Maclear regarding the resignation of his assistant, George Childe. Would like JH's opinion on the appointment of Maclear's sons to succeed Childe.
ALS 3pp† RS:HS 3.401 & ACC: Hydrographic Office Letter Book

9416. 1852–6–7 From Alvan Clark Boston (Mass.)
Has found an error in the printed list he sent recently. Wonders if a certain star listed has ever been observed as a double.
ALS 1p RS:HS 5.342

9417. 1852–6–7 From Henry Drury Harness Sherborne, Dorsetshire
Correspondence between HH and George Arbuthnot regarding comparative advantages of exporting coins versus bullion. Requests JH's opinion. Quick sketch of HH's views about air pump apparatus.
ALS 3pp TxU:H/M–0263.10; Reel 1087

9418. 1852–6–9 To Julia Margaret Cameron Harley St.
JH, now laboring in the vast 'Mammonium' [the Mint], missed a visit from JC and was unable to deliver JH's daughter Julia into JC's arms.
ALS 3pp DMC.695A.885.21

9419. 1852–6–10 From Henry Drury Harness Sherborne, Dorsetshire
Leonard Wyon's private work at Mint should have terminated on 30 Apr. HH warned Wyon not to use Mint apparatus for Punjabi or Exhibition medals after May. [H. C. C.] Owen requested reasonable extension. JH should close Punjabi account then ask Exhibition committee to set number of medals that they need and hold Wyon to that. Further explanation of air pump apparatus.
ALS 4pp TxU:H/M–0263.11; Reel 1087

9420. [1852–6–10] To [Henry Drury Harness] [Royal Mint]
Moneyer's former policy has resulted in two pay scales for Mint employees. Asks help in standardizing pay scale. Repair cost for air pump may be double original estimate. Treasury rejected JH's appeal on behalf of F. R. Brande; plans to appeal to Chancellor of Exchequer.
Adraft 2pp TxU:H/L-0141; Reel 1054

9421. 1852–6–11 To Francis Beaufort 32 Harley St.
Urges that Thomas Maclear's son be appointed to assist Maclear in the observatory [at the Cape].
ALS 3pp† Hydrographic Office Misc F26F1.5

9422. 1852 6 11 From Henry Drury Harness Sherborne, Dorsetshire
Thanks JH for reviewing HH's recommendation for improving air pump. Suggests way to increase work output of Sizing room. Advises JH to simplify pay scale for Mint workers. Promotion of young workers.
ALS 4pp † TxU:H/M–0263.12; Reel 1087

9423. 1852–6–11 From [A. J. F.] Yvon Villarceau Paris
Is sending a mathematical work on the stability of a moving locomotive, and hopes it will interest JH. YV would like it presented to the R.S.L. Refers to YV's work on double stars.
ALS 2pp (French) RS:HS 17.422

9424. 1852–6–12 To Margaret Brodie Herschel 32 Harley St.
Analogy between MH's love and action of light. All is ready for MH's arrival next week. Daughter Julia will spend month with J. M. Cameron. Miss Beaumont reports that JH's son John is recovering well from illness [at Clapham Grammar School?]. Cannot come to Collingwood Saturday; must work at 'Mammonium' [Mint]. Bring photography by JH for Mr. Hilhouse.
ALS 4pp TxU:H/L-0543.5; Reel 1053

9425. 1852–6–12 To Charles Edward Trevelyan Royal Mint
Recommends hiring Mr. Brande Junior [F. L. Brande] as

supernumerary clerk. Pressure from Bank [of England] to increase gold coinage.
AdraftS 4pp TxU:H/L-0482.19; Reel 1055

9426. 1852–6–12 To Charles Edward Trevelyan Royal Mint
Proposes hiring clerk to assist Master of Mint with correspondence and computations. [JH annotation of Nov. 1853: Sorry that this letter was sent. Enormous error.]
Adraft 4pp †(by JH) TxU:H/L-0482.18; Reel 1055

9427. 1852–6–13 From William Henry Bateson St. John's College, Cambridge
WB read JH's note to Cambridge University Commissioners on 10 June. Commission will meet on 1 July.
ALS 2pp TxU:H/M–0057.20; Reel 1087

9428. 1852–6–14 From Charles Edward Trevelyan Treasury
Supports JH's plan to hire F. R. Brande as supernumerary clerk and Henry Finch to replace Brande, but tell them explicitly that these are not permanent positions.
LS 3pp TxU:H/M–0552.31; Reel 1093

9429. 1852–6–14 From Charles Edward Trevelyan Treasury
Believes JH's handwriting is sufficiently clear for official correspondence. Denies request for additional worker to serve as JH's personal clerk.
LS 3pp TxU:H/M–0552.30; Reel 1093

9430. 1852–6–18 To Arthur Wellesley Royal Mint
'Not official.' Submits 'electrotype cast' of Waterloo Medal for AW's inspection.
ACC 1p TxU:H/L-0495; Reel 1055

9431. 1852–6–19 From George Biddell Airy Royal Observatory, Greenwich
A note accompanying the return of some papers and letters to JH.
ACCS 1p RGO 6.374.127

9432. 1852–6–21 From Henry Drury Harness Sherborne Dorsetshire
Sends Audit office letter. Address of HH's next destination.
ALS 1p TxU:H/M–0263.13; Reel 1087

9433. 1852–6–21 To [R.S.L. Council] n.p.
Reports on G. G. Stokes's 'most remarkable and important' paper on refrangibility of light, recommending highly its publication in full. Notes in postscript that JH's experimental results had come close to those of Stokes.
ADS 7pp RS RR.2.230

9434. 1852–6–21 From Charles Edward Trevelyan Treasury
Approves delay in JH's report on coinage until JH can determine how unusual present demand will affect Mint operations.
LS 2pp TxU:H/M–0552.32; Reel 1093

9435. 1852–6–22 From Adolph C. Petersen Altona
No decision has yet been made about closing or moving the observatory, but thanks JH for his letter [see JH's 1852–5–19], which may become useful support for retaining the observatory.
ALS 2pp (German) RS:HS 13.375

9436. 1852–6–24 From John R. Hind 22 Grove Rd., St. John's Wood
Takes the earliest opportunity of announcing the discovery of a new planet [Melpomene]. Gives readings.
ALS 3pp RS:HS 9.356

9437. 1852–6–24 From Charles Edward Trevelyan Treasury
Asks JH to call on CT when convenient. [JH annotation: Discussed objection of Lord C [Chandos?=Richard Grenville] to hiring Brande.]
LS 1p, † TxU:H/M–0552.33; Reel 1093

9438. 1852–6–29 From William Henry Bateson St. John's College, Cambridge
Will send WB's notes regarding Cambridge University Commissioners' discussion of tuition. Expects fuller treatment at next meeting [1 July].
ALS 3pp TxU:H/M–0057.21; Reel 1087

9439. 1852–6–29 From Margaret Seymour Park Lane
Invites JH to visit and meet the Austrian Minister 'who *delights* in the study of the heavens.'
ALS 2pp RS:HS 17.195

9440. 1852–7–1 From Henry Drury Harness St. Donats, Bridgend, Glamorganshire, South Wales
Encloses letter from W. G. Anderson regarding coins to be reserved for deputy paymaster general. Asks about JH's letter to W. T. Brande, who wrote to HH for advice about salary for son F. R. Brande. Returns signed receipt from Charles Elouis.
ALS 4pp TxU:H/M–0264.1; Reel 1087

9441. 1852–7–5 From Henry Drury Harness St. Donats, Bridgend, Glamorganshire, South Wales
Encourages JH to seek uniform system for hiring personnel and regulating rates for wages in Mint, but to carry out changes gradually. Describes in detail transition to new Mint system. Additional funds for Waterloo Medal. Ideas for improving air pump.
ALS 4pp TxU:H/M–0264.2; Reel 1087

9442. 1852–7–7 To Charles Edward Trevelyan Royal Mint
W. H. Barton concurs with JH on value of promoting F. R. Brande.
ACCS 4pp TxU:H/L-0482.20; Reel 1055

9443. 1852–7–9 From Henry Drury Harness St. Donats, Bridgend, Glamorganshire, South Wales
Problems with wages among workers in Melting house. More ideas for improving air pump.
ALS 4pp TxU:H/M–0264.3; Reel 1087

9444. 1852–7–12 From Francis Beaufort Admiralty
Asks JH for a copy of his *Cape Results* to be provided to the new Queen's College at Belfast.
ACC 2pp Hydrographic Office Letter Book

9445. 1852–7–12 To Margaret Brodie Herschel Royal Mint
Sons John and Alexander visited JH yesterday. Forwards son William's letter describing voyage. News from Cape of Good Hope about [Eighth Frontier War]. Silver panic at Bank of England. Trouble dislodging '*remoras*' (old officers) who retard changes at Mint.
AL 4pp TxU:H/L-0543.6; Reel 1053

9446. 1852–7–14 From Henry Drury Harness St. Donats Bridgend, Glamorganshire, South Wales
Note from Hunt & Roskell. HH gave instructions for sale of proof coins before HH left London. W. T. Brande does not

agree with wage system at Mint.
ALS 3pp TxU:H/M–0264.4; Reel 1087

9447. 1852–7–15 To Francis Beaufort 32 Harley St.
Will instruct Smith, Elder & Co. to provide a presentation copy of JH's *Cape Results* to Queen's College, Belfast. What has happened with JH's recommendation about Thomas Maclear's son [see JH's 1852–6–11]?
ALS 3pp Hydrographic Office Misc F26F1.6

9448. 1852–7–16 From Francis Beaufort n.p.
Thinks that a note must have miscarried as he is sure he informed him that the Admiralty had appointed Thomas Maclear's son [George Maclear] an assistant at the Cape and his other son [John Maclear] to a ship.
ALS 2pp RS:HS 3.402

9449. 1852–7–16 From William Smith Williams 65 Cornhill
Per JH's directions, sent copies [of JH's *Cape Results*] to Queen's College, Belfast. Awaiting JH's instructions for dispatching few remaining presentation copies to institutions abroad.
ALS 1p RAS:JH Archive 10/3.20; Reel 9

9450. 1852–7–18 To Thomas Maclear [32 Harley St.]
Approves of G. W. H. Maclear's appointment at Cape Observatory [Second Assistant]. JH's family will welcome G. W. H. Maclear to Collingwood during his England visit.
ALS 3pp Cape Archives/Maclear Papers File 100

9451. 1852–[7–18?] To Roderick Impey Murchison n.p.
Interested in RM's statements on various subjects, including gold diggings and the 'declension of California Produce.'[?]
ALS 2pp EUL:Gen 523/6

9452. 1852–7–19 To Henry Finch Royal Mint
JH made clear on 13 Mar. that HF's position in reorganized Mint would be that of newly entering junior clerk and that F. R. Brande's position warranted more seniority.
ACCS 2pp TxU:H/L-0153; Reel 1054

9453. 1852–7–19 From Henry Drury Harness St. Donats, Bridgend, Glamorganshire, South Wales
Surprised over the uproar caused by 'malcontents' at Mint. Advises JH not to concede anything. Ask Treasury for verbal authority for each step JH takes in hiring, dismissing, or paying employees. If JH needs help, HH will return immediately. P.S.: Learned that F. R. Brande is on salary now.
ALS 3pp TxU:H/M–0264.5; Reel 1087

9454. 1852–7–20 From George Arbuthnot Treasury
Asks JH for Christian names of colleague.
ALS 3pp RS:HS 1.403

9455. 1852–7–20 From Henry Drury Harness St. Donats, Bridgend, Glamorganshire, South Wales
Forwards note from Mr. [Greund?] referring to correspondence with which HH is not acquainted.
ALS 1p TxU:H/M–0264.6; Reel 1087

9456. 1852–7?–20 From William Henry Sykes India House
Encloses newly-discovered proof that John Forbes Watson's wet bulb depression was not a clerical error.
ALS 1p RS MM.16.166

9457. 1852–7–22 From William Henry Bateson [No. 10] Downing St.
[Cambridge University Commission] Board will meet tomorrow. Will print JH's papers and send proof to JH.
ALS 3pp TxU:H/M–0057.22; Reel 1087

9458. 1852–7–23 From William Henry Bateson [No. 10] Downing St.
Cambridge University Commissioners will meet 4 Aug.
ALS 2pp TxU:H/M–0057.23; Reel 1087

9459. 1852–7–23 From Heinrich Christian Schumacher Schleswig
Mentions his stay at Pulkowa Observatory in 1844 and 1845. Refers to [Alexander von] Humboldt's *Kosmos*. Discusses a function of prisms in telescopes.
ALS 3pp (German) RS:HS 17.179

9460. 1852–7–24 From Alexander John Ellis Corstorphine, Edinburgh
Describes second edition of AE's phonetic Latinis alphabet, referred to by JH in *Admiralty Manual*. Sent copy to C. K. J. Bunsen. This will be discussed at Society of Arts's Education Exhibition on 31 July. Encloses printed circular. Hopes JH will attend.
ALS 4pp TxU:H/M–0178; Reel 1087

9461. 1852–[7]–24 To William James Herschel n.p.
Draws attention to expenses for WH's education at Haileybury College. Urges use of regular system of accounts. JH sent request for extension of J. F. L. P. Maclear's leave from ship *Castor*. Assumes that W. H. B. Hamilton is still secretary to Admiralty. Hopes WH, Maclear, and [William] Waterfield will avoid mischief. How is WH's 'broken back'?
AL 4pp damaged TxU:H/L-0543.7; Reel 1053

9462. 1852–7–26 From Henry Drury Harness St. Donats, Bridgend, Glamorganshire, South Wales
Happy to hear that JH settled difficulties with [W. T.] Brande. Talents of W. H. Barton and Charles Elouis. Reward for R. F. Suft. Advises exchanging Mr. Cumberland for more useful clerk from some other public office.
ALS 4pp TxU:H/M–0264.7; Reel 1087

9463. 1852–7–26 To Margaret Brodie Herschel Royal Mint
Major fire in Refinery last night. Forwarded letter from Miss Meyer to MH. Used Mr. Love's help for unpleasant task. Glad that J. F. L. P. Maclear enjoys his naval profession. How is son William's back?
ALS 4pp TxU:H/L-0543.8; Reel 1053

9464. 1852–7–26 From John R. Hind 22 Grove Rd., St. John's Wood
Regrets the trouble JH has been caused, but it was due to the non-arrival of his letter. Cannot claim fellowship of the R.S.L. as he did not attend any of the meetings following his election. Gives a list of his planetary discoveries.
ALS 3pp RS:HS 9.357

9465. 1852–7–26 To Charles Edward Trevelyan Royal Mint
Crisis averted. W. T. Brande [Senior] withdrew letter of resignation submitted by his son F. R. Brande, who went away to country to calm down.
ACCS 4pp TxU:H/L-0482.21; Reel 1055

9466. [1852–7–27] To Margaret Brodie Herschel n.p.
Encloses note from W. H. B. Hamilton. Request for extension of leave for J. F. L. P. Maclear should go to captain of *Meander*, but JH does not know where it is, so JH wrote to captain of *Victory*. If Maclear receives no word by tomorrow, tell him to report to commanding officer to whom

he *thinks* he is assigned. Problems arranging instructions in assaying for Mr. Wooley. JH and assay master H. W. Field are not on good terms. More likely that Thomas Graham and [W. A.] Miller will permit a visitor, but neither has time to train apprentices.
ALS 3pp TxU:H/L-0536.17; Reel 1053

9467. 1852–7–28 To George Hamilton-Gordon 32 Harley St.
Points out deviations from policies recommended in 1848 by Mint Commission. Supports consolidation of supervisory positions. Protests C. E. Trevelyan's idea that JH's scientific pursuits in spare time as Master of Mint must be directed to public projects.
ALS 5pp TxU:H/L-0003; Reel 1054

9468. 1852–7–28 From Charles Edward Trevelyan Treasury
Introduces Deputy Assistant Commissary General Bower, who wishes to learn minting processes before leaving for Australia.
LS 3pp TxU:H/M–0552.35; Reel 1093

9469. 1852–7–31 To Edward Sabine n.p.
Stopped compilation of comet orbits because it had already been done. Would provide money from donation fund to distribute copies of this information.
C 1p RS:HS 23.114

9470. 1852–7–31 To [?] 32 Harley St.
A note accompanying a repayment of expenses incurred on behalf of some of JH's children.
ALS 1p WT 68134.2

9471. 1852–8–4 From Henry Drury Harness St. Donats, Bridgend, Glamorganshire, South Wales
Explains wages of Mint workers doing piece work. Discourages plan to pay equal wages in all departments. Policy for promotions. Problem of gold ingots of non-standard weight and alloy being imported by Bank of England.
ALS 7pp TxU:H/M–0265.1; Reel 1087

9472. 1852–8–6 From I. H. Jephson Downing St.
Next meeting of Cambridge [University] Commission will be on 13 Aug.
ALS 1p TxU:H/M–0297; Reel 1087

9473. 1852–8–8 To Margaret Brodie Herschel 32 Harley St.
Describes ship *Earl of Aberdeen* and son William's departure [for Scotland]. JH got lost returning from London Docks; describes squalid neighborhoods. Visited Elizabeth Baily, who is ill, and Mrs. Stewart, who is composed and well. J. C. Stewart and Rose Stewart are both ill. University Commission is finishing its report. Attitude toward dissenters is more liberal that JH expected. MH's concertina lessons with Mrs. Jones.
ALS 4pp TxU:H/L-0543.9; Reel 1053

9474. 1852–8–12 To Margaret Brodie Herschel [London]
Son Alexander has recovered and went with his brother John to Clapham School today. Will send J. F. L. P. Maclear's parcel to ship *Meander*. J. M. Cameron sent her translation of Part II of Dr. Heinemann's '[*Pialiss?*] *van Actevelde*.'
ALS 3pp TxU:H/L-0543.10; Reel 1053

9475. 1852–8–12 From William Henry Sykes India House
Plans date and time for first ascent of balloon experiment. WS's daughter has fallen in love with JH's son William.
ALS 3pp RS:HS 17.150

9476. 1852–8–14 To Margaret Brodie Herschel [London]
MH's birthday. Country needs bad harvest or expensive war to counteract 'monstrous plethora of gold' accumulated in England. Americans calling for naval war with England.
ALS 4pp TxU:H/L-0543.11; Reel 1053

9477. 1852–8–18 From James A. Gordon Stocks, Tring
Encloses the last gazette of the Princess. Seem to be prospering. Brother-in-law is off to Australia in the *Great Britain*.
ALS 4pp RS:HS 8.199

9478. 1852–8–18 To Margaret Brodie Herschel Royal Mint
MH's illness. Note from son Alexander at Clapham Grammar School. Martha [?] sends purse for MH's birthday. Balloon ascended to 19,000 feet yesterday with Charles Green, landed near Cambridge.
AL 4pp TxU:H/L-0543.12; Reel 1053

9479. 1852–8–19 To Caroline Emily Mary Herschel n.p.
CH's news that Mrs. Jones is with Margaret Herschel relieves JH's anxiety. Questions treatment of Mr. Wolf's horse.
ALS 4pp TxU:H/L-0524; Reel 1053

9480. 1852–8–21 From Henry Drury Harness St. Donats, Bridgend, Glamorganshire, South Wales
JH must deal with difficulty concerning Mr. Forbes, who will be foolish not to retire instead of letting case go to the Treasury. Work of current month is excellent.
ALS 2pp TxU:H/M–0265.2; Reel 1087

9481. 1852–8–21 From Charles Edward Trevelyan Treasury
Approves JH's action in transferring £2000 of half crowns to Bank [of England]. Will restrain Commissariat demand for silver coins until present pressure from Bank for coinage is over. Instructions for colonial coinage are obsolete; only *new* coins should be sent overseas. Plan to produce reserve stock of coins for Commissariat. [JH annotation: calculates minimum reserve stock.]
LS 4pp † TxU:H/M–0552.36; Reel 1093

9482. 1852–8–22 From William Henry Bateson St. John's College, Cambridge
All Cambridge University Commissioners, including JH, have approved report. Adam Sedgwick will finalize it with Bishop of Chester [John Graham] tomorrow. Board will meet on 24 Aug. to review report.
ALS 4pp TxU:H/M–0057.24; Reel 1087

9483. 1852–8–22 From John R. Hind 22 Grove Rd., St. John's Wood
Has found another planet [Fortuna]; gives readings. Finds that the famous eclipse of Thales took place in 585 B.C., not 610 B.C., as stated by Francis Baily and Mr. Altmann.
ALS 3pp RS:HS 9.358

9484. 1852–8–24 To Margaret Brodie Herschel [London]
Send bond that secures Peter Stewart's debt to JH; insurance company needs it. Let Mr. Wolf and Mr. Parkhurst arrange transfer of horse. Richard Jones seems well but complains about his feet. Forgot to mail Parkhurst's letters.
AL 4pp TxU:H/L-0543.13; Reel 1053

9485. 1852–8–24 From Charles Edward Trevelyan n.p.
Permission granted for issue of £50,000. Please return CT's

note to JH instructing that only new coins be sent to Colonies.
AL 2pp TxU:H/L-0482.22; Reel 1055

9486. 1852–8–25 From Augustus De Morgan 7 Camden St.
Has been examining and sorting all Francis Baily's correspondence with a view to its preservation. Suggests Greenwich as a suitable repository for most of it. Will be returning JH's letters for his selection of important letters to be preserved.
ALS 4pp RS:HS 6.261

9487. 1852–8–25 From John Hearn North Shields
Thanks him for the testimonial for his son who has recently died, leaving two children unprovided for. Can he assist with a place at Christ's Hospital for one?
ALS 2pp RS:HS 10.174

9488. 1852–8–25 To Margaret Brodie Herschel 32 Harley St.
Difficult to accept C.'s [daughter Caroline's?] choice. JH had other hopes for her. Mr. Wolf's mare is beautiful and well trained. Instructed Mrs. Wolf to meet both trains today. Leaves rest to MH. Five major issues at Mint and University Committee take up all of JH's time. Will dine at Lady Bryant's tomorrow. Railroad investments. Send Peter Stewart's bond quickly.
ALS 4pp TxU:H/L-0543.14; Reel 1053

9489. 1852–8–25 To [James Calder] Stewart Harley St.
Asks advice about investments. Received photographs from John [?].
ALS 3pp TxU:H/L-0438; Reel 1055

9490. 1852–8–26 To Margaret Louisa Herschel [London]
Recalls ghost story told by Louisa. Answers question about primate hands. Recommends study of mathematics in order to understand art. Regards to [Mary C.] Lipscombe.
ALS 4pp TxU:H/L-0534; Reel 1053

9491. 1852–8–27 From William Henry Bateson [No. 10] Downing St.
Sends final edition of report by Cambridge University Commissioners, who will meet on 30 Aug. to sign it.
ALS 2pp TxU:H/M-0057.25; Reel 1087

9492. 1852–8–27 To Margaret Brodie Herschel Royal Mint
Sad that MH's health is poor while facing 'deplorable affair' with MH's mother. JH's strength has returned. University Commission concluded its report this week. Attended Lady Bryant's dinner, met E. M. [Eneas Mackintosh?], who considers 'it all arranged.'
ALS 4pp TxU:H/L-0543.15; Reel 1053

9493. 1852–8–28 To Augustus De Morgan 32 Harley St.
Comments on astronomical discoveries, and depressing life in London.
C 2pp RS:HS 23.124

9494. 1852–8–28 To Margaret Brodie Herschel [Royal Mint]
Describes [G. H.] Gordon, who visited JH at Mint yesterday. Sends copy of University Commission report. Recounts struggle between Adam Sedgwick and George Peacock over report's conclusions.
AL 4pp TxU:H/L-0543.16; Reel 1053

9495. 1852–8–29 From Augustus De Morgan 7 Camden St.
The job [of sorting Francis Baily's correspondence] may be done at leisure. All his family are at Herne Bay. Regarding chronology. Annibal De Gasparis will be equal with J. R. Hind when he locates another planet. Is astonished how well books last.
ALS 3pp RS:HS 6.262 S. De Morgan, *Memoir of Augustus De Morgan*, 216–17

9496. 1852–8–31 From Henry Drury Harness Cowbridge, Glamorganshire, South Wales
Dispute over seniority between R. F. Suft and Mr. Cumberland. Neither of these junior clerks qualifies for vacancy of senior clerk, who should be competent to stand in during comptroller's absence. Ask W. H. Barton for recommendations.
ALS 4pp TxU:H/M-0265.3; Reel 1087

9497. 1852–9 From William Macquarie Molle Corinbatore, Madras
Recalls meeting JH in 1840, introduced by R. C. Stewart of Bombay. Saw JH's letter on artificial production of ice in *Year-Book of Facts* for 1851 (p. 157). Describes WM's unsuccessful experiments with commercial powders in tropical climate. Asks for information about portable air compressors. Local surgeon, Dr. Porteus, whom JH met at Cape of Good Hope, asks to be remembered to JH's family.
ALS 12pp TxU:H/M-0374; Reel 1087

9498. 1852–9–1 To James Calder Stewart n.p.
Questions about investments.
ALS 3pp TxU:H/L-0439; Reel 1055

9499. 1852–9–2 From George Arbuthnot Treasury
Demand for silver coins by Commissariat department [in Malta] must wait until Mint has met demand by Bank [of England]. Address all official letters to C. E. Trevelyan, even if responding to letters signed by GA.
ALS 3pp TxU:H/M-0009; Reel 1062

9500. 1852–9–2 From Thomas Maclear Royal Observatory, Cape of Good Hope
William Mann goes to England to receive transit circle. TM describes plans for the instrument.
ALS 4pp Cape Archives/Maclear Papers File 100

9501. 1852–9–2 To Roderick Impey Murchison Harley St.
Cannot introduce a Mr. Grant to overly 'pressed' Mint assayers. Will help Grant as far as possible, but returns 'excellent' testimonials, having no part in 'manning' possible Australian mint.
ALS 3pp EUL:Gen523/4/43

9502. 1852–9–3 From William O'Shaughnessy Brooke 8 Bentinck Terrace
Would like to visit the Mint with his family.
AL 2pp RS:HS 13.186

9503. 1852–9–3 To James Calder Stewart n.p.
Payment from insurance office. Thanks for advice on India bonds. JH's investments.
ALS 3pp TxU:H/L-0440; Reel 1055

9504. 1852–9–4 From George Arbuthnot Treasury
[Marked 'Private.'] Governor of Bank [of England] is in 'great fuss' about not receiving silver coins. What is Mint doing? Suspects some 'nonsense' behind this urgent demand for silver. In C. E. Trevelyan's absence, GA will not act precipitously.
ALS 4pp TxU:H/M-0010; Reel 1062

9505. [1852]–9–4 To Henry Drury Harness Royal Mint
Housing accommodations for Mint officers.
AdraftS 4pp TxU:H/L-0183.3; Reel 1054

9506. [1852]–9–4 To Henry Drury Harness 32 Harley St.
Grateful for HH's advice on dealing with Mr. Cumberland. JH spoke to [W. H. Barton] about Cumberland's claim to seniority over R. F. Suft. Encloses letter from Bank of England that indicates extraordinary work for Mint in near future. Recent production of coins. Suggests stronger series of checks on piecework of all men and officers at Mint.
ACCS 2pp † TxU:H/L-0183.2; Reel 1054

9507. 1852–9–6 To George Biddell Airy 32 Harley St.
Is seeking clarification about some papers GA sent to JH.
ALS 1p RGO 6.466.502

9508. 1852–9–6 From Henry Drury Harness Lillington, Sherborne, Dorsetshire
Considers Thomson Hankey's note ridiculous; pressure by Bank of England for more silver coinage is unnecessary. Reassignment of Mint residences to Mint officers. Discourages plan to extend piece work wages to include Mint officers.
ALS 4pp TxU:H/M–0265.4; Reel 1087

9509. 1852–9–7 From George Biddell Airy Royal Observatory, Greenwich
Offers some explanations for the reports in the papers [see JH's 1852–9–6] of seeing strange objects in the heavens; GA calls it 'hemiplegia.'
ACCS 4pp RGO 6.466.503

9510. 1852–9–10 To [?] Royal Mint
Cannot accede to the request for a visit to the Mint at this time. JH thanks for the cards of exchange rates with other currencies, and offers a suggestion for an improvement in them.
ALS 2pp WT B.8

9511. 1852–9–14 To Richarda Airy 32 Harley St
Reports significant improvement in Margaret's [JH's wife] health, after a long period of illness.
ALS 2pp JHS 2.9

9512. 1852–9–14 From George Arbuthnot Treasury
[Marked 'Private.'] Ordered Commissariat officer in Malta to send home £51,000 of silver coin for use by Mint in meeting demand of Bank [of England] for coins to export. Wishes government would take steps to reduce export of silver. Concern over public reaction to reducing silver content of coins.
ALS 8pp TxU:H/M–0011; Reel 1062

9513. 1852–9–16 To James Calder Stewart Royal Mint
Personal finances. Needs information on East India Company investment.
ALS 2pp TxU:H/L-0441; Reel 1055

9514. 1852–9–17 To Henry Drury Harness Royal Mint
Transfer of refinery from Mint to Anthony de Rothschild is delayed by failure to determine value of platinum vessels. Bank of England's demand for silver coins and half sovereigns is becoming urgent. Questions organization of new Weighing room. Staff of Mint clerks must be increased during intense coining activity. Effectiveness of weighing machine.
ACCS 4pp TxU:H/L-0184.2; Reel 1054

9515. 1852–9–20 From Henry Drury Harness Lillington, Sherborne, Dorsetshire
Procedures for evaluating refinery leased to Anthony de Rothschild. Problem of accurately weighing small coins. Size of permanent staff in Mint office should be determined by *ordinary* work, not by times of pressure. Consider borrowing temporary clerks from other government offices. Returns draft of JH's letter to Rothschild.
ALS 4pp TxU:H/M–0265.5; Reel 1087

9516. 1852–9–28 From Henry Drury Harness Lillington, Sherborne, Dorsetshire
Wages for a weighing room assistant. There is no extra copy of Coinage Act. Believes Treasury would approve of bonus for delivery of more silver or gold than planned.
ALS 2pp TxU:H/M–0265.6; Reel 1087

9517. 1852–9–28 From John R. Hind 22 Grove Rd., St. John's Wood
Angelo Secchi has found the second head of Biela's comet. Gives the elements of J. R. Hind's last planet (Fortuna).
ALS 4pp RS:HS 9.359

9518. 1852–9–28 From Thomas Maclear Royal Observatory, Cape of Good Hope
Sends parallax measurements for Alpha Centauri.
ALS 1p Cape Archives/Maclear Papers File 100

9519. 1852–10–3 To Margaret Louisa Herschel 32 Harley St.
Thanks for portrait of Louisa. Plans series of pictures of all JH's children. Margaret Brodie Herschel moved her mother [Mrs. Stewart] into new lodgings, but is now ill again. JH and London household are quarantined. No one may enter or leave. John Herschel [son of JH] is also ill, but under Mrs. Pritchard's care [at Clapham Grammar School]. Lists books and papers to be sent to JH.
ALS 4pp TxU:H/L-0535; Reel 1053

9520. 1852–10–7 From William George Anderson Pay Office
Shortage of silver coins at Bank of England. Explains recurring monthly drain of silver when ships crews, families, and pensioners must be paid. Begs JH to send £5000 of silver coin to Bank.
ALS 4pp TxU:H/M–0918; Reel 1083

9521. 1852–10–7 From Henry Drury Harness Lillington, Sherborne, Dorsetshire
Encloses pamphlets about [W. T.] Denison's experiments to determine accuracy of gold specimens by use of specific gravity. [JH annotation: Wrote to HH on 12 Oct. regarding Mr. Forbes and H. W. Field.]
ALS 1p † TxU:H/M–0266.1; Reel 1087

9522. 1852–10–8 To Henry W. Field Royal Mint
Instructions for assaying Indian coins.
ACCS 2pp TxU:H/L-0150; Reel 1054

9523. 1852–10–13 From Henry Drury Harness Lillington, Sherborne, Dorsetshire
Doubts [W. T.] Denison's claim that specific gravity is dependable for testing gold coins. Recommends dismissing Mr. Forbes at once. H. W. Field cannot be selective about assays but must perform these whenever JH requires.
ALS 3pp TxU:H/M–0266.2; Reel 1087

9524. 1852–10–13 To Henry Drury Harness [Royal Mint]
Pressure for coining gold, silver, and copper is becoming urgent. Complaints by Bank of England are reflecting badly on new system at Mint. Received two officers from Commissariat department to assist during crisis. Hopes HH

will return soon.
AAbsS 1p TxU:H/L-0184.3; Reel 1054

9525. 1852–10–14 From Henry Drury Harness Lillington, Sherborne, Dorsetshire
Arranges to meet JH at a place other than the Mint.
ALS 2pp TxU:H/M–0266.3; Reel 1087

9526. 1852–10–15 To My dear Sir Royal Mint
Arranges to meet [?] at JH's residence tomorrow to discuss business.
ACCS 1p TxU:H/L-0665; Reel 1089

9527. [1852]–10–16 To Henry Drury Harness 32 Harley St.
Real emergency is occurring. Assured C. E. Trevelyan that Mint would increase its efforts. Will see Mr. Mocatta on Monday. Hopes silver becomes available soon.
ACCS 3pp †(by JH) TxU:H/L-0184.4; Reel 1054

9528. 1852–10–16 From Henry Drury Harness 3 Hyde Park Terrace, Kensington Gore
In response to Treasury's declaration of 'emergency' condition, HH promises to assist in any way JH needs.
ALS 1p TxU:H/M–0266.4; Reel 1087

9529. 1852–10–16 From Charles Edward Trevelyan Treasury
Encloses letter from War Office. Current crisis in coin production has focused public scrutiny upon reorganized Mint system. '[W]e ... are on trial.' [JH annotation: Pensioners are complaining to War Office that they are not paid, due to lack of silver coins.]
ALS 3pp † TxU:H/M–0552.37; Reel 1093

9530. 1852–10–17 To Henry Drury Harness [32 Harley St.]
Call on JH as soon as possible.
ACCS 1p TxU:H/L-0184.6; Reel 1054

9531. 1852–10–17 To Henry Drury Harness 32 Harley St.
Directs HH to return to Mint immediately.
ACCS 3pp †(by JH) TxU:H/L-0184.5; Reel 1054

9532. 1852–10–18 From Henry Drury Harness 3 Hyde Park Terrace, Kensington Gore
Will not come to the Mint tomorrow.
ALS 1p TxU:H/M–0266.5; Reel 1087

9533. 1852–10–19 From Henry Drury Harness 3 Hyde Park Terrace, Kensington Gore
JH's description yesterday of events at Mint led HH to stay away. Surprised that JH wishes to see HH. No one should interfere between W. T. Brande and William Buckle.
ALS 2pp TxU:H/M–0266.6; Reel 1087

9534. 1852–10–19 To Henry Drury Harness 32 Harley St.
C. E. Trevelyan directed that Mr. Smith not be under direction of William Buckle. JH must devote entire morning to settling this delicate matter at Treasury. Please draw up working plan for Coining department. Contingency plan for coining if silver supply is interrupted.
ACCS 2pp TxU:H/L-0184.7; Reel 1054

9535. 1852–10–20 From Henry Drury Harness 3 Hyde Park Terrace, Kensington Gore
Forwards note from unknown author. Direct HH's servants to come to HH's residence.
ALS 2pp TxU:H/M–0266.7; Reel 1087

9536. 1852–10–20 To My dear Sir Royal Mint
Measures indicated in annexed document are absolutely necessary in present circumstances.
ACCS 2pp TxU:H/L-0666.2; Reel 1089

9537. 1852–10–21 From Henry Drury Harness London
Silver coinage. If Mr. Mocatta cannot deliver silver, perhaps the Rothschilds in Paris can. Did JH obtain authority yet to equalize subsistence wages for workers in Melting and Coining departments?
ALS 3pp TxU:H/M–0266.8; Reel 1087

9538. 1852–10–23 From George Biddell Airy Royal Observatory, Greenwich
Happened to meet JH's wife, Margaret; was very pleased to see her in such good health. Has heard a rumor about a Bank of England resolution on decimalization [see GA's 1852–3–15].
ACCS 2pp RGO 6.341.25

9539. 1852–10–23 To [William Thomas Brande] Royal Mint
Make immediate arrangements to accept extra workers into Die and Coining departments to bring coining operations up to maximum capacity.
AdraftS 2pp TxU:H/L-0666.1; Reel 1089

9540. 1852–10–23 To Henry Drury Harness [32 Harley St.]
Requires that HH return. 'The storm is coming on thicker.' Danger of coining operations breaking down. Report of coin production from W. T. Brande.
ACC 1p TxU:H/L-0184.8; Reel 1054

9541. 1852–10–23 To Charles Edward Trevelyan Royal Mint
Has not received reply from Anthony de Rothschild. Foresees great demand soon for gold and silver. Gives W. T. Brande's estimate of maximum possible output by Mint.
ACCS 4pp TxU:H/L-0482.23; Reel 1055

9542. 1852–10–25 From Thomas Bell Royal Society, Somerset House
Requests JH's opinion of H. P. Brougham's paper and reports of G. B. Airy and G. G. Stokes thereon.
ALS 2pp RS MM.16.155

9543. 1852–10–25 To [Mint officers] Royal Mint
[Formal notice] Work hours and pay rates for temporary workers.
ACCS 1p TxU:H/L-0667; Reel 1089

9544. 1852–10–26 To George Biddell Airy Royal Mint
The Bank of England will start to deal with weights of precious metals in decimals only [see GA's 1852–10–23]. JH responds to a letter in the *Times* about the shortage of silver coins.
ALS 3pp RGO 6.341.26

9545. 1852–10–26 From Charles Edward Trevelyan to Henry Drury Harness Treasury
New emergency at Mint, created by demand for gold and silver coins by people and Bank of England. JH unable to cope alone. Directs HH to cancel vacation and return to assist.
C 3pp TxU:H/M–1060; Reel 1083

9546. 1852–10–27 To Charles Edward Trevelyan Harley St.
Suggestions how to prepare for increasing public demand for silver. Mint workers began working 12-hour days on

Monday.
AdraftS 4pp TxU:H/L-0482.24; Reel 1055

9547. 1852–10–27 From Charles Edward Trevelyan Treasury
[Marked 'Confidential.'] CT will respond to the *Times* query regarding Mint [production of silver coins]. Plan to meet with CT when H. D. Harness returns. Glad to hear that Mint workmen began 12-hour workdays on 25 Oct.
LS 2pp TxU:H/M–0552.38; Reel 1093

9548. 1852–10–27 From Leonard Charles Wyon 22 Blomfield Road, Maida Hill
Thanks JH for processing payment of LW's compensation allowance. Will stop at Mint next week to examine letter from Treasury.
ALS 2pp TxU:H/M–0604; Reel 1086

9549. 1852–10–28 To William Parsons (Lord Rosse) Royal Mint
Will attend R.S.L. council today to present B.A.A.S. request for £300 to support balloon ascents for meteorological research.
ACCS 3pp TxU:H/L-0343; Reel 1054

9550. 1852–10–29 From Charles Edward Trevelyan Treasury
[Marked 'Confidential.'] Agrees with JH that assay of freshly minted coins may be eliminated [to reduce delivery time]. Keep official copy of JH's letter and send copy to CT. Anticipates no delay in new appointments at Mint when Mr. Forbes retires.
(A)LS 1p TxU:H/M–0552.39; Reel 1093

9551. 1852–10–30 To George Peacock 32 Harley St.
Asks GP to participate in the marriage ceremony of JH's daughter Caroline.
C 1p RS:HS 23.125

9552. 1852–11–1 To [Thomas Bell] Harley St.
Agrees with G. B. Airy's and G. G. Stokes's critique of H. P. Brougham's paper, finding paper unacceptable for publication in *R.S.P.T.*
draft RS MM.16.156

9553. 1852–11–1 To Samuel Hunter Christie 32 Harley St.
Report on a paper by Henry Brougham on light experiments, particularly undulatory theory, which JH believes undeserving of publication in *R.S.P.T.*
ALS 3pp RS RR.2.37 (draft: RS MM 16.156)

9554. [1852–11] From Henry Hart Milman Deanery, St. Paul's
Offering seats for the funeral of Arthur Wellesley (1st Duke of Wellington) in St. Paul's.
ALS 1p RS:HS 12.351

9555. [1852–11–3] To William Richard Hamilton [Royal Mint]
Will give every consideration to case of Benedetto Pistrucci, but Engraving department is long overdue for changes to bring it into conformity with all other departments in reorganized Mint.
Adraft 2pp TxU:H/L-0179; Reel 1054

9556. 1852–11–4 From George Gabriel Stokes Pembroke College, Cambridge
Has been studying the spectral lines of 'invisible' rays.
ALS 4pp RS:HS 17.28

9557. 1852–11–5 From Charles Hay Cameron East Sheen
Sends additions to the list of books, which Edward Ryan and himself have prepared. Congratulations on the forthcoming marriage of his daughter.
ALS 3pp RS:HS 5.146

9558. 1852–11–5 To Richard Sheepshanks 32 Harley St.
Needs information about the late Thomas Glanville Taylor, astronomer at Madras Observatory, as JH is trying to help one of Taylor's sons obtain a post in the Royal Navy.
ALS 3pp RS:HS 16.120 (C: RS:HS 25.9.75 & Cdraft: 25.15.12)

9559. 1852–11–6 To Richard Jones n.p.
Thanks for RJ's 'unwearied vigour and activity of mind.'
C 1p RS:HS 23.126

9560. 1852–11–7 To George Gabriel Stokes 32 Harley St.
Thanks GS for 'canary' glass. Pleased with GS's 'photological' research, extending the spectrum.
ALS 1p Cambridge University Library 7656/H693A
Stokes, *Correspondence*, *1*, 27–8

9561. 1852–11–8 From Michael Faraday Royal Institution
Sending a letter he has received from [Rudolf?] Wolfe. Will send it to the R.S.L. or the R.A.S after JH has seen it, whichever he thinks best.
ALS 1p RS:HS 7.197 Williams, *Selected Correspondence, 2*, 671

9562. 1852–11–9 From George Gabriel Stokes Pembroke College, Cambridge
Asks about problems in GS's paper that JH is reviewing. Remarks on spectral lines paper presented by Antoine Becquerel to Académie des sciences.
ALS 4pp RS:HS 17.29

9563. 1852–11–10 To Michael Faraday Harley St.
On the periodicity of sun spots.
C 2pp RS:HS 23.127 Williams, *Selected Correspondence, 2*, 671–2

9564. 1852–11–11 From Charles Edward Trevelyan Treasury
Colonial 'Chest' contains excess of silver coins. Requests that £30,000 be sent from Malta to England.
LS 2pp TxU:H/M–0552.40; Reel 1093

9565. 1852–11–12 From Michael Faraday Royal Institution
Has not read up the different treatises on elementary chemistry but gives three authors: George Fowne, Thomas Graham, and W. T. Brande. Has sent [Rudolf?] Wolfe's letter to Edward Sabine.
CC 1p RS:HS 7.196 Williams, *Selected Correspondence, 2*, 673–4

9566. 1852–11–13 From William Johnson Patent Office, 166 Buchanan St., Glasgow
Sends four issues of *Practical Mechanic's Journal* and unpublished article, all referring to JH's and to Mr. Trevithick's contributions to subject of ice making. John Gorrie has attacked WJ's view. Asks JH to confirm facts in articles.
ALS 3pp † TxU:H/M–0302; Reel 1087

9567. 1852–11–13 From Edward Sabine Woolwich
Hopes JH will present application of Kew Committee for financial aid at next council meeting. Includes suggestions

for maximizing credibility of application.
ALS 4pp RS MM.16.158

9568. 1852–11–15 From Henry Drury Harness Royal Mint
No news from Mr. Mocatta. Discrepancy of £126 on Mint books will be quickly settled. Thanks for gift of tickets to HH.
ALS 1p TxU:H/M–0266.9; Reel 1087

9569. 1852–11–17 From John R. Hind 22 Grove Rd., St. John's Wood
Has discovered a seventh planet [Kalliope]. Gives readings.
ALS 1p RS:HS 9.360

9570. 1852–11–22 From George Biddell Airy Royal Observatory, Greenwich
Giving his views on an observatory being linked with a university.
ALS 4pp RS:HS 1.223 (C: RGO 6.337.123)

9571. 1852–11–22 To George Biddell Airy 32 Harley St.
Is sorry that Sydney University was not too happy with one of the professors chosen, but feels the Sydney Trust made the right choice [see GA's 1852–11–22].
ALS 3pp RGO 6.337.125

9572. 1852–11–22 From George Biddell Airy Royal Observatory, Greenwich
A note to accompany the forwarding of some papers.
ACCS 2pp RGO 6.374.128

9573. 1852–11–22 From Charles Edward Trevelyan Treasury
Begin copper coinage for Ireland as soon as pressure for gold and silver coinage slackens. Quantity is not known, but it will be considerable. Mr. Watt is needed in Bahamas; plan to get new clerk from Customs House to replace Watt. CT is arranging for Mr. Julyan to remain as long as JH needs Julyan. Encloses extract from letter by Rev. Walker of Liverpool.
LS 3pp TxU:H/M–0552.41; Reel 1093

9574. 1852–11–22 To Charles Edward Trevelyan n.p.
Please return JH's letter of 20 Nov. Wants to correct error regarding men employed by G. F. G. Mathison.
ACCS 2pp TxU:H/L-0482.25; Reel 1055

9575. 1852–11–22 To [Charles Edward Trevelyan] Royal Mint
Charles Elouis discovered errors in Mint accounts since 1837 that explain earlier discrepancies. JH requests authority from Treasury commissioners to adjust previous accounts to correct those errors.
AdraftS 2pp TxU:H/L-0668; Reel 1089

9576. 1852–11–24 From Francis Beaufort Admiralty
A response to JH's actions in seeking a position for [?] Taylor.
ACC 1p Hydrographic Office Letter Book

9577. 1852–11–24 From Charles Edward Trevelyan Treasury
Glad that [Mr. Watt and Mr. Julyan] are working out well. CT will arrange for them to remain as long as needed. Delaying Watt's transfer to Bahamas. Acquaint them both with all aspects of Mint business, especially assaying.
LS 3pp TxU:H/M–0552.42; Reel 1093

9578. 1852–11–30 From Charles Edward Trevelyan Treasury
Lord Chandos [Richard Grenville] wants to know superannuation deductions and pension expectations of Mint workmen.
ALS 2pp TxU:H/M–0552.43; Reel 1093

9579. [1852–11–31 or later] To Margaret Brodie Herschel [London]
Arrangements about MH coming to London, and JH sending some things to Collingwood.
ALS 3pp JHS 1.80

9580. 1852–12–1 From H. R. Reynolds Treasury
HR can do nothing about [Horiton?] until 'Major of that place' writes to JH. Has turned this case over to HR's assistant for Mint affairs.
ALS 2pp TxU:H/M–0440; Reel 1093

9581. 1852–12–2 From Henry Drury Harness Royal Mint
Surprised by visit of Treasury people to attend examination of diets from Sheffield and Bringham. Waiting for authority to continue issuing coins previous to pyx assay.
ALS 2pp TxU:H/M–0267.1; Reel 1087

9582. 1852–12–2 From William Johnson Patent Office, 166 Buchanan St., Glasgow
Asks permission to publish parts of JH's 21 Nov. 1852 letter to WJ to refute John Gorrie's denial of JH's contribution to ice making. Sends issue of *Practical Mechanic's Journal* containing Gorrie's paper.
ALS 3pp TxU:H/M–0303; Reel 1087

9583. 1852–12–3 To Charles Babbage n.p.
Invitation to attend the wedding ceremony at St. James' Piccadilly [for the wedding of Caroline E. M. Herschel and Col. Alexander H. Gordon]
ALS 1p RS:HS 2.326

9584. 1852–12–3 From Henry Drury Harness Royal Mint
Sends messenger to JH with checks for wages and enclosed correspondence.
ALS 1p TxU:H/M–0267.2; Reel 1087

9585. 1852–12–3 To [Henry Drury Harness] [Royal Mint]
Encloses Treasury letter authorizing issuance of coin in anticipation of pyx assay. Please open and process all letters addressed to Master of Mint, except those marked 'private.' Procedures for hiring new assayer, Mr. Westwood, to replace Mr. Shaw. Return Shaw's bond to him. Encloses official papers regarding Indian pyx. H. W. Field was prepared for visit by Treasury officers.
AdraftS 4pp TxU:H/L-0669; Reel 1089

9586. 1852–12–4 From Henry Drury Harness Royal Mint
Returned Mr. Shaw's bond to Mr. Regland and asked Regland to clarify relationship between two Mr. Westwoods. Sends papers for JH's signature. Reports have returned on 50 gold ingots. Compares errors among assays by W. H. Field, Thomas Graham, and W. A. Miller.
ALS 2pp TxU:H/M–0267.4; Reel 1087

9587. 1852–12–4 From Henry Drury Harness Royal Mint
Sends assay book as JH requested. Cannot find HH's notes on silver coinage; will continue to search.
ALS 2pp TxU:H/M–0267.3; Reel 1087

9588. 1852–12–4 From George Peacock Deanery, Ely
Will be pleased to officiate at wedding of JH's daughter Caroline [see JH's 1852–10–30].
ALS 3pp RS:HS 13.304

9589. 1852–12–6 From Henry Drury Harness Royal Mint
Gold assay reports by Thomas Graham contain too many errors. Distrusts all previous assays performed at Graham's laboratory.
ALS 2pp TxU:H/M–0267.5; Reel 1087

9590. 1852–12–7 To the *Athenaeum* 32 Harley St.
Requests publication of a letter [see John Stewart's [1852]] that JH received from Stewart and in which Stewart explains an effective photographic process Stewart uses.
L *Athenaeum,* #1311 (Dec. 11, 1852), 1363

9591. 1852–12–7 From Henry Drury Harness Royal Mint
Inland Revenue office does not understand that punches and dies for coins may be produced outside Mint. Anthony de Rothschild claims to have lost large sum due to improper assay [of gold], but still wants to sell £10,000 more to Mint. Mint can comply with only first part of order from House of Commons; second part requires information kept secret by Bank of England.
ALS 4pp TxU:H/M–0267.6; Reel 1087

9592. 1852–12–8 From Henry Drury Harness Royal Mint
Letters to Inland Revenue office regarding [Leonard] Wyon's work are ready for JH to sign. Accuracy of coins.
ALS 2pp TxU:H/M–0267.8; Reel 1087

9593. 1852–12–8 From Henry Drury Harness Royal Mint
According to H. R. Reynolds, Mr. Westwood Sr., though a bondsman to the Crown, may still serve as surety for Mr. Westwood Jr. Should HH notify Mr. Regland that JH accepts two recommended sureties?
ALS 1p TxU:H/M–0267.7; Reel 1087

9594. 1852–12–8 From George Peacock Athenaeum
Has arrived in London, ready for the wedding on following day [see GP's 1852–12–4].
ALS 1p RS:HS 13.305

9595. 1852–12–9 From Henry Drury Harness Royal Mint
Forwards letter containing extraordinary proceedings relative to copper coinage for Ireland. Will send check for weekly wages to JH for signature.
ALS 3pp TxU:H/M–0267.9; Reel 1087

9596. 1852–12–9 From Charles Edward Trevelyan Treasury
Order for 'Return of Coin' melted by Bank [of England] was sent to Bank's governor [Thomson Hankey] and deputy governor. Congratulates JH on [marriage of daughter Caroline to son of next prime minister, Lord Aberdeen].
(A)LS 2pp TxU:H/M–0552.44; Reel 1093

9597. 1852–12–10 From Henry Drury Harness Royal Mint
Gave Mr. Baxter's letter to H. W. Field, so Field may return relics. Forwards letter from J. W. Morrison to JH. Advises JH not to recommend machinery for Sydney branch mint unless Treasury asks JH to do so. Congratulations on wedding of JH's daughter Caroline.
ALS 2pp TxU:H/M–0267.10; Reel 1087

9598. 1852–12–10 From William Lassell Valletta
Has been given every facility to erect his telescope in Malta. Atmosphere has been excellent for observations. Has studied the satellites of Uranus and the ring of Saturn; the latter is puzzling. Has been astonished with the views of Orion.
ALS 4pp RS:HS 11.148

9599. 1852–12–10 From John William Lubbock n.p.
Lady Lubbock has received a letter from a friend of JH's and it seems desirable that JH should know the contents. Can only communicate it verbally. Sends their congratulations on the forthcoming marriage of Miss Caroline Herschel.
ALS 4pp RS:HS 11.389

9600. 1852–12–10 To My dear Sir n.p.
Embarrassed that JH's servant sent away messenger who brought papers for JH's signature. Does not consider Mint to be bound by Ralph Heaton's terms for copper coinage. JH prefers contract with Heaton similar to former contract with Boulton & Watt. Dies should be supplied by Mint, not made by contractor.
ACCS 2pp TxU:H/L–0670; Reel 1089

9601. 1852–12–11 From Adolphe Quetelet Brussels
Lists volumes for JH sent to R.S.L. One includes observations of transit instrument of 1836–1839. Also follow-up on climate of Belgium. Thanks for wonderful reception in London.
ALS 3pp (French) RS:HS 14.251

9602. 1852–12–13 From Henry Drury Harness Royal Mint
C. E. Trevelyan wants JH to meet with governor of Bank of England. Congratulations on success of JH's son.
ALS 2pp TxU:H/M–0267.11; Reel 1087

9603. 1852–12–15 From John R. Hind 22 Grove Rd., St. John's Wood
Has discovered an eighth planet. Gives readings. Proposes to call it Thalia.
ALS 2pp RS:HS 9.361

9604. 1852–12–15 From Charles Edward Trevelyan Treasury
JH and H. D. Harness will meet with governor and deputy governor of Bank [of England] at three today in Treasury office.
LS 2pp TxU:H/M–0552.45; Reel 1093

9605. 1852–12–17 From Thomas Maclear Royal Observatory, Cape of Good Hope
The Australian ship carrying the gold for England has visited the Cape. Many of the Cape people have moved to Australia. Comments on the gold.
ALS 1p RS:HS 12.158

9606. 1852–12–18 From Henry Drury Harness Royal Mint
Apologizes for unclear language in a recent letter.
ALS 1p TxU:H/M–0267.12; Reel 1087

9607. 1852–12–21 From James A. Gordon n.p.
Was pleased to hear the good news. There have been movements in the political world. Had a long talk with George H. Gordon (4th Earl of A[berdeen]). Hopes Lady Herschel has recovered.
ALS 4pp RS:HS 8.200

9608. [1852–12–23?] To Margaret Brodie Herschel Royal Mint
Will try to get away tomorrow in time to join Gordons for trip to Collingwood [for Christmas]. Request from MH's milliner. Happy that MH patched things over with Stevensons. MH's letter to Mrs. Maclear was returned.
ALS 4pp TxU:H/L–0536.4; Reel 1053

9609. 1852–12–27 From Henry Drury Harness Royal Mint
Forwards for JH's signature two orders from Treasury

approving rates for piece work in Mint.
ALS 1p TxU:H/M–0267.13; Reel 1087

9610. 1852–12–29 From Henry Drury Harness Royal Mint
Forwards papers from Treasury. Advises JH to arrange for ample supply of copper coins for colonies at same time arrangements are made for coins for Ireland.
ALS 1p TxU:H/M–0267.14; Reel 1087

9611. 1852–12–29 From William Shanks Houghton-le-Spring
Describes WS's forthcoming volume, and asks JH to subscribe to it.
ALS 3pp RS:HS 17.202

9612. 1852–12–31 To Margaret Brodie Herschel [Royal Mint]
Colonel G.'s [Gordon's?] servant brought two pheasants from Collingwood. JH sent one to Mrs. Stewart. Encloses postal orders. W. T. Hooper will send certificate from Haileybury College for JH's son William. [Servant] James will stay on. Ends year at Mint with temporary cessation of gold coinage and 'huge battle of Silver.' Lists daily family expenses in London for 20–31 Dec.
ALS 4pp TxU:H/L–0543.17; Reel 1053

9613. [1853?] To Charles Babbage n.p.
Send the names and addresses of any friends who may wish to see the Mint. Thanks for pamphlet on income tax.
ALS 2pp RS:HS 2.393

9614. [1853 or later] To Charles Babbage n.p.
Number of half-crowns minted in the years 1832–53.
ALS 1p RS:HS 2.377

9615. 1853 To Augustus De Morgan n.p.
JH looking for some writings by Isaac Newton while at the Mint, but most of that seems to have vanished.
C 1p RS:HS 23.131

9616. [1853] To [Augustus De Morgan] n.p.
Has been reading AD's article on Mrs. Benton; agrees that AD has made it probable that she was involved in a private marriage. Comments on changing place of women in society.
ALS 3pp Univ. of London Library

9617. 1853 From James A. Gordon 7 Grafton St.
Saw Mrs. A. Gordon yesterday. Finds that JH dines alone; would be pleased if he would join him anytime when convenient. Had a long letter from Mr. Pellew today.
ALS 3pp RS:HS 8.201

9618. [1853–1] To Margaret Brodie Herschel n.p.
Family finances. Plans to let Charlotte go on Friday, to keep peace with regular cook. Will go to Naish on Saturday and bring daughters back on Monday. Will dine on Tuesday with new Chancellor of Exchequer [W. E. Gladstone]. Concern for daughter Amelia; with good nursing, she will grow new skin. Leonhard Schmitz's new Roman History is probably reproduction of B. G. Niebuhr's lectures.
ALS 2pp TxU:H/L–0536.12; Reel 1053

9619. [1853] To [Edward Sabine?] n.p.
Postscript regarding the brothers [Adolphe, Hermann, and Robert] Schlagintweit and funding for their proposed expedition.
AL 1p inc RS Sa.685

9620. [1853?] To Charles [Edward Trevelyan] [Royal Mint]
Reports gold coin production.
AdraftS 3pp TxU:H/L–0483.2; Reel 1055

9621. 1853–1–2 To Alexander Hamilton-Gordon 32 Harley St.
AH's father, Lord Aberdeen [G. Hamilton-Gordon] misunderstood JH, who never meant to suggest that office of deputy master of Mint was unnecessary. JH intended to stress that *some* responsibilities could be borne by only the Master.
ACCS 3pp TxU:H/L–0167; Reel 1054

9622. [1853–1–2?] To Margaret Brodie Herschel n.p.
Thanks for MH's New Year's blessings. Divine help is sustaining JH in adversity. Family finances. Summary of 1852 accounts. Grandma [Mrs. Stewart] is ill. Mint business includes opening two Australian branches, large order for copper coinage, and decision about Mint's relationship to Bank of England. H. D. Harness is doing well.
ALS 4pp TxU:H/L–0536.16; Reel 1053

9623. 1853–1–3 To Margaret Brodie Herschel [London]
Talks about things being sent out to Collingwood, about bills and the shortness of money, and the pressure of work at the Mint.
ALS 3pp JHS 1.90

9624. 1853–1–3 From George Gabriel Stokes Pembroke College, Cambridge
Surprised by length of spectrum produced by the combustion of metals, using electricity.
ALS 3pp RS:HS 17.30

9625. 1853–1–4 To [James Calder] Stewart n.p.
Asks for advice on JH's investments.
ALS 4pp TxU:H/L–0442; Reel 1055

9626. 1853–1–4 To George Gabriel Stokes 32 Harley St.
Thanks GS for his paper on fluorescent spectra; JH is interested in having these rays used on quartz, hoping to find his 'thermic' rays acting.
C 2pp RS:HS 25.10.3

9627. 1853–1–6 From J. F. Julius Schmidt Sternwarte zu Bonn
Sends work on shooting stars. Requests all information on telescopic meteors. Realizes that the frequency (nombre horaire) and the variation (variation horaire) of the shooting stars remains unknown. Hopes for global cooperation in this matter. Sends work on total solar eclipse.
ALS 4pp (German) RS:HS 17.177

9628. 1853–1–7 From Charles Edward Trevelyan Treasury
What quantity of copper coins for Ireland and colonies should be called for in bids? Estimate at least 300 tons for Ireland. What quantity in sterling is equivalent to one ton of copper pence, halfpence, and farthings? What colony ordered half- and quarter-farthings? [JH annotation: Ton of copper = £224. Ionian Isles and Ceylon use small copper coins.]
LS 2pp † TxU:H/M–0553.1; Reel 1093

9629. 1853–1–8 To Margaret Brodie Herschel Royal Mint
Frightened by MH's account of daughter Amelia's serious burns. Encloses revised account sheet that shows £196 missing. Son William will come to Collingwood soon. Saw daughter Caroline yesterday.
ALS 4pp TxU:H/L–0544.1; Reel 1053

9630. 1853–1–9 To Richard Sheepshanks 32 Harley St.
Asks for William Lassell's address in Malta, as JH's son, William, may be able to stop in to visit Lassell.
C 1p RS:HS 25.9.76

9631. 1853–1–10 From Charles Hay Cameron East Sheen
Sending letters of introduction for JH's son [William] when he visits India and wishing him success in his appointment.
ALS 2pp RS:HS 5.147

9632. [1853–1–10?] To Margaret Brodie Herschel n.p.
Thank daughter Isabella for account of Amelia Herschel's condition. Bank [of England] is in trouble again. Lists [Christmas] cards received. Attended Michael Faraday's wonderful lecture last night.
ALS 3pp TxU:H/L-0536.15; Reel 1053

9633. 1853–1–11 From Richard Sheepshanks Reading
Address of William Lassell in Malta. Congratulations on appointment of JH's son William to H.E.I.C. and on marriage of JH's daughter Caroline. Testing of standard measure bars at Greenwich.
ALS 4pp TxU:H/M–0541; Reel 1093

9634. 1853–1–12 From Henry Drury Harness Royal Mint
Forwards coinage return, assay book, and one letter to JH during JH's illness.
ALS 1p TxU:H/M–0268.1; Reel 1087

9635. 1853–1–13 From Thomas Bell Royal Society, Somerset House
Is having letters regarding 'Southern Telescope' sent to JH. Requests JH's remarks, to be read at committee meeting.
ALS 2pp RS MM.16.133

9636. 1853–1–14 From Charles Edward Trevelyan Treasury
Please comment on enclosed letter from Thomson Hankey.
LS 1p TxU:H/M–0553.2; Reel 1093

9637. 1853–1–17 To Miss Cornwallis Collingwood
Gratitude for nursing JH's daughter Amelia after severe burns. Taking JH's son William to Folkstone. Regards to Miss Lipscombe.
ALS 4pp TxU:H/L-0131; Reel 1054 (C: RS:HS 23:128)

9638. [1853–1–17 or later] To Margaret Brodie Herschel n.p.
Helped son William board ship to France. Instructions for sending mail to William in India. Recipe for croton pills. Letter from Matilda [Grahame Stewart]. What shall JH tell J. E. Mayall about 'J. S.'s photographs'?
ALS 4pp TxU:H/L-0536.13; Reel 1053

9639. 1853–1–17 To Charles [Edward Trevelyan] Harley St.
Explains JH's recent absence. In Collingwood to visit seriously burned daughter [Amelia] and to send son [William] off to India.
ACCS 1p TxU:H/L-0483.1; Reel 1055

9640. 1853–1–18 To Thomas Bell n.p.
Has received papers from TB on Southern telescope [see TB's 1853–1–13]. Wishes to avoid hasty judgment of the matter.
ALS 2pp RS MM 19.106

9641. 1853–1–19 From William Thomson Queen's College, Oxford
Sends JH a copy of the third edition of his *Laws of Thought*.
ALS 2pp RS:HS 17.436

9642. 1853–1–23 To Margaret Brodie Herschel [London]
Changes in daughter Caroline's life. Will send note received from son William in Paris. German poetry. Visited Elizabeth Baily. Received visit from Eve Gordon. Financial panic in Paris is spreading to Bank of England. Louis Napoleon's 'love match.' P.S.: Doubts if daughters Isabella or Louisa will understand [B. G.] Niebuhr's first volume.
ALS 4pp TxU:H/L-0544.2; Reel 1053

9643. 1853–1–23 From Adolphe Quetelet Brussels
Congratulations on marriage of JH's daughter Caroline Herschel. Expects JH has received works sent 1852–12–11. Sends more but assures JH he must not read everything.
AC 2pp RS:HS 14.252

9644. 1853–1–24 From Thomas Maclear Royal Observatory, Cape of Good Hope
Has just heard that Caroline Herschel is to marry the Hon. Mr. Gordon. Sends her his best wishes. This week TM hopes to commence the alteration to the circle room.
ALS 2pp RS:HS 12.159

9645. 1853–1–25 From James Maclaren Pashley House
Hardly likes to trouble him to read the enclosed. Gives his views on the gold question.
ALS 2pp RS:HS 12.27

9646. 1853–1–27 From Charles Edward Trevelyan Treasury
Found missing papers, including [James] Pennington's memorandum. Sends them to JH. Please send H. D. Harness's memorandum.
LS 1p TxU:H/M–0553.3; Reel 1093

9647. 1853–1–28 To Margaret Brodie Herschel [London]
Expects daughter Caroline and husband and Eneas Mackintosh on Sunday. Must hire seven supervisors and 28 workers at Mint. Bank of England requires £600,000 per week of new coins. Family finances. Engraving of William Whewell is 'wretched.' Mr. Dudgeon visited yesterday to learn JH's new photographic method of coating glass with gluten and acetic acid.
ALS 4pp TxU:H/L-0544.3; Reel 1053

9648. 1853–2–1 To George Biddell Airy 32 Harley St.
Argues for the decimalization of all weights, measures, and money [see JH's 1852–10–26].
ALS 4pp RGO 6.341.28

9649. 1853–2–1 From George Biddell Airy Royal Observatory, Greenwich
A note accompanying a book of observations being forwarded to JH by GA, and coming from Georgetown, USA.
ACCS 1p RGO 6.59.184

9650. [1853–2–2] To Charles Edward Trevelyan [Royal Mint]
Review enclosed letter from JH to Lord Aberdeen [G. Hamilton-Gordon]. Henry Finch seems willing to accept position in reorganized Mint.
ACC 1p TxU:H/M–0553.5; Reel 1093

9651. 1853–2–2 From Charles Edward Trevelyan Treasury
Prefers that Henry Finch and Mr. Hill, Jr. be assigned to temporary positions in Mint. Send official letter to Lord Aberdeen [G. Hamilton-Gordon] about this. That leaves one patronage appointment for W. G. Hayter to fill.
ALS 3pp TxU:H/M–0553.4; Reel 1093

9652. 1853–2–3 From Charles Edward Trevelyan Treasury
Return George Arbuthnot's final memorandum on Australian mint.
LS 1p TxU:H/M–0553.6; Reel 1093

9653. 1853–2–3 To James Wilson Royal Mint
Reports average output of gold coinage up to 'these changes in the production of Gold.' Demand is now increasing.
Adraft 4pp TxU:H/L-0502; Reel 1055

9654. 1853–2–5 From James Maclaren Pashley House
Comments on the currency. Thinks it would be feasible to share tutors for their children as the name of JH would attract men with superior ability.
ALS 4pp RS:HS 12.28

9655. 1853–2–6 From George Biddell Airy Flamsteed House, Greenwich
Has heard that one of JH's daughters [Amelia] has been hurt by burning; expresses concern.
ALS 1p TxU:H/M-0636.2; Reel 1084

9656. 1853–2–7 From George Biddell Airy Royal Observatory, Greenwich
Concerning GA's eclipse lecture and [G.] Seyffarth's claims regarding eclipse observations.
ALS 3pp RS:HS 1.224

9657. 1853–2–7 From George Biddell Airy Royal Observatory, Greenwich
The standardization of weights and measures is moving ahead; the question of storage of standards must be addressed.
ALS 2pp TxU:H/M-0636.1; Reel 1084 (C: RGO 6.341.30)

9658. 1853–2–7 From George Biddell Airy Royal Observatory, Greenwich
Proposes rules of procedure for use of secondary standards [see GA's 1853–2–7].
ALS 4pp TxU:H/M-0636.3; Reel 1084 (C: RGO 6.341.31)

9659. 1853–2–7 To George Biddell Airy n.p.
Question about whether the effect of Venus on the motion of the moon's node is measurable; this was brought on by reading J. H. Seyffert's writings.
LCC 1p inc RGO 6.81.4

9660. 1853–2–7 From George Biddell Airy Royal Observatory, Greenwich
Believes that the observations that concern JH [see JH's 1853–2–7] with respect to the moon's orbit do not argue for Venusian effects, but GA finds J. H. Seyffert's writings disturbing, too.
ACCS 3pp RGO 6.81.5

9661. 1853–2–7 To George Biddell Airy n.p.
Thank you note upon receipt of the book of observations [see GA's 1853–2–1].
LCC 1p RGO 6.59.186

9662. 1853–2–8 To William Goodenough Hayter Royal Mint
Junior clerk C. B. Piers resigned from Mint. Recommends Henry Finch to succeed Piers. Recalls sacrifice made by Finch and F. R. Brande when Moneyers were disbanded. Asked Lord Aberdeen [G. Hamilton-Gordon] that Mr. Hill Jr. be hired for temporary duty in Mint during coinage crisis.
AdraftS 4pp TxU:H/L-0188; Reel 1054

9663. 1853–2–9 To George Biddell Airy n.p.
Comments on accidental burns to JH's daughter Amelia.
draft 1p RS:HS 23.129

9664. 1853–2–9 To George Biddell Airy Royal Mint
Comments on GA's proposals about the storage of primary standards [see GA's 1853–2–7, i.e., RGO 6.341.30], and about secondary standards [see GA's 1853–2–7, i.e., RGO 6.341.31].
ALS 2pp RGO 6.341.34

9665. 1853–2–[10] From James A. Gordon Naish House, Wrangle, near Bristol
Testimonial to good character of Mr. Ottley Jr., to whose father JG delegated all JG's powers while on Antigua. Asks JH to assist young Ottley.
ALS 4pp TxU:H/M–0932; Reel 1083

9666. 1853–2–11 From George Biddell Airy Royal Observatory, Greenwich
Raises a problem for the Standards Commission to consider, namely, that the Bank of England finds it difficult to deal with the decimalization of troy ounces [in the bullion trade] [see JH's 1853–2–9].
LS 3pp TxU:H/M-0636.4; Reel 1084 (C: RGO 6.341.35)

9667. 1853–2–12 From George Biddell Airy Royal Observatory, Greenwich
Is very much in favor of JH's coinage ideas [see JH's [1851]–1–4], but thinks they may be too bold.
ACCS 3pp RGO 6.349.11

9668. 1853–2–14 To George Biddell Airy Royal Mint
Talks about the means to introduce JH's radical reform ideas for coinage [see GA's 1853–2–12]; would like the Bank of England to go further in decimalizing weights of coins [see JH's 1853–2–11].
ALS 4pp RGO 6.349.14 & 6.341.37 (C: TxU:H/M-0036.4v)

9669. 1853–2–14 To 'My dear Sir' Royal Mint
Suggests regulations regarding bullion. Resolution of [Treasury] commissioners was forwarded to Sir Charles Wood.
ACCS 1p TxU:H/M–0636.4.2; Reel 1084

9670. 1853–2–14 From H. R. Reynolds Treasury
Encloses draft [contract] for coinage of copper.
ALS 1p TxU:H/M–0441; Reel 1093

9671. 1853–2–15 From George Biddell Airy Royal Observatory, Greenwich
Believes that the increasing use of decimal weights should mean that no further evidence need be taken by the Standards Commission [see JH's 1853–2–14].
ALS 2pp TxU:H/M-0636.5; Reel 1084 (C: RGO 6.341.38)

9672. 1853–2–15 From [John] M. Field Assay Office
Notes change of bars on pieces 26 and 27 prior to their arrival at assay office.
ALS 1p TxU:H/M–1128.3 [p. 52]; Reel 1060

9673. 1853–2–16 To George Biddell Airy Royal Mint
Responds about a meeting date [see GA's 1853–2–15]; further thoughts on the introduction of a new coinage system [see JH's 1853–2–14].
ALS 2pp RGO 6.349.15 & 6.341.39

9674. 1853–2–16 From Charles Edward Trevelyan Treasury

Call on CT when convenient.
LS 1p TxU:H/M–0553.7; Reel 1093

9675. 1853–2–17 From H. R. Reynolds Treasury
Encloses amended draft proclamation regarding florin for JH's approval.
ALS 3pp TxU:H/M–0442; Reel 1093

9676. 1853–2–18 From George Biddell Airy Royal Observatory, Greenwich
A notice of meeting of the Standards Commission.
LS 1p TxU:H/M-0636.6; Reel 1084

9677. 1853–2–18 To Henry Finch Royal Mint
C. E. Trevelyan has confirmed JH's recollection that HF agreed to annual salary of £100 when entering reorganized Mint.
ACCS 1p TxU:H/L-0154; Reel 1054

9678. 1853–2–18 To Charles Edward Trevelyan [Royal Mint]
Matters discussed in meeting with CT: hiring of Henry Finch on probation; denial of Robert Mushet's claim for raise, except through promotion (JH annotation: Now Mushet will *expect* promotion); [Chancellor of Exchequer] W. E. Gladstone's alternative to shortage of copper coins.
[JH note] 1p TxU:H/M–0553.8; Reel 1093

9679. 1853–2–19 To Henry Finch Royal Mint
Formal announcement of HF's nomination by Treasury to fill office of junior clerk in Mint Office vacated by [C. B.] Piers. HF is assigned to W. T. Brande in Coining department.
ACCS 1p TxU:H/L-0155; Reel 1054

9680. 1853–2–19 To Charles Edward Trevelyan Royal Mint
Forwards request of 'Mr. M' as promised.
ALS 2pp TxU:H/L-0483.3; Reel 1055

9681. 1853–2–20 To Margaret Brodie Herschel [London]
Is trying to arrange for Mrs. Gordon to go to Collingwood, and a few other bits of news.
ALS 4pp JHS 1.95

9682. 1853–2–20 From Joseph Whitworth Manchester
Encloses copy of JW's letter to G. B. Airy, chair of Standards Commission. Invites JH to come and observe what JW has accomplished regarding standard measures.
ALS 2pp TxU:H/M–0581; Reel 1093

9683. 1853–2–21 To Margaret Brodie Herschel [London]
Commiserates with MH and her 'horrid headache'; sends on a letter from daughter Caroline.
ALS 3pp JHS 1.96

9684. 1853–2–24 From George Biddell Airy Royal Observatory, Greenwich
Seeks JH's approval, and signature, on a paper; does not wish to talk about copper coinage.
ACCS 3pp RGO 6.349.24

9685. 1853–2–25 From George Biddell Airy n.p.
Agenda items for 2 Mar. 1853 meeting of Commission for Restoration of Standards of Lengths and Weights.
LS 4pp ††(by JH) TxU:H/M–0656; Reel 1084

9686. 1853–2–27 To George Biddell Airy 32 Harley St.
The meeting date is fine, but no hour or place has been specified [see JH's 1853–2–16].
ALS 1p RGO 6.341.40

9687. 1853–2–28 From George Biddell Airy Royal Observatory, Greenwich
Clarifies meeting details [see JH's 1853–2–27].
ACCS 1p RGO 6.341.41

9688. 1853–2–29 To Margaret Brodie Herschel [London]
About MH's continuing illness [see JH's 1853–2–21] and the illness of Mrs. Knowles; some of their children are spending the weekend with JH.
ALS 4pp JHS 1.97

9689. [1853]–3–5 To Charles Edward Trevelyan Harley St.
Has not yet shown to H. D. Harness the paper referred to in JH's note to CT yesterday.
ACC 1p TxU:H/L-0484.13; Reel 1055

9690. 1853–3–5 From Charles Edward Trevelyan Treasury
[Marked 'Confidential.'] JH's reaction to paper was precisely what CT felt. Plans to communicate it to no one else.
LS 2pp TxU:H/M–0553.9; Reel 1093

9691. 1853–3–7 To Thomas Bell 32 Harley St.
Recommendations for design and construction of new telescope proposed by R.S.L. for Southern Hemisphere. Need to coordinate observations in both hemispheres.
AdraftS 10pp TxU:H/L-0090; Reel 1054

9692. 1853–3–8 To Charles [Edward Trevelyan] Royal Mint
Custom House authorities want return of two clerks loaned to Mint. W. T. Brande chose Mr. Harding as replacement clerk. JH does not believe that Harding will work out. John Beckwith claims new record this week for number of assays.
Adraft 1p TxU:H/L-0483.4; Reel 1055

9693. 1853–3–9 From Charles Piazzi Smyth Hillside Crescent, Edinburgh
Sends photographs of Great Dragon and the Ice Cavern in Teneriffe. Notes that images are finer when glass plates are used.
ALS 4pp RS:HS 16.233

9694. 1853–3–10 From Henry Piddington Calcutta
Of the general ignorance of meteorological matters, especially among sailors; wider distribution of HP's works would help, but HP is too poor to do that.
ALS 8pp RS:HS 13.408

9695. [1853]–3–10 From Frederick Waymouth Gibbs n.p.
Arranges visit to Royal Mint for Prince Albert Edward and Prince Alfred.
ALS 2pp TxU:H/M–0205; Reel 1087

9696. 1853–3–15 From George Biddell Airy Royal Observatory, Greenwich
Does not believe that the technical weights of the Bank of England should be given the same legal status as primary, and official, weights, measures, and coinage [see GA's 1853–2–11].
ALS 4pp TxU:H/M-0636.7; Reel 1084 (C: RGO 6.349.16)

9697. 1853–3–16 To William Henry Barton Royal Mint
Instructions for collecting 'Gold and Silver sweep' from coining and melting rooms and arranging annual sale of 'sweep and old stores' at end of March.
AdraftS 2pp TxU:H/L-0081; Reel 1054

9698. [1853–3–16] From Frederick Waymouth Gibbs n.p.
Prince Albert has postponed visit to Royal Mint for two princes until after Easter.
ALS 3pp TxU:H/M–0206; Reel 1087

9699. 1853–3–17 To Charles [Edward Trevelyan] Royal Mint
Wages for Henry Finch's temporary services.
ACCS 3pp TxU:H/L-0483.5; Reel 1055

9700. 1853–3–21 To George Biddell Airy [Royal Mint]
Details of how coins are tested and stamped for weight. Procedures are not regulated, and variances are common. Encloses Mr. Miller's letter on decimal coinage. [JH annotation: 'Substance sent to Airy March 21 / 53 but altogether diff in form and wording.']
Adraft 4pp †(by JH) TxU:H/L-0595; Reel 1089

9701. 1853–3–21 To George Biddell Airy Royal Mint
Responds with return of letter part and an invitation to the Royal Mint [see GA's 1853–3–21]; expands further on the coinage/weight problem.
ALS 4pp RGO 6.349.21

9702. 1853–3–21 From George Biddell Airy Royal Observatory, Greenwich
Wants to visit the Royal Mint; the coinage/weight question is more complex than GA imagined [see GA's 1853–3–15]; seeks JH's opinion on part of a letter to the Chancellor of the Exchequer.
ACCS 2pp RGO 6.349.18

9703. 1853–3–21 To [?] Harley St.
Regrets will be unable to be steward at a 'Public dinner.'
ALS 1p St. John's College, Cambridge

9704. [1853–3–22 or later] To Margaret Brodie Herschel London
Talks about JH's continuing cough, refers to some bills that need to be settled, and tells MH that JH is committed to going out on Sunday, much against his will.
ALS 1p inc JHS 6.63

9705. 1853–3–25 From Augustus De Morgan 7 Camden St.
The book on the coinage of England in the 1670s is by one William Jeake. Comments on the diversity of coins available at that date and how their values changed by proclamation. Should the present coinage be decimalized he hopes the half crown will be withdrawn.
ALS 3pp RS:HS 6.263

9706. 1853–3–25 To Margaret Brodie Herschel Collingwood
Winter scenes at Collingwood. Legal information about civil days beginning at midnight, while astronomical days begin at noon. Daughters Francisca, Mathilda Rose, and Amelia are well and active. Enjoying quiet pace in country.
ALS 4pp TxU:H/L-0544.4; Reel 1053

9707. 1853–3–26 To George Biddell Airy Collingwood
Supports the report, but JH believes as an employee of the Treasury, he cannot sign the report on coinage to the Chancellor of the Exchequer; does not anticipate swift action on the report.
ALS 3pp RGO 6.349.26

9708. 1853–3–28 From George Biddell Airy n.p.
Understands JH's position [see JH's 1853–3–26]; expects the report will be taken to Parliament.
ACCS 1p RGO 6.349.30

9709. 1853–4–2 To [James Calder] Stewart 32 Harley St.
Letter of recommendation, praising JS's character and talent.
ALS 3pp TxU:H/L-0443; Reel 1055

9710. 1853–4–2 From Charles Edward Trevelyan Treasury
Delays question of hiring Mr. Pilcher. Proceed with planning for Sydney branch mint.
ALS 2pp TxU:H/M–0553.10; Reel 1093

9711. 1853–4–3 From Henry Piddington Calcutta
Note accompanying more meteorological writings.
ALS 3pp RS:HS 13.409

9712. 1853–4–5 From Robert Main Greenwich
Wishes to apply for the vacancy that has occurred in the Superintendance of the Nautical Almanac, and would welcome JH's support.
ALS 4pp RS:HS 12.276

9713. 1853–4–6 From Matthew Collins Liverpool
Encloses pamphlets and would be pleased with any remarks of approbation.
ALS 2pp RS:HS 5.261

9714. 1853–4–6 From John Allen Slater Middlewich
Asks about precise link between rotation and revolution of satellites, and its relation to belief in existence of God.
ALS 4pp RS:HS 17.214

9715. 1853–4–11 To Augustus De Morgan Royal Mint
Asks for the experience of other countries in introducing decimal coinage.
C 1p RS:HS 23.130

9716. 1853–4–11 From Urbain J. J. Leverrier London
Thanks for his kind invitation. Returns to Paris on Thursday or Friday. Best wishes to Lady Herschel.
ALS 1p (French) RS:HS 11.206

9717. 1853–4–18 To James Wilson Royal Mint
Reports quantities of gold and silver issued by Mint in recent months.
ALS 3pp TxU:H/L-0503; Reel 1055

9718. 1853–4–25 From Joseph Burnley Hume Baltimore, Maryland
Describes new machinery at U.S. mint in Philadelphia. Refers JH to G. F. Dunning (visiting England) and Franklin Peale (chief coiner) for more information. U.S. mint uses humid assay of silver. Facts about gold assays.
ALS 4pp † TxU:H/M–0284; Reel 1087

9719. 1853–4–25 To [?] Harley St.
In response to a request for information, JH says that his father, William, never had a residence in London.
ALS 1p WT B.9

9720. [1853]–4–27 From Frederick Waymouth Gibbs n.p.
Gratitude for receipt of medals struck by Prince Albert Edward and Prince Alfred during their enjoyable visit to Royal Mint.
ALS 3pp TxU:H/M–0207; Reel 1087

9721. 1853–4–27 From Charles Edward Trevelyan Treasury
Calls JH's attention to articles on mint operations by J. T. Smith.
(A)LS 1p TxU:H/M–0553.11; Reel 1093

9722. 1853–5–5 To Charles [Edward Trevelyan] Royal Mint

Returns letters respecting [John] Goldsmith [Board of Works in Ireland], who has not yet appeared. Comments on colonial coins.
ACC 4pp TxU:H/L-0483.6; Reel 1055

9723. 1853–5–10 From Charles Piazzi Smyth Hillside Crescent, Edinburgh
Will send JH meridian observations that will be published; asks for his advice on them. Applies to government for an equatorial telescope. Discusses Teneriffe site.
ALS 4pp RS:HS 16.232

9724. 1853–[5?]–12 To William Thomas Brande Royal Mint
Instructions for transfer of dies and production of coins for Ionian [Islands] and Ceylon.
AdraftS 4pp TxU:H/L-0104; Reel 1054

9725. 1853–5–14 From George Arbuthnot Treasury
Legal problems with copper coins for New Brunswick.
ALS 5pp TxU:H/M–0012; Reel 1062

9726. 1853–5–14 To William Henry Fox Talbot 32 Harley St.
Thanks for examples of WT's photographic engraving; sends a photo in return.
ALS 2pp Science Museum Talbt 1/42 (C: RS:HS 23.132)

9727. 1853–5–16 To William Henry Barton Royal Mint
Instructions for melting gold and silver and casting coins.
ACCS 1p TxU:H/L-0082; Reel 1054

9728. 1853–5–16 From Adam Sedgwick Norwich
Asks JH to consider Reginald Octavius Day for an appointment to the new mint in Australia. Sends regards of his nieces and self.
ALS 4pp RS:HS 15.443

9729. 1853–5–17 From Admiralty Admiralty Office
Informing him that William Mann, assistant at Cape Town Observatory, has requested a salary increase, and would JH give his views.
ALS 1p RS:HS 1.18

9730. 1853–5–17 To Wilhelm Struve London
Thanks for dedication of WS's publication on double stars to JH's father and to F. W. Bessel. JH's 'present position' cuts JH off from active pursuit of scientific research. Compliments WS's accomplishments and those of WS's son, whom JH expects in England come summer.
C 1p RS:HS 23.133

9731. 1853–5–19 To Charles Edward Trevelyan [Royal Mint]
Proposes two Mint officers: Mr. Watt to be sent to Birmingham to take charge of copper coinage, and Mr. Julyan to be trained in Mint operations for future assignment either as comptroller of Mint or second-in-command to E. W. Ward at Sydney mint.
Adraft 4pp TxU:H/L-0483.7; Reel 1055

9732. 1853–5–20 From Charles Edward Trevelyan Treasury
Call on CT tomorrow before 2:30 pm.
LS 1p TxU:H/M–0553.13; Reel 1093

9733. 1853–5–20 From Charles Edward Trevelyan Treasury
Concurs with JH's choices of financial officers for minting operations in Birmingham and Sydney. Prefers Mr. Julyan for Sydney branch mint, but will not object to Mr. Watt. One guinea per week is appropriate pay for H. A. Severn.
LS 3pp TxU:H/M–0553.12; Reel 1093

9734. 1853–5–23 To William Hunter 32 Harley St.
Accepts invitation to Lord Mayor's conversazione on 8 June.
ALS 1p TxU:H/L-0748.1; Reel 1089

9735. 1853–5–25 To William Henry Barton Royal Mint
Gives instructions for the provision of certain kinds of metal bars for Mint use.
ALS 1p DMC.695A.885.18

9736. 1853–5–25 To Maria Sophia Herschel 32 Harley St.
To JH's daughter to inform her of the activities of her sisters, and to encourage her to obtain the maximum benefit from her travels in France and Germany with William Whewell and his wife.
ALS 4pp JHS 2.12

9737. 1853–5–28 From Henry Drury Harness Royal Mint
[Marked 'Private.'] Hopes that none of HH's actions led JH to resign as master of Mint. Plans to announce HH's own retirement from Mint at end of month.
ALS 2pp TxU:H/M–0268.2; Reel 1087

9738. 1853–5–28 To Henry Drury Harness n.p.
Private. Compliments HH's letter accompanying report by E. W. Ward, which JH forwarded to Treasury. Informally announced JH's resignation as master of Mint, but delayed formal letter until completion of JH's duties with Decimal Committee and Mint contract with Ralph Heaton & Sons. Needs information on silver assays.
AdraftS 2pp TxU:H/L-0185.1; Reel 1054

9739. 1853–5–30 To George Hamilton-Gordon 32 Harley St.
Requests permission to retire from position as Master of Mint.
AdraftS 3pp TxU:H/L-0004; Reel 1054

9740. 1853–5–30 To Henry Drury Harness 32 Harley St.
Assurance that JH approves of HH's conduct. Printer may send JH's report to House of Commons to Mint. Pay bills from Mr. Mocatta and goldsmith.
ACCS 3pp TxU:H/L-0185.2; Reel 1054

9741. 1853–5–30 From Henry Drury Harness Royal Mint
Grateful for JH's assurance [that HH did not upset JH]. No draft for Mr. Mocatta is required today. E. W. Ward and HH will meet JH at Treasury tomorrow.
ALS 1p TxU:H/M–0268.3; Reel 1087

9742. 1853–5–30 To Mary Somerville n.p.
The year has been peaceful for the Herschels and JH is doing little scientific work. Discusses the 'delusion' about 'turning tables,' which JH finds ridiculous.
ALS 4pp RS:HS 16.357 (C: RS:HS 23.134)

9743. 1853–5–30 From Charles Edward Trevelyan Treasury
Bring H. D. Harness and E. W. Ward to meet with CT tomorrow to discuss report on Sydney branch mint.
LS 2pp TxU:H/M–0553.14; Reel 1093

9744. [1853–6] To Joseph Burnley Hume n.p.
Differing opinions among manufacturers of minting machinery. Compares fly-press at Royal Mint with equipment at Philadelphia. Over nine million coins struck in past six months, with £630,000 per week at maximum capacity, three times more than under old Mint system.

Problems with introducing new machinery while production pressure continues. Five new coin weighing machines are 'greatest improvement' at Mint.
AdraftS 4pp TxU:H/L-0200; Reel 1054

9745. 1853–6–2 From Henry Drury Harness Royal Mint
Forwards bank documents and letter from Charles Elouis containing last month's finances. Asks JH to sign check for £10,000 for Mr. Mocatta, who does not want to be paid [for silver deliveries] through Bank of England.
ALS 2pp TxU:H/M–0268.4; Reel 1087

9746. 1853–6–3 From Henry Drury Harness Royal Mint
HH gave instructions to prepare for coining 500,000 half sovereigns. JH may inform Robert Mushet of this.
ALS 1p TxU:H/M–0268.5; Reel 1087

9747. 1853–6–7 From Henry Drury Harness Royal Mint
Sends papers that JH requested. H. U. Addington requested permission for Russian officer to visit Royal Mint. Mr. Mocatta wants second payment but has not yet delivered silver. Charles Elouis will investigate paying Mocatta through bullion office of Bank of England, as J. W. Morrison used to do. Reports 1797–99 copper coinage by Matthew Boulton.
ALS 3pp TxU:H/M–0268.7; Reel 1087

9748. 1853–6–7 From Henry Drury Harness Royal Mint
Forwards two notes from Mr. Mocatta, who wants to be paid for silver before delivering it. HH objects; there should be no more payments prior to delivery.
ALS 1p TxU:H/M–0268.6; Reel 1087

9749. 1853–6–8 To William Hallowes Miller Royal Mint
Spells out procedure for letting WM have access to standard weights.
CC(S) 2pp RGO 6.341.42

9750. 1853–6–8 To Charles Edward Trevelyan Royal Mint
E. W. Ward has called for bids on machinery for Sydney mint but requests written authorization before proceeding with contracts.
ACCS 2pp TxU:H/L-0483.8; Reel 1055

9751. 1853–6–9 From Henry Drury Harness Royal Mint
Forwards recent letters. The *Times* article today points out deficiency in supply of silver coins.
ALS 2pp TxU:H/M–0268.8; Reel 1087

9752. 1853–6–9 To Charles Pritchard [32 Harley St.?]
Expresses grave concern at interest in ideas of table turning among many of the boys at Clapham school. Writes of importance of the education of the next generation. Reiterates his claim that all force is associated with will.
ALS 3pp RS:HS 14.77 (C: RS:HS 23.135)

9753. 1853–6–10 To William Thomas Brande Royal Mint
Gives instructions as to how to prepare a piece of copper for coining.
ALS 1p WT B.10

9754. 1853–6–11 From George Biddell Airy Royal Observatory, Greenwich
Asks JH to let an examination be made of the standard pound kept at the Royal Mint [see WM's 1853–6–8].
ACCS 2pp RGO 6.339.62a

9755. 1853–6–11 From John Stewart Pau, Pyrenees
Describes a method developed by [J. J.] Heilmann of 'taking from glass negatives positive impressions of different dimensions.' Praises JH's suggestion that public records be preserved on microscopic negatives.
L *Athenaeum,* #1341 (July 9, 1853), 83 & *Journal of the Photographic Society, 1:8* (Aug. 22, 1853), 101–3

9756. 1853–6–13 To Herbert Spencer 32 Harley St.
Thanks HS for his 'highly interesting' paper on the 'Universal Postulate,' which will appear in the *Westminster Review.*
ALS 1p Univ. of London Library

9757. 1853–6–14 From Charles Edward Trevelyan Treasury
Approves JH's plan to pay E. W. Ward's sappers while they are employed at Mint.
ALS 1p TxU:H/M–0553.15; Reel 1093

9758. 1853–6–15 From Richard Jones Haileybury
Has begged Edward Ryan to see JH and have a consultation about some of RJ's affairs, which he would like to settle. Pleased to hear that Willy [Herschel] is doing so well.
ALS 3pp RS:HS 10.388

9759. 1853–6–15 From Charles Edward Trevelyan Treasury
Cannot find JH's official letter regarding appointment of bookkeeper at Mint. Please write another, proposing appointment of John Goldsmith to that office.
LS 3pp TxU:H/M–0553.16; Reel 1093

9760. 1853–6–16 To George Arbuthnot Royal Mint
Quality of silver for use in British coins. Coins issued to colonies.
AdraftS 4pp TxU:H/L-0038; Reel 1054

9761. [1853–6–16 or later] To Richard Jones n.p.
Invites RJ to visit. Wishes to discuss matter [of political economy] presented to RJ by [Edward] Ryan. Comments on examination system at Haileybury College, advising RJ to 'teach boldly.'
ALS 4pp TC, Camb. Add. Ms.c.53[62]

9762. 1853–6–17 To [Sir] Edward Ryan 32 Harley St.
Considering character of Lord A. [Aberdeen?], JH and wife believe that any letter on behalf of James [C. Stewart?] should come to Lord A. from Sir J. Graham or another public figure and should describe recent events with Lord Derby. Comments on examination system in India.
AdraftS 4pp TxU:H/L-0359; Reel 1054 (Cdraft: RS:HS 25.15.16)

9763. 1853–6–18 To John Donaldson Harley St.
JH's interpretation of John Reid's bequest supporting professorship of music at University of Edinburgh.
C 4pp RS:HS 19.144 (C: RS:HS 23.136)

9764. 1853–6–20 To George Biddell Airy Royal Mint
Informs GA of the discovery of an earlier standard pound weight found at the Royal Mint.
ALS 2pp RGO 6.341.43

9765. 1853–6–21 From George Biddell Airy Royal Observatory, Greenwich
Some difficulty about locating a particular standard pound weight, and the letter requesting the use of it [see JH's 1853–6–20].
ALS 2pp TxU:H/M-0636.8; Reel 1084 (C: RGO 6.341.44)

9766. 1853–6–22 From A. R. Drummond Charing Cross
Has started a fund for the Miss Grettons.
ALS 4pp RS:HS 6.505

9767. 1853–6–23 From Richard Jones Haileybury
Wished Edward Ryan to consult JH regarding [G. H. Gordon (4th Earl of Aberdeen)]. as he knows him better. Will not make any move until he has consulted Sir James Graham. Unable to manage Monday, but any day after will suit him. Mrs. Jones is very depressed and needs a change of air.
ALS 2pp RS:HS 10.389

9768. 1853–6–23 From Charles Edward Trevelyan Treasury
Write official letter to CT proposing wages for E. W. Ward's sappers and miners temporarily assisting at Mint. Letters for £150,000 of silver coins were sent.
LS 2pp TxU:H/M–0553.17; Reel 1093

9769. 1853–6–25 To Henry Drury Harness 32 Harley St.
JH remains at home today. HH should conduct assay. Arrange with W. H. Barton and W. T. Brande to suspend silver coinage for two days at end of month.
ACCS 2pp TxU:H/L-0185.3; Reel 1054

9770. 1853–[6]–25 To Ralph Heaton & Sons Royal Mint
Send 68 boxes of copper coins next week to Bank of Ireland and to Provincial Bank of Ireland.
Adraft 2pp TxU:H/L-0721; Reel 1089

9771. [1853]–6–30 From Joseph Hume [Bry?] Square
Gives mailing address in America for son J. B. Hume.
ALS 1p TxU:H/M–0283; Reel 1087

9772. 1853–7–2 From James Wilson Treasury Chambers
How many sovereigns and half sovereigns could be coined at Mint annually without interfering with other business?
ALS 2pp TxU:H/M–0587; Reel 1086

9773. 1853–7–4 From Alexander Hamilton-Gordon Argyll House
Lord Aberdeen [George Hamilton-Gordon] will see JH at 11 o'clock this morning.
ALS 1p TxU:H/M–0210; Reel 1087

9774. 1853–7–4 To George Hamilton-Gordon 32 Harley St.
Withdraws resignation [see JH's 1853–5–30] as Master of Mint. [JH annotation of 24 Feb. 1855: Letter not sent. Now regrets letting friends convince him to remain. JH has lived the past two years in 'the Shadow of death.']
ALS 2pp †† TxU:H/L-0005; Reel 1054

9775. 1853–7–6 To the *Athenaeum* H[ar]ley St.
Requests publication of a letter [see John Stewart's 1853–6–11] that JH received from Stewart and in which Stewart explains a method of 'taking from glass negatives positive impressions of different dimensions.'
L *Athenaeum,* #1341 (July 9, 1853), 83 & *Journal of the Photographic Society, 1:8* (Aug. 22, 1853), 101

9776. 1853–7–14 From Henry Finch [Royal Mint]
HF and F. R. Brande received similar appointments, but HF's annual salary is £100 while Brande's is £150. Asks JH to raise HF's salary to £150. [JH annotation: Replied on 19 July.]
ALS 3pp † TxU:H/M–0197; Reel 1087

9777. 1853–7–16 From John R. Hind Nautical Almanac Office
What would JH like done with the column 'Equinoctial Time' in the Nautical Almanac?
ALS 2pp RS:HS 9.362

9778. 1853–7–19 From Henry Finch Royal Mint
Disappointed by JH's response. Believes HF's case is identical to that of F. R. Brande. Lists benefits accruing to Brande that HF does not have. Asks JH to reconsider raise in HF's salary.
ALS 6pp TxU:H/M–0198; Reel 1087

9779. 1853–7–20 To William Buckle Royal Mint
JH cannot recommend anyone of the sort requested in letter from Mr. Wallenstein.
ACCS 2pp TxU:H/L-0116; Reel 1054

9780. 1853–7–20 From Francis Lawley Downing St.
Chancellor of Exchequer [W. E. Gladstone] will meet with JH next Friday.
ALS 3pp TxU:H/M–0328; Reel 1087

9781. 1853–7–20 To E. [Wallenstein?] [Royal Mint]
It is against Mint policy for JH to recommend any employee to act privately, as in EW's request for someone to inspect minting machinery.
ACCS 2pp TxU:H/L-0488; Reel 1055

9782. 1853–7–21 To William Rowan Hamilton 32 Harley St.
Eloquently, metaphorically congratulates WH on finally publishing *Lectures on Quaternions.*
ALS 4pp TCD 7762–72:257 Graves, *Hamilton, 2,* 681–2

9783. 1853–7–21 To John R. Hind 32 Harley St.
Outlines the changes he would like to see in the column for Equinoctial Time in the Nautical Almanac.
draft 4pp RS:HS 9.363 & 23.137

9784. 1853–7–21 From James Wilson Treasury Chambers
H. D. Harness will be leaving Mint to assume duties as Commissioner of Works in Ireland.
ALS 3pp TxU:H/M–0588; Reel 1086

9785. 1853–7–22 From Caroline Emily Mary Herschel Argyll House
Will spend weekend at Windsor Castle with Lord Aberdeen's family.
ALS 3pp TxU:H/M–0794; Reel 1084

9786. 1853–7–23 To William Ewart Gladstone 32 Harley St.
Advises against abolishing office of deputy master of Mint after H. D. Harness leaves at end of this week. Gives reasons.
Adraft 4pp TxU:H/L-0162; Reel 1054

9787. 1853–7–23 From Charles Edward Trevelyan Treasury
Thanks for catching CT's error. Returns letter, with £200,000 substituted for £150,000.
ALS 1p TxU:H/M–0553.18; Reel 1093

9788. 1853–7–28 From George Hamilton-Gordon Downing St.
Come to Argyll House tomorrow. Encloses reply by W. E. Gladstone to JH's letter of 23 July.
ALS 1p TxU:H/M–0001; Reel 1062

9789. 1853–7–30 From Henry Drury Harness Royal Mint
Notified [James] Wilson that HH's service at Mint will end today, but HH does not yet know what official date Treasury will record for HH's retirement.
ALS 2pp TxU:H/M–0268.9; Reel 1087

9790. [1853–8?] To Margaret Brodie Herschel 37 Tavistock Place
Postpones trip to Collingwood to dine with [Elizabeth] Baily on Thursday. Send coach to Cranbrook on Friday instead. Queen is going to Ireland. Annual revenue is good. Thick gloom and darkness over London today.
ALS 4pp TxU:H/L-0536.1; Reel 1053

9791. 1853–8–1 From Richard Jones Cathedral [Commission]
Still holds opinion that RJ expressed last night, but would speak to Lord A. [Aberdeen?] soon if RJ does not hear from Lord A.
ALS 1p TxU:H/M–0308; Reel 1087

9792. 1853–8 To Charles Edward Trevelyan Royal Mint
Retirement of H. D. Harness. Recommends W. H. Barton, comptroller since 1830, to become either new deputy-master or combination of comptroller and deputy-master. Repeats request to hire special clerk to assist Master. [JH annotation: No reply from Treasury.]
AdraftS 3pp †(by JH) TxU:H/L-0483.9 & -0483.10; Reel 1055

9793. 1853–8–3 From Augustus De Morgan 7 Camden St.
Has he heard the names of the coins decided on by the Commons when the Pound is decimalized? W. R. Hamilton has informed him that he has heard from JH. Is going to the seaside tomorrow.
ALS 3pp RS:HS 6.264

9794. 1853–8–8 To Charles Edward Trevelyan Royal Mint
Encloses draft of [JH's 1853–8] to commissioners of Treasury. Asks CT's permission to inform W. H. Barton of plan to merge offices of comptroller and deputy-master. Emphasizes need for special clerk to assist Master. New policy for signing Mint checks.
ACCS 3pp TxU:H/L-0483.11; Reel 1055

9795. 1853–8–9 To Augustus De Morgan n.p.
Thanks for AD's report on coinage [see JH's 1853–4–11], and for AD's puns.
C 1p RS:HS 23.138

9796. 1853–8–10 From Augustus De Morgan 7 Camden St.
Intends to send on Francis Baily's letters soon. Would like to see the coinage report.
ALS 4pp RS:HS 6.265

9797. 1853–8–10 From Charles Edward Trevelyan Treasury
Discusses salary for W. H. Barton in new combined office of deputy master and comptroller. Chancellor of Exchequer [W. E. Gladstone] may decide differently. Agrees that Master needs private secretary, but select one from among clerks already employed at Mint. Procedure for countersigning Mint checks.
LS 6pp TxU:H/M–0553.19; Reel 1093

9798. 1853–8–11 To Charles Edward Trevelyan Harley St.
R. L. Sheil led W. H. Barton to expect promotion to deputy-master, and Barton is not yet aware of Treasury plan to combine this office with duties of comptroller. Proposes giving Barton *both* salaries for performing both duties. Salaries for junior clerks.
AdraftS 4pp TxU:H/L-0483.12; Reel 1055

9799. 1853–8–12 To Charles Edward Trevelyan Harley St.
[Marked 'Confidential.'] Considering heavy workload from Jan. 1852 to June 1853, JH suggests bonuses for junior clerks and Mint officers. Surprised to learn that Charles Elouis felt passed over for deputy-master position at Sydney branch mint.
ACCS 4pp TxU:H/L-0483.13; Reel 1055

9800. 1853–8–13 To Margaret Brodie Herschel [London]
Cannot come home on weekend. Will try next weekend if [W. H.] Barton's affair can be settled.
ALS 4pp TxU:H/L-0544.5; Reel 1053

9801. 1853–8–13 To Margaret Brodie Herschel Mains [?]
About the smoky atmosphere in London, and a letter from son Alexander about his chemical work.
ALS 4pp JHS 1.21

9802. 1853–8–13 From Charles Edward Trevelyan Treasury
Forwarded to Chancellor of Exchequer [W. E. Gladstone] JH's letter regarding retirement of H. D. Harness. Does not approve of departure, in JH's plan, from practice of fixed pay for public officers. Surprised by grieved tone of Henry Finch, who turned down earlier generous offer to remain in reorganized Mint.
(A)LS 4pp TxU:H/M–0553.20; Reel 1093

9803. 1853–8–14 To Margaret Brodie Herschel n.p.
Received order for immediate report on state of coinage in preparation for debate in House of Lords next week. No progress on W. H. Barton's case. Permissiveness in Treasury administration. News of Elizabeth Baily, John and Catherine Bell, and Mrs. Gordon. P.S.: Three-volume novel *Adam Graeme at Mossgray* [by Margaret Oliphant] is 'beautiful.' JH has read first two volumes.
AL 8pp TxU:H/L-0544.6; Reel 1053

9804. 1853–8–15 To Margaret Brodie Herschel Royal Mint
Birthday greetings. Longs to be with MH. Comments on poet P. B. Shelley.
ALS 4pp TxU:H/L-0544.7, Reel 1053

9805. 1853–8–15 To Robert Mushet Royal Mint
Procedures for separating gold from silver and iridium by subsidence. Encloses volume of experiments by Charles Hatchett.
AdraftS 2pp TxU:H/L-0275; Reel 1054

9806. 1853–8–16 From George Biddell Airy Royal Observatory, Greenwich
Is arranging a meeting to seal up and deposit one set of standards [see GA's 1853–6–21].
ACCS 1p RGO 6.341.45

9807. 1853–8–16 From Frederick Fitzgerald Treasury Chambers
JH may call on James Wilson this afternoon.
ALS 1p TxU:H/M–0201; Reel 1087

9808. 1853–8–16 To Charles Edward Trevelyan Harley St.
Error in CT's letter to Chancellor of Exchequer concerning record keeping in Mint. Repeats request to hire special clerk to assist Master.
ACCS 3pp TxU:H/L-0483.14; Reel 1055

9809. 1853–8–17 From Robert Mushet Melting House, [Royal Mint]

Returns JH's book. Describes alloy of gold and silver. Copper-iridium alloy would work, but it would influence assays adversely. Procedure for separating gold and iridium.
ALS 3pp TxU:H/M–0386; Reel 1087

9810. 1853–8–17 To My dear Sir Royal Mint
C. E. Trevelyan reports that W. E. Gladstone agreed to 'subject of our conversation yesterday.' Does [?] still want to meet with JH at Treasury chambers today?
ACCS 1p TxU:H/L-0664; Reel 1089

9811. 1853–8–19 To Margaret Brodie Herschel Royal Mint
Postpone paying bills until end of month. Questions MH's birthday festivities. Congratulates Hayley family on remarriage. Poor Mr. Paynder is gone. JH finished third volume of *Adam Graeme at Mossgray* [by Margaret Oliphant].
AL 4pp TxU:H/L-0544.8; Reel 1053

9812. 1853–8–19 To Mary Elizabeth Lyell Harley St.
Thankful for the Lyells' return and accepts their invitation to dinner.
ALS 1p DMC.695A.885.7

9813. 1853–8–19 To Charles Edward Trevelyan Royal Mint
Submits E. W. Ward's report on feasibility of establishing branch mint in Sydney. Encourages Treasury to proceed with Ward's plan.
AdraftS 3pp TxU:H/L-0483.15; Reel 1055

9814. 1853–8–20 To George Biddell Airy n.p.
Explains why JH could not attend meeting [see GA's 1853–8–16]; hopes GA and family had a pleasant holiday.
ALS 2pp RGO 6.341.46

9815. [1853–8–22?] To Margaret Brodie Herschel [Royal Mint]
Mint business. W. T. Brande now has two jobs. Lyells are back. Mr. Newham reports legal problem determining who will pay for construction at 204 'OH^{d}' Street. Bank balance is low; check records to see when [Jesse] Piper's rent is due. Alexander Herschel's artistic ability. Urges MH to read Margaret Oliphant's *Adam Graeme of Mossgray*.
ALS 4pp TxU:H/L-0536.10; Reel 1053

9816. 1853–8–22 From [John] Stewart to Margaret Brodie Herschel Eaux Bonnes
Travels in France. Note of JS's discussion with Arthur Gordon about recent problems at Mint and JH's need for intelligent secretary. Possibility of reducing Mint to subsidiary of Treasury. Feels anxiety for JH's sufferings.
ALS 4pp TxU:H/M–1061; Reel 1083

9817. [1853–8–23] To Margaret Brodie Herschel [Royal Mint]
Deeply disappointed by Treasury's failure to pay an adequate salary to W. H. Barton for combined duties of deputy master and comptroller. C. E. Trevelyan granted only a junior and inexperienced clerk to assist JH. Feels 'sickened and disgusted' by Trevelyan's pettiness, which is crippling the effectiveness of the Mint staff.
C 1p TxU:H/M–0772.2; Reel 1084

9818. 1853–8–23 From Charles Edward Trevelyan Treasury
Surprised by JH's voluminous report on assays, which CT showed to Chancellor of Exchequer [W. E. Gladstone]. Will send copy of Gladstone's decision on several Mint matters.
P.S.: Gladstone read last letter that JH wrote to CT.
ALS 3pp TxU:H/M–0553.21; Reel 1093

9819. [1853]–8–23 To Charles Edward Trevelyan Harley St.
Recommends Mr. Aberdein as clerk and private secretary to Master of Mint. Suggests that Mr. Hill be promoted from temporary clerk to replace Aberdein in melting house.
ACCS 2pp TxU:H/L-0483.16; Reel 1055

9820. [1853]–8–24 From Margaret Brodie Herschel to [Alexander Hamilton-]Gordon Collingwood
Reviews vexations caused to JH by dealings with C. E. Trevelyan. Notes H. D. Harness's belief that Bank of England is maneuvering to take control of Royal Mint. Fears JH may become bitter about public offices. Encloses copy of JH's letter to MH [see JH's [1853–8–23]].
ALS 4pp TxU:H/M–0772.1; Reel 1084

9821. 1853–8–24 From Charles Edward Trevelyan Treasury
Meet CT and [James] Wilson at end of week. Hopes Lord Aberdeen [G. Hamilton-Gordon] will appoint Mr. Hill Jr. to vacant clerkship at Mint.
LS 2pp TxU:H/M–0553.23; Reel 1093

9822. 1853–8–24 From Charles Edward Trevelyan Treasury
Treasury intended that junior clerks in Mint should be available for training in *all* departments, to prepare them for higher positions. Suggestion for appointing supernumerary clerk. Formal conference will be held Saturday.
LS 4pp TxU:H/M–0553.22; Reel 1093

9823. 1853–8–24 To Charles Edward Trevelyan Royal Mint
[John] Goldsmith, temporarily assigned from Board of Works in Ireland to bookkeeper in coining department, is working out very well. Suggests offering Goldsmith permanent position and using him to train other accountants.
ACCS 3pp TxU:H/L-0483.17; Reel 1055

9824. 1853–8–26 To Julia Margaret Cameron Harley St.
Thanks JC for what she sent. Has been reading Alfred Tennyson's *Princess*. Praises it.
ALS 4pp Getty Center

9825. 1853–8–26 To Margaret Brodie Herschel [London]
Organizational changes at Mint. Deputy master declared unnecessary. W. T. Brande was promoted to combined office of deputy master and comptroller. JH protested to C. E. Trevelyan yesterday about low pay of junior clerks.
AL 4pp TxU:H/L-0544.9; Reel 1053

9826. 1853–8–27 To Margaret Brodie Herschel [London]
Is slightly ill, so JH decided not to come home; a few items about members of the family.
AL 4pp JHS 1.93

9827. 1853–8–30 From Augustus De Morgan 7 Camden St.
Is sending the packet of Francis Baily's letters today. Sees that Thomas Wright, the Milky Way man, kept a shop in Fleet St. and was a mathematical instrument maker. Warren de La Rue has doubtless sent him his picture of Saturn.
ALS 3pp RS:HS 6.266

9828. 1853–8–30 From Charles Edward Trevelyan Treasury
Sends draft of Treasury minutes for JH's inspection. E. W. Ward may immediately accept bid by Messrs. Taylor &

Allen.
LS 2pp TxU:H/M–0553.24; Reel 1093

9829. 1853–8–31 To William Thomas Brande Royal Mint
Asks WB to recommend candidates for supernumerary clerk to be appointed soon to Mint office. Has also asked W. H. Barton for recommendations.
AdraftS 4pp TxU:H/L-0105; Reel 1054

9830. 1853–9–1 To Margaret Brodie Herschel [London]
[Richard] Jones's application was successful, but his health is failing. JH's weekend trip to Collingwood is cancelled by problems with W. T. Brande, 'the most immanagable of men.' Brande and C. E. Trevelyan are JH's biggest problem. JH is sure of Trevelyan's '*ill offices.*'
AL 4pp TxU:H/L-0544.10; Reel 1053

9831. 1853–9–1 To Margaret Brodie Herschel Harley St.
Informs MH about the death of someone [Miss Maria Tunno?] dear to both of them.
ALS 3pp JHS 1.94

9832. 1853–9–1 From William Law [Downing St.]
Appointment of H. D. Harness to be commissioner of Public Works in Ireland was dated 16 Aug. [Reply to JH's letter to Mr. Fitzgerald.]
ALS 1p TxU:H/M–0326; Reel 1087

9833. 1853–9–2 From William Thomas Brande Durleigh near Bridgwater
Detained in travel; will return Monday. Recommends John Goldsmith as supplementary clerk. Advises against transferring Henry Finch or F. R. Brande from Coining to Melting departments.
ALS 4pp TxU:H/M–0119; Reel 1087

9834. 1853–9–2 To Margaret Brodie Herschel [London]
Instruct all daughters to date letters to JH. Loss of dear friend [Miss Maria Tunno?] leaves JH with no desire but to come home and be with family. Send carriage to railway station on chance that JH can get away from work tonight.
AL 2pp TxU:H/L-0544.11; Reel 1053

9835. 1853–9–3 From Charles Edward Trevelyan Treasury
Read enclosed draft of Treasury minutes to see if it correctly states JH's recommendations regarding signing of Mint checks. Please return CT's draft of minutes on Sydney branch mint. Encloses amusing letter from [W. N.] Forbes.
LS 2pp TxU:H/M–0553.25; Reel 1093

9836. 1853–9–7 From Charles Edward Trevelyan Treasury
Encloses copy of act prohibiting defacing of coins. Advises JH to make use of Treasury solicitor in all legal matters affecting Mint.
ALS 2pp TxU:H/M–0553.26; Reel 1093

9837. 1853–9–8 From George Biddell Airy n.p.
Drafting the Standards Commission report is taking up all of GA's spare time.
ACCS 1p RGO 6.341.47

9838. 1853–9–8 From George Biddell Airy Royal Observatory, Greenwich
Seeks instructions about return of the measures to the Royal Mint.
ACCS 2pp RGO 6.339.72

9839. 1853–9–8 To Margaret Brodie Herschel [London]
Visit to Collingwood lifted JH's spirits. Finances for Mrs. Rennie. Received copy of [William?] Platt's latest work from Miss Lipscombe. Sends condolences to Miss E. Tunno. Hopes loss will set young gentlemen thinking about 'brainless ardours.' Returns books to Collingwood.
ALS 4pp TxU:H/L-0544.12; Reel 1053

9840. 1853–9–8 From Charles Edward Trevelyan Treasury Chambers
Treasury commissioners seek JH's advice on feasibility of proposed observatory at Nottingham, to be funded by public money.
LS 3pp † TxU:H/M–0553.28; Reel 1093

9841. 1853–9–8 From Charles Edward Trevelyan Treasury
Reply to enclosed 'Reference,' then return it to 'Memorialists' by 20 Sept. Papers must be sent to [C. Rivers] Wilson at Boulogne.
LS 2pp TxU:H/M–0553.27; Reel 1093

9842. 1853–9–9 To George Biddell Airy Royal Mint
Asks GA about the best way to store standard measures [see GA's 1853–9–8].
ALS 2pp RGO 6.339.77

9843. 1853–9–9 To Miss Lipscombe Ludney
Thanks for transcribing Mr. Platt's humor. Miss Cornwallis's introduction of Greek into JH's family. Unable to visit Miss. L. at Goudhurst. Regards to Miss Cornwallis and Mrs. Cartwright.
ALS 4pp TxU:H/L-0226; Reel 1054 (C: RS:HS 23:140)

9844. 1853–9–10 To Charles Edward Trevelyan 32 Harley St.
Responds to CT's request for opinions on establishing astronomical and meteorological observatory at Nottingham. Distinguishes roles of royal, public, and private observatories for astronomy, and how these differ from meteorological observatories.
Adraft 4pp TxU:H/L-0483.18; Reel 1055

9845. [1853–9–10] To [Charles Edward Trevelyan] [Royal Mint]
JH's opinion on feasibility of establishing new astronomical and meteorological observatory at Nottingham. Compares this climate to other English sites.
ACCS 4pp TxU:H/L-0605; Reel 1089

9846. 1853–9–13 To William Thomas Brande Royal Mint
Concern for waste of gold generated in cutting of coins. Fears government enquiry if waste continues. [JH annotation: Handed to Brande in person.]
ACC 4pp †(by JH) TxU:H/L-0106; Reel 1054

9847. 1853–9–13 From Charles Edward Trevelyan Treasury
Approves JH's recommendation of John Goldsmith to position of supernumerary clerk in Mint office. Notify Lord Aberdeen [G. Hamilton-Gordon].
LS 2pp TxU:H/M–0553.29; Reel 1093

9848. 1853–9–14 From George Biddell Airy Royal Observatory, Greenwich
Writes to JH to re-arrange a time for a meeting.
ACCS 3pp RGO 6.339.82

9849. 1853–9–14 To George Hamilton-Gordon Royal Mint
Recommends John Goldsmith for position of supernumerary clerk in Mint office.
AdraftS 2pp TxU:H/L-0006; Reel 1054

9850. 1853–9–14 To Matthew Marshall Royal Mint
Henceforth, checks drawn on account of Master of Mint at Bank of England will not require initials of Charles Elouis.
ALS 2pp TxU:H/L-0246; Reel 1054

9851. 1853–9–15 To George Biddell Airy Royal Mint
Agrees to new meeting time [see GA's 1853–9–14].
ALS 1p RGO 6.339.88

9852. 1853–9–21 To J. Buckley 32 Harley St.
Thanks JB for a description of a fairly unique meteor and urges public notice of it be given.
ALS 3pp APS B.H435p.16

9853. 1853–9–23 From George Arbuthnot Treasury
JH should receive authority soon for appointing John Goldsmith. George Rennie is angry that E. W. Ward gathered valuable information from Rennie regarding machinery for Sydney branch mint, but then gave contract to another manufacturer. Expect Treasury Board to ask JH for report on this. Perhaps Ward can 'pacify' Rennie.
ALS 5pp TxU:H/M–0013; Reel 1062

9854. 1853–9–29 From Robert Mushet Melting House, [Royal Mint]
Withdraws RM's request for leave. Complains of noxious fumes in Melting room and from nearby refinery. Objects to frequent changes in RM's assistants; none remain long enough to learn Melting room procedures. Recent loss of Mr. Aberdein.
ALS 4pp TxU:H/M–0387; Reel 1087

9855. 1853–9–30 To Robert Mushet Royal Mint
Frequent change in RM's chief assistants is unfortunate, but junior officers must rotate in order to learn duties in every branch of Mint.
ACCS 1p TxU:H/L-0276; Reel 1054

9856. 1853–10–1 To Mr. Harris Royal Mint
H. D. Harness no longer resides at Mint. Explains division of JH's taxes between Collingwood and London residence at Harley Street.
ACC 1p TxU:H/L-0186; Reel 1054

9857. 1853–10–3 To Richarda Airy Royal Mint
Making arrangements to send two of JH's sons to Greenwich, so that G. B. Airy can take them to Woolwich [Arsenal] to watch mortar practice.
ALS 3pp JHS 2.10

9858. 1853–10–12 To John William Lubbock Harley St.
Hopes JL will call application of JH's brother-in-law for position of Secretary of Bank of Asia to attention of a director, who is JL's relative.
ALS 4pp RS LUB.H.361

9859. 1853–10–17 From Augustus De Morgan 7 Camden St.
Has found a seventeenth-century mathematical manuscript amongst Francis Baily's papers with JH's handwriting on it; can he explain the mystery? There is also a letter from P. L. M. Maupertius to James Bradley, which he proposes sending to the R.S.L.
ALS 3pp RS:HS 6.267

9860. 1853–10–19 To Charles Edward Trevelyan Royal Mint
Forwards letter from E. W. Ward to commissioners of Treasury, proposing that J. Triskett and ten royal engineers accompany Ward to Australia to supervise construction of branch mint in Sydney.
AdraftS 2pp TxU:H/L-0483.19; Reel 1055

9861. 1853–10–21 From George Biddell Airy Royal Observatory, Greenwich
A note accompanying the first draft of the report of the Standards Commission, with a request for comment.
LS 2pp TxU:H/M-0636.9; Reel 1084

9862. 1853–10–21 From Henry Finch Royal Mint
Still believes that HF's experience and competence are equal to those of F. R. Brande. Plans to appeal directly to Treasury.
ALS 2pp TxU:H/M–0199.1; Reel 1087

9863. 1853–10–21 To Henry Finch [Royal Mint]
Explains why JH considers HF's case to be different from that of F. R. Brande. Does not believe that HF's appeal to Treasury will be seriously considered.
ACC 1p inc TxU:H/M–0199.2; Reel 1087

9864. 1853–10–24 To Margaret Brodie Herschel [London]
News of friends and of daughter Caroline; is sending out letters, papers, money, and JH's 'likeness.'
ALS 3pp JHS 1.98

9865. 1853–10–26 From William Henry Sykes Brie Maritime
Describes travels abroad and their educational value, news on balloon ascents, barometer corrections, Henri Regnault's hygrometer, invention of electric weaver and other new applications of electricity throughout Europe.
ALS 8pp RS:HS 17.151

9866. 1853–10–27 To Margaret Brodie Herschel [London]
Comments on the weather, daughter Margaret Louisa's sunny disposition, what magazine to send son Willy, and JH's shortage of shirts.
ALS 4pp JHS 1.106

9867. 1853–10–28 To Augustus De Morgan Collingwood
JH's copy of Apollonius's work on conic sections was given to Francis Baily. [P. L. M. de] Maupertuis's letter to [James] Bradley is 'real prize.' Comment on AD's puns. Finds W. R. Hamilton's *Quaternions* 'horribly metaphysical.'
ALS 4pp RAS: MSS Add.119

9868. 1853–10–31 To Maria Sophia Herschel [London]
Full of messages to JH's wife, Margaret, and bits of family news.
AL 4pp JHS 1.99

9869. 1853–10–31 To Robert Mushet Royal Mint
Deliver bars produced today to Mr. Hill, who will commence work immediately in anticipation of assay.
ACCS 1p TxU:H/L-0277; Reel 1054

9870. 1853–10–31 From Henry Thoby Prinsep East India House
Discusses Addiscombe nominations; says he will nominate Margaret Herschel's nephew.
ALS 2pp RS:HS 14.188

9871. 1853–11–1 To Margaret Brodie Herschel [London]
Asks for personal linen to be sent; comments on a letter from daughter Margaret Louisa in Paris, and on the French and Americans generally.
AL 3pp JHS 1.103

9872. 1853–11–3 To Henry Drury Harness n.p.
Approves financial reports. Production of sovereigns and half sovereigns.
ACCS 1p TxU:H/L-0185.4; Reel 1054

9873. 1853–11–4 To Charles Edward Trevelyan n.p.
Requests increase of wages for Mint employees, in preparation for resuming gold coinage on extensive scale.
Adraft 2pp TxU:H/L-0483.20; Reel 1055

9874. 1853–11–6 To Margaret Brodie Herschel [London]
A shipment of special china for MH has arrived from Paris; some news about daughter Caroline, and several stories JH heard at a literary society meeting.
AL 4pp JHS 1.102

9875. 1853–11–6 From William Parsons (Lord Rosse) The Castle, Parsonstown
Would like JH to write the obituary of François Arago for the R.S.L.
ALS 3pp RS:HS 13.231

9876. 1853–11–7 To William Thomas Brande [Royal Mint]
Cease coining florins and commence production of shillings and sixpence. [JH annotation: Saw Brande today and gave these instructions verbally.]
ACC 2pp †(by JH) TxU:H/L-0671; Reel 1089

9877. 1853–11–7 To Margaret Brodie Herschel London
Comments on the Crimean War and the reports of Russian successes against Turkey; JH gives his opinion that England is more American than it is European.
ALS 1p inc JHS 6.45

9878. 1853–11–7 To William Henry Fox Talbot 32 Harley St.
May need WT's permission for the publication in a book on coinage of a photograph of JH prepared using WT's patented process.
ALS 4pp Science Museum Talbt 1/43

9879. 1853–11–9 To William Thomas Brande Royal Mint
Calls attention to tendency of recently cut silver coins to be too light in weight.
ACCS 1p TxU:H/L-0107; Reel 1054

9880. 1853–11–9 To William Parsons (Lord Rosse) 32 Harley St.
Declines WP's request that JH write an obituary notice of François Arago. Agrees R.S.L. should publish such. Suggests [Charles] Wheatstone as author.
ALS 3pp Rosse Papers K2.3 (C: RS:HS 13.231bis & 23.141)

9881. 1853–11–9 To [Charles Edward Trevelyan] Royal Mint
Encloses letter, with JH's approval, from E. W. Ward at Sydney branch mint requesting authority to appoint W. S. Jevons and F. B. Miller as assayers, H. A. Severn and John MacCutcheon as clerks, two foremen of mechanics, and one engineer in Coining department.
AdraftS 2pp TxU:H/L-0672; Reel 1089

9882. 1853–11–[10] To Margaret Brodie Herschel [London]
Comments on son John and daughter Caroline; wants some material to deal with a possible infringement of W. H. Fox Talbot's patent.
ALS 4pp JHS 1.67

9883. 1853–11–10 From Henry Thoby Prinsep East India House
Nominates Margaret Herschel's nephew for Addiscombe.
ALS 1p RS:HS 14.189

9884. 1853–11–11 From Edward Joseph Lowe Beeston
Has heard of a projected attack on the Henry Lawson instruments at the Observatory. Would like a line from JH on the value of the instruments. Has written to W. R. Dawes, [John?] Lee, and W. H. Smyth.
ALS 3pp RS:HS 11.326

9885. 1853–11–13 To Isabella Herschel [London]
Asking for a photograph of himself [see JH's 1853–11–[10]]; complains about the binders putting too much material in one book; talks about daughter Caroline spending much time at Windsor.
AL 4pp JHS 1.68

9886. 1853–11–13 From William Henry Fox Talbot Greta Bank, Keswick
Willingly gives JH permission to engrave a portrait photograph taken with WT's process. WT asks JH if he would be willing to testify on WT's behalf if necessary, relating to infringement of patent rights. WT announces that John Hind has discovered another asteroid [Euterpe].
ALS 4pp RS:HS B26.48

9887. 1853–11–18 To Margaret Brodie Herschel [London]
Comments on, and forwards, a letter from son Willy; also talks about the departure of a servant [?], the health of Uncle James [James Calder Stewart], and daughter Caroline's situation.
ALS 4pp JHS 1.107

9888. [1853–11–18 or later] To [?] [Royal Mint]
Refers to [James] Wilson's letter of 18 Nov. Queen must proclaim certain coins to be legal tender in colonies before Mint may issue them.
Adraft 1p TxU:H/L-0504; Reel 1055

9889. 1853–11–21 To Margaret Brodie Herschel [London]
About the death of a cousin and attendant funeral arrangements, the status of family finances, and some preparations for Christmas.
ALS 6pp JHS 1.100

9890. 1853–11–21 From William Lassell Starfield, Liverpool
If the Herschels intend visiting the B.A.A.S. meeting at Liverpool he will be pleased to accommodate them.
ALS 2pp RS:HS 11.149

9891. 1853–11–22 To W. Graves Royal Mint
Details of next week's shipment of copper coins from Mint to four branches of Bank of Ireland.
AdraftS 2pp TxU:H/L-0171; Reel 1054

9892. 1853–11–22 From Charles Edward Trevelyan Treasury
Confer with CT on rates of piecework for gold coinage before making decision. Did JH receive Treasury's official approval of remaining appointments to Sydney branch mint?
LS 2pp TxU:H/M–0553.30; Reel 1093

9893. 1853–11–25 To W. Graves Royal Mint
Modifies method of accounting for copper coins shipped from Mint to Bank of Ireland.
AdraftS 2pp TxU:H/L-0172; Reel 1054

9894. 1853–11–28 From John R. Hind 22 Grove Rd., St. John's Wood
Regrets he has been wrongly styled F.R.S. Comments on the projected Lawson Observatory at Nottingham and the high value placed on the instruments. Outlines his plan for the Equinoctial Time column in the Nautical Almanac.
ALS 7pp RS:HS 9.364

9895. 1853–11–28 To Charles Richard Weld 32 Harley St.
J. R. Hind [whose election to the R.S.L. had been declared void] has been referred to as 'F.R.S.' several times. This is not the fault of Hind, who has objected to it.
ALS 3pp RS MC.5.128

9896. 1853–11–29 To Margaret Brodie Herschel [London]
A close friend has died and JH is concerned in the arrangements for supporting the remaining family member; JH has submitted various papers for son John [to enter the Navy?]; JH dreams of himself at home.
AL 4pp JHS 1.108

9897. 1853–12–1 To [Charles Edward Trevelyan] Royal Mint
Forwards to Treasury commissioners JH's correspondence with managers and directors of Bank of Ireland and Provincial Bank of Ireland regarding introduction of new copper coin. Requests authorization to proceed with distribution.
AdraftS 2pp TxU:H/L-0673; Reel 1089

9898. 1853–12–4 To Margaret Brodie Herschel [London]
Mostly about sick people—their daughter Caroline, Fanny Bailey, and James Adam Gordon; son John passed his entrance examination [to the Navy?], and JH is still worried about the financial arrangements [see JH's 1853–11–29].
AL 4pp JHS 1.109

9899. 1853–12–5 From Henry Halford Jones and Robert Angus Smith Manchester
[Form letter] Announces next meeting of Literary and Philosophical Society, with presentation by [Peter] Spence on atmospheric electricity.
LS 1p TxU:H/L-0719.1; Reel 1089

9900. 1853–12–5 To Charles Edward Trevelyan Royal Mint
Recommends increasing salaries of Mint officers assigned to Sydney, to match salary increases announced by governor of New South Wales.
AdraftS 2pp TxU:H/L-0483.21; Reel 1055

9901. 1853–12–8 From G. B. n.p.
Instead of index of all printed papers relating to Mint requested by JH, GB suggests set of indexes from 1800 to 1852 by Mr. Hausard. [Note from 'm' to JH on verso: 'Have you not the "railiamutard" Indexes already at the Mint?']
ALS 2pp TxU:H/L-0724; Reel 1089

9902. 1853–12–10 From George Biddell Airy Royal Observatory, Greenwich
Does not know of the details of French or American coinage, but Augustus de Morgan has undertaken to find someone to help JH with this problem.
ACCS 2pp RGO 6.349.31

9903. 1853–12–12 From Sir Robert Harry Inglis 7 Bedford Square
Invitation to dinner for Friday next.
ALS 1p RS:HS 10.253

9904. 1853–12–14 To Charles Edward Trevelyan Royal Mint
Forwards request for review of Charles Elouis's right to superannuation pay. Repeats JH's earlier request (18 Mar. 1852) that Master of Mint be officially named among government officers listed in Superannuation Act.
ACCS 3pp TxU:H/L-0483.22; Reel 1055

9905. 1853–12–21 To C. Barnard Royal Mint
Returns list of articles needed by assay office at mint in [Melbourne,] Victoria, with list of recommended English suppliers.
AdraftS 3pp TxU:H/L-0080; Reel 1054

9906. 1853–12–22 From George Biddell Airy Royal Observatory, Greenwich
Is waiting for JH's comments on the first draft of the Standards Commission report [see GA's 1853–9–8].
ALS 1p TxU:H/M-0636.10; Reel 1084 (C: RGO 6.341.50)

9907. 1853–12–22 To Charles Babbage Harley St.
Hopes CB will dine with JH at Hawkhurst next week as JH hears CB will be in the neighborhood.
ALS 1p RS:HS 2.327

9908. 1853–12–22 To Julia Margaret Cameron 32 Harley St.
Makes arrangements to accompany Miss Julia Cameron [daughter of JC] to Collingwood. Thanks for JC's intervention on behalf of JH's nephew, Mr. Stewart. Expresses interest in Charles Hay Cameron's new undertaking.
ALS 4pp Getty Center

9909. 1853–12–22 To Julia Margaret Cameron [London]
Realizes that JH's 1853–12–22 will not reach JC. Makes arrangements for JC to send JH's daughter Julia to meet JH and to return to Collingwood with him on the weekend.
ALS 3pp DMC.695A.885.22

9910. 1853–12–22 From Sir Richard Owen The Cottage, Sheen
Mr. Chadwick, who is now RO's guest, believes that one of JH's sons is at a preparatory school near Clapham. Would welcome JH's advice on a suitable school.
ALS 2pp RS:HS 13.195

9911. 1853–12–22 To [Charles Edward Trevelyan] Royal Mint
Regarding future improvements at Mint that would involve Board of Works, JH prefers to postpone major renovations that JH suggested in Dec. 1852. Notes two minor repairs needed in Melting Room.
Adraft 2pp TxU:H/L-0714; Reel 1089

9912. 1853–12–26 From Charles Fulbrook Dallington, Hurstgreen
Believes that the moon has some influence on the amount of rainfall and sends some of his observations on this subject. Welcomes any comments.
ALS 3pp RS:HS 7.386

9913. [1854 or later] To the Board of Trade n.p.
Response to the Board of Trade's refusal to supervise the adjustment of compasses on sea-going ships. [This letter was prepared on behalf of the Council of the R.S.L.]
draft 11pp RS:HS B26.70

9914. 1854 To Augustus De Morgan 35 Bedford Place
Comments on circulation of new coinage, and on another

integral of Henry Warburton's.
C 2pp RS:HS 23.152

9915. [1854 or later] To Caroline Emily Mary Herschel [London]
Tells CH about a Crimean War medal being ordered, the Mint not being allowed to design it, but is expected to produce it. JH feels placed in a very difficult position.
AL 4pp inc JHS 1.69

9916. 1854 To Isabella Herschel 32 Harley St.
Talks about crystal rocks in JH's collection, which may be used, with great care.
ALS 3pp JHS 1.62

9917. [1854?] From James Calder Stewart to Margaret Brodie Herschel Athenaeum
Visited JH at Mint today. Notes JH's health and recovery. JH reminisced of childhood episodes with William Herschel.
ALS 7pp TxU:H/M–1050; Reel 1083

9918. [1854] To Margaret Brodie Herschel [London]
About the health of JH's servant [?], Knowles.
ALS 2pp JHS 1.26

9919. [1854] To Margaret Brodie Herschel [London]
About the school progress of sons John and Alexander; JH is in a very depressed state, looks at his current life [at the Mint?] with 'loathing,' and cannot imagine surviving it for more than a few months.
ALS 4pp JHS 1.14

9920. [1854 to 1855] To Richard Jones n.p.
Is taking RJ's admonitions to heart, and along with taking quinine and staying at home, JH feels that his nervous system is being restored.
C 2pp RS:HS 25.14.17

9921. [1854 to 1855] To My Dear Sir Collingwood
Forwards letter by E. W. Ward reporting progress at Sydney branch mint and assay results of Australian gold.
AdraftS 1p TxU:H/M–0154.2; Reel 1087

9922. [1854 to 1855] To 'Sir' [Royal Mint]
Progress of production on [?] medal, with ribbons and clasp, by [Heat & Roskell,] jewellers.
Adraft 1p inc TxU:H/L-0617; Reel 1089

9923. [1854?] From Adam Sedgwick [Cambridge, England]
'Delighted' to hear that JH is in 'such good spirits.' Has been giving regular class lectures.
ALS 1p inc Harvard: Houghton AAH 67m–67 (63)

9924. 1854 To James Calder Stewart n.p.
Approves Mr. Adam's use of logarithms. Sends set of tables. Glad that little Emily is better.
ALS 2pp TxU:H/L-0444; Reel 1055

9925. [1854] To [Charles Edward Trevelyan] [Royal Mint]
Requests permission to close Mint for three months to repair and replace machinery that has deteriorated under heavy workload of past two years. Plans to begin at end of this month. Please inform JH of any pending orders for coinage.
AdraftS 1p TxU:H/L-0625; Reel 1089

9926. [1854] To Benjamin Wyon [Royal Mint]
Changes in Crimean Medal.
C 4pp TxU:H/L-0510; Reel 1055

9927. [1854 to 1855] To [?] [Royal Mint]
Outlines administrative organization, duties, and jurisdiction of Sydney branch mint. Procedure for sending sample coins to London for examination and approval.
ACCS 2pp TxU:H/L-0627; Reel 1089

9928. 1854–1–2 To William Henry Barton Royal Mint
Make arrangements with Robert Mushet to transfer Mr. Hill Jr. to duty in Coining department under direction of W. T. Brande.
ACCS 1p TxU:H/L-0676.2; Reel 1089

9929. 1854–1–2 To Mint Staff Royal Mint
Orders Stronghold no. 3 in Coining Department to be cleaned daily instead of weekly.
ACCS 2pp TxU:H/L-0676.1; Reel 1089

9930. 1854–1–3 To [Ralph Heaton & Sons] Royal Mint
Directions for dispatching copper coins to eleven branches of Provincial Bank of Ireland.
ACCS 2pp TxU:H/L-0677; Reel 1089

9931. 1854–1–3 From William Whewell Lowestoft
Asks JH's opinion of an anonymous book [WW's *Of the Plurality of Worlds: An Essay*]. Describes it written 'very fairly' but as presenting views very different from JH's.
ALS 4pp RS:HS 18.213 Todhunter, *Whewell, 2,* 399–400

9932. [1854–1–5] To 'Dear Sir' [Royal Mint]
Analysis of quantities of coins in circulation. Explains ways in which gold and silver are diverted from coins to other purposes.
AdraftS 5pp TxU:H/L-0596; Reel 1089

9933. 1854–1–5 From Henry T. De La Beche Museum, Jermyn Square
Would he look at the accompanying work by O. P. A. P. Dufrénoy. Feels that Dufrénoy would advocate only that which was right.
ALS 2pp RS:HS 6.135

9934. 1854–1–5 To Charles [Edward Trevelyan] [Royal Mint]
Comments on various kinds of scales for measuring distances.
Adraft 2pp TxU:H/L-0484.1; Reel 1055

9935. 1854–1–5 To [Charles Edward Trevelyan] Royal Mint
Reports total coinage from 1814 to 1853 and silver coinage from 1801 to 1815. Gives reasons why gold coins in circulation 'fall far short' of those minted. History of copper coinage since 1790.
Adraft 3pp TxU:H/L-0678.2; Reel 1089

9936. 1854–1–6 To William Thomas Brande Royal Mint
Differences between scales in Mint office and Rolling Room. Concern for gold wasted by abrasion.
ALS 3pp TxU:H/L-0108.2; Reel 1054

9937. 1854–1–7 To William Ewart Gladstone Royal Mint
Forwards copies of new coins to be minted for Australia, and asks WG [Chancellor of the Exchequer] to seek the Queen's approval for them.
ALS 3pp BL Aaa 44377.25

9938. 1854–1–9 From Charles Edward Trevelyan Treasury
Forwards letter from Henry Finch to CT. It may help JH in considering Finch's case.
LS 1p TxU:H/M–0554.1; Reel 1093

9939. 1854–1–11 To Charles Edward Trevelyan Harley St.
CT's memorandum on 'Superannuation Question.' JH argues that money or rights given to any Mint employee for his services should not be revocable but 'should be absolutely his property.'
ACCS 3pp TxU:H/L-0484.2; Reel 1055

9940. 1854–1–11 To [Charles Edward Trevelyan] Royal Mint
Received [CT]'s letter of 9 Jan. 1854 conveying Duke of Newcastle's request for artist to execute Crimean Medal, and Treasury commissioners' directive for JH to include cost of dies for medal in 1855 Mint expenses. JH feels obligated to employ Mint's own Medaller and Engraver, but will accept dies from another artist if Queen Victoria so desires.
AdraftS 2pp TxU:H/L-0679; Reel 1089

9941. 1854–1–12 To Ralph Heaton & Sons Royal Mint
Directions for shipping copper coins to ten branches of Bank of Ireland.
ACCS 1p TxU:H/L-0190; Reel 1054

9942. 1854–1–13 From Charles Edward Trevelyan Treasury
Grateful for JH's paper on '*Deduction from Salaries*' question, but notes precedent in common law for exceptions in cases of dismissal from public service.
LS 4pp TxU:H/M–0554.2; Reel 1093

9943. 1854–1–15 From François N. M. Moigno St. Germain de Pres
Compliments to the Herschels. Was interested in JH's article in the last number of *Good Words*. A long time since he has seen Alexander Herschel.
ALS 2pp (French) RS:HS 12.352

9944. 1854–1–16 From Charles Edward Trevelyan Treasury
[Form letter] Treasury solicits recommendations on which of three decimal distance scales should be adopted by Ordnance Survey for maps.
LS 2pp TxU:H/M–0554.11; Reel 1093

9945. 1854–1–16 To Charles Edward Trevelyan Royal Mint
Comments on which rights should be forfeited when public servants are dismissed.
ACCS 3pp TxU:H/L-0484.3; Reel 1055

9946. 1854–1–16 To [Charles Edward Trevelyan] [Royal Mint]
Now that pressure for gold and silver coinage has abated, JH wants to eliminate some temporary clerks assigned to Mint. Mr. Watt and H. A. Severn were assigned to Sydney branch mint. [John] Goldsmith was appointed supernumerary clerk in Mint office. JH informed Mr. Welsh that Welsh's services will end soon.
AdraftS 2pp TxU:H/L-0680; Reel 1089

9947. 1854–1–17 To ? n.p.
Specifies procedure for combining the contents of two pots.
ALS 1p Adler Planetarium

9948. 1854–1–17 To Alexander James Beresford Hope Royal Mint
A letter of regret that JH is unable to attend the funeral of [William Carr Beresford].
ALS 3pp APS B.H435p.17

9949. 1854–1–17 To Robert Mushet Royal Mint
Notes striking increase recently in gold wasted in melting process. Calls for inquiry and report of cause by RM.
Adraft 2pp TxU:H/L-0278; Reel 1054

9950. 1854–1–18 From William Thomas Brande Royal Mint
Number of coin dies required by Sydney branch mint and lifetime of each die. WB would need six months to produce these.
ALS 2pp TxU:H/M–0121; Reel 1087

9951. 1854–1–18 From William Thomas Brande Royal Mint
James Wyon has proven to be competent engraver for two years.
ALS 2pp TxU:H/M–0120; Reel 1087

9952. 1854–1–18 To Margaret Brodie Herschel [London]
Directions for sending mail to son Willy [in India?]; about daughter Margaret Louisa's health, the family bills, some stories about people JH met at a dinner, and finally, JH's despair about his work.
ALS 4pp JHS 1.101

9953. 1854–1–21 From Charles Edward Trevelyan Treasury
Acknowledges CT's error in writing name of 'Rose & Crown.' Returns JH's unsigned letter on copper coinage with two errors noted. Concurs with JH on coins proposed for adoption.
LS 3pp TxU:H/M–0554.3; Reel 1093

9954. 1854–1–23 To Charles Edward Trevelyan Harley St.
Distance scales to be used on Ordnance Survey maps.
ACC 3pp TxU:H/L-0484.4; Reel 1055

9955. 1854–1–25 To Charles Edward Trevelyan Royal Mint
Requests permission to pay Mint expenses from special account until Parliamentary grant becomes 'available.'
ACCS 3pp TxU:H/L-0484.5; Reel 1055

9956. 1854–1–27 To Miss Cornwallis London
Rapid changes in science leave JH unable to recommend newest things in organic chemistry. Suggests works by H. V. Regnault or Justus von Liebig, and Lyon Playfair's work on chemistry of foods. Notes error in Liebig's work on body chemistry.
ALS 4pp † TxU:H/L-0132; Reel 1054 (C: RS:HS 23:142)

9957. 1854–1–27 From François N. M. Moigno St. Germain de Pres
Thanks for his articles on the prediction of time; comments on it. Has been translating an article of John Tyndall.
ALS 2pp (French) RS:HS 12.353

9958. 1854–1–27 To [Charles Edward Trevelyan] [Royal Mint]
Requests permission from Treasury to draw from Master's account at Bank [of England] to cover expenses that cannot be paid from depleted Exchequer account before 31 Mar., when Parliamentary grant for next fiscal year becomes available.
ACCS 2pp TxU:H/L-0681; Reel 1089

9959. 1854–1–28 From Charles Babbage Dorset St., Manchester Square
Saw in the *Times* that JH is working on universal alphabet. Sends Wolfgang Kempelen's book (1791) on mechanism of

speech. Speculates on machine that could produce phonetic sounds for universal language.
ALS 4pp TxU:H/M–0031; Reel 1062

9960. 1854–1–30 To [Christian C. J.] Bunsen [32 Harley St.]
Unable to attend today's conference on replacing present system of English writing with phonetic system such as 'Missionary Alphabet.' Hopes no final conclusions will be drawn without opportunity for further discussion. Gives JH's opinions and objections. Contrasts systems proposed by [Charles] Trevelyan, [J. B.] Gilchrist, and [Thomas] Young.
AdraftS 5pp & ACCS 13pp TxU:H/L-0117 & -0018; Reel 1054

9961. 1854–1–30 To Isabella Herschel Royal Mint
Family news and comments.
ALS 3pp JHS 1.70

9962. 1854–1–31 From John Banks Hastings
Requests donation of copy of JH's *Cape Results* to library of Mechanics Institution.
ALS 3pp † RAS:JH Archive 10/5.81; Reel 9

9963. 1854–1–31 To William Thomas Brande Royal Mint
Notes deficit in Coining Room between 23 and 28 Jan.. Following enquiries in December into similar shortages, JH now requires that WB submit written reports of daily accounts.
AdraftS 2pp TxU:H/L-0108.1; Reel 1054

9964. [1854–2]. To Margaret Brodie Herschel [London]
JH feels harassed by life and finds solace in MH's letters; thinks electricity could operate the Jacquard loom and the pianoforte. Argues against a recent book [William Whewell's *Of the Plurality of Worlds*], which claims the earth is the only inhabited body in the universe.
ALS 4pp JHS 1.110

9965. 1854–2–1 to 15 To Margaret Brodie Herschel [London]
Has recovered from his depression [see JH's 1854–2–1]; sees a possibility of a new position as a Civil Service appointment examiner.
AL 4pp JHS 1.57

9966. 1854–2–1 To Margaret Brodie Herschel [London]
Very depressed and concerned about the family finances, both short and long term.
AL 4pp JHS 1.56

9967. 1854–2–1 From Joseph Jopling 6 Budge Rd., St. Johns Wood
Sending diagrams of Halley's Comet traced from a work, *The Orbs of Heaven.*
ALS 3pp RS:HS 10.416

9968. [1854–2] To William Whewell n.p.
Comments on *Of the Plurality of Worlds: An Essay.* Suspects it is by WW. Disagrees with many parts, but finds some sections 'striking.' Laments that he has no time for intellectual matters, punning that he must concentrate on 'making money.'
ALS 4pp TC, Camb. Add. Ms.a.207[90] (C: RS:HS 23.140)

9969. 1854–2–4 To William Thomas Brande Royal Mint
Send books from Cutting Room. JH wants to question W. H. Barton about deficits in Coining Room. [Two assistants] will be in on Monday.
ALS 2pp TxU:H/L-0108.3; Reel 1054

9970. 1854–2–4 To Henry Anthony Jeffreys 32 Harley St.
Considers the education of the lower orders an honorable object, but has never given a public lecture and feels unable to assist.
draft 2pp RS:HS 10.294 & 23.143

9971. 1854–2–9 To William Thomas Brande Royal Mint
Send written evaluation of Henry Finch, whose year of probation expires 14 Feb. Should Finch be hired permanently to Mint staff?
ALS 4pp TxU:H/L-0108.4; Reel 1054

9972. 1854–2–9 To William Ewart Gladstone [Royal Mint]
Requests the addition of one wage increase, as a matter of fairness, to a previously agreed list for submission to the Treasury Lords.
ALS 3pp BL Aaa 44382.120

9973. 1854–2–9 To W. Graves Royal Mint
Second payment by Bank of Ireland for copper coins received from Mint.
ACC 2pp TxU:H/L-0173; Reel 1054

9974. 1854–2–9 To Ralph Heaton & Sons Royal Mint
Appoints Mr. Hill, senior clerk in Mint office, to replace Mr. Julyan as superintendent of copper coinage at Birmingham.
ACCS 2pp TxU:H/L-0682; Reel 1089

9975. 1854–2–9 From Margaret Seymour Park Lane
Invites JH to dine quietly with MS and the [11th] Duke of Somerset [E. A. Seymour], who is in 'delicate health.'
ALS 2pp RS:HS 17.196

9976. 1854–2–9 To Charles Edward Trevelyan [Royal Mint]
Please confirm that duties of appointment, which JH and CT discussed today, will permit residing outside London.
ACCS 2pp TxU:H/L-0484.6; Reel 1055

9977. 1854–2–10 From William Henry Stephenson Treasury
[Marked 'Private.'] Suggests that JH delay request to include office of master of Mint under Superannuation Act until after government revises Act.
ALS 2pp † TxU:H/M–0502; Reel 1093

9978. 1854–2–11 To 'Dear Sir' Royal Mint
Returns [?]'s letter, dated January, for correction. Believes that [?] intended it to be dated February.
AdraftS 2pp TxU:H/L-0683; Reel 1089

9979. 1854–2–11 To Edward Wolstenholme Ward Royal Mint
Mr. Watt requests leave from duties of new Sydney branch mint to join Commissariat department in Malta. If war is declared [in Crimea], Watt will resign from Mint. Difficulty in finding replacement so close to EW's departure for Australia.
ALS 1p TxU:H/L-0489; Reel 1055

9980. 1854–2–13 From George Biddell Airy Royal Observatory, Greenwich
Please send comments on the first draft of the Standards Commission report [see GA's 1853–12–22]; GA wants to prepare a second draft.
ALS 1p TxU:H/M-0637.1; Reel 1084 (C: RGO 6.341.49)

9981. 1854–2–13 To [Charles Edward Trevelyan] [Royal Mint]
Forwards report by W. T. Brande on performance of Henry

Finch during one year probation in coining department. JH recommends advancing Henry Finch and F. R. Brande to permanent positions.
Adraft 2pp TxU:H/L-0684.2; Reel 1089

9982. 1854–2–14 To George Biddell Airy 32 Harley St.
Suggests a reference to the need to reduce the variety of strange measures being used in the country [see GA's 1854–2–14]; wants to know who the examiners were at Cambridge University in Dec. 1853.
ALS 4pp RGO 6.341.50

9983. 1854–2–14 To Henry Finch [Royal Mint]
HF's year of probation in Coining department ends today. JH will recommend appointing HF to permanent position, but HF must take responsibility to acquaint himself with operations of all other departments if HF wishes further promotion. Admonishes HF to behave with dignity and adapt to new Mint organization without 'harking back after the old state.'
Adraft 2pp TxU:H/L-0156; Reel 1054

9984. 1854–2–14 To Charles Edward Trevelyan Royal Mint
E. W. Ward's assistant Mr. Watt, proposed supervisor of bullion office in Sydney mint, volunteered for military expedition to Constantinople. JH asks Treasury to choose replacement.
ACCS 1p TxU:H/L-0484.7; Reel 1055

9985. 1854–2–15 To Margaret Brodie Herschel [London]
Daughter Caroline is preparing for the problems that the Crimean War may bring; JH has applied for an Examinership [see JH's 1854–2–1 to 15].
ALS 4pp JHS 1.55

9986. 1854–2–15 or later To Margaret Brodie Herschel [London]
The effect of the Crimean War on the Mint, and on the family, for example, daughter Caroline will want to follow her husband as close to the war area as possible [see JH's 1854–2–15].
AL 2pp JHS 1.54

9987. 1854–2–15 To Sir John Romilly Collingwood
Comments on Dr. Dodd's views of study of law versus civil law. Wants university study to be as broad as possible. Understanding civil law requires knowledge of general jurisprudence.
ALS 1p RS:HS 14.501 (C: RS:HS 23.144)

9988. 1854–2–15 To Charles Edward Trevelyan Harley St.
[Marked 'Private.'] List of Cambridge men that JH considers eligible to be moderators and examiners for Mathematical Tripos.
AdraftS 2pp TxU:H/L-0484.8; Reel 1055

9989. 1854–2–15 To Charles [Edward Trevelyan] Harley St.
[Marked 'Private.'] Asks CT to submit JH's name to Lord Aberdeen [George Hamilton Gordon] for position on Cambridge board of examiners, provided this duty does not require residence in London.
ACCS 2pp TxU:H/L-0484.9; Reel 1055

9990. 1854–2–16 From George Biddell Airy Royal Observatory, Greenwich
Doubts that proscribing certain weight usages will be effective, unless the users feel a need to change [see JH's 1854–2–14]
ALS 2pp TxU:H/M-0637.2; Reel 1084 (C: RGO 6.341.52)

9991. 1854–2–16 From Charles Edward Trevelyan Treasury
Doubts that examiner could live outside London and still fulfill duties.
LS 2pp TxU:H/M–0554.4; Reel 1093

9992. [1854]–2–17 From Mrs. Mushet Royal Mint
Will act according to JH's advice in future, but believes police at Mint gate were also at fault for not adhering to consistent policy of inspecting parcels carried out by servants and visitors.
ALS 4pp TxU:H/M–0385; Reel 1087

9993. 1854–2–18 From William Nairn Forbes Calcutta Mint
Appreciation of JH's son [John]. Admired JH's letter of 2 Feb. 1852 to Treasury, in which WF's most important recommendations [to 1848 committee for Mint reform] have been carried out. Dependence on steam for transportation has raised price of coal. Proposes two methods to replace coal: (a) use electrical conductivity of seawater in presence of dissimilar metals to propel ships, and (b) generate heat in 'Earth Batteries,' similar to dungheaps. Asks JH to approve request by John Gilbert, retired mechanical engineer at Calcutta mint, who wishes to visit Royal Mint, but was refused several years ago.
ALS 3pp RS:HS 7.374

9994. 1854–2–18 To Margaret Brodie Herschel [Royal Mint]
Duties of Examiner of superannuation claims for Treasury will not permit JH to live in Collingwood and may be more burdensome than those of Mint. Upset at losing Charles Elouis, who will leave Mint to replace Mr. Watt as superintendent of bullion at mint in Sydney. Concern for daughter Caroline, whose husband, [A. H.] Gordon, leaves before March, leaving her alone in Malta when 'her hour is due.'
AL 4pp TxU:H/L-0762; Reel 1089

9995. 1854–2–18 To Mrs. Mushet Royal Mint
Regrets confusion in signing of gate passes that created awkwardness for Mrs. M. Hopes police arrangements will not trouble Mint residents too much.
ACC 2pp TxU:H/L-0279; Reel 1054

9996. 1854–2–20 From William Allen Miller King's College, London
Offers possible explanations to account for discrepancies between assays of WM and of Thomas Graham.
ALS 3pp † TxU:H/M–0368; Reel 1087

9997. 1854–2–21 To 'Sir' [Royal Mint]
JH paid salary for Mr. Watt from 22 June 1853 to 17 Feb. 1854, when Watt was officially assigned to Sydney branch mint. Mr. Hill will replace Mr. Julyan, who was recalled from Birmingham to Commissary Department on 10 Feb. 1854. Mr. Nash will take Mr. Hill's duties as senior clerk in Mint office and senior officer in charge of dies.
Adraft 2pp TxU:H/L-0685; Reel 1089

9998. 1854–2–21 From C. T. von Kersten Cité Bergère, Hôtel du Rhin, Paris
Introduces self to JH and describes CK's modified English alphabet, which is both phonetic and orthographical, and CK's 'philosophical universal alphabet' that would work for all languages. Criticizes phonotypy of Isaac Pitman and A. J. Ellis. Asks JH for loan of £8 with which CK can publish

manuscripts. CK made similar appeals to C. K. J. Bunsen and C. E. Trevelyan.
ALS 5pp TxU:H/M–0316; Reel 1087

9999. 1854–2–22 From John Banks Hastings
Acknowledges receipt of JH's *Cape Results* for library of Mechanics Institution.
ALS 3pp RAS:JH Archive 10/5.82; Reel 9

10000. 1854–2–23 From Alfred Smee 7 Finsbury Circus
Note to accompany a second edition of work on vision.
ALS 2pp RS:HS 16.172

10001. 1854–2–23 To Charles Edward Trevelyan Royal Mint
Proposes eliminating small annual sum paid for rent of Victualling office by making one payment for thirty-year lease. [JH note on verso: Calculations relating to 'State of Gold in Mint.']
ACCS 1p [JH note] TxU:H/L-0484.10; Reel 1055

10002. 1854–2–24 From Charles Edward Trevelyan Treasury
Explain why JH wants to credit five years of previous service to F. R. Brande and Henry Finch at time of their promotions.
LS 3pp TxU:H/M–0554.5; Reel 1093

10003. 1854–2–25 From William Spence Lower Seymour St.
Explains and encloses paper on eyeless insects and their situation in caves.
AL 4pp RS:HS 17.223

10004. 1854–2–25 To Charles Edward Trevelyan Royal Mint
Five-year apprenticeships of Henry Finch and F. R. Brande qualify them to succeed [William] Buckle in coining department. Salaries for Mint employees.
ACCS 4pp TxU:H/L-0484.11; Reel 1055

10005. 1854–2–27 To William Thomas Brande Royal Mint
JH still waiting for report on balances used in coin production, in order to explain deficit of 23–28 Jan.
ACC 2pp inc TxU:H/L-0108.5; Reel 1054

10006. 1854–2–27 From Adam Sedgwick Cambridge
William Conybeare has asked AS to assist in the certification of George Wilson. Mentions William Whewell's book [*Of the Plurality of Worlds: An Essay*], but has been too ill to study it.
ALS 8pp RS:HS 15.444

10007. 1854–2–28 To Margaret Brodie Herschel Royal Mint
About Willy's success, and about the weather and JH's health.
ALS 4pp JHS 1.53

10008. 1854–3–1 To Margaret Brodie Herschel [London]
Is looking forward anxiously to coming to Collingwood on Friday; the Mint had a surprise visit from the Queen this day.
ALS 4pp JHS 1.52

10009. 1854–3 To [Charles Edward Trevelyan] [Royal Mint]
In response to letter from Sir Emerson Tennant to Mr. Lawley, JH intends to apply for public pyx trial as soon as coining of gold is concluded.
ACCS 1p TxU:H/L-0686; Reel 1089

10010. 1854–3–2 From Henry Peter Brougham Grafton St.
Would like, if he can lay his hands on it, a paper of John Dalton's on the defect of vision.
ALS 2pp RS:HS 4.307

10011. 1854–3–2 To Mrs. James A. Gordon n.p.
A letter of love and condolence to a woman and her husband, the latter being very near death.
C 1p RS:HS 23:145

10012. 1854–3–2 To C. T. von Kersten n.p.
Object of CK's modified English alphabet is different from object intended by conference recently held at C. K. J. Bunsen's. Those of A. J. Ellis and Isaac Pitman have appeared in print and are favorably accepted; doubts that CK's characters would supplant them. See Bunsen regarding CK's universal alphabet.
ACCS 4pp TxU:H/L-0212; Reel 1054

10013. 1854–3–2 From Charles Edward Trevelyan Treasury
Crediting F. R. Brande with five years of previous service does not conform to Treasury commissioners' intent at time Brande was offered new position in Mint. Hold to original schedule of salaries and promotions for Brande and Henry Finch.
LS 4pp TxU:H/M–0554.6; Reel 1093

10014. 1854–3–3 To Henry Peter Brougham 32 Harley St.
[Responding to HB's 1854–3–2], JH states that no manuscript accompanied the collection of colored silks sent to JH by John Dalton. Describes the collection and discusses 'idioptic vision.'
ALS 4pp Univ.Coll.London Brougham

10015. 1854–3–3 From Charles Edward Trevelyan [Treasury]
Now understands JH's reason for promoting F. R. Brande. Will now deal with cases of Henry Finch and James Wyon.
ALS 1p TxU:H/M–0554.7; Reel 1093

10016. 1854–3–3 To Charles [Edward Trevelyan] Royal Mint
Further discussion of salary for F. R. Brande after promotion.
ACCS 3pp TxU:H/L-0484.12; Reel 1055

10017. 1854–3–3 From George Waldegrave 4 Harley St.
GW's cousin J. A. Gordon approaching death, but calm. GW going to Naish.
ALS 3pp RS:HS 18.365

10018. 1854–3–4 From George Waldegrave Naish, near Bristol
J. A. Gordon died quietly last night. Widow and Henrietta are thankful for JH's letter.
ALS 4pp RS:HS 18.366

10019. 1854–3–8 To William Thomas Brande Royal Mint
Questions about WB's 1851 report on gas apparatuses in Mint, and feasibility of repairing versus removing them.
AdraftS 2pp TxU.H/L-0108.6, Reel 1054

10020. 1854–3–8 From Charles Edward Trevelyan [Treasury]
Forwards suggestion from CT's cousin, Ernest Perceval.
ALS 1p TxU:H/M–0554.8; Reel 1093

10021. 1854–3–9 To Margaret Brodie Herschel Royal Mint

Comments on the health of MH and daughters Margaret Louisa and Francisca, and then JH complains of his own health problems.
ALS 1p JHS 6.44b

10022. [1854–3–10] To Charles Edward Trevelyan [Royal Mint]
Discourages suggestion to issue both milled and unmilled florins. This would confuse the public and promote counterfeiting.
ACC 2pp TxU:H/M–0554.9; Reel 1093

10023. 1854–3–10 To Mr. Wilkins Royal Mint
Sends cheque for £14, annual rent for land occupied by former Victualling office of Mint. Please send draft for 'conveyance.' JH will submit it to solicitor of Mint.
ACCS 2pp TxU:H/L-0499; Reel 1055

10024. 1854–3–11 To Adam Sedgwick Harley Street
Has signed George Wilson's certificate. Says William Whewell's book is 'clever but wrong.' Says moral and intellectual life is possible on Jupiter.
ALS 4pp RS:HS 15.445 (C: RS:HS 23.146)

10025. 1854–3–14 From George Biddell Airy Royal Observatory, Greenwich
Wants to meet with JH and W. H. Miller at the Exchequer to examine their standard equipment.
ALS 2pp TxU:H/M-0637.4; Reel 1084 (C: RGO 6.341.53)

10026. 1854–3–14 From Edward Joseph Lowe Highfield House
Comments on zodiacal light. Would like JH's opinion on his theory. Has been experimenting further with the propagation of plants with collodion. Money for the Henry Lawson Observatory has been subscribed and hopes soon to appoint trustees.
ALS 4pp RS:HS 11.327

10027. 1854–3–15 To George Biddell Airy 32 Harley St.
Congratulates GA on the discovery of an 'old British encampment' [probably to do with British Roman battles]. Agrees to meeting time and place [see GA's 1854–3–14].
ALS 2pp RGO 6.341.54 & 6.468.246

10028. 1854–3–15 To Charles Edward Trevelyan Royal Mint
Governor of Malta [Sir William Reid] requests £500 in copper coins. Mint does not have enough.
AdraftS 2pp TxU:H/L-0484.14; Reel 1055

10029. 1854–3–16 To Griffith George Lewis Royal Mint
Unable to strike honorary medal from dies of Royal Military Academy. Under reformed Mint system, those dies are private property of engraver L. C. Wyon.
Adraft 2pp TxU:H/L-0225; Reel 1054

10030. 1854–3–17 To Margaret Brodie Herschel [London]
Mostly about a Mrs. [J. A.] Gordon, who has suffered a personal loss [death of her husband], and a young man who seems to be lost.
ALS 4pp JHS 1.51

10031. 1854–3–18 From George Biddell Airy Royal Observatory, Greenwich
A note accompanying a copy of the second draft of the Standards Commission report [see GA's 1854–2–14].
ALS 1p TxU:H/M-0637.5; Reel 1084 (C: RGO 6.341.55)

10032. 1854–3–21 To [?] Royal Mint
Formal testimonial to competence of J. L. Drury, temporary clerk in Mint office and melting department since 31 Jan. 1853.
AdraftS 1p TxU:H/L-0688; Reel 1089

10033. 1854–3–22 From George Biddell Airy Royal Observatory, Greenwich
A notice of meeting of the Standards Committee supervising the standardized weighing and measuring, indicating the need to bring to the meeting members expense sheets.
LS 3pp TxU:H/M-0637.6; Reel 1084

10034. 1854–3–22 From George Biddell Airy Royal Observatory, Greenwich
A note accompanying a letter for T. S. Rice [Lord Monteagle], which JH is to sign and forward.
ACCS 1p RGO 6.341.56

10035. 1854–3–22 To Margaret Brodie Herschel [London]
Mrs. Newton, MH's servant, is being sued by a man who recommended her for the job, and JH is to be a witness in MH's stead.
AL 4pp JHS 1.50

10036. 1854–3–22 To Charles Edward Trevelyan [Royal Mint]
[Marked 'Extract.'] Asks Treasury to define position of J. [J. S.?] Wyon at Mint.
ACCS 1p TxU:H/L-0484.15; Reel 1055

10037. 1854–3–23 From George Biddell Airy Royal Observatory, Greenwich
There has been some confusion about sending packages of material; GA wants to know what JH has received.
ALS 2pp TxU:H/M-0637.8; Reel 1084

10038. 1854–3–23 From George Biddell Airy Royal Observatory, Greenwich
Urges JH to stay at Collingwood [see JH's 1854–3–23], but send in any comments on the second draft of the report and authorize GA to act on JH's behalf.
ALS 2pp TxU:H/M-0637.7; Reel 1084 (C: RGO 6.341.59)

10039. 1854–3–23 To George Biddell Airy n.p.
Is reluctant to attend meeting [see GA's 1854–3–14], as JH had intended to spend a few days at Collingwood.
ALS 3pp RGO 6.341.57

10040. 1854–3–23 From George Biddell Airy to Margaret Brodie Herschel Royal Observatory, Greenwich
Writes to MH to insist that JH not come to the meeting [see JH's 1854–3–23].
ACCS 3pp RGO 6.341.60

10041. 1854–3–23 To Charles Edward Trevelyan Royal Mint
Encourages Treasury to maintain distinction between temporary and permanent workers at Mint.
AdraftS 2pp TxU:H/L-0484.16; Reel 1055

10042. 1854–3–24 To George Biddell Airy [London]
Sends some suggestions for additions to draft report of the Standards Commission [see GA's 1854–3–23]; thanks GA for the report on the eclipse of Thales.
ALS 4pp RGO 6.341.64

10043. 1854–3–24 From Margaret Brodie Herschel to George Biddell Airy Collingwood
Profusely thanks GA for his letter [see GA's 1854–3–23].

The family will keep JH at home. Talks about her interest in a nearby mound as a possible ancient site.
ALS 3pp RGO 6.341.62

10044. 1854–3–24 To William Thomas Brande Royal Mint
Gold and silver coinage will be suspended during 'Enquiry week.' Use this time to produce copper coins.
ACCS 1p & AdraftS TxU:H/L-0108.7 & 0108.8; Reel 1054

10045. 1854–3–24 To Edward Joseph Lowe n.p.
Comments on EL's paper on zodiacal light; JH disagrees with EL's arguments and suggests rethinking them.
ALS 3pp BL Aaa 43829.21

10046. 1854–3–24 To William Parsons (Lord Rosse) 32 Harley St.
Regrets inability to write [obituary] memoir of D. F. J. Arago for R.S.L. Anniversary Meeting.
ALS 3pp RS MC.5.158

10047. 1854–3–24 To C[harles] Sterry London
Confirms arrangement to meet next day.
ALS 1p DMC.695A.885.32

10048. 1854–3–24 To [Audit Office] [Royal Mint]
Will submit statement if [Audit Office] insists, but does not believe this will afford any check on correctness of silver bullion inventory.
Adraft 2pp TxU:H/L-0687; Reel 1089

10049. 1854–3–25 To William Thomas Brande Royal Mint
'In Re Napier's Model,' WB is not to exceed the approved limit of £25.
Adraft 1p TxU:H/L-0108.9; Reel 1054

10050. 1854–3–29 To Margaret Brodie Herschel [London]
Mostly about the court case [see JH's 1854–3–22]; it appears the plaintiff lost.
ALS 4pp JHS 1.49

10051. 1854–3–30 To Robert Mushet Royal Mint
Deliver gold, silver, and copper scrap and unusable equipment to W. H. Barton to be sold. [JH annotation: Similar notices sent to W. T. Brande and to H. W. Field.]
ACCS 2pp † TxU:H/L-0280; Reel 1054

10052. [1854–4–1 or later] To 'Sir' [Royal Mint]
Requests that £30,000, granted by Parliament, be issued to replenish expenditures from Master's cash account during fiscal year ending 31 Mar. 1855.
ACCS 1p TxU:H/L-0619; Reel 1089

10053. 1854–4–1 To [Augustus De Morgan] n.p.
Discusses the pros and cons of AD's suggestion of a three penny piece.
ALS 2pp Univ. of London Library

10054. 1854–4–3 To James Calder Stewart Collingwood
[Dictated to Isabella Herschel] Transfer of funds from Thomas Baldwin's estate. [Isabella's note:] JH is still 'weak & very quickly fevered.'
L 3pp TxU:H/L-0445; Reel 1055

10055. 1854–4–3 To Charles Edward Trevelyan Royal Mint
Charles Elouis requests return of contributions (£180) to Superannuation Fund over 19 years of Mint service, in preparation for Elouis's appointment as superintendent of bullion office in Sydney mint.
ACCS 1p TxU:H/L-0484.17; Reel 1055

10056. 1854–4–4 From George Biddell Airy Royal Observatory, Greenwich
A note accompanying papers related to Melbourne University.
ACCS 1p RGO 6.335.34

10057. 1854–4–4 To [Charles Edward Trevelyan] Royal Mint
Recommends R. F. Suft to replace Charles Elouis as registrar and accountant in Mint office, and Charles Sterry to replace Suft as senior clerk. Seeks Treasury commissioners' approval for hiring properly qualified assistant to resident assayer, to replace Sterry.
Adraft 2pp TxU:H/L-0662; Reel 1089

10058. 1854–4–4 To Charles Edward Trevelyan Royal Mint
Recommends R. F. Suft to succeed Charles Elouis as registrar and accountant of Royal Mint. Recommends Charles Sterry to replace Suft as senior clerk in Mint office.
ACCS 2pp TxU:H/L-0484.18; Reel 1055

10059. 1854–4–5 To 'Sir' Royal Mint
Explains to Treasury why copper needed for coinage is supplied by private contract, rather than by public competition, following 1849 transfer of copper manufacturing from Boulton & Watt in Soho to Ralph Heaton & Sons in Birmingham.
AdraftS 3pp TxU:H/L-0690; Reel 1089

10060. 1854–4–5 From George Biddell Airy Royal Observatory, Greenwich
JH and GA together with Robert Lowe and Henry Malden are to find four professors for Melbourne University [see GA's 1854–4–4].
ACCS 2pp RGO 6.335.35

10061. 1854–4–5 To [Charles Edward Trevelyan] [Royal Mint]
Paymaster General's office requires special approval from Treasury to pay allowance for quarters to fireman George Busher, transferred from Queen's assay office to resident assay office.
ACCS 1p TxU:H/L-0689; Reel 1089

10062. 1854–4–6 To George Biddell Airy Royal Mint
Does not really want to be involved in this venture [see GA's 1854–4–5 and JH's 1854–4–10].
ALS 2pp RGO 6.335.36

10063. 1854–4–7 From George Biddell Airy Royal Observatory, Greenwich
Suggests a meeting to discuss whether to agree to act for Melbourne University [see JH's 1854–4–6].
ACCS 2pp RGO 6.335.37

10064. 1854–4–7 To Margaret Brodie Herschel [London]
Describes how to use the telescope at Collingwood, and comments on a number of people, including some of their children.
AL 4pp JHS 1.48

10065. 1854–4–7 To Margaret Louisa Herschel [London]
About arrangements for a few days at Collingwood, an accident to sister Caroline's 'W.L.' [?], and news about brother Willy, together with some moralizing.
ALS 4pp JHS 1.63

10066. 1854–4–8 From George Biddell Airy Royal Observatory, Greenwich
Makes a correction in the weight of an item reported in the Standards Commission report.
ALS 1p TxU:H/M-0637.9; Reel 1084

10067. 1854–4–8 From George Biddell Airy Royal Observatory, Greenwich
Sends a revision of the weight of the standard pound deposited at the Royal Mint [see JH's 1853–9–9].
ACCS 1p RGO 6.339.181

10068. 1854–4–10 From George Biddell Airy Royal Observatory, Greenwich
May not be able to make meeting [see JH's 1854–4–10].
ACCS 2pp RGO 6.335.39

10069. 1854–4–10 To George Biddell Airy n.p.
Agrees to an informal meeting and will organize it [See GA's 1854–4–7].
ALS 2pp RGO 6.335.38

10070. 1854–4–10 To Margaret Brodie Herschel [London]
Complains about the work involved in selecting professors for Melbourne University [see JH's 1854–4–6].
ALS 4pp JHS 1.47

10071. 1854–4–10 To Charles Edward Trevelyan Royal Mint
Recommends that former gas plant at Mint be maintained and kept in reserve, in case of interruptions in commercial supply from Ratcliffe Gas Company. Current prices of gas and coal.
Adraft 3pp TxU:H/L-0484.19; Reel 1055

10072. 1854–4–10 From Henry Warburton 45 Cadogan Place
Heard JH plans new edition of *Examples of Finite Differences* [1820]. Will send list of errors HW found in original. Results of HW's investigation of self-repeating series of numbers.
ALS 6pp RS:HS 18.62

10073. 1854–4–11 To 'Sir' Royal Mint
Advises [?] to transfer copper coins from Wexford to Dublin. Regrets that farthing is not better known and used by humbler classes in Ireland. Should JH send small remittance to [?]'s manager in Ballyshannon?
ACCS 2pp TxU:H/L-0691; Reel 1089

10074. 1854–4–11 From George Biddell Airy Flamsteed House, Greenwich
Making some additional arrangements along with meeting [see JH's 1854–4–10].
ACCS 2pp RGO 6.335.40

10075. 1854–4–21 To Mint Staff Royal Mint
Declares 26 Apr. a holiday at Mint, in keeping with Queen Victoria's proclamation that this be observed as 'a day of Humiliation and Prayer.'
ACCS 1p & C TxU:H/L-0692; Reel 1089

10076. 1854–4–22 To Henry Thomas De La Beche Royal Mint
Asks HD to recommend a young man to fill a vacancy at junior clerk, and who can do assaying as well. JH provides details about the position.
ALS 4pp WT B.11

10077. 1854–4–22 To Margaret Brodie Herschel Royal Mint
Comments on JH's health, daughter Margaret Louisa's trip to John William Lubbock's, and on the Preston Strike.
ALS 4pp JHS 1.46

10078. 1854–4–22 From John William Lubbock n.p.
Astronomer Royal and his wife are visiting the Lubbocks on Saturday; will JH join them for Saturday and Sunday? Will collect him if he likes. Louisa Herschel was the belle of the Bromley Ball.
ALS 2pp RS:HS 11.390

10079. 1854–4–22 To [Leonard Charles Wyon] Royal Mint
Office of Chief Medalist was abolished on 1 Dec. 1853. [LW]'s salary from then until 31 Mar. 1854 was drawn from Mint under title of 'Medalist and Engraver.' After 1 Apr. 1854, that salary will be issued by Paymaster General.
ACCS 3pp TxU:H/L-0693; Reel 1089

10080. 1854–4–24 To John P. Gassiot? 32 Harley St.
Suggests method for obtaining daily photographic record of appearance of sun and its spots.
ALS 3pp RS MC.5.164

10081. 1854–4–24 To Anthony de Rothschild Royal Mint
Has scrap gold that needs refining, but it is less than 100-pound minimum set by AR in contract with Mint. Requests cost and conditions under which AR will accept this.
ACCS 3pp TxU:H/L-0345; Reel 1054

10082. 1854–4–24 To Anthony de Rothschild Royal Mint
[Marked 'Private.'] JH discovered new metal while examining residues of iridium. Suspects that it is impurity in Australian gold. Asks AR to send waste residues from refinery to JH for tests.
ALS 4pp TxU:H/L-0344; Reel 1054

10083. 1854–4–26 To Margaret Brodie Herschel 32 Harley St.
About a church service JH attended, some foolish university rules, and the benefits of studying chemistry in a medical faculty.
AL 4pp JHS 1.45

10084. [1854–4–28 ?] To John William Lubbock [Royal Mint ?]
JH is trying to arrange where to meet JL.
ALS 1p Oxford History of Science Museum 58/6

10085. 1854–4–29 To Margaret Brodie Herschel [London]
A strong letter of approbation for a position taken, and a letter written, by MH.
AL 4pp JHS 1.42

10086. 1854–4–29 To John William Lubbock Royal Mint
As JH's [1854–4–28 ?] did not elicit a response, JH is again trying to arrange at what station to meet JL.
ALS 1p Oxford History of Science Museum 58/3

10087. 1854–5 To George Peacock London
Writes to ask GP to help publicize the need for candidates for professorships at the University of Melbourne.
draft 1p RS:HS 13.306a

10088. 1854–5–3 To Charles Edward Trevelyan Royal Mint
Proposes changes in dies for New Brunswick coins. Awaits Treasury authorization to announce contract with coin suppliers.
ACC 2pp TxU:H/L-0484.20; Reel 1055

10089. 1854–5–4 To Mint Staff Royal Mint
Announces appointment of Charles Elouis to superintendent of bullion office at Sydney branch mint, effective 1 May 1854. R. F. Suft will replace Elouis as registrar and accountant in Mint office. Charles Sterry will take Suft's former position as senior clerk. E. L. J. Ridsdale is appointed, subject to [one year] probation, to junior clerk and assistant assayer.
ACCS 1p TxU:H/L-0694; Reel 1089

10090. 1854–5–4 To Charles Edward Trevelyan [Royal Mint]
Appointed E. L. J. Ridsdale, recommended to JH by College of Science on Jermyn Street, to replace Charles Sterry as junior clerk of Mint office and assistant assayer.
Adraft 2pp TxU:H/L-0484.21; Reel 1055

10091. 1854–5–5 To Margaret Brodie Herschel Royal Mint
Of the health of [Mr.] Knowles [servant], a visit to the Baden Powells by JH, and family financial arrangements.
ALS 8pp JHS 1.41

10092. 1854–5–6 To Isabella Herschel [London]
Tells a silly story, and then complains about Charles Babbage stopping by and talking about CB's lighthouse.
ALS 4pp JHS 1.40

10093. 1854–5–8 To Margaret Brodie Herschel [London]
Concerned about the state of health of Knowles [a servant], and how to deal with him and his wife; comments on the educational prospects of son Alexander.
ALS 4pp JHS 1.39

10094. 1854–5–9 From John P. Gassiot 77 Mark Lane
The Kew Committee has requested estimates from Thomas Cook[e] and Andrew Ross for an instrument to carry out JH's suggestions. Is grateful for these suggestions.
ALS 2pp RS:HS 8.59

10095. 1854–5–10 To George Biddell Airy n.p.
Is organizing a meeting to begin narrowing the field of candidates for the Melbourne University professorships [see JH's 1854–4–7].
ALS 1p RGO 6.335.41

10096. 1854–5–11 To [Sophia Beck?] London
Has caught a bad cold and is unfit for human society, and so JH begs to be excused from the evening's social gathering.
ALS 1p WT 52792

10097. 1854–[5]–11 From Henry Warburton 45 Cadogan Place
Recalls JH's predecessor [Isaac Newton] at Mint was also interested in Bernoulli's 'Isoperimetrical Problems.' HW plans paper on new expression for coefficients of differences of zero in self-repeating series.
ALS 6pp RS:HS 18.63

10098. 1854–5–12 To George Biddell Airy n.p.
As GA cannot attend meeting, JH seeks his opinions on certain questions [see GA's 1854–4–11].
ALS 2pp RGO 6.335.42

10099. [1854]–5–12 To Margaret Brodie Herschel Royal Mint, London
Talks about his loneliness, his wish to be at home with MH, and his concern about MH's contemplated trip to Aberdeenshire.
AL 2pp JHS 1.38

10100. 1854–5–12 To Edward Wolstenholme Ward Royal Mint
Sends to EW official documents relating to new branch mint in Sydney, and copy of document regarding salvage of goods lost on ship *Hannibal*.
ALS 2pp TxU:H/L-0490.1; Reel 1055

10101. 1854–5–13 To Margaret Brodie Herschel [London]
Mostly about attending James Adam Gordon's funeral at Hadham in Hertfordshire.
ALS 4pp JHS 1.37

10102. 1854–5–15 From Wilhelm Hansen Pentonville
Has been commissioned by Guillaume Weber to present a paper about an instrument for measuring the earth's magnetism, to JH. Would like to present it personally; when can he call?
ALS 1p RS:HS 10.153

10103. 1854–5–15 From William Henry Fox Talbot Athenaeum
Asks JH to sign an affidavit that WT is the inventor of the calotype or talbotype as some are trying to break WT's patent hold.
ALS 5pp RS:HS B26.49

10104. 1854–5–16 To Charles Edward Trevelyan Royal Mint
Sent 27 cases of coins to [Sir William Reid], governor of Malta, via G. Baillie, agent for colonies.
ACCS 1p TxU:H/L-0484.22; Reel 1055

10105. 1854–5–16 To Charles [Edward Trevelyan] Royal Mint
Treasury has not yet confirmed appointment of E. L. J. Ridsdale as junior clerk and assistant assayer. Superannuation for J. French and S. Bradstock. Status of revised die for New Brunswick coins. Budget for new machinery.
Adraft 1p TxU:H/L-0484.23; Reel 1055

10106. 1854–5–18 From William Henry Fox Talbot Athenaeum
Sends JH a copy of the affidavit [see WT's 1854–5–15] and asks for JH's emendations, if any, and his signature. Sends JH a photograph engraved on steel.
ALS 3pp RS:HS B26.50

10107. [1854–5–18 or later] To [Charles Edward Trevelyan] [Royal Mint]
[Following Treasury commissioners' instructions given in [?]'s letter of 18 May 1854,] £3000 of New Brunswick copper coins are ready for delivery to Kay Rutherford & Co. in Liverpool, agents for colonial government.
Adraft 1p inc TxU:H/L-0623; Reel 1089

10108. 1854–5–19 To [William Thomas] Brande Royal Mint
Treasury commissioners approved designs for New Brunswick penny and halfpenny. Prepare dies, and send these to Ralph Heaton & Sons for production.
ACCS 2pp TxU:H/L-0695; Reel 1089

10109. 1854–5–19 To William Henry Fox Talbot [London]
Comments on the affidavit supplied by WT; will sign most of it but objects to one section.
ALS 6pp RS:HS B26.51 R. Derek Wood, 'The Involvement of Sir John Herschel in the Photographic Patent

Case, *Talbot v. Henderson*, 1854,' *Annals of Science*, *27* (1971), 249–51

10110. 1854–5–19 To [Charles Edward Trevelyan] Royal Mint
JH has studied petition for pension by John Austin, dismissed from Mint in 1834. Finds no reason to overturn earlier rulings of Mint Board denying that claim.
AdraftS 1p TxU:H/L-0696; Reel 1089

10111. 1854–5–20 To Margaret Brodie Herschel [London]
About where, and how, the older children are to spend their holidays, and about the health of [Mr.] Knowles.
ALS 3pp JHS 1.35

10112. 1854–5–20 From William Henry Fox Talbot Athenaeum
Explains the way an affidavit works in the Court of Chancery. Because JH feels uncomfortable about certain aspects of the affidavit, WT will not trouble him about it again.
ALS 4pp RS:HS B26.52

10113. 1854–5–[21] To Margaret Brodie Herschel [London]
Feels overwhelmed by the unreasonable expectations of his political masters, and at the same time the new university at Melbourne, Australia wants JH to help find professors.
AL 4pp JHS 1.36

10114. 1854–5–21 To William Henry Fox Talbot [London]
Clarifies JH's position, by saying that he clearly supports WT as the inventor of the calotype or talbotype, but has reservations about the collodion process.
ALS 4pp RS:HS B26.53 R. Derek Wood, 'The Involvement of Sir John Herschel in the Photographic Patent Case, *Talbot v. Henderson*, 1854,' *Annals of Science*, *27* (1971), 252–3

10115. 1854–5–22 To Edward Sabine Collingwood
Discusses auroral observation. Will reply to [John] Lefroy's letter soon. Wants grant proposals sent to outlying R.S.L. members for comments. Cannot attend Kew Committee meeting.
draft 4pp RS:HS 15.238 (C: RS:HS 23.147)

10116. 1854–5–23 From William Henry Fox Talbot Athenaeum
WT's opponents have named JH in one of their affidavits against WT. WT is therefore asking JH to respond to what is stated in that affidavit.
ALS 4pp RS:HS B26.54

10117. 1854–5–24 From Thomas Woods Parsonstown [Offaly, Ireland]
Submits TW's chemical process for making photographs.
ALS 3pp TxU:H/M–0600; Reel 1086

10118. 1854–5–25 From George Biddell Airy Royal Observatory, Greenwich
Agrees to a procedural matter for the selection committee for Melbourne University [see JH's 1854–5–12].
ACCS 2pp RGO 6.335.43

10119. 1854–5–25 To Margaret Brodie Herschel [London]
Compares his work day with MH's, and feels that hers is more worthwhile and more pleasant.
AL 4pp JHS 1.34

10120. 1854–5–26 To William Parsons (Lord Rosse) Harley St.
Arrangements JH has made for admission [to Mint?] for WP and some of WP's friends.
ALS 2pp Rosse Papers K2.4

10121. 1854–5–26 From Price and Bolton Lincoln's Inn
Write to inform JH of the successful conclusion of the patent case brought by W. H. Fox Talbot.
AL 1p RS:HS B26.56

10122. 1854–5–26 From William Henry Fox Talbot Athenaeum
Reports to JH that the patent infringement case had been decided in WT's favor, and tells of one incident in court.
ALS 3pp RS:HS B26.55

10123. 1854–5–30 To George Biddell Airy n.p.
The circulation of material begins [see GA's 1854–5–25].
ALS 2pp RGO 6.335.44

10124. 1854–5–30 To Henry W. Field [Royal Mint]
Asks HF to report on 'six assay pieces' enclosed.
ALS 1p Royal Mint

10125. 1854–5–31 To Margaret Brodie Herschel [London]
Comments about times some of the children have been with JH, and about family financial accounts.
ALS 4pp JHS 1.32

10126. 1854–5–31 To Roderick Impey Murchison n.p.
Sending an account of a 'certain Dr [David] Livingstone.' JH agrees with Thomas Maclear that Livingstone deserves encouragement. Has received RM's testimonial on behalf of [Frederick] McCoy.
ALS 2pp NatLibScot MS.2522, f.43

10127. 1854–6–2 To Mrs. [Richard] Jones n.p.
Sends her a newly minted florin.
C 1p RS:HS 23.148

10128. 1854–6–4 To Margaret Brodie Herschel [London]
About Willy being in some kind of trouble, articles by friends in the *Westminster Review*, the Crimean War, and the health of MH and JH.
AL 4pp JHS 1.28

10129. [1854]–6–4 To [Alexander James Beresford Hope] Collingwood
Thanks for inquiring as to the welfare of JH's wife, Margaret, and the latest child, both of whom are well. JH is still doing poorly, but appreciates the use of Marshal Beresford's armchair.
ALS 3pp WT 63313.4

10130. 1854–6–5 To E. A. Hoffay [Royal Mint]
Reports sums dispatched from Mint to colonies. Requests reimbursement from paymaster general before 30 June.
AdraftS 1p TxU:H/L-0196; Reel 1054

10131. 1854–6–8 To Margaret Brodie Herschel [London]
Of the heavy work load dealing with Melbourne University appointments [see JH's 1854–5–[21]], and of Charles Babbage's unacceptable behavior at a meeting of the Royal Observatory's Board of Visitors.
ALS 4pp JHS 1.33

10132. 1854–[6]–9 To Ralph Heaton & Sons Royal Mint
Directions for shipping copper coins to eight banks in Ireland.

Complains that shipping boxes are not strong enough.
Adraft 3pp TxU:H/L-0189; Reel 1054

10133. 1854–6–10 To George Gabriel Stokes London
Writes to acknowledge receipt of a testimonial on behalf of [George] Wilson.
C 1p RS:HS 25.10.4

10134. 1854–6–11 To George Biddell Airy n.p.
Many applications and testimonials regarding Melbourne University professorships have arrived. How soon would a meeting to reduce the numbers be useful [see JH's 1854–5–30]?
ALS 2pp RGO 6.335.46

10135. 1854–6–13 From George Biddell Airy Royal Observatory, Greenwich
Agrees to a meeting [see JH's 1854–6–11], and comments on one possible candidate for a Melbourne University professorship.
ACCS 3pp RGO 6.335.48

10136. 1854–6–13 To Margaret Brodie Herschel [London]
News of some of the children with JH or near him, such as Caroline and Margaret Louisa; comments about son Willy.
AL 3pp JHS 1.31

10137. 1854–6–13 To Charles Edward Trevelyan Royal Mint
Availability of silver for making Turkish coins.
AdraftS 2pp TxU:H/L-0484.24; Reel 1055

10138. 1854–[6]–14 To Margaret Brodie Herschel [London]
Commenting on the health of some mutual acquaintances, and on son Willy's poor memory.
AL 2pp inc JHS 1.4

10139. 1854–6–14 To James Calder Stewart Harley St.
Rejoices at JS's new position. Asks how to send two official documents safely to India.
ALS 3pp TxU:H/L-0446; Reel 1055

10140. 1854–6–14 To [Charles Edward Trevelyan] Royal Mint
Requests reimbursement from Treasury for coins sent to Falkland Islands, Constantinople, and Hong Kong.
Adraft S 2pp TxU:H/L-0697; Reel 1089

10141. 1854–6–15 To Margaret Brodie Herschel [London]
About Willy Herschel's health, and JH's heavy work load and his desire for a relaxing time away from work.
AL 4pp JHS 1.30

10142. 1854–6–16 To George Biddell Airy Royal Mint
Is swamped with the demands of the Melbourne University professor search [see JH's 1854–6–16] and major upheavals at the Mint.
ALS 4pp RGO 6.335.52

10143. 1854–6–16 To George Biddell Airy n.p.
Assesses most of the candidates for the mathematics professorship at Melbourne University [see GA's 1854–6–13].
ALS 3pp RGO 6.335.50

10144. 1854–6–16 To Charles Edward Trevelyan Royal Mint
Compares quality of silver in dollars and in [Turkish] piasters.
ACCS 1p TxU:H/L-0484.25; Reel 1055

10145. 1854–6–16 To [Charles Edward Trevelyan] Royal Mint
As requested by Treasury commissioners, JH will have Turkish coins sent by Commissary General William Filder weighed and assayed to determine their value relative to English gold standard.
ACCS 1p TxU:H/L-0698; Reel 1089

10146. [1854–6–16 or later] To [?] [Royal Mint]
Reports assay results of British coins minted in Turkey.
Adraft 2pp TxU:H/L-0597; Reel 1089

10147. 1854–6–17 From George Biddell Airy Royal Observatory, Greenwich
Sends acknowledgment for meeting date and receipt of letters [see JH's 1854–6–17].
ACCS 1p RGO 6.335.55

10148. 1854–6–17 To George Biddell Airy n.p.
Clears up confusion about one candidate for the Melbourne University professorship by realizing there are two with the same last name [see JH's 1854–6–16].
ALS 2pp RGO 6.335.54

10149. 1854–6–20 To George Biddell Airy n.p.
Sends some new, negative information about one of the prime candidates for a Melbourne University professorship [see JH's 1854–6–17].
ALS 2pp RGO 6.335.56

10150. 1854–6–20 To [J. F. Julius ?] Schmidt n.p.
Has received [J]S's letter and testimonial
ALS 1p NSUB Cod.Ms.Philos.182

10151. 1854–6–21 From George Biddell Airy Royal Observatory, Greenwich
Must use the information available, and so need to select the next candidate in line for the Melbourne University professorship [see JH's 1854–6–20].
ACCS 2pp RGO 6.335.58

10152. [1854?]–6–21 To Mocatta & Goldsmid Royal Mint
Orders £20,000 of [silver] pieces, to be delivered Saturday or Monday.
AdraftS 2pp TxU:H/L-0265; Reel 1054

10153. 1854–6–21 To Charles Edward Trevelyan Royal Mint
Failure of Treasury to notify Bank [of England] about appointment of R. F. Suft as new registrar and accountant to Mint is creating problems with signing checks. Further analysis of silver in dollars.
ACC 1p TxU:H/L-0484.26; Reel 1055

10154. 1854–6–21 To [Charles Edward Trevelyan] Royal Mint
Returns letter of 5 June 1854 from Commissary General William Filder to Treasury. Reports value of five Turkish coins relative to English gold standard.
AdraftS 2pp TxU:H/L-0699; Reel 1089

10155. 1854–6–22 To Margaret Brodie Herschel [London]
Family news such as the death of Mrs. Knowles [servant?)] JH took Margaret Louisa dancing, the school progress of John and Alexander, and comments about the Crimean War.
ALS 3pp JHS 1.29

10156. 1854–6–23 From John R. Hind 22 Grove Road, St. John's Wood
Computation of Alpha Centauri's angle and distance, derived with JH's 'graphical method.' Compares this with results by other astronomers.
ALS 3pp † RAS:JH Archive 14/3.18; Reel 11

10157. 1854–6–23 To 'Sir' Royal Mint
Large accumulation of coins since gold coinage was suspended. Recommends pyx trial to certify new coins.
AdraftS 1p TxU:H/L-0700; Reel 1089

10158. 1854–6–24 To Charles C. F. Greville Royal Mint
Requests official copy of Privy Council order of 19 Aug. 1853 establishing Sydney branch mint. This will travel to Australia with deputy master and remain there.
ALS 1p WT B.12 (AdraftS TxU:H/L-0174; Reel 1054)

10159. 1854–[6]–26 To James Wilson Harley St.
Wants to put stop to unauthorized private work being performed for Board of Inland Revenue by Mint employees using Mint equipment.
AdraftS 2pp TxU:H/L-0505.2; Reel 1055

10160. 1854–7–1 To [Mint Staff] Royal Mint
[Formal announcement] Announces departure from Coining Department of Stephen Bradstock Sr., Joseph French, Henry Godfrey, Henry Stevens, Alexander Arthur, and George Arthur. Approves promotions of Richard Files Jr., William Bradstock, Alfred Bradshaw, and Vincent Shelley to fixed salaries in Coining and Die department.
AdraftS 1p TxU:H/L-0701; Reel 1089

10161. 1854–7–3 To Margaret Brodie Herschel [London]
About the payment of bills, JH's depression about his life [at work?], and his pleasure in his family.
ALS 3pp JHS 1.24

10162. 1854–7–3 To Charles Edward Trevelyan [Royal Mint]
Requests permission of Treasury to send sample from Gold Trial Plate of 1829 to E. W. Ward, to be used at Sydney branch mint for comparison of standards.
Adraft 1p TxU:H/L-0484.27; Reel 1055

10163. 1854–7–5 From George Robert Waterhouse British Museum
Answers JH's request about Borneo snakes. Forty reported species, contrary to opinion that Borneo has none. None in Iceland or Tierra del Fuego. New Zealand is warm enough, but none reported there.
ALS 3pp RS:HS 18.100

10164. 1854–7–7 To Charles Edward Trevelyan [Royal Mint]
Report to Treasury on quality of Turkish coins sent by commissary general [William Filder] for analysis.
Adraft 1p TxU:H/L-0484.28; Reel 1055

10165. 1854–7–12 To George Biddell Airy London
Problem of late applicants for a position, especially as they may be good [see GA's 1854–6–21].
ALS 4pp RGO 6.335.59

10166. 1854–7–15 From George Biddell Airy The Grange, Borrowdale, Keswick
Believes that deadlines must be adhered to regarding search for Melbourne University professors [see JH's 1854–7–12]; adds comments on other candidates.
ALS 4pp RGO 6.335.61

10167. 1854–7–19 To Charles Edward Trevelyan Royal Mint
E. W. Ward requests that formal authorization for Sydney branch mint be sent from Treasury to [Sir Charles Fitzroy] governor of New South Wales, and that Queen promulgate an order for issuing Australian sovereigns.
ACCS 1p TxU:H/L-0484.30; Reel 1055

10168. 1854–7–19 To Charles Edward Trevelyan Royal Mint
E. W. Ward will depart for Australia on 4 Aug. with year's supply of dies and assay equipment. This created great demand on die department, and J. Wyon requests extra compensation. JH concurs.
ACCS 3pp TxU:H/L-0484.29; Reel 1055

10169. 1854–7–24 To Charles William Pasley [Royal Mint]
Reaffirms JH's earlier statement that complete conversion to decimal coinage would require twenty years.
ACCS 2pp TxU:H/L-0296; Reel 1054

10170. 1854–7–24 To Charles Edward Trevelyan Royal Mint
Have Bank of England deposit £20,083 in account of Master of Mint to pay for 80,000 five-franc pieces sent to Constantinople on 28 June.
ACCS 1p & Adraft 1p TxU:H/L-0484.31 & -0484.31 verso; Reel 1055

10171. 1854–7–29 From Augustus De Morgan n.p.
Sending a comic squib.
AL 1p RS:HS 6.433

10172. [1854–7–29 or later] To [Charles Edward Trevelyan] [Royal Mint]
Proposes new Mint position called Scalesman, to be filled by William Bradshaw, and to dismiss Bradshaw's present helper. Pay of lamplighter, Mr. Holland, is inadequate for services Holland renders in maintaining gas plumbing at Mint, formerly performed by Mr. [Love].
Adraft 1p inc TxU:H/L-0624.2; Reel 1089

10173. [1854–7–29 or later] To [Charles Edward Trevelyan] [Royal Mint]
Calls Treasury's attention to Mint employee Henry Godfrey, disabled by heart condition. JH sees no alternative but to dismiss Godfrey, but superannuation fund is not yet sufficient to support Godfrey. Asks permission to pay Godfrey £20 on dismissal.
Adraft 2pp TxU:H/L-0624.1; Reel 1089

10174. 1854–7–29 To [Edward Wolstenholme Ward] [Royal Mint]
Receipt for one trial plate of gold, to be given to Sydney branch mint.
Adraft 1p inc TxU:H/L-0624.3; Reel 1089

10175. 1854–7–29 To Edward Wolstenholme Ward Royal Mint
Forwards authority from Treasury for EW to proceed to New South Wales and superintend completion of branch mint in Sydney.
ACCS 2pp TxU:H/L-0491; Reel 1055

10176. 1854–7–31 To [Charles Edward Trevelyan] [Royal Mint]
Pay rate for policemen assigned to Mint has been increased by Metropolitan Police. Requests authorization from Treasury for this new expense.
AdraftS 1p & C TxU:H/L-0702; Reel 1089

10177. 1854–8 From J. Triskett to
Cottans & Hallen, Engineers Royal Mint, Sydney
Specifications for construction of iron gates and railings to be used in construction of new mint in Sydney. Send cost estimates to JH, who will appoint a construction superintendent.
Abs (by JH) 1p TxU:H/L-0713.3; Reel 1089

10178. 1854–8 To Charles Edward Trevelyan Royal Mint
Policy regarding redemption of mutilated silver coins.
AdraftS 2pp TxU:H/L-0484.32; Reel 1055

10179. 1854–8–12 From George Biddell Airy n.p.
Seeks current information about the selection process for the Melbourne University professorships [see GA's 1854–7–15].
ACCS 2pp RGO 6.335.63

10180. 1854–8–14 From George Biddell Airy Royal Observatory, Greenwich
Agrees to meeting date [see JH's 1854–8–14].
ACCS 1p RGO 6.335.66

10181. 1854–8–14 To George Biddell Airy n.p.
Informs GA of the candidates for the Melbourne University professorships; wants to meet with the committee [see GA's 1854–8–12].
ALS 4pp RGO 6.335.64

10182. 1854–8–14 From Alexander Hamilton-Gordon Camp, Gerrickler
Soldiers are suffering in this unhealthy country for want of occupation. Hopes he has been able to lighten his duties at the Mint. Administration seems inefficient in the supply to the troops.
C 2pp RS:HS 8.147

10183. 1854–8–15 To Charles Edward Trevelyan [Royal Mint]
Heavy demand for silver coins by Commissariat Service in Turkey. Mint supply was recently exhausted. Ask Treasury to borrow more from Bank [of England].
Adraft 1p TxU:H/L-0484.33; Reel 1055

10184. 1854–8–16 To 'Sir' Royal Mint
Earliest date for pyx trial will be November, leaving ample time to summon jury. Contact wardens soon. JH approves oath for jurymen, but suggests changes in wording.
AdraftS 4pp TxU:H/L-0703; Reel 1089

10185. 1854–8–16 To George Biddell Airy 32 Harley St.
Some concern about a lack of communication from Melbourne University, especially about money [see JH's 1854–8–14].
ALS 4pp RGO 6.335.67

10186. 1854–8–16 From Henry Warburton 45 Cadogan Place
Cambridge [Philosophical] Society will publish HW's paper on self-repeating series. Suggests JH clarify some equations in new edition of [*Examples of Finite Differences* (1820)].
ALS 4pp RS:HS 18.64

10187. 1854–8–16 To [Mint Staff] Royal Mint
[Formal announcement] Assigns personnel from each department to order and to inspect coal deliveries for coming year.
AdraftS 1p TxU:H/L-0704; Reel 1089

10188. 1854–8–17 From George Biddell Airy Royal Observatory, Greenwich
Agrees that JH should write to the Melbourne University authorities and point out some concerns [see JH's 1854–8–16].
ACCS 1p RGO 6.335.69

10189. 1854–8–17 From Henry Warburton 45 Cadogan Place
Additional suggestions for improving clarity of equations in new edition of JH's [*Examples of Finite Differences* (1820)].
ALS 3pp RS:HS 18.65

10190. 1854–8–18 To George Biddell Airy 32 Harley St.
Further interim arrangements for one of the Melbourne University professors [see GA's 1854–8–18]; comments about timing by electric telegraph.
ALS 2pp RGO 6.335.72

10191. 1854–8–18 From George Biddell Airy Royal Observatory, Greenwich
Rumor has it that Australia is short of cash; can only wait and see [see GA's 1854–8–17].
ACCS 3pp RGO 6.335.70

10192. 1854–8–18 To Margaret Brodie Herschel [London]
Mostly about family health, bills, and the war in the Crimea.
ALS 4pp JHS 1.19

10193. 1854–8–19 From George Biddell Airy Royal Observatory, Greenwich
Results of observation of transits at Greenwich and Paris.
ALS 4pp RS:HS 1.225

10194. 1854–8–19 From George Biddell Airy Royal Observatory, Greenwich
Responds to JH's interim arrangements [see JH's 1854–8–18].
ACCS 1p RGO 6.335.73

10195. 1854–8–19 To [Charles Edward Trevelyan] Royal Mint
Reports assayers's analyses of Chinese imitations of Spanish coin sent to Treasury by John Bowring [governor of Hong Kong].
ACCS 1p & Adraft 2pp TxU:H/L 0598 & 0705; Reel 1089

10196. 1854–8–19 From Henry Warburton 45 Cadogan Place
Additional improvements for new edition of JH's [*Examples of Finite Differences* (1820)].
ALS 4pp RS:HS 18.66

10197. 1854–8–20 To Henry Warburton 32 Harley St.
Discusses points raised about equations [see HW's 1854–8–19].
AdraftS 4pp RS:HS 18.67

10198. 1854–8–22 From Charles Edward Trevelyan Treasury
Inspect enclosed Treasury minutes regarding Robert Mushet, and inform CT if JH has objections. Encloses several papers on Turkish currency.
LS 2pp TxU:H/M–0554.10; Reel 1093

10199. 1854–8–23 To [Charles Edward Trevelyan] Royal Mint
Reports departure of E. W. Ward, with officers, staff, and machinery for Sydney branch mint. Submits statement of expenses and requests reimbursement for outfitting Sydney mint.
AdraftS 3pp TxU:H/L-0706; Reel 1089

10200. 1854–8–24 From L. Caggiati 44 Warwick St, Regent St.
LC, from Parma [Italy], and Baron de Luttteroth wish to visit Royal Mint and meet JH. [JH annotation of 27 Aug.: Ordered Smith, Elder & Co. to give copy of JH's *Cape Results* to L. Caggiati for delivery to 'Prof. [Cobla Dinester?] of [Royal?] Observatory in Parma.]
L 1p (French), JH note RAS:JH Archive 10/6.4; Reel 9

10201. 1854–8–24 To Margaret Brodie Herschel [London]
A collection of household and family matters.
ALS 4pp JHS 1.18

10202. 1854–8–24 From George Gabriel Stokes Pembroke College, Cambridge
Just received JH's report on GS's paper on refrangibility of light. Elaborates on lavender and brown rays.
ALS 15pp TxU:H/M–0504; Reel 1093

10203. 1854–8–24 To Charles Edward Trevelyan Harley St.
Melters' contract officially ended on 7 Mar. [1851], but they continued to profit from Mint until 31 July. Therefore, compensation for Robert Mushet should begin on 31 July. Returns paper on Turkish coins. Mr. Stickney's value for grain of pure gold is accurate.
ALS 3pp TxU:H/L-0484.34; Reel 1055

10204. 1854–8–25 To [Charles Edward Trevelyan] Royal Mint
Submits design for Australian sovereign and half-sovereign for approval by Queen Victoria. Requests permission to begin producing these coins as soon as Sydney mint is operational.
AdraftS 1p TxU:H/L-0710; Reel 1089

10205. 1854–8–25 From Henry Warburton 45 Cadogan Place
Further discussion of need for two constants [see HW's 1854–[5]–11]. Sends first installment of errata HW found in JH's [*Examples of Finite Differences* (1820)]. Realizes JH's duties [at Mint] afford little time to publish new edition. Suggests printing only errata list to first edition.
ALS 7pp RS:HS 18.68

10206. 1854–8–27 To George Biddell Airy n.p.
Arranging with GA for the provision of regular R.A.S. publications to be supplied to a foreign member.
ALS 1p inc RGO 6.234.389

10207. 1854–8–27 To Margaret Brodie Herschel [London]
Very discouraged about the current state of the Crimean War.
ALS 4pp JHS 1.17

10208. 1854–8–27 From Thomas Maclear Royal Observatory, Cape of Good Hope
The Cape Government is to employ a geologist to survey the mineral resources. Copper mines are all the rage at the moment and the value of shares is increasing rapidly. No further news of David Livingstone. Mounting of the transit circle is progressing.
ALS 4pp RS:HS 12.160

10209. 1854–8–27 To George Gabriel Stokes London
Comments on color as a sensation, and on GS's thoughts on 'lavender rays.'
C 2pp RS:HS 25.10.5

10210. 1854–8–28 To James Ross Snowden Royal Mint
Thanks for 1853 report of U.S. mint. Pleased that U.S. assays show improved purity of English gold coins. Seeks information about new coin weighing machine by Baron [A. P.] Seguire.
Adraft 2pp TxU:H/L-0707; Reel 1089

10211. 1854–8–28 To Charles Edward Trevelyan Royal Mint
Letter from Heaton's [Birmingham mint]. Delay in production of copper coins for New Brunswick. Coins ordered by G. Baillie for Ceylon are ready.
AdraftS 1p TxU:H/L-0484.35; Reel 1055

10212. 1854–8–29 To Charles Edward Trevelyan [Royal Mint]
Describes new machinery used in coining process and duties of workmen and clerks.
Adraft 1p TxU:H/L-0484.36.1; Reel 1055

10213. 1854–8–30 To L. Jones Royal Mint
Informs LJ that JH has a copy of the work of Signor Ferrari, who provided it, and JH will place his copy in the Mint library.
ALS 1p BL Aaa 44382.100

10214. 1854–9 To G. Baillie Royal Mint
Delivered £3000 of copper coins for shipment to Ceylon today. Previous invoice for earlier shipment of equal amount of coins has not yet been paid.
AdraftS 1p TxU:H/L-0084; Reel 1054

10215. [1854–9 or later] To [?] [Royal Mint]
Notes sent by W. H. Barton to parties with delinquent bills.
Adraft 1p (on JH's 1854–8–29) TxU:H/L-0484.36.2; Reel 1055

10216. 1854–9–5 From William Ewart Gladstone Broadstairs
Agrees with JH's opinion regarding allowances for those Mint officers mentioned in JH's letter of 1 Sept. Accepts proposals made by JH. Sent curious Italian work on decimal system to JH.
ALS 3pp TxU:H/M–0208; Reel 1087

10217. 1854–9–5 To Matthew Marshall [Royal Mint]
Mint will be ready to receive new gold [from Bank of England] and to begin coining early next month.
AdraftS 1p TxU:H/L-0247; Reel 1054

10218. 1854–9–6 To George Biddell Airy Royal Mint
Is to have a visit from Bank of England representatives to discuss making an adjustment in the weight of a gold sovereign; seeks GA's comments on the proposed changes.
ALS 3pp RGO 6.341.66

10219. 1854–9–7 From George Biddell Airy Royal Observatory, Greenwich
Encourages JH to view the change being proposed [see JH's 1854–9–6] as an improvement.
ACCS 4pp RGO 6.341.68

10220. 1854–9–9 To William Ewart Gladstone [Royal Mint]
Requests salary raise of £10 for assistant assayer.
ACCS 2pp TxU:H/L-0163; Reel 1054

10221. 1854–9–14 From Jean-Bapt[iste] Frener Lucerne
Applies for employment as engraver at the Royal Mint. Summarizes his past experience and provides a seal of his

own design.
ALS 2pp (French) TxU:H/M–0929

10222. 1854–9–19 From George Biddell Airy Royal Observatory, Greenwich
Sends on a letter from George Peacock, who is quite opposed to the proposed changes [see JH's 1854–9–6]; GA would be interested in JH's response.
ACCS 2pp RGO 6.341.70

10223. 1854–9–20 To George Biddell Airy [London]
The proposed change would make little difference to the Royal Mint, but as GA points out, new coins would have an effect on banking transactions [see GA's 1854–9–19].
ALS 4pp RGO 6.341.71

10224. 1854–9–23 From Giovanni Antonio Plana Turin
Apologizes for not sending JH copies of GP's recent writings. Is now beginning a memoir on action–at–a–distance [?], and wonders if the R.A.S. or R.S.L. would be interested.
ALS 3pp (French) RS:HS 13.439

10225. 1854–9–26 From Henry Warburton 45 Cadogan Place
Sends copy of HW's paper, with manuscript listing properties of differences of zero. Is there a published table of such differences like that in JH's [*Examples of Finite Differences* (1820)]?
ALS 4pp RS:HS 18.69

10226. 1854–10–5 From Henry Warburton 45 Cadogan Place
Learned to simplify expression for Bernoulli's number B=2n+1 using JH's formula.
ALS 3pp RS:HS 18.70

10227. 1854–10–7 From William Rutter Dawes Wateringbury, Maidstone
Will be happy to place his diagrams of the sun at JH's disposal should he wish to include them in the new edition of his book. Gives recent observations made by his Clarke object glass.
ALS 4pp RS:HS 6.99

10228. 1854–10–9 To [William Benjamin Carpenter] Collingwood
Thanks for sending 4th edition of WC's Physiology [*Principles of General and Comparative Physiology,* 4th ed., 1854].
ALS 2pp BostonPubLib Ms. Eng. 204

10229. 1854–10–9 From [Henry Warburton] Ore Place [Hastings]
Offers improved version of formulas sent in previous letter.
ALS 3pp inc RS:HS 18.71

10230. 1854–10–14 To 'Sir' Royal Mint
As directed by Treasury commissioners, in [James] Wilson's letter of 30 Sept. 1854, JH sends draft of royal proclamation, approved by Treasury solicitor, announcing new coins for New Brunswick.
Adraft 1p TxU:H/L-0709; Reel 1089

10231. 1854–10–15 To Margaret Brodie Herschel High Elms, Farnborough
About JH's social activities in London.
ALS 4pp JHS 1.20

10232. 1854–10–17 To George Biddell Airy n.p.
Finally concluding the Melbourne University business by asking for GA's expenses; the money finally came from Australia [see GA's 1854–8–18].
ALS 2pp RGO 6.335.74

10233. 1854–10–17 From Charles William Goodwin Die Office, Royal Mint
After 25 years service in Mint, CG is still not eligible for raise. Asks to be considered for position in Die department that will permit increases in salary.
ALS 3pp TxU:H/M–0209; Reel 1087

10234. [1854–10–17 or later] From William Thomas Brande Royal Mint
Advises hiring C. W. Goodwin as resident employee.
ALS 2pp TxU:H/M–0116.1; Reel 1087

10235. 1854–10–18 From George Biddell Airy Royal Observatory, Greenwich
Responds to JH's request [see JH's 1854–10–17].
LACC 2pp inc RGO 6.335.75

10236. 1854–10–18 From George Biddell Airy n.p.
Informs JH that GA (and his assistants) have been making pendulum observations in a coal pit.
ACCS 1p inc RGO 6.212.342

10237. 1854–10–18 From James [Struthers] Stewart Cape of Good Hope
Describes the Cape Botanical Garden, which is finally opened. Congratulates JH on the marriage of JH's daughter [Caroline Emily]. Copper mining in the colony successful. Will move to a new cottage.
ALS 8pp RS:HS 17.13

10238. 1854–10–21 To Isabella Herschel [London]
Strong desire for letters. Misses family. Emptiness of 'civilized' life. Encourages IH's expression of feminine feelings. Concern about Napoleon III, England's national debt, and America's 'vain school-boy-ism.'
ALS 4pp TxU:H/L-0527; Reel 1053

10239. 1854–10–28 To William Whewell 35 Bedford Place, Russell Square
Describes the abilities and special interests of JH's son Alexander, who, JH hopes, will enter Trinity College, Cambridge, in Oct. 1855.
ALS 4pp TC, Camb. Add. Ms.a.207[91] (C: RS:HS 23.149)

10240. 1854–10–29 To James Calder Stewart n.p.
Accepts dinner invitation to meet Mr. [Haufmann?]. Problems with JH's back.
ALS 2pp TxU:H/L-0447; Reel 1055

10241. 1854–10–31 To Margaret Louisa Herschel [London]
A letter of sympathy for MH, as she and her brother Willy have lost a good friend in Edward Wrench [?].
ALS 4pp JHS 1.65

10242. 1854–11 To Bankers Royal Mint
Seeks information on the reaction of the public to the florin or two shilling piece.
L *Preliminary Report of the Decimal Coinage Commissioners* (British Parliamentary Papers 1857, Session 2, vol. 19), pp. 125–6

10243. 1854–11 To John Gellibrand Hubbard Royal Mint
Writes to the Governor of the Bank of England to inquire as to the preference of the public for particular coins, especially the florin [two shilling] piece.
LS 1p DMC.695A.885.17

10244. 1854–11–1 From Augustus De Morgan 7 Camden St.
Sending an equation for JH to solve. Thanks for sending details of the weight of the florin, and the song from Punch. If he wishes JH may borrow books from the College library.
ALS 4pp RS:HS 6.268

10245. 1854–11–2 To Augustus De Morgan 35 Bedford Place
Sends Henry Warburton's equation and integral; thanks AD for humorous story sent to Collingwood.
C 2pp RS:HS 23.150

10246. 1854–11–2 To William Ewart Gladstone Royal Mint
Describes current state of coinage of silver coins.
ALS 4pp BL Aaa 44382.252

10247. 1854–11–3 From Augustus De Morgan 7 Camden St.
Sending one of his equations for JH's comments.
ALS 2pp RS:HS 6.269

10248. 1854–11–6 To Augustus De Morgan Royal Mint
Accepts AD's correction regarding Henry Warburton's writings [see JH's 1854–11–2].
C 1p RS:HS 23.151

10249. 1854–11–7 To William Ewart Gladstone Royal Mint
Suggests a circular to the banks in Great Britain to get the information that WG wants.
ALS 2pp BL Aaa 44382.262

10250. 1854–11–9 From George Biddell Airy Royal Observatory, Greenwich
Inviting JH, together with other members of the R.A.S., for dinner on the following day.
ACCS 1p RGO 6.234.475

10251. 1854–11–9 From Annibal de Gasparis Naples
Sending his theory of the orbits of comets. Comments on this.
ALS 3pp (French) RS:HS 8.47

10252. 1854–11–10 To Leonard Charles Wyon Royal Mint
New South Wales Government wishes to have medals struck. JH has recommended LW. Is LW interested? LW must not, however, use Mint machinery in this process.
ALS 4pp New South Wales State Library

10253. 1854–11–15 From Henry Warburton 45 Cadogan Place
Happy that JH may be leaving Mint and returning to scientific pursuits. HW resumed work on equations in finite differences. Hopes to reconcile discrepancy between JH's and P. S. Laplace's solutions. JH's solution renders only particular, not general, solutions.
ALS 4pp RS:HS 18.72

10254. 1854–11–17 From Henry Warburton 45 Cadogan Place
Approves of JH sharing HW's findings. HW on excellent terms with Augustus De Morgan. Suspects two integrals exist, one real and one imaginary, for P. S. Laplace's equation.
ALS 4pp RS:HS 18.73

10255. 1854–11–21 To William Henry Barton Royal Mint
Recent gold from Bank of England exhibits unusual properties in melting. Add 0.0005 extra alloy to each ingot.
ACCS 1p & C TxU:H/L-0711; Reel 1089

10256. 1854–11–25 From C. Bulstrode 2 Park Place, Regent's Park
Send inventory of [furnishings at] 32 Harley Street. New occupant is waiting to move in.
ALS 1p TxU:H/L-0083.1; Reel 1054

10257. 1854–11–27 To William Ewart Gladstone Royal Mint
Deals with the question of standard weights in relation to testing the accuracy of the weights of coins.
ALS 2pp BL Aaa 44382.324

10258. 1854–11–28 To J. H. Jennings Royal Mint
JJ's letter to Chancellor of Exchequer [W. E. Gladstone] regarding JJ's discovery of new alloy that makes gold strong as steel was forwarded to JH. Send samples to JH.
ACCS 1p & AdraftS TxU:H/L-0204 & -0203; Reel 1054

10259. [1854–12 to 1855–4] To [Edward Wolstenholme Ward] Royal Mint
Forwards copies of Treasury minutes and duplicate original authorizing Sydney branch mint to begin circulating Australian sovereigns and half sovereigns. Please advise JH of dies needed for 1856. [JH annotation to printer concerning arrangements of documents.]
Adraft 3pp †(by JH) TxU:H/L-0674; Reel 1089

10260. 1854–12–3 To Margaret Brodie Herschel [London]
Discusses rumors about the war on the continent, news of some visits JH had, and JH's dream life.
ALS 4pp JHS 1.10

10261. 1854–12–8 To Margaret Louisa Herschel [London]
Tells his daughter about a formal occasion for weighing gold in the presence of various officials and guilds.
ALS 4pp JHS 1.9

10262. 1854–12–10 From Henry Warburton 45 Cadogan Place
Testing recent work on equations analogous to P. S. Laplace's but of higher order. Solution so simple that HW expects to find it published elsewhere. Corrects error in previous letter regarding generality of arbitrary constants.
ALS 4pp RS:HS 18.74

10263. 1854–12–11 From Henry Warburton 45 Cadogan Place
Demonstrates easier way to treat equations analogous to P. S. Laplace's.
ALS 3pp RS:HS 18.75

10264. 1854–12–12 From Henry Warburton 45 Cadogan Place
Please correct error in HW's earlier calculations.
ALS 1p RS:HS 18.76

10265. 1854–12–13 From Charles Babbage Dorset
Sends JH a copy of a recent issue of the *Athenaeum* in which falsehoods, copied from a Richard Sheepshanks pamphlet, have been printed. CB wants JH to respond to, and correct, these falsehoods quickly.
ALS 3pp RS:HS B27.19

10266. 1854–12–15 To Charles Babbage Royal Mint
Refuses to become a partisan in this conflict [see CB's 1854–

12–13]. JH would only become involved if he felt he could be a peacemaker.
(A)CC(S) 2pp RS:HS B27.20

10267. 1854–12–15 From Richard Bentley New Burlington St.
Asks JH for the manuscript of Lieut. Burton's travels, which was forwarded by way of JH.
ALS 2pp RS:HS B27.80

10268. 1854–12–16 From George Biddell Airy n.p.
Appear to be answers to specific queries about examinations and tutors [at Cambridge?].
ACCS 1p RGO 6.375.308

10269. 1854–12–18 To Richard Bentley 35 Bedford Place
In response to RB's letter of 1854–12–15, JH states that the manuscript of Lieut. Burton's travels was not given to JH for transmission, JH has never heard of RB in connection with this manuscript, nor had JH received an earlier letter, and finally JH does not have the manuscript.
ACCS 1p RS:HS B27.81

10270. 1854–12–18 To Margaret Brodie Herschel [London]
Talks about their son Alexander's nature, and JH's busy weekend and headaches.
ALS 4pp JHS 1.7

10271. 1854–12–20 To [Charles Edward Trevelyan] [Royal Mint]
Mint salaries and bills have exhausted present funds. Please ask Treasury for another £12,000.
AdraftS 1p TxU:H/L-0712; Reel 1089

10272. 1854–12–21 To Cottans & Hallen, Engineers Royal Mint
Forwards letter from J. Triskett. Send to JH engineering design and cost estimate for iron gates and railings to be used for security at new mint building in Sydney. Then deliver materials to W. O. Young for shipment to Australia.
ACCS 1p TxU:H/L-0713.1; Reel 1089

10273. 1854–12–21 To Henry Drury Harness Royal Mint
Results of recent pyx trial.
ACCS 2pp TxU:H/L-0185.5; Reel 1054

10274. 1854–12–23 From William Rutter Dawes Wateringbury, Maidstone
Would like his opinion on a divided glass, double image micrometer before he submits it to an optician. George Dollond has lent him some divided glasses of long focus. Would like to borrow G. B. Amici's micrometer if JH could spare it.
ALS 4pp RS:HS 6.100

10275. 1854–12–29 From William Rutter Dawes Wateringbury, Maidstone
Is much obliged for the loan of G. B. Amici's micrometer. Comments on some of his readings with it. Is sorry to hear of the ill-health of Lady Herschel.
ALS 4pp RS:HS 6.101

10276. [1855?] To William Whewell Trinity Lodge
Urges WW to write to 'S' [Adam Sedgwick?]. Only in this way can there be a restoration of good relations after the quarrel between S. and WW.
ALS 2pp TC, Camb. Add.Ms.a. 216[34]

10277. 1855–1–2 From William Rutter Dawes Wateringbury, Maidstone
Is grateful for the gift of the instrument and hopes to put it to good use. Will report on it when he has mastered its use. Hopes the country air is improving JH's health.
ALS 3pp RS:HS 6.102

10278. 1855–1–7 From Henry Warburton Ore Place [Hastings]
Corrects calculations relating to arbitrary constants in HW's previous letters. Encourages JH to avail himself of this information in second edition [of *Examples of Finite Differences* (1820)] .
ALS 7pp RS:HS 18.77

10279. 1855–1–8 To Margaret Brodie Herschel [London]
Mostly about the health of MH and several of the children, and about JH's health.
ALS 4pp JHS 1.5

10280. 1855–1–8 To William Parsons (Lord Rosse) 35 Bedford Place
Would WP accept the presidency of the Photographic Society? Describes that society.
ALS 4pp Rosse Papers K2.5

10281. 1855–1–8 To Charles Edward Trevelyan n.p.
Submits estimates of silver losses expected by Mint from Apr. 1855 to Mar. 1856.
AdraftS 2pp TxU:H/L-0485.1; Reel 1055

10282. 1855–1–9 From Henry Warburton Ore Place [Hastings]
Relative to previous letter [see HW's 1855–1–7], HW found *two* possible equations, due to ambiguity of signs.
ALS 1p RS:HS 18.78

10283. 1855–1–11 To Charles [Edward Trevelyan] Royal Mint
Fears wounding feelings of Benjamin Wyon if JH is seen as ordering dies for Crimean medal from Wyon's relative and rival. Sends official letter that relieves JH of responsibility for choosing artist for medal.
ACCS 1p TxU:H/L-0124; Reel 1054

10284. 1855–1–12 To Charles Edward Trevelyan Royal Mint
Delivery of dies for Crimean medal to Master of Mint. JH ordered 60,000 ounces of silver for its production.
ACCS 4pp TxU:H/L-0485.2; Reel 1055

10285. 1855–1–13 To Ralph Heaton & Sons Royal Mint
Directions for shipping copper coins to Bank of Ireland, in same quantities stated in JH's directive of 11 Dec. 1854 to RH.
ACC 1p TxU:H/L-0191; Reel 1054

10286. 1855–1–13 To Mrs. [Richard] Jones n.p.
Regrets that JH's health does not allow him to visit his 'dear suffering friend' [Richard Jones], who JH hopes may yet rally.
C 1p RS:HS 23.153

10287. 1855–1–15 From Henry Roberts to Charles Edward Trevelyan War Department
Explains mistake that made JH angry. War Department presumed to commission medal of Queen without consulting Master of Mint first. Benjamin Wyon has been instructed to cease work until JH is informed and approves of project.
ALS 3pp TxU:H/M-1040; Reel 1083

10288. 1855–1–16 From Charles Edward Trevelyan Treasury

[Marked 'Private.'] Encloses letter from Henry Roberts, assistant secretary to War Department. Will convey JH's reply to Roberts.
ALS 2pp TxU:H/M–0555.1; Reel 1093

10289. 1855–1–16 To Charles Edward Trevelyan Royal Mint
Any of JH's letters to CT may be shown to [Henry] Roberts.
ACCS 1p TxU:H/L-0485.4; Reel 1055

10290. 1855–1–16 To Charles Edward Trevelyan Royal Mint
[Marked 'Private.'] Did not wish to offend [Henry] Roberts, but JH is in awkward position regarding L. C. Wyon, who, as Mint employee, is competing with rival artist for design of Crimean medal. Silver supply is adequate for medal production, but machinery is limited. Expects three-month delay.
ACCS 3pp TxU:H/L-0485.5; Reel 1055

10291. 1855–1–18 To Benjamin Hawes Royal Mint
Dies for [?] medal will be ready next week. How many medals should be struck?
Adraft 2pp TxU:H/L-0187; Reel 1054

10292. 1855–1–23 To George Hamilton-Gordon Royal Mint
Deterioration of JH's health. Requests permission to retire as master of Mint. Recommends W. H. Barton to succeed JH.
ACCS 3pp & AdraftS 3pp TxU:H/L-0007 & -0008; Reel 1054

10293. 1855–1–23 To William Ewart Gladstone Royal Mint
In reply to WG's 8 Dec. 1854 note, JH analyzes all Mint operations and labor expenses, and reports favorable balance. Advises against Treasury's plan to have money minted by private contractors.
AdraftS 6pp TxU:H/L-0164; Reel 1054

10294. 1855–1–23 To William Ewart Gladstone Royal Mint
JH wishes to resign for health reasons. Has already informed Lord Aberdeen [George Hamilton-Gordon]. Recommends W. H. Barton to succeed JH as Master of Mint.
ACCS 3pp & AdraftS TxU:H/L-0166 & -0165; Reel 1054

10295. 1855–1–24 From Caroline Emily Mary Herschel Argyll House
Arthur [Gordon?] visited to reveal that Lord John Russell just resigned, expected others will follow, and urged JH to follow their example and 'give up the Mint.' Lord Aberdeen [G. Hamilton-Gordon] wants to give newly created position of 'President of the Board of Examiners' to JH as a 'parting gift.' Arthur assured CH that JH could perform all its duties from Collingwood. Relates 'abuse' CH overheard in House of Commons yesterday.
ALS 3pp TxU:H/M–0791; Reel 1084

10296. 1855–1–24 To Caroline Emily Mary Herschel [London]
Sent JH's resignation proposal to both Lord Aberdeen [G. Hamilton-Gordon] and W. E. Gladstone yesterday. JH would have welcomed presidency of Board of Examiners one year earlier, but now desires only rest and recovery. Feels 'broken down.'
ALS 4pp TxU:H/L-0525; Reel 1053

10297. 1855–1–25 To Peter le Neve Foster Royal Mint
Delcines invitation to chair Society of Arts meeting because of poor health; may attend as visitor.
ALS 2pp St. John's College, Cambridge

10298. 1855–1–27 To J[ames] Heaviside n.p.
Thanks for having been informed of the death of an acquaintance.
C 1p RS:HS 23.154

10299. 1855–1–28 From A. B. Orlebar Bombay Magnetic Observatory
Has been nominated Professor of Maths at the Native College and is in charge of the Observatory, now being erected. Would welcome suggestions from JH as to observations that would be useful.
ALS 2pp RS:HS 13.183

10300. 1855–1–30 To George Gabriel Stokes 35 Bedford Place
Advises GS against spending £800 to verify the available drawings of Saturn; instead JH suggests alternative methods.
C 1p RS:HS 25.10.6

10301. 1855–1–31 To Caroline Emily Mary Herschel [London]
Recalls asking [Richard] Jones years ago if JH should seek public office. Jones was horrified, but suggested mastership of Mint as only suitable position. After four years in office, JH longs to escape it and return to family.
ALS 3pp TxU:H/L-0526; Reel 1053

10302. 1855–1–31 From [?] n.p.
Lord Aberdeen [G. Hamilton-Gordon] must first refer JH's resignation to Queen Victoria. [?] encouraged Aberdeen to do it quickly, while Aberdeen is still in office.
AL 1p † TxU:H/L-0715; Reel 1089

10303. [1855–2] From Caroline Emily Mary Herschel n.p.
Relates CH's and Arthur [Gordon?]'s conversation yesterday with Lord Aberdeen [George Hamilton-Gordon] on how to enable JH to resign without public 'mess.' Three weeks ago, Aberdeen would have accepted JH's resignation and appointed William Henry Barton as Master, but now Aberdeen's successor must decide.
ALS 3pp TxU:H/M–0792.1; Reel 1084

10304. 1855–2 To Isabella Herschel [London]
Writes to answer a question about the expansion of water on freezing.
AL 2pp JHS 1.43

10305. 1855–2–2 To George Biddell Airy Royal Mint
Having been asked for copies of standards, it seems to JH that the Royal Mint is not the office to decide on such requests.
ALS 3pp RGO 6.341.73

10306. 1855–2–3 From George Biddell Airy Royal Observatory, Greenwich
The requests should be made to the Exchequer office [see JH's 1855–2–2]; complains about the incompetence of the Chancellor of the Exchequer [W. E. Gladstone].
ACCS 2pp RGO 6.341.75

10307. 1855–2–6 To William Sharpey 35 Bedford Place
Discusses proposal to fix up the aerial telescope of Christiaan Huygens so as to compare his drawings of Saturn with observations made by contemporary observers using Huygens's telescope. Expresses skepticism about the project.
ALS 4pp RS MM 19.113

10308. 1855–2–9 From George Hamilton-Gordon Downing St.
Received JH's letter of resignation during ministerial crisis but elected not to present it to Queen Victoria. Send it to GG's successor [H. J. Temple], together with testimony to merits of W. H. Barton as JH's successor. [JH's daughter] Caroline continues well.
ALS 2pp TxU:H/M–0002; Reel 1062

10309. 1855–2–11 To George Hamilton-Gordon 35 Bedford Place
Chancellor of Exchequer [W. E. Gladstone] will arrange with Lord Palmerston [H. J. Temple] for JH's retirement as Master of Mint. Thanks for GG's support against 'external influences' during JH's term.
AdraftS 4pp TxU:H/L-0009; Reel 1054

10310. 1855–2–11 To George Hamilton-Gordon 35 Bedford Place
Writes to inform GG that JH has seen the Chancellor of the Exchequer and tendered his resignation as Master of the Mint.
ALS 3pp BL Aaa 43255.100

10311. 1855–2–12 From Charles Edward Trevelyan Treasury
Call on CT when convenient.
LS 1p † TxU:H/M–0555.2; Reel 1093

10312. 1855–2–13 To Charles Edward Trevelyan Royal Mint
Estimates expenses and problems in production of medals for Kaffir war veterans. Asks for authority to determine wages for workmen employed on Kaffir and Crimean medals.
Adraft 2pp TxU:H/L-0485.6; Reel 1055

10313. 1855–2–15 To [James Calder] Stewart n.p.
Returns 'table.' Thank grandma for loan of it.
ALS 1p TxU:H/L-0448; Reel 1055

10314. 1855–2–21 To William Thomas Brande Royal Mint
Urgency for delivery of gold coins is reduced. Begin rolling and cutting silver bars.
AdraftS 2pp TxU:H/L-0109.1; Reel 1054

10315. 1855–2–21 From Frederick Robert Brande and Henry Finch Royal Mint
[Copy in JH's hand of memorial to Treasury from F. R. Brande and Henry Finch. JH marked this copy 'Strictly private....'] Recount improvements that FB and HF have made in Mint operations, but complain that they do not receive the consideration, encouragement, or salary that is due them. [After reading this memorial, JH opened his finished letter (also 21 Feb 1855) to Brande and added a P.S. that led Brande and Finch to withdraw this memorial.]
C 3pp, JH note 1p TxU:H/M–0920; Reel 1083

10316. 1855–2–22 To Charles Edward Trevelyan [Royal Mint]
Forwards to Treasury, *without* JH's approval, request by F. R. Brande and Henry Finch, junior clerks in Mint office, for higher salary. [JH note: Brande disclaimed knowledge of this request. Brande and Finch retracted it, but wanted to send another. JH refused.]
ACCS 4pp, JH note 1p TxU:H/L-0485.7; Reel 1055

10317. 1855–2–23 From Thomas Maclear Royal Observatory, Cape of Good Hope
Regrets to hear of the ill health of JH. Believes that a sea voyage to the Cape would work wonders. The transit circle is in full working order. Comments on some of the work connected with it. Has heard that an accident had occurred to William Parsons' (3rd Earl of Rosse) 6-foot speculum. No news of David Livingstone.
ALS 4pp RS:HS 12.161

10318. 1855–2–25 From Thomas Graham 4 [Gordon Square]
TG volunteers to succeed JH as master of Mint if JH resigns. Willing to replace JH in summers when TG's college duties are suspended. [JH annotation to wife Margaret: '...there are greater fools in the world' than JH.]
ALS 3pp † TxU:H/M–0218.1; Reel 1087

10319. 1855–2–25 To George Gabriel Stokes 35 Bedford Place
Declines to serve on experimental committee for R.S.L.
ALS 2pp Cambridge University Library 7656/H694

10320. 1855–2–26 From James J. Anderson n.p.
Thanks for JH's letter to Mr. McCulloch on JA's behalf. Hopes for permanent appointment.
ALS 2pp TxU:H/M–0005; Reel 1062

10321. 1855–2–27 To Thomas Graham n.p.
Has no power to determine JH's successor to mastership of Mint. Make application to Lord Palmerston [H. J. Temple].
ACCS 1p TxU:H/M–0218.2; Reel 1087

10322. 1855–2–28 To George Gabriel Stokes n.p.
Agrees to help in other ways than as committee member [see JH's 1855–2–25].
ALS 2pp Cambridge University Library 7656/H695

10323. [1855–3] To 'My Lord' [Royal Mint]
Reports status of Crimean medal and Cape medal [Kaffir medal?]. Benjamin Wyon expects to have models ready soon.
Adraft 1p TxU:H/M–0946.2; Reel 1083

10324. 1855–3–2 To [?] Royal Mint
Extraordinary demand for gold and silver coin in 1852–54 has subsided. Reports disposition of twelve Custom House employees who were assigned to Mint as temporary clerks for this period.
Adraft 3pp TxU:H/L-0717; Reel 1089

10325. 1855–3–6 To William Thomas Brande Bedford Place
Reminds WB that Mint rules do not permit paying skilled laborer's wages to casual employees such as Patrick Shanahan.
AdraftS 3pp TxU:H/L-0109.2; Reel 1054

10326. 1855–3–6 From Leonard Charles Wyon 22 Blomfield Road [Maida Hill]
Will visit JH tomorrow at Bedford Place to submit LW's wax model for Kaffir medal.
ALS 2pp TxU:H/M–0606; Reel 1086

10327. 1855–3–7 To William Henry Barton 35 Bedford Place
Expects difficulty with wages listed for Patrick Shanahan and one other worker at Mint. Find documents that authorized JH to pay additional overtime wages during coinage pressure of last two years.
ACCS 2pp TxU:H/M–0512.2; Reel 1093

10328. 1855–3–7 From William Henry Barton Royal Mint
Encloses two Mint documents relating to wages for Patrick

Shanahan and another worker. JH failed to return papers on *Hannibal* claim.
ALS 2pp TxU:H/M–0044.2; Reel 1062

10329. 1855–3–7 From William Henry Barton Royal Mint
Delayed paying [W. O.] Young until JH approves papers on *Hannibal* claim [for ship's salvage].
ALS 2pp TxU:H/M–0044.1; Reel 1062

10330. [1855–3–7] From William Thomas Brande Royal Mint
Gives former reason for retaining Charles Stiles, who may now be dismissed.
ALS 4pp TxU:H/M–0116.2; Reel 1087

10331. 1855–3–7 From William Thomas Brande Royal Mint
Wages and overtime pay for Patrick Shanahan.
ALS 4pp TxU:H/M–0122; Reel 1087

10332. 1855–3–7 From Alexander Shaw to Henry John Temple n.p.
Copy [in JH's hand] of doctor's prescription for four to five weeks of bed rest and country air for JH before resuming duties at Mint.
C 1p TxU:H/M–1047; Reel 1083

10333. 1855–3–7 To George Gabriel Stokes 35 Bedford Place
Asks GS about ordering instruments from [George] Wilson.
C 1p RS:HS 25.10.7

10334. 1855–3–7 From Robert Francis Suft Royal Mint
Sends audit form with covering letter for JH's signature.
ALS 1p TxU:H/M–0512.1; Reel 1093

10335. 1855–3–7 To Henry John Temple 35 Bedford Place, Russell Square
JH's health is worse. Please expedite JH's request to resign from mastership of Mint. Seeks HT's approval to turn JH's duties over to W. H. Barton.
ALS 3pp TxU:H/L-0294; Reel 1054

10336. [1855–3–8] To [William Henry Barton] [32 Harley St.]
Procedure for E. W. Ward to recover expenses for construction of Sydney branch mint. Interview [W. O.] Young for particulars on shipments of materials. Discusses justification of Patrick Shanahan's wages. L. C. Wyon submitted wax model of Kaffir medal, but JH recalls that Treasury asked for drawing.
ACC 4pp TxU:H/L-0629; Reel 1089

10337. 1855–3–8 From William Thomas Brande Royal Mint
Notified Patrick Shanahan that his pay requires 12 full hours attendance. Gave Charles Stiles one week notice of termination. Forwarded 'Parisian Memorandum to Mr. Napier.'
ALS 2pp TxU:H/M–0123; Reel 1087

10338. 1855–3–8 To [William Thomas Brande] n.p.
Agrees to pay Patrick Shanahan five shillings per day, but JH believes this to be 'very high wages' for unskilled labor and requires twelve hours attendance each day. Leonard Wyon had prepared wax model of Kaffir medal. Dismiss Charles Stiles.
Adraft 3pp TxU:H/L-0718; Reel 1089

10339. 1855–3–9 From Frederick McCoy University of Melbourne
Asking JH to pay the bill for the books FM took to Australia with him.
ALS 2pp RGO 6.336.47

10340. 1855–3–10 From William Henry Barton Royal Mint
Sends letter from Mr. Roberts, memo on silver, and other papers. R. F. Suft ordered boxes to be made.
ALS 2pp TxU:H/M–0045.2; Reel 1087

10341. 1855–3–10 From William Henry Barton Royal Mint
In JH's report on state of Crimean Medal to Lord Panmure [Fox Maule], be aware that Benjamin Wyon may not have working dies ready when promised.
ALS 2pp TxU:H/M–0045.1; Reel 1087

10342. 1855–3–10 To George Gabriel Stokes n.p.
Inquiring about the cost of instruments being shipped.
C 1p RS:HS 25.10.8

10343. 1855–3–10 To Charles Edward Trevelyan 35 Bedford Place
Unable to perform Mint duties. Doctor directed JH to country for four to five weeks of complete rest. JH notified Lord Palmerston [H. J. Temple] and asked to be relieved of Mastership.
ACCS 2pp TxU:H/L-0485.8; Reel 1055

10344. 1855–3–10 From Charles Edward Trevelyan Treasury
Appointment of JH's successor is delayed by changes in Cabinet. CT spoke to George Lewis, who promised to speak to Lord Palmerston [H. J. Temple]. Will forward JH's latest note as reminder to Lewis. JH is free to leave London; W. H. Barton is competent to handle JH's duties.
LS 3pp TxU:H/M–0555.3; Reel 1093

10345. 1855–3–12 From William Henry Barton Royal Mint
Policy on adding alloy to gold coins. Reviews JH's instructions for WB during JH's absence. Hopes JH will heal. Should WB hunt up G. Baillie again regarding Ceylon coinage?
ALS 3pp TxU:H/M–0046; Reel 1087

10346. 1855–3–12 From William Thomas Brande Royal Mint
Will produce 6,000 Crimean medals each week, as soon as Benjamin Wyon provides reverse dies.
ALS 3pp TxU:H/M–0124; Reel 1087

10347. [1855–3–13] To [William Henry Barton] [Royal Mint]
Mint debt is too great. If G. Baillie cannot liquidate soon, inform Treasury about Mint's shortage of cash. Instructions regarding reports and duties that must be completed before end of March.
ACC 4pp inc TxU:H/L-0604; Reel 1089

10348. 1855–3–13 From William Law Downing St.
Directs JH to call on Lord Palmerston [H. J. Temple] tomorrow.
ALS 2pp TxU:H/M–0327; Reel 1087

10349. 1855–3–14 From Henry W. Field [Royal Mint]
Bank of England will begin importing gold again next week. Asks permission to be relieved of attending the pyx, in order to assay gold ingots. W. H. Barton will still have assistance

of E. L. J. Ridsdale. Notes £600 savings made by HF in Assay department.
ALS 3pp TxU:H/M–0192; Reel 1087

10350. 1855–3–16 From William Henry Barton Royal Mint
E. L. J. Ridsdale is qualified for Assay office and will return on probation. Production [of Crimean medal]. Distribution of copper coins. Hopes JH's health is improving.
ALS 6pp TxU:H/M–0047; Reel 1087

10351. 1855–3–17 From William Henry Barton Royal Mint
New details of gold shipment to Australia aboard vessel *Hannibal*. Will order more copper from Hays & Son for use as gold alloy at Sydney branch mint.
ALS 3pp TxU:H/M–0048; Reel 1087

10352. 1855–3–17 To [William Henry Barton] Collingwood
Approves delegating E. L. J. Ridsdale to represent [WB] in 'attending the Pyx' during assay process. Does not approve of transferring Ridsdale before his probationary year as assayer is completed.
ACCS 4pp TxU:H/L–0720; Reel 1089

10353. 1855–3–18 From William Thomas Brande Royal Mint
Estimates cost of producing a single [Crimean] medal at 35 shillings. Die Department now working 13 hours daily on medal.
ALS 3pp TxU:H/M–0125; Reel 1087

10354. 1855–3–18 From Augustus De Morgan 7 Camden St.
Hopes to hear how he is progressing. Looks as though he wanted rest and air. Sends a theorem for his consideration.
ALS 4pp RS:HS 6.270

10355. 1855–3–21 From Leonard Charles Wyon 22 Blomfield Road [Maida Hill]
Must know soon if LW's wax model for Kaffir medal has been approved.
ALS 2pp TxU:H/M–0605; Reel 1086

10356. 1855–3–23 To George Biddell Airy [Collingwood]
JH is very ill; would GA please deal with the disbursement of funds from Melbourne University [see JH's 1854–10–17].
L 3pp RGO 6.336.35

10357. 1855–3–25 To 'Dear Sir' Collingwood
[?]'s model was forwarded. [W. H.] Barton will communicate results back to [?].
C 1p TxU:H/L–0716; Reel 1089

10358. 1855–3–25 To John Wrottesley from Margaret Brodie Herschel Collingwood
Writing on JH's behalf, MH begs JH be excused from writing a memoir of C. F. Gauss, as JH is very ill.
C 1p RS:HS 25.12.1

10359. 1855–3–26 From Henry W. Field Royal Mint
Concern for careless work of E. L. J. Ridsdale, who is still on probation as assayer. Explains precedent for letting HF's assistant attend pyx trial.
ALS 3pp TxU:H/M–0193; Reel 1087

10360. 1855–3–29 From William Henry Barton Royal Mint
Sudden appointment of Thomas Graham to succeed JH as Master of Mint. Told C. E. Trevelyan that WB should have been considered. Details of gold shipment to Australia aboard *Hannibal*. Conflict with Lord Panmure [Fox Maule] over Crimean Medal. Report that Queen has approved Kaffir medal ribbon, but War department lost Benjamin Wyon's wax model for medal.
ALS 8pp TxU:H/M–0049; Reel 1087

10361. 1855–3–29 To Henry W. Field Collingwood
HF's proposal for new position of assistant resident assayer deviates from procedures for reorganized Mint by avoiding usual course of promotion. Use services of E. L. J. Ridsdale instead. If Ridsdale is incompetent, he must be notified that his probation year is unsatisfactory.
C 1p TxU:H/L–0151; Reel 1054

10362. [1855–3–30] To [William Henry Barton] [Royal Mint]
JH is paralyzed and in pain. Appreciates [WB]'s efforts to keep [Mint] running smoothly. Comments on JH's successor. Decision to contract out Crimean medal simplifies Mint's responsibilities. Review Mint structure and systems. Invite bids for Kaffir medal. Disappearance of L. C. Wyon's wax model of Kaffir medal.
ACC 2pp TxU:H/L–0611; Reel 1089

10363. 1855–3–31 From William Henry Barton Royal Mint
Mint business. Should WB continue to intervene in hunt for Benjamin Wyon's model [of Kaffir medal lost at War Department]? Still no official letter of Queen's approval for Kaffir Medal.
ALS 4pp TxU:H/M–0050; Reel 1087

10364. 1855–3–31 To Henry John Temple Collingwood
Approves HT's choice of H. D. Harness to succeed JH as Master of Mint. Characteristics for which HT should look in selecting JH's successor if Harness is not available.
(A)draft 2pp TxU:H/L–0295; Reel 1054

10365. 1855–4–4 From William Henry Barton Royal Mint
Sends Mint accounts of last three months for JH's approval. Mr. Ramsay [War Department] still silent on disappearance of Benjamin Wyon's model for Kaffir Medal. Will ask Lord Panmure [Fox Maule] for permission to have Wyon make medal from new model *not* approved by Queen. Has not yet seen JH's successor, Thomas Graham.
ALS 4pp TxU:H/M–0051; Reel 1087

10366. 1855–4–5 To William Henry Barton Collingwood
Signs and returns Mint accounts. Notes that two are missing.
C 1p (on C. Bulstrode's 1854–11–25) TxU:H/L–0083.2; Reel 1054

10367. 1855–4–10 From James Hill Brightling
Hill introduces courier Thomas Adams, who will pick up JH's parcels. Hopes JH's health is improving. [On verso, signed at Collingwood] Adams's receipt for parcels and one letter, all addressed to officials at Royal Mint.
ALS 1p, Thomas Adams note 1p TxU:H/M–0937; Reel 1083

10368. 1855–4–13 To George Biddell Airy Collingwood
A letter of information, together with a list of all material, sent with the material, to GA [see JH's 1855–4–13].
LCC 5pp RGO 6.336.41

10369. 1855–4–13 To George Biddell Airy [Collingwood]
A note of information about the impending forwarding of all

Melbourne University materials to GA [see JH's 1855–3–23].
L 3pp RGO 6.336.39

10370. 1855–4–14 From George Biddell Airy Royal Observatory, Greenwich
Acknowledges receipt of a parcel of material [see JH's 1855–4–13].
ACCS 1p RGO 6.336.44

10371. [1855]–4–15 From George Peacock Deanery of Ely
Upset by Cambridge Bill now before Parliament. It contains changes contrary to report from [Cambridge University] Commission. W. H. Bateson will write to John Romilly, and GP writes to JH, both seeking letters to Lord Chancellor explaining this problem.
ALS 7pp TxU:H/M–0412; Reel 1087

10372. 1855–4–16 From Edward Joseph Lowe Beeston
Henry Lawson has presented all his meteorological instruments and his transit instrument to this Observatory. Hopes to study the relation of the atmosphere to the prevalence of disease. Wonders if the Government or some society would be willing to finance assistant in the observatory.
ALS 4pp RS:HS 11.328

10373. 1855–4–16 From Lyon Playfair Department of Science and Art, Marlborough House, Pall Mall
Discusses possibility of JH taking position of a judge for Britain at a scientific exposition in France.
ALS 4pp RS:HS 14.170

10374. 1855–4–17 To George Peacock Collingwood
Comments on the action of the Council of the Senate of Cambridge University.
draft 1p RS:HS 13.306 & 23.155

10375. 1855–4–19 From Elias Loomis New York University
Has recently published a volume on *Practical Astronomy* and dispatched a copy via Messrs. Trubner for JH. Believes it fills a need as there has been no suitable work available for beginners.
ALS 2pp RS:HS 11.308

10376. 1855–4–19 From George Peacock Deanery of Ely
JH's views coincide with GP's views about constitution of [Cambridge] University. Adam Sedgwick, GP, and John Romilly will compose joint letter to Lord Palmerston [H. J. Temple], pointing out that present parliamentary bill [to make new statutes for Cambridge] is totally opposed to recommendations [of Cambridge University Commission] and to liberal constitution granted to Oxford University.
ALS 3pp TxU:H/M–0421; Reel 1087

10377. [1855]–4–20 From George Peacock Athenaeum
Will send copies tomorrow to JH and to Bishop of Chester [John Graham] of letter that GP, W. H. Bateson, and Adam Sedgwick wrote to Lord Palmerston [H. J. Temple], after consulting with John Romilly. Please sign and return it quickly, so it can be presented to [House of] Lords before Tuesday. Encloses copy of bill.
ALS 3pp TxU:H/M–0413; Reel 1087

10378. 1855–4–21 From William Henry Bateson London
Sends copy of letter by Dean of Ely [George Peacock], Adam Sedgwick, and 'Master of the Rolls' [John Romilly] to be sent to Lord Palmerston [H. J. Temple] and to Lord Chancellor. Please sign and return it.
ALS 2pp TxU:H/M–0057.26; Reel 1087

10379. 1855–4–24 To Edward Sabine n.p.
Recommends use of photography to prepare records of sunspot activity. Makes suggestions on how this could be done most effectively.
L *R.A.S.M.N., 15* (1855), 158–9

10380. 1855–4–27 From Sir Robert Harry Inglis 7 Bedford Square
Hastens to acknowledge his letter and regrets the Literary Society will not be able to count JH among its members.
ALS 2pp RS:HS 10.254

10381. 1855–5–1 From William Henry Barton Royal Mint
Final pay approved for JH. Thomas Graham has not yet received full authority [as JH's successor]. Production of Crimean Medal. No success in finding Benjamin Wyon's model of Kaffir Medal in War Department. Encloses reply from Philadelphia Mint regarding new weighing machine. Shipment of copper to Sydney branch mint. Cost of producing Kaffir ribbon. Wyon's dies are breaking.
ALS 8pp TxU:H/M–0052; Reel 1087

10382. 1855–5–6 From Alexander von Humboldt Berlin
Sends a letter of praise for the family Herschel, including especially JH and his father, William.
ALS 3pp (German) RS:HS 10.74

10383. 1855–5–8 From George Peacock 58 Trumpington St., {Cambridge]
Told John Romilly that JH cannot attend meeting to review amendments to Cambridge [University] bill. Large protest meeting by university residents yesterday. Believes Lord Chancellor [R. M. Rolfe] will abandon bill. GP's view of how amended bill should distribute authority at university.
ALS 4pp TxU:H/M–0422; Reel 1087

10384. 1855–5–10 From William Henry Bateson London
Sign and return three enclosed letters from Cambridge University Commissioners to Lord Palmerston [H. J. Temple], John Russell, and Lord Chancellor [R. M. Rolfe] regarding amended Cambridge University bill.
ALS 3pp TxU:H/M–0057.27; Reel 1087

10385. 1855–5–11 To [William Henry Bateson] Collingwood
Returns three letters [related to Cambridge University Commission], approved and signed. JH is 'utterly incapacitated from writing.'
draft 1p TxU:H/M–0130.2; Reel 1087

10386. 1855–5–13 From Urbain J. J. Leverrier to Margaret Brodie Herschel Paris
Was in London last week, but was unable to see JH as he had returned to Collingwood.
C 1p (French) RS:HS 11.207

10387. 1855–5–16 To [Thomas Graham] Collingwood
Encloses papers sent to JH from Treasury. R. F. Suft has responses from principal bankers in England; send copies of these at once to [James] Wilson.
C 1p TxU:H/M–0173.2; Reel 1087

10388. 1855–5–16 To G. Russell Collingwood
Circular regarding coinage of florins, approved by W. E. Gladstone, plus enquiry sent to English bankers and JH's abstract of replies from bankers, are all stored at Mint. [James] Wilson may obtain copies there. Will forward GR's letter of 14 May 1855 to Mint and ask that these papers be located.
C 2pp TxU:H/M–0172.2; Reel 1087

10389. 1855–5–17 From Edward Joshua Cooper Markree
Regarding the non-arrival of his star catalogues. Thinks that four volumes will finish the observations.
ALS 3pp RS:HS 5.291

10390. 1855–5–23 To George Biddell Airy Collingwood
A letter to accompany Frederick McCoy's 1855–3–9, indicating that it seems appropriate to pay the bills. One of the professors just appointed has died in Australia.
LS 4pp RGO 6.336.45

10391. 1855–5–24 From George Biddell Airy Royal Observatory, Greenwich
Will communicate with the other committee members about a replacement for the Melbourne University professor who died [see JH's 1855–5–23].
ACCS 1p RGO 6.336.49

10392. 1855–5–24 From George Peacock Deanery of Ely
Appeals to Lord Chancellor [R. M. Rolfe] are producing great changes in [parliamentary bill for new statutes at Cambridge University]. But Heads of Houses are angry. William Whewell's responses are 'arbitrary and despotic.' Extreme agitation at Cambridge.
ALS 4pp TxU:H/M–0423; Reel 1087

10393. 1855–5–25 To Edward Wolstenholme Ward Collingwood
JH, 'completely broken in health,' ceased to be Master of Mint in Apr. 1855. Successor is Professor [Thomas] Graham of University College. Keep Master informed of stock of dies for coins in Sydney. Glad that EW's office as deputy-master is independent of colonial government. Regards to Charles Elouis and J. Triskett.
C 2pp TxU:H/L-0492.1; Reel 1055

10394. 1855–5–28 From William Rutter Dawes Wateringbury, Maidstone
Results of observations using the G. B. Amici micrometer.
ALS 4pp RS:HS 6.103

10395. 1855–5–30 From George Biddell Airy Royal Observatory, Greenwich
About transferring the Melbourne University funds into a trust account managed by the rest of the committee [see JH's 1855–3–23].
ACCS 3pp RGO 6.336.50

10396. [1855–6 or later] To [James Calder] Stewart n.p.
Transfer of annuity to Mrs. Baldwin from JH's account as executor for Thomas Baldwin. Julia and Constance [Herschel] doing well at Anstey Cottage. News of [Isabella] is not good. Margaret had bad time but is now comfortable.
ALS 4pp TxU:H/L-0406; Reel 1055

10397. 1855–6–5 To A. J. Stotherd Royal Mint
JH has no knowledge of government's plan to issue Benedetto Pistrucci's Waterloo medallion to public.
ALS 2pp TxU:H/L-0767; Reel 1089

10398. 1855–6–16 To Isabella Herschel London
Writes to JH with family news, because JH's wife, Margaret, is ill.
AL 4pp JHS 1.44

10399. 1855–6–16 To [James Calder] Stewart n.p.
Thanks for report of JS's position. [Lucy?] is at Mr. Buillier's.
ALS 2pp TxU:H/L-0449; Reel 1055

10400. 1855–7–2 To William Henry Smyth Collingwood
Thanks WS for 'kind expressions on my behalf.' Believes these may be the last words he may write.
C 1p RS:HS 23.156

10401. 1855–7–6 From John R. Hind 22 Grove Rd., St. John's Wood
Has heard from W. H. Smyth that JH is suffering from serious and protracted indisposition. Hopes he will soon be restored to health as he himself has been.
ALS 4pp RS:HS 9.365

10402. 1855–7–8 To George Biddell Airy [Collingwood]
Has had a letter from Melbourne University, with instructions to transfer all the money to Melbourne's agent in London [see GA's 1855–5–30]; JH is slowly recovering.
ALS 4pp RGO 6.336.52

10403. 1855–7–15 From Charles Pritchard Rome
Visits Angelo Secchi at Rome observatory, where they are comparing information in *Cape Results* to their observations.
ALS 1p inc RS:HS 14.78

10404. 1855–7–20 From Melbourne University Melbourne
Would like an opinion on the question of attendance fees paid by students attending their lectures.
ALS 2pp RS:HS 13.32

10405. 1855–7–23 From Robert Francis Suft Royal Mint
Sends Mint accounts of Apr. 1855 for JH's signature.
ALS 1p TxU:H/M–0513; Reel 1093

10406. 1855–7–29 To [Sir] Edward Ryan Collingwood
Thanks for ER's help in obtaining appointment for Duncan Stewart. JH has been unable to stand up for five months; must write in supine position.
ALS 4pp TxU:H/L-0765.5; Reel 1089

10407. 1855–8–1 From John Henry Nelson to Alexander Herschel Courtfield, Ryde
Grateful for informing JN regarding JH's health. Ten years since they met, but hopes to renew his acquaintance as opportunity arises.
ALS 3pp RS:HS 13.95

10408. 1855–8–1 From John Henry Nelson Courtfield, Ryde
Best wishes for JH's health and convalescence.
ALS 3pp RS:HS 13.94

10409. 1855–8–2 From John Barlow 5 Berkeley St.
Regrets JH's name is not to be on their list. Lady Herschel's request will be attended to. Hopes he will attend any meeting he feels able to when in London.
ALS 3pp RS:HS 5.18

10410. 1855–8–2 From William Henry Fitton 18 Norfolk Crescent
Congratulations on JH's election as one of the eight foreign members of the French Institute.
ALS 4pp RS:HS 7.240

10411. 1855–8–4 From Augustus De Morgan to Miss Herschel 7 Camden St.
Has not heard from her father for some time. Would she inform her father that Richard Sheepshanks died this morning of apoplexy.
ALS 2pp RS:HS 6.271

10412. 1855–8–6 To Augustus De Morgan Collingwood
Expresses great sorrow at the death of Richard Sheepshanks; JH has been ill for many months.
C 2pp RS:HS 23.157

10413. 1855–8–9 From Augustus De Morgan 7 Camden St.
Was pleased to see his handwriting again especially as he had heard that JH had died. Elizabeth Baily was upset by the death of Richard Sheepshanks. Will be going down for the funeral tomorrow and hopes to meet G. B. Airy and William Simms. Both of JH's daughters have corresponded with him over his health.
ALS 3pp RS:HS 6.272

10414. 1855–8–13 From George Biddell Airy Royal Observatory, Greenwich
A note accompanying a copy of the act of Parliament legalizing the new standards, the end product of the work of the Standards Commission.
ACCS 1p RGO 6.341.76

10415. 1855–8–13 From George Biddell Airy Royal Observatory, Greenwich
Is trying to clear up matters related to Melbourne University [see JH's 1855–7–8].
ACCS 2pp RGO 6.336.54

10416. 1855–8–24 To Jean Baptiste Elie de Beaumont Hawkhurst
Expresses elation and gratitude at JH's election as a foreign associate of the Académie des sciences. Is especially honored because JH has been elected in succession to C. F. Gauss.
ALS 3pp Académie des sciences (Paris)

10417. 1855–8–31 To William Hallowes Miller 32 Harley St.
Is trying to arrange for a group of those interested to examine a glass disc.
C 1p RS:HS 23.158

10418. 1855–9–4 From James Nash 40 George St., Euston Sq.
JN's duties at Mint ended with successful completion of copper coinage at Birmingham. Asks JH for testimonial to help JN find other government employment.
ALS 3pp TxU:H/M–0397; Reel 1087

10419. 1855–9–23 From Augustus De Morgan 7 Camden St.
Is glad to hear that his health is improving. Has been preparing the wording of a tablet to be erected in memory of Richard Sheepshanks. Sends a proof of one of his theorems.
ALS 4pp RS:HS 6.273

10420. 1855–9–26 To Warren de La Rue Collingwood
Thanks for engraving of Saturn; comments on other aspects of observing that planet. JH notes he is confined to a wheelchair.
C 2pp RS:HS 23.159

10421. 1855–9–27 To George Biddell Airy Collingwood
Further instructions from Melbourne about the transfer of funds [see JH's 1855–7–8]; will be happy to get rid of this problem.
LS 4pp RGO 6.336.55

10422. [1855]–9–27 To James Nash Collingwood
Attests to JN's competence at Mint and in superintending copper coinage at Birmingham.
C 1p TxU:H/L-0282; Reel 1054

10423. 1855–9–28 From George Biddell Airy Royal Observatory, Greenwich
The end of the Melbourne University business seems happily now at hand [see JH's 1855–9–27].
ACCS 3pp RGO 6.336.57

10424. 1855–10–4 From J[ohn] Cooper Cambridge
Has secured lodgings for JH's son [Alexander]. Gives details of requirements.
ALS 3pp RS:HS 5.367

10425. [1855?]–10–4 From Elizabeth White 5 Hubert Terrace, Dover
EW writes for Elizabeth Baily, who communicates what she can remember to help JH commemorate her brother Francis Baily, who left a journal of American travels that Miss Baily hopes to make public. Found list of Francis's works that ought to have been forwarded to JH. Miss Baily would like to meet with JH in October or November.
ALS 4pp RS:HS 18.225

10426. 1855–10–16 From Thomas Hindle Tarleton
Describes an unusual double rainbow with the second bow extending from an angle to the main arc.
ALS 3pp diag Harvard: Houghton AAH 67m–67 (84)

10427. 1855–10–20 From Augustus De Morgan 7 Camden St.
Was pleased to hear from Elizabeth Baily that JH is improving in health. Knows a doctor who can produce jaundice. Plan is on foot to publish Francis Baily's travels in America. Hopes JH will allow his life of Baily to be included. Monument to Richard Sheepshanks is under consideration.
ALS 4pp RS:HS 6.274

10428. 1855–10–24 From George Biddell Airy Royal Observatory, Greenwich
Requesting comments on enclosed papers concerning Greenwich Observatory.
ALS 2pp RS:HS 1.226 (C: RGO 6.12.148)

10429. 1855–10–25 From Hermann Schlagintweit Cherra Poonjee
Happy to have met JH. Discusses journey in Himalayas. Would be glad to send JH any information he would like about the tropical climate.
ALS 4pp RS:HS 15.352

10430. 1855–10–31 To George Biddell Airy n.p.
Encourages GA to make requests to ensure that the Royal Observatory instruments are first rate [see GA's 1855–10–24].
ALS 4pp RGO 6.12.185

10431. 1855–11 From Achille Brachet Paris
On telescopes and their fittings.
ALS 3pp RS:HS 4.211

10432. 1855–11 To [James Calder] Stewart n.p.
Reports death of [Richard Jones]. John Herschel (son) returned from Scotland and goes to Addiscombe Thursday. 'Our governors' have a mess to deal with. Margaret is still weak. What is usual fee for London surgeon?
ALS 4pp TxU:H/L-0450; Reel 1055

10433. 1855–11–2 From Henry Anthony Jeffreys to Margaret Brodie Herschel Hawkhurst Parsonage

Has difficulty in providing Mr. Loyd with a pew; would she object if he moves the pew occupied by her servants to a different part of the church? Hopes JH did not catch a chill yesterday.
ALS 4pp RS:HS 10.295

10434. 1855–11–6 To Augustus De Morgan Collingwood
Comments on Francis Baily's travel writings and on Richard Sheepshanks's epitaph.
C 1p RS:HS 23.160

10435. 1855–11–8 From Achille Brachet Paris
Concerning telescopes and their fittings, and queries relative to them. Intends to publish a work on them.
ALS 3pp RS:HS 4.210

10436. 1855–11–9 From John Phillips B.A.A.S., Oxford
JH is appointed to a committee that will consider repeating the Balloon Experiments of 1852. William Whewell will head the committee.
L 1p TxU:H/M–0428.21; Reel 1093

10437. 1855–11–10 From Augustus De Morgan 7 Camden St.
Was pleased to see his signature again and to hear that he is in good spirits. He will turn into a chemist. Will send him a proof of his article and appendices upon Francis Baily.
ALS 3pp RS:HS 6.275

10438. 1855–11–10 From Henry Anthony Jeffreys Hawkhurst Parsonage
Outlines the arrangements for allocating the pews in the church.
ALS 4pp RS:HS 10.296

10439. 1855–11–15 To Achille Brachet Collingwood
Answer to his queries on telescopes and rays of light. Is sorry he rejects Joseph Fraunhofer's fixed lenses.
draft 4pp RS:HS 4.212 & 23.161

10440. 1855–11–17 To [James Calder] Stewart n.p.
Many cases of disease in Hawkhurst parish. JH's children are scattered. JH and Margaret Herschel are regaining strength. James [?] is ill with fatal stomach disorder.
ALS 3pp TxU:H/L-0451; Reel 1055

10441. 1855–11–21 From Augustus De Morgan 7 Camden St.
Thinks of putting JH's notes at the end of the life [of Francis Baily].
ALS 1p RS:HS 6.276

10442. 1855–12–5 To Henry Peter Brougham Collingwood
Expresses thanks and praise for HB's *Analytical View of Sir Isaac Newton's* Principia [London, 1855].
ALS 1p Univ.Coll.London Brougham

10443. 1855–12–5 To Augustus De Morgan Collingwood
On AD's note on Richard Sheepshanks's writing on the standard yard.
C 1p RS:HS 23.162

10444. 1855–12–6 From Augustus De Morgan 7 Camden St.
JH's handwriting is improving, which shows he is on the mend. Does not propose to show the failure of the scale. Quotes an error in his Latin.
ALS 3pp RS:HS 6.277

10445. 1855–12–7 From Augustus De Morgan 7 Camden St.
Submitting a vote for JH's comment. Quotes various points of Latin grammar.
ALS 2pp RS:HS 6.278

10446. 1855–12–12 To Edward William Brayley Collingwood
Concerning exchange of some papers on meteorology.
C 1p RS:HS 23.164

10447. 1855–12–12 To Augustus De Morgan n.p.
On proper and bad Latin.
C 1p RS:HS 23.163

10448. 1855–12–14 From George Biddell Airy Royal Observatory, Greenwich
Is sending his speculations on the attraction of mountains.
ALS 3pp RS:HS 1.227 (C: RGO 6.376.358)

10449. 1855–12–16 From Amelia Matilda Murray Brome Hall, Scole
Pleased to hear that JH's daughter is to be her successor. Thinks the Queen is right though AM's letters will not turn out to be so pro-slavery as given out to be. [Postscript gives reasons for her resignation.]
ALS s RS:HS 12.442–3

10450. 1855–12–17 From Adam Sedgwick Cambridge
Has JH received the second or third of AS's Cambridge Paleozoic? Asks about William Whewell's pamphlets regarding the University. Describes trip to Scotland.
ALS 4pp RS:HS 15.446

10451. 1855–12–18 From Edward William Brayley London Institution
Part I of Howard's Barometrographia will be ready on Thursday. Is pleased to hear his health has improved. A new edition of JH's *Prelim. Discourse* would be welcome.
ALS 4pp RS:HS 4.216

10452. 1855–12–22 From B. Anderson Royal Mint
Thanking him for his recommendation of BA's son for service at the Stationery Office.
ALS 2pp RS:HS 1.396

10453. 1855–12–24 From Amelia Matilda Murray Merton Hall, Thetford
Regrets to hear that JH has been ill. Gives the title of her book and the reasons for her views on slavery.
ALS 4pp RS:HS 12.444

10454. 1855–12–25 From Augustus De Morgan 7 Camden St.
Is glad to hear his health is improving. Has heard that JH is holding a large reception for all his descendents. Is progressing well with the book [on Francis Baily]. Sends a puzzle in spherical trigonometry.
ALS 3pp RS:HS 6.279

10455. [1856 or later?] To William Stanley Jevons Collingwood
Not aware of any experiments to ascertain the amount of personal error in the estimate of time of a star transit. JH then comments on monetary matters including the value and gold equivalence of sovereigns issued by the Mint.
C 2pp RS:HS 25.14.16

10456. [1856] To Adam Sedgwick Collingwood
Discusses AS's new book [*Synopsis of the Classification of the British Palaeozoic Rocks...*]. Wishes to have the completed work sent to him and sends congratulations on its completion. JH's health is improving.
ALS 4pp RS:HS 15.448 (C: RS:HS 23.185)

10457. [1856 or later] To ? n.p.
A note indicating JH's willingness to say a few words.
ALS 1p DMC.695A.885.13

10458. 1856–1–7 From Augustus De Morgan 7 Camden St.
Wishing him well for the new year. Hears JH has visited the East India College. Sends a theorem. H. P. Brougham (Baron Brougham and Vaux) amuses himself by finding the laws of central force for curves. Has been finding information on a nephew of Isaac Newton. George Stanhope (6th Earl of Chesterfield) and T. A. W. Parker (9th Earl of Macclesfield) were pupils of Abraham De Moivre.
ALS 8pp RS:HS 6.280

10459. 1856–1–11 From Thomas Spring-Rice Exchequer
TS will be chairman of Decimal Coinage commission. Do majority of bankers queried by JH favor this decimal system? Death of William Whewell's wife, Cordelia, sister-in-law of TS.
ALS 2pp TxU:H/M–0376; Reel 1087

10460. [1856]–1–15 From Charles De la Pryme Clarendon Mansion, Brighton
Encourages JH to contest Cambridge University seat in Parliament, following death of Henry Goulburn.
ALS 3pp TxU:H/M–0162; Reel 1087

10461. 1856–2–5 To Thomas Maclear Collingwood
Sends regrets over TM's son not passing exam. Tries unsuccessfully to arrange a second testing. Fears war in Europe.
ALS 4pp Cape Archives/Maclear Papers File 100

10462. 1856–2–9 From Edward Sabine 13 Ashley Place
Glad JH is recovering and free from official duties [at the Mint]. Discusses his upcoming papers regarding magnetism.
ALS 4pp RS:HS 15.239

10463. 1856–2–11 From Alexander Hamilton-Gordon Horse Guards
Would like a letter from JH giving him authority to have the box to open. Is busy with the Report.
C 2pp RS:HS 8.148

10464. 1856–2–11 From Roderick Impey Murchison 16 Belgrave Square
Asks JH's opinion about distribution of government fund for advancement of science handed over to R.S.L.
ALS 4pp RS MM.16.126

10465. [1856–2–11 or later] From Roderick Impey Murchison 16 Belgrave Square
Thanks for resolutions regarding distribution of government aid, particularly JH's dwelling on good example of B.A.A.S. Discusses proceedings of distribution committee meeting. [Marked 'Private'.]
ALS 2pp RS MM.16.127

10466. 1856–2–11 To Adam Sedgwick Collingwood
Praises AS's new book [*Synopsis of the Classification of the British Palaeozoic Rocks...*]. All is well at Collingwood.
ALS 2pp RS:HS 15.447 (C: RS:HS 23.165)

10467. 1856–2–13 To Alexander Hamilton-Gordon Collingwood
Sorry to have given him the trouble over the box; sends the required letter. Unfair of the *Times* to blame AH for the disasters in the Crimea.
draft 4pp RS:HS 8.149 (C: RS:HS 23.166)

10468. 1856–2–14 To H. J. Luff[?] Collingwood
Deals with the computation of a tax matter [regarding Observatory House in Slough?].
ACCS 4pp BostonPubLib Ms.Eng. 196(3)

10469. 1856–2–17 To William Macarthur Collingwood
JH, having examined the accounts, finds the charges for the medals fit with Leonard C. Wyon's agreement.
ALS 2pp New South Wales State Library

10470. 1856–2–18 From Edward Vivian Torquay
Discusses observations of color and light.
ALS 4pp RS:HS 17.464

10471. 1856–2–28 From Edward William Brayley London Institution
Is sending C. T. Beke's certificate for his signature. Hopes his health is still improving.
ALS 4pp RS:HS 4.217

10472. 1856–3–2 From Roderick Impey Murchison 16 Belgrave Square
Describes proceedings of government aid distribution committee meeting. JH's letter was so well-received JH was elected committee member. Asks JH to write example-illustrated report of resolutions and to be committee's orator.
ALS 6pp RS MM.16.128

10473. 1856–3–6 From James Booth Society of Arts
Council of Society of Arts would like JH's support for their new scheme of examinations.
ALS 2pp RS:HS 4.194

10474. 1856–3–9 To James Booth Collingwood
Declares his confidence in the certificates of proficiency about to be awarded by the Society of Arts.
L *Journal of the Society of Arts, 4* (April 19, 1856), 342

10475. 1856–3–12 From James Booth Society of Arts
Thanks for his support for their scheme of examinations.
ALS 3pp RS:HS 4.195

10476. 1856–3–15 From Roderick Impey Murchison 16 Belgrave Square
Discusses JH's resolution regarding distribution of government grants, which Committee adopted but R.S.L. Council opposed. Relays in great detail proceedings and

decisions of meeting.
ALS 10pp RS MM.16.129

10477. 1856–3–18 From G. R. Stevenson Tangwood
Hopes JH will speak well of GS and help secure appointment that will not force GS to 'desert my flock,' as GS's current militia assignment will.
ALS 4pp RS:HS 17.234

10478. 1856–3–19 To James Booth Collingwood
Conveys his confidence in the appropriateness of the certificate of proficiency awarded by the Society of Arts to men who have done well on the Society's examination.
ALS 3pp Columbia Univ. Library (ACCS: RS:HS 4.196)

10479. 1856–3–19 To John William Lubbock Collingwood
Is sending a letter of support for a neighbor for the position of commander of constabulary about to be established in Kent.
C 2pp RS:HS 23.167

10480. 1856–3–25 From Harry George Wakelyn Smith Manchester
Note asking JH to patronize an artist's exhibition in London.
ALS 4pp RS:HS 16.202

10481. 1856–3–28 From Alexander Wight Jr. 6 Clarenden Terrace, Notting Hill
New method for constructing circular lenses up to 10-feet diameter for £50. Focusing power sufficient to melt metals. Wants to send 3-inch sample to JH.
ALS 3pp RS:HS 18.245

10482. 1856–3–28 To John Wrottesley Collingwood
Is able to sit up in a wheel chair only a few hours a day; JH's illness has left him out of the sphere of scientific activity so he is unable to make any useful suggestions about medals.
C 2pp RS:HS 25.12.2

10483. 1856–4–3 From Alexander Wight Jr. 6 Clarenden Terrace, Notting Hill
Sends small lens. Its components neither parallel nor concentric; contains water. Suggests new polishing method. Estimates 4-foot lens would cost £7–£8. [JH note: Tested and returned lens. Noted advantages and deficiencies and suggests improvements. Listed others who built similar lenses.]
ALS 3pp, JH note 1p RS:HS 18.246

10484. [1856–4–6 or earlier] From George Biddell Airy n.p.
Regarding experiments on the Mohamet's coffin theory. His son Wilfrid has won a scholarship at Trinity College, Cambridge.
AL inc 4pp RS:HS 1.330

10485. [1856]–4–6 To George Biddell Airy Collingwood
[Reply to GA's [1856–4–6 or earlier] on Mohamet's coffin theory.] Congratulations on GA's son winning a scholarship. JH's health is better, but he cannot walk yet.
draft 4pp RS:HS 1.331

10486. 1856–4–9 From Alexander John Ellis Edinburgh
Sending his completed universal alphabet with comments.
ALS 4pp RS:HS 7.39

10487. 1856–4–9 From Alexander Wight Jr. 6 Clarenden Terrace, Notting Hill
Lens safely returned. Not aware of lenses in 1851 Exhibition or of C. F. Sturm's solid lenses. Notes advantages of plate glass. Offers to make three foot by three foot square lens. Recognizes limitations for use in astronomy; hopes for other applications.
ALS 4pp RS:HS 18.247

10488. 1856–4–12 From William Benjamin Carpenter Univ. Hall, Gordon Square
Is considering becoming a candidate for the position of Registrar at London University and would welcome support from JH.
ALS 6pp RS:HS 5.190

10489. 1856–4–14 From George Biddell Airy Royal Observatory, Greenwich
Requesting advice on his reflex zenith tube. Regarding progress on 'Standards.'
AL 4pp RS:HS 1.228 (C: RGO 6.376.360)

10490. 1856–4–14 From Few & Co. Covent Garden
Thanks for the check. Will search for the deeds.
AL 1p RS:HS 7.211

10491. 1856–4–16 To George Biddell Airy Collingwood
A number of options are proposed by JH to help GA solve his problem with the mercury level [see GA's 1856–4–14]; JH is feeling much better but fears he may 'never again be able to walk.'
ALS 4pp RGO 6.376.364

10492. 1856–4–16 From William Benjamin Carpenter Univ. Hall, Gordon Square
Application for the position of Registrar at [London] University.
C 4pp RS:HS 5.191

10493. 1856–4–17 From Mary Somerville Florence
Needs help understanding Percades [?] quantity. Inquires as to well-being of family and sends regards.
ALS 3pp RS:HS 16.358

10494. 1856–4–18 From William Benjamin Carpenter Univ. Hall, Gordon Square
Is grateful for JH's support for his application as Registrar. Encloses a copy of his application. Gives details of a binocular microscope he has received from Paris.
ALS 3pp RS:HS 5.192

10495. 1856–4–23 From George Biddell Airy Royal Observatory, Greenwich
Wants one or more of JH's actinometers. Notes his suggestion about the zenith tube.
ALS 1p RS:HS 1.229 (C: RGO 6.216.114)

10496. 1856–4–24 To George Biddell Airy Collingwood
Has no actinometers available [see GA's 1856–4–23], but tells GA where he can get one; rhapsodizes about the beauty of nature around him, and the silliness of a person.
ALS 3pp RGO 6.216.115

10497. 1856–4–24 To Edward Wolstenholme Ward Collingwood
Happy to hear that [new Sydney mint] is going well. JH confined with painful illness since Mar. 1855; still unable to walk or stand upright without support. Aluminium may be used in coins, but JH doubts it will replace iron in pots and pans. War is ended, but people worry about further Russian aggression.
C 4pp RS:HS 18.374 (C: 23.168)

10498. 1856–4–28 From Henry Warburton 45 Cadogan Place
Heard JH not well. Offers him 'a nosegay, gathered from

[JH's] own garden of the Differences of the Powers of Zero.' More notes on the self-repeating series. Developed two equations that HW may submit to Cambridge [Philosophical] Society.
ALS 8pp RS:HS 18.79

10499. 1856–4–29 To Mr. Holloway Collingwood
Plans to sell JH's property at Upton, including tenement Holloway occupies, to neighbors.
Adraft 1p inc RS:HS 18.85a

10500. 1856–5–5 To 'Sir' Collingwood
JH's recommendations to Admiralty regarding experiments to be conducted by expedition to Teneriffe led by C. P. Smyth.
ACCS 4pp TxU:H/L-0733; Reel 1089

10501. 1856–5–5 To B. Osborne Collingwood
The Admiralty has given a grant for an astronomical expedition to the Peak of Teneriffe, and JH is now providing a list of details that should be observed in such an expedition.
C 4pp RS:HS 23.169

10502. 1856–5–7 From Henry Warburton 45 Cadogan Place
HW neglected to state principle from which he derived equation for eliminating differences. Expresses it in formula.
ALS 3pp RS:HS 18.80

10503. 1856–5–8 From John Gorham Tonbridge
Has he repeated his experiments with the photographic representation of the prismatic structure? Regarding a numerical nomenclature for the colors.
ALS 3pp RS:HS 8.219

10504. 1856–5–11 From Augustus De Morgan 7 Camden St.
Has not heard from him for a long time. Thought that the Government was going to turn the R.A.S. out of Somerset House. There is someone to come to the rescue of Richard Sheepshanks. Has not seen Charles Babbage for many years. Their views are incompatible. C. P. Smyth is off to Teneriffe so now is the time for JH to make suggestions. Has set up a committee to make by-laws.
ALS 4pp RS:HS 6.281

10505. 1856–5–12 To Thomas Andrews Collingwood
Formal note of thanks for TA's paper on ozone.
C 1p RS:HS 23.171

10506. 1856–5–13 From John Gellibrand Hubbard Bank of England
Where in JH's evidence to Parliamentary Decimal Commission did JH give number of coins that require recoinage?
ALS 1p TxU:H/M–0278; Reel 1087

10507. 1856–5–14 From T[?] P[?] Admiralty
Writes on behalf of the Lords of the Admiralty to thank JH for his suggestions relating to C. P. Smyth's trip to the peak of Teneriffe.
ACC 1p Hydrographic Office Letter Book

10508. 1856–5–15 From Henry Warburton 45 Cadogan Place
Glad that JH shows renewed interest in mathematics. Besides HW, JH, Augustus De Morgan, and Mr. Gerard, there appear to be no others interested in researching the differences of the powers of zero. Gave outline of HW's paper at Cambridge Philosophical Society. Plans to revise it according to suggestions from JH and De Morgan.
ALS 4pp RS:HS 18.81

10509. 1856–5–16 To John Gellibrand Hubbard Collingwood
Responds to Hubbard's inquiry about coinage.
AdraftS 4pp TxU:H/L-0197; Reel 1054

10510. 1856–5–17 From George Biddell Airy Royal Observatory, Greenwich
Relating his experiments with the zenith tube.
ALS 4pp RS:HS 1.230

10511. 1856–5–17 From Henry Warburton 45 Cadogan Place
Quick way to determine coefficients in tables of differences of powers of zero.
ALS 2pp RS:HS 18.82

10512. 1856–5–19 To Augustus De Morgan Collingwood
Comments on the state of JH's health, and on things astronomical and mathematical.
C 2pp RS:HS 23.172

10513. 1856–5–20 From Augustus De Morgan 7 Camden St.
Is glad to hear his account of facts. Always thought he would recover his health. Gives one of his own mathematical formulae. Has no idea how the Decimal Coinage Commission is progressing.
ALS 4pp RS:HS 6.282

10514. 1856–5–21 From Thomas Andrews Queens College, Belfast
Thanks for JH's kind comments on TA's paper on ozone.
ALS 4pp inc RS:HS 1.344

10515. 1856–5–22 From William Benjamin Carpenter Univ. Hall, Gordon Square
Was elected yesterday to the Registrarship of London University, and thanks JH for his support.
ALS 1p RS:HS 5.193

10516. 1856–5–25 From Henry Warburton 45 Cadogan Place
Sends law of coefficients, for developing summation theorem.
ALS 3pp RS:HS 18.83

10517. 1856–5–26 From [F. Soring?] Paris
[Form letter] Offers to send *Commercium epistolicum J. Collins et aliorum...* (1856), by J. B. Biot and Francisque Lefort, a critical edition of works by mathematician John Collins.
(A)LS 1p (French) RAS:JH Archive 14/1.227 verso; Reel 10

10518. 1856–5–26 From Henry Warburton 45 Cadogan Place
Coefficients of differences of odd powers of zero observe same law as those of even powers. Example of table of coefficients of differences of 0^{15}. Table took 15 minutes to compute by HW's process of summation. Notes error in table for 0^{12} sent earlier.
ALS 3pp RS:HS 18.84

10519. 1856–5–27 To Henry Warburton Collingwood
Compliments HW's expression for the formula for coefficients of differences of the odd powers of zero. JH busy, but hopes soon to develop demonstration.
ACCS 2pp RS:HS 18.85b

10520. 1856–5–30 From Henry Warburton 45 Cadogan Place

Glad for JH's renewed interest in mathematics. At 72, HW expects to do no more. Reviewed JH's letter of 14 May [1856]. Fears JH misunderstood meaning of symbol HW employed. Quotes explanation from HW's paper on self-repeating series.
ALS 4pp damaged RS:HS 18.86

10521. 1856–6–2 To Margaret Brodie Herschel [London]
Complaining about the recent wet weather.
ALS 2pp JHS 1.66

10522. 1856–6–2 From Henry Warburton 45 Cadogan Place
Reflects on relations between differences of same power of zero. Reviews procedure developed for determining terms and law of coefficients. JH's critique expanded role for these, but HW disagrees with JH's latest comments.
ALS 3pp RS:HS 18.87

10523. 1856–6–3 From George Biddell Airy Royal Observatory, Greenwich
Has procured an actinometer and would like details of the scale.
ALS 3pp RS:HS 1.231

10524. 1856–6–4 To George Biddell Airy Collingwood
Some more information on the use of the actinometer [see JH's 1856–4–24 & GA's 1856–6–3]; some news of JH's wife, Margaret.
ALS 3pp RGO 6.216.117

10525. 1856–6–4 From Augustus De Morgan 7 Camden St.
Would he answer some of his queries when he has leisure? Queries on the law of beats for unison sounds. Met JH's daughter on Monday, who gave a fair account of JH and his recent work.
ALS 3pp RS:HS 6.283

10526. 1856–6–6 To Alexander Stewart Herschel Collingwood
Lectures AH on the 'examination paper of life' after AH had complained that he was disgusted by a college examination.
ALS 3pp JHS 1.64

10527. [1856–6–7 or earlier?] To Augustus De Morgan n.p.
Wants to make a new musical instrument on the principle of resonance; JH also suggests some improvements in the construction of the organ.
C 2pp RS:HS 25.14.7

10528. 1856–6–8 From Augustus De Morgan 7 Camden St.
Has long held the view that a pipe underneath the strings of a piano would increase the resonance. Had not conceived such an instrument as JH suggested. Why are 3rds and 6ths more agreeable to the ear than 4ths and 5ths?
ALS 3pp RS:HS 6.284 S. De Morgan, *Memoir of Augustus De Morgan*, 287

10529. 1856–6–14 From Baden Powell 6 Stanhope St.
Regrets letter read the night before at the R.A.S. of JH's wish to resign and hopes JH's health allows him to come occasionally and be a nominal member. Discusses Piazzi Smyth's preparations for voyage to Teneriffe.
ALS 4pp RS:HS 14.48

10530. 1856–6–15 To Margaret Brodie Herschel and Isabella Herschel Collingwood
Celebration planned for arrival of MH and IH. Son John is home from Addiscombe. JH's health. News of Hawkhurst. Regards to Mr. and Mrs. Greig.
ALS 3pp TxU:H/L-0545; Reel 1053

10531. 1856–6–16 From Hannah M. Bouvier Crosswick, N.J.
Sending him a copy of her *Familiar Astronomy* and inviting his comments on this work. His works are well known in America.
ALS 2pp RS:HS 5.90

10532. 1856–6–24 To Edward Sabine Collingwood
Comments on a proposed grant to publish T. R. Robinson's Armagh observations.
ALS 4pp Cambridge University Library 7656/RS8

10533. 1856–6–27 From George Gabriel Stokes Laurel Lodge, Barnet
Writes to clarify his understanding of the word 'fluorescence.' Asks JH to review a paper on color-blindness.
ALS 8pp RS:HS 17.31

10534. 1856–7–1 To George Gabriel Stokes Collingwood
Comments on GS's ideas [see GS's 1856–6–27] about the nature of fluorescence. Asks for strong horseshoe magnets from R.S.L. to try to solve the problem of 'Mahomet's coffin.'
ALS 7pp RS:HS 17.32 (C: RS:HS 23.173 & 25.10.9)

10535. 1856–7–2 From Sir William Pole to Cha[rle]s May 8 Great George St.
Claims himself perfect case of di-chromic vision, only seeing blue and yellow. Gave paper on his condition to R.S.L., quoting and supporting JH's theory. Hears JH is ill, or would send letter directly to JH.
ALS 3pp RS:HS 14.17

10536. 1856–7–3 From Joseph Bullar Southampton
Death of his sister a blow to John Bullar [father of writer].
ALS 2pp RS:HS 4.336

10537. 1856–7–4 From Joseph Baxendell Stocks St., Manchester
Recent observations of specified stars and their variations.
ALS 2pp RS:HS 3.305

10538. 1856–7–8 From George Gabriel Stokes Laurel Lodge, Barnet
Further thoughts [see GS's 1856–6–27] on the nature of fluorescence, including that it is a transient phosphorescence.
ALS 6pp RS:HS 17.33

10539. 1856–7–9 From George Gabriel Stokes Laurel Lodge, Barnet
Continues comments on fluorescence from letter of 1856–7–8; then shows mathematically that JH's solution for the problem of Mahomet's coffin [see GS's 1856–7–1] will not work.
AL 8pp RS:HS 17.34

10540. 1856–7–14 To George Gabriel Stokes Collingwood
Thanks for the magnetic theory [see GS's 1856–7–9] and agrees it argues against JH's ideas about Mahomet's coffin. Describes some interesting magnetic experiments JH has seen.
C 4pp RS:HS 17.35 (C: RS:HS 23.174 & 25.10.10)

10541. 1856–7–15 From Joseph Bullar Southampton
Their whole family is grateful for JH's letter of sympathy to

JB's father.
ALS 2pp RS:HS 4.337

10542. 1856–7–15 To Alexander Stewart Herschel Collingwood
Is concerned about the receipt of AH's accounts; lectures AH about the danger of moving into someone else's room [more luxurious] for the vacation; finally asks AH to discover who among the 'university-men' is color-blind, and to what degree.
ALS 4pp JHS 1.112

10543. 1856–7–15 From George Gabriel Stokes Laurel Lodge, Barnet
Comments on one of the magnetism experiments JH described in his letter of 1856–7–14.
ALS 4pp RS:HS 17.36

10544. 1856–7–15 From John Welsh Kew Observatory, Richmond
Describes R.S.L.'s compound magnet. Recommends cleaning and remagnetizing. Awaits JH's instructions.
ALS 4pp RS:HS 18.139

10545. 1856–7–16 To James Calder Stewart Collingwood
Deposit enclosed money at Cadet House for John Herschel's readmission [to Addiscombe].
ALS 1p TxU:H/L-0452; Reel 1055

10546. 1856–7–18 From John McClelland Lewes
Will give him great pleasure to call at Collingwood in the near future. Grieved to hear of JH's illness. Begs acceptance of a set of papers of Dr. William Griffith.
ALS 3pp RS:HS 12.7

10547. 1856–7–19 From Augustus De Morgan 7 Camden St.
Sending a pamphlet, which may give JH some amusement.
ALS 2pp RS:HS 6.285

10548. 1856–7–21 From Warren de La Rue St Mary's Road, Cannonbury
At the recent A.G.M. of the R.A.S. Club, JH was elected an Honorary Life Member. Hopes he will attend many more meetings.
ALS 2pp RS:HS 6.136

10549. 1856–7–23 From Charles May Great George St.
Recently came across some instructions JH had drawn up for persons who are color blind. Sends a note explaining the work of William Pole on this same subject. Has left the manufactory at Ipswich.
ALS 2pp RS:HS 12.313

10550. 1856–7–24 From Charles Elouis Sydney
Pleased to hear from Capt. E. W. Ward's letter that JH still remembered him. Grieves to hear of his bad health. Mint at Sydney is progressing. Rich gold fields at Victoria. Looks forward to the re-establishment of steam communication.
ALS 6pp RS:HS 7.130

10551. 1856–7–26 To Thomas Maclear Collingwood
Sends copy of *Admiralty Manual*, which includes JH article on meteorology. Asks TM what to do with Bokkeveld meteorite specimen still in his possession.
ALS 4pp Cape Archives/Maclear Papers File 100

10552. 1856–7–27 To John Tyndall Collingwood
Thanks for papers on diamagnetism and lecture on 'slaty cleavage.' Compares with phenomenon of 'ringing coins' in mint. Explains with examples conformity of JT's views with JH's own on geological stratification. Praises simplicity yet completeness of JT's work.
C 4pp RS:HS 17.383b (C: 23.176; C: RI 499)
Philosophical Magazine, 4th ser., *12* (1856), 197–8

10553. 1856–7–28 From John Bullar Basset Wood near Southampton
Is grateful for JH's remembrances and sympathy in his affliction [death of his sister]
ALS 4pp RS:HS 4.326

10554. 1856–7–30 From John Tyndall Royal Institution
Agrees with and asks permission to publish JH's letter in the *Philosophical Magazine.* Is going to Switzerland and Tyrol, where hopes to see phenomenon of glacial descent [see JH's 1856–7–31].
ALS 4pp RS:HS 17.384 (C: RI 501)

10555. 1856–7–31 To John Tyndall Collingwood
Clarifies letter on motion and structure modification of glaciers. JH thinks glacier ice unfavorable for exhibiting JT's phenomenon [see JT's 1856–7–30].
C 1p diag RS:HS 23.175 (C: RI 502)

10556. 1856–8–2 To Woronzow Greig Collingwood
Returns Mary Somerville's note and the papers. Does not feel competent to give his opinion on the papers as a whole as there are parts outside his field of study. Comments on some of the points raised, and on the whole he disagrees with the conclusions of the papers.
draft 7pp RS:HS 9.38 & 23.177

10557. 1856–8–6 From Sir William Pole 8 Great George St.
Thanks JH for his note. States R.S.L. is sending his paper to JH as a referee. Gives details of his own case of di-chromic vision.
ALS 4pp RS:HS 14.18

10558. 1856–8–7 To Woronzow Greig Collingwood
Has kept Mary Somerville's papers a little longer. If these papers are to be published they would require considerable revision.
draft 3pp RS:HS 9.39 & 23.178

10559. 1856–8–14 From John Bullar Basset Wood near Southampton
The money presented by his son is to be used to enlarge the Infirmary.
ALS 2pp RS:HS 4.327

10560. 1856–8–15 From Augustus De Morgan 7 Camden St.
JH's last letter was written in a good strong hand. Recommends suitable drink for hot weather. Gives his views about the possible marriage of Isaac Newton's niece. Has had a letter from a Mr. Wither on mathematical books.
ALS 4pp RS:HS 6.286 S. De Morgan, *Memoir of Augustus De Morgan*, 287–9

10561. 1856–8–18 To George Biddell Airy n.p.
A note to accompany the forwarding of a communication to GA.
ALS 1p RGO 6.376.366

10562. 1856–8–19 From George Biddell Airy Flamsteed House, Greenwich
Relating his experiences on his recent tours abroad.
ALS 3pp RS:HS 1.232 (C: RGO 6.376.367)

10563. [1856]–8–19 From C. De La Pryme Marine Parade, Brighton
Inviting JH to become a candidate at Cambridge.
ALS 4pp RS:HS 6.487

10564. 1856–8–21 From Walter Calverley Trevelyan Wallington, Newcastle-Upon-Tyne
Inquires after JH's interest on changing the spelling of the English language to a phonetic system.
ALS 3pp RS:HS 17.446

10565. 1856–8–25 From Augustus De Morgan 7 Camden St.
Many thanks for his invitation, but he does not intend going to Eastbourne. Further remarks on his own theory regarding the marriage of Isaac Newton's niece.
ALS 4pp RS:HS 6.287

10566. 1856–8–30 To John Murray, [Jr.] Collingwood
Has no changes to make in JH's *Admiralty Manual.*
ALS 1p Gerald M. Friedman

10567. 1856–9–6 From Henry Clifton Sorby Sheffield
In response to JH's letter to John Tyndall, explains views on the mechanical nature of geological phenomenon of slaty cleavage.
ALS 4pp RS:HS 17.220

10568. 1856–9–9 From William Henry Smyth St. John's Lodge
Visited Liverpool for inspection of *School Frigate*. Observed Triton. C. P. Smyth will install large telescope at Alta Vista; WS sends excerpt from son's letter. C. P. Smyth also works on spectral line observations.
ALS 3pp RS:HS 16.300

10569. 1856–9–10 From Augustus De Morgan 7 Camden St.
Has fallen from a ladder in his library and dislocated his shoulder, which accounts for the form of his letter. Has he any recollections of Humphry Davy rubbing his hands together?
ALS 3pp RS:HS 6.288 S. De Morgan, *Memoir of Augustus De Morgan*, 289

10570. 1856–9–11 To William Henry Smyth Collingwood
Success of C. P. Smyth's Teneriffe expedition pleases JH; anxious for declination of nebulae taken from high altitudes. Upset over G. J. Stoney's reproduction of JH's collimating telescope without giving him credit.
C 2pp RS:HS 23.180

10571. 1856–9–11 To Henry Clifton Sorby Collingwood
Thanks HS for response to and explains further JH's letter to John Tyndall regarding explanation of origin of geological slaty cleavage.
C 1p RS:HS 23.179

10572. 1856–9–12 To Augustus De Morgan Collingwood
Thought that AD's last letter was supposed to be written in verse. Hopes his injury will soon improve. Does not recollect any habit of Humphry Davy rubbing his hands together. A professor, [G. J.?] Stoney, has re-invented JH's collimating telescope and not improved it.
draft 3pp RS:HS 6.289 & 23.181

10573. 1856–9–13 From Augustus De Morgan 7 Camden St.
Quotes a rhyme. May have been injured even if he had been at Collingwood. Possibly [G. J.?] Stoney has not read JH on the telescope. Is thriving in health but sleeping very well.
ALS 4pp RS:HS 6.290

10574. 1856–10–9 From Francis Charles Knowles Belvedere House, Ryde
Thinks it may be a good idea to utilize aluminium to coat the speculum of reflecting telescopes. Is shortly to take command of iron works and furnaces and hopes to carry out experiments to produce barytic glass.
ALS 4pp RS:HS 11.65

10575. 1856–10–10 From Thomas Greenwood Brighton
Found his letter acknowledging the gift of his book awaiting him on his return from the continent. Reasons why he wrote the book. Has had letter from William Whewell. Hears that JH's health has not been good.
ALS 4pp RS:HS 9.29

10576. 1856–10–10 From George Gabriel Stokes Royal Society, Somerset House
Sends [William?] Pole's paper on color blindness, in case JH wishes to examine it.
ALS 2pp RS MM.16.157

10577. 1856–10–12 From Warren de La Rue St Mary's Road Cannonbury
Returned late on Thursday. Is gratified by JH's praise of his drawings of Saturn. Includes some further drawings and comments on JH's queries.
ALS 11pp RS:HS 6.137

10578. 1856–10–12 To George Gabriel Stokes Collingwood
Acknowledges receipt of a paper by [William] Pole, which JH will review.
C 1p RS:HS 25.10.11

10579. 1856–10–13 From Warren de La Rue St Mary's Road, Cannonbury
Made some observations of Jupiter after his last letter. Thought JH might be interested in seeing them. Would be grateful for any comments, which he will incorporate into his final drawing.
ALS 2pp RS:HS 6.138

10580. 1856–10–13 From Manuel John Johnson Oxford
Pleased to hear that JH was engaged on his old pursuits. Is grateful for pointing out his mistake. Regarding meteorological registration by means of photography.
ALS 8pp RS:HS 10.336

10581. 1856–10–16 To Francis Charles Knowles Collingwood
Thinks that aluminum coating on telescope mirrors may work [see FK's 1856–10–9]; also suggests looking into flouric glass containing aluminum for lenses of low dispersion power.
C 1p RS:HS 23.182

10582. 1856–10–25 From Thomas Maclear Royal Observatory, Cape of Good Hope
Sends some photographs with comments. Pleased to hear JH is in much better health. Received a circular from U. J. J. Leverrier requesting meteorological information. Sends meteorological observations made at the Observatory during the last fourteen years.
AL 4pp inc RS:HS 12.162

10583. 1856–11–5 From Peter le Neve Foster Society of Arts, Adelphi
Thought that the book presented by JH had come from

Algernon Percy (4th Duke of Northumberland); regrets the error.
ALS 2pp RS:HS 7.323

10584. 1856–11–8 From Augustus De Morgan 7 Camden St.
Hopes his health is improving as his own is. Saw Henry Warburton last week. Gives one of his own theorems. Sends a lecture on the decimal coinage.
ALS 4pp RS:HS 6.291 S. De Morgan, *Memoir of Augustus De Morgan*, 292

10585. 1856–11–9 To George Gabriel Stokes Collingwood
Needs more time for reviewing Pole's paper [see JH's 1856–10–12].
C 1p RS:HS 25.10.12

10586. 1856–11–11 To Augustus De Morgan n.p.
Comments on the state of JH's health, color blindness, missing R.A.S. *Notices,* and decimal coinage.
C 2pp RS:HS 23.183

10587. 1856–11–12 From Sir William Pole 8 Great George St.
Gives particulars of his vision, which WP characterizes as dichromic. Recounts experiments with red glass sent by JH to WP. Offers to visit if JH wants personal interview.
ALS 4pp RS:HS 14.19

10588. 1856–11–13 From Augustus De Morgan 7 Camden St.
Pleased to hear his health is improving, but he should rest in the morning. Thinks it will be a long time before the decimal coinage can be introduced. Comments on cooking in the next century. Has been looking at a 10" bar of aluminum, which [Thomas?] Graham has sent him.
ALS 4pp RS:HS 6.292 S. De Morgan, *Memoir of Augustus De Morgan*, 291–2

10589. 1856–11–13 From William Sharpey Royal Society, Somerset House
Sends photograph of drawing of lunar crater Copernicus made with reflector at Rome, to aid B.A.A.S.'s effort in comparing moon's surface with that of earth.
ALS 3pp RS MM.16.163

10590. 1856–11–14 From T. Dudgeon Cargen near Dumfries
Hears that JH is working on the theories of color blindness and encloses some China silks that may prove useful.
ALS 2pp RS:HS 6.509

10591. 1856–11–14 To George Gabriel Stokes Collingwood
Is sending in JH's report on Pole's paper [see JH's 1856–11–9].
C 1p RS:HS 25.10.13a

10592. 1856–11–14 To George Gabriel Stokes Collingwood
Thanks the council of the R.S.L. through GS for the gift of a portrait of Nicholas Copernicus.
C 1p RS:HS 25.10.13b

10593. 1856–11–17 From Sir William Pole 8 Great George St.
Thanks JH for note explaining colored glasses and 'positive' colors, saying WP only had a chance to do experiments with 'negative' colors thus far. Accepts JH's invitation to visit.
ALS 2pp RS:HS 14.20

10594. 1856–11–21 From Achille Brachet Paris
Sends a few explanations of G. B. Amici's third and sixth series microscopes.
ALS 1p RS:HS 4.213

10595. 1856–11–24 From Hannah M. Bouvier Crosswick, N.J.
Sent him a few months ago a copy of her book *Familiar Astronomy*. Wonders if it has miscarried as she has heard nothing.
ALS 1p RS:HS 5.91

10596. 1856–11–[30] From Achille Brachet Paris
Was pleased to receive his letter. Further concerning G. B. Amici's microscopes.
ALS 2pp RS:HS 4.214

10597. 1856–12–2 To James Booth Collingwood
Compliments JB on publication of 'Trigonometry of the Parabola.' Recalls JH's papers on catenaries.
AdraftS 3pp TxU:H/L-0101; Reel 1054

10598. 1856–12–4 From James Maclaren Pashley House
Sends another pamphlet on the gold question. On the last page he will find JM has adopted JH's suggestion, though he does not give the source. Thinks that these suggestions should prove useful to France.
ALS 2pp RS:HS 12.29

10599. 1856–12–5 From Francis Charles Knowles Belvedere House, Ryde
Has received news of a cargo of cryolite from Greenland. Outlines the various possibilities of using it for the manufacture of chemicals and glass. Sends his method of manufacturing aluminium.
ALS 7pp RS:HS 11.66

10600. 1856–12–5 From Patrick Francis Robertson 8 Old Jewry
Needs advice concerning Egyptian money. Proposes coining money there as a solution to some of the problems.
ALS 2pp RS:HS 14.368

10601. 1856–12–6 To Francis Charles Knowles Collingwood
Would be inclined to try a combination of cryolite and silex to obtain a glass. Comments on his process for decomposing alumina by cyanide of sodium. Sorry to see that FK uses the current forms of chemical notation.
C 4pp RS:HS 11.67 & 23.184

10602. 1856[?]–12–11 To James David Forbes London
Letter of introduction for the Drs. Hermann and Robert Schlagentweit, whose scientific interests parallel JF's.
ALS 2pp St. Andrews 109

10603. 1856–12–15 From Henry W. Field Royal Mint
Congratulations on the success of his son at Addiscombe.
ALS 4pp RS:HS 7.215

10604. 1856–12–15 From Robert M. Morrell 4 Beaufort Bldgs., Strand
Forwards the draft of the words of the Memorial to be presented to Her Majesty. Nonmembers of the League have signed. On receipt of JH's reply will forward the written headings for JH's signature.
ALS 1p RS:HS 13.52

10605. 1856–12–23 From Henry W. Field Royal Mint
Pleased to hear of his improved health. Has sent some of the substance he requested. Affairs at the Mint not very

satisfactory.
ALS 8pp RS:HS 7.216

10606. 1856–12–28 From Edward Wolstenholme Ward Sydney
Sad to hear of JH's suffering and retirement from Mint. Describes production rates, problems, equipment, personnel, and quality of gold at Sydney mint. Starting an observatory. G. B. Airy promises to send astronomer. New South Wales legislature is not friendly to scientific research. Describes Australian society.
ALS 12pp RS:HS 18.375

10607. 1856–12–30 From John McClelland to Margaret Herschel London
Will always remember the happy day he spent at Collingwood. Spent the evening with Mr. Grant. Has ordered a sea picture for JH. Has obtained the remainder of the Griffith papers.
ALS 3pp RS:HS 12.8

10608. [1857 or later] From Sophia Beck n.p.
All is well. Hope JH hears 'good accounts from India.'
ALS 1p inc RAS:JH Archive 12/1.3.7 verso; Reel 10

10609. [1857–1] From Jean Baptiste André Dumas n.p.
Discusses an edition of Antoine Lavoisier's works [3 vols., 1862–65] edited by JD. Comments on the edition and on the arrangements for JH receiving it.
ALS 2pp (French) RS:HS 6.511

10610. 1857 From W. G. Penny Nottingham Terrace
Sends a copy of his *Astronomical Investigations.* Claims there is an 'inequality in the motions of Mars and the Earth.'
ALS 2pp RS:HS 14.153

10611. 1857 From Henri Victor Regnault n.p.
Discusses his itinerary for the British Isles. Gives address where he may be reached. Wishes to visit JH at Collingwood. Congratulates JH on his improved health.
ALS 2pp (French) RS:HS 14.320

10612. [1857?] From Edward Sabine n.p.
Returns polar maps. Advises on methods of map projection, favoring polar projection. Working on new maps.
ALS 2pp RS:HS 15.330

10613. 1857–1–2 From Sir Edward Ryan Garden Lodge
Sends information concerning India in response to letter of JH's son. [Sir Charles Edward] Trevelyan has printed such information occasionally. Many do not see the importance of India.
ALS 4pp RS:HS 14.463

10614. [1857]–1–6 From W. B. Clegram Gloucester
J. C. Symons is spreading locally false theories about the rotation of the moon. Would like JH's views on this.
ALS 9pp RS:HS 5.349

10615. 1857–1–8 To W. B. Clegram Collingwood
Regrets that he thinks the faith of the public in astronomy should be shaken by the errors of J. C. Symons. Gives the theory of the rotation of the moon.
draft 3pp RS:HS 5.350 & 23.186

10616. 1857–1–14 To James David Forbes Collingwood
Thanks JF for his dissertation on the Progress of Mathematical and Physical Science. Discusses health of JF and JH. Suggests that some actinometric measurements are not as seriously flawed as at first thought [see JH's 1847–3–1].
ALS 5pp St. Andrews 55a (C: RS:HS 23.187)

10617. 1857–1–16 From Augustus De Morgan 7 Camden St.
Thanks for the magic square, which he is now returning. Seems no end to the possibilities of such squares.
ALS 4pp RS:HS 6.293

10618. 1857–1–16 To William Henry Sykes [Collingwood?]
Interprets [John] Dalton's theory of the nonelasticity of gas particles, expressing its limitations, particularly in regard to a Mr. Paton's paper.
ALS 4pp RS:HS17.152 (C: RS:HS 23.188)

10619. 1857–1–24 From George Peacock Deanery, Ely
About the state of GP's health, and plans for a vacation,
ALS 4pp RS:HS 13.307

10620. 1857–1–26 From G. F. Daniell near Chichester
Forwarding a copy of a letter written by GD's father in which he quotes from JH's writings. Is grateful for the tribute he paid to his father at the B.A.A.S. meeting.
ALS 1p RS:HS 6.469

10621. 1857–1–27 From William Henry Smyth St. John's Lodge
Asks JH to recommend C. P. Smyth for R.S.L. fellowship. Will avoid R.S.L. meetings due to declining health. Considers a second edition of *Cycle of Celestial Objects.*
ALS 2pp RS:HS 16.301

10622. 1857–1–29 To William Henry Smyth [Collingwood]
Agrees to support C. P. Smyth for membership in R.S.L. Objects to C. P. Smyth's statement regarding his discoveries observing at high altitudes. Stresses importance of variable star work.
ALS 7pp RS:HS 16.302 (C: RS:HS 23.189)

10623. 1857–2 To Augustus De Morgan Collingwood
Offers AD a puzzle and JH's solution to AD's problem about a body revolving in the evolute of an ellipse. On sending Francis Baily's correspondence to Greenwich.
C 2pp RS:HS 23.190

10624. 1857–2 From William Henry Smyth St. John's Lodge
Thanks JH for discovering an important error in a work WS asked him to proof read. Asks JH about a variable in Canis Minor over which there is some debate.
ALS 2pp RS:HS 16.303

10625. 1857–2–2 To Secretary, Académie des sciences Collingwood
Thanks for having sent various publications; believes that the Academy's intention has been to send JH complete series; states which issues JH has not received.
ALS 4pp Académie des sciences (Paris)

10626. 1857–2–6 From George Biddell Airy London
Writes to arrange a visit to Collingwood.
ACCS 4pp RGO 6.377.429

10627. 1857–2–7 From William Hopkins Cambridge
Has been absent from home, which accounts for his belated reply. Comments on JH's difficulties in relation to W. H. Hopkins's memoir on the external temperature of the earth. JH's son is progressing well.
ALS 8pp RS:HS 9.478

10628. 1857–2–8 From Augustus De Morgan 7 Camden St.
JH has got hold of the explanation about central forces. Need not worry about Elizabeth Baily; he will see to that. Do not reject any letters; they may be important in the future. Gives one of his own theorems.
ALS 4pp RS:HS 6.294

10629. 1857–2–27 From Michel Chevalier Paris
Has received the agreement noted on his ticket of 14 Feb.
ALS 2pp (French) RS:HS 5.252

10630. 1857–3–2 To Hannah M. Bouvier Collingwood
Thanks for HB's *Familiar Astronomy*. Praises it as a good introduction to astronomy.
ALS 2pp Huntington Library

10631. 1857–3–2 From John Tyndall Royal Institution
Sends sketch of lecture on structure and motion of glaciers. Wants to conduct experiments on physical properties of ice, and applying polarized light. Beyond paper, is interested in cleavage of ice.
ALS 4pp RS:HS 17.385 (C: RI 503)

10632. 1857–3–3 From William Henry Smyth St. John's Lodge
JH's letter of recommendation for C. P. Smyth misdirected. Little time remains for nomination.
ALS 2pp RS:HS 16.304

10633. 1857–3–10 From Samuel Tyler Trevenick City, Md.
Informs JH that he wrote a book on Baconian philosophy inspired by JH's *Prelim. Discourse*.
ALS 3pp RS:HS 17.455

10634. 1857–3–15 To Edward Sabine Collingwood
Writing article [for *Encyclopaedia Britannica*] on meteorology. Asks questions regarding work of [H. W.] Dove and [Thomas?] Taylor regarding cyclone and storm theory. Discusses barometric fluctuation.
C 2pp RS:HS 23.191

10635. 1857–3–15 To John Wrottesley Collingwood
Comments on a number of resolutions of the R.S.L., as always encouraging keeping government activity at a distance.
C 2pp RS:HS 25.12.3

10636. 1857–3–17 To George Biddell Airy Collingwood
Expresses unease about P. S. Laplace's formulation of the barometric determination of height; in a postscript comments that Michael Faraday's lecture on conservation of force is a 'strange production.'
ALS 4pp RGO 6.251.265

10637. 1857–3–17 To John Tyndall Collingwood
Thanks for lecture on glacier cleavage. Wonders how 'regelation' occurs when two masses of ice at 32° are placed in contact.
C 4pp RS:HS 17.386

10638. 1857–3–18 To Roderick Impey Murchison Collingwood
JH's health is improving. Sends a portion of a letter from Thomas Maclear regarding Dr. [David] Livingstone's observations. Suggests the portion be read at the Geographical Society. Asks RM to invite Livingstone to visit JH.
ALS 4pp encl Royal Geographical Society

10639. 1857–3–19 From Edward Sabine 13 Ashley Place
Discusses [H. W.] Dove's works regarding rotating storms. New information from Russian observatories gives a more complete understanding of the phenomenon. Has not seen Keith Johnson's new edition of *Physical Atlas*.
ALS 7pp RS:HS 15.240

10640. 1857–3–19 From Charles Piazzi Smyth Hillside Crescent, Edinburgh
Sends atmospheric observations to JH before transmission to Admiralty.
ALS 2pp RS:HS 16.234

10641. 1857–3–20 From George Biddell Airy Royal Observatory, Greenwich
Regarding P. S. Laplace's barometric measurements. Michael Faraday.
ALS 4pp RS:HS 1.233 (C: RGO 6.251.167)

10642. 1857–3–20 From John Tyndall Royal Institution
Remarks on physical properties of ice and their bearing on glacial phenomena. Asks for help inventing term for idea of 'fracture and renewal' of glacial bendings.
ALS 4pp RS:HS 17.387 (C: RI 504)

10643. 1857–3–26 From Adalbert Adolf Mühry Göttingen
Would like to dedicate to JH a book on climatology that AM is preparing.
ALS 3pp (German) RS:HS 12.375

10644. 1857–3–29 From Augustus De Morgan 7 Camden St.
Sees by the announcement that he is to be out in April. Is there anything he can do in the way of proof reading? Is preparing answers to 65 queries of S. J. Loyd (1st Baron Overstone) on the decimal coinage.
ALS 2pp RS:HS 6.295

10645. 1857–4–1 From Hannah M. Bouvier Crosswick, N.J.
Thanks for his favorable comments on her book. Sends proof sheets as part of her forthcoming book on authors.
ALS 2pp RS:HS 5.92

10646. 1857–4–1 To Augustus De Morgan Collingwood
Reading proofs of JH's *Essays Q. E. R.*; working on an article on meteorology.
C 2pp RS:HS 23.192

10647. 1857–4–2 From Augustus De Morgan 7 Camden St.
Sending a little joke on a political theme. Further regarding a query of JH's on a rabbit. Gives a problem on a polygon. Meteorology is peculiar as weather does not appear to be governed by any laws.
ALS 4pp RS:HS 6.296

10648. 1857–4–2 From Henry James Ordnance Map Office
Is glad he likes the idea of projecting more than a hemisphere. Compares his own projection and a stereoscopic projection and comments on the results. Is now having another projection made that should give an accurate map of America.
ALS 4pp RS:HS 10.271

10649. 1857–4–3 To Henry James n.p.
Thanks for his letter and projection. Compares it with one of his own devising and encloses a chart of the isothermal lines of the Northern hemisphere. Has used a different system for

his *Cape Results*.
C 4pp RS:HS 10.272 & 23.194

10650. 1857–4–3 To Edward Sabine Collingwood
Thanks ES for his paper. Sun spot activity should increase in 1857. Discusses his own earlier ideas concerning relationship between sun spots and Aurora Borealis.
C 4pp RS:HS 15.241 (C: RS:HS 23.193)

10651. 1857–4–3 To the London *Times* n.p.
Urges that observations be made of sunspots. Notes that sunspots are much more evident now than in 1856. [Written by JH under the pseudonym 'Helioscopus']
L London *Times* (April 6, 1857), p. 12, col. 4

10652. 1857–4–6 From Henry James Ordnance Map Office
Is grateful for his valuable remarks on the various projections of the sphere. Thinks the projection devised by JH to be well suited for isothermal lines. Has constructed a chart of the polar regions for the same purpose. Will try JH's projection.
ALS 3pp RS:HS 10.273

10653. 1857–4–6 From Edward Sabine 13 Ashley Place
Mentions [P.] Plantamour's memoir on barometer and a criticism by Mr. Dilams[?]. Comprehending why variations of magnetic disturbances correspond in period with terrestrial year is difficult.
ALS 3pp RS:HS 15.243

10654. 1857–4–6 From Edward Sabine 13 Ashley Place
Glad to know which of ES's papers JH has. Sends another along with maps from [Keith] Johnson's *Physical Atlas*. Discusses JH's ideas regarding relationship between solar spots and magnetic disturbances.
ALS 4pp RS:HS 15.242

10655. 1857–4–8 To Edward Sabine Collingwood
Discusses advantages of different magnetic charts. Grateful for [P.] Plantamour's observations. Satisfied with [J. R.] Wolf's period of 11.11 years for solar spots.
C 2pp RS:HS 23.195

10656. 1857–4–11 From Thomas Spring-Rice Park St
Questions from Lord Overstone [S. J. Loyd] about decimal coinage system.
ALS 3pp TxU:H/M–0377; Reel 1087

10657. 1857–4–12 To William Whewell Collingwood
Looking forward to a visit from WW. Glad to hear that WW is editing Richard Jones's unpublished works. Sad to hear that George Peacock is seriously ill. Comments negatively on a book by Herbert Spencer.
ALS 4pp TC, Camb. Add. Ms.a.207[92] (C: RS:HS 23.196)

10658. 1857–4–13 From John Welsh Kew Observatory, Richmond
Reliability of barometer and thermometer measurements from balloons. Compares results of several European observers. Suggests new expedition to Teneriffe [Canary Islands].
ALS 4pp RS:HS 18.140

10659. 1857–4–14 From Mary Somerville Florence
Thanks JH for critiquing her paper during his illness. MS's *On the Connexion of the Physical Sciences* is to be published in translation at Cesna.
ALS 4pp RS:HS 16.359

10660. 1857–4–15 From James Glaisher Lewisham
Has the blocks for the snow crystals. Can easily take some photographs if he desires them. Also has an article on crystals.
ALS 4pp RS:HS 8.128

10661. 1857–4–16 To George Biddell Airy Collingwood
Suggests ways of eliminating transmission of vibration to a piece of sensitive equipment.
C 3pp RS:HS 23.197

10662. 1857–4–16 From George Peacock 4 Grand Parade, St. Leonards-on-Sea
Is too ill to visit JH as he had intended.
ALS 4pp RS:HS 13.308

10663. 1857–4–24 To F[rederick] J[ames] Furnivall Collingwood
FF having suggested some time ago the publication of a collection of JH's essay, JH now offers to send FF JH's *Essays Q.E.R.*
ALS 1p Huntington Library

10664. 1857–4–25 From Frederick J[ames] Furnivall Lincolns Inn
Will be delighted to receive a copy of his collected essays. These are very valuable and enhance his reputation.
ALS 2pp RS:HS 7.387

10665. 1857–4–26 From James Glaisher Lewisham
Sending him engraved single copies, also photographic copies of the larger ones. Comments on his researches into snow crystals.
ALS 4pp RS:HS 8.129

10666. 1857–4–28 From Sir Joseph D. Hooker Kew
JH's letter of resignation was read at a full meeting of the [R.S.L.] club last evening and it was the unanimous wish of those present that JH should not resign.
ALS 4pp RS:HS 10.205

10667. 1857–4–28 From Smith, Elder & Co. 65 Cornhill
Note clarifying items of account.
ALS 2pp RS:HS 16.203

10668. 1857–5–1 From John Welsh Kew Observatory, Richmond
Cannot answer JH's question about electricity in heavy showers. Observations were discontinued years ago. Refers to Francis Ronalds's past experiments. Cites 1849 *B.A.A.S. Rep.* for map of storm observations. Notes changes in polarity of electricity during storms.
ALS 4pp RS:HS 18.141

10669. 1857–5–5 To William Whewell [Collingwood]
Invites WW to Collingwood, informing him of other visitors coming and new rail schedule.
ALS 3pp TC, Camb. Add. Ms.a.207[93]

10670. 1857–5–6 To Warren de La Rue Collingwood
Thanks for engravings of Saturn and Jupiter; would like to see an astronomical picture book produced.
C 1p RS:HS 23.198

10671. 1857–5–8 From Thomas Spring-Rice Exchequer
Encloses report by Lord Overstone [S. J. Loyd, on decimal coinage]. Will answer JH's note tomorrow.
ALS 1p TxU:H/M–0378; Reel 1087

10672. 1857–5–9 From Thomas Spring-Rice Exchequer
Just now received 'blue book' on decimal coinage from printer. Forwards it to JH. Apologizes for putting this burden on JH, but TS wants to be fully aware of all issues relating to decimal coinage. Answers JH's questions about

report.
ALS 2pp TxU:H/M–0379; Reel 1087

10673. 1857–5–11 From Warren de La Rue 110 Bunhill Row
Is pleased that JH approves of his drawings of the planets. Hopes to publish some drawings of the other planets in the future. Has moved his observatory but hopes soon to start observations. Unable to comment on Otto Struve's observations.
ALS 3pp RS:HS 6.139

10674. 1857–5–12 From Thomas Spring-Rice Exchequer
Quotes 1682 tract by William Petty advising Marquis of Halifax to employ decimal system of currency.
ALS 2pp TxU:H/M–0380; Reel 1087

10675. 1857–5–15 To Edward Sabine Collingwood
Heard ES recommended JH write an article for *Edinburgh Review* on terrestrial magnetism. Asks what he should read to write this article. Photographs ES sent of sun not promising.
C 2pp RS:HS 23.199

10676. 1857–5–15 From Smith, Elder & Co. 65 Cornhill
Further account clarification [see SE's 1857–4–28].
ALS 2pp RS:HS 16.204

10677. 1857–[5?]–15 To Thomas Spring-Rice Collingwood
Responds to questions by Lord Overstone [on decimal coinage]. TS may edit these as needed.
ALS 2pp TxU:H/L–0264; Reel 1054

10678. 1857–5–16 To Augustus De Morgan Collingwood
On some odd results JH has arrived at about P. S. Laplace's barometric formula.
C 1p RS:HS 23.200

10679. 1857–5–20 From Augustus De Morgan 7 Camden St.
Returns his cancel, which he found very amusing. Mrs. De Morgan is now at Hampstead suffering from an incurable complaint. Further comments on the beats of musical consonances and Robert Smith's theories.
ALS 4pp RS:HS 6.297

10680. 1857–5–21 From George Biddell Airy Royal Observatory, Greenwich
Note accompanying forwarding of a paper by Otto Struve, and other minor matters.
ALS 3pp RS:HS 1.233a (C: RGO 6.377.431)

10681. 1857–5–22 From Charles Hay Cameron Putney Heath
Thanks for the valuable gift of his *Essays Q.E.R.* Has just moved into this cottage and would welcome a visit from JH when in the neighborhood.
ALS 3pp RS:HS 5.148

10682. 1857–5–23 From Adam Sedgwick Lowestoft
JH's *Essays* [*Essays Q.E.R.*] has reached Lowestoft, but AS has been too ill to read them. Sends regards to family.
ALS 4pp RS:HS 15.450

10683. 1857–5–26 From George Peacock Deanery, Ely
Thanks JH for a copy of his *Essays Q.E.R.*, and adds some family news.
ALS 3pp RS:HS 13.309

10684. 1857–5–26 From Edward Sabine 13 Ashley Place
Asks JH to read ES's publications for the R.S.L. proceedings concerning colonial magnetic observations. Wishes JH could see R.S.L.'s recent accomplishments in magnetism.
ALS 4pp RS:HS 15.244

10685. 1857–5–27 From William Rutter Dawes Haddenham, Thame
Is grateful for the copy of his *Essays*. Query about measuring the focal length of a double object glass. Gives results of his recent observations. Hopes accounts of JH's son are good.
ALS 4pp RS:HS 6.104

10686. 1857–5–28 To Humphrey Lloyd Collingwood
Regrets that he cannot attend B.A.A.S. meeting and stay with HL; thanks HL for his paper on the undulating theory.
C 1p RS:HS 23.202

10687. 1857–5–31 From Alfred Gautier Lowestoft
Was pleased to receive his letter of the 28th. Proposes to visit JH on 3 June and will travel via Etchingham. Looks forward to meeting Lady JH and the family.
ALS 2pp (French) RS:HS 8.89

10688. 1857–6–8 From Richard Christopher Carrington Observatory, Redhill
Has heard that he is now residing at Collingwood and sends two papers for his perusal. Is the sketch correct? Also encloses two specimen pages of his forthcoming catalogue.
ALS 3pp RS:HS 5.200

10689. 1857–6–9 To Edward Sabine Collingwood
Willing to write article if *Quarterly* [*Review*] will publish it, if he may take his time writing, and if ES will provide aid regarding meteorology and magnetism.
C 1p RS:HS 23.203

10690. 1857–6–10 To Augustus De Morgan Collingwood
On magic squares.
C 2pp RS:HS 23.204

10691. 1857–6–13 From Alexander Bridport Becher Admiralty
Sends copy of proposed warrant of appointments to Board of Visitors of Royal Observatory. Asks JH to make alterations and forward these to John Wrottesley by [17 June].
ALS 2pp TxU:H/M–0072; Reel 1087

10692. 1857–6–16 From Annibal de Gasparis Naples
Requesting JH's interest in the work of Oscar Capocci.
ALS 1p (French) RS:HS 8.48

10693. 1857–6–17 From Arthur Swain Howard Lowe Highfield House near Nottingham
Sends barometer readings.
ALS 5pp encl Harvard: Houghton AAH 67m–67 (92)

10694. 1857[?]–6–20 From Christian F. Schönbein n.p.
Print of [Leonhard] Euler's portrait has been made. Copy being sent to JH.
ALS 1p RS:HS 15.366

10695. 1857–6–25 From K. B. Martin Harbour Office, Ramsgate
Sends data from Harbour Office; discusses observation of the tides.
ALS 2pp diag Harvard: Houghton AAH 67m–67 (95)

10696. 1857–6–25 From Edward Sabine 13 Ashley Place
Glad JH is reviewing advancement of science by R.S.L. members. Disappointed that little was accomplished in

certain areas where more was expected.
ALS 3pp RS:HS 15.245

10697. 1857–7 From Sir William Pole 3 Bird Cage Walk
Explains why his paper on color vision took the artist, dyer, etc., point of view rather than that of scientific chromatology. Willing to investigate a more scientific approach and asks permission for another visit.
ALS 2pp RS:HS 14.21

10698. 1857–7–2 From Archibald Smith Decimal Coinage Commission
Encloses, for JH's corrections, proofs of JH's reply to Lord Overstone's questions.
ALS 2pp TxU:H/M–0485; Reel 1093

10699. 1857–7–3 From Rudolf Wolf Zürich
A note of thanks upon receipt of a copy of JH's *Cape Results.*
ALS 1p (German) RS:HS 19.298

10700. 1857–7–11 From Archibald Smith Lincoln's Inn
Encloses proof of revision of JH's reply to Lord Overstone's questions. Will print 25 to 50 copies. Does JH's report on Canadian currency to House of Commons contain anything new that commissioners should know about?
ALS 3pp TxU:H/M–0486; Reel 1093

10701. 1857–7–13 From Michel Chevalier Paris
On gold and other monetary matters.
ALS 2pp (French) RS:HS 5.253

10702. 1857–7–13 From Edward Joseph Lowe Beeston
Does not quite understand JH's note; can he clarify the reading he requires? Comments on his readings on thunderstorms.
ALS 3pp RS:HS 11.329

10703. 1857–7–28 From John Bullar Basset Wood near Southampton
Hopes that reports he has heard of his health are unfounded. The wing of the infirmary has been completed.
ALS 3pp RS:HS 4.328

10704. 1857–8–7 From Augustus De Morgan 7 Camden St.
Hopes his health is improving. Where can he find the article expressing adjectives in algebraic symbols? Went to the R.A.S. in July and found much as usual. Mrs. De M. now at Hythe with the children. Elizabeth Baily appears to be in excellent health.
ALS 4pp RS:HS 6.298

10705. 1857–8–10 From Henri Victor Regnault Sèvres
Regrets deeply having missed seeing JH. Sends greetings to JH's family. Enjoyed trip around British Isles. Will now continue scientific work.
ALS 2pp (French) RS:HS 14.314

10706. [1857–8–11 or later] From Henri Victor Regnault to Margaret Herschel Sèvres
Apologizes profusely for not having written earlier. Thanks for a wonderful stay in England. Has been pressured into new and undesired positions. Sends porcelain tea service from Sèvres.
C 4pp (French) RS:HS 14.315

10707. 1857–8–15 From Sir William Pole 3 Bird Cage Walk
Plans to visit JH next week [to discuss color blindness].
ALS 1p RS:HS 14.22

10708. 1857–8–17 From Sir William Pole 3 Bird Cage Walk
Will bring [M. E.] Chevreul's color diagrams and new spectrum representation when he visits JH; confirms mode and date of arrival.
ALS 2pp RS:HS 14.23

10709. 1857–8–18 To John Herschel (son) Collingwood
The letter contains financial arrangements with son John while still in England, extensive comments on the fighting and restructuring of the colonial system in India, some matters relating to earth strata, and finally an extended discussion of the 'Cavendish experiment' [for measuring the gravitational attraction between lead balls].
ALS 7pp JHS 5.6

10710. 1857–8–20 From W. H. Barton Royal Mint
Thanks for the gift of the book. Regarding some financial affairs of JH.
ALS 3pp RS:HS 3.297

10711. 1857–8–27 To Abbé François Moigno Collingwood
Invites FM to visit him at Collingwood on his return trip to Paris from the B.A.A.S. meeting in Dublin.
ALS 4pp Dunedin Public Lib.

10712. 1857–8–30 From Sir William Pole Clérmont, France
Thanks Herschels for their hospitality and tells about layers of volcanic ash encountered on a walk.
ALS 4pp RS:HS 14.24

10713. 1857–9–7 From Jean B. L. Foucault Dublin
Is honored by his kind invitation to visit him, which he hopes to accept. Has great respect for JH's work.
ALS 2pp (French) RS:HS 7.377

10714. 1857–9–7 From François N. M. Moigno Dublin
Has received the letter and invitation. Will travel to Etchingham station and walk to Collingwood.
ALS 2pp (French) RS:HS 12.354

10715. 1857–9–7 To [R.S.L. Council] Collingwood
Is returning C. P. Smyth's paper ['Report on the Teneriffe Astronomical Experiment of 1856,' [*R.S.P.T., 148,* 165–], believing it now 'unobjectionable for publication.'
ALS 1p RS RR.3.251

10716. 1857–9–17 To William Whewell Collingwood
Thanks WW for letters. Has been visited by [J. B. L.] Foucault; impressed by Foucault's instruments. Notes large number of sunspots.
ALS 4pp TC, Camb. Add. Ms.a.207[99] (C: RS:HS 23.205)

10717. 1857–9–25 To Warren de La Rue Collingwood
Thanks for photograph of the moon.
C 1p RS:HS 23.206

10718. 1857–9–25 From J. H. Symons Bristol
Will send JS's treatise; wishes could offer something 'more intrinsically valuable' as token of gratitude for everything.
ALS 4pp RS:HS 17.254

10719. 1857–9–28 From Warren de La Rue 110 Bunhill Row
Has not been able to go to Kew Observatory yet. The photoheliograph has to await John Welsh's return from Scotland. A reflector with a diagonal mirror would be suitable for celestial objects. Has now removed to Cranford.
ALS 2pp RS:HS 6.140

10720. 1857–9–29 To George Biddell Airy Collingwood
Asks GA for clarification of the value for the 'mean equatorial horizontal parallax' of the moon, since two different values are given by different authors.
ALS 4pp RGO 6.377.433

10721. 1857–9–30 From George Biddell Airy n.p.
Offers some thoughts on JH's problem [see JH's 1857–9–29], but needs additional time before commenting more fully.
ACCS 3pp RGO 6.377.435

10722. 1857–10–1 From George Biddell Airy Royal Observatory, Greenwich
Regarding errors in Francis Baily's tables.
ALS 3pp RS:HS 1.234

10723. 1857–10–1 From Augustus De Morgan 7 Camden St.
Decimal Association will not publish answers to questions by Lord Overstone [S. J. Loyd] until next session. Seeks consent of JH and other members to publish these sooner.
ALS 2pp TxU:H/M–0163; Reel 1087

10724. 1857–10–1 From Alexander Hamilton-Gordon Argyll House
Has been very busy; otherwise he would have acknowledged his cheque before. Is glad to hear better news from India. Bad organization for the arming of the British troops in India.
ALS 4pp RS:HS 8.150

10725. 1857–10–3 From George Biddell Airy Royal Observatory, Greenwich
Regarding orbits of comets. Lunar elements JH requires are given in P. A. Hansen's tables.
ALS 3pp RS:HS 1.235 (C: RGO 6.377.437)

10726. 1857–10–5 From James Stark Edinburgh
Asks JH to look over JS's paper on correcting tables for barometric readings.
ALS 4pp RS:HS 17.1

10727. 1857–10–6 From R[obert] Few Covent Garden
Has been unable to find the copy of the marriage settlement [for Margaret Louisa Herschel?].
ALS 3pp RS:HS 7.212

10728. 1857–10–8 To Augustus De Morgan Collingwood
Regarding the possibility of decimal coinage. Washes his hands of adjectives expressed in algebraic form. Has received good news from India.
draft 2pp RS:HS 6.299 & 23.207

10729. 1857–10–8 To James Stark Collingwood
Discusses JS's paper on barometric table corrections in relation to mercury readings.
ALS 2pp RS:HS 17.2 (C: RS:HS 23.208)

10730. 1857–10–9 From Augustus De Morgan 7 Camden St.
S. J. Loyd's (1st Baron Overstone) queries and the work of the Commission on the Decimal Coinage. Is sure he read the algebraic adjectives in the *Memoirs of the Analytical Society*. May be related in some way. His wife is slowly recovering. Has JH read about the Devil's Elixir?
ALS 7pp RS:HS 6.300 S. De Morgan, *Memoir of Augustus De Morgan*, 293

10731. 1857–10–11 From Augustus De Morgan 7 Camden St.
Has received a list of witnesses whose answers have been printed for Thomas Spring-Rice (1st Baron Monteagle).
ALS 1p RS:HS 6.301

10732. 1857–10–14 From Augustus De Morgan 7 Camden St.
Hopes the Indian mail has taken a weight off JH's mind. George Peacock is anxious that JH should be in their plot regarding the introduction of the decimal coinage.
ALS 3pp RS:HS 6.302

10733. 1857–10–14 From James Stark Edinburgh
Thanks JH for proofing his paper on barometric reading corrections.
ALS 1p RS:HS 17.3

10734. 1857–10–19 From Humphrey Lloyd Paris
At the last meeting of the B.A.A.S. a resolution was adopted calling for the continuance of the Magnetical Observations. Would work through the joint R.S.L. and R.A.S. committees as before. Would JH agree to continue as chairman? Let him know at the Athenaeum where he returns in a few days.
ALS 4pp RS:HS 11.287

10735. 1857–10–21 To Augustus De Morgan Collingwood
Comments about JH's ancestor, Hercules, and replies to S. J. Loyd's (1st Baron Overstone) queries [see AD's 1857–10–9].
C 1p RS:HS 23.210

10736. 1857–10–21 To William Lassell Collingwood
Is preparing a new edition of *Outlines Astr.*, and would like to know from WL the status of each of the satellites of Uranus and Neptune.
C 1p RS:HS 23.209

10737. 1857–10–23 From Augustus De Morgan 7 Camden St.
Will send the proofs. Has heard all about JH's ancestor [Hercules]. Miss Louisa has committed manslaughter.
ALS 4pp RS:HS 6.303

10738. 1857–10–24 To Humphrey Lloyd Collingwood
Begs not to be chairman of the proposed Magnetic Conference Committee [see HL's 1857–10–19], but would serve on the committee.
C 1p RS:HS 23.211

10739. 1857–10–24 From Woodbine Parish St. Leonards-on-Sea
Places JH's name in a volume containing an account of the meteorite that struck ground [in North America]. WP also includes a description of the coca trees of Peru [in which JH took an interest].
ALS 1p RS:HS 14.134

10740. 1857–10–29 From William Lassell Sandfield Park near Liverpool
Was indebted to JH for the names of the satellites of Uranus. Will find details of his observations in the *R.A.S.M.N.* Gives volumes. Is thinking of returning to Malta and erect a larger telescope.
ALS 4pp RS:HS 11.150

10741. 1857–11–1 From Michel Chevalier Paris
Has finished his essay and returns to the subject of gold. Cites articles.
ALS 3pp (French) RS:HS 5.254

10742. 1857–11–1 To William Lassell Collingwood
Wishes all success to his proposed visit to Malta. Has been able to find the volumes of *R.A.S.M.N.* Comments on WL's

system of levers for his telescope.
draft 2pp RS:HS 11.151 & 23.212

10743. 1857–11–1 To George Gabriel Stokes Collingwood
Is willing to have his review sent on to the author, [C. P.] Smyth; asks GS to send the R.S.L.'s copy of Keith Johnson's *Physical Atlas* to JH.
C 1p RS:HS 25.10.14

10744. 1857–11–3 From Charles Babbage Dorset St.
Regarding a suitable engraving of William Herschel to be added to William Walker's engraving of a meeting at the Royal Institution.
ALS 2pp RS:HS 2.328

10745. 1857–11–8 To William Henry Smyth Collingwood
Pleased that WS included François Arago's lectures in his compilation. Believes the Italian edition could have been better edited. Remarks on the *Leviathan* problem.
C 1p RS:HS 23.213

10746. [1857–11–8 or earlier?] To William Henry Smyth n.p.
Has heard WS is publishing an English translation of François Arago's astronomical writings. Urges against including Arago's lectures if the Italian translation is indicative of their level of accuracy. Lists numerous errors in that edition. [Marked 'not sent.' Note added to CDraft (RS:HS 25.13.13) states: 'A letter differently worded but noting some of these points, but not all (for particular reasons) if I remember right was sent.'
AL 3pp inc RS:HS 16:323

10747. 1857–11–9 From William Keith Murray Crieff
Is pleased that JH approves of the photographs. Previously purchased W. H. Smyth's equatorial, the tube and object-glass made by [Thomas?] Cooke. Had a visit from Smyth recently and he advised WM to prepare a paper for the R.A.S.
ALS 4pp RS:HS 13.62

10748. 1857–11–11 From Augustus De Morgan 7 Camden St.
Has the proofs. Gives some queries concerning them.
ALS 1p RS:HS 6.304

10749. 1857–11–11 From William Lassell Sandfield Park near Liverpool
Thanks for his suggestions respecting the mode of observation by collimator; will give it careful consideration. Sends two photographs of the method of mounting his observation tower.
ALS 2pp RS:HS 11.152

10750. 1857–11–11 To [James Calder] Stewart n.p.
Sends £100 check for JS.
ALS 1p TxU:H/L-0453; Reel 1055

10751. 1857–11–12 From William Henry Smyth St. John's Lodge
Thanks JH for advice on several persons and for suggestion concerning the *Leviathan.* Remarks on R. C. Carrington's circumpolar star work. Writes to R.S.L. Council concerning François Arago's nomination for medal.
ALS 2pp RS:HS 16.305

10752. 1857–11–13 From Augustus De Morgan 7 Camden St.
S. J. Loyd (1st Baron Overstone) has agreed that the papers be presented to the Crown. Pray let him have his proof as soon as possible.
ALS 2pp RS:HS 6.305

10753. 1857–11–15 To Margaret Brodie Herschel [London]
Sends reports about sons Alexander and John at [Charles] Pritchard's school; there follow a number of minor matters of household information; JH will be very busy in the coming week.
ALS 4pp JHS 1.111

10754. 1857–11–17 From Ann Smyth St. John's Lodge
Leviathan successfully launched. Forwards papers from W. H. Smyth. Congratulates JH on coming 'happy event' [marriage of Margaret Herschel].
ALS 2pp RS:HS 16.216

10755. 1857–11–22 From John Bullar Basset Wood near Southampton
Is grateful for the copies of his essays, which he treasures. Hopes his anxiety for his son in India will be groundless.
ALS 4pp RS:HS 4.329

10756. 1857–11–26 To Augustus De Morgan n.p.
A proposition in perspective, and some nonsense.
C 1p RS:HS 23.215

10757. 1857–11–27 From Augustus De Morgan 7 Camden St.
Write to Mr. Richerly for as many as he wants. Has sent proof to George Peacock; might send one to the Dean of Salisbury. County persons are capital people for their purposes.
ALS 2pp RS:HS 6.306

10758. 1857–12 From Mary H. Lubbock High Elms, Farnborough, Kent
Preparing for wedding. Thanks JH for present. Happy that JH's daughter 'Lou' [Margaret Louisa Herschel?] will be bridesmaid.
ALS 3pp RS:HS 19.287

10759. 1857–12–3 To William Henry Smyth Collingwood
Makes suggestion on freeing the *Leviathan* from its trapped position.
C 2pp RS:HS 23.214

10760. 1857–12–7 From George Peacock Deanery, Ely
Family news and comments on some of JH's writings.
ALS 4p RS:HS 13.310

10761. 1857–12–26 To James Prescott Joule Collingwood
Thanks JJ for the counterfeit coin he sent; JH comments on the possibility of oval coinage.
C 1p RS:HS 23.216

10762. 1857–12–31 From Adalbert Adolf Mühry Göttingen
Letter accompanying AM's book on climatology.
ALS 4pp (German) RS:HS 12.376

10763. 1858 From Charles Babbage n.p.
Wishes to introduce an American friend, Professor Stephen Alexander, to JH.
ALS 1p RS:HS 2.349

10764. 1858–1 From Julia Margaret Cameron Putney Heath
Sends him photograph of his godchild. Would like some

lines from JH that he (the godchild) might keep.
ALS 3pp RS:HS 5.153

10765. 1858 To Augustus De Morgan n.p.
Is AD interested in the Lowndean Professorship at Cambridge?
C 1p RS:HS 23.251

10766. 1858–1–1 From Augustus De Morgan 7 Camden St.
Too busy to answer his last letter until now. His theorem on perspective is pretty and easy. Quotes one he uses. Has been busy finding the proof that every algebraic equation has a root. Has been organ tuning. Comments on the method of tuning using beats.
ALS 3pp RS:HS 6.307

10767. [1858 or later] From Sophia E. De Morgan 41 Chalcot Villas NW
Sending a magazine containing a notice of JH's paper on sensorial vision.
ALS 2pp RS:HS 6.435

10768. [1858 or later] From Sophia E. De Morgan n.p.
Sending some examples of sensorial vision
ALS 4pp RS:HS 6.434

10769. [1858] To William Robert Grove n.p.
Thanks for and comments on new edition of WG's *Correlation of Physical Forces*, particularly the subject of transformation of heat into motion.
ALS 2pp RI

10770. [1858?] To Margaret Brodie Herschel n.p.
JH spent most of the day before in a meeting, and with a man selling an engraving of scientists of 1806.
ALS 4pp JHS 4.10

10771. [1858?] To Margaret Brodie Herschel 2[?] Howard Place, Norfolk St.
Describes lodgings taken in London; JH is working hard on his *Physical Geography*.
ALS 4pp JHS 4.7

10772. [1858?] From Henri Victor Regnault n.p.
Expects that JH has received instructions sent through Mr. Royer. Informs JH that HR is sick and will visit some friends to recuperate.
ALS 1p (French) RS:HS 14.313

10773. [1858?] From Henri Victor Regnault n.p.
Will ask Parisian doctors for information JH wanted and transmit it immediately. Hopes change of air will improve JH's health. Regrets JH is not closer to the capital so that the best doctors could help. Is writing Mr. Royer to ask him to write to JH.
ALS 2pp (French) RS:HS 14:312

10774. [1858 to 1859] To Edward Sabine [Collingwood]
Thanks ES for receipt of information about magnetic curves and the work of [A. D.] Bache. JH is glad to hear ES is back to active work again, but JH says he is too ill to go to the B.A.A.S. meeting [in Aberdeen].
C 1p RS:HS 25.14.29

10775. 1858–1–1 From Charles Edward Trevelyan Treasury
Reports misplacement of letters of William Herschel after publication of extracts for *The Times*.
ALS 3pp RS:HS 17.444

10776. 1858–1–13 To John Henry Griesbach Collingwood
Sorry to hear he has been laid up, but his recovery has been quicker than his own. Unable to help him as his own lands have to be sold, but encloses a check for £20, which please acknowledge with a promissory note.
draft 3pp RS:HS 9.63

10777. 1858–1–17 From James Maclaren Pashley House
Is obliged for JH's observations on his statement of [Joseph?] Hume's doctrine. These errors are discreditable and will diminish the utility of the book.
ALS 2pp RS:HS 12.30

10778. 1858–1–18 From Charles Babbage Dorset St., Manchester Square
Sorry that CB's nephew bothered JH with request for money from fund for CB's sister. Nephew must deal directly with Reversionary Interest Society. Recalls mutual vow of JH and CB forty years ago to pursue many intellectual endeavors.
ALS 4pp TxU:H/M–0032; Reel 1062

10779. 1858–1–18 From Henry Anthony Jeffreys Parsonage, Hawkhurst
Requests an address of a Mr. Macintosh in order to inquire about a lecture topic.
ALS 1p Harvard: Houghton AAH 67m–67 (107)

10780. 1858–1–18 From H. C. Marshall Leeds
Regrets he will be unable to attend the forthcoming wedding.
ALS 2pp RS:HS 13.20

10781. 1858–1–20 To Mary Somerville Collingwood
Feeling better, but writing still painful. Writing articles on meteorology and geology for *Encyclopaedia Britannica.* May use information from her *Physical Geography.*
ALS 3pp RS:HS 16.360 (C: RS:HS 23.217)

10782. 1858–1–22 From John Bullar Basset Wood near Southampton
Sending him local paper so that he can see how the Oxford news has been received. Corporation of Southampton has come into a legacy.
ALS 4pp RS:HS 4.330

10783. 1858–1–22 From Leonard Horner 17 Queens Rd., West
Has had a volume of essays of JH's and wonders if the first one [address to the subscribers of Windsor Public Library] has been published separately; if not, he would like to issue it in a cheap form.
ALS 2pp RS:HS 9.495

10784. 1858–1–23 From William Sharpey R.S.L., Somerset House
On 10 Dec., JH, George Peacock, and William Whewell were named to R.S.L. committee to cooperate with B.A.A.S. for purpose of continuing observations of terrestrial magnetism.
ALS 3pp TxU:H/M–0479; Reel 1093

10785. 1858–1–25 To George Biddell Airy Collingwood
Thanks GA for a letter of introduction to friends in Rome [used by JH's daughter Margaret Louisa and her new husband, Reginald Dyke Marshall]; JH is willing to serve on a committee with GA and others if he does not have to go to London for meetings.
ALS 4pp RGO 6.694.471

10786. 1858–1–26 To Elizabeth Colling Collingwood
Thanks for the gift of a book of verse.
C 1p RS:HS 23.220

10787. 1858–1–26 From James Maclaren Pashley House
Has been thinking over the names of solicitors and believes he can recommend Messrs. Lewin of Southampton St. The senior partner is a brother of JM and JH should keep his papers when he concludes his business with his present solicitors.
ALS 3pp RS:HS 12.31

10788. 1858–1–29 From George Biddell Airy Royal Observatory, Greenwich
Wants someone else to take responsibility for committee on magnetic observations [see JH's 1858–1–25]; GA proposes Edward Sabine.
ALS 4pp TxU:H/M-0638.1; Reel 1084 (C: RGO 6.694.474)

10789. 1858–1–30 To [Roderick Impey Murchison?] Collingwood
Has been appointed, with William Whewell, to committee to help ensure continuance of observations on terrestrial magnetism. Needs names of those on B.A.A.S. committee with whom they will work.
ALS 4pp EUL:Gen 784/1/13

10790. 1858–1–30 To William Whewell Collingwood
Discusses B.A.A.S. business and asks for WW's views on some magnetic observations.
ALS 3pp TC, Camb. Add. Ms.a.207[95] (C: RS:HS 23.218)

10791. 1858–1–31 From P. O'Callaghan Cookridge Hall, Leeds
Sends a copy of [John?] Dalton's letter for which he is obliged. Expresses appreciation of his recent visit to Kent and Sussex.
ALS 3pp RS:HS 13.156

10792. 1858–1–31 To John Phillips Collingwood
Send names of members of B.A.A.S. committee appointed to cooperate with R.S.L. committee for purpose of procuring continuance of observations of terrestrial magnetism.
CS 3pp TxU:H/L-0302; Reel 1054

10793. 1858–2–1 From Augustus De Morgan 7 Camden St.
Congratulations and best wishes for JH's daughter recently married. If he has any material for the annual report, please send it.
ALS 4pp RS:HS 6.308

10794. 1858–2 To [John Herschel (son)] [Collingwood]
Louisa and her husband have arrived safely at Rome. JH poked his right eye with a stick, so now it is red.
ALS 2pp JHS 1.122

10795. [1858?]–2–1 From Edward Sabine 13 Ashley Place
Sends section X and earlier paper on Kew and Nertschinsk disturbances, which supply key to theory of magnetic disturbances.
ALS 3pp RS:HS 15.315

10796. 1858–2–2 To Augustus De Morgan Collingwood
Questions the exact beginning of the year 1857, and offers 'Old King Cole' in Latin.
C 1p RS:HS 23.221

10797. 1858–[2–4] From John Phillips B.A.A.S., Dublin
Humphrey Lloyd, Thomas Robinson, and Edward Sabine will form a committee to continue studying magnetic observations.
(A)LS 1p TxU:H/M–0428.23; Reel 1093

10798. 1858–2–4 From John Phillips B.A.A.S., Magdalen Bridge, Oxford
Forwards copy of the resolution of the General Committee of the B.A.A.S. concerning the continuation of study of magnetic observations.
ALS 2pp TxU:H/M–0428.22; Reel 1093

10799. 1858–2–4 From William Whewell Trinity Lodge
Approves G. B. Airy's suggestion concerning magnetic observations.
ALS 2pp TxU:H/M-5071

10800. [1858–2–6 or later] To Augustus De Morgan n.p.
Still worrying about where does the day begin?
C 1p RS:HS 23.219

10801. [1858]–2–6 From William Henry Fitton 4 Sussex Gardens
Dined yesterday in company with David Livingstone who would like to visit JH at Collingwood. Has been much impressed with Livingstone.
LS 4pp RS:HS 7.242

10802. 1858–2–6 To Edward Sabine n.p.
[Extract] Learned that JH was appointed by R.S.L. to committee to cooperate with B.A.A.S. to promote continuance of terrestrial magnetism survey. Asks ES to send committee summary of results already obtained and ES's opinion of how best to conduct future observations.
C 3pp TxU:H/L-0370; Reel 1054

10803. 1858–2–6 To Edward Sabine Collingwood
JH's improved health would allow him, if asked, to chair the Chemical Section at the B.A.A.S meeting. Notes that JH, G. B. Airy, William Whewell, and George Peacock have been appointed to a committee to cooperate with the B.A.A.S. committee dealing with the continuation of terrestrial magnetism observations.
C 2pp RS:HS 23:222

10804. 1858–2–7 From Augustus De Morgan 7 Camden St.
Queries on what meridian the year began. Returned a book to the R.S.L., which had been away for many years. Believes the R.S.L. Library was expurgated between 1734 and 1740.
ALS 4pp RS:HS 6.309 S. De Morgan, *Memoir of Augustus De Morgan*, 293–4

10805. 1858–2–11 From Augustus De Morgan 7 Camden St.
Further regarding the meridian on which the year began and various times throughout the world.
ALS 4pp RS:HS 6.310 S. De Morgan, *Memoir of Augustus De Morgan*, 294–5

10806. 1858–2–11 From David Livingstone 18 Hart St.
Regrets he will be unable to visit JH. Had a great desire to see him after the success of JH's suggestions for his last expedition.
ALS 4pp RS:HS 11.486

10807. 1858–2–18 From Thomas I. M. Forster Brussels
Would like a copy of his 'Peep into Other Worlds' as he has written a similar pamphlet. Has a pamphlet for JH about the philosophy of playthings and future life. Has been trying to obtain J. P. Espy's work on circular storms without success; does he know of a copy?
ALS 4pp RS:HS 7.341

10808. 1858–2–22 From Joseph Henry Smithsonian Institution

Is grateful for the encouragement he has given L. W. Me[e]ch. This Institution is endeavoring to collect information on the climate of North America and he is entering the field of meteorology for the first time. Sends some papers and would be glad of his comments.
ALS 4pp RS:HS 10.180

10809. 1858–2–22 From David Rawson 41 Nelson St., Liverpool
Has read JH's article 'with great pleasure.'
ALS 1p Harvard: Houghton AAH 67m–67 (107)

10810. 1858–2–25 To President, B.A.A.S. Collingwood
Requesting clarification of nature of work of joint committee of R.S.L. and B.A.A.S.
C 1p RS:HS 23.223

10811. 1858–2–25 To Edward Sabine n.p.
Is glad to sign certificate. Complains about tendency of Alexander von Humboldt, whose volume Mrs. Sabine is translating, to ignore other scientists' work.
ALS 4pp RS Sa.666

10812. 1858–2–26 From Warren de La Rue Royal Astronomical Society
Announces that the R.A.S. will print a map showing the path of the 115 Mar. 1858 eclipse.
L 1p Harvard: Houghton

10813. 1858–2–26 From Edward Sabine 13 Ashley Place
Discusses ellipticity and results of [G. B.] Airy and [F. W.] Bessel. Also, [Alexander von] Humboldt's magnetic results.
ALS 4pp RS:HS 15.246

10814. 1858–2–28 To T. P. Anderson Collingwood
Finds no inaccuracies in statement made to TA about Hollier's trust fund. Knows of no incumbrances on reversionary interests of W. H. B. Hollier.
ACCS 1p TxU:H/L-0028; Reel 1054

10815. 1858–2–28 To Mary Somerville Collingwood
Lists facts from various astronomical catalogues, such as the number of fixed, double, and binary stars. Discusses parallax. Thanks her for second edition of *Physical Geography*.
ALS 4pp RS:HS 16.361 (C: RS:HS 23.224)

10816. 1858–3 To Augustus De Morgan n.p.
Riddles, Latin nursery rhymes, and an eclipse description.
C 2pp RS:HS 23.227

10817. 1858–3–1 From Humphrey Lloyd Trinity College, Dublin
Calls attention to vol. 3 of Toronto observations, where Edward Sabine, [T. R.] Robinson, and HL expressed their views on continuing magnetic observations. Emphasizes importance of discovering physical and mathematical laws that govern magnetic phenomena. Sufficient data already exist for this. Recommends continuing hourly observations for five more years at seven foreign observatories (listed). [JH note: Routing slip to B.A.A.S. committee, instructing each member to read and forward HL's recommendations.]
ALS 4pp †, JH note 1p TxU:H/M–0337; Reel 1087

10818. 1858–3–1 From William Henry Smyth St. John's Lodge
'Making perhaps my last attack' on Gamma Virginis. Night air is increasingly detrimental to health.
ALS 3pp RS:HS 16.306

10819. 1858–3–2 To William Henry Smyth Collingwood
Wishes to put WS's final version of Gamma Virginis orbit in new edition of his catalogue. Asks questions about Mediterranean.
C 2pp RS:HS 23.225

10820. 1858–3–3 To Robert Grant Collingwood
About possible changes in the second edition of JH's *Outlines Astr.* due to the writings of another astronomer, Ignatio Porro.
C 1p RS:HS 23.226

10821. 1858–3–4 From James Booth Wandsworth Vicarage
Requesting JH's support for his application for the position of Professor of Geometry at Gresham College.
ALS 3pp RS:HS 4.197

10822. 1858–3–4 From William Henry Smyth St. John's Lodge
Sends JH copy of his work on Mediterranean; answers some of JH's questions.
ALS 3pp RS:HS 16.307

10823. 1858–3–10 From Richard Christopher Carrington Observatory, Redhill
Mr. Grant has just given him JH's note. Regarding Col. Parres's [?] helioscope.
ALS 4pp RS:HS 5.201

10824. 1858–3–10 From William Henry Smyth St. John's Lodge
Elaborates on JH's questions concerning the Mediterranean. Refers him to Capt. [T. A. B.] Spratt.
ALS 3pp RS:HS 16.308

10825. 1858–3–13 From George Biddell Airy Royal Observatory, Greenwich
Would support the establishment of Colonial magnetic and meteorological observatories if Edward Sabine were to continue to study the observations made [see GA's 1858–1–29].
ALS 4pp † TxU:H/M-0638.2; Reel 1084 (C: RGO 6.694.477)

10826. 1858–3–15 From Mary Somerville Florence
Enjoyed visit of Louisa [Herschel Marshall]. Wants copies of JH's new articles. Is witnessing a solar eclipse, in which half the disc was covered, as she writes.
ALS 4pp RS:HS 16.362

10827. 1858–3–16 From Robert Grant Royal Astronomical Society
Carrington will be communicating with JH regarding Ignatio Porro's instruments. Pleased to hear of the projected new edition of JH's *Outlines Astr.* Is engaged translating D. F. J. Arago's *Astronomy*. Points out omissions in this work. Comments on the planet Neptune.
ALS 12pp RS:HS 9.16

10828. 1858–3–20 From Augustus De Morgan 7 Camden St.
Regarding the recent eclipse of the sun. Has not been able to guess all his riddles.
ALS 3pp RS:HS 6.311

10829. 1858–3–21 To Robert Grant Collingwood
About viewing the solar eclipse, and the many errors in G. B. Airy's *Lectures*.
C 2pp RS:HS 23.228

10830. 1858–3–22 From Few & Co. Covent Garden
Have looked up the deeds and will make a schedule of them.
ALS 2pp RS:HS 7.213

10831. 1858–3–25 From William Whewell London
Expresses his views, generally supportive, on the question of the continuation of magnetic observations at various colonial stations.
ALS 3pp TxU:H/M-0572

10832. 1858–3–26 From Henry Anthony Jeffreys Parsonage, Hawkhurst
Asks JH to sign a petition to allow a parishioner to marry his late wife's sister.
ALS 1p Harvard: Houghton AAH 67m–67 (107)

10833. 1858–4–8 From Robert FitzRoy Meteorological Dept., B.o.T.
Sending some publications to show the kind of work he is doing. Has read JH's *Admiralty Manual* and congratulates him on its authority. Has been reading J. P. Espy's Fourth Report. Note of W. J. S. Pullen's soundings in the South Atlantic.
ALS 5pp RS:HS 7.251

10834. 1858–4–10 To Robert FitzRoy Collingwood
Comments on wind and current charts JH has received from RF.
C 2pp RS:HS 23.229

10835. 1858–4–16 From Edward Sabine 13 Ashley Place
Replies to specific requests by committee for summary of benefits from research in terrestrial magnetism and meteorology and for ES's opinion on continuance of observatories. [JH annotation: Routing list to G. B. Airy, George Peacock, and William Whewell.]
ALS 4pp † TxU:H/M–0525.1; Reel 1093

10836. 1858–4–21 From Edward Sabine 13 Ashley Place
Sends proofs of editor's notes from vol. 4 of [Elizabeth J. Sabine's] translation of Alexander von Humboldt's] *Cosmos*. Assumes that JH and magnetic committee members all have copies of 'Introduction' to vol. 3 of Toronto observations. Managed to clear up 'mystification' in J. B. Biot's account of pendulum experiments.
ALS 4pp TxU:H/M–0525.2; Reel 1093

10837. [1858 or later]–4–23 From Henry Holland 25 Brook St.
Comments on JH's paper on *Sensorial Vision* (1858). Covered some of this ground in HH's own book, *Mental Physiology* (1852).
ALS 4pp RS:HS 9.440

10838. 1858–4–23 From Joseph Wolff Vicarage, Ile Brewers, Taunton, Somersetshire
After JW's 1843–45 mission to find Charles Stoddart and Arthur Conolly in Bukhara [Uzbek], JW was given poor parish in Dorsetshire. Attempting to raise £1,000 for new church and school. Would JH contribute?
ALS 6pp RS:HS 18.432

10839. 1858–4–24 From Andres Poey Paris
Has read with much pleasure JH's dissertation on meteorology published in the *Encyclopaedia Britannica.* Requests an English version. Sends several memoires on meteorology. Regrets not being able to send a complete set. Asks JH to send other works for the Observatory of Havana.
ALS 4pp RS:HS 14.171

10840. 1858–4–27 From Edward Sabine 13 Ashley Place
ES's paper on magnetic observatories in colonies is same as 'Introduction' to vol. 3 of Toronto observations. Publisher has delayed [E. J. Sabine's] translation of vol. 4 of *Cosmos*; please send JH's copy of 'Editor's Notes' to other members of magnetic committee. J. B. Biot intentionally subordinated English pendulum experiments to French experiments, and Alexander von Humboldt was misled.
ALS 4pp TxU:H/M–0525.3; Reel 1093

10841. 1858–4–28 To Edward Sabine Collingwood
Must hold to 1/290 as measure of ellipticity. Committee agrees to recommend continuance of magnetic observations if JH will continue to do what he has done in the past.
C 1p RS:HS 23.230

10842. 1858–5[?] To James David Forbes n.p.
Asks if JF's article on glaciers is for *Encyclopaedia Britannica*. Requests that JF change a computational error in his copy of JH's article on meteorology.
ALS 2pp St. Andrews 49

10843. 1858–5–4 From George Biddell Airy Royal Observatory, Greenwich
Sends JH some responses to a letter from Edward Sabine on Colonial observatories [see GA's 1858–3–13].
ALS 5pp † TxU:H/M-0638.3; Reel 1084 (C: RGO 6.694.479)

10844. 1858–5–4 From Robert FitzRoy Meteorological Dept., B.o.T.
Entirely agrees with his views on M. F. Maury. Is surprised at David Brewster's review of Maury's work. Would JH comment on RF's work on weather forecasting.
ALS 4pp RS:HS 7.252

10845. 1858–5–7 To John Herschel (son) Collingwood
The family is waiting to hear from son John; JH comments on the end of the East India Company's rule in India, and on another newspaper item.
ALS 4pp JHS 5.47

10846. 1858–5–8 From Edward Sabine 13 Ashley Place
Current activities at magnetic observatories at Cape of Good Hope, Hobart [Tasmania], and Toronto. Future avenues for research in terrestrial magnetism include correlation with sunspot cycles, annual seasons, parallels of latitude, and lunar position. Urges establishment of many more observing stations. [JH annotation: Routing list to G. B. Airy, George Peacock, and William Whewell.]
ALS 9pp † TxU:H/M–0525.4; Reel 1093

10847. 1858–5–8 To Edward Sabine Collingwood
Describes in detail proceedings of a committee meeting, particularly noting remarks made about ES and stressing committee's wish to see evidence of progress.
ALS 5pp RS Sa.667

10848. 1858–5–11 From William Whewell Trinity Lodge
Comments briefly on the positioning of magnetic observatories. Unsure whether he understands Humphrey Lloyd's proposal.
ALS 3pp TxU:H/M-5073

10849. 1858–5–12 From George Biddell Airy Royal Observatory, Greenwich
Reacts to JH's suggestions [see JH's 1858–5–[10]] for altering GA's responses [see GA's 1858–5–4] to Edward Sabine's report on magnetic observatories [includes a silly riddle, apparently passed on from Augustus De Morgan].

ALS 2pp, GA note 1p TxU:H/M-0638.4; Reel 1084 (C: RGO 6.694.494)

10850. 1858–5–12 or earlier To George Biddell Airy n.p.
Suggests GA alter some phraseology and tone in his response [see GA's 1858–5–4] to Edward Sabine's report on magnetic observatories.
ALS 3pp RGO 6.694.492

10851. 1858–5–13 From George Biddell Airy Royal Observatory, Greenwich
Provides a more detailed response to Edward Sabine's report on magnetic observatories [see GA's 1858–5–12].
ALS 4pp † TxU:H/M-0638.5; Reel 1084 (C: RGO 6.694.496)

10852. [1858]–5–15 From Edward Sabine n.p.
Forwards report, just received, of three years of lunar diurnal variation at Cape of Good Hope. [JH annotation: Routing list to G. B. Airy, George Peacock, and William Whewell.]
ALS 2pp † TxU:H/M–0525.5; Reel 1093

10853. 1858–5–18 From George Biddell Airy Royal Observatory, Greenwich
Has looked at the magnetic observations made at the Cape, and comments on their usefulness.
ALS 1p TxU:H/M-0638.6; Reel 1084 (C: RGO 6.694.498)

10854. 1858–5–19 From James David Forbes Edinburgh
Thanks for JH's article on meteorology from the *Encyclopaedia Britannica*. Has been investigating climatology lately. His own health has not been good. Sends some of his own reprints that may be of interest.
ALS 3pp RS:HS 7.314

10855. 1858–5–19 From S. Cleveland Benenden
Sending description of effects seen at Kirkby Lonsdale during the recent eclipse.
ALS 2pp RS:HS 5.351

10856. 1858–5–22 To Augustus De Morgan Collingwood
About insects JH's children caught and photographed; on a book on harmonics.
C 1p RS:HS 23.231

10857. 1858–5–22 To James David Forbes Collingwood
Expresses thanks for several of JF's writings and relief at JF's good health. Comments on JH's own writing.
ALS 4pp St. Andrews 57a

10858. 1858–5–24 From William Whewell Trinity Lodge
Has received papers relevant to the Magnetic Committee. Comments on letters by G. B. Airy and Edward Sabine, siding mainly with Airy.
ALS 4pp TxU:H/M-5074

10859. 1858–5–25 From Augustus De Morgan 7 Camden St.
Has examined the animalcules JH sent. Sends some of his own obtained from ink! Regarding the variations of spelling his own name.
ALS 4pp RS:HS 6.312 S. De Morgan, *Memoir of Augustus De Morgan*, 295–6

10860. 1858–5–27 From William Whewell Trinity Lodge
Believes that it is not worth the cost to sustain many permanent magnetic observatories, but a few for a limited time would make sense. Unclear which observatories these should be.
ALS 1p TxU:H/M-5075

10861. 1858–5–28 From Henry Anthony Jeffreys Hawkhurst Parsonage
Comments on his plans for re-seating the church, and other matters.
ALS 4pp RS:HS 10.297

10862. 1858–6 From J[ames] G[arth] Marshall Headingley, Leeds
Is grateful to JH for sending him a copy of the *Essay on Meteorology*. Takes an interest in the subject himself and gives his views on cyclones and rotary storms.
ALS 8pp RS:HS 12.290

10863. 1858–6–5 From Hermann Schlagintweit Jermyn St.
Sends vol. 4 of [Alexander von] Humboldt's *Cosmos*. Would like to share information gathered in India.
ALS 4pp RS:HS 15.353

10864. 1858–6–8 From James David Forbes Edinburgh
Article on glaciers is from the *Encyclopaedia Britannica* and was restricted because of space. Mistakes by John Tyndall and others have arisen because his own *Letters on Glaciers* has been overlooked. William Whewell is to be married.
ALS 2pp RS:HS 7.315

10865. 1858–6–10 From Edward Sabine 13 Ashley Place
Will leave London for several weeks when ES's wife is well. Alexander von Humboldt is pleased with English translation of *Cosmos*. Preparations for magnetic research in Australia, Bavaria, [British] Guiana, and Mauritius. [JH annotation: Routing list to G. B. Airy, George Peacock, and William Whewell.]
ALS 5pp †(on cover) TxU:H/M–0525.6; Reel 1093

10866. 1858–6–12 From George Biddell Airy Royal Observatory, Greenwich
If magnetic observation stations are to be set up and staffed, then personnel should be considered soon.
ALS 4pp † TxU:H/M-0638.7; Reel 1084(C: RGO 6.694.499)

10867. 1858–6–18 To Edward Sabine Collingwood
Sends G. B. Airy's remarks concerning ES's last communication. Discusses future of observatories. Asks ES to indicate which observatories he feels are most important.
C 2pp RS:HS 23.232

10868. 1858–6–19 To Thomas Romney Robinson Collingwood
Thanks TR for his paper on magnetism and his article on speculae. Discusses [J. P.] Gassiot's work on galvanic stratification. JH's health is better.
C 1p RS:HS 23.245

10869. 1858–6–21 From Charles Edward Trevelyan to Thomas Graham Treasury Chambers
Encloses copy of Treasury Board's ruling that disbands the 'Compassionate Fund for the Mint Workmen,' established in 1851 by JH. Return all funds to Exchequer, and discontinue regular payments to widows and children.
C 1p, encl C 2pp TxU:H/M–1062; Reel 1083

10870. 1858–6–21 To William Whewell Collingwood
Congratulates WW on his planned marriage to Lady Affleck. Forwards some letters from Edward Sabine.
ALS 4pp TC, Camb. Add. Ms.a.207[94] (C: RS:HS 23.233)

10871. 1858–6–23 From Roderick Impey Murchison Geological Survey of Great Britain Office
Will JH add his signature to the slip to add to the memorial to

prevent the Natural History collections being removed from the British Museum.
ALS 3pp RS:HS 12.418

10872. 1858 [?]–6–24 To William Parsons (Lord Rosse) Collingwood
Compares in regard to the awarding of the R.S.L.'s Copley medal the contributions of three scientists: Friedrich Wöhler, Wilhelm Weber, and [illegible].
ALS 4pp Rosse Papers K2.6

10873. 1858–6–25 From George Biddell Airy Royal Observatory, Greenwich
Regarding a possible successor to U. J. J. Leverrier at the Paris Observatory.
ALS 4pp RS:HS 1.236 (C: RGO 6.145.185)

10874. [1858]–6–25 From Edward Sabine St. Leonards-on-Sea
Prospects for making magnetic and meteorological observations at Peking, which is more desirable than Shanghai or Hong Kong. Hopes for popular review of subject in *Quarterly* [*Review*].
ALS 3pp TxU:H/M–0525.7; Reel 1093

10875. 1858–6–26 To George Biddell Airy Collingwood
Intends to set up a meeting with appropriate members of the B.A.A.S., in order to prepare presentation about magnetic observatories for the B.A.A.S. meetings in Leeds.
ACC 2pp inc RGO 6.694.500

10876. 1858–6–26 To George Biddell Airy Collingwood
Responds to GA's 1858–6–25, and essentially agrees with GA's assessment.
ALS 4pp RGO 6.145.187

10877. 1858–6–26 From Edward Sabine St. Leonards
States his detailed recommendations concerning the continuation, extension, and location of magnetic researches at various locations throughout the world.
L *B.A.A.S. Report 1858* (1859); 303–5

10878. 1858–6–26 From Edward Sabine St. Leonards
Summarizes for JH the more extensive report appended, which recommends to the R.S.L. and the B.A.A.S. the establishment of magnetic observatories in several Colonial locations.
CC 5pp RGO 6.694.508v

10879. 1858–6–30 From George Biddell Airy Royal Observatory, Greenwich
Supports JH's intentions [see JH's 1858–6–26] and Edward Sabine's proposals about magnetic observations, but displays little faith in the effectiveness of the B.A.A.S.
ALS 4pp TxU:H/M-0638.8; Reel 1084 (C: RGO 6.694.501)

10880. 1858–7 From J[ames] G[arth] Marshall Headingley, Leeds
It was good of JH to reply so fully to his letter. Now has a better understanding of the subject than before. Read a paper on the geology of the Lake District at the local Philosophical Society. Comments on his own theories of rock foundations.
ALS 12pp RS:HS 12.291

10881. [1858–7–1 or earlier] To Edward Sabine Collingwood
Invites ES and his wife [at St. Leonards] to visit . JH is impressed with ES's ability to develop a working plan for terrestrial magnetism observations. Is prepared to write a review article [about magnetism?] if invited, but JH refuses to write such articles and offer them 'for acceptance or rejection.'
C 1p RS:HS 25.14.28

10882. 1858–7–2 From George Biddell Airy Royal Observatory, Greenwich
Regarding an observer for the forthcoming eclipse.
ALS 3pp RS:HS 1.237 (C: RGO 6.115.156)

10883. 1858–7–2 From Edward Sabine St. Leonards-on-Sea
Wife not well enough to visit Collingwood. ES may come next week. University of Kasan Observatory may assist British endeavors. Mr. Bolyani is very capable.
ALS 4pp RS:HS 15.247

10884. [1858–7–3 or later] To Edward Sabine [Collingwood]
Has incorporated ES's suggestions into the draft report about [the University of] Kasan.
C 1p RS:HS 25.14.27

10885. 1858–7–7 From Robert Brown Fairlie by Greenock
Has written a book on elementary harmony and would like to send it to JH for his views on its worth.
ALS 5pp RS:HS 4.309

10886. 1858–7–9 From H. C. Marshall Leeds
Believes it may suit JH to be HM's guest at Westwood during the meeting of the B.A.A.S. at Leeds. Hopes this is possible.
ALS 2pp RS:HS 13.21

10887. 1858–7–10 From Henry Darvill Windsor
Sees that JH is entitled to have his name on the register of voters. Would he fill up the form and return it; then his name will be entered on the new Register.
ALS 1p RS:HS 6.472

10888. 1858–7–16 From James David Forbes Pitlochry
Has he read John Tyndall's paper on the plasticity of ice and the theory of glaciers? Indebted to Michael Faraday for some excellent observations and the term 'Regulation.'
ALS 2pp RS:HS 7.316

10889. 1858–7–18 To James David Forbes Collingwood
Comments on JF's article about glaciers and particularly on the solid, liquid, and other states of matter.
ALS 3pp, encl 2pp St. Andrews 76a (encl 76b)

10890. 1858–7–18 From William Whewell Trinity Lodge
Generally approves of Edward Sabine's plan concerning magnetic observatories.
ALS 2pp TxU:H/M-5076

10891. 1858–7–20 From George Smith London
Received JH's letter of 17 July. Sent woodcut to W. S. Smyth as requested.
ALS 1p TxU:H/L-0755; Reel 1089

10892. 1858–7–20 To William Whewell Trinity Lodge
Forwards magnetic materials written by George Peacock. Hopes to meet with JH.
ALS 3pp TxU:H/M-5077

10893. 1858–7–22 From George Peacock Deanery, Ely
Confer with William Whewell on magnetic observations, then notify GP. Health is improving. Cambridge residents are angry over proposed changes but ignore similar changes at Oxford.
ALS 4pp TxU:H/M–0426; Reel 1093

10894. 1858–7–23 From Edward Sabine Llanfair, Llandovery, South Wales
Forwards latest computations of lunar diurnal variation in horizontal force of terrestrial magnetism at Cape of Good Hope. Describes method used by Archibald Smith to distinguish lunar from terrestrial magnetism.
ALS 3pp TxU:H/M–0525.8; Reel 1093

10895. 1858–7–30 From P. O'Callaghan Cookridge Hall, Leeds
Heard from Louisa [JH's daughter] that JH was interested to hear about the [Leeds] Literary and Philosophical Society. Gives details and recent lectures. Would be pleased for JH to give a lecture.
ALS 4pp RS:HS 13.157

10896. 1858–8–7 From Augustus De Morgan 7 Camden St.
Has not heard from him for a long time. Hopes all is well. Elizabeth Baily has had a stroke but is slowly mending. Arthur Baily has committed suicide. His own wife and children are at the seaside. Hopes JH or one of his daughters will write soon.
ALS 4pp RS:HS 6.313

10897. 1858–8–8 To Humphrey Lloyd Collingwood
R.S.L. committee recommends continuing three magnetic observatories at Falkland Islands, Newfoundland, and Vancouver Island.
ACCS 2pp TxU:H/L-0755.2; Reel 1089

10898. 1858–8–8 To Humphrey Lloyd Collingwood
R.S.L. committee for considering continuance of magnetic observations has recommended that hourly observations be carried out at Falkland Islands, Newfoundland, and Vancouver Island. Alternate stations are Ceylon and Bermuda. Will not print all observations. Hopes colonial governments will participate. JH and William Whewell will be at [B.A.A.S. meeting in] Leeds.
C 2pp TxU:H/L-0224; Reel 1054

10899. 1858–8–15 From P. O'Callaghan Cookridge Hall, Leeds
All are exceedingly obliged by JH's kind thought for the [Leeds] Literary and Philosophical Society. Subject JH mentions will be most interesting. Paper should not be too long. News from Ingwell most comforting.
ALS 3pp RS:HS 13.158

10900. 1858–8–20 From Humphrey Lloyd Limerick
Forwarded JH's letter of 8 Aug. 1858 to members of magnetic committee, who will respond before B.A.A.S. meeting in Leeds next month. Asks that both committees meet jointly on first day. Edward Sabine is willing to superintend new series of observations proposed by JH.
ALS 4pp † TxU:H/M–0338; Reel 1087

10901. 1858–8–26 From W. H. Barton Royal Mint
Regarding the financial affairs of JH.
ALS 4pp RS:HS 3.298

10902. 1858–8–26 From Joseph Baxendell Stocks St., Manchester
Regarding a difference in their observations of a certain star.
ALS 3pp RS:HS 3.306

10903. 1858–8–26 From Augustus De Morgan 7 Camden St.
Sees that JH was the author of the article on mathematics in Brewster's Encyclopaedia. Comments on this and Isaac Newton's authorship of the notes and review of the *Commercium Epistolicum.*
ALS 7pp RS:HS 6.314

10904. 1858–8–27 To Patrick Francis Robertson Collingwood
Thanks PR for the sewerage book. Suggests a method of disposing of sewage in the river. Concerned about solid waste, which he thinks should be used as fertilizer.
(S) copy 6pp RS:HS 14.497 (C: RS:HS 23.234)

10905. 1858–8–29 From Patrick Francis Robertson Halton
Thanks JH for his comments on PR's plan for London drainage. Asks another question regarding the system.
ALS 1p RS:HS 14.369

10906. 1858–8–30 To John P. Gassiot Collingwood
Thanks for his valuable paper on galvanic stratification. Hopes to meet him at the B.A.A.S. meeting at Leeds. Encloses some copper salts and would be grateful for his comments on them. Hopes his health has improved.
ALS 2pp RS:HS 8.60 & 23.235

10907. 1858–9–1 From Augustus De Morgan 7 Camden St.
JH keeps his astronomy right up to date. Regarding the significance of the word 'month' in legal phraseology. Gives two riddles.
ALS 3pp RS:HS 6.315

10908. 1858–9–6 To Charles Babbage Collingwood
Received another notice from General Reversionary & Trust Society regarding H. J. Hollier. JH asks again to be replaced as trustee for Mary Anne Hollier trust.
ALS 4pp TxU:H/L-0054; Reel 1054

10909. 1858–9–6 From William Henry Smyth St. John's Lodge
Thanks JH for a copy of the new edition of *Outlines Astr.*
AL 1p inc Harvard: Houghton

10910. 1858–9–8 To George Biddell Airy Collingwood
Humphrey Lloyd wants R.S.L. and B.A.A.S. committees on magnetism to meet together at B.A.A.S. meeting in Leeds. JH suggests inviting Edward Sabine.
ALS 4pp TxU:H/L-0020; Reel 1054

10911. 1858–9–8 From John P. Gassiot Clapham Common
Has been trying to obtain information on JH's query, which accounts for the delay in replying to his letter. Would like to show him his own experiments with electricity in a vacuum should he come to town.
ALS 4pp RS:HS 8.61

10912. 1858–9–9 From George Biddell Airy Royal Observatory, Greenwich
Indicates GA's intention of being at the Leeds meeting [of the B.A.A.S.].
ALS 2pp TxU:H/M-0638.10; Reel 1084

10913. 1858–9–9 From George Biddell Airy Royal Observatory, Greenwich
Is unhappy with one aspect of Edward Sabine's report [see ES's 1858–6–26], specifically the number of days each year when simultaneous observations will be made at all stations.
ALS 3pp TxU:H/M-0638.9; Reel 1084 (C: RGO 6.694.512)

10914. 1858–9–9 From Alfred Barry Leeds
Would like JH's presence at the distribution of prizes at the Oxford Middle Class Examination at Town Hall. His support

of the scheme, which he thinks the most important event in his memory, will be very welcome.
ALS 4pp RS:HS 5.22

10915. 1858–9–11 From Warren de La Rue 110 Bunhill Row
Has just dispatched to Hawkhurst station a positive copy in glass of a photograph of the moon. Has many photographs but this is the best one. Has constructed a reflecting stereoscope. Angelo Secchi visited him but the weather was not suitable for observations.
ALS 2pp RS:HS 6.141

10916. 1858–9–12 From Charles Babbage Dorset St., Manchester Square
Complications with JH's trusteeship for CB's sister [Mary Anne Hollier]. Details of how other trusts are managed. Will consult Edward Ryan about JH's desire to resign from trust. CB's brother-in-law Wolryche Whitmore died.
ALS 4pp TxU:H/M–0033; Reel 1062

10917. 1858–9–12 To Augustus De Morgan Collingwood
Errors to be corrected in a new edition of one of JH's writings, including spelling AD's name in the French way.
C 2pp RS:HS 23.236

10918. 1858–9–13 To Warren de La Rue Collingwood
Thanks for the lunar photographs; comments on other observations.
C 1p RS:HS 23.237

10919. 1858–9–13 From William Henry Smyth St. John's Lodge
Finds mistake in JH publication [*Outlines Astr.*] that assigned discovery of Hyperion to W. R. Dawes and G. P. Bond without crediting William Lassell.
ALS 3pp RS:HS 16.309

10920. 1858–9–15 To Elizabeth Colling Leeds
Acknowledges receipt of a communication about a comet/meteor from EC.
C 1p RS:HS 23.240

10921. 1858–9–15 From Augustus De Morgan 7 Camden St.
Does not intend writing a life of Isaac Newton or G. W. Leibniz. Comments on the apparent delusions of Newton regarding other men's inventions.
ALS 4pp RS:HS 6.316

10922. 1858–9–15 To Margaret Brodie Herschel Collingwood
Mostly about arrangements which MH is making for JH's travel to, and accommodation in, Leeds for the B.A.A.S. meetings; JH wants extremely detailed instructions of just what to do. Goes on to comment on the health of the family and JH's own poor health.
ALS 4pp JHS 5.32

10923. 1858–9–15 To William Lassell Collingwood
Is dismayed to find in the new edition of his own *Outlines Astr.* that he has attributed the discovery of *Hyperion* to W. R. Dawes without crediting WL; will have this corrected in a future edition.
draft 8pp RS:HS 11.153–4 (C: RS:HS 23.239)

10924. 1858–9–15 From Warren de La Rue 110 Bunhill Row
Is pleased the photograph arrived safely. Would be pleased to show him his collection of original negatives. Comments on the difficulties of using a telescope with a longer focus. Regarding improvements to the photoheliography at Kew. Encloses sketch of the comet.
ALS 3pp RS:HS 6.142

10925. 1858–9–15 To William Henry Smyth Collingwood
Sends apologies to William Lassell for omission of discovery credit.
C 1p RS:HS 23.238

10926. 1858–9–15 To William Whewell Collingwood
Is trying to arrange a magnetic conference during the first few days of other meetings.
ALS 3pp WT 66602

10927. 1858–9–16 From W. H. Barton Royal Mint
Arrangements for JH to meet him to arrange the sale of stock.
ALS 2pp RS:HS 3.299

10928. 1858–9–16 From George Peacock Deanery, Ely
Forwarded JH's and Humphrey Lloyd's letters to G. B. Airy and William Whewell. Cannot attend B.A.A.S. meeting at Leeds. Does JH concur with Edward Sabine's latest suggestions?
ALS 3pp TxU:H/M–0427; Reel 1093

10929. 1858–9–17 From William Lassell Sandfield Park near Liverpool
Do not let the error in *Outlines Astr.* disturb him; he quite understands. The 4-foot telescope proceeds. The speculum has been polished and awaits trial. If JH is attending the B.A.A.S. meeting hopes he will visit Lassell.
ALS 3pp RS:HS 11.155

10930. 1858–9–17 From William Henry Smyth St. John's Lodge
Writes that he was not aware of W. R. Dawes's hand in discovery of Hyperion; will note this in R.S.L. records.
ALS 2pp RS:HS 16.310

10931. 1858–9–24 From Alfred Barry Leeds
Hopes he will be attending the prize-giving. Sends a resolution to be passed at the meeting.
ALS 3pp RS:HS 5.23

10932. 1858–10–2 From Charles Robinson Leeds
Asks for JH's autograph. Aspires to be an astronomer. Was impressed by JH's feeling for humanity at the Leed's Young Men's Christian Institute.
ALS 3pp RS:HS 14.498

10933. 1858–10–6 From William Hallowes Miller 7 Scroope Terrace, Cambridge
Initial results of WM's crystallographic examination of copper salts sent by JH. Awaits results of electrical discharge experiments.
ALS 4pp TxU:H/M–0369; Reel 1087

10934. 1858–10–8 From William Crookes 20 Mornington Road
Sending a copy of the *Photographic News* in which an extract from JH's speech at the B.A.A.S. appears. Would like further details of JH's theories regarding the chemistry of photography.
ALS 3pp RS:HS 5.305

10935. 1858–10–8 To Margaret Brodie Herschel 2 Howard St.
Gives a detailed account of the wanderings and visits of JH and son John; among other things they observed a comet at

night.
AL 4pp JHS 5.21

10936. 1858–10–8 From Henry James Ordnance Map Office
Sending a copy of the Account of the Principal Triangulation of the United Kingdom.
ALS 1p RS:HS 10.274

10937. 1858–10–9 From William Hallowes Miller 7 Scroope Terrace, Cambridge
Further results from tests that WM conducted on copper salts sent by JH.
ALS 3pp TxU:H/M–0370; Reel 1087

10938. 1858–10–10 To William Crookes Collingwood
Comments on various specimen photographic papers JH is sending; JH believes he has isolated a metal he wants to call 'Junonium.'
C draft 2pp RS:HS 23.242

10939. 1858–10–10 To Warren de La Rue Collingwood
Has now had time to examine the stereoscopic photograph of the moon. Thinks it is a wonderful effect and opens up a new field for terrestrial objects.
draft 3pp RS:HS 6.143 (C: RS:HS 23.241)

10940. 1858–10–10 To E. L. Withers Collingwood
Assures EW that earth is spheroidal and that sun and planets follow laws of gravitation, contrary to absurd claims of Mr. Parallax. Regrets that Parallax is using JH's name to support 'ridiculous opinions.'
ALS 4pp TxU:H/L-0498; Reel 1055

10941. 1858–10–10 To E. L. Withers Collingwood
Dismisses the lectures of a Mr. Parallax, who is going about the country lecturing against the basic views of science, and using JH's ideas in support.
Cdraft 1p RS:HS 25.15.17

10942. 1858–10–12 From Charles Robinson Leeds
Acknowledges JH's letter and thanks him for it. Praises JH profusely.
ALS 1p RS:HS 14.499

10943. 1858–10–13 From John Bullar Basset Wood near Southampton
Is pleased to hear of his support for the Young Men's Association at Leeds. Is reading the new edition of JH's *Outlines Astr*.
ALS 4pp RS:HS 4.331

10944. 1858–10–13 To William Crookes Collingwood
Has just returned home and found the *Photographic News* and the queries regarding chemical experiments and photography. Gives details, but regrets he has not published much on it yet as he is still experimenting.
draft 4pp RS:HS 5.306

10945. 1858–10–13 To Augustus De Morgan Collingwood
On the definition of an island, and an invitation to lecture about a comet.
C 2pp RS:HS 23.243

10946. 1858–10–13 From [John Griffith?] Clapham [Commons?]
Regrets being absent when JH and sons called at [JG]'s home. Hopes to show JH some peculiar electrical phenomena that [JG] observed. Encloses two letters from [W. H.] Miller.
ALS 2pp TxU:H/M–0933; Reel 1083

10947. 1858–10–13 To the *Photographic News* Collingwood
Discusses the use in photography of a metallic substance named Junonium.
L *Photographic News, 1* (Oct. 29, 1858), 86

10948. 1858–10–15 From William Crookes 20 Mornington Road
Thanks for letter. Will insert the information in a forthcoming issue of the *Photographic News*. Awaits with interest news of the new metals discovered by JH. Would like a small quantity of Junoniate of soda.
ALS 4pp RS:HS 5.307

10949. 1858–10–15 From Augustus De Morgan 7 Camden St.
Sending a query regarding JH's philosophy.
ALS 2pp RS:HS 6.317 S. De Morgan, *Memoir of Augustus De Morgan*, 297

10950. 1858–10–18 From Augustus De Morgan 7 Camden St.
JH has turned round his question. Samuel Earnshaw has professed to have integrated the pipe equation. Gives an equation of his own.
ALS 4pp RS:HS 6.318

10951. 1858–10–19 To William Whewell Collingwood
Discusses various arrangements of the chemical elements, including those of [J. P.] Cooke and [John] Mercer. Asks WW about capillary action.
ALS 3pp TC, Camb. Add. Ms.a.207[96] (C: RS:HS 23.244)

10952. 1858–10–23 From William Crookes n.p.
Invites remarks from JH. Except for W. H. F. Talbot's 'new discovery,' the subject of copying paintings draws the most attention from readers.
ALS 1p inc RAS:JH Archive 12/1.4; Reel 10

10953. 1858–10–23 From William Crookes 20 Mornington Road
Letter and bottle of solution arrived safely. His latest photographic experiments. Is pleased to hear that JH approves of his efforts to chronicle photographic experiments.
AL 3pp inc RS:HS 5.308

10954. 1858–10–24 To Edward Sabine Collingwood
Regarding report on magnetic observatories to be laid before R.S.L. president and council.
ALS 2pp RS MC.5.364

10955. 1858–10–25 From Julia Margaret Cameron Putney Heath
Her daughter is shortly to be married to Charles Norman. News of the defects of vision in her children and her anxieties regarding it.
ALS 32pp RS:HS 5.154

10956. 1858–10–26 From Henry James Ordnance Map Office
Sends a list of the principal mountains in the United Kingdom and their altitudes. Has just returned from a tour in Ireland and Scotland; otherwise he would have replied earlier.
ALS 1p RS:HS 10.275

10957. 1858–10–31 To Sir William J. Hooker Collingwood

Please distinguish between the genus *Wellingtonia* and the genus *Sequoia*.
ALS 2pp Royal Botanic Gardens

10958. 1858–11–1 From Sir William J. Hooker Royal Botanic Gardens, Kew
Regarding the reasons for the confusion in the nomenclature of the genus *Wellingtonia* and *Sequoia*. Comments on the aquatic *Anacharis*.
ALS 8pp RS:HS 9.462

10959. 1858–11–4 From Robert Brown Fairlie by Greenock
Had no idea that JH had written on sound. Sends him an abstract of his book on harmony. Explains various aspects of chords and their notation.
ALS 7pp RS:HS 4.310

10960. 1858–11–6 From Robert Brown Fairlie by Greenock
Has sent him the abstract on harmony. Further remarks on this.
ALS 2pp RS:HS 4.311

10961. 1858–11–6 To the *Photographic News* Collingwood
Discusses the use of the stereoscope, including stereoscopic views of the moon and sun. Also discusses the appropriate stereoscopic angle.
L *Photographic News, 1* (Nov. 12, 1858), 110

10962. 1858–11–8 From Matthew Collins Dublin
Sending one of his tracts. Comments on the subject of duplicate equalities.
ALS 2pp RS:HS 5.262

10963. 1858–11–10 From William Selwyn Cambridge
WS's brother-in-law [George Peacock] has died after an illness.
ALS 4pp RS:HS 15.462

10964. 1858–11–10 From Henry Augustus Severn Mint, Sydney
Having outgrown his self-built telescope, hopes to borrow a larger 'metal' from JH, or to obtain from JH William Herschel's method for polishing telescopes.
ALS 4pp RS:HS 17.192

10965. 1858–11–11 From Johann Franz Encke Berlin
On some of the elements of the orbit of a newly appearing comet.
ALS 3pp (German) RS:HS 7.99

10966. 1858–11–13 From Arthur Cayley Stone Bldgs., Lincoln's Inn
Is thinking of applying for the Lowndes professorship at Cambridge, and would welcome a testimonial from him.
ALS 2pp RS:HS 5.226

10967. 1858–11–15 From Augustus De Morgan 7 Camden St.
Does not intend trying for the Lowndean Professorship. Hopes that JH has been asked to do a memoir of George Peacock. Hears he has been traveling about the country. Sends a riddle.
ALS 4pp RS:HS 6.319 S. De Morgan, *Memoir of Augustus De Morgan*, 297–8

10968. 1858–11–15 From William Ellis Hoddesden
Sending a volume dealing with his latest travels in Madagascar.
AL 1p RS:HS 7.129

10969. 1858–11–17 To John Wrottesley Collingwood
Agrees to write memoir [of George Peacock], but this will take some time as he is taking son [John] to Southampton to leave for India.
C 1p RS:HS 25.12.4

10970. 1858–11–18 From William Benjamin Carpenter Univ. Hall, Gordon Square
Would like his views on binocular vision. Regarding David Brewster's theories on this subject.
ALS 5pp RS:HS 5.194

10971. 1858–11–19 From George Biddell Airy Royal Observatory, Greenwich
Encloses copy of a letter addressed to GA from Toronto University (dated 30 October 1858 and stating that the Toronto Observatory has not been closed in spite of what the R.S.L. and B.A.A.S. joint committee have stated). Difficulties because of Edward Sabine's illness.
ALS 3pp RS:HS 1.238 (C: RGO 6.694.517)

10972. 1858–11–19 To Margaret Brodie Herschel Kelways Hotel, Southampton
Describes JH's visit to an old, bedridden friend [Bullar?]; JH has seen son John on board ship at Southampton, and then comments on the beauty of the country, which JH enjoyed as he traveled.
ALS 4pp JHS 4.11

10973. 1858–11–19 To [George Gabriel Stokes] n.p.
Thanks for paper on magnetism, and other minor matters.
ALS 3pp Cambridge University Library 7656/TR63

10974. 1858–11–20 To Margaret Brodie Herschel Kelway's Hotel, Southampton
Has put son John on board ship and they both looked for Mrs. Russell [?] but could not find her; JH reports on the health of daughter Amelia and about the time of their return home.
ALS 4pp JHS 5.33

10975. 1858–11–20 From William Henry Sykes 47 Albion St.
Before meeting of Balloon Committee, requests JH's opinion on value of renewing balloon ascents.
ALS 3pp RS:HS 17.153

10976. 1858–11–23 To George Biddell Airy Collingwood
Giving reply for GA to send to Toronto. Is grieved to hear of Edward Sabine's illness.
draft 4pp RS:HS 1.239 (C: RS:HS 23.248 & C: RGO 6.694.519)

10977. 1858–11–23 To William Benjamin Carpenter n.p.
Has not read article on binocular vision. Gives his views on this subject. Sends a copy of a paper he read at Leeds on visual impressions.
draft 5pp RS:HS 5.195 & 23.247

10978. 1858–11–23 To William Henry Sykes Collingwood
Urges renewal of high balloon ascents with periodic observations both to further knowledge of law of decrement of temperature and pressure and for 'ulterior investigations.'
C 2pp RS:HS17.154 (C: RS:HS 23.246)

10979. 1858–11–24 From Richard Christopher Carrington Observatory, Redhill
Methods of taking photographs of the sun to a smaller scale,

at Kew. This should be made the exclusive duty of one person.
ALS 4pp RS:HS 5.202

10980. 1858–11–30 From George Robert Waterhouse British Museum
Responds to JH's questions about discovery sites of extinct species of saurians, mammoths, snakes, giant sloths, etc.
ALS 3pp RS:HS 18.102

10981. 1858–12–4 From Sir Richard Owen British Museum
Asks JH to review R.S.L.'s reply to Prince Albert regarding extension of B.A.A.S. magnetic and meteorological observations. RO enjoyed photo[micro]graphs of aquatic animalcules; these give clear insight into 'the spontaneous production of the species....'
ALS 3pp TxU:H/M–0403; Reel 1087

10982. 1858–12–9 From John Wrottesley Wrottesley [Staffordshire]
Government refuses to 'entertain the Magnetic Proposals' this year, claiming cost overruns. Probably a mistake, but Edward Sabine's illness precludes easy reply. Expect discussion at next R.S.L. Council meeting.
ALS 4pp RS:HS 18.302

10983. 1858–12–13 From John Daniel Runkle Cambridge Mass.
Sends *Mathematical Monthly*. Asks for a list of JH's publications. Would also like a photograph and biographical sketch.
ALS 2pp RS:HS 14.511

10984. 1858–12–15 To Robert Brown Collingwood
Regrets he has kept his theory on harmony so long; returns it and comments on various aspects.
draft 4pp RS:HS 4.312 & 23.249

10985. 1858–12–17 To Augustus De Morgan Collingwood
About biographical information on George Peacock.
C 1p RS:HS 23.252

10986. 1858–12–17 From Balfour Stewart Alva St., Edinburgh
Hopes that JH will express opinion of BS's paper on radiant heat, as recommendation for BS's appointment to the Chair of Natural Philosophy at St. Andrews.
ALS 1p RS:HS 17.236

10987. 1858–12–17 To William Whewell Collingwood
Thanks for WW's views on capillary action. Asks for facts about George Peacock on whom JH was writing an obituary.
ALS 4pp TC, Camb. Add. Ms.a.207[97] (C: RS:HS 23.250)

10988. 1858–12–18 From Sir Richard Owen British Museum
[Marked 'Private.'] Treasury replied unfavorably to request for extension of magnetic observations. C. E. Trevelyan suggests that joint letter from JH and presidents of R.S.L. and B.A.A.S. proposing only one observer (no observatory) at Peking would be acceptable, if estimated cost is included. No other locations will be approved. This concession was made possible by influence of Prince Albert.
ALS 4pp TxU:H/M–0404; Reel 1087

10989. 1858–12–18 To George Biddell Airy Collingwood
Asks GA to help in determining cost of one observation station in Peking [see Richard Owen's 1858–12–18].
ALS 3pp RGO 6.694.530

10990. 1858–12–18 From Augustus De Morgan 7 Camden St.
Regarding George Peacock and his mathematical work. Gives some of his own theories, which he intends publishing. Includes a riddle.
ALS 4pp RS:HS 6.320

10991. 1858–12–18 From Sir Richard Owen British Museum
The Treasury has, with great reluctance, agreed to establish one magnetic observatory at Peking, rather than the extensive plans proposed [see Edward Sabine's 1858–6–26].
ALS 2pp RGO 6.694.532

10992. 1858–12–19 To Sir Richard Owen Collingwood
If only one observer is assigned to Peking, observations could not be conducted. Describes staff and expenses at other magnetic observatories. Edward Sabine's plan to adapt all instruments to photographic self-registering instruments.
C 6pp TxU:H/L-0289; Reel 1054

10993. 1858–12–20 From Robert Brown Fairlie by Greenock
Is grateful for his comments on his book on harmony. Explains one of his theories more fully.
ALS 3pp RS:HS 4.313

10994. 1858–12–21 From George Biddell Airy n.p.
Given the Treasury stance [see Richard Owen's 1858–12–18], GA sees little point in setting up one station [see JH's 1858–12–18].
ACCS 4pp RGO 6.694.533

10995. 1858–12–22 To Sir Richard Owen Collingwood
Sends RO a note from G. B. Airy on the inadvisability of having a meteorological observatory in Peking. JH agrees with Airy. When William Whewell arrives for a visit, JH will seek his advice.
ALS 3pp Dunedin Public Lib. (C: TxU:H/L-0290; Reel 1054 & RS Sa.668)

10996. 1858–12–22 To Sir Richard Owen Collingwood
Discusses proposal to establish magnetic and meteorological observatories at Peking, Newfoundland, Vancouver, and Falkland.
C 2pp RS MC.5.381

10997. 1858–12–25 From Balfour Stewart Alva St., Edinburgh
Thanks JH for testimonial toward BS's appointment to Chair of Natural Philosophy at St. Andrews. Encloses other testimonials.
ALS 1p RS:HS 17.237

10998. 1858–12–28 From Augustus De Morgan 7 Camden St.
Would JH extract the dates from his memoir of George Peacock as AD must write something for the R.A.S. Encloses some meteorological queries.
ALS 2pp RS:HS 6.321

10999. 1858–12–30 From Peter le Neve Foster Society of Arts, Adelphi
Circular letter inviting comments on the proposal of the Council of the Society of Arts to hold another exhibition in 1861.
ALS 3pp RS:HS 7.324

11000. 1858–12–30 From Annibal de Gasparis Naples
Sending details of his astronomical theories.
ALS 1p (French) RS:HS 8.49

11001. [1859] From Sir Benjamin Collins Brodie [the elder] Broome Park
Is indisposed so will not be coming to London; leaves it to him and Edward Sabine to send answer to H. P. F. P. Clinton (5th Duke of Newcastle). Would like an up-to-date citation for W. E. Weber as he has been awarded the Royal Medal.
ALS 3pp RS:HS 4.292

11002. 1859–1–1 From Augustus De Morgan 7 Camden St.
Many thanks for the dates for the memoir of George Peacock. Queries the name of his father. Comments on various church matters.
ALS 4pp RS:HS 6.322 S. De Morgan, *Memoir of Augustus De Morgan*, 298–9

11003. [1859?] To Isabella Herschel London
Tells about missing a meeting with HRH Prince Albert, about a paper read at a R.S.L. meeting, and describes how to figure out how strong a reading glass JH needs to buy for his wife, Margaret. JH saw Charles Wheatstone, who has devised an alphabetic telegraph.
AL 4pp JHS 5.18

11004. [1859?] To Margaret Brodie Herschel Collingwood
About an expected death, an accident involving an omnibus, and MH's return arrangements for the following day.
ALS 3pp JHS 4.16

11005. [1859?] To Margaret Brodie Herschel London
Talks of a committee attempting a massive collection of knowledge from the colonies; JH is considering exchanging some furniture between London and Collingwood.
ALS 4pp JHS 4.6

11006. [1859?] To Margaret Brodie Herschel London
JH decides not to exchange furniture [see JH's 1859?, i.e., #11005] on daughter Caroline's recommendation; some financial matters.
AL 2pp JHS 4.8

11007. [1859 or later] To [John Herschel (son)] Collingwood
About the cost of sending a telegram to India.
AL 2pp inc JHS 6.61

11008. 1859 To [Margaret Brodie Herschel] [Collingwood]
Sons William and John have brought photographs from India; also writes news of family and friends.
AL 2pp inc JHS 4.29

11009. 1859 To Woodbine Parish Collingwood
Comments on WP's account of a meteor striking ground in America.
C 1p RS:HS 23.290

11010. [1859 or later?] From Hensleigh Wedgwood 6 Adelphi Terrace
Recalls JH's letter from Cape regarding study of etymology. Encloses HW's own effort to develop this.
ALS 1p RS:HS 18.134

11011. 1859–1–6 From Augustus De Morgan 7 Camden St.
Many thanks for the information. Finds that George Peacock wrote a book entitled the *Measurer*. Someone has sent him a geometrical question with abominable diagrams.
AL 2pp RS:HS 6.323

11012. 1859–1–6 From Edward Sabine 36 Eversfield Place, St. Leonards[-on-Sea]
Sends copy of ES's reply to Richard Owen, who reported Owen's interview with C. E. Trevelyan. Suggests waiting until summer to apply for government funds to support observatories in Peking and Vancouver Island. ES's recent illness.
ALS 4pp TxU:H/M–0526.1; Reel 1093

11013. 1859–1–10 From Langham Dale Kingsclere near Newbury
Has been employed in London studying the way in which Government aid is granted to further popular education. Will be moving on to Glasgow soon to do similar work. Has been offered an L.L.D. if he can obtain a recommendation from someone in this country.
ALS 4pp RS:HS 6.462

11014. 1859–1–10 From Herbert Spencer St. John's Wood
Sending JH his *Essays* [*Essays: Scientific, Political, and Speculative,* 1857] and HS's 'Recent Astronomy and the Nebular Hypothesis,' the latter just published in the *Westminster Review*. Requests JH's comments on the latter publication.
ALS 4pp RS:HS 16.483

11015. 1859–1–12 From Richard Christopher Carrington Observatory, Redhill
Did not expect to see U. J. J. Leverrier's prediction of an inferior planet verified so soon. Observations of E. M. Lescarbault meritorious. Is writing to Charles Tennant and W. R. Birt to join in the observations of the planet.
ALS 4pp RS:HS 5.203

11016. 1859–1–16 To Herbert Spencer Collingwood
Thanks HS for sending [see HS's 1859–1–10] HS's *Essays* and his 'Recent Astronomy and the Nebular Hypothesis.' Regarding the latter, points out a geometrical error, expresses reservations about HS's and Auguste Comte's views of the nebular hypothesis, and argues against HS's theory of sunspot structure. Agrees with HS's idea that some nebulae may be nearer than commonly thought.
CC 8pp RS:HS 16.485 (C: 23.253)

11017. 1859–1–18 From Langham Dale Kingsclere near Newbury
Is grateful for the kind manner in which he answered his letter. Gives news of the College in South Africa. Details of Shaw House, Newbury.
ALS 2pp RS:HS 6.463

11018. 1859–1–22 From Richard Christopher Carrington Observatory, Redhill
Capt. W. S. Jacob has supplied him with enclosed letter as a basis for the Committee to work on. Would like his opinion on the scheme. Hopes Poonah will be the base for new observations. Size and quality of the Kew solar photographs.
ALS 4pp RS:HS 5.204

11019. 1859–1–25 To Richard Christopher Carrington Collingwood
On the desirability of one good observatory in India. Thinks Poonah would be the best site. Time not suitable to press for this establishment.
draft 4pp RS:HS 5.205 & 23.254

11020. 1859–1–25 From Michel Chevalier Paris
Is working on a volume on the gold standard.
ALS 2pp (French) RS:HS 5.255

11021. 1859–1–27 From James Booth Wandsworth Vicarage
Sending parts of his paper on the Lagocyclic curve.
ALS 2pp RS:HS 4.198

11022. 1859–1–27 From Charles Edward Trevelyan Grosvenor Crescent
Returns a letter from JH.
ALS 2pp RS:HS 17.445

11023. 1859–1–29 From Sir Edward Ryan Garden Lodge
Has shown JH's son's letter to [C. E.] Trevelyan and encloses his reply. Affairs in India may become more dangerous.
ALS 2pp RS:HS 14.464

11024. 1859–2–2 From M. Felix Billet Dijon
Sending his work on optics. Comments on their mutual work in this field.
ALS 3pp (French) RS:HS 5.63

11025. 1859–2–4 From Joseph Ellison Portlock Queen's Gardens, Hyde Park
Asks JH to sign a document relating to Col. Gelland [?].
ALS 2pp RS:HS 14.178

11026. 1859–2–5 To T. P. Anderson Collingwood
JH's only knowledge of Hollier trust fund was explained in earlier letters. Signs and returns annual report for trust fund.
ACCS 1p TxU:H/L-0029; Reel 1054

11027. 1859–2–6 To M. Felix Billet Collingwood
Thanks for his book on optics. Comments on various points.
draft 4pp (French) RS:HS 5.64 & 23.255

11028. 1859–2–11 From Augustus De Morgan 7 Camden St.
Sends a riddle. Has added a stanza. Does not credit the story of [T. C.?] Jansen's children. Comments on Roger Bacon's telescope.
ALS 3pp RS:HS 6.324

11029. 1859–2–24 From Augustus De Morgan 7 Camden St.
Hopes he is well as he has not heard from him for a long time. Sends his own theories regarding the composition of force.
ALS 6pp RS:HS 6.325

11030. 1859–2–24 From Alexander John Ellis Colney Hatch
Would like JH's support for his candidature for the R.S.L.
ALS 2pp RS:HS 7.40

11031. 1859–3–1 From Alexander John Ellis Colney Hatch
Will bring down his R.S.L. form to Collingwood. Regarding the phonetics of Hindustani.
ALS 7pp RS:HS 7.41

11032. 1859–3–3 From Edward William Brayley London Institution
Has used JH's writings for his lectures, but would like a simplification of statements dealing with the relation of temperature and pressure in the atmosphere.
ALS 6pp RS:HS 4.218

11033. 1859–3–6 To Edward William Brayley Collingwood
Tries to help EB understand JH's writings on meteorology; some comments on location of writings.
C 1p RS:HS 23.256

11034. 1859–3–7 From Edward William Brayley London Institution
Thanks for his suggestion for a lecture on graphical processes. Has had little communication with W. R. Birt since the latter left Kew. Has returned the book by Karl Kreil. Would be pleased to receive the Russian Observations.
ALS 4pp RS:HS 4.219

11035. 1859–3–7 To Augustus De Morgan Collingwood
Responds to AD's 1859–2–24 on forces, which degenerates into nonsense; comments on James Kemplay's writing on comets.
C 2pp RS:HS 23.257

11036. 1859–3–13 From Richard Christopher Carrington Observatory, Redhill
Is thinking of applying for the position of Observer at the Radcliffe Observatory and would like JH's support.
ALS 4pp RS:HS 5.206

11037. 1859–3–14 From George Biddell Airy Royal Observatory, Greenwich
Wants information for the proposed visit to Auvergne. Regarding the motion of the sun.
ALS 4pp RS:HS 1.240 (C: RGO 6.377.440)

11038. 1859–3–16 From Alexander K. Johnston Edinburgh
His letter and accompanying packet arrived for which accept his best thanks for the valuable contribution to the literature of physical geography. Is gratified by JH's commendation of his own *Physical Atlas*.
ALS 2pp RS:HS 10.448

11039. 1859–3–18 From Alexander J. B. Hope Arklow Ho. Connaught Place
Intends to stand for the University of Cambridge and would be pleased to receive his vote and interest. [Note on JH's reply: Shall have his vote; in any case will not vote against him.]
ALS 3pp† RS:HS 9.464

11040. 1859–3–18 From P. Kellom Edinburgh
Royal Society of Edinburgh would like JH to referee a paper on the mean temperature of the earth by J. D. Forbes. Would JH be willing? Sends a little biography of the late Professor Thomas Henderson, who is now nearly forgotten in his native city.
ALS 4pp RS:HS 11.77

11041. 1859–3–19 From James David Forbes Edinburgh
Has requested his publishers to send JH his recent volume on glaciers. Thanks for JH's *Physical Geography*.
ALS 2pp RS:HS 7.317

11042. 1859–3–21 From Richard Christopher Carrington Observatory, Redhill
Agrees with what he proposes to express to the Radcliffe Trustees. Capt. W. S. Jacob has retired to this country from India and will also give him his support. Gives address of the secretary of the Radcliffe Trustees.
ALS 4pp RS:HS 5.207

11043. 1859–3–22 To James David Forbes Collingwood
Expresses appreciation for JF's publicationon glaciers.
ALS 2pp St. Andrews 18 (C: RS:HS 23.258)

11044. 1859–3–25 To Alexander K. Johnston Collingwood
Has just received the first part of Johnston's *General Atlas* for which accept his very best thanks. It is a beautiful work.

Has been working at cartography himself and produced a projection of 97–100ths of the whole surface of the globe in one chart. Comments on this.
C 4pp RS:HS 10.449 & 23.259

11045. 1859–3–25 To Roderick Impey Murchison Collingwood
Sending details of his idea for a new projection of the sphere. If printed will need the maps re-engraved.
draft 2pp RS:HS 12.419

11046. 1859–3–25 To [Philip Kelland] Collingwood
Willing to referee paper by J. D. Forbes. Just received [PK]'s biography of Thomas Henderson, whom JH wishes had remained longer at Cape of Good Hope.
ALS 2pp TxU:H/L-0754; Reel 1089

11047. 1859–3–26 From Roderick Impey Murchison n.p.
Thanks for his memoir on the projection of the sphere; it will be read at their meeting after next Monday. Up to theirs neck in the Yang-ste-kiang.
ALS 4pp RS:HS 12.420

11048. 1859–3–26 From Edward Sabine n.p.
Received order for self-recording photographic instruments for magnetic observations from Mr. Bolyani, who plans to improve observatory at University of Kazan. ES quoted prices for making these in London and invited Bolyani to inspect observatory at Kew. ES now prefers Shanghai to Peking as site for observatory. Geographic distribution of global observatories. Reading JH's 'Physical Geography' [1859]. Sends two sunspot photographs from Kew.
ALS 6pp TxU:H/M–0526.2; Reel 1093

11049. 1859–3–27 To G. Bramwell Collingwood
Testimonial in favor of R. C. Carrington to succeed late M. J. Johnson as director of Radcliffe observatory. Understands that W. S. Jacobs is also a candidate.
ACCS 1p RS:HS 19.145

11050. 1859–4–1 From Alexander J. B. Hope Trafalgar Hotel, Spring Gds.
Would be grateful if JH would allow his name to be added to AH's Committee.
ALS 1p RS:HS 9.465

11051. 1859–4–6 From Robert Main Greenwich
Is a candidate for the position of Radcliffe Observer and would be pleased if JH would send a testimonial to the Radcliffe Trustees.
ALS 4pp RS:HS 12.277

11052. 1859–4–8 From Edward William Brayley London Institution
Would like JH's support for his application for a position of examiner at East India Company's College. Cannot find out what he has done with Karl Kreil's book. Will send him a copy of his lecture.
ALS 4pp RS:HS 4.220

11053. 1859–4–10 From Thomas Webster Cranbrook
Not at home today when JH's son called. Conveys Mr. Allan's gratitude for JH's promise to inaugurate literary society.
ALS 2pp RS:HS 18.131

11054. 1859–4–16 From Robert Arthington Leeds
Saw him at the B.A.A.S. meeting at Leeds. Raises query regarding the sun and Thomas Paine's views on deism.
ALS 4pp RS:HS 1.414

11055. 1859–4–16 To Henry Anthony Jeffreys Collingwood
Regrets delay in replying. Explains the various editions of Sir William Herschel's works.
ACCS 3pp RS:HS 10.436 (C: RS:HS 23.262)

11056. 1859–4–18 To Robert Arthington, Junior Collingwood
[Responding to RA's 1859–4–16,] discusses the sun, Thomas Paine's theories, and the writing of Genesis.
draft 4pp RS:HS 1.415 & 23.261

11057. 1859–4–21 From Sir Benjamin Collins Brodie [the elder] 14 Saville Row
Would like nominations for the Copley and Royal medals.
ALS 1p RS:HS 4.281

11058. 1859–4–21 From George Gabriel Stokes n.p.
Says that Council of R.S.L. will reconsider whether William Pole's paper on color-blindness should be published.
C inc 3pp Cambridge University Library 7656/RS107

11059. 1859–4–22 From Roderick Impey Murchison 16 Belgrave Square
Thanks for the memoir on his ingenious projection. Now sends a short theoretical view of Robert Stephenson on the astronomical reasons for fossil plants in the Arctic regions. Would like JH's comments.
ALS 4pp RS:HS 12.421

11060. 1859–4–24 To Roderick Impey Murchison Collingwood
Comments on Robert Stephenson's theory regarding the presence of fossil plants in Arctic regions and on possible astronomical causes of such fossilization; gives some of his own theories.
ALS 4pp diag RS:HS 12.422 (C: 23.263 and Geological Society)

11061. 1859–4–24 To George Gabriel Stokes Collingwood
Approves changes made by William Pole in his paper [see GS's 1859–4–21]; comments on appropriate presentation of JH's report on it.
ALS 4pp Cambridge University Library 7656/RS108

11062. 1859–4–25 From John P. Gassiot Clapham Common
Would like JH to read the enclosed and if agreeable, to add his name to those supporting the scheme [Scientific Relief Fund].
ALS 2pp RS:HS 8.62

11063. 1859–4–25 To [?] Collingwood
Comments on paper on [Strati?-]friction of electric light.
ALS 3pp St. John's College, Cambridge

11064. 1859–4–29 From John P. Gassiot Clapham Common
Would like to add JH's name to the first published list [of Scientific Relief Fund]. Regarding donations to the fund.
ALS 3pp RS:HS 8.63

11065. 1859–5–2 From Augustus De Morgan 7 Camden St.
Did his work on the Southern Constellations bring him into contact with a Frederic Houtmann? Regarding a manuscript stated to be by Galileo.
ALS 3pp RS:HS 6.326 S. De Morgan, *Memoir of Augustus De Morgan*, 299–300

11066. 1859–5–3 From Alexander Bridport Becher and Captain R[obert] Lee Admiralty
[Form letter] Announces 4 June visitation of Royal Observatory by Board of Visitors.
(A)LS 1p RAS:JH Archive 12/1.4; Reel 10

11067. 1859–5–6 From Augustus De Morgan 7 Camden St.
Regarding one of JH's riddles. Sees he must write an article on Frederic Houtmann. Will be glad of the information on Galileo.
ALS 2pp RS:HS 6.327

11068. 1859–5–8 From Augustus De Morgan 7 Camden St.
What does he make of the accompanying quotation from Amerigo Vespucci?
ALS 3pp RS:HS 6.328

11069. 1859–5–10 From Matwei Matwejewitsch Gussew London
Had a card to visit Greenwich Observatory the same day as he received JH's letter so was unable to visit him.
ALS 2pp (French) RS:HS 9.151

11070. 1859–5–12 From Thomas Nelson, Jr Edinburgh
Sending a section of their new atlas of the world, which includes some new features. Would like JH's comments. Will send the complete work when published if JH would like to see it.
ALS 3pp RS:HS 13.148

11071. 1859–5–13 To George Gabriel Stokes Collingwood
On some changes to the report on paper by William Pole [see JH's 1859–4–24]. Describes some experiments carried out with colored 'prismatic' light.
ALS 3pp Cambridge University Library 7656/H696

11072. 1859–5–14 From Edward Sabine 13 Ashley Place
Asks permission to include JH on R.S.L. committee, headed by C. J. Selwyn (relative of George Peacock) and Dr. Beale, to purchase portrait of Peacock from artist Douglas Blakiston. Self-recording magnetic instruments have reached Washington safely.
ALS 4pp TxU:H/M–0526.3; Reel 1093

11073. 1859–5–15 To Augustus De Morgan Collingwood
Comments on 'cause' and 'will.'
C 2pp RS:HS 23.264

11074. 1859–5–16 From W. Bonsey Slough
Regarding the faculty pew in the new church. Is sorry he was out when JH called. Hears he is contemplating a change in his property.
ALS 3pp RS:HS 4.190

11075. 1859–5–17 From George [Hamilton]-Gordon Argyll House
The B.A.A.S. meeting is to be held at Aberdeen and would be pleased to accommodate JH should he be attending the meeting. Will not be going himself.
ALS 3pp RS:HS 9.104

11076. 1859–5–18 From Augustus De Morgan 7 Camden St.
Thanks for his pamphlet; will comment on it later when he has properly read it. Regarding Maine de Biranand his philosophy. Gives some of his own theories.
ALS 3pp RS:HS 6.329 S. De Morgan, *Memoir of Augustus De Morgan*, 300

11077. 1859–5–19 To Thomas Nelson Jr. Collingwood
Is much obliged for the gift of the new atlas. Has never seen an atlas so clearly and beautifully executed. Comments on various points. Wishes that the usual parallels of latitude and longitude had been included.
draft 4pp RS:HS 13.149

11078. 1859–5–21 To George Gabriel Stokes Collingwood
Additional comments on William Pole's paper [see JH's 1859–5–13]; adds comments lost from JH's paper on sound.
ALS 3pp Cambridge University Library 7656/H697

11079. 1859–5–23 From George Gabriel Stokes Lensfield Cottage, Cambridge
Comments on the effect of intensity of illumination on the distinctness of the spectrum.
ALS 7pp RS:HS 17.37

11080. 1859–5–31 From Edward Sabine Bournemouth
Will assist JH in preparing review of terrestrial magnetism. Progress in publishing vol. 2 of observations from St. Helena. Prince Albert favors new observatories and hopes to see JH [at B.A.A.S. meeting] in Aberdeen. Research in British colonies. Demonstration of Mr. Gossert's apparatus by [T. R.] Robinson.
ALS 4pp TxU:H/M–0526.4; Reel 1093

11081. 1859–6–3 From John P. Gassiot Clapham Common
Called at Burlington House and sees that more subscriptions have been added to the fund.
ALS 2pp RS:HS 8.64

11082. 1859–6–8 To Thomas Maclear Collingwood
Makes arrangements for TM's visit.
ALS 3pp Cape Archives/Maclear Papers File 100

11083. 1859–6–10 From Matthew Fontaine Maury Observatory, Washington
Thanks for JH's article on *Physical Geography* from the new edition of the *Encyclopaedia Britannica*. Comments on this and gives details of some of his own researches into the causes of ocean currents.
ALS 21pp RS:HS 12.308

11084. 1859–6–11 To George Biddell Airy Collingwood
The Radcliffe Trustees are asking for assistance in naming a new director of the Oxford Observatory; JH sends a draft response.
ALS 2pp RGO 6.146.266

11085. 1859–6–11 To Sidney Herbert Collingwood
Draft of letter to be sent to Radcliffe Trustees, now being sent to G. B. Airy for his information [see JH's 1859–6–11].
draft 4pp RGO 6.146.267

11086. 1859–6–14 From George Biddell Airy n.p.
Responds to same request as JH identifies in his 1859–6–11.
ACCS 3pp RGO 6.146.283

11087. 1859–6–14 To Chairman, Society of Arts Collingwood
Unable to attend the most recent meeting of the Society of Arts, where the idea of having a fixed musical pitch was discussed. As a committee has been struck to study this, JH sends in his thoughts on the topic.
ACCS 4pp RS:HS B27.82 (draft: RS:HS B27.83)

11088. 1859–6–14 To William Whewell, Chair of the Musical Pitch Committee of the Society of Arts

Collingwood
Discusses the issue [then under debate] of a fixed musical pitch, urging that C be fixed at 512 vibrations per second and explaining in detail his reasons.
L *Journal of the Society of Arts, 7* (July 8, 1859), 581 & *Athenaeum,* #1654 (July 9, 1859), 56–7

11089. 1859–6–15 To George Biddell Airy Collingwood
Returns GA's letter copy [see GA's 1859–6–14] with thanks, noting it is more complete than JH's own.
ALS 2pp RGO 6.146.285

11090. 1859–6–15 From James Duncan Denholm near Hawick
Pointing out an error in JH's statement in an encyclopaedia that there are no serpents in Borneo.
ALS 4pp RS:HS 6.512

11091. 1859–6–25 From François N. M. Moigno Paris
Sending two volumes of his own *Cosmos*. Regrets he did not visit him in 1857.
ALS 1p (French) RS:HS 12.355

11092. 1859–6–27 From James Duncan Denholm near Hawick
Was pleased to receive JH's note. Further regarding the existence of serpents in Borneo. Suggests form of correction in encyclopaedia.
ALS 3pp RS:HS 6.513

11093. 1859–7–2 From Edward Sabine Falmouth
A. D. Bache reports that self-recording instruments are installed at Washington observatory and ready to begin observations. ES is making field observations to confirm changes in magnetic lines in England since ES's 1837 survey. [JH annotation: Summary of Bache's letter. Joint survey by Smithsonian Institution and U.S. Coast Survey will begin when Bache receives full instructions from Sabine.]
ALS 4pp † TxU:H/M–0526.5; Reel 1093

11094. 1859–7–12 To Augustus De Morgan Collingwood
Comments on decimal coinage, the weather, and politics.
C 2pp RS:HS 23.265

11095. 1859–7–12 From Augustus De Morgan 7 Camden St.
Is suffering from the heat. Was grateful for his information on Galileo's handwriting. Outlines the facts of the recent controversy on a reputed manuscript of Galileo. Will be moving house shortly.
ALS 8pp RS:HS 6.330

11096. 1859–7–13 From Edward Sabine 13 Ashley Place
Does JH wish to alter ES's reply to A. D. Bache? Origins of names for American rivers. Hopes JH is well enough to accept the invitation of Lord Aberdeen [G. Hamilton-Gordon].
ALS 4pp TxU:H/M–0526.6; Reel 1093

11097. 1859–7–16 To Samuel Austin Allibone Collingwood
Has not yet received SA's *Dictionary of English Literature,* which SA indicated he was sending.
ALS 1p Huntington Library

11098. 1859–7–17 From Johann Franz Encke Berlin
Sends some invitations to participate in an Alexander von Humboldt Foundation.
ALS 1p (German) RS:HS 7.100

11099. 1859–7–17 To Edward Sabine Collingwood
Discusses desirability of middle latitude stations in North America. Impressed by the work of [A. D.] Bache.
C 1p RS:HS 23.266

11100. 1859–7–20 From William Crookes 20 Mornington Road
Thanks for his interesting communication to the *Photographic News* and for the spectra. Gives details of his own photographic experiments with Iodide of silver.
ALS 4pp RS:HS 5.309

11101. 1859–7–20 From Woodbine Parish St. Leonards-on-Sea
Sends newspaper article about the meteor fall in Oswego County. Compares the account to a meteorite that fell in the Santiago Plains of Chile. Discusses authenticity of suspected meteorite specimens.
ALS 3pp RS:HS 14.135

11102. 1859–7–24 To Elizabeth Colling Collingwood
Note accompanying something being discussed in the music world.
C 1p RS:HS 23.267

11103. 1859–7–24 To [?] Collingwood
Sees no need for changes to the plates for *Outlines Astr.*
ALS 1p APS B.H435p.18

11104. [1859]–7–27 To William Empson Collingwood
Thanks for sending the note of A. J. Ellis. Comments on this and his own apparent mistake. Pleased to hear that the invalid is progressing. Their potato crop has an attack of the blight.
draft 4pp RS:HS 7.75

11105. 1859–7–28 From Edward Sabine 13 Ashley Place
Replied to J. A. da Silva that Kew Observatory would gladly train Portuguese observers and superintend manufacture of instruments, but that ES could not speak for English government. [JH annotation: Summary of letter from da Silva, director of meteorological observatory in Lisbon, requesting training for Portuguese observers at Kew, British influence on Portuguese government to join magnetic survey, and advice on instruments.]
ALS 2pp †† TxU:H/M–0526.7; Reel 1093

11106. 1859–7–30 From William Crookes 20 Mornington Road
His Daguerreotype arrived safely. Hopes to get some better photographs of the spectra before the end of the summer. Sees why their results differ.
ALS 3pp RS:HS 5.310

11107. 1859–7–31 To Matthew Fontaine Maury Collingwood
Acknowledges the proof sheets of the new edition of the *Physical Geography of the Sea.* Will read them with the attention they deserve. Does not at present agree with the theories of MM.
C 4pp RS:HS 12.309 & 23.268

11108. 1859–8–1 From John Gorham Tonbridge
Has been working on the rotation of colored discs. Would be pleased to show him some of the novel effects. Is surgeon to the local Grammar School.
ALS 3pp RS:HS 8.220

11109. 1859–8–9 From John Gorham Tonbridge
Proposes to be at Collingwood on Tuesday next. Intends coming early so as to be able to carry out his experiments by

daylight.
ALS 3pp RS:HS 8.221

11110. 1859–8–11 From John Gorham Tonbridge
Hopes to be at Collingwood on Saturday at 2 o'clock.
ALS 1p RS:HS 8.222

11111. 1859–8–13 From George Biddell Airy Royal Observatory, Greenwich
Relating to his holiday experience in France.
ALS 4pp RS:HS 1.241 (C: RGO 6.377.442)

11112. 1859–8–13 From Henry Stevens 4 Trafalgar Square
Will send a package for JH through a third party.
ALS 1p Harvard: Houghton AAH 67m–67 (107)

11113. 1859–8–15 From Sir Benjamin Collins Brodie [the elder] Broome, Betchworth, Surrey
Response to JH's article on 'Sensorial Vision' (1858). Notes on physical and metaphysical elements of perception and consciousness.
ALS 4pp TxU:H/M–0128; Reel 1087

11114. 1859–8–18 To Augustus De Morgan Collingwood
Congratulates AD on a successful move to a new house.
C 1p RS:HS 23.269

11115. 1859–8–20 To the *Photographic News* Collingwood
Recounts experiments JH conducted on the action of the solar spectrum on various silver salts. Explains how JH produced his solar spectrum.
L *Photographic News, 3* (Sept. 9, 1859), 2–3

11116. 1859–8–23 From George Wilson n.p.
Received JH's 'Remarks on Colour-Blindness' [1859]. Describes lack of success in urging Army, Navy surgeons to continue GW's investigations. Comments on statistical distribution of color-blindness, hereditary factors, and safety implications for military and railroads. Reports specific cases among contemporaries. Asks JH's advice on seeking assistance from B.A.A.S.
ALS 16pp RS:HS 18.268

11117. 1859–8–24 To George Biddell Airy Collingwood
Responds to GA's account of his tour [see GA's 1859–8–13] with interest, and adds a description of JH's poor health.
ALS 4pp RGO 6.377.444

11118. 1859–8–24 From Thomas Maclear Spring Gardens
Relates his experiences while visiting the Paris Observatory with G. B. Airy. Later travelled to Brussels to meet L. A. J. Quetelet. Is now preparing for a visit to Ireland and Mrs. Maclear will go to Bedfordshire. The meteorological observations [of the Cape] are to be printed in England under the direction of Robert FitzRoy.
ALS 8pp RS:HS 12.163

11119. 1859–8–25 From [Paul Sinnis?] B.A.A.S., 6 Queen Street Place, Upper Thames St.
Announces meeting of Council at Marishal College, Aberdeen on 14 Sept.
(A)LS 1p TxU:H/M–0921.9; Reel 1083

11120. 1859–8–25 From [Paul Sinnis] B.A.A.S., 6 Queen Street Place, Upper Thames St.
Announces meeting of Kew Committee at Observatory on 29 Aug.
(A)LS 1p TxU:H/M–0921.8; Reel 1083

11121. 1859–8–26 From George Biddell Airy Royal Observatory, Greenwich
Regarding the geographical position of places in France. Comments on JH's paper on light and colors.
ALS 4pp RS:HS 1.242 (C: RGO 6.377.446)

11122. 1859–8–26 From William Crookes 20 Mornington Road
Regrets that the article was received too late to appear in the current number. Will send a proof in a day or so. Would like to print some illustrations for JH's articles.
ALS 3pp RS:HS 5.312

11123. [1859–8–27 or later] To Margaret Brodie Herschel Collingwood
Talks about the beauty of the flowers at Collingwood, and then comments on his *Essays Q.E.R.*
ALS 4pp JHS 4.17

11124. 1859–8–27 To Margaret Brodie Herschel Collingwood
Writes about the activities of several family members, and then comments on the location of a railway extension, which will come closer to Collingwood.
ALS 4pp JHS 4.18

11125. 1859–8–27 To William Lassell Collingwood
Asks for details of WL's system of speculum polishing and telescope construction as JH is preparing an article on telescopes for the *Encyclopaedia Britannica.*
C 1p RS:HS 23.271

11126. 1859–8–27 To Warren de La Rue Collingwood
Requests information about specula for telescopes, especially silvered glass ones.
C 2pp RS:HS 23.272

11127. 1859–8–28 To Elizabeth Colling Collingwood
Thanks for some verses; comments on the current state of Indian religious society.
C 1p RS:HS 23.273

11128. 1859–8–29 From Edward Sabine 13 Ashley Place
Notified Richard Owen that Kew observatory is ready to receive trainees for magnetic observations. Stewart Balfour estimates six weeks to train observers, who will then take instruments and leave for British Columbia or Shanghai. New observatory at University of Mississippi in U.S. American Association for the Advancement of Science has expressed interest in magnetic survey. Success in photographing sunspots, which R. C. Carrington and Warren de La Rue will supervise. Recent observations.
ALS 6pp TxU:H/M–0526.8; Reel 1093

11129. 1859–8–29 To Norton Shaw Collingwood
Returns proof for JH's ['On a New Projection of the Sphere']. Requests 25 copies when printed.
ALS 1p Royal Geographical Society

11130. 1859–8–30 To William Crookes Collingwood
Sends spectra obtained by Ioduret of potassium. Gives details of his latest photographic experiments.
draft 4pp RS:HS 5.311 (C: RS:HS 23.270)

11131. 1859–8–30 From Warren de La Rue Cranford, Middlesex
Discusses a telescope ordered for C. A. von Steinheil of Munich and the design of WR's glass polishing machine. Mentions visit by Thomas Maclear. Kew Telescope produces photos of sunspots.
ALS 8pp Harvard: Houghton AAH 67m–67 (48)

11132. 1859–8–31 From William Lassell Bradstowe near Liverpool
Discusses installation of a 4-ft. aperture equatorial telescope. Describes plans for a glass polishing machine in great detail.
ALS 10pp Harvard: Houghton AAH 67m–67 (48)

11133. 1859–[9?] To T. P. Anderson Collingwood
Names three nephews of Charles Babbage and their financial status relative to Hollier fund, for which JH is trustee.
ACCS 1p TxU:H/L-0030; Reel 1054

11134. [1859–9] From Mary Grantham Camden Hill
Is pleased that JH has been appointed to succeed Elizabeth Baily at Tavistock Place. Regarding a situation for Miss Baily's servant.
ALS 4pp RS:HS 9.117

11135. 1859–9–1 To Edward Sabine Collingwood
JH cannot attend Aberdeen [B.A.A.S.] meeting due to severe illness. Will try to sketch draft of response to application to British and foreign governments for cooperation in magnetic projects.
C 1p RS:HS 23.274

11136. [1859]–9–3 From Edward Sabine 13 Ashley Place
Returns JH report with one correction, substituting Kazan for North Cape, because [Christopher] Hansteen cannot find Norwegian observers. Will make copies of its final form and take one to B.A.A.S. meeting in Aberdeen. Mr. Bolyani has visited every magnetic observatory in Europe. 1854 was the year of minimum magnetic disturbances in Peking.
ALS 4pp TxU:H/M–0517.8; Reel 1093

11137. 1859–9–4 To Humphrey Lloyd Collingwood
Sends procedures for preparation of the report of the Magnetic Committee for the meeting of the B.A.A.S., which JH cannot attend due to illness; asks some questions about achromatic lens theory.
C 2pp RS:HS 23.276

11138. 1859–9–4 To William Whewell Collingwood
Asks WW to check JH's draft report for the Meteorological and Magnetic Committee. Illness will keep him from attending the B.A.A.S. meeting
ALS 2pp TC, Camb. Add. Ms.a.207[98] (C: RS:HS 23.275)

11139. 1859–9–5 From Edward Sabine 13 Ashley Place
Encloses ES's résumé of magnetic surveys accomplished by U.K., written to encourage similar surveys in China and North America. Perhaps JH may append this to JH's report.
ALS 4pp TxU:H/M–0526.9; Reel 1093

11140. 1859–9–6 From Augustus De Morgan 41 Chalcot Villas NW
Elizabeth Baily died this morning, quietly and painlessly. Her last few weeks had been saddened by delusions.
ALS 3pp RS:HS 6.331

11141. [1859]–9–7 From Edward Sabine 13 Ashley Place
Sends Council report to be read next week. Instructed printers Taylor & Francis to accept any changes that JH may request. Sent copies of JH's report to Humphrey Lloyd and G. B. Airy. Gives ES's address in Scotland.
ALS 3pp TxU:H/M–0517.9; Reel 1093

11142. 1859–9–9 From George Biddell Airy Royal Observatory, Greenwich
Encloses a paper and gives his comments on it.
ALS 3pp RS:HS 1.243 (C: RGO 6.694.535)

11143. 1859–9–9 From George Biddell Airy n.p.
About GA's plans to attend the B.A.A.S. meeting in Leeds.
ACCS 2pp RGO 6.407.194

11144. 1859–9–9 From Augustus De Morgan 41 Chalcot Villas NW
All arrangements have been made for the burial of Elizabeth Baily. When did JH first become acquainted with Francis Baily? Gives his own earliest recollections. Has now placed all his books on their shelves.
ALS 4pp RS:HS 6.332 S. De Morgan, *Memoir of Augustus De Morgan*, 301–2

11145. 1859–9–13 To Humphrey Lloyd Collingwood
Thanks HL for the response to JH's queries [see JH's 1859–9–4].
C 1p RS:HS 23.277

11146. 1859–9–14 From J. B. Kearney Canterbury
Has no objection to his letter being communicated to the R.A.S. Comments on his observations. Thanks for his paper on musical pitch. Sees that new copper coinage is to be introduced and hopes that JH will use his influence to introduce the decimal system.
ALS 4pp RS:HS 11.73

11147. [1859]–9–15 To Margaret Brodie Herschel Collingwood
JH is reorganizing house while MH is away; other family news; MH is comforting someone very ill.
ALS 4pp JHS 4.3

11148. 1859–9–15 From E. Moorsom 7 Cambridge Road
Thanks for fine basket of venison. Hopes that Lady Herschel is stronger after her visit to the North.
ALS 3pp RS:HS 13.50

11149. 1859–9–16 From Thomas Maclear Kings Cross Station
Relates experiences while on a 10 day visit to Ireland. Now waiting for a train to Aberdeen. Professor [Philip?] Kelland injured in rail accident at Hitchin. Comments on the carelessness of railway workers. Agitation again at the Cape over the recall of Sir George Grey. Jack [TM's son?] has received orders for service, perhaps China.
ALS 8pp RS:HS 12.164

11150. 1859–9–18 From Augustus De Morgan 41 Chalcot Villas NW
Hears that JH is to inherit Baily's house. Recalls how he first met W. H. Wollaston and Francis Baily.
ALS 4pp RS:HS 6.333

11151. 1859–9–22 From John Bullar Basset Wood near Southampton
Shares JB's experiences with visual abnormalities similar to those described in JH's 'Sensorial Vision' (1858). Encloses copy of related subject that JB found in Section 45 of John Locke's 'On the Conduct of the Understanding.'
ALS 4pp TxU:H/M–0922; Reel 1083

11152. 1859–9–23 To Margaret Brodie Herschel Collingwood
About the impending death of someone they both hold dear; the nature of the Scots, and an inheritance from Elizabeth Baily.
AL 4pp JHS 4.1

11153. 1859–9–23 To William Lassell Collingwood
Comments on WL's information on telescope construction, and adds some of JH's own experience.

C 2pp RS:HS 23.278 (extract 1p Harvard: Houghton AAH 67m–67 {48})

11154. 1859–9–26 From Adam and Charles Black Edinburgh
Will allow more pages for the text of JH's *Telescope.*
ALS 1p Harvard: Houghton AAH 67m–67 (48)

11155. 1859–9–29 From William Whewell Athenaeum
Asks JH for his impressions of the Aberdeen B.A.A.S. meeting and of U. J. J. Leverrier's claim that there may be an intermercurial planet.
ALS 3pp RS:HS 18.214

11156. 1859–10–6 From Adam and Charles Black Edinburgh
Concerning the publication of JH's *Telescope.*
ALS 1p Harvard: Houghton AAH 67m–67 (48)

11157. 1859–10–10 From Robert Grant Royal Astronomical Society
Intends applying for the chair of Practical Astronomy at Glasgow University. Outlines the work he has already done in astronomy. Would like a testimonial from JH.
ALS 7pp RS:HS 9.17

11158. 1859–10–15 From Charles Manley 25 Great George St., Westminster
Announces funeral of Robert Stephenson.
ALS 2pp TxU:H/M–0347.1; Reel 1087

11159. 1859–10–17 To Elizabeth Colling Collingwood
Apology for delay in returning a biographical note; some concern about state of Indian society.
C 1p RS:HS 23.280

11160. 1859–10–18 From Robert Grant Royal Astronomical Society
Is grateful for the testimonial JH has sent, and will be pleased to use it in support of his application to Sir G. C. Lewis.
ALS 3pp RS:HS 9.18

11161. 1859–10–19 From S. Bax Tavistock Place
Miss B. Baily is delighted with the account of JH, and of his son in India. She has recovered from the cold and hopes that Lady Herschel is feeling better.
ALS 3pp RS:HS 5.28

11162. 1859–10–20 From Henry W. Field Royal Mint
Would like a testimonial from JH which would prove useful when he retires. Points out his value to JH when he analyzed the metals used.
ALS 3pp RS:HS 7.217

11163. 1859–10–22 From John Phillips B.A.A.S., Oxford
JH appointed to a committee including George Wilson, David Brewster, Clerk Maxwell, William Thomson, and William Pole to study color blindness. They will be granted £10.
L 1p TxU:H/M–0428.24; Reel 1093

11164. 1859–10–22 To Elizabeth White Collingwood
Encloses note from R. J. Kane. JH observed oscillating wave created by small waterfall in EW's garden at Holybourne. Requests diagram and measurements of it.
C 5pp RS:HS 19.146 (C: 23.92)

11165. 1859–10–26 From Edward Sabine 13 Ashley Place
Extract of letter from Charles Grey, asking B.A.A.S. to inform Prince Albert of estimated expenses for proposed five-year survey of terrestrial magnetism. As president of joint committee of B.A.A.S. and R.S.L., JH should answer this. Entry of three U.S. observatories may eliminate need for observatory in Newfoundland. Recent visit to Kew by A. T. Kupffer suggests that Russian stations may soon have self-recording instruments. Proposes suspending observatory in Falkland Islands until colonies decide to join survey. Dutch observatory in Java. Describes buildings needed for observatories. Estimates expenses for six years.
ALS 7pp, encl 1p TxU:H/M–0526.11; Reel 1093

11166. 1859–10–[27] To Henry W. Field Collingwood
Testimony to the value of HF's work when JH was Master of the Mint.
draft 3pp RS:HS 7.218 & 23.279

11167. 1859–10–27 From Samuel Fry Brighton
Sending a stereoscopic photograph of the gibbous moon and requesting information on various points. Has recently taken many photographs of the moon in various states.
ALS 4pp RS:HS 7.384

11168. 1859–10–28 To Edward Sabine Collingwood
Requests information concerning a letter to the Treasury, which ES had asked JH to write, regarding funding for [meteorological] stations in Vancouver, Shanghai, Newfoundland, and the Falkland Islands.
C 2pp RS:HS 23.281

11169. 1859–10–29 From Elizabeth White Holybourne [near Alton]
Wavelet in garden stream is gone due to scarce water, but EW describes it from memory. [Husband] Walpole adds measurements. Returns note from R. B. Kane. EW hurt that her word about Elizabeth Baily was doubted by Lady K. B. Kane. Thanks JH for accepting EW's claim.
ALS 3pp RS:HS 19.297

11170. 1859–10–31 From Henry W. Field Royal Mint
Is grateful for JH's testimonial, which will doubtless be of great value.
ALS 2pp RS:HS 7.219

11171. 1859–10–31 From Edward Sabine 13 Ashley Place
Proper procedure for communicating with Charles Grey. Cost and effectiveness of self-registering instruments versus older visual instruments. ES's method of estimating observatory expenses. Importance of observatory in Shanghai.
ALS 4pp TxU:H/M–0526.10; Reel 1093

11172. 1859–11–1 To 'Sir' Collingwood
Responds to [?]'s 21 Oct. 1859 letter to John Phillips expressing Prince Albert's interest in the impact that proposed B.A.A.S. magnetic and meteorological observations will have on Treasury funds.
ACC 4pp TxU:H/L-0729; Reel 1089

11173. 1859–[11] To Margaret Brodie Herschel Collingwood
Reports to MH on the contents of Elizabeth Baily's will [JH is given the house in Tavistock Place], and adds some family news, concluding by commiserating with MH's situation.
ALS 4pp JHS 5.17

11174. 1859–[11] To Margaret Brodie Herschel Collingwood
Talks about the death of Elizabeth Baily, and the beauty of the fall flowers; comments on correspondence that has arrived, and includes some family news.
ALS 4pp JHS 4.28

11175. 1859–11–2 To Charles Grey Collingwood
Outlines B.A.A.S.'s proposals regarding magnetic and meteorological observations, to be laid before Prince Albert.
ALS 12pp inc RS Sa.669

11176. 1859–11–3 From James David Forbes Edinburgh
Is a candidate for the position of Principal at St. Andrews United College in succession to David Brewster, who has moved to Edinburgh. Would be pleased to receive a testimonial from JH. Has already received one from Brewster.
ALS 3pp RS:HS 7.318

11177. 1859–11–4 From Augustus De Morgan Univ. College London
Someone has experimented with sound and finds that the pitch of a note is altered when transmitted through solids; is this true in JH's experience? Supposes that Lady Herschel has now returned.
ALS 2pp RS:HS 6.334

11178. [1859–11–4] To Royal Geographical Society n.p.
Requests that proofs of his paper ['On a New Projection of the Sphere'] be sent to him at 37 Tavistock Place.
ALS 1p Royal Geographical Society

11179. 1859–11–7 To Augustus De Morgan 37 Tavistock Place
Comments on fluorescence and the contents of Francis Baily's desk drawer.
C 2pp RS:HS 23.282

11180. 1859–11–7 To Charles Dickens 37 Tavistock Place
Sorry to trouble CD with note today regarding occupying the premises.
Adraft 1p inc RS:HS 19.195

11181. 1859–11–7 From Charles Dickens Tavistock Square
Would give him great pleasure to comply with JH's request, but the trees are in the garden of a Mr. Powell.
AL 1p RS:HS 6.492

11182. 1859–11–7 To Royal Geographical Society 37 Tavistock Place
Asks that the diagrams alluded to be printed with JH's paper ['On a New Projection of the Sphere'].
ALS 2pp Royal Geographical Society

11183. [1859–11–7 or earlier]. From Edward Sabine n.p.
Must postpone meeting at Kew Observatory because of a severe cold. Asks JH to choose a later date.
LS 3pp RS:HS 15.311

11184. 1859–11–7 From Edward Sabine 13 Ashley Place
Would be glad to accompany JH to Kew to see instruments.
ALS 4pp RS:HS 15.248

11185. 1859–11–8 To Warren de La Rue 37 Tavistock Place
Further questions about the physical optics of telescopes [see JH's 1859–8–27].
C 2pp RS:HS 23.284

11186. 1859–11–8 To Edward Sabine 37 Tavistock Place
Will accompany ES to Kew Observatory. Serving as legatee for the recently deceased Elizabeth Baily, sister of Francis Baily.
C 1p RS:HS 23.283

11187. 1859–11–9 To James David Forbes 37 Tavistock Place
Wishes JF success and rest in his new position [at St. Andrews].
ALS 3pp St. Andrews 167

11188. 1859–11–9 To George Cornewall Lewis 37 Tavistock Place, Russell Square
Recommending J. D. Forbes as a suitable person for the appointment of Principal at the University of St. Andrews.
draft 3pp RS:HS 11.480 & 23.285

11189. 1859–11–10 From Adam and Charles Black Edinburgh
Questions concerning the manuscript of JH's *Telescope.* Will send proofs to London.
ALS 1p Harvard: Houghton AAH 67m–67 (48)

11190. 1859–11–10 From Maurice Drummond London
Announces appointment of James D. Forbes to the Principalship of the United College at St. Andrews.
ALS 1p St. Andrews 174

11191. 1859–11–10 From Samuel Fry Brighton
Fears his last letter may have gone astray owing to incomplete address. Will send off the photo of the moon on Saturday. Explains what he means by seven days of difference. Has also taken many photographs of the planets, which he will be pleased to send for his perusal.
ALS 4pp RS:HS 7.385

11192. 1859–11–10 To Margaret Brodie Herschel 37 Tavistock Place
Of the comings and goings of some of their daughters in London; other family matters, including finances.
AL 4pp JHS 4.9

11193. 1859–11–10 From Warren de La Rue Observatory, Cranford
Offers to show JH his observatory and the polishing apparatus he uses. Comments on various telescope makers, especially A. Ross.
ALS Paul Luther

11194. 1859–11–10 From George Gabriel Stokes Athenaeum
Further comments on subject of GS's 1859–11–10; comments on several experiments involving light and color.
ALS 8pp RS:HS 17.39

11195. 1859–11–10 From George Gabriel Stokes Royal Society
Comments about reflective power of metals.
ALS 3pp RS:HS 17.38

11196. [1859]–11–11 From Charles Robert Darwin Down, Bromley, Kent
Sending CD's new book, *The Origin of Species.* Acknowledges CD's deep obligation to JH's *Prelim. Discourse.* 'Scarcely anything in my life made so deep an impression on me....'
ALS Xpp TxU Darwin, *Correspondence, 7,* 370–1

11197. 1859–11–11 From Edward Sabine 13 Ashley Place
Will travel to Kew Tuesday. Encloses answer to Duke of Newcastle regarding proposal for colonies. Welcomes any comments from JH.
ALS 4pp RS:HS 15.249

11198. 1859–11–12 From Edward Sabine to Benjamin Collins Brodie [the elder] 13 Ashley Place
Plan of Colonial Office submitted to R.S.L. does not yet have a 'definite or practical shape.' Recommends JH as best person to aid in matter.
ALS 4pp RS:HS 15.251

11199. 1859–11–12 To Alexander Stewart Herschel 37 Tavistock Place
Reports on a meeting of the R.A.S., mostly talking about sun spots.
C draft 2pp RS:HS 23.286

11200. 1859–11–12 From Edward Sabine 13 Ashley Place
Encloses letter from Colonial Office to R.S.L., which was referred to ES because [B. C.] Brodie [the elder] was out of town. Brodie hopes to see JH when JH comes to London.
ALS 3pp RS:HS 15.250

11201. 1859–11–12 From Thomas Spencer 192 Euston Road
Explains and encloses abstract of paper on action of magnetic oxides under various conditions. Discusses their purifying influence on water.
ALS 4pp RS:HS 17.224

11202. 1859–11–12 From George Gabriel Stokes Lensfield Cottage,Cambridge
Partial repeat of letter of 1859–11–10, together with some considerations about the molecular behavior producing fluorescence.
ALS 7pp RS:HS 17.40

11203. 1859–11–13 To Edward Sabine 37 Tavistock Place
Reluctant to recommend undertaking major scientific projects in colonies at government expense. Believes situation there will work itself out.
ALS 4pp RS:HS 15.252 (C: RS:HS 23.287)

11204. 1859–11–16 From Baden Powell 6 Stanhope St.
Regarding JH's question about formulae for intensities of reflected light, BP encloses a paper concerning G. G. Stokes's conclusions about the direction of the plane of vibrations.
ALS 3pp RS:HS 14.49

11205. 1859–11–17 From Robert Grant Royal Astronomical Society
Has received a letter from Sir George Lewis stating that he has been appointed to the chair of Practical Astronomy at Glasgow University. Is grateful for his support.
ALS 2pp RS:HS 9.19

11206. 1859–11–17 To Thomas Romney Robinson 37 Tavistock Place
Thanks for several papers; JH describes article on telescopes he has prepared for the *Encyclopaedia Britannica.*
ALS 3pp Cambridge University Library 7656/TR64 (C: RS:HS 23.288)

11207. 1859–11–18 From Adam and Charles Black Edinburgh
Receives additions to JH's *Telescope.*
ALS 1p Harvard: Houghton AAH 67m–67 (48)

11208. 1859–11–18 From Edward Joshua Cooper Leamington
Is sending the third volume of Ecliptic stars. Has arrived in Leamington for the winter and for his daughter's marriage.
ALS 3pp RS:HS 5.292

11209. 1859–11–18 To William Rowan Hamilton 37 Tavistock Place
Again admires WH's quaternions but begs that WH make them more accessible. Suggests book with rules clearly stated and examples relating them to ordinary problems.
ALS 4pp TCD 1493:1129 (C: TCD 1493:1130, TCD 7762–72:1460) Graves, *Hamilton, 2,* 121

11210. 1859–11–18 From Edward Sabine 13 Ashley Place
Discusses [Angelo] Secchi's letter. Can supply the Jesuits in Cuba with magnetic apparatus they need. Discusses Secchi's views on cause of earthquakes and volcanic eruptions.
ALS 4pp RS:HS 15.253

11211. 1859–11–21 From William Rutter Dawes Hopefield Lodge, Thame
Discusses his telescopes and their dimensions. Wants to obtain an object glass of an 8-in. aperture.
ALS 3pp Harvard: Houghton AAH 67m–67 (48)

11212. 1859–11–23 From Charles Babbage Dorset St.
Advises him to put his house in the hands of a temporary tenant so that he can rest.
ALS 4pp RS:HS 2.329

11213. 1859–11–23 To George Biddell Airy 37 Tavistock Place
Is preparing a list of telescopes with their focal lengths and quality, and asks GA to correct JH's list.
ALS 3pp RGO 6.170.195

11214. 1859–11–23 To Frederick Howlett London
Sending FH drawings of sunspots made by W. R. Dawes. Asks FH to inspect them and forward them to R.A.S. secretary R. C. Carrington along with any sunspots drawings made by FH on the same dates.
ALS 2pp Adler Ms. 13

11215. 1859–11–24 From Edward Joshua Cooper Leamington
Gives details of his various telescopes and cameras.
ALS 4pp RS:HS 5.293

11216. 1859–11–24 From Henry Simms [London]
Provides information on glass made by HS.
AL 1p Harvard: Houghton AAH 67m–67 (48)

11217. 1859–11–25 From George Biddell Airy Royal Observatory, Greenwich
Regarding telescopes in various places.
ALS 2pp RS:HS 1.244 (C: RGO 6.170.197)

11218. 1859–11–25 To William Rowan Hamilton 37 Tavistock Place
Reports favorably on sections of WH's book [*Elements of Quaternions*] that JH read. Predicts its importance and wide circulation. Is too busy to read more now.
ALS 3pp TCD 1493:1389 (C: TCD 1493:1135, TCD 7762–72:?)

11219. 1859–11–26 From Henry A. Fletcher Whitehaven
Provides information on a telescope being built by [Thomas] Cooke for HF's brother, [Isaac] Fletcher.
ALS 3pp Harvard: Houghton AAH 67m–67 (48)

11220. 1859–11–26 From William Allen Miller King's College London

The yellow substance turns out to be pentasulfide of arsenic, with a slight excess of sulfur; comments on its properties.
ALS 3pp RS:HS 12.344

11221. 1859–11–27 From David T. Ansted Bath
Informing him that he has just seen an original portrait of William Herschel in a dealer's shop in Bath.
ALS 4pp RS:HS 1.399

11222. 1859–11–27 or later To Margaret Brodie Herschel 37 Tavistock Place
Is writing to MH about Charles Pritchard's letter about work for son Alexander [see CP's 1859–11–27]; JH is now asking MH for her thoughts on the matter and adds other possibilities for Alexander. Comments on the superior meat available in London, and on his health.
ALS 4pp JHS 5.29

11223. 1859–11–27 From Charles Pritchard Clapham
Writes to JH to propose hiring JH's son Alexander for a year to work in CP's laboratory and make observations in CP's observatory.
C(S) 3pp JHS 5.28

11224. 1859–11–27 From William Whewell Trinity Lodge
Asks JH about two publications: [Felix Eberty's anonymous] *The Stars and the Earth* and J. C. Maxwell's theory of compound colors, WW recommending the latter.
ALS 4pp inc RS:HS 18.215

11225. 1859–11–28 From Edward Joshua Cooper 42 Clarendon Square
Confirms measurements of a transit circle.
ALS 1p Harvard: Houghton AAH 67m–67 (48)

11226. 1859–11–28 From John Henry Dallmeyer 2 Featherstone Buildings
Answers JH's inquiry concerning JD's telescope manufacturing process.
ALS 4pp Harvard: Houghton AAH 67m–67 (48)

11227. 1859–11–29 From Thomas Cooke & Sons Buckingham Works, York
Sends dimensions of an object glass.
2pp Harvard: Houghton AAH 67m–67 (48)

11228. 1859–11–29 To Augustus De Morgan 37 Tavistock Place
Tells AD how to deal with JH's letters to Francis Baily.
C 1p RS:HS 23.289

11229. 1859–11–29 From Charles May Great George St.
Sends a small fragment chipped from the second great bell. Comments on the metal of the bell and the causes of the cracks. Unable to attend the R.S.L. dinner tomorrow.
ALS 3pp RS:HS 12.314

11230. 1859–11–30 From Thomas Graham Royal Mint
Was gratified to see his handwriting, and has pleasure in sending the order. Duties were oppressive but have now eased. The Pyx has been delayed for two years but must take place soon.
ALS 3pp RS:HS 8.228

11231. 1859–11–30 To Margaret Brodie Herschel 37 Tavistock Place
Mostly news of what JH has been doing the last day or so, including a visit to daughter Caroline, whose daughter is better, and about JH's accommodations and food, together with comments about [William Henry] Fitton.
ALS 4pp JHS 5.19

11232. 1859–12–1 From Richard Christopher Carrington Observatory, Redhill
[Frederick] Howlett's additional drawings have been revised and placed with the first ones. Reminded editor they should be noticed. Perhaps JH would inform S. B. Howlett that the Society appreciates the drawings. Is about to examine C. J. G. Pastorff's manuscript.
ALS 4pp RS:HS 5.208

11233. 1859–12–1 From John Henry Dallmeyer 19 Bloomsbury St.
Informs JH that he is taking over the optical workshop of his late father-in-law, Andrew Ross.
L 2pp Harvard: Houghton AAH 67m–67 (48)

11234. 1859–12 From Richard Griffin & Co. London
Is thinking of publishing a biographical dictionary and encloses an article on JH, which they would like him to revise if necessary.
ALS 1p RS:HS 9.141

11235. 1859–12–1 From Thomas Grubb Dublin
Discusses his efforts making reflecting telescopes.
ALS 4pp Harvard: Houghton AAH 67m–67 (48)

11236. 1859–12–3 From Roderick Impey Murchison 16 Belgrave Square
Sending the syllabus of the Government School of Mines, which should give JH all the information. Will be pleased to see him at the Museum if he has any further queries.
ALS 6pp RS:HS 12.423

11237. 1859–12–4 To Charles Babbage n.p.
Has found a temporary tenant for his house.
ALS 3pp RS:HS 2.330

11238. 1859–12–5 From W. Bonsey Slough
Will be pleased to see him on Wednesday. Hopes he will stay the night.
ALS 1p RS:HS 4.191

11239. 1859–12–5 From Richard Christopher Carrington Observatory, Redhill
Queries regarding C. J. G. Pastorff's observations. Believes author is now dead, but can he inform him concerning his mode of observation.
ALS 4pp RS:HS 5.209

11240. 1859–12–5 From Warren de La Rue 110 Bunhill Row
Sending R. C. Carrington's certificate for JH's signature.
ALS 2pp RS:HS 6.144

11241. 1859–12–5 To Charles May 37 Tavistock Place, Russell Square
Thanks for the fragment of poor [Ben?]. Comments on this and the composition of the metal of bells.
draft 4pp RS:HS 12.315 & 23.291

11242. 1859–12–6 From Thomas Grubb Dublin
Informs JH which observatories received the object lenses from his workshops.
ALS 2pp Harvard: Houghton AAH 67m–67 (48)

11243. 1859–12–7 From Joanna B. Horner 17 Queens Rd., West
Charles Lyell has given her JH's translation of J. W. Goethe's

poem. Comments on it.
ALS 3pp RS:HS 9.491

11244. 1859–12–8 From [Marc] Secrétan 73 Rue du Faubourg Saint-Jacques, Paris
Lists dimensions of the equatorial telescope at Paris Observatory.
ALS 2pp (French) Harvard: Houghton AAH 67m–67 (48)

11245. 1859–12–10 From David T. Ansted Guernsey
Further regarding the portrait of William Herschel, and his advice to the present owner.
ALS 3pp RS:HS 1.400

11246. 1859–12–10 From William Allen Miller King's College London
Thanks for the photographic notes and for the remarks on color blindness. The remarks on the composition of colors interested him as he has difficulty in distinguishing shades of brown and green.
ALS 2pp RS:HS 12.345

11247. 1859–12–12 To Norton Shaw Hawkhurst
Any materials to be sent to JH should be sent to JH's Hawkhurst address.
ALS 1p Royal Geographical Society

11248. 1859–12–13 From Richard Christopher Carrington Observatory, Redhill
The Council of the R.A.S. will be pleased to receive any further drawings of S. B. Howlett, but will not undertake to publish them owing to the expense.
ALS 3pp RS:HS 5.210

11249. 1859–12–14 From Richard Christopher Carrington Observatory, Redhill
Answer to JH's query concerning the drawings of S. B. Howlett. Thanks for his information regarding C. J. G. Pastorff.
ALS 2pp RS:HS 5.211

11250. 1859–12–14 From Thomas Greenwood 14 Westbourne Terrace
Circular letter informing him that the third volume of his *Cathedra Petri* will be published shortly.
AL 1p RS:HS 9.30

11251. 1859–12–15 From Charles Pritchard Clapham Grammar School
Searches for asteroids. Measures double stars and variables; Alexander Herschel assists. Asks JH's opinion of parallax measuring methods. R.S.L. funds offered to M. J. Johnson's family in his memory.
ALS 4pp RS:HS 14.79

11252. 1859–12–21 From A. H. Brotherton 14 Woburn Place
Sending a proposition which he thinks will be worth his notice.
ALS 1p RS:HS 5.108

11253. 1859–12–27 From Thomas Greenwood 14 Westbourne Terrace
Sending the prospectus for the third volume of his *Cathedra Petri*. Would like his friends to talk about it, and a competent person to review it. Has just had a friendly letter from William Whewell, the only one of his friends now resident at College.
ALS 4pp RS:HS 9.31

11254. 1859–12–31 From Herbert Spencer Oakley Square
[Responding to JH's 1859–1–16,] HS defends the nebular hypothesis, HS's theory of comets, and sets out HS's theory of sunspot formation, noting that John Tyndall supports it.
ALS 5pp RS:HS 16.486

11255. 1860 or later From Julia Margaret Cameron Fresh Water Bay, Isle of Wight
Family news, wedding of a friend's daughter, suffering with jaundice.
ALS 6pp inc RS:HS 5.176 & 177

11256. 1860 To Augustus De Morgan 37 Tavistock Place
A story about Voltaire; arrangements to visit AD.
C 2pp RS:HS 23.292

11257. [1860] From Edward Joseph Lowe Highfield House
Sends meteorological observation charts from Highfield House Observatory.
ALS 1p encl 8pp Harvard: Houghton AAH 67m–67 (42)

11258. [1860?] To Herbert McLeod n.p.
Reasons why he is unable to sign the Memorial to the bishops regarding their attitude to *Essays and Reviews*.
C 4pp RS:HS 13.12 (C: RS:HS 23.385 & 25.13.14 & 15)

11259. [1860 to 1866?] From Samuel Newington n.p.
Hopes JH will return when fruit is ripe to harvest.
ALS 1p inc RAS:JH Archive 12/1.3.5 verso; Reel 10

11260. 1860–1–4 From John Barlow Torquay
Has just received his note and written to the Royal Institution so that he may get an answer without delay. Sends his best wishes to the family.
ALS 1p RS:HS 5.19

11261. 1860–1–4 From Alexander J. B. Hope Bedgebury Park Cranbrook
Thanks for the additional subscription to the Rifle Corps. Regrets that he could not attend due to indisposition.
ALS 3pp RS:HS 9.466

11262. 1860–1–4 From Robert Walker Harley, near Bath
Asks about other portraits of William Herschel like one RW owns, which National Portrait Gallery wants to buy. Will send miniature of WH's portrait to JH.
ALS 4pp RS:HS 18.367

11263. 1860–1–5 From Samuel Newington 10 Upper George St.
Is grateful for JH's opinion. Can hardly imagine that such a simple contrivance should answer the purpose. Saw the patent agent who has offered to send a list of all the writings on the subject. Also brought the matter before B. C. Brodie (the elder), who now seems totally blind.
ALS 4pp RS:HS 13.123

11264. 1860–1–6 From Sir Benjamin Collins Brodie [the elder] Broome Park
Regarding increasing the payment to the Secretaries of the R.S.L. Would like JH's comments.
ALS 6pp RS:HS 4.282

11265. 1860–1–7 To Alexander Stewart Herschel Collingwood
A letter of great sadness: someone is at death's door; not much hope is held out. [This letter may refer to the imminent death of daughter Margaret Louisa, because of its reference to

poor Reginald, which was the name of MLH's husband].
ALS 3pp JHS 1.113

11266. 1860–1–8 To [Mr.] Bell Collingwood
Gratitude for sending [John] Taylor's book. Believes dimensions of Great Pyramid exhibit knowledge of geometry but not of calculus among ancient Egyptians. Suggests way to modify British standard of length to produce 'perfect metrical system' superior to that of French.
CS 7pp TxU:H/L-0089; Reel 1054

11267. 1860–1–8 To Sir Benjamin Collins Brodie [the elder] Collingwood
Agrees with BB that the honorarium paid to the secretaries should be increased. Further comments on this point.
draft 4pp RS:HS 4.283 & 23.293

11268. 1860–1–8 To William Rowan Hamilton Collingwood
Sends back manuscript of WH's [*Elements of Quaternions*] with comments. Says it's excellent but somewhat distracting to a beginner.
ALS 4pp TCD 1493:1138

11269. 1860–1–9 From Robert Brown Free Manse, Dunoon
Would like to dedicate his 'Elements of musical science' to JH.
ALS 2pp RS:HS 4.314

11270. [1860?]–1–9 From Ann Smyth St. John's Lodge
Sends W. H. Smyth's history of R.S.L. Club [*Sketch of the Rise and Progress of the Royal Society Club,* 1860].
Mentions W. H. Smyth's retiring from visiting R.S.L. Club.
AL 1p inc RS:HS 16.221

11271. 1860–1–11 From Isaac Fletcher Tarn Bank near Cockermouth
Regrets the delay in answering JH's letter but he has been abroad. Will supply any details if not too late for his purpose. The mounting of his large telescope is peculiar and is similar to that used by Smyth.
ALS 3pp RS:HS 7.272

11272. [1860]–1–12 From Sir Benjamin Collins Brodie [the elder] 14 Saville Row
Has interpreted the word 'honorary' quite correctly.
ALS 1p RS:HS 4.290

11273. [1860]–1–12 To Herbert Spencer Collingwood
[Responding to HS's 1859–12–31,] JH presents additional problems in the nebular hypothesis, critiques in detail HS's views of the distribution of cometary orbits, and argues against HS's theory of sunspot formation.
CC 4pp RS:HS 16.484 (C: 23.294)

11274. 1860–1–14 From Robert Brown Free Manse, Dunoon
Is grateful for being allowed to dedicate his book to JH; also for his information on harmonics.
ALS 1p RS:HS 4.315

11275. 1860–1–14 From Robert Walker Harley, near Bath
Miniature of William Herschel's portrait was lost and recovered. At same sale where RW bought WH's portrait, was sold 7-foot reflector, inside which were several papers in WH's handwriting. New owner would not accept £40. Does JH advise offering more?
ALS 4pp RS:HS 18.368

11276. 1860–1–16 From Richard Griffin & Co. London
Is grateful for his assistance and will send him a proof. Can nothing be done about a new edition of his articles on Light and Sound from the *Metropolitana*?
ALS 2pp RS:HS 9.142

11277. 1860–1–18 From Herbert Spencer Oakley Square
[Responding to JH's 1860}–1–12,] HS admits some of JH's objections to HS's views on distribution of cometary orbits, but attempts to salvage HS's cyclonic theory of sunspot formation.
ALS 2pp RS:HS 16.487

11278. 1860–1–18 From Robert Walker Harley, near Bath
Please send directions to Collingwood, so RW may send miniature of William Herschel's portrait to JH.
ALS 1p RS:HS 18.369

11279. 1860–1–20 From Bernard Cracroft to Margaret Brodie Herschel Nackthorn, Lincoln
Sending mathematical papers of his friend for JH. Is on the point of going soldiering again. Would welcome position of tutor if she hears of one going vacant.
ALS 4pp RS:HS 5.295

11280. 1860–1–27 From Thomas Greenwood 14 Westbourne Terrace
Was pleased to see JH's writing once more. Is still active though not quite so energetic as in previous times. Has had appreciative letters about his books, though the general public ignores them. Comments on the papacy. Has two more books ready for the press.
ALS 4pp RS:HS 9.32

11281. 1860–1–27 From Alexander J. B. Hope Bedgebury Park Cranbrook
Has just returned from town and found JH's letter in Greenwood's volumes. Looks forward with pleasure to perusing the volumes.
ALS 3pp RS:HS 9.467

11282. 1860–1–27 From Edward Joseph Lowe Beeston
Has experienced an extraordinary meteorological year and been kept busy observing all the changes. Is to submit another certificate to the R.S.L. and would be pleased if JH would sign it.
ALS 4pp RS:HS 11.330

11283. 1860–1–27 From George Gabriel Stokes London
Referring to some experiments in sound, GS proceeds to provide the theoretical basis for them.
ALS 12pp RS:HS 17.41

11284. 1860–1–28 From Warren de La Rue 110 Bunhill Row
Sending Robert Main's certificate for JH's signature.
ALS 1p RS:HS 6.145

11285. 1860–2–2 From Edward Joshua Cooper Leamington
Regarding modifications to the water eye piece of his telescope.
ALS 6pp RS:HS 5.294

11286. 1860–2–3 To William Henry Smyth [Collingwood]
Claims if he had known certain informations concerning discovery of Neptune, he would have ranked J. C. Adams's claim equal with U. J. J. Leverrier's.
ALS 2pp inc RS:HS 16.311 (C: RS:HS 23.295)

11287. 1860–2–4 From George Smith 65 Cornhill
Received JH's letter and enclosure of yesterday. Gave instructions to refuse newspapers arriving for JH without postage from Cape of Good Hope.
ALS 1p TxU:H/M–0487.1; Reel 1093

11288. 1860–2–8 To Edward Joshua Cooper Collingwood
On the usefulness of Stephen's Writing Fluid to stop the sun's excess of light at large apertures in the telescope.
C 1p RS:HS 23.296

11289. 1860–2–8 From William Lassell Sandfield Park near Liverpool
Sends a lithograph of the 4-ft. aperture equatorial. Has made no observations with it yet. Had difficulty with the surface of the speculum. Comments on the drawing and on the use of the instrument. Cannot recollect seeing any of the irregularities pointed out by JH in his letter.
ALS 4pp RS:HS 11.156

11290. 1860–2–8 From William Henry Smyth St. John's Lodge
Asks JH's opinion of placement of some of William Herschel's documents. Agrees with JH on Neptune case, although he claims to have difficulty formulating an argument.
ALS 2pp RS:HS 16.312

11291. 1860–2–9 From William Crookes 20 Mornington Road
Would like a letter from JH supporting him against the charge of incompetency in his editing of the *Photographic News*.
ALS 4pp RS:HS 5.313

11292. 1860–2–10 From Thomas P. Thompson Eliot Vale, Blackheath
Asks JH's opinion of a tuning fork design; includes plan of the instrument.
C 1p diag RS:HS 17.347

11293. 1860–2–11 From Neil Arnott Cumberland Terrace
Is sending pamphlet which he hopes will persuade Aberdeen University authorities to retain the chair of Natural Philosophy. Would like JH's comments on this.
ALS 2pp RS:HS 1.409

11294. 1860–2–11 To William Lassell Collingwood
Comments on WL's 4-foot equatorial telescope of which WL sent JH a lithograph [see WL's 1860–2–8].
C 1p RS:HS 23.297

11295. 1860–2–13 From William Lassell Sandfield Park near Liverpool
Pointing out the reasons for some apparent discrepancies in the drawing of his telescope. The drawing was hurriedly done by his daughter.
ALS 3pp RS:HS 11.157

11296. 1860–2–13 From Thomas P. Thompson Eliot Vale, Blackheath
Thanks JH for advice on tuning fork. Further enquires about the instrument.
L 3pp RS:HS 17.348

11297. 1860–2–14 From B. Birkenthal Oxford
Is asking JH for assistance.
ALS 3pp (German) RS:HS 5.71

11298. 1860–2–15 From Edward Joseph Lowe Beeston
Thanks for signing his certificate for the R.S.L. Has promise of support from members of Council. Has just contrived a new ozone box; gives details. Carrying out observations with underground thermometers.
ALS 4pp RS:HS 11.331

11299. 1860–2–16 From George Paxton Young Toronto, Canada West
Sends published paper containing GY's proof that roots of algebraic equations above fourth degree cannot be represented by finite functions, and showing how roots of fifth degree equations may be found. GY's method may be extended to all higher degrees.
ALS 3pp RS:HS 18.445

11300. 1860–2–17 From B. Birkenthal Oxford
A follow-up letter [see BB's 1860–2–14] asking for an immediate response to BB's request for money.
ALS 2pp (German) RS:HS 5.72

11301. 1860–2–18 From William Crookes 20 Mornington Road
Is grateful for the help JH's letter gave him in his legal dispute with the *Photographic News*.
ALS 2pp RS:HS 5.314

11302. 1860–2–19 From T[homas] Radford Oxford
Has a foreigner staying at his house. Has seen letter by this man addressed to JH. TR wants to know if the foreigner is respectable. Thinks his name is [B.] Birkenthal.
ALS 3pp RS:HS 14.467

11303. 1860–2–20 From John Fremlin Streatfield 15 Upper Brook St.
Sends his personal scientific journal to JH. Charles Babbage has already given JS much help.
ALS 1p † Harvard: Houghton AAH 67m–67 (107)

11304. 1860–2–25 From Richard Christopher Carrington Observatory, Redhill
Regarding the sun's equation. Sun spots.
ALS 4pp RS:HS 5.212

11305. 1860–3–1 From George Biddell Airy Royal Observatory, Greenwich
Passes on information on stars received from a Bengal pilot.
ALS 2pp RS:HS 1.245 (C: RGO 6.195.250)

11306. 1860–3–1 To William Rowan Hamilton Collingwood
Apologizes for not returning sheets of *Elements of Quaternions* sooner, but was ill. Again cautions WH that method might be too advanced for elementary readers.
ALS 3pp TCD 1493:1144 (C: TCD 1493:1145)

11307. 1860–3–2 To George Biddell Airy n.p.
Is arguing for the development of a British metrical system based on the length of the polar axis of the earth.
ALS 4pp inc RGO 6.457.147

11308. 1860–3–2 To George Biddell Airy Collingwood
Offers some possible explanation for what the unnamed observer saw [see GA's 1860–3–1].
ALS 2pp RGO 6.195.251

11309. 1860–3–5 From Herbert Spencer Oakley Square
Asks JH to become a subscriber to a work that HS is bringing out. Lists prominent persons who have already done so.
ALS 1p RS:HS 16.488

11310. 1860–3–9 From Samuel Newington Ridgeway, Ticehurst

The writer of the enclosed production has given lectures at SN's establishment. Introduces him to JH. Hopes to give a lecture at Hawkhurst. Would be pleased for any of the Herschel family to view the orchid house, now in full bloom.
ALS 4pp RS:HS 13.124

11311. 1860–3–12 From Samuel Newington Ridgeway, Ticehurst
Miss Newington will have pleasure in seeing Lady Herschel and Miss Power on Monday or Tuesday next. Will send over a copy of the Orchid House tomorrow.
ALS 2pp RS:HS 13.125

11312. 1860–3–12 From William Williams Goudhurst
Offers his services as a coachman.
ALS 1p Harvard: Houghton AAH 67m–67 (107)

11312a. 1860–3–17 To Herbert Spencer Collingwood
See 13143, which is misdated.

11313. 1860–3–17 From Norman Pogson Hartwell, Aylesbury
Dissatisfied with [F. W. A.] Angelander's nomenclature of variable stars; seeks JH's approval of NP's nomenclature for the new 'Hartwell Atlas of Variable Stars.'
ALS 4pp RS:HS 14.1

11314. [186?]–3–19 From Cuthbert W. Johnson Croydon
Pointing out an error by JH in assigning the authorship of a weather table to Sir William Herschel.
ALS 3pp RS:HS 10.447

11315. 1860–3–19 To Norman Pogson Collingwood
Agrees with NP's proposed method of designating variable stars. Offers some suggestions from JH's own experience.
ALS 4pp RS:HS 14.2 (C: RS:HS 23.299)

11316. 1860–3–20 From Thomas Goodchild Guildford
Sending a descriptive pamphlet of an optical invention of his own for simple experiments with color.
ALS 2pp RS:HS 9.102

11317. 1860–3–21 From Julia Margaret Cameron Bromley Common
News of the birth of her grandchild. Husband and two sons are gone back to Ceylon. Hazards and illness while travelling there.
ALS 25pp RS:HS 5.155

11318. 1860–3–23 From Joseph Jopling 6 Vassall Terrace
Outlining his geometrical laws for the Great Pyramid.
ALS 4pp RS:HS 10.417

11319. 1860–3–26 From Josiah Parsons Cooke London
Sending letter of Prof. Asa Gray. American Academy would like to acquire some of the back numbers of the publications of the R.A.S. Is grateful for JH's mention of his work on the chemical elements. Will be in London for a week or two.
ALS 3pp RS:HS 5.363

11320. 1860–3–26 From William Selwyn Cambridge
Mentions a passage of Aristotle regarding comets. Proposes that Origen's theory of the Magi may be correct. Perhaps the Magi saw a comet.
ALS 4pp RS:HS 15.463

11321. 1860–3–28 From William Henry Smyth St. John's Lodge
May include JH's position on Neptune discovery in a new edition of *Cycle of Celestial Objects*. Will review objects catalogued in the volume with refractor.
ALS 2pp RS:HS 16.313

11322. 1860–3–29 To Edward Sabine Collingwood
Against changing location of Royal Observatory [from Greenwich], but will defer to opinion of [G. B. Airy] Astronomer Royal. Airy is most competent to judge situation. Still ill.
C 2pp RS:HS 23.300

11323. 1860–4–1 From Josiah Parsons Cooke Highbury
Is grateful for JH's assistance in obtaining back numbers of the publications of the R.A.S. Would like the chance to discuss his chemical experiments before the Royal Institution.
ALS 4pp RS:HS 5.364

11324. 1860–4–3 From Norman Pogson Hartwell, Aylesbury
Thanks JH for supporting his proposed method of designating variable stars, adding that G. B. Airy, J. R. Hind, and W. H. Smyth also accept it. Hopes to succeed M. J. Johnson as director of Oxford's Radcliffe Observatory.
ALS 8pp RS:HS 14.3

11325. 1860–4–9 From Thomas Greenwood 14 Westbourne Terrace
Is grateful for JH's intervention on his behalf. Comments on A. J. Beresford-Hope's review of his own book, and would be pleased if JH would pass on his comments.
ALS 4pp RS:HS 9.33

11326. 1860–4–11 From Robert Brown Free Manse, Dunoon
Would like his opinion on the form to be adopted for the dedication.
ALS 3pp RS:HS 4.316

11327. 1860–4–11 From Alexander J. B. Hope Arklow Ho. Connaught Place
Is obliged for letting him see Greenwood's letter about the book, but he did not himself write the review; it was written by an eminent scholar in that field.
ALS 3pp RS:HS 9.468

11328. 1860–4–14 From M. Felix Billet Dijon
Comments on his own work in optics and related fields.
ALS 4pp (French) RS:HS 5.65

11329. 1860–4–15 From Norman Pogson Hartwell, Aylesbury
Has heard that JH's son [Alexander] is observing variable stars at Charles Pritchard's school. Gives latest data on these stars. Suggests other variables to observe. Has adopted JH's proposal of using Julian day numbers for dating variables in his catalogue of variables.
ALS 6pp RS:HS 14.4

11330. 1860–4–16 To [Sir Edward] Ryan Collingwood
Comments on, and encloses, a letter from JH's son William James, who has accepted responsibility for trying to resolve a grave situation [in India].
ALS 3pp APS B.H435p.19

11331. 1860–4–17 From Augustus De Morgan 41 Chalcot Villas NW
Has a copy of J. R. Young's investigation. Gives one of his own theorems.
ALS 4pp RS:HS 6.335

11332. 1860–4–18 From Herbert Spencer 17 Wilmot St., Derby

Thanks JH for agreeing [see #13143] to be a subscriber to HS's project and for JH's critique of HS's use of the term 'Absolute.' Admits that it has difficulties.
ALS 3pp RS:HS 16.490

11333. 1860–4–21 From George Biddell Airy Royal Observatory, Greenwich
Regarding the printing of JH's British Metrical scale. Comparison between the chimes of Big Ben and those of St. Mary's, Cambridge.
ALS 4pp RS:HS 1.246 (C: RGO 6.379.138)

11334. 1860–4–21 From Bartholomew Price St. Giles St., Oxford
Invites JH to stay at BP's home for the B.A.A.S. meeting 27 June. Notes the new museum opens at the same time.
ALS 3pp RS:HS 14.55

11335. 1860–4–23 From George Biddell Airy Royal Observatory, Greenwich
Further details on the chimes of Big Ben.
ALS 2pp RS:HS 1.247

11336. 1860–4–23 To the *Athenaeum* Collingwood
Expresses his views on how the British should select their standard of length. Also discusses John Taylor's views of the Great Pyramid.
L *Athenaeum,* #1696 (April 28, 1860), 581–2 & John Taylor, *The Great Pyramid,* 2nd. ed. (London, 1864), pp. 39–43

11337. 1860–4–23 From Richard Christopher Carrington Observatory, Redhill
Sending a short paper which is to be printed, and would be glad with any comments. Currents indicated are deduced for a small number of cases only.
ALS 4pp RS:HS 5.213

11338. 1860–4–24 From Norman Pogson Hartwell, Aylesbury
Thanks JH for 'testimonial' to the Radcliffe Observatory on his behalf, but the response is not encouraging. Describes the tribulations of his career in astronomy. Makes further suggestions regarding the variable stars being observed by JH's son [Alexander].
ALS 8pp RS:HS 14.5

11339. 1860–4–25 To Anne Sheepshanks Collingwood
Does not wish to forward some letters of AS's brother, Richard, as JH has made marginal notes he would not wish others to read.
ALS 3pp RS:HS 16.122 (C: RS:HS 25.9.77)

11340. 1860–4–27 From Edward Sabine 13 Ashley Place
Sent proof for vol. 2 of St. Helena observations to JH. JH's promised article in *Edinburgh Review* on progress in science of terrestrial magnetism. Embarrassing delay by government in responding to Prince Albert's request for five-year magnetic survey is resulting in loss of interest among prospective directors. Proposal by Americans to assume leadership of survey in North America. Dutch observatory in Java and Jesuit observatory in Cuba.
ALS 12pp TxU:H/M–0527.1; Reel 1093

11341. 1860–4–30 From Henry W. Field Royal Mint
Sending some crystals of a substance he obtained by the separation of gold from iridium. Would be glad of his comments.
ALS 4pp RS:HS 7.220

11342. 1860–5–1 To the *Athenaeum* Collingwood
[Following up on JH's 1860–4–23,] notes another numerical relationship between the dimensions of the Great Pyramid and those of the earth.
L *Athenaeum,* #1697 (May 5, 1860), 617 & John Taylor, *The Great Pyramid,* 2nd. ed. (London, 1864), p. 44

11343. 1860–5–1 From J. Jackson Rochdale
Sending a printed circular on his proposed change in the currency.
ALS 1p RS:HS 10.432

11344. 1860–5–2 From Richard Christopher Carrington Observatory, Redhill
Thanks for his comments on the recent paper. Further points on the sun's equation. Will read his article on meteorology. Radcliffe Trustees have advertised the post again as there was only three applicants.
ALS 8pp RS:HS 5.214

11345. 1860–5–3 To Edward Sabine Collingwood
Too ill to work on anything besides nebula catalogue. Discusses recent meteorological phenomena, period of solar spots, and great pyramid.
C 2pp RS:HS 23.301

11346. 1860–5–5 From Joseph Jopling 6 Vassall Terrace
Pleased to hear that he has received his statements representing curves; comments on these. Further regarding his views on the Great Pyramid.
ALS 4pp RS:HS 10.418

11347. 1860–5–7 To the *Photographic News* Collingwood
Discusses the possibility of stereoscopic photography of action scenes. Also speculates about color photography.
L *Photographic News, 4* (May 11, 1860), 13

11348. 1860–5–8 From Woronzow Greig Surrey Lodge Lambeth
William King (Earl of Lovelace) was in Somerset when JH's letter arrived, but he placed it before him on his return. Thinks the G. Hartnell affair is the responsibility of J. R. Townshend (3rd Vicount Sydney). His own mother is doing well considering her age. Hopes JH has had good news from India.
ALS 4pp RS:HS 9.40

11349. 1860–5–8 From Norman Pogson to Alexander Herschel Hartwell, Aylesbury
Describes efforts to secure position at Oxford's Radcliffe Observatory, noting JH's testimonial. Suggests AH observe four circumpolar variables and makes other recommendations concerning AH's observation program.
ALS 8pp RS:HS 14.6

11350. 1860–5–12 From George Biddell Airy Royal Observatory, Greenwich
[Form Letter] GA's address, as Astronomer Royal, to Board of Visitors. Progress report on F. G. W. Struve's proposal for joint French-English-Belgian triangulation survey.
ACCS 4pp TxU:H/M–0650; Reel 1084

11351. 1860–5–14 From Thomas Milner Gibson Board of Trade
[Form letter] Wishes to nominate JH as member of commission to prepare for International Statistical Conference meeting in London in July.
LS 1p TxU:H/L-0161.1; Reel 1054

11352. [1860]–5–14 From Baden Powell 6 Stanhope St.
Glad to hear JH feeling better. Enclosing a 'shorter and simpler' explanation of some experiment by Léon Foucault, asking for comments.
ALS 3pp RS:HS 14.50

11353. 1860–5–16 To Thomas Milner Gibson Collingwood
Declines nomination to commission to prepare for International Statistical Conference meeting in London.
ACCS 1p (on TG's 1860–5–14) TxU:H/L-0161.2; Reel 1054

11354. 1860–5–16 From Arthur Penrhyn Stanley Oxford
Offers JH and Lady Herschel lodging during Oxford B.A.A.S. meeting.
ALS 1p RS:HS 17.228

11355. [1860]–5–19 From Friedrich Max Muller All Soul's, Oxford
Would like a testimonial from JH as he is applying for the Professorship of Sanskrit, left vacant by the death of H. H. Wilson.
ALS 3pp RS:HS 13.58

11356. 1860–5–22 From Thomas Maclear Royal Observatory, Cape of Good Hope
F. E. Wilmot leaves for command in Canada. Requests JH's advice in finding a schoolmaster to prepare his son Harold for Woolwich.
ALS 7pp Cape Archives/Maclear Papers File 100

11357. 1860–5–25 To John Herschel (son) Collingwood
Comments extensively on the calculation, and elimination, of error in geodesic measurements; JH concludes with a few family news items.
ALS 4pp JHS 5.3

11358. 1860–[6]–6 To Alexander Stewart Herschel [Collingwood]
Mostly taken up with JH advising AH about changing his course of studies at Cambridge, together with some family news from home.
ALS 3pp JHS 1.114

11359. 1860–6–6 From Alexander J. B. Hope Arklow Ho. Connaught Place
Regrets to hear that he proposes to retire from the captaincy of the Hawkhurst Company of the Rifle Volunteers.
ALS 1p RS:HS 9.469

11360. 1860–6–9 From William Radcliffe Birt Victoria Park
Would like confirmation of his London address before returning the books. Has borrowed from the Society the barometric observations made at Cracow.
ALS 4pp RS:HS 4.142

11361. 1860–6–11 From Alexander Bridport Becher Admiralty
Announces 18 June meeting of Board of Visitors of Royal Observatory at R.S.L. Subject will be proposed grant by [E. A. S. Seymour, 12th] Duke of Somerset to [Peter A.] Hansen.
ALS 1p TxU:H/M–0073; Reel 1087

11362. 1860–6–12 From Sir Benjamin Collins Brodie [the elder] Savile Row
Hopes JH will attend [18 June] meeting of Visitors of [Royal] Observatory to discuss letter of [E. A. S. Seymour, 12th] Duke of Somerset, belittling value of [P. A.] Hansen's observations.
ALS 4pp TxU:H/M–0129; Reel 1087

11363. 1860–6–13 From George Biddell Airy Royal Observatory, Greenwich
Urges JH to attend the next meeting of the Board of Visitors of the Royal Observatory, especially as GA believes some of the members of the Board do not understand the scientific problem [?].
ALS 1p TxU:H/M-0639.1; Reel 1084 (ACCS: RGO 6.83.136)

11364. 1860–6–13 From Joseph Jopling 6 Vassall Terrace
Saw J. S. Perring for only a short time. Explained his own theories and models of the Great Pyramid at the Royal Institution. Enlarges on his theories. Has been invited to lecture to the Sino-Egyptian Society.
ALS 4pp RS:HS 10.419

11365. 1860–6–13 From [Harriet?] Lubbock High Elms, Farnborough, Kent
Sends 'something about the Moon.' It is all the printer has produced so far.
ALS 1p TxU:H/M–0938; Reel 1083

11366. 1860–6–14 From Edward Loyd Hawkhurst
Is giving a dinner for the Hawkhurst Rifle Volunteers and would welcome JH's presence.
ALS 4pp RS:HS 11.345

11367. 1860–6–15 From Alexander Bridport Becher Admiralty
18 June meeting of Board of Visitors of Royal Observatory will begin one hour earlier. Encloses copy of letter of [E. A. S. Seymour, 12th] Duke of Somerset, and response from Astronomer Royal [G. B. Airy].
ALS 1p TxU:H/M–0074; Reel 1087

11368. 1860–6–15 To Thomas Stevenson 37 Tavistock Place
Thanks TS for TS's publication on lighthouse illumination; adds some of JH's own ideas on the subject.
ALS 3pp NatLibScot MS.785, f.23

11369. 1860–6–15 From George Gabriel Stokes Royal Society
Enquires whether JH wants a paper published in *R.S.P.T.* or in R.S.L. *Proceedings.*
ALS 2pp RS:HS 17.42

11370. 1860–6–19 From Norman Pogson to Alexander Herschel Hartwell, Aylesbury
Describes efforts to obtain position at Oxford's Radcliffe Observatory. Advocates that AH and all variable star observers follow a uniform plan. Laments confusion over indefiniteness of term 'magnitude' and his lack of proper equipment and adequate recognition. Invites AH to visit.
ALS 12pp RS:HS 14.7

11371. 1860–6–21 From Francis Galton Athenaeum
Intends going to Spain for the eclipse to make observations on solar radiation. Would like advice on the type of actinometer to use.
ALS 4pp RS:HS 9.88

11372. 1860–6–23 From Charles Pritchard Clapham Grammar School
Asks JH's opinion on CP's eclipse observations plans, especially on which equipment to use. Intends to use a 'diagonal solar eyepiece' during the eclipse. Plans to observe carefully and to sketch solar flares.

ALS 4pp inc RS:HS 14.80 Ada Pritchard, *Charles Pritchard: Memoirs of His Life,* 222–3

11373. 1860–6–25 From Thomas Stevenson Bridge of Allan
Believing JH gives TS credit for a proposal not TS's own, clarifies TS's work in lighthouse illumination.
ALS 2pp RS:HS 17.235

11374. [1860–6–25 or later] To [?] n.p.
Comments on the behavior of Sirius with its companion, and on the likelihood of the existence of the inter–Mercurial planet.
AL 1p inc JHS 6.64

11375. 1860–6–30 To *Photographic News* n.p.
[Writing under the pseudonym 'Redde Suum Cuique'], JH notes that a recently published process for recovering silver from old plate was in fact anticipated by James Keir in a 1790 *R.S.P.T.* paper. Attacks a recent misuse of the word 'actinometer.'
L *Photographic News, 4* (1860), 118

11376. 1860–6–31 To Alexander J. B. Hope Collingwood
Comments on the publication of several volumes.
ALS 2pp WT 63313.1

11377. 1860–7–5 From John Hawkins Hitchin
Thanks for his letter about Thomas Maclear's boy. Thought he was in the care of his uncle. Will do whatever JH thinks is best.
ALS 2pp RS:HS 10.170

11378. 1860–7–6 From Augustus De Morgan 41 Chalcot Villas NW
His strength is improving. The Airys are off to Bilboa.
ALS 5pp RS:HS 6.336

11379. 1860–7–12 From Norton Shaw [London]
Sends copy of paper for corrections.
ALS 1p Harvard: Houghton AAH 67m–67 (107)

11380. 1860–7–15 From Edward Sabine St. Mary's Cottage, Llandovery, South Wales
A. D. Bache reports that U.S. will establish magnetic observatory in Washington Territory. ES is preparing instructions for North American survey. Suggests change to JH's proposed article on terrestrial magnetism in *Edinburgh Review*.
ALS 4pp TxU:H/M–0527.2; Reel 1093

11381. 1860–7–18 From J. G. Thompson Madras
Sends JH pamphlets on the representation of Oriental languages using the Roman alphabet. Outlines the advantages of this system.
ALS 4pp RS:HS 17.434

11382. 1860–7–23 Sir William Pole Bilbao, Spain
Describes, using a diagram, how the solar protuberances appeared to him as a color-blind person when he observed a solar eclipse.
ALS 4pp RS:HS 14.25

11383. 1860–7–24 From George Robert Waterhouse British Museum
On behalf of GW's wife, thanks for JH's gift of table and bedcurtains.
ALS 1p RS:HS 18.103

11384. 1860–7–27 From Edward Sabine Llandovery
[Government refusal to support five-year magnetic survey] has eliminated services of T. W. Blakiston and Thomas Hull as directors of two magnetic observatories. Instruments for Montreal observatory. Recent magnetic changes in North America. Success of elder B. C. Brodie's eye operation.
ALS 4pp TxU:H/M–0527.3; Reel 1093

11385. 1860–7–28 From Charles Pritchard Clapham Grammar School
Sends observations of eclipse; asks JH to check accuracy. Details observations with diagonal solar eyepiece. Describes solar corona and lists bright stars visible during totality.
ALS 7pp inc RS:HS 14.81 Ada Pritchard, *Charles Pritchard: Memoirs of His Life,* 223–7

11386. 1860–7–29 From John Hawkins Hitchin
Is grateful for his suggestion of Charles Pritchard as tutor for Thomas Maclear's sons; has written to him and to Mr. Knight, who is the present tutor. Knight will not be well pleased.
ALS 3pp RS:HS 10.171

11387. [1860?]–7–31 From Charles Pritchard Clapham Grammar School
Problems making actinometer observations during solar eclipse. Reports he did not observe the supposed planet Vulcan, but a fixed star. Sends micrometer observations. Describes shape of prominences.
ALS 4pp RS:HS 14.110 Ada Pritchard, *Charles Pritchard: Memoirs of His Life,* 227–9

11388. 1860–8–2 To Edward Sabine Collingwood
Occupied with matters besides magnetism lately. Interest in science of Prince Consort [Albert] is praiseworthy. Discusses several observatories and letter of [T. W.?] Blakiston.
C 2pp RS:HS 23.302

11389. [1860]–8–6 To Thomas Maclear [Collingwood]
Congratulates TM on receiving knighthood. Sends recommendations and makes inquiries concerning Harold Maclear's schooling.
ALS 4pp Cape Archives/Maclear Papers File 100

11390. 1860–8–6 From Sir William Pole 3 Storey's Gate
Thanks for note of 29 [July]. Expresses doubt over accuracy of his observations of a solar eclipse. Stresses the necessity of careful comparison of various observations. Asks JH's opinion of a photograph with prominences greater and corona less than supposed.
ALS 3pp RS:HS 14.26

11391. 1860–8–10 From Patrick Francis Robertson 12 Leadenhall Street
Sends pamphlet on repaying the national debt. Discusses various problems and possibilities concerning the debt.
ALS 2pp RS:HS 14.370

11392. 1860–8–11 To [T. P. Anderson] Collingwood
Has no knowledge of transactions affecting trust fund for W. H. B. Hollier, other than what JH noted in previous letters.
ALS 1p TxU:H/L-0757.1; Reel 1089

11393. 1860–8–16 To William Parsons (Lord Rosse) Collingwood
Sends JH's 'On Atoms' to WP.
ALS 1p Rosse Papers K2.7

11394. 1860–8–17 From William Lockhart Blackheath
Has today dispatched a parcel to JH containing a copy of a Chinese translation of JH's work on astronomy; thinks this

work will be of great service to the Chinese.
ALS 3pp RS:HS 11.487

11395. 1860–8–18 From Edward Sabine Llandovery
Bibliography of works on terrestrial magnetism published in past four years.
ALS 8pp TxU:H/M–0527.4; Reel 1093

11396. 1860–8–20 From Thomas Maclear Royal Observatory, Cape of Good Hope
George Grey's wife sent back to England after her affair was discovered.
ALS 2pp Cape Archives/Maclear Papers File 100

11397. 1860–8–21 To Henry Samuel Boase Collingwood
Thanks HB for a copy of his work on the philosophy of nature.
C 1p RS:HS 23.303

11398. 1860–8–22 To Robert FitzRoy Collingwood
Asks for meteorological data that RF may have collected.
C 2pp RS:HS 23.304

11399. 1860–8–23 To George Biddell Airy Collingwood
Asks for details about a strange drawing of Jupiter JH had seen at the Royal Observatory some months earlier, and offers some comments about Warren de La Rue's eclipse photographs.
ALS 3pp RGO 6.379.140

11400. 1860–8–29 From W. Paterson Glossop
Has just heard a report that JH had predicted that there would be a series of heavy floods during the month of August. Wonders if this originated from JH.
ALS 2pp RS:HS 13.239 *Glossop Record* (Sept. 8, 1860)

11401. 1860–8–30 To Norton Shaw Collingwood
Please send copies for distribution to JH's friends of JH's paper ['On a New Projection of the Sphere'].
ALS 2pp Royal Geographical Society

11402. 1860–8–31 From James Dimmock Darwen
Sending suggestions for improvements in the submarine cable and would like JH's comments.
ALS 3pp RS:HS 6.493

11403. 1860–9–1 To W. Paterson Collingwood
Thanks for letter regarding his prediction of the floods. May be partly to blame as he did mention certain happenings that could be the result of solar phenomenon. Comments on this. If WP communicates this to a paper would like two or three copies.
draft 3pp RS:HS 13.240 & 23.305 *Glossop Record* (Sept. 8, 1860)

11404. 1860–9–2 From Augustus De Morgan 41 Chalcot Villas NW
Is he thinking of letting Francis Baily's house at Tavistock Place as he knows someone who is desirous of renting it. Was able to walk two miles at a quick pace yesterday. Rest of his family are at the seaside.
ALS 4pp RS:HS 6.337

11405. 1860–9–4 To Jean Baptiste Biot Collingwood
Was very pleased to receive another of JB's works. Glad to be associated in his revelations concerning molecular and polarization powers. Recall their past meetings and hopes to renew acquaintance in Paris at some future date.
draft 4pp RS:HS 4.94a & 23.306

11406. 1860–9–4 To Augustus De Morgan Collingwood
Selling house [in London]; comments on H. S. Boase's writing on annuities.
C 1p RS:HS 23.307

11407. 1860–9–4 From W. Paterson Glossop
Thanks for valuable and interesting letter; has sent copies to the *Glossop Record* and *Manchester Examiner* and will send copies of the papers should the letter be printed. Will be glad to know the results of his valuable observations.
ALS 1p RS:HS 13.241

11408. 1860–9–5 From Augustus De Morgan 41 Chalcot Villas NW
Knew he would get a speedy answer. Quick, the Bristol bookseller, has sent him an edition of Francis Walkingame's *Arithmetic*. Thinks JH is right to sell Francis Baily's house. G. B. Libri may wish to buy. His puns are unavoidable. Hopes JH will have his house valued.
ALS 7pp RS:HS 6.338

11409. 1860–9–8 From W. Paterson Glossop
JH's letter has appeared in the *Glossop Herald* and *Manchester Examiner*. Has ordered copies and will send some to JH. Has sent a copy to Sir Henry James.
ALS 2pp RS:HS 13.242

11410. 1860–9–10 From Reginald Stuart Poole 1 West Terrace, Worthing
Thanks JH for reading his preface to *The Genesis of the Earth and Man,* and expresses the author's gratitude that JH believes the book will be interesting.
ALS 3pp RS:HS 14.41

11411. 1860–9–10 From Edward Sabine Cowstone Rectory, Norfolk
Sent packet to JH last week containing publications listed in ES's bibliography of recent magnetic surveys.
ALS 2pp TxU:H/M–0527.5; Reel 1093

11412. 1860–9–11 To John Herschel (son) Collingwood
Begins with a discussion of the probability of error in a series of measurements and comments on astronomical observations including the observation of a comet.
ALS 4pp JHS 5.4

11413. 1860–9–11 From Joseph Jopling 6 Vassall Terrace
Unable to give in a letter the practical geometrical results he has worked out from the laws of the king's chamber of the Great Pyramid. It would grace a history of Egyptian astronomy.
ALS 3pp RS:HS 10.420

11414. 1860–9–12 From Augustus De Morgan 41 Chalcot Villas NW
Thanks for the newspaper cutting repudiating his wisdom on weather matters. Has had erroneous charges laid at his own door. What arms has he taken?
ALS 4pp RS:HS 6.339

11415. 1860–9–12 From James Dimmock Darwen
Thanks for his letter, but regrets he should have misconstrued the reason for his suggestions. Has no commercial motives.
ALS 3pp RS:HS 6.494

11416. 1860–9–14 From George Bentley New Burlington St.
Would like the opportunity of publishing a work for him.
ALS 2pp RS:HS 5.54

11417. 1860–9–14 From W. E. Hickson Wrotham
Sending an account of an earth shock at Wrotham on 3 Sept. and inquiring if it was noticed at Hawkhurst.
ALS 3pp RS:HS 10.189

11418. 1860–9–16 From Jean Baptiste Biot Paris
Regarding the missing publications of the Academy that JH has not received. Further regarding polarization in molecules.
ALS 3pp (French) RS:HS 4.95

11419. 1860–9–17 From John Murray, [Jr.] Albemarle St.
Has recently read in the papers JH's correspondence concerning the weather. Would like an article on this subject from JH for the *Quarterly Review*. Comments on the influence of the moon on the weather.
ALS 3pp RS:HS 12.448

11420. 1860–9–17 To Edward Sabine Collinwood
Lists magnetic books received from ES. It will be some time before JH can direct his attention to magnetic issues.
C 2pp RS:HS 23.308

11421. 1860–9–18 From Joseph Jopling 6 Vassall Terrace
Thanks for his letter. Any astronomical observations can be recorded on a plane or spherical surface. 108 equal cubes will exactly fill the coffer in the king's chamber of the Great Pyramid.
ALS 2pp RS:HS 10.421

11422. 1860–9–18 From Charles Manley 24 Great George St., Westminster
According to JH's instructions, portrait of late Robert Stephenson will be sent to 37 Tavistock Place. Thanks for JH's approval of paper on color blindness by CM's friend Mr. Pole. Would Pole be eligible for R.S.L. membership?
ALS 2pp TxU:H/M–0348.1; Reel 1087

11423. [1860–9–18 or later] To [?] n.p.
Criticizes point in [recipient's] scheme for issuing bank notes.
Adraft 1p inc TxU:H/M–0348.2; Reel 1087

11424. 1860–9–20 From George Bentley New Burlington St.
Is grateful for his suggestions. Would he have leisure to edit a new edition of the letters?
ALS 2pp RS:HS 5.55

11425. 1860–9–20 From John Bell Rolvenden Park
Hopes that JH will attend the luncheon when J. R. Townshend (3rd Viscount Sydney) inspects the 37th Kent R.V. battalion.
ALS 2pp RS:HS 5.43

11426. 1860–9–21 From Mary Somerville La Spezia Piemonte
Thanks JH for scientific papers. Is writing about relations between inorganic matter, physical and vital forces, and microscopic plants. Praises [Giuseppe] Garibaldi and his performance.
ALS 4pp RS:HS 16.363

11427. 1860–9–22 From Augustus De Morgan 41 Chalcot Villas NW
Sending a moral song. Sent one to W. H. Smyth.
ALS 2pp RS:HS 6.340

11428. 1860–9–22 From James Wylde 10 Stationers Hall Court
As editor for Griffin & Co., JW encloses that portion of book on 'Cont[emporary] Biography' that relates to JH. Asks JH for alterations.
ALS 1p RS:HS 18.438

11429. 1860–9–25 From William MacPherson 50A Albermarle St.
John Murray has shown him JH's note. Is pleased to have an article from JH for the *Quarterly Review* on his observations. Hopes he will show the influence of the moon on the weather.
ALS 1p RS:HS 12.266

11430. 1860–9–25 From James Marsh [Llanstead?], Essex
Requests autographs of William Parsons (Lord Rosse) and William Herschel for JM's collection.
ALS 1p TxU:H/MW-0246.4; Reel 15

11431. 1860–9–26 From John Hind 5 Orchard St., Kensington
Has read some extracts from JH's *Prelim. Discourse* and would be grateful if he has an old or soiled copy to dispose of.
ALS 3pp RS:HS 9.316

11432. 1860–9–28 From Augustus De Morgan 41 Chalcot Villas NW
Can JH come in on Sunday evening? Where do the authors of drinking songs get their facts?
ALS 2pp RS:HS 6.341

11433. 1860–10–1 From Joseph Jopling 6 Vassall Terrace
Outlining a method for making star maps with a pyramid.
ALS 1p RS:HS 10.422

11434. 1860–10–2 From James Wylde 10 Stationers Hall Court
As editor for Griffin & Co., JW forwards latest proof for JH to check [see JW's 1860–9–22].
ALS 1p RS:HS 18.439

11435. 1860–10–5 From Norman Pogson Hartwell, Aylesbury
Regrets that JH's son Alexander is quitting astronomy. Asks JH to support his effort to obtain position at Madras Observatory. Announces first project would be survey of southern heavens in completion of [F. W. A.] Argelander's new atlas.
ALS 4pp RS:HS 14.8

11436. 1860–10–6 To Norman Pogson Collingwood
Supplies what is in effect a testimonial to NP's abilities for use in NP's quest to become director of the Madras Observatory.
draft 3pp RS:HS 14.9 (C: RS:HS 23.309)

11437. 1860–10–7 To William Henry Smyth Collingwood
Asks WS opinion of attaining a civil pension for N. R. Pogson in honor of his astronomical accomplishment.
C 1p RS:HS 23.310

11438. 1860–10–8 From John Bell Rolvenden Park
Hopes that JH will reply for the 'visitors' at the forthcoming luncheon.
ALS 2pp RS:HS 5.44

11439. 1860–10–9 From Charles W. Tuttle Boston, Mass.
Requests a copy of JH's memoir on Caroline Herschel. Points out an error in the fifth edition of *Outlines Astr.*
ALS 4pp RS:HS 17.451

11440. 1860–10–11 From Henry Samuel Boase Claverhouse near Dundee
Is gratified by his kind remarks on his *Philosophy of Nature*. Regrets he failed to give due acknowledgement to JH's work, but will redress this in the future. Regarding the probable constitution of nature.
ALS 3pp RS:HS 5.80

11441. 1860–10–11 From William Henry Smyth St. John's Lodge
N. R. Pogson appointed to Madras Observatory. Appointment includes a pension, so a civil pension will not be necessary.
ALS 2pp RS:HS 16.314

11442. 1860–10–12 From Robert Brown Fairlie by Greenock
His book on musical science is ready and a copy will be forwarded to him later when published.
ALS 1p RS:HS 4.317

11443. 1860–10–12 From William Henry Smyth St. John's Lodge
Refers to a discovery made at the R.S.L., which, despite being based on 'circumstantial evidence,' convinces WS of its truth.
ALS 1p RS:HS 16.318

11444. 1860–10–13 From John Hind 5 Orchard St., Kensington
Is grateful for the gift of his *Prelim. Discourse*. Had to pay 4/8 excess postage.
ALS 3pp RS:HS 9.317

11445. 1860–10–15 From Benjamin Duprat Paris
Has dispatched *Memoirs* and *Comptes rendus* of the Academy. Regrets the delay, which is due to negligence on the part of the railway.
ALS 2pp RS:HS 6.516

11446. 1860–10–18 From John Hind 5 Orchard St., Kensington
Thanks for the 4/8 postage. Query on William Hopkins's theory of the Earth's formation. Would be grateful for any scientific works.
ALS 3pp RS:HS 9.318

11447. 1860–10–20 From George Biddell Airy Royal Observatory, Greenwich
Circular letter advising of the availability of back copies of various Royal Observatory printed observations.
(A)LS 3pp TxU:H/M-0639.2; Reel 1084

11448. 1860–10–21 From Thomas Maclear Royal Observatory, Cape of Good Hope
Details of Lady Grey's shipboard affair with the Admiral.
ALS 2pp Cape Archives/Maclear Papers File 100

11449. 1860–10–21 From Thomas Maclear Royal Observatory, Cape of Good Hope
Grateful for JH's congratulations. News of the education of his own sons. Pleased to hear of Warren de La Rue's success with photographing the pink projections during the total darkness. Recent astronomical observations. News of the Cape Electric Telegraph. Mr. Wollaston (a nephew of W. H. Wollaston) is about to make a survey for the telegraph wire. Gives details of the electric clock in connection with the transit circle.
ALS 8pp RS:HS 12.165

11450. 1860–10–22 To Robert FitzRoy n.p.
Asks for information about wrecks at sea and lives lost.
C 1p inc RS:HS 23.311

11451. 1860–10–23 From Robert FitzRoy Board of Trade and Admiralty Meteorological Department
Tells JH he will search for the requested weather data. Encloses 'Weather Reports in the Newspaper.'
ALS 4pp encl 8pp Harvard: Houghton AAH 67m–67 (66)

11452. 1860–10–23 From James Glaisher Blackheath
Sending the quarters reports. Comments on these reports.
ALS 4pp RS:HS 8.130

11453. 1860–10–24 To Robert FitzRoy Collingwood
Thanks for a number of RF's writings; asks for more rainfall data to try to establish a relationship between rainfall and the solar cycle.
C 2pp RS:HS 23.312

11454. 1860–10–25 From George Biddell Airy n.p.
Comments on Jupiter's appearance and on the eclipse photographs [see JH's 1860–8–23].
ACCS 4pp RGO 6.379.142

11455. 1860–10–25 From James Glaisher Blackheath
Has received his note. Occurs to him that the best plan would be for JH to borrow his manuscripts on rain. If he will notify him of the Railway address, he will dispatch them at once.
ALS 3pp RS:HS 8.131

11456. 1860–10–25 From Joseph Jopling 6 Vassall Terrace
Encloses [a diagram] of the transverse section of the entrance of the Great Pyramid, with details of the travelling platform for making observations.
ALS 2pp RS:HS 10.423

11457. 1860–10–27 From Robert FitzRoy Board of Trade and Admiralty Meteorological Department
Reports that rainfall information is not as complete as should be. Describes available data from Meteorological Department. Thanks JH for comments on his work.
ALS 20pp Harvard: Houghton AAH 67m 67 (67)

11458. 1860–10–27 From James Glaisher Blackheath
Has sent his rain papers from Lewisham station this morning. Is in no hurry for their return. Will send on his manuscript when he can find it.
AL 2pp inc RS:HS 8.132

11459. 1860–10–28 From Norman Pogson Hartwell, Aylesbury
Announces success in securing directorship of Madras Observatory, thanks to JH and G. B. Airy. Cannot now change the names for asteroids Hestia or Isis. Asks for suggestions on names in the case of future discoveries.
ALS 4pp RS:HS 14.10

11460. 1860–10–29 From Robert FitzRoy Board of Trade and Admiralty Meteorological Department
Discusses data gathering from 'the colonies.' Forwards a Scottish publication on meteorology.
ALS 4pp Harvard: Houghton AAH 67m–67 (68)

11461. 1860–10–29 From John R. Hind 22 Grove Rd., St. John's Wood
JH's copy of the *Nautical Almanac* for 1864 will be forwarded to the R.A.S. J. H. Mädler has solved the problem of one of William Herschel's observations.
ALS 4pp RS:HS 9.366

11462. 1860–10–29 From James Stark Edinburgh
Sends rainfall data. Resigned secretaryship of the Meteorological Society. Discusses possible influence of sunspots on rainfall.
ALS 6pp Harvard: Houghton AAH 67m–67 (98)

11463. 1860–10–30 From Robert FitzRoy Board of Trade and Admiralty Meteorological Department
Sends more data and papers on meteorological matters. Discusses weather observations by C. P. Smyth.
ALS 8pp Harvard: Houghton AAH 67m–67 (69)

11464. 1860–10–30 To William MacPherson Collingwood
Will not be able to produce his article in time for the January issue, but could write a much better article on meteorology if given longer, till July.
draft 1p RS:HS 12.267 & 23.313

11465. 1860–10–30 To Norman Pogson Collingwood
Congratulates NP on appointment to Madras Observatory. States that he will use both Isis and Hestia nomenclature without misgiving. Furnishes names from Greek mythology for NP to use for asteroid discoveries.
ALS 4pp RS:HS 14.11 (C: RS:HS 23.314) Note:KAA

11466. 1860–10–31 To Robert FitzRoy Collingwood
Thanks for all the material received; comments on some of RF's meteorological ideas.
C 2pp RS:HS 23.315

11467. 1860–10–31 From Norton Shaw [London]
Forwards copies of maps to accompany JH's paper.
ALS 1p Harvard: Houghton AAH 67m–67 (63)

11468. 1860–11–2 From Robert FitzRoy Board of Trade and Admiralty Meteorological Department
More rainfall data.
ALS 7pp Harvard: Houghton AAH 67m–67 (70)

11469. 1860–11–2 To John Wrottesley Collingwood
Thanks JW for his double star catalogue; JH comments on a few items contained therein.
C 1p RS:HS 25.12.5

11470. 1860–11–3 From William MacPherson 48 Inverness Terrace
Thanks for his letter. Will look forward to receiving his article in time for the July issue of the *Quarterly Review*.
ALS 2pp RS:HS 12.268

11471. 1860–11–5 From Elias Loomis London
Is grateful for JH's kind invitation to Collingwood, but regrets he cannot find time to come as he sails on the day after tomorrow and has much business to fit in before then.
ALS 2pp RS:HS 11.309

11472. 1860–11–6 To Robert FitzRoy Collingwood
Comments on RF's writing on drifting ice; comments extensively on RF's ideas about air circulation.
C 7pp RS:HS 23.316

11473. 1860–11–7 From Robert FitzRoy Board of Trade and Admiralty Meteorological Department
Thanks JH for his advice. Comments on various aspects of Iceland and Greenland.
ALS 4pp Harvard: Houghton AAH 67m–67 (71)

11474. 1860–11–7 From John Wrottesley Wrottesley [Staffordshire]
Gives probability formula used to determine weights of stars 3 & 35 [see JH's 1860–11–2]. Explains calculation procedures previous to 1856 and those used since 1857.
ALS 4pp RS:HS 18.303

11475. 1860–11–8 From Robert FitzRoy Board of Trade and Admiralty Meteorological Department
Discusses James Stark's weather observations.
ALS 4pp Harvard: Houghton AAH 67m–67 (72)

11476. 1860–11–13 To William Lassell Collingwood
Comments on WL's nebulae diagrams and asks for more.
C 1p RS:HS 23.317

11477. 1860–11–13 From Edward Joseph Lowe Observatory Beeston near Nottingham
Sends copies of observatory's printed reports, 1859–60.
ALS 1p RAS:JH Archive 12/1.3.16 verso; Reel 10

11478. 1860–11–13 From George Gabriel Stokes Lensfield Cottage Cambridge
Further about the paper referred to in GS's 1860–6–15.
ALS 2pp RS:HS 17.43

11479. 1860–11–15 From Robert FitzRoy Board of Trade and Admiralty Meteorological Department
Describes a starfish brought alive to England from Iceland by an amateur naturalist.
ALS 4pp Harvard: Houghton AAH 67m–67 (73)

11480. 1860–11–17 From T. Friend Dublin
Sending weather returns that he has kept at Valencia Island. There is a possibility of error in the thermometer readings.
AL 3pp RS:HS 7.362

11481. [1860]–11–17 From Ann Smyth St. John's Lodge
W. H. Smyth wants another copy of JH's dialogue on atoms. W. H. Flower, Smyths' son-in-law, will have a paper published in *R.S.P.T.*
ALS 4pp RS:HS 16.220

11482. 1860–11–21 From Alexander Wylie Singapore
Received *Outlines Astr.* before leaving China. Describes 'comet' sighting [27 May 1860]. Sent details of it to William Lockhart [see Lockhart's RS:HS 19.243]. Another witness, 'Rev. G. John,' will send report to JH.
ALS 9pp RS:HS 19.303

11483. 1860–11–24 From Robert FitzRoy Board of Trade and Admiralty Meteorological Department
Thanks JH for critiques of his paper; discusses Matthew Maury's work *Physical Geography of the Sea*.
ALS 8pp Harvard: Houghton AAH 67m–67 (74)

11484. 1860–11–24 From William Henry Smyth St. John's Lodge
Asks JH's opinion of Charles Shadwell; wishes to elect him a Fellow of the R.S.L. Includes copy of letter of testimonial to Robert FitzRoy. Believes Shadwell's lunar work of great merit.
ALS 2pp encl (RS:HS 16.315) RS:HS 16.316

11485. 1860–11–28 To Robert FitzRoy Collingwood
Comments on difficulties JH finds in some of RF's writings.
C 4pp RS:HS 23.318

11486. 1860–11–28 To William Henry Smyth Collingwood
Signs Charles Shadwell's recommendation. Sends new maps of earth, which include more surface with less distortion. Sends copy of JH's *Telescope*.
C 2pp RS:HS 23.319

11487. 1860–11–29 From Augustus De Morgan 41 Chalcot Villas NW
Thanks for his article on the telescope. Comments on this. Sends another extract from a drinking song. Hopes that G. B. Airy will be the next president of the R.A.S.
ALS 3pp RS:HS 6.342

11488. 1860–11–29 From Robert FitzRoy Board of Trade and Admiralty Meteorological Department
Discusses atmospheric wave theory in reference to an experiment in JH's *Meteorology*. Describes weather patterns observed on the voyage of the *H.M.S. Beagle*.
ALS 7pp Harvard: Houghton AAH 67m–67 (75)

11489. 1860–11–29 To Edward Sabine Collingwood
Asks for information concerning the magnetic observations and compilations made in various countries and regions.
C 2pp RS:HS 23.320

11490. 1860–12–1 From William Parsons (Lord Rosse) The Castle, Parsonstown
Many thanks for JH's article on the telescope; will read it with much interest. Has employed an assistant to prepare drawings of the nebula. The six-foot reflector requires repolishing.
ALS 4pp RS:HS 13.232

11491. 1860–12–1 From Warren de La Rue Observatory, Cranford Mx.
Please accept his thanks for his article on telescopes. Comments on JH's views. Would like 50 copies of the article to distribute to opticians.
ALS 2pp RS:HS 6.146

11492. 1860–12–5 From James Smith Brazier Aberdeen University
Wishes to apply for a position as professor and would like JH's recommendation.
ALS 4pp RS:HS 5.98

11493. 1860–12–6 To George Biddell Airy n.p.
A note with a copy of JH's article on telescopes for the *Encyclopaedia Britannica*, together with some papers to be forwarded to James Glaisher
ALS 3pp RGO 6.379.144

11494. 1860–12–6 From Edward Sabine 13 Ashley Place
Summary of research on terrestrial magnetism in other countries. Will send paper on magnetic storms to JH.
ALS 4pp TxU:H/M–0527.6; Reel 1093

11495. 1860–12–6 From William Selwyn Cambridge
Sends pictures of sun. Thanks JH for sparking his interest in the sun. Wants JH to direct the B.A.A.S.'s use of the Kew telescope to take pictures.
ALS 4pp RS:HS 15.464

11496. 1860–12–7 From George Biddell Airy Royal Observatory, Greenwich
Thanking him for his paper on telescopes. Observations of JH's semicircular map of the world.
ALS 3pp RS:HS 1.248 (C: RGO 6.475.233)

11497. 1860–12–8 To George Biddell Airy Collingwood
Has tried a number of different fractional projections, some more interesting than others [see GA's 1860–12–7].
ALS 3pp RGO 6.475.235

11498. 1860–12–10 From James Smith Brazier Aberdeen University
Is grateful for his prompt reply and kind remarks about his own qualifications [i.e., JB's].
ALS 1p RS:HS 5.99

11499. 1860–12–11 From William Lassell Sandfield Park near Liverpool
Was grateful for the copy of JH's article on the telescope. Sends parcels containing two specimens of 4-ft.specula; comments on these. Weather has been unsuitable for observations.
ALS 4pp RS:HS 11.158

11500. 1860–12–12 From Sir William Pole 3 Storey's Gate
Sends his certificate for the R.S.L. to JH to sign.
ALS 1p RS:HS 14.27

11501. 1860–12–12 To the London *Times* Collingwood
[It having been asserted in the *Times* that the Russians have not communicated about the climate at Peking], JH notes that in fact wonderfully detailed meteorological observations for Peking from 1850 to 1855 have been widely distributed by the Russians.
C 1p RS:HS 23:321 London *Times* (Dec. 14, 1860), p. 7, col. 5

11502. 1860–12–13 To [James Calder] Stewart Collingwood
Please withdraw [?]'s name from ballot until JH can confer with him when [he] comes home on leave next year. Thanks for care that JS is showing to Alexander S. Herschel.
ALS 2pp TxU:H/L-0454; Reel 1055

11503. 1860–12–14 From Arthur Cayley Stone Bldgs., Lincoln's Inn
Returning the corrected proof of the note to be added to one of JH's papers.
ALS 1p RS:HS 5.227

11504. 1860–12–14 From John Murray, [Jr.] Albemarle St.
Has had a paper placed in his hands from an Icelandic savant regarding drifts from Greenland. Thinks that JH may find this paper useful.
ALS 2pp RS:HS 12.449

11505. 1860–12–17 From Samuel Newington Ridgeway, Ticehurst
Can JH spare a few minutes one morning as he wishes to show him an invention to overcome seasickness; would like his comments.
ALS 2pp RS:HS 13.126

11506. 1860–12–18 From Edward Sabine 13 Ashley Place
Sends sample sheets of magnetic reductions from continual observations photographed at Kew. Explains reduction formulas. Will send to JH new paper by ES on lunar diurnal variation. Carlo Matteucci reports interest by new government in Naples in reviving meteorological observatory there.
ALS 4pp TxU:H/M–0527.7; Reel 1093

11507. 1860–12–20 To Edward Sabine Collingwood
Thanks for declination readings from photograms at Kew. Thinks meteorological observatory on Vesuvius is good idea, but not sure a magnetic one is. Includes two charts of world.
C 1p RS:HS 23.322

11508. 1860–12–21 From Edward Sabine R.S.L., Burlington House
Asks JH to comment on ES's paper, which will be read to

R.S.L. on 10 Jan. Corrects errors in two earlier letters to JH. Will adopt 'Photograms' instead of 'Photographs.'
ALS 3pp TxU:H/M–0527.8; Reel 1093

11509. 1860–12–21 From John Gardner Wilkinson Rhossili, Swansea
Thanks for copy of JH's ['Remarks on Colour-Blindness' (1859–60)].
L 2pp RS:HS 18.409

11510. 1860–12–22 From John Ball Bedgebury Park
Is obliged for the maps and articles on physical geography. Comments on the heights of various mountains and encloses a list of queries.
ALS 19pp RS:HS 5.11

11511. 1860–12–22 To Samuel Newington Collingwood
Is interested to hear of SN's anti-seasickness invention, especially as JH has just devised a swing-cot for the same purpose.
C 1p RS:HS 23.323

11512. 1860–12–23 From Joseph Jopling 6 Vassall Terrace
Has received colored tracing from Italy of drawings by Col. Beek. Compares them with what may be done using the Great Pyramid.
ALS 2pp RS:HS 10.424

11513. 1860–12–25 From Samuel Newington Ridgeway, Ticehurst
Will not detain JH more than half an hour; if his invention is not similar to JH's it will amuse him. Gives his reasons for seasickness. Will call tomorrow.
ALS 4pp RS:HS 13.127

11514. 1860–12–26 From William Rutter Dawes Hopefield, Haddenham, Thame
Announcing the sudden death of his wife. The Taylors are with him.
ALS 3pp RS:HS 6.105

11515. 1860–12–27 From Samuel Newington Ridgeway, Ticehurst
Since they met yesterday he has thought of a further modification in the apparatus to prevent seasickness and would like JH's opinion.
ALS 3pp RS:HS 13.128

11516. 1860–12–29 From Samuel Newington Ridgeway, Ticehurst
Has overcome pitch and roll with his inventions. Instrument is quite simple. What is he to do with it now?
ALS 1p RS:HS 13.129

11517. 1860–12–31 From Samuel Newington Ridgeway, Ticehurst
Has made a bad sketch of the instrument to prevent seasickness; comments on it.
ALS 3pp RS:HS 13.130

11518. 1860–12–31 From William Henry Smyth St. John's Lodge
Awaits copy of JH's *Telescope.* Problems with by-laws of R.A.S.
ALS 3pp RS:HS 16.317

11519. 1861 From Augustus De Morgan 41 Chalcot Villas NW
Is pleased JH can look his trial in the face. Regarding logic. Has seen a neat construction by JH's son.
ALS 6pp RS:HS 6.343

11520. [1861 or later] To Frederick Howlett n.p.
Returns a paper of Mr. Stobart, which deals with astronomical knowledge of the early Egyptians.
C 1p RS:HS 25.14.13

11521. 1861 From Henri Victor Regnault Sèvres
Talked to [L.-F.-C.?] Breguet for JH. Discusses work [on hygrometer?/metallic thermometer?] in detail, including diagrams. Regrets not having seen JH's daughters in Paris. Says he is recuperating from the 'catastrophe' and can still use his left eye.
ALS 4pp (French) RS:HS 14.316

11522. [1861?] To [T. R. Robinson?] n.p.
Discusses the project of erecting a large telescope in the southern hemisphere [Melbourne telescope] for observation of nebulae. Expresses reservations and makes suggestions about the project.
ALS 6pp? RS MM 12.77

11523. [1861] From William Selwyn n.p.
Detached postscript discussing sunspots, comparing them to the low barometric pressure center in cyclones. J. S. Henslow is dying.
AL 2pp inc RS:HS 15.515

11524. [1861 to 1864] From [?] n.p.
Letter of introduction for William Walker, who wants to show Walker's nearly finished engraving, *Distinguished Men of Science Living 1807–8*, to JH.
AL 1p TxU:H/L-0761; Reel 1089

11525. 1861–1–6 To James Calder Stewart Collingwood
Alexander S. Herschel's education goals and job prospects. [Lucy?] has published 'Marshall's Essays.' Question about investments.
ALS 4pp TxU:H/L-0455; Reel 1055

11526. 1861–1–9 To [Adam and Charles Black, Publishers] Collingwood
Sending first 64 pages of JH's *Physical Geography* with some insertions and corrections. Requests proofs in duplicate.
ALS 2pp BostonPubLib Ms. Eng. 204

11527. 1861–1–10 From Frederick Howlett St. Augustine's, Hurst Green
Expressing sympathy on the illness of Margaret Louisa Marshall (JH's daughter). Would like to hear the latest news by the bearer. Would JH append his name to the enclosed form.
ALS 4pp RS:HS 10.14

11528. 1861–1–11 From John Ball 18 Park St.
Sends him latest edition of the book brought out by the Alpine Club. A member has found the height of one of the mountains lower than that usually stated. Has written an article on glaciers for this month's *Edinburgh Review*.
ALS 4pp RS:HS 5.12

11529. 1861–1–[12] To John Herschel (son) [Collingwood]
Writes to inform son John of the sudden death [after only a very brief illness] of John's sister Margaret Louisa. [The letter is quiet and accepting, praising God for the happy and blameless life that 'Looey' had led and what joy she had brought to the family.]
ALS 3pp JHS 5.5

11530. 1861–1–14 From William Rutter Dawes Hopefield, Haddenham, Thame
Expressing condolence at loss of some member of JH's family [JH's daughter Margaret Louisa].
ALS 2pp RS:HS 6.106

11531. 1861–1–17 From Smith, Elder & Co. 65 Cornhill
Acknowledgement of fulfilling JH's requirements.
ALS 1p RS:HS 16.205

11532. 1861–1–18 From P. W. D. Baddeley Bonn
Is pleased that JH has received his book on whirlwinds. Comments on this and would welcome any criticism from JH on this work.
ALS 4pp RS:HS 5.3

11533. 1861–1–19 From Augustus De Morgan 41 Chalcot Villas NW
Has heard of the heavy blow that has fallen on the Herschels [death of JH's daughter Margaret Louisa]. Hopes they are all bearing up.
ALS 3pp RS:HS 6.344

11534. 1861–1–24 From George Biddell Airy Flamsteed House, Greenwich
Offering sympathy on the death of JH's daughter [Margaret Louisa].
ALS 3pp RS:HS 1.249

11535. 1861–1–24 From Joseph Jopling 6 Vassall Terrace
Encloses another design for permanent lines on the faces of a pyramid to observe and record astronomical facts; comments on his methods.
ALS 4pp RS:HS 10.425

11536. 1861–1–25 From Andrew Graham Markree Observatory
Has been urged to apply for the Headmastership of Sir Joseph Williamson's School at Rochester and would like JH's opinion on his suitability. Encloses a note from Edward Cooper.
ALS 2pp RS:HS 9.111

11537. 1861–1–26 From John Ball 18 Park St.
Is sorry he wrote his last letter on the day of the sad event. Hopes JH will use the table of heights in the Alpine volume. Comments on the accuracy of various heights given by authorities.
ALS 4pp RS:HS 5.13

11538. 1861–1–27 From William Whewell Trinity Lodge
Offers consolation to JH on the death of his daughter Margaret Louisa.
AL 4pp RS:HS 18.216

11539. 1861–1–28 From Mary Somerville Florence
Condolences on death of JH's daughter [Margaret] Louisa.
ALS 1p RS:HS 16.364

11540. 1861–1–29 From John William Lubbock Mansion House St.
Thinks there is nothing JH's son should be afraid of if the Directors will appoint him. Sorry to hear JH has been suffering. Lady Lubbock has been to St. Leonards.
ALS 2pp RS:HS 11.391

11541. 1861–1–29 To Edward Sabine Collingwood
Has been too ill to write terrestrial magnetism paper for *Edinburgh Review*. Suggests other people to write it.
draft 1p RS:HS 15.254 (C: RS:HS 23.324)

11542. 1861–1–30 From John William Lubbock Mansion House St.
Mr. Higham has had to retire due to ill health. If appointed JH's son will have nothing to do with the duties of actuary. Hopes to send something about the moon soon.
ALS 4pp RS:HS 11.392

11543. 1861–1–30 From Edward Sabine [13 Ashley Place]
Like JH, Smyths have been suffering bronchial problems. Urges JH not to give up [see JH's 1861–1–29] plan of JH writing a review on terrestrial magnetisim.
ALS 4pp RS:HS 15.255

11544. 1861–1–31 From John Murray, [Jr.] Albemarle St.
Regrets that JH's ill health has been the cause of his abandonment of his paper on the weather. Editor of the *Quarterly Review* will be pleased to receive it should JH resume his writing.
ALS 1p RS:HS 12.450

11545. 1861–2 To *Daily News* n.p.
Regarding the question of forged documents by the Indigo planters of Bengal and the advantages they hold over the Ryotts, who cannot read or write.
ALS 4pp RS:HS 6.461

11546. 1861–2–2 From Edward Loyd Hawkhurst
Is sorry to hear that JH has been so unwell. The Rifle Association is to compete for prizes. Is JH in favor of the Association?
ALS 3pp RS:HS 11.346

11547. 1861–2–2 From N. Newbon 1 Wardrobe Place
Thanks for letter. Mrs. Moorsam was interred at Norwood Cemetery last Wednesday. NN's health is bad so son represented her. Bequests to Isabella and Caroline Herschel.
ALS 2pp RS:HS 13.152

11548. 1861–2–3 From Johann Franz Encke Berlin
On the orbits of short period comets, and the shortening of such orbits.
ALS 4pp (German) RS:HS 7.101

11549. 1861–2–4 From John William Lubbock Mansion House St.
The Governors find that many of the senior members favor Mr. Stute for the appointment, so in the circumstances thinks it best that JH's son should not appear as a candidate.
ALS 2pp RS:HS 11.393

11550. 1861–2–6 From Roderick Impey Murchison Geological Survey of Great Britain Office
Regarding the geology of the Highlands. Will send him his own memoir. Will try at the next Council to obtain what JH wants.
ALS 4pp RS:HS 12.424

11551. 1861–2–6 From Norton Shaw [London]
Informs JH that a printer will make multiple copies of a paper for him.
ALS 1p Harvard: Houghton AAH 67m–67 (107)

11552. 1861–2–10 From William Whewell to Margaret Herschel Trinity Lodge
Offers consolation to Margaret Herschel on the death of her daughter Margaret Louisa.
ALS 4pp RS:HS 18.217

11553. 1861–2–11 From Andrew Graham Markree Observatory
Is grateful for his prompt reply; it is just what he required. Has not heard the result of his application yet.
ALS 1p RS:HS 9.112

11554. 1861–2–12 From Henry Anthony Jeffreys Hawkhurst Parsonage
Obliged for his note regarding the church rates. Comments on JH's views and gives summary of the work carried out by the local incumbent.
ALS 4pp RS:HS 10.299

11555. 1861–2–12 To Henry Anthony Jeffreys Collingwood
Returns the petitions. Regarding the one for the maintenance of the church rates. Has qualms about signing this as he does not entirely agree with some of the statements.
C draft 4pp RS:HS 10.298 & 23.325

11556. 1861–2–13 From Joseph Jopling 6 Vassall Terrace
Is obliged for his note and statement on the necessity of pegs at the ends of the lines of the Great Pyramid; comments on this. Has seen Augustus De Morgan and explained his theories. Urges the need for a permanent practical geometrical gallery.
ALS 4pp RS:HS 10.426

11557. 1861–2–15 From *Daily News* London
Agrees with him on the subject of his note, but thinks it best not to print it as names would be demanded.
ALS 2pp RS:HS 6.459

11558. 1861–2–20 From *Daily News* London
If he has an opportunity he will use the information in his letter, but he will not disclose his identity.
ALS 2pp RS:HS 6.460

11559. 1861–2–25 From Richard Christopher Carrington Observatory, Redhill
James Challis has proposed retiring from Cambridge Observatory. Would like JH's views on the matter of his chances of succeeding Challis.
ALS 4pp RS:HS 5.215

11560. 1861–2–25 From Tiburce Parmentier Gand
Realizes that they have never been introduced. Nonetheless, requests JH's personal opinion on a matter of high importance to science. Asks JH to critique a memoir.
ALS 2pp RS:HS 14.138

11561. 1861–2–25 From Matthew Digby Wyatt 54 Guildford St.
Thanks for JH's memoir of Francis Baily. Will take down and store the hatchment put up for the late [Elizabeth] Bailey. To whom does it belong?
ALS 1p RS:HS 18.437

11562. 1861–2–28 To Richard Christopher Carrington Collingwood

Has no means of knowing what course the University of Cambridge will take in the event of the resignation of James Challis. Hopes that RC will not cease from his astronomical labors.
draft 3pp RS:HS 5.216 & 23.326

11563. 1861–3–1 From P. W. D. Baddeley Bonn
Points out error in his own book and enlarges on that point.
ALS 6pp RS:HS 5.4

11564. 1861–3–1 From Alexander K. Isbister College of Preceptors
Would like to reprint JH's paper 'On the Projection of the Sphere' in their *Educational Times*. Can he supply a reduced diagram?
ALS 4pp RS:HS 10.270

11565. 1861–3–9 From Richard Christopher Carrington Redhill
James Challis has informed him that J. C. Adams is to be the new Observer at Cambridge. Does not see how he himself can ever obtain the post. Seems likely that he may give up the profession.
ALS 4pp RS:HS 5.217

11566. 1861–3–11 From John Henry Nelson Hyde Park Gardens
Thanks for the case of wine. Saw Alexander Herschel this morning and is getting some information from him. Has been on board the *Warrior*.
ALS 3pp RS:HS 13.96

11567. 1861–3–14 From Bartholomew Price St. Giles St., Oxford
Thanks JH for a memoir on the projection of a sphere. Expresses the wish that map-making, so neglected, be taken up thoroughly.
ALS 1p RS:HS 14.56

11568. 1861–3–14 From George Gabriel Stokes Royal Society
About a request of JH for some palladium from the R.S.L.
ALS 3pp RS:HS 17.44

11569. [1861?]–3–15 From Thomas Collins Simon n.p.
Tells JH that he can be ready to see him at any time.
ALS 1p inc Harvard: Houghton AAH 67m–67 (97)

11570. 1861–3–17 From Augustus De Morgan 41 Chalcot Villas NW
Many thanks for his new projection of the sphere. J. L. Lagrange has two papers of 1779 on the subject. John Lee elected president at the R.A.S. He himself has resigned from the council.
ALS 3pp RS:HS 6.345

11571. 1861–3–17 From Alexander Hamilton-Gordon Argyll House
Forgot to reply to his letter as he locked it away. Regarding the supply of oxygen. Does not think it applicable in the case of a confinement.
ALS 3pp RS:HS 8.151

11572. 1861–3–18 From Robert Charles Winthrop Boston, Mass.
Encloses pamphlet on atmospheric electricity by Joseph Henry and two pages about 'Carpet experiments' in New York by Joseph Lovering. U.S. going through 'sad political trial.' Does astronomy offer examples of severed systems or broken planets being re-united?
ALS 4pp RS:HS 18.427

11573. 1861–3–19 From William Selwyn Vine College, Cambridge
Discusses 'autograph of the sun' he sent earlier. Discusses measuring relative temperature of sun. Has an unpublished letter of Galileo to Cardinal Barberini regarding sunspots.
ALS 4pp RS:HS 15.465

11574. 1861–3–22 From Richard Lewis Royal National Life Boat Institution
Appeals for JH's support of the Life Boat Institution.
ALS 1p Harvard: Houghton AAH 67m–67 (63)

11575. 1861–3–22 To George Gabriel Stokes Collingwood
Thanks for the palladium [see GS's 1861–3–22].
ALS 1p Cambridge University Library 7656/H698

11576. 1861–3–22 From George Gabriel Stokes Royal Society
Note to accompany sending of palladium [see GS's 1861–3–14].
ALS 1p RS:HS 17.45

11577. [1861 to 1865]–3–23 From Frederic W. Ragg Malin's Lee, Shifnal [Shropshire]
Claims to have demonstrated that light intensity increases as distance from light source *increases*. Applies this to inner planets, where sun should appear dimmer than on Earth. [JH annotation: 'Frivolous & Vexatious.']
ALS 3pp † TxU:H/M–0432; Reel 1093

11578. 1861–3–27 From George Biddell Airy Royal Observatory, Greenwich
Is sending a tract on 'Probable Errors.' Further observations on JH's article on telescopes.
ALS 2pp RS:HS 1.250

11579. 1861–3–27 From George Biddell Airy n.p.
Reminds JH that the algebraic formulation of the theory of the achromatic telescope eyepiece was formulated by GA.
ACCS 1p RGO 6.476.390

11580. 1861–3–27 To H[enry] Stevens Hawkhurst
Expresses thanks for sending *Smithsonian Contribution to Knowledge,* a report by A. D. Bache, and materials from the American Philosophical Society.
ALS 1p Huntington Library

11581. 1861–3–29 To George Biddell Airy Collingwood
Recognizes the validity of GA's claim to the eyepiece theory [see GA's 1861–3–27]; JH did not have space to deal adequately with the subject of telescopes.
ALS 4pp RGO 6.476.391

11582. 1861–3–29 or earlier From G. P. Gutehingt[?] Queen's Head Hotel, Hawkhurst
The Churchwardens of Hawkhurst are inviting JH to the Easter dinner.
LS 1p RS:HS B27.87j

11583. 1861–4–1 From Alexander Hamilton-Gordon Colchester
Cannot lay his hands on JH's paper at the moment. Family is going on well. Will send details of the secret code of Army signals.
ALS 1p RS:HS 8.152

11584. 1861–4–3 From George Boole Blackrock near Cork
Thanks for JH's paper 'On a New Projection of the Sphere.' Wonders why no one has discussed this important problem

before.
ALS 1p RS:HS 5.83

11585. 1861–4–6 From Thomas Watson 16 Henrietta St., Cavendish Square
Sends photographic copy of mural tablet TW placed in chapel of St. John's College to honor TW's late uncle, Thomas Catton.
ALS 3pp RS:HS 18.385

11586. 1861–4–8 From George Boole Blackrock near Cork
Questions arising from JH's 'On a New Projection of the Sphere.' Will send his solution if JH is interested.
ALS 4pp RS:HS 5.84

11587. 1861–4–8 From John Daniel Runkle Cambridge, Mass.
Sends *Mathematical Monthly* with JH's portrait and catalog of his writings. Smithsonian will send rain data JH wants. Discusses politics and hopes England will help break the cotton monopoly and thereby end slavery.
ALS 5pp RS:HS 14.511a

11588. 1861–4–10 From John Henry Nelson n.p.
Thanks for JH's kindness and the good things sent. Now sends a specimen of what photography can achieve on a large scale.
ALS 2pp RS:HS 13.97

11589. 1861–4–12 To Mary Somerville Collingwood
Thanks MS for sympathy letter. Discusses present situation in Italy. Discusses his children and their scattered locations, e.g., a son fighting in the Indigo affair. Discusses his articles for *Encyclopaedia Britannica.*
ALS 4pp RS:HS 16.365 (C: RS:HS 23.328)

11590. 1861–4–15 From Norman Pogson Madras
Replies to JH's inquiries about Messier 80, noting that a bright star, since vanished, obliterated the cluster. Describes poor but improving instrumentation at the Madras Observatory. Asks about JH's son Alexander.
ALS 4pp RS:HS 14.12

11591. 1861–4–22 From George Biddell Airy Royal Observatory, Greenwich
Asks for his comments on the results of his Magnetic Register 1848–1857.
ALS 4pp RS:HS 1.251 (C: RGO 6.694.319)

11592. 1861–4–22 From Charles G. B. Daubeny Botanic Garden, Oxford
Is sending one of his own pamphlets which may interest JH. Hopes he is enjoying better health now he has resigned from the Mint.
ALS 4pp RS:HS 6.21

11593. 1861–4–23 From George Biddell Airy Royal Observatory, Greenwich
Further remarks concerning his Magnetic Registers [see GA's 1861–4–22].
ALS 2pp RS:HS 1.253

11594. 1861–4–23 To George Biddell Airy Collingwood
Asks GA many questions about his magnetical explanation [see GA's 1861–4–22].
ALS 4pp RGO 6.694.321 (drafts RS:HS 1.252 & 23.329)

11595. 1861–4–24 From George Biddell Airy Royal Observatory, Greenwich
Regarding the evaporation of water as a source of electricity. Magnetic results in relation to the North Atlantic.
ALS 4pp RS:HS 1.254 (C: RGO 6.694.324)

11596. 1861–4–24 From Thomas Hamilton Belfast
Requesting JH's autograph.
ALS 1p RS:HS 10.151

11597. 1861–4–27 From William Selwyn Cambridge
Discusses Galileo's unpublished letter on sunspots. Outlines comparisons between tropical storms and sunspots as discussed at Cambridge Philosophical Society. Wishes to meet [Frederick] Howlett to discuss sunspots.
ALS 4pp RS:HS 15.466

11598. 1861–5–2 From Thomas Collins Simon n.p.
Asks for advice on matters in which JH is 'the highest authority in this or any other country.'
ALS 1p inc Harvard: Houghton AAH 67m–67 (107)

11599. 1861–5–3 From W. H. Barton Royal Mint
Regarding the Pyx accounts. Has seen JH's son. What does he think of the new coins.
ALS 3pp RS:HS 3.300

11600. 1861–5–6 To Robert FitzRoy Collingwood
On the effect of the earth passing through a comet's tail.
C 1p RS:HS 23.330

11601. 1861–5–8 From Guillaume de Gruyer Gand
Has had no reply to his letter of 29 March requesting his opinion on a memoir on the solar system. Would be grateful for a reply.
ALS 1p (French) RS:HS 9.149

11602. [1861]–5–10 From J. H. Baillie Upper Harley St.
Was grateful for his letter. News of her aunt's death.
ALS 4pp RS:HS 5.5

11603. 1861–5–10 From Robert FitzRoy Meteorological Dept., B.o.T.
Thanks for sending his *Physical Geography.* Wonders whether a rainbow could be reflected to the eye from water. Startled to read J. R. Hind's letter about the comet. Encloses a letter from a railway surveyor about an earthquake in Mendoza. This person has been responsible for the railway between Chile and Peru. Comments on the gradients and working of the line.
ALS 12pp RS:HS 7.253

11604. 1861–5–11 From Frederick C. Fitton 4 Sussex Gardens
His father has had an attack of a most severe kind and is not expected to recover.
ALS 2pp RS:HS 7.369

11605. 1861–5–11 From Peter le Neve Foster Society of Arts, Adelphi
Query regarding the musical pitch question. [Contains draft of Herschel's reply]
ALS 2pp RS:HS 7.325

11606. 1861–5–12 To Robert FitzRoy Collingwood
Comments on whether a rainbow could be reflected to the eye from water. Comets approach quite near to the earth on many occasions. Regarding the development of South America.
C 5pp RS:HS 7.254 & 23.331

11607. 1861–5–12 To *The London Review* Collingwood
Discusses the theory of rainbow formation, especially for rainbows seen as reflections from bodies of water.
L *London Review, 2* (May 18, 1861), 568–9

11608. 1861–5–13 From George Boole Blackrock near Cork
After he started his calculation on the spheroid he found that several persons had already worked on it. Is sending his own paper on this subject under separate cover.
ALS 1p RS:HS 5.85

11609. 1861–5–13 From James Joseph Sylvester Woolwich Common
Thanks for memoirs sent.
ALS 2pp RS:HS 17.161

11610. 1861–5–14 From Robert FitzRoy Meteorological Dept., B.o.T.
Is grateful for JH's views on the rainbow question. Intends to send his comments to the *London Review*, if he has no objections. What effects would the Comet have if it hits the Earth? Mr. Wheelwright appears to be in error on his earthquake views.
ALS 4pp RS:HS 7.255

11611. 1861–5–14 From John Lee Doctors' Commons
A Chinese translation of JH's book on astronomy was presented to the R.A.S. at the last meeting. Comments on this edition. Admiral W. H. Smyth is now in London. Received an agreeable letter from Alexander Herschel.
ALS 3pp RS:HS 11.179

11612. [1861 or later]–5–15 From Joseph Bullar Basset Wood near Southampton
His letter arrived just before his father's death.
ALS 4pp RS:HS 4.341

11613. 1861–5–16 From James Nasmyth Penshurst
Had a visit from James Maclaren, who now conveys JN's drawings of the spots on the sun. Comments on the effect of the 'willow leaves.'
ALS 4pp RS:HS 13.77

11614. 1861–5–17 From Robert FitzRoy Meteorological Dept., B.o.T.
Has checked over JH's letter on rainbows and will have it signed Herschel. Encloses a scrap about rain. Mr. Wheelwright's note is curious. Will send a copy of the *London Review*.
ALS 4pp RS:HS 7.256

11615. [1861]–5–23 From Charles Robert Darwin Down
Is grateful for the present of JH's *Physical Geography*. Is pleased also with JH's note on his own book on species. Comments on some of the theories expressed in this work.
ALS 4pp RS:HS 6.17 Darwin, *Correspondence, 9,* 135–6

11616. 1861–5–25 From Joseph Baxendell Stocks St., Manchester
Thanks for his book on meteorology. On the reliability of hygrometers.
ALS 3pp RS:HS 3.307

11617. 1861–5–25 From John R. Hind 22 Grove Rd., St. John's Wood
Has been studying the star discovered by Anthelm [Voituret Anthelme] in 1670. The star he discovered in *Ophiuchus* has not disappeared as stated by some astronomers. Bishop is seriously ill.
ALS 3pp RS:HS 9.367

11618. [1861?]–6–8 From Ann Smyth 27 Victoria St.
Regrets missing JH's visit. W. H. Smyth having trouble walking.
ALS 3pp RS:HS 16.222

11619. 1861–6–10 To T. P. Anderson n.p.
Not aware of any changes in trust fund status of T. H. Hollier since Jan. 1860.
ACCS 1p TxU:H/L-0031; Reel 1054

11620. 1861–6–10 From John R. Hind 22 Grove Rd., St. John's Wood
Has received several letters from German astronomers on U. J. J. Leverrier's suggestions for the re-naming of the asteroids. Comments on this. Is preparing a report for the R.A.S. and would like to hear JH's views on the subject.
ALS 6pp RS:HS 9.368

11621. 1861–6–10 From Charles Pritchard [Clapham Grammar School]
Observes and measures sunspots. Takes spectrum of solar photosphere; detects sodium and magnesium absorption lines. Pleased with Alexander Herschel's work on the solar spectrum; wishes he would begin work on stellar spectra.
ALS 4pp inc RS:HS 14.82

11622. 1861–6–10 From Marc Seguin Montbard
Important office in French Institute has been passed on to MS. Compliments JH and many other members of R.S.L., and suggests exchange of ideas, asking JH for delivery of latest edition of JH's astronomy text, discusses idea of heat as motion, and other concepts. Expresses high scientific aspirations.
ALS 4pp (French) RS:HS 17.184

11623. 1861–6–12 From Andres Poey London
Sends some memoirs to JH. Is keeping the most recent ones. Has always cited JH in his works. Requests JH's critique of AP's work on photo-electric images. Leaves London in an hour and regrets not having met JH.
ALS 3pp RS:HS 14.172

11624. 1861–6–15 From Alexis Caswell London
Will be leaving for America shortly so regrets he will be unable to visit him. Has happy memories of his interview at Greenwich.
ALS 1p RS:HS 5.327

11625. 1861–6–15 From John R. Hind 22 Grove Rd., St. John's Wood
Is very pleased he agrees with the present system of naming the asteroids; so does G. B. Airy. George Bishop died yesterday, and the Observatory is now in the hands of Bishop's son.
ALS 2pp RS:HS 9.369

11626. 1861–6–21 To Sir William J. Hooker Collingwood
Found some old packets of seeds of *Protea* yesterday and wonders if he would like them to be sown to see if they would germinate. Comments on some of the plants in his garden. Tried some Peruvian potatoes.
draft 4pp RS:HS 9.463 & 23.332

11627. 1861–6–23 To Alexander Stewart Herschel Collingwood
AH's college exams. Prefers that AH come home, rather than JH coming to Clapham and using Charles Pritchard's astronomy instruments. Describes 'revolving reflector' that JH designed for studying star spectra. Margaret and Duncan Stewart are coming to Clapham tomorrow for races. Many visitors at Collingwood.
ALS 4pp TxU:H/L-0523; Reel 1053

11628. 1861–6–24 From Alexander Bridport Becher Admiralty

Regrets delay in paying JH's expenses for visitation of Royal Observatory.
ALS 1p TxU:H/M–0075; Reel 1087

11629. 1861–6–24 From John R. Hind 22 Grove Rd., St. John's Wood
Sends his calculation of the number of days from the commencement of the Julian period to 1 Jan. 1801; the result agrees with that of JH.
ALS 2pp RS:HS 9.370

11630. 1861–6–24 From Sir Joseph D. Hooker Kew
Father is unable to write so he has to thank him for his letter and the offer of Leucadendron seeds. Will try some experiments with them. Comments on some of C. R. Darwin's theories. Has he tried the Peruvian potato experiment? Does not remember a Camellia fruiting.
ALS 6pp RS:HS 10.206

11631. 1861–6–24 To Walter White Collingwood
Asks that various papers by JH be sent along with the *R.S.P.T.* to four foreign locations.
ALS 2pp Columbia Univ. Library

11632. 1861–6–26 From Warren de La Rue 110 Bunhill Row
Gives diameter of the sun's image. Used an ordinary eye-piece. There must be some mistake about the penumbra. It is possible now to take daily records of the surface of the sun.
ALS 4pp RS:HS 6.148

11633. 1861–6–30 From Charles Babbage Dorset St.
Is sending a book of travels by Mr. Tyler, which gives an account of the Mexican stone and flint implements. Alexander Herschel called. The latter's experiments with hydrogen are useful.
ALS 3pp RS:HS 2.331

11634. 1861–7–1 From Augustus De Morgan 41 Chalcot Villas NW
Has not heard from him for a long time. Did he receive his red and black logarithms? JH's son Alexander has sent him a mathematical problem. George Bishop has died and the observatory and instruments are to be sold. Sends two riddles. Sees that JH has proved that the sun is liquid fire.
ALS 4pp RS:HS 6.346

11635. [1861–7 or later] To Marc Seguin n.p.
Thanks for MS's works, communicated through [Picot] Tremblay. Worries that JH has insufficient time and lags behind in the physical sciences. Comments on ideas put forward in 1824 by JH and discussed by MS. Has found MS's name connected with suspension bridge over Rhone River.
ALS 4pp RS:HS 17.185

11636. 1861–7–6 To *The London Review* Collingwood
Reports JH's observations of a comet [Comet II Tebbutt ?] first seen from England ca. 29 June 1861. Remarks that it 'far exceeded in brightness any comet I have before observed.'
L *London Review, 3* (July 13, 1861), 46–7

11637. 1861–7–10 From James Challis Cambridge Observatory
Encloses sketch of the comet seen on night of 30 June. Comments on various aspects of the comet.
ALS 3pp RS:HS 5.237

11638. 1861–7–12 To Elizabeth Colling Collingwood
Sends a work on physical geography in return for a receipt of some lines of poetry.
C 2pp RS:HS 23.333

11639. 1861–7–14 To Augustus De Morgan Collingwood
Raised from his lethargy by the appearance of the comet; comments on the nature of the sun; can AD recommend a book on surveying for JH's son John.
C 2pp RS:HS 23.334

11640. 1861–7–15 From Augustus De Morgan 41 Chalcot Villas NW
Sends a sonnet on the comet. Regarding works on spheroidal triangles. Supposes he may use the Feldhausen anecdote. Regarding the law of facility.
ALS 4pp RS:HS 6.347

11641. 1861–7–17 From W. H. Barton Royal Mint
Further regarding the Pyx accounts.
ALS 6pp RS:HS 3.301

11642. 1861–7–21 From William Stanley Jevons University College London
Sending two papers he published some years ago, which agree with JH's theories regarding clouds. JH does not appear to know of their existence.
ALS 2pp RS:HS 10.319 Black (ed.), *Papers and Correspondence of...Jevons*, 2:432

11643. 1861–7–21 To John Tyndall Collingwood
Thanks for and comments on paper and lectures on 'diathermancy &c' of gases, scientific teaching, and solar chemistry. Regrets that JT mentioned all names connected with light research except JH's. Remarks on absorption of light, molecular vibrations.
C 3pp RS:HS 23.335 (ACCS: RS:HS 17.388a; C: RI 505–6)

11644. 1861–7–23 To S[amuel] P[ickworth] Woodward Collingwood
Thanks SW for a copy of the reviews of JH's *Meteorology* and *Physical Geography.* Comments on related matters such as importing English birds into Australia, observations made by Alexander von Humboldt, and catching specimens behind a sailing ship.
ALS 4pp diag BL Aaa 42580.137

11645. 1861–7–24 To Michael Faraday Collingwood
More on lacunae in JH's set of MF's writings [see JH's 1838–10–3].
C 1p RS:HS 23.336

11646. 1861–7–24 From John Tyndall [Royal Institution]
Apologizes for not having seen JH's paper on spectra of colored flames and thus having overlooked JH's prior claims [see JH's 1861–7–21]. Will try to make amends.
ALS 4pp RS:HS 17.388 (C: RI 507)

11647. 1861–7–25 From Adolphe Quetelet Brussels
Sends list of works received on an official form. Followed by letter. Suffers deeply from death of his wife, daughter, and her son. Had hoped to see JH in London, but JH was absent.
ALS 3pp (French) RS:HS 14.253

11648. 1861–7–26 From William Lassell Sandfield Park near Liverpool
Is on the point of departure for Malta and would welcome letters of introduction from JH.
ALS 3pp RS:HS 11.159

11649. 1861–7–26 From George Smith 45 Pall Mall
About repayment of a loan made from JH.
ALS 1p RS:HS 16.184

11650. 1861–7–29 From Michael Faraday Hampton Court
Found the paper JH required and has sent off the quarto version. Wishes he had something else to send as well.
ALS 2pp RS:HS 7.198

11651. 1861–7–29 From Peter le Neve Foster Society of Arts, Adelphi
His son Clement will be working in the neighborhood of Hawkhurst for the Geological Survey and would be pleased if JH would give any local advice as required.
ALS 2pp RS:HS 7.326

11652. 1861–7–29 From Edward Sabine Llandovery
Sends letter regarding meteorology from the Smithsonian to JH. Pleased with magnetic work of [A. D.] Bache at Girard College.
ALS 2pp RS:HS 15.256

11653. 1861–7–30 To William Stanley Jevons Collingwood
Thanks WJ for his papers on clouds [see WJ's 1861–7–21]. JH had not seen these, but praises their contents.
ALS 3pp John Rylands University Library of Manchester (C: RS:HS 23.337) Black (ed.), *Papers and Correspondence of...Jevons*, 2:433–4

11654. 1861–8–3 To George Gabriel Stokes Collingwood
Asks R.S.L. Council to request Colonial Office to provide official recommendation to authorities in Malta for William Lassell, going there to observe.
ALS 4pp Cambridge University Library 7656/H699 (C: Univ.Coll.London MS.Add.227)

11655. 1861–8–6 From George Gabriel Stokes Lensfield Cottage, Cambridge
Responds to a letter from JH, which GS has forwarded to Edward Sabine.
ALS 3pp RS:HS 17.46

11656. 1861–8–7 From Edward Sabine Llandovery
Certain the Colonial Secretary [Duke of Newcastle] would write to governor of Malta on behalf of [William] Lassell.
ALS 4pp RS:HS 15.257

11657. 1861–8–13 From John Henry Dallmeyer 19 Bloomsbury St.
Thanks for his note on the performance of the telescope. Gives suggested reasons for faults. Sends a print of a photograph of a map recently taken by one of his achromatic lenses.
ALS 3pp RS:HS 6.1

11658. 1861–8–14 From Sir Benjamin Collins Brodie [the elder] Bonchurch, I.o.W.
Has had no reply from H. P. F. P. Clinton (5th Duke of Newcastle); will write to H. J. Temple (3rd Viscount Palmerston) later if no answer is received.
ALS 2pp RS:HS 4.284

11659. 1861–8–14 To John William Lubbock Collingwood
Explains JH's objections to JL's shadow-extinction hypothesis of meteors. Believes that several sorts of meteors—magnetoelectric, stony, vaporous, etc.—exist.
ACC 4pp TxU:H/L-0234; Reel 1054

11660. 1861–8–15 To John William Lubbock Collingwood
Has got his paper on shooting stars. Comments on JL's theories of meteors. Pleased to hear a favorable account of Montague Lubbock.
C 4pp RS:HS 11.394, 23.338, & 25.15.18 *Letter from Sir. J. F. W. Herschel, Bart., to Sir. J. W. Lubbock, Bart., on Shooting Stars* (London: Goodhall & Dinsdale, 1861)

11661. 1861–8–15 From Roderick Impey Murchison Tunbridge Wells
Sends a copy of the Address of the Geographical Society, and also a few copies of the sketch of G. J. J. Hamilton Gordon (5th Earl of Aberdeen). Taken a house at Tunbridge Wells for the benefit of their health. Preparing for the Manchester meeting.
ALS 3pp RS:HS 12.425

11662. 1861–8–16 From G. C. B. n.p.
Reports band of cloud resembling zodiacal light near sunset on 8 Aug.
ALS 1p inc TxU:H/L-0753.2; Reel 1089

11663. 1861–8–17 From Sir Benjamin Collins Brodie [the elder] Bonchurch, I.o.W.
Has written to E. A. S. Seymour (12th Duke of Somerset) and will send William Lassell's letter of introduction to Admiral Edward Codrington.
ALS 2pp RS:HS 4.285

11664. 1861–8–18 From John William Lubbock High Elms, Farnborough
Read his letter with great pleasure. Would like to publish it with JH's permission. Incapable of mental exertion himself. Son Neville just married and set up in business. The Lubbocks are going abroad for 3 or 4 weeks. Looks forward to visiting JH in September.
ALS 4pp RS:HS 11.395

11665. 1861–8–19 From Thomas Maclear Royal Observatory, Cape of Good Hope
Describes last days of Mary Maclear. Burial on Observatory grounds.
ALS 4pp Cape Archives/Maclear Papers File 100

11666. 1861–8–20 From Sir Benjamin Collins Brodie [the elder] Bonchurch, I.o.W.
Has just received the enclosed.
ALS 1p RS:HS 4.286

11667. 1861–8–21 From Sir Benjamin Collins Brodie [the elder] Bonchurch, I.o.W.
Has received the enclosed from E. A. S. Seymour (12th Duke of Somerset). Has sent the memorandum to William Lassell. Can also send the Duke's letter if he thinks it worthwhile.
ALS 2pp RS:HS 4.287

11668. 1861–8–21 From John William Lubbock Mansion House St.
Looking forward to their visit to the JH's on the ninth. His own children are becoming versed in astronomy. Has sent JH's letter to the printer and will send him a proof later.
ALS 2pp RS:HS 11.396

11669. 1861–8–22 From Sir Benjamin Collins Brodie [the elder] Bonchurch, I.o.W.
Sends the official answer from the Admiralty about the William Lassell affair. Suggests all the letters about this subject should be placed in the archives of the R.S.L. for future reference.
ALS 2pp RS:HS 4.288

11670. 1861–8–22 From Sir Charles Lyell 53 Harley St.
Thanks for the valuable present, which he is reading as fast as possible. Comments on the work in the light of his own researches. Is glad he put in a note about C. R. Darwin.
ALS 3pp RS:HS 11.425

11671. 1861–8–23 From William Lassell Sandfield Park near Liverpool
Thanks JH for his efforts in having Sir Benjamin Brodie provide letters of introduction for WL's time in Malta.
ALS 2pp RS:HS 11.160

11672. 1861–8–24 From Warren de La Rue 110 Bunhill Row
Sending him his first photograph of the sun's spot. Has caused a lot of preparation. Comments on various points.
ALS 3pp RS:HS 6.147

11673. 1861–8–26 From Samuel Highley 70 Dean St.
Believes that David Brewster's instrument for examining the lines of the spectrum was based on a formula of JH. Would be pleased if he could supply him with the details and he will then request J. H. Dallmeyer to construct one.
ALS 3pp RS:HS 9.313

11674. 1861–8–26 From William Pollard 11 High St., Hastings
'Lot and parcel' arrived on 20 Aug.
ALS 1p RAS:JH Archive 12/1.3.9 verso; Reel 10

11675. 1861–8–27 From John R. Hind 22 Grove Rd., St. John's Wood
[James?] Ferguson wants *Titania* changed to *Echo,* which he has now inserted in the forthcoming *Nautical Almanac*. H. M. S. Goldschmidt has recovered *Pseudo-Daphne*. U. J. J. Leverrier does not wish to name No. 59.
ALS 2pp RS:HS 9.371

11676. 1861–8–29 From Eustace Button Clevedon
Has he published any further discussions on trade winds since those in his articles 'Physical Geography' and 'Meteorology'?
ALS 2pp RS:HS 5.130

11677. 1861–8–29 From John William Lubbock High Elms, Farnborough
Sends a proof, which please return with the necessary alterations and additions. JH's daughter is quite well.
ALS 1p RS:HS 11.397

11678. 1861–8–31 From Sir Benjamin Collins Brodie [the elder] Bonchurch, I.o.W.
Please send William Lassell the original letters to take with him when he goes to Malta, as he has only the copies at present.
ALS 3pp RS:HS 4.289

11679. 1861–8–31 From William Lassell Sandfield Park near Liverpool
Details about WL's preparation for departure for Malta.
ALS 4pp RS:HS 11.162

11680. [1861–8–31 or later]. To Edward Sabine n.p.
Advises on the construction of the Melbourne reflecting telescope. Thinks [W. P.] Wilson should come to watch construction.
draft 2pp RS:HS 15.334

11681. 1861–8–31 From Edward Sabine 13 Ashley Place
Encloses letter from [W. P.] Wilson of Melbourne and ES's response for JH's perusal. Discusses [Warren] de La Rue's work on solar photography.
ALS 2pp RS:HS 15.258

11682. 1861–9–1 To W[illiam] P[arkinson] Wilson Collingwood
Outlines advantages of erecting great reflector in Melbourne [WW's colony] for observation of southern nebulae.
ALS 3pp RS Sa.670

11683. 1861–9–4 From Eustace Button Clevedon
Thanks for his information. Has been trying to obtain a copy of JH's *Outlines Astr.* Further queries regarding trade-winds.
ALS 4pp RS:HS 5.131

11684. 1861–9–4 From Samuel Highley 70 Dean St.
Is grateful for his note. Will take an early opportunity of placing JH's communication before his readers. Curious the oversight should not have been noticed before. Has no more details of David Brewster's instrument.
ALS 2pp RS:HS 9.314

11685. 1861–9–5 From *The Educational Times* Fleet St.
Asks JH to contribute to *The Educational Times.*
AL 1p inc Harvard: Houghton

11686. 1861–9–6 From Samuel Highley 70 Dean St.
On Monday he should receive the proof of his communication. Comments on this communication and how it will be presented.
ALS 3pp RS:HS 9.315

11687. 1861–9–11 From John William Lubbock High Elms, Farnborough
Has been looking at JH's treatise on meteorology, especially the part relating to the barometrical formula. It seems that JH missed JL's paper on refraction published in *Transactions of the Astronomical Society,* and is therefore not aware of the true nature of P. S. Laplace's atmosphere; comments on this.
ALS 6pp RS:HS 11.398

11688. 1861–9–14 To John William Lubbock Collingwood
About the barometric formula for the measurement of heights.
draft 4pp RS:HS 11.400 (C: RS:HS 23.339)

11689. 1861–9–14 To John William Lubbock Collingwood
Does not possess the *Transactions of the Astronomical Society,* which contains JL's paper. Had to limit the scope of his article so could not deal with the point mentioned in JL's letter. Comments on P. S. Laplace's formula.
C 5pp RS:HS 11.399–400 & 23.339

11690. 1861–9–17 From Alfred P. Wire Training College, Battersea
Compares JH's 'stereoscopic phenakistiscope' (in 'Instantaneous Photography,' 1860) with AW's invention called 'Thaumatrope.' Suggests how to make moving pictures. 'Government Certificated [School] Masters' in England, Ireland, and Scotland should be enlisted to make meteorology observations.
ALS 3pp RS:HS 18.428

11691. 1861–9–18 From John William Lubbock High Elms, Farnborough
Of course JH is right; regrets he gave him so much trouble. Hopes his own construction of P. S. Laplace's atmosphere is correct. Hopes JH's daughter had a prosperous journey.
ALS 2pp RS:HS 11.401

11692. 1861–9–23 From George Biddell Airy Royal Observatory, Greenwich
Encloses printed copies of letters received from Johann Lamont regarding apparatus for examining Galvanic currents in the earth; would like his comments on them. Gave lecture in Manchester on the eclipse.
ALS 4pp RS:HS 1.255 (C: RGO 6.476.55)

11693. 1861–9–26 To Alfred P. Wire Collingwood
AW's 'Thaumatrope' is ingenious, but applies only to periodic movement. JH aims to reproduce *non*-periodic motion. Commends AW's suggestion to employ government schoolmasters as meteorology observers.
C 1p RS:HS 23.340

11694. 1861–9–28 To George Biddell Airy Collingwood
Encourages GA to develop some experiments to study the electrical phenomenon [see GA's 1861–9–23] more carefully.
ALS 4pp RGO 6.476.57

11695. 1861–9–30 From Alfred P. Wire Training College, Battersea
JH's reply was first AW heard that 'Stereoscopic Thaumatrope' had already been made. Lists two requirements for workable system of enlisting government schoolmasters for meteorological observations. AW and fellow students are ready to assist.
ALS 3pp RS:HS 18.429

11696. 1861–10–1 To Thomas Maclear Collingwood
Expresses condolences on death of Mary Maclear.
ALS 4pp Cape Archives/Maclear Papers File 100

11697. 1861–10–4 From Jean Baptiste Biot Paris
Has pleasure in informing JH that a parcel of publications of the Academy is being forwarded to him.
ALS 1p (French) RS:HS 4.96

11698. 1861–10–5 From Robert FitzRoy Meteorological Dept., B.o.T.
Sending some reports of the Meteorological Dept. Would also like to send a wind-glass and a 'watch=aneroid' but has refrained as he knows JH is encumbered with many such things. Encloses a note from Charles Green (Rush's air pilot). Sends a few more notes for his rain collection.
ALS 4pp RS:HS 7.257

11699. 1861–10–6 From Augustus De Morgan 41 Chalcot Villas NW
Has sent the packet of letters written to Francis Baily to Greenwich, but kept back one or two including the one on the R.S.L. Secretaryship. Is told that James South is now suffering from deafness. Has sent a paper on errors of observation to Cambridge. Comments on this.
ALS 4pp RS:HS 6.348

11700. 1861–10–7 To Robert FitzRoy Collingwood
Thanks for RF's daily barometric reports in the *Times;* comments on RF's investigation of the 'Royal Charter' storm.
C 2pp RS:HS 23.341

11701. 1861–10–14 From Augustus De Morgan 41 Chalcot Villas NW
Regarding the unsuitableness of the weather for astronomers. Sends an answer to one of JH's equations. Regarding the quarrels of Sir James South. Sends some nursery rhymes that seem appropriate.
AL 8pp RS:HS 6.349

11702. 1861–10–14 From John R. Hind Nautical Almanac Office
Will incorporate his suggestion respecting the introduction of the current Julian date in the *Nautical Almanac*. The name 'Baucis' has been proposed for No. 59. Has sent U. J. J. Leverrier his own views on nomenclature. Bishop's Observatory may be moved to Twickenham.
ALS 3pp RS:HS 9.372

11703. 1861–10–17 To Robert Main Collingwood
Thanks for the receipt of the Radcliffe catalogue of stars, but wishes observations had been equatorial rather than circumpolar.
C 1p RS:HS 23.342

11704. 1861–10–20 From A[lphonse] D'Orsan Birkenhead
Sending a copy of his work 'Our satellite,' and would be grateful for any comments.
ALS 3pp RS:HS 6.499

11705. 1861–10–21 From Robert Main Radcliffe Observatory
Is obliged for JH's letter acknowledging the receipt of the *Radcliffe Catalogue*; regrets the delay and will inquire into the cause. Has purchased R. C. Carrington's transit circle for the Radcliffe and would welcome JH's advice as to its employment.
ALS 4pp RS:HS 12.278

11706. 1861–10–22 From Joseph Bullar Southampton
Regarding electro-magnetic forces. His father's health is good.
ALS 4pp RS:HS 4.338

11707. 1861–10–24 To William Henry Sykes Collingwood
Stresses importance for Balloon Committee of obtaining corresponding temperatures and pressures; describes two highly sensitive thermometers that might be built for this purpose.
ALS 4pp RS:HS17.155 (C: RS:HS 23.343)

11708. 1861–10–25 To Robert Main Collingwood
Suggests an observational program for RM's newly purchased transit circle [see RM's 1861–10–21].
C 1p RS:HS 23.344

11709. 1861–10–28 From William Henry Sykes 47 Albion St.
Updates JH on activities of Balloon Committee, and thanks him for suggestion regarding temperature measurement. Mentions improving health of Mrs. Sykes.
ALS 4pp RS:HS 17.156

11710. 1861–11–1 From John R. Hind 22 Grove Rd., St. John's Wood
Will send the *Nautical Almanac* to Smith, Elder and Co. Is pleased they agree on the naming of the asteroids. Gives readings for the approaching transit of Mercury.
ALS 3pp RS:HS 9.373

11711. 1861–11–2 To John R. Hind Collingwood
About the naming of asteroids.
C 2pp RS:HS 23.352

11712. 1861–11–5 From Isaac Todhunter St. John's, Cambridge
Thanks JH for compliments on his book [*Theory of Equations*]. Requests JH's signature for his application to the

R.S.L.
ALS 2pp RS:HS 17.352

11713. 1861–11–7 To Warren de La Rue Collingwood
Thanks for pictures of Mars; speculates on the atmospheres of some planets.
C 1p RS:HS 23.346

11714. 1861–11–10 From Warren de La Rue Observatory, Cranford
Is pleased that JH likes his plates of Mars. Has a great collection of drawings of Mars but only a few of Venus. Hopes by the Spring to have the photoheliograph working. Has just completed his volume on the eclipse for 1860.
ALS 3pp RS:HS 6.149

11715. 1861–11–10 To John Tyndall Collingwood
Reply to JT's 'Remarks on Radiation and Absorption.' Calls non-absorption of air and vapor key to understanding meteorological phenomena. Remarks on 'radiant heat' from moon.
C 2pp RS:HS 23.347 (C: RI 508)

11716. 1861–11–11 From Gabriel Auguste Daubrée Paris
Asking JH to accept his book on the rock crystals.
ALS 3pp (French) RS:HS 6.29

11717. 1861–11–16 From William Whewell to Margaret Herschel Trinity Lodge
Expresses happiness that JH's health has improved and comments on various family matters.
ALS 4pp RS:HS 18.218

11718. 1861–11–20 From William Selwyn Cambridge
Sends series of solar heliographs. If further suggestions for improvement of these are noted, please send them. Is preparing to photograph Dec. 31 eclipse.
ALS 4pp RS:HS 15.567

11719. 1861–11–21 From John Tyndall Royal Institution
Gives account of observations on lunar radiation, including detailed description of experiments.
ALS 6pp RS:HS 17.389 (C: RI 509–11)

11720. 1861–11–22 To William Selwyn Collingwood
Has no suggestions to improve photographs. Wants to determine whether the degradation of light from center to edge of sun is 'real.' Does not think sun's size affects focus.
C 1p RS:HS 23.349

11721. 1861–11–22 To John Tyndall n.p.
Does not believe surface of moon can radiate cold; perhaps new moon can. Explains this geometrically.
C 3pp diag RS:HS 23.348 (C: RI 512–13)

11722. 1861–11–23 From John Tyndall Royal Institution
Clarifies that JT does not believe moon radiates cold itself, but effects are 'as if' it did. Relates this to atmosphere.
ALS 4pp RS:HS 17.390a (C: 514)

11723. 1861–11–24 To John Tyndall Collingwood
Finds JT's work on thermotics 'highly interesting,' although criticizes measurements of 'tenuity' of vapor, and explanation of scattering of sunlight and formation of corona around moon during eclipse.
C 2pp RS:HS 23.350 (C: RI 515)

11724. 1861–11–26 From William Lassell 9 Piazza Sliema, Malta
About WL's activity in setting up his observatory.
ALS 4pp RS:HS 11.163

11725. 1861–11–28 To Gabriel Auguste Daubrée Collingwood
Thanks GD for memoir on the subject of geological transformations.
C 1p RS:HS 23.351

11726. [1861 or later]–11–29 From Joseph Bullar Basset Wood near Southampton
Is sending him an engraving of his late father. Sees that JH has finished his work he told him about.
ALS 3pp RS:HS 4.342

11727. 1861–11–29 From John R. Hind 22 Grove Rd., St. John's Wood
Has received a letter from U. J. J. Leverrier requesting a name for No. 59; sees difficulties; so would JH suggest a name. Has sent the *Nautical Almanac* supplement and a memoir by Yvon Villarceau to Smith, Elder and Co.
ALS 3pp RS:HS 9.374

11728. 1861–12–2 To John R. Hind Collingwood
Suggests the name 'Clio' for asteroid No. 59. Gives a list of proposed names for asteroids.
draft 4pp RS:HS 9.375

11729. 1861–12–2 From Henri Victor Regnault Sèvres
Forwards letter to HR from [L.-F.-C.?] Breguet, who desires to construct a metallic thermometer for JH. HR requests quick reply from JH.
ALS 2pp (French) RS:HS 14.317

11730. 1861–12–4 From John R. Hind 22 Grove Rd., St. John's Wood
Is grateful for the ample list of suggested names for asteroids that JH has sent. Outlines the difficulties over the name 'Clio'. Would prefer 'Olympia.'. Believes 'Pseudo-Daphna' will be called 'Galatea'.
ALS 4pp RS:HS 9.376

11731. 1861–12–5 From Edward Sabine Burlington House
Requests JH sign certificate for [W. F.] Hook. Asks JH to forward it to William Whewell for his signature.
ALS 2pp RS:HS 15.259

11732. 1861–12–5 To William Whewell Collingwood
JH's 'scientific activity' has long been at zero, but JH has of late been preparing a lecture on the sun and translating Homer's *Iliad.*
ALS 4pp TC, Camb. Add. Ms.a.207[101] (C: RS:HS 23.353)

11733. 1861–12–6 To Edward Sabine Collingwood
Has signed certificate for [W. F.] Hook. [See ES 1861–12–5] Congratulates ES [on presidency of R.S.L.]. JH on B.A.A.S. balloon committee. Proposes design for metallic thermometer.
C 2pp RS:HS 23.354

11734. 1861–12–7 To Joseph Barclay Pentland Collingwood
Recalls mutual friends, Georges Cuvier and Mary Somerville. Gratitude for JP's kindness in Rome to JH's daughter, Margaret Louisa, now deceased. Asks JP to assist newlyweds David and Mary Power, who are caught in Rome by David's failing health and need advice on medical treatment. [Annotation by Constance Anne (Herschel) Lubbock identifies couple as David Power and bride Mary Lipscompe. David died soon after, and his widow married JH's son John.]
ACC 4pp †(by C. A. Lubbock) TxU:H/L-0299; Reel 1054

11735. 1861–12–9 From John Lee Doctors' Commons
Would like JH's approval to submit the enclosed certificate on behalf of JH's son. No. 29 of the *Record of Recreative Science* contains a good article on Light by E. J. Lowe, who mentions JH's name.
ALS 2pp RS:HS 11.180

11736. [1861]–12–9 From Edward Sabine 13 Ashley Place
A metallic thermometer will be constructed at Kew Observatory to test JH's proposal. Experiments on Index of Friction of gases to be resumed at Kew.
ALS 2pp RS:HS 15.310

11737. 1861–12–10 To John Lee n.p.
Grateful for JL's support of JH's son [?] for R.A.S. membership, but JH advised [son] not to apply until [son] had made some significant contribution to astronomy worthy of membership. [JH annotation: How different from today, when election is so easy.]
ALS 4pp † TxU:H/L-0223; Reel 1054

11738. 1861–12–10 To [husband of Lady Mildred ?] Collingwood
Due to poor health of both JH and his wife, Margaret, JH writes to decline an invitation to dinner.
ALS 2pp DMC.695A.885.14 (C: DMC.695A.885.16)

11739. 1861–12–11 To Henri Victor Regnault Collingwood
Describes objections to proposed metallic thermometer. Describes another simpler design for a metallic thermometer. Will ask [William] Sykes to consider HR's design.
C 4pp RS:HS 14.318 (C: RS:HS 23.356)

11740. 1861–12–11 To Edward Sabine Collingwood
Pleased ES and [Balfour?] Stewart like plan for metallic thermometer. Suggests way to obviate effect of pendulous movement of the suspended weight.
C 1p RS:HS 23.355

11741. 1861–12–12 To William Whewell Collingwood
Asks WW's opinion of JH's hexameter translation of the opening section of Homer's *Iliad.* Comments on the value of hexameter verse.
ALS 4pp TC, Camb. Add. Ms.a.207[100] (C: RS:HS 23.357)

11742. 1861–12–17 From Thomas F. Jamieson Ellon, Aberdeenshire
Thanks for his note and explanations. Comments on the variation in the height of the Caspian Sea at various periods.
ALS 1p RS:HS 10.435

11743. 1861–12–21 From Thomas Maclear Royal Observatory, Cape of Good Hope
Is grateful for JH's expression of sympathy over the death of his wife. One of his daughters has left on a ship for England. Sends a work by S. Salomon describing the visit of Prince Alfred to the Cape.
ALS 3pp RS:HS 12.166

11744. 1861–12–21 To Bartholomew Price Collingwood
Thanks for sending vol. 4 of BP's *Infinitesimal Calculus* and for earlier sending vols. 1 and 2. Praises parts of vol. 4. Will another volume be forthcoming?
C 1p RS:HS 23.358

11745. 1861–12–22 From Reginald Dyke Marshall Cookridge Hall, Leeds
Has not been able to obtain a return of the rainfall for Bassenthwaite or Seathwaite, but sends one for Keswick, which may be of interest. Was at Keswick on 25 Nov. and has never seen heavier rain than fell on that day. Children are well. Sad loss [death of Prince Albert] to the Queen and nation.
ALS 4pp RS:HS 12.292

11746. [1861–12–23] From Matthew Collins Dublin
Hopes the almanac he sent him has arrived safely. Intends it for the use of country schoolteachers. Has found a new and easy proof for one of W. R. Hamilton's theorems.
ALS 2pp RS:HS 5.263

11747. 1861–12–26 From Roderick Impey Murchison 16 Belgrave Square
Thanks for the copy of JH's *Physical Geography*. Comments on some of the sections of this work, especially the Weald. Sends his own paper on the 'Permian.'
ALS 4pp RS:HS 12.426

11748. [1861?]–12–27 From Ann Smyth St. John's Lodge
W. H. Smyth publishes essay. Sends family photographs.
ALS 4pp RS:HS 16.223

11749. 1861–12–28 From Angelo Secchi Rome
Sends bulletins of the Observatory, which contain both meteorological and astronomical information. Asks JH if experiments are worth pursuing.
ALS 2pp RS:HS 15.415

11750. [1862?] From ? Abinger Royal Orthopaedic Hospital, 315 Oxford St.
Requests donation to hospital.
LS 1p TxU:H/M–0917.1; Reel 1083

11751. [1862–1 to 1862–8] From George Biddell Airy Royal Observatory, Greenwich
Not surprised that actual costs [of publishing JH's catalogue of nebulae] departed from estimates. Reports vote by [R.S.L.] committee. Does not believe publication will be expensive.
ALS 2pp RAS:JH Archive 12/1.1.13; Reel 10

11752. [1862 1] From Matthew Collins n.p.
Is glad his article on central forces pleased him. Sends one of his latest solutions. Would like a few lines by way of a testimonial. Has been unfortunate when applying for positions as he has lacked support.
ALS 4pp RS:HS 5.265

11753. 1862 From Henry C. Criswick Greenwich
Is grateful that he has been allowed to dedicate his book to JH. Will incorporate the suggestions.
ALS 2pp RS:HS 5.376

11754. 1862 To William Robert Grove n.p.
Thanks for WG's *Correlation of Physical Forces*; JH is still uncertain about the nature of heat.
C 1p RS:HS 23.379

11755. [1862 to 1868] From Henry Michael Jenkins Somerset House
[Printed notice] Solicitation by Geological Society of contributions for Charles Nichols, 22-year member who has 'fallen into...difficulty.'
L 1p TxU:H/L-0232.1; Reel 1054

11756. [1862?] From John Lee [Wolverhampton]
Notes on trapping air at different elevations [during balloon ascent]. G. B. Airy and JH will be absent. W. H. Sykes will

arrive this evening. Will make another ascent Monday.
ALS 1p inc RAS:JH Archive 12/1.3.14 verso; Reel 10

11757. [1862] From John Lee n.p.
Receives report from Crystal Palace concerning balloon test.
ALS 1p inc Harvard: Houghton AAH 67m–67 (107)

11758. [1862] From John Lee n.p.
Discusses plans for a hot-air balloon.
ALS 1p inc Harvard: Houghton AAH 67m–67 (107)

11759. [1862?] From William Watkiss Lloyd n.p.
Brings coincidence to JH's attention. Believes WL's essay was 'worthy of a better fate.'
ALS 1p inc RAS:JH Archive 12/1.2.12 verso; Reel 10

11760. [1862] From D. C. E. van Monckhoven n.p.
Sending an incomplete copy of the fourth edition of his own book on photography [*Traité général de photographie*], which he would like to dedicate to JH.
ALS 1p RS:HS 12.358

11761. [1862–1–4] From Matthew Collins Dublin
Sorry to hear that the almanac did not reach him. Would be glad for any of JH's comments. Has put previous address.
ALS 2pp RS:HS 5.264

11762. 1862–1–6 From William Rutter Dawes to Margaret Brodie Herschel Hopefield, Haddenham, Thame
Sending a copy of a book he has found of great benefit. News of his household. Mrs. Gordon (JH's daughter Caroline) must have been shocked at the death of the Prince Consort.
AL 5pp inc RS:HS 6.107

11763. 1862–1–9 From William Henry Sykes 47 Albion St.
Invites JH to meeting of Balloon Committee and updates everything in previous letter.
ALS 3pp RS:HS 17.157

11764. 1862–1–11 From William Selwyn Foxton Royston
Discusses a depression on a heliograph, which he cannot reconcile with [Frederick] Howlett's observations. Conjectures over-exposure in printing may be the culprit.
ALS 4pp RS:HS 15.468

11765. 1862–1–13 From John R. Hind 22 Grove Rd., St. John's Wood
Regarding the apparent disappearance of a nebula. Comments on this.
ALS 4pp RS:HS 9.377

11766. 1862–1–14 From George F. Chambers Eastbourne
Will be pleased to furnish details of instruments used in a future edition. What does JH think of his attempts to copy his originals?
ALS 3pp RS:HS 5.330

11767. 1862–1–18 From Edward Joseph Lowe Beeston
Requesting JH to propose EL as a candidate for the R.S.L. The B.A.A.S. meeting will be in Nottingham in 1863. Has been to Spain to observe the eclipse.
ALS 3pp RS:HS 11.332

11768. 1862–1–18 From William Henry Sykes 47 Albion St.
Asks for JH's opinion on balloon ascents, to be presented at next Balloon Committee meeting. Explains enclosed letters and diagrams regarding ascents; mentions that new thermometer suggested by JH is under construction.
ALS 4pp RS:HS 17.158

11769. 1862–1–22 From John R. Hind 22 Grove Rd., St. John's Wood
Cannot see how there can be two opinions regarding JH's catalogue of nebulae. Suitable work for the R.A.S. to publish. Further regarding the lost nebula in Taurus.
ALS 2pp RS:HS 9.378

11770. 1862–1–22 To William Whewell Collingwood
Thanks for volume three of WW's Plato edition. Responds to WW's comments on JH's translation of the first section of Homer's *Iliad.*
ALS 4pp TC, Camb. Add. Ms.a.207[102] (C: RS:HS 23.359)

11771. 1862–1–23 To George Biddell Airy n.p.
Is preparing a list of nebulae and their positions; does GA think anyone would print them?
ALS 2pp inc RGO 6.200.2

11772. 1862–1–23 To George Biddell Airy Collingwood
Has sent an R.S.L. membership application certificate for GA to sign and send on.
ALS 2pp RGO 6.392.381

11773. 1862–1–24 From George Biddell Airy Royal Observatory, Greenwich
[Responding to JH's 1862–1–23], states that R.A.S. finances not sufficient to publish JH's 'Catalogue of Nebulae,' but R.S.L. is 'rolling in wealth.' Work could be done at [Greenwich Observatory], where GA is already printing 'Skeleton Forms for Precessions.' Forwarded R.S.L. certificate for [E. J.] Lowe to W. H. Smyth.
ALS 3pp RAS:JH Archive 12/1.1.3; Reel 10 (ACCS: RGO 6.200.3)

11774. 1862–1–24 From Robert FitzRoy Meteorological Dept., B.o.T.
Weather has been so interesting lately that he is emboldened to send the enclosed chart. Department is progressing in spite of U. J. J. Leverrier's protest. Picked up one of JH's pamphlets published at the Cape. Sends one of the Meteorological Manuals issued by the Dept.
ALS 4pp RS:HS 7.258

11775. 1862–1–25 From Edward Joseph Lowe Beeston
Is grateful to JH for proposing EL as a candidate for the R.S.L. Has been experimenting with ozone; comments on his observations.
ALS 4pp RS:HS 11.333

11776. 1862–1–27 To Robert FitzRoy n.p.
Comments on barometric readings and the signaling process.
C 2pp RS:HS 23.360

11777. 1862–1–28 To George Biddell Airy Collingwood
Suggests preparing an estimate of the cost of preparing and printing a nebula catalogue so JH can approach the R.S.L. [see GA's 1862–1–24].
ALS 2pp RGO 6.200.5

11778. 1862–1–28 From Henry Pearse St. John's Rectory, Bedford
A daughter of Thomas Maclear is to spend several days with JH and his family before returning to the Cape.
ALS 4pp RS:HS 13.331

11779. 1862–1–29 From John P. Gassiot Clapham Common
Sending him [Balfour] Stewart's certificate in the hopes that

he will add his name to it. Hopes to show him his experiments when he has the time.
ALS 3pp RS:HS 8.65

11780. 1862–1–29 To Thomas Maclear Collingwood
Arranged for Maggie Maclear to be met after her arrival in England and for her trip back to Cape Town.
ALS 4pp Cape Archives/Maclear Papers File 100

11781. 1862–1–30 From George Biddell Airy Royal Observatory, Greenwich
Estimates seven weeks and £10 to compute JH's nebulae data. Suggests JH use new 'Skeleton Form.'
ALS 3pp RAS:JH Archive 12/1.1.2; Reel 10

11782. 1862–1–31 From John R. Hind 22 Grove Rd., St. John's Wood
Has been considering how to incorporate JH's suggestions regarding the insertion of the Julian dates in the *Nautical Almanac*. Would like his views. Quotes extracts from a letter of U. J. J. Leverrier.
ALS 4pp RS:HS 9.379

11783. 1862–2–3 From Jacques Babinet Paris
On a static measure for gravity.
AL 3pp inc (French) RS:HS 3.4

11784. 1862–2–4 To Jacques Babinet Collingwood
Notes ingenuity of JB's gravimetric balance. Astonished that it did not occur to anyone before. [Letter continues 5 Feb.:] Suggestion for improving torsion thread arrangement.
(A)CCS 6pp RS:HS 19.147

11785. 1862–2–4 From Frederick Drew 13 Edbury St.
Is going to Kashmir for a geological survey. Wonders if while he is there he can carry out any meteorological research.
ALS 4pp RS:HS 6.504

11786. 1862–2–6 From George Biddell Airy Royal Observatory, Greenwich
Returned JH's 'original sheets of nebulae.' Kept sheets of calculations. Will write about money accounts later. Lord Palmerston's communication about Thomas Maclear's pension.
ALS 3pp RAS:JH Archive 12/1.1.14; Reel 10

11787. 1862–2–6 From Thomas Davidson Weardale, Co. Durham
Many years ago he read a paper to the Royal Scottish Society of Arts on the merits of JH's telescope compared with that of William Parsons (3rd Earl of Rosse); this paper has never been printed. Gives details of his own optical work.
ALS 6pp RS:HS 6.35

11788. 1862–2–6 To Alexander Stewart Herschel Collingwood
Comments on AH's chemical results and on an apparently anomalous sample of common salt being sent by JH; sends AH £5 for the month.
ALS 3pp JHS 1.116

11789. 1862–2–7 From John R. Hind 22 Grove Rd., St. John's Wood
Glad that JH approves introducing Julian dates. Notes their use in American lunar tables. Positions of three new nebulae.
ALS 3pp † RAS:JH Archive 12/1.6.3; Reel 10

11790. 1862–2–7 From Edward Joseph Lowe Beeston
Hears that his chance of success at the R.S.L. is good. [John?] Lee will forward the certificate to JH for presentation to the R.S.L.
ALS 3pp RS:HS 11.334

11791. 1862–2–9 To Thomas Davidson Collingwood
Outlines the differences between the telescope of his own father and that of Lord Oxmantown [William Parsons]. Comments on various lenses. Sends him a little work of his own on telescopes.
draft 2pp RS:HS 6.36 (C: RS:HS 23.361)

11792. [1862]–2–11 To Margaret Brodie Herschel Collingwood
Mostly family news, concluding with concerns about family finances.
ALS 4pp JHS 4.2

11793. [1862?]–2–11 From Edward Sabine n.p.
JH will be pleased with Mr. de Souza's and [Charles] Smallwood's letters. American and Havana instruments were dispatched. [Hermann] Schlagintweit has made interesting magnetic survey of India.
ALS 1p RS:HS 15.329

11794. 1862–2–12 From Thomas Davidson Weardale, Co. Durham
Is grateful for JH's letter and pamphlet. Worked on portraiture before Mr. Baird and assisted him later. Outlines the various lenses he used and made for other people. Is pleased JH has a good opinion of the work of [W. B.?] Rogers. Is he familiar with the works of [Thomas?] Dicks in astronomy?
ALS 6pp RS:HS 6.37

11795. 1862–2–13 To Alexander Stewart Herschel Collingwood
Comments on a number of chemical phenomena, and asks AH to analyze several solid samples; further, JH discusses meteor falls, with emphasis on the application of Newton's laws and terminal velocities.
ALS 7pp JHS 1.117

11796. 1862–2–22 From Jacques Babinet Rue Servandoni N. 15, Paris
Describes procedures, apparatus, and calculations for determining the static gravity of the earth. Thanks JH for some of his writings, and comments that JH, like Newton, has discovered the mind of the Creator and passed this on to humanity.
ALS 3pp diag (French) RS:HS 19.202

11797. 1862–2–24 To Jacques Babinet Collingwood
JB's solution to torsion thread arrangement is simpler and more ingenious than JH's. Enquires about details of gravimetric balance. Pendulum measurements. Density of continents. Clarification of JH's privileges as foreign associate of Institute. Death of J. B. Biot.
C 4pp RS:HS 19.148

11798. 1862–2–24 From Henri de Senarmont Paris
Asks JH's help in efforts to publish a complete edition of writings of physicist Augustin Fresnel.
ALS 3pp (French) RS:HS 17.188

11799. 1862–2–28 To Warren de La Rue Collingwood
Thanks for fine engravings; is concerned about the 'miraculous phenomena' depicted on some other people's engravings, 'especially American ones.'
C 1p RS:HS 23.364

11800. 1862–2–29 From Thomas Davidson Newcastle-upon-Tyne

Has now had ample time in which to study JH's pamphlet. Comments on various points. Has constructed a 4 1/2 inch aperture achromatic telescope, but the buyer went bankrupt. Would welcome assistance from JH.
ALS 4pp RS:HS 6.38

11801. 1862–[3] To the Manchester Literary and Philosophical Society n.p.
Responds to some erroneous claims made at the Society regarding JH's views in meteorology, e.g., JH asserts his support for the Hadleian theory of winds.
L *Manchester Philosophical Society Proceedings, 2* (1862), 215–7

11802. 1862–3–3 To Edward Sabine Collingwood
Asks help of R.S.L. in reducing all observations to 1870 in JH's catalog of nebulae. Estimates cost at £10.
C 2pp TxU:H/L-0371; Reel 1054

11803. 1862–3–3 To George Gabriel Stokes Collingwood
Has completed catalogue of nebulae and star clusters observed by Charles-Joseph Messier, JH's father, and JH. Requests R.S.L.'s help in 'reducing [them] to an epoch.'
ALS 4pp RS MC.6.210

11804. 1862–3–5 To Alexander Stewart Herschel Collingwood
Encourages AH in his intention to apply for a position at the University of Glasgow; gives AH advice about fulfilling his university responsibilities.
ALS 4pp JHS 1.105

11805. 1862–3–5 From George Gabriel Stokes Lensfield Cottage, Cambridge
Will present to R.S.L. Council JH's application for money for reducing JH's catalogue of nebulae.
ALS 2pp RAS:JH Archive 12/1.10.4; Reel 10

11806. 1862–3–6 From Joseph Baxendell Stocks St., Manchester
Thanks for his letter, which he will communicate to the Lit. and Phil. Soc. Regarding the views of William Hopkins. JH's theories regarding the effects of the sun spots are similar to a recent Frenchman's.
ALS 3pp RS:HS 3.308

11807. 1862–3–11 To Augustus De Morgan Collingwood
Asks what AD knows about the Andersonian Institution in Glasgow, which JH's son Alexander is considering as a place of employment. JH then deals with a problem in probability theory, and concludes the letter with a riddle.
ALS 4pp WT 45498.44

11808. 1862–3–11 From Henry P. Finlayson [Dover]
A theory of the construction of the solar system [see also HF's other letter of 1862–3–11].
ALS 2pp inc RS:HS 19.216

11809. 1862–3–11 From Henry P. Finlayson Dover
Sending two very singular empirical laws relating to the solar system and inviting JH's comments on them.
ALS 2pp RS:HS 7.368

11810. 1862–3–11 To Alexander Stewart Herschel Collingwood
Is giving advice about AH's application for a position at the University of Glasgow, and indicating what kind of help JH can provide.
ALS 6pp JHS 1.120

11811. 1862–3–11 From Henry Anthony Jeffreys Hawkhurst Parsonage
Board of Guardians have invited the Parish to consider the proposed new Assessment Bill. Sends a copy and would like JH's comments.
ALS 4pp RS:HS 10.300

11812. 1862–3–12 From Heinrich Louis d'Arrest Copenhagen
Comments on JH's work on nebulae and offers some of his own nebula observations. HA would be greatly honored if JH would include HA's observations in the catalog JH is preparing.
ALS 3pp (German) RAS:JH Archive 12/1.8.11; Reel 10

11813. 1862–3–12 From Thomas P. Kirkman Croft Rectory near Warrington
Thanks for acknowledging his mathematical memoir. Intended his paper on polyhedra for the French prize, but was annoyed they would not recognize his work, so submitted it to the R.S.L. instead. Is sending a paper to William Cavendish (7th Duke of Devonshire), who is a mathematician.
ALS 3pp RS:HS 11.43

11814. 1862–3–13 From Robert Richardson 36 Great George St.
Asks for JH's help in defending current patent laws in Parliament.
ALS 3pp RS:HS 14.487

11815. 1862–3–14 From A. R. Drummond Belvoir Castle
Has received a letter from the Misses Gretton and in consequence has opened an account on their behalf.
ALS 3pp RS:HS 6.506

11816. 1862–3–14 To Robert Richardson Hawkhurst
States that the current patent laws are unjust and would prefer to see them repealed rather than maintained.
ALS 1p RS:HS 14.487b

11817. 1862–3–15 To Robert Harley Collingwood
About the solution of algebraic equations and JH's previous work on this topic.
C 1p RS:HS 23.365

11818. 1862–3–17 From Herbert Spencer 29 Bloomsbury Square
Recalls JH having written that an 'annulus of stars might be in equilibrium.' Asks where this claim appears in JH's writings.
ALS 2pp RS:HS 16.491

11819. 1862–3–18 From Robert Harley Brighouse
Is grateful to JH for calling attention to his paper on analysis. Has forwarded his letter to George Boole. Hopes he can quote it in his paper for the Manchester Philosophical Society. Sends a copy of his memoir from the *R.S.P.T.*
ALS 3pp RS:HS 9.238

11820. [1862]–3–19 From Henri de Senarmont Paris
Thanks for help [see HS's 1862–2–24]; requests JH's further assistance in publishing a complete edition of writings of Augustin Fresnel.
ALS 2pp (French) RS:HS 17.189

11821. 1862–3–20 From John R. Hind 22 Grove Rd., St. John's Wood
Encloses slips to show how he has incorporated the Julian dates. Agrees with the American lunar tables. The nebula in

Taurus is invisible to William Lassell.
ALS 3pp RS:HS 9.380

11822. 1862–3–22 From Thomas Maclear Royal Observatory, Cape of Good Hope
Maggie Maclear returned to Cape Town after attempting to run away to England. Herschel family instrumental in her safe return.
ALS 2pp Cape Archives/Maclear Papers File 100

11823. 1862–3–24 From Thomas P. Kirkman Croft Rectory near Warrington
Regarding the report of the polyedra prize question of the French Academy.
ALS 3pp RS:HS 11.44

11824. 1862–3–25 From Walter White R.S.L.
Does JH want £20 sent directly or deposited in JH's account?
ALS 1p RAS:JH Archive 12/1.10.15; Reel 10

11825. [1862?]–3–28 From Henry Holland Brook St.
Hopes JH will accept enclosed little volume.
ALS 2pp RS:HS 9.443

11826. 1862–3–28 From George Gabriel Stokes R.S.L.
R.S.L. Council voted £20 for cost of reducing JH's catalogue of nebulae.
ALS 3pp RAS:JH Archive 12/1.10.5; Reel 10

11827. 1862–3–29 From Thomas P. Kirkman Croft Rectory near Warrington
Further regarding the polyedra prize of the French Academy.
ALS 2pp RS:HS 11.45

11828. 1862–3–30 To George Biddell Airy Collingwood
Ready to send nebula data to GA for reduction; question of dates need decisions [see GA's 1862–1–30].
ALS 4pp RGO 6.200.11

11829. 1862–3–31 From George Biddell Airy Royal Observatory, Greenwich
Send quickly JH's manuscript of 'Catalogue of Nebulae.' It will provide work for computer whom GA was about to lay off. Prefers reducing JH's data to 1860's. GA's improvement on F. W. Bessel's reduction formula.
ALS 4pp RAS:JH Archive 12/1.1.4; Reel 10

11830. 1862–4–2 To George Biddell Airy Collingwood
Forgot, in earlier note [see JH's 1862–4–2], to indicate agreement with GA's suggestions [see GA's 1862–3–31].
ALS 1p RGO 6.200.13

11831. 1862–4–2 To George Biddell Airy Collingwood
Is sending first part of observational data for nebula catalogue [see GA's 1862–3–31].
ALS 2pp RGO 6.200.10

11832. 1862–4–3 From George Biddell Airy [Royal Observatory, Greenwich]
Material sent by JH [see JH's 1862–4–2] arrived safely.
ACCS 2pp RGO 6.200.14

11833. 1862–4–3 From Walter White R.S.L.
Please send receipt for £20.
ALS 1p RAS:JH Archive 12/1.10.16; Reel 10

11834. 1862–4–4 To John R. Hind Collingwood
Reports the disappearance of a nebula seen by William Herschel in Coma Berenices, but which H. L. d'Arrest did not locate in 1850.
C 4pp TxU:H/L-0193; Reel 1054 *R.A.S.M.N.*, 22 (1862), 248–50

11835. 1862–4–7 To Sir Henry Taylor Collingwood
Thanks for HT's *St. Clement's Eve.* Has been reading it with much interest. Praises it.
ALS 2pp Bodleian Library, Oxford

11836. 1862–4–8 From George Julius P. Scrope 25 Hyde Park Gate
Thanks for volume on physical geography and for encouragement at start of GS's career. Hopes JH will approve of GS's latest work.
ALS 3pp RS:HS 17.183

11837. 1862–4–9 From George Biddell Airy [Royal Observatory, Greenwich]
Acknowledgment of receipt of more material from JH [see GA's 1862–4–3].
ACCS 1p RGO 6.200.15

11838. 1862–4–9 From John R. Hind Nautical Almanac Office
Was interested to receive the information regarding another missing nebula; would like to mention it at the R.A.S. The late George Bishop's observatory is now being rebuilt at Twickenham. Will return N. R. Pogson's table of Julian dates tomorrow. Has adopted the first of JH's suggestions.
ALS 2pp RS:HS 9.381

11839. 1862–4–10 From John R. Hind London
Returns N. R. Pogson's tables. Has received a communication from [W.?] Lummis concerning a dark spot on the sun's disc. Has been unable to make any calculations yet.
ALS 2pp RS:HS 9.382

11840. 1862–4–10 To William Whewell Collingwood
Responds to WW's comments on JH's *Iliad* translation. Is preparing a 'General Index Catalogue of Nebulae,' with help from computists supplied by G. B. Airy. Asks WW to recommend JH's son Alexander for a professorship at the Andersonian University of Glasgow.
ALS 4pp TC, Camb. Add. Ms.a.207[103] (C: RS:HS 23.366)

11841. 1862–4–12 From George Biddell Airy [Royal Observatory, Greenwich]
Needs to know, from JH, the date of the nebula observations [see GA's 1862–4–9].
ACCS 3pp RGO 6.200.16

11842. 1862?–4–12 From Adolphe Quetelet n.p.
Informs JH of his grief over losing his son and his wife. His other son has left the military and now is helping AQ. Talks of Prince Albert's death. Asks JH not to forget AQ.
ALS 1p (French) inc Académie belgique: #26

11843. 1862–4–12 To John Tyndall Collingwood
Remarks on JT's papers on radiant heat and related role of moisture. Will send requested sample of rock salt. Explains difficulties using it and hopes JT will have better success, perhaps even repeating JH's experiments.
CC 4pp RS:HS 17.390b (C: 23.367; C: RI 516–17)

11844. 1862–4–13 To George Biddell Airy Collingwood
Responds carefully to the question of observational, and reduction, dates [see GA's 1862–4–12].
ALS 4pp RGO 6.200.18

11845. 1862–4–15 To Adolphe Quetelet Collingwood
Condolences on death of AQ's wife and son. Having lost a daughter, JH sympathizes. Discusses merits of Prince Albert. Lists Belgian Academy memoirs possessed to date. Mentions work on a general index catalogue of nebula. [J. R.] Hind's

nebulae and others are disappearing.
ALS 4pp Académie belgique: #27

11846. 1862–4–15 From John Tyndall Royal Institution
Finds JH's approval incredibly gratifying. Thanks for gift of rock salt. Is sending unscientific book; trusts JH's interests are broad enough to appreciate it.
ALS 4pp RS:HS 17.391 (C: RI 518)

11847. 1862–4–16 From George Biddell Airy [Royal Observatory, Greenwich]
Acknowledges receipt of more material from JH [see JH's 1862–4–13].
ACCS 1p RGO 6.200.20

11848. 1862–4–17 From George Biddell Airy Royal Observatory, Greenwich
Has received notice from the Foreign Office that the Russian Government has sanctioned the establishment of a magnetic and meteorological observatory at Pekin.
ALS 2pp RS:HS 1.256 (C: RGO 6.695.215)

11849. 1862–4–17 To George Biddell Airy Collingwood
Has now sent the last of the nebula observations to be reduced for printing [see GA's 1862–4–16].
ALS 2pp RGO 6.200.22

11850. 1862–4–17 To Robert FitzRoy Collingwood
Thanks for a report by RF; comments on barometric readings.
C 1p RS:HS 23.369

11851. 1862–4–17 To William Whewell Collingwood
Sends, along with comments, JH's translation of part of Book II of Homer's *Iliad.* Is avoiding consulting other translations.
ALS 3pp TC, Camb. Add. Ms.a.207[104] (C: RS:HS 23.368)

11852. 1862–4–19 From Augustus De Morgan 41 Chalcot Villas NW
Sent his paper on probabilities a few days ago. Shows how his own method is different from that of L. A. J. Quetelet. Sends a poem.
ALS 4pp RS:HS 6.350 S. De Morgan, *Memoir of Augustus De Morgan*, 308–9 inc

11853. 1862–4–21 From William Cavendish Lismore Castle
Would like to include JH's name among those proposed for honorary degrees at Cambridge when he becomes Chancellor of the University.
ALS 2pp RS:HS 5.329

11854. [1862–4–21 or later] To William Cavendish [Collingwood]
JH's reply to WC's 1862–4–21.
draft 1p RS:HS 5.329verso

11855. 1862–4–21 From Robert FitzRoy Meteorological Dept. B.o.T.
Regarding water in the atmosphere. Effect of the moon on clouds. Some of the curiosities seen by his son while at the Embassy in Japan, including a device for giving warnings of earthquakes. Intends to show combinations of wind currents.
ALS 13pp RS:HS 7.259

11856. 1862–4–21 From Thomas W. Richardson Brede near Northiam
Sends sketch of reflecting camera and describes its composition. Can now make reflectors under 7 inches very inexpensively.
ALS 4pp RS:HS 14.488

11857. 1862–4–23 To Robert FitzRoy Collingwood
Comments on the pressure of the atmosphere; is convinced of the effect of the full moon on weather.
C 3pp RS:HS 23.370

11858. 1862–4–24 From George Biddell Airy Royal Observatory, Greenwich
Printing 1000 copies of JH's Catalogue [of Nebulae]. Please transfer funds to GA to cover expenses.
ALS 1p RAS:JH Archive 12/1.1.5; Reel 10

11859. 1862–4–24 From George Biddell Airy [Royal Observatory, Greenwich]
Sends JH a copy of the form to be used for reduced observations [see JH's 1862–4–17].
ACCS 1p RGO 6.200.27

11860. 1862–4–24 From Chalmers Guthrie London
Two casks of wine will be sent today to JH at Collingwood.
ALS 1p RS:HS 19.149a

11861. 1862–4–25 To George Biddell Airy Collingwood
Approves of form for reduced observations [see GA's 1862–4–24] and sends GA payment against expenses.
ALS 1p RGO 6.200.28

11862. 1862–4–25 From Robert FitzRoy Meteorological Dept. B.o.T.
Further comments regarding the pressure of the atmosphere. His sketch was to demonstrate the currents of air. JH's comments on the moon were interesting.
ALS 8pp RS:HS 7.260

11863. 1862–4–26 From George Biddell Airy Royal Observatory, Greenwich
Received £10 for cost of reducing JH's 'Catalogue of Nebulae.' Send description of catalogue.
ALS 1p RAS:JH Archive 12/1.10.10; Reel 10

11864. 1862–4–26 From George Biddell Airy Royal Observatory, Greenwich
Acknowledges receipt of payment [see JH's 1862–4–25]; asks for final approval of printing structure for the nebula catalogue.
ACCS 1p RGO 6.200.28a

11865. 1862–4–26 From John Eustace Prescott n.p.
Sending JH some crystals for examination. The quartz crystals show spirals. Invites JH to supply information concerning the crystals.
ALS 3pp RS:HS 14.51

11866. 1862–4–27 From Julia Margaret Cameron Little Holland House
Sending him R. J. Mann's letter. Mann could call to see him if convenient. News of her children.
ALS 16pp RS:HS 5.156

11867. 1862–4–28 From Edward Joshua Cooper 9 Neville Park, Tunbridge Wells
Disappointed that EC could not visit Collingwood in February while in Tunbridge Wells to visit daughter, Mrs. Arthur Warre. Leaving for London to see Exhibition.
ALS 3pp RAS:JH Archive 12/1.5.2; Reel 10

11868. 1862–4–28 From Sir Henry Taylor East Sheen
Thanks JH for favorable notice of his play. Sends sample of D. G. Rossetti's translation of Dante's *Vita nuova.*
ALS 5pp RS:HS 17.325

11869. 1862–4–29 From Augustus De Morgan 91 Adelaide Road
Many thanks for the hexameters. Comments on these and the meter of English poetry.
ALS 3pp RS:HS 6.351 S. De Morgan, *Memoir of Augustus De Morgan*, 309

11870. 1862–4–29 From John William Lubbock High Elms, Farnborough, Kent
Acknowledges receipt of [JH's *Cape Results*].
ALS 1p RAS:JH Archive 10/5.83; Reel 9

11871. 1862–4–30 To George Biddell Airy Collingwood
Provides final intended structure of the nebula catalogue page [see GA's 1862–4–26].
ALS 3pp RGO 6.200.30

11872. 1862–5 From William Wesley Washington, D.C.
[Written on back of printed notice from Smithsonian Institution, asking British correspondents to forward literary and scientific materials via WW in London.] Received JH's note and vol. 7 of Smithsonian Collections. Sent JH's parcel to Washington.
(A)L 1p TxU:H/MW-0246.2; Reel 15

11873. 1862–5–2 To Augustus De Morgan Collingwood
JH's inability to concentrate; comments in reply to AD's 1862–4–19 and1862–4–29.
C 2pp RS:HS 23.372

11874. 1862–5–2 To Robert FitzRoy Collingwood
On the dynamics of atmospheric pressure and the pressure on clouds and birds.
C 5pp RS:HS 23.371

11875. 1862–5–3 From Edward Joshua Cooper Holloway's Hotel, [Dover St.]
Will send records of cometic nebulae EC found in Nice during winter of 1844–1845. These did not appear in JH's catalogue. Does JH have all nebulae found by William Parsons?
ALS 3pp RAS:JH Archive 12/1.5.3; Reel 10

11876. 1862–5–3 From Augustus De Morgan 91 Adelaide Road
Many years ago JH had said his time was nearly up; now he is doing it again. Hopes it is a good omen. Regarding Greek hexameters and English translations of heroic poetry.
ALS 4pp RS:HS 6.352 S. De Morgan, *Memoir of Augustus De Morgan*, 310

11877. 1862–5–3 From Smith, Elder & Co. 65 Cornhill
Note of thanks to accompany payment for JH's first installment of the *Iliad* translation.
(A)LS 1p RS:HS 16.206

11878. 1862–5–4 To William Lassell Collingwood
As JH is preparing a catalogue of all known nebulae, would WL please provide a complete listing of all he has seen.
C 2pp RS:HS 23.373

11879. 1862–5–6 From George Smith 45 Pall Mall
About disposal of extra copies of JH's printed observations.
ALS 1p RS:HS 16.185

11880. 1862–5–8 From Edward Joshua Cooper Holloway's Hotel, [Dover St.]
Sending sketches of nebulae made in Nice and Naples in 1843–1845 using K. L. Harding's maps. Please return them.
ALS 2pp RAS:JH Archive 12/1.5.4; Reel 10

11881. 1862–5–8 To William Whewell Collingwood
Sends another section of JH's *Iliad* translation. Regrets he cannot come to the Cambridge B.A.A.S. meeting. Sends lines of the *Iliad* translated by Augustus DeMorgan.
ALS 4pp TC, Camb. Add. Ms.a.207[105] (C: RS:HS 23.374)

11882. 1862–5–10 From William Parsons (Lord Rosse) Birr Castle, Parsonstown
Offers assistance for JH's proposed catalogue of nebulae. Will send estimated positions of recently discovered nebulae.
ALS 6pp RAS:JH Archive 12/1.2.13; Reel 10

11883. 1862–5–10 From Edward Sabine Burlington House
Wants suggestions as to whom R.S.L. should award medals.
ALS 4pp RS:HS 15.260

11884. 1862–5–11 From Mary Somerville Turin
Cannot travel to London. Wishes she could return to her country and friends. Thankful for her health. Wants Herschels to visit Italy.
ALS 3pp RS:HS 16.366

11885. 1862–5–12 From Thomas Greenwood 14 Westbourne Terrace
Was pleased to receive JH's version of Book I of the *Iliad*. Comments on the difficulties of translating ancient works into modern language. Hopes his health is sounder.
ALS 4pp RS:HS 9.34

11886. 1862–5–13 To the London *Times* Collingwood
Times of 12 May called for remedy to industrial pollution. In 1838 JH visited soda production plant near Newcastle and suggested profitable modification to fume chimneys that removed harmful vapors and generated useful compounds.
AdraftS 2pp RS:HS 19.149b

11887. 1862–5–14 To the London *Times* Collingwood
Calls public attention to an inexpensive process JH devised in 1838 to eliminate noxious acid fumes coming from manufacturers of soda.
L London *Times* (May 15, 1862), p. 11, col. 4

11888. 1862–5–15 From P. O'Callaghan Leeds Philosophical & Literary Society
Sends drawings of masses of ice that fell in the neighborhood on the 7th. Would like drawings returned to deposit them in the museum. Hopes grandchildren reached Collingwood safely.
ALS 3pp RS:HS 13.159

11889. 1862–5–16 From John Eustace Prescott n.p.
Thinks application of heat will not harm carbonate of lead crystal, and urges JH to 'boil, torture, or otherwise put it to the question in whatever way' JH sees fit.
ALS 2pp RS:HS 14.52

11890. 1862–5–20 From Robert Moffat to Thomas Maclear Kuruman
Would like an introduction to JH so that RM can send him the final proofs of a book he has been working at, which has been returned unopened from the Institut de France.
ALS 4pp RS:HS 13.45

11891. 1862–5–20 From Robert Moffat Kuruman
Has instructed the printers to send JH proofs of RM's book on the phonetic alphabet. [Note by JH: 'sent the first proofs to France.']
ALS 5pp RS:HS 13.46

11892. 1862–5–22 From Thomas Webster Cranbrook
Thanks for copy of JH's translation of *Iliad.*
ALS 2pp RS:HS 18.132

11893. 1862–5–23 From William Lassell 9 Piazza Sliema, Malta
Recently observed nebulae. Changes in island climate reduce observation time. Happy that JH plans to publish all nebulae in one catalogue. Has no new ones to offer. Observations of Saturn's rings. WL's telescope.
ALS 8pp RAS:JH Archive 12/1.7.2; Reel 10

11894. 1862–5–25 From Julia Margaret Cameron Freshwater, I.o.W.
His gift has given her great pleasure. Will show his translation of Homer to the Tennysons. Has lost her neighbor Mr. Clough. Would like a visit from the Herschels. News of Alfred Tennyson's current work.
ALS 12pp RS:HS 5.157

11895. 1862–5–26 From J[ames] M[osey] Cranswick Manchester
Would be glad to know in which work he argued against the eternal existence of matter.
ALS 3pp RS:HS 5.372

11896. 1862–5–26 From Augustus De Morgan 91 Adelaide Road
Is meditating on a paper on the infinity of magnitude. Does he recollect a paper by J. J. Sylvester on the passage of a bullet in a gun barrel?
ALS 3pp RS:HS 6.353

11897. 1862–5–27 From George Biddell Airy [Royal Observatory, Greenwich]
Seeks clarification of the position of one of the listed nebulae [see JH's 1862–4–30].
ACCS 1p RGO 6.200.34

11898. 1862–5–27 To J[ames] M[osey] Cranswick Collingwood
[Responding to JC's 1862–5–26], discusses JH's argument against the eternal self existence of matter. Gives reference to where argument appears in JH's *Prelim. Discourse.*
ALS 4pp Science Museum Archives MS 1007 (ACCS: RS:HS 5.373 & C: 23.375)

11899. 1862–5–27 From John R. Hind Nautical Almanac Office, Gray's Inn
[Jean] Chacornac found nebula in Coma Berenices that JH thought had disappeared. It is highly variable. Notes on nebulae of variable magnitude.
ALS 2pp RAS:JH Archive 12/1.6.4; Reel 10

11900. 1862–5–27 From Thomas Parsons Stroud, Gloucestershire
Sends a cutting from the *Stroud Journal* concerning JH's father, which may interest JH.
ALS 2pp RS:HS 13.223

11901. 1862–5–28 To George Biddell Airy n.p.
Provides the information in response to GA's request for the position of one nebula [see GA's 1862–5–27].
LCC 1p RGO 6.200.36

11902. 1862–5–28 To George Biddell Airy Collingwood
Comments on several nebulae, and on GA's receiving an honorary degree [from Cambridge University].
ALS 3pp RGO 6.380.263

11903. 1862–5–28 From Robert Brown Fairlie by Greenock
Acknowledges his letter and his invitation to Collingwood. Was pleased to have the company of the young ladies on their journey from Paris.
ALS 2pp RS:HS 4.318

11904. 1862–5–29 From J[ames] M[osey] Cranswick Manchester
Thanks for his long and valuable reply to his query. States how the query originated.
ALS 3pp RS:HS 5.374

11905. 1862–5–29 From Charles G. B. Daubeny Botanic Garden, Oxford
Thanks for sending him a copy of his translation of the first book of the *Iliad.* Although his health is better he spent some time near Naples and saw the eruption of Vesuvius. Comments on volcanic activity.
ALS 4pp RS:HS 6.22

11906. 1862–5–29 To Augustus De Morgan Collingwood
Hopes AD will be successful in his projected paper on the infinity of magnitude. Does not recollect the paper by J. J. Sylvester. Has had a letter from Manchester on his own theory of eternal existence.
draft 2pp RS:HS 6.354 & 23.376 S. De Morgan, *Memoir of Augustus De Morgan*, 311

11907. 1862–5–30 From [Thomas] Melvill [India Office]
Of five institutions in India named by JH, three no longer exist. Suggests alternate recipients for *Cape Results.* TM will forward copies to India.
ALS 3pp RAS:JH Archive 10/5.99; Reel 9

11908. 1862–5–30 From Augustus De Morgan 91 Adelaide Road
Suspects that J. J. Sylvester and JH agree in their theories. Is inclined to agree with them as his third paper on logic shows. Has read JH's book and will return it soon. Has 30,000 life policies to calculate.
ALS 3pp RS:HS 6.355

11909. 1862–5–31 From Edward Joseph Lowe Beeston
Please accept the enclosed book, which he has dedicated to JH. Has been unsuccessful at the R.S.L. Hopes some day to have ample means with which to pursue science.
ALS 4pp RS:HS 11.335

11910. 1862–5–31 To Thomas Parsons Collingwood
Outlines the early life of his father, Sir William Herschel, and shows that the cutting received from the *Stroud Journal* was inaccurate. [Marked 'not sent.']
draft 4pp RS:HS 13.224 & 23.377

11911. 1862–6–2 From George Biddell Airy Royal Observatory, Greenwich
More corrections of errors [in reduction of JH's catalogue of nebulae].
ALS 3pp † RAS:JH Archive 12/1.11.9; Reel 10

11912. 1862–6–2 To Thomas Parsons Collingwood
Gives details of William Herschel's youth, to correct errors in TP's anecdote about WH in *Stroud Journal.*
C 6pp RS:HS 13.225 (C: RS:HS 19.172b) *Stroud Journal,* June 7, 1862

11913. 1862–6–2 From Adam Sedgwick Cambridge
Writes to point out an error in a pamphlet he had written. Discusses plans for upcoming trip to the north.
ALS 4pp RS:HS 15.451

11914. 1862–6–3 To [Alexander Stewart Herschel] Collingwood
The family is complaining about not having any news from AH; comments about AH's explanation of the color of copper sulfate by twilight.
ALS 1p JHS 1.104

11915. 1862–6–3 From Sir Charles Lyell Florence
Came to Florence on account of the sudden death of Mrs. [Joanna B.] Horner. Thanks for the translation of a book of the *Iliad*, but thinks only Greek scholars should read it. Has some queries regarding the migration of hippopotomi; can JH supply any facts or references?
ALS 3pp RS:HS 11.426

11916. 1862–6–5 From George Biddell Airy Royal Observatory, Greenwich
More corrections of errors [in reduction of JH's catalogue of nebulae].
ALS 2pp, encl Amem 2pp RAS:JH Archive 12/1.11.11; Reel 10

11917. 1862–6–4 From Thomas Parsons Stroud
Thanks JH for his 1862–6–2. Apologizes for misstatements TP had made [*Stroud Journal*, 10 May 1862] about William Herschel.
L RS:HS 19.172c *Stroud Journal* 7 June 1862

11918. 1862–6–5 From Smith, Elder & Co. 65 Cornhill
Per JH's directions, forwarded presentation copies [of JH's *Cape Results*] to J. C. Melvill for distribution to libraries in India.
ALS 1p RAS:JH Archive 10/3.21; Reel 9

11919. 1862–6–7 From Edward William Brayley London Institution
Is applying for professorship at Addiscombe Military College and would like JH's support.
ALS 5pp RS:HS 4.221

11920. 1862–6–9 From [Thomas] Melvill India Office
Received five copies of JH's *Cape Results*. Will forward them to designated institutions in India.
ALS 2pp RAS:JH Archive 10/5.100; Reel 9

11921. 1862–6–9 To Thomas Parsons Collingwood
Has just received the *Stroud Journal* and TP's letter. Thinks it a fair and proper thing that TP has done. His own letter had no sinister insinuation.
C 2pp RS:HS 13.226 & 23.378

11922. 1862–6–12 From George Smith 45 Pall Mall
Checking the accuracy of a writer's expression of JH's views.
ALS 2pp RS:HS 16.186

11923. 1862–6–15 To George Gabriel Stokes Collingwood
JH regrets he will not have time to review paper on Indian meteorology.
ALS 3pp Cambridge University Library 7656/H700

11924. 1862–6–17 From Thomas Maclear Royal Observatory, Cape of Good Hope
Enclosed is from Mr. Moffat, son of the missionary. Thanks for the essay on Meteorology. The Cape Meteorological Observations for 1841–7 were printed under the direction of Edward Sabine. The mss. for the later series are in possession of Robert FitzRoy, who is trying to find the means to print them. Serious gales at the Cape. Vessel wrecked on Sunday night attempting to enter Table Bay.
ALS 4pp RS:HS 12.167

11925. 1862–6–18 From William G. McKay Plumstead
Is an amateur astronomer, interested especially in sidereal astronomy. Has constructed a small telescope. Can JH advise him on good books dealing with sidereal astronomy?
ALS 3pp RS:HS 13.5

11926. 1862–6–20 From George Gabriel Stokes Royal Society
Requests that JH return a paper sent to him.
ALS 3pp RS:HS 17.47

11927. 1862–6–21 To [R.S.L. Council] n.p.
Reports on and deems fit for publication [William] Hopkins's paper [*R.S.P.T., 153*,. 677–] on theory of motion of glaciers.
ADS 10pp RS RR.5.117

11928. 1862–6–23 To William Parsons (Lord Rosse) Collingwood
Thanks WP for WP's paper on nebulae and on WP's giant reflecting telescope. Comments on and/or raises questions about various nebulae.
ALS 4pp inc Rosse Papers K2.8

11929. 1862–6–25 From George Biddell Airy Royal Observatory, Greenwich
As GA is about to leave on holiday, he brings JH up to date on the state of the nebula calculations [see JH's 1862–5–28].
ALS 2pp TxU:H/M-0640; Reel 1084 (C: RGO 6.200.38)

11930. 1862–6–26 To Alexander Stewart Herschel Collingwood
Thanks AH for the chemical analyses [see JH's 1862–2–13]; gives fatherly advice about spending, or not spending, money.
ALS 4pp JHS 1.118

11931. 1862–6–26 To John Herschel (son) Collingwood
Expresses concern about son John's loneliness in India, and so JH proceeds to tell him of the Great Exhibition of 1862, and about scientific work being done such as improvements in meteorology, solar photography, and the work being done on JH's catalogue of nebulae. No decision has been made about the Edinburgh professorship for which son Alexander has applied.
ALS 6pp JHS 5.7

11932. 1862–6–27 To George Biddell Airy Collingwood
The R.S.L. grant is almost used up; how much more money will be needed for the nebula catalogue [see GA's 1862–6–25]?
ALS 3pp RGO 6.200.39 (draft TxU:M/H-06404)

11933. 1862–6–27 To George Biddell Airy Collingwood
Sends £10 to GA, exhausting grant from R.S.L. GA's estimate of 5600 nebulae was too high. Recommends asking R.S.L. for additional £50.
ACCS 2pp TxU:H/M–0640.2; Reel 1084

11934. 1862–6–28 From Edward William Brayley Southgate Road
Has been ill, which accounts for the delay in answering his letter. Has applied for the position at Addiscombe. Remarks concerning volcanoes and his own and JH's queries regarding them.
ALS 12pp RS:HS 4.222

11935. 1862–6–29 From George Biddell Airy Royal Observatory, Greenwich
Attempts a more accurate estimate of work to be done on nebula catalogue [see JH's 1862–6–27].
ALS 2pp RGO 6.200.41

11936. 1862–7–1 From William Henry Smyth Athenaeum Club
Sends letter from Mr. Rüppell, who wishes to get a copy of R.A.S. medal.
ALS 2pp RS:HS 16.319

11937. 1862–7–2 To George Biddell Airy Collingwood
In view of the additional costs, should the nebula catalogue be reduced to a 'working list', or should JH go back to the R.S.L. for more money [see GA's 1862–6–29]?
ALS 6pp RGO 6.200.42

11938. 1862–7–2 To William Henry Smyth Collingwood
Suggests Mr. Rüppell apply to Council of R.A.S. for copy of medal. Mentions the head on the medal is [Isaac] Newton's, not William Herschel's.
C 1p RS:HS 23.380

11939. 1862–7–3 To Edward William Brayley Collingwood
Sorry to hear of his illness. Answers to his queries regarding the colloid state of ice. Cannot agree about metals of alkalis and earths uncombined in the earth.
draft 4pp RS:HS 4.223 & 23.381

11940. 1862–7–5 From Edward William Brayley n.p.
Was pleased to obtain JH's news regarding ice. Agrees with him about metals of alkalis in the earth. Further regarding volcanoes.
ALS 4pp RS:HS 4.224

11941. 1862–7–12 From John William Lubbock High Elms, Farnborough
JH's conjectures are natural but Augustus De Morgan had nothing to do with JL's *Treatise on Probability*. JL wrote it in conjunction with J. E. Drinkwater Bethune, although the title page does not show the names of the authors. Understands what JH means about the moon but cannot see what will be gained. Sorry to hear of JH's mishap.
ALS 4pp RS:HS 11.402

11942. 1862–7–14 From George Smith 45 Pall Mall
Relating to writing for the *Cornhill Magazine*, and spare copies of the *Cape Results*.
ALS 4pp RS:HS 16.187

11943. 1862–7–15 From John Lee Wolverhampton
Outlining events in the projected balloon ascent by Messrs. James Glaisher and Henry Coxwell.
ALS 3pp RS:HS 11.181

11944. 1862–7–18 From John Lee Wolverhampton
Stormy night followed by calm morning. Balloon is now filling for 9 a.m. liftoff.
ALS 1p RAS:JH Archive 12/1.3.6 verso; Reel 10

11945. 1862–7–22 To George Biddell Airy Collingwood
After collecting and reducing all nebula observations by JH and William Herschel, JH objects to GA's suggestion that JH should pay for calculations needed to prepare catalogue of nebulae. Hesitates to request additional funds from R.S.L.
C 5pp inc TxU:H/L-0021; Reel 1054

11946. 1862–7–23 From Adam Sedgwick Cambridge
Thanks JH for translation of *Iliad*. AS's health has been poor. Invites the Herschels to Norwich.
ALS 4pp RS:HS15.452

11947. 1862–7–24 To E. Loyd Collingwood
Writes to ask EL a series of questions about probabilities related to shooting at targets, with variations in the size and shape of the targets.
C 3pp RS:HS 23.382

11948. 1862–7–24 From William Parsons (Lord Rosse) Brighton
Assistant Samuel Hunter will send WP's observations to JH. Original intent was to confirm JH's nebulae, but WP soon found new nebulae. Trouble finding good assistants. Note on new transit instrument.
ALS 8pp RAS:JH Archive 12/1.2.15; Reel 10

11949. 1862–7–25 From Edward William Brayley London Institution
Sends articles he has written for the English Encyclopaedia. Still awaiting results of the professorship at Addiscombe.
ALS 2pp RS:HS 4.225

11950. 1862–7–25 From Augustus De Morgan 91 Adelaide Road
Should not imagine that he is declining when a hot day comes along. Regarding Thomas Harriot and his theological views.
ALS 4pp RS:HS 6.356

11951. [1862?]–7–26 From Henry Holland 72 Brook St.
Thanks JH for mentioning HH's paper. Attributes electrical phenomena to 'a *material element*,' perhaps a modified form of ether, distinct from other matter.
ALS 3pp RS:HS 19.277

11952. 1862–7–29 From William Henry Bateson Whitby
If he intends attending the B.A.A.S. meeting in October he hopes that he will stay at St. John's Lodge. Lady and Miss Herschel are welcome.
ALS 3pp RS:HS 5.26

11953. 1862–7–30 To Augustus De Morgan Collingwood
Discusses various etymologies, some coming out of JH's efforts to translate Homer's *Iliad*. Also discusses a book sale and some anagrams.
ALS 4pp Bodleian Library, Oxford

11954. 1862–7–31 From Elizabeth White [Holybourne, near Alton]
Lady K. B. Kane will erect tablet to her aunt Elizabeth Baily in Thatcham Church. Comments on International [Exhibition].
ALS 3pp RS:HS 18.223

11955. 1862–8–1 From George Biddell Airy Pitlochrie, North Britain
Ferry delayed by 19 July hurricane. Studying classical granite veins in Glen Tilt under [J. D.] Forbes's tutelage. Not disturbed by cost overrun of JH's catalogue of nebulae, but sorry that JH is 'so much vexed.' Simply tell R.S.L. committee that JH needs more money; do not worry about scruples.
ALS 4pp † RAS:JH Archive 12/1.1.6; Reel 10

11956. 1862–8–2 From John R. Hind 22 Grove Road, St. John's Wood
Positions of new comet, calculated from 'Florence observations.'
ALS 2pp RAS:JH Archive 12/1.6.5; Reel 10

11957. 1862–8–3 From Augustus De Morgan 91 Adelaide Road
His family are now by the sea in Wales. Comments on translations.
ALS 4pp RS:HS 6.357

11958. 1862–8–5 From George Smith 45 Pall Mall
Paper prepared by JH is judged as being 'too deep' for the readers of the *Cornhill Magazine*.
ALS 1p RS:HS 16.188

11959. 1862–8–6 To George Smith n.p.
Withdraws article that JH submitted to *Cornhill Magazine* because, as GS noted, it was meant for different class of readers. But asks for 250 copies to be printed for JH's private circulation.
Adraft 1p TxU:H/L-0386; Reel 1054 (Cdraft: RS:HS 25.15.19)

11960. 1862–8–7 From Elizabeth White [Holybourne, near Alton]
Sends Lady K. B. Kane's Dublin address. Describes former residence and neighborhood of Elizabeth Baily. EW's husband [Walpole] returned from Yorkshire Monday.
ALS 4pp RS:HS 18.224

11961. 1862–8–9 From Augustus De Morgan 91 Adelaide Road
Returns JH's manuscript on algebra. Wishes parts of it had been published. Wrote a treatise on mechanics, which he submitted to the University when applying for his present position. Regarding J. L. F. Bertrand and their papers on series.
ALS 10pp RS:HS 6.358 S. De Morgan, *Memoir of Augustus De Morgan*, 311–14

11962. 1862–8–9 From William Parsons (Lord Rosse) Yacht *Titania*, Ryde
Still searching for journal of WP's observations. Notes on micrometer used in measuring nebulae positions. WP's assistants are not qualified to make new measurements.
ALS 7pp RAS:JH Archive 12/1.2.17; Reel 10

11963. 1862–8–9 From Eyre B. Powell Madras
Thanks JH for copy of *Cape Observations*; says it will be useful to students of Madras College. Remarks on double star observations of his own confirmed in *Outlines Astr.* Wants to start a double star catalogue.
ALS 4pp RS:HS 14.184

11964. 1862–8–11 To George Biddell Airy Collingwood
Seeks state of the nebula catalogue so an accurate estimate of the remaining costs may be made [see GA's 1862–8–1].
ALS 1p RGO 6.200.47

11965. 1862–8–12 From Adolphe Quetelet Brussels
Hopes he has not lost JH's friendship. AQ's son left military to help AQ. Death of Prince Albert was a shock.
ALS 3pp (French) RS:HS 14.254

11966. 1862–8–12 From George Biddell Airy Royal Observatory, Greenwich
Proposes continuing work on JH's [Catalogue of] Nebulae, and not worrying about funding. Gives details of money and time spent so far. Work is about half done.
ALS 4pp RAS:JH Archive 12/1.1.7; Reel 10

11967. 1862–8–12 From William Selwyn Foxton Royston
Sending autographs. Discusses the faculae on the eastern limb and the reception of JH's note at the R.A.S. [Frederick] Howlett is drawing a group of sunspots.
ALS 4pp RS:HS 15.469

11968. 1862–8–12 From George Smith 45 Pall Mall
Relating to a paper JH sent in, and copies of another of JH's papers that have been printed.
ALS 2pp RS:HS 16.189

11969. 1862–8–13 From Titus Armellini Rome
Giving information on his invention of the floating barometer.
ALS 4pp RS:HS 1.405

11970. 1862–8–13 From James Inglis Cochrane Largs
Thanks for the first book of the *Iliad*. Did not realize that it had come from JH. Comments on the translation of Greek verses.
ALS 2pp RS:HS 5.353

11971. 1862–8–14 To Augustus De Morgan Collingwood
On preparing convenient small tables to be carried with one; comments on concept formation.
C 2pp RS:HS 23.383

11972. 1862–8–14 From G. Sabler Vilna
Encloses GS's address in hopes of meeting esteemed JH.
ALS 2pp RS:HS 17.165

11973. 1862–8–15 From Augustus De Morgan 91 Adelaide Road
Does not use tables for rough and ready work, but backs his tables against the slide rule. Thanks for the dialogue ['On Atoms']. Regarding blind atoms.
ALS 4pp RS:HS 6.359 S. De Morgan, *Memoir of Augustus De Morgan*, 314–15 inc

11974. 1862–8–15 From Henry Pearse St. John's Rectory, Bedford
Is trying to make arrangements for Thomas Maclear's son Harry to find a means of survival, and is seeking JH's advice.
ALS 6pp RS:HS 13.332

11975. 1862–8–15 From Edward Sabine 13 Ashley Place
Forwarded JH's request for additional £16 [to publish map of nebulae] to [R.S.L. secretaries] William Sharpey and G. G. Stokes. Expects approval.
ALS 2pp RAS:JH Archive 12/1.10.2; Reel 10

11976. 1862–8–15 To Edward Sabine Collingwood
JH's catalog of nebulae, reduced to 1880 as originally planned, is nearly complete and consumed all of £20 granted by R.S.L. Seeks £16 more to expand catalog and reduce data to 1860, as recommended by G. B. Airy.
ALS 4pp RS MC.6.244 (Adraft: TxU:H/L-0372; Reel 1054 & C: RS:HS 23.384)

11977. 1862–8–16 To Edward Sabine Collingwood
Needs to know soon if R.S.L. will approve grant to expand JH's catalog of nebulae. Otherwise G. B. Airy will dismiss man employed to do those computations. [JH annotation: Sent similar letter to G. G. Stokes on same date.]
AdraftS 3pp †(by JH) TxU:H/L-0373; Reel 1054

11978. 1862–8–19 From Titus Armellini Rome
Further information on his invention of the floating barometer.
ALS 4pp RS:HS 1.406

11979. 1862–8–20 From George Gabriel Stokes Malahide, Ireland
Edward Sabine sent JH's letter to GS with instructions to forward it to William Sharpey. Sees no difficulty in proceeding.
ALS 2pp RAS:JH Archive 12/1.10.6; Reel 10

11980. 1862–8–21 From J. J. Ewer & Co. 87 Regent St.
Regarding a purported relation of Caroline Herschel and her right to any property.
ALS 2pp RS:HS 7.139

11981. 1862–8–21 From Edward A. Seymour Admiralty
Unable to appoint a Mr. Maclea [Harry Maclear?], on whose behalf JH had applied, to Cadetship in the Royal Marines, in justice to many older candidates.
ALS 2pp RS:HS 17.197

11982. 1862–8–25 From Henry Pearse St. John's Rectory, Bedford
Informs JH of the effectiveness of everyone's activity on behalf of Harry Maclear [see HP's 1862–8–15], and thanks JH for his part in this.
ALS 2pp RS:HS 13.333

11983. 1862–8–25 From Edward Sabine [St. Mary's Cottage], Llandovery
Officially approves JH's request for additional £16.
ALS 2pp RAS:JH Archive 12/1.10.3; Reel 10

11984. 1862–8–28 From George Smith Interlaken
Relating to publication of JH's work in the *Cornhill Magazine*.
ALS 1p RS:HS 16.190

11985. 1862–8–31 To George Biddell Airy Collingwood
Expects R.S.L. to provide additional money, but JH will finish work himself rather than go over the amount granted [see GA's 1862–8–12].
ALS 2pp RGO 6.200.51

11986. 1862–9–1 From Johannes von Gumpach Hauteville, Guernsey
Has published a pamphlet on the *True Figure and Dimensions of the Figure of the Earth*, which he sends for his comments. Has had no success with G. B. Airy. Has found an error in J. F. Encke's work on the comet.
ALS 6pp RS:HS 9.72

11987. 1862–9–2 From George Biddell Airy Royal Observatory, Greenwich
Sorry to hear about JH's pain. Encloses notes from GA's testimony to parliamentary committee investigating changing standards of weights and measures. GA's work continues on 'veritable precessions.'
ALS 3pp RAS:JH Archive 12/1.1.8; Reel 10

11988. 1862–9–3 From John Bullar Basset Wood near Southampton
Is grateful for the ? which made him laugh. Hopes that Lady Herschel will accept the enclosed.
ALS 2pp RS:HS 4.332

11989. 1862–9–3 To Johannes von Gumpach Collingwood
Thanks for the pamphlet on the Figure of the Earth, but regrets he cannot agree with its conclusions. Points out some of the more serious errors.
draft 4pp RS:HS 9.73 & 23.386

11990. 1862–9–11 From John C. Sikes n.p.
Confirms through personal experience a passage in JH's *Physical Geography* regarding an 1860 earthquake.
ALS 4pp RS:HS 17.204

11991. 1862–9–13 From Smith, Elder & Co. 65 Cornhill
Note accompanying proofs of a contribution by JH to 'Survey of Literature, Science, and Art'.
ALS 1p RS:HS 16.207

11992. 1862–9–14 From Warren de La Rue Observatory, Cranford
Regarding the pictures of the moon by M. M. Gussew. Gives details of the eclipse of 1860. Heliograph has been working at his observatory since February. G. Sabler has now left and will doubtless do good work when he receives his instrument from J. H. Dallmeyer.
ALS 4pp RS:HS 6.150

11993. 1862–9–15 From Johannes von Gumpach Hauteville, Guernsey
Thanks for his frank statement on his pamphlet, but would like his opinion later when he has studied it more thoroughly.
ALS 3pp RS:HS 9.74

11994. 1862–9–16 From Smith, Elder & Co. 65 Cornhill
Note accompanying some material being returned to JH.
ALS 1p RS:HS 16.208

11995. 1862–9–17 From John Lee Hartwell House
James Glaisher has made two more balloon ascents and reached six miles high. W. R. Birt has been observing at Hartwell during the present lunation and discovered several new spots, a sea, and craters. Sends details. Is sending a petition to H. J. Temple (3rd Viscount Palmerston) for a pension for Birt and would be glad if JH would look it over and sign it.
ALS 4pp RS:HS 11.182

11996. 1862–9–17 To *The London Review* Collingwood
Congratulations on the balloon ascent to 30,000 feet by James Glaisher and Henry Coxwell. Suggests far greater heights are possible using compressed oxygen to breathe and the 'Peruvian *coca* leaf' [cocaine] as a source of strength.
L *London Review, 5* (Sept. 20, 1862), 264

11997. 1862–9–26 From William Lassell 9 Piazza Slierna, Malta
Describes galaxy-like gaseous nebula with interior elliptical ring.
ALS 4pp RAS:JH Archive 12/1.7.5; Reel 10

11998. 1862–9–26 From William Lassell 9 Piazza Slierna, Malta
Delighted with performance of WL's telescopes in Maltese climate. Invites JH to visit. Awaiting JH's catalogue of nebulae. Seeking assistant.
ALS 4pp RAS:JH Archive 12/1.7.4; Reel 10

11999. 1862–9–26 From William Parsons (Lord Rosse) 80 Marine Parade, Brighton
Encloses letter from G. J. Stoney, who had journal of WP's nebulae observations. Hopes WP's observations will be helpful for JH's proposed catalogue of nebulae.
ALS 4pp, encl ALS 7pp RAS:JH Archive 12/1.2.19; Reel 10

12000. 1862–9–27 To Joseph Whitworth Collingwood
Quotes report of limitations in JW's new artillery shell for penetrating ironclad vessels. Recounts JH's [Apr. 1836] experiments at Cape of Good Hope to increase explosive power of gunpowder by adding [Luke] Howard's fulminating mercury.
ALS 4pp RS:HS 18.403 (C: 23:387)

12001. 1862–9–29 From Heinrich Louis d'Arrest Copenhagen
More about the nebula catalog JH is preparing [see HA's

1862–3–12], and still wondering if there is some way HA's observations might be used.
ALS 3pp (French) RAS:JH Archive 12/1.8.12; Reel 10

12002. 1862–9–30 From George Smith 45 Pall Mall
Further to publication of JH's work [see GS's 1862–8–28 & 1862–8–12].
ALS 1p RS:HS 16.191

12003. 1862–10 From George Phillips Bond Cambridge, [Mass.]
Forwarding account of the great comet of 1858. Regarding illustrations for astronomical monographs. Includes several publications of Harvard Observatory.
ALS 2pp RS:HS 4.179

12004. 1862–10–1 To George Smith Collingwood
Sends JH's translation of Book II of Homer's *Iliad* (1862). Hopes to publish it in *Cornhill Magazine*. Three reasons for wanting it to appear soon.
AdraftS 3pp RS:HS 19.150 (C: RS:HS 23.388)

12005. 1862–10–3 To [William Lassell] Collingwood
Thanks for superb new information about nebulae in William Herschel's list. Wishes Alexander Herschel could serve as [WL]'s assistant in Valetta. Success of JH's son Alexander in chemistry and meteor research.
AdraftS 4pp TxU:H/L-0732; Reel 1089

12006. 1862–10–7 From William Henry Sykes British Association, Cambridge
Communicates Balloon Committee's feeling that balloon should not go as high as JH suggested, despite Henry Coxwell's having safely ascended higher. Reports successful presentation of previous results, hoping that Committee will find sufficient funding to continue.
ALS 4pp RS:HS 17.159

12007. 1862–10–8 From Henry Hart Milman Bedgebury Park, Cranbrook
Has been amused and interested by the paper that JH has sent him. Is sure the world would welcome the pronouncements of JH on the atomic theory.
ALS 2pp RS:HS 12.349

12008. 1862–10–8 From William Selwyn Foxton Royston
Discusses problems with darkness on solar autographs and also the appearance of a white spot or 'ghost.' Cannot explain the 'notch' in the Aug. 4 autographs.
ALS 8pp RS:HS 15.470

12009. 1862–10–16 To William Selwyn Collingwood
Has photographs of September 23 . Desires half pictures 'with the sun's limb in the middle of the view.' Cannot yet draw conclusions about the notches of the limb.
ALS 3pp RS:HS 15.471 (C: RS:HS 23.389)

12010. 1862–10–17 To T. P. Anderson Collingwood
Not aware of any changes in trust fund status of T. H. Hollier since 1861. Returns letter to bank accountant.
ACCS 1p TxU:H/L-0032; Reel 1054

12011. 1862–10–17 To Heinrich Louis d'Arrest n.p.
Reduction of JH's catalog of nebulae will not be completed before Christmas. HA's success in discovering nebulae and novae. Has compared positions of all nebulae observed by both JH and William Herschel; very few show evidence of shift. Thanks for offering HA's observations, but space in catalog is too limited.
Adraft 2pp inc TxU:H/M–0347.2 & TxU:H/M–0350.2; Reel 1087

12012. 1862–10–17 To Heinrich Louis d'Arrest n.p.
Nebulous nature of some stars. Recent reports on variability of some nebulae.
ACCS 2pp TxU:H/L-0135; Reel 1054

12013. 1862–10–17 From Johannes von Gumpach Hauteville, Guernsey
Has had no further communication from JH and would be pleased to receive any further comments if JH has now read the pamphlet more thoroughly.
ALS 4pp RS:HS 9.75

12014. 1862–10–18 From William Selwyn Foxton Royston
Will send requested half pictures. Notes that the 'ghost' always appears in the center of the plate. Discusses position lines and notches.
ALS 4pp RS:HS 15.472

12015. 1862–10–21 From James Dalmahay to Mr. Stewart 9 Flores St.
Believes he has solved the mystery of rain gauges holding more water when at the top of a building than when lower down. Would like JH's address so that he can communicate with him.
ALS 4pp RS:HS 6.464

12016. 1862–10–22 From William Lassell 9 Piazza Slierna, Malta
Received JH's list of nebulae. Will observe and confirm as many as possible. Honored by Alexander Herschel's offer to be WL's assistant. Defines WL's expectations of assistant.
ALS 4pp RAS:JH Archive 12/1.7.8; Reel 10

12017. 1862–10–24 To William Lassell Collingwood
Apologizes for returning WL's letter unopened. JH receives many letters with postage due from strangers and does not accept them. Did not recognize WL's address in Malta. Gives summary of JH's 3 Oct. 1862 letter to WL that seems to have been lost.
ACC 1p TxU.H/M–0383.2; Reel 1087

12018. 1862–10–24 From W. Paterson Glossop
Is giving a lecture on the atmosphere, and wonders if JH has written anywhere on the meteorology of the year 1860.
ALS 3pp RS:HS 13.243

12019. 1862–10–25 To Henry Wentworth Acland Collingwood
Discusses standards of weight and measure, especially the idea of a proposed unit called the 'Medical Grain.' JH strongly opposes this.
C 3pp Bodleian Library, Oxford

12020. 1862–10–25 From William Sharpey R.S.L.
R.S.L. Council voted additional £20 for reduction of JH's catalogue of nebulae. Inform Walter White where money should be sent.
ALS 1p RAS:JH Archive 12/1.10.19; Reel 10

12021. 1862–10–27 From Adalbert Adolf Mühry Göttingen
Brings JH news of AM's recent activity, including another book on climatology.
ALS 2pp (German) RS:HS 12.377

12022. 1862–10–27 From Walter White R.S.L.
Deposited £20 in JH's account at Drummond's Bank.
ALS 1p RAS:JH Archive 12/1.10.17; Reel 10

12023. 1862–10–28 From Henry Wentworth Acland Oxford
Informing him of a projected change in the scale of weights to be used in the new British Pharmacopaeia. Would like JH's opinion on this.
ALS 7pp RS:HS 1.383

12024. 1862–10–29 To Johannes von Gumpach Collingwood
Does not agree with the conclusion of JG's pamphlet, nor does he think an expedition necessary. Suggests termination of correspondence.
draft 1p RS:HS 9.76 & 23.390

12025. 1862–10–29 From Sir Richard Owen British Museum
Expressing thanks for the gift of one of JH's memoirs.
ALS 1p RS:HS 13.196

12026. 1862–10–29 To William Whewell Collingwood
Is sending his essay on atoms. Responds to WW's comments on JH's *Iliad* translation. Comments on controversy between G. B. Airy and Edward Sabine.
ALS 3pp TC, Camb. Add. Ms.a.207[106] (C: RS:HS 23.391)

12027. 1862–10–31 From Sir Charles Lyell 53 Harley St.
Has been requested by Edward Twisleton of the Public Schools Commission to give evidence in favor of introducing the teaching of the elements of science in schools. Would like to know JH's opinion, and also if he is willing to give evidence.
ALS 4pp RS:HS 11.427

12028. 1862–[11–1] To the *Athenaeum* n.p.
Calls attention to a report of a lake in Central America that 'gives rise to *two rivers.*'
L *Athenaeum,* #1881 (Nov. 14, 1863), 652 inc

12029. 1862–11–1 From Johannes von Gumpach Hauteville, Guernsey
Comments on JH's objections. Hopes to print some of the correspondence in his forthcoming book.
ALS 6pp RS:HS 9.77

12030. [1862]–11–1 From Henry Holland Brook St.
Will send copy of second edition of HH's *Essays on Scientific...Subjects* (1862). Received letter from HH's friend and former patient Louis Napoleon, who will send copy of Napoleon's book on Gallic campaigns of Caesar.
ALS 3pp RS:HS 9.436

12031. 1862–11–1 From William Lassell 9 Piazza Slierna, Malta
Best candidate for WL's assistant seems to be [Albert] Marth of Durham. Describes new planetary nebula.
ALS 4pp RAS:JH Archive 12/1.7.10; Reel 10

12032. 1862–11–4 From John R. Hind 22 Grove Road, St. John's Wood
Compares positions for variable nebula in Taurus reported by several astronomers. Plans to incorporate JH's suggestion regarding Julian calendar into more convenient table of dates.
ALS 4pp RAS:JH Archive 12/1.6.6; Reel 10

12033. 1862–11–5 To George Biddell Airy Collingwood
Is working on the descriptions of the nebulae for the catalogue [see GA's 1862–9–2].
ALS 4pp RGO 6.200.54

12034. 1862–11–6 From George Biddell Airy Royal Observatory, Greenwich
Little work remaining in calculations for JH's nebulae. Likes JH's plan to represent 'the quantification of different qualifications' by numbers. Hold payment until work is done. Astounding disturbances in 'magnetic curves' (1841–1857). GA now accepts usefulness of magnetic observatories.
ALS 4pp RAS:JH Archive 12/1.1.9; Reel 10

12035. 1862–11–6 From Edward Sabine 13 Ashley Place
Tells who received R.S.L. medals. Discusses possibility that [Henri] Regnault's recent work deserves Copley Medal. Wants JH's opinions. JH's letter concerning Melbourne telescope was sent to Lord Rosse.
ALS 4pp RS:HS 15.261

12036. 1862–11–7 From George Biddell Airy [Royal Observatory, Greenwich]
The reductions are essentially complete; encourages JH in his part of the work and concludes 'It will be a noble work' [see JH's 1862–11–5].
LACC 2pp RGO 6.200.60

12037. 1862–11–7 From George Biddell Airy n.p.
About some major variations in terrestrial magnetism readings.
ACCS 2pp RGO 6.695.490

12038. 1862–11–7 From R. Freke Dublin
Regrets the delay in his reply, but JH's letter was misdirected. It appears that his own theories are neither original nor correct and thanks him for his opinions.
ALS 4pp RS:HS 7.380

12039. 1862–11–8 From Charles Babbage Dorset St., Manchester Square
Marquis of Bute [J. P. C. Stuart] purchased property of CB's sister [Mary Anne Hollier] in Cardiff. [Charles] Few & Co., managers of property, sent title for CB's and JH's signatures. Sorry that JH was troubled again by pressure from CB's nephews for money.
ALS 3pp † TxU:H/M–0034; Reel 1062

12040. 1862–11–8 To Charles Babbage Collingwood
Regarding financial affairs. Is having trouble with his eyes. Has done little of late.
ALS 4pp RS:HS 2.332

12041. 1862–11–8 From Edward Twisleton 3 Rutland Gate
At the suggestion of Charles Lyell, requests JH's support for his plan for public school reform. Concerns include teaching of modern languages and time allotted for natural science. Calls for firm grounding in mathematics, including a thorough study of Books I–IV of Euclid's *Elements of Geometry*.
ALS 2pp RS:HS 17.452a

12042. 1862–11–9 To George Biddell Airy Collingwood
Accepts an invitation to dine with GA, and adds comments about problems JH is having with his vision.
ALS 4pp RGO 6.380.265

12043. 1862–11–10 From Alfred William Bennett 5 Bishopsgate St.
Sends a copy of the first part of A. D'Orsan's new work on *Our Satellite*. Intends to bring it to the notice of the R.A.S. at

their next meeting.
ALS 2pp RS:HS 5.47

12044. 1862–11–10 From Norman Macleod Adelaide Place, Glasgow
Is the editor of *Good Words* and is trying to build up the quality of its contributors. Would be grateful if JH could be induced to contribute some articles.
ALS 4pp RS:HS 12.260

12045. 1862–11–11 To James Dalmahay Collingwood
Comments on some papers of JD relating to the theory of rain.
C 1p RS:HS 23.392

12046. 1862–11–11 From R. Freke Dublin
Recognizes the force of JH's arguments but wrote his paper not for a reputation in science but for protection of his character as a gentleman.
ALS 4pp RS:HS 7.381

12047. 1862–11–11 To William Lassell Collingwood
Suggests Albert Marth of Durham observatory as possible assistant to WL. JH's notes and diagram of nebular ring.
ACCS 2pp TxU:H/L-0221; Reel 1054

12048. 1862–11–11 From Edward Sabine 13 Ashley Place
Lord Rosse agrees with JH regarding Melbourne telescope. His only objection to proposed increase in size is because of the added expense.
ALS 1p RS:HS 15.262

12049. 1862–11–12 To Edward Twisleton Collingwood
Claims 'I really have no definitely formed opinion' concerning public school curricula. Approves of ET's plan for teaching mathematics and increased study of foreign language.
ACCS 2pp RS:HS 17.452b (C: RS:HS 23.393)

12050. 1862–11–13 From Edward Twisleton 3 Rutland Gate
Wants JH to write to the Public School Commissioners to express his views on what direction education should take in the public schools, especially JH's views on the teaching of the physical sciences.
ALS 8pp RS:HS B27.85

12051. 1862–11–14 From James Dalmahay 9 Forres St.
Is grateful for his comments on his paper. Will adapt them to his own hypothesis.
ALS 2pp RS:HS 6.465

12052. 1862–11–14 From G. Mühry Hamelin
Writes to ask JH to contribute to the living expenses of GM's brother, Adalbart Adolf Mühry.
ALS 6pp (German) RS:HS 12.378

12053. 1862–11–14 From Edward Sabine 13 Ashley Place
Sends [William] Lassell's reply and JH's own letter regarding telescope at Melbourne for review. All will be forwarded to Lord Rosse. Also sends lithograph.
ALS 3pp RS:HS 15.263

12054. 1862–11–16 To Harriet Lubbock n.p.
Thanks for the perusal of the papers; these show a maturity of thought unexpected in a youth of eighteen. Comments on the moral questions included.
C 4pp RS:HS 11.349 & 23.394

12055. 1862–11–17 From George Biddell Airy Royal Observatory, Greenwich
Astronomische Nachrichten 1392 compares positions of forty nebulae by G. F. J. A. von Auwers, H. L. d'Arrest, and P. E. Laugier. Calculations for JH's [Catalogue of] Nebulae are complete and ready for inspection.
ALS 2pp RAS:JH Archive 12/1.1.10; Reel 10

12056. 1862–11–17 From Walter White R.S.L.
Please send receipt for £20.
ALS 1p † RAS:JH Archive 12/1.10.18; Reel 10

12057. 1862–11–18 To George Biddell Airy Collingwood
Asks GA for a copy of G. F. J. A. Auwers's nebula catalogue [see GA's 1862–11–17], and a sample of the nebula listing for JH's catalogue.
ALS 3pp RGO 6.200.62

12058. 1862–11–18 To Edward Sabine Collingwood
Discusses proposed Melbourne telescope, including [William] Lassell's proposed mounting and its implications for observing nebulae.
L draft 2pp RS:HS 15.264 (C: 23.395)

12059. 1862–11–19 From George Biddell Airy Royal Observatory, Greenwich
Accounts for money spent in reducing JH's catalogue of nebulae. Encloses sample page from catalogue and *Astronomische Nachrichten No. 1392* [see JH's 1862–11–18].
ALS 2pp RAS:JH Archive 12/1.10.13; Reel 10 (ACCS: RGO 6.200.64)

12060. 1862–11–19 From Alfred William Bennett 5 Bishopsgate St.
Thanks for letter. Further regarding *Our Satellite* by A. D'Orsan and its similarity to a work of Warren de La Rue.
ALS 3pp RS:HS 5.48

12061. 1862–11–20 To Alfred William Bennett Collingwood
Comments on the recent meeting at the R.A.S. when A. D'Orsan's book was shown.
draft 2pp RS:HS 5.49 & 23.396

12062. 1862–11–20 From William Hallowes Miller 7 Scroope Terrace, Cambridge
Writes to invite JH to publish, in the *R.S.P.T.*, his report on William Hopkins's paper on glaciers.
ALS 2pp RS:HS 13.41

12063. 1862–11–20 To George Gabriel Stokes Collingwood
Requests return of proofs, in triplicate, of JH's report on a paper on stopping cracks in cement.
ALS 1p Cambridge University Library 7656/H701

12064. 1862–11–24 From James Glaisher Blackheath
Sending the results of some recent balloon flights. Comments on these. Would be pleased to see Alexander Herschel when he is free.
ALS 8pp RS:HS 8.133

12065. 1862–11–24 To Henry de Worms Collingwood
Comments on HW's *The Earth and Its Mechanism.* HW erred by taking at face value François Arago's 'enormous blunder' of proving rotation from diurnal aberration.
C 4pp RS:HS 18.433 (C: 23.397)

12066. 1862–11–29 From Edward Joseph Lowe Beeston
Encloses account of a meteor seen on the 27th; pass it over to

JH's son as it may be useful. Has been making observations on ozone.
ALS 4pp RS:HS 11.336

12067. 1862–12–1 To Norman Macleod Collingwood
Regrets delay in answering letter but has been away from Collingwood and letter was not forwarded. Is unwilling to write article for another journal as he is already producing articles for an editor. Would editor of *Good Words* be interested in printing three lectures he gave at Hawkhurst on volcanoes, comets, and the sun? Could work this up for an article.
draft 4pp RS:HS 12.264 & 23.398

12068. 1862–12–3 From W[illiam] G[eorge] Spencer Derby
In light of recent discussions of metric system, which WS dislikes for 'lack of divisibility,' asks where to find JH's article on advantages of British measurement system.
ALS 2pp RS:HS 17.225

12069. 1862–12–4 From Mr. Schaeffer 29 St. Swithins Lane
Is sending from the Navy Board a new publication.
ALS 1p RS:HS 17.171

12070. 1862–12–5 From Edward Sabine 13 Ashley Place
Discusses proposed Melbourne telescope, opinions of various scientists, and projected costs of construction.
ALS 4pp RS:HS 15.265

12071. 1862–12–6 From Edward Sabine 13 Ashley Place
Sends [William] Lassell's letter regarding Melbourne telescope. Pleased with scientific memoirs received from Harvard College.
ALS 2pp RS:HS 15.266

12072. 1862–12–8 From John R. Hind Nautical Almanac Office, Gray's Inn
E. W. L. Tempel's 1860 discovery of nebula near Merope.
ALS 3pp RAS:JH Archive 12/1.6.7; Reel 10

12073. 1862–12–10 To Edward Sabine Collingwood
Extensive comments on the type and construction of telescope to be supplied to Melbourne University.
C 4pp Cambridge University Library 7656/H702 (C: Rosse Papers K2.)

12074. 1862–12–15 From Sir Isaac L. Bell Washington Co. Durham
Invitation to JH to visit him if he is attending the B.A.A.S. meeting.
ALS 3pp RS:HS 4.40

12075. 1862–12–16 To Henry [Holland] Collingwood
Discusses some of HH's criticisms of JH's translation of Homer's *Iliad*; JH also remarks on the cause of the Gulf Stream.
ALS 4pp APS B.H435p.20

12076. 1862–12–17 From Alexander J. B. Hope Bedgebury Park
Announces a meeting of the Ball Committee to be held at Cranbrook.
ALS 1p RS:HS B27.87d

12077. 1862–12–17 From William Lassell 9 Piazza Sliema, Malta
About the procedures WL uses when observing, and the responsibilities of an assistant.
ALS 4pp RS:HS 11.164

12078. 1862–12–17 To William Henry Smyth Collingwood
Concerned over the fact the retiring pension for John Russell does not provide for his wife and children.
C 3pp RS:HS 23.399

12079. 1862–12–20 From Sir Isaac L. Bell Newcastle-upon-Tyne
Thanks for the gift of volumes on meteorology and the telescope. Hopes that JH will be able to visit Newcastle with the B.A.A.S.
ALS 2pp RS:HS 4.41

12080. 1862–12–23 From James Challis Cambridge
Has read JH's brochure with pleasure. Regarding the alleged obstruseness of some of his own mathematical theories.
ALS 2pp RS:HS 5.238

12081. 1862–12–23 From G[oodwin] Kilburne Hawkhurst
Informs JH of a meeting of the finance committee of the Hawkhurst Volunteers.
ALS 1p RS:HS B27.87c

12082. 1862–12–23 From William Henry Smyth St. John's Lodge
Directs JH to inquire regarding a pension for John Russell. Note C. P. Smyth's use of *Outlines Astr.* Mentions Isaac Fletcher as a promising amateur astronomer.
ALS 3pp RS:HS 16.320

12083. 1862–12–23 From George Gabriel Stokes Lensfield Cottage, Cambridge
Comments on matters relating to the provision of a reflecting telescope to Melbourne University [see JH's 1862–12–10].
ALS 4pp RS:HS 17.48

12084. 1862–12–24 From Robert FitzRoy 1 Parliament St.
Sending a gift of his *Weather Book*. Hopes JH will send his comments on chapter 18. Has scarcely seen M. F. Maury.
ALS 4pp RS:HS 7.261

12085. 1862–12–24 From George Gabriel Stokes Cambridge
Trying to resolve disagreement between author of a paper, William Hopkins, and JH, a referee.
ALS 2pp encl RS:HS 17.49

12086. 1862–12–25 To Robert FitzRoy Collingwood
Thanks for gift of a book [RF's *Weather Book*].
C 1p RS:HS 23.400

12087. 1862–12–25 To William Parsons (Lord Rosse) Collingwood
Discusses plans for a memorial for Thomas Maclear. Requests change in manner in which WP records positions of nebulae observed at WP's observatory.
ALS 4pp Rosse Papers K2.9

12088. 1862–12–27 To William Whewell Collingwood
Sends translation of Book III of Homer's *Iliad* and a memorial for Thomas Maclear.
ALS 4pp TC, Camb. Add. Ms.a.207[107] (C: RS:HS 23.401)

12089. 1862–12–28 To John Wrottesley Collingwood
Asks JW to sign the memorial for Thomas Maclear, indicating that JH has corrected the matter related to Maclear's pension.
C 1p RS:HS 25.12.6

12090. 1862–12–29 From Edward Joseph Lowe Beeston
Ozone powders are in dry state. Intends tomorrow to make experiments with ozone using a kite. Can JH inform him about the amount of nitric acid in the air.
ALS 3pp RS:HS 11.337

12091. 1862–12–30 To Edward Sabine Collingwood
Objects to Board of Visitors acting as 'self-constituted Court-Martial' to examine personal conduct of Board's chairman [James South], whom G. B. Airy no longer recognizes as competent. [JH notes that he did not send this letter, but kept it because it gave the *reasons* behind the shorter form [TxU:H/L-0374], which he preferred.]
ACCS 3pp TxU:H/L-0375; Reel 1054

12092. 1862–12–30 To Edward Sabine Collingwood
[Rough draft (1p) crossed out, followed by:] Confidential propostion that G. B. Airy submitted to Board of Visitors is inappropriate. Queen's warrant does not empower Board to consider such matters.
ACCS 1p TxU:H/L-0374; Reel 1054

12093. 1862–12–31 From Edward William Brayley London Institution
Would like to pass on JH's views on ice to A. R. Abbott, who is giving a lecture on glaciers at the Friend's Institute. Remarks on storms. Confusion of Thomas Young's views with JH's. Regarding JH's paper on earthquakes and volcanoes.
ALS 4pp RS:HS 4.226

12094. 1862–12–31 From G. Mühry Hamelin
Is wondering if his letter [see GM's 1862–11–14] reached JH.
ALS 1p (German) RS:HS 12.379

12095. 1862–12–31 From Smith, Elder & Co. 65 Cornhill
Note accompanying payment for JH's writing [see SE's 1862–9–13].
(A)LS 1p RS:HS 16.209

12096. 1863–1–1 From George Biddell Airy Royal Observatory, Greenwich
Invites JH to R.A.S. dinner on 9 Jan. to meet Sir G. C. Lewis.
ALS 1p RAS:JH Archive 13/3.98 verso; Reel 10

12097. 1863–1–1 To Edward William Brayley Collingwood
JH criticizes [Robert] Mallet's views on the nature of earthquakes and the upheaval of the earth.
ALS 3pp APS B.H435p.21 (C: RS:HS 24.1)

12098. 1863 From Robert FitzRoy n.p.
Ideas respecting gravitating lunisolar action on the atmosphere—as apart from that of heat or other influence.
RS:HS 19.217

12099. 1863 or later To Robert FitzRoy n.p.
Remains unconvinced by RF's ideas on gravitating lunisolar action on the atmosphere.
C 1p RS:HS 25.14.10

12100. [1863?] To John Herschel (son) [Collingwood]
A short note to comment on photographs son John has sent; also inquires about prospects for the India survey John has discussed.
ALS 2pp inc JHS 5.8b

12101. [1863] From Henry Holland Brook St.
Received JH's translation of fourth book of *Iliad.* Will return it with comments in a few days. Attended two or three excellent lectures by JH's son. HH needs to balance mental efforts with more physical activity.
ALS 3pp RS:HS 9.435

12102. [1863?] To [R.S.L. Council] n.p.
Reports on and believes fit for publication paper [*R.S.P.T., 153,*. 309–] by G. B. Airy on diurnal irregularities of terrestrial magnetism.
C 5pp diag RS RR.5.7

12103. 1863–1–2 From Camille Flammarion Paris
Sending a little work on the philosophy of astronomy [Pluralité des mondes habités].
ALS 1p (French) RS:HS 7.267

12104. [1863]–1–2 From Henry Holland Brook St.
Received manuscript of JH's translation of third book of *Iliad.*
ALS 2pp RS:HS 9.434

12105. 1863–1–2 To D. C. E. van Monckhoven Collingwood
Is honored by DM's letter and proposal to dedicate his book [*Traité général de photographie*] to JH. Points out some errors of attribution he would like incorporated if not too late. Pass on his good wishes to J. A. F. Plateau.
C 7pp RS:HS 12.359 & 24.2

12106. 1863–1–3 From George Biddell Airy Playford, near Ipswich
Concerning Thomas Maclear.
ALS 2pp RS:HS 1.257

12107. 1863–1–3 To Elizabeth Colling Collingwood
Thanks for a number of items of poetry and sends EC a translation of some poetry by Friedrich Schiller.
C 2pp RS:HS 24.3

12108. 1863–1–4 From John Tyndall Royal Institution
Asks permission to print extract of JH's letter on heat spots of spectrum of rock salt sample. Hopes JH's son [Alexander] will speak at Royal Institution; JT would help with illustration. Wonders how blue color of water relates to sky color and polarization.
ALS 4pp RS:HS 17.392a (C: RI 520)

12109. 1863–1–5 To John Tyndall Collingwood
Likes JT's explanation of sky's blue color. Has difficulty with JT's explanation of polarization of skylight. Comments on absorptive powers of vapor and liquid.
CC 4pp RS:HS 17.392b (C: 24.4; C: RI 521)

12110. 1863–1–6 From William H. Callcott Kensington
Is pleased that JH has used Lady Callcott's account of the 1822 earthquake at Valparaiso. She was pained by the doubts of her accuracy by G. B. Greenough.
ALS 3pp RS:HS 5.141

12111. 1863–1–6 From A[lexander Henley] Grant St. Bride's Avenue
Would like some contributions from JH for the *Churchman's Family Magazine.*
ALS 2pp RS:HS 9.116

12112. 1863–1–6 From William Lassell Malta
After receiving testimonials of the highest order he has appointed [Albert?] Marth to be his assistant. Encloses some observations of nebulae. If JH's sons are visiting Malta he would be pleased to see them.
ALS 3pp RS:HS 11.165

12113. 1863–1–8 From John Bullar Basset Wood near Southampton
Has been reading his work on volcanoes and earthquakes ['About Volcanoes and Earthquakes'] with interest. Regarding his own health.
ALS 4pp RS:HS 4.333

12114. 1863–1–9 From William H. Callcott Kensington
Is grateful for his reply. Where can he find a statement of Isaac Newton regarding the relationship between colors and music?
ALS 4pp RS:HS 5.142

12115. 1863–1–10 From Thomas P. Kirkman Croft Rectory near Warrington
Many thanks for his information. Comments again on the prize being offered by the Academy. The answer is now being published by the R.S.L.
ALS 3pp RS:HS 11.46

12116. 1863–1–12 From Joseph Henry Washington, D.C.
Official receipt for copy of JH's *The Telescope* (1861) by library at Smithsonian Institution.
(A)LS 1p TxU:H/L-0339.1; Reel 1054

12117. 1863–1–13 From D. C. E. van Monckhoven Gand
Received his letter and comments with pleasure. Asks some questions on photographic experiments.
ALS 3pp (French) RS:HS 12.360

12118. 1863–1–13 From William Parsons (Lord Rosse) Birr Castle, Parsonstown
Preparing classified index of observatory journals, but information will not be accurate enough to help JH. G. J. Stoney is too busy to help remeasure nebulae positions.
ALS 4pp RAS:JH Archive 12/1.2.20; Reel 10

12119. 1863–1–13 To George Biddell Airy Collingwood
Anxious to receive final sheets of nebula catalogue from GA [see GA's 1862–11–19], as JH wants to enter the descriptions and complete the work.
ALS 2pp RGO 6.200.69

12120. 1863–1–14 To Elizabeth Colling Collingwood
Thanks for a collection of poems; comments on lunar eclipses.
C 1p RS:HS 24.6

12121. 1863–1–14 From Richard Craven Whitby
Has read JH's publication on volcanoes ['About Volcanoes and Earthquakes']. Comments on the formation of rocks and the creation of the world.
ALS 3pp RS:HS 5.375

12122. 1863–1–14 From Edward James Stone Royal Observatory, Greenwich
Gives details of calculations [for JH's catalogue of nebulae]. Three more weeks of work needed.
ALS 1p RAS:JH Archive 12/1.1.17; Reel 10

12123. 1863–1–14 From Edward James Stone [Royal Observatory, Greenwich]
In GA's absence, ES reports on the state of completion of the final copy of the nebula catalogue [see JH's 1863–1–13].
ACCS 1p RGO 6.200.70

12124. 1863–1–15 From Anthropological Society London
Requesting him to become a member of the newly formed Anthropological Society.
ALS 3pp RS:HS 1.401

12125. 1863–1–15 From Robert FitzRoy Meteorological Dept., B.o.T.
Enjoyed reading the dialogues of Hermogenes and Hermione [JH's 'On Atoms']. Found William Higgins's book some years ago and it appeared the basis for John Dalton's views. Finds astronomy and geology the basis for Old Testament inspiration. How absurd is the modern notion of circuits.
ALS 6pp RS:HS 7.262

12126. 1863–1–19 To Augustus De Morgan Collingwood
Austrian consul, Mr. Schaeffer, sent JH one copy of *Voyage of the Navara* and map, care of R.A.S. If parcel arrives, send it to Smith, Elder & Co.
ALS 2pp RAS MSS Add.26.9

12127. 1863–1–20 From Jean Baptiste Elie de Beaumont Paris
Regarding their journals and publication dates.
ALS 1p RS:HS 5.31

12128. 1863–1–21 From William Rutter Dawes Hopefield, Haddenham, Thame
Is grateful for the note and brochure. Is grieved to hear of the illness in his family. His own health is not good. Comments on the article on Telescopes by JH in the *Encyclopaedia Britannica.*
ALS 4pp RS:HS 6.108

12129. 1863–1–22 To William Whewell Collingwood
Thanks WW for and comments on WW's *Lectures on Political Economy.* Has learned that the theory of rent is exploded. JH's daughter Julia is seriously ill.
ALS 3pp TC, Camb. Add. Ms.a.207[109] (C: RS:HS 24.7)

12130. 1863–1–26 From Asa Gray Cambridge, Mass.
The missing parts will be sent in the Spring. The Academy would be pleased to receive any of JH's publications. Thanks for the pamphlet 'On Atoms.'
ALS 2pp inc RS:HS 9.120

12131. 1863–1–26 From William Selwyn College, Ely
WS's photographer may get a grant from the R.S.L. Object glass for new telescope was stolen. They are still using the old one.
ALS 4pp RS:HS 15.473

12132. 1863–1–27 To William Parsons (Lord Rosse) Collingwood
Has received letter from Lord Palmerston [Henry John Temple] approving JH's request for a memorial for Thomas Maclear.
ALS 1p Rosse Papers K2.10

12133. 1863–1–27 To John Wrottesley Collingwood
Informs JW that JH has received notice from H. J. Temple [Lord Palmerston] that Thomas Maclear is to be put on the next list to receive a pension.
C 1p RS:HS 25.12.7

12134. 1863–1–28 From Herbert Spencer Hinde St., Manchester Square
Thanks JH for sending HS a copy of JH's dialogue ['On Atoms']. Comments on it.
ALS 2pp RS:HS 16.492

12135. 1863–1–29 From William Rutter Dawes Hopefield, Haddenham, Thame
Pleased to hear that the memorial on behalf of Thomas Maclear has been successful. Comments on the mild weather. Hopes the health of his daughter is improving.
ALS 2pp RS:HS 6.109

12136. 1863–1–29 From Robert Main Radcliffe Observatory
Is pleased that the Government have put Sir Thomas Maclear on the Civil list. Has read JH's paper on atoms with much interest and amusement. Hopes that JH will call at the Observatory when in Oxford. Finds the R. C. Carrington transit circle a valuable instrument. Comments on his observations.
ALS 4pp RS:HS 12.279

12137. 1863–1–29 From D. C. E. van Monckhoven Gand
Further inquiries on photographic processes.
ALS 1p (French) RS:HS 12.361

12138. 1863–1–29 From John Henry Whitaker Manchester
Eighteen months ago, JH sent to JW autographs of William and Caroline Herschel and Baron von Humboldt. Now seeks another of each.
ALS 2pp RS:HS 18.397

12139. 1863–1–30 From Amédée Victor Guillemin 17 Quai d'Anjou, Paris
Is preparing a popular work on astronomical phenomena and would welcome copies of JH's *Outlines Astr.* and *Cape Results*.
ALS 1p (French) RS:HS 9.67

12140. 1863–1–30 From John Lee Hartwell House
Is pleased H. J. Temple (3rd Viscount Palmerston) has replied so promptly to JH's memorial regarding Thomas Maclear. Would like JH's sanction for JL's support of George Rümker as the new Observer at Hamburg.
ALS 3pp RS:HS 11.183

12141. 1863–1–30 To George Gabriel Stokes Collingwood
Further to dealing with report on William Hopkins's paper [see GS's 1862–12–24].
ALS 2pp Cambridge University Library 7656/RS332

12142. 1863–1–31 From Smith, Elder & Co. 65 Cornhill
Note accompanying payment for 'Notes on Science' for *Cornhill Magazine*
(A)LS 1p RS:HS 16.210

12143. 1863–2–1 From Heinrich Louis d'Arrest Copenhagen
Is sending JH a list of 38 nebulae, their positions reduced to 1830, and described using the terms that JH uses. Comments on some of the difficulties in identifying the nebulae, and includes reference to other astronomer's work.
ALS 4pp (French) RAS:JH Archive 12/1.8.13; Reel 10

12144. 1863–2–2 From Edward Sabine 13 Ashley Place
Wrote to [A.] Sawitsch regarding Russian pendulum experiments. Encloses Sawitsch's reply. Discusses experimental possibilities.
ALS 2pp RS:HS 15.267

12145. 1863–2–3 From William Radcliffe Birt Victoria Park
Warren de La Rue has suggested that WB apply for a government grant to further his researches on the lunar surface. Would be grateful for his support.
ALS 3pp RS:HS 4.143

12146. 1863–2–3 From G. Mühry Hamelin
Thanks JH for the efforts made on behalf of his brother, Adalbart Adolf Mühry. [The London Medical Relief Fund was only intended for residents of England, and so could not help.]
ALS 2pp (German) RS:HS 12.380

12147. 1863–2–3 From Thomas Newbon Hammersmith
Informing JH that TN's father died this morning.
ALS 1p RS:HS 13.151

12148. 1863–2–5 To Edward Sabine Collingwood
Agrees that the pendulum should be observed at major Indian stations because Russians did not avail themselves of offered vacuum apparatus and pendulums.
C 1p RS:HS 24.8

12149. 1863–2–6 From George Biddell Airy [Royal Observatory, Greenwich]
Is sending JH the completed sheets of nebula reductions for printing [see JH's 1863–2–6].
ACCS 2pp RGO 6.200.73

12150. 1863–2–6 To George Biddell Airy Collingwood
Once again anxious about completion of nebula catalogue, and about cost over run [see Edward Stone's 1863–1–14].
ALS 3pp RGO 6.200.71 (C: TxU:H/L-0022)

12151. 1863–2–6 From William Hopkins Cambridge
Has received the report on his paper on glaciers. Comments on one or two points. Hopes it will create interest in the subject.
ALS 3pp RS:HS 9.480

12152. 1863–2–7 From George Biddell Airy Royal Observatory, Greenwich
Sends a final accounting for the calculating work done at the Royal Observatory on JH's nebula catalogue [see JH's 1863–2–7].
ALS 3pp RAS:JH Archive 12/1.10.14; Reel 10 (ACCS: RGO 6.200.76)

12153. 1863–2–7 To George Biddell Airy Collingwood
Acknowledges receipt of nebula catalogue sheets from GA [see GA's 1863–2–6]; apologizes for JH's anxious letter [see JH's 1863–2–6].
ALS 3pp RGO 6.200.74

12154. 1863–2–8 From Amédée Victor Guillemin 17 Quai d'Anjou, Paris
Is grateful for JH's kind action [see AG's 1863–1–30]. Address the parcels to the London address of Hachette.
ALS 1p (French) RS:HS 9.68

12155. 1863–2–9 From George Biddell Airy Royal Observatory, Greenwich
Giving results (including diagram) of a comparison of earth currents and magnetometer disturbances.
ALS 6pp RS:HS 1.258 (C: RGO 6.476.118)

12156. 1863–2–10 To William Radcliffe Birt Collingwood
Is satisfied with WB's work on meteorology. Comments on WB's intention to diagram the moon.
C 1p RS:HS 24.9

12157. 1863–2–13 To George Biddell Airy Collingwood
Has found a few minor errors in the calculations checked so far [see GA's 1863–2–6]; comments on magnetic tracings JH received from GA.
ALS 3pp RGO 6.200.79

12158. 1863–2–13 To George Biddell Airy n.p.
Offers a few ideas to GA [see GA's 1863–2–9], but does not

have a coherent explanation.
LCC 1p inc RGO 6.476.126

12159. 1863–2–14 From George Biddell Airy n.p.
Will take into account JH's suggestions [see JH's 1863–2–13] and see where this leads.
ACCS 1p RGO 6.476.127

12160. 1863–2–15 From Jacques Babinet Paris
Regarding the gravimetric balance.
ALS 4pp (French) RS:HS 3.5

12161. 1863–2–17 From Edward Sabine 13 Ashley Place
Sends [Warren] de La Rue's letter on Southern[?] Telescope for JH's review. Swedish Academy working on reviving project of measuring arc of meridian.
ALS 2pp RS:HS 15.268

12162. 1863–2–19 To Jacques Babinet Collingwood
Suggestions on improving JB's gravimetric balance, constructed on principle of 'Bifilar suspension.' Refers to 1861 report of similar invention by J. A. Brown and another article on this topic.
C 4pp RS:HS 19.151 (C: RS:HS 24.10)

12163. 1863–2–19 From William Corbett Burder Bristol
Sending him a correction for his article on meteorology relating to the rainfall of Bristol. Intends publishing a book on climate and will send him a copy.
ALS 2pp RS:HS 5.128

12164. 1863–2–20 From George Biddell Airy Royal Observatory, Greenwich
Did JH receive G. F. J. A. von Auwers's catalog of JH's and William Herschel's nebulae?
ALS 1p RAS:JH Archive 12/1.1.11; Reel 10

12165. 1863–2–20 From George Biddell Airy [Royal Observatory, Greenwich]
Is JH interested in seeing G. F. J. A. Auwers's catalogue of JH's and William Herschel's nebulae?
ACCS 1p RGO 6.200.81

12166. 1863–2–20 From George Gabriel Stokes Royal Society
Asks JH to comment on the scientific merit of a large number of plates on solar spots sent to R.S.L. by Richard Carrington.
ALS 8pp RS:HS 17.50

12167. 1863–2–21 From John R. Hind Nautical Almanac Office
Would like JH to add his signature to Hind's certificate for the R.S.L., and then forward it to G. B. Airy.
ALS 3pp RS:HS 9.383

12168. 1863–2–22 To George Biddell Airy Collingwood
Might have saved JH some work if he had known what G. F. J. A. Auwers was doing [see GA's 1863–2–20]; has found some more errors in nebula calculations.
ALS 4pp RGO 6.200.82

12169. 1863–2–22 To Edward Sabine n.p.
Expresses interest in Swedish plan to survey Spitzbergen and to measure meridian of arc. Discusses gravitometer being developed by [Jacques] Babinet.
C 1p RS:HS 15.270 (C: RS:HS 24.12)

12170. 1863–2–22 To Edward Sabine Collingwood
Returns [Warren] de La Rue's letter and discusses its contents, especially the making, mounting, and mirror for de La Rue's reflecting telescope.
(S) 2pp RS:HS 15.269 (C: RS:HS 24.11)

12171. 1863–2–23 From George Biddell Airy Royal Observatory, Greenwich
Expect some errors in calculations of positions for JH's catalogue of nebulae. Suggestions on format.
ALS 2pp RAS:JH Archive 12/1.1.16; Reel 10

12172. 1863–2–23 From George Biddell Airy [Royal Observatory, Greenwich]
Refraction spectra of six stars exhibit lines corresponding with 'Solar Line F.'
ALS 1p RAS:JH Archive 12/1.1.15; Reel 10

12173. 1863–2–23 From George Biddell Airy Royal Observatory, Greenwich
Sends copy of G. F. J. A. von Auwers's catalog of JH's and William Herschel's nebulae. Observatory has another copy. Notes odd lines seen in 'Refraction Spectra of 11 stars.'
ALS 1p RAS:JH Archive 12/1.1.12; Reel 10

12174. 1863–2–23 From Adolphe Quetelet Brussels
Cruel illness has kept him from work for a while. Just finished *Histoire des sciences mathématique et physique chez les belges*. AQ's son works with proper motion, double and multiple stars, and absolute positions.
ALS 4pp (French) RS:HS 14.255 (ALS?: Académie belgique #28)

12175. 1863–2–23 From Alexander Wylie 18 Murray St., Camden Town
Heard of JH's 'domestic affliction.' Postpones visit till next week.
ALS 2pp RS:HS 19.305

12176. 1863–2–23 From Alexander Wylie 18 Murray St., Camden Town
Recalls translating *Outlines Astr.* into Chinese. Wishes to visit JH before returning to China.
ALS 2pp RS:HS 19.304

12177. 1863–2–24 From William Corbett Burder Bristol
Is grateful for his reply and anything may be quoted from his works. JH's *Prelim. Discourse* is just the work for the general public and should whet the appetite for more.
ALS 1p RS:HS 5.129

12178. 1863–2–26 To Augustus De Morgan Collingwood
Comments on family health; asks AD if he has read James Challis on the topics of the ether and of zodiacal light.
C 1p RS:HS 23.363

12179. 1863–2–27 From Charles Babbage Dorset St., Manchester Square
Apologizes for JH's further troubles with trust fund for CB's sister [Mary Anne Hollier], whose three sons are selling off their inheritances. Solicitor Charles Few will take care of legal issues. Not likely that JH can resign as trustee.
ALS 3pp TxU:H/M–0035; Reel 1062

12180. 1863–3–1 From John R. Hind Nautical Almanac Office
Explaining the discrepancy in the *Nautical Almanac;* will endeavor to make a clearer explanation in the next edition. Returns JH's son's memorandum.
ALS 4pp RS:HS 9.384

12181. [1863?]–3–2 From Edward Sabine 13 Ashley Place
Describes pendulum design. Believes [Henry] Kater's is the

best. Regrets Russians want to make own pendulum. Discusses nominees, including [G. P.] Bond, for R.S.L. foreign membership.
ALS 4pp RS:HS 15.313

12182. 1863–3–2 To George Gabriel Stokes Collingwood
Responds to GS's 1863–2–20, strongly recommending the support of the R.S.L. to defray part of the cost of publishing Richard Carrington's work.
draft 2pp RS:HS 17.51 (C: RS:HS 24.13)

12183. 1863–3–4 From Richard Christopher Carrington Isleworth
Has heard that JH has to decide if the illustrations to his own paper on sun spots are necessary. Thinks they are of vital importance.
ALS 4pp RS:HS 5.218

12184. 1863–3–5 To Heinrich Louis d'Arrest n.p.
JH received and will insert HA's list of nebulae into JH's catalogue. Progress of catalogue. Wishes JH had earlier knowledge of [G. F. J. A.] Auwer's reduction of William Herschel's nebulae, which JH just received.
ACCS 1p TxU:H/L-0136; Reel 1054

12185. 1863–3–5 From Augustus De Morgan 91 Adelaide Road
Is sorry to hear that the JHs have fever in the house. Hopes it will soon move away. Fever has been prevalent in Kent this year. Believes London is more healthy. Sends a problem.
ALS 3pp RS:HS 6.360

12186. 1863–3–5 From Joseph Jopling 6 Vassall Terrace
Several of the illustrations in JH's *Outlines Astr.* have a familiar look. Doubtless all based on eternal laws. Comments on his own theories.
ALS 3pp RS:HS 10.427

12187. 1863–3–5 From George Gabriel Stokes Royal Society
Acknowledges receipt of JH's 1863–3–2, and another dealing with disagreement with William Hopkins [see GS's 1862–12–24 & JH's 1863–1–30].
ALS 3pp RS:HS 17.52

12188. 1863–3–6 From Edward Joseph Lowe Beeston
Sending a paper he has written on his ozone experiments; please communicate it to the R.S.L. if JH thinks it worthy. Would JH request the R.S.L. to re-suspend EL's certificate as he seems certain of some support from Council. Will soon complete his work on ferns.
ALS 4pp RS:HS 11.338

12189. 1863–3–6 To Edward Sabine Collingwood
Supports G. P. Bond of Harvard College for foreign membership of R.S.L.; proposes H. E. Sainte-Claire Deville for the Rumford Medal, for his development of a high temperature laboratory furnace applied to metallurgy and the 'disseverance' of the hydrogen and oxygen of water.
ALS 4pp Cambridge University Library 7656/RS374A

12190. 1863–3–7 From Augustus De Morgan 91 Adelaide Road
Hopes all is well. Has not seen the point about James Challis to which JH refers, but Challis has produced strange theorems.
ALS 4pp RS:HS 6.361

12191. 1863–3–8 From A. V. Guillemin 17 Quai d'Anjou, Paris
Is grateful for the Memoirs. Hopes that JH will think his own work worthwhile when it appears.
ALS 2pp (French) RS:HS 9.69

12192. 1863–3–9 From Richard Christopher Carrington Isleworth
Is grateful for JH's report on his paper. Regarding the plates for this paper.
ALS 3pp RS:HS 5.219

12193. 1863–3–10 From Robert FitzRoy Meteorological Dept., B.o.T.
Book he intended to write has had to be postponed as he has been asked to prepare a second edition of his *Weather Book.* Sends a proof of the revision of the 18th chapter.
ALS 4pp RS:HS 7.263

12194. 1863–3–11 From Robert FitzRoy n.p.
Speculates on gravitational effect of sun and moon on earth's atmosphere.
ALS 4pp RS:HS 19.217

12195. 1863–3–13 To Robert FitzRoy Collingwood
Comments on RF's ideas on the gravitational effect of the sun and moon on the earth's atmosphere.
C 2pp RS:HS 23.327

12196. 1863–3–13 From [Pridler?] Sons & Co. Newenden [Kent]
Requests payment of JH's coal bill.
ALS 1p RS:HS 19.176a

12197. 1863–3–15 To John Tyndall Collingwood
Thanks for lectures on heat motion theory and radiation through earth's atmosphere, and for brochure on 'conformation' of the Alps. Asks about upcoming lecture of JH's son [Alexander].
C 1p RS:HS 24.14 (C: RI 522)

12198. 1863–3–16 From John Tyndall Royal Institution
Remarks on upcoming lecture of JH's son [Alexander], and on glacial movement.
ALS 4pp RS:HS 17.393a (C: RI 523)

12199. 1863–3–20 From Augustus De Morgan 91 Adelaide Road
Sends an extract from J. J. Lalande's article on planets from the *Encyclopédie méthodique.* Sends two of his own theorems.
ALS 4pp RS:HS 6.362

12200. 1863–3–20 From Robert FitzRoy n.p.
Thanks for his opinion; it will save him from exposure, but it is too late to get the book altered. Comments on some parts of JH's book on meteorology. The atmosphere appears to be smaller in extent than John Dalton conceived.
ALS 4pp RS:HS 7.265

12201. 1863–3–22 From John B. N. Hennessey Hanover Square
Was grateful to receive JH's kind letter. Will give him great pleasure to speak of JH's son John. Wonders if his other son has returned from India.
ALS 3pp RS:HS 10.178

12202. 1863–4–4 From Thomas P. Kirkman Croft Rectory near Warrington
Has lately resumed the study of the theory of groups. Further comments on the French Academy and the award of the mathematical medal.
ALS 3pp RS:HS 11.47

12203. 1863–4–6 From Adolphe Quetelet Brussels
Sent information on meteor spotted in Belgium and in Germany. Has heard differing opinions on heights of meteors. Talks about statistics of population.
ALS 4pp (French) RS:HS 14.256

12204. 1863–4–7 To William James Herschel Halton in Hastings
Family eagerly awaits WJH's return home after ten years in India. JH is visiting cousins in Halton. Locations of WJH's brothers and sisters. If possible, stop in Malta and visit William Lassell's equatorial reflector. Alexander Herschel will lecture on meteors at Royal Institution.
ALS 4pp TxU:H/L-0558; Reel 1053

12205. 1863–4–10 From Edward Twisleton Athenaeum
Reminds JH that the Public School Commissioners are still awaiting JH's response to the invitation to comment on the direction of education [see ET's 1862–11–13].
ALS 4pp RS:HS B27.86

12206. 1863–4–11 or later To Montague Bernard [Collingwood]
In response to the request [see Edward Twisleton's 1863–4–10] to comment on mathematics and physical science education for the Public School Commissioners, JH urges that more teaching be done in these areas.
draft 9pp RS:HS B27.87a

12207. 1863–4–11 To James Calder Stewart Collingwood
Thanks for help in Mrs. Baldwin's affairs. JH's family's health is improving. Major house repairs in progress.
ALS 3pp TxU:H/L-0456; Reel 1055

12208. 1863–4–13 To Sir Charles Lyell Collingwood
Received some proof sheets this morning of the Appendix to CL's book on the *Antiquity of Man*. Has read through the book with great interest. Comments on various points. Will have to revise their ideas on the length of human existence. Has had another letter from Twisleton so supposes he must try and write on science in schools. Have had sickness for the first three months of the year.
ALS 8pp APS B.D25.L.4 (C: RS:HS 11.428 & 24.15)

12209. 1863–4–20 From Henry W. Field Royal Mint
Would like photographs of JH, one for himself and one for the American Philosophical Society. Busy at the Mint. Issue of bronze coinage continues. Supplying foreign Mints with his gas apparatus.
ALS 4pp RS:HS 7.221

12210. 1863–4–20 From Roderick Impey Murchison Geological Survey of Great Britain Office
Outlining the requirements for the new Assistant Secretary at the Geographical Society. Gave Frank Buckland encouragement last week.
ALS 4pp RS:HS 12.427

12211. 1863–4–24 To Heinrich Louis d'Arrest n.p.
[JH's summary of letter] Instructions and corrections relative to data on nebula for JH's catalogue.
Amem 1p RAS:JH Archive 12/1.8.9; Reel 10

12212. 1863–4–25 From Roderick Impey Murchison n.p.
Attended the lectures given by JH's son at the [Royal Institution] and Michael Faraday and Bence Jones think he should not be considered for the post at the Geographical as he is cut for more scientific things. Would like JH's views.
ALS 4pp RS:HS 12.428

12213. 1863–4–28 From John Henry Nelson Hyde Park Gardens
Grateful for letters and enclosures. Has been ill himself, but is pleased to hear those at Collingwood are well. Hopes the prospect of Alexander [Herschel]'s appointment will be realized.
ALS 4pp RS:HS 13.98

12214. 1863–4–28 From Jean Reynaud Neuilly near Paris
Talks about variations in climate, glaciers, and compares climates of various historical periods. JR is writing about such matters in a second edition of one of his works.
ALS 4pp (French) RS:HS 14.479

12215. [1863?]–4–28 From John Tyndall Royal Institution
Thanks for a gift.
ALS 1p RS:HS 17.411 (C: RI 519)

12216. 1863–5–2 To George Gabriel Stokes Collingwood
Returns paper of G. B. Airy, Astronomer Royal, and submits report on it.
ALS 1p Cambridge University Library 7656/H703

12217. 1863–5–2 From William Trant Leeds Astronomical Society
Requests JH send an essay for the Leeds Astronomical Society lecture series.
ALS 2pp RS:HS 17.361

12218. 1863–5–6 To Count Paul Edmond de Strzelecki Collingwood
Introduces JH's son Alexander to PS, and hopes PS will support Alexander, who is seeking to become Assistant Secretary of the Royal Geographical Society.
ALS 3pp WT 62824.1

12219. 1863–5–8 From Heinrich Louis d'Arrest Copenhagen
Is sending JH several lists of nebulae, and comments about some of the items. [Almost illegible.]
ALS 2pp (French) RAS:JH Archive 12/1.8.14; Reel 10

12220. 1863–5–9 From Count Paul Edmund de Strzelecki 20b Savile Row
Will help JH or JH's son [Alexander], whose mind PS admires, in any way possible.
ALS 4pp RS:HS 17.246

12221. 1863–5–10 From Augustus De Morgan 91 Adelaide Road
Like Professor Olearius Schneiderhausen, AD has had only one coat in two years. JH will go on for many more years. Sends one of his theorems.
ALS 4pp RS:HS 6.363 S. De Morgan, *Memoir of Augustus De Morgan*, 320–1

12222. 1863–5–12 From Joseph D. Everett Windsor N.S.
Is applying for a post at Queen's College, Belfast and would be glad if JH would send a testimonial. Is sending some of his papers. Has been carrying out research on atmospheric electricity.
ALS 3pp RS:HS 7.137

12223. 1863–5–14 From George Biddell Airy Royal Observatory, Greenwich
Does JH need the original calculation sheets to deal with the errors in the nebula catalogue [see GA's 1863–2–23]?
ALS 2pp TxU:H/M-0641; Reel 1084 (C: RGO 6.200.85)

12224. 1863–5–15 To George Biddell Airy Collingwood
Explains how to detect errors in catalogued positions of stars. List of new nebulae in [G. F. J. A.] Auwers's catalogue.
ACCS 2pp TxU:H/M–0641.2; Reel 1084

12225. 1863–5–15 To George Biddell Airy Collingwood
Sends GA a list of errors found in the sheets for the nebula catalogue [see GA 1863–5–14].
ALS 5pp RGO 6.200.86

12226. 1863–5–15 From Alexander Bridport Becher Admiralty
Annual visitation of Royal Observatory by Board of Visitors will be held 6 June.
(A)LS 1p TxU:H/M–0076; Reel 1087

12227. 1863–5–15 From George Phillips Bond 162 New Bond St.
Sending through H. M. Parker proof of the engraving of the Nebula of Orion. Comments on this.
ALS 1p RS:HS 4.180

12228. 1863–5–16 From Robert FitzRoy Meteorological Office
Sending a paper to replace the one sent previously. Comments on the effect of the moon on the atmosphere. Wrote his *Weather Book* in a great hurry. Regrets the expression 'lines of force' but had the approbation of several practical electricians.
ALS 8pp RS:HS 7.264

12229. 1863–5–17 From Heinrich Louis d'Arrest Copenhagen
Is writing to clarify some earlier information about a list of nebulae that he had sent to JH.
ALS 2pp† (French) RAS:JH Archive 12/1.8.15; Reel 10

12230. 1863–5–17 From Thomas Maclear Royal Observatory, Cape of Good Hope
Grateful for the efforts of JH and his fellow astronomers in England for the continuation of TM's pension. Busy unpacking valuable library of Sir George Grey presented to the Cape. Details of his recent observations of stars. Weather has been bad. Surface of the object glass of transit circle is badly corroded. Sailor son is well again.
ALS 4pp RS:HS 12.168

12231. 1863–5–19 From George C. Hodgkinson The Lodge, Louth
Would like his views on an actinometrical scheme. Has been busy over the affairs of the late Sir James C. Ross.
ALS 7pp RS:HS 9.413

12232. 1863–5–23 From William Trant Leeds Astronomical Society
Hopes JH will present the essay prepared for the Leeds Astronomical Society lecture series in person. Describes format of the series and background of the society members.
ALS 3pp RS:HS 17.362

12233. 1863–5–26 From George Biddell Airy Royal Observatory, Greenwich
Addresses each error JH discovered in reduction of JH's catalogue of nebulae [see JH's 1863–5–15].
ALS 4pp † RAS:JH Archive 12/1.11.5; Reel 10 (ACCS: RGO 6.200.93)

12234. 1863–5–26 From Henry Anthony Jeffreys [Hawkhurst] Parsonage
Gratitude for copy of JH's *Iliad*. Hopes [Margaret] Herschel is [recovering from] her journey and that accounts from Ireland are still good.
ALS 1p RAS:JH Archive 12/6.12 verso; Reel 10

12235. 1863–5–28 From John Bullar Basset Wood near Southampton
The delay in thanking him for his gift of the *Iliad* has been occasioned by ill health.
ALS 2pp RS:HS 4.334

12236. 1863–5–30 From George C. Hodgkinson The Lodge, Louth
Is grateful for JH's notice of his ideas. Outlines his ideas for an actinometer.
ALS 4pp RS:HS 9.414

12237. 1863–5–31 To George Biddell Airy Collingwood
Provides last list of errors and explains JH's role in detecting them [see GA's 1863–5–26]; is worried about the task of proofreading.
ALS 10pp RGO 6.200.95 (C: TxU:H/L-0023)

12238. 1863–6–1 From Sir Isaac L. Bell Newcastle-upon-Tyne
Hopes that JH will be able to visit Newcastle.
ALS 2pp RS:HS 4.42

12239. 1863–6–2 To George Biddell Airy Collingwood
A letter of introduction asking GA to admit Major Robertson and his sisters to the observatory.
ALS 2pp RGO 6.709.131

12240. 1863–6–2 From George Biddell Airy Royal Observatory, Greenwich
Is working on the correction of calculation errors [see JH's 1863–5–31]; GA and his assistants will deal with the problem of proofreading.
ACCS 3pp RGO 6.200.100

12241. 1863–6–[2?] To John Herschel (son) [Collingwood]
Saw J. B. N. Hennessey [who had just come from India] at a Greenwich visitation day, but did not have time to speak to him there. JH will invite him to Collingwood.
ALS 3pp JHS 5.8a

12242. 1863–6–3 From George Biddell Airy n.p.
Responds to JH [see JH's 1863–6–2] about basic rules for admission to the observatory [for JH's future guidance]; GA discourages the visit of ladies 'who understand nothing and learn nothing.'
ACCS 2pp RGO 6.709.132

12243. 1863–6–3 To Richarda Airy Collingwood
Passes on recent reports of good news that daughter Caroline is improving, and that JH's wife, Margaret, survived the journey well. [Margaret went to Dublin to be with Caroline when she became very ill at the birth of her daughter Kathleen.]
ALS 3pp JHS 2,8

12244. 1863–6–3 From Negretti & Zambra London
Have made several of JH's actinometers but have always experienced difficulty with the fluid, ammonia sulphate of copper. Can some other fluid be used? They are now making two actinometers for a foreign observatory and would like to make them as accurate as possible.
ALS 2pp RS:HS 13.147

12245. 1863–6–3 From John Taylor Leonard Place, Kensington
Asks whether JT can reprint JH's letters on the British

modular standard of length, which letters were published in the *Athenaeum*.
ALS 1p RS:HS 17.328

12246. 1863–6–4 To George Biddell Airy Collingwood
Comments on GA's activity in correcting the calculation errors [see GA's 1863–6–2].
ALS 3pp RGO 6.200.104

12247. 1863–6–4 To [John Taylor] Collingwood
Gives permission to reprint letters to *Athenaeum* on standard of length, in light of recent bill to metricize. Requests as many copies as possible.
ALS 2pp St. John's College, Cambridge

12248. 1863–6–5 From George Biddell Airy [Royal Observatory, Greenwich]
Correcting work is gradually drawing to a close [see JH's 1863–6–4].
ACCS 2pp RGO 6.200.106

12249. 1863–6–5 From George C. Hodgkinson The Lodge, Louth
Finds there is not one of JH's actinometers at Kew; would be grateful if he could inform him who has been working with one.
ALS 4pp RS:HS 9.415

12250. 1863–6–6 From John Taylor Leonard Place, Kensington
Will send as many copies as possible of JT's reprinting of JH's letters to the *Athenaeum*. Asks if JH has seen Thomas Rawson Birks's book on matter and ether.
ALS 1p RS:HS 17.329

12251. 1863–6–8 From Sir Isaac L. Bell Newcastle-upon-Tyne
Is sorry to hear that JH is unable to attend the B.A.A.S. meeting. Has been working on a new formula for thallium, and will forward a sample if required.
ALS 2pp RS:HS 4.43

12252. 1863–6–10 From George Phillips Bond 162 New Bond St.
As he has been absent on the continent he has been unable to accept any of JH's kind invitations. His stay in Europe has been short, but full of gratification.
ALS 1p RS:HS 4.181

12253. 1863–6–10 From George C. Hodgkinson The Lodge, Louth
Thanks for the information. Has written to Henry Barrow, and to Negretti and Zambra. The latter can supply actinometers for 14 guineas; too expensive for him. Barrow has just replied that he can supply for 6 guineas.
ALS 2pp RS:HS 9.416

12254. 1863–6–10 From Roderick Impey Murchison 16 Belgrave Square
Sends the appendix to the volume being printed and hopes he can quote the extract from JH's letter. Comments on James Croll's theories. Points out a mistake in the report from the *Morning Post*.
ALS 8pp RS:HS 12.429

12255. 1863–6–10 To William Whewell Collingwood
Seeks WW's signature for the Thomas Maclear Memorial. JH's daughter Julia has been ill.
ALS 4pp TC, Camb. Add. Ms.a.207[108]

12256. 1863–6–11 To Sir Isaac L. Bell Collingwood
Would be pleased to receive a sample of thallium. Queries regarding the manufacture of pure nickel and cobalt.
draft 4pp RS:HS 4.44 & 24.16

12257. 1863–6–11 To [John Taylor] Collingwood
Discusses dispersal of copies of letter against adoption of 'French metre' [see JH's 1863–6–4].
ALS 3pp St. John's College, Cambridge

12258. 1863–6–13 From Mary Somerville Spezia
Has begun to write about science again after a long absence. Intended to revise small part of her *On the Connexion of the Physical Sciences,* but it became quite long. Wants JH's advice on it.
ALS 2pp RS:HS 16.367

12259. 1863–6–17 From Sir Isaac L. Bell Newcastle-upon-Tyne
Is sending a sample of thallium. Answers to JH's queries regarding cobalt.
AL 3pp inc RS:HS 4.45

12260. 1863–6–18 To Sir Isaac L. Bell Collingwood
Thanks for the thallium. Further regarding his 'cobalt steel' and its uses. Fluorin glass.
draft 3pp RS:HS 4.46 & 24.17

12261. 1863–6–19 From John Taylor Leonard Place, Kensington
Suggests JH send copies of JH's letters from the *Athenaeum* to members of Parliament because JH has more influence than JT.
ALS 1p RS:HS 17.330

12262. 1863–6–23 From John B. N. Hennessey Bath
Letter did not reach him as he had left town for Clifton. Hopes to reach town in time to see JH's son [John] before he leaves. Unable to meet him at Greenwich as he had to return with Waugh.
ALS 3pp RS:HS 10.179

12263. 1863–6–25 From Thomas Humphrey National School
Announces quarterly meeting of National School Committee on 27 June.
ALS 1p RAS:JH Archive 12/1.9.4 verso; Reel 10

12264. [1863]–6–27 From Mary Elizabeth Lyell 73 Harley St.
Received JH's translation, which is better than any blank verse can do. Invites JH to visit Lyells.
ALS 2pp TxU:H/M–0342; Reel 1087

12265. 1863–6–27 From Adolphe Quetelet Brussels
Thanks for article by JH's son Alexander Herschel on meteors. Has printed it in *Bulletin* of the academy.
ALS 1p (French) RS:HS 14.257

12266. 1863–6–27 From Edward Sabine 13 Ashley Place
R.S.L. Council has decided to publish correspondence regarding the Melbourne telescope. Sends early pages for JH's corrections.
ALS 2pp RS:HS 15.271

12267. 1863–6–29 From Titus Armellini Rome
New invention for measuring barometrical variations on an increased scale.
ALS 2pp RS:HS 1.407

12268. 1863–6–29 From Alexander B. Becher Admiralty
Has been very busy, which will account for his error. Will adopt JH's plan.
ALS 3pp RS:HS 5.32

12269. 1863–6–29 From Alexander Hamilton-Gordon Curragh Camp
Hopes to move Caroline (JH's daughter) to the seaside; then she will have a better chance of making a recovery. Probably expecting Willy home soon.
ALS 3pp RS:HS 8.153

12270. 1863–7–7 From Woronzow Greig Surrey Lodge, Lambeth
Received the enclosed [manuscript] some time ago with instructions to forward it to JH. Delayed sending it as he was uncertain if JH could spare the time to look through it. Regarding R. A. C. Godwin-Austen and his discoveries of human fossils at Abbeville. WG will be going to Bath for a few months for health reasons.
ALS 8pp RS:HS 9.41

12271. 1863–7–8 To George Biddell Airy n.p.
Was surprised at the large vote in support of a parliamentary bill to abolish the British Standard of weights and measures; JH hopes bill will not pass.
CC(S) 1p inc RGO 6.358.259

12272. 1863–7–8 To George Biddell Airy Collingwood
Has spent much time on figures and annotations for JH's nebula catalogue [see GA's 1863–6–5]; JH needs information from a paper in GA's hands.
ALS 3pp RGO 6.200.107

12273. 1863–7–9 From Edward James Stone [Royal Observatory, Greenwich]
As G. B. Airy is in Scotland on vacation, ES sends the requested information [see JH's 1863–7–8].
ACCS 1p RGO 6.200.109

12274. 1863–7–12 From Mary Somerville Spezia
Introduces Signore Capellini, professor of geology at Bologna, to JH in letter. Asks JH to assist him in any way possible.
ALS 1p RS:HS 16.368

12275. 1863–7–14 To John Herschel (son) Collingwood
Comments on zero and observational errors when using a telescope, and adds some comments about family relationships and JH's nebula catalogue. [Part of this letter is indecipherable.]
ALS 4pp JHS 5.9

12276. 1863–7–15 From George Phillips Bond Cambridge [Mass.]
Sending back number of their 'Annals' to complete JH's set. Remarks on the parts already published.
ALS 1p RS:HS 4.182

12277. 1863–7–15 From Augustus De Morgan 91 Adelaide Road
Regarding translations of Walter Mapes regarding the origin of the word theodolite.
ALS 4pp RS:HS 6.364

12278. 1863–7–17 To Augustus De Morgan n.p.
Agrees that the weather is hot. The word theodolite. Where did he obtain his verses by Walter Mapes? Why not come and visit them?
draft 4pp RS:HS 6.365 & 24.18

12279. 1863–7–18 To François Napoleon Marie Moigno Collingwood
Writes to FM to show several ways in which a series of prisms may be used in a spectroscope to have the light exiting in the same line as it entered the spectroscope. [Appends 4 pages of diagrams.]
ALS 3pp (French) DMC.695A.885.37 *Les Mondes, 3* (1863), 403–5

12280. 1863–7–21 From Thomas Maclear n.p.
Glad to hear of the recovery of JH's daughters. Encloses a copy of a letter of thanks written to H. J. Temple (3rd Viscount Palmerston). Error in the sun's distance is peculiar. Has been elected corresponding member of the French Academy. Family has been attacked by the influenza.
ALS 2pp RS:HS 12.169

12281. 1863–7–26 From [Antoine C. G.] Jobert 15 Portland Crescent, Leeds
Thanks JH for his order.
ALS 1p Harvard: Houghton AAH 67m–67 (63)

12282. 1863–7–30 To Mary Somerville Collingwood
Is sending manuscript with comments. Congratulates her 'activity of mind.' Critiqued her paper according to the effect it will have when published. Working on catalogue of nebulae.
ALS 4pp RS:HS 16.369 (draft: RS:HS 16:370)

12283. 1863–7–31 From Woronzow Greig York House Hotel, Bath
Has just received JH's letter and note for his mother. The manuscript has arrived safely at Surrey Lodge. Is deeply grateful for JH's comments and will follow out his instructions. Has also just heard from his mother.
ALS 3pp RS:HS 9.42

12284. 1863–8 From R. S. Brown Edinburgh
Sending papers to JH and would be glad if he would communicate them either to the B.A.A.S. or some influential journal.
ALS 1p RS:HS 5.112

12285. [1863–8] To Adolphe Quetelet n.p.
Describes in vivid detail the account of an 'extraordinary meteor' in October 1854. Given by a person from Hurworth.
AL 4pp Académie belgique: #30 *Bulletin de l'académie royale des sciences, des lettres et des beaux-arts de belgique,* 2nd ser., *16* (1865), 315–9

12286. 1863–8–2 From John Capellini London
Has received his letter and will visit him on Tuesday, arriving at Etchingham station at 3:35 p.m.
ALS 1p RS:HS 5.324

12287. 1863–8–4 From F. Gough Hastings
His note and the maps have arrived safely. Has made the necessary arrangements for the night of the 10th.
ALS 4pp RS:HS 9.109

12288. 1863–8–8 From Augustus De Morgan 91 Adelaide Road
Wife and family are on holiday at Portmadoc. Obtains his rhymes from Thomas Wright's *Political Songs of England*. Relates a tale of how a priest tricked a farmer in medieval times.
ALS 4pp RS:HS 6.366

12289. 1863–8–9 From John Wrottesley Wrottesley [Staffordshire]
Invites Herschels to visit Wrottesley [Hall, Staffordshire] to

witness James Glaisher's '*last high* Balloon Ascent' from Wolverhampton on 13 Aug.
ALS 3pp RS:HS 18.304

12290. 1863–8–14 From George Biddell Airy Royal Observatory, Greenwich
Has climbed Ben Nevis.
ALS 1p RS:HS 1.260 (C: RGO 6.380.267 inc)

12291. 1863–8–14 From George Biddell Airy Royal Observatory, Greenwich
Regarding William Lassell's nebulae drawings in the R.A.S.'s Memoirs. House of Commons action concerning 'Weights and Measures.'
ALS 3pp RS:HS 1.259

12292. 1863–8–14 George Biddell Airy [Royal Observatory, Greenwich]
Additional information is provided to that sent by E. J. Stone [see ES's 1863–7–9].
ACCS 2pp RGO 6.200.110

12293. 1863–8–14 From John Wrottesley Wolverhampton [Staffordshire]
Invites Herschels to view James Glaisher's next balloon ascent, sometime after 20 Aug. Glaisher's report to B.A.A.S. on eight ascents is just out. JW identifies rapid ascent as cause of observation errors. Predicts adoption of imported [metric] 'Decimal Scale' from Continent.
ALS 4pp RS:HS 18.305

12294. 1863–8–15 From Charles Babbage Dorset St., Manchester Square
Charles Few will send deed for JH's signature. It transfers property of CB's sister [Mary Anne Hollier] in Cardiff to Marquis of Bute [J. P. C. Stuart]. Expects no further changes in trust and no more troubles for JH. Saw Alexander Herschel's work on 'shooting stars' published by B.A.A.S. Illness of CB's son [Henry P. Babbage] in India. CB's latest book.
ALS 4pp † TxU:H/M–0036; Reel 1062

12295. 1863–8–16 From Adolphe Quetelet Brussels
Has received letter on meteors of 1868–8–10. Map was very interesting. Is going to Berlin for congress on statistics. Discusses phenomena of 1854. Asks for JH's opinion on whether meteors are terrestrial in origin.
ALS 4pp (French) RS:HS 14.258

12296. [1863–8–18] From Few & Co. 2 Henrietta St., Covent Garden
Instructions for signing and returning deed [of conveyance for Woodlands property in Hollier Trust]. [JH annotation, 19 Aug. 1863: Received and executed deed. It was witnessed by JH's son William, now recently returned from Bengal and resident at Hawkhurst for next three years. Returned deed to Few & Stamford.]
ALS 1p †† inc TxU:H/M–0188; Reel 1087

12297. 1863–8–18 To Adolphe Quetelet Collingwood
Tells AQ that last letter on meteors was written not by JH but by A. S. Herschel. JH attributes phenomena to cosmic origins. Gives further details.
ALS 4pp (French) Académie belgique: #32

12298. [1863–8–19 or later] To Adolphe Quetelet n.p.
Has talked to the lady who wrote account of the meteor. Finds her reliable. Has discovered other sightings of it. Sends report in French for printing in AQ's notices on meteors.
ALS 3pp Académie belgique: #33

12299. 1863–8–20 From John C. Broun Wynberg
Thought he may be interested in the news of the proposed lectureship at the Cape.
ALS 1p RS:HS 5.109

12300. 1863–8–20 From [Rawson William] Rawson Cape Town
RR's brother-in-law values proximity to Collingwood. Discusses flora and the state of botanical science at the Cape. Sends letter from colonial botanist.
ALS 4pp RS:HS 14.472

12301. 1863–8–22 To William Crookes Collingwood
Thanks WC for a copy of a paper published by WC in the *P.M.*
ALS 1p WT 69293.2

12302. 1863–8–22 From Few & Co. 2 Henrietta St., Covent Garden
Received deed of conveyance for Woodlands property, in Hollier Trust, executed by JH. Will advise JH when money has been received.
ALS 1p TxU:H/M–0189; Reel 1087

12303. 1863–8–25 From Mary Somerville Spezia
Thanks JH for criticism of manuscript. Sends revised introductory paragraph. Thanks him for kindness to Signore Capellini. Scientific interest in Italy grows as politics settle.
ALS 3pp RS:HS 16.371

12304. 1863–9–1 From Few & Co. 2 Henrietta St., Covent Garden
Sign enclosed authorization for Few & Co. to obtain Hollier Trust deeds from bank. Deeds will be exchanged for £1000.
ALS 2pp TxU:H/M–0190; Reel 1087

12305. [1863–9] To Adolphe Quetelet Collingwood
Three maps of meteors observed in 1863 by JH[?] and Alexander Herschel
maps 3pp Académie belgique: #34, 35, 36

12306. 1863–9–2 From Thomas P. Kirkman Croft Rectory near Warrington
Outlining the events leading to the rejection of his claim for the French Academy medal for mathematics.
ALS 6pp RS:HS 11.48

12307. 1863–9–3 To George Booth Tyndale Collingwood
Responds with thanks to GT, a fellow trustee of the British Museum, who has written to welcome JH and congratulate him on his election. JH is concerned about having enough time to fulfill his obligations.
ALS 3pp WT B.13

12308. 1863–9–7 From Sir Isaac L. Bell Washington Co. Durham
Has received his note with the queries. Will be forwarding a copy of last week's proceedings to him later.
ALS 2pp RS:HS 4.47

12309. 1863–9–7 From François N. M. Moigno London
Will come and visit him on Friday afternoon.
ALS 1p (French) RS:HS 12.356

12310. 1863–9–8 From Charles Babbage Eywood, Kington, Herefordshire
Please sign enclosed legal document and return to Messrs. 'Praed.' CB overworked, but hopes to complete *Passages from the Life of a Philosopher* before Christmas.
ALS 2pp TxU:H/M–0037; Reel 1062

12311. 1863–9–17 From William Rutter Dawes Hopefield, Haddenham, Thame
Thomas Cooke of York, a maker of object glasses, is about to visit Paris and would like an introduction from JH in order to visit the Paris Observatory. Hopes JH's son is back at Collingwood. Own health is poor, but is improving.
ALS 3pp RS:HS 6.110

12312. 1863–9–25 From Longman and Co. 39 Paternoster Row
Does JH want any corrections made in plates for *Outlines Astr.*?
L 1p RAS:JH Archive 12/6.11 verso; Reel 10

12313. 1863–9–25 From Alexander B. McGrigor St. Vincent St., Glasgow
Inviting JH to give one of the lectures at Glasgow.
ALS 7pp RS:HS 13.4a

12314. 1863–9–28 From James Nasmyth Penshurst
Is pleased that JH is to mention JN's 'willow leaves' on the sun in his new edition of *Outlines Astr.* Will endeavor to make a more perfect photographic copy of the original drawing. Gives a drawing and description of apparatus he uses for solar observations.
ALS 3pp RS:HS 13.78

12315. 1863–9–29 To Alexander B. McGrigor n.p.
Health makes it impossible for JH to travel to Glasgow.
draft 1p RS:HS 13.4b

12316. 1863–9–29 To [James Nasmyth] Collingwood
Gives advice on observing and photographing the sun and moon. Invites JN to visit JH.
ALS 4pp Massachusetts Historical Society

12317. 1863–10–1 From James Nasmyth Penshurst
Sends by book post two photographs of the original drawings of the sun's observations, also a photograph of a model of the S.E. portion of the moon; comments on them. Will be pleased to accept JH's invitation to visit Collingwood.
ALS 4pp RS:HS 13.79

12318. 1863–10–1 From Adolphe Quetelet Brussels
Received documents from JH just before leaving for Berlin. Discussed periodic stars in Berlin. Stayed in Berlin longer than expected, but was received well by old friends.
ALS 3pp (French) RS:HS 14.259

12319. 1863–10–2 From William Trant Leeds Astronomical Society
Thanks JH upon receiving the essay for the Leeds Astronomical Society lecture series ['The Yard, the Pendulum, and the Metre, Considered in Reference to the Choice of a Standard of Length'].
ALS 2pp RS:HS 17.363

12320. 1863–10–3 To George Biddell Airy Collingwood
Thanks GA for receipt of some Royal Observatory publications; is still pushing the British metrical standard based on the length of the earth's polar axis [see JH's 1860–3–2].
ALS 4pp RGO 6.358.261

12321. 1863–10–5 From William Crookes 20 Mornington Road
Is co-editor of a new quarterly scientific journal [*Quarterly Journal of Science*] and would be pleased to receive any contributions from JH, which would be liberally paid for.
ALS 2pp RS:HS 5.315

12322. 1863–10–6 From William Trant Leeds Astronomical Society
Leeds Astronomical Society will publish the essay written by JH for their lecture series after its presentation. WT will forward the proof for JH's corrections.
ALS 3pp RS:HS 17.364

12323. 1863–10–7 From George Biddell Airy Royal Observatory, Greenwich
Concerning the 'Standards.'
ALS 4pp RS:HS 1.261

12324. 1863–10–8 To George Biddell Airy n.p.
JH wants to avoid using the meter in England; asks GA for the results of A. R. Clarke's calculations on the figure of the earth [see GA's 1863–10–7].
CC 2pp inc RGO 6.358.263 & 6.423.661

12325. 1863–10–8 To George Biddell Airy Collingwood
Comments on the ease of calculation in geodesy in JH's version of the English system of units, and asks GA for some clarifications as JH prepares the seventh edition of his *Outlines Astr.*
ALS 4pp RGO 6.380.268

12326. 1863–10–9 From George Biddell Airy Royal Observatory, Greenwich
Regarding A. R. Clarke's paper on Geodesy. Benjamin Peirce's Lunar Table. The sun's parallax. JH's nebulae catalogue. Regarding the English inch.
ALS 6pp RS:HS 1.262 (C: RGO 6.85.105 & 85.269 & 200.112)

12327. 1863–10–9 From George Biddell Airy n.p.
Is still not happy with the earth's axis as a source of standard length; sends JH the results of A. R. Clarke's calculations on the figure of the earth [see JH's 1863–10–8].
ACCS 2pp RGO 6.358.266 & 6.423.662

12328. 1863–10–9 From William Trant Leeds Astronomical Society
Points out minor errors in the manuscript of JH's 'The Yard, the Pendulum, and the Metre.'
ALS 2pp RS:HS 17.365

12329. 1863–10–11 To Alexander R. Clarke Collingwood
Thanks for sending him his memoir on the *Figure of the Earth.* Comments on this. His son John would find a copy of this memoir of great assistance in his work in India; would he send him one.
draft 2pp RS:HS 5.343 & 24.19

12330. 1863–10–12 From Alexander R. Clarke Southampton
Thanks for his comments on his memoir. Sends a few additions, which he will be submitting to the R.A.S. in due course. Will send a copy of the memoir to JH's son.
ALS 4pp RS:HS 5.344

12331. 1863–10–12 From John R. Hind Nautical Almanac Office
Sees that JH mentions 'the longitude of perihelion on the ecliptic' as one of the elements he would like to have in the case of the small planets. Would like his advice.
ALS 2pp RS:HS 9.385

12332. 1863–10–12 From Longman and Co 29 Paternoster Row
Informs JH that his photographs of the moon's surface and diagrams of the sun's envelope were sent to the engraver.
L 1p inc Harvard: Houghton AAH 67m–67 (107)

12333. 1863–10–13 From Camille Flammarion Paris
Is very grateful for his work on the plurality of worlds. Comments on his own recent work and the work of JH on nebulae.
ALS 3pp RS:HS 7.268

12334. 1863–10–13 From William Trant Leeds Astronomical Society
Notes reception of corrections to JH's 'The Yard, the Pendulum, and the Metre.' Will send JH 100 copies of the essay.
ALS 2pp RS:HS 17.366

12335. 1863–10–14 From John R. Hind Nautical Almanac Office
Comments on JH's advice regarding the 'longitude of perihelion in the ecliptic.'
ALS 3pp RS:HS 9.386

12336. 1863–10–15 From William Crookes 20 Mornington Road
WC's goals in founding *Quarterly Journal of Science.* Invites JH to submit article for first issue.
ALS 2pp RAS:JH Archive 13/1.1; Reel 10

12337. 1863–10–15 From John R. Hind Nautical Almanac Office
Explaining some of the signs in the enclosed table. Has sent the *Nautical Almanac* for 1867.
ALS 3pp RS:HS 9.387

12338. 1863–10–15 From James Nasmyth Penshurst
Fine weather is getting rare at this time of year and would therefore like to induce the Herschels to visit him while it lasts.
ALS 3pp RS:HS 13.80

12339. 1863–10–16 From William Trant Leeds Astronomical Society
Announces JH's essay, 'The Yard, the Pendulum, and the Metre' will be read to the Leeds Astronomical Society on 10 Oct. 1863.
ALS 1p RS:HS 17.367

12340. 1863–10–18 From James Nasmyth Penshurst
Regrets that JH's engagements prevent them visiting Penshurst. JN too is unable to manage the dates quoted. Hopes his own Herschelian helioscope will be in action by the time the Herschels visit Penshurst.
ALS 2pp RS:HS 13.81

12341. 1863–10–19 From Charles Pritchard Freshwater
Warren de La Rue observes 'willow leaves' [mottling on solar surface]. Diagonal solar eyepiece works with great success. Praises de La Rue's work as having astounding accuracy.
ALS 4pp RS:HS 14.83 Ada Pritchard, *Charles Pritchard: Memoirs of His Life,* 240–1

12342. 1863–10–20 To George Biddell Airy Collingwood
Comments on the possibility of employment in India for GA's son, Hubert.
C 2pp RS:HS 24.20

12343. 1863–10–20 From William Crookes 20 Mornington Road
Would JH's son favor him with a resume of astronomical discovery for his new quarterly; if so, could he have his address?
ALS 1p RS:HS 5.316

12344. 1863–10–20 From P. O'Callaghan Leeds Philosophical & Literary Society
Sends a portion of JH's lecture for the Leeds Astronomical Society, which has just been received from the printer. PO's wife's health is improving and JH's grandchildren are flourishing.
ALS 2pp RS:HS 13.160

12345. 1863–10–21 From William Rutter Dawes Hopefield, Haddenham, Thame
Thanks to JH's note Thomas Cooke was able to visit the Paris Observatory. His own health is slowly improving.
ALS 2pp RS:HS 6.111

12346. 1863–10–21 To James Prescott Joule Collingwood
Thanks for his paper on amalgams. Regarding the amalgam of iron, he pointed it out many years ago in his Bakerian lecture. Would be interesting to experiment further with iron in solution of mercury.
C 3pp RS:HS 10.429

12347. 1863–10–21 From William Trant Leeds Astronomical Society
Informs JH that the Leeds Astronomical Society will print 'The Yard, the Pendulum, and the Metre' the day after its presentation.
ALS 2pp RS:HS 17.368

12348. 1863–10–22 To Gustave Augustin Quesneville Collingwood
Received four issues of GQ's *Moniteur Scientifique*. Saw mention in No. 163 of JH's proposed modification to British system of weights and measures. JH's paper on this topic is now before Astronomical Society of Leeds.
C 4pp TxU:H/L-0320; Reel 1054

12349. 1863–10–22 From Edward Sabine n.p.
R.S.L. Club locale changed to St. James's Hotel. Wants JH's advice on two papers ES is writing on results from Kew Observatory.
LS 4pp RS:HS 15.272

12350. 1863–10–24 From George Phillips Bond Harvard Observatory, Cambridge, Mass.
Read of JH's intention to prepare general catalogue of nebulae. Encloses list of unpublished nebulae found at Harvard over past 15 years. Finds similarity in form but difference in theory between JH's equations for curve of object glasses ['On Aberrations of Compound Lenses and Object Glasses' (1821)] and those of K. A. Steinheil and C. F. Gauss. Asks about Josef Fraunhofer's theory.
ALS 2pp TxU:H/M–0110; Reel 1087

12351. 1863–10–24 From James Prescott Joule Old Trafford
Much obliged for his letters regarding the amalgam of iron. Cannot account for his ignorance of JH's Bakerian lecture.
ALS 2pp RS:HS 10.430

12352. 1863–10–24 From Warren de La Rue to [Charles] Pritchard Hotel du Louvre, Paris
Regarding the 'willow leaves' seen by James Nasmyth on the surface of the sun.
ALS 4pp RS:HS 6.151

12353. 1863–10–25 To Edward Sabine Collingwood
Accepts invitation to dine with ES. Working on sixth edition of *Outlines Astr.* Has sent argument against converting to metric system to Leeds.
C 1p RS:HS 24.21

12354. 1863–10–27 From George Biddell Airy Royal Observatory, Greenwich
Regarding employment for GA's son in the Educational Departments in India.
ALS 2pp RS:HS 1.263

12355. 1863–10–27 To Augustus De Morgan Collingwood
Comments on Greek fire lead JH into a variety of recollections.
C 2pp RS:HS 24.22

12356. 1863–10–29 From John R. Hind Nautical Almanac Office
Sending the revision of the first half sheet for the 1867 supplement containing the latest elements of the small planets.
ALS 1p RS:HS 9.388

12357. 1863–10–29 From Edward Sabine Burlington House
JH to read his paper on nebulae Thursday, 19 Nov. Dinner will be at St. James's Hotel.
ALS 2pp RS:HS 15.273

12358. 1863–10–29 From William Trant Leeds Astronomical Society
Sends regrets to JH over the error in *The Leeds Mercury* claiming that the Leeds Astronomical Society had the copyright to 'The Yard, the Pendulum, and the Metre.'
ALS 6pp RS:HS 17.369

12359. 1863–11–3 From Roderick Impey Murchison Geological Survey of Great Britain Office
President of the R.S.L. has suggested that RM should propose Sir Henry Barkly as a fellow. Comments on Barkly's work in geography and geology.
ALS 2pp RS:HS 12.430

12360. 1863–11–4 From Edwin Dunkin 22, Circus St., Greenwich
Is sorry the publication of his own paper on solar motion was too late for JH to notice it in his new edition of the *Outlines Astr.*
ALS 4pp RS:HS 6.514

12361. 1863–11–9 From Walter White Burlington House
Notice of annual election of R.S.L. council and officers on 30 Nov. 1863.
LS 1p TxU:H/L-0339.2; Reel 1054

12362. 1863–11–13 From Thomas P. Kirkman n.p.
Hopes he glanced at the last words of TK's article on the polyhedra in the R.S.L. Proceedings.
AL 2pp inc RS:HS 11.53

12363. 1863–11–13 From Thomas P. Kirkman Croft Rectory near Warrington
Sending a demonstration of the last theorem of Pierre Fermat, which has been the subject of a prize of the Academy many times. Would like his views on the logic. Arthur Cayley has done everything possible to prevent TK from winning the Academy medal.
ALS 8pp RS:HS 11.49, 54, 55

12364. 1863–11–13 From William Trant Leeds Astronomical Society
Reports that 100 copies of the Leeds essay JH requested will be ready the next day. Thanks JH for his understanding concerning the error in *The Leeds Mercury* [see WT's 1863–10–29].
ALS 4pp RS:HS 17.370

12365. 1863–11–16 From Thomas P. Kirkman Croft Rectory near Warrington
Finds part of his proof was defective, but it makes no difference to the final result.
ALS 2pp RS:HS 11.50

12366. 1863–11–16 From William Trant Leeds Astronomical Society
Sends copies of 'The Yard, the Pendulum, and the Metre' published by Leeds Astronomical Society.
ALS 3pp RS:HS 17.371

12367. 1863–11–17 From Charles Grey Tunbridge Wells
At Turners Hill near East Grinstead is a pond, which feeds the Rivers Medway and Ouse.
AL 3pp RS:HS 9.128

12368. 1863–11–19 From Augustus De Morgan 91 Adelaide Road
Regarding a suitable length for the proposed metric system. Hopes the metrical people will continue to agitate.
ALS 2pp RS:HS 6.367 S. De Morgan, *Memoir of Augustus De Morgan*, 321–2

12369. 1863–11–19 From Charles Grey Tunbridge Wells
Would be pleased to show JH the pond, but thinks it should be carefully examined before any reference is made to it.
AL 3pp RS:HS 9.129

12370. 1863–11–20 From Alexander R. Clarke Southampton
Thanks for sending him his essay on the yard, pendulum and metre. Hopes it will engage the attention of the public. A measure of 50" would be very convenient.
ALS 2pp RS:HS 5.345

12371. 1863–11–21 From George Biddell Airy Royal Observatory, Greenwich
Regarding the Proceedings of the Southern Telescope Committee.
ALS 1p RS:HS 1.264 (C: RGO 6.255.232)

12372. 1863–11–22 To George Biddell Airy Collingwood
Sends GA [see GA's 1863–11–21] JH's copy of the Melbourne telescope correspondence and reports; needs information from GA and Edwin Dunkin about the sun's motion, although JH questions some of Dunkin's work.
ALS 4pp RGO 6.255.92 (C: RGO 6.255.233 inc)

12373. [1863]–11–23 From Emma Buckle Royal Mint
Would like JH's support for sending a memorial to the Lords of the Treasury for a Civil List pension for herself.
ALS 3pp RS:HS 5.115

12374. 1863–11–23 To Frederick Howlett Collingwood
Response to some papers sent to R.A.S.
draft 1p RS:HS 24.23

12375. 1863–11–23 To George Griffith Collingwood
Has learned JH has been named to the B.A.A.S. 'Committee of Weights and Measures.' Requests addresses for various members of the committee so that JH can send each a copy of a paper JH published.
ALS 3pp Manchester Central Library

12376. 1863–11–23 From Warren de La Rue Observatory, Cranford
Thinks it possible that a letter he wrote to Charles Pritchard

may have been passed to JH. Thinks there is some substance in James Nasmyth's claim on seeing 'Willow leaves' on the surface of the sun.
ALS 2pp RS:HS 6.152

12377. 1863–11–24 From John Henry Nelson Hyde Park Gardens
Will call on W. Bonsey on his way to Devon and may be able to give a valuation of the house and grounds at Slough.
ALS 3pp RS:HS 13.99

12378. 1863–11–25 From Thomas P. Kirkman Croft Rectory near Warrington
Having sent a bad proof, now sends a sound one.
ALS 3pp RS:HS 11.51

12379. 1863–11–27 From George Biddell Airy Royal Observatory, Greenwich
Requesting JH to write to Thomas Maclear to get him to finish his work on the Trigonometrical Survey of the Cape.
ALS 4pp RS:HS 1.265

12380. [1863]–11–27 From Emma Buckle Royal Mint
Is grateful for his letter. Thomas Graham will forward a letter of recommendation with the memorial and she would also like a note from JH.
ALS 4pp RS:HS 5.116

12381. 1863–11–27 To George Griffith Collingwood
Does not personally know any members of the committee [Committee for the Restoration of the Standards] except Prof. [William Hallowes] Miller. Consequently, asks GG to distribute JH's paper 'officially.'
ALS 1p Manchester Central Library

12382. 1863–11–30 From Thomas Graham Royal Mint
Regarding Mrs. Emma Buckle's (wife of William) application for financial assistance. Would JH address a letter to him stating the facts. Does not hold out much hope of obtaining financial assistance from the government.
ALS 2pp RS:HS 8.229

12383. 1863–12–1 To George Biddell Airy Collingwood
Thanks GA for the papers sent; asks GA to review enclosed note on solar motion [see JH's 1863–11–22].
ALS 2pp inc RGO 6.255.94

12384. 1863–12–1 To George Biddell Airy n.p.
Has seen colored fringes on clouds described, and explained, in a journal. Does GA accept the explanation?
ALS 2pp inc RGO 6.703.114

12385. 1863–12–2 From George Biddell Airy Royal Observatory, Greenwich
Thanking him for his statement about solar motion.
ALS 1p RS:HS 1.269 (C: RGO 6.255.95)

12386. 1863–12–2 From George Biddell Airy Royal Observatory, Greenwich
Regarding JH's remarks about colored fringes in the clouds.
ALS 1p RS:HS 1.268 (C: RGO 6.703.115)

12387. 1863–12–2 From George Biddell Airy Royal Observatory, Greenwich
Regarding the work of Thomas Maclear.
ALS 1p RS:HS 1.267

12388. 1863–12–2 From George Biddell Airy Royal Observatory, Greenwich
Is returning correspondence on Southern Telescope Committee and the Melbourne telescope. Comments on the working of the scheme.
ALS 4pp RS:HS 1.266 (C: RGO 6.255.234)

12389. 1863–12–2 From George Knott Cuckfield
Believes there is to be a new edition of JH's *Outlines Astr.* Points out some queries regarding some of the stars listed in the previous edition.
ALS 4pp RS:HS 11.87

12390. 1863–12–3 To Thomas Maclear Collingwood
Britain's change to the metric system displeases JH. Anxious for completion of TM's work, on arc of meridian measurements. Completes 'Catalogue of Nebulae.'
ALS 4pp Cape Archives/Maclear Papers File 100

12391. 1863–12–4 From George Phillips Bond Harvard Observatory, Cambridge, Mass.
Unpublished Harvard list contains nine nebulae published in [G. F. J. A.] Auwers's catalogue of novae, because those were first discovered at Harvard. Thanks JH for pointing out two others already published, which GB will exclude from list. Admits possibility of 'illusions' in some. Asks JH to compare results in C. F. Gauss's article on achromatic object glass. Sent copy of T. H. Safford's article on companion of Sirius.
ALS 2pp TxU:H/M–0111; Reel 1087

12392. 1863–12–4 From William Rutter Dawes Hopefield, Haddenham, Thame
Regarding the spots on the sun known as James Nasmyth's 'Willow leaves.'
ALS 6pp RS:HS 6.112

12393. 1863–12–4 To George Gabriel Stokes Collingwood
Comments on some communications with G. B. Airy, Astronomer Royal; complains he has received no formal acknowledgement of large number of manuscripts deposited with R.S.L.
ALS 4pp Cambridge University Library 7656/RS422

12394. 1863–12–5 From John Henry Nelson n.p.
Regrets that the weather has delayed his reply. Has now been able to get W. Bonsey's opinion on the valuation of JH's property at Slough; comments on this. Building land not much in demand at present.
ALS 3pp RS:HS 13.100

12395. 1863–12–7 From George Gabriel Stokes Lensfield Cottage, Cambridge
Acknowledges receipt of comments regarding Melbourne University telescope.
ALS 3pp RS:HS 17.53

12396. [1863]–12–8 From Emma Buckle Royal Mint
With his permission she hopes to send his letter with the memorial to the Treasury. Will let him know the outcome.
ALS 3pp RS:HS 5.117

12397. 1863–12–8 From Charles Rogers 20 Caledonia [St.?], Liverpool
Observations of cloud patterns after storms.
ALS 2pp TxU:H/M–0460; Reel 1093

12398. 1863–12–12 From William Rutter Dawes Hopefield, Haddenham, Thame
Is obliged for the perusal of Warren de La Rue's paper. Further comments on James Nasmyth's 'Willow leaves.'
ALS 8pp RS:HS 6.113

12399. 1863–12–15 To Elizabeth Colling Collingwood
About a great meteor seen earlier in the year; JH is slowly translating Homer's *Iliad.*
C 2pp RS:HS 24.25

12400. 1863–12–15 To Richard Hodgson Collingwood
Presents eight-volume copy of William Herschel's 'Sweep,' transcribed by late Caroline Herschel. Reviews her career.
AdraftS 4pp RS:HS 152 (C: RS:HS 24.24)

12401. 1863–12–15 From Charles Pritchard Freshwater
Describes debate at R.A.S. over the nature of the 'willow leaves.' Different parties argue that the 'leaves' are solid, precipitate, or non-luminous gas.
ALS 4pp RS:HS 14.84 Ada Pritchard, *Charles Pritchard: Memoirs of His Life,* 241–2

12402. 1863–12–16 From George Smith 45 Pall Mall
Asks JH to review paper on Aristotle's physics by George Henry Lewes, whose letter to GS is enclosed.
ALS 1p encl RS:HS 16.192 & 16.193

12403. 1863–12–16 From John Taylor Leonard Place, Kensington
Is sending JH copies of recent book [JT's *The Great Pyramid*].
ALS 1p RS:HS 17.331

12404. 1863–12–17 To [John Taylor] Collingwood
Discusses JT's 'Battle of the Standards,' especially JH's letters against adoption of metric system and another lecture JH encloses. Comments on error in JT's frontispiece on the great pyramid.
ALS 4pp St. John's College, Cambridge

12405. 1863–12–18 From Edward William Brayley London Institution
[John P.] Gassiot has just informed him of his experiments on the sun's light. Importance of this in relation to present theories.
ALS 3pp RS:HS 4.227

12406. 1863–12–21 From William Rutter Dawes Hopefield, Haddenham, Thame
Is sorry he has caused so much trouble with his 'Solar views.' Sends a recent paper he has written on the subject. Is appealing to James Nasmyth. Encloses sketch of an ear-shaped disc.
ALS 6pp RS:HS 6.114

12407. 1863–12–22 From Henry Enfield Roscoe and Joseph Baxendell Manchester
[Printed notice] Literary and Philosophical Society will meet on 29 Dec.
L 1p TxU:H/MW-0246.3; Reel 15

12408. 1863–12–24 From 'Velocity' [Hope Bank?], Stockport
Proposes hypothetical problem in physics of gravitation involving three unequal bodies falling equal distances. Asks JH to respond in *Good Works*.
ALS 4pp TxU:H/M–0949; Reel 1083

12409. 1863–12–24 From Robert FitzRoy Meteorological Office
Sends three pamphlets by Lieut. John Maclear. Was very interested in JH's 'Words' for January.
ALS 4pp RS:HS 7.266

12410. 1863–12–24 To James Thomas Walker Collingwood
Received tables of heights recorded in 'Sind, the Punjab, N.W. Provinces, and Central Asia.' Sees scientific interest in JW's described effect of refraction on levelling. Work is important to irrigation.
C 1p RS:HS 18.12

12411. 1863–12–26 From George Smith 8 Eastern Terrace, Brighton
Thanks JH for reviewing George Henry Lewes's paper [see GS's 1863–12–16].
ALS 3pp RS:HS 16.194

12412. 1863–12–28 From George Henry Lewes North Bank, Regents Park
Is grateful for the trouble JH has taken in annotating the proofs. Will send him a copy of the book when printed. Thanks also for the atomic dialogue.
ALS 3pp RS:HS 11.478

12413. 1863–12–29 From Henry W. Field Royal Mint
Is grateful for the photographs. Recent happenings at the Mint. Has had a visit from Henry Bingley, who looks well.
ALS 4pp RS:HS 7.222

12414. 1863–12–29 From Algernon Percy (4th Duke of Northumberland) to Alexander Herschel Alnwick Castle
Expresses thanks for receipt of photographs of JH.
ALS 1p RS:HS 13.363

12415. 1863–12–29 From William Edward Hickson Fairseat, Wrotham
Submits WH's ideas on moon's influence on weather. JH's son Alexander owes WH a letter.
ALS 3pp TxU:H/M–0275; Reel 1087

12416. 1863–12–29 From John Charles Sloan 42 Wilton Square, Islington
Proposes modification to balloon gondola for scientific research.
ALS 2pp TxU:H/M–0483; Reel 1093

12417. [1864?] To H.R.H. [?] Collingwood
Asks that consideration be given to 'wishes of the Cadet' in selecting a regiment for him.
AdraftS 1p inc RS:HS 19.155

12418. [1864?] To John Herschel (son) Collingwood
About geodesic books and JH's encouragement to John to keep asking questions; indeed JH is almost insistent on son John's writing to JH with questions, book needs, and confidences. JH also talks about finding the arithmetical mean of a number of observations. [Also included is the 1st page of a letter to John from one of his sisters.]
ALS 3pp JHS 5.14

12419. [1864 to 1866] To [G. Wharton Simpson?] n.p.
If enclosed item is accepted for publication in *Photographic News*, JH wants to inspect its press, especially 'as respects the Cipher,' and to receive 25 copies.
ALS 1p TxU:H/L-0745; Reel 1089

12420. [1864–1 or later] To [?] n.p.
Encloses copy of JH's essay entitled *The Yard, the Pendulum, & the Metre*... (1863). Speculates on causes for deviation in pendulum measurements. Recommends using torsion gravimeters in 'proposed operation.'
Adraft 1p inc TxU:H/L-0602.2; Reel 1089

12421. [1864?] To ? n.p.
In response to the gift of a book, JH comments on the

philosophy of knowledge, and goes on to say that JH agrees with [John Stuart ?] Mill's rejection of the syllogism as a means of argument. In the postscript, JH comments on the nature of heat.
draft 4pp JHS 5.2

12422. [1864?] To [R.S.L. Council] n.p.
Reports on G. B. Airy's analysis [*R.S.P.T., 153,*. 617–] of 177 magnetic storms.
AD 5pp inc? RS RR.5.8

12423. 1864–1–2 From James Thomas Walker 24 Blessington Road, Lee, Blackheath
Received JH's two letters acknowledging receipt of 'Volume of Tables of Heights' of the Great Trigonometrical Survey of India. Hopes to visit JH and discuss phenomenon of terrestrial refraction. Received valuable assistance on tables from JH's son John.
ALS 3pp RS:HS 18.13

12424. 1864–1–3 From Gabriel Auguste Daubrée Paris
Correspondence relating to the fall of a meteorite in South Africa on 13 Oct.1838, and the provision of samples from Thomas Maclear by way of JH. [Letter illegible in parts.]
ALS 4pp (French) RS:HS 6.30

12425. 1864–1–4 From William Edward Hickson Fairseat, Wrotham
Replies to opinions of astronomers [regarding moon's influence on weather]. Still waiting to hear from JH's son Alexander whether microscopists have searched for foraminifera in meteorites.
ALS 8pp TxU:H/M–0276; Reel 1087

12426. 1864–1–4 From Smith, Beck & Beck 31 Cornhill
Sending photographs of moon's surface.
ALS 1p RS:HS 17.219

12427. 1864–1–4 From John Taylor Leonard Place, Kensington
Discusses aspects of JT's new book [*The Great Pyramid*], especially its front matter.
ALS 2pp RS:HS 17.332

12428. 1864–1–5 From Alfred Yeomans 3 Foster St., Stonebridge
Poses question of sun's differential gravitational attraction on opposite sides of earth, inspired by JH's article ['Sun,' 1863] in *Good Words*.
ALS 3pp TxU:H/M–0611; Reel 1086

12429. 1864–1–7 To George Gabriel Stokes Collingwood
Note accompanying report on paper by G. B. Airy.
ALS 2pp Cambridge University Library 7656/RS428a

12430. 1864–1–8 From Walter White Burlington House
G. B. Airy's paper has been received at R.S.L.
ALS 1p TxU:H/M–0579.1; Reel 1093

12431. 1864–1–13 From James Samuelson 11 New Burlington St.
At request of William Crookes, JS [editor of new *Quarterly Journal of Science*] invites JH to submit original article on sunspots. Promises liberal honorarium . See article on sunspots by C. A. F. Peters in *Zeitschrift für populäre Mittheilungen aus dem Gebiete der Astronomie und verwandter Wissenchaften*.
ALS 2pp RAS:JH Archive 13/1.2; Reel 10

12432. 1864–1–15 From Richard Christopher Carrington Isleworth W.
RC's book [*Observations of the Spots on the Sun*] is done. First copy, exhibited at R.A.S. last week, was well reported in *R.A.S.M.N.* Will send copy to JH.
ALS 4pp RAS:JH Archive 13/2.1; Reel 10

12433. 1864–1–15 From James Samuelson 11 New Burlington St.
Encloses copy of C. A. F. Peters's article in *Zeitschrift für populäre Mittheilungen aus dem Gebiete der Astronomie und verwandter Wissenschaften*. Please submit JH's article on sunspots by February. Audience of *Quarterly Journal of Science* are '*intelligent students* of other branches and intelligent laymen.'
ALS 2pp RAS:JH Archive 13/1.3; Reel 10

12434. 1864–1–16 From Charles Pritchard Freshwater
Thanks JH for solar observations. Discusses 'willow leaves' observations made at Greenwich Observatory after following JH's suggestions for solar eyepiece.
ALS 4pp RS:HS 14.85 Ada Pritchard, *Charles Pritchard: Memoirs of His Life*, 242

12435. 1864–1–16 From George Shepherd 26 Throgmorton St.
Argues against JH's position that weather forecasts cannot be made for more than 48 hours in advance.
ALS 4pp RS:HS 16.141

12436. 1864–1–17 To George Shepherd Collingwood
Responds to GS's 1864–1–16.
C 3pp RS:HS 16.142

12437. 1864–1–18 From Richard Christopher Carrington Isleworth W.
Formulas for determining rotation rates of sunspots relative to latitude on solar surface. Believes equatorial parts of photosphere are dragged faster than others. Speculates on rotation rate of solid body below (if such exists).
ALS 4pp † RAS:JH Archive 13/2.2; Reel 10

12438. 1864–1–19 From George Shepherd 26 Throgmorton St.
The argument continues [see JH's 1864–1–17 & GS's 1864–1–16].
ALS 3pp RS:HS 16.143

12439. 1864–1–20 From M. Hume 3 Berck St., Regent St.
A note to say that MH is returning JH's dress coat by train.
ALS 1p RS:HS B27.25b

12440. 1864–1–20 From Charles Pritchard Royal Astronomical Society
Gratitude for JH's gift of Caroline Herschel's autograph manuscripts to R.A.S.
ALS 2pp RAS:JH Archive 15/5.2; Reel 11

12441. 1864–1–20 From Henry Enfield Roscoe and Joseph Baxendell Manchester
[Printed notice] Next meeting of Literary and Philosophical Society will be 26 Jan.
L 1p TxU:H/L-0305.1; Reel 1054

12442. 1864–1–20 From James Samuelson 8 Jubilee St., Everton, Liverpool
Needs *Zeitschrift* volume; then JH may have it back [see JS's 1864–1–15]. JH's article should be 10–12 pages.
Compliments JH's writings as popular without being 'popularising.'
ALS 2pp RAS:JH Archive 13/1.4; Reel 10

12443. 1864–1–21 From Thomas Spring-Rice Exchequer
Platinum pound-weight standard is deteriorating. Wants Treasury to bring it and the yard standard under observation of surviving committee members.
ALS 2pp RS:HS 14.481

12444. 1864–1–23 From George Biddell Airy n.p.
Replies to JH about the date of the transit of Mercury [see JH's 1864–1–[26]].
ACCS 2pp RGO 6.85.161

12445. 1864–1–24 From Charles Pritchard Freshwater
Asks JH for information on the Beaufoy Clock owned by R.A.S., which is now missing. Requests JH to read R.A.S. Annual Report.
ALS 3pp RS:HS 14.86 Ada Pritchard, *Charles Pritchard: Memoirs of His Life,* 246–7

12446. 1864–1–25 From George Alexander Hamilton Treasury Chambers
Treasury Commissioners ask that surviving members of 1853 Standards Committee reexamine Parliamentary standards of length and measure preserved at office of Exchequer.
LS 2pp TxU:H/M–0226; Reel 1087

12447. 1864–1–25 From Frederick Howlett St. Augustine's, Hurst Green
Comments on the sun spots he has been observing, and some new phenomenon. Any chance of a visit from JH?
ALS 4pp RS:HS 10.15

12448. 1864–1–26 To George Biddell Airy Collingwood
Has been asked to reweigh the damaged Parliamentary standard pound weight; thinks that W. H. Miller would be a more suitable person to do the job.
AL(S) 2pp RGO 6.358.267

12449. 1864–1–[26] To George Biddell Airy n.p.
Raises with GA the likelihood of an error in a date of a transit of Mercury in a paper by F. A. Winnecke.
ALS 2pp inc RGO 6.85.158

12450. 1864–1–26 To Frederick Howlett Collingwood
On the observation of sun spots.
C 1p RS:HS 24.26

12451. 1864–1–27 From George Biddell Airy Royal Observatory, Greenwich
Regarding Thomas Spring-Rice (1st Baron Monteagle), the Treasury and the Commission on Weights and Measures.
ALS 4pp RS:HS 1.270

12452. 1864–1–28 To F. W. A. Argelander Collingwood
Thanks for FA's star atlas series, one of which is a duplicate; reports apparent variation of a nebula.
C 2pp RS:HS 24.27

12453. 1864–1–28 To George Alexander Hamilton Collingwood
Received report by Lord Monteagle [Thomas Spring-Rice] on Standard Platinum Pound. Agrees to participate with members of 1853 Standards Committee in reexamining standards.
AdraftS 1p TxU:H/L-0177; Reel 1054

12454. 1864–1–29 From Hermann Kindt Yarm, Yorkshire
Expresses thanks for the writings JH has sent him; asks for other writings by JH.
ALS 4pp (German) RS:HS 11.24

12455. 1864–1–30 To F. W. A. Argelander Collingwood
Mistaken about seeming duplicate [see JH's 1864–1–28].
C 1p RS:HS 24.28

12456. 1864–1–30 To François N. M. Moigno Collingwood
FM has published, in translation, JH's article ['Weather and Weather Prophets' (1864)], already published in *Good Words.* JH was not aware that this was FM's intent, and JH warns him that some editors may be sensitive about such matters.
ALS 3pp DMC.695A.885.29 (Cdraft: RS:HS 25.15.20 & ACCS: TxU:H/L-0244; Reel 1054)

12457. 1864–2–1 From Edward Joseph Lowe Beeston
Encloses a paper on a hail and snow storm, which hand over to the British Meteorological Society when he has read it. Would JH request the R.S.L. Secretary to re-suspend his certificate as he understands he has a good chance of being elected this year.
ALS 3pp RS:HS 11.339

12458. 1864–2–2 From Richard Christopher Carrington Royal Brewery, Brentford
Copy of RC's book will be sent to JH from R.S.L. Thanks JH's son for letter.
ALS 3pp RAS:JH Archive 13/2.3; Reel 10

12459. [1864}–2–2 From François N. M. Moigno St. Germain de Pres
Regarding JH's article in *Good Words.*
ALS 1p (French) RS:HS 12.357

12460. 1864–2–4 From John C. Broun Wynberg
Further regarding the lectureship at the Cape of Good Hope.
ALS 2pp RS:HS 5.110

12461. 1864–2–4 From George Alexander Hamilton Treasury Chambers
Committee to reexamine Parliamentary standards of length and measure will meet on 11 Feb. in office of comptroller-general of Exchequer [Thomas Spring-Rice]. If committee desires to compare three copies of Standards, those copies will be brought to Exchequer office.
LS 2pp TxU:H/M–0227; Reel 1087

12462. 1864–2–4 From George Gabriel Stokes Royal Society
Sends R.S.L. Council's formal acknowledgement of star catalogue manuscript [see JH's 1863–12–4].
ALS 3pp RS:HS 17.55

12463. 1864–2–4 From George Gabriel Stokes Royal Society
Discusses printing of JH's catalogue of nebulae.
ALS 4pp RS:HS 17.54

12464. 1864–2–4 To Strahan & Co. Collingwood
Comments and will consider writing on article subjects proposed by Dr. [Norman?] McLeod. Requests Hebrew version of Genesis, chapter 1, with pronunciation and translation guides.
C 1p inc RS:HS 24.29

12465. 1864–2–5 To George Biddell Airy Collingwood
Is trying to discover the train connections that will allow JH to come to a meeting of the Standards Committee without coming to London overnight. 'Catalogue of Nebulae' is now in proof state [see GA's 1863–10–9] and needs to be checked.
ALS 4pp RGO 6.358.270 & 6.200.111

12466. 1864–2–5 From Hermann Kindt Yarm, Yorkshire
Thanks for further translations from JH and could HK have an extra copy for a dear friend.
ALS 4pp (German) RS:HS 11.25

12467. 1864–2–5 From James Samuelson 8 Jubilee St., Everton, Liverpool
Received JH's drawings. Will send proof of plate.
ALS 1p RAS:JH Archive 13/1.5; Reel 10

12468. 1864–2–5 To George Gabriel Stokes Collingwood
Responds to GS's letters of 1864–2–4 about procedures for printing JH's 'Catalogue of Nebulae' and receipt of William Herschel's manuscripts .
ALS 4pp Cambridge University Library 7656/H704

12469. 1864–2–6 From George Biddell Airy Royal Observatory, Greenwich
Further regarding the forthcoming meeting of the Standards Commission. Regarding JH's 'Catalogue of Nebulae.'
ALS 3pp RS:HS 1.271

12470. 1864–2–6 From George Biddell Airy n.p.
Informs JH about train service [see JH's 1864–2–5], and encourages JH to stay with GA from the night before the meeting.
ACC 2pp inc RGO 6.358.271

12471. 1864–2–6 or later To George Biddell Airy n.p.
As JH has made minor changes in many calculated results, the proof sheets cannot be easily compared with the calculation sheets [see GA's 1864–2–6].
ALS 4pp RGO 6.200.114

12472. 1864–2–6 From George Biddell Airy [Royal Observatory, Greenwich]
The practice at the Royal Observatory is to compare proof copies with calculation sheets [see JH's 1864–2–5].
ACCS 1p RGO 6.200.113

12473. 1864–2–6 From John Henry Nelson n.p.
The contents of JH's note has not surprised him. Encloses report of the trial for the better understanding of JH. Will be at Upton in a day or so.
ALS 2pp RS:HS 13.101

12474. 1864–2–7 To *Photographic News* Collingwood
Asserts that a process for obtaining 'Turnbull's blue' recently published in *PN* was discovered and published in 1842 by JH. Adds a second case in which JH anticipated another author's result.
ALS 7pp † TxU:H/L-0303; Reel 1054 *Photographic News, 8* (Feb. 12, 1864), 82

12475. 1864–2–7 To Felix Taylor Collingwood
JH declines to negotiate further about FT's continuing occupancy of Herschel House.
ACCS 2pp Oxford History of Science Museum 58/11

12476. 1864–2–8 From George Biddell Airy Royal Observatory, Greenwich
Regarding JH's forthcoming visit. On the platinum in the pound weight becoming browned.
ALS 2pp RS:HS 1.272

12477. 1864–2–8 From Charles Pritchard Freshwater
Thanks JH for reading R.A.S. records. G. P. Bond will not receive R.A.S. medal. The Beaufoy Clock is recovered.
ALS 3pp RS:HS 14.87

12478. 1864–2–9 To George Biddell Airy Collingwood
Is now too ill to come to the meeting of the Standards Committee; replies to GA's queries about how to clean up the pound [see GA's 1864–2–8].
ALS 4pp RGO 6.358.273

12479. 1864–2–10 To William Whewell Collingwood
Responds to WW's comments on parts of JH's translation of Homer's *Iliad.* Is sending his rendering of Book V. Has been ill.
ALS 5pp TC, Camb. Add. Ms.a.207[110] (C: RS:HS 24.30)

12480. 1864–2–10 From Edwin Richard Windham Wyndham-Quin n.p.
Augusta Wyndham-Quin and EW enjoyed visit by JH's daughter, who delivered JH's essay on pendulum (1863) and 'one of the best photographs' EQ ever saw. Send another.
ALS 2pp RS:HS 18.440

12481. 1864–2–11 From Hermann Kindt Yarm, Yorkshire
Thoughts on translation of literature, with comments specifically on JH's *Iliad* translation.
ALS 4pp (German) RS:HS 11.26

12482. 1864–2–11 From Charles Rodgers Charity School, Dewsbury
CR's hypothesis regarding effects of rotation of large masses, like sun and Jupiter, on rotation and orbits of small planets and satellites. Quotes *Genesis* for support.
ALS 3pp TxU:H/M–0457; Reel 1093

12483. 1864–2–12 From Augustus De Morgan 91 Adelaide Road
Hopes all are well. Comments on Charles Babbage's latest book. Intends telling the truth about James Ivory's delusions. Sends proof that ten is an even number.
ALS 4pp RS:HS 6.368

12484. 1864–2–13 From George Biddell Airy Royal Observatory, Greenwich
Expecting him soon. Experiments on the National pound.
ALS 4pp RS:HS 1.273

12485. 1864–2–15 From Charles Babbage Dorset St.
Will JH and his son [William] come and dine with him this week?
ALS 1p RS:HS 2.333

12486. 1864–2–16 From Richard Hodgson Grove House, Highgate Road
Correction to R.A.S. report regarding eyepiece that RH invented for viewing sun, but which was attributed to JH by Charles Pritchard.
ALS 2pp † (by Charles Pritchard) RS:HS 19.278

12487. 1864–2–18 To Richard Hodgson Collingwood
Charles Pritchard was wrong in ascribing priority for diagonal reflecting telescope in Pritchard's address to R.A.S. As early as 1825, JH was first to use diagonal reflector. Describes its construction.
C 3pp RS:HS 19.153 (C: RS:HS 24.31)

12488. 1864–2–20 From Charles Pritchard Freshwater
Richard Hodgson claims to have invented the diagonal solar eyepiece despite JH's description of it in *Cape Results.*
ALS 3pp RH:HS 14.88

12489. 1864–2–20 From James Samuelson 8 Jubilee St., Everton, Liverpool
Send JH's manuscript immediately, or JS must postpone JH's

article for next issue. Received Henry Draper's article on 'Silverised Glass Telescopes & Celestial Photo[graphy]' mentioning JH and 'Miss Herschel.'
ALS 2pp RAS:JH Archive 13/1.6; Reel 10

12490. 1864–2–21 From Charles Pritchard [Freshwater]
Richard Hodgson pursues claim to invention of the diagonal solar eyepiece while claiming that JH disclaimed the invention.
ALS 3pp RS:HS 14.89

12491. [1864–2–22 or later] From William Stronach Ardmellie, Banffshire
Forwards WS's observations of cloud cover in Banffshire during full moons of Dec. 1863, Jan. 1864, and Feb. 1864, as suggested by JH in *Good Words*.
ALS 1p, encl 3pp TxU:H/M–0511; Reel 1093

12492. 1864–2–23 From George Biddell Airy Royal Observatory, Greenwich
Regarding the transit of Mercury. Error in calculating the pound weights.
ALS 3pp RS:HS 1.274

12493. 1864–2–23 To George Gabriel Stokes Collingwood
Further about nebulae catalogue [see JH's 1864–2–5].
ALS 4pp Cambridge University Library 7656/H705

12494. 1864–2–24 From W. K. von Haidinger Vienna
Has sent a collection of his own communications for JH via the Foreign Secretary of the R.S.L.
ALS 1p RS:HS 9.152

12495. 1864–2–24 From John R. Hind Nautical Almanac Office
JH's letter went to the old address, which accounts for the delay. Regarding his query, gives date of the conjunction of the planets.
ALS 2pp RS:HS 9.389

12496. 1864–2–24 From James Samuelson 8 Jubilee St., Everton, Liverpool
JH's request for three proofs contradicts printer's policy of issuing only one of each. Assurance that JH's manuscript will be examined carefully for errors.
ALS 2pp RAS:JH Archive 13/1.7; Reel 10

12497. 1864–2–25 From Edward James Stone Royal Observatory, Greenwich
Intrigued by discovery of 'willow leav[e]s' on sun's photosphere, bought new telescope eyepiece. Reports observations on 'leav[e]s' and overall 'rice-like' texture of sun under telescope.
C 4pp RS:HS 13.82

12498. 1864–2–26 From Julia Margaret Cameron Freshwater, I.o.W.
Photographic news. Has heard about the JHs.
ALS 12pp RS:HS 5.159

12499. 1864–2–26 From William Rutter Dawes Hopefield, Haddenham, Thame
Would like account of the telescope of Warren de La Rue which JH has. E. J. Stone of the Royal Observatory has seen objects on the sun's surface like grains of rice.
ALS 3pp RS:HS 6.115

12500. 1864–2–26 From James Samuelson 8 Jubilee St., Everton, Liverpool
Corrections on JH's manuscript on sunspots. Please keep alterations to minimum. Will send complimentary copies of *Quarterly Journal of Science*. Please write brief note on 'vaporous ring' and its bearing upon sunspots.
ALS 4pp RAS:JH Archive 13/1.8; Reel 10

12501. 1864–2–27 From Richard Hodgson Brighton
Thanks for his letter. Has no copy of JH's Cape Observations so cannot refer to them. Comments on articles on telescopes by JH; they do not lead him to adopt the arrangement he adverted to 10 years ago. Regarding priorities in discoveries.
ALS 3pp RS:HS 10.199

12502. 1864–2–27 To Richard Hodgson Collingwood
Did not insinuate that RH plagiarized JH's idea of 'diagonal reflector,' but feels justified in defending JH's claim to being first to apply it.
ACCS 3pp RS:HS 19.154 (C: RS:HS 24.32)

12503. 1864–2–29 From M. & N. Hanhart, Lithographers 64 Charlotte St., Rathbone Place W.
Sends proofs and originals of plates [for JH's catalogue of nebulae], as directed by Mr. Churchill.
L 1p RAS:JH Archive 13/3.99; Reel 10

12504. 1864–3–1 From Charles Pritchard to Richard Hodgson Freshwater
Cites quotation from *Cape Results* proving that JH invented the diagonal solar eyepiece. Discusses RH's claim. Note to JH accompanies this letter.
CCS(A) 7pp RS:HS 14.90

12505. 1864–3–1 From Herbert Spencer 29 Bloomsbury Square
Doubts the claim that Auguste Comte has significantly influenced English scientists. Asks whether JH believes Comte has influenced JH or others. Asks whether it is more difficult to calculate the moon's motion or the motion of a projectile through a resisting medium.
ALS 4pp RS:HS 16.493

12506. 1864–3–2 From James Samuelson 8 Jubilee St., Everton, Liverpool
As requested, sent six copies of *Quarterly Journal of Science, No. II* to JH and two to Russia and Switzerland. Publisher's policy prohibits sending reprints. Sends proof of JH's plate with suggested alterations to 'Willow leaves.'
ALS 4pp RAS:JH Archive 13/1.9; Reel 10

12507. 1864–3–5 From John Henry Griesbach Campden Hill Villas
Regarding a musical phenomenon pointed out to him by F. A. Gore-Ouseley.
ALS 10pp RS:HS 9.64

12508. 1864–3–5 From James Samuelson 8 Jubilee St., Everton, Liverpool
Alterations to proofs of JH's plate [on sunspots].
ALS 1p RAS:JH Archive 13/1.10; Reel 10

12509. 1864–3–6 To [Frederick Howlett] n.p.
Comments on observations of sun and sunspots, especially R. C. Carrington's book on the subject.
ALS 4pp St. John's College, Cambridge (C: RS:HS 24.33)

12510. [1864?]–3–7 From [Augusta] Wyndham-Quin n.p.
Just received photograph of JH, because Mrs. Gordon forgot to forward it. [E. R. W. Wyndham-Quin] Lord Dunraven is jealous and wants one too. Enjoyed visit of Mr. [Juetty?] and Mrs. Gordon.
ALS 4pp RS:HS 18.441

12511. 1864–3–8 To Elizabeth Colling Collingwood
Coincidences between observations of a meteor seen by EC and another account.
C 2pp RS:HS 24.34

12512. 1864–3–9 From George Biddell Airy Royal Observatory, Greenwich
Meeting of the Standards Commission on 16 March.
ALS 1p RS:HS 1.275

12513. 1864–3–9 From Charles Rodgers Charity School, Dewsbury
Corrects error in sun's velocity that CR stated in earlier letter to JH today.
ALS 1p TxU:H/M–0459; Reel 1093

12514. 1864–3–9 From Charles Rodgers Charity School, Dewsbury
Demonstrates mathematically that sun's orbit cannot be elliptical.
ALS 2pp TxU:H/M–0458; Reel 1093

12515. 1864–3–9 To Herbert Spencer Collingwood
In response to HS's query [1864–3–1] about the influence of Auguste Comte on English scientists, JH refers HS to JH's 1845 B.A.A.S. presidential address. States that calculating the motion of a projectile through a resisting medium is extremely difficult.
CC 4pp RS:HS 16.494 (C: 24.35)

12516. 1864–3–12 From Richard Hodgson Chingford
[Charles?] Pritchard has embittered the subject so he must just acknowledge the receipt of JH's letter.
ALS 1p RS:HS 10.200

12517. 1864–3–12 From James Samuelson 8 Jubilee St., Everton, Liverpool
Alterations to proofs of JH's plate [on sunspots].
ALS 2pp RAS:JH Archive 13/1.11; Reel 10

12518. 1864–3–13 From James Samuelson 8 Jubilee St., Everton, Liverpool
Alterations to proofs of JH's plate [on sunspots]. Explains tinting process.
ALS 2pp RAS:JH Archive 13/1.12; Reel 10

12519. 1864–3–15 To George Biddell Airy Collingwood
Is unable to attend another meeting of the Standards Committee, due to severe illness; describes the best way to pack the standards away, and talks about the strange behavior of gallic acid.
ALS 4pp RGO 6.358.278

12520. 1864–3–16 To John Herschel (son) Collingwood
Announces son William's engagement to Emma Hardcastle; warns son John against the Anglo–Indian ladies; JH comments at length on a problem in surveying, and concludes with the news that daughter Francisca's health is improving.
ALS 8pp JHS 5.13

12521. 1864–3–16 From James Samuelson Everton, Liverpool
Apologizes for printer's misspelling of JH's name on plate. Encloses letter from Hanhart Lithographers with suggestions on tinting JH's plate of solar spots.
ALS 1p, encl ALS 1p RAS:JH Archive 13/1.13; Reel 10

12522. [1864–3–19] From Frederick Howlett n.p.
Drawing of 'the singular mottled appearance of portions of the Sun's disk ... March 19th 1864'.
DS 1p RS:HS 19.227

12523. [1864]–3–20 From Julia Margaret Cameron Freshwater, I.o.W.
Hopes his children have recovered their health. Comments on her recent photographic work. Wishes she could photograph JH.
ALS 16pp RS:HS 5.158

12524. 1864–3–22 From Frederick Howlett St. Augustine's, Hurst Green
Has returned James Nasmyth's letter and has never seen Warren de La Rue's letter. Comments on Nasmyth's 'willow leaves.' Have been observed at Greenwich Observatory. Thanks for his remarks on the best kind of eye-piece.
ALS 8pp RS:HS 10.16

12525. 1864–3–23 From Thomas Francis Wade 58 Upper Seymour St., Portman Square
Lady Herschel is valuable friend, having written to TW often in China. TW was to have 'bowed before John Steward' but postponed. Heard of JH's son's intentions from [Alexander Hamilton?] Gordon.
ALS 8pp RS:HS 18.4

12526. 1864–3–24 From James Samuelson 8 Jubilee St., Everton, Liverpool
'Private & confidential.' Asks JH to recommend reliable writer for 'Chronicles of Astronomy' section in *Quarterly Journal of Science*. Has purchased and will amalgamate with *Edinburgh [New] Philosophical Journal*. [JH annotation: Recommended [Edward] Brayley and [Robert] Grant].
ALS 2pp † RAS:JH Archive 13/1.14; Reel 10

12527. 1864–3–28 From Edward John Jennings Hawkhurst
Hopes that JH will be able to attend the dinner of the 38th Kent Rifle Volunteers on 5 April.
ALS 1p RS:HS 10.439

12528. 1864–3–29 From W. Bonsey Slough
Would like a photograph of JH. Recent and forthcoming events connected with Slough.
ALS 3pp RS:HS 4.192

12529. 1864–3–29 From James Samuelson 8 Jubilee St., Everton, Liverpool
Thanks JH for recommendations [see JS's 1864–3–24]. Plans new section in *Quarterly Journal of Science* to review recent events in science. Shall avoid 'craving for novelty' evident in other publications.
ALS 2pp RAS:JH Archive 13/1.15; Reel 10

12530. 1864–3–31 From Charles Pritchard Freshwater
Forwards letter concerning 'solar bright particles' by W. R. Dawes. Inquires if diagonal solar eyepiece used in JH's solar spectrum observations.
ALS 2pp RS:HS 14.91

12531. [1864–4] From John Curwen Forset Gate
Is about to provide a tuning fork for C and would like JH's permission to call it the 'Herschel C.'
ALS 1p RS:HS 5.383

12532. 1864–4–2 From W. Bonsey Belle Vue, Slough
Regarding a tenant for Herschel House. Family news.
ALS 3pp RS:HS 4.193

12533. 1864–4–2 From Richard Christopher Carrington Isleworth
Thanks for sending the paper in the *Quarterly Journal*. Comments on the importance of some of the points raised.
ALS 3pp RS:HS 5.220

12534. 1864–4–2 To William Whewell Collingwood
Thanks WW for comments on Book V of JH's *Iliad* translation; sending Book VI. JH's son William and JH's daughters enjoyed visiting WW. Has suffered for three months from severe bronchitis.
ALS 3pp TC, Camb. Add. Ms.a.207[111] (C: RS:HS 24.36)

12535. 1864–4–3 From James Samuelson 8 Jubilee St., Everton, Liverpool
Forwards paper by Henry Draper of New York.
ALS 1p RAS:JH Archive 13/1.16; Reel 10

12536. 1864–4–4 From Alexander Elder Lancing
Would like JH's recommendation to obtain a place in Greenwich School for a foster son of his own.
ALS 2pp RS:HS 7.124

12537. 1864–4–6 From Hermann Kindt Yarm, Yorkshire
HK has a number of letters from prominent Continental scientists, which he offers to JH for his perusal; adds further comments on translation.
ALS 4pp (German) RS:HS 11.27

12538. 1864–4–6 From Isaac Todhunter St. John's, Cambridge
Printing *History of the Mathematical Theory of Probability.* Inquires about a probability calculation of Wilhelm Struve's in JH's *Outlines Astr.*
ALS 4pp RS:HS 17.353

12539. 1864–4–7 To Isaac Todhunter Collingwood
Unable to find the calculation of Wilhelm Struve about which IT inquired. Will investigate the matter further.
C 2pp RS:HS 24.37

12540. 1864–4–11 To William Whewell Collingwood
Responds to some of WW's comments on JH's *Iliad* translation. May publish Books I–VI. Comments on Alexander Pope's translation.
ALS 4pp TC, Camb. Add. Ms.a.207[112] (C: RS:HS 24.38)

12541. 1864–4–12 To Richarda Airy Collingwood
Sends congratulations on the Airy's daughter Hilda's engagement to Edward J. Routh; also thanks for the congratulations sent to JH's son William James on his engagement to Emma Hardcastle.
ALS 4pp JHS 2.11

12542. 1864–4–12 From Bewicke Blackburn 25 Hans Place
Describes convection phenomenon observed in hot varnish boiler. This may explain sunspots.
ALS 6pp TxU:H/M–0106; Reel 1087

12543. 1864–4–15 To [James] M. Gilliss Collingwood
Thanks JG for the receipt of astronomical and meteorological observations made at Washington Observatory.
ALS 1p DMC.695A.885.12

12544. 1864–4–16 From William Walker 64 Margaret St., Cavendish Square
WW's engraving [*Distinguished Men of Science Living 1807–8*] is selling well. Plans another for scientists of 1846–47. Requests that JH sit for photograph.
ALS 2pp RS:HS 18.18

12545. 1864–4–18 To William Walker Collingwood
Agrees to sit for photographic portrait by WW. Thanks for reserving copy of WW's successful [*Distinguished Men of Science Living 1807–8*] for JH.
ALS 1p Bedgebury

12546. 1864–4–19 To Maria Sophia Herschel Collingwood
Reluctantly agrees to review a manuscript for a Mr. Adams.
C 2pp RS:HS 24.39

12547. 1864–4–20 From William Rutter Dawes Hopefield, Haddenham, Thame
Is ready to hear that he is to have a visit from Willie Herschel. Regrets he cannot entertain friends as he used to. Has been laid up with ill-health. Would like information on how to prepare a chemical precipitate as that forming the sun's surface. Warren de La Rue has written to him.
ALS 4pp RS:HS 6.116

12548. 1864–4–21 To Augustus De Morgan Collingwood
Comments on metrification.
C 1p RS:HS 24.40

12549. 1864–4–22 From Mrs. Gardiner Orsett Terrace, Hyde Park
Hears that Herchell (sic) House is for sale and would like details of the terms.
AL 1p RS:HS 9.89

12550. 1864–4–22 From William Hallowes Miller Cambridge
Cases have been ordered by Mr. Graham. Final meeting of the Committee for the Restoration of the Standards is fixed for 29 Apr. Will notify him should it be postponed.
ALS 2pp RS:HS 13.42

12551. 1864–4–23 From George Biddell Airy Royal Observatory, Greenwich
Sending him the map of the world.
ALS 1p RS:HS 1.277 (C: RGO 6.480.332)

12552. 1864–4–23 From George Biddell Airy Royal Observatory, Greenwich
Meeting of the Standard Commission on 29 April.
ALS 1p RS:HS 1.276

12553. 1864–4–25 To Mrs. Gardiner Collingwood
Has not advertised Herschel House as yet and does not know whether to let or sell it, but would be prepared to discuss the matter with principals only.
draft 2pp RS:HS 9.90

12554. 1864–4–26 From Rycroft Reeve Picadilly
Invites JH to become a fellow of the Genealogical and Historical Society.
ALS 1p RS:HS 14. 475

12555. 1864–4–27 From Frederick Howlett St. Augustine's, Hurst Green
Hopes the JH's are all well. Would like an eye-piece made by Mr. Cook[e] or any leading optician on the principles of JH as laid down in his paper in *R.S.P.T.* Shall he refer the optician to this paper?
ALS 4pp RS:HS 10.17

12556. 1864–4–28 To George Biddell Airy Collingwood
Will miss yet another meeting of the Standards Committee due to illness [bronchitis]; W. H. Miller has brought JH up to date on the activity of the committee, and JH agrees with what is being done.
ALS 3pp RGO 6.358.280

12557. 1864–4–28 From Henry W. Cookson Cambridge
Will he allow his name to be put forward to receive the honorary degree of L.L.D. in June?
ALS 1p RS:HS 5.365

12558. 1864–4–29 To Edward Sabine Collingwood
Had hoped to attend ES's soiree, but has influenza. Praises ES's report on Mr. McClintock's magnetic observations at Port Barrow.
ALS 3pp RS Sa.671

12559. 1864–4–30 To Henry W. Cookson Collingwood
Declining the offer of a honorary degree by the University of Cambridge on the grounds of ill health.
draft 3pp RS:HS 5.366 & 24.41

12560. 1864–4–30 To Frederick Howlett Collingwood
More on the construction of a good optical eyepiece for solar observations [see FH's 1864–4–27].
C 2pp RS:HS 24.42

12561. [1864–5] To Sir Charles Lyell n.p.
Introducing his eldest son and his son's newly married wife.
draft 2pp RS:HS 11.429

12562. 1864–5–1 From Warren de La Rue Observatory, Cranford
Regarding the solar photographs. They are not magnified originals but taken direct. Comments on these photographs. Has had correspondence with W. R. Dawes regarding the 'willow leaves' of James Nasmyth.
ALS 3pp RS:HS 6.153

12563. 1864–5–2 From Augustus De Morgan 91 Adelaide Road
Returns a book of poetry. His family is at Hastings. Sends a theorem.
ALS 6pp RS:HS 6.369

12564. 1864–5–3 From Walter M. Adams 55 Montagu Square
Is sending a manuscript for JH's comments and opinion.
ALS 3pp RS:HS 1.385

12565. 1864–5–3 To Augustus De Morgan Collingwood
Comments on some mathematical problems; asks if AD knows about a German geometer by the name of [Bernhard] Riemann; suggests some new names for trigonometry.
C 2pp RS:HS 24.43

12566. 1864–5–4 To Walter M. Adams Collingwood
Acknowledging receipt of the manuscript. JH will be unable to give an opinion as the work is so voluminous that he cannot spare the time, unless WA is prepared to loan it for a long period.
draft 2pp RS:HS 1.386 & 24.45

12567. 1864–5–4 To Augustus De Morgan Collingwood
Comments on health of family, and on parliament considering a bill to change weights and measures.
C 2pp RS:HS 24.44

12568. 1864–5–5 From Walter M. Adams 55 Montagu Square
Apologizing for his mistake in expecting an opinion of his manuscript, but would be glad if JH would read the first section only and comment on it.
ALS 3pp RS:HS 1.387

12569. 1864–5–6 From Thomas Humphrey Hawkhurst National School
Announces a meeting of the Committee of Management for the Hawkhurst National School.
ALS 1p inc Harvard: Houghton AAH 67m–67 (62)

12570. 1864–5–9 From Frederick Howlett St. Augustine's, Hurst Green
Is grateful for his communication on the proposed solar eyepiece. Will call round on his way to town to ascertain if he can leave the manuscript with [Thomas?] Cooke, the instrument maker, so that he can have some idea of the work involved. James Nasmyth will pay attention to screen projection.
ALS 4pp RS:HS 10.19

12571. 1864–5–10 From Walter M. Adams 55 Montagu Square
Thanking him for the return of the manuscript and for his comments.
ALS 2pp RS:HS 1.388

12572. 1864–5–12 From George Gabriel Stokes Royal Society
Asks JH to review 'magnetical' paper by Edward Sabine.
ALS 3pp RS:HS 17.56

12573. 1864–5–13 To George Gabriel Stokes Collingwood
Agrees to review paper [see GS's 1864–5–12].
ALS 2pp Cambridge University Library 7656/RS436

12574. 1864–5–14 From Fedor Thomas 5 Quai d'Anjou, Paris
Not too familiar with JH's work, as FT's work has taken him out of the area of JH's interest. Has very limited time for mathematics, but indicates his current interest. Refers to some writings of FT on topics of weights, measures, and monetary matters.
ALS 3pp (French) RS:HS 17.432

12575. 1864–5–16 To William Whewell Collingwood
Sends Book VII of JH's *Iliad* translation. Son William to be married Thursday. JH continues to be ill.
ALS 3pp TC, Camb. Add. Ms.a.207[113] (C: RS:HS 24.46)

12576. 1864–5–17 From Bewicke Blackburn 25 Hans Place
Regarding his own experiments with the froth of boiling oil.
ALS 4pp RS:HS 5.74

12577. 1864–5–19 From James Nasmyth Penshurst
Hopes JH and Lady Herschel will be able to visit them on some days named. Will write respecting trains as soon as they hear the answer.
ALS 2pp RS:HS 13.83

12578. 1864–5–21 To Merchants Taylor Co. Collingwood
Compliments to the company.
ALS 1p Harvard: Houghton AAH 67m–67 (62)

12579. 1864–5–22 From James Nasmyth Penshurst
Regrets Lady Herschel cannot visit them, but is pleased they are to receive JH and his daughter. Mrs. Nasmyth is writing regarding trains and they will be met at Penshurst.
ALS 2pp RS:HS 13.84

12580. 1864–5–24 From Alfred Potiquet Paris
Discusses work on biographies of scientific men.
ALS 1p (French) RS:HS 14.179

12581. 1864–5–24 To George Gabriel Stokes Collingwood
Reports on Edward Sabine's paper [*R.S.P.T., 154,* 227–] comparing magnetic disturbances at Kew and Nertschinsk. Makes suggestion regarding period of solar spots and other

matters, but deems paper worthy of publication.
ALS 4pp RS RR.5.234

12582. 1864–5–25 To John Herschel (son) Collingwood
Positively responds to son John's idea of geodesic measurements of much larger than usual triangles; suggests that John is not likely to be able to correct the whole world's English; JH reports about sickness in the family including his own seemingly permanent bronchitis.
ALS 4pp JHS 5.11

12583. 1864–5–30 From William Rutter Dawes to Margaret Brodie Herschel Hopefield, Haddenham, Thame
Was pleased to receive her note when Willie [William James Herschel] visited him. Hopes Willie will be happy in his married life. Was unable to accompany Willie to Aylesbury. Sorry to hear she has been unwell.
ALS 4pp RS:HS 6.117

12584. 1864–6–1 From Warren de La Rue Observatory, Cranford
Agrees with him that J. Müller's drawings do not show the entities around the sunspot clearly. Has nearly completed the apparatus for holding one of JH's sun prisms to his reflector. Hopes to do something with it soon.
ALS 2pp RS:HS 6.154

12585. [1864–6] To [Edward Sabine?] n.p.
Pendulum experiments are already approved at principal stations in Russian trigonometrical survey. Recent communication from J. H. Pratt to R.S.L. about pendulum observations.
Adraft 2pp inc TxU:H/M–0579.2; Reel 1093

12586. 1864–6–2 To Sir Charles Lyell Collingwood
Writes to introduce JH's son William James and his new bride to the Lyell's. Both JH and his wife, Margaret, are suffering from illness.
ALS 4pp APS B.D25.L.5

12587. 1864–6–2 From James Nasmyth Penshurst
Concerned to hear that JH's health suffered as the result of the visit to JN. Happy that Warren de La Rue and [J.?] Müller are giving their attention to the objects on the sun. Comments on a peculiar sun spot. Appreciates the promise of the new edition of JH's *Outlines Astr.* Thanks for the cuneiform signature.
ALS 4pp RS:HS 13.85

12588. 1864–6–3 To George Biddell Airy n.p.
About observing a grain-shaped spot on the sun; greetings to Friedrich Winnecke.
ALS 2pp inc RGO 6.255.44

12589. 1864–6–5 To Charles Babbage Collingwood
Thanking him for his book [probably *Passages in the Life of a Philosopher*].
ALS 2pp RS:HS 2.334

12590. 1864–6–6 From James Thomas Walker 24 Blessington Road, Lee, [Blackheath]
Returned from Gloucestershire. Will visit Collingwood on Monday.
ALS 2pp RS:HS 18.14

12591. 1864–6–6 From Friedrich August Theodor Winnecke Royal Observatory, Greenwich
Became interested in astronomy as a result of reading the papers of William Herschel. FW read these at Hanover in the collection of Caroline Herschel. FW is now trying to arrange a visit to JH.
ALS 3pp (German) RS:HS 18.269

12592. 1864–6–8 From Charles T. Beke Lower Norwood
Sending a draft of a memorial for JH to sign.
ALS 3pp RS:HS 4.32

12593. 1864–6–9 From Edward Sabine R.S.L., Burlington House
Encloses description of Col. [J. T.?] Walker's experiments. Asks for JH's opinion of these.
ALS 2pp TxU:H/M–0528; Reel 1093

12594. [1864–6–9 or later] To Edward Sabine n.p.
Pendulum experiments to be conducted at stations of Great Indian Trigonometrical Survey.
Adraft 2pp TxU:H/L-0360.1; Reel 1054

12595. 1864–6–9 From Friedrich August Theodor Winnecke Royal Observatory, Greenwich
Needs to make some adjustment about FW's visit to JH, as FW was not aware that Sunday trains in Britain run differently from other days.
ALS 2pp (German) RS:HS 18.270

12596. 1864–6–11 To Edward Sabine Collingwood
Plans to confer with [J. T.] Walker and F. A. T. Winnecke from Pulkowa to learn of Russian pendulum experiments before reporting to R.S.L. council. J. H. Pratt's measurement of polar axis and theory about earth's center of gravity.
ACCS 4pp TxU:H/L-0376; Reel 1054 (C: RS:HS 24.47)

12597. 1864–6–16 To Edward Sabine Collingwood
Gives reasons for and discusses technicalities of supporting pendulum experiments at astronomical and geodesical stations in Indian trigonometrical survey.
ALS 4pp RS MC.7.99

12598. 1864–6–18 To the London *Times* Collingwood
Supports, for scientific and commercial reasons, the retention of the British system of measures, arguing against adoption of the metric or decimal system.
C 1p JHS 5.1 [London] *Times* (1864–6–21), 7, col. 6

12599. 1864–6–26 From Friedrich August Theodor Winnecke Elze
Thanks JH for the day spent with him. FW tells JH about the situation of the grave of Caroline Herschel. Includes silhouette pictures found at Mrs. Groskopf's.
ALS 2pp (German) RS:HS 18.271

12600. 1864–6–29 From Charles T. Beke Lower Norwood
Sending the memorial for JH to sign.
ALS 2pp RS:HS 4.33

12601. 1864–6–29 From James Haig 4 Old Square, Lincoln's Inn
Sending another copy of a new edition of one of his books. Thinks he stated too strongly his objections to other people's work in his previous edition. Hopes JH will find time to read it.
ALS 2pp RS:HS 9.159

12602. 1864–6–29 To R.A.S. Collingwood
Learned that R.A.S. plans to use aplanatic lenses for solar studies. Submits set of unpublished tables by W. L. Newman [see Newman's 1845–2–18] for calculating radii of such lenses. Refers to work of this kind by Josef Fraunhofer, [G. P.] Bond, K. A. Steinheil, and C. F. Gauss.
ALS 7pp RAS: MSS Add.29 [suppl.]

12603. 1864–7–1 To James Haig [?] Collingwood
Thanks for his books. Encloses one of his own pamphlets on Atoms, which will show his personal view of mind versus matter. Comments on some of Haig's terminology.
C 5pp RS:HS 9.160 & 24.50

12604. 1864–7 From Francis Cranmer Penrose Cathedral Surveyor's Office, St Paul's
Is sending JH a paper on standardization of measurement, using decimalization but not the meter.
ALS 4pp RS:HS 13.334

12605. 1864–7–1 To the London *Times* Collingwood
Responds to objections published in the *Times* of 30 June to JH's 1864–6–18 letter to the *Times*.
L London *Times* (July 4, 1864), p. 11, col. 2

12606. [1864–7] To William Whewell [Collingwood]
Responds to WW's comments on JH's *Iliad* translation. Sending Book X.
ALS 3pp TC, Camb. Add. Ms.a.207[114] (C: RS:HS 24.65)

12607. 1864–7–2 From William Henry Smyth St. John's Lodge
Studies sidereal chromatics. Asks if William Herschel recorded colors of Beta Cygni in a 1779 observation.
ALS 2pp RS:HS 16.321

12608. 1864–7–3 From Warren de La Rue Observatory, Cranford
Appreciates the difference between Robert Hooke's and JH's suggestions for a helioscopic telescope. Comments on these suggestions. Has recently constructed a reflecting prism. Comments on this.
AL 4pp inc RS:HS 6.155

12609. 1864–7–4 From James Haig 4 Old Square, Lincoln's Inn
Thanks for his letter and amusing pamphlet on atoms. Comments on their use of certain terms. Does not think their views differ radically. Thinks heat is a form of movement.
ALS 3pp RS:HS 9.161

12610. 1864–7–4 To William Henry Smyth Collingwood
Sends WS William Herschel's chromatic observations of Beta Cygni. Completing 'Catalogue of Nebulae.' Works on translating Book IX of *Iliad*; fears he will not complete translation of *Iliad*.
C 2pp RS:HS 24.51

12611. 1864–7–4 To the London *Times* Collingwood
Corrects typographical errors in JH's recent letters [1864–6–18 & 1864–7–1] to the *Times*.
L London *Times* (July 6, 1864), p. 7, col. 2

12612. 1864–7–6 From William Henry Smyth St. John's Lodge
Questions dates in William Herschel's observations. Wishes JH luck on completing translation of *Iliad*.
ALS 2pp RS:HS 16.322

12613. 1864–7–11 From Warren de La Rue 110 Bunhill Row
Was not the person who sent him some photographs of the sun. Was glad the Astronomical Soiree was well spoken of. James Nasmyth was conspicuous and is pleased he was mentioned in JH's *Outlines Astr.*
ALS 2pp RS:HS 6.156

12614. 1864–7–11 From Isaac Todhunter St. John's, Cambridge
Asks again about Wilhelm Struve calculation in *Outlines Astr.*
ALS 3pp RS:HS 17.354

12615. [1864]–7–12 From Harry W. Maclear Hitchin
Much obliged for JH applying to the Horse Guards on HM's behalf. Has now received his appointment to the 28th Regiment.
ALS 2pp RS:HS 13.10

12616. 1864–7–14 To Augustus De Morgan Collingwood
Comments on the state of JH's health; about ciphers and about decimalization.
C 3pp RS:HS 24.52

12617. 1864–7–15 From Augustus De Morgan 91 Adelaide Road
Does not like the sound of the bronchitis. Gives a remedy that he found effective. Gives one of his own ciphers.
ALS 4pp RS:HS 6.370

12618. 1864–7–18 From Warren de La Rue 110 Bunhill Row
Encloses a letter from William Simms, which contains a suggestion for JH's reflecting prism. Inclines to the view that a reflector is not the most suitable instrument for sun observations.
ALS 2pp RS:HS 6.157

12619 1864–7–18 To Augustus De Morgan Collingwood
Comments on remedies suggested by AD [see AD's 1864–7–15]; on ciphers.
C 2pp RS:HS 24.53

12620. 1864–7–18 To Isaac Todhunter Collingwood
Cannot find the source of supposed Wilhelm Struve probability calculation; believes he has made a mistake. Regrets the error was not noticed in time to correct the seventh edition of *Outlines Astr.*
C 1p RS:HS 24.54

12621. 1864–7–18 To John Tyndall Collingwood
Thanks for and comments on papers on scientific history and molecular physics, particularly on a magnetic experiment and on dynamical efficiency of coal relative to motion theory of light and heat. Comments on nature of heat, 'vis viva.' Deplores laxity with which physicists use and confuse terms 'force,' 'motion,' and 'energy.'
C or CC 12pp diag RS:HS 17.393b (C: 24.48; C: RI 524–7)

12622. 1864–7–20 From Warren de La Rue 110 Bunhill Row
Sending the tracing again with a few additions. Comments on this and the one sent yesterday.
ALS 3pp RS:HS 6.158

12623. 1864–7–20 From Thomas Maclear Royal Observatory, Cape of Good Hope
The [Samuel] Parlby case is a hard one. Has put JH's letter and enclosures into the hands of Reid [TM's son-in-law]. Will arrange for the case to be brought before the Governor. Arrangements are being made to erect a statue of Sir George Grey in the Botanic Garden in front of the new library and museum. Is proud of his own sailor son [John Maclear].
ALS 4pp RS:HS 12.170

12624. 1864–7–21 From Isaac Todhunter St. John's, Cambridge
Thanks JH for looking into the Wilhelm Struve probability

calculation; makes note of the change in his work.
ALS 3pp RS:HS 17.355

12625. 1864–7–22 From W[illiam] A[rthur] Darby Manchester
Sending a specimen of a list of celestial objects which he is in process of publishing. Would like his comments.
ALS 3pp RS:HS 6.471

12626. 1864–7–23 From William Lassell Malta
Was grateful for a copy of the proof sheets of JH's forthcoming 'Catalogue of Nebulae.' Comments on the Colonial book post. May discover a few hundred new nebulae but they are faint. Regrets to learn that JH's health is not good. Temperature has been very warm.
ALS 4pp RS:HS 11.166

12627. 1864–7–24 From Augustus De Morgan 91 Adelaide Road
Has been in a state of collapse all day, but a dose of '02H' has worked wonders. Gives anecdotes of Diogenes and Plato.
ALS 4pp RS:HS 6.371

12628. 1864–7–25 From William Rutter Dawes Hopefield, Haddenham, Thame
A long letter dealing with the reports of various observers of James Nasmyth's 'Willow leaves' on the sun.
ALS 6pp RS:HS 6.118

12629. 1864–7–25 From William E. Hearn Melbourne
Thanks for his letter acknowledging *Plutology* and for the proof of the article in which JH refers to it. Authorship is difficult in these parts away from the proper books. Gives reasons why he chose the particular title and spelling.
ALS 3pp RS:HS 10.176

12630. 1864–7–27 To Henry Bence Jones Collingwood
His son informs him of the projected series of lectures at the Royal Institution on weather forecasting and meteorology. Comments on this.
draft 2pp RS:HS 10.337 (C: RS:HS 24.55)

12631. 1864–8–5 From George Biddell Airy Royal Observatory, Greenwich
Sending him the draft of the Report of the Standards Committee for his signature.
ALS 2pp RS:HS 1.278

12632. 1864–8–6 From Warren de La Rue 110 Bunhill Row
Sending a photograph of the moon 38" in diameter. Begs him to accept it as a mark of appreciation.
ALS 3pp RS:HS 6.159

12633. 1864–8–6 To Margaret Brodie Herschel Collingwood
About family finances, JH's health, and activities around the house.
ALS 4pp JHS 5.24

12634. 1864–8–8 To the R.A.S. n.p.
Describes a revolving eyepiece that could protect the eye from excess light in solar observation.
L *R.A.S.M.N., 24* (1864), 220–1

12635. 1864–8–9 To William Henry Smyth Collingwood
Believes he will complete *Iliad.*
C 1p RS:HS 24.56

12636. 1864–8–10 To Margaret Brodie Herschel Collingwood
Some friction has arisen between JH and MH over finances, and JH apologizes; JH is describing the newly-acquired horse, and reports on mail received and the activities of the children.
ALS 4pp JHS 5.25

12637. 1864–8–18 From John C. Broun Wynberg
Regrets the expense incurred by the enclosed letter. Sends the report from the Colonial Office.
ALS 1p RS:HS 5.111

12638. 1864–8–18 To Richard Christopher Carrington Collingwood
If RC's formulas for sunspots are correct, then 'trade-wind theory' is challenged and sun's photosphere may behave as envelope circulating around sun according to laws of planetary motion. Questions formula relating to sunspot velocity.
ACCS 4pp RAS:JH Archive 13/2.4; Reel 10

12639. 1864–8–18 From Augustus De Morgan 91 Adelaide Road
Did JH know when he was Master of the Mint that Isaac Newton had introduced a golden quarter-guinea? Most of his family are at Aldeburgh. Does he know Tristram Shandy's sense of proportion? Charles Babbage has become a public benefactor by the passing of the act against barrel organs.
ALS 3pp RS:HS 6.372 S. De Morgan, *Memoir of Augustus De Morgan*, 323–4

12640. 1864–8–[18] To James Calder Stewart from Isabella Herschel [Collingwood]
News of Herschel family. JH asks for information about 'India Stock' [investments].
ALS 8pp TxU:H/L-0457; Reel 1055

12641. 1864–8–19 To Augustus De Morgan n.p.
Comments on minting of coins [see AD's 1864–8–18], the state of JH's health, and the need for an elocution teacher for one of JH's sons.
C 4pp RS:HS 24.57

12642. 1864–8–21 From Augustus De Morgan 91 Adelaide Road
Has a quarter sovereign given him by Thomas Graham. Will mention the coin in *Notes and Queries*. Gives a pun on his own name. Anecdote of rain in India when his father and mother were travelling.
ALS 4pp RS:HS 6.373

12643. 1864–8–22 From Thomas Maclear Royal Observatory, Cape of Good Hope
Congratulations on marriage of JH's son William. Hopes Willy and his bride will call at the Cape on their way to India. Attorney General has just returned from Grahamstown with the [Samuel] Parlby papers. Has finished work on measurement of arc of meridian.
ALS 5pp RS:HS 12.171

12644. 1864–8–23 To Francis Cranmer Penrose Collingwood
Finds FP's paper [see FP's 1864–7] interesting, but does not believe his proposal has any hope of acceptance.
draft 4pp RS:HS 13.335 & 24.58

12645. 1864–8–23 From William Trant, Jr. 22 Bedford Place, Leeds
Asks for JH's advice on astronomy as a profession.
ALS 2pp RS:HS 17.372

12646. 1864–8–25 To William Thomson Collingwood
Thanks WT for paper on rigidity of the earth; agrees tidal action affects not only fluid motion of earth, but the solid as well. Mentions JH's paper of the previous year suggesting that tidal waves cause earthquakes. Raises questions about the idea of the earth as a magnet and about WT's dynamical theory of heat.
ALS 2pp RS:HS 17.349b (C: RS:HS 24.59) F. James, 'Between Two Scientific Generations: John Herschel's Rejection of the Conservation of Energy in His 1864 Correspondence with William Thomson,' *Notes and Records of the Royal Society*, *40* (1985), 55–7

12647. 1864–8–30 To [William Trant, Jr.] Collingwood
Responds to [WT]'s 'passionate desire to become an astronomer.' Distinguishes between doing astronomy and enjoying its discoveries. Explains training procedures.
ALS 4pp damaged TxU:H/L-0728; Reel 1089

12648. 1864–9–1 From William Trant, Jr. 22 Bedford Place, Leeds
Thanks JH for advice on the pursuit of astronomy, especially concerning the importance of mathematics. Desires to know astronomy not just as a 'pleasurable companion' but as a 'bosom friend.'
ALS 2pp RS:HS 17.373

12649. 1864–9–5 From Capel H. Berger Sion House, Lower Clapton
Requests that JH join many other scientists in signing a declaration in support of scripture.
L *Athenaeum,* #1925 (Sept. 17, 1864), 375 & London *Times* (Sept. 20, 1864), p. 7, col. 6

12650. 1864–9–5 To [John Benjamin Smith] Collingwood
Thanks JS for JS's speech. Discusses in detail JS's views on the introduction of a gold currency into India. Believes that this is a necessity and explains why. Notes problems and advantages involved in the circulation of money based on two standards, silver and gold.
ALS 7pp Manchester Central Library

12651. 1864–9–5 To [James Calder] Stewart Collingwood
JH's India [H.E.I.C.] investments. Looking forward to seeing John [Herschel; son of JH] when he returns on leave.
ALS 3pp TxU:H/L-0458; Reel 1055

12652. 1864–9–6 To Editor of the *Athenaeum* Collingwood
Sends a circular and JH's response to it for possible publication.
ALS 1p WT B.14

12653. 1864–9–6 To Capel H. Berger Collingwood
Gives his reasons for declining to join with over two hundred other scientists in signing the 'Declaration' regarding science and scripture [which was at that time being circulated].
L *Athenaeum,* #1925 (Sept. 17, 1864), 375 & London *Times* (Sept. 20, 1864), p. 7, col. 6

12654. 1864–9–7 To J[ohn] B[enjamin] Smith Collingwood
Since writing JH's 1864–9–5 on JS's ideas regarding the introduction of a gold standard for India, JH has learned that 'the Sovereign *is* to be circulated at 10 Rupees.' JH discusses what he foresees as the results of this action. Offers suggestions for the future.
ALS 5pp Manchester Central Library

12655. 1864–9–8 From George C. Hodgkinson The Lodge, Louth
Tried Kew and the Admiralty for actinometers but had to obtain one from Henry Barrow in the end, which was broken en route. Thinks these are too fragile for mountain climbing.
ALS 4pp RS:HS 9.417

12656. 1864–9–10 From William Hepworth Dixon Collingwood
Wants to send JH's reply to the circular [see JH's 1864–9–6] to the author of the circular, and allow him to reply.
ALS 1p WT B.15

12657. 1864–9–11 To John Herschel (son) Collingwood
Is pleased with the new wife of son William; JH is very supportive of son John's developing interest in language [an interest of JH's]; JH has not heard of any proposed Himalayan observatory; is sending a copy of his nebula catalogue.
ALS 4pp JHS 5.10

12658. 1864–9–12 or later To John Herschel (son) Collingwood
Refers son John to JH's *Meteorology* for help with the mechanics of dynamic wave curves; encourages John to do good whenever he has the opportunity.
ALS 4pp JHS 5.12

12659. 1864–9–13 From Augustus De Morgan 91 Adelaide Road
Regarding the antiquity of the + and - signs. Origin of *quid vides.*
ALS 6pp RS:HS 6.374

12660. 1864–9–15 To Augustus De Morgan Collingwood
On algebraic signs, riddles, and the state of JH's health.
C 3pp RS:HS 24.60

12661. 1864–9–16 From Augustus De Morgan 91 Adelaide Road
Hears that JH sits in his study with a stove. This is unhealthy for anyone with bronchial tendencies.
ALS 2pp RS:HS 6.375

12662. 1864–9–17 From George C. Hodgkinson The Lodge, Louth
Is grateful for JH's reply. Gives reasons why he was looking for smaller and more portable instruments to take with him to the Alps.
ALS 6pp RS:HS 9.418

12663. 1864–9–17 From John Tyndall Royal Institution
Regrets 'stupidity' of Italian post office, which lost note sent to JH suggesting reprinting of passage from *Treatise Astr.* of 1833 on operations of sun in second edition of JT's book. Also wants to publish JT's 'extremely philosophical views' on muscle contraction.
ALS 3pp RS:HS 17.394a (C: RI 528)

12664. 1864–9–19 From Smith, Elder & Co. 65 Cornhill
Acknowledgement of order from JH.
(A)LS 1p RS:HS 16.211

12665. 1864–9–20 From Robert Addams York Ter., Regents Park
Thanking him for his letter to the *Times* [apparently on freedom of religious opinion].
ALS 1p RS:HS 1.391

12666. 1864–9–20 From Augustus De Morgan 91 Adelaide Road

JH expressed himself with great precision in the old *Discourses*. Concerning one of the publications of the B.A.A.S.
ALS 4pp RS:HS 6.376 S. De Morgan, *Memoir of Augustus De Morgan*, 326 inc

12667. 1864–9–20 From Alfred Potiquet Paris
Asks for information on William Herschel.
ALS 1p (French) RS:HS 14.180

12668. 1864–9–21 From Roderick Impey Murchison Bath
Introducing two Russians, N. de Khanikoff and P. H. A. Moritz, who would like to visit JH. They are interested in making a magnetic survey of Georgia and RM has written a letter supporting their claims to the Grand Duke Michael. Meeting was marred by the death of J. H. Speke. Now on their way to Bristol.
ALS 4pp RS:HS 12.431

12669. 1864–9–21 From John Benjamin Smith Ascot
Outlines disadvantages of establishing a gold currency in India.
ALS 3pp RS:HS 17.215 (draft: Manchester Central Library)

12670. 1864–9–23 From William Rutter Dawes Hopefield, Haddenham, Thame
Is returning JH's paper on 'solar spots.' Comments on George Wilson of Glasgow and his suggestions on the use of telescopes. Encloses copy of a letter on the Moon's rotation. Was pleased to see JH's letter in the *Times* on the agreement between Scripture and Science.
ALS 3pp RS:HS 6.119

12671. 1864–9–23 From N. de Khanikoff London
Would like to visit JH, together with his friend Dr. P. H. A. Moritz, Director of the Observatory at Tiflis.
ALS 1p (French) RS:HS 11.80

12672. 1864–9–24 From Warren de La Rue Observatory, Cranford
Has just returned from a short tour of Russia. Visited Pulkowa and Nijny Novgorod. Had difficulties with couriers. JH's suggestion for a rotating disc with slit to reduce sunlight sounds feasible. Weather was unsuitable at Pulkowa for any observations.
ALS 4pp RS:HS 6.160

12673. 1864–9–25 From George Griffith 5 Park Villas, Oxford
Will be very glad to send his paper to the gentlemen concerned, but encloses their names and addresses in case JH wishes to send direct.
ALS 3pp RS:HS 9.143

12674. 1864–9–25 To Joseph A. F. Plateau n.p.
Apologizes for not having answered earlier. JH's health has been 'indifferent.' Received 'Sur un Problème curieux de Magnetisme' yesterday. Discusses 'alleged free suspension of Mahomet's coffin.' Agrees with JP that 'establishment of a stable equilibrium under the influence of mere magnetic forces is impossible.' Elaborates on an experiment relating to the 'problem.' Lists the works by JP that he has.
ALS 4pp RS:HS 14.167 (C: RS:HS 24.61)

12675. 1864–9–27 To George Biddell Airy Collingwood
Is very upset that JH's name has been attached to a group that wishes to import the 'French metrical system' into Britain; JH urges petitions or whatever is needed to retain the British system.
ALS 4pp RGO 6.358.288

12676. 1864–9–27 From Camille Flammarion Paris
Sending a new edition of his own work on the *Plurality of Worlds*.
ALS 1p (French) RS:HS 7.269

12677. 1864–9–27 From William Stanley Jevons Owens College, Manchester
JH's statement regarding scientific inquiry and religion seems to him invaluable, and he agrees with it entirely.
ALS 3pp RS:HS 10.320 Black (ed.), *Papers and Correspondence of...Jevons*, 3:60

12678. 1864–9–28 From George Biddell Airy Royal Observatory, Greenwich
Daughter has returned from Switzerland.
ALS 1p RS:HS 1.280

12679. 1864–9–28 From George Biddell Airy Royal Observatory, Greenwich
Regarding the opposition to the proposed introduction of the French Metrical system.
ALS 4pp RS:HS 1.279

12680. 1864–10–2 From Augustus De Morgan 91 Adelaide Road
Has long advocated the right to contract. Has no fear that the metrical system will become compulsory. American named J. W. Nystrom wishes to introduce a unit of 16. Comments on the Bible being 'every word of it is God's word.'
ALS 6pp RS:HS 6.377

12681. 1864–10–3 From Thomas P. Kirkman Croft Rectory near Warrington
Was pleased with JH's weighty and judicious rebuke. Sends a copy of a letter he wrote to the *Times* concerning religion and the natural sciences.
ALS 3pp RS:HS 11.52

12682. 1864–10–3 From James Robert Napier Glasgow
Professor W. J. M. Rankine lent JN JH's paper on the meter, yard, and foot. Would JH present a copy to the Institution of Engineers in Scotland? JN is president at the moment and hopes to introduce this subject. Hopes that JH will publish any further ideas he has on the subject.
ALS 2pp RS:HS 13.144

12683. 1864–10–6 From William Lassell Malta
Encloses lithographs of some of his drawings of nebulae, and two tracings of [?] 2241. Has presented his large equatorial to Melbourne and invited the Astronomer Royal [G. B. Airy] to Malta to observe through the telescope before it is dispatched to Australia. Would also welcome a visit from JH.
ALS 4pp RS:HS 11.167

12684. 1864–10–6 From Joseph A. F. Plateau Gand
Submits some observations concerning the experiment mentioned in JH's last letter [1864–9–25]. Discusses in detail his results of an experiment concerning the impossibility of a stable equilibrium. Decides stable equilibrium exists only in vertical position, not in horizontal. Sends another copy of his latest paper. Mentions a printing error in his memoir concerning molecular attraction.
ALS 3pp RS:HS 14.168

12685. 1864–10–11 From James South Observatory, Kensington
Would like JH's permission to make public JH's letter of

1826–11–14 in defence of the conduct of the late Humphry Davy.
(A)CC(S) 2pp RS:HS B27.21

12686. 1864–10–14 From Georg F. J. A. Auwers Gotha
Thanking him for his catalogue of nebulae. Gives amendments to his own readings of the observations of nebulae.
ALS 6pp RS:HS 1.424

12687. 1864–10–14 From Hermann Kindt Yarm, Yorks
Mrs. Polko was gratified by JH's translation of Frederick Schiller's poem. She would like a photograph and autograph of JH now, and wonders if he would like a photo of her brother Edoard Vogel? HK would like some additional copies of JH's translations.
ALS 4pp RS:HS 11.28

12688. 1864–10–14 To James South Collingwood
States unequivocally that the letter referred to in JS's 1864–10–11 has not 'a syllable' that can in any way affect the question of the conduct or character of the late Humphry Davy. JH could not conceive of publication of the letter without the express consent of the other person mentioned in it [Charles Babbage].
ACCS 2pp RS:HS B27.22

12689. 1864–10–15 To [Heinrich Louis d'Arrest?] Collingwood
Encloses photograph of JH for Madame [Polko?]. Would like to receive photograph of [Edoard] Vogel. Will send six copies of JH's translations of J. C. F. Schiller's *The Walk.* JH corrected Schiller's error in line 85.
ALS 3pp TxU:H/L-0749.6; Reel 1089

12690. 1864–10–15 From James South Observatory, Kensington
Received JH's letter of 1864–10–14 'with grief almost inexpressible.'
ALS 1p RS:HS B27.23

12691. 1864–10–17 From Hermann Kindt Yarm, Yorks
Is grateful for the photograph (which he considers bears a resemblance to Edoard Vogel) and also the extra copies of his translation, which he will transmit to friends in Germany.
ALS 4pp RS:HS 11.29

12692. 1864–10–21 To William Whewell Collingwood
Sends Book XI of JH's *Iliad* translation. Comments on Book XI. Severe bronchitis continues.
ALS 4pp TC, Camb. Add. Ms.a.207[115] (C: RS:HS 24.62)

12693. 1864–10–22 To *Photographic News* Collingwood
Notes introduction of 'Worthytype' process into commercial photography. Calls attention to process discovered by JH in 1832 for purifying uranium.
ALS 1p † inc & AdraftS 1p (on Roscoe's and Baxendell's 1864–1–20) TxU:H/L-0304 & -0305.2; Reel 1054

12694. 1864–10–22 To *Photographic News* Collingwood
Describes a method created by JH [see JH's 'Lettre ... sur ... un nouveau procédé pour effectuer la purification complète de l'oxide d'urane,' *Annales de chimie, 49* (1832), 306–11] for obtaining pure uranium.
L *Photographic News, 8* (Oct. 28, 1864), 520 & *American Journal of Photography,* n.s., *7* (Dec. 1, 1864), 250–1

12695. 1864–10–23 To William Lassell Collingwood
Comments on the several drawings of the same nebulae; thanks WL for the invitation to Malta but JH is too ill to accept.
C 2pp RS:HS 24.63

12696. 1864–10–25 From Warren de La Rue Observatory, Cranford
Has just received JH's 'Catalogue of Nebulae.' Has placed his skeleton tube in the cradle of his equatorial. Comments on this and gives sketches.
ALS 3pp RS:HS 6.161

12697. 1864–10–25 From Hermann Kindt Yarm, Yorks
Is grateful for his kind and flattering notice of his sonnet. Encloses Edoard Vogel's photograph.
ALS 4pp RS:HS 11.30

12698. 1864–10–26 From George Griffith B.A.A.S., Oxford
Resolution reappointing JH and others to Balloon Committee for further experiments.
L 1p TxU:H/M–0921.1; Reel 1083

12699. 1864–10–26 From Frederick W. Holland Portsea Place, Connaught Square
Sir Henry James does not wish to be connected with the new expedition to Sinai until the work of the previous one has been cleared. Will approach the Royal Geographical Society with his scheme. [H. S.] Palmer is the man to go on this expedition.
ALS 3pp RS:HS 9.423

12700. 1864–10–27 From Alfred Brothers Manchester
Is preparing a lecture on photography and wonders if JH was the inventor of the hyposulfite fixing process.
ALS 3pp RS:HS 4.296

12701. 1864–10–29 To Alfred Brothers Collingwood
Outlines his claim as being the inventor of the hyposulfite fixing process.
draft 4pp RS:HS 4.297 & 24.64 *British Journal of Photography, 13* (May 19, 1866), 236 & *Photographic News,* 10 (May 18, 1866), 230–1

12702. [1864–11 or later] To [?] n.p.
Committee pamphlet expressed incorrect representation of JH's views on changes in British standards of measurement.
Adraft 1p inc TxU:H/M–0935.2; Reel 1083

12703. 1864–11–2 From Alfred Brothers Manchester
Is grateful for his letter regarding the claim for the discovery of the hyposulfite fixing process.
ALS 2pp RS:HS 4.298

12704. 1864–11–7 From Alfred Brothers Manchester
Regarding the differences between portraits and photographs. Is sending him some magnesium for photographic purposes.
ALS 4pp RS:HS 4.299

12705. 1864–11–7 From G. Wharton Simpson 18 Canonbury Park South
Photographic News received copies of papers from *R.S.P.T.* giving evidence of [JH's] use of hyposulfite of soda. Will refer to W. H. F. Talbot on use of bromine, as JH suggested. P. B. Goddard of Philadelphia claims to have used bromine in Daguerreotype process in 1839, but did not announce it publicly until 1843.
ALS 2pp RAS:JH Archive 13/3.110; Reel 10

12706. 1864–11–9 From Charles Pritchard Freshwater
Compares W. L. Newman's and CP's calculations of optical coefficients to those in JH's work. Sends JH formulae and tables for calculating telescopic lens thickness and shape in

relation to the dimensions of the telescope. Wishes JH luck on *Iliad*.
ALS 9pp RS:HS 14.92

12707. 1864–11–10 From Charles Pritchard Royal Astronomical Society
Compares JH's equation for lens curvature with that of C. F. Gauss.
ALS 2pp RS:HS 14.93

12708. 1864–11–12 From William Hartley 45 Windham St., Bury, Lancashire
Encloses prints of specimens produced by WH's new lathe for manufacturing telescope specula. Asks JH's opinion of results.
ALS 1p † TxU:H/M–0935.1; Reel 1083

12709. 1864–11–12 From William Thomson 2 College, Glasgow
Reads JH's earthquake article and believes the effects of the transferences of matter of which JH speaks must be very small. Discusses the dynamical theory of heat. Will send to JH volume WT is editing.
ALS 12pp RS:HS 17.350a

12710. 1864–11–13 To John Tyndall Collingwood
Thanks for and comments on paper on luminous and obscure radiation, particularly on extra spectral rays, incandescence of platinum, and experiment using transcalescence of iodine.
ALS 3pp RI 305 (C: 529–30); draft 4pp RS:HS 17.394b (C: 24.66)

12711. 1864–11–14 From John Tyndall Royal Institution
Is not 'at liberty' to speak of experiments on change of refrangibility of 'extra-red' rays. Discusses other experiments on extra-red regions, having used various prisms but not obtaining JH's results.
ALS 4pp RS:HS 17.395 (C: RI 531)

12712. 1864–11–15 To T. P. Anderson and J. H. Shonbridge Collingwood
Signed and returned accountant's form. Delays reply until JH can confer with other trustees [of Hollier Trust Fund]..C 1p TxU:H/L-0033; Reel 1054

12713. 1864–11–16 To William Hartley Collingwood
Thanks for the beautiful engine-turned patterns; they exceed in beauty anything he has seen before. Should think he would have no difficulty in polishing reflectors.
C 4pp RS:HS 10.166 (C: RS:HS 24.67)

12714. 1864–11–16 To John Tyndall Collingwood
Surprised that JT cannot obtain heat spots [see JT's 1864–11–14]. Describes glasses and lenses used.
C or CC 3pp RS:HS 17.396 (C: RI 532 & RS:HS 24.68))

12715. 1864–11–17 To Few & Co. Collingwood
As instructed by Charles Babbage on 27 Feb. 1863, JH forwards to Few & Co. recent letter from solicitors T. P. Anderson and J. H. Shonbridge requesting information about funds in Hollier Trust.
ACCS 1p TxU:H/L-0149; Reel 1054

12716. 1864–11–17 To T. P. Anderson and J. H. Shonbridge Collingwood
Charles Babbage instructed JH to send TA's and JS's enquiry to [Charles] Few & Co., solicitors to Hollier Trust Fund.
ACCS 1p TxU:H/L-0034; Reel 1054

12717. 1864–11–17 From John Tyndall Royal Institution
JT does not have 'luxury' of using sunshine; uses electric light. Perhaps this adds to inability to obtain heat spots [see JH's 1864–11–16].
ALS 2pp RS:HS 17.397 (C: RI 533)

12718. 1864–11–17 From [A. J. F.] Yvon Villarceau Paris
Thanks for JH's 'Catalogue of Nebulae.' Comments on the problems of accuracy in YV's meridian instrument due to fluctuating temperatures and other problems. Also refers to geodetic and trigonometric considerations.
ALS 4pp (French) RS:HS 17.423

12719. 1864–11–18 From Few & Co. 2 Henrietta St., Covent Garden
Received JH's letter of 17 Nov. and full instructions from [Charles] Babbage. Will reply to solicitor of General Reversionary & Investment Company regarding Hollier Trust.
AL 1p TxU:H/M–0191; Reel 1087

12720. 1864–11–18 From John Tyndall Royal Institution
Asks JH to be a shareholder in venture to publish new weekly scientific journal.
ALS 2pp RS:HS 17.398a (C: RI 534)

12721. 1864–11–19 To Roderick Impey Murchison Collingwood
Asks RM whether RM thinks any of the geographical work of Karl Ritter should be translated into English.
ALS 4pp Geological Society

12722. [1864–11–20] From Norman Macleod Adelaide Place Glasgow
[Alexander] Strahan would like an article from JH on the [Great] Pyramid, but NM thinks that Piazzi Smyth is the man. All the same he would appreciate an article from JH, perhaps on a chemical theme.
ALS 4pp RS:HS 12.263

12723. 1864–11–21 From Roderick Impey Murchison Geological Survey of Great Britain Office
Regarding the feasibility of a translation of Carl Ritter's geographical works. Will put the matter to the Council of the Royal Geographical Society and let JH know their views. The Russians were grateful for JH's help.
ALS 6pp RS:HS 12.432

12724. 1864–11–21 To John Tyndall Collingwood
Wishes luck but expresses inability to help with new science journal [see JT's 1864–11–18].
ACCS 2pp RS:HS 17.398b (C: 24.69; C: RI 535)

12725. 1864–11–22 From Warren de La Rue 110 Bunhill Row
Encloses a letter from Angelo Secchi. Sent him a Herschelian eye piece and is having a prism prepared for him. Secchi has seen James Nasmyth's 'willow leaves.'
ALS 2pp RS:HS 6.162

12726. 1864–11–22 To William Thomson Collingwood
Thanks WT for pageproofs of his work on analytical dynamics. Compliments work on Isaac Newton's view of nature of luminous ether. Gives own views on ether and motion theory of heat.
draft 2pp RS:HS 17.350b (C: RS:HS 24.70) F. James, 'Between Two Scientific Generations: John Herschel's Rejection of the Conservation of Energy in His 1864 Correspondence with William Thomson,' *Notes and Records of the Royal Society*, *40* (1985), 59–61

12727. 1864–11–24 From George Biddell Airy Royal Observatory, Greenwich
Informing him that he has just heard of the death of Wilhelm Struve.
ALS 1p RS:HS 1.281 (C: RGO 6.381.439)

12728. 1864–11–24 From Charles Pritchard Freshwater
Praises JH's *Iliad.* Asks R.S.L. for funds to cover expenses incurred by spherical observations. Details information to be included in optical tables.
ALS 6pp RS:HS 14.94

12729. 1864–11–25 To George Biddell Airy Collingwood
Compares earlier and more recent pound determinations and finds interesting results; wants GA's opinion on the accuracy of his work, as JH wishes to send it to the India committee.
ALS 4pp RGO 6.359.540

12730. 1864–11–25 To Norman Macleod Collingwood
Declines the opportunity of writing an article on the Great Pyramid for *Good Words,* especially as he disagrees significantly with Charles Piazzi Smyth's views. Might write an article on another subject.
C 2pp RS:HS 24.71

12731. 1864–11–29 From Edward William Brayley London Institution
Sends his article on the sun and the British Almanac for 1865. Regarding various current theories relating to the sun. The professorship he applied for has not materialized.
ALS 4pp RS:HS 4.228

12732. 1864–12 From Charles Thomas Longley[?] Croydon
An invitation to a meeting.
ALS 1p RS:HS 11.455

12733. 1864–12 To Peter Guthrie Tait Collingwood
Thanks PT for papers on energy. Discusses topics of the papers, including quaternions and the concept of heat equalling motion.
ACCS 3pp RS:HS 17.425b (C: RS:HS 24.72)

12734. 1864–12–4 To Julia Margaret Cameron Collingwood
Has copied out some extracts from his translation of Homer and is wondering whether to publish the first twelve books. Comments on his views and poetry of Alfred Tennyson (1st Baron Tennyson).
draft 6pp RS:HS 5.160 & 24.74

12735. 1864–12–4 To Warren de La Rue Collingwood
Thanks for the diary and pocket book. Sees that Amédée Victor Guillemin has published an illustrated Astronomy [*Le ciel* (Paris, 1864)]. Thinks there is scope for publishing a volume of engravings of astronomical objects. Gives his own idea for the construction of a photometer.
draft 4pp RS:HS 6.163 & 24.73

12736. 1864–12–7 To Edward William Brayley n.p.
Calling his attention to one of his own articles. Concerning Mr. Magers and Roger Boscovitch's views. Concerning Magellanic Clouds.
draft 1p RS:HS 4.229 & 24.75

12737. 1864–12–7 From Julia Margaret Cameron n.p.
Is grateful for JH's extracts from Homer. Hopes JH will publish before he has finished all the translation. Will write again soon.
ALS 12pp RS:HS 5.161

12738. 1864–12–9 From T. P. Anderson and J. H. Shonbridge 4 Lincoln's Inn Fields
H. J. Hollier asked General Reversionary and Investment Company (Anderson's clients) for another advance from Hollier Trust Fund. As trustee, JH is asked to sign and return enclosed receipt.
ALS 1p TxU:H/M–0006; Reel 1062

12739. 1864–12–9 From Peter Guthrie Tait Edinburgh
Sends, at the request of William Thomson, articles from *North British Review* on the topic of energy.
ALS 1p RS:HS 17.425a

12740. 1864–12–9 From James Thomas Walker 1 Cromwell Road, South Kensington
Sent JH 'Index Chart' of Great Trigonometrical Survey of India. Compares five triangulation methods: longitudinal series, George Everest's, G. B. Airy's, that of JH's son John, and minimum squares. JH's son and bride avoided cyclone that destroyed Calcutta Observatory.
ALS 4pp RS:HS 18.15

12741. 1864–12–10 To Elizabeth Colling Collingwood
Comments about translating Homer and about other translations appearing.
C 2pp RS:HS 24.76

12742. 1864–12–10 To James Thomas Walker Collingwood
Received Indian survey chart. JH's son John wrote of higher-order ellipsoidal triangles with sides based on geodesical triangles. JH objects that these violate principle of diagonal bracing. Proposes different triangles and names sites in India as apexes. Received G. B. Airy's paper on reduction.
C 4pp RS:HS 18.16 (C: 24.77)

12743. 1864–12–12 To Friedrich Wilhelm Rudolph Engelmann Collingwood
Thanks for the valuable series of measures of double stars. Regarding JH's method of calculating the orbit of a double star, which RE seems to have misunderstood.
draft 2pp inc RS:HS 7.131 (C: RS:HS 24.78)

12744. 1864–12–12 From John Gibson Macvicar Manse of Moffat
Enclosing a paper for JH's perusal.
ALS 1p RS:HS 13.14

12745. 1864–12–12 From Warren de La Rue Observatory, Cranford
Has received a letter from Angelo Secchi who states that he has seen James Nasmyth's 'willow leaves.' Remainder of letter relates to the JH prism, which he sent him through Cardinal N. P. S. Wiseman.
ALS 2pp RS:HS 6.164

12746. 1864–12–13 From Amédée Victor Guillemin 17 Quai d'Anjou, Paris
Was pleased to receive JH's letter and criticisms on his book *Le ciel.* Comments on some of JH's criticisms.
ALS 3pp (French) RS:HS 9.70

12747. 1864–12–15 From James Thomas Walker 1 Cromwell Road, South Kensington
Indian Triangulation chart showed progress to May 1862. Explains why arcs of meridians and parallels were chosen over diagonal braces. G. B. Airy's method of determining errors helped. Will take note and package from Maria [Herschel] to JH's son John.
ALS 5pp RS:HS 18.17

12748. 1864–12–19 From Charles Babbage Dorset
Believes that John Davy's letter to the *Philosophical Magazine* was suggested by James South. To defend himself, CB now needs JH's permission to publish some of their correspondence.
ALS 3pp RS:HS B27.28 (draft: BL Aaa 32309.172)

12749. 1864–12–19 From James Smith 58 High St., Ayr
Offers JS's wind observations, as requested in JH's ['Weather and Weather Prophets' (1864)] in *Good Words*.
ALS 1p TxU:H/M–0488; Reel 1093

12750. 1864–12–19 From Walter White Royal Society
Will distribute copies of a JH paper.
ALS 1p † Harvard: Houghton AAH 67m–67 (100)

12751. 1864–12–20 To Charles Babbage Collingwood
Gives CB permission to publish the portions of letters requested by him. JH tells CB that he has responded to John Davy's letter in the *PM.*
ALS 3pp BL Aaa 32309.184 (ACCS: RS:HS B27.26)

12752. 1864–12–20 From George L. Paulin and J. S. L. Paulin The Walls, Berwick upon Tweed
Enclose table of GP's and JP's observations, inspired by JH's article in Jan. 1864 *Good Words*. Apologize for disputing JH's authority, but GP and JP found that full moon has no influence on local cloud cover.
ALS 2pp TxU:H/M–0405; Reel 1087

12753. 1864–12–20 To *Philosophical Magazine* Collingwood
Attempts to clarify claims published by John Davy concerning whether his brother Humphry Davy had in 1827 promised Charles Babbage a position as secretary of the R.S.L.
ACCS 4pp RS:HS B27.24 *Phil. Mag.*, 4th ser., *29* (Jan. 1865) 77–8

12754. 1864–12–20 To James South Collingwood
As JH has been approached by Charles Babbage for permission to publish certain portions of JH's correspondence, JH is writing to say that he gives both Babbage and JS permission to publish what they will.
ACCS 1p RS:HS B27.25a

12755. 1864–12–20 From W. H. Wood Sea View House, Weston-super-Mare, Somersetshire
Encloses WW's table of meteorological data for Jan.-Dec. 1864, made in response to JH's claim that full moon influences cloud cover.
ALS 1p, encl 1p TxU:H/M–0598; Reel 1086

12756. 1864–12–21 From Edward William Brayley London Institution
[John] Davy's letter mentioning JH is published in the current *P.M.* Regarding JH's and Mr. Magers's psychic views.
ALS 4pp RS:HS 4.230

12757. 1864–12–21 From Charles Pritchard Freshwater
Requests JH's reflections on the deceased Wilhelm Struve. Discusses making of telescope object glass. Continues work on the optical tables.
ALS 4pp RS:HS 14.95

12758. 1864–12–22 From George Gabriel Stokes Royal Society
Sends paper on earth's temperature for JH's opinion.
ALS 2pp RS:HS 17.57

12759. 1864–12–23 From Gabriel Auguste Daubrée Paris
Further details about exhibiting samples of the South African meteorite [see GD's 1864–1–3]. [Letter illegible in parts.]
ALS 4pp (French) RS:HS 6.31

12760. 1864–12–23 From William J. Edge Benenden Vicarage
Sending a weather almanac by a person called Orlando Whistlecroft. Comments on his weather forecasting.
ALS 6pp RS:HS 7.119

12761. 1864–12–23 To William Henry Smyth Collingwood
Comments on WS's publication on sidereal chromatics. Believes validity of doctrine of three primary colors proved.
C 3pp RS:HS 24.79

12762. 1864–12–24 From H. S. Bryant Portugal St.
Has to forward a copy of the Golden Calendar to him; how shall he send it?
ALS 3pp RS:HS 5.114

12763. 1864–12–24 From James Liddell Bodmin
Encloses JL's weather observations over past year. Asks if uniform style exists for all observers to follow. JL's views about moon's effect on climate. Wishes Robert FitzRoy would abandon weather prognostication and concentrate on valuable daily reports.
ALS 4pp, encl 12pp TxU:H/M–0331; Reel 1087 & TxU:H/M–1165; Reel 1060

12764. 1864–12–27 From Butterworth and Heath Firm London
Sends a proof of an engraved print.
ALS 4pp Harvard: Houghton AAH 67m–67 (87)

12765. 1864–12–29 From N. de Khanikoff London
Sends the first volume of his memoir on the Perse. Has contacted Jacques Babinet regarding experiments on variations of gravity. Comments on the work of the various researchers in this field.
ALS 3pp (French) RS:HS 11.81

12766. 1864–12–29 From William Selwyn College, Ely
Discusses translating the *Iliad.* Discusses Emile Gautier's theory of the maculae, which does not seem to explain what they see.
ALS 4pp RS:HS 15.474

12767. 1864–12–31 From Julia Margaret Cameron Fresh Water Bay, Isle of Wight
Is sending JH and Lady Herschel a collection of photographs JC has taken. Discusses them, and her philosophy of photography. Thanks JH for being 'her first Teacher.'
ALS 8pp Colin Ford, *The Cameron Collection,* 140–1

12768. 1865 or later From Julia Margaret Cameron n.p.
Note with some photographs.
AL 2pp inc RS:HS 5.180 & 181

12769. 1865–1 From Arthur Cayley Blackheath
Has kept the petition and will forward it after the meeting. His brother is now in the 16th book of the *Iliad.*
ALS 2pp RS:HS 5.228

12770. 1865 From Mr. Eyre Newbury
Will be pleased for JH to see the house and grounds at any time.
AL 1p RS:HS 7.140

12771. [1865] To Editor of *Saturday Review* n.p.
Passionately protests bill before Parliament for compulsory adoption of French metric system, 'a piece of French national vanity.'
Adraft 3pp TxU:H/L-0382; Reel 1054

12772. 1865–1–1 To William Selwyn Collingwood
Thanks WS for his *Iliad* translation. JH has just finished Book 3 of his own translation of the *Iliad.*
C 1p RS:HS 24.80

12773. [1865] To ? n.p.
Fragment of a letter discussing diffraction, JH's son's journey to observe eclipse and strong polarization of sun's corona, evidence of vast solar atmosphere.
AL 2pp inc diag RS Sa.684

12774. 1865–1–2 From Thomas L. Dames Edenberry
Will be very pleased to assist him in obtaining the fossils, but can he give him the location of the quarries.
ALS 3pp RS:HS 6.467

12775. 1865–1–[2] To John Herschel (son) Collingwood
Discusses at length the means of calculating the mean daily rainfall in India, and comments on son John's work in geodesy.
ALS 8pp JHS 6.59

12776. 1865–1–3 To Charles Thomas Longley Collingwood
Has received the notice of the meeting, but regrets his ill health will prevent his attendance at the meeting.
C 1p RS:HS 11.456

12777. 1865–1–4 From James Challis Cambridge
Thanks for his friendly letter written on Christmas day. Regarding the Cambridge Observations just published. His own theories of hydrodynamics.
ALS 8pp RS:HS 5.239

12778. 1865–1–7 From John Ryall Queen's College, Cork
Asks for JH's opinion of the mathematical work of Professor [George] Boole, who recently died. Needs testimony to get pension for his widow and child.
ALS 4pp RS:HS 14.514

12779. 1865–1–9 To John Herschel (son) Collingwood
Works out the derivation for a mathematical problem, and JH suggests that son John should use this derivation in his paper; it is christening day for William, son of John's sister Caroline. [Also included is a one page letter of family news from sister Julia.]
ALS 3pp inc; encl JHS 5.16

12780. 1865–1–10 To John Ryall Collingwood
At JR's request [see JR's 1865–1–7], attests to merits of late George Boole as mathematician, in support of petition for government pension for Boole's widow.
AdraftS 3pp TxU:H/L-0358; Reel 1054 (C: RS:HS 25.15.21)

12781. 1865–1–12 From John Ryall Queen's College, Cork
Thanks JH for his opinion of [George] Boole. It may help his widow and children in their application to Lord Palmerston [H. J. Temple].
ALS 3pp RS:HS 14.515

12782. 1865–1–12 To [R.S.L. Council] Collingwood
Reports on H. G. Hennessey's paper [R.S.L. *Proceedings, 13,* 312–] on synchronous distribution of temperature over earth's surface. Feels bewildered by paper, and believes it could be condensed.
ADS 8pp RS RR.5.116

12783. 1865–1–13 To [Sir] Edward Ryan Collingwood
Encloses published letter of 20 Dec. 1864 from JH to editors of *Philosophical Magazine and Journal* concerning Charles Babbage, who 'revived this miserable subject' of 1826 dispute between W. H. Wollaston and Humphry Davy regarding Babbage's candidacy for R.S.L. secretaryship.
ALS 4pp, encl 2pp TxU:H/L-0765.6; Reel 1089

12784. 1865–1–14 From Warren de La Rue 110 Bunhill Row
Would JH look through the proof of WD's speech on the presentation of a [R.A.S.] medal to Professor G. P. Bond and make any comments or suggestions.
ALS 1p RS:HS 6.165

12785. 1865–1–14 From George Gabriel Stokes Lensfield Cottage, Cambridge
Asks JH to support a petition to the government for a pension for the widow and children of George Boole.
ALS 4pp RS:HS 17.58

12786. 1865–1–16 From Sir Edward Ryan Garden Lodge
Thanks JH for copy of [John] Davy's letter. Is sorry to see Charles Babbage pursuing old disputes in his recent publication. Wife's health is failing.
ALS 5pp RS:HS 14. 465

12787. 1865–1–16 From W. H. Wood Sea View House, Weston-super-Mare, Somersetshire
Sends additional meteorological data for 10–12 Jan. 1865, showing effect of full moon on cloud cover.
ALS 2pp TxU:H/M–0599; Reel 1086 (C: RS:HS 24.81)

12788. 1865–1–17 From William Allen Miller Burlington House
WM's friend William Huggins is a candidate for the fellowship of the R.S.L. and would be pleased for JH's support. Please return the certificate to WM when he has signed.
ALS 2pp RS:HS 12.346

12789. 1865–1–20 From Gabriel Auguste Daubrée Paris
Says JH will receive an official letter of thanks from the administration of the Muséum d'Histoire Naturelle for the meteorite samples [see GD's 1865–1–23].
ALS 2pp (French) RS:HS 6.34

12790. 1865–1–21 To Adolphe Quetelet Collingwood, Hawkhurst Kent
Lists works received. Has been suffering severely from bronchitic attacks. Is translating *Iliad* into hexameter verse.
ALS 4pp Académie belgique: #37 (Transcription inc: #38)

12791. 1865–1–23 From Gabriel Auguste Daubrée Paris
Officially writes on behalf of the Muséum d'Histoire Naturelle to thank JH for his assistance in providing samples of the South African meteorite.
LS 1p (French) RS:HS 6.33

12792. 1865–1–23 From Henry Herman India Office, Westminster
Received seven copies [of JH's *Cape Results*]. Will forward these to India.
ALS 1p RAS:JH Archive 10/5.101; Reel 9

12793. 1865–1–23 From Edward John Jennings Hawkhurst
Regrets that illness prevented JH's attendance at the Finance Committee. Gives the reasons why a change of uniform is necessary for the Rifle Volunteers.
ALS 4pp RS:HS 10.440

12794. 1865–1–24 From William Rutter Dawes to Margaret Brodie Herschel Hopefield, Haddenham, Thame
Is grateful for her letter of condolence on the death of his niece. Is thankful to hear of the safe arrival of her son. His own health is better.
ALS 4pp RS:HS 6.120

12795. 1865–1–25 From Charles Babbage Dorset
Commiserates with JH in his illness, and says CB is ill also and confined to his house. Seeing JH's letter in the *Philosophical Magazine* [see JH's 1864–12–20] seems to CB to obviate completely the need for any further response.
ALS 4pp RS:HS B27.29

12796. 1865–1–28 From Bernard Cracroft Davies St.
Sending two pamphlets which he thinks are of great value, the one on the cure for sea-sickness being especially noteworthy. Would be glad of JH's support for Dr. John Chapman's claims.
ALS 8pp RS:HS 5.296

12797. 1865–1–29 To Sir Charles Lyell Collingwood
Sending a lump of a substance quarried near Edenderry; has found some very similar in a quarry of his own. Can send some more specimens. Congratulations on his Baronetcy. Has been prostrated with bronchitis.
C 4pp RS:HS 11.430 & 24.83

12798. 1865–1–29 To William Whewell Collingwood
Sends Book XIV of JH's *Iliad* translation. Fears he may not live to finish it and may not find a publisher. Glorious winter in England. Son William and wife have arrived in Calcutta.
ALS 4pp TC, Camb. Add. Ms.a.207[116] (C: RS:HS 24.82)

12799. 1865–1–30 From Warren de La Rue 110 Bunhill Row
Encloses two certificates and hopes that he will give the candidates his support. Has just received a letter from Angelo Secchi on the Herschelian prism.
ALS 1p RS:HS 6.166

12800. 1865–1–31 From Bernard Cracroft Davies St.
Thanks for kind letter. Pamphlets should have reached him by now. Discourses on the effects of cold and heat on the spinal cord.
ALS 8pp RS:HS 5.297

12801. 1865–1–31 To Warren de La Rue Collingwood
Returns the certificates duly signed and also Angelo Secchi's letter, which he was pleased to see. Elaborates on the possible causes of James Nasmyth's 'willow leaves.'
draft 3pp RS:HS 6.167 & 24.84 *R.A.S.M.N., 25* (1865), 152–3 inc

12802. 1865–1–31 From Sir Charles Lyell 53 Harley St.
Origin of siliceous minerals sent by JH. Planning tenth edition of *Principles of Geology*. Convinced that changes in position of land and sea were principal causes of climate changes. Questions [James] Croll's theory of astronomical influences on climate, based on data from U. J. J. Leverrier. Compares earth-pillars in Rhone valley with those in Tyrol. Wants tenth edition to include woodcut of JH's [1821?] drawings of earth-pillars.
ALS 12pp RS:HS 11.431 Katherine Lyell, *Lyell, 2,* 386–9

12803. [1865–2 or earlier] From Helen Underhill Summerfield, Wolverhampton
Thanks for 'valuable addition' to HU's book.
ALS 1p inc RS:HS 19.176b

12804. 1865–2–2 From William Radcliffe Birt Aylesbury
Notifying him of a forthcoming meeting of the Lunar Committee of the B.A.A.S. Hopes he will attend.
ALS 3pp RS:HS 4.144

12805. 1865–2–3 To Sir Charles Lyell Collingwood
Comments on the revolution of the apsides of the earth's orbit and its effect on glacier theory; further comments on the earth pyramids of Botzen [see CL's 1865–1–31], and associated phenomena.
AL 4pp APS B.D25.L.6 (C: RS:HS 24:85)

12806. 1865–2–3 From W[illiam] J. Millar Glasgow
Having read JH's paper in *Good Words* on the weather he commenced a series of observations of the state of the sky during three days of each lunation. Will be pleased to continue sending his observations if JH is willing to accept them.
ALS 3pp RS:HS 13.40

12807. 1865–2–4 From James Croll Andersonian University, Glasgow
Sending his paper on the causes of change of climate, and would be glad of JH's comments and opinion on its worth.
ALS 3pp RS:HS 5.299

12808. 1865–2–5 To Phillipp Ludwig von Seidel Collingwood
PS's results on light and magnitude of stars agree with JH's results at the Cape. Comments on PS's research on light of planets.
ALS 4pp RS:HS 17.186 (C 2pp: 24.86)

12809. 1865–2–6 From Charles Babbage Dorset
Draws JH's attention to an article in the most recent *Philosophical Magazine.*
ALS 2pp† RS:HS B27.30 (draft?: BL Aaa 32309.209)

12810. 1865–2–6 To James Croll Collingwood
In response to comments in a paper by JC, JH writes to establish his priority with respect to the significance of the revolution of the apsides of the earth's orbit and the eccentricity of the earth's orbit in affecting temperatures on the earth. JH had spelled this out in an 1830 paper for the *Transactions of the Geological Society,* and repeated it in his *Outlines Astr.*
L(S) 3pp inc APS B.D25.L.7; draft: RS:HS 5.300 & C: 24.87–8 J. C. Irons, *Autobiographical Sketch of James Croll* (London, 1896), 120–2

12811. 1865–2–6 To Sir Charles Lyell Collingwood
Writes to CL about JH's response to James Croll's paper. Includes an extract from JH's letter to James Croll [see JH's 1865–2–6].
ALS 3pp APS B.D25.L.8 (C: RS:HS 24.89)

12812. 1865–2–7 From John Tyndall Royal Institution
Sends copy of JT's '3rd Memoir,' along with specimens of liquid mercury ethyl and mercury methyl from discoverers Edward Frankland and B. F. Duppa.
ALS 2pp RS:HS 17.399a (C: RI 536)

12813. 1865–2–10 From Warren de La Rue 110 Bunhill Row
Was pleased to receive JH's letter requesting him to

communicate the subject to the Society. Has been busy preparing the Annual Report so has not had time to follow up Mr. Traye's papers.
ALS 1p RS:HS 6.168

12814. 1865–2–10 To [Walter] White Collingwood
Asks WW to check the R.S.L. council book for 23 Nov. 1826 to inform JH as to who was present.
ALS 1p BostonPubLib Ms. Eng. 204

12815. 1865–2–11 From Sir Charles Lyell 53 Harley St.
Regarding the editions in which he cited JH's paper. Further comments on James Croll's paper, and on the effects of glacial and astronomical events. Is the quantity of heat radiated by the sun in the course of the year always the same? Thanks for the drawings of the Botzen columns.
ALS 22pp RS:HS 11.432

12816. 1865–2–11 To W[illiam] J. Millar Collingwood
On the significance of the full moon and its influence on the weather.
C 1p RS:HS 24.90

12817. 1865–2–11 From Charles Pritchard Freshwater
R.S.L. approves of CP's optical work, believing it will be helpful at observatories abroad. CP comments on Alexander Herschel's prism experiments.
ALS 4pp RS:HS 14.96

12818. 1865–2–11 From Walter White Royal Society
At JH's request, WW sends JH a list of members present at the 'irregular' council meeting on 23 Nov. 1826, together with the business recorded in the minutes, which does not include any mention of the matter of the secretaryship of the R.S.L.
ALS 3pp RS:HS B27.31

12819. 1865–2–12 To *Philosophical Magazine*
Collingwood
Attempts further clarification of the matter discussed in JH's 1864–12–20 to the *P.M.* Reiterates JH's recollection of a conversation between W. H. Wollaston and Humphry Davy, which JH had described in a letter [see JH's 1826–11–25 to Francis Baily] and which JH refers to in his letter of resignation to Davies Gilbert [see JH's 1827–6–27], which latter letter JH quotes.
ACCS 4pp RS:HS B27.34 *Phil. Mag.*, 4th ser., *29* (Mar. 1865) 246–7

12820. 1865–2–14 From James Croll Andersonian University, Glasgow
Thanks for his letter concerning his own paper on the changes in climate. Points out parts where JH seems to have been mistaken in his views as expressed.
ALS 3pp RS:HS 5.301

12821. 1865–2–15 From Thomas L. Dames Edenberry
By all means retain the specimens and send them to Lyell or anyone else as he sees fit.
ALS 1p RS:HS 6.468

12822. 1865–2–15 To Sir Charles Lyell Collingwood
Does not gainsay the influence of distribution of land and sea on the climate. The change of eccentricity is also a powerful influence. Comments on this. Sees that CL's book is published. Is it too late to send his drawings of the earth pyramids?
ALS 4pp APS B.D25.L.9 (C: RS:HS 11.433 & 24.91 & 92)

12823. 1865–2–16 From Sir Charles Lyell 53 Harley St.
Has ordered that a copy of his new 6th edition of the *Elements* [*of Geology*] should be sent to JH. He wants the drawings of the earth pillars for the 10th edition of his *Principles* dealing with the effect of rain. JH's drawing made a great impression on him.
ALS 3pp RS:HS 11.434

12824. 1865–2–17 To James Croll Collingwood
Thanks for his letter and comments. Is always pleased to have any omission pointed out. Comments on his theories of the changing of the climates.
draft 5pp RS:HS 5.302 & 24.93 J. C. Irons, *Autobiographical Sketch of James Croll*, 125–7

12825. 1865–2–17 From H[enry Drury] Harness Chatham
Regarding the assaying at the Mint. Hopes JH will use his influence to assist W. H. Barton's son.
ALS 4pp RS:HS 10.161

12826. 1865–2–19 To Sir Charles Lyell Collingwood
Thanks for his *Elements* [*of Geology*]. Evidence from the Arctic shells is decisive. Sends drawings by book post. Comments on these drawings. Near Arezzo are similar phenomena in a soft sandy limestone. Measured the tallest column at Riten and found it 74' 10" from stone cap to base.
ALS 4pp APS B.D25.L.10 (draft: RS:HS 11.435 & 24.94 & 95)

12827. 1865–2–20 From George Gabriel Stokes
Burlington House
A letter from the Secretary of R. S. L. asking JH to review several papers. [Appended is a copy of JH's review in draft form.]
LS 1p DMC.695A.885.34

12828. 1865–2–21 From Sir Charles Lyell 53 Harley St.
Impressed by JH's [1824] drawings of earth-pillars in Botzen. Plans to visit them this year. Causes of earth-pillars. Speculates on climate changes on planet Mars.
ALS 7pp RS:HS 11.436 Katherine Lyell, *Lyell, 2*, 386–9

12829. 1865–2–22 To William Hallowes Miller
Collingwood
Is translating Homer's *Iliad* and has heard a verse is missing from the standard versions. Can WM help?
C 3pp RS:HS 24.96

12830. 1865–2–23 From James Croll Andersonian University, Glasgow
Gives various references in answer to JH's queries. Further regarding climatic changes.
ALS 4pp RS:HS 5.303

12831. 1865–2–23 From Johann Galle Berlin
A note of thanks for copies of recent writings by JH and his son Alexander.
ALS 2pp (German) RS:HS 9.86

12832. 1865–2–24 From W. H. Barton Bushey Park, Teddington
Regarding a position at the Mint for his own son.
ALS 4pp RS:HS 3.302

12833. 1865–2–24 From Bernard Cracroft Davies St.
Speculations on the causes of hydrophobia. Has he seen the work of J. B. Luys on the spinal system?
ALS 5pp RS:HS 5.298

12834. 1865–2–26 To Sir Charles Lyell Collingwood
Some comments on the variation of the eccentricity of the earth's orbit, and the geography of Mars.
ALS 4pp APS B.D25.L.11 (C: RS:HS 24.97)

12835. 1865–2–28 From W. H. Barton Bushey Park, Teddington
Proposes to visit JH at Hawkhurst.
ALS 2pp RS:HS 3.303

12836. [1865–3?] To [Admiralty?] n.p.
Objections by Board of Visitors to proposed new railway tunnel near Greenwich observatory.
Adraft 4pp RS:HS 19.176c

12837. 1865–3–1 From Thomas MacKereth Eccles
Doubts that JH put his approval on an article about shooting stars in the 'upper atmosphere' in *The Intellectual Observer.*
ALS 4pp † Harvard: Houghton AAH 67m–67 (94)

12838. 1865–3–3 From David Brewster Edinburgh
Is sending an unpublished pamphlet on dioptric lights and would be glad of his opinions as he is claiming public recognition.
ALS 3pp RS:HS 4.268

12839. [1865]–3–3 From Andrew Scott Waugh Athenaeum Club, Pall Mall
Requests JH sign R.S.L. certificate attesting to qualifications of H. E. L. Thuillier, AW's deputy in India. Hopes to see [JH's son] John when John returns from India.
ALS 2pp RS:HS 18.125

12840. 1865–3–4 From Sir Charles Lyell 53 Harley St.
Took JH's box of specimens and had a discussion with William Logan, A. C. Ramsay, and some young surveyors, one of whom had examined JH's neighborhood. Similar rhomboids to the Irish ones have been located in Wales. Hopes to have a discussion on this subject later. Can JH give, or refer him to data, the temperature of celestial space without the sun? Returns the drawings with many thanks.
ALS 5pp RS:HS 11.437

12841. 1865–3–5 To Sir Charles Lyell Collingwood
Comments on why JH gave 239°F as the temperature of space. Geological specimens may be given to the Geological Society or to the Jermyn Street Museum.
ALS 9pp APS B.D25.L.12 (C: RS:HS 11.438 & 24.98)

12842. 1865–3–9 To David Brewster Collingwood
Thanks for his letter regarding the dioptric lights. Gives his own views as to their importance.
draft 4pp RS:HS 4.269 & 24.100

12843. 1865–3–9 From Thomas Mackereth Eccles
Thanks for kind information regarding Christopher Hansteen. Will obtain a copy of the *Admiralty Manual* to read JH's paper on meteorology.
ALS 2pp RS:HS 13.6

12844. 1865–3–14 From Eliza Susan Quincy Boston, Mass.
Sends tributes to memory of her father [Josiah Quincy]. Discusses [James] Grahame and [George] Bond, who both have died. Also mentions Harvard Observatory and American Civil War.
ALS 4pp RS:HS 14.279

12845. 1865–3–15 From John R. Hind 8 Matson Villas, Richmond
Regarding the comet seen at Melbourne (Australia), may be the 1843 comet discussed by JH in his *Outlines Astr.* Comments on this.
ALS 4pp RS:HS 9.390

12846. 1865–3–15 From Edward Sabine Burlington House
Asks whether Greenwich Board of Visitors should meet to answer communication from Admiralty or if a response could be circulated among Visitors and sent if it meets with approval.
ALS 4pp RS:HS 15.274

12847. 1865–3–18 From Adolphe Quetelet Brussels
Lists works JH should have received. Asks JH to take another look in his library. Has received more on meteors. Sees no objection to a higher terrestrial atmosphere. Will send *Histoire des sciences mathématique et physique chez les belges.*
ALS 4pp (French) RS:HS 14.260

12848. 1865–3–18 To Edward Sabine Collingwood
Encloses proposed resolution to controversy over whether to build railway tunnel under park. Will be unable to attend meeting because of chronic bronchitis.
ALS 4pp encl RS Sa.672

12849. 1865–3–19 From Adolphe Quetelet Brussels
Lists works sent to JH. Discusses hypothesis concerning meteors and the atmosphere. Mentions his work 'Sur la physique du globe.' Cites Mr. Bravais. Attaches note sent to AQ by Royal Academy of Brussels, listing works JH should have received. Encloses list of errata for the *Bulletin de l'académie royale de belgique,* ser. 2, *15,* No. 6.
ALS 2pp (French) Académie belgique: #39 q 39a

12850. 1865–3–21 From John R. Hind 8 Matson Villas, Richmond
Further regarding the southern comet; does not think it can be the 1843 one until he receives further observations.
ALS 3pp RS:HS 9.391

12851. 1865–3–25 From Edward William Brayley London Institution
Regarding certain passages in JH's Cape Observations.
ALS 4pp RS:HS 4.231

12852. 1865–3–25 From Edward Sabine 13 Ashley Place
Visitors Board will meet to discuss railway companies' proposal to remove Greenwich Observatory to another site. Includes copy of Warren de La Rue's letter to ES.
ALS 3pp encl RS:HS 15.275

12853. 1865–3–25 From James Joseph Sylvester Woolwich Common
Thanks for note explaining refusal to undertake another new subject, JS's mathematical paper. Updates on algebra books.
ALS 3pp RS:HS 17.162

12854. 1865–3–26 To Edward William Brayley Collingwood
On the physical structure of the sun's surface.
C 2pp RS:HS 24.101

12855. 1865–3–28 From Roderick Impey Murchison 16 Belgrave Square
Sends a Memoir of W. E. Hickson for JH's views so that he can communicate the paper to the Royal Geographical Society. Regarding a maritime expedition to the North Pole. There never has been a North Pole Expedition.
ALS 4pp RS:HS 12.433

12856. 1865–3–29 From Thomas Humphrey Hawkhurst National School
Announces a National School quarterly meeting.
ALS 1p Harvard: Houghton AAH 67m–67 (90)

12857. 1865–3–29 To Edward Sabine Collingwood
Objects to plan to move Royal Observatory to site other than Greenwich, but will defer to G. B. Airy's opinion. JH's deteriorating health.
C 5pp TxU:H/L-0377; Reel 1054

12858. 1865–3–31 From Edward William Brayley London Institution
His views regarding the constitution of the sun coincides with JH's. Is giving a lecture for J. P. Gassiot in which he would like to quote some of JH's views.
ALS 3pp RS:HS 4.232

12859. 1865–3–31 To Eliza Susan Quincy Collingwood
Thanks her for photograph of EQ's late father [Josiah Quincy]. Glad [George] Bond knew he was awarded R.A.S. Medal before his death. Notes the generations of mankind are 'unequal.'
C 2pp RS:HS 14.280 (C: RS:HS 24.102)

12860. 1865–4–1 To John Herschel (son) Collingwood
Most of the letter is devoted to a mathematical derivation to show that part of son John's work on errors in geodetic observations is in error; however, part is highly praised by JH. Remainder of letter is clearly intended as a morale booster for John.
ALS 5pp JHS 6.58

12861. 1865–4–3 From Roderick Impey Murchison 16 Belgrave Square
Thanks for letter, but has not received W. E. Hickson's memoir. Regrets JH sees no use in a survey of the North Pole region. The scheme seems to be one that the President of the Geographical Society should encourage. The ribald buffoonery of the *Times* has done much good amongst the scientific world.
ALS 4pp RS:HS 12.434

12862. 1865–4–4 From Elizabeth Drummond Portobello, Edinburgh
A friend of hers is writing a memoir of her brother Thomas Drummond, and would be glad for a few minutes conversation with JH.
ALS 1p RS:HS 6.449

12863. 1865–4–4 From Martha Somerville Spezia
Written for Mary Somerville. Requests information on sunspots and on William Huggins's observations of nebulae. Mary is recovering from illness. They send regards to Herschels.
ALS 4pp RS:HS 16.324

12864. 1865–4–5 To Augustus De Morgan Collingwood
Longs to hear from AD; comments on JH's health and on the weather.
C 2pp RS:HS 24.103

12865. 1865–4–5 From Elizabeth Drummond Portobello, Edinburgh
Would like a copy of JH's letter that he wrote many years ago in support of an invention by her brother.
ALS 5pp RS:HS 6.450

12866. 1865–4–6 From Augustus De Morgan 91 Adelaide Road
The winter has not killed them, though his wife is a convalescent. Knows Mr. Peters as he is a member of the Athenaeum. Airy has fun in him and a love of Greek Drama. Charles Babbage and James South are at war again. Sends jokes.
ALS 4pp RS:HS 6.378

12867. 1865–4–6 To Margaret Brodie Herschel 41 Wania [?] Square
Reports on accommodations and the activities of those family members with JH.
ALS 4pp JHS 5.22

12868. 1865–4–10 From Herbert Spencer 88 Kensington Gardens Square
Recounts efforts to establish *The Reader* as a respected weekly journal of science. Asks that JH submit material for publication in it.
ALS 4pp RS:HS 16.495

12869. 1865–4–11 From W. K. von Haidinger Vienna
Was grateful for his kind letter. Doubtless he has received further of his papers by now, including the one on the meteor of Oct. 1863. Sends a note on two lines of the *Iliad.* Has heard interesting news of Alexander Herschel's labors.
ALS 2pp RS:HS 9.153

12870. 1865–4–11 To Mary Somerville Collingwood
Comments on the political change within Italy. Answers questions about spectrum analysis, spectra of nebulae, and [solar] 'willow leaves.' Says the view of universe as a collection of billiard balls is problematic. Includes postscript from JH's daughter Bella describing family events.
ALS 5pp RS:HS 16.372 (C: RS:HS 24:104)

12871. 1865–4–12 From George Biddell Airy Royal Observatory, Greenwich
Is returning the Proceedings of the Manchester Society, and The Magnetical and Meteorological Observations for 1851–58. Regarding the registration of magnetical disturbances.
ALS 4pp RS:HS 1.282 (C: RGO 6.679.452)

12872. 1865–4–12 To Margaret Brodie Herschel [Collingwood]
Tells MH about the contents of letters from sons John and William, and about daughter Isabella's illness.
ALS 4pp JHS 5.23

12873. 1865–4–12 From John F. McLennan Cambray Villa, Reigate
Doubtless JH has heard from Miss Elizabeth Drummond JM's desire to meet JH. Would like to call at Collingwood on Saturday.
ALS 1p RS:HS 12.250

12874. 1865–4–17 From John F. McLennan Hare Ct, Inner Temple
Thanks for his kind reception on Saturday. Will communicate with Miss Elizabeth Drummond. Gives list of letters he borrowed from JH. Intends to try to meet William Yolland.
ALS 3pp RS:HS 12.251

12875. 1865–4–17 To Editor of the *Reader* Collingwood
Appreciates invitation to publish some writings by JH in *The Reader,* but does not at present have suitable manuscripts. Is no longer active in astronomical observation.
ALS 2pp Sydney Ross Collection

12876. 1865–4–18 From Herbert Spencer n.p.
The Reader does not plan to publish a list of contributors, but would welcome letters by JH, which would imply JH's support for the new weekly journal of science.
ALS 4pp RS:HS 16.496

12877. 1865–4–19 From Edward Loyd Hawkhurst
The Archbishop of Canterbury will be lunching with EL on

Tuesday; will JH join them?
ALS 2pp RS:HS 11.347

12878. 1865–4–23 From John R. Hind 8 Matson Villas, Richmond
Sending the elements of 79, 81 and 82; is not certain if N. R. Pogson has sent the orbit for No. 80. It is reckoned in the same way as in the last edition of JH's *Outlines Astr.*
ALS 2pp RS:HS 9.392

12879. 1865–4–24 From G[eorge] C. Renouard Swanscombe Rectory
Tells several anecdotes about people William Herschel knew. Discusses William Herschel and his work.
ALS 3pp RS:HS 14.328

12880. 1865–4–25 To Sir Charles Lyell Collingwood
Comments on a table of U. J. J. Leverrier on the eccentricities of the earth's orbit.
ALS 4pp APS B.D25.L.13 (C: RS:HS 24.105)

12881. 1865–4–25 To G[eorge] C. Renouard Collingwood
Thanks GR for letter and anecdotes. Discusses Uranus and Jupiter and the long years the inhabitants of those planets, if there are any, must experience.
C 2pp RS:HS 24.106

12882. 1865–4–26 To Elizabeth Colling Collingwood
Comments on the unevenness of generations; speculations about life on Uranus; expects to finish translation of the *Iliad* by the end of the year.
C 2pp RS:HS 24.107

12883. 1865–4–27 From Thomas Graham Royal Mint
Thanks for his letter and also the enclosed one of B. H. Babbage. Regarding appointments to the proposed Mint at Melbourne. The difficulties involved.
ALS 4pp RS:HS 8.230

12884. 1865–4–28 From Sir Charles Lyell 53 Harley St.
Discussion on JH's geological specimens took place at the Geological Society on Wednesday. Gives some of the points raised at the meeting. Regarding the length of time that separates the extremes of eccentricity of the earth's orbit. Has raised the matter with G. B. Airy.
ALS 10pp RS:HS 11.439

12885. 1865–4–28 From G[eorge] C. Renouard Swanscombe Rectory
Thanks JH for amusing letter. Discusses William Herschel, the great reflector from Slough, his family, and his health.
ALS 4pp RS:HS 14.329

12886. 1865–4–30 To Alexander Herschel from Henry Bence Jones London
HJ writes to AH to inform him that HJ has nominated AH for the vacancy left by Robert FitzRoy's death [see Henry Holland's 1865–5–1].
CC 2pp JHS 6.56b

12887. 1865–4–30 To Roderick Impey Murchison Collingwood
Criticizes the manuscript of Mr. Hickson [?] about meteorology, the diameter of the sun, and conditions at the North Pole. JH recommends against publishing the work without considerable editing.
ALS 3pp BL Aaa 46126.402

12888. 1865–5–1 From Henry Holland 15 Brook St.
HH has nominated JH's son Alexander to fill a post become vacant by the death of Robert FitzRoy.
CC 1p JHS 6.56a

12889. 1865–5–1 From Sir Charles Lyell 53 Harley St.
Has just received the accompanying letter from James Croll so that JH can save himself the trouble of answering CL's query.
ALS 2pp RS:HS 11.440

12890. 1865–5–2 To Henry Holland Collingwood
Writes to thank HH for his kindness in nominating son Alexander for the meteorological post [see HH's 1865–5–1].
CC 3pp JHS 6.56c

12891. 1865–5–4 From William Radcliffe Birt Victoria Park
Notifying him of a forthcoming meeting of the Lunar Committee of the B.A.A.S.
ALS 1p RS:HS 4.145

12892. 1865–5–4 From John F. McLennan 22 Hill St., Edinburgh
Thanks for copies of his interesting and conclusive letters. Has just returned to Edinburgh and has not yet seen Miss Elizabeth Drummond. Hopes the delightful weather has relieved JH's bronchitis.
ALS 2pp RS:HS 12.252

12893. 1865–5–5 From A. B. Power Blackheath
Accepts position at Blackheath Preparatory School.
ALS 7pp RS:HS 14.186

12894. 1865–5–5 To Rudolf Wolf Collingwood
Has sent RW a copy of JH's catalog of nebulae and also a biographical sketch of JH's father. Thanks RW for and praises RW's works on sun spots and on the aurora borealis. Notes that G. B. Airy is now disposed to accepting the existence of solar 'willow leaves.'
C 3pp RS:HS 24.155

12895. 1865–5–6 To Sir Charles Lyell Collingwood
Further comments on the theory of glacier formation, urging causes beyond merely astronomical ones.
ALS 4pp APS B.D25.L.14 (C. RS:HS 24.108)

12896. 1865–5–6 To the London *Times* n.p.
Signing himself 'A.B.C.D.,' asks whether a priest is guilty of perjury for having refused to testify regarding a murder on the grounds that this would violate the seal of the confessional.
C 2pp RS:HS 24.109

12897. 1865–5–7 or later To Roderick Impey Murchison [Collingwood]
JH's son Alexander is up for membership in the R.A.S., and JH would appreciate RM speaking in favor if the occasion arises. Introduces JH's son John to RM and his wife. Comments on the possibility of still finding [David] Livingstone alive in Africa.
ALS 4pp BL Aaa 46126.406

12898. 1865–5–7 To Roderick Impey Murchison Collingwood
Thanks RM for his interest in a possible position for JH's son Alexander. It now appears that the position will likely go to someone else, but there is so much work to be done in the meteorological office that another person, such as Alexander, would be ideal. Perhaps RM could put in a word for him.
AL 4pp BL Aaa 46126.404

12899. 1865–5–7 To Adam Sedgwick Collingwood
Is glad that J. C. Maxwell has received the Aberdeen Professorship. Has forwarded 'your last' to the Principal. Is feeling better, but still needs crutches.
ALS 4pp RS:HS 15.449 (C: RS:HS 23.170)

12900. 1865–5–13 From Walter White Burlington House
Forwarding JH's parcel to Jean Chacornac in Paris. Explains R.S.L.'s postage policy for international parcels to and from members.
ALS 2pp RS:HS 18.230

12901. 1865–5–14 To Harriet Lubbock Collingwood
John Lubbock shall have JH's vote. Comments on the duties of an M.P. Cannot promise to canvas but will speak to anyone with influence he meets.
C 7pp RS:HS 11.350–1 & 24.110–11

12902. 1865–5–16 From J. Park Harrison Emhurst, Guildford
Sending a curve of mean temperature [of the moon]. Comments on this. Is to be printed in the proceedings of the R.S.L.
ALS 3pp RS:HS 10.165

12903. 1865–5–16 or later To J. Park Harrison n.p.
Replies to JH's 1865–5–16, expressing thanks for materials sent. Comments on temperature.
draft 1p RS:HS 10.165 verso

12904. 1865–5–17 From Hermann Kindt Yarm, Yorks
JH's translation of Frederick Schiller's poem has given great pleasure to many of his German friends. Quotes extract from the comments of Ferdinand Freiligrath.
ALS 4pp RS:HS 11.31

12905. [1865]–5–17 From Harriet Lubbock High Elms, Farnborough
Thanks for his kind note. Is convinced that John will benefit the House by becoming an M.P.
ALS 4pp RS:HS 11.352

12906. 1865–5–21 To William Matthews Collingwood
Results of calculations of an equation for the relation of temperature and altitude.
draft 4pp Harvard: Houghton AAH 67m–67 (101)

12907. 1865–5–22 From W. Herschel Halberstadt
Introduces WH to JH, explains where WH fits into the Herschel family, and tells a little about WH's life.
ALS 3pp (German) RS:HS 10.184

12908. 1865–5–22 From John Lubbock Chislehurst
Thanks for his letter. Sorry to hear that JH's name is not on the Register, but can they add his name to JL's Committee? His father is very ill.
ALS 3pp RS:HS 11.354

12909. 1865–5–25 To William Matthews Collingwood
Sends results of calculations of an equation for the relation of temperature and altitude.
ALS 4pp Harvard: Houghton AAH 67m–67 (102)

12910. 1865–5–25 From William Mathews Jr. Carpenter Rd., Edgbaston
Thanks for letter. Is expecting JH's son to dinner on Sunday. James Glaisher's observations all appear to point in the same direction. Will visit Collingwood when that way. Will they see JH at Birmingham for the B.A.A.S. meeting?
ALS 3pp RS:HS 12.298

12911. 1865–5–25 From William Walker 64 Margaret St., Cavendish Square
Thanks for photograph of JH, but current outcry of photographers about copyright forces WW to request that JH sit for portrait in WW's own studio.
ALS 1p RS:HS 18.19

12912. 1865–5–26 To John Lubbock Collingwood
Cannot make up his mind to support either one or other of the two great political divisions at the moment.
C 4pp RS:HS 11.355 & 24.112

12913. 1865–5–26 From James Nasmyth n.p.
Has great pleasure in receiving a visit from W. C. Morland. Will be pleased to visit the Herschels. Has been travelling in Italy; visited Vesuvius and also called on Angelo Secchi, who was engaged in making a model of the sun's disc and using oat corn grains to illustrate the willow leaf effects.
ALS 4pp RS:HS 13.86

12914. [1865–5–27] To [Editor of *Daily News*] n.p.
Responds to article by Mr. Yates in *Daily News* of the 18 May that misrepresented JH's views on standard units of measure.
Adraft 1p inc TxU:H/L-0735; Reel 1089

12915. 1865–5–27 To Henry James Collingwood
Concurs with HJ's proposal to assign Sinai survey party to H. S. Palmer. Cost of survey will be £1220. Encloses letter printed in *Daily News* in reply to Mr. Yates on British standards of measure, with calculations based on Capt. Clarke's results.
ACCS 2pp TxU:H/L-0201; Reel 1054

12916. 1865–5–30 From Isaac Todhunter St. John's, Cambridge
Thanks JH for compliments on his work [*History of the Mathematical Theory of Probability*]. Suggests changes in the Wilhelm Struve probability problem.
ALS 3pp RS:HS 17.356

12917. 1865–5–31 To Humphrey Lloyd Collingwood
Writes to ask for clarification of HL's views on the dispersion of light.
ALS 3pp APS B.H435p.22 (C: RS:HS 24.113)

12918. 1865–6–1 From George Gabriel Stokes Lensfield Cottage, Cambridge
Responds to a paper by Dr. Barnard, in which is propounded a view of dispersion of light being related to its intensity.
ALS 12pp RS:HS 17.59

12919. 1865–6–2 To James Calder Stewart Collingwood
Thanks for the memorandum about the lion. Recounts tale about Emperor Hadrian; mentions Joseph Butler, Samuel Butler, and family health problems. Glad that JS recovered from 'ugly accident.'
ALS 4pp BostonPubLib Ms.Eng.334 (C: TxU:H/L-0459; Reel 1055)

12920. 1865–6–[3] To James Calder Stewart n.p.
Reviews 'Hughes Map of the World.' Questions on India [H.E.I.C.] investments.
ALS 4pp TxU:H/L-0460; Reel 1055

12921. 1865–6–3 To George Gabriel Stokes Collingwood
Returns a paper on Indian meteorology, as he cannot devote time to it.
ALS 2pp Cambridge University Library 7656/H707

12922. 1865–6–3 To George Gabriel Stokes Collingwood
Comments on GS's 1865–6–1.
ALS 2pp Cambridge University Library 7656/H707

12923. 1865–6–4 From William Mathews Jr. Carpenter Rd., Edgbaston
Grateful for JH's welcome present of his article on meteorology, particularly as it contained the author's manuscript additions. Hopes that JH's son was pleased with the reception at Birmingham.
ALS 2pp RS:HS 12.299

12924. 1865–6–5 From Humphrey Lloyd Kilcroney, Bray
Has not seen the paper in the Smithsonian Institution, but believes the author is in error. Has looked at A. L. Cauchy's Dispersion Memoir, but thinks Baden Powell's paper presents a more distinct formula. Hopes to send in a few weeks the first volume of HL's [*Observations Made at the Magnetic and Meteorological Observatory at Trinity College, Dublin*].
ALS 4pp RS:HS 11.288

12925. 1865–6–6 From George C. Hodgkinson The Lodge, Louth
Has now applied to the R.S.L. for a Government Grant and would like JH's support for his application. Wishes sunfall as well as rainfall was recorded.
ALS 4pp RS:HS 9.419

12926. 1865–6–6 From George Gabriel Stokes Cambridge
GS believes he erred in his 1865–6–1; now writes to correct it.
ALS 11pp RS:HS 17.60

12927. 1865–6–13 From George James Symons 136 Camden Road
Sends JH complete set of rain papers and monthly circulars. Is grateful for JH's interest in meteorology.
ALS 3pp RS:HS 17.256

12928. 1865–6–14 From Smith, Elder & Co. 65 Cornhill
Acknowledgement of order from JH, and clarification of instructions.
ALS 2pp RS:HS 16.212

12929. 1865–6–15 To Harriet Lubbock Collingwood
Has been kept informed of the illness of Sir John William Lubbock by JH's daughter [Matilda] Rose. Hears things are now more serious and proposes to bring Rose home to Collingwood.
C 4pp RS:HS 11.353 & 24.114

12930. 1865–6–16 From George Gabriel Stokes Royal Society
Asks JH to consider revising report on a paper.
ALS 3pp RS:HS 17.61

12931. 1865–6–23 To [?] Collingwood
Comments on the preparation of an article on light by JH for publication in *Good Words.*
ALS 4pp APS B.H435p.23

12932. 1865–6–26 From George C. Hodgkinson Hôtel du Louvre, Paris
Found it impossible to answer JH's letter before his departure. Delighted with JH's contrivance. Hopes to make some observations on Mont Blanc.
ALS 3pp RS:HS 9.420

12933. 1865–6–27 To Edward Sabine Collingwood
Discusses negotiations between government and R.S.L. regarding Meteorological Departments of Board of Trade.
ALS 4pp RS Sa.674

12934. 1865–7–4 To Augustus De Morgan Collingwood
On concept formation.
C 2pp RS:HS 24.115

12935. 1865–7–4 To Isaac Todhunter Collingwood
Sends copy of paper on 'projection of the sphere in which the problem is to make any infinitely small figure on the projection similar to that on the sphere,' which he has presented to the Royal Geographical Society.
C 1p RS:HS 24.116

12936. 1865–7–11 To Henry Malden Collingwood
Requesting information as to where he can find details of the principle that regulates the application of the Greek accents.
C 3pp RS:HS 13.16 & 24.117

12937. 1865–7–14 To Bishop of Kingston Collingwood
Is pleased that the bishop found nothing religiously objectionable in JH's two papers, censored by the editor of the *Fortnightly Review,* in which JH suggested the spiritual origination of force, an idea that JH states has repeatedly appeared in JH's publications.
C 2pp RS:HS 24:118

12938. 1865–7–15 From Hermann Kindt Bampton, Penrith
Requests that JH write a few lines to an old friend of HK's in Germany.
ALS 4pp (German) RS:HS 11.32

12939. 1865–7–22 From Edmund Wheeler 48 Tollington Road, Holloway
Offers to lecture to JH's 'association.' Describes microscope equipment designed by EW and son, who now manufactures portable telescopes. Diagrams and questions about new object glasses and compound lenses. Offers free telescope to Alexander Herschel. Progress of Atlantic Cable offers good lecture topic for next session.
ALS 4pp RS:HS 18.394

12940. 1865–7–24 From Edmund Wheeler 48 Tollington Road, Holloway
Thanks JH for letter and explanation. Will contact Charles Pritchard for details. Please forward 'enclosed' to [Alexander Herschel].
ALS 1p RS:HS 18.395

12941. 1865–7–26 From William Walker 64 Margaret St., Cavendish Square
Will expect JH on Monday.
ALS 1p RS:HS 18.20

12942. 1865–7–26 Johann Karl Friedrich Zöllner Leipzig
A note to accompany the sending of some papers to JH and the R.A.S.
ALS 1p (German) RS:HS 18.363

12943. 1865–7–28 From William Matthews Egbaston, Birmingham
Replies to JH's letter of 1865–5–25 concerning the altitude/temperature equation.
ALS 3pp Harvard: Houghton AAH 67m–67 (96)

12944. 1865–[7?]–29 To William James Herschel n.p.
Grateful for letter from WJH's wife Emma. Maria Herschel's engagement to Henry Hardcastle. WJH's report on improvements in British administration of India is good news. News and whereabouts of Herschel family. JH paid

WJH's annual taxes, but not dues to Amateur Photographic Association requested by J. Melhuish. Alexander Herschel leaves Tuesday for Pyrenees.
ALS 4pp TxU:H/L-0559; Reel 1053

12945. 1865–7–29 From Humphrey Lloyd Kilcroney, Bray
Comments that A. J. Fresnel's papers in the *Annales de chimie* are really incomplete sketches of theories, which he developed later. Theory clearly stated in [Felix] Billet's *Traité d'optique physique.* Friend of his at Trinity College has produced a peculiar liquid.
ALS 3pp RS:HS 11.289

12946. 1865–8–5 From F. Charsley Slough
Send anyone interested in the telescope over at any time. Mr. Deverill is the best man to paint it.
ALS 2pp RS:HS 5.248

12947. 1865–8–7 From John Lee Hartwell House
W. R. Birt is engaged sending out notices to the members of the Moon Committee. Wishes JH could attend. Has commenced a large map of the moon, which members can compare with actual observations. Admiral W. H. Smyth is well in mind, but suffers much from illness.
ALS 2pp RS:HS 11.184

12948. 1865–8–10 From William Mathews Jr. Carpenter Rd., Edgbaston
Hopes that JH will be attending the forthcoming meeting of the B.A.A.S., and offering the hospitality of one of their members.
ALS 2pp RS:HS 12.300

12949. 1865–8–11. To [James Nasmyth] Collingwood
Declines offer of models of the moon's structure. Does not plan to present a lunar paper to the R.A.S.
ALS 4pp BostonPubLib Ms.196(1)

12950. 1865–8–12 From Robert Stirling Newall Gateshead
Has two pieces of crown and flint glass suitable for making an object glass but they are not thoroughly annealed. Can JH give him any advice?
ALS 3pp RS:HS 13.150

12951. 1865–8–14 From Thomas Maclear Royal Observatory, Cape of Good Hope
Will attend hearing of J. S. Parlby inheritance case.
ALS 4pp Cape Archives/Maclear Papers File 100

12952. 1865–8–14 From Adolphe Quetelet Brussels
Sends JH other copies of works not received. Also work on international statistics. Has found out that JH has written an article on probability. Would like to use it.
ALS 3pp (French) RS:HS 14.261

12953. 1865–8–14 To Adolphe Quetelet n.p.
Assumes JH has not received the volumes sent over a month ago. Mentions international statistics briefly. Wishes to add an article about probability to a French translation of JH's 1850 review of AQ's book.
draft by JH 2pp (French) Académie belgique: #40

12954. 1865–8–16 From Adolphe Quetelet Brussels
Notes for a letter to JH and son. Intends to discuss the 'phenomenon of October 1854.' Will continue observation of proper motion of stars. Will mention 'savants' who disagree and who agree with his opinions. Means to request that JH continue corresponding.
C 1p (French) inc Académie belgique: #31 *Bulletin de l'académie royale des sciences, des lettres et des beaux-arts de belgique,* 2nd ser., *16* (1865), 319–20

12955. 1865–8–22 To Adolphe Quetelet Collingwood
Has consulted with publisher of JH's 1850 review of AQ's book. Is flattered and sees no reason why AQ's enlarged work cannot prefix or append a French translation of JH's review. Publisher requests written acknowledgement. Asks AQ to attach a note, correcting his archery example.
ALS 4pp Académie belgique: #41 (C: #42)

12956. 1865–8–26 From James Syme St Vincent Place, Glasgow
Having read JH's article on light, sends observations on peculiar effects of light, particularly 'terrestrial refraction,' on rifle-shooting targets.
ALS 3pp RS:HS 17.252

12957. 1865–8–28 From James Glaisher Blackheath
Do not send the whole parcel; one form for each month is all he requires. Has not received the report of the Luminous Committee yet. Has sent Harrow's bill to JH's son.
ALS 4pp RS:HS 8.134

12958. 1865–8–29 From R. Arthur Kinglake Weston-super-Mare
Sending a photograph of John Locke the philosopher. Trying to raise a fund to provide a memorial for Thomas Young. Would be pleased if JH would subscribe to it. RK's brother is the author of a work on the Crimean War.
ALS 4pp RS:HS 11.85

12959. 1865–9–1 From Peter Martyn Aberdeen
Is now reading JH's article on Light in the May number of *Good Words*. Gives his own physiological and phrenological cause of color blindness and cites an example in the Rev. F. Drew.
ALS 4pp RS:HS 13.23

12960. 1865–9–10 To Adolphe Quetelet Collingwood
Confirms receipt of journals sent to JH by LQ. Calculations of probabilities for shots to hit targets.
Adraft 4pp TxU:H/L-0731; Reel 1089

12961. 1865–9–12 To William Henry Smyth Collingwood
Explains, according to the laws of probability, how the arrows should have been distributed on the archery target at St. Leonard's, and compares that with the actual distribution.
ALS 4pp WT 67430.2

12962. 1865–9–13 From Augustus De Morgan 91 Adelaide Road
W. R. Hamilton and W. H. Smyth are dead. Corresponded with Hamilton for 30 years though only met him once. Has before him the proofs of his own paper on the origin of the + and - signs.
ALS 4pp RS:HS 6.379 S. De Morgan, *Memoir of Augustus De Morgan*, 333

12963. 1865–9–13 From Thomas Maclear Royal Observatory, Cape of Good Hope
Friction between Eastern Province and Western Province in Parliament causes delay in J. S. Parlby case.
ALS 4pp Cape Archives/Maclear Papers File 100

12964. 1865–9–14 To Augustus De Morgan Collingwood
About old friends, alive or dead; comments on a puzzle in probability; returns to JH's old haunts.
C 4pp RS:HS 24.119

12965. 1865–9–[14?] To Lady Helen Bayley Hamilton Collingwood
Expresses condolences on death of William Hamilton, mentioning his work and character.
ALS 4pp TCD 7762–72:1684 Graves, *Hamilton, 3,* 214–5

12966. 1865–9–14 To Adolphe Quetelet Collingwood
Has received packet mentioned in 1865–8–14. Lists works received. Gives lengthy explanation of the error made in the target example. Asks AQ to append a note, stating that JH is aware of the fallacious answer.
ALS 4pp Académie belgique: #43

12967. 1865–9–14 From James Syme St Vincent Place, Glasgow
Discusses several causes of light disturbing aim in rifle-shooting. Hopes JH will be able to enlighten JS and others on this subject.
ALS 1p RS:HS 17.253

12968. 1865–9–16 From Ellen Toynbee St. John's Lodge
Expresses gratitude over JH's sympathy for ET's father's death. Describes his last days.
ALS 4pp RS:HS 17.441

12969. 1865–9–19 From William Edwin Hamilton [Dunsink] Observatory
On behalf of WH's mother, thanks JH for condolences on death of WH's father, W. R. Hamilton. Recalls boyhood visits to Collingwood.
ALS 2pp RS:HS 19.274

12970. 1865–9–21 From Augustus De Morgan 91 Adelaide Road
Has no information about H. T. Colebrooke. Has been corresponding with T. B. [?] Macaulay (1st Baron Macaulay) regarding the marriage of Isaac Newton's niece. Has JH heard the story of George III and his madness.
ALS 3pp RS:HS 6.380 S. De Morgan, *Memoir of Augustus De Morgan*, 333 inc

12971. 1865–9–25 To Augustus De Morgan Collingwood
Comments further on probability [see JH's 1865–9–14]; JH's daughter Maria is to be married; fills letter with much nonsense.
C 3pp RS:HS 24.120

12972. 1865–9–27 From Augustus De Morgan 91 Adelaide Road
Regarding the story of George III and his reason for addressing Parliament in this way. Pleased to hear the stove is to be removed. Great need for a weather theory. Regarding the identities of the Thomas Streets.
ALS 7pp RS:HS 6.381S. De Morgan, *Memoir of Augustus De Morgan*, 334 inc

12973. 1865–9–27 To William Whewell Collingwood
Marriage of daughter Maria set for 12 October. Daughter Amelia will visit Whewells in November. JH's *Iliad* translation nearly complete.
ALS 4pp TC, Camb. Add. Ms.a.207[117] (C: RS:HS 24.121)

12974. 1865 9–29 From William Rutter Dawes Hopefield, Haddenham, Thame
Is pleased to hear of the forthcoming marriage of JH's daughter [Maria Sophia]. Was shocked to hear of the sudden death of his neighbor Admiral W. H. Smyth.
ALS 2pp RS:HS 6.121

12975. 1865–9–30 From Thomas Hill Glasgow
The Directors of the Andersonian University (Glasgow) have made a mistake in not appointing JH's son [Alexander]. What is his son doing now as they would like him to lecture.
ALS 2pp RS:HS 10.192

12976. 1865–9–30 To Henry James Collingwood
Regrets delay in acknowledging his *Notes on the Great Pyramid*; it was due to a temporary loss of the work. Finds HJ's theory on the inclination of the passages very practical. Comments on the accuracy of the pyramid builders.
C 3pp RS:HS 10.276 & 24.122

12977. 1865–10–2 From Edward Sabine 13 Ashley Place
Sends, for JH's signature, request from daughter of the late W. R. Hamilton for continuance of Hamilton's pension.
ALS 1p RS:HS 15.276

12978. 1865–10–3 To Editor of the *Intellectual Observer,* Collingwood
A French civil engineer having proposed in the previous issue of the *IO* the idea of supporting roofs on the principle of the suspension bridge, JH reports that JH came up with this idea in 1836 and provides a sketch of such that JH made in that year.
draft 3pp diag DMC.695A.885.35 *Intellectual Observer, 8* (1866), 275–6

12979. 1865–10–3 From John Lee Hartwell House
Sends a copy of the address of the President of the B.A.A.S. at Birmingham, also an Aylesbury newspaper containing the funeral sermon on Admiral W. H. Smyth. Misses Smyth very much. Hopes JH is well.
ALS 2pp RS:HS 11.185

12980. 1865–10–3 To Adam Sedgwick Collingwood
JH's daughter Maria is getting married. JH has been ill. The *Iliad* translation is almost finished.
ALS 4pp RS:HS 15.453 (C: RS:HS 24.123)

12981. 1865–10–4 From Adolphe Quetelet Brussels
Was in Paris when JH's letter arrived. Will print JH's article. Asks if there are changes to be made. Discusses future project *Physique sociale,* which will include theory of probability applied to ethics and political science.
ALS 4pp (French) RS:HS 14.262

12982. 1865–10–5 From Jacob Tomlin Edengale, Tamworth
Requests JH's opinions on his theories of origins of language. Believes Hebrew is the root of all languages.
ALS 3pp RS:HS 17.439

12983. 1865–10–6 From Hermann Kindt Bampton, Penrith
Requests that JH send a few lines for a friend of HK's who is trying to develop a facsimile reproduction system.
ALS 3pp (German) RS:HS 11.33

12984. 1865–10–6 From Charles Pritchard Freshwater
Asks if JH anticipated the results of W. L. Newman's work on lens curvature.
ALS 4pp RS:HS 14.97

12985. 1865–10–6 From Adam Sedgwick Lowestoft
Thanks AS for his letter. Offers congratulations to newly married Maria Herschel. Will be returning to Cambridge to give his 46th course of lectures.
ALS 4pp RS:HS 15.454

12986. 1865–10–10 From George Biddell Airy Royal Observatory, Greenwich
Concerning JH's memorial to W. R. Hamilton's daughter.

Holiday in Keswick. Comments on JH's papers on optics and force.
ALS 4pp RS:HS 1.283 (C: RGO 6.381.442)

12987. 1865–10–10 From Helen E. A. Hamilton Dollymount near Dublin
Her mother is suffering from inflammation of the eyes so is unable to reply to his kind letter. Her father had a warm regard for his old friends.
ALS 3pp RS:HS 10.149

12988. 1865–10–13 From Thomas Maclear Royal Observatory, Cape of Good Hope
Appeal in J. S. Parlby case fails.
ALS 3pp Cape Archives/Maclear Papers File 100

12989. 1865–10–13 From Robert Francis Suft Royal Mint
Asks JH to send memorandum about position of senior clerks at Royal Mint.
ALS 3pp JH note RS:HS 17.247

12990. 1865–10–14 From George Griffith B.A.A.S., Oxford
Resolution reappointing JH and others to Balloon Committee for further experiments.
L 1p TxU:H/M–0921.4; Reel 1083

12991. 1865–10–14 From George Griffith B.A.A.S., Oxford
Resolution reappointing JH and others to Lunar Committee to continue mapping surface of moon.
L 1p TxU:H/M–0921.3; Reel 1083

12992. 1865–10–14 From George Griffith B.A.A.S., Oxford
Resolution reappointing JH and others to examine K. L. C. Rümker's astronomical observations in southern hemisphere and determine if these should be published.
L 1p TxU:H/M–0921.2; Reel 1083

12993. 1865–10–16 To George Biddell Airy Collingwood
Is working on a problem in polarized light, and finds his result disagreeing with F. J. D. Arago's third law; leads JH to think light cannot be undulatory in nature.
ALS 4pp RGO 6.478.479 (drafts RS:HS 1.284 & 24.12)

12994. 1865–10–18 From George Biddell Airy Royal Observatory, Greenwich
Explaining JH's polarization difficulty [see JH's 1865–10–16].
ALS 7pp RS:HS 1.286 (C: RGO 6.478.481)

12995. 1865–10–18 From George Biddell Airy Royal Observatory, Greenwich
Writes about a long standing, very old, member of the Royal Observatory's Board of Visitors [W. H. Smyth] having been summarily dismissed from the Board.
ALS 3pp RS:HS 1.285 (ACCS: RGO 6.15.75)

12996. 1865–10–18 From Thomas Humphrey Hawkhurst National School
Announces a quarterly meeting of the Hawkhurst National School Board.
ALS 1p Harvard: Houghton AAH 67m–67 (63)

12997. 1865–10–18 From J. B. Kearney Fareham
Congratulations on the marriage of one of JH's daughters [Maria Sophia]. Well remembers the hours he spent with JH. Has had to struggle for a living. Gives some of his theories on music and light that he will not be able to transmit to print.
ALS 8pp RS:HS 11.74

12998. 1865–10–18 To the Manchester Literary and Philosophical Society Collingwood
Suggests a method employing compressed air for cooling the working area in a deep mine.
L *Manchester Literary and Philosophical Society Proceedings, 5* (1866), 11–12

12999. 1865–10–18 From John Eustace Prescott Hawkhurst
Thanks JH for his expressions of goodwill and friendship. Asks for introductions to JH's friends in Cambridge.
ALS 3pp RS:HS 14.53

13000. 1865–10–19 From John Lee Hartwell House
Has seen the announcement of the marriage of one of JH's daughters [Maria Sophia] to Henry Hardcastle. Wishes them well, and would like to see them if they are likely to be in that area at all.
ALS 2pp RS:HS 11.186

13001. 1865–10–21 From George Biddell Airy Royal Observatory, Greenwich
Printed address of the Astronomer Royal to individual members of the Board of Visitors of the Royal Observatory.
DS 3pp RS:HS 1.287

13002. 1865–10–22 To William Whewell Collingwood
Asks WW to welcome Mr. Prescott, a minister taking a position in Cambridge. Comments on a new *Iliad* translation.
ALS 4pp TC, Camb. Add. Ms.a.207[118] (C: RS:HS 24.125)

13003. 1865–10–26 From Joseph Baxendell Lit. & Phil. Soc., Manchester
Thanks for the paper which will be communicated to the next meeting of the Lit and Phil Soc. Regarding Mr. Greave's paper on the ventilation of mines.
ALS 2pp RS:HS 3.309

13004. 1865–10–27 To Arthur James Melhuish Collingwood
JH's son William now lives in Bengal. JH cannot explain why William's membership to Amateur Photographic Association is unpaid. Will notify William in JH's next letter.
C 4pp TxU:H/L-0258; Reel 1054

13005. 1865–10–28 From Bartholomew Price 11 St. Giles St., Oxford
Apologizes for JH not getting copy of vol. 3 of BP's *Infinitesimal Calculus.* Promises to send one. Hopes to add another volume. Asks JH about some problems in probability theory.
ALS 3pp RS:HS 14.57

13006. 1865–10–29 To George Biddell Airy n.p.
Saddened by the poor treatment of W. H. Smyth [see GA's 1865–10–18].
LCC 1p RGO 6.15.77

13007. 1865–10–29 To George Biddell Airy Collingwood
Comments on GA's explanations about polarized light, and discusses GA's idea of the nature of light [see GA's 1865–10–18].
ALS 8pp RGO 6.478.487

13008. 1865–10–29 To John Herschel (son) Collingwood
A 'pep talk' from father to son on the occasion of son John's birthday.
ALS 2pp JHS 6.57

13009. 1865–10–30 From George Biddell Airy Royal Observatory, Greenwich
Further regarding polarized light and its interferences.
ALS 4pp RS:HS 1.288 (C: RGO 6.478.491)

13010. 1865–10–30 To George Biddell Airy n.p.
Passes along a report that someone took a 4-foot telescope to the top of Snowdon [mountain], from where he could see the eight satellites of Saturn and the rings of Neptune.
ALS 1p inc RGO 6.256.299

13011. 1865–10–31 To George Biddell Airy Collingwood
Realizes that GA is quite correct [see GA's 1865–10–30]; JH had neglected a basic principle about interference.
AL 2pp inc RGO 6.478.493

13012. 1865–11–1 To Bartholomew Price Collingwood
Offers solutions to some of the problems in BP's *Infinitesimal Calculus* about which BP asked [see BP's 1865–10–28]. Raises a problem in probability theory regarding archery targets for BP.
ALS 8pp diagram RS:HS 14.58 (C: RS:HS 24.126)

13013. [1865–11] To William Whewell [Collingwood]
Thanks WW for and comments on his suggestions on JH's translation of Book XXIV of Homer's *Iliad.*
ALS 3pp TC, Camb. Add. Ms.a.207[122] (C: RS:HS 24.130)

13014. 1865–11–2 From Joseph Baxendell Lit. & Phil. Soc., Manchester
Is sending the proofs of his recent communication for his correction.
ALS 1p RS:HS 3.310

13015. 1865–11–3 To William Whewell Collingwood
Sends Book XXIV of JH's *Iliad* translation. Pleased with his translation, but worries whether a publisher will take it. JH's daughter Maria enjoyed visiting WW; daughter Amelia arriving soon.
ALS 3pp TC, Camb. Add. Ms.a.207[119] (C: RS:HS 24.127)

13016. 1865–11–3 From William Whewell Trinity Lodge
Delighted by visit from JH's daughter Maria and her new husband. Looks forward to meeting Mr. Prescott [see JH's 1865–10–22].
ALS 3pp RS:HS 18.219

13017. 1865–11–7 From Neil Arnott Cumberland Terrace
Thanking him for his note. Relates events while in India.
ALS 2pp RS:HS 1.410

13018. 1865–11–10 To Adam Sedgwick Collingwood
Asks that AS show JH's son Alexander around Norwich when he arrives there. Reflects on changes in geology. Has finished hexameter *Iliad* translation.
ALS 4pp RS:HS 15.455 (C: RS:HS 24.128)

13019. 1865–11–12 From William Mathews Jr. to Alexander Herschel Carpenter Rd., Edgbaston
Thanks for his paper on the Pyrenees. Has he made any calculations on the depth from which thermal springs derive? The observations made with the spectroscope are interesting. Sent JH two memoirs by Count de S. Robert; would like them back and any comments JH has to make on them.
ALS 3pp RS:HS 12.301

13020. 1865–11–13 From Thomas Maclear Royal Observatory, Cape of Good Hope
J. S. Parlby's sister receives estate.
C 2pp Cape Archives/Maclear Papers File 100

13021. 1865–11–14 From William Whewell Trinity Lodge
Is returning Book XXIV of JH's *Iliad* translation. Asks him about his plans for publication. Has written letter of introduction for JH's daughter Maria.
ALS 3pp RS:HS 18.220

13022. 1865–11–15 To George Gabriel Stokes Collingwood
A letter of introduction on behalf of the Prescott family.
ALS 2pp Cambridge University Library 7656/H708

13023. 1865–11–16 From Joseph Bullar Basset Wood near Southampton
Wished to quote JH's name in connection with one of his theories.
ALS 3pp RS:HS 4.339

13024. 1865–11–18 From John Rate Peakridge, Staffordshire
Is trying to procure a church clock. Asks for advice from JH on choosing a clockmaker and a clock.
ALS 4pp RS:HS 14.470

13025. 1865–11–18 From B. W. S. Vallack St. Budeaux, near Plymouth
Comments on George Witt's work on silvering glass mirrors and on JH's *Telescope.* States problems with his Newtonian telescope. Works on designs for a binocular telescope.
ALS 15pp diag RS:HS 17.415

13026. 1865–11–20 From Isaac Fletcher Tarn Bank near Cockermouth
Is engaged drawing up a memoir of the late Admiral W. H. Smyth for the R.A.S. Would be glad of any of JH's observations on his work as an astronomer and as a member of the Board of Visitors of Greenwich.
ALS 3pp RS:HS 7.273

13027. [1865–11–20 or earlier] From Andrew Scott Waugh 7 Petersham Terrace, Queensgate, South Kensington
Requests JH sign certificate to R.S.L. attesting to Indian historian J. W. Kaye. Recalls 1861 visit with JH at Greenwich Observatory. Hopes JH's sons are doing well in India.
ALS 2pp † RS:HS 18.126

13028. 1865–11–23 To William Stanley Jevons Collingwood
Is grateful for his valuable work on the Coal question. Such a work has been long wanted. Used to have a theory himself that the tides might be utilized, but working it out found it was not economic.
ALS 4pp John Rylands University Library of Manchester (C: RS:HS 10.321 & 24.129) H. A. Jevons, *Letters and Journals of W. Stanley Jevons* (London, 1886), 216 & Black (ed.), *Papers and Correspondence of...Jevons*, 3:77–8

13029. 1865–11–23 From William Mathews Jr. Carpenter Rd., Edgbaston
Is greatly obliged for his two letters and postscript containing his criticism of the atmospheric law suggested by Count de S. Robert. Hopes soon to make a careful examination of the subject.
ALS 2pp RS:HS 12.302

13030. 1865–11–24 From Edward William Brayley London Institution
Sending comparison to Almanac of 1866. Concerning meteorites. Now principal librarian and professor at the

Institution.
ALS 3pp RS:HS 4.233

13031. 1865–11–25 From Joseph Bullar Basset Wood near Southampton
Thanks for referring him to one of his essays.
ALS 3pp RS:HS 4.340

13032. 1865–11–28 From Sir Charles Lyell 53 Harley St.
Sending the results of his comparison of JH's drawing and the actual earth pillars of Ritten and Stalden.
ALS 10pp RS:HS 11.441 Katherine Lyell, *Lyell, 2*, 404

13033. 1865–11–29 From Evan Lewis Accrington
Sending a copy of his own book, *God's Week of Work,* which he hopes JH will accept; would welcome any comments on it.
ALS 3pp RS:HS 11.479

13034. 1865–11–29 From Jacob Tomlin Edengale, Tamworth
Thanks JH for contributions to his book on language.
ALS 3pp RS:HS 17.440

13035. 1865–11–30 To George Biddell Airy Collingwood
Acknowledges receipt of Royal Observatory's astronomical, meteorological, and magnetic observations for 1863.
ALS 1p TxU:H/L-0024; Reel 1054

13036. 1865–11–30 To George Biddell Airy Collingwood
Approves of GA's development of a new type of eyepiece for use with the Royal Observatory's transit circle telescope.
ALS 2pp RGO 6.15.94

13037. 1865–11–30 To George Biddell Airy n.p.
Has noted that Josef Stefan of the Vienna Academy of Sciences holds views on light and its polarization similar to GA's [see GA's 1865–10–30]. Agrees with GA's correction of an essay on Julius Caesar's landing in Britain.
ALS 3pp RGO 6.478.504 & 6.480.585

13038. 1865–12–1 From Isaac Fletcher Tarn Bank near Cockermouth
Is grateful for his copy of the catalogue of nebulae.
ALS 1p RS:HS 7.274

13039. 1865–12–4 To Edward William Brayley Collingwood
Questions EB's use of terms 'ponderable' and 'imponderable' matter in EB's almanac.
C 1p RS:HS 24.131

13040. 1865–12–4 From Francis R. Drew Malvern College
Outlining the state of his own perception in regard to colors and harmony. Where can he find JH's published views on heat?
ALS 4pp RS:HS 6.503

13041. 1865–12–6 From Patrick Francis Robertson 12 Leadenhall St.
Found article with good description of Bank act. PR's ideas regarding issuance of currency notes.
ALS 3pp TxU:H/M–0451; Reel 1093

13042. 1865–12–7 From Henry Wentworth Acland Oxford
Is sending a copy of HA's Harveian Oration for JH's perusal.
ALS 4pp RS:HS 1.384

13043. 1865–12–7 To Sir Charles Lyell Collingwood
Describes exactly the route JH took to get to the point where he made his drawings of the earth pyramids at Botzen.
ALS 3pp APS B.D25.L.15 (C: RS:HS 24.132)

13044. 1865–12–7 To William Whewell Collingwood
Describes JH's search for a publisher of JH's *Iliad* translation. Sending a sample to [Matthew] Arnold. Hopes WW is better; feels very old.
ALS 4pp TC, Camb. Add. Ms.a.207[120] (C: RS:HS 24.133)

13045. 1865–12–8 To Patrick Francis Robertson n.p.
Insists that artificial representations of capital–currency, checks, bills, titles, etc.–are *not* capital. Examines arguments on both sides of issue.
Adraft 2pp damaged TxU:H/L-0339.3; Reel 1054

13046. 1865–12–10 From Wilhelm H. I. Bleek Cape Town
Forwarding a chapter of his comparative grammar of the South African language for JH's comments.
ALS 6pp RS:HS 5.76

13047. 1865–12–11 To 'dear Sir' Collingwood
Just read [?]'s article on [John?] Blackwood. JH's views on effectiveness of [Robert] Peel's act and on relative roles of government and Bank of England in nation's economy.
ACC 5pp TxU:H/L-0734; Reel 1089

13048. 1865–12–11 To William Whewell Collingwood
Asks WW to recommend JH's *Iliad* translation to Macmillan. Discusses possibility of publication of selections.
ALS 4pp TC, Camb. Add. Ms.a.207[121] (C: RS:HS 24.134 & draft: RS:HS 18.221)

13049. 1865–12–13 From George Biddell Airy Royal Observatory, Greenwich
Regarding I. K. Brunel's bridges and JH's letter on suspension roofs. Effects of a thunderstorm on a water barometer.
ALS 4pp RS:HS 1.289

13050. 1865–12–14 To Augustus De Morgan Collingwood
Comments on a mathematics paper AD sent. Inquires about W. S. Jevons. Is still working on translation of the *Iliad.*
C 2pp RS:HS 24.135

13051. 1865–12–14 From George C. Hodgkinson The Lodge, Louth
Has been voted a grant of £30 from the R.S.L. Committee. Results of his recent observations in the Alps. Comments on these and gives indexes for instruments.
ALS 10pp RS:HS 9.421

13052. 1865–12–14 From William Stanley Jevons Owens College Manchester
JH's letter has taken a long time to reach him, which accounts for his delay in replying. Is pleased that JH likes his pamphlet on the coal question. Has been puzzled by the character of Francis Bacon.
ALS 4pp RS:HS 10.322 Black (ed.), *Papers and Correspondence of...Jevons*, 3:80

13053. 1865–12–19 From Augustus De Morgan 91 Adelaide Road
Regarding the identity of J. P. [W. S.] Jevons. Hopes JH will now set to work on the *Odyssey.* Sends four riddles.
ALS 4pp RS:HS 6.382

13054. 1865–12–19 To Edwin Richard Windham Wyndham-Quin Collingwood
Thanks for sending [Charles] Graves's elegy on W. R.

Hamilton, reflecting on Hamilton's great, 'high-souled,' 'enthusiastic' character.
C inc 2pp TCD 7762–72:1696 Graves, *Hamilton, 3,* 226

13055. 1865–12–28 To the *Photographic News* Collingwood
Discusses recent progress in color photography. Stresses that what is most needed is a way of making negatives from which colored positives can be produced. Recounts JH's recent experiments relating to the action of light on solutions of platinum.
L *Photographic News, 10* (Jan. 5, 1866), 5–6 & *American Journal of Photography,* n.s., *8* (Feb. 1, 1866), 342–5

13056. 1865–12–29 From William Stanley Jevons Clynnog, Caernarvon
Encloses a note he has received from Macmillan (publishers). Would be grateful if JH could spare the time to write a short notice of the work as he feels an obligation towards macMillan for publishing his work.
ALS 4pp RS:HS 10.323 Black (ed.), *Papers and Correspondence of...Jevons*, 3:83

13057. 1865–12–30 From [J. Gomez Banley?] 54 Lombardy St.
Thanks for gift of JH's Catalog to library of [?] observatory.
ALS 1p TxU:H/L-0725.1; Reel 1089

13058. 1865–12–30 To E. Sheldon and Co. Collingwood
Acknowledges receiving a shipment.
ALS 1p Harvard: Houghton AAH 67m–67 (62)

13059. [1866] From Henry Coxwell n.p.
Thanks for letter. Regarding the use of oxygen when ballooning. Comments on other aspects of high altitude flight.
ALS 4pp RS:HS 5.369

13060. 1866 To Charles Pritchard Collingwood
Discusses opinions of Charles Darwin's theories. Doubtful that the day has shortened due to tidal forces since the earth's beginning. Alexander Herschel helps in the revision of William Herschel's double star catalogue. Alex is preparing to lecture at Glasgow University.
ALS 7pp RS:HS 14.101 (C: RS:HS 24.182)

13061. 1866–1–4 From James Liddell Bodmin
Sending his observations on the effect of the moon on the earth's atmosphere.
ALS 2pp RS:HS 11.483

13062. 1866–1–7 From Louis Marie Joseph, Comte de Chevigny n.p.
Has received his letter and the sheets of the translations. Comments on this and outlines further plans for translating.
ALS 3pp (French) RS:HS 5.337

13063. 1866–1–8 To William Stanley Jevons Collingwood
Expresses concern that the present rate of use of natural resources and the increase in population may create serious problems in the future.
ALS 3pp John Rylands University Library of Manchester
Black (ed.), *Papers and Correspondence of...Jevons*, 3:83–4

13064. 1866–1–9 From Samuel Birch British Museum
Regarding the structure and surface of the moon as revealed by stereoscopic photography.
ALS 4pp RS:HS 5.68

13065. 1866–1–11 From Henry Coxwell Tottenham
Has been reading JH's letter of 1862 concerning aerial flight and would welcome any hints or suggestions from JH on this subject.
ALS 4pp RS:HS 5.368

13066. 1866–1–12 To [T. P. Anderson] Collingwood
Has no knowledge of transactions affecting trust fund for W. H. B. Hollier, other than what JH noted in previous letters.
ALS 1p TxU:H/L-0757.2; Reel 1089

13067. 1866–1–13 To [Matthew Arnold?] Collingwood
Thanks for remarks on early portions of JH's *Iliad* translation. Hopes MA can go over 2800 particularly 'shaky' lines.
ALS 3pp EUL:Gen1731/22

13068. 1866–1–13 To Samuel Birch Collingwood
Is not convinced by his arguments that the moon's surface consists of snow and glaciers. Quotes examples relating to his experiences at the Cape.
draft 4pp RS:HS 5.69 & 24.136

13069. 1866–1–13 To Elizabeth Colling Collingwood
Has sent his translation of Homer's *Iliad* to the publisher and is having a collection of the best portions sent to EC.
C 2pp RS:HS 24.137

13070. 1866–1–14 To [Editor of the *Photographic News*] [Collingwood]
Dr. [T. L.] Phipson has proposed a solution to a cipher JH published in *Photographic News,* but JH did not receive the issue in which it appeared. Please send that issue to JH.
ALS National Library of Australia, Petherick, MS 760/18/127

13071. 1866–1–14 To William Whewell Collingwood
Macmillan is publishing JH's *Iliad* translation. Wishes to dedicate it to WW. [Matthew] Arnold wrote an encouraging letter on the sample translation sent him. Daughter Isabella appreciates your hospitality.
ALS 4pp TC, Camb. Add. Ms.a.207[123] (C: RS:HS 24.138)

13072. 1866–1–15 From Henry Anthony Jeffreys Parsonage
JH's solution is perfectly satisfactory.
ALS 1p TxU:H/M–0296.1; Reel 1087

13073. 1866–1–15 From Angelo Secchi Rome Observatory
Rome College honors JH as one of the most important figures in science. AS praises the contributions of the Herschel family to science. Will send observations of Rome Observatory for 1865.
ALS 1p (Italian) RS:HS 15.416

13074. 1866–1–16 From Samuel Birch British Museum
Further regarding his own theories on the surface of the moon as revealed by photography.
ALS 4pp RS:HS 5.70

13075. 1866–1–16 To the *Photographic News* Collingwood
Reveals the solution to a cipher given in JH's 1865–12–28 letter to *PN.*
L *Photographic News, 10* (Jan. 19, 1866), 35

13076. 1866–1–16 From George Rümker Hamburg
Sends observations of circumpolar nebulae originally published in the *Astronomische Nachrichten.* Hopes calculating their motion will eventually be possible. Wants JH's opinion on GR's work.
ALS 4pp RS:HS 14.509

13077. 1866–1–17 From John Lubbock 15 Lombard St.
Has seen JH's nephew and will be pleased to sign his certificate. His instinct and reason are against dating Stonehenge as post-Roman.
ALS 2pp RS:HS 11.356

13078. 1866–1–19 From John Couch Adams Cambridge Observatory
Asking JH to send a few encouraging lines to George Rümker, director of the Hamburg Observatory, on his memoir 'Circumpolar nebeln.'
ALS 3pp RS:HS 1.9

13079. 1866–1–19 To Isaac Todhunter Collingwood
Describes the problem of 'probability of three points on a plane forming an acute triangle.' Notes Augustus De Morgan has the same solution to the problem.
C 2pp RS:HS 24.139

13080. 1866–1–25 From B. W. S. Vallack St. Budeaux, near Plymouth

Thanks JH for corrections to the mirrors of his telescope. Says the telescope finally gives 'a satisfactory view of Saturn.' Discusses the lenses for BV's binocular telescope. Thinks photography could benefit from binocular vision.
ALS 11pp diag RS:HS 17.416

13081. 1866–1–26 From William E. H. Lecky Berne
Has just written to Longman's to say how pleased he will be if Miss Kerth translates his book into French. Regrets delay, but has been travelling in Italy.
ALS 2pp RS:HS 11.467

13082. 1866–1–28 From Julia Margaret Cameron Freshwater, I.o.W.
Sending some of her photographs. Has won medals for her efforts but would like JH's opinion so it can be quoted. Has been very busy with photographic work. Her son has gone to Ceylon and she now has no children in the house.
ALS 26pp RS:HS 5.162

13083. 1866–1–29 From James Challis Cambridge
Sending a certificate proposing Professor William Selwyn; would JH add qualifications, sign it and send it on to G. B. Airy.
ALS 3pp RS:HS 5.240

13084. 1866–1–30 From John R. Hind 8 Matson Villas, Richmond
Thinks that E. W. L. Tempel's comet may be connected with that of Biela. Comments on this.
ALS 4pp RS:HS 9.393

13085. [1866?]–2 From John Tyndall [Royal Institution?]
Corrects misinformation about refraction and dispersion of mercury ethyl and mercury methyl.
ALS 1p JH note RS:HS 17.399c

13086. 1866–2–2 From Isaac Todhunter St. John's, Cambridge
Makes suggestions for the solution to the three point probability problem using integrals of infinity.
ALS 4pp RS:HS 17.357

13087. 1866–2–5 To Julia Margaret Cameron Collingwood
Cannot give advice on the artistic merits of her photographs as he is not qualified, but offers suggestions and praise on the mechanics of the art.
draft 3pp RS:HS 5.163 & 24.140

13088. 1866–2–8 From Ernst F. W. Klinkerfues Göttingen
Sending a pamphlet on the quality of refracted light. Hopes JH will communicate it to the R.S.L.
ALS 4pp (German) RS:HS 11.56

13089. 1866–2–12 From B. W. S. Vallack St. Budeaux, near Plymouth
Responds to JH's criticisms of his binocular telescope plans.
ALS 4pp RS:HS 17.417a

13090. 1866–2–16 To Elizabeth Colling Collingwood
Main object in translating Homer was to 'wipe off the stigma cast on English hexameters by such people as Tennyson.'
C 1p RS:HS 24.142

13091. 1866–2–16 To B. W. S. Vallack Collingwood
Suggests weaknesses in BV's plans for a binocular telescope.
ACCS 4pp RS:HS 17.417b (C: RS:HS 24.141)

13092. 1866–2–18 From Julia Margaret Cameron Freshwater, I.o.W.
Is grateful for the praise JH has given her photographic efforts. Intends to do a series of life-sized head photographs. Is anxious about Henry Taylor, who has been overworking in the colonial cause.
ALS 12pp RS:HS 5.164

13093. 1866–2–19 From Charles Pritchard Freshwater
Informs JH, Charles Babbage, and James South that they are the surviving original members of the R.A S. Requests JH write an account of the R.A.S.'s founding. Discusses the evidence of the 'personal will' of God in creation.
ALS 3pp RS:HS 14.98

13094. 1866–2–20 From Goodwin Kilburne Hawkhurst
Announces a committee meeting for a volunteer group.
ALS 1p Harvard: Houghton AAH 67m–67 (63)

13095. 1866–2–22 From John Lubbock High Elms, Farnborough
Thanks for his letter. Agrees that the stones of Stonehenge must have come from the neighborhood. The chips in the barrows must have come from the hewn stones of Stonehenge.
ALS 3pp RS:HS 11.357

13096. 1866–2–23 From William Lloyd Birbeck 7 Vaughton Terrace, Birmingham
Offers to purchase scientific manuscripts for JH at a sale in London.
ALS 1p Harvard: Houghton AAH 67m–67 (62)

13097. 1866–2–23 To [George Gabriel Stokes] n.p.
Reports on paper [*R.S.P.T., 156,* 181–] by [William] Huggins on spectra of certain nebulae with mode of determining their brightness. Doubts some presuppositions, but regards paper as fit for publication in *R.S.P.T.*
ADS 3pp RS RR.6.149

13098. 1866–2–23 From George Thurnell Bramhope near Otley
Praises JH's essay 'The Yard, the Pendulum, and the Metre.' Believes C. P. Smyth's theories concerning the Great Pyramid being a standard of measurement.
ALS 4pp RS:HS 17.437

13099. 1866–2–24 To [Jean Baptiste André Dumas] Collingwood
Thanks for having sent a copy of the works of Antoine Lavoisier [3 vols., 1862–65, ed. by JD].
ALS 2pp (French) Académie des sciences (Paris)

13100. [1866–2–24 or earlier] From Robert Hind Hawkhurst
Asks to visit JH today.
ALS 1p RS:HS 19.157a

13101. 1866–2–24 To Ernst F. W. Klinkerfues n.p.
Has now received his pamphlet. Does not agree with its findings, but will submit it to the R.S.L., though it cannot be printed in *R.S.P.T.* as it has already appeared elsewhere. Comments on some of EK's theories.
draft 4pp RS:HS 11.57

13102. 1866–2–24 To George Gabriel Stokes Collingwood
Forwarding copy of paper by Ernst F. W. Klinkerfues; EK purports to show that movement of a star toward or away from an observer will affect its refraction in an achromatic prism.

ALS 3pp Cambridge University Library 7656/H709 (ACCS: RS:HS 19.157b)

13103. 1866–2–25 From William Selwyn Vine Cottage, Cambridge
Says William Whewell has fallen from his horse and has been injured. Thanks JH for a paper he signed for the R.S.L.
ALS 4pp RS:HS 15.475

13104. 1866–2–26 To William Selwyn Collingwood
Has heard about William Whewell's accident. Hopes he will recover, but knows his advanced age may hinder this. Wishes WS to send news of progress.
C 1p RS:HS 24.143

13105. 1866–2–27 To George Biddell Airy Collingwood
About William Whewell's accident, and the ideas of E. F. W. Klinkerfues on the behavior of light due to the motion of a star source.
ALS 4pp RGO 6.381.444

13106. 1866–2–27 From William Selwyn Vine Cottage, Cambridge
William Whewell has gotten up to walk several times. The left side of his body and face is still 'not quite right.'
ALS 4pp RS:HS 15.476

13107. 1866–2–27 From George Gabriel Stokes Lensfield Cottage, Cambridge
Comments against the paper of E. F. W. Klinkerfues about observations of dispersed star light [see JH's 1866–2–24].
ALS 12pp RS:HS 17.62

13108. 1866–2–28 To George Gabriel Stokes Collingwood
Agrees with GS's assessment of E. F. W. Klinkerfues's paper [see GS's 1866–2–27].
ALS 4pp Cambridge University Library 7656/H710

13109. 1866–3–1 From George Biddell Airy Royal Observatory, Greenwich
Has sent the 1854 Greenwich Magnetical and Meteorological Observations. Illness of the Master of Trinity College, Cambridge (William Whewell). Undulating theory of stars in motion.
ALS 3pp RS:HS 1.290

13110. 1866–3–1 From George Biddell Airy n.p.
Comments on the state of William Whewell's health, and about the theories of E. F. W. Klinkerfues [see JH's 1866–2–27].
ACCS 3pp RGO 6.381.446

13111. 1866–3–1 From Helen E. A. Hamilton 15 Blessington St., Dublin
Expressing the thanks of her mother for signing the memorial.
ALS 1p RS:HS 10.150

13112. 1866–3 To Robert Harley Collingwood
Comments on analytic symbolism in mathematics.
C 2pp RS:HS 24.149

13113. 1866–3–1 From William Selwyn Vine Cottage, Cambridge
William Whewell slightly improved. Dr. [George] Humphry has been elected professor of Anatomy. Is indebted to JH because of the paper he sent to the R.S.L.
ALS 3pp RS:HS 15.477

13114. 1866–3–2 From William Selwyn Vine Cottage, Cambridge
William Whewell seems to have permanent damage to his brain. Whewell has an article in *MacMillan's Magazine* regarding Auguste Comte's philosophy.
ALS 3pp RS:HS 15.478

13115. 1866–3–2 From George Smith 45 Pall Mall
Asks JH to recommend a reviewer for a popular book on astronomy.
ALS 2pp RS:HS 16.195

13116. 1866–3–2 From George Gabriel Stokes Lensfield Cottage, Cambridge
Further comments on Ernst Klinkerfues's paper, in response to JH's 1866–2–28.
ALS 4pp RS:HS 17.63

13117. 1866–3–3 From J. Fitzmaurice Army & Navy Club
A relation of JF has a handsome portrait of a William Herschel dressed in Jewish costume, to dispose of and wonders if JH would be interested in it. [Note by Herschel that this is Rabbi Herschel and no relation to the family.]
ALS 2pp RS:HS 7.370

13118. 1866–3–3 From William Selwyn Vine Cottage, Cambridge
William Whewell's symptoms are worse, but his mind is still active.
ALS 4pp RS:HS 15.479

13119. 1866–3–3 To [George Smith] Collingwood
Is trying to assist in finding someone to review a book [*The Heavens*] by [Amédée] Guillemin.
ALS 2pp APS B.H435p.24

13120. 1866–3–3 To John Tyndall n.p.
Describes experiment involving viewing spectrum of a light flash as it advances and recedes. [Marked 'not sent on further consideration.']
ALS? 3pp RS:HS 17.399b (C: 24.144)

13121. 1866–3–4 From William Selwyn Vine Cottage, Cambridge
William Whewell's nieces have little hope for his recovery. Whewell is anxious about his article on Grote's Plato for *MacMillan's Magazine.*
ALS 4pp RS:HS 15.480

13122. 1866–3–4 To George Gabriel Stokes Collingwood
Comments further on Ernst Klinkerfues's work [see JH's 1866–2–28]; on the behavior of light emitted by an electric spark.
ALS 4pp Cambridge University Library 7656/H711 (draft RS:HS 17.64 & C: RS:HS 24.145)

13123. 1866–3–5 From James Glaisher Blackheath
Editor of the *Leisure Hour* would like an article on the Correlation of the Physical forces. Would JH be prepared to write such an article?
ALS 4pp RS:HS 8.135

13124. [1866–3–5?] From William Selwyn n.p.
Detached postscript that refers to the behavior of William Whewell before his death.
ALS 1p inc RS:HS 15.516

13125. 1866–3–5 To William Selwyn Collingwood
Thinks William Whewell's excitement may be a 'precursor to exhaustion.' The hope of recovery is very slight. JH is

interested in Whewell's article on Plato.
C 1p RS:HS 24.146

13126. 1866–3–5 From William Selwyn Vine Cottage, Cambridge
William Whewell is still alive, but struggling. Dr. [Gilbert] French has died suddenly.
ALS 4pp RS:HS 15.481

13127. 1866–3–6 From John Eustace Prescott Cambridge
The Master of Trinity [William Whewell] died this afternoon. Regrets that his first letter from Cambridge brings such news.
ALS 3pp RS:HS 14.54

13128. 1866–3–6 From William Selwyn Vine Cottage, Cambridge
Relays message from Dr. [George] Humphry. William Whewell is conscious but knows he is dying.
ALS 2pp RS:HS 15.482

13129. 1866–3–6 From William Selwyn Cambridge
William Whewell died peacefully that afternoon. The place of burial has not been determined. WS will speak about him on Friday. Takes comfort in his writings.
ALS 5pp RS:HS 15.483

13130. 1866–3–6 From George Smith 45 Pall Mall
Thanks JH for recommendation [see GS's 1866–3–2].
ALS 2pp RS:HS 16.196

13131. 1866–3–6 From Francis Diedrich Wackerbarth Uppsala
Bemoans lack of complete collection in English of William Herschel's works. Questions quality of J. W. A. Pfaff's 1826 German work on WH, whose English nomenclature for stellar astronomy is not yet adopted in other languages. Suggests English government or R.A.S. underwrite cost of publication.
ALS 2pp RS:HS 18.1

13132. [1866]–3–7 From Katherine A. S. Gibson Trinity Lodge, Cambridge
Announcing the death of William Whewell and plans for his funeral. Hopes JH will not come in his present state of health.
ALS 4pp RS:HS 8.107

13133. 1866–3–7 From Robert Harley The Manse, Brighouse
Is writing an article on the life and work of George Boole and would like JH's advice on one of D. F. Gregory's references.
ALS 3pp RS:HS 9.239

13134. 1866–3–7 From William Stanley Jevons 9 Birch Grove, Rusholme
Intends to apply for a professorship at Owens College and would like to use JH's letter as a testimonial. Would also like to send a copy of this letter to the Chancellor of the Exchequer.
ALS 2pp RS:HS 10.324 Black (ed.), *Papers and Correspondence of...Jevons*, 3:85

13135. 1866–3–7 To William Selwyn Collingwood
Thanks WS for communicating the events of William Whewell's last days.
C 1p RS:HS 24.147

13136. 1866–3–8 From Edward William Brayley London Institution
Has answered Smith's request by letter this evening. Thanks for the specimen of the meteorite.
ALS 2pp RS:HS 4.234

13137. 1866–3–8 To William Stanley Jevons Collingwood
Agrees to WJ's request to make use of an earlier letter of JH's [see WJ's 1866–3–7].
ALS 2pp John Rylands University Library of Manchester (C: RS:HS 24.148) Black (ed.), *Papers and Correspondence of...Jevons*, 3:85–6

13138. 1866–3–8 From Ernst F. W. Klinkerfues Göttingen
Is grateful for his friendly letter. Gives a fuller explanation of his own theories on light. Would be glad if he would show this letter to Professor G. G. Stokes.
ALS 8pp RS:HS 11.58

13139. 1866–3–8 From William Selwyn Cambridge
William Whewell will be buried in Trinity College Chapel. The thought of speaking about Whewell depresses WS.
ALS 3pp RS:HS 15.484

13140. 1866–3–13 From Robert Harley The Manse, Brighouse
Has not had time to finish the George Boole paper, but the biographical part is in print. Will send him the complete article when it is finished.
ALS 2pp RS:HS 9.240

13141. [1866]–3–13 From Wilfred Heeley to William James Herschel Calcutta
Is Secretary of a committee for advising on Indian weights and measures and would welcome any reports and pamphlets by WH's father [note on verso stating the letter has been passed to his father].
ALS 3pp RS:HS 9.275

13142. 1866–3–15 From Thomas Maclear to Margaret Brodie Herschel Cape of Good Hope
[Samuel] Parlby papers are in the hands of [John] Steuart, who will hand them over to the new Attorney General when he arrives. Several epidemics have visited the Observatory and William Mann is now convalescing in Natal. [James] Adamson has caused unpleasantness with the reduction of the meteorological journals. Railroad building has stopped at the Cape. Difficulties with the electric telegraph.
ALS 5pp RS:HS 12.172

13143. 1866–3–17 To Herbert Spencer Collingwood
Will subscribe to HS's work as requested [see HS's 1860–3–5], but dissents 'from *very large* portions' of HS's views. Especially objects to HS's adoption of the 'Shibboleth of the Hegel & Schelling School of German Philosophy—"the Absolute."' [This letter misdated; correct date: 1860–3–17.]
ALS 4pp Univ. of London Library (CC: RS:HS 16.489 & C: 23.298)

13144. 1866–3–21 To Francis Diedrich Wackerbarth Collingwood
FW's suggestion to collect William Herschel's published works into one volume was considered earlier, but publishers found it not profitable. All works except two are in *R.S.P.T.*, 1780–1818. [Letter continues 22 Mar.:] JH almost completed identifying and assembling all of WH's double stars and micrometrical measures. Thanks for calling attention to 'unpublished series of 400 'Guages [sic] ('hausters')' noticed by [Maximilian Weisse]. JH was not aware of these. [Letter continues 23 Mar.:] Does not approve of FW's suggestion to request subscriptions to fund collection of WH's works. [Note by copyist: List of

differences between original letter and copy.] [Unknown annotation: 'Original in No. 2 Corresp. Tin box B.8.'] C 2pp, note (by ?) 1p TxU:H/L-0487; Reel 1055 (C: RS:HS 25.15.27)

13145. 1866–3–22 From Thomas Hill Merrylea
Has his note of the 19th and also saw JH's son [Alexander] for a few minutes yesterday. Agrees that JH's son should not become a candidate until he has given the lectures. Hopes to see him at Merrylea.
ALS 2pp RS:HS 10.193

13146. 1866–3–23 From Edmund B. Denison 33 Queen Anne St.
Sending a small book he has written on astronomy. Owes all his knowledge to JH's *Outlines Astr.* Comments on the changes that could be made should any new edition of the *Outlines* be contemplated.
ALS 4pp RS:HS 6.489

13147. 1866–3–25 From Rudolf Wolf Zürich
Is sending JH a copy of some of RW's writings on sunspots. RW has been considering writing about the work of William and Caroline Herschel before the discovery of Uranus.
ALS 2pp (German) RS:HS 19.299

13148. 1866–3–26 To François N. M. Moigno Collingwood
Corrections to reports in 12 Mar. 1866 issue of *Les Mondes* regarding ecclesiastical titles in England, William Whewell's career, and W. T. Brande's role at Royal Mint. Also, English do *not* eat cats; 'catsup' is made of *vegetables.* Nor was JH's daughter [Maria Sophia] ever acquainted with late [Ludovico] Calandrelli.
ACCS 3pp TxU:H/L-0245; Reel 1054 (Cdraft: RS:HS 25.15.22)

13149. 1866–3–26 From George Gabriel Stokes London
Comments on a letter GS received from Ernst Klinkerfues, which leaves GS convinced of the basic correctness of the views of GS and JH.
ALS 6pp RS:HS 17.65

13150. 1866–3–29 From William Selwyn Foxton Royston
Compiling three year series of helioautographs. Wants suggestions for noting the position of planets supposedly influencing sunspots. Discusses a passage from one of Galileo's letters concerning the influence of the planets on the sun's face.
ALS 4pp RS:HS 15.485

13151. 1866–4–1 From Augustus De Morgan 91 Adelaide Road
Has not heard from him for a long time; hopes his health is still good. Was sorry to hear of the death of Whewell. Sends a proof that has been puzzling him for a long time, and a French riddle.
ALS 4pp RS:HS 6.383

13152. 1866–4–2 To William Selwyn Collingwood
Discusses how to denote the positions of the celestial bodies in solar photographs. Advocates a system that shows differences of heliocentric longitude on the sides of the sun.
C 2pp RS:HS 24.150

13153. 1866–4–4 From William Selwyn Foxton Royston
Limited space precludes use of JH's suggestions for the photographic display. Will continue the series of solar autographs up to 11 years if he can.
ALS 4pp RS:HS 15.486

13154. 1866–4–6 To O. A. Hodgson Collingwood
As executor for estate of late [Thomas] Baldwin, JH received OH's certificate and order for payment to Mrs. Baldwin. Concern for Mrs. Baldwin's advanced age and her lax method of writing checks. Disposition of estate after Mrs. Baldwin dies.
ACCS 4pp TxU:H/L-0195; Reel 1054

13155. 1866–4–6 To William Selwyn Collingwood
Helioautographs are beautiful. Advises keeping original plan for presentation. Says 'Clarke's' [Harvey Carlisle's] article on William Whewell in *MacMillan's* is satisfactory. Describes an 'absurd paragraph' regarding Whewell in François Moigno's *Mondes.*
C 1p RS:HS 24.151

13156. 1866–4–7 From M[ichael] Connal Mauritius
Regarding his own theory of the determination of heights by the barometer, which has been modified by reading JH's pamphlet on Meteorology.
ALS 7pp RS:HS 5.360

13157. 1866–4–10 From Alfred Brothers Manchester
Would like to publish his letter on the hyposulfite fixing process.
ALS 2pp RS:HS 4.300

13158. 1866–4–10 To Augustus De Morgan Collingwood
Comments on JH's health, the death of William Whewell, and extended comments on a mathematical proof provided by AD [see AD's 1866–4–1].
C 3pp RS:HS 24.152

13159. 1866–4–11 From Augustus De Morgan 91 Adelaide Road
Considers JH's account of his own health tolerably good. The *Odyssey* will set him up entirely. Sends proof of one of his theorems. New professorship of International Law. Should jot down biographical details of William Whewell for the R.S.L.
ALS 4pp RS:HS 6.384

13160. 1866–4–16 To Alexander Stewart Herschel Collingwood
Mostly talks about having been offered some money to pay for the reduction of William Herschel's [JH's father] double star observations in preparation for publication.
AL 2pp inc JHS 1.123

13161. 1866–4–16 From John Wrottesley 18 Chapel St., Park Lane
Criticizes Charles Rümker's catalogue of southern stars.
ALS 4pp RGO 6.412.155

13162. 1866–4–17 From Francis Charles Knowles Belvedere House, Ryde
Would like JH's comments on one of his own theories regarding screw propulsion of ships. Sent it to the Admiralty, who rejected it. Can JH's son recommend a good resume of the physical and chemical composition of aerolithes?
ALS 12pp RS:HS 11.68

13163. 1866–4–17 From Charles Pritchard Freshwater
R.A.S. wants JH to direct the preparation of William Herschel's catalogue of double stars for publication. CP suggests Alexander Herschel prepare the catalogue with JH's supervision. G. B. Airy observes the 'willow leaves.'
ALS 4pp RS:HS 14.99 Ada Pritchard, *Charles Pritchard: Memoirs of His Life,* 254–5

13164. 1866–4–18 From John R. Hind 8 Matson Villas, Richmond
Regrets his reply [to Hind's 1866–1–30?] is late owing to absence on an excursion. Has not heard that either comet has been perceived. Has not seen Otto Struve's remarks.
ALS 2pp RS:HS 9.394

13165. 1866–4–20 From George Biddell Airy Royal Observatory, Greenwich
Further regarding the possible introduction of the metrical system for weights and measures.
ALS 2pp RS:HS 1.291

13166. 1866–4–20 To George Biddell Airy n.p.
Inquires of GA as to the current status of parliamentary moves to introduce the metric system into Britain [see GA's 1864–9–28].
LCC 1p RGO 6.359.240

13167. 1866–4–20 To George Biddell Airy n.p.
Comments on William Whewell's replacement on the Royal Observatory's Board of Visitors.
LCC 1p RGO 6.15.110

13168. 1866–4–20 To George Biddell Airy Collingwood
Comments on the quality of the star catalogue of Charles Rümker [see John Wrottesley's 1866–4–16].
ALS 2pp RGO 6.412.158

13169. 1866–4–21 From George Biddell Airy Royal Observatory, Greenwich
Regarding JH's remarks on Robert Willis. Comments on the usefulness of W. H. Miller of Cambridge.
ALS 2pp RS:HS 1.293 (C: RGO 6.15.111)

13170. 1866–4–21 From George Biddell Airy Royal Observatory, Greenwich
Regarding the measurement of the heat and light of the sun by means of the movements of a comet.
ALS 2pp RS:HS 1.292

13171. 1866–4–21 From Warren de La Rue 110 Bunhill Row
Has sent on the correspondence about K. L. C. Rümker's Southern Catalogue. Has recorded his vote in favor of the £150 for the reduction. Proposes sending his eldest son to Paris to study French and Art.
ALS 3pp RS:HS 6.169

13172. 1866–4–23 To Francis Charles Knowles Collingwood
In response to FK's 1866–4–17, JH says that he knows nothing about screw propulsion in ships and so cannot comment on FK's paper.
C 2pp RS:HS 24.154

13173. 1866–4–23 To Charles Pritchard Collingwood
Thanks CP and R.A.S. Council for funds to help with JH's catalogue of William Herschel's double stars. Alexander Herschel will begin work soon on the project.
L Ada Pritchard, *Charles Pritchard: Memoirs of His Life,* 255–6

13174. 1866–4–25 To George Biddell Airy Collingwood
Has heard that the Indian government may adopt the metric system; wants to gather, and send to India, information against metrication; JH still wants to bring the earth's axis length into the measuring system.
ALS 4pp RGO 6.359.243

13175. 1866–4–25 To Margaret Brodie Herschel Collingwood
Responds to MH's report on the health of [?]; JH and family went out nightingale hunting last night. [Also enclosed is a short letter from daughter Amelia to MH.]
ALS 3pp encl JHS 5.36

13176. 1866–4–27 From George Biddell Airy Royal Observatory, Greenwich
Regarding the documents for the Standards Commission.
ALS 4pp RS:HS 1.294

13177. 1866–4–29 To George Biddell Airy Collingwood
Avails himself of GA's offer of assistance to obtain information [see GA's 1866–4–27].
ALS 3pp RGO 6.359.247

13178. 1866–4–29 To [?] Collingwood
Please thank Dr. Gardner for his 'Calendar for the Correction of Dates.'
ALS 1p BostonPubLib Ms. Eng. 204

13179. 1866–4–30 From M. S. Bell Gt. Cumberland Place
Is grateful for MB's *Ode to Sensibility.* Regarding the Lords of the Level.
ALS 3pp RS:HS 5.46

13180. [1866–5 or later] From Francis Diedrich Wackerbarth [Uppsala]
Sends children's book to JH. Believes JH, who reads German, will understand Swedish.
ALS 1p RS:HS 18.3

13181. 1866–5–2 From Francis Diedrich Wackerbarth Uppsala
Grateful to learn that [collection of William Herschel's works] is being addressed.
ALS 1p RS:HS 18.2

13182. 1866–5–3 From Emily Hardcastle n.p.
Found the enclosed in a bookseller's shop, found it amusing, and sends it to JH to explain the mystery.
ALS 3pp RS:HS 10.156

13183. 1866–5–4 From George Biddell Airy Royal Observatory, Greenwich
Is sending a packet of seven papers on metrical and decimal systems.
ALS 1p RS:HS 1.295

13184. 1866–5–4 To Emily Hardcastle Collingwood
Comments on EH's 'Magic pictures', which JH said he produced and described in a paper twenty–six years earlier; JH is however unable to explain the process chemically.
C 1p RS:HS 24.154

13185. 1866–5–11 From Agnes Zimmermann to Caroline Emily Mary Herschel 13 Dorchester Place, Blandford Square
Sends information requested by JH about musical keys.
ALS 2pp TxU:H/M–1074; Reel 1060

13186. 1866–5–12 From G. Wharton Simpson Canonbury
Wishes to ascertain certain facts for article on history and aspects of photography. Especially, asks where to find first record of JH and [Louis] Daguerre using hyposulfite of soda for fixing photographs.
ALS 3pp RS:HS 17.210

13187. 1866–5–12 From John Wrottesley 18 Chapel St., Park Lane
Need to meet to discuss the question of publishing Charles Rümker's star catalogue [see JH's 1866–4–20].
ALS 2pp RGO 6.412.177

13188. 1866–5–14 From J. Melvill India Office
The packet for Wilfred Heely has just been delivered and JM will arrange for it to be forwarded to Calcutta by the Mail steamer.
ALS 1p RS:HS 12.320

13189. 1866–5–14 From G. Wharton Simpson Canonbury
Thanks JH for helping to clarify origins of Louis Daguerre's use of hyposulfite of soda for fixing photographs. Provides further information on photography, particular solar microscope photography.
ALS 4pp RS:HS 17.211

13190. 1866–5–15 From [Samuel Fisher?] Merchant Taylor's Hall, E.C.
[Form letter] Invitation to company dinner at Merchant Taylor's Company on 11 June to meet Duke of Cambridge [George William Frederick Charles].
LS 1p RAS:JH Archive 14/1.199 verso; Reel 10

13191. 1866–5–16 From Alexander Henning 8 Orsett Place
Will give him great pleasure if he can assist J. Parkhurst in any way, but regulations in Messrs. Green's ships will preclude him from their service. Met Margaret Brodie Herschel at the Stewarts a few days ago.
ALS 2pp RS:HS 9.306

13192. 1866–5–16 From M. Hume 2 Beak St.
Sends patterns for a suit. Cannot obtain the exact pattern JH wanted.
ALS 1p Harvard: Houghton AAH 67m–67 (63)

13193. 1866–5–16 From John Wrottesley 18 Chapel St, Park Lane
Sends documents requested by G. B. Airy. Hopes JH will attend visitation to hear minority view of Warren de La Rue and James Glaisher.
ALS 2pp RS:HS 18.306

13194. 1866–5–18 From Sir William Huggins Upper Tulse Hill
Giving observations of a new star observed by him and some of his correspondents.
AL 4pp RS:HS 10.40

13195. 1866–5–19 To Sir William Huggins Collingwood
Comments on new star reported by WH [see WH's 1866–5–18]; includes diagram of that portion of the sky.
C 2pp RS:HS 24.156

13196. 1866–5–19 To the *Photographic News* Collingwood
Quotes from an 1840 publication by JH to show JH's priority over a recent request for a patent for 'Magic pictures.' Recounts some recent photographic experiments by JH on the action of light on platinum.
L *Photographic News, 10* (May 25, 1866), 244

13197. 1866–5–20 From John F. South Kensington
Tells JH that JS's brother, James, is very ill and infirm, almost blind and with his memory failing, but wishing to be reconciled to those with whom he was in conflict in earlier times. Charles Babbage has visited. Would JH come or write?
ALS 4pp RS:HS B27.38

13198. 1866–5–21 To George Biddell Airy Collingwood
Comments on Charles Rümker's observations and agrees to a meeting [see John Wrottesley's 1866–5–12].
ALS 3pp RGO 6.412.186

13199. 1866–5–21 From Sir William Huggins Upper Tulse Hill
Much obliged for his observations of June 1842. Probably the same star that is now being observed. Gives observations sent him by G. B. Airy.
ALS 3pp RS:HS 10.41

13200. 1866–5–21 To John Wrottesley Collingwood
Mostly about arranging a meeting with both JW and G. B. Airy at the Greenwich Visitation.
C 1p RS:HS 25.12.8

13201. 1866–5–23 From John Wrottesley 10 Eaton Terrace
Unable to meet in Greenwich at time JH proposed. JW suggested later hour to G. B. Airy.
ALS 2pp RS:HS 18.307

13202. 1866–5–24 From George Biddell Airy Royal Observatory, Greenwich
Regarding the visit by E. A. S. Seymour (12th Duke of Somerset) to the Board of Visitors.
ALS 1p RS:HS 1.296 (C: RGO 6.412.192)

13203. 1866–5–24 From M. Hume 2 Beak St.
Sends suit that JH ordered.
ALS 1p Harvard: Houghton AAH 67m–67 (63)

13204. 1866–5–25 To George Biddell Airy Collingwood
Not well enough to attend Visitation Day at the Royal Observatory, but would be pleased if GA would invite JH's son John.
ALS 2pp RGO 6.15.605

13205. 1866–5–25 To George Biddell Airy Collingwood
Responds to meeting arrangements [see GA's 1866–5–24].
ALS 1p RGO 6.412.196

13206. 1866–5–26 From George Biddell Airy Royal Observatory, Greenwich
Wants a paper for the R.A.S. on the Corona star [T Coronae].
ALS 2pp RS:HS 1.299

13207. 1866–5–26 From George Biddell Airy Royal Observatory, Greenwich
Regarding visitation card for the Royal Observatory for JH's son John.
ALS 1p RS:HS 1.298 (C: RGO 6.15.607)

13208. 1866–5–26 From George Biddell Airy Royal Observatory, Greenwich
New date for proposed meeting on the Charles [=K. L. C.] Rümker business.
ALS 1p RS:HS 1.297 (C: RGO 6.412.198)

13209. 1866–5–26 From John Tyndall Royal Institution
Comments on excellence of the lecture at the Royal Institution of JH's son [Alexander].
ALS 2pp RS:HS 17.400 (C: RI 539)

13210. 1866–5–28 To the R.A.S. Collingwood
Reports JH's 1842 observation of a star near Epsilon Coronae, which may be tne new variable star [T Coronae].

L *Royal Astronomical Society Monthly Notices, 26* (1866), 299–300

13211. 1866–5–28 To John Tyndall Collingwood
Thanks for note [see JT's 1866–5–26] on improvement of the lecturing style of JH's son [Alexander]. JH's son has studied lecture's subject deeply and loves science; JH hopes will be physics professor at a 'considerable institution.'
C 1p RS:HS 24.157 (C: RI 540)

13212. 1866–5–29 From Edward James Stone Royal Observatory, Greenwich
In response to a letter from JH [*R.A.S. Monthly Notices, 26* (1866), 299–300] giving eye estimations of magnitudes of stars about Epsilon Coronae, suggests that the star JH observed may be 'the new variable' [T Coronae]. Reports Greenwich observations of this new variable star.
ALS 4pp diag RS:HS 17.240

13213. 1866–5–30 To James Croll Collingwood
Thanks JC for his papers on 'ice-cap and eccentricities [of planets].' Comments on the papers, noting the interactions between geology and astronomy.
L J. C. Irons, *Autobiographical Sketch of James Croll* (London, 1896), 128

13214. 1866–5–31 From George Biddell Airy Royal Observatory, Greenwich
Fixing a new time for the meeting to discuss Charles Rümker.
ALS 1p RS:HS 1.300 (C: RGO 6.412.208)

13215. 1866–5–31 To James South Collingwood
Writes a very kind letter, encouraging JS to remember the good days when JS and JH worked together in astronomy, and to forget less happy events. [This letter was enclosed in JH's 1865–5–31 to John F. South.]
ACCS 3pp RS:HS B27.39

13216. 1866–5–31 To John F. South Collingwood
As JH rarely comes to London. and is himself in poor health, it is unlikely that JH can visit James South. JH wants, however, also to wipe bitter recollections from his mind.
ACCS 3pp RS:HS B27.41

13217. 1866–6–3 From John F. South Kensington
Thanks JH for his kind letter [see JH's 1866–5–31]. James South has heard that his godson, William James Herschel, has returned from India. Could JH visit? [A greeting is appended by James South.]
ALS 3pp RS:HS B27.40

13218. 1866–6–4 From Frederick Levick Blaina, Newport
Has seen JH's letter regarding the use of compressed air for working machinery in deep mines. Encloses two papers on the subject by one of his sons. Comments on the use FL has made of compressed air in his mines.
ALS 3pp RS:HS 11.477

13219. 1866–6–12 From John Wrottesley 10 Eaton Terrace
Review enclosed document, then send it to Warren de La Rue, James Glaisher, and back to JW, who will forward it to B.A.A.S.
ALS 2pp RS:HS 18.308

13220. 1866–6–16 From William Radcliffe Birt B.A.A.S. Lunar Committee
Announces meeting of B.A.A.S. Lunar Committee.
ALS 1p Harvard: Houghton AAH 67m–67 (63)

13221. 1866–6–18 To ? Collingwood
[Writing to a correspondent in Manchester], apologizes that because of illness JH unable to take chair at 'your College Meeting.'
ALS 3pp Adler Planetarium

13222. 1866–6–22 From Wilfred Heeley Calcutta
Is grateful for the pamphlets on Standards and will inform him of any decisions made by the Committee. Comments on the various national standards and systems used.
ALS 4pp RS:HS 9.276

13223. 1866–6–22 From James Nasmyth to Alexander Herschel Penshurst
Sends a sketch and details of the great meteor seen on 20 June. Hopes that JH is enjoying the fine weather.
ALS 4pp RS:HS 13.87

13224. 1866–6–23 From William Stanley Jevons 9 Birch Grove, Rusholme
Was flattered by JH's message and regrets JH was unable to be present at the Prize-giving. Thinks he may like to see the enclosed extract from *Engineering* on the tides. Thinks good will come from the Royal Commission on Coal.
ALS 6pp RS:HS 10.325 Black (ed.), *Papers and Correspondence of...Jevons*, 3:122–3

13225. 1866–6–30 To William Stanley Jevons Collingwood
Many thanks for the Preface to the second edition of WJ's pamphlet, and for the extract on tides. Comments on this and the use of tidal power. Was disappointed at not being able to visit Manchester but suffered a severe fall.
ALS 3pp John Rylands University Library of Manchester (draft: RS:HS 10.326) Black (ed.), *Papers and Correspondence of...Jevons*, 3:123–4

13226. 1866–7–13 To George Biddell Airy Collingwood
A note to accompany forwarding of papers and letters relating to Charles Rümker's observations [see GA's 1866–5–31].
ALS 1p RGO 6.412.220

13227. 1866–7–16 From Wilfred Heeley Calcutta
Thanks for the parliamentary papers, which the Committee have found very useful. Will soon forward his report, which recommends the metric system.
ALS 1p RS:HS 9.277

13228. 1866–7–19 To Robert Harley Collingwood
Expresses thanks for RH, in a paper on George Boole in the *British Quarterly Review,* giving JH credit for having published 'my introduction of the mode of separation of the symbol of differentiation from the differentiated function and executing on it operations analogous to those of common algebra.'
C 1p RS:HS 24.158

13229. 1866–7–28 To Wilfred Heeley Collingwood
Is sending WH papers and material on weights and measures. Proceeds to argue strongly against any introduction of the metric system into India.
ALS 4pp DMC.695A.885.8 (C: RS:HS 9.278 & 24.159)

13230. 1866–8–2 To [?] Collingwood
Declines to be involved in another [scientific ?] undertaking, as JH's health is too precarious.
ALS 2pp WT 67694

13231. 1866–8–4 To Samuel Newington Collingwood
Comments that the sound of the waves on the seashore seems

louder as the rhythm matches the pulse of the heart.
C 1p RS:HS 24.160

13232. 1866–8–4 From Frederick Rixon Clapham
Asks JH to give 2 or 3 lectures on astronomy in the upcoming course at his school. JH declines the offer in an attached draft of a response.
ALS 2pp † RS:HS 14.492

13233. 1866–8–11 To Edward Sabine Collingwood
Opposes sending expensive scientific equipment for J. B. N. Hennessey to use in India; suggests instead a variety of useful observations needing to be made, using inexpensive equipment. Discusses son's [Alexander] spectroscope observations of meteors.
ALS 4pp Cambridge University Library 7656/RS534 (draft: RS:HS 15.277 & C: RS:HS 24.161)

13234. 1866–8–12 To Augustus De Morgan Collingwood
Comments on JH's health, and on items that are to go into JH's *Familiar Lectures*.
C 3pp RS:HS 24.162

13235. 1866–8–14 To James Croll Collingwood
Thanks JC for his paper on the 'Influence of the Tidal Wave on the Moon's Motion.' Expresses disagreement with a number of aspects of JC's argument.
L J. C. Irons, *Autobiographical Sketch of James Croll* (London, 1896), 236–9

13236. 1866–8–15 To Margaret Brodie Herschel Collingwood
Pleased to hear of the birth of a daughter to their daughter Maria [Hardcastle]; JH has been breathing carbolic acid fumes for the treatment of bronchitis.
ALS 4pp JHS 4.15

13237. 1866–8–17 From Augustus De Morgan 91 Adelaide Road
Thanks for letter. Has just turned 60 himself. Does not like the sound of the bronchitis. Wife thanks him for the pamphlet on Force. It is a dreadful puzzle. Philosophers must deny the existence of things they do not understand. Family members are at Walmer. Gives examples of mistakes being copied. Encloses two riddles.
ALS 4pp RS:HS 6.385

13238. 1866–8–17 From William Selwyn College, Ely
Has devised a method of exhibiting the photographs of the configuration of the planets. WS's sister, Mrs. George Peacock, will marry the new Master of Trinity [William H. Thompson].
ALS 4pp RS:HS 15.487

13239. 1866–8–18 From James Croll Glasgow
Thanks for letter and criticism of his own paper. Seems to have misunderstood most of his points. Explains some of his theories on the movement of the earth and tides.
ALS 8pp RS:HS 5.304

13240. 1866–8–18 From Edward Sabine Llandovery
Discusses [John] Hennessey's proposal [see ES's 1866–8–11]. Suggests sending achromatic along with a qualified observer. Thinks Alexander Herschel might be suitable for such a post. Melbourne telescope progressing rapidly.
LS 4pp RS:HS 15.278

13241. 1866–8–19 From Edward Livingston Youmans 16 Little Britain
Plans to compile and republish articles on claims of scientific education by famous scientists. Names nine who have consented. Wishes to include JH's letter to [J. C.] Adamson.
ALS 2pp RS:HS 18.443

13242. 1866–8–20 To Margaret Brodie Herschel Collingwood
News of the family servants, correspondence both JH and MH need to deal with, and prospects for son John.
ALS 4pp JHS 4.12

13243. 1866–8–20 To William Selwyn Collingwood
Says the plan for the exhibition is ingenious. Discusses the 11 year cycle of sunspots and says the sun was spotless the previous day.
C 1p RS:HS 24.164

13244. 1866–8–24 From Thomas Harris Highgate
Settles a matter of debt.
ALS 1p Harvard: Houghton AAH 67m–67 (60)

13245. 1866–8–26 To Roderick Impey Murchison Collingwood
RM has inadvertently sent letter intended for 'S,' a chemist, to JH. Having read letter, JH comments that he esteems 'S,' but thinks it advisable to tell 'S' that B.A.A.S. as a body will not recommend 'S''s invention to government.
ALS 4pp EUL:Gen523/4/36

13246. 1866–8–28 From George Rümker Hamburg
Sends observations of additional circumpolar nebulae. Notes differences between his observations and those of JH. Hopes to determine positions of very faint nebulae in the future.
ALS 4pp RS:HS 14.510

13247. 1866–8–30 To Edward Sabine Collingwood
Did not realize [John] Hennessey was considering constructing an achromatic telescope. Thanks ES for considering his son, Alexander, for the Melbourne post.
draft 1p RS:HS 15.279 (C: RS:HS 24.163)

13248. 1866–8–31 From Augustus De Morgan 91 Adelaide Road
Thought that JH and Charles Babbage were of the same year, but Cambridge Calendar shows differently, can JH solve the puzzle? Visited H. Crabb Robinson yesterday who gave him some sonnets that had been addressed to Wordsworth. Regarding the 1st edition of the *Eikon basilike*. Reason for Sir James South's knighthood.
ALS 4pp RS:HS 6.386

13249. 1866–9–1 To Augustus De Morgan Collingwood
Comments on Charles Babbage's entry into Cambridge; strange stories from old letters.
C 1p RS:HS 24.165

13250. [1866–9?] From Agnes Greig n.p.
Discusses plans for a visit to JH and Lady Herschel.
ALS 1p inc Harvard: Houghton AAH 67m–67 (107)

13251. 1866–9–1 From Edward Sabine Llandovery
B.A.A.S. appointed Committee to explore astronomical and meteorological possibilities in India. Considers [G. G.] Stokes's recommendation among the best. Suggests meteorological experiments.
LS 4pp RS:HS 15.280

13252. 1866–9–3 From Augustus De Morgan 91 Adelaide Road
Returns his letter. Recommends a biographical dictionary by Charles Hale. Paradoxes connected with theology are endless.
ALS 3pp RS:HS 6.387

13253. 1866–9–4 To Edward Sabine Collingwood
Further recommendations regarding useful equipment and observations to be made in India [see JH's 1866–8–11]. Agrees that meteorological experiments there are desirable. Offers JH's son John's experience of India to ES.
ALS 4pp Cambridge University Library 7656/RS536 (draft: RS:HS 15.281 & C: RS:HS 24.166)

13254. 1866–9–6 From Mary Somerville Spezia
Thanks him for papers he sent. Has finished her last work. Disappointed that science is often devoted to war and weaponry.
ALS 4pp RS:HS 16.373

13255. 1866–9–10 From Charles C. Nelson Upton Court near Slough
Giving news of the health of his brother, who will shortly be moving to St. Leonard's for health reasons.
ALS 3pp RS:HS 13.89

13256. 1866–9–11 To George Biddell Airy Collingwood
Has been asked to serve on a new Commission on Weights and Measures; indications are that the metric system is being recommended to the Indian government.
ALS 2pp RGO 6.359.250

13257. 1866–9–11 From Daniel Mackintosh Malvern Link
Has been travelling around lecturing on geology since he left the London Polytechnic in 1861. Wishes that someone like JH would consider geological features from a dynamical point of view.
ALS 3pp RS:HS 13.7

13258. 1866–9–13 From George Biddell Airy Royal Observatory, Greenwich
Is glad JH is going to assist with the Standards. Outlines his scheme for the Standards.
ALS 3pp RS:HS 1.301

13259. 1866–9–19 From Agnes Greig to Margaret Brodie Herschel Cranley Place, Onslow Square
Will be pleased to come on Friday and receive Sir John's counsel
ALS 3pp RS:HS 9.126

13260. 1866–9–24 From George Biddell Airy Royal Observatory, Greenwich
Wants JH's opinion on the question of maintaining observatories at Bombay and Madras.
ALS 4pp RS:HS 1.302 (C: RGO 6.149.7)

13261. 1866–9–24 From John Henry Dallmeyer 19 Bloomsbury St.
Outlining phenomena in the refraction of light. Has been experimenting with wide angle lenses for cameras and gives some of their defects. Would like JH's opinion on how to get rid of the 'secondary focus.'
ALS 12pp RS:HS 6.2

13262. 1866–9–25 To Julia Margaret Cameron Collingwood
Thanks JC for sending a batch of JC's photographs; praises them. Does not know the author of 'Man & his Dwelling' or of 'Ecce Homo.' Dislikes the latter.
ALS 5pp Colin Ford, *The Cameron Collection*, 142

13263. 1866–9–26 To George Biddell Airy [Collingwood]
Comments on the work of the Commission on Weights and Measures as outlined by GA [see GA's 1866–9–13].
ALS 4pp RGO 6.359.254

13264. 1866–9–26 To George Biddell Airy Collingwood
JH responds with an assessment of the Indian observatories [see GA's 1866–9–24].
ALS 4pp RGO 6.149.14

13265. 1866–9–27 From Augustus De Morgan 91 Adelaide Road
Tells a story of the translation of Arabic by Englishman who turned out to be Richard Baily, the brother of Francis. Where is it set forth that a leech is a barometer? What is the truth about the relations between Charles Babbage and James South? Has he a catalogue of Ulugh Beg?
ALS 3pp RS:HS 6.388 S. De Morgan, *Memoir of Augustus De Morgan*, 335

13266. 1866–9–28 From Charles T. Beke Bekesbourne, Canterbury
Is sending another memorial for JH to sign.
ALS 1p RS:HS 4.34

13267. 1866–9–29 From Charles Babbage Dorset St.
Regarding the signing of a new document for CB's sister. Has been reading JH's paper on the sun. Wishes he would print a list of all his papers.
ALS 2pp RS:HS 2.335

13268. 1866–9–29 From Hermann Kindt 344 Stretford Rd., Manchester
Asks JH to autograph a copy of *Essays Q.E.R.,* which HK wants to send to a friend.
ALS 2pp (German) RS:HS 11.34

13269. 1866–9–29 From Sir Charles Lyell 53 Harley St.
Hopes to send a copy of his new edition [of *Principles of Geology*] in a few weeks. Hopes he will be satisfied with the manner in which he used JH's drawings. Comments about variations in the obliquity of the ecliptic. Quotes correspondence he has had with E. J. Stone of Greenwich Observatory. Would like JH's opinion on the matter.
ALS 7pp RS:HS 11.442

13270. 1866–9–29 From Adolphe Quetelet Brussels
Has received works on meteors. Has been working for years with [H. A.] Newton on getting help from observatories in southern hemisphere on meteors. Resumes work on *Physique sociale.*
ALS 3pp (French) RS:HS 14.263

13271. 1866–9–30 To John Henry Dallmeyer Collingwood
Provides a physical explanation for a white patch in the center of JD's photograph.
C 2pp RS:HS 24.167 *Photographic News, 11* (June 21, 1867), 295–6

13272. [1866–10] To Charles Babbage n.p.
Is returning the trust deed duly signed. A list of his papers was printed in an American periodical. Hopes his yachting improved his health. He is suffering from bronchitis. Trouble with the Irish.
ALS 4pp RS:HS 2.336

13273. 1866–10–1 To Sir Charles Lyell n.p.
Comments on the relationship between the plane of the earth's orbit and the obliquity of the ecliptic.
C 3pp inc RS:HS 24.168

13274. 1866–10–2 From Charles T. Beke Bekesbourne, Canterbury
Thanks JH for promptly signing memorial [see CB's 1866–9–28].
ALS 1p RS:HS 4.35

13275. 1866–10–2 From Agnes Greig Cranley Place, Onslow Square
Has taken Mary Somerville's manuscript [*On Molecular and Microscopic Science*] to John Murray. Would like JH's opinion on the scientific value of her manuscript, and if he could assist her in seeing it through the press.
ALS 4pp RS:HS 9.127

13276. 1866–10–2 To Daniel Mackintosh Collingwood
Agrees that a large part of the hills and valleys are due to tidal action. Comments further on this, especially relating to parts of France. The Hawkhurst lectures have ceased and are not likely to be resumed.
C 4pp RS:HS 13.8 & 24.169

13277. 1866–10–2 From James Joseph Sylvester Woolwich Common
Wonders if mathematical result recently presented by [J. A.] Serret was published many years ago by JH. Comments on recurrence of identical problems in mathematical study.
ALS 4pp RS:HS 17.163

13278. 1866–10–4 To Elizabeth Colling Collingwood
Thanks for EC's poetry; comments on other translators of Homer, and on mythical beasts.
C 2pp RS:HS 24.170

13279. 1866–10–5 From George Biddell Airy Royal Observatory, Greenwich
Would like account of shooting stars by JH's son [Alexander]. Gives his observations on the properties of shooting stars.
ALS 3pp RS:HS 1.303 (C: RGO 6.478.336)

13280. 1866–10–5 From Hermann Kindt Fallowfield, Manchester
Letter of general adulation.
ALS 3pp (German) RS:HS 11.35

13281. 1866–10–9 To Augustus De Morgan Collingwood
Complains about Arabic star names; comments on squabble between Charles Babbage and James South.
C 2pp RS:HS 24.171

13282. 1866–10–12 From Henry Francis Blanford Calcutta
As directed by Meteorological Committee, HB sends report on 1864 Calcutta cyclone.
(A)LS 1p TxU:H/M–0107.1; Reel 1087

13283. 1866–10–13 To Sir Charles Lyell Collingwood
Comments on some of CL's writings; further argues that a canal at Suez and/or Darien would have a major effect on European climate due to the effect of ocean currents coming through such canals.
ALS 4pp APS B.D25.L.16 (C: RS:HS 24.172)

13284. 1866–10–19 From Sir Charles Lyell 53 Harley St.
Did not intend to trouble him with his proofs, but as he has cited JH in the chapter dealing with the astronomical effect on climate he sends proofs for this chapter. Would like to quote JH's two letters on the subject. The whole book is set up and hopes to send the volume in November.
ALS 4pp RS:HS 11.443

13285. 1866–10–19 From John Rate Peakridge, Staffs
Trying to ascertain the time for the eclipse of Jupiter's satellites. Wants to know the proper correction for his own telescope.
ALS 4pp RS:HS 14.471

13286. 1866–10–20 To Sir Charles Lyell Collingwood
Some minor suggestions regarding CL's proofs [see CL's 1866–10–19].
ALS 2pp† APS B.D25.L.17 (C: RS:HS 24.173)

13287. 1866–10–20 From William C. Morland Lamberhurst
Is grateful for JH's translation of Homer, which he values. Regrets to see JH's son so fatigued.
ALS 3pp RS:HS 12.368

13288. 1866–10–22 From Sir Charles Lyell 73 Harley St.
Was pleased to get his twelve pages of slips returned this morning. In one he proposes an alteration of 31 instead of 1003 to 1000. Which is correct?
ALS 2pp RS:HS 11.446

13289. [1866?]–10–23 From Julia Margaret Cameron Freshwater Bay
Is grateful for the volumes. Hopes his poetry endeavors will succeed. John Taylor is now at East Sheen. She has won a medal for photography at Southampton.
ALS 12pp RS:HS 5.173

13290. 1866–10–23 From Mary Somerville Spezia
Thanks for correcting proof sheets. Discusses JH's and [William] Grove's views on correlating physical forces. Tends to trust JH's views.
ALS 2pp RS:HS 16.374

13291. 1866–10–26 From George Griffith B.A.A.S., Oxford
Resolution reappointing JH and others to committee to persuade Russian government to establish magnetic observatory at Tiflis [Tbilisi, Georgia] under direction of Prof. [?] Moritz.
L 1p TxU:H/M–0921.5; Reel 1083

13292. 1866–10–29 From F. W. A. Argelander Bonn
Observations on a new variable star [T Coronae]. Enquiry regarding one of JH's observations.
ALS 4pp (German) RS:HS 1.375

13293. 1866–10–29 From Henry Hardcastle 6 Crown Office Row
Regrets he should have felt so much anxiety about Maria (HH's wife and JH's daughter), but she is progressing well. Have settled down well in their new house.
ALS 2pp RS:HS 9.227

13294. [1866]–11–5 To John Herschel (son) Collingwood
Worried by not having heard from son John, especially as there is cholera in India; JH sends news of the family, especially of how bright the younger girls are; also comments on world events and about books sent to John.
ALS 4pp JHS 5.15

13295. 1866–11–5 From Mary Anne Wilson 35 Norland Square, Notting Hill
Recalls happy days JH spent with MW's relatives. MW is now 76 and in good health, but experiencing financial difficulties. Will JH help her? Reports news of her brothers, Alexander and Patrick.
ALS 4pp RS:HS 18.420

13296. 1866–11–6 To F. W. A. Argelander Collingwood
Regarding his observations. Sending a copy of his original observations. Regarding the new variable star [T Coronae].
draft 6pp RS:HS 1.376 & 24.174

13297. 1866–11–6 From John F. McLennan 22 Hill St., Edinburgh
Sends four sheets of his *Memoir of Thomas Drummond* and would be pleased if JH would look over them and correct any errors. Miss Elizabeth Drummond continues an invalid. Is also sending the proof sheets to Sir T. A. Larcom.
ALS 2pp RS:HS 12.253

13298. 1866–11–14 To the British Meteorological Society Collingwood
Presents in some detail and with an illustration his observations of the meteor shower 13–14 Nov. 1866.
L *Proceedings of the British Meteorological Society, 3* (1867), 273–5

13299. 1866–11–14 From India Office London
R. A. T. Gascoyne-Cecil (Viscount Cranborne) would like JH's opinion on the value of the investigations carried out by Messrs. Schlagintweit in 1854; four volumes have already been published.
ALS 3pp RS:HS 10.263

13300. [1866]–11–15 From Charles G. B. Daubeny Botanic Gardens, Oxford
[Printed letter] Collecting poems related to natural history and physical science. Lists 16, including JH's 'On the Herschelean Telescope.'
L 1p TxU:H/M–0156; Reel 1087

13301. 1866–11–15 To Daniel Mackintosh Collingwood
Has been asked to comment on the writings of Hermann von Schlagintweit, but JH has not yet had time to read them carefully and so cannot comment.
C 1p RS:HS 24.175

13302. 1866–11–16 From John F. McLennan 22 Hill St., Edinburgh
Many thanks for his prompt attention to the proof sheets; sends some more. Miss Elizabeth Drummond is gratified by JH's interest.
ALS 2pp RS:HS 12.254

13303. 1866–11–16 To [India Office] Collingwood
Corrects earlier statement that JH has not received vol. 4 of the study of India written by [Robert and Hermann] Schlagentweit
ALS 2pp BostonPubLib Ms.Eng.231(1)

13304. 1866–11–17 From Charles Pritchard [Freshwater]
Compares observations of meteor shower in Orion with JH. Describes methods and observations in detail.
ALS 4pp RS:HS 14.100 Ada Pritchard, *Charles Pritchard: Memoirs of His Life,* 257

13305. 1866–11–18 To John F. McLennan Collingwood
Regarding the inventor of the compensation bars. Would not attribute their invention to Thomas Drummond in the light of J. E. Portlock's evidence. Thanks for the paper on ancient Greek kinship.
C 6pp RS:HS 12.255 & 24.176

13306. 1866–11–19 From Charles G. B. Daubeny Botanic Garden Oxford
If his own collection of poems is printed he hopes JH will allow him to include the one on the telescope. His own health has improved. Outlines a proposition for JH to work on if he has time.
ALS 4pp RS:HS 6.23

13307. 1866–11–19 From Thomas Maclear to Margaret Brodie Herschel Cape of Good Hope
Outlining the chances and difficulties of bringing the [Samuel] Parlby case before the Cape Government. Alexander Herschel would have been delighted with the display of shooting stars and volcanic eruptions seen at the Cape last Thursday.
ALS 4pp RS:HS 12.173

13308. 1866–11–19 To George Gabriel Stokes Collingwood
Carefully describes how to pack actinometers being shipped to India.
C 2pp RS:HS 25.10.15

13309. 1866–11–20 From John F. McLennan 22 Hill St., Edinburgh
Was not prepared for the facts regarding the invention of the compensation bars as expressed in JH's letter. Will investigate the various claimants further.
ALS 4pp RS:HS 12.256

13310. 1866–11–21 From William Rutter Dawes Hopefield, Haddenham, Thame
Hears that JH is preparing his father's measurements of double stars for the R.A.S.; would like to publish his own work on double stars in the same volume. Would like to know when the work will be ready for publication.
ALS 4pp RS:HS 6.122

13311. 1866–11–22 From Augustus De Morgan 91 Adelaide Road
Many thanks for his book [*Familiar Lectures*], which he is reading from the beginning. Comments on JH's views. Is Sidees a star or a constellation? Congratulations on his completion of Homer. Everyone is translating Homer now, and gives one of his own examples.
ALS 5pp RS:HS 6.389

13312. 1866–11–22 From Adolphe Quetelet Brussels
Apologizes if AQ has missed answering one of JH's letters. Talks about meteorology of Belgium worked on over 30 years. Wants JH's photo for his work on statistics. Congratulates JH on his son's good work. Has received work from Collingwood and would love to know author.
ALS 4pp (French) RS:HS 14.264

13313. 1866–11–22 From Mary Anne Wilson 35 Norland Square, Notting Hill
Deep appreciation for JH's donation.
ALS 2pp RS:HS 18.421

13314. 1866–11–24 From William Rutter Dawes Hopefield, Haddenham, Thame
Is grateful for his prompt reply about the catalogue of double stars. His own paper will run to some 250 pages. Otto Struve has offered to send him the observations of some of the stars he is interested in.
ALS 4pp RS:HS 6.123

13315. 1866–11–26 To Robert Gascoyne–Cecil, Lord Cranborne Collingwood
Comments on a multi-volume work on India by the Schlaginweits [Hermann and Robert], suggesting that the work contains much extraneous computation and detail and should be compressed. Questions whether further publication on the present scale is justified.
C 7pp RS:HS 24.177

13316. 1866–11–26 From Bartholomew Price 11 St. Giles St., Oxford
Thanks JH for his lectures [*Familiar Lectures*] and praises his style. Has encouraged Clarendon Press to ask JH to write a short treatise on inductive logic.
ALS 3pp RS:HS 14.59

13317. 1866–11–27 From J. Park Harrison Ewhurst, Guildford
Notes regular maxima and minima in 50-year temperature curves. Offers theory relating these to phases of moon. Compares it to J. H. Mädler's theory.
ALS 4pp RS:HS 19.276

13318. 1866–11–27 From Henry Enfield Roscoe Manchester Literary and Philosophical Society
Acknowledges receipt of JH's letter and fungi spore specimens. The letter was read and specimens exhibited at last meeting. Offers thanks.
ALS 1p RS:HS 14.424

13319. 1866–11–28 From James Challis Cambridge
Would be grateful if JH would add his name to the enclosed certificate for Edward Walker. Large number of readings of meteors taken at Cambridge Observatory. Comments on these observations.
ALS 3pp RS:HS 5.241

13320. 1866–11–28 To [Drummond's Bank] Collingwood
Grants permission for JH's son John to remove one bundle of papers from JH's storage box.
Adraft 1p TxU:H/L-0182.1; Reel 1054

13321. 1866–11–28 From India Office London
Lord Cranborne is grateful for JH's full and complete report on the scientific investigations of Messrs. Schlagintweit.
ALS 2pp RS:HS 10.264

13322. 1866–12–1 From John F. McLennan 22 Hill St., Edinburgh
Sends proof sheet on the mathematics of representation. Intends to print in full JH's letter to Mrs. Drummond. Has not yet received information regarding the measuring bars as T. A. Larcom is busy.
ALS 4pp RS:HS 12.257

13323. 1866–12–3 From Edward Sabine n.p.
Informs JH that his comments against sending a major telescope to India were decisive [see JH's 1866–8–11 & 1866–9–4]; biggest cost impediment was salary of observer.
LS 4pp RS:HS 15.282 (C: Cambridge University Library 7656/RS552)

13324. 1866–12–4 To Sir Charles Lyell Collingwood
Appreciates the receipt of volume one of the tenth edition of CL's *Principles of Geology.*
ALS 2pp APS B.D25.L.18 (C: RS:HS 24.178)

13325. 1866–12–4 From Bartholomew Price Oxford
Supports G. W. Kitchin's request that JH write treatise on inductive logic for benefit of Oxford students.
ALS 4pp TxU:H/M–0410; Reel 1087

13326. 1866–12–5 To Edward Sabine Collingwood
Maintains that a first class refractor is not needed at Masouri in India [see ES's 1866–9–4]. No special characteristics there justify the great expense of such a project.
ALS 3pp RS Sa.675 (C RS:HS 15.283 & RS:HS 24.179)

13327. 1866–12–6 From George William Kitchin Walton Manor, Oxford
Clarendon Press plans to publish series on logic. Invites JH to write treatise on inductive logic. Expect letters on this from John Phillips and Bartholomew Price.
ALS 4pp TxU:H/M–0318; Reel 1087

13328. 1866–12–6 From Edward Sabine n.p.
Discusses current debate over placement of telescope in India. Discusses various possibilities for telescope.
LS 4pp RS:HS 15.284

13329. 1866–12–6 From John Williams Somerset House
Acknowledges receipt by R.A.S. of papers relating to William Herschel's observations of double stars.
ALS 1p RS:HS 18.410

13330. 1866–12–7 From John Phillips n.p.
Requests JH's views on inductive logic for a publication by Cambridge.
ALS 3pp TxU:H/M–0428.25; Reel 1093

13331. 1866–12–8 To Edward Sabine Collingwood
In response to ES's 1866–12–6, JH sees no benefit in a great equatorial telescope in India being an itinerant instrument.
ALS 3pp Cambridge University Library 7656/RS554 (ACCS: TxU:H/L-0378; Reel 1054 & Cdraft: RS:HS 25.15.23)

13332. 1866–12–9 To George William Kitchin n.p.
Needs time to consider GK's request. What connection is expected between [Clarendon] Press and each author, or among all authors in GK's proposed series?
AdraftS 2pp TxU:H/L-0213; Reel 1054

13333. 1866–12–10 To Elizabeth Colling Collingwood
On the 'Great Phenomenon' of 13 and 14 Nov. [meteor shower]; JH appears to believe that there is an annular ring of very minute planets circling the sun.
C 2pp RS:HS 24.180

13334. 1866–12–11 From Edward Joseph Lowe Beeston
Sends some queries regarding meteor showers.
ALS 8pp RS:HS 11.340

13335. 1866–12–13 To Edward Joseph Lowe Collingwood
Did not mean to say that meteors describe an exact circle round the sun. Comments on the phenomena of meteors.
C 5pp RS:HS 11.341

13336. 1866–12–13 From John F. McLennan 22 Hill St., Edinburgh
Sends the last sheet of the Memoir that will be necessary for JH to see. Sheets on the measuring bars are held in abeyance.
ALS 1p RS:HS 12.258

13337. 1866–12–13 From William Selwyn Vine Cottage, Cambridge
Thanks JH for his sympathy regarding WS's fall. Describes the meteors he saw from his window on December 4.
ALS 4pp RS:HS 15.488

13338. 1866–12–14 From George William Kitchin Walton Manor, Oxford
Grateful that JH is considering [writing treatise on inductive logic]. Each book in series is independent of other authors. Conveyed JH's request to Bartholomew Price.
ALS 3pp TxU:H/M–0319; Reel 1087

13339. 1866–12–14 To William Selwyn Collingwood
Congratulates WS on his recovery. Thanks him for the meteor observations. The evidence points to an 11 rather than a 10 year period.
C 1p RS:HS 24.181

13340. 1866–12–15 From Clement M. Ingleby Ilford
Was interested to read in JH's *Familiar Lectures* the chapter on sensorial vision; has been subject himself to the phenomenon described by JH. Queries one of the statements and would like verification from JH on this point.
ALS 4pp RS:HS 10.267

13341. 1866–12–17 From Elias Loomis Yale College
Is indebted to JH for a number of scientific memoirs, including the 'Catalogue of Nebulae' and the tract on *Physical Geography,* both of which he highly esteems. Sends via the Smithsonian his own pamphlet on the *Aurora Borealis.*
ALS 1p RS:HS 11.310

13342. 1866–12–18 From Thomas Maclear to Margaret Brodie Herschel Royal Observatory, Cape of Good Hope
Reports a second failure over the [Samuel] Parlby affair. Buried Ellen last Thursday. Unemployment and lack of food at Cape owing to the financial difficulties of the government. Difficulties over the standards now.
ALS 2pp RS:HS 12.174

13343. 1866–12–18 From Richard Hodgson Royal Astronomical Society
Gratitude for JH's gift to R.A.S. of William Herschel's manuscript series on observations of double stars. Encloses two pages of proofs for JH's approval.
ALS 2pp RAS:JH Archive 15/5.3; Reel 11

13344. 1866–12–20 From W. Masters Kishnagur
Giving details of the meteors that fell recently.
ALS 2pp RS:HS 13.25

13345. 1866–12–24 From Adolphe Quetelet Brussels
Expresses good wishes for JH. Talks about aging. Notes JH's interest in poetry. Thanks for 'Of the Estimation of Skill in Targetshooting.' Has given it to [Edouard] Mailley for translation and printing in annual bulletin of Brussels observatory. Asks if JH wishes to be named as author.
ALS 4pp (French) RS:HS 14.265

13346. 1866–12–27 To Mary Baldwin Collingwood
Returns letter of George Bryandt. Dispute over auction of property belonging to MB's late brother [Thomas], of whose will JH is executor.
AdraftS 4pp † TxU:H/L-0072; Reel 1054

13347. 1866–12–31 From Thomas Maclear Royal Observatory, Cape of Good Hope
Sending a further section to be added to the Arc of Meridian. Comments on Nicolas Lacaille's work on the meridian. Has cherished memories of thirty years previous to this time of the year.
ALS 2pp RS:HS 12.175

13348. [1867 to 1869] To William James Herschel [Collingwood]
Avoid forcing metric system on India. Hopes Commission will adopt 'geometrical system of [A. T.] Kupffer and others. Winter weather. Family health and news. Alexander S. Herschel gave lecture at Leeds, reports many Japanese students at Glasgow. JH predicts that industrious Japanese may supplant Europe and America in next century. Details of JH's humorous plan for 'telegraphic personal transfer' [teleportation] to accelerate travel. Rash of bombings by mail in London. Attacks on police.
ALS 4pp TxU:H/L-0557; Reel 1053

13349. [1867?] From James Samuelson New Burlington St.
Has decided to accept papers treating of new researches. These will be inserted prominently in journal [*Quarterly Journal of Science*]. Would be happy to hear from JH.
ALS 1p RS:HS 15.341

13350. [1867 or after] From William Selwyn n.p.
Discusses displaying solar autographs to the best advantage. Invites JH's son Alexander to read his meteor paper at the Cambridge Philosophical Society.
ALS 4pp inc RS:HS 15.518

13351. [1867 or later] To George Gabriel Stokes n.p.
On a correction to a report [JH's obituary on William Whewell].
ALS 2pp Cambridge University Library 7656/H732

13352. 1867–1–1 To George Gabriel Stokes Collingwood
Acknowledges receipt of paper on nebulae.
ALS 1p Cambridge University Library 7656/H716

13353. [1867] To [Longman & Green Publishers] Collingwood
Sends a correction for the ninth edition of JH's *Outlines Astr.*
ALS 2pp Harvard: Houghton

13354. 1867–1–10 To Mary Baldwin [Collingwood]
Advice on choosing executors, drawing will, and disposing of MB's property. Sends semiannual dividend from Drummond's.
AdraftS 4pp TxU:H/L-0073; Reel 1054

13355. 1867–1–15 From William Selwyn Foxton Royston
Discusses Aristotle's and Richard Whately's descriptions of the sensitivity of the eye's lateral areas. Asks JH to review cover sheet and opine whether the predictions described are miracles.
ALS 4pp RS:HS 15.489

13356. 1867–1–16 From Edward Sabine n.p.
Sends copy of B.A.A.S. resolution regarding telescope in India. ES and R.S.L. colleagues agree to take no further steps in matter.
LS 4pp encl RS:HS 15.285

13357. 1867–1–17 To Margaret Brodie Herschel Collingwood
Tells MH about the wedding plans of son John and Mary Power; comments on the health of daughter Constance, and sends a silly sketch by Emma Hardcastle. [Part of the letter is illegible.]
ALS 4pp JHS 5.37

13358. 1867–1–17 To William Agei Collingwood
Thanks WS for describing Aristotle's and Richard Whately's observation of the great sensitivity of the eye's lateral portion. Congratulates WS on becoming Dean of Norwich. Draft discusses miracles and lists possible arguments against WS's idea that miracles are not a violation of nature
C 1p RS:HS 24.183 & draft: RS:HS 15.490

13359. 1867–1–24 From William Selwyn Foxton Royston
Dr. [Edward] Goulburn, not WS, is Dean of Norwich. Is trying to refute the position of David Hume that miracles are

violations of nature.
ALS 9pp RS:HS 15.491

13360. 1867–1–26 From Alexander J. B. Hope Bedgebury Park Cranbrook
Has arranged for JH to receive a couple of plants of *Fitzroya Patagonia*. Hopes that the inclement weather has caused no distress.
ALS 3pp RS:HS 9.470

13361. 1867–1–26 From Urbain J. J. Leverrier Paris
Has received the *R.A.S.M.N.* containing the article on the movements of the asteroids. Has addressed a letter on the subject to the *Moniteur universel* and would be glad of JH's comments.
ALS 2pp (French) RS:HS 11.208

13362. 1867–1–27 To Alexander J. B. Hope Collingwood
About some flowers, including a new variety being advertised.
ALS 3pp WT 63313.2

13363. 1867–1–27 To Urbain J. J. Leverrier Collingwood
Comments on UL's paper on meteors. Outlines the nebular theory of the solar system.
draft 3pp RS:HS 11.209 (C: RS:HS 11.210 & 24.184)
'Etoiles filantes de novembre,' *Moniteur universel* (Feb. 4, 1867), p. 105, col. 1 inc

13364. 1867–1–29 From George Biddell Airy Royal Observatory, Greenwich
Thanking him for lecture papers. Has sent his paper on partial differential equations.
ALS 4pp RS:HS 1.304 (C: RGO 6.381.448)

13365. 1867–1–30 From Camille Flammarion Paris
Sending the first volume of a new series on astronomy entitled *Studies and Lectures....* Is continuing his work in the *Times*.
ALS 1p (French) RS:HS 7.270

13366. 1867–2 From Jacques Babinet n.p.
Regarding the work of Emmanuel Liais (astronomer).
ALS 2pp (French) RS:HS 3.6

13367. 1867–2–2 To Joseph Baxendell Collingwood
Regarding gales of wind and their relation to violent changes in barometric pressure.
ALS 4pp RS:HS 3.311 & 24.186

13368. 1867–2–2 From Christopher Cooke Oxford
Regarding the doubtful actions of a Lieut. R. J. Morrison falsely obtaining names purporting to support his scheme for observations. Intends to lay the matter before William Parsons [3rd Earl of Rosse].
ALS 4pp RS:HS 5.361

13369. 1867–2–2 To Manchester Literary and Philosophical Society Collingwood
Notes that JH had pointed out the perpendicularity of the direction of wind to the direction of a wavefront's advance. Encourages research on the causes of such meteorological phenomena.
L *Manchester Literary and Philosophical Society Proceedings, 6* (1867), 91–3 & *Revue Maritime et coloniale, 19* (1867), 927–8

13370. 1867–2–3 From Urbain J. J. Leverrier Paris
Sends an extract from JH's letter, which has been printed in the *Moniteur*.
ALS 2pp (French) RS:HS 11.211

13371. 1867–2–4 To Elizabeth Colling [Collingwood]
Is in poor health; laments poor reception of his translation of the *Iliad*.
C 2pp RS:HS 24.187

13372. 1867–2–7 From Christopher Cooke Oxford
Thanks for his prompt reply to his query about Lieut. R. J. Morrison. He has discovered the horoscope of Edmond Halley.
AL 3pp RS:HS 5.362

13373. 1867–2–10 To Mary Baldwin Collingwood
Alterations to MB's will. Objects to MB's plan to return to live in Mr. Binsted's house at Anstey.
AdraftS 4pp TxU:H/L-0074; Reel 1054

13374. 1867–[2–12] To James Calder Stewart Collingwood
Anecdote about William Whewell. Religious beliefs in England. Foolish opinion about moon expressed in the *Times*.
ALS 4pp TxU:H/L-0461; Reel 1055

13375. 1867–2–14 From John Henry Dallmeyer 19 Bloomsbury St.
Has not had time until now to carry out experiments on 'the central spot' as seen in photographs as related in JH's last letter. His recent trials with various lenses have verified JH's theories.
ALS 7pp RS:HS 6.3

13376. 1867–2–16 To John Henry Dallmeyer Collingwood
Is he acquainted with his ordinary reading glass? Gives details. Always carries a small one around with him for reading small print; thinks there would be a good market for them.
draft 1p RS:HS 6.4 & 24.188

13377. 1867–2–18 From John Henry Dallmeyer 19 Bloomsbury St.
Regarding various types of glass for lenses. Will make two of JH's reading glasses if he will send details.
ALS 6pp RS:HS 6.5

13378. 1867–2–19 To John Henry Dallmeyer Collingwood
Outlining suitability of various types of glass for optical work. Gives details of theories of his reading lenses and calls attention to one of his own papers in the *R.S.P.T.* for 1821.
draft 2pp RS:HS 6.6 & 24.189

13379. 1867–2–21 From John Henry Dallmeyer 19 Bloomsbury St.
Thanks for letter and details of his paper in the *R.S.P.T.* Wishes there was an index of authors and subjects for students. Theories regarding the action of light on glass. Has interested Messrs. Chance in his theories. Gives reasons why he wants a reading glass of 3 1/2 inches.
ALS 8pp RS:HS 6.7

13380. 1867–2–21 From Norman Macleod [South Kensington Museum]
Council on Education requests loan of several portraits owned by JH for use in Second National Portrait Exhibition.
AL 1p inc RAS:JH Archive 14/1.192 verso; Reel 10

13381. 1867–2–25 From William H. Callcott Kensington
Would be grateful if he would inform him where his paper on concert pitch is published.
ALS 1p RS:HS 5.143

13382. 1867–2–26 From W[illiam] Cave Thomas 49 Torrington Square
Discusses matters of ascetics; praises JH's ideas.
ALS 3pp RS:HS 17.433

13383. 1867–3–2 From C. H. D. B. Ballot Utrecht
Is honored that JH should have noticed one of his articles. Quotes further articles, which may be of interest. Regarding JH's views on air and his own theories.
ALS 4pp RS:HS 5.132

13384. 1867–3–5 To Jacques Babinet Collingwood
No mention of Emmanuel Liais's *L'espace céleste* in English journals. Contact London publishers to find translator for it. Alexander Herschel is now professor of natural philosophy at Andersonian Institution in Glasgow. Hopes JB's elegant method of measuring gravity by torsion will not die before being put into practice.
AdraftS 4pp RS:HS 19.158

13385. 1867–3–5 From James Joseph Sylvester Woolwich Common
Sends JH proceedings of Mathematical Society of London and asks JH to join.
ALS 3pp RS:HS 17.164

13386. 1867–3–7 To Jacques Babinet Collingwood
Not willing to be involved in translation of French work on astronomy [see JB's 1867–2].
C 2pp (French) RS:HS 24.190

13387. 1867–3–10 To Mary Baldwin Collingwood
Gives more reasons for objecting to MB, 80 years old, leaving Stubbington and moving to Anstey [see JH's 1867–2–10].
ACCS 4pp & AdraftS TxU:H/L-0076 & L-0075; Reel 1054

13388. 1867–3–12 From [Norman Macleod] [South Kensington Museum]
Grateful for JH's contribution of portraits [to National Portrait Exhibition]. Packing and shipping instructions. Portraits must arrive by 30 Mar.
AL 1p inc RAS:JH Archive 14/1.206 verso; Reel 10

13389. 1867–3–14 From William Fairbairn Manchester
Invitation to stay with them should he visit the B.A.A.S. meeting at Manchester.
ALS 1p RS:HS 7.363

13390. 1867–3–14 From Norman Macleod South Kensington Museum
[Form letter] Safe receipt of William Herschel's portrait.
(A)LS 1p RAS:JH Archive 14/1.194 verso; Reel 10

13391. 1867–3–15 From Norman Macleod [South Kensington Museum]
Will carefully follow directions in JH's 13 Mar. letter.
ALS 1p RAS:JH Archive 14/1.193 verso; Reel 10

13392. 1867–3–20 To Augustus De Morgan Collingwood
Comments on mathematical comments AD has sent JH; JH is sorry to see AD has retired from his professorship; comments on the hard winter.
C 3pp RS:HS 24.191

13393. 1867–[3]–21 To [?] Collingwood
Received notice to transfer one-fifth of funds from [Mary Anne Babbage's] 1823 marriage settlement to W. H. B. Hollier.
ACCS 1p TxU:H/L-0636; Reel 1089

13394. 1867–3–25 From Augustus De Morgan 91 Adelaide Road
How is he during this spell of cold weather? Gives reason why he resigned from the College. Gives a theorem. Has no news of his own successor at the College.
ALS 4pp RS:HS 6.390S. De Morgan, *Memoir of Augustus De Morgan*, 369 inc

13395. 1867–3–27 From Thomas Humphrey National School
Announces quarterly meeting of National School Committee on 30 March.
ALS 1p TxU:H/MW-0246.7; Reel 15

13396. 1867–3–27 From H. Temple Humphreys 28 Westbourne Park Rd.
Would like JH's advice on the best shape for a rolling magnet with application for railway rolling stock.
ALS 8pp RS:HS 10.76

13397. 1867–3–27 From Charles Pritchard [Freshwater]
Asks if JH still considering editing William Herschel's *R.S.P.T.* publications. Suggests that Alexander Herschel edit the work. Speculates that William Herschel observed the 'willow leaves.' JH's star catalogue will be printed by the R.A.S.
ALS 4pp RS:HS 14.102 Ada Pritchard, *Charles Pritchard: Memoirs of His Life*, 257–8

13398. 1867–3–28 From John R. Hind London
[John?] Farley has just given him the readings for the minor planets. Encloses these and also readings for the major planets from U. J. J. Leverrier.
ALS 3pp RS:HS 9.395

13399. 1867–3–29 From C. H. Gordon-Lennox Board of Trade
Has been considering the subject of standard weights and measures and thinks a Royal Commission is the best idea. Regrets that JH's health prevents him from being a member.
ALS 3pp RS:HS 9.108

13400. 1867–4 From Norman Macleod South Kensington Museum
Second National Portrait Exhibition will open to public on 3 May. Invites JH to private viewing on 2 May.
ALS 1p RAS:JH Archive 14/1.200 verso; Reel 10

13401. 1867–4–2 From H. Temple Humphreys 28 Westbourne Park Rd.
Is grateful for his information. Believes he has found a way of overcoming the difficulties. Outlines his scheme for railway axles. Believes it will prove a great boon to the railways.
ALS 10pp RS:HS 10.77

13402. [1867?]–4–2 From Edward Sabine 13 Ashley Place
Asks JH to send an actinometer to Kew. It will be forwarded with other meteorological instruments to the Paris Exhibition.
ALS 2pp RS:HS 15.312

13403. [1867]–4–10 From Julia Margaret Cameron [Freshwater Bay]
Has been very busy. His poetry has inspired her to renewed efforts. Health not good.
ALS 8pp RS:HS 5.165

13404. 1867–4–10 [?] From Augustus De Morgan 91 Adelaide Road
Has heard a rumor that he is not well. One of his daughters

has been recuperating at Hastings. Sends a paradox. Comments on the editing of Isaac Newton's book on Daniel by Benjamin Smith, his nephew. R.S.L. has produced Vol. 1 of the list of scientific papers. Blaise Pascal affair is in a lull.
ALS 6pp RS:HS 6.391

13405. 1867–4–10 To Charles Pritchard Collingwood
JH is busy correcting first proofs of pages on double stars. Thanks for binding JH's star [allineations?]. CP's suggestion [see CP's 1867–3–27] to JH's son Alexander, to collect and edit William Herschel's papers, entails too much work for one editor. JH dreads thought of such work. Doubts CP's claim that WH observed fixed star in Corona.
ACCS 4pp TxU:H/L-0319; Reel 1054 (Cdraft: RS:HS 25.15.24)

13406. 1867–4–11 From Charles Hay Cameron Freshwater, I.o.W.
Is grateful for his gift of *Familiar Lectures* and has read them with delight. Is pleased that J. M. Cameron's photographs have given pleasure.
ALS 3pp RS:HS 5.149

13407. 1867–4–13 To [Mrs. Thomas Romney Robinson] Collingwood
Delighted to receive *Memoir of Maria Edgeworth.* Praises it and expresses thanks for it having been sent. Regards to Dr. Robinson.
ALS 2pp Bodleian Library, Oxford

13408. 1867–4–13 To Mrs. Socker[?] Collingwood
Delighted to receive *Memoir of Maria Edgeworth* [ed. by F. A. Edgeworth, 1867]. Praises Edgeworth. Whom should JH thank for this gift?
ALS 4pp Bodleian Library, Oxford

13409. 1867–4–14 From Edward Sabine Burlington House
Glad to have met JH's son [John]. Describes benefits of [R.S.L.] meetings for cultivation of science.
ALS 2pp RS:HS 15.286

13410. 1867–4–18 From Rudolf Wolf Zürich
Had written previously to JH to obtain more information about JH's father, William. Is now asking JH for a response to the same requests.
ALS 2pp (German) RS:HS 19.301

13411. [1867]–4–20 From Julia Margaret Cameron Freshwater, I.o.W.
Has only just received his letter. Feared for the safety of her parcel. Wishes he would write a poem on photography. John Taylor and the Camerons think that JH's poetry is beautiful. Writes in haste to catch the post.
AL 8pp inc RS:HS 5.166

13412. 1867–4–20 From Augustus De Morgan 91 Adelaide Road
Sends a theorem, which beats Blaise Pascal's by points.
ALS 2pp RS:HS 6.392

13413. 1867–4–21 To Rudolf Wolf Collingwood
Thanks RW for sending RW's *Mittheilungen* and RW's *Neue Untersuchungen.* Replies to RW's queries about JH's ancestry and about the current state of JH's father's largest reflecting telescope.
C 2pp RS:HS 24.192

13414. 1867–4–24 From Spottiswoode & Co. New St. Square
Received JH's insertions. Will send proof on Friday.
ALS 1p TxU:H/L-0463; Reel 1055

13415. 1867–4–25 To Augustus De Morgan Collingwood
Comments on AD's theorem [see AD's 1867–4–20].
C 2pp RS:HS 24.193

13416. 1867–4–25 From Rudolf Wolf Zürich
Thanks JH for his letter, and RW sends on additional writings in astronomy.
ALS 1p (German) RS:HS 19.300

13417. 1867–4–26 From Emilien Frossard Pasteur Bagnieres de Biguerre
What does JH think of the idea of establishing a meteorological observatory on the Pic du Midi?
ALS 4pp (French) RS:HS 13.237

13418. 1867–4–27 From Frederick Howlett St. Augustine's, Hurst Green
Regrets he was out when JH called. Thanks for the elegant verses. Will be moving shortly to a place nearer London, which promises much. Gives notes on some more sunspots which he has recently observed.
ALS 4pp RS:HS 10.21

13419. 1867–4–29 To Margaret Brodie Herschel [Collingwood]
About traveling to Halton; JH is finding working on his double star catalogue fairly severe drudgery.
AL 4pp JHS 4.14

13420. 1867–4–29 To Frederick Howlett n.p.
Would like FH to study a phenomenon that JH has noticed on the sun's disk and that has no connections with sun-spots.
C 4pp RS:HS 10.18 & 24.194

13421. 1867–5 From R. Calvin Clapham and Robert Spence Watson Newcastle upon Tyne
Invite JH to give lecture series at Literary and Philosophical Society next winter.
ALS 1p RS:HS 19.160a

13422. 1867–5–1 From H. Hotham Tunbridge Wells
Is puzzled by two apparent contradictions in JH's recent book *Familiar Lectures.* Are they printer's errors? Was related by marriage with the late Sir John William Lubbock.
ALS 4pp RS:HS 10.208

13423. 1867–5–2 To H. Hotham Collingwood
Thank you for his letter. There are no misprints or conflicting statements in his *Familiar Lectures.* Explains the various points.
draft 2pp RS:HS 10.209 & 24.195

13424. 1867–5–3 From George Gabriel Stokes R.S.L., Burlington House
Preparing for total eclipse of sun in 1868. Conferred with William Huggins about making spectroscopic analysis of red protuberances. Asks JH's opinion.
ALS 8pp TxU:H/M–0505; Reel 1093

13425. 1867–5–3 From George Gabriel Stokes n.p.
Asks JH's opinion on matter of purchase of expensive telescope for solar eclipse (1868) and other observations by William Huggins.
C 2pp Cambridge University Library 7656/H712

13426. 1867–5–4 From Frederick Howlett St. Augustine's, Hurst Green
Is uncertain if he has detected the lightlines on the solar

surface to which JH has called attention, but there does seem to be something unusual there. No sunspots are visible just now.
ALS 4pp RS:HS 10.20

13427. 1867–5–4 To Rudolf Wolf Collingwood
Thanks RW for sending him a photograph of RW; sends photograph of JH in return.
C 1p RS:HS 24.196

13428. 1867–5–5 To George Biddell Airy Collingwood
Is assisting in communication between G. G. Stokes and GA, about stereoscopic observation of eclipses; JH is not mobile enough to attend the next Board of Visitors meeting.
ALS 3pp RGO 6.122.177

13429. 1867–5–5 To Frederick Howlett Collingwood
Proposes to travel to visit FH next day to observe the sun with him.
C 1p RS:HS 24.198

13430. 1867–5–5 To George Gabriel Stokes n.p.
Describes problem with spectral lines in telescope while trying to understand William Huggins's results.
Adraft 1p TxU:H/L-0463 verso; Reel 1055

13431. 1867–5–5 To George Gabriel Stokes Collingwood
Comments on impossibility of increasing the intrinsic illumination of a source with a telescope; how to obtain the spectrum of red flames of the sun.
ALS 4pp Cambridge University Library 7656/H713 (draft RS:HS 17.66 & C: RS:HS 24.197) Stokes, *Correspondence, 1,* 211–2

13432. 1867–5–6 From George Biddell Airy Greenwich
Response to a letter from JH to G. G. Stokes [1867–5–5] seen by GA, on the problem of the effect of the telescope on illumination from a light source, especially related to a solar eclipse.
ALS 4pp Cambridge University Library 7656/A3339

13433. 1867–5–6 From George Biddell Airy Royal Observatory, Greenwich
Comments on the use of different telescopes and their effect on the albedo of the sun, related to solar eclipses [see JH's 1867–5–5].
ALS 4pp RGO 6.479.162

13434. 1867–5–6 From Charles C. Nelson 30 Hyde Park Gardens
His brother [John H. Nelson] is at Brighton, and may be coming to London soon. Bodily health improves but mental health shows no improvement.
ALS 3pp RS:HS 13.90

13435. 1867–5–7 To George Biddell Airy Collingwood
A note with G. G. Stokes' letter, which JH forgot to enclose [see JH's 1867–5–5].
ALS 4pp RGO 6.122.175

13436. 1867–5–8 From George Biddell Airy Royal Observatory, Greenwich
Returns JH's letter from G. G. Stokes about eclipse spectroscopy; GA has written to Stokes as well and encloses a copy of this letter.
ALS 1p, encl C(S) 3pp TxU:H/M-0642; Reel 1084 (C: RGO 6.122.179)

13437. 1867–5–8 From J. Melvill India Office
Sends a pamphlet by a friend of his, Mr. Thornton, on the squaring of the circle; would like JH's opinion on this pamphlet.
ALS 3pp RS:HS 12.321

13438. 1867–5–9 From Roderick Impey Murchison 16 Belgrave Square
Regrets he was out when JH's son called, but has sent him a card for Saturday. Pleased to hear that JH supports RM over his David Livingstone appeal. Suspects that Livingstone will turn up after a year.
ALS 4pp RS:HS 12.435

13439. 1867–5–9 To George Gabriel Stokes Collingwood
Believes that expensive telescope wanted by William Huggins is unnecessary for the intended purposes [see GS's 1867–5–3]; JH offers a telescope of his own to R.S.L.
ALS 4pp Cambridge University Library 7656/H714 (draft RS:HS 17.67 & C: RS:HS 24.199)

13440. 1867–5–10 To J. Melvill n.p.
Does not think Mr. Thornton's pamphlets solve the problem of squaring the circle as the circle is 3/4 of the circumscribed square. Comments further on these theories.
C 4pp RS:HS 12.322 & 24.200

13441. 1867–5–11 From J. Melvill India Office
Please accept his grateful thanks for his comments on Mr. Thornton's squaring of the circle. Will send these comments on to Thornton.
ALS 2pp RS:HS 12.323

13442. 1867–5–12 From Augustus De Morgan 91 Adelaide Road
The theorem is one of the consequences of the Julius Plücker system. Comments on JH's proposed method.
ALS 3pp RS:HS 6.393

13443. 1867–5–12 or earlier To George Gabriel Stokes Collingwood
Agrees reluctantly to write obituary notice of William Whewell; wishes he had been asked earlier.
ALS 3pp Cambridge University Library 7656/H733

13444. 1867–5–13 From George Gabriel Stokes Lensfield Cottage, Cambridge
Sends list of tutors at Trinity College contemporary with William Whewell; comments on telescope needed for eclipse viewing.
ALS 8pp RS:HS 17.68

13445. 1867–5–16 From Arthur J. Joyce 36 Gower St.
Outlining a series of experiments based on JH's work on *Light*.
ALS 5pp RS:HS 10.451

13446. 1867–5–16 From Charles C. Nelson 30 Hyde Park Gardens
Read JH's letter to his brother and it did not seem to affect him one way or another. Mental health is no better.
ALS 2pp RS:HS 13.91

13447. 1867–5–17 From L. Adamson Colonial Office, Cape Town, Cape of Good Hope
[Form letter] At governor's request, sends JH copy of 'Trigonometrical Survey of Southeast Coast of this Colony, by Captain W. Bailey, R.E.'
(A)LS 1p RAS:JH Archive 14/1.204 verso; Reel 10

13448. 1867–5–17 From George Gabriel Stokes R.S.L., Burlington House
Will not apply to government for money to purchase telescope for observing [1868] eclipse. William Huggins

believes that larger telescope is needed. Would JH's son [John] make these observations [for R.S.L.]?
ALS 4pp TxU:H/M–0506; Reel 1093

13449. 1867–5–18 From Charles G. B. Daubeny Botanic Garden Oxford
His collection of poetry has made good progress. Outlines the proposed contents of his book. Has not thought of including Latin compositions.
ALS 3pp RS:HS 6.24

13450. 1867–5–18 To John Herschel (son) Collingwood
The R.S.L. proposes sending telescopic and spectroscopic equipment to India to observe the solar eclipse of 1868. They want to know if son John would be prepared to make the observations. John could use some of his leave time in England to prepare.
AL 3pp inc JHS 6.54

13451. 1867–5–18 To George Gabriel Stokes Collingwood
Comments on the time needed by JH's son [John] to perform observations requested by R.S.L.; thanks for list of tutors [see GS's 1867–5–13].
ALS 3pp Cambridge University Library 7656/H715

13452. 1867–5–19 From John Herschel (son) Castle Hotel, Hastings
Wrote to G. G. Stokes and agreed to seek the permission requested by R.S.L. Will arrange for personal interview with Stokes. Is this 'independent of [Charles] Tennant's proposal'? Outlines survey plans if given charge of large instrument.
ALS 4pp TxU:H/M–0799; Reel 1085

13453. [1867]–5–20 or earlier To Margaret Brodie Herschel Collingwood
Family news; JH is very proud of the accomplishments of young daughter Constance Anne.
ALS 4pp JHS 4.4

13454. 1867–5–24 From Sir Benjamin Collins Brodie, Jr. Cowley House Oxford
Would be pleased if after reading BB's paper JH would let BB know any criticisms.
ALS 3pp RS:HS 4.293

13455. 1867–5–30 To Margaret Brodie Herschel Collingwood
News of many of the children, as well as news from London.
ALS 4pp JHS 4.5

13456. 1867–5–31 To George Gabriel Stokes Collingwood
Note accompanying JH's obituary notice of William Whewell.
ALS 1p Cambridge University Library 7656/RS581

13457. 1867–6–2 To Mr. Hutton Collingwood
Returns his Australian friend's speculations. Comments on these theories regarding atoms, etc., and gives other books and articles showing that his theories are not unique.
C 4pp RS:HS 10.220 & 24.185

13458. 1867–6–3 To Thomas Maclear Collingwood
Requests observations of Zeta Argus. Margaret Herschel travels on the continent. Liberal Party under William Gladstone unsuccessful with reforms.
ALS 4pp Cape Archives/Maclear Papers File 100

13459. 1867–6–6 From Robert Grahame Jr. to Margaret Brodie Herschel Paris
Thanks for the photographic portrait of JH, which he prizes greatly. Comments on various points of the photograph. No need to be uneasy about the rooms or cabs in Paris. Matilda (RG's sister) must rest her eyes; then she will recover quickly.
ALS 3pp RS:HS 9.14

13460. 1867–6–8 From Roderick Impey Murchison 16 Belgrave Square
Is bewildered in his old age by the new theories of James Croll on the glacial epochs. What does JH think of these theories? Sees no proof of glacial action in Paleozoic times.
ALS 4pp RS:HS 12.436

13461. 1867–6–12 To John Herschel (son) Collingwood
Writes to son John [who is now in England] further about the solar eclipse observations [see JH's 1867–5–18]; comments on family matters, and talks about making improvements in photographic processes, so that JH is able to print on both sides of the paper.
ALS 4pp JHS 6.53

13462. 1867–6–13 To Sir Benjamin Collins Brodie, Jr. Collingwood
Comments on a paper recently submitted to him by BB on chemical notation.
draft 7pp RS:HS 4.294 & 24.201

13463. 1867–6–17 From John Chapman 24 & 25 Oxford St.
British Stationery Warehouse received JH's instructions today. Gives prices and delivery time of 'despatch box.'
ALS 1p TxU:H/L-0392.1; Reel 1054

13464. 1867–6–17 From C[harles] Crowden Grammar School, Cranbrook
Is grateful that JH will preside at the prize giving. Gives details of the various prizes.
ALS 2pp RS:HS 5.377

13465. 1867–6–25 From Robert L. J. Ellery Melbourne Observatory
Notification that three volumes of the Melbourne Astronomical Observatory have been dispatched.
ASD 1p RS:HS 7.125

13466. 1867–6–26 From Sir Benjamin Collins Brodie, Jr. Cowley House Oxford
Thanks for his letter regarding chemical calculus. Further comments on his theories.
ALS 11pp RS:HS 4.295

13467. 1867–6–27 From Charles Tennant 62 Russell Square
Sends a copy of the third edition of his book, *The Bank of England*. Requests JH's support in his study of the subject.
ALS 8pp RS:HS 17.430

13468. 1867–6–30 To Mary Baldwin Collingwood
Sends MB's will, drawn up by JH, with instructions for completing and signing it. Sends MB's semiannual dividend.
AdraftS 3pp TxU:H/L-0077; Reel 1054

13469. 1867–7–1 From Charles Tennant 62 Russell Square
Expresses gratitude for JH's considerations of his work on the Bank of England. Claims the Bank needs major reform. Wants an influential voice to back his claim; asks JH to be

that voice.
ALS 7pp RS:HS 17.431

13470. 1867–7–2 From Rudolf Wolf Zürich
Because RW has had no response from JH, he is worried that if the letter and printed material were not lost, then JH must be dissatisfied with what RW had written about JH's father, William.
ALS 1p (German) RS:HS 19.302

13471. 1867–7–6 From Samuel Charles Wilks Nursling Rectory, Southampton
Expresses awe at scientific advances during SW's lifetime. Concludes that scientists since Francis Bacon failed to account for human variation when concluding experimental results.
ALS 8pp RS:HS 18.251

13472. 1867–7–7 To Rudolf Wolf Collingwood
Thanks RW for sending copies of RW's memoir on JH's father. States that JH has now revised a catalogue of all JH's father's double stars.
C 1p RS:HS 24.202

13473. 1867–7–8 To William Stanley Jevons Collingwood
Acknowledges receipt of WJ's treatise on logic, and comments on it.
ALS 3pp John Rylands University Library of Manchester (C: RS:HS 24.203)

13474. 1867–7–10 From Joseph Wilson Swan Newcastle-upon-Tyne
In light of JH's interest in and contribution to photography, encloses with explanation photographs produced by 'the Carbon process' using Indian ink coloring.
ALS 3pp RS:HS 17.251

13475. 1867–7–12 From J[ames] McDowell Cambridge
Thanks for kind reply and favorable opinion on his Geometry. Arthur Cayley also expressed a favorable opinion on his writings. Would like a testimonial from JH for the Chair of Mathematics at Cork.
ALS 2pp RS:HS 13.1a

13476. 1867–7–13 From John Bell Rolvenden Park
Regarding a suitable coachman for JH.
ALS 3pp RS:HS 5.45

13477. 1867–7–15 To J[ames] McDowell n.p.
Cannot supply testimonial as he has no knowledge of any other of JM's writings, nor does he know him personally.
draft 1p RS:HS 13.1b

13478. 1867–7–15 From Arthur B. Simpson Bexhill
Asking for answers to specific questions in hydrodynamics, which may be related to a future patent application.
ALS 6pp RS:HS 16.147

13479. 1867–7–19 To George Gabriel Stokes Collingwood
Agrees to allow Lord Oxmantown to add JH's remarks to Lord Oxmantown's paper.
C 1p RS:HS 25.10.16

13480. 1867–7–21 From Urbain J. J. Leverrier n.p.
Has lost Alexander Herschel's Paris address. Otto Struve will be visiting the Observatory on 10 Aug.
ALS 1p (French) RS:HS 11.212

13481. 1867–7–22 From John F. McLennan 22 Hill St., Edinburgh
Sends the remaining papers belonging to JH. Is grateful for his assistance.
ALS 2pp RS:HS 12.259

13482. 1867–7–22 From William Parsons (Lord Rosse) 4 Connaught Place
Laurence Parsons (Lord Oxmantown) has gone to Ireland to discharge his duties of High Sheriff. Sends finished proof of the engraving of Orion; comments on this. Has not worked on the nebula himself for many years. [J. K.?] Hunter was a good artist.
ALS 4pp RS:HS 13.233

13483. 1867–7–22 From Samuel Charles Wilks Nursling Rectory, Southampton
Thanks JH for response. Continues to criticize Francis Bacon. Quotes Scriptures: fallen nature accounts for human errors in observations. [Emphatic JH annotation: *No answer.*]
ALS 6pp † RS:HS 18.252

13484. 1867–7–23 To William Parsons (Lord Rosse) Collingwood
Thanks WP for having sent chart of the Orion nebula. Comments on its quality.
ALS 2pp Rosse Papers K2.11

13485. 1867–8–2 From Henry Wadsworth Longfellow Nahant near Boston
Expresses thanks for JH sending JH's translation of Homer's *Iliad.* Endorses JH's use of hexameter verse. Looks forward to reading it.
ALS 4pp BostonPubLib Ms.Am.230

13486. 1867–8–4 From Leonor Fresnel St. Cheron
Engaged in bringing out a new edition of A. J. Fresnel's works and has some queries.
ALS 4pp RS:HS 7.382

13487. 1867–8–5 From Matthew Arnold Chester Square
Thanking him for sending the translation of Friedrich Schiller [*Schillers Spaziergang, Translated into Latin Verse,* 1867].
ALS 2pp RS:HS 1.408

13488. 1867–8–7 To Leonor Fresnel Collingwood
Answers LF's queries of 4 Aug.
ALS 4pp (French) RS:HS 7.383 & 24.204

13489. 1867–8–7 From Robert Harley The Manse, Brighouse
Intends applying for the chair of mathematics at Queen's College, Cork and would be grateful for a testimonial from JH.
ALS 3pp RS:HS 9.241

13490. 1867–8–7 From Patrick Francis Robertson 7 Pall Mall
Thanks JH for translation of [Friedrich] Schiller's 'The Walk.' Reform bill will again be before the House tomorrow. Denounces mischief of the Radicals.
ALS 2pp RS:HS 14.371

13491. 1867–8–8 From Augustus De Morgan 91 Adelaide Road
Many thanks for the Latin translation of Frederick Schiller's poem. Comments on various points of poetry. Further about the Pascal-Newton letters. Regarding William Pearson and the founding of the R.A.S.

ALS 4pp RS:HS 6.394 S. De Morgan, *Memoir of Augustus De Morgan*, 371–2 inc

13492. 1867–8–11 From Angelo Secchi London
In London for a short time. Wants to meet JH and will come to see him the following day.
ALS 2pp RS:HS 15.417

13493. 1867–8–14 From George Biddell Airy Trossachs Hotel, Achray
Regarding earth currents and galvanic registers. Results of recent experiments on these.
ALS 8pp RS:HS 1.305

13494. 1867–8–15 From Augustus De Morgan 91 Adelaide Road
Further concerning the Newton-Pascal letters. Comments on the various line endings for poetry. Why in Runic Almanacs the days of the week begin with Monday.
ALS 4pp RS:HS 6.395 S. De Morgan, *Memoir of Augustus De Morgan*, 372–3 inc

13495. 1867–8–19 From Roderick Impey Murchison Purley Hall, Reading
Is grateful for a copy of 'Spaziergang' in Latin verses. Envies JH's varied accomplishments. Fourth edition of his *Siluria* will be issued in October. Relates how James Croll was appointed a geological surveyor in Scotland.
ALS 4pp RS:HS 12.437

13496. 1867–8–20 To Elizabeth Colling Collingwood
Thanks EC for her sonnet; talks about solar photography.
C 1p RS:HS 24.205

13497. 1867–8–20 From Norman Macleod South Kensington Museum
Portrait of William Herschel will be returned after Second National Portrait Exhibition closes 31 Aug. Gratitude for JH's loan.
ALS 2pp RAS:JH Archive 14/1.195–196 verso; Reel 10

13498. 1867–8–21 To Caroline Emily Mary Herschel Collingwood
Sending various publications by JH on light. [Responding to CH's queries], explains the nature of color and the action of the sun and stars in producing heat and light.
ALS 4pp Bodleian Library, Oxford

13499. 1867–8–21 To James Calder Stewart Collingwood
Apologizes for hurried note sent earlier. Sends correction for JH's version of the 'Well.'
ALS 1p TxU:H/L-0462; Reel 1055

13500. 1867–8–22 From William Selwyn Broomfield, Salop
Thanks JH for a 'corrected edition of Deambulatz.' [?] Angelo Secchi has been at Cambridge. Is making composite photographs showing the activity of the sun and planets in the last year.
AL 4pp inc RS:HS 15.492

13501. 1867–8–27 From James Samuelson Liverpool
[Charles] Darwin and [William] Fairbairn have started a certificate describing JS's contributions to science. Asks whether JH would sign it.
ALS 1p RS:HS 15.338

13502. 1867–8–29 From Augustus De Morgan 91 Adelaide Road
Regarding David Brewster and the inaccuracies in his life of Isaac Newton.
ALS 4pp RS:HS 6.396

13503. 1867–8–29 To William James Herschel Collingwood
Considering severe famine in Orissa province, JH believes that WJH's request for leave to return to England is untimely, particularly because WJH's superior is also leaving. Letter from [C. F.] Montresor speaks highly of WJH's expertise in irrigation. Suggests that WJH stay and help.
ALS 4pp TxU:H/L-0561; Reel 1053

13504. 1867–8–30 From Alexander J. B. Hope Bedgebury Park Cranbrook
Thanks for the gift of his poem; it has afforded him great pleasure. Is starting on a tour of South Wales and the West Country.
ALS 3pp RS:HS 9.471

13505. 1867–8–30 From Edward James Stone Royal Observatory, Greenwich
The Berlin Academy has decided to publish Charles Rümker's observations, thus removing the problem from the hands of the B.A.A.S. [see JH's 1866–7–13].
ACCS 4pp RGO 6.412.242

13506. 1867–9–2 From James Samuelson Worsley Villa, St. Domingo Grove, Liverpool
Asks JH to sign enclosed certificate over [Charles] Darwin's signature. JS will take it to B.A.A.S. meeting in Dundee next week.
AL 1p inc TxU:H/L-0760.3; Reel 1089

13507. 1867–9–3 From Gathorne Gathorne -Hardy Staplehurst
A note to thank JH for sending some Latin verses of his own.
ALS 2pp RS:HS 10.160

13508. 1867–9–7 From Augustus De Morgan 91 Adelaide Road
In one of JH's letters he mentions a name of a forger; was it G. B. Libri? It would not be easy to detect a forgery.
ALS 2pp RS:HS 6.397

13509. 1867–9–9 From Augustus De Morgan 91 Adelaide Road
Further points on the Michel Chasles forgeries.
ALS 4pp RS:HS 6.398

13510. 1867–9–9 From Eliza Susan Quincy Quincy, Mass.
Wants to send JH and [John] Stewart memoirs of her father, [Josiah] Quincy. Praises JH's *Iliad* translation) and his *Cape Results*. Introduces her nephew, General [?] Quincy.
ALS 7pp RS:HS 14.281

13511. 1867–9–11 From Julia Margaret Cameron Freshwater, I.o.W.
Her husband is well again and her own health is much improved. Would be interested to know which photographs he likes best and why.
ALS 8pp RS:HS 5.167

13512. 1867–9–12 To George Biddell Airy n.p.
It appears that the Indian government has now approved the introduction of the metric system; JH is very upset.
AL 2pp inc RGO 6.359.534

13513. 1867–9–12 To George Biddell Airy Collingwood
Is very pleased with resolution about the observations of

Charles Rümker [see Edward Sabine's 1867–8–30].
ALS 1p RGO 6.412.244

13514. 1867–9–12 To George Biddell Airy Collingwood
Comments on GA's theory of the thermo-electric origin of terrestrial magnetism.
ALS 4pp RGO 6.695.491

13515. 1867–9–12 To John Herschel (son) Collingwood
Has been observing sunspots and talks about sunspot cycles; JH is glad to see that son John has been practicing observing with the spectroscope in preparation for the Indian solar eclipse [see JH's 1867–6–12].
ALS 4pp JHS 6.52

13516. 1867–9–14 From George Biddell Airy Royal Observatory, Greenwich
Regarding the theory of magnetic currents.
ALS 2pp RS:HS 1.307 (C: RGO 6.695.493)

13517. 1867–9–14 From George Biddell Airy Royal Observatory, Greenwich
Thanking him for his paper on Indian Standards and systems of measurement.
ALS 2pp RS:HS 1.306

13518. 1867–9–14 From George Biddell Airy n.p.
Expresses GA's contempt for those who legislate for others without knowing what they are doing; GA is talking about the Indian legislation [see JH's 1867–9–12].
ACCS 2pp RGO 6.359.535

13519. 1867–9–18 From Augustus De Morgan 91 Adelaide Road
T. A. Hirst is trying to get Michel Chasles to give up his authority.
ALS 3pp RS:HS 6.399

13520. 1867–9–19 From Thomas Romney Robinson Armagh
Asks JH to sign memorial for [John?] Stevelly. Discusses excited reaction to [Michel] Chasles claims regarding Blaise Pascal and Isaac Newton.
ALS 3pp RS:HS 14.380

13521. 1867–9–26 From David Brewster Allerly by Melrose
Has heard that he has been suffering from the cramp and suggests a few remedies that he himself has found beneficial.
ALS 4pp RS:HS 4.270

13522. 1867–9–27 From James Parker Broad St., Oxford
Sends JH poems for possible publication in a selection edited by Charles Daubeny; asks JH to identify the one which he wants in the volume.
ALS 4pp RS:HS 14.137

13523. 1867–9–29 To John Herschel (son) Collingwood
Provides sources of information about polarization, spectroscopy and solar physics for son John, who is continuing his preparation for observing the solar eclipse [see JH's 1867–9–12].
ALS 4pp JHS 6.51

13524. 1867–9–30 To Eliza Susan Quincy Collingwood
EQ may forward packages for JH to [John] Stewart in London. Discusses [James] Grahame. Pleased EQ liked his translation of the *Iliad*.
C 2pp RS:HS 14.282 (C: RS:HS 24.206)

13525. 1867–10 To Richard Strachey Collingwood
Offers advice to president of committee for adoption of uniform system of weights and measures for India.
AdraftS 8pp TxU:H/L-0468 (C: 0469 & 0470); Reel 1055

13526. 1867–10 To Richard Strachey Collingwood
Offers advice to president of committee for adoption of uniform system of weights and measures for India.
Adraft 1p inc TxU:H/L-0467; Reel 1055

13527. 1867–10–12 From William Selwyn Cambridge
Lack of circularity in some solar autographs is due to clouds. Discusses a quote regarding stars in the Southern Hemisphere. Also, Aristotle's examination of the eye.
ALS 4pp RS:HS 15.493

13528. 1867–10–15 From Eliza Susan Quincy Quincy, Mass.
Sent her father's [Josiah Quincy] memoirs to London as JH directed. Praises JH's *Iliad* translation, *Cape Results,* and astronomical career.
ALS 3pp RS:HS 14.283

13529. 1867–10–15 From Henry Enfield Roscoe Owens College, Manchester
Will forward information on spectrum analysis when he receives it. Asks if JH is interested in studying the photochemistry of the sun during the upcoming eclipse.
ALS 4pp RS:HS 14.425

13530. 1867–10–16 To Augustus De Morgan Collingwood
Requests AD send JH an astronomical drinking song.
C 1p RS:HS 24.207

13531. 1867–10–18 From Charles G. B. Daubeny Botanic Garden Oxford
Editor cannot complain of JH's decision not to allow him to insert his poem on the telescope as that of Augustus De Morgan substituted is so good. Book will include poems of both dead and living men. Has unfortunately damaged the copy of De Morgan.
ALS 4pp RS:HS 6.25

13532. 1867–10–18 From Augustus De Morgan 91 Adelaide Road
His son George has died and was buried yesterday. Wife is bearing up well. Is ready to assent to the publication of the song.
ALS 4pp RS:HS 6.400 S. De Morgan, *Memoir of Augustus De Morgan*, 376–7 inc

13533. 1867–10–23 From Augustus De Morgan 91 Adelaide Road
Has managed to write out the song and insert the Blaise Pascal verse. Gave some advice to an Assurance Office yesterday, so he must be improving. Illness in his family this season has taken the form of great prostration. Gives curious confirmation of Isaac Newton's lack of knowledge of French. Blaise Pascal affair is growing into an epic.
ALS 4pp RS:HS 6.401

13534. 1867–10–25 From Henry Enfield Roscoe Owens College, Manchester
Pleased that JH's son [John] will examine the chemical intensity of the sun during the eclipse. Offers suggestions for successful experiment. Sends paper regarding chemical intensity at the tropics.
ALS 4pp RS:HS 14.426

13535. 1867–10–27 To the *Photographic News* Collingwood
Maintains that JH did not invent the thaumsacope [thaumatrope], which some have ascribed to him. JH does note that he proposed moving pictures in an 1860 publication, five years before Alonzo G. Grant sought patent protection for this idea.
L *Photographic News, 11* (Nov. 8, 1867), 541–2

13536. 1867–10–28 From Robert Hunt St. Agnes, Cornwall
Has been requested by the Editor of the *Quarterly Journal of Science* to write a paper on JH and modern astronomy. Would like JH's sanction, and also a few notes on the most important aspects.
ALS 3pp RS:HS 10.133

13537. 1867–10–29 To George Gabriel Stokes Collingwood
Asks GS to explain to R.S.L. Council why JH's son John cannot appear at a meeting of the Council, as he is due to sail for India.
ALS 4pp Cambridge University Library 7656/RS617

13538. 1867–10–30 To Robert Hunt Collingwood
In response to RH's 1867–10–28, JH sends a list of his writings on astronomy.
ALS 4pp George Eastman House, 17.17 (C: RS:HS 24.208)

13539. 1867–10–30 To Adolphe Quetelet Collingwood
Ashamed not to have written sooner. Has received AQ's meteorology of Belgium. Wishes success with his work on social physics. Enjoyed AQ's *Histoire des sciences mathématique et physique chez les belges*. Disgusted that the French insist Isaac Newton stole his ideas on the system of gravitation from Blaise Pascal. Hopes to send AQ his 'synopsis of all micrometrical measures' made by William Herschel on double stars. Sends Latin version of Friedrich Schiller's 'Spaziergang.'
AL 1p inc Académie belgique: #45 (C: #46)

13540. 1867–10–30 From John Tyndall Royal Institution
Asks whether Michael Faraday's work with glass resulted in improved manufacturing or if rather its main scientific value was in discovery of diamagnetism.
ALS 2pp JH note RS:HS 17.401 (C: RI 541)

13541. 1867–11–1 From James Forrest 25 Great George St., Westminster
[Form letter] Schedule of meetings for Institution of Civil Engineers. As honorary member, JH is invited to attend.
(A)LS 1p RAS:JH Archive 14/3.7.2 verso; Reel 11

13542. 1867–11–1 From Robert Hunt Penzance
Is grateful for his letter and will study the memoirs to which he refers. The Editor would also like a photograph; can JH oblige?
ALS 3pp RS:HS 10.134

13543. 1867–11–2 To Elizabeth Drummond n.p.
JH is 'shaky & feeble.' Pleased with biography of ED's brother Thomas. JH's son John and bride departed today for India. Deaths of JH's contemporaries. Describes Constance Herschel's whooping cough.
AL 4pp inc RS:HS 19.159 (C: RS:HS 24.209)

13544. 1867–11–3 To George Biddell Airy [Collingwood]
Writes to inquire as to the exact definition of the gallon and the pound, and the circumstances under which the appropriate measures would be made.
ALS 2pp RGO 6.359.536

13545. 1867–11–3 To John Tyndall Collingwood
Believes Michael Faraday's boro-silicate of lead has not been used for telescopes, but JH and others have tried using it for other optical purposes. Describes method for producing veinless flint glass.
C 2pp RS:HS 24.210

13546. 1867–11–4 From George Biddell Airy Royal Observatory, Greenwich
Provides JH with best definition that GA has available for the gallon [see JH's 1867–11–3], noting that methods for establishing standards are revised, even if the standards supposedly are not.
ALS 4pp TxU:H/M-0643.1; Reel 1084 (C: RGO 6.359.537)

13547. 1867–11–4 From Thomas Whitehouse Seager St., West Bromwich
Plans for refracting telescope TW is constructing. Diagrams and questions about lens construction.
ALS 3pp † RS:HS 18.402

13548. 1867–11–5 To The East India Company Collingwood
Feels that he has been quoted out of context, making it appear that JH favors the introduction of the metric system in India. This being not the case, JH then proceeds to show how the British system of measures is related to terrestrial measurements, especially the length of the earth's polar axis. In so doing, JH argues for the greater accuracy in the definition of the British units, as compared to the metric.
ALS 8pp APS 509.L56.20

13549. 1867–11–5 To Richard Strachey Collingwood
Offers advice to president of committee for adoption of uniform system of weights and measures for India.
C 8pp TxU:H/L-0469 & 0470; Reel 1055

13550. [1867]–11–5 To J[ohn] H[enry] Whitaker Collingwood
Has nothing of Mendelsohn to send to JW. Has R.S.L. letters, but JH does not want to part with them.
ALS 2pp TxU:H/L-0769.1; Reel 1089

13551. 1867–11–8 From Rodolphe Falb Graz
Sending details of his theory on meteorology and cometography and inviting JH's comments.
ALS 4pp (French) RS:HS 7.364

13552. 1867–11–8 From Henry Samuel King 65 Cornhill
Soliciting articles for *People's Magazine*.
ALS 4pp RS:HS 11.38

13553. 1867–11–11 From William Selwyn Vine Cottage, Cambridge
Sends JH a copy of 'Enoch.' Thanks JH for his helpful criticisms.
ALS 3pp RS:HS 15.494

13554. 1867–11–18 From George Gabriel Stokes Lensfield Cottage, Cambridge
About some instruments to be taken to India by JH's son John.
ALS 4pp RS:HS 17.69

13555. 1867–11–20 To George Gabriel Stokes Collingwood
Does not know if equipment from R.S.L. reached JH's son

John at Southampton [see GS's 1867–11–18].
ALS 4pp Cambridge University Library 7656/H717

13556. 1867–11–21 To J. Melvill Collingwood
Please transmit enclosed recommendations by JH to Richard Strachey or other members of committee to establish uniform system of weights and measures for India.
AdraftS 1p TxU:H/L-0763; Reel 1089

13557. [1867–11–22] To Charles Babbage n.p.
News of the forthcoming marriage of his daughter. Everyone seems to be mad on making ice.
ALS 4pp RS:HS 2.389

13558. 1867–11–23 From J. Melvill India Office
Impressed by JH's argument against French metric system. Will forward JH's letter to Richard Strachey or, if Strachey is in Abyssinia, to committee for weights and measures in India.
ALS 3pp TxU:H/M–0355; Reel 1087

13559. 1867–11–26 To *People's Magazine* Collingwood
Describes a method JH devised of producing 'autographic representations of fungi on glass.'
L *People's Magazine,* new ser., *1* (Jan., 1868), 62–3

13560. 1867–11–28 From William James Herschel Midnapore
Reports his opposition, shared by JH, to Britain and India adopting the metric system. Discusses WH's wife, Emma, WH's brother John, a coming solar eclipse, and a theory of gravity.
ALS 8pp Sydney Ross Collection

13561. 1867–11–29 To George Biddell Airy Collingwood
Concerned about 'Commercial weight' of new standard of weights and measures.
ALS 4pp TxU:H/L-0025; Reel 1054

13562. 1867–11–29 From Pierce Butler Ulcombe Rectory, Staplehurst
[Last page only] Seeks financial support for exploratory survey of Sinai Peninsula.
L 1p inc TxU:H/MW-0246.5; Reel 15

13563. 1867–11–29 To William Hallowes Miller Collingwood
Suspects that discrepancy between old and new commercial weights of Standard Pound may be due to trapped air in new standard. Needs WM's reply soon, so JH may inform India Committee on weights and measures.
C 4pp TxU:H/L-0261; Reel 1054

13564. 1867–11–30 From George Biddell Airy Royal Observatory, Greenwich
Writes to acknowledge receipt of letter [see JH's 1864–11–25], but it will take time to work through JH's calculations carefully.
ALS 2pp TxU:H/M-0643.2; Reel 1084 (C: RGO 6.359.542)

13565. 186[7]–12–1 To Edwin Dunkin Collingwood
Deals with several problems related to several papers.
ALS 3pp WT 67469

13566. 1867–12–2 To George Biddell Airy Collingwood
Is still trying to bring the various definitions of the pound into agreement [see GA's 1867–11–30]; the question of weighing in air or vacuum remains as a problem.
ALS 4pp RGO 6.359.543 (C: TxU:H/L-0026)

13567. 1867–12–2 From Alexander Hamilton-Gordon Bombay
Is pleased to hear his own sons are progressing well. Hopes Caroline (JH's daughter) will soon be able to come out. Is temporarily in charge of the Bombay army, which gives him plenty of work.
C 1p RS:HS 8.154

13568. 1867–12–2 To Richard Strachey Collingwood
'Not Sent.' Clarifies values quoted in JH's letter of 5 Nov. to RS.
C 4pp TxU:H/L-0471.1 & -0471.2; Reel 1055

13569. 1867–12–3 From George Biddell Airy Royal Observatory, Greenwich
Thanks JH for his useful letter [see JH's 1867–12–2]; it will be of value to the Commission on Weights and Measures.
ALS 1p † TxU:H/M-0643.3; Reel 1084 (C: RGO 6.359.545)

13570. 1867–12–3 To George Biddell Airy Collingwood
Having helped to create confusion surrounding the pound weight [see GA's 1867–12–3], JH now understands it simply and clearly; JH wonders if he is 'growing elderly.'
ALS 4pp RGO 6.359.546 (C: TxU:H/L-0027)

13571. 1867–12–[3] From William Hallowes Miller 7 Scroope Terrace, Cambridge
Compares values obtained by various observers in measuring new standards of weight and length.
ALS 4pp TxU:H/M–0371; Reel 1087

13572. 1867–12–4 From George Biddell Airy Royal Observatory, Greenwich
Thanks JH for his clear letter [see JH's 1867–12–3] about the pound weight standards.
ALS 2pp TxU:H/M-0643.4; Reel 1084 (C: RGO 6.359.548)

13573. 1867–12–4 From Adolphe Quetelet n.p.
Apologizes for not writing sooner. Busy with trip to Italy. Announces intention to dedicate entire next year to his *Physique sociale*. Thanks JH for his comments on AQ's son's work on temperature. Also grateful for commendation on AQ's history of science. Brings up AQ's son's work on proper motion.
C 2pp (French) Académie belgique: #47

13574. 1867–12–5 To Pierce Butler Collingwood
Describes JH's theodolite by G. F. Reichenbach and offers to loan it for PB's survey [of Sinai Peninsula].
Adraft 4pp TxU:H/L-0119; Reel 1054

13575. 1867–12–5 From F. A. Herschel Griesbach Hull
Was grateful for his kind note and check. Family is still in difficulties. Employed by a Liverpool firm to sell engineering articles.
ALS 4pp RS:HS 9.136

13576. 1867–12–5 To William Hallowes Miller Collingwood
Discusses WM's reply regarding commercial weight of new Standard Pound.
ACCS 3pp TxU:H/L-0262; Reel 1054

13577. 1867–12–6 To [William Hallowes Miller] Collingwood
Measurements by A. T. Kupffer afford greater accuracy for [specific] gravity of water than values used by Henry Kater as quoted in postscript of JH's 5 Nov. [1867] letter to [WM].
AdraftS 1p inc TxU:H/M–0107.2; Reel 1087

13578. 1867–12–6 From Adolphe Quetelet Brussels
Discusses coming year. Asks JH how to deal with JH's

review [of AQ's *Lettres sur la théorie des probabilités*]. Talks about his friend [Michel] Chasles. Also about AQ's son. Mentions JH's translation of [J. C. F. von] Schiller's 'Spaziergang.'
ALS 3pp (French) RS:HS 14.266

13579. 1867–12–6 To Richard Strachey Collingwood
Clarifies information in postscript of JH's letter of 5 Nov. to RS.
C 3pp TxU:H/L-0472; Reel 1055

13580. 1867–12–7 From William Selwyn Vine Cottage, Cambridge
Discusses translation of 'Enochus.' Thanks JH for the 'Rex Colius,' which amused Adam Sedgwick and WS.
ALS 4pp RS:HS 15.495

13581. 1867–12–9 To Elizabeth Colling Collingwood
Comments on reports of meteor sightings from the United States and Italy; explains why one may see a satellite of Jupiter where there is none.
C 2pp RS:HS 24.211

13582. 1867–12–10 From Joseph Alfred Hardcastle 4 Chesham [St.?]
Please clarify differences between French metric system and JH's adaptation of British system of measurements. Willing to raise issue in Parliament, but wants to understand it well. Hopes to see Emma soon.
ALS 3pp TxU:H/M–0241; Reel 1087

13583. 1867–12–17 From William Rutter Dawes Hopefield, Haddenham, Thame
Is exceedingly grateful for the Synopsis of William Herschel's double-star observations. Comments on some of the readings. Has read his translation of Schiller's 'Spaziergang' with pleasure. Will be writing to Margaret Brodie Herschel.
ALS 5pp RS:HS 6.124

13584. 1867–12–20 From Edward William Brayley London Institution
Sending the syllabus of Ellis's lecture. Has sent to JH's son [Alexander] a paper on meteors, which contains his views in cosmical philosophy. Elected Laurence Parsons (4th Earl of Rosse) to the R.S.L. yesterday.
ALS 3pp RS:HS 4.235

13585. 1867–12–23 From Henry Hudson Glenville, Fermoy
As JH has approved Dr. W. C. Wells's *Theory of Dew* he encloses extracts from the *Gardeners' Chronicle* to show that Wells based his paper on a false theory.
ALS 1p RS:HS 10.28

13586. 1867–12–28 From William Selwyn College, Ely
Discusses the translation of 'Enochus' and various suggestions he has been given for its improvement.
ALS 4pp RS:HS 15.496

13587. 1867–12–29 From Eliza Susan Quincy Boston, Mass.
Asks if JH received a copy of the memoirs of her father [Josiah Quincy]. They will be a 'valuable addition to the history of the U.S.'
ALS 4pp RS:HS 14.284

13588. 1867–12–29 From Alfred Smee 7 Finsbury Circus
Note accompanying a copy of a paper by AS.
ALS 1p RS:HS 16.173

13589. 1867–12–31 From William Rutter Dawes Hopefield, Haddenham, Thame
Is trying to obtain parts of *R.S.P.T.* containing the original communication of William Herschel on changes in double stars. Would be glad of JH's assistance. Printer has made mistakes in spite of his proof corrections. Weather most unsuitable for observations due to fog. Is pleased JH is making a synopsis of all the double-star measurements.
ALS 6pp RS:HS 6.125

13590. 1867–12–31 From John Henry Griesbach 48 Bedford Gardens
Would like him to accept his little work on musical sounds. F. A. Gore-Ouseley has invented some new organ pipes.
ALS 2pp RS:HS 9.65

13591. [1868 or 1869] From Charles G. B. Daubeny n.p.
Printed prospectus for volume of poems on Natural History and Physical Science.
D 2pp RS:HS 6.28

13592. [1868 or later] To Georg A. Erman n.p.
Comments on the bad habits creeping into the language of the German Academies. Sends list of papers and also a transcript of GE's paper for the *Astronomische Nachrichten*.
draft 3pp RS:HS 7.113

13593. 1868–1–4 From Henry Hudson Glenville, Fermoy
Makes no claim to originality for his experiments; thinks highly of Dr. W. C. Wells but would like to know whether [Pierre] Prévost's theory is tenable.
ALS 4pp RS:HS 10.29

13594. 1868–1–8 From William Rutter Dawes Hopefield, Haddenham, Thame
Is grateful for his prompt reply about his father's papers on double-stars. Had no idea that Angelo Secchi's Means were so inaccurate. Gives the principle on which his own Means are based. The weather has been wretched and most unsuitable for any observations.
ALS 5pp RS:HS 6.126

13595. 1868–1–8 From Friedrich August Theodor Winnecke 17 Kriegstrasse, Karlsruhe
Needs some information about the astronomical sweeps carried out by JH's father, William. FW comments about the location and shape of some nebulae, and what seem to be differences with other catalogues.
ALS 4pp (German) †† RS:HS 18.272

13596. 1868–1–11 From Henry Samuel King 65 Cornhill
Did not contemplate gratuitous contributions from JH.
ALS 4pp RS:HS 11.39

13597. 1868–1–12 To William Henry Fox Talbot Collingwood
Asks WT whether WT made a certain claim about the spectrum of strontium. Recounts observations of the strontium spectrum made by JH in 1822.
ALS 4pp Science Museum Talbt 1/44

13598. 1868–1–13 To Eliza Susan Quincy Collingwood
Received memoirs of [Josiah] Quincy. Has not had time to peruse the work in its entirety. Thanks EQ for the book.
C 2pp RS:HS 14.285 (C: RS:HS 24.5)

13599. 1868–1–14 From John Gorham Tonbridge
Some years ago JH witnessed his experiments on the rotation of discs. Has written a paper on this subject and would be glad of his comments on whether he has chosen the right

forms for the colors of the circle.
ALS 2pp RS:HS 8.223

13600. 1868–1–25 To *Photographic News* [Collingwood]
Points out that [Joel?] Spiller's article, 'Hyposulphite of Ammonia for fixing,' is 'copied verbatim' from a work of JH's.
ALS 4pp RS:HS 14.160

13601. 1868–1–26 To John Herschel (son) [Collingwood]
Has heard from Emma, son William's wife, that the solar eclipse will occur in the rainy season, and thus will not likely be visible [see JH's 1867–9–29]; JH is wondering if the equipment from the R.S.L. arrived safely.
ALS 4pp JHS 6.50

13602. 1868–1–31 To the *Photographic News* Collingwood
Discusses the action of hyposulfite of ammonia on muriate of silver. Gives directions for preparing hyposulfite of silver.
L *Photographic News, 12* (Jan. 31, 1868), 57

13603. 1868–2–5 From William Selwyn Vine Cottage, Cambridge
Will assume trusteeship if he is named in the document. Discusses trying to find Neptune with his telescope.
ALS 4pp RS:HS 15.497

13604. 1868–2–8 From Secretaries to Beresford-Hope's Committee British Hotel, Cockspur St.
[Printed notice] Ask election support for A. J. Beresford-Hope, representing Cambridge University.
L 1p TxU:H/L-0232.5; Reel 1054

13605. 1868–2–8 From Alexander J. B. Hope British Hotel Cockspur St.
Hopes he can have JH's vote. Would like to add JH's name to his committee, which would not involve him in any work.
ALS 3pp RS:HS 9.472

13606. 1868–2–8 From Frederic J. Jackson Bawtry
Sends copy of FJ's new theory of musical scale. Asks JH's opinion.
ALS 2pp TxU:H/M–0291; Reel 1087

13607. 1868–2–10 From James Graham London
Sending the dimensions of the Great Stone in the quarry near Baalbek. Has sent his paper to the Consul in Egypt. Greatly enjoyed his stay at the Herschels'.
ALS 6pp RS:HS 9.114

13608. 1868–2–10 From Alexander J. B. Hope British Hotel Cockspur St.
Many thanks. There is no name he prizes more than JH's.
ALS 2pp RS:HS 9.473

13609. 1868–2–19 From Thomas R. Jones Hopefield, Haddenham, Thame
Death of TJ's uncle, W. R. Dawes. Funeral is tomorrow.
ALS 2pp TxU:H/M–0309; Reel 1087

13610. 1868–2–21 To William Selwyn Collingwood
Discusses the sudden motion of some sunspots. Does not think planetary action is a very probable cause. Discusses upcoming conjunction of Venus and Jupiter.
C 2pp RS:HS 24.212

13611. 1868–2–29 From Hubert Airy Flamsteed House, Greenwich
Sends article on temporary color-blindness. G. B. Airy's book on sound [*On Sound and Atmospheric Vibrations*] is finished.
ALS 2pp RS:HS 19.306

13612. 1868–3–1 To Elizabeth Colling Collingwood
Comments on the death of a number of friends, his own poor health, and how he spent the winter working through his double-star observations.
C 2pp RS:HS 24.213

13613. 1868–3–3 From Julia Margaret Cameron Little Holland House
Sending him photographs and mounts for him to sign and return. Can keep one for himself. Remembers the day she spent with him.
ALS 5pp RS:HS 5.168

13614. 1868–3–3 From William Selwyn Vine Cottage, Cambridge
Has autographs showing the spots' motion. Will send them soon. Expresses regret at the death of W. R. Dawes. Refers to recent article on the optician George Merz.
ALS 4pp RS:HS 15.498 GMB

13615. 1868–3–5 To Frederic J. Jackson Collingwood
Difficulty understanding notation used in [FJ]'s pamphlet on musical scales. JH's comments on [?]'s formulas and conclusions.
ALS 7pp TxU:H/L-0738; Reel 1089

13616. 1868–3–7 From Edward William Brayley London Institution
Giving some lectures and asks JH for permission to quote from his writings,
ALS 4pp RS:HS 4.236

13617. 1868–3–7 From A. J. Simpson Slough
Requests financial assistance in building a new church in JH's parish.
ALS 4pp RS:HS 17.208

13618. 1868–3–9 To A. J. Simpson Collingwood
Replies regretfully that other urgent financial demands prevent JH from helping to build another church in AS's parish. Already is helping with one such church.
ALS 1p RS:HS 17.209

13619. 1868–3–10 From H. Turberville Aberavon, Glam.
Asks for JH's advice over a defective pair of glass discs. Discusses possibility of manufacturing optical discs in conjunction with the artist.
ALS 10pp RS:HS 17.378a

13620. 1868–3–12 From Henry Enfield Roscoe and Joseph Baxendell Manchester
[Form letter] Notice of next meeting of Manchester Literary and Philosophical Society.
L 1p RS:HS 19.161a

13621. 1868–3–12 From Mary Somerville Naples
Thanks JH for saving her from publishing a fault-filled book. Describes eruptions of Vesuvius. Agnes [Greig] enjoyed visiting the Herschels. MS wishes she could see them again.
ALS 2pp RS:HS 16.375

13622. [1868]–3–14 From Arthur Cayley Cambridge
Has returned sheets one and two to the printers and is sending the corresponding part of the manuscript. Please send his corrections to J. Henderson.
ALS 2pp RS:HS 5.229

13623. 1868–3–14 From Edward Loyd Hawkhurst
The Archbishop [of Canterbury, C. T. Longley] will be lunching with EL on the 24th and if JH is able will he join them for lunch.
ALS 3pp RS:HS 11.348

13624. 1868–3–14 From William C. Mathison Guildford
Has received with great pleasure JH's obituary of their dear friend [William Whewell]. Such a clear account was much needed.
ALS 1p RS:HS 13.27

13625. 1868–3–16 From Adam Sedgwick Kendal
Eyesight is failing. Discusses William Whewell's death and pays him tribute. Says Professor [W. H.] Thompson will replace Whewell.
ALS 10pp RS:HS 15.456

13626. 1868–3–17 To H. Turberville Collingwood
Explains HT's problem with the defective telescope glass. Advises caution about entering optical glass manufacturing. Urges HT to take back his will from JH's possession and entrust it to the care of a legal advisor.
ACC 2pp inc RS:HS 17.378b (C: 24.214)

13627. 1868–3–23 From Frederic J. Jackson Bawtry
Responds to JH's critique of FJ's theory of construction of musical scale.
ALS 6pp TxU:H/M–0292; Reel 1087

13628. 1868–3–24 From Arthur Cayley Cambridge
Has asked J. Henderson to send him the revised proof of JH's paper 'Additional Identification of Double Stars.' It will appear in the *Monthly Notes* and later in the *Memoirs*.
ALS 2pp RS:HS 5.230

13629. 1868–3–24 From William Lassell Ray Lodge, Maidenhead
Sending his impression, formed during a recent visit to Dublin, of the Great Telescope destined for Melbourne. Considers it a triumph of mechanical invention.
ALS 7pp RS:HS 11.168

13630. 1868–3–27 From George Biddell Airy Royal Observatory, Greenwich
Has received the bill of William Ewart concerning the metrical scale.
ALS 1p RS:HS 1.309

13631. 1868–3–27 From George Biddell Airy Royal Observatory, Greenwich
Further regarding his theories on sound.
ALS 2pp RS:HS 1.308

13632. 1868–3–27 From H. Turberville Barnstaple
Explains conditions of his will and the distribution of his estate.
AL 4pp inc RS:HS 17.379

13633. 1868–3–29 To George Biddell Airy Collingwood
Suggests reducing the number of Greenwich astronomical observations printed, but not reducing the magnetic and meteorological observations.
ALS 4pp RGO 6.16.5

13634. 1868–3–30 From Roderick Impey Murchison n.p.
Death of Pierce Butler. Recommended F. W. Holland, who has explored Palestine, to Henry James. RM has been too busy to keep track of survey fund.
ALS 4pp TxU:H/M–0384; Reel 1087

13635. 1868–4–1 From Russell Henry Manners Henrietta St.
Forwards a drawing of the great nebula of Orion, which he has just received from Angelo Secchi for JH. Quotes Secchi's remarks.
ALS 3pp RS:HS 12.286

13636. 1868–4–3 From George Biddell Airy Royal Observatory, Greenwich
Proposes a principle for the establishment of musical scales to satisfy musicians, not mathematicians, as JH had apparently done.
ALS 2pp TxU:H/M-0644.1; Reel 1084

13637. 1868–4–3 To Roderick Impey Murchison Collingwood
In Dec. 1867, JH offered small theodolite to Pierce Butler for survey of Sinai peninsula, and agreed to serve as co-trustee of Sinai Survey Fund with Henry James. Heard nothing until news of Butler's death and announcement that F. W. Holland and George Williams would take leadership and that RM was involved. Renews JH's offer, but health prevents JH from attending meetings.
AdraftS 4pp TxU:H/L-0271; Reel 1054

13638. 1868–4–3 From William Selwyn Foxton Royston
Has made arrangements to have copies of solar autographs sent. Discusses sunspots. Inquires whether JH knows of the University of Lund in Sweden.
ALS 4pp RS:HS 15.499

13639. 1868–4–6 To Alexander J. B. Hope n.p.
Opposes parliamentary bill to adopt French metric system. Explains weaknesses of system.
ACCS 4pp TxU:H/L-0091; Reel 1054

13640. 1868–4–8 From Alexander J. B. Hope Bedgebury Park, Cranbrook
Thanks JH for calling AH's attention to 'mischievous & preposterous metric measure bill.' Will work to defeat it [in Parliament].
ALS 4pp TxU:H/M–0082; Reel 1087

13641. 1868–4–12 From Eliza Susan Quincy Boston, Mass.
Thanks JH for his *Familiar Lectures*. Sends meteorological information.
ALS 3pp RS:HS 14.286

13642. 1868–4–15 From William Selwyn Foxton Royston
Is graphically charting the sun and planets to compare them. Explains the lines and positions of bodies in his diagrams.
AL 4pp inc RS:HS 15.500

13643. 1868–4–17 To Margaret Brodie Herschel St. Leonards
About arrangements for JH's return home, together with information about paying bills and the state of JH's finances.
ALS 4pp JHS 5.26

13644. 1868–4–18 To Augustus De Morgan Collingwood
Comments on a number of mathematical matters, on a book on positivism, and increased sunspot activity.
C 3pp RS:HS 24.215

13645. 1868–4–21 To [Henry W.] Field[?] Collingwood
Hopes HF and his wife can join the Herschels for lunch with [P. F.] Robertson at Collingwood. Gives information on transportation.
ALS 4pp BostonPubLib Ms.196(1)

13646. 1868–4–24 To [?] n.p.
Relation of British imperial units of weight to their 'geometrical synonyms.'
Adraft 1p inc TxU:H/M–0296.2; Reel 1087

13647. 1868–4–25 To [?] n.p.
Relation of imperial measures of capacity to their 'geometrical synonyms.'
Adraft 1p TxU:H/M–0341.2; Reel 1087

13648. [1868]–4–28 From Joseph Alfred Hardcastle 27 Pall Mall
Problem of Irish church has forced postponement of Parliamentary discussion of metric bill until 13 May.
ALS 2pp TxU:H/M–0242; Reel 1087

13649. 1868–4–28 From Alexander J. B. Hope Arklow House, Connaught Place
Working on rejection of metric bill [in Parliament]. Grateful for information provided by JH.
ALS 1p TxU:H/M–0083; Reel 1087

13650. 1868–4–28 To the Manchester Literary and Philosophical Society Collingwood
Attempts to clear up a misstatement made before the Society by J. C. Dyer concerning JH's views in regard to optics, especially in relation to the luminiferous ether.
L *Manchester Literary and Philosophical Society Proceedings, 7* (1868), 193

13651. 1868–4–28 From Samuel Tyler Washington, D.C.
Sends copy of ST's *Progress of Philosophy in the Past and in the Future*.
ALS 3pp RS:HS 17.456

13652. 1868–4–29 From Patrick Francis Robertson n.p.
Received JH's note. Will vote as JH advised. Bill is scheduled [to be read] on 13 May.
ALS 1p TxU:H/M–0452; Reel 1093

13653. 1868–4–30 From Hubert Airy Flamsteed House, Greenwich
Observations on various forms of [hemianopsia], which afflicts HA, his father [G. B. Airy], and JH. Detailed description of recent attack. Seven sketches with captions.
ALS 8pp RS:HS 19.307

13654. 1868–4–30 From Augustus De Morgan 91 Adelaide Road
JH's account of his own health is much better than he hoped for. Has he seen AD's contribution to English?
ALS 4pp RS:HS 6.402

13655. [1868]–4–30 From Joseph Alfred Hardcastle 27 Pall Mall
Thanks for JH's notes on metric system. Will study these before Parliamentary debate on 13 May. Calculates that 63 percent of exports go to British possessions, while only 18 percent go to Continent. Sees this as argument against introducing metric system.
ALS 3pp TxU:H/M–0243; Reel 1087

13656. [1868–4–30 or earlier] From Clements R. Markham and Richard Henry Major 15 Whitehall Place, S.W.
[Printed notice] Invitation to meetings of Royal Geographical Society.
L 1p TxU:H/L-0232.2; Reel 1054

13657. 1868–4–30 To the London *Times* Collingwood
Addresses problems in metric bill introduced to Parliament by William Ewart. Argues for the retention of the British system of weights and measures. Shows that the inch can be defined as one five hundred millionth of the length of the earth's polar axis.
Adraft 5pp (on TxU:H/L-0232.1 through L-0232.5) TxU:H/L-0232.6; Reel 1054 London *Times* (May 2, 1868), p. 12, col. 6

13658. 1868–5–1 From Patrick Francis Robertson 12 Leadenhall St.
Encloses papers from J. B. Smith, supporter of weights and measures bill. [JH's daughters] Maria and Amelia visited House [of Commons] yesterday, but ladies's gallery was full and they could not observe proceedings. Trial of American president [Andrew Johnson] is political farce.
ALS 2pp TxU:H/M–0453; Reel 1093

13659. 1868–5 To James Samuelson n.p.
JH is bound by promise to J. M. Cameron not to sit for portraits by other photographers. Will not sign photograph of JH sent by JS if this is intended for publication. Offers to write article on musical scale for *Quarterly Journal of Science*.
ACCS 4pp TxU:H/L-0380; Reel 1054 (Cdraft: RS:HS 25.15.25)

13660. 1868–5–2 From Joseph Alfred Hardcastle 27 Pall Mall
Why did French chose arc from Dunkirk to Barcelona as basis for metric measurements? Explain meridian lines. Why are linear metric measurements founded on arc, rather than on straight line? Criticizes William Ewart for introducing metric bill.
ALS 4pp TxU:H/M–0244; Reel 1087

13661. 1868–5–5 From Patrick Francis Robertson House of Commons
Returns petition to JH. Read JH's paper. Venomous propositions [in House of Commons].
ALS 1p TxU:H/M–0454; Reel 1093

13662. 1868–5–8 From Hubert Airy Flamsteed House, Greenwich
JH's approval encourages HA to publish account [of hemianopsia]. JH may keep HA's drawings. Notes on attack suffered by JH.
ALS 3pp RS:HS 19.308

13663. 1868–5–8 To Elizabeth Colling Collingwood
Comments on several aspects of poetry, including EC's; JH has been quite ill; talks about walking on water with a water velocipede.
C 2pp RS:HS 24.216

13664. 1868–5–10 From Charles Pritchard Freshwater, [Isle of Wight]
Informal request for JH to write biography of W. R. Dawes for R.A.S. Recent marriages in CP's family.
ALS 3pp TxU:H/M–0411; Reel 1087

13665. 1868–5–14 From Alexander J. B. Hope Arklow Ho. Connaught Place
Sees what a majority William Ewart's bill had yesterday. Comments on the events leading to this. So glad of the marriage of Amelia Herschel.
ALS 4pp RS:HS 9.474

13666. 1868–5–20 To *Daily News* Collingwood
JH's views on relation of British Imperial measures to length of Earth's axis were misrepresented in 18 May article. Explains them in detail. Notes scientific basis of British system.

C 4pp TxU:H/L-0134; Reel 1054 *Daily News* (May 27, 1868), p. 6, col. 4

13667. 1868–5–21 To John Herschel (son) Collingwood
Speaks of the awkwardness of some telescopic instruments, but urges son John to persevere; comments favorably on daughter Amelia's prospective husband, Thomas Wade, and comments disparagingly about the government's introduction of a metrication bill.
ALS 4pp JHS 6.49

13668. 1868–5–22 From George Biddell Airy Royal Observatory, Greenwich
More on musical scales [see GA's 1868–4–3]; GA here proposes the use of hyperbolic logarithms to help establish the scales.
ALS 2pp † TxU:H/M-0644.2; Reel 1084

13669. 1868–5–23 To George Biddell Airy Collingwood
Not well enough to attend Visitation Day at the Royal Observatory; fears JH will not likely attend any future meetings and should perhaps withdraw from the Board of Visitors.
ALS 1p RGO 6.16.31

13670. 1868–5–25 From Henry James War Office, 4 New St., Spring Gardens
Asks JH's approval for naming George Williams to take charge of Sinai survey fund following death of Pierce Butler. Will ask Secretary of State for War [J. S. Pakington] to place H. S. Palmer in charge and assign draftsmen and photographers to assist. Suggests survey goals. F. W. Holland agreed to join survey party. Estimated cost of survey is £1220.
ALS 4pp TxU:H/M–0293; Reel 1087

13671. 1868–5–26 From George Biddell Airy Royal Observatory, Greenwich
Stating that the Board of Visitors could not afford to lose the services of JH.
ALS 2pp RS:HS 1.311 (C: RGO 6.16.32)

13672. 1868–5–26 From George Biddell Airy Royal Observatory, Greenwich
Is sending William Pole's protest against the system of equal temperament in music [abstract of protest given].
ALS 1p RS:HS 1.310

13673. 1868–5–26 From Philip Kelland Wardie, [Edinburgh]
Just received JH's paper. Will need two weeks to review it.
ALS 2pp TxU:H/M–0315; Reel 1087

13674. 1868–5–26 From Sir William Pole 3 Storey's Gate
Thanks JH for his note. Asks that JH's paper on scales [1868 'On Musical Scales'] be made available to him.
ALS 1p RS:HS 14.28

13675. 1868–5–27 From Thomas R. Jones Trinity Parsonage, Huddersfield
Encloses biographical notes on W. R. Dawes.
ALS 2pp TxU:H/M–0310; Reel 1087

13676. 1868–5–29 From Alexander J. B. Hope Arklow Ho. Connaught Place
Sending a copy of his speech on Weights and Measures. Will be spending the recess in Dorset.
ALS 1p RS:HS 9.475

13677. 1868–5–29 From William Selwyn Foxton Royston
Discusses his diagram of planetary distances to the sun and earth. Will soon have autographs of sun showing the 'change-up' of a spot.
ALS 4pp RS:HS 15.501

13678. 1868–6–1 From James Samuelson Liverpool
Will publish JH's article on musical scales in October issue[?] of his journal (*Quarterly Journal of Science*). Discusses JH's opinions on copyrights.
ALS 3pp RS:HS 15.339

13679. 1868–6–2 From Thomas R. Jones Trinity Parsonage, Huddersfield
Additional biographical notes on W. R. Dawes.
ALS 4pp TxU:H/M–0311; Reel 1087

13680. 1868–6–6 From Thomas R. Jones Trinity Parsonage, Huddersfield
W. R. Dawes's only publications were one or two sermons and astronomical observations at George Bishop's observatory. Sent Dawes's measurements of double stars to William Huggins. Dawes's diaries did not begin until 1852.
ALS 3pp TxU:H/M–0312; Reel 1087

13681. 1868–6–10 From Roderick Impey Murchison School of Mines
Tyndall wishes to resign his lectureship at the Royal Institution. If Alexander [Herschel] would like this position, RM can recommend his appointment to the Governors.
ALS 4pp RS:HS 12.438

13682. 1868–6–12 From A. & C. Black Edinburgh
The three volumes will be dispatched from the binders next week. Should he require any more he will be charged at trade price. They wish the sales would improve.
ALS 1p RS:HS 5.73

13683. 1868–6–12 From Roderick Impey Murchison School of Mines
Outlines the constitution of the Royal School of Mines. Now finds he has to consult other professors before he can recommend Alexander [Herschel] for a Lectureship.
ALS 4pp RS:HS 12.439

13684. 1868–6–15 From George Knott Cuckfield
Regarding JH's comments on a paper of Francis Abbott read at the R.A.S.'s meeting on 12 June. Believes the paragraph in question came from a paper of Sir William Herschel. Understands that Lieut. John Herschel has just examined this nebula.
ALS 2pp RS:HS 11.88

13685. 1868–6–20 From St. John Vincent Day Buchanan St., Glasgow
Is grateful for JH's examination of his paper on the Pyramids. Comments on various aspects of this paper. Thanks for his two papers on weights and measures.
ALS 5pp RS:HS 6.479

13686. 1868–6–21 From Adolphe Quetelet Brussels
Offers translation of JH's review [of AQ's *Lettres sur la théorie des probabilités*] to look over. Has been updating his own text to reflect progress.
ALS 3pp (French) RS:HS 14.267

13687. 1868–6–21 From Adolphe Quetelet n.p.
Has worked much on revision of his *Physique sociale*. Asks for JH's changes to JH's review in *Edinburgh Review*. Puts

translation at JH's disposal.
draft (French) inc 1p Académie belgique: #48

13688. 1868–6–22 To Adolphe Quetelet Collingwood
Flattered that AQ wants JH's remarks to preface AQ's new work. No need to send French translation of JH's work. Asks AQ to omit certain passages.
ALS 3pp Académie belgique: #49 (draft: #50)

13689. 1868–6–23 From Adam Sedgwick Suffolk
Has printed a Memorial about Norwich. Discusses this pamphlet. Has been ill. Discusses solitude and says almost none of his friends are still living.
ALS 4pp RS:HS 15.457

13690. 1868–6–25 To George Biddell Airy Collingwood
Thanks for gift sent to JH's daughter Amelia; comments on William Petrie's number mysticism with the Great Pyramid.
C 2pp RS:HS 24.217

13691. 1868–6–26 From Mary Somerville Naples
Congratulations on fine Dante translation. Met [John] Tyndall and enjoyed his company. Has made several changes in her book [*Molecular and Microscopic Science*]. Is 'quite well for her age.'
ALS 3pp RS:HS 16.376

13692. 1868–6[?]–28 From Adolphe Quetelet Brussels
Apologizes for not writing. Has devoted all his time to *Physique sociale*. Has just finished last page.
draft? 1p (French) inc Académie belgique: #51

13693. 1868–6–30 To Sir Charles Lyell Collingwood
Sends a paper, which may help to clarify the effect of the Sahara Desert on certain European winds.
ALS 4pp APS B.D25.L.19 (C: RS:HS 24.218)

13694. [1868–7 or later] From Elizabeth Drummond Portobello, Edinburgh
Has received his translation of *Inferno* safely, and it has been greatly admired by competent judges. Weather has been very bad. He can keep the extract from the *Examiner*.
ALS 1p RS:HS 6.452

13695. 1868–7–2 From Sir William Huggins Upper Tulse Hill
Has just found that the bands of light from the comet are resolved by the spectroscope into bands that constitute a modified form of carbon. The spectrum of the comet was compared directly with a current of olefiant gas.
ALS 2pp RS:HS 10.42

13696. 1868–7–3 From Smith, Elder & Co. 65 Cornhill
Note accompanying shipping of a specially bound volume of JH's *Cape Results*. JH has noted on back of letter that volume will be returned as it arrived without plates or frontispiece.
ALS 1p RS:HS 16.213

13697. 1868–7–6 From William B. Clarke St. Leonards near Sydney, NSW
Sending details of the atmosphere being affected by the new moon. Has not regularly received his astronomical notices owing to a fault in the address. News of scientific happenings in Australia.
ALS 8pp RS:HS 5.348

13698. 1868–7–6 To Sir William Huggins Collingwood
Comments on WH's spectroscopic examination of cometary tails [see WH's 1868–7–2].
C 1p RS:HS 24.219

13699. [1868]–7–6. From Henry Bence Jones 84 Brook St.
John Tyndall left London in a hurry, but he has urged the claims of JH's son [Alexander] as a professor at the School of Mines.
ALS 2pp RS:HS 10.340

13700. 1868–7–6 From Sir Charles Lyell 73 Harley St.
Much obliged for calling his attention to M. C. E. Du Four's paper, though he cannot reconcile the general reasoning. Comments on the effect of the Sirocco on the Föhn of the Alps.
ALS 5pp RS:HS 11.444

13701. 1868–7–6 From Smith, Elder & Co. 65 Cornhill
Apology for errors in binding [see SE's 1868–7–3]; new volume being bound.
ALS 1p RS:HS 16.214

13702. 1868–7–6 From Sir Henry Taylor Bournemouth
Thanks JH for sending JH's work on Dante's *Divine Comedy*. Confesses not to be a Dante expert, but believes *terza rima* best for the poem in English.
ALS 8pp RS:HS 17.326

13703. 1868–7–8 To John Herschel (son) Collingwood
Comments on son John's spectroscopic and telescopic observations, and encourages observing a particular nebula, as it has been reported as having changed shape since JH observed it in the 1830's.
ALS 2pp JHS 6.48

13704. 1868–7–11 From Sir Charles Lyell 73 Harley St.
Sends Captain [P. F.] Shortland's account of his recent soundings and temperatures of the sea. Comments on these readings in the light of theory. What does JH think of the matter.
ALS 5pp RS:HS 11.445

13705. 1868–7–11 From John Tyndall Griesbach
JT has suggested that JH's son [Alexander] be appointed to position in Royal School of Mines, but colleagues, unwisely, JT thinks, want someone more well-known.
ALS 3pp RS:HS 17.402

13706. 1868–7–13 From Georg F. J. A. Auwers Greenwich
Would like to visit JH at Collingwood.
ALS 1p RS:HS 1.425

13707. 1868–7–13 To Sir Charles Lyell Collingwood
On the temperatures of water in the oceans, and the freezing temperature of sea water.
ALS 4pp APS B.D25.L.20 (C: RS:HS 24.220)

13708. [1868]–7–14 From Henry Holland Brook St.
JH's translation of first canto of Dante's *Inferno* (1868) is best HH has ever seen. Thanks for JH's commendation of HH's own translations [?]. More observations of soap bubbles.
ALS 3pp RS:HS 9.431

13709. 1868–7–14 To [George Gabriel Stokes] n.p.
Reports on and recommends publication of paper [*R.S.P.T.*, *159*, 1–] by Warren de La Rue, Balfour Stewart, and Benjamin Loewy containing heliographic positions and areas of sunspots observed in 1862 and 1863.
ADS 3pp RS RR.6.101

13710. 1868–7–15 From Georg F. J. A. Auwers Greenwich

Will visit JH at Collingwood on Friday.
ALS 1p RS:HS 1.426

13711. 1868–7–15 To Charles Babbage n.p.
Invitation to attend his daughter's wedding.
ALS 3pp RS:HS 2.337

13712. [1868]–7–22 From Marianne Spring-Rice Hallsteads, Penrith [Cumbria]
Gratitude for gift of JH's translation [Dante]. Pending marriage of Amelia Herschel to Thomas Wade.
ALS 4pp TxU:H/M–0907; Reel 1083

13713. [1868]–7–25 From John Murray, [Jr.] 50 Albemarle St.
Has heard from Mrs. Greig [wife of Woronzow Greig] that JH has consented to read the proof sheets of Mrs. Mary Somerville's work on molecular science. Sends the first 44 pages and would be pleased for any comments. Mrs. Somerville is in Naples.
ALS 2pp RS:HS 12.456

13714. 1868–7–27 From Sir William Huggins Upper Tulse Hill
Is grateful for his note. Has sent a paper to the R.S.L. on cometary phenomena and encloses a copy of the part in which he refers to JH's views. Would be glad if it were possible to explain the phenomena of tails.
ALS 2pp RS:HS 10.43

13715. [1868]–7–27 From John Murray, [Jr.] 50 Albemarle St.
Acknowledging the return of the proofs of Mrs. Mary Somerville's book. Is grateful for the revision and will forward the sheets to Mrs. Somerville for her to arrange and incorporate the suggestions.
ALS 1p RS:HS 12.457

13716. 1868–7–29 From William Selwyn Foxton Royston
Thanks JH for the Dante translation. He and Charles Jenyns commend the translation. Sends a translation [into Latin] of [S. T. Coleridge's] 'Genevieve' for JH's comments.
ALS 4pp RS:HS 15.502

13717. 1868–7–31 To Sir William Huggins 4 Chesham St.
Has been occupied with the wedding of his daughter [Amelia], so unable to read his paper before. Comments on the various theories regarding the tail of the comet. Regrets his statements do not agree with those of WH.
C 3pp RS:HS 10.44 & 24.221

13718. 1868–8–1 From Sir William Huggins Upper Tulse Hill
Much obliged for his suggestions regarding the comet; comments on these suggestions. His own observations were made on the comet, not its tail. Would be pleased for him to visit his observatory when in London.
ALS 4pp RS:HS 10.45

13719. 1868–8–1 To William Selwyn Chesham Street
Thanks WS for 'Genevieve' translation. Will not continue the Dante because a Terza Rima translation has been done previously. The sun is behaving oddly.
C 2pp RS:HS 24.222

13720. 1868–8–5 To Sir William Huggins Collingwood
Comments on WH's finding the carbon line in WH's spectroscopic examinations of cometary tails.
C 1p RS:HS 24.223

13721. 1868–8–11 To William Thomson n.p.
Comments on WT's paper on geological time. Unsatisfied with the efficiency of tidal friction in retarding the rotation of the earth. Considers effect of external attraction on a rotating body to relate to momentum.
ALS 4pp RS:HS 17.351a (C: RS:HS 24.224)

13722. 1868–8–18 From Frederic J. Jackson Bawtry
Thanks for his article on musical scales. Misses a description of the elements of the simple scale. Comments on various theories.
ALS 6pp RS:HS 10.431

13723. 1868–8–19 From Augustus De Morgan 91 Adelaide Road
Reasons why the sun has been so hot recently. Thanks for the paper on the scale. Comments on this. Sends a mathematical problem.
ALS 4pp RS:HS 6.403

13724. 1868–8–25 From Sir Edward Ryan Brechin
Wants information concerning Sir Edward Colebrooke's involvement in the formation of the Royal Astronomical Society. Cannot ask Charles Babbage because his memory is poor.
ALS 4pp RS:HS 14.466

13725. 1868–8–27 From Rider Garrett Huddersfield
JH's notes and version of the first canto of Dante's *Inferno* has made him attempt a version in the original rhyming meter. Would like his comments.
ALS 3pp RS:HS 9.93

13726. [1868?]–8–28 From Lady M[ary] Egerton Mountfield Court
Many thanks for his article on the musical scale. Will try to understand it when she has more leisure.
ALS 4pp RS:HS 7.123

13727. 1868–8–28 From Adolphe Quetelet Brussels
Sends first pages of JH's work on theory of probability. Asks for modifications.
ALS 2pp (French) RS:HS 14.268

13728. 1868–8–29 To Rider Garrett n.p.
Has received her version of the first canto of Dante's *Inferno*. Did she know it has already been done in an almost perfect way by Mr. Ford, prebend of Exeter? Comments on various points of her translation.
draft 6pp RS:HS 9.94 & 24.225

13729. 1868–8–29 From Charles Sterry Royal Mint
Death of William H. Barton.
ALS 5pp TxU:H/M–0272; Reel 1087

13730. 1868–9–1 To Elizabeth Colling Collingwood
Given up idea of translating Dante; comments on meteor shower report; suggests EC write an ode on poverty.
C 3pp RS:HS 24.226

13731. 1868–9–1 To Adolphe Quetelet Collingwood
Sends back excellent translation of his review with minor corrections. Mentions 'fine display of meteors with long trains' on 10 August. Gives news of his son Lieutenant John Herschel's observations. Acknowledges works received.
ALS 3pp Académie belgique: #52

13732. [1868]–9–3 From Rider Garrett Huddersfield
Is very grateful for his comments on her translation of Dante's *Inferno*. Will obtain Mr. Ford's version. Intends to continue with her translation though she sees no prospect of it

being printed.
ALS 4pp RS:HS 9.95

13733. 1868–9–3 From Charles Sterry Royal Mint
Hoped to succeed late William H. Barton as comptroller and deputy master of Mint, but learned that government plans to make this office a political appointment. Seeks testimonial from JH on CS's behalf.
ALS 4pp TxU:H/M–0273; Reel 1087

13734. 1868–9–4 From George Gabriel Stokes Lensfield Cottage, Cambridge
Comments on communications to R.S.L. by JH's son John in India.
ALS 4pp RS:HS 17.70

13735. 1868–9–7 To Mary Baldwin Collingwood
News of JH's son William and family, visiting England. JH and wife suffer from hot weather. JH's son John observed 'great eclipse.'
ALS 4pp † TxU:H/L-0078; Reel 1054

13736. 1868–9–7 To Charles Sterry n.p.
Recalls CS's promotion to senior clerk in Mint office in May or June 1854, with full approval by W. H. Barton. Attests to CS's merits then, but advises CS to seek recommendation now from current master of Mint.
AdraftS 1p (on J. Chapman's 1867–6–17) TxU:H/L-0392.2; Reel 1054

13737. 1868–9–8 From Charles Sterry Royal Mint
Gratitude for JH's recollection of circumstances that led to CS's appointment at Royal Mint. Death of W. H. Barton, whose duties CS took over. Lists famous persons who have written to Treasury on CS's behalf, when Treasury is planning to replace CS with a political appointee.
ALS 4pp TxU:H/M–0939; Reel 1083

13738. 1868–9–12 From Adolphe Quetelet Brussels
Thanks for comments on AQ's *Sur l'homme moyen*. Translator of JH's work is too timid to allow himself to be known, but appreciates JH's compliments. Says practical statistics has progressed whereas philosophical statistics remains stationary.
ALS 3pp (French) RS:HS 14.269

13739. 1868–9–14 To John Herschel (son) Collingwood
Is anxiously awaiting details of son John's observations of the solar eclipse in India; JH sends details of sunspots observed by JH on 18 Aug., and concludes with family news.
ALS 4pp JHS 6.47

13740. 1868–9–16 From William Mathews Jr. Carpenter Rd., Edgbaston
While in Switzerland requested his wife to send to JH a little tract on Count de S. Robert's hypsometrical investigations. Found that Robert's formulae worked correctly for ordinary European heights. Realizes that JH demolished Robert's theories, but Robert also showed observations on P. S. Laplace's theories.
ALS 3pp RS:HS 12.303

13741. 1868–9–18 From Emilien Frossard Pasteur Bagnieres de Biguerre
Remembers when JH introduced him to the Museum of Geology in Jermyn St.. Would like to exchange some rocks and minerals for some English fossils. Can JH advise him to whom to write?
ALS 3pp RS:HS 13.238

13742. 1868–9–19 To Robert Main Collingwood
Thanks RM for the double star observations he has sent.
C 1p RS:HS 24.227

13743. 1868–9–21 From Robert Main Radcliffe Observatory
Comments on double star observations and some papers on the subject.
ALS 3pp RS:HS 12.280

13744. 1868–9–24 From William Thomson Largs by Greenock
Replies in detail to JH's objection to explaining the retardation of the earth's rotation as due to tidal friction, JH having urged that this effect should be analyzed in terms of momentum rather than of *vis viva* [kinetic energy], as WT maintained.
ALS 12pp RS:HS 17.351b

13745. 1868–9–26 To William Selwyn Collingwood
Describes the peculiar pairs of sunspots he has recently witnessed. Thinks the earth affected their appearance somehow. Is using the Julian calendar for dating observations.
C 3pp RS:HS 24.228

13746. 1868–9–26 From Mary Somerville Naples
Congratulations on marriage of JH's daughter [Amelia]. Is writing her autobiography. Interested in solar eclipses, and says more will be known through photography and spectrum analysis.
ALS 4pp RS:HS 16.377

13747. 1868–9–26 From Edward James Stone Royal Observatory, Greenwich
Reports on examination of observations of 1769 of Venus's transit over the sun's disc. Discusses variations at different sites in 'breaking and formation of the black drop,' and possible causes of error in original observations. Will send copy of paper.
ALS 8pp RS:HS 17.241

13748. 1868–9–29 To Robert Main Collingwood
Sends RM a copy of Angelo Secchi's double star observations pointing to a very large number of calculation errors.
C 2pp RS:HS 24.229

13749. 1868–9–29 From Edward Sabine 13 Ashley Place
Wants information regarding undulatory theory of light controversy and the role of Lord [Henry] Brougham for his obituary. Congratulates good performance of JH's son [John?] in India.
ALS 4pp RS:HS 15.287

13750. 1868–9–29 From William Selwyn Foxton Royston
Discusses the great reflector at Lord Rosse's in Ireland. Will send solar autographs JH requested. Thanks JH for the hints about the Julian Calendar.
ALS 3pp RS:HS 15.503

13751. 1868–9–30 From Robert Main Radcliffe Observatory
About the location of several different sets of double star observations.
ALS 3pp RS:HS 12.281

13752. 1868–10–1 From Robert Main Radcliffe Observatory
Comments on the quality of some of the double star

observations in print.
ALS 3pp RS:HS 12.282

13753. 1868–10–1 From Robert Mushet Royal Mint
Gratitude for JH's generous testimonial on RM's behalf.
ALS 2pp TxU:H/M–0388; Reel 1087

13754. 1868–10–1 To Edward Sabine Collingwood
Discusses Lord [Henry] Brougham and his optical papers. Thanks ES for kind remarks regarding JH's son [John?]. Discusses polarization of corona.
AL 3pp Cambridge University Library 7656/RS678 (draft: RS:HS 15.288 & C: RS:HS 24.230)

13755. 1868–10–2 From Adolphe Quetelet Brussels
Thanks for proofed pages. Received *Astronomische Beobachtungen auf der Sternwarte zu Bonn.* Finished determining by electric means the different observational longitudes of Leiden and Brussels.
ALS 4pp (French) RS:HS 14.270

13756. [1868]–10–2 From Edward Sabine 13 Ashley Place
Thanks JH for information regarding Henry Brougham [see ES's 1868–9–29 & 1868–10–8]. Also for comment on polarization of aurora. Encloses Captain [John?] Hay's letter, but not plate.
ALS 2pp RS:HS 15.314

13757. 1868–10–5 From Thomas P. Thompson Eliot Vale, Blackheath
Thanks JH for his pamphlet on musical scales.
ALS 1p RS:HS 17.349a

13758. 1868–10–8 To John Herschel (son) n.p.
Congratulations on success in observing solar spectrum and corona. JH's views on sun's corona [with diagram]. Agrees that communications in India are abominable.
ALS 2pp TxU:H/L-0529; Reel 1053

13759. 1868–10–8 From Edward Sabine Burlington House
Discusses [William] Sharpey's view of JH's opinion of [Henry] Brougham's optical writings. Asks for JH's advice on presenting meteorological observations. On observations of an eclipse.
ALS 4pp RS:HS 15.289

13760. 1868–10–10 To Frederick Howlett Collingwood
Details about sunspot observations.
C 2pp RS:HS 24.231

13761. 1868–10–11 To Edward Sabine Collingwood
Discusses papers concerning light on [William] Sharpey's list. Discusses, very critically, Henry Brougham's optical researches.
draft 2pp RS:HS 15.290 (C: RS:HS 24.232)

13762. 1868–10–12 To Edward Sabine Collingwood
Reviews additional work of H. P. Brougham [see JH's 1868–10–1] as has been requested in preparation for an eloge . JH is critical of poorly described and poorly understood work, unrelentingly tied to Isaac Newton's corpuscular ideas.
ALS 5pp Cambridge University Library 7656/H718

13763. 1868–10–13 From Frederick Howlett St. Paul's, New Beckenham
Much obliged for his kind notes on sunspots; there has been a wonderful variety lately. Will make a study of the Julian dates. Can see the Crystal Palace fireworks splendidly from his house.
ALS 3pp RS:HS 10.22

13764. 1868–10–14 To Sir William Huggins Collingwood
A note to accompany JH's biographical sketch of William Dawes.
C 1p RS:HS 24.233

13765. [1868?]–10–15 From Henrietta G. M. L. Chatterton Henley-in-Arden
Was pleased to receive JH's translation of Dante. Wished he lived near enough to pay them a visit.
ALS 4pp RS:HS 5.333

13766. 1868–10–16 From George Biddell Airy Royal Observatory, Greenwich
Further regarding the theory of sound.
ALS 2pp RS:HS 1.312

13767. [1868?]–10–19 From S. Langley King William St.
Is preparing a sketch of Sir John Lubbock for a weekly, and would be glad of JH's opinion regarding Lubbock as a man of science and his advanced political views.
ALS 2pp RS:HS 11.96

13768. 1868–10–20 From Henry James Ordnance Survey Office, Southampton
Subscriptions for Sinai survey fund are nearly complete. Survey party will embark for Alexandria aboard *Ripon* on 24 Oct. Will send account books to JH and R. I. Murchison. Will not need JH's astronomical instrument.
ALS 4pp TxU:H/M–0294; Reel 1087

13769. 1868–10–21 From William H. Callcott Kensington
Is grateful for his article on *Musical Scales*. Queries regarding musical sounds. Has had to leave his old abode due to building operations.
ALS 4pp RS:HS 5.144

13770. 1868–10–23 From Frederick W. Holland Portsea Place Connaught Square
R. I. Murchison has forwarded JH's letter about the noises at Nakkus in Sinai. Has heard the noises several times and thinks they are caused by the movement of sand. There is much static electricity in sand.
ALS 3pp RS:HS 9.424

13771. 1868–10–23 To William Selwyn Collingwood
Some photographs support the sudden disappearance of the spot on 17 Sept. Some photographs are misdated. The relation of Jupiter and sunspots suggests previously unknown interconnections in the solar system.
C 3pp RS:HS 24.234

13772. 1868–10–24 From William Selwyn Cambridge
Sent JH's remarks to Mr. Titterton at Ely. Will endeavor to maintain accuracy. The remarks about Jupiter's action are contrary to the general consensus that planets in opposition close up spots.
ALS 4pp RS:HS 15.504

13773. 1868–10–25 To S. Langley Collingwood
Answer has been delayed as SL's letter was directed to MacMillans. Can be no two opinions on the contribution to scientific knowledge of Sir John Lubbock. Has full confidence in his political opinions.
C 3pp RS:HS 11.97 & 24.235

13774. 1868–10–28 From Alexander John Ellis Argyll Road Kensington
Thanks for his article on musical scales. Will read it later when he has time. Comments on the system of equal temperament. Will send some of his own papers later.
ALS 4pp RS:HS 7.42

13775. 1868–10–30 To Elizabeth Colling Collingwood
Thanks for verses on the transit of Mercury; comments on observation of an auroral arch and eclipse observations of the solar corona.
C 2pp RS:HS 24.236

13776. 1868–10–30 From Alexander Strahan 56 Ludgate Hill
Hopes that JH will contribute series of papers under heading, 'A Life's Retrospect in Science,' from various periods of JH's career, for publication in *Good Words*.
ALS 3pp RS:HS 17.243

13777. 1868–11–1 From P. O'Callaghan Tennis Ct. Club
This town and surrounding area were visited by an earthquake on Friday night. Gives time and details. Alexander Herschel is to give a lecture at Leamington in the ensuing season.
ALS 4pp RS:HS 13.161

13778. 1868–11–3 From George A. MacFarren 7 Hamilton Terrace
Has read with great interest JH's paper on the musical scales. Comments on JH's paper and expounds some of his own theories on the musical scales. Would like JH's comments on his paper.
ALS 9pp RS:HS 12.12

13779. 1868–11–3 To Charles Pritchard Collingwood
Solar spectrum observations of John Herschel (son of JH) may have produced detection of the photosphere and corona at a time other than a total solar eclipse.
ALS 3pp RS:HS 14.103 *R.A.S.M.N., 29* (1869), 5–6 inc

13780. [1868?]–11–4 From William C. Mathison Dickleburgh, Scole
Accept his best thanks for JH's paper on musical scales. It seems to dispose of the whole matter in the most satisfactory manner. Hopes his own pupil is well.
ALS 1p RS:HS 13.28

13781. 1868–11–5 From S. Langley King William St.
Can he publish JH's statement regarding John Lubbock? Outlines the wishes of other scientists regarding the printing of their views.
ALS 3pp RS:HS 11.98

13782. 1868–11–5 From Alexander Strahan 56 Ludgate Hill
Regrets that JH declines offer [see AS's 1868–10–30] to publish series of papers in *Good Words*. Hopes JH will nevertheless occasionally send short papers.
ALS 2pp RS:HS 17.244

13783. 1868–11–6 From Frederick W. Holland Alexandria
Had no time to write before he sailed [see FH's 1868–10–23, which is similar in content]; forward letters to the Ordnance Office, Southampton.
ALS 3pp RS:HS 9.425

13784. 1868–11–6 From George Gabriel Stokes Royal Society
Questions about JH's report on a paper; is sending on copies of JH's son John's papers.
ALS 3pp RS:HS 17.71

13785. 1868–11–8 To George Gabriel Stokes Collingwood
Comments on JH's son's reports [see GS's 1868–11–6]; recalls idea of JH's father, William Herschel, about nature of solar light; transit of Mercury 'beautiful'.
ALS 4pp Cambridge University Library 7656/H719

13786. 1868–11–9 From John Henry Griesbach 48 Bedford Gardens
Is grateful for his paper on musical scales. Comments on the characteristics of wind instruments. Some friends of his saw a brilliant meteor fall last Tuesday.
ALS 4pp RS:HS 9.66

13787. 1868–11–9 From Thomas Henry Huxley Royal School of Mines
Has placed Mr. Frosmard's letter before the Geological Society, and also discussed the matter with R. I. Murchison, but regrets he is unable to help.
ALS 4pp RS:HS 10.221

13788. 1868–11–10 To George Johnstone Stoney Collingwood
Thanks for paper on physical constitution of sun and stars. Discusses possibility of vast atmosphere for sun.
ACCS 1p RS:HS 17.242 (C: 24.237)

13789. [1868–11–14?] From Hubert Airy Flamsteed House, Greenwich
Writing about [hemianopsia] for doctor's thesis at Cambridge. Asks permission to quote report of JH's attack [see JH's 1868–5–4].
ALS 2pp † RS:HS 19.309

13790. 1868–11–14 From Edward Livingston Youmans [Athenaeum Club]
Encloses note from John Tyndall. Asks for interview. Wants to publish some of JH's papers in America. Purpose of EY's journal is to give double audience to scientific writers.
ALS 2pp RS:HS 18.444

13791. 1868–11–16 To George A. MacFarren Collingwood
Is obliged for GM's letter and for the trouble he has taken over JH's paper on musical scales. Object in writing this paper was to clear the ground as it were. Elucidates some of the points in his own paper and comments on parts of GM's paper.
draft 8pp RS:HS 12.13 & 24.238

13792. 1868–11–18 From Hubert Airy Flamsteed House, Greenwich
Received JH's permission to quote JH [in HA's thesis on hemianopsia] and describe JH's latest attack, including 'patches of coloured checquer work.'
ALS 1p RS:HS 19.310

13793. 1868–11–20 From St. John Vincent Day Dowanhill, Glasgow
Intends publishing a second edition of his paper on the Great Pyramid entitled 'Papers and Correspondence on the Pyramid.' Will respect JH's wishes on any points.
ALS 2pp RS:HS 6.480

13794. 1868–11–20 From Sir William Huggins Upper Tulse Hill
Sending a photograph of a solar prominence he has received from Major J. F. Tennant. Comments on various features. At

the Transit of Mercury he saw the phenomenon described by J. H. Schröter.
ALS 3pp RS:HS 10.46

13795. 1868–11–21 From George A. MacFarren 7 Hamilton Terrace
Appreciates JH's beautiful simplification of the numerals in musical arithmetic. Wishes JH would give up the point of the variability of the supertonic and transfer it to the submediant. Agrees with JH over the need for a keyboard instrument possessing a perfect chromatic scale in any particular key.
ALS 4pp RS:HS 12.14

13796. 1868–11–21 From [Osborne?] Turner Bristol
Requests permission to include extract from *Prelim. Discourse* in *The Sixth English Reading Book.*
L 2pp RS:HS 17.450

13797. 1868–11–22 To Sir William Huggins Collingwood
Thanks to WH for the photograph of solar protruberances; comments on polarization of light from the sun.
C 2pp RS:HS 24.239

13798. 1868–11–23 From John Herschel (son) Bangalore
Reports his observations of the nebula around Eta Argus. Provides a sketch. Has observed fifty other nebulae.
L *R.A.S.M.N., 29* (1869), 82–4

13799. 1868–11–24 To Robert Main Collingwood
Seems to JH to be some systematic error in some of RM's observations; JH includes examples.
C 4pp RS:HS 24.240

13800. 1868–11–26 To John Tyndall Collingwood
Thanks for paper on action of solar and electric light on vapors, which should give JT further insight into blue color of sky and polarization of skylight. Comments on the latter. JH's son [Alexander] is working hard at Glasgow.
C 2pp RS:HS 24.241 (C: RI 547–8)

13801. 1868–11–27 From John Tyndall Royal Institution
Is studying vesicles. JH's son [Alexander] may yet get Royal School of Mines appointment.
ALS 2pp RS:HS 17.403

13802. 1868–11–28 From Robert Main Radcliffe Observatory, Oxford
Notes strengths and weakness of heliometer for observing double stars. Calls it 'laborious, troublesome instrument.'
ALS 6pp RAS:JH Archive 14/3.17; Reel 11

13803. [1868]–11–28 From Edward Sullivan 13 Grosvenor Place
Their mutual friend James Sivewright had given ES plan for boat swing to combat Lady Sullivan's seasickness. Having sold boat in which swing was built, but planning to sail to Mediterranean, ES hopes JH will send another copy of plan.
ALS 2pp RS:HS 17.248

13804. 1868–11–29 From John Herschel (son) Bangalore
Provides further [see JH's son's 1868–11–23] observations, along with drawings, of the nebula around Eta Argus.
L *R.A.S.M.N., 29* (1869), 85–6

13805. 1868–11–30 From Patrick Shanahan 9 Fair St., Stepney
Is losing position as Royal Mint messenger, which JH had helped him obtain. Asks JH to tell Mint Master that position was intended as permanent.
ALS 3pp RS:HS 17.200

13806. 1868–11–30 From James Sivewright St. Helens
Hopes to obtain JH's plan for building a swing to counteract seasickness. Comments on mutual acquaintances.
ALS 3pp RS:HS 17.212

13807. 1868–11–30 From John Tyndall Royal Institution
Observations on subjecting carbonic acid to concentrated beam of an electric lamp. Records experimental result so far. Interested in ordinary dust particles under electric light.
ALS 4pp RS:HS 17.404 (C: RI 550)

13808. 1868–12–1 From George Gabriel Stokes Royal Society
Organizational matters regarding paper reports; comments on light from sun's corona [see JH's 1868–11–8].
ALS 4pp RS:HS 17.72

13809. 1868–12–1 From John Tyndall Royal Institution
Continues experiments with carbonic acid and electric light [see JT's 1868–11–30], hoping to 'explode' idea that atmosphere's polarization is due to reflection by air particles.
ALS 7pp RS:HS 17.405a (C: RI 551–2)

13810. 1868–12–2 From James Sivewright St. Helens
Thanks JH for sending plan of swing to counteract seasickness. This will enable his long suffering friend Lady Sullivan to escape hay fever by going to sea but without thereby being beset by seasickness.
ALS 2pp RS:HS 17.213

13811. 1868–12–2 To George Gabriel Stokes Collingwood
Further comments on solar light [see GS's 1868–12–1].
ALS 2pp Cambridge University Library 7656/H720

13812. [1868]–12–2 From Edward Sullivan 13 Grosvenor Place
Thanks JH for sending plan for swinging cot to prevent seasickness [see ES's [1868]–11–28] and asks for more information.
ALS 2pp RS:HS 17.249

13813. 1868–12–2 To John Tyndall Collingwood
Thinks JT's work on polarization will lead to remarkable discovery. Discusses production of rainbow, parallels to Isaac Newton's explanation of black spot on a soap bubble, and problems in JT's undulatory theory explanation of reflection. In JT's experiments, what are nebulous particles produced by light in gas or vapor?
CC 4pp RS:HS17.405b (C: 24.242; C: RI 553–4)

13814. 1868–12–3 From Charles Sterry [Royal Mint]
Disappointed that Benjamin Disraeli's secretary, Charles Fremantle, was appointed to vacancy left at Mint by death of William H. Barton. Thanks for JH's efforts on CS's behalf.
ALS 3pp TxU:H/M–0274; Reel 1087

13815. [1868]–12–3 From John Tyndall Royal Institution
Continues experiments with carbonic acid and electric light [see JT's 1868–11–30], suggesting possible cause of blue cloud color and eventual whitening of light.
ALS 4pp RS:HS 17.412 (C: RI 555)

13816. 1868–12–4 From James Scotson Prestwich
Asks JH where to find further information about a salt road and cliffs mentioned in JH's *Physical Geography.*
ALS 3pp RS:HS 17.182

13817. 1868–12–5 From Frederick Pollock Clapham
Discusses an experiment, described by JH in JH's *Familiar Lectures,* involving applying colors to paper. FP finds that

his results differ from JH's.
ALS 4pp RS:HS 14.175

13818. [1868–12–9?] To Patrick Shanahan n.p.
To JH's knowledge, PS's Royal Mint position bore no absolute guarantee of permanence unless he had been paying a 'superannuation' fund.
ALS 1p RS:HS 17.201

13819. 1868–12–10 From Redmond Barry Melbourne
Sends him a vocabulary of Aboriginal dialects and University calendar. The response to his questionnaire has been very disappointing.
ALS 5pp RS:HS 5.24

13820. 1868–12–10 From George Gabriel Stokes Royal Society
Asks JH to look at a paper on musical scales; comments on a paper by A. Prazmosky on polarization of light from the solar corona.
ALS 8pp RS:HS 17.73

13821. 1868–12–12 From Sir Frederick Pollock Clapham
Thanks JH for reply to letter questioning color experiment in JH's *Familiar Lectures*. Will conduct the experiment that JH suggests in the book.
ALS 2pp RS:HS 14.176

13822. 1868–12–12 From Edward Sabine n.p.
Asks JH to sign [C. P. B.] Walker's certificate for the R.S.L.
ALS 1p RS:HS 15.291

13823. 1868–12–12 To George Gabriel Stokes Collingwood
Returns paper on musical scales; agrees with GS's comments on A. Prazmosky's paper [see GS's 1868–12–10].
ALS 4pp Cambridge University Library 7656/H721

13824. [1868]–12–12 From John Tyndall Royal Institution
Asks JH to review paper on blue color of sky, polarization of skylight, and polarization of light by cloudy matter.
ALS 1p RS:HS 17.413 (C: RI 556)

13825. 1868–12–13 To John Tyndall n.p.
JT's paper on polarization of sky light suggests that neutral points are functions of cloud density. This verifies JH's earlier explanation of blue sky color. Incomplete polarization. Cause of blue color in water. Corrects note on W. A. Miller's observations of rainbows.
Adraft 4pp RS:HS 19.160b

13826. 1868–12–14 To John Tyndall Collingwood
Thanks for paper verifying JH's prediction that explanation of blue sky color carries with it that of polarization of skylight. Involves ultimate link between chemical and analytical dynamics. Notes that 'neutral points' in sky polarization have yet to be explained; offers tentative explanation based on clues in JT's work.
C 4pp RS:HS 24.243 (C: RI 557–8)

13827. 1868–12–15 From Sir William Huggins Upper Tulse Hill
Has arranged for the last *R.A.S.M.N.* to be sent to him. Lieut. John Herschel has detected a fourth line in the spectrum of the Nebula in Orion.
ALS 2pp RS:HS 10.47

13828. 1868–12–17 From D. McNeill Warwick Square
Will be sending some Pyretheum roseum and some birdlime with instructions for their use. Wrote a few lines to JH's son.
ALS 4pp RS:HS 13.13

13829. 1868–12–22 To the R.A.S. Collingwood
There having been reports of changes in the nebula around Eta Argus, JH sends the R.A.S. a letter from JH's son John [see JH's son's 1868–11–23] reporting his observations of that object. JH adds his comments.
L *R.A.S.M.N., 29* (1869), 84–5

13830. 1868–12–24 From Russell Henry Manners Henrietta St.
Thanks for his communication regarding the reported changes in Eta Argus, and also Lieut. John Herschel's letter from Bangalore. These will form a valuable addition to the next *R.A.S.M.N.*
ALS 2pp RS:HS 12.287

13831. 1868–12–25 To George Gabriel Stokes Collingwood
Comments on a paper on the effects of lateral movement in diminishing the intensity of sound.
ALS 2pp Cambridge University Library 7656/H722
Stokes, *Correspondence, 1,* 193 inc

13832. 1868–12–26 To Elizabeth Colling Collingwood
Comments on observations of meteors, comets, and the transit of Mercury.
C 2pp RS:HS 24.244

13833. 1868–12–28 From Adolphe Quetelet Brussels
Announces first volume of *Physique sociale* will appear soon. Discusses humans and law of nature. Says science is neglected in France.
ALS 3pp (French) RS:HS 14.271

13834. 1868–12–29 To Sir Henry Taylor Collingwood
Has received HT's pamphlet on crime and its prevention and punishment. Praises it and questions some particulars.
ALS 4pp inc Bodleian Library, Oxford

13835. [1869] From Edward William Brayley Hinckley
Encloses copies of papers referred to in his previous letter.
ALS 3pp RS:HS 4.238

13836. [1869] To Alexander Hamilton-Gordon n.p.
Discusses the 'strange proceedings on the part of the Bombay Authorities.' Mentions actions in this regard by Carry [JH's daughter Caroline].
draft 2pp RS:HS 8.155 (C: RS:HS 24.272)

13837. 1869–1 To Christian August Friedrich Peters Collingwood
Thanks CP for sending volumes of *Astronomische Nachrichten*. JH notes missing copies in his collection, and asks how he might best obtain these volumes.
ALS 4pp (French) RS:HS 13.368 (C: RS:HS 24.245)

13838. 1869–1–1 From Sir Henry Taylor East Sheen
Distributed a pamphlet on his theory of criminal punishment. Believes present system not effective. Draws favorable response.
ALS 4pp RS:HS 17.327

13839. 1869–1–2 To [Secretary of R.A.S.] Collingwood
JH sends and comments on drawings and a letter [see JH's son's 1868–11–23] from JH's son John regarding the nebula around Eta Argus. JH seems disinclined to accept the claim that Eta Argus has changed.
AdraftS 4pp, encl 2pp TxU:H/L-0739, 0740, 0741; Reel 1089 *R.A.S.M.N., 29* (1869), 82–8

13840. [1869–1–5 or later] To ? [Collingwood]
Is a listing of some of JH's papers in an attempt to clarify how many were distributed, and then JH indicates how many he will send to addressee.
AL 2pp inc JHS 6.62

13841. 1869–1–9 From Arthur Cayley Cambridge
Has sent the figures accompanying his second communication to the lithographers. Thinks his letter of 22 Dec. intelligible without them. Will send the proofs of the paper.
ALS 3pp RS:HS 5.231

13842. 1869–1–11 To George Gabriel Stokes Collingwood
Requests sending of a copy of recent R.S.L. *Proceedings* to JH's son in India.
ALS 2pp Cambridge University Library 7656/H723

13843. 1869–1–13 To George Biddell Airy Collingwood
Has received GA's paper on spectral lines and is amazed that it is possible to see a spectrum in the light from faint nebulae.
ALS 3pp RGO 6.381.451

13844. 1869–1–13 From Henry W. Field Royal Mint
Asks JH to confirm that JH appointed Charles Sterry as HF's assistant assayer on 25 Apr. 1851 after Henry Bingley declined position. Original letter of appointment is lost, and HF is accused of falsifying records.
ALS 4pp TxU:H/M–0195; Reel 1087

13845. 1869–1–14? To Adolphe Quetelet Collingwood
Was going to write in French, but switched languages. Honored by AQ's insertion of JH's review of AQ's *Physique sociale* as introduction. Has been very ill and has no strength for attention to anything beyond bodily sufferings. Lists works received.
ALS 4pp Académie belgique: #55

13846. 1869–1–14 From George Gabriel Stokes Royal Society
Information about the distribution of papers.
ALS 3pp RS:HS 17.74

13847. 1869–1–16 To Henry W. Field Collingwood
Cannot recall details or find documents regarding promotion of Charles Sterry to HF's assistant assayer in 1851. Remembers that Henry Bingley declined that position; it was then proper for HF to recommend someone.
AdraftS 2pp TxU:H/L–0152; Reel 1054

13848. 1869–1–18 From Elizabeth Drummond Portobello, Edinburgh
Sends an extract from the *Examiner*. Would be pleased if he would acknowledge it.
ALS 1p RS:HS 6.451

13849. 1869–1–21 From Arthur Cayley Cambridge
Send two copies of the Eta Argus paper. Send any additional corrections to J. Henderson. Has written for the mss. of Lieut. John Herschel's first letter.
ALS 2pp RS:HS 5.232

13850. 1869–1–22 From Sir William Huggins Upper Tulse Hill
Sends a proof of JH's obituary of W. R. Dawes. The concluding part of the manuscript was mislaid; would he therefore re-write the missing portion. Hopes to attend Professor Alexander Herschel's lecture this evening.
ALS 2pp RS:HS 10.48

13851. 1869–1–23 To Sir William Huggins Collingwood
Note with return of proof of JH's biographical sketch [see WH's 1869–1–22] of William Dawes.
C 1p RS:HS 24.246

13852. 1869–1–24 To William Benjamin Carpenter Collingwood
Thanks for his report on dredging operations. Comments on the findings.
draft 4pp RS:HS 5.196 & 24.247

13853. 1869–1–25 From Sir William Huggins Upper Tulse Hill
Professor Joseph Winlock of Harvard agrees with all WH's observations of Nebula of Orion. The observations at Bishop's Observatory between 1844 and 1851 appear to be the work of W. R. Dawes. Comments on the reasons why the bright lines are so easily seen.
ALS 2pp RS:HS 10.49

13854. 1869–1–26 From William Benjamin Carpenter 56 Regents Park Road
Thanks for his kind letter. Is surprised that the formation of flint attracted him. Comments on this and similar matters.
ALS 8pp RS:HS 5.197

13855. 1869–1–26 From Henry W. Field Royal Mint
Lists dates in 1851 on which steps were taken in appointment of Charles Sterry as HF's assistant assayer. HF is still under accusation.
ALS 3pp TxU:H/M–0196; Reel 1087

13856. 1869–1–26 From Robert Main Radcliffe Observatory
Offers to send on to JH any notices of double star observations that he receives.
ALS 2pp RS:HS 12.283

13857. 1869–1–27 From Martha Somerville Naples
Mary Somerville is sending her autobiography and desires comments and suggestions. She is busy with mathematics.
ALS 4pp RS:HS 16.325

13858. 1869–1–28 From Arthur Cayley Cambridge
Has altered two of the plates in one of JH's papers. Hopes that he will agree with the alteration.
ALS 2pp RS:HS 5.233

13859. 1869–1–30 To Robert Main Collingwood
Thanks RM for his offer [see RM's 1869–1–26]; wishes astronomers would develop an accepted system of indicating the quality of observations.
C 2pp RS:HS 24.248

13860. 1869–2–1 From Alexander V. Bewley York
Sending a query regarding the colors seen on clouds at sunset.
ALS 3pp RS:HS 5.58

13861. 1869–2–2 To 'dear Sir' Collingwood
Received [?]'s paper no. 10, which completes JH's collection. Returns duplicates of [?]'s papers on Kew and Nerchinsk 'disturbances' and on 'Residual Laws of disturbance.'
ALS 2pp TxU:H/L–0742; Reel 1089

13862. 1869–2–2 From Christian August Friedrich Peters Altona
Is supplying JH with copies of his missing journals [see JH's 1869–1], and will send more later.
ALS 3pp (French) RS:HS 13.369

13863. 1869–2–3 From Adolphe Quetelet Brussels
Obtained and sent JH 2 copies of *Physique sociale*. Thanks JH for observations but they are too scientific for his lectures.
ALS 3pp (French) RS:HS 14.272

13864. [1869]–2–3 From Martha Somerville Naples
Thanks JH for his letters to her mother (Mary Somerville), who is not well enough to write. Describes eruption of Vesuvius.
ALS 4pp RS:HS 16.326

13865. [1869]–2–5 From John Murray, [Jr.] 50 Albemarle St.
Has today sent by book post the manuscript of Mrs. Mary Somerville's autobiography, which JH is to report on. Understands that Mrs. Somerville is amenable to good advice.
ALS 2pp RS:HS 12.453

13866. 1869–2–7 To Joseph Barclay Pentland [Collingwood]
Receives Mary Somerville's autobiography; explains poor health will prohibit him from detailed scrutiny of the writing.
ALS 3pp RS:HS 14.155 (C: RS:HS 24.249)

13867. 1869–2–7 From Joseph Barclay Pentland Rome
Sends Mary Somerville's autobiography; visited her in Naples. JP will return to England in the summer.
ALS 3pp RS:HS 14.154

13868. 1869–2–8 From Robert P. Graves 33 Wellington St., Dublin
Has been requested by the relatives of the late W. R. Hamilton to write a memoir. Would be glad if JH would send him any suitable letters of Hamilton in his possession.
ALS 8pp RS:HS 9.26

13869. 1869–2–8 From William C. Morland Lamberhurst
Many thanks for sending the seeds of Wellingtonia. Has two Cryptomerias also grown from seed. Was much interested in Alexander's lecture.
ALS 3pp RS:HS 12.369

13870. 1869–2–14 From Robert P. Graves 33 Wellington St., Dublin
Agrees with him about the impossibility of sending W. R. Hamilton's letters by post. Thinks the best plan would be to send them to his brother at the Athenaeum, who will bring them to Ireland.
ALS 4pp RS:HS 9.27

13871. 1869–2–14 To Sir William Huggins Collingwood
Comments on some incorrect attributions to JH.
C 1p RS:HS 24.250

13872. 1869–2–22 To G. W. Harris Collingwood
Expresses thanks for receiving the 1868 volume of the *Transactions of the Society of Engineers.*
ALS 1p Owen Gingerich

13873. 1869–2–23 To William James Herschel Collingwood
Comments on WJH's appointment as judge in India. Contrasts legal practices in England and India. Reviews calamities WJH witnessed in India. Alexander Herschel is lecturing about eclipse at Royal Institution and Leamington. Family news. P.S.: Please write to W. E. Hearn in Melbourne regarding Hearn's book *Government of England.*
ALS 4pp TxU:H/L-0562; Reel 1053

13874. 1869–2–24 From Robert P. Graves 33 Wellington St., Dublin
Package of letters has arrived safely via the Globe Parcel Express Co., for which accept his best thanks. Would like a page or two of JH's recollections of W. R. Hamilton if possible. Has sent a pamphlet on William Wordsworth and the Lake Country.
ALS 4pp RS:HS 9.28

13875. 1869–2–26 From Mary Elizabeth Lyell Baston Hall, Bury St. Edmunds
Does JH have any letters from H. C. Oersted to lend to Oersted's daughter for publication?
ALS 3pp RS:HS 19.289

13876. 1869–3–1 From J[ohn] Jones Brynhyfryd, Wrexham
Sending some tables to show the relation between the mean distances of the primary and secondary planets.
ALS 4pp RS:HS 10.342

13877. 1869–3–4 From William Henry Bateson St. John's Lodge, Cambridge
Would be pleased if JH would come to the consecration of the new chapel in May.
ALS 2pp RS:HS 5.27

13878. 1869–3–10 From S. Alexander Renshaw New Basford
Desires advice on construction of glass prisms for Newtonian reflecting telescopes. Also asks how to test the parabolic figure of a glass speculum before silvering.
ALS 4pp RS:HS 14.477

13879. 1869–3–11 From Charles Pritchard Freshwater
Has discovered William Herschel's memoirs in R.A.S. library; arranged and read them. Asks JH whether any copies exist so the originals can be checked for completion before binding. Questions JH on William Herschel's opinion of the nature of nebulae and boundaries of the galaxy.
ALS 4pp RS:HS 14.104 Ada Pritchard, *Charles Pritchard: Memoirs of His Life,* 258–9

13880. 1869–3–14 To Mary Somerville n.p.
Commentary on manuscript [*Personal Recollections*?] that [MS] sent to JH one month ago. Career of scientific learning and domestic happiness will inspire future generations. Suggests publishing it posthumously for greater impact. Corrects passages about Charles Babbage. Gives purpose, history, and membership of each of *two* Standards committees, one of 1819 and one of 1838. Suggests avoiding topic of bitter controversy over invention of electric telegraph.
Adraft 6pp (on G. B. Airy's 1840–10–13 & 1840–5–18 & on [?]'s 1867–9–2) TxU:H/L-0760.4; Reel 1089 (C: RS:HS 16.378 & 24.251)

13881. 1869–3–16 From William Allen Miller King's College, London
The Meteorological Committee of the R.S.L. is about to commence publishing the summary of the observation collected from the seven observatories. Would like JH's advice on method of publishing the barometric observations.
ALS 3pp RS:HS 12.347

13882. 1869–3–18 From Charles Pritchard Freshwater
Suggests remedy for bronchitis problem in JH family. Young John Herschel's work on nebulae pleases CP. Comments on William Huggins's work on solar spectra.
ALS 4pp RS:HS 14.105

13883. 1869–3–18 To William Allen Miller Collingwood
Is fully satisfied with the objections in WM's letter regarding the individual records of barometric pressure. Gives advice on how the records should be shown.
draft 3pp RS:HS 12.348 & 24.252

13884. 1869–3–18 From Sir William Pole 3 Storey's Gate
Notices in letter of last May JH's postscript about the rate of color blindness occurring to overworked or ill artists, and doubts it becomes di-chromic, but merely a 'weakness' of vision.
ALS 2pp RS:HS 14.29

13885. 1869–3–18 To Balfour Stewart Glasgow
Agrees with BS's objection against performing daily analyses of individual barometric pressure records. Believes monthly analyses of importance for observing laws of annual pressure.
C 3pp RS:HS 24.253

13886. 1869–3–19 From Christian August Friedrich Peters Altona
Thanks JH for the memoirs he sent to CP, who will reciprocate with some of his own.
ALS 1p (French) RS:HS 13.370

13887. 1869–3–23 From Mary Somerville Naples
Has made the changes JH suggested. Autobiography will be published posthumously. Met [Henry W.] Longfellow. Wishes the Herschels would visit because it might improve JH's health.
ALS 3pp RS:HS 16.379

13888. 1869–3–25 From George Robert Waterhouse British Museum
Consulted A. C. L. G. Günther, museum ichthyologist. Answers JH's question about poisonous herring *Clupea thryssa*. [A. K.] *Johnston's Physical Atlas* attributes Ruminants article to GW, but GW wrote only Rodentia article.
ALS 3pp RS:HS 18.104

13889. 1869–3–26 From Mathilde Oersted Copenhagen
It was a great pleasure to receive JH's letter. Sees that he still remembers the day at Portsmouth. Hopes to collect and publish her father's letters and now returns those written to JH.
ALS 3pp RS:HS 13.198

13890. 1869–3–27 To Sir William Huggins Collingwood
Comments on the radiant heat of heavenly bodies and the motion of Sirius.
C 2pp RS:HS 24.254

13891. 1869–3–29 From George Robert Waterhouse British Museum
Extensive discussion of comparative anatomy, taxonomy, and distribution of rodents. Encourages JH to use term *Hystricidae* for porcupines in JH's [*Physical Geography* (1861)].
ALS 6pp RS:HS 18.105

13892. 1869–3–30 From George Robert Waterhouse British Museum
Reptile genera in JH's [*Physical Geography*] are consistent with those of [Edward] Forbes cited in *Johnston's Physical Atlas* [1854], but new genera have been discovered. Questions JH's claim that Elgin sandstone is Devonian. Notes Richard Owen's estimates on pterodactyl size, and distinct tail feathers [illustration] on recent specimen from Bavaria. Corrects JH's names for beaver genera.
ALS 4pp RS:HS 18.106

13893. 1869–4–1 From James Challis Cambridge
Is pleased with the interest JH is taking in his volume. Understands the difficulty of comprehending it at once. Further comments on his own dynamical theory of dispersion.
ALS 7pp RS:HS 5.242

13894. 1869–4–1 To George Robert Waterhouse Collingwood
Thanks for answers and for notice on *Apteryx* [*Maerurus* ?], which JH remembers watching Richard Owen inspect. Interested in W. B. Carpenter's and C. W. Thomson's ['Bathybius?]' as origin of chalk-flint. Discusses inorganic chemical allotropes. Organic allotropes may exist, serving as 'agents of vital action' and subject to a higher power.
Adraft 6pp RS:HS 18.107 (C: 24.255)

13895. 1869–4–5 From Christian August Friedrich Peters Altona
Comments on exchange of astronomical works with JH. CP expects to improve the observatory library now that the Prussians have taken over Altona. Is making arrangements to carry out some pendulum experiments.
ALS 4pp (German) RS:HS 13.371

13896. 1869–4–10 From Amédée Victor Guillemin Orsay, near Paris
Is preparing a new edition of his book *Le ciel,* which J. N. Lockyer is translating into English. Would like to reproduce new information on shooting stars and meteors and would like JH's assistance.
ALS 2pp (French) RS:HS 9.71

13897. [1869]–4–19 From John Murray, [Jr.] 50A Albemarle St.
Is prepared to take charge of Mrs. Mary Somerville's memoir. Found her wonderfully well when he met her at Naples.
ALS 1p RS:HS 12.454

13898. [1869]–4–22 From John Murray, [Jr.] 50A Albemarle St.
Has safely received the packet containing Mrs. Mary Somerville's manuscript. She has been awarded the Victoria medal of the Geographical Society.
AL 1p RS:HS 12.455

13899. 1869–4–24 From William E. Hearn Melbourne
Is pleased that his book, the *Government of England,* meets with JH's approval.
ALS 1p RS:HS 10.175

13900. 1869–4–25 From Wilhelm Schur Berlin
Is happy that his discovery of 70 Ophiuchi was included in JH's catalogue of double stars. Thanks for correction on a logarithm of the focal point equation; intends to change this in *Astronomische Nachrichten.*
ALS 1p (German) RS:HS 17.180

13901. 1869–4–26 From Russell Henry Manners Henrietta St.
Thanks for the additional information regarding Argus by Elias Loomis and J. Tebutt; it will appear in the next *R.A.S.M.N.*
ALS 1p RS:HS 12.288

13902. 1869–4–27 or earlier To John Herschel (son) Collingwood

Talks about problems of observing nebulae—one of which seems to have changed—and one of which is in the catalogue and does not seem to exist.
AL 1p inc JHS 6.46

13903. 1869–5 To Elizabeth Colling Collingwood
Comments on 'rubbish' published by Poet Laureate, EC's report of the weather of 1849, and the flowers and plants growing in JH's garden.
C 3pp RS:HS 24.257

13904. 1869–5–5 From Neil Arnott Cumberland Terrace
Is sending a pamphlet. Has given the profits of his recent book to various universities.
ALS 2pp RS:HS 1.411

13905. 1869–5–6 From George Biddell Airy Royal Observatory, Greenwich
Is sending his lectures on magnetism he gave at Cambridge.
ALS 2pp RS:HS 1.313

13906. 1869–5–6 To William James Herschel Collingwood
Glad that Emma's jaundice is over. Criticises metric system of measure. Encloses JH's response to Mr. Yates's letter in the *Times* on Geometrical British System. Upset by present resistance to anything English. This is what 'lost us...Neptune.' Would like to see Arabic translation of *Outlines Astr.* Relieved that 'shabby' French translation by [L. M. J.] Chevigné was stopped. Failure of gold standard in India. French and English coins. Suggestions on WJH's effort to introduce oaths into Indian justice system. Popularity of velocipedes [bicycles]. Family news.
ALS 4pp TxU:H/L-0564; Reel 1053

13907. 1869–5–8 To George Biddell Airy Collingwood
Has read the syllabus of GA's lectures on magnetism; it reminds JH of his attempt to solve the problem of 'Mahomet's coffin', which JH still believes would work.
ALS 4pp RGO 6.695.494 (drafts RS:HS 1.314 & 24.256)

13908. 1869–5–9 From Augustus De Morgan 91 Adelaide Road
Has not heard from him for a long time. He himself has been suffering with debility. Has done no maths. Is moving at midsummer to a new house at Merton Road.
ALS 3pp RS:HS 6.404

13909. 1869–5–9 From Thomas Webster to Margaret Brodie Herschel Cranbrook, Staplehurst
Recommends artisan to restore portrait of William Herschel. [Henry] Hardcastle and wife [Maria Sophia Herschel] visited TW's lodgings, lately occupied by [Thomas] Wade and wife [Amelia Herschel].
ALS 2pp RS:HS 18.133

13910. 1869–5–10 From William Benjamin Carpenter University of London
Is glad he is reconsidering the question of temperature. Regarding the deep sea thermometers.
ALS 4pp RS:HS 5.198

13911. 1869–5–11 To Augustus De Morgan Collingwood
On AD's and JH's illnesses; asks AD about Latin verses of the medieval period.
C 2pp RS:HS 24.258

13912. 1869–5–15 From Augustus De Morgan 91 Adelaide Rd.
Regarding middle Latin poets and similar English ones. Hopes JH's health improves with the advent of Spring.
ALS 4pp RS:HS 6.405b

13913. 1869–5–18 From George Biddell Airy Royal Observatory, Greenwich
Concerning JH's Mahomet's coffin theory.
ALS 2pp RS:HS 1.315 (C: RGO 6.695.495)

13914. 1869–5–19 From W. R. Greenhill Hawkhurst
Is grateful for the testimonial and for the favorable way in which he is regarded.
ALS 2pp RS:HS 9.124

13915. 1869–5–20 From Redmond Barry Melbourne
Is grateful for his courteous criticism of his research into Aboriginal dialects. Casts have been made of the heads of several of the aborigines.
ALS 4pp RS:HS 5.25

13916. 1869–5–21 From Joseph Winlock Cambridge, Mass.
Acknowledges receipt by Harvard College Observatory of JH's 'A Synopsis of all Sir William Herschel's Micrometrical Measurements...' (1867).
ALS 1p RS:HS 18.425

13917. 1869–5–24 From William Benson 147 Albany St.
Has sent a copy of his book on color to the R.S.L. library and would be glad of JH's comments on it.
ALS 4pp RS:HS 5.50

13918. 1869–5–26 From Thomas Francis Wade 58 Seymour St.
Delayed answering JH's letter. Describes new baby son. Thanks Margaret Brodie Herschel for note.
ALS 4pp RS:HS 18.5

13919. 1869–5–27 To William Benson Collingwood
Thanks for the gift of the paper on color. Comments on this and the subject in general.
draft 2pp RS:HS 5.51 & 24.259

13920. 1869–5–28 To Sir William Huggins Collingwood
About an increase in sunspot activity.
C 1p RS:HS 24.262

13921. 1869–5–28 To Henry Enfield Roscoe Collingwood
Thanks HR for sending HR's book on spectrum analysis [*Spectrum Analysis,* 1869]. Praises that publication.
C 1p RS:HS 24.261

13922. 1869–5–28 To John Tyndall Collingwood
Thanks for paper on comets. Offers questions, problems to be resolved, particularly on termination of comet.
CC 4pp RS:HS 17.406a (C: RI 559–60 & RS:HS 24:260)

13923. 1869–5–29 From John Tyndall Royal Institution
Thanks for JH's suggestions on JT's outline of comet speculations. [G. B.] Airy, too, has replied. Includes more speculations on 'envelopes' and nucleus of comet.
ALS 4pp RS:HS 17.406b (C: RI 562)

13924. 1869–5–29 To [John Tyndall] Collingwood
'Great Comet of 1843' presents difficulties for JT's explanation of comets' tails [continuation of JH's 1869–5–28].
ALS 1p DMC.695A.885.11 (C: RS:HS 24.263 & RI 561)

13925. 1869–5–30 To George Biddell Airy n.p.
Is considering a new edition of *Outlines Astr.*[?] and wants to provide the latest determinations of the periods of rotation of

the planets.
ALS 3pp RGO 6.257.356

13926. 1869–5–30 To John Tyndall Collingwood
Suggests experiments with regard to polarization and vaporization, and comets' tails.
CC 4pp diag RS:HS17.407a (C: 24.264; C: RI 563–4)

13927. 1869–5–31 From George Biddell Airy Royal Observatory, Greenwich
Gives information on the rotational periods of Mars, Jupiter, and Saturn [see JH's 1869–5–30].
ALS 2pp RS:HS 1.316 (ACCS: RGO 6.257.359)

13928. 1869–5–31 From William Benson 147 Albany St.
Is grateful for the paper and comments. Has referred to JH's papers many times in the lectures he has given. Hopes to produce a cheap book on color.
ALS 4pp RS:HS 5.52

13929. 1869–5–31 From Sir William Huggins Upper Tulse Hill
Has written to Lieut. John Herschel suggesting points on which solar observations might be valuable. Is pleased he has seen the bright lines. Saw some bright lines in the middle of the sun's disk.
ALS 2pp RS:HS 10.50

13930. [1869–6] From James Haig Beechwood, Bournemouth
Thanks for his kind note and his tracts on *Atoms* and *Forces*. Thinks J. S. Mill's *Logic* is dangerous. Hopes he will read his book thoroughly and not be too critical of his astronomy.
ALS 4pp RS:HS 9.158

13931. 1869–6 From Hermann Kindt Neustrelitz
Asks JH for copies of both Latin and English translations of Schiller's 'Walk'.
ALS 2pp (German) RS:HS 11.36

13932. 1869–6–1 From John Tyndall Royal Institution
In response to JH's comments on JT's explanation of comets says faintness of head and nucleus does not indicate non-existence; eye may not see all. Experiments with polarized light produce 'gorgeous' 'residual blue.'
ALS 4pp RS:HS 17.407b (C: RI 565)

13933. 1869–6–3 From Frederick J. O. Evans Admiralty
Will JH be able to attend the forthcoming meeting of the Board of Visitors? If not would he inform the Board of his intentions.
ALS 2pp RS:HS 7.132

13934. 1869–6–3 To Sir William Huggins Collingwood
Note to accompany the reported observations of JH's son John.
C 1p RS:HS 24.267

13935. 1869–6–4 To Frederick J. O. Evans Collingwood
Unable to attend the forthcoming meeting of the Board of Visitors. Would like to resign from the Board as his enfeebled health will prevent him from attending future meetings.
draft 2pp RS:HS 7.133 & 24.265

13936. 1869–6–4 To James Haig [?] Collingwood
Has received his book on symbolism from the publishers. Comments on some of its views. Feels he has been treated a little harshly.
C 4pp RS:HS 9.157 & 24.266

13937. 1869–[6]–6 To George Gabriel Stokes Collingwood
Correction to be made in JH's biographical sketch of William Whewell.
ALS 1p TxU:H/L-0464; Reel 1055

13938. 1869–6–7 From Sir William Huggins Upper Tulse Hill
Would like his advice on how to publicize Lieut. John Herschel's letter; shall he send it to G. G. Stokes? Comments on the method for viewing the forms of prominences.
ALS 4pp RS:HS 10.51

13939. 1869–6–8 From Frederick J. O. Evans Admiralty
Submitted JH's wishes to the Board, but the Board wishes him to remain a member as they value his advice, even if they cannot count on a personal appearance.
ALS 1p RS:HS 7.134

13940. 1869–6–9 To Sir William Huggins Collingwood
Thinks it would be best to send his son's letter to G. G. Stokes at the R.S.L. Cannot give a positive opinion on WH's plan for seeing the prominences. Finds some discrepancies in the Report of WH's paper on Sirius and the paper itself.
C 4pp RS:HS 10.52 & 24.268

13941. 1869–6–10 To John Herschel (son) Collingwood
Forwarded both of John's papers on solar observations to William Huggins. Comments on solar prominences and spots, and on Eta Argus. Concern for Esther's fever. Report of 'great meteor' seen at Collingwood. Alexander Herschel came home from Glasgow with new device called 'bicycle' that amuses neighbors. Explains optics problem in telescopes. English system of measures.
ALS 2pp TxU:H/L-0530; Reel 1053

13942. 1869–6–11 From Sir William Huggins Upper Tulse Hill
Has sent Lieut. John Herschel's letter to G. G. Stokes. Gives reasons for the discrepancies in the report of his paper and the paper itself. Is wondering whether to set up his large telescope further from London.
ALS 2pp RS:HS 10.53

13943. 1869–6–12 From Adam Sedgwick Cambridge
Wilberforce Clarke, the nephew of an old friend, has been appointed to a meteorological observatory. He needs a list of useful books and instruments for his job from JH.
ALS 4pp RS:HS 15.458

13944. 1869–6–16 From Julie Laurence Levi Canterbury
Outlining the sad financial case of Dr. C. T. Beke. Would JH join with some more well-wishers and subscribe towards the upkeep of Dr. Beke?
ALS 4pp RS:HS 11.475

13945. 1869–6–16 To Adam Sedgwick Collingwood
Praises AS for commencing his fifty-second course of lectures at age 84. Laments his own frailty. Recommends an array of meteorological books . Lists 'useful' meteorological instruments. Mentions family matters.
ALS 4pp RS:HS 15.459 (C: RS:HS 24.269)

13946. 1869–6–17 From Henry Bence Jones 84 Brook St.
Is to be a public meeting at the Royal Institution regarding a testimonial to Michael Faraday. Will JH oblige by doing the honors?
ALS 2pp RS:HS 10.338

13947. 1869–6–17 To Julie Laurence Levi Collingwood
Sympathizes with Dr. C. T. Beke but regrets he is unable to

contribute towards his assistance.
C 2pp RS:HS 11.476

13948. 1869–6–19 To Henry Bence Jones Collingwood
Grieves to say that he will be unable to attend the meeting at the Royal Institution to honor Michael Faraday, as his health is so poor. Thinks Faraday was a great man in all ways.
C 2pp RS:HS 10.339 (C: RS:HS 24.270)

13949. [1869–6–27?] From John Tyndall Royal Institution
Encloses reference to one of JH's letters to JT that has already been printed.
ALS 1p RS:HS 17.414 (C: RI 566)

13950. 1869–6–27 To [?] n.p.
Develops mathematical theorem.
ACCS 2pp inc TxU:H/M–0306.2 & TxU:H/M–0307.2; Reel 1087

13951. 1869–6–28 To Elizabeth Colling Collingwood
Comments on poetry, EC's and Robert Browning's; talks about the bicycle.
C 2pp RS:HS 24.271

13952. 1869–6–30 From Augustus De Morgan 6 Merton Road
Has been suffering with congestion of the brain. Moved house yesterday.
ALS 3pp RS:HS 6.405

13953. 1869–7–1 To Augustus De Morgan Collingwood
About AD's health and the cold summer.
C 2pp RS:HS 24.273

13954. [1869–7–7 or earlier] From Thomas Warner 47 Sussex Square, Brighton
Please review TW's enclosed new book on civil time-keeping: *How to Keep the Clock Right*. [JH annotation: Answered 7 July.]
ALS 2pp † RS:HS 18.91

13955. 1869–7–7 To Thomas Warner Collingwood
Received TW's book and approves simplicity and economy of its method of keeping time by using small fixed telescope to reduce star observations. Disapproves of using 'Dominical Letters.' Suggests using Julian dates.
L 'Preface to Improved Edition' of Thomas Warner, *How To Keep the Clock Right* (1876)

13956. 1869–7–9 From William Stanley Jevons Dover
Was very happy to receive JH's remarks on his little logical work. Comments on this work. Is not satisfied with the logic of J. S. Mill. Has not seen [James] Haig's work yet, but hopes to read it on return from the Continent.
ALS 4pp RS:HS 10.327

13957. 1869–7–14 To John Herschel (son) n.p.
Glad for Mary's recovery. Congratulations on John's promotion to captain. Suggests method to study solar prominences without spectroscope. [Letter continues 15 June:] Alexander Herschel reports that G. B. Airy tried this method unsuccessfully long ago, but JH is sure it will work.
ALS 3pp TxU:H/L-0531; Reel 1053

13958. 1869–7–15 To George Gabriel Stokes Collingwood
Reports on paper [*R.S.P.T., 159,*. 575–] by [Thomas] Andrews on continuity of gaseous and liquid states of matter, believing it fit for publication.
ALS 3pp RS RR.6.14

13959. 1869–7–18 To George Gabriel Stokes Collingwood
Note to accompany an extract from JH's son John's letter, reporting on his observations of solar prominences.
ALS 3pp Cambridge University Library 7656/H724

13960. 1869–7–20 From George Gabriel Stokes Lensfield Cottage, Cambridge
Extended comments on John Herschel's [JH's son] observations of the spectrum of the solar corona.
ALS 11pp RS:HS 17.75

13961. 1869–7–22 To Augustus De Morgan Collingwood
Concerned about AD's health. Offers theory of the constitution of matter.
C 2pp RS:HS 24.274

13962. 1869–7–24 From Hermann Kindt Neustrelitz
Thanks JH for his letter and package of his writings.
ALS 1p (German) RS:HS 11.37

13963. 1869–7–24 From Richard Anthony Proctor 5 Trigon Road, Kennington
Cites various celestial appearances as a basis for concluding that the Milky Way takes the form, not of a disk or ring, but of a series of convolutions. Urges that the nebulae form part of the Milky Way rather than being separate universes, as many had suggested. Asks JH to critique these ideas.
ALS 8pp RS:HS 14.112 Proctor, *Other Suns,* 392–402

13964. 1869–7–26 From Augustus De Morgan 6 Merton Road
Health is not good. Is not up to the theory of atoms. Have had broiling heat for many weeks.
ALS 2pp RS:HS 6.406

13965. 1869–7–26 From Francis Cranmer Penrose Colbyfield, Wimbledon
A note to accompany a paper by FP on the prediction of occultations and eclipses.
ALS 3pp RS:HS 13.336

13966. 1869–7–30 To Mary Somerville Collingwood
Attached comments [not included] convey JH's views on MS's new manuscript [*On Molecular and Microscopical Science,* 1869]. Expresses reservations about MS's endorsement of current ideas of forces and their correlation and conservation.
C 3pp RS:HS 24:275

13967. 1869–8–1 To Richard Anthony Proctor Collingwood
Encourages RP in his speculations about the nebulae and the structure of the Milky Way, which RP suggested is formed of a 'system of convolutions,' but raises objections to RP's views. Discusses idea that the Milky Way contains miniatures of itself and that beyond it may be a hierarchy of universes comparable to the Milky Way.
ALS 5pp RS:HS 24.276 Proctor, *Other Suns,* 402–5

13968. 1869–8–9 From Richard Anthony Proctor 5 Trigon Road, Kennington
Elaborates on his hypothesis that all forms of nebulae are part of the Milky Way and on his 'twisted-stream theory' of the Milky Way's structure. Favors view that dark matter is relatively common in space and that it extinguishes light from distant objects. Proposes method of measuring stellar diameters.
ALS 5pp RS:HS 14.113 Proctor, *Other Suns,* 406–14 inc

13969. 1869–8–10 From Adolphe Quetelet Brussels
Is trying to get JH 2 copies of *Physique sociale.* Discusses additions. Addresses JH's comments concerning power of a supreme being. Sends 19th volume of annals of observatory and AQ's work on international congress on statistics.
ALS 3pp (French) RS:HS 14.273 (C: Académie belgique: #56)

13970. 1869–8–14 From W[ilhelm] Veltmann Oberhausen
Apologizes for not acknowledging JH's work, Asks for copy of Royal Geographical Society's papers.
Discusses'[François] Arago's work on light beams and prisms. Requests JH's opinion concerning [E. F. W.] Klinkerfues's theory of light, which seems to make [Augustin] Fresnel's hypothesis superfluous.
ALS 4pp (German) RS:HS 17.461a

13971. 1869–8–16 To Margaret Brodie Herschel Collingwood
Of the fruit in the garden and the arrival of various of their daughters; JH seems to be feeling a little sorry for himself, but concludes with a riddle in French.
AL 4pp JHS 5.30

13972. 1869–8–18 To Adolphe Quetelet Collingwood
Thanks for second volume of *Physique sociale.* Happy to see AQ refuting 'the results of statistics of life, accident, crime...[as] indicative of an absence of free agency in human beings and the presence of some sort of impelling necessity.' Mentions JH's daughter's marriage and her learning Chinese.
ALS 3pp Académie belgique: #53

13973. 1869–8–19 From Thomas Graham Great Malvern
His letter has been forwarded. JH's correspondence with the Mint to which J. G. Hubbard refers has been published in the Report of the Commissioners on International Coinage 1868. Thinks expansion of mercurial amalgam is due to hydrogen. Hopes JH will accept a small medallion.
ALS 3pp RS:HS 8.231

13974. 1869–8–20 To Richard Anthony Proctor Collingwood
Discusses Olbers's Paradox, raises objections to RP's argument for the existence of dark celestial matter, and points out problems in RP's method of measuring stellar diameters.
C 3pp RS:HS 14.114 & RS:HS 24.277 Proctor, *Other Suns,* 414–15 inc

13975. 1869–8–20 From Thomas Romney Robinson Exeter
Sends 'first fruits' of Melbourne telescope in form of a letter from [A.] LeSueur and [R. L. J.] Ellery. Includes discussion of nebulae.
LS 6pp RS:HS 14.381

13976. 1869–8–21 To Thomas Romney Robinson Collingwood
Thanks for notice on first performance of Melbourne telescope and its contributions to knowledge about nature and role of nebulae. Comments on relationship of nebula and apparently associated stars. Looks forward to report on Magellanic Clouds.
ALS 4pp Cambridge University Library 7656/TR65 (C: RS:HS 24.278 & RS Sa.676)

13977. 1869–8–22 From Richard Anthony Proctor 5 Trigon Road, Kennington
Accepts JH's objection to his method of measuring stellar diameters. Quotes an argument he had published suggesting the existence of a hierarchy of systems in an infinite universe. Discusses his studies of the relation of proper motions to the sun's motion. Asks JH's permission to publish JH's idea that some nebulae may be 'miniature' of the Milky Way.
ALS 7pp RS:HS 14.115

13978. 1869–8–22 From Adolphe Quetelet Brussels
Thanks JH for his positive judgement of his work. Mentions that [François] Moigno could only say that it was anti-religious. Another work on statistics has emerged: *Versuch einer Sozialethik,* 2nd. vol., by Alexander von Ochingen. In few days leaves for international conference on statistics.
ALS 3pp (French) RS:HS 14.274 (C: Académie belgique: #54)

13979. 1869–8–22 From Thomas Warner 47 Sussex Square, Brighton
Requests permission to print JH's response approving of fixed-telescope system for time-keeping set out in TW's new book on civil time-keeping, *How to Keep the Clock Right.*
ALS 2pp RS:HS 18.89

13980. 1869–8–24 From William Vernon Harcourt Nuneham Park, Abingdon
Is very pleased that JH has accepted his little volume. Is sure JH would have made a success of a similar work. Behavior of animals.
ALS 5pp RS:HS 9.221

13981. [1869]–8–30 To the London *Times* [Collingwood]
Discusses in detail the then currently debated issue of a 'mintcharge or seigniorage' on gold coinage. Also discusses the idea of an international coinage.
Adraft 1p inc TxU:H/M–0950.2; Reel 1083 London *Times* (Sept. 2, 1869), p. 10, col. 4–5 & *Speeches, Letters, Articles, &c on the Gold Coinage Controversy of 1869* (London: Bank of England, 1870), 116–23

13982. [1869–9] To William Stanley Jevons Collingwood
Is unaware of experiment measuring amount of absolute error in transit measurements. Discusses matters relating to coinage and the standards underlying it.
ALS 4pp John Rylands University Library of Manchester Black (ed.), *Papers and Correspondence of...Jevons,* 3:216–7

13983. 1869–[9 or 10] From J[ohn] Jones Brynhyfryd, Wrexham
Sending tables showing the periodical occurrences of solar spots, earthquakes, volcanic eruptions, etc.; comments on his theories regarding these.
ALS 4pp RS:HS 10.344

13984. [1869–9 to 1870–2] To the London*Times* Collingwood
Encloses letter regarding coinage under discussion in Parliament.
ACCS 1p TxU:H/L-0129; Reel 1054

13985. 1869–9–2 From George Gabriel Stokes Torquay
Still worried about the meaning of some terms in John Herschel's [JH's son] paper [see GS's 1869–7–20].
ALS 4pp RS:HS 17.76

13986. 1869–9–3 To George Gabriel Stokes Collingwood
Agrees with GS's clarification of 'continuous spectra' and other terms [see GS's 1869–9–2]. Includes a copy of part of a letter from John Herschel [JH's son] about solar prominences.
ALS 2pp encl Cambridge University Library 7656/H725

13987. 1869–9–4 From George Biddell Airy Royal Observatory, Greenwich

Regarding JH's letter to the *Times* on the gold coinage question.
ALS 2pp RS:HS 1.317

13988. 1869–9–4 From C[harles] Rivers Wilson [Treasury]
In recent letter to the *Times* regarding international coinage, JH attributed viewpoint to Col. [J. T.] Smith that was in fact expressed by J. B. Smith, M.P. for Stockport. CW's remarks on valuation of coins.
ALS 10pp † TxU:H/M–0582; Reel 1093

13989. [1869–9–6] To the London *Times* n.p.
Corrects a misstatement JH made in his 1869–8–30 letter to the *Times*.
L London *Times* (Sept. 9, 1869), p. 4, col. 6

13990. 1869–9–7 To C[harles] Rivers Wilson Collingwood
Thanks for catching JH's error. Will write to the *Times* and substitute 'Mr. Smith, M.P.' for 'Col. Smith.' JH does not understand 'tariff of conversion' or how it will simplify calling in old coins. Prefers 'Binary system' to 'absolutely demonetizing silver.'
C 4pp TxU:H/L–0500; Reel 1055

13991. 1869–9–8 From George Gabriel Stokes Torquay
Clarifications of some confusion in the proof copy of a paper by John Herschel [JH's son].
ALS 8pp RS:HS 17.77

13992. 1869–9–10 From Augustus De Morgan 6 Merton Road
Is moving slowly. Regarding an edition of William Spence's *Mathematical Essays* edited by JH.
ALS 3pp RS:HS 6.407

13993. 1869–9–11 To Augustus De Morgan Collingwood
Is delighted to see him astride one of his old hobbies. Regarding the edition of William Spence's *Mathematical Essays*. Only remembers an 1819 one. His own health is not good.
draft 4pp RS:HS 6.408 & 24.279

13994. 1869–9–12 From A. Lincke Golssen, Prussia
AL claims to be a relative; could JH help financially with the education of AL's sons?
ALS 4pp RS:HS 11.485

13995. 1869–9–13 From John Thomas Smith The Mythe, Tewksbury
Encloses copy of JS's letter to the *Times* in response to JH's recently published objections to currency system in India. Wishes to meet JH and discuss issue in private, then publish joint statement on subject.
ALS 4pp TxU:H/M–0489; Reel 1093

13996. 1869–9–14 From James Maclaren Constable Burton
Read JH's letter in the *Times* with great pleasure. Would like to see the gold question remain as at present. Has not been successful with the Wellingtonia seed; none has come up. Regrets to hear the change in the weather has brought on JH's bronchitis.
ALS 7pp RS:HS 12.32

13997. 1869–9–15 From J[ohn] Jones Brynhyfryd, Wrexham
Is grateful for his letter. Is also pleased that the proportions between groups of planets and satellites merits his attention. Hopes to forward some tables on sunspots.
ALS 3pp RS:HS 10.343

13998. 1869–9–16 From Bartholomew Price 11 St. Giles St., Oxford
JH having declined offer from Clarendon Press to write a treatise on inductive logic, [Thomas] Fowler of Oxford, who had earlier written a book on deductive logic, drafted an inductive logic, relying heavily of JH's *Prelim. Discourse* and *Familiar Lectures*. Asks JH to make suggestions on the manuscript.
ALS 4pp RS:HS 14.60

13999. 1869–9–16 To John Thomas Smith Collingwood
Sees no likelihood that private meeting would resolve issue between JH and JS. JS misunderstood JH's argument on valuation of British coins and currency. Encloses JS's note to Editor of the *Times*.
C 5pp TxU:H/L–0387; Reel 1054

14000. 1869–9–20 To Sir William Huggins Collingwood
Suggests the planet Mercury may be a good subject for his spectroscopic experiments. Gives comments and explanations on his proposals. May like to communicate the results to the R.A.S.
C 6pp RS:HS 10.54 & 24.280

14001. 1869–9–21 From Robert L. J. Ellery Melbourne Observatory
Notification that a volume of the Melbourne Observatory papers has been dispatched.
ASD 1p RS:HS 7.126

14002. [1869]–9–21 From Thomas Fowler Lincoln College, Oxford
Hears that JH has agreed to look through the sheets of his book *Inductive Logic*. Is very pleased about this and would be grateful for any comments. Is intended as an elementary textbook.
ALS 3pp RS:HS 7.353

14003. 1869–9–21 To Sir William Huggins Collingwood
Note to correct calculations in an earlier letter [see JH's 1869–9–20].
C 1p RS:HS 24.281

14004. 1869–9–22 To Sir William Huggins Collingwood
Burn his [last] letter as he made a blunder in his geometry. Cannot imagine how he made the blunder. Gives new formulae for observing Mercury.
draft 1p RS:HS 10.56 & 24.282

14005. 1869–9–22 From Sir William Huggins Upper Tulse Hill
Unable to make any observations at the moment as his observatory is being enlarged to receive his new large telescope. Thinks it would be a good idea to carry out observations on Mercury. Is not sorry to be able to rest his eyes from his trying spectrum experiments.
ALS 3pp RS:HS 10.55

14006. 1869–9–23 To Thomas Maclear Collingwood
Writes of children's accomplishments. Discusses possible changes in nebula near Eta Argus.
ALS 3pp Cape Archives/Maclear Papers File 100

14007. 1869–9–26 From Augustus De Morgan 6 Merton Road
Regarding a correction to Francis Baily's epitaph. Health is improving slowly.
ALS 2pp RS:HS 6.409

14008. 1869–9–26 From Eliza Susan Quincy Quincy, Mass.
Describes celebration of [Alexander von] Humboldt's centennial in Boston. Sends photograph of total eclipse of sun, from Henry Morton, and various other articles. Praises science.
ALS 4pp RS:HS 14.288

14009. 1869–9–27 From F. Charsley Slough
Proposals for new footpath at Slough near his property; would JH object?
ALS 3pp RS:HS 5.249

14010. [1869–9–27 or later] To F. Charsley [Collingwood]
Reply to FC's 1869–9–27.
draft 1p RS:HS 5.249verso

14011. [1869]–9–27 From Thomas Fowler Lincoln College, Oxford
Consider his note on empirical laws as unwritten as he intends introducing a paragraph on this subject. Gives a list of the contents of his book.
ALS 2pp RS:HS 7.354

14012. 1869–9–27 From Sir William Huggins Upper Tulse Hill
Is much obliged for the more correct determination of Mercury. As soon as he gets his new apparatus he will try to obtain observations of the planet.
ALS 2pp RS:HS 10.57

14013. 1869–10 To Thomas Fowler n.p.
Comments on the proofs of his book that he has received so far. Has been ill and not feeling like intellectual work, so would he request his printer to send no more sheets for the time being.
draft 6pp RS:HS 7.356 & 24.283

14014. 1869–10 To Urbain J. J. Leverrier n.p.
JH responds to UL's concern about the Isaac Newton–Blaise Pascal forgeries [see UL's 1869–10–4].
C 1p RS:HS 24.290

14015. 1869–10 To William Selwyn Collingwood
Sends [William Whewell's] 'Isle of Sirens' and an acknowledgement to it in 'skimble skamble Hexameters.' Is eager to receive solar photographs and will send others by Professor [George?] Morton.
C 1p RS:HS 24.289

14016. 1869–10–2 From Edward William Brayley London Institution
Regarding Dr. Thomas Andrews's Bakerian lecture and the gaseous and liquid states. Comments on this and his previously stated theory. Has seen Alexander Herschel and thinks he should receive more credit for his work on the Meteor Committee. Scientific education of young people.
ALS 10pp RS:HS 4.237

14017. 1869–10–3 From Thomas Warner Royal Hotel, College Green, Bristol
Sends proof sheet of JH's approval letter before it goes to printer.
ALS 2pp RS:HS 18.90

14018. 1869–10–4 From Urbain J. J. Leverrier Paris
Regarding Michel Chasles and the Isaac Newton-Blaise Pascal forgeries.
ALS 2pp (French) RS:HS 11.213

14019. 1869–10–5 To Edward William Brayley Collingwood
Comments on liquid and gaseous states of matter and on colloids; appreciates EB's kind words about JH's son Alexander.
C 2pp RS:HS 24.284

14020. 1869–10–6 From Thomas Fowler Lincoln College, Oxford
Is extremely grateful for his comments on his book and will try to incorporate them. His book is intended as an introduction only. Delegates prefer s instead of z.
ALS 5pp RS:HS 7.355

14021. 1869–10–10 To the *Athenaeum* Collingwood
Calls attention to James Grahame's *History of the United States of North America*. Criticizes the *Saturday Review* for having slighted that work.
L *Athenaeum*, #2190 (Oct. 16, 1869), 498

14022. 1869–10–10 From H. Turberville Barnstaple
Describes his visit to Exeter. Hopes JH received a package sent the previous week. Asks question concerning lens making.
ALS 8pp RS:HS 17.380

14023. 1869–10–15 To Augustus De Morgan Collingwood
Discusses an exposed case of forgery involving supposed manuscripts of Blaise Pascal and the mathematician Michel Chasles.
C 2pp RS:HS 24.285

14024. 1869–10–15 From Frederick W. Holland Portsea Place, Connaught Square
Sinai expedition did not complete its work. Would like to send out Messrs. E. H. Palmer and C. F. T. Drake again. Can they continue to use JH's name as a trustee?
ALS 4pp RS:HS 9.426

14025. 1869–10–16 From William Selwyn Foxton, Royston
Has been translating into Latin 'your Dean's "Kentish Fire."' Has good Latin versions of [Oliver Goldsmith's] 'Edwin and Angelina' by Lord Stratford de Recliffe and of [Thomas] Gray's 'Elegy' by 'Chief Justice [Henry Thomas?] Cockburn.' Sends his 'Genevieve.'
ALS 4pp RS:HS 15.505

14026. 1869–10–18 From John R. Hind Meadow Lodge, Twickenham
Sends photograph of a medal lately struck in Paris to commemorate the discovery of 100 asteroids, on which Hind appears.
ALS 1p RS:HS 9.396

14027. 1869–10–18 To Frederick W. Holland n.p.
Takes it that the expense of the Sinai Expedition so far carried out has been met from the funds. Sees no objections to a continuation of the expedition.
C 1p RS:HS 9.427 & 24.287

14028. 1869–10–18 From J. Peter Lesley American Philosophical Society, Philadelphia
The Society will be happy to make JH's set of the Proceedings as complete as possible.
ALS 2pp RS:HS 11.472

14029. 1869–10–18 To Eliza Susan Quincy Collingwood
Thanks EQ for report of [Alexander von] Humboldt's centenary and the photographs from Henry Morton [see EQ's

1869–9–26]. Discusses neglect of James Grahame's work in England.
C 4pp RS:HS 14.289 (C: RS:HS 24.286)

14030. 1869–10–20 From Augustus De Morgan 6 Merton Road
Regarding a mistake in his old address. Hopes U. J. J. Leverrier will arrive safely after all. Cold weather is bracing him up.
ALS 3pp RS:HS 6.410 S. De Morgan, *Memoir of Augustus De Morgan*, 395–6

14031. 1869–10–22 From Augustus De Morgan 6 Merton Road
U. J. J. Leverrier arrived safely yesterday. Michel Chasles has a terrible time ahead of him. Cold weather is doing him good.
ALS 2pp RS:HS 6.411

14032. 1869–10–23 To George Biddell Airy Collingwood
Continues to rail against the imposition of the metric system in India; argues against any change in the British sovereign coin.
ALS 2pp RGO 6.361.631

14033. 1869–10–23 From John R. Hind Nautical Almanac Office
His last note on the Julian period puzzled him. Has not taken many double star measurements, but will send on a few later. Has not finished the ecliptical charts yet. Regarding a planet more distant than Neptune.
ALS 4pp RS:HS 9.397

14034. 1869–10–23 To William Selwyn Collingwood
Is working on a compilation of the measures of double stars. If [Johann] Wolf's period of 11.11 years is correct, why should 1810-11, a minimum, mark the first observations of solar spots?
C 2pp RS:HS 24.288

14035. 1869–10–25 To Elizabeth Colling Collingwood
Thanks for the gift of a sonnet; JH too much in figures to allow the muse to repay the gift.
C 1p RS:HS 24.291

14036. 1869–10–27 From Thomas Fowler Lincoln College, Oxford
Is extremely grateful for his comments. Agrees with him about the medical examples. All the sheets have now been sent.
ALS 2pp RS:HS 7.357

14037. 1869–10–27 From William Selwyn Foxton Royston
Has autographs that display the activity of the sun. Sends sonnets he has written. Asks JH to mention distribution of nebulae at R.A.S.
ALS 4pp RS:HS 15.506

14038. 1869–10–30 To the *Journal of the Society of Arts* Collingwood
Discusses methods of diminishing the effect of a ship's rolling and pitching on a person on a cot or chair on the ship.
L *Journal of the Society of Arts, 17* (Nov. 5, 1869), 907

14039. 1869–10–31 To Duncan [Stewart] Collingwood
Explains cause of DS's double vision. Suggests corrective lenses [diagrams].
ALS 3pp TxU:H/L-0142; Reel 1054

14040. 1869–11–1 From E. S. Newall St. Rogue, Clifton
Is a collector of autographs and would be pleased to receive an example of JH's.
ALS 4pp RS:HS 13.117

14041. [1869–11?] To William Selwyn n.p.
Discusses discrepencies between his observations for September and WS's photograph. It seems the new spots result from Jupiter being in opposition.
draft 4pp diag RS:HS 15.514

14042. 1869–11–2 From E. S. Newall St. Rogue, Clifton
Is grateful for JH's reply. Wishes she had obtained the autographs of Sir William and Caroline Herschel.
ALS 4pp RS:HS 13.118

14043. 1869–11–3 To William Selwyn Collingwood
Describes the similarities between JH's sketches and WS's photographs of sunspots. Has greater faith in own observations. Wishes WS would add Julian dates. Sends Professor [George?] Morton's photographs.
C 2pp RS:HS 24.292

14044. 1869–11–5 To the *Journal of the Society of Arts* Collingwood
Proposes a method of diminishing the effect of a ship's rolling and pitching on a person on a cot or chair on the ship.
L *Journal of the Society of Arts, 17* (Nov. 12, 1869), 917

14045. 1869–11–6 From Robin Allen Trinity House
Is sending three Mexican Gulf shells, and a paper-weight made from a piece of the Wolf Rock off Penzance.
draft RS:HS 1.392

14046. 1869–11–6 From Richard Anthony Proctor 5 Trigon Road, Kennington
Asks JH's views on a system of constructing and annotating a set of celestial maps RP was preparing. Requests JH's public endorsement of those maps.
ALS 4pp RS:HS 14.116

14047. 1869–11–8 From Augustus De Morgan 6 Merton Road
Kept indoors by the rain, so returns U. J. J. Leverrier with many thanks. Regarding Michel Chasles and the Pascal forgeries. Has lost his friend G. B. Libri. Had a visit from Robert Turner over the Pascal affair.
ALS 7pp RS:HS 6.412 S. De Morgan, *Memoir of Augustus De Morgan*, 396–7

14048. 1869–11–9 From Eliza Susan Quincy Boston, Mass.
Pleased JH wants to defend [James] Grahame's work. Describes [Henry] Morton's work and sends additional publications. Intrigued by JH's prediction that a great discovery concerning the sun will be made soon.
ALS 4pp RS:HS 14.290

14049. 1869–11–11 To William James Herschel Collingwood
Proud of WJH for introducing improvements into Indian judiciary system. Sorry to hear that WJH may have to leave Dacca.
ALS 2pp TxU:H/L-0565; Reel 1053

14050. 1869–11–13 From Richard Anthony Proctor 5 Trigon Road, Kennington
Discusses his progress preparing star maps. Stresses that such maps can suggest theories of the arrangement of stars and nebulae. Mentions various papers and maps RP was then

publishing
ALS 4pp RS:HS 14.118

14051. 1869–11–13 From Richard Anthony Proctor 5 Trigon Road, Kennington
Thanks JH for suggestions concerning a star atlas RP was preparing. Informs JH of his progress in making that atlas.
ALS 4pp RS:HS 14.117

14052. 1869–11–16 From Peter le Neve Foster Society of Arts, Adelphi
Council of the Society of Arts has offered prizes. Would be pleased if JH would serve on the Committee to decide the prizewinners.
ALS 1p RS:HS 7.327

14053. 186[9]–11–16 To John Herschel (son) Collingwood
Asks son John to stop at the British Museum and get the completion of a quotation JH needs; also asks John to pick up JH's *R.S.P.T.* copies not picked up recently.
ALS 3pp JHS 6.55

14054. 1869–11–16 From John Tyndall Clifton
Is delighted with JH's exposition of colors of thin plates and of measurements of a wave of light in *Familiar Lectures*. Asks permission to quote from it.
ALS 3pp RS:HS 17.408 (C: RI 569)

14055. 1869–11–18 From Peter le Neve Foster Society of Arts, Adelphi
Council of the Society of Arts has decided to form a committee to watch the interests of telegraphy. Invites JH to become a member.
ALS 2pp RS:HS 7.328

14056. 1869–11–18 From Thomas Maclear Royal Observatory, Cape of Good Hope
Will learn with regret that Feldhausen has had to be sold because R. J. Jones, the owner, was in debt. Outlines the alterations to the house and to the neighborhood now that the railroad has appeared. Has a tribe of grandchildren now.
ALS 6pp RS:HS 12.176

14057. 1869–11–18 From R. C. Phillips Lower Broughton, Manchester
Inquires about the mathematical nature of the chords of music, offering some ideas of his own.
ALS 2pp RS:HS 13.389

14058. 1869–11–22 From Henry Anthony Jeffreys Hawkhurst Parsonage
No such fast has been proclaimed. Sends two extracts from the *Times*.
ALS 2pp RS:HS 10.301

14059. 1869–11–24 To R. C. Phillips Collingwood
Comments on the mathematical ratios in chords at some length, in answer to RP's 1869–11–18.
draft 4pp RS:HS 13.390 & 24.293

14060. 1869–11–24 From B. F. Sandy Naval Observatory, Washington
Regrets that JH has not received annual volume of astronomical observations for 1865, sent months ago. Is sending another.
ALS 1p RS:HS 17.168

14061. 1869–11–25 To John Herschel (son) Collingwood
Is not prepared to come to any distinct conclusions about the effect of local attractions in geodetic surveying, but talks about the problem for the whole letter.
ALS 4pp JHS 5.46

14062. 1869–11–25 To William James Herschel Collingwood
Questions about Indian civil law regarding Bengal promisory notes and duties on probate.
ALS 2pp TxU:H/L-0566; Reel 1053

14063. 1869–12–[2] From T. P. Anderson and J. H. Shonbridge [4 Lincoln's Inn Fields]
[Summary in JH's hand:] Enquiry about Henry John Hollier's request for advance on mortgage. Enclosed form to be signed by JH [as co-trustee] or chief accountant of Bank of England. [JH note: signed form and returned it to Anderson and Shonbridge. Forwarded letter to Few & Co.]
Amem 1p TxU:H/M–0917.2; Reel 1083

14064. 1869–12–3 From J[ohn] Browning Clapham Old Town
Sending JH a set of stereograms of prisms.
ALS 2pp RS:HS 5.113

14065. 1869–12–9 From Pierce Butler Ulcombe Rectory, Staplehurst
Grateful for JH's interest in survey of Sinai Peninsula and loan of JH's theodolite. [H. S.] Palmer will visit JH and receive instruction in its use. Asks to name JH as co-trustee of survey with Henry James.
ALS 4pp TxU:H/M–0131; Reel 1087

14066. 1869–12–12 From Pierce Butler Long's Hotel, Bond St.
JH and R. I. Murchison have agreed to be trustees [of Sinai survey; see PB's 1869–12–9]. Subscriptions to support survey. Sends description of expedition. Royal Geological Society unanimously approved survey.
ALS 4pp TxU:H/M–0132; Reel 1087

14067. 1869–12–14 From George Setten Plymouth
Admiring 'working man' requests signature to be placed under JH's portrait as 'a remembrance of a great astronomer.'
ALS 3pp RS:HS 17.191

14068. 1869–12–15 To the London *Times* n.p.
Calls attention to sunspots then visible. [Written by JH under the pseudonym 'Helioscopus']
L London *Times* (Dec. 17, 1869), p. 7, col. 6

14069. 1869–12–18 From Thomas A. Hirst Athenaeum
Asks JH to sign a certificate for Fellowship of the R.S.L. of [William Stanley] Jevons.
ALS 2pp RS:HS 10.196

14070. 1869–12–18 From Henry Samuel King 65 Cornhill
Sending a little volume of poetry written by his wife.
ALS 3pp RS:HS 11.40

14071. 1869–12–18 From Mary Somerville Naples
Sends Christmas greetings to family. Still has her intellect and is staying busy. Thanks Margaret Brodie Herschel for visiting Agnes [Greig].
ALS 2pp RS:HS 16.380

14072. 1869–12–19 To Mrs. [Richard] Jones Collingwood
Best wishes on Mrs. Jones's birthday.
C 2pp RS:HS 24.294

14073. 1869–12–20 To Henry Samuel King Collingwood
Many thanks for Mrs. King's volume of poetry. Comments on some of the poems.
C 4pp RS:HS 11.41 & 24.295

14074. 1869–12–22 From T[heseus] A. Cheney Watkins, N.Y.
Thanks for his kind letter regarding his book. Would be very grateful if JH would send him his photograph and autograph signature, also signatures of any famous Europeans for his album.
ALS 4pp RS:HS 5.334

14075. 1869–12–24 To Sir Isaac L. Bell Collingwood
Many thanks for the papers on the chemistry of the blast furnace. Remarks concerning this.
draft 4pp RS:HS 4.48 & 24.296

14076. 1869–12–27 From R. C. Phillips Lower Broughton, Manchester
More on the mathematical ratios of musical chords, and of dissonance [see JH's 1869–11–24].
ALS 4pp RS:HS 13.391

14077. 1869–12–30 From Walter White Burlington House
Thanks for news of Mrs. Birtwhistle. Hopes JH returns to good health.
ALS 2pp RS:HS 18.231

14078. [1870] From Thomas Brassey St Leonards-on-Sea
Is grateful for his comments on the pamphlet on Trade Unions. Hopes to produce some more on similar subjects.
ALS 2pp RS:HS 5.97

14079. [1870]. From G. W. Royston-Pigott Lansdowne Crescent
Asks permission to send JH several articles he wrote concerning optics.
ALS 3pp RS:HS 14.506

14080. [1870] From William Selwyn Foxton, Royston
Thanks JH for drawing of solar maculae from 1854-61. Some hesitancy about the accuracy of [Alexander?] Wilson's observations and hypotheses. Sixty people wish to travel to Spain to see eclipse of the sun.
ALS 4pp RS:HS 15.513

14081. [1870 or earlier] From Edward Vivian n.p.
Mrs. W. Vivian is visiting south of France.
ALS 1p inc TxU:H/M–0950.1; Reel 1083

14082. 1870–1–3 To Hugh Martin Collingwood
Grateful for HM's sending J. H. Stirling's essay 'As Regards Protoplasm' (1869) to JH. T. H. Huxley overlooked allotropes of organic chemicals as possible origin of life. In this 'Biocratical state,' allotropes would not *be* alive, but would be 'designedly...fitted' to act as agents of vital processes under a causative power.
C 4pp TxU:H/L-0248; Reel 1054 (Cdraft: RS:HS 25.15.26)

14083. 1870–1–8 From William Stanley Jevons Wittington
Received copy of JH's 'Half Dozen Propositions Regarding the Gold Coinage.' Does not agree with JH's last proposition. England will soon be only European nation not using international coinage. Expects that new difficulties for 'Mr. [Robert] Lowe's scheme' will be caused by France.
ALS 4pp TxU:H/M–0300; Reel 1087

14084. 1870–1–10 From Alexander J. B. Hope Bedgebury Park, Cranbrook
Hopes JH's argument favoring pound sterling will be read by members of Parliament.
ALS 2pp TxU:H/M–0084; Reel 1087

14085. 1870–1–11 From Edward Hardcastle n.p.
Compliments JH's 'Half Dozen Propositions on the Gold Coinage.' Hopes these produce changes in handling of gold. Belgian calico is competing with English calico and producing economic distress in Manchester.
ALS 3pp TxU:H/M–0240; Reel 1087

14086. 1870–1–11 From Joseph Alfred Hardcastle 54 Queen's Gate Terrace
Just received another copy of JH's 'Half Dozen Propositions on the Gold Coinage' from son Henry Hardcastle. These should create problems for Chancellor of Exchequer. Hopes to understand these before Parliamentary discussion on gold currency begins.
ALS 2pp TxU:H/M–0245; Reel 1087

14087. 1870–1–11 From Russell Henry Manners Somerset House
Thanks for the note and communication sent to the Assistant Secretary.
ALS 1p RS:HS 12.289

14088. 1870–1–11 From J[ames] H[utchison] Stirling to Rev. Hugh Martin Piershill, Edinburgh
JS thanks HM for having copied out for JS a letter of JH's apparently regarding JS's paper. JS admits 'It never occurred to me even in dream to get the opinion of Sir John Herschel on my poor essay.'
ALS 7pp RS:HS 17.238

14089. 1870–1–13 To Elizabeth Colling Collingwood
Thanks for the volume of EC's verse; comments on various aspects of poetry.
C 4pp RS:HS 24.297

14090. 1870–1–13 From John Gellibrand Hubbard Addington Manor, Winslow
Wants to print JH's 'Half a Dozen Propositions Respecting the Gold Coinage' as conclusion to coinage controversy being published by governor of Bank of England. Asks for clarification on several points in Propositions. Hubbard is working to improve assay results at Bank.
ALS 8pp TxU:H/M–0279; Reel 1087

14091. 1870–1–13 From Thomas Maclear Royal Observatory, Cape of Good Hope
Sending photographs and plans of the Feldhausen estate. Comments on the recent sale of this property. Is anxious about the safety of David Livingstone. Comments on Livingstone's plans and the observational work he has been doing.
ALS 4pp RS:HS 12.177

14092. 1870–1–14 From J[ames] H[utchison] Stirling Piershill, Edinburgh
Thanks JH for favorable opinion of JS's paper, as expressed to Hugh Martin. Comments on JH's remarks on allotropy, and on connection between physics and life. Believes Germans preeminent in physiology.
ALS 4pp RS:HS 17.239

14093. 1870–1–15 To John Gellibrand Hubbard Collingwood
Has no objection to including JH's 'Half Dozen Propositions on the Gold Coinage' into Bank of England's publication

about gold controversy. Offers revision to JH's original article. Responds to Hubbard's comments on six propositions. Remarks on minting by private individuals.
Adraft 4pp TxU:H/L-0198; Reel 1054

14094. 1870–1–18 From Thomson Hankey 45 Portland Place
Agrees with JH's propositions regarding gold currency, but offers minor corrections.
ALS 4pp TxU:H/M–0238; Reel 1087

14095. 1870–1–18 From Henry Hardcastle n.p.
Will put JH's paper ['Half Dozen Propositions on the Gold Coinage'] at Reform Club, [United] Oxford Club, and Cambridge [University] Club. Father [J. A. Hardcastle] will do same at Brooks's Club. Suggests that JH also send copy to Athenaeum.
ALS 1p TxU:H/M–0246; Reel 1087

14096. 1870–1–18 From Patrick Francis Robertson Carlton Club
Enthusiastically approves of JH's 'Half Dozen Propositions Respecting Gold Coinage.' Will distribute copies at Carlton Club.
ALS 2pp TxU:H/M–0455; Reel 1093

14097. 1870–1–19 To Thomson Hankey Collingwood
Asks TH to reconsider JH's second Proposition on converting gold coins to pounds sterling. Offers examples. Sends copies of JH's 'Propositions' to TH and to [J. G.] Hubbard. Notes testimony to integrity of British coinage given by U.S. assay master [W. E. Du Bois].
AdraftS 3pp [on JH's 1866–11–28 (TxU:H/L-0182.1), on T. E. Selwyn's undated (TxU:H/L-0182.2), and on Pierce Butler's undated (TxU:H/L-0182.3)] TxU:H/L-0182.4; Reel 1054

14098. 1870–1–19 From John Gellibrand Hubbard Addington Manor, Winslow
Suggests alterations to JH's 'Half Dozen Propositions Regarding the Gold Coinage' before this is added to Bank of England's publication on coinage controversy. Pleased by testimony of U.S. assay master [W. E. Du Bois].
ALS 6pp TxU:H/M–0280; Reel 1087

14099. [1870–1–19 or earlier] From T. E. Selwyn n.p.
Reports on mental and physical condition of [?].
ALS 1p inc TxU:H/L-0182.2; Reel 1054

14100. 1870–1–20 From George Biddell Airy Playford near Ipswich
Responds to proposals of JH [?], and comments on the problems of the Royal Mint charges for coinage, especially coinage for foreign authorities.
ALS 2pp TxU:H/M-0645.1; Reel 1084

14101. 1870–1–21 From Hugh Martin Lasswade near Edinburgh
Grateful for JH's valuable letter. Has sent a copy to Dr. J. H. Stirling to show the coincidence of their views.
ALS 3pp RS:HS 13.22

14102. 1870–1–21 From Dr. Ricketts Birkenhead
Asks where JH's opinion that subsidence of land is caused by the pressure of material accumulation is stated in detail.
L 1p RS:HS 14.489

14103. 1870–1–23 From John S. Bourne Claremont, Exeter
Appreciation of JH's arguments in favor of pound sterling.
ALS 1p TxU:H/M–0115; Reel 1087

14104. 1870–1–23 From Henry Hardcastle 4 Gresham St.
Regarding the various methods of levying the gas rate for street lighting.
ALS 2pp RS:HS 9.228

14105. 1870–1–24 To John Gellibrand Hubbard Collingwood
Responds to Hubbard's queries about JH's 'Half a Dozen Propositions Respecting the Gold Coinage.' Quantity of sterling required to conduct wholesale and retail transactions.
C 2pp TxU:H/L-0199; Reel 1054

14106. 1870–1–26 From Henry Hardcastle Temple
Further regarding the various ways of levying gas rates. It appears that the method adopted at Hawkhurst is incorrect.
ALS 3pp RS:HS 9.229

14107. 1870–1–27 From John Gellibrand Hubbard Addington Manor, Winslow
Secretary to Bank of England will soon write to JH for permission to include JH's revised 'Half Dozen Propositions Regarding the Gold Coinage' at end of Bank's collection of papers on coinage controversy.
ALS 3pp TxU:H/M–0281; Reel 1087

14108. 1870–2–3 From Hammond Chubb Bank of England
Bank intends to publish letters and papers relating to 'Gold Coinage Controversy.' Does JH object to including JH's 'Half a dozen Propositions respecting the Gold Coinage'? [JH annotation: Answered 4 Feb.; approved, and enclosed corrected copy.]
(A)LS 1p † TxU:H/M–0137; Reel 1087

14109. 1870–2–5 From George Biddell Airy Royal Observatory, Greenwich
Has JH seen R. A. Proctor's theory regarding the Milky Way?
C 1p RS:HS 1.318 (C: RGO 6.258.499)

14110. 1870–2–6 From Julia Margaret Cameron Freshwater, I.o.W.
News of her sons. Has sold many of her photographs. Husband is in Ceylon.
ALS 36pp RS:HS 5.170

14111. 1870–2–7 To Richard Anthony Proctor [Collingwood]
Discusses various telescopes of his father and his father's [erroneous] announcement of his discovery of four additional satellites of Uranus.
draft 4pp inc RS:HS 14.119 Constance Lubbock, *Herschel Chronicle* (Cambridge, England, 1933), 290

14112. 1870–2–8 From F. Charsley Slough
Has made a mistake in the lease. Will fall in with any suggestion of JH.
ALS 2pp RS:HS 5.250

14113. [1870–2–8 or later] To F. Charsley [Collingwood]
Reply to FC's 1870–2–8.
draft 1p RS:HS 5.250verso

14114. 1870–2–8 From Hammond Chubb Bank of England
Sends proof of Bank's article on gold coinage. Tell HC of any corrections that JH may want.
ALS 2pp TxU:H/M–0138; Reel 1087

14115. 1870–2–8 From Edward Hardcastle Fern Hill, Cheetham Hill, Manchester

Prospects for investment in Great Eastern Railway for JH's son William.
ALS 2pp RS:HS 19.275

14116. 1870–2–9 From George Biddell Airy Royal Observatory, Greenwich
Has he heard about the resignations of the Assistants at the Paris Observatory? Further regarding the present situation at the Paris Observatory.
ALS 2pp RS:HS 1.319

14117. 1870–2–9 To George Biddell Airy Collingwood
Responds that R. A. Proctor's theory [see GA's 1870–2–5] is possible, and explains why.
ALS 4pp RGO 6.258.500

14118. 1870–2–9 From George Biddell Airy n.p.
About the dismissal of U. J. J. Leverrier at Paris observatory.
ACCS 2pp RGO 6.150.62

14119. 1870–2–9 To John Herschel (son) Collingwood
A long rambling statement against the metric system and its proposed introduction into India.
ALS 2pp JHS 5.45

14120. 1870–2–10 To Hammond Chubb Collingwood
Returns proofs with slight alterations. Encloses copy of letter JH sent to the *Times* on similar subject.
ACCS 1p TxU:H/L-0128; Reel 1054

14121. 1870–2–10 To Mary [Cornwallis Lipscomb Power Herschel] Collingwood
Thanks MH for copying reply from JH's son John to Col. Tennant's letter about 'ruinous' metric system. T. F. and Amelia [Herschel] Wade arrived in China. News of storms and shipping disasters. Denounces papal infallibility. Reviews Pierre Lanfrey's life of Napoleon. Tell JH's son John about changes in southern stars. Compares lectures of Chandra Kesub-den to sermons of John Wesley.
ALS 4pp TxU:H/L-0227; Reel 1054

14122. 1870–2–10 To the London *Times* [Collingwood]
Proposes a method of defraying the cost of coinage by means of seigniorage involving silver coinage.
L London *Times* (Feb. 12, 1870), p. 10, col. 5

14123. 1870–2–11 From Hammond Chubb Bank of England
Printer can still include JH's corrections to article [on gold coinage], but not JH's enclosed letter to the *Times*.
ALS 2pp TxU:H/M–0139; Reel 1087

14124. 1870–2–11 From Henry James 4 New St., Spring Gardens
Received his note on the Great Pyramid when travelling in Scotland. Quoted in his reply to C. P. Smyth. Encloses a note from F. D. Wackerbarth of Uppsala. Will be sending him the volumes containing the photographs of the Sinai survey.
ALS 4pp RS:HS 10.277

14125. 1870–2–14 From F. Charsley Slough
Is satisfied with the alteration he made in the lease and his bankers will be forwarding the premium. Thanks for the photograph of the contrivance for the ladder.
ALS 2pp RS:HS 5.251

14126. 1870–2–15 From Augustus De Morgan 6 Merton Road
Progressing slowly. Knows nothing about C. M. Hall or the whereabouts of his telescopes. Regarding the London Institution. Sees that U. J. J. Leverrier has been dismissed.
AL 4pp RS:HS 6.413

14127. 1870–2–16 To Henry H. H. Cameron Collingwood
Advice and guidance on the path he is to pursue when he goes to university.
draft 4pp RS:HS 5.317 & 24.298

14128. 1870–2–18 From Henry H. H. Cameron Bonchurch, I.o.W.
Thanks for his kindly advice. News of his own progress.
ALS 4pp RS:HS 5.318

14129. 1870–2–23 From Hammond Chubb Bank of England
Sends copy of Bank's published volume on gold coinage controversy.
(A)LS 1p TxU:H/M–0140.1; Reel 1087

14130. 1870–2–24 To Hammond Chubb [Collingwood]
Gratitude for copy of Bank's published volume on gold coinage controversy. Regrets that volume does not contain letter to the *Times* containing U.S. assay master W. E. Du Bois's testimony to integrity of British gold coinage, 'a most satisfactory answer to Mr. Segal's charge against it.'
ACC 1p TxU:H/M–0140.2; Reel 1087

14131. 1870–2–25 From E. S. Newall St. Rogue, Clifton
JH's kind letter and enclosures came when she was ill, which accounts for the delay in replying. Is exceedingly grateful for all the autographs. Her brother-in-law [R. S. Newall] has erected a 25" telescope at Gateshead and doubtless would be pleased to show it to JH. Would be grateful for any photographs to go with the autograph collection.
ALS 10pp RS:HS 13.119

14132. 1870–2–26 From George Biddell Airy Royal Observatory, Greenwich
Is sending extract of a letter [included] received from R. L. J. Ellery, Melbourne[deals with observations of Eta Argus].
ALS 1p RS:HS 1.320 & 1.321 (C: RGO 6.172.498)

14133. 1870–2–28 From Julia Margaret Cameron Freshwater, I.o.W.
Sends a copy of the *Graphic*. Is grateful for JH's letter to her son.
ALS 4pp RS:HS 5.171

14134. 1870–2–28 To Elizabeth Colling Collingwood
Much on poetry; for a sonnet on the sun by EC, JH sends some photographs of the sun.
C 4pp RS:HS 24.299

14135. 1870–3–8 From E. S. Newall St. Rogue, Clifton
Is grateful for the two portraits. Is sure that her brother would be pleased to show Professor Herschel his telescope. Let him know when in Gateshead.
ALS 4pp RS:HS 13.120

14136. 1870–3–8 From James Samuelson Liverpool
Asks JH to write to Earl de Grey [G. F. S. Robinson] stating JS's qualifications for position on Royal Commission 'for inquiring into the question of the Extension of State and to Sciences.'
ALS 1p RS:HS 15.340a

14137. 1870–3–9 From H. Temple Humphreys North Strand, Limerick
May be interested in seeing the enclosed [probably RS:HS 19. 229]. Outlines his proposals, which he hopes to get

adopted in Ireland. Trusts the east winds are not affecting him.
ALS 4pp RS:HS 10.78

14138. 1870–3–10 From Henry Denny Leeds
Inviting him to donate any copies of his works on astronomy, etc.
ALS 3pp RS:HS 6.490

14139. 1870–3–[13 or earlier] From R. S. White Southborough, Tunbridge Wells
Sports event set for first week in April. May JH's name be added to subscription list for prizes?
ALS 1p RS:HS 18.400

14140. 1870–3–13 To R. S. White Collingwood
JH will not join subscribers for prizes. Objects to so much emphasis on exhibitions of competitive sports, to the detriment of serious education. Feels young people benefit more from spontaneous exercise.
ACCS 4pp RS:HS 18.401 (C: RS:HS 24.300)

14141. 1870–3–16 To James Samuelson Collingwood
Cannot address Earl de Grey [G. F. S. Robinson] concerning members of Royal Commission without a request from the Earl. Cannot, therefore, write on behalf of JS [see RS:HS 15.340a].
copy 2pp RS:HS 15.340b

14142. [1870]–3–17 From Edward Hardcastle Manchester
The paper JH sent relates to a sale of stock belonging to Willy (JH's son). Hopes he does not suffer from the return of winter.
ALS 2pp RS:HS 10.157

14143. 1870–3–21 From Henry Denny Leeds
Thanks for his kind reply and transmittal of books. Has he ever written anything on volcanoes?
ALS 2pp RS:HS 6.491

14144. 1870–3–22 From H. Temple Humphreys North Strand, Limerick
Many thanks for his kind note. Pleased to find he is in good health and taking an interest in the work of the youngsters. Hopes Alexander (JH's son) showed him his plan for a binocular reflector.
ALS 4pp RS:HS 10.79

14145. 1870–3–25 From Walter White Burlington House
Encloses copy of R.S.L. statutes. G. G. Stokes cannot attend eclipse committee. Thanks for information about Mrs. Birtwhistle.
ALS 3pp RS:HS 18.232

14146. 1870–3–26 From Richard Anthony Proctor 5 Trigon Road, Kennington
Reports finding, while preparing a star atlas, that bright stars appear to be very dense in a region associated with the Milky Way but centered on the greater Magellanic Cloud. Because of a coming lecture on stellar distribution, 'I ... am saturating myself with Herschelian astronomy.'
ALS 12pp RS:HS 14.120

14147. 1870–3–27 From W. K. von Haidinger Vienna
Sending the continuation of one of his own reports he submitted to the Academy. Comments on various meteorites and the results.
ALS 4pp RS:HS 9.154

14148. 1870–3–28 From C. De La Pryme Reform Club
Is engaged in bringing out a diary of his ancestor. Would like JH's view on whether Isaac Newton's mind was ever deranged.
ALS 2pp RS:HS 6.486

14149. 1870–4–1 To Richard Anthony Proctor Collingwood
Suggests reasons for doubting the distribution of bright stars that RP had reported. Responds to RP's query concerning a statement in *Outlines Ast.* Encourages RP's hypothesizing on star distribution.
C 4pp RS:HS 24.301 Proctor, *Other Suns,* 417 inc

14150. 1870–4–1 From Patrick Francis Robertson 7 Pall Mall
Informs JH of plans for JH receiving 5 pounds of coca he had ordered.
ALS 1p RS:HS 14.372

14151. 1870–4–2 To Augustus De Morgan Collingwood
Asks AD's advice on an application JH received the previous day. Comments on JH's health.
C 2pp RS:HS 24.302

14152. 1870–4–3 From Richard Anthony Proctor 5 Trigon Road, Kennington
Reports on his studies of stellar and nebular distributions, discussing especially whether the Magellanic Clouds show a relation to the Milky Way stars. Speculates on whether recent studies on meteors bear on stellar formation and change. Stresses need for statistical studies in stellar astronomy.
ALS 12pp RS:HS 14.121

14153. 1870–4–4 From Nicholas Michell Queen's Road, St. Johns Wood
Sends a copy of his own book, the *Immortals,* which he would like JH to glance at as the Herschels are mentioned.
ALS 2pp RS:HS 13.38

14154. 1870–4–4 From Patrick Francis Robertson 26 Old Broad Street
Encloses letter giving charges for 5 pounds of coca ordered by JH [see PR's 1870–4–1]. Asks how PR should pay.
ALS 1p RS:HS 14.373

14155. 1870–4–8 To Duncan [Stewart] Collingwood
Comments on ability of eye to adapt to prismatic glass in corrective spectacles for double vision [see JH's 1869–10–31]. Heard of bad health of DS and John [Stewart]. JH is fighting bronchitis.
ALS 4pp TxU:H/L-0143; Reel 1054

14156. 1870–4–9 To [W. K. von Haidinger] Collingwood
Thanks WH for materials WH sent, including information about scientific activity in Vienna and papers on meteorites. Recounts some of the history of the Analytical Society. Discusses meteorites and also Homeric references to iron.
ALS 4pp BostonPubLib Ms. Eng. 204 (C: RS:HS 24.303)

14157. 1870–4–11 From G. Walmly Trevor House, 2, Warwick Gardens
Is publishing photographic volume of eminent scientists. Lists twelve. Would like to include JH.
ALS 2pp † RS:HS 18.372

14158. 1870–4–13 From Augustus De Morgan 6 Merton Road
JH's letter lays terrible responsibilities at his door. Should

not neglect the offer. Regarding the Julian calendar.
ALS 5pp RS:HS 6.414

14159. 1870–4–14 To Thomas Brassey Collingwood
Thanks for his pamphlet on 'Trade Unions and the Cost of Labour.' Comments on this.
draft 2pp RS:HS 5.96

14160. 1870–4–15 From Miss B. S. Clarke Boxted, Colchester
Discusses her theory of aerolites and the zodiacal light. Used to visit his house at Slough.
ALS 4pp RS:HS 5.346

14161. 1870–4–20 To Editor of *Pall Mall Budget* n.p.
Examples from *Iliad* and other sources to show that military offensive tactic called 'Vertical Fire' did not originate with [L. N. M.] Carnot, as Mr. Mallet claimed in recent issue of *Budget*.
C 2pp TxU:H/L-0293; Reel 1054

14162. 1870–4–21 From Miss B. S. Clarke Boxted, Colchester
Thanks for his letter about her theory regarding aerolites. Has another query on this question.
ALS 3pp RS:HS 5.347

14163. 1870–4–23 To Warren de La Rue Collingwood
Inquires about other sunspot observations made by JH.
C 1p RS:HS 24.305

14164. 1870–4–23 To William Selwyn Collingwood
Asks whether WS has photographs of the sun for 21 and 22 April. Notes how curious sunspot activity has been. Doubts that such is reconcilable with the notion of 'meteoric in-falls.'
C 1p RS:HS 24.304

14165. 1870–4–23 To Duncan [Stewart] Collingwood
No permanent optical solution to DS's double vision. Remarks on description by DS's son of cavalry duty in Ireland and Canada.
ALS 4pp TxU:H/L-0144; Reel 1054

14166. 1870–4–24 To Frederick Howlett Collingwood
Writes to compare notes on sunspot observations.
C 2pp RS:HS 24.306

14167. 1870–4–24 From Edward Sabine n.p.
Note to accompany forwarding of a letter; remarks on T. R. Robinson's upcoming cataract surgery.
C 1p encl Cambridge University Library 7656/H702

14168. 1870–4–25 From Warren de La Rue Observatory, Cranford
Has written to Kew Observatory to send on the photographs.
ALS 1p RS:HS 6.170

14169. [1870–4?–25?] From William Selwyn n.p.
Detached postscript that discusses the margins of the two photographs, which seem to show that all planets are in the near hemisphere.
AL 4pp inc RS:HS 15.517

14170. 1870–4–25 From William Selwyn Foxton, Royston
Is sending solar autographs. Notes that in groups of spots the larger precede the smaller spots. Can JH form any theories from this?
ALS 4pp RS:HS 15.507

14171. 1870–4–26 From Henry Hudson Glenville, Fermoy
Would like him to accept the enclosed paper on the cause and effect of dew. Comments on various points.
ALS 3pp RS:HS 10.30

14172. 1870–4–26 From Henry James Ordnance Survey Office
The volumes of the Sinai photographs will be dispatched tomorrow. The cost of the survey has been covered by subscriptions and he will forward a statement of the account later.
ALS 2pp RS:HS 10.278

14173. 1870–4–28 From Georg A. Erman Berlin
Sends two of his publications, one on terrestrial magnetism at Berlin. Comments on the results, and wonders if JH would enlist the help of the B.A.A.S. to finance further studies by H. J. R. Petersen.
ALS 6pp RS:HS 7.104

14174. 1870–4–28 From Frederick Howlett East Tisted, Alton
Has no idea whether the sunspots appeared suddenly or gradually. The Kew people may be able to help in this matter. Comments on some more sunspots.
ALS 4pp inc RS:HS 10.23

14175. 1870–4–28 To Edward Sabine Collingwood
ES has been forwarded letter explaining why 'great telescope' at Melbourne will not work and suggesting construction corrections, on which JH, who believes such 'a priori condemnation' unjustified, comments herein.
ALS 3pp RS MC.9.63 (C: RS:HS 24.308)

14176. 1870–4–28 To Edward Sabine Collingwood
Comments on the effect of great temperature variations on the speculum of a telescope.
C 3pp Cambridge University Library 7656/H727 (C: RS:HS 307)

14177. 1870–4–30 From Richard Anthony Proctor 5 Trigon Road, Kennington
Sends map from his new atlas. Reports finding a region rich in bright stars in the northern hemisphere but distinct from the Milky Way. States that in a Royal Institution lecture, he will claim that telescopes cannot reach the limits of the sidereal system and that it is far more complex than traditionally assumed.
ALS 4pp RS:HS 14.122

14178. 1870–4–30 From Warren de La Rue Observatory, Cranford
Has just learnt that some negatives were sent to his former place of residence. Will forward them when he goes to the City. A vote has been made from the Donation Fund of the R.S.L. for six copies of the Kew Solar negatives.
ALS 3pp RS:HS 6.171

14179. 1870–4–30 To William Selwyn Collingwood
Is returning March heliographs. Discusses the appearance of the spots. Has not received the April heliographs.
C 1p RS:HS 24.309

14180. [1870–5] From Henry Hardcastle Nather Hall
Will endeavor to obtain information on the Dock shares. Maria (HH's wife and JH's daughter) and infant are doing well.
ALS 2pp RS:HS 9.232

14181. 1870–5–2 From Henry Hudson Glenville, Fermoy
Is grateful for his note. Would like to be certain of W. C. Wells's theory before he revises his opinion. His original

paper was much larger. Gives some examples where he thinks Wells is in error.
ALS 3pp RS:HS 10.31

14182. 1870–5–2 To William Selwyn Collingwood
Wants to compare Mr. Titterton's photographs of the spot groups to [Warren] de La Rue's. Has tried to determine if some localities regularly have spots.
C 2pp RS:HS 24.310

14183. 1870–5–3 From J. T. Raymond Cowley St. John
Cannot afford fees he must pay to be ordained. Asks for financial assistance. Includes letters from a tutor and a professor.
ALS 3pp RS:HS 14.473

14184. 1870–5–3 From Walter White Burlington House
Sends report of Edward Sabine's 'conversaziones' of 12 Mar. and 23 Apr. Mrs. Birtwhistle wrote from Peking on 2 Feb.
ALS 2pp inc RS:HS 18.233

14185. 1870–5–4 From Adolf Bernhard Meyer 12 Victoria Road, Kensington
Forwards letter from G. A. Erman to JH and two papers by Erman. AM wants to visit JH to learn of prospects for Erman before AM leaves soon for zoological and magnetic observations in Malay Archipelago.
ALS 2pp TxU:H/M–0363; Reel 1087

14186. 1870–5–5 To Warren de La Rue Collingwood
Thanks for the photographs WD sent for the dates requested in JH's 1870–4–23.
C 2pp RS:HS 24.311

14187. 1870–5–6 To Adolf Bernhard Meyer Collingwood
Thanks for sending G. A. Erman's two papers on terrestrial magnetism and letter to JH. Will forward these to G. G. Stokes with recommendation that B.A.A.S. continue funding for reduction work of H. J. R. Petersen. Awaits AM's visit to Collingwood.
ALS 3pp TxU:H/L-0260; Reel 1054

14188. [1870]–5–6 To George Gabriel Stokes Collingwood
Forwards letter from G. A. Erman of Berlin requesting that B.A.A.S. fund [H. J. R.] Petersen's study of terrestrial magnetism.
C 3pp TxU:H/L-0465; Reel 1055

14189. 1870–5–6 To George Gabriel Stokes Collingwood
Writes to support grant request from Georg Erman of Berlin to allow Erman to do calculations on magnetism observations from circumnavigation of world.
C 4pp Cambridge University Library 7656/H728

14190. 1870–5–9 From Hubert Airy Flamsteed House, Greenwich
Describes difficulties in producing glass pen and attaching it to pendulum to produce precise curves for aligning equatorial [telescope mount]. Sent artistic sample to JH's wife.
ALS 4pp RS:HS 19.311

14191. 1870–5–9 From Eyre B. Powell Madras
Thanks JH for mention of him in *Outlines Astr.* Sends detailed observations of Alpha Centauri system.
ALS 7pp RS:HS 14.185

14192. 1870–5–10 From George Gabriel Stokes Lensfield Cottage, Cambridge
Acknowledges receipt of letter and papers from G. A. Erman.
ALS 2pp TxU:H/M–0507; Reel 1093

14193. 1870–5–11 To Richard Anthony Proctor Collingwood
Thanks RP for sending RP's *Other Worlds than Ours* and supplies comments, some favorable, on numerous claims made therein, e.g., on RP's theory that Jupiter and Saturn are hot and to some degree luminous.
ALS 4pp RS:HS 14.123 (C: RS:HS 24.312) Proctor, *Other Suns,* 415–17 inc

14194. 1870–5–11 From William Selwyn Foxton, Royston
Desires permission to present JH's 'diagram of "macularity" to the R.A.S.
ALS 4pp RS:HS 15.508

14195. 1870–5–12 To George Biddell Airy Collingwood
About a grant to support [H. J. R.] Petersen while he carries out the reduction of terrestrial magnetic observations.
ALS 3pp RGO 695.462

14196. 1870–5–12 From Richard Anthony Proctor 5 Trigon Road, Kennington
Thanks JH for commentary on RP's *Other Worlds than Ours.* Responds to JH's comments, accepting most objections. Acknowledges his major debt to JH's writings.
ALS 8pp RS:HS 14.124

14197. 1870–5–12 To William Selwyn Collingwood
The research shows no particularly 'maculiferous' area on the sun, but WS may, as WS requested, present 'data' to the R.A.S. Has noticed two new spots on helioautographs on 20 and 27 April.
C 2pp RS:HS 24.313

14198. 1870–5–15 From George Biddell Airy Royal Observatory, Greenwich
Refers to several items of correspondence, and then comments on the calculation of Gaussian earth-constants to a higher power than has been done to date.
ALS 2pp TxU:H/M-0645.2; Reel 1084

14199. 1870–5–15 From George Biddell Airy n.p.
Would support the application for a grant [see JH's 1870–5–12], but would like to see some changes in the way calculations are made.
ACCS 2pp RGO 6.695.464

14200. 1870–5–16 From Sir William Huggins R. Astronom. Soc.
The Council of the R.A.S. would like JH's opinion on the accompanying paper and drawing by Francis Abbott, and whether it should appear in the *R.A.S.M.N.*
ALS 1p RS:HS 10.58

14201. 1870–5–16 To Richard Anthony Proctor Collingwood
Assures RP that he is not upset at RP for publicly questioning some of JH's ideas. Continues discussion of point in perturbation theory for Uranus.
C 1p RS:HS 24.314

14202. 1870–5–17 From Richard Anthony Proctor 5 Trigon Road, Kennington
Admits JH is correct in suggested correction of RP's statement regarding perturbations of Uranus. Suggests that JH urge astronomers to observe the solar corona carefully.
ALS 4pp RS:HS 14.125

14203. 1870–5–18 To Sir William Huggins n.p.
Giving his opinion on a paper and drawing of [Francis]

Abbott.
draft 3pp RS:HS 10.59

14204. 1870–5–19 To Sir William Huggins n.p.
Further report on a paper and drawing of [Francis] Abbott [very similar to WH's 1870–5–18].
CC 4pp RS:HS 10.60

14205. 1870–5–19 To the R.A.S. Collingwood
Expresses skepticism on the claims made by F. Abbott, observing from Tasmania, of the detection of changes in the Eta Argus nebula. Compares Abbott's drawings with drawings made by JH and and by JH's son John.
L *R.A.S.M.N., 31* (1871), 228–30

14206. 1870–5–24 To Georg A. Erman Collingwood
About an application to the B.A.A.S. for a grant to fund the calculations related to reducing the magnetic observations of 1847–48.
C 4pp RS:HS 24.315

14207. 1870–5–24 From Adolf Bernhard Meyer 12 Victoria Road, Kensington
Will visit JH on 27 May.
ALS 1p TxU:H/M–0364; Reel 1087

14208. 1870–5–25 From Adolf Bernhard Meyer 12 Victoria Road, Kensington
Apologizes for delay in responding to JH. Wrote to JH yesterday about AM's plan to visit Collingwood.
ALS 1p TxU:H/M–0365; Reel 1087

14209. 1870–5–27 From J. Ryle Eastbourne
Appreciates JH's opinion regarding variability of climate, although it disagrees with his own. His views are not provable and the question remains one of probability.
ALS 3pp † RS:HS 14.516

14210. [1870–5–27 or after] From J. Ryle Eastbourne
[Richard?] Proctor [?] agrees with JR's views on Cosmical Climates. Re-presents his theory because he feels he did not state it fairly in his earlier [1870–5–27] letter.
ALS 4pp RS:HS 14.517

14211. 1870–5–30 From W. K. von Haidinger Dornbach, near Vienna
Is grateful for his letter. Sends a report of the Geological Institute, which includes JH's notes on the founding of the Cambridge Analytical Society. Sends his own report on Homeric irons and on the R.S.L. catalogue of scientific papers. Austrian science was at a low ebb but is now improving.
ALS 4pp RS:HS 9.155

14212. 1870–5–30 From Henry Hardcastle 5 Paper Bldgs. Temple
Sending letter from a firm of stockbrokers giving a reply to JH's query regarding the East India Dock shares.
ALS 2pp RS:HS 9.230

14213. 1870–6–1 To Richard Anthony Proctor Collingwood
Continues to explain matters in perturbation theory. Stresses need for careful observations in solar eclipses. Doubts the existence of the hypothetical intermercurial planet Vulcan.
C 2pp RS:HS 24.316

14214. 1870–6–2 From Richard Anthony Proctor 5 Trigon Road, Kennington
Admits his misstatement in RP's *Other Worlds than Ours* concerning perturbation of outer planets. Shares JH's doubts about existence of Vulcan. Speculates on the extent of meteoric material in the solar system. Comments negatively on J. Norman Lockyer's views on meteors.
ALS 4pp RS:HS 14.126

14215. 1870–6–6 From George Biddell Airy Royal Observatory, Greenwich
Regarding Alexander Herschel's paper on acoustical oscillations recently sent to G. A. Erman.
ALS 1p RS:HS 1.322 (C: RGO 6.382.355)

14216. 1870–6–7 To John Tyndall Collingwood
Thanks for work on diamagnetism and magne-crystallic action [*Researches on Diamagnetism* (1870)], which JH is excited to read. Mentions long-abandoned plans for magnetization experiments. Envies those who can see JT's 'magic' experiments at Royal Institution.
C 2pp RS:HS 24.317 (C: RI 571)

14217. 1870–6–9 From John Tyndall Royal Institution
Gratified by JH's 'sympathy and approbation,' and that JH's writing and spirit remain 'firm' and 'fresh.'
ALS 1p RS:HS 17.409 (C: RI 572)

14218. 1870–6–11 From William Selwyn Foxton Royston
Discusses the different strata of the ocean and [William] Carpenter's lecture on deep sea dredging. Is digging an artesian well.
ALS 4pp RS:HS 15.509

14219. 1870–6–13 To 'Chairman of the Eclipse Committee of the Royal Society' Collingwood
Cannot attend Eclipse Committee meeting but suggests that an observer carefully look for planets interior to Mercury's orbit, which would, if existing, reach conjunction during total solar eclipses.
ALS 3pp RS MC.9.96

14220. 1870–6–15 To Elizabeth Colling Collingwood
Of toads in rocks and stones, and martins in blocks of ice under rivers.
C 3pp RS:HS 24.318

14221. 1870–6–16 From Joseph Gurney Barclay Knotts Green, Leyton
Would he accept a volume of observations carried out at the Leyton Observatory?
ALS 1p RS:HS 5.16

14222. 1870–6–16 From Henry Hardcastle 5 Paper Bldgs. Temple
Cannot give a definite answer regarding liability on the Dock stock without knowing the Act of Parliament under which the East India Company was incorporated.
ALS 2pp RS:HS 9.231

14223. 1870–6–16 From William Selwyn Foxton Royston
Sends two copies of JH's 'macular conspectus.' Mr. Titterton was very pleased that JH approved of his work.
ALS 3pp RS:HS 15.510

14224. 1870–6–17 To Augustus De Morgan Collingwood
Writes about JH's health; justifies writing by sending some curves prepared by JH's son Alexander.
C 1p RS:HS 24.319

14225. 1870–6–17 From John Knox Laughton Portsmouth

Sending JL's book *Physical Geography* (1870)
AL 1p RS:HS 11.464

14226. 1870–6–17 To William Selwyn Collingwood
Sends diagrams showing three new couplets of sunspots. The sun is entering a new phase of activity. Sends drawings made with a glass pen by his son Alexander.
C 1p RS:HS 24.320

14227. 1870–6–17 From Thomas Francis Wade [Peking]
Received JH's note. TW's wife Amelia [Herschel] and baby fared poorly in Panama and California. China better now than in 1853–65. Constitution flimsy but still vital. Describes 'Mahometan insurrection.'
ALS 4pp RS:HS 18.6

14228. 1870–6–24 From William Selwyn Foxton Royston
Thanks JH for the drawings. Will send new solar autographs showing macula. Is working on the autographic series of the activity of the sun during its 11 year period.
ALS 4pp RS:HS 15.511

14229. 1870–6–25 From Augustus De Morgan 6 Merton Road
Is creeping along slowly like Alcuin's snail. Curvers are very curious. Informs him where there is an account of Coca. Comments on this as a tonic.
ALS 4pp RS:HS 6.415

14230. 1870–6–26 From Julia Margaret Cameron Freshwater, I.o.W.
Has had a visit from Alick (JH's son Alexander) and his friends. Sends her latest photograph.
ALS 4pp RS:HS 5.172

14231. [1870]–6–29 From Edward Hardcastle Manchester
Has been intending to write to inform him of the successful meeting on Friday, but the four day holiday has caused extra work. JH's consent to preside provided extra stimulus.
ALS 4pp RS:HS 10.158

14232. 1870–7–4 From Joseph Gurney Barclay 54 Lombard St.
Is grateful for his observations on the Leyton work. Will go through them and let him know the answers later.
ALS 1p RS:HS 5.17

14233. 1870–7–7 From Charles George Talmage Barclay Observatory, Leyton
Acknowledges JH's comparison of Barclay Observatory results with those of other observatories. Sends more observations.
ALS 7pp RS:HS 17.426

14234. 1870–7–10 To Otto Wilhelm Struve Collingwood
Thanks for many works, especially double star catalogs, about which JH requests further information, and which will help JH's attempt to compile digest of *all* measures of *all* known double stars.
C 2pp RS:HS 24.321

14235. 1870–7–13 From Warren de La Rue Observatory, Cranford
Would like to incorporate JH's suggestions in his paper and he has revised the numbers. Made an attempt to photograph the moon while totally eclipsed, but was unsuccessful.
ALS 3pp RS:HS 6.172

14236. 1870–7–14 To Warren de La Rue Collingwood
Comments on sunspot activity; agrees to inclusion of some paragraphs in a paper WD is writing.
C 1p RS:HS 24.322

14237. 1870–7–22 From George Frederick Watts Little Holland House, Kensington
Please send GW information about a wondrous coca [cocaine] reported by Mrs. Gordon.
ALS 1p RS:HS 18.388

14238. 1870–7–25 From Thomas Marsden Burstow Rectory, Crawley
Sending a work intended for publication on the common force of the universe. Would be glad of JH's comments.
ALS 3pp RS:HS 13.19

14239. 1870–7–25 From Richard Anthony Proctor 5 Trigon Road, Kennington
Asks JH's advice on whether a method devised by RP of charting bright stars in isographic projections based on star gauges seems best. Describes the method and urges that the charts will produce useful suggestions about the distribution of stars.
ALS 4pp RS:HS 14.127

14240. 1870–7–26 From George Frederick Watts Little Holland House, Kensington
Thanks for prompt response and JH's sharing such a scarce commodity [cocaine; see GW's 1870–7–22]. Hopes it will heal depression that interferes with GW's work.
ALS 2pp RS:HS 18.389

14241. 1870–7–28 To Richard Anthony Proctor Collingwood
Approves RP's method of preparing isographic projection charts of the distribution of bright stars, noting that JH's gauges of the southern stars are incomplete.
C 1p RS:HS 24.323 Proctor, *Other Suns,* 418–19 inc

14242. 1870–7–31 From George Biddell Airy Tunbridge Wells
Regarding GA's proposed visit to Hawkhurst. Is losing the services of his first assistant. Book on magnetism almost printed. Is revising his article on sound. Is trying out a water telescope.
ALS 4pp RS:HS 1.323 (C: RGO 6.382.353)

14243. 1870–7–31 From George Biddell Airy n.p.
About a title change for the Astronomer Royal's chief assistant.
ACCS 1p RGO 6.6.109

14244. 1870–7–31 To Elizabeth Colling Collingwood
Of talking fishes, a great meteor falling in Tripoli, and the 'madness of man' in the wake of the Franco–Prussian War.
C 2pp RS:HS 24.324

14245. 1870–8–1 From Arthur Cayley Cambridge
Sends the mss. and two copies of the proof for JH to check. Queries one point. Please return papers to him when they have been checked.
ALS 3pp RS:HS 5.234

14246. 1870–8–1 or earlier To John Herschel (son) Collingwood
Talks about signatures needed for a fellowship certificate for the R.S.L.
ALS 4pp JHS 5.44

14247. 1870–8–1 From Richard Anthony Proctor 5 Trigon Road, Kennington
Thanks JH for advice on planned chart; explains method of dealing with unguaged areas. RP has adopted a somewhat different method of charting.
ALS 1p RS:HS 14.128

14248. 1870–8–2 To George Biddell Airy Collingwood
Is preparing a catalogue of all double star observations, and is having much difficulty with the observations of J. H. Mädler.
ALS 4pp RGO 6.197.291

14249. 1870–8–2 To George Biddell Airy n.p.
Some comments about the proposed title change [see GA's 1870–7–31].
LCC 1p RGO 6.6.109a

14250. 1870–8–4 To John Herschel (son) Collingwood
Discusses the signatures needed for a fellowship certificate for the R.S.L. [see JH's 1870–8–1 or earlier]; comments on the arrival of son Alexander home from a holiday in Norway.
ALS 4pp JHS 5.43

14251. 1870–8–7 From Frances Annie Griesbach Hull
Wishes to take up a theatrical position at Aberdeen, and would be grateful for any financial assistance. Has already received £5 from the Queen.
ALS 4pp RS:HS 9.133

14252. 1870–8–15 From W. Paterson Manchester
Is working on what he hopes will be the true theory of the earth, but would like advice from an astronomer on one point. Believes JH has been asked this question before, according to Charles Lyell.
ALS 3pp RS:HS 13.244

14253. 1870–8–18 To Elizabeth Colling Collingwood
Comments about some pieces of German poetry.
C 1p RS:HS 24.325

14254. 1870–8–18 To Sir William Huggins Collingwood
Requests that WH sign the certificate for the election of JH's son John as a Fellow of the R.S.L.
C 1p RS:HS 24.326

14255. 1870–8–21 To Sir William Huggins Collingwood
Apologizes for forgetting that since WH is on the Council of the R.S.L. he is precluded from signing [see JH's 1870–8–18].
C 1p RS:HS 24.327

14256. 1870–8–22 From Frances Annie Griesbach Aberdeen
Is grateful for JH's assistance. Is now residing in Aberdeen but still in financial difficulties so would be pleased with any assistance.
ALS 4pp RS:HS 9.134

14257. 1870–8–25 From Emily Beke Bekesbourne, Canterbury
Relating the dire condition of Dr. C. T. Beke and requesting financial assistance from JH.
ALS 6pp RS:HS 5.38

14258. 1870–8–25 From Frederick Howlett East Tisted, Alton
Intends bringing the subject of sunspots before the B.A.A.S. next September and would therefore like JH's views on whether in his opinion these spots are mechanical or magnetic. Also any other comments would be welcome.
ALS 4pp RS:HS 10.24

14259. 1870–8–29 From F. Austen Capel Manor, Horsmonden
Sends extract from Eduard Fournier's *Le Vieux-Neuf.*
ALS 6pp RS:HS 1.423

14260. 1870–8–29 To Frederick Howlett Collingwood
Comments on remarkable solar activity. JH admits he has no theory to explain sunspots.
ALS 4pp Adler Ms. 13 (C: RS:HS 24.328)

14261. 1870–8–29 From Robert Main Radcliffe Observatory
Asks JH if he wants to make any changes in his article on meteorology for a new edition of the *Admiralty Manual.*
ALS 2pp RS:HS 12.284

14262. 1870–8–30 From W. K. von Haidinger Dornbach, near Vienna
Regrets the delay in sending his paper on Homeric irons, but he has been suffering from ill health. Comments on JH's translation of one of Homer's stanzas.
ALS 3pp RS:HS 9.156

14263. 1870–8–31 To William Selwyn Collingwood
Discusses the movement of the 'great group' of sunspots over the last six months. Asks WS to sign a certificate for his son [John].
C 2pp RS:HS 24.329

14264. 1870–9–1 To J. Peter Lesley Collingwood
Has received, through the good offices of the Smithsonian Institution, a number of back issues as well as current issues of the *Proceedings* of the American Philosophical Society of Philadelphia. The best way to send these volumes is by way of the Smithsonian.
ALS 2pp APS (no i.d.)

14265. 1870–9 From Archibald Campbell Tait Addington Park
Expresses concern over Church of England schools. Suggests appeal to raise funds.
ALS 4pp RS:HS 17.424

14266. 1870–9–2 From Robert Main Radcliffe Observatory
Sends some remeasured double star positions from the Radcliffe Observatory, together with comments about other double star observations.
ALS 2pp RS:HS 12.285

14267. 1870–9–2 From William Selwyn Foxton, Royston
Has been trying to make pendulum curves. Believes they may help explain the different sunspot forms. Thanks JH for the spectroscopic observation memoirs.
ALS 4pp RS:HS 15.512

14268. 1870–9–4 To William Selwyn Collingwood
Thanks WS for August 29 photographs. Is sending JH's son's [John's] certificate. Pities the 'fallen Emperor' [Napoleon III].
C 1p RS:HS 24.330

14269. 1870–9–7 To John Herschel (son) Collingwood
Gives advice to son John about whom to ask for a signature for the certificate for fellowship in the R.S.L. [see JH's 1870–8–4]; talks about the war [Franco-Prussian ?], and announces the birth of a son to daughter Amelia and Thomas Wade in Peking.
ALS 2pp JHS 5.42

14270. 1870–9–7 From Frederick Howlett East Tisted, Alton
Is grateful for JH's note. Agrees with him about the mystery of the sunspots. Is awaiting some heliographs from Ely. Is inquiring from Kew as to any recent magnetic disturbances that may coincide with the appearance of the sunspots. Is grateful for Alexander Hershel's information.
ALS 4pp RS:HS 10.25

14271. 1870–9–12 From Edward Sabine Llandovery
Sends packet from [John] Hennessey to JH for his perusal. Discusses recent weather. Wants JH's opinion of self-recording meteorological instruments. Mentions wreck of the *Hansa*.
ALS 4pp RS:HS 15.292

14272. 1870–9–14 To Edward Sabine Collingwood
Approves of ES's actinometric methods, but wonders about absorbent power of alcoholic liquid. Laments 'sad conclusion' to career of 'Hansa,' and hopes her companion is more successful.
ALS 4pp RS Sa.678

14273. 1870–9–17 From Edward Sabine Llandovery
Sends telegram from the *Germania*. Discusses crew of *Hansa* [see 15.292]. Asks if JH wants certain experiments done at Kew. Troubled by [W. A.] Miller's sudden illness.
LS 4pp RS:HS 15.293

14274. 1870–9–20 To Edward Sabine Collingwood
Thanks for telegram and has dispatched papers about absorption power of actinometer liquid. Hopes letter written in Lady Sabine's handwriting does not indicate serious illness of ES.
ALS 2pp RS Sa.679

14275. 1870–9–20 To George Gabriel Stokes Collingwood
In commenting on a paper on actinometric measurements varying with depth of atmosphere, JH questions the accuracy of the equipment used.
ALS 4pp Cambridge University Library 7656/H729

14276. [1870?]–9–22 From Norman Macleod n.p.
Has been away from home, which accounts for the delay in replying to JH's note. Send the manuscript to NM. Hopes for an article from JH.
ALS 1p RS:HS 12.265

14277. 1870–9–22 From George Gabriel Stokes Blackrocks, Bush Mills, Ireland
B.A.A.S. in Liverpool approved £50 grant to G. A. Erman, to be administered by Erman and JH, for computation of Gaussian constant for 1839.
ALS 6pp TxU:H/M–0508; Reel 1093

14278. 1870–9–24 To Norman Macleod Collingwood
Will send manuscript of article by JH's daughter [?] on history of Alsace and Lorraine for publication in NM's *Good Words*. JH is busy compiling general catalog of all known double stars.
ACC 3pp TxU:H/L-0243; Reel 1054

14279. 1870–9–27 To George Gabriel Stokes Collingwood
Thanks for grant of £50 from B.A.A.S. for calculations related to request from Georg Erman [see JH's 1870–5–6].
ALS 3pp Cambridge University Library 7656/H726

14280. 1870–9–28 From Sir William Huggins Upper Tulse Hill
Has had a letter from Cleveland Abbe inquiring when JH's Catalogue of Double Stars can be expected. Howard and Thomas Grubb have just erected his telescope but the eye end is not finished. Instrument will be very 'handy.'
ALS 3pp RS:HS 10.61

14281. 1870–9–28 From George Gabriel Stokes Armagh
Comments on JH's concerns about actinometric measurements [see JH's 1870–9–20]; recommends the study of periodicity of waves on Atlantic Coast.
ALS 14pp RS:HS 17.78

14282. 1870–9–29 From Robert Brown Fairlie by Greenock
Confined by accident but is grateful for his paper on musical scales and offers his comments on it.
ALS 4pp RS:HS 4.319

14283. 1870–9–29 From Edward Sabine Llandovery
Sends notice from [C. P. B.] Walker regarding pendulum experiments in India. Would welcome suggestions for new experiments in India.
ALS 4pp RS:HS 15.294

14284. 1870–9–30 To Sir William Huggins Collingwood
JH is preparing a digest of all double star observations; also comments on sunspot activity.
C 3pp RS:HS 24.331

14285. 1870–10–3 To Edward Sabine Collingwood
Discusses significance of Col. Napier's pendulum observations to research on earth's constitution and gravitation.
ALS 3pp RS MM.10.86A (C: RS MM.10.86B)

14286. 1870–10–5 From Sir William Huggins Upper Tulse Hill
Has sent JH's letter to Cleveland Abbe with a request to send JH the information. Has not seen the sun through his new telescope yet.
ALS 1p RS:HS 10.62

14287. 1870–10–6 To Augustus De Morgan Collingwood
Asks a whole series of silly riddles, and JH complains of not having heard from AD.
ALS 1p APS B.H435p.25

14288. 1870–10–6 From George Griffith B.A.A.S., Harrow
Notice of £50 grant from B.A.A.S. to JH and G. A. Erman to support H. J. R. Petersen in recomputing Gaussian constant for 1839. [JH annotation: £40 remitted 13 Mar. 1871; £10 remaining.]
ALS 4pp † TxU:H/M–0220; Reel 1087

14289. 1870–10–7 To Georg A. Erman Collingwood
Writes to clarify the actions of the B.A.A.S. [see JH's 1870–5–24].
C 2pp RS:HS 24.332

14290. 1870–10–7 To George Griffith Collingwood
Having been placed jointly in charge with G. A. Erman of a B.A.A.S. grant expenditure to support H. J. R. Petersen's recomputation of the Gaussian constant for 1829, JH asks for corrections in the statement of the actions funded.
ALS 2pp Manchester Central Library (C: TxU:H/L-0176; Reel 1054)

14291. 1870–10–7 To George Gabriel Stokes Collingwood

Corrects date on publication by G. A. Erman.
ALS 1p TxU:H/L-0466; Reel 1055

14292. 1870–10–7 To George Gabriel Stokes Collingwood
Requests some rewording of B.A.A.S. resolution [see JH's 1870–9–27].
ALS 1p Cambridge University Library 7656/H730

14293. 1870–10–8 From George Gabriel Stokes Armagh
Reports on status of John Robinson's eye following cataract surgery.
ALS 3pp RS:HS 17.79

14294. 1870–10–9 To Adolphe Quetelet Collingwood
Apologizes for not writing often. Has been ill. 'Moderate' use of 'Coca de Peru' [cocaine] helps. Is working on a 'General Catalogue of double stars reduced to 1830.0.' Acknowledges works received.
ALS 3pp Académie belgique: #57

14295. 1870–10–10 From George Griffith 4 Woodridge, Harrow
Sorry to learn of error in B.A.A.S. resolution regarding computation of Gaussian constant. JH's proposal, explained by G. G. Stokes, was passed by Committee of Recommendations. Errors occurred later, in transcription. Will resubmit JH's proposal to Council.
ALS 4pp TxU:H/M–0221; Reel 1087

14296. 1870–10–10 From George Gabriel Stokes Observatory Armagh
Explains error in B.A.A.S. report on grant for G. A. Erman. Will notify George Griffith.
ALS 10pp TxU:H/M–0509; Reel 1093

14297. 1870–10–12 To Georg Adolph Erman Collingwood
Corrects a misunderstanding about the date to which magnetic observations are to be reduced.
ACCS 1p RS:HS 25.7.7

14298. 1870–10–12 From John R. Hind Twickenham
Has calculated the ecliptic of Thucydides. Comments on this and the reading given by E. W. Hartwig. Regarding the 7 eclipses of the sun during the past 30 years. G. Bishop's assistant (W. Plummer) is investigating the orbit of the great comet of 1861. *Nautical Almanac* for 1874 delayed due to extra work on the transit of Venus.
ALS 8pp RS:HS 9.398

14299. 1870–10–15 From Edward Joseph Lowe Highfield House
Would like to become a member of the Board of Directors of the Midland Railway for which there is a vacancy. Would welcome a testimonial from JH.
ALS 3pp RS:HS 11.342

14300. 1870–10–16 To John P. Gassiot Collingwood
Gives JG a journal reference.
C 1p RS:HS 24.334

14301. 1870–10–17 From Cleveland Abbe Cincinnati
Regarding the publication of the Cincinnati catalogue of double stars. Is sending the Annual Report of the Cincinnati Observatory.
ALS 3pp RS:HS 1.378

14302. 1870–10–17 From John P. Gassiot Clapham Common
Many thanks for his reference. Will check it.
ALS 1p RS:HS 8.66

14303. 1870–10–19 To Elizabeth Colling Collingwood
Recounts various reports of phenomena in the sky related to the 'great Moscow phenomenon' of 'mock Suns and inverted Arches.'
C 3pp RS:HS 24.335

14304. 1870–10–20 From Georg A. Erman Pleishammer, Stenberg
Did not receive JH's letters as he was away from the University. Rejoices at his news of the B.A.A.S. grant.
ALS 1p RS:HS 7.105

14305. 1870–10–23 To John Tyndall Collingwood
Thanks for paper on action of rays 'of high refrangibility' on gases and vapors. Attempts to understand reflection of wave motion on particles with diameters 'incomparably smaller' than length of wave.
ALS 4pp RI 459 (draft: RS:HS 17.409a; C: RS:HS 24.336; C: RI 573–4)

14306. 1870–10–24 From Alexander [Hamilton-]Gordon to Margaret Brodie Herschel Eaton Square
Would like an order from Sir John to allow him to read in the British Museum.
AL 1p inc RS:HS 8.156

14307. 1870–10–29 From Augustus De Morgan 6 Merton Road
Thanks for his enquiry, but has no energy for conundrums at the moment. Finds he has a tendency to tumble as his strength returns.
ALS 3pp RS:HS 6.416

14308. 1870–10–29 From George Griffith 4 Woodridge, Harrow
Clerical error of '1839' instead of '1829' in JH's proposal has occurred in many other reports. No need to wait for Council approval. Will also inform G. A. Erman.
ALS 2pp TxU:H/M–0222.1; Reel 1087

14309. 1870–10–29 From William Stanley Jevons Parsonage Rd., Withington
Scientific men at Manchester, J. P. Joule, Balfour Stewart, and H. E. Roscoe especially, feel that there is not enough influence on the Government regarding scientific affairs. Thinks the leading scientific men should join together. Would JH lend his name and influence to such an undertaking?
ALS 4pp RS:HS 10.328 Black (ed.), *Papers and Correspondence of...Jevons*, 3:236–7

14310. 1870–10–30 To Julia Margaret Cameron Collingwood
Sends a copy of JH's version of the German Rhine Song, which JC requested. Comments favorably on French losses in the Franco-Prussian War. Thanks her for visiting.
ALS 2 pp encl Dunedin Public Lib.

14311. 1870–11–1 From William H. Attree East Grinstead
Question regarding auroral streams. [Part of this letter is missing; digest was obtained by consulting JH's reply of 1870–11–3]
ALS 2pp RS:HS 1.417

14312. 1870–11–3 To William H. Attree Collingwood
Relationship of electricity to magnetism, and relevance to

'auroral streamers'.
C 2pp RS:HS 24.337

14313. 1870–11–3 From William B. Davis 61 Charlotte St.
Is looking for employment as a calculator and if JH has the superintendence of the reduction of C. F. Gauss's *Constants* he would like the position.
ALS 2pp RS:HS 6.475

14314. 1870–11–3 To George Griffith Collingwood
Agrees with GG and G. G. Stokes that the 'Magnetic Reduction Grant' should be approved. G. A. Erman also agrees.
ALS 1p Manchester Central Library (ACCS: TxU:H/M–0222.2; Reel 1087)

14315. 1870–11–4 To Elizabeth Colling Collingwood
Thanks for the clippings about an aurora; thoughts on the relationship of poetry to music.
C 1p RS:HS 24.338

14316. 1870–11–6 To John Lubbock Collingwood
Returns, with thanks, a copy of JL's father's [John William Lubbock] work on lunar theory, and compliments JL on his own writings on prehistoric man.
C 1p RS:HS 24.339

14317. 1870–11–8 From William H. Attree East Grinstead
Thanking him for his reply to his query of 1870–11–1.
ALS 4pp RS:HS 1.418

14318. 1870–11–10 To William Stanley Jevons Collingwood
[Responding to WJ's 1870–10–29], JH declines participating, suggesting that too many lobbying groups already exist and that such groups frequently end up agitating for ends others than those that led to their creation.
ALS 2pp John Rylands University Library of Manchester
Black (ed.), *Papers and Correspondence of...Jevons*, 3:237–8

14319. 1870–11–15 From Robert Brown Fairlie by Greenock
Further comments on JH's paper on the musical scales.
ALS 2pp RS:HS 4.320

14320. 1870–11–15 From Richard John Nelson Stoke, Devonport
JH may recall meeting him at the Cape in March 1836. Now had a query regarding the size and limits of zodiacal light. Quotes case of errors in observations made by the Captain and Mate of a ship on which RN was returning from the Cape. Has now retired from the service.
ALS 4pp RS:HS 13.115

14321. 1870–11–17 To John Herschel (son) Collingwood
Asks son John whom he would like to have sign the certificate for fellowship in the R.S.L. [see JH's 1870–9–7]; JH talks about ways of dealing with local irregularities in geodetic surveying [see JH's 1869–11–25]; is wishing Amelia and family were out of China and into India.
AL 4pp inc JHS 5.41

14322. [1870]–11–23 From Andrew Scott Waugh [7 Petersham Terrace, Queensgate Garden, & Athenaeum Club]
Acknowledges return of R.S.L. certificate signed by JH. Anticipating visit with JH's son John, whom AW appointed to Great Trigonometrical Survey of India.
ALS 2pp † RS:HS 18.127

14323. 1870–11–25 From Katherine A. S. Gibson 52 Marina, St. Leonards
What are JH's views on a suitable person to write a memoir of William Whewell.
ALS 7pp RS:HS 8.105

14324. 1870–11 -[26 or later] To Katherine A. S. Gibson n.p.
Outlines the difficulty of selecting a person to write a memoir of William Whewell.
draft 1p RS:HS 8.106 (C: RS:HS 24.341)

14325. 1870–11–28 From Cleveland Abbe New York
Acknowledging receipt of JH's Slough catalogues. Further news regarding the publication of Professor O. M. Mitchel[l]'s Observations.
ALS 3pp RS:HS 1.379

14326. 1870–11–30 From Thomas G. Bunt Bristol
Sending calculations on the relative proportions of land to sea on the surface of the globe. Would welcome any comments from JH.
ALS 4pp RS:HS 5.126

14327. 1870–12–1 From Richard John Nelson Stoke, Devonport
Obliged for JH's kind and complete reply to his query on zodiacal light. Hopes the diagram will be included in the next edition of JH's book. Would like to include JH's visiting card in his album.
ALS 3pp RS:HS 13.116

14328. 1870–12–3 From George Biddell Airy n.p.
Is sending on G. G. Stokes instructions about polarization; several expeditions have left for more southerly locations to observe the upcoming solar eclipse.
ACCS 2pp RGO 6.382.357

14329. 1870–12–3 To George Biddell Airy Collingwood
About the health of Richarda Airy and that of JH.
ALS 2pp RGO 6.382.356

14330. [1870?]–12–3 From John Tyndall Royal Institution
Thanks for suggestion-filled letter. Remarks on blue color of water.
ALS 2pp RS:HS 17.410 (C: RI 575)

14331. 1870–12–5 From George Biddell Airy Royal Observatory, Greenwich
Is sending G. G. Stokes's instructions about polarization. Observation parties are off to Oran, Gibraltar and Cadiz.
ALS 2pp RS:HS 1.324

14332. 1870–12–5 From Thomas G. Bunt Bristol
Sending his apologies for some errors he made in his calculations in his letter of 30 Nov.
ALS 3pp RS:HS 5.127

14333. 1870–12–7 From Henry Samuel King 65 Cornhill
His eldest son, Henry Seymour King, will deliver the annual oration at Charterhouse on 12 Dec. Would give him great pleasure if JH could be present.
ALS 3pp RS:HS 11.42

14334. 1870–12–9 To Charles Babbage Collingwood
Thanks CB for receipt of a meteorological work by CB's son [B. H. Babbage], now in Australia. JH says he grows stupider every day and is a 'prisoner in the bonds of bronchitic tyranny.'
ALS 4pp BL Aaa 37199.524

14335. 1870–12–12 From G. W. Royston-Pigott Lansdowne Crescent
Discusses his thoughts on high powered definition and residual aberration in microscopes. Thinks they may be applicable to telescopes.
ALS 6pp RS:HS 14.507

14336. 1870–12–20 From Georg A. Erman Berlin
Sends an article from the *Astronomische Nachrichten* acknowledging the assistance of the B.A.A.S. to two Prussian scientists.
ALS 1p RS:HS 7.106

14337. 1870–12–20 To Roderick Impey Murchison Collingwood
Is pleased to hear of signs of recovery for RM from his ill health, and all wish him a complete recovery.
ALS 2pp BL Aaa 46126.408

14338. 1870–12–21 From George Biddell Airy Flamsteed House, Greenwich
Is sending two copies of his book on magnetism [*Treatise on Magnetism* (London, 1870)].
ALS 1p RS:HS 1.325

14339. 1870–12–22 To George Biddell Airy Collingwood
Thanking him for copies of books. People at Collingwood are gazing at the eclipse. Concerning the astatic needle.
draft 4pp RS:HS 1.326 (C: RS:HS 24.342 & draft: RGO 6.132.110)

14340. 1870–12–23 From John Henry Dallmeyer 19 Bloomsbury St.
Has made a trial of JH's lens but it was too large to be satisfactory. Regarding the 'graphoscope,' a new toy. Has been concerned with the manufacture of an improved form of opera glass. Present forms have many imperfections. Gives details of his own theories and sends one for JH's notice.
ALS 7pp RS:HS 6.8

14341. 1870–12–24 From Alexander J. B. Hope Bedgebury Park, Cranbrook
Hopes he will accept a Christmas box, *viz,* a portrait of Sir William Herschel by Thomas Lawrence. It was given him by his friend Sir John Kirkland, who is sure of its authenticity.
ALS 2pp RS:HS 9.476

14342. 1870–12–25 To John Henry Dallmeyer Collingwood
Thanks and a detailed description of the functioning of a pair of binoculars sent by JD to JH.
C 2pp RS:HS 24.340

14343. 1870–12–26 From Charles H. Weston Ensleigh Observatory, Lansdown near Bath
Moon not uniformly dark during last eclipse of sun. In 1858, Mr. Stuart saw same phenomenon, but JH, W. R. Dawes, and [James] Breen reported no such partial illumination. Suggests that this observation was due to refrangibility in medium, not to actual illumination of moon.
ALS 5pp RS:HS 18.393

14344. 1870–12–27 From John Henry Dallmeyer 19 Bloomsbury St.
Thanks for his letter. Is glad the binoculars give a satisfactory performance, but regrets the width between the tubes does not suit JH's eyes. Will alter this and send a new pair later.
ALS 3pp RS:HS 6.9

14345. 1870–12–29 From Adolphe Quetelet Brussels
Is working on humans and their proportions. Encloses brochures. Wishes to dedicate *L'Anthropométric* to JH.
ALS 3pp (French) RS:HS 14.275 (C: Académie belgique: #58)

14346. 1870–12–29 From Adolphe Quetelet [Brussels]
Thanks JH for writing. Complains of obstacles to his work. Happy to see JH recognizes work begun by Blaise Pascal and continued by famed mathematicians. Asks JH to respond to AQ's wish to dedicate his book [*L'Anthropométric*] to JH with only one word: 'J'accepte.'
draft 2pp (French) Académie belgique: #60

14347. 1870–12–29 To Adolphe Quetelet Collingwood
Inconceivable that JH merely writes 'J'accepte' (I accept) in response to the great honor AQ bestows upon him by dedicating his book to JH. Congratulates AQ on his 'fresh contribution to statistical knowledge.' Has just completed his catalogue of approximately 10,000 double and multiple stars.
ALS 4pp (French) Académie belgique: #58 (draft: AB: #59) A. Quetelet, 'Notice sur Sir John Fred. Will. Herschel,' *Académie royale des science, des lettres et des beaux-arts annuaire* (1872), 194–6

14348. 1871–1–6 From Julia Margaret Cameron Freshwater, I.o.W.
Is grateful for the interest he takes in her son and for his letter. Comments on the German war. Has a house full of visitors. It has been very cold and they have been without water.
ALS 12pp RS:HS 5.169

14349. 1871–1–9 From Daniel Vaughan Cincinnati
Sends an abstract of his research entitled 'The Effects of Tide on Planetary Motion.'
ALS 2pp encl RS:HS 17.460

14350. 1871–1–11 From Charles Davies Fishkill-on-Hudson
Has sent a work on the metric system in France, which includes JH's own valuable lecture.
ALS 1p RS:HS 6.474.

14351. 1871–1–13 From B. G. Jenkins 4 Moreton Place
Thanks for JH's remarks on his own 'What is matter'; comments on these. Also regarding JH's theory of the sun's repulsive force and William Herschel's observations on the strange shape of Saturn.
ALS 67pp diag RS:HS 10.315

14352. 1871–1–13 From B. G. Jenkins 4 Moreton Place
Sending a long letter proving one of JH's theories regarding the sun and confirmation of one of William Herschel's observations concerning Saturn.
ALS 2pp RS:HS 10.314

14353. 1871–1–14 To William John Macquorn Rankine Collingwood
Writes to thank WR for the paper on the thermodynamic theory of waves he sent; JH is too ill to respond or even to deal with the analysis in the paper. Nonetheless, JH does make some suggestions for WR's consideration.
(S) 4pp RS:HS 14.468 (C: RS:HS 24.343 & CC: JHS 5.40)

14354. 1871–1–20 From William John Macquorn Rankine 51 Avenue Road
Will mention JH's subject for investigation to Sir William Thomson. Is busy with Committee on Ships of War.
ALS 3pp RS:HS 14.469

14355. 1871–1–21 To *Nature* Collingwood
Requests publication of a notice of the recalculation by H. J. R. Petersen of the Gaussian constants of terrestrial magnetism.
AdraftS 1p inc RS:HS 19.161b *Nature, 3* (Jan. 26, 1871), 249

14356. 1871–1–22 To Robert Ellis Dudgeon Collingwood
Comments on use of 'air' lens rather than glass in some circumstances.
ALS 3pp APS B.H435p.26 (C: RS:HS 24.348)

14357. 1871–1–22 To Georg A. Erman n.p.
Has sent a note to *Nature* stating the details of the grant paid by the B.A.A.S. to H. J. R. Petersen for magnetic observations.
draft 1p RS:HS 7.107

14358. 1871–1–23 From Robert Ellis Dudgeon 53 Montagu Square
Is grateful for his comments on his own paper on vision. Would he refer him to the work by Mr. Chopal on lens.
ALS 3pp RS:HS 6.507

14359. 1871–1–24 To Thomas Andrews Collingwood
Thanking him for his paper on heat developed in combination of acids and bases. Commenting on the various makes of calorimeters.
ALS 3pp RS:HS 1.345 & 24.344

14360. 1871–1–26 To Georg A. Erman Collingwood
Note to accompany a translation of one of GE's writings; also comments on the 'reform' of the German language.
C 2pp RS:HS 24.349

14361. 1871–1–31 From Thomas Andrews Queens College, Belfast
Replying to JH's letter of 24 Jan. 1871. Thanking him for his suggestions. Has not seen JH's Cape Observations. Has seen report of R. W. Bunsen's experiments. Comments on these experiments.
ALS 6pp RS:HS 1.346

14362. 1871–1–31 To Elizabeth Colling Collingwood
Comments on spelling reform being attempted in Germany and the United States, and being proposed in England.
C 3pp RS:HS 24.350

14363. 1871–1–31 From Richard Anthony Proctor 5 Trigon Road, Kennington
Has sent JH a copy of his new book [*The Sun*].
ALS 1p RS:HS 14.129

14364. 1871–2–2 From John S. Harris Cork
Queries regarding the red glow seen during an eclipse of the sun.
ALS 2pp RS:HS 10.162

14365. 1871–2–2 From H. E. Hubbard Nottingham
Would like his views on the correct definition of a billion, to settle an argument.
ALS 1p RS:HS 10.212

14366. 1871–2–2 To Richard Anthony Proctor Collingwood
Thanks RP for sending RP's *Sun.* Suggests a theory that the solar corona, rather than originating in the earth's atmosphere, is produced by reflection from meteoric dust. Informs RP that he has completed a catalogue of all observations of double stars.
C 5pp RS:HS 24.351

14367. 1871–2–3 From Richard Anthony Proctor 5 Trigon Road, Kennington
Raises objections to JH's theory of the solar corona. Discusses possible existence of extensive meteoric dust in the solar system. Requests permission to dedicate a book on sidereal astronomy to JH. Asks JH whether his father in later years always used a front focus for his large telescopes and whether he saw the supposed four additional Uranian satellites with his 40-foot reflector.
ALS 10 pp RS:HS 14.130

14368. 1871–2–5 To Thomas Andrews Collingwood
Regarding patents, and quoting the results of experiments by foreigners without acknowledging the sources.
draft 3pp RS:HS 1.347 & 24.345

14369. 1871–2–7 To Richard Anthony Proctor Collingwood
Replies to RP's objections to his theory of the solar corona. Informs RP that William Herschel's larger telescopes were used as front focus. Suggests explanation of his father's disconfirmed discovery of four additional Uranian moons.
C 4pp RS:HS 24.352 Proctor, *Other Suns,* 419 inc & Constance Lubbock, *Herschel Chronicle*, 290

14370. 1871–2–8 From Richard Anthony Proctor 5 Trigon Road, Kennington
Discusses JH's response to his comments on JH's theory of the solar corona. Discusses great difficulty of writing his planned book on sidereal astronomy.
ALS 3pp RS:HS 14.131

14371. 1871–2–9 To Elizabeth Colling Collingwood
Comments on the attitude of the French towards the Germans in light of the war.
C 2pp RS:HS 24.353

14372. 1871–2–9 From John R. Hind Nautical Almanac Office
Forgot to call his attention to the column regarding the Julian period in the *Nautical Almanac.* Something strange about the neighborhood of *R. Leporis.* He has no doubt seen the table of Julian dates in the *Connaissance des temps.*
ALS 3pp RS:HS 9.399

14373. 1871–2–11 From Thomas Andrews Queens College, Belfast
Has notified J. C. Poggendorf of JH's notes on calorimeters. Discussion of Michael Faraday's indebtedness to JH's *Prelim. Discourse.*
ALS 4pp RS:HS 1.348

14374. 1871–2–11 To Warren de La Rue Collingwood
Comments on sunspot activity.
C 2pp RS:HS 24.354

14375. 1871–2–13 From John P. Gassiot Clapham Common
B.A.A.S. has announced its intention of giving up Kew Observatory. What does he think of the possibility of the R.S.L. taking it over for Magnetic Observations? Sends a copy of the Kew report, also a copy of a letter from [Balfour] Stewart.
ALS 4pp RS:HS 8.67

14376. 1871–2–15 From Warren de La Rue Observatory, Cranford
Is glad he is pleased with the work of Balfour Stewart and himself. Has applied to R.S.L. for a further grant to carry on the observations and computations for another year. Intends to finance the last year himself. Since the death of his brother

he has returned to business.
ALS 3pp RS:HS 6.173

14377. 1871–2–15 From George Robert Waterhouse British Museum
Turned JH's mineral specimen over to N. S. Maskelyne. Poor health of GW's wife.
ALS 2pp RS:HS 18.108

14378. 1871–2–16 From Nevil Story-Maskelyne British Museum
His colleague, [G. R.] Waterhouse, has passed an inquiry from JH to NM regarding a mineralogical character. It is a specimen of Obsidian, very high in iron content.
ALS 2pp RS:HS 13.24

14379. 1871–2–17 To John P. Gassiot Collingwood
Comments on R.S.L. acquiring Kew Observatory building for creation of magnetic observatory [see JG's 1871–2–13].
draft 1p RS:HS 8.68 (C: RS:HS 24.355 & 356)

14380. 1871–2–18 From Alfred Brothers Manchester
Encloses photographs taken of the last eclipse. Draws his attention to certain aspects of the eclipse.
ALS 3pp RS:HS 4.301

14381. 1871–2–18 From George Matthew Whipple Kew Observatory, Richmond, Surrey
Sends prints of solar photographs taken last week at Kew, as requested by Warren de La Rue. Regrets that picture on 11 Feb. was unsatisfactory.
ALS 1p RS:HS 18.396

14382. 1871–2–19 To John Herschel (son) Collingwood
Mostly about how to deal with the implications of local attractions in geodetic surveying [see JH's 1870–11–17]; JH has had the mineral son John sent analyzed and sends the results. JH commiserates with daughter-in-law Mary's illness.
ALS 4pp JHS 5.39 (C: 5.38 omitting the last paragraph)

14383. 1871–2–20 From Thomas Andrews Queens College, Belfast
Has received a letter from J. C. Poggendorf. Further reports on his experiments with gas.
ALS 6pp RS:HS 1.349

14384. 1871–2–20 To Alfred Brothers Collingwood
Thanks for, and comments on, some photographs of the solar corona.
C 2pp RS:HS 24.357

14385. 1871–2–21 From Alfred Brothers Manchester
Explaining why the moon's disc is circular.
ALS 2pp RS:HS 4.302

14386. 1871–2–22 To Thomas Andrews n.p.
Thanks TA for information on ice calorimetric work.
C 1p RS:HS 24.346

14387. 1871–2–23 From Robert Ellis Dudgeon 53 Montagu Square
Relating his recent experiments in dioptrics. Encloses a model of his air-lens. Is grateful for the reference.
ALS 4pp RS:HS 6.508

14388. 1871–2–24 From John P. Gassiot Clapham Common
Is obliged for his comments on the future of Kew Observatory. They differ from the others received, but they will be read out. Has he ever thought of publishing his reports to the R.S.L.? Thinks it would be a good idea if they could be published.
ALS 3pp RS:HS 8.69

14389. 1871–2–27 From Richard Anthony Proctor 5 Trigon Road, Kennington
Reports on recent solar eclipse photographs of the solar corona which jeopardize JH's meteoric theory of the corona's origin. Concludes the corona must be 'after all a phenomenon of eruption!!'
ALS 4pp RS:HS 14.132

14390. 1871–2–28 From John R. Hind Twickenham
Sends calculation of the Transits of Venus in 2004 and 2012. Cannot find any record of new double stars. Is in doubt about *R. Leporis*. Quotes an historical incident regarding an eclipse of the sun and the removal of Mahomet's pulpit. Did F. X. von Zach make his calculations himself?
ALS 6pp RS:HS 9.400

14391. 1871–3–1 To Richard Anthony Proctor Collingwood
Admits that recent heliographs of the corona favor the eruption theory over JH's meteoric theory. Notes that this raises question of the nature of the photosphere. Cites recent results showing link between sunspot formation and terrestrial magnetic disturbances.
C 2pp RS:HS 24.358

14392. 1871–3–2 From Georg A. Erman Berlin
Outlines the recent work in terrestrial magnetism by H. J. R. Petersen and himself. Comments on their results.
ALS 15pp (German) RS:HS 7.108–9

14393. 1871–3–3 From John P. Gassiot Clapham Common
His object in suggesting that JH's reports should be purchased was to preserve them from oblivion.
ALS 2pp RS:HS 8.70

14394. 1871–3–4 From George Griffith B.A.A.S., Harrow
Announces business meeting of Council on 11 Mar.
L 1p TxU:H/M–0921.6; Reel 1083

14395. 1871–3–4 From Richard Anthony Proctor 5 Trigon Road, Kennington
Suggests that meteors come sometimes from the sun, more typically from other stars.
ALS 4pp RS:HS 14.133

14396. 1871–3–9 To William Spottiswoode Collingwood
Needs address of [H. J. R.] Petersen of Kiel, to whom JH must send grant of £50 from B.A.A.S. for reduction of Gaussian constant. Gives address of G. A. Erman. JH's poor health.
ACCS 2pp TxU:H/L-0389; Reel 1054

14397. 1871–3–10 From William Benson 147 Albany St.
Sends a copy of his book on the science of color. Their theories regarding color seem to becoming more popular.
ALS 2pp RS:HS 5.53

14398. 1871–3–11 From William Spottiswoode 50 Grosvenor Place
Will send B.A.A.S. grant for £40 directly to G. A. Erman for reduction of Gaussian constant unless WS hears differently from JH.
ALS 3pp TxU:H/M–0497; Reel 1093

14399. 1871–3–12 To William Spottiswoode Collingwood

Thanks WS for volunteering to forward B.A.A.S. grant to G. A. Erman while JH is ill.
AdraftS 1p TxU:H/L-0390; Reel 1054

14400. 1871-3-13 From George Biddell Airy Royal Observatory, Greenwich
Regarding a reflector for observation of the forthcoming eclipse.
ALS 2pp RS:HS 1.327

14401. 1871-3-13 From George Biddell Airy n.p.
Is looking for an observer for a newly-mounted telescope in India for the December 1871 eclipse. Would JH's son John be available?
ACCS 2pp RGO 6.134.127

14402. 1871-3-14 From James Nasmyth Penshurst
Is gratified and encouraged by JH's approval of JN's illustrations of the lunar surface. The improvement in printing processes makes it possible to use such photographic illustrations.
ALS 4pp RS:HS 13.88

14403. 1871-3-16 To George Biddell Airy Collingwood
Does not know what commitment of John Herschel (JH's son) would be for December 1871, but encourages GA to communicate with him [see GA's 1871-3-13].
ALS 1p RGO 6.134.128

14404. 1871-3-17 From William Mann to the Admiralty Royal Observatory, Cape of Good Hope
Outlining his work at the Royal Observatory since 1839 and requesting an addition to his income.
C 7pp RS:HS 13.17

14405. [1871]-3-18 From Sophia E. De Morgan 6 Merton Road
Announcing the death of her husband Augustus.
ALS 3pp RS:HS 6.436

14406. 1871-3-19 From Adolphe Quetelet Brussels?
Offers JH the first exemplar of AQ's work [*Physique sociale*]. Realizes no one will ever appreciate years of work that it has taken.
C 1p (French) Académie belgique: #62

14407. 1871-3-21[?] To Adolphe Quetelet n.p.
Reiterates his congratulations to AQ for his work on statistics. Believes AQ will have many readers. Mentions that JH is now eighty years old.
ALS 4pp Académie belgique: #63 A. Quetelet, 'Notice sur Sir John Fred. Will. Herschel,' *Académie royale des science, des lettres et des beaux-arts annuaire* (1872), 197-9

14408. 1871-3-22 To Elizabeth Colling Collingwood
More remarks relating to French/German relations [see JH's 1871-2-9]. Thanks EC for more poetry and an account of an earthquake she experienced. JH is in poor health.
C 2pp RS:HS 24.359

14409. 1871-3-25 To Georg A. Erman Collingwood
No doubt he has received some of the money granted by the B.A.A.S. Has sent it all to GE as he did not know the address of H. J. R. Petersen. Unable to follow GE's scientific arguments. Require simplification.
draft 2pp RS:HS 7.110 (C: RS:HS 24.360)

14410. 1871-3-27 From Georg A. Erman Berlin
Thanks JH for assistance in settling matters with the B.A.A.S.; comments about travel plans and magnetic observations.
ALS 1p (German) RS:HS 7.111

14411. 1871-3-29 From Henry W. Field Royal Mint
Sends [T. F.?] Fremantle's first annual report. [Richard?] Roberts memo was drawn up from information obtained from himself. Has now retired, but with no great financial assistance.
ALS 4pp RS:HS 7.223

14412. 1871-3-29 From Frederick Herschel Waterhouse Geological Society, Somerset House
Requests testimonial from JH to assist FW in obtaining better appointment on Society's library staff.
ALS 1p TxU:H/M-0560; Reel 1093

14413. 1871-3-31 From Georg A. Erman Berlin
Regrets that his bad handwriting caused JH to misread some of his remarks. Will try to produce a clearer exposition of his results. Comments on his expositions.
ALS 1p RS:HS 7.112

14414. 1871-3-31 To Frederick Herschel Waterhouse Collingwood
Please clarify FW's request for testimonial from JH to assist FW in obtaining better appointment. JH does not know FW's capabilities and can testify only to good character of FW's family.
AdraftS 2pp TxU:H/L-0493; Reel 1055

14415. 1871-4-5 From Alexander J. B. Hope Arklow House, Connaught Place
[J. B.] Smith plans second reading [of metric bill] on 18 [Apr.] before Parliament adjourns. AH will approve it again. Philip Harwood, editor of *Saturday Review*, asks for article by JH on [metric system] before House votes on it.
ALS 4pp TxU:H/M-0085; Reel 1087

14416. 1871-4-5 From Frederick Howlett East Tisted, Alton
Is anxious to have an astronomical eye-piece based on JH's principles. Encloses communication from Messrs. Horne and Thornthwaite, who would like details for their employee, [W.?] Ackland, to work from.
ALS 4pp RS:HS 10.26

14417. 1871-4-7 To Frederick Howlett Collingwood
Is at present too ill to involve himself in any calculations necessary for making a good optical eyepiece.
C 1p RS:HS 24.361

14418. 1871-4-10 From Frederick Howlett East Tisted, Alton
Is grateful for his reply. Hopes he is feeling better. Escaped worst of the winter by expedition to Algeria, though the results were disappointing. Comments on some of his observations. Feels sure [W.?] Ackland will be able to construct a suitable eyepiece.
ALS 7pp RS:HS 10.27

14419. 1871-4-11 From Thomas Andrews Queens College, Belfast
Encloses notice which J. C. Poggendorf has published regarding JH. Details of further modifications in his experimental apparatus.
ALS 4pp RS:HS 1.350

14420. 1871-4-12 From W. Ackland Newgate St.
Has been unable to locate [W. L.] Newman's tables (astronomical) at the R.A.S. Would he please send further

details.
ALS 3pp RS:HS 1.380

14421. 1871–4–13 To W. Ackland Collingwood
Is distressed to hear that [W. L.] Newman's tables cannot be found at the R.A.S. Please make a further search. Does the R.A.S. have C. J. G. [=J. W.] Pastorff's volume of drawings of sunspots?
draft 3pp RS:HS 1.381 & 24.362

14422. 1871–4–13 To Thomas Andrews Collingwood
Thanks TA for further information [see JH's 1871–2–22] on TA's work on ice calorimeter.
C 1p RS:HS 24.347

14423. 1871–4–13 From William Spottiswoode Her Majesty's Printing Office
Acknowledges receipt of 'present' from [H. J. R.] Petersen.
ALS 1p TxU:H/M–0498; Reel 1093

14424. 1871–4–14 From W. Ackland Newgate St.
He has seen C. J. G. Pastorff's drawings at the R.A.S. Still no trace of Mr. Newman's tables.
ALS 3pp RS:HS 1.382

14425. 1871–4–15 From William Benjamin Carpenter University of London
Hopes that he found the report of the Porcupine Expedition worth his consideration. Sends a paper in which he examines the question of the Gulf Stream.
ALS 4pp RS:HS 5.199

14426. 1871–4–15 To Elizabeth Colling Collingwood
Comments on several aspects of poetry; further comments on French/German relations [see JH's 1871–2–9 & 1871–3–22].
C 2pp RS:HS 24.363

14427. 1871–4–15 From Henry Anthony Jeffreys Hawkhurst Parsonage
Thanks for his Easter offering. Sends a little book which he may find useful.
ALS 2pp RS:HS 10.302

14428. 1871–4–19 To William Benjamin Carpenter Collingwood
Discusses WC's ideas on the causes of oceanic circulation. Notes that wind currents are easier to study than water currents. Glad WC got his specimens of Mediterranean water.
L 5pp Royal Geographical Society *Proceedings of the Royal Geographical Society, 15* (1871), 211–12 & *Nature, 4* (May 25, 1871), 71

14429. 1871–4–20 To Alexander J. B. Hope Collingwood
Has despatched a letter on a subject of concern to JH to the editor of the *Saturday News,* but now feels it is too feeble. Is mostly feeling too ill to write or think for long.
ALS 2pp WT 63313.3

14430. 1871–4–21 To John Williams Collingwood
Encloses copy of JH's letter to R.A.S. [see JH's 1864–6–29] that accompanied JH's submission of W. L. Newman's tables for determining radii of aplanatic lenses. Suspects that there was more than one volume of tables.
ALS 2pp RAS: MSS Add.29 [suppl.]

14431. 1871–4–24 From John Williams Somerset House
In Nov. 1864, R.A.S. council instructed secretary Charles Pritchard to write to JH regarding tables by late [W. L.] Newman of York, but JW finds no evidence that Pritchard complied. Cannot find Newman's tables. Pritchard may have taken them.
ALS 3pp RS:HS 18.411

14432. 1871–4–25 From John Banks Bleak House, Hastings
Opposes bill on decimalization of [British] weights and measures.
ALS 4pp TxU:H/M–0041; Reel 1062

14433. 1871–4–25 From C[harles] Crowden Grammar School, Cranbrook
Is a candidate for the headship of Reading Grammar School and would be grateful if JH would give him a testimonial.
ALS 2pp RS:HS 5.378

14434. 1871–4–25 From John Henry Nelson Upton Court
Many thanks for his comprehensive explanations for the use of that word. Regrets JH's tenant at Slough is so unpunctual. Has JH any plan or map of his property at Slough and Upton?
ALS 4pp RS:HS 13.102

14435. 1871–4–27 From Charles Pritchard Oxford
Tests several possible rules for curving lenses including W. L. Newman's and [Thomas] Cooke's. Tries to put W. L. Newman's rules for lens thickness-to-curvature ratios in a simpler equation.
ALS 4pp RS:HS 14.106

14436. 1871–4–28 To Sir William Huggins Collingwood
Thanks WH for informing JH of his son John being selected as a Council nominee for election to F.R.S.
C 1p RS:HS 24.364

14437. 1871–4–29 From C[harles] Crowden Grammar School, Cranbrook
Is grateful for the testimonial JH has given him; it will be useful.
ALS 1p RS:HS 5.379

14438. 1871–5–2 From John Banks Bleak House, Hastings
JH's 25 Apr. 1871 reply encouraged JB to oppose introduction of French metric system. [Thomas] Brassey and Mr. Shuttleworth will vote against it [in House of Commons].
ALS 2pp TxU:H/M–0042; Reel 1062

14439. 1871–5–4 From Alexander J. B. Hope [Arklow House], Connaught Place
Philip Harwood, editor of *Saturday Review,* wants to use JH's letter as part of article on metric system on 13 May [1871] in anticipation of 16 May debate in House [of Commons]. Second reading of bill was postponed until 6 June.
ALS 4pp TxU:H/M–0086; Reel 1087

14440. 1871–5–9 From Philip Harwood G 1, The Albany, Piccadilly
Second reading of metric bill in Parliament was postponed to 8 June. PH will delay publication of JH's article in *Saturday Review* to coincide with public interest in bill.
ALS 2pp TxU:H/M–0248; Reel 1087

14441. 1871–5–10 From Margaret Brodie Herschel to [Duncan Stewart] Collingwood
Provides details of JH's health, which is very tenuous, but seems to be improving.
ALS 4pp JHS 2.3

14442. 1871–5–11 From Margaret Brodie Herschel to Duncan Stewart Collingwood
Writes of the gentle, peaceful nature of JH's death that

morning in the presence of many of the family.
ALS 4pp JHS 2.4

14443. 1871–6–13 From William Huggins to Robert Main R.A.S., Somerset House
Directs RM and Charles Pritchard to examine late JH's manuscript on history of double stars, sent to Radcliffe Observatory.
ALS 3pp RAS:JH Archive 14/4.2.2; Reel 11

14444. 1871–7–14 From Adolphe Quetelet Brussels
Thanks JH for accepting dedication. This has brought AQ utmost pleasure. Has dedicated few works. Expresses his esteem for JH's character and talent.
ALS 2pp (French) RS:HS 14.276 (C: Académie belgique: #61)

14445. 1871–8–12 From Sophia E. De Morgan to Margaret Brodie Herschel 6 Merton Road
Has been trying to collect letters of her husband as she intends writing a memoir on him [*Memoir of Augustus De Morgan* (London, 1882)]. Would be grateful if she could borrow any of his letters to JH.
ALS 2pp RS:HS 6.437

14446. [1871]–8–19 From Sophia E. De Morgan to Margaret Brodie Herschel Whitby
Was just leaving home when her note arrived. If the letters have not been already dispatched send them in a few weeks' time.
ALS 4pp RS:HS 6.438

14447. [1871]–9–19 From Sophia E. De Morgan to Margaret Brodie Herschel 6 Merton Road
Sending 35 letters, has 20 more that she will send later if she would like to see them. Wishes she had the photographs of insects. Thinks an appendix of all the riddles would be amusing.
ALS 3pp RS:HS 6.439

14448. 1871–9–28 From Sophia E. De Morgan to Margaret Brodie Herschel 6 Merton Road
Has received the packet of letters and has chosen 28 of them. Quite understands that some of the more personal ones should not be included in the memoir. Her strength does not seem equal to the task. Hopes to show the true events connected with University College. *Budget of Paradoxes* is ready for the press.
ALS 5pp RS:HS 6.440

14449. 1871–11–8 From Robert Main and Charles Pritchard to R.A.S. Oxford
Report on late JH's manuscript on *General History of Double Stars*, bequeathed to R.A.S.
ALS 2pp RAS:JH Archive 14/4.2.3; Reel 11

14450. 1873–2–1 From Sophia E. De Morgan to Margaret Brodie Herschel 6 Merton Road
Will send a large number of the remaining letters next week. Would be grateful if she would put them in order. The *Budget of Paradoxes* [of Augustus De Morgan] has been pleasant but difficult work owing to the printers' strike.
ALS 4pp RS:HS 6.441

14451. 1873–2–19 From Sophia E. De Morgan to Margaret Brodie Herschel 6 Merton Road
Letters arrived on Monday and will be kept safe. Feels envious at Miss Herschel's forthcoming visit to Rome. Mentions a friend she may meet there. Work progresses slowly.
ALS 3pp RS:HS 6.442

14452. 1873–5–6 From Sophia E. De Morgan to Margaret Brodie Herschel 6 Merton Road
Has sent off the last of the letters. Anne Sheepshanks is 83 and still clear in the head. If she visits the Royal Academy, she will see the bust of Augustus De Morgan there.
ALS 3pp RS:HS 6.443

14453. 1874–1–18 From F. W. A. Argelander to Col. John Herschel (son) Bonn
FA is sending JH's letters to Col. John Herschel. Has also asked a number of JH's European correspondents to do the same.
ALS 5pp (German) RS:HS 1.377

14454. 1874–5–14 From Robert Grahame Jr. to Margaret Brodie Herschel Bristol
Will look for letters from JH in RG's father's papers.
ALS 2pp inc RS:HS 9.14a

14455. 1874–6–6 From Alexander Stewart Herschel to William Lassell Collingwood
Encloses inventory of JH's unpublished manuscripts on *General History of Double Stars*, which AH forwarded to R.A.S. Estimates final total of JH's double star observations.
ALS 7pp RAS:JH Archive 14/4.1.1; Reel 11

14456. 1875–10–6 From George Biddell Airy to Margaret Brodie Herschel Flamsteed House, Greenwich
What should he do with the letters received previously from her?
ALS 2pp RS:HS 1.328

14457. 1875–10–8 From George Biddell Airy to Margaret Brodie Herschel Flamsteed House, Greenwich
Is sending a packet of letters. Relates about the loss of his wife.
ALS 2pp RS:HS 1.329

14458. n.d. To 'Dear Sir' n.p.
Some papers to be presented at [B.A.A.S.] meeting do not fit subjects of established sections. Suggests possible arrangements. Returns Mr. Ellise's report; cannot condense it any further.
ALS 3pp TxU:H/L-0764; Reel 1089

14459. n.d. To 'dear Sir' n.p.
Asks for clarification regarding legal units of measurement established by Act of 1824.
ACC 2pp inc TxU:H/L-0737; Reel 1089

14460. n.d. To 'Sir' [Royal Mint]
Thanks for 'parcel D W & M' of 26 Apr. Notes regarding British weights and measures, relative to [?]'s plan to circulate petition.
Adraft 1p TxU:H/L-0620; Reel 1089

14461. n.d. To George Biddell Airy n.p.
About the health of Richarda Airy, JH, and William Parsons [Earl of Rosse].
ALS 3pp RGO 6.381.440

14462. n.d. To George Biddell Airy n.p.
Is defending himself against attacks for having become involved in the succession for Surveyor-General of India; JH further comments on photographic experiments, using various parts of the spectrum.
ALS 8pp RGO 6.368.624

14463. n.d. From François J. D. Arago Paris
Has sent a bronze bust of George Cuvier to the R.S.L.
ALS 2pp (French) RS:HS 1.367

14464. n.d. From S[olomon] Atkinson China Terrace
Concerning a delicate matter and a young lady's name.
ALS 2pp RS:HS 1.416

14465. n.d. From Charles Babbage n.p.
Giving details of the secretary of an Italian organization. Is sending a letter of David Brewster.
ALS 1p RS:HS 2.338

14466. n.d. From Charles Babbage Devonshire St.
The hour of meeting of council is to be altered. Bring his paper on notation.
ALS 1p RS:HS 2.343

14467. n.d. From Charles Babbage n.p.
Is forwarding parcels. Getting ready for some experiments.
ALS 2pp RS:HS 2.344

14468. n.d. From Charles Babbage n.p.
Remarks on various equations.
ALS 3pp RS:HS 2.348

14469. n.d. From Charles Babbage n.p.
General family news.
ALS 2pp RS:HS 2.351

14470. [?–8?] From Charles Babbage n.p.
Family news.
ALS 2pp RS:HS 2.352

14471. n.d. To Charles Babbage n.p.
[Letter entirely 'in cipher' (code) except for 'Dear Babbage.']
AL 1p † TxU:H/L-0046; Reel 1054

14472. n.d. To Charles Babbage Norfolk St.
Copy formula and send it to [Richard?] Taylor.
ALS 1p RS:HS 2.359

14473. n.d. To Charles Babbage n.p.
Bad news: a Frenchman has submitted a design for a mathematical machine.
ALS 1p RS:HS 2.363

14474. n.d. To Charles Babbage n.p.
Unable to attend the Astronomical Society meeting. Encloses two papers for him to deal with.
ALS 3pp RS:HS 2.364

14475. n.d. To Charles Babbage n.p.
Hopes to see him in town and then they can go to see Joseph Clement. His observations are being spoilt by bad weather.
ALS 2pp RS:HS 2.368

14476. n.d. To Charles Babbage n.p.
News concerning Edward Ryan.
AL 2pp RS:HS 2.370

14477. n.d. To Charles Babbage n.p.
Will call on him tomorrow.
ALS 2pp RS:HS 2.372

14478. n.d. To Charles Babbage n.p.
Introducing a Mr. Wartam of the Armenian church at Vienna, who knows CB's work on mathematics.
ALS 2pp RS:HS 2.373

14479. n.d. To Charles Babbage n.p.
Will call for him on his way to the river.
ALS 1p RS:HS 2.374

14480. n.d. To Charles Babbage n.p.
Will he meet him tomorrow.
ALS 1p RS:HS 2.375

14481. n.d. To Charles Babbage n.p.
Regrets that he will be unable to come and see the experiments as he is laid up with rheumatism.
ALS 3pp RS:HS 2.376

14482. n.d. To Charles Babbage n.p.
Is unable to dine with him as his rheumatism is too bad.
ALS 4pp RS:HS 2.378

14483. n.d. To Charles Babbage n.p.
Can he lend him Thomas Young's lectures?
ALS 1p RS:HS 2.380

14484. n.d. To Charles Babbage n.p.
Will visit him on Sunday if the weather is suitable.
ALS 1p RS:HS 2.381

14485. n.d. To Charles Babbage n.p.
Invitation to dine with him.
ALS 1p RS:HS 2.382

14486. n.d. To Charles Babbage n.p.
Will he send the drawing of St. Olave's? Has lost his red pocket book.
ALS 2pp RS:HS 2.383

14487. n.d. To Charles Babbage n.p.
Sending an 'ill matched pair.' Has got a cold.
ALS 2pp RS:HS 2.384

14488. n.d. To Charles Babbage n.p.
Has some certificates for him to sign. Would he read the translation of C. F. Gauss's paper on Friday?
AL 2pp RS:HS 2.385

14489. n.d. To Charles Babbage n.p.
Will be pleased to dine with him. Has been unlucky of late.
ALS 1p RS:HS 2.386

14490. n.d. To Charles Babbage n.p.
Is returning his porisms. Comments on his geometrical language.
ALS 2pp RS:HS 2.387

14491. n.d. To Charles Babbage n.p.
Invitation to dine with him.
ALS 1p RS:HS 2.390

14492. n.d. To Charles Babbage n.p.
Is sorry to hear of his misfortunes. Regarding the Ryans.
ALS 2pp RS:HS 2.391

14493. n.d. To Charles Babbage n.p.
Many thanks for the Adelaide sovereign. More will be welcome if they send them.
ALS 1p RS:HS 2.392

14494. n.d. To Charles Babbage n.p.
Will be spending a quiet weekend near Tring. Unable to attend CB's *Soirée*.
ALS 2pp RS:HS 2.394

14495. n.d. To Charles Babbage n.p.
Thanks for some article.
ALS 1p RS:HS 2.395

14496. n.d. To Charles Babbage n.p.
Directory of specified shops. List of chemical apparatus.
AL 3pp RS:HS 2.398

14497. n.d. To Charles Babbage n.p.
Will not be able to come to town until Saturday. Has solved one of his chemical problems. Has not done anything to his mathematical works.
ALS 1p RS:HS 2.401

14498. n.d. To Charles Babbage n.p.
Will join him on Wednesday morning and go together. Has new result in astronomy.
ALS 2pp RS:HS 2.402

14499. n.d. From Georgiana Babbage 1 Dorset St., Manchester Square
Would like to bring one of her children when they visit him [at Christmas?].
ALS 3pp RS:HS 2.354

14500. n.d. From Georgiana Babbage 1 Dorset St., Manchester Square
Thanking him for the gift of china.
ALS 2pp RS:HS 2.355

14501. n.d. From Georgiana Babbage 1 Dorset St., Manchester Square
Thanking him for gift.
ALS 3pp RS:HS 2.356

14502. n.d. From Agnes Baillie [Honiton]
Thanks for the packet. Remarks on Devon and recent events.
ALS 2pp RS:HS 3.13

14503. [?]–7–8 From Agnes Baillie Hampstead
Family news.
ALS 3pp RS:HS 3.14

14504. [?]–7–19 From Agnes Baillie Hampstead
Called last Tuesday. Unable to see him as he was away.
ALS 2pp RS:HS 3.15

14505. n.d. From Francis Baily n.p.
Has received 50 copies of the enclosed work [?]. Stresses its value.
ALS 1p RS:HS 3.276

14506. n.d. From Francis Baily n.p.
Regarding the observations of JH and Giuseppe Piazzi on 'Comae.'
ALS 2pp RS:HS 3.277

14507. n.d. From Francis Baily n.p.
Will be happy to accommodate him.
ALS 1p RS:HS 3.278

14508. n.d. From Francis Baily n.p.
Meeting with J. C. Spencer (Viscount Althorp) next Wednesday at Downing St.
ALS 1p RS:HS 3.279

14509. n.d. To Francis Baily n.p.
Query regarding positions of stars in the various star catalogues.
AL 2pp RS:HS 3.284 (C: RS:HS 25.8.64)

14510. n.d. To Mary Baldwin n.p.
Thanks for news. Those 'assembled' speak often of MB. Love to JH's mother.
ALS 2pp TxU:H/L-0057; Reel 1054

14511. n.d. To Mary Baldwin n.p.
Questions about property lines around Baldwin family's Anstey Cottage. Mentions several family members and neighbors.
ALS 2pp TxU:H/L-0060; Reel 1054

14512. [?]–10–12 From Francis Beaufort Edgeworthtown
Encloses account of an explosion against the side of the *Royal George*, which he first showed to Gilbert Elliot (2nd Earl of Minto). Presents Mrs. Beaufort's best wishes to Margaret Brodie Herschel.
ALS 2pp RS:HS 3.410

14513. [?]–7–15 From Frederick W. Beechey Westbourne Crescent
Has been absent down the Bristol Channel so did not receive his communication until his return to Portishead. Does not wish to delay publication but would like to see the manuscript before it goes to press. Encloses answers to some of JH's queries. Has the chart of currents ready.
ALS 4pp RS:HS 4.25

14514. [?]–3–28 From Catherine Bell 14 Old Burlington St.
Thanks for the gift. Family news.
ALS 4pp RS:HS 4.37

14515. n.d. From Catherine Bell Kirkwalls, near Maidenhead
Thanks for Margaret Brodie Herschel's note. Hopes to see

them in the near future.
ALS 4pp RS:HS 4.38

14516. n.d. From George Bishop Regents Park
Encloses Merz's catalogue of instruments.
ALS 1p RS:HS 4.152

14517. n.d. From Margaret Power Blessington n.p.
Will be pleased to accompany him to the observatory on Thursday. Hopes that by then he and C. G. B. Daubeny will have recovered.
AL 1p RS:HS 5.77

14518. n.d. From Margaret Power Blessington n.p.
Unable to receive him on Wednesday; hopes he can manage Thursday.
AL 2pp RS:HS 5.78

14519. n.d. From Matthew R. Boulton n.p.
Has delayed writing in order to assist JH's friend. Is sending the medal to his agent in London.
ALS 2pp RS:HS 5.87

14520. [?]–5–17 From William Thomas Brande Royal Mint
Regrets he was not there when JH called at the Mint; otherwise his reception would have been different.
ALS 2pp RS:HS 5.95

14521. n.d. From Edward William Brayley Hinckley
Regarding sun spots and the phrase 'cloudy stratum.'
ALS 4pp RS:HS 4.239

14522. n.d. From Edward William Brayley Hinckley
Regarding the site of an ancient battle. Has received an actinometer for his lectures. Regarding the compressibility of liquids.
ALS 3pp RS:HS 4.240

14523. [?]–8–23 From Philothea M. Brodie Reigate
Has just seen her brother who will be able to find the color-blind French artist's address if JH can send her his name.
ALS 2pp RS:HS 5.107

14524. n.d. From Sir Benjamin Collins Brodie [the elder] Broome Park
Has received the enclosed from Edward Sabine. May not be able to come to London owing to a cold.
ALS 2pp RS:HS 4.291

14525. n.d. From John Bullar Basset Wood near Southampton
Regarding poetry and JH's gift of some of his own.
ALS 3pp RS:HS 4.335

14526. n.d. From Henry H. H. Cameron Freshwater Bay, I.o.W.
Is grateful for his kind advice and will bear it in mind when at Oxford.
ALS 3pp RS:HS 5.319

14527. n.d. From Julia Margaret Cameron n.p.
Sending her last series of photographs. Comments on her photographic work.
ALS 4pp RS:HS 5.174

14528. n.d. From Julia Margaret Cameron n.p.
Part of a love poem.
AL 1p inc RS:HS 5.175

14529. n.d. From Julia Margaret Cameron n.p.
Family news.
AL 1p inc RS:HS 5.182

14530. n.d. From Henrietta G. M. L. Chatterton Finchden, Tenterden
Is grateful for his kind gift of the book. Hopes he will be able to visit them.
ALS 3pp RS:HS 5.332

14531. [?]–1–20 From Edward Codrington n.p.
Will take charge of the chronometers for Capt. Richard Copeland. Will convey anything else JH has for the Mediterranean. Too busy to reply to his letter yesterday.
ALS 2pp RS:HS 5.355

14532. n.d. To Thomas Edward Colebrooke n.p.
Reports on considerations that led to the election in 1824 of Henry T. Colebrooke to the presidency of the Astronomical Society.
L Xpp inc T. E. Colebrooke, *Life of H. T. Colebrooke* (London, 1873), 297

14533. n.d. From Augustin Louis Cauchy n.p.
Sending to the R.S.L. his memoir on the theory of light. Comments on his own theories.
ALS 3pp (French) RS:HS 5.328

14534. [?]–3–3 From Benjamin D'Urban Government House
Introducing a German gentleman who is on his way to South Australia.
ALS 1p RS:HS 6.517

14535. n.d. From Benjamin D'Urban Wynberg
Is grateful for the news from Natal. Congratulates him on his decision to send troops so promptly.
C 2pp RS:HS 6.518

14536. n.d. From Charles G. B. Daubeny n.p.
Would be grateful if he would deliver the enclosed [note?] when he arrives at Florence as it concerns his baggage. Hopes he experiences no ill effects from his stay in Rome.
ALS 2pp RS:HS 6.26

14537. n.d. From Charles G. B. Daubeny Botanic Garden Oxford
Many thanks for his Latin translation of the poem. H. M. Parker is publishing them and will advise him.
ALS 2pp RS:HS 6.27

14538. n.d. From Humphry Davy Cassiobury
Intended calling on him after he sent the papers, but was prevented. Has had discussions with W. T. Brande about the duties of the secretaries and will see JH about this at the club.
ALS 2pp RS:HS 6.40

14539. n.d. From Humphry Davy n.p.
Is returning James South's papers. Will discuss publication matters on Thursday. Encloses a communication from [Thomas?] Weaver. What shall he do about [William?] Nicol?
ALS 2pp RS:HS 6.41

14540. n.d. From Humphry Davy n.p.
JH's second note has prevented him writing to [William?] Nicol. Question of communication must rest with John Barlow and S. H. Christie. Sends a note on the extraction of silver. J. G. Children has a method for the South American Bubble companies. Will have a long paper on electro-

chemical phenomena connected with copper sheathing.
ALS 3pp RS:HS 6.42

14541. n.d. From Humphry Davy n.p.
Let meeting be on Saturday. May visit glass house on the way to Greenwich. Has sent W. T. Brande's paper to the R.S.L. Has JH tried magnetic effects on resin and sulfur?
ALS 2pp RS:HS 6.43

14542. n.d. From Humphry Davy n.p.
Hopes to see R. S. Dundas (2nd Viscount Melville) before the next meeting of Council. Approves of JH's views. Everard Home's paper will be read this evening; hopes JH's paper on parallax will follow.
ALS 1p RS:HS 6.44

14543. n.d. From Humphry Davy n.p.
Is still confined to his room so would JH look over the enclosed papers for mistakes. Hopes Council will consider the position of James Hudson.
ALS 3pp RS:HS 6.45

14544. n.d. From Humphry Davy n.p.
Will have great pleasure in going down on Sunday and will take his carriage. Hopes Mr. Carter will be in the party.
ALS 1p RS:HS 6.46

14545. n.d. From Humphry Davy n.p.
Waited to see W. H. Wollaston who wishes to be called Honorary Commissioner of Longitude. Hopes to see him tomorrow.
C 1p RS:HS 6.47

14546. n.d. From Augustus De Morgan 5 Upper Gower St.
Introducing an uncle-in-law who is residing at the Cape for health reasons. Normally lives at Madras and may prove useful to JH should he visit Madras.
ALS 3pp RS:HS 6.417

14547. n.d. From Augustus De Morgan 69 Gower St.
Sending various tracts and a miscellaneous collection of lives. Would he look at the life of William Herschel and make any alterations or corrections as he sees fit.
ALS 2pp RS:HS 6.418

14548. n.d. From Augustus De Morgan 69 Gower St.
Sending an algebraic problem and its solution.
ALS 3pp RS:HS 6.419

14549. n.d. From Augustus De Morgan 69 Gower St.
Pointing out an error in one of JH's definitions in the treatise on sound.
ALS 2pp RS:HS 6.420

14550. n.d. From Augustus De Morgan 7 Camden St.
Saw W. S. Stratford yesterday. Has everything ready. Richard Sheepshanks will assist with the pendulums. If he wants more on the interest works, write and he will send a complete list.
ALS 2pp RS:HS 6.421

14551. n.d. From Augustus De Morgan 7 Camden St.
Regarding a citation for J. W. Lubbock's award.
ALS 7pp RS:HS 6.422

14552. n.d. From Augustus De Morgan 41 Chalcot Villas NW
Regarding the task of sorting his books. All W. H. Smyth's daughters are married; wait until the last one of JH's is married. Has had a letter from W. R. Hamilton. Heat has been dreadful.
ALS 4pp RS:HS 6.423

14553. n.d. From Augustus De Morgan n.p.
Regarding JH's new theory of perspective.
ALS 3pp RS:HS 6.424

14554. n.d. From Augustus De Morgan n.p.
Regarding a proof of L. A. J. Quetelet and the proposition that arises from it. Wife is nursing her mother.
AL 4pp detached postscript of letter RS:HS 6.425

14555. n.d. From Augustus De Morgan n.p.
Anne Sheepshanks would like JH's copy of Richard Sheepshanks's pamphlet in answer to James South and Charles Babbage, to bind with other things and return it to JH.
AL 1p detached postscript of letter RS:HS 6.426

14556. n.d. From Augustus De Morgan n.p.
Did the German edition of William Herschel's works ever materialize? Was Francis Baily's Synopsis of Laplace's Principal elements of Astronomy published? Gives a query regarding the possibility of the latter.
AL 4pp detached postscript of letter RS:HS 6.427

14557. n.d. From Augustus De Morgan n.p.
Sending some tracts which JH does not possess. Elizabeth Baily is rather better, though still weak.
AL 3pp detached postscript of letter RS:HS 6.428

14558. n.d. From Augustus De Morgan n.p.
What does he think of [Joseph] Liouville's journal in which 13 of the 14 communications are by the editor? Sends another riddle.
ALS 2pp detached postscript of letter RS:HS 6.429

14559. n.d. From Augustus De Morgan n.p.
Sends a sonnet to express thanks for one of JH's books. Regarding an answer to S. J. Loyd (1st Baron Overstone). Mr. Vaughan and the question of gold.
ALS 2pp inc RS:HS 6.430

14560. n.d. From Augustus De Morgan n.p.
Sending proposed prospectus for a 'fun^{2}y algebra.'
AL 3pp inc RS:HS 6.431

14561. n.d. From Augustus De Morgan n.p.
Sending proposed prospectus for a 'fun^{2}y algebra.'
AL 1p inc RS:HS 6.432

14562. [?]–10–29 To Augustus De Morgan 10 Hanover Terrace, Regents Park
Gratitude for books sent by AD. Encloses AD's biography of William Herschel with JH's comments.
ALS 1p, encl 7pp RAS: MSS De Morgan 3

14563. [?]–12–17 From John T. Delane Printing House Square
Regrets that any letters of JH should have been neglected, but cannot undertake to open and read all letters addressed to him.
AL 3pp RS:HS 6.485

14564. n.d. From William Hepworth Dixon St. John's Wood
Sending a privately printed tract on the confession and submission of Chancellor Bacon. Invites comments.
ALS 2pp RS:HS 6.496

14565. [?]–10–1 From James Douglas Leith
Received his present this morning for which he is grateful.

Comments on William Whewell's philosophy.
ALS 3pp RS:HS 6.500

14566. n.d. From J. W. Dudgeon 63 Pall Mall
Gives details of his experiments with glutenesed plates.
ALS 3pp RS:HS 6.456

14567. n.d. From J. W. Dudgeon 63 Pall Mall
Received his note and the glass last night. In case his plates have not been treated with iodide of potassium he has put some on one or two.
ALS 3pp RS:HS 6.457

14568. n.d. From J. W. Dudgeon 63 Pall Mall
When he last saw him he was carrying out experiments on colors. Sends a sample of Chinese natural green dye, which has just come into his hands.
ALS 2pp RS:HS 6.458

14569. n.d. From F. M. Eardley-Wilmot n.p.
Will be pleased to come to dinner on Friday next.
AL 1p RS:HS 7.8

14570. [?]–8–24 From Emma Edge Benenden Vicarage
Is grateful for the present of leaves. Will try some to see if it effects a cure. Busy working for the relief of the wounded soldiers.
ALS 4pp RS:HS 7.118

14571. n.d. From Maria Edgeworth n.p.
Come at once.
ALS 4pp RS:HS 7.31

14572. n.d. From William Empson n.p.
Sending notes of a few passages he intends to alter in JH's article.
AL 4pp inc RS:HS 7.77

14573. n.d. From William Empson n.p.
Sending an article that may interest him. Regarding the 'water controversy' of the late eighteenth century. George Brown is writing a life of Henry Cavendish.
AL 4pp inc RS:HS 7.78

14574. n.d. From William Empson n.p.
If he looks at A. K. Johnston's Physical Atlas he will find the measurements he requires, also references to [John?] MacDonald's Andrometer.
AL 2pp detached postscript of a letter. RS:HS 7.79

14575. n.d. From William Empson n.p.
Regarding the spelling of various words. The college is being disinfected. News of Richard Jones.
ALS 2pp inc RS:HS 7.80

14576. n.d. From William Empson n.p.
Queries regarding the printing of JH's paper.
ALS 2pp inc RS:HS 7.81

14577. n.d. From John Fairbairn n.p.
The Papers have arrived safely. The frontier is quiet.
ALS 1p RS:HS 7.153

14578. [?]–3–7 From Michael Faraday n.p.
Cites six journal articles relating to chemical separation of titanium and iron. Sent report to Henry Kater.
ALS 1p TxU:H/M–0187; Reel 1087

14579. n.d. To Michael Faraday n.p.
Expresses JH's regret at not having been able to come to MF's experimental demonstration.
C 1p RS:HS 25.14.8

14580. n.d. To Michael Faraday Slough
Discusses dispersive power of enclosed prism.
ALS 3pp RI 487–8 F3B

14581. n.d. To Michael Faraday n.p.
Maintains opinion that would be improper to append remarks from JH's note to MF's paper.
ALS 2pp RI 367 F3B

14582. [?]–4–22 From Cyrus West Field Hastings
Accepts his invitation to lunch, but his wife and daughter will be unable to come.
ALS 3pp RS:HS 7.367

14583. n.d. From Henry W. Field Much Hadham
Regrets he wrote at such an awkward time. Regarding suitable portraits of JH. Will wait for a more suitable one. HF has just had his photograph taken.
ALS 4pp RS:HS 7.224

14584. n.d. From William Henry Fitton n.p.
News of his own children.
AL inc 2pp RS:HS 7.244

14585. n.d. From Sir John Franklin Devonshire St.
Finds that Joseph Paris [?] has left both his addresses so has written requesting him to call on JH.
ALS 1p RS:HS 7.361

14586. n.d. From Princess Galitzine n.p.
Would like him to accept a small work on the analysis of force.
AL 3pp (French) RS:HS 9.84

14587. [?]–7–2 From John P. Gassiot Clapham Common
William Parsons (3rd Earl of Rosse), Edward Sabine and a few members of the Philosophical Club are coming to JG's house on Monday, and it would give much pleasure if JH would join them.
ALS 2pp RS:HS 8.71

14588. n.d. To [Joseph Goodall?] [London?]
Accepts an invitation to dinner.
AL 1p WT B.18

14589. [?]–8–9 From George [Hamilton]-Gordon Argyll House
Would like to see him at dinner tomorrow.
ALS 1p RS:HS 9.105

14590. n.d. From James A. Gordon 8 Park St.
Many thanks for the 'light reading,' which he returns together with the paragraph from the *Observer*. Won't be able to return to the country yet as he is not fit enough. Knew the family of Reginald Poole at Bridgewater.
ALS 2pp RS:HS 8.206

14591. n.d. From James A. Gordon n.p.
Still in a recumbent position and finds writing awkward. Sympathizes with JH's position in the col d'aspin, as he is afraid of precipices himself. Has had a letter from Captain Stevenson. Last quarterly was interesting on Louis Philippe.
ALS 4pp RS:HS 8.207

14592. n.d. From James A. Gordon n.p.
Believes the enclosed will give him satisfaction. Four of the Horaces are now provided for. Strange way the English have of going against people's express wishes. Regarding Russia's effrontery.
ALS 2pp RS:HS 8.209

14593. n.d. From James A. Gordon n.p.
An optician named Davis would like JH's opinion on an instrument he has designed for observing the sun's atmosphere. Would he look in on his way to the City? What a glorious day the 1st of May was.
ALS 3pp RS:HS 8.211

14594. n.d. From James A. Gordon n.p.
Sending this message by JH's two daughters, who he hopes have enjoyed their stay.
ALS 3pp RS:HS 8.212

14595. n.d. From James A. Gordon n.p.
His own health does not seem to improve. Would like to visit the German Spas he mentions but could not undertake the journey involved.
ALS 8pp inc RS:HS 8.213

14596. [?]–12–15 From James A. Gordon Naish, Bristol
Feeling worn out with his illness. Hopes JH will be able to go to Collingwood for Christmas. Thanks for all the trouble he took over the Mint-Clark affair.
AL 4pp inc RS:HS 8.214

14597. n.d. From James A. Gordon n.p.
Will have great pleasure in escorting Margaret Brodie Herschel. His address will be the University Club, Suffolk St. Will make the Parliamentary experiment if JH comes.
ALS 2pp inc RS:HS 8.216

14598. n.d. From James A. Gordon n.p.
Effect of crystals on mesmerized persons. Brompton Square was blocked by people waiting to see Marguerite Gardiner (Countess of Blessington). Gas lighting by electricity.
AL 2pp inc RS:HS 8.217

14599. n.d. From James A. Gordon n.p.
Regrets he could not help JH. Thought Mr. Townley had spoken to JH. His brother-in-law has seen Robert Houdini in Paris.
AL 2pp inc RS:HS 8.218

14600. n.d. From Lewis Conway-Gordon Southsea
Arrangements for the 12th suits him well, and looks forward to staying with him.
ALS 1p RS:HS 9.106

14601. n.d. From Lewis Conway-Gordon 6 St. George's St.
Would be very pleased to spend a day with him before his return.
ALS 1p RS:HS 9.107

14602. n.d. To Mrs. Gordon [London ?]
Accepts an invitation to tea.
ALS 1p WT B.17

14603. [?]–8–20 From Murial[?] Gordon Naish, Bristol
James A. Gordon has been moved to Naish, though it will take some time before he will be able to move about. Has had a letter introducing a Mr. Agevado, who wishes to see one of the farms, but unfortunately left no address. Can JH advise Mr. Stewart?
ALS 4pp RS:HS 8.215

14604. n.d. From James Grahame Kensington Square
Is unable to be of assistance to JH in the matter of a military appointment.
ALS 2pp RS:HS 8.443

14605. n.d. From James Grahame London
About JG's intended marriage.
ALS 1p RS:HS 8.444

14606. [?]–11–28 From James Grahame Hatchett's Hotel, Piccadilly
Has briefly come into London on distressing business.
ALS 1p RS:HS 8.445

14607. n.d. From Robert Grahame, Sr. London
Will be pleased to breakfast with him tomorrow, but must leave before ten in order to keep an appointment.
ALS 1p RS:HS 9.8

14608. n.d. From Cath[erine] Gretton n.p.
Has heard nothing of JH since mutual friend died two years ago. This May is cold and cheerless.
ALS 1p inc RAS:JH Archive 12/1.3.12 verso; Reel 10

14609. n.d. From Basil Hall n.p.
Further explanation of a matter in gunnery. Can send further sketches if he is still interested.
ALS 7pp inc RS:HS 9.193bis

14610. n.d. From Basil Hall n.p.
Has accepted an invitation to visit the *Victory* and hopes to meet the JH's at the same party. Is off to the dockyard to see about an anchor.
ALS 2pp RS:HS 9.194

14611. n.d. From Basil Hall n.p.
Has been staying at Rome, where the climate did not suit him, but is now on a small ship touring the Mediterranean. Sends a letter of Feliciano Scarpelini, who has a man working a specula made of marble. Palermo Observatory is being put on a secure footing.
ALS 3pp inc RS:HS 9.195

14612. n.d. From Basil Hall n.p.
Further about the experiments of T. T. Grant. Man presented him with a sealed packet on the subject of the precession of the equinoxes by means of the libration of the moon.
ALS 2pp inc RS:HS 9.197

14613. n.d. From Alexander Hamilton-Gordon Argyll House
Short letter in code.
ALS 1p RS:HS 8.157

14614. n.d. From John Bradley Harbord n.p.
Proposes the distinguishing terms for types of navigation: 'geo-navigation' and 'celo-navigation.'
ALS 1p Harvard: Houghton

14615. n.d. From Mary Harcourt St. Leonards [near Windsor]
Was delighted to receive news of the birth of JH's son. Has been suffering from headache herself.
ALS 2pp RS:HS 10.154

14616. [?]–10–21 From William Vernon Harcourt Bolton Percy, Tadcaster
Where can he obtain details of F. W. Bessel's experiments to which JH refers? Comments on experiments with pendulums.
ALS 4pp RS:HS 9.224

14617. [?]–10–13 From William Vernon Harcourt Bolton Percy, Tadcaster
Would like his opinion on one of the experiments of Isaac Newton, described in the *Principia*. Has this experiment ever

been repeated?
ALS 3pp RS:HS 9.225

14618. n.d. From Colonel [Arthur Edward] Hardinge Pinder Castle
JH is being subtly chastised for missing a meeting that the Prince of Wales attended.
ALS 3pp RS:HS 10.159

14619. n.d. From Joseph Hartnell Elfords
Has just received his note. Will consider JH's views in regard to the new road.
ALS 2pp RS:HS 10.168

14620. n.d. From William Henry Harvey n.p.
Sending a number of W. J. Hooker's Icones in which he will find a figure of the little green orchis. Can find nothing like his Satyrium. Would he send the bulb trowel he left at the Herschels'.
ALS 4pp RS:HS 9.252

14621. n.d. From William Henry Harvey n.p.
Many thanks for the bulb; succeeded also in digging one up. Has collected a sackload of bulbs from Table Mountain and sends a few orchidae. Hopes the orobanche will not wither. The composite from JH's garden is *Zinnia elegans*, a Mexican plant.
ALS 4pp RS:HS 9.253

14622. [?]–10–13 From William Henry Harvey n.p.
JH's plant is a *Serruria,* a very large genus. Is not certain of the name of the little yellow bulb. Working on a glossary of botanical jargon and a *Genera of S. African Plants*. Hopes he will send him any odd looking plant when he goes out bulbing. Is working on the Compositae at the moment. Alphonse de Candolle has issued a beautiful arrangement of this difficult family.
ALS 4pp RS:HS 9.254

14623. n.d. From William Henry Harvey n.p.
Further regarding vegetable colors. Advises him to write to Mrs. Griffiths (Torquay) about algae.
ALS 4pp RS:HS 9.255

14624. [?]–7–16 From Eduard Heis Aachen
Astronomical matters.
ALS 3pp (German) RS:HS 10.177

14625. n.d. From Alexander Henning George's Hotel
Just going on board to bring back Lady Ryan (wife of Sir Edward?) and will then call to see him. The furniture will be landed immediately. Proposes to sail Saturday morning.
ALS 3pp RS:HS 9.307

14626. [?]–6–24 From Alexander Henning 7 Chester Place
Gives an example of lunar observations. Considers this a more accurate measurement than that obtained by the chronometer.
ALS 2pp RS:HS 9.308

14627. n.d From Alexander Henning n.p.
Sends regards to Herschel family.
ALS 1p inc Harvard: Houghton AAH 67m–67 (107)

14628. n.d. To Caroline Emily Mary Herschel [London]
A light hearted letter to CH, together with a few lines for JH's wife, Margaret.
ALS 4pp JHS 1.89

14629. n.d. From G. H. Ford to Margaret Herschel n.p.
Is grateful for the gift of Sir John's latest volume and would she express her thanks to Sir John.
ALS 2pp RS:HS 7.375

14630. n.d. From Henry Anthony Jeffreys to Margaret Brodie Herschel Hawkhurst Parsonage
Accepts her kind invitation for today with pleasure. Regarding the number of vowel sounds in the English and French languages.
ALS 3pp RS:HS 10.309

14631. n.d. From Henry Anthony Jeffreys to Margaret Brodie Herschel Hawkhurst Parsonage
Is obliged for her note. Further regarding the distribution of custom to local tradesmen.
ALS 4pp RS:HS 10.311

14632. n.d. From John Henry Nelson to Isabella Herschel n.p.
Only on Friday was the paragraph concerning Alexander Herschel brought to the notice of JN. Comments on Alexander's work.
ALS 3pp RS:HS 13.114

14633. [?]–3–27 From John Henry Nelson to Margaret Brodie Herschel n.p.
Uncle Duncan and his protégé were launched on their northern expedition this morning. Was unable to act on the kind suggestion owing to a visit to the dentist and the cold weather.
ALS 3pp RS:HS 13.108

14634. [?]–9–5 From John Henry Nelson to Margaret Brodie Herschel The Queens, Hastings
Much obliged for her thoughtfulness. Called and met Mrs. Gordon [JH's daughter?], who gave a good account of all at home. Has been in severe pain from a chill.
ALS 4pp RS:HS 13.109

14635. n.d. To Margaret Brodie Herschel [London]
Commenting on aspects of family life, obviously raised in an earlier letter by MH.
ALS 2pp JHS 1.3

14636. n.d. To Margaret Brodie Herschel [London]
Instructions about ordering wine, etc., for a [Christmas?] party.
ALS 3pp JHS 1.6

14637. n.d. To Margaret Brodie Herschel [London]
Writes to commiserate with MH during a difficult time for her; is coming home on the weekend and bringing daughter Isabella with him.
ALS 3pp JHS 1.11

14638. n.d. To Margaret Brodie Herschel [London]
Mostly about bonnets, and letters to be directed to the family at Collingwood.
ALS 4pp JHS 1.12

14639. n.d. To Margaret Brodie Herschel [London]
Concerned about son Willy, who seems very lonely in India.
AL 2pp inc JHS 1.13

14640. n.d. To Margaret Brodie Herschel [London]
Concerned about their daughter Caroline, who is considering going to Constantinople.
ALS 2pp JHS 1.16

14641. n.d. To Margaret Brodie Herschel [London]
Recounts rushing around the previous day to get some business done.
ALS 3pp JHS 1.22

14642. n.d. To Margaret Brodie Herschel [London]
Comments on domestic matters, his pleasure in the accomplishment of his children, and on the death of a family friend.
AL 4pp JHS 1.23

14643. [?]–6–29 To Margaret Brodie Herschel [London]
Mostly about family business.
AL 4pp JHS 1.25

14644. n.d. To Margaret Brodie Herschel [London]
About the coal tax and the building of St. Paul's, political struggles, and a visit from daughters Caroline and Margaret Louisa.
AL 3pp JHS 1.27

14645. n.d. To Margaret Brodie Herschel [London]
The 'boys' have arrived from school. Comments on some household matters.
AL 2pp JHS 1.60

14646. n.d. To Margaret Brodie Herschel London
Thanks for letters from MH and daughter Isabella; some matters of bills and wages.
ALS 2pp JHS 1.79

14647. n.d. To Margaret Brodie Herschel [London]
About their brave daughter Isabella suffering physical hurt quietly; also about an aging family acquaintance, and about buying a horse and carriage.
ALS 4pp JHS 1.87

14648. n.d. To Margaret Brodie Herschel [London]
Praises MH for her knowledge. Discusses worsening conditions at the Mint and the effect this may have on daughter Isabella, who is staying with JH.
ALS 4pp JHS 1.88

14649. n.d. To Margaret Brodie Herschel London
Speaks of JH's pride in son John, and about arrangements for coming out to Collingwood.
ALS 4pp JHS 4.13

14650. n.d. To Margaret Brodie Herschel [Collingwood]
JH is writing to Peter Stewart through MH, and is trying to offer financial assistance to MH's brother in clearing up the accounts of Smith, Elder & Company.
ALS 1p JHS 4.22

14651. n.d. To Margaret Brodie Herschel n.p.
Sends news of the day, since MH left that morning.
ALS 2pp JHS 4.26

14652. n.d. To Margaret Brodie Herschel [Collingwood]
Asks MH's advice about spending money to renovate and improve the house at Slough so it will be easier to get a long term tenant, or whether to sell the house.
ALS 4pp JHS 4.30

14653. n.d. To Margaret Brodie Herschel London
Indicates intentions regarding JH's return to Collingwood, and talks about Mrs. Grundy, pianist and singer.
ALS 4pp JHS 5.20

14654. [?]–5–19 To Margaret Brodie Herschel Collingwood
Sending mail, money, and news to MH; redecorating is going ahead at Collingwood.
ALS 4pp JHS 5.27

14655. n.d. To Margaret Brodie Herschel Collingwood
Expresses grief over the loss of friend/companion [?] to daughter Maria.
AL 1p inc JHS 6.44a

14656. n.d. To Margaret Louisa Herschel [London]
About a rather disastrous dinner party JH gave, and reports on a visit to JH's sons together with some messages for MH's mother.
ALS 4pp JHS 1.92

14657. n.d. From Richard Jones to Margaret Brodie Herschel 14 Arlington St.
Mrs. Jones has asked him to finish the note for her as she is going to Brighton. Hopes to hear from JH when all is settled about Hawkhurst.
ALS 3pp RS:HS 10.398

14658. n.d. From [Alexander Stewart] Herschel n.p.
Matlock received [JH's] account, but bill was not enclosed. Is bill in papers that [JH] sent to [AH]?
ALS 1p TxU:H/M–1128.2 [p. 45 verso]; Reel 1060

14659. n.d. To John Herschel (son) [Collingwood]
About changes in metals due to frequent expansions and contractions, and about the health of JH's son William.
AL 1p inc JHS 6.65

14660. n.d. To [Margaret Brodie Herschel] [London]
About son Willy's school habits and about daughter Caroline, as reported by her father-in-law[?].
AL 4pp JHS 1.71

14661. n.d. From John R. Hind 3 Allsop's Terrace, New Rd.
Discovered a telescopic comet on the confines of *Camelopardes* [sic] and *Cassiopeia*. Sends observations.
ALS 1p RS:HS 9.334

14662. n.d. From William Hirley n.p.
Requests data.
ALS 1p inc Harvard: Houghton AAH 67m–67 (107)

14663. n.d. From George C. Hodgkinson n.p.
Comments on his observations made that year in the Alps. Hopes to return again with more instruments.
AL 4pp inc RS:HS 9.422

14664. [?]–7–10 From Henry Holland 72 Brook St.
Sends unpublished volume written at the request of HH's children. Experiments with soap bubbles. Heard that JH translated first canto of Dante's *Inferno*.
ALS 3pp RS:HS 9.433

14665. [?]–5–23 From Alexander J. B. Hope Arklow Ho. Connaught Place
Has written privately to the Chancellor of the Exchequer enclosing JH's letter. Thinks this preferable to asking the question publicly. Will be spending Whitsun at Bedgebury.
ALS 2pp RS:HS 9.477

14666. n.d. From William Hopkins Cambridge
Has written to Edward Sabine about the unauthorized use of JH's name on a B.A.A.S. petition for uniformity of weights and measures.
ALS 3pp RS:HS 9.481

14667. [?]–2–16 From R. W. Howard-Vyse Stoke
Is grateful that JH has allowed him to insert his opinion in the

passages in his own work. Regrets the trouble he has been put to.
ALS 1p RS:HS 10.11

14668. [?]–3–13 From R. W. Howard-Vyse Stoke
Would like to quote his authority in matters connected with the passages in the pyramids.
ALS 2pp RS:HS 10.12

14669. n.d. From R. W. Howard-Vyse Stoke
Returns his letter, which he finds so interesting he would like to call on him personally respecting it. Has written to Mr. Perning concerning the longitude and latitude of the pyramids. Has arranged to send vol. 1 of the great pyramid at Gizeh.
ALS 3pp RS:HS 10.13

14670. n.d From Francis J. Hughes Penally, Tenby
Has read his article on the Musical Scales. Has been working at a theory regarding the scales based on laws derived from the Bible. Submitted it to Dr. [H. J.?] Gauntlett, who was impressed. Would like JH's opinion.
ALS 7pp RS:HS 10.213

14671. n.d. From Robert Hunt n.p.
Has a work on *Elementary Physics* almost through the press; it is intended for the young. Hopes to construct an instrument to measure amount of chemical action in the solar rays. Would like JH to speak to a member of the Government about a position RH would like.
ALS 4pp inc RS:HS 10.135

14672. n.d. To [Robert Hunt?] n.p.
Doubts that there is any significant difference of density between the northern and southern hemisphere.
ALS 2pp WT 68037.15

14673. n.d. From A. M. Hussey Bromley
His first series of illustrations are temporarily bound up and if JH likes he can borrow them for six months. Would like to retain Mr. Corda's book for a little longer.
ALS 4pp RS:HS 10.219

14674. n.d. From James Rose Innes Cape Town
Was gratified to hear that JH had agreed to devote Wednesday to the examination of the physical and mathematical classes. Is grateful for JH's help and sacrifices.
ALS 2pp RS:HS 10.268

14675. n.d. From James Ivory n.p.
Has examined the papers JH left with him and has no objections to the calculations; comments on these. Will forward an additional remark to Francis Baily.
ALS 4pp RS:HS 10.259

14676. [?]–6–11 From Henry Anthony Jeffreys Hawkhurst Parsonage
Regarding the rates for JH's house.
ALS 4pp RS:HS 10.303

14677. n.d. From Henry Anthony Jeffreys Hawkhurst Parsonage
Has read the lecture with great delight and returns it as requested; would like to borrow it again at some future date to show it to other people.
ALS 3pp RS:HS 10.304

14678. [?]–2–18 From Henry Anthony Jeffreys Hawkhurst Parsonage
Has been away from home since he sent the last books. Now sends the Secretary's and Treasurer's books; comments on these books and their makeup. Presumes the person who collects the money should have a small account book.
ALS 4pp RS:HS 10.305

14679. n.d. From Henry Anthony Jeffreys Hawkhurst Parsonage
Tidd Pratt has given his permission for the rules and they can be enrolled as soon as the tables have been added. Will arrange a meeting date with Mr. Harrison. Leaves home tomorrow.
ALS 3pp RS:HS 10.307

14680. [?]–12–24 From Henry Anthony Jeffreys Hawkhurst Parsonage
Regarding Mrs. Tollemache and her views conflicting with those of HJ.
AL 4pp inc RS:HS 10.308

14681. n.d. From Henry Anthony Jeffreys Hawkhurst Parsonage
Regarding the principles of trade and business in the parish.
ALS 4pp RS:HS 10.310

14682. n.d. From Henry Anthony Jeffreys Hawkhurst Parsonage
Confesses his error about Mr. Fowler. Regarding the problem of the number of hairs on a human head.
ALS 2pp RS:HS 10.312

14683. n.d. From Henry Anthony Jeffreys Hawkhurst Parsonage
Returns the paper JH sent him. Found it quite a mathematical study. Comments on its theories. Gives a theory of his own.
ALS 3pp RS:HS 10.313

14684. n.d. From Henry Anthony Jeffreys [Hawkhurst] Parsonage
Invites JH to dinner to meet Mr. Wales, rector of Woodchurch, who will assist in pre-Christmas examinations at 'one of the Classical & Commercial' schools in HJ's parish.
ALS 1p TxU:H/M–0295.1; Reel 1087

14685. [?]–8–28 From C. A. Jensen 36 Great Marlborough St.
Writes to thank JH and his wife, Margaret, for their kind hospitality. CJ also mentions several other people he met.
ALS 1p (German) RS:HS 10.443

14686. n.d. From Thomas Best Jervis 17 Oxford Terrace
Called on him the other morning to consult him on several points. Has just been appointed Surveyor General of India to succeed George Everest, and would welcome JH's advice on several matters.
ALS 3pp RS:HS 10.318

14687. n.d. From George Johnston 5 Oak Place
Asks for support in gaining a pension from the army.
ALS 1p Harvard: Houghton AAH 67m–67 (63)

14688. [?]–7–18 From Henry Bence Jones Royal Institution
Is anxious that JH's son should have every chance of making himself known. Suggests dates for lectures and fees.
ALS 2pp RS:HS 10.341

14689. n.d. To Mr. Jones n.p.
A note accompanying a key that may hopefully open a bookcase in a room in Harley St. [?].
ALS 3pp WT A-371

14690. n.d. From Richard Jones Brasted
John Musgrave will deliver this. If JH is visiting James South's, Musgrave would like to go as he hopes to set up an observatory in Waterford. Mr. Wilding would like to be an F.R.S.
ALS 2pp RS:HS 10.391

14691. n.d. From Richard Jones Haileybury
Has just returned from Scotland. Did not call on James Graham but would like to know how he is progressing. Travelled back via the border country and can understand Walter Scott's enthusiasm. Liked the people in Edinburgh. Would like to know how the Herschels are and hopes to come down to see them soon.
ALS 3pp RS:HS 10.392

14692. n.d. From Richard Jones Tithe Commission
Since Mrs. Jones returned she has been suffering from a severe cold. Hopes to see her brother James before he sails. Is pleased JH will wage war against the Puseyites. Will read T. B. Macaulay again.
ALS 3pp RS:HS 10.393

14693. n.d. From Richard Jones Tithe Commission
Can get away the time JH mentions. Expects [Edward?] Ryan any minute. Sanscrit would be very useful for Willy [Herschel] as it forms the basis of most Indian dialects. Mrs. Jones has just lost her eldest sister, and is at Brighton.
ALS 4pp RS:HS 10.394

14694. n.d. From Richard Jones Haileybury
Encloses a letter he has received from Mr. Chapman, who thinks JH has done right. Has caught influenza on their return to Haileybury. Has over 80 grafted roses, which JH must come to see.
ALS 4pp RS:HS 10.395

14695. n.d. From Richard Jones Tithe Commission
Was very pleased to see his handwriting again and also to hear that he likes his new abode. William Whewell is coming on Saturday and RJ will propose JH's scheme to him. There are many nightingales at Haileybury. Has a new garden with 250 different rose trees. Will try and bring Mrs. Jones down to see him.
ALS 4pp RS:HS 10.396

14696. n.d. From Richard Jones Tithe Commission
Blue salvias will come by Thursday's coach. Is he to send back the review? Has lost Margaret Brodie Herschel's last letter.
ALS 1p RS:HS 10.397

14697. n.d. From Richard Jones Tithe Commission
Jones and Willy [Herschel] took the coach at Tunbridge and arrived home before six. Willy [Herschel] is about to start for Hertford and hopes to show him the Assizes later in the week.
ALS 3pp RS:HS 10.399

14698. n.d. From Richard Jones Tithe Office
Glad to hear he is back to his original idea. Can reach London by 10 a.m. if he wishes to stay longer at Haileybury. Fears the case of W--- is beyond all hope. His poor roses are all prostrated and disfigured.
ALS 4pp RS:HS 10.401

14699. n.d. From Richard Jones Tithe Commission
Giving details of the Chartist unrest in the City. [Also contains note from William Empson to JH.]
ALS 3pp RS:HS 10.402

14700. n.d. From Richard Jones Tithe Office
Health is improving. Had not seen F. P. G. Guizot's book, but thinks it a good one. Sees JH has got among the philosophers again. Cannot see why U. J. J. Leverrier is hurt. Will be staying in town for a few days.
ALS 4pp RS:HS 10.403

14701. n.d. From Richard Jones Haileybury
Sends the parliamentary papers about the observatory and some of the Dryopteris fern. How to use guano on roses.
ALS 2pp RS:HS 10.404

14702. n.d. From Richard Jones Tithe Office
Take Roundell Palmer by all means. Do not get any signatures to the address until he hears again. Regarding the difficulty of separating the Queen from the State. Mrs. Jones will travel to JH's after London.
ALS 4pp RS:HS 10.405

14703. n.d. From Richard Jones Tithe Commission
Sending down some copies of a military magazine, which contains articles on contouring that may interest him. Overrated the science of military drawing. Don't overwork Johnny [Herschel]. Persian is a useful language in India.
ALS 4pp RS:HS 10.406

14704. n.d. From Richard Jones Haileybury
Wakeford Attree, who is bringing this, will give a full account of the health of RJ, which is not good. Rejoices in Willy's (JH's son) change of Presidency as Bengal offers the widest field.
ALS 3pp RS:HS 10.407

14705. n.d. From Richard Jones Tithe Commission
Cannot send a decisive opinion about Johnny [Herschel]'s commission. Comments on the pros and cons of service in India or England. He starts for Geneva and plans to saunter around Switzerland.
ALS 4pp RS:HS 10.408

14706. n.d. From Richard Jones Charity Commission
Unable to answer JH's letter as soon as he received it as he was suffering from an attack of flatulence. Rejoices to hear JH has a prospect of peace. His eye is gone, but does not miss it as much as he expected. Will come to town in the Spring.
ALS 3pp RS:HS 10.409

14707. n.d. To Henry Kater [Slough]
Thanks HK for a chronometer, and sends some polishing papers.
ALS 2pp WT 57468

14708. n.d. From Mrs. Mary Frances Kater n.p.
Would like JH to come and have tea to discuss the idea of a fluid object glass, with which her husband is at present engrossed.
ALS 1p RS:HS 11.23

14709. n.d. To Mrs. Mary Frances Kater n.p.
Unable to come to the party last night as he was suffering from severe cold.
C 1p RS:HS 11.23bis

14710. n.d. To Johann Lamont n.p.
Has been waiting for an opportunity to thank him for his communication. This letter will be delivered by Mr. H. Bell, son of an eminent chemical manufacturer.
draft 1p RS:HS 11.453

14711. [?]–12–2 From [Miss] C. Langton Farnham Lodge
Requesting JH to start a subscription list to relieve their penury.
ALS 4pp RS:HS 11.457

14712. n.d. From John Lee n.p.
Discusses experiment.
ALS 1p inc Harvard: Houghton AAH 67m–67 (107)

14713. n.d. From John Lee n.p.
Discusses books loaned to JH.
ALS 1p inc Harvard: Houghton AAH 67m–67 (107)

14714. n.d. From Princess Lieven Salthill
Is grateful for his offer but unable to avail herself of it as she is leaving Salthill shortly.
AL 3pp (French) RS:HS 11.484

14715. n.d. From Joseph Johann Littrow n.p.
Describes, in considerable detail, the arrangement of an [?] observatory.
AL 2pp inc (German) RS:HS 11.262

14716. n.d. From Humphrey Lloyd Manchester
Pointing out one aspect of JH's report with which he does not agree, *viz.*, the advantage of very small magnets in observations.
ALS 4pp RS:HS 11.290

14717. [?]–12–15 From Edward Joseph Lowe Highfield House
Would like to print in his own *Treatise on Atmospheric Phenomena* JH's account of the Autora Borealis of May 1843. Had a large batch of papers from Sir Thomas Brisbane. Sends one of his weekly accounts.
ALS 3pp RS:HS 11.343

14718. [?]–12–3 From John William Lubbock Mansion House St.
Many years ago JH questioned him about the calculation of the perturbations in an inclined orbit, the satellites of Herschel [Uranus] for example. Has only recently seen the light. Is now confident that he can calculate the perturbations of any planet in any orbit. Gives details and would be glad of JH's comments.
ALS 12pp RS:HS 11.404

14719. [?]–12–23 From John William Lubbock Mansion House St.
Can he recollect the origin of Table V in the Report of the Committee on Meteorology showing the elastic force of aqueous vapor? James Apjohn says JL calculated it but JL has no recollection. JL's mother is in a hopeless state.
ALS 3pp RS:HS 11.403

14720. n.d. From Carl F. H. von Ludwig n.p.
Went to see J. C. Chase and Dr. Andrew Smith, and a meeting has been arranged at the Public Library tomorrow at 1 p.m. Smith had hoped to call on JH this morning, but was detained by the Governor.
ALS 1p RS:HS 11.408

14721. [?]–7–11 From Sir Charles Lyell 9 Crown Office Row, Temple
Scrope does not mention the depth to which the current has cut the granite in his memoir. In CL's revision can he quote JH on the pebbles?
ALS 2pp RS:HS 11.447

14722. n.d. From Sir Charles Lyell 73 Harley St.
Sends today the remainder of his astronomical chapter. JH's comments or corrections will be welcome.
ALS 2pp RS:HS 11.449

14723. [?]–8–17 From George A. MacFarren 7 Hamilton Terrace
Is flattered by receiving a copy of JH's interesting pamphlet. When he has read it will offer some of his own thoughts on the subject. Would be grateful for JH's comments.
ALS 2pp RS:HS 12.15

14724. n.d. From E[neas] MacKintosh Montagu Square
There is no stated time for Willie's [JH's son] appearance at India House; any day suitable for him and Charles Pritchard will do. Suggests Wednesday next, the tenth, if suitable.
ALS 2pp RS:HS 12.25

14725. n.d. From James Maclaren n.p.
Has received the Wellingtonia seeds safely and they seem all right. Glad to hear that the Paulownias have grown so well. Still has two camellias. Sorry to hear a poor account of his health. Hears reports of John Herschel [JH's son]. Own son has been called to the Bar.
ALS 3pp RS:HS 12.33

14726. n.d. From Thomas Maclear n.p.
Received a note yesterday giving the news that the French fleet is contemplating an action against Jonathan. The comet showed itself last night, but not long enough to obtain an observation.
ALS 1p RS:HS 12.209

14727. n.d. From Thomas Maclear n.p.
Mrs. Maclear's letter is not ready so it will have to travel by the *Aerolite*. Children have still escaped the measles.
ALS 1p RS:HS 12.244

14728. n.d. To Thomas Maclear n.p.
Comments and queries on various astronomical observations.
draft 2pp RS:HS 12.178bis

14729. n.d. From Norman Macleod Manchester
Is not certain what has happened to JH's paper. Is on his way to London to try and clear up the mystery.
ALS 3pp RS:HS 12.261

14730. n.d. From Norman Macleod Adelaide Place Glasgow
Please excuse the delay in replying to JH's letter. He has been busy opening churches and travelling. [Alexander] Strahan will write about the article.
ALS 3pp RS:HS 12.262

14731. [?]–8–27 From Reginald Dyke Marshall Keswick
Thanks for sending him the copy of the articles on meteorology; has been away from home grouse shooting. The day he arrived in Northumberland he saw a peculiar fan-like arrangement of clouds. Encloses a photograph of Louisa [RM's wife and JH's daughter].
ALS 4pp RS:HS 12.293

14732. [?]–12–8 From Reginald Dyke Marshall Leeds
Glad to send a good account of Louisa [RM's wife and JH's daughter] and her daughter.
ALS 2pp RS:HS 12.294

14733. n.d. From Carlo Matteucci n.p.
Objections to one of JH's theories.
AL 2pp inc (French) RS:HS 12.307

14734. n.d. To Sir James McGrigor n.p.
Has recently received JM's letter and is pleased to find

recognition of scientific merit. Welcomes his promised help. Does not feel justified in approaching Benjamin D'Urban on the subject.
draft 1p RS:HS 12.19 & 21.211

14735. n.d. To John Stuart Mill n.p.
Two problems verifying a theory of P. S. Laplace.
draft 4pp RS:HS 12.340 & 22.270

14736. n.d. From Mrs. [W. T.] Miller 7 Scroope Terrace, Cambridge
Describes JH's 1826 rescue of a mountain guide caught in an avalanche on Monte Rosa.
C 4pp inc & C 1p TxU:H/M–1023; Reel 1083

14737. n.d. From Edward Moore n.p.
Discusses work to be finished before the end of the year.
ALS 1p inc Harvard: Houghton AAH 67m–67 (63)

14738. n.d. From Thomas J. Moore Starfield, Ill.
Has read the first edition of JH's *Treatise Astr.* published in America. Has several queries regarding statements in this book, for which he would be pleased if JH would supply the answers.
ALS 1p RS:HS 13.49

14739. [?]–8–26 From William C. Morland Lamberhurst
Many thanks for his kind inquiries after the health of the General. Appreciates JH's kindness to his son when at Slough. Did JH see John Dawson's (2nd Earl of Portarlington) prophecy of fine weather?
ALS 4pp RS:HS 12.370

14740. n.d. From Roderick Impey Murchison n.p.
[Marked 'Private & Confidential.'] Seeks help from JH and George Peacock in persuading 'His Eminence' to change date of meeting.
ALS 3pp TxU:H/M–0382; Reel 1087

14741. n.d. From Roderick Impey Murchison n.p.
Mr. Simonoff desires to see [JH] and 'distinguished vicar.'
ALS 1p inc TxU:H/M–0383.1; Reel 1087

14742. n.d. To My dear Sir n.p.
Will not comment on 'Extracts of paper &c;' [?]'s 'Suggestions' to [Standards] Committee are thorough and sufficient. Reluctant to introduce changes in units of measurement. Suggestions for security of new standards.
ACC 4pp inc TxU:H/L-0593; Reel 1089

14743. n.d. To My dear Sir n.p.
Send books for JH to Patrick Stewart at 65 Cornhill. Will return 'Balloon papers' of J. W. Lubbock soon. Grateful for [?]'s efforts in Macedoine Melloni's cause.
AL 2pp TxU:H/L-0594; Reel 1089

14744. n.d. To My dear Sir n.p.
Finally acknowledges that public will not accept decimal currency. Agrees to suppress article 103 and corresponding table. Sorry to see [Indian] rupee sacrificed.
AL 2pp inc TxU:H/L-0614; Reel 1089

14745. n.d. To My dear Sir n.p.
Instructions for positioning new lenses in telescope do not correspond to actual labels on lenses. Suspects lenses were switched after leaving William Simms's shop. Please advise.
ACC 3pp inc TxU:H/L-0618; Reel 1089

14746. [?]–6–13 From George Thomas Napier Warren's Hotel, Regent St.
Has been so busy visiting various people and relatives that he has had no time to answer JH's letter. Leaves early in September for Rome and Naples and will be unable to visit Collingwood.
ALS 4pp RS:HS 13.73

14747. n.d. From William Napier n.p.
Discusses business matters.
ALS 1p Harvard: Houghton AAH 67m–67 (62)

14748. n.d. From John Henry Nelson n.p.
A year has passed since he sent news of local events and people. Hopes JH's work is progressing well. Gives news of local happenings at Slough.
AL 4pp inc RS:HS 13.103

14749. [?]–9–30 From John Henry Nelson Belle Vue Lodge, Ramsgate
The circumstance to which he alluded last week is thought to have taken place in 24 or 25 when the Slingsby family was at Farnham. Intends going to Paris for a few days at the end of this week.
ALS 4pp RS:HS 13.104

14750. [?]–2–21 From John Henry Nelson Upton Court
Has not answered his letter as he was called to town. Has seen the solicitor who drew up the agreements for the letting of JH's property at Slough. The tenant can now be given notice. Would JH like JN to make it known that the property is now open to offer again?
ALS 3pp RS:HS 13.105

14751. [?]–11–19 From John Henry Nelson Hyde Park Gardens
The works JH directed to be sent to JN from the Board of Trade had arrived before Nelson returned to town. Hopes that JH will accept the contents of the enclosed packet. Can make any alteration to the size or power as desired.
ALS 2pp RS:HS 13.107

14752. [?]–10–10 From John Henry Nelson 43 Sussex Gardens, Hyde Park
Probably heard that JN has been over the channel for health reasons. Encloses a seed of a chestnut from Hougoumont. Hears good reports of JH's health.
ALS 4pp RS:HS 13.110

14753. [?]–7–25 From John Henry Nelson Hyde Park Gardens
Will be traveling abroad for health reasons, but will call at Collingwood on Wednesday.
ALS 2pp RS:HS 13.111

14754. [?]–7–30 From John Henry Nelson Upton Court
Gives more details of F. Charsley, the future tenant of JH's property at Slough. Hopes that Alexander Herschel will not strive to scale the skyscrapers of Switzerland.
ALS 3pp RS:HS 13.112

14755. [?]–2–12 From John Henry Nelson Upton Court
Regarding the late trial and the unsuitable tenant for JH's property.
ALS 2pp RS:HS 13.113

14756. n.d. From John Henry Nelson to [Margaret Brodie] Herschel Hyde Park Gardens
Is grateful for the news. His own health is poor and is about to travel to Switzerland on doctor's advice.
ALS 3pp RS:HS 13.106

14757. [?]–5–5 From Sir Richard Owen Royal College of Surgeons

Has sent a note of his titular condition to the printer. Has been busy with a series of lectures on development.
ALS 2pp RS:HS 13.197

14758. n.d. From George Peacock n.p.
Note to say the matter of Mr. Lax's examination is not yet organized.
AL 1p RS:HS 13.311

14759. n.d. From George Peacock n.p.
Questions of publishing, including JH's considering the publication of William Herschel's writings.
ALS 3pp RS:HS 13.312

14760. [?]–5–1 From George Peacock Deanery, Ely
About the possibility of attending a meeting.
ALS 2pp RS:HS 13.313

14761. [?]–2–20 From George Peacock [Ingasy], Devon
Encouragement to JH for republication of some earlier works, together with news of GP's health.
ALS 4pp RS:HS 13.314

14762. [?]–3–31 From George Peacock Bedgebury
Arrangements about GP visiting JH.
ALS 2pp RS:HS 13.315

14763. [?]–3–29 From George Peacock 9 [Carlisle] Parade, Hastings
GP's poor health prevents him from visiting JH as had intended.
ALS 4pp RS:HS 13.316

14764. n.d. From George Peacock Athenaeum
Arrangements about meeting JH in London.
C 1p RS:HS 13.317

14765. n.d. From Thomas Peers n.p.
Notes how density of earth and moon are important to theories of light.
ALS 1p inc RAS:JH Archive 12/1.3.10 verso; Reel 10

14766. n.d. From C. Petit Observatoire de Toulouse
Writes for her husband, who is very ill and receives no one. Would have made an exception for JH's son, if he had been forwarned. Expresses her husband's deepest regrets and regards.
ALS 2pp RS:HS 14.157

14767. n.d. From Henry Piddington Calcutta
A note accompanying additional writings on the behavior of hurricanes.
ALS 3pp RS:HS 13.410

14768. n.d. From Claude S. M. Pouillet n.p.
Regrets being unable to visit JH at Slough while CP was in England.
ALS 2pp (French) RS:HS 14.182

14769. n.d. From Charles Pritchard n.p.
Experiments with lens curvature and prisms. Wants to discover set rules for lens thickness-to-curvature ratio.
AL 4pp inc RS:HS 14. 111

14770. n.d. To Adolphe Quetelet n.p.
Tells AQ of his daughters's fevers. Rejoices in the power of science to distract him temporarily from worries. Hopes to complete his *Catalogus generalis nebularum* by end of R.S.L. session.
ALS 1p Académie belgique: #29

14771. n.d. To Chairman of R.S.L. Physical Committee n.p.
Recommends unreservedly for publication in *R.S.P.T.* papers by G. G. Stokes and Baden Powell.
ALS 2pp RS RR.1.234

14772. n.d. From Henri Victor Regnault n.p.
Sends photographs from study of trees. Has news from Mr. Stewart. Is attempting to complete his works on gases and vapors. Also working on treatise on physics. Wishes to see his work also published in English.
ALS 2pp (French) RS:HS 14.319

14773. n.d. From James Clark Ross n.p.
Engaged at the Lightning Commission. Will assist JH when he needs help. Used chart of the Antarctic region he mentioned.
ALS 1p cov RS:HS 14.431

14774. n.d. To Edward Sabine Collingwood
Thanks for gelatine [?] paper. Hopes to photograph the sun.
ALS 2pp RS Sa.677

14775. n.d. To Edward Sabine n.p.
Invites to dinner. Will try to arrange meeting between ES and Danish hydrographer Captain Lahrtman[?].
ALS 1p RS Sa.682

14776. n.d. From Richard Sheepshanks n.p.
Sorting out some business with printing, and has done the calculations on his chronometrical observations made in 1838.
ALS 3pp RS:HS 16.123

14777. n.d. From Richard Sheepshanks Athenaeum
About several items of R.A.S. business, especially the *R.A.S.M.N.*
ALS 5pp RS:HS 16.132

14778. n.d. From Richard Sheepshanks Reading
Sends a letter to JH (related to Altona Observatory business?); is working on another *Notice*.
ALS 3pp RS:HS 16.135

14779. n.d. From Richard Sheepshanks n.p.
Has widely distributed prints (of whom?).
AL 1p inc RS:HS 16.136

14780. n.d. From Richard Sheepshanks n.p.
Asks JH to burn letter.
ALS 1p inc RS:HS 16.137

14781. n.d. To Richard Sheepshanks n.p.
Agrees to provide some written work which will go into the report of the weights and measures commission[?].
C 1p RS:HS 25.9.79

14782. n.d. To Richard Sheepshanks n.p.
A note to accompany a Russian work on weights and measures.
ALS 2pp RS:HS 16.139 (C: RS:HS 25.9.78)

14783. n.d. From Algernon O. Simon University College, London
Requests JH's autograph and 'carte de visite' for a collection.
ALS 2pp RS:HS 17.207

14784. n.d. From Arthur B. Simpson Bexhill
Asking for an appointment to meet with JH.
ALS 3pp RS:HS 16.148

14785. n.d. From Arthur B. Simpson Bexhill
Missed one opportunity to meet JH; now seeks another.
ALS 3pp RS:HS 16.149

14786. n.d. To Dear Sir n.p.
Notes peculiar spectrum emitted from [?]'s 'ingenious lamp.' Compares spectra from incandescent lime and from hydrated salts of lime.
ACCS 2pp TxU:H/L-0630; Reel 1089

14787. n.d. From Alfred Smee n.p.
About etymology.
ALS 1p inc RS:HS 16.174

14788. n.d. To William Sotheby [56 Devonshire St.]
Replies positively to WS's invitation to a party.
D 1p RS:HS 17.222b

14789. n.d. To James Calder Stewart n.p.
Opinions on proper representation of 'majority' in Parliament.
Adraft 2pp inc TxU:H/L-0397 verso; Reel 1055

14790. n.d. To James Calder Stewart n.p.
Details for paying tuition of John Herschel (son) at Addiscombe.
ALS 2pp TxU:H/L-0399; Reel 1055

14791. n.d. To James Calder Stewart Anstey near Alton
Will arrive in Portsmouth tomorrow.
ALS 1p TxU:H/L-0400; Reel 1055

14792. n.d. To James Calder Stewart n.p.
Please read JH's enclosed response to 'J.G.'s [James Grahame's?] unjust, harsh...letter;' then show it to Mr. Whalley, before JH sends it. JS's exposé was excellent.
ALS 2pp TxU:H/L-0402; Reel 1055

14793. n.d. To James Calder Stewart n.p.
Sends deposit to be put into JH's account at Drummonds [Bank] as executor of Thomas Baldwin's estate. Margaret will send longer letter soon.
ALS 3pp TxU:H/L-0403; Reel 1055

14794. n.d. To [James Calder] Stewart n.p.
Applied to Admiralty for 'instruments' for the African [Committee?]. Will bring Henry Kater's theodolite and copies of Taylor's and Hutton's tables for Duncan Stewart.
ALS 2pp TxU:H/L-0408; Reel 1055

14795. n.d. To [James Calder] Stewart n.p.
Encloses note to Sir Roderick, who is ill. Margaret's strength is returning, but she faces painful confinement 'down here.' Asks JS to carry out one of JH's duties as executor of JH's cousin Thomas Baldwin.
ALS 4pp TxU:H/L-0398; Reel 1055

14796. n.d. To [James Calder] Stewart n.p.
JH's time to depart is not far off. What to do with additional lenses? Enquires about winter weather in Crimea. Will go to Collingwood Friday.
ALS 2pp TxU:H/L-0405; Reel 1055

14797. n.d. To [James Calder] Stewart [Slough]
Impending struggle will be 'most formidable we have ever had.' Both sides are wrong. Wishes military were better armed and English people less 'Sybaritious.' Hails 'freedom of the Press.' Send document on 'faculty pew' to Slough church.
ALS 4pp TxU:H/L-0407; Reel 1055

14798. n.d. To [James Calder] Stewart n.p.
Invites JS to Sunday dinner with JH's sons Alexander and John.
ALS 2pp TxU:H/L-0409; Reel 1055

14799. n.d. To [George Gabriel Stokes] n.p.
JH requests that J. H. N. Hennessey be allowed to attend meeting of R.S.L. that evening.
ALS 1p Cambridge University Library 7656/H731

14800. [?]–11–29 From H. Symons Malvern
Is ill, but encloses revised copy of letter on rotation, from which HS hopes he has removed all possibly offensive statements.
ALS 3pp RS:HS 17.255

14801. n.d. To Thomas Francis Wade n.p.
Unexpected turn; nothing to do but pray it may lead to happiness. One so singularly fitted by talent should be offered position he deserved.
Adraft 2pp RS:HS 18.7

14802. n.d. From Henry Warburton n.p.
Sends hints. Will give explanation at Wednesday's meeting of 'Committee on Charles Babbage.'
ALS 1p RS:HS 18.88

14803. n.d. From [?] n.p.
A lady is trying to make an appointment with JH.
AL 1p RS:HS 1.397

14804. n.d. From [?] n.p.
Appeals to [JH] for expert opinion on author's 'true theory of our planetary System,' which could initiate a revolution in science but was met by ridicule.
AL 1p inc RAS:JH Archive 12/1.2.7 verso; Reel 10

14805. n.d. To [?] n.p.
Postscript: returns on investments.
AL 1p inc TxU:H/L-0411; Reel 1055

14806. n.d. To [?] n.p.
Explains JH's reasons for compiling list of discrepancies between JH's and [?]'s catalogs of stars.
ACCS 2pp TxU:H/L-0606; Reel 1089

14807. n.d. To [?] n.p.
Proposes modification to telescope to accommodate spectroscope for study of sun. Will write to Astronomer Royal [G. B. Airy] about it.
Adraft 1p inc TxU:H/L-0631; Reel 1089

14808. n.d. To [?] n.p.
Sends congratulations to some one who has received a Doctor of Medicine degree.
ALS 2pp WT 69294

14809. n.d. To [?] n.p.
Excuses himself from a social evening due to pressure of work [observing].
ALS 1p WT B.16

14810. n.d. To ? n.p.
Comments on the use of a particular phrase and its italicization.
ALS 1p inc DMC.695A.885.3

14811. n.d. To ? n.p.
Fragment of a letter, discussing unity of an R.S.L. committee.
AL 2pp inc RS MM.3.6

14812. n.d. To ? n.p.
Gives permission for a publisher to copy an engraving for a work by Richard Sheepshanks.
ALS 3pp Harvard: Houghton fMS Eng 945 (112)

14813. n.d. To Mr. [Holdeness?] n.p.
Asks for loan of H's largest 'coated Electrical fan' tomorrow.
ALS 1p TxU:H/L-0744; Reel 1089

14814. n.d. From [J.?] Herbert to Lord Norhampton [Northampton?] Admiralty
[Extract, in JH's hand, of letter:] Ask JH to determine exact price of telescope and time of its delivery. Commissioners are aware that observatory will need equatorially mounted telescope as soon as 'measurements of Lacaille's base' are completed.
C 1p TxU:H/M–1003; Reel 1083

14815. n.d. To [R.S.L. Council] n.p.
Reports on and recommends for publication [William?] Baxter's paper detailing experiments 'on living and recently dead animals' to determine signs of current electricity manifested during 'organic process of secretion.'
CDS 2pp RS RR.[1?]

APPENDIX I

BIBLIOGRAPHY OF PUBLICATIONS BY JOHN HERSCHEL

Prefatory Statement

The sources for this bibliography (an earlier version of which was published as M. J. Crowe, "Bibliography of the Publications of Sir John Herschel," in Brian Warner (ed.), *John Herschel 1792–1992: Bicentennial Symposium* (Royal Society of South Africa, 1994), pp. 125–140) are numerous. The two most important of these are the bibliography Herschel himself compiled and published in 1861 ("A Complete Catalogue of the Writings of Sir John Herschel," *Mathematical Monthly, 3* (1861), 220–7), and the list of his papers compiled in the *Royal Society Catalogue of Scientific Papers.* The latter bibliography suffers from a number of limitations, e.g., no books or monographs were included. Its limitations are further indicated by the fact that whereas it listed 155 publications, the present bibliography, even without counting reprintings, provides information on nearly four hundred publications. The former bibliography, that compiled by Herschel himself and including 221 items, also suffers from limitations. In addition to the obvious fact that it listed no post-1861 publications, its citations are in many cases inaccurate and/or overly cryptic, e.g., page references were in most cases not provided.

In addition to these two invaluable sources, the present bibliography is based on various lists made by Herschel of his publications and on his six volume compilation of his publications, these items being preserved at the Harry Ransom Humanities Research Center at the University of Texas. The use of modern bibliographical finding sources, e.g., *The Wellesley Index of Victorian Periodicals,* has also expanded the present compilation. The lists compiled by Herschel as well as the bibliography of his writings he prepared provide in most cases the evidence for ascribing to him the publications listed below that originally appeared anonymously.

In nearly every instance, the references in this bibliography have been checked against the original publication. No attempt has been made in this bibliography to include references to reprintings of Herschel publications that have appeared in the twentieth century. No doubt inaccuracies and omissions appear in the present bibliography. Its compiler will greatly appreciate receiving information about these.

Abbreviations Employed

Persons and Books

JH = John Herschel
EEQR = JH's *Essays from the Edinburgh and Quarterly Reviews* (London, 1857)
FL = JH's *Familiar Lectures on Scientific Subjects* (London, 1866)

Journals

Note: The following abbreviations for journals appearing repeatedly in this list have been adopted from the *Royal Society Catalogue of Scientific Papers 1800–1900.*
A. C. = *Annales de Chimie*
A. Ps. C. = *Annalen der Physik und Chemie*
As. S. M. Not. = *Royal Astronomical Society Monthly Notices*
As. S. Mm. = *Royal Astronomical Society Memoirs*
As. Nr. = *Astronomische Nachrichten*
B. A. Rp. = *British Association for the Advancement of Science Report*
Edinb. J. Sc. = *Edinburgh Journal of Science*
Edinb. Ph. J. = *Edinburgh Philosophical Journal*
Ph. Mg. = *Philosophical Magazine*
Phil. Trans. = *Royal Society Philosophical Transactions*
R. S. P. = *Royal Society Proceedings*

1812

Lover of the Modern Analysis, A [JH], "Analytical Formulae for the Tangent, Cotangent, &c.," *Journal of Natural Philosophy, Chemistry, and the Arts, 31* (1812), 133–6.

Analyticus [JH], "Trigonometrical Formulae for Sines and Cosines, &c.," *Journal of Natural Philosophy, Chemistry, and the Arts, 32* (1812), 13–16.

1813

JH, "On Equations of Differences and Their Application to the Determination of Functions from Given Conditions," *Memoirs of the Analytical Society* (1813), 65–114.

[JH and Charles Babbage], "Preface," *Memoirs of the Analytical Society* (1813), i–xxii.

JH, "On a Remarkable Application of Cotes's Theorem," *Phil. Trans.* (1813), 8–26.

[JH], "On Trigonometrical Series, Particularly Those Whose Terms Are Multiplied by the Tangents, Cotangents, Secants, &c. of Quantities in Arithmetic Progression, Together with Some Singular Transformations; with Notes Relating to a Variety of Subjects Connected with the Preceding Memoir," *Memoirs of the Analytical Society* (1813), 33–64.

1814

JH, "Considerations of Various Points of Analysis," *Phil. Trans.* (1814), 440–68.

JH, "New Properties of the Conic Sections," [Leybourn's] *Mathematical Repository,* new ser., *3,* part 2 (1814), 58–9.

1816

JH, "On the Development of Exponential Functions; Together with Several New Theorems Relating to Finite Differences," *Phil. Trans.* (1816), 25–45.

JH, Charles Babbage and George Peacock (translators): Silvestre François Lacroix, *An Elementary Treatise on the Differential and Integral Calculus,* with Appendix and Notes (Cambridge: J. Smith, 1816).

1817

JH, "Isoperimetrical Problems," *Edinburgh Encyclopaedia,* vol. 12 (1817), 320–7.

1818

JH, "On Circulating Functions, and on the Integration of a Class of Equations of Finite Differences into Which They Enter as Coefficients," *Phil. Trans.* (1818), 144–68.

JH, "Mathematics," *Edinburgh Encyclopaedia,* vol. 13, Parts 1, 2 (1818), 358–83.

1819

JH, "Additional Facts Relative to the Hyposulphurous Acid," *Edinb. Ph. J., 1* (1819), 396–400.

JH, "On the Hyposulphurous Acid and Its Compounds," *Edinb. Ph. J., 1* (1819), 8–29; also in *A. C., 14* (1820), 353–63; *Annales générale des sciences physiques, 3* (1820), 305–6; *Neues Journal der Pharmacie für Aerzte, Apotheker, und Chemiker, 5* (1821), 308–21. Note: the last item was reprinted from *A. C.*

JH, "Preface to the Essays" and "Note on Essay II" in *Mathematical Essays by the Late William Spence, Esq.*, ed. by JH (London, 1819), pp. xxvii–xxxii and 151–70; see also p. 295.

H, J. F. W. [JH], "[Questions 362–9 and Their Solutions]," [Leybourn's] *Mathematical Repository,* new ser., *4,* pt. 1 (1819), 61–71.

JH, "[Questions 406–9 and Their Solutions]," [Leybourn's] *Mathematical Repository,* new ser., *4,* pt. 1 (1819), 152–9.

[JH], "Sir William Herschel's Researches Respecting the Distances of the Fixed Stars," *Edinb. Ph. J., 1* (1819), 201–2.

1820

JH, "On the Action of Crystallized Bodies on Homogeneous Light, and on the Causes of the Deviation of Newton's Scale in the Tints Which Many of Them Develope on Exposure to a Polarised Ray," *Phil. Trans.* (1820), 45–100; also in *Annals of Philosophy, 1* (1821), 115–32; 161–81.

JH, "On the Application of a New Mode of Analysis to the Theory and Summation of Certain Extensive Classes of Series," *Edinb. Ph. J., 2* (1820), 23–33.

JH, "On Certain Optical Phenomena Exhibited by Mother-of-Pearl, Depending on Its Internal Structure," *Edinb. Ph. J.,* 2 (1820), 114–21.

JH, *A Collection of Examples of the Applications of the Calculus of Finite Differences* (Cambridge: J. Smith, 1820), 171 pp.; German translation by C. H. Schnuse (Braunsweig: E. Leibrock, 1859).

JH, "The Refractive Power of Hyposulphite of Soda and Silver [and of] Nitrite of Lead," *Edinb. Ph. J., 2* (1820), 184.

JH, "Some Additional Facts Relating to the Habitudes of the Hyposulphurous Acid and Its Union with Metallic Oxides," *Edinb. Ph. J., 2* (1820), 154–6.

1821

JH, "On the Aberrations of Compound Lenses and Object-glasses," *Phil. Trans.* (1821), 222–67.

[JH], *Address of the Astronomical Society of London, Explanatory of Their Views and Objects* (London, 1821), 14 pp. Republished in *Memoirs of the Astronomical Society, 1* (1822), 1–7.

JH, "On the Separation of Iron from Other Metals," *Phil. Trans.* (1821), 293–9; also in *A. C., 20* (1822), 304–6; *Annals of Philosophy, 1* (1821), 389–90; *3* (1822), 95–9; *Journal für Chemie und Physik, 32* (1821), 452–3; *Ph. Mg., 59* (1822), 86–90.

1822

JH, "On Certain Remarkable Instances of Deviation from Newton's Scale in the Tints Developed by Crystals with One Axis of Double Refraction on Exposure to Polarized Light," *Cambridge Philosophical Society Transactions, 1* (1822), 21–41; also in *Edinb. Ph. J., 4* (1820–1), 334–44; *5* (1821), 334–41.

JH, "On a New Method of Computing Occultations of the Fixed Stars," *As. S. Mm., 1* (1822), 325–8. Note: Despite the date of this issue of the *As. S. Mm.,* this paper was read Jan. 10, 1823.

JH, "Practical Rules for the Determination of the Radii of a Double Achromatic Object-glass," *Edinb. Ph. J., 6* (1822), 361–71; also in *Journal für Chemie und Physik, 38* (1823), 370–87.

JH, "On the Reduction of Certain Classes of Functional Equations to Equations of Finite Differences," *Cambridge Philosophical Society Transactions, 1* (1822), 77–87.

JH, "On a Remarkable Peculiarity in the Law of the Extraordinary Refraction of Differently-Coloured Rays Exhibited by Certain Varieties of Aphophyllite," *Cambridge Philosophical Society Transactions, 1* (1822), 241–7.

JH, "On the Rotation Impressed by Plates of Rock Crystal on the Planes of Polarization of the Rays of Light, as Connected with Certain Peculiarities in Its Crystallization," *Cambridge Philosophical Society Transactions, 1* (1822), 43–52; summary in *Edinb. Ph. J., 3* (1820), 397–8.

JH, "Subsidiary Tables for Facilitating the Computation of Annual Tables of the Apparent Places of Forty-Six Principal Fixed Stars, . . . to Which Is Prefixed a Statement of the Formulae Employed, and the Elements Adopted in Their Construction," *As. S. Mm., 1* (1822), 421–96. Despite the date for this issue of *Memoirs,* this paper was read Dec. 12, 1823.

JH and Charles Babbage, "Barometrical Observations Made at the Fall of Staubbach," *Edinb. Ph. J., 6* (1822), 224–7; *Notizen aus dem Gebiete der Natur- und Heilkunde, 2* (1822), 308–9.

1823

JH, "On the Absorption of Light by Coloured Media, and on the Colours of the Prismatic Spectrum Exhibited by Certain Flames; with an Account of a Ready Mode of Determining the Absolute Dispersive Power of Any Medium, by Direct Experiment," *Edinburgh Royal Society Transactions, 9* (1823), 445–60.

[JH], "To the Lark," "The Sailor's Departure," and "The Lament" [three poems] in Joanna Baillie (ed.), *A Collection of Poems, Chiefly Manuscript, and from Living Authors* (London: Longman, 1823), pp. 81–3, 142, and 192–3; "To the Lark" was reprinted in *EEQR,* pp. 734–6.

JH, "[Letter on a Table Contained in the Nautical Almanac]," *As. Nr., 1* (1823), 30–1.

1824

JH, "On Certain Motions Produced in Fluid Conductors When Transmitting the Electric Current," *Phil. Trans.* (1824), 162–96; also in *A. C., 28* (1825), 280–318; *A. Ps. C., 1* (1824), 351–6; *Journal für Chemie und Physik, 44* (1825), 177–216; *Annals of Philosophy, 8* (1824), 170–6; 271–86.

JH and James South, "Observations of the Apparent Distances and Positions of 380 Double and Triple Stars, Made in the Years 1821, 1822, and 1823, and Compared with Those of Other Astronomers; Together with an Account of Such Changes as Appear to Have Taken Place in Them since Their First Discovery. Also a Description of a Five-feet Equatorial Instrument Employed in the Observations," *Phil. Trans.* (1824), pt. 3, 1–412; summary in *Edinb. J. Sc., 3* (1825), 281–8; *4* (1826), 66–70.

1825

[JH], "[Abstract of a] Memoir . . . by M. Plana . . . Relating to the Theory of the Perturbations of the Planets Expounded in the *Mécanique Celeste,*" *Ph. Mg., 66* (1825), 468–71.

Anonymous [JH], "Analysis of Sir D. Brewster's 'Description of a Monochromatic Lamp,' and Sir J. F. W. Herschel's Paper 'On the Absorption of Light by Coloured Media and the Colours of Certain Flames,'" *Edinb. J. Sc., 2* (1825), 344–8.

JH, "Extract of a Letter to Dr. Ritchie Explaining a New Process of Actinometry," *Edinb. J. Sc., 3* (1825), 107.

JH, "On the Mechanical Effect Produced When a Conducting Liquid Is Electrified in Contact with Mercury," *Edinb. J. Sc., 2* (1825), 193–9; also in *Journal für Chemie und Physik, 48* (1826), 246–9.

JH, "Notice of a Remarkable Occurrence of Serpentine at the Junction of Sienite with the Dolomite of the Tyrol," *Edinb. J. Sc., 3* (1825), 126–9.

JH and Charles Babbage, "Account of the Repetition of M. Arago's Experiments on the Magnetism Manifested by Various Substances during the Act of Rotation," *Phil. Trans.* (1825), 467–96; summary in *Ph. Mg., 66* (1825), 97–8.

1826

JH, "Account of a Series of Observations, Made in the Summer of the Year 1825, for the Purpose of Determining the Difference of Meridians of the Royal Observatories of Greenwich and Paris," *Phil. Trans.* (1826), 77–126.

JH, "Account of Some Observations Made with a 20–feet Reflecting Telescope: Comprehending, 1. Descriptions and Approximate Places of 321 New Double and Triple Stars. 2. Observations of the Second Comet of 1825. 3. An Account of the Actual State of the Great Nebula in Orion, Compared with Those of Former Astronomers. 4. Observations of the Nebula in the Girdle of Andromeda," *As. S. Mm., 2* (1826), 459–97.

JH, "[Letter on Fraunhofer's Claims for the Inferiority of Reflecting as Compared to Refracting Telescopes]," *As. Nr., 4* (1826), 231–6.

H. [JH], The notes signed H. in James South, "Observations of the Distances and Positions of 458 Double and Triple Stars, Made in the Years 1823, 1824, and 1825; Together with a Re-examination of 36 Stars of the Same Description, the Distances and Positions of Which Were Communicated in a Former Memoir," *Phil. Trans.* (1826), 1–391.

JH, "Observations on Mr. Fraunhofer's Memoir on the Inferiority of Reflecting Telescopes When Compared with Refractors," *Quarterly Journal of Science, 20* (1826), 288–93.

JH, "On the Parallax of the Fixed Stars," *Phil. Trans.* (1826), 266–80.

1827

JH, "Address Delivered by J. F. W. Herschel, Esq., President of the Astronomical Society of London, on the Occasion of the Distribution of the Honorary Medals of That Society . . . to Francis Baily, Esq., Lieutenant W. S. Stratford, R. N., and Colonel Mark Beaufoy," *Ph. Mg.,* 2nd ser., *1* (1827), 455–66; also in *As. S. Mm., 3* (1829), 123–36; reprinted in *EEQR,* pp. 466–88.

JH, "Correction of an Error in a Paper Published in the *Philosophical Transactions,* Entitled 'On the Parallax of the Fixed Stars,'" *Phil. Trans.* (1827), 126–7.

JH, "Nature of the Light Emitted by Lime, in a High State of Incandescence," *Edinb. J. Sc., 6* (1827), 176.

JH, "Paper . . . Read at a Meeting of the Board of Longitude [on the Reform of the Nautical Almanac]," (Printed by Order of the Board, 1827), 7 pp.

1828

JH, "[Letter on] Calculating Machinery," London *Times* (August 19, 1828), p. 3, col. 2.

1829

JH, "Account of an Experiment Made on the Composition of Oil of Cassia, to Determine the Cause of Its High Dispersive Power," *Edinb. J. Sc., 10* (1829), 308–9.

JH, "Account of Two Remarkable Cases of Insensibility in the Eye to Particular Colours," *Edinb. J. Sc., 10* (1829), 153–9.

JH, "An Address . . . on February 8, 1828, on Presenting the Honorary Medals to Lieutenant-Gen. Sir Thomas MacDougal Brisbane . . ., and James Dunlop, Esq.," *As. S. Mm., 3* (1829), 399–412; reprinted in *EEQR,* pp. 489–503.

JH, "Approximate Places and Descriptions of 295 New Double and Triple Stars, Discovered in the Course of a Series of Observations with a 20–feet Reflecting Telescope; Together with Some Observations of Double Stars Previously Known," *As. S. Mm., 3* (1829), 47–63; summary in *As. S. M. Not., 1* (1827), 25–6.

JH, "Sur les couleurs de différentes flammes et sur les spectres qu'elles produisent quand on les analise au moyen du prisme," *Correspondance mathématique et physique, 5* (1829), 254; *A. Ps. C., 16* (1829), 186–7.

JH, "Observations with a 20–feet Reflecting Telescope.—Third Series; Containing a Catalogue of 384 New Double and Multiple Stars; Completing a First Thousand of Those Objects Detected in Sweeps with that Instrument; Together with Observations of Some Previously Known,"

As. S. Mm., 3 (1829), 177–213; summaries in *As. S. M. Not., 1* (1828), 45–7 and *Edinb. J. Sc., 9* (1828), 90–3.

JH, "Physical Astronomy," *Encyclopaedia Metropolitana,* 2nd Division: Mixed Sciences, vol. 1 (1829), pp. 647–729; 733–4; reprinted in *Cyclopaedia of Astronomyy* (London, John Joseph Griffin and Company, 1848).

JH, "Table of the Refractive Powers of Several Bodies," *Edinb. J. Sc., 10* (1829), 296–301.

1830

JH, *Letter from J. F. W. Herschel [Containing Remarks on Dr. Young's Mathematical and Physical Investigations]* (London, 1830), 15 pp.

JH, "Light," *Encyclopaedia Metropolitana,* 2nd Division: Mixed Sciences, vol. 2 (1830), pp. 341–586. Note that this article is dated Dec. 12, 1827) and that a two volume French translation made by P. F. Verhulst and A. Quetelet was published (Paris, 1829–1833); a German translation by J. E. Eduard Schmidt was also published (Stuttgart and Tubingen, 1831).

JH, *A Preliminary Discourse on the Study of Natural Philosophy* (London: Longman, 1830), vii + 372 pp.; continued to be published at late as 1867; French translation (Paris, 1834); German translation by F. C. Henrici (Göttingen: Vandenhoeck and Ruprecht, 1836); Italian translation by Gaetano Demarchi (Torino, 1840); Polish translation (Warsaw, 1955).

JH, "Sound," *Encyclopaedia Metropolitana,* 2nd Division: Mixed Sciences, vol. 2 (1830), pp. 747–824. Note: This article is dated Feb. 3, 1830.

1831

JH, "An Address . . . on February 13, 1829 on Presenting the Honorary Medals to the Rev. William Pearson, Professor Bessel, and Professor Schumacher," *As. S. Mm., 4* (1831), 215–24; reprinted in *EEQR,* pp. 504–18.

JH, "Einige Beugungs-Erscheinungen," *A. Ps. C., 23* (1831), 281–8.

JH, "Fourth Series of Observations with a 20–feet Reflector; Containing the Mean Places and Other Particulars of 1236 Double Stars, as Determined at Slough, in the Years 1828 and 1829, with That Instrument, (the Greater Part of Them Not Previously Described)," *As. S. Mm., 4* (1831), 331–78.

JH, "The General Nature and Advantages of the Study of the Physical Sciences" (from Herschel's *Preliminary Discourse. . .*) in *The American Library of Useful Knowledge* (Boston: Stimson and Clapp, 1831), vol. 1, pp. 267–320. This consists of Chs. 1–3 of Part I of JH's *Preliminary Discourse. . . .*

1832

JH, "On the Action of Light in Determining the Precipitation of Muriate of Platinum by Lime-water," *Ph. Mg.,* 2nd ser., *1* (1832), 58–60; also in *Annalen der Pharmacie, 3* (1832), 337–9; *A. Ps. C., 26* (1832), 176; *Journal für Chemie und Physik, 65* (1832), 262–3.

JH, "On the Cause of the Subterranean Sounds Heard at Nakoos, near Tor, in Arabia," *Geological Society Proceedings, 1* (1832), 388.

[JH], "The Comet [of Biela]," London *Times* (Sept. 26, 1832), p. 2, col. 6; reprinted in the *Evening Mail* for Sept. 25–6, 1832.

[JH], "The Comet [of Biela]," London *Times* (Sept. 27, 1832), p. 2, col. 6; reprinted in the *Evening Mail* for Sept. 27, 1832.

JH, "Lettre . . . sur la séparation de l'oxide de fer, et sur un nouveau procédé pour effectuer la purification complète de l'oxide d'urane," *A. C., 49* (1832), 306–11; also in *A. Ps. C., 25* (1832), 627–8.

JH, "Sur une propriété optique des rayons colorés dans le borax, et sur les révolutions périodiques de quelques étoiles doubles," *Correspondance mathématique et physique, 7* (1832), 77–9.

[JH], "[Review of] *Mechanism of the Heavens.* By Mrs. Somerville [and of] *Mécanique celeste.* By the Marquis de la Place, &c. Translated, with a Commentary, by Nathaniel Bowditch, LL. D., &c., Volume I," *Quarterly Review, 47* (1832), 537–59; reprinted in *EEQR,* pp. 21–62.

1833

JH, *An Address to the Subscribers to the Windsor and Eton Public Library and Reading Room* (London: Smith, Elder, 1833), 32 pp.; 2nd ed. (London, 1834); reprinted in *EEQR,* pp. 1–20 and in: Anonymous, *The Importance of Literature to Men of Business: A Series of Addresses Delivered at Various Popular Institutions* (London, 1852), pp. 31–48.

JH, "Description of a Machine for Resolving by Inspection Certain Important Forms of Transcendental Equations," *Cambridge Philosophical Society Transactions, 4* (1833), 425–40.

JH, "Fifth Catalogue of Double Stars Observed at Slough in the Years 1830 and 1831 with the 20–feet Reflector; Containing the Places, Descriptions, and Measured Angles of Position of 2,007 of Those Objects, of Which 1,304 Have Not Been Found Described in Any Previous Collection; the Whole Reduced to the Epoch 1830.0," *As. S. Mm., 6* (1833), 1–73.

JH, "On the Investigation of the Orbits of Revolving Double Stars; Being a Supplement to a Paper Entitled 'Micrometrical Measures of 364 Double Stars,' &c., &c.," *As. S. Mm., 5* (1833), 171–222; summary in *Ph. Mg.,* 2nd ser., *10* (1831), 443–5.

JH, "[Letter Reporting Observations of δ Cygni, μ Aquarii, α Capricorni, and Biela's Comet]," *As. Nr., 10* (1833), 317–20.

JH, "Micrometrical Measures of 364 Double Stars with a 7–feet Equatorial Achromatic Telescope, Taken at Slough, in the Years 1828, 1829, and 1830," *As. S. Mm., 5* (1833), 13–91; summary in *Ph. Mg.,* 2nd. ser., *10* (1831), 443–5.

JH, "Note sur la manière d'agir de l'acide nitrique sur le fer," *A. C., 54* (1833), 87–94; reprinted in *Annalen der Pharmacie, 10* (1834), 250–5; *A. Ps. C., 32* (1834), 211–16.

JH, "Notice of the Elliptic Orbit of ξ Boötis, with a Second Approximation to the Orbit of γ Virginis. To Which Is Appended, a Notice of the Elliptic Orbit of η Coronae," *As. S. Mm., 6* (1833), 149–57; summary in *As. S. M. Not., 2* (1831–33), 191–2.

JH, "Notice of a Remarkable Deposition of Ice round the Decaying Stems of Vegetables during Frost," *Ph. Mg.,* 3rd ser., *2* (1833), 110–11; also in *A. Ps. C., 28* (1833), 231–3.

JH, "Observations of Biela's Comet," *As. S. Mm., 6* (1833), 99–109; for a slightly shortened version, see *As. S. M. Not., 2* (1831–33), 117–24; see also *Ph. Mg.,* 3rd ser., *2* (1833), 222–7.

JH, "Observations of Nebulae and Clusters of Stars, Made at Slough, with a 20-feet Reflector between the Years 1825 and 1833," *Phil. Trans.* (1833), 359–505.

JH, "Remarks on a Fifth Catalogue of Double Stars, Communicated to the Royal Astronomical Society, June 7, 1832," *As. S. Mm., 6* (1833), 74–81; summary in *As. S. M. Not., 2* (1833), 137–40.

JH, "Sir John Herschel's Report [on the Fluid-Lens Telescope, Constructed for the Royal Society on Mr. Barlow's Principles]," *R. S. P., 3* (1833–4), 245–6.

JH, *A Treatise on Astronomy* (London: Longman, 1833), viii + 422 pp.; went through numerous English and American editions and printings until at least 1851; German translation as *Populäre Astronomie* by Julius Michaelis (Leipzig: L. Voss, 1838); there seem to be three French translations, one by Peyrot and published in 1834; another by A. Cournot and published in 1835; and a third by A. D. Vergnaud (1837). A Danish translation made by P. Pedersen appeared in 1836.

1834

JH, " On the Absorption of Light by Coloured Media, Viewed in Connexion with the Undulatory Theory," *B. A. Rp. for 1833* (1834), 373–4; *Notizen aus dem Gebiete der Natur- und Heilkunde, 40* (1834), 276–80; *Ph. Mg.*, 3rd ser., *3* (1833), 401–12; *A. Ps. C., 31* (1834), 245–61; *Correspondance mathématique et physique, 8* (1834), 114–27; reprinted in *FL*, pp. 476–94.

JH, "Explanation of the Principle and Construction of the Actinometer," *B. A. Rp. for 1833* (1834), 379–81.

JH, "Register of Atmospherical Phenomena and the Temperature of the Sea," *Athenaeum*, #364 (Oct. 18, 1834), 767.

1835

JH, *Annual Report of the South African and Scientific Institution* (Read July 11, 1835).

JH, "On the Astronomical Causes Which May Influence Geological Phenomena," *Geological Society Transactions, 3* (1835), 293–299; summaries in *Geological Society Proceedings, 1* (1830–1), 244–5 and *Ph. Mg.*, 2nd ser., *9* (1831), 136–8.

JH, "Barometrical and Other Observations," *Athenaeum*, #391 (April 25, 1835), 320.

JH, "Extract of a Letter . . . to Francis Baily [That Consists of an Account of Some Double Stars and Nebulae Observed at the Cape of Good Hope]," *As. S. M. Not., 3* (1835), 75–7.

[JH], *Instructions for Making and Registering Meteorological Observations in Southern Africa, and Other Countries in the South Seas, and Also at Sea* (London: Bradbury and Evans, 1835), 17 pp.; also published by G. Greig, Cape Town.

JH, "Instructions for Making and Registering Meteorological Observations," *As. S. M. Not., 3* (1835), 121–2.

JH, "[Extract from J. Herschel's] *Instructions for Making and Registering Meteorological Observations,*" *Athenaeum*, #401 (July 4, 1835), 509–10.

JH, "[Letter Reporting Some Observations Made at the Cape of Good Hope]," *As. Nr., 12* (1835), 273–6.

JH, "A List of Test Objects, Principally Double Stars, Arranged in Classes, for the Trial of Telescopes in Various Respects as to Light, Distinctness, &c.," *As. S. Mm., 8* (1835), 25–32; see also *As. S. M. Not., 3* (1834), 18–19.

JH, "On the Satellites of Uranus," *As. S. Mm., 8* (1835), 1–24; summary in *As. S. M. Not., 3* (1834), 35–6.

JH, "A Second Series of Micrometrical Measures of Double Stars, Chiefly Performed with the 7–feet Equatorial, at Slough, in the Years 1831, 2, and 3," *As. S. Mm., 8* (1835), 37–59.

1836

JH (translator) of "The Death Song of a Nadowessee Chief" by Schiller, in James Grahame, *The History of the United States of North America,* vol. 3 (London, 1836), 418–19; reprinted in *EEQR*, pp. 698–702.

JH, "Extract of a Letter . . . on the Present Appearance of γ Virginis," *As. S. M. Not., 3* (1836), 197–8.

JH, "[Extract of a Letter to William Rowan Hamilton]," *Athenaeum*, #462 (Sept. 3, 1836), 627–8.

JH, "Hourly Meteorological Observations for the Summer Solstice, Made at . . . Cape of Good Hope, on the 21st and 22nd of December, 1835," *Athenaeum*, #446 (May 14, 1836), 345.

JH, "Hourly Meteorological Observations for the March Equinox, 1836, Made at . . . Cape of Good Hope," *Athenaeum*, #466 (Oct. 1, 1836), 705.

JH, "Hourly Meteorological Observations for the June Solstice of 1836, Made at . . . Cape of Good Hope," *Athenaeum*, #470 (Oct. 29, 1836), 769.

JH, "Observations de la comète de Halley, faites au Cap," *Comptes rendus, 3* (1836), 505–6.

JH, "Pretendues découvertes dans la Lune," *Comptes rendus, 3* (1836), 505.

[JH], *Report of the Meteorological Committee of the South African Literary and Philosophical Institution, Read July A. D. 1836* (Cape Town: the Institution, 1836), 18pp.; reprinted in *South African Quarterly Journal,* 2nd ser., *1* (Dec., 1836), 1–18.

JH, "Second Report of the Meteorological Committee of the South African Literary and Scientific Institution," *Edinburgh New Philosophical Journal, 21* (1836), 239–46.

JH, "Sixth Catalogue of Double Stars, Observed at Slough, in the Years 1831 and 1832, with the 20–feet Reflector; Containing the Places, Descriptions, and Measured Angles of Position of 286 of Those Objects, of Which 105 Have Not Been Previously Described. Reduced to the Epoch 1830.0," *As. S. Mm., 9* (1836), 193–204; summary in *As. S. M. Not., 3* (1835), 137–8.

JH, "Views on Scientific and General Education, Applied to the Proposed System of Instruction in the South African College," *Ph. Mg.*, 3rd ser., *8* (1836), 432–8.

1837

JH, "Extract from a Letter from Sir John F. W. Herschel to C. Lyell [on the Theory of Volcanic Phenomena]," *Ph. Mg.*, 3rd ser., *11* (1837), 212–14.

JH, "Extract from a Letter from Sir John F. W. Herschel to R. I. Murchison . . . in Explanation of the Former [Letter], to C. Lyell [on the Theory of Volcanic Phenomena]," *Ph. Mg.*, 3rd ser., *11* (1837), 214–15.

JH, "[Extract of a Letter to Charles Lyell Dated Feb. 20th, 1836, 'On the Primary Cause of Geological Upheavals and Disturbances]'" in Charles Babbage, *The Ninth Bridgewater Treatise* (London: John Murray, 1837), 202–13.

JH, "[Extract of a Letter to Roderick I. Murchison on the *Primum Mobile* of Geological Disturbances]" in Charles

Babbage, *The Ninth Bridgewater Treatise* (London: John Murray, 1837), 214–17.

JH, "Hourly Meteorological Observations Made . . . at the Cape of Good Hope, on the 20th, 21st, and 22nd September, 1836," *Athenaeum*, #483 (Jan. 28, 1837), 64.

JH, "Hourly Meteorological Observations for the March Equinox, 1837, Made . . . at the Cape of Good Hope," *Athenaeum*, #505 (July 1, 1837), 482.

JH, "Hourly Meteorological Observations Made at . . . Cape of Good Hope, on the 21st and 22nd of June, 1837," *Athenaeum*, #525 (Nov. 18, 1837), 847.

JH, "Meteorological Observations . . . Made at the Cape of Good Hope, on the 20th, 21st, and 22nd December, 1836," *Athenaeum*, #499 (May 20, 1837), 360.

JH, "On the Prepared or Peculiar Voltaic Condition of Iron," *Ph. Mg.*, *11* (1837), 329–33.

[JH]. *Report of the Meteorological Committee [of the South African Literary and Philosophical Institution]: Part 1, Read July 17, 1837* (Cape Town, 1837), 20 pp.

1838

JH, "Account of a Remarkable Increase of Magnitude of the Star η in the Constellation *Argo*, Observed . . . at the Cape, December 16–17, 1837," *As. S. M. Not.*, *4* (1838), 121–2.

JH, "Extracts from a Letter . . . to C. Lyell [on Phenomena Connected with the Internal Temperature of the Earth]," *Geological Society Proceedings*, *2* (1838), 548–52; 596–8.

JH, "Extraits de deux lettres addressées du Cap de Bonne-Espérance," *Bibliothèque universelle des sciences, belles-lettres, et arts*, *14* (1838), 116–28; *Notizen aus dem Gebiete der Natur- und Heilkunde*, *6* (1838), 321–9.

JH, "Herschel's Telescope," London *Times* (Oct. 10, 1838), p. 5, col. 4.

JH, "Hourly Meteorological Observations Made at Feldhausen on the 21st and 22nd of September, 1837," *Athenaeum*, #539 (Feb. 24, 1838), 143.

JH, "[Letter on the Sudden Change of Magnitude of the Star η Argus]," *As. Nr.*, *15* (1838), 311–12.

JH, "Observations of the Comet of Halley, after the Perihelion Passage in 1836: Made at Feldhausen, Cape of Good Hope," *As. S. Mm.*, *10* (1838), 325–36; for an abbreviated version, see *Uebersicht der Arbeiten und Veränderungen der schlesischen Gesellschaft für vaterländische Kultur im Jahre 1839* (1840), 32–5.

JH, "[Remarks at] The Herschel Dinner," *Athenaeum*, #555 (June 16, 1838), 423–7:424–5.

1839

JH, "On Halley's Comet," *B. A. Rp. for 1838* (1839), pt. 2, 19–20.

JH, "[Letter on the Action of the Dissevered Rays of Light in the Solar Spectrum]," *Athenaeum*, #621 (Sept. 21, 1839), 724–5.

JH, "[Letter on the Variable Stars η Argus, α Hydrae, and α Cassiopeiae; and on the Comparative Brightness of the Stars]," *As. Nr.*, *16* (1839), 187–190; *Bibliothèque universelle des sciences, belles-lettres, et arts*, *20* (1839) 177–9.

JH, "Note on the Art of Photography, or the Application of the Chemical Rays of Light to the Purposes of Pictorial Representation," *R. S. P.*, *4* (1839), 131–3; summary in *Athenaeum*, #535 (March 23, 1839), 223; *Notizen aus dem Gebiete der Natur- und Heilkunde*, *10* (1839), 260–1.

JH, "Notice of a Chemical Examination of a Specimen of Native Iron, from the East Bank of the Great Fish River, in South Africa," *Ph. Mg.*, 3rd ser., *14* (1839), 32–4; *American Journal of Science*, *36* (1839), 213–4.

JH, "Notice of the Meteoric Phenomena at Feldhausen, South Africa, on the 13th and 14th November, 1837," *Meteorological Society Transactions*, *1* (1839), 77.

JH, "Observations on Stars and Nebulae at the Cape of Good Hope," *B. A. Rp. for 1838* (1839), pt. 2, 17–19.

JH, "On the Structure of the Vitreous Humour of the Eye of a Shark," *B. A. Rp. for 1838* (1839), pt. 2, 15–16.

JH, "On the Suspected Variability of the Star α Cassiopeiae," *As. S. M. Not.*, *4* (1839), 195–6.

1840

JH, "An Address . . . on February 14, 1840 on Presenting the Honorary Medal to M. Jean Plana," *As. S. Mm.*, *11* (1840), 331–9; also in *As. S. M. Not.*, *5* (1840–43), 32–8; reprinted in *EEQR*, pp. 519–31 but misdated as Feb. 19, 1840.

JH, "On the Chemical Action of the Rays of the Solar Spectrum on Preparations of Silver and Other Substances, Both Metallic and Non-metallic, and on Some Photographic Processes," *Phil. Trans.* (1840), 1–59; *Notizen aus dem Gebiete der Natur- und Heilkunde*, *14* (1840), 129–34.

JH, "A Letter to the Rev. William Whewell, President of the Section[,] on the Chemical Action of the Solar Rays" *B. A. Rp. for 1839* (1840), pt. 2, 9–11.

JH, "The Old Telescope," with a translation into German by Madame Mädler, *As. Nr.*, *17* (1840), 323–6.

JH, *Report of the Committee of Physics, Including Meteorology, on the Objects of Scientific Inquiry in Those Sciences* (Royal Society, 1840).

JH, "Report on the Subject of a Series of Resolutions Adopted by the British Association at Their Meeting in August, 1838, at Newcastle," *B. A. Rp. for 1839* (1840), pt. 2, 31–42.

[JH], "[Review of] 1. *Allgemeine Theorie des Erdmagnetismus* [by] C. F. Gauss and W. Weber. . . . 2. *Intensitas vis Magneticae* [by] Gauss. . . . 3. *Lettre de M. de Humboldt . . . sur les moyens propres à perfectionner la connaissance du Magnétisme Terrestre* . . . . 4. *Report of the Committee of Physics*. . . .," *Quarterly Review*, *66* (1840), 271–312; reprinted in *EEQR*, pp. 63–141.

JH, "On a Simple Mode of Obtaining from a Common Argand Oil Lamp a Greatly Increased Quantity of Light," *Ph. Mg.*, 3rd. ser., *16* (1840), 194–5.

JH, "On the Variability and Periodical Nature of the Star α Orionis," *As. S. Mm.*, *11* (1840), 269–78; summaries in *As. S. M. Not.*, *5* (1840), 11–16 and *Bibliothèque universelle des sciences, belles-lettres, et arts*, *26* (1840), 189–92.

1841

JH, "[Letter on Meteors]," *Athenaeum*, #721 (Aug. 21, 1841), 643.

JH, "[Letter on the Question of the Moon's Influence on the Weather]," *As. Nr.*, *19* (1842), 175–8.

JH, "Report of a Committee for the Reduction of Lacaille's Stars in the Coelum Australe Stelliferum," *B. A. Rp. for 1840* (1841), 427.

JH, "Report of the Committee . . . Appointed to Superintend the Scientific Cooperation of the Association in the

Researches Relative to Terrestrial Magnetism," *B. A. Rp. for 1840* (1841), 427–34.

JH, "Report on the Reduction of Meteorological Observations Made at the Equinoxes and Solstices, on the Part of a Committee Appointed by the British Association at Newcastle," *B. A. Rp. for 1840* (1841), pt. 1, 432–4.

JH, "[Review of] 1. *History of the Inductive Sciences* . . ., 2. *Philosophy of the Inductive Sciences* . . . by Rev. William Whewell," *Quarterly Review, 68* (1841), 177–238; reprinted in *EEQR,* pp. 142–256.

JH, "Second Report of a Committee of the British Association . . . for Revising the Nomenclature of the Stars," *B. A. Rp. for 1840* (1841), 426–7.

1842

JH, "On the Action of the Rays of the Solar Spectrum on Vegetable Colours, and on Some New Photographic Processes," *Phil. Trans.* (1842), 181–214; *Ph. Mg.,* 3rd. ser., *22* (1843), 5–21, 107–116, 135, 170–80, 246–52, 505–11; *Daguerreian Journal, 2, no. 3–6* (June 16–Aug. 1, 1851), 65–72, 97–106, 129–36, 161–5; *R. S. P., 4* (1842), 397–8; *Notizen aus dem Gebiete der Natur- und Heilkunde, 23* (1842), 294–6.

JH, "An Address . . . [on] February 12, 1841, on Presenting the Honorary Medal to M. Bessel," *As. S. Mm., 12* (1842), 442–54; also in *As. S. M. Not., 5* (1841), 89–98; reprinted in *EEQR,* pp. 532–51.

JH, "On the Advantages to Be Attained by a Revision and Re-arrangement of the Constellations, with Especial Reference to Those of the Southern Hemisphere, and on the Principles upon Which Such Re-arrangement Ought to Be Conducted," *As. S. Mm., 12* (1842), 201–24; summary in *As. S. M. Not., 5* (1841), 116–18.

JH, "Extracts from a Letter Addressed . . . to Mr. Murchison, Explanatory of the Phaenomena of the Freezing Cave of Illetzkaya Zatchita," *Geological Society Proceedings, 3* (1842), 697–9; *Edinburgh New Philosophical Journal, 34* (1843), 14–16; *Ph. Mg.,* 3rd. ser., *21* (1842), 359–61.

JH, "Letter [Submitting Various Photographs]," *B. A. Rp. for 1841* (1842), pt. 2, 40.

JH, "[Letter 'On Certain Photographic Effects']," *Athenaeum,* #773 (Aug. 20, 1842), 748.

JH, "Report of a Committee . . . Appointed for the Purpose of Superintending the Scientific Co-operation of the British Association in the System of Simultaneous Observations in Terrestrial Magnetism and Meteorology," *B. A. Rp. for 1841* (1842), pt. 2, 38–41.

JH, "Report of a Committee . . . to Superintend the Reduction of Meteorological Observations," *B. A. Rp. for 1841* (1842), pt. 1, 42–6.

JH, *Revised Instructions for the Use of the Magnetic and Meteorological Observatories and for the Magnetic Surveys. Prepared for the Committee of Physics and Meteorology of the Royal Society* (Royal Society, 1842).

JH, "On Some Phaenomena Observed on Glaciers, and on the Internal Temperature of Large Masses of Ice or Snow, with Some Remarks on the Natural Ice-caves Which Occur below the Limit of Perpetual Snow," *Geological Society Proceedings, 3* (1842), 699–702; *Edinburgh New Philosophical Journal, 34* (1843), 17–21; *Ph. Mg.,* 3rd. ser., *21* (1842), 362–4.

[JH (translator)], Friedrich Schiller, *The Walk. Translated in the Original Metre from the German* (Printed for Private Circulation, 1842), 23 pp.; reprinted in *EEQR,* pp. 683–97.

1843

JH, "On the Action of the Rays of the Solar Spectrum on the Daguerreotype Plate," *Ph. Mg.,* 3rd. ser., *22* (1843), 120–32.

JH, "On Certain Improvements on Photographic Processes Described in a Former Communication, and on the Parathermic Rays of the Solar Spectrum," *Phil. Trans.* (1843), 1–6.

JH, "The Comet and the Zodiacal Light," *Annals of Philosophical Discovery, 1* (1843), 187–9

JH, " The [Great] Comet [of 1843]," London *Times* (March 21, 1843), p. 5, col. 3.

JH, "The [Great] Comet [of 1843]," London *Times* (April 4, 1843), p. 6, col. 1.

JH, "The [Great] Comet [of 1843]," *Morning Herald* (March 24, 1843), p. 5, col. 3.

JH, "The [Great] Comet [of 1843]," *Morning Herald* (March 29, 1843), p. 6, col. 4.

JH, "On the Increase in Magnitude of the Star η Cygni," *As. S. M. Not., 6* (1843), 23–4.

JH, "[Letter on an] Aurora Borealis," *Athenaeum,* #811 (May 11, 1843), 465.

JH, "[Letter on] The Great Explosion at Dover," *Athenaeum,* #797 (Feb. 4, 1843), 111.

JH, "[Letter] On a Remarkable Photographic Process," *Athenaeum,* #829 (Sept. 16, 1843), 847.

JH, "Notice of an Extraordinary Luminous Appearance Seen in the Heavens on the 17th of March, 1843," *R. S. P., 4* (1843), 450–3.

JH, "Progress of the [Great] Comet [of 1843]," London *Times* (March 23, 1843), p. 6, col. 2.

JH, "On the Reflecting Telescope of the Late Sir William Herschel," *Athenaeum,* #831 (Sept. 30, 1843), 884. Note: This is a reply to some remarks made by T. R. Robinson at the 1843 B. A. A. S. meeting as reported in the *Athenaeum,* #830 (Sept. 23, 1843), 866–7.

JH, "Reply to Dr. Robinson [on the Reflecting Telescopes of the Late Sir William Herschel]," *Athenaeum,* #836 (Nov. 4, 1843), 983–4. Note: For Robinson's comments, see *Athenaeum,* #834 (Oct. 21, 1843), 945–6 and for Robinson's reply to JH, see *Athenaeum,* #838 (Nov. 18,1843), 1027–8.

JH, "Report of the Committee . . . Appointed to Conduct the Co-operation of the British Association in the System of Simultaneous Observations in Terrestrial Magnetism and Meteorology," *B. A. Rp. for 1842* (1843), 1–11.

JH, "Report of the Committee for the Reduction of Lacaille's Stars," *B. A. Rp. for 1842* (1843), 205.

JH, "The Variability of η Argus," *As. S. M. Not., 6* (1843), 9.

1844

JH, "Contributions to Actino-Chemistry. On the Amphitype, a New Photographic Process," *Athenaeum,* #886 (Oct., 19, 1844), 954.

JH, "Further Remarks on the Revision of the Southern Constellations," *As. S. M. Not., 6* (1844), 60–2.

JH, "[Letter Commenting on Mr. Clerihew's Observations in India of the Bifurcation of the Tail of the Great Comet of 1843]," *As. Nr., 21* (1844), 199–202.

JH, "Notice of a Remarkable Photographic Process by Which Dormant Pictures Are Produced, Capable of Development by the Breath or by Keeping in a Moist Atmosphere," *B. A. Rp. for 1843* (1844), pt. 2, 8.

JH, "Observations on the Entrance Passages in the Pyramids of Ghizeh," *Ph. Mg.,* 3rd. ser., *24* (1844), 481–4.

JH, "Report of a Committee . . . Appointed to Conduct the Co-operation of the British Association in the System of Simultaneous Magnetical and Meteorological Observations," *B. A. Rp. for 1843* (1844), 54–60.

JH, "Report of the Committee Appointed for the Reduction of Meteorological Observations," *B. A. Rp. for 1843* (1844), 60–103.

JH, "Report of the Committee . . . for Revising the Nomenclature of the Stars," *B. A. Rp. for 1843* (1844), 292.

1845

JH, "'Αμόρφωτα, No. I. On a Case of Superficial Colour Presented by a Homogeneous Liquid Internally Colourless; No. II. On the Epipolic Dispersion of Light, Being a Supplement to [No. I]," *Phil. Trans.* (1845), 143–53; *A. C.*, 3rd. ser., *38* (1853), 378–81; *R. S. P.*, *5* (1843), 547.

JH, "Contributions to Actino-Chemistry. On the Amphitype, a New Photographic Process," *B. A. Rp. for 1844* (1845), 12–13.

Capensis [JH], "On Dibbling Seed," *Gardener's Chronicle and Agricultural Gazette* (March 22, 1845), 197.

JH, "[Letter on Mr. Houzeau's Theory on the Effect of Aberration Arising from Proper Motion in Changing the Apparent Orbits of Double Stars]," *As. Nr.*, *22* (1845), 249–54.

JH, "Report of a Committee . . . Appointed to Conduct the Co-operation of the British Association in the System of Simultaneous Magnetical and Meteorological Observations," *B. A. Rp. for 1844* (1845), pt. 1, 143–55.

JH, "Report of a Committee . . . for Revising the Nomenclature of the Stars," *B. A. Rp. for 1844* (1845), pt. 1, 32–42.

1846

JH, "[Letter on the] Electric Telegraph," London *Times* (Sept. 21, 1846), p. 6.

JH, "[Letter on] Le Verrier's Planet," *Athenaeum*, #988 (Oct. 3, 1846), 1091.

JH, "Memoir of Francis Baily, Esq.," *As. S. Mm.*, *15* (1846), 311–58; also in *As. S. M. Not.*, *6* (1844), 89–121; reprinted in *EEQR*, pp. 552–620 and in Francis Baily, *Journal of a Tour in Unsettled Parts of North America in 1796 & 1797* (London: Baily Brothers, 1856), pp. 1–69.

JH, "[Model for the Globe of the Moon]," *B. A. Rp. for 1845* (1846), pt. 2, 4–5.

JH, "[Presidential] Address," *B. A. Rp. for 1845* (1846), xxvii-xliv; reprinted in *EEQR*, pp. 634–82.

JH, "[Report on Observations of Biela's Double Comet, 28 Jan. 1846]," *As. Nr.*, *24* (1846), 19–20.

JH, "Seventh Report of the Committee . . . Appointed to Conduct the Cooperation of the British Association in the System of Simultaneous Magnetical and Meteorological Observations," *B. A. Rp. for 1845* (1846), 1–73.

1847

JH, "On the Diminution of Brightness of β Ursae Minoris," *As. S. M. Not.*, *7* (1847), 95–6.

JH, "Extract of a letter from Sir John Herschel," *As. S. M. Not.*, *7* (1847), 24.

JH, "[Obituary of] Friedrich Wilhelm Bessel," *As. S. M. Not.*, *7* (1847), 200–14.

JH, "Preface" to Nicolas Louis de Lacaille, *A Catalogue of 9766 Stars in the Southern Hemisphere, for the Beginning of the Year 1750, from the Observations of the Abbé de Lacaille, Made at the Cape of Good Hope in the Years 1751 and 1752. Reduced at the Expense of the British Association for the Advancement of Science under the Immediate Superintendence of the Late Professor Henderson* (London, 1847), pp. iii-xi.

JH, *Results of Astronomical Observations Made during the Years 1834, 5, 6, 7, 8 at the Cape of Good Hope; Being the Completion of a Telescopic Survey of the Whole Surface of the Visible Heavens, Commenced in 1825 by Sir J. F. W. Herschel* (London: Smith and Elder, 1847), xx + 452 pp.

[JH], "Separation of Phosphoric Acid from Urine," *Gardener's Chronicle and Agricultural Gazette* (April 10, 1847), 244.

JH, "Solar Spots," *As. S. M. Not.*, *8* (1847), 14–15.

JH, W. W., and J. C. H. [JH, William Whewell and J. C. Hare], *English Hexameter Translations from Schiller, Göthe, Homer, Callinus, and Meleager* (London, 1847), vii + 277. JH's translation of Schiller's "The Walk" (pp. 4–28) was his contribution to this volume.

1848

JH, "[Extract of a Letter on the Nomenclature of the Small Planets]," *As. Nr.*, *26* (1848), 254.

JH, "[Letter on a] Lunar Rainbow," *Athenaeum*, #1099 (Nov. 18, 1848), 1149–50.

JH, "[Letter on] The Planet Neptune," *Athenaeum*, #1100 (Nov. 25, 1848), 1176.

Geometer [JH], "[Letter on] A Problem in Perspective," *Athenaeum*, #1100 (Nov. 25, 1848), 1179.

JH, "[Obituary Notice of] Miss Caroline Lucretia Herschel," *Athenaeum*, #1065 (Jan. 22, 1848), 84.

JH, "Remarks on the Subject of the Government Free Schools, and Other Means of Promoting Education among All Classes of the Inhabitants of the Cape Colony. . ." in Charles J. F. Bunbury, *Journal of a Residence at the Cape of Good Hope; with Excursions into the Interior* (London, 1848), pp. 281–97.

[JH], "[Review of] *Kosmos*. . . [by] Alexander von Humboldt. Ersten Band. . . [and] *Cosmos*. . . . By Alexander von Humboldt, Vol. I. . .," *Edinburgh Review*, *87* (Jan., 1848), 170–229; reprinted in *EEQR*, pp. 257–364.

[JH], "[Synopsis of] *Results of Astronomical Observations . . . at the Cape of Good Hope. . .*," *Philosophical Magazine*, ser. 3, *32* (1848), 378–89; 518–26; ser. 3, *33* (1848), 145–54; 231–7; 296–309.

1849

JH, "An Address . . . [on] February 11, 1848, on the Subject of the Award of the Testimonials [to Messrs. Hansen, Hencke, Hind, Bishop, Lubbock, Leverrier, Adams, Argelander, Airy, Everest, and Weisse]," *As. S. Mm.*, *17* (1849), 162–85; same paper in *As. S. M. Not.*, *8* (1848), 102–19.

JH (ed.), *Admiralty Manual of Scientific Enquiry; Prepared for the Use of Her Majesty's Navy: and Adapted for Travellers in General* (London: John Murray, 1849), xi + 488 pp.; 2nd ed. in 1851; 3rd in 1859; 4th 1871; 5th in 1886. JH's publications in this volume are the preface (pp. v–ix), the article "Meteorology" (pp. 268–322), as well as the "Appendix" (pp. 11–13) to the article "Astronomy" and the "Appendix" (pp. 441–4) to the article "Ethnology." The last item was reprinted in *EEQR*, pp. 745–50.

JH, "On the Determination of the Most Probable Orbit of a Binary Star from the Assemblage of a Great Number of Observed Angles of Position," *As. S. Mm., 18* (1850), 47–68; summary in *As. S. M. Not., 9* (1849), 166–72.

JH, "Extrait d'une lettre à M. Le Verrier sur sa méthode pour la détermination de l'orbite des étoiles doubles," *Comptes rendus, 28* (1849), 421–2.

JH, *Outlines of Astronomy* (London: Longman, 1849), xiv + 661 pp.; 3rd. ed. appeared in 1850; 4th ed. in 1851; the 5th in 1858; the 6th in 1859; the 7th by 1864; the 8th in 1865; the 9th in 1867; the 10th in 1869; the 11th in 1871; and it continued to be published at least until 1902. Arabic and Chinese translations also appeared.

1850

JH, "An Address . . . [on] February 9, 1849, on Presenting the Honorary Medal to William Lassell, Esq. of Liverpool," *As. S. Mm., 18* (1850), 192–200; also in *As. S. M. Not., 9* (1849), 87–92; reprinted in *EEQR*, pp. 621–33.

JH, "On the Algebraic Expression of the Number of Partitions of Which a Given Number Is Susceptible," *Phil. Trans.* (1850), 399–422.

[JH], "In Cupidinem flentem," *Athenaeum*, #1162 (Feb. 2, 1850), 132.

JH, "[Letter], *Bedford Mercury* (May 18, 1850).

JH, "[Letter on de Gasparis's discovery of the asteroid Parthenope] New Planet," *Athenaeum*, #1179 (June 1, 1850), 589.

[JH], "[Letter on the Gregorian Reform of the Calendar]," *Athenaeum*, #1162 (Feb. 2, 1850), 132.

[JH], "[Review of] 1. *Lettres sur la Théorie des Probabilitités* . . . . Par M. A. Quetelet. . . . 2. *Letters . . . on the Theory of Probability*. . . . By M. A. Quetelet . . .," *Edinburgh Review, 92* (July, 1850), 1–57. Reprinted in *EEQR*, pp. 365–465.

1851

JH and others, *Suggestions to Astronomers for the Observation of the Total Eclipse of the Sun on July 28, 1851* (London: R. Taylor, 1851?), 14 pp.

1852

JH, "[Evidence on Examination on May 26, 1853 before the Select Committee of the House of Commons on the Decimal Coinage]" in *Report from the Select Committee on Decimal Coinage* (British Parliamentary Papers 1852–53, vol. 22), pp. 45–62 of report.

JH, "[Letter on] Photographic Landscapes on Paper," *Athenaeum*, #1311 (Dec. 11, 1852), 1363.

1853

JH, "[Letter on the] New Photographic Process," *Athenaeum*, #1341 (July 9, 1853), 83.

JH, "[Letter on] Photographic Pantograph," *Journal of the Photographic Society, 1:8* (Aug. 22, 1853), 101.

JH, "On the Substitution of Bromine for Iodine in Photographic Processes," *Journal of the Photographic Society, 1* (1853), 70.

1854

JH, "Anthotype: Photography with Vegetable Juices," *Humphrey's Journal of Photography, 6:5* (June 15, 1854), 80.

JH, "The Colouring Matter of Flowers," *Humphrey's Journal of Photography, 5:24* (April 1, 1854), 380–4.

1855

JH, "On the Application of Photography to Astronomical Observation," *Athenaeum*, #1435 (April 28, 1855), 493.

JH, "On the Application of Photography to Astronomical Observations," *As. S. M. Not., 15* (1855), 158–9.

1856

JH, "[Letter on the] Society of Arts Examinations," *Journal of the Society of Arts, 4* (April 19, 1856), 342.

JH, "Remarks on Slaty Cleavage, and the Contortions of Rocks," *Ph. Mg.*, 4th ser., *12* (1856), 197–8.

1857

JH, "Circular to Bankers [on the Circulation of the Florin]" and "Analysis of Replies," in *Preliminary Report of the Decimal Coinage Commissioners* (British Parliamentary Papers 1857, Session 2, vol. 19), pp. 125–6 of the report.

JH, *Decimal Association. Answers to the Questions Communicated by Lord Overstone to the Decimal Coinage Commissioners* (London: M. S. Rickerby, 1857), 38 pp.

JH, *Essays from the Edinburgh and Quarterly Reviews, with Addresses and Other Pieces* (London: Longman, 1857), 750 pp.; (reprinted New York: Arno, 1981).

JH, "Meteorology," *Encyclopaedia Britannica*, 8th ed., vol. 14 (1857), pp. 636–90.

JH, "[Replies to Lord Overstone's Queries on the Decimal Coinage]" in *Questions Communicated by Lord Overstone to the Decimal Coinage Commissioners, with Answers* (British Parliamentary Papers 1857–58, vol. 23), pp. 94–119 of the report.

Helioscopus [JH], "Solar Spots," London *Times* (April 6, 1857), p. 12, col. 4.

1858

JH, "On the Photographic Properties of Junonium—A New Metallic Element," *Photographic News, 1* (Oct. 29, 1858), 86.

JH, *Sensorial Vision: A Paper Read . . . before the Philosophical and Literary Society of Leeds* (Leeds: E. Baines, 1858), 16 pp.; reprinted in *FL*, pp. 400–18.

JH, "The Stereoscopic Angle," *Photographic News, 1* (Nov. 12, 1858), 110.

1859

JH, "The Action of the Solar Spectrum upon Certain Compounds of Silver," *Photographic News, 2* (July 22,1859), 229–30.

JH, "The Action of the Solar Spectrum upon Certain Compounds of Silver," *Photographic News, 3* (Sept. 9,1859), 2–3.

JH, " Address [to the Chemical Section] by Sir J. F. W. Herschel, Bart, President of the Section," *B. A. Rp. for 1858* (1859) 41–5.

JH, *Manual of Meteorology* (extracted from the *Admiralty Manual of Scientific Enquiry*, 3rd ed.), (London: W. Clowes,1859), 52 pp.

JH, "On a New Projection of the Sphere," *Royal Geographical Society Journal, 30* (1860), 100–6; also in

Proceedings of the Royal Geographical Society, 3 (1859), 174–7.

[JH], "[Obituary Notice of] The Late Very Rev. Dr. Peacock, Dean of Ely," *R. S. P., 9* (1858), 536–43.

JH, "Physical Geography," *Encyclopaedia Britannica,* 8th ed., vol. 17 (1859), pp. 569–647.

JH, "Remarks on Colour-Blindness," *R. S. P., 10* (1859–60), 72–84.

JH, "Report of the Joint Committee of the Royal Society and the British Association for Procuring a Continuance of the Magnetic and Meteorological Observatories," *B. A. Rp. for 1858* (1859), 295–305.

JH, "Uniform Musical Pitch," *Athenaeum,* #1654 (July 9, 1859), 56–7.

JH, "Uniform Musical Pitch," *Journal of the Society of Arts, 7* (July 8, 1859), 581.

1860

JH, "Instantaneous Photography," *Photographic News, 4* (May 11, 1860), 13.

JH, "[Letter on] British Modular Standard of Length," *Athenaeum,* #1696 (April 28, 1860), 581–2; reprinted in John Taylor, *The Great Pyramid. Why Was It Built? & Who Built It,* 2nd. ed. (London, 1864), pp. 39–43 and in *Journal of the Franklin Institute, 40* (1860), 47–51.

JH, "[Letter on the] Climate of Pekin," London *Times* (Dec. 14, 1860), p. 7, col. 5.

JH, "[Letter on] Standard of Measure," *Athenaeum,* #1697 (May 5, 1860), 617; reprinted in John Taylor, *The Great Pyramid. Why Was It Built? & Who Built It,* 2nd. ed. (London, 1864), p. 44.

Redde Suum Cuique [JH], "[Letter on] Obtaining Pure Silver from Old Plate—Actinometer," *Photographic News, 4* (1860), 118.

JH, "Sir John Herschel and the Weather," *Glossop Record* (Sept. 8, 1860).

JH, "Telescope," *Encyclopaedia Britannica,* 8th ed., vol. 21 (1860), 117–45.

1861

JH, "A Complete Catalogue of the Writings of Sir John Herschel," *Mathematical Monthly, 3* (1861), 220–7.

JH, "On the Forms of Lenses Suited for Destroying Spherical Aberration," *British Journal of Photography, 8* (Nov., 1, 1861), 381–5.

JH, "On the Formula Investigated by Dr. Brinkley for the General Term in the Development of Lagrange's Expression for the Summation of Series and for Successive Integration," *Royal Society Philosophical Transactions, 150* (1861), 319–21; see also *R. S. P., 10* (1859–60), 500–1.

JH, *Letter from Sir. J. F. W. Herschel, Bart., to Sir. J. W. Lubbock, Bart., on Shooting Stars* (London: Goodhall & Dinsdale, 1861), 4 pp.

JH, "[Letter on] The Comet," *London Review, 3* (July 13, 1861), 46–7.

JH, *Meteorology. From the Encyclopaedia Britannica* (Edinburgh: A. and C. Black, 1861), vii + 288 pp.; 2nd. ed. 1862.

JH, *Physical Geography: From the Encyclopaedia Britannica* (Edinburgh: A. and C. Black, 1861), vii + 441; 2nd ed., 1862; 4th ed., 1872; 5th ed., 1875.

JH, "Projection of the Sphere," *Educational Times, 14* (Oct., 1861), 149.

JH, "[Letter on] Rainbows," *London Review, 2* (May 18, 1861), 568–9.

JH, "[Replies to the Queries Circulated by the Royal Commission for Lights, Buoys, and Beacons]" in *Report of the Commissioners Appointed to Inquire into the Condition and Management of Lights, Buoys, and Beacons* (British Parliamentary Papers 1861, vol. 25), vol. 2, pp. 593–6.

JH, "Swing Cot for Sea Voyages," *Journal of the Society of Arts, 9* (Jan. 4, 1861), 97–8.

JH, *The Telescope* (Edinburgh: A. & C. Black, 1861), vii + 190 pp.

1862

JH, "Book I. of the Iliad, Translated in the Hexameter Verse," *Cornhill Magazine, 5* (May, 1862), 590–609; reprinted as JH, *Book I. of the Iliad Translated in the Hexameter Metre* (Privately printed, London, 1862), 22 pp.

JH, *Book II. of the Iliad Translated in the Hexameter Metre* (Privately printed, London, 1862), 23 pp.

[JH], "The Comet," *Cornhill Magazine, 6* (Oct., 1862), 550–1.

[JH], "Figure of the Moon and of the Earth," *Cornhill Magazine, 6* (Oct., 1862), 548–50.

JH, "The Last Scientific Balloon Ascent," *London Review, 5* (Sept. 20, 1862), 264.

JH, "[Letter to Thomas Parsons on] The Late Dr.[William] Herschel," *Stroud Journal,* June 7, 1862.

JH, "Letter from Sir John Herschel to Mr. Hind on the Disappearance of a Nebula in Coma Berenices," *As. S. M. Not., 22* (1862), 248–50.

JH, "[Letter on a Paper 'On the Present State of Meteorology,' by Mr. T. Hopkins]," *Manchester Philosophical Society Proceedings, 2* (1862), 215–7.

JH, "Noxious Vapours," London *Times* (May 15, 1862), p. 11, col. 4.

[JH], "On Shooting Stars," *Cornhill Magazine, 6* (Nov., 1862), 713.

JH, "Some Remarks Appended to a Report on Mr. Hopkins's Paper 'On the Theory of the Motion of Glaciers,'" *R. S. P., 12* (1862–3), 676–9; *Ph. Mg.,* 4th ser., *27* (1864), 539–41.

[JH], "Sun Spots and the Magnetic Needle," *Cornhill Magazine, 6* (Sept., 1862), 407.

[JH], "Velocity of Light," *Cornhill Magazine, 6* (Dec., 1862), 855–6.

1863

JH, "On Atoms," *Smithsonian Institution Annual Report 1862* (Washington, D.C., 1863), 413-15; reprinted in *Fortnightly Review, 1* (1865), 81–4 and in *FL,* pp. 452–9.

JH, "Combinaisons nouvelles de prismes," *Les mondes, 3* (1863), 403–5.

JH, "On Comets," *Good Words, 4* (1863), 476–82; 549–557; reprinted in *FL,* pp. 91–141.

JH, *An Essay Entitled the Yard, the Pendulum, & the Metre, Considered in Reference to the Choice of a Standard of Length* (written for the Leeds Astronomical Society), (London: Longman, 1863), 24 pp.; reprinted in *FL,* pp. 419–51.

JH, "On the History of Earthquakes and Volcanos [sic]," *Good Words, 4* (Feb., 1863), 141–8.

JH, "[Letter on] Lakes with Two Outlets," *Athenaeum,* #1881 (Nov. 14, 1863), 652.

[JH and G. H. Lewes], "Notes on Science," *Cornhill Magazine, 7* (Feb., 1863), 276–80.

[JH and G. H. Lewes], "Notes on Science," *Cornhill Magazine, 7* (March, 1863), 412–6.

JH, "The Sun," *Good Words, 4* (April, 1863), 273–84; reprinted in *FL,* pp. 47–90.

[JH], "Variability of Nebulae," *Cornhill Magazine, 7* (Jan., 1863), 43–4.

JH, "About Volcanoes and Earthquakes," *Good Words, 4* (Feb., 1863), 53–8; reprinted in *FL,* pp. 1–46.

1864

JH, "The Battle of the Standards," London *Times* (June 21, 1864), p. 7, col. 6.

JH, "The Battle of the Standards," London *Times* (July 4, 1864), p. 11, col. 2.

JH, "Celestial Measurings and Weighings," *Good Words, 5* (June, 1864), 489–500; reprinted in *FL,* pp. 176–218.

JH, "To the Editor of the Times," London *Times* (July 6, 1864), p. 7, col. 2.

JH, "[Letter on] The Supply and Preparation of Uranium," *Photographic News, 8* (Oct. 28, 1864), 520; reprinted in *American Journal of Photography and the Allied Arts and Sciences,* n.s., *7* (Dec. 1, 1864), 250–1.

JH, "[Revolving Eyepiece for Solar Observation]," *As. S. M. Not., 24* (1864), 220–1.

JH, "Science and Scripture" *Athenaeum,* #1925 (Sept. 17, 1864), 375.

JH, "Science and Scripture," London *Times* (Sept. 20, 1864), p. 7, col. 6.

JH, "Sir J. F. W. Herschel's Cyanotype Processes.—A Reclamation," *Photographic News, 8* (Feb. 12, 1864), 82.

JH, "On the Solar Spots," *Quarterly Journal of Science, 1* (1864), 219–35.

JH, "The Weather, and Weather Prophets," *Good Words, 5* (Jan., 1864), 57–64; reprinted in *FL,* pp. 142–75.

1865

JH, "Catalogue of Nebulae and Clusters of Stars," *Phil. Trans., 154* (1865), 1–137.

JH, "Sur les étoiles filantes," *Bulletin de l'académie royale des sciences, des lettres et des Beaux-arts de belgique,* 2nd ser., *16* (1865), 319–20.

JH, "Extract of a Letter . . . to Mr. De. La Rue [Respecting Mr. Nasmyth's 'Willow Leaves']," *As. S. M. Not., 25* (1865), 152–3.

JH, "A Letter from Sir J. F. W. Herschel, Bart., to the Editors of the Philosophical Magazine and Journal, in Reference to a Recent Communication to That Work of Dr. J. Davy, etc.," *Ph. Mg.,* 4th ser., *29* (1865), 246–7.

JH, "On Light," *Good Words, 6* (April-Dec., 1865), 320–6; 358–64; 496–502; 653–9; 815–22; 892–901; reprinted in *FL,* pp. 219–399.

JH, "Sur une météore extraordinaire, observé à Hurworth (comté de Durham), en octobre 1854," *Bulletin de l'académie royale des sciences, des lettres et des Beaux-arts de belgique,* 2nd ser., *16* (1865), 315–9.

JH, "Notes Accompanying Some Rhomboidal Specimens of Iron-Sandstone, &c., Presented to the Society," *Geological Society Quarterly Journal, 21* (1865), 421; *Ph. Mg.,* 4th ser., *29* (1865), 551; possibly in *Geological Society Proceedings,* no. 132 (1865).

JH, "On the Origin of Force," *Fortnightly Review, 1* (1865), 435–42; reprinted in *FL,* pp. 460–75.

JH, "Remarks on the Letter Published by Dr. J. Davy in the December Number of the Ph. Mg.," *Ph. Mg.,* 4th ser., *29* (1865), 77–8, 246–7.

1866

JH, "Chain Suspension Roofs," *Intellectual Observer, 8* (1866), 275–6.

JH, *Familiar Lectures on Scientific Subjects* (London: A. Strahan, 1866), xii + 507 pp.; numerous reprintings up to at least 1886; 1877 reprinting as *Popular Lectures on Scientific Subjects* (London).

JH (translator), *The Iliad of Homer, Translated into English Accentuated Hexameters* (London: Macmillan, 1866), vii + 560 pp.

JH, "[Letter on a Method of Cooling the Workings of Deep Coal Mines]," *Manchester Literary and Philosophical Society Proceedings, 5* (1866), 11–12.

JH, "[Letter headed] Note on the First Use of Hyposulphite of Soda in Photography," *British Journal of Photography, 13* (May 19, 1866), 236.

JH, "[Letter of Oct. 29, 1864 in Anonymous,] The Origin of the Fixing Process," *Photographic News, 10* (May 18, 1866), 229–31:230–1.

JH, "The Magic Photograph Patent.—Action of Light on Platinum," *Photographic News, 10* (May 25, 1866), 243–4.

JH, "Photography in Natural Colours, Etc." *Photographic News, 10* (Jan. 5, 1866), 5–6; *American Journal of Photography and the Allied Arts and Sciences,* n.s., *8* (Feb. 1, 1866), 342–5.

JH, "On a Supposed Observation of the New Variable near ε Coronae," *As. S. M. Not., 26* (1866), 299–300.

JH, "Sir J. Herschel's Cipher," *Photographic News, 10* (Jan. 19, 1866), 35. See also pp. 6, 23, and 47 for relevant materials.

1867

JH, "Etoiles filantes de novembre," *Moniteur universel* (Feb. 4, 1867), p. 105, col. 1.

JH, "The Invention of the Thaumsacope [Thaumatrope]," *Photographic News, 11* (Nov. 8, 1867), 541–2.

JH, "[Letter on Barometric Waves]," *Manchester Literary and Philosophical Society Proceedings, 6* (1867), 91–3.

JH, "[Letter to Mr. Dallmeyer of Sept. 30, 1866 on Flare in Photography]," *Photographic News, 11* (June 21, 1867), 295–6.

JH, "On the Meteoric Shower of 1866, Nov. 13–14," *As. S. M. Not. 27* (1867), 19–21; *Ph. Mg.,* 4th ser., *33* (1867), 156–7.

JH, "Meteoric Shower of Nov. 13–14, 1866," *Proceedings of the British Meteorological Society, 3* (1867), 273–5.

JH, "Notice Explanatory of a Series of MS. Charts, Containing the Estimated Magnitudes of Stars Visible to the Naked Eye in Both Hemispheres, Presented by Him to the Royal Astronomical Society," *As. S. M. Not., 27* (1867), 213–16.

JH, *Schillers Spaziergang, Translated into Latin Verse* (London: R. Barrett, 1867), 8 pp.

JH, "A Synopsis of All Sir William Herschel's Micrometrical Measurements and Estimated Positions and Distances of the Double Stars Described by Him, Together with a Catalogue of Those Stars in Order of Right Ascension for the Epoch 1800.0, So Far As They Are Capable of Identification," *As. S. Mm., 35* (1867), 21–136.

JH, "Le vent et le baromètre," *Revue Maritime et coloniale, 19* (1867), 927–8.

1868

JH, "[Addendum dated May 11, 1852 to] Memorandum [dated May 5, 1852] by the Master of the Mint [Relating to the Expense of Gold Coinage]," *Report from the Royal Commission on International Coinage; Together with the Minutes of Evidence and Appendix* (London, 1868), Appendix: pp. 326–7.

JH, "Additional Identifications of Double Stars in the Synoptic Catalogues of Sir William Herschel's Micrometrical Measurements &c, with a List of Errata," *As. S. M. Not., 28* (1868), 151–4.

JH, "Hyposulphite of Ammonia," *Photographic News, 12* (Jan. 31, 1868), 57.

JH, "L'Inferno of Dante, Canto I," *Cornhill Magazine, 18* (July, 1868), 38–42.

JH, "[Letter dated April 30, 1852 from] The Master of the Mint to the Treasury [Relating to the Expense of Gold Coinage]," *Report from the Royal Commission on International Coinage; Together with the Minutes of Evidence and Appendix* (London, 1868), Appendix: pp. 324–5.

JH, "Letter [on a Statement in Mr. Dyer's Paper 'On the Mutations of Imponderable Elements]," *Manchester Literary and Philosophical Society Proceedings, 7* (1868), 193.

JH, "Memorandum [dated May 5, 1852] by the Master of the Mint [Relating to the Expense of Gold Coinage]," *Report from the Royal Commission on International Coinage; Together with the Minutes of Evidence and Appendix* (London, 1868), Appendix: pp. 325–6.

JH, "On Musical Scales," *Quarterly Journal of Science, 5* (July, 1868), 338–352.

[JH], "[Obituary Notice of] William Whewell," *R. S. P., 16* (1868), li-lxi.

JH, "To the Rainbow" (a poem), *People's Magazine,* new ser., *1* (Feb. 1, 1868), 91.

JH, "Sir John Herschel and British Imperial Measures," *Daily News* (May 27,1868), p. 6, col. 4.

JH, "Some Remarks on Mr. Arbuthnot's Observations of May 17, 1852, on Sir J. Herschel's Statement, by the Master of the Mint," *Report from the Royal Commission on International Coinage; Together with the Minutes of Evidence and Appendix* (London, 1868), Appendix: pp. 328–9.

JH, "On Variable Star η Argûs and Its Surrounding Nebula," *As. S. M. Not., 28* (1868), 225–9.

JH, "Vegetable Autographs," *People's Magazine,* new ser., *1* (Jan., 1868), 62–3.

JH, "Weights and Measures," London *Times* (May 2, 1868), p. 12, col. 6.

1869

JH, "The Channel Passage," *Journal of the Society of Arts, 17* (Nov. 5, 1869), 907.

JH, "The Channel Passage," *Journal of the Society of Arts, 17* (Nov. 12, 1869), 917.

JH, "The Coinage," London *Times* (Sept. 9, 1869), p. 4, col. 6.

Helioscopus [JH], "To the Editor of the Times [on Solar Spots]," London *Times* (Dec. 17, 1869), p. 7, col. 6.

JH, "Introduction" to Adolphe Quetelet, *Physique sociale: Essai sur le developpement des facultés de l'homme,* 2 vols., vol I (Bruxelles, 1869), pp. 1–89. This is a translation of JH's review of Quetelet on probabilities in the *Edinburgh Review* of 1850.

JH, "[Letter on] Gold Coinage," London *Times* (Sept. 2, 1869), p. 10, col. 4–5.

JH, "[Letter on James Grahame's writings on] American History," *Athenaeum,* #2190 (Oct. 16, 1869), 498.

JH, "On a Possible Method of Viewing the Red Flames without an Eclipse" *As. S. M. Not., 29* (1869), 5–6.

JH and John Herschel [son], "The Great Nebula round η Argûs" *As. S. M. Not., 29* (1869), 82–8.

1870

JH, "[Letter of September 2, 1869 to the London *Times* on] Gold Coinage," *Speeches, Letters, Articles, &c. on the Gold Coinage Controversy of 1869* (London: Bank of England, 1870), pp. 116–23.

JH "Half a Dozen Propositions Respecting the Gold Coinage," *The Gold Coinage Controversy, 1969* (London: Bank of England, 1870), pp. 308–10.

H., J. F. W. [JH], "Our Silver Coinage," London *Times* (Feb. 12, 1870), p. 10, col. 5.

1871

JH, "The Gaussian Constants of Terrestrial Magnetism," *Nature, 3* (Jan. 26, 1871), 249.

JH, "Remarks on Mr. Abbott's . . . Paper on η Argûs," *As. S. M. Not., 31* (1871), 228–30.

JH, "Seventh Catalogue of Double Stars, Observed at Slough, in the Years 1823–1828, Inclusive, with the 20–feet Reflector; 84 of Which Have Not Been Previously Described," *As. S. Mm., 38* (1871), 1–16.

JH, "Sir John Herschel on Ocean Currents," *Nature, 4* (May 25, 1871), 71.

1872

JH, "Queries Relative to Double Stars," *As. S. M. Not., 32* (1872), 29–31.

1874

JH, "A Catalogue of 10,300 Multiple and Double Stars Arranged in the Order of Right Ascension," ed. by Rev. R. Main and Rev. C. Pritchard, *As. S. Mm., 40* (1874), i-xii + 144.

1887

JH, "[Correspondence with Richard Proctor]" in Proctor's *Other Suns than Ours* (London, 1887), pp. 393–418.

1938

JH, *Poems & Pastimes* (Cambridge, England: privately printed), 28 pp.

1994

JH, "Address to the Young Gentlemen of the Mathematical Classes in the South African College at the Cape of Good Hope—Dec. 1834" in Brian Warner (ed.), *John Herschel 1792–1992: Bicentennial Symposium* (Royal Society of South Africa, 1994), pp. 121–24.

APPENDIX II

BIBLIOGRAPHY OF WORKS CONTAINING PRINTED HERSCHEL CORRESPONDENCE

Note: Included in this list are publications that contain printed letters to or from John Herschel. This list is far from complete. Moreover, time and funding restrictions have necessitated that less time be given to the search for printed versions of Herschel letters than we might have wished. Letters written by John Herschel and published during his life, for example, in scientific or popular journals or in newspapers are not included in this list because they have been referenced in the bibliography of publications by Herschel.

Babbage, Charles, *The Ninth Bridgewater Treatise: A Fragment* (London: John Murray, 1837), Appendix, Note I.

Black, R. D. Collison (ed.), *Papers and Correspondence of William Stanley Jevons*, vol. 2 (Clifton, New Jersey: Augustus M. Kelley, 1973).

Black, R. D. Collison (ed.), *Papers and Correspondence of William Stanley Jevons*, vol. 3 (London: Macmillan, 1977).

Brock, W. H., "The Selection of the Authors of the Bridgewater Treatises," *Notes and Records of the Royal Society, 21* (1966), 162–79.

Buckland, Gail, *Fox Talbot and the Invention of Photography* (London: Scholar Press, 1980).

Bunbury, Charles J. F., *Journal of a Residence at the Cape of Good Hope; with Excursions into the Interior* (London, 1848).

Cannon, Walter F., "The Impact of Uniformitarianism: Two Letters from John Herschel to Charles Lyell, 1836–1837," *Proceedings of the American Philosophical Society, 105* (1961), 301–14.

Chant, C. A., "Two Letters of a Famous Astronomer," *Royal Astronomical Society of Canada Journal, 38* (1944), 225–7.

Clark, John Willis, *The Life and Letters of the Reverend Adam Sedgwick*, vol. 2 (Cambridge, England, 1890).

Colebrooke, T. E., *Life of H. T. Colebrooke* (London, 1873).

Darwin, Charles, *The Correspondence of Charles Darwin* (Cambridge: Cambridge University Press, 1984–).

De Morgan, Sophie Elizabeth, *Memoir of Augustus De Mogran* (London: Longmans, Green, and Co., 1882].

Evans, David S., "Sir John Herschel to Captain Smyth," *Quarterly Bulletin of the South African Library, 25* (September, 1970), 3–17.

Evans, David S., "The Astronomical Work of Sir John Herschel at the Cape," *South African Public Library Quarterly Bulletin, 12:2* (December, 1957), 44–58.

Evans, David S., Terence J. Deeming, Betty Hall Evans, and Stephen Goldfarb (eds.), *Herschel at the Cape: Diaries and Correspondence of Sir John Herschel, 1834–1838* (Austin, Texas: Univ. of Texas Press, 1969).

Ferguson, W. T., and R. F. M. Immelman, *Sir John Herschel and Education at the Cape, 1834–1840* (Cape Town: Oxford University Press, 1961).

Ford, Colin, *The Cameron Collection: An Album of Photographs by Julia Margaret Cameron, Presented to Sir John Herschel* (London: Van Nostrand Reinhold, 1975).

Gernsheim, Helmut, "Talbot's and Herschel's Photographic Experiments in 1839," *Image, 8:3* (Sept., 1959), 132–7.

Graves, Robert Perceval, *Life of Sir William Rowan Hamilton,* vol. 1 (Dublin: Hodges, Figgis, & Co., 1882); vol. 2 (Dublin: Hodges, Figgis, & Co., 1885); vol. 3 (Dublin: Hodges, Figgis, & Co., 1889).

Harding, M. C. (ed.), *Correspondance de H. C. Örsted avec divers Savants*, vol. 2 (Copenhagen, 1920).

Herschel, John, "Extraits de deux lettres addressées du Cap de Bonne-Espérance,"*Bibliothèque universelle des sciences, belles-lettres, et arts, 14* (1838), 116–29.

Herschel, Mrs. John [Mary Cornwallis], *Memoir and Correspondence of Caroline Herschel,* 2nd ed. (London: John Murray, 1879).

Hoskin, Michael, "Astronomical Correspondence of William Rowan Hamilton," *Journal for the History of Astronomy, 15* (1984), 69–73.

Hyman, Anthony, *Charles Babbage: Pioneer of the Computer* (Oxford: Oxford Univ. Press, 1984).

Irons, James Campbell, *Autobiographical Sketch of James Croll* (London: Edward Stanford, 1896).

James, Frank A. J. L. (ed.), *The Correspondence of Michael Faraday,* Vol. 1, (London: Institution of Electrical Engineers, 1991).

James, Frank A. J. L. , "Between Two Scientific Generations: John Herschel's Rejection of the Conservation of Energy in His 1864 Correspondence with William Thomson," *Notes and Records of the Royal Society, 40* (1985), 53–62.

Larmor, Joseph (ed.), *Memoir and Scientific Correspondence of the Late Sir George Gabriel Stokes*, 2 vols. (Cambridge, England: Cambridge University Press, 1907).

Lubbock, Constance, *The Herschel Chronicle* (Cambridge: Cambridge Univ. Press, 1933).

Lyell, Katherine M. (ed.), *Life, Letters, and Journals of Sir Charles Lyell, Bart.* 2 vols. (London, 1881).

Maclear, Thomas, "Sir John Herschel at the Cape," *Monthly Magazine*, new ser., *3* (Sept., 1871), 128–40.

McLennan, John F., *Memoir of Thomas Drummond* (Edinburgh, 1867).

Moore, Sarah, "A Newly-Discovered Letter of J. F. W. Herschel Concerning the Plumian Professorship," *Journal for the History of Astronomy, 25* (1994), 142–3.

Morrell, Jack, and Arnold Thackray (eds.), *Gentlemen of Science: Early Correspondence of the British Association for the Advancement of Science*. Camden Fourth Series, Vol. 30. London: Royal Historical Society, 1984.

O'Brien, R. Barry, *Thomas Drummond, Under-Secretary in Ireland, 1835–40: Life and Letters* (London: K. Paul Trench, 1889).

Pritchard, Ada, *Charles Pritchard: Memoirs of His Life* (London: Seeley and Co., 1897).

Proctor, Richard, *Other Suns than Ours* (London, 1887).

Quetelet, L. A. J., "Notice sur John Fred. Will. Herschel," *Académie royale des sciences, des lettres et des beaux-arts de Belgique. Annuaire,* (1872), 161–99.

Return to the House of Commons (Feb. 17, 1852).

Ross, Sydney, "Unpublished Letters of Faraday and Others to Edward Daniel Clarke," *Bulletin for the History of Chemistry 11* (1991), 79–86.

Schultze, R. S., "Re-discovery and Description of Original Material on the Photographic Researches of Sir John F. W. Herschel, 1839–1844," *Journal of Photographic Science, 13* (1965), 57–68.

Shairp, J. C., P. G. Tait, and A. Adams-Reilly, *Life and Letters of James David Forbes, F.R.S.* (London, 1873).

Somerville, Martha, *Personal Recollections from Early Life to Old Age of Mary Somerville* (London, 1874).

Speeches, Letters, Articles, &c on the Gold Coinage Controversy of 1869 (London: Bank of England, 1870).

Todhunter, Isaac, *William Whewell, D.D., Master of Trinity College: An Account of His Writings,* 2 vols. (London: Macmillan, 1876).

Warner, Brian, *Charles Piazzi Smyth: Astronomer-Artist, His Cape Years 1835–1845* (Cape Town: A. A. Balkema, 1983).

Warner, Brian and Nancy (eds.), *Maclear and Herschel: Letters and Diaries at the Cape of Good Hope, 1834–1838* (Rotterdam: A. A. Balkema, 1984).

Warner, Thomas. "Preface to Improved Edition," *How to Keep the Clock Right*, Improved Edition (1876).

Williams, L. Pearce, *The Selected Correspondence of Michael Faraday,* 2 vols. (London: Cambridge Univ. Press, 1971).

Wood, R. Derek, "Latent Developments from Gallic Acid, 1839," *Journal of Photographic Science, 28:1* (1980), 36–42.

APPENDIX III

BIBLIOGRAPHY OF PUBLICATIONS ON JOHN HERSCHEL

Achinstein, Peter, "Light Hypotheses," *Studies in History and Philosophy of Science, 18* (1987), 293–337.

Agassi, Joseph, "Sir John Herschel's Philosophy of Success," *Historical Studies in the Physical Sciences, 1* (1969), 1–36.

Airy, George Biddell, "Address on Presenting the Honorary Medal to Sir J. F. W. Herschel," *Royal Astronomical Society Memoirs, 9* (1836), 303–12.

Aitken, Robert Grant, *The Binary Stars*, 2nd ed. (London, 1935).

Alborn, Timothy L., "The Business of Induction: Industry and Genius in the Language of British Scientific Reform, 1820–1840," History of Science, 34 (1996), 91–121.

Alborn, Timothy, "The 'End of Natural Philosophy' Revisited: Varieties of Scientific Discovery," *Nuncius: Annali di Storia della Scienza, 3(2)* (1988), 227–50.

Anonymous, "The Cape Sketches of Sir John Herschel," *South African Public Library Quarterly Bulletin, 7* (1952), 37–8.

Anonymous, *Catalogue of Valuable Printed Books, Autograph Letters, Historical Documents, etc. [Including] The Library, Papers & Scientific Apparatus of Sir William, Caroline & Sir John Herschel* (London: Sotheby and Co., 1958).

Anonymous, "British Association for the Advancement of Science [Meeting in 1845 at Cambridge]," *Illustrated London News, 6*, no. 165 (June 28, 1845), 405. With portrait of John Herschel.

Anonymous, [Comment on Death of John Herschel], *Nature, 4* (May 1871), 89.

Anonymous, [Death Notice of John Herschel], *American Journal of Science, 110* (1871), 479.

Anonymous, [Death Notice of John Herschel], *Nature, 4* (May 1871), 50.

Anonymous, "Funeral of Sir John Herschel," *Astronomical Register, 9* (1872), 140–1.

Anonymous, "Herschel's 'Nebulous Doubles' Revisited," *Sky and Telescope, 87* (June, 1994), 15.

Anonymous, "The Herschel Dinner," *Athenaeum*, #555 (June 16, 1838), 423–7.

Anonymous, "The Late Sir John Herschel," *Astronomical Register, 9* (1872), 138–40.

Anonymous, "The Late Sir John Herschel," *Every Saturday* (Boston), *10* (June 24, 1871), 593.

Anonymous, "The Late Sir John Herschel," *Illustrated London News, 58*, no. 1652 (May 27, 1871), 511–2. With portrait.

Anonymous, "Natural Philosophy," *Fraser's Magazine, 3* (1831), 697–702.

Anonymous, "Note on the Camera Lucida," *South African Public Library Quarterly Bulletin, 12:2* (December, 1957), 72–81 with plates.

Anonymous, [Obituary], *American Academy of Arts and Sciences Proceeding, 8* (1872), 461–71.

Anonymous, [Obituary], *Nature, 4* (1871), 60.

Anonymous, [Obituary], London *Times* (May 13, 1871), 5.

Anonymous, "Obituary Notice," *Photographic World, 1*, no. 7 (July, 1871), 201–2.

Anonymous, "[Review of] *An Elementary Treatise on the Differential and Integral Calculus.* By S. F. Lacroix. Translated from the French [by Charles Babbage, John Herschel, and George Peacock," *Monthly Review, 87* (1818), 179–85.

Anonymous, "[Review of] *Outlines of Astronomy.* By Sir John W Herschel ...," *Eclectic Review, 90* (1849), 576–98.

Anonymous, "[Review of] *Outlines of Astronomy,*" *Eclectic Review, 47* (1959), 33–9; reprinted from Westminster Review.

Anonymous, "[Review of] *Outlines of Astronomy.* By Sir John W Herschel ...," *Philosophical Magazine, 34* (1849), 452–5.

Anonymous, "[Review of] *A Preliminary Discourse on the Study of Natural Philosophy.* By J. F. W. Herschel," *Athenaeum*, #167 (Jan. 8, 1831), 38–9.

Anonymous, "[Review of] *A Preliminary Discourse on the Study of Natural Philosophy.* By J. F. W. Herschel," *Monthly Review*, n. s., *1* (1831), 286–93.

Anonymous, "[Review of JH's *Preliminary Discourse*]," *Revue encyclopédique ou analyse raisonée, 49* (1831), 369–70.

Anonymous, "[Review of JH's] *Results of Astronomical Observations ... at the Cape of Good Hope*," *Athenaeum*, #1034 (Aug. 21, 1847), 885.

Anonymous, "Science [including a review of the 5th edition of John Herschel's *Outlines of Astronomy*]," *Westminster Review*, n. s., *15* (1859), 184–91.

Anonymous, "Scientific Gossip," *Athenaeum*, 1155 (Dec. 15, 1849), 1278.

Anonymous, "Sir J. F. W. Herschel, Bart., K. H., etc.," *Living Age, 61* (1859), 515–22.

Anonymous, "Sir J. F. W. Herschel, Bart.," *Leisure Hour, 14* (Oct. 7, 1865), 631–5.

Anonymous, "Sir John Herschel," *Appleton's Journal of Literature, Science, and Art, 6* (Aug. 5, 1871), 141–2.

Anonymous, "Sir John Herschel," *Saturday Review* (May 20, 1871), 624–5.

Anonymous, "Sir John Herschel and Modern Astronomy," *Quarterly Journal of Science, 5* (1868), 186–96.

Anonymous, "Sir John Herschel and the New Test," *Saturday Review of Politics, Literature, Science, and Art, 18* (Sept. 24, 1864), 386–7.

Anonymous, "[On the Slab Placed above the Grave of John Herschel in Westminster Abbey]," *Nature, 5* (Feb. 1, 1872), 272.

Anonymous, *The Story of the Herschels, a Family of Astronomers* (London, n. d. but probably ca. 1875). (This book was written for young people)

Anonymous, "Study of Natural Philosophy," *Quarterly Journal of Education, 3* (1832), 60–73:69–73. This is a review of JH's *Preliminary Discourse*.

Anonymous, "The Will of the Late Sir William Herschel," *Gentleman's Magazine, 92* (1822), 650.

Arnold, Harry John Phillip, *William Henry Fox Talbot: Pioneer of Photography and Man of Science* (London, 1977).

Ashbrook, Joseph, "John Herschel's Expedition to South Africa" in Ashbrook's *The Astronomical Scrapbook: Skywatchers, Pioneers, and Seekers in Astronomy*, ed. by L. J. Robinson (Cambridge, Mass.: Sky Publishing Co., 1984), pp. 37–41.

Ashworth, William J., "Memory, Efficiency, and Symbolic Analysis: Charles Babbage, John Herschel, and the Industrial Mind," *Isis, 87* (1996), 629–653.

Ashworth, William J., "The Calculating Eye: Baily, Herschel, Babbage and the Business of Astronomy,"

British Journal for the History of Science, 27 (1994), 409–41.

Augustus Frederick, Duke of Sussex, "[Presentation of Royal Society's Royal Medal to John Herschel]," *Royal Society Proceedings, 3* (1830–37), 222–4.

Augustus Frederick, Duke of Sussex, "[Presentation of Royal Society's Royal Medal to John Herschel]," *Royal Society Proceedings, 3* (1830–37), 443–4.

Ball, Robert S., "John Herschel" in Ball's *Great Astronomers* (London, 1895), pp. 246–71.

Ball, Robert S., "The Great Astronomers. V.-John Herschel," *Good Words, 35* (1894), 835–43; also in *Eclectic Magazine, 124* (1894), 205–.

Barnard, Keppel H., "Sir John Herschel and the S. A. Literary and Scientific Institution," *South African Public Library Quarterly Bulletin, 9* (1954), 10–11.

Barr, E. Scott, "Sir John Frederick William Herschel (1792–1871)," *Optics and Photonic News, 1* (May, 1990), 23–9.

Bartholomew, C. F., "The Discovery of the Solar Granulation," *Royal Astronomical Society Quarterly Journal, 17* (1976), 263–89.

Becher, Harvey, "Radicals, Whigs and Conservatives: The Middle and Lower Classes in the Analytical Revolution at Cambridge in the Age of Aristocracy," *British Journal for the History of Science, 28* (1994), 405–26.

Becher, Harvey, "William Whewell and Cambridge Mathematics," Historical *Studies in the Physical Sciences, 11* (1980), 1–48.

Beck, Daniel A., "Life on the Moon: A Short History of the Hansen Hypothesis," *Annals of Science, 41* (1984), 463–70.

Bennett, James, "Catalogue of the Archives and Manuscripts of the Royal Astronomical Society," *Royal Astronomical Society Memoirs, 85* (1978), 1–90. Contains a directory to the Herschel Archive, the materials by Caroline, William, and John Herschel held by the Royal Astronomical Society.

Bennett, J. A., "'On the Power of Penetrating into Space': The Telescopes of William Herschel," *Journal for the History of Astronomy, 7* (1976), 75–108.

Blackwell, Richard J., "The Inductivist Model of Science: A Study in Nineteenth Century Philosophy of Science," *Modern Schoolman, 51* (1974), 197–212.

Bolt, Marvin, *John Herschel's Scientific Method in Context: Natural Knowledge in Early Ninetennth–Century Britain* (a doctoral dissertation at the Univ. of Notre Dame; nearing completion).

[Brewster, David], "Drawings by the Agency of Light," *Edinburgh Review, 76* (1843), 275–344.

[Brewster, David], "Recent History of Astronomy [Review of Herschel and South on Double Stars]," *Quarterly Review, 38* (1828), 1–15.

[Brewster, David], "[Review of] *Results of Astronomical Observations ... at the Cape of Good Hope....* By Sir John Herschel," *North British Review, 8* (Feb., 1848), 491–533; reprinted in *Littell's Living Age, 16* (1848), 577–96.

[Brewster, David], "[Review of] *A Treatise on Sound.* By J. F. W. Herschell [sic]," *Quarterly Review, 44* (1831), 475–511.

Brock, W. H., "The Selection of the Authors of the Bridgewater Treatises," *Notes and Records of the Royal Society, 21* (1966), 162–79.

Brock, W. H., and R. M. MacLeod, "The Scientists' Declaration: Reflexions on Science and Belief in the Wake of Essays and Reviews," *British Journal for the History of Science, 9* (1976), 39–66.

Buchdahl, Gerd, "[Review of] John Herschel's *A Preliminary Discourse on the Study of Natural Philosophy,*" *British Journal for the History of Science, 4* (1968), 173–5.

Buckland, Gail, *Fox Talbot and the Invention of Photography* (London: Scholar Press, 1980).

Buttmann, Günther, "John Frederick William Herschel: Aum 200. Geburtstag des Astronomen und Physikers," *Sterne und Weltraume*, 31 Jahrg. (Juli, 1992), 452–3.

Buttmann, Günther, *The Shadow of the Telescope: A Biography of John Herschel*, trans. by B. E. J. Pagel (New York: Charles Scribner's Sons, 1970). Translation of Buttmann's *John Herschel: Lebensbild eines Naturforschers* (Stuttgart, 1965).

Calvert, H. R., "The Presentation of Caroline Herschel's Telescope to the Society by Sir John Herschel," *Royal Astronomical Society Quarterly Journal, 11* (1970), 105–8.

Cannon, Susan Faye, *Science in Culture: The Early Victorian Period* (New York: Science History Publications, 1978).

Cannon, Walter Faye, "Charles Lyell, Radical Actualism, and Theory," *British Journal of the History of Science, 9* (1976), 104–20.

Cannon, Walter Faye, "John Herschel," *Encyclopedia of Philosophy*, ed. by Paul Edwards, vol. 3 (New York, 1967), 490–1.

Cannon, Walter Faye, "John Herschel and the Idea of Science," *Journal of the History of Ideas, 22* (1961), 215–39.

Cannon, Walter Faye, "The Impact of Cultural Uniformitarianism. Two Letters from John Herschel to Charles Lyell, 1836–1837," *American Philosophical Society Proceedings, 105* (June 1961), 301–14.

Cannon, Walter Faye, "The Problem of Miracles in the 1830's," *Victorian Studies, 4* (1960), 4–32.

[Carlyle, J. A.], "Natural Philosophy [A Review of John Herschel's *Preliminary Discourse on the Study of Natural Philosophy*], *Fraser's Magazine, 3* (July, 1831), 698–702.

Casini, Paolo, "Herschel, Whewell, Stuart Mill e l''analogia della natura,'" *Rivista di Filosofia, 72* (1981), 372–91.

Cawood, John, "The Magnetic Crusade: Science and Politics in Early Victorian Britain," *Isis, 70* (1979), 493–518.

Cawood, John, "Terrestrial Magnetism and the Development of International Collaboration in the Early Nineteenth Century," *Annals of Science, 34* (1977), 551–87.

Chambers, G. F., "Sir John Herschel's Drawings of Nebulae," *Astronomical Register, 4* (1866), 220.

Chant, C. A., "Two Letters of a Famous Astronomer," *Royal Astronomical Society of Canada Journal, 38* (1944), 225–7.

Chapman, Allan, "An Occupation for an Independent Gentleman: Astronomy in the Life of John Herschel," *Vistas in Astronomy, 36* (1993), 71–116.

Charpa, Ulrich, "John F. W. Herschels Methodologie der Erfahrungswissenschaft," *Philosophia naturalis, 24* (1987), 121–48.

Chen, Xiang, *Theories, Experiments, and Human Agents: The Controversy between Emissionists and Undulationistss in Britain, 1827–1859* (A 1992 doctoral disseration at Virginia Polytechnic Institute and State University).

C[lerke], A[gnes] M., "Herschel. Sir John Frederick William," *Dictionary of National Biography,* vol. IX (London, 1922), pp. 714–9.

Clerke, Agnes M., *The Herschels and Modern Astronomy* (London, 1901); see esp. 142–219.

Cockrell, Peter R., "The Meteorological Society of London," *Weather, 23* (1968), 357–61.

Colebrooke, T. E., *Life of H. T. Colebrooke* (London, 1873).

Craig, John, *The Mint: A History of the London Mint from A.D. 287 to 1948* (Cambridge, England, 1953), pp. 317–23.

Crossley, Edward, Joseph Gledhill, and James M. Wilson, *A Handbook of Double Stars* (London, 1879).

Crowe, Michael J., "Alexander Herschel," in *History of Astronomy: An Encyclopedia,* ed. by John Lankford (New York: Garland), 255.

Crowe, Michael J., "Bibliography of the Publications of Sir John F. W. Herschel" in Brian Warner (ed.), *John Herschel 1792–1992: Bicentennial Symposium* (Royal Society of South Africa, 1994), pp. 125–140.

Crowe, Michael J., "Introduction," in M. J. Crowe (ed.), *The Letters and Papers of Sir John Herschel: A Guide to the Manuscripts and Microfilm* (in the series: Collections from the Royal Society), (Bethesda, Maryland: University Publications of America, 1991), pp. v–xxvii..

Crowe, Michael J., "John F. W. Herschel," in *History of Astronomy: An Encyclopedia,* ed. by John Lankford (New York: Garland), 256–7.

Crowe, Michael J., "John Herschel: Britain's First Modern Physical Scientist" in Brian Warner (ed.), *John Herschel 1792–1992: Bicentennial Symposium* (Royal Society of South Africa, 1994), pp. 57–69.

Crowe, Michael J., "The John Herschel Correspondence Project" in Brian Warner (ed.), *John Herschel 1792–1992: Bicentennial Symposium* (Royal Society of South Africa, 1994), pp. 117–19.

Crowe, Michael J., (ed.), *The Letters and Papers of Sir John Herschel: A Guide to the Manuscripts and Microfilm* (in the series: Collections from the Royal Society), (Bethesda, Maryland: University Publications of America, 1991).

Crowe, Michael J., *Modern Theories of the Universe from Herschel to Hubble* (New York: Dover, 1994).

Crowe, Michael J., *The Extraterrestrial Life Debate 1750–1900: The Idea of a Plurality of Worlds from Kant to Lowell* (Cambridge: Cambridge Univ. Press, 1986).

Crowe, Michael J., "Sir John Herschel and the Leeds Astronomical Society," *British Astronomical Association Journal,* in press (1997).

D., G. M., "Sir William Herschel," *English Mechanic, 13* (June 2, 1871), 263.

Day, Archibald, *The Admiralty Hydrographic Service* (London, 1967).

Deacon, Margaret, *Scientists and the Sea 1650–1900: A Study of Marine Science* (London, 1971).

De Morgan, Sophia, *Memoir of Augustus De Morgan* (London, 1882).

Dodge, N. S., "Memoir of Sir John F. W. Herschel," *Annual Report of the Smithsonian Institution for 1871* (Washington, D. C., 1872), 109–35.

Dreyer, J. L. E. and H. H. Turner (eds.), *History of the Royal Astronomical Society 1820–1920* (London, 1923).

Dubbey, J. M., "The Introduction of the Differential Notation to Great Britain," *Annals of Science, 19* (1963), 37–48.

Dubbey, J. M., *The Mathematical Work of Charles Babbage* (Cambridge, England, 1978).

Ducasse, C. J., "John Herschel's Methods of Experimental Inquiry" in *Theories of Scientific Method: The Renaissance through the Nineteenth Century,* ed. by Edward H. Madden (Seattle, 1966), pp. 153–82. Reprint of Ducasse's "John Herschel's Philosophy of Science," *Studies in the History of Culture* (Menash, Wis.: American Council of Learned Societies, 1942), pp. 279–309.

Dunkin, Edwin,"Sir John F. W. Herschel" in *Obituary Notices of Astronomers* (Edinburgh, 1879), pp. 47–85.

Dyer, G. P., "'One of the best men of business': Master of the Royal Mint" in *John Herschel 1792–1871: A Bicentennial Commemoration*, ed. by D. G. King-Hele (London: The Royal Society, 1992), pp. 105–13.

Eder, Joseph Maria, "John Frederick William Herschel," *Photographische Correspondenz*, no. 316 (1887), 7–9.

Elkhadem, H., "La correspondence d'Adolphe Quetelet avec Sir John Herschel: Un example de la richesse du fonds Quetelet" in *Adolphe Quetelet 1796–1874: Hommages et contributions*, published by the Académie royale de Belgique (Brussels, 1975), 46–57.

Egami, Fuyuko and Ryuichi Yasugi, "The Influence of the Writings on the Philosophy of Science [of Herschel and Whewell] on the Works of Ch. Darwin," *Kagakusi Kenkyu, 11* (1972), 57–64. (In Japanese.)

Enros, Philip C., *The Analytical Society: Mathematics at Cambridge during the Early Nineteenth Century* (a 1980 doctoral dissertation at the University of Toronto).

Enros, Philip C., "The Analytical Society (1812–1813): Precursor of the Renewal of Cambridge Mathematics," *Historia Mathematica, 10* (1983), 24–47.

Enros, Philip C., "Cambridge University and the Adoption of Analytics in Early Nineteenth-Century England" in *Social History of Nineteen Century Mathematics*, ed. by H. Mehrtens, H. Bos, and I. Schneider (Boston: Birkhäuser, 1981), 135–47.

Evans, David S., "Dashing and Dutiful: Herschel and Maclear Made a Strange If Effective Team in Their Astronomical Work at the Cape of Good Hope," *Science, 127* (1958), 935–48.

Evans, David S., "Herschel, John Frederick William," *Dictionary of Scientific Biography*, ed. by C. C. Gillispie, vol. VI (New York, 1972), pp. 323–8.

Evans, David S., "Homage to Herschel," *Contemporary Review, 219* (1971), 299–306.

Evans, David S., "Sir John Herschel to Captain Smyth," *Quarterly Bulletin of the South African Library, 25* (September, 1970), 3–17.

Evans, David S., "The Astronomical Work of Sir John Herschel at the Cape," *South African Public Library Quarterly Bulletin, 12*:2 (December, 1957), 44–58.

Evans, David S., "Two Items of Astronomical Africana: I. Lacaille Writes to His Friend de la Condamine; II. Observations at the Cape by Sir John Herschel," *South African Public Library Quarterly Bulletin, 7* (1952), 3–9.

Evans, David S., *Under Capricorn: A History of Southern Hemisphere Astronomy* (Bristol: Adam Hilger, 1988).

Evans, David S., Terence J. Deeming, Betty Hall Evans, and Stephen Goldfarb (eds), *Herschel at the Cape: Diaries and Correspondence of Sir John Herschel, 1834–1838* (Austin, Texas: Univ. of Texas Press, 1969).

Ferguson, W. T., "Sir John Frederick William Herschel," *Dictionary of South African Biography*, vol. 1 (Johannesburg, 1968), pp. 361–4.

Ferguson, W. T., and Immelman, R. F. M., *Sir John Herschel and Education at the Cape, 1834–1840* (Cape Town: Oxford University Press, 1961).

Fernie, Donald, *The Whisper and the Vision: The Voyages of Astronomers* (Toronto, 1976), 57–105. This chapter is entitled: "'... in the Country of the Hottentots': Maclear and Herschel in South Africa."

Fernie, Donald, "The Last Universalist [John Herschel]," *American Scientist, 80* (Sept./Oct., 1992), 422–4.

Field, Henry W., "Obituary Notice of Sir J. F. W. Herschel, Bt. ...," *American Philosophical Society Proceedings, 12* (1871), 217–23.

Fine, Arthur, "Foreword" to John Herschel, *Preliminary Discourse on the Study of Natural Philosophy* (Chicago, 1987 reprint of the 1830 original), pp. v–xxi.

Fisch, Menachem, "'The Emergency Which Has Arrived': The Problematic History of Nineteenth-Century British Algebra," *British Journal for the History of Science, 27* (1994), 247–76.

Fleming, James Rodger, *Meteorology in America, 1800–1870* (Baltimore: Johns Hopkins Univ. Press, 1990).

[Forbes, J. D.], "[Review of Herschel's *Astronomical Observations ... at the Cape of Good Hope]*," *Quarterly Review, 85* (June, 1849), 1–31.

Forbes, James David, "Sir John Frederick William Herschel" in Forbes's "Dissertation Sixth" in *Encyclopaedia Britannica*, 8th ed., vol. 1 (1860), pp. 861–2.

Ford, Colin, *The Cameron Collection: An Album of Photographs by Julia Margaret Cameron, Presented to Sir John Herschel* (London: Van Nostrand Reinhold, 1975).

Friede, E. P., "The First Photographer in South Africa," *Africana News and Notes, 6* (1948–9), 86–8.

[Galloway, Thomas], "Sir John Herschel's Astronomy," *Edinburgh Review, 58* (1833–34), 164–98.

[Galloway, Thomas], "[Review of] *Traité de la lumière*. Par. J. F. W. Herschel [and] *Mémoire sur la Diffraction de la Lumière* [and] *Mémoire sur la Double Refraction. ...* Par M. A. Fresnel," *Foreign Quarterly Review, 7* (April, 1831), 283–306.

Garber, Elizabeth, "Aspects of the Introduction of Probability into Physics," *Centaurus, 17* (1973), 11–39.

Garland, Martha McMackin, *Cambridge before Darwin: The Ideal of a Liberal Education 1800–1860* (Cambridge, England, 1980).

Garland, Martha M., *A Liberal Education: The Development of an Ideal at the University of Cambridge 1800–1860* (a 1975 Ohio State University doctoral dissertation).

Gates, J. W. C., and J. Maxwell, "The Herschel Condition" in *John Herschel 1792–1871: A Bicentennial Commemoration*, ed. by D. G. King-Hele (London: The Royal Society, 1992), pp. 101–3.

[Gauss, C. F.], "[Review of] *A Preliminary Discourse of [sic] the Study of Natural Philosophy* by J. F. W. Herschel," *Göttingische gelehrte Anzeigen* (Dec. 16 and 19, 1833), 2005–21.

[Gauss, C. F.], "[Review of] A *Treatise on Astronomy* by Sir by J. F. W. Herschel," *Göttingische gelehrte Anzeigen* (July 5, 1834), 1059–76.

Gautier, A., *Société de physique et d'histoire naturelle de Genève mémoires, 21* (1872), 586–9.

Gernsheim, Helmut, "Talbot's and Herschel's Photographic Experiments in 1839," *Image, 8*:3 (September, 1959), 132–7.

Gernsheim, Helmut and Alison, *The History of Photography from the Earliest Use of the Camera Obscura in the 14th Century up to 1914* (New York, 1955).

Gilbert, G., *Photography: The Early Years* (New York, 1980).

Gill, David, "Dr. Gill on Sir John Herschel," *Nature, 58* (1898), 356.

Gleason, Mary Louise, *The Royal Society of London: Years of Reforn 1827–1847* (New York: Garland, 1991).

Good, Gregory Alan, *J. F. W. Herschel's Optical Researches: A Study in Method* (a 1982 University of Toronto doctoral dissertation).

Good, Gregory, "John Herschel's Optical Researches and the Development of His Ideas on Method and Causality," *Studies in History and Philosophy of Science, 18* (1987), 1–41.

Gooding, David, "'He Who Proves, Discovers': John Herschel, William Pepys and the Faraday Effect," *Notes and Records of the Royal Society, 39* (1985), 229–44.

Gore, J. E., "The Herschels and the Nebulae," *Gentleman's Magazine*, n. s., *65* (1900), 78–87.

Gosser, H. Mark, "Further Notes on Herschel and Talbot: The Term 'Photography'," *History of Photography, 5* (1981), 269.

G[ould], B[enjamin] A., "Herschel's *Outlines of Astronomy,*" *Christian Examiner, 47* (1849), 268–95.

Gower, Barry, "Astronomy and Probability: Forbes versus Michell on the Distribution of the Stars," *Annals of Science, 39* (1982), 145–60.

Grattan-Guinness, I., "Dark Star in a New Light," [London] *Times Higher Education Supplement* (Sept. 4, 1992), 14.

Grattan-Guinness, I., "Mathematics and Mathematical Physics from Cambridge, 1815–40: A Survey of the Achievements and the French Influences" in *Wranglers and Physicists: Studies on Cambridge Physics in the Nineteenth Century*, ed. by Peter Harman (Manchester, 1985), pp. 84–111.

Grattan-Guinness, I., "The Young Mathematician" in *John Herschel 1792–1871: A Bicentennial Commemoration*, ed. by D. G. King-Hele (London: The Royal Society, 1992), pp. 17–28.

Hall, A. Rupert, *The Cambridge Philosophical Society: A History 1819–1969* (Cambridge: Cambridge Philosophical Society, 1969).

Hall, Marie Boas, *All Scientists Now: The Royal Society in the Nineteenth Century* (Cambridge, England, 1984).

Hall, Marie Boas, "The 'Distinguished Man of Science'" in *John Herschel 1792–1871: A Bicentennial Commemoration*, ed. by D. G. King-Hele (London: The Royal Society, 1992), pp. 115–23.

Hammond, J. H., and Jill Austin, *The Camera Lucida in Art and Science* (Bristol: Adam Hilger, 1987).

Harley, Timothy, *Lives Which Live: A Series of Biographical Reviews. 1. Sir John Herschel* (London, 1884).

Hearnshaw, J. B., *The Analysis of Starlight: One Hundred and Fifty Years of Astronomical Spectroscopy* (Cambridge, England, 1986).

Heimann, P. M., "Molecular Forces, Statistical Representation and Maxwell's Demon," *Studies in History and Philosophy of Science, 1* (1970), 189–211.

Henroteau, F. C., "Double and Multiple Stars," *Handbuch der Astrophysik, 6* (Berlin, 1928), pp. 299–468:303–4.

Herschel, Alexander, "On Some Original Glass Photographs Produced and Preserved by the Late Sir John Herschel, Bart., from His Earliest Experiments on the Chemical Properties of Light," Photographic Journal, 15:234 (June 15, 1872), 153–5.

Herschel, John, "A Complete Catalogue of the Writings of Sir John Herschel," *Mathematical Monthly, 3* (1861), 220–7.

Herschel, John, "A Course of Readings in 1834," *South African Libraries, 7* (1940), 138–144; 145–154.

Herschel, Mrs. John [Mary Cornwallis], *Memoir and Correspondence of Caroline Herschel* (London, 1876).

Hetherington, Barry, "The Herschel Arms," *British Astronomical Association Journal, 82* (1972), 132–3.

Hoffleit, Dorrit, "A History of Variable Star Astronomy to 1900 and Slightly Beyond," *Journal of American Association of Variable Star Observers, 15* (1986), 77–106.

Holden, Edward S., "The Three Herschels," *The Century Illustrated Magazine, 30* (1885), 179–85.

Hollander, S., "William Whewell and John Stuart Mill on the Methodology of Political Economy," *Studies in the History and Philosophy of Science, 14* (1983), 127–68.

Hoskin, Michael, "Astronomical Correspondence of William Rowan Hamilton," *Journal for the History of Astronomy, 15* (1984), 69–73.

Hoskin, Michael, "John Herschel and Astronomy: A Bicentennial Appraisal" in Brian Warner (ed.), *John Herschel 1792–1992: Bicentennial Symposium* (Royal Society of South Africa, 1994), pp. 1–17.

Hoskin, Michael, "John Herschel's Cosmology," *Journal for the History of Astronomy, 18* (1987), 1–34.

Hoskin, Michael, *Stellar Astronomy: Historical Studies* (Chalfont St. Giles, 1982).

Hughes, David W., "Sir John F. W. Herschel, Meteoroid Streams and the Solar Cycle," *Vistas in Astronomy, 39* (1995), 335–46.

Hull, David, "Charles Darwin and Nineteenth-Century Philosophies of Science" in Ronald N. Giere and Richard Westfall (eds.), *Foundations of Scientific Method: The Nineteenth Century* (Bloomington, Ind., 1970), 115–32.

Hussey, William J., "Notes on the Progress of Double Star Astronomy," *Publications of the Astronomical Society of the Pacific, 12* (1900), 91–103.

Hyman, Anthony, *Charles Babbage: Pioneer of the Computer* (Oxford, 1984).

Irons, James Campbell, *Autobiographical Sketch of James Croll* (London: Edward Stanford, 1896).

Jain, Chaman Lal, *Methodology and Epistemology: An Examination of Sir John Frederick William Herschel's Philosophy of Science with Reference to His Theory of Knowledge* (a 1975 doctoral dissertation at Indiana University).

Jaki, Stanley L., *The Milky Way: An Elusive Road for Science* (New York, 1972).

Jaki, Stanley L., *The Paradox of Olbers' Paradox* (New York, 1969).

James, Frank A. J. L., "The Conservation of Energy, Theories of Absorption and Resonating Molecules, 1851–1854: G. G. Stokes, A. J. Ångström and W. Thomson," *Notes and Records of the Royal Society, 38* (1983), 79–107.

James, Frank A. J. L., "The Creation of a Victorian Myth: The Historiography of Spectroscopy," *History of Science, 23* (1985), 1–24.

James, Frank A. J. L., "The Debate on the Nature of the Absorption of Light 1830–1835: A Core-Set Analysis," *History of Science, 21* (1983), 335–68.

James, Frank A. J. L., "The Discovery of Line Spectra," *Ambix, 32* (1985), 53–70.

James, Frank A. J. L., *The Early Development of Spectroscopy and Astrophysics* (a 1981 doctoral dissertation at the University of London, Imperial College.)

James, Frank A. J. L., "Spectro-Chemistry and Myth: A Rejoinder," *History of Science, 24* (1986), 433–7.

James, Frank A. J. L., "Thermodynamics and Sources of Solar Heat, 1846–1862," *British Journal for the History of Science, 15* (1982), 155–81.

James, Frank A. J. L., "Between Two Scientific Generations: John Herschel's Rejection of the Principle of the Conservation of Energy in His 1864 Correspondence with William Thomson," *Notes and Records of the Royal Society, 40* (1985), 53–62.

Jenkins, Reese, *Some Interrelations of Science, Technology, and the Photographic Industry in the Nineteenth Century* (a 1966 doctoral dissertation at the University of Wisconsin).

Jones, Gordon, "The Scientific Publications of Henry Kater," *Centaurus, 11* (1965), 152–89.

Kidwell, Peggy Aldrich, "Prelude to Solar Energy: Pouillet, Herschel, Forbes, and the Solar Constant," *Annals of Science, 38* (1981), 457–76.

Kidwell, Peggy Aldrich, *Solar Radiation and Heat from Kepler to Helmholtz (1600–1860)* (a 1979 doctoral dissertation at Yale University).

King-Hele, D. G. (ed.), *John Herschel 1792–1871: A Bicentennial Commemoration*, ed. by D. G. King-Hele (London: The Royal Society, 1992).

King-Hele, D. G., "Introduction" to *John Herschel 1792–1871: A Bicentennial Commemoration*, ed. by D. G. King-Hele (London: The Royal Society, 1992), pp. 1–2

Knight, David, "Introduction" to reprinting of John Herschel (ed.), *Admiralty Manual of Scientific Enquiry* (Folkestone, 1974), pp. v–xxii.

Kondor, Gusztáv, "Emlékbeszéd Herschel János ... felett," *Ertekez´sek a methematikai osztály köréböl* (Pest: Magyar Tudományos Akademia) *3* (1874).

Koppelman, Elaine, "The Calculus of Operations and the Rise of Abstract Algebra," *Archive for History of Exact Science, 8* (1971), 155–242.

Laudan, Larry, *Science and Hypothesis: Historical Essays on Scientific Methodology* (Dordrecht, 1971).

Leech, James, *John Herschel and Victorian Science: Exhibition of Manuscripts, Photographs, Drawings, and Books from the History of Science Collection* (Austin: Humanities Research Center, 1966).

Leuckfeld, Paul, "Zur logischen Lehre von der Induction. Geschichtliche Untersuchungen. ... John Herschel," *Archiv für Geschichte der Philosophie, 10* (1897), 340–60.

Lewis, G. G., "Report of the Construction and Fixing the Herschel Obelisk at Feldhausen," *Royal Astronomical Society Memoirs, 15* (1846), 169–70.

Longwell, Chester R., "Herschel's View of Isostatic Adjustment," *American Journal of Science, 16* (1928), 451–3.

Lubbock, Constance, *The Herschel Chronicle* (Cambridge, Cambridge Univ. Press, 1933)

Lubbock, Constance A., *A Short Biography of Sir John F. W. Herschel, Bt.* (Cambridge, England: privately printed, c. 1938).

Lubbock, Sir John, "[Presentation of the Royal Society's Royal Medal to JH]," *Royal Society Proceedings, 4* (1837–43), 255.

Macfarlane, Alexander, "Sir John Frederick William Herschel" in Macfarlane's *Lectures on Ten British Physicists of the Nineteenth Century* (New York, 1919), pp. 131–41.

McIntyre, Donald, "The Herschel Obelisk," *South African Public Library Quarterly Bulletin, 8* (1954) 87–92.

Maclear, Thomas, "Account of the Erection of the Herschel Obelisk at the Cape of Good Hope," *Royal Astronomical Society Memoirs, 15* (1846), 165–9.

Maclear, T., "Sir John Herschel at the Cape," *Cape Monthly Magazine*, n. s., *3* (1871), 129–140.

McGann, Guy W., "Sir John Herschel and the Birth of Celestial Photography," *Griffith Observer, 58* (Sept., 1994), 2–11.

McLennan, John Ferguson, *Memoir of Thomas Drummond* (Edinburgh, 1867).

MacLeod, Roy M., "Whigs versus Savants: Reflections on the Reform Movement in the Royal Society" in Ian Inkster and Jack Morrell (eds.), *Metropolis and Province: Science and British Culture 1780–1850* (London: Hutchinson, 1983), 55–90.

Mädler, J. H., "Die 40 füssige Herschelsche Telescop," *Astronomische Nachrichten, 17* (1840), 323–6.

Mailly, E., *Tableau de l'astronomie dans l'hemisphère austral et dans l'Inde* (Bruxelles, 1872); published under this title in *Académie royale des sciences, des lettres et des beaux-arts de Belgique. Mémoires couronnés et autres mémoires, 23* (1873), see ch. VIII, pp. 109–24.

Malherbe, Ernst G., *Education in South Africa (1652–1922),* (Cape Town, 1925), esp. pp. 75–7.

Malin, David, "The Splendor of Eta Carinae," *Sky and Telescope, 73* (Jan., 1987), 14–18.

May, Walther, "Darwin und John Herschel," *Wissenschaftliche Rundschau,* (1911–12), 347–8.

[Mill, J. S.], "[Review of] *A Preliminary Discourse on the Study of Natural Philosophy.* By John Frederick William Herschel," *Examiner* (March 20, 1831), 179–80.

Miller, David Philip, "The Revival of the Physical Sciences in Britain 1815–1840," *Osiris*, 2nd ser., *2* (1986), 107–34.

Miller, David Philip, "Method and the 'Micropolitics' of Science: The Early Years of the Geological and Astronomical Societies of London" in J. A. Schuster and R. R. Yeo (eds.), *The Politics and Rhetoric of Scientific Method: Historical Studies* (Dordrecht, 1986), 227–57.

Millman, Peter M., "The Herschel Dynasty: Part II: John Herschel," *Royal Astronomical Society of Canada Journal, 74* (1980), 203–15.

Millman, Peter M., "The Herschel Dynasty: Part III: Alexander Stewart Herschel," *Royal Astronomical Society of Canada Journal, 74* (1980), 279–90.

Mitchell, Maria, "Maria Mitchell's Reminiscences of the Herschels," *The Century Illustrated Magazine, 38* (1889), 903–9.

Moore, Keith, "'Space and time forgot': John Herschel as a Poet" in *John Herschel 1792–1871: A Bicentennial Commemoration*, ed. by D. G. King-Hele (London: The Royal Society, 1992), pp. 29–36.

Moore, Patrick, *Alexander Herschel: The "Meteor Man" 1836–1907* (Bath: The William Herschel Society, 1993).

Moore, Patrick, "John Herschel and the 'Solitary One [Alpha Hydrae]'" in Moore's *Fireside Astronomy* (Chichester: John Wiley & Sons, 1992), 140–1.

Moore, Patrick, "A Look Back to 1833" in Moore's *Fireside Astronomy* (Chichester: John Wiley & Sons, 1992), 14–15.

Moore, Patrick, *Sir John Herschel: Explorer of the Southern Skies* (Bath: William Herschel Society, 1992).

Moore, Patrick and Pete Collins, *The Astronomy of Southern Africa* (London, 1977).

Moore, Sarah, "A Newly-Discovered Letter of J. F. W. Herschel Concerning the Plumian Professorship," *Journal for the History of Astronomy, 25* (1994), 142–3.

Morell, Jack, "William Whewell: Rough Diamond," *History of Science, 32* (1994), 345–359.

Newhall, Beaumont, "Sir J. F. W. Herschel," *The Complete Photographer, 5* (New York, 1943), 1963–5.

Nightengale, Howard C., "Joseph Alfred Hardcastle FRAS," *Bulletin of the William Herschel Society, 41* (Summer, 1993), 3–4.

Northampton, Spencer Joshua Alwyne Compton, Marquis of, "[Presentation of the Royal Society's Copley Medal to JH]," *Royal Society Proceedings, 5* (1843–50), 702–3.

[Olmsted, Denison]. "[Review of John Herschel's] *Results of Astronomical Observations ... at the Cape of Good Hope*," *American Journal of Science and Arts*, 2nd ser., *5* (May, 1848), 86–100.

Olson, Richard, *Scottish Philosophy and British Physics 1750–1880* (Princeton,1975).

Ostroff, Eugene, "Herschel and Talbot: Photographic Research," *Journal of Photographic Science, 27* (1979), 73–80.

Pancaldi, Giuliano, "Conoscenza fine a se stessa, tecniche e pubblico della scienza nel 'Preliminary discourse' di J. Herschel" in Antonio Santucci (ed.) *Eredità dell'Illuminismo* (Bologna, 1979), pp. 435–62.

Panteki, M., "William Wallace and the Introduction of Continental Calculus to Britain: A Letter to George Peacock," *Historia Mathematica, 14* (1987), 119–32.

Panter, John Raspin, *Sir John F. W. Herschel and Scientific Thought in Early Nineteenth-Century Britain* (a 1976 doctoral dissertation at the University of New South Wales in Australia).

Partridge, Michael, "Introduction" to John Herschel, *A Preliminary Discourse on the Study of Natural Philosophy* (New York, 1966), pp. vii–lvi.

Patterson, Elizabeth Chambers, *Mary Somerville and Cultivation of Science, 1815–1840* (Boston: Martinus Nijhoff, 1983).

[Peacock, George], "Herschel's Survey of the Southern Heavens," *Edinburgh Review, 88* (1848), 104–43.

Pells, E. G., "Sir John Herschel's Contribution to Educational Developments at the Cape of Good Hope," *South African Public Library Quarterly Bulletin, 12*:2 (December, 1957), 58–65.

Pettoello, Renato, "Lettere di F. E. Beneke a J. F. W. Herschel e a W. Whewell," *Rivista di storia della filosofia, 44* (1989), pp. 691–724.

Plackett, Robin L., "Herschel on Estimation," *Journal of the Royal Statistical Society*, Ser. A, *155* (1992), 29–35.

Porter, Theodore M., *The Rise of Statistical Thinking 1820–1900* (Princeton, 1986).

Porter, Theodore M., "A Statistical Survey of Gases: Maxwell's Social Physics," *Historical Studies in the Physical Sciences, 12* (1981), 77–116.

Pritchard, Ada, *Charles Pritchard: Memoirs of His Life* (London: Seeley and Co., 1897).

P[ritchard], C[harles], "Herschel, Sir John Frederick William," *Encyclopaedia Britannica*, 9th ed., vol. XI (New York, 1880), 768–70.

P[ritchard], C[harles], "J. F. W. Herschel," *Royal Astronomical Society Monthly Notices, 32* (1872), 122–42.

[Pritchard, Charles], "John Herschel," *Kaiserlichen Akademie der Wissenschaften* (Wien) *Almanach* (1873), 147–56.

P[ritchard], C[harles], "Sir John Herschel," *Nature, 4* (May 25, 1871), 69–71.

[Proctor, Richard A.], "The Herschels and the Star Depths," *Cornhill Magazine, 34* (1871), 48–62; also published in: *Eclectic Magazine*, 77 ( ), 426– ; in Living Age, 110 ( ), 323– ; and in Proctor's, *The Borderland of Science* (London, 1882), 1–30.

[Proctor, Richard A.], "Sir John Herschel," *English Mechanic, 13* (May 19,1871), 195; reprinted in Proctor's *Essays on Astronomy* (London, 1872), pp. 1–7

Proctor, Richard A., "Sir John Herschel as a Theorist in Astronomy," *St. Paul's Magazine, 8* (June,1871), 326–39; reprinted in Proctor's *Essays on Astronomy* (London, 1872), pp. 8–28.

Proctor, Richard, "The Study of Astronomy," *Fraser's Magazine, 84* (September, 1871), 282–292 reprinted in Proctor's *Essays on Astronomy* (London, 1872), pp. 29–48.

Quetelet, L. A. J., "Notice sur John Fred. Will. Herschel," *Académie royale des sciences, des lettres et des beaux-arts de Belgique. Annuaire*, (1872), 161–99.

Quetelet, L. A. J., "Sir John-F.-W. Herschel," *Académie royale des sciences de Belgique Bulletin, 31* (1871), 478–84.

Ring, Francis J., "John Herschel and His Heritage" in *John Herschel 1792–1871: A Bicentennial Commemoration*, ed. by D. G. King-Hele (London: The Royal Society, 1992), pp. 3–15.

Roberts, Chris, "The Photographic Pioneer" in *John Herschel 1792–1871: A Bicentennial Commemoration*, ed. by D. G. King-Hele (London: The Royal Society, 1992), pp. 67–76.

Robinson, A. M. Lewin, "The Cape in the 1830's," *South African Public Library Quarterly Bulletin, 12*:2 (December, 1957), 65–72.

R[obinson], T. R[omney], "[Obituary Notice of] John Frederick William Herschel," *Royal Society of London Proceedings, 20* (1872), xvii–xxiii.

Ronan, Colin A., "The Astronomer: Follower of a New Tradition" in *John Herschel 1792–1871: A Bicentennial Commemoration*, ed. by D. G. King-Hele (London: The Royal Society, 1992), pp. 39–49.

Ronan, Colin, A. "John Herschel (1792–1871)," *Endeavour*, new ser., *16* (Dec., 1992), 178–81.

Ronan, Colin A., "John Herschel and His Contemporaries" in Ronan's *The Astronomers* (London, 1964), 198–211.

Ross, Sydney, "Herschel and Hypo" in Sydney Ross, *Nineteenth-Century Attitudes: Men of Science* (Dordrecht: Kluwer, 1991), pp. 173–93.

Ross, Sydney, "John Herschel on Faraday and on Science" *Notes and Records of the Royal Society, 33* (1978), 77–82; reprinted in Sydney Ross, *Nineteenth-Century Attitudes: Men of Science* (Dordrecht: Kluwer, 1991), pp. 194–202.

Ross, Sydney, "Sir John Herschel's Marginal Notes on Mill's On Liberty, 1859," *Journal of the History of Ideas, 29* (1968), 123–30; reprinted in Sydney Ross, *Nineteenth-Century Attitudes: Men of Science* (Dordrecht: Kluwer, 1991), pp. 203–13.

Ross, Sydney, "Unpublished Letters of Faraday and Others to Edward Daniel Clarke," *Bulletin for the History of Chemistry 11* (1991), 79–86.

Rourke, J. P., "John Herschel and the Cape Flora, 1834–1839" in Brian Warner (ed.), *John Herschel 1792–1992: Bicentennial Symposium* (Royal Society of South Africa, 1994), pp. 71–86.

Ruse, Michael, "Charles Lyell and the Philosophers of Science," *British Journal for the History of Science, 9* (1976), 121–31.

Ruse, Michael, "Darwin and Herschel," *Studies in History and Philosophy of Science, 9* (1978), 323–31.

Ruse, Michael, "Darwin's Debt to Philosophy: An Examination of the Influence of the Philosophical Ideas of John F. W. Herschel and William Whewell on the Development of Charles Darwin's Theory of Evolution," *Studies in History and Philosophy of Science, 6* (1975), 159–81. For a critique, see Paul R. Thagard, "Darwin and Whewell," *Studies in History and Philosophy of Science, 6* (1975), 159–81, and a reply by Ruse, pp. 323–31.

Ruse, Michael, "The Relationship between Science and Religion in Britain, 1830–1870," *Church History, 44* (1975), 505–22.

Schaaf, Larry J., "L'amour de l'lumière: Herschel, Talbot, et la photographie" in *Les multiples inventions de la photographie* (Paris: Association française pour la diffusion du patrimoine de photographique, 1989), 114–24.

Schaaf, Larry J., "Correspondence," *History of Photography, 5* (1981), 269–70.

Schaaf, Larry J., "Herschel, Talbot, and Photography: Spring 1831 and Spring 1839," *History of Photography, 4* (1980), 181–204.

Schaaf, Larry J., "John Herschel, Photography and the Camera Lucida" in Brian Warner (ed.), *John Herschel 1792–1992: Bicentennial Symposium* (Royal Society of South Africa, 1994), pp. 87–102.

Schaaf, Larry J., *Out of the Shadows: Herschel, Talbot, & the Invention of Photography* (New Haven: Yale Univ. Press, 1992).

Schaaf, Larry J., "The Poetry of Light: Herschel, Art, and Photography" in *John Herschel 1792–1871: A Bicentennial Commemoration*, ed. by D. G. King-Hele (London: The Royal Society, 1992), pp. 77–99.

Schaaf, Larry J., *Sir John Herschel the Invention of Photography* (London?: Royal Photographic Society, 1981).

Schaaf, Larry J. (ed.), *Selected Correspondence of William Henry Fox Talbot 1823–1874* (London: Science Museum and National Museum of Photography, Film & Television, 1994).

Schaaf, Larry J., "Sir John Herschel's 1839 Royal Society Paper on Photography," *History of Photography, 3* (1979), 47–60.

Schaaf, Larry J., *Tracings of Light: Sir John Herschel & The Camera Lucida* (San Francisco: Friends of Photography, 1989).

Schultze, R. S., "Re-discovery and Description of Original Material on the Photographic Researches of Sir John F. W. Herschel, 1839–1844," *Journal of Photographic Science, 13* (1965), 57–68; also published as "Wiedergefundenes Versuchsmaterial zu Sir John Herschels Photographischen Experimenten aus den Jahren 1839–1844," *Photographische Korrespondenz, Jubiläumsfestschrift*, (1965), 41–54.

Schweber, S. S., "John Herschel and Charles Darwin: A Study in Parallel Lives," *Journal of the HIstory of Biology, 22* (1989), 1–71.

Schweber, S. S., "John F. W. Herschel: A Prefatory Essay" in S. S. Schweber (ed.), *Aspects of the Life and Thought of Sir John Frederick Herschel*, vol. I (New York: Arno, 1981), 1–158.

Schweber, S. S., "Scientists as Intellectuals: The Early Victorians" in J. Paradis and T. Postlewait (eds.), *Victorian Science and Victorian Values: Literary Perspectives* (New Brunswick, N. J., 1985), 1–37; also published as vol. 360 of *New York Academy of Sciences Annals* (1981), 1–37.

Shorland, Eileen, "The Last of the Philosophers: Sir John Herschel, Bart., 1792–1871," *British Astronomical Association Journal, 83* (1973), 335–40.

Sims, Alan, V., "Celestial Cartography—Sir John Herschel," *Bulletin of the William Herschel Society, 41* (Summer, 1993), 8.

Smith, Crosbie, "Geologists and Mathematicians: The Rise of Physical Geology" in *Wranglers and Physicists: Studies on Cambridge Physics in the Nineteenth Century*, ed. by Peter Harman (Manchester, 1985), pp. 49–83.

Smith, Robert, "The Cambridge Network in Action: The Discovery of Neptune," *Isis, 80* (1989), 395–422.

Sol Obscuratus [C. N. Hastie], *The Newtonian and Herschelian versus the Harringtonian Theory of the Universe* (Ryde, 1866).

Somerville, Martha, *Personal Recollections from Early Life to Old Age of Mary Somerville* (London, 1874).

Spargo, P. E., "Foundations Strong and Lasting—Herschel's Work in Education at the Cape" in Brian Warner (ed.), *John Herschel 1792–1992: Bicentennial Symposium* (Royal Society of South Africa, 1994), pp. 103–15.

Stanley, A. P., "Science and Religion. A Sermon Preached in Westminster Abbey on May 21, 1871, Being the Sunday

Following the Funeral of Sir John Herschel," *Good Words for 1871*, 453–9.

Stoy, R. H., "Sir John Herschel (1792–1871)," *South African Public Library Quarterly Bulletin, 12*:2 (December, 1957), 41–4.

Strong, John V., "John Stuart Mill, John Herschel, and the 'Probability of Causes,'" *PSA* (1978), 31–41.

Strong, John V., *Studies in the Logic of Theory Assessment in Early Victorian Britain, 1830–1860* (a 1979 doctoral dissertation at the University of Pittsburgh).

Struve, Otto, "Milestones in Double Star Astronomy," *Sky and Telescope, 24* (July, 1962), 17–19.

Sutton, M. A., "Sir John Herschel and the Development of Spectroscopy in Britain," *British Journal for the History of Science, 7* (1974), 42–60.

Sutton, M. A., "Spectroscopy, Historiography and Myth: The Victorians Vindicated," *History of Science, 24* (1986), 425–32.

Tait, P. G., [Obituary of John Herschel], *Royal Society of Edinburgh Proceedings, 7* (1872), 543–6.

Thomson, Sir William, "Inaugural Address [to B.A.A.S.]," *Nature, 4* (Aug. 3, 1871), 262–270:262–263; also appears in *British Association for the Advancement of Science Report ... 1871* (London, 1872), lxxxiv–cv:lxxxv–lxxxvi.

Todhunter, Isaac, *William Whewell, D.D., Master of Trinity College: An Account of His Writings*, 2 vols. (London: Macmillan, 1876).

Turner, A. J., "Some Comments by Caroline Herschel on the Use of the 40–ft Telescope," *Journal for the History of Astronomy, 8* (1977), 196–8.

Turner, Stephen P., *The Search for a Methodology of Social Science* (Dordrecht, 1986).

Turvey, Peter J., "Sir John Herschel and the Abandonment of Charles Babbage's Difference Engine No. 1," *Notes and Records of the Royal Society, 45* (1991), 165–76.

Varley, D. H., "Sir John Herschel at the Cape," *South African Public Library Quarterly Bulletin, 12*:2 (December, 1957), 39–41.

Walborn, Nolan R., and Martha H. Liller, "The Earliest Spectroscopic Observations of Eta Carinae and Its Interaction with the Carina Nebula," *Astrophysical Journal, 211* (1977), 181–3.

Wallace, William, *Causality and Scientific Explanation*, vol. II (Ann Arbor, 1974).

Walls, Gordon L., "Sir John Herschel and Contact Lenses," *Science, 111* (1950), 501.

Warner, Brian, *Astronomers at the Royal Observatory, Cape of Good Hope: A History with Emphasis on the Nineteenth Century* (Cape Town: A. A. Balkema, 1979).

Warner, Brian, *Charles Piazzi Smyth: Astronomer-Artist, His Cape Years 1835–1845* (Cape Town: A. A. Balkema, 1983).

Warner, Brian, "Herschel Bicentenary," *South African Journal of Science, 88* (Sept./Oct., 1992), 458–61.

Warner, Brian, "The Herschel Obelisk," *South African Public Library Quarterly Bulletin, 32* (1978), 56–69.

Warner, Brian, "Herscheliana," *Vistas in Astronomy, 30* (1987), 85–96.

Warner, Brian (ed.), *John Herschel 1792–1992: Bicentennial Symposium* (Royal Society of South Africa, 1994).

Warner, Brian (ed.), *Margaret Herschel: Letters from the Cape 1834–1838* (Cape Town: Friends of the South African Library, 1991).

Warner, Brian, "Portrait of a 40–foot Giant," *Sky and Telescope, 71* (March, 1986), 253–4.

Warner, Brian, "The Remnants of the Cape's Astronomical Heritage," *South African Museums Association Bulletin, 18* (1988), 103–5.

Warner, Brian, *Royal Observatory: Cape of Good Hope: The Founding of a Colonial Observatory* (Dordrecht: Kluwer, 1995).

Warner, Brian, "Sir John Herschel 1792–1871," *South African Journal of Science, 88* (Sept./Oct.) 1992), 458–61.

Warner, Brian, "Sir John Herschel at the Cape of Good Hope" in Brian Warner (ed.), *John Herschel 1792–1992: Bicentennial Symposium* (Royal Society of South Africa, 1994), pp. 19–55.

Warner, Brian, "Sir John Herschel's Description of His 20–feet Reflector," *Vistas in Astronomy, 23* (1979), 75–107.

Warner, Brian, "The Years at the Cape of Good Hope" in *John Herschel 1792–1871: A Bicentennial Commemoration*, ed. by D. G. King-Hele (London: The Royal Society, 1992), pp. 51–66.

Warner, Brian and Nancy Warner, *Maclear and Herschel: Letters and Diaries at the Cape of Good Hope, 1834–1838* (Rotterdam: A. A. Balkema, 1984).

Whewell, William, "[Review of] *A Preliminary Discourse on the Study of Natural Philosophy*. By J. F. W. Herschel," *Quarterly Review, 45* (1831), 374–407.

Whitrow, Gerald J., "Some Prominent Personalities and Events in the Early History of the Royal Astronomical Society," *Royal Astronomical Society Quarterly Journal, 11* (1970), 89–104.

Wilkes, M. V., "Herschel, Peacock, Babbage and the Development of the Cambridge Curriculum," *Notes and Records of the Royal Society, 44* (1990), 205–19.

Williams, L. Pearce, "The Royal Society and the Founding of the B.A.A.S.," *Notes and Records of the Royal Society, 16* (1961), 221–33.

Williams, Mari, "Beyond the Planets: Early Nineteenth-Century Studies of Double Stars," *British Journal for the History of Science, 17* (1984), 295–309.

Wilson, David B., "Concepts of Physical Nature: John Herschel to Karl Pearson" in U. C. Knoepflmacher and G. B. Tennyson (eds.), *Nature and the Victorian Imagination* (Berkeley, 1977), pp. 201–15.

Wilson, David B., "The Educational Matrix: Physics Education at Early-Victorian Cambridge, Edinburgh and Glasgow Universities" in *Wranglers and Physicists: Studies on Cambridge Physics in the Nineteenth Century*, ed. by Peter Harman (Manchester, 1985), pp.12–48.

Wilson, David B., "Herschel and Whewell's Version of Newtonianism," *Journal of the History of Ideas, 35* (1974), 79–97.

Wilson, David B., *The Reception of the Wave Theory of Light by Cambridge Physicists (1820–1850): A Case Study in Nineteenth Century Mechanical Philosophy* (a 1968 doctoral dissertation at Johns Hopkins University).

Wolfendale, Arnold, "The Optical Dimension" in *John Herschel 1792–1871: A Bicentennial Commemoration*, ed. by D. G. King-Hele (London: The Royal Society, 1992), p. 37.

Wood, R. Derek, "Latent Developments from Gallic Acid, 1839," *Journal of Photographic Science, 28*:1 (1980), 36–42.

Wood, R. Derek, "The Involvement of Sir John Herschel in the Photographic Patent Case, **Talbot v. Henderson**," *Annals of Science, 27* (1971), 239–64.

Woods, John J., "John Frederick William Herschel" in *Victorian Britain: An Encyclopedia* (New York: Garland, 1988), 360–1.

Wright, D. C., "Double Star Astronomy in the UK 1832–1932," *Quarterly Journal of the Royal Astronomical Society, 34* (1993), 423–40.

Yeo, Richard, *Defining Science: William Whewell, Natural Knowledge, and Public Debate in Early Victorian Britain* (Cambridge: Cambridge Univ. Press, 1993)

Yeo, Richard, "An Idol of the Marketplace: Baconianism in Nineteenth Century Britain," *History of Science, 33* (1985), 252–98.

Yeo, Richard, "Reviewing Herschel's Discourse," *Studies in the History and Philosophy of Sciene, 20* (1989), 541–52.

Yeo, Richard, "Scientific Method and the Image of Science 1831–1891" in Roy MacLeod and Peter Collins (eds.), *The Parliament of Science: The British Association for the Advancement of Science 1831–1891* (Northwood, Middlesex, 1981), 65–88.

Yeo, Richard, "Scientific Method and the Rhetoric of Science in Britain, 1830–1917" in J. A. Schuster and R. R. Yeo (eds.), *The Politics and Rhetoric of Scientific Method: Historical Studies* (Dordrecht, 1986), 259–97.

APPENDIX IV

BIOGRAPHICAL REGISTER AND INDEX TO CORRESPONDENTS

Letters to or from a correspondent are listed after the correspondent's name and are arranged by year, the year number appearing in bold type. The letters are designated by the numbers assigned in this Calendar. Calendar numbers of letters to the correspondent are in roman type. Calendar numbers of letters from the correspondent (typically these are letters to John Herschel) are in italic type.

The source of the information in the biography of the correspondent is listed in parentheses after the identification. A bibliography of the sources used along with a key to the abbreviations employed is provided at the end of the register.

Abbe, Cleveland (1838–1916). American meteorologist. One of the pioneers of modern meteorology. (*DAB, DSB*)
1870 *14301 14325*

Aberdeen, Lord (see Hamilton-Gordon, George)

Abinger, Mr. Chairman of the Committee of Management at Royal Orthopaedic Hospital in London. ([1862?])
1862 *11750*

Academy of Science (Berlin), Royal
1847 *7342*

Academy of Sciences in Göttingen, Royal
1840 4521

Académie des sciences, Secretary
1857 10625

Ackland, William (1820–95). English optician and writer. Wrote article on stereoscopic photography. (*MEB, AlliSup*)
1871 *14420* 14421 *14424*

Acland, Henry Wentworth Dyke (1815–1900). British physician. Professor of medicine at Oxford. (*DNB*)
1862 12019 *12023* **1865** *13042*

Adam and Charles Black
1859 *11154 11156 11189 11207* **1861** 11526 **1868** *13682*

Adams, John Couch (1819–92). English mathematician and astronomer. Calculated the location of the planet Neptune, 1845. (*DNB, DSB*)
1847 *7147 7367 7474* **1848** *7708 7856* 7875 *7880* **1851** *8715 9131* **1866** *13078*

Adams, Walter Marsham (fl. 1865–70). English dramatist. (*AlliSup*)
1864 *12564* 12566 *12568 12571*

Adamson, James Constantine (1797–1875). Minister in South Africa. Administrator of South African College. Lived in U.S., 1853–60. (*DSAfB*)
1835 3218 **1840** *4748* **1846** *6772 6823* **1847** *7203* **1848** *7518 7522*

Adamson, John. Honorary secretary to Literary and Philosophical Society of Newcastle-on-Tyne in 1847. (1847–10–13)
1847 *7288*

Adamson, L. Clerk for Colonial Secretary at Cape of Good Hope in 1867. (1867–5–17)
1867 *13447*

Addams, Robert
1864 *12665*

Addington, Henry Unwin (1790–1870). English civil servant. Under-secretary of State for foreign affairs, 1842–1854. (*DNB, MEB*)
1845 *6203*

Admiralty
1839 3904 **1848** *7491 7501 7585* **1853** *9729* **1865** 12836 **1871** 14404

Airy, George Biddell (1801–1892). English astronomer. Lucasian and Plumian professorships at Cambridge. Astronomer Royal 1835–81. Close friend of JH. (*DNB, DSB*)
1824 *955* 958 *966 967* **1826** *1317* 1320 *1322 1350* 1352 *1353* 1388 1411 *1412* **1827** *1466* 1468 *1475* 1479 *1482* 1489 *1494* 1495 *1496 1502 1504* 1506 *1507 1515* 1517 1594 *1599* **1828** *1682* 1685 *1854* 1861 *1879* 1897 **1829** *2042* 2058 **1830** *2091 2109* 2116 2117 *2167* 2170 *2190* 2194 *2253* 2264 **1831** *2276 2285* 2286 *2387* *2465* **1832** *2520 2564* 2566 *2567* 2581 *2686 2701 2710* **1833** *2740 2771* 2842 *2846 2852* 2853 2855 2883 *2885* *2886* 2887 *2907* **1834** *2973 2987 3018* **1835** *3072* 3074 *3153* 3183 **1836** *3262 3401* **1838** *3653* 3661 *3674 3677 3707 3712* 3725 3830 *3840 3846* 3848 3851 3857 *3861* 3862 *3865* **1839** *3927 3948 3973* 3979 4050 *4070 4071 4130 4160* 4183 *4212 4252* 4255 *4273 4304* 4309 4332 *4333* **1840** *4399* 4543 4544 *4546 4547 4548* 4549 *4551* 4552 4612 4613 *4622 4628 4629 4631* 4632 *4643 4644* 4646 *4655 4656 4686* 4696 4697 *4706 4714* **1841** *4784* 4791 *4794 4808 4809 4810 4815 4816* 4817 *4851* 4852 *4856* 4884 *4885* 4886 *4892* 4894 *4898 4899* 4900 *4928 4932* 4933 *4954 4955* 4956 *4966* 4968 4969 *4970* 4971 4972 *4974* 4975 *4976* 4979 *5046 5047 5048* **1842** 5055 *5075 5077* 5078 5086 5089 *5090* 5094 *5097* 5098 *5099* 5117 *5123* 5126 *5128* 5131 *5135* 5149 *5154* 5157 *5159 5171 5180* 5243 *5246* 5249 5250 *5254* 5258 5266 *5270* 5276 5277 *5279 5280* 5282 *5283 5302* 5303 **1843** *5360* 5361 *5363 5365* 5366 *5371 5372 5405* 5406 5416 *5425 5452 5468* 5469 5489 *5493* 5497 *5500 5514* 5517 *5520 5521 5531* 5538 *5543 5564* 5566 *5574* 5575 *5582* 5583 *5586 5592* 5602 5606 *5608* 5614 5619 *5628* *5629* 5630 *5659* **1844** *5713 5721 5754* 5762 5763 *5765* *5828* 5829 5830 5842 *5852* 5868 *5881* 5899 *5900 5901* *5905* 5907 *5915 5941 5942* 5971 *5972* 5976 *5978* 5993 *5994* 5997 *6001* 6002 *6016* 6021 6025 6033 *6043* 6044 **1845** 6141 *6165 6168* 6172 *6174* 6175 *6176 6184 6185* *6186 6192* 6204 *6208 6209 6210* 6214 *6224 6227* 6228 *6234 6244* 6246 *6266 6267* 6287 *6341* 6348 *6355 6377* *6386* 6387 6389 6390 6392 *6393* 6396 *6398 6399* 6403 *6405* 6408 6409 6412 *6413 6423* 6424 *6430 6431 6433* *6435 6445 6466 6482 6485* 6487 *6488 6489* 6493 *6494* 6500 *6506 6518 6519* 6527 **1846** 6631 *6661 6676* 6677 *6685 6686* 6687 *6693 6768 6769* 6773 *6774 6775 6776* *6779* 6782 6783 6795 6796 *6798 6799* 6803 *6804* 6818

6828 *6829 6830 6837* **1847** *6862* 6875 *6876 6877 6878* 6891 *6896 6897* 6899 6900 *6904 6906 6941* 6942 6987 *6992* 6993 *6998 6999 7000 7006* 7007 7015 7018 *7045* 7050 7051 *7235* 7237 *7239 7244* 7245 7246 *7265* 7266 *7287 7309* 7330 *7335 7343* 7347 *7358* 7375 *7377* 7379 *7384 7400 7413* 7442 *7454 7455* **1848** *7500 7502 7508* 7509 7510 *7533* 7534 7535 *7546 7547 7548* 7557 *7558* 7560 7561 *7563 7590 7595* 7597 *7600 7661 7662* 7663 *7694 7695* 7696 7698 *7772* 7773 *7774 7776 7803 7828* *7835* 7839 7840 7841 *7843 7844 7845 7846* 7847 *7848* *7850 7853* 7854 *7862 7872* 7876 7877 *7881* 7882 *7884* 7885 7886 7887 7889 7892 7893 *7894 7895 7899 7921* *7928 7929* 7931 **1849** *7998* 8000 *8016* 8017 8018 *8019* *8020 8026* 8027 *8029* 8030 *8033 8034* 8035 *8043 8064* 8065 *8109 8121* 8122 8126 *8146* **1850** *8253* 8259 *8263* 8265 8316 *8318* 8375 *8376* 8377 *8464* 8465 8466 *8468* *8469* **1851** 8548 *8554 8555 8556 8564* 8565 *8568 8570* *8651* 8655 *8659 8709 8714* 8718 8719 *8720* 8722 *8736* 8737 *8769 8771 8785 8789* 8790 *8821 8839 8943* 8944 8968 *8971 8973 8982* 8990 8991 *8992* 9009 9010 *9024* *9025 9027 9028* 9035 *9036* 9040 *9044 9053* 9063 9064 *9067 9077* 9080 *9081 9091 9092 9096* 9097 *9099* 9114 *9115 9123 9130 9133* 9134 *9139 9147* 9149 9152 9153 *9154* **1852** 9155 9167 *9168* 9171 *9173* 9179 *9193 9198* *9199* 9203 9223 *9227 9228 9241 9242* 9245 *9278* 9285 *9287* 9292 9408 *9431* 9507 *9509 9538* 9544 *9570* 9571 *9572* **1853** 9648 *9649 9655 9656 9657 9658* 9659 *9660* 9661 9663 9664 *9666 9667* 9668 *9671* 9673 *9676 9684* *9685* 9686 *9687 9696* 9700 9701 *9702* 9707 *9708 9754* 9764 *9765 9806* 9814 *9837 9838* 9842 *9848* 9851 *9861* *9902 9906* **1854** *9980* 9982 *9990 10025* 10027 *10031* *10033 10034 10037 10038* 10039 *10040* 10042 10043 *10056 10060* 10062 *10063 10066 10067 10068* 10069 *10074* 10095 10098 *10118* 10123 10134 *10135* 10142 10143 *10147* 10148 10149 *10151* 10165 *10166 10179* *10180* 10181 10185 *10188* 10190 *10191 10193 10194* 10206 10218 *10219 10222* 10223 10232 *10235 10236* *10250 10268* **1855** 10305 *10306* 10356 10368 10369 *10370* 10390 *10391 10395* 10402 *10414 10415* 10421 *10423 10428* 10430 *10448* **1856** *10484* 10485 *10489* 10491 *10495* 10496 *10510 10523* 10524 10561 *10562* **1857** *10626* 10636 *10641* 10661 *10680* 10720 *10721* *10722 10725* **1858** 10785 *10788 10825 10843 10849* 10850 *10851 10853 10866 10873* 10875 10876 *10879* *10882* 10910 *10912 10913 10971* 10976 10988 *10994* **1859** *11037* 11084 *11086* 11089 *11111* 11117 *11121* *11142 11143* 11212 *11217* **1860** *11305* 11307 11308 *11333 11335 11350 11363* 11399 *11447 11454* 11493 *11496* 11497 **1861** *11534 11578 11579* 11581 *11591* *11593* 11594 *11595 11692* 11694 **1862** *11751* 11771 11772 *11773* 11777 *11781 11786* 11828 *11829* 11830 11831 *11832 11837 11841* 11844 *11847 11848* 11849 *11858 11859* 11861 *11863 11864* 11871 *11897* 11901 11902 *11911 11917 11929* 11932 11933 *11935* 11937 11945 *11955* 11964 *11965* 11985 *11987* 12033 *12034* *12036 12037* 12042 *12055* 12057 *12059* **1863** *12096* *12106* 12117 *12149* 12150 *12152* 12153 *12155* 12157 12158 *12159 12164 12165* 12168 *12171 12172 12173* *12223* 12224 12225 *12233* 12237 12239 *12240 12242* 12246 *12248* 12271 12272 *12290 12291 12292* 12320 *12323* 12324 12325 *12326 12327* 12342 *12354 12371* 12372 *12379* 12383 12384 *12385 12386 12387 12388* **1864** *12444* 12448 12449 *12451* 12465 *12469 12470* 12471 *12472 12476* 12478 *12484 12492 12512* 12519 *12551 12552* 12556 12588 *12631* 12675 *12678 12679* *12727* 12729 **1865** *12871 12986* 12993 *12994 12995* *13001* 13006 13007 *13009* 13010 13011 13035 13036 13037 *13049* **1866** 13105 *13109 13110 13165* 13166 13167 13168 *13169 13170* 13174 *13176* 13177 *13183* 13198 *13202* 13204 13205 *13206 13207 13208 13214* 13226 13256 *13258 13260* 13263 13264 *13279* **1867** *13364* 13428 *13432 13433* 13435 *13436 13493* 13512 13513 13514 *13516 13517 13518* 13544 *13546* 13561 *13564* 13566 *13569* 13570 *13572* **1868** *13630 13631* 13633 *13636 13668* 13669 *13671 13672* 13690 *13766* **1869** 13843 *13905* 13907 *13913* 13925 *13927 13987* 14032 **1870** *14100 14109 14116* 14117 *14118 14132* 14195 *14198 14199 14215 14242 14243* 14248 14249 *14328* 14329 *14331 14338* 14339 **1871** *14400 14401* 14403 **1875** *14456 14457* **undated** 14461 14462

Airy, Hubert (1838–?). Fourth-born son of G. B. Airy. (Airy, *Autobiography*)
1868 *13611 13653 13662 13789 13792* **1870** *14190*

Airy, Richarda (née Smith) (1804–75). Wife of G. B. Airy, married 1830. (Airy, *Autobiography*)
1834 *2992* **1839** 4262 **1841** 4823 **1843** *5631* **1852** 9511 **1853** 9857 **1863** 12243 **1864** 12541

Albany Institute (Albany, New York), The
1836 3389

Albert, Prince (1819–61). Prince Consort of England. (*DNB*)
1839 4145

Alexander, James Edward (1803–85). British army officer. Aide-de-camp to Benjamin D'Urban at the Cape. (*DNB*)
1837 3539

Allen, Robin (1820–99). British civil servant. Secretary of Trinity House from 1867–81. (*MEB*)
1869 *14045*

Allen, William H. Probably a stationer operating as William H. Allen & Co., at 7 Leadenhall St. in London.
1842 *5210*

Allibone, Samuel Austin (1816–89). American bibliographer. Compiled *Dictionary of English Literature* (3 vols., 1858–1870) (*DAB*)
1859 11097

Amici, Giovanni B. (1786–1868). Italian microscopist and astronomer. Director of astronomical observatory and museum at Florence, 1831–59. (*DSB*)
1825 1099 *1117 1225* **1826** *1303 1330 1375* **1827** *1511 1529 1535* **1834** *3004*

Amici, Vincenzo. Son of Giovanni Battista Amici.
1827 *1595*

Ampère, André Marie (1775–1836). French mathematician, astronomer, and experimental physicist. Pioneer in electromagnetism. (*DSB*)
1825 *1241* **1827** *1576*

Anderson, B.
1855 *10452*

Anderson, James J.
1855 *10320*

Anderson, T. P. Lawyer, with partner, J. H. Shonbridge. Advised JH about marriage settlement trust for Mary Anne (Babbage) Hollier, sister of Charles Babbage. (1864–11–17)

1858 10814 **1859** 11026 11133 **1860** 11392 **1861** 11619 **1862** 12010 **1864** 12712 12716 *12738* **1866** 13066 **1869** *14063*

Anderson, William George (1804–97). British civil servant. Auditor for Admiralty and Exchequer. (*MEB*)
1852 *9520*

Andrews, Thomas (1813–85). Irish academic. Professor of chemistry at several institutions in Belfast. (*DSB*)
1843 5450 **1856** 10505 *10514* **1871** 14359 *14361* 14368 *14373* *14383* 14386 *14419* 14422

***Annales de chimie et de physique*, Editors of the**
1832 2559 2562

Ansted, David T. (1814–80). English geologist. Professor of geology at King's College, London. (*Alli, DNB*)
1859 *11221* *11245*

Anthropological Society
1863 *12124*

Apelt, Ernst Friederich (1813–59). German philosopher, historian of science. (*DSB*)
1845 *6198*

Arago, D. François J. (1786–1853). French physicist and astronomer. Director of Paris Observatory. (*DSB*)
1822 669 670 693 **1830** 2098 **1831** *2292* *2340* **1832** 2563 *2605* *2642* *2687* **1833** 2836 2845 *2927* **1836** 3234 *3310* **1837** 3461 **1839** 4131 4137 4150 4272 **1843** *5433* **1846** *6806* 6842 **1848** 7480 *7655* **undated** *14463*

Arbuthnot, George (1802–64). British civil servant. Senior civil servant in the Treasury. (*DNB, MEB*)
1852 *9390* *9412* *9454* *9499* *9504* *9512* **1853** *9725* 9760 *9853*

Archer, Frederick Scott (1813–1857). Sculptor, chemist, and pioneer photographer. Photographic shop at 105 Great Russell St. (*DNB, MEB*)
1844 *5963*

Argelander, F. W. A. (1799–1875). German astronomer. Professor at Helsinki and at Bonn. Produced star catalogues. (*DSB*)
1830 *2217* **1842** 5229 *5268* *5296* 5300 *5312* **1844** 5787 **1848** 7949 **1849** *7963* 7972 **1864** 12452 12455 **1866** *13292* 13296 **1874** *14453*

Armellini, Titus. Italian inventor, living in Rome. (1862–8–13)
1862 *11969* *11978* **1863** *12267*

Arnold, Matthew (1822–88). English poet and academic. Headmaster at Rugby, inspector of schools. (*DNB, MEB*)
1866 13067 **1867** *13487*

Arnott, Neil (1788–1874). Scottish physician and natural philosopher. (*MEB, DSB*)
1860 *11293* **1865** *13017* **1869** *13904*

Artaud, William (fl. 1776–1822). British portrait painter. (*DNB*)
1819 *336*

Arthington Jr., Robert (1823?–1900). English benefactor. Contributed significantly to English missions in Africa. (*MEB*)
1859 *11054* 11056

Astronomische Nachrichten
1823 841 **1825** 1221 **1832** 2667 **1835** 3079 **1837** 3598 **1838** 3624 **1841** 5032 **1843** 5634 **1844** 5959

***Athenaeum*, the**
1841 4973 **1842** 5237 **1843** 5380 5464 5603 5626 5627 **1846** 6754 **1848** 7908 7911 7920 **1850** 8205 8206 8313 **1852** 9590 **1853** 9775 **1860** 11336 11342 **1862** 12028 **1864** 12652 **1869** 14021

Atkinson, Jasper (1790–1856). Royal Mint employee. Provost of Company of Moneyers. (*MEB*)
1851 *8641* *8644* *8667* *8691* 8694 *8695* 8730 8777 *8804* 8814 *8815* 8827 *8828* *8829* 8833 8844 *8848* 8856 *8857* *8864* *8868* 8870

Atkinson, Solomon (?–1865). English barrister and writer about the law. (*MEB*)
undated *14464*

Attree, William H. Nephew of Richard Jones. (1870–11–1)
1870 *14311* 14312 *14317*

Auckland, Lord (see Eden, George)

Audit Office
1854 10048

Augustus Frederick (1773–1843). Baron Arklow, Duke of Sussex, 1801. President of R.S.L. 1830–38. (*DNB*)
1838 3668 3669 3693 3694 *3751*

Austen, F. A friend of JH's family, living at Capel Manor, Horsmonden.
1870 *14259*

Auwers, Georg F. J. A. (1838–1915). German astronomer at Gotha. Extended JH's nebular observations. Completed new reductions on James Bradley's stars. (*Who's*)
1864 *12686* **1868** *13706* *13710*

Ayern, Thomas. Coal company official in Newcastle. (1845–8–4)
1845 *6349* *6468*

B[?], E. L.
1846 *6659* *6777*

B., G.
1853 *9901*

B., G. C.
1861 *11662*

Babbage, Benjamin Herschel (1815–?). Oldest son of Charles Babbage. (Hyman, *Babbage*)
1844 *5826*

Babbage, Charles (1792–1871). English mathematician, classmate and close friend of JH. Developed calculating machinery. Lucasian Professor. (*DNB, DSB*)
1812 27 28 29 *41* 42 *43* 44 45 48 *54* **1813** 58 *59* 61 *65* *67* 68 *71* *73* 77 78 83 87 89 91 92 93 **1814** 102 *103* 104

105 *110 113 115 116* 117 118 119 *120* 123 126 127 128 *130 131* **1815** *135* 136 *138* 139 140 143 *145* 146 147 *149* 151 *152 153 154* 155 156 *158 159* 160 *162* **1816** 168 169 *171* 172 *173* 174 *178 179 181 182* 183 *184 186 196* 197 207 **1817** *210 214* 215 *218 219 222 228* 229 230 232 *235* 236 *238* 251 *253 257* 261 *265 273* 275 **1818** 282 *284 285* 288 293 *299 304* 307 *308 312 313* 314 315 *316* 317 318 320 *321* 322 *323* **1819** 324 *325 330* 339 *340* 341 *342 343* 344 345 *347* 350 356 357 *361 363* 364 *365* 369 376 *377* **1820** *395* 396 *400* 401 *403 414* 415 *419 421 422* 425 429 432 434 *437* 446 *453* 454 *461* 463 *472 473* 475 *485* 486 487 488 *490* 492 *500* 501 504 **1821** *505* 519 *522 523 532* 533 548 559 *561 571 572* 580 584 *586* 587 *594 616* 617 622 635 637 *643* **1822** *646* 681 698 *702* 707 *730* 747 758 *767 794 800* 815 819 *822* 834 **1823** 861 *885 887* 888 *889 893* 895 *899 922* **1824** 952 *974* 989 996 998 *1004* 1023 *1045* 1057 *1060* **1825** 1092 *1265* **1826** 1332 *1368* 1384 1408 *1419* 1421 1425 *1432* 1433 *1436* **1827** 1440 *1441* 1461 1538 *1549* 1562 *1619* 1636 **1828** 1676 1694 1695 *1710 1735* 1843 **1829** *1930* 1933 *1968* 2043 *2061* 2063 **1830** 2071 2108 *2112* 2113 *2126 2131* 2133 *2135* 2165 *2200* 2203 2221 2227 *2230* 2244 2254 **1831** *2300* 2307 *2341 2378* 2380 2411 *2412 2440* 2445 *2449 2467* 2473 *2474* **1832** 2490 2507 *2509 2540* 2698 *2699* 2700 *2704* **1833** *2754 2773* 2778 2779 *2806 2809* 2820 *2822* **1834** 2949 *3008* 3043 **1835** 3057 **1836** 3276 **1837** 3548 **1838** 3687 3747 *3766* **1839** *3918 3987* **1840** 4382 **1841** 4760 4987 *5025* 5027 **1842** 5056 **1843** *5385* 5386 *5578 5641* 5648 *5652* 5671 *5672* **1844** *5678* 5927 **1845** *6173* 6177 *6191 6501* 6507 **1846** 6642 **1847** 6887 6991 7037 *7043 7163* **1848** 7481 7587 *7601* **1851** 8491 *8621* 8625 8779 8921 8927 *9011* 9014 **1852** 9156 9266 9583 **1853** 9613 9614 9907 **1854** *9959 10265* 10266 **1857** *10744* **1858** *10763 10778* 10908 *10916* **1859** *11213* 11237 **1861** *11633* **1862** *12039* 12040 **1863** *12179 12294 12310* **1864** *12485* 12589 *12748* 12751 **1865** *12795 12809* **1866** *13267* 13272 **1867** 13557 **1868** 13711 **1870** 14334 **undated** *14465 14466 14467 14468 14469 14470* 14471 14472 14473 14474 14475 14476 14477 14478 14479 14480 14481 14482 14483 14484 14485 14486 14487 14488 14489 14490 14491 14492 14493 14494 14495 14496 14497 14498

Babbage, Elizabeth [Betty] Plumleigh (?–1844). Mother of Charles Babbage and Mary Anne Babbage Hollier. (Hyman, *Babbage*)
1827 1573 *1574 1581 1583* **1828** *1684 1693 1718* 1719 *1764 1810* 1811 *1813 1815 1816 1831 1846 1863* **1839** 4263 **1841** *4818* **1842** *5271*

Babbage, Georgiana (née Whitmore) (?–1827). Sister of Woolrych Whitmore and Louisa Whitmore Ryan (wife of Sir Edward). Married Charles Babbage, 1814. (Hyman, *Babbage*)
1821 585 **1826** *1409* **1827** 1442 *1460 1462 1469* 1470 *1471* **undated** *14499 14500 14501*

Babinet, Jacques (1794–1872). French physicist and meteorologist. Librarian at Bureau des Longitudes. (*DBF, DSB*)
1832 *2622* 2628 **1833** *2811* **1849** *7951* **1862** *11783* 11784 *11796* 11797 **1863** *12160* 12162 **1867** *13366* 13384 13386

Bache, Alexander Dallas (1806–67). American physicist. Established first U. S. magnetic and meteorological observatory. (*NCAB, DSB*)
1844 6034 **1848** *7646*

Bache, Franklin (1792–1864). American chemist and physician. Officer in American Philosophical Society. (*DAB*)
1826 *1346*

Baddeley, P. W. D.
1861 *11532 11563*

Baillie, Agnes (1760–1861). Elder sister of Joanna Baillie. (*MEB*)
1820 *418 426 436* **1839** 4284 **1844** *5967 5981* **1845** *6354* **1850** *8267 8294* **undated** *14502 14503 14504*

Baillie, G. Coinage agent for the government of Ceylon. (JH's 1854–9–?)
1854 10214

Baillie, J. H. Niece of Agnes Baillie. ([1861]–5–10)
1861 *11602*

Baillie, Joanna (1762–1851). Scottish poet and playwright. Youngest child of Rev. James Baillie. (*DNB, MEB*)
1820 *428* **1822** *672 708 713* **1842** *5342* **1847** *7067* 7072 *7076* **1850** *8268 8335 8393*

Baillie-Hamilton, William A. (see Hamilton, William A. Baillie–)

Baily, B. Cousin of Francis Baily, and involved in executing FB's will. (1844–9–14)
1844 *5926* 5928 **1845** *6091 6095*

Baily, Edward Hodges (1788–1867). English sculptor. Executed many celebrated portrait statues and groups. (*DNB, MEB*)
1849 *7982 8103 8111 8113 8120 8123*

Baily, Francis (1774–1844). English astronomer and stockbroker, and close friend of JH. Cofounder of the Astronomical Society. (*DNB, DSB*)
1819 326 **1820** *397 398 411 413* 456 *460* **1821** *530 543 555* 573 *575 625* **1822** *658 673 678 679 686 687 695 717 743 755* **1823** *859* **1824** *969* 1030 **1825** *1124 1135 1167 1179 1235 1243* **1826** *1298 1311* 1312 1318 *1414* 1422 **1827** *1459 1571 1613* 1635 *1652* **1828** *1663 1670* 1696 *1698 1720* 1724 *1737* 1739 *1757* 1761 1772 *1776 1798 1818 1833 1850* **1829** *1949 2046* 2048 **1830** 2258 **1831** 2289 *2297* 2301 *2313 2333* 2334 2346 *2349 2350* 2365 *2370 2389 2395* 2482 *2486* **1832** 2491 2547 *2548 2557* 2560 2574 *2576* 2645 *2705* **1833** *2733* 2833 *2871* 2900 2908 2922 **1834** 2943 *2950* 2991 *3000* 3019 **1835** 3090 3092 3118 *3155* 3169 3196 3223 **1836** *3245* 3283 *3317 3350 3368 3416 3419* **1837** 3450 3463 *3509* 3599 **1838** 3657 *3760* 3762 3771 3772 3804 *3806* **1839** 3905 *3920* 3922 *3928* 3964 *3965 3974* 3978 *3988* 4054 *4063 4066* 4086 4087 4207 *4214 4245* 4247 *4281 4286* 4291 *4308* 4310 *4313 4318* **1840** 4345 *4371* 4373 *4375 4418 4428* 4431 *4444 4468 4474 4488* 4499 4513 *4523* 4537 4574 4616 *4619* 4711 *4716 4727* 4730 *4731 4732 4735* 4737 4741 *4746* 4756 **1841** *4770 4776* 4793 *4795* 4862 4863 *4866* 4869 4870 *4872 4897* 4901 4908 *4911* 4914 *4946* 4947 *4951 4957* 4958 4963 *4988* 4999 *5002 5018* 5019 *5021* **1842** 5057 *5091* 5095 5118 *5218 5251* 5252 *5253* 5259 5267 *5269 5304* 5306 **1843** *5351* 5393 *5395* 5396 5399 5400 *5402* 5410 5412 *5414 5418 5421* 5423 *5429 5438* 5439 *5455* 5466 5470 *5473* 5475 *5477 5479 5519* 5526 *5549* 5550 *5551 5552 5560* 5565 *5568 5570 5587* 5589 *5595* 5609 *5610 5632 5643* 5649 5666 *5668 5669* **1844** *5706 5708 5717 5730*

5733 *5735 5737 5770* **1847** 6848 **undated** *14505 14506 14507 14508* 14509

Baldwin, Mary (1787?–?). Daughter of Thomas Baldwin, niece of Lady Mary Herschel, and cousin of JH.
1824 *1003* 1033 *1038* **1825** 1219 **1828** 1716 **1829** 1914 *1965* 2023 **1830** 2072 **1842** 5272 **1844** 5860 5883 5893 **1850** 8256 8443 **1851** 9015 **1866** 13346 **1867** 13354 13373 13387 13468 **1868** 13735 **undated** 14510 14511

Baldwin, Miss Probably Mary Baldwin. (Robt. Grahame Sr.'s 1822–8–30)
1822 779

Baldwin, Robert (c. 1780–1858). English printer and bookseller. Started *London Magazine.* (*MEB*)
1828 *1713*

Baldwin, Sophia
1812 *49 51*

Baldwin Sr., Thomas (?–1821). Brother of Mary Baldwin Pitt Herschel and uncle of JH.
1808 8

Ball, John (1818–89). Irish naturalist who contributed many papers to learned societies. (*MEB, AlliSup*)
1860 *11510* **1861** *11528 11537*

Ballot, C. H. D. Buys (see Buys Ballot, Christoph Hendrik Diederik)

Bancroft, George (1800–91). American historian and diplomat. Published *History of the United States* (10 vols.). U.S. minister to England. (*DAB, NCAB*)
1845 *6264* **1847** 7193 *7240* 7247 *7272* 7337 7344 **1848** 7646

Bank of England
1851 8543

Bankers
1854 10242

Banks, John. Secretary to Mechanics Institution of Hastings in 1854. (1854–1–31)
1854 *9962 9999* **1871** *14432 14438*

Banks, Joseph (1743–1820). English botanist. Travelled with Captain James Cook. President of the Royal Society. (*DNB, DSB*)
1815 *157*

Banley, J. Gomez
1865 *13057*

Barclay, Joseph Gurney (1816–98). Partner in bank. Set up observatory at his residence at Knott's Green. (*MEB, Alli*)
1870 *14221 14232*

Barlow, John (1799–1869). Chaplain at Kensington Palace. Secretary to Royal Institution, 1842–60. (*MEB*)
1852 *9229* **1855** *10409* **1860** *11260*

Barlow, Peter (1776–1862). Mathematician at Royal Military Academy, Woolwich, 1801–47. Invented mechanism for reducing errors in ships' compasses. (*DNB, DSB*)
1823 *881* **1824** *1089* **1825** *1119* **1827** *1560* **1828** *1671 1673* **1830** *2123 2130* 2138 *2144* 2158 *2205*

Barnard, C. Agent acting on behalf of the Assay Office in Victoria, Australia. (JH's 1853–12–21)
1853 9905

Barrett, C. P. Member, 1851, of planning committee for improving Slough. (1851–4–11)
1851 *8743* 8747

Barrow, Henry
1845 *6109*

Barrow, John (1764–1848). Secretary to the Admiralty, 1804–45. Cofounder of the Royal Geographical Society. (*DNB*)
1825 *1247* **1830** *2243*

Barry, Alfred (1826–1910). English chaplain-in-ordinary to the Queen. Son of prominent architect, Sir Charles Barry. (*AlliSup*)
1858 *10914 10931*

Barry, Redmond (1813–80). Irish-born colonial judge at Victoria, Australia. (*DNB, MEB*)
1868 *13819* **1869** *13915*

Barthes & Lowell
1846 *6551*

Bartlett, John Russell (1805–86). American antiquarian and public servant. (*NCAB, DAB*)
1847 *7348*

Barton, William Henry (1802–68). Royal Mint employee, rising to deputy-master, while JH was master of the Mint. (*MEB*)
1853 9697 9727 9735 **1854** 9928 10255 **1855** 10327 *10328 10329* 10336 *10340 10341 10345* 10347 *10350 10351* 10352 *10360* 10362 *10363 10365* 10366 *10381* **1857** *10710* **1858** *10901 10927* **1861** *11599 11641* **1865** *12832 12835*

Basire, James (1796–1869). British engraver, the last of several generations of engravers. (*DNB*)
1845 6056 *6391*

Bateman, William (1826–1901). Lord lieutenant of the County of Hereford. 2nd Baron Bateman, 1845. (*Peerage*)
1845 *6421*

Bateson, William Henry (1812–81). English academic administrator at Cambridge University. Master of St. John's College. (*DNB*)
1850 *8410 8418 8451 8453* **1851** *8599 8622 8761 8787 8835 8842 8871 9006 9093 9118 9145* **1852** *9239 9261 9272 9276 9317 9348 9374 9427 9438 9457 9458 9482 9491* **1855** *10378 10384* 10385 **1862** *11952* **1869** *13877*

Bavarian Academy of Science, Royal
1848 *7948*

Bax, S. A friend of the Baily family. (1859–10–19)
1859 *11161*

Baxendell, Joseph (1815–87). British meteorologist and astronomer. Secretary to the Manchester Literary and Philosophical Society. (*Who's*)
1848 *7935* **1856** *10537* **1858** *10902* **1861** *11616* **1862** *11806* **1863** *12407* **1864** *12439* **1865** *13003 13014* **1867** 13367 **1868** *13620*

Beale, A.
1840 *4514*

Beattie, William (1793–1875). English physician and later private secretary for the Duke of Clarence (afterwards William IV). (*DNB, MEB*)
1847 *7349 7364*

Beaufort, Francis (1774–1857). British naval officer. Uncle of novelist Maria Edgeworth. Helped build telegraph line in Ireland. Hydrographer for the Admiralty. (*DNB, MEB*)
1827 1539 *1542* **1828** 1773 *1786 1864* **1830** *2094 2147 2154* 2245 **1831** 2322 *2325 2403* 2404 *2405* 2408 *2410* 2413 *2421* 2425 2427 *2437* 2441 2468 *2479* 2481 2484 *2485* **1832** *2504 2510* 2513 *2542 2543 2569 2577 2618* **1833** *2727 2752* 2801 *2807 2863 2872 2875 2898 2909 2916* **1834** *2990* **1835** *3050 3061 3071* 3091 *3099 3158* 3189 3230 **1836** 3259 *3277 3306 3321 3335 3344* 3349 *3358 3375 3381* **1837** *3448 3605* **1838** *3670 3702* 3704 3709 *3710* 3726 *3792 3800 3823 3863 3895 3898* **1839** *3931 3944 3982 3992 4027 4162 4167 4170 4179* 4184 *4223 4226 4238 4280 4300 4341* **1840** *4376 4380 4569 4583 4752* **1841** *4785* 4792 *4799 4819 4828 5051* **1843** 5476 *5478* **1844** *5863 5873* **1846** *6646* **1847** 7136 *7144 7173 7232 7378 7394 7414 7416* 7438 *7443 7456* **1848** *7528 7559 7570 7571 7580 7629 7634 7641 7672* 7676 *7679 7738 7744 7777 7800 7805 7842 7918* **1849** 7952 *7969 8037 8086* 8089 8102 *8104 8106* **1850** 8287 *8289* 8295 **1851** *8897 9136* 9150 **1852** *9240 9262* 9302 9303 9304 *9305 9333 9337* 9351 *9368 9407* 9409 *9415* 9421 *9444* 9447 *9448 9576*
undated *14512*

Becher, Alexander Bridport (1796–1876). British naval officer. Worked in Hydrographic office, and edited the *Nautical Magazine*. (*MEB*)
1838 *3866* **1844** *5918* **1848** *7806 7923* **1857** *10691* **1859** *11066* **1860** *11361 11367* **1861** *11628* **1863** *12226 12268*

Beck, Sophia
1854 10096 **1857** *10608*

Beckedorff, Sophie. Friend of Caroline Lucretia Herschel after Caroline's return to Hanover. Visited JH at Collingwood. (1848–4–4)
1838 *3711 3815* **1846** *6651 6653 6658* **1847** *7350* **1848** 7514 *7552 7659* **1849** *7954* **1850** *8214*

Beckwith, E. S. [Mrs.] JH is a trustee for EB's marriage settlement.
1833 *2714* **1848** *7718*

Beckwith, Thomas. Friend of JH's family. (1822–8–23)
1822 *777*

Beckwith, Thomas P. Son of E. S. Beckwith.
1847 *7055 7061*

Becquerel, Antoine Cesar (1788–1878). French physicist. One of the founders of electrochemistry. (*DSB, Who's*)
1851 *8492*

Beechey, Frederick W. (1796–1856). Naval officer and polar explorer. Surveyed parts of the African and Irish coastlines. (*DNB, MEB*)
1848 *7676 7714 7720 7730* 7732 7735 *7739 7742 7750* 7752 7753 *7757 7827* **1849** *7960* **1851** *8639 8648*
undated *14513*

Beer, Wilhelm (1797–1850). German astronomer. Mapped principal features of the moon. (*Who's*)
1835 *3166* **1837** *3522* **1838** *3666*

Beke, Charles T. (1800–74). English explorer of Abyssinia. Toured the alleged Mount Sinai in Egypt. (*DNB, MEB*)
1834 *2981* **1840** *4639* **1846** *6720 6732* **1847** *7095* **1849** *7976* **1864** *12592 12600* **1866** *13266 13274*

Beke, Emily Alston. Second wife of Charles Tilstone Beke. (*DNB*)
1870 *14257*

Belfour, Edward Secretary to the Trustees of the Hunterian Collection. (1840–1–30)
1840 4408

Belime, de
1848 *7613*

Bell, [Mr.]
1860 11266

Bell, Catherine (1780–1855). Married John Bell, 1821.
1834 *2930* **1844** *5831* **undated** *14514 14515*

Bell, Sir Charles (1774–1842). British anatomist. Studied nerve functions. Professor of surgery at Edinburgh. (*DNB, DSB*)
1832 *2675* 2680

Bell, Charles Davidson (1813–82). Artist and civil servant. Nephew of Sir John Bell. Draughtsman and surveyor at the Cape. (*SADNB, DSAfB*)
1835 *3200* **1848** *7609 7665* **1851** *9008*

Bell, Sir Isaac L. (1816–1904). British metallurgical chemist and industrialist. (*DNB*)
1862 *12074 12079* **1863** *12238 12251* 12256 *12259* 12260 *12308* **1869** 14075

Bell, James. Tuscan consul general in London in 1847. (1847–10–11)
1847 *7283 7306*

Bell, John. A neighbor of JH.
1860 *11424 11438* **1867** *13476*

Bell, Sir John (1782–1876). British army officer and civil servant at the Cape. Interested in educational reform. (*DNB, DSAfB*)
1834 *2971 3021 3022* 3026 **1835** *3137 3159* **1838** *3618 3628* 3638 *3639 3875* **1846** *6662* **1850** *8454*

Bell, Marion Shaw (1787–1876). Wife of Sir Charles Bell. (*MEB, AlliSup*)
1866 *13179*

Bell, Thomas (1792–1880). British comparative anatomist and dental surgeon. Professor of zoology at King's College, London. (*DNB, MEB*)
1852 9360 *9373* 9398 *9542* 9552 **1853** *9635* 9640 9691

Bellew, Richard M. (1803–80). British M.P. and a lord of the Treasury. (*MEB*)
1851 *8680*

Beneke, Friedrich Eduard (1798–1854). German philosopher. Professor at Göttingen and Berlin. (*NDB*)
1842 *5190* **1843** *5675* **1845** *6187 6336 6524* **1850** *8273 8357*

Bennett, Alfred William (1833–1902). British botanist, bookseller and publisher. Lecturer in botany and editor of *Nature*. (*DNB*)
1862 *12043 12060* 12061

Benson, William (fl. 1865). English chromatologist, writer. (*AlliSup*)
1869 *13917* 13919 *13928* **1871** *14397*

Bentley, George (1828–95). English publisher, of numerous works of fiction, travel, history, and biography. (*MEB, AlliSup*)
1860 *11416 11425*

Bentley, Richard (1794–1871). English publisher. Printed two novels of Charles Dickens, *Oliver Twist* and *Barnaby Rudge*. (*DNB, MEB*)
1854 *10267* 10269

Beresford, William Carr (1768–1854). British army officer and 1st Viscount Beresford. Governor of Madiera. M.P. for Cambridge University. (*DNB, Peerage*)
1845 *6453 6461*

Beresford-Hope, Alexander James (see Hope, Alexander J. B.)

Beresford-Hope's Committee, Secretaries to
1868 *13604*

Berger, Capel H.
1864 *12649* 12653

Bernard, Montague (1820–1882). English scholar of international law. (*DNB*)
1863 12206

Berzelius, Jöns Jakob (1779–1848). Swedish electrochemist and mineralogist. Professor of medicine in Stockholm. (*DSB*)
1838 *3695*

Bessel, Friedrich Wilhelm (1784–1846). German mathematician and astronomer. Made first measurement of stellar parallax, 1838. (*DSB*)
1820 442 *468* **1821** 631 633 **1822** 666 *709 746* 770 **1828** *1760* **1833** 2776 **1834** *2989* **1835** 3116 **1838** *3671 3788* 3807 *3831* **1839** *4138* 4343 **1840** *4519* **1842** 5181 *5211* 5212 *5215* 5217 *5226 5227* 5230 *5310* **1843** *5388* 5490 *5506* **1844** *5709* 5879 *5964*

Best, Baron George (1756–1823).
1821 *553*

Bethune, John Elliott Drinkwater (see Drinkwater-Bethune, John Elliott)

Bewley, Alexander V.
1847 7297 **1869** *13860*

Bianchi, Joseph (1791–1866). Professor of astronomy at Milan in 1821. (*Houzeau*)
1828 *1664* **1829** *1944* **1838** *3651*

Bidwell, J. E.
1841 5044

Billet, M. Felix (1808–82). French physicist. Most of his work was in optics. (*Who's*)
1859 *11024* 11027 **1860** *11328*

Bingley, Henry. Employed at Mint, 1825–51. (Challis, *Mint*)
1851 8602 *8675 8676 8677 8687 8707* 8710 8716 *8748 8750* 8751 *8762 8763* 8764 8810 *8816 8873 8909* 8918 *8936* 8937 *8954* 8958 *8961* 8962

Biot, Jean Baptiste (1774–1862). French physicist. Made balloon ascent (with Joseph Louis Gay–Lussac). Studied rotary polarization of light. (*DSB, Who's*)
1819 *348 378* 387 **1820** 435 *438 480* **1821** 515 **1836** *3338* **1838** 3680 **1839** *4100* **1841** *4913 4927* **1860** 11405 *11418* **1861** *11697*

Biot, Madame. Wife of Jean Baptiste Biot, who taught her science and mathematics. (*DSB*)
1839 *3906 4094*

Birbeck, William Lloyd
1866 *13096*

Birch, Samuel (1813–85). Keeper of Oriental Antiquities at the British Museum, 1861–85. (*MEB, AlliSup*)
1866 *13064* 13068 *13074*

Birkenthal, B.
1860 *11297 11300*

Birkett, Daniel. Executor, with JH and Robert Spottiswoode, of the estate of E. S. Beckwith's husband. (1830–8–30)
1830 *2220 2222 2225* **1831** *2331*

Birt, William Radcliffe (1804–81). Meteorologist and astronomer. Employed by JH to reduce barometric observations. (*DSB, MEB*)
1838 3776 **1840** 4554 4623 4627 4633 4671 4745 4754 **1841** 4761 4800 4831 4843 4875 4906 4939 4952 5033 **1842** 5193 5200 5281 5284 **1843** 5345 5367 5369 *5409 5544* 5545 5547 5554 5559 5561 *5588* 5591 *5598* 5607 *5612* **1844** *5701 5736 5740* 5807 5889 5912 5916 *5921* 5929 **1845** *6089 6098* 6253 *6269* 6294 *6450* 6454 *6486* 6490 *6508* **1847** *6946* 6947 *7065* 7068 *7216* 7269 7444 *7457* **1848** 7496 *7503 7517* 7594 *7598* 7603 7625 *7647 7653* 7654 7656 *7715 7721* 7724 *7804 7807* **1849** *7973 7977 7987 7991 8031 8061* 8078 8084 *8085 8090 8168 8174* **1850** *8194 8196 8221 8274* 8278 *8296 8340* 8343 *8345* 8347 **1860** *11360* **1863** *12145* 12156 **1865** *12804 12891* **1866** *13220*

Bishop, George (1785–1861). Vintner and astronomer. Erected a private observatory near Regent's Park. (*MEB*)
1845 *6317* **1846** *6602* **1847** 7117 *7310 7323* **1848** *7705* **1851** 8797 *8805* **undated** *14516*

Blackburn, Bewicke. Friend of George Biddell Airy. (1864–5–17)
1864 *12542 12576*

Blagden, Charles (1748–1820). Prominent English physician and chemist. Secretary of the R.S.L., 1784–1820. (*DNB, Alli*)
1819 *337*

Bland, Miles (1786–1867). Mathematician and divine. Church official at Wells Cathedral. (*DNB, MEB*)
1818 *283* **1821** *535* 538 **1822** *731* 732 **1831** 2266 *2275* **1832** *2528*

Blane, Gilbert (1749–1834). Scottish physician. Advised the British government on sanitary and medical matters. (*DNB, Alli*)
1831 *2267 2293* 2450 *2451*

Blanford, Henry Francis (1834–93). Meteorologist and geologist. Held numerous scientific positions in India. (*DNB, MEB*)
1866 *13282*

Bleek, Wilhelm H. I. (1827–75). German-born expert on South African philology. (*DNB, AlliSup*)
1865 *13046*

Blessington, Margaret Power (1789–1849). English author who was one of the great socialites of her day. (*DNB, Alli*)
undated *14517 14518*

Blewitt, Octavian (1810–84). Secretary to Royal Literary Fund in London, 1839–84. Editor of *Gardener's Chronicle* and travel writer. (*MEB*)
1843 *5445*

Board of Trade, the
1854 9914

Boase, Henry Samuel (1799–1883). English geologist, physician, and industrialist. (*DNB, AlliSup*)
1860 11397 *11440*

Bode, Johann Elert (1747–1826). German astronomer. Director of the Berlin Observatory and member of the Berlin Academy. (*DSB, Who's*)
1820 443 *483* **1821** 544 *565 604 626 640* **1822** *722* **1825** *1108*

Boguslawski, Palm Heinrich Ludwig (1789–1851). German astronomer and director of Breslau [Wroclaw] observatory. (*DSB, NDB*)
1838 *3774 3818* **1844** *5759 5938* **1845** *6131* **1848** *7529*

Boileau, John Theophilus (1805–86). Officer in Bengal Engineers and director of magnetic observatory in Simla, India. (*MEB*)
1839 *4257* **1840** *4402 4455*

Bold, [S?]. JH's host at the B.A.A.S. meeting in Southampton in 1846. (1846–9–8)
1846 *6728*

Bond, George Phillips (1825–65). American astronomer. Discovered the Saturnian satellite Hyperion. Pioneer of photographic astronomy. (*DAB, DSB*)
1851 *8847 8904* **1862** *12003* **1863** *12227 12252 12276 12350 12391*

Bond, William Cranch (1789–1859). Chronometer maker and astronomer. Pioneer in meteorology and magnetism. (*DAB, DSB*)
1839 *3996* 4132 **1847** *7337* **1849** *7993 8066*

Bonne, Charles Rigobert Marie (1771–1839). French army officer and engineer. (*DBF*)
1825 *1127 1184 1192 1194 1195 1217*

Bonsey, W. A neighbor of JH at Slough.
1859 *11074 11238* **1864** *12528 12532*

Boole, George (1815–64). Mathematician and logician. Professor of mathematics at Queen's College, Cork. (*DNB, MEB*)
1861 *11584 11586 11608*

Booth, James (1806–78). Mathematician, divine and academic administrator. (*DNB, MEB*)
1839 4074 **1856** *10473* 10474 *10475* 10478 10597 **1858** *10821* **1859** *11021*

Borden, Jr., Gail (1801–74). American surveyor and inventor. Patented a process for evaporating milk. (*Who's, DAB*)
1851 *8851*

Bosse, M. French publisher. (JH's 1844–3–17)
1844 5760

Boulton, Matthew Robinson (1770–1842). Co-owner of Soho Engineering Works. Developed scientific management techniques. (*Who's*)
1829 1986 *1989* 1990 **undated** *14519*

Bourke, Richard (1777–1855). Governor of eastern district of the Cape of Good Hope, 1825–8. (*DNB, MEB*)
1839 *4213*

Bourne, John S.
1870 *14103*

Bouvard, Alexis (1767–1843). French astronomer. First pointed out irregularities in Uranus' orbit, 1821. (*Who's*)
1820 444 *457* **1821** 525 *557* 576 **1824** 950

Bouvier, Hannah M. (1811–?). American-born astronomer. Wrote acclaimed work *Familiar Astronomy.* (*Alli*)
1856 *10531 10595* **1857** 10630 *10645*

Bowdich, Sarah (née Lee) (1791–1856). British artist and author of works of natural history and travel. Married T. E. Bowdich (1791–1824). (*MEB*)
1828 *1703 1712* **1830** *2156* **1833** *2721 2746* **1838** *3656*

Bowditch, Henry Ingersoll (1808–92). American physician and abolitionist. Professor of medicine at Harvard. (*NCAB, DAB*)
1847 *7338*

Bowditch [Jr.], Nathaniel. Son of Nathaniel Bowditch, Sr.
1838 *3779* **1839** *4097 4120*

Bowditch [Sr.], Nathaniel (1773–1838). American navigator, astronomer and mathematician. Translated most of Laplace's *Mécanique céleste.* (*Who's*)
1829 *2050* **1830** 2114 **1832** *2586*

Boye, [I. I.]
1844 *5745*

Brachet, Achille
1855 *10431 10435* 10439 **1856** *10594 10596*

Bramwell, G. Secretary to trustees of Radcliffe Observatory at Oxford in 1859. (JH's 1859–3–27)
1859 11049

Brande, Frederick Robert (1829–?). Apprentice moneyer at Mint. Son of W. T. Brande. (Craig, *Mint*)
1855 *10315*

Brande, William Thomas (1788–1866). Chemistry professor and mint employee. (*DNB, MEB*)
1833 *2817* **1850** *8419* 8441 **1851** *8773 8928* 8951 **1852** *9296* 9539 **1853** 9724 9753 9829 *9833* 9846 9876 9879 **1854** 9936 *9950 9951* 9963 9969 9971 10005 10019 10044 10049 10108 *10233* **1855** 10314 10325 *10330 10331 10337* 10338 *10346 10353* **undated** *14520*

Brass, J[ohn]
1812 *52*

Brassey, Thomas (1836–1918). British nautical engineer. M.P. for Hastings, 1868–86. (*DNB*)
1870 *14078* 14159

Brayley, Edward William (1802–70). English scientific author. Disputed JH's thermotic theory of plutonic and volcanic action. (*DNB, Alli*)
1845 *6157* **1846** *6678* **1855** 10446 *10451* **1856** *10471* **1859** *11032* 11033 *11034 11052* **1862** *11919 11934* 11939 *11940 11949 12093* **1863** 12097 *12405* **1864** *12731* 12736 *12756* **1865** *12851* 12854 *12858 13030* 13039 **1866** *13136* **1867** *13584* **1868** *13616* **1869** *13835 14016* 14019 **undated** *14521 14522*

Brazier, James Smith (1825–89). Professor of chemistry at Aberdeen University. (*MEB*)
1860 *11492 11498*

Breda, Jacob Gijsbertus Samuel van (1788–1867). Permanent secretary to Dutch Academy of Sciences, and museum director. (Williams, *Faraday*)
1848 *7657*

Brewster, David (1781–1868). Natural philosopher, divine and editor. University administrator at St. Andrew's and Edinburgh. (*DNB, DSB*)
1817 *242* 246 *247* 255 **1818** *290* 310 **1819** 331 332 352 359 *360 367 371* 373 *375* 380 *382* 383 *386* 388 389 **1820** 399 402 *404* 431 **1822** *667* 751 753 771 *813* **1823** 842 858 883 928 931 **1824** 940 959 *976* 1078 **1825** 1264 *1267* **1827** 1458 *1509* **1828** 1745 *1896* **1831** 2457 2458 *2469* 2471 *2480* **1832** 2492 *2587* 2588 2650 *2666* **1847** *7226* **1850** *8358* **1851** 8956 **1865** *12838* 12842 **1867** *13521*

Briarly, Mr.
1832 2532

Briggs, John Thomas (1781–1865). Accountant-general of the British navy. (*DNB*)
1850 *8287*

Briguola, Comte Jarsana de. Commissioner general for England of the General Shipwreck Relief Society. (1839–4–13)
1839 *4064*

Brinkley, John (1763–1835). British astronomer, mathematician, and divine. First Astronomer Royal for Ireland, 1792. (*Who's*)
1821 *520* **1822** *826* **1824** *964* 1085 *1088* **1825** *1109* **1826** *1304*

Brisbane, Thomas Makdougall (1773–1860). British astronomer and army officer. Governor of New South Wales, established Parramatta Observatory there. (*DNB, DSB*)
1827 1483 **1828** *1704* **1841** *5052* **1842** 5071 **1847** *7327*

B.A.A.S.
1840 4675 **1844** 5948

B.A.A.S. Committee
1848 7798

B.A.A.S., Physical Section of the
1841 4961

B.A.A.S., President
1858 10810

British Meteorological Society, the
1866 13298

Brockedon, William (1787–1854). English painter, author, and inventor. Published a journal of his travels in the Alps. (*DNB, Alli*)
1851 *8703 8717*

Brodie [the elder], Sir Benjamin Collins (1783–1862). Physician and surgeon. Surgeon to George IV and William IV. (*DNB, MEB*)
1859 *11001 11057 11113* 11198 **1860** *11264* 11267 *11272 11362* **1861** *11658 11663 11666 11667 11669 11678* **undated** *14524*

Brodie, Jr., Sir Benjamin Collins (1817–80). Chemist and professor at Oxford. Son of Benjamin Collins Brodie, the elder. (*DNB, MEB*)
1867 *13454* 13462 *13466*

Brodie, Philothea M.
undated *14523*

Bromley, Richard Madox (1813–65). English civil servant. Commissioner of Greenwich Hospital, 1863–5. (*DNB, MEB*)
1852 9246

Brooke, William O'Shaughnessy. British physician, who served in India and retired to England. Assay master at Calcutta mint. (1852–9–3)
1843 *5373 5384* **1852** *9502*

Brothers, Alfred. British pioneer photographer. Experimented with magnesium light. (Eder, *Photography*)
1864 *12700* 12701 *12703 12704* **1866** *13157* **1871** *14380* 14384 *14385*

Brotherton, A. H.
1859 *11252*

Brougham, Henry Peter (1778–1868). English barrister and politician. M.P. and in cabinet, especially as Lord Chancellor. Baron Brougham & Vaux (*DNB*)
1827 *1548* 1550 1551 *1552* **1850** 8175 *8176* **1854** *10010* 10014 **1855** 10442

Broun, John C. Lecturer in botany at King's College, Aberdeen, and then an itinerant lecturer in botany at the Cape. (1864–7–4)
1863 *12299* **1864** *12460 12637*

Brown, R. S.
1863 *12284*

Brown, Robert (c. 1789–1873). Scottish musicologist and writer on musical theory. (*Brown*)
1858 *10885 10959 10960* 10984 *10993* **1860** *11269 11274 11326 11442* **1862** *11903* **1870** *14282 14319*

Browning, John (fl. 1867–88). English astronomer and writer. (*AlliSup*)
1869 *14064*

Brunnov, Filipp Ivanovich (1797–1875). Russian diplomat and assistant to K. V. Nesselrode. Envoy to London. (*MERSH, GSE*)
1847 *7294 7370*

Bryant, H. S.
1864 *12762*

Buchanan, Thomas
1840 *4477*

Buckland, William (1784–1856). Geologist and divine. Served in a number of church positions. (*DNB, DSB*)
1828 *1655 1668* **1829** 1932 **1832** 2620 **1837** *3510* 3567 **1839** *4010*

Buckle, Emma. Wife of William Buckle. (1863–11–23)
1863 *12373 12380 12396*

Buckle, William (1794–1863). Employee of Bolton and Watt in Soho, Birmingham, and then at Royal Mint. (*MEB*)
1851 8739 8776 **1853** 9779

Buckley, J.
1853 9852

Buist, George (1805–1860). Anglo-Indian newspaper editor in Scotland and India. Licensed as a preacher in Scotland. (*DNB*)
1847 *7458*

Bullar, John (ca. 1774–1864). English schoolmaster, biographer, and religious writer. (*BBI*)
1846 *6749* **1847** *7196* **1856** *10553 10559* **1857** *10703 10755* **1858** *10782 10943* **1859** *11151* **1862** *11988* **1863** *12113 12235* **undated** *14525*

Bullar, Joseph (fl. 1841). English physician and travel writer. (*BBI*)
1856 *10536 10541* **1861** *11612 11706 11726* **1865** *13023 13031*

Bulstrode, C. Owner of, or agent for, house at 32 Harley St., where JH resided while Master of Mint, 1851–55. (1854–11–25)
1854 *10256*

Bunbury, Charles James Fox (1809–86). English botanist and natural historian. Wrote a book about the Cape. (*MEB*)
1847 *7290* 7298 *7311*

Bunbury, Henry (1778–1860). British army officer, historian, and M.P. (*DNB*)
1838 3672 **1840** 4634 4645

Bunsen, Christian C. J. (1791–1860). Prussian theologian and diplomat. (*NDtB*)
1834 *2940 2945* **1847** *7222* **1848** *7722* **1851** *9045* **1854** 9960

Bunt, Thomas G.
1870 *14326 14332*

Burder, William Corbett (1822–65). Brother to physician George J. Burder. Member of the R.A.S. (1863–2–19)
1863 *12163 12177*

Burke, John (1787–1848). Genealogist, and author of *Genealogical and Heraldic Dictionary of the Peerage and Baronetage of the United Kingdom*, 1826. (*DNB*)
1844 *5955*

Butler, Pierce (1826–68). British divine and explorer. Organizer of Sinai Peninsula Survey. (*MEB*)
1867 *13562* 13574 **1869** *14065 14066*

Butterworth and Heath Firm
1864 *12764*

Button, Eustace
1861 *11676 11683*

Buys Ballot, C. H. D. (1817–90). Dutch geologist. Professor of mathematics and physics at Utrecht. (*DSB*)
1867 *13383*

Byam, Mr. On staff at Lithographic Press, 36 Frith St., Soho, in 1822. (JH's 1822–9–2)
1822 762 769

C., Miss
1815 132

Cacciatore, Niccolo (1780–1841). Assistant to Giuseppi Piazzi, and director of Royal Observatory at Palermo. (*DBIt*)
1824 *1087* **1825** 1111 *1129 1136* **1826** *1431* **1827** *1451* **1828** 1656 **1834** *2975* **1836** *3251*

Caggiati, L. Professor of clinical medicine at Parma. (1854–8–27)
1854 *10200*

Calais, Postmaster at
1822 798

Caldwell, Captain. American naval officer who visited JH at the Cape. (Evans, *Herschel*)
1836 3239

Callcott, William H. (1807–82). English composer. Arranged numerous piano works. (*DNB, AlliSup*)
1863 *12110 12114* **1867** *13381* **1868** *13769*

Cameron, Charles Hay (1795–1880). English lawyer. President of the Council of Education for Bengal, India, 1843–8. (*DNB, AlliSup*)
1850 *8284* 8298 **1852** *9557* **1853** *9631* **1857** *10681* **1867** *13406*

Cameron, Henry H. H. Son of Charles Hay Cameron.
1870 14127 *14128* **undated** *14526*

Cameron, Julia Margaret (née Pattle) (1815–79). Indian-born portrait photographer. Took two portraits of JH, 1867. Wife of Charles Hay Cameron. (*DNB, MEB*)
1839 4125 **1841** 4922 **1846** 6700 **1847** *7019* **1848** *7482* **1849** *8128 8141* **1850** 8303 **1851** *8571* 8578 **1852** 9418 **1853** 9824 9908 9909 **1858** *10764 10955* **1860** *11255 11317* **1862** *11866 11894* **1864** *12498 12523* 12734 *12737 12767* **1865** *12768* **1866** *13082*

13087 *13092* 13262 *13289* **1867** *13403 13411 13511* **1868** *13613* **1870** *14110 14133 14230* 14310 **1871** *14348* **undated** *14527 14528 14529*

Candolle, Alphonse (1806–93). Swiss botanist and phytogeographer. An avid historian of science. (*Who's*) **1836** *3252* **1837** 3526 **1838** *3734*

Cantor, Theo E. (Theodore Edward) (fl. 1836–54). Irish physician, zoologist, and writer. (*BBI*) **1838** *3872* **1841** *5028*

Capellini, John [Giovanni] (1833–1922). Italian geologist and paleontologist. President of the Italian Geological Society. (*GDE*) **1863** *12286*

Capocchi, Ernesto (1798–1864). Italian astronomer. Observer at Naples Observatory. (*BHGW*) **1852** *9157*

Carpenter, William Benjamin (1813–85). British naturalist and professor of forensic medicine. Explored deep ocean sediments. (*DNB*) **1838** *3730* **1839** *4311* **1840** *4429* **1842** *5096* 5106 *5196* **1851** *8791 8811* 8818 **1854** 10228 **1856** *10488 10492 10494 10515* **1858** *10970* 10977 **1869** 13852 *13854 13910* **1871** *14425* 14428

Carrighan, Arthur Judd (1781?–1845). English divine. Fellow of St. John's College, Cambridge. (*Cantab*) **1827** *1644* 1648

Carrington, Richard Christopher (1826–75). British astronomer who built several private observatories. (*DNB, DSB*) **1857** *10688* **1858** *10823 10979* **1859** *11015 11018* 11019 *11036 11042 11232 11239 11248 11249* **1860** *11304 11337 11344* **1861** *11559* 11562 *11565* **1863** *12183 12192* **1864** *12432 12437 12458 12533* 12638

Castelcicala, Prince di. Probably belonged to the Ruffo family who were the princes of Castelcicala. Attached to Naples embassy in London. (*GDE*) **1847** *7331*

Caswell, Alexis (1799–1877). American astronomer. Made precise measurements on weather patterns. (*DAB, Who's*) **1861** *11624*

Catalan, Eugène Charles (1814–94). French mathematician. Lecturer in mathematics in Liége. (*DBF*) **1847** *7362*

Catton, Thomas (1760–1838). English astronomer and divine. Director of St. John's College observatory. (*DNB*) **1818** *281* **1820** *423* **1821** *638* **1822** 714 **1823** *933* **1827** *1640*

Cauchoix, Robert A. (1776–1845). French optician. Invented achromatic glass with long focal length. (*LGE*) **1833** 2856

Cauchy, Augustin Louis (1789–1857). French mathematician and physicist. Studied hydrodynamics and astronomy. (*DSB, Who's*) **undated** *14530*

Cavendish, William (1808–91). English politician, industrialist, and scientific patron. Chancellor of Cambridge University. 7th Duke of Devonshire. (*DNB*) **1862** *11853* 11854

Cayley, Arthur (1821–95). English mathematician and astronomer. Professor of pure mathematics at Cambridge. (*DSB, AlliSup*) **1858** *10966* **1860** *11503* **1865** *12769* **1868** *13622 13628* **1869** *13841 13849 13858* **1870** *14245*

Cecil, Robert Gascoyne- (see Gascoyne-Cecil, Robert A. T.)

Cello, Baron de. Bavarian minister to England in 1847. (1847–10–4) **1847** *7256*

Challis, James (1803–82). English astronomer and divine. Director of Cambridge Observatory, 1836–61. (*DNB, DSB*) **1840** 4684 **1844** 5794 **1847** *7162* **1849** *8050* **1850** *8379* **1861** *11637* **1862** *12080* **1865** *12777* **1866** *13083 13319* **1869** *13893*

Chambers, George F. (1841–?). English lawyer. Wrote *A Handbook of Descriptive and Practical Astronomy* (1861). (*Alli, AlliSup*) **1862** *11766*

Champnes, Edward. Probably the vicar at Upton, near Slough. (1849–10–15) **1849** *8112*

Chandos, Lord (see Grenville, Richard)

Chantrey, Francis Legatt (1781–1841). English sculptor. Famous for his portrait sculptures and his statues of children. (*DNB*) **1823** *857 863* **1829** *1915* **1830** *2206* **1831** *2374*

Chapman, John (1822–94). British watchmaker, bookseller, and publisher. Proprietor and editor of *Westminster Review*, 1851–94. M.D. 1857. (*MEB*) **1845** *6364* **1867** *13463*

Charlesworth, Edward (1813–93). Proprietor and editor of London's *Magazine of Natural History*, 1828–40. Curator at museum of Yorkshire Philosophical Society, 1844–58. (*MEB*) **1847** *7229*

Charsley, F. Tenant at Observatory House at Slough. (1865–8–5) **1865** *12946* **1869** *14009* 14010 **1870** *14112* 14113 *14125*

Chatterton, Henrietta G. M. L. (1806–76). English author. Published many tales, notes of travel, and poems. (*DNB, Alli*) **1868** *13765* **undated** *14531*

Cheney, Theseus A. (1830–78). American scholar. Studied local history and founded the Georgic Society and Library. (*AlliSup*) **1869** *14074*

Chesney, Francis Rawdon (1789–1872). Irish-born officer in the Royal Artillery. Made expeditions through Syria and the Euphrates valley. (*MEB, AlliSup*) **1841** *5039*

Chester, Bishop of (see Graham, John)

Chevalier, Charles Louis (1804–59). French optician. Perfected solar microscope. Invented new instruments for the camera. (*Who's*)
1840 *4681*

Chevalier, Michel (1806–79). French economist at the College of France. Wrote on the gold standard and other monetary matters. (*DBF*)
1857 *10629 10701 10741* **1859** *11020*

Chevallier, Temple (1794–1873). English divine and professor of astronomy, mathematics and Hebrew. (*DNB, DSB*)
1847 *7164 7174*

Chevigny, Louis Marie Joseph, Comte de (1793–1876). French author. Proposed French translation of JH's *Outlines Astr.*, but it was not completed. (*DBF*)
1866 *13062*

Chevreul, Michel Eugène (1786–1889). French chemist and gerontologist. A founder of modern organic chemistry. (*DSB, Who's*)
1839 *4103*

Children, John George (1777–1852). British chemist. Wrote on mineralogy and electricity. Employed at British Museum. (*DNB, Who's*)
1827 *1543* **1831** 2444 **1832** 2644 **1850** *8367*

Christian VIII (Christian Frederik) (1786–1848). King of Denmark, 1839–48. (*ScandHist*)
1840 4479 **1842** 5087

Christie, Samuel Hunter (1784–1865). British mathematician and student of magnetism. Secretary to R.S.L., 1837–54. (*DNB, MEB*)
1840 4604 *4728* **1841** 5022 **1843** 5415 **1847** *7404*
1851 8619 8758 **1852** 9553

Chubb, Hammond. Secretary to the Bank of England. (1870–2–8)
1870 *14108 14114* 14120 *14123 14129* 14130

Clapham, R. Calvin. Honorary secretary to Literary and Philosophical Society at Newcastle-upon-Tyne in 1867. (1867–5)
1867 *13422*

Clark, Alvan (1804–87). American portrait painter, astronomer, and manufacturer of astronomical instruments. (*DAB, DSB*)
1852 *9399 9416*

Clark, Richard. President of Metropolitan Light Company in 1848. (WS's 1848–8)
1848 *7799*

Clarke, Alexander R. (1828–1914). Scottish-born geodesist. Fellow of the Royal Society. (*Who's, AlliSup*)
1863 12329 *12330 12370*

Clarke, Edward Daniel (1769–1822). British mineralogist and University librarian at Cambridge. (*DNB, DSB*)
1816 *163* 164 165 166 167 176 198 *202* 208 **1817** *211 225* **1822** *647 648*

Clarke, Miss B. S. Daughter of Edward Daniel Clarke. (1870–4–15)
1870 *14160 14162*

Clarke, William B. (1798–1878). British geologist and divine. Discovered gold in Australia, 1841. (*MEB*)
1868 *13697*

Clarke-Whitfeld, John (1770–1836). British composer, organist, and choirmaster at St. John's College, Cambridge, 1797?–1821. (*DNB*)
1817 *233* **1820** *450 465 470*

Clegram, W. B.
1857 *10614* 10615

Clement, Joseph (1779–1844). Skilled draftsman and developer of machine tools. Supervisor of construction on Charles Babbage's calculating machine. (Lindgren, *Glory*)
1827 *1624* **1828** *1672 1705 1754 1787* 1793

Clerk, George (1787–1867). British M.P., Secretary to the Treasury, and Master of the Mint, 1845–6. (*DNB*)
1843 *5508*

Cleveland, S.
1858 *10854*

Clift, William (1775–1849). English naturalist. Curator of the Hunterian Museum. (*DSB, Who's*)
1836 *3257*

Cloete, Abraham Josiah. South-African officer in British army. Officer in charge of Cape Town when JH was at the Cape. (Warners, *Maclear*)
1835 3120

Clough, Arthur Hugh (1819–61). British poet and academic administrator. (*DNB*)
1851 9103

Coates, J.
1842 5235

Cochrane, James Inglis (fl. 1860). Translated Homer's *Iliad.* (*AlliSup*)
1862 *11970*

Coddington, Henry (?–1845). English mathematician. Wrote essays on optical theory. (*DNB*)
1828 *1780*

Codrington, Edward (1780–1851). English naval officer. (*DNB*)
undated *14532*

Colby, Thomas (1784–1852). British officer in Royal Engineers. Director of ordnance triangulation survey of England. (*DNB, MEB*)
1827 *1596*

Cole, James H. Secretary to the Chancellor of the Exchequer. (1851–11–25)
1851 *9029 9068 9078*

Cole, Thomas
1852 *9338 9354*

Colebrooke, Thomas Edward (1813–90). English M.P., lord lieutenant of Lanarkshire, and biographer. (*MEB*)
undated 14533

Colladon, Jean-Daniel (1802–93). Swiss engineer. Professor of mechanics at Paris and Geneva. (*Who's*)
1851 *8978*

Collen, Mr.
1842 5233 5240

Colling, Elizabeth (fl. 1856–70). English poet. Exchanged poetry with JH. (*AlliSup*)
1858 10786 10920 **1859** 11102 11127 11159 **1861** 11638 **1863** 12107 12120 12399 **1864** 12511 12741 **1865** 12882 **1866** 13069 13090 13278 13333 **1867** 13371 13496 13581 **1868** 13612 13663 13730 13775 13832 **1869** 13903 13951 14035 **1870** 14089 14134 14220 14244 14253 14303 14315 **1871** 14362 14371 14408 14426

Collingwood, Sarah Newnham- (see Newnham-Collingwood, Sarah)

Collins, Matthew
1853 *9713* **1858** *10962* **1861** *11746* **1862** *11752 11761*

Collins, Thomas. Publisher at 4 Church Place, Covent Garden. (*Brit. Mus. Cat.*)
1847 7052

Compton, Marianne. Daughter of Spencer Joshua Alwyne Compton. (1838–12–18)
1838 *3888*

Compton, Spencer J. A. (1790–1851). R.S.L. president, 1833–48. 2nd Marquis of Northampton. (*DNB*)
1838 3757 3896 *3902* **1839** *3999 4000* 4185 **1840** 4609 4617 4618 **1841** 5023 5042 **1843** 5507 **1844** 5848 **1845** 6261 *6262* 6404 *6406 6427* 6446 *6464* **1846** *6755 6784* 6810 **1847** 7038 *7138 7345 7397* **1848** *7632* 7635 *7643* 7890 7897 **1849** *8003* 8005 8105 *8107* **undated** 14814

Configliachi, Pietro (1777–1844). Italian mathematician and physcist. (*DBIt*)
1821 608

Connal, Michael (1817–93). Scottish education administrator. Chairman of Glasgow School Board, 1876–85. (*MEB*)
1866 *13156*

Conway-Gordon, Lewis (1838–95). Army officer in Bengal Engineers. Director general of railways in India. (*MEB*)
undated *14600 14601*

Cook, William
1838 *3889*

Cooke, Christopher (fl. 1850). English astrological and travel writer. (*AlliSup*)
1867 *13368 13372*

Cooke, Josiah Parsons (1827–94). American chemist. Made early attempt to classify the elements according to atomic weight. (*DSB, Who's*)
1860 *11319 11323*

Cookson, Henry W. (1810–76). English divine. Master of Peterhouse, Cambridge. (*MEB, DNB*)
1864 *12557* 12559

Cooper, Edward Joshua (1798–1863). Irish astronomer and world traveller. Established Markree Observatory. (*DNB, MEB*)
1833 *2763* 2768 *2866* **1834** *2963* **1836** 3315 **1838** 3777 *3799* **1839** *4227 4269* 4274 **1840** *4392* **1843** *5459* 5539 *5622* **1844** *5805* **1847** *7207 7217* 7223 *7359* 7363 *7405* **1848** *7690* 7691 *7701* **1852** 9217 9243 *9256* 9294 **1855** *10389* **1859** *11208 11215 11225* **1860** *11285* 11288 **1862** *11867 11875 11880*

Cooper, John (1812?–96). English divine and academic administrator. (*MEB*)
1855 *10424*

Cooper, John Thomas (1790–1854). British chemist. Involved with JH and Michael Faraday in glass making experiments. (James, *Faraday*)
1827 1480 *1493*

Cornwallis, Miss. Lived in Gaudhurst, near Collingwood. Nursed Amelia Herschel after severe burns in Jan. 1851. (1853–1–17)
1853 9637 **1854** 9956

Cottans & Hallen, Engineers
1854 *10177* 10272

Courtney, R. Probably an editor for Spottiswoodes & Shaw, printers, in 1850. (1850–2–20))
1850 *8227*

Cowie, Benjamin Morgan (1816–1900). English mathematician and divine. Professor of geometry at Gresham College. (*DNB, MEB*)
1847 *7033*

Cowper, William Francis (1811–88). M.P. for Hertford, 1835–68, serving in many political appointments. Assumed name of Cowper-Temple. (*DNB*)
1838 *3836*

Coxwell, Henry (Tracey) (1819–1900). British aeronaut. Made many balloon ascents, including one to a record height. (*DNB*)
1866 *13059 13065*

Cracroft, Bernard (1828–88). English barrister, stockbroker and political writer. (*AlliSup*)
1860 *11279* **1865** *12796 12800 12833*

Craeyvanger, L. Secretary to Council of University of Utrecht, Netherlands, in 1847. (1847–10–4)
1847 *7257*

Craig, William Gibson (1797–1878). British M.P. and a lord of the Treasury. (*MEB*)
1851 *8680*

Cranborne, Lord (see Gascoyne-Cecil, Robert A. T.)

Cranch, Joseph. Brother-in-law of William Cranch Bond; sent by Harvard to investigate the purchase of a telescope. (Jones, *Harvard*)
1842 *5260* 5261

Cranswick, James Mosey (fl. 1855). English religious writer. (*AlliSup*)
1862 *11895* 11898 *11904*

Craven, Richard
1863 *12121*

Crick, Thomas (1801–76). English divine, and fellow of St. John's College, Cambridge. (*MEB*)
1845 *6149*

Criswick, Henry C. (fl. 1860). Prepared an agriculturalist's weather guide and manual. (*AlliSup*)
1862 *11753*

Croll, James (1821–90). Scottish natural philosopher and geologist. Published many articles on physical geology and climatology. (*DSB, AlliSup*)
1865 *12807* 12810 *12820* 12824 *12830* **1866** 13213 13235 *13239*

Crookes, William (1832–1919). British chemist and physicist. President of Society for Psychical Research, 1897. (*DNB, DSB*)
1858 *10934* 10938 10944 *10948 10952 10953* **1859** *11100 11106 11122* 11130 **1860** *11291 11301* **1863** 12301 *12321 12336 12343*

Crowden, Charles. English minister and educator. Headmaster of Cranbrook Grammar School, 1866–88. (*AlliSup*)
1867 *13464* **1871** *14433 14437*

Cruickshank, John (1787–1875). British mathematician, and librarian at Aberdeen University. (*MEB*)
1843 *5401* **1847** *7351*

Cubitt, H.
1843 *5370 5374 5448*

Curwen, John (1816–80). English minister. Gave up the ministry in 1864 in order to promote the tonic sol-fa system of teaching music. (*DNB, AlliSup*)
1864 *12531*

Custance, R.
1844 5802

Cuvier, Georges C. L. D. (1769–1832). French naturalist. Famous for his paleontological reconstructions of fossils. (*DSB, Who's*)
1826 *1376* **1831** 2448 *2452*

D'Abbadie, Antoine Thompson (1810–97). Irish-born astronomer and explorer in Brazil and Ethopia. Improved geodetic instruments. (*Who's*)
1849 *8170*

Daguerre, Louis (1789–1851). French inventor and photographer. Developed the Daguerreotype. (*Who's*)
1839 4186

Daily News
1861 11545 *11557 11558* **1865** 12915 **1868** 13666

Dale, Langham (1826–98). English-born professor of English and classics at South African College, Cape Town. (*DSAfB*)
1859 *11013 11017*

Dallmeyer, John Henry (1830–83). German-born optician, working in London. Improved telescope and microscope lenses. (*MEB, Who's*)
1859 *11226 11233* **1861** *11657* **1866** *13261* 13271 **1867** *13375* 13376 *13377* 13378 *13379* **1870** *14340* 14342 *14344*

Dalmahay, James
1862 *12015* 12045 *12051*

Dalton, John (1766–1844). English physicist, chemist, and meteorologist. Developed law of partial pressures and atomic theory. (*DSB, Who's*)
1832 *2493* **1833** *2751* 2795

Dames, Thomas L.
1865 *12774 12821*

Daniell, Edmund R. (?–1854). Barrister, brother of John Frederic Daniell. (Williams, *Faraday*)
1827 1521

Daniell, G. F. Daughter of John Frederic Daniell. (1857–1–26)
1857 *10620*

Daniell, John F. (1790–1845). British physicist, meteorologist, and chemist. Invented hygrometer. (*DNB*)
1831 2456 *2459* 2463 **1845** *6100*

Darby, William Arthur (fl. 1860–70). English rector, astronomer and writer. (*AlliSup*)
1864 *12625*

d'Arrest, Heinrich Louis (1822–75). German astronomer. Contributed to the discovery of Neptune and to the comet that bears his name. (*DSB, Who's*)
1847 *7273* **1848** *7782* **1862** *11812 12001* 12011 12012 **1863** *12143* 12184 12211 *12219 12229* **1864** 12689

Darvill, Henry (1812–83). Solictor at Windsor, 1834–83. (*MEB*)
1858 *10887*

Darwin, Charles Robert (1809–82). English naturalist and author of *Origin of Species*. (*DSB*)
1838 *3826* **1848** *7567 7633 7697 7703 7733* **1849** *8059* **1859** *11196* **1861** *11615*

Daubeny, Charles G. B. (1795–1867). British physician, chemist, and botanist. Traveling companion to JH in Italy, 1824. (*DNB, DSB*)
1832 2529 **1837** *3506* **1850** 8461 **1861** *11592* **1862** *11905* **1866** *13300 13306* **1867** *13449 13531* **1868** *13591* **undated** *14536 14537*

Daubrée, Gabriel Auguste (1814–96). French geologist. Researched the origin and artificial production of minerals. (*DSB, Who's*)
1861 *11716* 11725 **1864** *12424 12759* **1865** *12789 12791*

Daussy, Pierre (1792–1860). French hydrographic engineer. Member of French Bureau des longitudes. (*DBF*)
1848 *7575 7762*

Davenport, Mr. Tenant farmer renting property owned by Lady Mary Herschel at time of William Herschel's death. (1822–9–26, 1822–10–4, 1822–10–20)
1822 802 809 820

Davidson, Thomas (1817–85). Scottish paleontologist. Especially interested in brachiopods. (*Who's, AlliSup*)
1862 *11787* 11791 *11794 11800*

Davies, Charles
1871 *14350*

Davies, John. Secretary to Literary and Philosophical Society of Manchester. (1847–10–5)
1847 *7262 7263*

Davis, William B.
1870 *14313*

Davy, Humphry (1778–1829). English natural philosopher. Professor of chemistry at the Royal Institution. (*DNB, DSB*)
1822 *829* **1826** 1415 **undated** *14538 14539 14540 14541 14542 14543 14544 14545*

Davy, Martin (1763–1839). English physician and master of Caius College, Cambridge. (*DNB*)
1828 *1677* 1683 *1689*

Dawes, William Rutter (1799–1868). British astronomer. Observed at, and constructed, several private observatories. (*DNB, DSB*)
1831 2343 2355 *2470* **1832** *2584 2678* 2690 **1833** *2786* 2789 *2841 2848 2899* **1835** *3177* **1836** *3426* **1839** *4042 4055* **1842** *5119 5150 5307* **1843** *5548* **1844** *5702 5705 5711 5741 5866* **1845** *6096 6286 6410* **1846** *6601 6609 6716 6750 6762 6781 6794 6797* 6800 *6801* **1847** *6952 6956 6963 7270 7301 7368 7369 7401 7430 7439* **1848** *7492 7564 7740 7822 7849 7878* **1849** *8063 8150* **1850** *8351* **1851** *8493 8572 8579 9121* **1852** *9233* **1854** *10227 10274 10275* **1855** *10277 10394* **1857** *10685* **1859** *11211* **1860** *11514* **1861** *11530* **1862** *11762* **1863** *12128 12135 12311 12345 12392 12398 12406* **1864** *12499 12547 12583 12628 12670* **1865** *12794 12974* **1866** *13310 13314* **1867** *13583 13589* **1868** *13594*

Dawson, Ira.
1827 *1484*

Day, St. John Vincent (fl. 1869–77). English archaeologist. Interested in Egyptian pyramids. (*AlliSup*)
1868 *13685 13793*

Decuppis, Pompilio
1839 *4095*

De Kupffer, Adolf Theodor von (see Kupffer, Adolf Theodor de)

De la Beche, Henry T. (1786–1855). English geologist. Did extensive fieldwork in continental Europe and Jamaica. (*DSB, Who's*)
1846 *6654 6743* **1847** *7460 7475* **1848** *7620 7621 7941* **1854** *9933* 10076

Delambre, Jean-Baptiste (1749–1822). French astronomer. Prepared tables for Uranus, Jupiter and its satellites, Saturn, and the sun. (*DSB, Who's*)
1820 *458* **1821** 526 **1822** *711* 720 723

Delane, John T. (1817–79). Editor of the *Times*, 1841–77. (*DNB, Alli*)
undated *14563*

De La Pryme, Charles (1815–99). Member of Reform Club. One of original founders of the Devonshire Reform Club. (*MEB, AlliSup*)
1856 *10460 10563* **1870** *14148*

De La Rive, Arthur-Auguste (1801–73). Swiss physicist. Studied electricity and heat theory. (*Who's*)
1845 *6293*

de La Rue, Warren (1815–89). British astronomer, chemist, and inventor. Pioneer in lunar photography. (*DNB, MEB*)
1855 10420 **1856** *10548 10577 10579* **1857** 10670 *10673* 10717 *10719* **1858** *10812 10915* 10918 *10921* 10939 **1859** 11123 *11131* 11185 *11190 11240* **1860** *11284 11490* **1861** *11632 11672* 11713 *11714* **1862** 11799 *11992* **1863** *12352 12374* **1864** *12561 12584 12608 12613 12618 12622 12632 12672 12696 12725* 12735 *12743* **1865** *12784 12799* 12801 *12813* **1866** *13171* **1870** 14163 *14168 14177* 14186 *14235* 14236 **1871** 14374 *14376*

De La Vane, C.
1833 *2831*

De Morgan, Augustus (1806–71). English mathematician. Active in R.A.S., and close friend of JH. (*DNB*)
1831 2308 2312 *2433 2453* **1832** 2494 *2515 2611 2637 2643* **1833** 2894 *2929* **1834** *2951* **1836** *3352* 3415 3418 **1837** *3445* **1840** *4572* 4599 **1841** *4811* 4903 *4995* **1842** 5313 *5316 5343* **1843** *5674* **1844** 5679 5680 5718 *5720* 5723 *5726 5731* 5739 5746 5747 5748 5749 *5750* 5752 *5753* 5769 *5771 5772 5773* 5774 *5775* 5776 *5777 5778 5781* 5783 5790 *5791* 5810 *5812* 5819 *5821 5849* 5913 5914 5917 *5919 5943 5954 5957 5980 6003* 6022 *6031 6046* 6049 **1845** *6069 6121 6142 6144 6156 6162 6166* 6232 *6255 6265 6339 6359 6473* 6475 *6479* **1846** 6576 *6577 6679* 6681 *6815* 6819 *6831* 6834 **1847** 6865 *6867 6871* 6874 6882 6892 *6928* 6937 *7056 7083* 7468 **1848** *7539* 7544 *7545 7553* 7565 *7568* 7572 7604 *7674* 7684 *7685* 7686 *7689* 7692 *7702* 7900 7909 *7912* 7932 *7937* **1849** 7981 *7983* 7990 *7994* 7999 *8004* 8008 8011 8012 8014 *8015* 8021 *8042 8051* 8073 8094 *8098* 8171 **1850** 8177 *8198* 8202 *8207 8213* 8249 *8257 8382 8397* **1851** 8557 **1852** *9486* 9493 *9495* **1853** 9615 9616 *9705* 9715 *9793* 9795 *9796 9827 9859* 9867 **1854** 9915 10053 *10171 10243* 10245 *10247* 10248 **1855** *10354 10411* 10412 *10413 10419 10427* 10434 *10437 10441* 10443 *10444 10445* 10447 *10454* **1856** *10458 10504* 10512 *10513 10525* 10527 *10528 10547 10560 10565 10569* 10572 *10573 10584* 10586 *10588* **1857** *10617* 10623 *10628 10644* 10646 *10647* 10678 *10679* 10690 *10704 10723* 10728 *10730 10731 10732* 10735 *10737 10748 10752* 10756 *10757* **1858** 10765 *10766 10793* 10796 10800 *10804 10805* 10816 *10828* 10856 *10859 10896 10903 10907* 10917 *10922* 10945 *10949 10950 10967* 10985 *10989 10998* **1859** *11002 11011 11028 11029* 11035 *11065 11067 11068* 11073 *11076* 11094 *11095* 11114 *11140 11144 11150 11177* 11179 11228 **1860** 11256 *11331 11378 11404* 11406 *11408 11414 11427 11432 11487* **1861** *11519 11533 11570 11634* 11639 *11640 11699 11701* **1862** 11807 *11852 11869* 11873 *11876 11896* 11906 *11907 11950* 11953 *11957 11961* 11971 *11973* **1863** 12126 12178 *12185 12190 12199 12221 12277* 12278 *12288* 12355 *12368* **1864** *12483* 12548 *12563* 12565 12567 12616 *12617* 12619 *12627 12639* 12641 *12642 12659* 12660 *12661 12666 12680* **1865** 12864 *12866* 12934 *12962* 12964 *12970* 12971 *12972* 13050 *13053* **1866** *13151* 13158 *13159* 13234 *13237 13248* 13249 *13252 13265* 13281 *13311* **1867** 13392 *13394 13404 13412* 13415 *13442 13491 13494 13502 13508 13509 13519* 13530 *13532 13533* **1868** 13644 *13654 13723* **1869** *13908* 13911 *13912 13952* 13953 13961 *13964 13992* 13993 *14007* 14023 *14030 14031 14047* **1870** *14126* 14151 *14158* 14224 *14229* 14287 *14307* **undated** *14546 14547 14548 14549 14550 14551 14552 14553*

14554 14555 14556 14557 14558 14559 14560 14561 14562

De Morgan, Sophia E. (née Frend). Married Augustus De Morgan, 1837. (*Who's*)
1858 *10767 10768* **1871** *14405 14445 14446 14447 14448* **1873** *14450 14451 14452*

Denison, Edmund B. (1816–1905). English lawyer and author. (*MEB, Alli*)
1866 *13146*

Denny, Henry (1803–71). English entomologist. Curator for the Literary and Philosophical Society in Leeds for forty-five years. (*DNB, Alli*)
1870 *14138 14143*

Dewar, Daniel (1787–1867). British divine and academic, mostly in Aberdeen. (*MEB*)
1847 *7258*

De Worms, Henry (1840–1903). English barrister and politician. 1st Baron Pirbright. (*DNB, Peerage*)
1862 12065

Dickens, Charles (1812–70). English novelist [a.k.a. John Huffham]. (*DNB, Alli*)
1859 11180 *11181*

Dietrichstein, Count Franz Joseph J. N. (1767–1854). Austrian diplomat. (*NDtB*)
1847 *7302*

Dilke, Charles Wentworth (1789–1864). British antiquary and critic. Editor of the *Athenaeum*, 1830–46. (*DNB*)
1848 7914

Dimmock, James
1860 *11402 11415*

Dixon, Robert V. (?–1885). Irish deacon. (*MEB*)
1849 *8134* 8135

Dixon, William Hepworth (1821–79). English historian and traveller. Prolific writer for periodicals. (*DNB, Alli*)
1864 *12656* **undated** *14564*

Dollond, George (1774–1852). Optician, and successor to his uncle, Peter Dollond. (*DNB, MEB*)
1820 *496* **1823** 854 **1824** *981* **1842** 5328 *5330 5333* 5336 *5338* 5340 **1843** *5353* 5354 *5357* **1847** *7171*

Donaldson, John (?–1865). Reid professor of music at University of Edinburgh, 1845–65. (*DNB, MEB*)
1853 9763

D'Orsan, Alphonse
1861 *11704*

Douglas, James
undated *14565*

Dovrell, E. R. Probably an editor with William Clowes & Son, printers. (1851–1–10)
1851 *8561*

Draper, John William (1811–82). American chemist, medical doctor, historian, and philosopher. Pioneer in photography. (*DAB, DSB*)
1840 *4630* **1842** *5285* 5322 **1844** *6010* **1845** 6140 6235

Drew, Francis R. (fl. 1859–81). Science master at Malvern College, 1864–81. (*AlliSup*)
1865 *13040*

Drew, Frederick (1836–91). Youngest son of the astronomer John Drew (1809–57). Associate of the Royal School of Mines. (*MEB, AlliSup*)
1862 *11785*

Drinkwater-Bethune, John Elliot (1801–51). Indian legislator and educationist. Barrister, served in a number of important positions in India. (*DNB, MEB*)
1832 *2651* 2679 *2681 2682 2683* **1838** *3678* **1842** *5084*

Drummond, A. R.
1853 *9766* **1862** *11815*

Drummond [Jr.], Elizabeth. Sister of Thomas Drummond; resided in Edinburgh in 1867. (O'Brien, *Drummond*)
1865 *12862 12865* **1867** 13543 **1868** *13694* **1869** *13848*

Drummond [Sr.], Elizabeth (née Somers). Mother of Thomas Drummond. (O'Brien, *Drummond*)
1840 4540 **1844** *5857*

Drummond, Maurice. An official in the Home Office. (1859–11–10)
1859 *11191*

Drummond, Thomas (1797–1840). British officer in Royal Engineers. Involved in ordnance survey. Governor of Ireland. (*DNB*)
1830 *2115* **1832** 2533 *2592* **1833** *2813*

Drummond's Bank
1847 *7437* **1866** 13321

Dudgeon, J. W.
undated *14566 14567 14568*

Dudgeon, Robert Ellis (1820–1904). Scottish-born homeopath. Inventor of a sphygmograph for registering pulse. (*Who's, AlliSup*)
1871 14356 *14358 14387*

Dudgeon, T.
1856 *10590*

Duff-Gordon, Alexander Cornewall (1811–72). British senior civil servant. (*MEB*)
1850 *8435*

Dulong, Pierre L. (1785–1838). French chemist and physicist. Formulated, with Petit, the law relating atomic weights and specific heats of chemical substances. (*DSB, Who's*)
1832 *2633*

Dumas, Jean Baptiste André (1800–84). French chemist; professor and administrator. (*DSB, Who's*)
1844 5788 **1857** *10609* **1866** 13099

Duncan, James (1826–90). Scottish-born gynecologist. Researched anesthetics. (*Who's, AlliSup*)
1859 *11090 11092*

Dundas, Robert Saunders (1771–1851). Politician and cabinet minister. 2nd Viscount Melville. (*DNB*)
1830 *2243*

Dunkin, Edwin (1821–97). English astronomer. Associated with the Royal Observatory at Greenwich, 1838–84. (*Who's, AlliSup*)
1863 *12360* **1867** 13565

Dunlop, James (1795–1848). British astronomer. Worked with Charles Rümker and T. M. Brisbane. (*DNB, DAusB*)
1828 *1726* **1837** 3597

Dunraven, Lord (see Wyndham-Quin, Edwin R. W.)

Dunraven, Lady Augusta (see Wyndham-Quin, Augusta)

Duprat, Benjamin. Librarian at the Bibliothéque Impérial et du Sénat in Paris. (1860–10–15)
1860 *11445*

D'Urban, Benjamin (1777–1849). English army officer. Governor and commander-in-chief of the Cape of Good Hope, 1833–7. (*DNB*)
undated *14534 14535*

Dutrochet, René J. H. (1776–1847). French animal and plant physiologist. Made first experiments on osmosis. (*DSB, Who's*)
1843 *5449*

Dyer, H. Moreton
1839 *3924*

Dyson, T.
1842 *5219*

E. Sheldon and Co.
1865 13058

Eardley-Wilmot, F. M. (1812–77). British army officer involved in Kaffir wars. Director of Cape meteorological observatory. (*MEB*)
1840 *4419* **1841** *4836 4873* **1842** 5152 *5238* **1843** *5501 5509* **1844** *6007* **1846** *6813* **1848** *7745* **undated** *14569*

East India Company
1867 13548

Eclipse Committee of the Royal Society', 'Chairman of the
1870 14219

Eden, George (1784–1849). English statesman, and Master of Mint, 1830–35. 1st Earl of Auckland. (*DNB*)
1847 *7386* 7387 *7389* 7461 *7466* **1848** *7554 7581 7588 7623 7810 7812 7916 7919*

Edge, Emma. Wife of William J. Edge. (EE's [?]–8–24)
undated *14570*

Edge, William J. Vicar of Benenden, Kent, 1858–77. (*Alli, AlliSup*)
1864 *12760*

Edgeworth, Maria (1767–1849). Novelist, personal acquaintance of JH. (*DNB*)
1822 *763 793* 799 **1826** *1301* 1305 **1827** *1492* **1828** *1727* 1743 *1744 1750* **1831** *2337* **1843** 5646 **1847** *7099* **1848** *7838* **undated** *14571*

Edgeworth, Michael Pakenham (1812–81). English botanist. Civil servant in India who wrote extensively about its plant-life. (*DNB, MEB*)
1843 *5462*

Edie, William. Army officer at the Cape. Travelled with Andrew Smith. (Warners, *Maclear*)
1835 3100

Edinburgh Philosophical Journal
1822 649

Educational Times
1861 *11685*

Edwards, William Frédéric (1776–1842). French physiologist, anthropologist, ethnologist. (*DSB, DBF*)
1820 *479* **1824** *1066*

Egerton, Lady M[ary]. Possibly the wife of Wilbraham Egerton, 2nd Baron Egerton of Tatton. (*Peerage*)
1868 *13726*

Elder, Alexander (1790–1876). Partner with George Smith as booksellers and publishers, first on Fenchurch St., then on Cornhill. (*MEB*)
1864 *12536*

Elie de Beaumont, Jean Baptiste (1798–1874). French geologist, professor, and administrator. Secretary of the Académie des sciences. (*Who's*)
1855 10416 **1863** *12127*

Ellery, Robert L. J. (1827–1908). English astronomer. Founded the Melbourne Observatory. (*Who's*)
1867 *13465* **1869** *14001*

Elliot, Second Earl of Minto, Gilbert (1782–1859). English M.P. and diplomat. 2nd Earl of Minto. Assumed additional surnames of Murray and Kynynmound. (*DNB, Peerage*)
1822 *697 725* 735 *739 744* **1831** 2428 2429 *2434* **1838** 3798 *3808 3809* 3833 *3837* **1839** *4294* 4299 *4306*

Elliot, James. Professor of mathematics at Liverpool. (JH's 1839–5–30)
1839 *4104* 4110

Ellis, Alexander John (1814–90). British mathematician and philologist. Reformed phonetic alphabet. (*DNB, MEB*)
1843 5480 **1851** *8664 8688* **1852** *9460* **1856** *10486* **1859** *11030 11031* **1868** *13774*

Ellis, William (1794–1872). English foreign missionary. Foreign Secretary of the London Missionary Society, 1831-41. (*DNB, MEB*)
1858 *10968*

Elouis, Charles. Employee at London Mint to 1854, and then at the mint in Sydney. (Challis, *Mint*)
1856 *10550*

Empson, William (1791–1852). British author and professor of law at Haileybury College. (*DNB, MEB*)

1847 *6849 7132 7155 7191 7241 7276 7291 7324 7336 7385 7390 7409 7424 7426 7469* **1848** *7483 7504 7519 7764* 7775 *7780 7818* **1849** *8087 8092 8108 8144* **1850** *8192 8195 8201 8209 8226 8237 8277 8325 8336* 8363 *8369* **1859** 11104 **undated** *14572 14573 14574 14575 14576*

Encke, Johann Franz (1791–1865). German astronomer and director of several observatories. (*NDtB, DSB*)
1824 *1044 1062* **1826** *1299* 1325 *1338* 1359 *1437* **1827** 1464 **1828** *1753* 1869 **1830** *2106* 2127 **1833** *2862* 2910 **1835** *3103* **1837** 3432 **1839** *3997* 4018 4129 **1840** *4486 4667 4674 4701* **1846** *6746* 6789 **1847** 6907 *6913 7391* **1850** *8260* **1858** *10965* **1859** *11098* **1861** *11548*

Enfield, Edward (1811–80). Mint employee, and academic administrator. (*DNB, MEB*)
1851 *8494 8656* 8671 *8681 8723* 8724 *8979*

Engelmann, Friedrich Wilhelm Rudolph (1841–88). German astronomer and bookseller. Built a private observatory. (*BHGW*)
1864 12744

Erman, Georg Adolph (1806–77). German physicist, meteorologist, and geologist. (*DSB*)
1845 *6129 6130* **1848** *7943* **1851** *8784* **1868** 13592 **1870** *14173* 14206 14289 14297 *14304 14336* **1871** 14357 14360 *14392* 14409 *14410 14413*

Evans, Frederick J. O. (1815–85). English hydrographer. Interested in terrestrial magnetism. (*DNB, MEB*)
1869 *13933* 13935 *13939*

Everest, George (1790–1866). English military engineer. Surveyor-general of India. Mount Everest is named after him. (*DNB, MEB*)
1844 *5973*

Everest, Robert
1828 *1791*

Everett, Edward (1794–1865). American clergyman, statesman, and professor at Harvard. (*DAB, NCAB*)
1845 *6305 6326* **1846** *6608* **1847** *7373* **1851** 8692

Everett, Joseph D. (1831–1904). English scholar. Taught mathematics and natural philosophy. (*AlliSup*)
1863 *12222*

Eyre, Mr.
1865 *12770*

Eyres, Charles (1809–76). English divine and academic administrator. (*Cantab*)
1850 *8391*

Fairbairn, John (1794–1864). Controversial newspaper editor in South Africa. Slavery abolitionist. (*DSAfB, SESA*)
1834 2953 *2954* **1835** *3081 3170* **1836** 3346 *3355 3391 3396* **1837** 3428 *3455 3458* 3551 *3552* **1838** *3646 3769* **undated** *14577*

Fairbairn, William (1789–1874). English engineer. Built and designed hundreds of bridges. (*DNB, MEB*)
1867 *13389*

Falb, Rodolphe (1838–1903). Austrian astronomer, meteorologist. Studied effect of the sun and the moon on weather. (*Who's*)
1867 *13551*

Falconer, Hugh (1808–65). Scottish paleontologist, botanist. Studied vertebrate fossils in Siwalik Hills, India. (*DSB, Who's*)
1844 *6042* **1846** 6529 6592 *6595*

Fallows, Fearon (1789–1831). English mathematician and astronomer. Classmate of JH. Director of Cape observatory. (*DNB*)
1816 205 **1820** *410* **1822** *724 729 741* 823 **1823** *865* **1831** 2277

Fallows, Mary Ann (née Hervey). The wife of English astronomer Fearon Fallows.
1831 *2472* 2475 *2477* **1832** 2495 *2561*

Faraday, Michael (1791–1867). English chemist and physicist. Pioneer in electromagnetism. (*DNB, DSB*)
1821 506 **1825** 1170 *1227* 1230 *1236 1248 1252* **1826** *1323 1333* 1334 1335 1336 *1339* 1340 1343 1351 *1367* **1827** 1485 1514 *1568* 1570 *1610* 1614 1615 **1828** 1678 1752 1788 *1794* 1852 *1875* **1832** 2658 *2661* 2662 **1838** 3784 *3789* **1839** *3960* **1840** *4393* **1845** 6279 6459 *6469* 6509 *6510* **1846** *6562* 6570 *6597 6605* **1847** *7175* **1848** *7582* **1851** *8613* 8665 *8666 8732 8749 9058* **1852** *9561* 9563 *9565* **1861** 11645 *11650* **undated** *14578* 14579 14580 14581

Faye, Hervé A. E. A. (1814–1902). French astronomer. Discovered comet, 1843. Improved astronomical techniques. (*DSB, Who's*)
1848 *7759*

Fenwick, John. Honorary secretary to Literary and Philosophical Society at Newcastle-upon-Tyne in 1847. (1847–10–13)
1847 *7288*

Few & Co.
1829 *1955* **1840** *4410 4668* **1841** *4825* **1843** *5424 5511 5515 5518* **1856** *10490* **1858** *10830* **1863** *12296 12302 12304* **1864** 12715 *12719*

Few, Charles. Solicitor for Few & Co. Administered marriage settlement for Charles Babbage's sister, Mary Anne, for which JH was trustee. (*MEB*)
1840 *4400 4412 4647* **1843** *5553 5556*

Few, Robert (1807–1887). British solicitor. (*MEB*)
1857 *10727*

Field, Cyrus West
undated *14582*

Field, Henry W. (1803–88). Mint employee during JH's mastership. (*MEB*)
1852 9522 **1854** 10124 **1855** *10349 10359* 10361 **1856** *10603 10605* **1859** *11162* 11166 *11170* **1860** *11341* **1863** *12209 12413* **1868** 13645 **1869** *13844* 13847 *13855* **1871** *14411* **undated** *14583*

Field, John M. Employee at Assay office during JH's mastership of the Mint. (Craig, *Mint*)
1853 *9672*

Field, Joshua (1787?–1863). British civil engineer and industrialist. Built marine engines. (*DNB*)
1851 8854 *8858* 8859

Field, Wardell, etc.
1845 *6451*

Finch, Henry (1831–?). Employee of the Mint during JH's mastership. (Brande and Finch's letter of 1855–2–21)
1852 9452 **1853** 9677 9679 *9776 9778 9862* 9863 **1854** 9983 **1855** *10315*

Finlayson, Henry P.
1862 *11808 11809*

Fisher, George (1794–1873). British astronomer and divine. Went as astronomer on naval exploratory expeditions. (*DNB, MEB*)
1847 *7236*

Fisher, Samuel. Clerk to the Merchant Taylor's Company. (1866–5–15)
1866 *13190*

Fitton, Frederick C. Son of William Henry Fitton. (1861–5–11)
1861 *11604*

Fitton, William Henry (1780–1861). British geologist and physician. (*DNB, DSB*)
1826 *1416* **1827** *1443 1578 1600 1641* **1828** 1741 1770 *1808 1812* **1829** *1992* **1830** 2231 *2256* 2261 **1831** 2420 *2432* **1832** *2541* **1836** *3423* **1838** *3610 3780* **1839** *4253* **1840** *4372 4577* **1843** 5346 **1844** *5885* **1847** 6909 *7259* **1855** *10410* **1858** *10801* **undated** *14584*

Fitzclarence, George Augustus Friederick (1794–1842). British army officer and orientalist. Oldest illegitimate son of William IV. (*DNB, Peerage*)
1828 *1838*

Fitzgerald, Frederick. Probably clerk in the Treasury office. (1853–8–16)
1853 *9807*

Fitzmaurice, J.
1866 *13117*

FitzRoy, Robert (1805–65). English meteorologist and naval officer. Commander of *HMS Beagle*, 1828–36. (*DNB, MEB*)
1836 *3342 3347 3351* 3392 **1841** 4802 *4820* **1852** 9158 **1858** *10833* 10834 *10844* **1860** 11398 11450 *11451* 11453 *11457 11460 11463* 11466 *11468* 11472 *11473 11475 11479 11483* 11485 *11488* **1861** 11600 *11603* 11606 *11610 11614 11698* 11700 **1862** *11774* 11776 11850 *11855* 11857 *11862* 11874 *12084* 12086 **1863** *12098* 12099 *12125 12193 12194* 12195 *12200 12228 12409*

Flammarion, Camille (1842–1925). French astronomer. Observed Mars extensively. Actively involved in popularization of science. (*DSB, Who's*)
1863 *12103 12333* **1864** *12676* **1867** *13365*

Fletcher, Henry A. Brother of Isaac Fletcher. (1859–11–26)
1859 *11219*

Fletcher, Isaac (1827–79). English coal owner and ironmaster. M. P. for Cockermouth, 1868–79. (*MEB*)
1849 *8124* **1860** *11271* **1865** *13026 13038*

Fletcher, Thomas Bell Elcock (1807–97). British physician and academic administrator. (*MEB*)
1847 *7212*

Flourens, Marie J. P. (1794–1867). French physiologist. Studied the central nervous system. (*DSB, Who's*)
1842 *5073*

Forbes, James David (1809–68). British experimenter and professor of natural philosophy at Edinburgh, 1833–60. (*DNB, DSB*)
1831 *2352* **1832** *2570 2575* 2579 *2582 2609* 2617 *2638 2684* 2691 *2697* 2702 **1833** *2737* 2745 *2770 2783* 2790 *2791 2889* 2903 **1835** *3063 3121* 3124 *3201* **1836** *3367 3405* 3406 **1837** *3433 3440* 3544 **1839** 4037 4118 *4152* **1840** 4466 *4475* 4480 *4571* 4578 *4580* 4641 *4651* **1841** *4858 5053* **1842** *5081 5146* 5148 5162 *5170 5173* 5174 5185 5311 **1843** 5368 *5375* **1844** 6040 **1845** 6073 6087 6281 *6288 6328* 6357 **1847** *6883* 6917 *6923* 7073 *7221* **1848** 7922 **1849** *8052* 8054 *8151* 8153 8154 **1850** 8364 8403 *8407* **1851** 8974 *9084* **1856** 10602 **1857** 10616 **1858** 10842 *10855* 10857 *10864 10888* 10889 **1859** *11041* 11043 *11176* 11187

Forbes, D.D., John (1801–74). British mathematician and Free Church minister. (*MEB*)
1847 *6932* **1848** *7675* 7706 *7719 7723*

Forbes, John S. (1804–1866). Scottish advocate. Brother of James David Forbes. (*MEB*)
1832 *2656* 2663 2695 2703 **1847** 6971

Forbes, Mrs.
1825 *1177*

Forbes, William Nairn (1796–1855). British officer in Royal Bengal Engineers. Master of Mint in Calcutta, 1836–55. (*MEB*)
1854 *9993*

Ford, G. H.
undated *14629*

Forman, Walter
1831 2377 2391 2418 **1832** *2612*

Forrest, James. Secretary to Institute of Civil Engineers in 1867. (1867–11–1)
1867 *13541*

Forshall, Josiah (1795–1863). Long term employee at the British Museum. (*DNB, MEB*)
1845 *6476* **1847** 7041 7104 **1848** *7615*

Forster, Thomas I. M. (1789–1860). English naturalist, astronomer. Cofounder of the short-lived Meteorological Society. (*DNB, MEB*)
1841 *5011* **1842** 5114 *5122 5136* **1843** *5355 5491 5635* **1844** *5756 5887 5944* **1845** *6155* **1846** *6555 6652* **1858** *10807*

Foster, Peter le Neve (1809–79). Secretary to the Society of Arts. Cofounder of the Photographic Society. (*DNB, MEB*)
1855 10297 **1856** *10583* **1858** *10999* **1861** *11605 11651* **1869** *14052 14055*

Foucault, Jean B. L. (1819–68). French physicist. Determined the absolute velocity of light. Constructed a

pendulum to demonstrate the earth's rotation. (*DSB, Who's*)
1857 *10713*

Fourier, Jean Baptiste Joseph (1768–1830). French mathematician and government administrator. Studied heat diffusion and theory of equations. (*DBF, DSB*)
1824 *941 960 985* **1825** 1158 *1178 1200* **1827** *1530 1534 1598* **1828** *1707 1762 1807* **1830** 2099 **1835** *3134*

Fowler, Thomas (1832–1904). English logician and author at Oxford University. (*DNB, Who's*)
1869 *14002 14011* 14013 *14020 14036*

Fox, Robert. A woodcraftsman by training. Became an optician because of interest. (1840–2–20)
1840 *4448*

Franklin, Sir John (1786–1847). English Arctic explorer and discoverer of the North-West Passage. Lieutenant-governor of Van Dieman's Land. (*DNB*)
1837 *3546* **1838** *3827* **1839** *3966* **undated** *14585*

Franklyn, Richard (1797–?). Member of company of Moneyers at London Mint, 1812–51. (Craig, *Mint*)
1851 *8673* 8682 *8696 8697 8753* 8786

Fraunhofer, Josef von (1781–1826). German physicist. Maker of achromatic lenses and optical instruments. Studied Fraunhofer lines. (*DSB, Who's*)
1825 1106 *1153*

Freke, R.
1862 *12038 12046*

Frener, Jean-Baptiste (1821–92). Swiss engraver. (*DHBS*)
1854 *10221*

Fresnel, Leonor. Brother of Augustin-Jean Fresnel. (*DBF*)
1867 *13486* 13488

Friend, T.
1860 *11480*

Fry, Samuel
1859 *11167 11192*

Fulbrook, Charles
1853 *9912*

Furnivall, Frederick James (1825–1910). Philologist, literary scholar and reformer. Established numerous literary societies. (*DNB, AlliSup*)
1857 10663 *10664*

Fuss, Nicholas [also Nikolai Ivanovich Fus] (1755–1826). Russian astronomer and mathematician. Active in St. Petersburg Academy of Sciences. (*DSB, GSE*)
1822 *752*

Fuss, Paul Heinrich (1797–1855). Secretary of St. Petersburg Academy of Science. Son of Nicolaus Fuss. (*BHGW*)
1844 *5710*

Fyers, A. British officer in Royal Engineers. Stationed on Mauritius in 1838.
1851 *8995*

Galbraith, William (1786–1850). British mathematician, writer. (*Alli*)
1849 *8053 8056 8152*

Galitzine, Princess
undated *14586*

Galle, Johann (1812–1910). German astronomer. Based on U. J. J. Leverrier's calculations, discovered the planet Neptune. (*DSB, Who's*)
1848 *7825* **1865** *12831*

Galloway, John (1804–94). English engineer. Cofounder of a large engineering enterprise in Manchester. (*MEB*)
1846 *6639*

Galloway, Thomas (1796–1851). British mathematician, and actuary of Amicable Life Assurance Co., 1833–51. (*DNB, MEB*)
1832 *2665* **1839** *4001 4078 4121* **1840** *4524 4530 4584 4712* **1844** *5755 5779 5784 5792 5793 6011 6018 6029 6038 6045 6054* **1845** *6169* **1846** *6824 6825* **1847** *6872 7121 7365 7371* **1848** *7515 7536 7555 7711 7879*

Galt, John (1779–1839). English novelist. Wrote over forty works. (*DNB, Alli*)
1816 *204* **1817** 258 *259 260* 264 *267 269*

Galton, Francis (1822–1911). English anthropologist. Introduced mental testing. Established fingerprinting for human identification. (*DSB, Who's*)
1860 *11371*

Gambart, Jean F. A. (1800–36). French astronomer. Discovered thirteen comets. (*DBF, Who's*)
1826 *1358 1366 1413 1417*

Gardiner, Mrs. Interested in buying Observatory House [in Slough]. (1864–4–22)
1864 *12549* 12553

Gardner, Thomas. English chemist on the council of the College of Chemists. (1844–2–1)
1844 *5722* **1845** *6356*

Garrett, Rider
1868 *13725* 13728 *13732*

Gascoyne-Cecil, Robert A. T. (1830–1903). British M.P., cabinet minister, prime minister. (*DNB*)
1866 13315

Gasparis, Annibal de (1819–1892). Italian astronomer, mathematician. (*DBIt*)
1850 *8307* **1851** *8807 8960* **1852** *9299 9300* **1854** *10251* **1857** *10692* **1858** *11000*

Gassiot, John P. (1797–1877). British scientific writer and researcher in chemical electricity. (*DNB*)
1839 *4155 4190 4191* **1847** *6915 6919 6926 6953* 6959 *6962* **1852** *9326* **1854** 10080 *10094* **1858** 10906 *10911* **1859** *11062 11064 11081* **1862** *11779* **1870** 14300 *14302* **1871** *14375* 14379 *14388 14393* **undated** *14587*

Gathorne-Hardy, Gathorne (1814–1906). English statesman. 1st Earl of Cranbook. (*DNB*)
1867 *13507*

Gauss, Carl Friedrich (1777–1855). German mathematician and theoretical astronomer. Developed fundamental theorems of algebra and arithmetic. (*DSB, Who's*)
1820 445 464 **1821** 527 563 574 *627* **1825** 1216 **1828** *1855* **1829** 1997 **1832** 2595 **1833** 2767 *2772* **1838** 3876 **1839** *4040 4101* 4127 **1842** 5264 *5295* **1844** 6035 **1845** *6137 6138* 6194 6250

Gautier, Alfred (also Jean Alfred) (1793–1881). Swiss astronomer and mathematician. Professor and director of observatory at Geneva. (*DHBS*)
1822 674 *684 808* **1828** *1728* **1833** *2797* **1836** *3246* **1841** *4937* **1842** *5132 5293* **1843** *5456* **1847** *6939* **1848** *7577* **1857** *10687*

Gemmellaro, Carlo (1787–1866). Sicilian naturalist and professor of geology. Hosted JH during his travels in Italy. (*DBU*)
1824 *1025*

Gemmellaro, Mario (1773–1839). Sicilian naturalist who studied volcanoes. Brother of Carlo Gemmellaro. (*GDE*)
1824 1028 *1080* **1826** *1300* **1827** 1453 *1611* **1832** 2601 *2621* **1833** *2895* **1834** *3029* **1835** *3066 3097 3135 3182* **1836** *3296 3322 3366*

Geoffroy Saint-Hilaire, Étienne (1772–1844). French naturalist, zoologist. Worked with Georges Cuvier on species classification. (*DSB, NBG*)
1839 *4106*

Georg Merz & Son
1841 *4962* **1843** 5576 **1845** 6455 6517 **1847** *6961 7201 7296 7433* **1848** *7627*

Gerling, Christian Ludwig (1788–1864). German astronomer, mathematician, physicist. Measured star magnitudes and solar parallax. (*Who's, NDtB*)
1843 *5394*

Gibbs, Frederick Waymouth (1821–98). British barrister, and tutor of Prince of Wales, 1852–58. (*MEB*)
1853 *9694 9698 9720*

Gibson, Katherine A. S. Friend of William Whewell.
1866 *13132* **1870** *14323* 14324

Gibson, Thomas Milner (1806–84). British yachtsman, M.P., and long term official of the Board of Trade. (*DNB, MEB*)
1860 *11351* 11353

Gilbert, Davies (1767–1839). Treasurer and President of R.S.L. and promoter of Cape observatory. (*DNB*)
1825 1209 1211 **1826** *1310 1313* **1827** 1444 1454 *1455* 1499 *1501* 1540 *1544* 1577 1589 *1645* 1649 **1828** 1657 1674 1789 *1796 1801* 1824 1826 1832 1835 1859 **1829** *1916* 1962 *1963* 1987 *2051* 2052 *2054* 2056 **1830** *2125* 2128 2129 *2187* 2188

Gilliss, James M. (1811–65). American astronomer. Superintendent of U.S. Naval Observatory from 1861. (*Who's*)
1864 12543

Gilpin, Mary Ann. Probably daughter of George Gilpin, clerk to the R.S.L. till he died in 1810. (Hall, *R.S.*)
1827 *1587*

Gipps, Elizabeth (née Ramsay). Married George Gipps in 1830. Visited JH at Cape. (*AusDB*)
1837 *3608*

Gipps, George (1791–1847). Colonial governor (N.S.W.) and officer in Royal Engineers. (*DNB, AusDB*)
1837 3606 3607 *3609*

Gladstone, William Ewart (1809–98). British M.P., cabinet minister, prime minister, and Master of the Mint. (*DNB, MEB*)
1853 9786 **1854** 9937 9972 *10216* 10220 10246 10249 10257 **1855** 10292 10293

Glaisher, James (1809–1903). British astronomer, meteorologist, and balloonist. Assistant to G. B. Airy at Greenwich. Set balloon record with Henry Coxwell. (*DNB, DSB*)
1857 *10660 10665* **1860** *11452 11455 11458* **1862** *12064* **1865** *12957* **1866** *13123*

Glenelg, Lord (see Grant, Charles)

Goddard, John Frederick. Owner/employee of Royal Gallery of Practical Science in London. (JH's 1839–7–10)
1839 4153 **1842** 5334

Goldingham, John (1767–1849). Civil engineer, and director of the observatory in Madras. (*BHGW*)
1827 1533 *1537*

Goldsmid, Isaac Lyon (1778–1859). British financier and philanthropist. Member of firm of Mocatta & Goldsmid, bullion brokers. (*DNB, MEB*)
1829 1942

Goltermann, H. G. Agent acting on behalf of Caroline Lucretia Herschel. (1823–8–27)
1823 *905*

Gompertz, Benjamin (1779–1865). English mathematician, astronomer. Compiled, with Francis Baily, a catalog of stars. (*DSB, Who's*)
1818 *291* **1823** *877* 878 **1830** *2179*

Goodall, Joseph (1760–1840). Provost of Eton. (*DNB*)
1820 *466* **1822** *650 651 835* **1823** *850* **1824** 970 **1828** *1790* **1839** *4169* **undated** 14588

Goodall, Mrs. Joseph. A neighbor of JH at Slough.
1838 3667

Goodchild, Thomas
1860 *11316*

Goodenough, Samuel (1743–1827). Bishop of Carlisle. Member of the Linnean Society and the Royal Society. (*DNB, Alli*)
1822 *795*

Goodwin, Charles William. Member of staff in Die department at London Mint in 1852–54. (1854–10–17)
1854 *10234*

Gordon, Alexander Cornewall Duff- (see Duff-Gordon, Alexander Cornewall)

Gordon, Alexander Hamilton- (see Hamilton-Gordon, Alexander)

Gordon, Caroline Emily Mary Hamilton- (see Herschel, Caroline Emily Mary)

Gordon, Emma Katherine (see Mrs. James A. Gordon. (*Cantab*)

Gordon, George Hamilton- (see Hamilton-Gordon, George)

Gordon, James A. (1791–1854). British M.P. and magistrate. Long time acquaintance of JH. Cousin to George Waldegrave. (*Cantab*)
1810 *15* **1818** *294* **1819** *354 381* **1820** 476 *478* **1821** 601 *614* **1829** *1966* **1832** *2496 2497 2498 2670 2692* **1833** *2732* **1843** *5535* **1844** *5703 6030* **1845** *6070* 6072 6079 *6081 6271 6282 6306 6452* **1849** *8077 8081 8088 8155* **1850** *8352 8365 8383 8385* 8389 *8394 8398 8405* 8426 *8427 8430* 8431 **1851** *8496 8605* 8612 *8615 8752 8988* **1852** *9279 9477 9607* **1853** *9617 9665* **undated** *14590 14591 14592 14593 14594 14595 14596 14597 14598 14599*

Gordon, L. D. Webster. Resident of Madiera. Introduced to JH by Thomas Murdoch. (TM's 1833–10–4)
1833 2890

Gordon, Lewis Conway- (see Conway-Gordon, Lewis)

Gordon, Mrs. Possibly Mrs. James A. Gordon.
undated 14602

Gordon, Mrs. James A. (née Woolley) Married to James A Gordon (*Cantab*)
1854 10011

Gordon, Murial. A relative of James A. Gordon.
undated *14603*

Gordon-Lennox, C. H. (1818–1903). English M.P. and cabinet member. 6th Duke of Richmond. (*DNB*)
1867 *13399*

Gore, John. British divine. (1832–1–17)
1832 2516

Gorham, John (fl. 1855–88). Surgeon, scientific writer. (*AlliSup*)
1856 *10503* **1859** *11108 11109 11110* **1868** *13599*

Goring, C. (fl. 1805). Religious writer. (*BBI*)
1823 *903* **1825** *1093 1104*

Gosset, Isaac (1783–1855). English divine. Chaplain to royal household at Windsor Castle. (*MEB*)
1830 *2136*

Gough, F.
1863 *12287*

Goulburn, Henry (1784–1856). English M.P. and cabinet member. Chancellor of the Exchequer, 1841–46. (*DNB*)
1842 5273

Gould, John (1804–81). English ornithologist. Taxidermist for the Zoological Society of London and author of several bird books. (*DSB, Who's*)
1846 *6556*

Graham, Andrew (ca. 1815–1908). Astronomer at Markree Observatory in Ireland. (1861–1–25)
1861 *11536 11553*

Graham, James
1868 *13607*

Graham, James Robert George (1792–1861). British M. P. and cabinet minister. First Lord of the Admiralty several times. (*DNB*)
1845 *6160*

Graham, John (1794–1865). British academic administrator and divine. Master of Christ's College, and Bishop of Chester. (*DNB, MEB*)
1850 *8178 8392* 8395 8396 *8401*

Graham, Thomas (1805–69). British chemist. Published *Elements of Chemistry*, 1842. Master of the Mint after JH. (*DNB, MEB*)
1845 *6201 6259* **1848** *7944* **1855** *10318* 10321 10387 **1858** 10869 **1859** *11230* **1863** *12382* **1865** *12883* **1869** *13973*

Grahame, James (1790–1842). Scottish barrister and historian of the U.S. Close personal friend of JH. Introduced JH to Margaret Brodie Stewart. (*Quincy*)
1810 *16 17* **1811** *19 20 22 24* **1812** *32 33 36 37 38 39 47 50 55* **1813** *57 63 64 66 72 75 79 82 84 90 94 99* **1814** *100 109 112 114 121 122* **1815** *133 134 137 148 150* **1816** *170 175 189 193 201* **1817** *212 217 220* 221 *231 234 248 256 263 278* **1818** *295 297 302 306 311* **1819** *328 349 368 370* 374 *379* **1820** *427 481 484 491* **1821** *529* 536 *540 588* **1822** *675 704 718 748 754* 785 *788 791 806* 810 *818* **1823** *851 874 894* **1824** *962 982* 995 1049 1064 *1091* **1825** *1118 1121 1164 1165 1181 1190* 1196 *1197 1201 1228 1231 1242 1259 1261* **1826** *1290 1293 1321 1347 1356 1361 1378 1391 1400 1403 1406 1428 1435* **1827** *1457 1463* 1467 *1472 1477 1497 1500 1518 1555 1572 1586 1590 1612* **1828** 1658 *1675 1679 1681 1699 1700 1717 1721 1742 1751 1803 1819 1844 1849* 1853 *1858 1860* 1865 *1876 1885 1905 1913* **1829** *1923 1938 1972* 1982 *2019 2025 2026 2027 2028 2029 2045 2068* **1830** 2073 *2092 2095 2204 2240 2242 2250* **1831** *2288 2336 2353 2384 2399 2442 2454 2489* **1832** *2553 2565 2593 2615 2694 2707* **1833** *2729 2742 2812 2828 2854 2876* **1834** *2931 2932 2938 3031* **1835** *3058 3190 3211 3214* **1836** *3261 3363* **1837** *3475* **1838** 3655 *3673 3858* **1839** *3923* **1840** *4387 4396 4682* **1842** *5189 5198 5203 5209* **undated** *14604 14605 14606*

Grahame, [Sir] James. Edinburgh University professor. (JH to C. P. Smyth on or about 1845–3–28)
1845 6150

Grahame, Jr., Robert. Son of James Grahame.
1843 *5644* **1844** 5843 *5850 5855 5892* **1867** *13459* **1874** *14454*

Grahame, Sr., Robert (1759–1851). Scottish politician, businessman. Lord Provost of Glasgow. Father of James Grahame. (*MEB*)
1819 *346* **1822** *779* **1825** *1212* **1830** *2152* **1832** *2544* **1833** *2808 2814* **undated** *14607*

Grant, Alexander Henley (fl. 1863–81). Editor, religious writer. (*BBI*)
1863 *12111*

Grant, Charles (1778–1866). English cabinet minister. Created Earl of Glenelg, 1835. (*DNB, BHF*)
1836 3301 **1838** 3643 **1839** *3945 3994*

Grant, James William (1788–1865). British astronomer in India and England. (*DNB*)
1837 *3527* **1843** *5532 5542* **1845** *6254 6480* **1847** *7204*

Grant, Robert (1814–1892). British astronomer. Author of *History of Physical Astronomy*, 1852. (*DNB*)
1852 *9334* 9343 **1858** 10820 *10827* 10829 **1859** *11157 11160 11205*

Grantham, Mary. Friend of Elizabeth Baily (sister of Francis). ([1849]–1–15)
1849 *7965* **1859** *11134*

Graves, John Thomas (1806–70). Irish-born jurist and mathematician. Worked on imaginary logarithms. (*DNB, Who's*)
1828 1659 1866 **1829** 1924 1925 1931 *1934* 1973

Graves, Robert P. (1810–93). English divine. Biographer of William Rowan Hamilton. (*MEB*)
1869 *13868 13870 13874*

Graves, W. Secretary to the Bank of Ireland. (JH's 1853–11–22)
1853 9891 9893 **1854** 9973

Gray, Asa (1810–88). American botanist and natural history professor. Corresponding secretary of A.A.A.S. (*DAB, DSB*)
1846 *6691* **1847** *7339* **1863** *12130*

Gray, John E. (1800–75). British naturalist and keeper of the zoological department at the British Museum. (*DNB, MEB*)
1846 *6665*

Greatheed, Bertie (1759–1826). English dramatist. (*DNB, Alli*)
1818 *309*

Green & Co.
1848 *7765*

Green, G. R. Librarian at Eton College, Windsor. (Francis Hodgson's 1847–9–17)
1847 *7219*

Greenhill, W. R. Neighbor of JH at Collingwood. (1869–5–19)
1869 *13914*

Greenough, George Bellas (1778–1855). English geographer and geologist. Cofounder of the Geological Society of London. (*DSB, Who's*)
1845 6502 *6513* 6515 **1850** 8299 *8301*

Greenwood, Thomas (1790–1871). English barrister. (*MEB, AlliSup*)
1856 *10575* **1859** *11250 11253* **1860** *11280 11325* **1862** *11885*

Gregory, Mrs. Wife of Olinthus Gilbert Gregory.
1841 4990

Gregory, Olinthus G. (1774–1841). British mathematician and editor. Cofounder of London University, 1827. (*DNB, DSB*)
1822 *733* **1827** *1523* **1835** *3215*

Greig, Agnes. Daughter-in-law of Mary Somerville, wife of Woronzow Greig. (1866–10–2)
1866 *13250 13259 13275*

Greig, Woronzow (1805?–1865). Barrister; eldest son of Mary Somerville. (*Women*)
1856 10556 10558 **1860** *11348* **1863** *12270 12283*

Grenville, Richard (1823–89). British statesman, M.P., and colonial governor (Madras). 3rd Duke of Buckingham and Chandos. (*DNB*)
1852 *9362 9382* 9383 *9386* 9391 9393

Gretton, [Catherine]. Possibly a relative of George Gretton.
undated *14608*

Gretton, George. British divine. Operated private school at Hitcham, which JH attended, 1800–09. Dean of Hereford. (JH's 1820–4–18)
1820 430 *433 441 447 448 452 462*

Greville, Charles C. F. (1794–1865). British political diarist. Senior civil servant. (*DNB, MEB*)
1851 *8652* **1854** 10158

Grey, Charles (1804–70). British army officer, M.P., and private secretary to Prince Albert and then to Queen Victoria. (*DNB, MEB*)
1859 11175 **1863** *12367 12369*

Grey, George (1799–1882). British barrister, M.P., and cabinet minister. (*DNB, MEB*)
1838 *3853* **1841** 5050 **1842** *5299* **1850** *8386* 8429

Griesbach, F. A. Herschel. Relative of JH, employed as a salesman by a Liverpool manufacturing firm. (FG's 1867–12–5)
1847 6948 **1867** *13575*

Griesbach, Frances Annie (1849–?). Eldest daughter of Francis A. Herschel Griesbach. Employed as an actress in Aberdeen. (1870–8–7)
1870 *14251 14256*

Griesbach, Frederick. Presumably a relative of JH.
1825 *1272*

Griesbach, George. Presumably a relative of JH.
1828 *1680 1692* **1829** *1998*

Griesbach, J. William. Probably a relative of JH, and former neighbor near Slough. (1842–11–23)
1842 *5317*

Griesbach, John Henry (1798–1875). Musician, composer, director. Amateur scientist in acoustics. (*MEB*)
1829 *2031* 2036 **1843** *5563 5596* **1844** *5853 5886 5930* **1845** *6309 6520* **1847** *6850 7234* **1850** *8384* **1858** 10776 **1864** *12507* **1867** *13590* **1868** *13786*

Griesbach, W. Herschel. Member of the German branch of the Griesbach family. Retired British army officer. (1823–5–8)
1823 *875* **1827** *1531* **1847** 7074

Griffith, George. Assistant general secretary to B.A.A.S., 1862–81. (Macleod, *Parliament*)

1863 12375 12381 **1864** *12673 12698* **1865** *12990 12991 12992* **1866** *13291* **1870** *14288* 14290 *14295 14308* 14314 **1871** *14394*

Griffith, John
1858 *10946*

Groombridge, Stephen (1755–1832). English astronomer. Compiled a catalog of circumpolar stars. (*DNB, Who's*)
1828 *1886 1907*

Groskopf, Dr. Concerned with the care of Caroline Lucretia Herschel after her return to Hanover. Likely related to the Herschels.
1846 *6821* **1847** *6974*

Grove, William Robert (1811–96). Welsh-born physicist, electrochemist. Invented the Grove gas voltaic battery. (*DSB, Who's*)
1842 5085 **1847** 6851 6960 6964 *6965 6968* **1852** 9349 **1858** 10769 **1862** 11754

Grover, Mrs. Member of tenant family on JH's property at Slough in 1842. (1842–10–28)
1842 5297

Grubb, Thomas (1800–78). British optician and engineer. Manufacturer of reflecting telescopes. (*DNB, MEB*)
1839 *4072* **1859** *11234 11242*

Gruyer, Guillaume de
1861 *11601*

Gubbins, C. In the Indian service in Bengal. (1850–4–6)
1850 *8269*

Guillemin, Amédée Victor (1826–93). French journalist, writer. Wrote a popular work on astronomical phenomena, *Le ciel.* (*DBF*)
1863 *12139 12154 12191* **1864** *12746* **1869** *13896*

Guinand, Aimé. French glassmaker. Son of Pierre Louis Guinand, glassmaker. (James, *Faraday*)
1827 *1591* 1597

Gumpach, Johannes von
1862 *11986* 11989 *11993 12013* 12024 *12029*

Gurney, Hudson (1775–1864). British antiquary and verse writer. Biographer of Thomas Young. (*DNB*)
1830 2134 2139 2183 2184 2193 2262 **1831** 2274 2298 2327

Gussew, Matwei Matwejewitsch (1826–66). Russian astronomer. Director of the observatory at Wilna. (*BHGW*)
1859 *11069*

Gutehingt, G. P. One of the church wardens at Hawkhurst church. (1861–3–29 or earlier)
1861 *11582*

Guthrie, Chalmers. London wine merchant in 1862. (1862–4–24)
1862 *11860*

Gwatkin, Richard. British divine. Friend and fellow student of JH at St. John's College. (*Cantab*)
1827 *1616* 1617 *1623 1627 1629 1642* 1646

Gye, James
1833 *2859*

H.[?], H. R.
1864 12417

H.E.I.C. Court of Directors, Chairman
1845 6310

Hagen, Jacob (1805–70). English-born landowner and politician in South Australia. (*AusDB*)
1849 *8040*

Haggard, William Debonaire (d. 1886). Head of the bullion office at the Bank of England in 1851–55. (Prescott's 1851–1–17)
1851 *8583 8584 8592*

Haidinger, W. K. von (1795–1871). Austrian mineralogist and geologist. Worked extensively in mineral optics. (*Who's*)
1864 *12494* **1865** *12869* **1870** *14147* 14156 *14211 14262*

Haig, James (1813–?). British barrister and philosophical writer. (*AlliSup*)
1864 *12601* 12603 *12609* **1869** *13930* 13936

Hall, Basil (1788–1844). British naval officer, explorer, and author. (*DNB*, Warners, *Maclear*)
1821 *570* **1824** *956* **1827** 1487 **1829** *1945* **1830** *2097 2215* **1831** *2357* 2369 *2478* **1832** *2613 2625* 2630 *2640* 2641 **1833** *2722 2780* **1834** *2978* 3044 *3046* **1835** *3104 3149 3191* **1836** *3237 3294 3323 3359 3409* **1837** *3442 3466 3467 3595* **1840** *4378 4707 4713* **undated** *14609 14610 14611 14612*

Hall, Margaret. Wife of Basil Hall. (1839–12–11)
1839 *4324*

Hall, Marshall (1790–1857). British physician and physiologist. R.S.L. council member. (*DNB, MEB*)
1839 4250 *4251 4258* 4259 *4296 4307* **1840** *4502* **1848** *7924*

Halle, H. Fraser (1808–86). Headmaster of South Lambeth grammar school for 33 years. (*MEB*)
1845 *6075* 6076

Halliwell, James O. (1820–89). English man of letters. Biographer of Shakespeare. (*DNB, MEB*)
1840 4603

Hamilton, Eliza Mary (1807–51). Irish poetess. Sister of William Rowan Hamilton. (*MEB, Alli*)
1841 *4840*

Hamilton, George Alexander (1802–71). British senior civil servant. (*DNB, MEB*)
1864 *12446* 12453 *12461*

Hamilton, Helen E. A. (1840–1870). Daughter of William Rowan Hamilton. (Graves, *Hamilton*)
1865 *12987* **1866** *13111*

Hamilton, Lady Helen Bayley (1804–1869). Wife of William Rowan Hamilton. (Graves, *Hamilton*)
1865 12965

Hamilton, Thomas
1861 *11596*

Hamilton, William A. Baillie- (1803–81). British naval officer and civil servant. (*MEB*)
1844 *5844* **1845** *6487 6500* **1848** *7700 7716* 7725 *7756 7760 7766 7778 7808* **1851** 8551

Hamilton, William Edwin (1834–?). Son of William Rowan Hamilton. Edited his father's *Elements of Quaternions*.
1865 *12969*

Hamilton, William John. Secretary to Geological Society of London. (1847–11–4)
1847 *7360*

Hamilton, William Richard (1777–1859). British antiquarian and diplomat. Brought Elgin marbles to England. (*DNB, MEB*)
1824 *1009 1013 1090* **1837** *3478* **1839** *4275* **1841** *4845* **1851** *8627 8820 9021 9104 9107 9119* **1852** *9319 9335* 9339 *9344* 9345 *9352* 9355 9555

Hamilton, William Rowan (1805–65). Irish mathematician and Astronomer Royal of Ireland. (*DNB*)
1827 *1593* 1603 *1625* 1630 **1828** 1746 1890 **1829** 1926 *1957* 2003 **1830** *2249* 2265 **1831** 2381 *2443* **1833** 2724 *2734* **1834** *3017* **1835** 3132 **1836** 3269 **1838** *3663* 3684 *3685 3859* **1839** *3950* 3959 4323 **1840** 4384 *4394* **1841** 4835 *4850* 4855 **1842** *5344* **1843** 5650 5653 *5663* **1846** *6647* 6747 6802 *6808* **1847** *6943* 6955 *6966* 7179 *7194 7251* 7267 *7316 7340* **1848** 7658 7747 7891 **1850** 8290 **1851** 8910 **1853** 9782 **1859** 11209 11218 **1860** 11268 11306

Hamilton-Gordon, Alexander (1817–90). British army officer. Second son of the 4th Earl of Aberdeen. Married JH's oldest daughter, Caroline. (*MEB*)
1853 9621 *9773* 9820 **1854** *10182* **1856** *10463* 10467 **1857** *10724* **1861** *11571 11583* **1863** *12269* **1867** *13567* **1869** 13836 **1870** *14306* **undated** *14613*

Hamilton-Gordon, Caroline Emily Mary (see Herschel, Caroline Emily Mary)

Hamilton-Gordon, George (1784–1860). English diplomat and statesman, prime minister. 4th Earl of Aberdeen. (*DNB, BHF*)
1851 8497 **1852** 9159 9467 **1853** 9739 9774 *9788* 9849 **1855** 10294 *10308* 10309 10310 **1859** *11075* **undated** *14589*

Hanhart, Lithographers, M. & N.
1864 *12503*

Hankey, Thomson (1805–93). British political economist, M.P., and senior partner in Thomson Hankey & Co., West India merchants. (*DNB, MEB*)
1870 *14094* 14097

Hankins, Richard. Clerk of securities at Treasury in 1850. (1850–12–2)
1850 *8444*

Hansen, Wilhelm (1832–1906). German mechanical engineer. Worked with steam engines. (*Who's, NDtB*)
1854 *10102*

Harbord, John Bradley
undated *14614*

Harcourt, Mary. Wife of William Venables Vernon Harcourt. (*DNB, Who's*)
1832 *2506* **undated** *14615*

Harcourt, William Vernon (1789–1871). British divine, and cofounder and active official of B.A.A.S. (*DNB, MEB*)
1831 *2268* 2409 2476 **1832** *2688* **1833** *2849* **1869** *13980* **undated** *14616 14617*

Hardcastle, Edward (1826–1905). British M.P. and Justice of the Peace for Hawkhurst. (*Cantab*)
1870 *14085 14115 14142 14231*

Hardcastle, Emily
1866 *13182* 13184

Hardcastle, Henry (1840–1922). English barrister, and director of Whittle Brewery, Chelmsford. Married Maria Sophia Herschel, 12 Oct 1865. (*Cantab*)
1866 *13293* **1870** *14095 14104 14106 14180 14212 14222*

Hardcastle, Joseph Alfred (1815–99). British M.P. Father of Henry and Emma, who married two of JH's children. (*MEB*)
1867 *13582* **1868** *13648 13655 13660* **1870** *14086*

Harding, Charles L. (1765–1834). German astronomer. Assistant at J. H. Schröter's observatory in Lilienthal. (*DSB*)
1820 *459* **1825** *1218* **1826** *1354* **1828** *1748* **1832** *2602*

Hardinge, Arthur Edward (1828–1892). British army officer and equerry to Prince Albert and then Queen Victoria. (*DNB*)
undated *14618*

Hardy, Gathorne Gathorne- (see Gathorne-Hardy, Gathorne)

Hare, Archdeacon Julius Charles (1795–1855). English scholar and divine. Wrote on theological topics. (*DNB*)
1851 8770

Harley, Robert (1828–1910). English mathematician. Researcher in higher algebra. (*DNBSupp*)
1862 11817 *11819* **1866** 13112 *13133 13140* 13228 **1867** *13489*

Harness, Henry Drury (1804–83). British officer in Royal Engineers. Deputy-master of Mint, 1850–52. (*DNB, MEB*)
1850 8445 *8449 8456 8459 8462 8470 8472 8473* **1851** *8498 8499* 8500 *8552 8558 8559 8567 8569 8573* 8585 *8586 8587 8590 8593 8597 8598 8606 8610 8616 8617 8620 8629 8630 8631 8633 8635 8640 8647 8649 8650 8653 8657 8668 8683 8689 8690 8698 8711 8721 8740 8744 8745 8767 8792 8794 8799 8808 8812 8817 8819 8822 8824 8826 8830 8836 8838 8841 8852 8853 8854* 8858 *8859 8882 8886 8890 8900 8903 8911 8914 8922 8940* 8941 *8953 8966 8984 8996 9012 9050 9051 9054 9055 9070 9120 9126 9142* **1852** *9176 9189* 9237 *9252 9267 9316 9375 9396 9400 9410 9417 9419* 9420 *9422 9432 9440 9441 9443 9446 9453 9455 9462 9471 9480 9496* 9505 9506 *9508* 9514 *9515 9516 9521 9523* 9524 *9525* 9527 *9528* 9530 9531 *9532 9533* 9534 *9535 9537* 9540 9545 *9568 9581 9584* 9585 *9586 9587 9589 9591 9592 9593 9595 9597 9602 9606 9609 9610* **1853** *9634 9737* 9738 9740 *9741 9745 9746 9747 9748 9751* 9769 *9789* 9872 **1854** 10273 **1865** *12825*

Harris, G. W. Secretary of the Society of Engineers. (JH's 1869–2–22)
1869 13872

Harris, John S.
1871 *14364*

Harris, Mr.
1853 9856

Harris, Mrs. C. Wife of C.(?) Harris, assistant to the treasurer of B.A.A.S. (1845–12–19)
1845 *6503*

Harris, Thomas
1866 *13244*

Harris, William Snow (1791–1867). British military surgeon and investigator of atmospheric electricity. (*DNB, MEB*)
1851 *9016 9082* 9094

Harrison, J. Park (fl. 1850–65). British scientist who studied the influence of the moon on air temperature. (*RSL Cat*)
1844 5766 **1865** *12902* 12903 **1866** *13317*

Hart, Joseph J. (fl. 1875). British writer. (*BBI*)
1840 *4346*

Hartley, William
1864 *12708* 12713

Hartnell, Joseph. Chairman of Hawkhurst Dispensary, providing medical relief to families in Hawkhurst parish. ([1845–11–25?])
1843 *5441* **1845** *6440* **undated** *14619*

Hartnup, John (1806–85). British observational astronomer. Assistant at J. Wrottesley's private observatory. (*MEB*)
1841 *4989* **1847** 7118 *7181 7187*

Harvey, William Henry (1811–66). English botanist. Colonial Treasurer at Cape Town, South Africa, 1836–42. (*DSB, Who's*)
1836 3235 3236 **1837** 3560 3561 *3562 3563 3575* **1838** *3752 3873* **1839** 4051 *4128* **1843** *5654* 5657 **1844** *5698 5782* **1845** *6368* **undated** *14620 14621 14622 14623*

Harwood, Philip (1809–87). British journalist and Unitarian minister. Editor for a number of British papers. (*DNB, MEB*)
1871 *14440*

Hatchett, Charles (1765–1847). English chemist. Discovered the element niobium (columbium). (*DSB, Who's*)
1821 *551*

Hausmann, Johann Friedrich Ludwig (1782–1859). German mineralogist and professor at Göttingen. (*Who's*)
1847 *7445*

Hawes, Benjamin (1797–1862). British soap manufacturer and M.P. Arranged partnership between Charles Wheatstone and William F. Cooke. (*DNB, MEB*)
1855 10291

Hawkins, Edward (1780–1867). British numismatist, antiquary, and banker. Employee of British Museum. (*DNB, MEB*)
1851 *8553*

Hawkins, John (1791–1877). English steward. Founder of the Hitchin Friendly Institution. (*MEB*)
1860 *11377 11386*

Hawkins, William (1758–1836). English agent for the East India Company at the Cape. Friend of JH. (Warners, *Maclear*)
1836 *3314* **1837** *3549* **1838** *3611 3612 3786*

Hawtrey, Edward Craven (1789–1862). British linguist and provost of Eton College, 1853–62. (*DNB, MEB*)
1844 *5888* **1847** *6967*

Hayday, James (1796–1872). Bookbinder, 1833–61. Exclusively contracted to Edward Gardner of Oxford Warehouse. (*DNB, MEB*)
1847 *7087 7096 7139*

Hayler, James. Agricultural laborer at Moor House (later Collingwood) prior to its purchase by JH in April 1840. (1840–1–9)
1840 *4381*

Hayter, William Goodenough (1792–1878). British M.P. and cabinet official. (*DNB, MEB*)
1853 9662

Headlam, Thomas Emerson (1813–75). English judge advocate-general, and M.P. for Newcastle-on-Tyne. (*DNB, MEB*)
1838 *3736* 3740 *3763* **1844** 5839 *5854*

Hearn, George W. (1812–1851). Professor of mathematics at Sandhurst College. (*MEB*)
1845 *6295* **1846** *6574 6579 6582* **1847** *6863 6868 6873 6879* 6881 *6885*

Hearn, John. Father of George Whitehead Hearn.
1851 *8976* **1852** *9487*

Hearn, William E. (1826–88). British political economist and barrister. Academic at Melbourne from 1854 on. (*DNB, DAusB*)
1864 *12629* **1869** *13899*

Heaviside, James (1808–97). British mathematician. Taught at Haileybury College, 1838–57. (*MEB*)
1855 10298

Hebeler, B. Prussian Consulate General in London in 1851. (1851–9–3)
1851 *8948*

Heeley, Wilfred. Secretary of the government commission on weights and measures in Calcutta. (1866–11–22)
1866 *13141 13222 13227* 13229

Heis, Eduard (1806–77). German professor of mathematics and astronomy in Münster. Editor of an astronomy journal. (*BE*)
1848 *7562* **undated** *14624*

Henderson, Ebenezer (1809–79). English scientist. Designed mechanism to check sidereal time. (*DNB, MEB*)
1849 *8138 8147* **1850** *8304 8309*

Henderson, J. Probably a bookseller or publisher, with offices at 46 Lime St., in 1822. (1822–7–30)
1822 756

Henderson, Thomas (1798–1844). British astronomer and Astronomer Royal for the Cape and for Scotland. (*DNB, DSB*)
1827 *1487* 1528 **1829** 1928 **1832** *2514* 2696 **1833** *2769 2782 2838* 2843 *2860 2888* 2906 **1834** *2986* **1835** *3164* **1836** *3414* **1837** *3464* **1840** *4669* **1841** *4944* **1842** *5104* 5108 *5113 5145 5164 5289* **1843** *5442 5451*

Hennessey, John B. N. (1829–1910). English surveyor. Deputy surveyor-general of India, 1883–4. (*DNB*)
1863 *12201 12262*

Henning, Alexander. British naval officer. Captain of the ship *Windsor* on which JH and family returned to England in 1838. (Evans, *Herschel*)
1843 *5516* **1866** *13191* **undated** *14625 14626 14627*

Henry, J.
1821 *508*

Henry, John. Assistant at Royal Observatory, Greenwich, in 1847. (Meadows, *Greenwich*)
1847 *7260*

Henry, Joseph (1797–1878). American physicist and mathematician. Studied electromagnetism. Director of the Smithsonian Institution, 1846–78. (*DAB, NCAB*)
1858 *10808* **1863** *12116*

Henry, William (1774–1836). English chemist. Discoverer of Henry's Law about gas absorption. (*DSB, Who's*)
1832 2508

Henslow, John Stevens (1795–1861). English botanist and professor at Cambridge. Improved the Cambridge Botanical Garden. (*DSB, Who's*)
1843 *5471* **1848** *7644 7938* **1849** *8137*

Herapath, John (1790–1868). British mathematician and journalist. Proprietor of several railway magazines. (*DNB, MEB*)
1822 816 **1825** 1213 **1830** 2232 **1848** *7831 7896*

Herbert, J.
undated *14814*

Herbert, Sidney (1810–61). British M.P. and cabinet minister. 1st Baron Herbert of Lea. (*DNB*)
1859 11085

Herman, Henry. Probably a clerk in India Office in 1865. (1865–1–23)
1865 *12792*

Herrick, Edward C. (1811–62). American entomologist. Librarian at Yale University, 1843–58. (*DAB*)
1838 *3824* **1839** 3983 *4180* **1840** 4659

Herschel, Alexander Stewart (1836–1907). English physicist, astronomer, and meteorologist. Fifth child and second son of JH. Did not marry. (*DNB*)
1855 10407 **1856** 10526 10542 **1859** 11199 **1860** 11265 11349 11358 11370 **1861** 11627 **1862** 11788 11795 11804 11810 11914 11930 **1863** 12414 **1865** 12886 13019 **1866** 13160 13223 **1874** *14455* **undated** *14658*

Herschel, Caroline Emily Mary (1830–1909). Oldest child of JH. Married Alexander Hamilton-Gordon, in 1852. Lady of the bedchamber to Queen Victoria, 1855–91.
1836 *3273* **1846** 6736 **1851** *8495* **1852** 9479 **1853** *9785* **1854** 9916 **1855** *10295* 10296 10301 *10303* **1866** 13185 **1867** 13498 **undated** 14628

Herschel, Caroline Lucretia (1750–1848). Sister of William Herschel. German singer and astronomer. Immigrated to England (1772) to assist JH's father William with his astronomical observations. (*DNB, DSB*)
1822 *839* **1823** *853 860 871 884 891* 897 *901* **1824** 1022 1048 *1053 1054* 1063 *1065* 1083 **1825** *1100 1128 1134* 1148 *1154* 1214 *1232* 1273 **1826** *1287* 1324 *1364* 1379 *1385 1401* 1410 *1427 1439* **1827** *1490* 1508 *1512 1554 1567 1585 1637* **1828** 1667 1733 1747 *1749 1759 1817 1841* 1902 **1829** *1917* 1927 1941 *1964* 1965 1977 *1984* 2044 **1830** *2085* 2164 *2181* 2189 *2235* **1831** 2281 *2314* 2351 *2359 2379* 2438 *2447 2487* **1832** *2517* 2523 2555 *2556* 2591 2607 **1833** 2720 2760 *2765 2767* 2772 2781 *2784* 2824 *2832* 2917 2921 2928 **1834** 2941 2959 *2974* 2982 *3006* **1835** 3076 3077 3198 **1836** 3273 3274 3291 *3341* 3364 3393 **1837** 3431 *3460* 3473 *3496* 3535 3580 **1838** 3658 3711 3741 3764 3870 *3886* 3903 **1839** *3919* 3984 4139 4198 4266 *4312* **1840** 4504 *4614 4615* 4635 4642 *4663* 4708 **1841** 4772 *4829* 4837 *4859* 4861 *4871* 5016 5017 **1842** 5112 5121 5192 5223 *5234* 5236 5265 5275 5290 5323 **1843** 5404 *5440* 5447 *5461* 5502 5590 5665 **1844** 5759 5820 *5833* 5861 **1845** 6131 6135 6351 6491 **1846** 6598 6610 6615 6820 **1847** 7103 7292

Herschel, Friedrich
1838 *3728*

Herschel, Isabella (1831–93). Second child of JH. Never married.
1846 6666 **1850** 8179 **1853** 9885 **1854** 9917 9961 10092 10238 **1855** 10304 10398 **1856** 10530 **1859** 11003 **1864** *12640* **undated** 14632

Herschel, Johann Dietrich (1755–1827). Youngest son of Isaac Herschel and brother of William and Caroline Herschel. Uncle of JH.
1810 *13* **1818** *300* **1822** 789 792 *831*

Herschel (son), John (1837–1921). English officer in the Royal Bengal Engineers. Appointed to Great Trigonometrical Survey of India, 1858–86. Sixth child and third son of JH. Married Mary Cornwallis Lipscomb, 1867. (*The Times*)**1857** 10709 **1858** 10794 10845 **1859** 11007 **1860** 11357 11412 **1861** 11529 **1862** 11931 **1863** 12100 12241 12275 **1864** 12418 12520 12582 12657 12658 **1865** 12775 12779 12860 13008 **1866** 13294 **1867** 13450 *13452* 13461 13515 13523 **1868** 13601 13667 13703 13739 13758 *13798 13804* **1869** 13902 13941 13957 14053 14062 **1870** 14119 14246 14250 14269 14321 **1871** 14382 **1874** 14453 **undated** 14659

Herschel, Margaret Brodie (née Stewart) (1810–84). Daughter of Rev. Alexander Stewart, Scottish minister. Married JH, 1829. (For pre-1829 letters, see Stewart, Margaret Brodie.)
1822 793 **1829** 1970 *2001* 2060 2062 2064 2065 2069 **1830** 2080 2081 2082 2096 2153 2191 2192 2195 2196

2197 2198 2199 2206 2207 2223 2236 2238 2241 **1831** 2337 2344 *2360* **1832** *2523* 2524 2527 2531 2561 2590 2596 2599 2603 **1833** 2796 2799 2802 2831 2877 2879 2880 2881 2912 2913 2919 **1834** 2992 **1835** 3054 *3077* **1836** 3237 3342 3361 **1838** 3613 3689 3719 3720 3721 3722 3727 3736 3746 3748 3749 3756 3758 3781 3815 *3829* 3884 **1839** 4002 4045 4140 4168 4324 **1840** 4389 4413 4420 4505 4507 4508 4512 4517 4526 4528 4687 **1841** 4762 4763 4814 4847 **1842** 5122 5182 **1843** 5347 5359 5661 **1844** 5728 5757 5857 6012 **1845** 6057 *6058* 6059 6060 6061 6062 6074 6077 6082 6084 6092 6093 6236 6248 6249 6297 6298 6301 6302 6303 6304 6307 6346 6350 6483 6484 6504 **1846** 6530 6531 6532 6533 6534 6554 6568 6572 6637 6638 6717 6718 6730 6733 6735 6739 6740 6741 6838 6844 **1847** 6852 7089 7090 7091 7114 7119 7145 7148 7196 7281 7284 **1848** 7484 7485 7511 *7537* 7583 7659 7687 7792 **1849** 7954 **1850** 8180 8214 8326 8339 8341 8344 8415 8416 8420 8421 8423 8485 8488 **1851** 8501 8502 8503 8504 8505 8506 8507 8508 8509 8510 8511 8546 8549 8607 8729 8780 8860 8894 8895 8905 8932 8945 8946 8947 8993 8999 9039 9121 **1852** 9160 9161 9162 9180 9184 9200 9204 9210 9211 9222 9230 9247 9274 9280 9291 9327 9363 9372 9401 9424 9445 9463 9466 9473 9474 9476 9478 9484 9488 9492 9494 9579 9608 9612 **1853** 9618 9622 9623 9629 9632 9638 9642 9647 9681 9683 9688 9704 9706 9790 9800 9801 9803 9804 9811 9815 9816 9817 *9820* 9825 9826 9830 9831 9834 9839 9864 9866 9871 9874 9877 9882 9887 9889 9896 9898 **1854** 9918 9919 9920 9952 9964 9965 9966 9985 9986 9994 10007 10008 10021 10030 10035 10040 *10043* 10050 10064 10070 10077 10083 10085 10091 10093 10099 10101 10111 10113 10119 10125 10128 10131 10136 10138 10141 10155 10161 10192 10201 10207 10231 10260 10270 **1855** 10279 *10358* 10386 10433 **1856** 10521 10530 10607 **1857** 10706 10753 **1858** 10770 10771 10923 10935 10972 10974 **1859** 11004 11005 11006 11008 11124 11125 11147 11152 11173 11174 11193 11222 11231 **1860** 11279 **1861** 11552 11717 **1862** 11762 11792 **1864** 12583 12633 12636 **1865** 12794 12867 12872 **1866** 13142 13175 13236 13242 13259 13307 13342 **1867** 13357 13419 13453 13455 13459 **1868** 13643 **1869** 13909 13971 **1870** 14306 **1871** *14441 14442* 14445 14446 14447 14448 **1873** 14450 14451 14452 **1874** 14454 **1875** 14456 14457 **1921** 3688 **undated** 14629 14630 14631 14633 14634 14635 14636 14637 14638 14639 14640 14641 14642 14643 14644 14645 14646 14647 14648 14649 14650 14651 14652 14653 14654 14655 14657 14660 14756

Herschel, Margaret Louisa (1834–61). Fourth child of JH. Married Reginald Dyke Marshall in 1858.
1850 8328 *8390 8395* **1852** 9281 9490 9519 **1854** 10065 10241 10261 **undated** 14656

Herschel, Maria Sophia (1839–1929). Seventh Child of JH. Married Henry Hardcastle, 1865.
1853 9736 9868 **1864** 12546

Herschel, Mary Cornwallis Lipscomb Power (?–1876). Married John Herschel (son of JH) in 1867.
1870 14121

Herschel, Mary Pitt (née Baldwin) (1750–1832). Sister of Thomas Baldwin and widow of John Pitt. Married William Herschel, 8 May 1788. Mother of JH.
1799 1 2 3 **1813** *95 97* **1819** 333 334 335 **1821** *507 541 545 558 560 587 589* 591 593 594 *598* 602 *603 606* 607 613 *623 624 628 642* **1822** *662 668 682 773* 774 *775 825* **1823** *855 862 910* **1824** 961 983 984 986 *987 991* 992 993 *997* 999 *1000* 1001 *1002* 1005 *1007* 1008 *1011* 1014 1015 1017 *1018 1019* 1021 *1026* 1027 1031 1032 1035 1037 *1039* 1040 *1042* 1043 1046 1050 *1052* 1055 *1056* 1058 *1059* 1061 **1825** *1233* **1826** 1370 1373 1377 1380 1382 1389 1390 **1827** *1476* 1536 **1828** *1898* 1899 1906 **1829** 1967 *1974 1979 1988 1993 1995* 1996 2000 2001 *2002 2005* 2007 2010 2013 *2014* 2015 *2016* 2017 2020 2021 2024 2033 *2041 2047* **1831** 2269 2360 2401

Herschel, Miss
1855 10411

Herschel, W. Son of Christian Friedrich Herschel. Relative of JH in Germany. (1865–5–22)
1865 *12907*

Herschel, William (1738–1822). First president of RAS, 1821. Discoverer of Uranus. Father of JH. (*Grove*)
1811 18 23 26 **1812** 34 40 53 **1813** *96 98* **1814** 107 **1817** *237* 242 **1818** 287 **1820** 467 **1821** *528* 592 600 605 609 611 **1822** 750 772

Herschel, William James (1833–1917). English-born Indian civil servant. Oldest son and third child of JH. Married Anne Emma Haldane Hardcastle, 1864. (*MOH*)
1844 *5743* **1851** 9030 **1852** 9163 9461 **1863** 12204 **1865** 12944 **1866** 13141 **1867** 13348 13503 *13560* **1869** 13873 13906 14049 14061

Heywood, James (1810–97). English university and library reformer. (*MEB, AlliSup*)
1843 *5524 5546*

Hickson, William Edward (1803–70). British educational writer and boot maker. Editor of *Westminster Review*, 1840–52. (*DNB, MEB*)
1860 *11417* **1863** *12415* **1864** *12425*

Highley, Samuel (fl. 1851–82). English writer of miscellaneous works. (*AlliSup*)
1861 *11673 11684 11686*

Hill, James. Appears to have assisted JH in his departure from the Mint. (1855–4–10)
1855 *10367*

Hill, Rowland (1795–1879). Post office reformer. Introduced penny postage. (*DNB, MEB*)
1838 *3659* **1852** *9411*

Hill, Thomas. An official of the Andersonian University in Glasgow.
1865 *12975* **1866** *13145*

Hind, John (1796–1866). English mathematician. Fellow and tutor of Sidney Sussex College. (*DNB, MEB*)
1860 *11431 11444 11446*

Hind, John R. (1823–95). British observational astronomer. Superintendent of the *Nautical Almanac*. (*DNB, MEB*)
1840 *4625* **1845** *6374 6382* 6401 *6402* 6415 *6417 6462 6525* **1846** *6571 6578 6596 6643 6650 6688 6753* **1847** *6888 7137 7146* 7149 *7156 7303 7325 7341* **1848** *7823* **1849** *7953 7970 7974* **1850** *8189 8215 8234 8258 8409* **1851** *8741 8795 8802 8831* **1852** *9346 9376 9436 9464 9483 9517 9569 9603* **1853** *9777* 9783 *9894* **1854** *10157* **1855** *10401* **1860** *11461* **1861** *11617 11620 11625 11629 11675 11702 11710*

11711 *11727* 11728 *11730* **1862** *11765 11769 11782 11789 11821* 11834 *11838 11839 11899 11956 12032 12072* **1863** *12167 12180 12331 12335 12337 12356* **1864** *12495* **1865** *12845 12850 12878* **1866** *13084 13164* **1867** *13398* **1869** *14026 14033* **1870** *14298* **1871** *14372 14390* **undated** *14661*

Hind, Robert
1866 *13100*

Hindle, Thomas
1855 *10426*

Hingston, John
1846 *6752 6764*

Hirley, William
undated *14662*

Hirst, Thomas A. (1830–92). English mathematician, writer. (*MEB*)
1869 *14069*

Hobhouse, John Cam (1786–1869). British political reformer, M.P., and cabinet minister. Created Baron Broughton de Gyfford, 1851. (*DNB*)
1839 *4109* 4114

Hodgkin, Thomas (1798–1866). English physician and professor of morbid anatomy, Guy's Hospital. Hodgkin's Disease is named after him. (*DNB, MEB*)
1824 *972 1024* **1826** 1284 **1830** *2168* 2169 **1831** *2356* 2371 *2431* **1832** *2519* 2521 *2522 2554* **1834** *2983 3039* **1836** *3408* **1840** *4374* 4425 **1851** 8512

Hodgkinson, George C. (1816–80). English meteorologist, astronomer, and writer on education. An avid mountain climber. (*DNB, MEB*)
1863 *12231 12236 12249 12253* **1864** *12655 12662* **1865** *12925 12932 13051* **undated** *14663*

Hodgson, Francis (1781–1852). English academic and divine. Provost of Eton College. (*DNB, MEB*)
1843 5348 *5533* **1847** *7219*

Hodgson, O. A. British divine. Executor for widow of [Thomas?] Baldwin in 1866. (1866–4–6)
1866 13154

Hodgson, Richard (1804–72). British publisher. Operated several private observatories. (*MEB*)
1863 12400 **1864** *12486* 12487 *12501* 12502 12504 *12516* **1866** *13343*

Hoffay, E. A. Agent in office of paymaster general in 1854. (1854–6–5)
1854 10130

Holdeness, Mr.
undated 14813

Holland, Frederick W. (1837–80). British geographer and divine. Involved in expeditions to the Sinai peninsula. (*MEB*)
1864 *12699* **1868** *13770 13783* **1869** *14024* 14027

Holland, Henry (1788–1873). Personal physician to Princess Caroline of Wales, to William IV, and to Prince Albert. (*DNB, MEB*)
1838 *3854* **1839** *4141* **1841** *4887 4930* **1842** *5241* **1843** *5356* **1850** *8191 8219 8222* **1858** *10837* **1862** *11825 11951 12030* 12075 **1863** *12101 12104* **1865** *12888* 12890 **1868** *13708* **undated** *14664*

Holland, P. H. A physician in Manchester.
1844 *5951 5982* 5988

Holloway, Mr. Tenant on JH's property in Upton in 1856.
1856 10499

Holman, James (1786–1857). Becoming blind in 1810, Holman (known as 'the blind traveler') journeyed around the world, 1827–32. (*MEB, Alli*)
1827 *1575* **1835** *3054*

Home, Sir Everard (1756–1832). English anatomist and surgeon to the king, 1808–32. Custodian of the Hunterian Collection. (*DSB, Who's*)
1822 *677 680*

Hooker, Sir Joseph D. (1817–1911). English botanist. Close friend of Charles Darwin. (*DSB, Who's*)
1857 *10666* **1861** *11630*

Hooker, Sir William J. (1785–1865). English botanist. Father of Joseph Dalton Hooker. Director of the Royal Botanic Gardens at Kew. (*DSB, Who's*)
1843 *5656* 5658 *5660* 5667 **1845** *6263 6280* 6289 *6361* 6363 *6365* 6378 6379 *6381* **1846** 6561 *6563 6573 6583* 6585 **1847** *7057* 7058 7446 *7459* **1848** 7497 7605 *7666 7748* 7749 **1850** 8181 **1858** 10957 *10958* **1861** 11626

Hope, Alexander J. B. (1820–87). English politician and author. Tory M.P., including for Cambridge University. (*DNB*)
1842 *5216* **1854** 9948 10129 **1859** *11039 11050* **1860** *11261 11281 11327 11359* 11376 **1862** *12076* **1867** *13360* 13362 *13504* **1868** *13605 13608* 13639 *13640 13649 13665 13676* **1870** *14084 14341* **1871** *14415* 14429 *14439* **undated** *14665*

Hope, James
1831 2309

Hopkins, Albert (1807–72). American astronomer, botanist, and preacher. Built the first American college observatory at Williams College. (*NCAB*)
1838 *3755*

Hopkins, William (1793–1866). British mathematician and geologist. Private tutor at Cambridge, 1827–66. (*DNB, MEB*)
1843 5458 **1857** *10627* **1863** *12151* **undated** *14666*

Horley, R. Young
1845 *6470*

Hornbuckle, Thomas W. (1775–1848). British academic administrator and divine. Offered JH position of subtutor at St. John's College, 1815. (*Cantab, Shadow*)
1815 *141* **1823** *912* **1824** *975 1077* **1826** *1392* 1394 **1827** *1488*

Horner, Joanna B. Friend of Mathilde Oersted. Probably daughter of Leonard Horner. (1852–5–4)
1852 *9364* 9380 **1859** *11243*

Horner, Leonard (1785–1864). Scottsh geologist and educational reformer. Promoted a comprehensive, science-based education. (*DNB, DSB*)
1827 1527 **1846** *6785* **1849** *8156* **1858** *10783*

Horsburgh, James (1762–1836). English hydrographer. Constructed navigational charts of the eastern seas. (*DNB, Alli*)
1833 *2810* **1834** *3032* **1835** *3136*

Hotham, H.
1867 *13421* 13423

Hotham, Henry
1830 *2243*

Howard, Luke (1772–1864). English meteorologist. Developed method of cloud classification. (*DNB, Who's*)
1839 4126 *4154* **1840** *4452* *4489* **1841** *4801* **1846** *6734* 6738 **1847** *7001*

Howard-Vyse, R. W. (1784–1853). English army officer. Explored many Egyptian pyramids. (*DNB, MEB*)
1839 *4067* **1843** *5403* **undated** *14667* *14668* *14669*

Howlett, Frederick. British divine. Vicar of St. Augustine's. A near neighbor to JH. (1861–1–10 & 1864–5–9)
1859 11214 **1861** 11520 *11527* **1863** 12376 **1864** *12447* 12450 12509 *12522* *12524* *12555* 12560 *12570* **1867** *13418* 13420 *13426* 13429 **1868** 13760 *13763* **1870** 14166 *14174* *14258* 14260 *14270* **1871** *14416* 14417 *14418*

Howlett, Samuel B. (1794–1874). British officer with military surveyors. Invented anemometer and improved mountain barometer. (*DNB, MEB*)
1838 *3801* **1839** *3989*

Howley, William (1766–1848). British divine. Archbishop of Canterbury and *ex officio* principal trustee of British Museum, 1829–48. (*DNB*)
1843 5571 5604 **1848** 7530

Hubbard, H. E.
1871 *14365*

Hubbard, John Gellibrand (1805–89). British M.P. and director of the Bank of England. 1st Baron Addington, 1807. (*DNB*)
1854 10244 **1856** *10506* 10509 **1870** *14090* 14093 *14098* 14105 *14107*

Hudson, Henry
1867 *13585* **1868** *13593* **1870** *14171* *14181*

Hudson, James (1804–59). Assistant secretary to R.S.L., 1829–38, and secretary to Royal Agricultural Society, 1838–59. (*MEB*)
1825 *1103* **1826** *1423* **1827** *1545* **1829** 2032 **1830** *2257* **1831** 2383 **1832** *2685* **1833** *2884* *2893* *2920* **1834** *2965* *3014* **1835** 3106 **1839** *3998* **1847** *7160* **1849** *8083*

Hügel, Baron Karl Alexander von (1796–1870). Austrian diplomat, geographer, and botanist. (*NDtB*)
1847 7242

Huggins, Sir William (1824–1910). British astronomer. Long term officer in the R.A.S. (*DNB, DSB*)
1866 *13194* 13195 *13199* **1868** *13695* 13698 *13714* 13717 *13718* 13720 13764 *13794* 13797 *13827* **1869** *13850* 13851 *13853* 13871 13890 13920 *13929* 13934 *13938* 13940 *13942* 14000 14003 14004 *14005* *14012* **1870** *14200* 14203 14204 14254 14255 *14280* 14284 *14286* **1871** 14436 *14443*

Hughes, Francis J. British divine. Vicar at Penally in Tenby.
undated *14670*

Hughes, Henry Hunter (1796–1884). Tutor at St. John's College, Cambridge. (*MEB*)
1829 *1983*

Hughes, Thomas Smart (1786–1847). English writer, historian, curate. (*DNB, Alli*)
1816 *190* **1829** *2066* **1830** 2088

Humboldt, Alexander von (1769–1859). German geologist, mineralogist, geographer, and world traveller. (*DSB*)
1828 *1847* **1829** 1975 **1838** *3665* 3733 **1842** 5058 *5244* 5247 **1844** *5799* 5835 **1845** *6278* 6447 **1850** *8197* 8220 *8366* 8370 *8373* *8489* **1851** *8912* 9041 **1852** *9330* **1855** *10382*

Hume, Joseph (1777–1855). British surgeon, M.P., and radical politician. (*DNB, MEB*)
1850 *8314* **1853** *9771*

Hume, Joseph Burnley. Secretary to the commission to inquire into abuses at the Mint. Son of Joseph Hume. (Craig, *Mint*)
1853 *9718* 9744

Hume, M. Tailor on Beak St.
1864 *12440* **1866** *13192* *13203*

Humphrey, Thomas. Secretary to National School Committee at Hawkhurst. (1865–10–18)
1863 *12263* **1864** *12569* **1865** *12856* *12996* **1867** *13395*

Humphreys, H. Temple. County surveyor in West Limerick, Ireland.
1867 *13396* *13401* **1870** *14137* *14144*

Hunt, Robert (1807–87). British science writer, chemist, and researcher in photography. (*DNB, MEB*)
1839 *4320* 4329 *4337* **1840** 4503 *4515* 4520 *4527* *4538* 4541 *4545* *4556* *4640* 4652 4715 **1841** *4764* *4765* *4787* 4813 4821 4895 *4907* **1842** 5129 5163 *5179* 5186 5262 *5291* *5305* *5315* 5319 *5324* 5325 5327 *5329* *5335* **1843** *5383* 5389 5392 *5426* 5430 *5453* 5485 *5495* *5527* *5584* 5593 *5637* *5664* **1844** 5707 *5714* *5744* *5767* *5796* 5803 *5813* *5815* *5824* *5935* *5952* 5956 *6032* **1845** *6148* *6193* 6200 *6318* 6428 *6521* **1846** 6550 6580 *6614* *6649* *6656* **1847** *6921* 6950 **1848** *7626* *7669* 7814 *7815* *7901* 7903 *7905* **1849** *8038* *8039* **1850** 8193 **1851** 8513 8580 **1852** 9224 9290 *9369* **1867** *13536* 13538 *13542* **undated** *14671* 14672

Hunt, W. H.
1843 *5487*

Hunter, William (1781–1856). London alderman, 1843–56, sheriff, 1844–45, and lord mayor, 1851–53. (*MEB*)
1853 9734

Hussey, A. M. Probably the wife of Thomas John Hussey. (Letter n.d. RS 10.219)
undated *14673*

Hussey, Thomas John (fl. 1840). British amateur astronomer and divine. Rector at Hayes, Kent. (*Dreyer*)

1825 1274 **1827** *1484* 1592 **1828** *1777* 1806 **1829** *1918 2035* **1830** *2214* 2226 **1831** 2394 2396 **1832** *2499* 2500 *2624 2627* **1833** 2792 **1847** *7346*

Hutton, Charles (1737–1823). English mathematician. Professor at the Royal Military Academy at Woolwich, 1773–1807. (*DSB, Who's*)
1822 *782*

Hutton, Mr.
1867 13457

Huxley, Thomas Henry (1825–95). British anatomist, physiologist, and naval surgeon. Academic from 1854–85. Strong supporter of Charles Darwin. (*DNB, DSB*)
1868 *13787*

Hymers, John (1803–87). English mathematician. President of St. John's College, Cambridge, 1848–52. (*DNB, MEB*)
1847 *7100* **1850** *8308*

India Office
1866 *13299* 13303 *13320*

Inghirami, Giovanni (1779–1851). Italian astronomer and geologist. Professor of mathematics and director of observatory in Florence. (*EnItal, DBU*)
1824 *1036* **1825** *1161*

Ingleby, Clement M. (1823–86). English writer, Shakespearean scholar. (*DNB, MEB*)
1866 *13340*

Inglis, Sir Robert Harry (1786–1855). English barrister and M.P. Trustee of the British Museum. (*DNB, MEB*)
1829 *2049* **1831** *2461* **1838** *3715 3819 3820* **1843** *5577 5673* **1844** *5697 5699* **1845** *6094 6104 6143 6343* **1846** *6640 6644* **1847** *7467* **1848** *7584 7607 7767 7930* **1849** *7975* **1850** *8271 8292 8432 8484* **1851** *8611 8774 8898* 9122 *9124* **1852** *9175 9212 9275* **1853** *9903* **1855** *10380*

Innes, James Rose (1799–1873). Scottish mathematician. Professor and academic administrator in South Africa. (*DSAfB, SESA*)
1840 *4409 4506 4626* **undated** *14674*

***Intellectual Observer*, Editor of the**
1865 12978

Isbister, Alexander K. (1822–83). English writer on education. (*DNB, MEB*)
1861 *11564*

Ivory, James (1765–1842). Scottish mathematician. Created Ivory's Theorem (for ellipsoids). Determined cometary orbits. (*DSB, Who's*)
1828 *1771* 1775 *1856* 1857 **1831** *2436* **1838** *3690 3706* **undated** *14675*

Ivory, William. Involved in settling affairs of Thomas Henderson, Astronomer Royal for Scotland.
1844 *5817 6047* **1845** *6083*

J. J. Ewer & Co.
1862 *11980*

Jackson, Frederic J. Music theorist. (1868-8-18)
1868 *13606* 13615 *13627 13722*

Jackson, J. Cashier to Bank of Ireland in Dublin in 1853.
1860 *11343*

Jackson, Julian (1790–1853). British geographer and officer in the British and Russian armies. Secretary to the Royal Geographical Society, 1841–47. (*DNB*)
1841 4804

James, Henry (1803–77). British army officer. Supervised several Ordnance and other surveys. (*DNB, MEB*)
1857 *10648* 10649 *10652* **1858** *10936 10956* **1865** 12914 12976 **1868** *13670 13768* **1870** *14124 14172*

Jameson, Robert (1774–1854). Scottish mineralogist, geologist, and natural historian. Professor at Edinburgh, 1803–54. (*DSB, Who's*)
1826 *1383* **1832** *2664*

Jamieson, Thomas F.
1861 *11742*

Jarnac, Phillipe (1815–75). French government minister, and envoy to England in 1847. (*LGE*)
1847 7157 *7165*

Jeans, George (fl. 1845). English astronomer. Wrote *Practical Astronomy for the Unlearned* (1848). (*Alli*)
1838 *3744* **1840** *4559 4679*

Jeffrey, Lord Francis (1773–1850). Scottish critic. Writer for the *Edinburgh Review*. Named to the Court of Session, 1834 (thus a law lord). (*DNB, Alli*)
1850 *8199* 8208

Jeffreys, Henry Anthony (1810–98). British divine and minister at Hawkhurst, 1839–97. Wrote on English decimal system. (*MEB*)
1840 *4347 4348 4349 4350 4351 4352 4353 4445 4460 4472* **1841** *4766* **1844** *5724 5729* **1845** *6101 6102 6367* 6369 *6372* **1849** *8023 8024* **1850** *8203* 8204 **1851** *8809 8879* **1854** 9970 **1855** *10433 10438* **1858** *10779 10832 10861* **1859** 11055 **1861** *11554* 11555 **1862** *11811* **1863** *12234* **1866** *13072* **1869** *14058* **1871** *14427* **undated** *14630 14631 14676 14677 14678 14679 14680 14681 14682 14683 14684*

Jenkins, B. G.
1871 *14351 14352*

Jenkins, Henry Michael (1840–86). British librarian, and functionary to Geological and Agricultural societies. (*MEB*)
1862 *11755*

Jenkins, Richard (1785–1853). Indian statesman, linguist, and officer of the East India Company. (*DNB, MEB*)
1839 4092

Jennings, Edward John
1864 *12527* **1865** *12793*

Jennings, H. C.
1845 *6308*

Jennings, J. H.
1854 10258

Jennings, Philip
1840 *4432*

Jensen, C. A. Professor at Copenhagen University; friend of Hans Christian Oersted. (1839–1–24)
1839 *4135* **undated** *14685*

Jephson, I. H. Secretary to the Cambridge Commission. (1852–8–6)
1852 *9472*

Jerdan, William (1782–1869). British journalist and editor of several papers and journals. (*DNB, MEB*)
1847 *7213*

Jervis, Thomas Best (1796–1857). Map maker and printer of maps. Officer in Bombay Engineers, 1813–44. (*MEB*)
1834 *2933* **1835** *3195* **1839** *4081* 4088 *4239* **undated** *14686*

Jevons, William Stanley (1835–82). British economist and logician. Assayer at Sydney mint, 1854–59. (*DNB, MEB*)
1856 10455 **1861** *11642* 11653 **1864** *12677* **1865** 13028 *13052 13056* **1866** 13063 *13134* 13137 *13224* 13225 **1867** 13473 **1869** *13956* 13982 **1870** *14083 14309* 14318

Jey, S. Servant/companion to Lady Watson, wife of William Watson, a friend of JH's father, William. (1816–9–26 & 1829–3–9)
1829 *1970* **1841** *4847*

Jobert, Antoine C. G.
1843 *5457 5463* **1846** *6663* **1863** *12281*

Johnson, Charles
1847 *6864*

Johnson, Christopher
1840 *4473*

Johnson, Cuthbert W. (1799–1878). English agricultural writer. Supported the Public Health Acts of 1848. (*DNB, MEB*)
1860 *11314*

Johnson, Manuel John (1805–59). British astronomer and East India Company artillery officer. Director of Radcliffe observatory, 1839–59. (*DNB, DSB*)
1839 *4020 4068* **1847** *6940 7205* **1848** *7785* **1856** *10580*

Johnson, Walter R. (1794–1852). American educator and scientist. (*NCAB*)
1847 *7352*

Johnson, William (1823–64). Patent agent in Glasgow. Editor of *Practical Mechanic's Journal*, 1848–64. (*MEB*)
1852 *9566 9582*

Johnston, Alexander K. (1804–71). Scottish-born geographer. Published the first physical atlas in England. (*MEB, Alli*)
1859 *11038* 11044

Johnston, George. Retired army officer of 6th Dragoon Guards. (See 14687.)
undated *14687*

Johnstone, John. An employee at the New Prison, Wandsworth, Surrey. (1849–3–8)
1849 *8001*

Jones, Henry Bence (1814–73). English physician, and chemist applying chemistry to questions in pathology. (*DNB, MEB*)
1864 12630 **1865** *12886* **1868** *13699* **1869** *13946* 13948 **undated** *14688*

Jones, Henry Halford. Secretary to Manchester Literary and Philosophical Society. (1853–12–5)
1853 *9899*

Jones, John (1800–75). English quartermaster-general and astronomer. (*MEB*)
1869 *13876 13983 13997*

Jones, L.
1854 10213

Jones, Mr. Probably a servant of JH.
undated 14689

Jones, Mrs. Richard. Wife of Richard Jones.
1854 10127 **1855** 10286 **1869** 14072

Jones, Richard (1790–1855). British political economist and divine. Taught at Haileybury College, 1835–55. (*DNB, MEB*)
1816 *194* **1822** *664* 757 *759 760 783* 796 **1826** *1365* 1372 **1827** *1481* **1830** *2255* **1831** *2278 2280 2294* 2296 **1832** *2545* **1833** *2730 2750 2816 2874 2892 2915 2919* **1834** 3015 **1837** *3480* 3587 **1838** 3778 **1839** *3993 4231* **1840** 4354 *4493 4662 4687 4759* **1841** *4778 4904 4940 4953 5003 5006 5013 5043* **1842** *5103 5139 5278* **1843** 5364 **1844** *6012* **1845** *6063 6362* **1846** *6667* 6756 **1847** *7113 7142* **1848** *7710 7783 7836 7852* **1849** *7958 8057* 8060 *8145* **1851** 8514 *8896* **1852** 9559 **1853** *9758* 9761 *9767 9791* **1854** 9921 **undated** *14657 14690 14691 14692 14693 14694 14695 14696 14697 14698 14699 14700 14701 14702 14703 14704 14705 14706*

Jones, Thomas. Librarian at Chetham's Library, Manchester. (1847–10–17)
1847 *7299* **1848** *7784*

Jones, Thomas R. Nephew of W. R. Dawes. (1868–2–19)
1868 *13609 13675 13679 13680*

Jones, William (1789–1834). British divine. Fellow of St. John's College, Cambridge. (*Cantab*)
1832 *2639 2668* **1833** *2711* 2725 *2788*

Jopling, Joseph (1831–84). British historical painter, clerk in Horse Guards, and prize-winning rifleman. (*DNB, MEB*)
1844 *5859* **1854** *9967* **1860** *11318 11346 11364 11413 11421 11433 11456 11512* **1861** *11535 11556* **1863** *12186*

Joseph, V.
1831 *2390*

Joule, James Prescott (1818–89). British physicist. Secretary of Literary and Philosophical Society of Manchester, 1842–89. (*DNB, DSB*)
1843 *5454* **1846** *6778* **1847** *7262 7263* **1857** 10761 **1863** 12346 *12351*

***Journal of the Society of Arts*, the**
1869 14038 14044

Joyce, Arthur J. (fl. 1853). English writer of miscellaneous works. (*AlliSup*)
1867 *13445*

Kaiser, Frederik (1808–72). German astronomer, professor, and director of Leiden observatory, 1837–72. (*DSB*)
1848 *7671*

Kane, Robert John (1809–90). Irish-born chemist, physician and industrialist. Married Katherine Baily, niece of Francis Baily. (*DNB, MEB*)
1846 *6729* **1847** 7218 *7230* 7285 7286 **1848** *7942* **1849** *7959* 7961 *7966* **1850** *8478*

Kater, Edward (?–1866). Son [?] of Henry Kater. Member of Royal Irish Academy. (*MEB*)
1840 *4485 4494*

Kater, Henry (1777–1835). British army officer and surveyor in Madras. Improved hygrometer and pendulum. (*DNB*)
1821 509 516 596 **1822** 652 691 696 715 *726 728* 742 764 **1823** 896 **1825** 1254 **1826** 1306 *1308* 1329 **1827** 1541 *1546* **1828** 1765 1782 *1783* 1804 **1829** 2053 *2055* **1830** *2100* 2101 2137 *2140* 2143 2172 2176 *2178* **1831** *2345* **1832** 2608 *2616 2671* **1833** 2747 **undated** 14707

Kater, Mrs. Mary Frances. Wife of Henry Kater.
1821 510 511 **1822** 780 **1824** *942* **1828** *1697* 1701 1702 **1830** *2084 2102* **undated** *14708* 14709

Kay, Joseph Henry (1814–75). British naval officer. Director of magnetic observatory at Hobart Town. (*MEB*)
1842 5153

Kearney, J. B.
1859 *11146* **1865** *12997*

Keating, William H. (1799–1840). American chemist and lawyer. Professor of chemistry at the University of Pennsylvania. (*NCAB*)
1825 1149

Keddie, William (1809–77). Secretary of the Glasgow Philosophical Society. (*MEB, Alli*)
1848 *7505*

Keely, Prof. American professor at Waterville College in Maine. (1849–1–31)
1849 *7980*

Keleher, William. Local arrangements committee secretary for B.A.A.S. meeting in Cork in 1843. (letter undated)
1843 *5488 5496* 5510

Kelland, Philip (1808–79). British mathematician and divine. Professor at Edinburgh, 1838–79. (*DNB, MEB*)
1859 *11040* 11046 **1868** *13673*

Kelvin, Lord (see Thomson, William)

Kemplay, James (1810–82). Barrister and author. (*MEB*)
1843 *5444*

Kersten, C. T. von. Amateur philologist. (1854–2–21)
1854 *9998* 10012

Keverbey de Kessel, Baron Charles L. G. J. Italian consul in Brussels in 1829. (CK's 1829–?)
1829 *2011*

Khanikoff, N. de (1819–78). Russian geographer and surveyor. (*BHGW*)
1864 *12671 12765*

Kidd, John (1775–1851). English chemist amd anatomist. Professor of medicine at Oxford. (*DSB, Who's*)
1825 *1168*

Kilburne, Goodwin. Secretary for Kent Rifle Volunteers. (1866–2–20)
1862 *12081* **1866** *13094*

Kindt, Hermann
1864 *12454 12466 12481 12537 12687 12691 12697* **1865** *12904 12938 12983* **1866** *13268 13280* **1869** *13931 13962*

King, Augusta Ada (née Byron) (1815–52). Lady Lovelace. Married William King, 1835. (*Peerage*)
1848 *7829 7855 7857 7860* 7865

King, Henry Samuel (1817–78). English banker, publisher, and bookseller. (*MEB*)
1867 *13552* **1868** *13596* **1869** *14070* 14073 **1870** *14333*

King, Joshua (1798–1857). President of Queen's College, Cambridge, 1832–57. Lucasian Professor, Cambridge, 1839–49. (*MEB*)
1825 *1122 1125*

King, William (1805–93). Lord lieutenant of County Surrey. Created Viscount Ockham and Earl of Lovelace, 1838. Added name -Noel. (*Burke's Peerage*)
1848 *7861* **1849** *8009*

Kinglake, R. Arthur
1865 *12958*

King-Noel, William (see King, William)

Kingston, Bishop of
1865 12937

Kirchoff, E. W. L. Solicitor in Hanover who handled legal affairs after the death of Caroline L. Herschel, 1848. (Annie Elise Knipping's 1848–1–29)
1848 *7516*

Kirkman, Thomas P. (1806–95). English mathematician. Noted for his polemic against Herbert Spencer's philosophy. (*DSB, Who's*)
1862 *11813 11823 11827* **1863** *12115 12202 12306 12362 12363 12365 12378* **1864** *12681*

Kitchin, George William (1827–1912). British divine and official at Clarendon Press, Oxford. (*DNB*)
1866 *13327* 13332 *13338*

Klingemann, C. Secretary to Hanover legation in London. (1847–8–23)
1847 *7176*

Klinkerfues, Ernst F. W. (1827–84). German astronomer and physicist. Studied cometary and planetary orbits. (*Who's, NDtB*)
1866 *13088* 13101 *13138*

Knight, Charles
1840 *4355*

Knightsee, John. Official at Bank of England, 1846. (1846–5–1)
1846 *6634*

Knipping, Adolph. Son of Anne Elise Knipping. (AEK's 1848–1–29)
1849 8110

Knipping, Anne Elise (née Herschel). Daughter of Johann Dietrich Herschel, elder sister of Mrs. Groskopf and cousin of JH. (1848–1–29)
1822 *814* **1847** *6944 7026* 7300 *7450* **1848** *7523* 7549 *7550 7650 7677 7819*

Knoblauch, Karl Herman (1820–95). German physicist. Professor at the University of Halle. (*BHGW*)
1847 *6936*

Knott, George (1835–94). English astronomer. Built observatory at his residence. (*MEB*)
1863 *12389* **1868** *13684*

Knowles, Charles Henry (1754–1831). English naval officer. Author of many technical pamphlets. (*DNB*)
1830 *2174*

Knowles, Francis Charles (1802–92). English barrister and writer. Only son of Charles Henry Knowles. Won Telford Prize for mathematics. (*MEB*)
1849 *8028* **1856** *10574* 10581 *10599* 10601 **1866** *13162* 13172

Knowles, [John] (1781–1841). British clerk in the Surveyor's Department of the Navy Office. (*DNB*)
1822 *712*

Koller, Baron. Attached to Austro-Hungarian embassy in London in 1847, along with Count Dietrichstein. (1847–10–18)
1847 *7304*

Koller, Marian (1792–1866). Austrian chemist and astronomer, and Benedictine monk. (*OBL*)
1847 *6949*

König, Charles D. E. (1774–1851). British mineralogist. Keeper of natural history department at the British Museum. (*DNB, MEB*)
1839 *4260* **1844** *5738*

Kreil, Karl (1798–1862). Austrian astronomer and geophysicist. Director of meteorological observatory at Prague. (*OBL*)
1838 *3761* **1839** *4264* 4325

Kupffer, Adolph Theodor [de] (1799–1865). German meteorologist. Director-general of magnetic observatories in Russia. (*NUC, NDtB*)
1840 *4664* **1844** *5977* **1845** *6116* **1849** *7979*

Lamb, William (1779–1848). English statesman and prime minister. 2nd Viscount Melbourne. (*DNB*)
1838 3759 *3765*

Lamont, Johann (1805–79). German astronomer and physicist. Involved in meteorological and magnetic observations. (*DSB*)
1846 *6822* **1852** *9340* **undated** 14710

Landeman, John
1838 *3675*

Langley, S.
1868 *13767* 13773 *13781*

Langton, C.
undated *14711*

Langton, Joseph. Tenant farmer on 'the Merke,' property inherited by JH's mother. (CL's 1833–10–11)
1830 *2086* **1831** 2361 *2362* 2366 *2367*

Langton, Martha. Possibly wife of Joseph Langton on 'the Merke'. (1833–10–10)
1833 *2896* **1839** *3921*

Langton, Miss C. Relative of Joseph Langton on 'the Merke.'
1833 2897

Laplace, Pierre Simon (1749–1827). French mathematician and physicist. Developed metric system. (*DSB*)
1822 665 *685 703 749* 838 **1823** *843 872* 904 **1824** 979 *1067* **1825** 1155 *1226*

Lardner, Dionysius (1793–1859). British mathematician and science writer. Edited *Cabinet Cyclopaedia,* 1829–49. (*DNB*)
1827 1503 **1828** *1784* 1797 *1799* 1805 1821 *1912* **1829** 1951 *1953 1956* 1960 **1830** *2090* 2093 2166 2186 2208 *2218 2224 2251* **1831** 2302 *2305* 2315 **1832** *2551* **1837** *3550* **1838** 3640

Largeteau, Charles Louis (1791–1857). French astronomer. Improved calculations for ecliptic conjunction of two celestial bodies. (*Who's, NBG*)
1826 *1315*

Lassell, William (1799–1880). British astronomer. Built several private observatories. (*DNB, DSB*)
1842 *5124* 5133 **1843** *5434* **1846** *6709* 6711 *6712 6727* 6737 6757 *6767* **1847** 6880 *6884 7128 7130 7150* **1848** 7524 *7538 7556* 7837 7858 *7863 7864* 7866 **1850** *8375* **1851** *9026* 9031 9032 *9085* **1852** *9306 9308 9324 9350* 9353 *9370 9598* **1853** *9890* **1857** 10736 *10740* 10742 *10749* **1858** 10924 *10929* **1859** 11126 *11132* 11153 **1860** *11289* 11294 *11295* 11476 *11499* **1861** *11648 11671 11679 11724* **1862** 11878 *11893 11997 11998* 12005 *12016* 12017 *12031* 12047 *12077* **1863** *12112* **1864** *12626 12683* 12695 **1868** *13629* **1874** 14455

Laughton, John Knox (1830–1915). English naval historian. Professor of modern history, King's College, London. (*AlliSup*)
1870 *14225*

Laugier, Paul Auguste Ernest (1812–72). French astronomer. Member of the Académie des sciences. (*NBG*)
1847 *7398* **1848** *7754*

Laurie, Joseph
1839 *4298*

Lavia, Gregorio Barnaba. Foreign secretary to Gioenian Academy of Natural Sciences of Catania, Sicily. (1825–10–6)
1825 1238

Law, Charles Ewan (1792–1850). British barrister and legal administrator. M.P. for Cambridge University. (*DNB*)
1835 *3047*

Law, William. Official in British prime minister's (Lord Palmerston's) office. (1855–3–13)
1851 *8733* **1853** *9832* **1855** *10348*

Lawley, Francis. Official in Exchequer office in 1853. (1853–7–20)
1853 *9780*

Lawrence, Thomas
1844 *5764*

Lawson, Henry (1774–1855). British astronomer. Equipped several private observatories. (*DNB, MEB*)
1847 *7180*

Learmonth, John (1789–1858). Scottish coach builder. Lord Provost of Edinburgh, 1832–3. (*MEB*)
1832 *2655*

Lecky, William E. H. (1838–1903). Irish-born historian and essayist. Wrote *History of England in the Eighteenth Century*. (*DNB, AlliSup*)
1866 *13081*

Lee, John (1783–1866). British lawyer, astronomer and antiquarian. Cofounder of the R.A.S. Built private observatory. (*DNB, MEB*)
1824 *1081* **1829** *1958* **1837** *3446 3453 3503* **1847** *6853 6918 7131* **1861** *11611 11735* 11737 **1862** *11756 11757 11758 11943 11944 11995* **1863** *12140* **1865** *12947 12979 13000* **undated** *14712 14713*

Lee, Robert. British naval officer, employed at the Admiralty. (1859–5–3)
1859 *11066*

Lee, Sarah (see Bowdich, Sarah)
1848 *7693 7709*

Lee, Stephen (fl. 1795 1825). English astronomer. Clerk and librarian to the R.S.L. (*Alli, Watt*)
1825 *1256* **1828** 1766 *1778*

Lefevre, Charles Shaw- (see Shaw-Lefevre, Charles)

Lefevre, John George Shaw- (see Shaw-Lefevre, John George)

Lefroy, John Henry (1817–90). British army officer and Arctic explorer. Colonial governor and director of several colonial magnetic observatories. (*DNB, DCB*)
1840 *4492* **1850** *8281*

Lemon, Sir Charles (1784–1868). British M.P. President of both Geological and Technological Societies in Cornwall. (*MEB*)
1844 5785 *5806 5816*

Lennox, Charles Henry Gordon- (see Gordon Lennox, Charles Henry)

Lerebours, Noel Paymal. French optician for the Paris Observatory. (*NBG*)
1840 *4555 4581*

Lesley, J. Peter (1819–1903). American geologist. Wrote *A Manual of Coal and Its Topography*. (*DSB, Who's*)
1869 *14028* **1870** 14264

Leslie, John (1766–1832). British mathematician and natural philosopher. (*DNB*)
1830 *2216*

Leverrier, Urbain J. J. (1811–77). French astronomer and meteorologist. Predicted the position of Neptune, 1846. (*NBG, DSB*)
1846 *6807* 6811 *6814* **1847** *6905 7332* **1848** *7832* **1849** *7985* 8006 *8007* 8010 *8025* **1853** *9716* **1855** *10386* **1867** *13361* 13363 *13370 13480* **1869** 14014 *14018*

Levi, Julie Laurence. Friend of Mrs. Beke (probably Emily Alston Beke). (1869–6–16)
1869 *13944* 13947

Levick, Frederick. Partner in Blaina iron works of Levick and Simpson. (1866–6–4)
1866 *13218*

Lewes, George Henry (1817–1878). English writer. Common-law husband of English novelist George Eliot. (*DNB, MEB*)
1863 *12412*

Lewin, Thomas (1805–79). English writer of works on law and on theology. (*DNB*)
1850 *8282*

Lewis, Evan (1828–69). English minister. Lectured on religious and scientific matters. (*DNB, MEB*)
1865 *13033*

Lewis, George Cornewall (1806–63). British barrister, M.P., cabinet minister, and editor of the *Edinburgh Review*. (*DNB, MEB*)
1851 *8788 8923* **1859** 11188

Lewis, Griffith George (1784–1859). British army officer. Commanded Royal Engineers at the Cape. (*DNB, MEB*)
1854 10029

Lewis, Richard. Secretary of Royal National Life Boat Institution. (1861–3–22)
1861 *11574*

Leybourn, Thomas (1770–1840). English mathematician. Professor at Sandhurst. (*DNB, Alli*)
1828 *1862*

Lichtenberg, Mr. Friend of Caroline Lucretia Herschel. (1837–9–15)
1837 *3538*

Liddell, Henry. Secretary of the Slough Mechanics Institute. (1848–4–15)
1848 *7673* 7681 *7699*

Liddell, James. British naval officer. Commanded the *Wellington* on many India voyages. (1864–12–24)
1864 *12763* **1866** *13061*

Lieven, Princess. The Lieven family was a noble family in Germany for centuries, with branches in Russia and Sweden. (*NDtB*)
undated *14714*

Lincke, A.
1869 *13994*

Linden, Baron de. The Linden family was a noble family in Germany for centuries. One member of that family was serving in London in the Württemburg legation in 1847. (*NDtB*)
1847 *7261* 7268

Lindenau, Bernard A. von (1779–1854). German astronomer, administrator. (*Who's, NDtB*)
1825 *1202* **1835** *3138* **1842** *5175*

Lindley, John (1799–1865). English botanist and horticulturist. Editor of *Gardners' Chronicle*. (*DNB, Who's*)
1838 *3787 3794 3855* **1839** 3932 *3940 3967 3995 4208* 4222 **1842** 5107 *5109* **1843** *5483* 5484 *5486* **1844** 5700 *6052*

Lipscomb, Mary Cornwallis (see Herschel, Mary Cornwallis Lipscomb Power)

Lipscomb, Miss
1853 9843

Lister, Joseph Jackson (1786-1869). Wine merchant and optical scientist. Work led to the modern microscope. (*DNB*)
1831 *2306*

Littrow, Joseph Johann [von] (1781–1840). Austrian astronomer and director of the observatory at Vienna. (*OBL*)
1821 *567 583* 641 **1822** *657 689* 740 *776* 836 **1823** *848 856 920 926 938* **1824** 954 957 *965 988* 1070 **1825** *1147 1193* **1826** *1438* **1827** 1516 *1559 1584* **1828** 1795 *1827* **1829** *2039* **1830** *2247 2248* **1831** *2330 2375* **1832** *2606* **undated** *14715*

Livingstone, David (1813–73). English explorer and missionary to Africa, physician. (*DNB, Who's*)
1858 *10806*

Lloyd, Humphrey (1800–81). Irish experimenter in optics and professor of natural philosophy. Organized global network of magnetic observatories. (*DNB, DSB*)
1838 *3679* 3681 *3738* 3742 3802 *3811* 3834 *3844* 3852 *3856* **1839** 4004 *4008 4073* 4077 *4192* 4199 *4205 4235* **1840** *4377* 4456 4597 *4600* 4648 *4698 4702* 4704 *4710 4734* **1841** *4938 4977* **1842** *5245* 5248 *5255* 5263 **1844** 5869 *5960* 5998 *6006* **1845** 6099 *6105 6106 6124* 6268 *6274* 6275 **1846** *6680 6770* **1848** 7736 **1850** *8297* **1857** 10686 *10734* 10738 **1858** *10817* 10897 10898 *10900* **1859** 11137 11145 **1865** 12917 *12924 12945* **undated** *14716*

Lloyd, John Augustus (1800–54). British engineer and surveyor. Surveyed parts of South and Central America. (*DNB, MEB*)
1835 3093 *3165* 3224 **1836** *3266 3270 3308 3400* **1837** *3474 3576*

Lloyd, William Watkiss (1813–93). British classical scholar and tobacco manufacturer. (*DNB, MEB*)
1862 *11759*

Lockhart, John Gibson (1794–1854). British author and classical scholar. Editor of the *Quarterly Review*, 1825–53. (*DNB, MEB*)
1832 *2648* 2652 **1840** *4620 4720 4722* 4723 **1841** *4865 4918*

Lockhart, William (1811–96). British physician and medical missionary to China. Translated medical works into Chinese. (*MEB*)
1860 *11394*

Lombardi, Antoine. Secretary of the Societa italienne des sciences at Modena. (1843–6–10)
1843 *5498* **1846** *6673*

London Bridge, Committee for Rebuilding
1824 1006

***London Review*, the**
1861 11607 11636 **1862** 11996

London, Lord Mayor of
1823 880

Longbottom, R. L.
1851 8813

Longfellow, Henry Wadsworth (1807–82). American poet and professor at Harvard. (*DAB*)
1867 *13485*

Longley, Charles Thomas (1794–1868). Archbishop of Canterbury, 1862–8. (*DNB, MEB*)
1864 *12732* **1865** 12776

Longman and Co.
1844 5789 **1863** *12312 12332*

Longman & Green Publishers
1867 13353

Loomis, Elias (1811–89). American mathematician, astronomer, and meteorologist. (*NCAB, DAB*)
1848 *7638* **1855** *10375* **1860** *11471* **1866** *13341*

Lovelace, Lord (see King, William)

Lovering, Joseph (1813–92). American mathematician and natural philosopher. Academic, and long term official with A.A.A.S. (*NCAB, DAB*)
1840 *4532* **1847** *7355*

Lowe, Arthur Swain Howard (1826–88). English meteorologist, illustrator. (*MEB*)
1843 *5505* **1857** *10693*

Lowe, Edward Joseph (1825–1900). British meteorologist and botanist. Cofounder of Meteorological Society. (*MEB*)
1844 *6004* **1845** *6277* **1847** *6935* **1848** *7639 7769 7790 7925* **1849** *7992 7997 8032 8129* 8136 *8139* 8142 **1850** *8311* **1853** *9884* **1854** *10026* 10045 **1855** *10372* **1857** *10702* **1860** *11257 11282 11298 11477* **1862** *11767 11775 11790 11909 12066 12090* **1863** *12188* **1864** *12457* **1866** *13334* 13335 **1870** *14299* **undated** *14717*

Lowry, Wilson (1762–1824). English engraver. Interested in philosophical, geological, and mineralogical topics. (*DNB, Alli*)
1820 405 *407* **1823** *873*

Loyd, Edward
1860 *11366* **1861** *11546* **1862** 11947 **1865** *12877* **1868** *13623*

Lubbock, Harriet (née Hotham) (?–1873). Married John William Lubbock, 1833.
1860 *11365* **1862** 12054 **1865** 12901 *12905* 12929

Lubbock, John (1834–1913). Scientist, banker, and member of Parliament. Son of John William Lubbock. (*DNB*)
1865 *12908* 12912 **1866** *13077 13095* **1870** 14316

Lubbock, John William (1803–65). British astronomer, mathematician, and banker. Long term R.S.L. officer. On Standards Commission (with JH?). (*DNB, DSB*)
1832 2501 *2676 2689* **1833** *2793* 2794 **1838** 3621 **1839** 3907 4093 4194 *4201 4256* **1840** 4356 *4388 4403* 4407 *4411 4430* 4476 *4478 4482* 4483 *4487 4495* 4500 4563 4564 *4594* 4601 *4602* 4606 *4610* 4611 4705 **1841** 4896 4916 4942 4996 *5004* 5034 **1844** 6048 *6050* 6051 *6055* **1845** 6078 *6080 6108* 6110 *6112 6113* 6118 *6122 6132 6145 6152 6158 6179* 6182 *6212* **1847** 7034 7077 *7124 7317 7402* 7417 7418 7431 7432 *7435* **1849** *8160* **1851** 9002 **1852** *9599* **1853** 9858 **1854** *10078* 10084 10086 **1856** 10479 **1861** *11540 11542 11549* 11659 11660 *11664 11668 11677 11687* 11688 11689 *11691* **1862** *11870 11941* **undated** *14718 14719*

Lubbock, Mary H. Daughter of J. W. Lubbock. JH and family attended her wedding in December 1857.
1857 *10758*

Luc, Antoinette de
1822 *778*

Ludwig, Carl F. H. von (1784–1847). German physiologist. Invented several instruments related to blood circulation. (*DSB, Who's*)
1836 *3258 3292* **undated** *14720*

Luff, H. J. A church warden at Slough. (W. Bonsey's 1859–5–16)
1856 10468

Lunn, Francis. British divine. Vicar at Butleigh. ([1823]–9–23)
1823 *890 908* **1824** *1012* 1071 *1084* **1825** 1098 *1101* **1829** 1919

Lutzardt de Wabern (see Wabern, Lutzardt de)

Lyell, Sir Charles (1797–1875). British geologist and academic. (*DNB, DSB*)
1825 *1159 1163* **1828** *1829* **1829** *2057* **1835** *3148* **1836** 3278 *3329 3333* **1837** *3435 3477* 3497 **1840** *4357* **1843** 5594 *5600* **1844** 6014 **1846** *6672* **1848** *7614* 7616 7617 **1851** *9113* **1861** *11670* **1862** *11915 12027* **1863** 12208 **1864** 12562 12586 **1865** 12797 *12802* 12805 12811 *12815* 12822 *12823* 12826 *12828* 12834 *12840* 12841 12880 *12884 12889* 12895 *13032* 13043 **1866** *13269* 13273 13283 *13284* 13286 *13288* 13324 **1868** 13693 *13700 13704* 13707 **undated** *14721 14722*

Lyell, Mary Elizabeth (née Horner) (1809–73). Conchologist. Married Charles Lyell, 1832.
1853 9812 **1863** *12264* **1869** *13875*

M., Mr.
1846 6621

M[?], A[?]
1847 *7319*

Macann, M.
1848 *7573 7586 7678 7682*

Macarthur, William (1800–1882). Australian-born member of the New South Wales legislative council. (*DNB*)
1856 10469

McBean, Thomas
1839 *4330 4338* **1840** *4414 4421* 4426 *4435*

McClelland, John (?–1883). English surgeon, geological writer and zoologist. Inspector General of Bengali hospitals. (*MEB, AlliSup*)
1837 *3577* **1842** *5337* **1856** *10546 10607*

McCoy, Frederick (1817–99). Irish-born physician and geologist. Professor and museum director in Melbourne, Australia. (*AusDB*)
1855 *10339*

McCullagh, James (1809–47). Irish mathematician and academic. (*DNB*)
1831 *2402* **1846** 6664 *6758*

Macculloch, James Ramsey (1789–1864). British economist, academic, and writer. (*MEB*)
1851 8832

Macdougal, Thomas
1839 *4108*

McDowell, James (?–1885). English mathematician, scientist and writer. (*MEB, AlliSup*)
1867 *13475* 13477

Macfarlane, Duncan (1771–1857). Principal of Glasgow University and moderator of the established church of Scotland. (*DNB, MEB*)
1838 3813 *3821* **1847** *7353*

Macfarren, George A. (1813–87). English musical composer and professor of music. (*DNB, MEB*)
1868 *13778* 13791 *13795* **undated** *14723*

McGrigor, Alexander B. Chair of a committee offering public science lectures in Glasgow. (1863–9–25)
1863 *12313* 12315

McGrigor, Sir James (1771-1858). Scottish-born army surgeon. (*DNB*)
1833 *2902* **1836** 3339 *3380* **undated** 14734

McIntyre, Mr.
1832 2534

McKay, William G.
1862 *11925*

Mackay, W. S.
1844 *5681*

Mackereth, Thomas
1865 *12837 12843*

Mackintosh, Daniel (1815–91). English lecturer on astronomy, geology, and physical geology. (*MEB, AlliSup*)
1866 *13257* 13276 13301

Mackintosh, Eneas. Guardian of Margaret Brodie Herschel before her marriage. Obtained appointments in East India Company for JH's sons John and William.

1830 *2149* **1843** 5567 *5569* **1849** *8071* 8072 *8074* *8095* **undated** *14724*

MacKintosh, James (1765–1832). English philosopher. Wrote *Vindiciae Gallicae* in response to Edmund Burke's *Reflections on the French Revolution.* (*DNB, Alli*)
1831 *2310*

McLachlan, Mr.
1845 *6342*

Maclaren, James
1853 *9645 9654* **1856** *10598* **1858** *10777 10787* **1869** *13996* **undated** *14725*

Maclean, C. Hope. Secretary to Statistical Society in 1834. (1834–5–19)
1834 *2977*

Maclear, Harry W.
1864 *12615*

Maclear, Mary (née Pearse) (?–1862). Wife of Thomas Maclear. Married in 1825. (Warners, *Maclear*)
1837 3566 **1846** *6844*

Maclear, Thomas (1794–1879). Irish-born surgeon and astronomer. Astronomer Royal at the Cape observatory, 1834–70. (*DNB, DSB*)
1833 *2748 2827 2837 2851 2861 2873 2901 2905* **1834** 2934 *2942* 2944 2947 *2948* 2955 *2956 2957 2958* 2961 2962 *2964 2966 2968 2969* 2970 *2979* 2980 2988 *2996 2997* 2998 *2999* 3002 *3003 3007* 3013 *3016 3023* 3033 *3036* 3037 3041 **1835** *3048* 3051 *3052 3053* 3055 *3056* 3060 3064 3067 *3070 3073 3075* 3078 3083 *3085* 3086 3087 3094 *3095 3098 3101* 3102 *3109 3112* 3122 *3123 3125 3127* 3129 *3130 3133 3139 3140* 3141 3142 *3143* 3144 *3146* 3147 3150 *3151* 3162 *3163* 3167 *3168* *3171* 3172 *3173 3175 3178* 3179 *3180 3181* 3184 *3185* 3186 *3187* 3188 *3193* 3194 3197 *3202* 3203 3204 *3205* *3206* 3207 3209 *3210* 3212 3213 *3216* 3217 3219 *3221* *3225* 3226 3227 *3228 3232* 3233 **1836** 3241 *3243* 3248 *3249* 3250 *3253 3254* 3255 *3256 3260* 3263 3264 *3271* 3279 3280 *3281 3282* 3284 *3285* 3286 3287 3288 *3289* 3290 3293 *3295 3297* 3298 *3299* 3300 3302 *3303 3309* 3311 *3312 3318* 3319 *3324* 3325 *3327 3330 3331* 3332 3348 *3354* 3356 *3357 3361* 3362 3369 *3370 3371* 3372 3376 *3377* 3378 3379 3382 *3383 3384 3386 3387* 3388 3390 3394 3395 *3402 3403 3410 3411* 3420 *3422* 3424 *3425 3427* **1837** *3430* 3436 3437 3439 3441 *3443* 3456 *3457 3459 3465 3468 3469* 3470 3471 *3472 3476 3479* 3494 *3498 3499 3500* 3501 3502 *3504* 3505 *3507* 3508 3514 *3515* 3516 3517 3519 *3520* 3524 *3525 3529* 3536 *3541* 3542 3543 *3545* 3554 *3555 3557 3558* 3559 *3568* 3569 *3570* 3571 *3573 3574* 3578 *3579* 3581 *3583 3584* *3585 3586 3588* 3589 3591 *3592 3593 3594* 3596 3600 *3604* **1838** *3614 3619* 3622 3625 *3626 3630 3631* 3632 *3633* 3636 3642 3645 3650 3696 *3775 3816 3817* 3825 **1839** 3908 *4002 4046* 4089 *4098* 4113 *4122 4123* 4133 *4209 4210* 4220 4267 *4271 4295* **1840** *4446* 4484 *4490* *4496 4724* **1841** *4838 4864* **1842** 5093 *5140* **1843** *5670* **1844** 5682 5867 5870 *5936 5937 6005* **1845** 6221 6233 6528 **1846** *6564 6606* 6625 *6645 6657 6838* **1847** *7030 7172* 7406 **1848** *7525 7593* 7743 *7820* *7950* **1849** *7984 8075* 8096 *8097 8101* 8157 *8159* **1850** *8231 8264 8359 8360* 8402 *8413* **1851** *8596* *8972 9088 9143* **1852** *9231* 9264 *9312* 9347 *9357* 9450 *9500 9518 9605* **1853** *9644* **1854** *10208* **1855** *10317* **1856** 10461 10551 *10582* **1859** 11082 *11118 11149* **1860** *11356* 11389 *11396 11448 11449* **1861** *11665* 11696 *11743* **1862** 11780 *11822* 11890 *11924* **1863** *12230 12280* 12390 **1864** *12623 12643* **1865** *12951* *12963 12988 13020* **1866** *13142 13307 13342 13347* **1867** 13458 **1869** 14006 *14056* **1870** *14091* **undated** *14726 14727* 14728

McLennan, John F. (1827–81). Scottish sociologist. Studied ancient marriage rites. (*DNB, MEB*)
1865 *12873 12874 12892* **1866** *13297 13302* 13305 *13309 13322 13336* **1867** *13481*

McLeod, Herbert
1860 11258

Macleod, Norman (1812–72). Scottish minister. Editor of *Good Words.* (*DNB, MEB*)
1862 *12044* 12067 **1864** *12722* 12730 **1867** *13380* *13388 13390 13391 13400 13497* **1870** *14276* 14278 **undated** *14729 14730*

McNeill, D.
1868 *13828*

MacPherson, William (1812–93). Scottish-born legal writer. Master of equity at Supreme Court of Calcutta, 1848–59. Editor of the *Quarterly Review,* 1860-67. (*DNB, MEB*)
1860 *11429* 11464 *11470*

Macvicar, John Gibson (1800–84). Irish author. Minister of Moffat, Dumfriesshire. (*DNB, MEB*)
1864 *12745*

Mädler, Johann Heinrich (1794–1874). German astronomer. Concerned with the moon, Mars, and double stars. (*DNB, Who's*)
1835 *3166* **1837** *3522* **1838** *3666* **1842** 5142 *5147* **1844** *5880* **1845** *6388*

Mädler, Minna. Wife of Johann Heinrich von Mädler.
1842 *5141*

Main, Robert (1808–78). British astronomer. Chief assistant to G. B. Airy at Royal Observatory, Greenwich, 1835–60. (*DNB, MEB*)
1844 *5874* 5876 **1846** *6698* **1847** *7125* **1853** *9712* **1859** *11051* **1861** 11703 *11705* 11708 **1863** *12136* **1868** 13742 *13743* 13748 *13751 13752* 13799 *13802* **1869** *13856* 13859 **1870** *14261 14266* **1871** 14443

Major, Richard Henry (1818–91). British geographer and keeper of maps at the British Museum. (*DNB, MEB*)
1868 *13656*

Malden, Henry (1800–76). English classical scholar. Professor of Greek at London University, 1831–76. (*DNB, MEB*)
1865 12936

Mallet, Robert (1810–81). Civil engineer, scientific investigator and industrialist. (*DNB, MEB*)
1846 *6724*

Manchester Literary and Philosophical Society, the
1862 11801 **1865** 12998 **1867** 13369 **1868** 13650

Manley, Charles. Secretary to the Institution of Civil Engineers, 25 Great George St., Westminster. (1847–8–5)
1847 *7126* **1859** *11158* **1860** *11422*

Mann, William (1817–73). British astronomer; chief assistant to Thomas Maclear at the Cape. (*DNB, MEB*)
1871 *14404*

Manners, Russell Henry (1800–70). English naval officer. President of R.A.S., 1868–70. (*MEB*)
1868 *13635 13830* **1869** *13901* **1870** *14087*

Mantell, Gideon (1790–1852). English geologist and paleontologist. Discovered fossil remains of dinosaurs. (*DNB, Who's*)
1825 *1263*

March, Mr. Undertaker for the funeral of William Herschel, JH's father. ([1822–9])
1822 *781*

Markham, Clements R. Secretary to Royal Geographical Society. (1868–4–30 or earlier)
1868 *13656*

Marsden, Thomas (1810–1890). English scholar and divine. (*MEB, AlliSup*)
1870 *14238*

Marsh, James
1860 *11430*

Marshall, H. C. Relative of R. D. Marshall, who married JH's daughter Margaret Louisa in 1858.
1858 *10780 10886*

Marshall, James Garth (1802–73). M.P. for Leeds, 1847–52. Sheriff of Yorkshire, 1860. (*MEB*)
1858 *10862 10880*

Marshall, Matthew (?–1873). First assistant cashier, 1829–35, and chief cashier, 1835–64, to Bank of England. (*MEB*)
1853 9850 **1854** 10217

Marshall, Reginald Dyke. Married Margaret Louisa, JH's daughter, in 1858.
1861 *11745* **undated** *14731 14732*

Martin, Hugh (1822–85). Minister of Free Church in Scotland, 1844–65. Examiner in mathematics at University of Edinburgh. (*DNB, MEB*)
1870 14082 14088 *14101*

Martin, James Ranald (1793–1874). British surgeon in India, 1826–40, and London, 1840–74. Studied tropical diseases. (*DNB, MEB*)
1845 *6499 6516* **1846** *6575*

Martin, K. B. Harbormaster at Ramsgate. (1857–6–25)
1857 *10695*

Martius, Karl von. Secretary of the Royal Bavarian Academy of Sciences. (JH's 1850–5–31)
1850 8320

Martyn, Peter. A physician in Aberdeen. (1865–9–1)
1865 *12959*

Maskelyne, Nevil Storey (see Storey-Maskelyne, Nevil)

Maskew, H. B. Librarian at South African Public Library, Cape Town, in 1848. (1848–12–20)
1848 *7947*

Masters, W. Professor at Kishangar College in India. (1866–12–20)
1866 *13344*

Mathews, Jr., William (fl. 1875). Possibly an English botanist. (*AlliSup*)
1865 12906 12909 *12910 12923 12943 12948 13019 13029* **1868** *13740*

Mathieu, Claude Louis (1783–1875). French astronomer. Professor at Collège de France and École polytechnique. (*Who's, NBG*)
1827 *1604* 1620

Mathison, Gilbert Farquhar Graeme (1803?–54). British clerk, and refiner of gold and silver at Mint, 1828–51. (Challis, *Mint*)
1851 *8636* 8700 *8704* 8759 *8760* 8772 *8913* 8914 *8969* 8970 *8977 9052 9061 9108*

Mathison, William C.
1868 *13624 13780*

Matteucci, Carlo (1811–68). Italian physicist and physician. Studied electricity and magnetism. (*Who's, EnItal*)
1839 *4003* **1846** *6710* **1851** *8708* **undated** *14733*

Maule, George (1776–1851). Solicitor to Treasury, 1817–51. (1851–4–22)
1851 8754 *8781*

Maury, Matthew Fontaine (1806–73). American oceanographer and naval officer. Superintendent of the Hydrographic Office. (*DAB, DSB*)
1847 *7354* **1859** *11083* 11107

Mauvais, Félix Victor (1809–54). French astronomer. Observer at Paris, and discoverer of four periodic comets. (*Who's*)
1847 *7410* **1848** *7761*

Maxwell, W. E. Nephew of William Pearson. (1848–8–27)
1847 *7185*

May, Charles (1800–60). British chemist, millwright, and inventor. Built private observatory. (*MEB*)
1832 *2672* 2673 **1833** *2755 2774 2911* 2923 **1849** 8163 *8166* **1856** 10535 *10549* **1859** *11229* 11241

Mayall, John Edwin (?–1867). British artist and photographer, living in London. (*MEB*)
1849 *7988*

Mayo, Herbert (1796–1852). English physiologist, anatomist. Discovered true function of facial nerves. (*DNB, Who's*)
1843 *5376*

Meadows, William. British naval officer, assistant to Thomas Henderson at the Cape observatory. (Warners, *Maclear*)
1831 2397

Medwell, Mr.
1851 *8881*

Melbourne, Lord (see Lamb, William)

Melbourne University
1855 *10404*

Melhuish, Arthur James (1829–95). British author and publisher. Involved with Amateur Photographic Association. (*MEB*)
1865 13004

Melloni, Macedonio (1798–1854). Italian physicist and director of meteorological observatory on Mt. Vesuvius. (*DSB, EnItal*)
1841 *4786* 4826 *4832 4860*

Melvill, J.
1866 *13188* **1867** *13437* 13440 *13441* 13556 *13558*

Melvill, James Cosmo (1792–1861). Long term employee of the East India Company, 1808–58. (*DNB, MEB*)
1839 *4124* **1847** *7166*

Melvill, Philip. Secretary of the East India Company. (JH's 1840–1–22)
1840 *4385* 4398

Melvill, Thomas. A clerk in the India Office of the British government. (1862–5–30)
1862 *11908 11920*

Melville, Lord (see Dundas, Robert Saunders)

Melville, W. Leslie. A relative of Margaret Brodie Herschel. Nominates John Herschel (JH's son) to a cadetship at Addiscombe. (E. Mackintosh's 1849–7–16 & 1848–9–11)
1848 *7826*

Merchants Taylor Co.
1864 12578

Merz, Georg (1793–1867). German optician. Cofounder of Munich Optical Institute. (*ADB*)
1843 *5599* **1845** *6474* 6477 **1846** *6559 6581* **1847** 7227

Meyer, Adolf Bernhard (1840–1911). German physician and friend of Georg Adolph Erman. (1870–5–4)
1870 *14185* 14187 *14207 14208*

Meyer, Van De (see Van De Meyer)

Michell, Nicholas (1807–80). English writer of several poems and novels. (*DNB, MEB*)
1870 *14153*

Micklethwaite, S. B. P. (1786–1853). British army officer, and Sheriff of Sussex. (*MEB*)
1841 *4780* 4781 *4782* **1843** *5474*

Mildred?], [husband of Lady
1861 11738

Miles, Alfred (1796–?). British naval officer. (*BBI*)
1835 *3231* **1847** *7318* **1848** *7578*

Mill, John Stuart (1806–73). English philosopher. Developed his father's (James Mill's) Utilitarianism. (*DNB, MEB*)
1843 *5460* **1845** *6322* 6327 6329 *6330* 6333 6334 *6337* 6338 *6505* 6511 *6522* **1846** *6587 6591 6607* 6611 6612 *6616* **undated** 14735

Millar, William J. (fl. 1870). English writer of miscellaneous works. (*AlliSup*)
1865 *12806* 12816

Miller, John Fletcher (1816–56). British meteorologist, who founded an observatory in the Lake district. (*MEB*)
1849 *8162 8169* **1850** *8240*

Miller, Mrs. W. T.
undated *14736*

Miller, William Allen (1817–70). English surgeon and chemist. Cofounder of the Chemical Society. (*DNB, MEB*)
1854 *9996* **1859** *11220 11246* **1865** *12788* **1869** *13881* 13882

Miller, William Hallowes (1801-80). English mineralogist and crystallographer. Constructed new standards of length and weight, 1843. (*DNB, MEB*)
1852 9234 9293 **1853** 9749 **1855** 10417 **1858** *10933 10937* **1862** *12062* **1864** *12550* **1865** 12829 **1867** 13563 *13571* 13576 13577

Milman, Henry Hart (1791–1868). Poet, dramatist, and dean of St Paul's Cathedral. (*DNB, MEB*)
1838 *3705* **1852** *9554* **1862** *12007*

Mint officers
1852 9543

Mint Staff
1851 8887 8975 **1854** 9929 10075 10089 10160 10187

Minto, Lord (see Elliot, Gilbert)

Mitchill, Samuel Lathan (1764–1831). American physicist, U. S. senator, and promoter of science. (*DAB*)
1822 653

Mitscherlich, Eilhard (1794–1863). Swedish chemist, mineralogist, and professor of chemistry at Berlin. (*DSB*)
1828 1911 **1830** *2162* **1831** 2406

Mocatta & Goldsmid
1854 10152

Moffat, Robert (1795–1883). English missionary to South Africa. Superintendent at Kuruman, 1825–70. (*DNB, MEB*)
1862 *11890 11891*

Moigno, François N. M. (1804–84). French physician and mathematician. Editor of scientific weekly *Cosmos*. (*NBG*)
1854 *9943 9957* **1857** 10711 *10714* **1859** *11091* **1863** 12279 *12309* **1864** 12456 *12459* **1866** 13148

Moll, Gerard (1785–1838). Dutch physicist. Professor of physics and mathematics at Utrecht. (*BHGW*)
1821 *517* 549

Molle, William Macquarie. British judge in India. Attended Haileybury College. (*MOH*)
1852 *9497*

Molyneux & Cope
1822 *832* **1823** 852

Molyneux, Robert (fl. 1800–40). Clock maker with office at 44 Devonshire St., Queen Square. (Warners, *Maclear*)
1830 2089

Monckhoven, D. C. E. van (1834–1882). Belgian chemist.
1862 *11760* **1863** 12105 *12118 12137*

Moneyers, Company of
1851 8618 8684 *8693*

Monteagle, Lord (see Spring-Rice, Thomas)

Monticelli, Teodoro (1759–1845). Italian naturalist, geologist, and professor in Naples. Investigated Mt. Vesuvius. (*EnItal*)
1825 1114 *1150* **1826** *1360* **1827** *1532* **1834** *3042* **1835** *3049* **1838** *3716*

Moore, Edward
undated *14737*

Moore, Thomas J.
undated *14738*

Moorsom, E. (?–1861). Companion to Lady Mary Herschel at Windsor in her later years.
1859 *11148*

Morison, William (?–1851). British army officer and M.P. Government official in India. (*MEB*)
1843 5645 **1850** *8283*

Morland, William C.
1866 *13287* **1869** *13869* **undated** *14739*

***Morning Post*, Editor of**
1822 811

***Morning Herald*, the**
1843 5419 5427

Morrell, Robert M. Honorary secretary of National Sunday League. (1856–12–15)
1856 *10604*

Morris-Stirling, T. D.
1846 *6744* 6748

Morton, J. W. A secretary in the Admiralty office. (1830–1–6)
1830 *2083* 2087

Mossotti, Ottaviano F. (1791–1863). Italian astronomer, mathematician, and physicist. University professor in several countries. (*DBU*)
1827 *1563* **1845** *6375*

Moxon, Charles
1844 *6036* **1846** 6839

Moyes, James. Printer. (*Brit. Mus. Cat.*)
1818 319

Mühry, Adalbert Adolf. Unofficial guardian of Caroline Lucretia Herschel's affairs after her return to Hanover. Writer on climate and disease. (1844–1–27)
1842 5308 **1843** *5413* **1844** 5696 *5715* **1857** *10643 10762* **1862** *12021*

Mühry, E. Probably a relative of Adalbert Adolf Mühry, and a relative of JH.
1841 *4890* **1843** *5623*

Mühry, G. Brother of Adalbert Adolf Mühry. (1862–11–14)
1862 *12052 12094* **1863** *12146*

Muirhead, James P. (1813–1898). English writer who wrote a book about James Watt's inventions. (*MEB, Alli*)
1847 6931 **1848** *7506*

Müller, G. W.
1826 *1374* **1843** *5391*

Max Müller, Friedrich (1823–1900) Boden Professor of Sanskrit at Oxford. (S. Langley's 1868–11–5)
1860 *11355*

Munby, C. Secretary in the Colonial Office. (1838–12–7)
1838 *3882*

Murchison, Charlotte. Wife of Roderick Impey Murchison.
1828 *1660*

Murchison, James P.
1847 6938

Murchison, Mrs. Kenneth. Her husband was Kenneth Murchison, a British naval officer resident for a time at the Cape, and brother of Roderick Impey Murchison.
1835 3088

Murchison, Roderick Impey (1792–1871). British geologist and naval officer. Director-general of Geological Survey, 1855–71. (*DNB, DSB*)
1829 2034 2067 **1830** *2259* **1831** *2283* 2284 **1832** 2502 2518 *2706* **1833** *2718* 2735 *2741* 2785 **1835** 3089 **1836** *3244* 3407 **1837** *3449 3486* **1838** *3652 3676* 3697 *3700 3735 3745* 3874 *3897* 3900 **1839** *3955 4023 4032* 4033 *4228 4340* **1840** 4358 **1842** *5059* 5060 5115 *5120 5125* 5134 *5204* **1844** 5786 *5797* 5798 *5800* 5801 *5804* 5961 **1845** *6240* 6247 **1846** 6535 6536 6593 *6594 6628* 6629 6630 6632 6635 *6670* 6703 *6708 6714* 6721 6723 *6793* **1848** *7904* **1850** 8223 8238 8246 8250 **1851** 8516 **1852** 9301 9451 9501 **1854** 10126 **1856** *10464 10465 10472 10476* **1857** 10638 **1858** 10789 *10871* **1859** 11045 *11047 11059* 11060 *11236* **1861** *11550 11661 11747* **1863** *12210 12212 12254 12359* **1864** *12668* 12721 *12723* **1865** *12855 12861* 12887 12897 12898 **1866** 13245 **1867** *13438 13460 13495* **1868** *13634* 13637 *13681 13683* **1870** 14337 **undated** *14740 14741*

Murdoch, Thomas. Fellow of R.S.L., and friend of JH.
1833 *2890*

Murray, Amelia Matilda (1795–1884). English writer. Maid of Honor to Queen Victoria, 1837–56. (*DNB, MEB*)
1855 *10449 10453*

Murray [Jr.], John (1808–92). Publisher, son of John Murray [Sr.]. Continued publishing firm. (*DNB*)
1848 *7933* **1849** *8055* **1850** *8378* **1856** 10566 **1860** *11419 11504* **1861** *11544* **1868** *13713 13715* **1869** *13865 13897 13898*

Murray [Sr.], John (1778–1843). Publisher, 1803–43. Established *Quarterly Review*, 1809. (*DNB*)
1832 *2598* **1840** *4637 4665*

Murray, William Keith (1801–61). Scottish army officer. Erected an observatory at Ochertyre, 1853. (*MEB*)
1851 *9065* **1857** *10747*

Mushet, Mrs.
1854 *9992* 9995

Mushet, Robert (1811–71). English Mint employee and author of works on coinage. (*DNB, MEB*)
1851 9046 9089 **1853** 9805 *9809 9854* 9855 9869 **1854** 9949 10051 **1868** *13753*

Napier, Francis (1819–98). English diplomat. Secretary of embassy at Constantinople, 1854–7. (*MEB, Alli*)
1851 *8614*

Napier, George Thomas (1784–1855). British army offficer, and governor at the Cape, 1837–43. (*DNB, MEB*)
1838 *3627 3634 3796* 3864 *3883* **1839** *4028 4249* 4326 **1840** 4433 *4509 4565* 4649 4650 **1841** *4796* 4964 **1847** *7264* **1849** *8091* **undated** *14746*

Napier, James Robert (1821–79). English ship builder, inventor. Wrote seventeen scientific papers. (*MEB*)
1864 *12682*

Napier, Richard. Brother of Sir George Thomas Napier.
1848 *7729*

Napier, William
undated *14747*

Narrien, John (1782–1860). English mathematician and astronomical writer. Compiled a series of mathematical texts. (*DNB, MEB*)
1833 *2844*

Nash, James. An employee of the Mint during JH's mastership. (1855–9–14)
1855 *10418* 10422

Nasmyth, James (1808–90). British astronomer and engineer. Invented the steam hammer. Built foundry near Manchester. (*DNB, DSB*)
1845 *6347* **1848** *7791* **1861** *11613* **1863** *12314* 12316 *12317 12338 12340* **1864** *12577 12579 12587* **1865** *12913* 12949 **1866** *13223* **1871** *14402*

Nature
1871 14355

Nayler, George (1764?–1831). English Garter King-of-Arms, 1815–31. (*DNB, Alli*)
1818 *287*

Negretti & Zambra
1863 *12244*

Nelson, Charles C. Probably brother of John Henry Nelson. (JHN's 1871–4–25)
1866 *13255* **1867** *13434 13446*

Nelson, John Henry. A friend of JH from Slough days.
1836 *3385* **1844** *5995* **1855** *10407 10408* **1861** *11566 11588* **1863** *12213 12377 12394* **1864** *12473* **1871** *14434* **undated** *14632 14633 14634 14748 14749 14750 14751 14752 14753 14754 14755 14756*

Nelson, Richard John (1803–77). British army officer and geologist. At Cape while JH was there. Published *Geology of the Bermudas.* (*DNB, MEB*)
1870 *14320 14327*

Nelson, Jr., Thomas (1822–92). Scottish publisher. Invented a rotary printing press. (*DNB, MEB*)
1859 *11070* 11077

Nesselrode, Karl Vasilevich (1780–1862). Russian diplomat. Accompanied Nicholas I on trips to London. (*MERSH, GSE*)
1847 *7305*

Neve, Mr. Agent for Sarah Newnham-Collingwood's property (Moor House), which JH bought and renamed Collingwood.
1839 4187

Newall, E. S.
1869 *14040 14042* **1870** *14131 14135*

Newall, Robert Stirling (1812–89). English engineer and astronomer. Improved the submarine telegraph cable. Gave a telscope to Cambridge. (*DNB, MEB*)
1865 *12950*

Newbold, Thomas John (1807–50). English writer and traveler. Knowledgeable about South India geology. (*DNB, Alli*)
1842 *5225* **1843** *5662*

Newbon, N. (?–1863). Friend of JH's family. (Th.N's 1863–2–3)
1861 *11547*

Newbon, Thomas. Son of N. Newbon. (NNN's 1861–2–2)
1863 *12147*

Newington, Samuel (1822?–82). Superintendent of mental home and horticulturist. (*Darwin*)
1860 *11259 11263 11310 11311 11505* 11511 *11513 11515 11516 11517* **1866** 13231

Newman, W. L. Amateur telescope builder. Actuary and secretary of Yorkshire Fire and Life Insurance Co. (1844–9–19)
1844 *5939 5945* **1845** *6111*

Newnham-Collingwood, Sarah (née Collingwood). Owner of Moor House, which JH purchased and renamed Collingwood. (*Herschel Soc.*)
1840 *4441* **1844** *5727*

Nichol, John Pringle (1804–59). British astronomer and academic. Rector of Montrose Academy. (*MEB*)
1838 3782 *3832* **1839** *3941 3972* **1847** *7214*

Nicholson, William (1753–1815). Scientist and inventor. Edited Nicholson's *Journal of Natural Philosophy*. (*DNB*)
1812 30 35

Nicollet, Joseph Nicolas (1786–1843). French-born explorer and mathematician. Immigrated to New Orleans, Louisiana, 1832. (*DAB, NBG*)
1821 *542* 556 569 **1822** 705 **1827** *1524 1565 1588 1631 1638* **1830** *2228*

Nobili, Leopold (1784–1835). Italian physicist. Studied electricity, including thermoelectric currents. (*DSB, Who's*)
1827 *1498*

Noehden, George Henry (1770–1826). German-born Director of the Department of Numismatics in the British Museum. (*Alli, Watt*)
1817 *244 252*

Normanby, Lord (see Phipps, George)

Northampton, Marquis of (see Compton, Spencer Joshua Alwyne)

Northumberland, Duke of (see Percy, Hugh (1817–12 Feb 47) *or* Percy, Algernon (1847–65))

O'Callaghan, P. Secretary of the Leeds Literary and Philosophical Society, and friend of Reginald Dyke Marshall and family. (1858–7–30)
1858 *10791 10895 10899* **1862** *11888* **1863** *12344* **1868** *13777*

Oersted, Hans Christian (1777–1851). Danish physicist and pioneer in electromagnetism and electrochemistry. (*DSB*)
1825 *1208* **1826** 1275 **1839** *4082 4115* 4177 **1840** 4405 **1841** *4909* 4915 *4929* 4997 *5024* 5038 **1846** *6599 6617* **1847** *7069* **1849** *8093 8116 8167 8173*

Oersted, Mathilde. Daughter of Hans Christian Oersted. (Dunsheath, *Giants*)
1869 *13889*

Ogilvie, Robert Annesley (1807–79). English Surveyor-General of Customs, 1863–76. (*MEB*)
1841 *5035*

Olbers, Heinrich Wilhelm Matthias (1758–1840). Physician and ophthalmologist. Built private observatory in Bremen. (*DSB*)
1820 391 **1821** 524 *539* **1827** *1650 1651* **1829** *2022* **1832** *2677* **1838** *3729*

Olmsted, Denison (1791–1859). American scientist with professorships in chemistry, mathematics, and astronomy. (*NCAB, DAB*)
1848 *7507*

Oriani, Barnaba (1752–1832). Italian astronomer at the Milan Observatory. Worked with Giuseppe Piazzi. (*Who's, EnItal*)
1820 *477*

Orlebar, A. B. Director of magnetic observatory at Bombay. (1845–2–12)
1846 *6565 6584* **1855** *10299*

Orloff, [Alekis F.?] (1786–1861). Russian army officer and diplomat. (*MERSH*)
1821 *609 615*

Osborne, B. Clerk in the Admiralty office. (JH's 1856–5–5)
1856 10501

Osborne, S. I. British divine, at one time at the church at Slough. (1840–12–1)
1840 4738

O'Shaughnessy, William Brooke (see Brooke, William O'Shaughnessy)

Osler, Abraham Follett (1808–1903). English meteorologist. Invented meteorological instruments. (*DNB*)
1838 *3753* **1839** *4244*

Ott, G. W. Agent for Jameson & Aders. (1824–3–3)
1824 *973*

Ouseley, Frederick A. G. (1825–89). English musician and composer. Professor of music at Oxford, 1855–89. (*DNB, MEB*)
1839 *4240*

Ouseley, Sir Gore (1770–1844). English diplomat and Oriental scholar. A founder of the Royal Asiatic Society of London. (*DNB, Alli*)
1839 *4254 4285* 4287

Ouseley, Harriet. Wife of Gore Ouseley. (*DNB*)
1839 *3910*

Owen, Sir Richard (1804–92). British surgeon and museum conservator. Professorships in anatomy and physiology. (*DNB, DSB*)
1844 *5732* **1847** *7462 7476* **1848** *7531 7652 7771 7779 7917* **1853** *9910* **1858** *10981 10990 10991* 10992 10995 10996 **1862** *12025* **undated** *14757*

P[?], T[?]
1856 *10507*

Paine, Robert Treat (1803–85). American astronomer and meteorologist. Bequeathed his fortune to Harvard Observatory. (Jones, *Harvard*)
1836 *3326* 3373 3374 **1837** *3454 3564* **1838** *3623* **1839** 4276 **1840** *4636* **1841** *4881* **1848** *7606*

***Pall Mall Budget*, Editor of**
1870 14161

Palmerston, Lord (see Temple, Henry John)

Paravey, Charles Hippolyte de (1787–1866?). French orientalist. (*NBG*)
1841 *5012 5029*

Parish, Woodbine (1796–1882). British statesman, geologist, and paleontologist. (*DNB*)
1857 *10739* **1859** 11009 *11101*

Park, Roswell (1807–69). American educator, army officer, and Episcopal clergyman. Founder of Racine College in Wisconsin. (*DAB, NCAB*)
1841 *5026*

Parker, James (1834–?). British publisher and historian. (*AlliSup*)
1867 *13522*

Parmentier, Tiburce. President of the Cercle philomatique in Gand, Belgium. (1861–3–25)
1861 *11560*

Parnell, C. I. Victim, along with JH, of an attempted fraud by an unnamed person.
1826 *1402* **1827** *1445*

Parry, Charles Henry (1779–1860). English physician. Traveled with Samuel Taylor Coleridge to Harz, 1799. (*DNB, MEB*)
1850 *8254*

Parry, Matilda
1822 *830*

Parry, William Edward (1790–1855). British Arctic explorer seeking Northwest Passage. Hydrographer to the Admiralty. (*DNB, AusDB*)
1821 *552* **1824** *947* **1826** *1314* **1828** *1736* **1839** *4021 4211*

Parsons, Thomas. Grandson of Thomas Parsons, a friend of William Herschel. (1862–5–27)
1862 *11900* 11910 11912 *11916* 11921

Parsons (Lord Rosse), William (1800–67). Irish astronomer, who erected a large telescope at Birr Castle. 3rd Earl of Rosse. (*DNB, DSB*)
1843 *5528* 5536 **1845** 6126 **1847** *7197* **1849** *7978 7989 8127* 8132 **1850** 8182 8272 *8329 8348 8417* 8487 **1851** 8517 *8701* 8725 **1852** 9549 **1853** *9875* 9880 **1854** 10046 10120 **1855** 10280 **1858** 10872 **1860** 11393 *11491* **1862** *11882* 11928 *11948 11962 11999* 12087 **1863** *12119* 12132 **1867** *13482* 13484

Partsch, Paul (1791–1856). Austrian geologist and mineralogist. (*OBL*)
1847 *7210*

Pasley, Charles William (1780–1861). British officer in Royal Engineers, 1798–1861. Published pamphlet on decimalizing measures and coinage. (*DNB, MEB*)
1846 6705 *6725* **1848** *7786 7789* **1854** 10169

Pasteur, Emilien Frossard
1867 *13417* **1868** *13741*

Pastorff, Charles. Son of Johann Wilhelm Pastorff.
1847 *6854* **1850** *8346*

Pastorff, Johann Wilhelm (1767–1838). German astronomer and civil servant. (*NDtB*)
1820 *392* 393 **1824** *963* 977

Paterson, W.
1860 *11400* 11403 *11407 11409* **1862** *12018* **1870** *14252*

Patterson, Robert M. (1787–1854). American chemist and mathematician. Professor, and director of the U. S. mint. (*NCAB*)
1823 868

Pattinson, Hugh Lee (1796–1858). English metallurgical chemist. Patented a process for desilvering lead. Built private observatory. (*DNB, Who's*)
1852 *9249 9270*

Paulin, George L.
1864 *12752*

Paulin, J. S. L.
1864 *12752*

Payne, E. W. Writer of books for children. (*Alli*)
1850 *8211 8218*

Peacock, George (1791–1858). English mathematician and Dean of Ely. Lowndean professor at Cambridge. Good friend of JH. (*DNB, DSB*)
1816 *187 199 203* **1817** *223 226 239 240 241 254 274 276 277* **1818** *279 280 286 289 292 296* **1819** *329 366 385 390* **1820** *408 424 439 449 495 498 499 502* **1821** *566 577 629 636 639* **1822** *671 692 784 797 827 840* **1823** *866* **1824** *980* **1827** *1653* **1828** *1690* **1838** *3649 3689* 3691 *3698* 3701 3793 **1841** *4773* **1842** *5137 5177* **1844** *5877 5950 5999* **1845** 6229 *6241 6245 6260 6291 6299* **1846** *6684* **1847** *7081* 7088 *7199* **1850** *8400 8404 8455 8458* **1851** *8755* **1852** 9551 *9588 9594* **1854** 10087 **1855** *10371* 10374 *10376 10377 10383 10392* **1857** *10619 10662 10683 10760* **1858** *10893 10928* **1921** *3688* **undated** *14758 14759 14760 14761 14762 14763 14764*

Pearse, Henry. Brother of Mrs. Mary Maclear, wife of Thomas Maclear. (see HP's 1850–12–2)
1850 *8446* **1851** *8628 8837* **1862** *11778 11974 11982*

Pearson, George (1751–1828). English physician and chemist. R.S.L. council member. (*DNB, DSB*)
1827 *1558*

Pearson, William (1767–1847). British astronomer and divine. Established several private observatories. (*DNB*)
1822 *660*

Peel, Robert (1788–1850). English statesman and Prime Minister, 1834–5, 41–6. (*DNB, MEB*)
1830 *2246* **1845** 6103 6147 6319 6323

Peers, Thomas
undated *14765*

Peirce, Benjamin (1809–80). American mathematician and astronomer. Professor at Harvard. (*NCAB, DAB*)
1842 *5239* **1847** *7344*

Pelouze, Théophile Jules (1807–67). French chemist and assayer for the French Mint. (*NBG, DSB*)
1851 8518

Penn, John (1805–78). English engineer. Manufactured the Perkins steam gun and designed marine steam engines. (*DNB, MEB*)
1829 *1971*

Penny, W. G. Amateur astronomer. (1857–9)
1857 *10610*

Penrose, Francis Cranmer (1817–1903). English architect, archaeologist and astronomer. Developed predictive methods for the moon and comets. (*DNB, Alli*)
1864 *12604* 12644 **1869** *13965*

Pentland, Joseph Barclay (1797–1873). Traveler and writer of travel guides. (*DNB, MEB*)
1861 11734 **1869** 13866 *13867*

People's Magazine
1867 13559

Pepys, William Hasledine (1775–1856). English inventor. Invented mercury gasometer, water gasholder. (*DNB, Who's*)
1821 *521* **1823** *864* **1845** 6460 *6467*

Percy (4th Duke of Northumberland), Algernon (1792–1865). British naval officer. (*DNB*)
1847 7003 *7004* 7035 7039 *7044 7047* 7049 *7062* 7066 *7070* 7092 *7098* 7107 7108 7109 *7110 7112 7115* 7116 *7123* 7129 *7133* 7143 *7254 7271 7477* **1848** 7493 7494 *7579* **1849** *8048* **1850** *8337* 8338 *8342* **1863** *12414*

Percy (3rd Duke of Northumberland), Hugh (1785–1847). British public servant. Subsidized publication of JH's *Cape Results.* (*DNB*)

1833 2823 *2825* 2835 *2839 2857 2882* **1835** *3059* 3131 *3174* **1836** *3345* 3413 **1837** 3601 **1838** *3692* 3766 **1839** 3937 4038 *4056* 4111 **1840** 4359 4360 *4699* **1841** 5000 5005 **1845** 6290

Pereira, Jonathan (1804–53). English pharmacologist and physician. Lecturer on materia medica. (*DNB, MEB*)
1843 *5573*

Perigal, Henry (1801–98). English astronomer, meteorologist. Believed the moon does not rotate. (*MEB*)
1839 *4099 4215* **1840** *4406 4463*

Perrey, M.
1833 2761

Peters, Christian August Friedrich (1806–80). German astronomer, who worked with F. G. W. Struve. Editor of *Astronomische Nachrichten*. (*DSB*)
1848 *7888* **1869** 13837 *13862 13886 13895*

Peters, W. C.
1852 9313

Petersen, Adolph C. (1804–54). German astronomer, observer at Altona. (*BHGW*)
1848 *7787* **1850** *8490* **1851** *8574* **1852** 9387 *9435*

Petit, C. Wife of Frédéric Petit.
undated *14766*

Petit, Frédéric (1810–65). French astronomer. Director of observatory at Toulouse. (*BHGW*)
1850 *8255*

Peyrot, J.
1833 *2926* **1835** *3065*

Pfaff, Johann Wilhelm A. (1774–1835). German scientist and mathematician. Son of the mathematician Johann Friederich Pfaff. (*DSB*)
1823 916 *934* **1824** 1020 **1827** *1643*

Phelps, Robert (1808–90). Cambridge University fellow and administrator. Wrote a treatise on optics. (*MEB*)
1847 *7312*

Phelps, Sackville
1845 *6432*

Philip, John (1775–1851). English missionary to South Africa. Prominent politician and administrator at the Cape. (*DNB, MEB*)
1834 *3009* 3034 **1835** *3069 3119* 3126 *3128* **1836** 3272 *3275*

Phillips, John (1800–74). British geologist and museum keeper. Contributed over a hundred papers to the scientific literature. (*DNB, DSB*)
1831 2393 **1838** *3754 3838* **1839** *4022* 4052 *4172 4173 4174* 4216 *4301 4302 4303* **1840** *4689 4690 4691 4692 4693* 4749 **1841** *4923 4924 4981 4982 4983 4984* **1842** *5166 5167 5168 5220 5221 5222* **1843** *5537 5611* **1844** *5965 5966* **1845** *6292 6300 6311 6312 6313 6314 6315 6316 6456* **1846** *6763* **1850** *8362* **1855** *10436* **1858** 10792 *10797 10798* **1859** *11163* **1866** *13330*

Phillips, R. C. A letter sorter at the Manchester post office. (1869–11–18)
1869 *14057* 14059 *14076*

Phillips, Sir Richard (1767–1840). English author, bookseller, publisher, schoolmaster, and patent-medicine vendor. (*DNB*)
1826 1276 **1827** *1446*

Phillips, Thomas
1843 5636

Philosophical Magazine
1833 2719 **1840** 4436 **1864** 12753 **1865** 12819

Philosophical Society of Philadelphia, the Secretary
1822 727

Phipps, George (1819–90). British army officer, M.P., and colonial governor. 2nd Marquis Normanby and 3rd Viscount Normanby. (*Peerage*)
1839 *3971 4015*

Photographic News
1858 10947 10961 **1859** 11115 **1860** 11347 11375 **1864** 12474 12693 12694 **1865** 13055 **1866** 13075 13196 **1867** 13535 **1868** 13600 13602

***Photographic News*, Editor of the**
1866 13070

Piazzi, Giuseppe (1746–1826). Director of Palermo Observatory in Sicily. (*GDE*)
1821 546 590 **1824** 943 **1826** *1297 1345*

Pickersgill, Henry William (1782–1875). Painter, especially of portraits. Painted the portrait of John Herschel at St. John's College, Cambridge. (*DNB, MEB*)
1833 2712 2805 **1845** *6064*

Pictet, Marc August (1752–1825). French physicist, meteorologist, geologist, and astronomer. (*DSB*)
1822 *828* **1823** *929* **1824** *994*

Piddington, Henry (1797–1858). English meteorologist. Studied storms of the Indian seas. (*DNB, MEB*)
1840 *4516 4725* **1845** *6370* **1848** *7713* **1850** *8190 8406* **1852** *9394* **1853** *9695 9711* **undated** *14767*

Pigott, Edward (fl. 1768–1807). English astronomer. Discovered a nebula, and a comet that bears his name. (*DSB, DNB*)
1821 562 *564*

Pillans, James (1778–1864). Scottish educational reformer. Professor of humanity and laws, Edinburgh University, 1820–63. (*DNB, MEB*)
1832 *2654* 2659

Piper, Jesse
1840 *4361* **1844** *6020*

Pirbright, Baron (see De Worms, Henry)

Pistrucci, Benedetto (1784–1855). Italian-born gem engraver. Became chief engraver at the Mint in London. (*DNB, MEB*)
1851 8519 *8544 8550 8581 8582* 8588 *8594 8662 8915 8934 8935* 8939 *9105 9110* 9111 *9116* 9127 *9135* **1852**

9172 9174 9177 9178 9181 *9213 9221 9225* 9282 *9284 9288 9309* 9320 *9328 9332 9341* 9384 *9385* 9414

Plackett, William. An official at the College of Heralds. (1838–7–2)
1838 *3713*

Plana, Giovanni Antonio (1781–1864). Italian mathematician and astronomer. Professor of astronomy at Turin, 1811–64. (*DSB, Who's*)
1824 *1016* **1825** 1102 *1133 1174* 1198 *1210 1240 1258 1268* **1826** *1283 1285 1355 1386 1399* **1828** *1730 1800* **1829** *1940 2018 2038* **1832** *2646* **1833** *2731* **1834** *3005* **1836** *3320* **1838** *3617* **1839** *4219* **1840** *4401 4557* **1850** *8353* **1854** *10224*

Planta, Joseph (1744–1827). Librarian at British Museum and paymaster of Exchequer bills. (*DNB*)
1822 *790* **1824** *944*

Plateau, Joseph A. F. (1801–83). Belgian physicist. Worked in physiological optics and invented a stroboscope. (*DSB, Who's*)
1864 12674 *12684*

Playfair, John (1748–1819). Scottish mathematician, physicist, and geologist. Supported James Hutton's theories. (*DSB, Who's*)
1812 *40*

Playfair, Lyon (1818–98). British chemist and mineralogist. Created Baron Playfair. (*DNB, MEB*)
1855 *10373*

Plowden, William Henry Chichele (1790–1880). British employee of the East India Company. Superintended British trade in China. (*MEB*)
1845 6139

Poey, Andres (1830?–?). Founder and director of the meteorological observatory in Havana. Later lived in Paris. (*BHGW*)
1858 *10839* **1861** *11623*

Pogson, Norman (1828–91). British astronomer, assisting at several observatories. Astronomer at Madras. (*DNB, MEB*)
1860 *11313* 11315 *11324 11329 11338 11349 11370 11435* 11436 *11459* 11465 **1861** *11590*

Poisson, Siméon Denis (1781–1840). French mathematician. Applied mathematics to physical problems. (*DSB, Who's*)
1826 *1319* **1832** *2674* **1833** *2744* **1838** *3683*

Pole, Sir William (1814–1900). English engineer and musician. (*MEB, Alli*)
1856 *10535 10557 10587 10593* **1857** *10697 10707 10708 10712* **1860** *11382 11390 11500* **1868** *13674* **1869** *13883*

Poletti, Luigi (1792–1869). Italian architect. (*EnItal, GDE*)
1824 *990*

Pollard, William
1861 *11674*

Pollock, Frederick (1815–74). Indian-born army officer in Bengal Engineers. (*MEB*)
1850 *8235* **1868** *13817 13821*

Pond, John (1767–1836). English astronomer, Astronomer Royal, 1811–35. Translated Laplace's *Système du monde*. (*DNB*)
1821 618 620 **1822** *737 738* **1825** 1207 **1827** *1447 1569*

Pons, Jean Louis (1761–1831). French astronomer and discoverer of comets. Director of several Italian observatories. (*DSB*)
1823 932 **1824** *1030*

Poole, Reginald Stuart (1832–95). English archaeologist and orientalist. Keeper for Coins and Metals, British Museum. (*DNB, MEB*)
1850 *8244 8251 8305* **1852** *9207 9235 9283* **1860** *11410*

Porter, George R. (1792–1852). English statistician. A founder of the Statistical Society. (*DNB, MEB*)
1848 *7869 7906 7913* **1849** *8148 8149*

Portlock, Joseph Ellison (1794–1864). British army officer and geologist. Worked on the Irish ordnance survey, 1825–43. (*DNB, MEB*)
1859 *11025*

Postmaster
1821 531

Potiquet, Alfred
1864 *12580 12667*

Potter, Richard (1799–1886). British scientific writer, merchant, and academic. Studied optics with John Dalton. (*DNB, MEB*)
1832 *2552* 2558

Potter, Sir John (1815–58). English M.P. Manchester merchant and draper. (*MEB*)
1852 9381

Pouillet, Claude S. M. (1790–1868). French physicist. Invented high-temperature thermometers. (*Who's, NBG*)
1822 *654* **undated** *14768*

Powell, Baden (1796–1860). English physicist. Savilian Professor of Geometry, Oxford, 1827–60. Studied radiation. (*DSB, DNB*)
1825 *1257* **1856** *10529* **1859** *11204* **1860** *11352*

Powell, Eyre B. (1819–1904). Director of Public Instruction, Madras, India, 1862–75. (*DIB*)
1862 *11963* **1870** *14191*

Power, A. B. A teacher. (1865–5–5)
1865 *12893*

Prescott, Henry James (1802–56). Director, deputy governor, and then governor of the Bank of England. (*MEB*)
1851 *8575* 8576

Prescott, John Eustace (1832–1920). Fellow of Corpus Christi College, Cambridge. (*Alli, AlliSup*)
1862 *11865 11889* **1865** *12999* **1866** *13127*

Price and Bolton
1854 *10121*

Price, Bartholomew (1818–98). British mathematician and divine. Professor and academic administrator. (*DNB, MEB*)
1860 *11334* **1861** *11567* 11744 **1865** *13005* 13012 **1866** *13316 13325* **1869** *13998*

Prichard, Constantine Estlin (1818–69). British divine. Academic and ecclesiatical administrator. Son of James Cowles Prichard. (*MEB*)
1849 *7962*

Prichard, James Cowles (1786–1848). British physician, ethnologist, and philologist. Published *Natural History of Man*, 1843. (*DNB*)
1848 *7624 7630 7631 7648 7651*

Pridler Sons & Co.
1863 *12196*

Prinsep, Henry Thoby (1792–1878). Indian civil servant of long and distinguished service. (*DNB*)
1853 *9870 9883*

Prinsep, James (1799–1840). British architect, orientalist, and assay master at both the Indian and British Mints. (*DNB*)
1834 2935 3035 *3040* **1835** *3108* **1836** *3242 3247 3337 3398* **1837** *3438 3553 3572* **1838** *3620*

Prinsep, Sara (née Pattle). Sister of Julia Margaret Cameron. Wife of Henry Thoby Prinsep.
1849 *8045*

Pritchard, Charles (1808–93). British astronomer and divine. Co-editor of JH's *Catalogue of Double Stars* and pioneer in stellar photography. Headmaster, Clapham Grammar School. (*DNB, MEB*)
1844 *5683* **1845** *6465* **1846** *6537* **1848** *7763* **1850** 8183 **1852** *9200* **1853** 9752 **1855** *10403* **1859** *11223 11251* **1860** *11372 11385 11387* **1861** *11621* **1863** *12341* 12352 *12401* **1864** *12434 12441 12445 12477 12488 12490 12504 12530 12706 12707 12728 12757* **1865** *12817 12984* **1866** 13060 *13093 13163* 13173 *13304* **1867** *13397* 13405 **1868** *13664* 13779 **1869** *13879 13884* **1871** *14435* **undated** *14769*

Privy Council
1851 8520

Proctor, Richard Anthony (1837–88). British astronomer and author. (*DNB, MEB*)
1869 *13963* 13967 *13968* 13974 *13977 14046 14050 14051* **1870** 14111 *14146* 14149 *14152 14178* 14193 *14196* 14201 *14202* 14213 *14214 14239* 14241 *14247* **1871** *14363* 14366 *14367* 14369 *14370 14389* 14391 *14395*

Prout, William (1785–1850). English chemist and physiologist. Created Prout's Hypothesis. (*DSB, Who's*)
1827 *1520* **1829** *1920*

Prussian Academy of Sciences
1828 1714

Pucci, Marquis
1820 *474*

Pulman, James (1783–1859). British official at the College of Heralds. (*MEB*)
1838 *3767*

'Q in a Corner'
1842 *5130*

Quesneville, Gustave Augustin (1810–?). French chemist and physician. Editor of *Revue scientifique* and of *Moniteur scientifique*. (*NBG*)
1863 12348

Quetelet, Adolphe (1796–1874). Belgian statistician, astronomer and meteorologist. Astronomer at Brussels Royal observatory. (*DSB*)
1826 *1429* **1827** 1525 *1526 1582 1601* **1828** *1669* 1851 **1829** *1921 2006 2012* **1830** *2157* 2160 *2163* **1831** *2382* 2483 **1832** 2600 **1833** *2759* 2924 **1834** *2952 3001* **1835** *3082* **1836** *3265* **1837** *3451* 3495 3534 *3547 3582* **1838** *3635* 3812 *3835* 3849 *3901* **1839** *4116* 4143 *4195 4314* **1840** 4362 *4447 4570 4657 4677* **1841** *4789 4910 4934* 4943 *4959 4967 4986 5020 5030* 5037 *5054* **1842** *5111 5127* 5160 *5165* 5176 *5309* **1843** *5585* **1844** *5725* 5734 *5962* 6037 **1845** 6251 *6425* **1846** 6620 **1847** *7161 7208* 7295 *7333* **1848** *7813* **1851** *8660 8916* **1852** *9601* **1853** *9643* **1861** *11647* **1862** *11842* 11845 *11966* **1863** *12174 12203 12265* 12285 *12295* 12297 12298 12305 *12318* **1865** 12790 *12847 12849 12952* 12953 *12954* 12955 12960 12966 *12981* **1866** *13270 13312 13345* **1867** 13539 *13573 13578* **1868** *13686 13687* 13688 *13692 13727* 13731 *13738 13755 13833* **1869** 13845 *13863 13969* 13972 *13978* **1870** 14294 *14345 14346* 14347 **1871** *14406* 14407 *14444* **undated** 14770

Quin, Augusta Wyndham- (see Wyndham-Quin, Augusta)

Quin, Edwin Richard Windham Wyndham- (see Wyndham-Quin, Edwin Richard Windham)

Quincy, Eliza Susan (fl. 1824–61). Daughter of Josiah Quincy. Student of early American history. (*Alli*)
1848 *7770* 7830 **1865** *12844* 12859 **1867** *13510* 13524 *13528 13587* **1868** 13598 *13641* **1869** *14008* 14029 *14048*

Quincy, Josiah (1772–1864). American politician, municipal reformer and educator. President, Harvard University, 1829–45. (*DAB, Alli*)
1843 *5621* **1844** *5832 5949* **1845** *6344 6345 6384* 6436 *6437*

Radford, Thomas
1860 *11302*

Ragg, Frederic W. Son of Thomas Ragg, curate of Malin's Lee, 1861–65. (*DNB*)
1861 *11577*

Ralph Heaton & Sons
1853 9770 **1854** 9930 9941 9974 10132 **1855** 10285

Ramage, John (1784–1835). Aberdeen merchant, interested in optics. (*BHGW*)
1826 *1337 1387* 1405 **1833** *2829*

Rankine, William John Macquorn (1820–72). Scottish engineer, physicist. Worked in railroad construction. Studied thermodynamics. (*DSB, Who's*)
1871 14353 *14354*

Raper, Henry (1799–1859). British naval officer and writer on navigation. (*DNB, MEB*)

1839 *4096* **1840** *4736* 4743 **1841** *4774 5008 5015* **1847** *7252* **1849** *8062*

Rate, John (fl. 1840–88). English vicar, writer. Vicar of Lamley, Staffordshire, 1848–81. (*AlliSup*)
1865 *13024* **1866** *13285*

Rawson, David
1858 *10809*

Rawson, Rawson William (1812–99). Colonial Secretary, Cape of Good Hope, 1854–64. (*MEB*)
1863 *12300*

Raymond, J. T. Graduate of St. John's College, Cambridge. (1870–5–3)
1870 *14183*

***Reader*, Editor of the**
1865 12875

Redfield, William C. (1789–1857). American geographer and meteorologist. Studied storms and hurricanes. (*DAB, Alli*)
1840 *4680 4753* **1846** *6566*

Redman, T. S.
1840 *4497*

Reeve, Rycroft. Secretary of the Geneological and Historical Society. (1864–4–26)
1864 *12554*

Regnault, Henri Victor (1810–78). French chemist and physicist. Studied specific heat. Discovered carbon tetrachloride. (*DSB, Who's*)
1857 *10611 10705 10706* **1858** *10772 10773* **1861** *11521 11729* 11739 **undated** *14772*

Rehausen, Johan Gothard, Baron de. Swedish minister to England. (1847–11–9)
1847 *7366*

Reid, William (1791–1858). British officer in the Royal Engineers, and colonial governor. (*DNB, Alli*)
1828 *1845* **1838** *3750* **1839** *3951 4107* **1849** *8082*

Rennie, John (1794–1874). British civil engineer. Built London Bridge, 1824-31. (*DNB, MEB*)
1821 568 **1834** *2995*

Renouard, George C. (1780–1867). English classical scholar. (*DNB, Alli*)
1817 *227 250* **1865** *12879* 12881 *12885*

Renshaw, S. Alexander (fl. 1875). English mathematician. (*AlliSup*)
1869 *13878*

Renwick, James (1792–1863). American engineer and writer. Professor of natural philosophy and chemistry at Columbia. (*DAB*)
1847 *7421*

Reslhuber, Augustin (1808–75). Austrian astronomer and Benedictine monk. Director of Kremsmünster observatory. (*OBL*)
1851 *8872*

Reugger, Mr. Councillor d'Stat, Berne, Switzerland. (1823–9–20)
1823 *907*

Revel, Count. The principality of Revel was in the control of the Broglie family, but the Count can not be identified. (*EnItal*)
1847 *7198*

Reventlow, Count Frederik (1791–1874). German politician from Scleswig-Holstein, representing the Danish king. (*BE*)
1847 *7167*

Reynaud, Jean (1806–63). French mining engineer and geologist. (*BHGW*)
1863 *12214*

Reynolds, H. R. Solicitor for the Treasury in charge of legal affairs for the Mint during JH's mastership. (1851–11–19)
1851 *9066 9071 9079* **1852** *9205 9226 9307 9310 9580* **1853** *9669 9675*

Rice, Charles Spring- (see Spring-Rice, Charles)

Rice, Marianne Spring- (see Spring-Rice, Marianne)

Rice, Thomas Spring- (see Spring-Rice, Thomas)

Rich, Elihu (1819–75). English writer, translator. Wrote a biography of Emanuel Swedenborg. (*MEB, Alli*)
1847 *7422*

Rich, Henry (1803–69). English M.P. and junior lord of the Treasury. (*MEB*)
1851 *8521 8522 8523 8524* 8525 *8642* 8726 *8731 8738* 8793 8796 8798 8823 8825 8834 8845 8846 8849 8855 *8865 8888 8899 8906* 8917 8924 8929 *8985* 8987 *9083 9098 9117*

Richard Griffin & Co.
1859 *11235* **1860** *11276*

Richard, John (1818–79). British stationer, partner in firm of Richard and Wilson. Built an observatory at West Hill, Wandsworth. (*MEB*)
1838 *3686* **1843** *5359 5379*

Richards, J.
1828 *1755*

Richardson, Charles (1775–1865). English lexicographer. (*DNB, Alli*)
1838 *3783 3841 3892*

Richardson, George. Captain of the *Mount Stuart Elphinstone* in 1834. (JH's 1834–1–17)
1834 2939 **1847** *7425*

Richardson, Mr. British practical astronomer. (JH's 1831–4–24)
1831 2342

Richardson, Robert (1812–91). English engineer, company director. (*MEB*)
1862 *11814* 11816

Richardson, Sir John (1787–1865). English explorer of the Arctic, naturalist and surgeon. Physician of the Royal Hospital at Haslar. (*DNB, MEB*)
1840 *4726* **1847** *7463* 7470 7473

Richardson, Thomas W.
1862 *11856*

Ricketts, Dr.
1870 *14102*

Riddle, Edward (1788–1854). British mathematician and astronomer. Taught at Royal Naval Hospital in Greenwich, 1821–51. (*DNB, MEB*)
1847 *7224*

Ridgway, Alexander F.
1845 *6385*

Riedl-Sevenstern, J. Viennese builder of moon globes. (1850–12–21)
1850 *8374 8480*

Rigaud, Stephen J. (1816–59). Bishop of Antigua. Eldest son of Stephen Peter Rigaud. (*DNB, Alli*)
1842 *5143 5161*

Rigaud, Stephen Peter (1774–1839). English mathematical historian and astronomer. Professor at Oxford. (*DNB, Alli*)
1824 *971* **1829** *1976* 1980 *2009* **1830** *2210* 2212 *2213 2219* 2234 *2239* **1831** 2335 *2338* 2400 **1832** 2530 *2537* **1833** *2738* **1837** 3518 3532 **1838** *3703*

Rintoul, Robert (1823–?). Member of company of Moneyers at Royal Mint, 1839–51. (Challis, *Mint*)
1851 *8699* 8702 8705 *8706* 8727 *8734*

Ritchie, William (1790–1837). Rector of Academy at Tain. (*DNB*)
1825 1107 *1110 1112 1113* 1120 *1126 1172* **1826** *1434* **1827** *1452 1478 1561 1626* **1828** *1665 1731 1738 1756* **1829** *1994* **1831** *2414 2435* **1832** *2657* 2660

Rive, Arthur-Auguste De La (see De La Rive, Arthur-Auguste)

Rixon, Frederick. A member of the lecture committee at Clapham school. (1866–8–4)
1866 *13232*

Robarts, W.
1845 *6213 6422*

Roberton, John David (?–1843). Assistant librarian, and assistant secretary, 1835–43, to R.S.L. (Hall, *R.S.*)
1838 3743 **1840** *4588 4595 4598 4605 4607 4624* 4695 *4703* **1841** *4941 5031* **1843** *5349 5580*

Roberts, Henry (1812?–74). British barrister and civil servant. (*MEB*)
1855 *10287*

Roberts, Martyn
1844 *5716*

Roberts, Mr. Probably a bookseller.
1830 2074

Roberts, Richard (1789–1864). English inventor. Patents in textiles and railways. (*DNB, MEB*)
1846 6751 *6761* 6771

Robertson, D.
1833 2743

Robertson, Patrick Francis (1807–85). British M.P. and deputy chairman of the Bank of Egypt. (*MEB*)
1848 7817 **1856** *10600* **1858** 10904 *10905* **1860** *11391* **1865** *13041* 13045 **1867** *13490* **1868** *13652 13658 13661* **1870** *14096 14150 14154*

Robinson, Charles
1858 *10932 10942*

Robinson, Mrs. Thomas Romney. Wife of Thomas Romney Robinson.
1867 13407

Robinson, Thomas Romney (1792–1882). Irish astronomer and mathematical physicist. Invented cup anemometer. (*DNB, MEB*)
1828 *1661* **1829** 2059 **1832** *2539* **1849** *8119* 8125 *8130* **1850** 8306 *8315* 8323 **1858** 10868 **1859** 11206 **1861** 11522 **1867** *13520* **1869** *13975* 13976

Robison, John (1778–1843). Scottish inventor. Son of scientific writer John Robison (1739–1805). (*DNB*)
1831 *2282 2339* **1839** *4044*

Robson, L. H.
1850 8266

Rodgers, Charles
1864 *12482 12513 12514*

Rodwell, Lionel. Secretary to the British section of the General Shipwreck Relief Society. (1839–8–3)
1839 *4193*

Rogers, Alexander. British mathematics teacher and amateur astronomer. Tutored the young JH.
1807 *4* **1808** *5 6 7 9* **1809** *10 11 12* **1810** *14* **1811** *25* **1812** *31* **1813** *56 74* **1814** *108 125* **1827** *1628* **1828** *1686 1715 1725 1785 1884* **1829** *1922 1948 1991 2030* **1830** *2075 2252* **1831** *2299* **1832** *2585 2623* **1833** *2728 2904*

Rogers, Charles (1825–90). Scottish minister, author, and historiographer. Published numerous periodicals. (*DNB, MEB*)
1863 *12397*

Rogers, Samuel (1763–1855). English poet. (*DNB, Alli*)
1842 *5341* **1843** *5358* **1850** *8275*

Roget, Peter Mark (1779–1869). British physician and science writer. Invented slide rule. Author of *Roget's Thesaurus.* (*DNB, MEB*)
1828 *1908* **1830** 2260 **1831** 2304 2323 **1841** 4827 **1842** *5194* **1843** 5615 **1845** *6090*

Romer, Leonard. Secretary (?) to the B.A.A.S. coal map committee. (1846–1–13)
1846 *6560*

Romilly, Sir John (1802–74). British barrister, M.P. and cabinet minister. Created 1st Baron Romilly. (*DNB*)
1854 9987

Ronalds, Sir Francis (1788–1873). British meteorologist and inventor. Devised instrumentation for meteorological observations. (*DNB, MEB*)
1847 *6893 7209* **1848** *7717* **1849** *8041*

Roscoe, Henry Enfield (1833–1915). English chemist and founder of the Manchester School of Chemistry. Active in many scientific organizations. (*DSB*)

1863 *12407* **1864** *12439* **1866** *13318* **1867** *13529* *13534* **1868** *13620* **1869** 13921

Rose, Henry John (1800–73). English theologian and divine. Rector of Haughton Conquest, Bedfordshire. (*MEB, Alli*)
1843 *5620*

Ross, James Clark (1800–62). British naval officer and explorer of both polar regions. Discovered north magnetic pole. (*DNB, MEB*)
1839 *4029* 4034 *4039* 4047 4164 4176 *4182* *4203* **1843** *5605* **1847** *6970* 6977 *7177* **1848** *7683* **undated** *14773*

Rosse, Lord (see Parsons, William)

Rothschild, Anthony de (1810–76). British financier, merchant, and diplomat. Leased refinery from Mint, 1852. (*DNB, MEB*)
1854 10081 10082

Rovell, G. A.
1843 *5352*

Roy, Rummohun
1832 *2550* *2583*

R.A.S., Secretary to
1869 13839

R.A.S., the
1836 3397 **1837** 3602 **1846** 6538 **1847** 6855 **1864** 12602 12634 **1866** 13210 **1868** 13829 **1870** 14205

Royal Geographical Society
1859 11178 11182

R.S.L., a member of the Glass Committee of the
1827 1448

R.S.L. Committee of Physics (including Meteorology), Chairman of
1844 *5837*

R.S.L. Committee of Physics, Chairman of
1846 6792

R.S.L. Council
1840 4368 4369 4370 4553 **1842** 5184 **1847** 7078 **1850** 8354 **1851** 9095 **1852** 9434 **1857** 10715 **1862** 11927 **1863** 12102 **1864** 12422 **1865** 12782 **undated** 14815

R.S.L. Joint Committee of Physics and Meteorology, Recipients of Report of
1840 4363

R.S.L. Physical Committee
1842 5116

R.S.L. Physical Committee, Chairman of
undated 14771

R.S.L. Physical Committee, Secretary of
1842 5061

R.S.L. Secretary
1843 5408 5617

***Royal Society Philosophical Transactions*, Editor of the**
1842 5301

Royston-Pigott, G. W. (1819–89). English surgeon and writer. Improved several scientific instruments. (*MEB, Alli*)
1870 *14079* *14335*

Rufenacht, Elise
1831 *2270*

Rümker, George. German astronomer. Illegitimate son of K. L. C. Rümker, whom he succeeded as director of Hamburg Observatory, 1857–97. (*DSB, AusDB*)
1866 *13076* *13246*

Rümker, Charles (Karl Ludwig Christian) (1788–1862). German-born astronomer, became British seaman. Observing in Australia, witnessed the return of Encke's comet, 1822. (*AusDB, DAusB*)
1827 *1449*

Rumsey, Lacy (?–1872). Clerk in the Treasury, retired 1856.
1850 *8471* *8475* *8476*

Rundell, W. W. (1816–97). British secretary to several scientific societies. (*MEB*)
1847 *7048* 7059 *7071*

Runkle, John Daniel (1822–1902). American mathematician and educator. President of Massachusetts Institute of Technology, 1870–8. (*DAB, Alli*)
1858 *10983* **1861** *11587*

Ruskin, John (1819–1900). British poet, artist, and social reformer. (*DNB, Bryan*)
1842 *5062* **1847** *7374*

Russell, E. Probably a printer, involved in the printing of Lacaille's and Lalande's star catalogs. (Stratford's 1846–8–14)
1846 *6701*

Russell, Frances A. M. (née Elliot-Murray-Kynynmound). Daughter of the 2nd Earl of Minto. Married Lord John Russell, 1841. (*Peerage*)
1851 *8591*

Russell, G. Probably a clerk at the Treasury. (JH's 1855–5–16)
1855 10388

Russell, Lord John (1792–1878). British M.P., cabinet minister, and prime minister. Known as Lord John Russell. Created 1st Earl Russell. (*DNB, BHF*)
1838 *3629* **1839** *4112* 4117 *4321* 4322 **1840** 4415 *4416* 4417 4422 *4437* **1850** *8319* 8321 8330 *8334* *8408* 8414 *8424* 8425 *8481* 8486 **1851** 8850 *8866* **1852** *9238*

Rutherford, J. H. Cape resident raising funds for legal defense of John Fairbairn. (JH's 1837–6–1)
1837 *3481* 3482 *3484* *3485* *3489* *3490* 3491 3492 *3493* 3511 3512

Rutherford, William (1798?–1871). Mathematician and master at Royal Military Academy, Woolwich. (*DNB, MEB*)
1847 *7372*

Ryall, John. Vice president of Queen's College, Cork, in 1865. (1865–1–10)
1865 *12778* 12780 *12781*

Ryan, Sir Edward (1793–1875). British barrister, and judge in India. College friend of JH and brother-in-law of Charles Babbage. (*DNB, MEB*)
1826 *1418* **1828** *1687* **1832** *2709* **1834** 3012 3024 *3028* **1835** *3096* *3229* **1836** *3421* **1842** 5195 *5197* **1847** *7040* **1849** *8172* **1850** *8200* **1851** *8874* *8876* *8883* *8892* 8986 **1852** *9208* **1853** 9762 **1855** 10406 **1857** *10613* **1859** *11023* **1860** 11330 **1865** 12783 *12786* **1868** *13724*

Ryle, J.
1870 *14209* *14210*

Sabine, Edward (1788–1883). British army officer, Arctic explorer, and pioneer in terrestrial magnetism. Long term officer in R.S.L. and B.A.A.S. (*DNB, DSB*)
1824 1068 **1825** 1131 *1138* 1139 *1140* 1141 1142 *1145* 1151 1157 1185 1187 *1188* 1199 *1203* *1204* *1206* *1255* **1826** *1277* *1288* 1291 *1292* **1838** *3739* *3805* *3810* *3822* *3842* *3847* *3867* *3877* 3879 *3887* *3899* **1839** 3911 *3912* 3913 *4005* *4006* *4012* *4013* *4024* *4048* *4053* *4057* *4059* 4061 *4065* *4069* *4075* *4076* *4079* *4134* *4142* *4146* *4149* *4156* *4161* 4165 *4175* *4178* *4181* *4188* *4189* *4196* *4197* *4202* *4206* *4217* *4218* *4221* *4229* *4234* *4270* *4277* *4278* *4279* *4282* *4283* *4288* *4289* *4292* *4293* *4297* *4305* *4317* *4327* *4328* 4331 *4334* 4335 *4339* *4342* *4344* **1840** 4364 *4379* *4390* *4424* *4438* *4440* *4449* *4469* *4491* *4510* *4518* *4525* *4529* *4531* *4533* *4535* *4539* *4550* *4558* *4560* *4562* *4566* *4567* *4573* *4575* *4576* *4579* *4582* *4589* *4590* *4591* *4593* *4658* *4660* *4670* *4672* *4676* *4678* *4683* *4685* *4709* *4718* *4729* *4733* *4739* *4740* *4742* *4744* *4750* *4757* *4758* **1841** *4767* *4777* *4779* 4783 *4790* 4797 *4798* *4803* *4805* *4806* *4807* 4812 *4822* *4830* *4833* *4834* *4839* *4841* *4848* *4853* *4878* 4879 *4880* *4882* 4925 *4935* *4936* *4945* *4980* *4985* *4994* *5001* *5036* **1842** *5063* *5064* 5065 5066 *5083* *5088* *5092* 5100 5101 5102 *5105* *5156* *5199* *5206* *5213* 5224 *5242* *5256* **1843** *5377* *5378* *5387* *5481* *5534* *5555* *5562* *5579* **1844** *5684* 5685 *5704* *5712* *5742* *5768* *5808* *5836* *5838* 5845 *5847* *5856* 5862 *5864* *5865* 5871 *5875* *5884* *5897* *5922* *5931* *5932* *5940* *5946* *5958* *5974* *5983* *5984* *5987* *6008* *6009* *6013* *6019* *6026* 6027 *6039* *6041* **1845** *6071* 6086 *6097* *6107* *6114* *6119* *6125* *6127* *6136* *6146* *6153* *6159* *6167* *6178* *6188* *6189* *6195* *6202* *6215* *6219* 6220 *6272* *6320* *6324* *6331* *6335* *6352* *6371* *6394* *6411* *6414* *6416* *6418* *6419* *6434* *6438* *6441* *6442* *6472* *6492* *6497* *6512* *6526* **1846** *6539* *6549* *6552* *6567* *6660* 6674 *6675* *6713* *6805* *6812* *6845* **1847** *6856* *6894* 6910 *6914* *6922* *6951* *7053* *7154* *7158* *7188* *7215* *7274* 7275 7277 7278 *7293* *7320* *7392* *7451* *7478* *7479* **1848** *7498* *7541* 7543 *7551* *7608* 7610 7618 *7619* *7622* *7628* *7636* *7649* *7670* 7726 *7727* 7728 *7731* *7741* *7746* *7751* *7755* *7758* *7781* *7788* *7793* *7794* *7795* 7796 *7824* 7915 **1849** *7955* 7980 8079 8131 **1850** *8224* *8229* *8241* *8242* 8243 8247 8310 *8312* 8349 *8350* *8412* **1851** *8930* 8933 *8983* *9047* 9075 **1852** 9214 *9286* *9295* *9325* *9329* 9469 *9567* **1853** 9619 **1854** 10115 **1855** 10379 **1856** *10462* 10532 **1857** *10612* 10634 *10639* 10650 *10653* *10654* 10655 10675 *10684* 10689 *10696* **1858** 10774 *10795* 10802 10803 10811 *10813* *10835* *10836* *10840* 10841 *10846* 10847 *10852* *10865* 10867 *10874* *10877* *10878* 10881 *10883* 10884 *10894* 10954 **1859** *11012* *11048* *11072* *11080* *11093* *11096* 11099 *11105* *11128* 11135 *11136* *11139* *11141* *11165* 11168 *11171* *11183* *11184* 11186 *11197* *11198* *11200* 11203 *11210* **1860** 11322 *11340* 11345 *11380* *11384* 11388 *11395* *11411* 11420 11489 *11494* *11506* 11507 *11508* **1861** 11541 *11543* *11652* *11656* 11680 *11681* *11731* 11733 *11736* 11740 **1862** *11793* 11802 *11883* *11975* 11976 11977 *11983* *12035* *12048* *12053* 12058 *12070* *12071* 12073 12091 12092 **1863** *12144* 12148 *12161* 12169 12170 *12181* 12189 *12266* *12349* 12353 *12357* **1864** 12558 12585 *12593* 12594 12596 12597 **1865** *12846* 12848 *12852* 12857 12933 *12977* **1866** 13233 *13240* 13247 *13251* 13253 *13323* 13326 *13328* 13331 **1867** *13356* *13402* *13409* **1868** *13749* 13754 *13756* *13759* 13761 13762 *13822* **1870** *14167* 14175 14176 *14271* 14272 *14273* 14274 *14283* 14285 **undated** 14774 14775

Sabine, Elizabeth (née Leeves) (1807–79). Translator of German scientific works. Married Edward Sabine, 1826. (*MEB*)
1850 8184 *8188*

Sabler, G.
1862 *11972*

Sadleir, Franc (1774–1851). Mathematician and language scholar. Provost at Trinity College, Dublin. (*DNB, MEB*)
1845 6325

Sadleir, William Digby (?–1858). Irish divine and poet. (*O'Donoghue*)
1852 *9170*

Salmon, John Drew (1802–59). British ornithologist and manager of ice company. (*MEB*)
1852 *9356*

Samuelson, James (1829–?). British publisher and science writer. (*Alli*)
1864 *12431* *12433* *12442* *12467* *12489* *12496* *12500* *12506* *12508* *12517* *12518* *12521* *12526* *12529* *12535* **1867** *13349* *13501* *13506* **1868** 13659 *13678* **1870** *14136* 14141

Sandy, B. F. Naval officer and superintendent of the U.S. Naval Observatory. (1869–11–24)
1869 *14060*

Santini, Giovanni (1787–1877). Italian astronomer, academic, and director of observatory at Padua. (*GDE, DBU*)
1825 *1175* *1260* **1828** *1792* **1832** *2669* **1833** 2775 **1839** *4232* **1840** *4755* **1844** *5895* **1850** *8372*

***Saturday Review*, Editor of**
1865 12771

Scarpellini, Feliciano (1762–1840). Italian astronomer. Founder of the Campidoglio Observatory. (*EnItal*)
1824 *1034* **1834** *2972*

Schaeffer, Mr.
1862 *12069*

Schaffheütl, Karl Emil (1803–90). German geologist and librarian. Professor at Munich and prolific author. (*BHGW*)
1845 *6252*

Schenck & Co.
1822 *690* **1823** *870*

Schimmelpenninck, Count Gerrit. Dutch minister to England. (1847–8–20)
1847 *7168 7195*

Schlagintweit, Adolph (1829–57). German naturalist, mountain climber, and geologist. Explored the western Himalayas with brother Robert. (*ADB, BE*)
1850 *8479*

Schlagintweit, Hermann (1826–1882). German physician, naturalist, and mountain climber. Explored the eastern Himalayas with brother Robert. (*ADB, BE*)
1850 *8447* **1851** *8547 8645* 8658 **1855** *10429* **1858** *10863*

Schmidt, J. K. Eduard (1803–32). German mathematician. Professor of mathematics, astronomy, and physics at Tübingen. (*BHGW*)
1828 *1781*

Schmidt, J. F. Julius (1825–84). German astronomer, geophysicist. Director of the German observatory in Athens. (*DSB, Who's*)
1853 *9627* **1854** 10150

Schomburgk, Robert H. (1804–65). German-born explorer. Traveled the West Indies and surveyed British Guyana. (*DNB, MEB*)
1845 *6237*

Schönbein, Christian F. (1799–1868). German chemist. Discovered guncotton, collodion. (*DSB, Who's*)
1839 *4241* **1840** *4365* **1843** *5465* **1845** *6296 6332* **1846** *6540 6603 6694 6696 6704 6706* **1850** *8371* **1857** *10694*

Schumacher, Christian Andreas (1810–54). Nephew of H. C. Schumacher. Assistant at Pulkowa Observatory, and then to Alexander von Humboldt. (Batten,*Struve*^)
1852 *9459*

Schumacher, Heinrich Christian (1780–1850). German astronomer and geodesist. Founded *Astronomische Nachrichten*. (*DSB*)
1821 550 *554 582* 595 *612* **1822** 837 **1823** *898 909* 917 *921* **1824** 951 *953* **1826** *1309* **1832** *2594 2597* **1835** 3080 *3084 3152* **1838** *3723 3724 3731 3737* **1839** *3925* 3926 3949 *3957* **1840** *4434 4461 4673* **1841** 4788 *5009 5045* 5049 **1842** *5076 5080* **1843** *5624 5651* **1844** *5858* **1845** *6283 6514* **1846** *6600* **1847** *6982 7027 7046* 7130 *7279 7307* **1848** *7640* **1849** *8058 8067 8158 8165* **1850** *8225 8276 8291 8317 8368*

Schur, Wilhelm (1846–1901). German astronomer. (*Who's*)
1869 *13900*

Schwabe, Samuel Heinrich (1789–1875). German astronomer and apothecary. Discovered the periodicity of sunspots. (*Who's*)
1847 *7440*

Schweigger, [J. S. C.] (1779–1857). German physicist and chemist. Invented electrometer and string galvanometer. (*DSB, Who's*)
1822 *655 734* 745 **1841** *4888*

Schwinck, G. Officer in the Corps of Engineers in Berlin. (1840–8–1)
1843 *5558*

Scotson, James
1868 *13816*

Scrope, George Julius P. (1797–1876). English geologist. Supported many of Charles Lyell's uniformitarian theories. (*DSB, Who's*)
1862 *11836*

Seake, S. M. Secretary to Henry Rich, Junior Lord of the Treasury. (1851–8–8)
1851 *8902 8931*

Secchi, Angelo (1818–76). Italian astronomer and physicist. Director of observatory at Collegio Romano. (*DSB*)
1861 *11749* **1866** *13073* **1867** *13492*

Secrétan, Marc (1804–67). French mathematician and optical engineer. Partner in Lerebours-Secrétan, Paris instrument makers. (*Who's*)
1859 *11244*

Sedgwick, Adam (1785–1873). British geologist and divine. Fellow of Trinity College, Cambridge, 1810–73. Founded Cambridge Philosophical Society. (*DNB, DSB*)
1820 *412 416* 482 *494* **1833** 2834 **1844** 5822 *5825* **1845** *6196 6197* 6205 6495 **1846** *6554* **1847** 7097 *7282* 7423 7447 *7452* **1848** 7486 *7489* 7495 7537 7540 *7660* 7664 *7680* 7688 **1850** 8185 **1853** *9728* **1854** *9923 10006* 10024 **1855** *10450* **1856** 10456 10466 **1857** *10682* **1862** *11913 11946* **1865** 12899 12980 *12985* 13018 **1868** *13625 13689* **1869** *13943* 13945

Sedgwick, John. Official on vestry board of Hawkhurst parish. (1840–6–24)
1840 *4587 4596*

Seguin, Marc (1786–1875). French engineer and physicist. Built first suspension bridge in France. Supported molecular theory of heat. (*DSB, Who's*)
1823 882 **1861** *11622* 11635

Seidel, Phillipp Ludwig von (1821–96). German astronomer and mathematician. Made photometric measurements of fixed stars and planets. (*DSB, Who's*)
1865 12808

Sellique, Mr.
1821 *578*

Selwyn, T. E.
1870 *14099*

Selwyn, William (1806–75). British divine and professor at Cambridge. (*DNB*)
1858 *10963* **1860** *11320 11495* **1861** *11523 11573 11597 11718* 11720 **1862** *11764 11967 12008* 12009 *12014* **1863** *12131* **1864** *12766* **1865** 12772 **1866** *13103* 13104 *13106 13113 13114 13118 13121 13124* 13125 *13126 13128 13129* 13135 *13139 13150* 13152 *13153* 13155 *13238* 13243 *13337* 13339 **1867** *13350 13355* 13358 *13359 13500 13527 13553 13580 13586* **1868** *13603* 13610 *13614 13638 13642 13677 13716* 13719 13745 *13750* 13771 *13772* **1869** 14015 *14025* 14034 *14037* 14041 14043 **1870** *14080* 14164 *14169 14170* 14179 14182 *14194* 14197 *14218 14223* 14226 *14228* 14263 *14267* 14268

Senarmont, Henri de (1808–62). French mineralogist. Professor in Paris. (*BHGW*)
1862 *11798 11820*

Seppings, Sir Robert (1767–1840). English naval architect and inventor. Improved shipbuilding methodology and invented improved lifting blocks. (*DNB, Who's*)
1822 *716*

Setten, George
1869 *14067*

Sevenstern, J. Riedl- (see Riedl-Sevenstern, J.)

Severn, Henry Augustus. Employed at the Mint in London, then at the Mint in Sydney, Australia. (1858–11–10)
1858 *10964*

Seymour, Edward A. (1775–1855). British scientist. 11th Duke of Somerset. (*DNB*)
1852 *9321*

Seymour, Edward A. S. (1804–85). British M.P., cabinet minister, and author. 12th Duke of Somerset. (*DNB*)
1862 *11981*

Seymour, Margaret. Wife of E. A. Seymour, 11th Duke of Somerset.
1852 *9361 9439* **1854** *9975*

Shadwell, Sir Charles (1814–86). English naval officer. Studied nautical astronomy. (*DNB, MEB*)
1847 *6945* **1849** *8047*

Shanahan, Patrick. An employee at the Mint during JH's mastership. (Brande's 1855–3–7)
1868 *13805* 13818

Shanks, William (fl. 1850). English mathematician. (*Alli*)
1852 *9611*

Sharp, William (1805–96). English physician. Promoted science teaching in schools. (*DNB, MEB*)
1849 *8115*

Sharpey, William (1802–1880). Scottish physician and professor of anatomy and physiology. Senior secretary to R.S.L., 1854–72. (*DNB, MEB*)
1855 10307 **1856** *10589* **1858** *10784* **1862** *12020*

Shaw, Alexander (1804–90). British surgeon at Middlesex Hospital, London. (*DNB*)
1855 *10332*

Shaw, Norton (fl. 1860). British geographer and secretary of the Royal Geographical Society. (*BBI*)
1859 11129 11247 **1860** *11379* 11401 *11467* **1861** *11551*

Shaw-Lefevre, Charles (1794–1888). English barrister and parliamentarian. 1st Viscount Eversley. (*DNB*)
1851 8515

Sheepshanks, Anne. Sister of Richard Sheepshanks.
1860 11339

Sheepshanks, Richard (1794–1855). British barrister and astronomer. Member of weights and measures commissions. (*DNB, MEB*)
1830 *2076* 2237 **1833** *2716* **1835** *3160* 3199 **1839** *4043* 4091 *4105* 4157 *4158* **1840** 4457 *4459 4464* 4465 4501 **1841** *4768 4917* 4919 *4920* **1842** *5067* 5082 *5187 5286 5292 5294* **1843** *5350* **1844** 5686 5687 5688 5689 *5690* 5691 *5923 5924 5925* 5953 5969 *5975* 5985 *5990 6015* 6017 6023 *6024* **1845** 6085 *6115 6117* 6123 *6133 6218* 6222 *6223* 6376 **1846** *6541 6542 6543* 6553 *6557* 6558 6569 6604 6619 6817 6826 6827 6832 6835 6836 6840 *6841* **1847** *6857 6858 6866* 6870 6889 6895 6901 6903 *6972 6978* 6980 6981 *6983* 6988 *6996 7002 7013* 7014 *7016 7021* 7022 7023 7082 *7084* 7085 7093 7094 *7101 7169* 7289 *7314* 7328 *7388* 7395 *7407* 7411 *7415 7419 7453* 7471 **1848** 7487 7512 7526 7527 7566 7576 7591 7611 7645 7707 *7809* 7811 7874 **1849** *7956 7964 7967* 7968 7971 *7995* 8002 8099 *8100* **1850** 8186 8236 8248 8261 8262 8270 8285 8286 8288 8331 **1851** *8566* 8712 9037 *9038 9100 9101* 9106 *9112* **1852** 9182 9206 *9209 9378* 9388 *9395* 9558 **1853** 9630 *9633* **undated** *14776 14777 14778 14779 14780* 14781 14782

Sheil, Richard Lalor (1791–1851). Irish dramatist, barrister, and politician. Public official, including master of the Mint. (*DNB, MEB*)
1850 8428 *8433 8452 8467*

Shepherd, George (fl. 1860). English engineer, meteorologist. (*Alli*)
1864 *12435* 12436 *12438*

Shepherd, John (?–1859). Member of the court of directors of the East India Company. Deputy master of Trinity House. (*MEB*)
1850 8187

Shepherde, William. General manager of Metropolitan Light Company in 1848. (1848–8)
1848 *7799*

Sherwood & Co.
1843 *5482* 5503 *5512*

Shonbridge, J. H. Lawyer, with partner, T. P. Anderson. Advised JH about marriage settlement trust for Mary Anne (Babbage) Hollier, sister of Charles Babbage. (1864–11–17)
1864 12712 12716 *12738* **1869** *14063*

Sikes, John C.
1862 *11990*

Silliman, Benjamin (1779–1864). American chemist, mineralogist, and geologist. Professor, Yale College. (*DSB, Who's*)
1843 *5397*

Simms, Henry
1859 *11216*

Simms, William (1793–1860). British scientific instrument maker. Partner with Edward Troughton in London. (*MEB*)
1840 *4404* **1847** *7356*

Simon, Algernon O.
undated *14783*

Simon, Thomas Collins. British author, writing on the relationship between science and religion.
1861 *11569 11598*

Simpson, A. J. British divine. Vicar of Upton Church near Slough. (1868–3–7)
1868 *13617* 13618

Simpson, Arthur B.
1867 *13478* **undated** *14784 14785*

Simpson, G. Wharton (1824–80). Proprietor and editor of *Photographic News*. (*MEB*)
1864 12419 *12705* **1866** *13186 13189*

Simson, John. Attached to Baden legation in London. (1847–10–1)
1847 *7248* **1848** *7898*

Sinnis, Paul. Assistant to the General Secretary of the B.A.A.S. (1859–8–29)
1845 *6242* **1859** *11119 11120*

Sivewright, James (1848–?). English politician, and friend of Edward Sullivan. (*BBI*)
1868 *13806 13810*

Skirrow, John (?–1846). British architect and land surveyor. Sent by the Admiralty to the Cape to build observatory, 1825. (*DSAfB*)
1836 *3412*

Slater, John Allen (fl. 1840). English poet. Wrote a collection of poetry entitled *Shadows of Thought* (1842). (*Alli*)
1853 *9714*

Sloan, John Charles
1863 *12416*

Smedley, Edward (1788–1836). English writer. Editor of the *Encyclopaedia Metropolitana*, 1822–36. (*DNB, Alli*)
1831 2303

Smee, Alfred (1818–77). British chemist, physiologist, and surgeon to the Bank of England. Invented several useful devices. (*DNB, MEB*)
1842 *5074 5158 5169* 5202 *5208 5257 5287 5314 5320 5331* **1843** *5398 5411* **1845** *6088 6170* 6171 *6199* **1847** *6869 6908 7280* **1849** *8044 8133* **1850** *8322 8361* **1854** *10000* **1867** *13588* **undated** *14787*

Smith, Andrew (1797–1872). British army surgeon, serving at the Cape, 1821–37. Involved in exploration of parts of Africa. (*DNB, DSAfB*)
1835 *3105* **1836** *3340* **1837** *3530 3603* **1838** *3717* **1839** *4171* **1841** *4889*

Smith, Archibald (1813–1872). British mathematician and lawyer. Secretary of Decimal Coinage Commission. (*DNB, MEB*)
1857 *10698 10700*

Smith, Beck & Beck
1864 *12426*

Smith, David William (1764–1837). British army officer. Canadian legislator and then administrator for the Duke of Northumberland. Cousin of Mary Pitt Herschel (*EnCan*)
1821 *619* **1822** 801 **1831** *2455* **1832** *2511*

Smith, Elder & Co.
1847 *7183 7192* **1848** *7596* **1857** *10667 10676* **1861** *11531* **1862** *11877 11918 11991 11994 12095* **1863** *12142* **1864** *12664* **1865** *12928* **1868** *13696 13701*

Smith, George (1824–1901). British publisher. Son of George Smith, partner of Alexander Elder. Founder and proprietor of *Dictionary of National Biography*. (*DNB*)
1848 *7939* **1858** *10891* **1860** *11287* **1861** *11649* **1862** *11879 11922 11942 11958* 11959 *11968 11984 12002* 12004 **1863** *12402 12411* **1866** *13115* 13119 *13130*

Smith, Harry George Wakelyn (1787–1860). English army officer. Governor of the Cape of Good Hope, 1847–52. Fought in the Kaffir wars. (*DNB, MEB*)
1837 3483 3487 *3488 3537* **1852** *9413* **1856** *10480*

Smith, J.
1839 *3914*

Smith, James (1782–1867). Scottish geologist and man of letters. Wrote many geological papers. (*DNB, MEB*)
1864 *12749*

Smith, John Benjamin (1794–1879). British M.P. and Manchester cotton merchant. (*MEB*)
1864 12650 12654 *12669*

Smith, John Thomas (1805–82). British officer in Madras Engineers. Master of Madras and Calcutta Mints. (*DNB, MEB*)
1869 *13995* 13999

Smith, Mary (née Taylor). Married David William Smith (2nd wife), 1803. (*EnCan*)
1838 *3654*

Smith, Mr. An agent involved with the purchase of Collingwood by JH. (1839–9–5)
1839 4230

Smith, Robert Angus. Secretary to Manchester Literary and Philosophical Society in 1853.
1853 *9899*

Smith, Robert Vernon (1800–73). British M.P. and cabinet minister. 1st Baron Lyveden. (*DNB*)
1840 *4383 4395 4439 4450 4453 4470 4471*

Smith, William. British M.P. for Norwich. Member of the Royal Society Club. (WS's 1833–9–17)
1833 *2762 2867* 2868

Smyth, Ann (née Warington). Married William Henry Smyth, 1815.
1842 *5110* **1843** *5616* **1844** *5692 6053* **1845** *6284* 6478 **1857** *10754* **1860** *11270 11481* **1861** *11618 11748*

Smyth, Charles Piazzi (1819–1900). British astronomer, Astronomer Royal for Scotland. Assisted Thomas Maclear at Cape, 1835–45. Son of W. H. Smyth. (*DNB, DSB*)
1842 *5068 5069* **1843** *5407* **1844** *5751* **1846** *6618 6726* **1847** *6927 7105* 7106 *7122 7140* **1850** *8217* **1853** *9693 9723* **1857** *10640*

Smyth, William Henry (1788–1865). British naval officer and surveyor. Built several private observatories. (*DNB, MEB*)

1827 1505 1547 1579 *1632* 1633 **1828** 1688 1722 *1767* 1774 *1779 1809 1828* 1836 *1878* 1882 **1830** 2171 *2185* 2202 2211 **1831** 2271 2272 *2364* 2372 2398 2439 *2446* **1832** *2568 2578* 2580 2610 *2626* 2635 *2636* 2653 *2708* **1833** 2713 2803 *2804 2815 2819* 2830 *2864* 2878 **1834** 2976 *2984 3011* 3045 **1835** *3062* 3113 *3154 3192* 3208 **1836** *3268 3316* 3328 *3360 3365* **1837** *3447 3556* **1838** *3791* 3795 **1839** *4083* 4090 **1840** *4700* **1842** 5144 5178 **1843** *5390 5467 5472 5494* 5499 5504 *5513* 5522 *5525* 5529 *5530* 5540 *5557* 5581 *5601 5638* 5642 *5655* **1844** 5814 *5991* 5992 **1845** 6065 6163 *6164 6180* 6216 *6225 6226 6238* 6239 *6373 6383 6429 6439* 6457 *6471* 6481 **1846** 6544 *6586* 6588 *6589 6622* 6623 6624 *6626 6683 6689* 6692 *6702 6780 6786* 6833 **1847** 6890 6920 6924 6929 6934 *6954* 6957 *6958 7134* 7326 7380 7381 7434 **1848** 7513 7833 *7883* 7902 7927 **1849** *7957* 8046 *8114* 8117 **1850** *8355* 8356 **1851** *8562* **1852** 9248 **1855** 10400 **1856** *10568* 10570 **1857** *10621* 10622 *10624 10632* 10745 10746 *10751* 10759 **1858** *10818* 10819 *10822 10824 10909 10919* 10925 *10930* **1860** 11286 *11290 11321* 11437 *11441 11443 11484* 11486 *11518* **1862** *11936* 11938 12078 *12082* **1864** *12607* 12610 *12612* 12635 12761 **1865** 12961

Snowden, James Ross (1810–78). American numismatist, lawyer, and director of the U.S. Mint. (*NCAB, DAB*)
1854 10210

Socker, Mrs. Wife of Professor Socker. (1867–4–13)
1867 13408

Society of Arts, Chairman
1859 11087

Somerset, Duchess of (see Seymour, Margaret)

Somerset, Duke of (see Seymour, Edward Adolphus (1793–1855) *or* Seymour, Edward Adolphus Seymour (1804–85))

Somerville, Martha. Daughter of Mary Somerville. (*Women*)
1865 *12863* **1869** *13857 13864*

Somerville, Mary (née Fairfax) (1780–1872). British mathematician and science writer. Married Russian naval officer Samuel Greig. Widowed, and married William Somerville. (*DNB, Women*)
1829 *1985 2040* **1830** *2077 2078* 2079 2104 *2110 2111 2118* 2119 2141 *2145 2161* 2180 *2233* **1831** *2291 2317* 2328 *2344 2354* 2373 **1843** *5633* **1844** 5761 *5882 5933* **1845** 6066 *6380* 6448 **1846** *6590* **1847** 7135 **1848** *7687* **1851** 8643 8742 **1853** 9742 **1856** *10493* **1857** *10659* **1858** 10781 10815 *10826* **1860** *11426* **1861** *11539* 11589 **1862** *11884* **1863** *12258 12274* 12282 *12303* **1865** 12870 **1866** *13254 13290* **1868** *13621 13691 13746* **1869** 13880 *13887* 13966 *14071*

Somerville, William (1771–1860). British physician and army surgeon. Served in several medical posts in the Empire. (*DNB*)
1826 *1357 1362* 1363 **1829** 2037 **1830** *2105 2132* **1831** *2385 2388* 2392 *2415* 2416 *2422 2423 2424* 2426 2430 *2460* **1832** 2549 **1833** 2757 2766 2918 **1835** 3156 3157 **1837** *3523* **1848** *7511* 7612 *7637* 7642 *7668*

Sorby, Henry Clifton (1826–1908). English chemist, geologist, biologist and metallurgist. (*DSB, Who's*)
1856 *10567* 10571

Soret, Frederick
1824 *1041*

Soring, F.
1856 *10517*

Sotheby, William (1757–1833). English author. Friend of many young authors, including Walter Scott and Samuel Taylor Coleridge. (*DNB, Alli*)
1826 *1327* **undated** 14788

South, James (1785–1867). English astronomer. Worked with JH on re-observing WH's double stars. (*DSB, Who's*)
1820 *467 497* **1821** *512 579 644* **1822** *656 661 701* 736 786 **1823** *844 886 900 906 911 918 919 935* 937 **1824** 948 949 967 968 *1010 1047* 1051 1082 *1086* **1825** 1094 1095 1096 *1123* 1130 *1144* 1146 *1152 1156 1160 1162* 1166 *1171 1173 1176 1182 1189 1191 1205* 1215 *1220* 1222 *1223 1229 1234 1239 1245* 1251 **1826** 1407 1420 **1827** *1622* **1828** 1662 *1723 1834* **1829** 1961 *1978 2008* **1830** *2103 2173* 2177 **1831** *2287 2462* 2464 *2466* 2488 **1832** *2572* 2573 *2647* **1833** 2818 *2821* 2826 **1835** 3134 **1838** *3843* 3845 *3868* 3869 *3871* **1864** *12685* 12688 *12690* 12754 **1866** 13215

South, John F. (1797–1882). English surgeon. Surgeon at St. Thomas Hospital. Half-brother of James South. (*DNB, MEB*)
1866 *13197* 13216 *13217*

Spahr, N. German nephew of Mrs. Gerlach. Resided in Geneva in 1843. (JH's 1843–12–31)
1835 3222 **1843** 5618 5676 5677

Spence, William (1783–1860). English entomologist. Wrote, with William Kirby, an *Introduction to Entomology*. (*DNB, MEB*)
1854 *10003*

Spencer, Herbert (1820–1903). English social philosopher. Supported laissez-faire economics and evolutionary theory. (*DSB, Who's*)
1853 9756 **1859** *11014* 11016 *11254* **1860** 11273 *11277 11309 11332* **1862** *11818* **1863** *12134* **1864** *12505* 12515 **1865** *12868 12876* **1866** 13143

Spencer, Thomas (fl. 1835-59). British chemist. (*RSL Cat*)
1859 *11201*

Spencer, William George (1790–1866). English mathematician. Skilled tutor of elementary mathematics. (*DNB, MEB*)
1862 *12068*

Spiers, Alexander (1807–69). English lexicographer. Compiled a *General English and French Dictionary*, 1846. (*DNB, MEB*)
1851 *8875*

Spottiswoode & Co.
1867 *13414*

Spottiswoode, Robert. Executor, with JH and Daniel Birkett, of the estate of E. S. Beckwith's husband. (1830–8–30)
1830 *2124*

Spottiswoode, William (1825–83). British mathematician, linguist, and physicist. Partner in Eyre & Spottiswoode, printers. (*DNB, MEB*)
1871 14396 *14398* 14399 *14423*

Spring-Rice, Charles
1845 *6270*

Spring-Rice, Marianne (née Marshall) (?–1889). Lady Monteagle. Married Thomas Spring–Rice, 1841. (*DNB*)
1868 *13712*

Spring-Rice, Thomas (1790–1866). British M.P. and cabinet minister. Created 1st Baron Monteagle. (*DNB, MEB*)
1838 *3662* **1856** *10459* **1857** *10656 10671 10672 10674* 10677 **1864** *12443*

St. Hilaire, Étienne Geoffroy (see Geoffroy Saint-Hilaire, Étienne)

St. John, John. British divine and sometime acting on behalf of St. John's College about JH's portrait. (1833–1–15)
1833 2726

Stanley, Arthur Penrhyn (1815–81). Dean of Westminster, 1863–81. Author of *History of Sinai and Palestine* (1856). (*DNB, MEB*)
1860 *11354*

Stanley, Edward (1793–1862). English surgeon and anatomy professor. Life member of council of the Royal College of Surgeons. (*DNB*)
1848 7542

Stanley, Owen (1811–?). English naval captain and surveyor. (*BBI*)
1846 *6666 6671*

Stark, James (1811–90). Scottish physician who practiced in Edinburgh. (*MEB, Alli*)
1857 *10726* 10729 *10733* **1860** *11462*

Starkey, Digby P. (1806–76). Irish writer of miscellaneous works. Contributed to many periodicals. (*MEB, Alli*)
1847 *7361*

Steinheil, C. A. von (1801–70). German physicist, astronomer. Discovered process of silvering glass mirrors. (*DSB, Who's*)
1842 *5172* 5205 **1845** *6340* **1847** *7079 7080*

Stephenson, William Henry (1811–98). English treasury clerk and secretary, 1827–62. (*MEB*)
1850 *8411 8434* 8439 **1851** *8685* **1854** *9977*

Sterry, Charles. Assayer and clerk in Mint office, 1851–[68?]. (JH's 1868–9–7)
1854 10047 **1868** *13729 13733* 13736 *13737 13814*

Stevelly, John. Professor at Belfast. (1850–11–26)
1845 *6366* **1846** *6765* **1850** *8436*

Stevens, Henry
1859 *11112* **1861** 11580

Stevenson, G. R. Officer in the West Kent militia. (1856–3–18)
1856 *10477*

Stevenson, Thomas (1818–87). Scottish engineer and meteorologist. Improved lighthouse illuminating systems. (*DNB, Who's*)
1860 11368 *11373*

Stewart, Balfour (1828–87). British physicist, meteorologist, and academic. Cofounder of Society for Psychical Research. (*DNB, DSB*)
1858 *10986 10997* **1869** 13885

Stewart, Duncan (1804–75). Eldest brother of Margaret Brodie Herschel. Medical doctor in service of East India Company, 1825–55. (Warners, *Maclear*)
1834 *2936* 2960 **1869** 14039 **1870** 14155 14165 **1871** 14441 14442

Stewart, Emilia. Mother of Margaret Brodie Herschel (née Stewart). London residence at 22 Beaumont St., Devonshire St., and at 57 York Terrace.
1828 1867 1870 *1871* 1873 *1874* 1877 1880 *1881* 1883 1888 1891 *1892* **1830** 2107 2142 2146 2151 **1831** 2363 2368 **1833** 2925 **1851** 9022

Stewart, James Calder. Brother of Margaret Brodie Herschel. Resided in Calcutta, 18[26?]–1833.
1831 2279 **1833** 2749 2777 **1834** 2967 2993 3038 **1835** 3117 3220 **1836** 3313 **1837** 3462 3533 **1838** 3644 3828 3829 3839 3850 **1839** 3930 3935 4017 *4163* 4265 **1844** 5693 5758 5780 **1849** 7996 **1851** 8526 9146 **1852** 9342 9379 9489 9498 9503 9513 **1853** 9625 9709 **1854** *9918* 9924 10054 10139 10240 **1855** 10313 10396 10399 10432 10440 **1856** 10545 **1857** 10750 **1860** 11502 **1861** 11525 **1863** 12207 **1864** 12640 12651 **1865** 12919 12920 **1867** 13374 13499 **undated** 14789 14790 14791 14792 14793 14794 14795 14796 14797 14798

Stewart, James Struthers (1791–1863). Scottish-born artist and M.P. in South Africa. Farmer at the Cape from 1834–63. (*DSAfB*)
1836 *3307* **1844** *5795* **1845** *6420 6458* **1854** *10237*

Stewart, John. Brother of Margaret Brodie Herschel. Accompanied sister and JH on 1829 trip through Europe. Husband of Matilda Grahame.
1839 4155 4159 4190 **1843** 5381 **1852** *9164* **1853** *9755 9816*

Stewart, Margaret Brodie (For post-1828 letters, see Herschel, Margaret Brodie.)
1828 1889 1893 1894 *1895* 1910

Stewart, Mr. One of Margaret Brodie Herschel's brothers. (Francis Beaufort's 1835–2–12)
1835 3071 **1862** 12015

Stewart, Peter. Brother of JH's wife, Margaret. Printer, in association with publishers Smith Elder & Co. Commonly called Peter, his formal name was Patrick. (JH's 1836–6–16)
1828 1885 **1830** 2150 **1833** 2787 **1834** 2994 3020 **1835** 3107 **1836** 3336 *3404* **1843** 5523 **1844** *5719 5728* **1845** *6181 6190 6395* **1847** *7334*

Stewart, Rose. Probably the wife of James Calder Stewart. (MBH & JH's 1839-10-18)
1839 4265

Stirling, James Hutchison (1820–1909). British philosopher and surgeon. Refuted T. H. Huxley's views on protoplasm. (*DNB*)
1870 *14088 14092*

Stirling, T. D. Morris- (see Morris-Stirling, T. D.)

Stockenström, Andries (1792–1864). Cape of Good Hope frontier soldier, administrator, and politician. Lieutenant governor of eastern districts, 1836–39. (*DSAfB, MEB*)
1838 *3893* **1839** *3975 3990 4140* 4144 *4147 4248* **1840** *4389 4397 4481 4653* 4654 **1841** *4771*

Stokes, Charles (1784–1853). British stockbroker and natural historian. (*MEB*)
1822 719

Stokes, George Gabriel (1819–1903). British mathematician and physicist. Lucasian professor, 1849–1903. Secretary of R.S.L., 1854–85. (*DSB*)
1849 8140 **1852** *9322* 9331 *9556* 9560 *9562* **1853** *9624* 9626 **1854** 10133 *10202* 10209 **1855** 10300 10319 10322 10333 10342 **1856** *10533* 10534 *10538 10539* 10540 *10543 10576* 10578 10585 10591 10592 **1857** 10743 **1858** *10973* **1859** *11058* 11061 11071 11078 *11079 11194 11195 11202* **1860** *11283 11369 11478* **1861** *11568* 11575 *11576* 11654 *11655* **1862** 11803 *11805 11826* 11923 *11926 11979* 12063 *12083 12085* **1863** 12141 *12166* 12182 *12187* 12216 12393 *12395* **1864** 12429 *12462 12463* 12468 12493 *12572* 12573 12581 *12758* **1865** *12785 12827 12918* 12921 12922 *12926 12930* 13022 **1866** 13097 13102 *13107* 13108 *13116* 13122 *13149* 13308 **1867** 13351 13352 *13424 13425* 13430 13431 13439 13443 *13444 13448* 13451 13456 13479 13537 *13554* 13555 **1868** 13709 *13734 13784* 13785 *13808* 13811 *13820* 13823 13831 **1869** 13842 *13846* 13937 13958 13959 *13960 13985* 13986 *13991* **1870** 14188 14189 *14192* 14275 *14277* 14279 *14281* 14291 14292 *14293 14296* **undated** *14799*

Stone, Edward James (1831–97). British astronomer. Chief assistant at Greenwich, Astronomer Royal at Cape, and Radcliffe observer. (*DNB, MEB*)
1863 *12122 12123 12273* **1864** *12497* **1866** *13212* **1867** *13505* **1868** *13747*

Stoney, George Johnstone (1826–1911). Irish astronomer and physicist. Assistant to William Parsons. (*DNB*)
1868 13788

Story-Maskelyne, Nevil (1823–1911) Professor of Mineralogy at Oxford and keeper of minerals at the British Museum. (*DNB*)
1871 *14378*

Stotherd, A. J.
1855 10397

Strachey, Richard (1817–1908). British army officer in Royal Bengal Engineers. Involved in botanical and geological exploration of the Himalayas and Tibet. (*DNB*)
1867 13525 13526 13549 13568 13579

Strahan & Co.
1864 12464

Strahan, Alexander (fl. 1870). English publisher [of *Good Words*?]. (*AlliSup*)
1868 *13776 13782*

Strangeways, John George Charles Fox (1803–59). British M.P. and official in Foreign Office. (*MEB*)
1839 *4204*

Stratford, William Samuel (1791–1853). British naval officer and long term council member of R.A.S. Superintendent of *Nautical Almanac*. (*DNB, MEB*)
1826 *1302* **1828** *1903* **1829** 1929 **1831** *2316 2318* 2319 2320 2321 2324 *2326* 2332 2407 **1838** 3641 **1844** *5890 5891 5894 5896 5898 5902 5903 5906 5908 5909 5910 5989* **1845** *6358 6498 6523* **1846** *6697 6699 6731* 6766 **1847** *6911 6912 6916 6973* 6975 *6979 6984 6985 6986 6989 6994 6995 6997 7005 7008 7009 7010 7011 7012 7017 7020 7024 7025 7028 7029 7031 7032 7036 7054 7060 7063 7064 7243 7315 7382 7383*

Streatfield, John Fremlin
1860 *11303*

Stronach, William
1864 *12491*

Struve, Wilhelm (1793–1864). German astronomer, geodesist, and academic. Director of Dorpat and Pulkowa observatories. (*DSB, GSE*)
1822 *710* **1823** 869 *892 902* **1825** 1105 *1143* 1180 **1826** 1296 1326 *1328 1404 1430* **1827** 1450 *1557* 1618 1634 *1639* **1828** *1734* 1768 1769 **1830** *2120* 2148 *2155 2182 2201 2209 2229* **1831** 2348 **1832** *2546* 2571 2631 **1833** *2715* 2758 *2840* 2869 *2891* 2914 **1834** 2985 **1836** *3267* **1838** 3637 *3880* **1840** *4561* 4621 **1841** *4775* 4998 **1842** *5191* **1844** *5818 5841* 5846 *5851* **1846** *6759 6760* 6843 **1847** *7120* **1848** 7834 *7946* **1853** 9730

Struve, Otto Wilhelm (1819–1905). German astronomer and geodesist. Son of Wilhelm Struve. Worked with his father at Dorpat and Pulkowa. (*DSB*)
1848 *7490* **1870** 14234

Strzelecki, Count Paul Edmund de (1793–1873). Polish-born explorer. Studied the Australian interior and discovered gold. (*DNB, Who's*)
1845 6256 **1863** 12218 *12220*

Styles, Mr.
1822 821

Suft, Robert Francis. Employed as a clerk at the Royal Mint, London, during JH's mastership. (1855-7-23)
1855 *10334 10405* **1865** *12989*

Sullivan, Edward (1826–99). English yachtsman and writer of miscellaneous works. (*MEB, Alli*)
1868 *13803 13812*

Sumner, Archbishop of Canterbury, John Bird (1780–1862). British divine; trustee of British Museum. (*DNB, MEB*)
1851 8661 9048

Sussex, Duke of (see Augustus, Frederick)

Swainson, William (1789–1855). English naturalist. Authored many zoological papers and books. (*DSB, Who's*)
1839 3961 *3962*

Swan, Joseph Wilson (1828–1914). English physicist and chemist. Improved photochemistry, and invented the incandescent electric lamp. (*Who's*)
1867 *13474*

Sykes, William Henry (1790–1872). British army officer and member of East India Company board of directors. On B.A.A.S. balloon committee. (*DNB*)
1839 *4060* 4062 **1844** *6028* **1845** *6321 6449* **1847** 6930 *7141* **1848** *7520* 7521 **1850** *8252 8280 8293 8450* **1851** *8861 9056 9074* **1852** *9165 9201* 9232 *9365 9456 9475* **1853** *9865* **1857** 10618 **1858** *10975* 10978 **1861** 11707 *11709* **1862** *11763 11768 12006*

Sylvester, James Joseph (1814–97). English mathematician. Developed theories of algebraic forms and algebraic determinants. (*DSB, Who's*)
1841 *4902* **1861** *11609* **1865** *12853* **1866** *13277* **1867** *13385*

Syme, James (1799–1870). Scottish surgeon. Improved many surgical methods. (*DNB, Who's*)
1865 *12956 12967*

Symons, George James (1838–1900). English meteorologist. Published annual volumes of weather statistics, 1860–99. (*MEB, Who's*)
1865 *12927*

Symons, H.
undated *14800*

Symons, J. H.
1857 *10718*

T.?], [J.
1839 4336

Tait, Archibald Campbell (1811–82). Headmaster at Rugby. Archbishop of Canterbury, 1868–82. (*DNB, MEB*)
1870 *14265*

Tait, Peter Guthrie (1831–1901). Professor of natural philosophy, Edinburgh University, 1860–1901. (*MEB, Alli*)
1864 12733 *12739*

***Talbot*, Commander of steam-packet**
1822 787

Talbot, William Henry Fox (1800–77). British mathematician and photochemist. Pioneer in photography. (*DNB, DSB*)
1826 *1278 1281 1282 1286* 1289 *1294 1307* 1331 *1348 1369* 1381 **1827** *1519 1553 1556 1564* 1566 *1606* 1608 **1829** 1935 **1830** 2263 **1833** *2753 2756 2764 2798 2800* **1836** 3238 **1837** 3434 *3531* **1838** 3647 3648 *3660* 3664 *3682 3718* **1839** *3934* 3936 *3938 3942* 3943 *3946 3952* 3953 3954 *3956* 3958 *3963 3968 3969 3980* 3981 *3985* 3991 *4007* 4009 *4014* 4025 *4030 4035 4058* 4080 4084 *4085* 4102 4136 *4148* 4151 *4166 4233* 4236 *4237 4290* 4315 *4319* **1840** 4458 *4462* 4467 4522 *4534* 4536 *4585* 4592 4661 *4666* **1841** 4844 *4846 4849* 4854 4857 4867 *4868* 4891 *4893 4912 4926* 4931 4948 *4949* **1842** *5151* 5155 *5188* **1843** *5422 5428* 5431 *5435* 5436 *5443* **1844** *5920 5934* **1847** *7313* 7321 *7329* **1848** *7569* 7602 **1850** *8437* 8448 **1851** *8782* 8783 8891 *8938* **1853** 9726 9878 *9886* **1854** *10103 10106* 10109 *10112* 10114 *10116 10122* **1868** 13597

Talmage, Charles George (?–1886). English astronomer. Director of private observatory of Joseph Gurney Barclay, 1865–86. (*MEB*)
1870 *14233*

Tatham, Ralph (1778–1857). Master of St. John's College and rector of Colkirk, Norfolk. (*MEB*)
1847 *6898 7127*

Taylor, Bridget
1842 5274

Taylor, Felix. Tenant of Observatory House in Slough. (JH's 1864–2–7)
1864 12475

Taylor, Sir Henry (1800–1886). English author of various works, including plays. (*DNB*)
1850 8216 **1862** 11835 *11868* **1868** *13702* 13834 **1869** *13838*

Taylor, [F.R.S.], John (1779–1863). British mine owner and chemical manufacturer. Long term officer of Geological Society and B.A.A.S. Brother of Richard Taylor, printer. (*MEB*)
1840 4391 *4451* **1841** *4824* **1845** *6353* **1848** 7704

Taylor, John (1781–1864). English publisher and author. Proprieter of the *London Magazine*. (*MEB*)
1863 *12245* 12247 *12250* 12257 *12261 12403* 12404 **1864** *12427*

Taylor, John Edward (1779–1863). British printer. Nephew and business partner of Richard Taylor. (*MEB*)
1846 *6695*

Taylor, John Ellor (1837–95). Editor, geologist and lecturer on science. Curator of Ipswich Museum, 1872–93. (*DNB*)
1840 4542

Taylor, John I. (?–1857). British astronomer, linguist, poet, and cotton merchant in Liverpool, 1807–57. (*MEB*)
1836 3240

Taylor, Richard (1781–1858). British printer, naturalist and editor of *Philosophical Magazine*. Brother of John Taylor (FRS). (*DNB, MEB*)
1842 *5298 5318 5326 5332* **1843** *5382 5492* **1844** *5968* **1845** *6206* 6211 **1846** *6695*

Taylor, Thomas Glanville (1804–48). British astronomer. Performed calculations for Stephen Groombridge's star catalogue. Director of Madras Observatory. (*DNB*)
1821 *632* **1836** *3353* **1837** *3565* **1838** *3885*

Temple, Henry John (1784–1865). English M.P. and prime minister. 3rd Viscount Palmerston. (*DNB*)
1820 420 **1826** *1295* **1848** 7867 *7871* **1855** 10332 10335 10364

Tennant, Charles (1823–1906). English merchant and art patron. Founder of chemical works at St. Rollox, Glasgow. (*DNB*)
1867 *13467 13469*

Thomas, Fedor
1864 *12574*

Thomas, William Cave (1825–?). English painter, writer. (*Alli, AlliSup*)
1867 *13382*

Thomas Cooke & Sons
1859 *11227*

Thompson, J. G.
1860 *11381*

Thompson, Thomas P. (1783–1869). English army and naval officer and M.P. Author of works on a variety of subjects. (*DNB, MEB*)
1860 *11292 11296* **1868** *13757*

Thomson, Thomas (1773–1852). British chemist and academic. Author of *System of Chemistry*, 7 eds. (*DNB, DSB*)
1827 *1486* 1510 **1844** 5840

Thomson, William (1824–1907). Irish-born physicist. Professor at Glasgow University. 1st Baron Kelvin. (*DNB, Who's*)
1853 *9641* **1864** 12646 *12709* 12726 **1868** 13721 *13744*

Thurnell, George
1866 *13098*

Timbs, John (1801–75). English author and editor. (*DNB*)
1838 3890

***Times*, the London**
1828 1814 **1832** 2632 2634 **1838** 3790 **1843** 5417 5420 5437 **1845** 6217 **1846** 6742 **1857** 10651 **1860** 11501 **1862** 11886 11887 **1864** 12598 12605 12611 **1865** 12896 **1868** 13657 **1869** 13981 13984 13989 14068 **1870** 14122

Tindal, Acton. Clerk of the Peace for Buckinghamshire (which included Slough). (1839–3–29)
1839 *4036*

Todhunter, Isaac (1820–84). British mathematician and fellow of St. John's College, Cambridge. (*DNB, MEB*)
1861 *11712* **1864** *12538* 12539 *12614* 12620 *12624* **1865** *12916* 12935 **1866** 13079 *13086*

Tomlin, Jacob (1793–1880). British missionary and divine. (*MEB*)
1865 *12982 13034*

Towson, John T. (1804–81). British chronometer and watch maker, scientific writer and pioneer in photography. Introduced to Admiralty by JH. (*DNB, MEB*)
1846 6707 *6719* **1847** *7441*

Toynbee, Ellen. Sister of Mrs. Baden Powell. (1865–9–16)
1865 *12968*

Tracy, V.
1849 *8080*

Trant, Jr., William (1844–1924). Founder and secretary of Leeds Astronomical Society. After working many years as a journalist became archivist to the province of Saskatchewan in Candada. (*DCB*)
1863 *12217 12232 12319 12322 12328 12334 12339 12347 12358 12364 12366* **1864** *12645* 12647 *12648*

Treasury, the
1852 9221 9358

Trechsel, F. (1776–1849). Swiss mathematician and physicist. (*BHGW*)
1822 *807* **1823** *930*

Tredgold, J. H. A friend of Andries Stockenström.
1839 *4242* 4243 *4246*

Trembley, Picot
1821 *599*

Trevelyan, Charles Edward (1807–86). Assistant secretary to the Treasury. Governor of Madras. (*DNB*)
1850 *8440* 8442 *8457* 8460 *8474* **1851** 8527 8528 8529 8530 8531 8532 8533 8545 8560 *8608 8623* 8624 8632 *8634 8637 8646* 8654 *8663 8669* 8672 *8674 8678 8679 8686 8713* 8728 *8735* 8746 *8756* 8757 *8765 8766* 8768 8778 *8800* 8801 *8840 8862 8869 8877 8878* 8884 *8889 8893* 8901 *8925* 8926 *8949* 8950 *8955* 8957 *8959* 8963 *8965 8967* 8980 8989 8994 *8998 9000* 9001 *9004* 9005 *9017 9018* 9019 9020 *9023* 9033 9034 *9049 9059* 9060 *9069 9072* 9073 *9086 9090 9102 9125 9137* 9138 *9140* 9144 *9148* **1852** 9166 *9185 9186* 9187 9188 *9190 9191* 9192 9194 *9195 9196* 9197 *9202 9215 9216 9218 9219* 9220 9236 9244 9250 9251 *9253 9254 9255 9257 9258 9259* 9260 *9263* 9265 *9269* 9271 9273 9297 *9298* 9311 *9318 9323* 9336 9359 *9366 9371 9377* 9389 *9392* 9397 *9402 9403* 9404 9406 9425 9426 *9428 9429 9433 9437* 9442 9465 *9468 9481 9485 9529* 9541 *9545* 9546 *9547 9550 9564 9573* 9574 9575 *9577 9578 9596 9604* **1853** 9620 *9628 9636* 9639 *9646* 9650 *9651 9652 9674* 9678 9680 *9689* 9690 9692 9699 *9710 9721* 9722 9731 *9732 9733 9743* 9750 *9757 9759 9768 9787* 9792 9794 *9797* 9798 9799 *9802* 9808 9813 *9818* 9819 *9821 9822* 9823 *9828 9835 9836 9840 9841* 9844 9845 *9847* 9860 9873 9881 *9892* 9897 9900 9904 9911 **1854** 9925 9934 9935 *9938* 9939 9940 *9942 9944* 9945 9946 *9953* 9954 9955 9958 9976 9981 9984 9988 9989 *9991* 10001 *10002* 10004 10009 *10013 10015* 10016 *10020* 10022 10028 10036 10041 10055 10057 10058 10061 10071 10088 10090 10104 10105 10107 10110 10137 10140 10144 10145 10153 10154 10162 10164 10167 10168 10170 10172 10173 10176 10178 10183 10195 *10198* 10199 10203 10204 10211 10212 10271 **1855** 10281 10283 10284 10287 *10288* 10289 10290 *10311* 10312 10316 10343 *10344* **1858** *10775 10869* **1859** *11022*

Trevelyan, Walter Calverley (1797–1879). English naturalist and antiquarian. President of United Kingdom Alliance. (*DNB, MEB*)
1856 *10564*

Triskett, J. Superintendent of coining department at Sydney mint. (1854–8)
1854 *10177*

Trumper, Joseph
1832 2538

Tudhope, Francis (1805–70). English schoolmaster and lay preacher at the Cape of Good Hope. (*DSAfB*)
1840 *4423 4442* **1841** *4769*

Tulley, Charles. Member of a London firm of instrument makers. (James, *Faraday*)
1827 *1602*

Tunno, M.
1839 *4268* **1842** *5228*

Turberville, H.
1868 *13619* 13626 *13632* **1869** *14022*

Turner, Osborne (1825–87). British publisher and bookseller. (*MEB*)
1868 *13796*

Tuttle, Charles W. (1829–81). American astronomer, lawyer and antiquarian. Explained Saturn's 'dusky' ring. (*DAB, Who's*)
1860 *11439*

Twisleton, Edward (1809–74). English politician. Writes to JH on behalf of public school commissioners. (*DNB, MEB*)
1862 *12041* 12049 *12050* **1863** *12205*

Tyler, Samuel (1809-77). American lawyer, writer. (*DAB, Alli*)
1857 *10633* **1868** *13651*

Tyndale, George Booth (1772–1855). English solicitor. (*MEB*)
1843 *5572* **1863** 12307

Tyndall, John (1820–93). British natural philosopher, mathematician, and railway engineer. (*DNB, DSB*)
1851 8603 *9042 9043* **1852** *9183* **1856** 10552 *10554* 10555 **1857** *10631* 10637 *10642* **1861** 11643 *11646* 11715 *11719* 11721 *11722* 11723 **1862** 11843 *11846* **1863** *12108* 12109 12197 *12198 12215* **1864** 12621 *12663* 12710 *12711* 12714 *12717 12720* 12724 **1865** *12812* **1866** *13085* 13120 *13209* 13211 **1867** *13540* 13545 **1868** *13705* 13800 *13801 13807 13809* 13813 *13815 13824* 13825 13826 **1869** 13922 *13923* 13924 13926 *13932 13949 14054* **1870** 14216 *14217* 14305 *14330*

Underhill, Helen. Probably the wife of Joseph Underhill, barrister of Wolverhampton. (*MEB*)
1865 *12803*

Utzschneider, Joseph von (1763–1840). German industrialist, associated with great optical works in Munich. (*de Clerq*)
1829 *2070*

Vallack, B. W. S. British divine. Vicar in village of St. Budeaux near Plymouth. (1865–11–18)
1865 *13025* **1866** *13080 13089* 13091

Vallot, [D. M.].
1821 *610*

Valz, J. E. Benjamin (1787–1867). French astronomer. Director of the observatory at Marseilles. (*BHGW*)
1839 *4031*

Van De Meyer, Mr. Belgian minister to England. (1847–8–25)
1847 *7182*

Vaughan, Daniel (1818–79). American astronomer, mathematician, and chemist. Worked on the rings of Saturn. (*DAB, NCAB*)
1871 *14349*

'Velocity'
1863 *12408*

Veltmann, Wilhelm (1832–1902). German mathematician and teacher. (*BHGW*)
1869 *13970*

Venn, Henry (1796–1873). English divine and honorary secretary to the Church Missionary Society. (*MEB*)
1846 6545 **1848** *7768* 7801 7802 *7816 7868*

Villarceau, A. J. F. Yvon (1813–83). French astronomer. Observer at Paris and member of Bureau des Longitudes. (*BHGW*)
1849 *8022* 8036 *8049 8143* **1851** *8880* **1852** *9423* **1864** *12718*

Vivian, Edward (1808–93). Science writer and poet. (*MEB*)
1856 *10470* **1870** *14081*

Vulliamy, Benjamin Lewis (1780–1854). Royal clockmaker, with a shop at 68 Pall Mall. (*DNB, MEB*)
1847 *7159 7186* 7190 **1848** *7712*

Vyse, Richard William Howard- (see Howard-Vyse, Richard William)

Wabern, Lutzardt de
1822 721

Wackerbarth, Francis Diedrich (1813–84). British mathematician, astronomer, linguist, and divine. Became Swedish citizen. (*NUC, MEB*)
1866 *13131* 13144 *13180 13181*

Wade, Thomas Francis (1818–95). British army officer, linguist, and diplomat. Married JH's daughter Amelia, 1868. (*DNB, MEB*)
1864 *12525* **1869** *13918* **1870** *14227* **undated** 14801

Wagner, Mr.
1840 4366

Walbeck, Henrik Johan (1793–1822). Astronomer, director of Åbo [now Turku, Finland] observatory in 1820–21. (*NUC*)
1820 440 *469* **1821** *513* 534 **1822** 706 *765*

Waldegrave, George G. (1786–1857). British naval officer. 2nd Baron Radstock. Cousin of James A, Gordon. (*DNB*)
1854 *10017 10018*

Walker, James Thomas (1826–96). British officer in the Royal Engineers, Superintendent of great trigonometric survey of India, 1861–83. (*DNB, MEB*)
1863 12410 **1864** *12423 12590 12740* 12742 *12747*

Walker, Robert. Owner of portrait of William Herschel. Attended auction of WH's effects in 1860.
1860 *11262 11275 11278*

Walker, Sears Cook (1805–53). American mathematician, astronomer, and actuary. (*NCAB, DAB*)
1847 *7464*

Walker, William (1791–1867). Scottish engraver and miniature painter to William IV. (*DNB, MEB*)
1864 *12544* 12545 **1865** *12911 12941*

Wallace, Alexander. Assistant to Thomas Henderson, Astronomer Royal for Scotland, 1834–44. (1847–5–8)
1847 *6990*

Wallace, W. C. American oculist in New York City. (1838–6–26)
1838 *3708*

Wallace, William (1768–1843). British mathematician and inventor. Devised eidograph and chorograph. (*DNB*)
1815 *142* 144 **1817** 266 *268* **1824** *978* **1825** 1132 *1237* *1266* 1270 **1829** *1939* **1840** *4694* 4747

Wallenstein, E.
1853 9781

Wallich, Nathaniel (1786–1854). Danish-born surgeon in British army. Botanist and explorer in Indian subcontinent. (*DNB, MEB*)
1828 *1802 1830 1868 1872 1900* **1836** *3399* **1850** *8438*

Wallis, John. Amateur astronomer. (1838–7–13)
1831 *2386*

Walmly, G.
1870 *14157*

Walter, Henry (1785–1859). British divine and professor of natural philosophy at Haileybury College. (*MEB*)
1852 *9315*

Warburton, Henry (1785–1858). British philosophical radical, timber merchant, and M.P. (*DNB, MEB*)
1816 *177 180* **1829** *1943* **1847** *7308 7376 7393 7396 7399 7403 7412 7420 7427 7428 7436 7448 7472* **1850** *8210 8212 8228 8230 8232 8233 8239 8324 8332 8333* **1854** *10072 10097 10186 10189 10196* 10197 *10205 10225 10226 10229 10253 10254 10262 10263 10264* **1855** *10278 10282* **1856** *10498 10502 10508 10511 10516 10518* 10519 *10520 10522* **undated** *14802*

Ward, Edward Wolstenholme (1823–90). British officer in Royal Engineers. Employed at mints in London, Sydney, and Melbourne. (*MEB, AusDB*)
1854 9979 10100 10174 10175 10259 **1855** 10393 **1856** 10497 *10606*

Ward, Henry George (1797–1860). Secretary to Admiralty, 1846–59. (*DNB*)
1848 *7532*

Warington, Robert (1807–67). British chemist. Cofounder of Chemical and Cavendish Societies. (*DNB*)
1841 *4842*

Warner, Thomas. Author of book on civil timekeeping: *How to Keep the Clock Right* . (*NUC*)
1869 *13954* 13955 *13979 14017*

Wartmann, Élie François (1817–86). Swiss electrochemist and physicist. (*NUC*)
1841 *4992* **1847** *7206 7429*

Waterhouse, Elizabeth Ann (née Griesbach). Daughter of G. L. J. Griesbach, likely a relative of JH. Married George Robert Waterhouse, 21 Dec 1834. (*DNB*)
1851 *8947*

Waterhouse, Frederick Herschel. Son of George Robert Waterhouse.
1871 *14412* 14414

Waterhouse, George Robert (1810–88). British architect and naturalist. Keeper of geological department at the British Museum. (*DNB*)
1837 *3540* **1843** *5597 5613 5639* **1845** *6397* **1851** *8964 9141 9151* **1854** *10163* **1858** *10980* **1860** *11383* **1869** *13888 13891 13892* 13894 **1871** *14377*

Watermeyer, J. Secretary for the Committee of South African Infant Schools. (1834–11–11)
1834 *3025 3027 3030*

Waterston, John James (1811–83). Physical chemist and astronomer. (*DSB*)
1846 *6682* 6690

Watson, Harrison (1798–1869). South African businessman and statesman. Neighbor to JH at the Cape. (*SADNB*)
1836 *3304 3343*

Watson, Robert Spence (1837–1911). British political, social, and educational reformer. Secretary to the Literary and Philosophical Society at Newcastle-upon-Tyne, 1862–93. (*DNB*)
1867 *13422*

Watson, Thomas (1792–1882). British physician and professor of medicine. Physician to the royal family, 1859–82. (*DNB*)
1844 *5979* **1861** *11585*

Watson, Sir William (1744–1825). British physician. Friend of William Herschel and godfather to JH. (*DNB*)
1816 *195 200 206* **1819** 338 *351 362* **1820** *406 471 489* **1821** *537 581 630* **1822** 803 **1824** 1029

Watt, James Duff. Deputy assistant commissary general at the Cape. (Warners, *Maclear*)
1837 3452

Watt, Jr., James (1769–1848). British engineer, and co-owner of Soho Engineering Works after the retirement of his father James Watt (1736–1819). (*DNB*)
1839 *4261*

Watts, George Frederick (1817–1904). British painter and sculptor. (*Bryan, DNB*)
1870 *14237 14240*

Wauchope, A. Probably the wife of Robert Wauchope.
1840 *4443*

Wauchope, Mrs. A. Probably the wife of Robert Wauchope.
1840 *4427*

Wauchope, Robert (1790?–1862). British naval officer, stationed at Cape, 1834–38. Invented time ball for evaluating chronometers. (*MEB*)
1835 *3110 3114* **1837** *3513* **1840** 4442 **1844** *5827*

Waugh, Andrew Scott (1810–78). British officer in Bengal Engineers. Surveyor-general of India, 1843–61. (*DNB, MEB*)
1865 *12839 13027* **1870** *14322*

Weber, Wilhelm Eduard (1804–91). German physicist. Initiated network of magnetic observatories. (*DSB*)
1838 *3699* **1846** *6546*

Webster, Matthew Henry. Corresponding secretary of Albany Institute in New York, 1837–40. (1837–2–16)
1837 *3444* **1838** *3785* **1840** *4608*

Webster, Thomas (1800–86). British painter. (*DNB, Bryan*)
1859 *11053* **1862** *11892* **1869** *13909*

Webster, W. H. B. Secretary to the Ipswich Philosophical Society. (1848–10–17)
1848 *7870 7926 7945*

Wedgwood, Hensleigh (1803–91). British barrister, philologist, and police magistrate. (*DNB, MEB*)
1844 *5694* **1848** *7488* **1859** *11010*

Weisse, Maximilian (1798–1863). Meteorologist and astronomer at observatory at Jagiellonian University, Cracow. (*NUC*)
1848 *7873* **1849** *8013*

Weld, Charles Richard (1813–69). British barrister and publisher. Official historian of the R.S.L. (*DNB, MEB*)
1844 *5809* **1845** *6243 6257 6426 6443 6463 6496* **1847** 6933 *7075 7102* **1852** *9277* **1853** 9895

Wellesley, Arthur (1769–1852). British army officer. 1st Duke of Wellington. (*DNB*)
1852 9430

Wellington, Duke of (see Wellesley, Arthur)

Welsh, John (1824–59). British meteorologist, balloonist, and astronomical observer. (*DNB, MEB*)
1856 *10544* **1857** *10658 10668*

Wesley, Charles (1757–1834). English musician, eldest son of Charles Wesley, the hymn writer. (*DNB*)
1828 1839 *1840 1842*

Wesley, William
1862 *11872*

Weston, Charles H. Astronomer at Ensleigh Observatory, near Bath. (1870–12–26)
1870 *14343*

Wheatstone, Charles (1802–75). British musical instrument maker, cryptographer, and inventor. (*DNB, DSB*)
1825 *1224* **1833** 2717 2847 *2850* **1835** *3115* 3176 **1838** 3894 **1839** 3915 *3976* **1840** 4751 **1841** 5040 *5041* **1842** *5201* 5207 5321 **1845** 6151 *6154* 6183 6207 **1846** *6633* **1847** *7465* **1848** *7599* 7667 7797 *7907* 7910 **1849** 8068 *8069* 8070 *8076*

Wheeler, Edmund (fl. 1855). English topical writer. (*Alli*)
1865 *12939 12940*

Whewell, William (1794–1866). English philosopher, mathematician, scientist, historian and philosopher of science. Close friend of JH. (*DSB, Who's*)
1817 216 *224* 245 249 **1818** *298* 301 303 *305* **1819** 384 **1821** 514 **1822** *663 817* 824 833 **1823** 913 914 *915* 925 936 **1824** 1079 **1826** *1316 1344 1349* 1371 *1395* 1398 **1827** 1456 1465 *1605 1621 1654* **1828** *1691 1711 1837 1848* **1829** 1936 *1950 1959 1969 1981* 2004 **1831** *2290* 2295 2347 *2358 2417* 2419 **1832** *2629 2649* **1833** *2723 2736* **1834** 2946 3010 **1835** 3068 3111 3145 3161 **1836** *3305 3334 3417* **1837** 3521 3528 **1838** *3768* 3770 3773 3803 3814 *3878* 3881 3891 **1839** 3939 3970 3986 4011 4041 4200 4224 4225 **1840** 4386 *4454* 4511 4568 4638 **1841** 4874 *4876* 4877 4905 *4921 4950* 4960 5007 *5010* 5014 **1842** *5072* 5079 5183 5214 5231 5232 5288 *5339* **1843** *5446* 5647 **1844** 5823 5872 *5878* 5904 *5911 6000* **1845** 6067 6120 6128 *6134* 6230 6231 6273 6276 6285 *6360* 6400 6407 6444 **1846** 6547 *6613* 6636 6641 6668 *6669* 6722 *6787* 6788 6790 *6809* 6816 6846 **1847** *6859* 6860 6861 *7255* 7449 **1848** 7499 7589 *7821* 7851 7934 **1849** 7986 8118 **1850** *8300* 8302 *8327* 8380 8387 8477 **1851** 8600 8601 8604 8638 8670 8803 8806 8885 **1854** *9931* 9968 10239 **1855** 10276 **1857** 10657 10669 10716 **1858** 10790 *10799 10831 10848 10858 10860* 10870 *10890* 10892 10926 10951 10987 **1859** 11138 *11155 11224* **1861** *11538 11552 11717* 11732 11741 **1862** 11770 11840 11851 11881 12026 12088 **1863** 12129 12255 **1864** 12479 12534 12540 12575 12606 12692 **1865** 12798 12973 13002 13013 13015 *13016 13021* 13044 13048 **1866** 13071

Whipple, George Mathews (1842–93). British physicist, meteorologist, and superintendent of Kew magnetic observatory. (*DNB*)
1871 *14381*

Whitaker, John Henry
1863 *12138* **1867** 13550

White, Elizabeth. Wife of Walpole White. Close friend of Elizabeth Baily, sister of Francis Baily.
1855 *10425* **1859** 11164 *11169* **1862** *11954 11960*

White, Henry. A nephew of Susan White (JH's godmother). (RS 18.227)
1833 *2858*

White, James William. A nephew of Susan White.
1833 *2865*

White, R. S. Honorary secretary for athletic event at Tunbridge Wells. (1870–3)
1870 *14139* 14140

White, Susan. JH's godmother, and aunt of James, Walpole, and Elizabeth White. Mistress of Sally, an elderly servant very fond of JH. (1814–9–25)
1814 *124* **1817** *270 271* **1823** *923 924* **1825** 1262 **1832** 2505 *2525 2526*

White, Walter (1811–93). British cabinetmaker and R.S.L. librarian. (*DNB, MEB*)
1861 11631 **1862** *11824 11833 12022 12056* **1863** *12361* **1864** *12430 12750* **1865** 12814 *12818 12900* **1869** *14077* **1870** *14145 14184*

Whitehouse, Thomas
1867 *13547*

Whitfeld, John Clarke- (see Clarke-Whitfield, John)

Whittaker, John William (1790?–1854). British philologist and divine. College chum of JH. (*DNB*)
1812 46 **1813** 60 *62 69* 70 *76* 80 81 85 86 88 **1814** 101 *106 111* **1816** *185 191* 192 **1817** 213 243 272 **1819** 353 **1822** 683 688

Whitworth, Joseph (1803–87). British mechanical engineer. Introduced uniform system of screw threads. (*DNB, MEB*)
1853 *9682* **1862** 12000

Wigan, Arthur Cleveland (1815–?). British writer. (*NUC*)
1849 *8164* **1850** *8399*

Wigginton, H. S. Secretary of Slough Mechanics' Institute. (1847–2–6)
1847 *6886*

Wight, Jr., Alexander
1856 *10481 10483 10487*

Wilcox, R. Officer in East India Company. Director of Lucknow Observatory. (*Dreyer*)
1840 *4688*

Wilke, C. M.
1847 *7200*

Wilkie, David (1785–1841). British painter to William IV and Queen Victoria. (*DNB, Bryan*)
1817 *209* **1820** *451* **1823** *867*

Wilkins, Mr. Owned land next to the Mint in London. (JH's (?)1854–3–10)
1854 10023

Wilkinson, Henry (1792–1838). British barrister and schoolmaster. College friend of JH. (*Cantab*)
1816 188

Wilkinson, John Gardner (1797–1875). Explorer and Egyptologist. (*DNB*)
1860 *11509*

Wilks, Samuel Charles (1789–1872). British evangelical minister, and editor of *Christain Observer*. Supporter of new geological views. (*DNB*)
1867 *13471 13483*

Willes, E. Eton graduate and nephew of Mrs. [Eardley?]-Wilmot. Close friend of JH.
1820 *503* **1828** *1763* **1829** *1947* **1838** *3797 3860* **1839** *3929 3933 4316* **1843** *5432*

Williams, John (1797–1874). Assistant secretary to R.A.S., 1847–74. Author of general indices to *R.A.S.M.N.* (*MEB*)
1866 *13329* **1871** 14430 *14431*

Williams, Thomas. Executor for the estate of the 3rd Duke of Northumberland. (1848–2–18)
1848 *7592*

Williams, William
1860 *11312*

Williams, William Smith (1800–75). British writer and bookseller. Literary adviser to Smith, Elder & Co. (*MEB*)
1824 *945* **1846** *6627* **1847** *6969 7086 7151* 7170 *7189 7202 7220 7228 7231 7238 7249 7253 7322 7357* **1848** *7734* **1852** *9449*

Wilmot, Frederick Marow Eardley- (see Eardley-Wilmot, Frederick Marow)

Wilson, Alexander. Keen currency reformer. (1843–10–31)
1843 *5625* 5640 **1844** *5947* 5970 *5986* **1850** *8463*

Wilson, Andrew. Husband of Helen Wilson. (1819–9–29)
1819 *372*

Wilson, Ann. A friend of JH, living with others known to JH.
1815 *161* **1819** *358*

Wilson, Charles Rivers (1831–1916). British civil servant and financier. Held positions in the Treasury, 1856–94. (*DNB*)
1869 *13988* 13990

Wilson, Francis Maria (née Edgeworth) (1799–1848). Sister of Maria Edgeworth. Wife of Lestock Peach Wilson. (Hare, *Edgeworth*)
1831 *2376*

Wilson, George (1818–59). British chemist, physician, scientific historian, and religious writer. Investigated color blindness. (*DNB*)
1859 *11116*

Wilson, Helen. Friend of Mrs. E. S. Beckwith and the Herschels.
1820 *493*

Wilson, James (1805–60). British political economist, M.P., and cabinet minister. (*DNB, MEB*)
1851 8534 *8577 8595 8907* 8908 *8919* 8920 **1853** 9653 9717 *9772 9784* **1854** 10159

Wilson, Mary Anne (1790–?). Sister of Alexander Wilson and of Patrick Wilson.
1866 *13295 13313*

Wilson, Miss
1817 209

Wilson, Pat. Friend of JH at Cambridge. (1811–6–6)
1811 *21*

Wilson, Thomas (1800–76). British educator and cofounder of B.A.A.S. Founded several technical societies. (*MEB*)
1850 *8388* 8390

Wilson, William Parkinson (1826?–1874). English-born mathematician. Professor of mathematics and astronomy at Melbourne, Australia. (*AusDB*)
1861 11682

Winlock, Joseph (1826–75). American astronomer and mathematician. (*NCAB*)
1869 *13916*

Winnecke, Friedrich August Theodor (1835–97). German astronomer and mathematician. Director of Strasbourg Observatory. (*Who's*)
1864 *12591 12595 12599* **1868** *13595*

Winther, H. T.
1844 *5834*

Winthrop, Robert Charles (1807–94). American politician, publisher, and philanthropist. (*NCAB*)
1861 *11572*

Wire, Alfred P.
1861 *11690* 11693 *11695*

Withers, E. L.
1858 10940 10941

Witte, Mrs. Resident of Hanover who designed globes of the moon. (1838–7–25)
1838 *3732* **1846** 6655

Witte, W. Husband of Mrs. Witte.
1847 *6976*

Wolf, Rudolf (1816–93). German astronomer and director of Zurich Observatory. *(BE)*
1857 *10699* **1865** 12894 **1866** *13147* **1867** *13410* 13413 *13416* 13427 *13470* 13472

Wolff, Joseph (1795–1862). British missionary, linguist, and divine. (*DNB, MEB*)
1858 *10838*

Wollaston, George Hyde. Probably the brother of William Hyde Wollaston.
1828 *1901 1904 1909*

Wollaston, William Hyde (1766–1828). British physiologist, chemist, physicist, and physician. (*DNB, DSB*)
1820 *455* **1821** *518* **1823** *845 846 847 879* **1825** *1097 1115 1137* **1826** *1393* **1827** *1473* 1474 *1513 1522*

Wood, Charles (1800–85). British civil servant and cabinet minister. 1st Viscount Halifax. (*DNB*)
1850 *8422* **1851** *8589 8609* 8626 8867 *8942 9003* 9007 9057 *9062 9076* 9087

Wood, James (1760–1839). British mathematician and divine. Master of St. John's College. (*DNB, Alli*)
1820 394 *417* **1826** 1396 *1397* 1424 *1426* **1829** *1946* **1831** *2311* **1832** *2512 2589* 2614

Wood, W. H.
1864 *12755* **1865** *12787*

Woods, Thomas
1854 *10117*

Woodward, Samuel Pickworth (1821–65). English naturalist. Professor of geology at Cirencester. (*DNB*)
1861 11644

Woolcott, George. Honorary secretary of Kent Opthalmic Hospital. (1847–6–9)
1847 *7042*

Woolhouse, Wesley Stoker Barker (1809–93). British mathematician and writer on music. (*MEB*)
1847 *7250*

Woollgar, John Webb (1795–1851). British solicitor. (*MEB*)
1846 *6791* **1847** *7184*

Woolsey, Theodore Dwight (1801–89). American minister, educator and president of Yale College. (*NCAB, DAB*)
1847 *7408*

Worms, Henry de (see De Worms, Henry)

Wrangham, E. Personal friend of Herschel and Beckwith families.
1822 *761*

Wrangham, Sarah. Personal friend of JH and probably daughter of E. Wrangham.
1827 *1580 1607*

Wrench, Mrs.
1844 5811

Wrottesley, John (1798–1867). British lawyer and cofounder of R.A.S. Built several private observatories. 2nd Baron Wrottesley. (*DNB, MEB*)
1839 *4016* 4019 *4026* **1847** *7111 7152 7153* **1855** 10358 **1856** 10482 **1857** 10635 **1858** 10969 *10982* **1860** 11469 *11474* **1862** 12089 **1863** 12133 *12289 12293* **1866** *13161 13187 13193* 13200 *13201 13219*

Wrottesley, Lord (see Wrottesley, John)

Wyatt, Matthew Digby (1820–77). British architect and writer on art. (*DNB, MEB*)
1861 *11561*

Wylde, James. Editor for Griffin & Co., publishers. (1860–9–22)
1860 *11428 11434*

Wylie, Alexander (1815–87). British translator and missionary to China. Translated JH's *Outlines Astr.* into Chinese. (*DNB, MEB*)
1860 *11482* **1863** *12175 12176*

Wyndham-Quin, Augusta (née Gould) (?–1866). Lady Dunraven. Married Edwin Wyndham-Quin, 1836. (*DNB*)
1864 *12510*

Wyndham-Quin, Edwin Richard Windham (1812–71). Irish archaeologist, astronomer, and spiritualist. 3rd Earl of Dunraven. (*DNB, Peerage*)
1847 *7178* **1864** *12480* **1865** 13054

Wyon, Benjamin (1802–58). Chief engraver of seals at the Mint. (*DNB*)
1854 9926

Wyon, Leonard Charles (1826–91). Engraver at the Mint. Succeeded his father, William. (*DN, MEB*)
1851 *9013 9128* 9129 *9132* **1852** *9548* **1854** 10079 10252 **1855** *10326 10355*

Wyon, William (1795–1851). Engraver at the Mint. (*DNB, MEB*)
1850 *8482* 8483 **1851** *8843*

Yapp, George Wagstaffe (1811–80). British journalist, associated with *Art Journal.* (*MEB*)
1847 *6902*

Yeomans, Alfred
1864 *12428*

Yinsly, J. Secretary(?) to the Royal Bavarian Academy of Sciences. (1848–11–29)
1848 *7936*

Yolland, William (1810–85). British officer in Royal Engineers. Published astronomical observations for Ordnance Survey. (*DNB, MEB*)
1847 *7211 7233*

Yorke, Lt. Col. Charles (1790–1880). British army officer. Fought in the Kaffir wars in 1852. (*DNB*)
1839 *3977*

Youmans, Edward Livingston (1821–87). American physician, editor, and promoter of scientific education. (*NCAB, DAB*)
1866 *13241* **1868** *13790*

Young, Eliza (née Maxwell) (1785–1859). Married Thomas Young, 1804. (*DNB*)
1830 *2121* 2122 *2159* 2175 **1833** *2739*

Young, F. A. Probably a neighbor of JH at Hawkhurst. ([1840]–11–12)
1840 *4719 4721*

Young, George Paxton (1819–89). Canadian philosopher, and school inspector. (*DCB*)
1860 *11299*

Young, Samuel (1776–1826). East India Company civil servant in Madras. (*Burke's Peerage*)
1819 *327 355* **1821** *547 597 634* **1825** *1183*

Young, Thomas (1773–1829). British physician, physicist, and Egyptologist. (*DNB*)
1820 *409* **1821** *645* **1822** *676 694 699 700* 766 *768* **1823** *849 927* **1824** *946* 1069 *1072 1073* 1074 *1075 1076* **1825** *1169* 1186 *1246 1249* 1250 *1253 1269* **1826** *1279 1280 1341 1342* **1827** *1647* **1828** *1706 1708 1709 1729 1732 1740 1820* 1822 *1823 1825 1887* **1829** 1937 *1952 1954 1999*

Zach, Franz Xaver von (1754–1832). German astronomer and surveyor. Editor of *Correspondance astronomique, géographique, hydrographique et statistique*. (*DSB*)
1822 *659* **1825** *1116*

Zantedeschi, Francesco (1797–1873). Italian priest and professor of physics at Padua. (*Who's*)
1846 *6847*

Zimmermann, Agnes
1866 *13185*

Zöllner, Johann Karl Friedrich (1834–82). German physicist, and inventor of light-measuring equipment. (*DSB*)
1865 *12942*

Zucchini, Zanobi
1841 *4993*

Unidentified
1814 129 **1817** 262 **1821** 621 **1822** 804 805 812 **1823** *876* **1824** 939 **1825** 1244 1271 **1827** 1491 1609 **1828** 1666 1758 **1831** 2273 **1832** 2503 2535 2536 2604 2619 2693 **1833** 2870 **1834** 2937 **1837** 3429 3590 **1838** 3615 3616 3714 **1839** 3909 3916 3917 3947 *4049* 4119 **1840** 4367 4498 4586 4717 **1841** 4883 *4965* 4978 4991 **1842** 5070 5138 **1843** 5362 5541 **1844** *5695* 5996 **1845** 6068 6161 6258 **1846** 6548 6648 6715 *6745* **1847** 6925 *7225* **1848** *7574* 7737 7859 7940 **1850** 8245 8279 8381 **1851** 8535 8536 8537 8538 8539 8540 8541 8542 8563 8775 8863 8952 8981 8997 9109 **1852** 9169 9268 9289 9314 9367 9405 9470 9510 9526 9536 9600 **1853** 9670 9703 9719 9810 9888 **1854** 9913 9922 9927 9932 9947 9978 9997 10032 10052 10059 10073 10146 10156 10184 10215 10230 **1855** *10302* 10323 10324 10357 **1856** 10457 10500 **1859** 11063 11103 11172 **1860** 11374 11423 **1861** *11524* **1864** 12420 12421 12702 **1865** 12773 12931 13047 **1866** 13178 13221 13230 **1867** 13393 **1868** 13646 13647 **1869** 13840 13861 13950 **undated** 14458 14459 14460 14742 14743 14744 14745 14786 *14803 14804* 14805 14806 14807 14808 14809 14810 14811 14812

BIBLIOGRAPHY OF SOURCES FOR BIOGRAPHICAL REGISTER

ADB: *Allgemeine deutsche Biographie.* Under the auspices of the Historical Commission of the Royal Academy of Sciences. 1st ed., Leipzig: 1875–1912. Reprint ed., 56 vols. Berlin: 1967–71.

Airy, *Autobiography*: Airy, George Biddell. *Autobiography of Sir George Biddell Airy.* Wilfred Airy, ed. Cambridge, England: 1896.

Alli: *A Critical Dictionary of English Literature, and British and American Authors, Living and Deceased.* By S. Austin Allibone. 3 vols. Philadelphia and London: 1859–71.

AlliSup: *A Critical Dictionary of English Literature, and British and American Authors, Living and Deceased.* Supplement by John Foster Kirk. 2 vols. Philadelphia and London. 1891.

AusDB: *Australian Dictionary of Biography: 1788–1850; 1851–1890.* Douglas Pike and Bede Nairn, general eds. 6 vols. Melbourne: 1966–76.

Batten, *Struve*: Batten, Alan H. *Resolute and Undertaking Characters: The Lives of Wilhelm and Otto Struve.* Dordrecht: 1988.

BBI: *British Biographical Index.* David Bank and Anthony Esposito, eds. London: 1990.

BE: *Brockhaus Enzyklopädie.* 19th ed., F. A. Brockhaus, ed. 24 vols. Mannheim: 1986.

BHF: *British Historical Facts, 1830–1900.* Compiled by Chris Cook and Brendan Keith. New York: 1975.

BHGW: *Biographisch-Literarisches Handwörterbuch zur Geschichte der Exacten Wissenschaften.* By Johann Christian Poggendorff. Vols. 1–4, Leipzig: 1863–1904; vol. 5, Leipzig und Berlin: 1926; vol. 6, Berlin: 1936–40.

Brit. Mus. Cat.: *British Museum General Catalogue of Printed Books.* Photolithographic ed. to 1955. 263 vols. London: 1965–66.

Brown: Brown, J. D., and Stratton, S. S. *British Musical Biography: A Dictionary of Musical Artists, Authors and Composers Born in Britain and Its Colonies.* Birmingham: 1897.

Bryan: Bryan, Michael. *Dictionary of Painters and Engravers, Biographical and Critical.* Revised and enlarged by R. E. Graves. 2 vols. London: 1886.

Burke's Peerage: *Burke's Peerage and Baronetage.* 105th edition, 4th impression. London: 1980.

Cantab: *Alumni Cantabrigiensis: A Biographical List of all Known Students, Graduates and Holders of Office at the University of Cambridge from the Earliest Times to 1900.* Compiled by John A. Venn. Part II. From 1752 to 1900. 6 vols. Cambridge:

Challis, *Mint*: Challis, C. E., ed. *A New History of the Royal Mint.* Cambridge, England: 1992.

Craig, *Mint*: Craig, John. *The Mint: A History of the London Mint from A.D. 287 to 1948.* Cambridge, England: 1953.

DAB: *Dictionary of American Biography.* Under the auspices of the American Council of Learned Societies. 20 vols., index. New York and London: 1928–37. Supplements 1–5, 1944–77.

Darwin: *A Calendar of the Correspondence of Charles Darwin, 1821–1882.* Frederick Burkhardt and Sydney Smith, eds. New York and London: 1985.

DAusB: *Dictionary of Australian Biography.* Percival Serle, ed. 2 vols. Sydney: 1949.

DBF: *Dictionnaire de biographie française.* Under the direction of J. Balteau, M. Barroux, M. Prevost et al. 18 vols., (A - Lacombe). Paris: 1933–94.

DBIt: *Dizionario biografico degli italiani.* Vols. 1–44 (A - Farina). Rome: 1960–94.

DBU: *Dizionario biografico universale.* Gottardo Garollo, ed. 2 vols. Milan: 1907.

DCB: *Dictionary of Canadian Biography.* 13 vols. (1000–1910). Toronto: 1966–94.

de Clercq: *Nineteenth-Century Scientific Instruments and Their Makers.* P. R. de Clercq, ed. Leiden and Amsterdam: 1985.

DHBS: *Dictionnaire historique et biographique de la suisse.* Compiled by the Administration du dictionnaire historique et biographique de la suisse. Neuchatel: 1926.

DIB: Buckland, C. E. *Dictionary of Indian Biography.* Reprint of 1st ed. New York: 1969

DNB: *Dictionary of National Biography.* Sir Leslie Stephen and Sir Sidney Lee, eds. Rept. ed., 21 vols. London: 1937–38. 13 supplements London: 1901–93.

Dreyer: *History of the Royal Astronomical Society, 1820–1920.* J. L. E. Dreyer and H. H. Turner, eds. London: 1923.

DSAfB: *Dictionary of South African Biography.* 3 vols. Cape Town: 1968–77.

DSB: *Dictionary of Scientific Biography.* C. C. Gillispie, ed. 14 vols., and 2 supps. New York: 1970–90.

Dunsheath, *Giants*: Dunsheath, Percy. *Giants of Electricity.* New York: 1967.

Eder, *Photography*: Eder, Josef Maria. *History of Photography.* Edward Epstean, trans. New York: 1945.

EnCan: *Encyclopedia Canadiana.* John E. Robbins, ed. 10 vols. Ottawa: 1963.

EnItal: *Enciclopedia italiana di scienze, lettere ed arti.* 35 vols. Rome: 1929–39.

Evans, *Herschel*: *Herschel at the Cape. Diaries and Correspondence of Sir John Herschel, 1834–1838.* David S. Evans, Terence J. Deeming, Betty Hall Evans, Stephen Goldfarb, eds. Austin and London: 1969.

Graves, *Hamilton*: Graves, Robert Percival. *Life of Sir William Rowan Hamilton, Knt., LL. D., ...* 3 vols. Dublin: 1882–89.

Grove: *The New Grove Dictionary of Music and Musicians.* Stanley Sadie, ed. 20 vols. London and Washington: 1980.

GSE: *Great Soviet Encyclopedia* (translation of 3rd ed. of *Bol'shaia Sovetskaia Entsiklopediia,* Moscow: 1970–8). 28 vols. New York and London: 1973–81.

Hall, *R.S.*: Hall, Marie Boas. *All Scientists Now: The Royal Society in the Nineteenth Century.* Cambridge and New York: 1984.

Hare, *Edgeworth*: Hare, A. J. C. *Maria Edgeworth: Life and Letters.* 2 vols. 1895.

Herschel Soc.: *Herschel Society Newsletter.* February, 1992. William Herschel Society, Bath, England.

Houzeau: *Bibliographie générale de l'astronomie.* Written and compiled by J. C. Houzeau and A. Lancaster. 1st ed., Brussels: 1880–89. New ed., 2 vols. London: 1964.

Hyman, *Babbage*: Hyman, Anthony. *Charles Babbage, Pioneer of the Computeer.* Princeton: 1982.

James, *Faraday*: *The Correspondence of Michael Faraday.* Frank A. J. L. James, ed. Vol. 1 (1811–1831). London: 1991.

Jones, *Harvard*. Jones, Bessie Zaban, and Lyle Gifford Boyd. *Harvard College Observatory: The First Four Directorships, 1839–1919*. Cambridge, Mass, 1971.

LGE: *La grande encyclopedie*. 31 vols. Paris: 1886–1902.

Lindgren, *Glory*: Lindgren, Michael. *Glory and Failure*. Trans. by Craig G. McKay. Cambridge, Mass.: 1990.

Meadows, *Greenwich*: Meadows, A. J. *Greenwich Observatory. The Royal Observatory at Greenwich and Herstmonceux, 1675–1975*. Vol. 2 (1836–1975). London: 1975.

MEB: *Modern English Biography*. Frederic Boase, ed. 3 vols. and supps., 3 vols. London: 1965

MERSH: *The Modern Encyclopedia of Russian and Soviet History*. Joseph L Wieczynski, ed. 46 vols. Gulf Breeze, Florida: 1976–87. 9 supps. Gulf Breeze, Florida: 1988–93.

MOH: Danvers, Frederick Charles, Sir M. Monier-Williams et al. *Memorials of Old Haileybury College*. Westminster: 1894.

NBG: *Nouvelle biographie générale*. 42 vols. (A - Saint-André). Paris: 1852–66.

NCAB: *The National Cyclopaedia of American Biography*. 63 vols. New York: 1898–1984.

NDtB: *Neue Deutsche Biographie*. Edited by the Historical Commission of the Bavarian Academy of Sciences. 17 vols., (A - Moller). Berlin: 1953–93.

NUC: *The National Union Catalog*. Pre-1956 Imprints. 686 vols. London and Chicago: 1968–80. Supplement, vols. 686–754, 1980–1.

O'Brien, *Drummond*: O'Brien, Richard Barry. *Thomas Drummond, Under-Secretary in Ireland, 1835–48: Life and Letters*. London: 1889.

O'Donoghue: O'Donoghue, David James. *The Poets of Ireland: A Biographical and Bibliographical Dictionary of Irish Writers of English Verse*. Dublin: 1912.

OBL: *Österreichisches biographisches Lexikon 1815–1950*. Leo Santifaller and Eva Obermayer-Marnach, eds. 10 vols., (A - Schobert). Vienna: 1957–1994.

Peerage: *The Complete Peerage of England, Scotland, Ireland, Great Britain and the United Kingdom (Extant, Extinct or Dormant)*. By George Edward Cokayne. Rev. ed., Vicary Gibbs, ed. 12 vols., (A - Towton). London: 1910–93.

Quincy: Quincy, Josiah. *Memoir of James Grahame, LL. D.* Boston: 1845.

RSL Cat: *Catalogue of Scientific Papers, 1800–1863*. Compiled by the Royal Society of London. 6 vols. London: 1868–73.

SADNB: *Southern African Dictionary of National Biography*. Compiled by Eric Rosenthal. London and New York: 1966.

ScandHist: *Dictionary of Scandinavian History*. Byron J. Nordstrom, ed. Westport, Conn.: 1986.

SESA: *Standard Encyclopaedia of Southern Africa*. 11 vols. and 1 supp. Cape Town: 1970–76.

Shadow: Buttmann, Günther. *The Shadow of the Telescope: A Biography of John Herschel*. B. E. J. Pagel, trans. Guildford and London: 1974.

The Times: *The Times of London*. London: 1788–1995.

Warners, *Maclear*: Warner, Brian and Nancy. *Maclear and Herschel: Letters & Diaries at the Cape of Good Hope, 1834–1838*. Cape Town: 1984.

Watt: Watt, Robert. *Bibliotheca Britannica: A General Index to British and Foreign Literature*. 4 vols. Edinburgh: 1824.

Who's: *World Who's Who in Science from Antiquity to the Present*. Allen G. Debus, ed. Chicago: 1968.

Williams, *Faraday*: *The Selected Correspondence of Michael Faraday*. L. Pearce Williams, ed. 2 vols. Cambridge and New York: 1971.

Women: Ogilvie, Marilyn Bailey. *Women in Science: Antiquity Through the Nineteenth Century*. Cambridge, Mass: 1986.

INDEX

Index references are to the numbers appearing in bold at the beginning of each entry in this Calendar. This index is both a name and a subject index for the summaries. Because the Biographical Register provides a listing by correspondent of the letters written to John Herschel or received from him, this function is not repeated in the index below. Information about a correspondent appearing in a letter to or from that correspondent is not typically listed under that correspondent's name because it is accessible through the Biographical Register.

Publications by John Herschel are listed alphabetically in the index rather than under Herschel's name. Publications by other authors are listed alphabetically and also under that author's name.